Lecture Notes in Computer Science 5073

Commenced Publication in 1973
Founding and Former Series Editors:
Gerhard Goos, Juris Hartmanis, and Jan van Leeuwen

Osvaldo Gervasi Beniamino Murgante
Antonio Laganà David Taniar
Youngsong Mun Marina L. Gavrilova (Eds.)

Computational Science and Its Applications – ICCSA 2008

International Conference
Perugia, Italy, June 30 – July 3, 2008
Proceedings, Part II

Springer

Volume Editors

Osvaldo Gervasi
University of Perugia, Italy
E-mail: osvaldo@unipg.it

Beniamino Murgante
Università degli Studi della Basilicata, Potenza, Italy
E-mail: beniamino.murgante@unibas.it

Antonio Laganà
University of Perugia, Italy
E-mail: lag@unipg.it

David Taniar
Monash University, Clayton, VIC , Australia
E-mail: david.taniar@infotech.monash.edu.au

Youngsong Mun
SoongSil University, Seoul, Korea
E-mail: mun@comp.ssu.ac.kr

Marina L. Gavrilova
University of Calgary, Canada
E-mail: marina@cpsc.ucalgary.ca

Library of Congress Control Number: 2008929608

CR Subject Classification (1998): F, D, G, H, I, J, C.2-3

LNCS Sublibrary: SL 1 – Theoretical Computer Science and General Issues

ISSN 0302-9743
ISBN-10 3-540-69840-X Springer Berlin Heidelberg New York
ISBN-13 978-3-540-69840-1 Springer Berlin Heidelberg New York

Springer is a part of Springer Science+Business Media

springer.com

Printed in Germany

Typesetting: Camera-ready by author, data conversion by Scientific Publishing Services, Chennai, India
Printed on acid-free paper SPIN: 12322664 06/3180 5 4 3 2 1 0

Preface

This two-volume set is assembled following the 2008 International Conference on Computational Science and Its Applications, ICCSA 2008, a premium international event held in Perugia, Italy, from June 30 to July 3, 2008.

The collection of fully refereed high-quality original works accepted as theme papers for presentation at ICCSA 2008 are published in this LNCS proceedings set. This outstanding collection complements the volume of workshop papers, traditionally published by IEEE Computer Society.

The continuous support of computational science researchers has helped ICCSA to become a firmly established forum in the area of scientific computing and the conference itself become a recurring scientific and professional meeting that cannot be given up.

The computational science field, based on fundamental disciplines such as mathematics, physics, and chemistry, is finding new computational approaches to foster the human progress in heterogeneous and fundamental areas such as aerospace and automotive industries, bioinformatics and nanotechnology studies, networks and grid computing, computational geometry and biometrics, computer education, virtual reality, and art. Due to the growing complexity of many challenges in computational science, the use of sophisticated algorithms and emerging technologies is inevitable. Together, these far-reaching scientific areas help to shape this conference in the areas of state-of-the-art computational science research and applications, encompassing the facilitating theoretical foundations and the innovative applications of such results in other areas.

The topics of the fully refereed papers span all the traditional as well as the emerging computational science realms, and correspond to the conference themes:

1) Computational Methods, Algorithms and Applications
2) High-Performance Technical Computing and Networks
3) Advanced and Emerging Applications
4) Geometric Modelling, Graphics and Visualization
5) Information Systems and Information Technologies

The present volume set also includes papers from more than 20 prominent workshops comprising ICCSA 2008 on the areas of embedded systems, geographical analysis, computational geometry, computational geomatics, computer graphics, virtual reality, computer modeling, computer algebra, mobile communications, wireless networks, computational forensics, data storage, information security, Web learning, software engineering, computational intelligence, digital security, biometrics, molecular structures, material design, ubiquitous computing, symbolic computations, Web systems and intelligence, and e-education.

We are very grateful to the Steering Committee and the Program Committee for their tremendous support in putting this conference together, the 500 referees for their diligent work in reviewing the submissions, the local committee members and all the sponsors, supporting organizations and volunteers of ICCSA for contributing their time, energy and resources to this event.

Finally, we thank all authors for their submissions making the ICCSA Conference year after year one of the premium events on the scientific community scene, facilitating the exchange of ideas, fostering new collaborations, and sharing the future of computational science.

June 2008

Osvaldo Gervasi
Marina L. Gavrilova

On behalf of the Co-editors
Beniamino Murgante
Antonio Laganà
David Taniar
Youngsong Mun

Organization

ICCSA 2008 was organized by the University of Perugia (Italy) and the University of Calgary (Canada).

Conference Chairs

Marina L. Gavrilova (University of Calgary, Canada), Scientific
Osvaldo Gervasi (University of Perugia, Italy), Program

Steering Committee

Marina L. Gavrilova (University of Calgary, Canada)
Osvaldo Gervasi (University of Perugia, Italy)
Eui-Nam Huh (Kyung Hee University, Korea)
Andres Iglesias (University of Cantabria, Spain)
Vipin Kumar (Army High Performance Computing Center and University of Minnesota, USA)
Antonio Laganà (University of Perugia, Italy)
Youngsong Mun (Soongsil University, Korea)
C.J. Kenneth Tan (OptimaNumerics, UK)
David Taniar (Monash University, Australia)

Program Committee

Jemal Abawajy (Deakin University, Australia)
Mir Atiqullah (Southern Polytechnic State University, USA)
Mark Baker (Portsmouth University, UK)
Brian A. Barsky (University of California, Berkeley, USA)
David Bell (The Queen's University of Belfast, UK)
Sergey Bereg (Duke University, USA)
Stefania Bertazzon (University of Calgary, Canada)
J.A. Rod Blais (University of Calgary, Canada)
Alexander V. Bogdanov (Institute for High Performance Computing and Data Bases, Russia)
Matt-Mouley Bouamrane (University of Manchester, UK)
John Brooke (University of Manchester, UK)
Martin Buecker (Aachen University, Germany)
Rajkumar Buyya (University of Melbourne, Australia)
Yves Caniou (INRIA, France)
Hyunseung Choo (Sungkyunkwan University, Korea)
Min Young Chung (Sungkyunkwan University, Korea)

Sponsoring Organizations

ICCSA 2008 would not have been possible without the tremendous support of many organizations and institutions, for which all organizers and participants of ICCSA 2008 express their sincere gratitude:

University of Perugia, Italy
University of Calgary, Canada
Innovative Computational Science Applications (ICSA)
MASTER-UP, Italy
SPARCS Laboratory, Universityu of Calgary, Canada
OptimaNumerics, UK

Table of Contents – Part II

Workshop on Information Systems and Information Technologies (ISIT 2008)

Workshop on Virtual Reality in Scientific Applications and Learning (VRSAL 2008)

Workshop on Ubiquitous Web Systems and Intelligence (UWSI 2008)

Workshop on Wireless and Ad-Hoc Networking (WAD 2008)

Workshop on Internet Communication Security (WICS 2008)

Workshop on Integrated Methods and Algorithms for Non-Destructive Evaluations of Concrete Structures (IMANDECS 2008)

General Track on Computational Methods, Algorithms and Applications

General Track on High Performance Technical Computing and Networks

General Track on Advanced and Emerging Applications

General Track on Information Systems and Information Technologies

Table of Contents – Part I

Workshop on Geographical Analysis, Urban Modeling, Spatial Statistics (GEO-AN-MOD 2008)

Workshop on Advances in Web Based Learning (AWBL 2007)

Workshop on Computer Aided Information (CAI) Technology (CAIT 2008)

Workshop on Computational GeoInformatics (COMPGEO 2008)

Workshop on Embedded Systems for Ubiquitous Computing (ESUC 2008)

Workshop on High-Performance Computing and Information Visualization (HPCIV 2008)

Workshop on Mobile Communications (MC 2008)

Workshop on Molecular Simulations Structures and Processes (MOSSAP 2008)

Workshop on PULSES – Logical, Technical and Computational Aspects of Transformations and Suddenly Emerging Phenomena (PULSES 2008)

General Track

A New Construction of Multivariate Public Key Encryption Scheme through Internally Perturbed Plus

Zhiwei Wang[1], Xuyun Nie[2], Shihui Zheng[1], Yixian Yang[1], and Zhihui Zhang[3]

[1]Information Security Center,
State Key Laboratory of Networking and Switching Technology,
Beijing University of Posts and Telecommunications, Beijing 100876, P.R.China
[2]School of Computer Science and Engineering,
University of Electronic Science and Technology of China,
Chengdu 610054, P.R.china
[3]Sony (China) Research Laboratory,Beijing 100080, P.R.China
zhwwang2007@yahoo.com.cn

Abstract. Public key cryptography is an important tool in our modern information society. Multivariate public key cryptosystem (MPKC) is a type of public key cryptosystem with high efficiency. In this paper, firstly, we propose a basic multivariate public key encryption scheme which is efficient but insecure. Secondly, we use the internally perturbed plus modifier to modify the basic scheme, and get a improved scheme which is secure against known types of attacks. Finally, we provide a practical implementation of the improved scheme.

Keywords: Multivariate Public Key Cryptosystem, Internally Perturbed Plus, Encryption Scheme, Quadratic Polynomial.

1 Introduction

Public key cryptography is an important tool in our modern information society. Traditional public key cryptosystems such as RSA and ElGamal rely on hard number theory based problems such as factoring or discrete logarithm. However, due to the quantum computer attack by Shor [1] and the demand for more efficient cryptosystems for small devices, there is a great challenge to build new public key cryptosystems, in particular the ones that could survive from future attacks utilizing quantum computers.

Multivariate public key cryptosystem (MPKC) is one of the promising alternatives. The security of MPKC relies on the difficulty of solving systems of nonlinear polynomial equations with many variables, and most of the time, the equations are chosen quadratic since solving quadratic systems is already an NP-complete problem in general. The public key of MPKC is mostly a set of quadratic polynomials, which are derived from composition of maps. Compared with RSA public key cryptosystem, the computation of MPKC can be very fast

O. Gervasi et al. (Eds.): ICCSA 2008, Part II, LNCS 5073, pp. 1–13, 2008.

because it is operated on a small finite field. So MPKC may be suitable for even low-end devices.

Let k be a finite field. The general form of MPKCs is of the form

$$Y = (y_1, \cdots, y_m) = F(x_1, \cdots, x_n) = L_2 \circ \phi \circ L_1(x_1, \cdots, x_n)$$

where F is a map from k^n to k^m, L_1 is a randomly chosen invertible affine transformation from k^n to k^n and L_2 is a randomly chosen invertible affine transformation from k^m to k^m. Define K is a degree l extension field of k. Let $\pi : K \to k^l$ be a natural k-linear isomorphism. Namely, we take a basis of K over k, $\theta_1, \cdots, \theta_l$, and define π by $\pi(a_1\theta_1 + \cdots + a_l\theta_l) = (a_1, \cdots, a_l)$ for any $a_1, \cdots, a_l \in k$. The map ϕ is called central map which is defined as $\phi = \pi \circ \tilde{\phi} \circ \pi^{-1}$. The map $\tilde{\phi}$ over K is the main difference between different MPKCs.

One of the basic ideas to design such a system was started by Matsumoto and Imai[2]. The map $\tilde{\phi}$ of MI scheme is of the form

$$\tilde{\phi}(x) = X^{1+q^\theta}$$

such that $gcd(q^\theta + 1, q^n - 1) = 1$. However, this scheme was proven insecure under an algebraic attack using the linearization equations by Patarin[3].

In 1996, Patarin extended the MI scheme and proposed the Hidden Field Equation cryptosystem(HFE)[4]. In this case, the difference from the original MI scheme is that $\tilde{\phi}$ is replaced by the map

$$\tilde{\phi} : X \mapsto \sum_{i,j}^{A} a_{ij} X^{q^i+q^j} + \sum_{i}^{B} b_i X^{q^i} + c$$

where coefficients are randomly chosen and the total degree of $\tilde{\phi}$ must be small, otherwise the decryption process will become too slow. However, a new algebraic attack using both the Minrank method and relinearization method by Kipnis and Shamir [5] shows that the number A can not be too small, but if A is big, the system is too slow due to the process of solving the polynomial equation in the decryption process. This is further confirmed by [6,7].

Another construction of central map is Stepwise Triangular System (STS). These schemes are TTM [8], MFE [11], TRMC [9], etc. But due to [10], [12], [13], [14], [15], they are all insecure. And we found that they have a common defect - there are some linearization equations (includes second order linearization equations) in these scheme.

Our motivation is to propose an efficient multivariate public key encryption scheme which can resist known attacks on MPKC, include linearization equation attack, Rank attack [10], XL & Gröbner basis algorithm attack, and differential attack.

For our construction, the idea is very simple. We construct $\tilde{\phi}$ with three high order equations over a "big" finite field K which is a degree l extension of k and k is $\mathbb{F}_2$. Due to the fact that squaring is a linear operation on a field of $\mathbb{F}_2$, the public key is a quadratic system over k.

The paper is organized as follows. We introduce the construction of the basic scheme in Section 2, provide its analysis in Section 3. In Section 4, we give an improved encryption scheme through internally perturbed plus. We present the security analysis in Section 5. In Section 6, we propose a practical implementation of the improved scheme. In the last Section, we conclude the paper.

2 Construction of the Basic Multivariate Public Key Encryption Scheme

Let $k = \mathbb{F}_2$ be a finite field with 2 elements. Let $K = \mathbb{F}_{2^l}$ be a degree l extension of k. Map $F : k^{2l} \to k^{3l}$ is a composition of three maps L_1, ϕ, L_2. Let

$$\begin{aligned}(x_1, \cdots, x_{2l}) &= L_1(m_1, \cdots, m_{2l}),\\ (y_1, \cdots, y_{3l}) &= \phi(x_1, \cdots, x_{2l}),\\ (z_1, \cdots, z_{3l}) &= L_2(y_1, \cdots, y_{3l})\end{aligned}$$

where L_1 and L_2 are invertible affine maps on k^{2l} and k^{3l}, respectively.

F can be expressed by

$$\begin{aligned}(z_1, \cdots, z_{3l}) &= F(m_1, \cdots, m_{2l})\\ &= L_2 \circ \phi \circ L_1(m_1, \cdots, m_{2l})\\ &= (F_1(m_1, \cdots, m_{2l}), \cdots, F_{3l}(m_1, \cdots, m_{2l}))\end{aligned}$$

Let $\pi : K \to k^l$ be a nature k-linear isomorphism. The central map $\phi = \pi \circ \tilde{\phi} \circ \pi^{-1}$, $\tilde{\phi}$ is defined over K as follows:

$$\begin{cases} Y_1 = (X_1^2 X_2)^2 + \alpha_1 (X_1^2 X_2);\\ Y_2 = (X_1 X_2)^2 + \alpha_2 (X_1 X_2);\\ Y_3 = X_1^2 X_2 + X_1 X_2.\end{cases} \tag{2.1}$$

where α_1, α_2 are randomly chosen in the finite field K, such that $\alpha_1, \alpha_2 \neq 0, 1$, $\alpha_1 \neq \alpha_2$, $\alpha_1 + \alpha_2 \neq 0$ to avoid decryption failure.

Utilizing the fact that squaring is a linear operation on the field $\mathbb{F}_2$, the three equations in (2.1) are a set of multivariate quadratic equations under the map π. So the public key in the basic scheme is a set of multivariate quadratic polynomials.

Public Key

The public key of the basic scheme includes the following:

1. The field k including its additive and multiplicative structure.
2. The $3l$ polynomials

$$F_1(m_1, \cdots, m_{2l}), \cdots, F_{3l}(m_1, \cdots, m_{2l}) \in k[m_1, \cdots, m_{2l}].$$

Private Key
The private key includes the two invertible affine transformations L_1, L_2 and their inverse, the random numbers α_1, α_2.

Encryption
Given a plaintext message $M' = (m'_1, \cdots, m'_{2l})$, the associate ciphertext is $Z' = (z'_1, \cdots, z'_{3l})$, that is

$$\begin{aligned} Z' = (z'_1, \cdots, z'_{3l}) &= F(m'_1, \cdots, m'_{2l}) \\ &= (F_1(m'_1, \cdots, m'_{2l}), \cdots, F_{3l}(m'_1, \cdots, m'_{2l})). \end{aligned}$$

Decryption
After received the ciphertext $(z'_1, \cdots, z'_{3l})$, we decrypt it by executing the following steps:
1. First compute

$$Y' = L_2^{-1}(Z') = (y'_1, \cdots, y'_{3l}).$$

And then derive Y'_1, Y'_2 and Y'_3 by map π^{-1} from $(y'_1, \cdots, y'_{3l})$.
2. Then we compute $\tilde{\phi}^{-1}$ to get the value of X'_1, X'_2 that is

$$(X'_1, X'_2) = \tilde{\phi}^{-1}(Y'_1, Y'_2, Y'_3).$$

3. At last, we derive $x'_1, \cdots, x'_{2l}$ from X'_1, X'_2 by map π. Denote $X' = (x'_1, \cdots, x'_{2l})$. Finally, we compute

$$M' = L_1^{-1}(X') = (m'_1, \cdots, m'_{2l}).$$

This is the correct plaintext corresponding to $Z' = (z'_1, \cdots, z'_{3l})$.

Remark: If one of X'_1, X'_2 is zero, then the decryption failure happens. For our scheme, we call a plaintext singular, which means at least one of its corresponding X'_1, X'_2 is zero; otherwise, the plaintext is called nonsingular. It is easily to show, the ratio of singular plaintext space is at most $2|K|^{-1}$. When we choose a proper l, the ratio is very small.

3 Analysis of the Basic Scheme

3.1 Computation Complexity

The encryption of our basic scheme like all the other MPKCs, is the evaluation of the public key polynomials. The encryption process is very efficient, which is the advantage of MPKC. So we mainly evaluate the decryption computation complexity of our basic scheme.

For decryption, we need first calculate L_2^{-1}, which can be neglected compared with the computation required for compute ϕ^{-1}. The same is true for compute L_1^{-1}.

The main step for computing ϕ^{-1} is to compute $\tilde{\phi}^{-1}$. For computing $\tilde{\phi}^{-1}$, we need to solve two univariate quadratic equations over K to get the values of

$X_1'^2X_2'$ and $X_1'X_2'$. At first, we transform the first two equations in (2.1) into the form $x^2 + x + \beta = 0$. This reduction could involve 2 multiplications. There is a simple method to solve this type equation, e.g.[21], pages 26 or Appendix A. According to Appendix A, when l is odd, the computation complexity of solve univariate quadratic equation can be neglected, because β^{2^i} can be done through shifting.

Then we derive two solutions from each equation. Hence we have 4 sets of $X_1'^2X_2', X_1'X_2'$. We randomly choose one and carry on decryption. We substitute the chosen set into the expression of Y_3, and compare the result to Y_3'. If equal, continue, else choose another set of $X_1'^2X_2', X_1'X_2'$. Repeat the proceed above till we derive the correct solution. The worst case, we need compute 4 times the value of Y_3. Fortunately, all the computations are additive, the complexity can be ignored.

Finally, we compute X_1', X_2' from $X_1'^2X_2', X_1'X_2'$. This process only requires 2 divisions.

Comparison with MI and HFE. Inverting the central map of MI cryptosystem costs one exponentiation with large exponent [22]. In the case of HFE, we need to execute d^2 exponentiations and d^3 multiplications [23] for d the total degree of the central map. For our basic scheme, when we choose l odd, inverting the central map only need 2 divisions and 2 multiplications. So our basic scheme is efficient in computation complexity.

3.2 Defect of Our Basic Scheme

Our basic scheme is insecure due to the linearization equations attack. We can construct some linearization equations as follows: From the first two equations of (2.1), we use the method in Appendix A to get the values of $X_1'^2X_2', X_1'X_2'$. We denote them as

$$\begin{cases} X_1'^2X_2' = f(Y_1'), \\ X_1'X_2' = f(Y_2') \end{cases} \tag{3.1}$$

Form (3.1), we can get

$$f(Y_1') = X_1' f(Y_2') \tag{3.2}$$

X_i' is linear in (3.2). So from (3.2), we can derive l linearizaion equations over finite field k. It is a big hidden trouble to our basic scheme. For practical use, we should use some modifications to improve our basic scheme.

4 An Improved Scheme

We use a method called *internally perturbed plus* to modify the basic scheme. The internal perturbed plus modifier is combined with two modifications *internal perturbation* and *plus*. The generic *plus* modifier adds $r \in \mathbf{Z}^+$ random equations n input variables each to the private key. Internal perturbation has been introduced for MI and HFE, as PMI and IPHFE respectively [24,25]. But PMI had been broken in [28] by differential attack. Ding and Gower [26] repaired the PMI scheme by *plus* modifier. They call this new scheme PMI+, which was proven to resist the differential attack. PMI+ is the first multivariate public key encryption scheme, which use the *internally perturbed plus* modifier. It still remain unbroken.

We now describe how to use the *internally perturbed plus* modifier to improve our basic scheme. Let r be a small positive integer and randomly choose r linear functions

$$z_1(x_1, \cdots, x_{2l}) = \sum_{j=1}^{2l} \alpha_{j1} x_j + \beta_1$$

$$\vdots$$

$$z_r(x_1, \cdots, x_{2l}) = \sum_{j=1}^{2l} \alpha_{jr} x_j + \beta_r$$

such that $z_i - \beta_i$ are linear dependant. Let $Z : k^{2l} \rightarrow k^r$ be the map defined by

$$Z(x_1, \cdots, x_{2l}) = (z_1(x_1, \cdots, x_{2l}), \cdots, z_r(x_1, \cdots, x_{2l})).$$

The map will be source of the internal perturbation of our basic scheme. Randomly choose $3l$ polynomials $\hat{f}_1, \cdots, \hat{f}_{3l} \in k[z_1, \cdots, z_r]$ of total degree 2. Let $\hat{F} : k^r \rightarrow k^{3l}$ defined by

$$\hat{F}(z_1, \cdots, z_r) = (\hat{f}_1(z_1, \cdots, z_r), \cdots, \hat{f}_{3l}(z_1, \cdots, z_r)),$$

and define the perturbation map $F^* : k^{2l} \rightarrow k^{3l}$ by

$$F^*(x_1, \cdots, x_{2l}) = \hat{F} \circ Z(x_1, \cdots, x_{2l}) = (f_1^*, \cdots, f_{3l}^*),$$

where $f_1^*, \cdots, f_{3l}^* \in k[x_1, \cdots, x_{2l}]$.

Let us replace $\bar{F}$ by

$$\bar{F} = L_2 \circ (\phi + F^*) \circ L_1 = (\bar{F}_1, \cdots, \bar{F}_{3l}).$$

Then, let $q_1, \cdots, q_a$ be a polynomials of total degree 2 in polynomial ring $k[x_1, \cdots, x_{2l}]$. we append this a polynomials to the map $\phi + F^* = (f_1, \cdots, f_{3l})$ to form

$$\bar{\bar{\phi}} = (f_1, \cdots, f_{3l}, q_1, \cdots, q_a).$$

We choose an invertible affine transformation $L_2 : k^{3l+a} \rightarrow k^{3l+a}$, and redefine $\bar{F}$ by

$$\bar{F} = L_2 \circ \bar{\bar{\phi}} \circ L_1 = (\bar{F}_1, \cdots, \bar{F}_{3l+a}).$$

Public Key

The public key of the improved scheme includes the following:

1. The field k including its additive and multiplicative structure.
2. The $3l + a$ total degree 2 polynomials $\bar{F}_1, \cdots, \bar{F}_{3l+a} \in k[x_1, \cdots, x_{2l}]$.

Private Key
The private key includes the following:
1. The map ϕ;
2. The set of linear functions $z_1, \cdots, z_r \in k[x_1, \cdots, x_{2l}]$;
3. The set of points in P defined by $P = \left\{(\mu, \lambda) | \hat{F}(\mu) = \lambda\right\}$, or equivalently, the set of the polynomials $\hat{f}_1, \cdots, \hat{f}_{3l} \in k[z_1, \cdots, z_r]$;
4. The two invertible affine transformations L_1, L_2 and their inverse.

Encryption
Given a plaintext message $M' = (m'_1, \cdots, m'_{2l})$, the associate ciphertext is $Z' = (z'_1, \cdots, z'_{3l+a})$, that is

$$\begin{aligned} Z' = (z'_1, \cdots, z'_{3l+a}) &= \bar{F}(m_1, \cdots, m_{2l}) \\ &= (\bar{F}_1(m_1, \cdots, m_{2l}), \cdots, \bar{F}_{3l+a}(m_1, \cdots, m_{2l})). \end{aligned}$$

Decryption
After derived the ciphertext $(z'_1, \cdots, z'_{3l+a})$, we decrypt it by executing the following steps:
1. First compute

$$Y' = L_2^{-1}(Z') = (y'_1, \cdots, y'_{3l+a}).$$

We set aside the last a coordinates that come from the polynomials $q_1, \cdots, q_a$.
2. For each $(\mu, \lambda) \in P$, compute

$$(x'_{\lambda 1}, \cdots, x'_{\lambda 2l}) = \phi^{-1}((y'_1, \cdots, y'_{3l}) + \lambda),$$

and check if

$$Z(x'_{\lambda 1}, \cdots, x'_{\lambda 2l}) = \mu.$$

If not, discard this $(x'_{\lambda 1}, \cdots, x'_{\lambda 2l})$; otherwise goto the next step.
3. Finally, we compute

$$M' = L_1^{-1}(x'_{\lambda 1}, \cdots, x'_{\lambda 2l}) = (m'_1, \cdots, m'_{2l}).$$

If there is only one $(m'_{\lambda 1}, \cdots, m'_{\lambda 2l})$, then this must be plaintext. However, it is very possible that we will have more than one $(m'_{\lambda 1}, \cdots, m'_{\lambda 2l})$. In this case, we can use the polynomials $q_1, \cdots, q_a$ to help decide which is the true plaintext.

5 Analysis of the Improved Scheme

5.1 Security Analysis

Currently known attacks on multivariate public key systems can be roughly classified into four kinds: linearization equation attack, Rank attack, XL or *Gröbner* Bases attack, and differential attack. In this section, we will discuss the security of our improved scheme against these attacks.

Linearization Equation Attack: This attack is first proposed by Patarin in 1995 to defeat the MI scheme [3]. The linearization equation is also called Patarin relation. Say a encryption scheme be satisfied the linearization equation means that plaintext variables x_i and public key polynomials or ciphertext variables y_j are satisfied the identical equation of the form

$$\sum_{i=1,j=1}^{n,m} a_{ij}x_iy_j + \sum_{i=1}^{n} b_ix_i + \sum_{j=1}^{m} c_jy_j + d = 0 \tag{5.1}$$

We can obtain linearization equations to attack our our basic scheme. In our improved scheme, we add the internal perturbation to the central map. So the formerly linearization equations need not be true. However, we are still not guarantee that linearization equation do not exist. Therefore, in order to test the improved scheme for the existence of linearization equations, we search directly for all the equations of the form of Equation (5.1), in the variables a_{ij},b_i,c_i,d, which hold for all plaintext/ciphertext pairs, for values of l and r. We choose $a = 0$, since the plus perturbation do not influence the existence of linearization equations. We selected 100 randomly chosen instances for each choices of parameters (l, r) with $44 \leq l \leq 48$ and $2 \leq r \leq 8$, but there does not exist linearization equations.

Ding et al. extended the idea of linearization equation attack to the high order linearization equation (HOLE) attack [15]. In high order linearization equation, only the plaintext variables x_i need to be linear, ciphertext variables y_j can be high order. We did not do experiment to test our improved scheme for the existence of high order linearization equation, because the computational complexity of these attack is more than 2^{80} when $l \geq 47$.

Rank Attack: This attack is firstly introduced to cryptography by Kipnis and Shamir [5] against HFE. Goubin and Courtois [10] used this attack for Triangular-Plus-Minus system. Yang and Chen generalized the idea of rank attack [16], which can be divided into low rank attack and high rank (dual rank) attack.

According to [16], the complexity of low rank attack is about $q^{vt}(m^2(nv/2 - m/6) + mn^2v)$. $q = |k|$, t is the smallest rank in linear combination of central equations, m is the number of equations in central map, n is the number of variables, and $v = \lceil \frac{m}{n} \rceil$. In our improved scheme, the minimum rank t is more than $2l$. Hence, when $l \geq 40$, the computational complexity of this type attack is more than 2^{80}.

Coppersmith et al. [17] first used the high rank attack against a multivariate scheme proposed by Shamir [29]. According to [16], the complexity of high rank is about $q^u(un^2 + n^3/6)$, where u is the minimal number of appearance in central map for any x_i. In our improved scheme, $u \geq 3l$. Therefore when $l \geq 30$, the computational complexity of this type attack is more than 2^{90}.

In a word, our improved scheme can resist the rank attack as long as $l \geq 40$.

Differential Attack: The basic idea of the differential attack is to use the distribution of the rank of the kernel of the differential (or difference) to detect which plaintext produce noise and which do not. If we only use the internally perturbed modifier to improve our basic scheme, this can be achieved, and then the attacker can "denoise" the system and use the linearization attack to break the underly basic scheme. Fouque et al. [28] use this method to break the PMI scheme. However, it is very easy to repair the PMI scheme using the plus method. The resulting scheme, PMI+ proposed by Ding and Gower [26], resists the differential attack by adding a few additional polynomials.

For our improved scheme, our experiment show that if $l = 47, r = 6$, we must take $a \geq 10$, so that it can resist the differential attack. Furthermore, to differential attack, the larger a is, the more secure the MPKC should be.

XL & Gröbner Basis Algorithm Attack: XL algorithm is a method to solve the system of multivariate polynomial equations. It proposed by Courtois et al. [19] on Eurocrypt 2000. However, according to Gwénolé Ars et al. [18], the XL algorithm is also a Gröbner basis algorithm which can be represented as a redundant variant of Gröbner basis algorithm $\mathbf{F_4}$ [27]. According to [19], when $m = n + C$, the complexity of algorithm is about

$$(\frac{n^{\sqrt{n}}}{\sqrt{n}!})^{\omega}$$

where $\omega = 3$ in the usual Gaussian reduction algorithm, and $\omega = 2.3766$ in improved algorithms. For our improved scheme, when $l \geq 47$, the computational complexity is more than 2^{115}.

To differential attack, it would seem that the larger a is, the more secure the MPKC should be. However, we must be careful to add too many extra polynomials since otherwise we may create a weakness to Gröbner bases attacks. Through computer experiments using the implementation of F_4 [27]algorithm in Magma, it is shown that if we choose $l \geq 40, r = 6$, then we can expect our improved scheme have a security of 2^{80} against a Gröbner attack using F_4. If we add a few *plus* polynomials to resist the differential attack, secure implementations against Gröbner attack using F_4 is $l \geq 40, r = 6$ and $a = 11$.

5.2 A Practical Implementation

We note that when $r = 0, a = 0$, the improved scheme reduced to the basic scheme. On the other hand, the system with $r = 3l$ would simply be system of $3l$ randomly chosen quadratic polynomials. Since the decryption is slower by a multiple of 2^r, we must not choose r too large. Although adding the more *plus* polynomials will achieve the more security against differential attack, we can not choose a too large, since these *plus* polynomials will increase successful probability Gröbner bases attack. $l = 47, r = 6, a = 11$ is a secure implementation. We will discuss the public/private key size and the computational complexity in this case.

Public Key Size

The public key contains $3l + a = 152$ quadratic polynomials. Each polynomials has $94 \times 93/2 = 4371$ quadratic terms, 94 linear terms and one constant terms. So the public key size is roughly 84kB.

Private Key Size

The private key includes the two invertible affine transformation L_1 and L_2 and their inverse, the perturbation vector $z_1, \cdots, z_r$, and the perturbation set P. The two affine maps and their inverse together require $94 \times 95 \times 2 + 152 \times 153 \times 2 = 64,372$ bits of storage, the six linear polynomial components of z require $95 \times 6 = 570$ bits of storage, and P requires $32 \times (6 + 2 \times 94) = 6208$ bits of storage. Therefore the private key requires roughly 8.9kB.

Encryption/Decryption Computational Complexity

For encryption, we need to compute the value of 152 quadratic polynomials for a given plaintext $M' = (m'_1, \cdots, m'_{94})$. Calculating the value of each polynomial need 8836 multiplications when we rewrite each quadratic polynomials as $\Sigma x_i(b_i + \Sigma a_{ij}x_j) + c$.

We select $r = 6$, so the decryption computational complexity of our improved scheme will be slower by multiple 2^6 than that of the basic scheme.

Table 1. comparison our original scheme with IPHFE and PMI+ with security level 2^{80}

scheme	Encryption costs	Decryption computational complexity	Parameters	Public key size (kB)	Private key size (kB)
Our scheme	8836 multiplications over $k = \mathbb{F}_2$	$2^6\mathcal{O}(2l^2 + 2l^3)$	$q = 2, l = 47, r = 6, a = 11$	84	8.9
IPHFE	10609 multiplications over $k = \mathbb{F}_2$	$2^6\mathcal{O}(l^3d^2 + 2l^2d^3)$	$q = 2, l = 103, r = 6, d = 129$	77	6.6
PMI^+	about 67081 multiplications over $k = \mathbb{F}_2$	$2^6\mathcal{O}((7l)^3)$	$q = 128, l = 37, r = 6, a = 11$	30	7.2

Comparison our scheme with PMI^+ and IPHFE.

Now, there are only two secure encryption scheme in MPKC, called IPHFE [25]and PMI+[26]. We make a comparison our original scheme with IPHFE and PMI^+. When we counting our costs in operations in the ground field k, we can have $\mathcal{O}(l^2)$ for multipling/squaring, $\mathcal{O}(l^3)$ for exponentiation/division. Here we have l the extension degree of the corresponding field $K = \mathbb{F}_{q^l}$ over the ground field $k = \mathbb{F}_2$. d is the total degree IPHFE. We keep this in mind when comparing our original scheme with IPHFE and PMI^+ (**Table 1**).

From the **Table 1**, we can see that the advantage of our scheme is the encryption/decryption computational complexity. Althrough, the key size of our scheme is increased.

6 Conclusion

In this paper, we propose a construction of multivariate public key encryption scheme through internally perturbed plus, which central map is different to the previous MPKCs like MI, HFE, TTM etc. Through comparison, the encryption/decryption computational complexity of our scheme is lower than the existing two schemes - IPHFE [25]and PMI+[26]. Through security analysis, we believe that if we choose a proper l, r, a, our scheme can resist the currently known attacks. We believe that our results, through experimental in nature, are promising and warrant further investigation.

Acknowledgement. This study is supported by SONY (China) Research Laboratory, National Basic Research Program of China (No 2007CB310704), National Natural Science Foundation of China (No 90718001) and the 111 Project (No. B08004).

References

1. Shor, P.: Polynomial-time algorithms for prime factorization and discrete logarithms on a quantum computer. SIAM Journal on Computing 26(5), 1484–1509 (1997)
2. Matsumoto, T., Imai, H.: Public quadratic polynomial-tuples for efficient signature verification and message encryption. In: Günther, C.G. (ed.) EUROCRYPT 1988. LNCS, vol. 330, pp. 419–453. Springer, Heidelberg (1988)
3. Patarin, J.: Cryptanalysis of the Matsumoto and Imai public key scheme of Eurocrypt 1988. In: Coppersmith, D. (ed.) CRYPTO 1995. LNCS, vol. 963, pp. 248–261. Springer, Heidelberg (1995)
4. Patarin, J.: Hidden field equations (HFE) and isomorphism of polynomials (IP): Two new families of asymmetric algoritms. In: Maurer, U.M. (ed.) EUROCRYPT 1996. LNCS, vol. 1070, pp. 33–48. Springer, Heidelberg (1996)
5. Kipnis, A., Shamir, A.: Cryptanalysis of the HFE public key cryptosystem by relinearization. In: Wiener, M.J. (ed.) CRYPTO 1999. LNCS, vol. 1666, pp. 19–30. Springer, Heidelberg (1999)
6. Courtois, N.: The security of hidden field equations (HFE) process in cryptology. In: Naccache, D. (ed.) CT-RSA 2001. LNCS, vol. 2020, pp. 266–281. Springer, Heidelberg (2001)
7. Faugere, J., Joux, A.: Algebraic cryptoanalysis Hidden Field Equation (HFE) cryptosystems using Grober bases. In: Boneh, D. (ed.) CRYPTO 2003. LNCS, vol. 2729, pp. 44–60. Springer, Heidelberg (2003)
8. Moh, T.: A fast public key system with signature and master key functions. Lecure notes at EE department of standford university (1999)
9. Wang, L., Chang, F.: Tractable rational map cryptosystems (revised on December 28, 2006), `http://eprint.iacr.org/2004/046`
10. Goubin, L., Courtois, N.: Cryptotanalysis of the TTM cryptosystem. In: Okamoto, T. (ed.) ASIACRYPT 2000. LNCS, vol. 1976, pp. 44–57. Springer, Heidelberg (2000)
11. Wang, L.-C., Yang, B.-Y., Hu, Y.-H., Lai, F.: A Medium-Field Multivariate Public Key Encryption Scheme. In: Pointcheval, D. (ed.) CT-RSA 2006. LNCS, vol. 3860, pp. 132–149. Springer, Heidelberg (2006)

12. Ding, J., Schmidt, D.: The new TTM implementation is not secure. In: Proceeding of Inernational Workshop on Coding, Cryptography and Combinatorics (CCC 2003), pp. 106–121 (2003)
13. Nie, X., Hu, L., Li, J., Updegrove, C., Ding, J.: Breaking a new instance of TTM cryptosystem. In: Zhou, J., Yung, M., Bao, F. (eds.) ACNS 2006. LNCS, vol. 3989, pp. 210–225. Springer, Heidelberg (2006)
14. Joux, A., Kunz-Jaquces, S., Muller, F., Ricordel, P.-M.: Cryptanalysis of the tractable rational map cryptosystem. In: Vaudenay, S. (ed.) PKC 2005. LNCS, vol. 3386, pp. 258–274. Springer, Heidelberg (2005)
15. Ding, J., Hu, L., Nie, X., Li, J., Wagner, J.: High order linearization Equation (HOLE) attack on multivariate public key cryptosystems. In: Okamoto, T., Wang, X. (eds.) PKC 2007. LNCS, vol. 4450, pp. 233–248. Springer, Heidelberg (2007)
16. Yang, B., Chen, J.: Building secure tame-like multivariate public key cryptosystems-the new TTS. In: Boyd, C., González Nieto, J.M. (eds.) ACISP 2005. LNCS, vol. 3574, pp. 518–531. Springer, Heidelberg (2005)
17. Coppersmith, D., Stern, J., Vaudeny, S.: The security of the birational permutation signature scheme. J. Cryptology 10(3), 207–221 (1997)
18. Ars, G., Faugere, J., Imai, H., Kawazoe, M., Sugita, M.: Comparision between XL and Grobner bases algorithms. In: Lee, P.J. (ed.) ASIACRYPT 2004. LNCS, vol. 3329, pp. 338–353. Springer, Heidelberg (2004)
19. Courtois, N., Klimov, A., Patarin, J., Shamir, A.: Efficient algorithms for solving overdefined systems of multivariate polynomial equations. In: Preneel, B. (ed.) EUROCRYPT 2000. LNCS, vol. 1807, pp. 392–407. Springer, Heidelberg (2000)
20. Yang, B., Chen, J., Courtois, N.: On asymptotic security estimats in XL and Grobner bases related algebraic cryptanalysis. In: López, J., Qing, S., Okamoto, E. (eds.) ICICS 2004. LNCS, vol. 3269, pp. 401–413. Springer, Heidelberg (2004)
21. Blake, I., Seroussi, G., Smart, N.: Elliptic curves in cryptography. Cambridge Unversity Press, Cambridge (1999)
22. Courtois, N., Goubin, L., Patrin, J.: Sflash: Primitive specification (second revised version), Submissions, Sflash, 11 pages (2002), https://www.cosic.east.kuleuven.be/nessie
23. Courtois, N., Goubin, L., Patrin, J.: Quartz: Primitive specification (second revised version), Submissions, Quartz, 18 pages (2001), https://www.cosic.east.kuleuven.be/nessie
24. Ding, J.: A new variant of the Matsumoto-Imai through perturbation. In: Bao, F., Deng, R., Zhou, J. (eds.) PKC 2004. LNCS, vol. 2947, pp. 305–318. Springer, Heidelberg (2004)
25. Ding, J., Schmidt, D.: Cryptanalysis of HEFV and the internal perturbation of HFE. In: Vaudenay, S. (ed.) PKC 2005. LNCS, vol. 3386, pp. 288–301. Springer, Heidelberg (2005)
26. Ding, J., Gower, J.: Inoculating multivariate schemes against differential attacks. In: Yung, M., Dodis, Y., Kiayias, A., Malkin, T.G. (eds.) PKC 2006. LNCS, vol. 3958, pp. 290–301. Springer, Heidelberg (2006)
27. Faugere, J.: A new efficient algorithm for computing Grobner bases(F4). Journal of Pure and Applied Algebra 139, 61–88 (1999)
28. Fouque, P.-A., Granboulan, L., Stern, J.: Differential cryptanalysis for multivariate schemes. In: Cramer, R.J.F. (ed.) EUROCRYPT 2005. LNCS, vol. 3494, pp. 341–353. Springer, Heidelberg (2005)
29. Shamir, A.: Efficient signature schemes based on birational permutations. In: Krawczyk, H. (ed.) CRYPTO 1998. LNCS, vol. 1462, pp. 257–266. Springer, Heidelberg (1998)

Appendix A: Solving Quadratic Equations in $\mathbb{F}_{2^n}$

This part is quoted from [21].

An equation form $x^2 + \beta = 0$ is trivially solved in $\mathbb{F}_{2^n}$ by writing its root explicitly as $x_0 = \beta^{2^{n-1}}$. Other non-trivial equations can always be brought into the canonical form

$$x^2 + x + \beta = 0 \tag{1}$$

This equation has solutions in $\mathbb{F}_{2^n}$ if and only if $Tr_{q|2}(\beta) = 0$. If x_0 is such a solution, then so is $x_0 + 1$.

The procedure for finding a solution varies according to the parity of n. If n is odd, an explicit solution is given by $x_0 = \tau(\beta)$, where τ, the *half-trace* function, is defined by

$$\tau(\beta) = \sum_{j=0}^{(n-1)/2} \beta^{2^{2^j}} \tag{2}$$

It can verified direct inspection that $\tau(\beta)^2 + \tau(\beta) = \beta + Tr_{q|2}(\beta)$, which verifies the solution when $Tr_{q|2}(\beta) = 0$.

When n is even, the half-trace will not do, and the solution is found using the following procedure. Let $\delta \in \mathbb{F}_{2^n}$ be such that $Tr_{q|2}(\delta) = 1$. Such a element can be obtained either by randomly drawing field elements until one of the right trace is found (with a probability of one half in each try), or by deterministically computing the traces of the basis element $\alpha_0, \alpha_1, \cdots, \alpha_{n-1}$. At least one basis element must have trace one. With δ at hand, a solution x_0 to equation (2) is given by

$$x_0 = \sum_{i=0}^{n-2} (\sum_{j=i+1}^{n-1} \delta^{2^j}) \beta^{2^i}. \tag{3}$$

To verify that x_0 is indeed a solution, we compute

$$\begin{aligned} x_0^2 + x_0 &= \sum_{i=1}^{n-1} (\sum_{j=i+1}^{n} \delta^{2^j}) \beta^{2^i} + \sum_{i=0}^{n-2} (\sum_{j=i+1}^{n-1} \delta^{2^j}) \beta^{2^i} \\ &= \delta(\beta^{2^{n-1}} + \beta^{2^{n-2}} + \cdots + b^2) + (\delta^{2^{n-1}} + \delta^{2^{n-2}} + \cdots + \delta^2)\beta \\ &= \delta\, Tr_{q|2}(\beta) + \beta, \end{aligned}$$

where the last equality follows from $\delta^{2^{n-1}} + \delta^{2^{n-2}} + \cdots + \delta^2 = Tr_{q|2}(\delta) + \delta = 1 + \delta$. Thus, $x_0^2 + x_0 = \beta$ if and only if $Tr_{q|2}(\beta) = 0$ as desired.

Using the Self-Organizing Map (SOM) Algorithm, as a Prototype E-Content Retrieval Tool

Athanasios S. Drigas and John Vrettaros

Department of Applied Technologies,NCSR "DEMOKRITOS"
Ag. Paraskeui, Greece
{dr,jvr}.imm.demokritos.gr

Abstract. SOM O.D.I.S.S.E.A.S is an intelligent searching tool using the Self-Organizing Map (SOM) algorithm, as a prototype e-content retrieval tool. The proposed searching tool has the ability to adjust and scale into any e-learning system that requires concept-based queries. In the proposed methodology, maps are used for the automatic replacement of the unstructured, the half structured and the multidimensional data of text, in a way that similar entries in the map are represented near between them. The performance and the functionality of the document organization, and the retrieval tool employing the SOM architecture, are also presented. Furthermore, experiments were performed to test the time performance of a learning algorithm used for the direct creation of teams of terms and texts enabling efficient searching and retrieval of the documents.

Keywords: e-content retrieval, e-learning, intelligent searching tool.

1 Introduction

With the advent of new technologies and the World Wide Web (WWW), enormous quantities of informative material are nowadays available on-line. Computers are increasingly changing from computing systems to portals, which permit to access big volumes of information. In parallel, due to the interest in reducing costs of education and stimulating people to never stop learning, e-learning applications have recently developed in educational, industrial and research institutions. However, e-learning platforms present some difficulties concerning the instructor and the student interaction. Effective mining and retrieval of the e-content is the major bottleneck of e-learning applications. The lack of metadata and classification of the e-content, force the development of powerful search engines.

The basic approaches concerning information retrieval and data mining in textual documents collections are: (1) searching using keywords or key documents, (2) exploration of the collection referring to "ome" organization or categorization of the documents, and (3) filtering of interesting documents from the incoming document stream. Keyword search systems can be automated rather easily whereas the organization of document collections has traditionally been carried out by hand. In a manual organization carried out for example, in libraries, classification schemes are defined and each document is positioned into one or several classes by a librarian.

O. Gervasi et al. (Eds.): ICCSA 2008, Part II, LNCS 5073, pp. 14–23, 2008.

Similarly, in the current hypertext systems the links between related documents are most often added by hand. One of the traditional methods of searching for texts that match a query is to index all the words (terms) that have appeared in the document collection. The query itself, typically a list of appropriate keywords, is compared with the term list of each document to find documents that match the query. In the existed applications, the educational content is multilingual and heterogeneous. Therefore, simple keyword queries are not capable for efficient mining of the available information. In order to bypass the aforementioned bottlenecks, we propose the use of Artificial Neural Networks (ANN). Due to their wide range of applications, ANNs have been an active research for the past decades [2]. A large variety of learning algorithms have been evolved and being employed in ANNS. A further categorization divides the network architectures into three distinct categories: feedforward, feed-backward and competitive [2]. The self-organizing maps or Kohonen's feature maps are feedforward, competitive ANN that employ a layer of input neurons and a single computational layer [7]. The neurons on the computational layer are fully connected to the input layer and are arranged on an N-dimensional lattice. In this paper, we shall use the SOM algorithm to cluster contextually similar documents into classes.

The ability of the SOM algorithm to produce spatially organized representations of the input space can be utilized in document organization, where organization refers to the representation and storage of the available data. An architecture based on the SOM algorithm that is capable of clustering documents according to their semantic similarities is the so-called WEBSOM architecture [4,5,6,7]. The WEBSOM consists of two distinct layers where the SOM algorithm is applied. The first layer is used to cluster the words found in the available training documents into semantically related collections. The second layer, which is activated after the completion of the first layer, clusters the available documents into classes that high probability contains relevant documents with respect to their semantic content. Due to that, the WEBSOM architecture regarded as a prominent candidate for document organization and retrieval. The structure of the paper is as follows. In the first section is presented the basic structure of the SOM architecture. The system architecture is presented in the second section. The software component constructed to illustrate the applicability of the proposed architecture is shown in the third section. Finally the performance of the training algorithm is illustrated in the final section.

2 Description of the SOM Algorithm

The basic Self-Organizing Map (SOM) can be visualized as a sheet-like-neural-network array (figure), the cells (or nodes) of which becomes specifically tuned to various input signal patterns or classes of patterns or classes of patterns in an orderly fashion. The learning process is competitive and supervised, meaning no teacher is needed to define the correct output (or actually the cell into which the input is mapped) for an input. In the basic version, only one map node (winner) at a time is activated corresponding to each input. The locations of the responses in the array tend to become ordered in the learning process as if some meaningful nonlinear coordinate system for the different input features were being created over the network.

Due to its competitive nature, the SOM algorithm identifies the best-matching, winning reference vector w(k) (or winner for short), to a specific feature vector xj with respect to a certain distance metric. The index s of the winning reference vector is given by:

$$s = \arg \min_{l=1}^{L} \| x_j - w_l(k) \|, \quad (1)$$

where ||.|| denotes the Euclidean distance.

A neighborhood updating, especially in the early iterations, is performed in order to achieve a global ordering of the input space onto the lattice, which is crucial for the good resolution of the map [4]. The term basic SOM will henceforth denote the on-line algorithm proposed by Kohonen [4] without any modification of speed-up techniques.

2.1 Salton's Vector Space Model

The vector space model [8] has been widely used in the traditional IR field. Most search engines also use similarity measures based on this model to rank web documents. The model creates a space in which both documents and queries are represented by vectors. For a fixed collection of documents a dimensional vector is generated for each document and each query from sets of terms with associated weights, where m is the number of unique terms in the document collection. Then a vector similarity function, such as the inner product, can be used to compute the similarity between a document and a query. (Fig. 1)

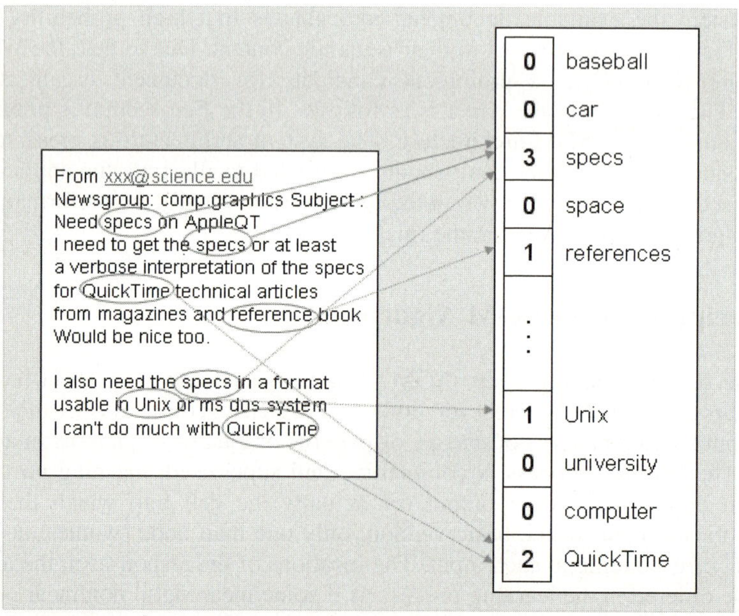

Fig. 1. The VSM. The text on the left is represented as the vector on the right. Each line of the vector is a different word of the text. Each line's record is the frequency of the word in the text.

The main problem of the vector space model is the large vocabulary in any sizable collection of free text documents, which results in a vast dimensionality of the document vectors.

In the following sections are presented, some methods for reducing the dimensionality. These are applicable to all cases where the documents are encoded using the vector space model, i.e. as the document-by-word matrix.

2.2 Latent Semantic Indexing

Latent Semantic Indexing (LSI) is one alternative to the original vector space model. LSI tries to make account to the co-occurrence of terms in documents when encoding the documents. One way of interpreting the LSI is that it represents the jth document by the vector

$$n'_j = \sum_k n_{jk} x'_k, \tag{2}$$

where n_{jk} denotes again the number of times the word k occurs in the jth document. The x'_k is the code that the LSI forms of the kth word by investigating the co-occurrence of the words within the documents. The term-by-document matrix, a matrix where each column is the word histogram corresponding to one document, is decomposed into a set of factors (eigenvectors) using the singular-value decomposition (SVD). The factors that have the least influence on the matrix are then discarded. The motivation behind omitting the smallest factors is that they most likely consist of noise. The vector x'_i can then be formed by using only the remaining factors, whereby the dimensionality is reduced.

2.3 Random Projection

A low-dimensionality representation for documents can be obtained by a random projection of the high-dimensional representation vector into a much lower-dimensional space [3]. The benefit compared with alternative methods such as the LSI is the extremely fast computation. The accuracy of the results is still comparable.

2.4 Word Clustering

Clustering methods can be used for reducing the number of data by grouping similar items together [3]. If similar words can be clustered together, documents can be represented as histograms of words clusters rather than of individual words. Various early approaches for categorizing words have been described [3]. In languages with rigid word order, such as English, the distribution of words in the immediate context of a word contains considerable amounts of information regarding the syntactic and semantic properties of the word. The SOM has been used to cluster words based on the distributions of words in their immediate contexts [3].

2.5 SOM Computation Complexity

The computational complexity of the SOM algorithm is only linear in the number of data samples. However, the complexity depends quadratically on the number of the map units. For document maps intended for browsing the document collection, the resolution (number of map units per number of documents) should be good, since browsing is easier if there are representations of only a few, about ten documents in a map unit on the average. Hence, for such resolution, the number of map units has to be proportional to the number of documents. For very large document collections the resulting computational complexity might become problematic.

3 SOM O.D.I.S.S.E.A.S Description

The authors determine the characteristics of the intelligent search based on the nature of the educational material (e-content) that interest the users (word docs, html pages, plain text). The system has the capability of document retrieval from databases aiming at the preparation and presentation of an e-learning course. The system is capable of retrieving certain educational texts by the users according to their "physical" questions. In the following paragraph we summarize the basic elements of the system. The descriptors are exported from the text of the multimedia material and transforming of these descriptors into compound search descriptors, in suitable vectors of characteristics. The authors concretize of the non supervised learning algorithm SOM for the successful information retrieval. The concrete methodology was peered against the method of simple equation of keywords as well as the one that makes use of the simple metric resemblance in the representation space of the texts (e.g. calculation of cosine between vectors and retrieval of those nearest in the vector that represents a question) because it provides better retrieval performance and releases the user from the need of creation of complicated educational components. The big advantage of the non supervised search models is that content managers are not obliged to create huge learning material (examples of questions with the connected answers). Taking into consideration that the user cannot as an expert in neural networks training, the software search module is supposed to supply him/her with the capability to search the database intelligently, via the combination of the automatic exported characteristics and his/her own keywords.

The proposed methodology for the creation of the intelligent search system is based on the SOM algorithm, described in the previous section. In the concrete application, the SOM maps are used fro the automatic placements of the unstructured or half structured and multidimensional data of text in such a way that similar entries in the map are represented near between them. Via a learning process, the t performance is illustrated in the following section, that final map allows the direct creation of teams of terms and teams of texts so that the distances between the different data can be directly used during the search and retrieval duration. An example of a previous successful application of the SOM networks in information retrieval is the web application WebSOM [2]. This application is based on the export of descriptors of texts from different SOMs, which replace the department of pretreatment, and representation of texts, in combination with a self-organized of the retrieved texts. It also provides the capability of a two dimension depiction of texts relative between them, so that the user has in his disposal a visual representation of the material in relevant categories. As recent researchers have shown that the functionalism of such visualization considering the help that it provides in the final user is arguable, in the concrete work, we do not use the depiction of map. On the contrary, we provide the capability of information retrieval from the database according to the content of texts and thus present the results in form of a list in a declining line of resemblance so that we decrease the difficulties faced by users.

3.1 Content Retrieval Using the SOM Algorithm

The SOM algorithm has been used to retrieve educational material from a database. The methodology of this operation follows. We export the descriptors from the text of the educational material. Responsible for this task are the designers of the database system. An automate method for transformation of the descriptors of the material as well as the compound search descriptors, in suitable vectors of characteristics is used. To perform this automatic operation we have used the VSM algorithm described previously. The VSM provides efficiency of the search results. The next step is to learn the SOM with the vectors of text characteristics. The result is the clustering of the used terms and texts in teams of relevant content. Following this step we must search for relevant texts with the use of questions. After the creation of the vector of characteristics of the question it is supplied in the entry of trained SOM network. The result is the calculation of the nearest Euclidean distance of teams of texts towards the question based on the activated neurons of the map. This team will contain texts with terms of approximate weight and consequently will present the highest affinity of content with the question. Search of relevant texts in neighbor teams is the following step. The attribute of the self-organization allows the user to search different relevant texts found in teams of neighbor neurons of map. A question that is placed to the system, formulated as word or as a combination of words, activates the processes of retrieval of texts relative with the question. The system seeks in the map of teams of terms thus neurons that correspond in the terms or in the combination of terms that exists in the question. Those texts represent in concrete neurons of the map of the teams of texts activated by the terms, are selected and thus are presented in the user. Additionally, it is possible to present texts by a concrete method of using metric resemblance between texts.

3.2 Training

After the creation of the vector maps, we perform training. The feature vectors are presented iteratively an adequate number of times to the neural network which perform clustering in an effort to build word classes containing semantically related words. This is based on empirical and theoretical observations that semantically related words have more or less the same preceding and succeeding words. The above process yields the so-called word categories map (WCM) [1]. After the computation of the document vectors the SOM method is used to cluster them. It is expected that the constructed documents classes contained contextually similar documents.

4 Experimental Results

The performance of the SOM algorithms in the proposed case study is illustrated in this section. The performance is measured using the Mean Square Error (MSE) between the reference vectors and the document vectors assigned to each neuron in the training phase. Furthermore the recall-precision performance is measured using query documents from a test set during the recall phase is used as an indirect measure of the quality of the document organization provided by the SOM algorithm.

To measure the effectiveness of a retrieval system, two widely used ratios are employed: the precision and the recall. Precision is defined as the proportion of retrieved documents that are relevant:

$$P = \frac{r}{n_2} \tag{3}$$

Recall is the proportion of relevant documents that are retrieved:

$$R = \frac{r}{n} \tag{4}$$

As the volume of the retrieved documents increases the above ratios are expected to change. The sequence of recall-precision pairs obtained yields the so-called recall-precision curve. Each query-document in the test set produces one recall-precision curve. An average over all the curves corresponding to query documents of the same topic obtained from the test set produces the average recall-precision curve. If the recall level does not equal to one, we proceed with the second best winner neuron and repeat the same procedure and so on. The comparison of the effectiveness between the retrieved documents utilizes that above-mentioned curve.

4.1 Software Component of Intelligent Search

The core of the system of the SOM algorithm and the VSM has been developed in ANSI C to ensure portability and compatibility in platforms of different type (Windows, UNIX, Linux, etc). the user interface can be designed overall round the basic and autonomous departments of the system in order to activate, through a GUI, the system operations and presentation of search results (standalone applications) via Visual Basic/C++, Delphi or Web based enabled as CGI scripting, ASP, JSP, PHP, etc. A basic issue of the final interface is the confidence estimation. The user tends to need a score of the results of its query in order to ameliorate his queering style. The following Fig. 4 presents an embodiment example of the search system into the e-learning platform O.D.I.S.S.E.A.S. [1] (Open Distance Interactive SyStem for Educational Applications) with the JSP (Java Server Pages) technology. (Table 1)

In order to prove the functionality of the proposed system, we analysed a collection of text and multimedia documents. Although the SOM algorithm can be applied only to text documents, in our collection we include multimedia documents. Multimedia

Table 1. Results obtained from the analysis of the multimedia digital library

Collection	Size (Mb)	Number of Documents	Average Document size(Mb)	Number of Lexical Form	Number of Terms
Textual	69	105	0,283	99,659	344,511
Audio (Audio Description)	90	35	2,1	35,024	140,832
Image (Image Description)	32	44	0,344	18,067	87,438
Video (Video Description)	930	16	65	3,967	23,898

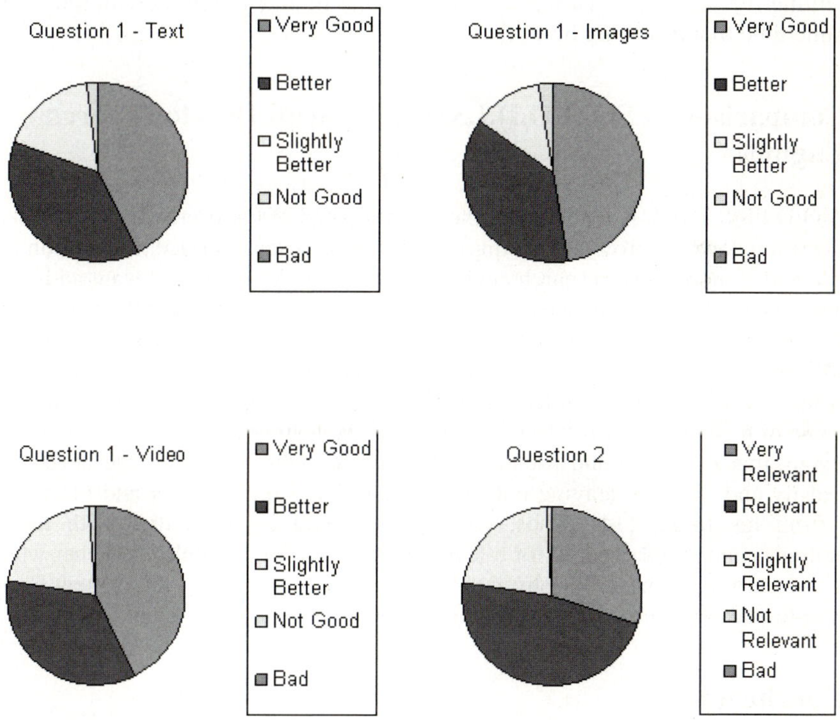

Fig. 2. The results of the two questions

documents are analysed through the annotation that is created for each picture, video and sound. In this all the collection is available in raw text format. The test collection consists of 200 documents, 95 are multimedia and 105 are textual. The average size of the documents is 2,3 Mb and the biggest document size is 180 Mb.

4.2 Discussion

The proposed system has been evaluated by means of its usability by the e-learning users. The e-learning users are the teachers who upload new material in the database and the students who download the teaching material. In order to measure the applicability of the system, we have set two questions to the users:

Question 1: how relevant (in percentage) are the retrieved documents, in collection categories, to the query compared to ordinary search engine?

Question 2: are you satisfied with the degree of correlation of the system?

The results of the questions are depicted in Fig. 2. The only disadvantage of the proposed system is that every time a new document is uploaded in the database, the learning process must run for the new document. This process is time consuming and costly. In the future work we are planning to improve the applicability of the system concerning the automatic learning process to tune in and watch the cultural event at the same time that it's being broadcast.

5 Comparison of SOM O.D.I.S.S.E.A.S with Desktop Search Engines

SOM-O.D.I.S.S.E.A.S is focusing on the specific need of the user, which is the search of e-learning content, providing a simple and friendly user interface, in addition to the Google and Yahoo desktop search engines which provide many irrelevant add-ins and information, concerning the need of the users. In comparison to the MS search engine, SOM-O.D.I.S.S.E.A.S. provides multiple search (e.g. picture and text), in addition to the MS search that provides advanced search in categories (e.g. text or video). Also the criteria of the search are based on the "free phrase" inputted by the user and not on the name of a file or other criteria. The interface is designed following the basic rules of design, that is simple and friendly user interface which satisfies the needs of the user easily and directly, leaving out information that might distract and disorient the user from his target [14]. Thus, even a user who isn't familiar with the new technologies can simply search for e-learning content. For example, a teacher who has created a learning scenario for physics' lesson wants to search for the content he used, in order to make a scenario for the mathematics' lesson.

6 Conclusion

In this paper is presented the use of the SOM algorithm along with a training algorithm, for document retrieval. The applicability of the algorithm is illustrated in an e-learning case study. A software component has been constructed to perform intelligent search in the educational documents. The performance of the training algorithm using the MSE measure has been presented. One of the basic issues concerning the Intelligent Systems is their ability to adjust and to be installed into any platform that requires methods of intelligent retrieval. The system was awarded after its application in various tests. Its pedagogical advantages were underlined not only by the students but also by the instructors handling and assessing the educational material. The instructors gained valuable time during their course as they could retrieve information using simple queries while students found the intelligent system necessary at their self-paced learning. Moreover, in a future expansion, the system is expected to provide reasons of the confidence estimation accompanied by the retrieved texts. That means that the user will be supplied with reasons of the certain search results and the scope of its query.

References

1. Drigas, A., Vrettaros, J., Kouremenos, S.: Open Distance Interactive System for Educational Applications O.D.I.S.S.E.A.S. In: Proceedings of the Technology and learning in Higher Education Conference, Samos (2001)
2. Honkela, J., Lagus, K., Kaski, S.: Selforganizing maps of large document collections. In: Deboeck, G., Kohonen, T. (eds.) Visual Explorations in Finance with self-organizing Maps, pp. 168–178. Springer, London (1998)
3. Lagus, K., Kaski, S.: Keyword selection method for characterizing text document maps. In: Proceedings of ICANN 1999, 9th international Conference on Artificial Neural Networks, vol. 1, pp. 371–376. IEE, London (1999)
4. Kohonen, T., Kaski, S., Lagus, K., Honkela, J., Paatero, V., Saarela, A.: Self organization of a massive text document collection. In: Oja, E., Kaski, S. (eds.) Kohonen Maps, pp. 171–182. Elsevier, Amsterdam (1999)
5. Lagus, K.: Text retrieval using self-organized document maps. Technical Report A61. Helsinki University of Technology, Laboratory of computer and information science (2000)
6. Kohonen, T.: Self-organization and Associative Memory, 3rd edn. Spriger, Heidelberg (1989)
7. Kohonen, T.: Self-Organization Maps. Springer, Heidelberg (1995)
8. Salton, A.: Automatic Text processing. Addition-Wesley Publishing Company, Inc., Reading (1995)
9. Kohonen, T.: Self-organization of very large documents collections. State of the art. In: Niklasson, L., Boden, M., Ziemke, T. (eds.) Proceedings of ICANN 1998, 8th international Conference on Artificial Neural Networks, vol. 1, pp. 65–74. IEE, London (1998)
10. Deerwester, G., Dumais, T., Furnas, W., Landauer, K., Harshman, R.: Indexing By Latent Semantic Analysis. Journal of the American Society of Information Science (1990)
11. Honkela, J., Kaski, S., Kohonen, T., Lagus, K.: Self-organizing maps of very large document collections: Justification for the WEBSOM method. In: Balderjahn, I., Mathar, R., Schader, M. (eds.) Classification, Data Analysis, and Data Highways, pp. 245–252. Springer, Berlin (1998)
12. Perfetti, R., Costantini, G.: Associative memories on BBS neural networks: a hardware-oriented learning algorithm. In: WSEAS NNA-FSFS-EC, pp. 109–454 (2003)
13. Triantafyllou, I., Carayannis, G.: Architectures and Techniques for Monolingual and Multilingual Information Retrieval Systems in a SOM Framework. In: WSEAS NNA-FSFS-EC, pp. 205–454 (2003)
14. Dix, A., Finlay, J., Abowd, G., Beale, R.: Human Computer Interaction, 3rd edn. Prentice Hall (2003)

Method of Extract Changing Scene in Animation

Young Jae Park[1], Youngsong Mun[1], Keun Su Lee[2], and Gye Young Kim[1]

[1] Dept. of Computer Soongsil Univ., Sangdo5dong, Dongjakgu
Seoul, Republic of Korea
webmaster@kuly.org, mun@ssu.ac.kr, gykim11@ssu.ac.kr
[2] Dept. of Computer Engineering, Hankyung Univ., Seokjungdong, Ansung City
Kyunggido, 456-749, Republic of Korea
kslee@hknu.ac.kr

Abstract. In this paper, we propose that method of extract changing scene in animation. A new method differed about attributes of animation based on already exist one. We mix histogram method and block method plus new things. We found it much improved than traditional method.

Keywords: animation, changing scene, detection, vision, hci.

1 Introduction

Already, there are many papers about Detection of transition scenes as we know. The method was extract changing from real scene. Generally, Detection of transition scenes means serial scene of the scene is different from the current scene as a transition point of detection. Movie can not changed by themselves unless people making a other scene into them. Because of it is real-time based. So if there is changed that kind of scene, it follows changed anything in the pixel too much, or if any scene is not changed, pixel is changing little bit. So it is the simplest method that Discrimination changing pixels between difference image through the frame to detection of transition scenes. But it could be happen unexpected if there is something changed kind of camera's size, rotate. Therefore, block-based comparison and histogram comparison is used by generally in many wheres.

1.1 Block Based on Extract Changing Scene

Block based method is divide one frame as 9 regions. And they have a represent 16x16 size blocks in center. Represent block is including lots of information about frame.

(1) is a each part divide. G is gray level, M is frame's number, K is pixel's location, P means measure part. Measure part is decide how much it would be measure pixel that around pixel's location

$$D\_sum^{i} = \sum_{k=0}^{size} | G_{m}^{i}(k+p) - G_{m-1}^{i}(k) \qquad (1)$$

O. Gervasi et al. (Eds.): ICCSA 2008, Part II, LNCS 5073, pp. 24–31, 2008.

One of min value is smaller than threshold value. It can be moving, or larger than threshold value, it can not be moving [4].

1.2 Histogram Comparison

Histogram method is compared histogram that pixel's frequency applies magnitude in color part. Hri(k), Hgi(k), Hbi(k) is color part(R, G, B)'s number(N) of frequency(k) in i'th frame. Alpha, beta, gamma, is constant for light level conversion follows NTSC standards. And using by each one is defined 0.299, 0.587, and 0.114. If D's value is larger than threshold value, the scene has changed. Or smaller than threshold value, the scene has not changed. [5][6]

$$D_{wr,wg,wb}(f_i, f_j) = \begin{cases} \sum_{k=0}^{N-1} (|H_i^r(k) - H_j^r(k)| \times \alpha \\ + |H_i^g(k) - H_j^g(k)| \times \beta \\ + |H_i^b(k) - H_j^b(k)| \times \gamma) \end{cases} \quad (2)$$

2 Changing Scene in Animation

In this section, Talking about method of extract changing scene in animation. Traditional method is based on real movie, but I suggest new ways to extract that in animation. Animation is creative by people. So there is variety Hero who has got specialty things by each one. But basically we have a common composition. So it looks similar. There are 3 animations. First is model animation. Second is 3D animation. Third is cell animation. Model animation is made a object by mud or clay and capturing there motion. And 3D animation is made by computer graphic. Last one, we can see cell animation many times. Like a TV or Theater. It's specialty is

Fig. 1. For example changing scene in animation

drawing many papers to expression background, Hero, object. Because of it follows lack of time, cost, and human efforts.

In the case of cell animation, it is express to a cell and then there is happening at the scene transitions if we change the background. And people drawing it by hands so there is not color enough to express everything unlike real scene. So I suggest detection of changed scenes using by histogram comparison and block-based comparison.

Fig. 2. The simple color animation. (Navy, Light Brown, Light Blue, Dark Blue, Light Yellow)

2.1 Extract Color Information

First, the judge in the case, one video scene change the color values in the next round of video footage of the same operation, will be whether the judge. This represents one of the same formulas.

$$SH = \sum_{x,y} S(x,y) \tag{3}$$

$$S(x,y) = \begin{cases} Sub(x,y) & if\, Sub(x,y) \le 180 \\ 360 - Sub(x,y) & if Sub(x,y) > 180 \end{cases} \tag{4}$$

$$Sub(x,y) = | H_1(x,y) - H_2(x,y) | \tag{5}$$

H1 is original image that was changed to color value. H2 is next image that was changed to color value. So sub image is made by gap of color value. There are 0° ~ 360° values. Because of using H part after changing color value for HSI Model. For example value 1's color is same as value 360's. If it is minus, the value is 359. It is large number expect. So, we use (4) to fix that value and make a S. Finally make a SH using by all of the result, and make a decision.

There changing scene existence by SH's value. (6), (7) is explains about H part of changing RGB model to HSI model to get color value [7].

$$H = \begin{cases} \theta & if\, B \le G \\ 360 - \theta & if\, B > G \end{cases} \tag{6}$$

$$\theta = \cos^{-1}\left(\frac{\frac{1}{2}[(R-G)+(R-B)}{\sqrt{(R-G)^2 + (R-B)(G-B)}} \right) \tag{7}$$

The above method is used when the video depending on the size of the operation is going to be the difference in speed. If the operation is high time resolution image geometric increase in order to have the security of the diagonal image processing pixel how to select and use.

2.2 Extract Histogram Information

If SH value is larger than threshold value, computer is starting to research because of SH value is going to be further part of changing scene.

Fig. 3. Recognition of the same scene as different scene

Scene was not changed animation is fixed color's number and color's position. And It's very easy to conscious. There is big gap of color to easy to conscious different as reality. Histogram comparison can be happen many times like that misunderstood same scene as different scene. that is why, even though, it is the same background, there are many parts of character, and moving between frame and frame like a Fig. 3, so we suggest 2 ways to solve them.

First, making a histogram like a Fig. 4, HSI model's feather is made of h value of color value, and h value is between 0° ~ 360°. make a histogram after divide 360° into 45°, and arrow's length is accumulate.

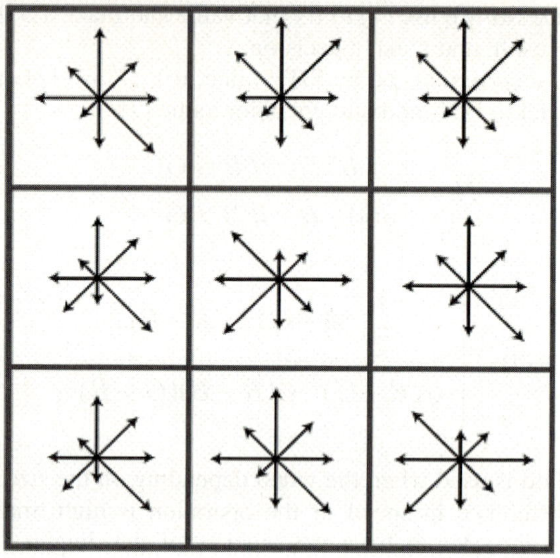

Fig. 4. Make a histogram after divide frame into nine's

3	2	3
2	1	2
3	2	3

Fig. 5. Each frame's weight

Generally, hero is take a center position, and background is take a outside. Apply on weight Fig. 5, so if the value is larger than threshold value, scene was changed, or smaller than threshold value, scene was not changed.

3 Experiments Result

In this chapter, proposed methods to the test, and to evaluate the performance. Performance measurement is accuracy and processing time.

We have used IBM Computer which is an Intel Pentium-4 3.0Ghz, 1GB Memory, NVIDIA Geforce 6600. OS is Microsoft Windows XP SP2, Language is Microsoft Visual Studio C++ 6.0. Experimental video is "Karekano" that production was "Gainax" in 1998. Resolution is 640x480 and 23.976 frame per second.

The result of using the traditional method shows [Table 1] and [Table 2]

Table 1. The result of block based on extract changing scene

Animation	Number of frame	Changing scene	Accuracy
1	34525	223	80.29%
2	34500	175	77.66%
3	54523	276	88.63%

Table 2. The result of histogram based on extract changing scene

Animation	Number of frame	Changing scene	Accuracy
1	34525	207	74.75%
2	34500	164	73.01%
3	54523	260	83.66%

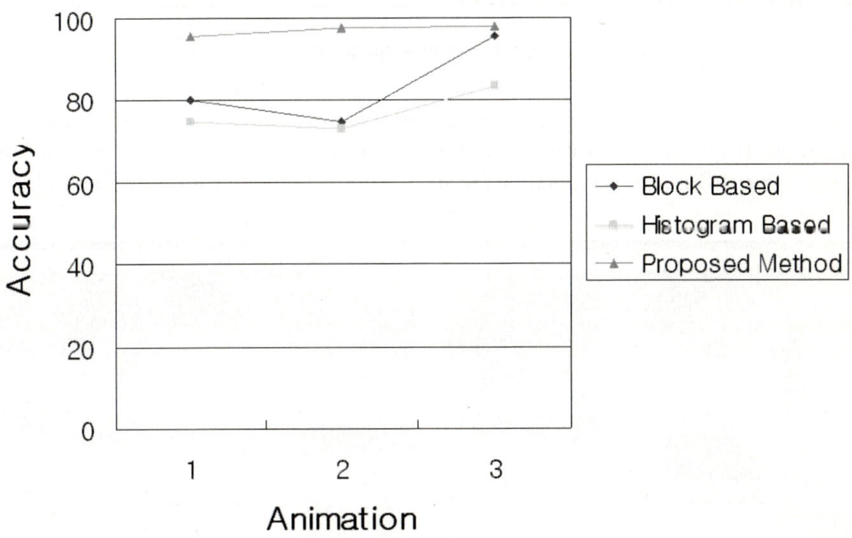

Fig. 6. The result of Block based, Histogram based and Proposed method

Table 3. The result of proposed method

Animation	Number of frame	Changing scene	Accuracy
1	34525	266	95.68%
2	34500	220	97.77%
3	54523	306	98.01%

The result of using the proposed method shows [Table 3].

We have tested the recognition performance for 103,548 frames. The result of experiment is 97% accuracy, and computing time is 2,984 sec. Fig. 6. is result that proposed method was better than traditional method.

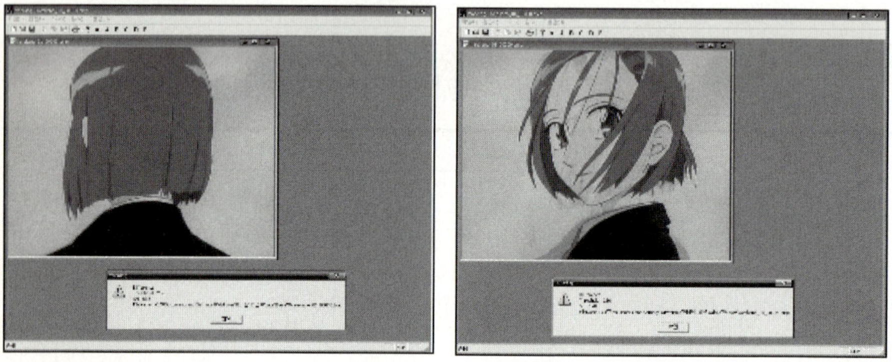

Fig. 7. Misrecognize same scene as difference scene

Fig.7. is an example that misrecognize same scene as difference scene using existing method. If character is fully rotated at one frame that it often malfunction. Fig.8. is a result that difference scenes to accurately recognize using proposed method.

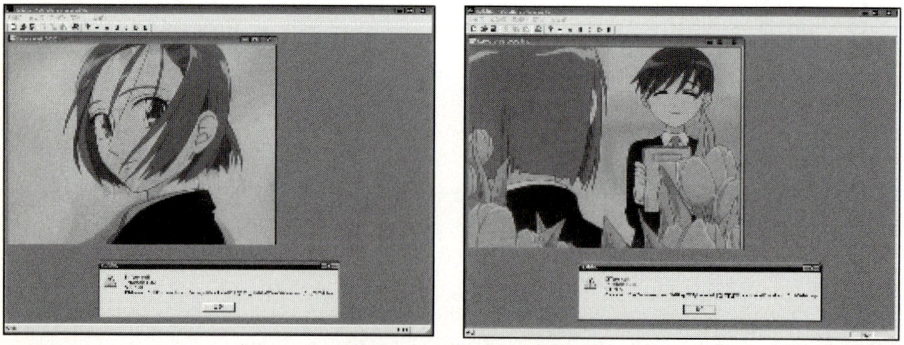

Fig. 8. Difference scenes to accurately recognize

4 Conclusion

Animation is artificial creations. So, it was difficult to extract changing scene using traditional method. The traditional method was not suitable for animation, because it was proposed by real world based. For the reason, we proposed new extract changing scene method and then satisfied.

Acknowledgement. This research is supported by Seoul R&BD Program. (10581C093113)

References

1. Lowe, D.G.: Distinctive Image Features from Scale-Invariant Keypoints. International Journal of Computer Vision 60(2), 91–110 (2004)
2. Se, S., Lowe, D., Little, J.: Mobile Robot Localization and Mapping with Uncertainty using Scale-Invariant Visual Landmarks. International Journal of Robotics Research 21(8), 735–758 (2002)
3. Lowe, D.G.: Object recognition from local scale-invariant features. In: International Conference on Computer Vision, Corfu., Greece, pp. 1150–1157 (1999)
4. Zhang, H.J., Kankanhalli, A., Smoliar, S.W.: Automatic Partitioning of Full-motion Video, Multimedia Systems. 1(1), 10–28 (1993)
5. Tonomura, Y.: Video handling based on structured information for hypermedia systems. In: Proc. ACM Int. Conf. Multimedia Information Systems, pp. 333–344 (1991)
6. Nagasaka, A., Tanaka, Y.: Automatic Video Indexing and Full-Video Search for Object Appearances. In: Knuth, E., Wegner, L. (eds.) Visual Database Systems II, pp. 113–127. Elsevier Science Publishers, Amsterdam (1992)
7. Gonzalez, R.C., Woods, R.E.: Digital Image Processing, 2nd edn., p. 793. Pearson Education (2002) ISBN 0-201-18075-8
8. Denzler, J., Niemann, H.: Active Rays: Polar-transformed Active Contours for Real-Time Contour Tracking. Academic Press, London (1999)
9. Fernando, W.A.C., Canagarajah, C.N., Bull, D.R.: A unified approach to scene change detection in uncompressed and compressed video. IEEE Transactions on Consumer Electronics 46(3), 769–779 (2000)
10. Seong, Y.K., Choi, Y.-H., Park, J.-A., Choi, T.-S.: A hard disk drive embedded digital satellite receiver with scene change detection for video indexing. IEEE Transactions on Consumer Electronics 48(3), 776–782 (2002)
11. Brown, M., Lowe, D.G.: Recognising panoramas. In: International Conference on Computer Vision (ICCV 2003), Nice, France, October 2003, pp. 1218–1225 (2003)

Improved Curvature Scale Space Based on Image Retrieval Using Dynamic Time Warping

Sang Hee Kim[1], Youngsong Mun[1], and Hyung Il Choi[2]

[1] School of computing, Soong-Sil University
hotdog4242@nate.com, Mun@ssu.ac.kr
[2] School of media, Soong-Sil University
hic@ssu.ac.kr

Abstract. In this paper, we suggest improved Improved Curvature Scale Space(ICSS) based on image retrieval using Dynamic Time Warping(DTW). This method consists of three process. At the first stage, we retrieved an image using ICSS method. We applied ICSS space to extract shape information. This method expressed contours of the object which is binarized through pre-process in ICSS space, then extract shape features of the object. At the second stage, we applied DTW for time series to determine similarities. DTW algorithm is good to find the optimal alignment between two time series. finally, merge ICSS shape feature with DTW information. This method is estimated with a database of 1000 images. With estimated result, we show that proposed method is better in accuracy of searching than without DTW.

Keywords: CSS, DTW, Image retrieval, Shape similarity.

1 Introduction

Advanced in memory technologies and processing speed have made it feasible to store a large number of images in computers. This has given rise to the problem of organizing them for a rapid access to their content. An image database system aims to help people in this regard and enable them to find their desired images as quickly as possible. In many applications, the user of image database remembers something about the content of his desired image or wishes to find similar images to an existing image. In content-based image database systems, intrinsic properties of images are captured in some feature vectors which are indexed or compared to one another during query processing to find similar images from the database[1].

Considerable amount of information exists in two dimensional boundaries of objects which enables us to recognize objects without using further information. As a result, shape similarity retrieval plays an important role in content database systems.

In conclusion, although the number of proposed methods is increasing rapidly, there are still a number of short-comings associated with each method. While the robustness of some methods is doubtful, other methods which exhibit a reasonable degree of robustness are often computationally expensive. In this paper, we introduce an efficient d robust shape representation method for shape similarity retrieval.[8]

The following is the organization of the remainder of this paper. In Sect. 2, the construction of the ICSS image and the properties of using its maxima as the shape

O. Gervasi et al. (Eds.): ICCSA 2008, Part II, LNCS 5073, pp. 32–42, 2008.

representation are explained. Section 3 describes the dynamic time warping algorithm and matching ICSS maxima image using DTW. In Sect. 4, merge ICSS shape feature with DTW information. In Sect. 5, Experimental and Result. And finally, Sect. 6 conclusions and future works.

2 ICSS(Improved Curvature Scale Space) Image

A number of shape representations have been suggested to recognize shapes even under affine transformation. Some of them are the extensions of well-known methods such as Fourier descriptors and moment invariants. The methods are then on a small number of objects for the purpose of pattern recognition. In both methods, the basic idea if to use parameterization which is robust with respect to affine transformation. In all cases, the methods are tested on a small number of objects and therefore the results are not reliable. Moreover, almost the same results can be achieved using conventional methods without modifications.

2.1 Contour Extract

A snake is initially placed near image contours under consideration, and then a procedure of energy minimization is applied to draw the snake to desirable image contours[2][3]. Such a model may yield different forms of image contours depending on an energy functional being minimized and a minimization algorithm used[4].

$$E^*_{snake} = \sum_{i=1}^{N} (\alpha E_{cont}(v_i) + \beta E_{curv}(v_i) + \gamma E_{image}(v_i))ds \tag{1}$$

$$v_i = (x_i, y_i)$$

Based on a greedy algorithm, our approach of minimization is iterative as in a greedy snake. In the discrete formulation of active contour models, the contour is represented as a set of snake points $v_i = (x_i, y_i)$ where x_i and y_i are the coordinates of the snake point i. At each iteration, the energy function is computed for the current location of v_i and each of its neighbors. The location having the smallest value is chosen as the new location of v_i. The iteration determines the convergence of energy minimization by detecting the number of points moved. Based on the greedy algorithm, our energy function has the form of Equation(1).

$$E_{image} = -\left|\Delta G_\sigma * SCDF(I(v))\right|^2 \tag{2}$$

$$SCDF(i,q) = e^{-\frac{1}{2}\left(\frac{(i-<i>^2}{\sigma_i^2}+\frac{(q-<q>^2}{\sigma_q^2}\right)} \tag{3}$$

The continuity term $E_{cont}(v_i) = \left| d^* - |v_i - v_{i-1}| \right| / \max_j \{ \left| d^* - |v_{ij} - v_{i-1}| \right| \}$ encourages even spacing of points. $|v_i - v_{i-1}|$ is the distance between two snake points under consideration, and d^* is the average of the length between points, where $\{v_{ij} \mid j = 0,2,..,7\}$ for a snake point $v_i = v_{i,0}$ and its eight-neighbor points $v_{ij} = (j \neq 0)$.

The curvature term $E_{cont}(v_i) = |v_{i-1} - 2v_i + v_{i+1}| / \max_j \{ |v_{i-1} - 2v_i + v_{i+1}| \}$ gives a reasonable and quick estimate of the local curvature of the snake. The image term $E_{image}(v_i) = (\min - mag(v_i))/(\max - \min)$ is to push a toward edge points, where max and min is maximum and minimum gradient in a neighborhood.

From Equation 2, SCDF shows a skin color distribution function, as a same Equation 3. Here I with q the average against I and Q from YIQ color space which the training pixels have, σ_i^2 with σ_q^2 it means a dispersion, i and q is I and Q of input pixel . Y elements the lucidity, I elements and Q elements show a hue form YIQ color space.

This space the top of color or change of chroma is few is a strong point to change of lucidity. The price of Equation 3, the price where the input pixel and training pixel are near in the similarity percentage recording 1, shows a near price in case 0 of opposition. Namely, when with Fig 1, skin color distribution function against a training pixel is composed, the pixel which is input and training pixel price of similarity percentage recording function were included in middle square territory all.

From guard part of face territory with the pixel which keeps a skin color the difference of skin color distribution function priced of two pixel for is big because like that the pixel is being contiguous. When like this quality about under using extracting the outline against a face territory, with image energy function is uses the lucidity price substitution skin color distribution price of Equation 3.

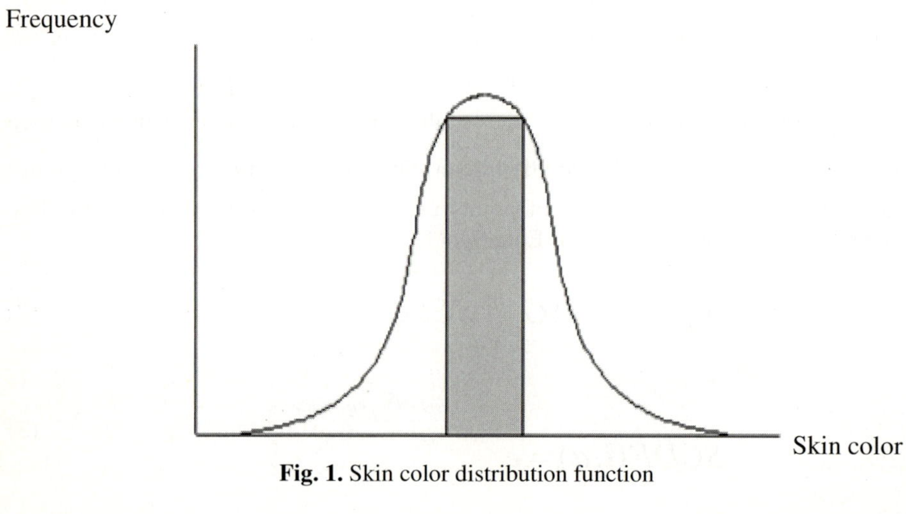

Fig. 1. Skin color distribution function

2.2 CSS(Curvature Scale Space)

Following the preprocessing stage, every object is represented by the x and y coordinates of its boundary points. The curve is then smoothed by a Gaussian function. The smoothed curve is called Γ_σ, where σ denotes the width of the Gaussian kernel, $g(u,\sigma)$. The locations of curvature zero-crossings on Γ_σ are determined at different levels of scale using Equation 4. The process starts with σ=1, and at each level, σ is increased by $\Delta\sigma$, chosen as 0.2 in our experiments, As σ increases, Γ_σ shrinks and becomes smoother, and the number of curvature zero crossing points on it decreases, Finally, when σ is sufficiently high, Γ_σ will be a convex curve with no curvature zero-crossings (see Fig.2). The process of creating ordered sequences of curves is referred to as the *evolution* of Γ.[1]

$$k(u,\sigma) = \frac{X_u(u,\sigma)y_{uu}(u,\sigma) - X_{uu}(u,\sigma)y_u(u,\sigma)}{(X_u(u,\sigma)^2 + Y_u(u,\sigma)^2)^{3/2}} \tag{4}$$

If we determine the locations of curvature zero-crossings of every Γ_σ during evolution, we can display the resulting points in (u,σ) plane, where u is the normalized arc length and σ is the width of the Gaussian kernel. The result of this process can be represented as a binary image called the CSS image of the curve. The intersection of every horizontal line with the contours in this image indicates the locations of curvature zero-crossings on the corresponding evolved curve. For example, by drawing a horizontal line at $\sigma = 10.0$, it is observed that there are 6 zero-crossing points of Γ_{10}. These points can also be found on the boundary of object in Fig.2 for $\sigma = 10$.

AS seen in Fig. 2, there are two curvature zero-crossings on every concave or convex part of the shape, and as the curve becomes smoother, these points approach each other and create a contour in the CSS image of the shape. When the segment is filled,

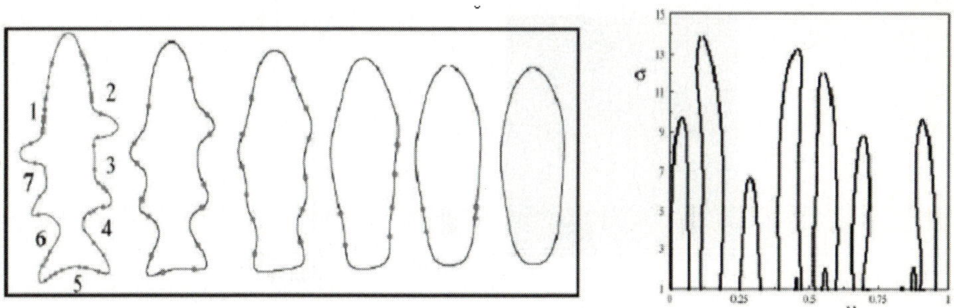

Fig. 2. Shrinkage and smoothing of the curve and decreasing of the number of curvature zero-crossings during the evolution, from left: $\sigma = 1,4,7,10,12,14$

the two points join and represent the maximum of the relevant contour. The height of this contour then reflects the depth and size of the concavity or convexity. The deeper and larger the segment, the higher the maximum. In other words, a contours maximum in the CSS image, represents a segment of the shape.

2.3 Noise Removed Method Using by Clustering

In Fig. 3, we seeing big 3 maximum points bigger than other things. Horizontal line lower part maximum points which it removes in noise from CSS method. But we can see the small 3 maximum points above the horizontal line, this maximum points appear from preprocessing or aliasing of contour extraction. Also which will not be removed, dropped the accuracy of similarity comparison and increased calculation time. In this paper, which complements this problem, we divides extract maximum point and removed maximum point using by clustering method. First, in standard point between that

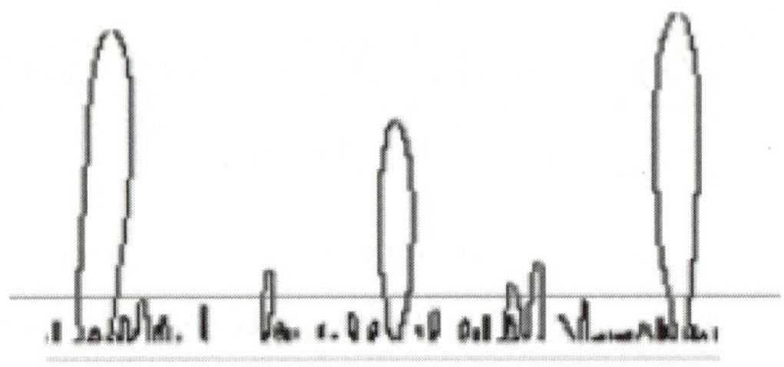

Fig. 3.

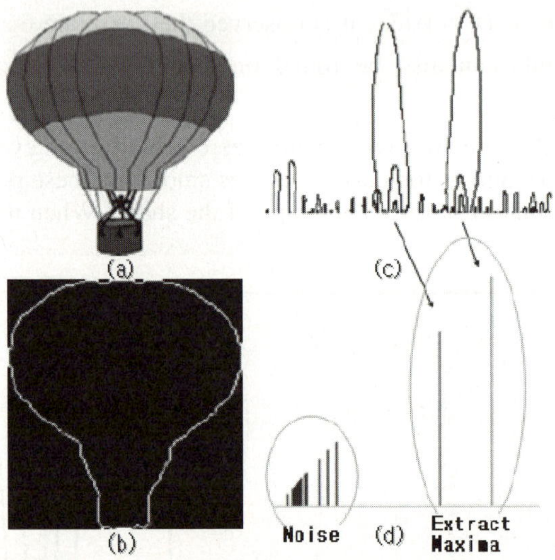

Fig. 4.

big maximum points and small maximum points which maximum point probably near clustering. After dividing in two groups, it calculates the mean value of each group, In standard calculated two mean value, clustering it does an existing maximum points, Until when does not being change of average of two groups, it repeats this process. Finally it comes to divide in two groups, we able to classify left side part is noise and right side part is maximum. Then the maximum point field which is being caught to 1/6 parts of the maximum, it will be able to classify with a noise.[9]

In Fig 4. (a) is input image, (b) is contour, (c) is CSS image, (d) image of after clustering, it is classify noise and maxima.

3 Dynamic Time Warping(DTW)

Dynamic time warping (DTW) is a technique that finds the optimal alignment between two time series if one time series may be "warped" non-linearly by stretching or shrinking it along its time axis. This warping between two time series can then be used to find corresponding regions between the two time series or to determine the similarity between the two time series.

A distance measurement between time series is needed to determine similarity between time series and for time series classification. Euclidean distance is an efficient distance measurement that can be used. The Euclidian distance between two time series is simply the sum of the squared distances from each nth point in one time series to the nth point in the other. The main disadvantage of using Euclidean distance for time series data is that its results are very unintuitive. If two time series are identical, but one is shifted slightly along the time axis, then Euclidean distance may consider them to be very different from each other. Dynamic time warping(DTW)was introduced[10] to overcome this limitation and give intuitive distance measurements between time series by ignoring both global and local shifts in the time dimension.

$$
\begin{aligned}
Q &= q_{1,} q_2, \dots, q_i, \dots, q_n \\
C &= c_{1,} c_2, \dots, c_i, \dots, c_n
\end{aligned}
\tag{5}
$$

Given two time series Q, and C of lengths $|Q|$ and $|C|$

Construct a warp path W

$$
\begin{aligned}
W &= w_{1,} w_2, \dots, w_i, \dots, w_n \\
\max(m,n) &\leq K \leq m+n-1
\end{aligned}
\tag{6}
$$

Where K is the length of the warp path and the kth element of the warp path is

$$w_k = (i.j)$$

The optimal warp path is the warp path is the minimum-distance warp path, where the distance of a warp path (Q,C) is

$$DTW(Q,C) = \min\{\frac{1}{K}\sqrt{\sum_{k=1}^{k} w_k} \tag{7}$$

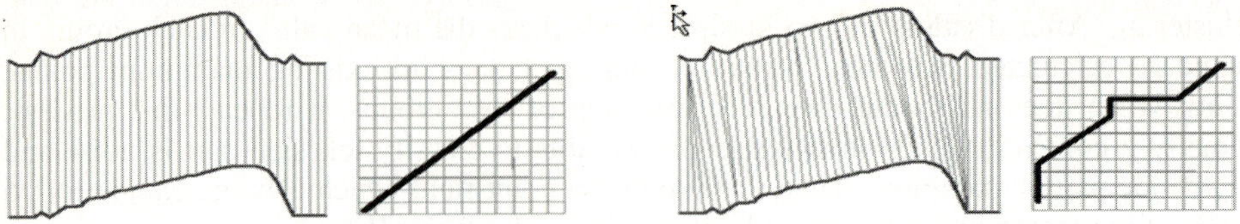

(a) (b)

Fig. 5. (a) is Euclidean distance(warping path), (b) is DTW(warping path)

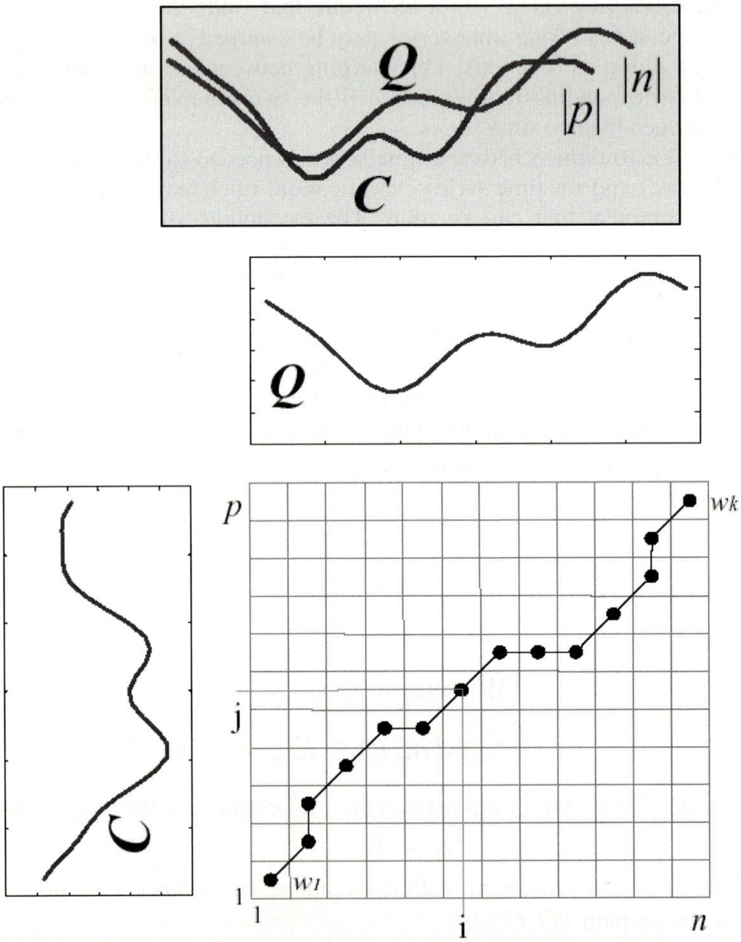

Fig. 6. Input data(Q,C) time series and warping path

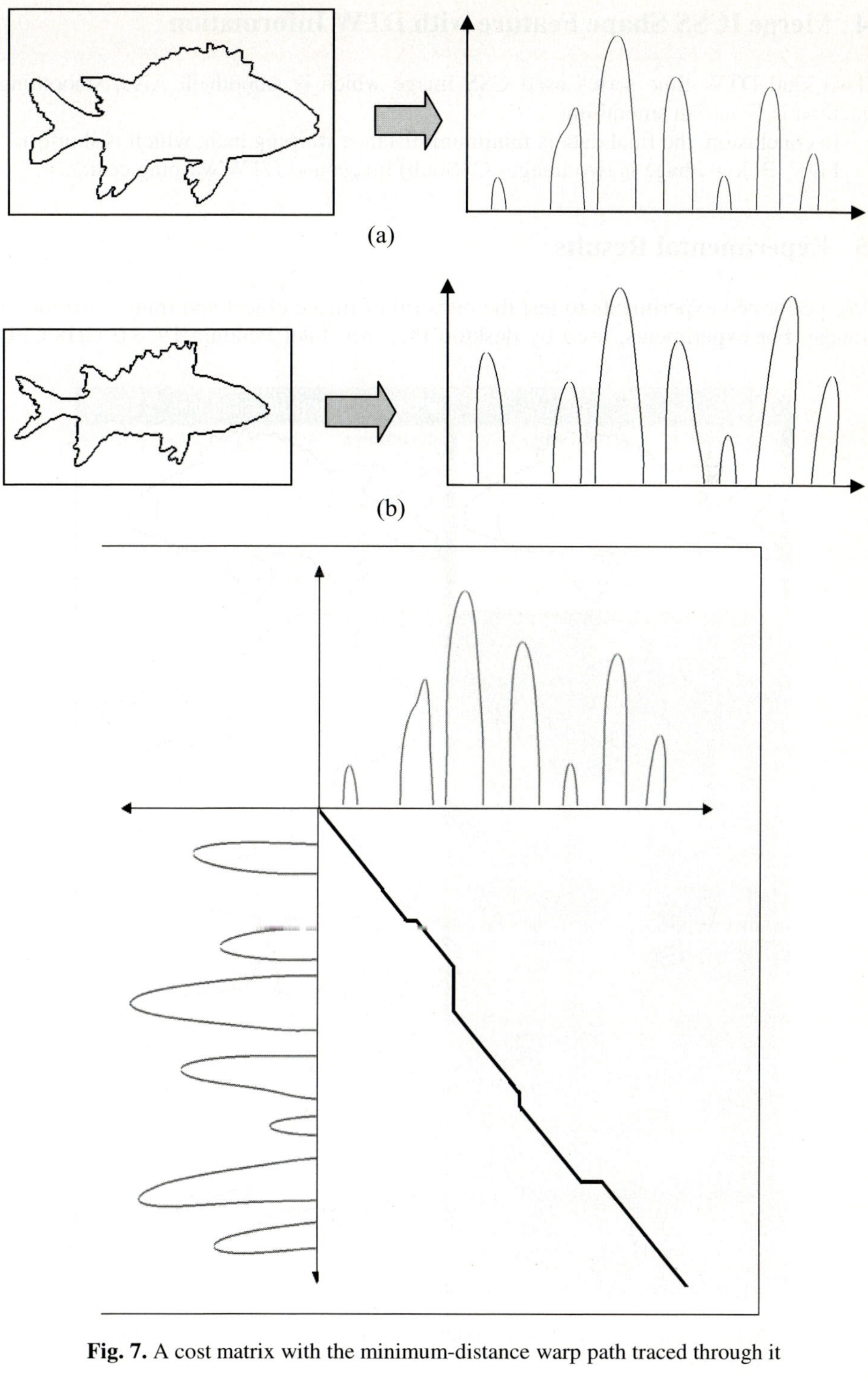

Fig. 7. A cost matrix with the minimum-distance warp path traced through it

4 Merge ICSS Shape Feature with DTW Information

Two kind DTW time series used CSS image which is smoothed. Also, smoothing method is Gaussian smoothing.

In conclusion, the final data is minimum-distance warping path, which is shortest.

Fig 7. Below image is two images CSS(a,b) image and DTW warping cost.

5 Experimental Results

We performed experiments to test the retrieval of image object and transformation of image. For experiments, used by desktop PC, spec Intel Pentium IV 3.0 GHz CPU

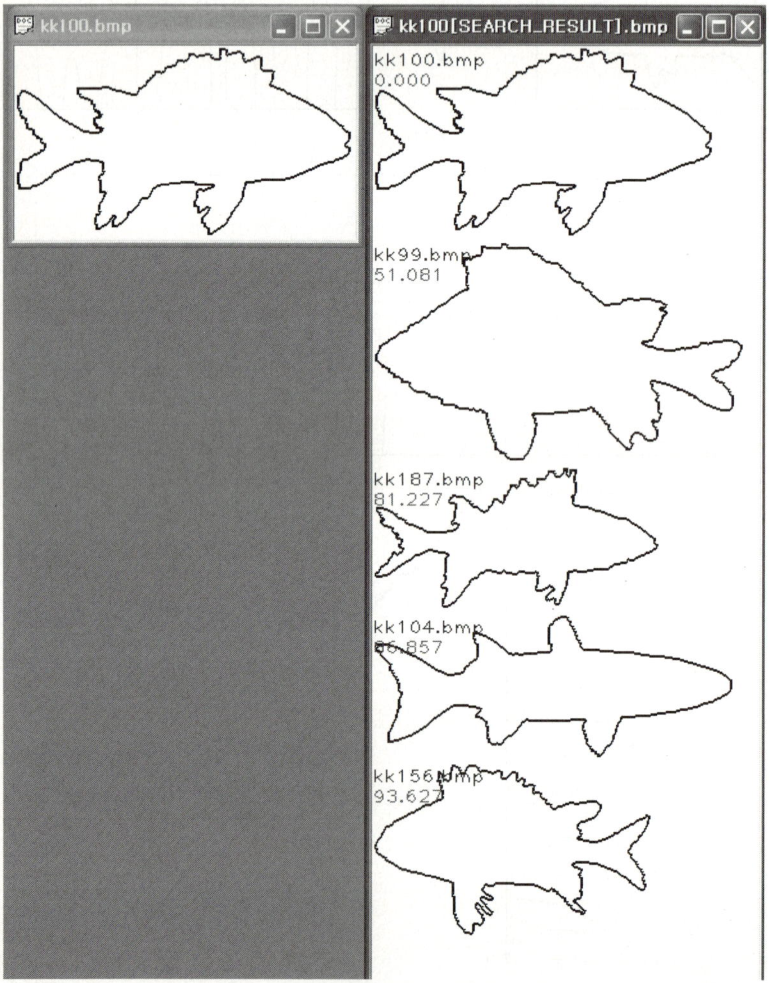

Fig. 8. Left : query image , Right : result image, Red number is DTW warping cost

and 1 GByte, OS is Windows XP professional, made a retrieval system by Visual C++ 6.0 and for DB, used Microsoft Access.

Fig. 8 query image and resultant image which is highest similarity.

6 Conclusions

We presented shape similarity retrieval method using ICSS image and DTW warping cost. This method has relatively high speed and good accuracy.

Our future work will focus on improving the accuracy as well as the speed of the techniques used here. Accuracy can be improved by using better punning techniques as well as using a larger feature set which discriminates images better from each other. Speed can be improved by optimizing our implementation of the dynamic time warping algorithm, as well as looking at related computational techniques to minimize the number of possible matches.

Acknowledgments. This research is supported by Seoul R&BD Program (10581C093113).

References

1. Apostol, N., Rajeev, R., Kyuseok, S.: WALRUS: A similarity retrieval algorithm for imagedatabase. In: Proceeding of the ACM SIGMOD International Conference on Management of Data, pp. 395–406. ACM Press, New York (1999)
2. Michael, K., Andrew, W., Demetri, T.: Snakes: Active Contour Models. Int. J. Computer-Vision. 1(4), 321–331 (1987)
3. Donna, J.W., Mubarak, S.: A Fast Algorithm for Active Contours and Curvature Estimation. CVGIP: Image Understanding 55(1), 14–26 (1992)
4. Chang, S.K., Yan, C.W., Dimitroff, C., Arndt, T.: An Intelligent Image Database System. In: Proc. Of IEEE Trans. On Software Engineering, IEEE Computer Society Press, Los Alamitos (1998)
5. Natsev, A., Rastogi, R., Shim, K.S.: WALRUS: A Similarity Retrieval Algorithm for Image Database. In: Proc. Of SIGMOD Conference, pp. 395–406 (1999)
6. Chua, T.S., Lim, S.K., Pung, H.K.: Content-Based Retrieval of Segmented Images. In: Proc. Of ACM Multimedia, San Francisco, pp. 211–218. ACM, New York (1994)
7. Stricker, M., Dimai, A.: Color Indexing with Weak Spatial Constraints, Storage and Retrieval for Image and Video Databases IV. In: SPIE Proceedings, vol. 2670 (1996)
8. Huang, J., Kumar, S.R., Mitra, M., Zhu, W.-j., Zabih, R.: Image Indexing Using Color Correlogram. In: International Conference on Computer Vision and Pattern Recognition, IEEE, Los Alamitos (1997)
9. Park, J.-Y., Kim, G.-Y., Choi, H.-I.: Image Retrieval Using By Skin Color and Shape Feature. In: Gervasi, O., Gavrilova, M.L. (eds.) ICCSA 2007, Part I. LNCS, vol. 4705, pp. 1045–1053. Springer, Heidelberg (2007)
10. Kruskall, J., Liberman, M.: The Symmetric Time Warping Problem: From Continuous to Discrete. In: time warps, String Edits and Macromolecules: The Theory and Practice of Sequence Comparison, pp. 125–161. Addison-Wesley Publishing Co., Reading (1983)

11. Abdulla, W., Chow, D., Sin, G.: Cross-words reference template for DTW-based speech recognition systems. In: Proc. IEEE TENCON, Bangalore, India (2003)
12. Salvador, S., Chan, P.: FastDTW:Toward Accurate Dynamic Time Warping in Linear Time and Space
13. Lee, D.-H., Kim, H.-J.: A fast content-based indexing and retrieval technique by the shape in large image database. The Journal of Systems and Software (2001)
14. Fuh, C.S., Cho, S.W., Essig, K.: Hierarchical color image region segmentation for content-based image retrieval system. IEEE Trans. Image Process 9(1), 156–162 (2000)
15. Kim, H.K., Kim, J.D.: Region-based shape descriptor invariant to rotation, scale and translation. Signal Processing: Image Communication 16, 87–93 (2000)
16. Mohanna, F., Mokhtarian, F.: An efficient Active Contour Model Through Curvature Scale Space Filtering. Multimedia Tools and Application (2003)

Recovery Schemes for Fast Fault Recovery on GMPLS Network

Kisu Kim[1], Won-hyuk Lee[2], Ki-sung Yu[2], and Jae-yong Lee[3],
Seongjin Ahn[1], and Jin-wook Chung[1]

[1] Dept. of Electrical and Computer Engineering, Sungkyunkwan Univ., Suwon, Korea
{kskim,sjahn,jwchung}@songgang.skku.ac.kr
[2] Korea Institute of Science and Technology Information, Daejeon, Korea
{livezone,ksyu}@kisti.re.kr
[3] Dept. of Internet Engineering, Hanseo Univ., Seosan, Korea
jylee@hanseo.ac.kr

Abstract. In this research, we proposed two types of fault recovery schemes designed by combining the protection and the restoration methods in order to increase the survivability of GMPLS networks. The first method is a transformation of the 1+1 protection method where it consumes more resources than the 1+1 protection method but has a faster fault recovery speed. The second method is a combination of the 1+1 protection method and the restoration method and can enhance the performance after recoveries. Although the two methods proposed in this paper consume more resources than the ones before, they have outstanding recovery time and performance after recoveries.

Keywords: GMPLS, fault recovery, protection, restoration, performance.

1 Introduction

The backbone network of the post-generation network is composed of optical telecommunication that provides broadband and many studies are being conducted in order to form an efficient optical network using IP. It is too difficult in reality for the existing optical network, characterized as passive and inflexible, to support the IP with a very active traffic property. The MPLS (Multi-Protocol Label Switching) applied GMPLS (Generalized Multi-Protocol Label Switching) was introduced in order to resolve this issue. Through this we were able to flexibly manage and control optical networks, which are passive and are unable to allocate active bandwidth according to the reconfiguration of network and traffic, by integrating them into a single control layer [1, 2, 3].

Since vast amounts of data are lost within a short frame of time when a fault occurs in a GMPLS network, the network survivability during cases of faults is becoming a critical issue. GMPLS, which is derived from the MPLS network, supports optical wavelength switching as well as standard packet switching thereby calling for a need of multilevel survivability strategy according to each data type [4, 5].

Survivability strategies used today are the protection method and the restoration method. The protection method is largely consisted of dedicated protection and shared

O. Gervasi et al. (Eds.): ICCSA 2008, Part II, LNCS 5073, pp. 43–55, 2008.

protection and can be classified as 1:1, 1+1, 1:N, or M:N according to the path mapping method.

While the protection method has a short recovery time, its disadvantage is that it wastes excessive resources. Shared protection methods such as the 1:N or M:N method intended to increase network resources efficiency are difficult to implement and they also cannot recover faults occurring in different paths simultaneously [1, 3].

The restoration method executes fault recovery through rerouting from the terminal node by notifying each terminal node from the fault detection node. Since it sets the path after the occurrence of a fault, it can set the recovery path as the optimal path but has a longer recovery time compared to the protection method [1, 3].

The network bandwidth today is increasing at an exponential rate as GMPLS, designed based on tens of gigabytes of optical network, can encounter huge data losses in a short period of time. Therefore, in this study, we propose two recovery methods that can enhance the performance after recoveries and minimize the recovery time but use more resources than the previous methods.

This paper is organized as follows. Chapter 2 describes the protection method and the restoration method. Chapter 3 proposes a recovery method that minimizes the recovery time and Chapter 4 introduces a recovery method that enhances the performance after a recovery. Chapter 5 analyzes performance of proposal methods. Chapter 6 discusses prospective researches and concludes this study.

2 Related Studies

2.1 Protection Method

In the protection method, which uses a preset recovery path, one or more backup paths are preset for one or more working paths and it can run under different levels for network survivability [7].

1+1 protection method. The 1+1 protection method sets a backup path that has identical resource attributes as the working path and transfers traffic identical to the one flowing through the working path through the backup path. The destinations of the paths receive the same traffic from two paths and that's where the traffic decision is made. Therefore, this requires twice as much network resources as setting a single working path thereby depreciating resource efficiency but is advantageous in terms of its recovery time for faults and that it has almost no traffic losses [3, 8].

1:1 protection method. Similar to the 1+1 protection method, the 1:1 protection method sets both the working path and the backup path simultaneously where traffic is transferred through the working path but not through the backup path. A fault message is received from the node that detected the fault in the working path and traffic is then sent through the backup path as this may result in a slight loss of traffic [3, 8, 9].

1+N protection method. As opposed to the 1+1 and 1:1 protection methods mentioned earlier, in the 1+N protection method, a single path provides a backup path to multiple working paths. While using a single backup path to support multiple working paths may increase the resource efficiency, the prevailing condition must be that the

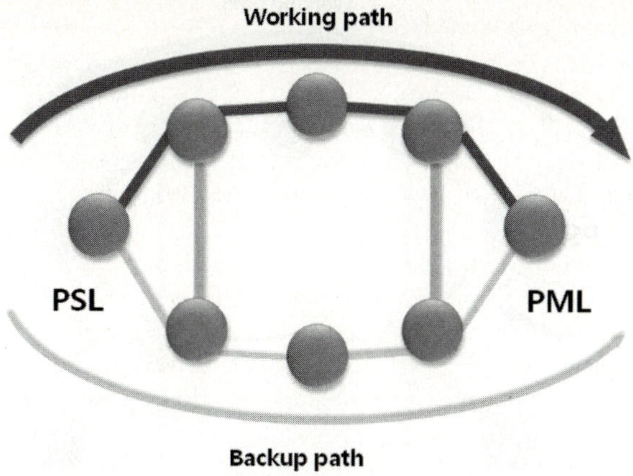

Fig. 1. Example of 1+1 protection method

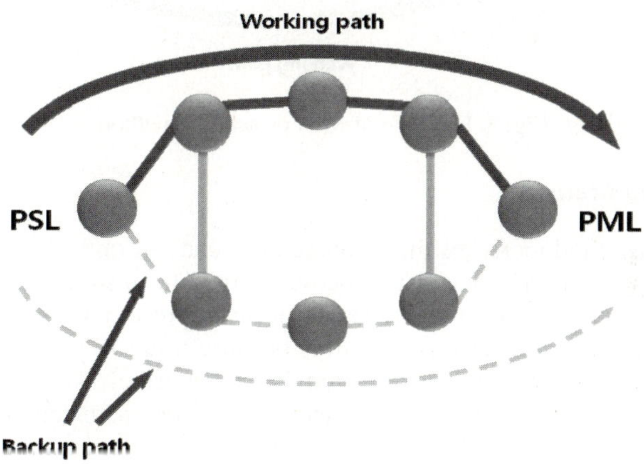

Fig. 2. Example of 1:1 protection method

transmitting LSR (Label Switch Router) and the receiving LSR must be the same and only one working path is recovered when a fault occurs in more than two working paths at the same time. Working paths should not be within the same SRLG (Shared Risk Link Group) in order to go around this issue [3, 8, 9].

M:N protection method. By setting M number of backup paths and N number of working paths, in the M:N protection method, any one of the M backup paths can support any one of the N working paths. Some resources may be wasted because backup paths are not used until a fault occurs and the efficiency may be increased by avoiding to set working paths within the same SRLG [3, 8, 9].

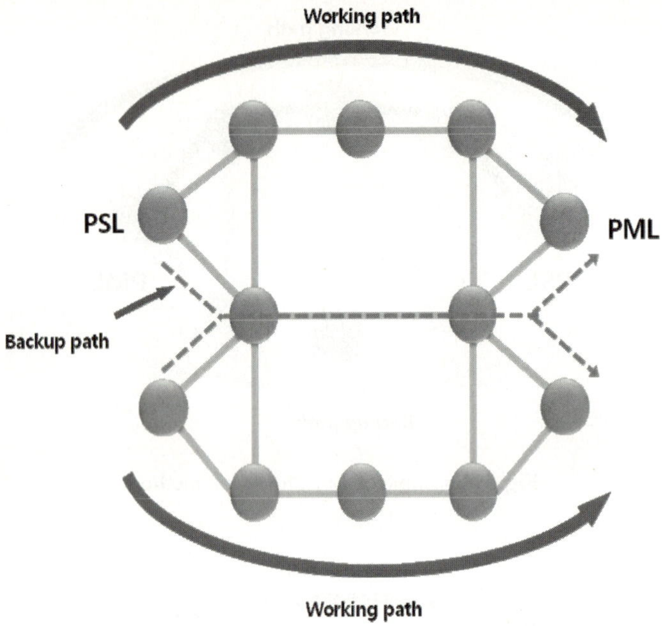

Fig. 3. Example of M:N protection method

2.2 Restoration Method

The restoration method increases the resource efficiency of optical network by setting a new backup path in case of a fault for recovery. The restoration mechanism has better resource efficiency because it does not have to preset backup paths in preparation of a fault but has a longer recovery time because it has to perform a backup path setting procedure after the fault has occurred. It can nevertheless optimize the network by having the router algorithm calculate and set recovery paths based on the current status and policy of the network after the recovery.

3 1+1+1 Protection Method

The 1+1 protection method sets a backup path that has identical resource attributes to the working path and transmitting traffic identical to the one transmitted through the working path through the backup path thereby reducing the recovery time for a fault and traffic losses [8].

The 1+1 protection method allows a relatively faster recover compared to other previously proposed methods where the recovery time is determined by the location and distance of the PML (Path Merge LSR) and the fault. If the distance between the PML and the fault is far like the one shown in [Figure 4], the recovery time may be long even for the 1+1 protection method [12].

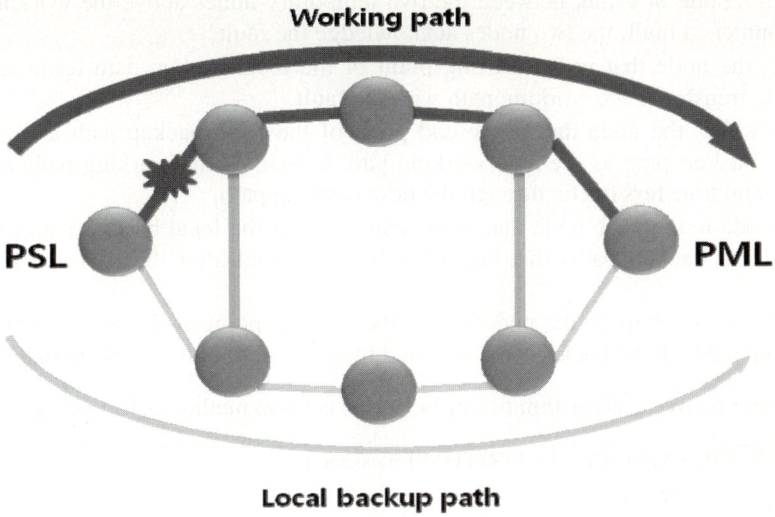

Fig. 4. This figure shows worst case of 1+1 protection method

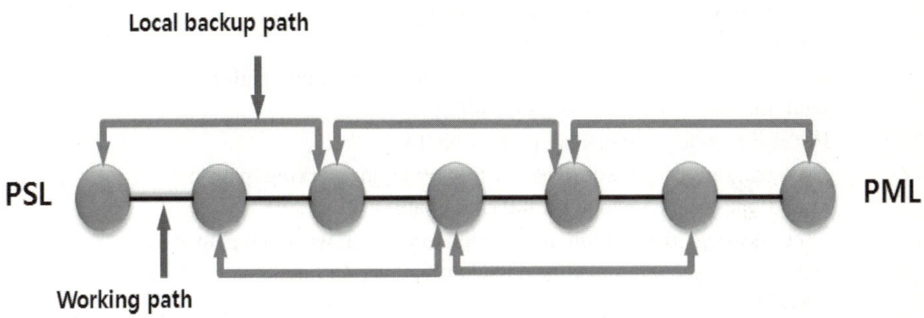

Fig. 5. This figure shows mechanism of 1+1+1 protection method

Therefore, we propose a recovery method in which each node above the working path has a local backup path as shown in [Figure 5] and that performs identical operations as the 1+1 protection method in order to minimize recovery time. We denote this method the 1+1+1 protection method because it consumes the same amount of resources as adding another backup path to the 1+1 protection method.

The procedure for the 1+1+1 protection method is following.

① Create a working path for transferring data.

② Create a local backup path for each node.

③ When transferring traffic through the working path, transmit identical traffic through the local backup path.

④ The two nodes that form the local backup path must be aware of the mapping relationship between the label used for the local backup path that they're associated in and the label used for the working path.

⑤ When a node or a link between the two temporary nodes above the working path encounters a fault, the two nodes acknowledge the fault.
⑥ Here, the node that is the starting point of the local backup path terminates the traffic transfer to the working path with the fault.
⑦ Meanwhile, the node that is the end point of the local backup path assumes the local backup path as the new working path instead of the working path with the fault and transfers traffic through the new working path.
⑧ The node next to the node that is the end point of the local backup path removes any local backup paths that are unlikely to be used after the occurrence of the fault.
⑨ The node previous to the node that is the starting point of the local backup path sets up a new local backup path required by the update of the working path.

The fault recovery algorithm of the 1+1+1 protection method is following.

```
FAULT RECOVERY ALGORITHM ( message )
SWITCH ( message )
  CASE 'left node or edge in working path has a fault' :
    working_path_receive_stop();
    IF ( local_backup_path is existed )
      set_working_path ( local_backup_path );
      working_path_receive_start ( );
      set_message_destination_node ( left_node_in_working_path );
      send_message ( 'update working path' );
      IF ( right_node_in_working_path != NULL )
        set_message_destination_node ( left_node_in_working_path );
        send_message ( 'change local backup path' );
        set_message_destination_node ( right_node_in_working_path );
        send_message ( 'remove local backup path by end node' );
      END IF
    ELSE
      set_message_destination_node ( right_node_in_working_path );
      send_message( 'left node or edge in working path has a fault' );
      working_path_send_stop ( );
    END IF
    break;
  END CASE
  CASE 'update working path' :
    IF ( left_node_in_working_path != NULL )
      set_message_destination_node ( left_node_in_working_path );
      send_message ( 'update end node in local backup path' );
    END IF
    break;
  END CASE
  CASE 'update end node in local backup path' :
    new_end_node = get_right_node_in_working_path( right_node_in_working_path );
```

```
        local_backup_path_remove ( );
        new_path = set_path ( present_node, new_end_node );
        set_local_backup_path ( new_path );
        break;
    END CASE
    CASE 'change local backup path' :
        working_path_send_stop ( );
        working_path_remove ( );
        set_working_path ( local_backup_path );
        new_end_node = get_right_node_in_working_path( right_node_in_working_path );
        new_path = set_path ( present_node, new_end_node );
        set_local_backup_path ( new_path );
        break;
    END CASE
    CASE 'right node or edge in working path has a fault' :
        working_path_send_stop ( );
        break;
    END CASE
    CASE 'remove local backup path by end node' :
        set_message_destination_node ( left_node_in_local_backup_path );
        send_message ( 'remove local backup path' );
        local_backup_path_receive_stop ( );
        break;
    END CASE
    CASE 'remove local backup path' :
        local_backup_path_send_stop ( );
        local_backup_path_remove ( );
        break;
    END CASE
    DEFAULT :
        return 'recovery is failed'
        break;
    END DEFAULT
END SWITCH
return 'recovery is success'
END FAULT RECOVERY ALGORITHM
```

An optimal recovery time can be reached for the 1+1+1 protection method regardless of the location of fault occurrences as it can recover not only faults on links but also faults on nodes. It can also reduce the amount of traffic lost by transmitting traffic through the backup path and the working path simultaneously as it is done in the 1+1 protection method. The 1+1+1 protection method can resolve any additional faults by removing unnecessary local backup paths and adding necessary local backup paths when the local backup path has been changed to a working path due to a fault.

4 Hybrid Recovery Method

As per the 1+1 protection method, it would not be correct to say that the backup path used when a fault occurs is the optimal path because the path is set prior to transferring the traffic. Therefore, in this study, we use the backup path set by using the 1+1 protection method as the working path for occurrence of a fault and we use the restoration method to create a new working path and propose a hybrid recovery that uses this as the working path. This is because the restoration method computes and configures backup paths based on the current status and policy of the network thereby choosing the optimal path.

The hybrid recovery procedure is following.

① Create a working path and a backup path for data transfer.

② When transferring traffic to the working path, transfer an identical traffic to the backup path setup by the 1+1 protection method.

② When a fault occurs in the working path, notify this to the PSL (Path Switching LSR) and the PML (Path Merge LSR).

③ The PML uses the backup path set by the 1+1 protection method as the working path and the PSL executes the same operation as the restoration method to create a new backup path.

④ When a backup path has been configured in the PSL, compare the backup path created by the restoration method and the backup path created by the 1+1 protection method.

⑤ If they are identical, the PSL creates a new backup path by finding the optimal path excluding the previous backup path.

⑥ If they are different, the backup path created by the restoration method becomes the new working path and the backup path created by the 1+1 protection method becomes a backup path again to transfer traffic.

The fault recovery algorithm of the hybrid recovery method is following.

```
FAULT RECOVERY ALGORITHM FOR HRM ( message )
SWITCH ( message )
  CASE 'left node or edge in working path has a fault' :
    IF ( present_node == PML )
      working_path_receive_stop ( );
      set_working_path ( backup_path );
      working_path_receive_start ( );
    ELSE
      set_destination_node ( right_node_in_working_path );
      send_message ( 'left node or edge in working path has a fault' );
      working_path_send_stop ( );
      working_path_remove ( );
    END IF
    break;
  END CASE
  CASE 'right node or edge in working path has a fault' :
```

```
        working_path_send_stop ( );
        working_path_remove ( );
          IF ( present_node == PSL )
            optimal_path = compute_optimal_path (PSL, PML);
            IF ( backup_path == optimal_path )
              set_working_path ( backup_path );
              new_path = set_path ( PSL, PML );
              set_backup_path ( new_path );
              backup_path_send_start( );
            ELSE
              new_path = set_path ( PSL, PML );
              set_working_path ( new_path );
              working_path_send_start ( );
              set_backup_path ( backup_path );
              set_destination_node ( PML );
              send_message ( 'change working path' );
            END IF
          ELSE
            set_destination_node ( left_node_in_working_path );
            send_message ( 'right node or edge in working path has a fault' );
          END IF
      break;
    END CASE
    CASE 'change working path' :
      temp_path = get_working_path ( );
      set_working_path ( backup_path );
      set_backup_path ( temp_path );
      break;
    END CASE
    DEFAULT :
      return 'recovery is failed'
      break;
    END DEFAULT
  END SWITCH
  return 'recovery is success'
  END FAULT RECOVERY ALGORITHM FOR HRM
```

By performing recovery using the 1+1 protection method, the hybrid recover method reduces traffic losses. Also, it optimizes the efficiency of traffic transfer by finding the optimal path after recovery to use as the working path. On the down side, it requires more control traffic compared to the previous protection or restoration method and requires more resources. However, the exponential increase in the bandwidth and the advancement of network tool technologies today mitigate the issue of overheads.

5 Analysis of Performance of Proposal Method

5.1 Analysis of Performance of 1+1+1 Protection Method

We compute average estimated time for fault recovery, when a network uses 1+1+1 protection method and has n nodes. n means the number of nodes on working path in the network. We also compute average estimated time for fault recovery, when a network uses 1+1 protection method and has n nodes, to compare performance of 1+1+1 protection method to 1+1 protection method.

Average estimated time for fault recovery of 1+1+1 protection method is followings. T_{trans} means transmission time between two nodes and *Tproc* means processing time in a node to deliver packets toward another node.

$$\begin{aligned} T_{avg} &= \frac{1}{n-1}\left\{(T_{trans}+T_{proc})+\sum_{i=1}^{n-1}\frac{T_{trans}}{2}\right\} \\ &= \frac{(n+1)}{2(n-1)}T_{trans}+\frac{1}{2(n-1)}T_{proc} \end{aligned} \tag{1}$$

Average estimated time for fault recovery of 1+1 protection method is followings.

$$\begin{aligned} T_{avg} &= \frac{1}{n-1}\sum_{i=1}^{n-1}\left\{(n-i-1)(T_{trans}+T_{proc})+\frac{T_{trans}}{2}\right\} \\ &= \frac{(n-1)\,T_{trans}}{2}+\frac{(n-2)\,T_{proc}}{2} \end{aligned} \tag{2}$$

[Figure 6], [Figure 7], [Figure 8] show difference between performance of 1+1+1 protection method and 1+1 protection method. These figures shows performance of 1+1+1 protection method is approximately fifth higher than performance of 1+1 protection methods if n is 5.

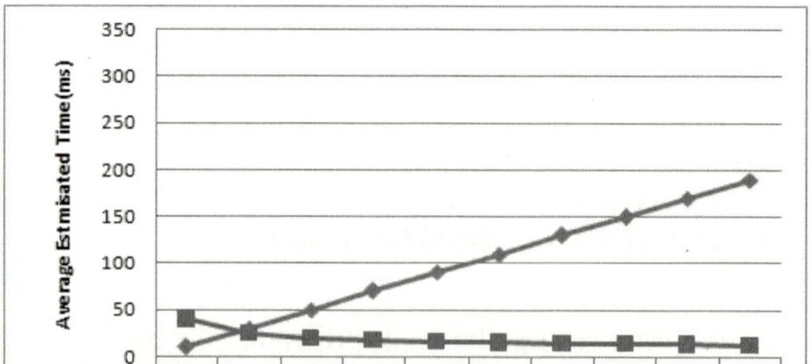

	2	3	4	5	6	7	8	9	10	11
1+1	10	30	50	70	90	110	130	150	170	190
1+1+1	40	25	20	17.5	16	15	14.2	13.7	13.3	13

Fig. 6. This figure shows performance of 1+1+1 protection method and performance of 1+1 protection method, when *Ttrans* and *Tproc* are 20ms

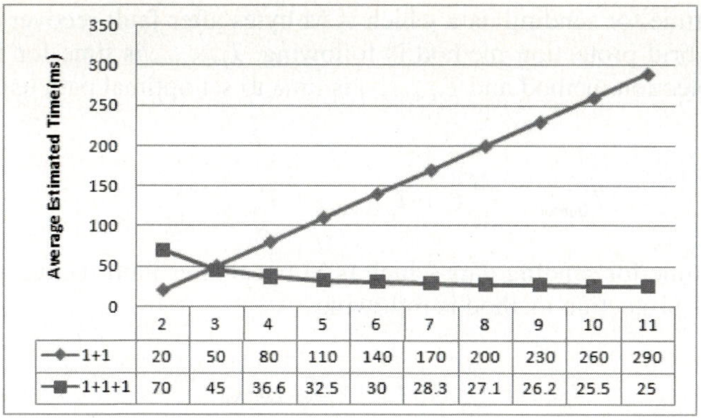

	2	3	4	5	6	7	8	9	10	11
1+1	20	50	80	110	140	170	200	230	260	290
1+1+1	70	45	36.6	32.5	30	28.3	27.1	26.2	25.5	25

Fig. 7. This figure shows performance of 1+1+1 protection method and performance of 1+1 protection method, when *Ttrans* is 40ms and *Tproc* is 20ms

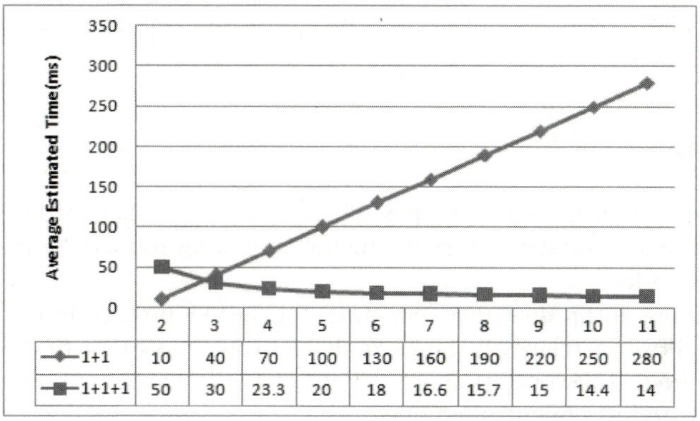

	2	3	4	5	6	7	8	9	10	11
1+1	10	40	70	100	130	160	190	220	250	280
1+1+1	50	30	23.3	20	18	16.6	15.7	15	14.4	14

Fig. 8. This figure shows performance of 1+1+1 protection method and performance of 1+1 protection method, when *Ttrans* is 20ms and *Tproc* is 40ms

These figures note that average estimated time of 1+1 protection method increased by 50% if *Trans* or *Tproc* becomes twice. Average estimated time of 1+1+1 protection method increased by approximately 90%, if *Ttrans* becomes twice. While *Tproc rarely affect* average estimated time of 1+1+1 protection time.

5.2 Analysis of Performance of Hybrid Protection Method

We compute estimated time for sending data which is M bytes after fault recovery, when a network uses Hybrid protection method. We also compute estimated time for sending data which is M bytes after fault recovery, when a network uses 1+1 protection method. t_1 is estimated time for sending data which is m bytes on optimal path and t_2 is estimated time for sending data which is m bytes on backup path of 1+1 protection method. *t2* is presented by (3).

$$t_2 = t_1 + \alpha \quad \alpha \tag{3}$$

Estimated time for sending data which is M bytes after fault recovery, when a network uses Hybrid protection method is following. $T_{protection}$ is time for fault recovery using 1+1 protection method and $T_{restoration}$ is time to set optimal path using restoration method.

$$T_{Hybrid} = \frac{M}{m} t_1 + T_{protection} + T_{restoration} \tag{4}$$

Estimated time for sending data which is M bytes after fault recovery, when a network uses 1+1 protection method is following.

$$\begin{aligned} T_{1+1} &= \frac{M}{m}(t_1 + \alpha) + T_{protection} \\ &= \frac{M}{m} t_1 + \frac{M}{m} \alpha + T_{protection} \end{aligned} \tag{5}$$

Therefore, performance of Hybrid protection method is higher than 1+1protection method if backup path in 1+1 protection method is not optimal path and $T_{restoration}$ is smaller than (M/m)α α

6 Conclusion

In this research, we proposed two types of fault recovery schemes designed by combining the protection and the restoration methods in order to increase the survivability of GMPLS networks.

By maintaining an optimal recovery time, the 1+1+1 protection method proposed in this study allows a relatively faster fault recovery compared to the existing 1+1 protection method. It also offers more variety of fault recoveries because it not only considers faults in links but also in nodes.

The hybrid recovery method proposed in this study is able to enhance the performance after recoveries by integrating the 1+1 protection method with the restoration method. In other words, it executes recovery using the 1+1 protection method and then it uses the restoration method to find the optimal path to use as the working path thereby optimizing the traffic transfer efficiency.

Although the two methods introduced in this study consume more resources than the ones in existence, they are more advantageous in terms of performance after recovery and recovery time. However, the exponential increase in the bandwidth and the advancement of network tool technologies today mitigate the issue of overheads.

There will be more studies along the way to examine the implementation of the two methods proposed in this paper because they are much more sophisticated than the ones before. It is also infeasible to compare these two methods with the ones before because they have not gone through any performance evaluation. Therefore, we will need to conduct a verification of their performances through simulations. Lastly, given the exponentially increasing bandwidth as of current, we will focus more on promoting recovery success rate or reducing recovery time more so than reducing resource consumption.

References

1. Kim, H.: A Study of Survivable Traffic Grooming in Broadband Convergence Networks, Thesis of Doctoral course, Sungkyunkwan Univ. (2005)
2. Grover, W.D.: Mesh-Based Survivable Networks - Options and Strategies for Optical, MPLS, SONET, and ATM Networking. Prentice-Hall, Englewood Cliffs (2004)
3. Nam, C., Go, K., Noh, M., Kim, S., Kim, H., Lee4, J., Chung, J.: LSA Expansion for Fault Recovery in GMPLS Network. IJCSNS 7(10), 114–122 (2007)
4. Mannie, E., et al. (eds.): Generalized MPLS Architecture, draft-ietf-ccamp-gmpls-architecture-03.txt, (August 2002)
5. Tomsu, P., Schmutzer, C.: Next Generation Optical Networks, pp. 161–180. Prentice Hall PTR (2002)
6. Papadimitriou, D., Mannie, E.: Analysis of Generalized Multi-Protocol Label Switching (GMPLS)-based Recovery Mechanisms (including Protection and Restoration), draft-ietf-ccamp-gmpls-recovery-analysis-04.txt, IETF (2004)
7. Makam, S., et al.: A Path Protection/Restoration Mechanism for MPLS Networks, Internet Draft, draft-makam-mpls-protection-00.txt (October 1999)
8. Mannie, E., Papadimitriou, D. (eds.): Recovery (Protection and Restoration) Terminology for Generalized Multi-Protocol Label Switching (GMPLS) RFC 4427 (2006)
9. Cisco co., MPLS Traffic Engineering: Shared Risk Link Groups (2007), `http://www.cisco.com/univercd/cc/td/doc/product/software/ios120/120newft/120limit/120s/120s29/fs29srlg.pdf`
10. Qin, Y., et al.: Study on a Joint Multiple Layer Restoration Scheme for IP over WDM Networks. IEEE Network, 43–48 (March 2003)
11. Hjalmtysson, G., Yates, J., Chaudhuri, S.: Restoration Services for the Optical Internet. In: Proceeding of SPIE, Terabit Optical Networking: Architecture, Control, and Management Issues, vol. 4213 (October 2000)
12. Oh, S., Kim, Y.: A Comparison of Restoration Schemes in MPLS Network, KICS vol 27(4C) (2002)

Efficient Facial Features Warping Using BSM (Bayesian Shape Model)

Joo-Hyung Kwon, Gye-Young Kim, and Youngsong Mun

School of Computing, Soongsil University, Seoul, Korea
joohyung3@ssu.ac.kr, gykim11@ssu.ac.kr, mun@ssu.ac.kr

Abstract. This paper proposes an efficient method for warping facial features. The existing methods use points, which are standard for facial features warping, without the reason or calculation. And existing methods have difficulties for facial feature warping. We estimate the standard points by using BSM(Bayeian Shape Model). From the experiment results for the various image, the proposed algorithm shows more natural results than the conventional algorithm and is more efficient than ASM(Active shape model).

Keywords: BSM, facial feature, warping.

1 Introduction

Recently one of subject, that computer visual field study, is the research about face. The research for face is various. Face detection, face recognition, facial feature extraction etc. This research must need face detection and feature point extraction. Specially feature point extraction is more important for face recognition and warping. This paper proposed other method. When we warp for facial expression deformation, more efficient by using BSM(Bayesian Shape Model). Chapter 2 explain the summary of system. And chapter 3 explain about BSM and compare passive method and ASM(Active shape model).

2 Summary of System

Fig.1 appears about the summary of system. We use BSM(Bayesian Shape Model) for getting feature points. After getting feature points, we can move feature points. Gap=(before position – after positon). Finally we warp image about gap.

3 Bayesian Model(BM)

3.1 Bayesian Rule

BSM(Bayesian Shape Model) originates from Bayesian Rule. So This chapter explains about that. Bayesian Rule relates the conditional and marginal probabilities of stochastic events A and B:

O. Gervasi et al. (Eds.): ICCSA 2008, Part II, LNCS 5073, pp. 56–66, 2008.

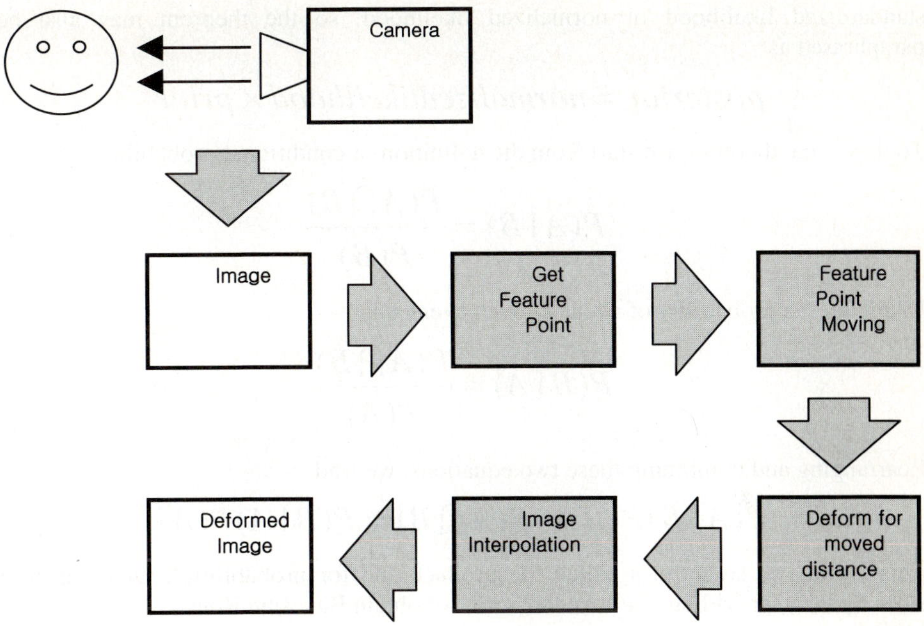

Fig. 1. Summary of System

$$P(A \mid B) = \frac{P(B \mid A)P(A)}{P(B)} \tag{1}$$

Each term in Bayesian Rule has a conventional name: P(*A*) is the prior probability or marginal probability of *A*. It is "prior" in the sense that it does not take into account any information about *B*. P(*A*|*B*) is the conditional probability of *A*, given *B*. It is also called the posterior probability because it is derived from or depends upon the specified value of *B*. P(*B*|*A*) is the conditional probability of *B* given *A*.

P(*B*) is the prior or marginal probability of *B*, and acts as a normalizing constant. Bayes' theorem can also be interpreted in terms of likelihood:

$$P(A \mid B) \propto L(A \mid B)P(A)$$

Here *L*(*A*|*B*) is the likelihood of *A* given fixed *B*. The rule is then an immediate consequence of the relationship *P*(*B* | *A*) = *L*(*A* | *B*). In many contexts the likelihood function *L* can be multiplied by a constant factor, so that it is proportional to, but does not equal the conditional probability *P*. With this terminology, the theorem may be paraphrased as

$$posterior = \frac{likelihood \times prior}{normalizingcons\tan t} \tag{2}$$

In words: the posterior probability is proportional to the product of the prior probability and the likelihood. In addition, the ratio *L*(*A*|*B*)/*P*(*B*) is sometimes called the standardized likelihood or normalized likelihood, so the theorem may also be paraphrased as

$$posterior = normalizedlikelihood \times prior$$

To derive the theorem, we start from the definition of conditional probability.

$$P(A \mid B) = \frac{P(A \cap B)}{P(B)} \tag{3}$$

Likewise, the probability of event *A* given event *B* is

$$P(B \mid A) = \frac{P(A \cap B)}{P(A)}$$

Rearranging and combining these two equations, we find

$$P(A \mid B)P(B) = P(A \cap B) = P(B \mid A)P(A)$$

This Lemma is sometimes called the product rule for probabilities. Dividing both sides by P(*B*), providing that it is non-zero, we obtain Bayesian Rule:

$$P(A \mid B) = \frac{P(B \mid A)P(A)}{P(B)} \tag{4}$$

3.2 Bayesian Framework and Energy Terms

The Bayesian Model (BM) formulates the matching of a deformable model to the object in a given image as a maximizing *a posteriori* (MAP) estimation problem. Denote the mean of the sample contours in the shape domain as $\overline{f_0}$ (the mean contour), the deformed version of $\overline{f_0}$ as $\overline{f}$ (the prototype contour), and the deformable contour in the image domain as f, where $\overline{f_0} \in R^{2\times N}$, $\overline{f} \in R^{2\times N}$ and $f \in R^{2\times N}$ are the matrices representing the corresponding contours formed by the coordinates of N landmark/boundary points. According to the Bayesian estimation, the joint posterior distribution of f and $\overline{f}$, $p(f, \overline{f} \mid d)$, is

$$p(f, \overline{f} \mid d) = \frac{p(d \mid f)p(f, \overline{f})}{p(d)} \tag{5}$$

Where $p(d, f) = p(d \mid f, \overline{f})$ is the likelihood distribution of input image data d.

$$p(f,\overline{f}) = p(f,\overline{f})p(\overline{f}) \tag{6}$$

is the joint prior distribution of f and $\overline{f}$. For a given image d, the MAP estimates, f_{MAP} and $\overline{f}_{MAP}$, can be defined as, $\{f_{MAP},\overline{f}_{MAP}\} = \arg\max_{f,\overline{f}}\{p(f,\overline{f} \mid d)\}$

$$= \arg\max_{f,\overline{f}}\{\frac{p(d \mid f)p(f \mid \overline{f})p(\overline{f})}{p(d)}\} \tag{7}$$

where $p(\overline{f})$ is the distribution of the prototype $\overline{f}$, $p(f \mid \overline{f})$ describes the variations between f and $\overline{f}$, and $p(d \mid f)$ indicates the matching between f and the salient features of the object in image d. It can be seen from Eq.7 that BM estimates the suitable prototype shape $\overline{f}$, and obtains a best deformable contour in the image domain f to match the object. Therefore, in the BM, not only the prototype, but also the deformation between f and the prototype $\overline{f}$ can be adaptively adjustable. Provided the densities in Eq.7 can be modeled as Gibb's distribution, i.e.

$$\begin{aligned} p(\overline{f}) &= Z_1^{-1}\exp\{-E_{con}(\overline{f})\} \\ p(f \mid \overline{f}) &= Z_2^{-1}\exp\{-E_{\text{int}}(f \mid \overline{f})\} \\ p(d \mid \overline{f}) &= Z_3^{-1}\exp\{-E_{ext}(d \mid f)\} \end{aligned} \tag{8}$$

where Z_1, Z_2, and Z_3 are the partition functions, maximizing the posterior distribution is equivalent to minimizing the corresponding energy function of the contour

$$\{f_{MAP},\overline{f}_{MAP}\} - \arg\min_{f,\overline{f}}\{E_{contour}\} \tag{9}$$

Where $E_{contour} = E_{con} + E_{\text{int}} + E_{ext} \cdot E_{con} = E_{con}(\overline{f})$ is the constraint energy term of the adjustable prototype contour $\overline{f}$, which limits the variations of . $\overline{f}$ and ensures that $\overline{f}$ is similar with $\overline{f_0}$ (in shape. $\overline{f_0}$ is the mean of all the aligned sample contours in the shape domain. $E_{\text{int}} = E_{\text{int}}(f \mid \overline{f})$ is the internal energy term that describes the global and local shape deformation between f and $\overline{f}$. The external energy term $E_{ext} = E_{ext}(d \mid f)$ defines the degree of matching between f and the salient features of image d.

3.3 The Constraint Energy of Prototype

The constraint energy term E_{con} of the prototype contour is caused by the prior distribution, $p(\overline{f})$, of the samples in the shape domain, which can be estimated from the normalized sample data and reflects the shape variations of the shapes in the shape domain. Linear or nonlinear methods can be used to estimate this prior distribution according to the application. In this paper, the method using PCA to estimate $p(\overline{f})$ is utilized, which reflects mainly the linear relationship of the shape variations. In cases where all the samples are aligned views of similar objects seen from a standard view, this distribution can be accurately modeled by a single Gaussian distribution [9].

$$p(\overline{f}) = \frac{\exp(-\frac{1}{2}\sum_{i=1}^{M}\frac{\omega_i^2}{e_i})}{(2\pi)^{M/2}\prod_{k=1}^{M} e_k^{1/2}} \tag{10}$$

Where

$$\omega = \Phi_M^T(\overline{f} - \overline{f_0}) \tag{11}$$

is the vector of the shape parameters, and $\overline{f} - \overline{f_0}$ is the deformation from $\overline{f_0}$ to $\overline{f}$. Φ_M is the matrix composed of the eigenvectors corresponding to the largest M eigenvalues e_i, $(1 \le i \le M)$, which is computed from the covariance matrix of all the sample contours. Therefore, using the PCA, a prototype contour can be reconstructed from $\overline{f_0}$ and a given ω,

$$\overline{f} = \overline{f_0} + \Phi_M \omega \tag{12}$$

Note in Eq.11 and Eq.12, $\overline{f}$ and $\overline{f_0}$ have been expanded to $2N \times 1$ vectors. The PCA representation preserves the major linear correlations of the sample shapes and discards the minor ones, hence provides an optimized approximation of $\overline{f}$ in the sense of least squares error. This representation describes the most significant modes of the shape variations or the global shape deformations subject to the prior distribution of the prototype contour. From Eq.10, the corresponding constraint energy is denoted as

$$E_{con} = \frac{1}{2}\sum_{i=1}^{M}\frac{\omega_i^2}{e_i} \tag{13}$$

The variations of the prototypes contour is limited by the plausible area of the corresponding shape parameters ω, which is defined as

$$\sum_{i=1}^{M} \frac{\omega_i^2}{e_i} \le M_t \tag{14}$$

The threshold, M_t, may be chosen using the x^2 distribution [6]. The constraint energy term ensures that the dynamically adjustable prototype contour remains similar with the mean shape during the matching process, and at the same time, large shape variations and deformations subject to the prior distribution of the samples can be tolerated.

3.4 Transformational Invariant Internal Energy

Since $p(f \mid \overline{f})$ or $E_{\text{int}}(f \mid \overline{f})$ measures the degree of matching between the deformable model f in the image domain and the prototype $\overline{f}$ in the shape domain, a transformational invariant internal energy term should be designed and incorporated in the BM to deal with the corresponding transformations (such as Euclidian, similarity and affine transforms) between the shape domain and the image domain. Mathematically, $\overline{f}$ and f are related by

$$f_i = T(\overline{f_i}) + \varepsilon, \ (1 \le i \le N), \tag{15}$$

where $T(\cdot)$ refers to the transformation operator, and ε represents the deformation. In this paper, the similarity or affine transformations are considered,

$$T(\overline{f_i}) = A\overline{f_i} + t, \ (1 \le i \le N) \tag{16}$$

where A is the corresponding transformational 2×2 matrix, t is a translation vector.

Since the point correspondence between f and $\overline{f}$ is known, by defining the least squares objective function, the transformational parameters can be easily determined [6, 10, 11]. Let $\hat{A}$ and $\hat{t}$ represents the alignment results of f and $\overline{f}$, the global internal energy term of the deformable contour is designed as,

$$E_{g\text{int}}(f \mid \overline{f}) = \frac{1}{N}\sum_{i=1}^{N}[(\overline{f_i} - \hat{A}^{-1}(f_i - \hat{t}))^T \cdot (\overline{f_i} - \hat{A}^{-1}(f_i - \hat{t}))] \tag{17}$$

It can be seen that the internal energy term is transformational invariant, which indicates the global degree of matching between the deformable contour f and the prototype $\overline{f}$.

In addition, to describe the local shape variations, a local internal energy term is also defined by the proportion of area,

$$E_{l\,\text{int}}(f_i \mid \overline{f}) = \frac{(S_1 + S_2)AREA_{proto}}{AREA_{aligned}} \tag{18}$$

$S_1 = S(\overline{f}_{i-1}, f_i^a, \overline{f}_i)$ and $S_2 = S(\overline{f}_i, f_i^a, \overline{f}_{i+1})$, where $S()$ is the area of the triangle formed by the three points inside the brackets. $AREA_{proto}$ and $AREA_{aligned}$ represent the interior areas formed by the hull of the prototype contour and the aligned deformable contour $(f_i^a = \hat{A}^{-1}(f_i - \hat{t}))$ in the shape domain respectively. When the areas of the triangles S_1 and S_2 are close to zero, the shape and position of $\overline{f}$ and f will also be close. In case the contour represents an open shape, the local internal energy at the end point of the contour will not be calculated.
In summary, the internal energy is composed of both the global and local terms,

$$E_{\text{int}}(f \mid \overline{f}) = E_{g\,\text{int}}(f \mid \overline{f}) + \frac{1}{N}\sum_{i=1}^{N} E_{l\,\text{int}}(f_i \mid \overline{f}) \tag{19}$$

which reflects the degree of fitting between $\overline{f}$ and f, and remains invariant under similarity and affine transformations.

3.5 External Image Constraints

The external energy term $E_{ext} = E(d \mid f)$ indicates the degree of matching between the deformable contour f and the salient image features. Minimizing E_{ext} adjusts f and moves it towards the object boundary in the image d. The external energy usually combines all the information of edge, texture, color and region, etc, so that it can provide an effective description of the matching. For example, the color information can be combined into the edge detection process, so that the edge maps will accurately stand for the boundaries of the interested objects.

Among various matching rules and external energy terms used in the literature, the energy term including both the gradient and directional edge information is utilized because of its simplicity and efficiency [3].

First, the image $d = \{d(x, y)\}$ is smoothed using Gaussian function, $G_\sigma(x, y)$, with the deviation σ.

$$d_\sigma(x, y) = G_\sigma(x, y) * d(x, y) \tag{20}$$

This convolution can be performed in frequency domain by using FFT/IFFT quickly.

Second, the normalized gradient of the smoothed image d_σ at each pixel location (x, y), denoted
$d^g_{\sigma x}(x, y), d^g_{\sigma y}(x, y)), (\| d^g_\sigma(x, y) \| \in [0,1])$, is computed.

At last, constraints on the deformable contour f ensure that f moves towards the object boundaries: when the image pixels along the contour have high gradient magnitude, and the direction of the image gradient along the contour is perpendicular to the orientation of the contour, the external energy is small. Therefore, the external energy function is defined as

$$E_{ext}(d \mid f) = \sum_{i=1}^{N} (1 - \| d^g_\sigma(x_i, y_i) \|)$$
$$\mid n(x_i, y_i) \cdot h(x_i, y_i) \mid \tag{21}$$

where "" is the dot product. $h(x, y)$ is the direction of the gradient $d^g_\sigma(x, y)$ and $h(x, y) = d^g_\sigma(x, y) / \| d^g_\sigma(x, y) \| \; and \; \| h(x, y) \| = 1$. $n(x_i, y_i)$ indicates the normal vector of the contour f at point $f_i = (x_i, y_i)$, with $\| n(x_i, y_i) \| = 1$ and $n(x_i, y_i) = [\begin{smallmatrix} 0 & -1 \\ 1 & 0 \end{smallmatrix}] V_i / \| V_i \|$, where $V_i = \frac{f_{i+1} - f_i}{\| f_{i+1} - f_i \|} + \frac{f_i - f_{i-1}}{\| f_i - f_{i-1} \|}$ is the tangent vector of contour f at point f_i.

3.6 Compare Others

This chapter shows the answer for "The work proposed is efficient. Why?". We compare three kinds of method. First, the passive method that we get standard points by mouse click. Second, the one of automatic methods by ASM. Third, the other of automatic methods by BSM. The first thing has very variable and needs lots of effort and time. Because the each standard point needs click by user. [4] verify that BSM is more efficient than ASM. ASM is frequent method in face recognition field. Next graph and chart are schematization of compared result in [4].

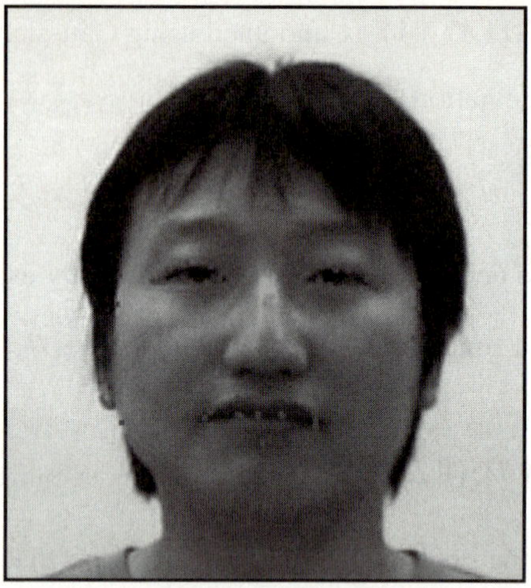

Fig. 2. Example of Feature Points

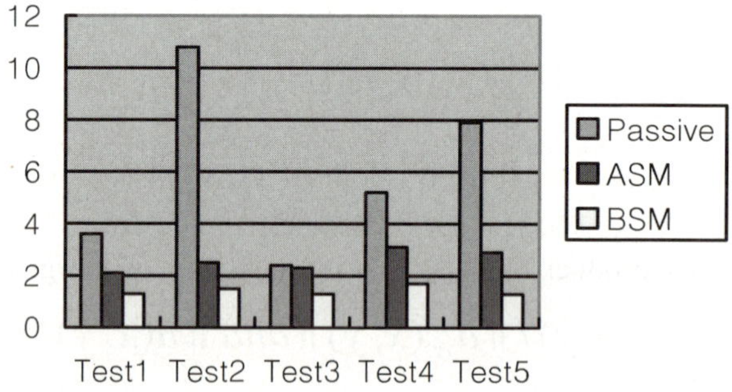

Fig. 3. Compared graph of passive, ASM, BSM

Table 1. The result of comparison (Passive, ASM, BSM)

	Test1	Test2	Test3	Test4	Test5
Passive	3.6	10.8	2.4	5.2	7.9
ASM	2.1	2.5	2.3	3.1	2.9
BSM	1.3	1.5	1.3	1.7	1.3

4 Experimental Results

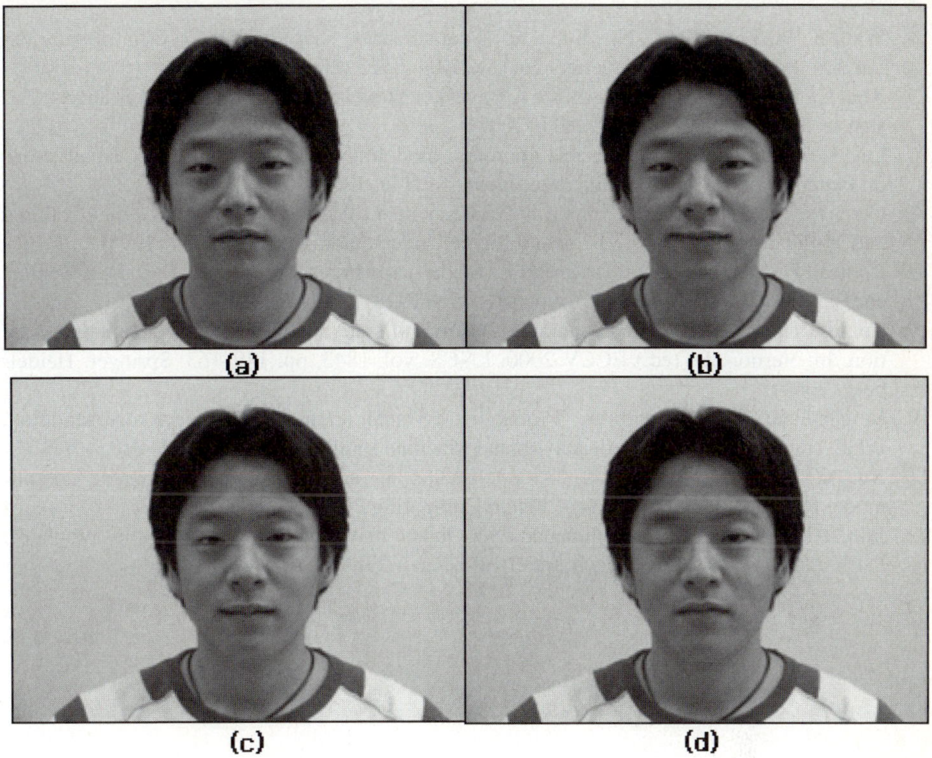

Fig. 4. Comparison between using BSM and not using BSM. (a)original image, (b)not using BSM, (c)(d)using BSM.

5 Conclusion

This paper proposes an efficient method for warping facial features. We use BSM(Bayesian Shape Model) for getting feature points. This method is more efficient than without BSM. And We compare two kinds of method. Recently the study about face is crowd and various. Next time we try topic that 3D facial warping for expression.

Acknowledgments

This research is supported by Seoul R&BD Program(10581C093113).

References

1. Knappmeyer, B., Thornton, I.M., Bulthoff, H.H.: The use of facial motion and facial form during the processing of identity. Vision research 43(18), 1921–1936 (2003)
2. Su, C., Zhuang, H., Huang, L., Wu, F.: Bayesian Network Enhanced Prediction Based Multiple Facial Feature Tracking. Journal of image and graphics 10(2), 175–180 (2005)

3. Xue, Z., Li, S.Z., Shen, D., Teoh, E.K.: A novel Bayesian shape model for facial feature extraction. IEEE Transactions on Patten Analysis and Machine Intelligence 1, 514–519 (2002)
4. Wan, K.W., Lam, K.M., Ng, K.C.: An accurate active shape model for facial feature extraction. Pattern recognition letters 26(15), 2409–2423 (2005)
5. Xue, Z., Li, S.Z., Lu, J., Teoh, E.K.: A Bayesian Model for Extracting Facial Features. Pattern recognition 36(12), 2819–2833 (2003)
6. Lai, K.F., Chin, R.T.: Deformable contours: modeling and extraction. IEEE Transactions on Patten Analysis and Machine Intelligence 17(11), 1084–1090 (1995)
7. Cootes, T.F., Taylor, C.J., Cooper, D.H., Graham, J.: Active shape models - their training and application. Computer Vision and Image Understanding 61(1), 38–59 (1995)
8. Cootes, T., Edwards, G.J., Taylor, C.J.: models. In: Proceeding of 5th European Conference on Computer Vision, vol. 2, pp. 484–498 (1998)
9. Cootes, T.F., Taylor, C.J.: Combining elastic and statistical models of appearance variation. In: Vernon, D. (ed.) ECCV 2000. LNCS, vol. 1843, pp. 149–163. Springer, Heidelberg (2000)
10. Moghaddam, B., Pentland, A.: Probabilistic visual learninig for object representation. IEEE Transactions on Patten Analysis and Machine Intelligence 19(7), 696–710 (1997)
11. Xue, Z., Li, S.Z., Teoh, E.K.: Ai-eigensnake: an affine-invariant deformable contour model for object matching. Image Vision Computing 20(2), 77–84 (2002)
12. Jain, A.K., Zhong, Y., Lakshmanan, S.: Object matching using deformable templates. IEEE Trans. Pattern Anal. Mach. Intell 18(3), 267–278 (1996)

Background Separation in Motion Picture Using Presumptive Eyes Position and Different Image

Se-Young Jang and Hoon-Sung Kwak

Department of Computer Engineering, Chonbuk National University,
Jeonju, 561-756, Korea
{sewise,hskwak}@chonbuk.ac.kr

Abstract. The paper used the method using the information of colors by converting RGB information, the color image information to HSI model and YC_bC_r model and the presumptive eyes position, as the detection method of the region of face and its minutiae. From the sample image, the study suggested the method to separate and extract a subject from its background even in the sample images where there are colors similar to a given skin colors and to replace the original background with another background. Then, by using the only 4.8% of the overall region, the initial region of interest to comprehend filtering and a moving object was created, which demonstrates that based on the overall region, the only 13% may be used to track an actually moving object.

Keywords: Moving picture, Image processing, Background.

1 Introduction

As mobile terminal has recently come into wide use, it is also attempted that various terminals capable of processing images are released in accordance with users' requirements. Therefore, application of image processing is becoming a matter of primary concern. While the terminals in the past mainly focused on voice call functions, the recent terminals support PDA functions and some support digital camera as image processing technology has been rapidly developed. However, due to the users' requirements for the mobile terminals with imaging function, especially for image processing, the terminals are being equipped with additional functions as well as the existing camera function. In addition, the camera function built on the terminals is rapidly developed, and its effective pixels are not significantly different with those of the exclusive stand-alone digital cameras, which demonstrate that mobile terminals, in the future, should consider or reinforce video messaging using digital camera function video share calling.

Based on the above background, the study attempted to separate a user's movement tracing or image as an object from its background. It is a necessary component in video share calling, in which a user becomes a subject in video, in order to process various additional functions. Despite of the fact that realizing object-tracing in video in real time by using computer video processing technology requires considerable

O. Gervasi et al. (Eds.): ICCSA 2008, Part II, LNCS 5073, pp. 67–77, 2008.

time and complexity, the study on object-in-video recognition and object tracing made actively progress as the computer performance has already led the related technologies[1].

Video information from mobile terminals basically message video and image information while being accompanied with the resultant information on a user's location and situation. That is to say, the information of a user's location and situation is contained in the background image. However, it may be often the cases that one does not wish to show the background image or it should be restricted due to security problem or other reasons. Therefore, the studies suggest the solution for the above situation by separating a user image from the entire image and then replacing the background image with a different image. Furthermore, eliminating background image also benefits by the reduction of data to transfer, increasing the compression ratio; assuming a same bandwidth, higher quality image may be transferred because the transmission speed increases more.

In case of the communication via a wire video phone transmitting a video including one's face, the technology to extract and separate a user's image from the background image consists of semi-automatic segmentation and background synthesis technology. The most currently used 'object segmentation technology' creates the segmented image information by either using blue screen, a special environment or manually designating the minutiae from its entire image and estimating the later movements. In the case, it may be embodied only in the limited conditions and moreover, it is not possible in mobile environment. As an alternative technology to separate a user from a background image in wire communication environments as well as to overcome the above limitations, comparing to the existing method to simply encode and transmit images input from the built-in digital camera of mobile terminals, the proposed method segments a target object to increase the efficiency of encoding in advance, increasing the coding efficiency and synthesizing the background that may violate one's privacy with another background.

2 Extraction of Motion Image Information

2.1 Detection of Face Using Color Models

In the past, black and white image was mainly adopted due to the limitations of image processing while the use of colorful image is increasing in order to use more video information [2]. Swain and Ballard reported that for the object discrimination and its position detection in color video, applying color histogram reiteratively had more performance in color video rather than in B&W video and that color histogram was more stable to express or discriminate an object, even in the cases of overlap image and in spite of a viewpoint to look at it.

Chang and Huang successfully extracted facial area and its minutiae by using a color facial image. First of all, they defined the range of RGB values of facial color and determined that the pixels within such RGB values would be facial region. To extract a facial region from a color image, Dai and Nakano converted RGB color image to YIQ model and used I element video. From the above experiments, the study defined the rage of I element colors that a face has in YIG model. With the range, the remainder was eliminated, except for the pixels of I element values, which are contained within the range of the converted I element image. Based on the quadratic probability density function to the resultant image, inertia, inverse difference and correlation features, that is, SGLD-based facial texture model was applied, extracting a facial area within an image.

2.2 Detection of Face Using Facial Minutia

The detection features that it may determine the position of face in a complicate background and find the eyes, nose and mouth by using low dimensional color or texture, irrespective of pose or lighting. However, it has a demerit that it may be significantly affected by noise. Many studies have been executed to recognize face by using the features of contour and shape of each facial component. First of all, it recognizes whether any facial element exist and determines the existence of face actually. The facial elements such as the eyes, nose and mouth are found by their contours. Then, the positional correlations of the eyes, nose and mouth are calculated, on which the existence of face would be determined.

The contour may be expressed as groups of contour for the eyes, eyebrow, nose and mouth while dividing planes. Zhang [3] and Qiiang [4] reported that the contours in a grey image and the linked cross points might be used recognize face or silhouette. In addition, Suzuki [5] found that face might be also recognized by drawing feature map by means of the contour distribution, measuring the angles of contours and comparing them to the angles of the contour representing the ordinary eyes, nose and mouth.

2.3 Motion Tracking in Motion Picture

The existing methods to find a moving object and trace its movement include region-based tracking, contour tracking, matching method and difference image-based method. The region-based tracking is the method to detect an object in a different image from its primary image, which is input by measuring the current background, by finding a connected region in continuous image and tracking a moving object.

The method may find a region connecting pixels that are beyond the thresholds in a different image but it faces a difficulty to segment each region in a complicate object status. The contour tracking method expresses a contour, which is the boundary of an object and tracks the object by dynamically updating the contour. It is advantageously superior to the former method in terms of no need of complicate calculation but it is not unfortunately able to track an object if the contour is partially hidden.

The matching method finds a matching point of which similarity is the most by extracting the pixel information or features of an object in a template image and moving the searching area [6]. The method has a demerit that it may not properly respond to brightness change, zooming or rotation of an object.

The different image-based method calculates a different image between adjacent images on a temporal axis and extracts the only moving region by using the different image. It acquires 3 consecutive images, calculates the difference between the middle reference image and the first image and then, the difference between the middle reference image and the third image. Then, it executed AND operation for the two different images, detecting the region, which is the contour of moving region. The different image is used as a template to track a user. However, it also requires additional operations because in reality, an object to detect may not be detected as a complete template.

3 Search the Position of Face in Temporary Image

In general, one may accurately detect a facial region regardless of one's facial size, shape, position or posture expressed in a visual image without difficulty and has the ability to discriminate who is expressed in an image among the people that the one knows. As such, in order to embody the man's ability in computer, it is first necessary to find a position of face located in a given image. That is, if a face should be found in a real time video, the position of face should be found in the video as soon as possible. The paper uses the presumptive eyes position and the presumptive skin color position to find the position of face in a given video [7].

For the searching method of eyes position, it uses the feature that the presumptive eyes region may be easily found by binary coding because the brightness of the pupil is very low. Also, by validating whether the only presumptive eyes region is the part of face, it may find the position of face in given video information. However, it adversely requires that the eyes acquired from the video information should be vivid. If the eyes are hidden by hair or the shade from a hat darkens the eyes area, the eyes may be also hidden. To overcome the shortcomings, the presumptive skin color position is also used.

The skin color is one of features discriminating one's face; that is, the regions to detect a face in the color model are skin colors (yellow, black, and white), black and white, which represent the other parts of faces (eyes, nose and mouse). The region of a subject for photography of interest is extracted from a still image by using the above skin color features with various conditions.

3.1 Search the Position of Face Using Facial Minutia

Blank region around the eyes and the eyes region are used in an image, in which the eyes position is found by converting the image to binary-coded image. That is, in a binary coded image, the region consisting of 4 blank regions and 1 eyes region is recognized as the eyes.

The method designates the left and right edges of pixels representing the eyes as the left/right boundaries of the eyes. The reason why the input image is converted to binary-coded image is because binary-coded image requires less data size than color or B&W image. In addition, less data may lead increased process speed and shorten the processing time. When converting to a binary-coded image, the threshold should be adjusted according to the adjacent brightness. If the adjacent environment is bright, adjusting the threshold higher changes the image to all white images. On the other hand, if adjusting it lower in a dark environment, the entire image may turn black. If so, the eyes region is hidden in the image, making it difficult to discriminate the eyes from the image. Thus, unless the skin color is not close to black, converting it into binary-coded image may contribute to easily discriminate the features of face.

Almost binary-coded image expressing the eyes and iris also shows the eyes, nose, hair, ears, mouth and eyebrows, and there are four blank regions around the eyes. The blank regions also exist between eyes, eye and ear, eyebrow and eye, and eye and nose. In addition, as if the region representing the eyes in an input image is converted to a binary-coded image, the number black pixels exist in proportion to the eyes' size.

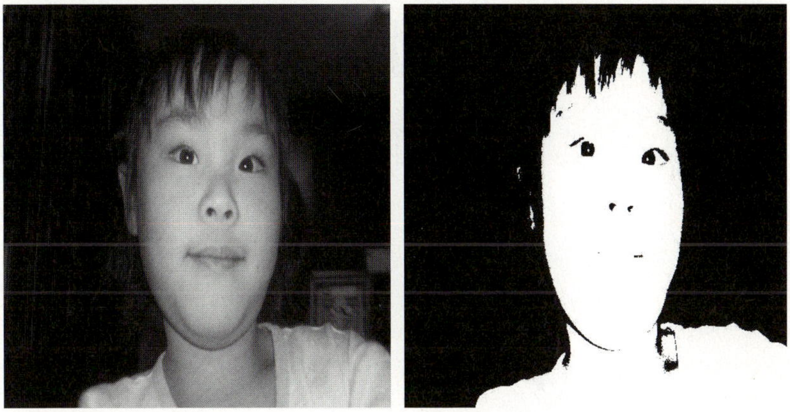

(a) (b)

Fig. 1. (a) Input image. (b) Converting to Binary-Coded Image.

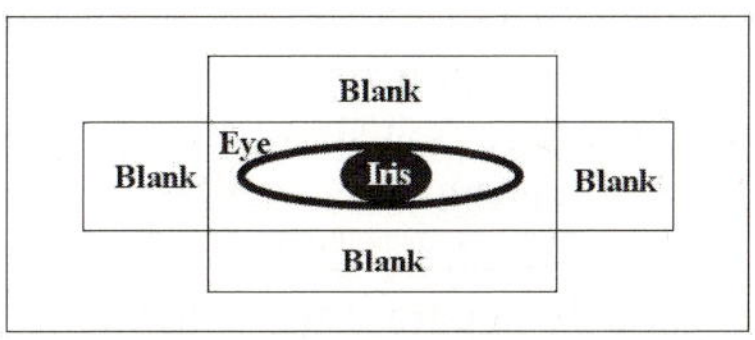

Fig. 2. Four blank directions

The regions are united to one region and mask-shaped, as shown in Figure 2. The mask is occupied as one region where four blank regions should exist in four directions and the number of black pixels in proportion to the eyes size should exist in the middle of it.

Figure 3 shows the method to apply the only region consisting 4 blank regions and 1 eyes region as the eyes in a binary-coded image by converting an input image into its binary-coded image. The region information is recognized by coordinates; in case clothes' color is similar to skin color in an image, it considers the region that is not statistically accurate in the skin color extraction and minimizes the cases that may not recognize it normally in the next step, facilitating the detection of face normally. Alternatively, it may reduce the misconception that hands or arms are recognized as face.

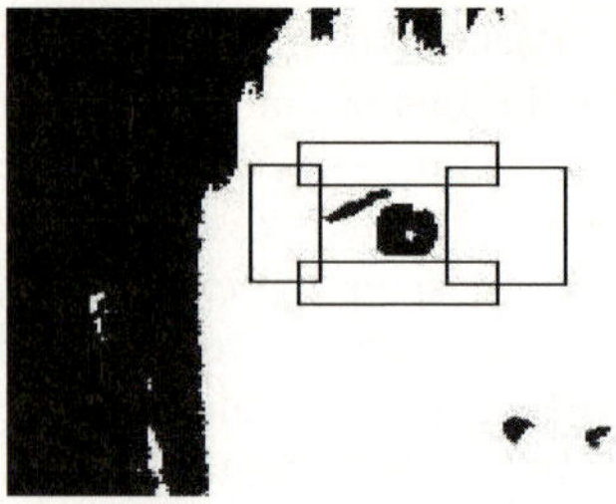

Fig. 3. Mask with 1 region in the center of eyes

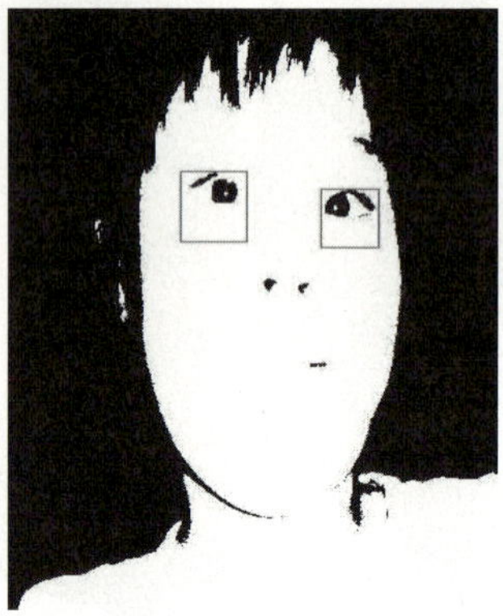

Fig. 4. Mask applied to input image

3.2 Search the Position of Face Using Color Models

A method using the information of colors should be capable of extremely modeling the skin color in the proper color region in order to separate facial region from an input image. Extracting a skin color region should be able to easily extract it under the conditions of searching and extracting skin region and not to be sensitive to the change of color impression in accordance with light's brightness.

HIS model may advantageously separate the information of colors and brightness from an image, being suitable for brightness-based technology. YC_bC_r model may separate light intensity from the information of colors, with which the data excluding the impact of light intensity from the information of colors may be utilized. Y represents light intensity and 'C_b' and 'C_r' mean blue color information and red color information respectively. If using the color models, the features using the color distribution for skin colors with various conditions are densely concentrated on a certain region, so the use of color distribution may be useful to detect the region of face. For input images, HIS model used the measurement results of skin color distribution region in the remaining color elements except for I element. It eliminates the remainder except for the presumptive skin region in the input images. The extracted region is converted to HIS information by using Formula (1).

$$
\begin{aligned}
I &= \frac{1}{3}(R+G+B) \\
S &= 1 - \frac{3}{(R+G+B)}[\min(R,G,B)] \\
H &= \cos^{-1}\left[\frac{\frac{1}{2}[(R-G)+(R-B]}{\sqrt{(R-G)^2+(R-B)(G-B)}}\right]
\end{aligned}
\tag{1}
$$

At the moment, the distributions of H and S were concentrated on a region, as a result of using H and S information except for I information, and it was used as the skin color distribution region by using the above results. Next, YC_bC_r model used the measurement results of skin color distribution range, especially for C_b and C_r color elements, except for Y information for brightness. The extracted region was converted to YC_bC_r information by using Formula (2).

$$
\begin{aligned}
Y &= 0.29900R + 0.58700G + 0.11400B \\
C_b &= -0.16874R - 0.33126G + 0.50000B \\
C_r &= 0.50000R - 0.41869G - 0.08131B
\end{aligned}
\tag{2}
$$

The model also shows that the distributions of color elements are concentrated on a specific range, and the results were used as the skin color distribution.

To obtain the desirable information in a color image, the skin color with various conditions, two color models, HIS and YC_bC_r were used to extract a densely concentrated region and set the region in an input image. It means the segmentation into blocks with a certain uniform size by using the skin color distribution. Next, by using the skin color distribution view, if a measurement is lower than the threshold, based on the number of pixels in a block, it determines that the region does not fall under skin color region and it is eliminated by regarding it as after-image or distortion.

(a) (b)

Fig. 5. (a) Input image. (b) Extraction region using skin color region.

Then, each region selected as skin color was compared to its up/down/right/left regions and if it is determined that it is not skin color region, it was also eliminated by regarding it as after-image or distortion.

3.3 Object Extraction and Movement Tracking Using Contour

To separate a subject for photography from its background in video information, it is firstly necessary to track an object. To track a moving object in video information, it is necessary to detect a moving object and separate it from the background. It extract image information in the first frame from the input image, executes masking operation only after the contour filtering and saves the current image into buffer for the different image process in the next step. From the second frame, the different images are calculated on the information on the previous image, and the block size calculated at the moment is set to $M * N$ while the candidate blocks for a moving object were extracted by setting the threshold. Formula (3) shows, as the method to discriminate the difference from the previous image in terms of pixels, the processing with Y element by converting an image to YC_bC_r color model.

$$D(x, y) = Threshold(\sum_{i=1}^{M}\sum_{j=1}^{N} Threshold(Y_k(x+i, y+j) - Y_{k-1}(x+i, y+j), t1), t2) \tag{3}$$

$$M : WindowSize(Width)$$
$$N : WindowSize(Height)$$

The contour block may be applied only to the initial blocks of interest through the above procedure. Assuming that the region designated by contour blocks is the region with the most information for the further movement, the up/down/right/left blocks of the existing contour block is recognized as the region that the existing contour blocks may be moved and set to the subject to detect accordingly. It may save the overall time by intensely analyzing the contour blocks set as the region of interest in order to extract a moving region at a high speed.

To extract the contour in the current image only from the initial blocks of interest, with the premise that when applying Sobel filter and Canny filter, the region with contour is the part with edge image, so it becomes the object of interest, it was filtered within a block and with the window size of (*N*, *M*) and it was registered as a block of interest only when the number of pixels is not less than the threshold, in which the blocks of interest may be divided into inner blocks and outer blocks by using block matching technique. The technique needs two premises; one is that the pixels in a same block move identically and the other is that blocks moves only vertically or horizontally.

The block matching technique divides an image into blocks of same size and expresses the pixels in a block as a motion vector. The discrimination may catch the movement of an object effectively faster than inspecting the overall region if using the contour block information in order to track the motion of an object in high speed video content. In the technique, contour filter is applied to extract contour from an input image and the blocks with changes more than the threshold in the previous different image are left while the remainder is eliminated as being regarded as the region with no or little change.

After the above process, the input image is turned to an image where the only human shaped contour is left while its background is almost eliminated, but at the same time, the inner region of the remaining image may be partially removed as it is wrongly recognized as part of background with little motion. It may be, however, separated into inner block and outer block if using the recovery method by determining whether it is background or inner block depending on the degrees of proximity to the blocks not less than the threshold in a different image. After then, image processing is labeled into the current inner block, outer block and adjacent block, which is adjacent to contour, and then, the adjacent block and ambient block are determined as the region to be further moved. Then, the overall operation efforts necessary for extracting contour may be reduced by reiterating the previous processes such as different image processing.

4 Experiment Result

The experiment was executed in two parts; still image and motion image. The region of face was detected by means of two methods to detect the eyes and the skin color. For the detection method using the eyes, the method to recognize the region with 4 blank regions and 1 eyes region as the eyes was applied. Then, for the skin color detection, it was detected by using the range of skin color values extracted. The experiment was executed to see the performance of the proposed method in a temporary image. If the eyes were found in each experiment, it was determined that it successfully detected the face. And, it was found that the shape or posture of a human and the size of face occupying in an image actually served as important elements to comprehend the minutiae. The existing object tracking algorithms have several restrictions; for instance and representatively, they secured a background with no object in advance to use it for different image or manually processing masking image in the first image, all of which seemed to be impossible to practice in reality.

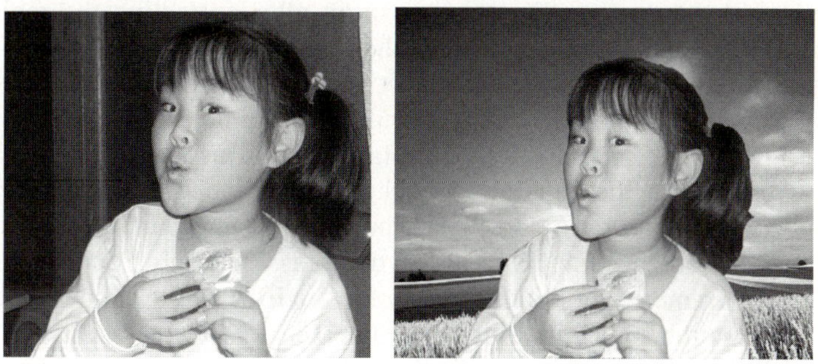

Fig. 6. Execution of program prepared for the experiment

Table 1. Shows the ratio of objects to search for the overall region(the overall region consists of 768 blocks)

Heading level	Example	Font size and style
0Frame~9Frame	450	5.85
10Frame~19Frame	821	10.96
20Frame~29Frame	898	11.69
30Frame~39Frame	901	11.73
40Frame~49Frame	826	10.75
50Frame~59Frame	933	12.14
60Frame~69Frame	817	10.63

In Table 1, a frame is divided into $M * N$, yielding totally 768 blocks and at the moment, the number of blocks selected as the initial blocks of interest in each frame was counted. Viewing the frames no. 0 through 9, the number of blocks based on each frame is 768, so the total number of blocks was 7680. Consequently, it means that the number of the initial blocks of interest was 385. Therefore, it shows that based on the overall region, the only region accounting for 4.8% was used as the basis to comprehend filtering and moving object. That the number of initial blocks increases early means searching a moving object and it could be found that the region was maintaining at the similar ratio. Figure 6 shows that the only 13% of the overall region was used to track a moving object in reality, and the actual ratio depends on the types of motion image or the size of moving object.

5 Conclusion

As the mechanism to extract the region of face and minutiae in still images, the study suggested the method using the information of color by converting RGB information, the color image information into HIS model and YC_bC_r model. Through the experiments, an effective method to extract skin color was suggested by separating the light intensity information from the information of color image from two models and it was possible to extract the minutiae of face by means of contour extraction filter, brightness analysis and the presumptive eyes position after extracting the region of skin color and to use it as the basis to determine the region of face.

In case of motion image, a moving object and its background could be tracked by using the information of the object and it could be separated from the background, on which the original background could be replaced with a user's desirable background. Then, it was found that the deletion of background reduces the data to transmit, increasing the compression ratio and furthermore, in a same bandwidth, the transmission speed may be increased, facilitating the transmission of high quality image.

As a result, a frame was divided into $M * N$ and the total number of blocks was compared to the number of initial blocks of interest. In conclusion, it was found that for the initial block of interest, the only about 4.8% of the overall region was used to comprehend filtering and moving object while the only 13% of the overall region was actually used to track a moving object.

References

1. Forsyth., D.A., Pone., J.: Computer Vision: A Modern Approach. Pearson Education (2003)
2. Srisuk, S., Kurutach, W.: A New Robust Face Detection in Color Images. In: IEEE International Conference on Automatic Face and Gesture Recognition, pp. 291–269. IEEE Press (2002)
3. Zhang, S., Qiao, H.: Face recognition with support vector machine. In: IEEE International Conference on Intelligent Systems and Signal Processing, pp. 8–13. IEEE Press (2003)
4. Gao, Q.: Extractiong object sihouettes by perceptual edge grouping. In: IEEE International Conference, pp. 2450–2454. IEEE Press (1997)
5. Suzuki, Y., Shibata, T.: Multiple-clue face detection algorithm using edge-based feature vectors. In: IEEE International Conference, pp. 737–740. IEEE Press, Los Alamitos (2004)
6. Karungaru, S., Fukumi, M., Akamatsu, N.: Face Recognition using edge-based feature vectors. In: IEEE International Conference, vol, IEEE Press, Los Alamitos (2004)
7. Park, J.I.: Object Separation from Moving image using Difference of Image and Image Boundary, Ph.D. Dissertation (2006)

An X3D Approach to Neuro-Rehabilitation

Mauro Zampolini[1], Riccardo Magni[2], and Osvaldo Gervasi[3]

[1] Department of Rehabilitation, ASL3 - Regione Umbria
Foligno Hospital, Via M. Arcamone, 06034, Foligno (PG), Italy
m.zampolini@als3.umbria.it
[2] Pragma Engineering,
Via della Pallotta, 5, I-06100 Perugia, Italy
riccardo.magni@pragmaeng.it
[3] Department of Mathematics and Computer Science, University of Perugia,
via Vanvitelli, 1, I-06123 Perugia, Italy
osvaldo@unipg.it

Abstract. We present an application based on X3D technologies for the telerehabilitation of patients affected by neurological diseases. In particular the virtual world is designed to integrate the activity made on NU!Reha[1] Desk product with the activities performed on the virtual world. The main target of our system are patients with heavy neurological disease (people not able to move a forced in static positions because of therapies, etc).

The system is designed to allow the remote monitoring and assessment of the patient's activities by the staff at the hospital.

1 Introduction

The development of telecommunication technologies allow to think to the telerehabilitation as a reliable system to provide a series of rehabilitation related services. One of the first use of tele-technologies in rehabilitation has been the evaluation of the outcome of rehabilitation programs[1] or the assessment of the needs of the patient[2,3,4]. The telerehabilitation may be useful in improving communication and access to health care over distance[5,6], offering the possibility of early diagnosis and start of therapy in acute conditions. Furthermore, telerehabilitation allows the continuous monitoring of people at risk cutting costs and time and it plays a crucial role shortening length of stay in hospital or specialized rehabilitation center.

In the present paper we describe the use of X3D technology to implement some applications able to integrate the activity of environments devoted to the rehabilitation of patients with very heavy neurological diseases, like NU!Reha produced by Pragma Engineering, with virtual and augmented reality applications. In fact, the NU!Reha system can be integrated with a virtual environment in which the patient can perform some exercises to improve her/his cognitive capabilities.

[1] NU!Reha is a trademark of Pragma Engineering srl.

O. Gervasi et al. (Eds.): ICCSA 2008, Part II, LNCS 5073, pp. 78–90, 2008.

The paper is structured as follows: in section 2 a review of the existing research related to neuro tele-rehabilitation and the adoption of virtual reality techniques is presented. In section 3 our methodology is discussed and the integration of the virtual reality application with the NU!Reha Desk is discussed. In section 4 a prototype application, related to the simulation of an exercise difficult to be implemented in a real context in case of patients with huge disease or unable to move autonomously, like accessing and using an elevator, is presented. In section 5 the architecture of the system is presented and discussed. In section 6 some conclusions are presented.

2 The Existing Research

Telerehabilitation approach of upper limb after stroke with a system including a Web-based library of status tests, therapy games, and progress charts, which can be used with a variety of input devices, including a low-cost force-feedback joystick capable of assisting or resisting in movement has been applied.[7]

Virtual reality techniques and high speed networks create an environment allowing clinicians and technical staff at different sites to interact in a 3-dimensional environment[8].

Virtual reality with haptic interfaces and networked PC's has been applied in orthopedic rehabilitation with the Rutgers & Stanford system for rehabilitation of hand, elbow, knee and ankle impairments to provide exercises while being monitored remotely by the therapists[9]. Preliminary results indicated that the system works well as a diagnostic tool and that the subjective evaluation by patients was very positive.[10]

A Cyber-Glove and Rutgers Master U-ND force feedback glove, allowing user-interaction with a virtual environment has been used in stroke rehabilitation. A pilot clinical trial using a system with a Cyber-Glove and Rutgers Master U-ND force feedback glove in three chronic stroke patients showed improvement on most of the hand parameters over the course of the training . The patients with greater cognitive impairment required more trials to acquire the skills requested. Also language processing appeared to be related to how effectively participants learned to use the virtual rehabilitation center[11].

The cognitive rehabilitation is a potential field for developing telerehabilitation, specially using virtual reality. The speech therapy has been provided in telerehabilitation too. The main findings in a pilot study are that the approach is feasible and the protocols should be improved in future studies[12].

The upper limb motor rehabilitation is one of the main field of application in stroke patients using gloves [13], facilitating systems or virtual reality.

A virtual reality system based on a 3D motion tracking system (Polhemus 3Space FastTrack, Vermont, U.S.A.) to capture the patients arm movements by a magnetic receiver and to transmit it to the computer screen has been recently developed. The receiver is attached to a real object (end effector) grasped by the patient, or to a glove, worn by the subject[14].

With a remote workstation it is possible to create the sequence of virtual exercises and supervise the patient's execution. The virtual tasks consisted mainly of simple movements which simulate activities of daily living, i.e. pouring water from a glass, using the hammer, posting an envelope into a mailbox slot, circumnavigating a donut, etc. The complexity of the task is determined by the physiotherapist, depending on a patient's motor deficit and the degree of motor recovery.

The subject moves the real object (envelope, carafe, hammer) on a virtual environment and views on the computer screen the trajectory of the corresponding virtual object. During performance, a patient could visualize on real time not only his movement but also the correct trajectory that he has to execute with the end effectors, to meet the constrains of the exercise, prerecorded by the physical therapist. With these system a pilot study in 5 stroke patients has been carried out. Subjects underwent the tele-rehabilitation program for 4 weeks. The therapy significantly improved the Fugl-Meyer mean score, the mean duration and the velocity of the movements, but not the Functional Independence Measure scale score[15].

H-CAD (Home Care Activity Desk) is an activity desk designed for the rehabilitation of the upper limb. This project deals with the development of a system capable to enable patients affected by multiple sclerosis, stroke or traumatic brain injury to perform rehabilitation treatment for the upper limb at home. The system has been designed to allow patients to perform exercises at home, monitoring from the hospital the execution of these exercises, allowing the patient and the therapist to interact through a teleconference system. The feedback mechanism and the parameter measurements have two purposes. One is to immediately inform the user about the performance of a specific task. The other is to allow the therapist in the hospital setting to check how many attempts have been made to perform a certain task and how many of these attempts have resulted in successful task completion. During the hospital experiment emerged that H-CAD system had really good acceptance and good general opinion. The worst scored characteristic of the H-CAD system was its aesthetic aspect.

The pilot study Nine Hole Peg Test values (show an improving trend from baseline to the end of home test. The analysis of data reveals an improvement either during A1 as B phases in all tested participants, whereas in the end of A2 phase NHPT tended back. Trial length does not allow knowing whether part of the improvement is sustained along time or motor function return to baseline. Further validation phase increasing home test length should be done.

A development of the HCAD system is the Hellodoc project. This project deals with the key activities to be implemented in order to validate the market in relation with the settlement of a home care based service capable to extend the rehabilitation treatment of patients affected by neurological diseases like TBI, Stroke or MS at home under the control and supervision of the hospital.

2.1 The Effectiveness of Telerehabilitation and Virtual Reality Approach

Telerehabilitation is a quite recent field for research. The first experimental studies started few years ago; we review 9 studies to analyze the evidences of the effectiveness of telerehabilitation. Unfortunately, the studies included a small number of patients and the design is limited since only 2 studies were randomized. One study was case-control and the other were observational with or without intervention. The first study analyzed is a telemonitoring to support frail elders compared to a control group with the same problems of health and same age. The effect seem relevant in term of function which is improved in the intervention group[16].

Another observational study without specific intervention and telemonitoring, emphasizes the reliability of remote evaluation of pressure sores aimed to indicate the strategy of treatment. The data confirm the good reliability of this approach[17].

The upper limb rehabilitation of stroke patient with virtual reality can be used locally to introduce new exercises and when applied remotely seems to be effective on improving the impairment without affect the function. The study is observational non randomized and without a group of control then the promising data should be confirmed with a larger trial[15]. Similar results with a similar VR paradigms has been obtained[18].

The remote intervention has been demonstrated at least as effective as the specific intervention of therapist at home setting[4]. Similar observation has been done in another study with different intervention.

Similar effect has been found with different protocol of exercise in other studies[19,20]. Using an haptic glove the recovery after hand surgery seems to be facilitated[21]. From the analysis of the studies we can conclude that there are not strong evidences about the efficacy of telerehabilitation due to the small size of the sample studied and the poor study design. However the telerehabilitation is a promising approach and the data show a trend toward a positive effect of this approach.

2.2 NeuroVR

A relevant approach of Virtual Reality techniques to the rehabilitation of neurologic disease has been carried out by the NeuroVR project [22]. NeuroVR is an open source virtual reality platform, providing the clinical professional with an editor of the Virtual Environment, which allows non-expert users to easily modify a virtual scene, to best suit the needs of the clinical setting.

The NeuroVR platform includes two main components, the Editor and the Player.The Editor is realized by customizing the User Interface of Blender, an integrated suite of 3D creation tools available on all major operating systems.Clinicians and researchers have the freedom to run, copy, distribute, study, change and improve the NeuroVR Editor software, so that the whole VR community benefits. Using the NeuroVR Editor, the psychological stimuli/stressors appropriate for any given scenario can be chosen from a rich database of 2D and 3D objects, and easily placed into the pre-designed virtual scenario by using an icon-based interface (no programming skills are required).

The other component of NeuroVR is the Player, which allows to navigate and interact with the VEs created using the NeuroVR Editor. When running a simulation, the system offers a set of standard features that contribute to increase the realism of the simulated scene. These include collision detection to control movements in the environment, realistic walk-style motion, advanced lighting techniques for enhanced image quality, and streaming of video textures using alpha channel for transparency. The player can be configured for two basic visualization modalities: immersive and non-immersive. The immersive modality allows the scene to be visualized using a head-mounted display, either in stereoscopic or in mono-mode; compatibility with head-tracking sensor is also provided. In the non-immersive modality, the virtual environment can be displayed using a desktop monitor or a wall projector. The user can interact with the virtual environment using either keyboard.

3 The Proposed Approach

The application is configured as the extension of the task-oriented therapy already developed for NU!Reha Desk product.

The experimental paradigm of the NU!Reha Desk has been the use of the technological features with the facilitation of the motor recovery. In fact, according to the most recent studies the motor action are facilitated by a significant contextual environment. Based on these observations in rehabilitation a therapy based on the task has been developed.

Taking into account this organization of the motor control a virtual reality approach seems to be not applicable in rehabilitation activity. Some studies already showed that the virtual reality should be useful for the recovery of motor activity.

A virtual environment allow to create a personalized exercise planning specific task not possible to realize with a real instrument. Virtual reality allow also to develop complex sequence of exercise improving the capacity of planning a series of action.

It is based on the usage of a purposely conceived "activity desk" for the execution of occupation therapy tasks and exercises, allowing different grasping actions and interactions with sensorised objects. The device is composed by a base unit and a list of components which have different functionality and scope of utilization as they are connected to the base unit. The virtual world related to the NU!reha desk is shown in Fig.1. At present, no components functionality is allowed in absence of the base unit, they are:

the base unit: the desk supporting the core structure of the system (both hardware and software). It is intended as the basic module of all the system: all other modules or subsystems are to be connected to this unit;

control pad: it allows the control of the system interface by part of the caregiver assisting the patient during the exercises;

shelf: shelf-like tool useful for the execution of vertical displacement based exercises, they are equipped of an array of IR proximity sensors allowing to detect the presence of a sensorised object (an object covered by the reflective labels);

keyboard module: module for the execution of exercises consisting in inputting a sequence of numbers with a phone/POS like keyboard;

RF Tool : a plastic cylinder containing an internal electronic allowing its horizontal localization;

video cameras and supports: a couple of webcams with articulated stands are supplied for the video recording of the exercises performed by the patient;

accessories: a list of accessories are supplied in order to keep simple and feasible the system usage; they include a sensorised pen, plastic jars and cups, a book-alike reproduction, foam cubes.

By means of accessories and sensors, a number of different exercises can be configured directly by the therapist allowing a full customizable environment to perform different grasp/grip actions and upper limb movements: writing and pre-graphism exercises, vertical and horizontal object displacements can be performed and monitored by the system.

The system is based on a sensorised LCD surface placed in front of the patient and acting as interface and working area for the execution of exercises. A SBC placed inside the table supports the execution of all tasks and acquisitions requested by each exercise. A WiFi PCMCIA interface allows the connection to external LAN or ADSL connections for data uploading and configuration

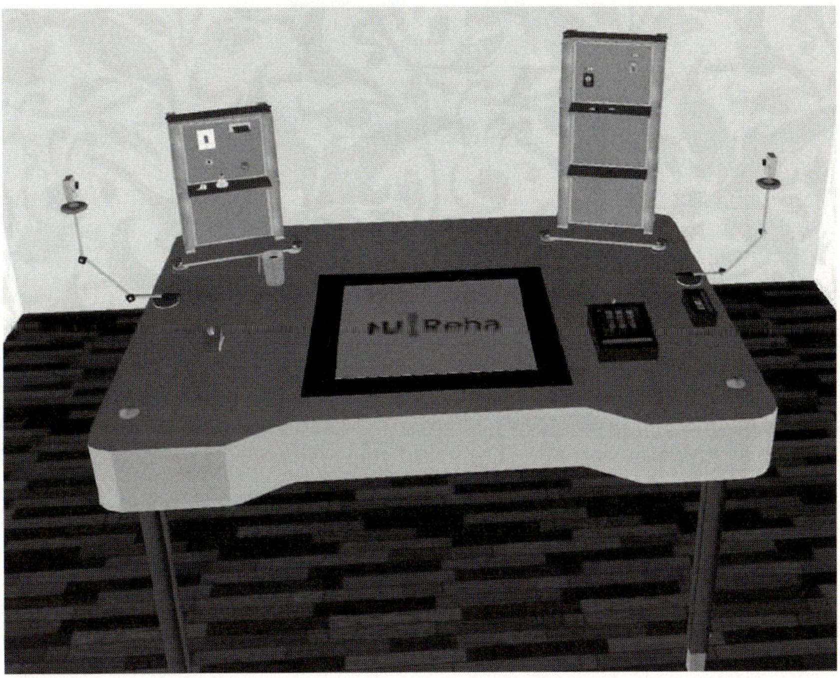

Fig. 1. The virtual world related to the full installation of the NU!Reha Desk

retrieving from an external server. A typical exercise includes the utilization of one or more sensorised accessories for the execution of contextual activities as they are provided by occupational therapy. For example, an exercise typology could include the interaction with the sensorised pen (similar in weight and dimension to a normal pen) on the plastic surface upon the screen. It understands a "writing" exercise (based on the writing performance) and/or "pre-graphism" exercise (based on trajectories like shown in Fig. 2). Another exercise typology utilizes a plastic jar with inside the RF tool: it is localized on the sensitized area allowing to monitor horizontal displacement through the positions suggested by the exercise, as shown in Fig.2.

Fig. 2. Two exercises on the NU!Reha Desk

4 Prototype Application

The experimental application introduces 3D visualization features through X3D technology as an extension and significant improvement of the present NU!Reha system. The aims of the application proposed are:

- to extend interaction capabilities of the present system by means of a 3D visualization of scenarios and/or represented objects;
- to propose new applications with respect to cognitive rehabilitation and the execution of complex tasks.
- to assess the ability of the patient performing the elementary tasks of the tele-rehabilitation

The first application is based on the need to associate the proposition of each exercise to an explanatory representation on screen. Considering the relative ability of patients (and the generic caregiver) to identify the expected task (action sequences, goals, etc), the vitalized representation of the exercise contributes to the effectiveness of the system.

Various animations show the type of movement of various objects. In particular the reduced number of objects in the scenario help the patient to identify the proper movement and illustrate the goal of the exercise and the expected movements. This is important because the presence of refinements could be perceived as 'distractors' with respect to the simple description of the exercise. The absence of scenario elements (like texture on backgrounds) which could confuse the perception of shapes and object recognition: 'neutral' backgrounds and lights are preferred in most cases if the system is utilized by a patient with TBI (Traumatic Brain Injury) exitus. The chance to choose different point of view and zooming factors, is then considered as a preferred feature for a better understanding of the amplitude and characteristics of displacements requested by the execution of each exercise.

The extension of the present application of the NU!Reha Desk is an important improvement derived from X3D technology. The system already includes a number of sensorised items and make it possible to sensorise objects of common use in daily life activities. The patient should be able to manipulate these objects, interfacing with the activity desk for the execution of a mimic of tasks requested in real life (such as moving the jar on the horizontal plane). The application is proposing the extension of this capabilities toward the representation of a 3D based "problems" displayed on the same sensorised LCD screen. The idea comes from the need to complete in the cognitive framework the application of the same telerehabilitation apparatus: this chance brings advantages not only on the economical basis (same system that can be exploited for an increasing number of exercises), but also in sense of flexibility and rehabilitative usefulness. In fact, the exitus of most of neurological patients is not referred solely to motor or cognitive impairments, normally they are linked together with different severity. The OT exercises proposed by NU!Reha Desk system are already understanding cognitive aspects, mainly referred to the action designs and coordination. The 3D visualization allows including in the exercise set some *mostly cognitive* exercise which can be mixed in the exercise session proposed to the patient. It means that with the utilization of such familiar interfaces (already utilized for contextual tasks), patient can complete his/her rehabilitation therapy including some exercises based on:

- problem solving;
- planning activities;
- categorizations;
- attention training.

The prototype application performs a planning exercise related to outdoor situation: utilizing an elevator. It allows the patient to explore the scene by means of the pen interface on the screen and to choose the correct sequence of actions to perform the expected task. The present application is targeted on a medium difficulty level reducing the number and typologies of components/signals outside of strictly needed for the task itself: the scenario decoding and interactions are almost simplified and easy to understand. The proposed exercise needs to plan and execute the following sequence:

a) to recognize the elevator;
b) to identify call push button;
c) to activate call button;
d) to await for doors open;
e) to enter into the elevator cabin;
f) to identify the button panel;
g) to activate the correct button (stated at the exercise beginning);
h) to await for door opening and exit.

This application includes feedback features which can optionally configured by the therapist himself: positive feedback (outlining the success in each action of the sequence), negative feedback, general feedback for the sequence completion.

The Fig.3 shows the sample application focused on the execution of some simple operations related to the usage of an elevator.

5 System's Architecture

The telerehabilitation service based on NU!Reha platform is managed by means of 3 software applications:

- configuration manager
- exercise manager
- data analysis manager

The configuration manager is a hospital-side application allowing to choose the exercise typologies and assemble them in a execution sequence. The patient would be asked to execute the exercise proposed following the sequence list and the repetition number set by the therapist during the configuration. The exercise manager is the application which is physically embedded into the portable unit at patient home. It is connected through a server connection to the configuration manager, downloading configuration files. This application is managing the following functions:

- *user interface* a simplified user interface controlled by means of 4 keys (control pad)
- *exercise management* starting from list and configuration options for each exercise
- *data recording* referred to camera views and data coming from sensors embedded in the portable unit
- *data encrypting and transmitting* toward the server unit.

Data analysis module retrieves data from server unit. It is a hospital-side application giving the chance to control and revise the information coming from the portable unit placed at patient home, it means for each exercise: execution

Fig. 3. Beginning of the exercise (upper panel); Call button activated (central panel); Lift control panel (lower panel)

(date and time), number of repetitions, overall duration, sensor data (delay from exercise start, id of activated sensor). All data can be analyzed by means of a kind of synoptic display, syncing them with the camera recording, revising patient posture and scenario.

The Virtual World is implemented in X3D[23], a ISO standard devoted to the realization of virtual environments on the Web. X3D is a powerful language that make it possible to interact with the user through various types of sensors. The Scene Access Interface (SAI) makes it possible to control and monitor the user behavior using programs written in Java or Javascript.

In the present case we use Java to implement several `Script` nodes that perform the data management, control and assess the patient's actions. The assessment results are passed to the NU!reha system so that the clinician can retrieve the results and monitor the patient's activity.

We used X3D also for training and visual help purposes. In fact, a virtual world has been implemented to show how to assemble the NU!reha system (for the clinicians) and which features are available to the user. The clinician can use the virtual world also to show to the patient how the various exercises available in the NU!reha system should be executed: several sensors and animations are able to show to the patient the correct sequence of actions to be performed and the expected behavior to be obtained.

6 Conclusions

This work is demonstrating the chances to apply the X3D technology in the field of newly developed solutions for the neurological rehabilitation. The virtual reality applied in rehabilitation is considered a promising approach to enhance the possibility to perform different exercises aimed to the recovery of the neurological desease. The problem is that the virtual reality failed to full emulate the real world loosing the strong facilitation of the contextual factors. The advantage of virtual reality system is the possibility of costumization. The originality of this work has been the introduction of the planning of a series of action that have a peculiar value to the series of exercises. This kind of approach is very important in the condition that involve the cognitive impairment (particularly the frontal lobe) and motor impairment.

During the development of this application, some predictor factors about feasibility and sustainability of this approach has been envisaged. They are referred both to the technical approach as well as to strategical issues. On the technical side:

- referring to effectiveness of neurological rehabilitation, the proposal of real-life 3D problems is preferred (instead of proposing just a 3D navigation in different environments), the chance to execute ordered list of tasks mimics outdoor real life situation;
- the chance to configure real 3D views even if simplified in textures and details is strongly adviced by the medical community, considering the actual cognitive ability of each patient and its evolution;
- the same concept of therapy imposes an "evolutionary" approach to the application developed: each exercise should be conceived in a modular way as it is referred to the cognitive complexity, accompanying patient in his/her recovery.

On the other side, some consideration could be placed in advance about the chance to propose these applications as a routinary approach in rehabilitation/telerehabilitation therapies:

- the role of operators (therapists, MD rehab responsible..) is changing rapidly, a telerehab service implies a number of different skills to be able to exploit the technological features offered, this fact understands the need to propose new applications and education jointly as part of a process aimed to develop a different way to intend the rehabilitation itself and therefore new processes to make it possible;

- the integration of this X3D application in an already available telerehabilitation product, demonstrates the chance to mix contestual and virtual exercises proposing therapy at home in effective way; hence at prototipe level the application would be tested with patients in telerehab setting.

In conclusion this system is a promising tool to be used in a rehabilitation center to integrate the normal rehabilitation activity. Furthermore there is the possibility to use it in telerehabilitation. A series of clinical studies are necessary to demonstrate the efficacy to facilitate the functional recovery of the system in the different conditions.

References

1. Russell, T.G., Wootton, R., et al.: Physical outcome measurements via the Internet: reliability at two Internet speeds. J. Telemed. Telecare 8(Suppl. 3), 50–52 (2002)
2. Ricker, J.H., Rosenthal, M., et al.: Telerehabilitation needs: a survey of persons with acquired brain injury. J. Head Trauma Rehabil 17(3), 242–250 (2002)
3. Robinson, S.S., Seale, D.E., et al.: Use of telemedicine to follow special needs children. Telemed. J. E. Health 9(1), 57–61 (2003)
4. Sanford, J.A., Griffiths, P.C., et al.: The effects of in-home rehabilitation on task self-efficacy in mobility-impaired adults: A randomized clinical trial. J. Am. Geriatr. Soc. 54(11), 1641–1648 (2006); Sanford, J. A., Jones, M., et al.: Using telerehabilitation to identify home modification needs. Assist. Technol. 16(1), 43–53 (2004)
5. Torsney, K.: Advantages and disadvantages of telerehabilitation for persons with neurological disabilities. NeuroRehabilitation 18(2), 183–185 (2003)
6. Hauber, R.P., Jones, M.L.: Telerehabilitation support for families at home caring for individuals in prolonged states of reduced consciousness. J. Head Trauma Rehabil. 17(6), 535–541 (2002)
7. Reinkensmeyer, D.J., Pang, C.T., et al.: Web-based telerehabilitation for the upper extremity after stroke. IEEE Trans. Neural Syst. Rehabil. Eng. 10(2), 102–108 (2002)
8. Ketelaer, P., De Wolf, L.: New Technologies in MS: present and future applications of virtual reality. The Int. MS. J. (2001)
9. Burdea, G., Popescu, V., et al.: Virtual reality-based orthopedic telerehabilitation. IEEE Trans. Rehabil. Eng. 8(3), 430–432 (2000)
10. Girone, M., Burdea, G., et al.: Orthopedic rehabilitation using the 'Rutgers ankle' interface. Stud. Health Technol. Inform. 70, 89–95 (2000)
11. Jack, D., Boian, R., et al.: Virtual reality-enhanced stroke rehabilitation. IEEE Trans. Neural. Syst. Rehabil. Eng. 9(3), 308–318 (2001)
12. Hill, A.J., Theodoros, D.G., et al.: An Internet-based telerehabilitation system for the assessment of motor speech disorders: a pilot study. Am. J. Speech Lang. Pathol. 15(1), 45–56 (2006)
13. Placidi, G.: A smart virtual glove for the hand telerehabilitation. Comput. Biol. Med (2006)
14. Piron, L., Cenni, F., et al.: Virtual Reality as an assessment tool for arm motor deficits after brain lesions. Stud. Health Technol. Inform. 81, 386–392 (2001)
15. Piron, L., Tonin, P., et al.: Motor tele-rehabilitation in post-stroke patients. Med. Inform. Internet Med. 29(2), 119–125 (2004)

16. Chumbler, N.R., Mann, W.C., et al.: The association of home-telehealth use and care coordination with improvement of functional and cognitive functioning in frail elderly men. Telemed. J. E. Health 10(2), 129–137 (2004)
17. Rintala, D.H., Krouskop, T.A., et al.: Telerehabilitation for veterans with a lower-limb amputation or ulcer: Technical acceptability of data. J. Rehabil. Res. Dev. 41(3B), 481–490 (2004)
18. Holden, M.K., Dyar, T.A., et al.: Telerehabilitation using a virtual environment improves upper extremity function in patients with stroke. IEEE Trans. Neural Syst. Rehabil. Eng. 15(1), 36–42 (2007)
19. Carey, J.R., Durfee, W.K., et al.: Comparison of Finger Tracking versus Simple Movement Training via Telerehabilitation to Alter Hand Function and Cortical Reorganization after Stroke. Neurorehabil. Neural Repair (2007)
20. Page, S.J., Levine, P.: Modified constraint-induced therapy extension: using remote technologies to improve function. Arch. Phys. Med. Rehabil. 88(7), 922–927 (2007)
21. Heuser, A., Kourtev, H., et al.: Telerehabilitation using the Rutgers Master II glove following carpal tunnel release surgery: proof-of-concept. IEEE Trans. Neural Syst. Rehabil. Eng. 15(1), 43–49 (2007)
22. You can find alla information about the project and the Software environment at the web site, `http://www.neurovr.org`
23. The X3D specifications and the documentation are available at the Web3D Consortium Web site, `http://www.web3d.org/x3d`

User Technology Readiness Measurement in Fingerprint Adoption at Higher Education Institution

Raudzatul Fathiyah Mohd Said, Shuzlina Abdul Rahman, Sofianita Mutalib, Marina Yusoff, and Azlinah Mohamed

Faculty of Information Technology and Quantitative Sciences,
Universiti Teknologi MARA, 40450 Shah Alam, Selangor, Malaysia
raudzatul.said@tricubes.com,
{shuzlina,sofi,marinay,azlinah}@tmsk.uitm.edu.my

Abstract. Due to difficulty in monitoring the attendance of members in the higher education institution (HEI), the HEI management has recognized the need to conduct a study on the adoption of fingerprint system. The study identified the appropriate indexing model and developed a questionnaire that suits for the environment. The higher education institution selected for this study is International Islamic University Malaysia (IIUM). The target groups were selected across a sample of the organisation's members. The returned questionnaires were analyzed using a standard statistical tool. Reliability and normality testing were deployed to ensure that the data is reliable and normal for further testing. Spearman's correlation and Kruskal Wallis were used to find the relation between elements of innovativeness, optimism, discomfort and insecurity. The findings have revealed that the innovativeness and optimism have a positive relation with the users' readiness. Meanwhile, discomfort and insecurity gave a negative relation to the faculty members'. This indicates that the faculty member is aware of the technology and optimist with the implementation of the technology in their organization. However, they still feel insecure and discomfort with the justification and impact of the implementation.

Keywords: Biometric, Fingerprint Technology, Technology Readiness Index Model, User Readiness.

1 Introduction

Biometric technology is widely used to detect and recognize individuals. Governments enforce their immigration laws by taking fingerprints of certain people entering their countries. Companies, hospitals and organizations protect the entrances of their buildings, interior rooms, computers and safes using fingerprint readers. Some companies keep track of working hours by automatically recognizing incoming and outgoing employees [1]. Current applications of biometric technologies range from controlling access to facilities, enabling voice recognition at call-centers, to control time and attendance of employees, to provide self-service access to safe deposit vaults at banks, or to cash checks in supermarkets [2]. There is a need to restrict the access and attendance of unauthorized persons for gaining access to the facilities and services as diverse as buildings,

O. Gervasi et al. (Eds.): ICCSA 2008, Part II, LNCS 5073, pp. 91–104, 2008.

rooms and computer labs. In Kulliyyah Information, Communication and Technology (ICT), there is no specific system in used to manage and monitor the attendance in the faculty. In this matter, Kulliyyah ICT may need a fingerprint verification system. With the intention of ensuring faculty members are ready to use the fingerprint attendance system in the faculty, specific study needs to be conducted. This study aims to measure the users' readiness level in implementing fingerprints attendance. Thus, past research and the relevant user readiness model are assessed. The model is expected to exhibit the relationship between contributors and inhibitors components related to technology readiness (TR).

2 Related Works

Biometric systems have been applied to a broad range of applications on several fields of our society such as forensic science, financial and trade security, physical access control check points, information systems security, customs and immigration, national identification cards, and driver licenses [3, 4], among other examples.

2.1 Biometric Technology

There are multitude of related theories, concepts, practices or strategies, and technologies which apply in fingerprints technologies. Personal identification or authentication using biometrics as defined by [3] applies pattern recognition techniques to measure physiological or behavioural characteristics. Generally, there are two types of biometric systems that enable the link between a person and his or her identity. The first type, identity verification or authentication that occur when a user claims who he is and the system accepts or declines his claim. The second type, identity identification occurs when the system establishes a subject identity or fails to do it without any prior claim.

Both [5] and [3] agreed that a biometric system can be based on any physiological or behavioural characteristics as long as the following properties are fulfilled;

i. Universality; Every person should have the characteristics
ii. Uniqueness; No two persons should be the same in terms of the biometric characteristics
iii. Permanence; The characteristics should be invariant with time
iv. Collectability; The characteristics must be measurable quantitatively and easy to acquire
v. Performance; The biometric technique accuracy level
vi. Acceptability; The level of user acceptance of the biometric system
vii. Circumvention; The level of difficulty in order to forge an identification/ authentication.

2.2 Fingerprint Technology

A fingerprint looks at the patterns found on a fingertip. There are a variety of approaches to fingerprint verification [2]. Some emulate the traditional police method by matching minutiae; others use straight pattern-matching devices; and still others are a bit more unique, including things like fringe patterns and ultrasonic. Some verification approaches can detect when a live finger is presented; some cannot. A greater variety of fingerprint devices is available than for any other biometric. As the prices of these devices and processing costs fall, using fingerprints for user verification is gaining acceptance [6].

There are some aspects that need to be considered regarding to fingerprint system [7] that includes robustness, acceptability, integration and speed storage requirement. It is important to consider how robust the system is to fraud and imitation. Fraud can occur at the enrolment stage as well as at the verification stage [8]. Therefore, the system should be robust to the users' fingerprints over time. For this, an adaptive system that gradually modifies the store templates may be used.

The technology must be easy to use during both the enrolment and comparison phases. It must also be socially acceptable. The users would not accept a system that may threaten their privacy and confidentiality or that might appear to treat them as potential suspects and criminals [9]. A clear public standpoint on the issue of privacy in relation to fingerprint technologies is required to ensure broad public acceptance. The time required to enroll, verify or identify a person is of critical importance to the acceptance and applicability of the system. Preferably, the acceptable verification time should be of the order of one second or faster. The storage requirement for the templates is also an important issue.

2.3 Fingerprint Application in Education Institution

These are among potential fingerprint applications which can be deployed in higher education institutions. This section mentions the universities and its biometric application which have been implemented fingerprints in their campuses. One example of deploying fingerprint is for payment systems whereby a specific user groups was monitored while using the services and facilities offered free or at subsidised rates [10]. University of Georgia for example uses fingerprint systems for university catering facilities which are only provided for registered students [11]. Also, food-service workers at Georgia punch in and out of their shifts with a time clock that scans their fingerprints so that a worker cannot clock in for an absent friend. The process, which measures the size and shape of the hand, takes only a few seconds. [10], the director of food services at Georgia expressed that there is no system is foolproof; however this is far more efficient for them than a photo-based system. Certainly, the university is among the first to use the fingerprint technologies widely, having relied on it in one form or another in its dining halls since 1974.

The fingerprint system was recently adopted by Campbell University for 1,200 voluntary users in the university's dining halls. Administrators at the institution claims that identification by fingerprint scan takes 1.6 seconds, compared to 3.5 seconds to swipe a student's ID card [12]. In addition, few school districts have begun using fingerprints for certain applications. The School District of Philadelphia began using finger scanning instead of time clocks to track the work hours of some of its employees [11].

2.4 Readiness Index Model

There are two (2) models that commonly used to measure user readiness which are Cloverleaf and Technology Readiness Index (TRI) Model. The Cloverleaf Model was developed by [13] as to identify the requirements for technology transfer successfulness. The tool has been customized for different environments. It can be used over time to compare technologies and adapt to the unique situation of users. Cloverleaf Model identified four main factors which give affect to the commercial success or failure of the new technology and its transferability and commercialization. The four (4) factors are the strength of the technology itself, the market attractiveness, commercialization avenues, and management support.

TRI was developed by [14] to measure a multiple scale for embracing new technologies. TRI construct refers to people's tendency to embrace and use new technologies for accomplishing goals in home life and at work. TRI consists of four (4) dimensions; optimism, innovativeness, discomfort and insecurity. The respondents will be classified as high, medium or low based on their scores on each of these dimensions also differed significantly in terms of their use of high-technology products and services.

The Cloverleaf model focused on the problem of commercialization within the private sector, and on the commercialization into the private sector of technology from government agencies, such as universities. For that the case, it relatively understudied arena of TR lies more on users behavior and perception. Indeed, the assessment process is particularly difficult when the organizations involve do not share the same culture, goals, modes of operation, time lines, and criteria for success.

Table 1. Summaries of studies using Cloverleaf and Technology Readiness Index (TRI) model

Author	Research Title	Model Used
Chang and Chen [15]	Assessing the Readiness of Internet-based Inter-organisational System (IOS) and Evaluating its Impact on Adoption	CM
Chang and Shaw [16]	Developing the Readiness Index of IT-enabled Supply Chain Collaboration	CM
Galbraith *et al.* [17]	"Spin-In" Technology Transfer for Small R&D Bio-Technology Firms: The Case of Bio-Defense	CM
Chang and Kannan [18]	Employee Technology Readiness and Adoption of Wireless Technology and Services	TRI
Matthing *et al.*[19]	Developing successful technology-based services: the issue of identifying and involving innovative users	TRI
Lin and Hsieh [20]	The role of technology readiness in customers' perception and adoption of self-service technologies	TRI
Lin *et al.* [21]	Consumer Adoption of e-Service: Integrating Technology Readiness with the Technology Acceptance Model	TRI
Lin and Peng [22]	The Cultural Dimension of Technology Readiness on Customer Value Chain in Technology-Based on Service Encounter	TRI

Whereby, positive and negative beliefs about technology may coexist, and people can be arrayed along technology beliefs variety anchored by strongly positive at one end and strongly negative at the other. Theoretically and empirically, people's TR correlates with their propensity to employ technology. Therefore, TRI caters for these factors. Besides, it has been proposed that consumers' TR has positive impacts on their online service quality perceptions and online behaviors. Table 1 summarized some studies that had been conducted using these two models. TRI appeared to be the most popular tool to measure users' readiness.

3 Methodology

There are four major phases involved in this phase and discussed in the following sections.

3.1 Data Collection

Questionnaire and survey method were used in this study. The questionnaire consists of twenty one (21) questions that were randomly distributed to one hundred and fifty (150) faculty members of Kulliyyah ICT, IIUM. The set of questionnaires had been developed based on contrast built by [14]. The survey was conducted randomly in Kulliyyah ICT that consists of measurement metrics based on the construct done during literature research. The objective of the measurement metrics is to relatively measure of readiness to implement fingerprint attendance system in Kulliyyah ICT, IIUM. Data analysis was done after surveys are completed. The metrics were measured using a SPSS tool. This is to develop a general theory on the readiness of Kulliyyah ICT in implementing fingerprint attendance systems.

3.2 Research Model

The research model has been chosen based on compatibility of the study with other studies which used the same model. Based on pool of studies which used the same model [18],[23], TRI model is the most compatibility to be used as a research model for this study. TRI consists of various technology beliefs, which are categorized into four components. The first component is optimism, where it is a positive view of technology and a belief that it offers people increased control, flexibility, and efficiency in their lives. Secondly, innovativeness which, means as a tendency to be a technology pioneer and thought leader. Further, discomfort, it is a perceived lack of control over technology and a feeling of being overwhelmed by it. Final component is insecurity. It is a distrust of technology and skepticism about its ability to work properly. The first two components, optimism and innovativeness were categorized as contributors. These components may increase a consumers' TR. The other two components, discomfort and insecurity were categorized as inhibitors, where these components may suppress TR [23]. This is shown in Figure 1.

Previous research by [14] showed that TRI is able to distinguish well between users and nonusers of high technology services. The model also able to identify specific groups of users for who discomfort and insecurity is likely to be very significant. In fact, TRI scores correlate well with consumers' ownership of technology-based products and services (people who own technology-based products and services have a significantly higher TRI score) and with people's use of technology-based services (as compared to those who had no plans to use the services, those who planned to use the services in the next 12 months or have use the services in the past 12 months, have

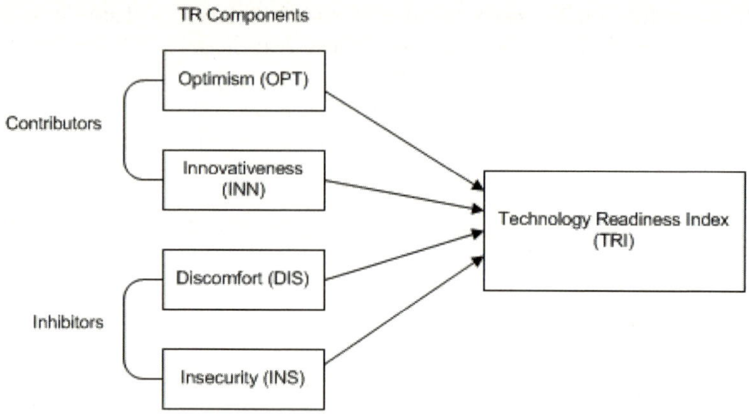

Fig. 1. Technology Readiness Index Model [18],[23],[14]

significantly higher TRI scores). Therefore, TRI becomes a research model of this research. This is due to the linkage between TRI scores and technology-related behaviors to embark new technologies.

3.3 Hypothesis Development

The hypotheses of the study were formulated as follows;

H_1: Innovativeness has a positive effect on readiness of users to use fingerprint attendance system

H_2: Optimism has a positive effect on readiness of users to use fingerprint attendance system

H_3: Discomfort has a negative effect on readiness of users to use fingerprint attendance system

H_4: Insecurity has a negative effect on readiness of users to use fingerprint attendance system

3.4 Data Analysis

Reliability and normality test analysis were employed to test the variables under optimism, innovativeness, discomfort and insecurity. For determining the relationship among the four (4) dimensions of TRI, Spearman's correlation was used due to the abnormality distribution within the constructs. The correlation testing will define whether the hypothesis is acceptable or reject.

Reliability Test

Before the actual analysis taken place, a reliability analysis had been done in order to gauge whether the items are reliable or vice versa. Based on the 100 respondents, the reliability analysis found that innovativeness score is 0.7979, optimism is 0.7324, discomfort is 0.7298 and insecurity is 0.7181. Since, the score of all the items is greater than 0.7 ($p>0.7$), therefore the reliability of all items for innovativeness, optimism, discomfort and insecurity had achieved. This is means that the constructs are reliable to be tested for further analysis. The results had been shown in Table 2.

Table 2. Reliability analysis

Components	Alpha	N of Items
Innovativeness	0.7979	5
Optimism	0.7324	5
Discomfort	0.7298	5
Insecurity	0.7181	6

Normality Test

The characteristic of data is explored with the aim to find out whether the distribution of variables is normal. Hence, normality test had been done to get the normality significance. The kolmogorov-Smirnov statistic with Lilliefors significance level is for testing normality with the normal probability. Table 3 showed that the significance level of all the constructs is less than 0.05 ($p<0.05$). Thus, it is proven from the output that all variables are relatively not normally distributed, meaning, the respondents' answer skewed to strongly disagree or strongly agree.

Table 3. Normality Test

Components	Kolmogorov-Smirnov[a] Sig.
Innovativeness	0.000
Optimism	0.000
Discomfort	0.000
Insecurity	0.004

a. Lilliefors Significance Correction

Spearman's Correlation

Since normality testing had showed that all constructs were not normally distributed, Spearman's correlation was the most appropriate analysis to be used to test the relationship between all constructs.

Table 4 showed that innovativeness 1 has no relation with innovativeness 2, 3 and 4. This is proved by the significant level which is less than 0.05. However, it has relation with innovative 5 where the significance level is greater than 0.05. Meanwhile, innovative 2, 3 and 4 are inter related between each other since its significance level is greater than 0.05.

Table 4. Spearman's Correlation of Innovativeness

		Inn1	Inn2	Inn3	Inn4	Inn5
Spearman's rho	Inn1		0.009	-0.087	-0.143	0.075
	Inn2			0.617**	0.606**	0.481**
	Inn3				0.759**	0.580**
	Inn4					0.415**

Table 5. Spearman's Correlation of Optimism

		Opt1	Opt2	Opt3	Opt4	Opt5
Spearman's rho	Opt1		0.112**	0.253*	0.222*	0.043
	Opt2			0.549**	0.345**	0.351**
	Opt3				0.237*	0.352**
	Opt4					0.067

For optimism construct, Table 5 showed that the significance level is greater than 0.05. Therefore, optimism constructs have strong relation between each construct.

Table 6. Spearman's Correlation of Discomfort

		Dis1	Dis2	Dis3	Dis4	Dis5
Spearman's rho	Dis1		0.661**	0.217*	0.519**	0.579**
	Dis2			0.261**	0.299**	0.596**
	Dis3				0.126	0.036
	Dis4					0.380**

Table 6 proved that all discomfort contrast had a strong element where the significance level is greater than 0.05. Nevertheless, significant level for discomfort 3 and 5 is 0.036, where it is less than 0.05. Therefore, the relation between these two (2) elements could not be assumed.

Table 7. Spearman's Correlation of Insecurity

		Ins1	Ins2	Ins3	Ins4	Ins5	Ins6
Spearman's rho	Ins1		0.340**	0.444**	0.105	0.218*	0.178
	Ins2			0.465**	0.110	0.310**	0.354**
	Ins3				0.281**	0.340**	0.227*
	Ins4					0.099	-0.119
	Ins5						0.534**

As shown in Table 7 the significance level of insecurity constructs is greater than 0.05, therefore the relationship between these constructs can be assumed. Nonetheless, relation between insecurity 4 and insecurity 6 is rejected since the significance level is – 0.119.

Kruskal-Wallis Test

The analysis found that all the variables were not normally distributed. For that reason, the Kruskal-Wallis test is found to be the most suitable method to examine possible differences between two (2) or more groups.

Table 8. Kruskal-Wallis Test between four (4) constructs of TRI and Occupation

	I	O	D	In
			Asymp. Sig.	
Occupation	0.002	0.089	0.000	0.011

Table 8 showed the result as occupation was associated with innovativeness, discomfort and insecurity. Yet, occupation is not significant with optimism where the value is greater than 0.05 ($p>0.05$). This means that the occupation influence the way how this group answered innovativeness, optimism, discomfort and insecurity variables.

Table 9. Kruskal-Wallis Test between four (4) constructs of TRI and User Experience

	I	O	D	In
			Asymp. Sig.	
User experience	0.006	0.046	0.000	0.003

In Table 9 output showed that user experience is associated with innovativeness, optimism, discomfort, and insecurity variables. This is verified by showing that the significance level for all four (4) variables is less than 0.05 ($p<0.05$). It shows how respondents with experience in using fingerprint system affect innovativeness, optimism, discomfort, and insecurity may differ with novice users of fingerprint system.

4 Findings

4.1 Hypotheses Testing

This section discusses the major findings of the data analysis pertaining to the four (4) hypotheses as mentioned in Section 3.3. The relationship is also being tested to prove that the construct has either positive or negative effect to TR. Since normality could not be assumed, Spearman's correlation is being adopted to test the relation between four (4) constructs.

Table 10. Spearman's correlation between Independent Variables

Spearman's rho		Optimism
	Innovativeness	0.273**

Hypothesis One, H_1: Innovativeness has positive effect on readiness of users to use fingerprint attendance system.

Hypothesis Two, H_2: Optimism has positive effect on readiness of users to use fingerprint attendance system

As shown in Table 10, the significant level of the relation between innovativeness and optimism is 0.273. The data showed sufficient evidence that innovativeness and optimism have positive effect to faculty's members to use fingerprint system. This is because the significant level is more than 0.05 ($p>0.05$), therefore hypothesis 1 (H_1) and 2 (H_2) are accepted.

Table 11. Spearman's correlation between Independent Variables

Spearman's rho		Insecurity
	Discomfort	-0.487**

Hypothesis Three, H_3: Discomfort has negative effect on readiness of users to use fingerprint attendance system.

Hypothesis Four, H_4: Insecurity has negative effect on readiness of users to use fingerprint attendance system.

The output in table 11 presented that the significant value between discomfort and insecurity is –0.487. Therefore, it was proven that there is a negative relationship between discomfort and insecurity. As a consequence, lower discomfort and insecurity scores are associated with readiness of the faculty to implement fingerprint attendance system. Based on this output, hypothesis 3 (H_3) and 4 (H_4) are proven.

4.2 Technology Readiness Index (TRI) Score

The score of TR for Kulliyyah ICT is calculated in order to determine whether the subject is ready to use fingerprint system or otherwise. From the descriptive analysis, the study shows the score of contributors and inhibitors variables. Previously, the analysis found the Innovativeness and Optimism scores were positively related to the faculty to adopt fingerprint attendance system. These scores were derived by getting the mean of each variable. Furthermore, the TRI elements of Discomfort and Insecurity scores were negatively related to faculty use, such that the readiness decreased as each of these elements increased.

Kulliyyah ICT had scored 19.50 for Optimism and 18.62 for Innovativeness. Whereby, the score for Discomfort is 11.20 and 17.45 for Insecurity. Since the score for Contributors element is higher than Inhibitors element, it means that Kulliyyah

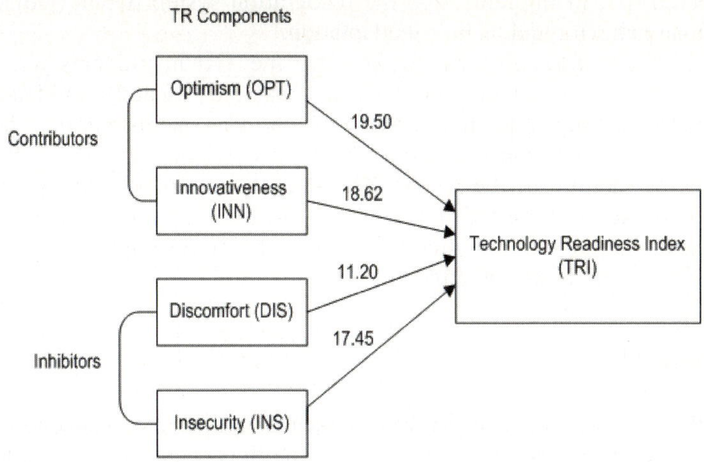

Fig. 2. Technology Readiness Index (TRI) score

ICT is ready to use fingerprint attendance system. Figure 2 summarized the score of TRI in Kulliyyah ICT. In the future, the result would help to facilitate the smoothness of the implementation and eventually the evaluation of the model.

5 Discussion

The study has identifed components that can be contributors for the faculty to use fingerprint attendance system. The analyzing of components of contributors is based on the TRI Model. The elements of contributors are Optimism and Innovativeness. Furthermore, only suitable constructs of optimism and innovativeness are being chosen as a factor. This is because certain constructs are not appropriate for the faculty's environment. The same model is being used in order to identify the components of inhibitors. As a result, Discomfort and Insecurity are the inhibitors components. Subsequently, only appropriate components of Discomfort and Insecurity are used in this study.

The analysis used TRI derived from the faculty responses to the twenty five (25) questions on the survey. A bivariate linear regression analysis was used to investigate each variable. Also elements had been tested and it showed that almost all are statistically significant ($p>0.05$) with the Kulliyyah ICT fingerprint attendance system used. Nonetheless, there are few elements which are insignificant. This might due to those elements were not suitable for this study, thus, removing the elements would be appropriate. Further, analysis was conducted on the four (4) elements of TR. Optimism and Innovativeness were considered contributors to TR, while Discomfort and Insecurity are considered inhibitors. As expected, the variables which contribute to TR were positively related to Kulliyyah technology use, and those elements that inhibit technology use were negatively related to faculty to use fingerprint attendance system. It is also observed that in implementing the recognition system would require further study on human behavior and technical adaptation.

The results of the normality test showed that the TRI instruments were not normally distributed. Abnormal distribution leads that the research could not test for certain analysis. Consequently, misleading or incorrect conclusions may be reached using the results. As with most research, limitations associated with sampling which could temper the validity of the results. The sample comprises entirely respondents who are expected to be more technology savvy than other faculty's member. Therefore, the findings require further attention. A more random sample is needed for getting a significant result in user readiness.

6 Conclusions

Findings of the study indicate that Innovativeness and Optimism components are the key drivers for gauging readiness of user to embrace the fingerprint technology. When both of these elements got high score it can be concluded that the faculty of Kulliyyah ICT is ready to implement fingerprint attendance system. Overall, the elements of TR, which are Optimism, Innovativeness, Discomfort and Insecurity, provide a better fit and exhibit stronger predictive power. In addition to that, from the results, it shows that all research objectives have been positively answered and TRI has shown to be an effective tool for identifying users' readiness for the implementation of a new technology.

There are several areas that need to be further investigated. Numerous factors actually affect Kulliyyah ICT members in using the system namely; the availability of technology resources, support or encouragement from management regarding technology, and the implementation strategies. Though, these factors are beyond the scope of this study, the investigation of these factors would be beneficial for further research.

Another key area for further analysis emerged from the results is TR. TR levels are derived from survey responses to questions that are groups into four (4) areas which are Optimism, Innovativeness, Discomfort and Insecurity. Insecurity and Discomfort are the main drivers of lower TR levels. Exposure to technology over time may have the strongest effect on the inhibiting components of readiness. It means that as students, lecturers or staffs are continually exposed to the system, their discomfort toward the technology may reduce, and their feelings of insecurity associated with using the fingerprint technology may reduce as well. This leads to the approach that user readiness is not static over a period of time and individual readiness can change. Therefore, relationship between technology exposure and TR need to be specifically researched and the results from this study could provide a basis for further investigation.

References

1. Bleumer, G.: Biometric Authentication and Multilateral Security (2006) (Retrieved February 26, 2006), `http://www.semper.org/sirene/people/gerrit/papers/bioauth.pdf`
2. Heracleous, L., Wirtz, J.: Biometrics – the New Edge in Service Industries Executive Briefings on Key Business & Management Issues (2006) (Retrieved June 26, 2006), `http://www.execed.oxford.edu`

3. Jain, A., Hong, L., Pankanti, S.: Biometric Identification. 43(2), 90–98 (2002)
4. Liu, S., Silverman, M.: A Practical Guide to Biometric Security Technology. IT Professional 3(1), 27–32 (2001)
5. Gamboa, H., Fred, A.: A Behavioural Biometric System Based on Human Computer Interaction (2006) (Retrieved March 4, 2006), http://www.lx.it.pt/~afred/anawebit/articles/AFredSPIE2004.pdf
6. Sukhai, N.B.: Access Control & Biometrics. In: Proceedings of the 1st annual conference on Information Security Curriculum Development, Georgia, pp. 124–127 (2004)
7. Lawson, W.J.: Enhancing Assistive Technologies: Through the Theoretical Adaptation of Biometric Technologies to People of Variable Abilities. Faculty of the School of Business Kennedy-Western University (2003)
8. Smith, C.L.: The Science of Biometrics Identification. Australian Science Teachers Journals 49(3), 34–39 (2003)
9. Harris, A.J., Yen, D.C.: Biometric authentication: assuring access to information. Journal: Information Management & Computer Security 10(1), 12–19 (2002)
10. Floyd, J.M.: Throwing Away the Keys (2006) (Retrieved March 4, 2006), http://chronicle.com/colloquy/2005/12/bio/
11. Kiernan, V.: Show Your Hand, Not Your ID. Information Technology. The Chronicle of Higher Education 52(15), 28 (2005)
12. Gale, D.: Biometrics Go Mainstream, Campus Technology. 101 Communications (2006) (Retrieved October 10, 2006), http://www.campus-technology.com/article.asp?id=1818
13. Heslop, L., McGregor, A.E., May, G.: Development of a Technology Readiness Assessment Measure: The Cloverleaf Model of Technology Transfer. Journal of Technology Transfer 26(4), 369–384 (2001)
14. Parasuraman, A.: Technology readiness index (TRI): A multiple-item scale to measure readiness to embrace New Technologies. Journal of Service Research 2(24), 307–320 (2000)
15. Chang, H.L., Chen, S.H.: Assessing the Readiness of Internet-based IOS and Evaluating its Impact on Adoption. In: Proceedings of the 38th Hawaii International Conference on System Sciences, pp. 1–10 (2005)
16. Chang, H.L., Shaw, M.J.: Developing the Readiness Index of IT-enabled Supply Chain Collaboration (2006) (Retrieved June 26, 2006), http://citebm.business.uiuc.edu/RN%20study/readiness%20index.pdf
17. Galbraith, C.S., DeNoble, A.F., Ehrlich, S.B.: "Spin-In" Technology Transfer for Small R&D Bio-Technology Firms: The Case of Bio-Defense. Journal of Technology Transfer 29, 377–382 (2004)
18. Chang, A.M., Kannan, P.K.: Employee Technology Readiness and Adoption of Wireless Technology and Services. In: Proceedings of the 39th Hawaii International Conference on System Sciences, pp. 1–9 (2006)
19. Matthing, J., Kristensson, P., Gustafsson, A., Parasuraman, A.: Developing successful technology-based services: the issue of identifying and involving innovative users. Journal of Services Marketing 20(5), 288–297 (2006)
20. Lin, J.S.C., Hsieh, P.L.: The role of technology readiness in customers' perception and adoption of self-service technologies. International Journal of Service Industry Management 17(5), 497–517 (2006)

21. Lin, C.-H., Shih, H.-Y., Sher, P.J., Wang, Y.-L.: Consumer adoption of e-service: integrating technology readiness with the technology acceptance model. Technology Management: A Unifying Discipline for Melting the Boundaries 3(31), 483–488 (2005)
22. Lin, C.H., Peng, C.H.: The Cultural Dimension of Technology Readiness on Customer Value Chain in Technology-Based Service Encounters. Journal of American Academy of Business 7(1), 176–180 (2005)
23. Massey, A.P., Khatri, V., Ramesh, V.: From the Web to the Wireless Web: Technology Readiness and Usability. In: Proceedings of the 38th Hawaii International Conference on System Sciences, pp. 1–10 (2005)

Web Usability on the Nintendo Wii Platform

Valentina Franzoni, Osvaldo Gervasi, Sergio Tasso, and Simonetta Pallottelli

Department of Mathematics and Computer Science, University of Perugia,
via Vanvitelli, 1, I-06123 Perugia, Italy
{osvaldo,sergio,simona}@unipg.it

Abstract. The aim of the present study is to propose a set of guidelines for designing Internet web sites usable and accessible with the Nintendo Wii console. After an accurate analysis of usability issues and the definition of the typical Wii Internet user, twelve usability guidelines will be described. The named guidelines are focused on the visibility, understandability, clickability and compatibility issues. To prove the effectiveness of the guidelines we have applied them to a test web site. We carried out a comparison test on a sample of forty individuals, selected among the various categories of the potential users of the Nintendo Wii console. The analysis of the resulting information confirms that the revised web site according to our guidelines is more usable and the improvement is more pronounced for weak categories (elderly and individuals with no experience with web browsing). Furthermore the adoption of the guidelines reduces the gap between users with different expertise, accessing a web site with the console.

1 Introduction

The development of computer graphics engines made available several commodity consoles produced by the major operators in the entertainment market. These consoles are enabled with the Internet access, the web surfing and other communication facilities that open new perspectives to facilitate the collaboration among people. The access to Internet through a console have some limitations respect to a personal computer, in particular the screen resolution, that impact the access to the information through the Web.

In the present paper we introduce twelve guidelines to adapt the content of web sites to the Nintendo Wii console in order to guarantee the compliance of the web content to the rules of usability and accessibility. The proposed guidelines will allow people to get information and use web services really usable and accessible, both using PC and the Nintendo Wii interface, offered as an alternative for PC web browsing.

The usability is defined by the ISO standard 9241-11 and in ISO/IEC standard 9126[1]. During the last 10 years some usability guidelines were also defined by J. Nielsen, D. Norman and other usability experts, concerning visibility, affordance, conceptual models, feedback, user protection and flexibility[2,3,4]. The usability of a web site is also a key factor to increase the user trust of a web site and the degree of web site loyalty[5].

O. Gervasi et al. (Eds.): ICCSA 2008, Part II, LNCS 5073, pp. 105–118, 2008.

To adapt our product to needs and to expectations of the final user, we have to understand the characteristics of the users, their activities and the organizational and social contexts in which they are living and operating. For what concerns the understanding and using friendliness, then, we can take advantage of the available consolidated usability principles and guidelines, based on the acquaintances of behavioral and cognitive aspects in human-computer interaction[6,7,8,9]. The spirit of ISO norms and usability guidelines is that the design of the interface and the contents planning of web sites must be user-centered. The users know their competences, culture, needs, limits and attitudes and it's necessary their involvement in planning, realization and management of a web site through usability tests.

Accessibility is defined and analyzed in several notes of the World Wide Web Consortium (W3C)[10]. The Italian Parliament has promulgated a law to define the rules to benefit disabled users in accessing computers instruments[11]. We want to guarantee accessibility to individuals belonging to categories experiencing difficulties in using web technologies: elderly, sick people, ipoliterates, users who operate in a low illuminated atmosphere, and/or using old instruments, users who have few competences and abilities to use Web and computers.

The paper is organized as follows: in section 2 a brief description of the Nintendo Wii console is given; in section 3 the users that performed the usability tests are described; in section 4 our twelve usability guidelines for the Nintendo Wii console are presented; section 5 discuss the usability tests made on a set of forty users and the results obtained; finally in section 6 some conclusions are illustrated.

2 The Nintendo Wii Console

The Nintendo Wii console makes possible to surf the Internet in an easy way. The browser Opera for Wii, based on Opera 9.10, is adapted for the innovative motion-sensitive controller, called WiiRemote: the user's actions are caught by the integrated motion sensors and sent via a bluetooth link to the SensorBar.The WiiRemote is designed to be ergonomic and easy to use. A few buttons, pushable with one hand, allow to use advanced controls. A symmetrical design and the possibility to adjust the relative settings, allow a full access to left-handed people. The WiiRemote includes also a built-in speaker and a rumble feature, to make the browsing interaction more realistic. An external socket is integrated to connect the Nunchuck, a second motion-sensitive controller with more buttons and a control stick, that expand the control system for two-hand actions. Movements are computed by a CPU Power PC, a 512 Mb flash memory and an ATI graphic accelerator. It's possible to connect up to 4 WiiRemotes, making possible to surf the Web in a very innovative way: together with family and friends. The users can participate together to the browsing sessions and are allowed to point on the web site elements, to increase the communicative interaction. During the web browsing sessions it's possible to zoom in and out (automatic or progressive zoom): the Outline Font feature focus the enlarged text elements. It's also possible to scroll the visible area, set the single-column mode (that reduces the layout, viewing the components in the same order in which they are written in the markup code and resizes the text areas), and other settings. The Internet channel supports Ajax with JavaScript 1.5, Flash up to 7, cookies, SSL, MP3, while ActiveX, PDF and Video Codecs are not supported and it's not possible to download any file from the Internet.

3 The User Experience

According to the studies on web usability made by the Nielsen Norman Group in the last years[2], a percentage of 66% of web sites are not usable (users can not complete the requested tasks during usability tests) for a generic PC web browsing. In most cases users abandon a not immediately useful or interesting web page after 35 seconds on average. This behavior and the subsequent research of new web pages where to find the requested information, is known as information foraging: the more new informative resources are easily available, the least users spend time on any one of them. So users, like opportunist predators, prefer to make quick use of immediately available portions of information rather than dwell on them to choose the best ones. Concerning Wii web browsing, it seems right to think that the percentage of unsuccess of web sites we talked about is raising, both depending on the lower precision of the pointing device and on the contrary on the extreme easiness of use of the same device. Therefore the range of users enlarges to a potentially less expert public, from professionals and keens of technologies to pure newbies, who approach the Web exclusively through the Nintendo Wii. It could be attractive where we can give a tasty accessible offer, designed to measure on user.

3.1 Teenagers

We can focus our attention on teenagers, still the main enjoyers of game consoles and guarantors of the success of the Web. The ability of new generations to handle interactive mechanism better than adults, in particular if they include the use of fingers like mouse, joystick and joypad, is due -wherever real- to a precocious formation and habit to use such instruments. Nevertheless, according to the studies and usability tests by the Nielsen Norman Group[2], teenagers are inclined to worry about Internet and its dangers, like virus, spam, malware and phishing and they avoid clicking on unknown elements, downloading and freely surfing the Web. We will focus for them on affordance and on simplification of contents and interfaces.

3.2 Elders

Another interesting typical user could be elderly people (over 65 years old), that represent a big percentage of the population. Nintendo gave a lot of publicity to the suitable of Wii for elderly users and in general who has not experience with web browsing and with the use of new technologies. The Wii could be right for them, because of the extreme simplicity of its interface and it could be used for physical and mental training of healthy seniors. We will plan to reduce for them any interaction, guaranteeing all the functionalities.

3.3 Sick and Disabled People

Wii can be used for training and rehabilitation for single persons or groups in public and private sanitary structures. In Padova University (Padova, Italy) Prof. Luciano Gamberini[12] of the Human Technology Lab of the faculty of Psychology started on September 2007 to develop systems for research and clinical intervention based on the Wii console. For this aim we invite web designers to respect all the guidelines about accessibility. Wii can be also used for simulation and education and for all the purposes of use of 3D reality.

4 Web Usability Guidelines on the Wii Interface

In the most economically advanced countries, Internet is replacing television programs. Several technologies have been exploited to surf the Internet using television as output device. The access of multimedial information through the TV needs interactive capabilities for which the TV interface is not enough specialized and several adjustments and integrations need to be introduced in order to make the interface efficient and comfortable.

The research in this area is going in the direction of "integration through simplification": the automatisms which are being developed and the corresponding services that gradually appears in our life go in the direction of a reduction of interactivity, because on TV, for example, it is difficult to scroll pages or browse hypertexts. So, the proposed medium on one side loses the entertainment capability of TV and on the other renounces to the interactive abundance of PCs. Wii inverts this trend, proposing a suitable interface for both group entertainment and interactive web browsing. Nevertheless, the majority of web sites is not designed to be surfed on Wii or any other device except computers. Furthermore a lot of them do not comply to accessibility and usability guidelines for the Web. So even on Wii Internet Channel they can not be properly visualized and their interactive features are not well operating.

The usability guidelines for the Wii interface are presented in table 1 and discussed in the following sections.

4.1 The Restoration of Visibility

The first and most important problem that we found on the Wii Internet Channel is related to TV resolution and image quality. To improve reading intelligibility of text elements, Wii developers have reduced the screen resolution so that the width of the page appears in 800 pixels. In spite of it, the visibility of the text is very low and the most used sizes of the text body - 10,12,14 - are clearly

Table 1. The 12 accessibility guidelines for the Nintendo Wii console

Guideline:	Text
First:	Balance the size and spacing of the fonts and all other components
Second:	Space out any component from the edges of the page
Third:	Test color matching and contrast using specific software
Fourth:	Simplify the design
Fifth:	Simplify the content
Sixth:	Reduce interaction to essentials, preserving all functionalities
Seventh:	Space clickable components, keeping them in the same visual area
Eighth:	Don't use components that don't allow spacing
Ninth:	Use only the supported formats
Tenth:	Use CSS for the design
Eleventh:	Guarantee light content
Twelfth:	Adopt the guidelines for web accessibility and usability

inadequate for a comfortable displaying, without making extreme use of the zooming and scrolling features. Because of the flicker effect of the screen, given by the low updating frequency of the TV image, text elements result not readable, inducing eye fatigue: the edges of the characters are not defined and they result hard and difficult to read. According to the first guideline we suggest to adopt a font size of at least 20 (26 recommended) for text and a line spacing of at least 1.5 lines. For the other components the same rules apply: all of them must be large and spaced enough. From our tests on resolution, it is evident that most TVs (especially old models) loose some pixels on the edges and, because of their rounded shape, the first visible pixels get deformed. It's necessary, then, to space out any component from the edges of the page. A margin of 15 pixels in every direction is an appropriate check point. We note that, because of low resolution, colors with a low contrast become illegible and cause tearing. The third guideline recommend to test color matching and contrast using specific software.

4.2 The Restoration of Understandability

The use of large fonts and components allows to see every element correctly, but doesn't guarantee a simple interpretation of it. To keep the same design with the same enlarged content is not possible, but even keeping the same elements disposed in a different design is not a good idea. In fact, if the number of elements is high, enlarging them we will obtain a too dense page or a too long one. In the first case, the web page becomes a complex and confusing scheme to understand. In the second, we force the user to a long search for the contents using the scroll feature, that increases the number of users abandoning the page before reaching the desired content, or unable to identify its position. We define the guidelines (fourth and fifth) to simplify the design and to simplify the content. The fourth guideline will make the design clearer and more pleasant and the fifth one will reduce the length of the web page and make it more readable. As a point of control for the fourth one we propose to verify that images, buttons, links, blocks of text are not flooding the page and that they are disposed in a maximum of 4 columns, better two in case of text blocks. The point of control for the fifth guideline is to readjust all text to a cognitive level of secondary school (that is right for teenagers, elders and people with higher levels of instruction but low capability of attention) and edit images so that they present levels of detail understandable with a maximum of two zooming operations. Bring the text at this level means both to simplify the language and syntax, both reduce the length. A good way to do so is to provide images that can communicate the proposed concepts. Furthermore, we want to remember that interactivity is another point that may create confusion using web pages. From our tests we see the seriousness of this issue: it's not always easy even for general users to identify and interpret the functionality of the various components. We propose the sixth guideline to reduce interaction to essentials, preserving all functionalities. A point of control for it can be not to include in the same page different components with the same interactive function, except when it's necessary to access them from very distant points of the page. An exception can be the logo in the header, that can be clickable as a link to the home page.

4.3 The Restoration of Clickability

In most web sites, the design includes lists and menus or other components with no vertical spacing. Due to the nature of the controller (sensible to very small movements) and considering the shape and size of the associated pointer (a little hand of 50x70 pixels with a slightly active area) it is rather difficult to stop and click on small or close components and the palm of the little hand may also cover adjacent links. To this end, we propose the seventh guideline, advising to space clickable components, keeping them in the same visual area. As point of control, a line spacing of 1.5 lines for text links and at least 10 pixels for buttons and similar elements, such as picture thumbnails, should be considered. Because especially combo boxes, drop-down and pop-up menus cause clickability problems, we specify not to use components that don't permit spacing.

4.4 The Restoration of Compatibility

We can see from the specification of the console that not all available formats for the web components are supported by the Wii Internet Channel, so we recommend to use only the supported formats.

The compatibility with Ajax includes the full support of Cascading Style Sheets (CSS). Using CSS for the design is fundamental to the usability of our web site, because the browser Opera for Wii provides a single-column mode that uses CSS for a correct view. The use of CSS is also a requirement for accessibility and allows to use functionalities like hovers, that otherwise could be designed only through the use of JavaScript, making the page heavier. Another problem that afflicts the Wii Internet Channel is the slowness of page loading, that induces the user to abandon the site and that hardly depends on the proposed content.

It's necessary, then, to guarantee light content, regulating the use, among others, of decorations and animations. Finally, we reaffirm the importance to maintain compatibility with the guidelines for web accessibility and usability.

5 Usability Tests

Our tests were conducted on a sample of 40 users. , according to suggestions given by J. Nielsen and M. Visciola.[13] selected among people who don't work in the fields of new technologies, marketing, web design or usability, because these people are not a significant user model. According to the suggestions of the Nielsen Norman Group[2], we assessed that the insiders have too many specific expertise and are not able to use a web site as anyone would do. We included in our study, on the other side, users without previous experience of the Internet. In general in web usability tests this class of users is not considered significant, but in our case the neophyte user is one of the most important user model to study, because the Wii console is proposed as an instrument to approach the Web even for the first time.

Our 40 users were selected also on the basis of their age, dividing them in 4 classes, according to the definitions established by the Italian Institute for Statistic (ISTAT): teenagers (12-19 years old), juveniles (20-39), adults (40-64), elders (over 65). The users were selected among the four categories is the same percentage.

We have chosen for our tests an exhisting web site, named site B, related to a restaurant, because we considered it a suitable target for our users.It was developed in 2003 using old methodologies, like tables for the graphic planning and the use of images for the text link presentation. That included a home page with very complex elements and some internal pages with one template. We redesigned it applying the twelve usability guidelines for the Wii interface. We used XHTML and CSS standards and some simple PHP techniques to optimize maintenance and to manage the contact form, and named it site A.

To keep usability both for Wii and PCs, we used JavaScript code for the platform identification and we created different CSS for the visualization respectively on Opera for Wii and on all other web browsers.

We carried out the usability tests on both websites. We made the test on a cathode ray tube TV, with a 22" display 4:3, to which we connected the Wii console and an ADSL line. We put our users at a distance of 2 meters, which we consider appropriate to this kind of TV and compatible with the furniture provision in a normal room; the selected distance is also the recommended one in Wii specifications. We made our tests in an adequately lit environment, with the light source put behind the user, in a position that avoids reflections on the screen.

Users were introduced in a small room for tests, separated from the waiting room so they did not perceive any disruption and to avoid anyone to listen to our tests from outside. In the room two other people were present to record data, at a distance of about 1 meter and outside view field, to avoid to distract the user.

We carried out the tests asking people to perform some tasks on both web sites: the redesigned web site and the original one. Our test users have not been previously trained for it. The purpose of the test was not explained to users before its conclusion, to avoid any influence during the evaluation of both test web sites. Once seated in our workstation, we introduced them to the Internet Channel fundamentals. Before beginning, we proposed a minute to try the different functions of the WiiRemote on the homepage of the site proposed, except for clicking, to avoid the beginning of navigation. We finally explained them they have to surf two different web sites, with similar content, in each of which they have asked to repeat the same 3 tasks, with an increasing level of difficulty.

We have based our tests on the "thinking aloud" method, in which we ask users to report aloud each thought while operating with our interface. Hearing the thoughts of users helps to understand why they are doing what they are doing, and to record any difficulty. One facilitator talked to the user, while the other only recorded times and operation numbers. The role of the first facilitator was to explain the features of the remote control and to give any clarification on it during the test, solving in real time any problem depending on the WiiRemote's functions and on the nature of the assigned tasks.

We didn't suggest how to complete tasks and didn't provide informations about the structure and the contents of both web sites. To count operations, we considered only clicks on clickable elements, both successful (as transitions from one page to another) and unsuccessful (as not accurate clicks and repeated ones). We also registered for each task if it was successful or not.

In figure 1 is shown the test environment and a user performing the test.

At the end of the practice of the tests, we briefly interviewed the users, recording data relating to their age, sex, experience with the Internet, and their evaluation of both sites, first A then B, with the following questions:

Q.1: Did you like to browse this web site?
Q.2: Did you like anything particular?
Q.3: Didn't you like anything particular?
Q.4: How would you define this experience of web browsing: funny, relaxing, difficult, boring, or other?
Q.5: Would you use again this web site on Wii if possible?

We also made two video recordings for each test: one camera was pointed on the screen, the other on the user, both with audio track, in order to view the user movement on the interface and on the WiiRemote and face expressions. The tasks have been assigned in sequence on site A and then repeated on site B. Both sites present roughly the same structure: we decided to test first the redesigned site, then the original, otherwise the user could be conditioned, knowing how to complete the tasks, altering the measured times. The three tasks assigned were at different levels of difficulty: find the address of the restaurant, find the price of defined dishes, find a way to book a table.

Levels were established on the basis of the depth of the links and the number of ways in which it was possible to end the task. The first task was easy, the

Fig. 1. A woman performing our test related to the usability of Nintendo Wii

second was of high difficulty and the last of medium difficulty. In this way the hardest test shouldn't be influenced by tiredness.

5.1 Results

We obtained from our study three types of data:

1. Objective data, relating to times and number of operations for each assigned task.
2. Subjective data, obtained with the feedback in the final interview.
3. Subjective data recorded by our facilitator, relating to any difficulty that users experienced during the practice of the tests, on the basis of the thinking aloud method.

All data were recorded for sites A and B separately and registered on a database. During the transcription, data have been classified by type (2 or 3).

The results showed that users consider the redesigned site better than the original and in particular more easy and pleasant to use, so that our guidelines are valid. In particular the improvement is more evident approaching to weak groups, namely elders and people without experience in the Internet. The use of our guidelines reduces the gap among the various ranges of people in the approach to the technology of the Wii web browsing.

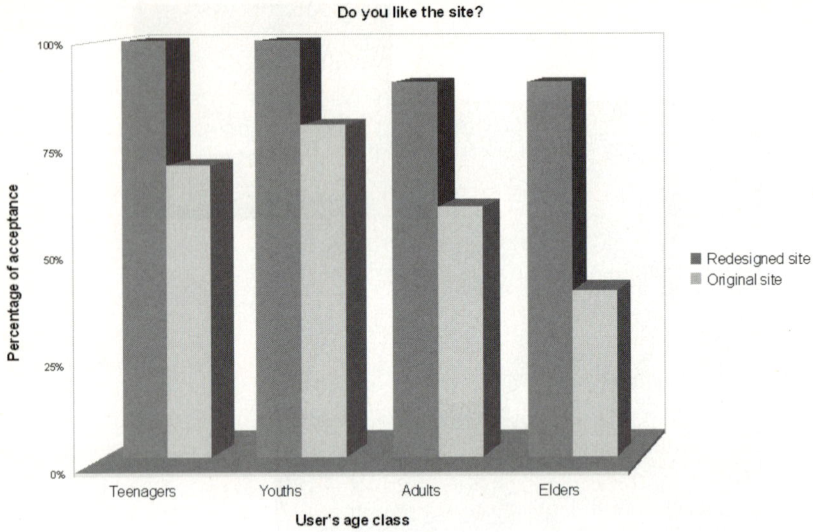

Fig. 2. Percentage of acceptance by the users of the redesigned and the original sites, as a function of the user's age class

5.2 Subjective Data from Feedback

Data recorded with interviews (type 2) revealed that the redesigned site had a higher approval. Both the general indicators (the answer to the questions Q.1 and Q.5 gave a much higher value in the redesigned site than in the original. The degree of acceptance of the two sites is shown as a function both of the user's age class (figure 2) and of the user's experience with the Internet (figure 3).

In addition, the intention of users to visit again the two sites is shown as a function both of the user's age class (figure 4) and of the user's experience with the Internet (figure 5). The figures show that the approval of the redesigned site is more evident when we get closer to the weak groups and the gap among the different groups declines considerably.

5.3 Subjective Data Recorded by the Facilitator

Each user reported a number of difficulties variable up to 2 (if none, we registered 'none'), for a total of 80 responses. The difficulties experienced by the users were: length of text, contents organization, frequent zooming and fear to click. Fear to click was registered when user kept pointer on a link without clicking on for over 5 seconds and reported aloud observations about it: this difficulty had the same attendance on both sites. The goodness of our guidelines is further confirmed by the collected responses, as shown in figure 6, where the problems experienced accessing the two sites are shown and it is evident that the problems with the redesigned site are almost solved.

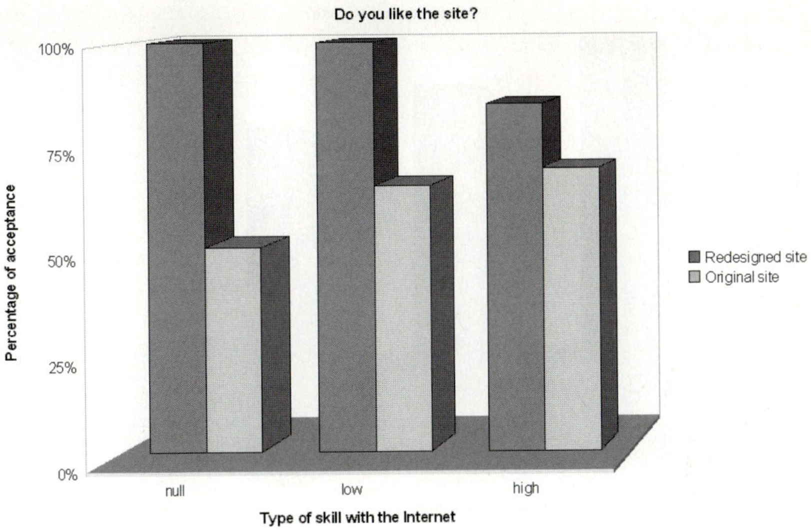

Fig. 3. Percentage of acceptance by the users of the redesigned and the original sites, as a function of the user's experience of the Internet

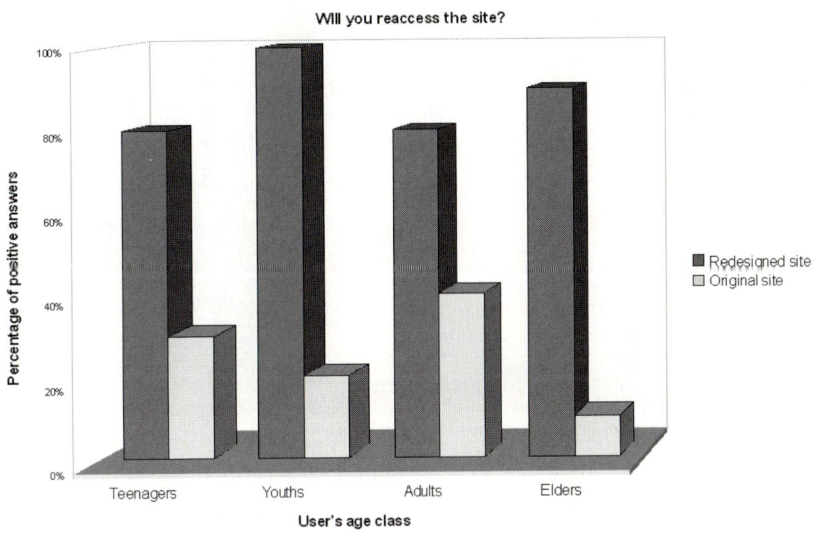

Fig. 4. Degree of intention of users to visit again the redesigned and the original sites, as a function of the user's age class

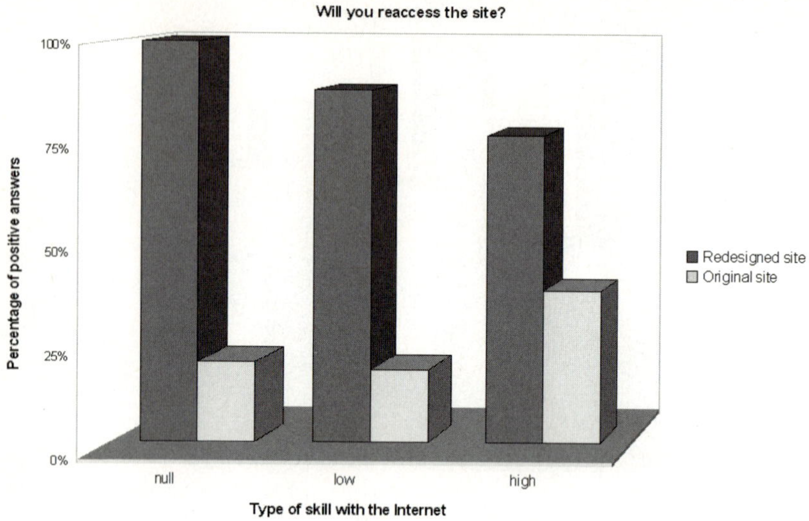

Fig. 5. Degree of intention of users to visit again the redesigned and the original sites, as a function of the user's experience with the Internet

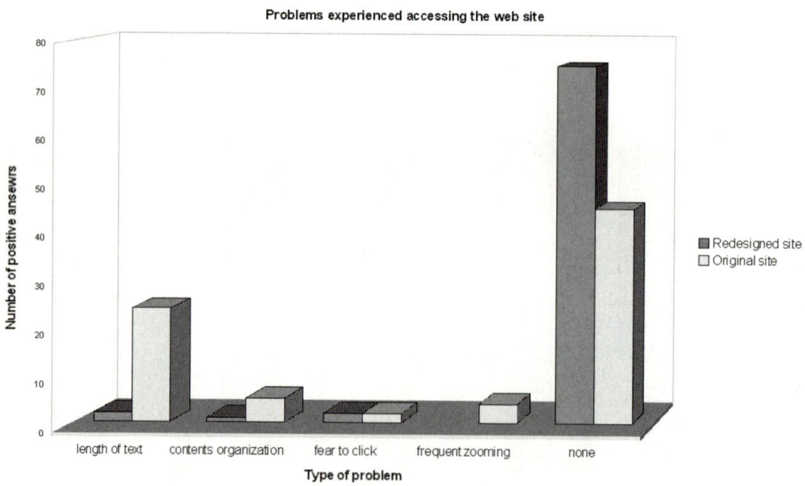

Fig. 6. Problems experienced by users accessing the redesigned and the original websites

5.4 Objective Data

The number of tasks completed is 111 for site A and 101 for site B. We notice that the solving method for tasks in both sites A and B was the same, so users that completed a task on site A knew how to complete it in site B. In spite of this, we have a higher percentage of success in site A than in site B. Even analyzing these data by age and experience in the Internet, we can notice the trend about the intergenerational gap and the difference among experience ranges. It confirms the subjective data from the feedback: the difference among the results in the various groups in site A is reduced.

6 Conclusions

The proposed twelve guidelines for the web designers in order to maximize the usability of the Nintendo Wii console have been used to restyle an existing web site. The results of the tests made on a set of 40 users grouped by 4 classes of age, related to the execution of three tasks on both web sites have been recorded. The data analysis clearly shows that the site following the guidelines is more usable and that the improvement is more pronounced for weak categories (elderly and individuals with no experience with Internet browsing), so the use of the proposed guidelines reduces the difference between the different users approaches to this technology. Indeed, the trend graphs for all of the data show that the redesigned site has received a larger rating and that the outcome is even more evident in weaker categories. It was also clear that the studied issues are actually those with the maximum weight on usability, and that the use of the guidelines solve the problems. We can conclude that the proposed guidelines maximize the usability of a website surfed with the Nintendo Wii console.

References

1. Abran, A., Khelifi, A., Suryn, W., Seffah, A.: Usability Meanings and Interpretations in ISO Standards. Software Quality Journal 11(4), 325–338 (2003)
2. Nielsen, J., Loranger, H.: Prioritizing Web Usability. New Riders Press, Berkeley (2006)
3. Lazar, J. (ed.): Universal Usability: Designing Computer Interfaces for Diverse User Populations. John Wiley & Sons (2007) ISBN 0-470-02727-4
4. Bias, R.G., Mayhew, D.J. (eds.): Cost-Justifying Usability. Academic Press, Boston (1994)
5. Flavian, C., Guinaliu, M., Gurrea, R.: The role played by perceived usability, satisfaction and consumer trust on website loyalty. Information and Management 43(1), 1–14 (2006)
6. Stewart, T.: Ergonomics standards concerning human-system interaction: Visual displays, controls and environmental requirements. Applied Ergonomics 26(4), 271–274 (1995)
7. Helander, M.: A Guide to Human Factors and Ergonomics, 2nd edn. CRC Press (2005) ISBN 0-415-28248-9
8. Sears, A., Jacko, J.A. (eds.): The Human-Computer Interaction Handbook, 2nd edn. CRC Press (2007) ISBN 0-8058-5870-9
9. Sharp, H., Rogers, Y., Preece, J.: Interaction Design: Beyond Human-Computer Interaction. John Wiley & Sons, Inc., West Sussex (2007)

10. Web content accessibility Guidelines 1.0 (1999), `http://www.w3.org/TR/WAI-WEBCONTENT/`; Techniques for Web Content Accessibility Guidelines 1.0 (2000), `http://www.w3.org/TR/WAI-WEBCONTENT-TECHS/`; Techniques for Web Content Accessibility Guidelines 2.0 (2007), `http://www.w3.org/TR/WCAG20/`
11. Legge 9 gennaio, n. 4, by Italian Parliament about the dispositions to benefit disabled users in accessing computers instruments (Disposizioni per favorire l'accesso dei soggetti disabili agli strumenti informatici) (2004)
12. Human Technology Laboratories, Università degli Studi di Padova, `http://psicologia.unipd.it/htlab`
13. Visciola, M.: Usabilità dei siti web, curare l'esperienza d'uso in Internet, Apogeo, Milano, Italy (2006)

Volterra Algorithm for Modelling Sea Surface Current Circulation from Satellite Altimetry Data

Maged Marghany, Mazlan Hashim, and Arthur P. Cracknell

Department of Remote Sensing
Faculty of Geoinformation Science and Engineering
Universiti Teknologi Malaysia
81310 UTM, Skudai, Johore Bahru, Malaysia
maged@fksg.utm.my,
magedupm@hotmail.com,
mazlan@fksg.utm.my,
cracknellarthur@hotmail.com

Abstract. This paper was utilized a new approach for modelling sea surface current from JASON-1 satellite altimetry data. This was based on utilizing of the Volterra series expansion in order to transform the time series satellite altimetry data into a real ocean surface current. Thus,the basic equation of hydrodynamic has been solved by second order Volterra model. Then, the Volterra kernel inversion used to obtain the sea surface current velocity. The finite element model of Lax-Wendorff schemes used which was based on triangular space-time elements to map the spatial current variation in the South China Sea over different monsoon periods. In situ sea surface current measurements were collected along the east coast of peninsular Malaysia by using electromagnetic current meters. The study shows that the maximum current magnitude of 1.2 m/s was occurred during the northeast monsoon period as compared to other monsoon periods. The main noticeable feature was an existence of anticlockwise eddy in the Gulf of Thailand. The results also shows a good correlation between in situ current measurements and the Volterra-Lax-Wendrof scheme with high R^2 of 0.91. It can be said that Volterra-Lax-Wendrof scheme can be used as numerical scheme for modelling sea surface current from altimetry data.

Keywords: JASON-1 satellite altimetry data, Volterra model, Lax-Wendorff schemes, Finite element model, Sea surface current.

1 Introduction

Satellite altimetry systems have been recognized as an excellent tool for coastal hydrodynmic investigations [8]. Indeed, satellite altimeters provide long-term and continuous coverage of wave and wind fields of the world oceans. One of the first satellites that used altimetry technology in the measurement of sea levels was the Seasat 1978-64A. In the 1990s with more complex and advanced imaging and altimetry systems, radar remote sensing has become more significant tool and has been

O. Gervasi et al. (Eds.): ICCSA 2008, Part II, LNCS 5073, pp. 119–128, 2008.

utilized by the ERS- 1, Geos-3 and TOPEX/Poseidon satellites. In 1992 TOPEX/Poseidon satellite altimetry mission was launched and its mission was ended in 2002. Since then it was replaced by Jason-1. Both satellite missions provide the most precise altimetry data when compared to others. Although satellite altimetry records are still quite short as compared to the tide gauge data sets, this technique appears quite promising for sea level change problem. In fact, it provides sea level measurement with large coverage. A precision of about 1 mm/year of measurement global change can be obtained. At present, few studies have been utilized different altimeter sensors for the sea surface investigations in the South China Sea [8]. Hu et al. [4] investigated the sea surface height with a period of 3-6 months using six years TOPEX/POSEIDON altimeter data. They reported that the sea surface height variations are associated strongly with current and eddies features. Imawaki et al. [3] introduces a new technique for acquiring high-resolution mean surface velocity by combined use of TOPEX/POSEIDON and ERS-1/2 altimeter data and drifter data obtained from 1992 through 2001 to detect the undulation of mean-surface dynamic topography down to the oceanic current. Indeed, the ocean current associated with geoid by the amount of sea surface dynamic topography is different compared to the mean–surface height acquired from satellite altimeter. In order to improve the hydrodynamic equation solutions of the mean sea-surface to invert accurate pattern of sea surface current we need to obtain altimeter data that is both more accurately and more widely distributed in time. Furthermore, we need to improve our knowledge of the long wavelength geoid so that its errors are substantially smaller than the corresponding values of the sea surface topography. The main contributions of this work is to design numerical scheme to reduce the impact of the Coriolis parameter in estimating sea surface current. As a matter of fact, at low latitude zone such as Malaysia where the existance of weak Coriolis parameter which leads to weak geostrophic current. According to this prospective, this study attempt to implement the second order of the Volterra model and Lax-Wendroff scheme to suppress the numerical solution of geostrophic continuity equation.

2 Data and Study Area

The study area is covered the South China Sea (SCS) where it is considered as an equatorial, semi-enclosed sea with a complex topography that includes large shallow regions. According to Wyrtki [9],the SCS is located between the Asian continent, Borneo, the Philippines, and Taiwan (Fig. 1). The northeastern part adjoins a deep sea basin, while the southern part is a shelf sea with depths less than 200 m. The SCS has two features that have very important and interesting effects on the general circulation: (1) the Coriolis force decreases to zero at the equator where both nonlinear and frictional forces become very important; and (2) the monsoon regime exerts a strong effect on the SCS circulation. Neither of the above phenomena are studied in coastal regions of the world ocean at the present time [9].

In situ sea surface current measurements were acquired on October 6 – 12 2003, April 6-12 2004, and March 1-8 2005, respectively during JASON-1 satellite altimetry overpass. The Merged Sea Level Anomaly (MSLA) data were acquired from JASON-1 satellite altimetry between 102° E to 114° E and 1 ° N to 10° N. The in

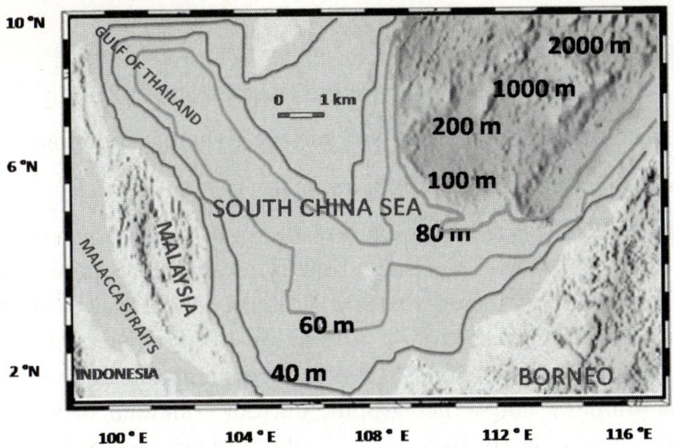

Fig. 1. Location of the South China Sea

situ sea surface current measurements howvever were collected between 102° 5'E to 105° 10'E and 2° 5'N to 6° 10'N (Fig. 2) using Valeport electro-magnetic current meter. In this context, the current meter was lowered down from the sea surface to water depth of 50 m.

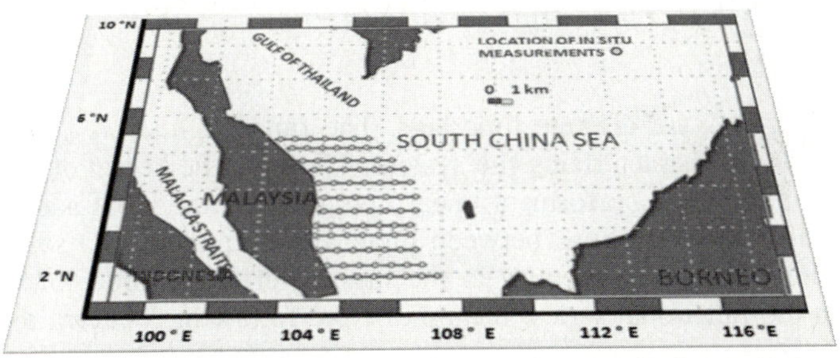

Fig. 2. Location of in situ measurements along east coast of Malaysia

3 Model

3.1 Volterra Model

According to Imawaki et al. [3] the instantaneous surface geostrophic velocity U_g can be obtained from temporal mean velocity $\bar{U}(x,t)$ and anomaly of sea-surface geostrophic velocity $\bar{U}(x,t) = (u', v')$ by following formula:

$$U_g(x,t) = \bar{U}(x) + \bar{U}(u',v') \tag{1}$$

where (u',v') are the sea surface geostrophic components in x and y directions which are estimated from the anomaly ξ of sea surface dynamic topography which practically equivalent to the sea-surface height anomaly ζ_s.

The Volterra model was used to express the geostrophic current velocity which were calculated from JASON-1 satellite altimetry as a series of nonlinear filters on the ocean surface current. This means that the Volterra model can be used to study the geostrophic current energy variation as a function of parameters such as the current direction, or the current waveform. A generalized, nonparametric framework to describe the input–output *x* and *y* geostrophic current components relation of a time-invariant nonlinear system. In discrete form, the Volterra series for input, $X(n)$, and output, $Y(n)$ as given by Inglada and Garello [2] can be expressed as:

$$\begin{aligned} Y(n) = {} & h_0 + \sum_{i_1=1}^{\infty} h_1(i_1)X(n-i_1) + \sum_{i_1=1}^{\infty}\sum_{i_2=1}^{\infty} h_2(i_1,i_2)X(n-i_1)X(n-i_2) + \\ & \sum_{i_1=1}^{\infty}\sum_{i_2=1}^{\infty}\sum_{i_3=1}^{\infty} h_3(i_1,i_2,i_3)X(n-i_1)X(n-i_2)X(n-i_3) + \dots\dots\dots + \\ & \sum_{i_1=1}^{\infty}\sum_{i_2=1}^{\infty}\dots\dots\dots\dots\sum_{i_k=1}^{\infty} h_k(i_1,i_2,\dots\dots\dots i_k)X(n-i_1)X(n-i_2)\dots\dots\dots\dots X(n-i_k) \end{aligned} \tag{2}$$

where, n, i_1 , i_2 ,...,i_k , are discrete time lags. The function $h_k(i_1 ,i_2 ,...,i_k)$ is the *k*th-order Volterra kernel characterizing the system. The h_1 is the kernel of the first order Volterra functional, which performs a linear operation on the input and h_2 , h_3 ,...,h_k capture the nonlinear interactions between input and output AVISO sea level variation. The order of the non-linearity is the highest effective order of the multiple summations in the functional series. It might be the following Volterra model can be used to express the instantaneous surface geostrophic velocity U_g as follows;

$$U_g(x,t) = N(\xi,\zeta_s) + \sum_{i=1}^{\infty} \int_R h_i(\bar{\tau},\xi) \prod_{j=1}^{i} \bar{U}_a(u'-\tau_j)(v'-\tau_i)d\tau \tag{3}$$

where $N(\xi,\zeta_s)$ is the geoid height which is estimated from sea surface dynamic topography ξ and the sea-surface height anomaly ζ_s as described by Imawaki et al [3]. Following, Inglada and Garello (1999), the mathematical expressions for second –order Volterra kernels (D_{2xx} and D_{2yy} of instantaneous surface geostrophic velocity U_g are as follows;

$$D_{2xx}(f_{1x},f_{1y},f_x-f_{1x},f_y-f_{1y})=\left[\xi_x\frac{\partial D_{1x}(f_{1x},f_{1y})}{\partial\zeta}(f-f_1)-\xi_x\bar{f}+\frac{\partial N(\vec{\xi})}{\partial\vec{\xi}}.(\bar{U}_{t_0}-\beta(\frac{\partial\zeta_s}{\partial y}-\frac{\partial\xi}{\partial y})\right]\times \beta\frac{\partial\xi}{\partial y}\bar{f} \quad (4)$$

$$D_{2yy}(f_{1x},f_{1y},f_x-f_{1x},f_y-f_{1y})=\left[\xi_y\frac{\partial D_{1y}(f_{1x},f_{1y})}{\partial\zeta}(f-f_1)-\left[\xi_y\bar{f}+\frac{\partial N(\vec{\xi})}{\partial\vec{\xi}}.(\bar{U}_{t_0}-\beta(\frac{\partial\zeta_s}{\partial x}-\frac{\partial\xi}{\partial x})\right]\right]\times \beta\frac{\partial\xi}{\partial x}\bar{f} \quad (5)$$

where $f_x, f_y, \bar{f}$ are the frequency domain in x and y dimensions while $\bar{f}$ is average frequency domain. Furthermore, β is the ratio of gravity acceleration to the Coriolis parameter expressing the effect rotation of the earth. In order to estimate the sea surface current from the altimeter data, we assume that there is a linear relationship between 2-D Fourier transform of mean surface slope variations obtained from JASON-1 satellite altimetry data ($F(f_x, f_y)$) and the second –order Volterra kernels D_{2xx} and D_{2yy} as follows:

$$F(f_x,f_y)=U_x(f_x,f_y).D_{2xx}(f_x,f_y) \quad (6)$$

$$F(f_x,f_y)=U_y(f_x,f_y).D_{2yy}(f_x,f_y) \quad (7)$$

Equations 6 and 5 are used to estimate the current at x and y directions where the inverse Volterra model is used to estimate the current flow as described by Inglada and Garello [2] by

$$D(f_x,f_y)=\begin{cases}\dfrac{1}{D_{2xx}(f_{1x},f_{1y},f_x-f_{1x},f_y-f_{1y})}\\[2ex]\dfrac{1}{D_{2yy}(f_{1x},f_{1y},f_x-f_{1x},f_y-f_{1y})}\end{cases} \quad (8)$$

The current direction can be obtained by using the traditional tan formula as follows:

$$\theta = \tan^{-1}\left(\frac{U_y}{U_x}\right) \tag{9}$$

3.2 Lax-Wendrof Scheme

The second-order accurate dispersive Lax-Wendrof scheme is used for sea surface current flow in JASON-1 altimeter data which can be written in predictor–corrector form in a staggered grid. The staggered grid consists of a primary grid where points are labeled with *(i, j, k)* and a dual grid where points are labeled with *(i* + 12,*j* + 12, *k* + 12*)* . In 2D, the centers of the cell edges in the primary grid are given by *(i* + 12, *j)*, etc. Following Liska and Wendroff [5], 2D, numerical flow for the two-step forms of Lax-Wendrof scheme is

$$U_{i,j}^{n+1} = U_{i,j}^{n} + \frac{\Delta t}{2\Delta x}\Big(v(U^{n+1/2}{}_{i+1/2,j+1/2}) + v(U^{n+1/2}{}_{i+1/2,j-1/2}) - v(U^{n+1/2}{}_{i-1/2,j+1/2}) - v(U^{n+1/2}{}_{i-1/2,j-1/2})\Big) + \frac{\Delta t}{\Delta y}\Big(G(U^{n+1/2}{}_{i+1/2,j+1/2}) + G(U^{n+1/2}{}_{i-1/2,j+1/2}) - G(U^{n+1/2}{}_{i+1/2,j-1/2}) - G(U^{n+1/2}{}_{i-1/2,j-1/2})\Big) \tag{10}$$

where *v* and *G* are smooth functions as described by Liska and Wendroff [5]. The values at the center of all faces of the primary cell on time level $n + 1/4$ are computed using the analog of 2D predictor by

$$U_{i,j+1/2,k+1/2}^{n+1/4} = \frac{1}{4}(U^{n}{}_{i,j,k} + U^{n}{}_{i,j+1,k} + U^{n}{}_{i,j,k+1} + U^{n}{}_{i,j+1,k+1}) + \frac{\Delta t}{4\Delta y}(G(U^{n+1/6}{}_{i,j,k+1/2})) \tag{11}$$

The finite difference scheme model was applied with traditionally rectangular grids due to most their simplicity. Model numerical solutions was enhanced by using the curvilinear nearly orthogonal, coastline-following grids. This technique was implemented to overcome the problems raised due to complicated SCS topography, and boundary conditions.

4 Results and Discussion

The geostrophic current patterns have been modeled by Volterra- Lax-Wendrof Scheme are shown in Figs 3, 4 and 5. Fig. 3 shows the typical water circulation during the inter-monsoon period of October 2003 of the South China Sea. In October 2003, non fixed current patterns basically was dominated at the northern and center parts of the South China Sea. The maximum current magnitude was 1.0 m/s was found in the northern and eastern coast of Malaysia. The minim current velocity was 0.4 m/s and observed in the Gulf of Thailand. It is interested to find that the stream current pattern was a dominant feature in Malacca Straits with current magnitude of 0.8 m/s. The east coast of Malaysia was dominated by southwards current flow which moving parallel to coastline. Part of these currents was turned south-east and moving parallel to Borneo coastline with current velocity of 1.0 m/s.

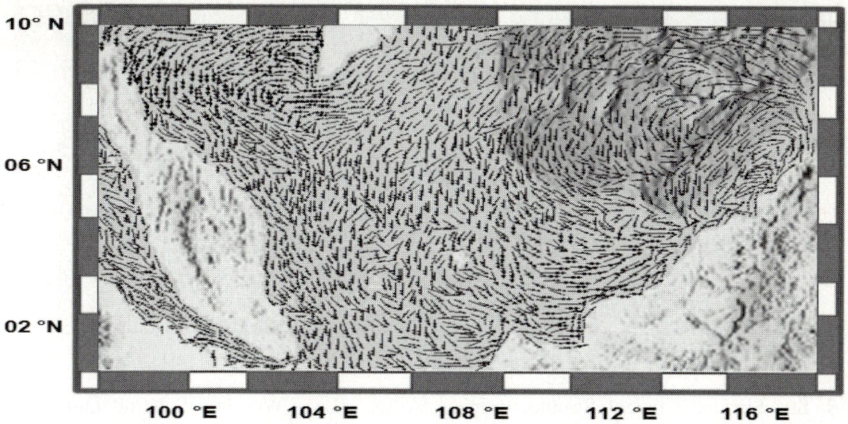

Fig. 3. Sea Surface Current Simulated from Volterra-Lax- Wendrof Scheme During October 2003

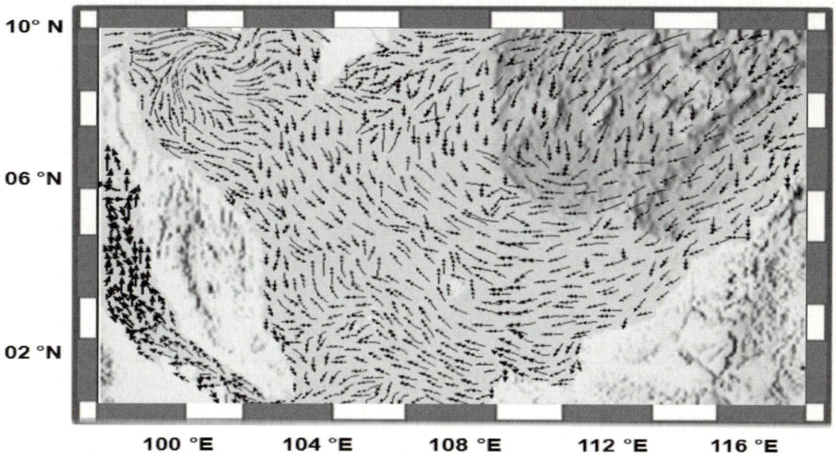

Fig. 4. Sea Surface Current Simulated from Volterra-Lax- Wendrof Scheme During April 2004

In April 2004, there was weak current flow along the east coast of Malaysia with magnitude 0.3 m/s. This current was moved towards southward direction. In Gulf of Thailand there were two eddies were moved opposite to each other with maximum velocity of 0.3 m/s. The southern eddy was moved anticlockwise while the north eddy was moved clockwise. It is obvious that there was southward current flow along the Borneo coastline with current velocity of 0.5 m/s. This current was turned to north and formed small scale eddy in the southern part of Vietnam coastline. The results confirm the study of Wryki [9], Maged [6] and Alejandro [1] . In March 2005, the current magnitude is increased and generate an eddy in the center of the South China sea. The maximum current magnitude was 1.2 m/s. the dominate current direction was northward. The Gulf of Thailand dominated by anticlockwise eddy with maximum velocity of 1.0 m/s.

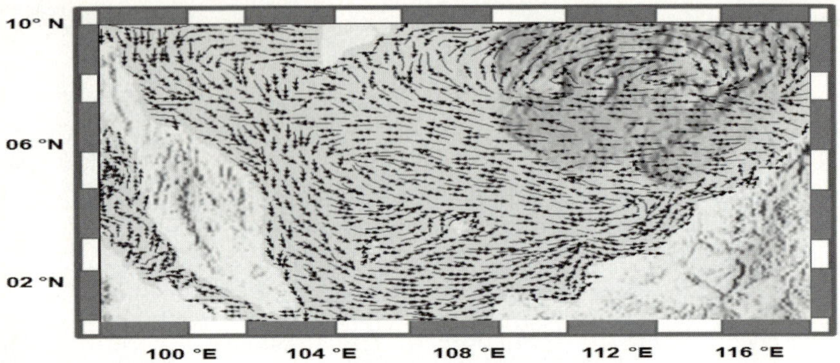

Fig. 5. Sea Surface Current Simulated from Volterra-Lax- Wendrof Scheme During March 2005

Fig. 5 shows the regression relationship between average in situ current measurements and the simulated current flow from Volterra- Lax-Wendrof Scheme. The scatter points are more closed to the regression line with R^2 value of 0.91 and RMSE of ±0.4 m/s which are indicated high correlation between in situ measured current and simulated one from Volterra-Lax-Wendrof Scheme (Table 1).

Table 1. Satistical Summary

Statistical Parameters	Values
r^2	0.91
p	0.002
RMSE	± 0.4 m

The Volterra algorithm was shown an overcome of the Coriolis force problem. It does mean that the Volterra algorithm has provided solution for low geostrophic current velocity always appears along Malaysian coastal water due to weakness of Coriolis parameter involves in continuity equation. In addition, Lax-Wendrof scheme can handle mixed sub and super-critical current flows directly, with no regard for directional nature of the computation (left to right or vice versa) [5].

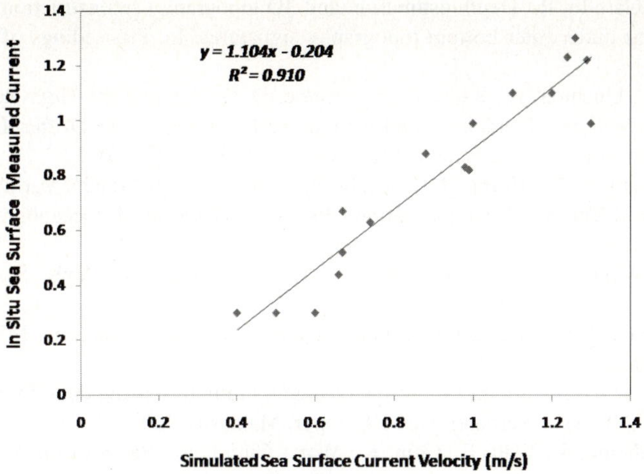

Fig. 6. Regression of Simulated Averaged Current and Measured In situ Current

5 Conclusions

This work has demonstrated the two-dimensional sea surface current flow modelling from JASON satellite altimetry data in the South China. The new technique was used to reduce the impact of Coriolis parameter in the continuity equation of geostrophic current. Two procedures were involved: second order Volterra Kernel and Lax-Wendrof Scheme. It is interested to name this algorithm as Volterra-Lax-Wendrof scheme. The results showed good correlation between in situ current measured and Volterra-Lax-Wendrof scheme with high r^2 of 0.91 and RMSE of ±0.4 m/s. It can be concluded that Volterra-Lax-Wendrof scheme could be used as good model to acquire fine resolution of sea surface current in such sea basin of the South China Sea.

Acknowledgements

We would like to express our appreciation to Universiti Teknologi Malaysia Research Management Center (RMC) for a great cooperation. In fact, RMC is providing excellent advice, plans and management for UTM researchers. We also acknowledge the financial e-science grants have received from the Malaysian Ministry of Science and Technology via its Intensified Research for Priority Areas programme.

Reference

1. Alejandro, C., Nasir, M.S.: Dynamic bahaviour of the upper layers of the South China Sea. In: Proceeding of the national conference on climate change, Universiti Pertanian Malaysia, Serdang, Selangor, Malaysia, August 12-13, 1996, pp. 135–140 (1996)

2. Inglada, J., Garello, R.: Depth estimation and 3D topography reconstruction from SAR images showing underwater bottom topography signatures. In: Proceedings of IGARSS 1999 (1999)
3. Imawaki, S., Unchida, H., Ichikawa, K., Ambe, D.: Estimating the High-Resolution Mean sea-Surface Velocity Field by Combined Use of Altimeter and Drifter Data for Geoid Model Improvement. Space Science Reviews 108, 195–204 (2003)
4. Hu, J., Kawamura, H., Hong, H., Kobashi, F., Wang, D.: 3-6 Months Variation of Sea surface Height in The South China Sea and Its Adjacent Ocean. J. Oceanography 57, 69–78 (2001)
5. Liska, L., Wendroff, B.: Composite schemes for conservation laws. SIAM J. Numer. Anal. 35(6), 2250–2271 (1998)
6. Maged, M.: Coastal Water Circulation off Kuala Terengganu, Malaysia. MSc. Thesis Universiti Pertanian Malaysia (1994)
7. Maged, M.: Wave spectra and shoreline change by remote sensing data. Ph.D. Thesis, Universiti Putra Malaysia, Serdang, Kuala Lumpur, Malaysia (2000)
8. Mohd, I., Ahemd, S., Wah, F., Chin, K.: Water Circulation Pattern from Sea Surface Current and Chlorophyll-A Derived Using Satellite Data in the South China Sea. In: Paper Presented at the 2nd International Hydrographic and Oceanographic Conference and exhibition 2005 (IHOCE 2005), Kuala Lumpur, Malaysia, July 5-7, 2005 (2005)
9. Wyrtki, K.: Physical Oceanography of the South-East Asian Waters. In: NAGA Report, vol. 2, Univ. Calif., Scripps Inst. Ocean, La Jolla (1962)

User-Centered Design of Web Pages

Chung-Hsing Yeh[1] and Yang-Cheng Lin[2]

[1] Clayton School of Information Technology, Faculty of Information Technology, Monash University, Clayton, Victoria 3800, Australia
ChungHsing.Yeh@infotech.monash.edu.au
[2] Department of Arts and Design, National Hualien University of Education, Hualien, 970, Taiwan
lyc0914@cm1.hinet.net

Abstract. This paper conducts an experimental study on university web pages to examine whether the hyperlink style affects users' feelings of a web page. We categorize 3 hyperlink styles, including text only, text and symbol, and text and icon. A grey relational analysis conducted suggests that the hyperlink style is a crucial element in user-centered web page design. The experimental study identifies 7 web page design elements and 33 representative webs pages as experimental samples for developing a neural network (NN) model. The NN mode is used to examine the complex relationship between 7 design elements and 3 users' feelings of web pages. With the NN model, a web page design database is built consisting of 3,645 different combinations of design elements, together with their associated feeling values. As a general rule, a web page with the "text only" hyperlinks has the highest possibility of being easiest and clearest. A web page with the "text and symbol" hyperlinks will have the most classic look, while the "text and icon" hyperlinks will make the web page have the most modern appearance. These design-supporting results provide useful insights in facilitating the user-centered web design process.

1 Introduction

A good web page design is convenient for users, while a bad web page design can be disorientating to users [17]. As the backbone of the Internet and the most important navigation facility [2], hyperlinks are the most essential ingredient of web pages. Displayed as a highlighted word, icon, or graphic, hyperlinks enable users to reach any intended web pages by clicking a mouse [19]. In other words, users can move directly from one web page to another by using hyperlinks. However, there is a significant usability [15] problem that hyperlinks often do not lead users to the information they expect [21]. Users often do not know where they are and what way to choose next [21]. For example, unclear hyperlinks of a web page may lead to users' operational failures, thus making the web page unusable [18]. In addition, users can even lose their orientation in a small hypertext system, if no orientation clues are given [16]. This is called "disorientation", meaning the tendency of the users to lose their sense of location and direction on the Internet [20].

O. Gervasi et al. (Eds.): ICCSA 2008, Part II, LNCS 5073, pp. 129–142, 2008.

To overcome potential problems of hyperlinks when designing web pages, quite a few studies have proposed solutions from various perspectives [1, 5, 18, 19, 20, 21]. For example, the proper use of iconic interfaces can reduce system complexity and help users interact with systems more easily [4]. However, the icons used on web pages often are ambiguous [4], resulted in another usability problem "cognitive overhead", meaning the additional effort and concentration necessary to maintain several tasks at one time [5]. Web designers must consider whether icons are easily identifiable as designing web pages, because non-identifiable icons may convey wrong meanings [4].

In this paper, we conduct an experimental study on university web pages to examine whether the hyperlink type or style used on a web page affects the users' feelings of the web page. We categorize 3 hyperlink styles, including "text only", "text and symbol", and "text and icon". The "text only" style means that the hyperlink contains text or words "only". The "text and symbol" style means that the hyperlink adds some symbols or graphics, such as small circles or square shapes without a specific meaning, while the "text and icon" style indicates the hyperlink with some icons, such as photographs or drawings having a specific meaning. In addition to the hyperlink style, there are other distinctive design elements of a web page that may also affect the users' overall feeling or perception of the web page. As such, we need to consider not only the hyperlink style but also how it interacts with other design elements in contributing to the users' overall feeling or perception of the web page.

In subsequent sections, we present the experiment study using a number of modeling and analytical techniques including morphological analysis [9, 10, 11], Kansei Engineering [8, 12], grey relational analysis [6], and neural networks (NN) [14].

2 Experimental Samples of Web Pages

We first used morphological analysis to extract common design elements of web pages. 96 homepages of university websites were collected as the experimental samples. The morphological analysis involved two steps. In the first step, 6 design experts with at least 3 years of web page design experience were asked to write down the influential design elements of the web pages that will affect users' perception of web pages, according to their knowledge and experience. In the second step, the 6 experts formed a focus group [3] to discuss the results and combine similar opinions or components. As a result of the morphological analysis, Table 1 shows the 7 influential design elements extracted from the 96 websites, together with their associated types.

Each design element has different types of its own, ranging from 3 to 5, as indicated by the type number 1, 2, 3, 4 or 5 in Table 1. For example, "Hyperlink Style (X_5)" has 3 types, including Text only, Text and symbol, and Text and icon, meaning that the hyperlink style is text "only", or "adding" some symbols (e.g. small circles or square shapes without a specific meaning), or "adding" some icons (e.g. photographs or drawings with a specific meaning). The design element "Ratio of Graphics to Text (X_1)" means that the ratio between the area used by graphics and the area used by text on a web page. This ratio takes 5 values, corresponding to 5 different types, including Above 3, Between 3-1, 1, Between 1-1/3, and Below 1/3.

Table 1. Extracted design elements of web pages

	Elements	Type 1	Type 2	Type 3	Type 4	Type 5
X_1	Ratio of Graphics to Text	Above 3	Between 3-1	1	Between 1-1/3	Below 1/3
X_2	Blank Ratio	0% - 20 %	20% - 40%	40% - 60%		
X_3	Layout Style	2 columns	3 columns	Multiple columns		
X_4	Frame Style	Up and down	Left and right	Compound style		
X_5	Hyperlink Style	Text only	Text and symbol	Text and icon		
X_6	Number of Colors	Below 4 colors	4-7 colors	Above 7 colors		
X_7	Background Color	Cold color	Warm color	Neutral color		

To facilitate the assessment process made by another group of subjects, we performed a cluster analysis on the 96 web page samples based on the result of morphological analysis. With the generation of a cluster tree diagram, we extracted 33 representative web page samples, including 27 training samples and 6 test samples for training and testing the NN model to be developed. Fig. 1 shows 6 representative experimental web page samples, which represent 3 types of "Hyperlink Style (X_5)" including "Text only", "Text and symbol" and "Text and icon" respectively. To collect numerical data for modeling and analyzing users' feelings on web pages, the degree to which the 33 web page samples matches a given set of users' feelings has to be assessed.

With Kansei engineering, a pair of image words is often used to describe users' feeling or perception about a specific image of a product. Based on our previous study [7], we used 3 feeling word pairs for describing users' feelings on web pages, including easy-difficult (E-D), clear-confusing (C-C), and classic-modern (C-M). 30 subjects were asked to assess the degree to which each of the 33 web page samples matches the E-D, C-C and C-M feelings respectively using 5 scales (1-5) of the semantic differential (SD) method [11]. For example, each subject assessed each web page sample for the E-D feeling on a scale of 1 to 5, where 1 is very easy to use and 5 is very difficult to use. For the C-C feeling, the value of 1 represents the web page is very clear to follow, and 5 represents the web page is very confusing. For the C-M feeling, the value of 1 represents the web page looks very classic, and 5 represents the web page looks very modern.

Table 2 shows the 33 experimental samples, including 6 test samples (asterisked), together with their corresponding assessment results. For each representative web page sample in Table 2, the first column shows the web page number and Columns 2-8 show the corresponding type number for each of its 7 design elements, as given in Table 1.The

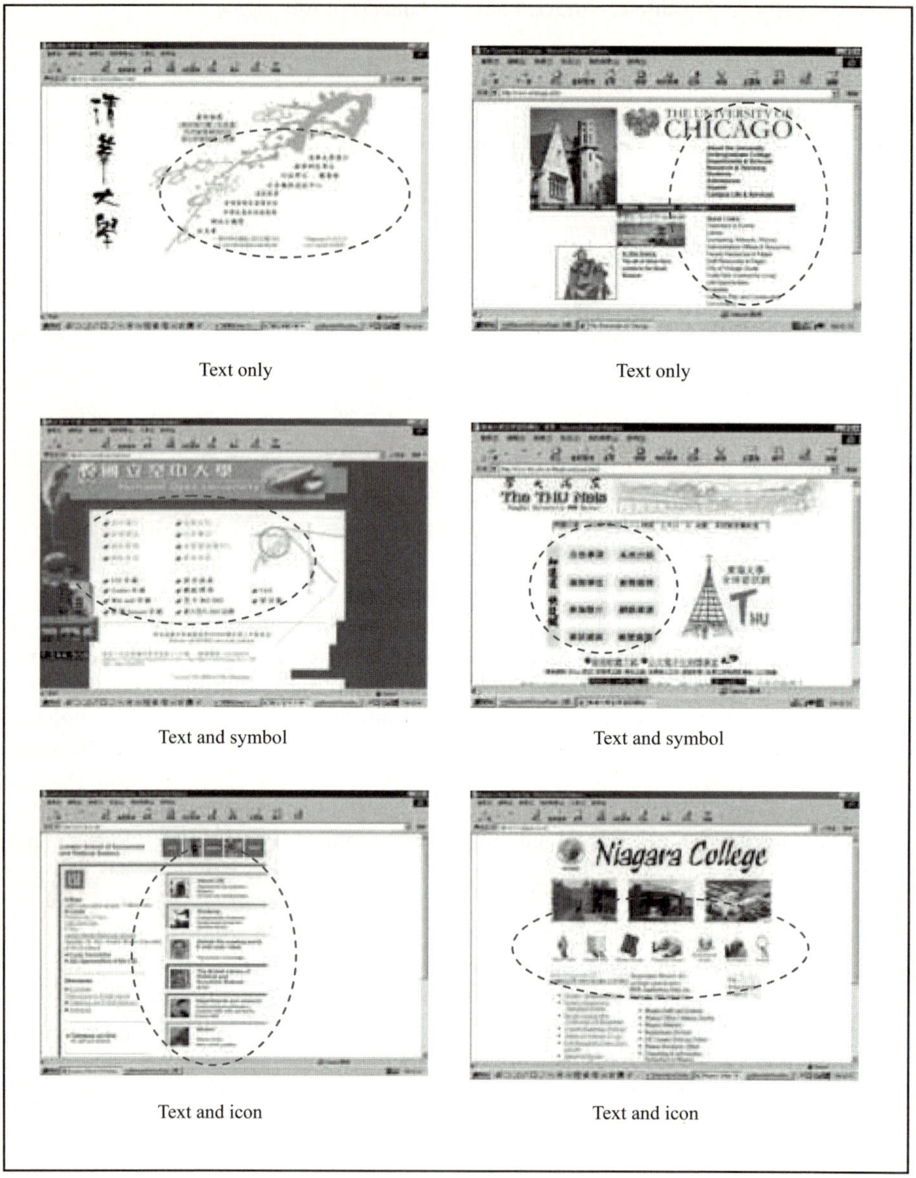

Fig. 1. Representative experimental samples of web pages

E-D, C-C, or C-M feeling value shown in the last three columns of Table 2 is the average of the values assessed by the 30 subjects. Table 2 provides a numerical data source for performing grey relational analysis and for training and testing neural network models, which are to be presented in the following section.

Table 2. The evaluation result for 33 selected web page samples, including 6 test samples (asterisked)

Web Page No.	X_1	X_2	X_3	X_4	X_5	X_6	X_7	E-D value	C-C value	C-M value
1	2	2	2	3	2	2	2	2.28	1.70	2.90
2	3	3	3	3	1	1	2	2.11	2.16	2.52
3	3	3	2	1	2	3	1	2.63	2.59	3.06
4	5	3	2	3	2	1	3	3.26	3.48	2.52
5	3	2	3	1	3	2	1	2.84	2.57	3.06
6	1	1	2	1	3	3	3	2.44	2.48	3.78
7	3	3	1	2	1	1	1	2.47	2.42	2.00
8	5	2	2	3	2	1	3	2.78	3.16	2.64
9	1	1	1	1	3	3	3	3.27	3.11	4.42
10	1	2	1	2	1	2	3	2.60	2.28	3.52
11	3	3	2	1	2	1	1	2.77	2.33	2.68
12	4	1	2	2	1	2	2	2.64	2.63	2.99
13	2	2	3	1	1	3	3	2.27	2.33	2.68
14	2	1	3	2	1	2	3	2.73	2.31	2.21
15	4	1	2	1	3	3	2	2.89	2.63	2.33
16	4	2	3	3	3	3	1	2.67	2.78	2.84
17	2	3	2	3	2	1	3	2.26	2.00	2.91
18	4	1	2	3	2	3	2	2.68	2.59	2.12
19	5	2	3	2	1	2	1	2.96	2.90	2.31
20	5	3	1	2	1	1	3	2.34	2.28	1.87
21	4	2	2	3	3	3	2	2.37	2.26	3.67
22	1	1	2	3	3	2	3	2.70	2.49	3.42
23	3	2	3	1	1	2	3	2.73	3.00	2.90
24	4	3	3	1	1	2	2	3.15	2.89	2.47
25	5	3	3	3	3	1	2	4.11	3.36	2.36
26	2	2	3	3	2	1	1	3.02	3.07	2.31
27	1	1	3	3	2	3	1	3.06	3.26	3.25
28*	2	2	2	1	3	2	2	2.80	2.20	3.27
29*	2	3	1	2	2	1	3	2.27	2.27	3.00
30*	1	2	2	2	3	2	3	2.80	3.07	4.00
31*	3	3	1	1	1	3	1	2.53	2.47	3.20
32*	2	1	3	3	3	3	1	3.60	3.40	3.00
33*	2	1	3	3	2	3	2	2.80	2.87	3.27

3 Experimental Analysis with GRA and NN

In this section, we apply the grey relational analysis (GRA) and the neural network (NN) techniques to analyze and model users' feelings on web pages. The GRA is used to determine the most influential design elements of web pages for a given set of users' feelings. The NN mode is used to examine the complex relationship between input variables (design elements of web pages) and output variables (users' feelings of web pages).

3.1 Grey Relational Analysis

The GRA is used to determine the relationship (similarity) between two series of stochastic data in a grey system. One is the reference series ($x_0 \in X$), and the other is the m comparison series ($x_i \in X$, $i = 1, 2, \ldots, m$), where $X = \{x_\sigma \square \sigma \square 0, 1, 2, \ldots, n\}$ is a given grey relational element set. The grey relational degree between the two series at a certain time point t is represented by the grey relational coefficient $r(x_0(k), x_i(k))$, defined as [6]:

$$r(x_0(k), x_i(k)) = \frac{\min_i \min_k |x_0(k) - x_i(k)| + \xi \max_i \max_k |x_0(k) - x_i(k)|}{|x_0(k) - x_i(k)| + \xi \max_i \max_k |x_0(k) - x_i(k)|},$$

$$k = 1, 2, \ldots, n; \; i = 1, 2, \ldots, m. \tag{1}$$

where $\xi \in [0,1]$ is a distinguishing coefficient for controlling the resolution scale, usually being assigned the value of 0.5. The grey relational degree of each comparison series x_i ($i = 1, 2, \ldots, m$) to reference series x_0 at all time points can be calculated by

$$r(x_0, x_i) = \frac{1}{n} \sum_{k=1}^{n} r\,(x_0(k), x_i(k)), \quad i = 1, 2, \ldots, m. \tag{2}$$

We performed the GRA to determine the most influential design elements of web pages for the E-D, C-C, and C-M feelings. The GRA involved the following seven steps:

Step 1: Present the original data series.
Step 2: Normalize the data series.
Step 3: Calculate $|x_0(k) - x_i(k)|$.
Step 4: Calculate $\max\{\max |x_0(k) - x_i(k)|\}$ and $\min\{\min |x_0(k) - x_i(k)|\}$.
Step 5: Set the distinguishing coefficient $\xi = 0.5$.
Step 6: Calculate the grey relational coefficient of the data series.
Step 7: Calculate the grey relational degree of the data series, and obtain the value of $r(x_0, x_i)$.

With the value ranging from 0 to 1, Table 3 shows the grey relational degree $r(x_0, x_i)$ between the web page feeling (x_0) and each of the 7 design elements (x_i, $i = 1, 2, \ldots, 7$) using the GRA. If $r(x_0, x_i) > r(x_0, x_j)$, then the design element x_i is closer to the web page feeling x_0 than the design element x_j. The higher the $r(x_0, x_i)$ value, the more influential the design element x_i [9].

As clearly indicated in Table 3, the "Hyperlink Style (x_5)" element has the highest GRA value of 0.66, affecting the E-D feeling the most, followed by the "Ratio of Graphics to Text (x_1)" element and the "Number of Colors (x_6)" element. This implies that the web page designers should focus their attention more on these three most influential elements, when the objective of designing web pages is to achieve a desirable E-D feeling. On the contrary, the designers may pay less attention to the less influential elements, as these elements contribute relatively little to users' E-D feeling. In addition, the C-M feeling has the same result as the E-D feeling that the "Hyperlink Style (x_5)"

element has the highest GRA value of 0.68 as the "Number of Colors (x_6)" element . For the C-C feeling, the "Ratio of Graphics to Text (x_1)" element has the highest GRA value of 0.70, while the GRA value of the "Hyperlink Style (x_5)" element is relatively low (0.60). However, the "Hyperlink Style (x_5)" element has the highest overall GRA value in terms of the three feeling values. This result clearly suggests that the choice of the hyperlink style is crucial in designing web pages for affecting user's E-D, C-C and C-M feelings as a whole.

Table 3. The result of GRA

$r(x_0, x_i)$	x_1	x_2	x_3	x_4	x_5	x_6	x_7
E-D feeling	0.63	0.60	0.60	0.57	0.66 *	0.61	0.54
C-C feeling	0.70 *	0.60	0.67	0.61	0.60	0.61	0.59
C-M feeling	0.60	0.58	0.60	0.53	0.68 *	0.68 *	0.60

3.2 The Neural Network Model

With their effective learning ability, NN models have been widely used to examine the complex and non-linear relationship between input variables and output variables [13]. In this study, we develop an NN model in order to examine the relationship between the 7 web page design elements and the 3 user feelings of web pages. With a single hidden layer, the NN model is a multilayered feedforward one trained with the backpropagation learning algorithm, as it is an effective and popular supervised learning algorithm [14]. In addition, the NN model uses the Delta learning rule and the Sigmoid transfer function. The number of neurons in the single hidden layer is 13, which is determined by (the number of input neurons + the number of output neurons) / 2. This is the most widely used rule for building NN models [22]. Table 4 shows the number of neurons used in the input, hidden and output layers of the NN model.

Table 4. Neurons of the NN model

Input layer	23 neurons for 23 types of 7 design elements.
Hidden layer	13 neurons, (23+3) / 2 = 13.
Output layer	3 neurons for the E-D, C-C, and C-M feeling values respectively.

The 23 types of the 7 design elements in Table 1 are used as the 23 input variables for the NN model. If the web page has a particular design element type, the value of the corresponding input neuron is 1; otherwise, the value is 0. The assessed average values of the E-D, C-C, and C-M feelings are used as the output neurons.

The 27 web page samples in the training set, given in Table 2, were used to train the NN model. Each model was trained 5,000 epochs at each run. When the cumulative training epochs were over 100,000, the training process was completed. Table 5 shows the training epochs of the model run and its corresponding root of mean square (RMS) errors.

Table 5. RMS errors of the NN model for the training set

Learning epochs	10000	20000	30000	40000	50000	60000	70000	80000	90000	100000
RMS error	0.0355	0.0206	0.0140	0.0127	0.0119	0.0112	0.0106	0.0105	0.0104	0.0104*

4 Performance Evaluation and Discussion

In this section, we used the 6 web page samples, indicated in Table 2, as the test set to evaluate the performance of the NN model in terms of its ability to predict the users' feeling values for a new web page with given design elements. Rows 2-4 of Table 6 show the average assessed values of the E-D, C-C and C-M feelings on the 6 test samples given by the 30 subjects, and Rows 5-7 show the predicted values for the three feelings by using the NN model trained in the previous section. The last column of Table 6 shows the RMS error of the NN model for the test set. If there is no difference or error between the predicted values and the expected values, the RMS error is 0.

Table 6. Predicted feeling values and RMS error of the NN model for the test set

	Web page feeling	Web page sample no.						RMS error
		28	29	30	31	32	33	
Average assessed value by subjects	E-D value	2.80	2.27	2.80	2.53	3.60	2.80	
	C-C value	2.20	2.27	3.07	2.47	3.40	2.87	
	C-M value	3.27	3.00	4.00	3.20	3.00	3.27	
Predicted value by the NN model	E-D value	2.43	3.00	2.36	2.69	2.79	2.82	
	C-C value	1.71	2.27	1.96	3.03	2.36	2.42	0.1868
	C-M value	3.15	2.43	3.85	2.94	2.17	2.08	

As indicated in Table 6, the RMS error of the NN model is 0.1868. This result indicates that the NN model has an accuracy rate of 81.32% (100%-18.68%) for predicting the values of the E-D, C-C, and C-M feelings of web pages. This suggests that the NN model is a promising approach for modeling users' feelings on web pages.

In design settings where the designers have to design the easiest and the clearest web page, what should they do? Or the designers design a web page with a lot of graphics (i.e. the ratio of graphics to text being above 3, Type 1 of X_1) and with the "text only" hyperlink style (Type 1 of X_5). If the designers want the web page to look the clearest to follow, and to look very modern, what should they do? These questions can be answered by using the NN model developed.

The NN model enables us to build a web page design database that can be used to help determine the best web page design for meeting specific web page feelings. The design database can be generated by inputting each of all possible combinations of design elements to the NN model individually for generating the associated feeling values. The resultant web page design database consists of 3,645 (=5×3×3×3×3×3×3) different combinations of design elements, together with their associated E-D, C-C, and C-M feeling values. As an illustration for the hyperlink style element, Table 7 shows the best and worst combinations of design elements respectively for the E-D and C-C feelings under the three hyperlink styles. That is, the combinations of design elements shown in the top half of Table 7 will produce the easiest and clearest web pages as perceived by users for the 3 different hyperlink styles respectively, while the bottom half of Table 7 shows the design element combinations of the most difficult and most confusing web pages.

Table 7. The best combination of design elements for the E-D and C-C feelings under three hyperlink styles

Feeling	X_5 ***Hyperlink Style***	X_1 Ratio of Graphics and Text	X_2 Blank Ratio	X_3 Layout Style	X_4 Frame Style	X_6 Number of Colors	X_7 Background Color	Total value of E-D and C-C feelings
Easiest and clearest web pages	***Text only*** ***(1)***	Between 3-1 (2)	40% - 60% (3)	3 columns (2)	Compound style (3)	4-7 colors (2)	Warm color (2)	3.51 (2.00, 1.51)
	Text and symbol ***(2)***	Between 3-1 (2)	20% - 40% (2)	3 columns (2)	Left and right (2)	Above 7 colors (3)	Warm color (2)	3.56 (2.09, 1.47)
	Text and icon ***(3)***	1 (3)	20% - 40% (2)	3 columns (2)	Left and right (2)	Above 7 colors (3)	Warm color (2)	3.65 (2.01, 1.64)
Most difficult and most confusing web pages	***Text only*** ***(1)***	Below 1/3 (5)	0% - 20% (1)	Multiple columns (3)	Up and down (1)	4-7 colors (2)	Warm color (2)	7.31 (3.72, 3.59)
	Text and symbol ***(2)***	Below 1/3 (5)	0% - 20% (1)	2 columns (1)	Up and down (1)	Below 4 colors (1)	Warm color (2)	7.90 (4.25, 3.65)
	Text and icon ***(3)***	Below 1/3 (5)	0% - 20% (1)	Multiple columns (3)	Up and down (1)	Below 4 colors (1)	Warm color (2)	7.86 (4.26, 3.60)

As evidenced in the experimental results presented above, the hyperlink style plays an important role in affecting users' feelings of web pages. When designing a web page, can designers choose a particular hyperlink style that best meets the users' perception of being the easiest to use and the clearest to follow? To address this question, we conducted a further experiment. Table 8 shows the result for each hyperlink style, which gives the minimum value, the maximum value, the average value and the standard deviation of a feeling (E-D, C-C or C-M) value for all the combinations of other design elements. The result suggests that a web page with the "Text only" hyperlink style will has the best possibility of being easiest and clearest. A web page with "Text and symbol" will have the most classic look, while "Text and icon" will have the most modern appearance.

Table 8. The E-D, C-C, and C-M feeling values for a given hyperlink style

	E-D				C-C				C-M			
Hyperlink Style	Min	Max	Ave.	Std. dev.	Min	Max	Ave.	Std. dev.	Min	Max	Ave.	Std. dev.
Text only	1.77	3.72	2.43	0.37	1.50	3.60	2.50	0.46	1.53	4.54	2.72	0.61
Text and symbol	1.98	4.25	3.13	0.46	1.46	3.76	2.80	0.49	1.33	4.57	2.66	0.71
Text and icon	1.96	4.29	3.10	0.52	1.45	3.70	2.56	0.51	1.42	4.75	2.92	0.76

It is commonly agreed that web pages should be easy to use and clear to follow. If the designers wish to design an easier and clearer web page, which looks very classic, is there any best combination of design elements for a given hyperlink style? On the contrary, if the designers want to get a very modern web page, is there any restriction? To answer these questions, we use the following settings respectively:

$$\text{(The E-D value + the C-C value)} \leqq 4, \text{ and the C-M value} \leqq 2.2 \quad (3)$$

$$\text{(The E-D value + the C-C value)} \leqq 4, \text{ and the C-M value} \geqq 3.8 \quad (4)$$

From the web page design database consisting of 3,645 different combinations of design elements, we use Eq. (3) to get the easier, clearer and very classic web pages, and Eq. (4) to obtain the easier, clearer and very modern web pages. Table 9 lists the common types (or the rules) of design elements under 3 hyperlink styles, whose corresponding web pages fit Eq. (3) or Eq. (4). As shown in Table 9, it indicates that if the designers want to design an easier and clearer web page, and simultaneously look very classic, they should design a web page with the ratio of graphics to text being between 3-1 (Type 2 of X_1) and with the "Left and right" frame style (Type 2 of X_4), no matter what hyperlink style is used. Furthermore, if the designers use the "Text and icon" hyperlink style (Type 3 of X_5), the designers then have to follow the restriction "Below 4 colors (Type 1 of X_6)" or "Above 7 colors (Type 3 of X_6)" to achieve the desirable web page feelings.

Table 9. Design element combinations for the very classic or very modern web pages with easier to use and clearer to follow

	X_5 Hyperlink Style	X_1 Ratio of Graphics	X_2 Blank Ratio	X_3 Layout Style	X_4 Frame Style	X_6 Number of Colors	X_7 Background Color
Easier, clearer, and very classic web pages	**Text only (1)**	Between 3-1 (2)	0% - 20% (1)	2 columns (1)	Left and right (2)	Below 4 colors (1)	
			40% - 60% (3)	Multiple columns (3)		Above 7 colors (3)	
	Text and symbol (2)	Between 3-1 (2)		3 columns (2)	Left and right (2)	Below 4 colors (1)	
						Above 7 colors (3)	
	Text and icon (3)	Between 3-1 (2)			Left and right (2)	Below 4 colors (1)	
						Above 7 colors (3)	
Easier, clearer, and very modern web pages	**Text only (1)**	Above 3 (1)	20% - 40% (2)	3 columns (2)			Warm color (2)
			40% - 60% (3)				
	Text and symbol (2)	Above 3 (1)	20% - 40% (2)	3 columns (2)	Left and right (2)	Above 7 colors (3)	Warm color (2)
		1 (3)			Compound style (3)		Neutral color (3)
	Text and icon (3)		20% - 40% (2)	3 columns (2)		Above 7 colors (3)	Warm color (2)
			40% - 60% (3)				Neutral color (3)

Table 9 provides useful insights in helping web designers design a new web page for meeting specific web page feelings. As a further study, Table 9 can be used as a basis for building multiple conditions if-then rules [11], as follows:

$$\text{Rule: IF } X_1 \text{ is } A_1 \text{ AND } X_2 \text{ is } A_2 \ldots \text{AND } X_n \text{ is } A_n,$$
$$\text{THEN } Y_1 \text{ is } B_1 \text{ AND } Y_2 \text{ is } B_2 \ldots \text{AND } Y_n \text{ is } B_n \quad (5)$$

where $A_1, A_2, \ldots, A_n$ and $B_1, B_2, \ldots, B_n$ are the linguistic values or terms, taken by the input linguistic variables $X_1, X_2, \ldots, X_n$ and the output linguistic variables $Y_1, Y_2, \ldots, Y_n$ respectively. The input linguistic variables $X_1, X_2, \ldots, X_n$ represent a set of the web page design elements and the output linguistic variables $Y_1, Y_2, \ldots, Y_n$ represent a set of web page feelings. For example, if the "Ratio of Graphics to Text" is "Above 3" (Type 1 of X_1), and "Blank Ratio" is "20% - 40%" (Type 2 of X_2) or "40% - 60%" (Type 3 of X_2), and "Layout Style" is "3 columns" (Type 2 of X_3), and "Hyperlink Style" is "Text only" (Type 1 of X_5), and "Background Color" is "Warm color" (Type 2 of X_7), then the "E-D feeling " is "Easier", and "C-C feeling" is "Clearer", and "C-M feeling" is "Very modern". As an illustration, Fig. 2 shows the homepage of a university website which has a look of easier to use, clearer to follow, and very modern.

Fig. 2. A web page with easier to use, clearer to follow, and very modern

5 Conclusion

Web pages consist of design elements such as hyperlinks that are essential to the functionality of web pages and affect users' feelings. In this paper, we have presented an experiment study on university homepages to address the issue of different hyperlink styles combined with other design elements of web pages for achieving the desirable users' feelings, using morphological analysis, Kansei Engineering, grey relational analysis, and neural network techniques. The experimental result has demonstrated that the grey relational analysis can help web designers focus their attention more on the influential design elements, and the NN model is a promising approach for modeling users' feelings of web pages. This study has shown the advantages of using these techniques for examining how the hyperlink style of a web page will affect users' feelings of the web page. Web designers should consider using the "Text only" hyperlink style on web pages in order to best match the users' feelings of easy-to-use and clear-to-follow. To make a web page look classic, the "text and symbol" hyperlink style can be used. The web pages with "text and icon" hyperlinks will have the most modern appearance. The result of this study provides useful insights in helping web designers design web pages that best meet desirable users' feelings.

References

1. Amichai-Hamburger, Y., Fine, A., Goldstein, A.: The Impact of Internet Interactivity and Need for Closure on Consumer Preference. Computers in Human Behavior 20, 103–117 (2004)
2. Amichai-Hamburger, Y., Kaynar, O., Fine, A.: The Effects of Need for Cognition on Internet Use. Computers in Human Behavior 23, 880–891 (2007)
3. Caplan, S.: Using Focus Groups Methodology for Ergonomic Design. Ergonomic 33, 527–533 (1990)
4. Cheng, H.-I., Patterson, P.E.: Iconic Hyperlinks on E-Commerce Websites. Applied Ergonomics 38, 65–69 (2007)
5. Conklin, J.: Hypertext: An Introduction and Survey. IEEE Computer 20, 17–40 (1987)
6. Deng, J.L.: Control Problems of Grey System. System and Control Letters 1, 288–294 (1982)
7. Guan, S.-S., Lin, Y.-C.: Building the Web Design System Using the Kansei Engineering. Journal of Design 17, 59–74 (2002)
8. Ishihara, S., Ishihara, K., Nagamachi, M.: An Automatic Builder for a Kansei Engineering Expert System Using Self-Organizing Neural Networks. International Journal of Industrial Ergonomics 15, 25–37 (1995)
9. Lai, H.-H., Lin, Y.-C., Yeh, C.-H.: Form Design of Product Image Using Grey Relational Analysis and Neural Network Models. Computers and Operations Research 32, 2689–2711 (2005)
10. Lai, H.-H., Lin, Y.-C., Yeh, C.-H., Wei, C.-H.: User Oriented Design for the Optimal Combination on Product Design. International Journal of Production Economics 100, 253–267 (2006)
11. Lin, Y.-C., Lai, H.-H., Yeh, C.-H.: Consumer-Oriented Product Form Design Based on Fuzzy Logic: A Case Study of Mobile Phones. International Journal of Industrial Ergonomics 37, 531–543 (2007)
12. Nagamachi, M.: Kansei Engineering: A New Ergonomics Consumer-Oriented Technology for Product Development. International Journal of Industrial Ergonomics 15, 3–10 (1995)
13. Negnevitsky, M.: Artificial Intelligence. Addison-Wesley, New York (2002)
14. Nelson, M., Illingworth, W.T.: A Practical Guide to Neural Nets. Addison-Wesley, New York (1991)

15. Nielsen, J.: Usability Engineering. Academic Press, United Kingdom (1993)
16. Nielsen, J.: Hypertext and Hypermedia. Academic Press, New York (1993)
17. Nielsen, J.: Top Ten Mistakes in Web Design. Sunsoft Distinguished Engineer (1996)
18. Ozok, A.A., Salvendy, G.: Measuring Consistency of Web Page Design and Its Effects on Performance and Satisfaction. Ergonomics 43, 443–460 (2000)
19. Qu, Z., Ghorbani, R.P., Li, H., Hunter, R.L., Hannah, C.D.: Advantages of Combined Touch Screen Technology and Text Hyperlink for the Pathology Grossing Manual: a Simple Approach to Access Instructive Information in Biohazardous Environments. Human Pathology 38, 420–425 (2007)
20. Ruddle, R.A., Howes, A., Payne, S.J., Jones, D.M.: The Effects of Hyperlinks on Navigation in Virtual Environments. International Journal of Human-Computer Studies 53, 551–581 (2000)
21. Weinreich, H., Lamersdorf, W.: Concepts for Improved Visualization of Web Link Attributes. Computer Networks 33, 403–416 (2000)
22. Yeh, C.-H., Lin, Y.-C.: Neural Network Models for Transforming Consumer Perception into Product Form Design. In: Wang, J., Yi, Z., Żurada, J.M., Lu, B.-L., Yin, H. (eds.) ISNN 2006. LNCS, vol. 3973, pp. 799–804. Springer, Heidelberg (2006)

Gesture Based Interface for Crime Scene Analysis: A Proposal

Andrea F. Abate[1], Maria De Marsico[2], Stefano Levialdi[2], Vincenzo Mastronardi[3], Stefano Ricciardi[1], and Genoveffa Tortora[1]

[1] Dipartimento di Matematica e Informatica, Università degli Studi di Salerno, 20186, Fisciano (SA), Italy
[2] Dipartimento di Informatica, Università degli Studi di Roma "La Sapienza" Via Salaria 113 - 00198 Roma
[3] Dipartimento di Scienze Psichiatriche e Medicina Psicologica, Università degli Studi di Roma "La Sapienza", Piazzale Aldo Moro, 5 -00185 Roma

Abstract. Within crime scene analysis, a framework providing interactive visualization and gesture based manipulation of virtual objects, while still seeing the real environment, seems a useful approach for the interpretation of cues and for instructional purposes as well. This paper presents a framework providing a collection of techniques to enhance reliability, accuracy and overall effectiveness of gesture-based interaction, applied to an interactive interpretation and evaluation of a crime scene in an augmented reality environment. The interface layout is visualized via a stereoscopic see-through capable Head Mounted Display (HMD), projecting graphics in the central region of the user's field of view, floating in a close-at-hand volume. The interaction paradigm concurrently exploits both hands to perform precise manipulation of 3D models of objects, eventually present on the crime scene, or even distance/angular measurements, allowing to formulate visual hypothesis with the lowest interaction effort. A real-time adaptation of interaction to the user's needs is performed by monitoring hands and fingers' dynamics, in order to allow both complex actions (like the above mentioned manipulation or measurement) and conventional keyboard-like operations.

1 Introduction

A number of well-established theories assert the possibility to identify crime culprits using a criminological database, built upon similar previously-solved cases. The main criminal profiles which could populate the database are the following (though further profiles might be also added) :

- Murders in general and Family murders
- Serial Killers
- Rapists
- Paedophiles
- Pyromaniacs
- Classic and Family mass murders
- Terrorists

O. Gervasi et al. (Eds.): ICCSA 2008, Part II, LNCS 5073, pp. 143–154, 2008.

Each profile usually corresponds to a specific modus operandi that can be investigated and reconstructed through a meticulous crime scene analysis. This approach may be combined with advanced crime scene investigation and simulation techniques. One of the expected outcomes is the ability to disambiguate complex scenarios more easily, by virtual on-the-site exploration of persons/objects/scene behaviour. In this way, for example, an help can be provided to solve one of the most tricky investigation problems, namely highlight staging or crime scene camouflage. The latter is defined as "the intentional forging of a crime or crime scene to sidetrack police inquiries and to make the criminal justice process fruitless". Notice that identifying the motive of a crime is one of the key issues in any criminal investigation. In staged crimes, understanding the motive will very often take to the culprit. In fact, experts suggest that in each staged crime the investigator will face the need to identify the motive for two different behaviours: 1) the original action that called for the staging and 2) the staging itself. In such crimes the investigator must be able to recognize not only the motive for the criminal act (greed, revengeful rage, game-playing, etc.), but also the motive for staging (instinct of self-preservation, embarrassment, shame, etc.). An early exploration of the real scene with advanced augmented reality tools may quickly highlight inconsistencies and contradictions in the scene setting, able to suggest staging, for example, the simulation of a virtual object trajectory may reveal the unnatural position of a real corresponding one.

Unfortunately, at present, no operational systematization of the above scenario exists worldwide. The aim of this study is to contribute in diminishing this gap by gathering deep criminological know-how with some of the most promising three-dimensional interaction and visualization technologies, to support an innovative investigation methodology based on augmented reality and multimodal gesture based interfaces. This approach will enable the possibility to visit an actual crime scene while displaying virtual objects and interacting with them to formulate hypothesis or measuring relevant features. The system is meant to support expert users in decision making and simulation of scenarios, but it is also usable for training purposes. This can provide a valuable help to actual and aspiring investigators as it could result in a significant contribution to the analysis of investigation hypotheses, allowing to visually verify the position and possible displacement of bodies, the position of the murder weapon, and the position and movements of the culprit, by comparing them with their virtual counterparts. We want to underline the differences of our proposal with a more traditional crime scene virtual reconstruction. On one hand, the latter provides the possibility to reproduce a crime scene in the most faithful way even at a distance and at any time, and the additional ability to perform different simulations. On the other hand, it requires a certain time to be completed, while our approach allows an early, advanced exploration of the fresh scene. The investigator is able to move in the real crime context and to inspect all its nooks: hypotheses can be virtually verified at the same time they are generated, with all evaluation elements still available. This fact must also be considered in relation with a peculiar characteristic of reconstructed scenes. Elements that are hidden in the components used for reconstruction, e.g. due to occlusion in scene pictures, remain hidden even in the most reliable virtual reconstruction. It could not be possible to consider such elements anymore, most of all if the scene has been later contaminated. In this way valuable cues might be lost. It comes out that the two approaches are complementary, as they provide different advantages. However, while virtual reconstruction of a real scene is a well-consolidated activity at present [1], our proposal of interaction is quite novel within the above mentioned activities. As a further consideration, a physical interface device, such as a keyboard or a mouse, might be difficult to position within a sensitive area such as a crime scene. This is one of the reasons, together with pure usability considerations that will be introduced later, that led us towards the choice of a virtual keyboard included within the augmented reality setting.

The rest of this paper is organized as follows. Multimodal gesture based interfaces and their characteristics are briefly presented in section 2., while the proposed framework is described in section 3, and briefly discussed in section 4. The paper concludes in section 5.

2 Multimodal Gesture Based Interfaces and Related Work

Recent advances in technology spur the investigation of novel interaction paradigms, able to exploit the increased technological power. Multimodal interfaces [2] are the natural candidates to get over the traditional WIMP (Windows, Icons, Menus, Pointers) interaction paradigm. They are characterized by the (possibly simultaneous) use of multiple human sensory modalities and can support combined input/output modes. A multimodal system is therefore expected to provide users with a more natural interaction. At present, multimodal interfaces have pervaded new application domains [3], like in advanced virtual reality systems which allow to simulate a real "physical" manipulation even when this is not possible due to logistical problems (e.g. remote operation) or to dangerous settings (e.g. radioactive materials and areas), or when it is convenient just to simulate a real operation (e.g. for training purposes). We must consider that the successful execution of direct manipulation tasks within an immersive virtual reality environment may critically depend on the abolition of any latency time between the moment an action is performed, e.g. by means of datagloves, and the moment the user can see the action effects. In this way the user perceives gesturing as an act that can be directly implemented at the interface. If this cannot be achieved, all usability benefits, coming from the direct manipulation paradigm, would be lost [4]. A further step is to definitively drop traditional "artificial" input devices (keyboard and mouse) to completely rely on "natural" channels such as voice and gestures.

Gesture based interfaces are surely not new in human-computer-interaction literature. They were originally presented as the most natural way to navigate through and interact with virtual reality environments, generated by military simulators or by entertainment applications [5]. More recently, thanks to technological progress and cost reduction, they have been proposed for advanced interaction in a broader spectrum of applicative fields, ranging from industrial processes to medical applications [6]. In this latter field the researches on gesture based interaction are often focused on the replacement of conventional input devices like keyboard and mouse with a contactless interface for sterile operating room interventions. For this kind of applications a visual approach to gesture recognition [7] is preferred because it does not require the user to wear any specific device as the recognition is based on video acquisition and processing of gestures. Along this line of research Graetzel et al. [8] exploit stereo cameras to capture both color and depth info to achieve reliable, high-speed hand detection and tracking within a user-specified workspace, and to interpret hand gestures as mouse commands (pointer movement and button presses). The Gestix system by Wachs et al. [9] is a further system designed for sterile HCI. It proposes a vision based gesture capture system which interprets user's gestures in real-time to navigate through and manipulate an image and data visualization environment. Dynamic navigation gestures are translated to commands based on their relative positions on the screen, while the gesture recognition relies on tracking of the user's hand based on color-motion cues; a state machine switches from navigation gestures to others such as zoom and rotate. Image viewing and browsing represents another relevant field of application [10] which can benefit from instrumented gloves and magnetic motion tracking technologies [11]. In their the recent work [12], Tani et al. discuss the use of a glove-driven interface in radiological workstations and present a prototype that aims at integrating common functions, such as virtual manipulation and navigation control, with a basic gesture interface. The user can control the mouse by simply pointing at the screen and moving the hand, or can perform gestures associated with specific commands in the given context. For example, by waving the hand over the keyboard the user calls the database browser. This paper proposes a gesture based interface targeted to interactive measurement and analysis of a crime scene. Though interaction following the desktop metaphor (e.g. windows to browse information and a mouse to move and manipulate objects) has become widespread and is claimed to be intuitive [13], this is not always true when it comes to 3D data manipulation. Common functions like image rotation with respect to a given point or landmark positioning, which can be easy to perform in two dimensional space, are more complex in three dimensions as the greater degree of freedom requires a more powerful interaction approach to make the system effective. As a result, approaches focusing on recognition accuracy, which tend to map mouse and keyboard functionalities through gestures, could not exploit the full potential of this interaction technique. The proposed framework combines one hand and two hands gestures with a virtual context adaptive graphic interface, which floats within the user's field of view by means of a see-through capable stereoscopic HMD. In particular, our work exploits six degrees of freedom head/wrists tracking plus gesture recognition for advanced human-system interaction. A more natural and familiar way of managing objects and situations is expected to improve global user performance and increase applications effectiveness. Nevertheless, in some situations the need arises to perform classical keyboard-and-mouse supported operations, such as selecting an object from a menu of models or entering parameters for a complex operation. On one hand, switching to a "real" device would break off the interaction flow in a disturbing way. On the other hand, we must consider the way the user is accustomed to perform such operations. Consistency with familiar interaction patterns is a very important guideline aiming at reducing the time and efforts needed to learn a new application [14]. For this reason, a virtual keyboard is projected onto the actual environment, so that the user can comfortably switch from manipulation operations to system interaction operations. Moreover, the familiar point-and-click pattern is re-proposed through fingers movements to extend the range of available functions while keeping the number of basic gestures to a minimum. The idea is allow the user to explore a real crime scene exploiting a small set of simple (user-wise) gestures and the visual interface to enable a wider set of commands and functionalities, improving both the efficacy and the accuracy of user-system interaction.

3 Description of Proposed Methodology

A schema of the proposed framework is in Fig. 1. After a brief calibration session, to initialize gloves sensors and check the trackers, the system enters in operating mode. Two separate data channels, respectively for the left and right hand, are pre-processed and then both feed the Gesture Recognition Engine. This module analyzes posture tokens outputted by datagloves, combined with motion data captured by the tracking system. As a result of this process a (one-hand or two-hands) gesture is recognized and passed to the Interaction Engine, which eventually selects a function of the virtual keyboard or translates the gesture in a transformation (rotation/traslation) of the 3D model sent to the Visualization Engine. Here two renderings (left and right) of the whole scene (interface plus anatomical model) are calculated and coherently displayed by the HMD. The main hardware and software components of system architecture are described in subsections 4.1 to 4.4.

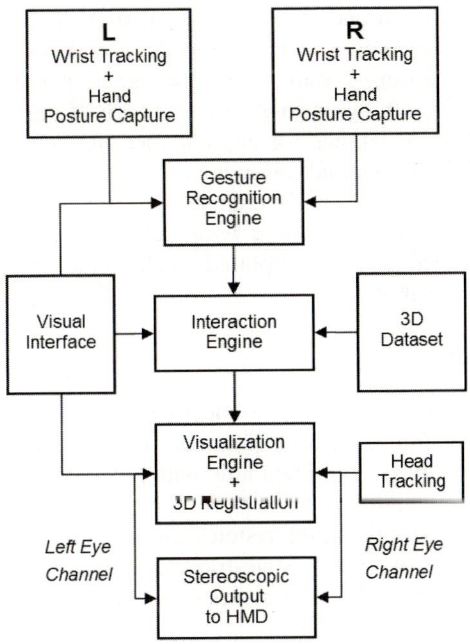

Fig. 1. Overall view of the proposed framework

3.1 User's Head and Hands Capture

The user input module is responsible for user's hands tracking within 3D space and hand posture acquisition. Here an accurate and reliable capture technique can be exploited, based on wireless instrumented gloves and ultrasonic tracking devices. Such choice simplifies the gesture recognition stage (e.g inter-hands and inter-fingers occlusions are not an issue anymore) as each single finger has individual sensors for flexion and abduction which are unaffected by any other fingers. In more detail, left and right hand posture acquisition is performed via a couple of wireless 5DT Datagglove 14 ultra, featuring fourteen channels for finger flexion and abduction measurement, with 12 bit of sampling resolution each. Additionally, a binary (open/closed) value for each finger, resulting from the comparison of the normalized joint flexion value to a threshold, leads to 2^4 (four fingers are considered excluding the thumb) different combinations or postures. Though it is possible to access a much more accurate finger status representation via each sensor's raw values, this built in classification is more appropriate for the posture recognition, as it simplifies both user training (partially flexed fingers do not compromise posture recognition) and the gesture recognition module design. The four hand postures used in this study (fist, index finger point, not index finger point and flat hand) are shown in Fig. 2a. They have been selected as they are the simplest to perform for most users and among the most used in natural interaction [15]. As datagloves do not provide any spatial information, the system relies on Ascension Flock of Birds magnetic motion tracking hardware, with six degrees-of-freedom, to detect head and wrists position in 3D space and their rotation on three axis (yaw, pitch and roll). The use of such data will be presented in the following. Among the advantages of this setup there is the wide capture volume (respect to video based solutions requiring the user to be positioned in a precise spot within camera field of view), an accuracy in the range of one millimeter for distance measurements and of 0.25 degrees for angular measurements, and a high sampling rate (up to 180 measures per second) allowing an accurate capture of fast movements. A preprocessing applied to each of six channels (for each hand) filters capture noise by means of a high frequency cut and a temporal average of sampled values. Then both left and right data streams are outputted, each including basic postures captured and positional/rotational information.

3.2 Gesture Recognition Engine

At this stage a specific software module checks for particular posture tokens which trigger associated interaction activities. The analysis is performed by means of a Gesture Recognition Engine based on timed automata [16], able to detect one-hand, two-hands and timed posture patterns which are associated to manipulation functions. Timed automata are labeled transition systems used to model the behavior of components in real-time systems over time. State-transition graphs are annotated with timing constraints. A timed automaton performs time-passage actions, in addition to ordinary input, output and internal actions. It accepts timed words, i.e. infinite sequences in which a real-valued time of occurrence is associated with each symbol. Timed automata so provide a feature that classical finite automata do not address in any way, namely timing. Embedding time allows changing the status of involved entities according to time-based events to enhance the quality of user-system interaction. The aim here is to augment the basic one-hand postures through timed patterns or via a combination of left and right hand for a simple yet more powerful interaction.

The use of timed automata offers a further key benefit for the proposed architecture, as it enables the application designer to formally verify the interaction model by means of well established model checking procedures [17]. In the proposed framework, eight gestures are used as shown in Fig. 2, two of which (Fig. 2b) are defined

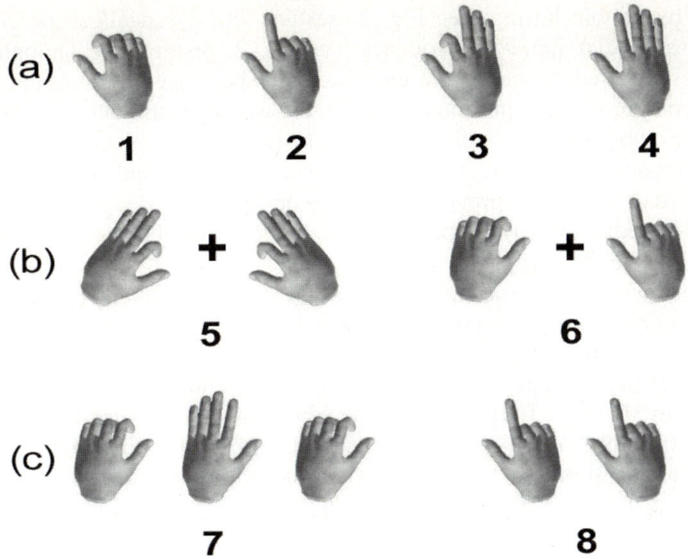

Fig. 2. One-hand (a), two-hands (b) and (c) time-based gestures

through a two hand combination of basic postures (Fig. 2a), while other two (Fig. 2c) are defined by a timed sequence of basic postures (for instance fist-flat_hand-fist, or double pointing). Recognized gestures are represented by a vector including gesture index, first hand x-y-z spatial coordinates, first hand yaw-pitch-roll angles, second hand x-y-z spatial coordinates, and second hand yaw-pitch-roll angles.

3.3 Interaction Engine

Once a valid gesture has been recognized the corresponding gesture vector is passed to the interaction engine, which exploits the same timed automata based design of recognition engine, and is responsible for any visual interaction allowed by the system, translating user's gestures into actions. Gestures are evaluated according to the current interaction status, so that the same gesture may trigger different actions in different operative contexts (rotation, measurements, landmark assignment, etc). Operational modes and manipulation functions are selected via a virtual interface, which is displayed within the field of view as a frame surrounding the 3D content, and including textual info related to the ongoing operations (numerical values for coordinates and angles, distances, etc.). The layout of such interface is visually perceived as a floating object, thanks to stereoscopic rendering. It is displayed in a close at hand position along the depth of visual field, according to a calibration procedure in which the user touches with the finger a sequence of small targets at various depths, thus allowing an adaptation of the parameters regulating the stereo effect. One of the main aims of interface design is to minimize the number of gestures required to operate it, yet preserving familiar interaction patterns. For this reason a gesture adaptation of the classical point and click interaction paradigm is adopted: selection is triggered by hitting an active area by index fingertip (see Fig. 2, gesture #2), an action or a confirm is triggered by double hitting (see Fig. 2, gesture #8), a cancel/escape command is triggered by a fist-flat_hand-fist sequence (see Fig. 2, gesture #7). Though only one-hand gestures are used to operate the interface, an experienced user could exploit both hands to operate in a faster and more comfortable way (for instance, typing characters in a text field by both hands results in a much faster operation than via a mouse based selection and would not require a physical keyboard). Visual and acoustical feedbacks are provided to confirm the "pressure" of a key or the acknowledgment of a particular command, thus reducing wrong operations. If required, interface layout can be hidden at any time via a gesture toggle.

At present, only a small set of functions has been implemented, allowing to rotate/move the object, to place landmarks over its surface and to take distance measurements between landmarks.

While object pan is more conventionally achieved with any of the two hands, object rotations in 3D space fully highlight the advantages of two-hand gestures, as they are addressed associating object rotation over three axis with the rotation of a vector connecting the two points of contact between index finger and thumb on each hand (see gesture #5 in Fig. 2 and Fig.3). This approach is much more precise than the typical one-hand based interaction, which is often seen for instance in VR applications. As a matter of fact, it allows a greater control of rotation (two hands better visualize the overall rotation), a more comfortable operation (it does not matter how each wrist rotate during interaction, as only the vector connecting the two hands is relevant) a less jerky interaction without losing responsiveness (a weighted average of both hands spatial info improves tracking). Once the rotation function has been selected, the user can set the rotation handle for each hand (the anchor points used to interact with the model) by gesture #3 simply moving the fingers along the object's surface. A valid handle's location is highlighted by a colored spot if it lies within a valid region. If the gesture #5 is recognized then the vector connecting the two handles is visualized and the interactive rotation is performed until this condition is true.

Landmark positioning over the crime scene volume may be accomplished according to two different interaction patterns. In the first case, a rotation of the object is performed, as explained before, in order to expose the location of interest, and then a landmark is placed by double hitting (gesture #8). A more intuitive, though less precise, pattern implies performing the rotation of the object by a single hand, grasping (see Fig. 2, gesture #1) the object in a position which acts as the pivot point and double pointing the location on which the landmark is to be assigned.

For any task involving positioning in 3D space, a precise calculation of actual finger positions is performed through a combination of a forward-kinematics applied to a 3D parametric hand model, which is adapted to the real user's hand measures during a calibration session. This setup is due only one time and may be saved and retrieved when the system is initialized. In this case the raw flexion values are exploited for each finger.

3.4 Visualization Engine

3D models and the virtual interface are both processed by the visualization engine, also responsible for AR related real time transformations and for the stereo rendering of 3D content. The engine is built on the Quest3D graphics toolkit and DirectX API.

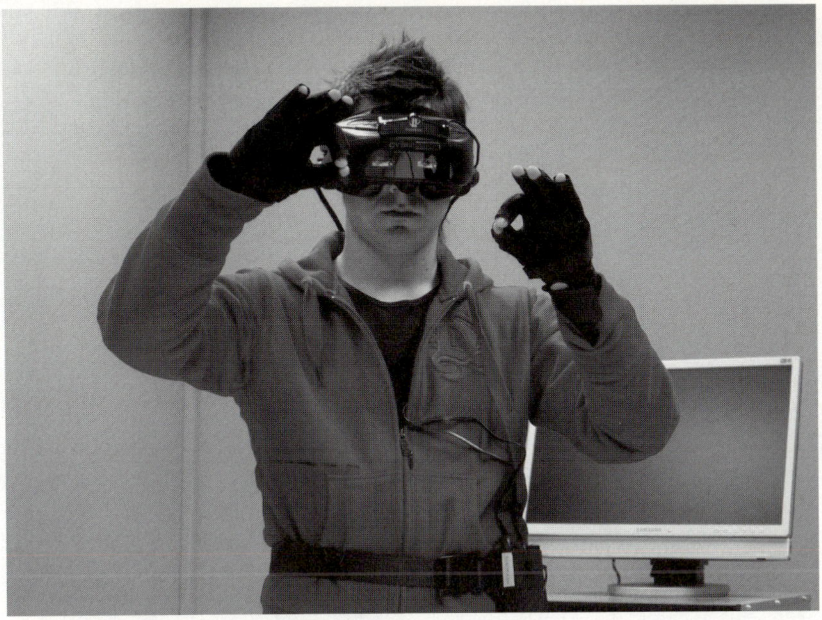

Fig. 3. User during object rotation by means of both hand. The rotation of a vector connecting the two points of contact between index finger and thumb on each hand is used to control the actual object rotation over three axis.

To provide the AR environment, the Visualization Engine exploits user's biometrics such as head's position and orientation, then processes these data to transform the virtual content as seen from user's point of view and coherently to a 3d model of surrounding environment, a crucial task referred as 3D registration. Actually, any AR architecture requires a precise registration of real and virtual objects. In other words, the objects in the real and virtual world must be properly aligned with respect to each other, or the illusion that the two worlds coexist will be compromised. To achieve this not trivial goal two main requirements have to be satisfied: a) the position and orientation of user's head have to be precisely tracked at a high sample rate; b) the physical world or at least the objects relevant to the application has to be precisely measured in the same 3D space in which the user operates. At runtime, two rendering cameras (one for each eye) are built, matching the exact position/orientation of user's eyes, transforming each vertex of each virtual object to be displayed onto the real scene accordingly. Two renderings (left and right) are then calculated and coherently displayed through an optical see-through Head Mounted Display (a Cybermind Visette SXGA), which is an helmet working by placing optical combiners in front of the user's eyes (see Fig. 4). These combiners are partially transmissive, so that the user can look directly through them to see the real world. The combiners are also partially reflective, so that the user sees virtual images bounced off the combiners from head-mounted LCD monitors. The rendering engine has been tailored to optical

Fig. 4. User wearing left and right hand 5DT- 14 dataglove ultra, plus left and right hand wireless 6DOF motion trackers from Intersense. A Cybermind see-through HMD is also worn to visualize the virtual keyboard and the virtual objects.

see through HMD, but it can be easily adapted to video see-through displays. Eventually, a selective culling of a virtual object may be performed whereas it is partially or totally behind a real object, but for the applications showcased in the presented case study this technique (and the overhead required to accurately model the real environment) could not be necessary.

4 Discussion

The framework described so far is only a part of a wider project which aims at improving the access to three dimensional data coming from crime scenes and its usage for crime scene investigation and reconstruction. As it is still in a early stage of development, the first concern is to assess if the proposed approach to 3D manipulation could match the requirements for a more accurate and intuitive way to deal with crime scene data. Today, modern investigation techniques are often based on crime scene 3D capture and analysis by means of conventional mouse and keyboard-based software tools. As the three dimensional data generated by these approaches can be quite complex, problems may easily arise, related to the way these data are accessible, and to how complex is interacting with them through a bi-dimensional display. Though stereoscopic displays have already been proposed for this kind of applications, the main issue is on the interface side. Indeed, operating on a 3D surface or volume requires a more powerful way to specify actions or locations in space which is not well addressed by usual 2D-based interaction paradigm, involving conventional interfaces. At the same time, many conventional functions have to be accessible and easy to use in a way similar to that possible with common devices (mouse and keyboard). To this regard, the first tests conducted on a small group of users are encouraging as all subjects reported an advantage in achieving the right landmark placement or in the object control, with an overall greater confidence feeling during interaction.

5 Concluding Remarks

In this paper a framework for crime scene interpretation has been presented, based on a gesture interface operating within an augmented reality environment,. The potential advantages of this approach may be summarized as follows: a common ground for discussion among police and scientific administrations, the possibility of "re-playing" the scene dynamically to observe details that may not be seen in a static picture (as in a photograph), the teaching value of showing different crime scenes to policemen during a specialized course, the evaluation of different cues and their relative importance in the scene interpretation. The proposed gesture-based interface approach exploits one-hand and two-hands timed interaction patterns, combined with a virtual floating interface. The preliminary tests conducted seem promising, as they show a user-wise perceptible improvement in performing 3D interaction tasks like the selection of a particular spot on a 3D surface or the distance measurement between features. As the work is in an early stage, more functions have to be implemented and more complete and accurate experiments are required to measure advantages in a real working environment. As a future development, we are planning to include a physical engine within our system, able to reproduce objects interaction and dynamics according to stored rules.

References

1. Gibson, S., Howard, T.: Interactive reconstruction of virtual environments from photographs, with application to scene-of-crime analysis. In: ACM symposium on Virtual reality software and technology, Seoul, Korea, pp. 41–48 (2000)
2. Jaimes, A., Sebe, N.: Multimodal human-computer interaction: A survey. Computer Vision and Image Understanding 108, 116–134 (2007)
3. Oviatt, S.: Multimodal Interfaces. In: Jacko, J., Sears, A. (eds.) The Human-Computer Interaction Handbook: Fundamentals, Evolving Technologies, and Emerging Applications, Lawrence Erlbaum, New Jersey (2003)
4. Reeves, L.M., Lai, J., Larson, J.A., Oviatt, S., Balaji, T.S., Buisine, S., Collings, P., Cohen, P., Kraal, B., Martin, J.-C., McTear, M., Raman, T.V., Stanney, K.M., Su, H., Ying Wang, Q.: Guidelines for Multimodal User Interface Design. Communications of the ACM 47(1), 57–59 (2004)
5. Brooks, F.: Grasping reality through illusion - Interactive graphics serving science. In: ACM CHI 1988, pp. 1–11 (1988)

6. Krapichler, C., Haubner, M., Lösch, A., Englmeier, K.: Human-Machine Interface for Medical Image Analysis and Visualization in Virtual Environments. In: IEEE conference on Acoustics, Speech and Signal Processing, ICASSP 1997, vol. 4, pp. 21–24 (1997)
7. Kohler, M., Schroter, S.: A Survey of Video-based Gesture Recognition - Stereo and Mono Systems. Technical Report 693, Informatik VII. Univ. of Dortmund (1998)
8. Graetzel, C., Fong, T., Grange, S., Baur, C.: A Non-Contact Mouse for Surgeon-Computer Interaction. Technology and Health Care 12(3), 245–257 (2004)
9. Wachs, J., Stern, H., Edan, Y., Gillam, M., Feied, C., Smith, M., Handler, J.: Gestix: A Doctor-Computer Sterile Gesture Interface for Dynamic Environments. In: Saad, A., et al. (eds.) Soft Computing in Industrial Applications. ASC 39, pp. 30–39. Springer (2007)
10. Dionisio, J.D.N., Bui, A.A.T., Ying, R., Morioka, C., Kangarloo, H.: A Gesture-Driven User Interface for Medical Image Viewing. Radiological Society of North America (RSNA), InfoRad Exhibit; Radiology, 807 (2003)
11. Wachs, J., Stern, H., Edan, Y., Gillan, M., Feied, C., Smith, M., Handler, J.: A Real-Time Hand Gesture Interface for Medical Visualization Applications. In: 10th Online World Conference on Soft Computing (2005)
12. Tani, B.S., Maia, R.S., von Wangenheim, A.: A Gesture Interface for Radiological Workstations. In: Twentieth IEEE International Symposium on Computer-Based Medical Systems CBMS 2007, pp. 27–32 (2007)
13. Duke, D.J.: Reasoning About Gestural Interaction. In: ACM/Eurographics 1995, vol. 14(3), pp. 55–66 (1995)
14. Dix, A., Finlay, J., Abowd, G., Beale, R.: Human-Computer Interaction, 3rd edn. Prentice Hall (2004)
15. Stern, H., Wachs, J., Edan, Y.: Optimal Hand Gesture Vocabulary Design Using Psycho-Physiological and Technical Factors. In: 7th International Conference on Automatic Face and Gesture Recognition, FG 2006, pp. 257–262 (2006)
16. Alur, R., Dill, D.L.: A theory of timed automata. Journal of Theoretical Computer Science 126(2), 183–235 (1994)
17. Alur, R., Courcoubetis, C., Dill, D.L.: Model checking for real-time systems. In: 5th Annual Symposium on Logic in Computer Science, pp. 414–425. IEEE Computer Society Press, New York (1990)

Script Language Generation Techniques for 3D Avatar Control Using User Annotation*

Teuk-Seob Song[1], Yoon-Chul Choy[2], Soon-Bum Lim[3], Won-Sung Shon[4], and Euyhyun Suh[1]

[1] Department of Computer Engineering, Mokwon University, Daejeon, Korea
teukseob@dreamwiz.com, ehsuh@mokwon.ac.kr
[2] Department of Computer Science, Yonsei University, Seoul, Korea
ycchoy@raibow.yonsei.ac.kr
[3] Department of Multimedia Science, Sookmyung Women's University, Korea
sblim@sookmyung.ac.kr
[4] Department of Computer Education, Gyeongin National University of Education, Incheon, Korea
sohnws@gin.ac.kr

Abstract. Generally, annotation in a paper document environment refers to underlining important sentences, drawing symbols such as stars or adding comments to the original document for additional explanation. In this paper, we propose automatic generation techniques for controlling a 3D avatar using user annotation. Via the proposed approach, even users who have no knowledge of markup language or script language can use the avatar easily.

Keywords: Avatar, Annotation, Script Language, 3D Avatar Control Technique.

1 Introduction

Embodied agents or humanoid avatars may effectively be used to communicate by human users, for example in an area of instructional VR(Virtual Reality). In recent years there has been a growing interest in embodied agents, in particular embodied conversational agents in the form of animated humanoid avatars. An interesting example is provided by avatar Steve [1] which is developed for educating complex manipulation process of an engine of a vessel also only operate in predefined process. WhizLow is developed for easier understanding of difficult courses such as internal processing procedure of computer using 3D avatar [2].

In this paper, we propose automatic script generation techniques which controls a 3D avatar based on the user's annotation. In general, an annotation on paper works means scoring under the important words, marking symbols such as an asterisk, or adding comments on the original document for the additional explanations. The annotation in this paper means the information including underlines, boxes and comments for a presentation except an avatar's behaviors saved

* "This work was supported by the Korea Research Foundation Grant funded by the Korean Government(MOEHRD)" (KRF-2007-331-D00429).

O. Gervasi et al. (Eds.): ICCSA 2008, Part II, LNCS 5073, pp. 155–167, 2008.

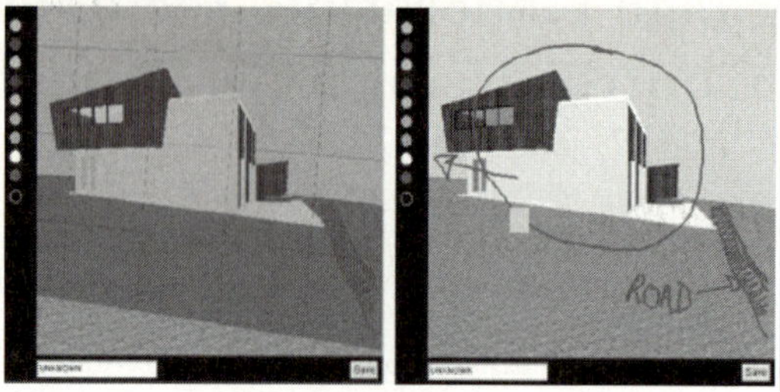

Fig. 1. Examples of a 3D Environment Annotation System Developed at the University of Washington [7]

in DB. Using annotation, users can express opinions [3,4]. They can also deal with large amount of information by grasping the details of a document through the use of the annotation rather than the entire document. In a multi-user environment, users can relay their opinions (important idea, additions, deletions, corrections, etc.).

An annotation system can be used in various areas of text-based web document environments such as information sharing [5], searching [6], re-editing, etc. Annotation is a useful tool for exchanging information among multiple users in not only document environments but also in 3D CAD design systems such as Spacepen [7] and Redliner [8]. In addition, due to wide distribution of high-speed internet and popularization of high performance personal computers, cyber space has become a familiar environment to a great number of people. An avatar in cyber space is typically an animated character that replaces the functions of users.

Avatars are used in commercial services of MSN, Yahoo, Freechal, MS Office, etc. There has been active study on both 2D and 3D avatars [9,10,11] and this is especially true at Pennsylvania State University [12,13] and North Carolina Sate University [1,2] and University of Twente of The Netherlands [14] and Tokyo Institute Technology [15,16,17]. 2D or 3D avatars of existing studies as well as commercialized avatars either perform limited actions or very complex script language is used. Steve [13] was developed for teaching the complex manipulation process of vessel engines and only operates in a predefined process. Furthermore, Steve and Whizlow [1] only handle simple actions such as pointing at target objects and there is no interaction between the target object and avatar. In most education sites, the avatar should perform various actions. In addition, a variety of marks made on presentation material using a mouse, pointer or board is needed, as shown in Figure 2.

Hence, there are limits to applying existing studies, which are based on predefined actions, to presentations or cyber education. Also, studies for controlling the actions of the avatar using existing script language focus on controlling the avatar itself such as its avatar or facial expressions and use complex script language.

Accordingly, there are limitations when end users, who are not professionals, try to control the avatar. The present study proposes a method whereby end users who are not familiar with script language can control a 3D avatar for

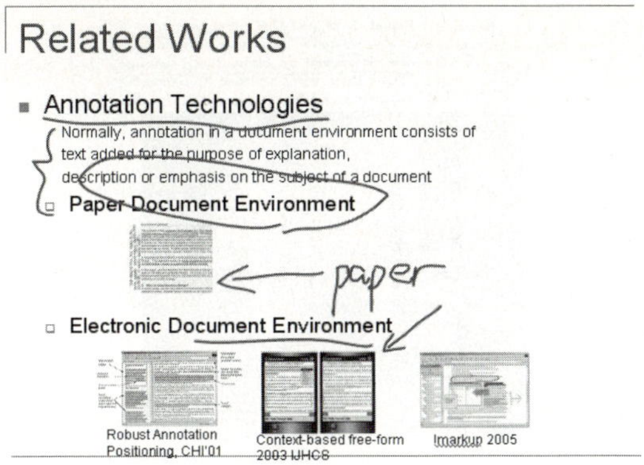

Fig. 2. Example of Presentation using MS PowerPoint

the purpose of making presentations. Because general end users are not familiar with script language, they can add various symbols such as underlining, checks, and arrows to presentation material through an interface based on a WIMP concept. The annotation created by users is saved in CVaML(Context Based Virtual Avatar Markup Language) and script language is automatically created via a context-based script language generator.

In this paper, Section 2 presents an analysis of related studies and Section 3 explains the avatar structure developed for the study as well as annotation DTD built for saving context information. Section 4 looks at the context-based avatar control method and Section 5 describes the user evaluation. Lastly, Section 6 presents conclusions and directions for future study.

2 Related Works

In this section, we present previous annotation systems in an electronic document environment and avatar control technology using script languages. The annotation interface on the electronic document provides different structures and functions depending on the realization space and its object. In most mobile reading equipment [18] and ebook PDAs [19,20], annotations are created using an interface that is similar to the interface for paper documents. To enable this, XLibris recognizes 5 types of marking and uses textual content simultaneously to support free marking creation and searching functions. Annotea is a LEAD (Live Early Adoption and Demonstration) project and it improves the W3C cooperation environment that shares annotation in web. It is made up of Amaya Editor and Browser. The original document uses an appropriate document or a structured document and annotation is defined using the RDP-structure.

Previous studies for controlling a 3D based avatar are introduced below, including work related to controlling the avatar via markup language such as TVML [21], AML [22], and MPML [16] and various application environments for avatars such as Steve [13] and WhizLow [1]. Recently, research has been

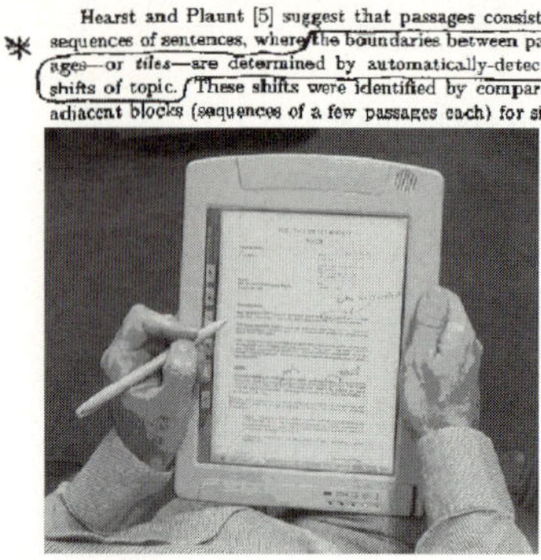

Fig. 3. Examples of Document Environment Annotation

conducted with the aim of increasing the convenience and reusability of the avatar. Studies for using 3D virtual environment for education or training purposes have proceeded since the early 1960s. In 1997 the 3D avatar Steve, developed at North Carolina University, became the first to be used for education or training purposes in a virtual environment. Steve was developed so as to facilitate manipulation of complex machineries such as engines of vessels used by the U.S. Navy. Students could follow the same process in a virtual environment. North Carolina University later developed WhizLow [1], which is similar to Steve but is for educating the internal (CPU, RAM, ALU operator) processing procedure using a 3D avatar.

Meanwhile, a research lab in Pennsylvania University presented Jack [2], an avatar designed for presentations in 3D virtual space and University of Twente [14] of The Netherlands is developing a realistic presentation room. However, like Steve, Jack does not used script language and hence there are limitations when trying to extend the application domain. The presentation room of University of Twente [14] does not provide interaction between the virtual environment and avatar.

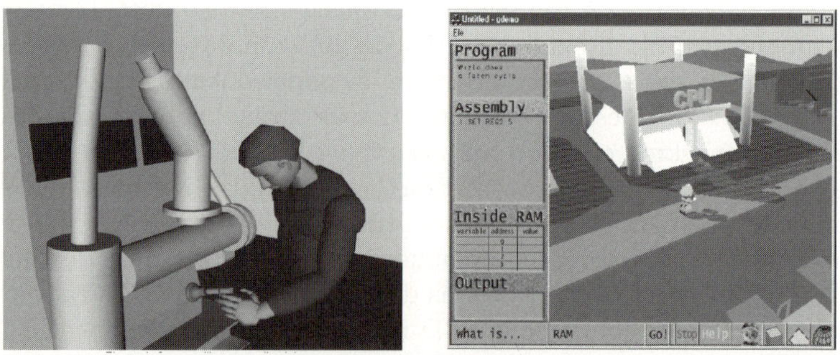

Fig. 4. 3D avatars Steve and WhizLow [1,2]

Table 1. Comparison of Previous Studies

	Input Method	Characteristics	Limitations
Steve [1], Jack[13]	No Editing tool	A variety of application environments were developed such as using 3D avatar in training or education	There are limitations when trying to implement the method to other domains because it does not use script language.
TVML[21], AML[22], MPML[17], CML[9]	Editor, Text based	A method for controlling avatar in a 3D virtual environment using script language was developed.	Because it is for specific purposes such as controlling facial expressions or behaviors, it cannot support interactivity with presentation material.

The following is a summary of studies that use script language for controlling the avatar in table 1. TVML is a language developed by NHK, a Japanese broadcast company. It uses text-based script language that automatically creates television programs for internet broadcast [16]. The TVML-based system can create in real-time a TV program using computer graphic, voice, and multimedia technologies on a desktop. TVML uses text-based language. Even though an editing tool exists for TVML, because it is text based much time is required for users to control the avatar since they have to indicate every avatar behavior.

CML (Character Markup Language)[9] is a XML-based script language and was developed for creating characteristics and motions of charactesr. The purpose of CML is fast and easy creation of life-like characters and agents in online applications and the virtual world. CML utilizes a method for embodying a character to agents and effective emotions depending on the given context.

3 Modeling of Avatar and CVaML

The avatar used in the present study was designed using H-Anim [23]. H-Anim is an ISO/IEC FCD 19774 International Standard and is a humanoid library developed for effective management of humans in 3D online virtual environments. Currently, version 200x is available. The tutor avatar used in the study is developed based on H-Anim and is composed of 15 segments and 11 joints as shown in following figure 5.

The following table 2 is a list of behaviors that the 3D avatar performs during a presentation in a virtual environment. Each behavior is saved in a DB in XML format and a behavior creation motion editor [11] is used.

Saving User Annotation Base on XML. In this work, for more precise application of user options in the 3D virtual environment, annotation information is created via a rule-based annotation system and this is used when creating script

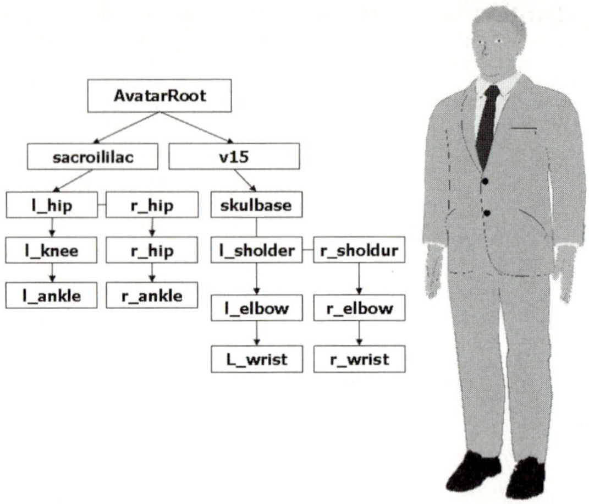

Fig. 5. Tutor Avatar Based on H-Anim

Table 2. Avatar's Behavior

Avatar's behavior	View Point	Description
Basic Behavior	Front	This involves presenting material by naturally using both hands.
Pointing	Presentation screen, Front	This involves pointing at a location of a presentation screen with a hand or a pointer.
Plain Symbol	Presentation screen, Front	This involves drawing plain symbols such as underline on the presentation material with the left or right hand.
Complex symbol	Presentation screen, Front	This involves drawing complex symbols such as a star. For a natural appearance, the right hand is used.
Comment	Front	This involves commenting and looking at the front of the presentation room or the presentation material.
Humor	Front	This is for facilitating user concentration on the lecture. Various behaviors can be created by users.
Movement	Front, left, right	This is when the length of the text that should be read or commented upon is long and there is spare time. During this time, the avatar moves from left to right.

language. Therefore, to express the annotation information that is created in the system in a format that can be easily understood by both human and machine, a XML based DTD, CVaML (Context based Virtual Avatar Markup Language)

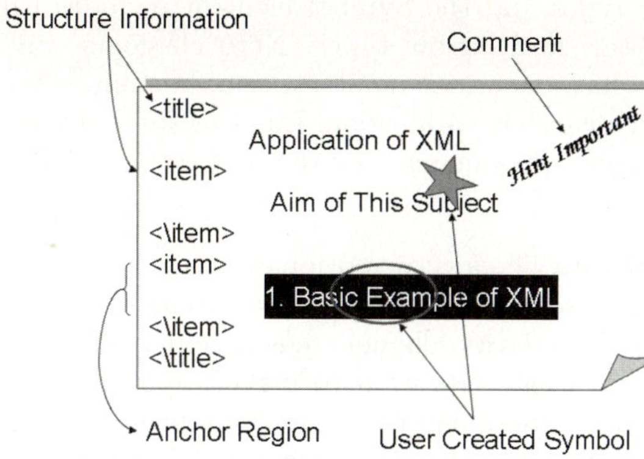

Fig. 6. Example of User Annotation

is defined. To support geometric information of annotation and a variety of content information in the proposed system, the basis of CVaML includes a variety of annotation types, semantic information, styles, and structure/non-structure information. The end user uses an intuitive interface that applies the WIMP concept and the saved information is stored in an Annotation file. The overall structure of the CVaML DTD, designed for storage of the user's annotation, is shown in Figure 6 and a detailed description is provided below.

Basic Structure of CVaML. As can be seen in Figure 7, CVaML DTD can put multiple annotations below UserAnnotation, the root element. Each annotation element is composed of AvatarAni, which is for supporting the style type of annotation created in the original document and certain behavior of the avatar designated by the user.

Type Element. Annotation type element is constructed with Lines, Symbol, Note, and Bracket elements depending on the style type. Line elements includes

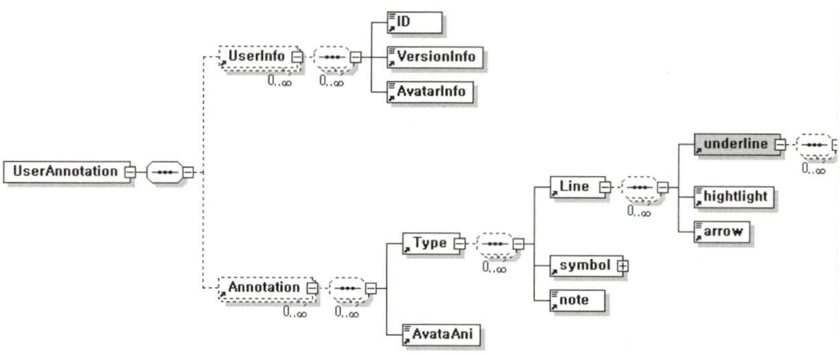

Fig. 7. The tree of CVaML Annotation DTD

Line and Highlight types and the Symbol element includes Ellipse, Rectangle, Crossout, Check, Star, and Arrow types. Note elements can be divided into Standalone_Note, which occurs alone at the top, bottom, left or right margins, and Add-Note, which can have additional Lines or symbol types in the content. These annotation types are influenced by the basic types of annotation proposed by Marshall [18].

Motion and Relation. The avatar motion element is composed of an element, which is acted by the user's annotation, and an element, which is controlled by use excluded type. The relation element has a semantic function of the annotation, and included a user's intention to make the annotation. It particularly referred to the semantic function of the annotation proposed by Marshall's study [18]. The relation element provides basic semantic features such as tip, remind, mark, interpretive, idea, and criticism while it provides user define features to reflect a user's original intention.

4 Script Language Generation Techniques

In this study, the annotation information created by users is managed separately from the XML file of lecture contents and using context information analysis module, path of elements, coordinates within the virtual space, existence or non-existence of avatar behavior, the length of comment are preprocessed and saved. The first process of this context information is through the creation of a logical tree, as illustrated in the example shown below.

In this study, the type elements included in CVaML, the avatar behavior designated by users, and the related contexts are analyzed. The analysis of the context information becomes the basis for applying rules. Each rule is processed through the rule model of the proposed system, which has a (If condition AND ⋯

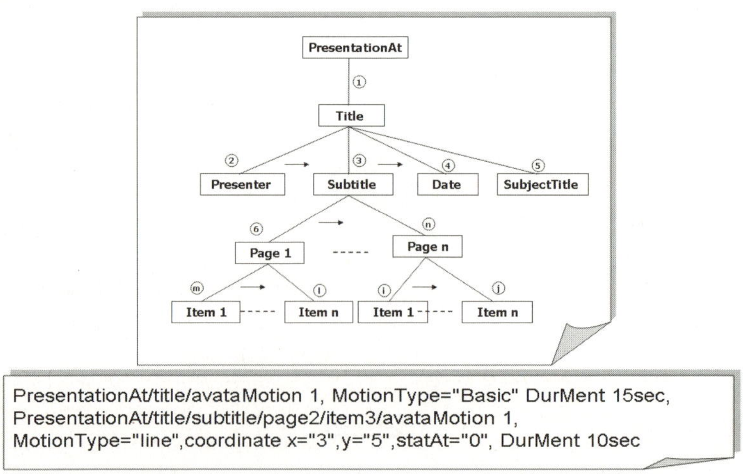

Fig. 8. Examples of a Logical Tree and Context Information

Table 3. An Example of Algorithm

IF (1) Type of Annotation is Underline
(2) The Annotation included in one item.
IF $ThSIS < C_1$
THEN
(1) Do annotation and ment at the same time
ELSE IF $ThSIS < C_2$
THEN
(1) Move presentation position and then do annotation while speaking

AND condition THEN application method) format. A summary of the method for analyzing context is provided in the following.

Simple Case where Annotation is Completely Included in One Structure Information. The simplest form is when a user's annotation is completely included in one structure information. The context information analysis module investigates the structure information and divides it into that having user's annotation and that which does not, and thereupon processes the information. When the annotation information is completely included in one structure information and it is a plain annotation (Line or Highlight), the avatar can use both hands for explanation. Hence, the usage of left and right hands is not considered but only the coordinate value of the user's annotation is saved.

$$ThSIS = \frac{\sum TotalTimeOfAvatarBehavior}{\sum TotalTimeOfAvatarTalk} \quad (1)$$

Development of a Prototype. In this part, we present the result of applying the proposed context-based avatar control system. The proposed system is based on Windows XP and Java3D and can support XML based lecture notes and 10 types of symbols and comments. As shown in Figure 9, the annotation information created by the user is managed in XML format and a logical tree is created after preprocessing. High-level script language is created after analysis

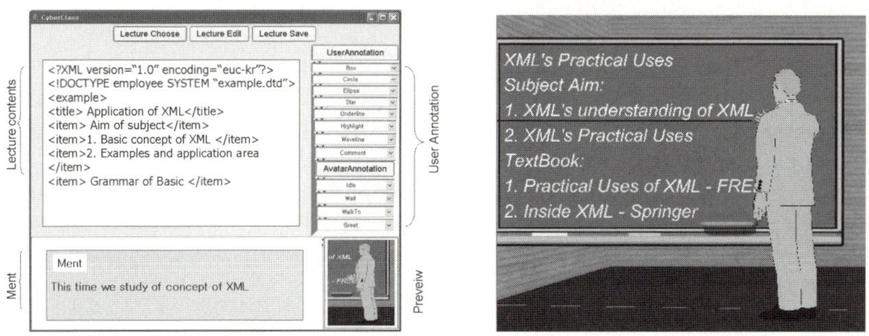

Fig. 9. Screen Shot of Prototype and Input Interface

of the logical tree. In the 1st step, the presence or the absence of user's annotation is distinguished by analyzing the structure information. When there are no annotations, the default value is performed. When there are annotations, rule-based high-level script language is created after considering the time it takes to perform and avatar behavior and the time it takes for reading the comments, spatial information, and the annotation type.

5 Experiment and Evaluation

In this section, the performance of the proposed system is evaluated. The experiment evaluated the naturalness of the avatar's behavior and accurate creation of underline and symbols for the annotation information created by one or multiple users. The proposed method was compared with an existing TVML editor. Because quantitative measurement of the proposed method is difficult, users were told to draw up a script that accompanies the avatar's behavior with underline or symbols and evaluated the results. The number of participants for the experiment was 20 and each user used each system and rated the degree of satisfaction on a scale from 0 to 10. 0 was very unsatisfactory and 10 was very satisfactory.

Question 1. A case where a comment is added to existing presentation material without any interaction between the presentation material and avatar such as a Symbol or Line. Question 2. A case where the behavior of the avatar is changed. There is some interaction between the presentation material and the

Table 4. Automatic Generated Sample Code

```
 <?xml version="1.0" encoding="UTF-8"?>
 <3DAvataControl>
 <AvatarInfo name="Tutor">
 <role>presentation</role>
 <Location>/Location/tutor.xml</Location>
 </AvatarInfo>
 <!-- No Annotation* -->
 <Presentation Title>
 <ment >
 This time we study of concept of XML.
 <DoMotionAvata = "Default">
 </ment>
 </Presentation Title>
 <Presentation presenter>
 <ment > My name is Gil Dong Hong, I am a cyber tutor. Nice to meet
every one.
 <DoMotionAvata = "Default" >
 </DoMotionAvata>
 </ment>
 </Presentationpresenter>
```

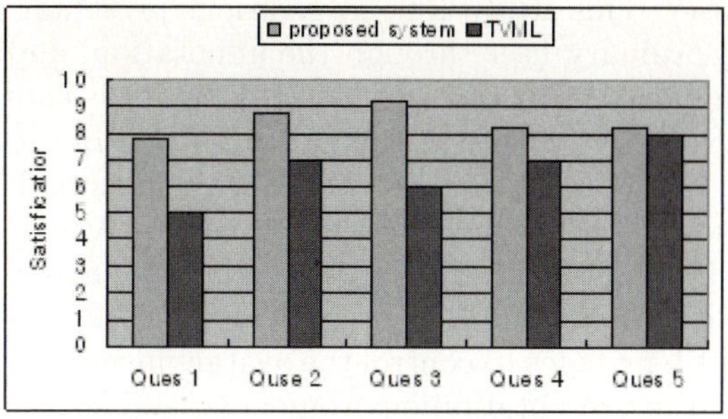

Fig. 10. Result of User Test

avatar but the interaction is very simple, such as pointing. Question 3. A case where a simple annotation such as Underline requires corresponding behavior from the avatar. Question 4. A case where the avatar performs simple behavior and an underline or an arrow is drawn on the presentation material and the comments of the avatar correspond to the behavior of the avatar. Question 5. A case where a multiple user environment is evaluated. Avatar behavior based on context information created by 10 people and a presentation is shown to 10 different users and they rate the naturalness and the effectiveness of the presentation. Our proposed system support a variety of symbol types that are not supported in existing systems. Also, users have to directly draw up text-based script language in existing systems whereas in the proposed system, script language is automatically created via an intuitive interface that is based on the WIMP concept.

Therefore, even non-professional users can easily control the 3D avatar and a more dynamic presentation is possible through a usage of a variety of annotations. From the user evaluation, it was demonstrated that the proposed system is superior to existing systems. In particular, the proposed system was proven effective in regard to Question 3, where various symbol types were created. In Question 3, because existing systems must use script language, much time is needed. However, in the proposed system, users use the intuitive interface to create symbols and do not have to write the script language, and as such, the proposed system was rated higher with respect to satisfaction. Question 5 tested the satisfaction rate when multiple users were involved and the rate of the proposed system was similar to that of other existing systems. In Question 5, the need for users to write script language was eliminated. Because of this, there was no noticeable difference between existing commercial systems and the proposed system.

6 Conclusion and Future Studies

In this paper, we study a control method of the avatar, which presents in 3D virtual spaces. When basic script languages or markup languages are used, there have been many difficulties to control a 3D avatar with the intention of the end user as a average user. This study is to save various presentation information of the end user as an ordinary user through the annotation, and to automatically generate script languages with the automatic generation module for the script based on the context. The avatar is designed by H-Anim in this study. The tutor avatar is designed by H-Anim composed of 15 segments and 11 joints. In this study, various presentation information to be used by CVaML, non-professional end users is saved in annotation and script language is created automatically in a script automatic creation module. It is applicable for the automatic generation method of the script languages to control the avatars in various fields. This paper limits the study to the presentation environments, and so has some limitations on the avatar behaviors and on the simple annotation. It requires to define the behaviors by the various avatars including those which forecast weather or report news, and to develop the types and the rules per annotation in the near future.

References

1. Rickel, J., Johnson, W.L.: Task-Oriented Collaboration with Embodied Agents in Virtual Worlds. Embodied Conversational Agents, 95–112 (2001)
2. Lester, J.C., Zettlemoyer, L.S., Gregoire, J.P., Bares, W.H.: Explanatory Lifelike Avatars: Performing User-Centered Tasks in 3D Learning Environments. In: International Conference on Autonomous Agents Proceedings of the third Annual Conference on Autonomous Agents, pp. 24–31 (1999)
3. Kantamneni Prasad, R., Narayanan, S.: Personalization of Information Retrieval through User Profiling. In: Proceedings of the IEEE International Conference on Systems, Man, and Cybernetics, Phoenix, AZ, pp. 3475–3478 (2001)
4. Web Annotations, `http://www.paulperry.net/notes/annotationns.asp`
5. Roscheisen, M., Mogensen, M., Winograd, T.: Shared Web Annotations as A Platform for Third-Party Value-Added Information Providers: Architecture, Protocols, and Usage Examples, Technical Report CSDTR/DLTR, Computer Science Department, Stanford University, `http://www-pcd.stanford.edu/`
6. Denoue, L., Vignollet, L.: An Annotation Tool for Web Browsers and Its Applications to Information Retrieval. In: Proceedings of RIAO 2000, Paris (April 2000)
7. Jung, T., Gross, M.D., Do, E.Y.: Annotating and Sketching on 3D Web models. In: Proceedings of International Conference on Intelligent User Interfaces(IUI 2002), San Francisco, California, USA, pp. 13–16 (2002)
8. Jung, T., Do, T.E.Y., Gross, M.D.: Redlining and Annotation of 3D Design Models on the Web. In: 8th International Conference on Computer Aided Architectural Design Futures 1999, pp. 81–98 (1999)
9. Arafa, Y., Mamdani, A.: Scripting Embodied Agents Behavior with CML: Character Markup Language. In: Proceedings of the 8th International Conference on Intelligent User Interfaces, pp. 313–316 (2003)
10. Kim, J.K., Sohn, W.S., Lim, S.B., Choy, Y.C.: A Layered Scripting Language Technique for Avatar Behavior Representation and Control. In: The 2004 AI Simulation and Planning, pp. 565–573 (2004)
11. Kim., G.Y., Kim, G.J., Choy, Y.C., Lim, S.B.: Modeling and Creating of Task-level Behavior. In: Motion, A. (ed.) Proceedings of Conference of KISS Fall, vol. 30(2), pp. 439–441 (2003)

12. Noma, T., Zhao, L., Badler, N.I.: Design of a Virtual Human Presenter. Computer Graphics and Applications 20(4), 79–85 (2001)
13. Noma, T., Badler, N.I.: A Virtual Human Presenter. In: Proc. IJCAI 1997 Workshop on Animated Interface Agents, pp. 45–51 (1997)
14. Nijholt, A., Welbergen, H.V., Zwiers, J.: Introducing an Embodied Virtual Presenter Agent in a Virtual Meeting Room. In: Proc. 23rd IASTED International Multi-Conference ARTIFICIAL Intelligence and Applications, pp. 14–16 (2005)
15. Shindo, Y., Matsuda, H.: Prototype of Cyber Teaching Assistant. In: IEEE International Conference on Advanced Learning Technologies, pp. 6–8 (2001)
16. Prendinger, H., Descamps, S., Ishizuka, M.: MPML: A Markup Language for Controlling the Behavior of Life-Like Characters. Journal of Visual Languages and Computing 15(2), 183–203 (2004)
17. Zong, Y., Dohi, H., Ishizuka, M.: Multimodal Presentation Markup Language MPML with Emotion Expression Functions Attached. In: Proceedings of the 2000 International Conference on Microelectronic Systems Education, pp. 11–13 (2000)
18. Marshall, C.C., Glovchinsky, G., Price, N.M.: Designing e-books for legal research. In: Proceedings of the first ACM/IEEE-CS Joint Conference on Digital Libraries, Roanoke, Virginia, pp. 41–48. ACM Press, NY (2001)
19. MS Reader. Microsoft (2001), `http://www.microsoft.com/reader`
20. RDF: Resource Description Framework, `http://www.w3.org/RDF/`
21. TVML (TV program Making Language), `http://www.nhk.or.jp/strl/tvml/`
22. Kshirsagar, S., Thalmann, N.M., Vuilleme, A.G., Thalmann, D., Kamyab, K., Mamdani, E.: Avatar Markup Language. In: Proceedings of the Workshop on Virtual Environments, pp. 169–177 (2002)
23. Humanoid Animation Working Group, The H-Anim Specification Version 200X, `http://www.h-anim.org`

Perception Based Representation of Female Avatar

Sergio Tasso, Osvaldo Gervasi, Francesco Scolastra,
and Giulianella Coletti

Department of Mathematics and Computer Science, University of Perugia,
via Vanvitelli, 1, I-06123 Perugia, Italy
{osvaldo,sergio}@unipg.it, coletti@dipmat.unipg.it

Abstract. We face the problem of building a model of a simulator to generate 3D virtual female body representation. To this purpose, we focus on some relevant indicators characterizing the definition of the measures of the various parts of a female body. We rely on a fuzzy modeling approach by using the interpretation (in terms of coherent conditional probability) of a membership function of a fuzzy set as a suitable likelihood. The components of the system are a simulator, implemented as a Web service, and a virtual reality environment based on X3D that allows the user to refine the body representation based on the users self-perception.

The avatar can be used for example both in e-commerce as a dummy which replaces one's own body for the measurements of virtual clothes and in crime investigation for the identikit construction based on a few clues.

1 Introduction

The study of the measures of the human body began in arts, in order to define some relationships among the various parts of the human body; nowadays it plays an important role in the design industry, in the realization of clothes, in ergonomics and architecture, where statistic data on the body size of the population are used to optimize the products. The modification of the lifestyles, eating habits and ethnic composition of the population, causes the modification of the averaged body measurements and therefore induce the systematic update of the statistical data.

In the present paper we describe the mathematical model, the simulator and the X3D environment designed to represent in a virtual world the female body according with her self-perception. The shape of the body is evaluated from few fuzzy expressions, like *"Height: tall"* and *"Waist: normal"*, and the measures of breast and hips, from which all remaining measures are evaluated by the simulator taking into account some statistical data related to the considered country and some medical information.

The simulator returns all evaluated measurements to the X3D application that enable the user to refine the X3D representation of the female body. The

O. Gervasi et al. (Eds.): ICCSA 2008, Part II, LNCS 5073, pp. 168–180, 2008.

refinement is made possible by some buttons that allow to increase or decrease some measures. In the virtual environment the action of moving the button right or left make an interpolation of the shape of the body from the current measure to the new measure evaluated from the button position. The animation produced is obtained interpolating some measures defined to classify the available categories for a given part of the body (i.e.: very lean, lean, normal, fat , very fat).

The paper is structured as follows: in section 2 the mathematical model adopted is described, in particular the insight of the coherent conditional probability and the expression allowing to combine probabilistic and fuzzy informations are illustrated. In section 3 the procedure followed for eliciting membership functions for the body measures of *height* and *waist* of girls is described. In section 4 the implementation of the simulator as a Web Service and the representation of the female body in a virtual world is described. Finally in section 5 some conclusions and future developments are described.

2 The Mathematical Model

Fuzzy set theory is a well known framework, introduced by Zadeh [1], for handling vague and imprecise information usually expressed in natural language.

One of most discussed problem to apply fuzzy set theory and fuzzy logic is the actual elicitation of the membership functions. The interpretation of the membership in terms of coherent conditional probability as a suitable likelihood give a good tool for solving this problem. Moreover it permits to combine in a natural way fuzzy information and statistical precise information and to update a membership function by using Bayes rule.

2.1 Fuzzy Sets and Conditional Probability

We refer to the state of information (at a given moment) of a real (or fictitious) person that will be denoted by "You". If X is a (not necessarily numerical) quantity with range C_X, let A_x be, for any $x \in C_X$, the event $\{X = x\}$. The family $\{A_x\}_{x \in C_x}$ is obviously a *partition* of the certain event $\Omega = C_X$.

Now, let φ_X be any *property* related to the quantity X : from a pragmatic point of view, it is natural to think that You have some information about possible values of X, which allows You to refer to a suitable *membership function* of the fuzzy subset of "elements of C_X with the property φ_X".

For example, if X is the numerical quantity expressing the diameter of balls in a box in cm and φ_X is the property "large", then the interest is in fact directed toward *conditional events* such as $E|A_x$, where x ranges over the possible value of the diameters, with

$$E = \{\text{You claim that } X \text{ has the property } \varphi_X\}\ , \quad A_x = \{\text{the value of } X \text{ is } x\}.$$

It follows that, while You may assign to each of these conditional events a degree of belief (subjective probability) $P(E|A_x)$, You must not assign a degree of belief $1 - P(E|A_x)$ to the event E under the assumption A_x^c (the value of X is not x), since an additivity rule *with respect to the conditioning events* does not hold. In other words, it seems sensible to identify the values of the membership function $\mu_{\varphi}(x)$ with suitable conditional probabilities.

Notice that this conditional probability $P(E|A_x)$ is *directly* introduced as a function on the set of conditional events (and without assuming any given algebraic structure). Is that possible? In the usual (Kolmorogovian) approach to conditional probability the answer is NO, since the introduction of $P(E|A_x)$ would require the consideration (and the assessment) of $P(E \wedge A_x)$ and $P(A_x)$ (assuming positivity of the latter). But this *could not be* a simple task: in fact in this context the only sensible procedure is to assign directly $P(E|A_x)$. For example, it is possible to assign the (conditional) probability that "You claim this number is small" knowing its value x, but not necessarily the probability that "The number has the value x" (not to mention that, for many choices of the quantity X, the corresponding probability can be equal to zero). These problems are easily by–passed in our framework.

2.2 Coherent Conditional Probability

The approach to probability used here (expounded, e.g., in [2], [3] and [4]: see also the book [5]) is based on *coherence* (a concept that goes back to de Finetti [7]). The starting point is a synthesis of the available information, expressed by one or more *events*: to this purpose, the concept of event must be given its more general meaning, that is expressed by a *proposition*. Moreover, events play a two–fold role, since we must consider not only those events which are the direct object of study, but also those which represent the relevant "state of information": in fact a set of *conditional* events, together with a relevant assessment of *conditional* probability, are the objects that allow to manage specific (conditional) situations and to update degrees of belief on the basis of the evidence.

A peculiarity (which entails a large flexibility in the management of any kind of uncertainty) of this approach to conditional probability is that, due to its *direct* assignment as a whole, the knowledge (or the assessment) of the "joint" and "marginal" unconditional probabilities $P(E \wedge H)$ and $P(H)$ is not required; moreover, the *conditioning* event H (which *must* be a *possible* event) may have *zero probability*.

Definition – Given a set $\mathcal{C} = \mathcal{G} \times \mathcal{B}^o$ of conditional events $E|H$ such that $\mathcal{G}$ is a Boolean algebra and $\mathcal{B} \subseteq \mathcal{G}$ is closed with respect to (finite) logical sums, and putting $\mathcal{B}^o = \mathcal{B} \setminus \{\emptyset\}$, then $P : \mathcal{C} \rightarrow [0, 1]$ is such that

(i) $P(H|H) = 1$, for every $H \in \mathcal{B}^o$,

(ii) $P(\cdot|H)$ is a (finitely additive) probability on $\mathcal{G}$ for any given $H \in \mathcal{B}^o$,

(iii) $P\big((E \wedge A)|H\big) = P(E|H) \cdot P\big(A|(E \wedge H)\big)$, for every E, $A \in \mathcal{G}$ and E, $E \wedge H \in \mathcal{B}^o$.

A conditional probability P is defined on $\mathcal{G} \times \mathcal{B}^o$: however it is possible, through the concept of *coherence*, to handle also those situations where we need to assess P on an *arbitrary* set of conditional events $\mathcal{C} = \{E_1|H_1, \ldots, E_n|H_n\}$.

The role of coherence is in fact that of ruling an extension process, starting from the classic *axioms for a conditional probability*.

Definition – The assessment $P(\cdot|\cdot)$ on $\mathcal{C}$ is *coherent* if there exists $\mathcal{C}' \supset \mathcal{C}$, with $\mathcal{C}' = \mathcal{G} \times \mathcal{B}^o$ ($\mathcal{G}$ Boolean algebra and $\mathcal{B} \subseteq \mathcal{G}$ closed with respect to logical sums) such that P can be extended from $\mathcal{C}$ to $\mathcal{C}'$ as a *conditional probability*.

Concerning coherence, a fundamental result is the following.

Theorem – *Let $\mathcal{K}$ be any family of conditional events, and take an* arbitrary *family $\mathcal{C} \subseteq \mathcal{K}$. Let P be an assessment on $\mathcal{C}$; then there exists a (possibly not unique) coherent extension of P to $\mathcal{K}$* if and only if *P is coherent on $\mathcal{C}$.*

Coherence of conditional assessments can be ruled by a fundamental *characterization theorem*, which is based on checking the compatibility of a suitable sequence of linear systems: for the sake of brevity, we avoid here any detail or deepening, and we just mention that a coherent conditional probability on an arbitrary family $\mathcal{C}$ can be characterized by suitably representing it (in any finite subset $\mathcal{F}$ of $\mathcal{C}$) by means of a *class* $\{P_\alpha\}$ of coherent *unconditional* probabilities giving rise to the so-called *zero-layers* (indexed by α). See, e.g., the book [5], and for a concrete application – to medical diagnosis – see [9].

Finally, notice that what is usually emphasized in the relevant literature – when a conditional probability $P(E|H)$ is taken into account – is only the fact that $P(\cdot|H)$ *is a probability for any given* H: this is a very restrictive (and misleading) view of conditional probability, corresponding trivially to just a modification of the so-called "sample space" Ω. It is instead essential to regard also the conditioning event H as a "variable", *i.e.* the "status" of H in $E|H$ is not just that of something representing a given *fact*, but that of an (uncertain) *event* (like E) for which the knowledge of its truth value is not required. In other words, even if beliefs may come from various sources, they can be treated in the same way, since the relevant *conditioning* events (including both *statistical data* and also – to use Zadeh's [6] terminology – *perception–based information*) can always be considered as being *assumed* propositions.

Roughly speaking, looking on a coherent conditional probability – acting as a membership function – as a general non-additive "uncertainty" measure $m(\cdot) = P(E|\cdot)$ of the conditioning events amounts to referring to what in the statistical jargon is called "likelihood", with its various "ad hoc" extensions from a point function to a set function. The problem is that without a clear, precise and rigorous mathematical frame, the likelihood "per se" is not a proper tool to deal with statistical inference and to manage partial and vague information (these aspects are discussed more deeply in [10]).

2.3 Taking into Account Hybrid Information

For a deepening of the above approach to fuzzy sets and for other relevant formal definitions, see [8], where not only how to define fuzzy subsets has been shoved, but also the counterparts of the basic continuous T-norms and the corresponding dual T-conorms have been introduced in a very natural way , bound to the former by *coherence*. In [11] the above approach has been used to realize, by focusing on some relevant aspects, a model which is suited for an effective evaluation of the quality of a health–care provider, e.g. clinical department, day hospital, etc..

Nevertheless the aim of that application is essentially descriptive, in fact it is not focused on the choice of one element of a class by taking into account a piece of information expressed in different ways (statistical data, fuzzy information and so on). For the application presented in this paper the final goal is exactly to find the element of a set (database), which has the highest probability to be very similar to an element of an other set (a real person for instance).

As discussed in [12], the above interpretation of fuzzy set as a coherent conditional probability provides a very natural method for choosing an element of C_X by using both probabilistic information and fuzzy information : precisely a possible probability on the elements of $\mathcal{C}_X$ (for instance their frequency or any distribution summarizing the available information) and fuzzy information, expressed by a membership function $\varphi_X(\cdot) = P(E_\varphi|\cdot)$. In fact we can choose the most probable element $x \in \mathcal{C}_X$ under the hypothesis E_φ. In fact by using Bayes theorem we can compute for every $x_i \in \mathcal{C}_X$) the value $P(A_{xi}|E_\varphi)$ as

$$P(A_{xi}|E_\varphi) = \alpha P(A_{xi})\varphi_X(x_i)$$

where $\alpha = \sum_i \varphi_X(x_i P(A_{xi})$.

So to obtain the goal it is sufficient to compute

$$max_i\{P(A_{xi})\varphi_X(x_i)\}. \qquad (1)$$

3 Assessing Membership Functions: An Exploratory Case Study

Considering the event E = "a girl claims a woman tall" and x ranging over all possible realizations of the variable X), we assign the membership function $\mu_E(x)$ by taking it equal to the probability of the conditional event $E|\{X = x\}$. This probability is assessed through the following procedure: the girl k of a chosen sample of girls is required to evaluate the degree of belief on the event E_k = "I claim a woman tall" (given $X = x$) by a number between 0 and 10, so we get a vector $(d_1, d_2, \ldots, d_n)$ (where n is the number of girls); then, putting

$$d = \frac{d_1 + d_2 + \ldots + d_n}{n} \,,$$

we assess $P(E|\{X = x\}) = d$. The collected information for normal height are shown in Table 1.

The same procedure can be applied to girls to find the analogous membership functions $\mu_F(x)$, $\mu_G(x)$, $\mu_H(x)$, $\mu_L(x)$ as a conditional probability, by considering, now, the event F = "a girl claims a women very tall" , G = "a girl claims a women short" H = "a girl claims a women very short" L = "a girl claims a women medium", as shown in Figure 1.

Similar procedures will be followed for eliciting membership function relate to waist values, as shown in Table 2 and Figure 2.

Table 1. Collected responses for medium height

	Height (cm)							
	145	150	155	165	170	175	180	185
d_1	0	3	6	10	9	6	0	0
d_2	0	1	7	10	10	6	1	0
d_3	0	2	5	10	9	6	0	0
d_4	0	3	6	10	10	5	0	0
d_5	0	2	7	10	10	5	1	0
d_6	0	2	6	10	9	6	0	0
d_7	0	3	6	10	9	4	2	0
d_8	0	1	5	10	9	5	1	0
d_9	0	2	8	9	10	6	1	0
d_{10}	0	1	6	9	10	5	1	0
d_{11}	0	3	6	10	10	4	1	0
d_{12}	0	2	8	10	9	4	1	0
d_{13}	0	1	8	9	9	5	2	0
d_{14}	0	2	6	9	10	6	4	0
d_{15}	0	3	7	9	9	6	1	0
d_{16}	0	1	7	9	10	6	3	0
d_{17}	0	2	7	10	10	3	2	0
d_{18}	0	0	5	10	9	5	1	0
d_{19}	0	1	6	9	10	6	1	0
d_{20}	0	2	6	9	9	5	2	0
d_{21}	0	1	7	10	9	6	3	0
$d/10$	0	0,18	0,64	0,96	0,94	0,52	0,13	0

Table 2. Collected responses for normal waist

	Waist (cm)							
	40	50	60	70	80	90	100	110
d_1	0	0	6	10	10	6	0	0
d_2	0	1	7	10	10	6	1	0
d_3	0	2	5	10	9	6	0	0
d_4	0	3	6	10	10	5	0	0
d_5	0	2	7	10	10	5	0	0
d_6	0	0	7	10	9	4	0	0
d_7	0	3	6	10	9	4	2	0
d_8	0	1	7	10	9	5	1	0
d_9	0	2	8	9	10	6	1	0
d_{10}	0	1	8	9	10	5	1	0
d_{11}	0	3	7	10	10	4	1	0
d_{12}	0	0	8	9	10	4	1	0
d_{13}	0	1	8	9	9	4	2	0
d_{14}	0	2	6	10	10	6	4	0
d_{15}	0	3	7	9	9	4	1	0
d_{16}	0	1	8	9	10	6	3	0
d_{17}	0	2	7	10	10	3	0	0
d_{18}	0	0	5	10	10	5	1	0
d_{19}	0	1	6	9	10	4	1	0
d_{20}	0	0	6	10	9	5	2	0
d_{21}	0	1	7	10	10	6	3	0
$d/10$	0	0,14	0,68	0,97	0,97	0,49	0,19	0

4 The Representation of a Female Body

We implemented an application based on the computational approach described so far in order to allow a user to represent in a virtual world a female body according to his/her perception of the body characteristics (how much tall how much lean, etc) expressed by the user in a fuzzy way. The detailed measurements of the various parts of the body are calculated by a simulator implementing the computational model as a Web Service. The complete set of the calculated measurements related to the various part of the body are returned to the (client) virtual world that represents the female body and enables the user to modify its shape until a satisfactory representation is achieved.

4.1 The Web Service Approach

A user can request the representation of a female avatar through a client connected to the Web Service server. The implemented Web Service is based on Java/SOAP technology. The server requires some basic information to the user, like height and waist. The user can choose whether to give in details to the system all measures or limiting them to a few hybrid information. Some values can be determined in a precise way (crisp, eg. age) others may be specified in a generic (fuzzy) way. In example the user to define the waist can express sentences like: "very lean", or "lean", "normal", "fat", "very fat". Higher will be the number of information given more precise will result the estimated measurements. According to the inputs entered, the simulator (based on both a statistical model

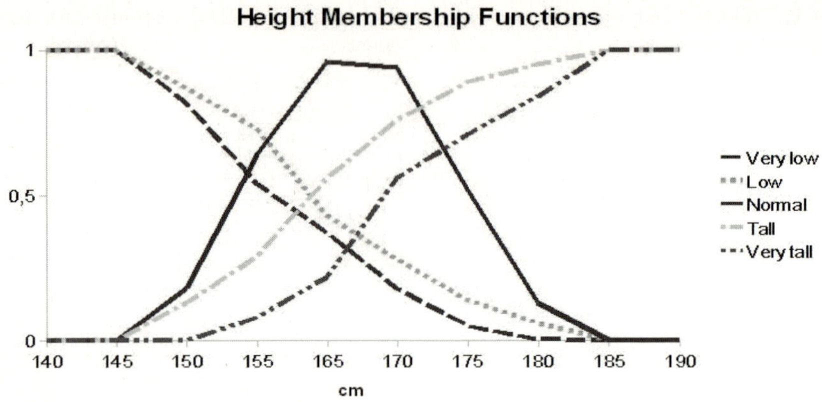

Fig. 1. Height Membership Functions

and on some formulas used in anthropometry and medicine), produces as output information a detailed list of measurements coded in XML format. The set of output information is passed to the client that through a virtual world implemented in X3D[13] will represent the avatar. If the user is not satisfied of the avatar representation, she/he can tune the various parts of the body through a set of controls. The relevant measurements of the body reshaped by the user will be passed back to the simulator to register how the evaluated measurements matched the user's perception of the body.

4.2 Calculation of Body Measurements

Many formulas are processed by the simulator because there are many different combinations of data that can be generated from the input values. For example, from only 4 input values (height, weight, age and a choice between hips, waist or breast), we can obtain more than 25 specific measures of the human body, largely sufficient for building a detailed three-dimensional model.

The following aspects must be taken into account:

- Data manually entered by the user will not be altered by formulas based on different data;
- If a female has some particular measure of her body, it might be wise to manually enter those values;
- The generated output data are to be considered valid only for women, being the taken proportions specific of a female body.

The most relevant formulas are those related to the calculation of size, age, height and waist. For example, knowing the size of the waist and shape, we can build developments in physical (hourglasslike, pearlike,etc.) or, knowing for example age, we can argue if the female is fit or sturdy. One of the most important formulas takes into account the Body Mass Index (BMI)[14].

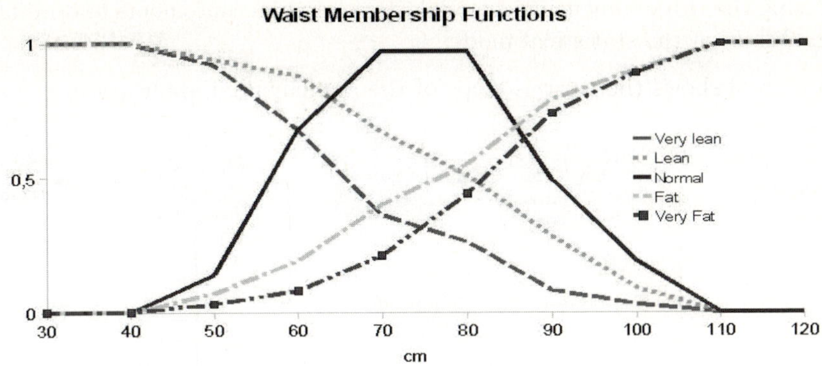

Fig. 2. Waist Membership Functions

Table 3. Ideal Weight formulas

Ideal Weight	Formula
Broca	$BW = h - 104$
Lorenz	$LW = h - 100 - ((h - 150)/4)$
Ottaviani-Travia	$0W = (0,672\dot{h}) - 50.95$
Van Der Wael	$VW = (h - 150)\dot{0}.75 + 50$

The weight (W) can be evaluated considering the ideal weights formulas shown in Table 3 as a function of the height h expressed in centimeters. We used the average value of these ideal weight values. The knowledge of the ideal weight is very useful because, when not provided directly by the user, we can use it to decide the development of the body, even to get more measurements. To calculate the approximate age is used the height and the circumference of the head (it is not necessary to provide the measure of the head, because this can be obtained by other values): in a adult person the head is approximatively the eighth part of the human body, in a teenager it is the seventh part, in a child only the fourth part. Other calculable measures are: the length of arms (always in proportion with the height and with the shoulders), the waist (derived by the measure of the hips or by the size of the breast), the size of the foot (and consequently of the shoe) and many others. Furthermore, given the foot length in centimeter, it is possible to evaluate the height, the size of the upper limbs and shoulders. The first proportion is: $height = 6 * foot$.

4.3 The Simulator Engine

The simulator program is composed of three main sections:

- Entering data and completion of the first model, starting with Fuzzy values
- Changing the data entered, with the possibility to choose between specific measures or directly from the X3D environment

- Saving the three-dimensional model, its related measurements and updating the tables of the statistical model.

The Figure 3 shows the various steps of the implemented application.

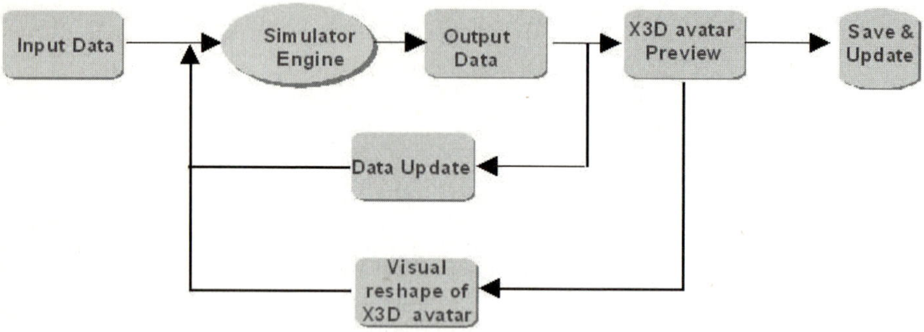

Fig. 3. Block diagram related to the implemented application

At the first input screen the user can choose his measurements into some Java input panel (implemented using `JcomboBox` classes) with the Fuzzy predetermined values. In particular, the user can indicate his height, waist, breast, and hips, as shown in Figure 4. After a fast computation, we evaluate the crisp values corresponding to the fuzzy information, which are shown in Table 4.

The crisp evaluation is based on various information. Firstly, the origin of the IP web client is used to determine the memberships related to the statistical data collected in the considered country. According to the frequency tables previously downloaded by international agencies the simulator detects the most probable value under the condition that one person claims "X has the property φ_X" (eg. "a female height 177 cm is tall"). The value is evaluated using the mathematical model discussed so far, choosing the maximum among the available values $P(X = x|E_\varphi) = \mu_\varphi(x) \cdot P(X = x)$, where P are the stored frequencies. Every goal of a specific session can be used to update the membership function by using a Bayesian procedure based on the interpretation of fuzzy subset as coherent probability shown in equation (1). The resulting values can be modified by the user editing directly the values into the related boxes.

Table 4. Crisp values corresponding to the fuzzy values specified in Figure 4

Measurement	Value
height (cm)	170
age (years)	26
weight (kg)	60
breast (cm)	84
waist (cm)	60
shoe size	38

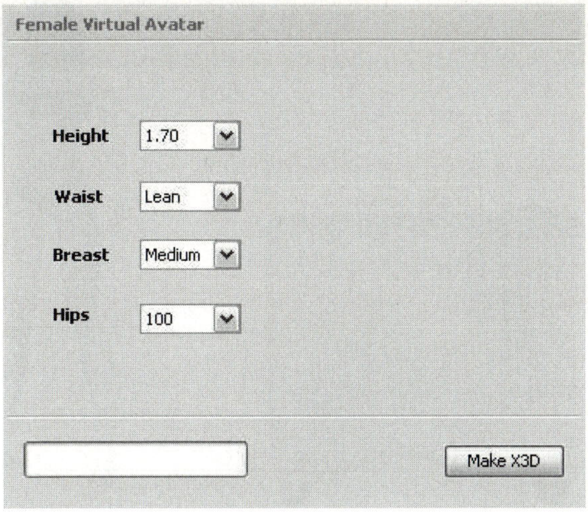

Fig. 4. Panel for the specification of the fuzzy input data

Measurement	Value	Measurement	Value
Height (cm)	170.00	Waist (cm)	60.00
Shoulders (cm)	42.50	Body	Regular
Head circumference (cm)	21.25	Broca Ideal weight	70.00
Age	26	Lorenz Ideal weight	65.00
Weight (Kg)	60.00	Ottaviani-Travia Ideal weight	63.29
Hips(cm)	108.00	Van Der Wael Ideal weight	62.00
Breast (cm)	84.00	Lorenz(2) Ideal weight	66.50
Bust (cm)	69.00	Weighted Ideal weight	65.00
Foot (cm)	28.33	Body Mass Index	20.76
Hand (cm)	21.25	Forearm (cm)	21.25
Arm (cm)	31.87	Upper limb (cm)	74.37
Wingspan (cm)	191.25	Shoe size	38

Fig. 5. Complete set of crisp measurements evaluated from the fuzzy information

When the values are confirmed, the user can see all possible output measurements both in crisp way (the available measurements are listed in Table 5) and/or in the preview X3D Virtual avatar representation (in Figures 6 and 7 the representations corresponding to two different body measurements are shown). These measurements are calculated by applying the formulas of anthropometry (presented in section 4.2). The avatar is shown with a chair for reasons of proportion.

The virtual world interact with the simulator through the Scene Access Interface (SAI)[15] implemented in Java.

The user is enabled to modify the virtual representation of the female body acting on some controls that interpolate the given part of the body among the ranges

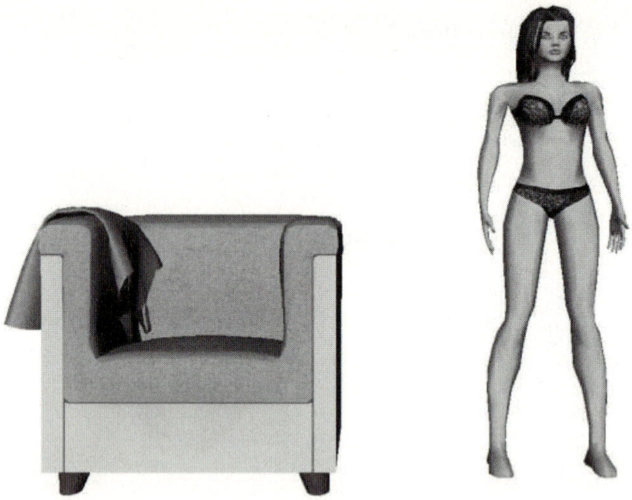

Fig. 6. X3D virtual world representing the avatar associated to a slim female

Fig. 7. X3D virtual world representing the avatar associated to a fat female

available for its measures (corresponding to the possible body sizes considered, according to the available membership functions). Once the user has reshaped the body in order to fulfill her self-perception, the resulting measures are returned to the simulator to update the statistics about the membership functions.

5 Conclusions

The implemented application for the representation of a female body is based on the user's perception of its characteristics specified by a membership function expressed in terms of coherent conditional probability as a suitable likelihood.

The representation, made using the X3D language, is suitable to be used through the web and can be used in various contexts: from the fashion, facilitating virtual fashion shows, to the identification of criminals and/or people under investigation, thanks to the reliability of the system to represent the female body.

Considering that the female body is much complicated to represent respect to the man body, the extention of our environment to be able to represent with the same degree of accuracy and friendliness the man body is an obvious and easy to implement development for the future.

Acknowledgments

Financial support from the the European Union through the COST action D37 (ELAMS working group) is acknowledged.

References

1. Zadeh, L.A.: Fuzzy sets. Information and Control 8, 338–353 (1965)
2. Coletti, G.: Coherent Numerical and Ordinal Probabilistic Assessments. IEEE Transactions on Systems, Man, and Cybernetics 24, 1747–1754 (1994)
3. Coletti, G., Scozzafava, R.: Characterization of Coherent Conditional Probabilities as a Tool for their Assessment and Extension. International Journal of Uncertainty, Fuzziness and Knowledge-Based System 4, 103–127 (1996)
4. Coletti, G., Scozzafava, R.: Conditioning and Inference in Intelligent Systems. Soft Computing 3, 118–130 (1999)
5. Coletti, G., Scozzafava, R.: Probabilistic logic in a coherent setting, Trends in Logic, n.15. Kluwer, Dordrecht (2002)
6. Zadeh, L.A.: Toward a perception–based theory of probabilistic reasoning with imprecise probabilities. Journal of Statistical Planning and Inference 105, 233–264 (2002)
7. de Finetti, B.: Teoria della probabilità, Einaudi, Torino (1970) - Engl. transl.: Theory of Probability, vol. 1 and 2. Wiley, Chichester (1974)
8. Coletti, G., Scozzafava, R.: Conditional probability, fuzzy sets and possibility: a unifying view. Fuzzy Sets and Systems 144, 227–249 (2004)
9. Coletti, G., Scozzafava, R.: The role of coherence in eliciting and handling "imprecise" probabilities and its application to medical diagnosis. Information Sciences 130, 41–65 (2000)

10. Coletti, G., Scozzafava, R.: Conditional Probability and Fuzzy Information. Computational Statistics & Data Analysis 51, 115–132 (2006)
11. Coletti, G., Paulon, L., Scozzafava, R., Vantaggi, B.: Measuring the Quality of Health-Care Services: A Likelihood-Based Fuzzy Modeling Approach. In: Mellouli, K. (ed.) ECSQARU 2007. LNCS (LNAI), vol. 4724, pp. 853–864. Springer, Heidelberg (2007)
12. Coletti, G., Scozzafava, R., Vantaggi, B.: Choosing on the basis of hybrid information (submitted for publication)
13. Brutzman, D., Daly, L.: X3D: 3D Graphics for Web Authors. Morgan Kaufmann Publishers (2007) ISBN-13: 978-0-12-088500-8, ISBN-10: 0-12-088500-X
14. Jackson, A.S., Stanforth, P.R., Gagnon., J., et al.: The effect of sex, age and race on estimating percentage body fat from body mass index: the Heritage Family Study. Int. J. Obes. Relat. Metab. Disord. 26(6), 789–796 (2002)
15. The Scene access interface (SAI) is defined in the Standard ISO/IEC, 19775-2:2004, Extensible 3D (X3D), Part 2. See the full text at the `http://www.web3d.org/x3d/specifications/ISO-IEC-19775-X3DAbstractSpecification_+_Amendment1_to_Part1/Part02/SAI.html`

Blended GIS Studies Model the Work of a Forest Expert

Markus Holopainen, Teppo Hujala, and Juhana Nieminen

University of Helsinki, Department of Forest Resource Management
P.O.Box 27, 00014 University of Helsinki, Finland
markus.holopainen@helsinki.fi, teppo.hujala@helsinki.fi,
juhana.nieminen@helsinki.fi

Abstract. Geoinformatics is a science involved with spatial data and spatial geographic information systems (GIS). At the Helsinki University Faculty of Agriculture and Forestry geoinformatics is widely practised in both the forest and agricultural sciences. Opportunities offered by the Web and the principles of research-orientated tuition have constituted the basis on which the planning and implementation of geoinformatics curricula have been carried out. The concept of blended tuition provides the means to combine these views in a consistent guideline. Tuition strives for pedagogical appropriateness by integrating local, remote and mobile studies into the learning environment, as well as activities resulting in research information. Here we present those blended tuition components and accumulative phases currently found in geoinformatics curricula. Our case further highlights activities for increasing student and tutor motivation and commitment to apply blended studies.

Keywords: Geographic information systems, remote sensing, tuition methods, blended tuition, forestry.

1 Preface

1.1 Blended Educational Methods at the Department of Forest Resource Management

Geoinformatics is a science involved with spatial data and spatial geographic information systems (GIS). At the Helsinki University Faculty of Agriculture and Forestry, geoinformatics is widely practised as an auxilliary science in both forestry (e.g. forest management, forest ecology, forest technology and forest economics) and the agricultural sciences. The Department of Forest Resource Management is responsible for the tuition of geoinformatics at the Faculty.

Opportunities offered by the Web and the principles of research-orientated tuition have constituted the basis on which the planning and implementation of geoinformatics tuition has been carried out at the Department in recent years. The concept of blended tuition provides the means to combine these views in a consistent guideline for developing tuition. Tuition strives for pedagogical appropriateness by integrating local, remote and mobile studyies into the learning environment, as well as activities resulting in research information. The methods and points of view to be integrated in a certain study period stem from the tuition goals of the period.

O. Gervasi et al. (Eds.): ICCSA 2008, Part II, LNCS 5073, pp. 181–190, 2008.

In the early stages, the students are automatically provided feedback regarding the GIS exercises carried out and given access to guided network discussions dealing with focused topics. Aquarium (Web-based) examinations provide a means to render studies more flexible, as well as support personal network tutelage (Heinimo and Hujala 2005). In the final stages, project-type studies are motivated by solving current practical problems of research interest. At best, the study course may result in a research article in addition to the study points gained.

Blended tuition components accumulate and are most prominent in those Master-level study projects where the authenticity of the topic evokes meaningfulness and the substances, methodologies and metaskills learnt during previous courses are utilized. These assignments offer a challenging but nonetheless safe environment to practise skills required by a researcher or other academic expert in a versatile manner.

Here we present those blended tuition components and accumulative phases currently found in geoinformatics curricula. Our focus lies in illustrating the principles and implementation of climax courses. In addition, an example is given on how Web opportunities and research-orientated tuition can be combined by blended tuition into a uniform principle for developing tuition. Our case further highlights activities for increasing student and tutor motivation and commitment to apply blended studyies.

1.2 Tuition of Geoinformatics at the Department – A Review of the Past Decade

Although the roots of remote-sensing-based forest inventory and drawing up of forestry maps stem from the early 20th century, forest geoinformatics was elevated to a field of research and tuition only in 1998 when the 'Forests in Geoinformation Systems' graduate school (Holopainen and Laasasenaho 2002) began functioning. At the same time, the Department was privileged to have Visiting Professor Jeremy Fried and Stuart Fish as his assistant for a year, who together founded and added an A-level geoinformatics curriculum to the Department's tuition offering. At this stage the study periods were largely technical base courses in nature, having the main purpose of training future researchers and tutors.

The next stage of development commenced in 2002 when a fixed-period, donation-funded professorship in geoinformatics was assigned to the Faculty of Agriculture and Forestry. As a result of this professorship, the geoinformatics curriculum was extended to the applied level and also included spatial applications related to the ecological and economical aspects of sustainable use of natural resources. In 2003 the Department was granted a tutorship in geoinformatics, which has made consistent development of tuition possible. This task was substantially aided during 2001 - 2004 by a half-time support person in Webtuition funded by the Centre of Tuition Technology.

Coinciding with the founding of the Department's new professorship was the founding of a second professorship in geoinformatics at the Faculty of Science, Department of Geography. A year later a fixed-time professorship in geoinformatics was established at the Helsinki University of Technology Department of Structural Engineering and Building. All these new professorships created, on their part, a basis for cooperation in this field in Finnlad. The 'Virtual University of Geoinformatics' project funded by the Ministry of Education has aided in orienting the tuition to blended and international direction. The third stage of development will be to extend forest geoinformatics tuition to the postgraduate level.

1.3 Developmental Steps of Web-Based GIS Tuition

From the very beginning, tuition in geoinformatics at the Department of Forest Resource Management was intended to utilize modern technology as much as possible. Since practically all students were campus students at that time, the initial role of the Web was to support local tuition. In practice, this meant delivering tuition material via the local network and public Web pages. The next step was the introduction of Web course tools (WebCT) which was, however, not widely used.

After this technology-driven start, emphasis was placed on developing the pedagogical aspects of GIS Web tuition, examples of which are the 'Illustrative spatial information on the Web' project (see Jäppinen 2002) which won the 2001 Tuition Technology Award at the University of Helsinki and the Mobile Studying Pilot (Hujala et al. 2003 a, b) which was devised in cooperation with the Centre of Tuition Technology.

Among others, flexible Web-based assessment or the so-called examination aquarium (Heinimo and Hujala 2005), guided tutorial Web discussions (see Hujala 2005) and preparing for the use of Wiki's and on-line messaging systems for pedagogically justified tuition are examples of the progress made in recent years in Web-based tuition. From using the web for supporting local tuition, a gradual transformation occurred, resulting in a combination of various forms of tuition. This mosaic of separate tuition components has since been developed from its initial holistic state to a more blended entity of meaningful study assignments and forms.

2 To a More Blended Tuition

2.1 Research-Oriented Tuition as a Foundation for Blended Curriculum

Flexibility and research orientation have been adopted as basic principles for developing GIS tuition at the Department. The aim has been to train students as experts who master the research-oriented way of working, the core of which encompasses critical reasoning and the ability to solve problems (Hujala et al. 2007). To achieve this, utilization of a colleague network in addition to the researcher's own expertise has been considered elementary (see also Beriter and Scardamalia 1993).

The principles for applying research-oriented tuition at the Department were previously defined as follows (Nieminen et al. 2005):

1. Studies involve solving of concrete research problems
2. The tutor's role is to act as a guiding colleague
3. Scientific reasoning is encouraged during the learning process
4. Interaction and peer assessment play a significant role in problem solving
5. Verbal presentations and research reviews are fundamental ways of working
6. Advanced level students may take part in real research projects

Research orientation was interpreted as comprising three components (Hujala et al. 2007): 1) tuition is based on methods developed or recommended by scientific investigations, 2) learning practices in research work are integrated into the tuition and 3) studies include processes resulting in research results. While considering blended tuition goals at the Department, Levonen et al. (2005) emphasized enabling (remote studyies, opportunity to choose) and to some extent rationalizing (distribution of material, automatic feedback, Web discussions). Modification of tuition practices was also considered, but the results have so far been achieved through courses integrated into research projects and renegotiations of the roles of both tutors and students (Brew 2003).

2.2 Fluent Integration of Blended Tuition and Research Orientation

Since research-oriented tuition is by nature a holistic approach to academic tuition, blended tuition can easily be considered as a concept and tool for promoting it. Blended tuition combines research orientation as well as information and communication technologies into a consistent directive for developing tuition. The concept of fluency (see Biggs 2003) is also a fundamental element in blending geoinformatics tuition. In other words, the focus has been on how well learning aims, tuition methods, learning assessment and the learning atmosphere have been in line with the blended tuition applications selected. Since the nature of study objectives deepens as studies progress, the implementation of blended tuition also changes when shifting from basic to advanced studies. At the curriculum level blended tuition in geoinformatics can be considered as a cumulative fusion in which the various forms of tuition and study are most fruitful and cooperative in courses at the most advanced level.

During the next study period, the students are guided into an actively theme-driven and time-scheduled educational Web discussion. Participation in the discussion is assessed as a part of the course performance, using criteria agreed on at the beginning of the period. Exercises involving the acquisition of spatial information in the field aid in adopting technical themes and link theoretically learned issues with practice. In addition to traditional lecture examinations, courses can be taken via a computerized examination aquarium system in which students are not bound to fixed examination dates. The opportunity offered to the tutor to give personal feedback can be considered as another major advantage of the examination aquarium system. This improves the learning outcome and meaningfulness of the examination decidedly (see Heinimo and Hujala 2005).

Merging of research and tuition is especially realized in forest inventory advanced-level courses involving geoinformatics. At its best, the outcome of a course has consisted of a research manuscript composed by both tutors and students, some of which have been published in refereed scientific series (e.g. Kangas et al. 2004, Leino et al. 2006). This has also motivated the students, since work carried out during the course has become meaningful. Experience has also shown that authentic of study tasks motivate. At the same time, students get an idea of what is expected from them when arranging their graduate thesis and what a researcher's work entails in practice. Courses in geoinformatics also offer a genuine opportunity to acquire research material for arranging their theses.

Some of the courses are arranged in cooperation with organizations such as forest centres, the Forest Service, and the cities of Helsinki and Espoo. Thus research information is linked with practical work. The principles of blended tuition can include all the geoinformatics tuition given by the Department. Blended tuition is therefore not a single technique, but a learning environment encompassing how the Department operates, the final objective of which is to develop students into critical experts in this field of science.

Development of blended tuition in geoinformatics has been influenced by research carried out by Boud and Brew (1995) and Brown et al. (1997), who emphasized the connection between applied methods of assessment and learning results. Several methods of assessment have, indeed, been utilized geoinformatics courses given by the Department, the purpose of which has been to aid in deeply directed learning. The methods used have included preliminary tests and examinations, assessment discussions, criticism and evaluation of all aspects of a course, and an interstudent feedback system.

Taras (2003) suggested that self-assessment included in a course should be combined with peer assessment and assessment of the tutor to aid in the development of metacognitive skills and good learning. The merging of these various methods of assessment was initially tested at the Department and development of these methods continues.

2.3 Blended Tuition in Geoinformatics

The blended tuition presented here is an example of blending at the curriculum level (Ross and Gage 2006). Although the planners have inserted a multitude of blended elements into the courses (Table 1), personal choices of the students finally decide on how blending is applied to each student. Student orientation is one of the major benefits stemming from the blended tuition resulting from a decade of development.

Tuition development has not been restricted to courses leading to a degree. Future degree tuition and more intensive international cooperation in tuition have been prepared by organizing specialized courses in which tuition opportunities provided by social media and advanced Web services have been demonstrated. These types of specialized courses have been organized together with the Viikki Tropical Resources Institute (VITRI), which is part of the Department of Forest Ecology. Another example is the GIS_AL3 course, which is a specialized geoinformatics course directed to international tutors and funded through EU Asialink-tuition cooperation.

During the courses, emphasis was placed on how remote and Web learning environments can be utilized to improve interuniversity tuition and to make it possible for 'virtual exchange students' to take part in courses given by cooperative universities. The courses encompassed the most recent dynamic Web tools for enhancing tuition; the techniques utilized included social bookmarks, dynamic maps (actively linked to the Google Maps service), Web presentations (SlideShare), Web video and e-conference services (YouTube / Adobe ConnectPro) and the BlackBoard learning environment as examples of flexible cooperative tools. In combining tuition with new

Table 1. Geoinformatics courses given by the Department of Forest Resource Management and the blended tuition methods included

Basic studies (annually)
GIS1 – Desktop GIS* (may be taken remotely)
GIS2 – GPS in agriculture and forestry*
GIS3 – Remote sensing 1 (may be taken remotely)
GIS4 – GIS analyses 1, ArcGIS* (may be taken remotely)
Advanced studies
GIS10 – Geostatistics and modelling
GIS11 – Applied GIS analysis*
GIS12 – Advanced remote sensing (may be taken remotely)
GIS13 – Logistics and business GIS*
GIS14 – Environmental GIS (in cooperation with the Faculty of Biosciences)*
GIS15 – WebGIS (Mapserver)*
MINV12 – Specialized remote sensing*
Marv4/2 – Forest inventory project
*The tuition language is either Finnish or English in all courses.
Basic study courses utilizing blended tuition methods
GIS1 – WebCT/Blackboard (exercises and Web essay)
GIS2 – OpenWWW (discussion – BSCW / Web knowledge searching / examination aquarium)
GIS – WebCT/Blackboard (exercises / examination aquarium)
GIS4 – WebCT/Blackboard (exercises / automatic correction system for map feedback / Web essay / examination aquarium)
Marv4/2 – Research project, distribution of material by the Web, open newsgroup discussions
Advanced study courses utilizing blended tuition methods
(all of which utilize the WebCT/Blackboard for distributing course material and the examination aquarium for taking exams)
GIS10 – Exercise wind-ups / Web commenting
GIS12 – Distribution of exercises / examination in the WebCT/Blackboard
GIS13 – Handing back and correcting of exercises via the WebCT/Blackboard
GIS14 – Webvideos / Handing back of exercises
MINV12 – Research Project, Web distribution of material, open newsgroup discussions

workgroup exercise tools, such as OpenSpace, WorldCafé or Full Circle, the role of the tutor has become more of a facilitator or mentor of the learning process in some courses.

By applying these techniques participants in the GIS1, GIS2, GIS4 and GIS_AL3 courses have produced information content for the courses by their own collective expertise. These new exercises have also given the students an opportunity to experience group work benefits at their best.

We conclude that the level of education offered in geoinformatics by applying the blended model forms a complete whole, from basic to advanced studies and ending at tutor training.

3 Examples of Basic and Advanced Forest GIS Courses

3.1 Sample Course GIS2 - GPS in Agriculture and Forestry

The objective of this basic couse in geoinformatics is to adopt the basics of satellite positioning systems and to learn how to acquire geographic data with such systems. In addition, the aim is to become acquainted with the most recent research in satellite positioning and its applications in agriculture and forestry. To pass the course, students are required to participate in exercise sessions and guided Web discussions, as well as pass the exam and carry out personal exercise tasks. Participation in the Web discussions can be substituted by writing an essay (but few have exercised this option). The motivation, implementation and participant assessment of guided Web discussions were described in further detail by Hujala (2005). In implementing the study period, issues such as a researchlike way of working, sharing of interstudent expertise and immediate application of learning theory in practice have emerged.

In course feedback, personal involvement has been appraised above all else and the complexity of the research articles studied the most criticized. From the blended tuition point of view, this study period is a highly typical course involving the application of local and Web tuition features.

3.2 Sample Course MINV12 - Advanced Remote Sensing; 3D Mensuration of the Forests

The objective of the Master and postgraduate level Advanced remote sensing course (MINV12) is to familiarize the student with currently on-going research in remote sensing, which can be utilized in the mensuration, inventory, monitoring and management of forest resources.

During the course, students become acquainted with the evolution of a small-scale research project. The course begins with a week-long local study period during which the students are oriented to a research project via lectures and exercises. After this, teams of two commence work on the projects which deal with tree-, plot- or compartment-level remote sensing. Field material acquired in the Forest resource inventory project course (Marv4/2), a prerequisite for this course, is used as research material. The work carried out is reported in accordance with a research report. Finally, a wind-up session is organized, during the which project results are presented by the teams. The project reports are later elaborated into research publications in which the students' input is also acknowledged when appropriate.

Web discussions carried out in the open newsgroup environment by the groups and tutors are fundamental to the course. During the course tutors are available for consultation via the Web in almost real-time. A common geoinformatics tuition electronic feedback system is also used during the course.

4 Summary and Deliberation

4.1 The Role of Blending Local Tuition and the Web

The objective of blending local tuition and the Web is to serve the (learning) objectives of the course. Completeness is also an aim: one blending, not separate segments.

Blending is also used to introduce flexibility, i.e. an attempt is made to tickle different study preferences. A versatile multitude of tuition methods also adds to and maintains student motivation and commitment. Research orientation is also stressed for, i.e. issues are found out in a way the researcher would. In addition to that of a researcher, blended tuition is used to model the work of a practical geoinformatics expert in a safe study environment.

4.2 Assessment Triangulation

Development of learning assessment has played a central role in the planning of blended tuition in geoinformatics. This has been supported on one hand by pedagogical assessment knowledge provided by the 'Flexible Web-based assessment practices' -project and on the other by focusing on the assessment methods as a part of the streamlining process. Another principle has been to utilize several methods of assessment, the procedure of which is called assessment triangulation.

On one hand the objective has been to assess skills in a streamlined manner and on the other to channel students into reaping the benefits of different study methods. Automatic feedback provided in the geographic analysis exercises has above all provided immediate feedback for the students, which is of great importance while learning technical basics and concepts. Continuous assessment provided by the exercises is a form of adaptive local tuition, in which the feedback given by the guiding tutor and peer discussions play a central role. Grading of Web discussions, in turn, motivates working in a Web community: a constructive, querying, dialoguelike and helpful participation is rewarded. Assessment of Web exams encourages tutors to give constructive feedback via the Web which, at best, can be personal. Assessment of project reports, carried out in some study periods, encourages students to adopt an investigative, holistic and cooperative way of working as well as a critical point of view.

4.3 Observations, Feedback and Challenges

Based on the feedback acquired during the geoinformatics courses given so far, we conclude that students above all appreciate concrete work. This arguably positive feedback has given rise to the challenge of how to motivate students in critical conceptual reasoning when the enthusiasm lies in the 'skills of the hand'. Student feedback concerning Web tasks has, in turn, been highly heterogeneous. The benefits and drawbacks associated with Web tasks are experienced differently, depending on the student's current life and study situation. Remotely carried out Web tasks clearly suit students arriving from other departments or minor degree students and students having either part-time or full-time jobs. On the other hand, the significance of local tuition is especially emphasized by the Department's own basic students.

A major future tuitional challenge is how to increase the meaningfulness of blended tuition methods and how the benefits achieved can be shown to the tutors as well as the students. Experiences gained so far has shown that combining research with blended tuition is a challenging task. The resulting workload for both students and tutors is greater than that of traditional courses. In addition, production of new knowledge always involves a risk of failure which may, in turn, affect the outcome of the course. It is, however, reasonable to further the merging of tuition and research, because it provides a means to offer students opportunities in research-based learning and also a way to motivate tutors to apply blended tuition, i.e. an investigative way of tutoring.

Development of geoinformatics tuition is also, in part, related to a Helsinki University internal project launched for developing the tuitional quality assessment system (e.g. the feedback system). A future web-based feedback system can be automated to such an extent that a certain questionnaire subset may be used to obtain useful information for all courses. Development of the feedback system and its linking with normal course and study registration processes enables a flexible way to gather, file and analyse feedback.

Helsinki University forestry departments have traditionally been internationally orientated. In recent years, these departments have increased their international activity, e.g. by taking part in the 'Finnish - Russian Cross Border University' project involving nine universities and by founding their own forest sciences Master-level study programme tutored in English (MSc in Forest Sciences and Business). In addition, the Department of Forest Resource Management has taken part in a geoinformatics Master-level study programme tutored in English and co-ordinated by the Department of Geography.

Blended tuition methods in geoinformatics are also useful from the international point of view. They support the community aspects of the English Master programmes. In addition, Web courses constructed in recent years provide foreign students with an opportunity to take part in geoinformatics studies, ranging from the basic level all the way to the Master and post-graduate levels. This type of entirety is probably quite unique in Helsinki University. Web tuition already enables cooperation at the Nordic level. The Asialink project has provided a base on which to build individual courses and even study entities for a much larger market.

Recently developed blended tuition in geoinformatics will play a decisive future role; flexibility, reusability and the opportunity to transform the roles of tutors and students among its benefits. Furthermore, blended tuition activities enhance students' own study efforts; especially desired are genuine situations, such as combining of multiple information sources, joint activities, sharing of expertise and skills for mediated interaction.

The need for new knowledge and research as well as the opportunities for developing new innovations are especially high in interfaces between different scientific disciplines. The Helsinki University campuses of Viikki and Kumpula together make up one of the most prominent centres for research, tuition and utilization of geoinformatics currently found in Finland. Future cooperation in research and tuition activities between these campuses should be furthered, thus ensuring the necessary qualifications for providing high-quality tuition. The work carried out so far in developing blended tuition in geoinformatics can be utilized for integrating various faculties. Web discussions, Web examinations and on-line seminars are especially applicable to cooperation in tuition between campuses (and universities). In fact, here their benefits may be better revealed than in traditional local tuition-based blended tuition.

References

Bereiter, C., Scardamalia, M.: Surpassing ourselves: An inquiry into the nature of expertise. Open Court, Chicago (1993)

Biggs, J.: Teaching for quality learning at university, 2nd edn. Society for Research into Higher Education & Open University Press (2003)

Boud, D., Brew, A.: Developing a typology for learner self assessment practices. Research and Development in Higher Education 18, 130–135 (1995)

Brew, A.: Teaching and Research: New relationships and their implications for inquiry-based teaching and learning in higher education. Higher Education Research and Development 22(1), 3–18 (2003)

Brown, G., Bull, J., Pendlebury, M.: Assessing Student Learning in Higher Education. Routledge, London (1997)

Heinimo, S.-M., Hujala, T.: Flexible Web-Based Assessment Methods - A Virtual University Project at the Department of Forest Resource Management at Viikki Campus. In: Szåcs, A., Ingeborg, B. (eds.) Proceedings of EDEN 2005 Annual Conference, Helsinki, Finland, European Distance and E-Learning Network, pp. 469–471 (2005)

Holopainen, M., Laasasenaho, J. (eds.): Metsät paikkatietojärjestelmissä –tutkijakoulu 1998-2002. Helsingin yliopiston metsävarojen käytön laitoksen julkaisuja, vol. 30 (2002)

Hujala, T.: Ohjattu BSCW-keskustelu osana kurssisuoritusta. Piirtoheitin 1/2005 (2005), `http://www.valt.helsinki.fi/piirtoheitin/bscw6.htm` ISSN 1795-3413

Hujala, T., Nieminen, J., Holopainen, M.: Blended Learning in a Research Intensive University – A Case Study from the Field of Geoinformatics. In: Proceedings of EDEN 2007 Annual Conference, Naples, Italy, European Distance and E-Learning Network (2007)

Hujala, T., Kynäslahti, H., Seppälä, P.: Creative Learning" – Mobility in action. In: Kynäslahti, H., Seppälä, P. (eds.) Mobile Learning, pp. 105–114. Edita Publishing Inc. (2003a)

Hujala, T., Kynäslahti, H., Seppälä, P.: Mobiiliopiskelua metsänarvioinnissa. Teoksessa Koski-Kotiranta, S. & Kynäslahti, H (toim.) Tukevasti verkko-opetukseen. Yliopistopaino, 189–201 (2003b)

Jäppinen, S.: Verkko-opetuksen haaste: Älä rakastu välineeseen! Yliopistolainen 4/2002, s. 6–7 (2002)

Kangas, A., Aakala, T., Alanen, H., Haavisto, M., Heikkilä, J., Kaila, A., Kankaanpää, S., Kämäri, H., Leino, O., Mäkinen, A., Nurmela, E., Oksa, S., Saari, A., Tikkanen, S.-M., Wathén, M.: Lahopuuinventoinnin menetelmien vertailu Nuuksion ulkoilualueilla. Metsätieteen aikakauskirja 1/2004, 43–51 (2004)

Levonen, J., Joutsenvirta, T., Parikka, R.: Blended Learning - Katsaus sulautuvaan yliopisto-opetukseen. Piirtoheitin 2/2005 (2005)

Leino, O., Holopainen, M., Mäkinen, A., Tuominen, R., Happonen, H., Kiviaho, T.: Pysty- ja maalahopuuston inventointi relaskooppiotannan avulla. Metsätieteen aikakauskirja 3/2006, 383–390 (2006)

Nieminen, J., Hujala, T., Holopainen, M.: Forestry Research as a basis for GIS related teaching at the University of Helsinki. In: Proceedings of Research-Based Teaching in Higher Education, University of Helsinki, Finland, pp. 73–76 (2005)

Ross, B., Gage, K.: Global Perspectives on Blended Learning. In: Bonk, C.J., Graham, C.R. (eds.) Handbook of blended learning: Global perspectives, local designs, Pfeiffer Publishing, San Francisco (2006)

Taras, M.: To Feedback or Not to Feedback in Student Self-assessment. Assessment & Evaluation in Higher Education 28(5), 449–565 (2003)

Towards Standards-Based Processing of Digital Elevation Models for Grid Computing through Web Processing Service (WPS)

Sandra Lanig, Arne Schilling, Beate Stollberg, and Alexander Zipf

Department of Geography, Chair of Cartography, University of Bonn,
Meckenheimer Allee 172, 53115 Bonn, Germany
lanig@geographie.uni-bonn.de, schilling@geographie.uni-bonn.de,
stollberg@geographie.uni-bonn.de, zipf@geographie.uni-bonn.de

Abstract. Digital Elevation Models (DEM) and 3D spatial data plays an important role in typical earth science applications. Numerous simulations, e.g. flood modeling, and spatial analysis, requires very exact terrain data. During the acquisition of these data, for an example by means of laserscanning, very large data sets results due to the high measuring point density (up to four points per square meter). Current classical Geo-Information-System (GIS) software cannot manage the demand of processing and analyzing these massive raw terrain data. A lack of computing power may appear. There is a need for sophisticated data processing techniques. For this purpose the use of Grid Computing is a good choice to accomplish high processing performance and storage capacity. To process these massive raw geodata we develop a range of terrain Web Processing Services (WPS) which are made available as Grid services.

Keywords: WPS, SDI, Grid Computing, OGC, Web services, GIS, processing, DEM.

1 Introduction

The need for computing power is steadily rising within Geosciences through the increasing volume of available geodata. Furthermore there is a range of applications requiring high performance computing, e.g. disaster management tasks. One example for the increasing demand is the processing of Digital Elevation Models (DEMs) and 3D spatial data which play an important role in typical earth science applications. Numerous simulations, e.g. flood modeling, are based on DEMs and 3D-models. For this purpose very computationally intensive preprocessing operations are necessary for the handling of massive amounts of raw geodata. This means that an efficient usage and optimization of current existing IT resources is necessary. Nowadays, these challenges are met by using Grid Computing technologies and development of so called Grid infrastructures for the combination of distributed computing power. Thus, the project Spatial Data Infrastructure-Grid (SDI-Grid) focuses on the efficient processing of spatial data by grid computing technologies and the integration of an SDI into a Grid environment.

O. Gervasi et al. (Eds.): ICCSA 2008, Part II, LNCS 5073, pp. 191–203, 2008.

Currently, distributed spatial data is accessed by services based on the specifications of the Open Geospatial Consortium (OGC). Visualization, vector and raster data access, along with the ability to search for spatial data is covered by the established OGC standards Web Map Service (WMS), Web Feature Service (WFS), Web Coverage Service (WCS), and Catalog Service for the Web (CSW), whereas a recognized standard for distributed spatial data processing was missing for a long time. This issue was addressed by the development of a Web Processing Service (WPS), which has been approved as official standard by the OGC in December 2007 [**Error! Reference source not found.**].

For offering a variety of DEM processing functionalities based on OGC Web Services (OWS) standards, we define and implement a set of such WPS processes. The huge amount of terrain data (especially laserscanner data) for the generation of a DEMs as well as modeling spatial processes based on the DEMs require an enormous demand on IT resources. Particularly during the laserscanning supported capture of three-dimensional data for modeling DEMs, very large data sets results in consequence of the high measuring point density (up to four points per square meter). The potential of this spatial data is increased by the design of services for the preparation, fusion and processing, based on grid technologies. Currently we are developing a range of terrain Web Processing Services (WPS) for generalization, geotessellation, spatial partitioning, converting and semantic streaming (sorted by relevance) of DEM data within the project SDI-Grid (www.gdi-grid.de). These services will be implemented into the German-Grid (D-Grid) Infrastructure. This kind of implementation accounts for various challenges concerning the combination of OGC and Grid standards. Especially the integration of OGC Web Services (OWS) into an Open Grid Services Architecture (OGSA) involves some obstacles which will be brought up later in this contribution.

The remainder of this contribution is structured as follows: Section 2 introduces basic concepts of both Grid Computing and SDIs and discusses how these can work together. Also a short overview of the project SDI-Grid is given. Section 3 gives an introduction of the OGC WPS specification, in particular aspects of processing DEMs and 3D-data with WPS. Section 4 demonstrates proposals for realizing DEM functions as WPS and section 5 illustrates a flood modelling simulation as an example scenario. The last section describes the broader context of related research activities.

2 Grid-Computing and SDI

The project SDI-Grid started in summer 2007 and is funded by the German Ministry for Education and Research (BMBF). The project focuses on solutions for efficient integration and processing of spatial data based on GIS and SDI technologies in a working SDI-Grid infrastructure. This connection of SDI and Grid standards is causing a couple of problems, which are for instance differences in service communication, the implementation of security mechanisms, and definition of service interfaces.

Concerning the service communication, there are differences in the service discovery, description, and messaging. The Grid community follows the IT mainstream. Grid Services are described by the Web Service Description Language (WSDL) and communicate by means of Simple Object Access Protocol (SOAP), both standards of the W3C. On the other side, SDI Web services are based on standards of the OGC. All OGC Web Services (OWS) implement the mandatory operation GetCapabilities for receiving a service description and communicate via the Hypertext Transfer Protocol (HTTP) methods GET and POST. This fundamental difference in the description and messaging methods leads to interoperability problems between OWS and "normal" Web Services. The OGC is aware of this drawback and recently started working on solutions. Lately approaches are developed to describe OGC Web Services with WSDL and messaging with SOAP.

Table 1. W3C and OGC Web Services

Web Service	W3C	OGC
Service description	WSDL	GetCapabilities
Messaging	SOAP/XML	HTTP GET/POST, KVP, GML

Furthermore there are two substantial differences between OGC Web Services and Grid Services. OWS are persistent and stateless, Grid services are transient and condition afflicted. To implement stateful Web Services, the Web Service Resource Framework (WSRF), a collection of specifications under the support of the Organization for the Advancement of Structured Information Standards (OASIS), is used.

For standardization of grid-based applications, the Global Grid Forum (GGF) [3] developed the Open Grid Service Architecture (OGSA). OGSA is a concept for the development of an open Grid infrastructure based on the principles of Service Oriented Architecture (SOA), taking the specifications of the WSRF into account. As a basic implementation of these concepts the software Globus Toolkit 4 (GT4) is used within the SDI-Grid project. The GT4 is an implementation of the Open Grid Service Infrastructure (OGSI), which is the first formal and technical specification of OGSA. It is a collection of software components which provide many of the building blocks that are necessary to create a Grid-based application. For developing Grid applications the GT4 primarily includes a number of high-level Grid infrastructure services that implement interfaces for managing processing power, storage, and other resources [19]

The first task is to achieve that OGC Web Services and the GT4-based Grid-infrastructure can work together. Thereby the SDI-Grid components are adapted to the D-Grid infrastructure. In the following we focus on the OGC Web Processing Service (WPS), since it represents the central component in SDIs for processing data and performing nearly arbitrary calculations. Therefore this Service is the first and major service, which will be connected to the Grid middleware, thus bridging the gap between the GRID world and the OGC world.

3 OGC Web Processing Service (WPS)

The OGC WPS standard was developed for offering any kind of GIS processing functionality by a standardized interface. A WPS may provide simple calculations (e.g. the calculation of a buffer) as well as complex computations (e.g. the simulation of a climate model), whereby a WPS is able to handle more than a single process. That means a WPS instance may provide a set of different processes. According to the specification, there are three mandatory operations performed by a WPS, namely GetCapabilities, DescribeProcess and Execute [14]:

- GetCapabilities returns a brief service metadata document describing the resources of the specific server implementation, and gives a short description of each process offered by the WPS instance.
- DescribeProcess returns a detailed description of a process including its required input- (including the allowed formats) and output-parameter.
- Execute finally run the offered process.

These operations can be requested by a client and performed by a WPS server implementation. Many of the metadata structures are based on the ISO 19115 international standard for geographic information metadata.

In general a client communicates with an OWS via the Hypertext Transfer Protocol (HTTP) by sending a request which is answered with a response by the OWS server. Requests can be executed by the two methods GET and POST. The GET request is simply transported via the Uniform Resource Locator (URL) where the URL has the following form:

```
http://<server_address>/<server_path>?<parameter_list>.
```

The <parameter_list> is a sequence of elements in the form <parameter_name>=<value>&. The combination of <parameter_name> and <value> is denoted as a Key Value Pair (KVP) encoding because the name of a parameter always links a pair with its value. The length of an URL is limited and therefore the GET method is only qualified for a handful of parameters. In order to send more complex requests, the POST method is used which allows the embedding of XML encoded documents inside the body of an HTTP message. The request encodings of the three WPS operations are listed in Table 2.

Table 2. Operation request encoding [14]

Operation name	Request encoding
GetCapabilities (mandatory)	KVP
DescribeProcess (mandatory)	KVP and optional XML
Execute (mandatory)	XML and optional KVP

As mentioned above, service communication in line with OWS differs from general computer industry standards. The OGC is aware of this fact and is working on solutions. So, e.g. the WPS is compatible with the Web Service Description Language (WSDL) and the Simple Object Access Protocol (SOAP), and definitions for how to use WPS with these standards have been defined in the specification.

A specific characteristic of the WPS is the passing of input parameters within the Execute request. The specification provides the possibility to hand over required parameters directly within the request or to pass a reference. This means a process requiring a specific dataset (e.g. GML encoded) as input should be able to retrieve this data from a web-accessible resource by using the reference sent within the Execute request. Especially regarding the used Grid Computing environment, this possibility of passing data references instead of raw data within a WPS request offers a huge advantage: required data can be directly stored within the Grid and retrieved by the WPS. So there is no need to send huge amounts of data around.

Due to the generic interface of the WPS, various GIS tasks of arbitrary complexity can be carried out. In our work package in the project SDI-Grid different tasks for processing large DEMs are implemented and integrated into the WPS. These are very calculation and memory intensive DEM-processing operations. Therefore the processing power of a Grid is necessary in order to perform these calculations more efficiently. Generally Grid Computing and OWS act on different interfaces, which prevent communication. This incompatibility problem needs to be solved. To link the two worlds Grid Computing and SDI, the coupling of WPS and Grid takes place via the transport of the processing services into the Grid. This "gridification" of the WPS infrastructure is realized within a special work package within the project. A first version of a generic WPS infrastructure has been developed within the deegree framework, a Java framework offering the main building blocks for SDI based on standards of the OGC and ISO/TC 211 (ISO Technical Committee 211 – Geographic Information/Geomatics) [6].

Within our work package we developed a set of processes for very computationally intensive geometrical preprocessing operations of massive geodata. As there are no OGC standardization proposals for DEM processing functionalities, we developed our own interfaces. These are then linked to the WPS interface and may serve as input for later standardization efforts concerning terrain data handling, especially they may be used for the development of so called WPS profiles which are standardized WPS processes. The developed interfaces and the implemented functionalities are described in detail in the following section.

4 Proposals for Realizing Preprocessing DEM Functions as WPS

In this section we present the terrain preprocessing services. In particular we are developing a range of services for geo-tessellation (WPS Generic Geotessellation Service Framework), spatial partitioning (WPS Spatial Partitioning Service), generalization (WPS Terrain Generalization Service), and streaming (WPS Relevance Sorting Service) of DEM data within the SDI-Grid project. Furthermore there are several services planned along with the Institute of Cartography of the University Hannover in our work package namely the WPS 3D Converter Service and the WPS 3D Generalization Service. Within this paper we will only focus on the first mentioned.

At first we extend the functionally of our existing terrain library. In the second step we make the developed operations available as WPS processes and third we integrate these WPS processes into the Grid infrastructure by using the GT4 as middleware. Finally these WPS terrain operations are integrated within selected use case scenarios by mapping them together with other services in a service chain workflow. After the grid-enablement of the individual components it is possible to send a request from a client to a WPS. The WPS instance will accomplish an authorization and execute the request by accessing the Grid infrastructure.

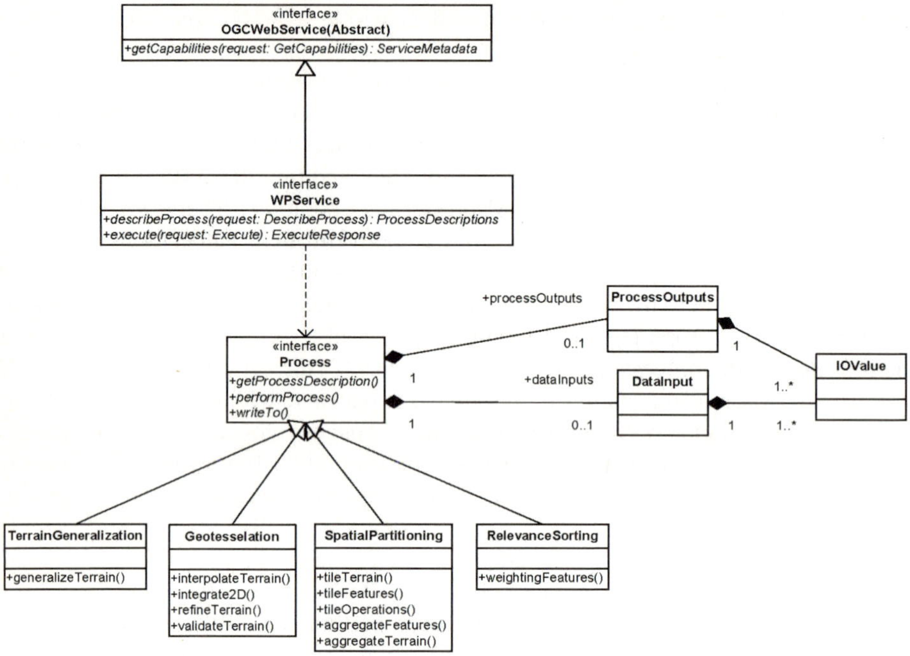

Fig. 1. WPS interface UML diagram with preprocessing services [14]

4.1 WPS Generic Geotessellation Service Framework

The developed Geotessellation Service generates DEMs from massive raw data, e.g. laserscanning data. The resulting surface geometry is stored as Triangulated Irregular Network (TIN). The default tessellation implementation is based on the Delaunay algorithm. This kind of tessellation divides the 2D space into triangles with specific properties, so that the resulting TIN is as regular as possible. Usually the height information is not taken into account during the computation, but attached as an attribute. Although this approach can be very fast, it sometimes yields imprecise results. Surface reconstruction techniques used in CAD software are far more efficient, but optimized for smaller, solid objects.

For geophysical computations like flooding or other hydrological simulations it is very important to reconstruct surface features like dikes and ditches very accurately. The apex line of dikes and banks and the bottom line of ditches must be appropriately represented by the triangulation. Another criterion is the volume of specific surface features, e.g. a river bed. Therefore we implemented a modified Delaunay algorithm, which incorporates the measurement of the surface curvature. In addition to the Delaunay criterion, that no other point may lie within the circum circle of a triangle, we also take the local curvature as an additional factor. The curvature is measured by multiplying the angle between the adjacent triangles at an edge with the edge length. When minimizing this curvature value, the TIN surface becomes smoother, and represents the original surface much better with the same amount of input points.

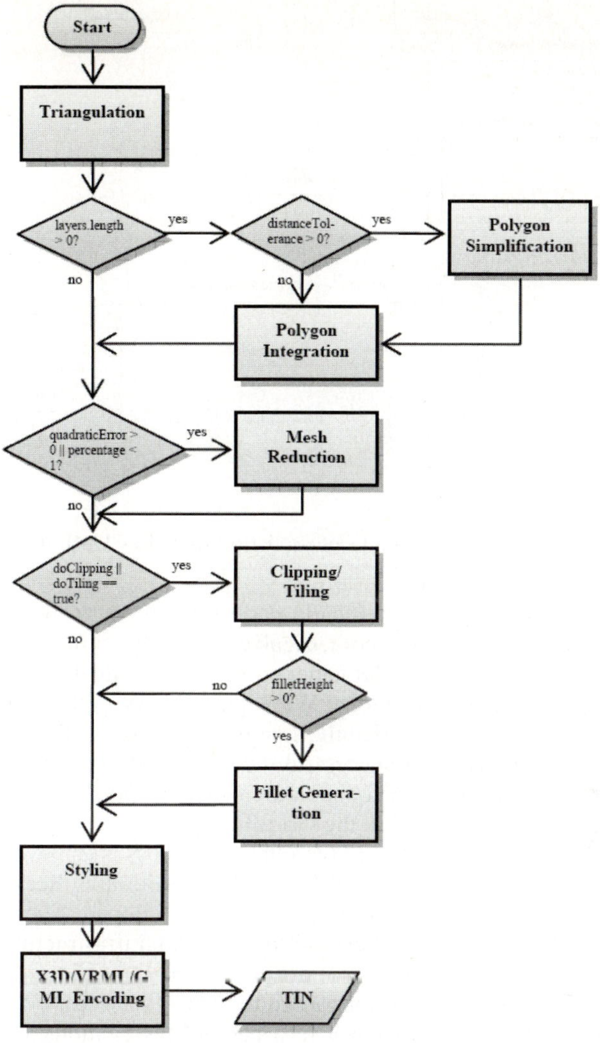

Fig. 2. Process chain of the Geotessellation Framework

The internal process chain of the tessellation functionality is shown in Fig. 2. The subsequent process is the integration of 2D polygon data; in our case areas of different soils, substrate, or land cover [17]. This data can be taken from the GIS of public authorities. The resulting TIN is divided into several segments containing possible additional information on the surface or substrate properties that is necessary for geo-simulations. Currently this segmentation is used for visualization purposes. Areas for vegetation, streets, water, and residential areas are cut into the terrain and different generic textures are applied to these segments.

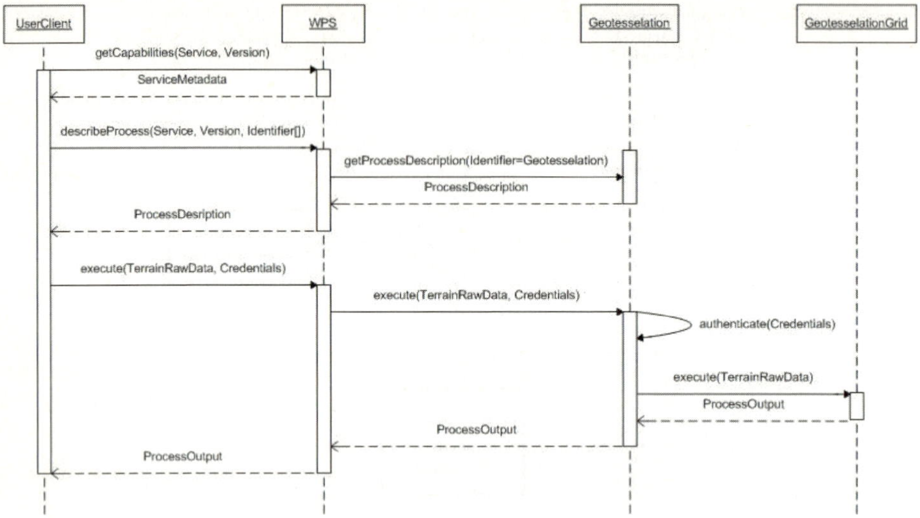

Fig. 3. Grid-enabled WPS Geotessellation workflow UML diagram

Other functions, which can be optionally part of this process chain, include a mesh reduction in order to reduce the geometrical complexity, tiling for cutting off not relevant areas at the borders, and fillet generation that is vertical faces at the borders.

A service client uses the conventional WPS interface. With the service request also credentials for the identification and authorizing to the Geotessellation Service will be submitted. If the authentication is successful the OGC WPS Geotessellation delegates the execution to an appropriate grid-enabled Geotessellation Service, which accomplishes the actual computation within the Grid (Fig. 3)

The parallelization of the whole geo-tessellation process is achieved by partitioning the input data spatially. Therefore the WPS Spatial Partitioning Service, which could be invoked by the WPS Geotessellation Service, is developed. Then it is possible to run many WPS processes in parallel within the Grid infrastructure. The resulting smaller fragments of the TIN afterwards are seamed together in order to reconstruct the final surface. These can then be forwarded to further services.

In order to guarantee the consistency in a parallel tessellation, the Geotessellation WPS parses all input parameters and is responsible for the interpolation of all input raw data and all subsequent steps. The WPS acts thus as central, coordinating component.

4.2 WPS Spatial Partitioning Service

In order to be able to accomplish a distributed processing in a Grid environment, the processes and the data management of the Web services must be parallelized. Depending upon functionality or structure of the data there are several partitioning versions, e.g. tiling datasets, parallelization by features or by operations.

In a preparing step of the parallelization for example, the Spatial Partitioning Service spatially partitions the dataset into smaller chunks or tiles and conducts other servers to execute the described job on the subset in parallel. Thus a parallel treatment, distributed on several Grid-nodes is possible. The spatial segmenting is carried

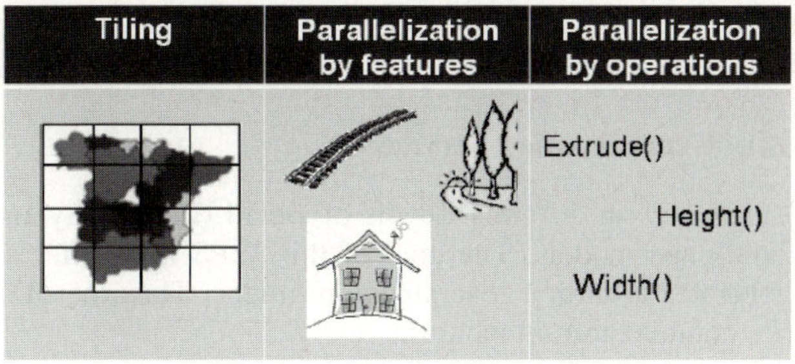

Fig. 4. Spatial Partitioning methods

out depending on the use case and according to different criteria, namely either simply via a raster or on the basis from geomorphologic, administrative or other criteria. The intermediate results are merged after the analysis of the individual tiles into a model.

For example a service call of the Partitioning WPS is made by the Geotessellation Service. The selected terrain area data that needs to be computed is partitioned and tessellated in parallel on the individual Grid nodes. The WPS Spatial Partitioning Service keeps the necessary parameters for partitioning from the inquiring Geotessellation Service as service request. The Partitioning Service makes the partitioning according to the request parameters of the Geotessellation Service and sends the result as service response back to the WPS Geotessellation Service. The service acts as so-called composed service which accomplishes the tessellation on different Grid nodes and administers it in an internal state.

4.3 WPS Relevance Sorting Service

Use-oriented boundary conditions, e.g. the constraint range at the streaming or the limitation of performance from the view of the client, can make it necessary that certain geodata objects or areas are processed with higher priority. Therefore a semantic weighting is to be computed in a preprocessing step dependent on application. This is used as basis for the treatment and the distribution of the geodata or the detail degree of the object with visualization.

Service requests must be treated with different priorities, so that the processes, with which e.g. a client waits for the answer preferentially is treated. In a preprocessing step a weighting of relevant geoobjects in dependence of the use case occurs. A goal is to produce a context sensitive prioritization for the processing of geodata. For the weighting of the geoobjects a standardized relevance function is defined. A classification that takes place which categories geoobjects into important and less important features by mapping corresponding relevance values. For the determination of the relevance values criteria for quantitative design must be specified. This takes place by a classification of the geoobjects depending on geometrical, semantic, visual and structural aspects. The sum of all classification factors results then in the relevance value of the geoobject.

The calculation of the semantic relevance of a geoobject and the assortment derived from it take over its own component, which is likewise put to the Grid over the

OGC WPS as service at the disposal. The relevance calculation takes place in the context of a generic framework.

4.4 WPS Terrain Generalization Service

Multi-scale 3D-models with different Levels of Detail (LOD) play an important role in terrain and landscape models. The goal of this WPS is to offer a service which automatically generates arbitrary generalization grades. Thus the 3D Generalization Service simplifies complex terrain models.

We distinguish components for the generalization of city models and of DEM. Within this paper we only consider the generalization of the terrain.

Based on the criteria given by the user, resolutions are assigned to parts of the described region. Furthermore the intersection of the generalized features with the terrain provided as constraints. For this purpose the Terrain Generalization Service supplies a terrain model according to these specifications.

In form of a database terrain models with low and higher resolution are administrated. The generalization of the based terrain model takes place in dependence of the application scenario as well as the topography. For example a relatively low resolution of the DEM is sufficient with smaller relief dynamics, without changing typical characteristics of the relief. The DEM should adopt to the relief by reducing the geometrical complexity by means of LOD handling. LOD generation is based on a mesh simplification algorithm, which removes elements of the original TIN that have a low significance until a certain error value is reached. To reduce large 3D models captured by laser-scanning, an edge contraction algorithm is used. The result of this procedure is a collection of quadratic surface tiles of different accuracy and size that can be put together in order to build a complete DEM.

The multi-resolution model makes different LODs of terrain geometry available. The administration of the terrain LOD takes place independently of the 3D city models. By the decoupled produced generalization grades of city and terrain model, the resolution and accuracy of the two models are independent. This makes it possible to adapt the resolution of the DEM to the use case. The geometrical consistency between city and terrain model must be guaranteed. This could take place by the definition of a Terrain Intersection Curve (TIC).

The Terrain Generalization Service supports usual formats for terrain models. Independent of the data format, it is important that the terrain model is examined for its consistency. Inconsistencies like gaps or different resolutions can occur as a result of the distributed computation.

5 Use Case `Flood Modeling´

For geophysical computations like flooding or other hydrological simulations it is very important to reconstruct surface features like dikes and ditches exactly. The apex line of dikes and the bottom line of ditches must be appropriately calculated by the WPS Geotesselation Service. Within our work package along with the Institute of River and Costal Engineering of the Technical University Hamburg further services are developed.

The Geotessellation WPS accomplishes an adapted interpolation of the raw data for the flood modeling scenario according to the following criteria:

- recognition and derivation of breaking edges (Feature detection)
- inserting of additional breaking edges
- consideration of area borders

Apart from measuring points of the DEM thus additionally breaking edges, whose altitude also must be given, in a terrain data base are charged with. The triangulation takes place only on request via a Delaunay algorithm, which generates an even triangle meshing. Additionally if terrain edges are present, these must be taken into account during the triangulation of the TIN. The influence of these is visible as hard ridges and sharply bordered terraces. In particular sharp transitions between plain areas and steep hillsides or walls must be preserved. Also contour or height lines could be used as a constraint.

For flood modeling and simulation it is therefore necessary to extend the pure triangulation of sample points (e.g. Delaunay) by the integration of linear feature like breaking edges, which is referred to as Constrained Triangulation (CT).

Also important is the segmentation of the surface into areas with different substrate or vegetation parameters. Rock and soil properties can be used in order to derive hydrological parameters. Especially the infiltration capabilities play an important role in hydrological simulations. Regarding the geometrical solution, the approach of [**Error! Reference source not found.**] can be used. We partition the terrain surface based on 2D areas/polygons which may come from geological, environmental, geo ecological, soil, or other data sets. The result is a DEM consisting of many smaller surface parts within the same TIN representing areas of different attributes, which can then be used as input for further flooding simulations.

To process the high-resolution terrain data, a parallelization takes place by the use of the WPS Partitioning Service. By means of tiling the massive raw data and tesselation at several grid nodes, an area in the context of the discretization process is divided into approximately equal large tiles and processes into dependence of available computing resources.

6 Conclusion and Future Work

The presented components and interfaces are currently being implemented within the German project SDI-Grid. With the now officially released WPS interface, the spectrum of SDI services has been extended, so that the task of SDIs is no longer to only share geo data but also to offer services and computation resources. We showed that the geo community shares some interests with the Grid computing community when it comes to the processing of large terrain data sets and that linking both can be beneficial. Since their development has been carried out in parallel in the past, different technologies have been adopted; however the ideas are very similar. Exploiting Grid computing for SDI geo processing requires a middleware that contains translations between protocols, e.g. HTTP POST to SOAP and additional services for the partitioning and parallelization, which is unfortunately domain specific.

References

1. Agrawal, A., Radhakrishna, M., Joshi, R.C.: Geometry-based Mapping and Rendering of Vector Data over LOD Phototextured 3D Terrain Models. In: WSCG 2006 - The 14th International Conference in Central Europe on Computer Graphics, Visualization and Computer Vision, Plzen -Bory, Czech Republic, January 30 – February 3 (2006)
2. Coors, V., Rossignac, J.: Delphi: geometry-based connectivity prediction in triangle mesh compression. The Visual Computer, International Journal of Computer Graphics 20(8-9) (2004)
3. Czajkowski, K., Ferguson, D., Foster, I., Frey, J., Graham, S., Maguire, T., Snelling, D., Tuecke, S.: From Open Grid Services Infrastructure to WS-Resource Framework: Refactoring & Evolution (2004), `http://www.ibm.com/developerworks/library/ws-resource/ogsi_to_wsrf_1.0.pdf`
4. Heckbert, P.S., Garland, M.: Survey of Polygonal Simplification Algorithms. In: Multiresolution Surface Modeling Course, SIGGRAPH 1997, Los Angeles, USA (1997)
5. Hoppe, H.: Smooth view-dependent level-of-detail control and its application to terrain rendering. In: IEEE Visualization, October 1998, pp. 35–42 (1998)
6. Heier, C., Kiehle, C.: Geodatenverarbeitung im Internet - der OGC Web Processing Service. GIS 2005 6, 39–43 (2005)
7. Global Grid Froum (GGF), `http://www.ggf.org/`
8. The Globus Alliance, http://www.globus.org/
9. Kiehle, C., Greve, K., Heier, C.: Standardized Geoprocessing – Taking Spatial Data Infrastructures one step further. In: Proceedings of the 9th AGILE International Conference on Geographic Information Science, Visegrád, Hungary (2006)
10. Kohout, Josef: Delaunay Triangulation in Parallel and Distributed Environment. PhD Thesis, University of West Bohemia, Pillsen, Czech Republic (2005)
11. Kolbe, T.H., Gröger, G., Plümer, L.: CityGML – Interoperable Access to 3D City Models. In: Proceedings of the Int. Symposium on Geoinformation for Disaster Management, Delft, The Netherlands, March 21-23 (2005)
12. OASIS, http://www.oasis-open.org/
13. OASIS WSRF Trechnical Committee, `http://www.oasis-open.org/committees/tc_home.php?wg_abbrev=wsrf`
14. OGC: Web Processing Service. OGC Standard, Document Reference, Number 05-007r7, Version 1.0.0 (2007)
15. Lindstrom, P., Pascucci, V.: Terrain Simplification Simplified: A General Framework for View-Dependent Out-of-Core Visualization. IEEE Transactions on Visualization and Computer Graphics 8(3), 239–254 (2002)
16. Schilling, A., Basanow, J., Zipf, A.: Vector Based Mapping Of Polygons On Irregular Terrain Meshes For Web 3D Map Services. In: 3rd International Conference on Web Information Systems and Technologies (WEBIST), Barcelona, Spain (March 2007)
17. Schilling, A., Zipf, A.: Generation of VRML City Models for Focus Based Tour Animations - Integration, Modeling and Presentation of Heterogeneous Geo-Data Sources. In: Web3D Conference, Saint Malo, France, 9-12.03.2003 (2002)
18. Schroeder, W.J.: A Topology Modifying Progressive Decimation Algorithm. In: IEEE Proceedings Visualization 1997, pp. 205–221 (1997)
19. Sotomayour, B., Childers, L.: Globus Toolkit 4: Programming Java services. Morgan Kaufmann, San Francisco (2006)

20. Stollberg, B., Zipf, A.: OGC Web Processing Service Interface for Web Service Orchestration - Aggregating Geo-processing Services in a Bomb Threat Scenario. In: Ware, J.M., Taylor, G.E. (eds.) W2GIS 2007. LNCS, vol. 4857. Springer, Heidelberg (2007)
21. Wartell, Z., Kang, E., Wasilewski, T., Ribarsky, W., Faust, N.: Rendering Vector Data over Global, Multi-resolution 3D Terrain. In: Proceedings of Joint EUROGRAPHICS – IEEE TCV Symposium on Visualization, Grenoble, France, May 26-28, 2003, pp. 213–222 (2003)
22. Weiser, A., Neis, P., Zipf, A.: Orchestrierung von OGC Web Diensten im Katastrophenmanagement - am Beispiel eines Emergency Route Service auf Basis der OpenLS Spezifikation. In: GIS - Zeitschrift für Geoinformatik. 09/2006, pp. 35–41 (2006)
23. Weiser, A., Zipf, A.: Web Service Orchestration (WSO) of OGC Web Services (OWS) for Disaster Management. In: Joined CIG/ISPRS Conference on Geomatics for Disaster and Risk Management, Toronto. Kanada, 23.-25.05.2007 (2007)
24. WSRF.NET, http://www.cs.virginia.edu/~gsw2c/wsrf.net.html

Identifying Spam Web Pages Based on Content Similarity

Maria Soledad Pera and Yiu-Kai Ng

Computer Science Department, Brigham Young University,
Provo, Utah, U.S.A.

Abstract. The Web provides its users with abundant information. Unfortunately, when a Web search is performed, both users and search engines are faced with an annoying problem: the presence of misleading Web pages, i.e., *spam* Web pages, that are ranked among legitimate Web pages. The mixed results downgrade the performance of search engines and frustrate users who are required to filter out useless information. In order to improve the quality of Web searches, the number of spam pages on the Web must be reduced, if they cannot be eradicated entirely. In this paper, we present a novel approach for identifying spam Web pages that have mismatched titles and bodies and/or low percentage of hidden content. By considering the content of Web pages, we develop a spam-detection tool that is (i) *reliable*, since we can accurately detect 94% of spam/legitimate Web pages, and (ii) *computational inexpensive*, since the word-correlation factors used for content analysis are precomputed. We have verified that our spam-detection approach outperforms existing anti-spam methods by an average of 10% in terms of *F-measure*.

1 Introduction

The Web is populated with all possible subject areas from personal health care to constitutional laws to religious beliefs as presented in online news articles, research papers, and customer-generated media (i.e., blogs), to name a few, and virtually all kinds of information can be found on the Web. With the huge amount of information to sort through, users turn to Web search engines for assistance in locating needed information. Hence, analyzing the content of relevant Web pages and ranking them accordingly is a crucial process in Web information retrieval (IR). As a result, a search engine spider crawling the Web not only entices gathering new information, but also imposes significant financial impact on the users, such as commercial transactions on the Web, that results from the ranking of (the contents of) Web pages determined by the search engine [19]. As mentioned in [1], there is an economic incentive for manipulating search engines' rankings by creating pages that score high independently of their real contents, even though the intention is unethical. Gradually, more pages are introduced on the Web that are considered legitimate when in fact they are not and ranked high when they should not be by existing search engines. As reported in [16], a significant portion of existing Web pages—14% in the year of 2006—are spam.

O. Gervasi et al. (Eds.): ICCSA 2008, Part II, LNCS 5073, pp. 204–219, 2008.

Spamming is a serious IR problem, since it (i) affects the quality of Web searches, (ii) damages the search engine's reputation, and (iii) weakens the user's confidence in the retrieved results. In general, Web spamming are treated as an attempt to receive an unjustifiably favorable relevance or high ranking for their Web pages, regardless of the true values of the pages. Spamming can also be defined as an attempt to deceive a search engine's relevancy ranking algorithm [18].

A number of existing spamming approaches rely on *links* within Web pages [8] to manipulate Web ranking algorithms such as PageRank [4], whereas others rely on infecting the *content* of Web pages [16], e.g., stuffing popular or concatenating words in Web pages, to increase their chance for matching Web queries. However, due to their complexity, neither considering links nor statistical features of Web page contents are effective approaches in identifying spam Web pages as the volume of Web pages is huge.

Fetterly et al. [9] use a semantic technique, i.e., actual word count, on Web pages for detecting spam. Instead, we consider the actual content (i.e., words) of Web pages and use the word-similarity values to identify and eliminate spam Web pages. We will show that by (i) computing the *degree of similarity* among the words in the title and the body of a Web page P, which is computed by using their word-correlation factors, (ii) using the percentage of *hidden content* within P, and/or (iii) considering the bigram or trigram *phrase-similarity* values of P, we can determine whether P is spam with high accuracy.

The remaining sections are organized as follows. In Section 2, we discuss existing anti-spam methods and address their differences with ours. In Section 3, we describe in detail our spam-detection approach. In section 4, we present the experimental results, which verify the accuracy of our spam-detection method. In Section 5, we give a conclusion and include the future directions of our work.

2 Related Work

Previous anti-spam work focus on applying two different strategies: *content analysis* and *link connection*. [16] introduce and combine several anti-spam heuristics based on the content of a Web page, which include (i) the number of words, (ii) the average length of the words, (iii) the amount of anchor text, (iv) the fraction of visible content, (v) the faction of globally popular words, and (vi) the likelihood of independent n-grams within the page. These heuristics are treated as features in classifying spam Web pages.

[11] also present several anti-spam heuristics according to page content that include (i) detecting the inclusion of terms in anchor text that are unrelated to the referenced Web pages, (ii) computing the amount of repetitive terms in a Web page P introduced by a spammer in an attempt to increase its relevance score, (iii) verifying the existence of a large number of unrelated terms, and (iv) identifying the presence of phrase stitching in P, which should increase its degree of relevance to several posted queries. Likewise, [9] analyze the statistical features of the host component of an URL and the excessive replication of content to establish content-based features that can be used by Web spam classifiers.

As a link-analysis approach, [2] introduce a damping function, which classifies spam and non-spam Web pages using (incoming and outgoing) links within a page without considering its content. Other well-known anti-spam techniques that rely on link analysis are described in [10,12]. [10] present a semi-automatic algorithm, TrustRank, which uses the link structure of the Web to discover pages that are likely to be legitimate with respect to a small set of (seed) legitimate pages that were manually identified by experts. Since this approach requires human assistance, it is not fully automated. [12], on the other hand, define a spam mass metric that reflects the impact of link spamming on the ranking of a page and use it to determine Web pages that are significantly beneficiaries of link spamming. As yet another link-based method, [3] introduce SpamRank, which is based on PageRank, to identify pages linked by a large number of other pages intended for misleading search engines to rank their target higher.

None of the approaches discussed above relies on the actual word semantic meaning in the content of a given Web page to detect spam Web pages, which is our spam-detection approach. In [17] we demonstrate that using the content of emails for detecting junk emails is effective, the same strategy that we adopt for finding spam Web pages in this paper.

3 A Content-Based Spam-Detection Approach

As discussed earlier, spam Web pages are a burden for (i) the Web servers, since (among others) they waste storage space and processing time for indexing and maintenance and (ii) the users who must deal with low-quality retrieved results (caused by spamming) when performing Web searches. In order to neutralize existing spamming tricks, we rely on the *content* (i.e., words and phrases and/or proportion of markup content) of a given Web page P to determine timely and accurately whether P should be treated as spam.

3.1 Our Web Spam-Detection Approach

[14] claim that the *title* of a document often reflects its *content*, and we are confident that this same concept applies to Web pages as well, since a legitimate Web page is a regular document with a title that describes its content, whereas the title of a spam Web page often does not. Consider the legitimate Web page (http://www.mothersbliss.co.uk/shopping/index) in Figure 1 in which the title reflects the content of the page, whereas Figure 2 shows a spam Web page (http://extc.co.uk) in which its content mismatches its title. We analyze the content of a Web page to determine how closely its *title* is related to its *body* (as discussed in Sections 3.2 and 3.4) and calculate the fraction of hidden content of a Web page (see Section 3.3), if necessary, in detecting spam Web pages.

3.2 Web Page Similarity

We rely on the *similarity measure* between the content of (represented by a sequence of words in) the title T and the body B of a Web page P to identify

<title> Baby, pregnancy, motherhood: discover the magic of being a mother at mothersbliss!</title>
<body>Baby names – Birth announcement. Are you ready for a baby yet? When is the best time to make a baby? Is my diet okay for my baby-to-be? Am I exercising enough ...? Mothersbliss™ has answers to all your concerns! ... </body>

Fig. 1. The title and (portion of the) content of a *legitimate* Web page

<title> Compare and find resources </title>
<body> Cars Computers Internet Electronics Loans Mortgages Credit Cards Hotels Flights Gadgets Furniture Health Beauty ... </body>

Fig. 2. The title and (portion of the) content of a *spam* Web page

spam Web pages. We determine the *degree of similarity* between T and B using the correlation factors of words in T and B as defined in a precomputed *word-correlation matrix*, which are generated by analyzing a set of 880,000 Wikipedia documents (downloaded from http://www.wikipedia.org/) to calculate the *correlation factor* (i.e., similarity value) of any two words[1] based on their (i) *frequency of co-occurrence* and (ii) relative *distance* as defined below.

$$c_{i,j} = \sum_{w_i \in V(w_i)} \sum_{w_j \in V(w_j)} \frac{1}{d(w_i, w_j)} , \tag{1}$$

where $d(w_i, w_j)$ denotes the *distance* between any two words w_i and w_j in any Wikipedia document D, and $V(w_i)$ ($V(w_j)$, respectively) is the set of stem variations of w_i (w_j, respectively) appeared in D. If w_i and w_j are consecutive words, then $d(w_i, w_j) = 1$, whereas if w_i and w_j do not co-occur in the same document, then $d(w_i, w_j) = \infty$, i.e., $\frac{1}{d(w_i,w_j)} = \frac{1}{\infty} = 0$. To avoid bias that occurs in documents of large size, we *normalize* the word-correlation factors as follows:

$$C_{i,j} = \frac{c_{i,j}}{|V(w_i)| \times |V(w_j)|} , \tag{2}$$

where $c_{i,j}$ is as defined in Equation 1.

Title-Body Similarity. Using the normalized word-correlation factors, we can compute the degree of similarity (between the words) of the title T and the body B of a Web page P, denoted $SimTB$. We focus only on T and B of P, since as shown in the experimental results (see Section 4), $SimTB$ of P can accurately determine whether P is spam or legitimate. In the case in which there is no title in P, we consider the *hidden content* of P (see Section 3.3).

To determine the $SimTB$ value of P, we calculate the similarity value of each word t in T with respect to each word b in B of P. The *higher* the correlation

[1] Words in the documents were *stemmed* (i.e., reduced to their grammatical roots, e.g., computer, computing, and computation are converted to compute) after all the *stopwords* (i.e., words that carry little meaning, e.g., articles, prepositions, and conjunctions) were removed, which minimized the number of words to be considered.

Table 1. Word-correlation factors and the μ-values among some of the words in the *title* and *body* of the legitimate Web page as shown in Figure 1

	names	announce	time	diet	baby	answer	...	μ-value
baby	8.9×10^{-8}	1.6×10^{-7}	9.8×10^{-8}	7.1×10^{-8}	1	7.3×10^{-8}	...	1
pregnancy	5.8×10^{-4}	4.1×10^{-5}	1.4×10^{-7}	6.1×10^{-7}	3.7×10^{-3}	2.5×10^{-10}	...	4.3×10^{-3}
discover	6.1×10^{-8}	7.1×10^{-2}	1.4×10^{-8}	3.3×10^{-8}	8.9×10^{-8}	8.1×10^{-9}	...	7.1×10^{-2}
...	...	...	...	...	...	...	...	...
Average	1.5×10^{-4}	1.8×10^{-2}	8.6×10^{-8}	1.9×10^{-7}	2.5×10^{-1}	3.6×10^{-8}	...	**0.3**

Table 2. Word-correlation factors and the μ-values among the words in the *title* and some of the words in the *body* of the spam Web page as shown in Figure 2

	computers	internet	electronics	mortgages	credit	flights	...	μ-value
compare	9.1×10^{-8}	2.5×10^{-7}	5.9×10^{-8}	2.3×10^{-8}	1.1×10^{-8}	6.1×10^{-8}	...	5.5×10^{-7}
find	2.2×10^{-7}	2.1×10^{-7}	1.4×10^{-8}	5.3×10^{-9}	1.6×10^{-8}	2.0×10^{-8}	...	5.9×10^{-7}
resources	6.2×10^{-8}	4.0×10^{-8}	5.4×10^{-8}	4.5×10^{-8}	1.7×10^{-7}	2.0×10^{-8}	...	4.4×10^{-7}
Average	1.3×10^{-7}	1.7×10^{-7}	4.2×10^{-8}	2.4×10^{-8}	6.6×10^{-8}	3.4×10^{-8}	...	**5.3**$\times10^{-7}$

factors among t and the words in B, the *higher* the similarity value between t and B, denoted $\mu_{t,B}$.

$$\mu_{t,B} = 1 - \prod_{b \in B} (1 - C_{t,b}) \ , \tag{3}$$

where $\mu_{t,B}$ is defined as the complement of a negated algebraic product, instead of the algebraic sum, and $C_{t,b}$ is as defined in Equation 2.

Once the μ-values of each word in T and all the words in B is calculated, we can determine the *degree of similarity* between T and B, which calculates the average similarity value of each word t in T and all the words in B, as

$$SimTB(T,B) = \frac{\mu_{1,B} + \mu_{2,B} + \ldots + \mu_{n,B}}{n} \ , \tag{4}$$

where n is the number of *words* in T. A *high* (*low*, respectively) $SimTB(T,B)$ value reflects a *high* (*low*, respectively) degree of similarity between T and B.

Example 1. Table 1 (Table 2, respectively) shows the correlation factors among the words in the title and the body of the Web page P_1 as shown in Figure 1 (P_2 in Figure 2, respectively). The degree of similarity, i.e., $SimTB$, of the title T_1 and the body B_1 of P_1 is 0.88, whereas the degree of similarity between the title T_2 and the body B_2 of P_2 is 3.2×10^{-7}. According to the computed $SimTB$ value, P_1 (P_2, respectively) is highly likely a legitimate (spam, respectively) page. □

Similarity Threshold Value. Having defined the $SimTB$ value of a given Web page P, we must determine an appropriate *word-similarity threshold value* V so that if $SimTB$ of $P \geq V$, then P is considered *legitimate*; otherwise, P is treated as *spam*. An ideal similarity threshold should (i) reduce to the

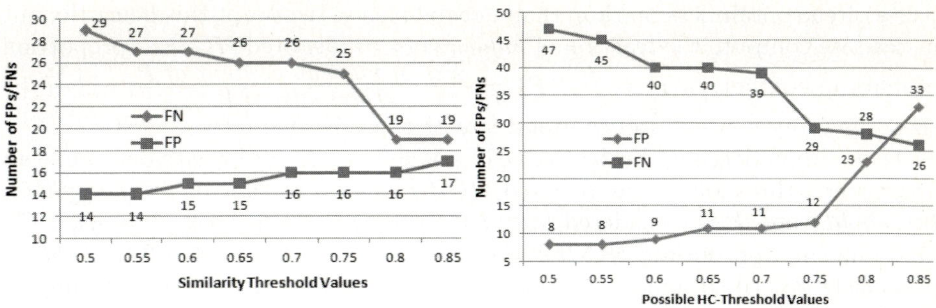

(a) *SimTB* threshold values (b) Hidden-content threshold values

Fig. 3. Number of False Positives (FPs) and False Negatives (FNs) computed by using different possible similarity threshold values on the Web pages in the *Threshold set*

minimum the number of spam Web pages identified as legitimate, i.e., *false negatives* (FNs), and (ii) avoid treating legitimate Web pages as spam, i.e., *false positives* (FPs). In order to determine the correct similarity threshold, we consider (i) Web pages in a test set, called *Threshold set*, and (ii) a number of possible similarity threshold values which yield different number of FPs and FNs. The *Threshold set* is a collection of 370 previously classified spam and non-spam Web pages (170 spam and 200 non-spam) randomly selected from the WEBSPAM-UK2006 dataset (http://www.yr-bcn.es/Webspam/datasets/), which is a well-known, publically available reference collection for Web spam research that consists of 77.9 million spam and non-spam pages. As shown in Figure 3(a), the optimal word-similarity threshold is **0.80**, since at 0.80 the total number of FPs and FNs are reduced to a minimum and neither the number of FPs nor FNs dominates the other.[2] Hence, we declare a Web page P as

$$Status(P) = \begin{cases} \textit{Legitimate} & \text{if } SimTB(P_T, P_B) \geq 0.80 \\ \textit{Spam} & \text{otherwise}, \end{cases} \quad (5)$$

where P_T (P_B, respectively) denotes the *title* (*body*, respectively) of P.

Using Equation 5, we classify the Web page P_1 in Figure 1 as legitimate, since $SimTB(T_1, B_1) = 0.88 \geq 0.80$, whereas the Web page P_2 in Figure 2 as spam, since $SimTB(T_2, B_2) = 3.2{\times}10^{-7} < 0.80$.

3.3 Fraction of the Hidden Content

Even if a Web page P lacks of a title, we can still determine whether P is spam by considering the fraction of *hidden content* in P. [16] define the visible content of a Web page P as the length (in characters) of all non-markup words in P divided by the total size (in characters) of P and claim that spam Web

[2] We verified the correctness of the *similarity* threshold value and other threshold values using another *Threshold set* S, which consists of 100 (38 spam and 62 non-spam) pages from WEBSPAM-UK2006, and S yields the *same* threshold values.

pages often contain less markup than normal pages. We adapt this heuristic, but instead we compute the fraction of *hidden content*, denoted HC, i.e., proportion of markup content, of P as $HC(P) = \frac{Size\ of\ markup\ content\ of\ P}{Total\ size\ of\ P}$, where the size of markup content and the total size of P are in characters.

Again, upon defining the HC value of a Web page P, we have to determine an appropriate threshold value, denoted *HC-threshold*, so that if $HC(P) \geq$ *HC-threshold*, then P is considered *legitimate*; otherwise, P is treated as *spam*. To determine an appropriate *HC-threshold* value, we used the same *Threshold set* previously described and computed the number of FPs and FNs according to different *HC-threshold* values. Figure 3(b) shows that the ideal *HC-threshold* value is **0.75**, since the number of FPs and FNs are reduced to a minimum at 0.75. Excluding the title in the HTML page P_1, a legitimate Web page (P_2, a spam Web page, respectively) as shown in Figure 1 (Figure 2, respectively), the fraction of hidden content of P_1 (P_2, respectively) is 0.87 (0.23, respectively). Using the chosen *HC-threshold* value, i.e., 0.75, we correctly classify P_1 and P_2.

3.4 The Use of Bigrams and Trigrams

We have observed that whenever there is at least one word[3] in the title T of a *spam* Web page P that also appears in the body B of P, then $SimTB(T, B)$ is *high*, which causes our spam-detection approach to misclassify P as *legitimate*. As a result, our spam-detection method yields higher than expected number of FNs. In order to further enhance our spam-detection approach, we consider *bigram* and *trigram*, instead of *unigram* (i.e., single-word as presented in Section 3.2), phrase-correlation factors of T and B in determining the content similarity between T and B. We consider *bigrams* and *trigrams*, since as claimed by [15] and verified by us, short phrases (i.e., bigrams and trigrams) increase the retrieval effectiveness, whereas using phrases of 4 or more words tends to retrieve unreliable results.

The Phrase-Similarity Value. In computing the *phrase-correlation factor*, denoted pcf, of any two bigrams or trigrams, p_1 and p_2, we apply the *Odds* [13] ratio, i.e., $Odds(H) = \frac{p(H)}{1-p(H)}$, on the normalized word-similarity factors as defined in Equation 2. *Odds* measures the prospective support based on a hypothesis H (i.e., n-grams) using prior knowledge $p(H)$ (i.e., the word-correlation factors of the n-grams) to determine the strength of a belief, which is pcf.

$$pcf_{p_1,p_2} = \frac{\prod_{i=1}^{n} C_{p_{1_i},p_{2_i}}}{1 - \prod_{i=1}^{n} C_{p_{1_i},p_{2_i}}}, \qquad (6)$$

where p_{1_i} and p_{2_i} are the ith $(1 \leq i \leq n)$ words in p_1 and p_2, respectively, and $C_{p_{1_i},p_{2_i}}$ is the *normalized word-similarity value* as defined in Equation 2. By using the computed phrase-correlation factors, we can replace $C_{t,b}$ in Equation 3 by pcf_{p_1,p_2} to determine (i) the μ-value between an n-gram $(2 \leq n \leq 3)$ in T and all the n-grams in B, as well as (ii) the *degree of similarity* between T

[3] After *stopwords* are removed and the remaining words are reduced to their *stems*.

Table 3. The *phrase-correction factors* and μ-values of (some of) the *bigrams* in the title with respect to the bigrams in the body of the legitimate Web page in Figure 1

	baby name	name birth	birth announce	announce ready	...	μ-value
baby pregnancy	4.1×10^{-5}	1.2×10^{-14}	1.3×10^{-11}	5.9×10^{-10}	...	4.1×10^{-5}
pregnancy motherhood	7.7×10^{-11}	2.4×10^{-11}	2.8×10^{-15}	3.7×10^{-12}	...	1.1×10^{-10}
motherhood discover	6.9×10^{-15}	8.1×10^{-13}	4.2×10^{-8}	2.9×10^{-13}	...	4.2×10^{-8}
...	...	...	...	...	...	...
Average	1.0×10^{-5}	6.3×10^{-12}	1.1×10^{-8}	9.1×10^{-10}	...	**1.0** $\times10^{-5}$

Table 4. The *phrase-correction factors* and μ-values of (some of) the *trigrams* in the title with respect to the trigrams in the body of the legitimate Web page in Figure 1

	baby name birth	name birth announce	birth announce ready	announce ready baby	...	μ-value
baby pregnancy motherhood	2.4×10^{-11}	2.4×10^{-22}	7.8×10^{-17}	6.7×10^{-17}	...	2.4×10^{-11}
pregnancy motherhood discover	4.7×10^{-16}	1.7×10^{-12}	3.9×10^{-20}	1.2×10^{-19}	...	1.7×10^{-12}
...	...	...	...	...	...	...
Average	6.1×10^{-12}	4.3×10^{-13}	1.0×10^{-15}	1.9×10^{-16}	...	**6.5** $\times10^{-12}$

and B, i.e., $SimTB$, using Equation 4 which overcomes the *unigram* problem that arises when an unigram in T appear in B. In adopting Equation 4 to compute the degree of similarity, n in the equation represents the number of *bigrams* (*trigrams*, respectively), instead of unigrams, in T. Table 3 (Table 4, respectively) shows (some of) the *phrase-correction factors* between the bigrams (trigrams, respectively) in the title and body of the legitimate page in Figure 1.

Example 2. Figure 4 shows a spam Web page P (http://khs.co.uk) in which the word KHS in its title T is repeated in its body B, yielding $SimTB(T, B)$ = 0.84 by Equation 4 on word-similarity measures. Using the word-similarity threshold value 0.80 as defined earlier, P is misclassified as *legitimate*. However, when considering the *bigrams* in T and B, $SimTB(T, B) = 0.57$ and using the threshold value (defined below), P is correctly classified as *spam*. □

The Phrase-Similarity Threshold Value. Prior to using the phrase-correlation factors, we define the *bigram-* (*trigram-*, respectively) *similarity threshold value* V so that for any Web page P, if $SimTB(T_P, B_P) \geq V$, where $SimTB(T_P, B_P)$ is computed by using the bigram- or trigram-correlation factors, then P is considered legitimate; otherwise, P is treated as spam.

In determining an ideal phrase-similarity threshold, we use the same *Threshold set* (in Section 3.2) to compute the number of FPs and FNs according to

<title> Compare KHS and find KHS resources on KHS </title>
<body> KHS Cars Computers Internet Electronics Loans Mortgages Credit Cards Hotels Flights Gadgets Furniture Heath Beauty ... Still looking for deals on KHS? ... </body>

Fig. 4. A sample spam Web page that is misclassified as *legitimate* when the (single-) *word*-similarity measure is applied, but is correctly classified as *spam* when the *phrase*-similarity value is considered

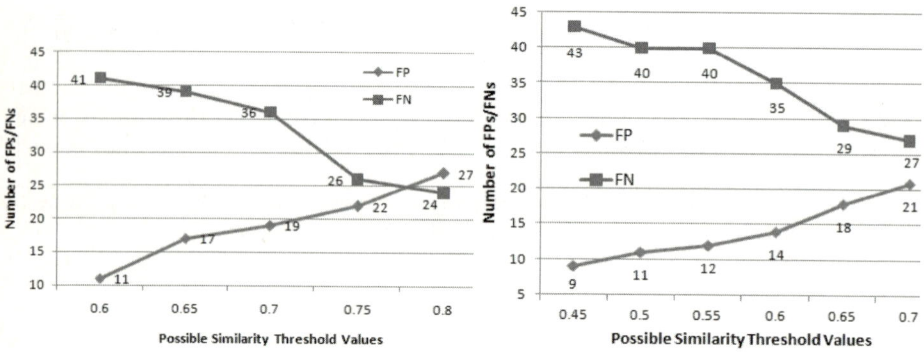

(a) *Bigram*-similarity threshold values

(b) *Trigram*-similarity threshold values

Fig. 5. Number of FPs and FNs computed by using different possible *bigram*- and *trigram*-similarity threshold values on the Web pages in the *Threshold set*

different phrase-similarity threshold values and choose the value V such that the total number of FPs and FNs at V are reduced to a minimum. Figure 5(a) shows that the optimal *bigram*-similarity threshold value is **0.75**, whereas Figure 5(b) indicates that the optimal *trigram*-similarity threshold value is **0.65**.

3.5 An Enhanced Similarity-Measure Method

We have considered alternative approaches to augment the use of phrase-correlation factors in computing the $SimTB$ value of a Web page that can further enhance the performance of our spam-detection approach. An alternative approach is to determine the similarity among n-gram phrases[4] ($1 \leq n \leq 3$) in the title T with respect to the ones in the body B of a Web page P and *penalize* P with a *lower* $SimTB$ value if B contains phrases that are similar to only a *few* phrases in T and *reward* P with a *higher* $SimTB$ value if B contains phrases which are related to a *number* of phrases in T.

[4] In the case when no bigrams or trigrams are available in the title T, i.e., after stopword removal and stemming on the words in T and only one word is available in T, then the (single) word-similarity will be considered.

The Enhanced Phrase-Similarity Approach. The enhanced phrase-similarity approach assures that if the phrases in the body B are closely related to most of the phrases in the title T, then the corresponding Web page P is more likely legitimate; otherwise, P is likely spam. We compute the enhanced similarity value between T and B of P by calculating the *sum* of the phrase-correlation factor of each bigram (trigram, respectively) *pt* in T with respect to each bigram (trigram, respectively) phrase in B.

$$spcf_{pt,B} = \sum_{j=1}^{m} pcf_{pt,j} \ , \tag{7}$$

where $pcf_{pt,j}$ is the *phrase-correlation factor* as defined in Equation 6 and m is the total number of the bigrams (trigrams, respectively) in B.

Once the $spcf$-value of each bigram (trigram, respectively) in T has been calculated, we can compute the *enhanced degree of similarity* between T and B, denoted $enSimTB$, as

$$enSimTB(T, B) = \sum_{i=1}^{n} Min(spcf_{i,B}, 1) \ , \tag{8}$$

where n is the total number of bigrams (trigrams, respectively) in T.

In calculating the $enSimTB$ value, we add the *minimal value* of 1 and the $spcf$-value of an n-gram phrase pt_i in T. We do so in order to *restrict* the similarity value of each pt_i in T with respect to the ones in B to 1, which is the similarity value for an *exact match*; otherwise, the $spcf$-value could be given too much weight over an exact match, which could raise the $enSimTB$ value much higher than necessary on a "few" good (or exact) matches. The $enSimTB$ measure should further enhance the $SimTB$ value defined in Equation 4 and its variation for phrase-correlation factors as discussed in Section 3.4, since neither one counts the *frequencies of occurrence* of related phrases in T and B.

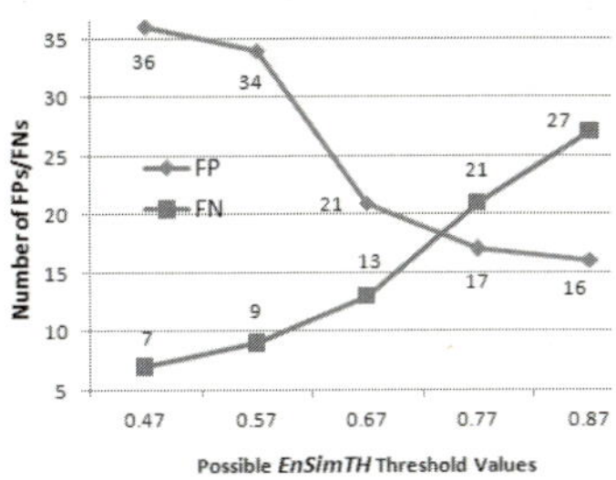

(a) $EnSimTB$ threshold values

<title>Loan Advice – your one stop loan advice site, loan broker, load lender, loan quote, load UK </title>
<body>... Cash Advance Payday Loans. Do you know the Benefits of Checking your Credit Report? The Effects of Consumer Debt. How To Eliminate Credit Card Debt. Credit and Debt Relief–A One Stop Solution. Credit Establishment 101. Comprehending a Credit Report... </Body>

(b) A sample *spam* Web page misclassified by the $SimTB$ value

Fig. 6. Determination of the ideal $EnSimTB$ threshold value and a classification example using the $SimTB$ versus $EnSimTB$ value based on the *bigram*-similarity

Table 5. Some of the bigram-similarity and *spcf*-values for the *bigrams* in the title T of the Web page in Figure 6(b) with respect to the *bigrams* in its body B

Bigrams in the Title	Bigrams in the Body						
	cash advance	advance payday	credit report	card debt	...	*spcf* value	*Min* (*spcf*, 1)
loan advice	1.5×10^{-9}	2.1×10^{-14}	5.0×10^{-16}	6.5×10^{-16}	...	2.14	1.00
advice site	2.7×10^{-16}	2.6×10^{-15}	4.5×10^{-15}	2.9×10^{-15}	...	7.2×10^{-12}	7.2×10^{-12}
site loan	4.5×10^{-13}	1.5×10^{-16}	1.3×10^{-15}	2.3×10^{-15}	...	9.7×10^{-10}	9.7×10^{-10}
...	...	...	...	...	...	...	...
				enSimTB	**4.0**	*EnSimTB*	**0.5**

To avoid the *length* bias in T, we normalize an *enSimTB* value as $EnSimTB(T, B) = \frac{enSimTB(T,B)}{n}$, where n is the total number of bigrams (trigrams, respectively) in T. Thus, $0 \leq EnSimTB(T, B) \leq 1$.

The *EnSimTB* Threshold Value. We define the appropriate *threshold value* for *EnSimTB*, which yields the cut-off value between spam and legitimate Web pages. Using the same *Threshold set* and different possible threshold values, we determine the number of FPs and FNs for each of the possible thresholds for *EnSimTB*. As shown in Figure 6(a), the *bigram EnSimTB-threshold* value should be **0.67**, which yields the minimal sum of FPs and FNs. (Note that the *trigram EnSimTB-threshold* value is not computed, since bigrams outperform trigrams in similarity measure and we only consider bigrams from here on.)

Example 3. Table 5 shows how closely related (some of) the *bigrams* in the title of the *spam* Web page P in Figure 6(b) are to the ones in its body. Using the bigram *SimTB* value of P, which is 0.75, P is misclassified as *legitimate*, since $SimTB(T_P, B_P) \geq 0.75$, the bigram *SimTB* threshold value. However, when the *EnSimTB* value is considered instead, P is correctly classified as *spam*, since $EnSimTB(T_P, B_P) = 0.5 < 0.67$, the bigram *EnSimTB* threshold value. □

The overall design of our spam-detection process is shown in Figure 7. Note that by considering the fraction of *hidden content* or the *EnSimTB* value of a Web page P after the *SimTB* value of P has been calculated, we are able to further reduce the number of FPs and FNs. (See details in Section 4.)

4 Experimental Results

In this section we discuss the dataset used for our empirical study and show the accuracy of our spam-detection approach in using n-gram ($1 \leq n \leq 3$) phrases, which verifies the effectiveness of our approach in detecting spam Web pages. In addition, we compare the performance of our spam-detection approach with other well-known, existing anti-spam methods.

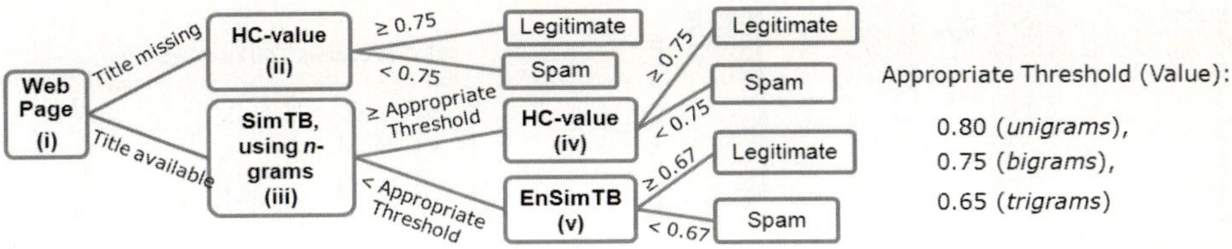

Fig. 7. The overall Web spam-detection process

4.1 Web Page Dataset

To show the *accuracy* of our spam-detection approach, which is measured by the number of FPs and FNs, we used the *Verification set*,[5] which consists of 1,040 randomly selected Web pages—370 labeled spam and 670 non-spam—from the WEBSPAM-UK2006 dataset. As stated in [5], WEBSPAM-UK2006 is appropriate and representative in establishing the accuracy of a given spam-detection approach, since the collection (i) includes a large variety of spam and non-spam Web pages, (ii) represents uniform random sample, (iii) consists of spam Web pages created by using different spam techniques, and (iv) is freely available to be used as a benchmark measure in detecting spam Web pages.

4.2 Accuracy of Our Approach in Using *SimTB* with(out) the *HC*-Value

We first verified (i) the effectiveness of our spam-detection approach in using n-grams ($1 \leq n \leq 3$) and (ii) the most accurate n-gram phrases in determining the $SimTB$ values between the title and the body of a Web page using $Accuracy = \frac{Correctly\ identified\ Web\ pages}{Total\ of\ Web\ pages}$ and $Error\ Rate$ = 1 - $Accuracy$, where *correctly identified Web pages* is the total number of Web pages minus FPs and FNs.

As shown in Figure 8(a), using *bigrams* and the $SimTB$ values on the Web pages in the *Verification set* yields the *accuracy* and *error rate* of 83% and 17% respectively, which outperforms the *unigram* and *trigram* approaches. Figure 8(b) shows the number of FPs and FNs of different n-grams ($1 \leq n \leq 3$) in misclassifying Web pages. According to the experimental results, *bigrams* significantly reduce the number of FPs and FNs as opposed to the number of FPs and FNs generated by using *unigrams* or *trigrams* in computing the $SimTB$ values.

We have observed that bigrams outperform trigrams because (closely) related 3-word phrases between the title and the body of a Web page occur less often than (closely) related 2-word phrases. As a result, the degree of similarity between the title and the body of a Web page is *lower* in using *trigrams* than *bigrams*, causing a *higher* number of FPs and FNs.

We further compared how well our spam-detection approach performs when considering both (i) bigram-similarity among the words in the title T and the

[5] Web pages in the *Verification set* are different from the ones in the *Threshold set*.

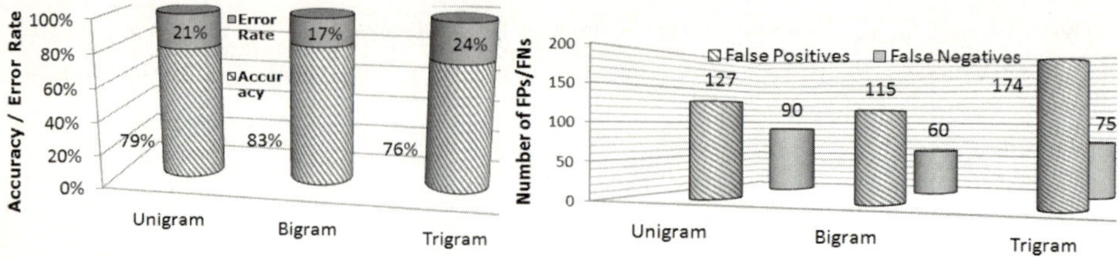

(a) The accuracy and error rates (b) The number of FPs and FNs

Fig. 8. Experimental results on using n-gram ($1 \leq n \leq 3$) phrases in determining the $SimTB$ values of the Web pages in the *Verification set*

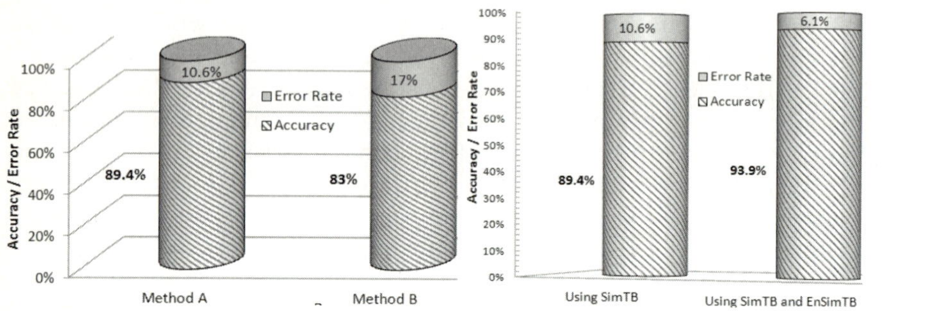

(a) *Accuracy* using Method A (bigram $SimTB$ + HC) versus using Method B (bigram $SimTB$ only)

(b) *Accuracy* and *Error Rates* of the $SimTB$ values with(out) the $EnSimTB$ values

Fig. 9. Experimental results computed on the Web pages in the *Verification set*

body B of a given Web page P and (ii) the fraction of *hidden content* of P (Method A), i.e., Steps (iii) and (iv) in Figure 7, as opposed to only considering the bigram-similarity measure between T and B of P (Method B), i.e., Step (iii) only, using the $SimTB$ values. Figure 9(a) shows that the accuracy of our approach is increased by close to 7% in applying Method A than Method B.

4.3 The Overall Accuracy of Our Spam Detection Approach

We have conducted further comparisons on using the bigram phrase-$SimTB$ measure with(out) the $EnSimTB$ values on the Web pages in the *Verification set*, i.e., Step (iii) with(out) Step (v) in Figure 7. As shown in Figure 9(b), we increase the accuracy by nearly 5%, yielding an accuracy ratio of 93.9% in detecting spam Web pages when using bigrams in computing the $SimTB$ and $EnSimTB$ values, instead of using solely the $SimTB$ values.

Even more so, by considering the HC-value as well as the $EnSimTB$ value (i.e., Step (iv) and (v) in Figure 7), in addition to the $SimTB$ value, we further reduce the number of FNs (as shown in Figure 10(a)) and obtain an *overall*

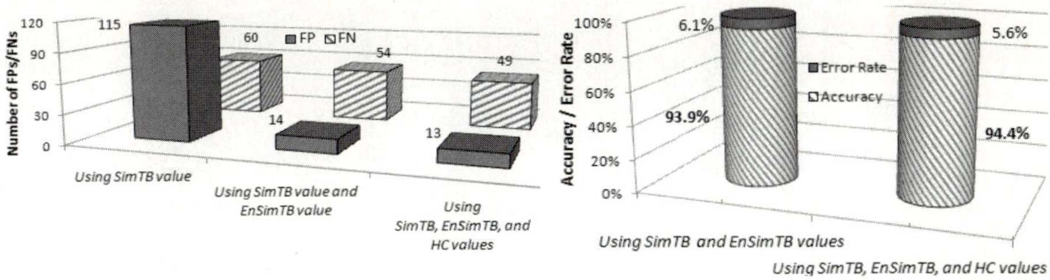

(a) Computed FPs and FNs (b) Computed Accuracy-Error Rates

Fig. 10. Experimental results of applying the previously described spam detection approaches on the Web pages in the *Verification set*

accuracy ratio of 94.4% (as shown in Figure 10(b)) without significantly increasing the computational complexity, since it requires only $\mathcal{O}(n)$ time in calculating the HC value, where n in the number of characters in a Web page P, and $\mathcal{O}(m^2)$ in computing the $EnSimTB$ value, where m is the number of bigrams in P.

4.4 Comparing the Performance of Our Spam-Detection Approach with Other Anti-spam Methods

We further compare the performance (in terms of *precision* and *recall*) of our spam-detection approach with other well-known anti-spam methods in [6], which consider link-based [1] and content-based [16] features, and the combination of both. The features described in [6], which include the degree-related measures, PageRank, TrustRank [10], and features described in [16], such as the number of words and average word length in a page, are served as inputs to the C4.5 decision-tree. Furthermore, [6] enhance the spam-detection accuracy by (i) implementing a graph clustering algorithm that evaluates whether the majority of hosts in a cluster C are spam and if so *all* the hosts in C are considered spam; (ii) applying the graph topology to smooth "spamicity" predictions by propagating them using random walks [20]; and (iii) using a stacked graphical learning scheme [7] to improve the quality of the original predictions.

In comparing the existing anti-spam methods listed above with ours, we consider the evaluation method defined in [6], which adopts the following matrix:

		Prediction	
		Non-Spam	Spam
True Label	Non-Spam	a	b
	Spam	c	d

[6] compute the *True Positive Rate* (or *recall*) $= \frac{d}{c+d}$, *False Positive Rate* $= \frac{b}{a+b}$, and *F-Measure* $= \frac{2 \times precision \times recall}{precission + recall}$, where *precision* is defined as $\frac{d}{b+d}$. *High* recall and precision translate into *high F-measure*, whereas *low* precision and

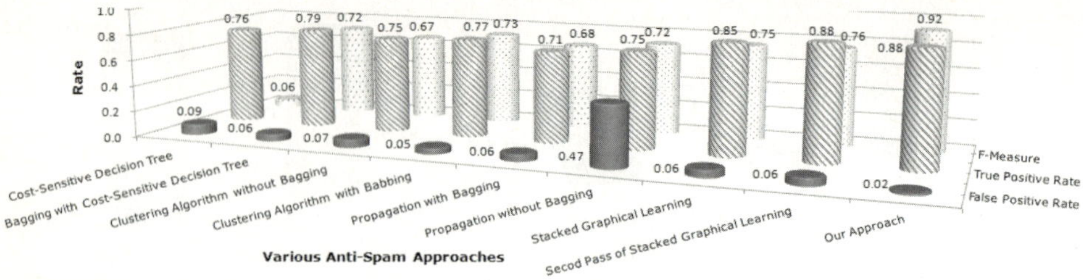

Fig. 11. The *False Positive Rate*, *True Positive Rate*, and *F-Measure* computed by using the WEBSPAM-UK2006 dataset applied to the approaches in [6] and ours

recall yield *low F-measure.* Furthermore, *high* (*low*, respectively) recall and *low* (*high*, respectively) precision generate *low F-measure.*

We used the Web pages in the WEBSPAM-UK2006 dataset to determine the *precision* and *recall* ratios, which dictate the *F-Measure*, of our spam-detection approach and the anti-spam methods in [6]. Figure 11 shows the results reported in [6] for different Web anti-spam detection methods using the classifier with the highest *F-Measure*, as well as the results generated by using our approach. Our spam-detection method clearly outperforms the other anti-spam methods by at least 10% (on the average) in terms of *F-Measure*, which indicates that we obtain *high* precision and recall in detecting spam Web pages, i.e., correctly identifying spam Web pages while avoiding misclassifying legitimate Web pages.

5 Conclusions and Future Work

In this paper, we present a spam-detection approach that can effectively identify spam Web pages to aid search engines in performing more adequate searches. Our anti-spam approach minimizes the user's time in looking through pages that are deceitful and do not contain useful information. In designing our anti-spam method, we consider (i) the (enhanced) *similarity* measures of phrases in the title with respect to the ones in the body of a Web page P, and (ii) the fraction of *hidden content* of P, if necessary, to determine whether P is spam. Experimental results show that by using our approach, we can classify spam Web pages with 94.4% accuracy. Even more so, our approach outperforms existing anti-spam approaches by close to 10% on the average in F-measure. Furthermore, our approach is computational inexpensive, since (i) the word-correlation factors used for computing the phrase-correlation factors are precomputed and (ii) the computational time to calculate the fraction of hidden content is insignificant.

We have observed that the use of *bigrams* significantly increases the performance of our spam-detection approach. Since the bigram-correlation values employed in our spam-detection tool are computed by using the unigram-correlation factors, we believe that constructing a phrase-correlation matrix directly from the Wikipedia documents could further enhance the performance of our approach in terms of (i) minimizing misclassified Web pages and (ii) reducing the computational time required to determine the $(En)SimTB$ values of Web pages.

References

1. Becchetti, L., Castillo, C., Donato, D., Leonardi, S., Baeza-Yates, R.: Link-Based Characterization and Detection of Web Spam. In: 2nd Intl. Workshop on Adversarial Information Retrieval on the Web (AIRWeb), pp. 1–8. ACM Press, New York (2006)
2. Becchetti, L., Castillo, C., Donato, D., Leonardi, S., Baeza-Yates, R.: Using Rank Propagation and Probabilistic Counting for Link-Based Spam-Detection. In: Workshop on Web Mining and Web Usage Analysis, pp. 1–8 (2006)
3. Benczur, A., Csalogany, K., Sarlos, T., Uher, M.: SpamRank-Fully Automatic Link Spam-detection. In: 1st AIRWeb Workshop, pp. 25–38. ACM, New York (2005)
4. Brin, S., Page, L.: The Anatomy of a Large-Scale Hypertextual Web Search Engine. In: 7th Intl. World Wide Web Conf., ACM, New York (1998)
5. Castillo, C., Donato, D., Becchetti, L., Boldi, P., Santini, M., Vigna, S.: A Reference Collection for Web Spam. SIGIR Forum 40(2), 11–24 (2006)
6. Castillo, C., Donato, D., Gionis, A., Murdock, V., Silvestri, F.: Know your Neighbors: Web Spam-detection Using the Web Topology. In: Intl. ACM SIGIR Conf., pp. 423–430. ACM, New York (2007)
7. Cohen, W.W., Kou, Z.: Stacked Graphical Learning: Approximating Learning Markov Random Fields Using Very Short Inhomogeneous Markov Chains. Technical Report, Machine Learning Department, Carnegie Mellon University (2006)
8. Davison, B.: Recognizing Nepotistic Links on the Web. In: Artificial Intelligence for Web Search, pp. 23–28. AAAI Press, Menlo Park (2000)
9. Fetterly, D., Manasse, M., Najork, M.: Spam, Damn Spam, and Statistics: Using Statistical Analysis to Locate Spam Web Pages. In: 7th Intl. Workshop on the Web and Databases (WebDB), pp. 1–6 (2004)
10. Gyongyi, Z., Garcia-Molina, H., Pedersen, J.: Combating Web Spam with TrustRank. In: 30th Intl. Conf. on VLDB, pp. 576–587. ACM, New York (2004)
11. Gyongyi, Z., Garcia-Molina, H.: Web Spam Taxonomy. In: 1st Intl. Workshop on Adversarial Information Retrieval on the Web (AIRWeb), pp. 39–47. ACM, New York (2005)
12. Gyongyi, Z., Berkin, P., Garcia-Molina, H., Pedersen, J.: Link Spam-Detection Based on Mass Estimation. In: Intl. Conf. on VLDB, pp. 439–450. ACM, New York (2006)
13. Judea, P.: Probabilistic Reasoning in the Intelligent Systems: Networks of Plausible Inference. Morgan Kauffman, San Francisco (1988)
14. Lam-Adesina, A., Jones, G.: Applying Summarization Techniques for Term Selection in Relevance Feedback. In: Intl. ACM SIGIR Conf., pp. 1–9. ACM, New York (2001)
15. Misjne, G., de Rijke, M.: Boosting Web Retrieval through Query Operations. In: European Conf. on Information Retrieval, pp. 501–516 (2005)
16. Ntoulas, A., Najork, M., Manasse, M., Fetterly, D.: Detecting Spam Web Pages through Content Analysis. In: Intl. Conf. on World Wide Web, pp. 83–92 (2006)
17. Pera, M.S., Ng, Y.-K.: Using Word Similarity to Eradicate Junk Emails. In: 16th ACM Conf. on Information and Knowledge Management, pp. 943–946 (2007)
18. Perkins, A.: The classification of Search Engine Spam (2001), `http://www.silverdisc.co.uk/articles/spam-classification/`
19. Svore, K.M., Wu, Q., Burges, J.C., Raman, A.: Improving Web Spam Classification Using Rank-Time Features. In: 3rd AIRWeb Workshop, pp. 9–16. ACM, New York (2007)
20. Zhou, D., Bousquet, O., Lal, T.N., Weston, J., Scholkopf, B.: Learning with Local and Global Consistency. Advance in Neural Info. Proc. Sys. 16, 321–328 (2004)

Categorical Data Clustering Using the Combinations of Attribute Values

Hee-Jung Do and Jae-Yearn Kim

Department of Industrial Engineering, Hanyang University
Sungdong-gu, Seoul 133-791, Korea
hyhjd4@hanyang.ac.kr
jyk@hanyang.ac.kr

Abstract. Clustering is an important technique for exploratory data analysis. While most of the earlier clustering algorithms focused on numerical data, real-world problems and data mining applications frequently involve categorical data. Here, we propose a new clustering algorithm for categorical data that is based on the frequency of attribute value combinations. Our algorithm finds all the combinations of attribute values in a record, which represent a subset of all the attribute values, and then groups the records using the frequency of these combinations. As our algorithm considers all the subsets of attribute values in a record, records in a cluster have not only similar attribute value sets but also strongly associated attribute values. We evaluated our algorithm with real and synthetic data sets, and the experimental results demonstrate the effectiveness of our algorithm.

Keywords: Data mining, Categorical clustering, Attribute value sets.

1 Introduction

Clustering is an important technique for exploratory data analysis [6]. Clustering problems occur in a wide range of applications and are an important component of data mining, statistical pattern recognition, machine learning, and information retrieval [14]. In data mining, clustering is widely used for grouping objects into classes or groups so that objects within one group are more similar to each other to than those in other groups.

Most of the earlier clustering algorithms focused on numerical data, where the clustering was based on a distance measure such as the Euclidean distance. However, data in the real world or in data mining applications frequently involve categorical data. The clustering of categorical data is a difficult, yet important task [5].

Categorical data consist of attribute sets whose domain is not numeric, and each attribute takes on a set of values that is not ordered. The most important feature of categorical data is that there is no natural distance between two objects, as there is no obvious order in their values [8]. Therefore, the traditional approach of converting categorical data into numeric values does not necessarily produce meaningful results [1]. Also, transforming categorical values to numeric values leads to loss of semantics and waste of storage when the domain of the categorical attribute is large. In addition,

O. Gervasi et al. (Eds.): ICCSA 2008, Part II, LNCS 5073, pp. 220–231, 2008.

Rock algorithm [4] showed that the Euclidean distance can be a poor measure of similarity for categorical attributes.

Therefore, similarity measures such as the Jaccard coefficient have often been used for clustering categorical data instead of the Euclidean distance. However, as the Jaccard coefficient computes the similarity between two objects only, it does not reflect the properties of the neighborhood of the objects. Thus, it merely provides local information involving only two objects, and cannot provide global information about the other objects. The Jaccard coefficient fails to capture the natural clustering of "not-so-well-separated" data sets with categorical attributes [4].

Here, we propose a new clustering algorithm for categorical data that is based on the frequency of attribute value combinations (FAVC). The FAVC algorithm examines all the combinations of attribute values in a record, and then groups records using the frequency of these combinations. That is, the combinations of the attribute value set represent a subset of attribute values. For categorical attributes, the more common attribute values that a cluster includes, the better the cluster is. Therefore, a cluster consists of attribute value sets whose frequency is higher than in any other cluster. However, even though a cluster has many common or similar attributes, it may be difficult to find associations among attribute values in this cluster. As the FAVC algorithm considers all the subsets of attribute values in a record, it can easily find associations among attribute values in a cluster. The FAVC algorithm helps to solve the problems mentioned above, and uses the fact that records in a cluster have not only similar attribute value sets but also strongly associated attribute values.

This paper is organized as follows. We review related works In Section 2. We establish the problem and describe our FAVC clustering algorithm in Section 3. We present the results of experimental studies on real and synthetic data sets in Section 4, and give our conclusions in Section 5.

2 Related Works

2.1 Clustering Algorithms for Categorical Data

Clustering algorithms can be divided into those with partitional and hierarchical clustering methods. A partitional clustering algorithm partitions the objects into k clusters based on a distance measure such as the Euclidean distance. A hierarchical clustering algorithm groups objects into trees of clusters [10].

Here, we review clustering algorithms for categorical data.

The k-modes algorithm [1] is an extension of the k-means algorithm for clustering categorical data that uses modes as the centroid of a cluster instead of means. For each attribute, the mode is the most frequent attribute value. However, choosing only one attribute value in each attribute domain, such as the highest frequency attribute value, can cause problems. For example, suppose that a gender attribute in a cluster is 49% male and 51% female. If only female is chosen as the centroid of this cluster, then the male information will be lost. Because males make up almost half of this cluster, there is just about as much male information as female [14].

The "robust clustering using links" (ROCK) algorithm [4] is a hierarchical clustering algorithm that treats market-basket data as categorical data. ROCK uses the notion of links between objects for clustering. A link is defined as the number of common neighbors between two objects, which can be neighbors if their similarity exceeds a user-defined threshold, ∂. The choice of threshold ∂ is the critical factor in the success of clustering in the ROCK algorithm, since the threshold is difficult to define in practice.

COOLCAT [5] is an entropy-based algorithm for clustering categorical data. Entropy is an effective tool for measuring the information and uncertainty of a random variable. COOLCAT uses the principle that the entropy of a set is inversely proportional to the number of common attribute values of the set. COOLCAT allocates objects into clusters to minimize the expected entropy of the overall cluster. In practice, the time complexity of COOLCAT can restrict the data sizes it can handle.

The "clustering data using summaries (CACTUS)" algorithm [3] is a fast summarization-based algorithm for clustering categorical data. CACTUS examines each pair of attribute values and detects the strongly connected pairs that exceed a threshold. The cluster consists of the strongly connected attribute values. CACTUS defines inter-attribute and intra-attribute summaries, which indicate how closely related values are from different and from the same values, respectively. However, CACTUS may lose information depending on the threshold; strongly associated items can often removed by a low threshold.

2.2 Evaluation of Clustering Results

A number of experimental models have been proposed to evaluate the quality of clustering results. The Davies-Bouldin (DB) index and the Calinski Harabasz (CH) index are among the many cluster evaluation models that are only suitable for numeric data [13]. In this paper, we use two techniques to evaluate the clustering effectiveness for categorical data.

The first is the category utility (CU) function. It attempts to maximize both the probability that two tuples in the same cluster have attribute values in common, and the probability that tuples from different clusters have different attributes. The function is shown in Eq. 1. This is the expression to calculate the expected value of the CU function, where $P(A_i = V_{ij} \mid C_k)$ is the conditional probability that attribute i has the value V_{ij} given cluster C_k, and $P(A_i = V_{ij})$ is the overall probability of the attribute i having the value V_{ij} in the entire data set. The higher the CU value, the better the clustering result.

$$CU = \sum_k \frac{C_k}{|D|} \sum_i \sum_j \left[P(A_i = V_{ij} \mid C_k)^2 - P(A_i = V_{ij})^2 \right] \quad (1)$$

The second measure of clustering quality is the expected entropy, which has often been used to represent clustering quality. For a discrete random variable X, D(X) is the set of values that X can take, and p(x) is the probability function of X. Then the entropy E(x) can be defined as shown in Eq (2).

$$E(X) = -\sum_{x \in D(X)} p(x) \log(P(x)) \quad (2)$$

The expected entropy of K clusters is defined Eq. (3),

$$E(\overline{C}) = \sum_k \left(\frac{|P(C_k)|}{|D|} E(P(C_k))\right) \quad (3)$$

where $E(P(C_k)$ represents the entropies of each cluster, and $|P(C_k)|$ denotes the points assigned to cluster C_k for $k = 1,..., K$. The symbol $\overline{C} = \{C_1,..., C_k\}$ represents the clustering.

3 FAVC Algorithm Model

In this section, we present our FAVC algorithm that is based on the frequency of attribute value combinations. We define the problem description in Section 3.1 and introduce the FAVC algorithm in Section 3.2.

3.1 Problem Description

The clustering problem we propose can be formulated as follows.

In this paper, data set D is a set of tuples(records) defined as $D = \{t_1,..., t_n\}$, where tuple t has a set of categorical attributes $A = \{A_1,..., A_k\}$, and each attribute A_i has v distinct values $\{a_1,..., a_v\}$. Tuple t is represented as $t = \{t.a_i,..., t.a_k\}$ and has a *k*-attribute value set. An example data set is described in Fig. 1 where there are 15 tuples, each of which has a tuple identification (TID). In the example data set of Fig. 1, tuple 1(TID_1) is represented as $t_1 = \{amz\}$ and has a 3-attribute value set. A clustering $C = \{C_1,..., C_j\}$ is a partition of the tuples into *j* clusters, where C_i is a cluster consisting of a set of tuples.

Definition 1. The combination of attribute value sets for tuple $t = \{t.a_1,...,t.a_k\}$ is denoted by $S(t)$. The combination of attribute value sets for each tuple $t_i = \{t_i.a_1,...,t_i.a_k\}$ is represented by $S(t_i)$. $S(t)$ is defined as the set of the subsets of attribute values containing each tuple except zero. The combination sets in $S(t)$ are the factors of $S(t)$.

In the example data set shown in Fig. 1, $S(t)$ of tuple 1 (TID_1), $t_1 = \{amz\}$, is $S(t_1) = [\{amz\},\{am\},\{mz\},\{az\},\{a\},\{m\},\{z\}]$ and {amz}, {am}, {mz}, {az}, {a}, {m}, and {z} are factors of $S(t_1)$.

Definition 2. The frequency of the $S(t_i)$ factors in each cluster $C_1,..., C_j$ is denoted by *Freq* $(S(t_i),\ C_{j=1,...,j})$. It is defined as the sum of the number of tuples containing the $S(t_i)$ factors in each cluster $C_1,..., C_j$.

TID	A_1	A_2	A_3	TID	A_1	A_2	A_3	TID	A_1	A_2	A_3
1	a	m	z	6	b	m	y	11	c	m	x
2	a	f	x	7	a	m	x	12	b	m	x
3	b	m	y	8	c	f	y	13	c	f	x
4	b	m	z	9	c	f	x	14	a	f	y
5	c	f	z	10	c	m	z	15	a	m	z

Fig. 1. An example of data set D

For example, suppose there are three clusters. The frequency of the $S(t_1)$ factors for tuple 1(TID_1) in each cluster $C_1, C_2,$ *and* C_3 is represented by $Freq(S(t_1), C_1)$, $Freq(S(t_1), C_2)$, and $Freq(S(t_1), C_3)$.

By computing all the combinations of the attribute value set for the tuple, we can cluster not only tuples consisting of the common or similar attribute value set, but also tuples with strongly associated attribute values.

A similarity measure, such as the Jaccard coefficient, computes the number of attribute values common to the two tuples. Although this measure can cluster tuples with common or similar attributes, it is less concerned with associations among attribute values. For example, consider the attribute value set, {Honda, August, New York}. The attribute sets which have a high Jaccard coefficient of 0.5 can be {Honda, September, New York} and {Honda, August, Chicago}. As these are very similar, they are grouped in the same cluster. However, it is difficult to find any association among these attribute value sets.

In the FAVC algorithm, attribute value sets, such as {Honda, August, New York}, {Honda, August, Chicago}, and {Honda, August, LA}, are in the same cluster. Here,

Structure of FAVC Algorithm

Step 1. Randomly partition tuples into k clusters from the dataset D.
Step 2. Read each tuple sequentially, compute $Freq(S(t_i), C_{j=1,\ldots,k})$, and allocate the tuple to the cluster with maximum $Freq(S(t_i), C_{j=1,\ldots,k})$.
Step 3. Update the cluster.
Step 4. Repeat Step 2 until no tuple is moved between clusters.
Step 5. Compute the expected entropy for the clustering result generated in Step 4.
Step 6. Repeat Steps 1–4 until finding the cluster with the minimum expected entropy within user-defined threshold ∂.
Step 7. Terminate FAVC when we output clustering result with good quality within threshold ∂.

Fig. 2. Overall structure of the FAVC algorithm

we can understand that common values, [Honda, August], in these attribute value sets represent co-occurrence. Thus, we can say that "Honda" and "August" are strongly associated in this cluster. This is because we compute all combinations of the attribute values for the tuples.

3.2 FAVC Algorithm Sequence

In this section, we describe the FAVC algorithm. Fig. 2 shows its overall structure. The contribution of the FAVC algorithm is found in Steps 1–4, while Steps 5–6 are for generating good quality clusters.

In Step 1, we randomly partition the tuples from the data set *D* into *j* clusters to form the initial clustering. We use a random method to form the initial clustering, similar to those used in other partitional clustering algorithms.

In Step 2, each tuple is read sequentially from data set *D*, and $Freq(S(t_i), \ C_{j=1,...,j})$ is computed with the random clusters generated in Step 1. A tuple is allocated to the cluster with maximum $Freq\ (S(t_i), \ C_{j=1,...,j})$, and then the cluster is updated. If two maximum values of $Freq(S(t_i), \ C_{j=1,...,j})$ are the same, we compare the frequency of $S(t_i)$ factors using a level-wise approach, where he next to *k*-attribute value set, the *k-1* attribute value set is compared in $S(t_i)$ factors. In the example shown in Fig. 4, if the sums of C_1 and C_2 were the same, we would first compare the value of the S(t) factor with the k-attribute value set . Here, k=3, and as the 3-attribute value set is {amz}, the first value compared is {amz}. As the {amz} value in C_1 is 1 and that in C_2 is 0, we select C_1. Again, if the values of {amz} were also the same, we would compare the values of {am}, {mz} and {az}. If the values of the 2-attribute value set were also the same, we would compare the 1-attribute value set.

Example. Consider the data set in Fig.1. We set j=3. We randomly partition the tuples from the data set *D* into 3 clusters to form the initial clustering, so each cluster contains 5 tuples. The random initial clustering is shown in Fig. 3. Each tuple is read sequentially from data set *D*, and $Freq\ (S(t_i), \ C_{j=1,...,j})$ is computed for the initial clustering. In Fig. 1, the first tuple TID_1 has the attribute value set {amz}. Therefore, the $S(t_1)$ of TID_1 is $S(t_1) = [\{amz\}, \{am\}, \{mz\}, \{az\}, \{a\}, \{m\}, \{z\}]$, and the factors of $S(t_1)$ are {amz}, {am}, {mz}, {az}, {a}, {m}, and {z}. Then, we compute $Freq\ (S(t_1), \ C_{1,2,3})$ for each cluster in Fig. 3. The results of $Freq(S(t_1), \ C_{1,2,3})$ are shown in Fig. 4. The {amz} factor occurs once in C_1, not at all in C_2, and once in C_3. The {am} factor occurs once in C_1, once in C_2, and once in C_3. The {mz} factor occurs twice in C_1, once in C_2, and once in C_3. The {az} factor occurs once in C_1, not at all in C_2, and once in C_3. The other factors are computed as above. Therefore, the sums of the number of tuples containing the $S(t_1)$ factors in C_1, C_2,

C1				C2				C3			
TID	A_1	A_2	A_3	TID	A_1	A_2	A_3	TID	A_1	A_2	A_3
1	a	m	z	6	b	m	y	11	a	m	z
2	a	f	x	7	a	m	x	12	b	m	x
3	b	m	y	8	c	f	y	13	c	f	x
4	b	m	z	9	c	f	z	14	a	f	y
5	c	f	z	10	c	m	x	15	a	m	z

Fig. 3. The initial clustering

and C_3 are $Freq(S(t_1),\ C_1) = 13$, $Freq(S(t_1),\ C_2) = 7$, and $Freq(S(t_1),\ C_3) = 10$, respectively. As C_1 is the maximum at 13, TID_1 is allocated to C_1. TID_1 does not move to another cluster, while TID_15 is moved from C_3 to C_1. As TID_1 and TID_15 have identical attribute values (amz), they are grouped in the same cluster.

In Step 3, because of the movement of the cluster in Step 2, the clusters are updated from their initial state as shown in Fig 5(a)

In Step 4, the algorithm repeats Step 2 with the updated clusters from Step 3, until no tuples move between clusters. Fig 5(b) shows the final clustering results.

In Step 5, we compute the expected entropy for the clustering results generated in Steps 1–4. The expected entropy is one of the measures capable of evaluating clustering results. As the FAVC algorithm involves partitional clustering, it takes the input parameter, *j*, and arbitrarily partitions the objects into *j* clusters for initial clustering. Depending on the initial random clustering, the clustering result can turn out differently.

attribute set	C 1	C 2	C 3
{amz}	1	0	1
{am}	1	1	1
{mz}	2	1	1
{az}	1	0	1
{a}	2	1	2
{m}	3	3	3
{z}	3	1	1
sum	13	7	10

Fig. 4. Results of Freq(S(t_1), $C_{1,2,3}$)

Therefore, in Step 6, we iterate Steps 1–4 to find an even better clustering result using the expected entropy after evaluating the quality of the clustering result generated by Steps 1-4.

In Step 7, we terminate our algorithm when the clustering results have the minimum expected entropy within a user-defined threshold ∂. The final clustering results in Fig. 5(b) are the clustering results generated by Steps 1-6.

In the final clustering results of Fig. 5(b), C_2 has the common attribute values [c and f], and these attribute values group only in C_2. Thus, common values [c and f] represent co-occurrence in C_2, and we can therefore say that [c and f] are strongly associated in C_2.

3.3 FAVC Complexity

The time complexity of FAVC is $O(rC(2^k - 1) \mid D \mid)$, where *r* is the number of iterations, *C* is the given cluster number, *k* is the number of attributes, and |*D*| is the data size. This is because the combination of *k*-attribute value sets for each tuple is calculated in each cluster, and is iterated until there is no movement between clusters.

C_1			
TID	A_1	A_2	A_3
1	a	m	z
2	a	f	x
3	b	m	y
4	b	m	z
5	c	f	z
6	b	m	y
15	a	m	z

C_2			
TID	A_1	A_2	A_3
8	c	f	y
9	c	f	x
10	c	m	x
13	c	f	x

C_3			
TID	A_1	A_2	A_3
7	a	m	x
11	c	m	x
12	b	m	x
14	a	f	y

(a)

C_1			
TID	A_1	A_2	A_3
1	a	m	z
2	a	f	x
3	b	m	y
4	b	m	z
6	b	m	y
10	c	m	z
15	a	m	z

C_2			
TID	A_1	A_2	A_3
5	c	f	z
8	c	f	y
9	c	f	z
13	c	f	z

C_3			
TID	A_1	A_2	A_3
7	a	m	z
11	c	m	z
12	b	m	z
14	a	f	y

(b)

Fig. 5. Clustering results: (a) updated and (b) final

4 Experimental Results

We conducted experiments with real and synthetic data sets to evaluate FAVC. We used real data sets to compare the quality of clustering between FAVC and COOLCAT. Our experiments were performed on a Sony laptop computer equipped with 1 GB of memory and a Pentium-M processor running at 2.13 GHz running the Windows XP Professional operating system.

4.1 Real Data Sets

We used two real data sets to evaluate clustering quality.

The "Mushroom" data set was obtained from the UCI Machine Learning Repository [11]. Each data record describes the physical characteristics (e.g., color, odor, shape) of a single mushroom. The data record has a poisonous/edible field for each mushroom, which was not used for clustering. All 22 attributes are categorical and the set contains 8124 records in total: 4,208 for edible mushrooms and 3,916 for poisonous ones.

The "Congressional Voting" data set was also obtained from the UCI Machine Learning Repository. Each record contains a Congressman's votes on 16 issues (e.g., education, spending, and crime). All attributes are Boolean with either Yes or No values. Each record has a Republican/Democrat classification label, which was one of the ones used for clustering. The set contains 435 records in total: 267 Democrats and 168 Republicans.

4.2 Real Data Set Results

Table 1 shows the clustering results for the Mushroom data set using FAVC and COOLCAT. We set the number of desired clusters for both algorithms to 20. We set ∂ to 20 as the number of iterations to find a good quality clustering result, i.e., the FAVC algorithm was iterated 20 times to find the minimum expected entropy. We used samples of various sizes. The expected entropy value obtained by FAVC was almost identical with that obtained by COOLCAT. In addition, the CU value of the FAVC result was almost identical to that of COOLCAT. However, even though the FAVC algorithm included the 20 entropy iterations, its running time was less than that of COOLCAT.

Table 2 shows the clustering result for the Congressional Votes set using both FAVC and COOLCAT. We set the number of desired clusters for both algorithms to 2. We set ∂ to 20, as for the Mushroom set. There was little difference in the results

Table 1. Clustering Results for FAVC(F) and COOLCAT(C) using Mushroom data set

Sample size	Expected entropy		CU		Running times(s)	
	F	C	F	C	F	C
10%	7.93	7.9	6.35	6.3	94	47
20%	10.8	10.7	6.45	6.38	267	1060
30%	11.0	10.91	6.39	6.43	765	2143
40%	9.85	9.86	6.5	6.5	828	2175

Table 2. Clustering Results for FAVC(F) and COOLCAT(C) using Congressional Voting data set

Sample size	Expected entropy		CU		Running times(s)	
	F	C	F	C	F	C
20%	13.86	13.85	2.9	2.9	1.3	0.32
40%	13.85	13.84	2.92	2.9	1.83	0.33
60%	13.87	13.85	2.92	2.92	3	0.36

for different sample sizes. The expected entropy of the FAVC result was almost identical to that produced by COOLCAT. In addition, the CU values showed that the FAVC clustering was almost identical to that of COOLCAT. The FAVC running time was longer than that of COOLCAT because of the iteration times.

4.3 Synthetic Data Sets and Results

We used synthetic data sets to evaluate the scalability of the FAVC algorithm. We used a synthetic data generator [12] to generate data sets in various data sizes from 1–10K, and with the number of attributes in the range of 5–30. Fig. 6 shows the scalability for some data sizes and numbers of attributes. We computed the running time of both FAVC and COOLCAT with these data. For FAVC, we only show the running time for Steps 1–4, and exclude the iterations to minimize the expected entropy (Steps 5–7) as the latter are not part of the main contribution of FAVC. By their nature, the iterations in Steps 5–7 may take a long time, depending on the threshold ∂. Thus, it is difficult to compare the running time with that of COOLCAT directly. The running time of FAVC increased linearly with the data size and the number of attributes. This was also true for COOLCAT, but the running time increased twice as fast as that of FAVC. However, in the case of the small data sets, the COOLCAT running time was better than that of FAVC.

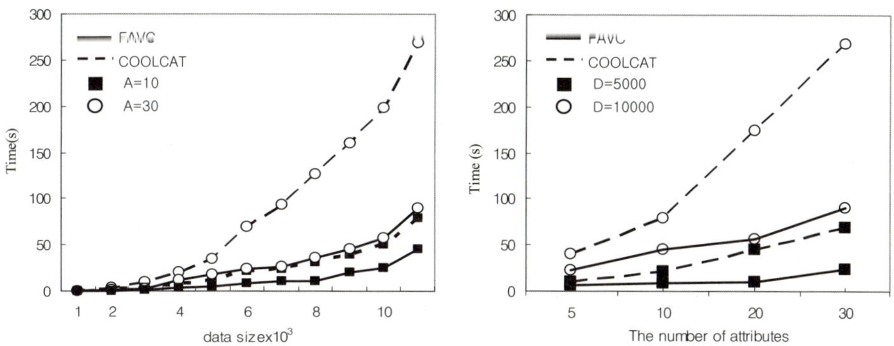

Fig. 6. Running times of FAVC and COOLCAT for synthetic data sets, where D is data size and A is the number of attributes

5 Conclusions

In this paper, we proposed a new clustering algorithm for categorical data based on the frequency of attribute value combinations. The FAVC algorithm calculates all the combinations of attribute values in a tuple, and then groups the tuples using the frequency of these combinations. In our method, objects in a cluster have not only similar attribute value sets but also strongly associated attribute value sets. This is because we perform the clustering, considering all the combinations of attribute value sets in a tuple. We evaluated our algorithm with both synthetic and real data sets.

Our experimental results showed that the FAVC algorithm is stable for various sample sizes and has a linear time complexity relative to data size. The clustering results for FAVC were better than or equivalent to those of COOLCAT, and the FAVC running time was superior to that of COOLCAT, especially for increased data sizes.

References

1. Huang, Z.: A Fast Clustering Algorithm to Cluster Very large Categorical Data Sets in Data Mining. In: Proceedings of ACM SIGMOD Workshop on Research Issues on data Mining and knowledge Discovery (1997)
2. Gibson, D., Kleinberg, J., Raghavan, P.: Clustering Categorical Data: An Approach based on Dynamical. In: Proceedings of the 24th International Conference on Very Large Databases (1998)
3. Ganti, V., Gehrke, J., Ramakrishnan, R.: CACTUS-Clustering Categorical Data Using Summaries. In: Proceedings of the fifth ACM SIGKDD International Conference on Knowledge Discovery and Data Mining, pp. 73–83 (1999)
4. Guha, S., Rastogi, R., Shim, K.: ROCK: A Robust Clustering Algorithm for Categorical Attributes. In: Proceedings of the 15th International Conference on Data Engineering (1999)
5. Barbara, D., Couto, J., Li, Y.: COOLCAT: An entropy-based algorithm for categorical clustering. In: Proceedings of the 2002 ACM CIKM International Conference on Information and Knowledge Management, pp. 590–599 (2002)
6. Yun, C.H., Chuang, K.T., Chen, M.S.: Adherence clustering: an efficient method for mining market-basket clusters. Information Systems 31, 170–186 (2006)
7. Hsu, C.C., Chen, Y.C.: Mining of Mixed data with application to catalog marketing. Expert Systems with Applications (2006)
8. Kim, M., Ramarkrishna, R.S.: Projected clustering for categorical datasets. Pattern Recognition Letters 27, 1405–1417 (2006)
9. Halkidi, M., Batistakis, Y., Vazirgiannis, M.: On Clustering Validation Techniques. Journal of Intelligent Information Systems (2001)
10. Han, J., Kamber, M.: Data Mining Concepts and Techniques. Morgan Kaufmann, San Francisco (2001)
11. UCI machine Learning Repository, http://www.ics.uci.edu/~mlearn/MLRepository.html
12. Dataset Generator (DatGen), http://www.datasetgenerator.com
13. Halkidi, M., Batistakis, Y., Vazirgiannis, M.: Clustering algorithms and validity measures. In: Thirteenth international conference on scientific and statistical database management, pp. 3–22 (2001)

14. Chen, H.L., Chuang, K.T., Chen, M.S.: Labeling Unclustered Categorical Data into Clusters Based on the Important Attribute Values. In: Proceedings of the 5th IEEE International Conference on Data Mining, pp. 106–113 (2005)
15. Mirkin, B.: Reinterpreting the Category Utility Function. Machine Learning, 1–11 (2001)
16. Gluck, A., Corter, J.: Information, Uncertainty, and the utility of categories. In: Proceedings of the Seventh Annual Conference of the Cognitive Science society (1985)
17. Ordonez, C., Omiecinski, E.: Efficient disk-based K-means clustering for relational database. IEEE Transactions on Knowledge and Data Engineering 16(8), 909–921 (2004)

Generating Fuzzy Equivalence Classes on RSS News Articles for Retrieving Correlated Information

Nathaniel Gustafson, Maria Soledad Pera, and Yiu-Kai Ng

Computer Science Department, Brigham Young University, Provo, Utah, U.S.A.

Abstract. Tens of thousands of news articles are posted on-line each day, covering topics from politics to science to current events. In order to better cope with this overwhelming volume of information, RSS (news) feeds are used to categorize newly posted articles. Nonetheless, most RSS users must filter through many articles within the same or different RSS feeds in order to locate articles pertaining to their particular interests. Due to the large number of news articles in individual RSS feeds, there is a need for further organizing articles to aid users in locating non-redundant, informative, and related articles of interest quickly. In this paper, we present a novel approach which uses the word-correlation factors in a fuzzy set information retrieval model to (i) filter out redundant news articles from RSS feeds, (ii) shed less-informative articles from the non-redundant ones, and (iii) cluster the remaining informative articles according to the fuzzy equivalence classes generated on the news articles. Our clustering approach requires little overhead or computational costs, and experimental results have shown that it outperforms other existing well-known clustering approaches.

1 Introduction

These days more information is being transmitted through the Internet on a daily basis than any individual could read or process in an entire lifetime. Besides entire libraries of news articles which have been digitally archived and are publicly available, there is a constant influx of new articles. Efforts to manually and regularly sort through this deluge of new articles posted on any Weblogs, news, and other Web sites to find particular items of interest not only are extremely time-consuming and labor-intensive, but it is also an impractical process. In solving this problem, recent efforts have been made on developing RSS[1] news feeds, which are XML documents that provide users with new, frequently updated news content automatically and allow users to subscribe to it.

Although RSS news feeds do save users considerable amounts of search time by classifying newly posted articles according to their subject areas, users still need to sort through the large number of articles posted within RSS files (of which a single RSS user may subscribe to many) in order to locate articles pertaining to their particular interests. Added to this problem is the presence of *redundant* articles, i.e., articles which contain information already included in other articles in different (or even the same) RSS feeds. In addition, *less-informative* articles, which are not entirely redundant, but which include significant overlapped information appeared in other articles, are common. For

[1] "RSS" refers to the following standards: Rich Site Summary (RSS 0.91), RDF Site Summary (RSS 0.9 and 1.0), and Really Simple Syndication (RSS 2.0).

O. Gervasi et al. (Eds.): ICCSA 2008, Part II, LNCS 5073, pp. 232–247, 2008.

example, a brief, breaking news article would become redundant or less-informative when its updated version is posted, with additional, detailed information on the same event. All of these increase the volume of articles which users must sort through in order to find useful information. Furthermore, there is a need for grouping newly posted articles by *content*, in order to aid users in seeking related information instantly. In this paper, we present an approach which employs the fuzzy set information retrieval (IR) model to cluster non-redundant, informative, and related RSS news articles.

While other existing clustering techniques (see details in Section 2) perform well under certain conditions, the conditions for such techniques are often restrictive. Some require user feedback, whereas others need a very large data set for analysis before effective clustering may be performed. Our redundant/less-informative article detection and relevant clustering technique has the advantage of (i) being context free, i.e., no preprocessing or user input is necessary and (ii) computationally effective. This, of course, is an enormous benefit for implementation in real-world systems.

The remainder of this paper is organized as follows. Section 2 discusses other work performed which pertains to either detecting replicated information or clustering (news articles). Section 3 describes the fuzzy set IR model, which is used as a basis for our informative-document detection approach. Section 4 introduces the fuzzy equivalence relation we use for grouping related news articles. Section 5 presents our clustering results and an analysis thereof. Section 6 gives our concluding remarks.

2 Related Work

Among the huge amount of related work in document/text/data analysis, we narrow our discussion on representative work closely related to detecting and/or clustering similar contents (topics) in either RSS news articles or Web documents.

The objective of topic detection and tracking (TDT) in [16] is to identify and label stories in several news streams that pertain to new or previously unidentified events to find those that discuss the same event as the one specified by a user. [2] suggest using a TF-IDF algorithm to recognize terms that explain the main topics in a weekly news archive and then cluster the sentences in the news articles with higher average weight according to the topics using a sentence vector. This approach, however, does not deal with the problem of information overload that should be minimized. [1] introduce a *shingling* approach for detecting similar Web documents. The detection method first represents the content of each document D by the subsequences of words in D, which are called *shingle*. Hereafter, [1] compare the set of shingles of size k belonged to any two documents to establish the overlap of the two for clustering purpose.

[9] capture the structure of on-line news events that make up a topic and the dependencies among them (i.e., event threading) through different event models. Even though the use of *cosine similarity* and *time-stamps* of news stories in [9] produces fairly good results when the events are provided, the performance deteriorates rapidly if the system has to discover the events itself. [4] utilize the R-measure, which is the normalized sum of the lengths of all *word suffixes* of the text repeated in documents to detect duplicates and plagiarism. Rather than using word suffixes to detect near-duplicate documents, [15] define *instance-level constraints* that determine to which class a document belongs

in order to identify duplicate and near-duplicate documents. In yet another clustering approach, [7] incorporate the user's prior knowledge, which indicate pairs of documents belonged to the same cluster, to obtain the desired cluster structures or to construct accurate clusters. This technique enables users to control the clustering process based on the *prior knowledge* specific to the target data set, which is, however, a constraint.

An incremental hierarchical text document clustering approach used for organizing documents from various on-line sources is presented in [11]. The approach depends on the frequency of occurrence and the contents of the words within documents, which is another *term-frequency* and *word-matching* approach, to determine the topic of a document for clustering. [6], who also consider RSS news articles, allow the user to find articles grouped by similar topics. In [6], the *k-nearest neighbor* algorithm locates the k nearest stories for each new story S so that the cosine similarity in the Vector Space Model computed for each of the k stories and S is not lower than the predefined threshold; otherwise, the content of S is treated as a new topic.

[10] use three different variations of the k-mean algorithm to find higher quality solutions in less time for clustering binary data streams. Even though the results of the incremental k-mean are good, dependence in initialization, sensitivity to outliers, and skewed distributions could affect the performance of the algorithms. [3] develop a *divide-and-merge* clustering methodology that combines top-down (divide) and bottom-up (merge) algorithms, which creates a tree structure T whose leaves are documents, and each cluster is a sub-tree rooted at a node of T. [12] describe a clustering algorithm that uses the contents of linked Web pages for clustering relevant Web pages. Thus, like ours, [12] rely on the content of a document to produce clusters of highly related documents. However, [12] only cluster documents containing a single topic, which is a restriction, since a document may cover more than one topic as in an RSS news article.

3 The Fuzzy Set IR Model and Redundant Articles Detection

Detecting non-redundant and informative RSS news articles is a challenging task, since RSS news feeds are *dynamic* in nature. RSS users subscribe to Web sites that typically add the content of news articles in the machine-readable, XML-formatted file regularly and rapidly. Two of the essential elements in an RSS file are the *title* and *description* of an *item* (i.e., a news article), since the former contains the headline of the article (story), whereas the latter often contains the first couple sentences of the article, and several items can appear in the same RSS file. We concatenate the title and description of each item and treat them as the *content descriptor* of the corresponding article and use a selective clustering approach based on a fuzzy equivalence relation to cluster the articles that possess (majority of the) information which is not included in other articles from either the same or different RSS feeds. This can be done by determining the degrees of similarity of any two articles using the lists of keywords[2] in their respective content

[2] Keywords denote words that are non-stopwords and stemmed. *Stopwords* are very common words, such as prepositions, demonstrative, interrogative, and indefinite pronouns, which do not provide useful information to distinguish the content of different articles. *Stemmed words* are words with common morphological and inflectional endings removed, which minimize the number of (semantically the same) keywords to be compared in different articles.

descriptors. The degrees of similarity can be computed by the *correlation factors* in the fuzzy set IR model among different keywords, which are predefined by using a set of more than 930,000 Wikipedia (http://wikipedia.org/) documents to determine the (i) *frequency* of co-occurrence and relative *distance* ($c_{i,j}$), (ii) normalized value ($nc_{i,j}$), and *keyword correlation factor* ($cf_{i,j}$) of each pair of keywords w_i and w_j as

$$c_{i,j} = \sum_{x \in V(w_i)} \sum_{y \in V(w_j)} \frac{1}{d(x,y)}; nc_{i,j} = \frac{c_{i,j}}{|V(w_i)| \times |V(w_j)|}; cf_{i,j} = \frac{\sum_{m=1}^{k} nc_{i,j}^{m}}{k}, \quad (1)$$

where $d(x, y)$ = $|Position(x)$ - $Position(y)|$ + 1 is the distance, i.e., the number of words, between words x and y in a Wikipedia document, $V(w_i)$ ($V(w_j)$, respectively) is the set of non-stop, stemmed words of w_i (w_j, respectively), $|V(w_i)|$ ($|V(w_j)|$, respectively) is the number of words in $V(w_i)$ ($V(w_j)$, respectively), and m is the mth out of the k ($1 \leq m \leq k$) Wikipedia documents in which both w_i and w_j, or their stemmed variances, occur.

According to the pre-computed keyword correlation factors, we define a fuzzy association $\mu_{i,j}$ of the *content descriptors* of an RSS news article pair A_i and A_j, along with their *degree of similarity* $Sim(i, j)$ as

$$\mu_{k_m,j} = 1 - \prod_{k_l \in A_j} (1 - cf_{m,l}), \forall_{k_m \in A_i}; Sim(i,j) = \frac{\mu_{k_1,j} + \mu_{k_2,j} + \cdots + \mu_{k_n,j}}{n}, \quad (2)$$

where $cf_{m,l}$ is given in Equation 1, $\mu_{k_m,j} \in [0, 1]$ reaches its maximum when $cf_{m,l} = 1$, i.e., $k_m = k_l$ for any $k_l \in A_j$, n is the number of keywords in A_i, and $Sim(i, j) \in [0, 1]$.

In general, $Sim(i, j) \neq Sim(j, i)$. If $Sim(i, j)$ = 0, then there are no keywords in A_i that is considered similar to any keyword in A_j. When $Sim(i, j)$ = 1 = $Sim(j, i)$, A_i and A_j are *identical*. If $Sim(i, j) \sim 1$ and $Sim(j, i) \ll 1$, where $\sim 1 \equiv \geq 0.93$ and $\ll 1 \equiv < 0.9$,[3] then A_i is *subsumed* by A_j, i.e., each keyword in A_i is (semantically) the same as some of the keywords in A_j. Using the correlation factors in *Sim(i, j)*, we determine whether A_i or A_j should be treated as (i) *redundant*, i.e., identical or one is "*subsumed*" by the other, (ii) one is *less-informative* than the other, or (iii) (un-)related.

Example 1. Consider the RSS news articles A_1 and A_2 in Figure 1(a). Since the keywords in the content descriptor of A_1 and A_2 are the same, $Sim(A_1, A_2) = 1 = Sim(A_2, A_1)$, and A_1 or A_2 is treated as *replicated*. Another two articles A_3 and A_4 are shown in Figure 1(b). The calculated similarity values between A_3 and A_4 are $Sim(A_3, A_4)$ = 0.69 and $Sim(A_4, A_3)$ = 0.93, which suggests that A_4 is subsumed by A_3.

Shown in Figure 1(c) are two other news articles A_5 and A_6. Both articles address similar event (i.e., Senator Barack Obama enters the presidential race). Since $Sim(A_5, A_6)$ = 0.36 and $Sim(A_6, A_5)$ = 0.28, the two articles contain some related information. Figure 1(d) shows another two news articles, A_7 and A_8. A_8 is *less-informative* than A_7, since $Sim(A_8, A_7)$ = 0.86 is high, which means that significant amount of information presented in A_8 is contained in A_7, whereas $Sim(A_7, A_8)$ = 0.42, which indicates that A_7 contains other information that is not available in A_8. During the clustering process of these articles, A_8 is an ideal choice to be eliminated. □

[3] The threshold values, 0.93 and 0.9, were determined by the conducted empirical study in 2006.

```
<item><title>Move Over, Botox: Wrinkle 'Filler' Is New Option</title>
<link><![CDATA[http://abcnews.go.com/Health/Cosmetic/story?id=2664045&CMP=...]]></link>
<pubDate>Thu, 30 Nov 2006 16:41:31 -0500</pubDate>
<description>Juvederm Injections Said to Fill Gaps in Aging Faces</description></item>

<item><title>Move Over, Botox: Wrinkle 'Filler' Is New Option</title>
<link><![CDATA[http://abcnews.go.com/Health/Cosmetic/story?id=2675805&CMP=...]]></link>
<pubDate>Thu, 23 Nov 2006 11:21:21 -0500</pubDate>
<description>Juvederm Injections Said to Fill Gaps in Aging Faces</description></item>
```

(a) A_1 (top) is *identical* to A_2 (bottom)

```
<title> Egyptian Ship With 1,300 Aboard Sinks </title>
<link>http://hosted.ap.org/dynamic/stories </link>
<description>An Egyptian cruise ship with 1,300 people aboard has sunk in the Red Sea off
   the Saudi coast during an overnight crossing. At least a dozen people have been rescued,
   but there are reports of dozens of bodies recovered. </description>

<title> Egyptian Ship With 1,300 Aboard Sinks </title>
<link>http://rssfeeds.usatoday.com/~r/UsatodaycomWorld-TopStories/~3/69844309/2007-... </link>
<pubDate>Tue, 2 Jan 2007 18:26:28 GMT</pubDate>
<description> An Egyptian cruise had 1,300 people aboard when it sank in the Red Sea off the
   Saudi coast during an overnight crossing. </description>
```

(b) A_3 (top) *subsumes* A_4 (bottom)

```
<title><![CDATA[Barack Starts 2008 Bid]]></title>
<link><![CDATA[http://www.washingtonpost.com/...]]></link>
<pubDate><![CDATA[Tue Jan 16 14:40 EST 2007]]></pubDate>
<description><![CDATA[Illinois senator files exploratory committee papers;
   official announcement is set for Feb. 10.]]></description></item>

<title>Barack Obama Jumps Into 2008 Race</title>
<pubDate>Tue, 16 Jan 2007 13:31:24 EST</pubDate>
<link>http://www.cbsnews.com/stories/2007/01/16/politics/main2361354...</link>
<description>Democratic Sen. Barack Obama of Illinois took the first step in
   a presidential bid, filing paperwork that will allow ...</description></item>
```

(c) A_5 (top) and A_6 (bottom) are *related*

```
<title>Clean penguins return to sea after spill</title>
<link>http://www.boston.com/news/science/articles/2006/08/01/clean_penguins_return_... </link>
<description>Dozens of freshly cleaned Magellanic penguins waddled into the ocean Monday ...,
   close to 200 goo-covered birds that were rescued and washed after an oil spill.</description>

<title>Clean Penguins Return to Sea After Spill</title>
<link><![CDATA[http://abcnews.go.com/Technology/wireStory?id=2261905&CMP=OTC-RSS..]]></link>
<pubDate>Tue, 01 Aug 2006 16:20:40 -0400</pubDate>
<description>Dozens of Cleaned-Up Penguins Return to Sea After Oil Spill Off Argentina's
   coast</description>
```

(d) A_8 (bottom) is *less-informative* than A_7 (top)

Fig. 1. Examples of identical, subsumed, related, and less-informative news articles

4 Fuzzy Equivalence Relation

After discarding *redundant* (i.e., identical or subsumed) RSS news articles based on their Sim values, we proceed to eliminate *less-informative* RSS news articles in a cluster being constructed. This task can be accomplished by first generating clusters of all the non-redundant RSS news articles (both informative and less-informative) that have a certain degree of similarity. A non-redundant cluster of RSS news articles is defined as

$$C_\alpha = \{d \mid Sim(d, e) \geq \alpha, \forall e \in C_\alpha\} \,, \tag{3}$$

where α is the *minimum degree of similarity* that any two articles in C_α must hold.

Clusters of news articles can be created by using a *fuzzy equivalence relation*, which is defined according to the degrees of similarity among different elements in a set, and is used to generate "crisp" subsets (i.e., clusters) [5]. R is a fuzzy equivalence relation on a fuzzy set Y, which defines a "crisp" relationship among the elements of Y, if R is *reflexive*, *symmetric*, and max-min *transitive*, i.e.,

$$R(x, x) = 1, \;\; \forall x \in Y \;, \;\; R(x, y) = R(y, x), \;\; \forall x, y \in Y \;, \tag{4}$$

$$R(x, z) \geq \max_{y \in Y}\{\min\{R(x, y), R(y, z)\}\}, \;\; \forall x, y, z \in Y \;. \tag{5}$$

The key for establishing a fuzzy equivalence relation is the definition of *transitivity*. The first definition for fuzzy transitivity was proposed by Zadeh [17], which is the max-min transitivity, as defined in Equation 5. However, the max-min transitivity is known to be a restrictive constraint, and is not applicable to the clustering problem that we deal with in this paper. This is because in order to apply the max-min transitivity to our clustering problem, it is required that for any two articles d_x and d_z in a cluster C, there cannot exist another article d_y in C which has similarities with both d_x and d_z that are greater than the similarity between d_x and d_z. Consider the following sentences:

S_1: *House passes new Medicare drug bill.* S_2: *On stem cell legislation, a new reprise.* S_3: *House passes bill relaxing limits on stem cell research.*
Since S_3 has higher similarity with both S_1 and S_2 than S_1 and S_2 have with each other, $max\text{-}min$ transitivity is not suitable to define a fuzzy equivalence relation in our work.

The $max\text{-}prod$ transitivity [18] (defined below) is not as restrictive as the $max\text{-}min$ transitivity and better suits the requirements of our fuzzy equivalence relation.

$$R(x, z) \geq \max_{y \in Y}\{R(x, y) \times R(y, z)\} . \tag{6}$$

The $max\text{-}prod$ transitivity can be easily satisfied if the involved values fall in the interval $[0, 1]$, such as our μ and Sim functions, since the product of two numbers $x, y \in [0, 1]$ is smaller than x and y, i.e., if $x, y \in [0, 1]$, then $x \geq x \times y$ and $y \geq x \times y$. Hence, we adapt the $max\text{-}prod$ transitivity in establishing our fuzzy equivalence relation.

In order to adapt the fuzzy equivalence relation with $max\text{-}prod$ transitivity for eliminating less-informative RSS news articles in a cluster, it is necessary to define a function that "combines" the similarity measures of any two news articles A_i and A_j, i.e., $Sim(i, j)$ and $Sim(j, i)$, into a single one, to satisfy the conditions of *symmetry* and *transitivity*. We first consider several combination functions and then choose the one that satisfies the $max\text{-}prod$ equivalent relation as the desired combination function.

4.1 Combination Functions

A commonly used combination function is *average*. However, the average of two pairs of significantly different similarity values, e.g., (0.5, 0.5) and (0.9, 0.1), can yield the same result, e.g., (0.5 + 0.5)/2 = (0.9 + 0.1)/2. Moreover, although the average function is fuzzy symmetric and reflexive, it is not $max\text{-}prod$ transitive. Two equations Q and Q' in [8], which combine two values, e.g., $Sim(i, j)$ and $Sim(j, i)$, can be defined as

$$Q(Sim(i, j), Sim(j, i)) = \frac{Sim(i, j) + Sim(j, i)}{1 - min(Sim(i, j), Sim(j, i))} , \tag{7}$$

$$Q'(Sim(i, j), Sim(j, i)) = (Sim(i, j) + Sim(j, i)) - (Sim(i, j) \times Sim(j, i)) . \tag{8}$$

Both Q and Q', are simple to compute; however, Q' has the same drawback as the average function, i.e., it yields the same result to significantly different pairs of values. For example, both (similarity) value pairs (0.9, 0.9) and (0.99, 0.1) are assigned the same value, i.e., 0.99, by Q'. In contrast, Q assigns a high value only when both similarity measures of $Sim(i, j)$ and $Sim(j, i)$ are high, as shown in Figure 2(a). Moreover, Q is *fuzzy-symmetric*, however, Q is neither fuzzy-reflexive nor fuzzy-transitive. We modify Q so that the modified Q function, FE, is a fuzzy equivalence relation.

$$FE(d_i, d_j) = \begin{cases} 1 & \text{if } i = j \\ 0.0001 & \text{if } Q(d_i, d_j) < 0.0001 \\ \frac{Q(d_i, d_j)}{\max(Q(d_x, d_y))} & \text{otherwise} . \end{cases} \tag{9}$$

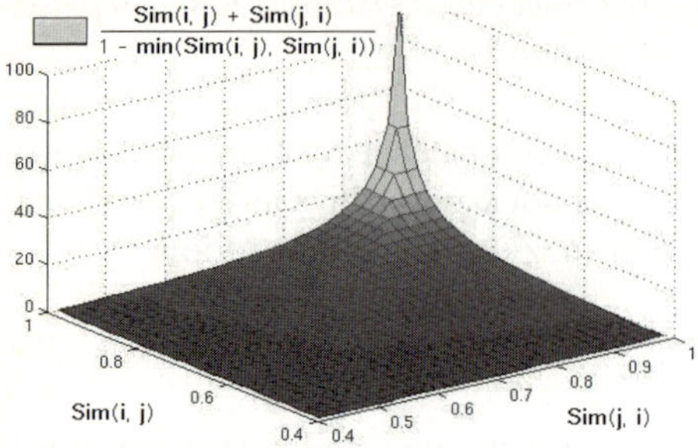

(a) A sample of values computed by using the Q function in Equation 7

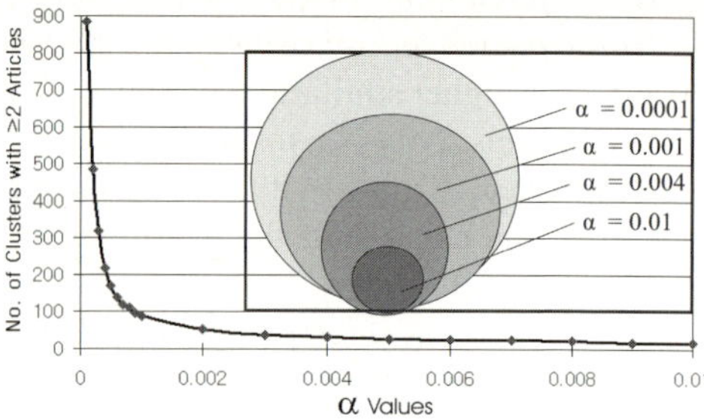

(b) Equivalence (chart) and subsummed classes (diagram) generated by 429 articles

Fig. 2. Q function and the numbers of equivalence classes generated by using different α values

The 1st condition in FE is introduced to satisfy *reflexivity*, whereas the 3rd condition, the normalized Q function, restricts the values of FE to [0, 1]. The 2nd condition guarantees *max-prod transitivity*, since 0.0001 limits the similarity value between any two articles so that the *max-prod* transitivity is satisfied,[4] a necessary and sufficient condition. Since Q is *symmetric* as defined earlier, FE is a fuzzy equivalence relation.

4.2 Clustering with α-Cuts

We apply the fuzzy equivalence relation FE to determine the equivalence classes (clusters) of non-redundant news articles posted under different RSS news feeds by setting the α value [5] of fuzzy sets as shown in Equation 3. An α value is used to generate an α-cut, which is a *set of clusters* such that every pair of news articles in the same cluster has a degree of similarity no less than α. The α-cut analysis, which is widely used in representing uncertainty information, to which the informativeness of RSS news articles belong, restricts degrees of fuzzy members in different classes. We adapt this analysis approach, instead of other clustering approaches, since α-cut is *seamless* for clustering news articles whose degrees of similarity are determined by the fuzzy set theory.

As α increases, the number of equivalence classes of the corresponding α-cut decreases, and the size of each equivalence class is reduced. (See the chart in Figure 2(b) for an example.) One important property of α-cuts is that the equivalence classes generated by some α-value x will always be subsumed by the equivalence classes generated by some α-value y, if $y < x$ (see the inserted diagram in Figure 2(b) for an example).

By using *smaller* α values, which generate *larger* clusters, we decrease (i) the chance of eliminating an *entire* cluster when discarding less-informative articles, and (ii) the number of singleton clusters created. However, the clusters generated may be *loosely related* if α is too small. Thus, we favor *larger* α values, which generate *fewer* and

[4] We verified the *max-prod transitivity* of FE using 400 randomly chosen news articles from 12 different RSS feeds, which include BBC News, PR Newswire, and Slashdot.

Table 1. Different α values and clusters (i.e., α-cuts) generated for a set of 4 news articles. If α is too low, A_2 is clustered with unrelated articles; however, if α is too high, A_4 is isolated from related articles A_1 and A_3. The α-value 0.0018 is effective.

A_1: Boy in Mo. kidnapping ready for school - Family of boy in Mo. kidnapping case recalls happily receiving word of his recovery

A_2: Embryo saved after Katrina is born - Parents of embryo rescued after hurricane Katrina celebrate birth of boy Tuesday

A_3: Abducted boy may have hard time leaving psychologically - He may be the latest of kidnap victims to suffer from Stockholm syndrome

A_4: Parents of missing boy say his life was threatened - Parents of Shawn Hornbeck, missing four years, say captor threatened his life

$\alpha = 0.0007$:$\{A_1 A_2 A_3 A_4\}$,...	$\alpha = 0.0018$:$\{A_1 A_3 A_4\}$,$\{A_2\}$,...	$\alpha = 0.0040$:$\{A_1 A_3\}$,$\{A_2\}$,$\{A_4\}$,...

Table 2. Various α values and their corresponding statistics on a set of 90 news articles. The α value 0.0019 is optimal due to its relatively low number of clusters and singletons.

Value of α	No. of Clusters	No. of Singletons	Average Number of Clusters an Article is in	Average Number of Articles Per Cluster
0.0001	182	0	8.32	4.38
0.0004	182	0	8.32	4.38
0.0007	182	0	8.32	4.38
0.0010	184	0	7.25	4.06
0.0013	140	1	5.46	4.01
0.0016	120	4	4.22	3.76
0.0019	**93**	**7**	**3.24**	**3.44**
0.0022	72	13	2.84	3.42
0.0025	60	20	2.51	3.50

smaller clusters in which articles are closely related. Table 1 shows the clusters generated by using different α values, whereas Table 2 provides the statistics for various α values on a sample set of 90 news articles. As the average number of articles per cluster in Table 2 decreases, α increases and clusters in the α-cut are tighter.

The formula for α has been empirically determined by using eight sample news articles sets (i.e., news feeds) ranging from 34 to 351 articles in size. A linear regression line was calculated using the principle of *least squares* on the test data. Optimal α values vary due to subjectivity, and are a question of both how *tight* clusters should be, and how many *singletons* should be allowed. Typically, an effective α value can be determined by the number of articles in the set, which we define as $\alpha = 15.23e^{-6} \times$ Number of Non-Redundant Articles + $1.49e^{-3}$, and the value of α not only determines what *clusters* will be generated, but also plays a key role, along with the ranking function (defined in Section 4.3), in deciding which news articles (from multiple RSS feeds) are *less-informative*. Moreover, the value of α determines the appropriate *number* of clusters to be generated, which affects the ranking on individual news articles.

Table 3. RSS news articles downloaded from New York Times (www.nytimes.com), Washington Post (www.washingtonpost.com), and CBS News (www.cbsnews.com) on January 16, 2007

ID	Title; (Portion of the) Description
0	The Warming of Greenland; Arctic melting accelerates, revealing uncharted islands and ...
1	Observatory: A Smelly Puzzle, Solved; The world's largest flower, a link between lobsters ...
2	Personal Health: 104 Teenagers Who Are Role Models for Weight Loss; A look at the ...
3	Q & A: Blood and History; Is there an evolutionary reason for human blood types?
4	Findings: The Voices in My Head Say 'Buy It!' Why Argue?; What is the difference ...
5	Ruins in Northern Syria Bear the Scars of a City's Final Battle; Excavation of ruins at Tell ...
6	Baghdad Explosions Kill Dozens, Wound Scores; Series of bomb blasts outside a Baghdad ...
7	Barack Obama Starts 2008 Bid; Illinois senator files exploratory committee papers; official ...
8	Trial Begins in CIA Leak Case; Defense lawyers argue publicity has damaged I. Lewis ...
9	Wide Berth Given on Sex Ed; Montgomery County schools invoke codes in defense of ...
10	Lawmaker Angers Blacks, Jews; Va. delegate says blacks "should get over" slavery, ...
11	More Than 100 Dead In Baghdad Attacks; More than 100 people were killed in attacks in ...
12	Barack Obama Jumps Into 2008 Race; Democratic Sen. Barack Obama of Illinois took the ...
13	Potential Libby Jurors Split On Iraq War; Potential jurors for the trial of a former White ...
14	Chills Across U.S. As Ice Storm Heads East; After weeks of unseasonably mild weather, ...
15	Oil Prices Plummet To 19-Month Low; Oil prices dropped by $2 a barrel to a 19-month low ...
16	Iran Buys Surplus U.S. Military Hardware; The Associated Press reports that U.S. military ...

4.3 The Ranking Approach

During the process of discarding less-informative articles generated by an α, we must retain at least one article in each cluster to ensure that the content or "story" of no cluster is lost. Thus, singleton clusters should not be eliminated.

Since we wish to reduce the total number of (sets of) articles, and the same articles may appear in different clusters,[5] we cannot treat every cluster separately while selecting less-informative articles to discard, since we could potentially discard all the articles in a cluster. Consider, for example, a set of articles $C = \{a, b, c, d, e, f\}$ and their FE values: $FE(a, b) = 0.055$, $FE(c, d) = 0.021$, $FE(e, f) = 0.014$, and the same FE value, 0.0025, for all the other possible pairs. If we set $\alpha = 0.1$, then three clusters $C_1 = \{a, b\}$, $C_2 = \{c, d\}$, and $C_3 = \{e, f\}$ are generated. Suppose we need to discard two articles from C. If we rank the articles only in the same cluster based on their similarity values, i.e., without considering the degrees of similarity with other articles in other clusters in which they reside, and discard the two with the highest similarity values, then a and b (which have the highest similarities values) are discarded, and consequently C_1 is eliminated, which is undesirable, since C_1 is the only cluster that contains the story of a and b. Thus, we rank each article A among all the clusters (generated by our α-cut equation) in which A resides, and discard those that have higher rankings (i.e., articles highly similar to others) among all the clusters, which are less-informative.

[5] A fuzzy equivalence relation based on max-$prod$ transitivity, which is differed from the max-min transitivity, does not always yield disjoint equivalence classes and is desirable, since articles may contain a variety of information, and should be allowed to reside in different classes (clusters) that include news articles with which they share some portion of information.

Table 4. News articles in each cluster (created by using the α-value 0.013) before and after eliminating less-informative ones, along with the ranking of the 17 articles in Table 3

Original Clusters	Clusters After Elimination	No. of Articles Eliminated	Final Set of Clusters
{ 0, 1 }	{ 0, 1 }	0	{ 0, 1 }
{ 2, 3 }	{ 2 }	1	
{ 0, 5 }	{ 0, 5 }	0	{ 0, 5 }
{ 4, 6, 14 }	{ 4, 14 }	1	{ 4, 14 }
{ 4, 8, 13 }	{ 4 }	2	
{ 8, 9 }	{ 9 }	1	{ 9 }
{ 6, 10, 11, 14 }	{ 10, 11, 14 }	1	{ 10, 11, 14 }
{ 7, 11 }	{ 11 }	1	
{ 0, 12 }	{ 0, 12 }	0	{ 0, 12 }
{ 2, 12 }	{ 2, 12 }	0	{ 2, 12 }
{ 7, 12 }	{ 12 }	1	
{ 12, 13 }	{ 12 }	1	
{ 6, 15, 16 }	{ 15, 16 }	1	{ 15, 16 }
{ 4, 6, 16 }	{ 4, 16 }	1	{ 4, 16 }
{ 11, 16 }	{ 11, 16 }	0	{ 11, 16 }

Ranking
7
8
3
6
11
13
12
4
1
10
0
9
14
5
16
2
15

We consider the function $Rank(d_i) = \frac{\sum_{k=1}^{N} \max\{\forall_{d_j \in C_k} Sim(i,j)\}}{N}$, where N is the number of non-singleton clusters C in which d_i appears, to rank articles in C created by an α-cut. $Rank$ computes the *average* of the *maximum* similarity values of an article d_i and each article d_j in each cluster C_k in which d_i appears.

According to the rankings, the top n ($n \geq 1$) ranked (i.e., less-informative) articles, are discarded. The value n can be determined by the *average* number of new articles that are accessed by the user on a regular basis. Furthermore, each generated cluster C can be represented by the top k (≥ 1) most frequently-occurred keywords in C.

Example 2. Consider the set of 17 non-redundant RSS news articles that were extracted from various RSS news feeds as shown in Table 3. Assume that 30% of the (less-informative) news articles (i.e., a total of five articles) are supposed to be deleted. Table 4 shows the non-singleton clusters generated by the α-value 0.013 using Equation 3, which along with the *rankings* of the 17 articles computed by using the $Rank$ function on their degrees of similarity, determine which five articles should be discarded.

The first article to be eliminated is Article 7, which is less-informative than Article 12 in the same cluster. (Both articles refer to Senator Obama's presidential bid, but Article 7 only mentions the fact, whereas Article 12 provides more detail.) Subsequently, Articles 8, 3, and 6 are eliminated, since each addresses political affairs that are mostly covered by other articles, e.g., Article 6 is covered in Article 11, and Article 8 is covered in Article 13. These articles were *manually examined* for their relative degrees of similarity. The next article in the ranking order to be eliminated is Article 11. This article is related to Article 6, which covers an accident in Baghdad and has already been eliminated. Since Article 11 is also grouped in a cluster with Article 7, in order to

Table 5. A portion of the 273 feeds used for verifying our analysis approach with areas **Bus**sines (Bs); **Ent**ertainment (ET); **H**ealth (Hl); **I**n**t**ernational (IT); **L**o**c**al (Lc); **Mi**scellaneous (Mi); **Po**litics (Po); **S**ports (Sp); **Te**chnology (Te); **T**ravel (Tr); **T**op **S**tories (TS); **U**nited **S**tates (US)

ID	Sources	RSS Feeds								
		No.	Number of Articles (in the Subject Area)							
1	1115.org	2	10 (TS)	10 (TS)						
2	abcnews.go.com	20	49 (Bs)	60 (Bs)	23 (ET)	32 (ET)	68 (Hl)	115 (Hl)	26 (IT)	43 (IT)
			17 (Po)	47 (Po)	30 (Sp)	39 (Sp)	54 (Te)	58 (Te)	25 (Tr)	67 (Tr)
			41 (TS)	47 (TS)	65 (US)	28 (US)				
3	adn.com	4	6 (Sp)	6 (Sp)	8 (TS)	8 (TS)				
4	blogs.zdnet.com	2	10 (Te)	10 (Te)						
5	bloomberg.com	10	32 each (Bs (3), ET, IT, Lc, Po, Sp, TS (2))							
6	boston.com	8	10 (Lc)	11 (Lc)	15 (Lc)	19 (Lc)	30 (Lc)	38 (Lc)	24 (Mi)	48 (TS)
7	businessweek.com	4	29 (Bs)	29 (Bs)	45 (TS)	45 (TS)				
8	cbsnews.com	8	56 (Bs)	50 (ET)	53 (Hl)	42 (IT)	52 (Po)	51 (Te)	7 (TS)	40 (US)
9	chinaview.cn	6	60 each (Bs, ET, Hl, IT, Mi, Po)							
10	chron.com	6	25 (Bs)	21 (ET)	23 (IT)	25 (Po)	21 (Sp)	20 (TS)		

avoid discarding the entire cluster {7, 11} after Article 7 has been removed, we retain Article 11. We eliminate the next article in the ranking, i.e., Article 13 (on politics).

After the elimination process is completed, we proceed to eliminate all the clusters that are either *duplicated* or *subsumed* by others (as shown in Table 4), which yields only informative clusters. □

5 Experimental Results

In this section, we present the conducted experiments and use their results to verify the accuracy of our approach on (i) detecting redundant and less-informative RSS news articles collected from various RSS news feeds and (ii) clustering informative articles. We analyzed different test sets and determined the number of *false positives*, i.e., news articles that were mistakingly treated as redundant or less-informative and discarded in each set, which was then converted into a percentage of *accuracy*. Table 5 and Figure 3 show the wide variation and subject areas of news articles from which each test case was drawn. The collections of articles in each test case was downloaded from multiple RSS feeds, often containing the same (or similar) news articles.

5.1 Verifying the Accuracy of Detecting Redundant RSS New Articles

Table 6 shows the size of each test case used for identifying redundant articles, along with its result verified by human judges. The results confirm the correctness of our redundancy-detection approach by the 11 test cases. After *manually* examining the articles in various test cases, we conclude that the overall success rate in detecting *redundant* articles (from either the same or different RSS feeds) is perfect and have achieved an average accurate rate of 86% in detecting all the *subsumed* articles. (The three missed subsumed articles are caused by our high threshold values, i.e., 0.93 and 0.9.)

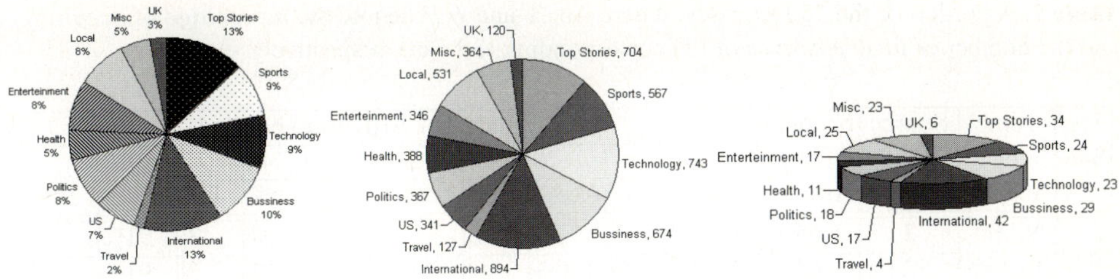

(a) Ratios of Areas (b) Number of Articles (c) Number of Distinct Feeds

Fig. 3. Statistical data on subject areas, number of articles, and distinct RSS feeds in Table 5

Table 6. Test cases for verifying the accuracy in detecting redundant (i.e., identical and subsumed) news articles, which were extracted from source data in Table 5

Test Case	Number of Articles	Identical Articles			Subsumed Articles		
		Existed	Eliminated	Accuracy Ratio	Existed	Eliminated	Accuracy Ratio
TC_1	11	0	0	100%	0	0	100%
TC_2	31	2	2	100%	5	4	80%
TC_3	62	3	3	100%	0	0	100%
TC_4	86	3	3	100%	1	1	100%
TC_5	103	8	8	100%	0	0	100%
TC_6	115	5	5	100%	0	0	100%
TC_7	128	7	7	100%	2	2	100%
TC_8	139	14	14	100%	5	4	80%
TC_9	179	2	2	100%	1	1	100%
TC_{10}	194	24	24	100%	3	3	100%
TC_{11}	388	34	34	100%	4	3	75%
AVG	131	9.3	9.3	100%	1.91	1.64	86%

5.2 Accuracy of Our Less-Informative News Articles Elimination Approach

We have considered 25 different test sets, each of which contains multiple non-redundant RSS news articles from various RSS news feeds. We *manually* evaluated the (less-informative) articles chosen automatically to be discarded and counted the number of false positives. We calculated the percentage of *accuracy* as $1 - \frac{FalsePositives}{Discarded}$.

The twenty-five test sets show a high degree of accuracy, even when the percentage of articles discarded is high. The percentages of discarded articles are in the range of 10% to 50%. Table 7 shows the accuracy percentage and the corresponding numbers of false positives for each distinct percentage of discarded less-informative news articles. The results show that the overall average of accuracy is 88%.

5.3 The Overall Accuracy and Observations of Our Clustering Approach

Consider the test set of articles as shown in Table 6 and their corresponding source as shown in Table 5 to further verify the accuracy in generating clusters of non-redundant

Table 7. A portion of the 25 test cases, where Acc% and FP denote the *percentage* of *accuracy* and the number of *false positives* of the corresponding test case, respectively

Test Set	Topic of Articles	Source of Articles	Number of Articles	Percentages of Articles to be shed									
				10%		20%		30%		40%		50%	
				Acc%	FP	Acc%	FP	Acc%	FP	Acc%	FP	Acc%	FP
1	World	3	65	100%	0	92%	1	85%	3	88%	3	91%	3
2	Techno	2	64	100%	0	100%	0	90%	2	88%	3	91%	3
3	USA	3	53	81%	1	91%	1	87%	2	91%	2	92%	2
4	World	2	51	100%	0	100%	0	87%	2	85%	3	84%	4
5	USA	2	50	60%	2	80%	2	87%	2	90%	2	88%	3
6	Sports	2	50	100%	0	90%	1	93%	1	90%	2	92%	2
7	USA	2	49	100%	0	100%	0	93%	1	95%	1	92%	2
8	World	2	46	78%	1	89%	1	93%	1	89%	2	91%	2
9	Sports	2	44	77%	1	89%	1	85%	2	89%	2	86%	3
10	World	2	40	100%	0	88%	1	83%	2	81%	3	85%	3
...	...	...	...	...	...	...	...	...	...	...	...	...	...
Average		1.84	36	**87%**	0.40	**90%**	0.68	**87%**	1.32	**86%**	1.84	**87%**	2.24

and informative news articles using 0.0019 as the α-cut value and discarding 30% of the less-informative news articles. The empirical results of the conducted experiments are shown in Table 8, which indicates an average of 96% accuracy on the test sets.

Using different values of α-cut and elimination percentages of articles in various test cases, we observe that the best clusters of a set of articles are generated when the recommended α-cut value (computed by using the α-equation in Section 4.2) is adapted, since (i) the number of singleton clusters is *low* (and in some cases there are none at all), (ii) a *good* number of clusters (i.e., the number of clusters do not exceed the number of articles) are generated, and (iii) *larger* clusters are reduced. According to the experimental results, the ideal number of clusters (with appropriate number of articles) is created when 20-30% of the articles are eliminated, since when the numbers of articles to be eliminated is *too small*, many similar articles could remain, whereas if we eliminate more than 30% of the articles, sometimes we not only eliminate similar ones, but also news stories entirely. Furthermore, when α-cut is *low* and the elimination percentage is small (10%), we often obtain clusters that include too many not-closely-related articles. When α-cut is *high* and the elimination percentage is also *high* (50%), we obtain more clusters, and most of them often consist of two articles. Hence, after the less-informative elimination process, all or most of the clusters are reduced to singleton clusters and in some cases, entire clusters might be eliminated (due to the requested number of articles to be deleted). However, when the number of articles is *small* and the elimination percentage is *high*, we are forced to eliminate some informative ones.

5.4 Performance Evaluation against Other Existing Clustering Approaches

We have verified the merit of our clustering approach by comparing its performance with other well-known clustering approaches, such as *k-means* and *Naive Bayes*. We

Table 8. Verifying the average accuracy (which is 95.8%) in clustering non-redundant and removing 30% of the less-informative RSS news articles using various test cases in Table 6

Test Case	No. of Articles	No. of Non-Redundant Articles	No. of Removed Less-Informative Articles	No. of Clusters Before Elimination	No. of Clusters After Elimination	Accuary of Our Clustering Approach
TC_1	11	11	3	5	3	66.7%
TC_2	31	25	8	10	9	100%
TC_3	62	59	18	46	31	100%
TC_4	86	82	25	110	81	96.0%
TC_5	103	95	28	100	77	96.4%
TC_6	115	110	33	128	82	97.0%
TC_7	128	120	36	145	114	97.3%
TC_8	139	122	36	268	251	100%
TC_9	179	176	53	274	225	100%
TC_{10}	194	168	51	282	254	100%
TC_{11}	388	351	105	914	751	100%

chose the Reuters corpora (Reuters-21578 distribution 1.0, http://kdd.ics.uci.edu/databases/reuters21578/reuters21578.html), which consists of 21,578 news articles manually assigned to one or more 135 established topics, for performance analysis, since the corpora is a popular benchmark data widely used for verifying the merit of any text categorization, IR, machine learning, knowledge discovery, or data mining method. We considered articles in the Reuters corpora that belong to a *unique* topic and ignored topics that have *less* than five articles to facilitate the comparisons, which is the same evaluation strategy used by [13], since it is more accurate to analyze articles that belong to a unique topic. As a result, 9,494 articles and 51 different topics were chosen. To quantify the performance among different clustering methods, including ours, we applied the mutual information (MI) metric [14] (given below), which defines how *similar* or *independent* any two given sets of news article clusters C and C' are, where $p(c_i)$ ($p(c'_j)$, respectively) is the *probability* that a randomly selected article from the reduced Reuters data set belongs to cluster c_i in C (c'_j in C', respectively) and $p(c_i, c'_j)$ is the *probability* that a randomly selected article belongs to both c_i and c'_j.

$$MI(C, C') = \sum_{\forall_{c_i} \in C, \forall_{c'_j} \in C'} p(c_i, c'_j) \times log_2 \frac{p(c_i, c'_j)}{p(c_i) \times p(c'_j)} \,. \tag{10}$$

Since $MI(C, C') \in [0..\max(H(C), H(C'))]$, where $H(C)$ and $H(C')$ denote the *entropy* of C and C', respectively, we normalize the MI metric value as $\widehat{MI}(C, C') = \frac{MI(C,C')}{max(H(C),H(C'))}$, where $\widehat{MI}(C, C') \in [0..1]$, $H(C) = -\sum_{i=1}^{k} p(c_i) log_2 p(c_i)$ measures the *purity* of the set of clusters in C, and $H(C')$ can be defined accordingly.

We evaluate our clustering approach according to the procedure described in [13], which randomly selects news articles from k (≥ 2) different clusters in the reduced Reuters corpora. The set of k clusters and the set of clusters C' generated by using our clustering approach (using α-cut = 0.0019 and discarding 30% of the articles) form the

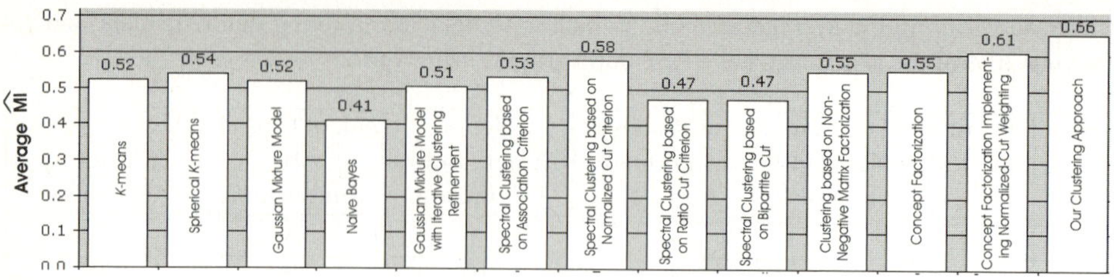

Fig. 4. Average normalized Mutual Information ($\widehat{MI}$s) computed for various clustering approaches given in [13] along with ours

two sets of clusters to obtain their $\widehat{MI}$ values. We repeated this process for different sets of k clusters with randomly selected articles, which are evaluated against C', to yield different test cases and results. The test cases include different numbers (15-50) of randomly chosen articles and topics, which generate (2-15) scalable clusters and yield various (0.39-0.88) $\widehat{MI}$s, a similar evaluation procedure as in [13].

Figure 4 shows the average $\widehat{MI}$ of the clusters generated by using our clustering approach and the other eleven clustering approaches mentioned in [13]. Among all the clustering approaches, *k-mean* selects k different seeds (i.e., documents) to determine the centroids of the clusters and iteratively attempts to find new clusters in order to minimize the total intra-cluster variance between the documents and their corresponding cluster centers. The *Gaussian Mixture Model* or *Naive Bayes* method defines a *probabilistic* cluster model and try to find the model by maximizing the likelihood of data to be clustered. The *Spectral-based clustering* approaches use eigenvectors of a *similarity* matrix derived from the data to be clustered, which establish a measure of similarity between any two points in the data in order to perform dimensionality reduction to reduce the dimensions in clustering. [14] use the *Non-Negative matrix*, which captures the topic of a document cluster and represents each document as an additive combination of the base topics, to determine the cluster any document D belongs by finding the base topic with which D has the largest projection value. The *Concept Factorization* technique models each concept as a linear combination of data points and computes the sets of linear coefficients for clustering and labeling the data points.

As illustrated in Figure 4, our clustering approach outperforms all the other clustering approaches based on averaging $\widehat{MI}$s computed for all the different test cases on the Reuters corpora. In fact, not only our clustering approach achieves a higher average $\widehat{MI}$ value than others, ours is not affected by initialization (i.e., the required training cases) as in other approaches. Furthermore, our clustering approach enhances its performance when the number of articles to be clustered is large, since before clustering we discard redundant and less-informative ones, which yields clusters with higher quality.

6 Conclusions

We have presented a method for clustering non-redundant and informative RSS news articles, which combines well-established techniques, such as α-cuts and fuzzy set logic, with other innovative approaches, such as using *distance correlation factors*, to filter and classify (i.e., cluster) accurately and efficiently information in RSS feeds. Moreover, we have developed a formula for computing the *optimal value* of α, resulting in α-*cuts* which contain an ideal number of clusters and very few singleton ones. We have also analyzed the results of our method for eliminating different percentages of redundant and less-informative RSS news articles, which achieve high accuracy (in the upper 80% in general). According to the experiments conducted, our clustering approach outperforms others. In fact, our clustering approach functions with little overhead. The keyword-correlation factors are precomputed, and the algorithms for removing redundant/less informative news articles and clustering informative ones are computationally inexpensive. Due to the *small overhead*, *high efficiency*, and without any *labor-intensive* user feedback, our clustering approach can be implemented in real-world RSS feeds to aid users in locating news articles of interest among tens of thousands posted daily.

References

1. Broder, A., Glassman, S., Manasse, M., Zweig, G.: Syntactic Clustering of the Web. Computer Networks and ISDN Systems 29, 8–13, 1157–1166 (1997)
2. Bun, K., Ishizuka, M.: Topic Extraction from News Archive Using TF*IDF Algorithm. In: Intl. Conf. on Web Information Systems Engineering (WISE), pp. 73–82. Springer (2002)
3. Cheng, D., Kannan, R., Vempala, S., Wang, G.: A Divide-and-Merge Methodology for Clustering. ACM TODS(31) 4, 1499–1525 (2006)
4. Khmelev, D., Teahan, W.: A Repetition-Based Measure for Verification of Text Collections and for Text Categorization. In: 26th Intl. ACM SIGIR Conf., pp. 104–110. ACM, New York (2003)
5. Klir, G.K., St. Clair, U., Yuan, B.: Fuzzy Set Theory, Foundations and Applications. Prentice Hall, New Jersey (1997)
6. Li, X., Yan, J., Deng, Z., Ji, L., Fan, W., Zhang, B., Chen, Z.: A Novel Clustering-Based RSS Aggregator. In: Intl. Conf. on World Wide Web, pp. 1309–1310. ACM, New York (2007)
7. Li, Y., Chung, S.: Document Clustering Based on Frequent Word Sequences. In: ACM Conf. on Information and Knowledge Management (CIKM), pp. 293–294. ACM, New York (2005)
8. Luger, G.: Artificial Intelligence, Structures and Strategies for Complex Problem Solving, 5th edn. Addison Wesley, San Francisco (2005)
9. Nallapati, R., Feng, A., Peng, F., Allan, J.: Event Threading within News Topics. In: ACM Conf. on Information and Knowledge Management (CIKM), pp. 446–453. ACM, New York (2004)
10. Ordonez, C.: Clustering Binary Data Streams with K-Means. In: ACM SIGMOD Workshop on Research Issues on Data Mining and Knowledge Discovery, pp. 10–17. ACM, New York (2003)
11. Sahoo, N., Callan, J., Krishnan, R., Duncan, G., Padman, R.: Incremental Hierarchical Clustering of Text Documents. In: ACM CIKM, pp. 357–366. ACM, New York (2006)
12. Wang, Y., Kitsuregawa, M.: Evaluating Contents-Link Coupled Web Page Clustering for Web Search Results. In: ACM CIKM, pp. 499–506. ACM, New York (2002)
13. Xu, W., Gong, Y.: News Article Clustering by Concept Factorization. In: 27th Intl. ACM SIGIR Conf, pp. 202–209. ACM, New York (2004)
14. Xu, W., Liu, X., Gong, Y.: News Article Clustering Based on Non-Negative Matrix Factorization. In: 26th Intl. ACM SIGIR Conf., pp. 267–273. ACM, New York (2003)
15. Yang, H., Callan, J.: Near-Duplicate Detection by Instance-Level Constrained Clustering. In: 29th Intl. ACM SIGIR Conf., pp. 421–428. ACM, New York (2006)
16. Yang, Y., Pierce, T., Carbonell, J.: A Study on Retrospective and On-Line Event Detection. In: 21th Intl. ACM SIGIR Conf., pp. 28–36. ACM, New York (1998)
17. Zadeh, L.: Similarity Relations and Fuzzy Orderings. Info. Sci(3), 177–200 (1970)
18. Zimmermann, H.: Fuzzy Set Theory and Its Applications. Kluwer Academic, Dordrecht (1991)

An On-the–Flight Proxy System for Multi-format Web Content Transcoding for Mobile Devices

Juan G. González Serna, Hugo Estrada Esquivel, Azucena Montes Rendon, Victor Sosa Sosa, and J. Carlos Olivares R.

Centro Nacional de Investigación y Desarrollo Tecnológico (CENIDET)
Cuernavaca, Morelos, Mexico
{gabriel,hestrada,amr,vjsosa,jcolivares04c}@cenidet.com.mx

Abstract. One of the trends in current information society is the visualization of Web sites using mobile devices. At present time, there are millions of mobile devices that are used by the owner to navigate in the web. However, not all the information placed in the web is appropriately adapted to be visualized in mobile devices. As a result, not all functionalities can be exploded by the users. To solve this situation, we need mechanisms to allow users of mobile devices to visualize Web sites independently if the Web sites were specifically designed or not for these types of devices. The problem of this scenario is that the contents of the Web site need to be adapted to the limitations of each mobile device. In this paper, an on-the-flight Web content transcoding mechanism is presented as solution to this problem. The mechanism proposes the inclusion of a new HTTP header in the HTTP package to describe a transcoding format command. This header is generated by a client-side Proxy named GAP, which runs on mobile devices with Windows CE operating system (Pocket PC, Smartphone, among others). The transcoding header is interpreted by a server-side Proxy, named GAT MT, that converts Web documents to WML, XHTML-MP, PDF, plain text, PostScript, XML and HTML adapted for mobile devices. To do this, the mechanism uses XSL technologies (XSLT and XSL-FO) to define Stylesheets according to a specific device.

1 Introduction

In current societies, where the information is extremely relevant to take the appropriated business decisions, the access to the information becomes to be a crucial factor. One of the media that has been used for accessing information in a very flexible manner is the Internet (more specific the Web). On the other side, the computer devices are getting smaller, which have lead to new paradigms to consider aspects such as mobility and information ubiquity.

One of the main issues of mobile devices is the lack of Web contents specifically designed for these types of devices (most of the content is designed for computer equipment with big screens, more colors, more processing power, etc). As a consequence, users can only visualize a very small part of the Web, which complicates the correct visualization of the content. The interaction schema (navigation, links selection, etc) is another factor affected by the small size of the current mobile devices.

O. Gervasi et al. (Eds.): ICCSA 2008, Part II, LNCS 5073, pp. 248–258, 2008.

Another mobile-device limitation is that browsers for mobile devices accept a very limited rate of formats. One of the consequences of the current lacks in this area is the cost associated to the Internet connection in mobile devices. The problems in visualization and interaction have direct influence on the time wasted to access a certain Web pages.

In order to try to guarantee that all Web resources can be available on mobile devices, we need some kind of mechanisms that transform the "normal" information into information adapted for mobile devices, which can be visualized better in these devices. This transformational mechanism is called transcoding. In this paper, an on-the-flight Web content transcoding mechanism is presented as solution of the visualization and interaction problems in mobile devices. The objective of transcoding mechanism is to allow users of mobile devices to visualize Web sites independently if the Web sites were specifically designed or not for these types of devices. To do this, the transcoding mechanism adapts the contents of the Web site according to the limitations of each mobile device. Specifically, the mechanism described in this work transform HTML resources to WML, XHTML-MP, PDF, PostScript, XML, plain text and a HTML reformatted version. With the proposed approach, we give a further step in providing the desired scenario where the information is available when, where and how the users decide it.

The paper is structured as follows: Section 2 presents the related works. Section 3 presents an overview of transcoding approach used in this research work. Section 4 describes the proposed approach, while Section 5 presents the results of evaluating the proposed approach. Finally, Section 6 concludes and briefly discusses future work.

2 Related Works

The idea behind Web content transcoding is a very old idea that has been addressed in different ways. Generally, transcoding has been applied to the transformation of multimedia contents (audio and video). That is the case of video television transcoding: PAL to NTSC, or in the case of audio: WAV to MP3.

At present, only a few on-the-flight transcoders exist that can be considered as multiplatform. This is the case of proxies such as the proxy Google wmlproxy that converts documents to WML. In other cases, there exist online transformers where users access to a Web site to introduce a URL they wish to transform and the system gives back the transformed page, which is the case of skweezer[3].

Some works have developed schemes that allow Web content transcoding "out-of-line". For example, a user downloads documents and when the documents are needed in another different format, the transformer is executed. This is the case of jTranscoder [4.

The current microbrowsers on mobile devices (e.g. Pocket Internet Explorer, Mozilla Minimo, Opera Mobile) perform a basic type of transcoding. They adapt documents to a single column to avoid the famous scrolling problem on mobile devices.

Another approach to solve this visualization problem is by having Web sites specially designed for a specific platform. That is the case of the Google finder, that has search interfaces for each platform and format (e.g. page for PDA, Cellular, iMode) or by having sites designed exclusively for this platform. This last idea has a main issue: the necessary investment in the administration of Web sites.

At the moment, some existing technologies offer solutions to guarantee that same information on the Web can be visualized on different forms depending on the device, which is the case of ASP.NET and Java struts. In ASP.NET, a site is designed with MobileWebForms, when a device accesses to the site, the Web server detects the type of device and generates a new page that can be visualized in that device according to its displaying characteristics. Struts are based on MVC (Model-View-Controller) [5] architectonic design, which basically separates the program logic (model), the way in which the events and the user interaction with the system are managed (controller), and finally the way in which the data are arranged (view). Making the interface independent of the data and the events, the interface can be adapted to other devices without affecting the controller and the model.

In our research group, we started to work with XML applied to the transformation of Web contents. Our first work [1] makes the transformation of Web documents, but it doesn't consider several target formats. Therefore, an extension has been proposed to this work to manage better the inclusion of several formats for mobile devices.

As a result of reviewing specialized literature in the area [7], [8], [9], [10], [11] y [12], we have detected the lack of mechanisms that manage the transformational process in a similar way we do.

3 Transcoding

Transcoding is the process to transform a document from a format to another [1]. In order to make the transformation possible, the following technologies were used:

XSL (eXtensible Stylesheet Language), which allows us to separate information of presentation. For example, in HTML, the same document shows the data and the schema to display the data. With XSL the presentation is structured in two forms: the data as must be visualized, which is represented by using CSS (Cascade StyleSheet) and the data as it needs to be transformed, which is represented using XSLT.

XSLT (eXtensible Stylesheet Language Transformation), which allows us transforming a document written in XML to another XML-based language. It makes the data portable, allowing, for example, that a XML file that represents a database scheme can be converted to HTML for making reports of the data contained in the document [2].

XSL-FO (eXtensible Stylesheet Language-Formmating Objects), which allows us making transformations to documents that are not XML-based (PDF, PostScript, plain text, etc.).

4 The Proposed Approach

The proposed approach is based on the definition of an intermediate XML model between the original Web site and the Web site reformulated according to a specific format for a mobile device. However, this is not enough to solve the problem, so we have implemented a client-side Proxy and a server-side Proxy. The client-side Proxy

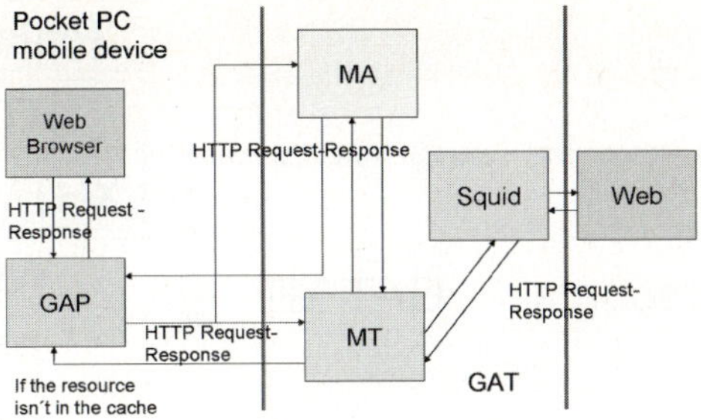

Fig. 1. System architecture

is called GAP (Hoarding Manager for Pocket PCs) and the server-side Proxy is called GAT (Hoarding and Transcoding Manager) as it is shown in figure 1.

The GAT is divided mainly in two mechanisms: MA (Hoarding Mechanism) and MT (Transcoding Mechanism). GAT MT was developed in Java using the following APIs: xalan-j and xerces for the transformation; JAX and DOM for the XML document path; FOP, Avalon and Batik for the transformation of objects, among others. In GAT ML tool, XML and XSLT technologies are used for document transcoding; on the other hand, JTidy is used to generate a XHTML document from a HTML source code.

The GAP, on the other hand, manages the disconnection events that could be presented in the mobile devices, reason why it hoards resources into the local cache of the devices. Also, it generates and sends the corresponding header to the GAT (see table 1) for content transcoding, which we have named X-Transform (this is because specific HTTP headers that have not been standardized must begin with the X capital letter).

GAT MT interprets the format that contains the X-Transform header (if any, otherwise it assumes HTML as the predetermined format) and if the device is a mobile device like Windows CE (Pocket PC, Smartphone), Palm OS (Palm, Treo, Clie) and Epoc (Cellular with Symbian OS) the page is transformed.

Table 1. Example of a HTTP header request generated by the GAP

GET http://www.cenidet.edu.mx/ HTTP/1.0
Accept: */*
UA-OS: Windows CE (Pocket PC) –Version 3.0
UA-Color: Color16
UA-Pixeles: 240x320
UA-CPU: ARM SA1110
UA-Voice: False
UA-Language: Mozilla/2.0
Accept-Encoding: gzip, deflate
User-Agent: Mozilla/2.0 (Compatible; MSIE 3.02; Windows CE; PPC; 240x320)
Host: www.itmorelia.edu.mx
Proxy-Connection: Keep-Alive
X-Transform: XHTML-MP; Complete

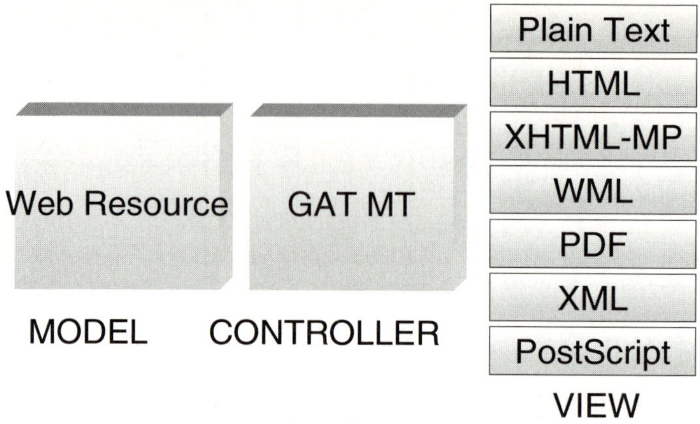

Fig. 2. MVC applied to the system

The transcoding mechanism is very similar to the MVC architecture; figure 2 shows the architecture of this method. The input model corresponds to the Web resource transformed to a XML file, the controller is the GAT MT that handles the way the document is transformed and finally, the view is represented by the document in all available formats.

The transformation process is the following: First, the system detects if the resource is within the cache of transformed documents and if so, it gives back the

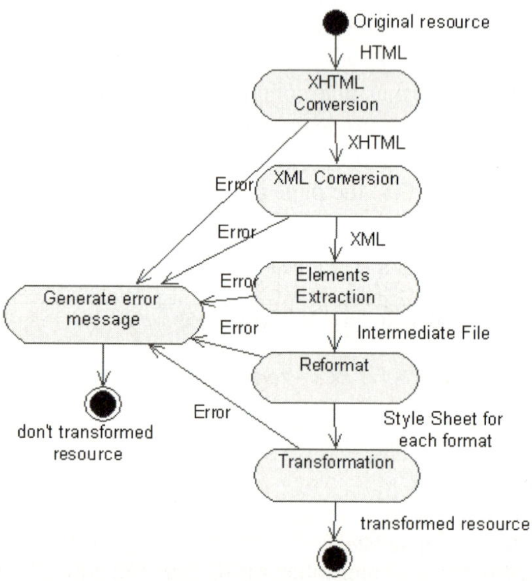

Fig. 3. Activity diagram of the transcoding general process

document immediately (this avoids the system to do the process several times, reason why the response time is improved). In case the cache doesn't contain the document, the system obtains it from the Internet, if the document is HTML; the JTidy tool denominated is used to transform it into a XHTML document. Based on this file, the system adds a special header taking this XHTML code to transform it into a XML document. Having the XML document, the processing can be applied to it for reformatting, which consists of distilling all those characteristics that are not available on mobile devices (tables, frames, forms, etc.). Later on, the system groups elements with the same type (text, images and links). Once the distilled document has been generated, the system generates the corresponding style sheet to the wished format. Figure 3 shows the activity diagram of the transcoding general process.

In table 2, there is an example of a style sheet in XSL-FO.

Table 2. Example of a style sheet in XSL-FO

```
<?xml version='1.0' encoding='iso-8859-1'?>
<xsl:stylesheet version="1.0" xmlns:xsl="http://www.w3.org/1999/XSL/Transform"
xmlns:fo="http://www.w3.org/1999/XSL/Format">
<xsl:output method="xml" indent="yes"/>
<xsl:template match="html">
<fo:root xmlns:fo="http://www.w3.org/1999/XSL/Format">
<fo:layout-master-set>....
        <fo:block font-family="Helvetica" font-size="18pt" text-align="end"
space-before="2cm">
GAT MT X-transform
Converter or resources Web to PDF
</fo:block>
<fo:block font-weight="bold" color='blue'>
Prueba del transcodificador</fo:block>
<fo:block font-size="12pt">
<xsl:value-of select="body" /><xsl:text> </xsl:text>
<fo:block font-size="12pt" color='red'> * Images *</fo:block>
<fo:block><fo:external-graphic src='cenidet.jpg' height='1.4cm' width='2.0cm' scal-
ing='uniform' /></fo:block>
<fo:block font-size="12pt" color='red'> * Links *
<fo:basic-link external-destination='http://www.cenidet.edu.mx/~wm-serna'
color='blue'>http://www.cenidet.edu.mx/~wm-serna/
</fo:basic-link>
...
</xsl:template></xsl:stylesheet>
```

Later on, the system continues with the transformation process using the xalan-j API if the formats are XHTML-MP, WML or reformatted HTML and the FOP API if the formats are PDF, PostScript, XML or plain text. Once the resource is generated, the cache mechanism gives back the resource to the client updating its cache list. In case of error, a page with the respective error is showed to the user (Figure 4).

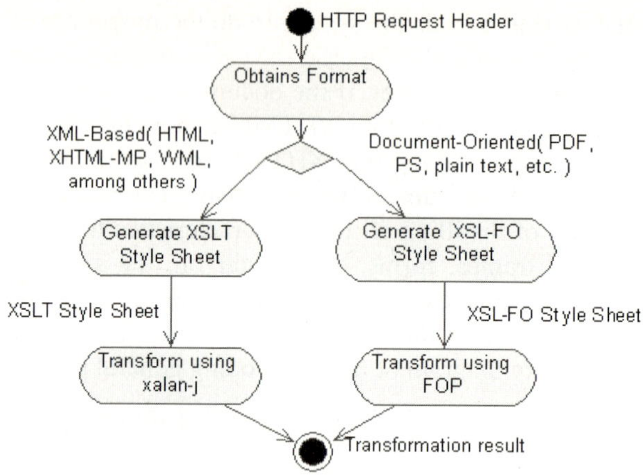

Fig. 4. Activity diagram of the transformation process

5 Results of Evaluating the Proposed Approach

The tests were made with the following equipment: a Pocket PC HP iPAQ rx3115 with Windows Mobile 2003, a Smartphone Motorola MPX 220 with Windows Mobile 2003. For the traditional devices where GAT MT was executed, an HP Pavilion Entertainment Notebook PC dv1025/a (Intel Centrino to 1,7 Ghz. with 512 Mb. RAM) was used.

The test consisted in the visualization of 100 sites in all the formats (the Web address corpus was obtained from interviews and preferences of the users). The average size of the analyzed pages was 30,476.81 bytes (29.76 Kb) and 56.99 objects embedded in the main page.

Only 29% of the sites used in the evaluation of the method were correctly transformed. The reason of this was that in most of the cases, it was not possible to correctly transform the Web sites into a XML document. It means that the immense

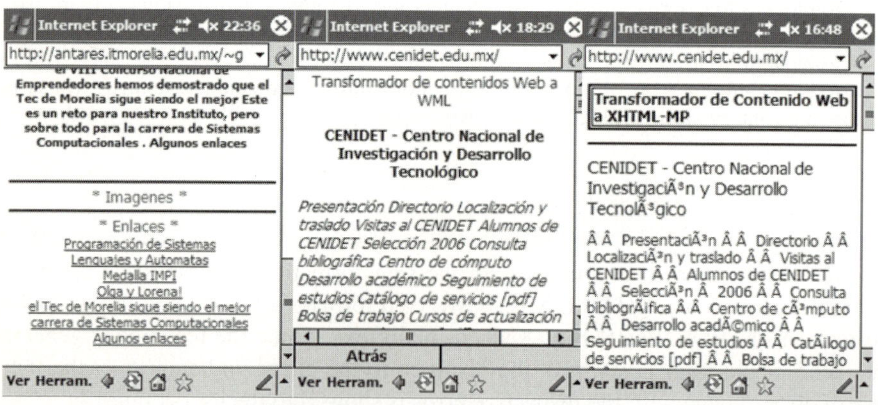

Fig. 5. Web resource transcoding using XSLT (reformatted HTML, WML and XHTML-MP respectively)

majority of Web sites are badly formed and structured. In figure 5, figure 6 and figure 7, some examples of transcoding results of the Web resources are shown. Figure 5 shows the transcoding using XSLT to generate HTML, WML and XHTML-MP documents. Figure 6 shows the transcoding using XSL-FO to generate PDF, XML and plain text documents. Finally, Figure 7 shows the result of transcoding a document into postscript file visualized in a desktop PC.

Fig. 6. Web resource transcoding using XSL-FO (PDF, XML and plain text respectively)

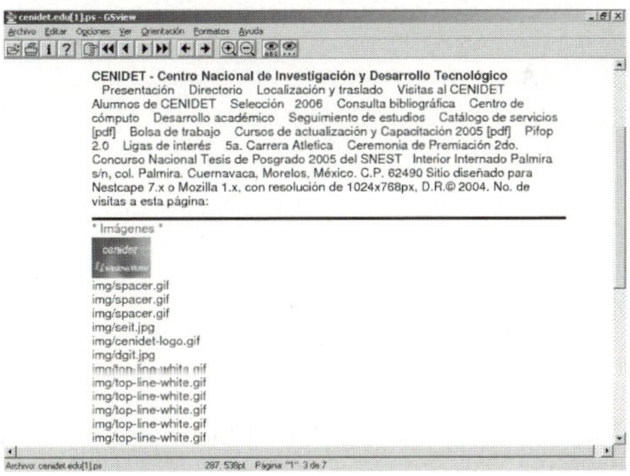

Fig. 7. Web resource transcoding in PostScript format visualized in a desktop PC

Table 4. Times obtained for the transcoding process

Format	Time in ms
PS	3,782.27
PDF	2,421.75
HTML	2,312.13
XML	1,794.17
TXT	1,767.06
XHTML-MP	1,574.34
WML	1,563.51

Table 4 shows the results of transcoding the 100 sites evaluated. Table considers the time wasted in the following activities: GAT MT receives the resource requested, obtains it, transforms it and gives it back to the GAP.

The average time of the transcoding process was 2,173.60 ms, whereas the average without transcoding was 1,589.53 ms; the transcoding process was 36.74% slower.

In the transformation process, we could observe that it is proportional to the size of the main page (a big document produces a high time for downloading and processing of the document). Also we observed that as more elaborated is the style sheet; greater is the time of the transcoding process. In general the use of XSL-FO (PDF and PostScript) is slower than XSLT, this due to the visualization format complexity (PDF and PS require greater amount of elements to display information), considering the fact that the elements extraction time, download time, etc. are similar.

Figure 8 shows the times for the transcoding process for each format taken into account.

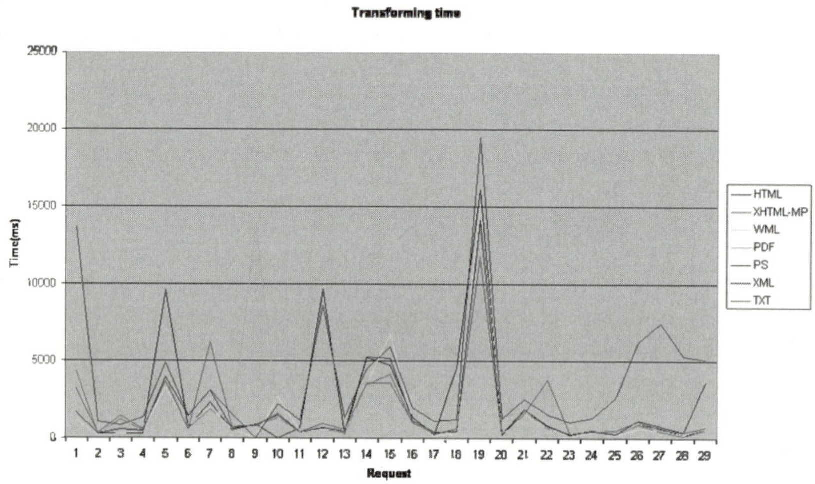

Fig. 8. Comparative graph of the transcoding times

6 Conclusions and Further Work

With the present work, we determined that it is possible to transform Web documents (HTML) "on-the-flight" to other formats like WML, XHTML-MP, PDF, XML, PostScript, plain text and reformatted HTML for mobile devices; obtaining acceptable response times. With the work presented here, we have developed a tool that allows Web content transcoding "on-the-flight" into different formats (the user could extend a new transformation format only specifying the code of the new style sheet) that can easily be adapted to other presentation formats for mobile devices.

This work has many areas of opportunity to grow. For example, in WML it is not possible to visualize images in the traditional image formats (jpeg, gif, etc.), these must be in a WBMP (Wireless BitMaP) format. At the moment, we are reviewing which changes can be made to transform these images.

On the other hand, in PDF files, the image paths are relative in most of the cases; reason why at the time of generating the PDF document these cannot be visualized. In this part, it would be needed to create a parser that brings the resources that are inherent to the documents, concretely the images, in this matter, the process would become slower and harder to implement.

Another improvement related to images would be to compress them, changing the size and the resolution so that they occupy less space. This is something difficult to implement but it would increase the navigation performance (perhaps it would make it slower). We must remember that images transference represents 52% of the Internet load.

Another activity that is possible for implementing is to separate the style sheet in the code, this is done in order to make the document conversion more extensile (also, we would have to modify the form to access to the style sheet). The advantage to make this is that the style sheet from the code wouldn't be directly modified, every time those changes are reflected. This activity is highly recommended to do.

Another possible activity is to generate a single intermediate file, a generic kind of metamodel to avoid the great amount of repeated resources due to the diversity of existing caches. The problem would be again to generate the document, perhaps it would save the problem of getting the document and its intermediate reformats, but every time the document needs to be re-generated losing therefore the original concept of the system cache in the GAT.

Another possible improvement is in the fact that if the document cannot be transformed, the system gives back the original document if it could not be reformatted and adapted (if the resource can't be obtained, then it is impossible to send this message). This is good because we detected that in some cases it is not possible to show the resource and the users get frustrated their navigation experience. This activity is widely recommended to do.

A substantial improvement would be to try to read any Web document and transform it into another one. This problem is n-complexity because it is needed to create tools that read and transcode in several formats. This is possible because transformation is made through XSL, which allows us to transform any XML document into any other document XML-based.

Another important modification consists of taking into account the devices restrictions and users preferences to make the transcoding, since the scheme proposed here is made thinking about Pocket PC devices.

References

1. Claudia Selene Uriarte Cabada. Transformador de Contenidos Web para Asistentes Personales Digitales, master thesis, cenidet, julio de (2004)
2. Gutiérrez, A., Martínez, R.: XML a través de ejemplos, Alfaomega Rama, Madrid, Spain (2001) ISBN: 970-15-0714-2
3. Skeezer [Last visited april 2006], `http://www.skweezer.net/`
4. Chanchaem, T.: A Survey on Internet Content Transcoding for Universal Access. Department of Computer Science, Kent State University (May 2003)

5. Burbeck, S.: Applications Programming in Smalltalk-80: How to use Model-View-Controller (MVC), 1987 and 1992
6. María del Rosario Vázquez Aragón, Generador Automático de Informes Web en Base al Esquema de Base de Datos Utilizando Tecnologías XML, master thesis, cenidet, November, p.112 (2003)
7. Schaefer, R., Dangberg, A., Mueller, W.: Fuzzy Rules for HTML Transcoding. In: Proceedings of the 35th Hawaii International Conference on System Sciences, IEEE, Los Alamitos (2002)
8. Laakko, T., Hiltunen, T.: Adapting Web Content to Mobile User Agents. In: IEEE Internet Computing Magazine, IEEE Computer Society, Los Alamitos (2005)
9. Dabas, A.: Proxy Server for Handhelds, master thesis, Czech Technical University at Praga, Faculty of Electrical Engineering (May 2005)
10. Alam, H., Rahman1, F.: Web Document Manipulation for Small Screen Devices: A Review. In: Web Document Analysis Workshop (WDA) (2003)
11. Bouillon, L., Vanderdonckt, J.: Retargeting Web Pages to other Computing Platforms with VAQUITA. In: van Deursen, A., Burd, E. (eds.) Proceedings of IEEE Working Conference on Reverse Engineering WCRE 2002 (Richmond, 28 Octobre-1 Novembre 2002), pp. 339–348. IEEE Computer Society Press, Los Alamitos (2002)
12. Maheshwari, A., Sharma, A., Ramamritham, K., Shenoy, P.: TranSquid: Transcoding and Caching Proxy for Heterogeneous E-Commerce Environments. In: Proceedings of the 12th IEEE Workshop on Research Issues in Data Engineering (RIDE 2002), San Jose, CA (February 2002)

A Semantic-Based Fully Visual Application for Context-Aware Matchmaking and Request Refinement in Ubiquitous Computing

Michele Ruta, Tommaso Di Noia, Eugenio Di Sciascio, and Floriano Scioscia

Politecnico di Bari
via Re David 200, I-70125
Bari, Italy
{m.ruta,t.dinoia,disciascio,f.scioscia}@poliba.it

Abstract. This paper presents a mobile visual application aimed at fully exploiting semantics of request descriptions to enable advanced services. Distinguishing aspects of its underlying framework include semantic ranking of request results and logic-based explanation of matchmaking outcomes as well as the context identification and exploitation to select suitable resources. The GUI has been designed and implemented to be effective for handheld devices with reduced screen capabilities. It requires no knowledge of any logic principle to be fully used. The framework is general-purpose and could be easily adapted to different mobile scenarios. Here it is motivated and presented in a tourism case study.

1 Introduction and Motivation

In ubiquitous computing scenarios, information technology can assist users in discovering resources examining alternatives and analyzing choices, thus aiding people to retrieve information satisfying their needs and/or giving more elements to make rational decisions. Especially in some "evanescent" scenarios, to discover a service/resource could be a very difficult issue: finding items encountering user needs often requires too much effort and time. This is particularly true when a user has a vague idea of what she wants, having limited knowledge of a specific domain.

When an effective network infrastructure is lacking, the process of supporting the user searching goods or services is an even more challenging subject. Basically, several issues concerning traditional service discovery are exasperated in the case of ubiquitous and pervasive approaches because of both the host mobility and limited capabilities of mobile devices. Small displays, uncomfortable input methods, reduced memory availability and computational capacities restrain the exploitation of such applications. Nevertheless the rising potentialities of wireless-enabled handheld devices provide the necessary basic requirements for implementing flexible discovery frameworks. The aim is to reduce the human effort in discovering resources and/or services, also granting an acceptable level

O. Gervasi et al. (Eds.): ICCSA 2008, Part II, LNCS 5073, pp. 259–274, 2008.

of accuracy and coping with user mobility and heterogeneity of pervasive scenarios. We present a "pocket" application enabling a semantic-based discovery in ubiquitous environments. The main goal we pursue is to allow users equipped with handheld devices to take advantage of semantic resource annotation and matchmaking as well as of logic-based ranking and explanation services, while hiding all technicalities from them and letting users interact with the system without requiring dependable wired infrastructures. The system we propose is a general-purpose mobile DSS (Decision Support System). It can be exploited by tourists performing a visit but also by a customer wanting to remotely retrieve goods or services getting them from a generic marketplace and so on. The knowledge domain is encapsulated within a specific ontology the user must select at the beginning of her interaction with the system.

A relevant feature of the proposed approach is the context awareness. It allows to perform a selective resource discovery based on proximity which is especially useful in the case of some specific mobile applications as the virtual guide presented here [1]. Recall that users equipped with PDAs or smartphones are dipped within a pervasive environment; hence they could be specifically interested to resources or services near them. Hence, during the discovery, resources/services close to the user should be ranked better than the ones far off (supposing an equivalent semantic distance from the request). In other words, the *semantic distance* between request and offered resource should be properly rectified taking into account the *physical distance* occurring between user and resource itself (if it has an environmental collocation). In the proposed virtual guide application, this feature has been obtained by means of the integration of a localization module within the tool. The application recognizes the user location and grades matchmaking outcomes according to vicinity criteria.

The retrieval process is accomplished across multiple steps. Request formulation is the first phase. It is critical especially in case of ontology-based systems as the one we present here. The query language has to be very simple but, at the same time, its expressiveness must allow to correctly interpret user requirements so retrieving only what the user is really looking for. In most cases users are unable to exploit logic formulas needed to use a formal ontology; they want a simple visual representation to manipulate the domain of interest. The overall system should be able to rapidly match resources to potential beneficiary interests and to present available resources in an appealing manner –compatible with restricted screens– facilitating examination and checking of their features.

We borrow techniques and ideas from the Semantic Web initiative and adapt them to volatile wireless scenarios. Semantic Web technologies applied to ubiquitous computing open extremely interesting new possibilities, including: formalization of annotated descriptions that become machine understandable so enabling interoperability; reasoning on descriptions and inference of new knowledge; validity of the Open World Assumption (OWA), overcoming limits of structured data models; possibility of going beyond physical boundaries of structured and fixed network infrastructures. By means of formal ontologies, expressed using OWL[1], knowledge about a specific domain can be modeled and exploited in order to derive new information from the one stated within metadata associated to each resource. Modeling the interest domain using an OWL ontology, the user is able to browse the related knowledge starting form "her vague idea" about the resource she wants to discover. By means of a preliminary selection of the reference ontology, the user fixes the scenario so conditioning the following interactions with the system. Different sessions in the application exploitation could refer to different ontologies and then could entail interaction with the system aimed at different purposes. For example a generic client could use the application as a pocket virtual guide for tourist purposes selecting a cultural heritage ontology and in a further phase after concluding her visit she can adopt it as a mobile shopping assistant to buy goods in a B2C m-marketplace: in that case she will select an e-commerce ontology. Once the request has been formalized w.r.t. the reference ontology, its formal relations are exploited in order to find resources able to satisfy user necessities. Based on the formal semantics of both the request and the returned resource/service descriptions, an explanation of the matchmaking outcome is then provided to the user to foster further interaction.

Main features of the proposed approach are: full exploitation of non-standard inferences to enable explanation services; semantic-based ranking of retrieved resources; fully graphical and usable interface with no prior knowledge of any logic principles; total absence of physical space-temporal bonds in system exploitation.

The framework we present here has been implemented within the Apulia project PE_082 "IC Technologies for tourist assistance through interactive consultancy of a virtual guide". It adopts the semantic-enhanced features described above and uses them to provide various user-oriented services, which benefit from a Knowledge Representation approach.

The remaining of the paper is structured as follows: in the next section, basics of matchmaking and exploited Description Logics inference services are briefly recalled. Section 3 outlines system features and architecture by means of a case study referred to a tourism scenario. In Section 4 we comment on related work. Finally conclusion and future work terminate the paper.

2 Matchmaking Principles

Main matchmaking issues emerge when the resource discovery is not limited to identity matches (which are probably too rare to be realistic in concrete applications), that is when the goal is to find resources suitable to grant also a not-complete satisfaction of the user request, in order to satisfy user needs "to the best possible extent". Semantic matchmaking is basically the process of finding best matches of a request among available resources, where both the request and the resource descriptions are semantically annotated w.r.t. a reference *ontology* (*i.e.*, a shared interpretation of the knowledge domain).

[1] OWL Web Ontology Language, W3C Recommendation 10 February 2004, http://www.w3.org/TR/owl-features/

Knowledge Representation (KR) approaches to matchmaking usually exploit classical deductive services, namely *classification* and *consistency*. Basically, given a request/resource pair opportunely annotated, classification allows to check whether all request specifications are included within the resource description, whereas consistency verifies whether some specification in the request contradicts (some of) the ones within the resource annotation. In both cases, the response is then a binary *true/false* value. Although these inference services are very useful in the early phases of a discovery process, they are not sufficient to rank a set of resources with respect to a request. A positive classification corresponds to a resource completely satisfying the request, yet a) such a match might not occur during a discovery process; b) various resources could be "almost" classified with respect to a request. A matchmaker may then order resources according to their "distance" from the request and it may hypothesize a semantic based explanation for the causes of mismatch. In an analogous way, when an inconsistency occurs, a matchmaker may: a) order resources, putting before the ones "less conflicting" with the request; b) provide an explanation whether the request was completely unsatisfiable or only some part of it was in conflict with available resources.

Given R (for Request) and O (for Offer) both consistent with respect to an ontology $\mathcal{T}$, logic-based approaches to matchmaking proposed in the literature [2,3] use classification and consistency to grade match results in five categories:

– **Exact.** All the features requested in R are exactly the same provided by O and vice versa –in formulae $\mathcal{T} \models R \Leftrightarrow O$.

– **Full-Subsumption.** All the features requested in R are contained in O –in formulae $\mathcal{T} \models O \Rightarrow R$.

– **Plug-In.** All the features offered in O are contained in R –in formulae $\mathcal{T} \models R \Rightarrow O$.

– **Potential-Intersection.** There is an intersection between the features offered in O and the ones requested in R –in formulae $\mathcal{T} \not\models \neg(R \wedge O)$.

– **Partial-Disjoint.** Some features requested in R are conflicting with some offered in O –in formulae $\mathcal{T} \models \neg(R \wedge O)$.

To better clarify differences among previous match types, let us suppose a tourist is performing a visit and she is interested in seeing "medieval palaces with courtyards" (this is the previously called **R**). If there is a resource $\mathbf{O_{exact}}$ annotated as "medieval palace with a courtyard", R and O coincide. From the matchmaking perspective, **Exact** match is obviously the best, because both R and O express the same preferences and, since all the resource characteristics requested in R are semantically implied by O (and vice versa), the user finds exactly what she is looking for. On the contrary, if we have $\mathbf{O_{full}}$ annotated as "medieval palace with a courtyard and frescoed roofs", all requirements in R are satisfied by O, but other non-conflicting characteristics are also specified in the returned resource. In a **Full** match all the interpretations for O are surely also interpretations for R and then O completely satisfies R. This means that all the resource characteristics requested in R are semantically implied by O but not, in general, vice versa. Then, in a full match, O may expose some unrequested characteristics. From the point of view of requester this is not a bad circumstance. In fact, characteristics she was looking for are already satisfied. If the provided resource is $\mathbf{O_{plug-in}}$ simply labeled as "medieval palace", all characteristics in O were required by R, but the requirement of a courtyard is not explicitly satisfied. **Plug-In** match expresses the circumstance when O is more generic than R, and then it is possible that the latter can be satisfied by the former. Some characteristics in R are not specified, implicitly or explicitly, in O. This is surely more appealing for the provider than for the requester. As said, here we adopt the OWA. In case the returned resource is $\mathbf{O_{potential}}$ annotated as a "medieval palace with a frescoed roof", neither all elements in R are in O nor vice versa. R and O are still compatible, since an explicit conflict does not occur.With **Potential** match we can only say that there is some similarity between O and R, and then O might potentially satisfy R. Probably some features of O are underspecified in its description. So the requester has to contact the provider if she wants to know something more about them. Finally, supposing $\mathbf{O_{partial}}$ is a "medieval church with a courtyard", a requirement in R is explicitly violated by O, making the provided resource incompatible with the request. **Partial** match states that R and O are conflicting, yet notice that the disjointness between them might be due only to some –maybe negligible from the requester's standpoint– incompatible characteristics. Hence, after a revision of opposed features, an agreement can be reached.

As previously discussed, usually logic-based approaches only allow a categorization within match types. But while exact and full matches can be rare, a user may get several potential and partial matches. Then a useful logic-based matchmaker should provide a –logic– ordering of available resources w.r.t. the request, but what we get using classification and consistency is a boolean answer. Also partial matches might be just "near miss", *e.g.*, maybe just one requirement is in conflict, but a pure consistency check returns a hopeless *false* result, while it could be interesting to order "not so bad" ads according to their similarity to the request.

2.1 DL Inference Services for Matchmaking

In the proposed approach we exploit Description Logics (DLs) –whose formal semantics is at the basis of the Ontology Web Language OWL-DL. Furthermore, there is a strict correspondence between the OWL-DL syntax and DIG 1.1 [4], but DIG is less verbose and more compact. Here we formalize examples by adopting DL syntax instead of OWL for the sake of compactness. DLs are a family of logic formalisms for Knowledge Representation [5] whose basic syntax elements are *concept names*, *role names*, *individuals*. Intuitively, concepts stand for sets of objects in the domain, and roles link objects in different concepts. Individuals are used for special named elements belonging to concepts.

Concepts can be used in *inclusion assertions* $O \sqsubseteq D$, and *definitions* $O \equiv D$, which impose restrictions on possible interpretations according to the knowledge elicited for a specific domain. The semantics of inclusions and definitions is based on set containment: an interpretation $\mathcal{I}$ satisfies an inclusion $O \sqsubseteq D$ if $O^{\mathcal{I}} \subseteq D^{\mathcal{I}}$, and it satisfies a definition $O \equiv D$ when $O^{\mathcal{I}} = D^{\mathcal{I}}$. A DL theory (a.k.a. TBox) is a set of inclusion assertions and definitions. A *model* of a TBox $\mathcal{T}$ is an interpretation satisfying all inclusions and definitions of $\mathcal{T}$.

DL-based systems usually provide at least two basic reasoning services:

– **Concept Satisfiability:** $\mathcal{T} \models R \not\sqsubseteq \bot$. Given a DL theory $\mathcal{T}$ and a concept R, does there exist at least one model of $\mathcal{T}$ assigning a non-empty extension to R?

– **Subsumption:** $\mathcal{T} \models R \sqsubseteq O$. Given a DL theory $\mathcal{T}$ and two concepts R and O, is R more general than O in any model of $\mathcal{T}$?

Nevertheless –given a DL– the matchmaking issues outlined above call for other, non-monotonic inference services that we briefly recall hereafter. Let us consider concepts R and O and an ontology $\mathcal{T}$. If a partial match occurs, *i.e.*, they are not compatible with each other with respect to $\mathcal{T}$, one may want to retract some specifications in R, G (for *Give up*), to obtain a concept K (for *Keep*) such that $K \sqcap O$ is satisfiable in $\mathcal{T}$.

In [6] the Concept Contraction problem was first defined as:

– **Concept Contraction:** Let $\mathcal{L}$ be a DL, R, O be two concepts in $\mathcal{L}$ and $\mathcal{T}$ be a set of axioms in $\mathcal{L}$, where both R and O are satisfiable in $\mathcal{T}$. A *Concept Contraction Problem* (CCP), identified by $\langle \mathcal{L}, R, O, \mathcal{T} \rangle$, consists of finding a pair of concepts $\langle G, K \rangle \in \mathcal{L} \times \mathcal{L}$ such that $\mathcal{T} \models R \equiv G \sqcap K$, and $K \sqcap O$ is satisfiable in $\mathcal{T}$. Then K is a *contraction* of R according to O and $\mathcal{T}$.

If nothing can be kept in R during the contraction process, we get the worst solution –from a matchmaking point of view– $\langle G, K \rangle = \langle R, \top \rangle$, that is give up everything of R. Conversely, if $R \sqcap O$ is satisfiable in $\mathcal{T}$, that is a potential match occurs, nothing has to be given up and the solution is $\langle \top, R \rangle$, *i.e.*, give up nothing. Hence, a Concept Contraction problem amounts to an extension of a Concept Satisfiability one. Since usually one wants to give up as few things as possible, some minimality criteria in the contraction must be defined [7]. If the offered resource O is a potential match for R, it is necessary to assess what should be hypothesized H in O in order to completely satisfy R and then move to a full match. In [8] the Concept Abduction problem was first defined as:

– **Concept Abduction:** Let $\mathcal{L}$ be a DL, R, O be two concepts in $\mathcal{L}$, and $\mathcal{T}$ be a set of axioms in $\mathcal{L}$, where both O and R are satisfiable w.r.t. $\mathcal{T}$. A *Concept Abduction Problem* (CAP), identified by $\langle \mathcal{L}, R, O, \mathcal{T} \rangle$, is to find a concept $H \in \mathcal{L}$ such that $\mathcal{T} \models O \sqcap H \sqsubseteq R$, and moreover $O \sqcap H$ is satisfiable in $\mathcal{T}$. We call H a *hypothesis* about O according to R and $\mathcal{T}$.

Obviously the definition refers to R and O consistent. If $O \sqsubseteq R$ then we have $H = \top$ as a solution to the related CAP. Hence, Concept Abduction amounts to extending subsumption. On the other hand, if $O \equiv \top$ then $H \sqsubseteq R$.

Referring to the matchmaking example of Section 2, let us consider a TBox $\mathcal{T}$ for the tourism domain containing the axioms (i) $Palace \sqsubseteq \exists hasAge \sqcap \exists hasRoof$ and (ii) $Palace \sqsubseteq \neg Church$. Then both request and all offers could be modeled in DL formalism, *e.g.*,

- $\mathbf{R} \equiv Palace \sqcap \forall hasAge.MiddleAge \sqcap \exists hasCourtyard$
- $\mathbf{O_{partial}} \equiv Church \sqcap \forall hasAge.MiddleAge \sqcap \exists hasCourtyard$

- $\mathbf{O_{potential}} \equiv Palace \sqcap \forall hasAge.MiddleAge \sqcap \forall hasRoof.FrescoedRoof$

Having two concepts whose conjunction is unsatisfiable, in the solution $\langle G, K \rangle$ to the CCP $\langle \mathcal{L}, R, O, \mathcal{T} \rangle$, G represents "why" R and O are not compatible. For instance, the CCP $\langle \mathcal{L}, \mathbf{R}, \mathbf{O_{partial}}, \mathcal{T} \rangle$ yields the pair $\langle G = Palace, K = \exists hasCourtyard \sqcap \forall hasAge.MiddleAge \rangle$. For Concept Abduction, having R and O such that $O \not\sqsubseteq R$, the solution H to the CAP $\langle \mathcal{L}, R, O, \mathcal{T} \rangle$ represents "why" the subsumption relation does not hold. H can be interpreted as *what is specified in R and not in O*. For example, the solution for the CAP $\langle \mathcal{L}, \mathbf{R}, \mathbf{O_{potential}}, \mathcal{T} \rangle$ is $H = \exists hasCourtyard$.

3 System Outline

Previous inference services have been exploited within a prototypical mobile client for semantic-based service/resource discovery. It is aimed at employing the semantic matchmaking approach outlined above, with a location-based resource filtering for mobile and ubiquitous applications.

3.1 Design and Development Guidelines

Design and development of the proposed application were driven by the following guidelines, taking the objective of maximizing efficiency, effectiveness and usability.

a. Limited computing resources of the target platform must be carefully taken into account. In particular for novel semantic-based mobile applications, from a performance viewpoint it is practically infeasible to reuse existing Semantic Web tools and libraries on current mobile devices. A compact and optimized implementation of the required features and technologies is thus needed.

b. Mobile computing platforms are much more heterogeneous than personal computers, with devices highly differentiating in form factor, computational and communication capabilities and operating systems. Cross-platform runtime environments have to be developed in order to overcome this fragmentation.

c. Graphical User Interface (GUI) design should endorse the peculiarities of mobile and pervasive computing. Unlike their desktop counterparts, mobile applications are characterized by a *bursting* usage pattern, *i.e.*, with frequent and short sessions. Hence, mobile GUIs must be designed so that users can satisfy their needs in a quick and straightforward way. A task-oriented and consistent look and feel is required, leveraging familiar metaphors which most users are accustomed to.

d. Finally, software design must be conscious of the inherent constraints of mobile ad-hoc networks. From an application viewpoint, the most important issues are unpredictable disconnections and low data rates. The former is mainly due to host mobility, higher transmission error rates of wireless links and limited battery duration; the latter is a typical concern of wireless networks w.r.t. wired ones and it is also due to energy saving requirements for small devices. Applications must be designed with built-in resilience against network failures and QoS degradation, so as to prevent unexpected behaviors.

For a greater compatibility with various mobile platforms, our tool was developed using Java Micro Edition (ME) technology, by far the most widespread cross-platform mobile environment. The compliance with one of the Java ME *profiles* ensures the compatibility with a broad class of mobile computing devices. The Java Mobile Information Device Profile[2] (MIDP) was selected, which is currently available for the majority of mobile phones and connected PDAs. Our tool is fully compliant with Java MIDP 2.0 specifications and API. All UI elements are accessible through the keyboard/keypad of the mobile device; additionally, MIDP transparently adapts to pointer-based interaction (*e.g.*, via touchscreen) on platforms where it is available.

In order to allow location-based service/resource provisioning, the application exploits the Java Location API JSR-179[3] to determine the device's location. JSR-179 provides a unified API to interact with all *location providers* –*i.e.*, real-time positioning technologies– that are available on the device. These include an internal GPS (Global Positioning System) receiver, an external GPS device connected *e.g.*, via Bluetooth and the mobile phone network (cell-based positioning). Accuracy depends on the positioning method, being typically higher for GPS than for cell-based techniques. Our tool requests a high-accuracy location determination firstly; if the accuracy requirement cannot be satisfied by available location providers on the device, the constraint is relaxed.

The proposed tool supports a subset of the DIG 1.1 interface (for its compactness DIG has been preferred to OWL) extended for *MAMAS-tng*[4] reasoner. This interface allows interaction with the state-of-the-art of Knowledge Representation Systems (KRS) through a classical request/response mechanism.

A lightweight implementation of the client-side DIG interface has been developed in Java. A specialized library was designed for efficient manipulation of knowledge bases. In order to minimize runtime memory consumption, *kXML*[5] Java streaming XML parser was adopted, which implements the open standard XML Pull API[6]. Streaming parsers allow an application to closely control the parsing process and do not build an in-memory syntax tree model for the XML document (as DOM parsers do). This increases speed and reduces memory requirements, which is highly desirable in resource-constrained environments. Moreover, streaming parsers are best choice for processing XML data incoming from network connections, since parsing can be pipelined with input reception.

3.2 Case Study: A Quick Tour in Bari

In order to better explain the approach, we consider the following case study: *Bob is in Bari for business. After his last meeting, he is near the old town centre with some spare time before reaching a colleague for dinner. He had never been in*

[2] http://java.sun.com/products/midp/
[3] http://jcp.org/en/jsr/detail?id=179
[4] http://sisinfab.poliba.it/MAMAS-tng/
[5] http://kxml.sourceforge.net/
[6] http://xmlpull.org/

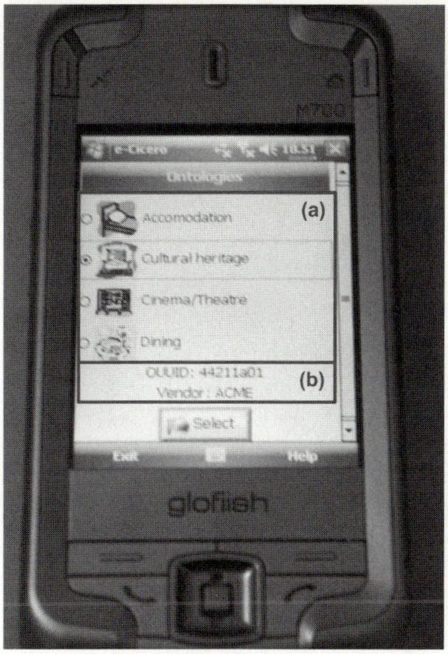

Fig. 1. Ontology selection screen

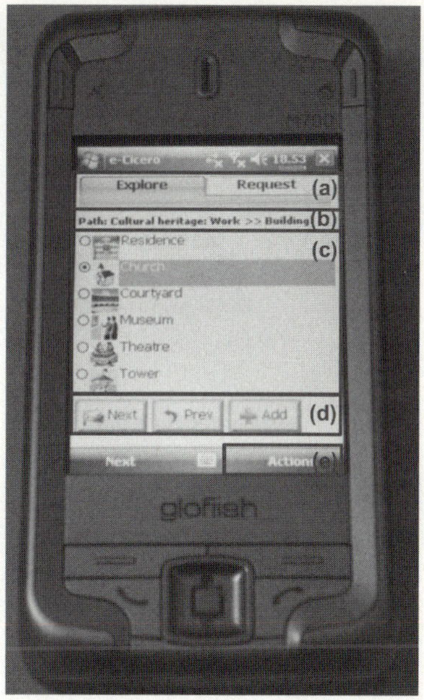

Fig. 2. Ontology browsing screen

Bari before and he knows very little about the city. Being interested in medieval art and particularly in churches, he would like to visit interesting places near his current location. Let us see how we model the discovery issue to meet Bob's needs.

Use your smartphone. *Bob is looking for medieval churches in a 1 Km range. Under GPRS/UMTS or Wi-Fi coverage, his GPS-enabled smartphone can connect to a service/resource provider in order to search for interesting items in the area.* The service provider keeps track of semantically annotated descriptions of different types of touristic points of interest in Apulia region together with their position coordinates. It interacts with mobile clients replying to their requests: in particular, it is equipped with a DL reasoner. In our current implementation we adopt the previously cited *MAMAS-tng*. Designed user interface is task-oriented. The provided mobile tool assists the user in service/resource discovery through the following three main tasks.

Ontology management. *Firstly, Bob selects cultural heritage as the resource category he is interested in.* Different domain ontologies are used to describe general resource classes (*e.g.,* accommodation, cultural heritage, movie/theatre shows). When the application is started up, the ontology selection screen is shown, as portrayed in Figure 1. A list of managed ontologies is displayed in **(a)**. When the user highlights an ontology, more detailed information is shown in the lower portion **(b)** of the screen. Each ontology is labeled by a universally unique identifier (OUUID) which is required to perform the ontology agreement between client

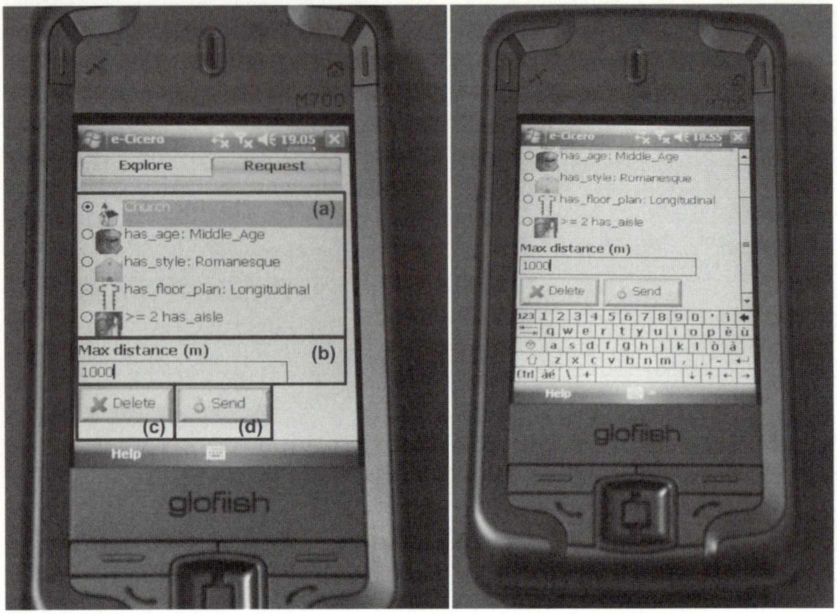

Fig. 3. Request confirmation screen

and resource provider before the matchmaking phase [9]; besides, ancillary information such as ontology domain name, vendor and an icon can be displayed.

As soon as it is started, the virtual guide begins device discovery in the background to find a suitable service provider.

Semantic request composition. *Bob composes his semantic-based request through a fully visual form. He browses resource features modeled in the domain ontology and selects desired characteristics, without actually seeing anything of the underlying DL-based formalism. Then he submits his request to the server.* Figure 2 shows the taxonomy browsing screen. A list-based browsing panel is displayed in **(b-c)**, adopting the same UI metaphors as file system navigation tools that are now familiar to PC users (*e.g.*, "My Computer" in Windows or "Nautilus" in GNOME). This avoids the complexity of typical tree or graph-based GUI representations of ontological knowledge, which can be cumbersome for exploring complex ontologies. In particular, the list **(c)** shows the current *focus* in the classification induced by terminological definitions and inclusions in the reference ontology. The directional keys of the mobile device or the buttons in **(d)** are used to "surf" the ontology by expanding an item or going back one level. The panel focus changes accordingly. Navigation *storyboard* is displayed in **(b)**, so that the user can orient himself even in deeper ontologies. The user can select desired features (corresponding to ontology's concepts and role constructors) to build his request, through either the keypad, the appropriate button in **(d)** or commands in the menu **(e)**. Finally, the tabs in **(a)** allow the user to switch from the browsing screen to the request confirmation screen and vice versa.

```
AD ⊑ Age                                   BC ⊑ Age
Middle_Age ⊑ AD                            Centralized ⊑ Floor_Plan
Longitudinal ⊑ Floor_Plan                  Square ⊑ Floor_Plan
Quadrangular ⊑ Square                      Byzantine ⊑ Style
Romanesque ⊑ Style                         Gothic ⊑ Style
Baroque ⊑ Style                            Portal ⊑ Architectural_Element
Cathedra ⊑ Architectural_Element           Aisle ⊑ Architectural_Element
Altar ⊑ Architectural_Element              Pulpit ⊑ Architectural_Element
Crypt ⊑ Architectural_Element              Apse ⊑ Architectural_Element
Window ⊑ Architectural_Element             Single_Light ⊑ Window
Double_Light ⊑ Window                      Triple_light ⊑ Window
Religious ⊑ Destination                    Private ⊑ Destination
Public ⊑ Destination                       Private ⊑ ¬Public
Private ⊑ ¬Religious
Building ⊑ ∃has_age ⊓ ∃has_floor_plan ⊓ ∃has_style
Residence ⊑ Building ⊓ ∃Destination ⊓ ∀Destination.Private
Church ⊑ Building ⊓ ∃Destination ⊓ ∀Destination.Religious ⊓ ∃has_altar ⊓ ∀has_altar.Altar
Castle ⊑ Residence
```

Fig. 4. Axioms in the toy cultural heritage ontology used in the case study

The request screen is shown in Figure 3. Currently selected features are shown in the list **(a)**; if the user changes his mind, he can remove a previously selected feature using the button **(c)**. Additionally, in **(b)** the user can specify a maximum distance D for resource retrieval. This range is submitted to the provider together with the current device coordinates –determined through the Location API. In this way, a circular area is identified, with its centre in the user's position: the service provider will return resources located in this area only. Eventually the user confirms his request and submits it to the resource provider through the appropriate command button **(d)**. *Bob would like to visit a Romanesque Middle Age church with longitudinal floor plan and two aisles.* W.r.t. our cultural heritage reference ontology (reported in Figure 4 only partially for the sake of brevity), this can be expressed formally as

R: $Church \sqcap \forall\ has\_age.Middle\_Age \sqcap \forall\ has\_floor\_plan.Longitudinal \sqcap \geq 2\ has\_aisle \sqcap \forall\ has\_style.Romanesque$

in DL classical notation. As it can be seen, requests are formulated as DL conjunctive queries. Each conjunct represents a requested resource feature; it can be an atomic concept selected from the ontology, a universal quantifier or an unqualified number restriction on roles. If the domain ontology defines an atomic concept as range for a particular role, the application moves the browsing panel focus to that concept when the role is expanded. The user can then select that concept or one of its subclasses as filler for the universal quantifier constructor. Otherwise, if no range is specified for a role, a number restriction constructor is assumed and the user is allowed to specify the cardinality constraint type and value (as in the case of *has_aisle*). The communications module was designed as a finite state machine to precisely retain knowledge about the progress of client-server interaction. By doing so, only failed operations are actually repeated, thus improving efficiency from both time and energy standpoints. Retry command, along with other less frequently used functions, is located in the menu so as not to cram the main screen area.

Results review and query refinement. *The reasoner filters all available resources, discarding the ones not located in the user-specified area. Then it performs the matchmaking between user request and each resource description. Results are ranked according to their semantic distance from the request and best matching resources are returned to Bob, along with their location.* For example, let us consider the following resources in the knowledge base of the provider:

S1: Basilica of St. Nicholas, Bari (distance from user: $d = 900$ m). A Romanesque Middle Age church, with longitudinal floor plant, two aisles, three portals and two towers. Other notable elements are its, crypt, altar, cathedra and Baroque ceiling. W.r.t. domain ontology, this is expressed as:

$Church \sqcap = 2\ has\_aisle \sqcap \forall\ has\_age.Middle\_Age \sqcap \forall\ has\_style.Romanesque \sqcap = 1\ has\_apse \sqcap = 3\ has\_portal \sqcap = 1\ has\_crypt \sqcap = 1\ has\_altar \sqcap = 2\ has\_tower \sqcap = 1\ has\_cathedra \sqcap \exists\ ceiling\_style \sqcap \forall\ ceiling\_style.Baroque \sqcap \forall\ has\_floor\_plan.Longitudinal$

S2: Norman-Hohenstaufen Castle, Bari ($d = 570$ m). It is described as a Middle Age castle, with Byzantine architectural style and a quadrangular plan with four towers.

$Castle \sqcap \forall\ has\_floor\_plan.Quadrangular \sqcap = 4\ has\_tower \sqcap \forall\ has\_style.Byzantine \sqcap \forall\ has\_age.Middle\_Age$

S3: Church of St. Scholastica ($d = 1100$ m). It is described as a Romanesque Middle Age church, with longitudinal floor plan, three aisles, an apse and a tower.

$Church \sqcap \forall\ has\_style.Romanesque \sqcap \forall\ has\_age.Middle\_Age \sqcap \forall\ has\_floor\_plan.Longitudinal \sqcap = 3\ has\_aisle \sqcap = 1\ has\_tower \sqcap = 1\ has\_apse$

S4: Church of St. Mark of the Venetians, Bari ($d = 430$ m). It is described as a Romanesque Middle Age church with two single-light windows and a tower.

$Church \sqcap \forall\ has\_style.Romanesque \sqcap \forall\ has\_window.Single\_Light \sqcap = 2\ has\_window \sqcap \forall\ has\_age.Middle\_Age \sqcap = 1\ has\_tower$

In order to make resource discovery really useful and effective, the user should receive some meaningful feedback about the outcome of his request. A simple list of results is then inadequate whereas a ranking score has more explanatory power. Moreover, looking at descriptions of retrieved resources the user might find other features he was not previously aware of, but he would be interested in. Table 1 reports results of the semantic-based matchmaking for the above example. **S3** is not processed since its distance from the user exceeds the limit, even though it would result in a full match. **S1** is a full match with the request, because it explicitly satisfies all user requirements. On the other hand, **S4** is described just as Romanesque Middle Age church, therefore due to Open World Assumption (what is not specified in a description has not to be interpreted as a constraint of absence) it is not specified whether it has a longitudinal floor plan with aisles or not: these characteristics become part of the *Hypothesis* computed through Concept Abduction. Finally, **S2** produces a partial match with user request, since it is a castle: this concept is incompatible with user request, so it forms the *Give Up* feature computed through Concept Contraction, while

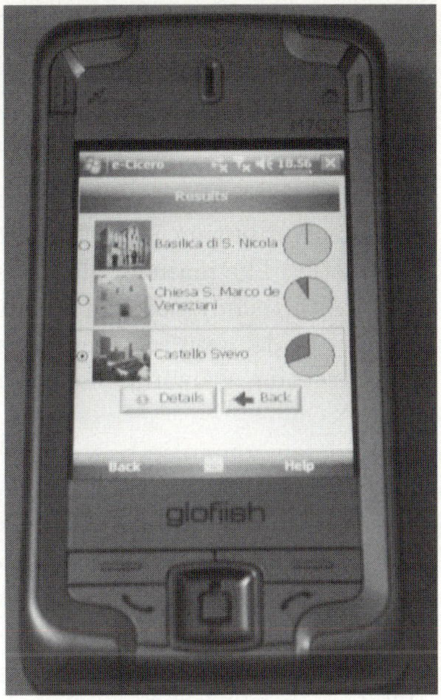

Fig. 5. Results screen

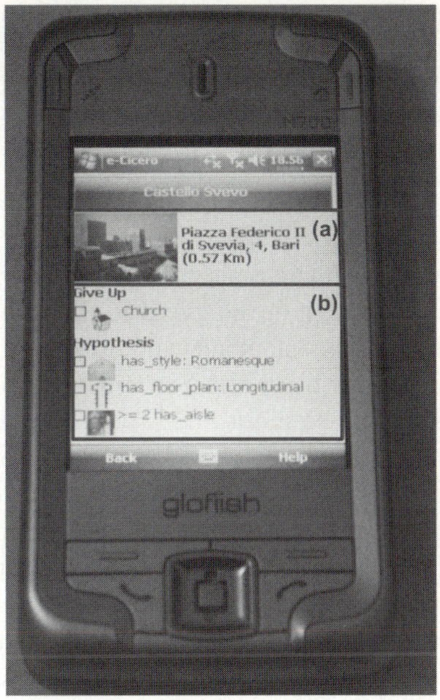

Fig. 6. Result details screen

the *Keep* expression includes the remaining of the request. For each advertised resource O_i, its overall score S is then computed using the *utility function*:

$$S(O_i, d_i) = 100[1 - \frac{s\_match(R, O_i)}{max(s\_match)}(1 + \frac{d_i}{D})]$$

where $s\_match(R, O_i)$ is the semantic distance from request to resource; $max(s\_match) \doteq s\_match(R, \top)$ is the maximum semantic distance from the request, which depends on axioms in the domain ontology; d_i is the physical distance of the resource from the user and D is the user-supplied distance limit. The purposes of the utility function are to weigh the result of semantic matching according to distance and to convert the score to a more user-friendly scale. Of course nearer resources are preferred, but in case of a full match $s\_match(R, O_i) = 0$ hence $S = 100$ regardless of distance, so that full matches will always be shown at the top of the result list.

The result screen is showed in Figure 5. Retrieved resources are listed –best matching first– along with their overall scores as pie charts. When the user selects a resource, a detail form screen is displayed, as depicted in Figure 6. A picture of the resource is shown in **(a)** together with its address and the distance from the user, while the semantically relevant resource properties contributing to the outcome are displayed in **(b)**.

If Bob is not satisfied with results, he can refine his semantic request and submit it again. The user can go back to the ontology browsing screen to modify his request. Furthermore, in case of partial match, he can select some elements of the *Give Up* list in the result details screen and they will be removed from the request automatically.

4 Related Work

Service/resource discovery in mobile and ubiquitous computing is a very active topic in both the research community and ICT industry. The main challenge is to provide effective and flexible discovery paradigms and techniques, yet intuitive enough to be of practical interest for a potentially wide user base. Ideas and technologies from the Semantic Web vision are attracting growing attention as an enabling element.

In [10] a prototypical mobile client is presented for semantic-based mobile service discovery, in which user preference specification plays a key role. An OWL ontology describes relevant properties of each service class. Ontology browsing is allowed through a graph-based representation, which is rendered adaptively. A pointing device and a large screen seem to be required in order to explore ontologies of moderate complexity with reasonable comfort. Moreover, preference specification requires a rather long interaction process, which can be impractical in mobile scenarios. Authors acknowledged this issue and introduced heuristic mechanisms to simplify interaction, *e.g.*, the exploitation of default values.

mSpace Mobile [11] is a location- and context-aware mobile Semantic Web client for tourism scenarios. Users can discover and combine services/resources in an area through an associative process, by browsing semantic relationships represented in an RDF knowledge base. Similarly to our ontology browsing, a "focus + context" paradigm is adopted in UI design. The desire to integrate multiple domains of information has led to a division of the user interface into many tiny sections, whose clarity and practical usability seem questionable. Moreover, knowledge is extracted from several independent sources to build a central RDF triple store, which can then be accessed by one or more servers in order to reply to client requests through the Internet. The proposed architecture is therefore hardly adaptable to MANET-based ubiquitous computing environments.

Table 1. Matchmaking results

Supply	**Match type**	*s_match*	**Outcome**	**Score**
S1: Basilica of St. Nicholas	Full	0	**Hypothesis:** $H = \top$;	100
S4: Church of St. Mark	Potential	3	**Hypothesis:** $\forall\ has\_floor\_plan.Longitudinal \sqcap \geq 2\ has\_aisle$	92
S2: Norman-Hohenstaufen Castle	Partial	11	**Give up:** *Church* **Keep:** $Building \sqcap \forall\ has\_age.Middle\_Age$ **Hypothesis:** $\forall\ has\_floor\_plan.Longitudinal \sqcap \geq 2\ has\_aisle \sqcap \forall\ has\_style.Romanesque$	68
S3: Cathedral of Trani	N.A.	N.A.	discarded due to distance	N.A.

Fully decentralized semantic-based resource/service discovery infrastructures in ubiquitous computing call for peer-to-peer interaction paradigms. Hence, mobile hosts themselves should be endowed with reasoning and matchmaking capabilities. To the best of our knowledge, Pocket KRHyper [12] is currently the only available reasoning engine for mobile devices. It provides satisfiability and subsumption inference services, which have been exploited by authors in a DL-based matchmaking scheme between user profiles and descriptions of resources/services [13]. Due to resource constraints, the proposed approach privileges simplicity over expressiveness and flexibility w.r.t. ours, and does not currently allow explicit explanation of outcomes. We conjecture that a native language optimized implementation can provide acceptable performance for more resource-intensive inference procedures, thus allowing to overcome those constraints at least partially.

5 Conclusion and Future Work

We have presented a fully visual mobile application for semantic-enabled resource discovery in ubiquitous computing. Non standard DL inference services are exploited to perform a matchmaking based on semantics of request/resource descriptions. The proposed tool is general purpose as it allows a mobile user to perform a discovery in multiple scenarios (from u-commerce [14] to u-tourism [1]) simply using her mobile phone. The final goal of our endeavor is to provide an m-DSS to retrieve resources/services through a fully dynamic wireless infrastructure, without relying on support facilities provided by more expensive information systems and wired networks. The approach has been motivated in a tourism case study. The proposed application aims to be really context-aware as it recognizes via GPS the user location and grades matchmaking outcomes according to vicinity criteria. Limits of the tool emerge because actually the time spent in performing the overall discovery and ranking procedure is still somewhat high to be definitively acceptable. From this point of view a significant bottleneck is represented by the matchmaker computation. The complexity of the matchmaker module claims for serious optimization and simplification of the reasoner features in sight of an explicit utilization for pervasive and ubiquitous applications. Future work will also focus on the enhancements of the user interface. A more immediate GUI also integrating a speech recognition engine is the most important effort we are now pursuing.

Acknowledgments

The authors are grateful to Saverio Ieva for useful implementations. They would also like to acknowledge the support of Apulia project PE_082 "IC Technologies for tourist assistance through interactive consultancy of a virtual guide".

References

1. Watson, R., Akselsen, S., Monod, E., Pitt, L.: The Open Tourism Consortium: Laying The Foundations for the Future of Tourism. European Management Journal 22(3), 315–326 (2004)
2. Paolucci, M., Kawamura, T., Payne, T., Sycara, K.: Semantic Matching of Web Services Capabilities. In: Horrocks, I., Hendler, J. (eds.) ISWC 2002. LNCS, vol. 2342, pp. 333–347. Springer, Heidelberg (2002)
3. Li, L., Horrocks, I.: A Software Framework for Matchmaking Based on Semantic Web Technology. International Journal of Electronic Commerce 8(4), 39–60 (2004)
4. Bechhofer, S., Möller, R., Crowther, P.: The DIG Description Logic Interface. In: Proceedings of the 16th International Workshop on Description Logics (DL 2003). CEUR Workshop Proceedings, vol. 81 (September 2003)
5. Baader, F., Calvanese, D., Mc Guinness, D., Nardi, D., Patel-Schneider, P.: The Description Logic Handbook. Cambridge University Press, Cambridge (2002)
6. Colucci, S., Di Noia, T., Di Sciascio, E., Donini, F.M., Mongiello, M.: Concept abduction and contraction in description logics. In: Proceedings of the 16th International Workshop on Description Logics (DL 2003). CEUR Workshop Proceedings, vol. 81 (September 2003)
7. Gärdenfors, P.: Knowledge in Flux: Modeling the Dynamics of Epistemic States. Bradford Books, MIT Press, Cambridge (1988)
8. Di Noia, T., Di Sciascio, E., Donini, F.M., Mongiello, M.: Abductive matchmaking using description logics. In: Proceedings of Eighteenth International Joint Conference on Artificial Intelligence IJCAI 2003, Acapulco, Mexico, MK, pp. 337–342 (August 2003)
9. Ruta, M., Di Noia, T., Di Sciascio, E., Donini, F.: Semantic based collaborative p2p in ubiquitous computing. Web Intelligence and Agent Systems 5(4), 375–391 (2007)
10. Noppens, O., Luther, M., Liebig, T., Wagner, M., Paolucci, M.: Ontology-supported Preference Handling for Mobile Music Selection. In: Proceedings of the Multidisciplinary Workshop on Advances in Preference Handling, Riva del Garda, Italy (August 2006)
11. Wilson, M., Russell, A., Smith, D., Owens, A., Schraefel, M.: mSpace Mobile: A Mobile Application for the Semantic Web. In: Proceedings of the End User Semantic Web Workshop at ISWC 2005 (2005)
12. Sinner, A., Kleemann, T.: KRHyper - In Your Pocket. In: Proc. of 20th International Conference on Automated Deduction (CADE-20), Tallinn, Estonia, July 2005, pp. 452–457 (2005)
13. Kleemann, T., Sinner, A.: User Profiles and Matchmaking on Mobile Phones. In: Bartenstein, O. (ed.) Proc. of 16th International Conference on Applications of Declarative Programming and Knowledge Management INAP 2005, Fukuoka (2005)
14. Watson, R., Pitt, L., Berthon, P., Zinkhan, G.: U-Commerce: Expanding the Universe of Marketing. Journal of the Academy of Marketing Science 30(4), 333–347 (2002)

A Profile Modelling Approach for E-Learning Systems

Corrado Mencar, Ciro Castiello, and Anna Maria Fanelli

Department of Informatics – University of Bari
Via E. Orabona, 4 – 70125 Bari ITALY
{mencar,castiello,fanelli}@di.uniba.it

Abstract. This paper presents a user profile modelling approach based on fuzzy logic techniques. The proposed approach is conceived to find application in the context of e-learning processes, with the aim of providing personalised contents to different categories of users. Several concepts are introduced and formalised within a peculiar mathematical framework which bases its working engine on an innovative scheme of metadata exploiting the expressive strength of fuzzy sets. Along with the formal presentation of the profile modelling approach, suitable examples of application are provided for the sake of illustration.

1 Introduction

Personalisation is one of the key issues pervading most of the technological applications designed for content providing, such as e-commerce systems, web portals, e-learning platforms and so on [1]. In the diverse contexts where they find application, personalisation mechanisms are mainly based on the definition of user profiles. These are formal structures representing different pieces of information related to the user, ranging from his expressed preferences or previous knowledge to his specific role within the area of interest. Usually, profiles are defined to represent categories of users sharing common features; in this way, user profiles stand as filters which favour the allocation of personalised contents.

Generally speaking, the definition of user profiles determines a specific granularity level to be introduced inside the area of interest. This kind of information granulation can be established on a range where the roughest case refers to the definition of a single profile for all the ensemble of users (no personalisation allowed), while the finest case refers to the definition of a distinct profile for each user (maximum level of personalisation). Inside this range, the choice for a particular granularity level is mainly driven by a number of factors concerning the trade-off between the involved costs and the produced benefits. More fine-grained profiles can be achieved through automated profiling systems, which rely on data mining and machine learning techniques [2]. Nevertheless, this kind of approach requires a significant learning time, during which the user behaviour must be monitored by the system in order to build customised profiles.

The work activity described in this paper refers to personalisation processes performed in the e-learning context: here the final aim consists in providing

O. Gervasi et al. (Eds.): ICCSA 2008, Part II, LNCS 5073, pp. 275–290, 2008.

Learning Objects (hereafter LOs) to the users, so as to satisfy their needs and goals at best [3]. E-learning platforms may profit from the adoption of automated user profiling mechanisms, since they could prove to be useful in improving the allocation of LOs to different categories of learners [4]. However, the management of the learning time necessary for monitoring the user behaviour appears to be somewhat problematic. In fact, in e-learning contexts the users should be addressed to LOs as soon as they enter the system, thus reducing the degree of freedom of a new user. For that reason, here a more convenient approach is proposed, aimed at combining the benefits of automated profiling mechanisms with the rigidity of e-learning systems. In practice, the users are assigned to pre-established user profiles, shared among the entire user community. At the same time, each user is also associated with individual profiles, which can be used to track his specific behaviour. The definition of complex profile structures (on the basis of simpler profile components) allows to build up any kind of user profile, responding to the articulated conditions of real world applications.

Additionally, a peculiarity of the proposed approach consists in the introduction of fuzzy logic for modelling the association of users and profiles. In typical real world situations, a user can be hardly characterised in terms of a single profile: the specification of fuzzy degrees of membership allows to associate a single user with multiple profiles. Moreover, the employment of fuzzy logic is useful also for defining a suitable metadata specification, to be adopted for the description of the LOs. Actually, metadata are largely used in profiling systems to characterise the objects involved in the personalisation process [5]. The usual association mechanisms based on the common metadata schemes produce the simple identification of a number of objects to be connected with the demanding users. Obviously, that means that a great deal of resources are left outside from the association process. Fuzzy logic allows for a more comprehensive metadata specification, including the description of imprecise properties of the LOs. Consequently, it can be realised a gradual association between users and LOs, configured as a ranking where degrees of compatibility are used to identify the most suitable LOs for each user, without excluding those characterised by lower degrees of compatibility.

The paper is organised as follows. In the next section the profile modelling approach is introduced. In section 4 the model for the description of the pedagogical resources is formalised, while section 5 is devoted to the formalisation of the model describing the profiles of the actors involved in the e-learning process. Section 6 closes the paper with some conclusive remarks.

2 Related Work

In the last decade soft computing techniques (including fuzzy logic, neural networks, probabilistic reasoning, genetic algorithms, etc.) have been successfully applied in user modelling [6,7].

Fuzzy logic is usually employed in user modelling for its intrinsic ability of representing and manipulating imprecise and graded concepts. The usefulness of fuzzy logic is generally appreciated when user models cannot be precisely defined without arbitrary approximations, which appears to be the most frequent case in many real world situations. For a survey on user modelling implemented by means of several paradigms, including fuzzy logic, see [8].

A noteworthy application of fuzzy logic for user modelling in e-learning systems is given in [9]. Here, fuzzy sets are used to model the user knowledge and are dynamically adapted while user learns new concepts from the e-learning platform. In [10] fuzzy rules are employed to register user actions and to refine the strength of the relationship between the user model attributes and the concepts of the knowledge domain. In [11] fuzzy sets are used to model beliefs about interactions that students make with LO and quizzes. Such a modelling capability is part of an educational system, which is able to evaluate how much a student plausibly studied her assigned resources.

Fuzzy Logic has been also used to perform user modelling in several areas other than e-learning systems. As an example, in [12] a fuzzy nearest neighbour approach is used in a collaborative filtering system to guess user preferences on the basis of historical records. In [13] a fuzzy approach for user modelling has been adopted to improve the interaction between the user and information retrieval systems. In [14] fuzzy multiple criteria analysis has been used as a tool for user modelling in a Sales Assistance software.

In most works the representation of a user models realised by means of fuzzy logic is flat, being usually based on a collection of fuzzy sets, or a vector of fuzzy values. On the contrary, some applications would require that users of an e-learning system should be associated to more complex representations, in order to better express the role, the knowledge, the preferences of each user in her professional context. Besides, it could be necessary to describe the roles of a user in complex terms, such as a composition of sub-roles. In the subsequent sections an approach is proposed that takes into account this kind of requirements by providing a highly flexible framework for representing user profiles.

3 Preliminaries on the Profile Modelling Approach

The activity described in this paper moves from the assumption that the proposed approach for profile modelling can find application in the context of a pedagogical platform developed for e-learning purposes. Precisely, our investigation is addressed to formalise the association process between Learning Objects (LOs) and users, on the basis of suitable metadata specifications.

The main concerns of our approach are related to two distinct categories of requirements. On the one hand, we advocate the independence of the proposed modelling strategy from the implementation details related to the e-learning platform. On the other hand, we propose a modelling approach endowed with the capabilities needed to overcome the most common difficulties related to the management of e-learning processes. As concerning the independence requirements, we intend to preserve:

- the independence from the actual representation of the LOs inside the e-learning platform;
- the independence from the actual representation of the users inside the e-learning platform;
- the independence from the specific technologies adopted for the representation of metada inside the e-learning platform.

By conforming to these independence requirements it is possible to devote peculiar attention to the management of the profile modelling process, regardless of the constraints related to the practical realisation of the e-learning platform. This kind of approach allows to set aside a proper definition of users and LOs: they can be simply acknowledged as class instances, without additional specifications.

As concerning the capability requirements, we intend to realise:

- the capability to employ metadata specifications allowing for the representation of imprecise properties;
- the capability to formalise profiles of high complexity;
- the capability to perform (possibly partial) associations of a single user with several profiles.

The key to association between users and LOs are metadata, which connect a LO with an attribute and its respective value. Our approach differentiates from usual metadata specifications since we assume that the value for an attribute, far from being simply an element inside the attribute domain, can be specified as a fuzzy set. The theory of fuzzy sets basically modifies the concept of membership: the classical binary membership leaves room for a more comprehensive variety of membership degrees, defined in terms of a mathematical function (as we are going to detail in the next section). In this way, fuzzy sets allow for a partial membership of their elements [15]. The employment of fuzzy metadata characterisation enables the definition of different properties related to a LO. In particular, we can distinguish among: simple properties (regarding the punctual evaluation of an attribute by determining a single value inside the set of infinite possible values); collective properties (regarding the extensional specification of a discrete set of values for an attribute); imprecise properties (regarding the intensional definition of a qualitative value for an attribute). It should be noted how this kind of approach produces a granulation of the attribute domains, where fuzzy sets are adopted to represent each information granule. This favours a mechanism of elaboration of concepts that is in agreement with the human reasoning schemes [16].

Actually, the formalisation of imprecise properties is included in our model to cope with the intrinsic difficulties related to some metadata characteristics, which can not be described in terms of simple or collective values. Attempts to formalise such properties by means of discretisation processes lead to arbitrariness, resulting in a poor management of the involved resources. The introduction of fuzzy sets is intended to overcome this kind of difficulties, together with the adoption of particular mathematical operators that are especially suitable for handling imprecise information. In this way, gradual associations can be realised between users and LOs, on the basis of a compatibility ranking. As a result, each user can be ultimately addressed to the most compatible LOs, without arbitrarily discarding those characterised by a lower degree of compatibility.

The user profiles are used to represent stereotypical categories of learners. In order to take into account stereotypes of high complexity, in this work the user profiles are formalised as collections of profile components. Analogously to the metadata specification, the profile components are characterised in terms of fuzzy sets: this homogeneity expedites the comparison process aiming at defining a compatibility degree between profile components and LOs. The aggregation of such compatibility degrees produces the final association of a profile with a LO.

Users are characterised by their corresponding profiles, however a single user hardly finds a full representation inside a single profile. For that reason, the proposed modelling approach allows a partial membership of users to different profiles, as in the case of real world situations. Therefore, the final association of a user with a LO is evaluated by considering the compatibility degrees related to the different profiles of belonging.

In the following sections we are going to detail the profile modelling approach by distinguishing the characterisation of the pedagogical resources from the description of the user profiling mechanisms. All the involved entities are formally defined in terms of mathematical concepts and suitable examples are provided to illustrate the working scheme of the modelling approach.

4 Modelling the Pedagogical Resources

4.1 Learning Objects and Attributes

A Learning Object (LO) is a reusable set of instructions inside an e-learning scenario. The model proposed in this paper leaves aside the peculiar structure of a LO which is simply defined as an element of a set. Let $\mathcal{O}$ be a non-empty set of physical objects, namely the LO space.

Definition 1. *A Learning Object (LO) is an element* o *in the LO space* $\mathcal{O}$*, i.e.* $o \in \mathcal{O}$.

With reference to a particular topic to be learned, a LO may be represented by a multimedia support, such as a document file, a presentation, a hypertext, and so on.

Each LO can be associated with a set of attributes. Generally speaking, an attribute may be numeric or symbolic and it is related to a (possibly infinite) number of distinct values. Let $\mathcal{A}$ be a non-empty set, namely the attribute space.

Definition 2. *An attribute is an element* A *in the attribute space* $\mathcal{A}$*, i.e.* $A \in \mathcal{A}$*. In particular, an attribute* A *is a set of values* $a \in A$.

Example 1. If we consider a LO represented by a digital resource, a list of related attributes may include:

1. the name of the LO;
2. the fruition time of the LO;
3. the creation date of the LO;
4. the complexity of the LO;
5. the scope of the LO.

The peculiarity of the proposed modelling approach consists in associating a particular LO with the imprecise values of its attributes. To manage these associations, the concept of fuzzy set is employed, standing as a generalisation of the classical concept of the mathematical set. By defining a fuzzy set over a domain, it is possible to extend the membership evaluation process to every element inside the domain, thus moving from binary membership values (0/1) to a gradation of membership values over a continuous range. For our purposes, we define a fuzzy set over each attribute of a LO as follows. Let $A \in \mathcal{A}$ be an attribute.

Definition 3. *A fuzzy set defined over* A *is a function:*

$$F_A : a \in A \rightarrow F_A(a) \in [0,1] \tag{1}$$

$F_A(a)$ *is called membership degree of value* a *in fuzzy set* F_A.

The definition of fuzzy sets enables the LO characterisation in terms of the correspondence between the attributes and their possible values. This kind of relationship can be defined in terms of a set of Attribute-Value pairs, specified as follows. Let $\mathcal{A}$ be the attribute space.

Definition 4. *An Attribute-Value pair is an ordered pair* (A, F_A), *being* $A \in \mathcal{A}$ *and* F_A *a fuzzy set defined over the attribute* A. *An Attribute-Value set is a set of Attribute-Value pairs:*

$$f = \{(A, F_A) | A \in \mathcal{A}\}.$$

Remark 1. An Attribute-Value set can be formalised as the function:

$$f : A \in \mathcal{A} \longrightarrow F_A \in \mathcal{F}_A,$$

being $\mathcal{F}_A$ the space of all the possible fuzzy sets which can be defined over the attribute A.

4.2 Metadata and Learning Objects Description

Attributes and values are strictly connected with a LO. Therefore, it is useful the introduction of the metadata concept (to be defined for every attribute), associating a LO with a fuzzy set which represents the attribute value. Let $A \in \mathcal{A}$ be an attribute of a LO $o \in \mathcal{O}$.

Definition 5. *A metadata* m_A *is a function associating the LO* o *with a fuzzy set defined on* A:

$$m_A : \mathcal{O} \rightarrow \mathcal{F}_A.$$

In order to obtain a thorough description for a LO, it is necessary to refer to its attributes and their related values. A straightforward mechanism to generate a LO description is the simple enumeration of the attributes, together with the fuzzy sets reporting the corresponding values. This kind of description is based on the set of metadata that it is possible to define for a LO. Let $\mathcal{A}$ be the attribute space.

Definition 6. *The description of a Learning Object $o \in \mathcal{O}$, with respect to $\mathcal{A}$, is the set of all the Attribute-Metadata pairs associated to o:*

$$D(o) = \{(A, m_A(o)) | A \in \mathcal{A}\}. \quad (2)$$

Remark 2. The description of a LO $o \in \mathcal{O}$ can be formalised as the function:

$$D(o) : A \in \mathcal{A} \rightarrow F_A \in \mathcal{F}_A.$$

Remark 3. The description $D(o)$ is an Attribute-Value set.

Remark 4. We admit the presence of attributes associated with the entire set of values, i.e. when $m_A(o) = A$. This condition is verified when no values are specified for the attribute A in the characterisation of the LO o.

Example 2. Inside the illustrative scenario introduced in the example 1, the LO description can be expressed by listing the attributes together with the fuzzy sets reporting the corresponding values:

1. Name $\rightarrow$ {"Introduction to word processing"/1};
2. Fruition time $\rightarrow$ about 10′ $= T\{8, 10, 15\}$;
3. Creation date $\rightarrow$ {"07-06-07"/1};
4. Complexity $\rightarrow$ {"Easy"/1, "Average"/0.3, "Expert"/0};
5. Scope $\rightarrow$ {"ICT"/0.7, "Word Processing"/1}.

It can be observed that the fuzzy sets reporting the values for the attributes «Name» and «Creation date» refer to simple properties of the LO. They assign the maximum membership degree (equal to 1) only to one of the infinite values the attributes may assume. All the other values are not reported inside the characterisations of the fuzzy sets, since their membership degree is equal to zero. This peculiar condition can be graphically represented by means of fuzzy singletons, as depicted in figure 1. The fuzzy sets reporting the values for the attributes «Complexity» and «Scope» refer to collective properties of the LO. They are defined over discrete sets and assign a membership degree to each one of the possible values, as depicted in figure 2. Finally, the fuzzy set reporting the value for the attribute «Fruition time» refers to an imprecise property of the LO. It is defined over a continuous set and assigns a membership degree to each one of the possible values by means of a triangular function, as depicted in figure 3. (It should be noted that different kinds of membership functions may be adopted, such as trapezoidal or Gaussian functions.)

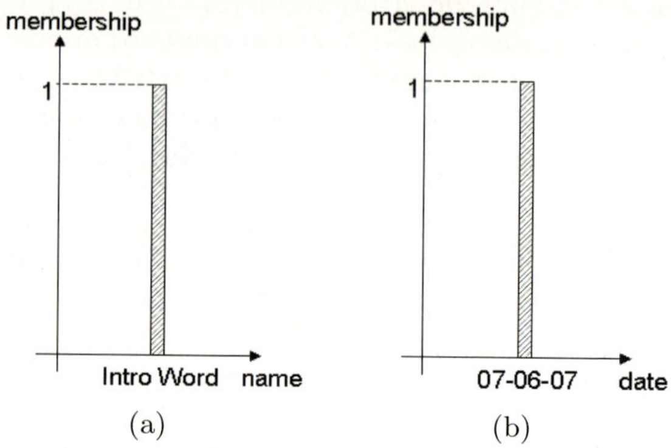

Fig. 1. Fuzzy singletons representing simple properties of a LO: the «Name» attribute (a) and the «Creation date» attribute (b)

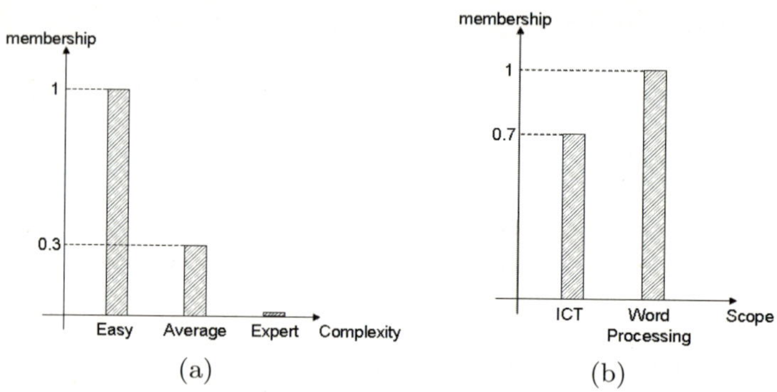

Fig. 2. Fuzzy sets representing collective properties of a LO: the «Complexity» attribute (a) and the «Scope» attribute (b)

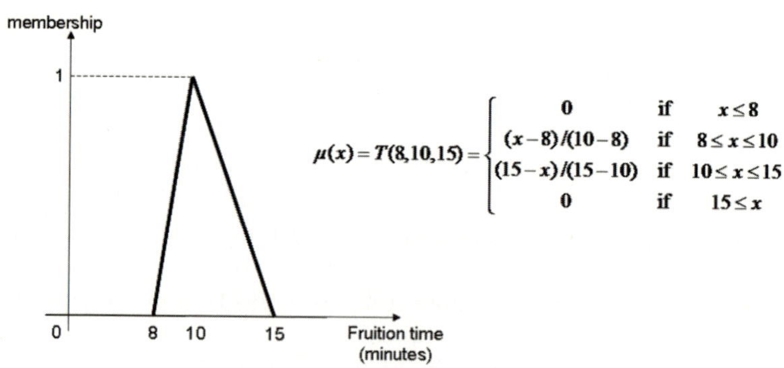

Fig. 3. Fuzzy set representing an imprecise property of a LO: the «Fruition time» attribute

Table 1. The matrix describing a sample LO Collection

	Name	Fruition time	Creation date	Complexity	Scope
LO1	{"Introduction to Word"/1}	about 10'	"07-06-07"/1	{"Easy"/1, "Average"/0.3, "Expert"/0}	{"ICT"/0.7, "WordProcessing"/1}
LO2	{"Introduction to Latex"/1}	$\frac{1}{2}$ hour	"22-05-07"/1	{"Easy"/0.4, "Average"/0.7, "Expert"/0.1}	{"ICT"/0.7, "WordProcessing"/1}
LO3	{"HTML for Dummies"/1}	about 40'	"22-04-07"/1	{"Easy"/0.3, "Average"/0.8, "Expert"/0.2}	{"ICT"/0.7, "Web"/1}

The mathematical formalisation of the LO description can be extended to manage several distinct LOs. In this case, the formula in the definition 6 should be properly generalised by means of a matrix representation, where rows and columns correspond to the LOs and their attributes, respectively. To this aim, we introduce the concept of LO Collection.

Definition 7. *A LO Collection O is a subset of the space $\mathcal{O}$, i.e. $O \subseteq \mathcal{O}$.*

The description of a LO Collection can be further specified with reference to the definition of the single LO description as follows.

Definition 8. *The description of a LO Collection O is the union set of all the Attribute-Value sets defined by the description of each LO:*

$$D(O) = \bigcup_{o \in \mathcal{O}} \{(D(o))\}. \tag{3}$$

Example 3. The previously described illustrative scenario can be expanded by involving a number of different LOs. The information reported in table 1 represents the matrix describing a sample LO Collection.

5 Modelling the Actors of the E-Learning Process

5.1 Profile Components and Compatibility Degrees

The user profiles are regarded as complex concepts whose analysis can be performed on the basis of simpler elements, that are the profile components. Each of them is formalised in terms of the previously introduced Attribute-Value pairs, so that the fuzzy valorisation of attributes can be replied. Let $\mathcal{A}$ be the attribute space.

Definition 9. *A profile component c is defined as the set of ordered pairs:*

$$c = \{(A, F_A) \; | A \in \mathcal{A}\}. \tag{4}$$

Remark 5. A profile component can be formalised as the function:

$$c : A \in \mathcal{A} \to F_A \in \mathcal{F}_A.$$

Remark 6. The ensemble of the profile components spans the set $\mathcal{C}$, namely the space of the profile components.

The formalisation of the profile components is useful to define the concept of user profiles.

Definition 10. *A user profile p is a set of profile components, i.e. $p \subseteq \mathcal{C}$.*

Remark 7. The ensemble of the user profiles spans the set $\mathcal{P} = 2^{\mathcal{C}}$ of user profiles.

Example 4. A specific user profile can be constituted by a number of profile components. As an example, we refer to a couple of profile components. The first one (c_1) is characterised by the following Attribute-Value pairs:

1. Fruition time $\to$ short $= TR\{0, 0, 15, 30\}$;
2. Complexity $\to$ {"Easy"/1, "Average"/1, "Expert"/0.5};
3. Scope $\to$ {"ICT"/0.5, "Word Processing"/0.8}.

The second profile component (c_2) is characterised by the following Attribute-Value pairs:

1. Complexity $\to$ {"Easy"/0.5, "Average"/1, "Expert"/1};
2. Scope $\to$ {"Management"/1}.

Such a user profile can be properly associated to a «secretary» profile and it is defined in terms of the same attributes employed for the LO descriptions reported in the previous examples. Here the «Fruition time», «Complexity» and «Scope» attributes refer to the characteristics of LOs that the user is supposed to be addressed to. The attributes not appearing in this example are not deemed useful for describing the profile components. The pieces of information reported in the example are quite illustrative of the usefulness of profile components. In fact, the first component c_1 is related to the ICT competence of the secretary, with special reference to the use of word processing software. This kind of competence can be reasonably regarded as a non-priority issue for the secretary profile; for that reason the related LOs are characterised by a low level complexity and a short fruition time. Conversely, the secretary profile is fully qualified in terms of management activities, as represented by the the maximum membership degree associated to the value of the «Scope» attribute in the second component c_2. As a consequence, more complex LOs are to be considered, without a specification for the «Fruition time» attribute: in this case the user should be addressed to LOs requiring any time of fruition.

The homogeneity between the LO description and the profile components is straightforward, as resulting from the comparison of definitions 6 and 9. The common structure of these elements allows the definition of a compatibility degree among them, which is actually evaluated between a couple of Attribute-Value sets. For this purpose, it is possible to exploit the possibility measure among fuzzy sets and the aggregation operators. Particularly, the possibility measure verifies the existence of an attribute value both in the profile component and in the LO description; the aggregation process, performed over the evaluated possibility measures, produces a compatibility degree between the component profile and the LO.

Definition 11. *The possibility degree between two fuzzy sets F'_A, F''_A, defined on the same attribute A, is defined as follows:*

$$\Pi(F'_A, F''_A) = \sup_{a \in A} \min\{F'_A(a), F''_A(a)\}.$$

The possibility degree provides a measure of the compatibility of two granular values defined on the same attribute. It is hence the basic operation for the definition of the compatibility degree between a LO and a profile component. The calculation of the possibility degree spans all the attributes in $\mathcal{A}$. As a consequence, given two Attribute-Value sets f_1, f_2 the related possibility degree can be specified.

Definition 12. *The possibility degree between two Attribute-Value sets f_1, f_2 is defined as follows:*

$$\Psi(f_1, f_2) : A \in \mathcal{A} \longrightarrow \Pi(f_1(A), f_2(A)) \in [0, 1].$$

The definition of the compatibility degree of the two Attribute-Value sets f_1, f_2 requires the aggregation of the possibility degrees attained for each attribute.

Definition 13. *The compatibility degree between f_1 and f_2 is defined as:*

$$K_\omega(f_1, f_2) = \omega(\Psi(f_1, f_2)).$$

Function ω is an OWA (Ordered Weight Average) aggregation operator:

$$\omega : [0, 1]^{|\mathcal{A}|} \to [0, 1],$$

defined as:

$$\omega\left(\pi_1, \pi_2, \ldots, \pi_{|A|}\right) = \sum_{j=1}^{|A|} \pi_{i_j} \cdot w_j,$$

where:

$$\pi_{i_1} \leq \pi_{i_1} \leq \cdots \leq \pi_{i_{|A|}}$$

and $w_1, w_2, \ldots, w_{|A|} \in [0, 1]$ are weight factors such that:

$$\sum_{j=1}^{|A|} w_j = 1.$$

Remark 8. By changing the weight factors, several OWA can be defined, such as the minimum function (by setting $w_1 = 1$ and $w_j = 0$ for $j > 1$) or the mean value function (by setting $w_j = 1/|A|$ for all j). The choice of a specific OWA is a matter of design.

Remark 9. The compatibility degree $K_\omega(c, D(o))$ between a profile component c and a LO description $D(o)$ can be defined in terms of the compatibility degree between a couple of Attribute-Value sets introduced by the definition 13.

Generally speaking, a user profile is compatible with a LO if at least one of its profile components is compatible with the LO. Since we are dealing with fuzzy evaluations, it is necessary to refer to the maximum compatibility degree evaluated for each profile component.

Definition 14. *The compatibility degree between a profile p and a LO o is defined as the maximum compatibility degree of the profile components:*

$$K_\omega(p, D(o)) = \max_{c \in p} K_\omega(c, D(o)).$$

Example 5. It is possible to evaluate the compatibility degree between the user profile defined in the example 4 and the LO description reported in the example 2. The compatibility degree is equal to the maximum compatibility degree between one of its profile components (that are c_1, c_2) and the LO description. By considering the profile component c_1, the evaluation of the possibility measures among the fuzzy sets defined for the attributes «Scope», «Complexity» and «Fruition time» is illustrated in figure 5, with the assistance of the graphical representations of the involved fuzzy sets. By adopting the minimum function as OWA aggregation function, the compatibility degree $K_\omega(c_1, LO)$ between the profile component c_1 and the LO can be properly evaluated as:

$$K_\omega(c_1, LO) = \omega(0.8, 1, 1) = 0.8.$$

An analogous process can be performed with reference to the profile component c_2, yielding the compatibility degree:

$$K_\omega(c_2, LO) = 0.$$

According to the definition 14, the final degree of compatibility between the «secretary» user profile and the LO is equal to:

$$\max(K_\omega(c_1, LO), K_\omega(c_2, LO)) = 0.8.$$

5.2 Users and User Profiles

The LOs are intended to be demanded by physical users. Each user can be associated with multiple profiles: these associations are characterised by fuzzy membership degrees. Three kinds of profiles have been conceived in our profile modelling approach:

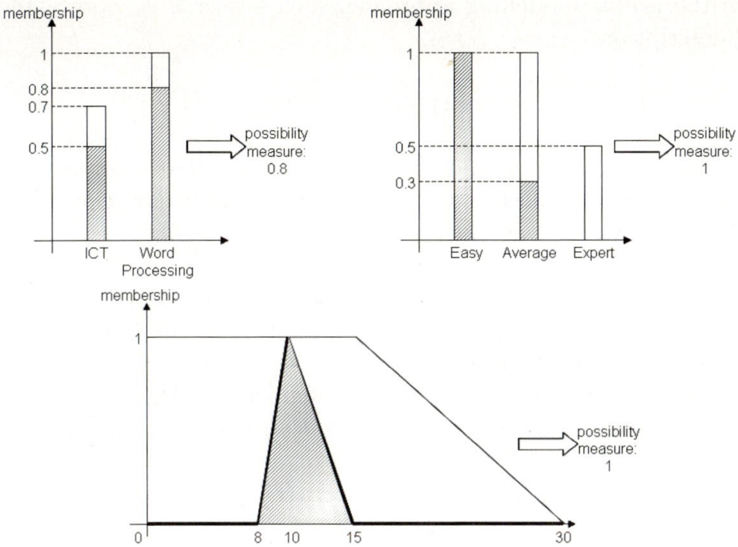

Fig. 4. Evaluation of the possibility measures among fuzzy sets

1. competence profiles (characterising the users in terms of their specific roles or working activities);
2. preference profiles (characterising the users in terms of their specific choices during the e-learning experience);
3. acquaintance profiles (characterising the users in terms of the specific information they have collected during the e-learning experience).

In any case, the structure of the profiles is the same as defined in the previous section for all the above specified categories. A physical user can be defined in terms of the membership degree with reference to a profile base. Let $\mathcal{U}$ be a non-empty set (set of users).

Definition 15. *A physical user is an element u in the set of users $\mathcal{U}$, i.e., $u \in \mathcal{U}$.*

Definition 16. *A profile base is a subset P of the profile space $\mathcal{P}$, i.e. $P \subseteq \mathcal{P}$.*

Let $P \subseteq \mathcal{P}$ be a profile base and let $u \in \mathcal{U}$ be a physical user.

Definition 17. *The description of the user u is defined by the fuzzy set:*

$$D_P(u) : p \in P \longrightarrow [0, 1].$$

Example 6. It could be possible to further detail the scenario illustrated in the example 4 by supposing that the «secretary» user profile may be compatible with some other user profile (possibly corresponding to some other working function). As an example, we could think of a person inside a company who plays the different roles of secretary and (to a lesser extent) of tax consultant. In the context of the profile modelling approach, such a user u is represented by the following description:

$$D(u) = \{\text{“secretary”}/0.8, \text{“tax consultant”}/0.2\}.$$

The above formalisation is based on the assumption that there exist both the user profile «secretary» and the user profile «tax consultant»: the latter may be described in a similar way as illustrated in the example 4.

The compatibility degree between a user and a LO can be defined on the basis of the compatibility degree between the description of the LO and the profiles associated with the user. In practice, several degrees of compatibility should be taken into account, weighted by the user membership degrees with respect to the profile base. Let $u \in \mathcal{U}$ be a user and let $P \subseteq \mathcal{P}$ be a profile base.

Definition 18. *The compatibility degree between the description of the user u and the description of the LO $o \in \mathcal{O}$ is defined as:*

$$K_\omega(D_P(u), D(o)) = \max_{p \in P} \min\{K_\omega(p, D(o)), D_P(u)(p)\}.$$

Example 7. With reference to the example 6, the compatibility degree between the user and the LO can be evaluated by the maximum compatibility degree between the LO and the user profiles (namely, the «secretary» and the «tax consultant» profiles). As concerning the «secretary» profile, we have already evaluated its compatibility degree with the LO, which is equal to 0.8. By supposing a compatibility degree equal to 0.1 for the (undefined) «tax consultant» profile, the ultimate compatibility degree between the user and the LO would be equal to 0.8, i.e. the maximum value of the profile compatibility degrees.

Finally, it is possible to formalise the different role of the previously specified profile categories for the user characterisation. Let $u \in \mathcal{U}$ be a user and let C (Competence), A (Acquaintance) and P (Preference) be three profile bases.

Definition 19. *The compatibility degree between the description of the user u and the description of the LO $o \in \mathcal{O}$ is defined as:*

$$\begin{aligned} K_\omega(u, o) = \min\{ &\max\{K_\omega(D_P(u), D(o)), K_\omega(D_C(u), D(o))\}, \\ &1 - K_\omega(D_A(u), D(o))\}. \end{aligned} \tag{5}$$

The relationship expressed by (5) represents the logical property associating a LO to a specific user on the basis of his competence, the preferences he has expressed during the e-learning process and the LOs he had the opportunity to get acquainted with. Specifically, relationship (5) expresses the logical property that associates a LO to a user if the latter has competence or preference on the LO, but he is not acquainted with it.

6 Conclusions

In this paper a profile modelling approach has been proposed to be applied in the context of e-learning processes. The peculiarity of the illustrated approach consists in the employment of fuzzy logic for modelling the descriptions of pedagogical contents and user profiles. In this way, it is possible to formalise a mathematical scheme of metadata conceived to describe the involved attributes both with respect to their simple properties and by considering their more complex collective and imprecise properties. That is done by defining a fuzzy set over each attribute, so that a fuzzy attribute valorisation can be determined. Moreover, the profiling mechanism benefits from the use of fuzzy membership values, since each user can be partially associated with more than a single profile. Finally, the adoption of fuzzy operators allows to improve the usual association mechanisms, enabling the evaluation of compatibility degrees, which constitute the basis for building up a ranking of resources to be associated with a specific user.

Future work is to be addressed to a more comprehensive study of the fuzzy operators involved in the association mechanisms, in order to define the most suitable functions for modelling the different semantics of the personalisation process.

References

1. Riecken, D.: Introduction: personalized views of personalization. Communications of the ACM 43(8), 26–28 (2000)
2. Eirinaki, M., Vazirgiannis, M.: Web mining for web personalization. ACM Transactions on Internet Technology 3(1), 1–27 (2003)
3. De Bra, P., Brusilovsky, P., Houben, G.: Adaptive hypermedia: from systems to framework. ACM Computing Surveys. Article No. 12 31(4) (1999)
4. Markellou, P., Mousourouli, I., Spiros, S., Tsakalidis, A.: Using semantic web mining technologies for personalized e-learning experiences. In: Uskov, V. (ed.) Web-based Education – WBE 2005, Acta Press (2005)
5. Neven, F., Duval, E.: Reusable learning objects: a survey of lom-based repositories. In: MULTIMEDIA 2002: Proceedings of the tenth ACM international conference on Multimedia, pp. 291–294. ACM Press, New York (2002)
6. Azvine, B., Wobcke, W.: Human-centred intelligent systems and soft computing. BT Technology Journal 16(3), 125–133 (1998)
7. Frías-Martínez, E., Magoulas, G., Chen, S., Macredie, R.: Recent soft computing approaches to user modeling in adaptive hypermedia. In: Adaptive Hypermedia and Adaptive Web-Based Systems, pp. 104–114. Springer, Heidelberg (2004)
8. Brusilovsky, P., Millán, E.: User models for adaptive hypermedia and adaptive educational systems. In: The Adaptive Web, pp. 3–53. Springer, Heidelberg (2007)
9. Kavcic, A.: Fuzzy user modeling for adaptation in educational hypermedia. IEEE Transactions on Systems, Man, and Cybernetics, Part C: Applications and Reviews 34(4), 439–449 (2004)
10. Martinovska, C.: A fuzzy-based approach to user model refinement in adaptive hypermedia systems. In: De Bra, P., Brusilovsky, P., Conejo, R. (eds.) AH 2002. LNCS, vol. 2347, pp. 411–414. Springer, Heidelberg (2002)

11. Kosba, E., Dimitrova, V., Boyle, R.: Using fuzzy techniques to model students in web-based learning environments. In: Knowledge-Based Intelligent Information and Engineering Systems, pp. 222–229. Springer, Heidelberg (2003)
12. Suryavanshi, B.S., Nematollaah Shiri, S.P.M.: A fuzzy hybrid collaborative filtering technique for web personalization. In: Proceedings of the 3rd Workshop on Intelligent Techniques for Web Personalization (ITWPŠ 2005), pp. 1–8 (2005)
13. John, R.I., Mooney, G.J.: Fuzzy user modeling for information retrieval on the world wide web. Knowl. Inf. Syst. 3(1), 81–95 (2001)
14. Popp, H., Lödel, D.: Fuzzy techniques and user modeling in sales assistants. User Modeling and User-Adapted Interaction 5(3), 349–370 (1995)
15. Zadeh, L.: Fuzzy sets. Information and Control 8, 338–353 (1965)
16. Zadeh, L.: A note on web intelligence, world knowledge and fuzzy logic. Data & Knowledge Engineering 50, 291–304 (2004)

Modeling and Analyzing Variability for Mobile Information Systems

Raian Ali, Fabiano Dalpiaz, and Paolo Giorgini

University of Trento - DISI, 38100, Povo, Trento, Italy
{raian.ali,fabiano.dalpiaz,paolo.giorgini}@disi.unitn.it

Abstract. Advances in size, power, and ubiquity of computing, sensors, and communication technology made possible the development of mobile or nomadic information systems. Variability of location and system behavior is a central issue in mobile information systems, where behavior of software has to change and re-adapt to the different location settings. This paper concerns modeling and analysis of the complementary relation between software and location variability. We use graphical and formal location modeling techniques, show how to elicit and use location model in conjunction with Tropos goal-oriented framework, and introduce automated analysis on the location-based models.

1 Introduction

Advances in computing, sensing and communication technology have recently led to the growth of interest in software mobility. Mobility emphasizes several concerns (space, time, personality, society, environment, and so on) often not considered by traditional desktop systems [1,2]. Besides computing ubiquity, the 21st century computing [3] is expected to have a core "mental" part: computing systems act on behalf of humans executing tasks without prompting them for and receiving their explicit requests, i.e. computing will realize the concept of agency, and interaction with computers is getting hence as minimal as possible [4]. Advances in technology do not necessarily imply the easiness of exploiting it, rather more challenges are introduced. Software systems can be given more responsibility, and they can now actively support several decision making processes. Consequently, appropriate software development methods and models need to be developed, or adapted, to cope with the new software scenarios.

Nomadic users expect smarter information systems, able to adapt their behavior without human intervention. Systems should be able to reason about the surrounding location, which includes users themselves, and adapt autonomously their behavior to location settings. What we need then is to model and analyze the variable locations the users can be part of, and define how location influences the behavior of the mobile information system (herafter MobIS). Changing behavior needs a kind of reasoning on what exists and what can be done based on criteria like user preferences, cost, time, priority, and so on. Besides modeling the variant locations, and the different behaviors the MobIS can adopt accordingly,

O. Gervasi et al. (Eds.): ICCSA 2008, Part II, LNCS 5073, pp. 291–306, 2008.

we need to define the process of analyzing and designing such systems that are distinguishably different from desktop driven information systems.

In this paper, we use Tropos [5,6] goal-oriented framework to analyze and model the different behaviors a MobIS can adopt. We use goal model to represent the alternative strategies for satisfying a goal, and then we attach location properties to a set of defined variability points on goal model. The integration between goal model and the location where the software operate, will allow us to support the decision making process when deriving a location-tailored MobIS instance as well as enabling different kinds of analysis. MobIS different strategies for satisfying a stakeholder goal, and location model are designed graphically by designers, and then formalized as a basic step for automated reasoning. Automated reasoning will allow to answer questions like: "*are all MobIS objectives achievable in a given location?*", "*what is the optimal alternative to achieve an objective in a given location?*", and "*what is the optimal set of modifications that have to be done in a given location to satisfy some MobIS objectives?*".

The paper is structured as follow: in Section 2, we review and discuss related work. In Section 3, we present a MobIS motivating example. In Section 4, we discuss how Tropos goal model can be adopted to analyze and represent the relation between goal satisfaction alternatives and location; we also use Tropos conceptualization to give a location definition. In Section 5, we show location-based goal modeling. In Section 6, we model location and location properties captured by the location-based goal model. In Section 7, we describe three kinds of automated analysis, and in Section 8 we draw conclusions and future work.

2 Related Work

Mobility is considered as the most important market and technological trend within ICT. Spatial mobility is about people moving around, having wireless access to information and services, but much more other factors imply mobility, like for instance the users's social, personal and task context [1]. Compared to traditional desktop driven information systems, MobIS has two main differences: limitation of hardware (energy, size, and computation power), and sensitivity to location changes. In this paper, we will focus on the second characteristic of MobIS. We need to model both the variable locations a MobIS is expected to operate in, and the different behaviors the MobIS can adopt accordingly.

In the area of context modeling, the relation between context and its use is not clearly considered (e.g. [7], [8] and [9]). We believe in the tight complementary relation between variable behavior (both human and software ones) and context. When the relation between context and its use is omitted, we cannot answer questions like " *how do we decide the relevant context?*", "*why do we need context?*" and "*how does context influence variable behavior?*". Modeling context information is not a standalone activity, rather context has be elicited in conjunction with the analysis we do for discovering the alternative software behaviors. Salifu et al. [10] investigate the use of problem descriptions to represent and analyze variability in context-aware software; the work recognizes the link between software requirements and context information as a basic step to design a context aware system.

Software variability is a term commonly used to represent software provided with different behaviors, whose variants can be produced guaranteeing low costs, short time, and high quality [11]. Feature modeling is a well known modeling technique exploited by product line engineering to derive a tailored product from a family of possible products [12]. A mobile software is expected to select *autonomously* among the different alternatives it supports depending on the location settings. Lapouchnian et al. [13] propose techniques to design autonomic software based on an extended goal modeling framework, but the relation with the surrounding location is not focused on. A variant of this approach is proposed by the same authors in [14], where the emphasis is on variability modeling under the requirements engineering perspective, and the classification of intentional variability when decomposing a goal. In our work, we focus on variability of location, i.e. the unintentional variability, which influences the applicability and the efficiency of each goal satisfaction alternative.

Software agents are motivated by the need to act on behalf of users. An agent can perceive the environment where it lives, reason about it, and take decisions autonomously according to its beliefs. Agent-based software is composed of a set of social interacting agents that together form a multi-agent system (MAS)[15]. Agent-oriented software engineering (AOSE) methodologies is concerned with finding models and methods that lead to agent-based systems [16]. We believe that agent paradigm fits well for MobIS development, for both of them perceive the environment and autonomously act on behalf of humans.

Goal models, mainly adopted by KAOS [17] and Tropos [5,6] agent-oriented methodologies, represent a paradigm shift from object orientation. While object-oriented analysis fits well to the late stages of requirement analysis, the goal-oriented analysis is more natural for the earlier stages [18], where we think it is also the natural place for modeling location. Goal models analyze, in a top-down way, a high level goal to discover the more specific sub-goals and tasks for satisfying that goal. Goal models allow for different alternatives to satisfy a goal, but do not specify in which locations each alternative can be adopted. We believe that alternative behaviors and location variability are complementary; supporting two alternative behaviors without specifying when to follow each of them rises the question *"why do we support two alternatives and not just one?"*. Conversely, considering location variability without supporting alternative behaviors rises the question *"what can we do if location changes?"*.

3 Motivating Example

To show how a MobIS can adapt its behaviour depending on the location, we describe an example concerning a client in a shopping mall. After entering the mall, the client inserts his/her smart-card into a card reader to get identified and authorized to take a PDA providing information about products. The company producing the PDA MobIS is required to develop it in a way it can adapt to different malls and support changes in a mall over time. One way of doing this is to consider all the variable mall settings, and design the software to autonomously choose an appropriate behaviour.

In the above introduced scenario, the PDA needs to get connected to the mall network, which can be either WiFi-based or a wired LAN whose access points are cable-based terminals. In the case of wired LAN the PDA has to show a demo that fits the client's language and expertise, and it should guide the client to the nearest free connection point. Positioning clients can be done in different ways, depending on the current location properties. An option is reading some positioning RFID tags located throughout the mall, using an RFID reader integrated to the PDA. Another way is using GPS positioning, if the client is in an open mall. If the mall contains some visible physical signs identifying the current zone, the client can be asked to provide what is written in the nearest visible sign. The scenario can also contain further variability aspects: when the connected client needs help about a product, the MobIS can either search for a suitable technician, query the mall database, redirect the PDA's browser to the mall website, or look for another client that can help.

The software company will consider different mall structures the MobIS supports. In each location the MobIS excludes some behaviours, and chooses one of the possible behaviours. The current location can also be analyzed to ensure that all MobIS goals are achievable. Furthermore, the existence of unachievable goals would motivate the location managers to change its properties (if possible), in order to overcome the problems that deny the satisfaction those goals.

4 Tropos for Location-Based Variability

Goal-based modeling is intended to explain *why* a requirement is needed in the system to be. Goal analysis consists of an iterative refinement of a root *goal* through *And-/Or- decomposition* into *subgoals*, until identifying the software requirements needed to satisfy that root goal. Two main frameworks characterize the research in this area, namely KAOS [17] and Tropos [5,6]. In this paper we refer to the latter, since some concepts it provides fit well to modeling location-based variability: social relations (*dependencies*) between system *actors*, *contributions* to non-functional requirements (*soft-goals*), and its support to all of software development life cycle phases.

Fig.1 shows a part of one Tropos goal model to clarify this framework main concepts. Actors (*Client MobIS* and *Mall Website*) have a set of top-level goals (*Provide Information to the Client*), which are iteratively decomposed into subgoals by and-decomposition (all subgoals should be achieved to fulfill the top goal) and or-decomposition (at least one subgoal should be achieved to fulfill the top goal). In Fig.1, the top-level goal is and-decomposed into *Establish Connection*, *Get product specification*, and *Provide Answer*; the goal *Provide answer* is or-decomposed into *Query Mall DB* and *Ask Website.* Soft-goals are goals for whose satisfaction there is no clear cut criteria (*Easy Connection* is a rather vague concept), and they are contributed either positively (0, +1] or

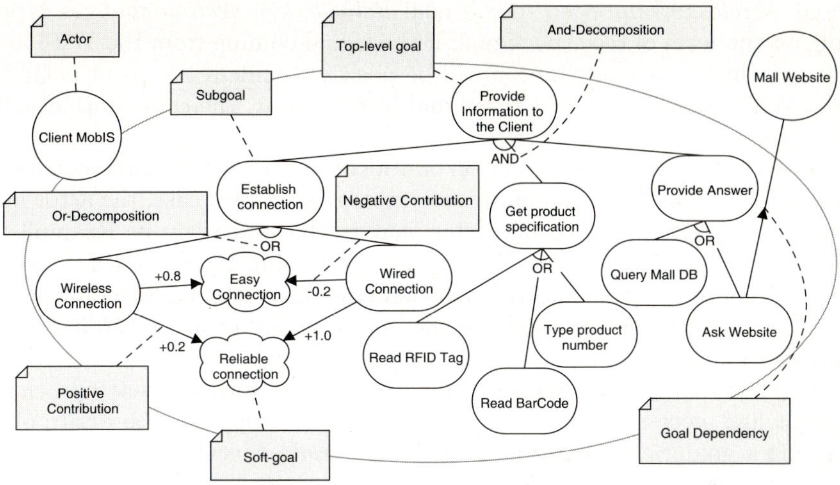

Fig. 1. A goal model labeled with the name of the basic concepts

negatively [-1, 0) by goals: *Wireless Connection* contributes positively (+0.8) to *Easy Connection*, while *Wired Connection* contributes negatively (-0.2) to *Easy Connection*). Goal dependencies represent situations where an actor cannot fulfill a goal by itself, but depends on another actor to fulfill it: actor *Client MobIS* depends on actor *Mall Website* for the achievement of goal *Ask Website*.

4.1 Location-Based Goal Model

In Tropos, the system is modeled as a set of social inter-dependent actors having goals, where actors can commit to strategy to satisfy their goals. Autonomous selection among alternative strategies requires representing the criteria an actor builds its decision upon. One alternative can be recommended in a certain location, while it can be even unapplicable in others. The selection criteria between alternatives is not explicitly modeled in the current Tropos goal model. We propose to integrate goal satisfaction intentional variability with location variability for deriving location-based software and enable automated reasoning.

When an actor tries to satisfy a goal, there can be several points where location specification is needed to take an appropriate decision between variable alternatives. We explain now our proposed list of goal model variability points:

1. *Location-based Or-decomposition*: Or-decomposition is the basic variability construct, but in current Tropos the choice of a specific Or-alternative is left to the actor intention, without explicitly considering location properties that can inhibit some alternatives. Example (from Fig.1): goal *Establish connection* can be achieved using *Wireless Connection* only if the mall has a wireless network and the client can access it.
2. *Location-based contribution to soft-goals*: the value of contributions to soft-goals can vary from one location to another. Example: the contribution from

goal *Wireless Connection* to soft-goal *Reliable Connection* changes depending on the level of received signal: if the signal coming from the WiFi access point is high, the contribution will be positive, while if the client is far from the WiFi access point and the signal level is poor, the contribution will be negative.

3. *Location-based dependency*: in certain locations an actor might not be able to satisfy a goal using any of its own strategies; in such case, the actor might delegate this goal to another actor that is able to satisfy it. Example: the MobIS can satisfy goal *Provide Answer* by fulfilling *Query Mall DB*; while if the database is offline and the mall website exists and has a mobile devices version, the MobIS can delegate the goal to another actor (*Mall Website*) browsing that website.
4. *Location-based goal activation*: an actor, when location settings change, might find necessary or possible triggering (or stopping) the desire of satisfying a goal. Self-activation is one of the main characteristics that distinguish an agent from an object [15], and Tropos needs to support it to better develop agent based systems. Example: if the MobIS has adopted the alternative *Wired Connection* for establishing connection, and while the client is getting to one cable-based terminal, the PDA detects a wireless signal, the goal *Wireless Connection* could be triggered to better satisfy the soft-goals.
5. *Location-based And-decomposition*: a sub-goal might (or might not) be needed in certain locations, that is some sub-goals are not always mandatory to fulfill the top-level goal in And-decomposition. Example: to satisfy the goal *Wired Connection*, the MobIS has first to show demo to client only if the client is using the system for the first time.

4.2 Location in Tropos

An information system designed to be self-reconfigurable needs to support different behaviors and to perform an autonomous selection among them according to some criteria, e.g. user preferences [19]. In our approach, the variability of MobIS is based on the different locations the system can adapt to, and the set of behaviors it can choose among. The autonomous location-based choice among alternatives and the applicability of each alternative requires a location model.

It is important to emphasize the subjective nature of defining a location. To clients, a mall employee is not more than a moving object if this employee does not satisfy any of the client objectives (like giving consultation about products). The mall network is not even noticeable if clients are not allowed to access it, or if the client has no need for it. Moreover, two different locations (L_1 and L_2) for an actor A_1 can be considered as a unique location $L_1 = L_2$ by an actor A_2. For instance, an actor with the objective of drinking coffee will see a mall with coffee machines different from a mall without them, while these two malls are equal for an actor who does not have such objective. Using Tropos concepts, we define location from the perspective of an actor as:

The set of available actors and resources that can be employed to achieve actor goals.

By "location", we basically refer to an environment with a high degree of commonality, e.g. shopping mall, museum, kids' room, airport, and so on. The commonality is based on the location social structure: the resources (physical and informational), the actors that exist together with the responsibilities, objectives, relations with other actors and resources each actor has, and the rules that coordinate the interaction between actors and the use of resources.

5 Location-Based Goal Analysis

Goal analysis will lead to discover alternative sets of software requirement that satisfy the analyzed goal. We propose to use goal analysis to also define location properties that are needed to decide between those alternatives. Each location property defines a set of locations, i.e. the locations where this property is evaluated to true. Location properties are associated to variability points on goal model as shown in Section 4.1.

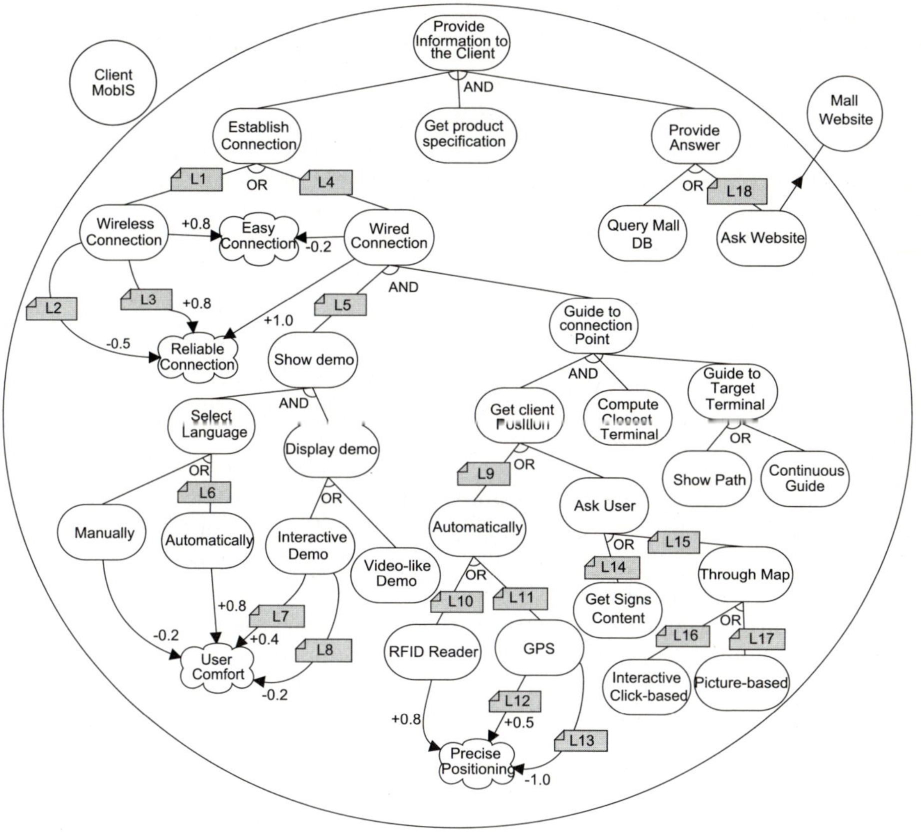

Fig. 2. A Location-based Goal Model

Fig.2 shows the goal model for a MobIS that is intended to operate in a shopping mall to help clients to get information about products. The top MobIS goal *"Provide Information to the Clients"* is top-down analyzed to get the MobIS requirements. We will mainly focus on the analysis of the sub-goal *"Establish Connection"*, and show how location properties can be defined on the variability points of the goal model. The location-based goal model is shown in Fig.2 and the location properties are expressed using labels on the variability points and will be formalized in the next section over the location model using Datalog¬.

The goal *"Establish Connection"* can be achieved by fulfilling *"Wireless Connection"* when the mobile client is in a mall providing a wireless network (L1) or by fulfilling *"Wired Connection"* if the mall provides USB Cable-Based connection (L4). Wireless connection contributes positively to the soft-goal *"Easy Connection"*, while cable connection is less easy and gives a negative contribution to that soft-goal. Cable-based connection is almost always reliable; adopting this choice will satisfy to high degree the soft-goal *"Reliable Connection"*, while wireless connection is not always reliable. If the mobile user is close to the WiFi access point, the signal is high (L3) and the contribution to the soft-goal *"Reliable Connection"* is positive; the same contribution is negative if the signal is weak (L2).

When the MobIS decides to establish a wired connection, the system has to show a demo to the client explaining the connection process, and then guide client to the nearest connection terminal. The demo is needed only if the client uses the system for the first time (L5), so *"Show Demo"* is an optional goal. In order to show a demo, the information system should decide the demo language: this information can be provided *"Manually"* by the client, or *"Automatically"* if the client mother language or the mall default language is known (L6). The automatic language selection contributes positively to soft-goal *"User Comfort"*, while the manual selection contributes negatively. The system will select then between displaying an interactive demo, or a video like one. The *"Interactive demo"* goal will contribute positively to the soft-goal *"User Comfort"* if the client has good expertise using PDAs and the used PDA has a touch screen (L7), while it contributes negatively in the other case (L8).

Now we briefly describe the rest of the model. The system will guide client to the nearest cable connection terminal. The client current position has to be identified: it can be obtained either automatically through GPS, or reading an RFID positioning tag, or manually by asking user to type the content of the nearest positioning physical sign, or by showing the user an interactive (or picture) map to specify his/her current position. Then, the path to the nearest free terminal will be computed and the client be guided to that terminal. After getting connected, the client might ask for different information about products. The client will specify the product, and then the system will provide the required information. In the case the mall provides accessible database, the MobIS might query it, otherwise it might delegate this goal to the mall website actor (if the website has a mobile devices version).

6 Location Modeling

Satisfiability of location properties on the goal model will determine the appropriate MobIS goals satisfaction strategy. In our broad vision, the MobIS will be instantiated according to the current location. Technology advances make possible generating and communicating location description beyond user awareness. Ubiquitous computing and communication technology provide data that enables the derivation of a specific location instance from a location model. The location properties will be evaluated on this location instance to derive an adequate instance of a multi-behavioral MobIS. This vision is shown in Fig.3.

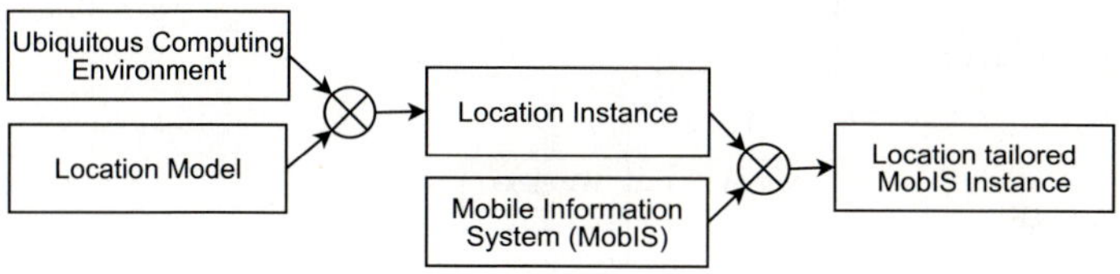

Fig. 3. The process of instantiating a location-tailored MobIS instance

Location models are the basis for any kind of reasoning on location. We propose both a graphical location model, which can be drawn by designers, and an equivalent formal location model that enables automated checking of the location properties against an instance of the location model. Location-based goal analysis helps us to discover the entities that compose location and the relationships among them. In other words, location model is elicited by the location-based goal analysis.

6.1 Graphical Location Model

In this work, we rely on UML class diagrams to describe graphically locations. Class diagrams allow us to have a clear drawing and simple description of the relevant classes and relationships describing the location, and they are often used in conceptual modeling, and location rules can be expressed over them using OCL constraints [20].

Fig.4 shows the class diagram for the mall scenario proposed in Section 3; here we just describe its peculiarities, its interpretation should be easy by reading the figure. The specialization (is-a) relation is used to identify classes that share a common part but differ for some attributes or relations: for instance, a `Network` can either be `Wireless` or `Wired` depending on the type of access points it is composed of. Attributes can be used to define required information characterizing classes or relations: a WiFi access point (`WiFiAP`) is characterized by its position (`WPos`, representing the X and Y coordinates in the mall topology); the association `ConnectedAP` defines the clients connected to a particular `WiFiAP`, and every `Client` has an attribute `SigLevel` defining the signal strength of the WiFiAP the client is covered by it. Association is used to depict the relations

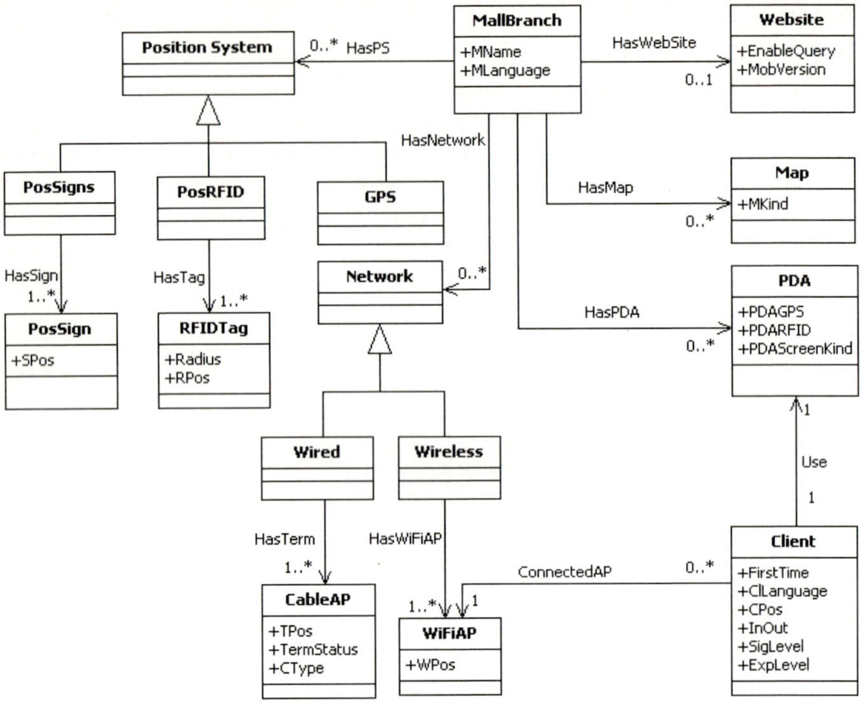

Fig. 4. Class diagram representing a location model

between different classes: a `MallBranch` has zero to n `HasMap` relations to `Map`, a `Client` uses one `PDA` (which is in turn used by one `Client`).

6.2 Formal Location Model

The second modeling technique defines a formal location model, which is the basis to verify which location properties hold in a particular location instance (i.e. an instance of the formal location model). The formalization we propose is based on Disjunctive Datalog (Datalog¬) [21], which satisfies our current need of formal knowledge representation. In particular, we use the syntax of the DLV solver [22]; it enables automated reasoning on knowledge, and its language adds some features like basic arithmetics support and cardinality check to the original Datalog¬ language. We show part of the translation of the class diagram of Fig.4 to DLV in Table 1.

Datalog¬ is not based on classes, relations, and attributes, and we have hence defined translation rules to derive an equivalent representation based on predicates. We defined a Datalog¬ predicate Class, which defines if a certain variable is a class in the class diagram meaning. We exploited the ♯count aggregate operator, which is used both to check one-to-one or many-to-one relations and

Table 1. Datalog¬ formalization of part of the location class diagram

MallBranch(A) :- Class(A), 1=♯count{B:MLanguage(A,B), Bool(B)}, 1=♯count{C:MName(A,C)}, 1=♯count{W : HasWebSite(A,W), WebSite(W)}.
MallBranchPDA(A,B) :- MallBranch(A), HasPDA(A,B), PDA(B).
PDA(A) :- Class(A), 1=♯count{B:PDAGPS(A,B)}, 1=♯count{C:PDARFID(A,C), Bool(C)}, 1=♯count{D:PDAScreenKind(A,D), ScreenKind(D)}.
Wired(A,B) :- Class(A), HasTerm(A,B), CableAP(B).
WiFiAP(A) :- Class(A), 1=♯count{X,Y:WPos(A,X,Y), ♯int(X), ♯int(Y)}.
Wireless(A,B) :- Class(A), HasWiFiAP(A,B), WiFiAP(B).
Network(A) :- Wired(A,B). Network(A) :- Wireless(A,B).
PDAConflict(A,F) :- Class(A), Class(B), PDA(F), Use(A,F), Use(B,F), A!=B
Client(A) :- Class(A), 1=♯count{X,Y:CPos(A,X,Y), ♯int(X), ♯int(Y)}, 1=♯count{C:InOut(A,C), Environment(C)}, 1=♯count{L:ClLanguage(A,L), Bool(L)}, 1=♯count{D:FirstTime(A,D), Bool(D)}, 1=♯count{F:Use(A,F), PDA(F)}, 1=♯count{S:SigLevel(A,S), SignalStrength(S)}, 1=♯count{P:ConnectedAP(A,P), WiFiAP(P)}, Use(A,G), not PDAConflict(A,G).

attributes; for example, a mall branch has exactly one attribute named `MName`, which represents the mall name. The representation of optional relations is based on the definition of auxiliary Datalog¬ predicates: for instance, to express that a `MallBranch` has an optional relation `HasPDA`, we defined a binary predicate `MallBranchPDA` that represents a `MallBranch` connected to a `PDA` by `HasPDA` relation. Specialization is represented by defining more predicates (one for each subclass) for the same superclass: `Network` can be implied either by `Wireless` and by `Wired`.

We describe now in details the definition of `Client`, using class diagrams terms *relation* and *attribute* even though they are expressed as predicates in Datalog¬. A `Client` is a `Class`, which has exactly one attribute `CPos` identifying its current position; in turn, `CPos` defines the X and Y position in the mall topology using two integers. A `Client` also has exactly one `InOut` attribute identifying its position between "outdoor" and "indoor" values, a boolean `ClLanguage` attribute which expresses if a default language has been set, a boolean `FirstTime` attribute that says if this is the first usage of the MobIS. Every `Client` is using one `PDA`, to which is connected by the relation `Use`. The `Client` also has an attribute `ExpLevel` that defines the expertise level in using a PDA. A certain `Client` cannot have any `PDAConflict` relation; it holds if there are two different users of the same `PDA`. A `Client` is associated to one `WiFiAP` through the relation `ConnectedAP`, and the signal strength is represented by the attribute `SigLevel`.

The location properties that were extracted informally by the location-based goal model of Fig.2, can be now formalized over the location model we already formalized. The formalization of the location properties (represented as labels in Fig.2) is shown in Table 2.

Table 2. The Location properties on the Goal model formalized in Datalog¬

(L1) A wireless network is available in the mall L1(A) :- Use(A,P), MallBranchPDA(M,P), NetworkedBranch(M,B), Wireless(B,C).
(L2-L3) There is a good (L2) or poor (L3) connection with a WiFi access point L2(A) :- Client(A), SigLevel(A,good). L3(A) :- Client(A), SigLevel(A,poor).
(L4) An USB connection can be established L4(A) :- Use(A,P), MallBranchPDA(M,P), NetworkedBranch(M,B), Wired(B,C), CType(C,usb), TermStatus(C,free).
(L5) The client uses the system for the first time L5(A) :- Client(A), FirstTime(A,true).
(L6) A default language has been chosen (by the mall or the client) L6(A) :- Client(A), ClLanguage(A,true). L6(A) :- Client(A), Use(A,P), MallBranchPDA(B,P), MallBranch(B), MLanguage(B,true).
(L7-L8) High (low, in L8) expertise in using the PDA and (not, in L8) touch screen L7(A) :- Client(A), Use(A,B), ExpLevel(A,low), PDAScreenKind(B,touch). L8(A) :- Client(A), not L7(A).
(L9) There is a positioning system in the mall L9(A) :- Client(A), Use(A,P), MallBranchPDA(B,P), MallBranchPS(B,Q).
(L10) The client's PDA is in a location covered by the RFID tag signal L10(A) :- Client(A), Use(A,B), inCoverage(B,C), PDARFID(B,true), MallBranchPDA(D,B), MallBranchPS(D,E), PosRFID(E,C).
(L11) The PDA has a connected/integrated GPS receiver L11(A) :- Client(A), Use(A,B), PDAGPS(B,true).
(L12-L13) The client is outdoor (L12) or indoor (L13) L12(A) :- Client(A), InOut(A,outdoor). L13(A) :- Client(A), InOut(A,indoor).
(L14-L15) There are position signs (L14) or a map (L15) in the mall L14(A) :- Client(A), Use(A,B), MallBranchPDA(C,B), MallBranchPS(C,D), PosSigns(D,E). L15(A) :- Client(A), Use(A,B), MallBranchPDA(C,B), MallBranchMap(C,M).
(L16-L17) Interactive map and PDA with touch screen (L16) or picture map L16(A) :- Client(A), Use(A,B), MallBranchPDA(C,B), MallBranchMap(C,M), MKind(M,interactive), PDAScreenKind(B,touch). L17(A) :- Client(A), Use(A,B), MallBranchPDA(C,B), MallBranchMap(C,M), MKind(M,picture).
(L18) The mall branch has a mobile version of its website L18(A) :- Client(A), Use(A,B), MallBranchPDA(C,B), HasWebSite(C,W), MobVersion(W,true).

7 Analyzing Location-Based Goal Models

This section presents three types of analysis for examining MobIS variability against a given location, and vice versa. A preliminary step consists of evaluating the validity of location properties at the variability points of the goal model (L1-L18 in Table 2) on an instance of the location model. This step can be done automatically using the DLV solver [22], and the result we get is those location properties that hold in the considered location instance.

We suppose the existence of two clients John and Mike, both located indoor and using a PDA with touch screen, GPS, and RFID reader. John has high expertise in using the PDA, while Mike's expertise level is low. Their positions in the mall are different; it is the first time Mike uses the system. Mike has set a default language, John has not. We assume them to be in a mall named *SuperMall*, provided with position signs for directing people and a wireless network, while mall map and mall region language are not known.

In the following we propose three types of analysis that can be executed on location-based Tropos models: location-based goal satisfiability (LGS), location properties satisfiability (LPS), and preference analysis (PA).

7.1 Location-Based Goal Satisfiability (LGS)

This kind of analysis is aimed to verify if a goal is achievable in the current location instance. In our example, with respect to client Mike, the goal *"Select Language"* cannot be achieved by choosing the sub-goal *"Automatically"*, because the evaluation of location property L6 is false.

The analysis can be performed using the goal reasoning algorithm proposed by Giorgini et al. [23] on the goal model restricted by the evaluation of the location properties. A strategy for evaluating satisfiability follows a top-down approach: starting from a top-level goal, we should check that all (at least one) sub-goals in and- (or-) decompositions can be achieved, or that the top-level goal can be achieved via a *makes* (+1.0) contribution from an achievable goal. In the mall MobIS, the top-level goal *"Provide Information to the Client"* can be fulfilled only if three subgoals are fulfilled; if the mall had neither a wireless connection nor a wired one, both L1 and L4 would be evaluated to false, and there wouldn't be any way of fulfilling the goal *"Establish connection"*.

7.2 Location Properties Satisfiability (LPS)

This analysis checks if the current location structure is compliant with the MobIS goals. It is exploited to identify what is missing in a particular location where some top-level goals have been identified as unsatisfiable by LGS. When a goal cannot be satisfied, LPS will identify the denying conditions and suggest ways for solving the problem.

If wireless network is unavailable because there are not wireless access points working, the goal *"Establish connection"* will be unsatisfiable, and the problem is that both L1 and L4 evaluate to false ($\neg L1 \wedge \neg L4$). Since there are not *makes* contributions to higher level goals, the only way of allowing the satisfiability of the top goal is change the location in such a way that L1 or L4 ($L1 \vee L4$) holds.

If we choose to enable L1, we have to examine the definition of L1 (Table 2); the first two predicates (*Use(A,P, _), MallBranchPDA(M,P)*) are true both for John and Mike, while the remaining two predicates are evaluated to false (*NetworkedBranch(M,B), Wireless(B,C)*). Going into the details of the fourth predicate, we find that there are not wireless access points B such that *HasWiFiAP(A,B)* and hence the problem is identified. That is, we need to establish a *HasWiFiAP* relation with at least one *WiFiAP*.

7.3 Preferences Analysis (PA)

This type of analysis requires the specification of preferences over alternatives. Preferences can be specified using contributions to soft-goals as in [19]. We need this analysis in two cases:

1. When there are several alternatives to satisfy a goal: in our example, satisfying the goal *"Select Language"* for client Mike can be achieved both *"Manually"* and *"Automatically"* (L6 is evaluated to true because Mike has a default language). When mall administrator preferences give the soft-goal *"User Comfort"* high priority, the analysis will suggest the alternative *"Automatically"*.
2. When there is no applicable alternative: in this case, LPS might provide several proposals about the needed location modifications. The choice of a specific option can base on the preferences over soft-goals. If the mall has no connection available, and the mall administration preferences give *"Easy Connection"* a higher priority than *"Reliable Connection"*, the preference analysis will suggest to introduce a wireless connection (make L1 true) alternative to the mall structure.

8 Conclusions and Future Work

In this paper we showed the importance of modeling location settings as a basis to derive location-based MobIS behaviour. We also showed the subjective nature of defining location. From the perspective of an Tropos actor, a location is a set of available resources and other actors that are needed for achieving the goals of that actor. Information about current location is not necessarily needed by end users, but by the MobIS to decide which behaviour it should adopt. We used a variant of Tropos goal model that supports variability handling on the basis of location properties.

We exploited Datalog¬ to formalize the location model and the location properties characterizing the variability points in the goal model. Formalization is needed because location properties are not generated to be simply shown, rather to enable the MobIS to reason about them. We presented three kinds of analysis that can be executed on the basis of location-based Tropos goal model.

Future work consists of expanding the proposed approach in various directions. We need to consider the meta-information the location information might have: for instance, some location entity attributes and relations are changeable over time, others are not. We also need to check how this can be presented formally and how the automated analysis will work with it. Moreover, we need to refine the analysis techniques by defining specific algorithms and test their efficiency.

Acknowledgement

This work has been partially funded by EU Commission, through the SERENITY project, by MIUR, through the MEnSA project (PRIN 2006), and by the Provincial Authority of Trentino, through the STAMPS project.

References

1. Krogstie, J., Lyytinen, K., Opdahl, A., Pernici, B., Siau, K., Smolander, K.: Research areas and challenges for mobile information systems. International Journal of Mobile Communications 2(3), 220–234 (2004)
2. Pernici, B.: Mobile information systems: infrastructure and design for adaptivity and flexibility. Springer, Heidelberg (2006)
3. Weiser, M.: The computer for the twenty-first century. Scientific American 265(3), 94–104 (1991)
4. Streitz, N., Nixon, P.: The disappearing computer. Commun. ACM 48(3) (2005)
5. Yu, E.: Modelling strategic relationships for process reengineering. Ph.D. Thesis, University of Toronto (1995)
6. Bresciani, P., Perini, A., Giorgini, P., Giunchiglia, F., Mylopoulos, J.: Tropos: An agent-oriented software development methodology. Autonomous Agents and Multi-Agent Systems 8(3), 203–236 (2004)
7. Yau, S., Liu, J.: Hierarchical situation modeling and reasoning for pervasive computing. In: Proceedings of 3rd Workshop on Software Technologies for Future Embedded and Ubiquitous Systems (SEUS), pp. 5–10 (2006)
8. Henricksen, K., Indulska, J.: A software engineering framework for context-aware pervasive computing. PerCom., 77–86 (2004)
9. Wang, X.H., Zhang, D.Q., Gu, T., Pung, H.K.: Ontology based context modeling and reasoning using owl. In: PERCOMW 2004: Proceedings of the Second IEEE Annual Conference on Pervasive Computing and Communications Workshops, pp. 18–22. IEEE Computer Society, Washington (2004)
10. Salifu, M., Nuseibeh, B., Rapanotti, L., Tun, T.: Using problem descriptions to represent variability for context-aware applications. In: First International Workshop on Variability Modelling of Software-intensive Systems (2007)
11. Pohl, K., Böckle, G., van der Linden, F.: Software Product Line Engineering: Foundations, Principles, and Techniques. Springer, Heidelberg (2005)
12. Kang, K., Kim, S., Lee, J., Kim, K., Shin, E., Huh, M.: Form: A feature-; oriented reuse method with domain-; specific reference architectures. Annals of Software Engineering 5, 143–168 (1998)
13. Lapouchnian, A., Yu, Y., Liaskos, S., Mylopoulos, J.: Requirements-driven design of autonomic application software. In: Proceedings of the 2006 conference of the Center for Advanced Studies on Collaborative research (2006)
14. Liaskos, S., Lapouchnian, A., Yu, Y., Yu, E., Mylopoulos, J.: On goal-based variability acquisition and analysis. In: Proc. 14th IEEE International Requirements Engineering Conference, September 2006, pp. 11–15. Minneapolis, USA (2006)
15. Jennings, N.: On agent-based software engineering. Artificial Intelligence 117(2), 277–296 (2000)
16. Henderson-Sellers, B., Giorgini, P.: Agent-oriented Methodologies. Idea Group Pub. (2005)
17. Dardenne, A., van Lamsweerde, A., Fickas, S.: Goal-directed requirements acquisition. In: Selected Papers of the Sixth International Workshop on Software Specification and Design table of contents, pp. 3–50 (1993)
18. Mylopoulos, J., Chung, L., Yu, E.: From object-oriented to goal-oriented requirements analysis. Commun. ACM 42(1), 31–37 (1999)
19. Liaskos, S., McIlraith, S., Mylopoulos, J.: Representing and reasoning with preference requirements using goals. Technical report, Dept. of Computer Science, University of Toronto (2006), `ftp://ftp.cs.toronto.edu/pub/reports/csrg/542`

20. Warmer, J., Kleppe, A.: The object constraint language: precise modeling with UML. Addison-Wesley Longman Publishing Co., Inc., Boston (1998)
21. Eiter, T., Gottlob, G., Mannila, H.: Disjunctive datalog. ACM Transactions on Database Systems (TODS) 22(3), 364–418 (1997)
22. Leone, N., Pfeifer, G., Faber, W., Eiter, T., Gottlob, G., Perri, S., Scarcello, F.: The dlv system for knowledge representation and reasoning. ACM Transactions on Computational Logic (TOCL) 7(3), 499–562 (2006)
23. Giorgini, P., Mylopoulos, J., Nicchiarelli, E., Sebastiani, R.: Reasoning with goal models. In: Spaccapietra, S., March, S.T., Kambayashi, Y. (eds.) ER 2002. LNCS, vol. 2503. Springer, Heidelberg (2002)

A Glance on Current XML Twig Pattern Matching Algorithms

Duy-Bo Dao[1] and Jinli Cao[2]

[1] Binh Dinh Teacher Training College, Binh Dinh, Vietnam
[2] La Trobe University, Bundoora, Vic 3083, Australia

Abstract. Extensible Markup Language (XML) has become a de facto standard for information representation and exchange over the Internet. The core operation of XML Query Processing is twig pattern matching, that is, finding in an XML document tree D all matches of a given twig query Q. Modern twig query matching algorithms often first decompose individual path matches and then merge them to form twig matches. In this paper, we review and evaluate two current efficient algorithms, namely *HolisticTwigStack* and *TwigList*. The *HolisticTwigStack* algorithm, which based on *TwigStack*, a notable and effective algorithm, uses a stack structure to avoid the storage of intermediate matches and the process of merging matches, whereas the *TwigList* algorithm use simple lists. Our analysis on experimental results show that the advantages and disadvantages of both *HolisticTwigStack* and *TwigList* for efficient XML Twig Pattern Matching in term of processing time and memory usage.

1 Introduction

XML has rapidly grown from a markup language to a standard for representation and exchange of heterogeneous data over the Web applications. Therefore, a vital concern is how to store and query potentially huge heterogeneous XML data efficiently.

XML employs as a tree structured model for representing data. The relationships between elements are specified as nested structures. Thus XML queries care not only the content or value but also the structure of XML data. Common XML query languages, such as XPath [1] and XQuery [2], issue structural queries over XML data. A XML query is typical defined as a pattern that opts subsets of elements satisfying certain structural relationships. One of the most common structural queries is the twig pattern query which specify patterns of selection predicates on multiple elements related by an ordered tree structure. XML twig queries are considered as the most common type of query processed over XML data.

Due to pattern matching is one of the most important types of XML queries to retrieve information from an XML database, finding all occurrences of the twig pattern is the core operation of XML query processing both in relational storage of XML databases and in native XML databases [3,4,5,6]. In the past years, many research of XML Query Processing principally concentrates on building

O. Gervasi et al. (Eds.): ICCSA 2008, Part II, LNCS 5073, pp. 307–321, 2008.

efficient structural indexes or numbering schemes. Because of digging the structural indexes, one can traverse XML data more efficiently by averting scanning unnecessary search space. Many reported researches [6,7,5,8,9,10] proposed many algorithms, such as *Stack-Tree* [11], *TJFast* [12], *PathStack* and *TwigStack* [6], *TwigStackList* [13], *Twig^{2}Stack* [4], *TSGeneric* [7], *TwigList* [3], and *HolisticTwigStack* [8], based on the labelling scheme that usually encodes the positional information of each element in an XML document. By using the region encoding to an XML document, it is easy to identify the structural relationship between any two elements without scanning the entire XML document tree. By discovering the advantageous properties of numbering schemes, these approaches are developed to process path and twig queries.

Most earlier algorithms [11,14,15] typically first decompose a twig pattern into many binary patterns or single paths and then search for matches the binary relationships. Finally, these matches are stitched together to get the final results to the twig query. The main drawback of such decomposition-based algorithms is that could generated a very large number of intermediate binary pattern matches even though the input and final results sizes are much more manageable. Aiming at the problem, Bruno *et al.* [6] propose a holistic twig join algorithm, namely *TwigStack* which generates only individual path matches that are ensured to contribute to the final results for only the ancestor-descendent type of queries. This approaches is proved to be I/O and CPU optimal for the ancestor-descendent type of queries among sequential algorithms that read the entire input for twigs. However, the *TwigStack* algorithm still can not control a large number of intermediate results for parent-child type of queries.

Many efficient algorithms for XML twig pattern matching have been proposed to amending the sub-optimal problem of the *TwigStack*. Among them, there are two current efficient algorithms, namely *HolisticTwigStack* [8] and *TwigList* [3]. The first algorithm based on a notable and effective algorithm *TwigStack* [6] with the region encoding scheme. This algorithm still preserves all the advantages of *TwigStack* and avoids the storage of intermediate matches and the process of merging matches. The second one, *TwigList* [3] algorithm using simple lists instead of using stacks as most previous algorithms. Both of them were evaluated better than both *TwigStack and Twig^{2}Stack* [4] in terms of time complexity, memory space and I/O complexity. However *HolisticTwigStack and TwigList* have never seen to compare together on efficient processing of XML Twig Queries.

The main contributions of this paper is summarized as follows:

- We review and analyze both of the *HolisticTwigStack and TwigList* on processing for XML Twig pattern matching. We examine on most advantages of two algorithms which compare to *TwigStack [6] and Twig^{2}Stack* [4]
- Then, we examine the results of two algorithms with the different approaches that avoids intermediate results to gain an efficient process for XML Twig Pattern matching.
- We present a summary from original experimental results on a range of real and synthetic data set, and query twig patterns. Our summaries validate our

analyses and show the superiority of *HolisticTwigStack and TwigList* over most previous algorithms.

The remainder of the paper proceeds as follows. Section 2 presents the related works in the area of efficient processing for XML Twig Pattern matching queries. Section 3 discusses two current efficient algorithms and outlines their problems. The paper gives analyses and evaluations on original experimental results of two algorithms in Section 4. Finally, Section 5 concludes this paper and propose some future research works.

2 Related Works

2.1 Data Model and Twig Pattern Query

XML document is commonly modelled as a node -labelled, ordered tree, where nodes represent elements, attributes and texts, and parent-child edges represent element-subelement, element attribute and element-text pairs. Most existing XML query processing algorithms [6,16,17,13,18,19,7,20,12,4], use a region code with 3-tuple *(leftPos, rightPos, levelNo)* to present the position of a node in the data tree. The region encodings support efficient evaluation of structural relationships. Typically, element u is an ancestor of element v if and only if $u.leftPos < v.leftPos < u.rightPos$. For parent-child relationship, if $u.level = v.level - 1$ then v is a child of u.

Queries in XML query languages make use of twig patterns to match relevant portions of element in an XML database. Twig pattern nodes may be elements, attributes and texts. Twig pattern edges are either parent-child relationships (denoted by /) or ancestor-descendant relationships (denoted by //).

Given a twig pattern T and an XML database tree D, a match of T in D is specified by a mapping from nodes in T to nodes in D, such that: *(i)* query node predicates are satisfied by the corresponding database tree nodes; and *(ii)* the parent-child and ancestor-descendant relationships between query nodes are satisfied by the corresponding database nodes. The answer to query T with m nodes can be conducted as a list of *m-ary* tuples, where each tuple $(t1,...,tm)$ consists of the database nodes that identify a distinct match of T in D.

2.2 Existing Twig Pattern Matching Algorithms

Pattern matching is one of the most important operation of XML query processing to retrieve information for an XML document. Since discovering the occurrences of twig (tree structure) query patterns is a core operation of XML Query Processing, A huge set of XML Twig Pattern Matching algorithms have been proposed in literatures.

Several XML Twig Join Algorithms are outlined as follows. First, based on containment join, prior research [11,14,16] decomposes a twig pattern into a set of binary structure relationships (Parent-Child or Ancestor-Descendent relationships). Then, each binary relationship is examined by using structural join

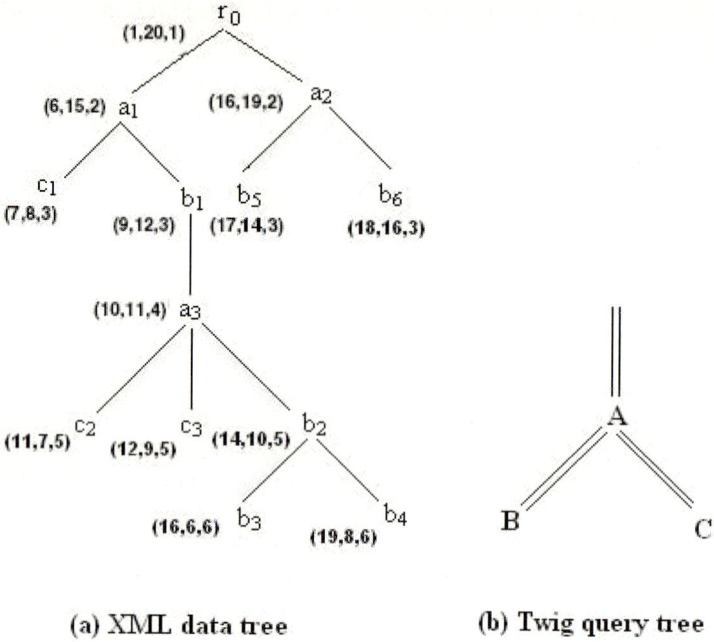

Fig. 1. Example XML data tree (a) and a Twig Pattern (b)

techniques and the final matches are gained by the merging process of individual binary join results. Among many reported studies, Zhang *et al.* in [14] proposed a multi-predicate merge join (MPMDJN) algorithm using inverted list and based on the region encoding (containment labelling) to process XML queries. Later, Al-Khalifa *et al.* in [11] gave a stack-based algorithm, called *Stack-Tree-Desc/Anc* which breaks the twig queries into a set of binary structure components. It is optimal for an Ancestor-Descendent and Parent-Child binary relationship. Then, Wu *et al.* in [16] researched on the problem of structural join order selection for complex queries. The main drawback of the above approaches is the large and unnecessary intermediate results generated by the algorithms.

Then, Bruno et al. in [6] proposed a novel holistic twig join algorithm *TwigStack* which avoids producing intermediate results unless they participate to the final result. However, the main limitation of *TwigStack* is that it may generate a large amount of useless intermediate results when query trees contain any Parent-Child edge (denoted as /-edges).

To accelerate XML Twig pattern matching, there is much research on the use of indexes. In particular, Jiang *et al.* in [19] proposed an *XML Region Tree (XR-tree)* which is a dynamic external memory index structure specially designed for nested XML data. With XR-Tree, they proposed a new algorithm, *TS-Generic+*, to effectively skip ancestors and descendants that do not contribute to the join. Recently, Lu *et al.* [13] also proposed *TwigStackList* algorithm to handle twig queries with Parent-Child relationships. Based on Dewey labelling scheme,

Lu *et al.* in [12] used an extended Dewey labelling scheme to propose a *TJFast* algorithm to traversal only leaf elements. However, almost the above algorithm can not avoid generating a large amount of unless path merges as shown in [9]. Most recently, Chen and his co-researchers in [4] proposed a *Twig*^{2}*Stack* algorithm using hierarchical-stacks instead of using enumeration of path matches. *Twig*^{2}*Stack* has proven that it outperforms both *TJFast* and *TwigStack*. But due to the complexity of hierarchical-stacks used in the algorithm, *Twig*^{2}*Stack* may conduct many random access and use a large space of memory. To averting the disadvantages of these algorithms, it some new approaches have to be developed to deal with those problems to gain an optimal for XML Query Processing.

In next section, the paper reviews and evaluates two current algorithms for efficient processing XML Twig Pattern Matching, namely *HolisticTwigStack* [8] and *TwigList* [3].

3 Current Efficient Algorithms on XML Twig Pattern Matching

Most recently, *HolisticTwigStack and TwigList* algorithms were evaluated better than both *TwigStack and Twig*^{2}*Stack* in terms of time complexity, memory space and I/O complexity. The rest of this section discusses these algorithms in details.

3.1 HolisticTwigStack Algorithm

The *HolisticTwigStack* algorithm was proposed by Jiang et al [8] is a one-phase holistic twig evaluation algorithm based on the *TwigStack* algorithm [6]. Due to the space limitation, interested readers referred to [8] for details and proofs of the algorithm.

Below, in brief, the paper discusses the algorithm via presenting an example. From the example in Figure 1, data structure of *HolisticTwigStack* is shown as Figure 2.

In Figure 2, first node a_1 is pushed onto the root stack S_A, followed by b_1 onto stack S_{B1} and b_1 onto stack S_{C1}. S_{C1} and S_{B1} are each linked to a_1, represented by the solid arrows, because a_1 is the closest pattern ancestor for those elements in S_{B1} and S_{C1}. Then a_3 is pushed onto the root stack S_A, it is a descendent of element a_1. Later, c_2 is pushed onto a new stack S_{C2} because it has a different closest pattern ancestor than c_1. S_{C2} is linked to a_3. When b_2 is encountered, because it is not a descendent of b_1, it pushed onto a new stack S_{B2}. S_{B2} is linked to a_3. Then b_3 is pushed onto the same stack with S_{B2} because it is a descendent of b_2 and has the same closest pattern ancestor a_3. b_4 is pushes onto a different stack S_{B3} because none of the top elements in the existing stacks of type B is an ancestor of b_4. But b_4 has the same closest pattern ancestor as elements in the stack S_{B2}. So S_{B2} and S_{B3} are linked together with a dotted arrow pointing from b_3 to its sibling element b_4. c_2 is placed in a new stack S_{C2} because it is not a descendant of c_1. S_{C2} is linked to a_3 with a solid arrow. Finally, c_3 is pushed onto a new stack because none of the top elements in the existing stacks of type C is an ancestor of

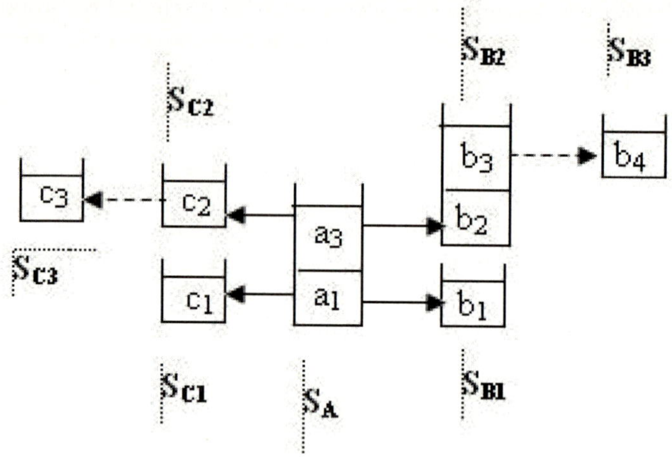

Fig. 2. Example Data Structure of HolisticTwigStack Algorithm

c_3. c_3 linked to element c_2 in S_{C2} because it has the same closest pattern ancestor as those elements in S_{C2}. In the structure, solid arrows are represented the closest pattern ancestor relationships and dotted arrows are used to link between the stack of the same type nodes that share the same closest pattern ancestor. Indeed, having a look at an example of stacks $\mathbf{S}_{B2}$ and $\mathbf{S}_{B3}$ when they are linked together by pointing a dotted arrow from b_3 to its sibling b_3.

The *HolisticTwigStack* is presented in Algorithm 1 (Figure 3). The authors of the algorithm have proposed several major lemmas and theorems for showing the correctness of the *HolisticTwigStack* and the analysis of its complexity.

In the algorithm, elements are checked for satisfaction of structural relationships as the same way done in *TwigStack* which call the function *getNext()* recursively. The algorithm try to hold the elements sharing the same *Top_Branch_Element* in memory until a disjoint element arrives. This disjoint element come to eliminate the participating of subtree rooted at the *Top_Branch_Element* in any further match. Like as in the *TwigStack*, an incoming element *el*is pushed into a stack only if it is certainly to take account for a twig match (*line 5*. Otherwise, it will advance to next element (*line 20*). After having reach the end of streams (*end*(***T***)), the algorithm output the remaining twig matches to root element left in *S_root* (*line 22-30*). For details and proofs of the algorithm, reading to [8].

Correctness and Complexity of the algorithm: *Suppose a twig pattern* ***D*** *comprises of n nodes and only A-D edges, and an XML data tree D. The worse case I/O and CPU processing time complexities linear in the sum of all sizes of n input lists and the output list will occur in the algorithm.*

3.2 TwigList Algorithm

Recently, Qin and his co-researchers in [3] proposed ***TwigList*** algorithm for processing twig pattern matching faster. *Twiglist* is independently developed and

Algorithm 1: HolisticTwigStack (Q)

```
begin
    while not end(Q) do
        q=getNext(Q);
        e=the current first element in the stream of q;
        if (q is of root type) or (at least one element in S_parent(q) covers e) then
            while (!Empty(S_TBN) and e is disjoint with S_TBN's top element) do
                if (no leaf stack is empty) then
                    ShowTwigSolution(S_root, S_root.size-1);
                end
                TopElem=s.pop();
                if Empty(S_TBN) then
                    Clean all child stacks of s;
                else
                    Update the links between stacks;
                end
            end
            Remove all the elements in each S_Upper_Node that are disjoint with e
            and update the links appropriately;
            MoveElementToStack(q, e);
        end
        AdvanceList(q);
    end
    while not empty(S_root) do
        ShowTwigSolution(S_root, S_root.size-1);
        S_root.Pop();
        if not Empty(S_root) then
            Update the links between stacks;
        else
            Clean all child stacks;
        end
    end
end
```

Fig. 3. Pseudo Code of *HolisticTwigStack* from [8]

inherited similarly characteristics with the *Twig*^{2}*Stack* algorithm. But, the *Twiglist* also substantial outperforms *Twig*^{2}*Stack*. The efficiency of the *Twig*^{2} *Stack* algorithm is obtained by using simple lists instead of using hierarchical-stacks as in *Twig*^{2}*Stack* to reduce the processing costs.

The algorithm focuses on efficient processing twig pattern matching queries. For determining ancestor-descendant relationships among nodes in an XML document tree, where a node of the XML tree is encoded with a regional encode as previous approaches.

The *TwigList* algorithm was proposed for processing twig pattern matching queries by maintaining a list for nodes in a query tree instead of using complicated hierarchical-stacks as in both *TwigStack* and *Twig*^{2}*Stack*. The list structure of the algorithm is briefly mentioned as follows.

Considering path $A//B$ against an XML Document tree $\boldsymbol{D}$. If an A-typed node is an ancestor of a set of B-typed nodes in XML document tree $\boldsymbol{D}$: *(i)* It has to be able to specify a minimal interval for the A-typed node to cover

Algorithm 2: *TwigList* (T,D)

Input: a query tree T with n nodes $\{V_1, \cdots, V_n\}$, and an XML tree D;
Output: all n-ary tuples as answers for Q;

1: let X_i be a sequence of V_i-typed XML tree nodes sorted in preorder, for all $1 \leq i \leq n$;
2: let X be the set of all X_i for $1 \leq i \leq n$;
3: obtain a set of lists $L = \{L_{V_1}, L_{V_2}, \cdots L_{V_n}\}$ by calling *TwigList-Construct* (T, X);
4: $R \leftarrow$ *TwigList-Enumerate* (T, L);
5: **return** R;

Fig. 4. Pseudo Code of the *TwigList* Algorithm [3]

all the B-typed nodes;*(ii)* It also must be in case that there does not exist any B-typed node in the interval which is not a descendant of the A-typed node. This property is not available to make use in existing algorithms *TwigStack* and *Twig*^{2}*Stack* due to push/pop operations together with stacks can not effectively hold this property. *TwigList* fully make use of the above property because of mainly using lists instead of stacks.

The *TwigList* algorithm is shown in Figure 4. The algorithm takes two input parameters, a query tree ***T(V,E)*** which presents a twig pattern matching query, and an XML document tree ***D***. The query tree ***T(V,E)*** has n nodes $\boldsymbol{V_1}$, $\boldsymbol{V_2}$, ..., $\boldsymbol{V_n}$. The algorithm conducts lists for all $\boldsymbol{V_i}$-typed nodes in the XML document tree ***D*** and then sort them following preorder. For a whole solution, the algorithm takes two steps to achieve the results. Firstly, it calls *TwigList-Construct* procedure to conducts a set of lists which compactly hold all twig patterns for answering ***T*** (*line 3*). Then, it calls *TwigList-Enumerate()* to obtain all n-ary tuples as answers for query tree ***T*** (*line 4*).

In here, the paper explain how the *TwigList* works using an example of the same twig query pattern matching query ***T***=*//A//B//C* (Figure 1(b)) against XML data tree ***D***(Figure 1(a)). In query tree ***T***, there are three types: *A, B, and C*, where *A* is the root node, and *B and C* are leaf nodes. According to *TwigList-Construct()* as inputs, X comprises of three sequences for three types, $X_A = <a_0\, a_1, a_2, a_3>$, $X_B = <b_1, b_2, b_3, b_4, b5, b_6>$, and $X_C = <c_1, c_2, c_3>$. *TwigList-Enumerate()* then will generate three lists, L_A, L_B, and L_C, to maintain all potential *n-ary*tuples for answering query ***T***. Here, every XML tree node, a_i in L_A will hold two pairs of pointers to specify the intervals for its B-typed descendant nodes (*leftPos*$_B$, *rightPos*$_B$), and its C-typed descendants, (*leftPos*$_C$, *rightPos*$_C$).

Figure 5(a) shows the stack S and the lists L_A, L_B, and L_C, after b_4 is pushed onto S. *Figure* 5(b) depicts S and the lists after b_5 is pushed into the stack S. When b_5 is pushed into the stack S, *toList()* enforces b_4, b_3, b_2, a_3, b_1, a_1 to be popped up, and then appended to the corresponding lists. Now is the timing for a_3 and $a_1$to fill in its *leftPos*$_B$ and *leftPos*$_C$ positions. As can be seen from *Figure* 5 (c), all patterns are held by the lists.

The main second step of the *TwigList*, it calls *TwigList-Enumerate()* with taking two input parameters , the n-node query tree ***T*** and a set of lists ***L***

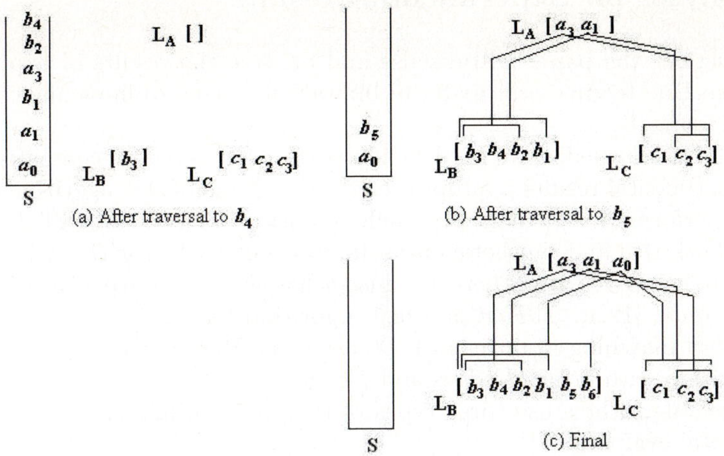

Fig. 5. Example of Using *TwigList-Construct()*

which is obtained in *TwigList-Construct()*. *TwigList-Enumerate()* inserts all *n-ary* tuples as results into a relation R to be returned.

Continuing with the above example, three lists are shown in *Figure* 5 (c). Initially, *start* = $[a_3,b_3,c_2]$ and *end* = $[a_0,b_6,c_3]$. Here, triple(a_3,a_1,a_0) specify the interval for all *A*-typed XML tree nodes. (b_3,b_4,b_2) specifies the interval where a_3's *B*-typed descendant exist, and (c_2,c_3) specifies the interval where a_3's *C*-typed descendant exist. At line 5, *move* = *start* = $[a_3,b_3,c_2]$ records the current positions for outputting answers. After calling *moreMatch()* to generate all *n*-ary tuples. The terminational condition is specified in *line 9* of the algorithm when there is no any tuple to be generated. Lastly, the final results from *TwigList-Enumerate()* are (a_3,b_3,c_2), (a_3,b_3,c_3), (a_3,b_4,c_2), (a_3,b_4,c_3), (a_3,b_2,c_2), and (a_3,b_2,c_3), followed by (a_1,b_3,c_1), ...

Time/space complexity of the *TwigList-Construct()* are both $\mathbf{O}(d.|X|)$ in the worse case, where d is the maximum degree of a node in the query tree $\boldsymbol{Q}$ and $|X|$ is the total number of nodes, v_i, in the XML data tree that is a V_i-typed node, $1 \leq i \geq n$.

3.3 Summary

Both *HolisticTwigStack, TwigList* has outperformed *TwigStack* and *Twig*^{2}*Stack* in terms of the time/space complexities which are lower bound to output all matchings of a *n*-node query tree.

In next section, the paper will analyze and evaluate on existing results from each experimental results which have been done by the original authors of both two current algorithms, *HolisticTwigStack, TwigList.*

4 Analyses on Experimental Results

In this section, the paper will discuss and analyze the results of experimental evaluation. Due to time and material limitations for us to implement the algorithms by ourself.

To our best knowledge, we tried to depict the evaluations on comparison of existing experimental results appropriately. The authors of the algorithms all evaluate the performance of their approach against *TwigStack* [6] and *Twig*^{2}*Stack* [4] using both real and synthetic datasets. Particularly, *Twig*^{2}*Stack* is chosen as the basic to compare with others, because it has shown to outperform *TwigStack* [6] and *TJFast* [12,4]. *TJFast* is a fast algorithm for efficient processing XML twig pattern matching in terms of I/O cost and CPU time among existing twig join algorithm with both //-edges and /-edges.

Both the algorithms use three types of twig pattern matching queries in the experimental evaluation:

- Type 1 (Q1) represents a set of wide and shallow twig queries (*e.g.* depth = 2 and width= 4 or 5 in the patterns)
- Type 2 (Q2) represents a set of deep and narrow twig queries with,(*e.g.* depth = 3 to 5 and width = 2)
- Type 3(Q3) represents a group of balanced queries ((*e.g.* depth = 2 to 3 and width = 2 to 3)

For *HolisticTwigStack* in [5,8], the queries chosen for DBLP, XMark, and Tree-Bank datasets are based on the same above ideas. For example, twig pattern matching queries are used in the algorithm as shown in Figure 6

For *TwigList* in [3], there are 15 query trees for three datasets as shown in Figure 7. Each set of 5 query trees have different combinations of /-edges and //-edges, some queries are selected from the queries used in [4].

In the rest of this section, the paper will review the results which measured the execution time in seconds and the use of memory space in bytes.

HolisticTwigStack: Figure 8(a) and Figure 8 (b) shown that the results on XMark. *TwigStack* uses 8% to 18% more time and 4.0 to 5.2 times more space than *HolisticTwigStack*. Usually, the larger the number of matches, the larger space ratio for a queries [5]. It is caused by the uniform tree structure of XMark dataset which is detailed as follows.

Dataset	Type	Twig Queries
XMark	2	// person[.//australia]//affica//city[.//country][gender]//education
XMark	1	// person[.//address][.//education][//gender]//age
XMark	3	// person[.//address[//city]//country]//profile[gender]//education
DBLP	2	//dblp[.//proceedings]//book[.//editor]//isbn
DBLP	2	//proceedings[.//editor][//publisher][//series]//isbn
DBLP	3	//dblp[.//proceedings]//book

Fig. 6. Twig queries using in *HolisticTwigStack* algorithm

Dataset	Type	Twig Queries
XMark	2	//item[location]/description//keyword
XMark		//people//person[.//address/zipcode][.//education][//gender]//age
XMark	2	//item[location][.//mailbox/mail//emph]/description//keyword
XMark	3	//people//person[.//address/zipcode][id]/profile[.//age]/education
XMark	3	//open_auction[.//reserve][.//annotation[.//person//parlist]//bidder//increase
DBLP	2	//dblp/proceedings[title]/author
DBLP	3	//dblp/article[author][./title]//year
DBLP	2	//dblp/proceedings[.//cite/label][title]/author
DBLP	2	//dblp/article[author][./title][.//url][.//ee]//year
DBLP	2	//article[.//mdate][.//volume][.//cite/label]//journal
Treebank	2	//S/VP//PP[.//NP/VBN/]/IN
Treebank	2	//S/VP/PP[IN]/NP/VBN
Treebank	2	//S/VP//PP[.//NN][.//NP[.//CD]/VBN]/IN
Treebank	2	//S[.//VP][.//NP]/VP/PP[IN]/NP/VBN
Treebank	2	//EMPTY[.//VP/PP//NMP][.//S[.//PP//JJ]//VBN/]//PP/NP//NONE

Fig. 7. Twig queries using in *TwigList* [3]

In XMark benchmark dataset, each person has a unique Address and profile fields and each address in turn has also a unique City and Country fields, each Profile has one and only one Education and Gender attributes. For the query *Type 3* in XMark, because of its balanced structure, each rooted element has only six descendant nodes. In the *HolisticTwigStack* algorithm, the six descendent elements will be pushed into stacks instantly after the root element (*Person*) is pushed onto the root stack to finish the twig match which is outputted instantly. Thus, at any time during execution, there are at most seven nodes in the stacks. Whereas, *TwigStack* has to store 4 single path matches for each suitable root element, which will takes 12 nodes in the stacks.

For comparison with *Twig*^{2}*Stack*, *HolisticTwigStack* much faster than which consumes time just 15%, 96% and 59% of *Twig*^{2}*Stack*. However, due to the uniform and balanced tree structure of XMark dataset, the space usage of both the algorithms are almost the same. In some cases, the space usage of these algorithm is quite small to be displayed (Figure 8(b)).

The results which are observed on TreeBank are also shown in (c) and (d). In TreeBank dataset, the execution time and space utilization for both *HolisticTwigStack* and *TwigStack* is nearly the same. But the *HolisticTwigStack* method is more faster than *Twig*^{2}*Stack*, it consumes about 30%, 90%, and 28% of *Twig*^{2}*Stack*'s time. It also uses an average 35% space usage of *Twig*^{2}*Stack*.

The results on DBLP dataset are shown in Figure 9. As shown in the chart, *HolisticTwigStack* uses much less time, 3.1, 4.4 and 1.4 times less, than *Twig*^{2}*Stack*. This is due to the complex stack manipulation and overhead for processing "non-optimality" nodes in the stacks. In this dataset, *Twig*^{2}*Stack* takes much more space for the algorithm's execution than *HolisticTwigStack* and event than in the DBLP and Xmark as well.

TwigList: As shown in Figure 10, *TwigList* outperforms *Twig*^{2}*Stack* in all experimental tests. Due to mainly *TwigList* used lists (linear structure) to

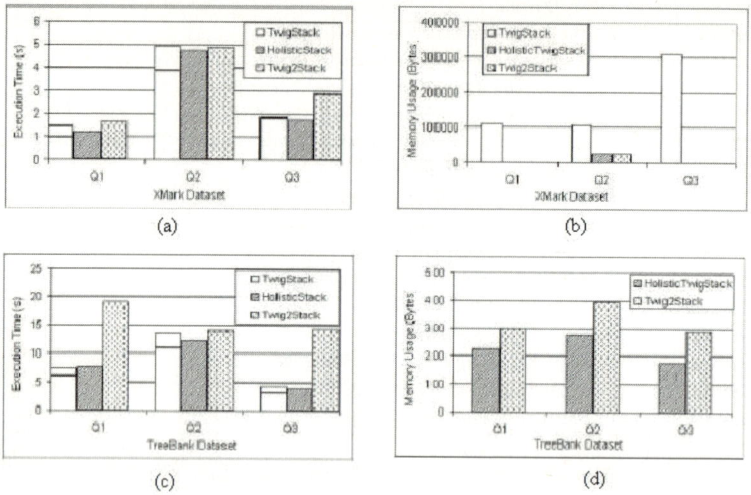

Fig. 8. Experimental Results from *HolisticTwigStack* [8]

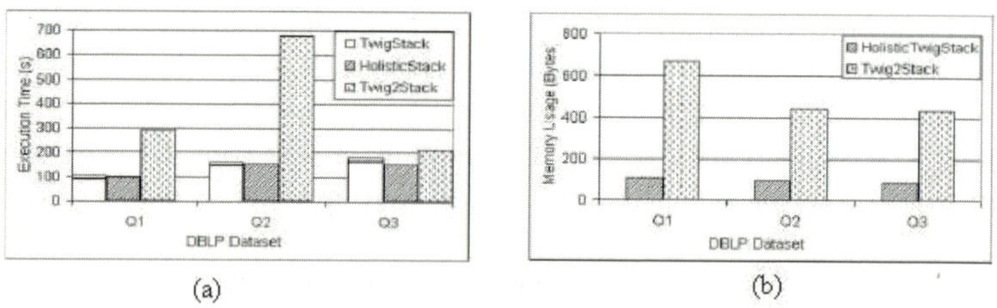

Fig. 9. Experimental Results from *HolisticTwigStack*(cont.) [8]

organize the elements instead of using complex hierarchical-stacks in *Twig*^{2}*Stack*. Furthermore, when enumerating final results *Twig*^{2}*Stack* uses a join technique which generate a lot of intermediate results, whereas *TwigList* does not produce any intermediate results.

For XMark dataset (Figure 10(a)) *TwigList* is much more faster than *Twig*^{2}*Stack*, it consumes time, on average, less 3-4 times than *Twig*^{2}*Stack*.For DBLP dataset (Figure 10(b)) *TwigList* is much more faster than *Twig*^{2}*Stack*, it consumes time, on average, less 4-8 times than *Twig*^{2}*Stack*. For TreeBank dataset (Figure 10(c)) *TwigList* is much more faster than *Twig*^{2}*Stack* when the query tree is became more complex, for example, TQ4 and TQ5.

In summary, *Twig*^{2}*Stack* based on lists (linear structure) replace a huge number of random accesses in both memory and disk. Figure 11 presents a summary for the analysis of all two algorithms.

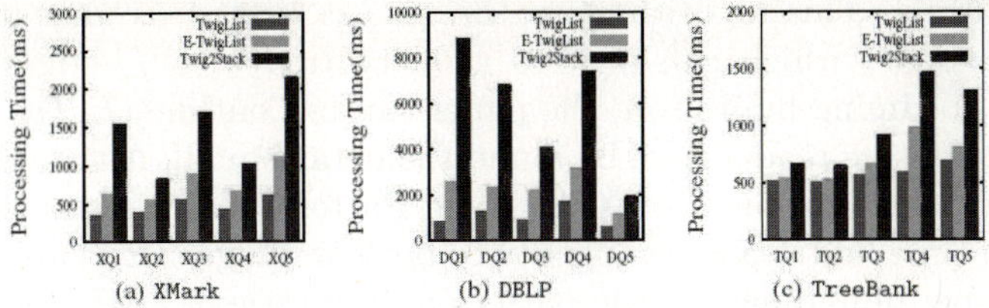

Fig. 10. Experimental Results from *TwigList* [3]

4.1 Summary and Evaluations

Experimental results of both algorithms have confirmed that the algorithms, *TwigList* and *HolisticTwigStack*, significantly outperform current efficient algorithms for XML Twig Pattern matching, specifically, *TwigStack* and *Twig*^{2}*Stack*. Indeed, *HolisticTwigStack* generally executes faster and requires less space than *TwigStack* and *Twig*^{2}*Stack*. The larger the query trees size and the complex the tree structures, the better the algorithm. The algorithm is proposed an effective and efficient one-phase holistic twig-pattern matching algorithm based on the stack-based *TwigStack*. By devising a novel stack structure to maintain matching patterns until entire twig patterns matches.

Tree Bank

	Twig²Stack			
	Execution Time			Memory Usage
	Type 1	Type 2	Type 3	
HolisticTwigStack	**~30%**	**~90%**	**~28%**	**35%**
TwigList	***n/a***	**~70%**	***n/a***	***n/a***

Xmark

	Twig²Stack			
	Execution Time			Memory Usage
	Type 1	Type 2	Type 3	
HolisticTwigStack	**~15%**	**~96%**	**~59%**	**100%**
TwigList	***n/a***	**~25-35%**	***n/a***	***NA***

DBLP

	Twig²Stack			
	Execution Time			Memory Usage
	Type 1	Type 2	Type 3	
HolisticTwigStack	**~24%**	**~18%**	**~42%**	**~20%**
TwigList	***n/a***	**12-25%**	***n/a***	***NA***

Fig. 11. Summary from original results

Meanwhile, the new *TwigList* algorithm using simple lists to compactly hold the twig patterns. This algorithm can be also extended as an external algorithm (*E-TwigList*) which can process a query tree with //-edges as well as /-edges. Due to timing limitation, the paper can not outline (*E-TwigList*) here, for details,interesting readers can be found the details in [3]. Although the above algorithms solve the problem of XML Twig Pattern Matching in different approaches, the time and space complexity of both two algorithms are linear w.r.t the total number of findings of twig pattern and the size of XML document tree.

In short, as statistics shown, *TwigList* seems to be chosen as a fast Twig Pattern Matching algorithm currently for efficient processing of XML Twig Query Matching on any complex native XML document database. XML document is modelled as an ordered tree with using containment labelling scheme (or *Region encoding*).

5 Conclusions and Future Works

XML twig pattern matching is a key issue for XML Query Processing and Query Optimization. In this paper, we discuss and evaluate two current efficient algorithms with using different techniques to address this problem. Our statistics from the analyses on original experimental results have confirmed that in comparison with *HolisticTwigStack*, the *TwigList* algorithm seems to be significantly more efficient in most case tested. As part of future work, we are planing to implement both two algorithms on the same test-bed for further comparison. Then we would like to study the pattern matching over using a novel labeling scheme called *extended Dewey* or *directed acyclic graphs* - ***DAGs*** to get a better performance in XML Query Processing and Query Optimization.

References

1. Berglund, A., Boag, S., Chamberlin, D., Fernadez, M.F., Kay, M., Robie, J., Simeon, J.: Xml path language (xpath). version 2.0 (2007), `http://www.w3.org/TR?Xpath20/`, `http://www.w3.org/TR/1999/REC-xpath-19991116`
2. Boag, S., Chamberlin, D., Fernandez, M.F., Florescu, D., Robie, J., Simeon, J.: Xquery 1.0: An xml query language. W3C Recommendation (January 2007), `http://www.w3.org/TR/2007/REC-xquery-20070123/`
3. Qin, L., Yu, J.X., Ding, B.: Twiglist: Make twig pattern matching fast. In: R.K.,, et al. (eds.) DASFAA 2007. LNCS, vol. 4443, pp. 850–862. Springer, Heidelberg (2007)
4. Chen, S., Li, H.-G., Tatemura, J., Hsiung, W.-P., Agrawal, D., Candan, K.S.: Twig2stack: Bottom-up processing of generalizedtree pattern queries over xml documents. In: VLDB 2006, September 2006, pp. 283–294. VLDB Endowment, Seoul (2006)
5. Jiang, Z., Luo, C., Hou, W.-C.: An efficient one-phase holistic twig join algorithm for xml data. In: CIKM 2006, pp. 786–787. ACM, New York (2006)

6. Bruno, N., Koudas, N., Srivastav, D.: Holistic twig joins: Optimal xml pattern matching. In: Proceedings of the 21st ACM SIGMOD International Conference Management of Data (SIGMOD 2002), pp. 310–321 (2002)
7. Jiang, H., Wang, W., Lu, H., Yu, J.X.: Holistic twig joins on indexed xml documents. In: Proceedings of the 29th VLDB Conference, Germany, VLDB (2003)
8. Jiang, Z., Luo, C., Hou, W.-C., Zhu, Q., Che, D.: Effiecient processing of xml twig pattern: A novel one-phase holistic solution. In: Wagner, R., Revell, N., Pernul, G. (eds.) DEXA 2007. LNCS, vol. 4653, pp. 87–97. Springer, Heidelberg (2007)
9. Choi, B., Mahoui, M., Wood, D.: On the optimal of holistic algorithms for twig queries. In: Mařík, V., Štěpánková, O., Retschitzegger, W. (eds.) DEXA 2003. LNCS, vol. 2736, pp. 28–37. Springer, Heidelberg (2003)
10. Aghili, S.A., Li, H.-G., Agrawal, D., Abbadi, A.E.: Twix: Twig structure and content matching of selective queries using binary labeling. In: Proceedings of the First International Conference on Scalable Information Systems (INFORSCALE 2006), Hong Kong, ACM, New York (2006)
11. Al-Khalifa, S., Jagadish, H.V., Koudas, N., Patel, J.M., Srivastava, D., Wu, Y.: Structural joins: A primitive for efficient xml query pattern matching. In: Proceedings of the 18th International Conference on Data Engineering (ICDE 2002) (2002)
12. Lu, J., Ling, T., Chen, T.: Tjfast: Effective processing of xml twig pattern matching. In: Proceedings of the 14th International Conference on World Wide Web (WWW), pp. 1118–1119 (2005)
13. Lu, J., Chen, T., Ling, T.W.: Efficient processing of xml twig patterns with parent child edges: A look-ahead approach. In: Grossman, D., Gravano, L., Zhai, C., Herzog, O., Evans, D.A. (eds.) Proceedings of the ACM Thirteenth Conference on Information and Knowledge Management (CIKM 2004), pp. 533–542. ACM, New York (2004)
14. Zhang, C., Naughton, J., DeWitt, D., Luo, Q., Lohman, G.: On supporting containment queries in relational database management systems. In: Proceedings of ACM SIGMOSD 2001, pp. 425–436. ACM, New York (2001)
15. Shanmugasundaram, J., Tufte, K., He, G., Zhang, C., DeWitt, D., Naughton, J.: Relational databases for query xml documents: Limitations and opportunities. In: Proceedings of the 25th VLDB Conference, pp. 302–314 (1999)
16. Wu, Y., Patel, J.M., Jagadish, H.V.: Structural join order selection for xml query optimization. In: Proceedings of the 19th International Conference on Data Engineering (ICDE 2003), pp. 443–454. IEEE, Los Alamitos (2003)
17. Chen, T., Lu, J., Ling, T.W.: On boosting holism in xml twig pattern matching using structural indexing techniques, Baltimore, Maryland, USA, pp. 455–466 (2005)
18. Chien, S.-Y., Vagena, Z., Zhang, D., Tsotras, V.J.: Efficient structural joins on indexed xml documents. In: Proceedings of the 28th International Conference on Very Large Data Bases (VLDB 2002), pp. 263–274 (2002)
19. Jiang, H., Lu, H., Wang, W., Ooi, B.C.: Xr-tree: Indexing xml data for efficient structural joins. In: Proceedings of the 19th International Conference on Data Engineering (ICDE 2003), pp. 253–264. IEEE, Los Alamitos (2003)
20. Lu, J., Ling, T.W., Chan, C.-Y., Chen, T.: From region encoding to extended dewey: On efficient processing of xml twig pattern matching. In: Proceedings of the 31st VLDB Conference, Norway, pp. 193–204. VLDB (2005)

Model-Based Evolution of an E-Learning Environment Based on Desktop Computer to Mobile Computing

Ana I. Molina[1], William J. Giraldo[2], Francisco Jurado[1],
Miguel A. Redondo[1], and Manuel Ortega[1]

[1] Department of Information Technologies and Systems,
Computer Science and Engineering Faculty,
Castilla – La Mancha University,
Paseo de la Universidad, 4. 13071 – Ciudad Real. Spain
{AnaIsabel.Molina,Francisco.Jurado,
Miguel.Redondo,Manuel.Ortega}@uclm.es
[2] Systems and Computer Engineering, University of Quindío, Quindío, Colombia
wjgiraldo@uniquindio.edu.co

Abstract. In the last years a great amount of collaborative applications have been developed. Advances in wireless technology and its integration on mobile devices offer support to user-to-user interaction on the move, becoming any place a potential collaborative scenario. With the aim of obtaining an appropiate support for the development of multi-plataform groupware applications we propose use a model-based approach for the development of interactive groupware applications (CIAM). This approach can be used for supporting the evolution of existing systems based on desktop metaphor towards mobile support. In this paper we show the application of this method to a case study, the teaching of Domotics. We will take as a starting point a collaborative e-learning environment called "Domosim-TPC".

Keywords: Mobile computing, Methodological framework, Graphical User Interfaces evolution process, Computer Supported Collaborative Learning, Model-Based Design.

1 Introduction

In the last years a great amount of collaborative applications have been developed. On the other hand some of them have been developed according to the paradigms of ubiquitous or mobile computing. Advances in wireless technology and its integration on mobile devices offer support to user-to-user interaction on the move, becoming any place a potential collaborative scenario.

Most of mobile collaborative systems are carried out in the same manner as other applications are developed, without taking into account the special characteristics of these paradigms. Therefore, the requirements that characterize these paradigms may not be considered in the most appropriate way; in special, we have to mention the aspects of user interface (UI) development and the perception of the context of the application. From our point of view we need appropriate frameworks and tools (Methodologies, processes, specifications techniques, CASE tools, etc.) to help in the

O. Gervasi et al. (Eds.): ICCSA 2008, Part II, LNCS 5073, pp. 322–334, 2008.

analysis and design processes of these complex applications [1, 2]. Clearly groupware activities can be improved in an important way by taking into account mobility aspects. However the current approaches do not offer an integrating and efficient solution that tackles jointly mobility issues and group work aspects.

The main goal of this paper is to incorporate the mobile computing paradigm in the teaching and learning of domains with a high experimental degree in order to take into account mobile computing possibilities. Also, the features of these domains provide an excellent framework to analyze the collaborative process.

With the aim of obtaining an appropiate support for the development of multi-plataform groupware applications we propose use a model-based approach for the development of interactive groupware applications (CIAM) [3]. This is a methodological framework supported by a set of notations for modeling and designing interactive and collaborative tools. This approach can be used for supporting the evolution of existing systems based on desktop metaphor towards mobile support [1]. We intend to identify common high-level task patterns in *Computer Supported Collaborative Learning* (CSCL) environments and guidelines that facilitate the creation of a complete semi-automatic environment that generates CSCL and mobile tools, independent of the study domain and of the platform.

In this paper we are going to show the application of our method to a case study, the teaching of Domotics. We will take as a starting point a collaborative e-learning environment called "Domosim-TPC" [4]. The paper is structured as follows. Section 2 describes the main features in the Domosim-TPC tool (used as a starting point environment). In the following section, the stages necessary to develop a mobile version of the aforementioned system are enumerated, that is, our model-based evolution process is described. This process is based on the use of conceptual specifications using the CIAN notation. This notation has been proposed in the context of a methodological approach called CIAM. In this section the CIAM methodological approach is shown, enumerating its several stages, and the aspects that are specified in each. Finally we will show the evolution process of the asynchronous tools in Domosim towards PDA support (individual workspace to design models) and we will draw some conclusions.

2 Domosim-TPC

The domain where our investigation is being applied is the learning of the design of automated control facilities in buildings and housing, also called Domotics. The term Domotics is associated to the set of elements that, when installed, interconnected and automatically controlled at home, release the user from the routine of intervening in everyday actions and, at the same time, provide optimized control over comfort, energetic consumption, security and communications. In this kind of training, the realization of practical experiments is especially important. In order to soften this problem by means of the use of technology, we have developed a distributed environment with support for distance learning of domotics design: DomoSim-TPC [4].

Using the DomoSim-TPC system (figure 1), the activities of practical learning of domotical design are structured in three clearly differentiated stages. In each of them diverse cognitive exercises are carried out and approached and representations of expert knowledge are used.

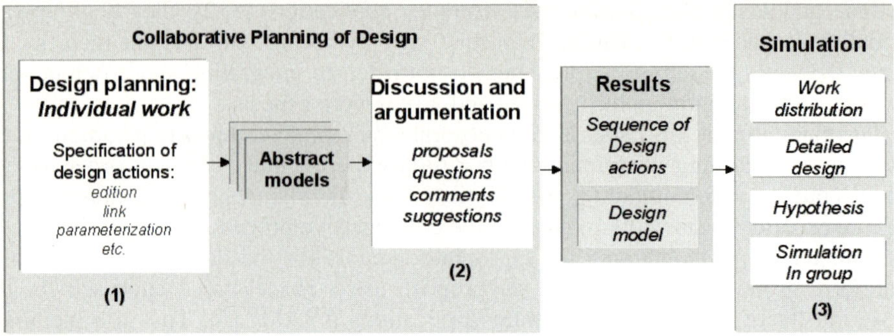

Fig. 1. Stages and tasks of domotical design learning carried out in an experience with DomoSim-TPC

Next, these stages are described concisely and the most outstanding tasks carried out in each one of them are pointed out:

- *Specification of models and planning of their design strategy.* In this stage the students, in an individual way, reflect and plan the steps to build a model satisfying the requirements proposed in the problem formulation. The strategy traced by the user is dynamically contrasted with an optimal plan of design for this problem.
- *Discussion, argument and search of consent* in the characteristics of the models individually built. In this stage, the participants discuss about the models built, about their types and about the steps carried out to obtain them. From this process a proposal (model) is obtained reflecting the viewpoint of each participant.
- *Detailed design and simulation in group.* Before checking the validity of the proposed solution, the apprentices should detail and organize the attributes associated to the objects that form the model. Later on, they consider the hypothesis and case studies that should be contrasted by means of the Collaborative Simulation of the behavior of this model.

We are interested in the effective use of such mobile computing devices for collaborative learning. There are tasks in the system Domosim-TPC which are susceptible of improvement through mobile computing. In particular, the Collaborative Planning of Design is a asynchronous and reflexive task which could be improved using mobile devices.

3 Model-Based Evolution Process

Our purpose entails improving the traditional classroom environment with the collaborative and the mobile computing paradigms. We intend to implement a prototype in a particular environment for collaborative learning (Domosim-TPC). We have to adapt the asynchronous tools of Domosim to the characteristics of mobile devices. To do this, it is necessary to restructure the user interface to adapt it to the constraints of size and utility of this kind of appliances.

In order to sustain a learning activity, we identify the concept of *space*. This is, a virtual structured place with resources and tools to perform a task (to solve a problem). There are three spaces: (1) *An individual workspace to design models satisfying a specification.* (2) *A shared space for discussion and argumentation* about the design actions that the learners have planned. (3) Another *shared space with the results of the discussion process.*

The process of evolution of Domosim-TPC towards mobile computing consists of several stages:

1. *Analysing tasks that can be improved by using mobile computing.*
2. *Design of tasks* taking mobile computing paradigm principles into account.
3. *Implementing a prototype* that applies proposed theories.
4. *Evaluating the prototype in real contexts.*
5. *Identifying the task patterns that could be common in CSCL environments,* based on the resolution of proposed problems and simulation of solutions contributed by students.
6. *Creating a tool that allows*, from a tasks model of a CSCL application, *obtaining in a semiautomatic way the equivalent interface for several mobile devices.*

If we want to generalize this process to the learning environment of other disciplines, we can automate the transformation process of the user interface. For CSCL tools developers, this introduces the problem of constructing multiple versions of applications for different devices. There are many dimensions to consider when designing context-dependent applications (environments, platforms, domain, users,...). With the aim of obtaining an appropiate support for the development of multi-plataform groupware applications we propose use a model-based approach for the development of interactive groupware applications (CIAM) [3]. This is a methodological framework supported by a set of notations for modeling and designing interactive and collaborative tools.

3.1 CIAM: A Methodological Approach for Modeling Interactive Groupware Applications

In this section the CIAM (*Collaborative Interactive Applications Methodology*) proposal is presented. CIAM is a methodological approach for the development of CSCW (*Computer Supported Cooperative Work*) applications that takes into account the modeling of group work and interaction issues. Unlike other existing proposals in the fields of conceptual modelling of CSCW systems and modeling of issues related with the Computer Human Interaction, CIAM considers the joint modeling of both issues, as well as the differentiation between the concepts cooperation and collaboration [5].

This approach consists of three main elements:

- A *conceptual framework* that clearly defines the concepts studied and modeled in each one of the phases in the methodological proposal [6].
- A *methodological framework* that defines the set of phases that compose the proposal, as well as the set of specification techniques to use in each of them [3]. In the figure 2 we can see the stages of the CIAM proposal. In each of them several collaborative and interactive systems issues are specified. The

information represented in each of the stages serves as basis for the modeling to be made in the following stage; so that this information is extended, related or specified with a greater level of detail in the following stage of the process.

- A *notation*, called CIAN (*Collaborative Interactive Applications Notation*), that allows expressing the peculiarities of the interactive groupware systems (figure 3).

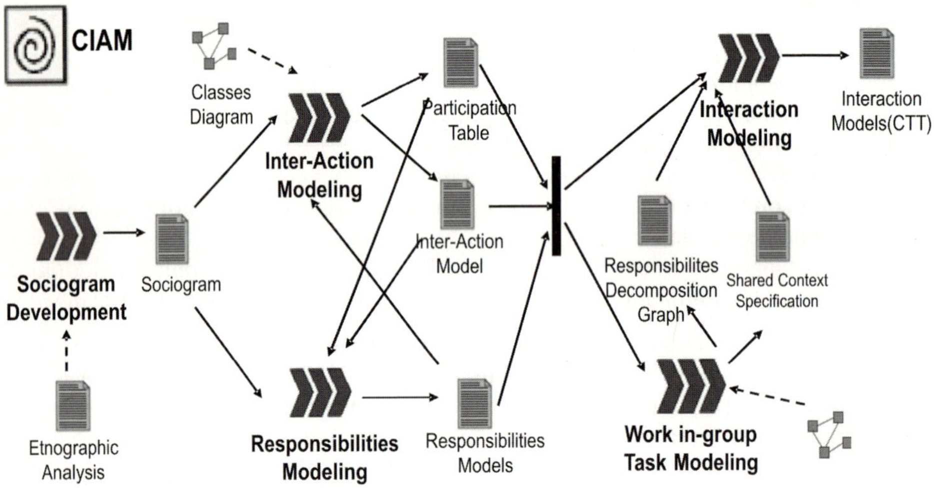

Fig. 2. CIAM methodological proposal stages

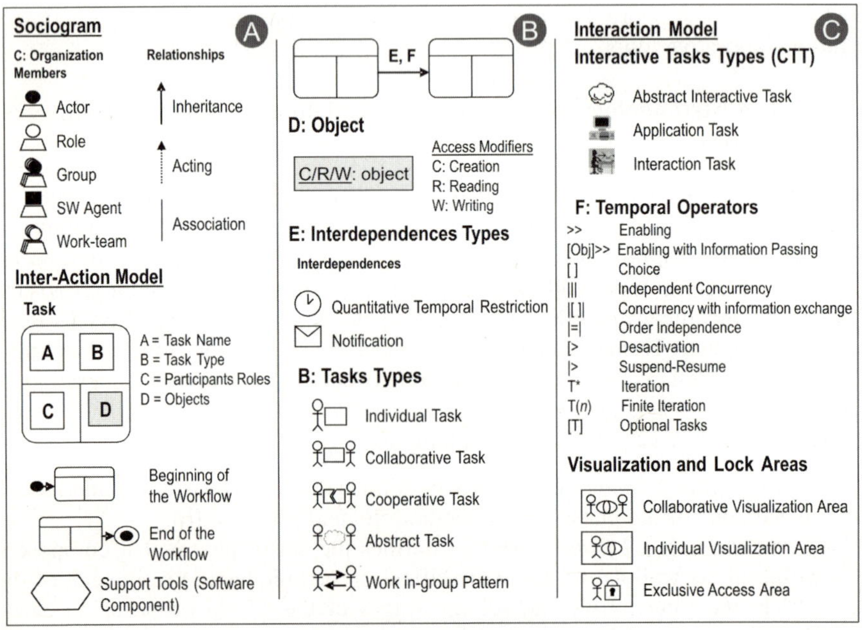

Fig. 3. CIAN Notation

In Table 1 we summarize the specification techniques used in the several stages of our methodological approach, the notation used and the product obtained in each.

Table 1. Specification techniques and result obtained in the several stages of CIAM

CIAM Methodology Stage	Specification Technique	Representation Type	Result obtained
Sociogram Developmen	Sociogram	Graphic	Organizacional Structure Specification *(roles, actors, software agents, etc)*
Responsibilities Modeling	Responsibilities Model	Textual	Detailed specification of the responsibilities of each role
Inter-Action Modeling	Participation Table	Textual	Relationships between the main tasks ans roles of the system
	Inter-Action Model	Graphic	Work Structure and workflow to be performed by the organization
Work in-group Tasks Modeling	Access Control Matrix	Textual	Relationships between data objects and roles at level of work in-group task
	Table of Operation Permissions on the Shared Context	Textual	Relationships between operations and roles at level of work in-group task
Collaborative Tasks Modeling	Responsibilites Decomposition Graph	Graphic	Responsibilities distrivution in a cooperative task specification
Cooperative Tasks Modeling	Shared Context Specification (classes diagram in standard UML notation)	Graphic	Specification of the Shared Context in a collaborative task, the division in visualization areas and the finalization policy
Interaction Modeling	Interactive Tasks Decomposition Tree (CTT notation enriched with icons for specifying visualization areas in collavorative tasks)	Graphic	Interaction Modeling at individual responsibility level; interaction with the shared context in collaborative tasks

4 Applying CIAN for Adapting Collaborative Tasks in Domosim-TPC to Mobile Computing Support

We intend to obtain the mobile version, and in particular, the PDA version of the individual plan edition space in Domosim-TPC (figure 4). The evolution process aforementioned is based on the use of conceptual models of the starting system. Next, we describe the analysis and modeling of the main tasks in asynchronous workspace of Domosim-TPC using the CIAN notation. The models, created in the context of CIAM, allows to adapt the user interface of Domosin-TPC to mobile computing support. Also, these conceptual models are used for automating and generating this evolution process.

Next we are going to explain the creation of the *sociogram* and the *inter-action* models associated with the system Domosim-TPC. In figure In Figures 5 and 6 we can see the appearance of both.

In the ***Sociogram*** *Development* stage, the organization structure is modeled, as well as the relationship between its members. In the example we have the following roles: *Teacher*, *Student*, *Observer* and *System* (that adds expert knowledge to the learning process). We identify some specialized students called *Planner* and *Designer* that take part in the two main tasks in the environment. The inheritance relationships

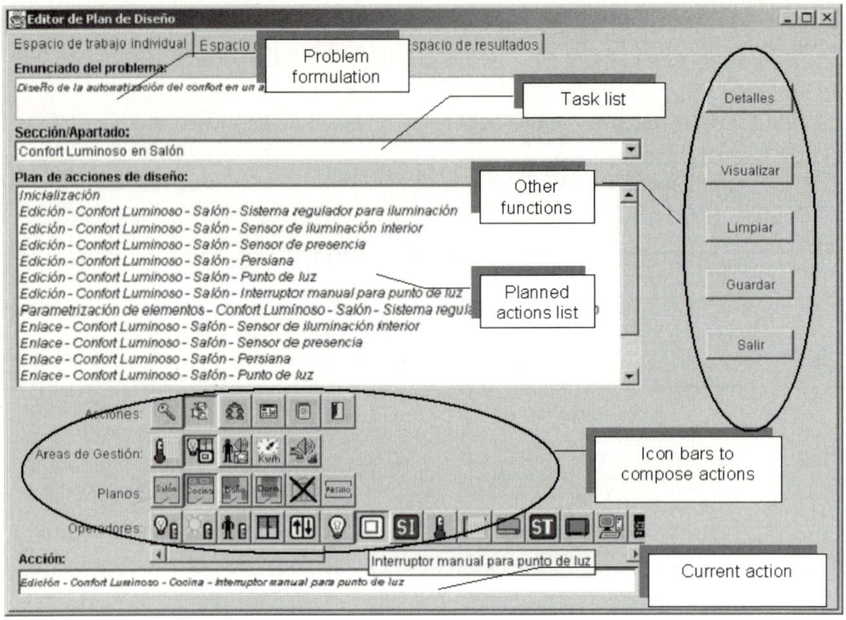

Fig. 4. Plan Editor User Interface

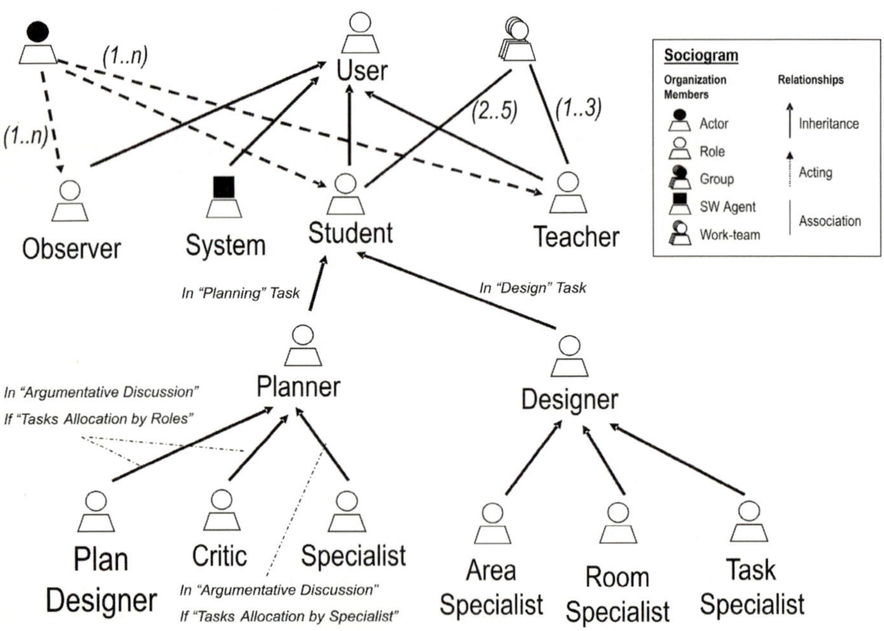

Fig. 5. Sociogram of Domosim-TPC

can be enriched with the definition of *conditions*. For example, we show that a *Student* can be specialized in the *Designer* role in the context of the *"Design"* Task. The *Planner* role has several subroles (*Plan Designer, Critic* or *Specialist*). A *Planner* can play any of them depending on the "*Task Allocation*" chosen for the activity. Once the *inheritance* (generalization / specialization) *relationships* among roles established, the *actor-role acting relationship* is added for the main roles in the diagram. This kind of relationship can be labelled when we want to express the cardinality (minimum and maximum) in the cases in which the specification establishes restrictions on the matter. The diagram can also show the relationships among roles that can, at a certain moment, work together. These relationships are expressed by means of *association relationships*. In figure 5 we can see that the *Student* role and the *Teacher* role are associated, creating work group. This indicates that there are tasks in which both with their respective responsibilities take part.

In the following two stages (***Responsibilities Modeling*** and ***Inter-Action Modeling*** stages) the abstract tasks (or processes) that define the work in-group developed in the organization, as well as the temporal and data dependencies that exist among them are described. It is created the so-called *participation table*, which provides an initial idea about the division of the work at the highest level of abstraction. Once the participation table has been constructed, the *Responsibilities Model* can be defined. It specifies in more detail and under an individual perspective the responsibilities associated to each one of the roles in the organization, adding to its shared responsibilities those that are exclusive to it. The information specified in these two stages is supplementary, being necessary for both models to be coherent.

Figure 6 shows the *inter-action model* associated to the system taken as example. In this example we use a symbol to express *Abstract Task* (6.d), that is, group work tasks that can be decomposed into others in a lower level of abstraction and of different kinds (6.g and 6.h). Collaborative and Cooperative tasks must specify the roles involved in its execution (6.a), whereas in the individual tasks only a role must appear. For all the tasks the objects manipulated and their access modifiers are indicated (6.b). For each task we can specify the so-called *Domain Independent Support Tool* (6.c). These are the supporting tools that implement well-known patterns or interaction protocols. Between the tasks *Configure Experiences* and *Planning* we can see temporal and data dependencies (6.c), indicating that the data *Activities* are transferred and the relation between these tasks is sequential (>>). Between task *Planning* and *Design and Simulation*, there is a period dependence (6.f) and a condition that must be checked (6.e).

In the ***Work in-Group Tasks Modelling*** stage the group tasks identified in the previous stages are described in more detail. Two different kinds of tasks are distinguished and modelled in a distinctive way: the *collaborative tasks* and the *cooperative tasks*. The *collaborative tasks* specification is based on the *shared context* definition [7], that is, the set of objects that are visible to the users and the actions that can be performed.

Finally, in the ***Interaction Modelling*** stage the interactive aspects of the application are modelled. For each task of individual nature detected in the previous stages of the process, an *interaction model* is created. The notation used for the interactive models is the *interactive tasks decomposition tree* from CTT [8]. As for the collaborative tasks, CIAM allows obtaining the interactive model directly from the *shared context* definition [9]. In particular, the division of the shared context made in the previous stage is used for organizing the UI in *workspaces*. The interaction model obtained facilitates the obtaining of the final UI.

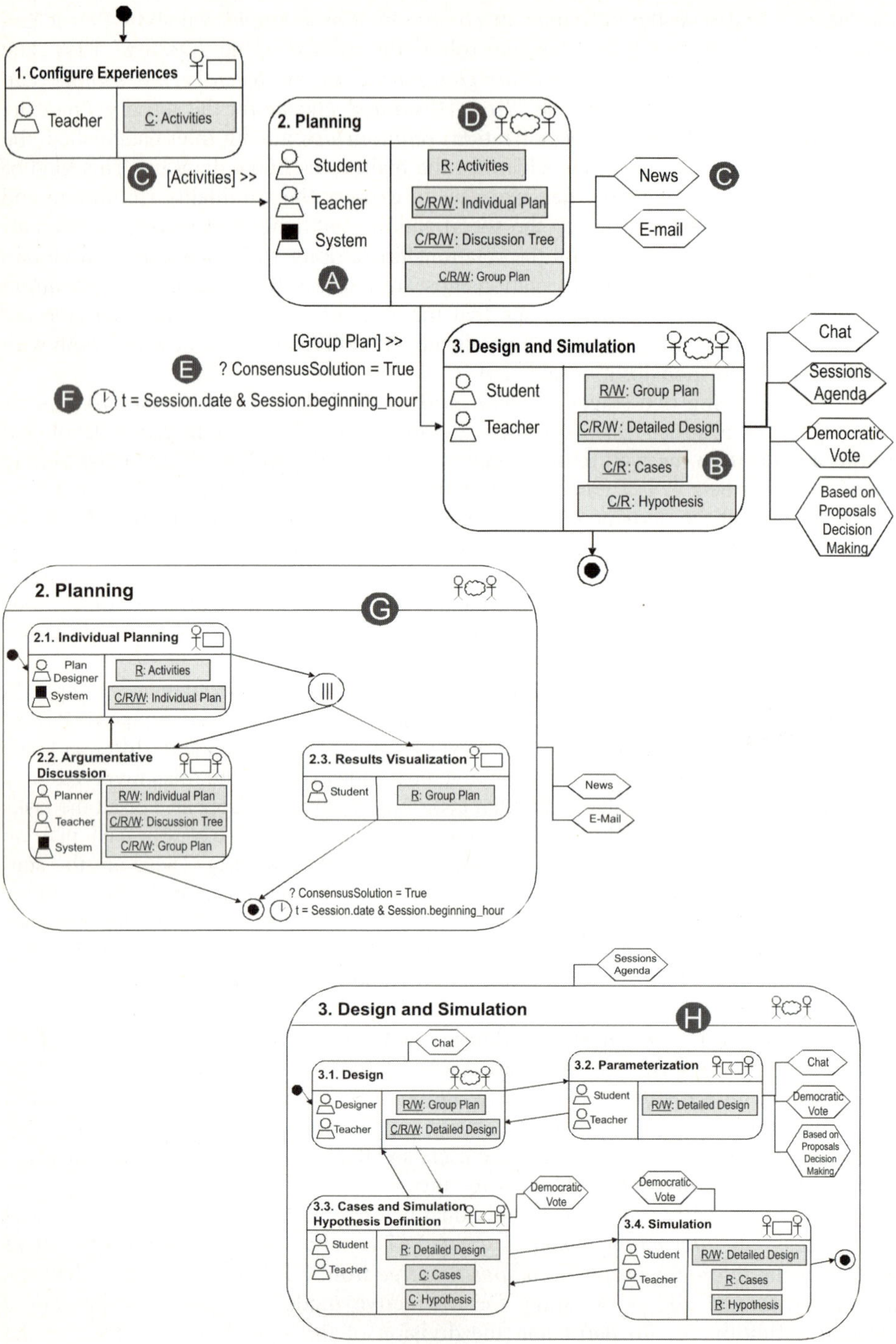

Fig. 6. Domosim-TPC Inter-Action Model

Figure 7 shows the task model in CTT notation for individual task *"Individual Planning"* (task 2.1 in figure 6). To obtain the version for PDA of the individual workspace, temporary relationships among the tasks and the domain-manipulated objects must be taken into account. This information allows creating the interface in which both the widgets (user interface objects) that show domain application objects (internal objects) and the widgets that allow executing certain actions applicable to these internal objects must appear together. In the design plan editor (individual workspace) two internal objects are handled: the *DESIGN_ACTION* and the *INDIVIDUAL_PLAN* (a collection of design actions). Figure 7 shows the names (in uppercase) of both objects. They are part of the name of the tasks that manipulate them.

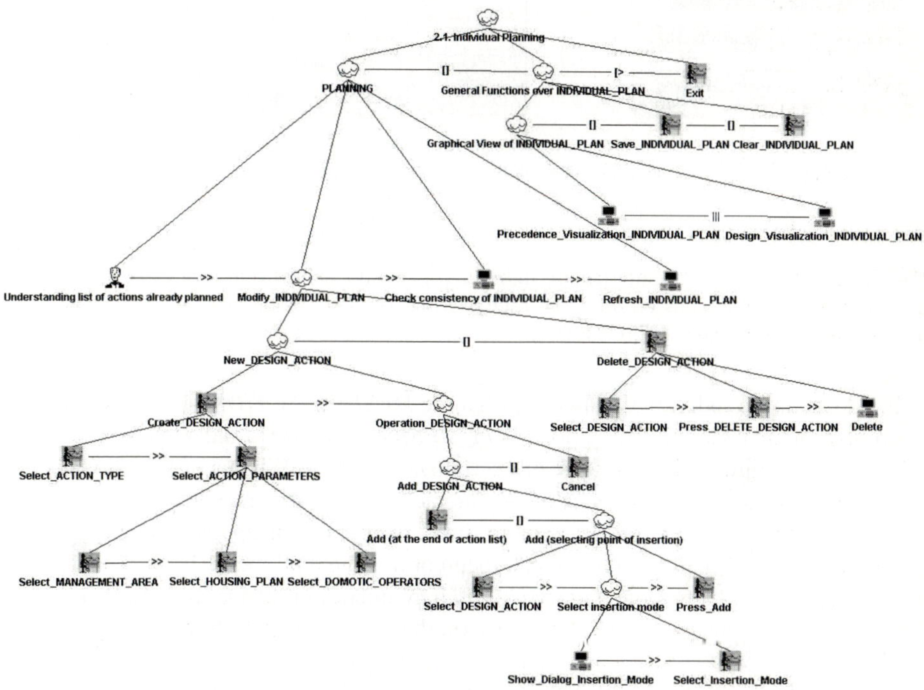

Fig. 7. Modeling of abstract task *"2.1 Individual Planning"*

Diagram in figure 7 shows the general functions that can be performed on the individual plan. It can be shown graphically. There are two modes of *visualization*: a list of nodes (a node represents an action) connected by arcs (representing precedence relationships); and the design of the scene that is created for executing the planned actions list. We can also *save* the individual plan. The option *Clear* eliminates all the information contained in the actions list. These actions are applicable to the *INDIVIDUAL_PLAN* object. These must appear in the user interface next to the object related (the list box that shows the sequence of steps in the plan). The resulting PDA interface of these subset of tasks are shown in figure 8 (a).

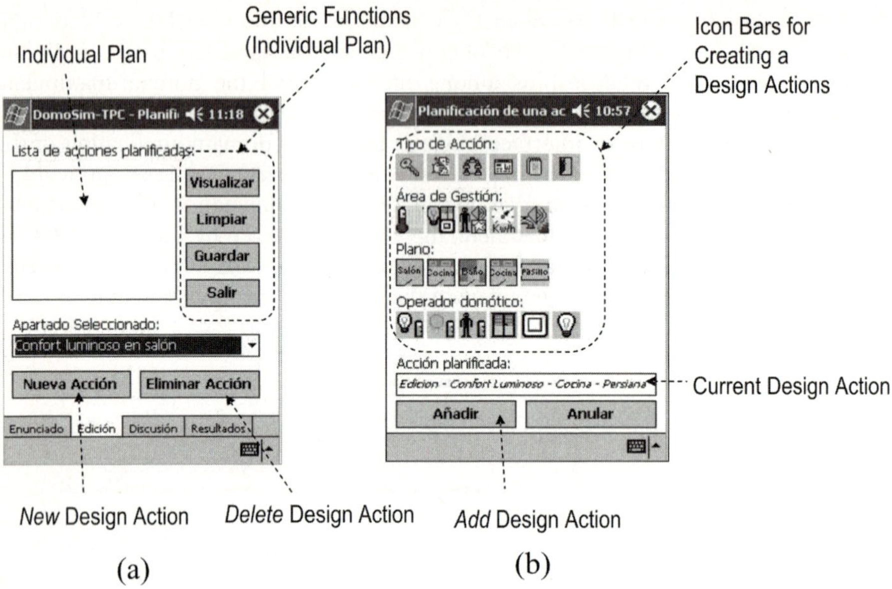

Fig. 8. PDA Version of the interface to individual plan edition. (a) Interface that allows showing and performing actions on the INDIVIDUAL_PLAN. (b) Dialog box that allows the creation of a new INDIVIDUAL_ACTION.

In addition, the individual plan editor handle *DESIGN_ACTION* objects. In the diagram shown in figure 7 the actions *Add_DESIGN_ACTION* and *Delete_DESIGN_ACTION* are included. The first one has certain complexity. When a task (that means an operation over a internal object) is of the interaction type, the mapping to a perceptible object (a widget in the interface) is more direct. This kind of operations can be represented by means of buttons, options in a menu or a contextual menu. It has been applied to the mapping of the *Delete_DESIGN_ACTION* operation, or the aforementioned generic functions, which the user can perform on the *INDIVIDUAL_PLAN* object.

However, when a task has a certain complexity, i.e., when a task is represented by an abstract task, with several abstraction levels and several interaction tasks (this occurs in the *New_DESIGN_ACTION* task), more complex visual components are necessary (a panel, in a PC version of the interface; or in a PDA, where there are display resolution constraints, a dialog box is a better choice). This occurs in the task that allows creating new design actions, as we can see in figure 8 (b). This dialog box appears whenever a new design action is created.

In figure 9 we can see the process followed by the CIAM proposal until the obtaining the interaction models. These models of interaction are expressed in CTT and works as an entry point to use it with development tools of user interfaces based on models as is the case of TERESA [10] or the Dygimes framework [11]. We have proposed a method for obtaining the final user interfaces that also considers the object manipulated in the context of interactive tasks. We have shown as applying our method an user interface is obtained starting from an interaction model that gathers the requirements to support collaborative tasks.

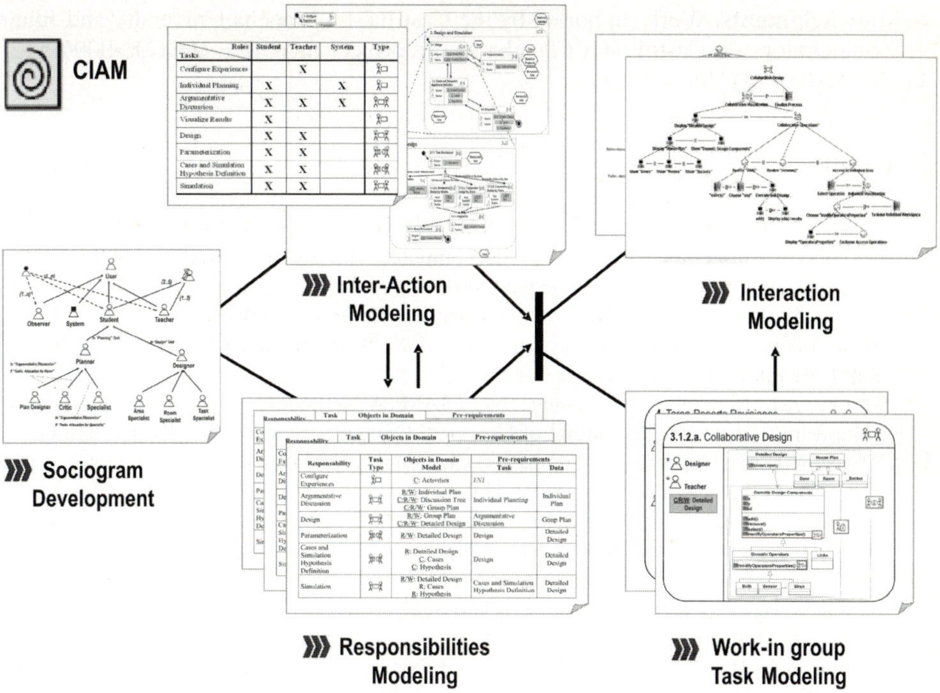

Fig. 9. Overall process followed by the CIAM proposal until the obtaining the interaction models

5 Conclusions and Future Work

In this article we have presented how to apply a methodological approach (CIAM) for the development and evolution of user interfaces to support collaborative and interactive activities that can be developed in a context of mobile computing. CIAM guides the designer following different phases from modeling to reaching interaction models. CIAM is based on the CIAN notation that allows users to accurately describe the features of a collaboration process (roles, responsibilities, tasks, shared context, etc).

We have shown the application of CIAM in the design and evolution process of the user interface that supports the task of collaborative planning of design in the context of the system Domosim-TPC.

Therefore, this has been a case of study really appropriate to discover the CIAM potential in helping to the development of collaborative user interfaces. Also it has been useful to show the necessities of extension of CIAM in order to add certain features of the mobile computing paradigm and especially some parameters that allow modeling the context. The results in the application of this Methodology show the necessity to include those aspects closely related with context modeling and the synchronization of contents; that is why we make an outline of the way to take into account these characteristics as a future work.

Acknowledgments. Work supported by the Castilla–La Mancha University and Junta de Comunidades de Castilla–La Mancha in the project mGUIDE (PBC08-0006-512) and M-CUIDE (TC20080552).

References

1. Eisenstein, J., Vanderdonckt, J., Puerta, A.: Applying model-based techniques to the development of user interfaces for mobile computers. In: ACM Conference on Intelligent User Interfaces IUI 2001, Alburqueque (2001)
2. Puerta, A., Eisenstein, J.: Towards a General a Computational Framework for Model-Based Interface Development Systems. In: IUI 1999: International Conference of Intelligent User Interfaces, Los Angeles (1999)
3. Molina, A.I., Redondo, M.A., Ortega, M.: CIAM: A methodology for the development of groupware user interfaces. Journal of Universal Computer Science (JUCS) (Designing the Human-Computer Interaction: Trends and Challenges) (2007)
4. Redondo, M.A., Bravo, C.: DomoSim-TPC: Collaborative Problem Solving to Support the Learning of Domotical Design. In: Computer Applications in Engineering Education, vol. 4(1), pp. 9–19. John Wiley & Sons, Chichester (2006)
5. Dillenbourg, P., et al.: The Evolution of Research on Collaborative Learning. In: Reimann, P., Spada, H. (eds.) Learning in humans and machines. Towards an interdisciplinary learning science, London, pp. 189–211 (1995)
6. Molina, A.I., Redondo, M.A., Ortega, M.: A conceptual and methodological framework for modeling interactive groupware applications. In: Dimitriadis, Y.A., Zigurs, I., Gómez-Sánchez, E. (eds.) CRIWG 2006. LNCS, vol. 4154. Springer, Heidelberg (2006)
7. Dourish, P., Bellotti, V.: Awareness and Coordination in Shared Workspaces. In: Proceedings of the Conference on Computer Supported Cooperative Work CSCW 1992, Toronto, Canada. ACM Press, New York (1992a)
8. Paternò, F.: ConcurTaskTrees: An Engineered Notation for Task Models. In: Diaper, D., Stanton, N.A. (eds.) The Handbook Of Task Analysis For HCI, LEA, Mahwah, NJ, pp. 483–501 (2004)
9. Grudin, J.: The Mechanics of Collaboration: Developing Low Cost Usability Evaluation Methods for Shared Workspaces. In: 9th IEEE International Workshops on Enabling Technologies: Infrastructure for Collaborative Enterprises (WETICE 2000) (2000)
10. Paternò, F.: CTTE: Support for Developing and Analyzing Task Models for Interactive System Design. IEEE Transanctions on Software Engineering 28(9) (2002)
11. Luyten, K.: Dynamic User Interface Generation for Mobile and Embedded Systems with Model-Based User Interface Development, PhD. thesis, Universiteit Limburg (2004)

An Effective Data Processing Method for Fast Clustering

Hyun-Joo Moon[1], Sangheon Kim[1], Jongbae Moon[2], and Eun-Ser Lee[3]

[1] Dept. of Cultural Contents, Hankuk University of Foreign Studies,
270 Imun 2-dong, Dongdaemun-gu,
Seoul, 130-082, Korea
hyunjoomoon@gmail.com, zoofoz@gmail.com

[2] Korea Institute of Science and Technology Information,
52-11, Eoeun-dong, Yuseong-gu,
Daejeon, 305-806, Korea
jbmoon@kisti.re.kr

[3] Dept. of Computer Engineering, Andong National University,
388 Songcheon-dong, Andong-city,
Gyeongsangbuk-do, 760-749, Korea
eslee@andong.ac.kr

Abstract. Because of the extensive diffusion of Internet usage, heterogeneous computing platforms, and ubiquitous computing technologies, Web data that are usually written in XML format are explosively increased. With the growth of Web data and the importance of their clustering, we need similarity detection method because it is a fundamental technology for efficient document management. In this paper, we introduce a similarity detection method that can check both semantic similarity and structural similarity between XML DTDs. For semantic checking, we adopt ontology technology, and we apply longest common string and longest nesting common string methods for structural checking. Our similarity detection method uses multi-tag sequences instead of traversing XML schema trees, so that it gets fast and reasonable results.

Keywords: XML, DTD, Similarity Detection, Ontology, Tag Sequences.

1 Introduction

As the usage of XML increases, the need for managing similar documents and detecting the difference and similarity among XML documents also increases. However, the detection method of the similarity in XML documents has difficulties because user defined tag names cause semantic ambiguity. For example, same tag names may have different meanings from documents to documents. In contrast, different tag names may have similar meanings.

For addressing this ambiguity problem, we first need to check XML DTDs for their elements' meaning and their internal structures [11]. In this paper, we introduce a similarity detection method in XML DTDs, which supports semantic and syntactic similarity among DTD documents. We adopt clustering strategies for efficient and fast detection.

This paper is organized as follows: In Section 2, we list related work. Section 3 discusses our system's architecture and similarity detection model. In Section 4, we will reveal our experiments, and finally we conclude our work in Section 5.

O. Gervasi et al. (Eds.): ICCSA 2008, Part II, LNCS 5073, pp. 335–347, 2008.

2 Related Works

Initial similarity detection methods use Tree Edit Distance Algorithms. These algorithms are designed to detect changes in structured documents or semi-structured documents using graph edit operations including delete, insert, and substitute. Various Tree Edit Distance Algorithms such as Valiente's algorithm, Tai's tree-to-tree correction algorithm, and Shasha and Zhang's algorithm are all based on atomic edit operations on nodes not considering subtree level operations such as subtree deletes and subtree moves [4][8].

Structural Document Differencing Algorithms such as LaDiff and MH-Diff generating edit scripts are based on the weighted matching problem. These algorithms do not make good results in semi-structured documents or in documents having node duplications [2][5][7].

Time Sequence-based Algorithms [9], such as Time Warping distance detection methods or DFT-based methods, totally differ from standard methods based on graph-matching algorithms that are generally computationally expensive, i.e., at least $O(N^2)$, where n is the number of elements in the two documents. This type of algorithms focuses on the fact that the computation of structural similarity is crucial to some clustering applications that can be of great value to the management of Web data, and allow a significant reduction of the required computation costs. They do not have enough semantic similarity detection mechanism, but they have very fast processing time.

Another interesting algorithm is XClust [10] that introduced ontological similarity measurement, and uses a tree traverse algorithm including PCC (Path Context Coefficient) concepts to capture the degree of similarity in the paths of two elements. However, XClust causes a serious velocity limitation, especially in large-scale databases or dissimilar document groups.

The comparison [7] among the existing detection methods is shown in Figure1.

Algorithm	Data Model	Mapping	Edit Operation				Run Time
			I	D	R	M	
Valiente	Ordered Unordered	Bottom-up	√	√	√	X	$O(n)$
Tai	Ordered	Ordered	√	√	√	X	$O(n^2 h^4)$
SZ	Ordered	Ordered	√	√	√	X	$O(n^2 h^2)$
LaDiff	Ordered	Bottom-up	√	√	√	√√	$O(ne+e^2)$
MH-Diff	Unordered	Induced graph	√	√	√	√√	$O(n^2)$
XyDiff	Ordered	Bottom-up Lazy-down	√√	√√	√	√	$O(nh)$
X-Diff	Unordered	Top-down	√√	√√	√	X	$O(n^2 d \log d)$
diffX	Ordered element Unordered attribute	Bottom-up Iterative-down	√	√√	√	√√	$O(n^2)$

I – Insert, **D** – Delete, **R** – Replace, **M** – Move
√ - Node, √√ - Node and Subtree, X – Not available
n – number of nodes, h – height, d – maximum degree, e – edit distance

Fig. 1. Comparison among Similarity Detection Methods

3 Similarity Detection Model

3.1 System Architecture

Our system named XJune uses XML DTDs for XML schema integration, and integration process consists of two phases; schema clustering and integrated schema extraction. Each

```
Algorithm NameSim(t1, t2, threshold1, threshold2, map)
Input:  t1 , t2 - two DTD trees to be compared with each other
        threshold1- The minimum ElementNameSim value to  have  the same tag index
        threshold2- The minimum ElementNameSim value to be included in LocalMatch
        map - the map that links each tag name to an index Number
Output:   the degree of name similarity between the two DTD trees

Begin
  a1 ← TreeTraverse(t1)
  a2 ← TreeTraverse(t2)
  next ← 0

  matrix ← {}
  for each e1   a1 do
     for each e2   a2 do
        n1 ← e1.name
        n2 ← e2.name
        n = NameSim(n1, n2, 3)

  if n > threshold1 then
    if both names are not exists in map then
       put (n1, next) into map
           put (n2, next) into map
   next ← next + 1
   else if n1 is not exist in map then put (n1, index of n2) into map
     else if n2 is not exist in map then put (n2, index of n1) into map
  else
       if n1 is not exist in map then
           put (n1, next) into map
           next ← next + 1
     if n2 is not exist in map then
           put (n2, next) into map
           next ← next + 1
   matrix ← matrix    (e1, e2,  n )
   MatchList ← LocalMatch(matrix, |a1|, |a2|, threshold2)
   return ∑sim / Max(|a1|, |a2|)
      sim    MatchList
End.

Function TreeTraverse(t)
Input:
  t - a  DTD tree
Output:
  an element array using visitor pattern which traverse tree
  in depth-first mechanism

ElementNameSim(w1, w2,  maxDepth) = OntologySim(w1, w2,  maxDepth)
```

Fig. 2. Element Name Similarity Detection

pair among DTDs is compared at once and the similarity is computed. After that, DTDs with higher similarity scores are integrated into a cluster, and then the cluster will have a new integrated DTD. Our system checks both semantic similarity and structural similarity in XML DTDs. We adopt ontology technology for semantic checking, and longest common string and longest nesting common string method for structural checking.

3.2 Ontological Element Similarity Detection

Ontological name similarity between nodes in two documents is identified using WordNet and our additional functions. The similarity values between two elements are from 0 to 1. 0 means they don't have any similarity, and 1 means they are exactly same. We limit the steps calls WordNet less than 3.

Based on the calculated each element name similarity `ElementName-Sim(`e_1, e_2`)`, the name similarity value between two DTD trees `Name Sim(`d_1, d_2`)` is calculated.

An element in DTD1 is $e_{1i} \in d_1$ and an element in DTD2 is $e_{2j} \in d_2$, In here, $1 \leq i \leq n$ and $1 \leq j \leq m$. The node number of DTD1, n, is defined as `NodeNumber(`d_1`)`, and m is defined as `NodeNumber(`d_2`)`. Algorithm for element name similarity detection is shown in Fig. 2.

3.3 Structural Similarity Detection

We adopt LCS (Longest Common String) and LNS (Longest Nesting common String) to detect structural similarity. LCS shows the same tag numbers in the same order,

```
                Algorithm LCS(seq1, seq2)

Input:
  seq1, seq2 – the tag sequences of DTD1 and DTD2
Output:
  LCS tag sequence of DTD1 and DTD2

Begin
  m = len(seq1)
      n = len(seq2)
      # An (m+1) times (n+1) matrix
      C = [[0] * (n+1) for i in range(m+1)]
      for i in range(1, m+1):
      for j in range(1, n+1):
                if seq1 [i-1] == seq2 [j-1]:
                   C[i][j] = C[i-1][j-1] + 1
                else C[i][j] = max(C[i][j-1], C[i-1][j])
      return C
  End.

Function Backtrack(C, seq1, seq2, i, j)
Begin
      if i == 0 or j == 0:
                return ""
      elif X[i-1] == Y[j-1]:
                return Backtrack(C, X, Y, i-1, j-1) + X[i-1]
      else if C[i][j-1] > C[i-1][j]:
                return Backtrack(C, X, Y, i, j-1)
      else return Backtrack(C, X, Y, i-1, j)
End.
```

Fig. 3. LCS extraction algorithm

which means the common nesting elements on both trees. Sometimes it includes some broken tags, which means that it can be include one of a start tag or its end tag, not both of them. LNS shows the common nesting elements on both trees without broken tags. From the information, we can extract all information of the tree structure to compare their meaning, and calculate the similarity with rapid processing speed. We use tag sequences and calculate DTD similarity very fast rather than tree traversing methods. The details are shown in algorithms on this paper. LCS and LNS extraction process use the encoding scheme named Invariant Direct Encoding Scheme and Depth-First Search tree traversing method.

LCS and LNS extraction algorithms are shown in Fig. 3 and Fig.4.

```
                    Algorithm LNS(seq1, seq2)

Input:
  seq1, seq2 – the tag sequences of DTD1 and DTD2
Output:
  LNS tag sequence of DTD1 and DTD2

Begin
for i in LCS
      // A stack S is used for check whether a tag has its pair tag.
      // If there is its pair tag, there are two same tags in the stack.
      // Therefore, they are popped together from the stack.
      // The remainders after putting in all tag sequences in LCS
      // are broken tags that don't have their pair tags.

      If there is the same index in S and their parent are the same,
         then  remove i from the stack S and
               record i to be a member of LNS
      else
         push i into the stack S
for i in LCS
         remove all i which do not recorded as members of LNS

LNS(LCS,seq1)
LNS(LCS,seq2)
End.
```

Fig. 4. LNS extraction algorithm

Our Structural Similarity algorithm using LNS and multi-level tag sequences to extract structural similarity between two DTD trees is described in Fig. 5.

3.4 DTD Similarity Detection Algorithm

The value of similarity between two DTD documents is calculated using semantic similarity and structural similarity and their weights. The calculated semantic similarity and structural similarity are added with their weights. We set the default weights of α and β as 0.4 and 0.6 for the simulation. Fig. 6 shows how to calculate document similarity algorithm in our method.

```
Algorithm NestingStructureSim(t1, t2, map)

Input: t1, t2 - DTD tree
   map  name to index map
Output: structural similarity between two trees

Begin
  length ← 0
  x1 ← 0
  x2 ← 0
  seq1 ← TagSequence(t1)
  seq2 ← TagSequence(t2)
   for (match ← NextBestMatch(x1, x2)) ≠ nil do
     length ← match.length
     x1 ← match.i + match.length
     x2 ← match.j + match.length
   return length / ((|seq1|+ |seq2|)- length)
End.

Function TagSequence(t)  using visitor pattern
Function NextBestMatch(seq1, seq2, m1, m2)
Input:
  seq1, seq2: tag sequences
  m1, m2: start index
Output:
  Next maximum-length match
   if there's no more match then nil

Begin
  if m1 >= |seq1| or m2 >= |seq2| then return nil
  if seq1[m1] = seq2[m2] then return (m1, m2, Length(seq1, seq2, m1, m2))
  else
    a ← Search(seq1, seq2, m1, m2)
    b ← Search(seq2, seq1, m2, m1)
  if a = nil and b = nil then return nil
  else if a = nil then return (b.j, b.i, b.length)
  else if b = nil then return (a.i, a.j, a.length)
  else  if a.length >= b.length then return (a.i, a.j, a.length)
  else  return (b.j, b.i, b.length)
End.

Function Search(seq1, seq2, m1, m2)
Begin
  for each p_{m1<=i<|seq1|} ∈ seq1 do
    for each q_{m2<=j<|seq2|} ∈ seq2 do
      if p_i = q_j then return (i, j, Length(seq1, seq2, i, j))
        return nil
End.

Function Length(seq1, seq2, m1, m2)
Input:
Output:
  The length matched from corresponding offset
  in each sequence
```

Fig. 5. Nesting Structure Similarity Detection

Algorithm DTDSim(t_1, t_2)

Input:
  t_1, t_2 - DTD tree
Output:
  DTD similarity

Begin
  initialize map<as name to index>
  return α * NameSim(t_1, t_2, map)
    + β* NestingStructureSim(t_1, t_2, map)
End.

Fig. 6. DTD Similarity Detection Algorithm

4 Experiments

4.1 Experimental Environment

For the experiment, we used Java SE 6.0 on Windows XP SP2 running on Pentium M 1.6 GHz RAM 1.56GB to implement our similarity detection scheme. We used JaxMe 2 DTD Parser to parse XML DTD documents. We also implemented a user interface to manipulate the threshold and weight values for similarity detection process.

4.2 Experimental Results

We show the simple experiment results. Fig. 7 shows two DTD trees with its encoding numbers, and Fig. 8 shows the similarity result with LCS, LNS, and other multi-level tag sequences. Two DTD trees have the same nesting structure and same data, but some different elements having same ontological meaning.

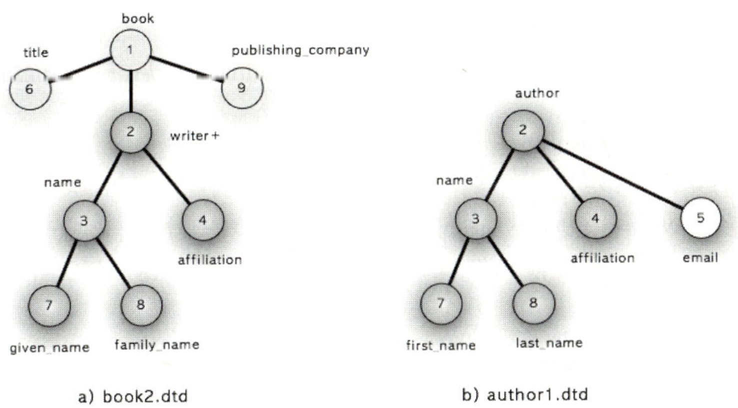

Fig. 7. Encoded DTD trees of `book2` and `author1`

As you can see in Fig.9, there are some differences. The second, seventh, and eighth elements have different element names but same meanings. The two DTDs are similar. The similarity between the two DTDs, therefore, is bigger than 0.5. Similarity has values from 0 to 1. 0 means the two element names are totally different, and 1 means they are exactly same. The other numbers mean they have similarity, and bigger number has higher similarity.

The similarity between two elements `book2` and `author1` is shown on the name matrix table in Fig.10. The similarity is calculated by `NameSim(t1, t2, threshold1, threshold2, map)` algorithm in Fig.4 that calls synonym function of WordNet ontology system to get the word's synonyms and.

	author	name	first_name	last_name	affiliation	email
book	0.0	0.0	0.0	0.0	0.0	0.0
title	0.0	0.0	0.0	0.0	0.0	0.0
writer+	0.81	0.0	0.0	0.0	0.0	0.0
name	0.0	1.0	0.0	0.0	0.0	0.0
given_name	0.0	0.0	0.81	0.0	0.0	0.0
family_name	0.0	0.0	0.0	0.81	0.0	0.0
affiliation	0.0	0.0	0.0	0.0	1.0	0.0
publishing_ company	0.0	0.0	0.0	0.0	0.0	0.0

Fig. 8. Name matrix of `book2` and `author1`

From the name matrix table, we can find the following match sets between the two DTDs. Each match set has the similarity value more than 0.7, which is the threshold value that we set the minimum similarity value to be considered as the same elements. The maximum call numbers of WordNet is defined 3 in the following experiment.

NameSim(given_name, first_name, 3) => 0.80999994
NameSim (affiliation, affiliation, 3) => 1.0
NameSim (writer, author, 3) => 0.80999994
NameSim (name, name, 3) => 1.0
NameSim (family_name, last_name, 3) => 0.80999994

From the name matrix table, we can calculate the *NameSim*(`book2`, `author1`).

NameSim (book2, author1, 0.7, 0.3, map)
= (0.81+1.0+0.81+1.0+0.81)/(16+12-5)
= 0.4629049

Using the match sets, two elements that their similarity is bigger than 0.7 have the same encoding number. Fig. 11 shows the encoding numbers of the two DTDs, which are numbers on each leaf.

Encoding Number	Element of DTD1	Element of DTD2
1	book	
2	writer	author
3	name	name
4	affiliation	affiliation
5		e-mail
6	title	
7	given_name	first_name
8	family_name	last_name
9	publishing_company	

Fig. 9. Encoding numbers of elements in two DTDs `book2` and `author1`

Now we can produce tag sequences of two DTDs. The tag sequence shows the structure of the DTD tree, and each element consists of two tags; start tag and end tag. Start tag and end tag of an element has the same encoding number. Additionally, we check the length of tag sequence to use for similarity calculation. The tag sequence of DTD1, `book2`, has length 16. It means that DTD1 has 8 elements, and consists of 16 tags; 8 start tags and 8 end tags. The tag sequence of DTD2, `author1`, has length 12.

Seq(book2) = (1,6,6,2,3,7,7,8,8,3,4,4,2,9,9,1)
Seq(author1) = (2,3,7,7,8,8,3,4,4,5,5,2)

For the additional information, we also produce depth level sequences of two DTDs. Each depth sequence shows the depth levels of corresponding elements in the tree. The information is used to remove some broken tags when the LCS has the broken tags.

DepthLevel(book2) = (1,2,2,2,3,4,4,4,4,3,3,3,2,2,2,1)
DepthLevel(author1) = (1,2,3,3,3,3,2,2,2,2,2,1)

LCS and LNS are derived from the tag sequences directly, and both depth level are both 10. It means that there are no broken tags.

LCS(Seq(book2), Seq(author1)) = (2,3,7,7,8,8,3,4,4,2)
LNS(Seq(book2), Seq(author1)) = (2,3,7,7,8,8,3,4,4,2)

As the next step, we calculate both depth level sequences of the LNS part in the two trees to figure out how much the two trees are different. At the moment, level average, level deviation of the two trees and cardinality similarity between the two trees are also calculated for the same reason.

LNS_DepthLevel(Seq(book2))= (2,3,4,4,4,4,3,3,3,2)
LNS_DepthLevelAverage(Seq(book2))= 3.2
LNS_DepthLevelDeviation(Seq(book2))= 0.7483314

LNS_DepthLevel(Seq(author1))=
(1,2,3,3,3,3,2,2,2,1)
LNS_DepthLevelAverage(Seq(author1))=2.2
LNS_DepthLevelDeviation(Seq(author1))= 0.7483314

LNS_CardinalitySimilarity = (0.7, 1.0, 1.0, 1.0, 1.0, 1.0, 1.0, 1.0, 1.0, 0.7)

From these calculation results, we can easily generate 4 similarity factors. The first factor, *sw1*, shows how mach parts are similar in the two trees. The second factor, *sw2*, explains the difference between the depth level average of LNS in both trees. The third factor, *sw3*, explains the difference between the depth level distributions of LNS in both trees. The last factor, *sw4*, is the cardinality similarity between LNS part in both trees. Note that *sw2* and *sw3* is difference, not similarity.

Sim1=0.5555556
Sim2=0.3125
Sim3=0.0
Sim4=0.93999994

Now, we can calculate the nesting structural similarity between two DTDs using the equation of 4 similarity factors. Refer the *NestingStructural- Similarity* algorithm in Figure 6.

NestingStructureSim(`book2, author1`)
= 0.5555556 * ((2-0.3125-0.0)+0.93999994
= 0.48657408

Finally, the similarity between two DTDs is calculated by semantic similarity and structural similarity.

getSemanticSim=0.4629049
getStructureSim()=0.48657408
getSim=0.47710645

Fig. 10 shows the similarity detection process on our detection system.

We tested 175 DTDs for the similarity detection and clustering. From the test result, we notice that XClust has an excellent approach to compute similarity, but it has some problems such as lack of generality and practicality. For example, if one of the DTDs is a subset of the other, it is different from the case that two DTDs are different but have some same elements. We are interested in similarity for clustering techniques; therefore, subsets of other DTDs should be more similarity. XClust also has a problem when it compares one small document with one big document. We concluded that XClust has a great quality when it compares similar sized documents, but it is not so good at compare totally different sized documents. Fig.11 shows the results of the experiments.

[Name Table]
1: book
2: writer author
3: name
4: affiliation
5: email
6: title
7: given_name first_name
8: family_name last_name
9: publishing_company

[Name Matrix]

	2	3	7	8	4	5
1	0.0	0.0	0.0	0.0	0.0	0.0
6	0.0	0.0	0.0	0.0	0.0	0.0
2	0.81	0.0	0.0	0.0	0.0	0.0
3	0.0	1.0	0.0	0.0	0.0	0.0
7	0.0	0.0	0.81	0.0	0.0	0.0
8	0.0	0.0	0.0	0.81	0.0	0.0
4	0.0	0.0	0.0	0.0	1.0	0.0
9	0.0	0.0	0.0	0.0	0.0	0.0

[Match Set]
(given_name,first_name) => 0.80999994
(affiliation,affiliation) => 1.0
(writer,author) => 0.80999994
(name,name) => 1.0
(family_name,last_name) => 0.80999994

Sequence1 = (1,6,6,2,3,7,7,8,8,3,4,4,2,9,9,1), len = 16
Level1 = (1,2,2,2,3,4,4,4,4,3,3,3,2,2,2,1)
Sequence2 = (2,3,7,7,8,8,3,4,4,5,5,2), len = 12
Level2 = (1,2,3,3,3,3,2,2,2,2,2,1)

LCS = (2,3,7,7,8,8,3,4,4,2), len = 10
LNS = (2,3,7,7,8,8,3,4,4,2), len = 10
LNS level1 = (2,3,4,4,4,4,3,3,3,2), average = 3.2, deviation = 0.7483314
LNS level2 = (1,2,3,3,3,3,2,2,2,1), average = 2.2, deviation = 0.7483314
LNS cardinality = (0.7, 1.0, 1.0, 1.0, 1.0, 1.0, 1.0, 1.0, 1.0, 0.7)
Sim1=0.5555556, x=18
Sim2=0.3125
Sim3=0.0
Sim4=0.93999994
getSemanticSim(threshold=0.3)=0.4629049 [1331ms]
getStructureSim()=0.48657408 [0ms]

getSim=0.47710645

Fig. 10. Similarity between `book2` and `author1`

	TEDAs	SDAs	XClust	XJune
Base Information	Structure	Structure	Structure + Semantic	Structure +Semantic
Ontology	N/A	N/A	Available	Available
Additional Ontology	N/A	N/A	N/A	Available
1:1 Mapping	Available	Available	Available	Available
M:M Mapping	N/A	N/A	Available	Available

Fig. 11. Comparison for Performance Capability

We think that clustering for huge amount of Web data should be generalized and proceeded in a short time. Using some metadata such as element level average and deviation, our method detects the difference of tree shapes and the distribution status between two documents. If two trees have the same level distribution but different deviation, it means that the two trees have the same parts which level depths are similar but the distribution of same nodes in the both trees is different. In the same manner, we can figure out how different the same parts in both trees are. Additional factors that detect more characteristics of DTD trees can be used for efficient clustering process.

5 Conclusions

In heterogeneous and dynamic environment such as mobile or ubiquitous network, fast and effective clustering methods are greatly useful for database management and fast search, especially in case of huge collections of web data.

We propose an efficient data processing method for fast clustering. Using the method, both flexible word compatibility and practicality can be achieved. If we consider further relationships between similarity detection algorithms and remarkably expensive additional ontological functions, this method can be one of the good solutions for the future.

For evaluation, we built a DTD database for similarity test set. The standard DTD set has 175 DTDs in several topics. The node numbers in the DTDs will be various from 5 to 100.

References

1. Extensible Markup Language (XML) 1.0 (1998), `http://www.w3c.org/TR/REC-xml`
2. Lian, W., Cheung, D.W., Yiu, S.: An Efficient and Scalable Algorithm for Clustering XML Documents by Structure. IEEE Transactions on Knowledge and Data Engineering 16(1) (January 2004)
3. Costa, G., Manco, G., Ortale, R., Tagarelli, A.: A Tree- Based Approach to Clustering XML Documents by Structure. In: Boulicaut, J.-F., Esposito, F., Giannotti, F., Pedreschi, D. (eds.) PKDD 2004. LNCS (LNAI), vol. 3202, pp. 137–148. Springer, Heidelberg (2004)

4. Klein, P., Tirthapura, S., Sharvit, D., Kimia, B.: A Tree-edit -distance Algorithm for Comparing Simple, Closed Shapes. In: Proceedings of the 11th Annual ACM SIAM Symposium of Discrete Algorithms, pp. 696–704 (2000)
5. Dalamagas, T., Cheng, T., Winkel, K.J., Sellis, T.: Clustering XML Documents Using Structural Summaries. In: Lindner, W., Mesiti, M., Türker, C., Tzitzikas, Y., Vakali, A.I. (eds.) EDBT 2004. LNCS, vol. 3268, pp. 547–556. Springer, Heidelberg (2004)
6. Borenstein, E., Sharon, E., Ullman, S.: Combining Top-down and Bottom-up Segmentation. In: Proceedings of the IEEE Conference on Computer Vision and Pattern Recognition (June 2004)
7. Ekram, R.A., Adma, A., Baysal, O.: Diffx: An Algorithm to Detect Changes in Multi-Version XML Documents. In: Proceedings of the 2005 Conference on the Centre for Advanced Studies on Collaborative Research (2005)
8. Zhang, K., Wang, J.T., Shasha, D.: On the Editing Distance between Undirected Acyclic Grahphs and Related Problems. In: Proceedings of the 6th Annual Symposium of Combinatorial Pattern Matching (1995)
9. Rafiei, D., Mendelzon, A.: Similarity-Based Queries for Time Series Data. In: Proceedings of the ACM International Conference on Management of Data, pp. 13–24 (May 1997)
10. Lee, M.L., Yang, L.H., Hsu, W., Yang, X.: XClust: Clustering XML Schemas for Effective Integration. In: Proceedings of the 11th ACM International Conference on Information and Knowledge Management, pp. 292–299 (2002)
11. Moon, H.J., Kim, K.J., Park, G.C., Yoo, C.W.: Effective Similarity Discovery from Semi-structured Documents. International Journal of Multimedia and Ubiquitous Engineering 1(4), 12–18 (2006)

MyGuide: A Mobile Context-Aware Exhibit Guide System

Jongmyung Choi[1] and Jongbae Moon[2]

[1] Dept. of Computer Engineering, Mokpo National University,
61 Cheonggye, Muan-gun, Jeonnam, 534-729, Korea
jmchoi@mokpo.ac.kr
[2] Korea Institute of Science and Technology Information (KISTI)
52-11, Eoeun-dong, Yuseong-gu, Daejeon, 305-806, Korea
jbmoon@kisti.re.kr

Abstract. Recently there have been researches on smart exhibition guide systems for visitor's convenience. In this paper, we introduce a context-aware, visitor-oriented guide system and some context related issues encountered during the development process. Our system provides a visitor with the most suitable guide information on the exhibit in both its content and its media format. For visitor-oriented guide service, we adopt four types of context information: visitor's location, visitor's background knowledge, media type supported by the terminal device, and visitor's preferred language. We also show how to model context in an object-oriented approach, how to apply ECAM software architecture to our system, and how to build our prototype system using RFID, MIDP APIs, and context-awareness.

Keywords: Context-aware, Guide system, Mobile, RFID.

1 Introduction

In museums and in exhibitions such as the Consumer Electronics Show (CES) [1], visitors need more customized guide systems rather than brochures or notes on the exhibit. For visitors' convenience, guide systems have advanced from brochures to digitalized systems. Recently, these digitalized guide systems have tended to adopt mobile devices, such as PDAs and mobile phones, and context-awareness for personalized services [2-7]. The main feature of these systems is that they focus on the visitor's convenience and needs. As the result, they have focused their interests on two kinds of technologies: services on mobile devices and location-awareness. However, they don't consider the visitor's background or profile, and provide the same information. In real world, some visitors may want more advanced and detailed information on the exhibits, while others may want simple and general information. In addition, some may want dynamic multimedia content, but others may want simple but quick text contents. In addition to the visitor's convenience, the visitor-oriented service that considers a visitor's background is also very important for the visitor's satisfaction. Therefore, smart guide system should be able to provide visitor-oriented guide services.

O. Gervasi et al. (Eds.): ICCSA 2008, Part II, LNCS 5073, pp. 348–359, 2008.

In this paper, we introduce a context-aware, visitor-oriented mobile guide system, named MyGuide. We adopt three kinds of technologies: RFID, mobile phone, and context-awareness to support the visitor's convenience. For the visitor-oriented guide service, our system considers three types of context information: the visitor's background knowledge, the media type supported by his/her mobile device, and his/her preferred language. With these three contextual elements, the visitor's location constitutes the context of our system. These context data determine the contents of the guide information and the media format.

We also introduce our prototype system and our experiences learned during the development process. In the early phase, we acquire the requirements from the stakeholders and identify the four contextual elements from these requirements. After that, we model context with the four contextual elements in object-oriented approach, and determine the software architecture. We adopt ECAM (Event, Control, Action, Model) architecture pattern [8] for our system because of its event-driven feature. We also design the system based on the software architecture and the database schema. In the implementation phase, we use MIDP [9] for the mobile phone programming, 13.56 MHz RFID reader and GNU java serial library [10] for RF tag reading, Oracle database for context information management, and Darwin Streaming Server [11] for multimedia services.

The main contributions of our work are four. First, we propose a visitor-oriented guide system that considers the visitor's background. It considers the visitor's knowledge level, media type, and preferred language, and determines the guide contents based on these data. Second, we adopt four types of context information for the guide services. Most of context-aware systems use single context information. However as the systems become more complex, they will use multiple types of context information more frequently. Third, we introduce how to model context with orthogonal context information in an object-oriented way. In the perspective of context-aware services, context is categorized into two: Multimedia and Text. Fourth, we show the development phase and the prototype system in detail. This will be a guideline or a sample for further context-aware system development.

The remainder of the paper is organized as follows. In Section 2, we discuss other works that are closely related to our work. Then we introduce some interesting issues on developing context-aware guide systems including system requirements, contextual elements, context modeling, and system architecture in Section 3. After that, in Section 4 we introduce our prototype system. Finally, we reveal the conclusions of our work in Section 5.

2 Related Works

Recently, there have been studies on smart guide system to serve museums and exhibitions. The main features of these systems are location-awareness and mobile devices. The representative studies are the researches of Keith [2,12], Sherry [3,4], and Jani [5].

Keith [2,12] introduces a city tour guide system that utilizes a wireless network and context-awareness. The system was initially intended to adopt three different types of context information: location, display device, and user profile [2], but its implementation uses only location [12]. The system determines the user's location using the location information from the base station.

Sherry [3,4] introduces an exhibition guide system using RFID. The main feature of this system is its bookmarking function. It means that visitors can bookmark any exhibit that draws his/her interests, and then he/she can make web pages with the information about the exhibits.

Jani [5] introduces an RFID-based system, mTag, which provides information to the pre-specified devices when the user carrying a tag approaches the RFID reader. The mTag system is similar to our system in that they both provide information when a tag approaches the fixed RFID reader.

Other representative systems in the field are the guide system of National Museum [6] and the system of the science exhibition [7] in Korea. These guide system use mobile phone and mobile RFID.

The existing researches and systems have some similarities with our system, but our system also has its own uniqueness. The differences and similarities of the systems are described in Table 1. MyGuide system uses multiple types of context information and supports the visitor-oriented guide services.

Table 1. Comparison of Smart Guide Systems

Attributes / Systems	Context Information	Location Sensing	Terminal Device	Visitor-oriented Service
Keith's [12]	Location	Base station	PDA	No
Sherry's [3][4]	Location	RFID	PDA	No
Jani's [5]	Location	RFID	PDA	No
Museum[6][7]	Location	Mobile RFID	Mobile Phone	No
MyGuide	Location, Knowledge, Media type, Language	RFID	Mobile Phone	Yes

3 Context-Aware Guide Systems

3.1 System Overview

Very diverse guide systems such as brochures, books, audio systems, and human guides have been utilized in the exhibition shows and museums. The main goal of MyGuide is to provide the most suitable information to the visitors in the most suitable format. In the requirements elicitation phase, we acquire the basic requirements as specified in section 3.2. In addition, we choose RFID, mobile phone, and context-awareness to meet the requirements. If we use these technologies, the visitors can receive the most suitable services as in the following scenario.

Scenario

When a visitor arrives at the museum, he/she registers his/her information (background knowledge, language) and his/her mobile phone information (facilities, phone number), and receives an RF tag. After registration, the visitor walks around the exhibits, finds an interesting exhibit, and accesses the RF tag to the reader attached to the exhibit. Then the system sends a message to the visitor. The message contains the information about the exhibit, or the URL for the multimedia information about the exhibit.

If we follow this scenario, we can generate the sequence diagram as shown in Fig. 1. We can figure out RFID reader, database for context information, context-server, and contents server from the scenario. The RFID reader sends the tag value to the context server, and it identifies the visitor. After that, it also can access context information stored in the context database using the tag value as a key. The contents server keeps the explanatory contents about the exhibits, and provides suitable contents to the visitor.

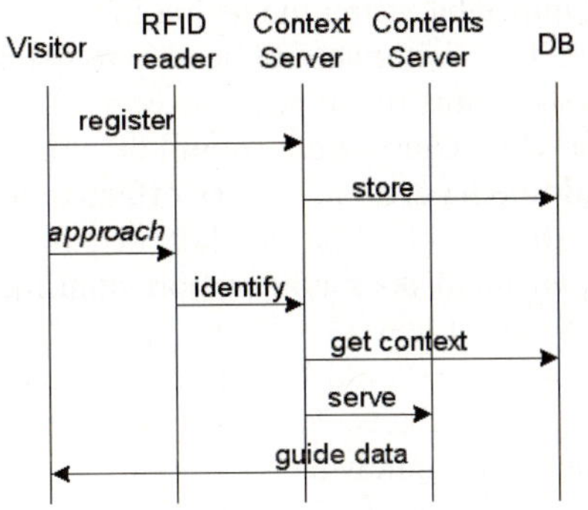

Fig. 1. Sequence Diagram for Scenario

From the basic scenario and the sequence diagram, we determine the overall architecture of the system. Fig. 2 shows the overall architecture of MyGuide system. When a visitor approached to the exhibit, the context server determines the context, and contents server provides guide information through CDMA network.

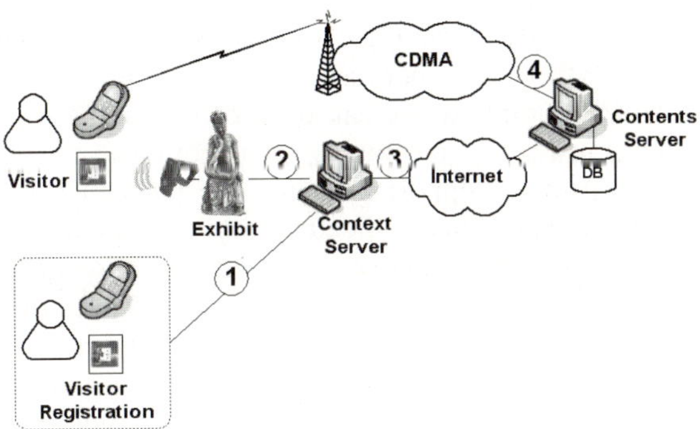

Fig. 2. Overall Architecture of MyGuide System

3.2 System Requirements and Development Strategy

From the goal and the scenario of MyGuide, we are able to draw the functional and non-functional requirements. The followings are the representative requirements for the system.

- (R1) Customized Services: Existing guide systems provide the same services to all visitors without considering the visitors' background knowledge or age. This can feel unsatisfactory to the visitors. Children and beginners don't understand the explanation with jargon, and experts are not satisfactory with an explanation that contains only general information. Therefore, our system considers the visitors' background and provides appropriate information according to their profiles.
- (R2) Easy Use: The end-system or terminal should be easy to use, and the visitors should feel comfortable using the technology.
- (R3) Personal Usage: The visitor should be able to get and control the services at any time when he/she wants in the museum.
- (R4) Multimedia Services: Our system should be able to deliver multimedia data to the visitors. Multimedia content is very effective to explain information to a visitor whether he/she is a child or an adult, or a beginner or an expert. On the other hand, if the terminal does not support multimedia service, it should be able to provide text data at least.
- (R5) Less Cost: Our system should be implemented with the least costs and efforts because of limited budgets. Therefore, we prefer to use the existing technologies without developing new technologies.
- (R6) Multiple Languages: One of the interesting requirements from stakeholders is for the system to support multiple languages as the number of visitors from other countries increases. Therefore, our system should support guide contents in multiple languages.

We searched the possible technologies available to meet the basic requirements, and we chose RFID, mobile phone, and context-awareness strategically. There are several reasons why we choose these technologies as follows;

- RFID: An RF tag is small enough to carry with the visitor and it is natural and easy to use. It is a stable technology because it has been used in various fields for a long time. Furthermore, there are ample system components and software modules for RFID systems, so we can reduce costs and efforts by using the existing components instead of developing them from the scratch.
- Mobile Phone: In order to support mobility, we need mobile devices including PDA, mobile phone, and other dedicated mobile terminals. From several kinds of devices, we choose mobile phone based on the following merits. First, most of the new products support multimedia display, and even older ones support at least a text message service without adding any hardware components or installing any software modules. Second, we can save cost and effort in developing end-systems for providing guide service. Third, the mobile phones are easy to carry and people tend to keep their phones with them. Fourth, they are user friendly. It is their own phone, and they use it every day. Fifth, a mobile phone is always connected to a CDMA or GSM network at any place. We don't need to consider a situation in which the end-system is disconnected.
- Context-awareness: Day[13] defines context as "any information that characterizes the entity". Context-aware systems adjust their services according to the context. We think the visitor's profile and location are good examples for context information. Based on these data, our system determines its service contents.

These technologies are stable and popular in their own fields. We only need to combine them to meet our functional requirements.

3.3 Contextual Elements

In the scenario of Section 3.1, we assume two things. First, the system database has a table that maps the exhibit and the RFID reader's identifiers. Therefore, we can get the visitor's location when a reader reads his/her RF tag. Second, the visitor registers his/her information with an RF tag value.

There are countless environmental elements around the context-aware system. However we cannot consider every contextual element, so that we limit the number of contextual elements in our system to the information that is necessary to meet the requirements. We identify four kinds of context information that affects the service of our system.

- The visitor's location
- The visitor's background knowledge
- The facilities of the visitor's mobile phone
- The visitor's preferred language

The visitor's location is primary context information. The system can determine which exhibit draws a visitor's attention according to his/her location. We use RF tags and RFID readers to identify the visitor's location. When the visitor with an RF tag approaches to an exhibit, the reader attached to the exhibit identifies his/her tag value. In the perspective of guide information, the location determines the subject of data to be served.

Considering the visitor's background knowledge is the most valuable feature of our system. The more we divide the level of the background knowledge, the more the service will be satisfactory. However, we should consider development costs and the maintenance problem of contents, so that we group the background knowledge into three levels.

- Beginning: basic information for children and those unfamiliar with the subject
- Normal: general information for ordinary adults and middle/high-school students
- Advanced: advanced information for experts or college students

The facilities of the visitor's mobile phone determine the media type of information. Some phones support multimedia data display, but others may have the only minimum functions for mobile phone. Therefore our system should consider the facilities of the visitor's mobile phone. We therefore categorize mobile phones into two types: Text or Multimedia.

The visitor's preferred language is also important context information to be considered. As the world becomes smaller, we need to serve the visitors from other countries in museum. We consider total three languages: Korean, English, and Chinese.

The four kinds of context information have their own set of possible values as specified above. A situation in our system is the value combination of four kinds of context information. And the value of context information determines the contents and the media format of the guide information.

3.4 Context Modeling

MyGuide adopts four kinds of context information for its services. The notable feature of them is that they are orthogonal and independent of each other. For example, the visitor's location has nothing to do with his/her preferred language, and the subject and the language of the contents have nothing in common. The four types of context information can be represented orthogonally in the four-dimensional coordinate as shown in Fig. 3.

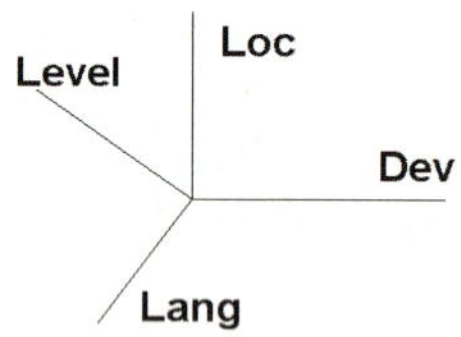

Fig. 3. Context Space of MyGuide System

According to Dey's definition [13], each of the four types of context information can be a context, but any one, two, or three of them cannot determine the guide service. Therefore, we view a context as the combination of all constituent context information. In our system, we define a context as the tuple of values of the four context information types. Then, a context may be represented as a dot in the coordinate system in Fig. 3.

In MyGuide system, we have too many contexts. Let C be the set of all possible contexts in MyGuide system. Then a context c (□ C) can be represented as a tuple: c = <loc, level, dev, lang>, where loc □ Loc, level □ Level, dev □ Dev, and lang □ Lang. We can imagine that Loc = { x | x is an exhibit. }, Dev = {multimedia, text}, Lang = {Korean, English, Chinese}, and Level = {beginning, normal, advanced}. Then we can calculate the number of contexts in MyGuide as follows:

$$|C| = |Loc| \times |Dev| \times |Lang| \times |Level|$$

Because this system has too many contexts to consider, we need to categorize them according to commonality and variability. In MyGuide system, media type determines the metadata and service server, so we categorize context information according to media type.

Fig. 4 shows the context model of MyGuide. The notation used in the figure follows our prior studies [14, 15]. The notation is similar to class diagram. According to this notation, a context may be implemented in multiple ways, so that it is represented as an interface. Similar contexts have common features, and the common features can be generated as context type. MyGuide has two contexts: Multimedia and Text.

We have four kinds of context information and two contexts. Therefore, when the system senses the contextual data from the external environment or from the internal source, it should be able to gather the data, interpret the data, and generate contexts. We call the component that plays the roles as the context reducer. Fig. 5 shows the context reducer.

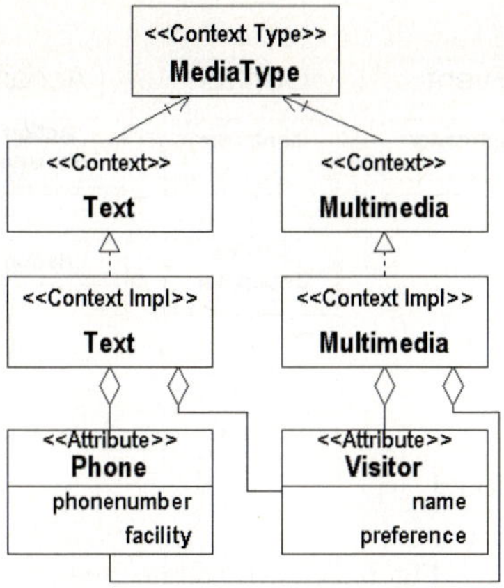

Fig. 4. MyGuide Context Model

Fig. 5. Context Reducer

3.5 Software Architecture

Software architecture is important enough to determine the system's framework and design. Compared to traditional systems, context-aware systems have some additional interesting features to consider in determining software architecture. Therefore, there have been interesting studies for architecture patterns for context-aware systems. We adopts ECAM (Event, Control, Action, Model) pattern [8] for MyGuide because it is event-driven and it uses multiple kinds of context information. ECAM is the extension of the ECA (Event, Control, Action) [16] pattern by adding Model for context information.

In ECAM pattern as shown in Fig. 6, TagReader reads the RF tag and generates a READ event. The event is transferred to the Controller in Control, and Context in Model. Context gathers context information from multiple sources and determines the current context of the system. Context in ECAM plays the role of context reducer in Fig. 5. Controller determines the service contents based on the Service Descriptor and the current context determined by Context. Then the Controller transfers the information about the service to ActionManager. ActionManager provides the service.

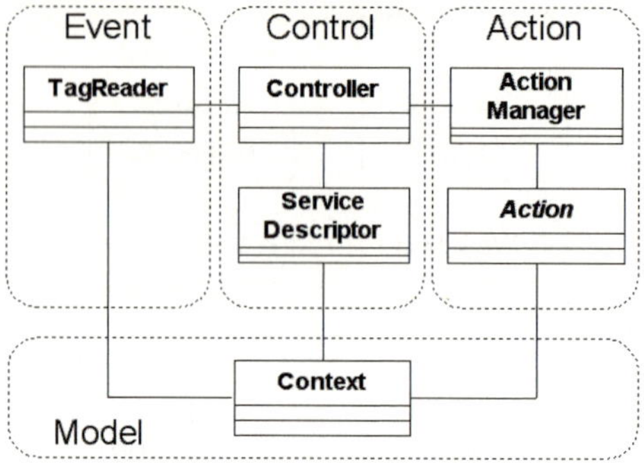

Fig. 6. ECAM Architecture

4 Prototype System

We implemented MyGuide prototype system on the Microsoft Window platform. We chose the Java language for software implementation, and MIDP [9] for mobile phone programming. We used 13.56 MHz RFID readers and card-type RF tags. We used GNU java serial communication library [10] to connect to the RFID readers. For multimedia service, we adopted Darwin Streaming Server from Apple [11]. We managed the context information and the guide data on the exhibits in Oracle 10g DBMS. Fig. 7 shows the architecture of the prototype system and its components.

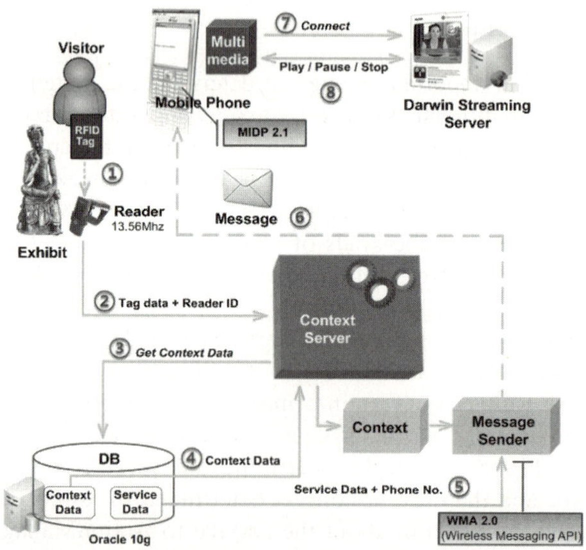

Fig. 7. Architecture of MyGuide Prototype System

MyGuide starts services on READ event from RFID reader. The visitor's profile and his/her mobile phone information are stored in DBMS. Fig. 8 shows the E-R diagram of the prototype system. Contents entity means the guide data for an exhibit. For an exhibit, there are several Contents records to provide visitors with suitable data. Visit entity means the trace data of a visitor's access to the exhibits.

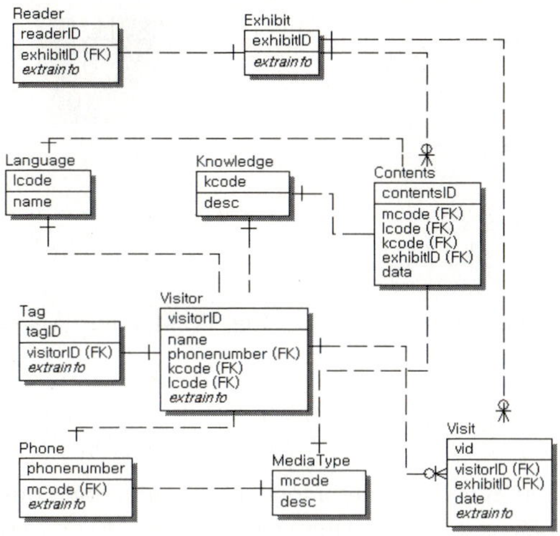

Fig. 8. E-R Diagram of MyGuide Prototype System

The class diagram of MyGuide is shown in Fig. 9. TagManager gets the values stored in the RF tag and the reader's identifier, and generates READ event. On the event, the Controller takes the control and coordinates the cooperation between CntxtReducer and MsgSender. CntxtReducer infers the current context from the values of contextual elements. It uses DbFacade for retrieving contextual data stored in DBMS. Controller determines the contents of service using the context, and asks MsgSender to generate Msg for the service and send it to the visitor. Msg consists of the explanation of the exhibit and the metadata.

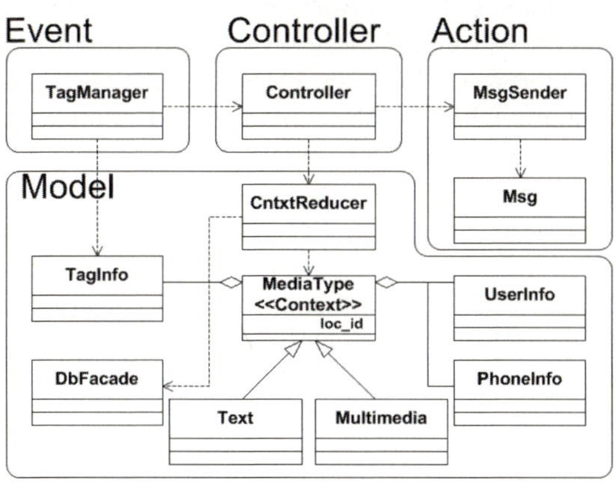

Fig. 9. Class Diagram of MyGuide Prototype System

Fig. 10 and Fig. 11 show the running examples of MyGuide prototype system. Fig. 10 shows the text service at beginning level in Korean. Fig. 11 shows the multimedia service at beginning level in Korean.

Fig. 10. Text Service

Fig. 11. Multimedia Service

5 Conclusions

Exhibitions and the museums actively adopt information and communication technologies to provide convenient services to the visitors in preparation for the ubiquitous computing era. However, until now existing systems haven't considered the visitor's background. Therefore, the guide services have not been truly visitor-oriented.

In this paper, we introduced a mobile guide system for exhibitions and museums. The most interesting feature of the system is context-awareness. It considers four kinds of context information: location, background knowledge, media type, and language. Based on the context information, the system provides the most suitable services to the visitor. For example, when a Korean visitor is an expert on the exhibit, and he/she has a mobile phone that supports multimedia, the system provides the multimedia type expert information in Korean.

We also introduced some interesting issues when developing context-aware systems: system requirements, context information, context modeling, software architecture, and its prototype system. We drew system requirements from stakeholders during the informal interview. From the requirements, we chose three basic technologies for the system: RFID, mobile phone, and context-awareness. We also found out the four types of context information. Because the context data are independent each other, there are too many possible contexts. Therefore, we model context from context information and determine two kinds of contexts. We chose ECAM architecture pattern for our system because it is event-driven and it can handle multiple kinds of context information. Our prototype system is implemented with 13.56MHz RFID reader, Darwin Streaming server, and MIDP APIs. The current version is running on the MIDP emulator, and we will port it to run on the mobile phone in the near future.

References

1. Consumer Electronics Show, http://www.cesweb.org/
2. Cheverst, K., et al.: Developing a Context-aware Electronic Tourist Guide: Some Issues and Experiences. CHI Letter 2(1), 17–24 (2000)
3. Hsi, S.: The Electronic Guidebook: A Study of User Experiences using Mobile Web Content in a Museum Setting. In: Proc. of the IEEE International Workshop on Wireless and Mobile Technologies in Education (2002)
4. Hsi, S., Fait, H.: RFID Enhances Visitors' Museum Experience at the Exploratorium. Comm. of the ACM 48(9), 60–65 (2005)
5. Korhonen, J., et al.: mTag - Architecture for Discovering Location Specific Mobile Web Services Using RFID and Its Evaluation with Two Case Studies. In: Proc. of AICT-ICIW, pp. 191–200 (2006)
6. Lee, E.: [Experience Mobile] PDA Video Guide System in National Central Musem, m-talk (December 2005), http://m-talk.inews24.com/
7. Kim, E.: Listing from mobile phone in Science Exhibition, mbn (November 14, 2007), http://news.mk.co.kr/
8. Choi, J., et al.: Context Model for RFID Systems. In: Proc. of 2nd International Conference on Ubiquitous Information Technologies and Applications (2007)
9. Mobile Information Device Profile (MIDP), http://java.sun.com/products/midp/
10. RXTX, http://www.rxtx.org/
11. Open Source Streaming Server, http://developer.apple.com/opensource/server/streaming/
12. Cheverst, K., et al.: Experiences of Developing and Deploying a Context-Aware Tourist Guide: The GUIDE Project. In: MOBICOM, pp. 20–31 (2000)
13. Dey, A.K.: Understanding and Using Context, Personal and Ubiquitous Computing. Special issue on Situated Interaction and Ubiquitous Computing (2001)
14. Choi, J., Park, J.-S.: Context Issues in RFID Systems. In: Proc. of ALPIT (2007)
15. Choi, J.: Context-Driven Requirements Analysis. In: Gervasi, O., Gavrilova, M.L. (eds.) ICCSA 2007, Part III. LNCS, vol. 4707, pp. 739–748. Springer, Heidelberg (2007)
16. Dockhorn Costa, P., Ferreira Pires, L., van Sinderen, M.: Architectural Patterns for Context-Aware Services Platforms. In: Proc. of the Second International Workshop on Ubiquitous Computing (2005)

Resource Management through Resource Virtualization in Distributed Network Environments

Jongbae Moon[1], Sangjoon Park[2], and Jongchan Lee[2]

[1] Korea Institute of Science and Technology Information,
52-11 Eoeun-dong, Yuseong-gu, Daejeon, 305-806, Korea,
[2] Department of Computer Information Science, Kunsan National University,
Jeollabuk-do, 573-701, Korea
jbmoon@kisti.re.kr, lubimia@hotmail.com, gederr@hanmail.net

Abstract. As distributed network has become a large-scale environment such as grid computing, software resource management is rising as the key issue. There are a lot of solutions to manage hardware and task resources in distributed computing environments. There also are many other types of resources such as software resources. However there are few solutions to manage these resources. In this paper, we propose to design, implement, and evaluate a new resource management system, which manages software resources effectively. This system uses application virtualization technology to manage software resources. This system greatly reduces traditional software deployment costs and associated installation problems in distributed computing environments. We conducted an experiment to show how the system works well. In this system, an added node can also execute a requested job without installation of applications.

1 Introduction

As computing technology advances, computing systems, which are geographically dispersed, are interconnected through networks and cooperate to achieve intended tasks. Moreover, not only system architecture but also software is more complicated. Such computing systems, which are called distributed systems, are composed of various types of resources and provide high throughput computing and high reliability. As distributed computing has evolved from a simple cluster to complicated grid computing, providing a reliable and efficient distributed computing environment largely depends on the effective management of these resources[1].

Resource management has always been a critical aspect, even in centralized computing environments and operating systems[2], [3]. Managing resources is much simpler in centralized systems than in distributed systems, since the resources are confined to a single location, and in general the operating system has full control of them. In distributed computing systems, however, these resources are scattered throughout the distributed computing environment and no single entity has full control of these resources. Thus, the management of resources in a distributed computing environment is inherently more difficult.

Various types of resources, which include both hardware resources and software resources, exist in distributed computing systems. Many researches focus on hardware

O. Gervasi et al. (Eds.): ICCSA 2008, Part II, LNCS 5073, pp. 360–370, 2008.

resource management[4], [5], [6], [7]. For example, some researches focus on monitoring work nodes, selecting available work nodes, allocating jobs to available work nodes, and balancing loads between work nodes. In this paper, we focus only on software resource management, because installing various kinds of software in many work nodes, keeping software up to date, monitoring software status, and performing routine maintenance require more efforts.

Recently, on-demand software streaming has been used to make managing PC Labs easy[8]. On-demand software streaming technology streams an application code itself to a client computer, which then executes it natively. On-demand software streaming provides the benefits of server-based computing. Server-based computing offers potential of reducing the total cost of computational service through reduced system management cost and better utilization of shared hardware resources[9]. Currently, only a few companies including [10], [11], [12], [13], and [14] are known to possess software streaming technology. On the other hand, software streaming is already used to manage software resources in many offices and school PC Labs, and its demand is increasing.

However, distributed systems do not support controlling streamed software resource. In this paper, we propose a new resource management system that can effectively manage software resources including streamed software in a distributed system. The proposed system adapts software streaming technology to a distributed system. We implemented a system including a job broker and a job management module. The proposed system greatly reduces traditional software deployment costs and associated installation problems.

The rest of this paper is organized as follows. Section 2 describes requirements for adapting software streaming to a cluster. Section 3 proposes a system model. Section 4 describes how to implement a system. Section 5 shows an experiment to test if the system works well. Section 6 summarizes the existing research regarding software streaming technologies, and Section 7 concludes the paper.

2 Requirements for Adapting Software Streaming to a Distributed System

In a distributed system, a resource broker provides mechanisms to effectively control job and system resources, but does not provide mechanisms to control software streaming. Therefore, a middleware for software streaming is necessary to control streamed software in a distributed system. Moreover, in order for software streaming to work well on a distributed system, the following requirements should be considered.

- Heterogeneity: as a distributed and cluster system could be composed of heterogeneous work nodes, we should implement and provide a common interface for streaming packages running in the heterogeneous system.
- Transparency: users would not necessarily to know if any applications needed are installed in work nodes or not. When applications are not installed, if we stream all applications to the client, it would waste time and money. Also, when preinstalled legacy application is exist, streaming application is not needed.
- Workflow: as requested jobs could be composed of several contiguous steps, workflow should support streaming tasks.

3 Proposed Resource Management System Model

To satisfy the requirements as described above, we propose a new resource management system that effectively manages software resources including streamed software in distributed environments. The system model is illustrated in Fig. 1, and consists of a job broker, a job management module, and software streaming package. In this system, users connect to the job broker and check the status of work nodes and then request jobs.

A distributed system can not control streamed jobs. To distribute streamed software on a distributed environment, a job broker, which is a middleware, is a necessity. The job broker schedules user requests, and then assigns jobs to a work node. Moreover, in this system, a batch job can be performed by using workflow service.

A job management module is necessary to execute both conventionally-installed applications and streamed software in a cluster system. The job management module, which is an intermediate between the job broker and the streaming client, executes streamed software without modification of the existing software streaming package. The job management module creates a process for streamed software and then executes it natively.

A streaming server makes an image by converting application into stream. Users do not need to consider system dependable information. A streaming client unpacks the image into the virtual file system, where the program is then run. Only a small part is needed at a time as it executes the application.

4 Implementation

We have implemented a system by using Z!Stream that was developed for Linux by SoftOnNet, Inc., and was implemented in C and PHP language. Also we used MSF (Meta Scheduling Framework)[15] system to build workflow. In this paper, we implemented a broker service and a job management module in Java language to support multiple platforms.

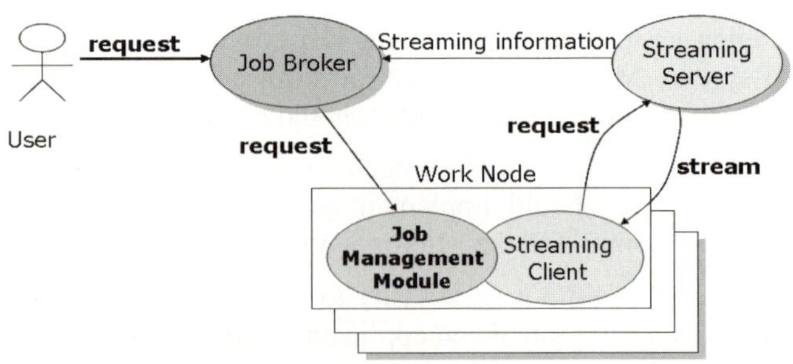

Fig. 1. A system model for adapting software streaming technology to cluster system

4.1 System Architecture

The architecture of the proposed system is shown in Fig. 2. The system consists of several modules and Z!Stream package. A job broker collects work node status information and requests jobs to work nodes using workflow. The job broker is composed of several modules, which are a workflow interpreter, a streaming map, work node monitor, and a job submission module. A job management module resides at a work node with Z!Stream client and executes requested jobs. The job management module consists of a status detector module, software detect module, and a job execution module. If the requested application does not exist, the job execution module requests the application to Z!Stream server. After Z!Stream server streams the application code to a work node, then the work node executes it natively.

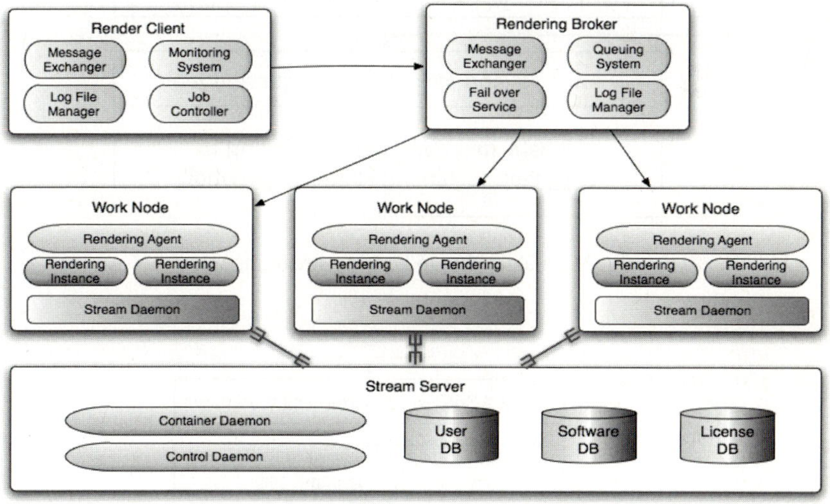

Fig. 2. Architecture of the proposed system

4.2 Implementation of Job Broker

A job broker consists of a streaming map, a workflow interpreter, a work node monitor, and a job submission module. The streaming map is a list of streamed software that a work node can stream and run. Fig. 3 shows an algorithm to keep a streaming map. The work node monitor collects status information of work nodes. The status information includes running job's status and system load information. The job status is one of the following: WAIT, RUNNING, and DONE. The workflow interpreter builds a workflow according to user requests, and selects work nodes to request jobs. The requested jobs may consist of one job step, but most of them are generally batch jobs which can consist of several job steps. To run batch jobs, workflow can be used. The workflow interpreter performs an algorithm described in Fig. 4, and selects nodes to run workflow jobs.

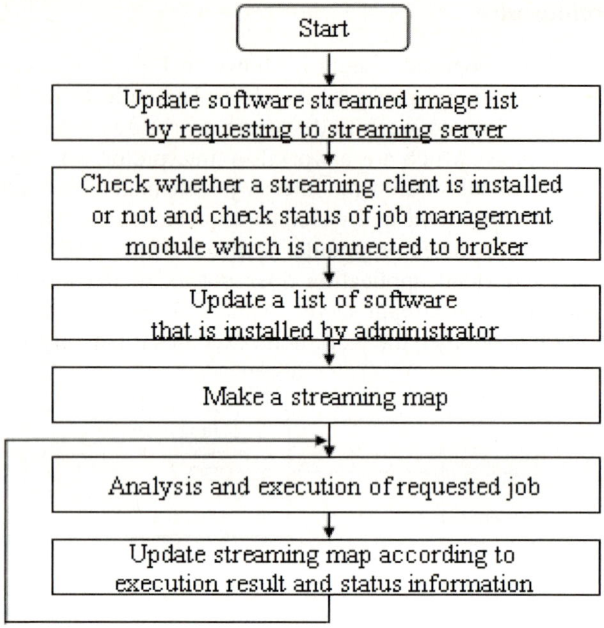

Fig. 3. An algorithm to keeping streaming map

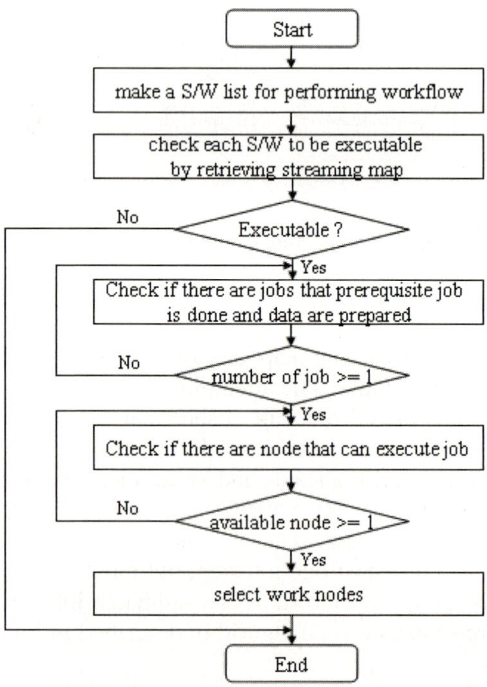

Fig. 4. Flowchart of the work node selection algorithm

XML schema and document for workflow are shown in Fig. 5, and XML schema and document for task supporting streamed software are shown in Fig. 6.

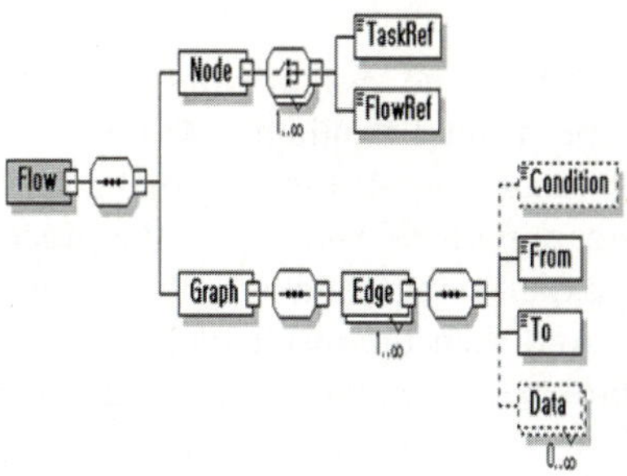


```xml
<?xml version="1.0" encoding="UTF-8"?>
<Flow xmlns:xsi="http://www.w3.org/2001/XMLSchema-instance"
xsi:noNamespaceSchemaLocation="msfflow.xsd">
    <Node>
        <TaskRef id="t1">seventy9@SSU:MSF_Task:blaststream</TaskRef>
        <TaskRef id="t2">seventy9@SSU:MSF_Task:modstream</TaskRef>
        <TaskRef id="t3">seventy9@SSU:MSF_Task:pharmomapstream</TaskRef>
        <TaskRef id="t3">seventy9@SSU:MSF_Task:pharmoscanstream</TaskRef>
    </Node>
    <Graph>
        <Edge>
            <From>t1</From>
            <To>t2</To>
        </Edge>
        <Edge>
            <From>t2</From>
            <To>t3</To>
        </Edge>
        <Edge>
            <From>t3</From>
            <To>t4</To>
        </Edge>
    </Graph>
</Flow>
```

Fig. 5. XML schema and document for workflow

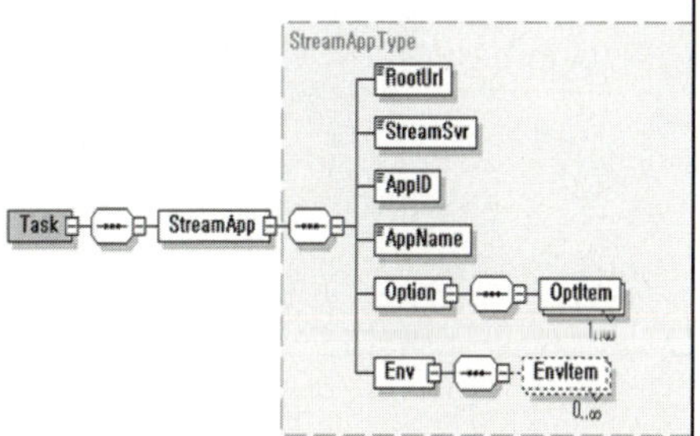


```xml
<?xml version="1.0" encoding="UTF-8"?>
<!--Sample XML file generated by XML Spy v4.3 U (http://www.xmlspy.com)-->
<Task xmlns:xsi="http://www.w3.org/2001/XMLSchema-instance"
xsi:noNamespaceSchemaLocation="C:₩msftask.xsd">
    <StreamApp>
        <RootUrl>http://203.253.23.56</RootUrl>
        <StreamSvr>203.253.23.56</StreamSvr>
        <AppID>mod2</AppID>
        <AppName>mod2</AppName>
        <Option>
            <OptItem name="inputfile" type="input" switch="" value="/moddata/1fdx_1durA.py"/>
        </Option>
        <Env>
            <EnvItem name="KEY_MODELLER8v1" value="MODELIRANJE"/>
            <EnvItem name="MODINSTALL8v1" value="/modeller"/>
            <EnvItem name="PYTHONHOME" value="/usr/lib/python2.3"/>
        </Env>
    </StreamApp>
</Task>
```

Fig. 6. XML schema and document for a task supporting streamed software

4.3 Implementation of Job Management Module

The job management module consists of status detector, software detector, and job execution module. The status detector collects conventionally-installed software list and process information, which the job execution module is performing. Then, the status detector forwards the list and information to a job broker. The job execution module uses a job queue. The OS control module controls conventionally-installed applications by calling system calls. The streaming job control module controls streamed jobs by calling interface, which the streaming client provides.

5 Experiment

In this section, we conducted an experiment to test how the system works well. We used four identically configured machines that have an Intel Pentium 4 3.0GHz CPU with 1GB Ram and 80GB E-IDE and run Linux Fedora Core 4 (Linux kernel 2.6.11). We used an application for 3D rendering as a streaming S/W because of the processor-intensive nature of 3D job. We conducted an experiment that a 3D job is submitted and the job is executed in distributed nodes. And then the result files which are resided in each node are sent to the broker. The broker gathers the result file and generates an animation file. After that a client can see an animation as a result.

Fig. 7 shows a client application which is responsible for connecting broker server.

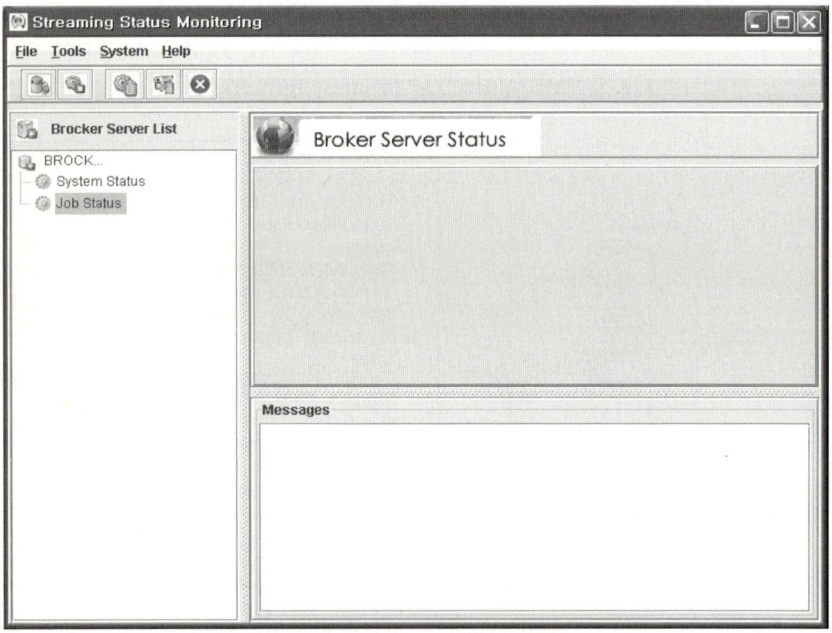

Fig. 7. GUI Interface for Clients

The broker is implemented in Java language, so the scalability is provided easily.

When a client submit a job, an input file and options are required. Fig. 8 shows a simple GUI for job submission.

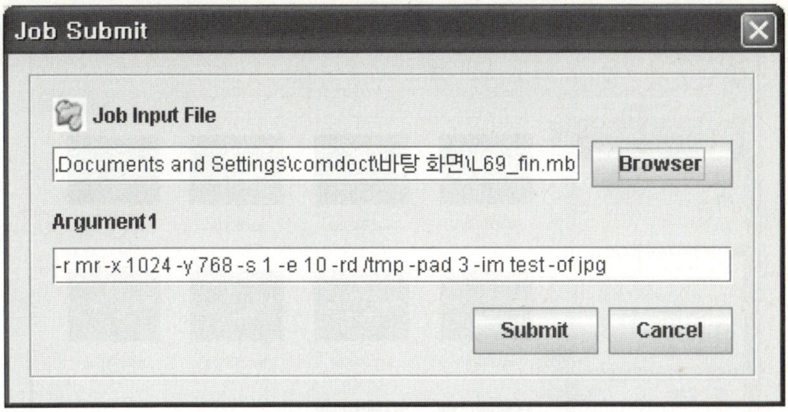

Fig. 8. Parameters for job submission

After a client submitted a job, systems and jobs are monitored in real time. The broker gathers all information about system information and job information. When a client access to the broker, a client can show CPU Load, Memory, and Job's status. This is shown in Fig. 9.

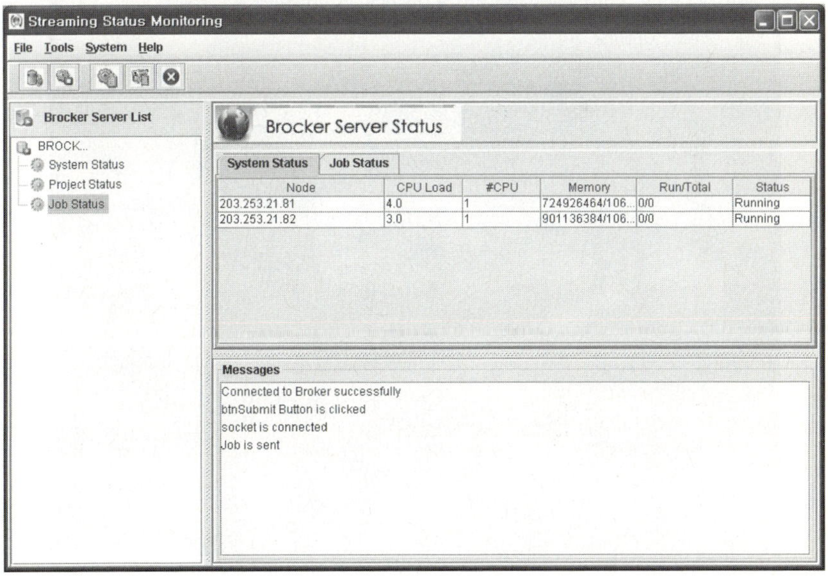

Fig. 9. Systems and jobs monitoring Interface

When a job is send to the broker, the job is assigned into many nodes in distributed network environments. Some of the nodes may have the application for the job, but some of these may not have. The nodes which don't have the application request to Streaming server and receive an appropriate application to execute the job.

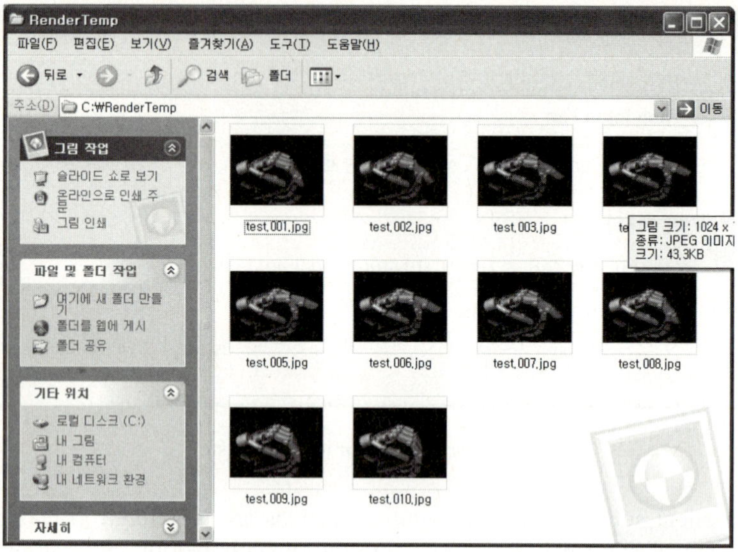

Fig. 10. Result files received from the broker

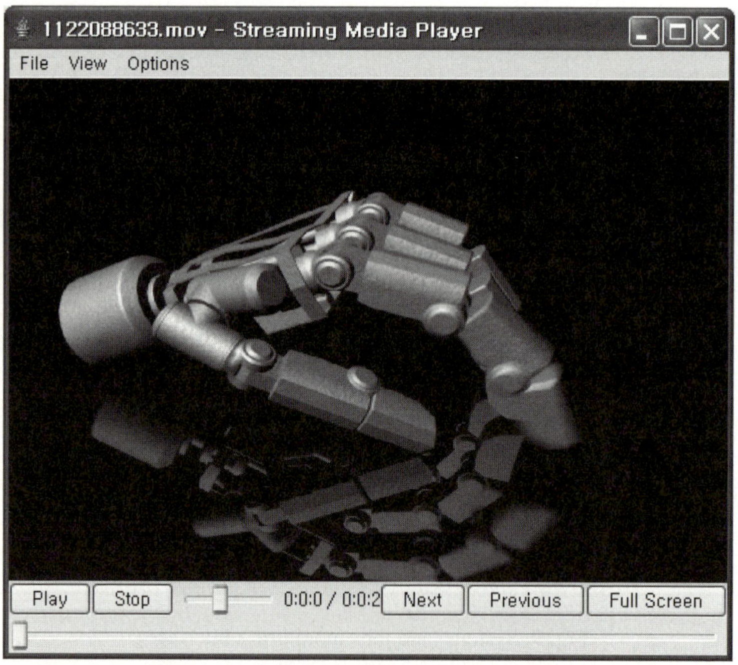

Fig. 11. Animation movie file made by the broker

Fig. 10 shows the result files and Fig. 11 shows the animation movie file which is generated by the broker. A client can see an animation movie as a result.

6 Related Works

On-demand software streaming technology executes an application with only a part of its image in a client; the client downloads only a part of the application from a streaming server through the network, and then executes the application. There are several companies that are known to possess on-demand software streaming technology.

AppStream platform[10] provides streaming applications that is executed by clients, and provides on-demand service including centralized management and control. AppStream's patented streaming technology segments applications' executable code and data into small parts called streamlets and sends the streamlets.

Softricty[11] provides SystemGuard which is a virtual run-time environment within which an application executes. SystemGuard maintains the integrity and reliability of the operating system by shielding it from change that is normally created by the application as it is installed and run. However, since the applications execute locally, access is still available to all local services including cut and paste, OLE, printing and network drives.

Stream Theory[12] provides software-streaming platforms which allow users to interact with fully functional Windows-based applications without installation on the desktop, making the service ubiquitous and transparent to the user. A client desktop software-on-demand capability provides a cost effective and user-friendly technology to access applications, accelerate software deployment and manage software resources.

Exent's EXEtender solution[13] offers cutting-edge technology with unparalleled performance to deliver applications via the Internet or enterprise network. Software applications are securely streamed to authorized end users. EXEtender enables the integration of Web-based authentication, authorization, and user profile management systems. EXEtender has a built-in security algorithm to ensure that all application software titles remain copy-protected.

SoftOnNet's Z!Stream[14] uses patent-pending Distributed Virtual Paging technology, and Z!Stream enables the most efficient delivery of any Windows software by sending the fewest needed executables to client computers. The software runs on the client side, freeing server space and dramatically reducing network demand compared to terminal emulation.

Most of solutions described above support only PC environments and only Windows OS. SoftOnNet, Inc., however, provides Z!Stream solution for Linux version.

7 Conclusions and Future Works

In this paper, we proposed a new resource management system which manages software resources effectively. The proposed system adapts on-demand software streaming technology to existing distributed systems. We implemented a broker and a job management module. The broker receives user requests and builds workflow if necessary, and then selects work nodes and requests jobs to work nodes. The job management module runs requested jobs and control the jobs including streamed jobs. This system can manage software resources effectively by using software streaming. Also, this system greatly reduces traditional software deployment costs and associated installation problems. A new added work node is not necessary to install any applications to run a job. Moreover, the system can perform workflow tasks.

There are some problems remaining in this system which require further research. The proposed system does not support batch systems such as PBS. Also, a job management module can not control completely streamed software because it gets only limited information from the streaming client. In future research, we shall take batch systems into consideration, and provide improved methods for controlling streamed software.

References

1. Goscinski, A., Brearman, M.: Resource Management in Large Distributed Systems. ACM SIGOPS Operating Systems Review 24, 7–25 (1990)
2. Tanenbaum, A.S.: Modern Operating Systems. Prentice Hall, Englewood Cliffs (1992)
3. Banga, G., Druschel, P., Mogul, J.C.: Resource containers: A new facility for resource management in server system. In: Proceedings of the 3rd USENIX Symposium on Operating Systems Design and Implementation (ODSI), New Orleans, LA (1999)
4. Eager, D.L., Lazowska, E.D., Zahorjan, J.: Adaptive Load Sharing in Homogeneous Distributed Systems. IEEE Trans. on Software Eng. SE-12(5) (May)
5. Fogg, I.S.: Distributed Scheduling Algorithms: A Survey, Technical Report No. 81, University of Queensland, Australia
6. Goscinski, A.: Distributed Operating Systems. The Logical Design. Addison-Wesley, Reading (1989)
7. Wang, Y.-T., Morris, R.J.T.: Load Sharing in Distributed Systems. IEEE Trans. On Computers C-34(3) (March)
8. AppStream Inc.: Solving the unique PC management challenges IT faces in K-12 schools, 2300. Geng Road, Suite 100, Palo Alto, CA 94303
9. Li, F., Nieh, J.: Optimal Linear Interpolation Coding for Server-based Computing. In: Proceedings of the IEEE International Conference on Communications (2002)
10. AppStream, http://www.appstream.com/
11. Softricty, Inc., http://www.softricity.com/
12. Stream Theory, http://www.streamtheory.com/
13. Exent Technologies, http://www.exent.com/
14. SoftOnNet, Inc., http://www.softonnet.com/

A Cost-Effective Mobile Video Streaming with an Adaptive Control of Inter-packet Delays

Youn-Sik Hong[1] and Jun-Sik Hong[2]

[1] Department of Computer Science and Eng., Univ. Of Incheon, 177 Dowha-dong Nam-gu 402-749 Incheon, Korea
yshong@incheon.ac.kr
[2] Department of Electrical and Electronics Eng., 36 Naedok-dong Sangdang-gu 360-764 Cheongju, Korea
jnskhong@cju.ac.kr

Abstract. In a recent VoD application, a stationary VoD server delivers video stream to a mobile device, particularly PDA, over wireless. Typically, the sending rate at the server is faster than the playback rate at the handheld device. Therefore, it can't receive streaming video as possible as the sending rate even if there is no congestion and extremely low BER over wireless. Our objective is to develop a simple cost effective rate control scheme using a buffer level at a mobile device. We classify the buffer at a mobile device into three distinct states depending on its level (occupancy). For each of these states a different rate control is to be applied.

Keywords: TFRC, video stream, minimum fill level, inter-packet delay.

1 Introduction

Delivering streaming video over wireless to mobile users is becoming a research topic of growing interests. With the emergence of small wireless handheld devices, it is expected that streaming video will be a major source of traffic to these mobile devices.

In a typical VoD (Video on Demand) application, a mobile device, particularly PDA (Personal Digital Assistant), is mainly used for receiving video streaming from its stationary server. A mobile device as a receiver needs some buffer space to smooth out fluctuations in the data receiving rate from such a network. At the start of a session or after a fast forward, the buffer needs to be filled up first before the video can be viewed. This results in a startup delay which depends on the buffer size at a mobile devices and the bandwidth available from the network. The buffer needs to be proper dimensioned so that the video display is continuous without underflow at the client buffer due to a reduced bandwidth available when the network is congested. Moreover, over-dimensioning the buffer results in an expensive client system and a large startup delay.

To sustain continuous video display at the mobile device, continuous monitoring of the network status is required to prevent transmission collapse. To address the concerns, real-time streaming needs more intelligent rate adaptation or rate control

O. Gervasi et al. (Eds.): ICCSA 2008, Part II, LNCS 5073, pp. 371–380, 2008.

mechanism. A widely popular rate control scheme for streaming in wired networks is equation-based rate control, also known as TCP-friendly rate control (TFRC) [1]. In TFRC the TCP-friendly rate is determined as a function of packet loss rate, round-trip time (RTT), and packet size, to mimic the long-term steady performance of TCP. One of the advantages to rate control using TFRC results in small rate fluctuation, making it attractive for streaming applications that require constant video quality.

A number of previous efforts improves the performance of TFRC over wireless but also brings up two concerns. First, mobile devices are often small and inexpensive with limited computation and buffer capacities. When a mobile client receives video stream from its stationary server the server and a mobile client act as a fast sender and a slow receiver, respectively, due to substantial differences in their computational capabilities. Therefore, even if there is no congestion and extremely low BER (bit error rate) over wireless, the mobile client can't receive streaming video as possible as the sending rate. The second concern is that a status of a mobile device can be determined as the level (occupancy) of the buffer. For example, if the buffer level is less than minimum fill level, the device stays in the state *underflow* which ensures that sending rate are too low. If the buffer level is quite larger than some specified value, it enters a state *overflow*. Then it prevents buffer overflow by ensuring that sending rates are not too aggressive.

Many of the works does not reflect the buffer status of a receiver. They have considered the network status only. Thus, our scheme is to reformulate the rate in the TFRC by considering the buffer status of a PDA. The advantages of our approaches are as follows: First, it does not require any modifications to network protocols, except at the application layer. To do this, inter-packet delay at the sender can be adjusted to control the sending rate. Second, it has the potential to fully consider processing capabilities of a mobile device. Our objective is to develop a simple cost effective scheme which offers an acceptable video quality over a noisy wireless channel using a limited buffer size of a PDA.

This paper consists of the following; we discuss the related works in Chapter 2. In Chapter 3, we present our proposed methods to improve performance. The experimental results are shown in Chapter 4. Finally, we conclude our works in Chapter 5.

2 Related Works

In this section, we briefly summarize some of the related works. *Padhye* et al [3] proposed the following model of the steady state throughput for a TCP flow, as a function of packet loss rate and round trip time.

$$T = \frac{S}{rtt\sqrt{\frac{2p}{3}} + rto(3\sqrt{\frac{3p}{8}})p(1+32p^2)} \tag{1}$$

where T represents the sending rate, S is the packet size, rtt is the end-to-end RTT (round-trip time), rto is the retransmission time-out (RTO), and p is the end-to-end packet loss rate. One of the most popular end-to-end streaming protocol is presented in [1]. It works well for wired network. But, in presence of wireless errors, it shows a degraded performance. Chen et al [2] uses a more simple model for TFRC as shown in Eq. 2, while it gives all the fundamental factors that affect the sending rate.

$$T = \frac{kS}{rtt\sqrt{p}} \tag{2}$$

where k is a constant factor. Chen et al[2] proposed multiple TFRC connections to fully utilize wireless channel.

Sisalem et al [4] proposed LDA (Loss Delay-based Adaptation) which is a video streaming scheme based on RTP/RTCP [8]. LDA uses RR (Receiver Report) message in RTP/RTCP to extract the network parameters such as RTT, packet loss, and RTO. Then LDA recalculates the sending rate by using Eq.1. SRTP [5] is similar to LDA. It attempts to smooth out the variation of the sending rate with network fluctuations. A typical problem of the above equation-based rate control approaches is a underutilization of the wireless channel. For example, the calculated sending rate is about 60% of the measured sending rate.

Zheng et al [7] derives a minimum handheld device buffer size. The minimum buffer size is proportional to the average system retransmission time. As the channel BER increases, the minimum buffer size increases abruptly in order to maintain a continuous display at the mobile device.

Some of the works focused on the issues to distinguish between packet loss caused by congestion and that caused by wireless channel error. *Cen* et al [8] proposed and evaluated several differentiation algorithms. *Yang* et al [7] proposed a video transport protocol (VTP) based on achieved rate estimation and loss discrimination algorithm.

3 A Buffer Controlled Adaptive Video Streaming

3.1 A Buffer Control Scheme

Basically, the client buffer must be filled to a minimum fill level before the client starts displaying. In other words, the amount of data retrieved for a stream is at least equal to the amount consumed by the stream's playback. The buffer level $B(t)$ at time t is computed the following equation:

$$B(t) = B(t-1) + T(t,t-1) - C(t,t-1) \tag{3}$$

where $B(t\text{-}1)$ represents buffered data at time t-1, $T(t,t\text{-}1)$ is the sending rate at the time interval $[t, t\text{-}1]$, and the $C(t, t\text{-}1)$ is the amount consumed the stream's playback at the time interval $[t, t\text{-}1]$. Thus, the buffer level varies with time.

Consider the illustration in Fig. 1. Assume that the sending rate is 1, 2, 2, 2 … and the amount consumed the stream's playback is 2 (constant). For this example, the minimum fill level and the overflow level is 2 and 7, respectively. Notice that every number represents the number of packets. At initial, the buffer has only one which is less than the minimum fill level. The client does not playing a video. At next step, after received 2 packets, 2 packets is consumed for stream's playback and 1 remains in the buffer. At the fourth step, the buffered data is 7 which are equal to the overflow level. So the sending rate must be decreased at some specified amount without considering the network status.

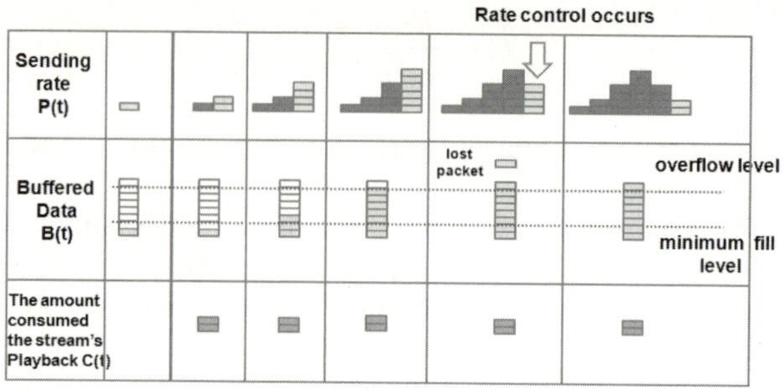

Fig. 1. An illustration of the proposed buffer controlled scheme

With our experimental results we can estimate the buffer level in Eq. 3. For simplicity, assume that the packet size *S* is constant. With the buffer size of 16,384 bytes, the measured data consumption rate per second is 1.56 *S*, provided the playback rate is also constant. The sending rate per second is approximately 10.7 *S*. Thus, Eq. 3 becomes Eq. 4.

$$B(t) \approx B_{\min} + 10.7 \times t \times S - 1.56 \times t \times S = B_{\min} + 9.14 \times t \times S \tag{4}$$

where $B_{\min}$ represents the minimum fill level of the buffer (*fixed*) and the unit of the time *t* is *second*. Notice that these empirical data has been obtained using the mobile device, PDA (with 400MHz CPU), in a wireless LAN based on IEEE 802.11b. Since the sending rate is faster than the data consumption rate, the buffer level increases rapidly with time.

The minimum buffer level required at the client to prevent underflow can be approximately expressed by Eq. 5.

$$B_{\min} \propto rtt \times C^{\max} \tag{5}$$

where $C^{\max} = \max_{0 \le i \le N}(C(i,i-1))$, *N* is the length of the video. To contrast with, buffers in routers are designed to be larger than the bandwidth-delay product [10].

If there is no user interaction and then the playback rate is be kept constant. In addition, the minimum fill level is pre-determined by Eq. 5 and remains constant. In that case, the buffer level can be affected by the sending rate. On the contrary, the sending rate can be controlled by the client buffer.

$$B(t) \propto T(t,t-1) \Leftrightarrow T(t,t-1) \propto B(t) \tag{6}$$

3.2 A Classification of the Buffer States

A buffer can be in one of the three states: *underflow*, *stable* (*normal*), or *overflow*. Since the buffer requirement for the client is related to its state, it is necessary to determine the stationary state probability of the client.

If the data production rate T (t-1) never lags the consumption rate C (t-1), there is never a net decrease in the amount of buffered data. Eq. 7 gives a constraint to prevent starvation at the mobile device (resulting in frozen display).

$$C(t) \le T(t), 0 \le t \le N \tag{7}$$

If this constraint is not satisfied, the status of the buffer remains the state *underflow*. However, in a typical situation, data production rate at the stationary server is higher than the consumption rate in the mobile device as shown in Eq. 4.

The maximum size of the buffer can be expressed as

$$T(t) - C(t) \le B_{max}, 0 \le t \le N \tag{8}$$

where B_{max} is the maximum buffer size. If this constraint is not satisfied, the status of the buffer remains the state *overflow*. That means some packets may be lost or dropped. The larger the buffer size, the longer the consumption rate. Finally, if the buffer level is greater than the minimum fill level and also is less than the overflow level, the status of the buffer remains the state *stable* (or *normal*).

We can summarize the above three states for the client buffer in Eq. 9. Note that α will be determined by empirically. Naturally, it can be proportional to the current sending rate.

$$\begin{aligned}
&\text{Underflow state: } 0 \le B(t) < B_{min} \\
&\text{Stable state: } B_{min} \le B(t) < B_{max} - \alpha \\
&\text{Overflow state: } B_{max} - \alpha < B(t) \le B_{max}
\end{aligned} \tag{9}$$

3.3 The Rate Control at the Application Layer

In an application layer, an inter packet delay (IPD) refers to the time interval between the transmission of two successive packets by any host. In general, the sending rate T increases as IPD decreases. In our test, we empirically chose IPD in the range of between 1ms and 20ms. Theoretically, IPD is chosen judiciously by considering network parameters, such as *rtt*, the size of the receive buffer, and the capacity of the link, to avoid the degradation of the overall performance. The following equation shows the resulting rate control scheme depending on the buffer state.

$$\textbf{Underflow state}: T(t) = \gamma \times T(t-1), \ where, 1.5 \le \gamma \le 2 \tag{10}$$

where k is used to adjust the resulting *IPD* (t) whose value falls into the range of $1ms \le IPD(t) \le 20ms$. Eq. 11 is applied to each of the following two states to find the IPD after computing the appropriate T (t). At the stable state, the conventional additive increase is applied. Δrtt is the *rtt increases* after a round [7].

$$\textbf{Stable state}: T(t) = \frac{T(t-1) \times rtt + 1}{rtt + \Delta rtt} \tag{11}$$

A parameter γ_2 between 0 and 1 is selected as the tolerable rate reduction ratio. When the overflow state is entered, the sending ratio should be reduced as

$$\textbf{Overflow state: } T(t) = \gamma_2 \times T(t-1), \text{ where } 0 < \gamma_2 < 1 \tag{12}$$

4 Performance Evaluations

4.1 Experiment Setup

We performed the two different experiments. One type of the experiments is carried out with the ns-2 simulator [11] to analysis the efficiency of the proposed rate control in different error rate conditions. The topology representing a mixed wired-cum-wireless network is used throughout this paper.

Another type of experiments is carried out on a test-bed which reproduces (on a small scale) a real prototype of the topology in Fig. 2. In this experiment, we trace the sending rate to see whether it reacts to buffer level changes quickly.

4.2 The Estimated Minimum Buffer Size and Maximum Buffer Size

First we measured *rtt* (in average) under different error rates through 10 runs to estimate the minimum buffer level and the maximum buffer level by using Eq. 5 and Eq. 8. These estimated parameters have been used to determine the adjustable constant α in Eq. 9 to classify the buffer levels.

In Fig. 2, the congestion window decreases as the error rate increases. That means the sending rate is lowest for the highest error rates. The calculated sending rates with respect to the error rates and the measured *rtt* (in average) by using Eq. 2 are summarized in Table 1. Notice that the packet size *S* is kept to be constant.

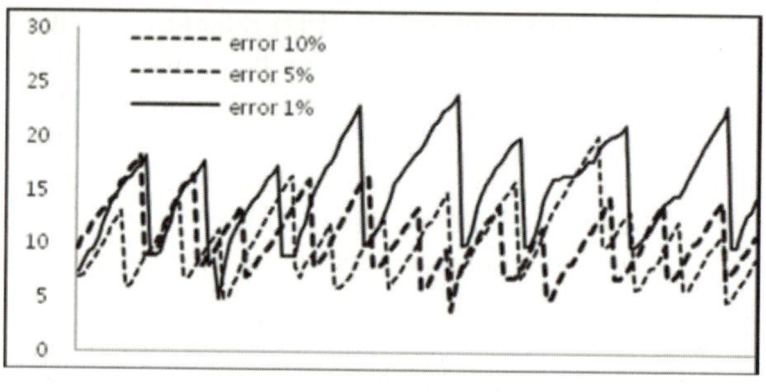

Fig. 2. The congestion window under different control error rates

Table 1. The calulated sending rate under different random error rate

Error rate	5%	10%	15%	20%
rtt (ms)	21.4	22.05	23.75	29.15
sending rate	313.47 x *S*	215.1209 x *S*	163.073 x *S*	115.0635 x *S*

The simulation results show that *rtt* has little changed within the error rate of less than 10%. In the absence of random errors, typical *rtt* is measured 19ms. Assume that the error rate is less than 10% and the data consumption rate is to be kept constant, $B_{\min}$ depends on the consumption rate; $B_{\min} > rtt \times C^{\max} \approx rtt \times C_{const} \approx 20ms \times C_{const}$

Our empirical results give us the following intuition. $\frac{C_{masured}}{T_{measured}} = \frac{1.56 \times S}{10.7 \times S} = 0.15 < 0.2$

Thus, the estimated lower bound on the buffer level is about to 20% of the fixed buffer size.

In a PDA, the socket buffer size is limited to 32,768 bytes. From the analysis of empirical results, the possible range of the constant α in Eq. 9 is 10~15% of the $B_{\max}$.

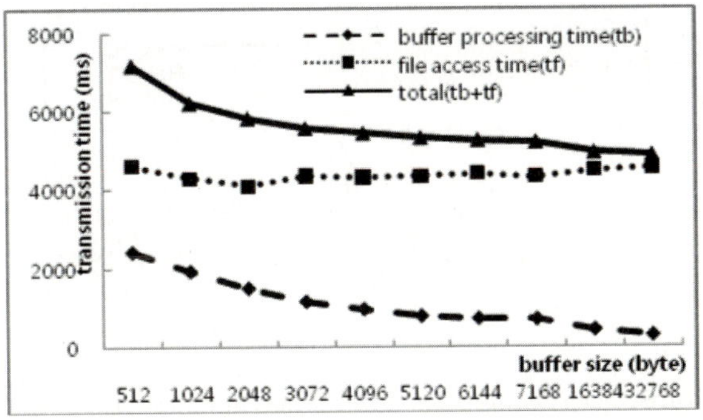

Fig. 3. The transmission time with respect to the buffer size

Because a larger buffer size results in a shorter processing time as shown in Fig. 3, the parameter α becomes smaller with a larger buffer size. That means the data consumption rate also increases as the buffer size increases.

4.3 The Rate Control Using the Inter-packet Delays

We test the rate variation to IPD changes under different packet sizes. This test is indicative of the agility of our proposed method and also of its efficiency. From the analysis of

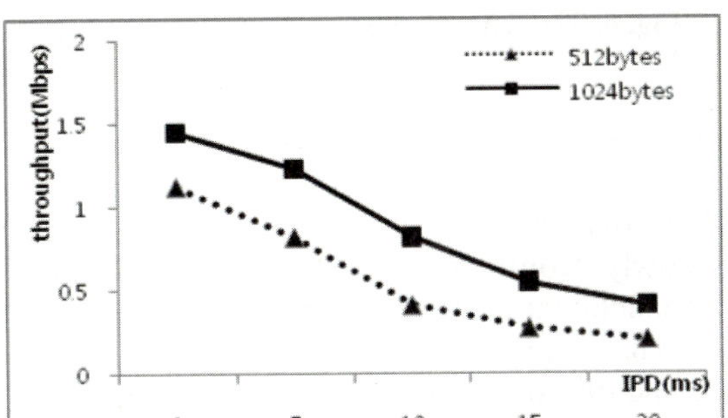

Fig. 4. The transmission time with respect to IPD

these empirical results the possible range of the IPD will be 1ms < IPD < 20ms. Notice that with smaller packet size the rate variation to IPD changes quickly. By setting the IPD to 20ms, then the rate will be drastically decreased to 24% of that of 1ms.

4.4 The Adaptive Rate Control and Its Effectiveness

We conducted simulations varying each of the parameters α, B_{min} , and B_{max} . Fig. 5 plots the rate with different B_{min} . B_{min} (20) indicates the 20% of the full buffer size. Fig. 6 plots the rate with different α. B_{max} (80) indicates the 80% of the full buffer size. That is, α is set to 20%. In both tests, the buffer size of 16,384 bytes was used.

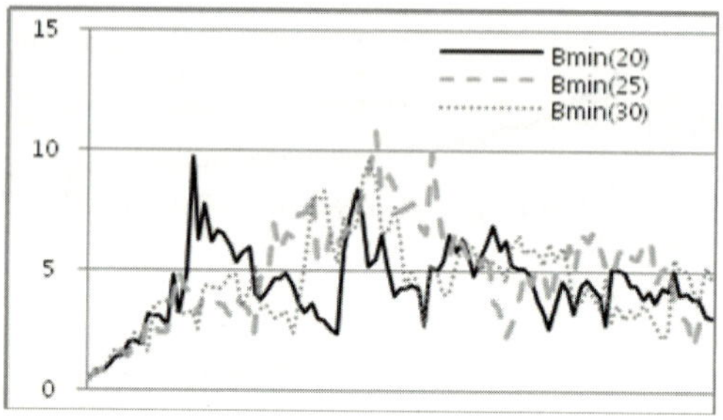

Fig. 5. The sending rates under different minimum fill level

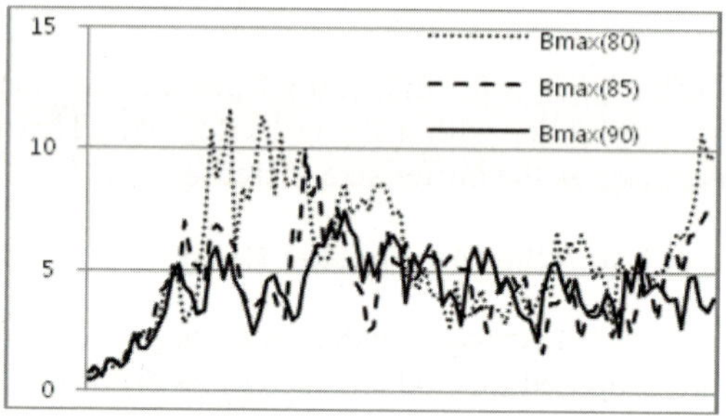

Fig. 6. The sending rate under different α in Eq.9

The standard deviation of B_{min} (20) and B_{max} (90) (i.e., α=10%) are the smallest. Thus, we continue to conduct simulations with these parameters. Fig.7 shows the rate with buffer sizes. With smaller buffer size it maintains a smoother rate and reacts to bandwidth changes slowly. However, it results in under-utilization of a wireless channel. We test the rate adaptation of the proposed method to bandwidth changes under different buffer size. The graph in Fig.8 presents the instantaneous sending rates of the proposed method and the conventional TFRC [1].

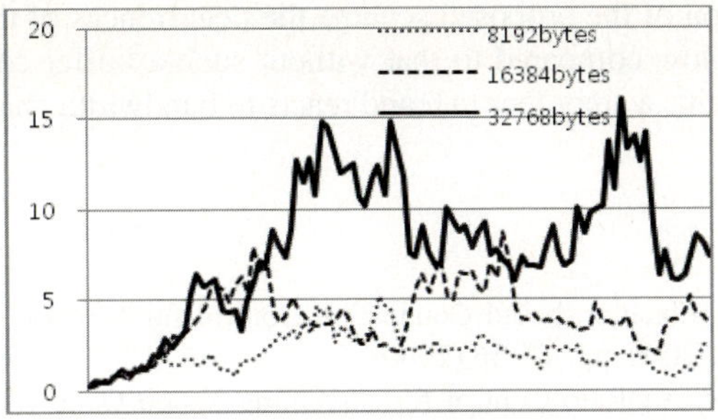

Fig. 7. The sending rate under different buffer size

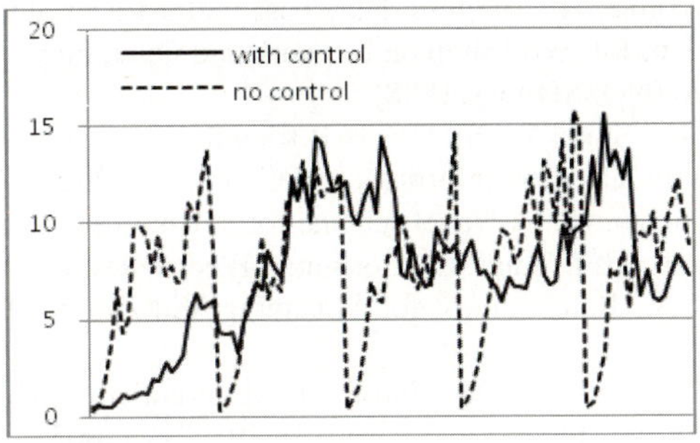

Fig. 8. The comparison result (with the buffer size of 32,768bytes)

With the adaptation of the proposed buffer control scheme the occurrences of entering the overflow state is relatively low compared to that without such a buffer control. In addition, since the sending rate is faster than the data consumption rate, the packet loss is occurred during the *overflow* state. With the overflowed socket buffer for a mobile device, it reacts to bandwidth changes slowly and then result in a performance degradation.

5 Concluding Remarks

In this paper, we proposed a buffer controlled adaptive video streaming scheme over wireless. In a prototype VoD system a VoD server delivers video stream to a mobile device, PDA. However, since a mobile device has limited computation and less buffer capacities, the socket buffer for it can be frequently overflowed even if there is no network congestion or low BER over wireless. So the rate at the server can be strongly affected by the receiver's state. We proposed a simple prediction for controlling the sending rate at the server by inspecting the current buffer level to reduce the possibility of packet loss. In addition, a simple cost effective rate control was presented by varying the inter-packet delays at the server. It does not require any modifications to network protocols.

With the adaptation of the proposed scheme the occurrences of the overflowed socket buffer are relatively low compared to that without such a buffer control. Thus, the proposed method maintains a smoother rate and reacts to bandwidth changes more quickly.

References

1. Floyed, S., et al.: Equation-based Congestion Control for Unicast Applications. In: Proc. ACM SIGCOMM 2000, pp. 43–56 (2000)
2. Chen, M., Zakhor, A.: Rate Control for Streaming Video Over Wireless. IEEE Wireless Communications, 32–41 (2005)
3. Padhye, J., Firoiu, V., Towsley, D., Kurpose, J.: Modeling TCP throughput: A simple model and its empirical validation. In: ACM SIGCOMM, pp. 303–314 (1998)
4. Sisalem, D., Schulzrinne, H.: The loss-delay based adjustment algorithm: A TCP-friendly adaptation scheme. In: Int. Workshop on Network and Operating System Support for Digital Audio and Video (NOSSDAV) (1998)
5. Song, B., Chung, K., Shin, Y.: SRTP: TCP-friendly congestion control for multimedia streaming. In: 16th Int. Conf. on Information Networking (2002)
6. Zheng, B., Atiquzzaman, M.: A Novel scheme for streaming multimedia to personal wireless handheld devices. IEEE Trans. On Consumer Electronics 49(1), 32–40 (2003)
7. Chen, M., Zakhor, A.: Rate Control for Streaming Video over Wireless. IEEE Infocom (2004)
8. Cen, S., Cosman, C., Voelker, G.M.: End-to-end differentiation of congestion and wireless losses. IEEE/ACM Trans. On Networking 11(5), 703–717 (2003)
9. Schulzrinne, H., Casner, S., Frederick, R., Jacobson, V.: RTP: A transport protocol for real-time applications. IETF, RFC 1889 (1988)
10. Andrew, L.L.H., et al.: Buffer Sizing for Non-homogeneous TCP Sources. IEEE Communication Letters 9(6), 567–569 (2005)
11. The network simulator ns-2, `http://www.isi.edu/nanam/ns/`

Scene Change Detection Using a Local Detection Tree and Clustering in Ubiquitous Environment

Shin Seong-Yoon[1], Lee Jong-Chan[1], Baek Seong-Eun[1], Kang Oh-Hyong[1], Shin Jung-Hoon[2], and Rhee Yang-Won[1]

[1] Dept. of Computer and Information Engineering, Kunsan Natl. Univ.
San 68, Miryong-Dong, Kunsan City, Jeonbuk 573-701, South Korea
{s3397220,chan2000,mepysto,ohkang,ywrhee}@kunsan.ac.kr
[2] College of Engineering Division of Applied Systems Engineering, Chonbuk Natl. Univ.
664-14, 1 Ga, Deokjin-Dong, Jeonju City, Jeonbuk, 561-756, South Korea
shinjh@chonbuk.ac.kr

Abstract. Processing of video data is embossed very importantly in ubiquitous environment. This paper proposes a Scene Change Detection method using the local decision tree and clustering. The local decision tree detects cluster boundaries wherein local scenes occur, in such a way as to compare time similarity distributions among the difference values between detected scenes and their adjacent frames, and group an unbroken sequence of frames with similarities in difference value into a cluster unit. In other words, the local decision tree method is used to detect local scenes from a cluster segmentation unit.

Keywords: Local Decision Tree, Cluster, Scene Change Detection.

1 Introduction

Processing of video data that is not structured in ubiquitous environment is very difficult. Videos feature large-capacity, non-linear and non-structured data. As such, much more research needs to be conducted in order to provide necessary information to users who are accustomed to dealing with structured data (i.e. text or image). In other words, more focus needs to be on developing various application technologies that allow for not only analysis of video content but also retrieval of semantic information, which goes beyond the mere video play-out function. To effectively meet the needs of such technologies, therefore, it is inevitably necessary to do research on video segmentation, which is the key to content-based video analysis.

Video segmentation refers to Scene Change Detection wherein images, texts, and audio in the media are analyzed according to their features, and are classified in a hierarchical manner to ensure the video is hierarchically structured. Over the past few decades, a lot of Scene Change Detection techniques have been proposed [1, 2]. One of the most common methods for Scene Change Detection is to compare the difference values for specific features among an unbroken sequence of frames. Despite its relatively high accuracy, this method is limited in its ability to ensure robustness [3].

O. Gervasi et al. (Eds.): ICCSA 2008, Part II, LNCS 5073, pp. 381–389, 2008.

2 Related Research

A faster and easier way to detect scene change is by comparing the number of pixels above a certain threshold [4, 5, 6]. However, this approach has the drawback of failing to distinguish between big transitions in a smaller region within a video, or between small transitions in a larger area. In other words, this method is highly sensitive to camera and/or object motions, and to brightness, so a greater number of scene changes are mistakenly detected.

The block-based comparison focuses on local features as a means of reducing sensitivity to camera and/or object motions. In the case of this approach, scene changes are detected in such a way as to segment each frame into blocks, and figure out the number of block pixels for two adjacent frames. Kast et al [7] performed a scene change detection task by calculating the likelihood ratio that was obtained from the mean and variance of different blocks. Shah et al [5] presented a method for detecting the blocks with the highest similarity to their preceding images in such a way as to segment the frames into 12 non-overlapping blocks, thereby obtaining gray-level values. Xion et al [8] proposed a net comparison scheme for detecting scene change by calculating only a small part of a video image. Xion et al [9] proposed a step-variable algorithm using time and spatial information. However, the above-mentioned methods have the disadvantage of showing high sensitivity to camera motions.

The most popular method for Scene Change Detection is through histogram comparison. Tono et al [4, 10, 11] proposed a simple method for detecting scenes according to a specific threshold through comparing gray-level histograms. Ueda et al [12] used the variation ratios of color histograms to detect scene change. Naga et al [11] performed a few simple statistical comparisons using gray-level and color histograms. Zhan et al [4] found that after comparing among pixel differences, statistics differences, and a few histograms, histogram techniques are a good exchange factor in terms of accuracy and speed. Naga et al [11] proposed an X^2-test method with focus on the difference value between the two frames and on camera and/or object motions. However, as a whole, the X^2-test method failed to outperform the linear histogram comparison method proposed by Tono et al [10], posing some disadvantages like incremental calculations. Garg et al [13] evaluated the performance of three histogram methods using six different color coordinate systems (RGB, HSV, YIQ, L*a*b*, L*u*v*, Munsell). Zhan et al [4] proposed a twin-comparison method for detecting gradual scene changes through the use of two thresholds. Bore et al [14] found that the twin-comparison method outperformed other video segmentation techniques in terms of ease of applicability to detecting scene change. The above-mentioned methods offer some advantages due to the nature of their simplicity and less sensitivity to camera and/or object motions. However, they are less likely to detect scene change, if provided with two different images, due to the absence of spatial information.

Recent research efforts involve the combination of color histograms with X^2-histograms [15], and/or the use of local, composite X^2-tests that prove to be robust due to their ability to detect all scenes from radical to gradual[16].

3 Determination of Scene Change Using a Local Decision Tree

In this paper, the Scene Change Detection method based on the local decision tree features the use of local X^2-histograms and normalization expressions. Local scenes are followed by the occurrence of global scenes - a process of detecting various scene changes following a scene change detection. Local scenes stay within the current scene until a new scene is detected. In addition, local scenes have multiple scene changes, so difference values need to be grouped by local thresholds into clusters. Figure 1 shows a graph of the overall numerical formula structure that is used for Scene Change Detection.

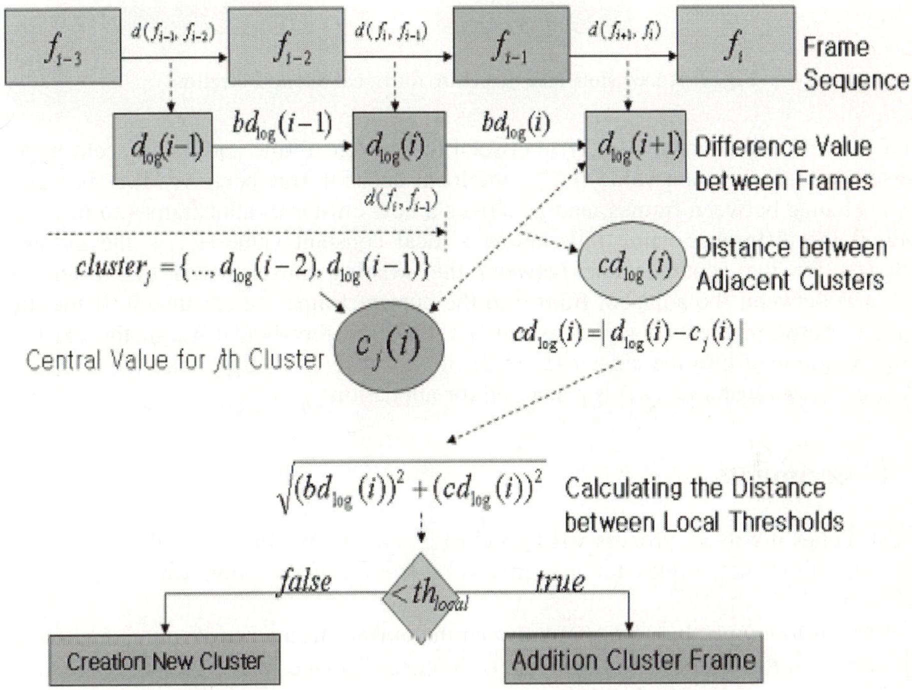

Fig. 1. A Numerical Structure Based on the Local Decision Tree

In the local decision tree, the ith difference value between the adjacent frames (f_i, f_{i-1})) is $d_{log}(i)$. In addition, $bd_{log}(i)$ refers to the absolute difference ($|d_{log}(i)$-$d_{log}(i$-$1)|$) between the ith and (i-1)th difference values. s_mc_j refers to the jth cluster in the mth scene, and $s_mc_j(i$-$1)$ refers to the central value for the frame (f_{i-1}). $cd_{log}(i)$ is the absolute difference value for the $s_mc_j(i$-$1)$ between the current frame(f_i) and the jth cluster. The local threshold distance (th_{local}) is obtained through the distance between ($bd_{log}(i)$ and $cd_{log}(i)$).

The decision tree for classifying local scenes is illustrated in Figure 2.

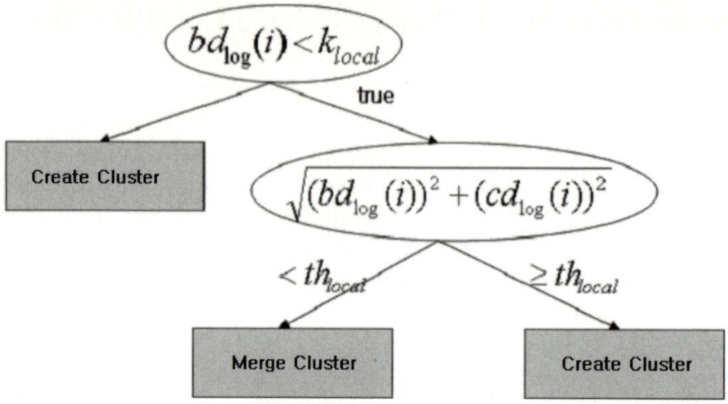

Fig. 2. A Decision Tree Structure for Local Scene Detection

If the difference value ($bd_{log}(i)$) between the current frame and its adjacent frame exceeds a local constant value (k_{local}), the local decision tree perceives it as having a scene change between frames, and generates a new cluster to allot frames to it. However, if the difference value falls below a local constant value (k_{local}), the distance ($cd_{log}(i)$) for the central value between the existing clusters, and the difference ($bd_{log}(i)$) between the adjacent frames in the current frame are calculated. If the distance () between the two values is smaller than the threshold (th_{local}), the existing frame is grouped into the cluster ($s_m c_j$). If the distance () is larger than the threshold (th_{local}), a new cluster ($s_m c_{j+1}$) is generated for allocation.

4 Experiments

Local scenes involve a process wherein clusters are created by the local decision tree from the difference values for a sequence of frames at the point where scenes are detected.

If the difference value ($bd_{log}(i)$) between the current frame and its preceding frame falls below a threshold (k_{local}), and the difference distance ($bcd_{\log}$) for the preceding cluster ($cd_{log}(i)$) exceeds a local threshold (th_{local}), a new cluster is created. Otherwise, local scenes are grouped into a preceding cluster. As illustrated in Figure 3, different thresholds are compared to detect global scene changes.

Figure 4 shows a selected portion of the distribution of the total difference values that are extracted from an advertisement video in order to do an experiment on local scene change.

The region where a lot of abrupt and gradual scene transitions occur due to the difference values obtained after scene change detection is determined as an experimental region through the use of the global decision tree. As shown in Figure 4, the experimental region includes a relatively large number of abrupt scene changes that are made due to the rapid motions of an object and a camera.

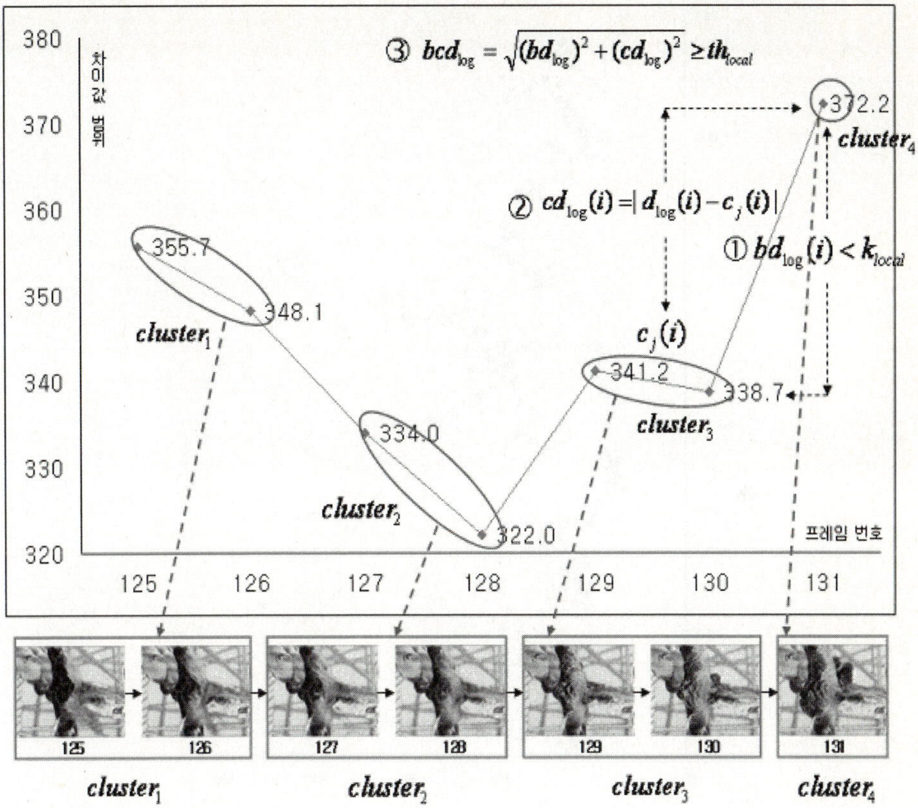

Fig. 3. A Comparison of Thresholds for Detecting Local Scene Changes

Figure 5 shows the experimental region classified by the local decision tree as a cluster unit.

Clusters can be created by the local decision tree in such a way as to group together the difference values for the frames with time similarity. As shown in Figure 5, if the distances ($bd_{log}(i)$) between the adjacent frames with time similarity meet a set threshold, and if the distances between the neighboring central values ($cd_{log}(i)$) are met, frames are grouped into the same cluster. Unless the said distances meet a set threshold, clusters are created by the local decision tree that create new clusters by calculating the difference values.

Figure 6 shows all the frames, central values, and numbers of the clusters ({c5, c6, c7}, {c21, c22}, {c24, c25, c26})) with time similarity in the case that the central values for the clusters meet a threshold (≥380).

As illustrated in Figure 6, the clusters organized by the frames with relatively high difference values include abrupt scene changes. This is because abrupt scene transitions lead to high difference values between adjacent frames.

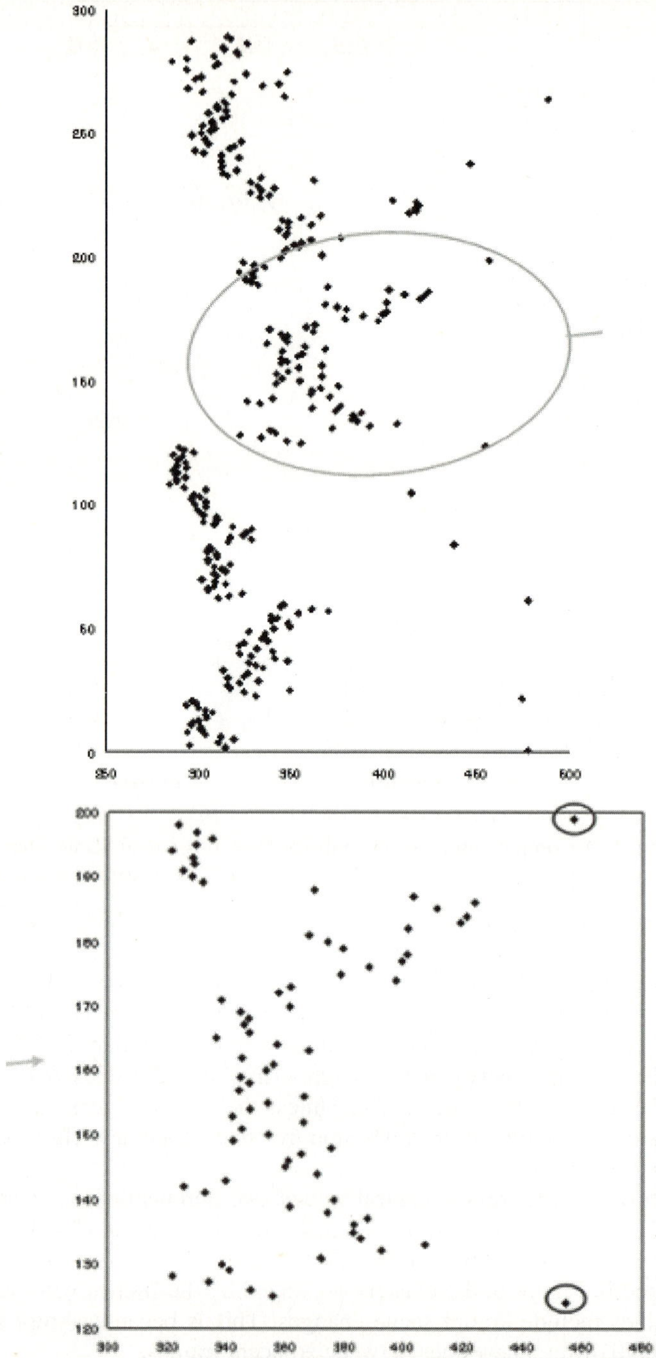

Fig. 4. A Selected Portion of the Distribution of the Total Difference Values for an Advertisement Video

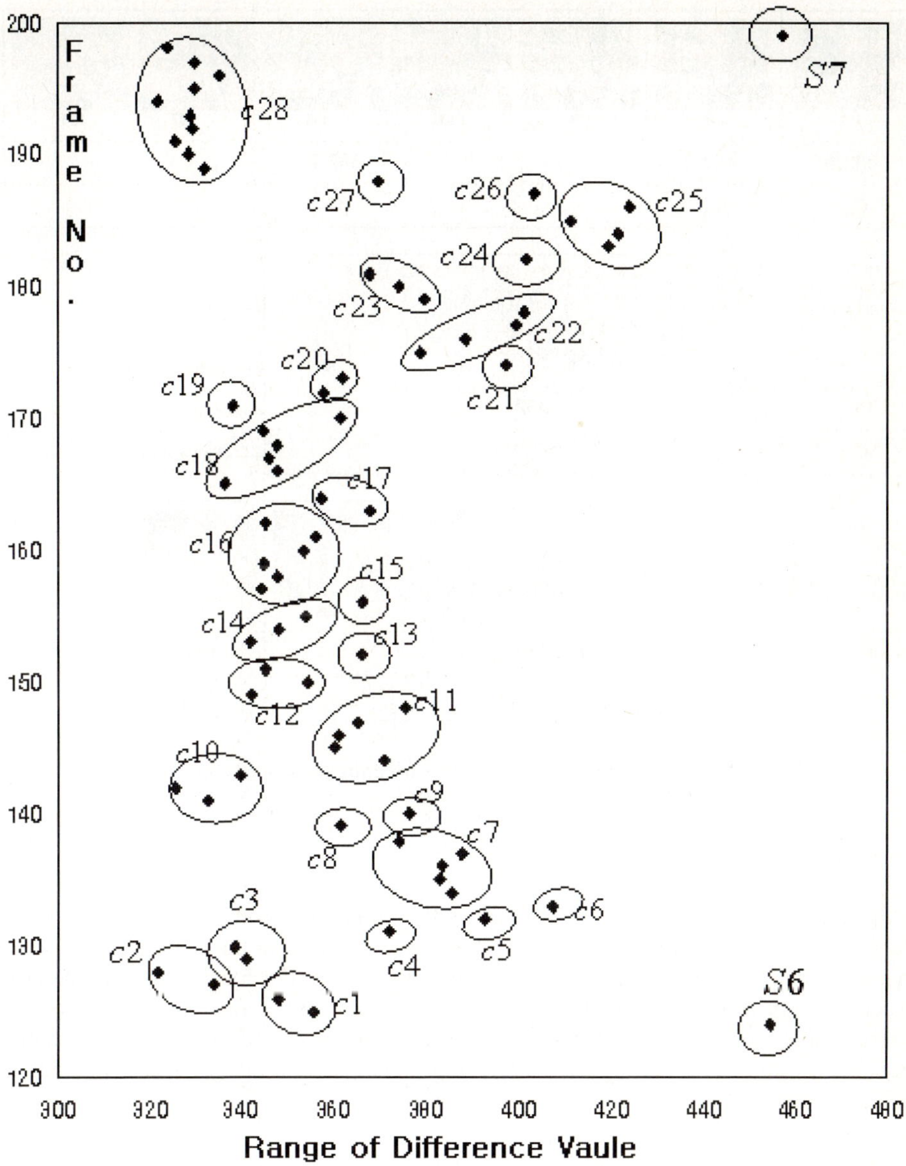

Fig. 5. The Distribution of the Clusters Created by the Local Decision Tree

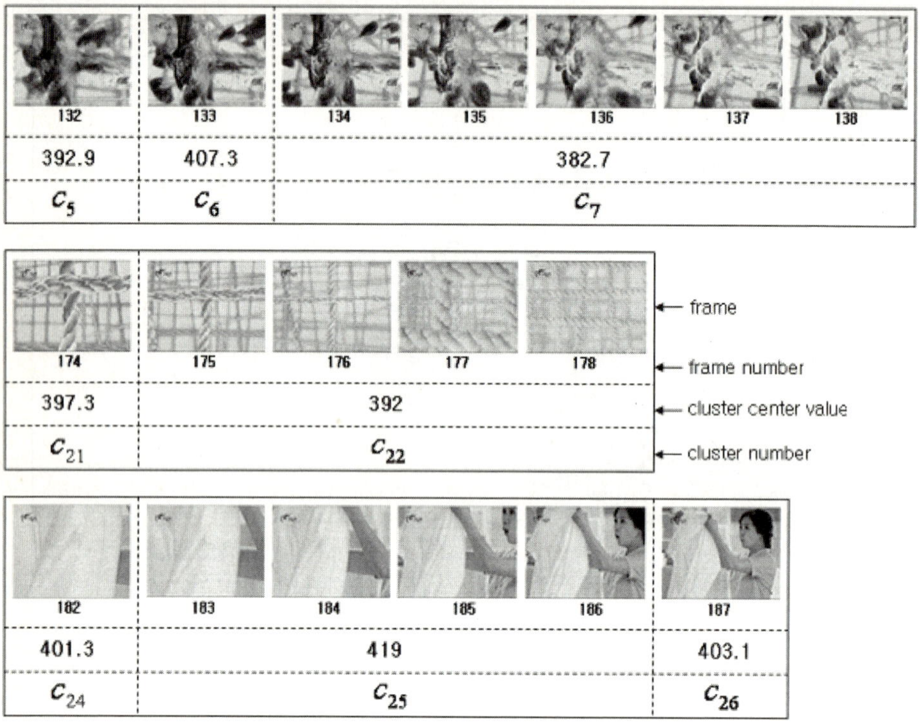

Fig. 6. The Construction of the Clusters Meeting a Central Value (≥380)

5 Conclusion

In this paper, scene changes are detected by the global decision tree in order to group clusters as local scene regions together. A scene is initiated at the point where a scene change is detected. At the point where the scene starts, the local decision tree constructs the clusters based on the similarity in frames' difference values, and on the central values for adjacent clusters. Clustering continues until another scene change sequence is detected. Through the recurring steps, several scenes and a number of clusters under the scenes are created from one video.

Acknowledgments. This work was supported in part by a grant from the Regional Innovation System of the Ministry of Commerce, Industry and Energy of Korea.

References

1. Koprinska, Carrato, S.: Temporal Video Segmentation: A Survey. In: Signal Processing Image Communication. Elsevier Science, Amsterdam (2001)
2. Ananger, G., Little, T.D.C.: A survey of technologies for parsing and indexing digital video. Journal of Visual Communication and Image Representation, 28–43 (1996)

3. Gargi, U., Kasturi, R., Strayer, S.H.: Performance Characterization of Video-Shot-Change Detection Methods. IEEE transaction on circuits and systems for video technology 10(1) (February 2000)
4. Zhang, H.J., Kankanhalli, A., Smoliar, S.W.: Automatic Partitioning of Full-motion Video. Multimedia Systems 1(1), 10–28 (1993)
5. Shahraray, B.: Scene Change Detection and Content-Based Sampling of Video Sequences. In: Digital Video Compression: Algorithms and Technologies, Proc. SPIE, vol. 2419, pp. 2–13 (1995)
6. Hampapur, A., Jain, R., Weymouth, T.: Digital Video Segmentation. In: Proc. ACM Multimedia, vol. 94, pp. 357–364 (1994)
7. Kasturi, R., Jain, R.: Dynamic Vision. In Computer Vision: Principles. IEEE Computer Society Press, Washington (1991)
8. Xiong, W., Lee, J.C.-M., Shen, D.-G.: Net Comparison: An Adaptive and Effective Method for Scene Change Detection. SPIE (1995)
9. Lee, J.C.-M., Li, Q., Xiong, W.: Automatic and Dynamic Video Manipulation. Research and Development in Information Retrieval (1998)
10. Tonomura, Y.: Video handling based on structured information for hypermedia systems. In: Proc. ACM Int. Conf. Multimedia Information Systems, pp. 333–344 (1991)
11. Nagasaka, A., Tanaka, Y.: Automatic Video Indexing and Full-Video Search for Object Appearances. In: Knuth, E., Wegner, L. (eds.) Visual Database Systems II, pp. 113–127. Elsevier Science Publishers, Amsterdam (1992)
12. Ueda, H., Miyatake, T., Yoshizawa, S.: IMPACT: An Interactive Natural-motion-picture Dedicated Multimedia Authoring System. In: Proceedings of CHI, 1991, pp. 343–350. ACM, New York (1991)
13. Gragi, U., Kasturi, R., Antani, S.: Evaluation of video sequence indexing and hierarchical video indexing. In: Proc. SPIE Conf. Storage and Retrieval in Image and Video Databases, pp. 1522–1530 (1995)
14. Boreczky, J.S., Rowe, L.A.: Comparison of Video Scene Change Detection Techniques. In: Proc, SPIE Conf. Storage and Retrieval in Image and Video Databases, pp. 170–179 (1996)
15. Shin, S.Y., Pyo, S.B.: Video Browsing Using An Efficient Scene Change Detection in Telematics. Journal of the Korea Society of Computer and Information 11(4), 147–154 (2006)
16. Kim, Y.L., Rhee, Y.W.: Scene Change Detection Using Local χ^2 Test. Journal of the Korea Society of Computer and Information 11(3), 193–201 (2006)

Shot Boundary Detection Using a Global Decision Tree in Ubiquitous Environment

Shin Seong-Yoon[1], Gao Cheng[1], Baek Seong-Eun[1], Park Sang-Joon[1],
Shin Jung-Hoon[2], and Rhee Yang-Won[1]

[1] Dept. of Computer and Information Engineering, Kunsan Natl. Univ.
San 68, Miryong-Dong, Kunsan City, Jeonbuk 573-701, South Korea
gaowen.net@gmail.com,
{s3397220,mepysto,lubimia,ywrhee}@kunsan.ac.kr
[2] College of Engineering Division of Applied Systems Engineering, Chonbuk Natl. Univ.
664-14, 1 Ga, Deokjin-Dong, Jeonju City, Jeonbuk, 561-756, South Korea
shinjh@chonbuk.ac.kr

Abstract. Role of video in Ubiquitous environment is growing. This paper presents a shot boundary detection method based on the global decision tree that allows for extraction of boundaries of high variations occurring due to camera breaks from frame difference values. For a start, difference values between frames are calculated through local X^2-histogram and normalization. Next, the distances between difference values are calculated through normalization. After the global threshold distance is computed based on the distances between the difference values calculated, a comparison between the distance value for two adjacent frames and the global threshold distance is performed to detect shot boundaries.

Keywords: Global Decision Tree, Shot Boundary Detection, Normalization, Global Threshold.

1 Introduction

Processing of video is increasing rapidly in Ubiquitous environment. Videos are non-structured data, and they need to be structured according to a certain rule to provide ease of access. To ensure its structurization, therefore, a video needs to be semantically classified.

As a basic unit in video segmentation, a shot refers to an unbroken sequence of frames from one camera. Based on the frame at the point where a shot occurs, a cluster refers to a set of adjacent frames that have a sequence of similar difference values; it is used to extract possible scene changes arising from camera and/or object motions. A group refers to a cluster set that is grouped together by a clustering algorithm according to the central value for clusters. The initial frame in a cluster is a basic unit used to extract representative frames.

This paper demonstrates how shot frames are extracted when, through normalization, the difference values between adjacent frames meet a specific threshold and are independently distributed. In addition, this paper shows how shot boundaries are detected according to a specific threshold obtained from converted difference values, through the use of the global decision tree as well as the proposed normalization

O. Gervasi et al. (Eds.): ICCSA 2008, Part II, LNCS 5073, pp. 390–398, 2008.

algorithm. The term 'global decision tree' refers to a method for extracting boundaries with high variations occurring due to camera breaks from the difference values between adjacent frames. Global decision trees are used to segment a video into shots by detecting shot boundaries where global shots occur. In other words, a global shot is the boundary between shots.

2 Related Work

One of the most common methods for video segmentation is to compare the difference values between two frames for shot detection. The shots detected are regarded as representative frames that are useful when structuring shots or scenes. Video classification methods that use difference values, however, have the problem of determining a threshold [1][2][3].

The clustering-based video structuring method involves combining relevant shots into a group thereby obtaining a higher-level video hierarchical structure, and then constructing a browsing tree that allows for a video structure, through the use of the clusters that are combined together. This method has the advantage of eliminating the necessity of determining a threshold due to the use of time and visual factors. However, it also has the drawback of failing to enable a content-based hierarchical structure [4].

The scene-based video structuring method includes a model-based approach and a general-purpose approach. The model-based scheme features a semantic video structurization and provides outstanding accuracy. However, this method poses some disadvantages given that it requires a preliminary model to be developed, as well as several components such as some time needed to set up the model, domain knowledge, and experience. On the contrary, the general-purpose scheme uses time constraint clustering wherein shots are grouped into clusters based on time-constraint clustering, thereby requiring no obvious domain. Based on the clusters, a scene transition graph is built where the scene flow makes up the structure of a video [5][6].

Furthermore, recent research efforts involve the combination of color histograms with χ^2-histograms[7], and/or the use of local, composite χ^2-tests that prove to be robust due to their ability to detect all scenes from radical to gradual.

In this paper, the X^2-histograms in [7][8] were used to detect scene changes. The Equation (1) below can be used to detect scene changes.

$$d(f_i, f_{i-1}) = \sum_{k=1}^{b} d_{\chi^2}(f_i, f_{i-1}, k) \tag{1}$$

$$\begin{aligned} d_{\chi^2}(f_i, f_{i-1}, k) = \sum_{j=1}^{n} (& \frac{(H_i^r(j) - H_{i-1}^r(j))^2}{\max(H_i^r(j), H_{i-1}^r(j))} \times \alpha \\ & + \frac{(H_i^g(j) - H_{i-1}^g(j))^2}{\max(H_i^g(j), H_{i-1}^g(j))} \times \beta \\ & + \frac{(H_i^b(j) - H_{i-1}^b(j))^2}{\max(H_i^b(j), H_{i-1}^b(j))} \times \gamma \end{aligned}$$

In this context, bl refers to the total number of blocks, and $H_i^r(k)$ refers to the difference value in the red-channel histogram between the block bl of the ith frame and the gray-level k. In addition, α, β, and γ are the constants used to convert grey-level intensity according to the NTSC standards where α=0.299, β=0.587, and γ=0.114.

As a means of normalization, the log functions and constants used in image processing to increase grey-level image values were modified and applied to the difference values. The Equation (2) below can apply to the proposed method.

$$d_{\log} = c \times \log(1+d^2)$$
$$c = \frac{\max(d_{\log})}{\max(\log(1+d^2))} \tag{2}$$

The log function d from above is the difference value extracted from the Equation (1), and the constant c is a multiplier constant obtained from d.

3 Global Scene Region Decision

A global scene region refers to the area of a video where an abrupt scene transition occurs due to a camera break. In the case of a global scene region, scene changes occur globally in the entire video flow according to certain rules.

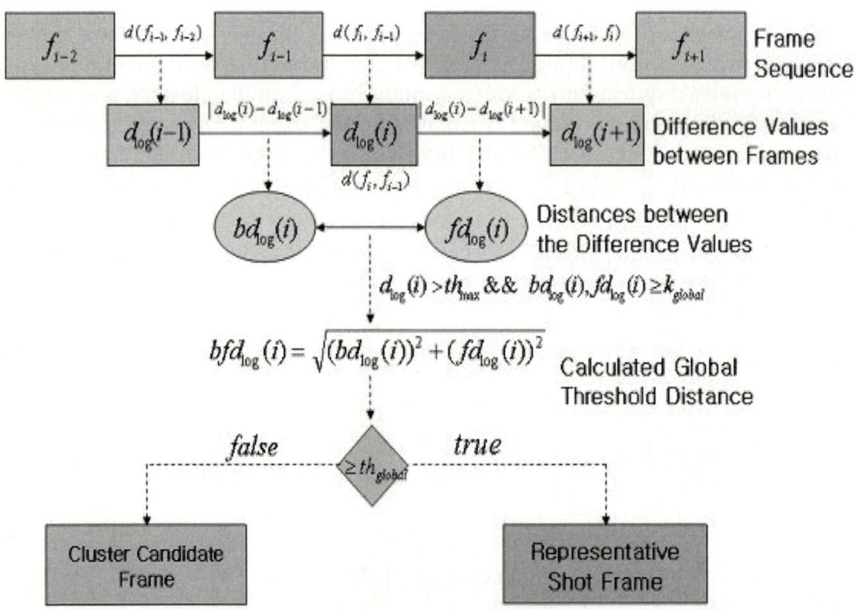

Fig. 1. The Global Decision Tree Structure

Generally, a shot boundary where a camera break occurs has wide deviations from the frames before and after the camera break. That's because the frames before and after a camera break have distribution information about the frames that are arranged in relatively close proximity to each other. In contrast, a shot boundary typically shows such marked difference values that it distinguishes itself from the frames before a camera break. Unless the difference values at the boundary exceed a specific threshold, scene changes due to the abrupt motions of an object or a camera, or a flash light can be predicted. Thus, such shot boundaries occurring globally due to camera motions can be used to detect scene changes.

Figure 1 illustrates the overall numerical expression that allows you to decide on a global scene. First off, $d(f_i, f_{i-1})$ is the expression for the X^2-histogram in Expression (1) obtained as a result of the first shot boundary detection. In addition, $d_{\log}(i)$ refers to the normalized difference value, which corresponds to the ith frame in the (f_i, f_{i-1}) frames that are arranged in sequence as a result of the first shot boundary detection. Also, $d_{\log}(i\text{-}1)$ refers to the normalized difference value for the $(i\text{-}1)$th frame; $bd_{\log}(i)$ refers to the difference between itself and the difference value for the preceding frame $(|d_{\log}(i)\text{-}d_{\log}(i\text{-}1)|)$; and $fd_{\log}(i)$ refers to the difference between itself and the difference value for the subsequent frame $(|d_{\log}(i)\text{-}d_{\log}(i+1))$.

Figure 2 shows various examples of scene changes that can be extracted from given conditions through the use of the global decision tree.

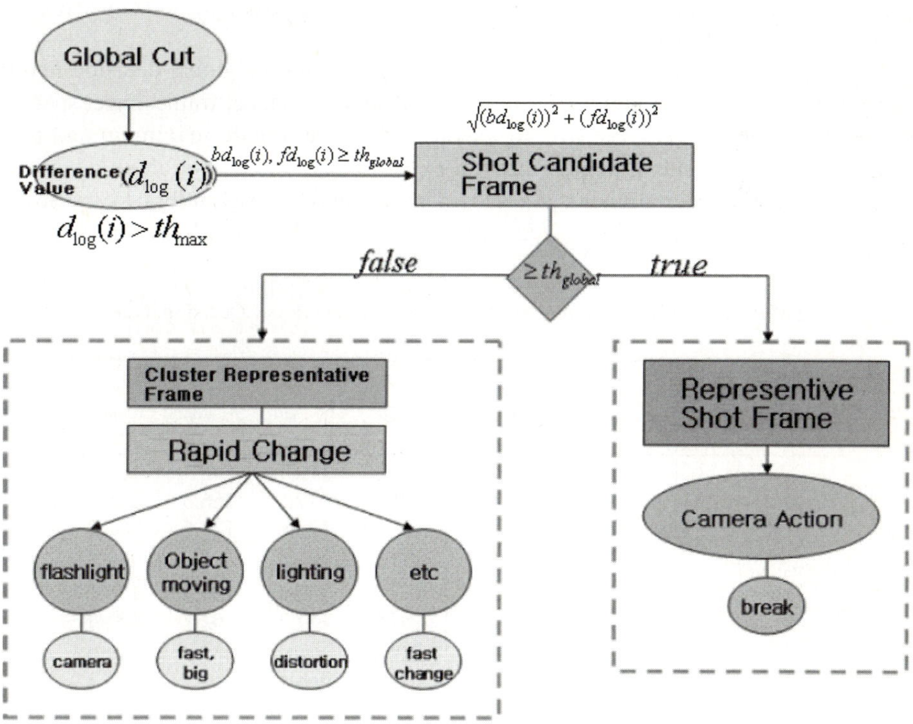

Fig. 2. Various Examples of Scene Changes

In order for the current frame to be regarded as the one where a camera break occurs, the difference value ($d_{\log}(i)$) should not just meet the maximum threshold (th_{max}), but the difference value between the present frame and its preceding frame ($bd_{\log}(i)$), or the difference value between the current frame and its subsequent frame ($fd_{\log}(i)$) should also achieve a specific constant value (k_{global}). Unless either of the two conditions is met, the current frame may be regarded as an abrupt scene tradition in which it can be grouped into a cluster, not as a shot frame. In such a case, the frame is classified into a cluster frame candidate, which is supposed to be determined by the local decision tree. In the meantime, the distance value ($bfd_{\log}(i)= \sqrt{(bd_{\log}(i))^2 + (fd_{\log}(i))^2}$) for adjacent frames should meet the global threshold (th_{global}). This is because the threshold conditions for different difference values as well as the distance obtained by adding two values should have a specific threshold range in such a way that the sensitivity of camera breaks to video features is considered to allow for adaptive decision. If the current difference value meets all of the four conditions, the frame corresponding to the current difference value is determined as the representative shot frame.

4 Experiments

Experiments were carried out using MS Visual C++ 6.0 and DirectX 8.1 running under Windows XP. Each video sequence was executed at a rate of 30 frame/sec. and at a resolution of 320 X 240.

Shots should be extracted from the inputted video using the global decision tree in a way that the distances among the difference values for adjacent frames meet specific thresholds. Table 1 shows the overall average of the minimum, maximum and total values for the thresholds obtained through experimental video data about advertisements, news, and entertainment. As shown in Table 1, $k_{global}1$ refers to the lower threshold limit for $bd_{log}(i)$, and $k_{global}2$ refers to the lower threshold limit for $fd_{log}(i)$.

Table 1. Threshold Ranges for Application of the Global Decision Tree

	Threshold	th_{max}	$K_{global1}$	$k_{global2}$	th_{global}
Advertisement	Minimum	415.2	118.1	98.7	162.3
	Maximum	488	177.6	166.3	228.1
	Total Average	456.3	145.6	128.6	195.2
News	Minimum	422.4	73.4	62.4	96.3
	Maximum	487.6	142.4	129.9	179.1
	Total Average	465.5	117.3	102.7	155.4
Entertainment	Minimum	447.5	132	88.1	158.7
	Maximum	501.2	205.2	208.3	292.4
	Total Average	472.9	172.4	167.8	241.3

If a video has a lot of scene changes as in advertisement or entertainment videos, the distance between adjacent frames is relatively long. On the other hand, a news video that has few abrupt scene changes and more gradual scene changes features a relatively shorter distance between adjacent frames. As a result of application of the thresholds available for a news video to an advertisement or entertainment video, there were few problems with the global shot detection. In this paper, therefore, the global thresholds were obtained from the distance between the two adjacent difference values, and the thresholds showing no sensitivity to video types were determined.

Figure 3 shows the global shots that were extracted from an advertisement video through the use of the global decision tree and thresholds.

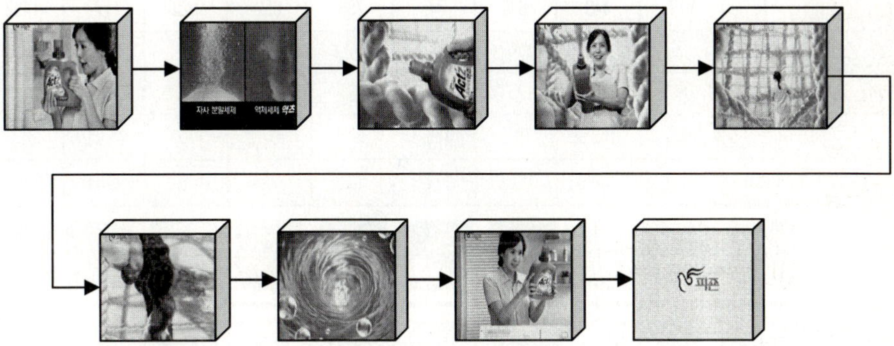

Fig. 3. The Global Shots Extracted from an Advertisement Video

Table 2 shows the comparative values that allow for a global shot extraction from the advertisement video, as illustrated in Figure 3.

Table 2. A Comparison of an Advertisement Video for Global Shot Extraction

	1shot	2shot	3shot	4shot	5shot	6shot	7shot	8shot	9shot
Frame Number	1	22	51	61	105	124	199	238	264
$d_{\log}(i\text{-}1)$	0	296.6	346.8	305.6	297	289.5	324.1	312.5	313.6
$d_{\log}(i)$	477.7	474.4	477.3	438.2	415.2	454.4	456.7	446.3	488
$d_{\log}(i+1)$	314.6	331	311	316	303.8	355.7	344.7	312.3	346.8
$bd_{\log}(i)$	477.7	177.6	130.5	132.6	118.1	164.9	132.6	133.8	174.3
$fd_{\log}(i)$	163.1	143.1	166.3	122.2	111.3	98.7	112	134	141.2
$bfd_{\log}(i)$	504.8	228.1	211.4	180.3	162.3	192.2	173.6	189.4	224.3

Advertisement videos have quite a few scene changes, so they are more likely to show big difference values that could be regarded as global shots when special photographs are taken or an object makes rapid movements.

Figure 4 illustrates the shots extracted from a news video with relatively few scene changes and more gradual scene changes, and the distances between difference values.

	1shot	2shot	3shot	4shot	5shot	6shot	7shot	8shot	9shot	10shot
Frame No	1	317	449	675	745	796	852	928	1177	1492
$d_{\log}(i)$	477.7	454.1	473.2	479.2	476.1	422.4	487.6	476.3	466.6	461.4
$bd_{\log}(i)$	477.7	106.7	129.7	125.0	126.6	73.4	142.4	130.7	129.0	110.6
$fd_{\log}(i)$	125.8	88.9	109.9	91.0	119.8	62.4	107.7	118.1	112.0	83.0
$bfd_{\log}(i)$	494.0	138.9	169.9	154.6	174.3	96.3	178.6	176.2	170.8	138.3

	11shot	12shot	13shot	14shot	15shot	16shot	17shot	18shot	19shot	20shot
Frame No	1577	1664	1778	2096	2211	2361	2427	2504	2550	2638
$d_{\log}(i)$	465.9	454.5	487.2	483.9	464.5	460.1	433.7	474.6	461.8	461.4
$bd_{\log}(i)$	118.5	108.9	139.3	136.5	123.6	113.4	93.4	129.4	117.1	74.4
$fd_{\log}(i)$	129.9	90.9	94.1	119.2	112.4	117.5	74.8	123.7	79.9	115.9
$bfd_{\log}(i)$	175.8	141.9	168.1	181.2	167.0	163.3	119.7	179.1	141.8	116.1

Fig. 4. A Comparison of the Distances between News Video Cuts

The proposed decision tree allows for removal of camera flashlights that lead to mistakenly extracting scene changes from a news video. The frames with flashlights are more likely to show relatively big difference values, so they are supposed to be determined by the decision tree as a shot. Figure 5 shows the difference values for different global shots extracted from an entertainment video.

Entertainment videos are highly likely to have a lot of relatively abrupt scene transitions, and their frames include many flashlights, especially in interview scenes. As a result, entertainment videos are more sensitive in determining a threshold, thereby making it quite difficult to identify a shot boundary. However, the proposed global decision tree allows global shots to be effectively extracted using given thresholds.

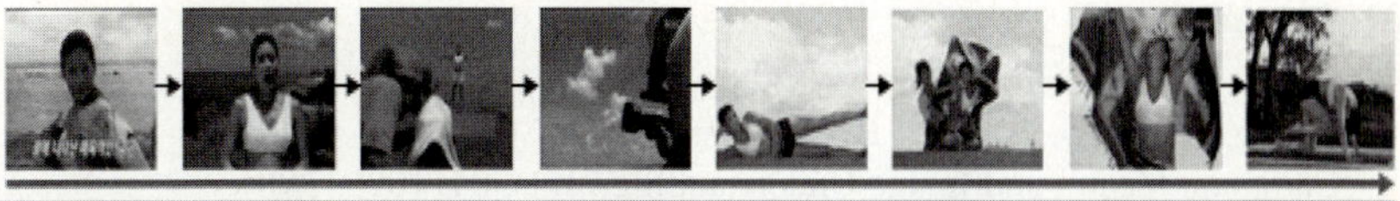

	1shot	2shot	3shot	4shot	5shot	6shot	7shot	8shot
Frame No	1	71	335	444	537	698	840	1210
$d_{log}(i)$	477.7	493.2	464.0	473.5	501.2	447.5	456.8	483.7
$bd_{log}(i)$	477.7	164.7	151.1	179.7	205.2	183.2	132.0	175.6
$fd_{log}(i)$	162.5	189.0	150.8	144.5	208.3	153.1	88.1	173.3
$bfd_{log}(i)$	504.6	250.7	213.5	230.6	292.4	238.7	158.7	246.7

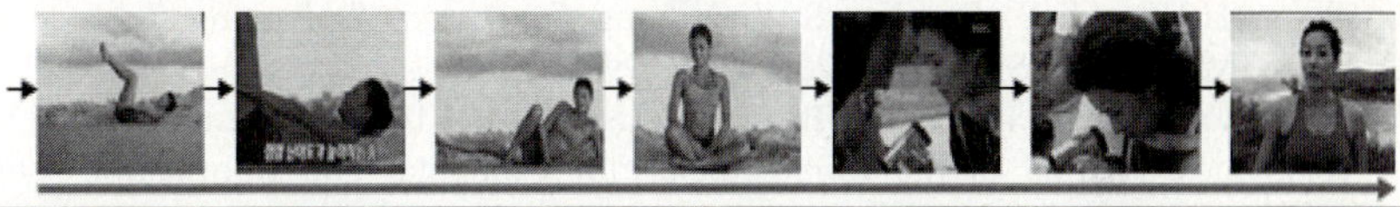

	9shot	10shot	11shot	12shot	13shot	14shot	15shot
Frame No	1372	1457	1531	1648	1775	1871	2116
$d_{log}(i)$	481.2	460.5	458.9	448.6	491.8	472.5	486.7
$bd_{log}(i)$	189.3	133.0	180.0	176.5	205.0	133.6	204.2
$fd_{log}(i)$	191.3	160.8	171.8	160.7	187.2	174.9	196.0
$bfd_{log}(i)$	269.2	208.7	248.8	238.7	277.6	220.1	283.1

Fig. 5. A Comparison of the Distances between Entertainment Video Cuts

5 Conclusion and Future Work

This paper demonstrated how abrupt scene changes like camera breaks were detected using the global decision tree. Shot frames were extracted in a way that scene changes were detected using the X^2-histogram from the inputted video, and that the difference values between adjacent frames were met through normalization. In addition, the proposed global decision tree allowed for extraction of scene changes, including the abrupt movements of an object and/or a camera, and flashlights.

Instead of text or voice, the color information of adjacent frames was only used in this paper. Therefore, there is a necessity to develop a scene change detection method that incorporates various media features. In addition, the focus of future research needs to be on video structurization, extracted scene analysis, and video representation and indexing.

Acknowledgments. This work was supported in part by a grant from the Regional Innovation System of the Ministry of Commerce, Industry and Energy of Korea.

References

1. Yeung, M., Yeo, B., Liu, B.: Segmentation of Video by Clustering and Graph Analysis. Computer Vision and Image Understanding 71(1), 94–109 (1998)
2. Yeo, B.-L.: Efficiency processing of compressed image and video. Technical report, PhD thesis. Princeton University (1996)
3. Zhang, H., Low, C.Y., Gong, Y., Smoliar, S.: Video Parsing Using Compressed Data. In: Proc. SPIE Conf. Image and Video Processing II, vol. 2182, pp. 142–149 (1994)
4. Zhong, D., Chang, S.F.: Video Object Model and Segmentation for Content-Based Video Indexing. In: IEEE International Conference on Circuits and Systems (June 1997)
5. Yeung, M.M., Yeo, B.-L., Wolf, W., Liu, B.: Video Browsing Using Clustering and Scene Transition Compressed Sequences. IS&T SPIE, Multimedia Computing and Networking (1995)
6. Zhang, H., Tan, S.Y., Smoliar, S.W., Yihong, G.: Automatic Parsing and Indexing of News Video. Multimedia Systems 2, 256–266 (1995)
7. Shin, S.Y., Pyo, S.B.: Video Browsing Using An Efficient Scene Change Detection in Telematics. Journal of the Korea Society of Computer and Information 11(4), 147–154 (2006)
8. Kim, Y.L., Rhee, Y.W.: Scene Change Detection Using Local x^2-Test. Journal of the Korea Society of Computer and Information 11(3), 193–201 (2006)

A Handover Scheme Supporting the Buffer Management in B3G Networks*

Sangjoon Park[1], Jongchan Lee[1], Changbok Lee[1], Ohhyung Kang[1], Seongyoon Shin[1], and Kwanjoong Kim[2]

[1] Dept. of Computer Information Engineering, Kunsan National University
[2] Dept. of Computer & Information Science, Hanseo University
{chan2000,lubimia}@kunsan.ac.kr

Abstract. In B3G Networks, service users can use heterogeneous networks to receive various services. We propose a buffer management scheme for reliable services. Service users can use mobile services in wireless networks such like WLAN and UMTS. In this paper, we present the network structure implementing policy-based functions based PDP-PEP to control heterogeneous access networks. Also, we propose the ISHO for specific traffic class by the buffer management of access network. By the buffer management scheme on ISHO, it can manage the resource usage as well as resource balancing, and packet loss rate can be lowered in burst traffic period.

1 Introduction

Network services are varied through wireless network technique evolution. In Beyond Three Generation (B3G) networks, mobile users can use various services by accessing heterogeneous network systems [1]-[3], and heterogeneous services as well as proper network service. From complicated service requests, b3G networks should provide the end-to-end QoS mechanism. For IP based network QoS IETF (Internet Engineering Task Force) considers two management scheme: IntServ and Diffserv [4][5]. Also, to advanced network services service users and service providers can negotiate the service contact to receive All-IP service by Service Level Agreement (SLA). By SLA, the network guarantees the reliable service providing. If an user receives low service quality, the service provider provides appropriate service management to an user. Once the SLA contact is determined, Service Level Specifications (SLS) [6] selects the QoS specification and parameters to providing service [7]. In B3G networks, access networks such like UMTS (Universal Mobile Telecommunication System) and WLAN (Wireless Local Area Network) are used for service users. The policy-based network management is implemented to control heterogeneous networks such like UMTS and WLAN in B3G networks. The IETF presents the two elements (PDP - Policy Decision Point and PEP - Policy Enforcement Point) for implementing policy-based function. By the PDP the service policy is determined to provide the

* This work was supported by the Korea Research Foundation Grant funded by the Korean Government (MOEHRD)(KRF-2007-314-D00184).

O. Gervasi et al. (Eds.): ICCSA 2008, Part II, LNCS 5073, pp. 399–408, 2008.

service, and the PEP implements the determined policy [8]-[11]. For the network service and management, Common Open Policy Service (COPS) is an applied protocol between the PDP and the PEP that it is proposed by the IETF [12]. In UMTS and WLAN, service users can receive network services provided from the PDP and the PEP on the SLA.

The SLA based B3G networks must provide various and reliable network service environment fro users. Hence, user terminals should support the multi-mode network system to access heterogeneous networks that mobile users are able to receive QoS guarantee to heterogeneous services. In this paper, we propose a buffer management scheme considering the traffic channel. The traffic class change is considered to implement the QoS management of the Diffserv. In the policy-based network structure, if available network resource is lowed to a threshold, specific traffic service is changed to another access network to provide active control for the buffer management. By the threshold of channel management, the network availability can increase, and the call blocking rate can be lowed for service users.

The rest of the paper is organized as follows. In Section II, we describe the SLA and the QoS management on the policy-based elements. Section III presents the proposed channel management scheme for the traffic class control. The simulation analysis results are presented in Section IV. Finally, we discuss conclusions in Section V.

2 Policy Based QoS Management

2.1 SLA Based QoS Management

In this paper, we consider the network architecture in which service users can contract the SLA with service providers. By the SLA the wireless service can be provided to implement end-to-end QoS management. The service users request the service quality as received service level so that the leveled service can be showed by the SLA. Service providers have the reliability to guarantee the QoS to each provided service. Hence, for reliable service purpose, the SLA is contracted between the users and providers (Fig. 1).

Service users can request the compensation if the network dose not guarantee the reliable service. Once the SLA is contract, and then the SLS can be determined. In the SLS, system parameters are defined to each service level. After successful SLA agreement, Inter-domain Policy Agents (IPAS) delivery the service request to each network [13]. By the SLA and SLS mapping, service provider continuously checks the service state for the reliable service and QoS management. After the SLS parameters are determined, in each networks actual network service will be presented.

2.2 SLA Structure

To provide the determined service by the SLA we consider the policy-based network architecture to support the Diffserv [5] for All-IP network service. For the UMTS and WLAN the policy-based function is adopted to control heterogeneous access networks. It is specific network service or network management mechanism that implements the network function by the network policy. In network management, service

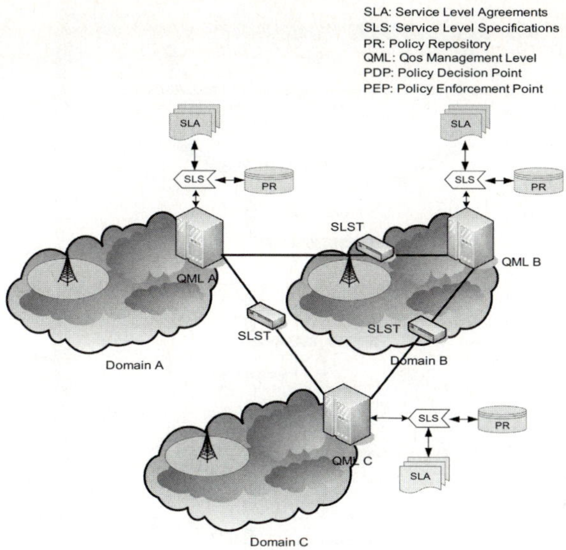

Fig. 1. SLA based service construction

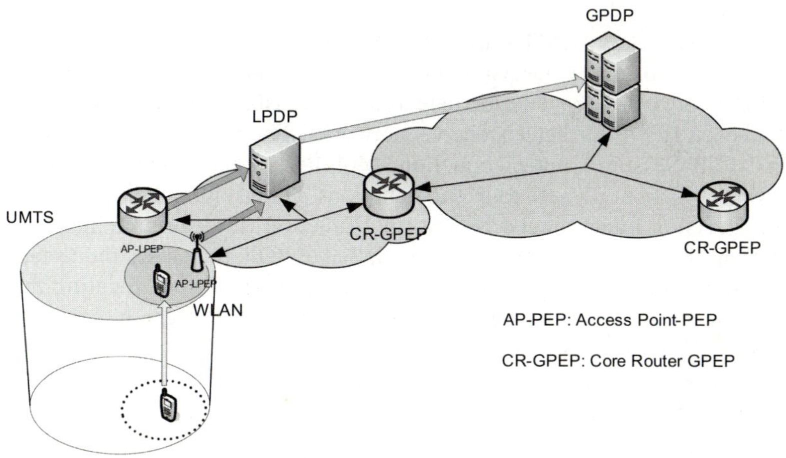

Fig. 2. Policy-based network structure

provider make new network policy for the resource and service, and then the management systems apply the new policy to the network systems. PDP and PEP are important elements to implement the policy-based service control. The policy can be determined by the PDP, and the PEP implements the network service by the policy. Also, the network stores, appends and deletes SLA policies in Policy Repository (PR). Fig. 2 shows the policy-based network architecture.

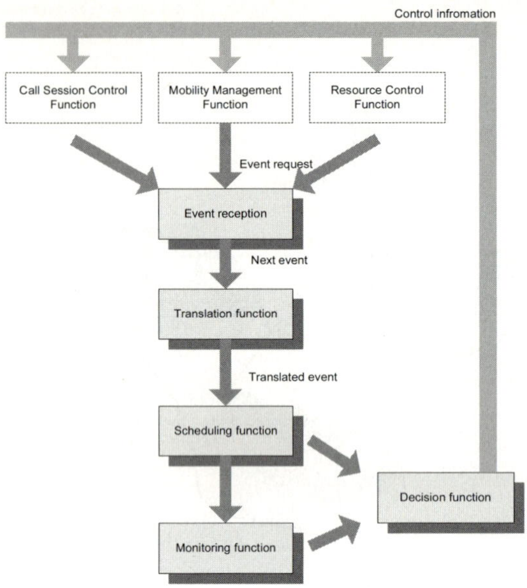

Fig. 3. PDP structure

The access networks (UMTS and WLAN) support the SLA function so that they operate their Local policy Decision Point (LPDP) independently. Also the Global PDP (GPDP) provides the intermediate role to translate heterogeneous services or policies to each LPDP for heterogeneous networks. The PDP consists of three control functions (Call Session Control Function, Mobility Management Function and Resource control Function) and four sub-functions (Translation Function, Scheduling Function, Decision Function and Monitoring Function) to determine the function implementation (Fig.3). In policy based model, the PDP translates requested service events from mobile users, and send instructions to the PEP for executing the system operation. The communication between the PDP and the PEP can be implemented as using the COPS protocol proposed in the IETF [5]. The COPS protocol provides a request/response model to exchange the policy information to determine actually the function implementation.

3 Handover Channel Management

In B3G networks supporting heterogeneous access networks, mobile users can receive the service mobility from the multi-networks service so that the Inter-system handover can be implemented to service mobility. Also, the mobile terminal must support the multi-modes system to provide multi-networks service. The terminal multi-modes can be implemented by multi-systems or software download. Service mobility is that the inter-system handover can occur by the network resource state without the terminal mobility. From the resource management the handover is able to occur to meet user request in the resource state.

3.1 System Structure

Fig. 4 presents the handover channel management to control traffic to each network level in the policy-based network. In this paper, we consider the network model to implement handover channel management. If a user use UMTS system and QoS management system knows the resource shortage state by the threshold, by the SLA the traffic class can be moved WLAN to provide appropriate service.

3.2 LPEP Resource Management

When mobile users receive network service in hot spot, the system cannot guarantee the end-to-end QoS caused by the high traffic load. In this paper, we consider the channel management scheme of LPEP (access router) to manage traffic buffer. Fig. 5 shows the relation between the service class and threshold in the buffer of access router. In Diffserv, generally three traffic class are defined, and each traffic session can be managed the Per Hop Behavior (PHB) group of traffic class. Access router

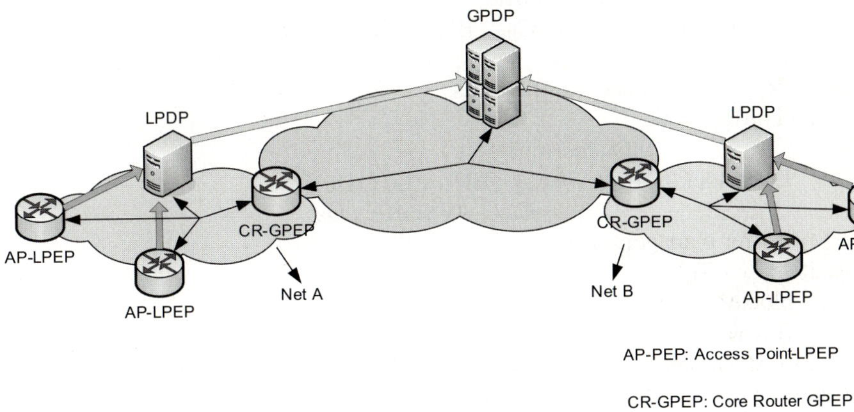

Fig. 4. Network structure

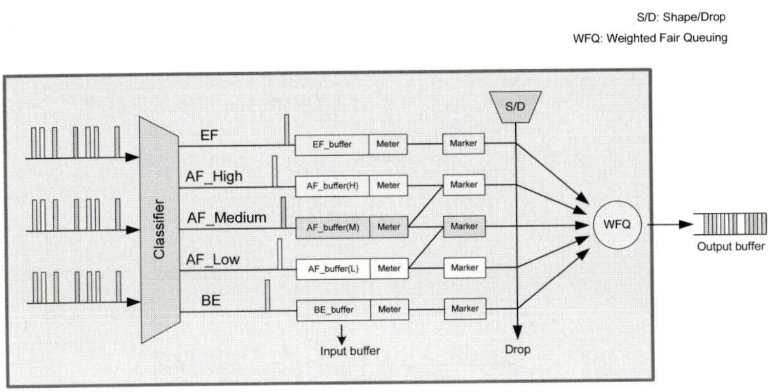

Fig. 5. Buffer management

buffer has three buffer threshold (T_H, T_M and T_L), and traffic transmission is determined by the buffer management. If the current buffer state is R_{CN}, the three buffer thresholds are T_L (Little threshold value about buffer of access router), T_M (Middle threshold value about buffer of access router) and T_H (High threshold value about buffer of access router), and three traffic classes are C_{EF} (Expedite Forwarding traffic class), C_{AF} (Assured Forwarding traffic class) and C_{BE} (Best effort traffic class), we present channel management algorithm in Fig. 6.

```
RCN - current buffer value LPEP, TL - Little threshold value
TM - Middle threshold value, TH - High threshold value
CEF - EF traffic class, CAF - AF traffic class
CBE - BE traffic class, InQueue - Insert the messages
Block - block traffic classes
If (RCN <= TL) InQueue(traffic)
   Else if (RCN <= TM) Check traffic class
        If (traffic == CEF || traffic == CAF)
                       InQueue(traffic)
        Else Block(traffic)   /* CBE : BE traffic class */
        Buffer management ()
Else if (RCN <= TH) Check traffic class
      If(Traffic == CEF)
                       InQueue(traffic)
        Else block(traffic)  /* CEF : EF traffic class
                                 CAF : AF traffic class */
        Buffer management ()
Else if (RCN > TH)
        block(traffic)
        Buffer management ()
/* LPEP transmits Queue_info message to LPDP traffic class control by policy
request handover about traffic class check policy mobility management Request handover to other LPDP handover decision */
```

Fig. 6. Channel management algorithm

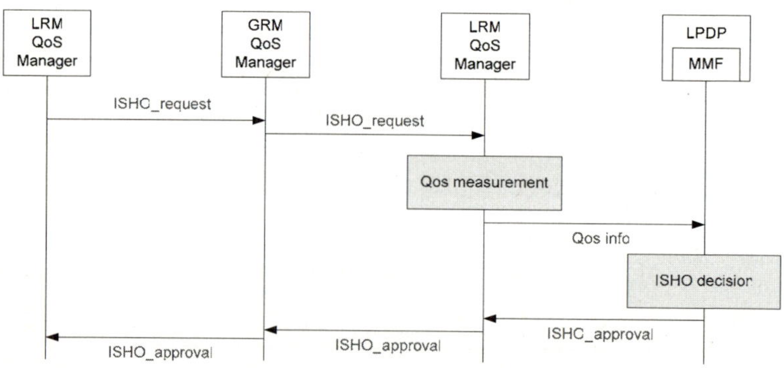

Fig. 7. Handover by service change

If $R_{CN} \leq T_L$, the router buffer can receive all class traffic. When $T_L \leq R_{CN} \leq T_M$, by the traffic check traffic, class EF and AF have transmission priority to BE traffic. Mobile user using low traffic service can change to heterogeneous network by the service handover (Fig. 7). The buffer information in the access router is transferred to LPDP (U) which controls the traffic class. By the class control of the available resource it request the handover to GPD that it checks the policy in the SLA. After the checking, it requires the handover to LPDP (W). Then, the LPDP (W) sends ack message to implement the handover. If $T_M < R_{CN} \leq T_H$, the EF traffic can be only accepted to the buffer instead other class services.

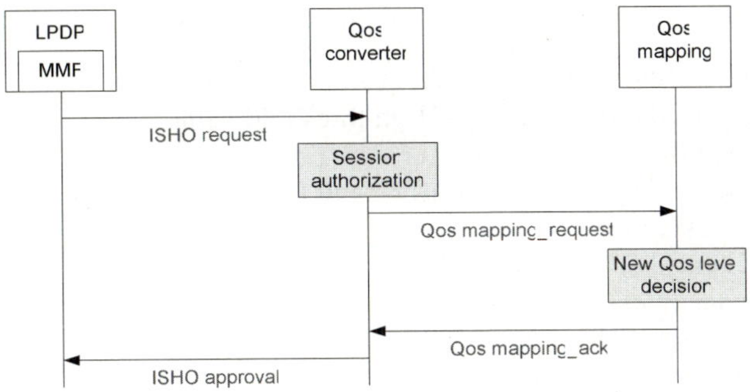

Fig. 8. Agreement process

Hence the handover occurs to heterogeneous network. After the MMF of LPDP (W) determines the handover, it requests the handover to the GPDP. After the GPDP does the agreement process, it allows the handover by sending ack message. The agreement process is shown in Fig. 8. When the MMF of GPDP receives the handover request, it sends the QoS information request message to the RCF of GPDP to manage lower level resource. Then the RCF of GPDP sends the resource information request message to the RCF of LPDP (U) to know the resource state of local area. By the monitoring, the RCF of LPDP (U) sends the ack message to the RCF of GPDP. Once the RCF of GPDP received ack message, it sends the ack message to the MMF of GPDP which also hears the resource state of local area. If for the handover the buffer state is $R_{CN} \leq T_L$, the MMF of GPDP sends the handover request to the MMF. When the MMF receives the request message, it sends the handover ack message to the MMF of GPDP by checking resource state. The GPDP implements the policy process for the handover, and the mobile terminal does the handover successfully after the process agreement. Hence, it can reduce the buffer overload by the handover to control the channel usage when the traffic congestion. If the threshold of buffer is checked by the overload, the load distribution to heterogeneous network can be implemented to guarantee the end-to-end QoS.

4 Simulation and Performance Evaluations

In this chapter, we describe the simulation environments to evaluates the proposed scheme, and discuss the results. We assume the UMTS and the WLAN as access networks. Fig. 9 shows the algorithm to the buffer management algorithm. After the LPEP checks the buffer state, it receives the traffic class as the threshold value, and the PDP manages the resource bye the RCF. If the outside event occurs, sub functions (translation, scheduling and decision) implements the policy function. The MMF also manages the handover and the terminal mobility. To UMTS and WLAN, the buffer size is 20 Mbytes, and system buffers can process 2 Mbps and 1 Mbps respectively. Table 1 presents system parameters to each class traffic transmission rate.

```
R_CN - current buffer value LPEP, T_L - Little threshold value
T_M - Middle threshold value, T_H - High threshold value
If (R_CN <= T_L) InQueue(traffic)
   Else if (R_CN <= T_M) InQueue(traffic) and Block(traffic)
      Else Block(traffic)
Resource_Control ()
Mobile_Management ()
If (ISHO request) ISHO_approve ()
   Else Buffer_State_Check()
Qos_check and ISHO_decision()
Reseource_Control();
/* Once ISHO is determined, the channel state is first checked for QoS manage-
ment. If the channel can be provided to a mobile terminal, the ISHO is finally ap-
proved. */
```

Fig. 9. Buffer management algorithm

Table 1. System parameters

UMTS	
Traffic class	Transmission rate
EF	256 – 128 Kbps
AF	128 – 64 Kbps
BE	64 - 0 Kbps
WLAN	
Traffic class	Transmission rate
EF	512 – 256 Kbps
AF	256 – 128 Kbps
BE	128 - 0 Kbps

We assume that the T_H, T_M and T_L of WLAN are 95%, 93% and 90%, and the T_H, T_M and T_L of UMTS are 90%, 88% and 85%. In UMTS the maximum transmission rate of EF class is 256kbps, when the AF class is 128kbs and the BE class is 64kbps.

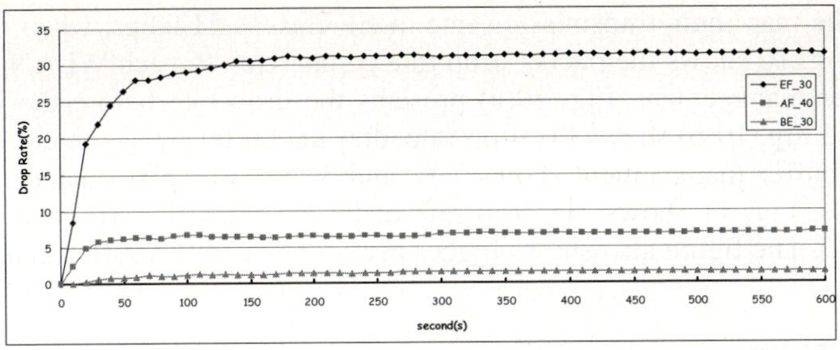

(a) Traffic drop rate before the buffer management

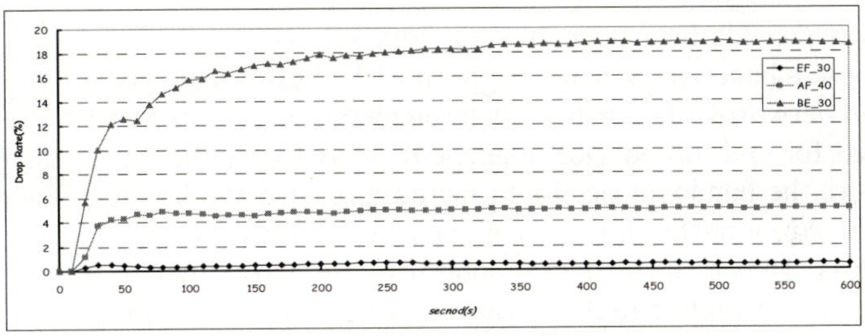

(b) Traffic drop rate after the buffer management

Fig. 10. Drop rate of buffer packet

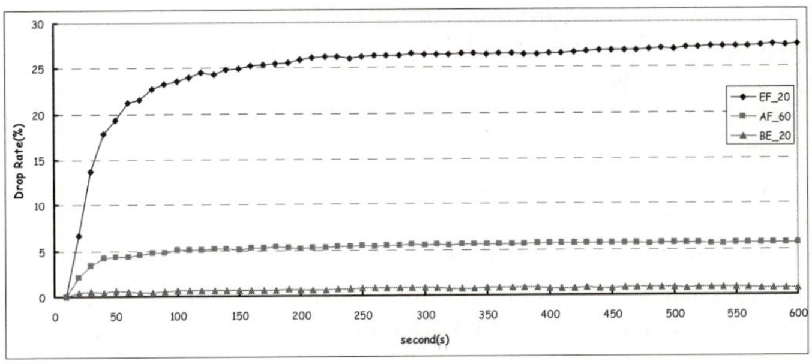

(a) Traffic drop rate before the buffer management

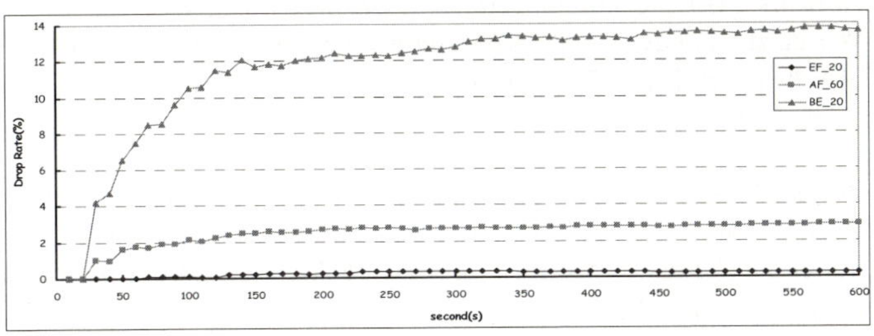

(b) Traffic drop rate after the buffer management

Fig. 11. Drop rate of buffer packet (AF Class ratio 60 %)

In WLAN the maximum transmission rate of EF class is 512kbps, when the AF class is 256kbs. Fig. 10 shows the packet drop rate comparison to each WLAN traffic class by the buffer management. Fig. 10(a) presents the drop rate before the buffer management, and Fig. 10(b) shows the drop rate after the buffer management. As figure is shown, the buffer management shows 1 % and 5 % lower performance to EF class and AF class. Fig. 11 shows the drop rate of buffer when the AF class traffic ratio increase 60%. The figure also shows higher performance of the buffer management.

5 Conclusions

B3G networks provide heterogeneous network services for mobile users. In this paper, we propose a handover scheme supporting the buffer management to increase the reliability of networks. To control heterogeneous access networks, the PDP-PEP model is used for end-to-end QoS management. Also, by the buffer management of access network the handover occurs to increase service availability. When the traffic overload occurs in a buffer, by the handover to the buffer management it can reduce the packet loss. Hence, instead user mobility, the active handover for the channel management can be applied to the advanced policy-based network service.

References

1. Marques, V., Aguiar, R.L., Garcia, C., Moreno, J.I., Beaujean, C., Melin, E., Liebsch, M.: An IP-based QoS architecture for 4G operator scenarios. IEEE Wireless Communications 10(3), 54–62 (2003)
2. Needham, M.L., Natarajan, N.: QoS in B3G Networks – an Overview. In: Proc. of ICCT 2003, pp. 1369–1372 (April 2003)
3. Politis, C., et al.: Cooperative Networks for the Future Wireless World. IEEE Communications Magazine 42(9), 70–79 (2004)
4. Mykoniati, E., et al.: Admission Control for Providing QoS in Diffserv IP Networks: The TEQUILA Approach. IEEE Communications Magazine 41(1), 38–44 (2003)
5. Blake, S., et al.: An Architecture for Differentiated Service. IETF RFC 2475 (December 1998)
6. Cortese, G., et al.: CADENUS: Creation and Deployment of End-User Services in Premium IP Networks. IEEE Communications Magazine 41(1), 54–60 (2003)
7. Zhuang, W., et al.: Policy-Based QoS Architecture in the IP Multimedia Subsystem of UMTS. IEEE Network 17(3), 51–57 (2003)
8. Marilly, E., et al.: Service Level Agreements: A Main Challenge for Next Generation Networks. In: Proc. of IEEE ECUMN 2002, pp. 297–304 (April 2002)
9. Zhuang, W., et al.: Policy-Based QoS Management Architecture in an Integrated UMTS and WLAN Environment. IEEE Communications Magazine 41(11), 118–125 (2003)
10. Yavakar, R., et al.: A Framework for Policy-based Admission Control. RFC2735 (January 2003)
11. Chan, K., et al.: COPS Usage for Policy Provisioning. RFC 3084 (March 2001)
12. Garcia-Macias, J.A., et al.: Quality of Service and Mobility for the Wireless Internet. Wireless Networks 9(4) (July 2003)
13. Mahbubul, et al.: Quality of Service among IP-based Heterogeneous Networks. IEEE Personal Communications (December 2001)

Weighted MDS for Sensor Localization

Duc Vo, Nhat Vo, and Subhash Challa

University of Technology, Sydney, Australia
dvo@eng.uts.edu.au
The University of Melbourne, Australia
n.vo@pgrad.unimelb.edu.au
subhash.challa@nicta.com.au

Abstract. Multidimensional Scaling (MDS) has been recently applied to node localization in sensor networks and gained some very impressive performance. MDS treats dissimilarities of pair-wise nodes directly as Euclidean distances and then makes use of the spectral decomposition of a doubly centered matrix of dissimilarities. However dissimilarities mainly estimated by Received Signal Strength (RSS) or by the Time of Arrival (TOA) of communication signal from the sender to the receiver used to suffer much errors when the distances between nodes are far. From this observation, Weighted Multidimensional Scaling (WMDS) is proposed in this paper. Different from MDS, WMDS incorporates weighting factors to account for the impact of pair-wise estimated dissimilarities in MDS framework. The further distance between two nodes is, the less "impact" weight should be considered. The experiment on real sensor network measurements of RSS and TOA shows the efficiency and novelty of WMDS for sensor localization problem in term of sensor location-estimated error.

Keywords: Sensor Localization, Sensor Network, Multidimensional Scaling, Nonmetric Multidimensional Scaling.

1 Introduction

Localization of the nodes is one of the main issues in a wireless sensor network. While many algorithms have been proposed to estimate the position of the nodes, there is still no algorithm that performs best in all networks [1]. In general, localization algorithms follow the following scheme: anchor unknown distance determination, deriving a nodes position given the anchor distances, and then refinement of the position estimates. Because mobile sensor networks are changing fast as time progresses, not much effort has been invested in researching the refinement phase. However, this phase can be successfully applied to static networks [2]. Localization algorithms can be divided into two categories: *range-based* and *range-free*. In *range-based* algorithms, nodes estimate their distance to seeds using some specialized hardware. Although the use of range measurements results in a fine grained localization scheme, range-based algorithms require the sensors contain hardware to make range measurements [3]. *Range-free* algorithms

O. Gervasi et al. (Eds.): ICCSA 2008, Part II, LNCS 5073, pp. 409–418, 2008.

estimate the location of sensor nodes by, either, exploiting the radio connectivity information among neighboring nodes, or exploiting the sensing capabilities that each sensor node possesses [4]. As a matter of fact, *range-free* techniques are more cost-effective because they do not require sensors to be equipped with any special hardware, but use less information than range-based algorithms. *Range-based* localization depends on the assumption that the absolute distance between a sender and a receiver can be estimated by *Received Signal Strength* (RSS) or by the *Time of Arrival* (TOA) of communication signal from the sender to the receiver. Bahl et al. [5] present RADAR, a radio-frequency (RF) based system for locating and tracking users inside buildings. RADAR operates by recording and processing signal strength information at multiple base stations positioned to provide overlapping coverage in the area of interest. Bischoff et al. [6] proposed a lightweight localization approach for supporting distance and range queries in ad hoc wireless sensor networks by using RSS to estimate distances. Ward et al. [7] used *Time of Arrival* of signals, while Priyantha et al. [8] and Savvides et al. [9] have used Time Difference of Arrival of signals to estimate distances. And recently Dil et al. [1] proposed a novel range-based algorithm based on the Monte Carlo approach. A review of *range-free* localization schemes from the perspectives of anchor-based and anchor-free solutions are well mentioned in [4]. We just name a few important recent algorithms as follow. He et al. [10] proposed a *range-free* algorithm called APIT in which all possible triangles of the seeds are formed. The location of a node is the center of intersection region of all triangles. In Gradient algorithm, proposed by Nagpal et al. in [11], the anchor nodes initiate a gradient that self-propagates and allows a sensor node to infer its distance from the anchor. After estimating distances to three anchors a sensor node infers its own location through multilateration. Niculescu and Nath [12] proposed DV-Hop which is similar to Gradient, but uses a different method for estimating the average distance per hop. In addition to the localization schemes mentioned above, many more have been proposed. In [13], Savvides et al. derive the Cramer-Rao lower bound (CRLB) for network localization, expressing the expected error characteristics for an ideal algorithm, and compare it to the actual error in an algorithm based on multilateration. Moore et al. [14] introduces the probabilistic notion of robust quadrilaterals as a way to avoid ip ambiguities that otherwise corrupt localization computations. The algorithm MAL (mobile-assisted localization) proposed in [15] is a method which employs a mobile user to assist in measuring distances between node pairs until these distance constraints form a "globally rigid" structure that guarantees a unique localization. Resilient LSS proposed in [16] is a localization scheme which is an extension of least squares scaling (LSS) with soft constraints to exploit deployment-specific requirements, such as minimum node spacing.

2 Related Work

We consider sensor location estimation when sensors measure received signal strength (RSS) or time-of-arrival (TOA) between themselves and neighboring sensors. We use an extensive TOA and RSS measurement campaign in an indoor office from the work in [17]. They derived Cramer Rao Bound (CRB) and maximum-likelihood estimators (MLEs) under Gaussian and log-normal models for the TOA and RSS measurements, respectively. Some related works are previous ones based on MDS [18][19][20]. MDS-MAP [18] uses a technique from mathematical psychology called Multidimensional Scaling (MDS). The intuition behind multidimensional scaling is simple. Suppose there are N points, suspended in a volume. We don't know the positions of the points, but we do know the distance between each pair of points. Multidimensional scaling is an $O(N^3)$ algorithm that uses the Law of Cosines and linear algebra to reconstruct the relative positions of the points based on the pairwise distances. Local MDS [19], a local version of MDS is used to compute maps of many local arrangements of nodes. These local maps are pieced together to obtain global maps. This method tends to perform better than the global MDS method when node density is non-uniform, or in coverage exist. The local calculations allow a distributed implementation, but weights are restricted to be either 0 or 1. In this paper, we focus on sensor location estimation using pair-wise RSS or TOA measurements in a wireless network. Consider a network of $N = m + n$ sensors with m reference and n blindfolded sensors ($m \ll n$). We also call those m known-location sensors anchor nodes. The sensor localization problem corresponds to the estimation of sensor coordinates $X = [x_1, x_2, ..., x_N]$, where $x_i \in \Re^2$ (extention of these results to $\Re^3$ is also possible). Suppose that $[x_1, x_2, ..., x_n]$ are unknown-location sensor coordinates and that $[x_{n+1}, x_{n+2}, ..., x_{n+m}]$ are known-location sensor coordinates, we aim to find $[x_1, x_2, ..., x_n]$ based on the estimated distance (or dissimilarity) δ_{ij} between sensor i and j. The estimation of δ_{ij} could be based on RSS or TOA. The next section of the paper describes Classical MDS and Weighted MDS in detail. Section 4 presents Procrustes Analysis which seeks the isotropic dilation and the rigid translation, reflection and rotation needed to match the estimated coordinate to the true coordinate of sensor locations. The proposed sensor localization algorithm is also presented in Section 4. Section 5 comes with the experiment on the real measured sensor network. Conclusions are stated in Section 6.

3 Multidimensional Scaling - MDS

Suppose a set of N objects is under consideration and between each pair of objects (i, j) there is a measurement δ_{ij} of the "dissimilarity" between the two objects. The typical goal of MDS is to find a low dimensional space, usually Euclidean, in which points in the space represent the objects, one point representing one object, and such that the distances between the points in the space $\{d_{ij}\}$ match as well as possible the original dissimilarities $\{\delta_{ij}\}$ [21].

3.1 Classical Multidimensional Scaling - CMDS

Classical Multidimensional Scaling treats dissimilarities $\{\delta_{ij}\}$ directly as Euclidean distances and then makes use of the spectral decomposition of a doubly centered matrix of dissimilarities. Formally, a symmetric $N \times N$ matrix $\Delta = [\delta_{ij}]$ is called a dissimilarity matrix if $\delta_{ij} \geq 0$ (nonnegative elements) and $\delta_{ii} = 0$ (zero diagonal elements). From a given dissimilarity matrix Δ, CMDS algorithm constructs a configuration of points in a Euclidean space of specified dimension p [22]. Because the distances in the configuration space are to be Euclidean, we can have

$$\delta_{ij}^2 = d_{ij}^2(X) = \|x_i - x_j\|^2 = (x_i - x_j)^T(x_i - x_j) \tag{1}$$

By writing the squared distances as $d_{ij}^2(X) = x_i^T x_i - 2x_i x_j + x_j^T x_j$ and defining $\Psi = \left[x_1^T x_1, .., x_N^T x_N\right]^T$ [20], the squared distance matrix $D = [d_{ij}^2(X)]_{i,j=1}^N$ now can be written as

$$D = \Psi e^T - 2X^T X + e\Psi^T \tag{2}$$

where $e \in \Re^n$ is a vector of all ones, and $X = [x_1, x_2, ..., x_N]$. Let us define the inner product matrix $B = X^T X$. Matrix B can be rewritten as

$$B = -HDH \tag{3}$$

where $H = I - ee^T/N$ is the centering operator. In order to recover the coordinates from B, CMDS needs to extract the eigenvectors and eigenvalues as follow

$$B = U\Lambda U^T \tag{4}$$

Let $\Lambda_p = diag(\lambda_1, .., \lambda_p)$ be the diagonal matrix of p largest non-zero eigenvalues of B, and $U_p = [u_1, .., u_p]$ is the corresponding eigenvectors. Then the coordinate matrix X is given by $X = U_p \Lambda_p^{1/2}$.

3.2 Weighted Multidimensional Scaling - WMDS

The objective of MDS is to find a matrix X such that $d_{ij}(X)$ matches δ_{ij} as closely as possible. This objective can be formulated in a another way by using the definition of the *Stress* function $s(X)$ as follow

$$X = \arg\min_X s(X) \tag{5}$$

where the *Stress* function $s(X)$ is the least-squares criterion to measure the distortion of the distances due to the embedding, i.e.

$$s(X) = \sum_{i<j} \left(\|x_i - x_j\|^2 - \delta_{ij}\right)^2 \tag{6}$$

Note that due to the symmetry of the dissimilarities and the distances, the summation only involves the pairs (i, j) where $i < j$. Different from the *Stress* function in MDS, the *Stress* function $s(X)$ in Weighted MDS contain nonnegative weights to weight the contribution of the corresponding distorted distances in computing and minimizing the *Stress* function and is defined as

$$s(X) = \sum_{i<j} w_{ij} \left(\|x_i - x_j\|^2 - \delta_{ij}\right)^2 \tag{7}$$

There are many ways that $s(X)$ could be minimized. For example, ones can use an iterative steepest descent approach. However, a significantly better (in terms of guarantees on, and rate of, convergence) method for minimizing stress was introduced by Jan de Leeuw [23]. De Leeuw's iterative majorization method at each step minimizes a simple convex function which both bounds s from above and touches the surface of s at a point Z, called the supporting point. In convex analysis such a function is called a majorizing function. This iterative majorization process is also referred to as the SMACOF algorithm (Scaling by majorizing a complicated function).

4 Procrustes Analysis and Proposed Algorithm

Procrustes analysis dilates, translates, reflects and rotates one of the configurations of points to match, as well as possible, the other, enabling a comparison of the two configurations to be made [21]. In this section, we will briefly review procrustes analysis and propose our sensor localization algorithm.

4.1 Procrustes Analysis

Procrustes analysis seeks the isotropic dilation and the rigid translation, reflection and rotation needed to match one configuration to the other [21]. Suppose a configuration of N points in a q-dimensional Euclidean space, with coordinates given by the $N \times q$ matrix $O_1 = [x_1, .., x_N]^T \in \Re^{N \times q}$, needs to be optimally matched to another configuration of N points in a p-dimensional Euclidean space ($p \geq q$), with coordinates given by the $O_2 = [y_1, .., y_N]^T \in \Re^{N \times p}$. In order to make both configurations be at the same p-dimensional space, we firstly add $p - q$ columns of zeros at the right of matrix O_1. It is assumed that the ith point in the first configuration is in a one-to-one correspondence with the ith point in the second configuration. Let the point x_i in the X space be dilated, translated, rotated to new coordinate $\tilde{x}_i = cA^T x_i + b$, the objective function $\Phi(c, A, b)$ needed to be minimized is the new sum of squared distances between points and presented as

$$\Phi(c, T, b) = \sum_{i=1}^{N} (y_i - cA^T x_i - b)^T (y_i - cA^T x_i - b) \tag{8}$$

where the matrix A is orthogonal giving a rigid rotation, vector b is a rigid translation vector, and c is the dilation. The solution to procrustes analysis is well described in [21] and can be summarized as following steps:

- Subtract the mean vectors for the configurations from each of the respective points in order to have the centroids at the origin.
- Find the rotation matrix $A = (O_1^T O_2 O_2^T O_1)^{1/2} (O_2^T O_1)^{-1}$ and rotate the O_1 configuration to $O_1 A$
- Scale the O_1 configuration by multiplying each coordinate by c, where $c = tr(O_1^T O_2 O_2^T O_1)^{1/2} / tr(O_1^T O_1)$

The value of minimized objective function is

$$\Phi_{opt} = tr(O_2^T O_2) - \left\{tr(O_1^T O_2 O_2^T O_1)^{1/2}\right\}^2 / tr(O_1^T O_1) \tag{9}$$

and the translation vector can be calculated as $b = \mu_y - cA^T \mu_x$, where μ_x and μ_y are the centroids of the two configurations.

4.2 Proposed Sensor Localization Algorithm

Our proposed sensor localization algorithm based on WMDS and procrustes analysis consists of three steps as follow:

Step 1. Construct the dissimilarity (or distance) matrix which contains the dissimilarities between all pairs of sensor nodes based on RSS or TOA.

Step 2. Apply WMDS to the dissimilarity matrix to get the two (or three) dimensional coordinates which is a 2D (or 3D) relative map (in our work, we only experiment on two dimensional map). The weights in WMDS are chosen as $w_{ij} = 1/\theta_{ij}$.

Step 3. Given sufficient anchor nodes , we use procrustes analysis to find the rigid rotation matrix A, rigid translation vector b, and the dilation c that match the configuration of absolute anchor nodes to the configuration of relative anchor nodes. And then we use A, b and c to transform the other relative nodes (location-unknown nodes) to form the final estimated map.

5 Experiment on a Measured Network

In our experiment, we use sensor network measurements of received signal strength (RSS) and time-of-arrival (TOA) were made publicly and originally reported in [17]. This data set includes the RSS and TOA range measurements from a network of 44 devices (4 of which are anchor nodes). The measurements were made in an open plan office building, within a 14 × 14m square area. Four devices near the corners are chosen as reference devices (node 3, 10, 35, 44). The remaining 40 devices are blindfolded devices (see Fig. 1). Before apply MDS or WMDS, we need to construct the dissimilarity (or distance) matrix which contains the dissimilarities between all pairs of sensor nodes based on RSS or TOA.

- *For Distance matrix based on RSS:* We use equation (7) in [17] to estimate range from given received power P_{ij} as:

$$\delta_{ij} = d_0 \left(\frac{P_0}{P_{ij}}\right)^{1/n_p} \tag{10}$$

  where P_0(dBm) is the received power in decibel milliwatts at a reference distance d_0 and The path loss exponent n_p is a function of the environment. More details can be referenced in [17].

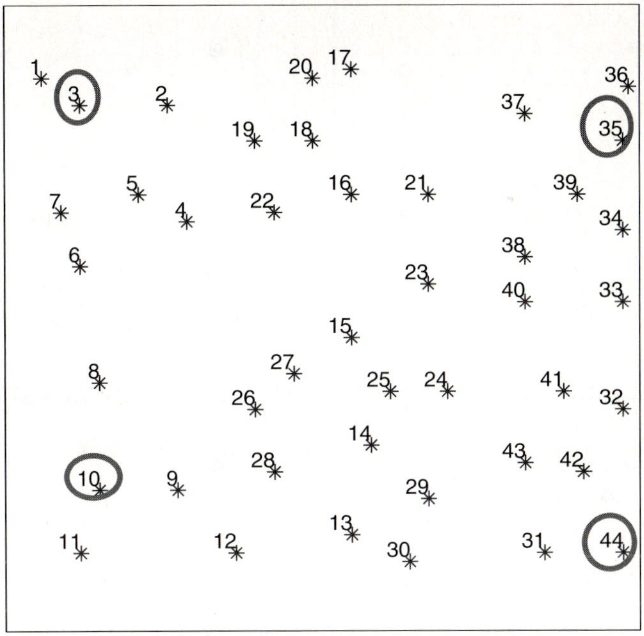

Fig. 1. Experimental sensor locations of 44 nodes with 4 circled anchor nodes (3, 10, 35, 44)

– *For Distance matrix based on TOA:* We use data matrix of time delay between sensors, in seconds (a 44 by 44 matrix). The (i, j) element is the T_{ij} reported in Section IV of [17], i.e., the average of 10 measured time delays, 5 with the transmitter at i and receiver at j, and 5 with the transmitter at j and receiver at i. The distance between sensor i and j is calculated as

$$\delta_{ij} = |T_{ij}| \times \rho \tag{11}$$

where ρ is speed of propagation (speed of light) in m/s.

In our experiments, the estimated device locations are compared with the actual locations with Root Mean Squared Error (RMSE) to measure location-estimated error. A comparison of sensor Localization based on MDS and WMDS using RSS and TOA range measurements from sensor network can be seen in Fig. 2 .

The true and estimated sensor positions are marked by '*' and 'x', respectively, where the lines represent the estimation errors and the anchor nodes are circled with red color. We also compare the performance of the WMDS algorithm to classical MDS, and the MLE based solutions from [17]. Table 1 summarizes the RMSE of the location estimates based on those methods. Next, in order to evaluate the performance of MDS and WMDS versus the number of anchor nodes, we choose randomly $m = 3, .., 10$ sensors as anchor nodes, then apply MDS and WMDS. 50 times of random selection were performed for each value

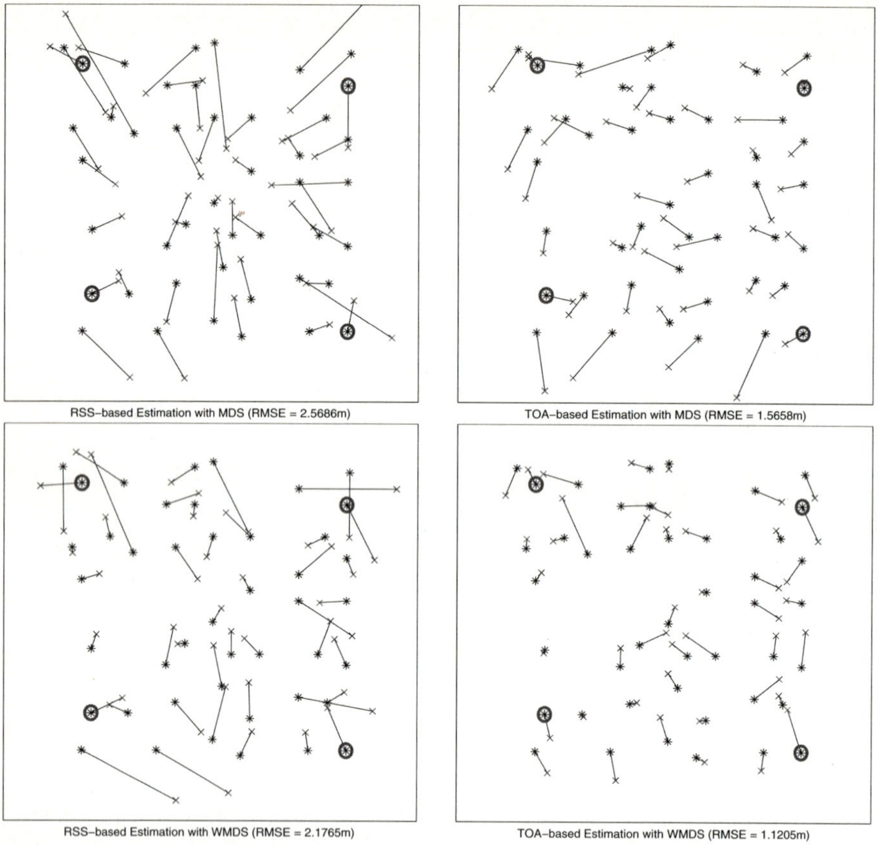

Fig. 2. Sensor Localization based on MDS and WMDS are performed using RSS and TOA range measurements from sensor network. True and estimated sensor locations are marked, respectively, by '*' and 'x', while anchor nodes are circled.

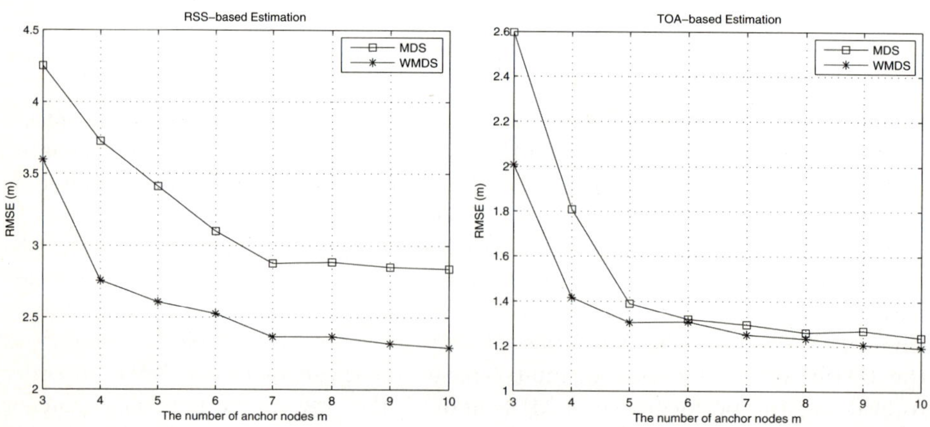

Fig. 3. The performance of MDS and WMDS versus the number of anchor nodes

Table 1. RMSE of location estimates based on MDS, MLE, and WMDS in experimental network

	MDS	MLE [17]	WMDS
RSS	2.57m	2.18m	**2.18m**
TOA	1.57m	1.23m	**1.12m**

of m and the average RMSE result was recorded. It can be seen from Fig. 3 that WMDS outperforms MDS, and the RMSE of both MDS and WMDS reduces when the number of anchor nodes increases.

6 Conclusion

We applied Weighted Multidimensional Scaling approach for sensor localization that outperforms MDS with real measurements of received signal strength (RSS) and time-of-arrival (TOA). Because WMDS includes the individual weights to reflect the impact of distances on the objective function, it is more effective for the problem of sensor localization than MDS. The impressive experiment results of WMDS on real measured sensor network exhibits the promising prospects of this approach for the problem of sensor localization. Our future work will be applying this approach for the sensor network with mere connectivity.

Acknowledgment

The authors would like to thank Dr. Neal Patwari for providing the measurements on RSS and TOA of sensor network.

References

1. Dil, B., Dulman, S., Havinga, P.J.M.: Range-based localization in mobile sensor networks. In: Römer, K., Karl, H., Mattern, F. (eds.) EWSN 2006. LNCS, vol. 3868, pp. 164–179. Springer, Heidelberg (2006)
2. Langendoen, K., Reijers, N.: Distributed localization in wireless sensor networks: a quantitative comparison. Comput. Networks 43(4), 499–518 (2003)
3. Rudafshani, M., Datta, S.: Localization in wireless sensor networks. In: IPSN, pp. 51–60 (2007)
4. Poovendran, R., Wang, C., Roy, S.: Secure Localization and Time Synchronization for Wireless Sensor and Ad Hoc Networks (Advances in Information Security). Springer, New York (2006)
5. Bahl, P., Padmanabhan, V.N.: Radar: An in-building rf-based user location and tracking system. In: INFOCOM, pp. 775–784 (2000)
6. Bischoff, U., Strohbach, M., Hazas, M., Kortuem, G.: Constraint-based distance estimation in ad-hoc wireless sensor networks. In: Römer, K., Karl, H., Mattern, F. (eds.) EWSN 2006. LNCS, vol. 3868, pp. 54–68. Springer, Heidelberg (2006)

7. Ward, A., Jones, A., Hopper, A.: A new location technique for the active office. IEEE Personnel Communications 4(5), 42–47 (1997)
8. Priyantha, N.B., Chakraborty, A., Balakrishnan, H.: The Cricket Location-Support System. In: 6th ACM MOBICOM, Boston, MA (2000)
9. Savvides, A., Han, C.C., Srivastava, M.B.: Dynamic fine-grained localization in ad-hoc networks of sensors. In: MOBICOM, pp. 166–179 (2001)
10. He, T., Huang, C., Blum, B.M., Stankovic, J.A., Abdelzaher, T.: Range-free localization schemes for large scale sensor networks. In: MobiCom 2003: Proceedings of the 9th annual international conference on Mobile computing and networking, pp. 81–95. ACM Press, New York (2003)
11. Nagpal, R., Shrobe, H.E., Bachrach, J.: Organizing a global coordinate system from local information on an ad hoc sensor network. In: Zhao, F., Guibas, L.J. (eds.) IPSN 2003. LNCS, vol. 2634, pp. 333–348. Springer, Heidelberg (2003)
12. Niculescu, D., Nath, B.: Ad hoc positioning system (aps). In: Proc. IEEE GlobeCom, San Antonio, AZ (2001)
13. Savvides, A., Garber, W.L., Adlakha, S., Moses, R.L., Srivastava, M.B.: On the error characteristics of multihop node localization in ad-hoc sensor networks. In: Zhao, F., Guibas, L.J. (eds.) IPSN 2003. LNCS, vol. 2634, pp. 317–332. Springer, Heidelberg (2003)
14. Moore, D., Leonard, J., Rus, D., Teller, S.: Robust distributed network localization with noisy range measurements. In: SenSys 2004: Proceedings of the 2nd international conference on Embedded networked sensor systems, pp. 50–61. ACM Press, New York (2004)
15. Priyantha, N.B., Balakrishnan, H., Demaine, E., Teller, S.: Mobile-Assisted Localization in Wireless Sensor Networks. In: IEEE INFOCOM, Miami, FL (2005)
16. Kwon, Y., Mechitov, K., Sundresh, S., Kim, W., Agha, G.: Resilient localization for sensor networks in outdoor environments. In: ICDCS 2005: Proceedings of the 25th IEEE International Conference on Distributed Computing Systems (ICDCS 2005), Washington, DC, USA, pp. 643–652. IEEE Computer Society Press, Los Alamitos (2005)
17. Patwari, N., Hero, A.O.I., Perkins, M., Correal, N., O'Dea, R.: Relative location estimation in wireless sensor networks. In: IEEE Transactions on Signal Processing, vol. 51, pp. 2137–2148 (2003)
18. Shang, Y., Ruml, W., Zhang, Y., Fromherz, M.P.J.: Localization from mere connectivity. In: MobiHoc, pp. 201–212 (2003)
19. Ji, X.: Sensor positioning in wireless ad-hoc sensor networks with multidimensional scaling. In: INFOCOM (2004)
20. Costa, J.A., Patwari, N., Alfred, O., Hero, I.: Distributed weighted-multidimensional scaling for node localization in sensor networks. ACM Trans. Sen. Netw. 2(1), 39–64 (2006)
21. Cox, T.F., Cox, M.A.A.: Multidimensional Scaling. Chapman and Hall, Boca Raton (1994)
22. Kearsley, A., Tapia, R., Trosset, M.: The Solution of the Metric STRESS and SSTRESS problems in Muldimensional Scaling Using Newton's Method. Computational Statistics 13(3), 369–396 (1998)
23. de Leeuw, J.: Applications of convex analysis to multidimensional scaling. Recent developments in statistics, 133–145 (1977)

A Comparative Study of Directional Node-Disjoint Multipath Routing in Wireless Ad Hoc Network*

Abhijit Saha[1], Shafqat-Ur-Rehman[2], Mikyung Kang[1], In-Hye Shin[1], Gyung-Leen Park[1,**], and Junghoon Lee[1]

[1] Dept. Of Computer Science and Statistics, Cheju National University,
[2] Dept. of Computer Engineering, Cheju National University,
690-756, Ara 1 Dong, Jeju-si, Jeju-do, Republic of Korea.
{abhijit298,shafqat,mkkang,ihshin76,glpark,jhlee}@cheju.ac.kr

Abstract. This paper compares the performance of on-demand Directional Node-Disjoint Multipath Routing (DNDMR) protocol with those of contemporary wireless ad hoc network routing protocols. DNDMR protocol especially performs well for grid-style network topology. Multipath routing inherently allows the establishment of multiple routes between source and destination. The important features of DNDMR are path accumulation with a novel directional method, reducing routing overhead by controlled propagation, and ensuring node-disjoint paths automatically. DNDMR improves the reliability of data communication (i.e., fault tolerance) over the wireless ad hoc network. It significantly reduces control traffic overhead and end-to-end delay, and improves throughput. We evaluate the comparative performance of DNDMR using ns-2 simulator. The performance is measured comparative to existing unipath routing (DSDV, DSR, AODV), and multipath routing (AOMDV) protocols. The simulation results reveal that DNDMR has better performance and more reliable than the contemporary unipath and multipath routing protocols.

Keywords: Wireless Network, Ad hoc Network, Routing Protocols, Unipath and Multipath Routing Protocols.

1 Introduction

Wireless networking is gaining large attention and becoming very popular in these years, due to its potential and possibilities. Infrastructure networks and mobile ad hoc networks are two main architectures of wireless networking. Infrastructure networks include cellular networks and wireless local area network. Users are connected via base stations/access points, and backbone networks. Ad hoc network is a recently developed part of wireless communication. The difference from traditional wireless networks is that there is no need for established infrastructure. Since there is no such

* "This research was supported by the Ministry of Knowledge Economy, Korea, under the ITRC (Information Technology Research Center) support program supervised by the IITA (Institute of Information Technology Advancement)" (IITA-2008-C1090-0801-0040).

** Corresponding author.

O. Gervasi et al. (Eds.): ICCSA 2008, Part II, LNCS 5073, pp. 419–432, 2008.

infrastructure and therefore no preinstalled routers, which can forward packets from one host to another, this task has to be taken over by mobile nodes of the network.

The main objective of ad hoc routing protocol is finding a short and optimized route from the source to the destination node without predetermined topology or centralized control. All nodes are mobile and topology changes frequently. Design issues in developing a routing protocol for wireless network is much complicated than those for wired network. Hence, routing protocol is a challenging issue in wireless ad hoc network communication. The major challenges for routing protocols are: mobility, bandwidth constraint, error-prone and shared channel, location-dependent contention, and other resource constraint. Minimum route acquisition delay, quick route reconfiguration, loop prevention, minimum control overhead, scalability, provisioning of Quality of Service, support for time-sensitive traffic are the major requirements of a routing protocol in ad hoc wireless networks.

Routing protocols can be classified either as proactive or reactive. Reactive also called on-demand routing protocols consume less bandwidth than proactive protocols. On-demand routing protocols have some limitation that all of them build and transmit on a unipath route for each data session. When there is a link break on the active route, all of them have to invoke a route discovery process. By establishing multiple paths between a source and destination in a single route discovery, on-demand multipath routing protocols can alleviate these problems. Path rediscovery needs to initiate only when there is no path available from source to destination. In the previous work, Directional Node-Disjoint Multipath Routing (DNDMR) protocol [7] is proposed and compared with on-demand unipath routing protocols. In this paper, a comparative study of DNDMR is presented with respect to contemporary unipath and multipath routing protocols in wireless ad hoc networks, as DNDMR reduces the routing overhead significantly and achieves two node-disjoint routing paths efficiently. To demonstrate the effectiveness of DNDMR, we have compared our protocol with Proactive unipath routing protocol (DSDV), Reactive or also known as On-demand unipath routing protocol (DSR, and AODV), and Multipath routing protocol (AOMDV). The comparative analysis shows that DNDMR has better performance, and more reliable compared with other mentioned protocols.

The rest of this paper is organized as follows: Section 2 introduces the related work. Section 3 describes the DNDMR protocol mechanism in detail. The performance evaluation is presented in Section 4, and Concluding remarks of this paper in Section 5.

2 Related Work

Based on the number of routers discovered between source and destination, the protocols can be either as unipath or multipath protocols. Most of the currently proposed routing protocols for ad hoc networks are unipath routing protocols. In this routing protocol, only a unipath or single route is used between a source and destination node. Destination-Sequenced Distance Vector (DSDV), Dynamic Source Routing (DSR), and Ad hoc On-demand Distance Vector (AODV) protocols are the most popular and widely used protocols belonging to the unipath protocols. On the other hand, Multipath routing consists of finding multipath or multiple routes between a source and destination node. These multiple paths between source and destination node pairs can be used to compensate for the dynamic and unpredictable nature of ad hoc networks.

Multipath routing is very useful technique to find out the multiple paths between source and destination by using a single route discovery. Multipath routing protocols can attempt to find node-disjoint, link-disjoint, or non-disjoint routes. Node-disjoint routes have no nodes or links in common. Link-disjoint routes have no links in common, but may have nodes in common. Non-disjoint routes can have nodes and links in common. The multipath routing is more effective than the single path routing because multipath can provide load balancing, fault-tolerance, and higher aggregated bandwidth. Multipath routing protocols in ad hoc networks have been proposed in [13-24]. Although these protocols build multiple routes on-demand, most of them just discuss link-disjoint or non-disjoint paths; very less work has been done for node-disjoint path, where node-disjoint routes offer the highest degree of fault-tolerance. In addition, all of them flood Route Request (RREQ) packets to the entire network in order to discover multiple routing paths.

2.1 Destination-Sequenced Distance-Vector Routing

The Destination Sequenced Distance Vector (DSDV) protocol [8] is a proactive hop-by-hop distance vector routing protocol, requiring each node to broadcast routing updates periodically. Here, every MH in the network maintains a routing table for all possible destinations within the network and the number of hops to each destination. Each entry is the marked with a sequence number assigned by the destination MH. The sequence numbers enable the MHs to distinguish stale routes from new ones, thereby avoiding the formation of routing loops. Routing table updates are periodically transmitted throughout the network in order to maintain consistency in the tables.

To alleviate potentially large network update traffic, two possible types of packets can be employed: full dumps or small increment packets. A full dump type of packet carries all available routing information and can require multiple Network Protocol Data Units (NPDUs). These packets are transmitted less frequently during periods of occasional movements. Smaller incremental packets are used to relay only the information that has changed since the last full dump. Each of these broadcasts should fit into a standard-size NPDU, thereby decreasing the amount of traffic generated. The MHs maintain an additional table where they store the data sent in the incremental routing information packets. New route broadcasts contain the address of the destination, the number of hops to reach the destination, the sequence number of the information received regarding the destination, as well as a new sequence number unique to the broadcast. The route labeled with the most recent sequence number is always used. In the event that two updates have the same sequence number, the route with the smaller metric is used in order to optimize (shorten) the path. MHs also keep track of settling time of the routes, or the weighted average time that routes to a destination could fluctuate before the route with the best metric is received. By delaying the broadcast of a routing update by the length of the settling time, MHs can reduce network traffic.

2.2 Dynamic Source Routing

The Dynamic Source Routing (DSR) protocol [10] is an on-demand routing protocol that is based on the concept of source routing: the source includes the whole route in every packet sent, and routes are discovered only when needed. DSR consists of two major phases: route discovery and route maintenance.

If a source node wants to send a packet to destination which it does not have a route in its route cache, it initiates route discovery procedure by broadcasting a Route Request (RREQ) message to its neighboring nodes. Upon receiving the RREQ, if a node knows of a route to the destination, it can send a Route Reply (RREP) message back to the source; otherwise it will append its own address to the list of the nodes in the RREQ and forward the request. Upon receiving the RREQ, the destination obtains a route from the source to the destination, and in the presence of bidirectional links it can simply reverse this route and use it for sending a RREP message along this route to the source. Sequence number mechanism ensures that a node only forwards each RREQ at most once. Since shorter routes require few hops, the first RREQ to reach the destination is likely to have taken a route that is minimum or close to minimum in terms of hop count.

If a source route breaks, the intermediate node sends a Route Error (RERR) message to the source node. When RERR packet is received, the hop in error is removed from the node's route cache and all routes containing the hops are truncated at that point. The source node may then resend the packet using an alternative route in its route cache, or initiate a new route discovery. If the intermediate node has a different route to the destination in its own cache, it can initiate packet salvaging and forward the packet using this alternative route.

2.3 Ad Hoc On-Demand Distance Vector Routing

The Ad Hoc On-Demand Distance Vector (AODV) routing [11] is a reactive protocol, even though it still uses characteristics of a proactive protocol. AODV takes the interesting parts of DSR and DSDV, in the sense that it uses the concept of route discovery and route maintenance of DSR and the concept of sequence numbers and sending of periodic hello messages from DSDV.

In AODV, when a source node has data traffic to send to a destination node, it first initiates a route discovery process. In this process, the source node broadcasts a Route Request (RREQ) packet. Neighbor nodes which do not know an active route for the requested destination node, forward the packet to their neighbors until an active route is found or the maximum number of hops is reached. When an intermediate node knows an active route to the requested destination, it sends a Route Reply (RREP) packet back to the source node in unicast mode. Eventually, the source node receives the RREP packet and opens the route.

When either destination or intermediate node moves, a Route Error (RERR) is sent to the affected source node. When source node receives the RERR, it can reinitiate route discovery if the route is still needed. Neighborhood information is obtained from broadcast Hello packet.

2.4 Ad Hoc On-Demand Multipath Distance Vector

Ad Hoc On-Demand Multipath Distance Vector (AOMDV) [17] is an extension to the AODV protocol for finding multiple loop-free and link disjoint paths. The protocol computes multiple loop-free and link-disjoint paths. Loop-freedom is guaranteed by using a notion of "advertised-hopcount". Link disjointness of multiple paths is achieved by using a particular property of flooding.

To keep track of multiple routes, the routing entries for each destination contain a list of the next-hops together with the corresponding hop counts. All the next hops have the same sequence number. For each destination, a node maintains the advertised hop count, which is defined as the maximum hop count for all the paths. This is the hop count used for sending route advertisements of the destination. Each duplicate route advertisement received by a node defines an alternative path to the destination. To ensure loop-freedom, a node only accepts an alternative path to the destination if it has a lower hop count than the advertised hop count for that destination. Because the maximum hop count is used, the advertised hop count therefore does not change for the same sequence number. When a node advertisement is received for a destination with a greater sequence number, the next-hop list and advertised hop count are reinitialized.

AOMDV can be used to find link-disjoint routes. To find disjoint routes, each node does not immediately reject duplicate RREQs. Each RREQ carries an additional field called first hop to indicate the first hop (neighbor of the source) taken by it. Also, each node maintains a first hop list for each RREQ to keep track of the list of neighbors of the source through which a copy of the RREQ has been received. In an attempt to get multiple link-disjoint routes, the destination replies to duplicate RREQs regardless of their first hop. To ensure link-disjointness in the first hop of the RREP, the destination only replies to RREQs arriving via unique neighbors. The trajectories of each RREP may intersect at an intermediate node, but each takes a different reverse path to the source to ensure link-disjoiness.

In the case of multipath routing, discovering the multiple routing paths from source to destination is very crucial task. As studied the related works, intermediate node(s) can take the duplicate RREQ over condition or can avoid the duplicate RREQ, depending on the routing discovery approach. In the case of DNDMR, intermediate nodes take the duplicate path and transmit the RREQ directionally, which results in the discovery of preeminent node-disjoint paths. We have reduced the routing overhead by controlled propagation of routing packet through the network. The routing packets thus received at the destination node automatically form node-disjoint paths. DNDMR is an approach with improved efficiency in terms of routing overhead, end-to-end delay, and fault tolerance. In the next section, Directional Node-Disjoint Multipath Routing (DNDMR) Protocol with low control traffic overhead is explained in details.

3 Directional Node-Disjoint Multipath Routing

3.1 Route Discovery

Directional Node-Disjoint Multipath Routing (DNDMR) Protocol is an on-demand routing protocol which builds the multiple node-disjoint routes from the source to the destination by using Route Request (RREQ) and Route Reply (RREP) messages. When a source requires a route to the destination but no route is known, it will initiate a route discovery by flooding a RREQ packet directionally throughout the entire network. Because of this flooding, several duplicate packets traverse through the multiple node-disjoint routes to the destination in direction-wise fashion. Once the RREQ reaches the destination, a RREP will reverse the route in the route record of the RREQ and traverse back through this route.

3.2 Route Request Propagation

The main goal of DNDMR is to build multiple node-disjoint paths with a very low routing overhead. To achieve this goal in on-demand routing, the destination node must know the entire path of all available routes so that it can select the node-disjoint paths. Therefore, we use the source routing approach where node information of route is included in the RREQ packet. When the source node has data packets to send to the destination, but does not have the route information, it broadcast the RREQ packet. Every node knows its one hop neighbor through neighbor discovery. Every node periodically broadcasts HELLO packets to its one-hop neighbors. This knowledge of one-hop neighbors is utilized while deciding the next hop of RREQ packet during the route discovery process. The packet contains source id and a sequence numbers that uniquely identify the packet. For propagation of RREQ in DNDMR, we have proposed a novel directional approach. Before broadcasting a packet, every node checks its one-hop neighbor nodes, and after that it floods the packet in its left, right, up, and down direction. After receiving RREQ from the source node, intermediate node(s) rebroadcast the packet respectively. Intermediate node can not rebroadcast the RREQ from which side it received the packet.

Moreover, when a node receives the same RREQ from a different incoming link than the link from first RREQ is received and if packets current hop count is not greater than the first received RREQ it is considered as the duplicate.

For finding the node-disjoint multiple paths in DNDMR, the intermediate node takes the duplicate RREQ and retransmits both RREQ in the entire network. Figure 1 shows the propagation process of DNDMR. When an intermediate node first receives a RREQ packet from its neighbor, it checks how many nodes there are in its one-hop neighborhood and then it decides whether duplicate RREQ can be received from another neighboring node or not. If intermediate node not received the duplicate RREQ, it retransmits the packet directionally (left, right, up, and down). If it finds that duplicate RREQ could be received from another neighboring node, the responding node, from which side it received the packet, retransmits the packet only to the opposite node (if the node is not available in opposite side, it will be discarded automatically).

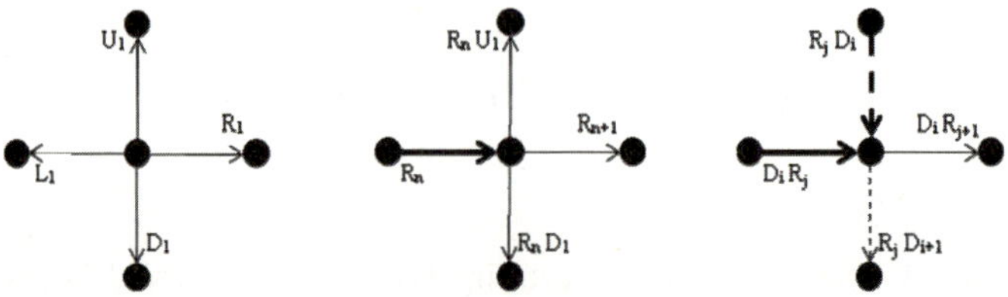

Figure 1(a) Source node Figure 1(b) Intermediate node

Fig. 1. Propagation Process in DNDMR

The direction of the propagation of RREQ is decided on the basis of relative position of the one-hop neighbors of the router node. Every node periodically broadcast its position to its one-hop neighbors and every node maintains a neighbor table. When RREQ is received by an intermediate node it decides the left, right, up, and down nodes in its neighborhood and based on this notion of left, right, up, and down, it sends the RREQ in the suitable direction. This controlled forwarding of RREQ packets is important in the sense that when RREQ packets reach the destination, node-disjoint paths are automatically ensured. Unlike the traditional node-disjoint protocols wherein node-disjoint paths are decided either at the source node or at the destination node, our protocol does not depend on any such mechanism. Instead control propagation of the RREQ packets automatically guarantees the formation of node-disjoint paths.

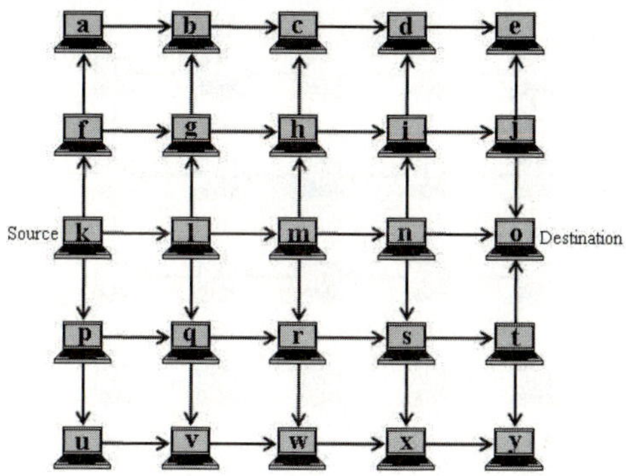

Fig. 2. Grid-style Network Topology

We mainly focus on the grid-style network which can be found in vehicle-roadside network [26]. The roadside units, such as 802.11 access point, are installed in the traffic lights, and communicate with each other. Traffic light deployment has typically grid-style topology in downtown area.

Figure 2 illustrates the entire RREQ propagation procedure of DNDMR. In this figure, *node q* could receive first RREQ from its up node (l) or left node (p). Suppose that, *node q* received the first RREQ from its up node (l). After receiving the RREQ, it found three (p, r and v) more neighboring nodes in its one-hop range. *Node q* can guess that it could receive the duplicate RREQ from its left node (p). Therefore, *node q* rebroadcast the receiving first RREQ to the opposite side that means from up node (l) to down node (v) and when *node q* will receive the duplicate RREQ from left node (p), it rebroadcast the packet to its opposite side that means its right node (r). With this process, DNDMR discovers the routing paths to the destination. By using this method, not only routing overhead is reduced methodically but multiple node-disjoint paths are also formed automatically at the destination.

3.3 Node-Disjoint Path Selection

In DNDMR, first received two route paths by destination are always node-disjoint route paths. For example of DNDMR, in Figure 3, three node-disjoint route paths are received by destination: k-l-m-n-o, k-f-g-h-i-j-o, and k-p-q-r-s-t-o. In most of the cases only two node-disjoint paths are received by the destination. Therefore, we limit the number of node-disjoint route paths to two in DNDMR. When the destination node first receives the RREQ, it records the entire path and sends the route reply (RREP) to the source by using that route. RREP packet includes the nodes id of the entire path and using this information, intermediate node can forward the packet. After this process, when the destination node receives the second RREQ, it sends another RREP to the source via the second route.

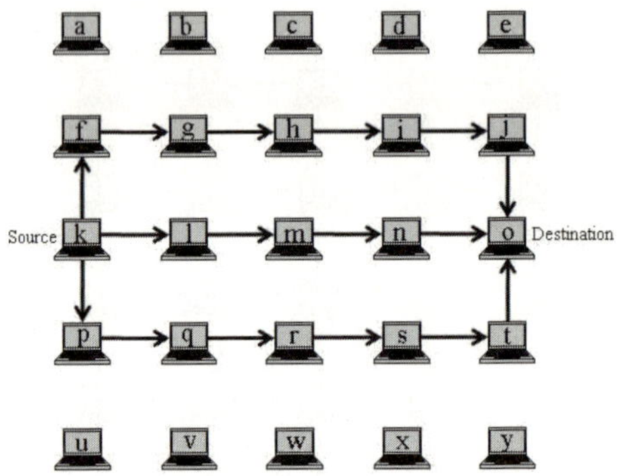

Fig. 3. Node-disjoint paths discovered

3.4 Route Maintenance

In the fault tolerance perspective, more reliable paths should be selected to reduce the chance of route failures. However because of the mobility, the congestion, and the packet collisions, route links can be broken often in ad hoc networks. In DNDMR, link can be disconnected when a node fails to send the packet to the next hop of the route. In this case responsible node will sends a Route Error (RERR) packet to the upstream direction of the route. The RERR message contains the route to the source and the immediate upstream and downstream nodes of the broken link. After receiving this RERR packet, the source node removes every entry from its route table that uses the broken link. If only one of the two route node-disjoint paths invalidated during the transmission, remaining valid route will be used for transmitting data packet. If all of the two routing paths are invalid, the source node initiates the route discovery process again. The summary of the pseudo code is revealed in the next section.

3.5 Pseudo Code

```
PROCEDURE SendDataPacket(Packet p)
BEGIN
      IF (this is source node)
            IF (no path available)
```

```
    Discover Node-Disjoint-Paths Routes(Dest)
        IF (primary path available)
            Call SendPacket(Primary Path)
        ELSE IF (secondary path 1 available)
            Call SendPacket(Secondary Path)
        ELSE
                Result = ForwardPacket(Path)
                IF (result is link failure)
                Send error info to source
END

PROCEDURE ForwardPacket(Path)
BEGIN
        IF (Dest is equal to Current Node)
            Call ReceivePacket(Packet)
        ELSE
            Send Packet to the next node
END

PROCEDURE DiscoverNeighbors(Node)
BEGIN
        Send Hello Packets to Neighbors
END

PROCEDURE Discover Node-Disjoint-Paths(Dest)

BEGIN
        IF (Current node is not Dest)
        {
            Decide Next hop from the Neighbor Table
            Send Packet to the Next hop
        }
        ELSE (Current node is Dest)
        {
            Send the route to the source
        }
END

PROCEDURE Main()
BEGIN
    Discover Neighbors
    Discover Node-Disjoint-Paths
    Transmit data on suitable path
END
```

4 Performance Evaluation

In order to demonstrate the effectiveness of DNDMR, we evaluate its comparative performance with that of DSDV, DSR, AODV, and AOMDV. We compare the performance in terms of routing traffic overhead, throughput, and end-to-end delay using ns-2 simulator [25]. Forty nine nodes are distributed according to a static grid network topology. The environment settings are explained in Table 1.

Table 1. The setup parameters used for the ns2 simulations

Antenna type	Omnidirectional
Propagation model	TwoRayGround
Transmission range	100m
MAC protocol	802.11 with RTS/CTS
MAC bandwidth	1 Mbit
Interface queue type	Drop-tail priority queue
Max. IFQ length	50
Propagation delay	1 ms
Node count	49
Network size	700m × 700m
Simulation time	130s

We have used TCP application as traffic generating source. The TCP packets are continuously transmitted and various performance metrics are measured in case of DSDV, DSR, AODV, AOMDV, and DNDMR.

Control traffic overhead or routing overhead illustrated the cumulative sum of routing packets generated in order to route data packets form the source to the destination throughout the duration of the simulation. Figure 4 shows the comparative routing

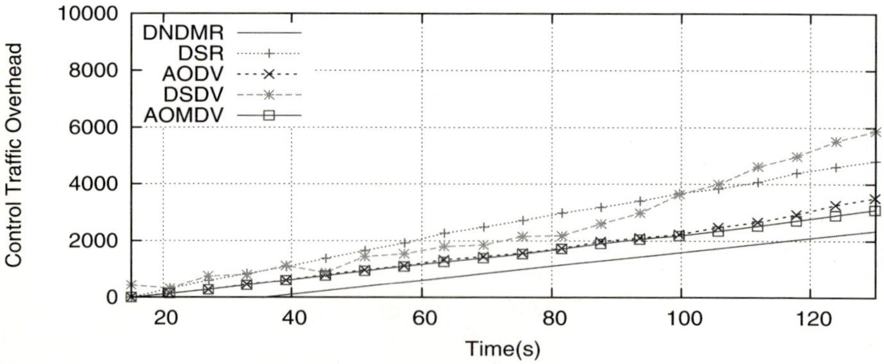

Fig. 4. Comparative Control Traffic Overhead

overhead. In case of DNDMR, it is observed that routing overhead significantly lower than DSDV, DSR, AODV, and AOMDV. The reason is that route discovery mechanism in DNDMR is highly optimized and the destination is discovered by the transmission of minimum packets throughout the network. As seen in Figure 4, the cumulative packets for all mentioned protocols at the end of simulation time are DSDV (6000 packets), DSR (5000 packets), AODV (3500 packets), AOMDV (3100 packets), and DNDMR (2300 packets), whereas DNDMR is comparatively less than the other protocols.

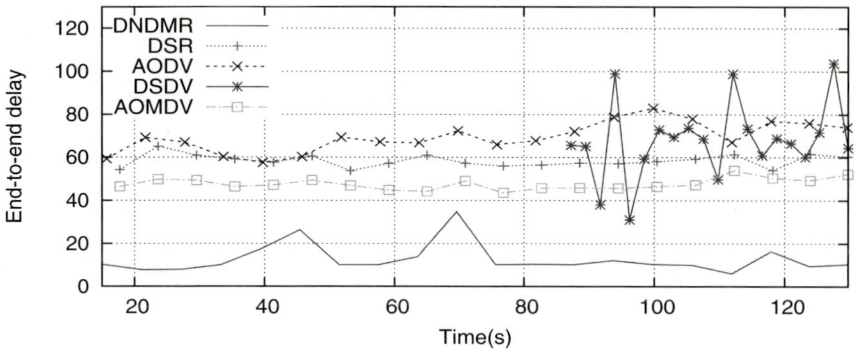

Fig. 5. End-to-End Delay Comparative

End-to-End delay characteristics of DNDMR are analyzed in comparison to DSDV, DSR, AODV, and AOMDV. DSDV, DSR, and AODV are unipath routing protocols. If the route fails, route needs to be rediscovered which results in longer end-to-end delays. On the other hand, AOMDV is disjoint multipath and DNDMR is node-disjoint multipath routing protocol. If one path fails, the traffic can be re-routed on the remaining path. Path rediscovery needs to initiate only when there is no path available from source to the destination. Hence, end-to-end delay is relatively less compared to DSDV, DSR, AODV, and AOMDV. As seen in Figure 5, the average end-to-end delays for all mentioned protocols are: DNDMR (15ms), AOMDV (50ms), DSDV (65ms), DSR (60ms), and AODV (70ms). Therefore, DNDMR is 3.3 times faster than AOMDV, 4 times faster than DSR, 4.3 times faster than DSDV, and 4.6 times faster than AODV.

We calculate the throughput as the number of bytes transferred per 5 seconds. The comparative analysis shows that DNDMR throughput is better than that of unipath routing protocols (DSDV, DSR, and AODV) and that of disjoint multipath routing protocol (AOMDV) as shown in Figure 6. The algorithm will perform better in dynamic environments. DNDMR is more fault-tolerant due to the presence of alternate node-disjoint routes. In case of node failure, DNDMR can re-route the traffic on any of the existing alternate paths instead of starting the process of new route discovery. This improves significantly the control traffic overhead, end-to-end delay, and throughput.

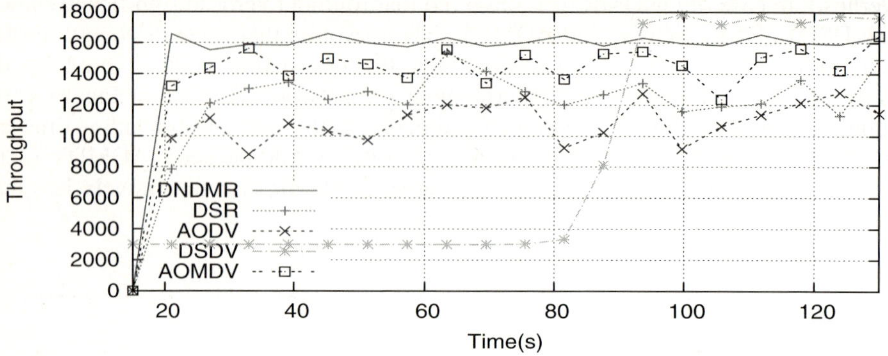

Fig. 6. Comparative Throughput

5 Conclusions

We have evaluated the comparative performance of Directional Node-Disjoint Multipath Routing (DNDMR) protocol which is an on-demand routing protocol for wireless ad hoc networks. DNDMR reduces the routing overhead significantly and achieves multiple node-disjoint paths efficiently. We evaluate the comparative performance of DNDMR using ns-2 simulator. Simulation results show that DNDMR route discovery and packet delivery is more efficient than the most popular unipath routing protocols (DSDV, DSR, and AODV), and the disjoint multipath routing protocol (AOMDV). The reason is that DNDMR discovers node-disjoint routes with minimum control traffic overhead. Also, the availability of more than one alternate path improves reliability. The improvement of the reliability further decreases the end-to-end delay. In the future work, DNDMR analysis will be done for the Quality-of-Service (QoS) in wireless ad hoc network.

References

1. Internet Engineering Task Force (IETF) Mobile Ad Hoc Networks (MANETs) Working Group Charter,
`http://www.ietf.org/html.charters/manet-charter.html`
2. Haas, Z.j., Tabrizi, S.: On Some Challenges, and Design Choices in Ad hoc Communications. In: The Proceedings of the IEEE, Military Communications Conference (MILCOM), Bedford, MA, October 1998, pp. 187–192 (1998)
3. Jubin, J., Tornow, J.D.: The DARPA Packet Radio Network Protocols. IEEE 75(1), 21–32 (1987)
4. Lin, Y.D., Hsu, Y.C.: Multi-Hop Cellular: A New Architecture for Wireless Communications. In: IEEE INFOCOM 2000, March 2000, pp. 1273–1282 (2000)
5. Zadeh, A.N., Jabbari, B., Pickholtz, R., Vojcic, B.: Self-Organizing Packet Radio Ad Hoc Networks with Overlay. IEEE Communication Magazine 40(6), 140–157 (2002)

6. Ananthapadmanabha, R., Manoj, B.S., Murthy, B.S., Siva Ram, C.: Multi-Hop Cellular Networks: The Architecture and Routing Protocol. In: IEEE PIMRC 2001, October 2001, vol. 2, pp. 78–82 (2001)
7. Saha, A., Ur Rehman, S., Park, G.-L., Lee, J.: Directional Node-Disjoint Multipath Routing in Wireless Ad Hoc Network. In: 19th IASTED International Conference on Parallel and Distributed Computing and Systems (PDCS 2007), Cambridge, Massachusetts, USA (November 2007)
8. Perkins, C.E., Bhagwat, P.: Highly Dynamic Destination-Sequenced Distance-Vector Routing (DSDV) for Mobile Computers. Comp. Commun. Rev., 234–244 (October 1994)
9. Murthy, S., Garcia-Luna-Aceves, J.J.: An Efficient Routing Protocol for Wireless Networks. ACM Mobile Networks and App. J. Special Issue on Routing in Mobile Communication Networks, 183–197 (October 1996)
10. Johnson, D.B., Maltz, D.A., Hu, Y.C.: The Dynamic Source Routing Protocol for Mobile Ad Hoc Networks (DSR). Technical Report, IETF 2003, IETF Draft, July 2004, Work in Progress (2004)
11. Perkings, C.E., Belding-Royer, E.M., Das, S.R.: Ad hoc On-Demand Distance Vector (AODV) Routing, IETF internet draft (July 2004), `http://www.ietf.org/internet-drafts/draft-ietf-manet-aodv-13.txt`
12. Park, V.D., Corson, M.S.: A Highly Adaptive Distributed Routing Algorithm for Mobile Wireless Networks. In: INFOCOM 1997 (April 1997)
13. Lee, S.-J., Gerla, M.: Split Multipath Routing with Maximally Disjoint Paths in Ad Hoc Networks. In: IEEE Int'l Conf. Comm., June 2001, pp. 3201–3205 (2001)
14. Wang, L., Shu, Y., Dong, M., Zhang, L., Yang, O.: Adaptive Multipath Source Routing in Ad Hoc Networks. In: IEEE ICC 2001, June 2001, vol. 3, pp. 867–871 (2001)
15. Wang, L., Shu, Y., Zhao, Z., Zhang, L., Yang, O.: Load Balancing of Multipath Source Routing in Ad Hoc Networks. In: IEEE ICCC 2002, April 2002, vol. 5, pp. 3197–3201 (2002)
16. Lee, S.-J., Gerla, M.: AODV-BR: Backup Routing in Ad hoc Networks. In: IEEE Wireless Communications and Networking Conference (WCNC), pp. 1311–1316 (2000)
17. Marina, M.K., Das, S.R.: On-Demand Multipath Distance Vector Routing in Ad Hoc Networks. In: IEEE Int'l Conf. Network Protocols (ICNP), November 2001, pp. 14–23 (2001)
18. Nasipuri, A., Das, S.R.: On-Demand Multipath Routing for Mobile Ad Hoc Networks. In: IEEE Int'l conf. computer comm. And networks (IC3N) (October 1999)
19. Ye, Z., Krishnamurthy, S.V., Tripathi, S.K.: A Framework for Reliable Routing in Mobile Ad Hoc Networks. In: IEEE INFOCOM 2003, April 2003, vol. 1, pp. 270–280 (2003)
20. Nasipuri, A., Castaneda, R., Das, S.R.: Performance of Multipath Routing for On-demand protocols in Mobile Ad Hoc Networks. J. ACM/Baltzer Mobile Networks and Applications J (MONET) 6, 339–349 (2001)
21. Wu, K., Harms, J.: On-Demand Multipath Routing for Mobile Ad hoc Networks. In: European Personal Mobile Communication Conference (EPMCC), Vienna, Austria (February 2001)
22. Xue, Y., Nahrstedt, K.: Fault Tolerant Routing in Mobile Ad Hoc Networks. In: IEEE Wireless Communications and Networking Conference, March 2003, vol. 2, pp. 1174–1179 (2003)
23. Li, X., Cuthbert, L.: Stable Node-Disjoint Multipath Routing with low overhead in mobile Ad Hoc Networks. In: IEEE Computer Society's 12th Annual Int'l Symposium on Modeling, Analysis and Simulation of Computer amd Tele- communicatons Systems, October 2004, pp. 184–191 (2004)

24. Mueller, S., Ghosal, D.: Multipath Routing in Mobile Ad Hoc Networks: Issues and Challenges. In: Calzarossa, M.C., Gelenbe, E. (eds.) MASCOTS 2003. LNCS, vol. 2965. Springer, Heidelberg (2004)
25. Fall, K., Varadhan, K.: ns Notes and documentation. Technical Report, VINT project, UC Berkeley and LBNL (1997)
26. Zhang, Y., Zhao, J., Cao, G.: On Scheduling Vechile-Roadside Data Access. In: The Fourth ACM Int'l Workshop on Vehicular Ad Hoc Networks (VANET) (2007)
27. Leung, R., Poon, E., Chan, A.C., Li, B.: MP-DSR: A Qos-aware Multi-path Dynamic Source Routing Protocol for Wireless Ad-Hoc Networks. In: 26th Annual IEEE International Conference on Local Computer Networks (LCN 2001), Tampa, FL, United States, November 2001, pp. 132–141 (2001)
28. Chen, J., Druschel, P., Subramanian, D.: An Efficient Multipath Forwarding Method. In: IEEE INFOCOM 1998, San Francisco, CA (March 1998)

Multipath Routing Scheme for Wireless Multihop Network

Binod Vaidya[1], Dong-You Choi[2], JongAn Park[2], and SeungJo Han[2]

[1] Dept. of Electronics & Computer Eng., Tribhuvan Univ., Nepal
bnvaidya@gmail.com
[2] Dept. of Information & Communication Eng., Chosun Univ., Korea
{dychoi,japark,sjbhan}@chosun.ac.kr

Abstract. Due to the increasing popularity and availability of portable wireless devices, which constitute of wireless mobile ad hoc network and sensor network, demand efficient routing protocols. On-demand routing protocols adapt well with dynamic topologies of the wireless ad hoc networks, due to their lower control overhead and quick response to route break. In this paper, we discuss on the design approaches for a typical multipath routing algorithm in mobile ad hoc network (MANET). And we propose a multipath routing for MANET. The proposed scheme is basically intended for highly dynamic ad hoc networks in which communication faults occur on a frequent basis. We conduct simulation for such a scheme, which shows that performance of the proposed scheme is better in term of various performance metrics.

Keywords: MANET, on-demand routing, AODV, multipath routing.

1 Introduction

Wireless ad hoc network, both mobile ad hoc network and sensor network, is emerging as an important area for new developments in the ubiquitous network. An ad hoc network is a system of wireless mobile nodes that dynamically self-organize in arbitrary and temporary network topologies allowing people and devices to internetwork without any preexisting communication infrastructure.

Due to the increasing popularity and availability of portable wireless devices, which constitute of wireless ad hoc network and wireless sensor network, demand efficient routing protocols. On-demand routing protocols adapt well with dynamic topologies of the wireless ad hoc networks, due to their lower control overhead and quick response to route break. AODV (Ad-hoc On-demand Distance Vector) [1] and DSR (Dynamic Source Routing) [2] are the most popular on-demand routing protocols in wireless ad hoc network.

However, in a single-path routing, previously created multi-hop route could frequently break because of node mobility and interference. New route discovery would initiate for each failure, in turn, inducing routing overheads and latency. A multipath routing protocol is a promising technique to overcome problems of frequent topological changes and link instability as the use of multiple paths

O. Gervasi et al. (Eds.): ICCSA 2008, Part II, LNCS 5073, pp. 433–445, 2008.

could diminish effect of possible node and link failures. Thus multipath routing protocols for wireless ad-hoc networks are deemed superior over conventional single-path ad-hoc routing protocols [3] as the former reduce end-to-end delay, increase reliability, provide robustness and load-balancing, and improve security [4].

This paper proposes a multipath routing protocol for wireless ad hoc network in which communication failures occur frequently. Selective RREQ forwarding and multiple alternative path discovery techniques are used to compute multiple paths comprising node-disjoints and alternate fail-safe route paths. The rest of this paper is organized as follows: Section 2 briefly discusses multiple path selection criteria. Section 3 presents existing work related to our scheme. Section 4 describes a proposed multipath scheme for MANET. Section 5 gives a performance evaluation of the proposed scheme. Section 6 concludes the paper and points out future works.

2 Multiple Path Selection Criteria

Eventhough multipath routing can be used effectively in improving the reliability, fault-tolerance, end-to-end delay, as well as in achieving load balancing, it is more complicated than single path routing.

In order to achieve performance benefits of multipath routing algorithm, it depends on the availability of the desired multiple paths and on the capability of the multipath finding technique as well as and routing technique. Creating multiple paths in the wireless ad hoc network efficiently with desired property still remains a key challenge in research.

2.1 Disjoint

Different multipath routing protocols have different approaches to create the multiple paths. Depending on how much the created paths are disjointed from each other, we can classify the paths in three categories: [4] non-disjoint path; node-disjoint path and partially disjoint path. Different types of multiple paths are shown in Figure 1.

In creating non-disjoint paths, we do not care of the joint points of the path. Instead, the algorithm focuses on finding the paths as more as possible.

Node-disjoint paths do not have any nodes in common, except the source and destination. In case of partially disjoint path, it depends on certain factors. One type of partially disjoint paths is link-disjoint path, which does not have common links, but may have nodes in common.

Another partially disjoint path is a fail-safe alternate path [5]. It is a path between source and destination if it bypasses at least one intermediate node on the primary path. Fail-safe multiple paths are different from node-disjoint and link-disjoint multiple paths, in the sense that fail-safe route paths can have both nodes and links in common.

The higher degree of disjunction, the probability of simultaneous failure of all paths decreases. It is due to rare chance of locating of two disjoint paths

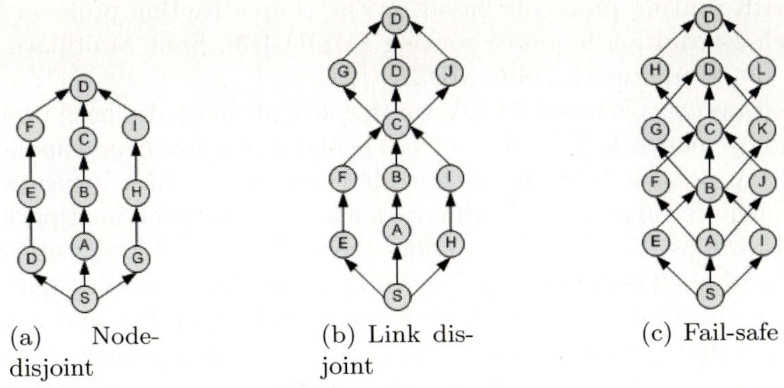

Fig. 1. Types of multiple paths

near the same source of noise or error. In contrast, the probability of an internal competition for acquiring the shared channel becomes more when the routes have more joint points.

Given the trade-offs between using node-disjoint and non-disjoint routing paths, partially disjoint path may offer a better compromise between two extremes.

2.2 Path Selection

In the case of single path routing, the destination is responsible for path selection. For multipath routing, one approach is that after receiving all RREQ packets, the destination chooses the desired routes and sends them towards source node by means of RREP packet [6,7,8]. Another approach is that the intermediate nodes can participate in the process of selecting the paths in a distribute manner [9,10,11,12].

2.3 Handling Duplicate RREQ

In order to obtain all potential paths, all RREQ copies should be processed. Thus one of important issues in multipath routing is to handle a duplicate RREQ by the intermediate node.

A duplicate RREQ has probably passed more hops to reach destination than the previous RREQ. In consequence, it has traveled longer distance and can be a good candidate for rejection. On the other hand, re-broadcasting of it could provide us by more alternate route paths which are basics of multipath routing. Hence, the intermediate nodes should handle the duplicate RREQ packets more carefully.

3 Related Works

We briefly present the related works to multipath routing. Some multipath routing protocols have been proposed for wireless ad hoc networks.

Multipath routing protocols based on the source routing protocol DSR are found such as Multipath source routing (MSR) [13], Split Multipath Routing (SMR) [14] and multipath DSR (MDSR) [15].

Multipath protocols based AODV routing scheme have also been proposed for wireless ad hoc networks [5,7,10,11,16,17]. Scalable multipath on-demand routing (SMORT) [5] uses the idea of fail-safe alternate path, which bypasses at least one intermediate node on the primary path. Node-disjoint multipath routing (NDMR)is proposed in [7], which applies the path accumulation feature of DSR to AODV. Ad hoc On-demand Multipath Distance Vector routing (AOMDV) is proposed in [10] that produces multiple routes with loop-free and link-disjoint properties. AODV-Multipath (AODVM) [11] is a routing framework to provide robustness to route breaks, which computes multiple node-disjoint paths between source and destination. AODV-BR [16] utilizes a mesh structure and provides multiple alternate routes without generating additional control messages. And Caching and Multipath Routing (CHAMP) is proposed in [17] that uses packet salvaging in multipath routing to provide improved fault tolerance.

4 Proposed Multipath Ad Hoc Routing

In the MANET paradigm routing is a challenging task due to mobility and the resulting inherent dynamic network topology. As mentioned earlier, according to the number of paths that are discovered from a route request, the routing protocols are divided into single path and multipath.

In this paper, we propose an efficient multipath routing protocol for mobile ad hoc network, which is a modification of a single path AODV routing protocol. The proposed framework is designed to compute not only node-disjoint paths but also fail-safe paths. So the combination of these path types allows the computation of more alternative paths than in node-disjoint or link-disjoint multipath routings. Thus the proposed scheme is named as AODV with Multiple Alternative Paths (AODV-MAP).

AODV-MAP has a main goal finding on-demand multiple paths between the source and the destination such that it not only lowers frequencies of costly route discoveries to keep end-to-end connection for longer time but also uses labeled multiple alternative paths for efficient traffic distribution among these paths. Thus it provides an easy mechanism to increase the likelihood of reliable data delivery by sending multiple copies of data along different paths.

AODV-MAP is basically intended for highly dynamic ad hoc networks in which communication failures occur frequently. In the proposed scheme, each node in MANET keeps and maintains tables - routing table, and neighbor node table. Neighbor node table is used to record neighborhood information, which is periodically updated.

As in AODV, there are three types of control messages - Route Request (RREQ), Route Reply (RREP) and Route Error (RERR). The proposed protocol has two basic phases, namely route discovery, and route maintenance.

4.1 Route Discovery

A source node S initiates route discovery process, when it wants to communicate to a destination D, for which it does not have a valid route. The source node S floods a route query packet to the entire network. On receiving a RREQ, intermediate nodes update their routing table for a reverse route to the source. Intermediate nodes are not allowed to send RREPs back to the source even when they have route information to the destination. This is essential to find desired multiple paths at the destination. AODV-MAP utilizes sequence numbers to ensure avoidance of routing loop and use of up-to-date route information.

Structures of RREQ and RREP packets of AODV-MAP are similar to those of AODV, with addition of path accumulation list in former and path accumulation and path label in latter. Path accumulation during route discovery is shown in Figure 2 and the path information ensures to send RREP to the source as well as enables to set path label at the destination.

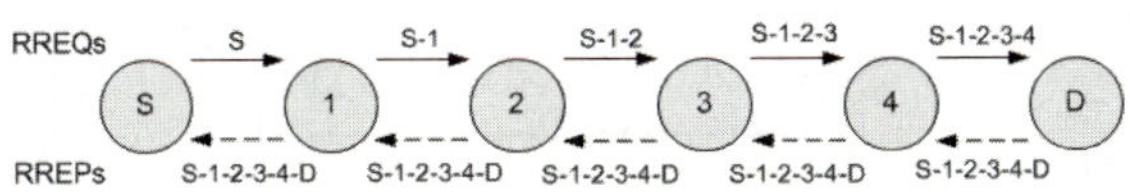

Fig. 2. Path accumulation during route discovery

AODV-MAP allows intermediate nodes to forward duplicate RREQ packets that come from different neighbors than previous RREQ and whose hop count is not larger than that of previous RREQ. This ensures to discover desired alternative paths. That is only if the following criteria of selective RREQ forwarding scheme are met:

- $hop\_cnt_{RREQ_i} = hop\_cnt_{prevRREQ}$
- $nn_{RREQ_i} \neq nn_{prevRREQ}$

where *hop_cnt* is number of hops from the source to the current node handling the request; and *nn* is neighbor node.

In this scheme, the destination is responsible for not only selecting node-disjoint and fail-safe paths from all the received routes but also defining the path with specific path label. When receiving the first RREQ, the destination records the route path of RREQ and sets it with path label 1. Then after copying route path of RREQ to a RREP packet, the destination node sends RREP to source node using path information in it. Hence the intermediate nodes can forward this packet using the path information in RREP.

When the destination receives a duplicate RREQ, it will compare route path of RREQ to that of the routing table. If source and destination nodes are same, the path is said to be a node-disjoint with the primary path so the destination determines it as path label 3. If at least one of intermediate nodes in the route path in the routing table is different from all of nodes in the route path of the

RREQ, the route is said to be a fail-safe with the primary path so destination determines it as path label 2. After setting proper path label in RREP, the destination sends it to the source along the path information in it. RREPs follow the reverse paths to reach the source node.

If any alternate path is received, the destination sends the RREP to the source along the route path of RREP. Otherwise, the received RREQ is discarded.

During route discovery, multiple alternative paths are discovered, which can be available for data transferring and shown in Figure 3.

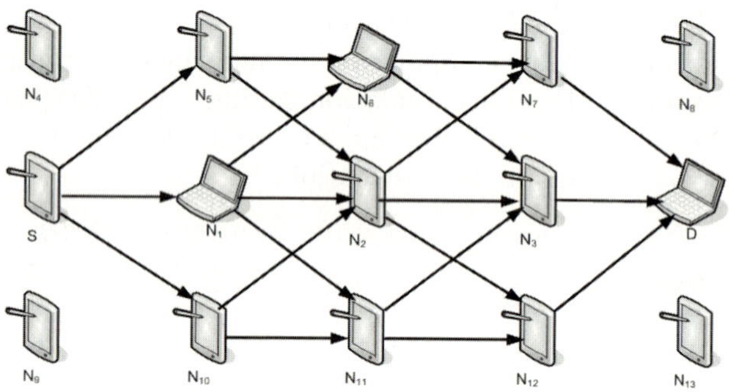

Fig. 3. Discovering multiple paths during route discovery

After completion of route discovery process, there will be a primary path $\langle S \rightarrow N_1 \rightarrow N_2 \rightarrow N_3 \rightarrow D \rangle$; two node-disjoint paths $\langle S \rightarrow N_5 \rightarrow N_6 \rightarrow N_7 \rightarrow D \rangle$, and $\langle S \rightarrow N_{10} \rightarrow N_{11} \rightarrow N_{12} \rightarrow D \rangle$; and a number of fail-safe paths $\langle S \rightarrow N_5 \rightarrow N_2 \rightarrow N_7 \rightarrow D \rangle$, $\langle S \rightarrow N_1 \rightarrow N_6 \rightarrow N_3 \rightarrow D \rangle$, $\langle S \rightarrow N_1 \rightarrow N_2 \rightarrow N_{12} \rightarrow D \rangle$, $\langle S \rightarrow N_{10} \rightarrow N_{11} \rightarrow N_3 \rightarrow D \rangle$.

It should be noted that a traffic distribution approach utilizing labelled multiple paths can be used for effective voice communication. This approach is out of the scope in this paper.

4.2 Route Maintenance

Normally route links in wireless ad hoc networks are broken frequently due to the mobility of nodes, congestion and packet collisions. The proposed scheme is capable of recovering broken routes immediately. When a node fails to deliver the data packets to the next hop of the route by receiving a link layer feedback from link layer or receives the RERR packet, it removes entries in its route table that uses the broken link and looks up its routing table if there is another entry for the destination.

If it has another entry for the destination, data packets therefore can be delivered through the alternate route and are not dropped when route breaks occur. If it has no another entry for the destination, it sends a RERR packet to the upstream node. When the source has no entry for the destination and the session is still active, it would initiate a new route discovery.

5 Performance Evaluations

In order to evaluate the performance of the proposed protocol and compare with node-disjoint multipath routing and AODV, we have simulated using OPNET Modeler [18]. In order to evaluate capability of these protocols for different node mobility, we changed node mobility by varying the maximum speed.

5.1 Simulation Environment

In the simulation, the network coverage areas are 1000m x 1000m and 1500m x 1500m for 50 mobile nodes 100 nodes respectively. Nodes communicate via radio signals with 250m of propagation range and the channel capacity of 2 Mbps. The IEEE 802.11 DCF is used as the MAC layer protocol in the simulation. DCF uses CSMA/CA technique to the packets.

The mobility and traffic models similar to those previously reported are used. The nodes are initially uniformly distributed throughout the network area and their movement is determined by the random waypoint mobility model [19]. According to this model, a node randomly selects a location with in the network area and moves towards it with a speed uniformly chosen between a predefined minimum and maximum values. The node stays in that position for a period, ie pause time, and then again selects another random location to move. The pause time is constant at 30 sec.

Constant bit-rate (CBR) sessions between source and destination pair in the network were occurred with each CBR session generating 10 packets / sec and the size of each data packet is 512 bytes. Each node buffers all data packets while waiting for a route. All packets (both data and routing) sent by the routing layer are queued at the buffer until the MAC layer can transmit them. Routing packets are given higher priority than data packets in the buffer. The sources and destinations are chosen randomly with uniform probabilities. Each run executes 800 seconds of simulation time.

5.2 Simulation Scenario

In order to evaluate performance of the AODV-MAP scheme and compare with other routing schemes such as AODV and node-disjoint multipath routing in different network conditions, two parameters are varied in the simulations:

- Number of nodes (50 nodes and 100 nodes)
- Maximum velocity of the nodes

In order to evaluate performances of the AODV and proposed protocols in mobility condition, the maximum velocity of the nodes parameter is varied in the simulation. 20 and 40 sources are modeled respectively to study the effect of varying mobility in networks of 50 and 100 nodes.

5.3 Performance Metrics

In our simulation, four performance metrics are used in the performance evaluation of the proposed scheme: route discovery frequency, average end-to-end delay, and packet delivery ratio [20].

- Route discovery frequency which is total number of route discoveries initiated per second.
- Average end-to-end delay includes all possible delays from the moment the packet is generated to the moment it is received by the destination node.
- Packet delivery ratio: The ratio of the data packets delivered to the destinations to those generated by the CBR sources.

5.4 Results and Analysis

A series of experiments were conducted with the maximum velocity for 20 sources of 50 nodes network and for 40 sources of 100 nodes network. The mobility was varied in order to see the effects of the route discovery frequency, average end-to-end delay, and packet delivery ratio. The packet sending rate is fixed at 10 packets / sec. A node can choose a speed between minimum (0 m/s) and maximum speed. The results were collected at constant speeds from 0 to 20 m/s.

Route discovery frequency (RDF) is a very important metric as smaller route discovery frequency of a given routing protocol implies that it can increase opportunities for setting up more alternative paths. Figures 4 and 5 show the route discovery frequency with respect to the maximum speed for networks of 50 nodes and 100 nodes respectively. In both cases, AODV-MAP has smaller route discovery frequency than AODV and the node-disjoint multipath routing. It can be seen that in case of AODV-MAP scheme and node-disjoint multipath routing, a new route discovery is invoked only when all paths fail.

In Figure 4, at the maximum speed of 10 m/s, route discovery frequency in AODV-MAP is lesser than AODV and node-disjoint multipath routing by about 34% and 20% respectively. Whereas in Figure 5, at the maximum speed of 10 m/s, route discovery frequency in AODV-MAP is lesser than AODV and node-disjoint multipath routing by about 62% and 40% respectively. AODV-MAP scheme maintains a low route discovery frequency compared to node-disjoint multipath routing and single path AODV. This indicates that the robust nature of the scheme to mobility of nodes and topological changes.

Average end-to-end delay of data packets which includes all possible delays caused by buffering during route discovery latency, queuing at interface queue, length of routing paths, retransmission delays at MAC, propagation and transfer times.

Figures 6 and 7 illustrate the change in the average end-to-end delay as a function of speed of nodes for networks with 50 nodes and 100 nodes respectively. It can be seen that the general trend of all curves is an increase in delay with the increase of velocity of nodes. The reason is mainly that high mobility of nodes

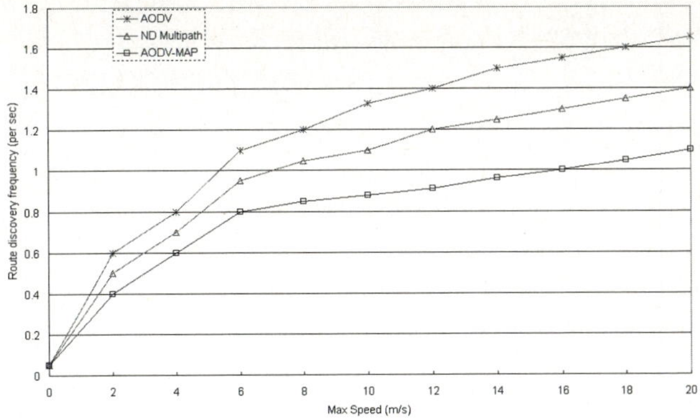

Fig. 4. Route Discovery Frequency for 50 nodes

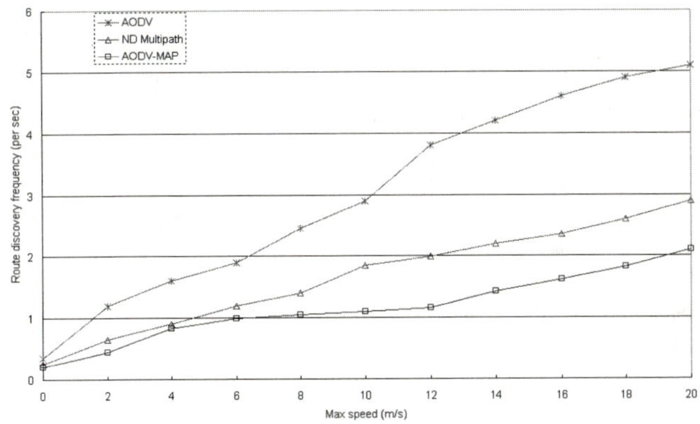

Fig. 5. Route Discovery Frequency for 100 nodes

results in an increased probability of link failure that causes an increase in the number of routing rediscovery processes.

The proposed protocol exhibits the smallest end-to-end delay whereas AODV incurs the largest delay. The delays of the proposed scheme and node-disjoint multipath routing are gradually increased after node velocity of 10m/s, while delay in AODV increases quickly as velocity increases. This is because availability of alternate routing paths in both AODV-MAP and node-disjoint multipath routing eliminate route discovery latency that contributes to the delay when active route fails. This result shows that the proposed scheme is effective especially in situations where nodes move frequently.

Packet delivery ratio (PDR) is one of the important metrics since it shows the loss rate, which in turn affects the maximum throughput of the network.

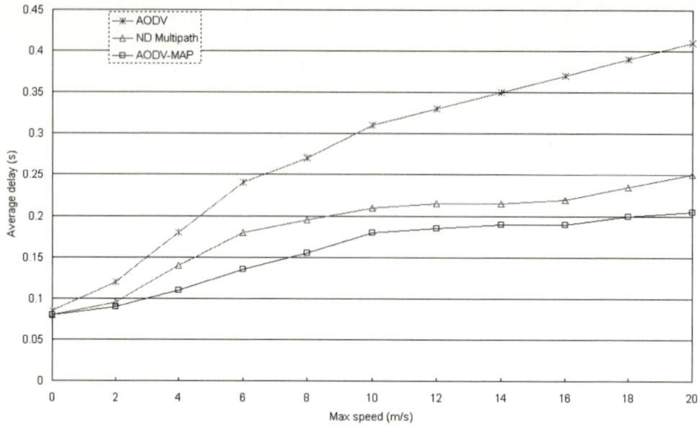

Fig. 6. Average end-to-end delay (n=50)

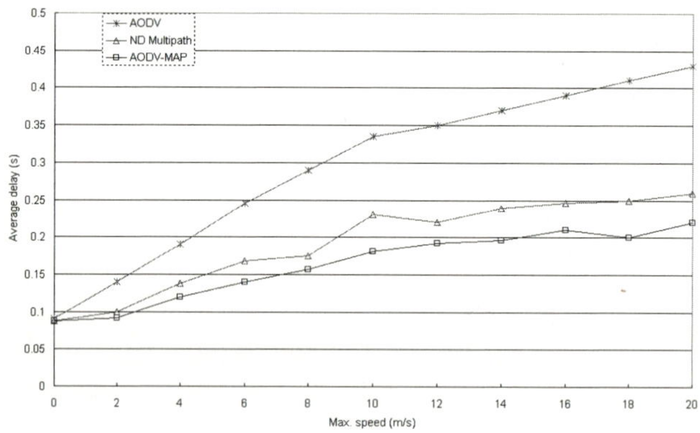

Fig. 7. Average end-to-end delay (n=100)

The packet delivery ratios of the three protocols are shown in Figures 8 and 9. They describe the variation of the packet delivery ratio as a function of the node speed. The probability of link failure increases with increase in the node speed, and hence the number of packet drops also increases. AODV-MAP scheme and node-disjoint multipath routing produce high packet delivery ratio due to presence of multiple paths. So when an active routing path is broken due to mobility of nodes, these protocols still can manage the communication between source and destination without pause or interrupt. Since AODV-MAP has both node-disjoint and fail-safe alternative paths, it can survive longer. Thus AODV-MAP scheme has much higher packet delivery ratio than both node-disjoint multipath routing and AODV. This mechanism of AODV-MAP assures high packet delivery ratio.

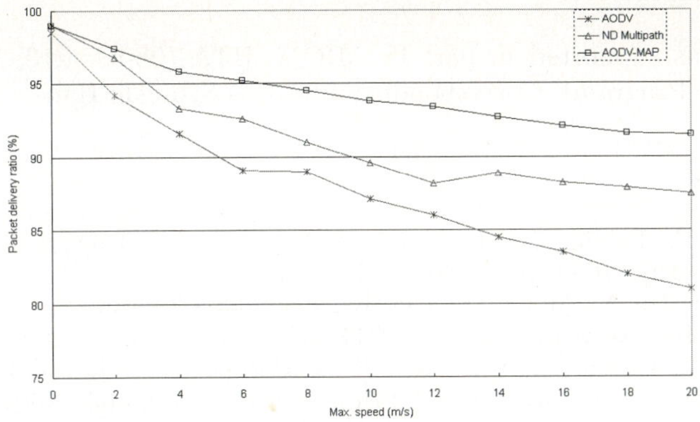

Fig. 8. Packet delivery ratio (PDR) (n=50)

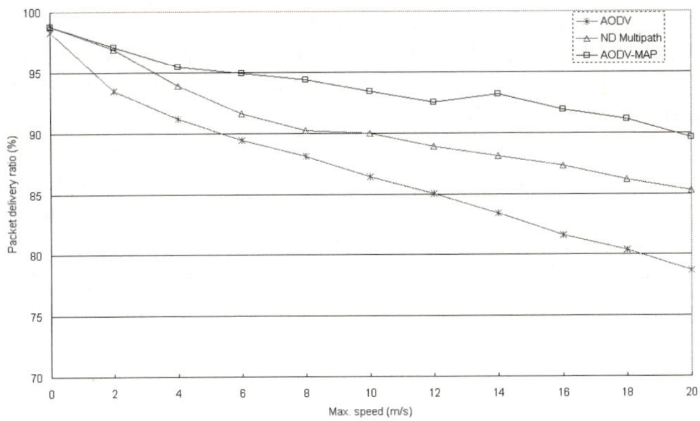

Fig. 9. Packet delivery ratio (PDR) (n=100)

6 Conclusions and Future Work

In this paper, an efficient multipath routing has been proposed to minimize frequent communication failures in MANETs. A simulation showed that the performance of the proposed scheme is much better than AODV and node-disjoint multipath routing in terms of route discovery frequency, end-to-end delay, and packet delivery ratio.

Currently AODV-MAP scheme has no power-ware mechanism. In future work, we will incorporate power-aware path selection algorithm for AODV-MAP routing protocol so that it can select longer-lived alternative paths. We will also extend our multipath routing protocol in hybrid ad-hoc networks and evaluate the global IP connectivity using hybrid mutlipath MANET.

Acknowledgement

This study was supported in part by MIC & IITA (08-infra-13, IT Research Infrastructure Program). Corresponding author is SeungJo Han.

References

1. Perkins, C.E., Belding-Royer, E., Das, S.R.: Ad hoc on-demand distance vector (AODV) routing. IETF RFC 3561 (July 2003)
2. Johnson, D., Hu, Y., Maltz, D.: The Dynamic Source Routing Protocol (DSR) for Mobile Ad Hoc Networks for IPv4. IETF RFC 4728 (February 2007)
3. Nasipuri, A., Castaneda, R., Das, S.R.: Performance of Multipath Routing for On-Demand Protocols in Mobile Ad Hoc Networks. In: Mobile Networks and Applications 6, pp. 339–349. Kluwer Academic Publishers, Dordrecht (2001)
4. Mueller, S., Tsang, R.P., Ghosal, D.: Multipath Routing in Mobile Ad Hoc Networks: Issues and Challenges. In: Calzarossa, M.C., Gelenbe, E. (eds.) MASCOTS 2003. LNCS, vol. 2965, pp. 209–234. Springer, Heidelberg (2004)
5. Reddy, L.R., Raghavan, S.V.: SMORT: Scalable multipath on-demand routing for mobile ad hoc networks. Elsevier Ad hoc Networks Journal 5(2), 162–188 (2007)
6. Nasipuri, A., Das, S.R.: On-Demand Multipath Routing for Mobile Ad Hoc Networks. In: Proceedings of the 8th International Conference on Computer Communications and Networks (1999)
7. Li, X., Cuthbert, L.: Stable node-disjoint multipath routing with low overhead in mobile ad hoc networks. In: Proc. of The IEEE 12th Annual International Symposium on Modeling, Analysis, and Simulation of Computer and Telecommunications Systems (MASCOT 2004), October 2004, pp. 184–191 (2004)
8. Chen, Y.S., Tseng, Y.C., Shue, J.P., Kuo, P.-H.: On-Demand, Link-State, Multi-Path QoS Routing in a Wireless Mobile Ad-Hoc Network. In: Proceedings of European Wireless (2002)
9. Wei, W., Zakhor, A.: Robust Multipath Source Routing Protocol (RMPSR) for Video Communication over Wireless Ad Hoc Networks. In: ICME 2004, Taiwan, pp. 1379–1382 (2004)
10. Marina, M.K., Das, S.R.: Ad hoc on-demand multipath distance vector routing. Wiley Wireless Communications and Mobile Computing 6(7), 969–988 (2006)
11. Ye, Z., Krishnamurthy, S.V., Tripathi, S.K.: A routing framework for providing robustness to node failures in mobile ad hoc networks. Elsevier Ad Hoc Networks Journal 2(1), 87–107 (2004)
12. Leung, R., Liu, J., Poon, E., Chan, A.C., Li, B.: MP-DSR: A QoS-aware Multipath Dynamic Source Routing Protocol for Wireless Ad hoc Networks. In: Proceedings of the 26th IEEE Annual Conference on Local Computer Networks, pp. 132–141 (2001)
13. Wang, L., Zhang, L., Shu, Y., Dong, M.: Multipath source routing in wireless ad hoc networks. In: Proc. of Canadian Conference on Electrical and Computer Engineering (CCECE 2000), March 2000, vol. 1, pp. 479–483 (2000)
14. Lee, S.J., Gerla, M.: Split multipath routing with maximally disjoint paths in ad hoc networks. In: Proc. of IEEE International Conference on Communications 2001, June 2001, vol. 3, pp. 867–871 (2001)
15. Mao, S., Lin, S., Panwar, S.S., Wang, Y., Celebi, E.: Video transport over ad hoc networks: multistream coding with multipath transport. IEEE Journal on Selected Areas in Communications 21(10), 1721–1737 (2003)

16. Lee, S.-J., Gerla, M.: AODV-BR: Backup Routing in Ad hoc Networks. In: Proceedings of the IEEE Wireless Communications and Networking Conference (WCNC 2000), September 2000, vol. 3, pp. 1311–1316 (2000)
17. Valera, A.C., Seah, W.K.G., Rao, S.V.: Improving protocol robustness in ad hoc networks through cooperative packet caching and shortest multipath routing. IEEE Transactions on Mobile Computing 4(5), 443–457 (2005)
18. OPNET Modeler Simulation Software, `http://www.opnet.com`
19. Broch, J., Maltz, D.A., Johnson, D.B., Hu, Y.C., Jetcheva, J.: A Performance Comparison of Multi-Hop Wireless Ad Hoc Network Routing Protocols. In: Proceedings of the Fourth Annual ACM/IEEE International Conference on Mobile Computing and Networking (MobiCom 1998), Dallas, USA, October 1998, pp. 1–13 (1998)
20. Corson, S., Macker, J.: Mobile ad hoc networking (MANET): Routing protocol performance issues and evaluation considerations. IETF RFC 2501 (January 1999)

Improvement of on Demand Multicast Routing Protocol in Ad Hoc Networks to Achieve Good Scalability and Reliability

MohammadReza EffatParvar[1], Mehdi EffatParvar[2], and Mahmoud Fathy[3]

[1] Informatics Services Corporation,Tehran, Iran
effatparvar@gmail.com
[2] Computer Eng Department, Islamic Azad University – Ardebil Branch, Ardebil, Iran
mehdi_effatparvar@yahoo.com
[3] Computer & IT Eng Department of Iran University of Science & Technology, Tehran, Iran
mahfathy@iust.ac.ir

Abstract. An ad hoc network is a dynamically reconfigurable wireless network with no fixed wired infrastructure. The primary concerns in ad hoc networks are bandwidth limitation and unpredictable dynamic topology. Reliable Multicast plays a significant role in many applications of Mobile Ad Hoc Networks (MANETs). The On-Demand Multicast Routing Protocol (ODMRP) was designed for multicast routing in ad hoc networks. In this paper, we propose a cluster-based on demand multicast routing protocol to the lack of extension of flat multicast routing protocols in ad hoc networks of large scale. Simulation results show that using clustering base ODMRP improves network performance in terms of end-to-end delay and control packets. Also we propose a local recovery approach to design a reliable multicast algorithm. By this approach nodes can join to multicast groups in short time and data delivery will be increase, so ODMRP transmits the data with good reliability.

Keywords: Ad hoc networks, Multicast, Scalability, Reliability, ODMRP, Cluster.

1 Introduction

Ad hoc networks consist of mobile nodes that autonomously establish connectivity via multi hop wireless communications. In many ad hoc applications, nodes need to collaborate to achieve common goals and are expected to communicate as a group rather than as pairs of individuals (point-to-point). Multicast is a very useful and efficient means of supporting group-oriented applications, especially in mobile/wireless environments where bandwidth is scarce and equipment has limited power [14]. Recently, wireless mesh networks have attracted much attention [3,4]. These networks typically have low maintenance overhead, high data rates, and are not energy constrained.

There are several problems which are specific to ad hoc networks. One of the problems is decrease of management capability as the size of ad hoc networks becomes

O. Gervasi et al. (Eds.): ICCSA 2008, Part II, LNCS 5073, pp. 446–457, 2008.

large. The other problem is frequent change of the topology of ad hoc networks due to node movement. In order to cope with these problems, the clustering scheme such as the autonomous clustering scheme [5] and Least Clusterhead Change (LCC) scheme has been proposed. The election of clusterheads has been a topic of many papers as described in [7]. In all of these papers the leader election guarantees that no node will be more than one hop away from a leader. Furthermore, their time complexity is O(n), where n is the number of nodes in the network.

Mesh-based multicast protocols [8], provide alternative paths and a link failure need not trigger the re computation of a mesh. Previous studies showed that mesh-based protocols are robust against topology change and are more suitable than tree-based protocols [9]. ODMRP is an ad hoc multicast routing protocol based on a multicast mesh [8].

Because importance of reliability there are some reliable protocols in MANET; that we can classify them in three categories: The categories are, Automatic Retransmission Request (ARQ) based, gossip-based and Forward Error Correction (FEC) based. The protocols are RMA [11], RALM [10] and ReAct [12] are belong to the ARQ-based category. In this paper, we propose a new method for clustering ODMRP; also we propose a new local recovery approach in ODMRP. Our clustering algorithm divides the network to some group, that these clusters are 1-hop clustering. By using clustering method protocol overhead will be decrease and we can establish scalable network. Local recovery method achieves fast recovery when route breakage happens, so destination can connect to source in new route or the same route.

Simulation results show that our approaches improve the performance of ODMRP, so we can use ODMRP as a scalable and reliable protocol.

The reminder of this paper is organized as follows. Sections 2 describes issues of ODMRP reliability and scalability. Section 3 describes our new clustering mechanism. In section 4 we show our local recovery approach. Section 5 follows with the simulation results and concluding remarks are made in section 6.

2 Issues of Reliability and Scalability in ODMRP

ODMRP [8] is a best-effort mesh-based multicast routing protocol. Hence, a major performance evaluation criterion is the average miss ratio, which is calculated as the number of packets that each receiver did not receive divided by the total number of packets sent by the source, and averaged over the total number of receivers in the multicast group. In this section, we study how the average miss ratio of ODMRP is affected by mobility, session size, and the transmission range. The simulation environment consists of 30 hosts in a 700 m * 700 m area. A single source multicasts 2000 512-byte DATA packet to a multicast group. Network bandwidth is 2 Mbit/sec. MAC protocol is 802.11. The random waypoint mobility pattern is used with equal minimum and maximum speeds of mobility (i.e., mobile hosts move with a constant speed). Several values of constant mobility speeds between 0 m/sec. (i.e., stationary hosts) and 25 m/sec. are considered. Figure 1 shows the miss ratio of ODMRP with different numbers of members.

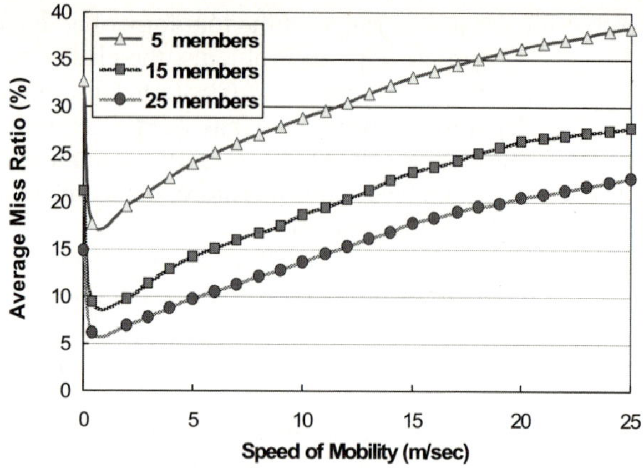

Fig. 1. Average miss ratio versus speed of mobility for different numbers of members, transmission range is 200m

At zero speed of mobility, the hosts are stationary and thus, the topology of the ODMRP multicast mesh is fixed. Hence, when a source multicasts a data packet, only the receivers that are in the mesh may receive it but other receivers that are far away from the mesh will not. As mobility increases, distant receivers may, while moving, reach the multicast mesh and hence start receiving some data packets causing the average miss ratio to decrease (as shown in Figure 1). However, as mobility increases further, there are more frequent disconnections in the multicast mesh and hence packet loss increases.

For larger session sizes, there are more forwarding nodes in the multicast mesh, and therefore more redundancy in the mesh, leading to falling back on alternate multicast links when a multicast link fails and thus, reducing the average miss ratio.

ODMRP always tries to include the shortest path within the forwarding group, so almost it has lower latency than the other algorithm. In this section, Figures show the performance metrics as functions of group size. With fixed pause time as 60 seconds, we have one multicast group of size from 20 to 200. Figure (2) shows the result for packet delivery ratio. As group sizes increases, ODMRP delivers more fractions of packets. The reason is that the forwarding mesh becomes more reliable with more redundant paths as it increases its size.

In this section, we study the performance behaviors with respect to the horizontal scalability. We consider the following 6 scenarios: 72 by 2, 48 by 3, 36 by 4, 24 by 6, 18 by 8 and 12 by 12. Here, "72 by 2" means 2 multicast groups, and 72 members per group. Thus, in all scenarios, the total number of receivers is fixed to 144. There is one source for each group. The traffic demand remains equal in all scenarios.

Figure-3(a) shows the packet delivery ratio and the variance among the groups in the network. As the number of groups' increases, performance of ODMRP shows quick drop to less than 10% for 12 groups. The size of meshes does not decrease proportional to the group sizes.

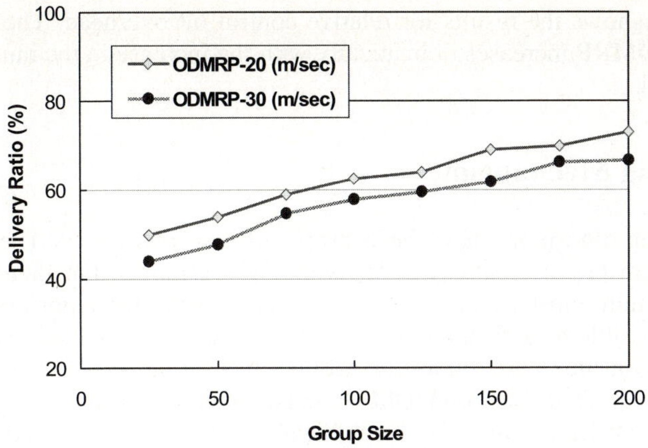

Fig. 2. Packet delivery ratio versus group size. (Pause time is 60s, 1 group, 1 source, speed 20 and 30 m/s).

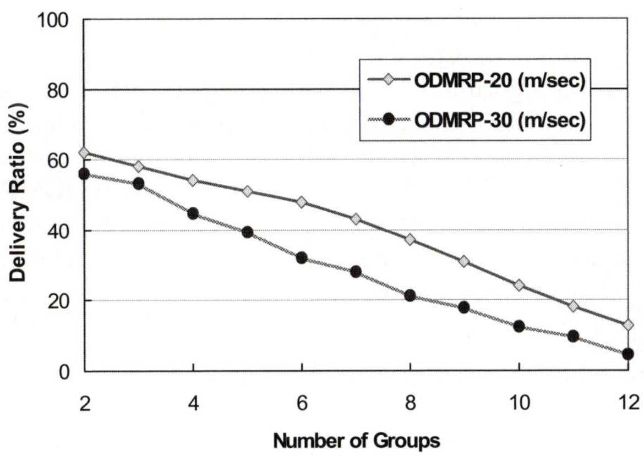

(a) Packet delivery ratio.

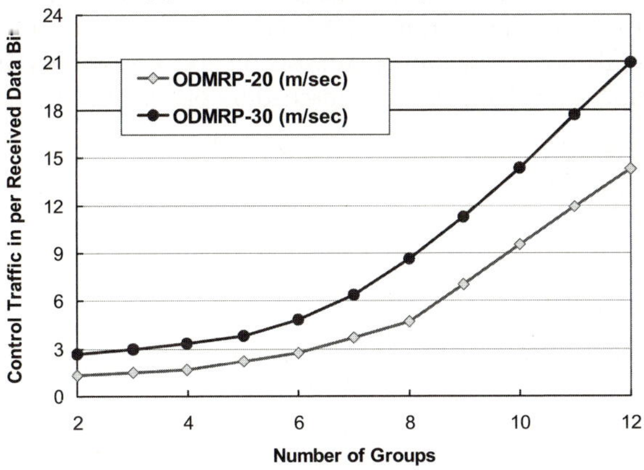

(b) Bit Overhead

Fig. 3. Performance versus group size. (Pause time is 60s, 1 source per group, speed 20 and 30 m/s).

Figure-3(b) shows the results for relative control bit overhead. The control traffic incurred by ODMRP increases dramatically with the increase in the number of groups and node speed.

3 Clustering Mechanism

Many clustering algorithms have been proposed, one of them is The Max-Min D-cluster algorithm [1], it creates clusters with a given radius, but there is no way to limit the maximum cluster size. On the other hand, using the algorithm presented in [2], it is not possible to limit the radius for the clusters. Newly many clustering algorithms worked on multicast routing in Ad hoc. In [15] we can see one of clustering approach that is implemented on ODMRP and adopted an enhanced weighted clustering algorithm (EWCA) to manage hierarchically its motion nodes, and forms the forwarding group mainly based on clusterheads. In this section we propose our clustering mechanism.

Our clustering mechanism has 2 stages, at first stage we divide the network in some 1-hop clusters and in second stage we define some condition and we determine the role of clusterheads and the cluster members.

Generally clustering algorithms have some disadvantages like overhead, needing the nodes stop during the algorithm execution and depending on underlying protocols. In our algorithm clusterheads do as gateways, so only the clusterheads transmit the Join-Request packets. In our algorithm all assumption are like ODMRP but we define some new fields in Join-Request packet in order to cluster the network at the beginning of simulation, also we define a new packet to find the new clusterhead during the simulation. With above assumption, our new algorithm, that we call SC-ODMRP is described as follows:

a) In the beginning of simulation each source sends the Join-Request packet (see figure 4).

b) On receiving the first Join-Request, each node determines the sender as the clusterhead of itself. Then the node puts the clusterhead address in Join-Request packet and broadcasts it (see figure 5).

c) When the Broadcasted Join-Request is received by the original sender, and it knows that has been selected as clusterhead by one node.

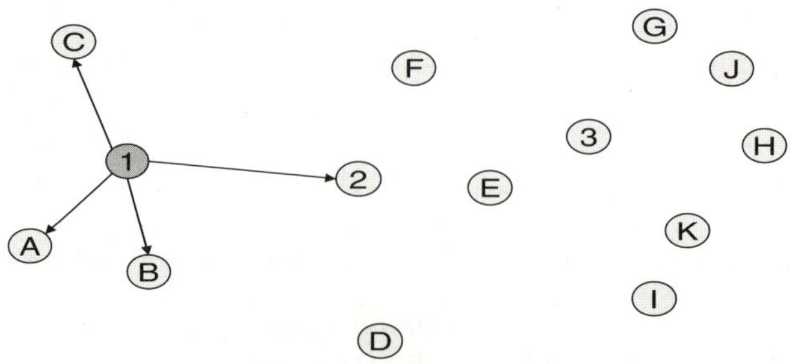

Fig. 4. Sending the first Join-Request by the source

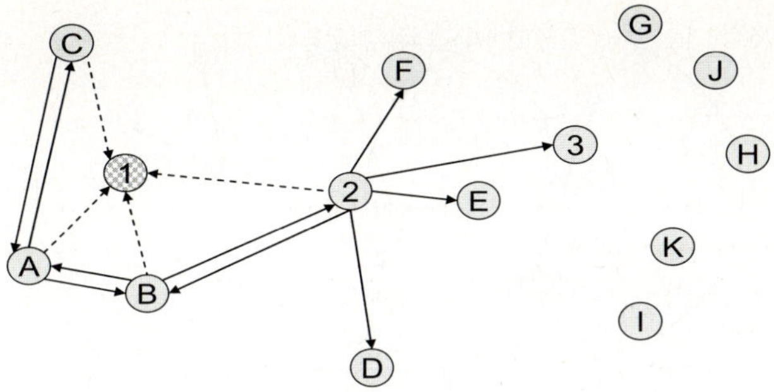

Fig. 5. Determine the clusterhead by first Join-Request

d) Stages b and c will be repeated by all nodes.

e) For the other Join-Request packets except the first, only two groups broadcast it: clusterhead and the nodes that receive the packet for the first time.

f) Each node can receive the packets from any clusterhead and can change its clusterhead.

g) If a node does not receive the Join-Request after a specific period (in our simulation 6s) sends a Find-Cluster packet to its neighbors. The node puts the multicast group address in the packet and sets the Find-Cluster status field. The status field may have three states:

- 0: determine that this Find-Cluster packet is a question packet to discover the clusterhead.
- 1: means that the Find-Cluster packet sender suggests itself as the clusterhead for the receiver.
- 2: it means that the sender has accepted the receiver as its clusterhead.

So finding a new clusterhead is done by sending three Find-Cluster packets with different status field. (see figures 6 and 7).

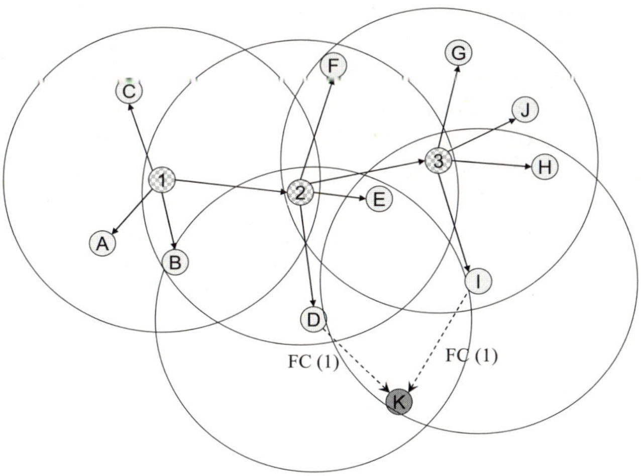

Fig. 6. node K does not have a clusterhead so after receiving the Find-cluster it can choice the clusterhead

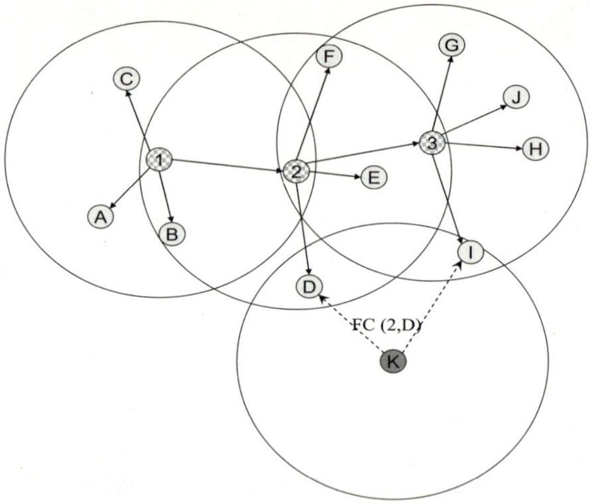

Fig. 7. Node K accepted D as its clusterhead

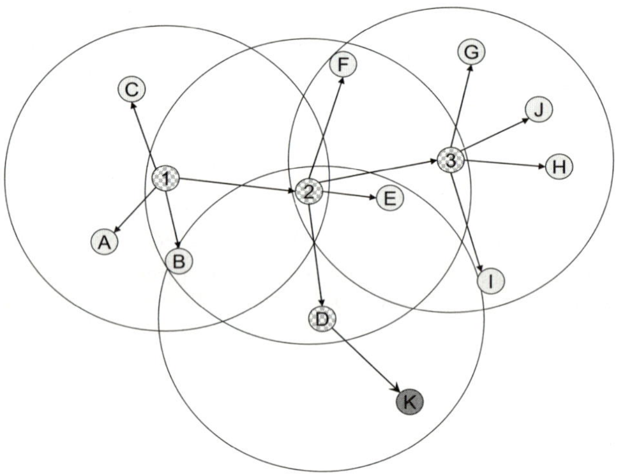

Fig. 8. Node D can send the packets to K

After determining the clusterhead, node can join to multicast group by sending Join-Reply packet like ODMRP. In our algorithm clusterhead is determining as forwarding group nodes; so they can transmit the data to destinations. In section 5 we show our simulation results completely. Simulation results show that SC-ODMRP improves the routing overhead and end to end delay.

4 Local Recovery Mechanism and Reliable Multicasting

After analyzing reliability and scalability on ODMRP, in this section we proposed a new recovery mechanism for ODMRP. Many routing algorithms use recovery method to increase the performance and decrease the delay between source and destination. Usually local recovery methods are used in tree based algorithm but sometimes because importance of data and our application we need to use this method in mesh based routing too. There is a number of works currently done on local route recovery, which include WAR (Witness-Aided Routing Protocol) [16], ABR (Associatively-Based Routing Protocol) and RDMAR (Relative Distance Micro discovery Ad-hoc Routing Protocol). In this section we introduce a new recovery approach while it keeping low complexity of the algorithm to a minimal. Complexity is an important attribute because nodes in MANET are generally small portable devices with little processing power, thus, the local recovery scheme must also cater to this property.

Because of many reasons in ODMRP it is possible that the discovered routes between source and destination break. Generally movement of node in MANET cause the route breakage, also if this movement occurred by one FG node (forwarding group node) it is harmful than the other cases because by movement of FG node one of the important connecting route between this FG and all other node connected to that FG (e.g. destination or FG nodes) will break so they can not receive any packet until other route is discovered. So mostly we considered this case in our algorithm.

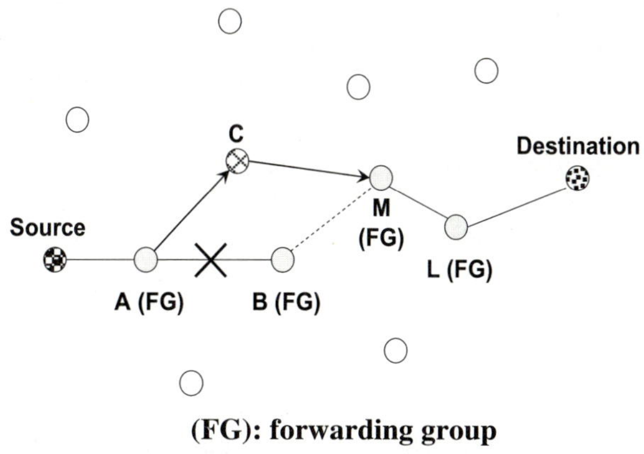

Fig. 9. Local Recover approach in our algorithm

With reference to Figure 9, if the direct link (AB) breaks off, there should exist, in most cases, some indirect route from A to the original next node B through some neighbor node C. In these situations, if a request packet is sent out to find the original next hop or other node which is at the further part of the original route with limited time-to-live (e.g. 2 hops), the possibility of repairing the current route should be high and the overhead should be much lower than using end-to-end global recovery. With the above assumptions, a new algorithm for the proposed this mechanism is described as follows:

When an intermediate FG node discovers that the link to the next hop has broken it would:

a) Save the data packet into local repair data buffer, set a timeout timer
b) Send out a two-hop wide local recovery request, containing a sequence of nodes including all the further nodes on the original route from A till the destination node D.

Any node, on receiving the local recovery request, will check whether it is on the node list, and if it is, would send back a local recovery reply

c) On receiving local recovery request from an intermediate node, A salvages the data packet at the local repair buffer and transmits it using the repaired route.

d) If no reply received after the local repair timer expires, the data packet is dropped, and an error message is send back to the source from A, the source may then initiate a new round of end-to-end route discovery.

For doing the local recovering method we changed the Join-Reply packet in ODMRP, so each receiver generates a Join-Table packet that is broadcast to its neighbors. The Join-Table packet contains the multicast groups address sequence of (source address, next hop address) pairs and a count of the number of pairs. In one pair the source address and the next hop address is the same and it is equal with receiver address. When a node receives a Join-Table, it checks if the next node address of one of the entries matches its own address. If it does the node realize that it is on the path to the source and thus becomes a part of the forwarding group for that source by setting its forwarding group flag. It then adds its own Join-Table to last received Join-Table and broadcast it. By this method each FG node knows the other node on the path between itself and destination. So FG node can recover the breakage route in suitable time with minimal complexity and without need the source. We call our algorithm LR-ODMRP and in next section we will compare it with ODMRP.

5 Simulation Results and Performance Evaluation

The simulation code was implemented within the Global Mobile Simulation (GloMoSim) library [13]. In the local recovery approach simulation, we modeled a network of 50 mobile hosts placed randomly within a 1000×1000m2 area. Radio propagation range for each node was 250 meters and channel capacity was 2Mbit/sec. Each simulation runs for 300 seconds of simulation time. The IEEE 802.11 Distributed Coordination Function was used as the medium access control protocol. We used Constant Bit Rate as our traffic. The size of data payload was 256 bytes. the source sends data at the rate of 20 packets/second, also We used Random-Way point as mobility model. For clustering mechanism we modeled the network by increasing the number of nodes and the speed. Nodes placed randomly within 2000×2000m2 area. Radio propagation range for each node was 376 meters and channel capacity was 2Mbit/sec.The size of data payload was 512 bytes and MAC protocol is IEEE 802.11 and simulation time is 300s.

Figure 10 show the protocol reliability varies with mobility (node speed) in our local recovery approach. This comparing shows that at lower speed the difference in packet delivery ratio is not outstanding.

In both case increased mobility requires that forwarding group member be updated more frequently also by increasing mobility causes frequent link change so by recovery method we can increase the packet delivery ratio. Figure 11 plots control overhead per data byte transferred as a function of mobility. Because of recovery packet in LR-ODMRP, this method slight increase overhead at higher speed.

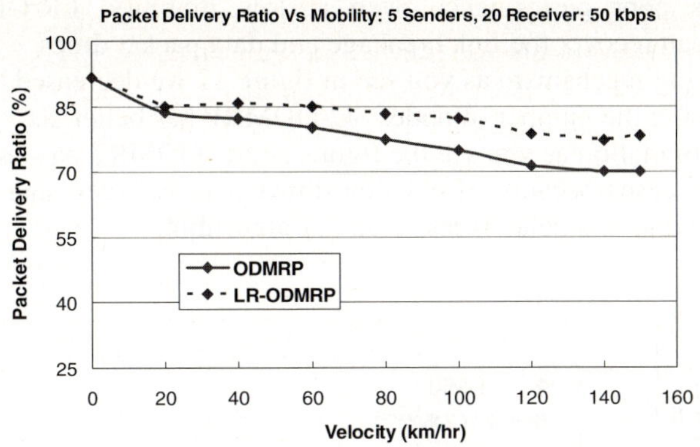

Fig. 10. Packet delivery ratio as a function of node mobility

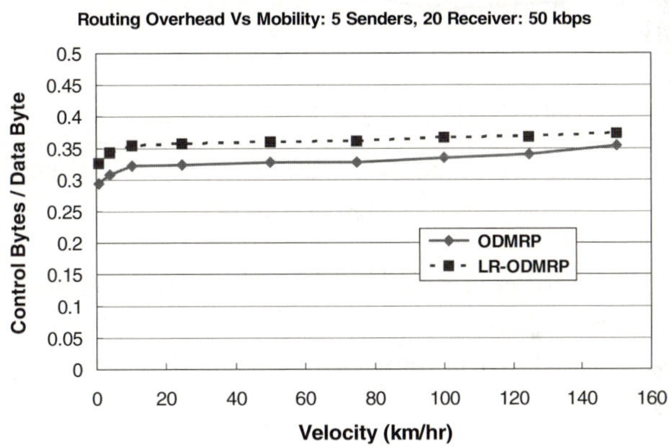

Fig. 11. Routing overhead as a function of node mobility

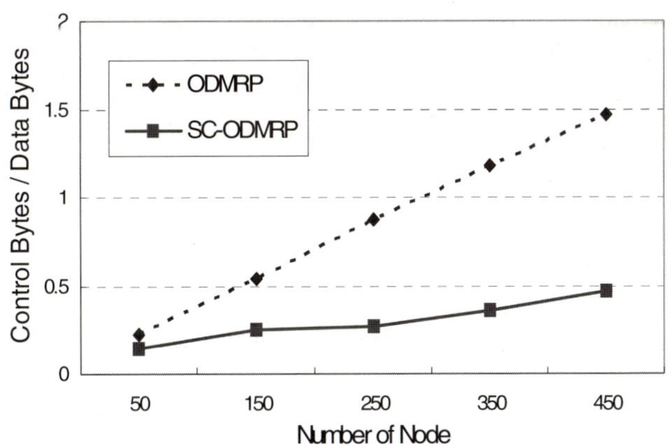

Fig. 12. Routing overhead as versus of number of nodes (2 groups, pause time 10s, speed 10m/s)

ODMRP has poor performance even at low mobility. LR-ODMRP act like ODMRP but it can recover the link breakage and data packet drops.

In our clustering mechanism as you see in figure 12 we decreased the routing overhead. By increasing the number of nodes SC-ODMRP has better act.

Also in our simulation as you see the figure 13 SC-ODMRP has less end to end delay, but in some cases because of existing direct path between sources and destinations ODMRP end to end delay is less than our algorithm.

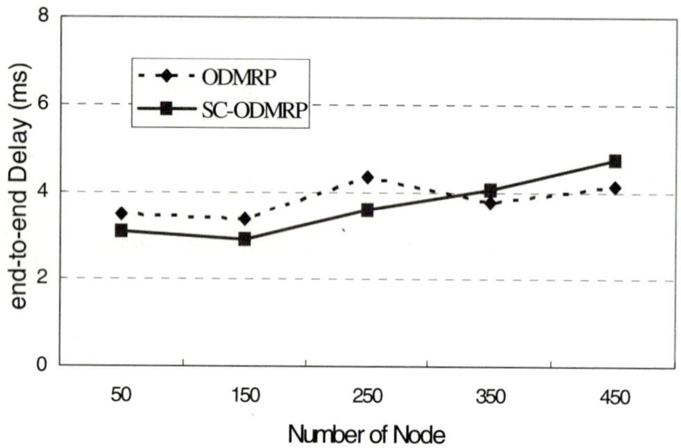

Fig. 13. End to end delay as versus number of nodes (2 groups, pause time 10s, speed 10m/s)

Our simulation showed that in most cases SC-ODMRP data delivery ratio is like ODMRP (figure 14).

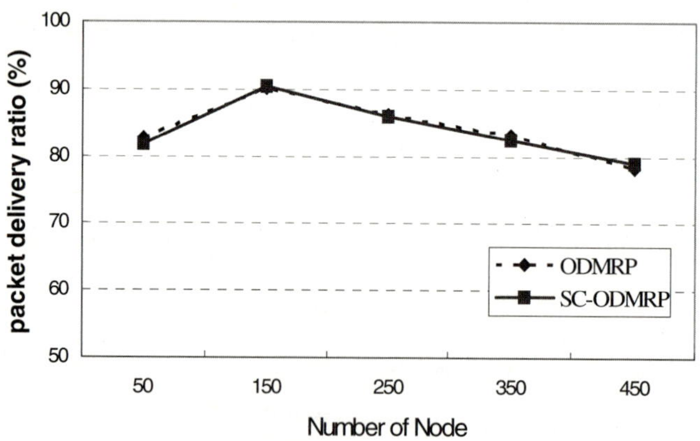

Fig. 14. Data delivery ratio as versus number of nodes (2 groups, pause time 10s, speed 10m/s)

But sometimes because of delay on sending the Find-cluster packet destination loss the data packets.

6 Conclusion

We introduced new techniques for clustering and local recovery. These techniques were applied to ODMRP and evaluated via several simulation scenarios. Our simulation showed that LR-ODMRP is more reliable than the ODMRP; also it has good performance when we increase the speeds of mobility and multicast group member, we can use LR-ODMRP in large scale topology and we can increase the packet delivery ratio and reliability. SC-ODMRP improves the network overhead in comparison of ODMRP by creating clusters and restricts the packet broadcasting. For the future works we intend to compact our local recovery method with clustering algorithm and we will extend it on other multicast routing protocols.

References

1. Amis, D., Prakash, R., Vuong, T.H.P., Huynh, D.T.: Max-Min D-Cluster Formation in Wireless Ad Hoc Networks. Proceedings of IEEE, 32–41 (March 2000)
2. Banerjee, S., Khuller, S.: A Clustering Scheme for Hierarchical Control in Multi-hop Wireless Networks, Technical Report CS-TR-4103, University of Maryland, College Park (February 2000)
3. MIT Roofnet, `http://www.pdos.lcs.mit.edu/roofnet`
4. Mesh Networks Inc., mesh networks technology overview, `http://www.meshnetworks.com`
5. Ohta, T., Inoue, S., Kakuda, Y.: An adaptive multihop clustering scheme for highly mobile ad hoc networks. In: Proc. 6th IEEE International Symposium on Autonomous Decentralized Systems (ISADS 2003), pp. 293–300 (2003)
6. Darehshoorzadeh, A., Dehghan, M., Jahed Motlagh, M.R.: Quality of Service Support for ODMRP Unicast Routing in Ad hoc Networks. In: International Conference on Digital Telecommunications (ICDT 2006), Côte d'Azur, France, August 29-31 (2006)
7. Baker, D.J., Ephremides, A.: The Architectural Organization of a Mobile Radio Network via a Distributed Algorithm. IEEE Transactions on Communications, 1694–1701 (1981)
8. Gerla, M., Lee, S.-J., Su., W.: On-demand multicast routing protocol (ODMRP) for ad hoc networks, IETF Internet Draft, draft-ietf-manet-odmrp-02.txt (2000)
9. Lee, S., Su, W., Hsu, J., Gerla, M., Bagrodia, R.: A performance comparison study of ad hoc wireless multicast protocols. In: INFOCOM 2000 (March 2000)
10. Tang, K., Obraczka, K., Lee, S.-J., Gerla, M.: A reliable, congestion-control led multicast transport protocol in multimedia multi-hop networks. In: The 5th International Symposium on Wireless Personal Multimedia Communications, pp. 252–256 (2002)
11. Gopalsamy, T., Singhal, M., Panda, D., Sadayappan, P.: A reliable multicast algorithm for mobile ad hoc networks. In: Proceedings of ICDCS (2002)
12. Rajendran, V., Yi, Y., Obraczka, K., Lee, S.J., Tang, K., Gerla, M.: Reliable, Adaptive, Congestion-Controlled Adhoc Multicast Transport Protocol: Combining Source-based and Local Recovery, UCSC Technical Report (2003)
13. UCLA Parallel Computing Laboratory, GloMoSim: A scalable simulation environment for wireless and wired network system
14. Iiyas, M.: The Handbook of Ad Hoc Wireless Networks. CRC Press, Boca Raton (2003)
15. Sun, X., Liu, W., Zhang, Z., Zhao, Y.: CODMRP: Cluster-Based On Demand Multicast Routing Protocol, WiCOM (September 2006)
16. Aron, I.D., Gupta, S.K.S.: A Witness-Aided Routing Protocol for Mobile Ad Hoc Networks with Unidirectional Links. In: Leong, H.V., Li, B., Lee, W.-C., Yin, L. (eds.) MDA 1999. LNCS, vol. 1748, pp. 24–33. Springer, Heidelberg (1999)

Cryptanalytic Flaws in Oh et al.'s ID-Based Authenticated Key Agreement Protocol*

Meng-Hui Lim[1], Sanggon Lee[2], and Hoonjae Lee[2]

[1] Department of Ubiquitous IT, Graduate School of Design & IT,
Dongseo University, Busan 617-716, Korea
menghui.lim@gmail.com

[2] Department of Information and Communication,
Dongseo University, Busan 617-716, Korea
{nok60, hjlee}@dongseo.ac.kr

Abstract. A key agreement protocol is designed for two or more entities to agree upon a shared secret key, which is used to preserve confidentiality and data integrity over an open network. In 2007, Oh et al. proposed an efficient ID-based authenticated key agreement protocol on elliptic curve pairings, which is believed to be able to generate two session keys securely after a protocol execution. However, we discover that their protocol is in fact susceptible to the basic impersonation attack as well as the key compromise impersonation attack. In this paper, we present the imperfections of Oh et al.'s scheme and subsequently we suggest a slight modification to the scheme which would resolve the problems.

1 Introduction

Key agreement protocols are essential in secure communications establishment across an insecure network. Specifically, two commmunication parties would exchange some public information in prior to create a mutual secret key that is known only to themselves. A secure key agreement protocol should be designed in such a way that no single protocol entity can predetermine or predict the secret key. This secret key (also known as a *session key*) if derived securely, can subsequently be used to create a confidential or integrity-protected communication channel among the entities.

Designing a secure key agreement protocol is not an easy task. A secure key agreement protocol would need to withstand both passive and active attacks which would most probably be launched by the adversaries when the messages are exchanged in a public channel. In the former case, it is assumed that the adversary, from time to time, eavesdrops on all the messages exchanged by the honest parties. Such an attack enables the adversaries to perform offline analysis on the captured data and extract any useful information whenever is possible. On contrary, the latter case involves a more powerful adversary who is assumed

* This work was supported by MIC of Korea, under ITRC support program supervised by IITA (IITA-2007-C1090- 0701-0026).

O. Gervasi et al. (Eds.): ICCSA 2008, Part II, LNCS 5073, pp. 458–467, 2008.

to be able to intercept, delete, inject, modify and replay messages in any online instance of the key agreement protocol. These attacks are usually preceived to be the capability of an outsider adversary in a two-party communication process.

To capture the notion of security, Wilson and Menezes [1,2] have defined a number of desirable security attributes which can be used to analyze a key agreement protocol. These security attributes are described as follows:

Known session key security. A protocol is considered to be *known session key secure* if it remains achieving its goal in the face of an adversary who has learned some previous session keys.

(Perfect) forward secrecy. A protocol enjoys *forward secrecy* if the secrecy of the previous session keys is not affected when the long term private keys of one or more entities are compromised. *Perfect forward secrecy* refers to the scenario when the long term private keys of all the participating entities are compromised.

Key-Compromise Impersonation Resilience. Suppose that a protocol entity A's long term private key has been disclosed. Obviously an adversary who knows this value can now impersonate A since it is precisely the value which identifies A. We say that a protocol is *key-compromise impersonation resilient* if this loss would not enable an adversary to masquerade as other legitimate entities to A as well or obtain other entities' secret key.

Unknown Key-Share Resilience. In an unknown key-share attack, an adversary convinces a group of entities that they share a session key with the adversary whereas in fact, the key is shared between the group and another party. This situation can be exploited in a number of ways by the adversary when the key is subsequently used to provide encryption or integrity.

Key Control Resilience. It should not be possible for any of the participants (or an adversary) to coerce the session key to a preselected or predicted value.

Since Boneh and Franklin [3] have initiated the development of identity(ID)-based encryption scheme using the notion of bilinear mapping based on elliptic curve pairings, many ID-based key agreement protocols have been proposed thereafter. In 2002, Smart [10] has proposed a notable ID-based authenticated key agreement protocol, which combines the ideas from Boneh-Franklin's work [3] and Joux's tripartite protocol [5]. However, this protocol does not provide the forward secrecy attribute, as pointed out by Shim in [9]. Shim has then proposed an alternative ID-based authenticated key agreement protocol and claimed it to be efficient and capable of satisfying most of the desired security features. Unfortunately, Sun-Hsieh [11] have demonstrated a valid man-in-the-middle attack on Shim's scheme few months later which renders her protocol to be insecure. In 2004, Ryu et al. [8] have proposed another efficient ID-based scheme, which requires every protocol participant to perform only one pairing computation and two point-multiplications in a protocol execution. Nevertheless, Boyd-Choo [4] and Yuan-Li [12] have discovered the insecurity of Ryu et al.'s scheme against the key compromise impersonation attack and the key reveal attack. Recently, Oh et al. [7] have proposed a new ID-based protocol which is able to generate two session keys in a protocol run. They claimed that their scheme is more secure and efficient as compared to Shim's protocol [9], Ryu et al.'s protocol [8] as well as Yuan-Li.'s protocol [12]. However, we strongly disagree with their claim as we have identified some fatal flaws in their scheme which would in fact endanger the protocol participants if it is employed.

Hence, the main purpose of this paper is to demonstrate the impersonation attacks on Oh et al.'s scheme. Besides that, we also suggest a slight modification to the original scheme so as to counter the defects. The structure of this paper is generally organized as follows. In the next section, we will illustrate the basic properties of bilinear pairings and some underlying assumptions explicitly. In Section 3, we will review Oh et al.'s ID-based authenticated key agreement protocol. In Section 4, we will demonstrate the impersonation attacks on their scheme. We will then illustrate our improvement in Section 5 and the subsequent security analysis in Section 6. Last but not least, we will conclude this paper in Section 7.

2 Preliminaries

Let $\mathbf{G}_1$ be an additive group of a large prime order, q and $\mathbf{G}_2$ be a multiplicative group of the same order, q. Let $P, Q \in \mathbf{G}_1$ and $\hat{e} : \mathbf{G}_1 \times \mathbf{G}_1 \longrightarrow \mathbf{G}_2$ be a bilinear pairing with the following properties:

- **Bilinearity:** $\hat{e}(aP, bQ) = \hat{e}(P, Q)^{ab} = \hat{e}(abP, Q)$ for any $a, b \in Z_q^*$.
- **Non-degeneracy:** $\hat{e}(P, Q) \neq 1$.
- **Computability:** There exists an efficient algorithm to compute $\hat{e}(P, Q)$.

A bilinear map which satisfies all three properties above is considered as *admissible bilinear*. It is noted that the Weil and Tate pairings associated with the supersingular elliptic curves or abelian varieties, can be modified to create such bilinear maps. Now, we describe some cryptographic problems:

Bilinear Diffie-Hellman Problem (BDHP). Let $\mathbf{G}_1$, $\mathbf{G}_2$, P and $\hat{e}$ be as above with the order q being prime. Given $\langle P, aP, bP, cP \rangle$ with $a, b, c \in Z_q^*$, compute $\hat{e}(P, P)^{abc} \in \mathbf{G}_2$. An algorithm α is deemed to have an advantage ϵ in solving the BDHP in $\langle \mathbf{G}_1, \mathbf{G}_2, \hat{e} \rangle$ based on the random choices of a, b, c in Z_q^* and the internal random operation of α if

$$Pr[\alpha(\langle P, aP, bP, cP \rangle) = \hat{e}(P, P)^{abc}] \geq \epsilon.$$

Computational Diffie-Hellman Problem (CDHP). Let P be an element of $\mathbf{G}_1$ as above. Given $\langle P, aP, bP \rangle$ with $a, b \in Z_q^*$, compute $abP \in \mathbf{G}_1$.

Discrete Logarithm Problem (DLP). Given two groups of elements P and Q, such that $Q = nP$. Find the integer n whenever such an integer exists.

For the remaining sections of this paper, we assume that BDHP, CDHP and DLP are hard such that there is no polynomial time algorithm to solve these cryptographic problems with non-negligible probability.

3 Review of Oh et al.'s Scheme

In this section, we revisit Oh et al's ID-based key agreement protocol [7], which consists of 3 phases, namely *Setup* phase, *Extract* phase and *Key Agreement* phase. We now describe each phases in detail.

Setup. Let E be a super-singular curve defined by $y^2 = x^3 + 1$ over F_p. The Key Generation Center (KGC) inputs a security parameter k into a BDH parameter generator $\mathcal{G}$, which returns two groups $\mathbf{G}_1$ and $\mathbf{G}_2$ of prime order q, a generator element P of $\mathbf{G}_1$, a bilinear map $\hat{e} : \mathbf{G}_1 \to \mathbf{G}_2$. KGC then chooses a random oracle $H : \{0,1\}^* \to \mathbf{G}_1$, selects a master key $s \in Z_q^*$, computes the master public key $P_{pub} = sP$ and publishes the system parameters $\{p, q, \mathbf{G}_1, \mathbf{G}_2, E, P, P_{pub}, H, \hat{e}\}$ while keeping s secret.

Extract. For an entity with identity information $ID \in \{0,1\}^*$, the public key is given as $Q_{ID} = H(ID)$. The KGC computes the corresponding private key as $S_{ID} = sQ_{ID}$ and issues S_{ID} to the entity via a secure channel. Hence, an ID-based key pair is defined as (Q_{ID}, S_{ID}), where $Q_{ID}, S_{ID} \in \mathbf{G}_1$.

Key Agreement. Suppose that two communicating entities, A and B wish to establish a session key by carrying out an instance of the protocol run. We denote A and B's public/private key pair to be Q_A/S_A and Q_B/S_B respectively. Then, the key exchange can be performed as follows:

1. A selects a random number $a \in Z_q^*$, computes $T_A = aP$ and sends T_A to B. In a symmetrical manner, B selects a random number $b \in Z_q^*$, computes $T_B = bP$ and sends T_B to A.

2. Upon receiving T_B, A computes the shared secrets

$$K_{AB} = \hat{e}(aQ_B, P_{pub}) = \hat{e}(Q_B, P)^{as} \tag{1}$$

and

$$K'_{AB} = \hat{e}(S_A, T_B) = \hat{e}(Q_A, P)^{bs}. \tag{2}$$

After receiving T_A, B computes the shared secrets

$$K_{BA} = \hat{e}(S_B, T_A) = \hat{e}(Q_B, P)^{as} \tag{3}$$

and

$$K'_{BA} = \hat{e}(bQ_A, P_{pub}) = \hat{e}(Q_A, P)^{bs}. \tag{4}$$

3. Subsequently, A and B compute two shared session keys, which are

$$\begin{aligned} K &= kdf(ID_A, ID_B, aT_B, K_{AB}) \\ &= kdf(ID_A, ID_B, bT_A, K_{BA}) \end{aligned} \tag{5}$$

and

$$K' = kdf(ID_A, ID_B, aT_B, K'_{AB}) \\ = kdf(ID_A, ID_B, bT_A, K'_{BA}), \quad (6)$$

where *kdf* is a key derivation function.

The authors claimed this protocol to be more efficient and secure than Shim's, Ryu et al.'s and Yuan et al.'s protocols by demonstrating a heuristic security and efficiency analysis. However, we discover that this is not the case where their protocol cannot withstand the basic impersonation attack and key compromise impersonation attack, and we will demonstrate these attacks in the next section.

4 Cryptanalysis on Oh et al.'s Scheme

In this section, we present two impersonation attacks on Oh et al.'s scheme in detail: the basic impersonation attack and the key compromise impersonation attack.

4.1 Basic Impersonation Attack

Similar to Joux's protocol [5], Oh et al.'s does not provide adequate authentication to prevent any unauthorized entity from establishing a session key with the legal entity. Indeed, as what Oh et al. claim, their scheme is able to produce two session keys efficiently at the end of a protocol run. However, they did not manage to secure both of the keys simultaneously. Eventually, one of the keys will be exposed to a malicious outsider adversary who impersonates another legal entity and engage with the target entity in a protocol execution. We now provide the details of the basic impersonation attack as follows:

Suppose that an adversary E, pretending as B, wants to establish a protocol run with A.

1. Initially, A selects a random number $a \in Z_q^*$, computes $T_A = aP$ and sends T_A to B. E intercepts the message, selects a random number $e \in Z_q^*$, computes $T_B^\star = eP$ and sends T_E to A on behalf of B.
2. Upon receiving $T_B^\star$, A computes the shared secrets

$$K_{AB} = \hat{e}(aQ_B, P_{pub}) = \hat{e}(Q_B, P)^{as} \quad (7)$$

and

$$K'_{AB} = \hat{e}(S_A, T_B^\star) = \hat{e}(Q_A, P)^{es}. \quad (8)$$

Then, as per protocol specification, A computes two shared session keys, which are

$$K = kdf(ID_A, ID_B, aT_B^\star, K_{AB}) \\ = kdf(ID_A, ID_B, aeP, \hat{e}(aQ_B, P_{pub})) \quad (9)$$

and

$$K' = kdf(ID_A, ID_B, aT_B^{\star}, K'_{AB}) \\ = kdf(ID_A, ID_B, aeP, \hat{e}(S_A, eP)). \tag{10}$$

3. Note that E does not know a, S_A and S_B. Hence, she does not have the knowledge in computing

$$K_{BA} = \hat{e}(S_B, T_A) = \hat{e}(aQ_B, P_{pub}) = K_{AB}.$$

However, she is able to calculate

$$K'_{BA} = \hat{e}(eQ_A, P_{pub}) = \hat{e}(Q_A, P)^{es} \tag{11}$$

which is equivalent to K'_{AB} in Eq. (8). E then proceeds to calculate

$$K' = kdf(ID_A, ID_B, eT_A, K'_{BA}) \\ = kdf(ID_A, ID_B, aeP, \hat{e}(eQ_A, P_{pub})). \tag{12}$$

Hence, E has successfully agreed on the second session key (K') with A by masquerading as B. Despite of being more efficient, Oh et al.'s scheme has failed to resist the basic impersonation attack, which may further results in other cryptanalytic attacks. In fact, this attack can also be carried out in a similar manner when E intends to cheat B by masquerading as A. If that is the case, E would be able to agree upon the first session key

$$K = kdf(ID_A, ID_B, aeP, \hat{e}(eQ_B, P_{pub})) \tag{13}$$

with A. Thus, we can conclude that Oh et al.'s scheme is insecure against the basic impersonation attack.

4.2 Key Compromise Impersonation Attack

Key Compromise Impersonation (KCI) attack is a kind of known-key attack, which can be performed by an adversary after compromising a protocol entity's private key. We often refer such entity as *corrupted*. Note that in practice, a malicious party may hack into an entity's computer in order to acquire the entity's private key, and even worse, the corrupted entity may be unaware of this intrusion. By learning this secret key, the adversary is now able to impersonate the corrupted party directly in communicating with other protocol entities. However, this is often not our interest to delve into it in a KCI analysis. On contrary, the general adversarial goal in the KCI attack is to impersonate any other legitimate party or parties and establish a common session key with the corrupted party in a protocol run.

In [7], Oh et al. have scrutinized the KCI resilience of their scheme and subsequently conjecture their scheme to be KCI secure. However, they only restrict their analysis in considering the adversary to be a passive eavesdropper, and thus they claim that the adversary is unable to compute the parameter

$aT_B = bT_A = abP$ (an instance of CDHP), which is required in computing the session keys K and K'. Regrettably they have left out the most important scenario in their analysis where the adversary would impersonate any legal entity to perform the attack. Recall that in Section 4.1, we have proven Oh et al.'s scheme to be susceptible to the basic impersonation attack. In the current KCI analysis where the adversary has acquired an additional secret key of the target entity, it is easy to see that their scheme cannot resist the key compromise impersonation attack as well since the adversary can employ the same strategy as in Section 4.1 to compute one of the session keys in the KCI attack. This significantly violates the KCI security of a key agreement protocol, and renders their scheme to be KCI insecure.

5 Improvements on Oh et al.'s Scheme

In the attacks which we have demonstrated previously, the adversary can only compute a shared secret out of two in each protocol execution. It is apparent that if we use the shared secrets separately to derive two distinct session keys, one of them will eventually be exposed. Hence, in order to secure the scheme, one direct way is to utilize the two shared secrets to derive only a session key, so that the adversary would not be able to compute it by learning merely a shared secret. With this, we suggest a minor modification to the session key computation of the scheme, where

$$\begin{aligned} K &= kdf(ID_A, ID_B, abP, K_{AB}, K'_{AB}) \\ &= kdf(ID_A, ID_B, abP, K_{BA}, K'_{BA}). \end{aligned} \tag{14}$$

We now provide a general overview of the enhanced protocol as follows:

1. Message Exchange:
   $A \rightarrow B : T_A = aP.$
   $B \rightarrow A : T_B = bP.$
2. Shared Secrets Computation:
   $A : K_{AB} = \hat{e}(aQ_B, P_{pub}), K'_{AB} = \hat{e}(S_A, T_B).$
   $B : K_{BA} = \hat{e}(S_B, T_A), K'_{BA} = \hat{e}(bQ_A, P_{pub}).$
3. Session Key Computation:
   $A : K = kdf(ID_A, ID_B, aT_B, K_{AB}, K'_{AB}).$
   $B : K = kdf(ID_A, ID_B, bT_A, K_{BA}, K'_{BA}).$

6 Security Analysis of the Improved Scheme

Now, we scrutinize the security of our improvement on Oh et al.'s scheme to ensure that the flaws have been overcome while the other desired security attributes are preserved.

Known session key security. In every protocol run, the ephemeral private key (a, b) that each protocol entity chooses would vary. Since these ephemeral values are involved in computing the session key of our enhanced protocol, the session key would also vary with every protocol execution. On top of that, a key deriving function (usually a one-way cryptographic hash function) is used to derive the session key. Hence, the knowledge of some previous session keys would not allow the adversary to gain any advantage in deriving any future and other previous session keys.

Perfect Forward secrecy. Suppose that A and B's long term private key S_A and S_B have been compromised. Besides, we also assume that the adversary has obtained some previous session keys established by A and B. However, the adversary is not able to derive any other previously established session keys since the adversary does not possess their respective ephemeral private keys which are needed in constructing those session keys. Hence our enhanced protocol is able to fulfill the perfect forward secrecy property. Furthermore, if the KGC's secret s is compromised at any point, our protocol is still able to retain the confidentiality of previous session keys since the adversary is not able to compute the component abP (which is a CDHP) in recovering the session key as shown in Eq. (14). Hence, our protocol is deemed providing the KGC forward secrecy too.

Key-Compromise Impersonation Resilience. Suppose that A's static private key S_A has been compromised and the adversary wishes to impersonate B in order to establish a session with A. However, she is unable to compute the first shared secret K_{BA} in Eq. (1) or (3) as she does not know a or S_B. Notice that in our enhanced protocol, the adversary is required to derive both of the shared secrets in order to compute the session key in Eq. (14). Hence, this significantly prevents the adversary from launching her KCI attack successfully. Generally, the same situation would result when the long term key S_B is compromised (the adversary fails to compute the second shared secret K'_{AB} in this case) as our enhanced protocol is symmetric. As a result, our enhanced protocol is able to withstand the KCI attack under all circumstances.

Key Replicating Resilience. The session key of our enhanced protocol is derived by using a key deriving function which takes in A and B's identity and the shared secrets. If the adversary carries out a key replicating attack on our enhanced protocol by means of choosing a random number $e \in Z_q^*$ and modifying the A's message to $e \cdot T_A$ as well as B's message to $e \cdot T_B$ during the message exchange phase, Alice's final session key $K = kdf(ID_A, ID_B, abeP, \hat{e}(Q_B, P)^{as}, \hat{e}(Q_A, P)^{bes})$ would be different from Bob's session key $K = kdf(ID_A, ID_B, abeP, \hat{e}(Q_B, P)^{aes}, \hat{e}(Q_A, P)^{bs})$ at the end of the protocol execution. With this, the adversary would not be able to force the establishment of non matching sessions to possess a common session key. As a result, if the adversary reveals A's session key, she would not be able to guess B's session key correctly with non-negligible probability and vice versa.

Basic Impersonation Resilience. Recall that our enhanced protocol mandates every protocol entity to compute both shared secrets (K_{AB} and K'_{AB} or K_{BA} and K'_{BA}) which are used subsequently to derive the session key.

This has in fact prevented the adversary's attempt in launching the basic impersonation attack as described in Section 4.1 since the adversary is only able to recover one of the shared secrets. Therefore, it is not difficult to see that with the knowledge of partial agreed secret, the adversary would not be able to distinguish the actual session key (which is the output of kdf) from a random string. Intuitively, we claim that our enhanced protocol can resist the basic impersonation attack perfectly against any unauthorized party.

Unknown Key-Share Resilience. In our enhanced scheme, the identity of the communicating parties have been included in the session key so as to prevent the attacker from launching the unknown key-share attack in various ways on our improved protocol. Hence, a stronger sense of authentication can be achieved explicitly.

Key Control Resilience. Apparently in our protocol, no single protocol participant could force the session key to a predetermined or predicted value since the session key is derived by using both A and B's long term and ephemeral private keys.

7 Conclusion

As a conclusion, we have rebuted Oh et al's claim about the security of their protocol by mounting a basic impersonation attack and a key compromise impersonation attack on their scheme. Based on these flaws, we have subsequently suggested a slight modification to the original scheme so as to secure the scheme. Besides justifying our safeguard against the impersonation attacks, we have presented a heuristic security analysis to ensure that our enhanced scheme remains preserving all the desirable security features that the original scheme possesses.

References

1. Blake-Wilson, S., Menezes, A.: Authenticated Diffie-Hellman Key Agreement Protocols. In: Tavares, S., Meijer, H. (eds.) SAC 1998. LNCS, vol. 1556, pp. 339–361. Springer, Heidelberg (1999)
2. Blake-Wilson, S., Johnson, D., Menezes, A.: Key Agreement Protocols and their Security Analysis. In: Proceedings of the 6th IMA International Conference on Cryptography and Coding. LNCS, vol. 1355, pp. 30–45 (1997)
3. Boneh, D., Franklin, M.: Identity-based Encryption from the Weil Pairing. In: Kilian, J. (ed.) CRYPTO 2001. LNCS, vol. 2139, pp. 213–229. Springer, Heidelberg (2001)
4. Boyd, C., Choo, K.-K.R.: Security of Two-Party Identity-Based Key Agreement. In: Dawson, E., Vaudenay, S. (eds.) Mycrypt 2005. LNCS, vol. 3715, pp. 229–243. Springer, Heidelberg (2005)
5. Joux, A.: A One-round Protocol for Tripartite Diffie-Hellman. In: Proceedings of the 4th International Algorithmic Number Theory Symposium (ANTS-IV). LNCS, vol. 1838, pp. 385–394. Springer, Heidelberg (2000)
6. McCullagh, N., Barreto, P.S.L.M.: A New Two-Party Identity-Based Authenticated Key Agreement. In: Menezes, A. (ed.) CT-RSA 2005. LNCS, vol. 3376, pp. 262–274. Springer, Heidelberg (2005)

7. Oh, J.-B., Yoon, E.-J., Yoo, K.-Y.: An Efficient ID-Based Authenticated Key Agreement Protocol with Pairings. In: Stojmenovic, I., Thulasiram, R.K., Yang, L.T., Jia, W., Guo, M., de Mello, R.F. (eds.) ISPA 2007. LNCS, vol. 4742, pp. 446–456. Springer, Heidelberg (2007)
8. Ryu, E.-K., Yoon, E.-J., Yoo, K.-Y.: An Efficient ID-Based Authenticated Key Agreement Protocol from Pairings. In: Mitrou, N.M., Kontovasilis, K., Rouskas, G.N., Iliadis, I., Merakos, L. (eds.) NETWORKING 2004. LNCS, vol. 3042, pp. 1458–1463. Springer, Heidelberg (2004)
9. Shim, K.: Efficient ID-based Authenticated Key Agreement Protocol based on Weil Pairing. Electronics Letters 39(8), 653–654 (2003)
10. Smart, N.P.: An Identity Based Authenticated Key Agreement Protocol Based on the Weil Pairing. Electronics Letters 38, 630–632 (2002)
11. Sun, H.M., Hsieh, B.T.: Security Analysis of Shim's Authenticated Key Agreement Protocols from Pairings, Cryptology ePrint Archive: Report (113) (2003)
12. Yuan, Q., Li, S.P.: A New Efficient ID-Based Authenticated Key Agreement Protocol, Cryptology ePrint Archive: Report (309) (2005)

AFPL, an Abstract Language Model for Firewall ACLs

S. Pozo, R. Ceballos, and R.M. Gasca

Department of Computer Languages and Systems
ETS Ingeniería Informática, University of Seville
Avda. Reina Mercedes S/N, 41012 Sevilla, Spain
Tel:+34 954559897 Fax: +34 954557139
sergiopozo@us.es, {ceballos,gasca}@lsi.us.es
http://www..lsi.us.es/~quivir

Abstract. Design and management of firewall rule sets is difficult and error prone, mainly because the translation of access control requirements to low level languages is difficult. Abstract languages have been proposed, but none have been adopted by the industry. We think that the main reason is that their complexity is close to many of the existing low level languages. Complexity is defined as the difficulty to express knowledge from the reality being modeled (access control requirements). In this paper, we analyze the most widely used firewall languages and different possibilities of abstraction. Based on this analysis, a model for Firewall languages is proposed, and a new simple yet expressive and powerful firewall abstract language, Abstract Firewall Policy Language (AFPL), is proposed. AFPL can then be translated to existing low level firewall languages, or be directly interpreted by firewall platforms. We expect that AFPL can fill the gap between requirements and low level firewall languages.

Keywords: acl, firewalls, model, language, access control, ruleset.

1 Introduction

A firewall is a network element that controls the traversal of packets across different network segments [1, 2], thus it is a mechanism to enforce an access control policy (ACP), represented as a finite access control list (ACL). A firewall ACL is a finite list of ordered (total order) condition/action rules, called the rule set. Each field of the condition part of a rule is called a selector. In firewall rule sets, selectors are for example IP addresses, ports and protocol.

Firewalls have to face many problems in modern networks. One of the most important ones is the high complexity of rule set design and management. Any error defining a rule set may compromise the system security by letting unwanted traffic pass or blocking desired traffic.

Networks have different access control requirements that must be translated by a network administrator to firewall rule sets. Writing and managing rule sets are a tedious, time consuming and error prone tasks for several reasons. Translation of requirements to rule sets is not a naïve task. One of the main reasons is that the gap

O. Gervasi et al. (Eds.): ICCSA 2008, Part II, LNCS 5073, pp. 468–483, 2008.

between the high level access control objectives or requirements and low level rule sets are too big. Firewall low level languages are, in general, difficult to learn, use and understand. Each firewall platform has its own low level language, which need to be known by the network administrator in order to implement the access control requirements. Changing from one vendor to another implies a complete rewrite of the rule set. Low level firewall languages are, in general, very different between them in syntax and semantics.

When a new model or language is going to be proposed, there is a compromise between expressivity and complexity. A very expressive model is generally more complex than a less expressive one. We define complexity as the difficulty to express knowledge from the reality being modelled. Existing language proposals for access control exist, some of them are specific to firewall domain and other are generic to access control. However, we think that these languages are too complex (in the sense of the previously stated definition), and that complexity is the main reason of not being widely adopted by the industry: it is very difficult to translate access control requirements *(the reality)* to these languages. In addition, the construction of models during the requirements analysis and system design phases can improve the quality of the resulting systems by providing a foundation for early analysis and fault detection, prior to final code generation. This approach has been taken in [18], where authors propose a firewall ACL model, and diagnose inconsistencies in it. Then, the consistent model can be used to generate particular low level rule sets.

Thus, a simple yet highly expressive and powerful language is needed to cover this need. In this paper, the most important and widely used firewall languages have been analyzed, as a starting point for the new language design. Then, different model possibilities for a new language are discussed, and an abstract firewall language model is proposed. Finally, a grammar based on XML Schema Definition (XSD) is proposed as a proof of concept. The language is called *AFPL, Abstract Firewall Policy Language.* This language can then be compiled to low level firewall languages or being interpreted directly by firewall platforms.

The paper is structured as follows. In section 2 related works are presented. Section 3 presents the analysis of firewall languages and the proposal of the model for AFPL. Section 4 presents a grammar discussion. Section 5 presents conclusions and future works for AFPL. Appendix I presents the full analysis of the selectors of the most used low level firewall languages. Appendix II presents a base AFPL model used in some parts of the paper. Finally, Appendix III presents the XSD grammar.

2 Related Works

Firewall languages tend to be very low level. Writing a rule set is a very difficult task [3] that usually requires in-depth knowledge of a particular firewall language and platform working internals. Recall that complexity is defined in this paper as the difficulty to express knowledge from the reality being modelled. Fortunately, some research groups have proposed languages to model access control policies. In [4] authors propose a high level firewall language, Firmato, and model it as an ERD which assign roles to services. This language can be compiled to several low level languages. They also provide a mechanism to completely separate the security policy from network topology. However, the complexity of this language is similar to many low level ones. In addition, the use of roles could cause confusion to many users, as RBAC models cannot be fully expressed with ACLs [14]. Finally, a major limitation of Firmato is that the language can only represent knowledge in positive logic (*allow* rules), which complicate the expression of exceptions. However, Due to this limitation, the rules expressed are always consistent and order independent.

FLIP [16] is a recently proposed firewall language which can also compile to several low level ones. Their authors claim that ACLs expressed in FLIP are consistent. However, they are because one of its limitations: it does not support overlapping between rule selectors. Prohibiting the use of overlaps is a major limitation, since it is impossible to express exceptions, which could result in the need to write tons of rules to express them. In addition, its syntax is even more complex than Firmato's one. However, due to this lack of expressiveness, FLIP ACLs are order independent.

NeTSPoC [17] is a description language for security policy and topology. It generates consistent rule sets resolving anomalies by giving *deny* rules a higher priority. This consistency resolution technique is not adequate, since the resulting rule set could not conform to requirements [19]. Again, the complexity of the syntax of NeTSPoC is similar to Firmato and FLIP.

HLFL [15] is another high level language which can compile to Linux, BSD and PIX. This language is very simple, but we think that it is too simple to express some scenarios. For example, it does not support port ranges in its syntax. In addition, its development stopped in 2003.

In [5] authors provide a language, Ponder, to represent routine network policies, which do not compile to any low level language. Unfortunately, its complexity surpasses the needs of the needed for firewall ACLs.

Some organisations have proposed languages to represent access control policies as XML documents, such as XACML [6], PCIM [7], Rule-ML [8], and SRML [9]. However, none of these languages are specific enough for firewall access control policies, which are a much simpler subset of access control policies. These languages tend to be very general and complex to be used in firewall domain, as they cover a wider spectrum of security policies and applications. Even UML has been proposed to model access control policies [13]. However, we consider that UML could be a valid language for the requirements definition phase, but then these models need to be translated to an intermediate level language. There are some good surveys about access control policy languages available [10, 11, 12].

For the reasons stated above, a simple yet highly expressive and powerful firewall language is needed. In this work, the most important and widely used firewall languages are analysed: Linux IPTables 1.3.7, Cisco PIX 7.0, FreeBSD 6.2 IPFilter, FreeBSD 6.2 IPFirewall, OpenBSD 3.7 Packet Filter, and Checkpoint Firewall-1 4.1. Then, different language models created in a bottom-up process are discussed. A model for the AFPL language created on the basis of them in a bottom-up approach is proposed. Finally, this model is used to construct the AFPL grammar in XSD. However, other grammars can be constructed if required, since one model is valid for all grammars. The provided XSD grammar should be taken as a proof of concept.

Finally, AFPL can be used to construct specifications of firewall ACLs, which could then be transformed to low level vendor specific rule sets. We expect that this language can fill the gap between requirements and low level firewall languages, and ease design and management of rule sets.

Firewall platforms are very different from one company to other, and even between the available Open Source platforms, which range from differences in the number and type of selectors, to severe differences in rule processing algorithms that can affect the design of the rule set. In addition, firewall platforms have other specific characteristics. These range from how does each platform treat stateful or stateless connection tracking, how the rule processing is done (forward, backward, with jumps…), etc. Characteristics not related with filtering parameters or selectors should be considered in the transformation algorithms. In this paper, only the AFPL model and a proof of concept grammar for it are proposed. Conversion algorithms to low level languages have been left for future works.

3 AFPL Model

The syntax and semantics of the most important and widely used firewall languages have been analyzed, in order to acquire sufficient knowledge of their characteristics and possibilities. Analysis result is presented in Appendix I. Explanations of the analysis are given in this section.

3.1 Model Creation Alternatives

When abstracting from reality in order to create a model in a bottom-up process, three main approaches can be taken. These approaches, applied to the firewall languages problem domain are described in the following sections (Fig. 1).

Factorized AFPL Model	Aggregated AFPL Model	Customized AFPL Model
- Simple	- Complex	- Balanced
- Minimal	- Maximal	- Customized
- Not complete	- Complete	- Not complete

Fig. 1. Characteristics of models for AFPL

3.1.1 Factorization of Selectors

The first approach is to analyze different firewall languages and factorize the common filtering selectors. Selectors can be expressed using different syntaxes, and thus a factorization of these syntaxes is also necessary. The resulting AFPL language model is composed of only the common filtering selectors and their common syntaxes (Appendix II).

The main benefit of this approach is that the model is minimal and very simple. A simple language with more typically used selectors is easier to understand and use for the end user than a language with rare selectors that are only used in a minority of situations. The main drawback of this approach is that, in general, the resulting model may not be sufficiently complete to represent reality. These selectors are not widely used because they are only applied in very specific situations. Even some of them are only used in one of the analyzed firewall languages. Note that, for example, Cisco

PIX platform does not support any of these selectors, but it is one of the most used firewalls in the world, proving that complexity is not a good approach to language design.

The same problem arises within each common selector. The syntaxes supported by the same selector of different firewall languages are usually different, having some common ones. Differences in syntax for a given selector could have some benefits: more compact rule sets that result in less memory consumption and/or fastest processing, etc. However, all these different syntaxes, if supported in AFPL model, need to be translated to the specific syntaxes supported in each specific low level language, which will complicate the translation process. A translation is possible *(necessary condition)* if properties exhibited by a specific firewall platform can be reproduced in another one with no extensions to its language or modifications to the matching algorithm. Thus, there could be cases where a translation may be impossible.

3.1.2 Aggregation of Selectors

This is just the opposite approach of factorization. The key idea is to take all filtering selectors with all their possible syntaxes, and construct AFPL model with them. Thus the resulting model will be composed of *all* common and uncommon parts of the low level language of each firewall platform.

The main benefit of this approach is that the resulting model is complete. This gives the users a high degree of flexibility (and also complexity) when using the language. The main drawback of this approach is complexity. The model and the resulting language could be similar in complexity to existing proposals of high level languages. When using the language for a specific problem, the user has to cope with all the complexity inherited from each firewall language (a lot of selectors, wide domains, etc.), which is clearly worse than if the user coded the solution directly in a low level language. We think that this is the main reason why existing proposals of languages have not been adopted by industry.

Having a model with selectors and syntaxes that are not supported by all firewall platforms has several disadvantages over the previous approach. Some selectors and their syntaxes present in some firewall languages would be impossible to represent in others. This implies that these selectors need to be emulated, which in some cases could be an impossible task, as it is going to be explained in the next section.

Under these circumstances, the resulting rule set could be very difficult to be generated from the AFPL language, or even impossible. We think that it is easier to automate optimization (as is the case of the factorized model) than emulation.

3.1.3 Customization of Selectors

Finally, this approach is the hybridization of the other two, it represents a trade-off. In this approach, each uncommon selector should be analyzed in detail. The main benefit of this approach is that usability and complexity of the resulting language is equilibrated. Some selectors and syntaxes used in the model need to be emulated in low level firewall languages, and others need to be optimized. This is the approach we have taken in order to design the AFPL model, and it is described in the next section.

3.2 AFPL Model

Our start point is the factorized AFPL model presented in Appendix II. We take it as a basic meta-model. In the next sections, more complexity is going to be added to this basic meta-model, in order to customize the selectors and their syntaxes.

3.2.1 Addition of Uncommon Selectors

An uncommon selector (Fig. 2) can be added if its behaviour can be emulated with the common selectors of *all* the considered firewall platforms, and it also adds new functionality to the model. A selector *adds new functionality* to the model if it cannot be emulated with rules which do not contain it. For example, the *Time To Live* selector cannot be emulated decomposing the rule that uses it in several other rules which do not use it.

A selector is emulable if its behaviour can be reproduced using a combination of one or more rules but using only the common selectors of firewall languages. Note that the use of interfaces for matching is optional in all analyzed platforms except in PIX. For this reason interface use in AFPL is optional. If the final rule set is for PIX, then the same rule set can be applied for all interfaces, as does FW-1. The same is applicable for interface direction, which is mandatory for PIX and IPFilter and optional for the rest of firewall languages.

The conclusion is that none of the uncommon selectors can be emulated, they cannot be incorporated into AFPL model. Not all matching engines of firewall platforms support them: they are impossible to emulate with the common selectors.

SELECTOR	EMULABLE	NEW FUNCTIONALITY	ADD TO AFPL
TCP Flags	x	√	x
TCP Options	x	√	x
IP Options	x	√	x
IP Version	x	√	x
Src/Dst MAC	x	√	x
ToS	x	√	x
TTL	x	√	x

Fig. 2. Analysis of uncommon selectors

3.2.2 Addition of Uncommon Syntaxes to Selectors

In this section, we analyze the possibility of supporting uncommon syntaxes in all the considered selectors. In general, we consider an uncommon syntax as a candidate for its addition to the group of supported syntaxes for that selector in the AFPL model, if it can be emulated with the common syntaxes of the same selector and it provides clear usability improvements for human users. A syntax *provides clear usability improvements* if its use by a human cannot introduce inconsistencies in the AFPL model and language and it provides compactness. This characteristic is subjective, and for that reason, there are syntaxes that could be considered for their inclusion in AFPL model (they are marked by an asterisk). However, since one of our main objectives is to keep AFPL model and language as small and simple as possible, we have decided to leave them out. The analysis is presented in Fig 3. Again, we base this analysis on results presented in Appendix I.

- **Source and Destination IP.** For these selectors there are several uncommon syntaxes to be analyzed. IP addresses can also be expressed as identifiers in FW-1, providing new functionality and a clear usability improvement. IPTables and FW-1 support the specification of groups of continuous IP addresses in these selectors, but their matching engines decompose a rule with this syntax in several rules with one IP per rule. A range of IP addresses can be expressed in several rules, thus we consider that this do not provides a clear usability improvement. IPFilter, IPFW, PF and FW-1 support domain names, but as a domain is the same as an identifier, no new functionality is provided. PF support tables, and PF and FW-1 support tables and lists, which are very similar to ranges, and thus will not be considered. IPFW permit logic expressions of IP addresses with OR and NOT logic operators, which again provides a new user oriented functionality which we consider that is too complex and error prone. PF can use an interface name (referring to the block of IP addresses statically assigned to that interface), a block of IPs of that interface and an interface which receives its IP from DHCP. Although this provides new functionality, in this case, the functionality is already in the actual meta-model, since it supports the specification of rules per interface. All firewall platform permit negation of address selectors, except PIX and IPFilter. Although this is new functionality, it provides no clear advantages over a syntax without negation, because a negated selector can be emulated with a combination of allow and deny rules. In conclusion, no new syntax is going to be added for Source and Destination IP address.
- **Interface.** Interfaces can be expressed in all firewall platforms as identifiers, representing the name of the interface of the underlying platform. All platforms except PIX and FW-1 can use groups or lists to represent a collection of interfaces, providing no new functionality but a minor usability improvement (firewalls typically have a few interfaces, this can be reproduced using several rules). IP addresses can be used in IPFW and FW-1, but it is the same as using directly the IP address of the network assigned to the interface, providing a new and easily emulable functionality and an improvement in usability, and thus will be considered (usually one knows the IP of an interface, but not the name of the interface). All platforms except PIX permit the use of a wildcard representing that a rule applies to all interfaces, since PIX rules must be assigned to a specific interface. Note that the use of interfaces for matching is optional in all analyzed platforms except in PIX. This represents a new functionality and a big usability improvement, since if not considered in the meta-model, the only way to represent a rule for all interfaces is to write the same rule but for each different interface. In addition, the wildcard can be easily emulated in PIX generating a rule for each interface. Thus for these reasons, we will consider it in the meta-model. Negation will not be considered for the same reasons explained in Source and Destination IP.

- **Interface Direction.** The only thing that is different in the analyzed firewall platforms when specifying the direction of the interface is that in PIX and IPFilter the use of direction is mandatory, and that they do not support wildcards. This is the same case of interfaces, as explained before. For the same reasons, wildcards will be incorporated in the meta-model. Again, for the same reasons stated above, negation is not considered.
- **Protocol.** IPFilter, IPFW and PF support the use of identifiers, but as explained above they are easily emulable user-centric improvements and thus will be considered. IPFilter support the specification of TCP plus UDP protocols at the same time, but again it can be decomposed in two rules, thus it is a user facility with no big improvement in ease of use. ICMPv6-based filtering is supported in PF, but its support depends on the matching algorithm implemented in the platform. Protocol lists are also supported in some firewalls, but as other lists, they can be decomposed in several rules. Logic OR/NOT expressions can also be used in IPFW, and negation is also possible as with other selectors in some firewall platforms, but for the same reasons stated above, they will no be considered in the model.
- **Source and Destination Ports.** Many range syntaxes are possible in many firewall platforms, as is the case of ranges ‘<p’, ‘<=p’, ‘>p’, ‘>=p’, ‘(p1, p2)’ and ‘)p1, p2(‘. These syntaxes provide no new functionality or a clear usability improvement, and can be easily emulated with the common “[p1, p2]” syntax without loss of functionality or performance. For this reason we will not include them in the meta-model. Lists of ports, logic OR/NOT expressions and negation are also possible in some platforms, but again we will not consider it for the same reasons stated before.
- **ICMP Type.** IPFW support lists as a way of representing various types of ICMP traffic, and many languages support negation, but will not be considered in the meta-model for the same reasons. The same reasons stated before are applicable here.
- **Action.** There are two common actions: allow and deny a packet. However, all analyzed firewall platforms except Cisco PIX support a third one, reject, that denies a packet, but with acknowledging it to the sender. This is a new functionality that cannot be emulated in PIX, but can be transformed into a deny action without a big loss in functionality (the packet will be denied, the only difference is the acknowledgement). For this reason, packet rejection will be permitted in the meta-model.

Fig. 3 presents a summary of the analysis and, as can be seen, nearly all the uncommon syntaxes are oriented to provide compactness of the rule set. However, we think that most of them also generate confusing rule sets, and can easily generate inconsistencies. Once the rule set has been created in AFPL, and prior to deployment, it is always possible to use one of the many existent rule set optimization algorithms. Fig. 4 represents the final AFPL model. The next point should be to define a simple grammar for it. I the next section, one is provided as a proof of concept.

SELECTOR	UNCOMMON SYNTAX	EMULABLE	NEW FUNCTIONALITY	USABILITY IMPROVEMENT	ADD TO AFPL
IP Address	Identifier/Domain	√	√	√	√
	Continuous range	√	x	x*	x
	Group	√	x	x*	x
	OR/NOT	√	x	x	x
	Negation	√	√	x	x
	Interface name	√	√	x	x
Interface	IP Address	√	√	√	√
	Group	√	x	x*	x
	Wildcard	√	x	√	√
	Negation	√	√	x	x
Interface Direction	Wildcard	√	x	√	√
	Negation	√	√	x	x
Protocol	Identifier	√	√	√	√
	TCP+UDP	√	x	x	x
	ICMPv6	x	√	x	x
	Group	√	x	x*	x
	OR/NOT	√	x	x	x
	Negation	√	√	x	x
Port	Range <p	√	x	x*	x
	Range >p	√	x	x*	x
	Range <=p	√	x	x*	x
	Range >=p	√	x	x*	x
	Range (p1,p2)	√	x	x*	x
	Range)p1,p2(	√	x	x*	x
	Group	√	x	x*	x
	OR/NOT	√	x	x	x
	Negation	√	√	x	x
ICMP Type	Group	√	x	x*	x

Fig. 3. Analysis of uncommon syntaxes

4 AFPL Grammar

As has been stated above, several different grammars can be constructed for the AFPL model. However, we have preferred to represent the proof of concept grammar as an XML Schema Definition (XSD) document, since perhaps it is most widely known than other languages, such as ABNF. Representing the AFPL model (Fig. 4) in XML is an easy task, since it is very simple. The XSD is presented in Appendix III. In addition, for illustrative purposes, Fig. 5 presents an example firewall ACL in AFPL which, due to space constraints, is only composed of two rules. Note that rule order is implicit in the XML, since rule matching algorithms follow a linear order in firewalls. Thus it is not necessary to explicitly represent rule priorities in the model. Comparing any of the reviewed high and low level languages with this XML representation, it can be seen how easy is to represent real life firewall knowledge in AFPL.

However, using XML instead of other languages has several advantages. First, as can be seen from the example of Fig. 5, the ACL is very well structured and it is easily understandable, even by a person without any particular knowledge about a concrete firewall language, XML syntax, or even the XSD schema. This also

Selector	Obligation	Dependencies	Common Syntax (Can be optimized)	Uncommon Syntax (Must be emulated)	Comments
Source and Destination IP Address	Mandatory		-IP -Block -Wildcard	-Identifier	
Interface	Optional		-Identifier	-Wildcard -IP	
Interface Direction	Optional		-In -Out	-Wildcard	IPTables can not represent direction, but can differentiate inbound from outbound interfaces
Protocol	Mandatory		-TCP -UDP -ICMP -Number -Wildcard*	-Identifier	*IPFilter and PF do not have wildcards, but the omission of this selector has the same effect
Source and Destination Port	Optional	Only if Protocol is TCP or UDP	-Number -Range: [p1,p2] -Identifier -Wildcard		
ICMP Type	Optional	Only if Protocol is ICMP	-Number -Identifier		
Action	Mandatory		-Allow -Deny	-Reject	Reject cannot be emulated in PIX, but deny can be used instead loosing the acknowledgement

Fig. 4. AFPL Model

```
<policy>
    <rule>
        <matches>
             <protocol>tcp</protocol>
             <ipsrc>140.192.37.20</ipsrc>
             <ipdst>all</ipdst>
             <prtdst>80</prtdst>
        </matches>
        <action>deny</action>
    </rule>
    <rule>
        <matches>
             <protocol>tcp</protocol>
             <ipsrc>140.192.37.0/24</ipsrc>
             <ipdst>all</ipdst>
             <prtdst>21</prtdst>
       </matches>
       <action>allow</action>
    </rule>
</policy>
```

Fig. 5. AFPL example

improves modification complexity to the ACL. Second, this XML can easily be parsed using available libraries, since most object oriented programming languages have an XML parsing library in their API by default. In addition, it is undefined in the schema (Appendix III) which value does an optional selector take if it is omitted in a rule. The parser is the responsible of valuating these selectors. In this paper, it is assumed that these selectors receive a wildcard value, as is the usual solution that most firewall languages take if a selector is omitted. As explained before, the AFPL model and language cannot represent characteristics other than the needed ones to take filtering decisions (selectors with their syntax and actions). Other characteristics related to how the ACL is going to be executed are not part of the model.

5 Conclusions and Future Works

Firewalls have to face many problems in modern networks. One of the most important ones is the high complexity of rule set design and management. Any error defining a rule set may compromise the system security by letting unwanted traffic pass or blocking desired traffic.

Different networks have different access control requirements that must be translated by a network administrator to firewall rule sets. Translation of requirements to rule sets is not a naïve task. One of the main reasons is that the gap between the high level access control objectives or requirements and low level rule sets is too big. Firewall low level languages are, in general, difficult to learn, use, and understand. A simple yet highly expressive and powerful language is needed to address this need.

In this paper we first analyzed the most important and widely used firewall languages. Then, different models for the new language have been discussed. Finally, a model and a proof of concept language created on the basis of them have been proposed. The language is called *AFPL, Abstract Firewall Policy Language*. The model can also be used to construct other grammars for the language. AFPL can be used to design firewall ACLs, which could then be transformed to low level vendor specific rule sets.

Our future research plan includes the design of a definitive grammar and the implementation of automatic translation algorithms to several low level firewall languages from AFPL.

Acknowledgements

This work has been partially funded by the Spanish *Ministerio de Educación y Ciencia* in the project DPI2006-15476-C02-01.

References

1. Chapman, D., Zwicky, E.: Building Internet Firewalls, 2nd edn. O'Reilly & Associates, Inc., Sebastopol (2000)
2. Cheswick, W., Belovin, S.: Firewalls and Internet Security, 2nd edn. Addison-Wesley, Reading (2003)

3. Wool, A.: A quantitative study of firewall configuration errors. IEEE Computer 37(6), 62–67 (2004)
4. Bartal, Y., Mayer, A., Nissim, K., Wool, A.: Firmato: A Novel Firewall Management Toolkit. ACM Transactions on Computer Systems 22(4) (2004)
5. Damianou, N., Dulay, N., Lupu, E., Sloman, M.: The Ponder Specification Language. In: Workshop on Policies for Distributed Systems and Networks (Policy2001), HP Labs Bristol, January 29-31 (2001)
6. OASIS eXtensible Access Control Markup Language (XACML), `http://www.oasis-open.org/committees/xacml/`
7. Moore, B., Ellesson, E., Strassner, J., Westerinen, A.: Policy Core Information Model (PCIM). IETF RFC 3060 (February 2001)
8. Rule Markup Language (RuleML), `http://www.ruleml.org/`
9. Simple Rule Markup Language (SRML): A General XML Rule Representation for Forward-chaining Rules. ILOG, S.A (May 2001)
10. De Capitani di Vimercati, S., Foresti, S., Jajodia, P.: Access control policies and languages. Int. J. Computational Science and Engineering (2007)
11. Ardagna, C.A., Damiani, E., De Capitani di Vimercati, S., Samarati, P.: XML-based Access Control Languages. Information Security Technical Report. Elsevier Science, Amsterdam (2004)
12. El-Atawy, A.: Survey on the Use of Formal Languages/Models for the Specification, Verification, and Enforcement of Security Policies", Technical reports, DePaul University, CTI 06-005 (April 2006)
13. Basin, D., Dorser, J., Lodderstedt, T.: Model Driven Security: from UML Models to Access Control Infrastructures. ACM Transactions on Software Engineering and Methodology 15(1) (January 2006)
14. Barkley, J.: Comparing simple role based access control models and access control lists. In: ACM Workshop on Role-Based access control (RBAC). ACM, New York (1997)
15. High Level Firewall Language, `http://www.hlfl.org`
16. Zhang, B., Al-Shaer, E., Jagadeesan, R., Riely, J., Pitcher, C.: Specifications of a High-level Conflict-free Firewall Policy Language for Multi-domain Networks. In: Proceedings of ACM Symposium on Access Control Models and Technologies (SACMAT 2007) (2007)
17. NeTSPoC: A nework Security Policy Compiler, `http://netspoc.berlios.de`
18. Pozo, S., Ceballos, R., Gasca, R.M.: Model Based Development of Firewall Rule Sets: Detecting and Diagnosing Errors. Information and Software Technology Journal (Spring 2008)
19. Pozo, S., Ceballos, R., Gasca, R.M.: CSP-based Firewall Rule Set Diagnosis using Security Policies. In: International Symposium on Frontiers in Availability, Reliability and Security (FARES), in International Conference on Availability, Reliability and Security (ARES), Vienna, Austria, April 2007, IEEE Computer Society Press, Los Alamitos (2007)

Appendix I. Analysis of Selectors of Firewall Languages

This analysis is related to the condition part of a rule, and refers to the conditions that may be matched against a packet that arrives at the firewall. Each firewall language has a different syntax for each selector. Note that, due to space constraints, uncommon selectors have been dropped.

I.1 Source and Destination Addresses

	iptables	*PIX*	*ipfilter*	*ipfw*	*pf*	*Fw-1*
Src/Dst IP	optional	mandatory	optional	mandatory	optional	optional
IP	√	√	√	√	√	√
Identifier						√
Block	√	√	√	√	√	√
Range	√					√
Domain			√	√	√	√
Table					√	
List					√	√
Logic OR/NOT expression				√		
Interface					√	
Block interface					√	
DHCP interface					√	
Negation	√			√	√	√
Wildcard	√	√	√	√	√	√
Default value	Any ip		Any ip		Any ip	Any ip

I.2 Interface

	iptables	*PIX*	*ipfilter*	*ipfw*	*pf*	*Fw-1*
Interface	optional	mandatory	optional	optional	optional	optional
Identifier	√	√	√	√	√	√
Group Identifier	√		√	√	√	
List					√	
IP Address				√		√
Negation	√			√	√	
Wildcard	√		√	√	√	√
Default value	Any interface		Any interface	Any interface	Any interface	Any interface

I.3 Interface Direction

	iptables	*PIX*	*ipfilter*	*ipfw*	*pf*	*Fw-1*
Direction	implicit	mandatory	mandatory	optional	optional	optional
In	√	√	√	√	√	√
Out	√	√	√	√	√	√
Negation				√		
Wildcard	√			√	√	√
Default value	Any direction			Any direction	Any direction	Any direction

I.4 Protocol

	iptables	*PIX*	*ipfilter*	*ipfw*	*pf*	*Fw-1*
Protocol	optional	mandatory	optional	mandatory	optional	optional
Number	√	√	√	√	√	√
Identifier			√	√	√	
TCP	√	√	√	√	√	√
UDP	√	√	√	√	√	√
TCP/UDP			√			
ICMP	√	√	√	√	√	√
ICMP6					√	
List	√				√	√
Logic OR/NOT expression				√		
Negation	√			√		√
Wildcard	√	√	√	√	√	√
Default value	Any protocol		Any protocol		Any protocol	Any protocol

I.5 Source and Destination Ports

	iptables	*PIX*	*ipfilter*	*ipfw*	*pf*	*Fw-1*
Src/Dst port	optional	optional	optional	optional	optional	optional
Number	√	√	√	√	√	√
Identifier	√	√	√	√	√	
Range ≠ p	using negation	√	√	using negation	√	using negation
Range < p	using negation	√	√		√	√
Range > p	using negation	√	√		√	√
Range ≤ p	√		√		√	using negation
Range ≥ p	√		√		√	using negation
Range [p1,p2]	√	√	√	√	√	√
Range (p1,p2)			√		√	
Range)p1,p2(	using negation		√	using negation	√	using negation
Block				√		
List	√			√	√	
Logic OR/NOT expression				√		
Negation	√			√		√
Wildcard	√	√	√	√	√	√
Default value	Any port	Any port	Any port	Any port	Any port	Any port
	iptables	*PIX*	*ipfilter*	*ipfw*	*pf*	*Fw-1*

Appendix II. Factorized AFPL Model

This analysis represents the result of the syntax factorization of the common selectors.

Selector	Obligation	Dependencies	Common Syntax (Can be optimized)
Source and Destination IP Address	Mandatory		-IP -Block -Wildcard
Interface	Optional		-Identifier
Interface Direction	Optional		-In -Out
Protocol	Mandatory		-TCP -UDP -ICMP -Number -Wildcard
Source and Destination Port	Optional	Only if Protocol is TCP or UDP	-Number -Range: [p1,p2] -Identifier -Wildcard
ICMP Type	Optional	Only if Protocol is ICMP	-Number -Identifier
Action	Mandatory		-Allow -Deny

Appendix III. AFPL XML Schema

The grammar is composed of a list of Condition/Action rules. The grammar should have at least one rule. This is because there is a required rule at the end of the ACL that represents the default policy that should be taken if no other previous rule matches. The condition part of the rule uses the selectors presented in Table 3, which also have specific syntaxes. These selectors are Source and Destination IP (mandatory), Source and Destination Ports (only if protocol is TCP or UDP), Protocol (mandatory), ICMP Type (only if protocol is ICMP), and finally interface and direction (optional).

```
<?xml version="1.0"?>
<xs:schema xmlns:xs="http://www.w3.org/2001/XMLSchema">
        <xs:element name="policy">
                <xs:complexType>
                        <xs:sequence>
                                <xs:element name="rule" minOccurs="1" maxOccurs="unbounded">
                                        <xs:complexType>
                                                <xs:sequence>
                                                        <xs:element name="matches" type="matchesType"/>
                                                        <xs:element name="action" type="actionType"/>
                                                </xs:sequence>
                                        </xs:complexType>
                                </xs:element>
                        </xs:sequence>
                </xs:complexType>
        </xs:element>
        <xs:complexType name="matchesType">
                <xs:all>
                        <xs:element name="interface" type="interfaceType" minOccurs="0"/>
                        <xs:element name="direction" type="directionType" minOccurs="0"/>
                        <xs:element name="ipsrc" type="ipType"/>
                        <xs:element name="ipdst" type="ipType"/>
                        <xs:element name="protocol" type="protoType"/>
                        <xs:element name="prtsrc" type="portType" minOccurs="0"/>
                        <xs:element name="prtdst" type="portType" minOccurs="0"/>
                        <xs:element name="icmptype" type="icmptType" minOccurs="0"/>
                </xs:all>
        </xs:complexType>
        <xs:simpleType name="actionType">
                <xs:restriction base="xs:string">
                        <xs:enumeration value="allow"/>
                        <xs:enumeration value="deny"/>
                        <xs:enumeration value="reject"/>
                </xs:restriction>
        </xs:simpleType>
        <xs:simpleType name="interfaceType">
                <xs:restriction base="xs:string">
                        <xs:pattern value="([0-2 ? 0-9 ? 0-9 \.[0-2 ? 0-9 ? 0-9 \.[0-2 ? 0-9 ? 0-9 \.[0-2 ? 0-
9 ? 0-9]) ( a-zA-Z0-9])+"/>
                </xs:restriction>
        </xs:simpleType>
        <xs:simpleType name="directionType">
                <xs:restriction base="xs:string">
                        <xs:enumeration value="in"/>
                        <xs:enumeration value="out"/>
                        <xs:enumeration value="both"/>
                </xs:restriction>
        </xs:simpleType>
        <xs:simpleType name="ipType">
                <xs:restriction base="xs:string">
                        <xs:pattern value="(([0-2 ? 0-9 ? 0-9 \.[0-2 ? 0-9 ? 0-9 \.[0-2 ? 0-9 ? 0-9 \.[0-
2 ? 0-9 ? 0-9])(/[0-3 ? 0-9])?  ( a-zA-Z0-9])+"/>
                </xs:restriction>
        </xs:simpleType>
        <xs:simpleType name="protoType">
                <xs:restriction base="xs:string">
                        <xs:pattern value="tcp udp icmp ( a-zA-Z])+ ( 0-2 ? 0-9 ? 0-9])"/>
                </xs:restriction>
        </xs:simpleType>
        <xs:simpleType name="portType">
                <xs:restriction base="xs:string">
                        <xs:pattern value="(([illegible] 0-9 ? 0-9 ? 0-9 ? 0-
9])?  ( a-zA-Z])+"/>
                </xs:restriction>
        </xs:simpleType>
        <xs:simpleType name="icmptType">
                <xs:restriction base="xs:string">
                        <xs:pattern value="([0-2 ? 0-9 ? 0-9]) ( a-zA-Z])+"/>
                </xs:restriction>
        </xs:simpleType>
</xs:schema>
```

The Design and Message Complexity of Secure Socket SCTP

Stefan Lindskog[1] and Anna Brunstrom[2]

[1] Centre for Quantifiable Quality of Service in Communication Systems
Norwegian University of Science and Technology, Trondheim, Norway
stefan.lindskog@q2s.ntnu.no

[2] Department of Computer Science, Karlstad University, Sweden
anna.brunstrom@kau.se

Abstract. This paper describes the design of secure socket SCTP (SS-SCTP). SS-SCTP is a new end-to-end security solution that uses the AUTH extension for integrity protection of messages and TLS for mutual authentication and key negotiation. Data confidentiality is in SS-SCTP provided through encryption at the socket layer. SS-SCTP aims to offer a high degree of security differentiation based on features in the base SCTP protocol as well as in standardized extensions. The flexible message concept provided in the base protocol plays a central role in the design of SS-SCTP. In the paper, a comparison of the message complexity produced by SS-SCTP, SCTP over IPsec, and TLS over SCTP is also presented. The main conclusion that can be drawn from the comparison is that, depending on the traffic pattern, SS-SCTP produces either less or similar message overhead compared to the standardized solutions when transferring user data.

Keywords: SCTP, end-to-end security, packet protection, security differentiation, message complexity, protocol design.

1 Introduction

Stream control transmission protocol (SCTP) [18] is a connection oriented transport protocol standardized by the signaling transport (SIGTRAN) working group within Internet engineering task force (IETF). SCTP has evolved to a mature, general purpose protocol on par with the traditional transport protocols user datagram protocol (UDP) and transmission control protocol (TCP). Compared to UDP and TCP, SCTP provides many new and interesting transport features that make it a strong candidate for next generation communication networks.

SCTP provides a reliable transport service, ensuring that data is transported across a network without error and in sequence. An unordered delivery service is also offered by the protocol. Prior to data transmission, an SCTP connection or association, as it is called in SCTP, is created between the endpoints. An association is maintained until all data has been successfully transmitted. Associations

O. Gervasi et al. (Eds.): ICCSA 2008, Part II, LNCS 5073, pp. 484–499, 2008.

in SCTP are initiated through a four-way handshake mechanism, which is aimed to protect against denial-of-service (DoS) attacks that try to block resources by sending a large number of forged setup requests. Furthermore, SCTP uses a flow and congestion control mechanism similar to TCP.

SCTP is message-oriented and supports framing of individual message boundaries based on the so-called chunk concept. Through the chunk concept, protocol extensions are easily implemented. In [19, 23, 24, 21], four such extensions that have reached a standard status are described.

Furthermore, SCTP provides advanced features for both multistreaming and multihoming. The idea with multistreaming is to decrease the impact of head-of-line (HoL) blocking. A stream in SCTP is a unidirectional channel within an association. Streams provide the ability to send separate sequences of ordered messages as independent flows. In particular, a packet loss in one stream does not inhibit the delivery of packets in other streams. Multihoming, on the other hand, is the ability for a single SCTP endpoint to support multiple IP addresses. In SCTP, multihoming was introduced as a mean to provide network fault tolerance. By introducing multihoming, additional security attacks could be envisioned. In [20], unique and well-known security threats associated with SCTP and especially its multihoming facility are described.

Neither SCTP nor UDP or TCP provide mechanisms for end-to-end (E2E) security provision. For SCTP, E2E security is recommended to be implemented through the use of either IP security (IPsec) [12] or transport layer security (TLS) [4]. Both SCTP over IPsec [1] and TLS over SCTP [8] have proven to introduce limitations and inefficiencies [25]. Two other solutions have therefore been proposed. These two are referred to as secure SCTP [25] and SCTP aware DTLS[1] [6], respectively. However, even these two new solutions have their limitations [13].

The design and message complexity analysis of an alternative E2E security solution, referred to as secure socket SCTP (SS-SCTP), is described in this paper. During the design of SS-SCTP, the aim has been to provide an E2E security solution that is efficient and, at the same time, can offer a high degree of security differentiation based on standardized features. SS-SCTP provides E2E security through the new authenticated chunks (AUTH) extension [24] for integrity protection of messages; TLS for mutual authentication and key exchange; and confidentiality protection of messages based on encryption and decryption at the socket layer. The general idea behind SS-SCTP together with a qualitative comparison with the four above mentioned E2E security solutions are presented in [13].

The remainder of the paper is organized as follows. In Section 2, the message concept in SCTP is described. A brief summary of the two standardized E2E security solutions is given in Section 3. The design of SS-SCTP is presented in Section 4. In Section 5, the message complexity for SS-SCTP, SCTP over IPsec, and TLS over SCTP is compared. Finally, concluding remarks are presented in Section 6.

[1] DTLS is an abbreviation for datagram TLS. The DTLS standard is described in [17].

2 SCTP Messages

As already mentioned, SCTP is message-oriented and supports framing of individual messages. One or more messages may be bundled into the same SCTP packet. An SCTP packet consists of a common header plus one or more chunks. A source and a destination port number (16 bits each), a verification tag (32 bits), and a checksum (32 bits) are included in the common header. The length of the common header is thus 96 bits (or 12 bytes) in total.

Chunks are used to frame messages within SCTP packets. Both control and user data are transferred as chunks. Each chunk has at least a type field (8 bits), a flag field (8 bits), and a length field (16 bits). Most chunk types may also contain a value field. The content in the value field is dependent on the chunk type. This implies that all chunks are at least 32 bits (4 bytes) in length.

An example of a packet with two control chunks and three data chunks is illustrated in Fig. 1. As can be seen in the figure the common header is placed first in the SCTP packet. After the common header, the control chunks are added followed by as many data chunks as will fit in the packet. The number of chunks that can be carried within the same SCTP packet is limited by the path maximum transmission unit (PMTU). If a single message is larger than the PMTU minus the IP header and the common header, SCTP will fragment the message and send the fragments within multiple SCTP packets.

Fig. 1. An example of an SCTP packet with the mandatory SCTP common header (white), two control chunks (black), and three data chunks (dark grey)

In the base SCTP protocol, 13 different chunk types are defined. These are described in Table 1. Note that only one chunk type, i.e., DATA, is used for transferring user data. The other 12 chunk types carry control information. INIT, INIT ACK, COOKIE ECHO, and COOKIE ACK are used when initializing a SCTP association. These are further described in Section 4. SHUTDOWN, SHUTDOWN ACK, and SHUTDOWN COMPLETE are transferred to gracefully terminate an association.

When designing SCTP a key feature was to make it extensible. The single most important feature for extensibility is the concept of chunks. By using the flexible chunk concept, four extensions have so far been standardized. These are partial reliability [19], padding chunk [23], authenticated chunks (referred to as the AUTH extension) [24], and dynamic address reconfiguration [21]. The extensions define new chunk types, new parameters for specific chunk types, and new error causes to be sent in ABORT and/or ERROR chunks. All in all, five additional chunk types have been defined in the four extensions—one new chunk type is defined in each extension except in the extension for dynamic address reconfiguration, where two new chunk types are defined. The additional chunk

Table 1. Chunk types defined in the base SCTP protocol [18]

Type	Name	Description
0	DATA	User data
1	INIT	Initiation
2	INIT ACK	Initiation acknowledgement
3	SACK	Selective acknowledgement
4	HEARTBEAT	Heartbeat request
5	HEARTBEAT ACK	Heartbeat acknowledgement
6	ABORT	Abort
7	SHUTDOWN	Shutdown
8	SHUTDOWN ACK	Shutdown acknowledgement
9	ERROR	Operational error
10	COOKIE ECHO	State cookie copy reply
11	COOKIE ACK	Cookie acknowledgement
14	SHUTDOWN COMPLETE	Shutdown complete

Table 2. Chunk types defined in the standardized SCTP extensions

Type	Name	Description
192	FORWARD TSN	Forward cumulative transmission sequence number (TSN) [19]
132	PAD	SCTP packet padding [23]
15	AUTH	Authentication chunk [24]
193	ASCONF	Address configuration change [21]
128	ASCONF ACK	Address configuration change acknowledgement [21]

types are summarized in Table 2. The only extension used in SS-SCTP is the AUTH extension and it is more thoroughly described in Section 4.

3 Standardized E2E Security Solutions

Two E2E security solutions for SCTP have been standardized until now, i.e., SCTP over IPsec and TLS over SCTP. A brief summary of these are given below.

3.1 SCTP over IPsec

The use of IPsec to secure E2E SCTP traffic is described in [1]. IPsec operates on the network layer and was designed to be independent of the transport protocol. The current version of IPsec has evolved to a fairly complex protocol. IPsec could in fact be seen as a number of protocols. When only integrity protection is needed the authentication header (AH) protocol [10] is used. If both integrity and confidentiality protection is needed, the encapsulating security payload (ESP) protocol [11] is used instead. To be able to dynamically establish security associations (SAs), a protocol referred to as the Internet key exchange (IKE) protocol [9] might be used.

The major problem with this solution, as described in [1], is the dynamic creation of SAs using IKE for multihomed hosts. When two multihomed hosts with n and m IP addresses, respectively, are establishing an SCTP association, $2 * m * n$ SAs must also be established. The creation of SAs is known to be an expensive operation and must be negotiated separately. When SCTP is used over IPsec all messages belonging to the same SCTP association are also protected in the same way. Hence, no security differentiation may be applied at run-time. This is mainly due to the fact that IPsec is neither aware of SCTP messages nor of SCTP streams. Securing all data transferred between two hosts may produce unnecessary computational burden at the endpoints.

3.2 TLS over SCTP

Another solution to implement E2E security is through the use of TLS over SCTP as described in [8]. TLS is a byte-oriented protocol that requires an underlying reliable and in-sequence delivery service, which is offered by SCTP as long as it does not use the unordered delivery service or the extension for partially reliable transport. Application data transfers are made over a TLS connection. Such a connection must be bidirectional. TLS provides key negotiation and endpoint authentication, data integrity through hashed message authentication codes (HMACs), and/or data confidentiality through encryption and decryption algorithms.

The most serious disadvantage with TLS over SCTP is that SCTP control chunks exchanged between two peers are completely unprotected. This is caused by the fact that TLS operates on a layer above SCTP and is thus not aware of the control chunks. Due to the requirement of TLS on reliable ordered delivery, the SCTP unordered delivery service as well as the extension for partially reliable transport is not at all supported. Note also that a bidirectional TLS connection is established using two unidirectional SCTP streams. Hence, a single unidirectional SCTP stream cannot be protected by TLS.

4 Design of SS-SCTP

The design of SS-SCTP is described in detail in this section. An overview of the protocol is first given. Then, the procedure for establishing a secure association is described followed by a description of user data transfers. The socket application programming interface (API) is summarized last in the section.

4.1 Overview

When designing SS-SCTP, the goal has been to provide as much security differentiation as possible using features defined in the base SCTP protocol and in its standardized extensions. A key feature used in SS-SCTP is the concept of messages and chunks introduced in the previous section. Another important feature in the base protocol is the multistreaming mechanism. With respect to the standardized extensions, only the AUTH extension is used. The structure of the involved components is illustrated in Fig. 2.

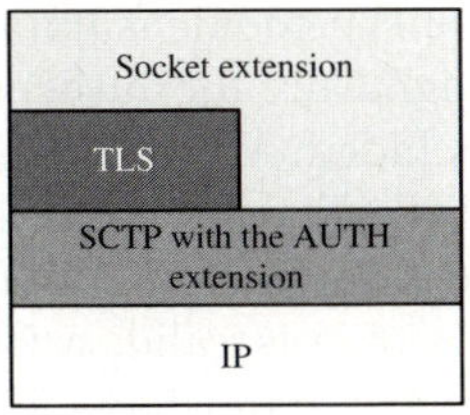

Fig. 2. Structure of SS-SCTP and its relation to other protocols

SS-SCTP provides data integrity protection of control and data chunks, and data confidentiality protection of data chunks. Data integrity protection is provided through HMACs by using the AUTH extension. Data confidentiality is, on the other hand, implemented through encryption and decryption at the socket layer using the OpenSSL crypt library [16]. A new socket API, which extends the base SCTP socket API, has therefore been defined. The API is further described in Subsection 4.4.

In order to provide data integrity and confidentiality protection as described above, a mechanism for mutual authentication and key exchange is needed. For this purpose, TLS is used. Note, however, that TLS is in SS-SCTP only used for mutual authentication and key exchange when establishing a new association, and for re-authentication and re-keying in an ongoing association. A similar usage of TLS is provided in the IEEE 802.11i standard [7] when open system authentication is used.

4.2 Secure Association Establishment

A secure association in SS-SCTP is essentially established in two steps. In step 1, an SCTP association is established. In this step, key parameters for the AUTH extension are exchanged. In step 2, mutual authentication and key exchange is performed using TLS. After these two steps are successfully completed, user data may be exchanged between the peers.

Step 1. In the first step of a secure association establishment, a four-way handshake is performed as defined in the base SCTP protocol. The message exchange in such a handshake is illustrated in Fig. 3. As can be seen in the figure, the first chunk to be sent is the INIT chunk. INIT contains information related to the association, such as an initiation tag, wished number of outbound streams, maximum number of inbound streams, and initial TSN. The initiation tag is a random number that will later be used as a verification tag in the SCTP common header. The INIT chunk may also contain pre-defined optional parameters, such as additional IP version 4 (IPv4) addresses, IP version 6 (IPv6) addresses, and hostnames. The second chunck exchanged is the INIT ACK chunk, which contains similar information to the INIT chunk except the mandatory state cookie parameter. The state cookie parameter is used as a defense against resource exhaustion attacks. The content of the cookie is implementation specific, but some guidelines are given in [18]. A signature for authentication when the state cookie comes back must at least be included. When the peer receives the INIT ACK, a COOKIE ECHO chunk is replied. The COOKIE ECHO contains a copy of the received state cookie. Upon reception of a COOKIE ECHO chunk, the included state cookie is verified. If the cookie is valid, a COOKIE ACK is sent. This is the final chunk sent in the four-way handshake. If, on the other hand, the state cookie is invalid, an ABORT chunk is replied. The COOKIE ACK chunk is an example of a chunk type that only contains the three mandatory fields, i.e., type, flag, and length, as described in Section 2. Note that both the COOKIE ECHO and the COOKIE ACK could be bundled with other chunks, e.g., DATA and SACK, in an SCTP packet.

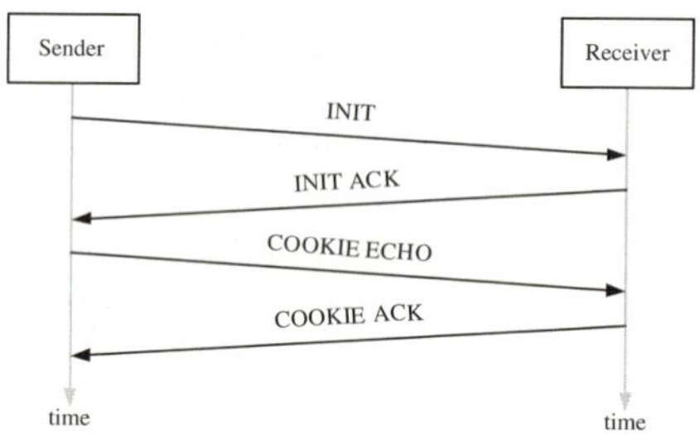

Fig. 3. Step 1 of the secure association establishment in SS-SCTP

When the AUTH extension is used, three additional parameters are exchanged in the INIT and/or INIT ACK chunks: a random parameter (RANDOM), a chunk list parameter (CHUNKS), and a requested HMAC algorithm parameter (HMAC-ALGO). RANDOM contains an arbitrary random number that is included once in the INIT or INIT ACK chunk. CHUNKS contains a list of chunk types that are required to be protected by an AUTH chunk. INIT, INIT ACK, SHUTDOWN COMPLETE, and AUTH are not allowed to be included in this list. HMAC-ALGO, finally, contains a list of supported HMAC algorithms in sender preference order. Two algorithms have so far been defined: secure hash algorithm number 1 (SHA-1) and SHA-256 [15]. These two produce a HMAC of 160 and 256 bits, respectively.

The actual integrity protection is provided through the AUTH chunk. The fields in an AUTH chunk type are illustrated in Fig. 4. Except the basic chunk fields, i.e., type, flag, and length, four additional fields are included. The shared key identifier field is used to notify the receiver which key to use. The HMAC identifier field specifies the HMAC algorithm that has been used to produce the following HMAC. The third field is the HMAC itself. The last field is a padding field, which is used to align the length of the chunk to a multiple of 32 bits. This implies that the length of the padding field could be 0, 8, 16 or 24 bits. For both SHA-1 and SHA-256, the length of the padding field will be 0, since the total length of an AUTH chunk with SHA-1 will be 224 bits and an AUTH chunk with SHA-256 will be 320 bits. Both 224 and 320 are multiples of 32.

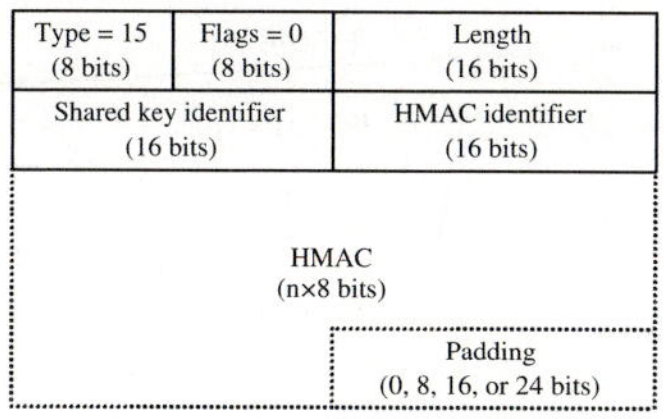

Fig. 4. Fields in an AUTH chunk

Step 2. Once an SCTP association is established, the next step is to perform a mutual authentication, an encryption algorithm negotiation, and a key exchange. In SS-SCTP, TLS is used for this purpose. Mutual authentication is achieved through certificate based methods using either RSA[2] or digital signature standard (DSS) certificates. For key exchange, three different algorithms could be used, i.e., RSA, Diffie-Hellman (DH), or ephemeral DH (DHE). As encryption algorithm, a symmetric block cipher according to the TLS standard may be selected.

In TLS, a combination of security algorithms is grouped together in so-called ciphersuites. When TLS was standardized it was designed to be extensible with respect to ciphersuites. For example, the addition of the AES ciphersuites is described in [3]. A ranking of certificate based ciphersuites with respect to security strengths is provided by the national institute of standards and technology (NIST) in [2]. Additional ciphersuites are described in [14] and [5], respectively.

The message flow in step 2 is depicted in Fig. 5. This message flow is according to the TLS standard. As depicted in the figure, the sender initiates the secure association establishment by sending a `client_hello` message containing protocol version and a list of preferred ciphersuites. The receiver will reply with a `server_hello` message which includes the preferred ciphersuite followed by `certificate`, `server_key_exchange`, `certificate_request`, and

[2] RSA is an abbreviation for Rivest, Shamir, and Adleman, who are the inventors of the algorithm.

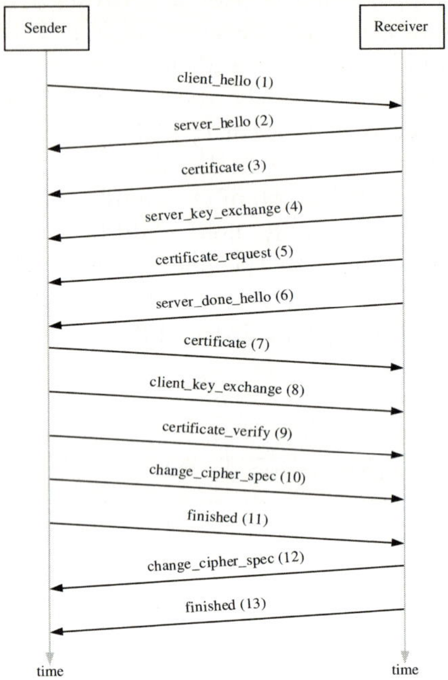

Fig. 5. Step 2 of the secure association establishment in SS-SCTP. Message numbers have been included in parenthesis after each message.

`server_done_hello` messages. Note that these five messages can be sent in either the same SCTP packet if they fit, or be divided into multiple SCTP packets. The sender replies with `certificate`, `client_key_exchange`, `certificate_verify`, `change_cipher_spec`, and `finished` messages, which can be sent in one or more SCTP packets. Finally, the receiver will respond with `change_cipher_spec` and `finished` messages.

After message 13 in Fig. 5, the next task is to establish the cryptographic keys. This is performed as follows. First, a pre-master secret is exchanged. Second, a 48-byte (384 bit) master secret is generated by each peer. From the master secret, a set of keys are derived. One key is referred to as the temporal key (TK). TK is the key used for data encryption of user messages.

As mentioned in the introduction, SCTP supports the concept of multistreaming in order to decrease the impact of HoL blocking. Each established stream has a stream identifier that identifies to which stream a message belongs. The number of streams within an association is negotiated in the first two messages, i.e., INIT and INIT ACK, sent in step 1 as was described above. In SS-SCTP, streams 0 in both directions are reserved for the TLS handshake procedure described above.

4.3 User Data Transfer

After successfully finishing steps 1 and 2, user data may be transferred. User data are transferred as DATA chunks. The fields in a DATA chunk type is illustrated in Fig. 6.

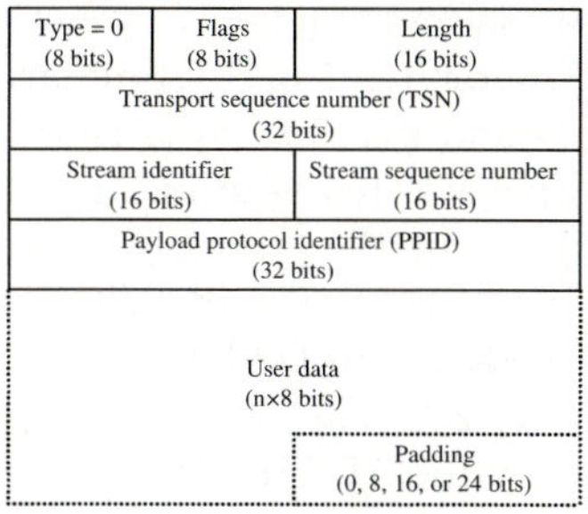

Fig. 6. Fields in a DATA chunk

For the DATA chunk type, three flags have been defined. These are the lower three bits in the flag field and are referred to as U, B, and E, respectively. The U bit is set if data carried in the chunk is unordered. The B bit is sent to indicate that the chunk is the first part of a fragmented message, while the E bit is set to denote that this is the last part of a fragmented message. If an entire message fits into a single chunk, both the B and the E bit is set. When transferring data pieces in the middle of a message, neither B nor E is set.

The TSN field is a sequence number for the association, while the stream sequence number is a sequence number for the stream. The stream identifier indicates which stream the chunk belongs to. The payload protocol identifier (PPID) field is a user supplied value. This field is ignored by SCTP, but is an important field in SS-SCTP. In SS-SCTP, the PPID value indicates whether the included user data is encrypted or not. A zero value denotes that the chunk carries plaintext data, whereas non-zero values denote encrypted data. In the latter case, the first 16 bits are used as a shared key identifier and the last 16 bits are used (or rather reserved) to denote encryption algorithm. In the current implementation of SS-SCTP, the encryption identifier field has no meaning and is therefore ignored by the receiver. The encryption algorithm is instead negotiated in step 2 as described above. This may however change in future versions of SS-SCTP. Note the similarities with the encoding scheme used in the AUTH chunk type.

The remaining two fields in the DATA chunk type is the user data field and the padding field. The user data field contains the actual user message, and the padding field is used to align data to 32 bit boundaries. When data confidentiality protection is needed, encryption is applied to user data. Encryption of data in SS-SCTP is performed at the application level on a per-message basis using the OpenSSL crypto library. For bulk data encryption, symmetric block ciphers such as data encryption standard (DES), triple DES (3DES), Blowfish, and advanced encryption standard (AES) are envisioned. When block ciphers are used, special treatment of the last block is needed. This is due to the fact that the algorithms operate on fixed size blocks, which implies that some form of padding is necessary. Fortunately, the OpenSSL crypto library has built-in support for handling padding of the last block. Except for the last block, blocks are encrypted using the `EVP_EncryptUpdate()` function. Encryption of the last block is performed with `EVP_EncryptFinal()`. Similarly, decryption is performed using `EVP_DecryptUpdate()` and `EVP_DecryptFinal()`.

An example of an SCTP packet that contains a SACK chunk and four DATA chunks is illustrated in Fig. 7. In the example it is assumed that the AUTH extension is used for integrity protection of SACK and DATA chunks. As is depicted in the figure, the messages in DATA chunks 1, 3, and 4 are encrypted, while the data in DATA chunk 2 is sent in plaintext. All four DATA chunks are, however, integrity protected by the AUTH chunk. Control chunks that have been requested to be integrity protected must be included after the AUTH chunk, but before the first DATA chunk[3]. All other control chunks are included before the AUTH chunk. The SACK chunk that is included in the SCTP packet depicted in Fig. 7 is placed after the AUTH chunk and is thus integrity protected.

Fig. 7. An SCTP packet with the mandatory SCTP common header (light grey), an AUTH chunk (black), an integrity protected SACK chunk (white), and four DATA chunks. DATA chunks 1, 3, and 4 (dark grey) are both integrity and confidentiality protected, while DATA chunk 2 (white) is only integrity protected.

4.4 New Socket API

A new socket API is provided by SS-SCTP, which extends the draft SCTP socket API described in [22]. The extension consists of the following four functions: `sctp_secureAssoc()`, `sctp_secureSend()`, `sctp_secureReceive()`, and `sctp_secureReassoc()`.

The `sctp_secureAssoc()` function establishes an SS-SCTP association. This implies that an ordinary SCTP association is first created with at least two outbound and two inbound streams. Then, mutual authentication and key negotiation is performed using TLS on streams 0 in both directions. This corresponds to steps 1 and 2 in the previous subsection.

The `sctp_secureSend()` function is used to send encrypted messages. The parameters are the same as in the `sctp_send()` function provided in [22]. The difference between these two is that the former encrypts messages, while the latter does not. Both functions are needed in order to provide security differentiation.

[3] In an SCTP packet, control chunks must be included before DATA chunks.

The `sctp_secureReceive()` function is used to receive and, if needed, decrypt messages at the receiver side. The decryption algorithm is, however, only applied on data in chunks with a non-zero PPID.

The `sctp_secureReassoc()` function is used to initiate a full TLS re-negotiation. This implies that both mutual authentication and key exchange is performed again. Such a re-negotiation is often referred to as re-authentication and re-keying.

5 Comparison of Message Complexity

A comparison of the message complexity when transferring user data using SCTP over IPsec, TLS over SCTP, and SS-SCTP is emphasized in this section. We consider cases when user messages are unbundled as well as bundled. For data integrity and confidentiality protection, the SHA-1 algorithm and the AES algorithm in cipher block chaining (CBC) mode with a key length of 128 bits are assumed. Since the focus is on the message overhead introduced when transferring user data, the secure association establishment is not considered. For the same reason, re-authentication and re-keying are not considered either. These events are also infrequent in comparison to data transfers.

5.1 Unbundled Data Transfers

Sending unbundled user messages means that each SCTP packet only carries a single user message. Such transfers will therefore result in a higher message overhead compared to a bundled transfer, simply due to the fact that some protocol headers must be duplicated in each SCTP packet. Unbundled data transfers are typically used to achieve as low latency as possible, which may be a requirement in signaling and multimedia applications.

A summary of the message complexity for the three different investigated E2E security solutions when transferring user messages unbundled is provided in Table 3. The figures in the table include all higher layer protocol headers starting with the IP layer and a user message of the specified size. IPv4 is assumed to be used here. Hence, when SS-SCTP is used, an IPv4 header (20 bytes), an SCTP common header (12 bytes), an AUTH chunk (28 bytes), and a DATA chunk header (16 bytes), and as much of the user message that fits is included in each SCTP packet. Similarly, when SCTP over IPsec is used an IPv4 header (20 bytes), an IPsec header/trailer (32 bytes in total independent of if AH or ESP is used), an SCTP common header (12 bytes), and a DATA chunk header (16 bytes), and as much of the user message that fits is included in each SCTP packet. TLS over SCTP, finally, includes an IPv4 header (20 bytes), an SCTP common header (12 bytes), a DATA chunk header (16 bytes), and as much of the user message that fits in each SCTP packet. In this case, the user message contains both the payload data, the TLS header (5 bytes for type, version, and length) and the TLS trailer (23 bytes for HMAC and padding).

As can be seen in the Table 3, SS-SCTP and TLS over SCTP produce the same amount of traffic for user messages that fit in a single SCTP packet. When

Table 3. Message complexity expressed in bytes for different sized user messages when transferred unbundled using IPv4. The PMTU is assumed to be 1500 bytes.

E2E security solution	**User message size [bytes]**				
	128	**256**	**1024**	**4096**	**16384**
SS-SCTP	204	332	1100	4324	17296
SCTP over IPsec	208	336	1104	4336	17344
TLS over SCTP	204	332	1100	4268	16988

message fragmentation is needed, i.e., when transferring user messages that do not fit in a single SCTP packet, TLS causes slightly less traffic compared to SS-SCTP. SCTP over IPsec results in the highest amount of network traffic although the difference with SS-SCTP is small.

The message overhead added for unbundled data transfers by the different solutions are depicted in Fig. 8. The message overheads introduced by all solutions are high when transferring small user messages, but rather small when transferring large messages.

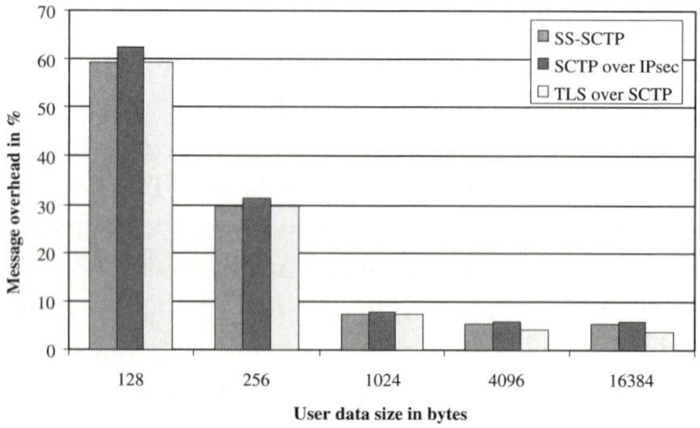

Fig. 8. Message overhead for different sized user messages

5.2 Bundled Data Transfers

When bundled data transfers are used, multiple user messages could potentially be carried by a single SCTP packet. Bundling is, however, only used when multiple chunks are ready to be sent and fit completely in a single SCTP packet. If that is not the case, each message is sent in one or more SCTP packets depending on the size of the data. By using bundling, the message overhead is reduced. Bundled data transfers could with advantage therefore be used by, e.g., notification and data log services. In such services, messages exchanged are typically small, i.e., a few hundred bytes. Slightly increased latency is also typically acceptable in such services.

The message complexity for the three studied E2E security solutions with different amounts of message bundling is depicted in Table 4. All messages are assumed to be of equal length, i.e., 256 bytes in this case. Such an assumption is fair when considering notification as well as data log services. When bundling is used, TLS causes the highest message complexity. This is due to the extra protocol overhead added by TLS for each message, i.e., the TLS header (5 bytes) and the TLS trailer (23 bytes) as described above. When TLS is used, five bundled user messages will therefore not fit into a single SCTP packet. Instead, the four first messages are sent in one SCTP packet and the fifth is sent in a separate one, assuming that no further messages arrive to the send queue before the second packet is sent. When comparing SS-SCTP and SCTP over IPsec, the message complexity is again similar.

Table 4. Message complexity expressed in bytes for different amount of message bundling using IPv4. Each user message is 256 bytes and the PMTU is assumed to be 1500 bytes.

E2E security solution	**# of user messages**				
	1	**2**	**3**	**4**	**5**
SS-SCTP	332	604	876	1148	1420
SCTP over IPsec	336	608	880	1152	1424
TLS over SCTP	332	632	932	1232	1564[†]

[†]For this case, two SCTP packets are needed to transfer these five user messages, since the total data size is greater than the PMTU.

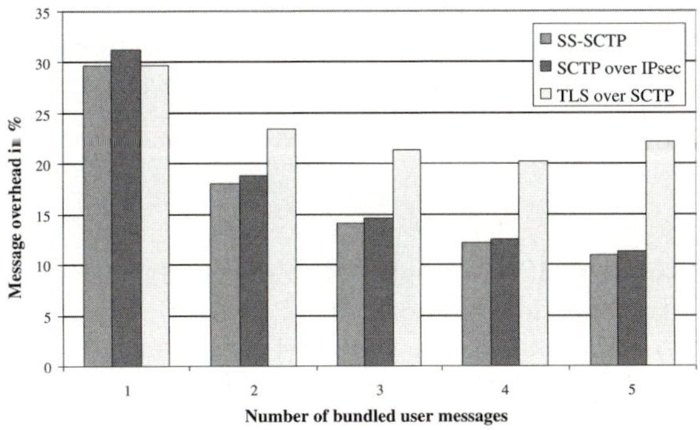

Fig. 9. Message overhead when user messages are bundled

Fig. 9 illustrates the high overhead introduced when TLS over SCTP is used to transfer bundled user messages. In the worst case illustrated in Fig. 9, i.e., five bundled messages, the overhead is more than twice as big as compared to SS-SCTP and SCTP over IPsec. This implies that TLS is not a preferable E2E security solution when transferring small messages that can be bundled together in SCTP packets. Finally, the difference between SS-SCTP and SCTP over IPsec with respect to message overhead is almost negligible.

6 Concluding Remarks

The design of SS-SCTP has been presented in this paper. The main focus has been on mechanisms for data integrity protection and data confidentiality protection. Data integrity protection is provided through the new AUTH extension, whereas data confidentiality protection is implemented at the socket layer. Mutual authentication and key exchange is conducted using certificate based methods provided by the TLS protocol.

In the paper, the message complexity of SS-SCTP is also investigated and compared to the two standardized E2E security solutions, i.e., SCTP over IPsec and TLS over SCTP. The conclusion is that SS-SCTP always produces slightly less message overhead compared to SCTP over IPsec. Compared to TLS over SCTP and when the bundling feature in SCTP is turned off, SS-SCTP adds the same message overhead when small messages are transferred and slightly more message overhead when transferring messages larger than the PMTU. When transferring small messages that can be bundled together in SCTP packets, SS-SCTP clearly outperforms TLS over SCTP.

The next step in the development of SS-SCTP is to finalize the prototype implementation. An experimental evaluation of SS-SCTP and a comparison with the two standardized E2E security solutions for SCTP will then be conducted.

Acknowledgment. The work at the Norwegian University of Science and Technology is financially supported by the research council of Norway. The work at Karlstad University is supported by grants from the Knowledge Foundations of Sweden with TietoEnator and Ericsson as industrial partners.

References

1. Bellovin, S., Ioannidis, J., Keromytis, A., Stewart, R.: RFC 3554: On the use of stream control transmission protocol (SCTP) with IPsec (July 2003)
2. Chernick, C.M., Edington III, C., Fanto, M.J., Rosenthal, R.: Guidelines for the selection and use of transport layer security (TLS) implementations. National Institute of Standards and Technology (NIST) (June 2005)
3. Chown, P.: RFC 3268: Advanced encryption standard (AES) ciphersuites for transport layer security (TLS) (June 2002)
4. Dierks, T., Rescorla, E.: RFC 4346: The transport layer security (TLS) protocol version 1.1 (April 2006)
5. Eronen, P., Tschofenig, H.: RFC 4279: Pre-shared key ciphersuites for transport layer security (TLS) (December 2005)
6. Hohendorf, C., Rathgeb, E.P., Unurkhaan, E., Tüxen, M.: Secure end-to-end transport over SCTP. Journal of Computers 2(4), 31–40 (2007)

7. Institute of Electrical and Electronic Engineers (IEEE). Amendment 6: Medium access control (MAC) security enhancements. IEEE Std 802.11i, July 23 (2004)
8. Jungmair, A., Rescorla, E., Tuexen, M.: RFC 3436: Transport layer security over stream control transmission protocol (December 2002)
9. Kaufman, C.: RFC4306: Internet key exchange (IKEv2) protocol (December 2005)
10. Kent, S.: RFC4302: IP authentication header (December 2005)
11. Kent, S.: RFC4303: IP encapsulating security payload (ESP) (December 2005)
12. Kent, S., Seo, K.: RFC 4301: Security architecture for the Internet protocol (December 2005)
13. Lindskog, S., Brunstrom, A.: An end-to-end security solution for SCTP. In: Proceedings of the Third International Conference on Availability, Reliability and Security (ARES 2008), Barcelona, Spain, March 4–7, 2008, pp. 526–531 (2008)
14. Medvinsky, A., Hur, M.: RFC 2712: Addition of Kerberos cipher suites to transport layer security (TLS) (October 1999)
15. National Institute of Standards and Technology (NIST). Secure hash standard (August 1, 2002), `http://csrc.nist.gov/publications/fips/fips180-2/fips180-2.pdf`
16. OpenSSL homepage (November 30, 2007), `http://www.openssl.org/`
17. Rescorla, E., Modadugu, N.: RFC 4347: Datagram transport layer security (April 2006)
18. Stewart, R.: RFC 4960: Stream control transmission protocol (September 2007)
19. Stewart, R., Ramalho, M., Xie, Q., Tuexen, M., Conrad, P.: RFC 3578: Stream control transmission protocol (SCTP) partial reliability extension (May 2004)
20. Stewart, R., Tuexen, M., Camarillo, G.: RFC 5062: Security attacks found against the stream control transmission protocol (SCTP) and current countermeasures (September 2007)
21. Stewart, R., Xie, Q., Tuexen, M., Maruyama, S., Kozuka, M.: RFC 5061: Stream control transmission protocol (SCTP) dynamic address reconfiguration (September 2007)
22. Stewart, R., Xie, Q., Yarroll, L., Poon, K., Tuexen, M.: Sockets API extensions for stream control transmission protocol (SCTP), draft-ietf-tsvwg-sctpsocket-15.txt (work in progress). Expires (January 10, 2008)
23. Tuexen, M., Stewart, R., Lei, P.: RFC 4820: Padding chunk and parameter for the stream control transmission protocol (SCTP) (March 2007)
24. Tuexen, M., Stewart, R., Lei, P., Rescorla, E.: RFC 4895: Authenticated chunks for stream control transmission protocol (SCTP) (August 2007)
25. Unurkhaan, E., Rathgeb, E.P., Jungmair, A.: Secure SCTP: A versatile secure transport protocol. Telecommunication Systems 27(2–4), 273–296 (2004)

Queueing Analysis for Networks Under DoS Attack

Amar Aissani

University of Science and Technology Houari Boumediene (USTHB),
BP 32 El Alia, Bab-Ez-Zouar, 16111, Algeria
amraissani@yahoo.fr

Abstract. In this paper, we consider a queueing model of computer network under DoS attacks. The arrival of SYN packets contains two types: the regular request packets and the attack packets that request for connections. We assume that the connection requests arrive according to a Poisson processes and the service times are general but different for the two types of requests. A maximum number of connections can be served at the same time. Each half-open connection is held for at most a deterministic or random period of time (timeout). We obtain the steady-state probability distribution of the stochastic process describing the evolution of such a system. Next, we show how to compute some security metrics such as the loss-probability or the buffer occupancy percentage of half-open connections for attack packets.

Keywords: Network security, DoS attack, Queueing, Steady-state distribution, Timeout.

1 Introduction

Networks attacks today (Denial of service (DoS) attack, worm, Trojan horse and virus) are increasing in frequencies, severity and sophistication. They exploit bugs in a specific operating system or vulnerabilities in TCP/IP implementation and can cause serious problems to normal business operations [1-5,7-9]. In this paper, we concentrate on DoS attaks which can be classified into several types. In flooding attack, malicious users (attackers) may flood a network with a large volume of data in order to deliberately consume the basic and limited resources of a victim, such as the process control blocs and the maximum allowed connections. In particular, DoS attacks may disrupt the normal operation of physical components in the network, and may also manipulate data in transit such as encrypted data. The impact of a DoS attack on a particular system will vary depending on the protocols and applications involved. Crippling DoS (CDoS) attack, may cause the degradation of Quality of Service(QoS) or render the service unavailable. Distributed Denial-of-service (DDoS) attacks are simply DoS attacks performed by multiple agents simultaneously. (Some statistics about DoS attacks can be found in [1] and some used networks security metrics in [1-5,7-9].

These works provide also references about some defence mechanisms which have been proposed in the literature to defend against DoS attacks (anomaly-detection, signature-scan including blocking attack packets, tracing and filtering). Many defence

O. Gervasi et al. (Eds.): ICCSA 2008, Part II, LNCS 5073, pp. 500–513, 2008.

mechanisms have been proposed [7,9] (anomaly-detection and signature-scan techniques, for example). These mechanisms include blocking attack packets to reduce the intensity of attacks, tracing the packets to locate the attacking source and using the proactive measures to filter the attack packets [4,5,7].

Most of the cited works are experimental studies trying to characterise the behaviour of the DoS attacks and quantitatively estimating their impact on the basis of the effective detection and filtering techniques. Most of these studies use simple queuing models to analyse various network security metrics (consumption metrics, complexity metrics,...) [3] Some others use more sophisticated queueing models, but the analysis is also based on experimental simulation [2]. There is some attempts to give more formal framework in order to analytically study the impact of DoS attacks using Queuing theory (see Chang [7]). The motivation is that when an attack has an impact on a network parameter, then such parameter can be used as an attack detection metric [3].

We can mention also in this research direction the work of Khan & Traore [3] which use a simple model M/M/1/K with round robin discipline to analyse the impact of DoS attacks on some parameters as response time or queue-growth-rate. Long et al [2] proposed an abstract queueing model showing the effect of DoS attacks on delay jitter and loss probability both at remote plant level and at controller level. C. Xenakis & al [8] develop an analytic model for an abstract model of UMTS mobile device . The analysis is carried out by modelling the IPsec processor and the transmitter as a system of two independent M/G/1 queues in tandem. Wang & al [4] use a two dimensional embedded Markov chain model to study the network under DoS attacks.

In this paper, we provide a next step in this direction in order to analytically characterise the security attributes of a network under DoS attacks such as the most prevalent SYN-flooding attack. We consider the model studied both in [4] and [5] by different methods. The goal is to quantify the damage that a successful attacker can have on the performance of the network such as the loss probability and buffer occupancy of half-open connections. The difference in our model is that the distribution of service times are arbitrarily distributed and we take into consideration the possibility of connection failures. On the other hand, it is obtained explicit formulas for the steady-state probabilities of the underlying stochastic process, which render the model more interesting from the computational point of view.

In the following section we describe the basic model of a network under flooding DoS attack and we derive the steady-state distribution of the network under such attack. In section 3, we consider some useful characteristics which can be used as security metrics. Section 4, is devoted to a discussion of the time-out , particularly the sensitivity of its distribution to the exponential assumption. In section 5 we consider numerical illustrations showing the effect of DoS attacks upon system performance. The accuracy of the model is tested by comparison with the similar model of [4] for different scenarios. Finally, in sections 7 and 8, we discuss the cases of Degradation of service DoS (DSDoS) attacks and non Poisson arrivals respectively.

2 The Basic Model

We consider the model described in [4,5] in which the input queue and service times are stochastic processes. In general, the arrival of SYN packets contains two types:

the regular request packets (first type) which describe the packet flow under network normal status and the attack packets (second type) that request for connections. The victim has a connection buffer of the backlog queue, in which at most N half-open connections are allowed simultaneously, so a maximum number of connections N can be served at the same time. We assume that the connection requests arrive according to a Poisson process with rate λ. This assumption is conform to some experimental studies [11], at least when the network is heavily used. Let $S^{(i)}$ be the service time of ith type of request packet with distribution function $H_i(x) = P(S^{(i)} < x)$, $i = 1,2$. We assume that $H_i(0+) = 1$. The arrivals of the regular request packets (from legitimate users) and the attack packets (from malicious users) are both Poisson processes with rate λ_1 and λ_2 respectively. The two arrival processes are independent of each other and of the holding times for half-open connections.

Let $\nu(t)$ and $\mu(t)$ be the numbers of the regular request packets and the attack packets at time t, respectively. It indicates the number of connections used by legitimate users and malicious users respectively. It is obvious that at any time $t \geq 0$, we have $\nu(t) + \mu(t) \leq N$, so that a maximum number of both legitimate and malicious connections N can be served in the same time. All connection requests that arrive when the server is in saturated state, $\nu(t) + \mu(t) = N$, will be rejected.

Let $\zeta(t) = \{\nu(t), \mu(t); \alpha_1(t), \dots, \alpha_{\nu(t)}; \beta_1(t), \dots, \beta_{\mu(t)}\}$ be the stochastic process where the discrete component is defined on the state space $IN \otimes IN$ and $\alpha_i(t) \in IR_+$ is the residual service time of a regular packet on the server N° i and $\beta_j(t) \in IR^+$ the residual service time of a malicious packet on the server N° j

$$F_{i,j}(t; x^{(i)}, y^{(j)}) = P\{\nu(t) = i, \mu(t) = j; \alpha^{(i)}(t) < x^{(i)}; \beta^{(j)}(t) < y^{(j)}\}$$

$$0 \leq i + j \leq N$$

Here we have denoted by $x^{(i)} = (x_1, \dots, x_i)$, $y^{(j)} = (y_1, \dots, y_j)$ and $\alpha^{(i)}(t) < x^{(i)}$ means that $\alpha_k(t) < x_k, \forall k = 1, \dots, i$.

If $\max(\tau_1, \tau_2) < \infty$, then the stochastic process $\{\zeta(t), t \geq 0\}$ has a stationary ergodic distribution. By using a method similar to that of Sebastianov [11] (see also [6] we can find that the steady-state probability distribution is solution of the following equilibrium system of partial differential equations

$$\sum_{p=1}^{i} \frac{\partial F_{i,j}}{\partial x_p} + \sum_{q=1}^{j} \frac{\partial F_{i,j}}{\partial y_q} - (\lambda_1 + \lambda_2)(1 - \delta_{i+j,N}) + \frac{\lambda_1}{i} \sum_{p=1}^{i} F_{i-1,j}(x(i,p); y^{(j)}) H_1(x_p) +$$

$$+ \frac{\lambda_2}{j} \sum_{q=1}^{j} F_{i,j-1}(x^{(i)}; y^{(j,q)}) H_2(y_q) = \sum_{p=1}^{i} \frac{\partial F_{i,j}(x_1, \dots, x_{p-1}, 0, x_{p+1}, \dots, x_i; y^{(j)})}{\partial x_p} +$$

$$\sum_{q=1}^{j} \frac{\partial F_{i,j}(x^{(i)};y_1,\ldots, y_{q-1},0, y_{q+1},\ldots, y_j)}{\partial y_q} + (i+1))\frac{\partial F_{i+1,j}(x_1,\ldots, x_i,0; y^{(j)})}{\partial x_{i+1}} +$$

$$+(j+1))\frac{\partial F_{i,j+1}(x^{(i)};y_1,\ldots, y_j,0)}{\partial y_{i+1}} ,$$

where $x^{(i,p)} = (x_1,\ldots, x_{p-1}, x_{p+1},\ldots, x_i)$ and $y^{(j,q)} = (y_1,\ldots, y_{q-1}, y_{q+1},\ldots, y_j)$

It is not difficult to see that the solution of this equation is of the form

$$F_{i,j}(x^{(i)}; y^{(j)}) = \frac{\lambda_1^i \lambda_2^j}{i!\, j!} F_{00} \times \Pi\left(H_1, x^{(i)}\right) \times \Pi\left(H_2, y^{(j)}\right), \tag{1}$$

$$\Pi\left(H, x^{(i)}\right) = \prod_{p=1}^{i} \int_0^{x_p} [1 - H(u)] du$$

Since

$$\lim_{\substack{x_p \to \infty, 1 \le p \le i \\ y_q \to \infty, 1 \le q \le j}} F_{i,j}(x^{(i)}; y^{(j)}) = \frac{(\lambda_1 \tau_1)^i (\lambda_2 \tau_2)^j}{i!\, j!} F_{00} ,$$

then, the unknown constant can be found from the normalisation condition

$$F_{00} \sum_{i=0}^{N} \sum_{j=0}^{N-i} \frac{(\lambda_1 \tau_1)^i (\lambda_2 \tau_2)^j}{i!\, j!} = 1 ,$$

Next, we get

$$F_{00} = \left[\sum_{\substack{i,j=0 \\ 0 \le i+j \le N}} \frac{(\lambda_1 \tau_1)^i (\lambda_2 \tau_2)^j}{i!\, j!} \right]^{-1} , \tag{2}$$

Finally, we obtain the probability that the server is in a specific state (i, j), indicating that i regular packets and j attack packets are in the system

$$p_{i,j} = \frac{(\lambda_1 \tau_1)^i (\lambda_2 \tau_2)^j}{i!\, j!} \left[\sum_{\substack{i,j=0 \\ 0 \le i+j \le N}} \frac{(\lambda_1 \tau_1)^i (\lambda_2 \tau_2)^j}{i!\, j!} \right]^{-1} , 0 \le i + j \le N , \tag{3}$$

Particular case. If $H_1(x) = 1 - e^{-\mu_1 x}, x \ge 0$ and $H_2(x) = 1 - e^{-\mu_2 x}, x \ge 0$, and $\rho_i = \lambda_i / \mu_i, i = 1,2$ then we obtain

$$p_{i,j} = \frac{(\rho_1)^i (\rho_2)^j}{i!\, j!} \left[\sum_{\substack{i,j=0 \\ 0 \le i+j \le N}} \frac{(\rho_1)^i (\rho_2)^j}{i!\, j!} \right]^{-1}, \tag{4}$$

3 Security Metrics

Now, from the above results, we can derive some security metrics which characterise the performance of the network under DoS attack. They are similar to that of [4,5].

Connection loss probability is a basic measure which can be obtained from the stationary distribution as follows

$$P_{loss} = \sum_{i=0}^{N} p_{i,N-i} \,,$$

We can use this characteristic as a detection intrusion metric by fixing threshold value $\varepsilon = \varepsilon(N)$ small enough to indicate the network security status; if $P_{loss} \ge \varepsilon$, then the network is under DoS attack.

Another performance measure which can be used as security metric is the buffer occupancy percentage of half-open connections for regular traffic. It is characterised by the mean ratio of the number of regular half-open connections to the maximum allowable number of half-open connections

$$P_r = \frac{1}{N} \sum_{i=0}^{N} i \sum_{j=0}^{N-i} p_{i,j} \,,$$

Similarly, the buffer occupancy percentage of half-open connections for attack packets is

$$P_a = \frac{1}{N} \sum_{i=0}^{N} \sum_{j=0}^{N-i} j p_{i,j} \,,$$

It represents the mean ratio of the number of attack packets to the maximum allowable number of half-open connections.

4 Time Out

In realistic situations [4,5] of SYN-flood attacks, each half-open connection is held for at most a deterministic (or random) period of time B which is time interval from the epoch that the half-open connection begins to the epoch that the connection is dropped. The event "*connection arrived*" represents a SYN message and the event "*connection completed*" corresponds to ACK message being received by the server. In the case of SSL connection depletion attack, N_1 and N_2 represent the number of legitimate and malicious completed TCP connections respectively, that have not yet established a secure channel and for which the negotiation phase is still in progress.

The *connection arrived* event represents the *Client hello* message being received by the server and the *connection completed* event represents the corresponding *Finished message* being sent by the server. It is even possible to consider a nested model, each level representing a different layer in the protocol stack [5].

It is possible to take into account these phenomena by modifying the equilibrium equations and then solve them. But, in this study we take a more rapid argumentation. According to our assumptions, connections completed events are generated at intervals of time from the *connection arrived* events, if the time_out has not elapsed. Otherwise, *connection expired* events are generated at time-out intervals from the *connection arrived* event, so that the effective legitimate service time S_B is

$$S_B = \begin{cases} S, \ if \ \ S \le B \\ B, if \ \ S > B \end{cases},$$

Then the probability distribution of S_B is

$$H_B(x) = H_2(x)H_2(B) + \chi(B < x)\overline{H}_2(B)),$$

It is not difficult to show that the mean time-out is

$$\tau_B = E(S_B) = B\overline{H}_2(x) + \int_0^B x dH_2(x) \quad , \tag{5}$$

$$\tau_b^{(2)} = B\overline{H}_2(B) = \int_0^B x^2 dH_2(x) \quad , \tag{6}$$

In particular, if $H_2(.)$ belongs to the class of NBUE (New Better than Used) probability distributions with mean $\overline{\tau} = \dfrac{1}{\mu}$, then $S_B \le_{st} S_{\exp}$ i.e. S_B is stochastically smaller than a random variable $S_{\exp}$ with distribution function $H_{\exp}(x) = e^{-\mu B}\delta(x - B) + \left[1 - e^{-\mu x}\right]\left[1 - e^{-\mu B}\right]$ and the mean time-out $\tau_B \le \tau_{\exp} = \dfrac{1 - e^{-\mu B}}{\mu}$ suggested in [5] for the effective mean service time of legitimate connections.

Now we can modify the probability of the event *connection failed* as follow $\phi = \phi_r + \phi_e$ where ϕ_r is the probability of the event *connection rejected* : it occurs when the server was not able to serve the connection because no more connection channels were available (the queue is full), and ϕ_e is the probability of the event *connection expired*: the server tried to serve the connection but the communication timed out and the connection was dropped.

If $\infty^{(i)}$ means that $x_k = \infty, 1 \le k \le i$, then in stationary regime, we have

$$\phi_e = 1 - \sum_{i=0}^{N} \sum_{j=0}^{N-i} \frac{\lambda_1^i \lambda_2^j}{i!\, j!} F_{00} \times \Pi\left(H_1, \infty^{(i)}\right) \times \prod_{q=1}^{j} \int_B^{\infty} [1 - H_2(u)] du \ ,$$

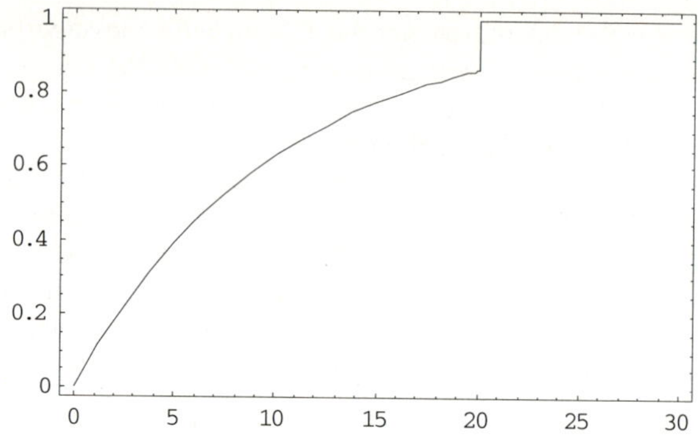

Fig. 1. Time-out distribution function for exponential service distribution ($\mu = 0.1$)

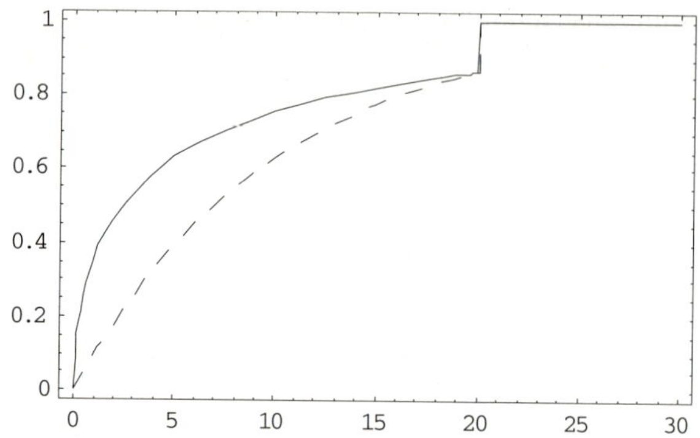

Fig. 2. Time-out distribution function for
- - - - exponential service distribution ($\mu = 0.1$)
——— Weibull service ($\lambda = 0.2, \alpha = 0.5$)

Consider now the problem of approximating analytically the time_out distribution. Consider two such distributions F and G, X and Y being the corresponding random variables. For example, G can be the exponential distribution used as approximating distribution in [4,5]. How we can estimate the deviation of the unknown F from the approximating one G? If for example $F <_* G$ and have the first two moments common, then $F = G$, where $<_*$ is the star stochastic order . The two corresponding random variables are ordered in this sense if and only if the random variable $Z = \Lambda_G(X)$ has an increasing failure rate average (IFRA) distribution.

It has been shown [13] that when the two distributions share the first moment, but the second moments are known, then the deviation of the two distributions can be estimated by

$$\sup_{x \geq 0} |F(x) - G(x)| \leq 3b^{2/3} [\zeta(X, Y)]^{1/3} ,$$

where b is the supremum of the density of either F or G and $\zeta(X,Y)$ is the average metric given by

$$\zeta(X,Y)=\int_0^\infty\left|\int_0^\infty[F(u)-G(u)]du\right|dx$$

From now, we assume that $X <_* Y$, $E(X)\leq E(Y)$ (if $E(X)<E(Y)$, then $F(x)<G(x)$ for some x and the second moments of F and G are finite). Then the average metric can be expressed through the two first moments $\zeta(X,Y)=\frac{1}{2}\left[E(Y^2)-E(X^2)\right]$. So, $F=G$ if and only if the two moments are equal. Now, if the approximating distribution G (or F) has a density uniformly bounded by b, then we have a family of bounds of the form

$$\sup_{x\geq 0}|F(x)-G(x)|\leq 3\left[\frac{b^2(\alpha)}{2}\left\{\upsilon(\alpha)-E\left(x^2\right)\right\}\right]^{1/3}, \tag{6}$$

where $b(\alpha)=\sup_{x\geq 0}\alpha g(x)\overline{G}^{\alpha-1}(x)$, $\nu(\alpha)=\int_0^\infty 2x\overline{G}^{\alpha}(x)dx$.

This bound comes from the fact that $F <_* G$ is equivalent to $F <_* (1-\overline{G}^{\alpha})$ for each $\alpha>0$. So the range of α in (6) is $\alpha_1<\alpha\leq\alpha_2$, where α_1 is the largest value of α such that $\overline{F}=1-F$ is completely dominated by $\overline{G}^{\alpha}$, and α_2 is such that the first moment of $1-\overline{G}^{\alpha_2}$ is the same as that of F. In some cases the "best possible approximation" can be obtained by minimizing the right-hand side with respect to α over the interval $(\alpha_1,\alpha_2]$.

For example, if H_2 is an IFRA distribution with first two moments μ_1 and μ_2, then H_D is also an IFRA distribution (at least for random time out which can be chosen as an IFRA random variable) and the approximating time-out distribution is exponential as in [4,5], with

$$\sup_{x\geq 0}\left|F(x)-e^{-\alpha x}\right|\leq\varepsilon(B)=3\left[1-\tau_B^{(2)}/\left(2(\tau_B)^2\right)\right]^{1/3},\quad \forall 0<\alpha\leq 1/\tau_B$$

Here $\alpha_1=0$, $\alpha_2=1/\tau_B$, and the minimal deviation is obtained for this last value of α.

This bound holds in fact for all upper non parametric distributions, in particular *NBU*, *NBUE*, *HNBUE*.

Consider for example, the following IFRA distribution $H_2(x)=(1-e^{-x})(1-e^{-2x})$, then figure 3 shows evolution of the error $\varepsilon(B)$ in (6) as a function of the mean timeout B.

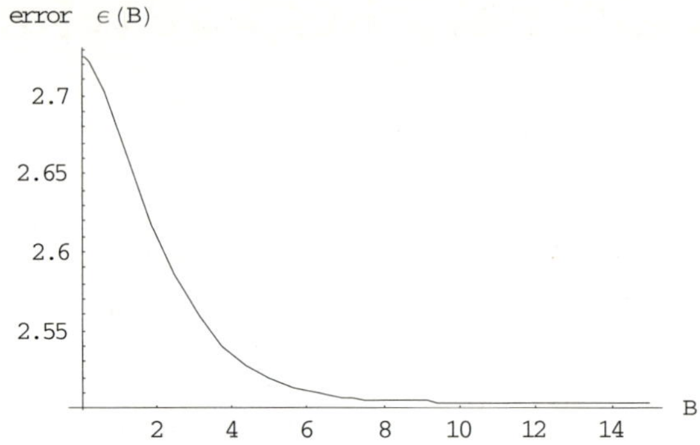

Fig. 3. Evolution of the upper error bound $\varepsilon(B)$ as a function of B

5 Numerical Examples

We consider now some numerical examples showing the effect of DoS attacks on some performance metrics.

First, we test the accuracy of our model (model 2) with the similar model of Wang (2007) [4] (model 1). In that work the basic numerical data are as follows. Let $\lambda_1 = 10$ packets/sec be the parameter of the Poisson arrival process of regular request packets ; $\lambda_2 = k\lambda_1$ is the parameter of the Poisson arrival process of attack request packets. In this case, the attack parameter k may be understood as the ratio of arrival rates between the attack packets and the regular request packets. The service time distribution is exponential $H_1(x) = H_2(x) = 1 - e^{-\mu x}$, with mean $\mu^{-1} = 100$ sec. Note that it is not difficult from our formula to take different distributions of service time for regular and attack requests, for example a Weibull distribution.

We consider the following scenarios:

Scenario 1. Let the mean time-out $B = 5$.

Scenario 2. Let the maximum allowable number of half-open connections $N = 20$.

For the first scenario, the values of the loss probabilities for both models 1 and 2 are given in table 1 for different values of the attack parameters k . Also, we considered different scenarios about the values of $N = 10,20,50$. As in [4], we observe that the loss probability increases with the increase of the attack traffic load. On the other hand, the values of P_{loss} for both models are similar, particularly for great values of k . The deviation between two models comes from the different nature of the assumptions about the mean time-out. Note however that our model 1 is more interesting from the computational point of view. Indeed, Wang & al [4] claim a computational complexity of $O(N^6)$ on a sequential machine , and $O(N^{5.7})$ on a parallel machine. Our method gives an explicit formula with complexity no greater than $O(N)$, in order to compute the normalisation constant.

Table 1. Comparison of loss probabilities for model 1 and model 2 (scenario 1: $B = 5$)

$k \downarrow$ $N \rightarrow$	$N = 10$		$N = 20$		$N = 40$	
	Model1	Model2	Model1	Model2	Model1	Model2
$k = 0$	0.01	--	--	--	--	--
$k = 0.2$	0.5	0.8324	--	0.6665	--	0.3456
$k = 0.4$	0.9	0.8559	0.45	0.7128	0.001	0.4321
$k = 0.7$	0.9	0.8809	0.9	0.7624	0.2	0.5279
$k = 1$	0.9	0.8986	0.9	0.7975	0.82	0.5966
$k = 1.25$	0.9	0.9097	0.9	0.8197	0.9	0.6404
$k = 1.6$	0.9	0.9218	0.9	0.8437	0.9	0.6880
$k = 1.9$	0.9	--	0.9	0.8597	0.9	0.7198
$k = 2.3$	0.9	--	0.9	0.8766	0.9	0.7534

The corresponding loss probabilities for the second scenario are given for both models in table 2, when the maximum number of allowed half-open connections is set to be $N = 20$. Other scenarios showing the mean ratio of the number of regular half-connections to the maximum allowable number of half-open connections versus the attack load result in similar tables and plots that are omitted here for brevity. In all scenarios , we observe small deviations between studied characteristics, while the computation time is considerably improved.

Table 2. Comparison of loss probabilities for model 1 and model 2 (scenario 2: $N = 20$)

$k \downarrow$ $B \rightarrow$	$B = 1$		$B = 10$		$B = 50s$	
	Model1	Model2	Model1	Model2	Model1	Model2
$k = 0$	0	--	--	--	--	--
$k = 0.2$	0.5	0.8324	--	0.6665	--	0.3456
$k = 0.4$	0.9	0.85596	0.45	0.7128	0.001	0.4321
$k = 0.7$	0.9	0.8809	0.9	0.7624	0.2	0.5279
$k = 1$	0.9	0.8986	0.9	0.7975	0.82	0.5966
$k = 1.25$	0.9	0.9097	--	0.8197	0.9	0.6404
$k = 1.6$	0.9	0.9218	--	0.8437	0.9	0.6880
$k = 1.9$	0.9	--	--	0.8597	--	0.7198
$k = 2.3$	0.9	--	--	0.8766	--	0.7534

Figure 4, shows the connection loss probability P_{loss} as a function of the attack traffic load with different settings for the maximum allowable number of half-open connections N . We use again the basic data of the work[4].

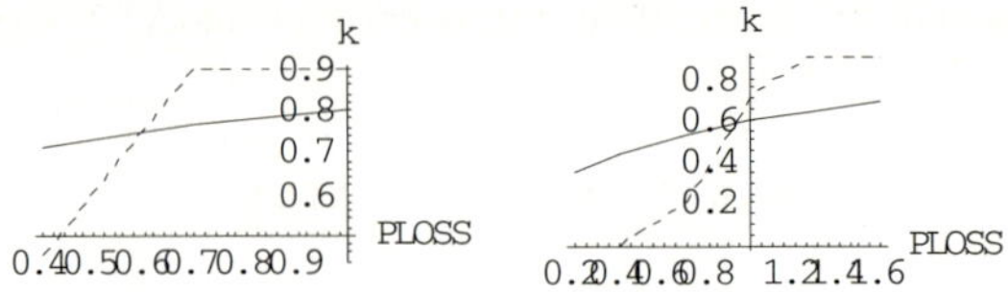

Fig. 4. Effect of the attack parameter k on the connection loss probability for different values of $N = 10,40$ ($H^{(i)}(x) = 1 - e^{-\mu x}$). The dashing line corresponds to model 1.

In a second set of experiments (Fig.5), we take a non exponential service distribution. We choose a Weibull distribution $H_i(x) = 1 - e^{-(\lambda x)^{\alpha}}$, with $\lambda = 0.02$ and $\alpha = 0.5$, so it has the same mean as the previous example, with a coefficient of variation greater than one. Figure 5, shows the effect of the attack for different scenarios about the maximum number of allowable half-open connections.

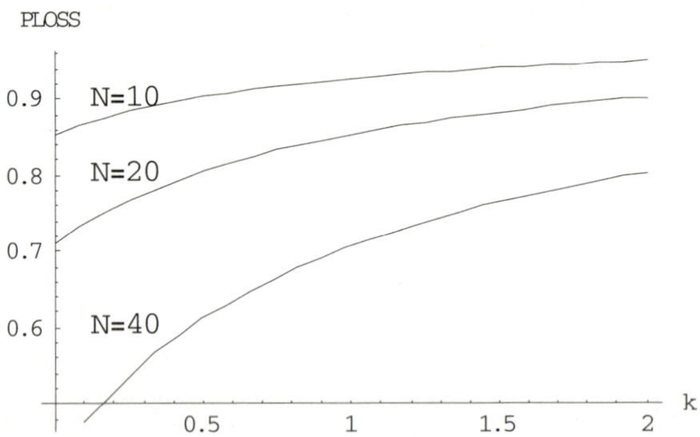

Fig. 5. Effect of the attack parameter k on the connection loss probability for different values of $N = 10,20,40$ ($H^{(i)}(x) = 1 - e^{-(\lambda x)^{\alpha}}$).

6 Degradation of Service DoS Attack

In this section we discuss the case of DSDoS attack where service becomes unavailable for a random restoration period of time. These DSDoS attacks occur according to the following process. If at any time, a connection is in course, then the life time of the connection (the period between beginnings of service until such an attack occurs) is a random variable ξ with distribution function $F(x)$. We assume that when such an attack occurs, a connection is interrupted with the loss of packet in service at this time. A restoration begins for a random period of time η with distribution function $G(x)$. On the other hand all arriving packets (regulars or attackers) finding all connections unavailable (busy or out of order) are lost. In this case we have again a solution of the form

$$F_{i,j,k}(x^{(i)};y^{(j)},u^{(k)},v^{(l)})=\frac{\lambda_1^i\lambda_2^j\gamma_1^k\gamma_2^l}{i!\,j!k!l!}F_0\times\Pi\big(H_1,x^{(i)}\big)\times\Pi\big(H_2,y^{(j)}\big)$$
$$\times\Pi\big(G_1,u^{(k)}\big)\times\Pi\big(G_2,v^{(l)}\big),$$

where $\gamma_i=\int_0^\infty[1-H_i(x)]dF(x)$ and $1\le i+j+k+l\le N$.

The underlying markovian process is ergodic if $\max(E(\xi),E(\eta),E(\tau_1),E(\tau_2))<\infty$.

7 Non Poisson Arrivals

Consider now the case of non Poisson arrival processes of regular or attack packets. It is indeed difficult to obtain an explicit solution for the modified basic process. However, we can obtain some useful assertions concerning the condition of non saturation of the system.

Let t_n^1 (resp. t_n^2) be the instant of the n-th regular (resp. attack) packet and $\xi_n^i=t_{n+1}^i-t_n^i,(i=1,2),n=0,1,\ldots$ Also, let τ_n^i be the service time of the n th packet of type $i,i=1,2$.

Now we consider the process $q_{((i,j))}(x)$ indicating that i legitimate connections and j malicious connections after a period of x units from the last arrivals, $q_{((i,j))}=q_{((i,j))}(0)$.

We assume that the sequence $\{\xi_n^i,\tau_n^i,-\infty<n<\infty\}$ is stationary (in the strict sense) and ergodic.

We follow the methodology of Borovkov [12] which consists to construct a judicious sequence of stationary events $\{A_{(i,j)}\}$ from a positive event A_0, by iteration $A_{(i,j)}=T^{i+j}A_0$. Here T is the shift transformation of measurable sets of the $\sigma-$algebra generated by the sequences of input random variables $\{\xi_n^i,\tau_n^i,-\infty<n<\infty\}$ corresponding to the measure preserving transformation on such measurable random variables. The event $B=\bigcup_{i+j\le -m+1}A_{(i,j)}$ is such that $TB\supset B$ and $P(TB)=P(B)$. Hence, $TB=B$ except perhaps on a negligible set (with zero probability). In view of the metric transitivity of our sequence, $P(B)=0$ or $P(B)=1$.

In order to have the last case, we need to choose a positive event A_0, so from the inequality $P(B)\ge P(A_0)>0$, we will deduce that $P(B)=1$. Finally, it follows that the sequence of processes $\{q_{n+k}^{(i,j)}(x)\}$ converges to a stationary process $\{q^{(i,j)}(x)\}$ satisfying $P(q^{(0,0)}(0)=1$. The convergence is understood here in a strong sense.

For example, a concrete sufficient conditions for $P(A_0) > 0$ to hold is that

$$P(\tau^{(i)} \leq m\xi^{(i)}) > 0, i = 1,2\text{, and } \max\left[E(\tau^{(1)}), E(\tau^{(2)})\right] < \infty$$

Another sufficient condition is the following: there exists x_0 such that for all $x \geq x_0$, $\Delta > 0$

$$P\left(\tau^{(i)} \in (x, x+\Delta) > 0\right), \max\left[E(\tau^{(1)}), E(\tau^{(2)})\right] < \infty$$

8 Conclusion

In this paper we proposed a new model for the analysis and security performance evaluation of computer networks. The model is interesting in the sense that its stochastic nature allows to capture attacker behaviour, system's response to attack and intrusion. The model takes into consideration the traditional failure processes in the Reliability sense So it can be included in more general simulation software environment. The accuracy of the model has been tested with a similar model for different scenarios about the input parameters. The developed model is particularly interesting from the computational point of view.

Further study may be evidently oriented to the case when the arrival processes are non Poisson although some experimental studies claim that the Poisson assumption is acceptable. Of course, an important effort need to be given to the simulation of real DoS attacks. However, the analytical nature of the model allows to conduct simulations by using some relevant data or few execution traces of a real system. So, we can draw conclusions under quite general assumptions about the parametric probability distributions. We also conjecture that the model in this form is insensitive to these distributions i. e. the loss probability depends only on the first moments of these distributions. An interesting problem is the optimisation of the above mentioned security metrics relatively to some controllable parameters and according to the envisaged technical architecture. Note finally, that a particular attention must be given to other types of DoS attacks as the crippling DoS attack or degradation Quality of Service (QoS) attacks [5].

Acknowledgements

The author is grateful to the referees for their comments. This work has been supported in parts by CNEPRU B00220060089 and Tassili 06 MDU 687 project grants.

References

1. Moore, D., Voelker, G., Savage, S.: Inferring Internet Denial-of-Service Activity. In: Proceedings of the 2001 USENIX Security Symposium, pp. 9–22. IEEE Press, New York (2001)
2. Long, M., Wu, C.-H., Hung, Y.: Denial of Service Attacks on Network-Based Control Systems: Impact and Mitigation. IEEE Transactions on Industrial Informatics 1(2), 85–96 (2005)

3. Khan, S., Traore, I.: Queue-based Analysis of DoS Attacks. In: Proceedings IEEE Workshop on Information Assurance and Security, United States Military Academy, West Point, pp. 266–273. IEEE Press, New York (2005)
4. Wang, Y., Lin, C., Li, Q.-L., Fang, Y.: A Queueing Analysis for the Denial of Service (DoS) Attacks. Computer Networks 51, 3564–3573 (2007)
5. Boteanu, D., Fernandez, J.M., McHugh, J., Mullins, J.: Queue Management as a DoS counter-measure? In: Garay, J.A., Lenstra, A.K., Mambo, M., Peralta, R. (eds.) ISC 2007. LNCS, vol. 4779, pp. 263–280. Springer, Heidelberg (2007)
6. Gnedenko, B.V., Kovalenko, I.N.: Introduction to Queueing Theory, Nauka, Moscow (1967)
7. Chang, R.K.C.: Defending Against Flooding-Based Distributed Denial-of-Service Attacks: A tutorial. IEEE Communications Magazine 40(10), 42–51 (2002)
8. Xenakis, C., Laoutaris, N., Merakos, L., Stavrakis, L.: A Generic Characterization of the Overheads Imposed by IPsec and Associated Cryptographic Algorithms. Computer Networks 5(17), 3225–3241 (2006)
9. Mirkovic, J., Reiher, P.: A Taxonomy of DDoS Defence Mechanisms. SIGCOMM Comput. Commun. Rev. 34(2), 39–53 (2004)
10. Sevastianov, B.A.: An Ergodic Theorem for Markov Processes and its Application to Telephone Systems with Refusals. Theory of Probability and it's applications, Tom II 1, 106–116 (1957)
11. Foster, L., Cao, J., Cleveland, W., Lin, D., Sun, D.: Internet traffic tends toward Poisson and independent as the load increases. In: Denison, D., Hansen, M., Holmes, C., Mallick, B., Yu, B. (eds.) Nonlinear estimation and Classification. LNCS, vol. 171, pp. 83–110. Springer, Heidelberg (2003)
12. Borovkov, A.A.: Ergodicity and Stability of Stochastic Processes. Wiley Series in Prob. & Stat. J. Wiley & Sons, Chichester (1998)
13. Sengupta, D., Deshpande, J.V.: Some Results on the Relative Ageing of Two Life Distributions. J.Appl. Prob. 31, 991–1003 (1994)

New E-Payment Scenarios in an Extended Version of the Traditional Model

Mildrey Carbonell, Joaquin Torres, Antonio Izquierdo, and Diego Suarez

Universidad Carlos III de Madrid
Leganés (Madrid) Spain
{mcarbone,jtmarque,aizquier,dsuarez}@inf.uc3m.es

Abstract. Most of the security proposals in commerce scenarios have been based on a classical e-payment system definition. This definition basically represents a client who sends a payment order to obtain some goods/services from the merchant, which the intentions of the real money transaction carry on between his financial institutions. Nevertheless, these definitions are not sufficiently robust when new aspects appear in the electronic payment transaction. We can identify some of those new aspects (such as: smart card with network capabilities, business mediator with advantage services, handheld devices with constrained connectivity, and multiparty scenarios) that could subordinate the design of current and future commerce scenarios. In this paper we extended the traditional e-payment system definition, in order to include these new aspects. Additionally, we describe two new payment models, where such aspects are involved, and where the secure solution needs to consider new security requirements.

Keywords: E-payment system, secure payment protocol, multi-party scenarios, m-commerce, smart card.

1 Introduction

Since the appearance of e-commerce, significant work has been done trying to find a standard definition of e-payment system describing the principal involved entities and the flow of transactions in which these entities are involved.

Common to all of these approaches is the description of the e-payment system as the interaction of five principal entities: the business entities *client (C)* and *merchant (M)*, the financial entities *issuer (I)* and *acquirer (A)* and a *payment system provider (PSP)* as a secure entity which performs electronic payment transactions on behalf of the issuer and the acquirer on the Internet side and on behalf of client and merchant on the private banking network side.

However, this definition presents some limitations when describing the trends in current e-payment systems in which new aspects emerge, such as: business mediator, smart card, multiparty scenarios, and mobile devices, all of them creating new scenarios and circumstances which need new secure solution. The main goals of our research activity are to show the necessity to extend the e-payment system definition, and the necessity of design new security solution to payment process. For that, we present two

O. Gervasi et al. (Eds.): ICCSA 2008, Part II, LNCS 5073, pp. 514–525, 2008.

new payment models where many aspects are involved generating new flow of message and new security requirements.

Following, we start with the consideration of the traditional definition of e-payment system and the principal flow of his messages, in section 2. Afterwards, in section 3, we present the new aspects, that we will be considered, in the extended e-payment scheme. In section 4, the two new payment models, with their new flow of messages are described. And next, in section 5 we present a brief description of the new security requirements that appears due to these new aspects. The paper is concluded in section 6.

2 E-Payment Model

In many works, the payment model [1][2][3][4] is defined as the interactions between five principal entities (C, M, I, A, PSP) (fig. 1).

- **Client (*C*):** Entity who wants to purchase goods or services from a merchant.
- **Merchant (*M*):** Entity who delivers the goods/services upon receipt of payment.
- **Issuer (*I*):** The financial organization issuing the valid electronic payment instrument (for example, credit/debit card, account and others). He will transfer funds from the customer and acquirer bank.
- **Acquirer (*A*):** The financial organization of the merchant. He verifies the validity of the deposited payment and forwards it to the *PSP*, in order to inform the merchant.
- **Payment system provider (*PSP*):** IN [1] is defined as the entity which performs payment interactions on behalf of *I* and *A* on the Internet side, and on behalf of *C* and *M* on the private banking network side. He receives the request of payment authorization from the merchants and communicates with the issuer in order to obtain a response. He could communicate with the customer (depending on the electronic payment instrument) for obtaining some information (account, password, etc.). If the *payment authorization* request is successful, he sends to the merchant this information and withdraws from the merchant to inform the acquirer. The function of this entity may be implemented by a credit-card company (such as VISA with 3D Secure [2]), a mobile operator (such as mobipay [5]), a secure gateway (such as paypal [6], authorized.net [7]) or a bank.

The principal flow of messages in the payment process is:

1. *Payment ordering:* PO is the interaction between *C* and *M* , there, *C* requests to purchase goods or services from *M* sending the required information (amount of purchases, issuer identification by payment instruments, etc.) to carry out the payment
2. *Authorization request: AR* It is the interaction by which *M* requests a payment authorization from the customer's issuer and waits for a response. The payment authorization process is handled through a PSP with the following functionalities:
   a) *Off-line client authentication:* Where the PSP check the client information, received from merchants, without need an online client authentication. In that way, the merchant receives all the private information of client, forward it to the PSP and this one connects to the issuer bank.

b) *Online client authentication:* Where the PSP checks the client information, which is received from merchants, by means an online authentication mechanism (PIN, password, certificates, etc.). With this purpose, the PSP establishes an *authentication channel* with the client.

3. *Authorization response:* Where the PSP sends the response (successful or unsuccessful depending on issuer decision) of the authentication process to the merchants. If it is successful, the response it is sent to the acquirer, in order to conclude the purchase.
4. *Payment sending:* Finally, the payment process conclude with an interaction between *I* and *A,* with the goals of transferring the requested amount. Normally, this type of transaction is performed under a private banking network, once the acquirer has received a successful authorization response. This interaction ends when *A* forwards the payment receipt to *M* through a PSP.

In figure 1 the arrows represent the directions of these transactions.

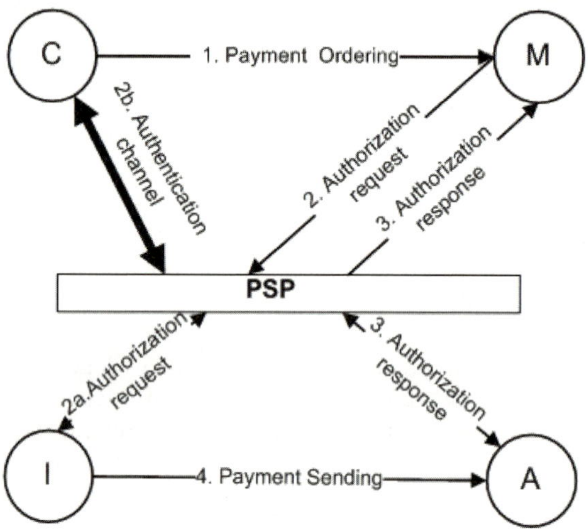

Fig. 1. Flow of messages in the traditional e-payment system

3 Extending Electronic Payment Model

3.1 New Technological Aspects Involved in Electronic Payment

Although, many scenarios have been described taking the traditional model as reference, we consider that it presents some limitations when describing actual electronic payment systems in which appear new technological aspects, such as: smart cards with network capabilities, handheld devices with reduced connectivity, business mediators with advantage services, multiplicity of involved entities.

Payment Network Smart Card (NSC)

Recently, many efforts have been made to transform the smart card into a device open to network interconnectivity. In order to do this, one must consider it as a fully functional Internet node in accordance, whenever possible, with established standards. Several works have been done to this field, such as that collected in [8], which has been oriented towards establishing a simplified TCP/IP protocol stack. Moreover, in [9] the authors aimed to provide a TCP-type protocol.

Although the resulting protocol did not fulfill all of the requirements established in the standard [10], it included the concept of agent-based Internet card. In [11], a review of the characteristics, steps and planning of what could be a new generation of smart cards with or without contacts is addressed. A clear evolution towards a networked smart card was quite evident. In [12] and [13], the card was already thought of as an Internet node which implemented standardized security and communication protocols, to be connected to a network via the host. The card was able to provide services or access Internet resources making use of protocol stacks in the same way as any other node on the network. Its use in security solutions was soon proposed. In this way, the network smart card was able to establish secure direct communication with remote Internet servers, as shown in [14][15]. This capacity allows the cards to guarantee online transactions.

A new remote authentication protocol architecture for network smart cards is described in [16] aiming to cover the authentication process globally with the maximum assurance of security. Her main characteristics are: 1) Stand-alone supplicant: The smart card adopts the functionality of standalone supplicant vs. split supplicant [17], 2) Remote authentication protocol architecture: Atomic smart card authentication protocol design and end-to-end mutual authentication schema on layer 2 of the OSI model.

These NSCs are powerful devices for being used as payment cards in electronic commerce. It could be used to establish a secure mechanism to authenticate the client.

Business mediator

Although, in electronic commerce seemed to suggest that e-commerce transactions would result in decreased costs for buyers and sellers alike, and would therefore ultimately lead to the elimination of mediator reducing this term to a digital shop (e-shop) like a mediator between virtual and real worlds [18]. A careful analysis of the structure and functions of electronic marketplaces reveals a different picture. The electronic business mediator have appeared [19][20][21][22] providing many value-adding functions that cannot be easily substituted or 'internalized' through direct supplier-buyer dealings, and hence mediating parties may continue to play a significant role even in the e-commerce world.

In the payment process, the business mediator is usually considered as entity which secure the payment transaction, this means; it is associated with the PSP. However, in others works it is represented, as entity capable of, among other functionalities, decrease the customer operations [23], simplifying the amount of transactions [24], provide a centralized the connection with the *PSP* [25], and else. In all the case, it is represented an entity with advantage services to client and to merchants, this means; in the *business process*, and out of the secure financial process such as: the PSP.

Consequently, the business mediator inside the payment process needs to be considered of as a new entity different to the PSP. His integration could modify the transaction between the involved entities and could generate some different scenarios.

Connectivity

Protocols designed in recent years for mobile payment systems are based on a scenario where all entities are directly interconnected to one another. This scenario (formally called "Full connectivity scenario") offers advantages to protocol designers because it allows them to simplify the design and development of payment protocols without losing security guarantees. Nevertheless, this scenario does not consider situations in which one of the engaging entities can not directly communicate with the other due to the impossibility of one of them to connect to Internet, maybe for restrictions of handheld device, geographic situation, etc. Considering previous works, these new scenarios of connectivity can be classified [26] as:

- *Disconnected*: *C* and *M* are disconnected from the *PSP* and directly communicate with each other using a local link.
- *Server-Centric:* The *PSP* is connected to both *C* and *M* but they are not directly connected with each other.
- *Client-Centric* [27]: *C* is connected to both *M* and the *PSP* but they are not directly connected with each other.
- *Kiosk-Centric* [28]: *M* is connected to both *C* and the *PSP* but they are not directly connected with each other.
- *Full-Connectivity:* All the entities are directly connected to one another.

Connectivity is one of the more important aspects which will determine the design of further security solutions for electronic payment, especially in mobile contexts. This concept implies in many cases new messages flow and new secure mechanisms to protect the traditional payment transactions.

Multiplicity of involved entities

Multiparty scenarios in electronic commerce [29] [30] are already well-know. Applications such as *virtual mall* or *market place, commercial search agent* describe a commerce scenario of multi-purchase, where one customer could interact with many merchants simultaneously; this means that it could obtain products from many merchants. In other business models, such as *demand aggregation*, appear scenarios where many customers need to form a partnership in order to increase their bargaining power and obtain discounts; in that way, finally this multiple customers could interact with one merchant of with many merchants, simultaneously.

Consequently, traditional topologies such as (one-to-many, ring, mesh, many-to-many) need to be considered in the payment model, and also, need to be integrated with the others previous aspects in order to represent real applications.

In that way, we propose to extend the electronic payment model to include these new aspects.

3.2 Extended E-Payment Model

Here, we will extend the electronic payment definition as the interactions between seven engaging parties (figure 2), three of which (I, A, PSP) remain unchanged from the classical model. Appear two new optional entities: (the dotted line around them represents that they may or may not be present) the business mediator (*Md*) and an enhanced payment card (NSC) with networking capabilities. The single merchant (M) is extended to one or more merchants (M+), and the client (C) is extended to one or more customers (C+).

The variety of connectivity conditions between the entities, are represented by arrows. It can be always direct connection, always indirect connection or one of them depending on the scenario.

The new entity Network smart card (*NSC*) needs a direct connection with the client (this means that the cardholder needs a mechanism to read the NSC). But although, the NSC could establish a direct authentication channel with his issuer bank, this connection could be by indirect communication channel.

And on the other hand, the mediator is located between *M+* and *C+*, since he could offer advantageous services, in the payment process, to both of them. Note that it is a different entity from *PSP*. In that way, his connection with all close entities (C+, M+, PSP) could by direct or indirect.

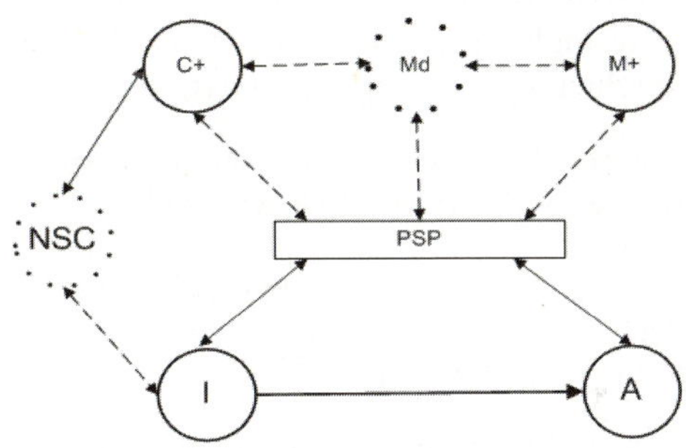

Fig. 2. Graphic representation of the extended e-payment model

In this case, inside the definition it is not possible to describe the flow of message because this one will depend on the involved entities and the type of connectivity (direct or indirect) between them. Due to that, in the next section, we describe two payment scenarios and the appropriate flow of messages. These scenarios could be derivated from the extended e-payment model, and could integrate all the new features.

4 New E-Payment Scenarios

Based on the new definition of the e-payment model we can generate a wide range of scenarios. We are interested in describing two novel scenarios where these new features are involved.

4.1 Client-Centric Payment with Multiple Merchants, Business Mediator, and Offline NSC (CC-M+ Payment)

This scenario (Figure 3.) could be described by the flow of messages and the participating entities as: one client (*C*), one business mediator (*Md*), multiple merchants

($M_i \in M$) and the *PSP* who provide the connection with the financial institutions (issuer and acquirer). The mediator plays the role of a virtual mall where the client can purchase products/services from different merchants. But especially, he acts as mediator between client and merchants, facilitating multi-purchases and multi-payment. In that way, *C* will delegate the multiple transactions with the merchants to *Md*.

However, the principal characteristic of this scenario is that the merchants and the *Md* are disconnected from the *PSP*, and only the *C* establishes a connection with this one. In that way, all the payment process, and the payment authorization process (requests and responses) need to flow through *C*. Due to that, the mediator could provide a single message as an appropriate combination of request/response messages. Finally, the client establishes a direct connection with the PSP, and he could directly send his private payment info to the PSP. The online authentication process could be completed with the authorization request process. In this scenario the payment process occurs in the following way:

1. The client sends the **payment order** *(PO)* to *Md*. In this case, this payment order could be composed by multiple payment orders PO_i to each merchant $Mi \in M+$, where $M+$ are the merchants selected by the client.

2. Once this payment order *PO* is received, the mediator split it in a number of PO_i and **distributes** it to each involved merchant $Mi \in M'$.

3. Afterwards, when merchants receive the payment order *POi,* they start with the **authorization request** *ARi* to the *PSP*. Since they have not connectivity with the *PSP*, they need to do the process through *Md*.

4. In this moment, the *Md* waits for all the authorization request *ARi* which is send by the involved merchants $Mi \in M'$. Once *Md* receives AR_i , it creates a single a single **authorization request** message and **forwards** it to *C***.**

5. *C* has the responsibility to connect to the *PSP* and to send all the **authorization requests** received from merchants through *Md*.

6. Upon reception of the message from *C*, the *PSP* could receive in this scenario the **authentication** information of C. And for that, he continues with the traditional responsibility of checking the authenticity of *C* by means of the issuer bank. After which, PSP sends the **authorization response** *ARs*. This will be composed by the responses *ARsi* to each involved merchants $Mi \in M+$. This *ARs* need to be sent through *C*.

7. The client receives the products/services when all involved merchants receive the **authorization response** *ARsi*. Due to that, *C* **forward** the *ARs* received from *PSP* to *Md*.

8. Finally, *Md* divides the message *ARs* and **distributes** the appropriate **authorization responses** *ARsi* to each involved merchants $Mi \in M+$.

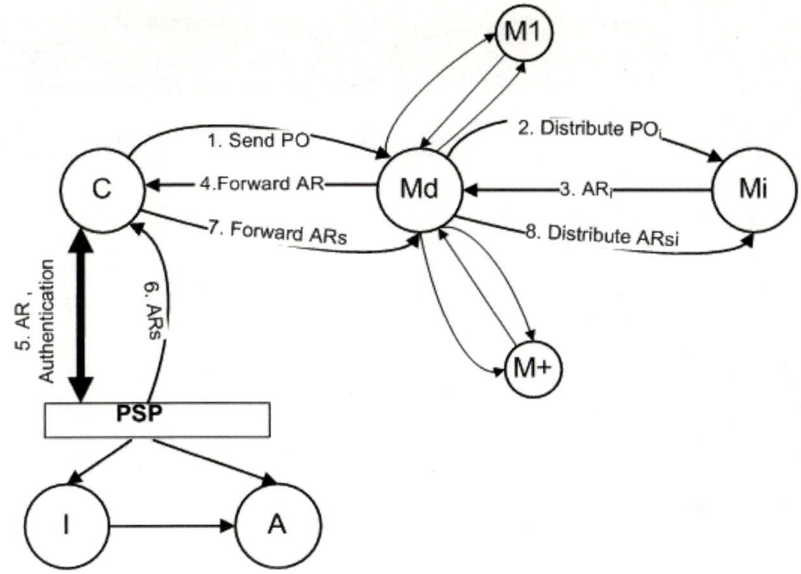

Fig. 3. Flow of messages in CC-M+ scenario

4.2 Client-Centric Payment with Multiple Merchants, Business Mediated, and Online NSC (CC-M+-NSC)

This new scenario (Figure 4.) could be described, as in the previous case as: a client (*C*), a business mediator (*Md*) and multiple merchants ($M_i \in M$) and the *PSP*, who provides the connection with the financial institutions (issuer and acquirer). The mediator plays the same role. And in the same way that in the previous one, the merchants and the *Md* are disconnected from the *PSP*. Only the *C* can communicate with the *PSP*. However in this scenario a new entity appears: an online NSC which has the possibility of directly authenticating the client with the issuer bank. In that way, the flow of payment process is transformed in the following way:

- The steps, from 1 to 4, are repeated in this scenario. Nevertheless, the step 5 introduce new variant.

5. *C* has the responsibility of connecting with the *PSP* to send all the **authorization requests** received from merchants through *Md*. The authentication channel is now provided by the NSC
6. Upon reception of a message from *C*, the *PSP* translate the authentication mechanism to the issuer bank...
7. Now the authentication process is handle by the NSC established an **authentication channel,** between NSC and I.
8. When all this process concludes (and the PSP receive the issuer response), the PSP sends the **authorization response** *ARs* (composed by the responses *ARsi*) to each involved merchants $Mi \in M'$ through *C*.

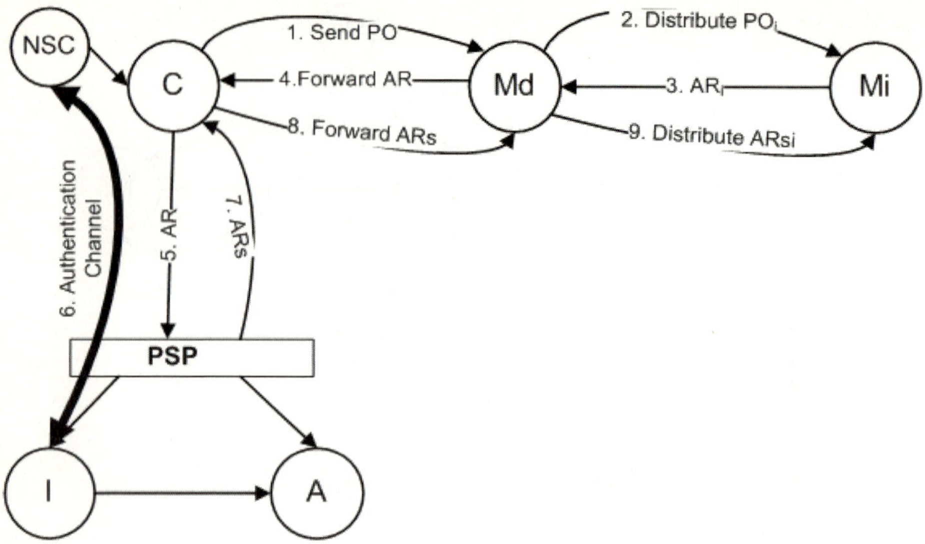

Fig. 4. Flow of message in the CC-M+-NSC payment scenario

9. *C* **forward** the *ARs* received from *PSP* to *Md*.
10. And Finally, *Md* split the message *ARs* and **distributes** the appropriate **authorization responses** *ARsi* to each involved merchants *Mi*∈*M'*.

5 New Security Requirements for the Extended Version of the Payment Model

Payment requires the greatest level of security in electronic commerce transactions [31][32]. For this reason, a solution is needed that guarantees confidentiality, authenticity, integrity and non-repudiation of the transactions. However, many new requirements need to be considered in our models. Now we enumerate those security requirements according to the new features.

Payment Network Smart Card

- A protected channel (confidential, authenticated and keeping integrity) should be built by own network smart card and the remote server (Access Control server controlled by the issuer bank).
- Standardized communication and transport protocols between the smart card and the issuer bank with security purpose.

Business mediator

- A secure architecture to create a trust relationship with the un-trusted *Md.*
- A secure mechanism to assign authorization to *Md*, for distributing the payment order, for joining the authorization request and, next for distributing the authorization responses for each involved merchant.

- A mechanism to create a non repudiation evidence of *Md's* participation in each step of the protocol.
- A mechanism to establish mutual authentication schemes between customer and merchants across untrustworthy mediator.

Connectivity

- A mechanism to guarantee the security of the message when going through one or many entities. For example all the authorization process messages through *Md*, and *C* to *PSP*.
- A secure mechanism to exchange cryptographic information, such as, a public key or a symmetric key, without online connectivity.

Multiplicity of involved merchants

- A mechanism to protect the confidentiality of the purchases and payment information between all involved merchants, in order to avoid that some malicious involved merchant could eavesdrop the information sent to other legitimate merchant.
- A mechanism to protect the integrity of the purchases and payment information, which is sent to the merchants. With the goal to of avoiding that some malicious merchant could modify the information sent to other merchants.
- A mechanism to avoid mutual coordination, between the involved entities, to carry malicious behaviors.

6 Conclusion

Security in traditional e-payment model is based on protecting the transaction flow between business entities (client and merchant) and their transactions with the financial institutions. However, the security architectures based on this model are not sufficiently robust when new participants with their respective functionalities and particularities engage the e-payment system.

In this paper we summarize four new aspects (payment network smart card, connectivity, business mediator, multiplicity of involved entities) to consider an extended version of traditional payment model.

Also, we describe two new payment scenarios (*Client-Centric payment with multiple merchants, business mediator, and offline NSC*, and, *Client-Centric payment with multiple merchants, business mediator, and online NSC*) in order to demonstrate how these new aspects could define new payment interactions and consequently requires new security solutions. A brief discussion of the security requirements of these scenarios allow us to how that it is not trivial to adapt the current payment solution to these new approach. In the future we are interested in developing a realistic secure payment solution for the proposed scenarios based on an extended version of the traditinal payment solutions.

References

1. Kungpisdan, S.: Modelling, design, and analysis of Secure Mobile Payment Systems., Thesis for Doctor of Philosophy, Faculty of Information Technology, Monash University (2005), http://beast.csse.monash.edu.au/~srini/theses/keng.pdf
2. 3-D Secure. System Overview. 70015-01 External Version Copyright © 2002-2003. Visa International Version 1.0.2 (May 01, 2003), http://partnernetwork.visa.com/pf/3dsec/download/trk_3dsec_system_overview_v102.pdf
3. Kou, W.: Payment technologies for e-commerce. Springer, Heidelberg (2003)
4. O'Mahony, D.: Electronic payment systems for e-commerce. Artech House (2001)
5. Secure system to pay with mobile phone charging the amount directly to a payment card, http://www.mobipay.com
6. Secure online payment gateway: PayPal, http://www.paypal.com
7. Payment gateway: Authorize.Net, http://www.authorizenet.com/
8. Rees, J., Honeyman, P.: Webcard: a Java Card web server. In: Proc. of 4th IFIP Smart Card Research and Advanced Application Conference, CARDIS 2000, Bristol, U.K (2000)
9. Urien, P.: Internet card, a smart card as a true Internet node. Computer Communications 23(17), 1655–1666 (2000)
10. Postel, J.: Transmission Control Protocol. IETF RFC 079 (September 1981)
11. IST Project RESET, Roadmap for European Research on Smartcard related Technologies, IST-2001-39046: Final Roadmap, v.5 (May 2003)
12. Montgomery, M., Ali, A., Lu, H.K.: Secure Network Card. Implementation of a Standard Network Stack in a Smart Card. In: Proc. of 4th IFIP Smart Card Research and Advanced Application Conference, CARDIS 2004, Toulouse, France, August 23-26, 2004. Kluwer Academic Publishers, Dordrecht (2004)
13. Lu, H.K.: New Advances in Smart Card Communications, International Conference on Computing. In: Communications and Control technologies (CCCT), Austin, TX, USA, August 14-17 (2004)
14. Lu, H.K., Ali, A.: Prevent On-line Identity Theft - Using Network Smart Cards for Secure On-line Transactions. In: Zhang, K., Zheng, Y. (eds.) ISC 2004. LNCS, vol. 3225. Springer, Heidelberg (2004)
15. Ali, A., Lu, K., Montgomery, M.: Network Smart Card: A New Paradigm of Secure Online Transactions. In: Proc. of Security and Privacy in the Age of Ubiquitous Computing, IFIP TC11 20th International Conference on Information Security (SEC 2005), Chiba, Japan, May 30 - June 1 (2005)
16. Torres, J., Izquierdo, A., Sierra, J.M.: Advances in network smart cards authentication. Computer Networks 51(9), 2249–2261 (2007)
17. Torres, J., Izquierdo, A., Sierra, J.M., Ribagorda, A.: Towards selfauthenticable smart cards. Computer Communications 29(15), 2781–2787 (2006)
18. Porter, M.: Strategy and the Internet. Harvard Business Review, 63–78 (March 2001)
19. Giaglis, G., Klein, S., O'Keefe, R.: The role of intermediaries in electronic marketplaces: developing a contingency model. Information Systems Journal 12(3), 231 (2002)
20. Dikaiakos, M.: Intermediary infrastructures for the World Wide Web. Computer Networks 45(4), 421–447 (2004)
21. Esparza, O., Muñoz, J., Soriano, M., Forné, J.: Secure brokerage mechanisms for mobile electronic commerce. Computer Communications 29(12), 2308–2321 (2006)
22. Bhargava, H., Choudhary, V.: Economics of an Information Intermediary with Aggregation Benefits. Information Systems Research 15(1), 22–36 (2004)

23. Bhargava, H., Choudhary, V.: Economics of an Information Intermediary with Aggregation Benefits. Information Systems Research 15(1), 22–36 (2004)
24. Wang, Y., Varadharajan, V.: A mobile autonomous agent-based secure payment protocol supporting multiple payments. In: IAT 2005, pp. 88–94 (2005)
25. Carbonell, M., Sierra, J., Torres, J., Izquierdo, A.: Security analysis of a new multi-party payment protocol with intermediary service. In: DEXA Workshops 2007, pp. 698–702 (2007)
26. Chari, S., Kermani, P., Smith, S., Tassiulas, L.: Security Issues in MCommerce: An Usage-Based Taxonomy. In: Proceedings of E-Commerce Agents, pp. 264–282 (2001)
27. Tellez Isaac, J., Sierra Cámara, J.M.: Anonymous Payment in a Client Centric Model for Digital Ecosystems. In: Proceedings of IEEE International Digital Ecosystems and Technologies (DEST), Australia (2007)
28. Tellez, J., Sierra, J.M., Izquierdo, A., Torres, J.: Anonymous Payment in a Kiosk Centric Model using Digital signature scheme with message recovery and Low Computional Power Devices. Journal of Theorical and Applied Electronic Commerce Research 1(2), 1–11 (2006)
29. Bartelt, A., Lamersdorf, W.: A multi-criteria taxonomy of business models in electronic commerce. In: Fiege, L., Mühl, G., Wilhelm, U.G. (eds.) WELCOM 2001. LNCS, vol. 2232, pp. 193–205. Springer, Heidelberg (2001)
30. Rappa, M.: Business models on the web. Managing the digital enterprise(October 2006), `http://digitalenterprise.org/models/models.html`
31. Stallings, W.: Cryptography and network security: principles and practices, ch. 16-20. Prentice Hall, Englewood Cliffs (2006)
32. Tsiakis, T., Stheohanidews, G.: The concept f security and trust in electronic payments. Computer & Security 24(1), 10–15 (2005)

Enhancing an Integer Challenge-Response Protocol

Robson de Oliveira Albuquerque[1], Luis Javier García Villalba[1], and Rafael Timóteo de Sousa Jr.[2]

[1] Grupo de Análisis, Seguridad y Sistemas (GASS)
Departamento de Ingeniería del Software e Inteligencia Artificial
Facultad de Informática, Despacho 431
Universidad Complutense de Madrid (UCM)
C/ Profesor José García Santesmases s/n
Ciudad Universitaria
28040 Madrid, Spain
{robson,javiergv}@fdi.ucm.es
[2] Departamento de Engenharia Elétrica
Faculdade de Tecnologia
Universidade de Brasília
Campus Universitário Darcy Ribeiro
Asa Norte – Brasília – DF
CEP 70.910-900, Brazil
desousa@unb.br

Abstract. In a decentralized network, such as a peer-to-peer or a spontaneous network, a significant trust factor for a peer is to gain a sufficient level of certainty on the other peers' real identity. In this paper we evaluate a proposed peer identification protocol that was designed for such environments and operates based on iterated challenge-response exchanges among peers. For this purpose, we introduce a new attack against this protocol and use the birthday paradox to model the number of operations until the proposed attack is successful. The modeling process, which results in the estimation of the upper bound effort for this successful attack, gives way to the definition of enhancements for the identification protocol. As a result, we define a new identification protocol based on multiple integer challenge-responses that, though not being a cryptographic protocol, represents for an attacker a challenge harder than breaking a symmetric cryptographic key by brute force attack. Our proposed attack shows how to break the GCP protocol without any previous knowledge on target secret information.

Keywords: Information Security, Spontaneous Self-Organized Networks, Peer Identification, Identification Protocol, Trust.

1 Introduction

The trust management in spontaneous self-organized networks is a mechanism used to decrease the complexity of the access control and authorization of operations, allowing an approach to cope with the decentralized structure of these networks, the dynamic of arrival and departure of users, the mobility and the possibility of a large number of participant entities.

O. Gervasi et al. (Eds.): ICCSA 2008, Part II, LNCS 5073, pp. 526–540, 2008.

In these spontaneous networks, the network control subsystem as well as the applications constitute distributed systems, which organize themselves autonomously in order to provide a communication and processing structure to the participant entities, even thought they do not have a previously determined structure neither a centralized control. These characteristics are commonly presented in mobile *ad-hoc* networks, computational grids, peer-to-peer systems, agent communities etc.

In these distributed self-organized systems the problem of identifying the entities that initially do not know or do not have the possibility to recognize each other is always present. In such a scenario, it is possible (and actually, very easy) that a remote entity presents more than one identity, which characterizes a Sybil attack [1]. Thus, in order to control the access or to authorize operations, an entity must be sure that different identities belong to different entities. Despite that, there is also the possibility of a group of remote entities to form an attack coalition. Trust [4], [8] has an important role in pointing out directions and tries to find real and effective solutions for identity and security problems.

The Green Card Protocol (GCP) [2] proposes a way to an entity, in a distributed system, for acquiring a sufficient level of trust in the real identity of another entity. GCP is based on an iterated challenge-answer process between two entities, with no restriction to the communication subsystem used, what tries to simplify the identification process.

Having the GCP way of operation in mind, the formal model of the environment presented in [1] for the multi-identity attack is considered adequate, in this article, for the analysis of an attack against the GCP. This formal model establishes the basic characteristics of a distributed self-organized system, which includes these reasonable assumptions:

- A group of entities, including a subgroup of well behaving entities and another sub group of misbehaving entities (possibly working in coalition).
- A broadcast communication medium that can be used by any entity, meaning that every entity may send a message that will be broadcasted to all other entities and every message can be observed by all entities.
- The messages are received by all other entities within a finite time interval. The delivery of the messages is guaranteed, but there is no guarantee that the entities will receive the messages in the same order.
- There is a minimum restriction of the computational resources available for each entity; however, this restriction allows the entities to establish point-to-point private and authenticated communication paths using public key cryptography.

Considering the context presented above, this article proposes a new attack plan against the GCP. This proposed plan is based on the mathematics behind the Birthday Paradox [3], a formulation known as the base of the birthday cryptographic attack. Then, the Birthday Paradox solution is used to estimate the upper bound effort for a successful attack against the GCP, using the attack plan presented, and this effort is shown to be much smaller than that it takes to accomplish a brute force attack completely. Moreover, the GCP protocol was implemented to allow a verification of the characteristics of the attack plan, through a distributed simulation.

As a result of this analysis, given the required enhancements for the GCP protocol, we define a new identification protocol based on multiple integer challenge-responses that, though not being a cryptographic protocol, represents for an attacker a challenge harder than breaking a symmetric cryptographic key by brute force attack.

The article is organized as follows. Section 2 describes the GCP and section 3, the Birthday Paradox. Section 4, 5 and 6 presents the attack plan against the GCP and the estimative of the upper bound effort for a successful attack, with some implementation data. Section 7 presents further discussion that is important about enhancing security in identification and gives the specification of a new approach for a challenge-based identification protocol. The conclusions, in section 8, address other possible security issues and discuss some necessities for strengthening this protocol. Also we point directions for further work with the integration of the notion of trust into an identification protocol.

2 Characteristics of the GCP Protocol

The GCP proposes a way to dynamically manage the identification of various entities (nodes) in completely decentralized environments, characterized by the absence of an infrastructure responsible for providing at all times the authentication of the nodes (in centralized or hybrid environments this authentication is usually provided by a central data base server, that verifies the logins and passwords or codes generated by cryptography functions).

Facing the absence of a centralized authentication entity, the responsibility of identifying other entities is given to the entity itself in an individual process. And that requires an exchange of information among the participants to provide some guarantee on the authenticity of the identity presented by each one. An entity A may verify the authenticity of the identity of B, by comparing the data provided by B with the data stored about B in a local database in A. To do so, A will send challenges to B, who must give the answers expected by A for each challenge.

Each entity is responsible for generating independently its identification data, which is its innate knowledge. For an entity A the innate knowledge $K=\{<q,r>\}$, is given by a set of tuples q e r, with the following characteristics:

1. The elements q and r are non negative integers. The field q represents the challenge and r the answer.
2. The possible values of q are limited to $q < k$, k being a constant value known by all entities. The tuples that form K can be represented in memory by tables with size k.
3. The value r is randomly generated. Although it is not specified in the protocol, it is assumed the r is within the interval [0,d}, with d=2^n, where n is the number of bits used to represent r. There is no loss of generality with this assumption, since n may be arbitrarily large and, can express in a simple way the characteristics of the protocol, like the memory usage by K (if the protocol is used by E entities, the memory usage in bits for each entity is E*k*n. For example, if there are 500 entities, using 256 tuples with 16 bits (2 bytes) for the values of r, it is necessary a database of 256 Kbytes for each entity).

The protocol uses this knowledge to exchange the challenges and answers in an iterated way between two entities. The content of the exchanged messages are the integer numbers *q* and *r*, which are the challenges and answers. For each challenge *q* sent by an entity A, an entity B must send back a correspondent answer *r*, as expected by A. Each correct answer provided by B, results in an increase of the trust A has in the authenticity of B's identity. The protocol is symmetric and has a piggybacking functionality, through which entity B, while responding a challenge from A, also sends a challenge to be answered by A. In each iteration, each entity chooses the challenge it will send randomly, that is, the challenge from A to B consists in a random choice from a position in the table of tuples which B must have previously acquired from A.

If an entity B is already identified by an entity A, the GCP also suggests a way for entity B to introduce a third entity C to A, thus creating a sort of presenting mechanism. This process brings up the issue of transitivity impossibility [4] in the trust relation from C to B to A, assuming that if A trusts in B, B trusts in C, it does not mean that A trusts in C. This is an important trust issue that is not explicit in the GCP specification.

It is also important to note that the GCP is not a cryptographic protocol, that is, there is no assumptions that the creation and communication of the tuples *<q,r>* demand cryptographic functions or operations. The GCP specification [2] presents the following arguments in favor of the protocol, especially concerning the robustness against impostors and the computational cost:

- Since *r* is a randomly generated number, it is possible, from a statistical point of view, to keep extremely low the probability of an imposter guessing the value of *r* correctly to a given challenge *q*, without knowing the tuple *<q,r>*.
- In an iteration of challenge-answer, a single incorrect reply results in the immediate rejection of the identity. But a correct reply does not mean the validation of the identity, but it only increases the probability of the identity being associated with the correct entity. Thus, it is possible to choose a number of iterations that provides an acceptable certainty of the identity.
- Considering that two entities A and B have already mutually validated their identities, the use of the protocol allows A and B to verify jointly if an entity C is an impostor. In case there are inconsistency between the data kept by A and B about C, then C may not be trustful.
- The protocol requires, in computational terms, processing power to compare integer numbers and storage capacity for databases containing the tuples of integers.

Considering all the aspects mentioned here, the only chance for an impostor not being discovered would be to answer correctly a sequence of challenges. This breach is exploited in the attack plan proposed in the following section.

3 Attack Plan Based on the Birthday Paradox

The attack plan is based on the use of an impostor with multiples identities, i.e. a Sybil class impostor SCI (or, equivalently, a coalition of imposters, since GCP provides the means for coalitions, due to the possibility of a third entity to be introduced between two already mutually trusted entities). In this manner, the imposter SCI can start an arbitrary number of iterative sections, presenting itself as B, trying to identify itself to an entity A.

The statistic analysis of the possibilities of SCI guessing correct answers indicates a possible attack against the GCP based on the probabilistic phenomenon called Birthday Paradox 3. It is interesting to note that such denomination refers to the fact that, to find a group of people in which (with 50% of possibilities) at least two people have the same date of birth, it is necessary to randomly gather a minimum of 23 people, which is a much lower value than 365 (the possible dates of the year), fact that goes against the intuition of the majority of the people. The underlying mathematics to this phenomenon is the base to the notorious Birthday Attack against cryptographic hash functions, given that the number of N-bits hash values that must be generated before getting a collision is not 2^N, but only $2^{N/2}$ [5].

In the attack plain against the GCP, the central question is to determine the probability of the impostor SCI generating random answers and these answers coincide with values of r in tuples $<q, r>$ relative to entity B that are in the challenge-answers table of the entity A, the attack's target. The knowledge that entity A has about B is stored in A, in the form of a table of k tuples:

$$K_B = \{<q_0, r_0>, <q_1, r_1> \ldots <q_i, r_i> \ldots <q_{k-1}, r_{k-1}>\} \quad (1)$$

In case the impostor SCI has the possibility to repeatedly present answers to one same challenge q, then SCI will have repeated chances to try to guess the corresponding value r, with the correctness possibilities increasing in each attempt, given that to each new attempt SCI decreases the space of values of attempt t, discarding the values already used in previous attempts. Thus, the probability to find a value t equals to r (collision) is defined by the solution of the birthday paradox.

Remembering that the values of the trials t are random integer numbers picked from an uniform discreet distribution among d possible values in the interval [0, d), it is simple to calculate first the probability pdif(t), of all the values of t are different from r. If $t > d$, this probability is 1 (*pigeonhole principle*). But, for $t \leq d$, this probability will be:

$$\text{pdif}(t) = 1 \times (1-1/d) \times (1-2/d) \ldots \times (1 - (t-1)/d) \quad (2)$$

This expression comes from that fact that the second trial cannot have the same value of the first one; the third cannot have the same value of the other two first ones, etc. An equivalent expression for this probability is:

$$\text{pdif}(t) = \text{product}(i = 1, i = t\text{-}1, 1 - i / d) \quad (3)$$

The event of at least two of the t trials result on same value (collision) is the complement to the event of all the values being different in the t trials. Thus, the collision probability is:

$$\text{pcol}(t) = 1 - \text{pdif}(t) \quad (4)$$

Fig. 1 shows, for an interval of 256 possible values, the evolution of this probability as a function of the number of trials. It is worth to notice that in this example in the twentieth trial the collision probability is greater than 50%.

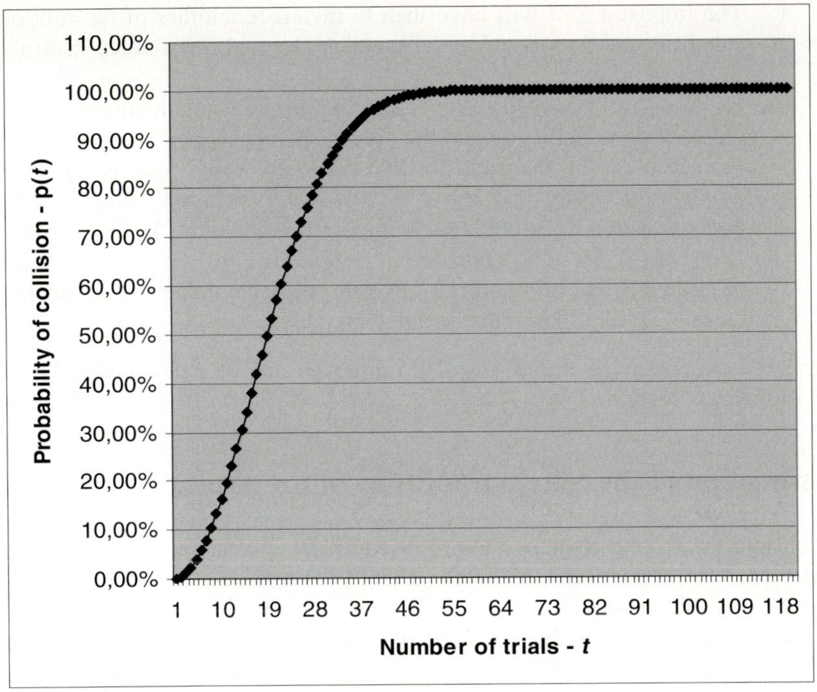

Fig. 1. Probability of collision of 2 values on t trials (d = 256)

Upon these perspectives, the plan of attack against the GCP can be formulated in the following terms:

- The impostor SCI intends to identify itself as entity B to entity A. Without having (yet) the knowledge on the identification of B in A, the impostor SCI keeps in memory a table of attempts TB = {position 0 – value [0, …, n], …, position k-1 – value [0, …, n] }, with k positions and k being a parameter of GCP known by all the entities and representing the amount of challenges that allows an establishment of trust in an identity.
- The impostor SCI introduces itself to A, as B.
- The impostor receives challenge *i* from A and, without having any information on what B would answer, SCI chooses a random reply and sends to A.
- If SCI guesses a correct answer (what may happen, since it is a random phenomenon), SCI writes down the reply in position *i* of table TB and passes to the next iteration, waiting the next challenge from A.
- If SCI guesses wrong, then its (fake) identity will immediately be refused by A, but SCI finds out that the value in position *i* of TB cannot be the attempted reply, what is a valuable and important information of the attacker's point of view.

- The impostor SCI will have then to incarnate another of its multiple personalities (as previewed in sybil attack [1]) and reintroduce itself to A, as B, one more time.
- As A is not interested in repeating the previous challenge, it chooses another random challenge and the process is repeated.
- In each iteration, the impostor SCI either answers correctly and continues or answers wrong and learns a little more, in a way that the table TB evolves with a negative knowledge on the identity of B, TB = {position 0 - ($!P_1$, $!P_2$ $!P_3$)},{…}, {position $_{k-1}$ - ($!Q_1$, !...,$!Q_N$). Then, for each position of table TB, the probability of SCI to guess the value that identifies B increases exactly like in the birthday paradox.

Hence an estimation of the effort of the imposter can be calculated, which is the object of the next section.

4 Estimation of the Maximum Effort of the Attack

Another approach to the issue of the proposed attack consists in considering that the impostor has all the possible tables TB. A table TB has k answers, and each answer can have d possible values.

The total number of possible tables is d^k. The impostor SCI could choose one of these tables iteratively and try the validation of identity with entity A, as B. In terms of the effort to succeed, the imposter would be facing a problem of the birthday paradox, in a space of possible d^k birthdays. However, given the attack plan above, a shortcut exists in which the impostor SCI has the possibility to attack each element of table TB separately, thus it will now face k problems of the birthday paradox, each one of these problems in a space of d possible birthdays.

The necessary effort in each case can be estimated by calculating the amount of attempts *t* to get, with more than 50% of possibilities, a randomly chosen integer number (of d possible values) that repeats the first choice (that is, one of the parts of the identification of B that is already in the knowledge base of A).

To calculate the amount of attempts, let's take the probability pesp(t, d) of in t attempts we get a value equal to a particularly determined value already chosen (equivalent to the probability of somebody in a group of t people to have the same anniversary that one determined person):

$$\text{pesp}(t, d) = 1 - ((d - 1) / d)^t \tag{5}$$

Inverting this expression, we get:

$$t = \log_{(d-1)/d} (1 - \text{pesp}) \tag{6}$$

To have a probability of collision greater than 50% (pesp > 1/2), the amount of attempts must be:

$$t_{1/2} \geq \log_{(d-1)/d} 1/2 \tag{7}$$

Fig.2 shows, for an interval of d = 256 possible values, the evolution of this probability in function of the number of attempts. Such result applied to the case of the plan of attack against the GCP indicates that "only" 177 attempts would be enough to the impostor generate a valid reply to one determined challenge of the attacked entity.

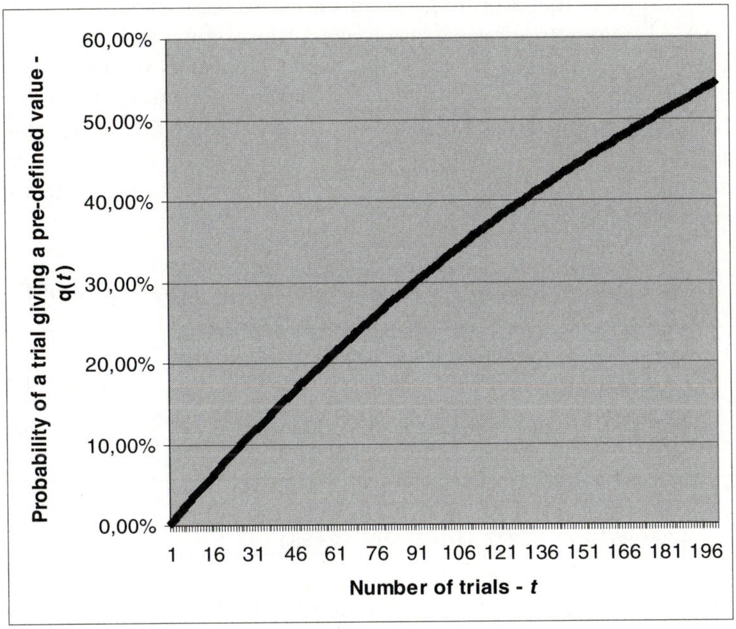

Fig. 2. Probability of a trial giving a pre-defined value (d = 256)

In consequence, considering that there are k answers to be guessed, the total effort of the attack is estimated as:

$$K\bullet t_{1/2} = k \bullet \log_{(d-1)/d} 1/2 \qquad (8)$$

Finally, we reiterate that such effort can be divided among multiple personalities of the imposter, what indicates this is the upper bound for a brute force attack of the Sybil class.

5 Numerical Analysis and Simulation

The analysis of the GCP illustrates the problem of identification of entities in spontaneous self-organized distributed environments. The existence of a shared secret between pairs of entities allows the mutual identification of the entities by the exchange of information concerning this secret.

The GCP proposes that such secret, or knowledge, concerning the identity of a peer entity, should be structured in the form of a table of integer numbers, without using cryptographic resources in the protection of these numbers. From an attacker point of view this is a puzzle of integers to be guessed.

Another way to see this knowledge K={<*q*, *r*>} of an identity is to consider that it is a non-cryptographic key with size k*n, formed by k parts with size n. This allows a comparison between the effort estimation for the considered attack against the GCP and the effort for a brute force attack against a cryptographic key of the same size. It is worth to remember that, using N bits for the cryptographic key, there are 2^N possible keys and, assuming that a brute force attack has to seek an average of half of this space of keys, $2^{N/2}$ attempts are necessary to get a successful attack [6].

Using the formula obtained in section 4, Table 1 presents the effort of the attack against the GCP, for answers *r* with size *n*, for chosen values of *n*. It is also shown the amount of attempts to break a knowledge K, and the occupation of memory to store knowledge K, in a system with 500 participant entities, using 256 challenges. At last, for comparison, it is shown the necessary effort for a successful brute force attack against a cryptographic key of size 256**n*.

Table 1. Estimated maximum effort for the birthday attack against GCP

Size (bits) of challenge answers (n)	Number of possible challenge answers (d)	Number of trials for a successful attack against 1 challenge (t)	Number of trials for a successful attack against 256 challenges (256 * t)	Memory occupation (bytes) in each entity (500 entities, 256 challenges)	Effort for breaking a 256*n bits cryptographic key
8	256	177	45.337	128.000	2^1024
16	65.536	45.426	11.628.991	256.000	2^2048
32	4.294.967.296	2.977.044.471	762.123.384.697	512.000	2^4096

Table 2 compares the effort to break knowledge K of the GCP with the effort to break keys of the DES cipher standard [6]. To get a comparison, in the same conditions of memory occupation, the size of 64 bits of DES key (56 bits of the key + 8 bits of parity) is used as limit of the size of knowledge K. Thus, in knowledge K, the answers with greater size imply in the use of a lesser amount of challenges. Being considered with either 4 16-bit or 2 32-bit answers, GCP compares favorably, representing a challenge (integer puzzle to be guessed) for an attacker as hard as, even much more hard than, breaking a DES key.

Table 2. Effort comparison between the birthday attack against GCP the brute force attack against DES keys

Size (bits) of challenge answers (n)	Number of possible challenge answers (d)	Number of trials for a successful attack against 1 challenge (t)	Number of challenges (k=(56+8)/n)	Quantidade de tentativas para sucesso do ataque (k * t)	Memory occupation (bytes) in each entity (500 entities)	Effort for breaking a 56 bits DES key (2^28)
8	256	177	8	1.417	4.000	268.435.456
16	65.536	45.426	4	181.703	4.000	268.435.456
32	4.294.967.296	2.977.044.471	2	5.954.088.943	4.000	268.435.456

The implementation of the protocol allowed us to obtain data from simulation of this protocol usage in a distributed environment. Table 3 shows results from an experiment with distributed software agents using GCP, with an impostor agent executing our attack

Table 3. Data from the simulation of the proposed attack against GCP

Size (bits) of challenge answers	Number of challenges	Estimated number of trials for a successful attack	Number of trials for a successful attack in protocol simulation
8	8	1.417	1.166
8	256	45.337	32.753
16	4	181.703	128.758
16	256	11.628.991	8.199.951

plan against a target agent. Shown results come from a large number of repetitions of the attack and contain medium values from these repetitions. As expected, the impostor effort is in every case under the estimated maximum effort obtained in section 4.

The estimative of the maximum effort of a successful attack against the protocol of identification GCP is not, however, sufficient for a characterization of the protocol, in particular with respect to other possible vulnerabilities and to the robustness of the protocol to attempts of exploration of these vulnerabilities. We comment on these issues in the next section.

6 Comments on GCP Vulnerabilities

One of the points that deserve attention is the initialization of knowledge K in each entity. Besides the issues of tuple indexation and memory occupation, it is necessary to define how the initialization of tables occur between peer entities and to analyze the consequences in terms of important parameters for the application environment, such as the expense of energy in the processing and the communications, as well as the communication channel bandwidth occupation.

Also, there is the issue of parameter k that all entity must know previously to configure knowledge K. The choice of parameter k seems to be determined by the necessity to choose the lesser possible k for a determined robustness against attacks, as greater k is, more storage space is consumed and more complexity in processing the answers is added. And it may be necessary to change the value of k dynamically, what would lead to an integral reset of the system.

The problem of the choice of a global parameter k has an equivalency in the use of the threshold cryptography [7], by coincidence a cryptography system where a global key is composed of p parts (each part is delivered to a present entity in the system), k of this p parts allow an reconstruction of enough information to accomplish operations of certification and authentication in a distributed system. The problem of the adequate choice of k is still without an optimized solution that takes in account the dynamics of entrance and exit of entities in the system, and consequently the dynamic change of the number of parts p.

The automation of the initialization process requires some cryptographic system among entities of the GCP. With a cryptographic system, the use of the GCP is questionable, since symmetrical or asymmetrical cryptography system also would allow, with equivalent effort and robustness, as indicated in Tables 1 and 2, the identification of peer entities, and other inexistent functionalities in the GCP, such as the confidentiality and the authentication of the messages. But, it is worth to comment that the usage of key exchange protocols and one time session keys, associated to an identification integer puzzle, are promising solutions for strengthening this initialization process.

Concerning the confidentiality necessity, for hypothesis, the GCP must operate in an environment susceptible to attacks, that is, in the presence of impostors. Then, for all practical effect the protocol must be ready to operate in a broadcast medium in the context of the distributed auto-organized system. Thus, the challenge-reply messages are spread out in the system and can easily be captured. Not being ciphered, such messages allow the impostors to acquire knowledge K, without performing active attacks. To avoid this possibility, it is rational to think of the hypothesis of a communication subsystem protected by cryptography, what leads us back to the question of the necessity of the GCP, in case there is already a cryptography system in action.

Therefore, some more study remains to be done regarding the configuration of the protocol and comparing it to cryptographic identification systems, thou it seems that non-cryptographic integer puzzles can be interesting for peer identification in decentralized networks.

7 Enhancing the Protocol

As result of the protocol analysis it is important to say that the estimation effort for a peer is very low in the number of attempts. This assumption takes in consideration the size that a tuple $<q,r>$ may have in bytes.

Assuming the kind of information that can be exchanged among peers in MANET, P2P or GRID systems, and the quantity of messages that goes through these kind of systems, the incorrect size of a challenge-response function and how it is formulated generates security flaws.

Having both assumptions in mind, the following paragraphs show how to enhance the strength of an identification protocol, like GCP.

The first enhancement concerns the initialization of the identification table, which must be built with *k* values. Each *k* corresponds to a tuple $t_i <q_i, r_i>$. So if *k* is of size 256, and *r* is an integer of size *s* in bytes, the table t could be built as:

$$t = \{ t_1 = <1, r_1>, \ldots, t_n = <n, r_n>, \ldots, t_{256} = <256, r_{256}> \} \tag{9}$$

Each element of this table is computed and exchanged among peers in the network using a Difie-Helman [9] approach. This will guarantee the confidentiality and the integrity of the identity table.

Another enhancement is defined for the exchange of challenge and responses during identification of peers. Instead of sending just one challenge at a time, one peer must send at least 3 challenges in only one message. The message M is a combination of three different challenges C. With this assumption the message can be expressed as:

$$M = \{(C_1, C_2, C_3)+ T_{DH}\} \tag{10}$$

Where each C is a tuple < q , r >, q is the position in the table, r is the value of the position and T_{DH} is the timestamp information exchanged with Diffie-Hellman. Extending the discussion, if the challenge-response mechanism is modified in order to send three or more challenges per message, the attacker now has to estimate not only one answer at a time, but the full message. This significantly increases the difficulty of breaking the challenge, because the probability now is much different of $2^{N/2}$.

Following this approach, the entity that sends the challenge may not confirm just one challenge C_1, C_2 or C_3, giving the opportunity to the attacker to built the challenge table by guessing. The sender must confirm the full challenge (not individual C_1, C_2, C_3) with just an answer of OK (M is correct) or ERR (M is incorrect). Thus not informing which tuple is incorrect.

Now, the only manner to initiate the communication process is that the attacker or the other entity really knows three or more answers of the identity table at one step. And if the protocol defines that is necessary at least three steps with correct response to message M to start the communication, the attacker can not efficiently built a challenge-response table as in GCP. In this case the attacker is forced to know at least three different answers at a time, so he can answer efficiently only one challenge (9 positions and the correct answer in total for a full handshake with this modification – three steps).

In the point of view of a legitimate entity (the one that has the correct table) in is only necessary pick up others values in its table. But in the attacker's point of view the probability of guessing three correct challenges at a time is much more difficult. And would take much more computing resources and time to solve the puzzle, thus making the challenge-response mechanism much more difficult to break by brute force, sybil attack or birthday paradox.

The size of k implies directly in more calculation to the attacker because is has to compute more high values in terms of probability. The higher the k value is the more difficult is to guess it, because the number of values would increase exponentially. The size of r in bytes is also important because the higher r is the higher is the set that an attacker must guess.

Then the following may be considered, a) the proper size of the challenge implies in more difficulty to the attacker (size of k and size of r), b) the use of Diffie-Hellman can help the confidentiality and the integrity of the protocol and also SSH [100] can be used to help exchange the tables.

To increase more the strength of the protocol, the answers to the message M should be built in terms of hash functions in conjunction with timestamp information exchanged with Diffie-Hellman T_{DH}. Let the response R of message M be:

$$R_M = \text{hash}\ \{M = [(C_1, A_1), (C_2, A_2), (C_3, A_3)] + T_{DH}\} \qquad (11)$$

Where A_1, A_2 and A_3 are the answers to challenges C_1, C_2 and C_3. The entity that sends the challenge knows the answers of each message M because he can also built the hash of the message M. He can also verify if the R_M corresponds to the given challenge using an internal table t_i (M, R_M, T_{DH}). This avoids interception of clear responses, because no value, but the hash is sent through the network.

Than the steps of the new protocol would be:

1. Built the table as: $\{t_1 = \langle 1, r_1\rangle, \ldots, t_n = \langle n, r_n\rangle, \ldots, t_k = \langle k, r_{256}\rangle\}$, where k is an integer of the size of the table and r an integer with the size of the challenge in bytes.
2. Exchange the table among peers in the network using a Difie-Helman or SSH approach.
3. Built challenges as $M = \{(C_1, C_2, C_3) + T_{DH}\}$, where each C is a tuple $\langle q,r\rangle$, q is the position in the table, r is the value of the position and T_{DH} is the timestamp information exchanged with Diffie-Hellman.
4. Response of challenge M as $R_M = \text{hash}\ \{M=[(C_1, A_1), (C_2, A_2), (C_3, A_3)] + T_{DH}\}$.
5. Starts the communication process if in local table (M, R_M, T_{DH}) the received hash corresponds to its pair in the table position after a three round handshake.

8 Conclusion and Future Work

One of the contributions shown in this paper comes from the analysis of an identification protocol, evaluation its strength against a brute force attack and the proposal of a new schema.

To do so, we formulated a novel attack based on the sybill attack and used the birthday paradox probabilistic phenomenon to deduce the maximum effort for this attack to be successful. We then developed the numerical analysis of the attack and implemented the protocol to gather data from a distributed simulation of its usage. Our proposed attack shows how to break the GCP protocol without any previous knowledge on target secret information. We propose enhancements to the identification protocol so that it can represent a hard challenge for an attacker. Then it seems to be interesting to continue research work on the utilization of integer puzzles for peer identification in decentralized networks.

The commented vulnerabilities and the possible solutions can be evaluated in numerical analysis, so a new protocol can be developed with solutions for peer knowledge initialization, confidentiality of challenge answers and resistance to answer replay. To avoid sending information in clear text, the negotiation of the communication process would be efficiently secured due to the Diffie-Hellman and SSH protocol characteristics. This point has already been proved its efficiency in terms of confidentiality and integrity once there is many implementations of both protocols.

Also the use of hash in the answers would avoid man-in-the-middle attacks. It protects the information and does not let an attacker know the correct answer to the message M, because it seems that hash functions, timestamp information and symmetric cryptography offer promising solutions for these issues and should increase strength against Sybil attacks. Besides all modifications proposed to the identification protocol, trust approach [8], [4], [11] can be considered as an approach to future work.

The trust approach can help the identification of peers in network, so it would try to address answers to questions like - how can one decide with which entities it may cooperate to reach its own objectives? To correct answer this question further research is needed because there is no specific formalism to model trust in groups and also represent the trust consensus.

Trust mechanisms can contribute to enhance the protocol in many different aspects because the trust vision can be observed as a modular scenario [111]. Considering this point of view, trust is built computing the combination of specific information (acquired trust, past situations, reputation consideration, and more). Also more problems may emerge in trust considerations thus making assumptions more difficult to evaluate and implement because of the intransitive characteristic of trust, that must be observed.

It is important to keep with further research because there is no assumption for a group trust model. This assumption, by its own characteristic, is not a trivial problem because an accepted formalism that represents trust consensus and group trust representation has not been achieved.

This problem represents new challenge in the trust research area because most of the known trust models represent the one to one (1:1) approach. But in distributed environments in many cases is necessary the one to many (1:N) or many to many (N:N) approach. In other words, it means that in distributed environments one may assume the communication process with the whole group (maybe without knowing all of its members).

Considering the same situation, the communication process may need the group to group view (N:N), what means that in distributed environments, the necessity to exchange or acquire information about groups may happen.

And without specific research in trust protocols, trust models, there is no short-time solution in distributed networks and identification schemas.

Acknowledgements

The authors thank MEC (Ministerio de Educación y Ciencia, Spain, under Project TEC2007-67129/TCM) and MITyC (Ministerio de Industria, Turismo y Comercio, Spain, under Project FIT 360000-2007-48) for their financial support.

References

1. Douceur, J.R.: The sybil attack. In: Druschel, P., Kaashoek, M.F., Rowstron, A. (eds.) IPTPS 2002. LNCS, vol. 2429, pp. 251–260. Springer, Heidelberg (2002)
2. Campadello, S.: The Green Card Protocol: an identification protocol for decentralized systems. In: Proceedings of the 2006 International Symposium on World of Wireless, Mobile and Multimedia Networks, pp. 647–651. IEEE Comp. Soc, Los Alamitos (2006)
3. Bloom, D.: A birthday problem. American Mathematical Monthly 80, 1141–1142 (1973)
4. Marsh, S.P.: Formalising Trust as a Computational Concept. PhD thesis, Dept. Comp. Science and Math., University of Stirling (1994)
5. Bellare, M., Kohno, T.: Hash Function Balance and Its Impact on Birthday Attacks. In: Cachin, C., Camenisch, J.L. (eds.) EUROCRYPT 2004. LNCS, vol. 3027, pp. 401–418. Springer, Heidelberg (2004)
6. Stallings, W.: Cryptography and Networks Security – Principles and Practices, pp. 82–83. Pearson Prentice Hall, Upper Saddle River (2006)

7. Shamir, A.: How to Share a Secret. Communications of the ACM 22(11), 612–613 (1979)
8. Gambetta, D.: Can We Trust Trust? In: Gambetta, D. (ed.) Trust: Making and Breaking Cooperative Relations, electronic edition, ch.13, pp. 213–237. Department of Sociology, University of Oxford (2000)
9. RFC 2631 - Diffie-Hellman Key Agreement Method
10. RFC 4251 - SSH Protocol Architecture
11. Patel, J.: A Trust and Reputation Model for Agent-Based Virtual Organizations. Thesis of Doctor of Philosophy. Faculty of Engineering and Applied Science. School of Electronics and Computer Science. University of Southampton (January 2007)

Non-Destructive Technique for Defect Localization in Concrete Structures Based on Ultrasonic Wave Propagation

Giuseppe Acciani, Girolamo Fornarelli, Antonio Giaquinto, Domenico Maiullari, and Gioacchino Brunetti

Politecnico di Bari, Dipartimento di Elettrotecnica ed Elettronica,
Via E. Orabona 4, 70125 Bari, Italy
{acciani,fornarelli,a.giaquinto,maiullari,
brunetti}@deemail.poliba.it

Abstract. Concrete structures require periodic inspections and quality control to assess their structural integrity. For this task several methodologies based on different technologies have been previously proposed. In particular, Non-Destructive Techniques, based on the use of ultrasonic wave propagation, have revealed attractive due to the possibility to perform reliable assessments of concrete structures. In this paper a method exploiting ultrasonic propagation characteristics is proposed. The aim of the method consists of determining the position of a defect by the computation of flight times related to signals reflected by anomalies in the structure. Such computation is based on a preliminary classification of defect positions that combines a genetic algorithm for a feature selection and a statistical approach for classification. The performances of the proposed method are evaluated in a specific case study, showing satisfactory numerical results, which show that this approach can be used to identify the position of small sized defects.

Keywords: NDT, Concrete Structure, Detection of Defect Position, Genetic algorithms, k-NN, Flight Time Computation.

1 Introduction

Concrete structures like bridges, tunnels, walls or infrastructures require periodic inspections and quality control to assess their structural integrity [1,2]. In order to track and evaluate the symptoms of operational anomalies, deterioration or damages that may compromise service quality or safety, many advanced sensing and measurements technologies have been developed and applied [3]. In addition to conventional destructive probing, Non-Destructive Techniques (NDT) have been developed with the purpose of performing easy, expressive and reliable assessment of concrete structures [2]. A large number of well known NDT methods for concrete structures, such as infrared termography, X-rays emission, ground-penetrating radar, electromagnetic induction techniques, acoustic impact method, ultrasonic pulse velocity method, ultrasonic pulse-echo method, impact-echo method have been proposed [1,3,4].

O. Gervasi et al. (Eds.): ICCSA 2008, Part II, LNCS 5073, pp. 541–554, 2008.

In last years, the methods based on the propagation of ultrasonic waveform have attracted researcher's interests due to their effectiveness in the localization of structural components and internal defects like cracks or voids [2]. In particular the pulse-echo method is a common, simple and suitable approach for field operation test. By driving a dynamic force via a piezo-electric element, elastic waves are generated and propagated in concrete structures where the stress waves are scattered, reflected, attenuated and resonated [2,5]. A receiver is used to detect the arrival of reflected waves (echoes), so that the flight time of these waves can be calculated. Wave reflections are produced by internal defects, interfaces between materials with different densities and elastic properties and boundaries of the solid. This method is used to determine the location of defects or interfaces taking into account the velocity of stress waves and measuring their flight time [1]. When the echoes have low amplitudes or noise is overlapped to the received signals, the detection of the instant in which echoes are received can be very difficult and the processing of ultrasonic waveforms needs advanced and refined techniques [5,6]. Such methods reveal effective in the evaluation of the flight time of ultrasonic signal echoes, but their application is focused on defects of large extension , like delamination or cracks.

In this paper an automatic inspection method to determine the position of defects of small dimensions inside a concrete structure is proposed. This method consists of two steps: the first step, based on a genetic algorithm and a statistical approach, provides a rough localization of defects; the second step determines a more accurate positioning of the defects starting from the information given by the first one.

The paper is organized as follows. In Section 2 an outline about the propagation of ultrasonic waves in concrete structure is given; Section 3 reports the proposed method for localizing the defects; in Section 4 experimental results and discussions are displayed.

2 Ultrasonic Wave Propagation in Concrete Structure

In order to evaluate the position of voids in a concrete structure a system based on the propagation of elastic waves in a homogeneous isotropic concrete specimen is developed.

In the considered test ultrasonic waves are generated by a transducer consisting of a piezoelectric thin layer. Such transducer is located at the centre of one specimen surface and works according to a *pulse-echo* methodology. In detail, pulses are transmitted and received on the same side of the specimen after being reflected by the opposite face and defects inside the media. Thus, a single sensor is exploited to transmit and receive signals, and the scheme of an indirect transmission method is realized. The excitation voltage applied to the piezoelectric layer is represented by a Hamming windowed sinusoid with a duration of three periods, whose analytical expression is given by:

$$
V_e(t)=\begin{cases} V_M\left\{\frac{1}{2}\left[1-\cos(\frac{2}{3}\pi ft)\right]\sin(2\pi ft)\right\} & \text{if } 0\leq t\leq \frac{3}{f} \\ 0 & \text{if } t\geq \frac{3}{f} \end{cases} \tag{1}
$$

being f the central frequency excitation and V_M the amplitude of the signal.

The governing equations for a homogeneous isotropic elastic solid in terms of displacements is represented by Navier's equations which, in absence of body forces, can be written as [7]:

$$
\rho\frac{\partial^2 \boldsymbol{u}}{\partial t^2}=(\lambda+\mu)\nabla(\nabla\cdot\boldsymbol{u})+\mu\nabla^2\boldsymbol{u} \tag{2}
$$

where:

$\boldsymbol{u}=[u_1 \quad u_2 \quad u_3]^T$ is the displacement field;
μ is the first Lamé constant;
λ is the second Lamé constant;
ρ is the density of the media.

In order to solve equations (2), Neumann's conditions at the free surfaces of the specimen have to be added. In particular, bonds on the derivative of the displacement with respect to spatial directions x_1, x_2, x_3 must be imposed. By considering that the traction vector has to vanish, boundary conditions can be written as:

$$
\boldsymbol{\sigma}\bullet\boldsymbol{n}=0 \tag{3}
$$

where $\boldsymbol{\sigma}$ is the stress tensor and $\boldsymbol{n}=[n_1 \quad n_2 \quad n_3]^T$ is the outward unit normal to the boundaries.

In a similar way, Dirichelet's conditions, regarding the displacement values, have to be imposed on each point of the contact surface between the concrete specimen and the piezoelectric emitter. Neglecting lateral deformations of the transducer, the displacements along the x_3 direction may be expressed by means of the following proportionality relationship:

$$
u_3(t)=\mathrm{d}\cdot V_e(t) \tag{4}
$$

where d is called piezoelectric constant. Transmitted elastic waves propagate in the media and provide a displacement field due to mechanical forces. This stress changes the material dimensions and an output resulting voltage is received by the sensor due to the piezoelectric effect. This voltage is proportional to the thickness changes $u_3(t)$ according to the expression:

$$
V_r(t)=\mathrm{h}\cdot u_3(t) \tag{5}
$$

where h is the strain-voltage constant. The obtained signal can be used to detect the presence and the positions of defects inside the concrete structure. To this aim, the ultrasonic waves propagation is analyzed on the horizontal plane containing the transducer axis as shown in Fig. 1. The defects inside the specimen are supposed included in this plane.

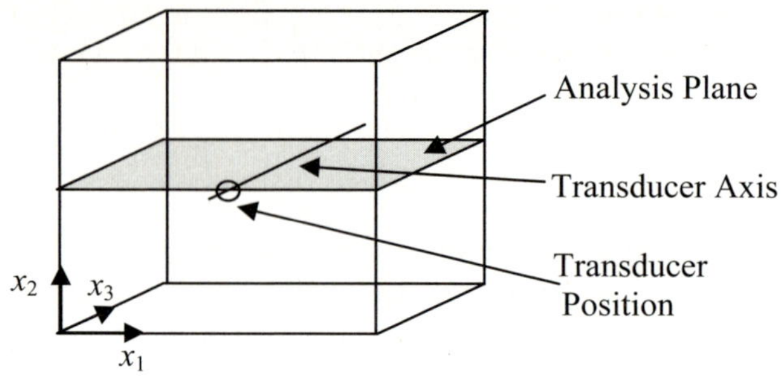

Fig. 1. Concrete specimen under test and diagnosis set up

It is worth to note that, given a particular wavefront, the displacement amplitude decreases as the distance from the transducer axis increases. When a defect is present inside the specimen, the reflected echo has different amplitude depending on the position of the defect. Therefore, there are some possible defect locations which provide echo amplitudes too low to be detected by the receiver. It follows that it is possible to identify a zone inside the media, called *radiation field*, defined by the set of defect positions that are detectable in the specimen under test. For these reasons, all the defects, considered in the present work, are placed in this area. In the bi-dimensional case the radiation field can be considered as a triangle whose upper vertex coincides with the emitter.

3 Method for the Identification of Defect Position

In this section an automatic inspection method to detect the position of defects inside a concrete structure is described. This method consists of two steps: the aim of the first step is to provide a rough localization of defects. Such information is successively exploited by the second step to supply a refinement of the positioning. In the following subsections the two steps are illustrated.

3.1 Classification of Defect Position

In order to classify a defect position, a set of 71 time and frequency parameters, successfully used in [8,9], is extracted from the waveforms received by the piezoelectric transducer with the aim of characterizing these signals. Then, the most significant features are chosen by a genetic selection algorithm [10].

Genetic algorithms (GAs) are a class of robust problem-solving techniques based on a population of possible solutions, which evolve through successive iterations by means of the application of three genetic operators: selection, crossover and mutation.

A solution is represented by a finite sequence of 0s and 1s called chromosome. In feature selection problems a chromosome has length c corresponding to the total number of features. A 0 represents a rejected feature, whereas a 1 stands for a selected one.

To start the solution process, the GA has to be initialized with a randomly generated population, which constitutes the initial solution. In order to produce the new possible solutions, two elements of a successive population are obtained by two chromosomes of a subset of the previous one. Their parts are randomly exchanged at a chosen point by means of a crossover operator. Moreover, in the chromosomes of the new population one element can be flipped according to a prefixed probability, on the base of a mechanism called mutation. Finally, a certain number of individuals is copied from the current generation to the next one with a mechanism of replication (or elitism). New populations are generated until a criterion is met; in the present case this criterion is given by a maximum number of iterations. The next generation is chosen from the new created individuals and from the previous one, according to a fitness function that allows to compute a score for each chromosome.

In the genetic feature selection, it is critical to design an appropriate fitness function to avoid local minima. Traditionally, in this kind of problems, a hit rate function is used, but this approach does not consider the selection performances which are the target to be pursued [11]. For this reason, in the present paper the following fitness function is used:

$$fitness = f(X_i) - \alpha \frac{|X_i|}{D} \tag{6}$$

where $f(\cdot)$ is the hit rate, $|X_i|$ is the number of selected features for the chromosome X_i, D is the number of all the measured features, α is a de-emphasis coefficient.

The classification performances are evaluated through the hit rate function and are estimated by a k-NN classifier. The α coefficient weights the contribution of the ratio between the used features and the number of all defined ones in the evaluation of the fitness function.

The selected number of features is obtained on the basis of the classification method. Therefore, this method achieves the feature reduction and the classification of voids position at the same time.

3.2 Defect Localization

The second step of the proposed method aims at computing a refined position of the defect. This target is performed by evaluating the time t_f that a defect echo takes to be registered by the sensor. The computation of t_f bases on the observation that two received signals reveal different energy distributions when defect is present or absent, respectively. In particular, these two signals are analyzed exploiting the information given by the method described in the previous section. This method provides the class of a defect. Each class identifies an area in which the defect is located, this implies that the defect lies in known spatial bounds. As a consequence, it is possible to compute the times t_{in} and t_{fin}, corresponding to the minimum and maximum time that a defect echo takes to be revealed by a sensor.

Let $s(t)$ and $d(t)$ be two signals acquired in the case of absence and presence of a defect, respectively. The energy difference between $s(t)$ and $d(t)$ shows maximum amplitude values corresponding to the beginning of a defect echo. On the base of this consideration, the energy functions $E(t)$ and $E_d(t)$ of the two signals are computed and the function $D(t) = |E(t) - E_d(t)|$ is studied. Function $D(t)$ is constant if $E(t)$ and $E_d(t)$ present similar trends, on the other hand $D(t)$ increases if the energy functions differ in shape. It follows that the time instants, in which the function $D(t)$ varies relevantly, are candidates to be the beginning of a defect echo sequence.

In order to single out such instants in which variations are maximum, the discrete derivative $\Delta D(t)$ of $D(t)$ is calculated and its maxima are searched. The choice of function $\Delta D(t)$ is due to the fact that the presence of noise makes the computation of instants, for which function $D(t)$ increases sharp, a hard task; in fact the selection of a threshold would be necessary. This choice could reveal difficult to be performed; on the other hand the computation of relative maxima of $\Delta D(t)$ allows to compute instants in which $E(t)$ and $E_d(t)$ most differ. Moreover, this methods minimize disturbs caused by numerical errors or noisy data.

In detail, if no interferences take place, the defect echo should present the same characteristics of the transmitted one, i.e. these two signals are identified by the same frequency and length. Consequently, the time instant t_M, for which maximum variation of $D(t)$ is given, coincides with the zero crossing placed before the maximum of the signal $d(t)$ in $[t_{in}, t_{fin}]$. The time t_f is given by the following expression:

$$t_f = t_M - 1.25 / f \, . \tag{7}$$

If the defect echo is overlapped to back wall echoes or transmitted signals, that is, if interferences present, the computation (7) of t_f constitutes an approximation of the initial instant of the defect echo. For this reason, different considerations have to be faced in order to identify t_f : in particular, the position of the time instant t_M depends on the amplitude and phase of defect echo which is overlapped to the transmitted signal or its replicas caused by board effects of the finite transmitting media.

For a sake of a better comprehension, in the following the possible cases and the corresponding method to compute t_f are described.

Case A) The defect echo is overlapped to the transmitted pulse: this condition is verified when the defect belongs to those classes which are close to the wall surface. Since the utilized sensor works in a pulse-echo way, the reception of the signal $d(t)$ can be performed after a time $t_s = 3/f$ corresponding to the length of the transmitted pulse and two subcases A_1 and A_2 could verify. The former corresponds to the situation in which the utmost part of the defect echo is overlapped to the transmitted pulse, the latter verifies if the defect echo develops in the time range $[t_s, t_{fin}]$.

Case A_1 is recognized when $M \in \left[t_s, \frac{t_s + t_{fin}}{2} \right]$, being M the absolute maximum of $\Delta D(t)$ restricted to $[t_{in}, t_{fin}]$, otherwise case A_2 is considered. If case A_1 is verified, then the time t_f is computed as:

$$t_f = t_M - 1.5 / f \, . \tag{8}$$

In this case a corrective factor modifies the computation in (7) because it is not possible to analyze the signal in $\left[t_{\mathrm{in}}, t_{\mathrm{s}}\right]$ due to the operating limits of the sensor in the receiving phase. If case A_2 holds, then this corrective factor reveals unnecessary, therefore t_{f} is given by (7).

Case B) In this case the defect echo can be separated from the back wall echo. Let v the speed of wave propagation in the media, th the thickness of the wall to be analyzed, then the transmitted signal takes $t_{\mathrm{b}} = th/v$ to be reflected by the bottom of the wall. Case B verifies if $t_{\mathrm{fin}} < t_{\mathrm{b}}$: in this case the time t_{f} is given by expression (7).

Case C) In this case the defect echo results overlapped to the back wall one. The following condition holds: $t_{\mathrm{fin}} \geq t_{\mathrm{b}}$.

Other two subcases, C_1 and C_2, can be distinguished. In case C_1 the defect echo mainly develops before the back wall one is received. On the other hand, in case C_2, it overlaps to the back wall echo. Case C_1 takes place if

$$\sum_{t_i \in \mathbf{T_L}} \Delta D(t_i) \geq \alpha \sum_{t_j \in \mathbf{T_R}} \Delta D(t_j) \qquad (9)$$

being:

$T_L = \left\{ t_i \in \left[t_{\mathrm{in}}, t_{\mathrm{b}}\right] \mid \Delta D(t_i) \text{ is a relative maximum} \right\};$

$T_R = \left\{ t_i \in \left[t_{\mathrm{b}}, t_{\mathrm{fin}}\right] \mid \Delta D(t_i) \text{ is a relative maximum} \right\};$

$\alpha \in [0,1]$.

If condition (9) is not satisfied, then case C_2 is verified. In both cases the time t_{f} is given by expression (7): in particular, $t_{\mathrm{M}} = t_{\mathrm{m}}$, where t_{m} is the instant such that $\Delta D(t_{\mathrm{m}})$ is the absolute maximum of $\Delta D(t)$ in $\left[t_{\mathrm{in}}; t_{\mathrm{b}}\right]$ or in $\left[t_{\mathrm{b}}; t_{\mathrm{fin}}\right]$ in case of C_1 or C_2, respectively.

The knowledge of the time t_{f} identifies a locus, which the bound of a defect should lie on. This locus coincides with a circumference whose radius is $r = \frac{1}{2} v t_{\mathrm{f}}$ and centre is the position of the transducer. Locus points are candidates to be a possible position for the defect. Therefore, the actual position of a defect can be obtained by intersecting two circumferences obtained by repeating the evaluation of t_{f} for two different positions of the transducer.

It should be noted that the intersection point between two circumferences are affected by errors, thus, three measures are conducted in order to enhance the accuracy in the computation of the final position. Let $P_0\left(x_{\mathrm{M}}, 0\right)$, $P_L\left(x_{\mathrm{M}} - l, 0\right)$ and $P_R\left(x_{\mathrm{M}} + l, 0\right)$ be three different points, in which the transducer is positioned. Each experiment allows to determine the time t_f of the defect echo signal, enabling to compute three circumferences γ, γ_L and γ_R whose centers are P, P_L and P_R. The intersections P_1 and P_2, between γ and γ_L and between γ and γ_R identify the position of the defect to be localized. It should be noted that P_1 and P_2 coincide if their computations are not affected by errors, but this condition is never satisfied. As a consequence, the proposed system computes the final position of the defect as the weighted sum of the coordinates of P_1 and P_2. To this purpose, after computing the class, which a defect belongs to, the intersections among γ, γ_L and γ_R could not satisfy spatial constraints of the class. Therefore, the coordinates of P_1 and P_2 are weighted by the factors $\alpha \in [0,1]$ and $1-\alpha$. If geometrical constraints are satisfied by the only point P_1, then $\alpha > 0.5$, whereas $\alpha < 0.5$ if constraints are satisfied by the only point P_2. In the following section experimental results are reported in order to show the performances of the proposed method.

4 Experimental Results and Discussion

The case study has been simulated by a commercial software tool based on the Finite Element Method (FEM). The simulations have been carried out considering an excitation frequency $f = 70\ \text{kHz}$, an analysis time $T = 4\times10^{-4}\ \text{s}$, a quad mapped mesh with dimension sides $m = 0.6\ \text{cm}$. Such value of m corresponds to 1/10 of the wavelength at the central frequency, in order to guarantee the numerical accuracy provided by the solution algorithm. The geometric and physical parameters of the considered specimen are summarized in Table 1.

The simulated piezoelectric layer is Piezo Zirconate Titanate (PZT 27), commonly employed in ultrasonic techniques having diameter $0.05\ \text{m}$, piezoelectric constant $d = 425\times10^{-12}\ \text{m/V}$, strain-voltage constant $h = 1.46\times10^{9}\ \text{V/m}$.

In order to test the proposed method for defect localization, 10 classes of defects with different positions inside the radiation field and 9 square voids (side 2 cm) for each class have been considered, as shown in Fig. 2.

Given a wavefront Γ, its centre C is aligned to the transducer axis. The displacement amplitude of a generic point P belonging to Γ reduces as the distance between P and C increases. Points, for which the displacement amplitude reduces to the 80% of the amplitude in C, represent bounds of the radiation field and are denoted by the oblique lines in Fig. 2.

A database of 90 simulated ultrasonic waveforms (echoes reflected by each defect) has been obtained with the specimen and the configuration analysis previously described. The collected signal set has been contaminated and augmented by additive white Gaussian noise, obtaining four noisy waveforms for each signal. The extended data set of 360 waveforms has been randomly divided into training subset (90%) and test subset (10%) with the aim to examine the robustness of the automatic interpretation scheme. According to the above described method, 71 time and frequency domain parameters for each waveform of the database have been evaluated. Then, the genetic feature selection based on k-NN algorithm has been used to select the optimal

Table 1. Geometric and physic characteristics of concrete specimen

Parameter	Value
Height (x_1)	1.0 m
Width (x_2)	1.0 m
Depth (x_3)	0.3 m
Young's Modulus	$E = 25\times10^9$ Pa
Density	$\rho = 2300$ kg/m^3
Poisson's ratio	$\nu = 0.33$
First Lamé constant	$\mu = 9.398496\times10^9$ Pa
Second Lamé constant	$\lambda = 1.824414\times10^{10}$ Pa

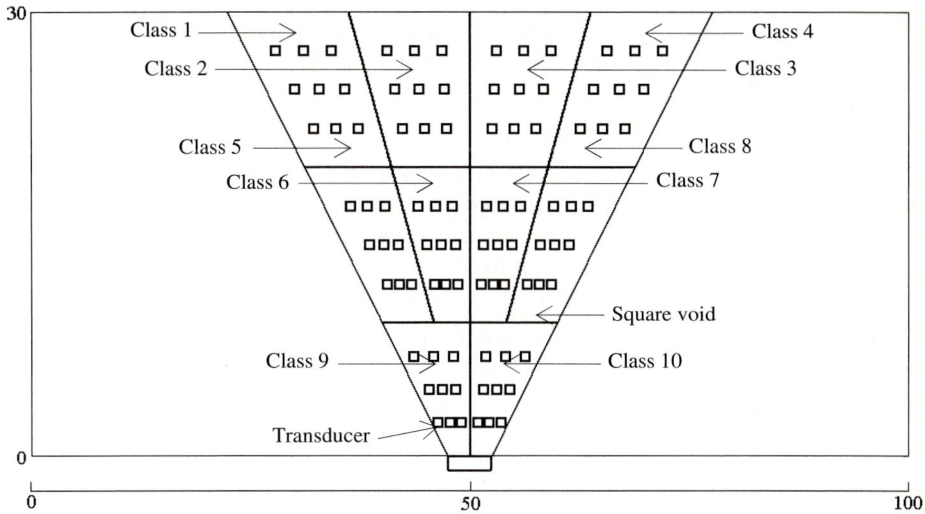

Fig. 2. Classes of defect in the radiation field

feature set for voids position classification. The performances of k-NN have been evaluated in terms of Mean Classification Error (MCE), calculated as the mean of the errors in the classification of the test set waveforms.

Some preliminary experiments have been carried out to try out an optimal set of operators and parameters of the GA. On the basis of the experience gained from these explorations, the parameters have been set as shown in the Table 2.

Using the reported parameters, the GA method and the number of neighbours k = 3 for the k-NN, 7 features have been selected by the initial 71:

Difference between 50% level and 25% level (spectrum CD);
Difference between 75% level and 25% level (spectrum CD);
Difference between 90% level and 25% level (spectrum CD);
Global rise frequency between 25% level and peak of spectrum;

Table 2. Description of genetic algorithm parameters

GA parameters settings	Value or type
Initial population choice	Random
Population size	100
Number of generations	100
Selection criterion	Roulette wheel
Elitism (no. of unchanged individuals)	10
Crossover (no. of selected individuals)	50
Crossover rule	Single point crossover
Crossover probability	1
Mutation probability	0.02
Termination criterion	Maximum no. of generations

Global rise frequency between 50% level and peak of spectrum;
Global fall frequency between peak of spectrum and 50% level;
Global rise variance between 25% level and peak of spectrum;
(CD = cumulative distribution)

This setting allows to achieve a classification rate of 98%. This result shows that the adopted approach provides a low error rate and a reliable method for the initial identification of voids position. The accurate value of the position is, then, evaluated in the second step of the proposed approach.

In Fig. 3 the trends of a signal $d(t)$ in the presence of a defect and the reference signal $s(t)$ in absence of defects are shown. It can be noted that signals present two main contributions: a signal corresponding to the transmitted pulse in the range $[0,t_s]$, being $t_s = 4.28\times10^{-5}$ s , and a back wall echo in the range $[t_b, t_b + 3/f]$, with $t_s = 1.49\times10^{-4}$ s . Other contributions are due to multiple reflections phenomena. In addition to these contributions, the signal $d(t)$ presents a spurious contribution between the transmitted pulse and the back wall echo. Such anomaly reveals the presence of a defect in the transmitting media.

Given the reported signals $d(t)$ and $s(t)$, their corresponding energies $E_d(t)$ and $E(t)$ are computed and shown in Fig. 4a (solid and dashed lines represent $E_d(t)$ and $E(t)$, respectively). Figs 4b and 4c report the functions $D(t) = |E(t) - E_d(t)|$ and its finite derivative $\Delta D(t)$. In Fig. 4c the vertical bars correspond to t_{in} and t_{fin} , respectively. It can be observed that in the reported experiment case B verifies. Moreover, experimental results confirm that, due to its strongly nonlinear shape, $D(t)$ makes the selection of a threshold for revealing time instants of maximum variations a hard task. For this reason, maxima of the finite derivative $\Delta D(t)$ are searched. In detail, its absolute maximum in the range $[t_{in}, t_{fin}]$ and its corresponding abscissa t_M are determined, then the time t_f is computed by expression (7), as reported in the previous

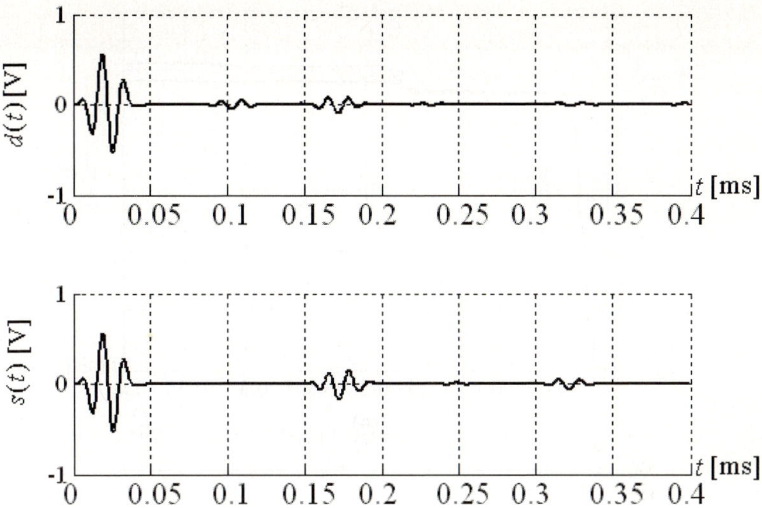

Fig. 3. Signal received in the presence of a defect and reference signal

section. It is worth to observe that numerical solutions are searched by considering the geometrical constraints deriving from the knowledge about the class of the defect.

Successively, the procedure is repeated by positioning the sensor at $P_0(x_M, 0)$, $P_L(x_M - l, 0)$ and $P_R(x_M + l, 0)$, being $x_M = 50$ cm and $l = 5$ cm. In this way, three times t_0, t_L and t_R are computed and three circumferences γ, γ_L and γ_R are determined. In Fig. 5 the loci γ (solid line), γ_L (dotted line) and γ_R (dotted/pointed line) are shown. In the same figure the defect to be detected is reported: the square corresponds to its bounds and the asterisk represents its centre. The intersections P_1 and P_2 between γ and the remaining circumferences are then calculated. Lastly, the final position of the defect is given by the barycentre of P_1 and P_2.

In order to quantify the performances of the proposed solution, the Mean Distance (MD) index has been computed as:

$$\text{MD} = \frac{1}{n}\sum_{k=1}^{n} \text{d}(P_k, D_k) \tag{10}$$

being P_k the position of the defect given by the proposed system, D_k the medium point of the defect side nearest to the sensor, $\text{d}(P_k, D_k)$ the Euclidean distance between P_k and D_k, and n the number of the elements in a specified class. The index MD represents the mean error, affecting the result, when evaluating positions of defects belonging to the same class.

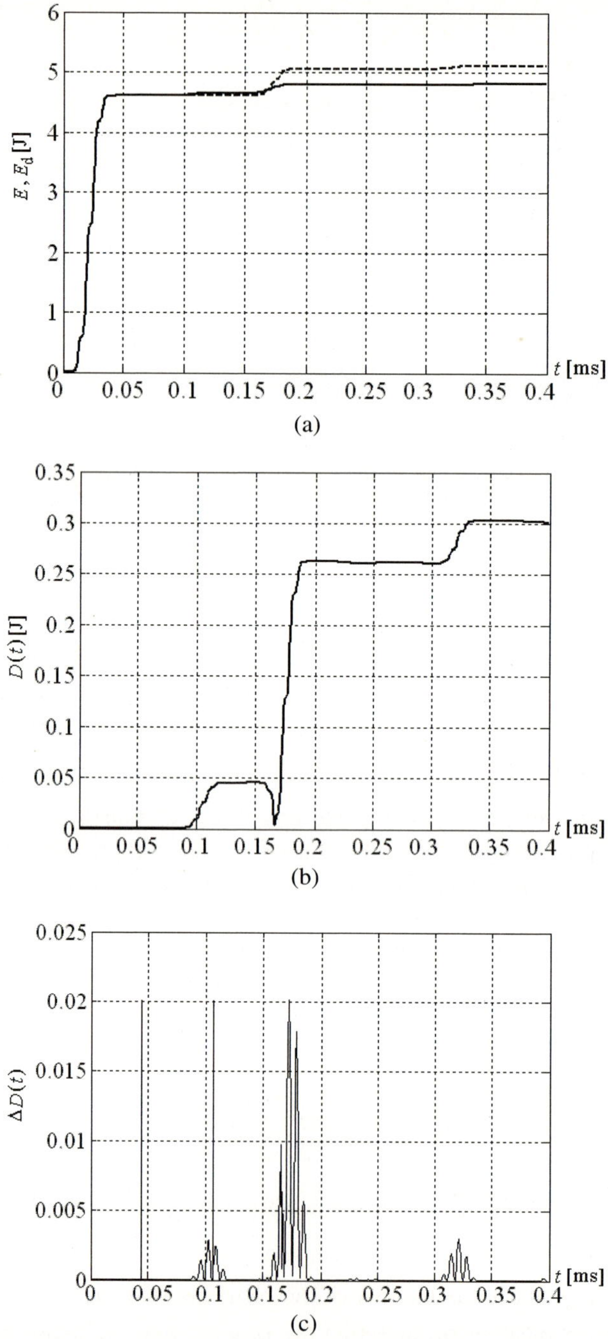

Fig. 4. (a) Functions $E(t)$ (dotted line) and $E_d(t)$ (solid line) (b) function $D(t)$ and (c) function $\Delta D(t)$

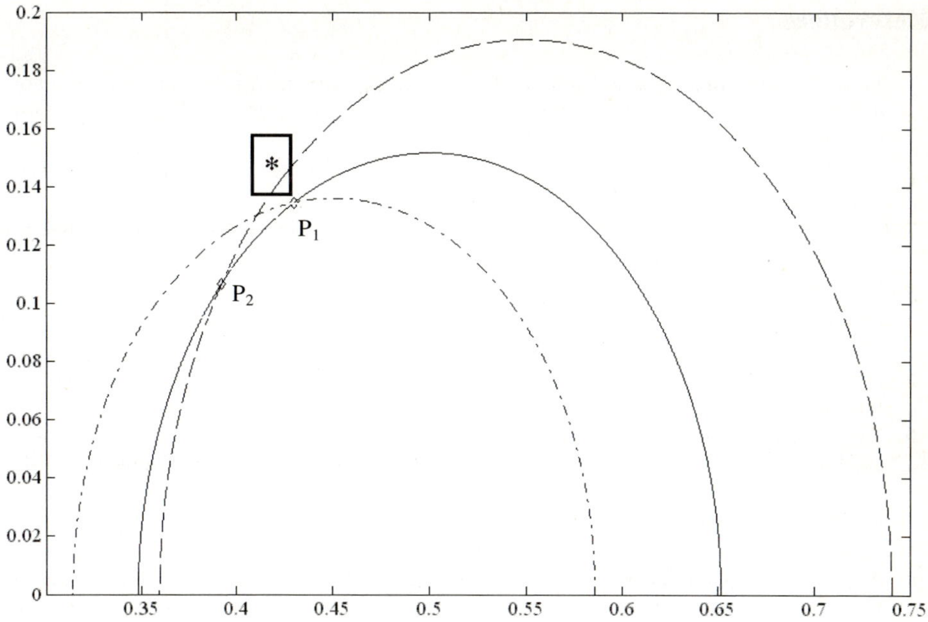

Fig. 5. Circumferences γ (solid line), γ_L (dotted line) and γ_R (dotted/pointed line)

Table 3 reports the values of MD with regard to the defects partitioned into the classes indicated in Fig.2.

Table 3. MD indices computed for the considered classes

	MD
Classes 1 and 4	3.21 cm
Classes 2 and 3	1.60 cm
Classes 5 and 8	1.76 cm
Classes 6 and 7	0.68 cm
Classes 9 and 10	1.66 cm
Average	1.78 cm

It can be noted that the MD values are common to two classes, because both the location of defects and the experimental scheme are symmetric.

Results are encouraging: an average error of 1.78 cm is achieved. This implies that the measure of a position is affected by a mean relative error equal to 5.93% of its dimension with a thickness $d = 30 \text{ cm}$. Moreover, the method has better performances when the central classes of the radiation field (classes 2, 3, 6, 7, 9 and 10) are considered: in this case the average error decreases to 1.31 cm, corresponding to 4.37% of the wall thickness.

References

1. Mori, K., Spagnoli, A., Murakami, Y., Kondo, G., Torigoe, I.: A new non-contacting non-destructive testing method for defect detection in concrete. NDT & E International 35(6), 399–406 (2002)
2. Schickert, M.: Ultrasonic Nde Of Concrete. In: IEEE Proceedings, IEEE Ultrasonics Symposium, 8-11, October 2002, vol. 1, pp. 739–748 (2002)
3. Xiang, S.K.T.: Detection and classification of flaws in concrete structure using bispectra and neural networks. NDT & E International 35(1), 19–27 (2002)
4. Berriman, J.R., Hutchins, D.A., Neild, A., Gan, T.H., Purnell, P.: The application of time-frequency analysis to the air-coupled ultrasonic testing of concrete. IEEE Transactions on Ultrasonics, Ferroelectrics and Frequency Control 53(4), 768–776 (2006)
5. Ohtsu, M., Watanabe, T.: Stack imaging of spectral amplitudes based on impact-echo for flaw detection. NDT & E International 35(3), 189–196 (2002)
6. Fan, X., Zuo, M.J., Wang, X.: Application of stationary wavelet transforms to ultrasonic crack detection. In: Canadian Conference on Electrical and Computer Engineering, May 2006, pp. 1432–1436 (2006)
7. Rose, J.L.: Ultrasonic Waves in Solid Media. Cambridge University Press, Cambridge (1999)
8. Acciani, G., Brunetti, G., Chiarantoni, E., Fornarelli, G.: Automatic Evaluation of Flaws in Pipes by means of Ultrasonic Waveforms and Neural Networks. In: 2006 International Joint Conference on Neural Networks Sheraton Vancouver Wall Centre Hotel, Vancouver, BC, Canada, 16-21 July (2006)
9. Case, T.J., Waag, R.C.: Flaw Identification from Time and Frequency Features of Ultrasonic Waveforms. IEEE Transactions on Ultrasonics, and Frequency Control 43(4), 592–600 (1996)
10. Muni, D.P., Pal, N.R., Das, J.: Genetic programming for simultaneous feature selection and classifier design. IEEE Transactions on Systems, Man and Cybernetics, Part B 36, 106–117 (2006)
11. Salcedo-Sanz, S., Camps-Valls, G., Perez-Cruz, F., Sepulveda-Sanchis, J., Bousono-Calzon, C.: Enhancing genetic feature selection through restricted search and walsh analysis. IEEE Transactions on Systems, Man and Cybernetics, Part C 34(4), 398–440 (2004)

Development of a Finite Element Code for Non-Destructive Evaluation of Concrete Structures by Means of Ultrasonic Waves

G. Aiello, E. Dilettoso, and N. Salerno

Dipartimento di Ingegneria Elettrica Elettronica e dei Sistemi
Università degli Studi di Catania
Viale Andrea Doria 6, I-95125 Catania, Italy
gaiello@diees.unict.it, edilettoso@diees.unict.it,
nsalerno@diees.unict.it

Abstract. Non-Destructive Evaluation techniques are of great relevance nowadays and are extensively employed in many areas of engineering to detect defects in products and structures, or to point out deteriorations in industrial plants. This paper focuses on techniques which make use of transient ultrasonic waves, the aim being numerical modelling and simulation of their propagation in concrete structures affected by defects, possibly in the presence of material inhomogeneities and/or voids. For this purpose a fully three-dimensional finite element code has been developed by the authors, who have validated and used it to perform numerous simulations, the results of which are presented and discussed.

1 Introduction

For several years, Non-Destructive Evaluation (NDE) techniques have acquired more and more relevance, as is witnessed by the progressive successful widening of their field of application and by the increase in the accuracy and flexibility they offer. Nowadays they are routinely and extensively employed in a great number of different areas of engineering and science to detect and characterize defects or corrosion phenomena in products and structures, to monitor the degree of safety of critical parts of industrial plants, to evaluate the state of conservation of the historical and artistic heritage, and for many other uses [1], [2], [3], [4]. In particular, in the field of protection of the architectural heritage there exists a growing activity aiming to search for, develop and use evaluation techniques for its preservation, featuring an increasing measure of accuracy and result reliability. Another requirement, also held in great consideration, is to reduce, if not to cancel out completely, the invasive nature of such techniques. In fact, the numerous NDE techniques available today, and more specifically those based on the use of electromagnetic and ultrasonic waves, are acquiring greater importance. As is to be expected, the two types of techniques present advantages and disadvantages, which render them to some extent complementary, as concerns both their sphere of use and the type of information obtainable. This paper deals with ultrasonic-wave NDE techniques which are applied to detect defects, such as voids and/or material irregularities, in concrete structures which are part, for example,

O. Gervasi et al. (Eds.): ICCSA 2008, Part II, LNCS 5073, pp. 555–565, 2008.

of the architectural heritage. Specifically, the main outcome of the present research consists of the development of a fully three-dimensional finite element code which was used to model and simulate the propagation of ultrasonic waves in such structures, and to present and discuss the results obtained by using the code. The computations were performed by assuming as an unknown the displacement vector field inside the structure, and a validation of the code has also been given.

It is worth pointing out that, due to the nature of the problem being examined, to model the concrete structures correctly the code must be able both to solve the equations in the presence of material inhomogeneities and/or geometrical disuniformities, of a size comparable to or greater than the wavelengths of the ultrasonic waves employed, and to describe accurately their transient propagation through a variety of solid materials of which the structures to be investigated are made [5], [6]. The code thus developed was then inserted into a wider finite element code, called ELFIN, entirely developed by the authors and their colleagues at the University of Catania for research aims only [7].

The paper is organized as follows. Sections 2 and 3 describe the mathematical and numerical formulation used to develop the code; in Section 4 a validation of the code is given and some results of the simulations performed are presented and discussed in detail; finally, in Section 5 the authors' conclusions are given.

2 Mathematical Formulation

According to the relevant literature [8], [9], [10], we study the propagation of ultrasonic waves in solids by means of the displacement vector **u.** In an isotropic homogeneous medium, under the standard assumptions of linear elasto-dynamics theory and the absence of viscosity, the governing equation is:

$$(\lambda+\mu)\nabla\nabla\cdot\mathbf{u}+\mu\Delta\mathbf{u}+\mathbf{f}_v=\rho\ddot{\mathbf{u}} \tag{1}$$

where $\mathbf{f}_v$ is the body force, λ and μ are the Lamé constants, ρ is the volume mass density and the dots denote the second-order time derivative.

Equation (1) has to be solved in a bounded domain Ω with boundary $\partial\Omega = \partial\Omega_g \cap \partial\Omega_m$ on which an assigned displacement $\mathbf{u}|_{\partial\Omega g}=\mathbf{u_g}$ (Dirichlet or geometrical condition) and/or an assigned traction $\mathbf{f}|_{\partial\Omega m}=\mathbf{f_m}$ (Newmann type or mechanical boundary condition) have to be imposed; the initial conditions $\mathbf{u}_0=\mathbf{u}(P,0)$ $\mathring{\mathbf{u}}_0=\mathring{\mathbf{u}}(P,0)$ must also be specified in each point P of Ω.

Making use of **u** it is then possible to compute the strain tensor ξ_{ij} i,j=1,2,3 in rectangular Cartesian coordinates as [8]:

$$\xi_{ij}=\frac{1}{2}\left(\frac{\partial u_i}{\partial x_j}+\frac{\partial u_j}{\partial x_i}\right) \tag{2}$$

Equation (1) can be derived by imposing a vanishing first-order variation $\delta^{(1)}F$ of the following functional F(**u**,t):

$$F(\mathbf{u},t) = \int_{\Omega} \left[-w_e + (-\rho\ddot{\mathbf{u}} + \mathbf{f}_v)\cdot\mathbf{u} \right] + \int_{\partial\Omega_m} \mathbf{f}_m \cdot \mathbf{u}, \tag{3}$$

where $w_e = \frac{1}{2}\sum_{i,j=1}^{3}\left(\lambda\xi_{ii}\,\xi_{jj} + 2\mu\xi_{ij}^2\right)$ is the total strain energy of the body [11]. Starting from (3) it is straightforward to write $\delta^{(1)}F$ as:

$$\delta^{(1)}F = \int_{\Omega} \left[-\delta w_e + (-\rho\ddot{\mathbf{u}} + \mathbf{f}_v)\cdot\boldsymbol{\delta}\mathbf{u} \right] + \int_{\partial\Omega_m} \mathbf{f}_m \cdot \boldsymbol{\delta}\mathbf{u} \tag{4}$$

Note that in deriving (4) the variation $\boldsymbol{\delta}\mathbf{u}(P)$ of $\mathbf{u}$ has to be considered as time-independent. Then, by means of (2) it is easy to deduce the following expression for δw_e, which will be used in the next section

$$\delta w_e = \sum_{i,j=1}^{3}\left[\lambda\frac{\partial u_j}{\partial x_j}\frac{\partial\delta u_i}{\partial x_i} + \mu\left(\frac{\partial u_j}{\partial x_i}\frac{\partial\delta u_j}{\partial x_i} + \frac{\partial u_j}{\partial x_i}\frac{\partial\delta u_i}{\partial x_j}\right)\right]. \tag{5}$$

In fact, as

$$\delta w_e = \sum_{i,j=1}^{3}\left(\lambda\xi_{jj}\,\delta\xi_{ii} + 2\,\mu\,\xi_{ij}\,\delta\xi_{ij}\right) \tag{6}$$

And

$$\xi_{jj}\,\delta\xi_{ii} = \frac{\partial u_j}{\partial x_j}\frac{\partial\delta u_i}{\partial x_i} \tag{7a}$$

$$\sum_{i,j=1}^{3}\xi_{ij}\,\delta\xi_{ij} = \frac{1}{2}\sum_{i,j=1}^{3}\left(\frac{\partial u_j}{\partial x_i}\frac{\partial\delta u_j}{\partial x_i} + \frac{\partial u_j}{\partial x_i}\frac{\partial\delta u_i}{\partial x_j}\right) \tag{7b}$$

through some algebraic manipulations expression (5) is deduced. Of course, making use of standard calculus tools, expression (4) can be recast in such a way as to prove that the stationary of (4) occurs when the solution of the boundary/initial value problem (1) is considered [11].

3 Numerical Formulation

In order to solve the boundary/initial value problem (1) numerically, we adopt a finite-element semi-discre-tization procedure in which the Raylegh-Ritz variatio-nal method is used to discretize the problem spatially with respect to the Cartesian coordinates x_1, x_2 and x_3, whereas a direct integration method is used to carry out the time-history analysis [12], [13], [14]. Making use of this approach, a system of linear second-order coupled differential equations of the form

$$\mathbf{M}\ddot{\mathbf{u}}_n + \mathbf{K}\mathbf{u}_n = \mathbf{R}_n \tag{8}$$

is derived, where $\mathbf{u}_n=\mathbf{u}(n\Delta t)$ and $\ddot{\mathbf{u}}_n=\ddot{\mathbf{u}}(n\Delta t)$ are the vectors containing the (unknown) nodal values of the displacement vector **u** and the acceleration vector $\ddot{\mathbf{u}}$, respectively; **M** and **K** are two square matrices, the former called the consistent mass matrix and the latter the stiffness matrix; $\mathbf{R}_n$ is the vector of known terms, its entries coming from both the boundary/initial conditions and the body (weight) force (however, in the present case, the contributions to $\mathbf{R}_n$ arising from the body force and the mechanical boundary conditions were discarded, as the influence of the former is quite negligible in the current context, and the latter does not concern the analysis being performed; in our formulation, therefore, $\mathbf{R}_n$ is only due to the inhomogeneous Dirichlet boundary/initial conditions and it is built by rearranging the known terms in the space-discretized equations by the standard per-element FEM assembly procedure); the integer n specifies the number of time iterations; and, finally, Δt is the size of the time step of the time integration scheme adopted, details of which will be given below. Let us now consider the criteria we adopted to number the equations and the unknowns of (8). Since the cardinality of the vectors, as well as the order of the matrices present in (8) is 3N, N being the number of nodes in the finite-element mesh with unknown displacement, each index h of a row in these arrays and each index k of a column (if any) can be written as $h=3(p-1)+i$ and $k=3(q-1)+j$, where p and q specify a pair of nodes and i and j a pair of components, respectively: for example, the entry $(\mathbf{M})_{hk}$ stands for the coefficient of the j-th component of the (unknown) displacement vector at the node q in the equation of the i-th component of the (unknown) displacement vector at node p As regards the expression of the entries of **M** and **K**, the first step in order to derive them is to expand the components of **u**, $\ddot{\mathbf{u}}$ and $\delta\mathbf{u}$ in terms of scalar shape functions α_p:

$$u_j = \sum_{q=1}^{N'} (\mathbf{u})_k \, \alpha_q \tag{9a}$$

$$\ddot{u}_j = \sum_{q=1}^{N'} (\ddot{\mathbf{u}})_k \, \alpha_q \tag{9b}$$

$$\delta u_i = \sum_{p=1}^{N'} (\delta\mathbf{u})_h \, \alpha_p \tag{9c}$$

$$\delta u_j = \sum_{p=1}^{N'} (\delta\mathbf{u})_l \, \alpha_p \tag{9d}$$

N' being the total number of nodes and $l=3(p-1)+j$, and then to introduce the expansions (9a), (9c) and (9d) into expression (5). However, in this way, in the fourfold sum over the indices i, j, p and q thus obtained both $(\delta\mathbf{U})_h$ and $(\delta\mathbf{U})_l$ appear. In order to transform it in such a way that only $(\delta\mathbf{U})_h$ appears, in the part of this sum containing $(\delta\mathbf{U})_l$, which comes from the expansion of the first term on the right-hand side of

(7b), it is necessary to exchange the indices i→ j and j→i; in fact, by so doing the index l transforms to h, and $(\boldsymbol{\delta}\mathbf{U})_l$ becomes $(\boldsymbol{\delta}\mathbf{U})_h$; note that by this exchange the index k takes the new value m=3(q-1)+i and thus some terms containing $(\mathbf{U})_m$ appear in the sum. After this, the following expression is deduced:

$$\delta w_e = \sum_{i=1}^{3}\sum_{p=1}^{N'}\left[\sum_{j=1}^{3}\sum_{q=1}^{N'}\left(\lambda\frac{\partial\alpha_p}{\partial x_i}\frac{\partial\alpha_q}{\partial x_j}+\mu\frac{\partial\alpha_p}{\partial x_j}\frac{\partial\alpha_q}{\partial x_i}\right)(\mathbf{U})_k\right](\delta\mathbf{U})_h \tag{10}$$

$$+\sum_{i=1}^{3}\sum_{p=1}^{N'}\left[\mu\sum_{q=1}^{N'}(\mathbf{U})_m\nabla\alpha_p\cdot\nabla\alpha_q\right](\delta\mathbf{U})_h \tag{10}$$

By imposing the stationary of functional (4) and taking into account the arbitrariness of $(\boldsymbol{\delta}\mathbf{U})_h$, one deduces from inspection of (10) an expression for the entries of **K**

$$(\mathbf{K})_{hk} = \int_\Omega\left[(\lambda+\mu)\frac{\partial\alpha_p}{\partial x_i}\frac{\partial\alpha_q}{\partial x_j}+\mu\nabla\alpha_p\cdot\nabla\alpha_q\right] \qquad i=j \tag{11a}$$

$$(\mathbf{K})_{hk} = \int_\Omega\left(\lambda\frac{\partial\alpha_p}{\partial x_j}\frac{\partial\alpha_q}{\partial x_i}+\mu\frac{\partial\alpha_p}{\partial x_i}\frac{\partial\alpha_q}{\partial x_j}\right) \qquad i\neq j \tag{11b}$$

In a similar way, discretizing the dynamic term $-\rho\ddot{\mathbf{u}}$ in (4) by means of the expansion (9b), the expression for **M** is deduced:

$$(\mathbf{M})_{hk} = \int_\Omega\rho\,\alpha_p\,\alpha_q \quad i=j,\ (\mathbf{M})_{hk}=0\ i\neq j \tag{12}$$

note that in the last two expressions the nodes p and q have be considered as belonging to the same finite element; moreover, by inspection it results from (11a), (11b) and (12) that the matrices **K** and **M** are both symmetric.

In the literature there are a number of methods to carry out the time integration of equation (8), and after careful examination we decided to use the standard central-difference method, for two main reasons [12]. Firstly, in the absence of dissipative effects, leading to the appearance of a first-order time derivative term in equation (8), and thus of a damping matrix, the method can be made explicit by diagonalizing the consistent mass matrix **M** by means of an effective and accurate lumping scheme, such as the HRZ one [12]; in this way, the disadvantage of the method being conditionally stable, thus requiring a sufficiently small time step to guarantee the numerical stability of the procedure, is widely compensated for by the conspicuous decrease in the computational effort needed to compute **u**, since it is no longer necessary in such a case to solve a linear algebraic system of a high order at each time iteration step. Secondly, the method is robust, second-order accurate and easy to implement.

Applying the central-difference method to equation (8), the following linear algebraic system is obtained [12]:

$$\mathbf{M}\mathbf{u}_{n+1}=(\Delta t)^2\mathbf{R}_n-(\Delta t)^2\mathbf{K}\mathbf{u}_n+\mathbf{M}(2\mathbf{u}_n-\mathbf{u}_{n-1}) \tag{13}$$

Then, making use of the HRZ mass lumping technique, a diagonal matrix **M'** is derived; its entries are computed by scaling those of **M** lying on its main diagonal by means of a per-element algorithm which preserves the global mass of the element [12]. After this, substituting **M** with **M'** in (13), and solving with respect to $\mathbf{u}_{n+1}$, the following relationship is deduced

$$\mathbf{u}_{n+1} = \Delta t^2 \left(\mathbf{M}'\right)^{-1} \mathbf{R}_n - \Delta t^2 \left(\mathbf{M}'\right)^{-1} \mathbf{K}\mathbf{u}_n + 2\mathbf{u}_n - \mathbf{u}_{n-1} \tag{14}$$

which was used to compute iteratively $\mathbf{u}_{n+1}$. However, the computation of $\mathbf{u}_1$ from (14) requires knowledge of the starting value $\mathbf{u}_{-1}$ which is related to the initial values $\mathbf{u}_0$, $\mathring{\mathbf{u}}_0$ and $\ddot{\mathbf{u}}_0$ by the following second-order approximate power series expansion

$$\mathbf{u}_{-1} = \mathbf{u}_0 - (\Delta t)\dot{\mathbf{u}}_0 + \frac{1}{2}(\Delta t)^2 \ddot{\mathbf{u}}_0 \tag{15}$$

By assuming $\mathbf{u}_0$=0 and $\mathbf{R}_0$=0 it results from (8) that $\ddot{\mathbf{u}}_0$ is zero, so that having assigned $\mathring{\mathbf{u}}_0$=0 in our computations, it follows that $\mathbf{u}_{-1}$ is also zero; as a consequence of this, $\mathbf{u}_1$ vanishes.

The stability of the central-difference method requires the following inequality to hold

$$\Delta t \le \frac{2}{\omega_{max}} \tag{16}$$

where ω_{max} is the highest natural frequency of the equation det(**K**-ω^2**M'**)=0.

In order to avoid the drawback of the great effort required by the direct numerical computation of ω_{max}, the upper bound of ω_{max} is commonly evaluated by making use of Gerschgori's theorem [12]. This criterion, for lumped mass matrices **M'**, states that

$$\omega_{max}^2 \le \max\left(k_{ii} + \sum_{j=1,\, j\ne i}^{n_e} \left|k_{ij}\right| \right) \backslash m'_{ii} \tag{17}$$

where i=1, 2, …, n_e, and n_e is the number of degrees of freedom per element.

4 Description of the Results

By using the formulation based on the finite element method (FEM) described in the previous section, we worked out a numerical code for analysis of the propagation of transient ultrasonic waves in elastic structures. Actually, this code is only one module of the much larger and more powerful FEM research code ELFIN, which has been developed by the authors and their colleagues at the University of Catania over the last twenty years for the computation of electromagnetic fields in many areas of electrical engineering [7].

In order to validate our code, a comparison between the results obtained by means of ELFIN with analogous ones derived by making use of the commercial FEM code ANSYS was made, and it will be shortly described in this Section. The structure taken into consideration was a concrete wall modelled geometrically as a homogeneous parallelepiped with thickness t = 0,25 m, length l = 6,00 m and height h = 6,00 m; the material parameters of interest have the following values: modulus of elasticity E = 22,00 GPa, Poisson's ratio $\nu = 0{,}15$ and volume density $\rho = 2400{,}00$ kg·/m^3; for the sake of completeness, the values of Lame's constant λ and μ are also given: $\lambda = 9{,}560$ GPa and $\mu = 4{,}099$ GPa.

The numerical results presented here were obtained by means of the ELFIN code and, for reasons of brevity, only one kind of defect consisting of a parallelepipedal shaped void is considered here. However, in order to analyze how the geometrical features of the defect influence the time variation of the components of the displacement vector **u**, in short hereafter termed response, an extensive numerical investigation intended to highlight if and how such a response depends on the size, orientation and position of the defect was carried out. To this ends, three different shapes of defect, referred to below as D1, D2 and D3, were examined and two different locations of their centres were taken into consideration.

In describing the results of the computations carried out, we will be referring to the Cartesian reference frame shown in Figure 1, which has the coordinate planes parallel to the surfaces of the wall and its origin coinciding with its geometrical centre. The sizes (in meters) of the defects D1, D2 and D3 along the axes x, y, and z are (0.25,0.25,0.15), (0.25,0.75,0.15) and (0.75,0.25,0.15) respectively. In the two series of analyses performed, the centres P1 and P2 of the defects and the point O where the response was observed (observation point) are located on the plane z =0, and their coordinates are, in order, (0.00,1.50, 0.00), (0.00, 2.25, 0.00) and (0.00, -1.50,0.00).

In the computations performed by means of ELFIN the structure was discretized by a regular first-order tetrahedral mesh, the number of subdivisions in the x, y and z directions being 48, 120 and 5 respectively; we need a finer discretization in the propagation direction (y axis) in order to obtain a better accuracy. The simulations

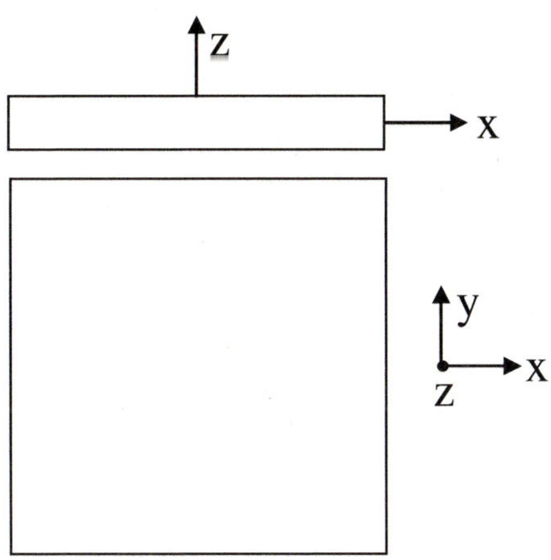

Fig. 1. Wall geometry and reference frame

performed demonstrated that this domain discretization guarantees an adequate resolution of the problem being investigated and a suitable relationship between the spatial and temporal steps. The mesh obtained contains 35.574 nodes and 144.000 finite elements.

The analyses of the structure were performed by setting the boundary conditions as follows: on the surfaces of the wall belonging to the planes of equations x=-3 and x=3 a homogeneous Dirichlet boundary condition was imposed on the y and z components of vector **u**, whereas at the points of the wall lying on the plane of equation y=-3 an inhomogeneous Dirichlet boundary condition was prescribed for the component x of **u**. Specifically, the former boundary condition, simulating the ultrasonic waves exciting the wall, assigns to its nodes belonging to the above plane an equal displacement along the x axis; the time dependence of this component was a sinusoidal signal modulated in amplitude by a Hanning window of six oscillations with the following waveform:

$$f(t) = A\sin(2\pi f t)\cdot\sin^2(2\pi f t/2n) \qquad 0 \le t \le t_0 \qquad (18)$$

where the amplitude A is 1 mm, the modulation frequency f is 2 kHz and the integer n is 3. The overall duration of the signal t_0 is 1,5 ms; outside this time interval the waveform is assumed to be identically zero and is plotted in Figure 2.

Taking into account the 1746 nodes with an assigned Dirichlet boundary condition, the global number of degrees of freedom was 104.976

As regards the simulations performed by means of the ANSYS code, the spatial and time discretizations employed were exactly the same as that just described, except for the finite elements adopted, which were eight-node bricks.

A crucial parameter of the simulations was the minimal time step required by the stability of the explicit numerical integration scheme. The upper bound of the time step computed according to (16) was 25 μs, the time step employed in the computations is 5 μs and the analyses were performed taking into consideration an overall duration of 30 ms.

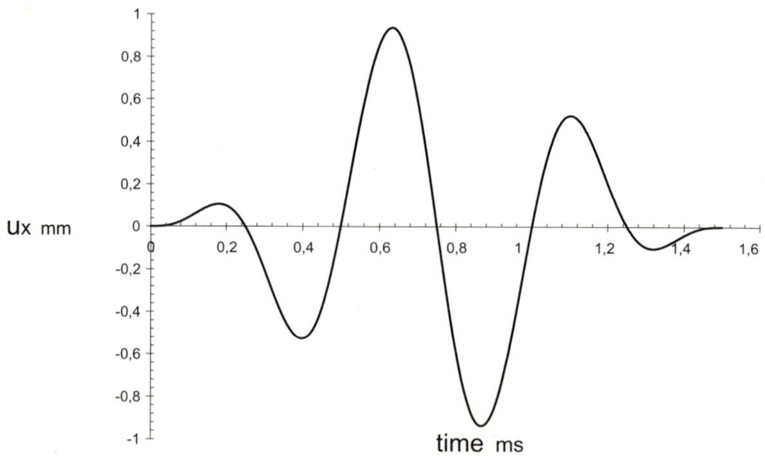

Fig. 2. Plot of the function f(t)

Figure 3 shows the x component of the displacement vector at the observation point O versus time without the presence of the defect: the dotted line refers to the analysis carried out by means of the ANSYS code, whereas the continuous line refers to that performed by means of the ELFIN code. By inspection of this it is quite evident that the two simulations are in good agreement with each other, so it can reasonably be stated that ELFIN was successfully validated.

In both plots in Figure 3 the ultrasonic wave generated by the device (direct wave) is followed after 5.25 ms by another clearly recognizable wave of identical amplitude produced by the reflection of the direct wave by the opposite end of the wall; it is worth noticing that, as expected, the amplitude of the reflected wave is equal to that of the direct one owing to the lack of dissipative phenomena in the model.

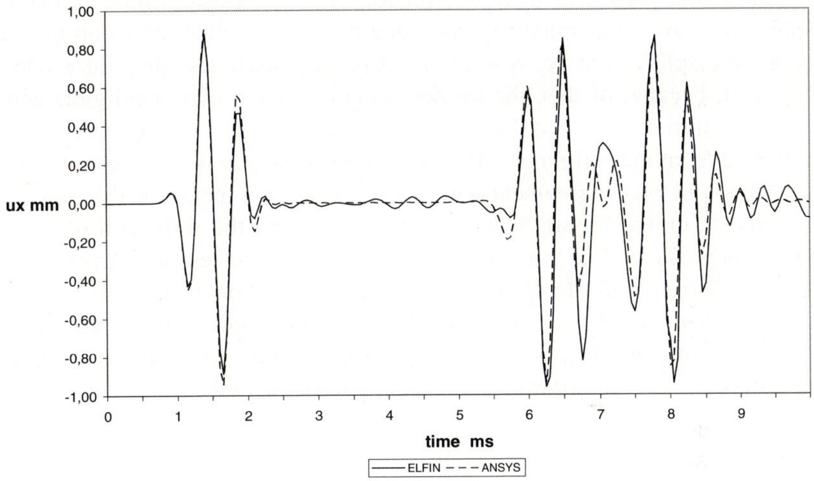

Fig. 3. Comparison between ELFIN and ANSYS without defect

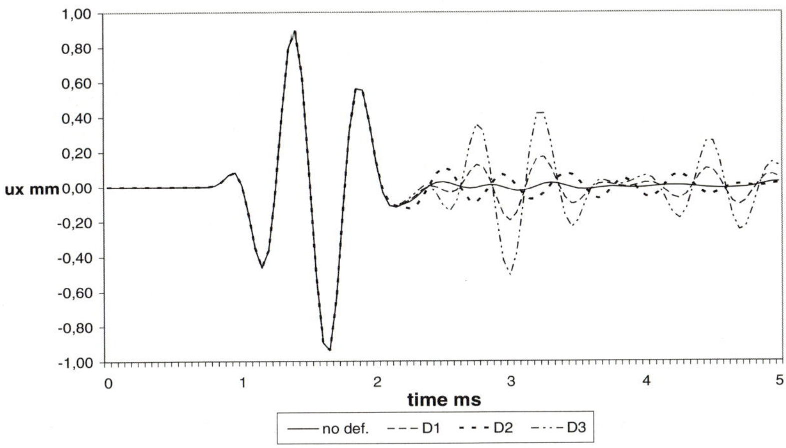

Fig. 4. Responses with the defects D1, D2 and D3 centred in P1

Figures 4 and 5 give the values versus time of the x component of the displacement vector **u** obtained by ELFIN at the observation point O in the presence of a defect; more specifically, Figure 4 refers to the cases in which the three defects D1, D2 and D3 are centred at the point P1, whereas Figure 5 regards only the defect D3 and takes into consideration the cases when it is located at the two different points P1 and P2. Moreover, in order to better hightlight the distorsions introduced by the defects D1, D2, and D3, in these figures the response with no defect is plotted with those occurring when only one defect is introduced in the mesh at the points previously specified. In these figures it is evident that the direct wave is followed, after a time interval depending on the distance of the defect from the boundary where the direct wave is generated, by another wave whose shape and amplitude depend essentially on the size of the defect in the wave propagation direction (y axes). Besides this, although it is not graphically shown, the numerical computations confirm that the y and the z components of the displacement are several orders of magnitude less than the x one, as is to be expected; because of this, the former components are quite negligible and they do not give useful indications to detect and recognize the defect. Another result which can be inferred from these figures is that the waves propagate with a velocity of about 2 km/s, which is in good agreement with that evaluated by means of [9]. Note that in Figures 4 and 5, unlike Figure 3, the wave produced by the reflection of the direct wave by the opposite end of the wall is not shown only for reasons of practical convenience (the figure would have been too large). Even though the results presented here and the others which were obtained require additional analyses, and more extended investigations need to be performed, on the whole they may be considered a promising starting point in order to detect and identify the defects.

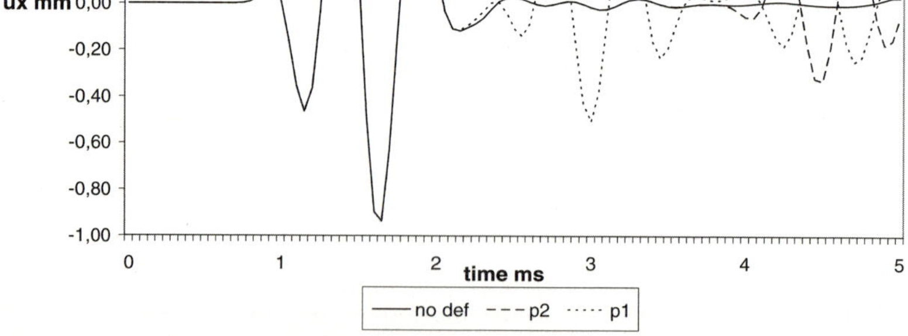

Fig. 5. Responses with the defect D3 centred in P1 and P2

5 Conclusions

In the paper we have presented the results of numerical computations carried out by means of a module of the large FEM research code ELFIN, specifically developed by the authors to analyze the propagation of ultrasonic waves in solid elastic media, in the context of the NDE technique. In particular, a concrete wall containing a void was considered, and several analyses were performed in order to investigate how the components of the displacement field are influenced by the position and dimensions of the void. The results demonstrate that the code can be usefully employed to detect and identify some kinds of defects.

In addition, the code has been validated and its results discussed in detail. In the near future we intend to make changes to the code so as to reduce considerably the computation time currently required.

References

1. Cawley, P., Alleyne, D.N.: The Use of Lamb Wave for Long Range Inspection of Large Structures. Ultrasonic 34, 287–290 (1996)
2. Alleyne, D.N., Cawley, P.: Long Range Propagation of Lamb Waves in Chemical Plant Pipework, Material Evaluation (1997)
3. Rose, J.L., Barshinger, J.: Using Ultrasonic Guided Wave Mode Cutoff for Corrosion Detection and Classification. In: 1998 IEEE Ultrasonic Symposium, pp. 851–854 (1998)
4. Binda, L., Saisi, A.: Non destructive testing applied to historic buildings: The case of some Sicilian Churches. In: Lourenço, P.B., Roca, P. (eds.) Historical Constructions, Guimarães (2001)
5. Ludwig, R., Lord, W.: A Finite-Element Formulation for thr Study of Ultrasonic NDT Systems. IEEE Transaction on Ultrasonic, Ferroelectric and Frequency Control 35(6), 809–819 (1988)
6. You, Z., Lusk, M., Ludwig, R., Lord, W.: Numerical Simulation of Ultrasonic Wave Propagation in Anisotro-pic and Attenuative Solid Materials. IEEE Transaction on Ultrasonic, Ferroelectric and Frequency Control 38(5), 436–445 (1991)
7. Aiello, G., Alfonzetti, S., Borzì, G., Salerno, N.: An Overview of the ELFIN Code for Finite Element Research in Electrical Engineering. In: Konrad, A., Brebbia, e.C. (eds.) Software for Electrical Engineering: Analysis and Design VI, pp. 143–152. WIT Press, Southampton (1999)
8. Fung, Y.C.: Solid Mechanics. Prentice-Hill (1965)
9. Graff, K.F.: Wave Motion in Elastic Solids. Dover Publications (1991)
10. Kolsky, H.: Stress Waves in Solids. Dover Publications (1963)
11. Washizu, K.: Variational Methods in Elasticity and Plasticity. Pergamon Press, Oxford (1972)
12. Cook, R.D., Malkus, D.S., Plesha, M.E.: Concepts and Applications of Finite Element Analysis. Prentice-Hill (1988)
13. Bathe, K.J., Wilson, E.L.: Numerical Methods in Finite Element Analysis. Prentice-Hall (1976)
14. Garnier, V., Chaix, J.F., Rossat, M., Payan, C.: Improvement in the Analysis of the Wave's Propagation in Concrete. In: ECNDT 2006, Berlin (2006)

Time, Wavelet and Hilbert-Huang Domain Analysis of Signals from Ultrasonic Based Equipment for the Non Destructive Evaluation of Concrete and Brick Masonry Walls

Antonino Musolino, Marco Raugi, Mauro Tucci, and Florin Turcu

Dipartimento di Sistemi Elettrici e Automazione, Università di Pisa.
Via Diotisalvi 2,56126, Pisa, Italy
{musolino,raugi,mauro.tucci,florin.turcu}@dsea.unipi.it

Abstract. In this paper we compare the effectiveness of the time, wavelet and Hilbert-Huang domain analysis of waveforms from a Non Destructive Test (NDT) equipment. The analysis of the signals from NDT systems is one of the most important activities in the recognition of the presence of possible defects that are usually associated with the presence of echoes from the defects themselves. Very often the echoes traveling back to the sensors are superimposed with the signal from the transducer. We analyze the waveforms obtained by couples of sensors in order to extract the echoes and to recognize the points were they originate. Data are obtained by a measurement campaign on a concrete wall and on masonry specimens of different texture and thickness.

Keywords: Wavelet transform, Hilbert-Huang transform, Non Destructive Test, Ultrasonic waves, Magnetostrictive transducers and sensors.

1 Introduction

The knowledge of the maintenance status of concrete and masonry structures is an important datum for the choice of specific consolidation techniques. Critical structures need to be evaluated during their service to ensure that they are not deteriorated and are free from defects. A number of non destructive techniques for the detection of defects, each with its advantages and drawbacks, have been developed in the past. Ultrasonic and sonic velocity test, stress wave transmission and impact echo are three important and widely used NDT methods based on the propagation of elastic waves. These methods are briefly discussed in [1] that contains an exhaustive and useful reference section.

Ultrasonic velocity testing units are commercially available from a number of sources. An ultrasonic pulser/receiver couple initiates a timing circuit as it sends an electrical signal to the source transducer, which in turn uses an internal piezoelectric crystal to generate a low-energy, high-frequency stress wave. The wave travels through the section where the receiving transducer converts the wave energy back to electrical energy.

O. Gervasi et al. (Eds.): ICCSA 2008, Part II, LNCS 5073, pp. 566–581, 2008.

Lower-frequency sonic stress waves are generated by an instrumented hammer whose constitution defines the energy and frequency content of the initial wave. The onset of the hammer pulse triggers an attached oscilloscope or digital data recorder to begin compiling data, as sensed by an accelerometer held against or attached to the wall at the receiver position. Converting wave trace data into velocity information is time consuming and must be conducted in a meticulous manner to obtain reliable results.

First developed for evaluating concrete, the impact-echo approach uses a transient stress wave typically generated at the face of the wall by a hammer or other mechanical impacter. Wave energy is reflected at points of elastic impedance variations within the wall, such as at an air void, the boundary between a masonry unit and adjacent grout, or at a crack. Multiple wave reflections set up vibrational frequencies within the section. The resulting waveform is then recorded and analysed in the frequency domain to identify dominant frequencies. The depth of the discontinuities can be identified as a function of frequency and of the compression characteristic velocity. Impact-echo approaches are attractive because access is required to only one wall face. One disadvantage of the technique is that no information can be obtained about materials beyond an initial void. The method also provides limited localized information.

Data from ultrasonic or sonic testing can be used as input to tomographic reconstruction algorithms to provide a cross-sectional representation of internal properties. In a homogeneous material, stress wave velocity is related to the material's dynamic stiffness, Poisson ratio, and material density.

In the evaluation of a wall, very little energy travels through air voids or gaps within the wall and, as a result, the apparent straight-line velocity decreases as the wave travels around internal voids or discontinuities. Measurement of pulse velocity is a point-by-point process using a gridwork set up on the wall surface. Through-wall velocity offers the most meaningful data. Semi direct transmission paths are also present in the gridwork. Data points can be interpreted individually or an entire data set can be used to generate a contour plot of through-wall velocity.

Tomographic analysis techniques have been shown to provide reasonable approximation of the size and extent of internal anomalies at the cost of considerable effort for acquiring the required large data set.

All these briefly described techniques are used to inspect in detail relatively small portions of concrete or masonry structures (typically walls or columns) located in correspondence of the equipment. As a consequence a great number of tests have to be performed to inspect the whole structure.

Ultrasounds based techniques have been introduced for long range NDT in different engineering fields [2]. In particular, in the case of plants with systems of pipes, ultrasonic torsional waves, guided by the walls of the pipes themselves, have been used to inspect portions of pipes up to 50 meters long. The presence of a defect in the structure produces an echo detected by properly positioned sensors; further investigation may by subsequently carried on to characterize the defect.

We propose a similar approach to the NDT of concrete or masonry walls; we firstly perform the inspection of a relatively large portion of the structure with the aim of detecting and locate possible defects. The extension and the shape of the defect can be subsequently investigated by the use of the former techniques on a narrow zone.

The feasibility of a long distance inspection technique for concrete structure based on the propagation of the elastic ultrasonic waves has been investigated by the authors in previous papers [3], [4]. We will now focus on the analysis of the signals detected by the sensors that is crucial to the practical feasibility of the proposed approach.

The presence of a defect produces an echo traveling back to the transducer. We consider the defect as a secondary source that activates when the wavefront from the transducer impacts the defect itself. The presence of superior modes across the thickness of the wall (when its thickness is comparable with the wavelength of the excitation) and the multiple reflections between the boundary surfaces makes difficult the identification of the echo. However, if the sensors are placed at two points that are one close to the other, it is possible to recognize the presence of a backward traveling wave by performing a proper comparison of the two waveforms. The knowledge of the arrival instants of the backward waves on an array of couples of sensors allows estimating the position of the defect. A correct estimate of the arrival instant is then crucial for a good performance of the triangulation algorithm that is usually used.

The paper is organized as follows. Section 2 will briefly discuss some aspects of the feasibility of the long range inspection method by considering a number of numerical simulations of the propagation of the elastic waves on a concrete and brick masonry walls with defects variously positioned.

Section 3 will show the test specimens and the NDT equipment used in the measurement campaign and will discuss a simple multipath model used to pre-process the sensors' waveforms. In the same section the main features of the wavelet and the Hilbert-Huang transforms that are relevant to the proposed application will be discussed. Section 4 will show some application examples of the proposed analysis technique to the data taken by a concrete wall and by a brick masonry wall with variously positioned defects.

2 Numerical Simulation

In order to assess the validity of the proposed long range inspection technique several numerical analysis of the propagation of elastic waves have been performed both on concrete and brick masonry walls. We simulated the action of the transducer by imposing tangential force on a small portion of the surface of the walls. Magnetostrictive transducers driven by known currents synthesized by electronic equipment have been used in the experimental campaign. The waveform shown in fig. 1 has been used in all the simulations. Force is zero for instants after 0.12 ms.

2.1 Concrete Walls

We considered a 4 meters high (y direction), 1.5 meters long (x direction) concrete wall with a thickness of 15 centimeters. The transducer is located in the centre of the upper surface. It is 2 centimeters high and 10 centimeters long and is able to transmit an x-directed force to the underlying portion of the wall surface.

We assumed the following physical constants for the concrete. Elasticity modulus $E = 2.8 \cdot 10^{10}\ N/m^2$; Poisson ratio $\upsilon = 0.5$; mass density $\rho = 2.4 \cdot 10^3\ kg/m^3$; damping parameters: $\alpha_M = 5.0 \cdot 10^{-4}\ s^{-1}$ and $\alpha_K = 1.0 \cdot 10^{-8}\ s$.

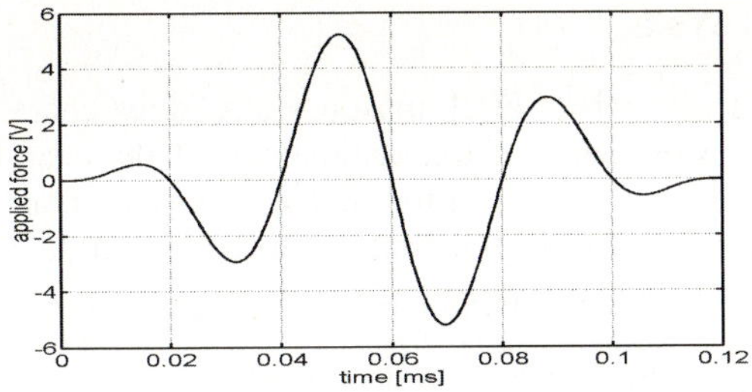

Fig. 1. Time waveform of the transducer driving current; the transmitted force is proportional

Because of the symmetries only the upper half of the structure has been simulated. We started performing the simulation of the propagation of an elastic wave on the wall without defects [5], [6]. Subsequently we examined three configurations characterized by the presence of a defect (a void long 10 cm and high 3 cm that passes through the wall on its entire thickness) in different positions. The first one at the coordinates x=0, y=1 (the middle of the shown portion of the geometry), the second at x=0, y=1.8 (20 cm below the upper edge of the wall) and the third at x= -0.4 and y=1.0.

Figure 2 shows the color maps of the computed x-displacement at properly selected instants. The effects of the presence of the defects are evident. In particular it is possible to recognize the presence of waves that travel from the defect backward to the transducer. Obviously these images are not available when performing a realistic measurement as they would require a huge number of sensors for their construction while the number of sensors seldom exceeds ten. As the thickness of the wall is greater (about twice) than the wavelength at 30 kHz, higher order modes appear in the propagation of the elastic waves. The response of the wall does not consist of a single pulse shaped similarly to the excitation force traveling from the transducer but lasts for a longer time than that shown in fig. 1. The wavefront of the echo reflected by the defect is then superimposed with the tail of the wave from the transducer and this may hide the echo itself. What it is clear at a first glance form fig. 2 it is very difficult to reconstruct by the information taken by a reasonable number of sensors.

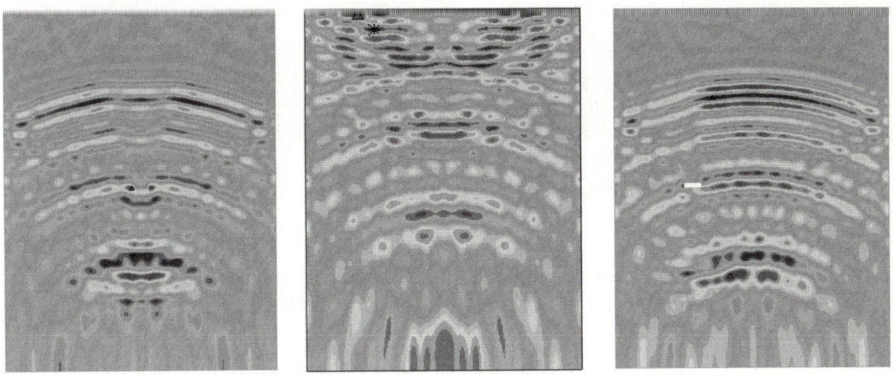

Fig. 2. X-axis component of the displacement at various times instant for the wall with the void. The first image refers to the wall with central void at 0.78 ms, the second to the void near the upper edge at 1.08 ms and the third to the void in the left hand portion of the wall at 0.78 ms.

2.2 Brick Masonry Walls

We investigated the two realistic brick masonry structures characterized by different texture shown in fig. 3. Because of the symmetries of the examined textures it was possible to analyze only one quarter of the first geometry and one half of the last one. The adopted boundary conditions, included those related to the symmetries, are shown in the figures. Physical parameters for the bricks are: $E_b = 1.4 \cdot 10^{10}\ N/m^2$, $\upsilon_b = 0.1$ and $\rho_b = 1.9 \cdot 10^3\ kg/m^3$, and for the mortar: $E_m = 0.9 \cdot 10^{10}\ N/m^2$, $\upsilon_m = 0.1$ and $\rho_m = 1.2 \cdot 10^3\ kg/m^3$. The damping parameters for both bricks and mortar are the same as those for concrete.

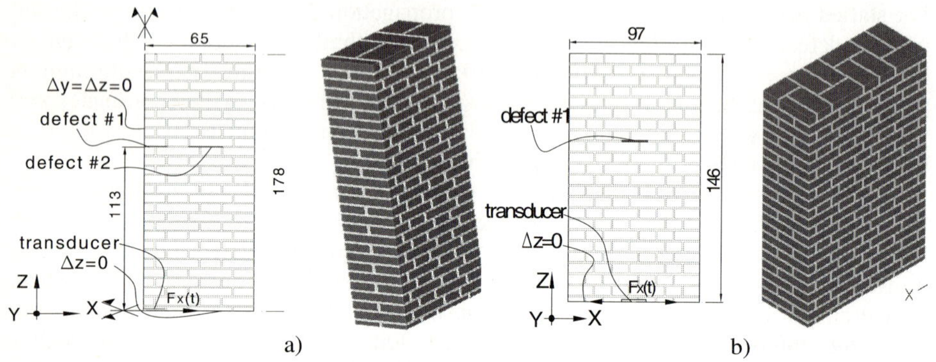

Fig. 3. Dimensions and texture of wall #1 and #2

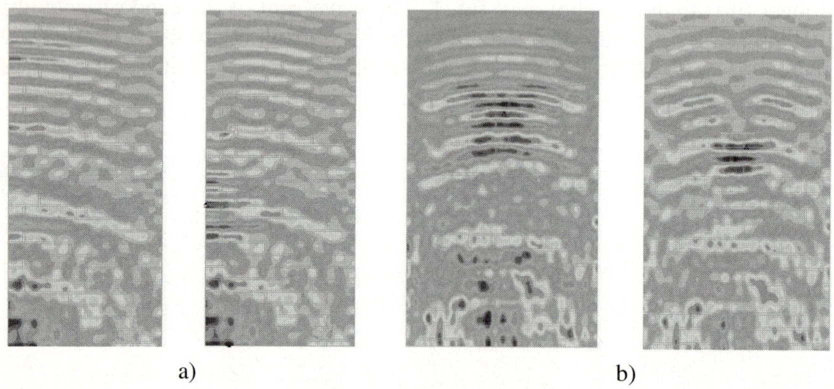

Fig. 4. Comparison of the X-axis component of the displacement for the intact and defective walls. The first two pictures refer to wall #1, the other to wall #2.

We firstly performed the numerical analysis on the intact walls; subsequently we introduced the defects in the positions shown in fig. 3. The defects for wall #1 are 0.25x0.25x0.01 (meters) and for wall #2 is 0.195x0.38x0.01. All consist in a missed layer of mortar along the entire wall thickness. The comparison of the X components of the displacements on the upper XZ face between the intact and the defective walls clearly evidences the effects of the presence of the defects. In all cases the defects act as secondary sources that produce reflected waves that propagate backward to the transducer.

Propagation in these structures is characterized by a pronounced directionality that is a consequence of the anisotropy of the system due to the presence of the mortar layers. The presence of the defect almost entirely reflects the corresponding portion of the incident wave front (this is well evidenced by fig. 4b that correspond to a texture preventing symmetry with respect the YZ plane) and substantially blocks the propagation; the directional characteristics of the propagation in the structures prevent the reconstitution of a compact wave front behind the defect, with the consequent formation of a region with very small displacements that extends far behind and in correspondence of the defect. The directionality is confirmed by the absence of the reflected waves from the defect in fig. 3a not aligned with of the transducer.

3 Signal Analysis

As discussed and shown in the previous section the recognition in the data taken from sensors of patterns similar to that shown in figure 1 is not possible because of the presence of higher order modes in the propagation of the elastic waves as the thickness of the wall is comparable with the wavelength of the excitation. However, if two points are close one to the other it is possible to recognize the presence of a backward traveling wave by comparison of the displacement waveforms at these two points.

3.1 The System MsM2020®

All the experimental data have been obtained by measurements campaign performed by using the diagnostic system MsS 2020® shown in fig. 5 with its transducer-sensor [7]. This apparatus, based on magnetostrictive transducers and sensors, is able to produce the propagation of transversal guided waves in the frequency range 4-128 kHz. A nickel strip is glued on the surface of the wall and is properly magnetized by using a rare hearts permanent magnet.

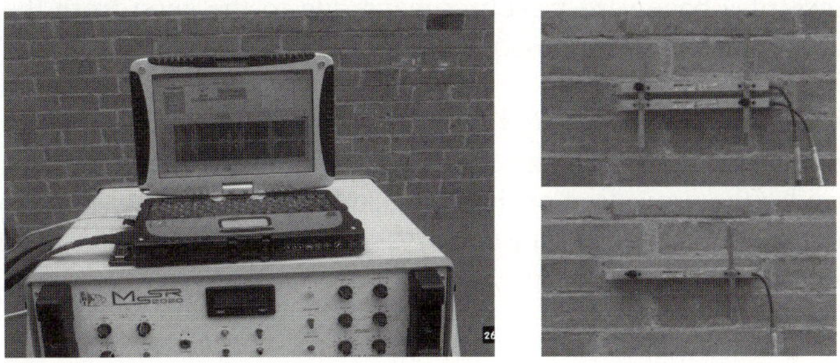

Fig. 5. The MsS 2020® system and its transducer-sensor

A coil driven by the system MsS 2020® is placed over the strip. Because of the strong magnetostrictive properties of the nickel, a current flowing on the coil is able to produce a deformation on the nickel strip that is transmitted to the wall. The operation of the can be reversed: a displacement of the magnetized nickel strip is able to induce an electromotive force at the terminals of the coil.

3.2 Multipath Model of Wave Propagation

Let us consider two points A and B as shown fig 6a. The position of the transducer is known as well as the thickness of the wall. We also assume the knowledge of the transverse velocity of the elastic waves. Propagation is not purely transversal and as a consequence the formula $v = \sqrt{E/(2\rho(\upsilon+1))}$ could not be used as it represents only a rough estimate. It is however possible to obtain a better estimate of the velocity by a contextual measurement. In the case of the presence of a forward traveling wave only it is possible to relate the waveforms at points A and B; by properly delaying the curve at point A it is expected to become very similar to that at point B.

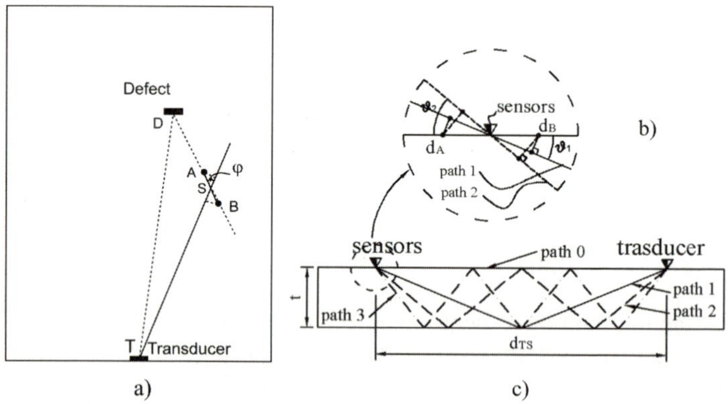

Fig. 6. a) Experimental setup for the concrete wall; b) schematic position of transducer, defect and sensors couple, c) multipath model

The presence of a backward traveling wave due to the echo from the defect prevents this occurrence and as a consequence may be used as a sign of the presence of the defect.

The delay times may be evaluated by using a simple model based on the different lengths of the multiple paths that exist between the transducer and the sensors.

Let us consider the projection of the distance between the two points on the direction connecting the center of the transducer with S, the midpoint of A and B.

$$\Delta s = \overline{AB} \cos \varphi \tag{1}$$

If we recall the shape of the wavefront in the figs. 2 and 4, Δs can be considered the distance between A and B as seen by the transducer. Let us now consider the cross section of the wall corresponding to the segment TS as shown in fig. 6b that shows the direct path and a few other paths between the transducer and the mid point of A and B. Let d_A and d_B be the distances form the transducer to the projections of A and B on TS respectively. The length of TS is $d_{TS} = 0.5 \cdot (d_A + d_B)$. The delay of the direct path is $\Delta t_0 = (d_A - d_B)/v$, ($v$ is the transversal velocity), that of the path #k is

$$\Delta t_k = ((d_A - d_B)\cos\theta_k)/v; \quad \theta_k = tg^{-1}\left((2kt)/d_{TS}\right) \tag{2}$$

as shown in fig. 6b.

These different (decreasing) delays are assigned to different portions of the waveform by the following rule. Δt_0 is assigned to the portion of curve that starts at time $t_0 = l_0/v$ where l_0 is the length of the direct path (see fig. 6b) and lasts $\Delta T = 0.12ms$ sec (the time interval where the excitation is not zero, as reported in fig. 2). Δt_k is assigned to the portion of curve that starts at time

$$t_k = \frac{l_k}{v} = \frac{2k}{v}\sqrt{t^2 + \left(\frac{d_A + d_B}{4k}\right)^2} \tag{3}$$

and lasts $\Delta T = 0.12ms$ (l_k is the length of the path #k).

The correspondence built by this rule is a piecewise constant function. Once a proper smoothing is performed, the curve of the displacements at point A is accordingly shifted to obtain a curve that is almost coincident with that at point B.

An example of the waveforms related to the concrete wall with the void in the upper position and detected by a couple of sensors at 70 and 73 cm from the transducer (aligned with the transducer and the defect) are shown in figs. 7 and 8. The curves in fig. 8 are coincident until the instant $\bar{t} = 1.3 \cdot 10^{-3}$ sec. This means that at this instant the reflected wave from the defect has reached the sensors couple after traveling from the transducer to the defect and back to the sensors: $d_{tot} = d_{TD} + d_{DS} \simeq 2.82 \simeq \bar{t} \cdot v = 1.3 \cdot 10^{-3} \cdot 2180$. The recognition of the arrival instant of the echo from the defect by comparison of the time waveforms above shown may lead to inaccurate results. The accuracy may be increased by performing the product of the waveforms in fig. 8.

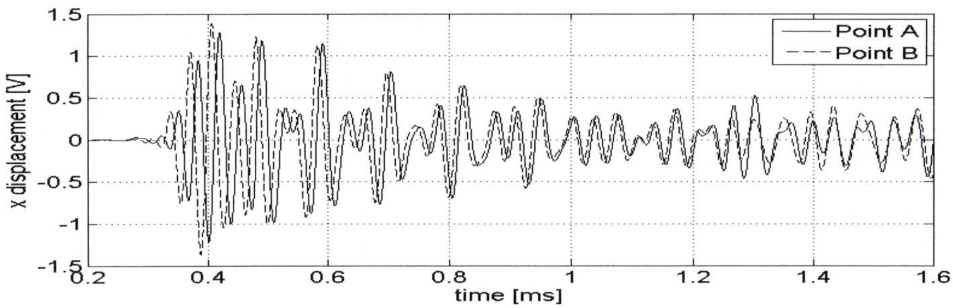

Fig. 7. *X*-displacements at points A and B (vertical axis unit is volt)

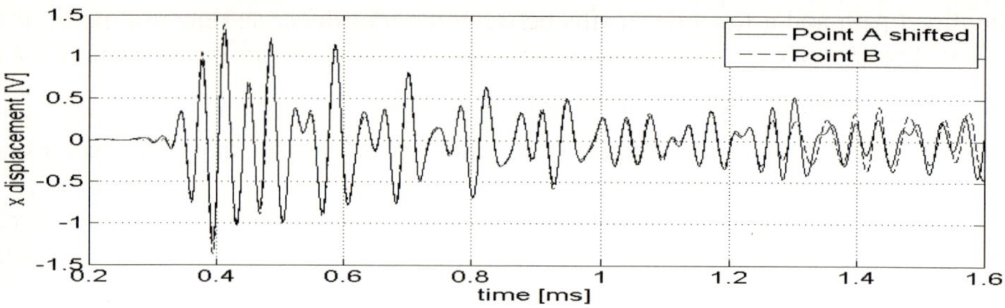

Fig. 8. Comparison of the x-displacements at point B and point A shifted

When the two curves are both of the same sign the product assumes positive values, while negative values are related to the existence of a phase difference between the factors. The improvement in the accuracy by using the product is not fully satisfactory. The analysis of the signals by the Hilbert-Huang transform and by the wavelet transform will be investigated too.

3.3 Hilbert-Huang Transform

As known the Hilbert-Huang [8] transform projects a signal on an adaptive basis. The basis functions named Empirical Mode Functions (EMF) have an oscillatory behavior characterized by the alternate presence of a minimum, a zero crossing and a maximum. This feature makes the EMF very attractive in detecting the phase difference between two signals by considering the zero crossing instants of the main EMF components, i.e. those with the higher frequencies.

3.4 Wavelet Transform

The wavelet analysis [9] is based on the introduction of a window function $\Psi(u)$, called mother wavelet. It is scaled and translated according to the expression: $\Psi_{s,t}(u) = |s|^{-p}\, \Psi\left((u-t)/s\right)$ where s is the scaling factor and p is chosen for normalization. The continuous Wavelet Transform of $x(u)$ is defined as:

$$\bar{x}(s,t) = \int_{-\infty}^{+\infty} \Psi_{s,t}(u) x(u) du = \left\langle \Psi_{s,t}, x \right\rangle \tag{4}$$

If the value of the WT of $x(u)$ is high for a determined scale and time, the scaled wavelet at the time t and the signal are strictly correlated.

The importance of the WT, if compared to the Short Time Fourier Transform (STFT) can be roughly explained by the following remarks. Let us suppose that the mother wavelet is centered on t_0 and has time width T. Then $\Psi_{s,t}$ is centered on st_0+t and has time width sT; consequently the WT gives information about the signal in the interval [st_0+t-sT, st_0+t-sT]. With the increase of s, the width of the window increases, while the resolution decreases. This is in accordance with the fact that low frequency components are better revealed by the use of wide time windows, while high frequency components need a narrow time window.

As a matter of fact, the WT studies the signal via a window characterized by a continuous variation of width avoiding the limit of the STFT, which is based on the use of a fixed window.

Furthermore it is possible to define a multi-resolution scheme, on bases of wavelet functions at different scales, that acts on a signal like a sub band filtering procedure. This leads to the possibility of decomposing a signal into a sum between a blurred version of the signal itself (x^M) and a series of details (d^i).

It is evident that this property is useful in pattern recognition applications, since the different details can enhance and/or show characteristics not visible in the original signal. In particular we will focus on the comparison of the zero crossing instants of signals as those in fig. 8.

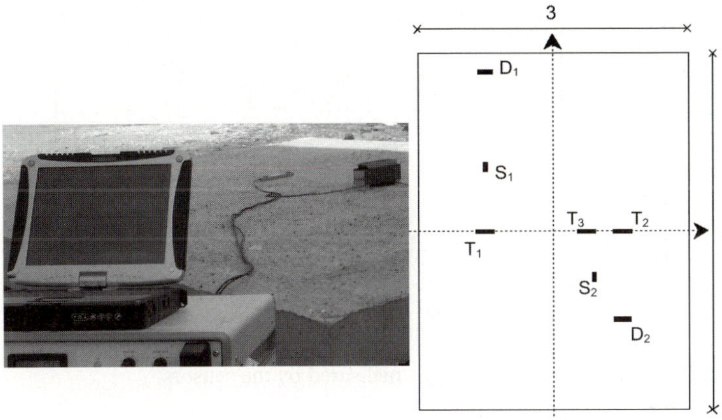

$$T_1 \equiv (-0.75, 0.0)$$
$$T_2 \equiv (0.75, 0.0)$$
$$T_3 \equiv (0.40, 0.0)$$
$$D_1 \equiv (-0.75, 1.80)$$
$$D_2 \equiv (0.75, -1.0)$$
$$S_1 \equiv \begin{pmatrix} -0.75, 0.70 \\ -0.75, 0.73 \end{pmatrix}$$
$$S_2 \equiv \begin{pmatrix} 0.45, -0.515 \\ 0.45, -0.555 \end{pmatrix}$$

Fig. 9. Dimensions of the concrete specimen and position of the transducers, defects and sensors used in the experimental measurements. The centre of the wall was assumed as the origin of the coordinates.

4 Experimental Results

An experimental campaign has been performed on a concrete wall shown in fig. 9 and on the 2 brick masonry walls whose feature are introduced in fig. 3. The numerical results shown in fig. 2a describe with good approximation the interaction between the couple transducer-defect D_2-T_2; fig. 2b is related to D_1-T_1 and fig. 2c is related to T_3-D_2. S_1 denotes the couple of sensors whose measured waveforms are shown in figs. 7 and 8.

Figure 10 shows the displacement at the sensors of the couple in S_2 when T_2 is active after that the shifting procedure described in 3.2 has been performed. An inspection of the curves shown in fig. 10 reveals a phase difference occurring for the first time around $t_0 \simeq 0.78\,ms$. This occurrence may be evidenced by considering the product of the curves as shown in fig. 11. We see that the product assumes for the first time negative values near t_0.

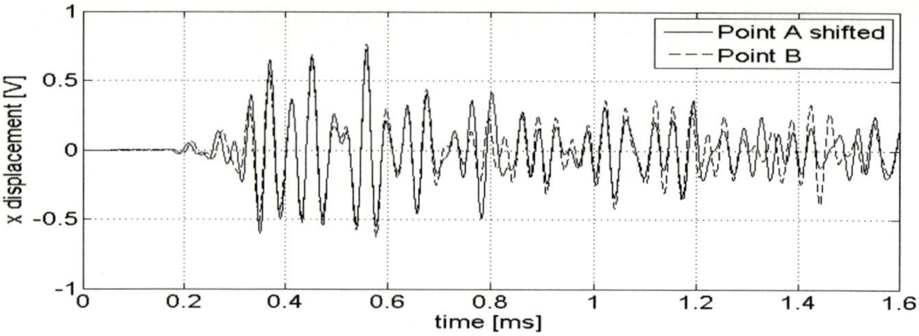

Fig. 10. X-displacements at sensors of the couple in S_2 (vertical axis unit is volt)

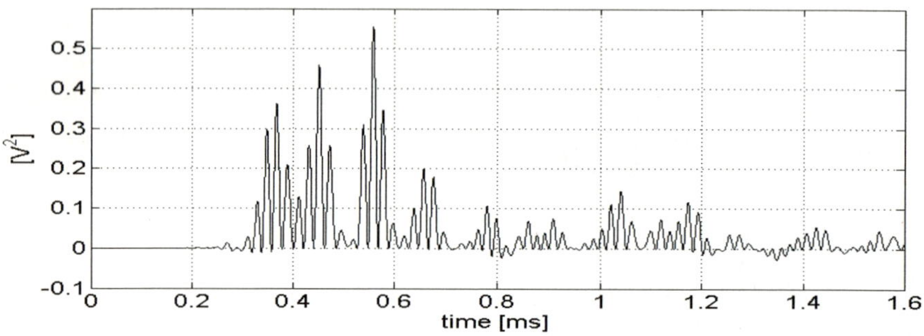

Fig. 11. Product of the displacements measured by the sensors

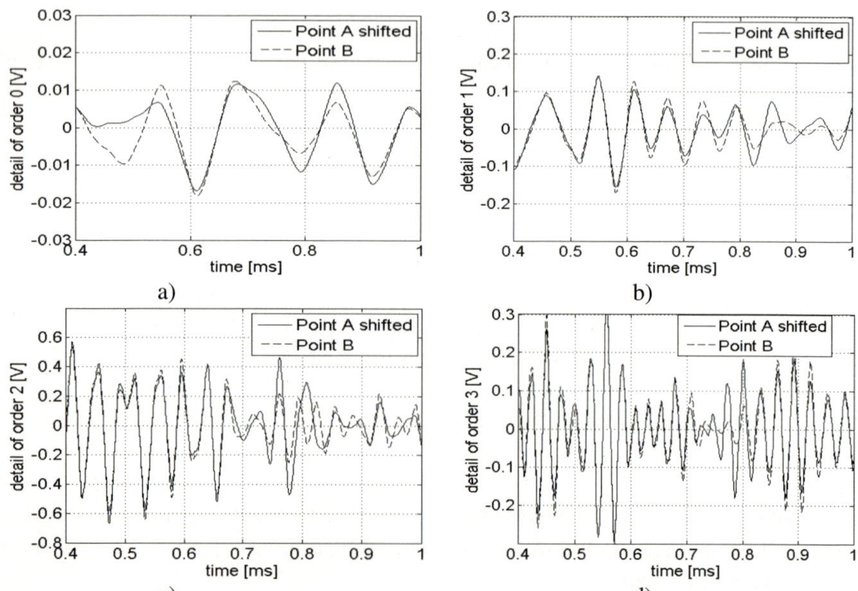

Fig. 12. Comparison of the blurred version (a) and of the details of the signals (b-d)

A better localization of the instant where the phase difference starts can be performed by comparing the sensors waveform in the wavelet and in the Hilbert-Huang domains. Let us begin with the wavelet transform. We used Daubechies wavelet on the interval with NV=6 vanishing moments [9] and compared the details the curves at different levels. For a better understanding of the figures we report only the portions of the curves in the interval *0.5-1.0 ms.*

The occurrence of the phase difference can be acknowledged at all the detail level with increasing resolution. In particular fig. 12 d) clearly shows the phase difference between the two signals in the interval $[0.73 \cdot 10^{-3}, 0.8 \cdot 10^{-3}]$.

The comparison in the Hilbert-Huang domain is carried on by comparison of the EMF for the two signals. As the EMF functions of the form $a(t)\sin(p(t))$ it is possible to define instantaneous phase and frequency and this allows an accurate location of the instants of phase variation of one signal with respect to the other.

Figure 13 shows the comparison of the waveforms of the first two EMFs of the signals that are those characterized by the highest frequency content. The other EMFs contain low frequency information and do not permit a good enough focusing on the time domain. Figure 14 shows the instantaneous frequency of the EMFs of fig. 13. The phase variation of one signal with respect to the other is clearly evidenced by the peaks.

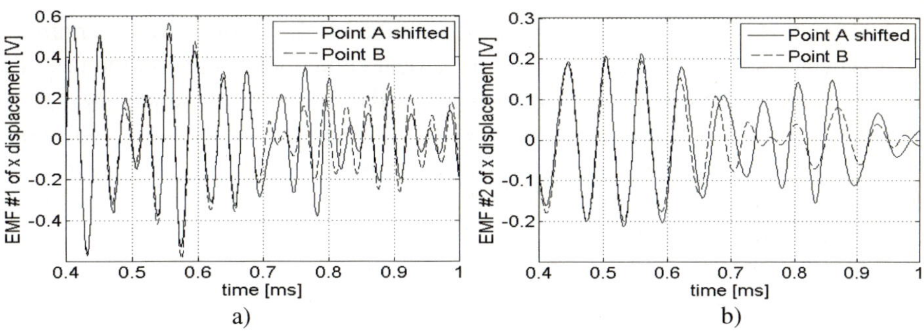

Fig. 13. Comparison of the first two EMFs

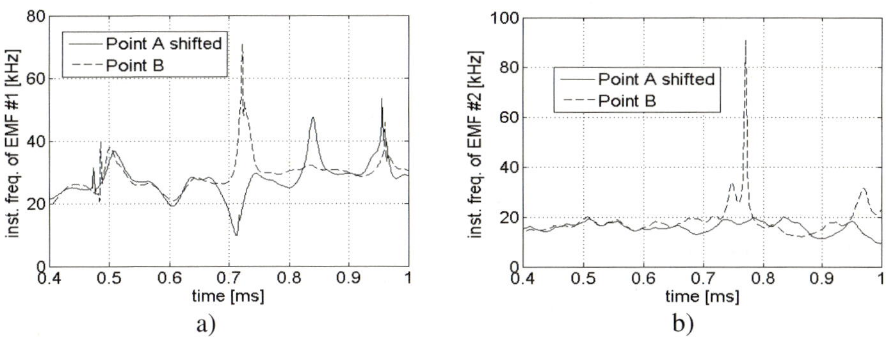

Fig. 14. Comparison of the instantaneous frequency of the first two EMFs

The rough estimate of $t_0 \simeq 0.78\,ms$ obtained by the analysis of the waveforms in the time domain has been refined by the use of the wavelet transform that allows a more precise evaluation by considering together the figs. 12 c) and d) that suggests the new estimate $t_0 \simeq 0.74\,ms$. The analysis of the instantaneous frequency content of the EMFs obtained by the Hilbert-Huang transform provides a further estimate $t_0 \simeq 0.72\,ms$ that is the time instant where in both figures 14 a) and b) a strong phase variation of one signal with respect to the other appears for the first time.

If, as shown in fig. 9, we consider the total distance $D_{tot} = \overline{T_2D_2} + \overline{D_2S_2} \simeq 1.55\,m$ and assume $v \simeq 2218\,m/s$ as the velocity of the elastic wave (contextually measured) we find by using the t_0 above estimated $D_{meas} = v \cdot t_0 = 1.60\,m$ that is very close to D_{tot}.

This confirms the validity of the proposed approach based on the Hilbert-Huang transform. Furthermore the interpretation of the data is much easier than those provided by time and wavelet analysis and the error probability is greatly reduced.

In an error free context the data relative to two sensors couple are sufficient to locate the defect by performing a triangulation. In this contest, where errors of several nature may affect the accuracy of the total distance D_{meas} obtained by the sensors, an array of sensors couple (from 5 to 10) is used and a least square approach is carried out.

Measurements have also been performed on the brick masonry walls reported in fig. 3. The analysis of the results of the numerical simulations suggests the use of an array of sensors aligned with the transducer in the direction of propagation of the transverse waves. The presence of the defect almost entirely reflects the corresponding portion of the incident wave front and substantially blocks the propagation; as a consequence, sensors positioned behind the defect will measure very small displacements if compared with sensors positioned between the transducer and the defect that will detect also the echo from the defect itself. Furthermore the analysis of waveforms from couples of sensors as those above described may locate the defect position.

In the wall # 1 we considered a sensor couple placed on the axis of the transducer and at a distance of 0.45 m from the transducer itself. In the wall #2 the distance form the transducer is 0.53m. The displacement of the sensors of both the couples is 2.75 cm. The following figure shows the results of the analysis in the time, wavelet and Hilbert-Huang domains for the waveforms of the sensors couples.

The comparative analysis of the data in fig. 15 and relative to the wall #1 suggests an instant $t_{arr,1} \simeq 1.05\,ms$ as the arrival instant of an echo from the defect. By using the measured values of the velocity $v_{meas} = 1780\,m/s$ we obtain for the distance between the transducer and the defect the value $D_{td,1} = 0.5 \cdot \left(v_{meas} \cdot t_{arr,1} + D_{ts,1}\right) = 1.16\,m$ that is an excellent estimate of the defect position. From curves in fig. 16 related to wall #2 we obtain $t_{arr,2} \simeq 0.70\,ms$ and $D_{td,2} = 0.5 \cdot \left(v_{meas} \cdot t_{arr,2} + D_{ts,2}\right) = 0.89\,m$; the effective distance between the transducer and the defect is 0.97m.

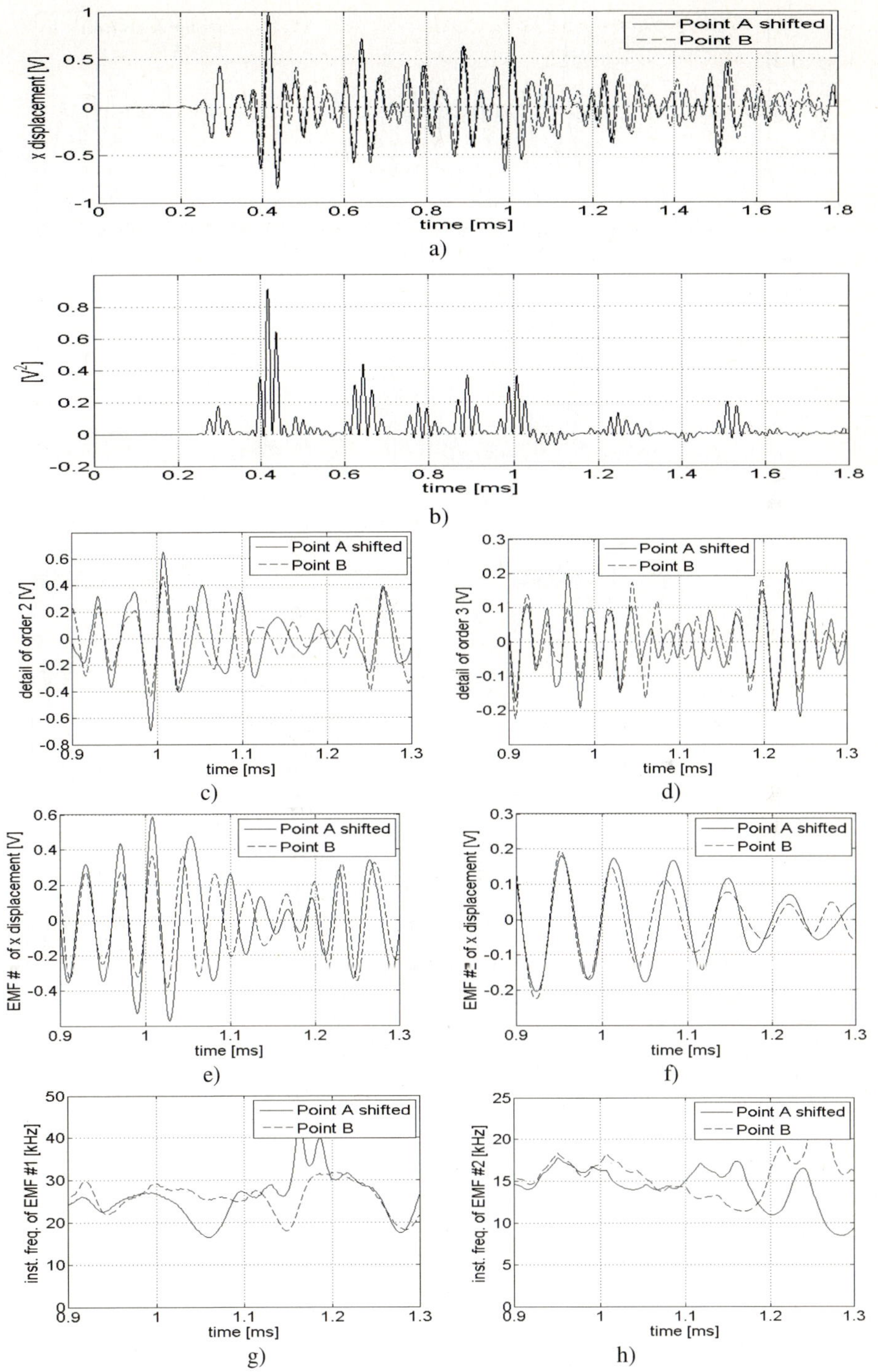

Fig. 15. Time: a), b); wavelet: c), d); Hilbert-Huang (HH) EMF e), f) and HH inst. freq. g), h)

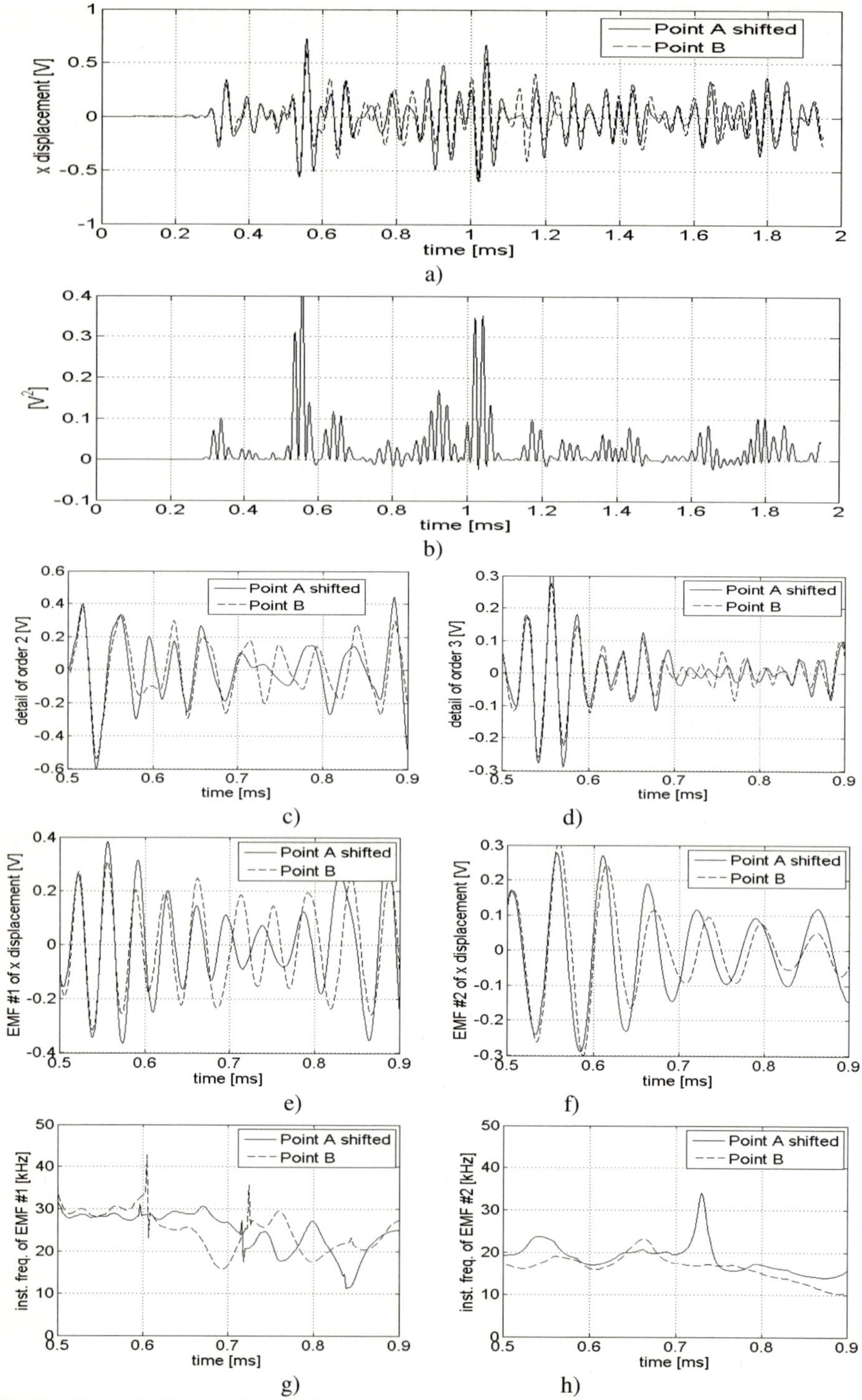

Fig. 16. Time: a), b); wavelet: c), d); Hilbert-Huang (HH) EMF e), f) and HH inst. freq. g), h)

5 Conclusion

A comparison between time, wavelet and Hilbert-Huang domain analysis of data from ultrasonic nondestructive test for concrete and masonry wall has been presented. The analysis performed by using the Hilbert-Huang transform has demonstrated as the more reliable one. Nevertheless a comparative analysis of the results obtained by all the three examined techniques can help in obtaining even more precise and reliable results.

Acknowledgments. This work was supported in part by the Italian Ministry of University (MIUR) under a program for the Development of Research of National Interest (PRIN GRANT 2005094532) and by the Fondazione Cassa di Risparmi di Livorno.

References

1. Shuller, M.P.: Nondestructive testing and damage assessment of masonry structures. Prog. Struct. Engng. Mater 5, 239–251 (2003)
2. Cawley, P., Alleyne, D.: Practical Long Range Guided Wave Inspection - Managing Complexity. Review of Quantitative Nondestructive Evaluation 22, 22–37 (2003)
3. Musolino, A., Raugi, M., Tucci, M., Turcu, F.: Feasibility of Defect Detection in Concrete Structures via Ultrasonic Investigation. In: Progress in Electromagnetic Research Symposium, Prague, Czech Republic, August 27-30, 2007, pp. 371–375 (2007)
4. Musolino, A., Raugi, M., Tucci, M., Turcu, F.: Voids Detection in Brick Masonry Structures by Using Ultrasonic Testing. In: IEEE Ultrasonic Symp., pp. 1377–1380. New York (October 2007)
5. Kaltenbacher, M.: Numerical Simulation of Mechatronic Sensors and Actuators. Springer, Berlin (2004)
6. Graff, K.F.: Wave Motion in Elastic Solids. Oxford University Press, Oxford (1975)
7. Operating and technical instructions for magnetostrictive sensor (MSS) instrumentation system, mod. MsSR2020, MsSR2020D, by Sensor Systems and NDE Technology Dept., Applied Physics Division, South West Research Institute, San Antonio, Texas (December 2004)
8. Huang, N.E., Shen, Z., Long, S.R., Wu, M.C., Shih, H.H., Zheng, Q., Yen, N.-C., Tung, C.C., Liu, H.H.: The empirical mode decomposition and the Hilbert spectrum for nonlinear and non-stationary time series analysis. In: Proc. R. Soc., London, Ser. A, vol. 454, pp. 903–995 (1998)
9. Daubechies., I.: Ten Lectures on Wavelets. SIAM, Philadelphia (1992)

Acoustic NDT on Building Materials Using Features Extraction Techniques

M. Camplani[1], B. Cannas[1], F. Cau[1], G. Concu[2], and M. Usai[1]

[1] Electric and Electronic Dept., Univ. of Cagliari, Cagliari, Italy
{cannas,massimo.camplani,musai,francesca.cau}@diee.unica.it
[2]Dept. of Structural Engineering, Univ. of Cagliari, Cagliari, Italy
gconcu@unica.it

Abstract. Ultrasonic materials analysis is based on the principle that the propagation of any wave is affected by the medium through which it travels. Thus, changes in measurable parameters associated with the passage of a wave through a material can be correlated with changes in physical properties of the material. In this paper, an ultrasonic technique has been experimented for non destructive testing of building structures. In particular, several parameters associated with acoustic waves propagating through a trachite masonry and a concrete pillar have been analyzed and discussed making reference to a Finite Element model of the structures.

Keywords: Non Destructive Testing, ultrasonic waves, feature extraction.

1 Introduction

The ageing of building heritage and the need of guarantee its safety, especially when reconstruction of its past life and of its previous restorations is not clear, has led to the creation, the evolution and the rapid diffusion of non-destructive diagnostic methodologies, in order to proper manage buildings rehabilitation. In fact, every work must be preceded by careful observation of the effective state of the building so as to ascertain the nature of possible structural inadequacies and be able to proceed with works that are suitable and sufficient for the rehabilitation of the building and its characteristics.

Among the different non-destructive diagnostic methods, ultrasonic inspections seem to be very suitable for evaluating a building condition, because they give information with immediacy, rapidity and relatively low cost. Ultrasonic materials analysis is based on a simple principle of physics: the propagation of any wave will be affected by the medium through which it travels. Thus, changes in measurable parameters associated with the passage of a wave through a material can be correlated with changes in physical properties of the material [1].

Traditional application of ultrasonic techniques is based on measurements of the velocity v of waves propagating through the material. The velocity is obtained from the ratio l/t, where t (travel time) is the time the wave needs to travel along the path of length l. It is a function of material elastic modulus E, Poisson's number ν and density ρ, thus it is directly related to structures elastic parameters, crucial for inspections of structures stability and lastingness.

O. Gervasi et al. (Eds.): ICCSA 2008, Part II, LNCS 5073, pp. 582–595, 2008.

These methods are preferentially carried out applying the Through Transmission Technique, in which the wave is transmitted by a transducer (Emitter) through the test object and received by a second transducer (Receiver) on the opposite side. This allows to obtain the average velocity *v* related to paths crossing the object's thickness [2]. The studies of ultrasonic methods have been centered on medical or materials engineering applications for laboratory testing. The valuable handmade analysis has begun in the first of '90 and nowadays it is performed using considerable approximations to the detriment of results precision.

The inaccurate results are also caused by the subjective methodology often used for interpretations. This contributed to form the common opinion that the use of such methodologies to analyze valuable buildings does not give reliable results. Conversely, the aforesaid techniques could show notable diagnostic properties if used in an appropriate way, that is providing them with innovative techniques, which employ and develop advanced computational tools and testing devices.

In this paper, the potentiality of a novel approach to ultrasonic investigation is analyzed. This approach involves the analysis of several parameters in time and in frequency domain, associated with acoustic waves propagating through the material.

2 Case Study

The analysis has been carried out on a stone masonry model and on a concrete pillar model. The acoustic wave is transmitted through the structure and received by a second transducer positioned on its opposite side, according to the Through Transmission Technique. Changes in received signal provide indications of variations in material continuity.

In order to simulate the wave propagation through the two structures, they have been both simulated through a finite element 3-D model, developed using the commercial code ANSYS [3]. In particular, the LS DYNA modulus, which solves structural and mechanical problems in a transient domain, has been used. The numerical analyses have been carried out simulating the use of piezoelectric transducers for emitting and receiving the signal. Simulations have been performed on a Hewlett-Packard

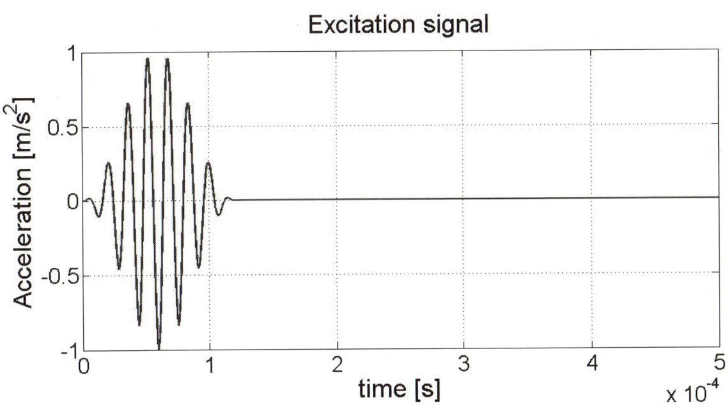

Fig. 1. Excitation signal

Workstation xw9300 with Opteron tm Processor 250, 2.39Ghz, 3.50GB of RAM Physical Address Extension. Each simulation (one ray path) lasted 35 minutes. Due to symmetry of the structures, it has been sufficient to analyse only one quarter of them.

The excitation y(t) is an acceleration signal applied in the grid's nodes on one side of the wall. It is a seven and half-cycle tone burst enclosed in a Hanning window (Fig.1) [4]

$$y(t) = \frac{1}{2}\sin(2\pi ft)\cdot\left[1 - \cos\left(\frac{2}{15}2\pi ft\right)\right] \tag{1}$$

2.1 Model of the Stone Masonry

The wall is 90cm wide, 62cm high and 38cm thick, and it is made of Trachite blocks sized $20{\times}38{\times}12\text{cm}^3$, settled as shown in Figures 2-3 and jointed with cement lime mortar. The block assigned to the central position of the wall was not settled, thus realizing a macro-cavity with the same size of the missing block (Fig. 3), and assumed as a known anomaly. Mortar joints have been assumed to be 1cm thick. Figure 4 shows the finite element model of the masonry.

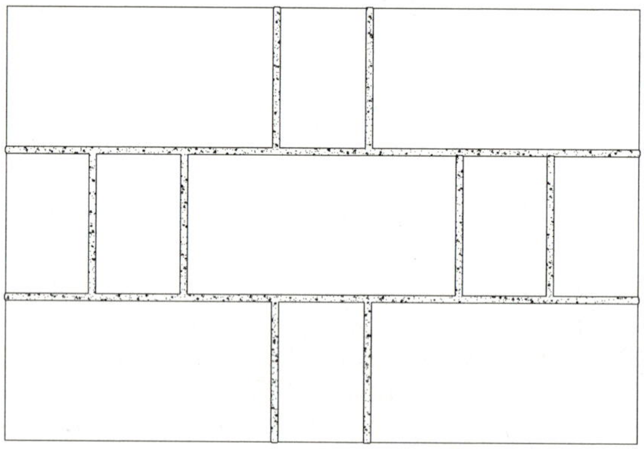

Fig. 2. Front view of the wall

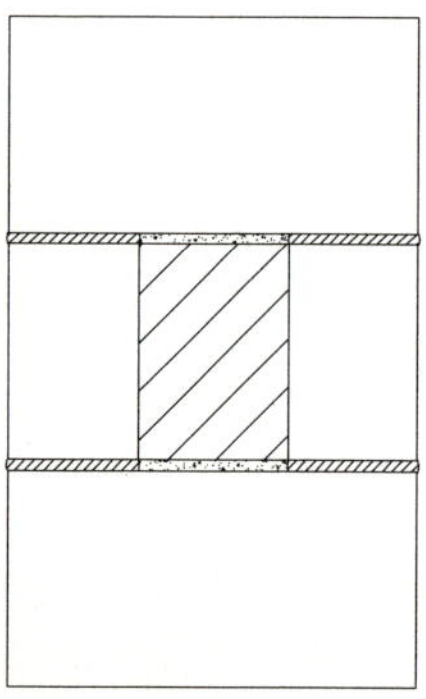

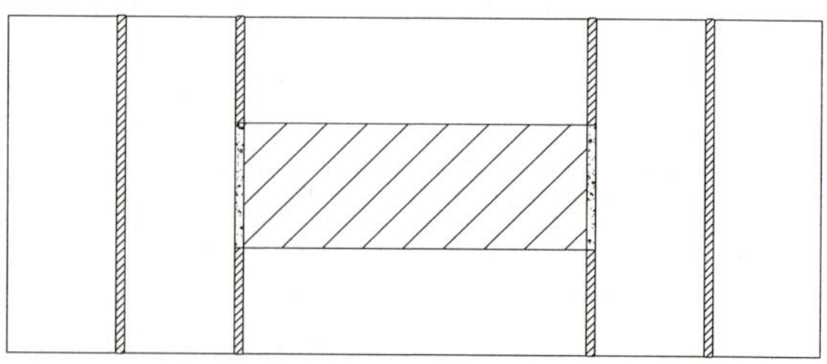

Fig. 3. Vertical plane section: position of the cavity (left) and horizontal plane section intercepting the cavity (right)

Table 1. Material properties

Properties	**Young modulus** [GN/m^2]	**Poisson coefficient**	**Density** [kg/m^3]
Trachite	6	0.2	1750
Mortar	15	0.1	1800
Concrete	33	0.3	2300
Air	1.5 10^{-4}	0.1	1.3

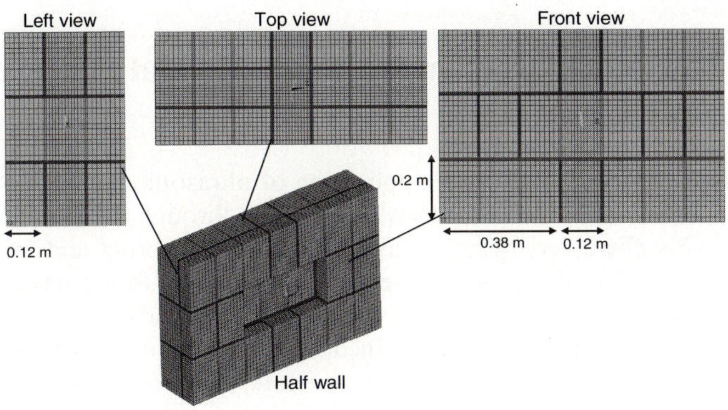

Fig. 4. FEM model of the wall

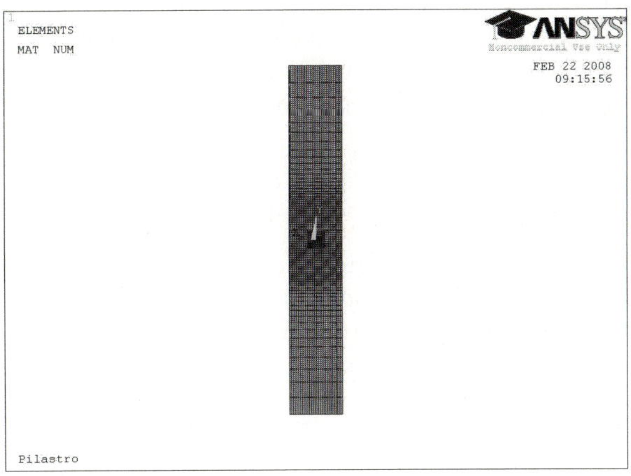

Fig. 5. FEM model of the concrete pillar

Trachite specimens have been prepared and then tested for the determination of compressive strength and elastic modulus, following the Italian Standards UNI EN 1926 and UNI EN 14580. The material properties assumed for the trachite, the mortar, and the cavity are reported in Table I.

2.2 Model of the Concrete Pillar

The concrete pillar is 30cm wide, 200cm high and 30cm thick. The ultrasonic testing analysis of the pillar model has been performed assuming the presence of a cavity of 10cm per side.

The material properties assumed for the concrete and the cavity are reported in Table I. Three simulation sessions have been performed, modifying the position of the defect in order to test the precision of the method. Figure 5 shows the finite element model of the pillar.

3 Waves Features Extraction for Anomalies and Material Classification

As previously mentioned, traditional application of ultrasonic techniques is based on measurements of the velocity v of waves propagating through the material. Nevertheless, other wave's characteristics such as attenuation, scattering and frequency content, primarily related to the elastic wave power, allow one to get more and relevant information about the material. In fact, wave power is absorbed or attenuated at different rates in different materials, determined in a complex manner by interactive effects of density, hardness, viscosity, and structure. Moreover, waves reflect from boundaries between dissimilar materials, thus changes in material structure, e.g. presence of discontinuities or defects, can affect the amplitude, direction, and frequency content of scattered signals. Finally, all materials tend to act to some degree as a low pass filter, attenuating or scattering the higher frequency components of a broadband wave more than the lower frequency components. Thus, analysis of changes in the remaining frequency content of a selected broadband pulse that has passed through the test material can keep track of the combined effects of attenuation and scattering as previously described [2, 5].

3.1 The Waves Features

Let $\mathbf{x}_n=[x(t_n)]$ $n=1,\ldots N$, be the time series of the wave recorded at the receivers and $\mathbf{X}(\mathrm{f})$ the frequency spectrum of $x(t)$. Several wave features can be utilized for the analysis both in the time domain and in the frequency domain, as shown in [4,6-13]. In this research only 21 features have been considered, discarding the features that cannot be correlated enough with changes in physical properties of the tested structures. In fact, a preliminary analysis of the discarded features highlighted that they do not give an evident contribution to the identification of the different materials of the structure, e.g. stones, mortar and void for the masonry, concrete and void for the pillar, because of their lack of sensitivity toward the variations in the continuity of those type of materials.

The 21 selected features are listed in the following.

Features in the Time Domain

1. Travel Time

Travel time is the difference between the arrival time of the received signal and its starting time.

2. Velocity of Propagation

$$v = \frac{\text{piece thickness}}{\text{travel time}} \tag{2}$$

3. Maximum peak-peak amplitude

4. Maximum absolute values of the amplitude

5. Minimum absolute values of the amplitude

6. Mean value of the normalized amplitude

7. Variance of normalized amplitude

8. Time reversibility

Time-reversibility is the skewness of the slopes normalized by standard deviation of the slopes taken to the third power.

$$t_{rev} = \frac{1}{\sigma^3 (N-1)} \sum_{i=2}^{N} \left(\frac{x(t_i) - x(t_i - 1)}{t_i - t_{i-1}} \right) \tag{3}$$

9. Third order auto-covariance

$$t_{C3} = \frac{1}{(N-2)} \sum_{i=3}^{N} \left(x(t_i) - x(t_i - 1) x(t_i - 2) \right) \tag{4}$$

The Third order autocovariance is the higher order extension of the traditional autocovariance (a slide of the third order moment estimation).

10. Mean value of the normalized envelope

11. Variance of the normalized envelope

12. Local rise time from 25% level to the peak of the envelope

13. Local rise time from 50% level to the peak of the envelop

14. Signal Power

Features in frequency domain

15. Spectrum principal frequency

$$f_{\max} : \left| \mathbf{X}(f_{\max}) \right| = \max_f \left| \mathbf{X}(f) \right| \tag{5}$$

16. Spectrum central frequency

$$f_c = \frac{f_1 + f_2}{2} \tag{6}$$

where f_1, f_2 are respectively the lower and upper cutoff frequencies that delimit the spectrum bandwidth.

17. Spectrum bandwidth

$$B = f_2 - f_1 \tag{7}$$

18. Spectrum normalized mean value
19. Spectrum normalized variance
20. Spectrum maximum amplitude

Moreover, the *attenuation of the input signal* α has been evaluated

$$\alpha = \frac{f_s - f_r}{\sigma_s^2} \tag{8}$$

where σ_s^2 is the spectrum variance of the input signal, and f_s and f_r are the input signal centroid frequency and the received signal centroid frequency. The input signal attenuation has been estimated from spectral centroid down-shift using a statistically based method.

4 Feature Extraction

As previously mentioned, the test method is the Through Transmission Technique. In this method, the signal is transmitted through the test object and received by a second transducer on the opposite side.

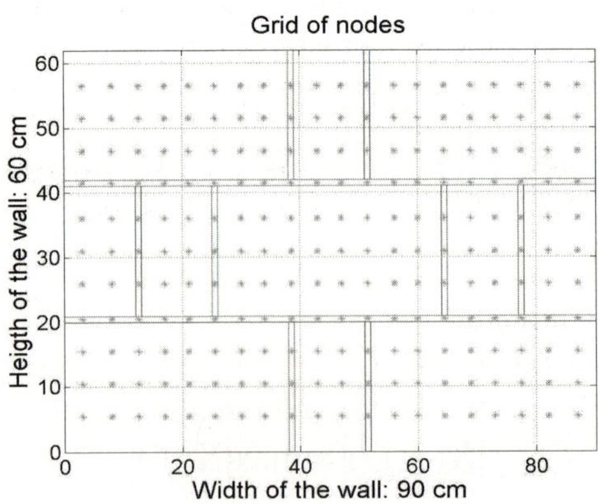

Fig. 6. Grid of test points

4.1 Features Extraction on the Stone Masonry Model

220 emitters and 220 receivers have been positioned on a grid of 11×20 nodes in the opposite surfaces of the wall (see Fig. 6). At the end of the simulation sessions, 220 signals have been obtained, one for each point of the grid of receivers. From each node of the grid, the 21 parameters cited in section 3 have been extracted in order to investigate the presence of anomalies and obtain some information on the material. Each parameter is associated with a matrix $\mathbf{A}_i(20x11)$, i=1,...21, containing the parameter values in each node of the grid. A map of the distribution of each feature has been derived visualising the data recorded in the grid's nodes.

In order to evaluate how each feature correlates with materials properties, a matrix $\mathbf{M}(20x11)$ has been built. The element m_{ij} identifies the type of path the wave, emitted in the i,j position of the grid, propagates through. In particular, the wave can propagate through four types of paths: trachite, mortar-trachite, mortar, air-trachite. Figure 7 shows the map associated to the M matrix.

As an example of the features maps, Figure 8 shows the distribution map of the signal power. In this figure, each square identifies one receiver and it is representative of its neighbourhood.

It can be noticed that this feature, as the amplitude variance, is able to well-define the contour of the defect, but not the mortar joints.

The map of the velocity, shown in Figure 9, is able to discriminate three main types of ray-paths. Only the paths through trachite and those through mortar-trachite are undistinguishable. In the centre of the map the presence of the cavity region can be clearly seen, although it is smaller than expected.

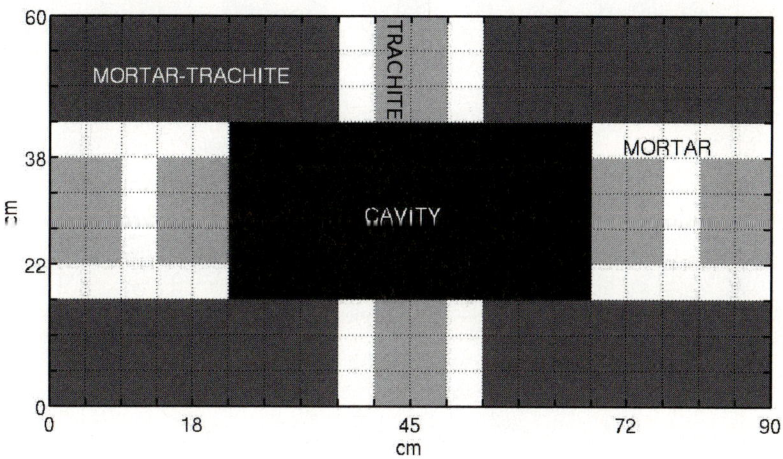

Fig. 7. Map associated to the M matrix

K-means algorithm has been used to partition the map of each feature into clusters corresponding to different types of ray paths [14]. K-means performs the grouping by minimizing some metric relative to the centroids of the clusters. In this case, the metric is the Euclidean distance between data and the corresponding cluster centroid.

In order to evaluate the goodness of the clustering, the error E has been evaluated for each map as the number of misclassified pixels. The features corresponding to the highest values of the error are the minimum absolute value of the signal, the spectrum maximum amplitude, and the variance of the normalized envelope.

Figure 10 reports the map obtained clustering the velocity field. As it can be noted three types of ray-paths can be identified and the cavity is still smaller than expected.

In order to exploit the complementary contribution of the information contained in the various features, the high-dimensional distribution obtained by the union of the different features has been considered.

The resulting distribution can be compared with the material distribution in order to identify different ray-paths. Also for this task, K-means algorithm has been used to partition the high-dimensional features space into the clusters corresponding to different ray paths.

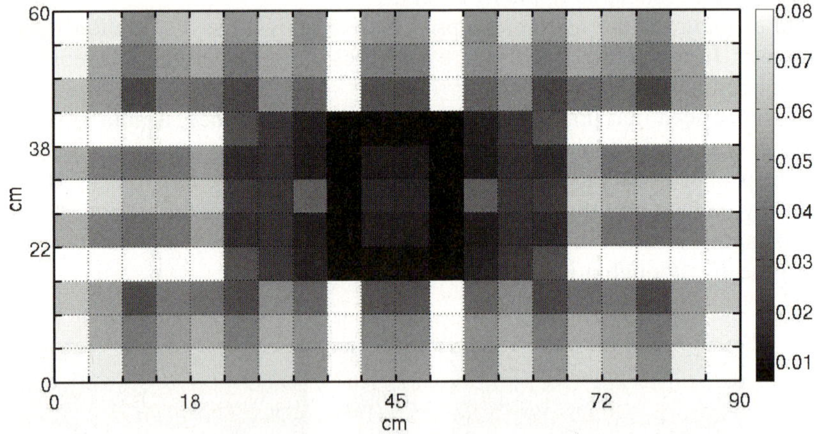

Fig. 8. Signal Power: distribution map

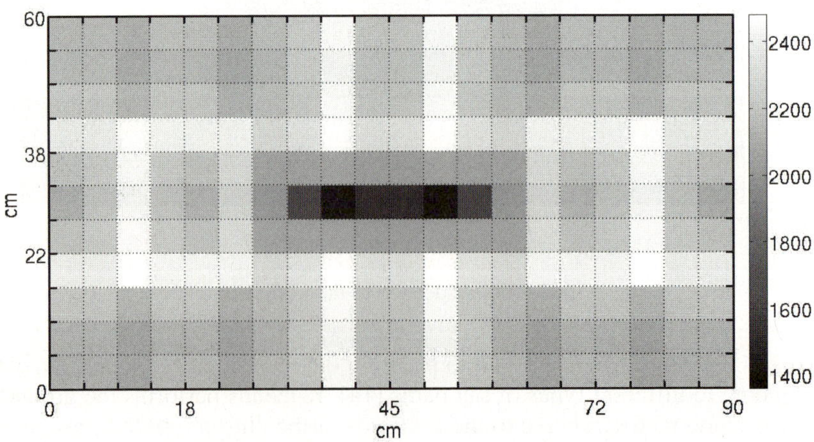

Fig. 9. Propagation velocity: distribution map

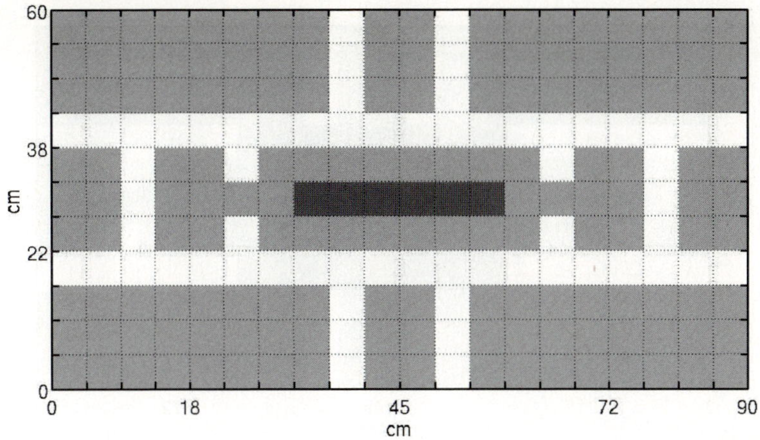

Fig. 10. Map of the clustered velocity field

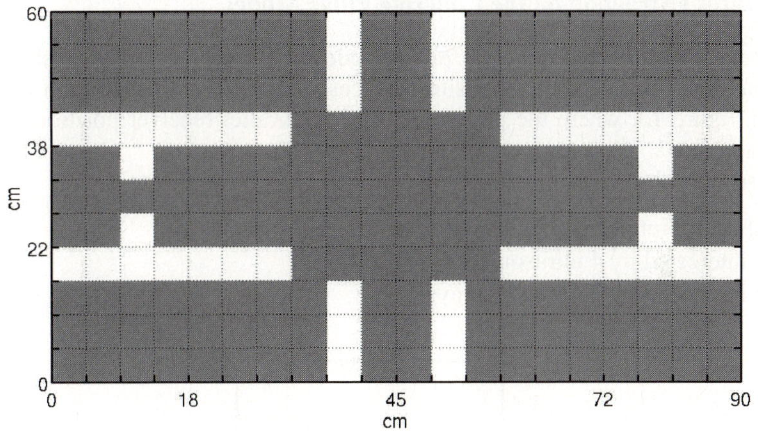

Fig. 11. Map derived from clustering of selected features (2 clusters)

By discarding the three features corresponding to the highest values of the error, an 18-dimensional feature space is obtained. Figure 11 and Figure 12 show the resulting maps .

Figure 11 reports the map obtained partitioning the feature space into 2 clusters. The paths crossing the trachite, mortar-trachite and the cavity cannot be distinguished. Only the mortar joints are accurately defined.

Figure 12 shows the map obtained partitioning the feature space into 3 clusters. In this case, the paths crossing the trachite and those crossing mortar-trachite cannot be still clearly distinguished. Nevertheless, the cavity inside the structure is accurately defined, as well as the mortar joints. The partitioning into 4 clusters does produces the same map obtained in the case of 3 clusters.

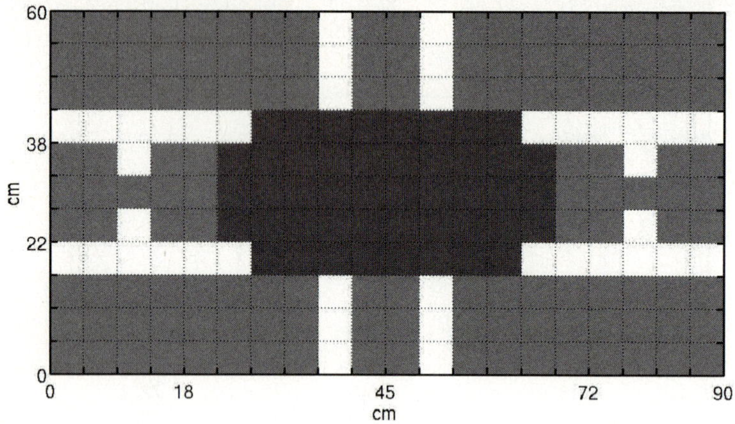

Fig. 12. Map derived from clustering of selected features (3 and 4 clusters)

4.2 Features Extraction on the Concrete Pillar Model

Features extraction on the concrete pillar model has been performed positioning 35 emitters and 35 receivers on opposite surfaces of the wall, 30cmx45cm spread, as shown in Figure 13, where the points identify the emitter/receiver positions and thus the ultrasonic rays.

Three cases have been examined:

- central defect interested by only one ultrasonic ray;
- defect interested by 4 ultrasonic rays;
- defect interested by 2 ultrasonic rays.

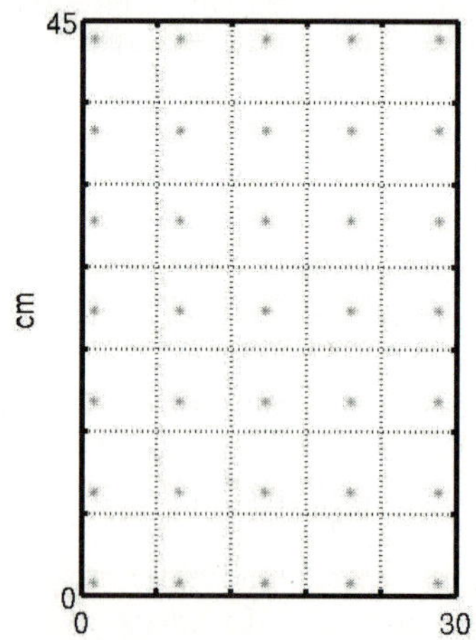

Fig. 13. Emitter and receiver positions

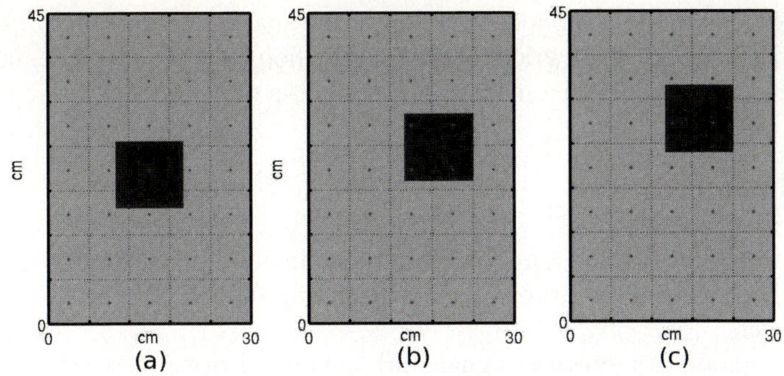

Fig. 14. Map associated to the M matrix

At the end of the simulation sessions, 35 signals have been obtained, one for each receiver point. From each of these points, the 21 features cited in section 3 have been extracted. Figure 14 (a-c) shows the maps associated to the M matrix. In particular, the wave can propagate through two types of paths: air-concrete and concrete. In each figure the dark square identifies the defect, the small points are the pixel centres.

Figure 15 (a-c) shows the velocity maps. The defect in the centre of the section (fig. 15 (a) dark square) is correctly positioned but its dimension is smaller than expected. This is due both to the dimension of the adopted grid (7cm) and to the position and dimension of the defect. In fact for the first defect, only one ray intercepts the defect area.

Analogue considerations can be done for the other defects.

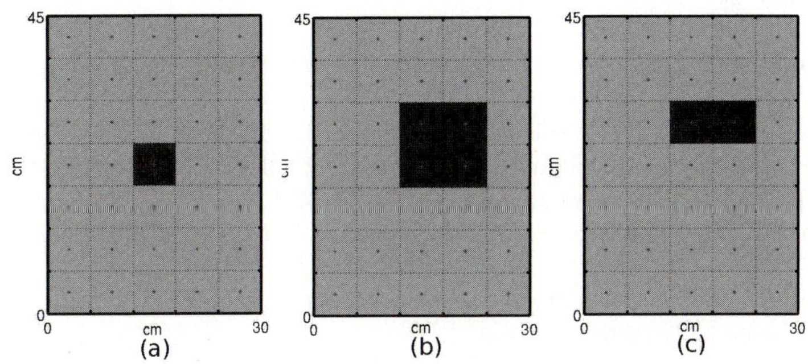

Fig. 15. Velocity map of the structure

The defect in figure 15 (b), which is interested by 4 rays, is identified with better precision. Thus, it can be said that the velocity map is always able to exactly center the defect, even if the defect size is underestimated with respect to the real dimension.

The analysis of the other features listed in section 3.1 does not provide further indications related to defects identification and sizing, thus their use in this case is redundant.

A sharper resolution in the identification of the shape and position of the defect can be achieved using more thick grids for data acquisition, On the other side, increasing the number of receivers results in a more time-consuming measurement session.

5 Conclusion

The effectiveness of a novel approach to ultrasonic buildings testing, based on the extraction of several features from waves propagating through the structures' thickness, has been tested.

The propagation of ultrasonic signals through both a stone masonry wall and a concrete pillar has been simulated using a finite element method, and various wave's features, in time and frequency domains, have been extracted from the received signals.

Results show that the proposed method is able to locate anomalies and to define them with a resolution that directly depends on the dimension of the acquisition grid and on the relative position of receivers and defect. For homogeneous structures, such as the concrete pillar, the defect localization can be achieved by simply processing of velocity data.

When the structure is made of different materials, such as the trachite wall (trachite and mortar), the effectiveness of ultrasonic testing is improved if the method is performed by resorting to the clustering of the high dimensional features space. Clustering procedures permit to compress the information related to several features in a single map, which represents a signature of the examined structure allowing one to clearly detect materials properties and structures' discontinuities.

References

1. Krautkramer, J., Krautkramer, H.: Ultrasonic testing of materials. Springer, New York (1990)
2. Cianfrone, F.: Indagini microsismiche e ultrasoniche. In: Seminario Sperimentazione su strutture; attualità ed affidabilità delle metodologie d'indagine. Venice (1993)
3. Ansys, Release 10. Documentation for ANSYS (2007)
4. Jackson, L.B.: Digital Filters and Signal Processing. Kluwer Academic Publishers, Boston (1995)
5. Nelligan, T.: An introduction to ultrasonic material analysis. Olympus NDT (2007), http://www.olympusndt.com
6. Cohen, L.: Time-frequency analysis. Prentice Hall, Englewood Clifts (1995)
7. Vergara, L., Gosalbez, J., Miralles, R., Bosch, I.: On estimating the center frequency of ultrasonic pulses. Ultrasonic 42, 813–818 (2004)
8. Salazar, A., Miralles, R., Parra, A., Vergara, L., Carrascosa, B.: Ultrasonic non- destructive testing of archaeological ceramics. In: European NDT Conference (2006)
9. Liu, L., Lane, J.W., Quan, Y.: Radar attenuation tomography using the centroid frequency downshift method. Journal of Applied Geophysics 40, 105–116 (1988)
10. Hinton, Y.L.: Problems associated with statistical patter recognition of acoustic emission signals in a compact tension fatigue specimen. NASA/TP-1999-209351 ARL-TR-1691 (1999)

11. Miralles, R., Gosalbes, J., Salazar, A.: Detection of non linearities caused by bubbles in ultrasonic signals. In: 13th European Signal Processing Conference, Antalya, Turkey (2005)
12. Schreiber, T., Schmitz, T.: Testing for non linearity in unevenly sampled time series. Phys. Rev. 59, 4044–4047 (1999)
13. Gautama, T., Van Hulle, M.M., Mandic, D.P.: On the characterization of the deterministic/stochastic and linear/nonlinear nature of time series. Technical report DPM-04-5, Imperial College, London (2004)
14. MacQueen., J.B.: Some Methods for classification and Analysis of Multivariate Observations. In: 5th Berkeley Symposium on Mathematical Statistics and Probability, University of California Press, Berkeley (1967)

Acoustic Tomography for Non Destructive Testing of Stone Masonry

M. Camplani[1], B. Cannas[1], S. Carcangiu[1], G. Concu[2], A. Fanni[1], A. Montisci[1], and M.L. Mulas[1]

[1] Electric and Electronic Dept., Univ. of Cagliari, Cagliari, Italy
{massimo.camplani,cannas,sara.carcangiu, fanni,amontisci}@diee.unica.it
[2] Dept. of Structural Engineering, Univ. of Cagliari, Cagliari, Italy
gconcu@unica.it

Abstract. Acoustic methods seem to be very suitable for evaluating a building condition because they give information with immediacy, rapidity and relatively low cost. They are based on measurements of the velocity of acoustic waves propagating through the material. Those methods are often carried out applying the Through Transmission Technique, in which the wave is transmitted by a transducer through the test object and received by a second transducer on the opposite side. In this paper, the sonic tomography imaging based on this technique has been used for non destructive testing of a stone masonry. The tomography implies that a ill-posed linear equations system has to be solved, in order to determine the velocities distribution inside the tested structure, thus highlighting the presence of anomalies. Different inversion algorithms, chosen between the most commonly used, have been implemented for determining the distribution of waves velocity in selected sections of the tested masonry.

Keywords: Non Destructive Testing, ultrasonic waves, stone masonry, tomography.

1 Introduction

In the last decades, extending the useful service life of structures and infrastructures has become of crucial importance, due to cultural, social, and economic factors. There is a strong interest in achieving this goal, and special care is reserved to the management of the increasing cost of building estate maintenance.

Management and maintenance systems for assets and structures have been developed in order to handle the information flow and store relevant data, to plan and organize the maintenance activities, and to prepare and manage maintenance budgets. In particular non destructive techniques (NDTs) for buildings diagnosis and evaluation have been experimented. NDTs are particularly suitable when dealing with structures and infrastructures because they allow one to not interfere with the state of the asset [1].

Among the non destructive techniques, sonic and ultrasonic methods, based on measurements of the velocity v of acoustic waves propagating through the material, seem to be very suitable for evaluating a building condition because they give information with immediacy, rapidity and relatively low cost [2].

O. Gervasi et al. (Eds.): ICCSA 2008, Part II, LNCS 5073, pp. 596–605, 2008.

Those methods are preferentially carried out applying the Through Transmission Technique, in which the wave is transmitted by a transducer (Emitter) through the test object and received by a second transducer (Receiver) on the opposite side. This allows to measure the time t that the wave needs to travel through the object's thickness, from the emitter to the receiver, along a path of length l; the average velocity of the wave is then obtained from the ratio l/t [3].

The major limit of the Through Transmission Technique consists of describing the velocity field in each section of the object using only one value of velocity for each path, i.e., hypothesizing that the mean is homogeneous along each wave path. The Tomographic Technique overcomes this limit using numerical analysis as a real measurement instrument, combining the results of several Through Transmission Technique applications for a sharper and reliable investigation of the object.

The present work aims to point out the reliability of acoustic tomography in the diagnosis and characterization of stone masonry buildings, deepening the analysis of the issues which affect the algebraic problem conditioning and highlighting the most suitable techniques for the algebraic problem solution.

2 The Tomographic Problem

An acoustic wave propagating through any object spends a definite time to travel from a point to another of the object. The wave covers the path between the two points spending the time t and propagating with a mean velocity v. When the distance between the two points reduces to zero a local velocity v_p, and a local slowness $s = 1/v_p$ can be defined for the point p.

The acoustic behavior of a selected section of the object is then defined when the slowness s is known continuously in every point of the investigated section. This function can be approximated dividing the section into a grid of N rectangular cells (pixels) in which v is supposed to be constant (Fig. 1).

The Tomographic Problem consists on obtaining the slowness of the N pixels starting from the knowledge of M travel times measured along a series of paths joining couples of transducers located on opposite or adjacent sides of the section.

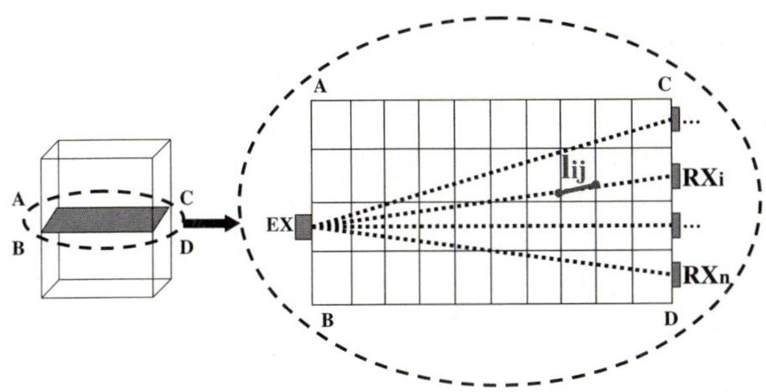

Fig. 1. Schema of Linear Tomographic Problem

The ray-paths of waves depend on the velocity distribution, and their sharp definition is a not easy problem to solve, especially when dealing with structures made of different elements such as stone masonry. In this study, the Linear Tomography has been applied, which considers the ray-paths to be rectilinear.

In order to obtain the values of the slowness in the grid, the following equations system has to be solved:

$$\mathbf{Ls} = \mathbf{t} \ . \tag{1}$$

where :
$\mathbf{t} = [t_1, t_2, \ldots t_M]$ is the vector of measured travel times;
$\mathbf{s} = [s_1, s_2, \ldots s_N]$ is the vector of slowness;
$\mathbf{L} = [l_{11}, l_{12}, \ldots l_{NM}]$ is the coefficients matrix, whose generic element l_{ij} is the length of the ith ray in the jth cell.
Thus, the Tomographic solution consists of determining the vector **s** as:

$$\mathbf{s} = \mathbf{L}^{-1}\mathbf{t} \ . \tag{2}$$

To avoid instability in matrix inversion, the number of cells must be smaller than the number of measured travel times. If the inverse of **L** exists it can be directly evaluated.

However, the inverse of **L** generally does not exist since **L** is not a square matrix, it is ill conditioned, and it has not full rank. Thus, other methods have to be used to solve the problem. In this paper three conventional inversion algorithms have been implemented. To compare their performance, the residual:

$$R = \frac{\|\mathbf{t} - \mathbf{L}\hat{\mathbf{s}}\|}{\|\mathbf{t}\|} \ . \tag{3}$$

has been evaluated, where $\hat{\mathbf{s}}$ is the obtained slowness distribution.

2.1 Singular Value Decomposition

A common method to obtain the solution of equations system (1) in the least square sense is the Singular Value Decomposition (SVD) [4-6]. The singular value decomposition can be used for computing the pseudo-inverse of the **L** matrix. Indeed, this method produces a diagonal matrix **D**, of the same dimension of **L** and with non-negative diagonal elements in decreasing order, and two unitary matrices **U** and **V** so that $\mathbf{L} = \mathbf{UDV}^{\mathrm{T}}$. Then, the pseudo-inverse of the matrix **L** with singular value decomposition is $\mathbf{L}^{+} = \mathbf{VD}^{-1}\mathbf{U}^{\mathrm{T}}$ and the solution of equations system (1) can be written as $\mathbf{s} = \mathbf{L}^{+}\mathbf{t}$.

This solution is the minimum norm solution and it is only mathematically suitable. The inverse problem is ill posed and ill conditioned, making the solution sensitive to measurement errors and noise. Regularization methods are needed to treat this ill posedness. It can be shown that the small singular values mainly represents the noise and can be discarded. Truncated SVD (TSVD) may be considered as having a filter, and hence it is less sensitive to high frequency noise in the measurements.

2.2 Iterative Methods

Because of the nonlinear relationship between the velocity and the travel time, it is almost impossible to find the solution by a single step algorithm using a linear approximation. Thus, iterative methods can be used, such as ART (Algebraic Reconstruction Technique) and SIRT (Simultaneous Iteration Reconstruction Technique). Both methods need a starting value of velocity, and then they modify iteratively this value by minimizing the difference between the measured travel times and the travel times calculated in the previous iteration. While ART goes on ray-path after ray-path, SIRT takes into account the effect of all ray-paths crossing each cell.

In the *N*-dimensional space each equation in (1) represents a hyperplane. When a unique solution exists, the intersection of all the hyperplanes is a single point. A computational procedure to locate the solution consists of starting with an initial solution, denoted by:

$$\mathbf{s}^{(0)} = \left(s_1^{(0)}, s_2^{(0)}, \ldots, s_N^{(0)}\right). \tag{4}$$

This initial solution is projected on the hyperplane represented by the first equation in (1) giving $\mathbf{s}^{(1)}$. $\mathbf{s}^{(1)}$ is then projected on the hyperplane represented by the second equation in (1) to yield $\mathbf{s}^{(i-1)}$ and so on. When $\mathbf{s}^{(i-1)}$ is projected on the hyperplane represented by the ith equation to yield $\mathbf{s}^{(i)}$, the process can be mathematically described by :

$$\mathbf{s}^{(i)} = \mathbf{s}^{(i-1)} + \frac{t_i - \mathbf{l}_i^T \cdot \mathbf{s}^{(i-1)}}{\mathbf{l}_i^T \mathbf{l}_i} \mathbf{l}_i . \tag{5}$$

where $\mathbf{l}_i$ is the ith raw of the matrix **L**.

For an over determined problem, $M > N$, ART does not give a unique solution, but this depends on the starting point. The tomographic system is often over determined and measurement noise is present. In this case an unique solution does not exist and the solution found by ART will oscillate in the neighborhood of the intersections of the hyperplanes.

The Simultaneous Iterative Reconstruction Technique (SIRT) algorithm [4-6] uses the same equations as in the ART algorithm; the difference is that SIRT modifies the slowness model taking into account at each iteration the effect of all ray-paths crossing each cell. The new value of each cell is the average value of all the computed values for each hyperplane. Then, using the SIRT algorithm, better solutions are usually obtained at the expense of slower convergence.

3 Case Study

The analysis has been carried out on a full scale stone masonry model. The wall is 90 cm wide, 62 cm high and 38 cm thick, and it is made of Trachite blocks sized $20 \times 38 \times 12$ cm^3, settled as shown in Figures 2-4 and jointed with cement lime mortar. The block assigned to the central position of the wall was not settled, thus

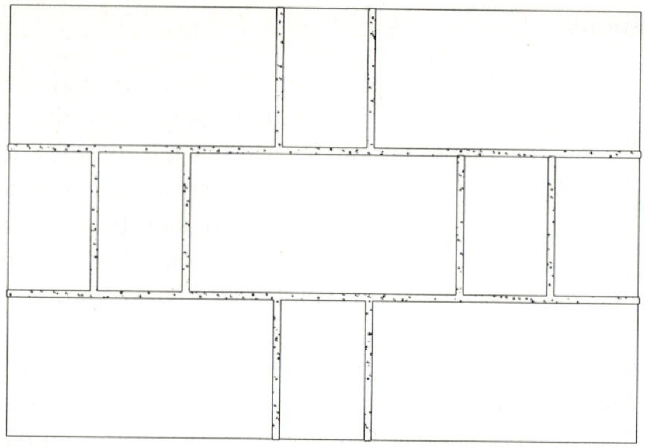

Fig. 2. Front view of the wall

realizing a macro-cavity with the same size of the missing block (Fig. 3), and assumed as a known anomaly. Mortar joints have been assumed to be 1cm thick, but since the wall was manually built by a builder, actual dimensions are not so precise.

Trachite specimens have been prepared and then tested for the determination of compressive strength and elastic modulus, following the Italian Standards UNI EN 1926 and UNI EN 14580. Results are shown in Table 3. The acoustic wave is transmitted through the test object and received by a second transducer positioned in one of the faces on the structure. Changes in received signal provide indications of variations in material continuity.

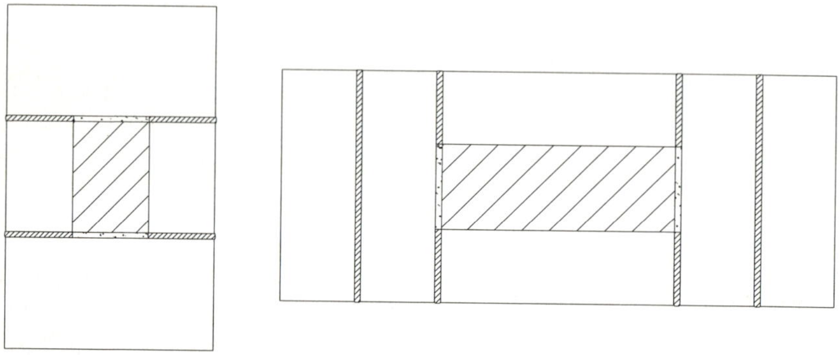

Fig. 3. Vertical plane section: position of the cavity (left) and Horizontal plane tomographic section intercepting the cavity (right)

Table 1. Trachite properties

Properties	Value (MPa)
Compressive strength	40.5
Static elastic modulus	6100

4 The Tomographic Analysis

The Tomography Technique has been applied to the real stone masonry. A horizontal plane section crossing the wall in order to intercept the central void (Fig. 3 right panel) has been chosen. The investigated section was 90 cm wide and 38 cm thick, and it has been divided in 40 cells 9 cm × 9.5 cm.

Emitters and receivers have been alternatively positioned along the side of the section, in the centers of the cells. Each acquisition has been run activating the receivers positioned in the sides of the wall parallel and adjacent to the side where the emitter was positioned (Fig. 4).

Using this measurements configuration, the section is crossed by 138 paths, whose relative travel times have been measured. Thus, system (1) consists of 138 equations and 40 unknowns.

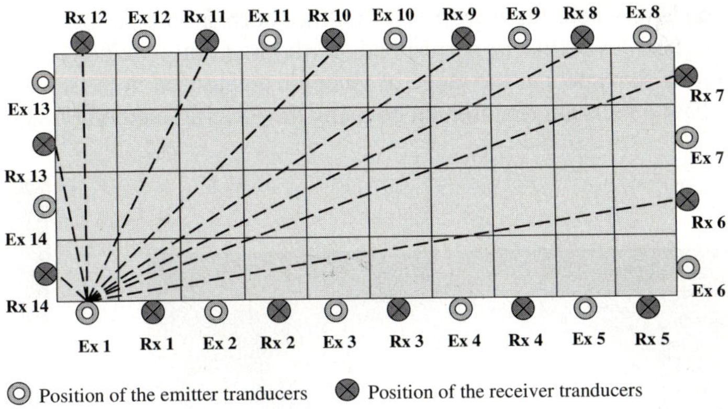

Fig. 4. Measures scheme

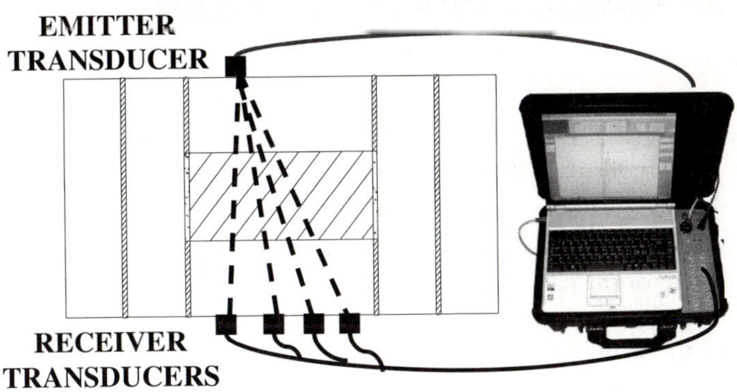

Fig. 5. Measurement set up

The testing equipment, developed by Boviar s.r.l., includes:

- a piezoelectric transducer for emitting signals;
- 4 piezoelectric transducers for receiving signals;
- a multi-channel acquisition system for signals generation and preliminary analysis;
- a PC for data storage and signal processing.

The measurement set-up and the operative procedure are shown in Figure 5.

The software for tomographic data analysis was developed in Matlab®. All the results in this section have been performed using an AMD Opteron Processor 250 2.39GHz with 3.5GB of RAM running under Windows XP Professional.
The technique described in Section 2 has been applied to the case study of Section 3.

Due to the different density of the cells, the velocity should be spread within the range between 10^3 and $3 \cdot 10^3$ m/s, which correspond to the velocity of the wave respectively in the air and in the trachite. The map of velocities is represented by a 256 levels gray scale diagram, where the lowest level (white) corresponds to the minimum velocity, and the highest level (black) corresponds to the maximum velocity. The 256 levels are normalized with respect to the range of velocities measured in the represented tomography. The 10 × 4 cells are represented by using the Kriging method [7-9].

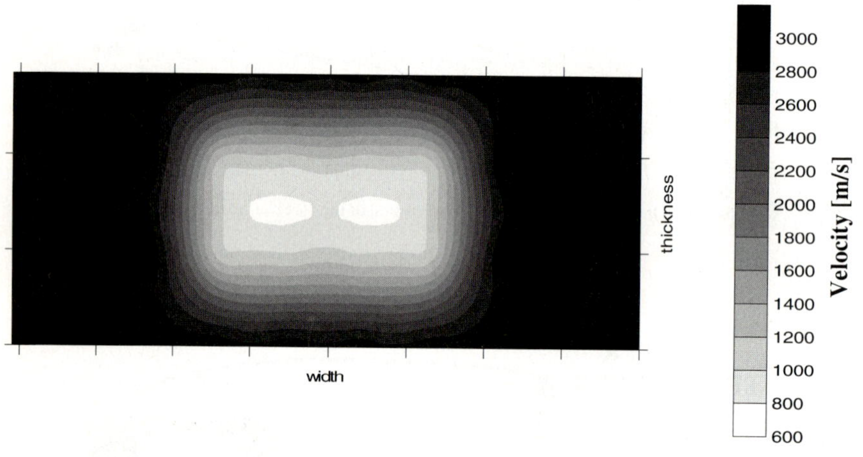

Fig. 6. Ideal tomographic map of the structure

The ideal velocity map of the selected section has been elaborated assuming a velocity of 10^3 m/s for the cavity and a velocity of $3 \cdot 10^3$ m/s for the stones, without taking into account the mortar (Fig. 6).

In Figure 7 the map obtained with the SVD method is shown. The calculation time has been of 16 ms, and the final residual was R = 0.10088. As it can be seen, the Kriging method identify a central region, corresponding to the position of the cavity, where the velocity of the wave is lower.

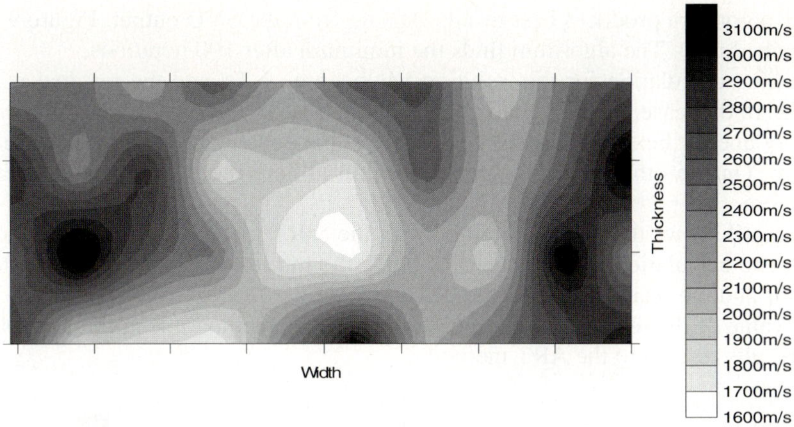

Fig. 7. Tomographic map obtained with SVD method

Figure 8 shows the map obtained with the Truncated SVD method.

The range [100 – 25]% of SVD matrix rank has been examined. No significant visual improvement was observed beyond 77%. To adopt the best image quality for the Truncated Singular Value Decomposition (TSVD) keeping the residual relatively low, the TSVD algorithm was set to 77% of original rank, i.e., by discarding singular values less than 0.2. The final residual was equal to 0.10328. As it can be noted, the regularization attenuates the noising components, so that the region associated to the central cavity is here wider, and the peripheral regions with low velocity are no longer present.

Iterative algorithms have been tested for 1000 iterations starting from different initial points, i.e., a vector of zeros, the SVD output, the TSVD output and the vector composed by the average slowness of the ray-paths crossing each cell. In order to obtain reasonable values of velocity, the slowness values are constrained in the range [0-6000 m/s].

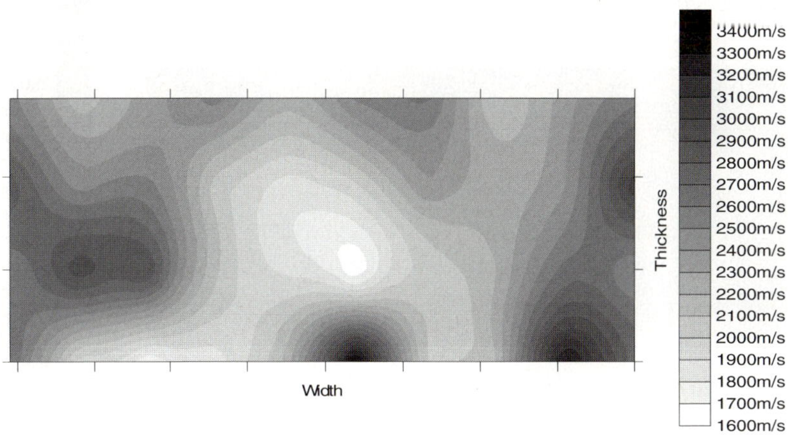

Fig. 8. Tomographic map obtained with SVDT method

ART algorithm produces best results starting from the SVD output. Figure 9 shows the obtained map. The algorithm finds the minimum after 100 iterations.

The time calculation for this solution has been of 1.2 s and the residual was R = 0.13222. In this case, the residual is bigger than those of the other algorithms, but this is mainly due to the high value of velocity in some cells placed in the left side of the structure. On the other hand, the central region is here more clearly distinguishable than before.

Figure 10 shows the results obtained with the SIRT method starting from the SVD output. The calculation time has been of 0.9 s; the minimum has been found after 100 iterations and the residual was R =0.10143.

The value of the residual shows that with the SIRT method it is possible to obtain a better results respect to the ART method.

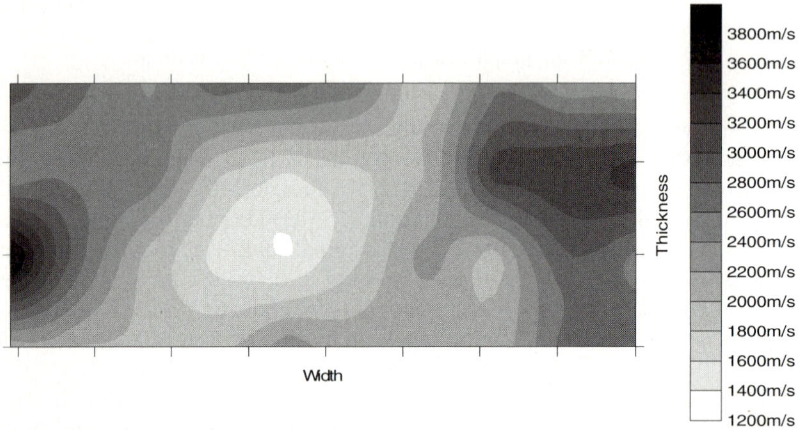

Fig. 9. Tomographic map obtained with ART method

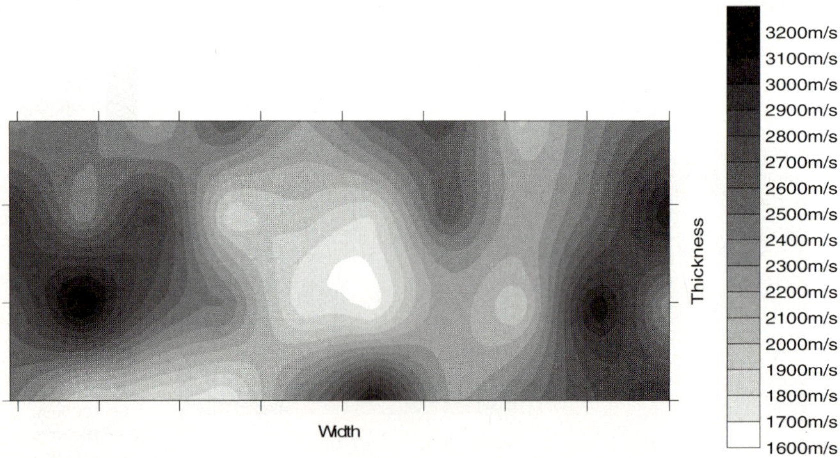

Fig. 10. Tomographic map obtained with SIRT method

5 Conclusion

The present paper illustrates the sonic tomography performed on a real full scale masonry. Due to the uncertainties of measurements, a set of redundant measures has to be acquired, so that an over determined equations system has been solved. Four different algorithms have been used to solve such equations system and the results have been compared.

The obtained results show that all the algorithms are able to identify the presence of an anomaly in the stone masonry. In spite of a bigger value of the residual, the SVD algorithm seems to be the most suitable one for the reconstruction of the tomography, because it shows more clearly than the other algorithms the presence of the defect inside the wall.

The results obtained with the tested approaches are valid to define with a good precision the contour of a macroscopic cavity inside the structure. Using more thick grids for data acquisition, it will be possible to define the shape of defect with smaller dimensions, thus achieving a sharper resolution.

References

1. Fib bulletin 17. Management, maintenance and strengthening of concrete structures. (2002)
2. Krautkramer, J. and Krautkramer H.: Ultrasonic testing of materials. Springer Verlag, New York (1990)
3. Cianfrone, F.: Indagini microsismiche e ultrasoniche. In: Seminario Sperimentazione su strutture; attualità ed affidabilità delle metodologie d'indagine. Venice (1993)
4. Berryman, J.G.: Non linear inversion and tomography. Lecture notes, Earth Resource Laboratory, MIT (1991)
5. Herman, G.T.: Image Reconstruction from Projection: the Fundamentals of Computerized Tomography. Academic, New York (1980)
6. Ivansson, S.: Seismic Borehole Tomography – Theory and Computational Methods. Proc IEEE 74, 328--338 (1986)
7. Deutsch, C.V. and Journal, A.G.: GSLIB Geostatistical Software Library and User's Guide. Oxford University Press, New York (1992)
8. Isaaks, E.H. and Srivastava, R.M.: An Introduction to Applied Geostatistics. Oxford University Press, Oxford (1989)
9. Watson, D.F.: Contouring: A Guide To The Analysis And Display of Spatial Data. Pergamon Press, New York (1992)

Time and Frequency Approaches to Non Destructive Testing in Concrete Pillars Using Neural Networks

Barbara Cannas, Sara Carcangiu, Francesca Cau, Alessandra Fanni, and Augusto Montisci

Department of Electrical and Electronic Engineering, University of Cagliari, Piazza d'Armi, 09123 Cagliari, Italy
{cannas,s.carcangiu,francesca.cau,fanni,amontisci}@diee.unica.it

Abstract. In this paper, Multi Layer Perceptron neural networks have been trained to identify the position of defects in concrete structures analyzed using an ultrasound technique. A diagnostic model obtained by means of Finite Elements techniques has been used to model the ultrasound transmission through a concrete pillar of specified size affected by defects in different positions. The obtained signals have been processed both in the time and frequency domains, in order to reduce data dimensionality and to compute suitable features. Results show good accuracy in the identification of the position of the faults.

Keywords: Non Destructive Testing, ultrasonic waves, neural networks.

1 Introduction

In recent years, several methods have been proposed that use the propagation of ultrasonic waves for Non Destructive Testing (NDT) tasks in concrete [1-12]. In fact, the wave velocity, amplitude and spectrum clearly depend on the state of the material.
An electro-mechanical transducer is used to generate an impulsive ultrasonic stress wave that propagates into the object being inspected as compression, shear, and surface waves. Reflection of the stress pulse occurs at boundaries separating materials with different densities and elastic properties.

In the *impact echo testing* the pulse traveling back to the transducer, which also acts as a receiver, is displayed on an oscilloscope, and the round trip travel time of the pulse is measured electronically. By knowing the speed of the stress wave, the distance to the reflecting interface can be determined.
In the transparency testing that requires access to two surfaces, the receiver is in the opposite side of the transmitter. The first pulse that arrives at the receiving transducer will be diffracted around the periphery of the eventual defect [1].

Literature reports different approaches to concrete strength estimation based on the correlation between physical parameters and concrete strength [2,3]. For example, the correlation between wave velocity and compression strength is a classical technique that can be applied to each concrete after a calibration campaign. The measurement of wave's velocity can be performed on cores extracted from the structure; thus, they are representative of a region with limited and supposedly homogeneous size. In [4] the authors analyzed the longitudinal speeds of propagation of ultrasound in various

O. Gervasi et al. (Eds.): ICCSA 2008, Part II, LNCS 5073, pp. 606–616, 2008.

samples of concrete, confirming the exponential relationship with the samples' compressive strength characteristics as evaluated during destructive testing.

Imaging and interpretation of the waves in the time domain as well as in the frequency domain can be used to measure the thickness and the location of the defect. In [5] acoustic parameters obtained from received signals generated by impact on concrete surface are analyzed to estimate defect sizes.

The attenuation of the transmitted [6] or backscattered waves [7], the spectrum degeneration [8,9] or the evolution of the signal energy [10] are useful ultrasonic indexes of the damaging evolution.

Generally speaking, in wave propagation based methods, the user must interpret the recorded signals. Qualitative interpretation requires a user experience in waveform recognition.

Artificial intelligence techniques such as neural networks, seem to be particularly suitable to make automatic the ultrasonic signals interpretation for their ability of self-learning and generalization. In [11] a Multi Layer Perceptron neural network is trained to recognize three damage conditions in a reinforced concrete beam. The beam is divided into eight regions as diagnosis location and the damage is simulated by the reduction of the material strength.

In [12] a concrete undamaged detection system is built using fuzzy neural networks to obtain the concrete strength, processing the rebounding value and the carbonation depth.

In this paper, a Multi Layer Perceptron (MLP) neural network is trained to assess the location of defects in a concrete pillar. The experimental test pillar is modeled using finite elements methods. Numerical analysis of a pillar of rectangular cross section is performed to generate training and testing samples for neural network assessing task. In particular, the wave source is due to an impact in the pillar surface. The received signal, recorded in the opposite face of the pillar, is used to feed the network after a suitable preprocessing.

The paper is organized as follows: section 2 describes the diagnostic technique, section 3 illustrates the data set. The adopted feature extraction techniques are reported in section 4. Section 5 describes the neural diagnostic models, whereas in section 6 the results are presented and discussed. Finally, section 7 reports the conclusions.

2 The Transparency Technique

The transparency technique is based on the propagation of ultrasonic waves in the test structure. A mechanical impact is applied on the surface of the concrete element and a receiver is located in the opposite surface of the structure. The elastic waves, detected at the receiver, are converted into electrical signals by directly coupling piezoelectric transducers in the surfaces of the structure.

Usually, the applied excitation frequency ranges from 15 to 75 kHz, and its propagation velocity in a mean is a function of the elastic characteristics of the mean itself (Young modulus, and Poisson coefficient) and of its density.

Dishomogeneities in the structures (cracks, degraded regions, cavities, etc.) produce variation of the propagation velocity; reflections, refractions, partial absorption and attenuation of the wave. The analysis of such phenomena can be useful to evaluate the presence of such dishomogeneities.

Fig. 1 shows an experimental set up to apply the transparency technique.

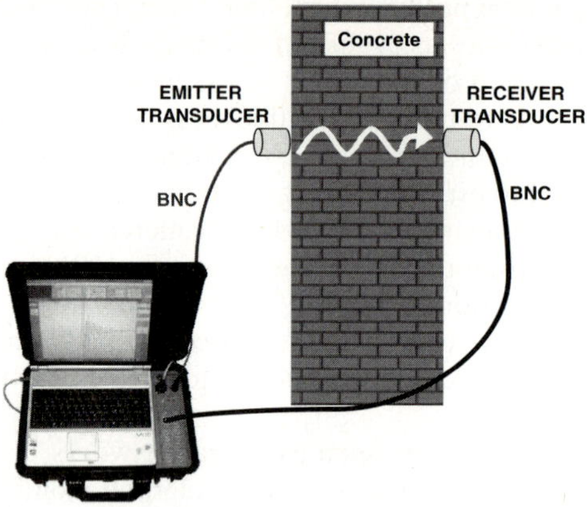

Fig. 1. Experimental set-up

3 Data Set

The propagation of an elastic wave through a concrete pillar has been simulated by means of the Finite Element Method (FEM), using LS-DYNA module of ANSYS [13].

The numerical computations have been carried out simulating an advanced multi-channel acquisition system. In the examined cases, the instrument has been simulated as made up of three excitation transducers and three receiving transducers located in

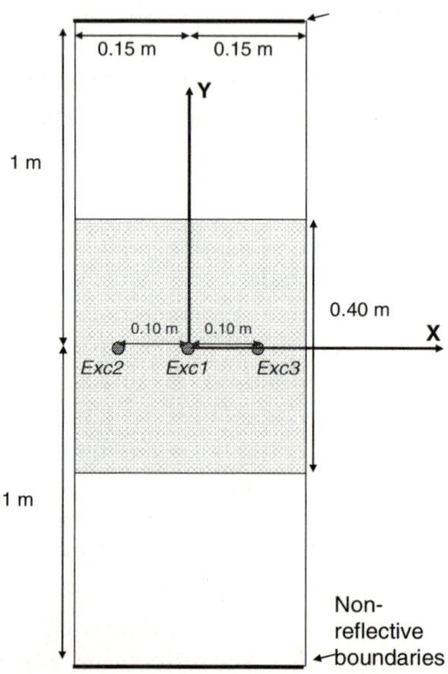

Fig. 2. Front view of the pillar and excitation points

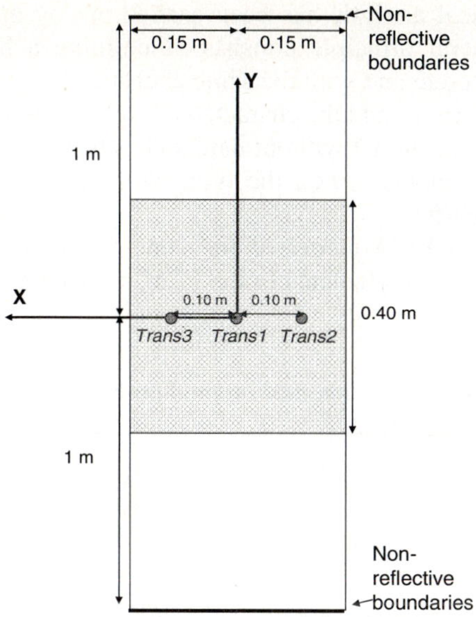

Fig. 3. Back view of the pillar and observing points

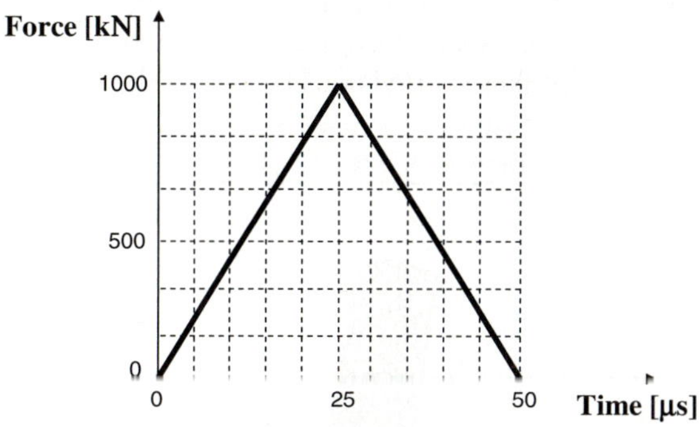

Fig. 4. Excitation signal

the opposite faces of a pillar, parallel to the z = 0 plane (see Fig. 2 and Fig. 3). The numerical computations simulate an experimental session where a portion of the pillar 40 cm high is inspected.

The excitation is a force characterized by a triangular shape, lasting 50μs, and whose amplitude is equal to 10^3 kN [5, 14]. The excitation signal is shown in Fig. 4.

The generated wave propagates in the pillar until it is received by the transducer in the opposite side.

Data have been generated by simulating propagation and scattering of elastic waves in the concrete structure with defects sized $8 \times 8 \times 8$ cm^3 in different positions.

The transient numerical analysis has been carried out by means of the finite element method. The general principle consists of defining a homogeneous medium equivalent to that of concrete and with the same acoustic characteristics.

The geometrical dimensions and the characteristics of the pillar are shown in Table 1.

An analysis of the pillar model without defects has been firstly performed in order to set the simulation parameters, i.e., the dimensions of the elements in the FEM model, and the sampling time.

The three dimensional FEM model of the concrete pillar is reported in Fig. 5, whereas the dimensions of the elements in the FEM model, and the sampling time are shown in Table 2.

Table 1. Geometrical dimensions and material properties of the concrete pillar

High [m]	2
Width [m]	0.3
Depth [m]	0.3
Young's modulus [GN/m^2]	33
Poisson's coefficient	0.300
Density [kg/m^3]	2300

Table 2. Model parameters

Number of x-axes subdivisions	30
Number of y-axes subdivisions	100
Number of z-axes subdivisions	30
Sampling time [μs]	5
Excitation frequency [kHz]	50

Fig. 5. 3-D FEM model of the concrete pillar

Non reflecting boundaries have been imposed on the upper and on the bottom surfaces of the pillar in order to simulate an infinite boundary. These constraints are typically used to limit the size of the model and to prevent artificial wave reflections at the boundary.

In a second phase, the same pillar has been considered, which presents different defects. Varying the location of the defect into the inspected portion of the pillar, several simulations have been performed.

The defects are modeled by simply varying the characteristics of a set of elements inside the pillar, in order to simulate cavities.

The defect positions and physical characteristics are reported in Table 3 and Table 4.

Table 3. Defect barycentre positions.

Range of x-axis [m]	$0 \div -0.10$
Range of y-axis [m]	$0 \div 0.12$
Range of z-axis [m]	$-0.08 \div 0.08$

Table 4. Physical Defect characteristics

Young's modulus [GN/m^2]	$1.5 \cdot 10^{-4}$
Poisson's coefficient	0.1
Density [kg/m^3]	1.3

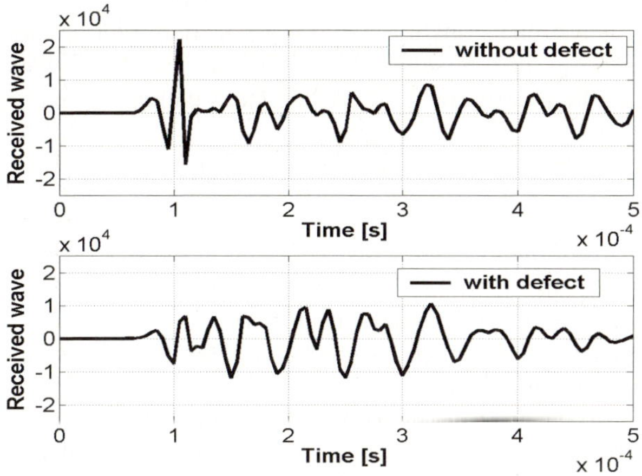

Fig. 6. Transmitted waves corresponding to pillar without and with defect

The resulting data set is composed by 80 defect cases plus the fault free case. The limited number of simulations is due to the high CPU time required for each of them.

For each defect, the used signals are the waves measured in the 3 observing points.

In order to reduce the data dimensionality, the acquired waves obtained from the numerical simulations have been pre-processed.

Two procedures have been proposed, dealing with time domain data and frequency domain data.

A time window of 0.05 ms has been chosen. This time window contains 100 samples. In Fig. 6, the received signals corresponding to pillar without and with a defect respectively, are reported.

4 Feature Extraction

In general, if D is the number of simulations ($D = 81$ in the present paper), S is the number of samples in the chosen temporal window ($S = 100$ in the present paper), and N is the number of receiving transducers ($N = 3$), D different matrices of dimension $S \times N$ can be built. Let $\mathbf{M_d}$, $d = 1,\ldots D$, be the matrix of dimension $S \times N$ characterizing a particular defective pillar or the pillar without defect. Each matrix $\mathbf{M_d}$, which represents the pattern describing the corresponding defect, has been processed (as detailed in the following) in order to generate input patterns suitable to feed the NN classifier. The corresponding output pattern will be constructed associating to the input pattern the corresponding coding of the defect, i.e., a flag representing the presence or the absence of defect and, in case of defective pillar, the barycentre position.

As outlined, some post processing are necessary on the $\mathbf{M_d}$ matrices, in order to extract the significant features and to reduce the data dimensionality.

4.1 Feature Extraction in the Frequency Domain

A two-dimensional Fast Fourier Transform (FFT-2D) has been applied to each of the $\mathbf{M_d}$ matrices, obtaining D new couples of matrices, $\mathbf{A_d}$ and $\mathbf{P_d}$, containing the amplitude and the phase components of the FFT-2D respectively.

To reduce the data dimensionality, a subset of K amplitude components has been selected from each $\mathbf{A_d}$. The K selected components are those whose amplitude is greater than k times the maximum value of the matrix elements, i.e.,

$$a_{ij} \geq k \cdot \max_{s=1,\ldots S;n=1,\ldots N} \left(a_{sn}\right), \quad k = 0.1,\ldots,0.6$$

The selected K amplitude components of all the D matrices have been stored in a unique matrix of dimension $K \times D$, and the corresponding phase components have been stored in another matrix of the same dimension.

The resulting matrices have been normalized so that they have mean equal to zero and standard deviation equal to 1.

The best value of k has been heuristically determined. In particular, six neural network training sessions have been performed by varying k between 0.1 and 0.6. The network with the minimum Mean Squared Error on the test set is retained.

4.2 Feature Extraction in the Time Domain

In the time domain, the signal acquired on the pillar without defects is used as offset for all the other acquisitions.

For all the signals, a temporal window [t_{start} - t_{end}] has been selected, where $t_{start} = 110$ µs, $t_{end} = t_{start} + (h + 4) \cdot 5$ µs, $h = 1,\ldots,20$.

In terms of $\mathbf{M_d}$ matrix, only a sub-matrix has been considered for the neural network training, whose rows refer to the samples belonging to the chosen window. The obtained matrix has been normalized so that it has the mean value equal to zero and standard deviation equal to 1.

The best value of h has been heuristically identified. In particular, twenty neural network training sessions have been performed by varying h between 1 and 20. The network with the minimum Mean Squared Error on the test set is retained.

5 Neural Network NDT System

A Multi-Layer Perceptron (MLP) neural network is constituted by an input layer, one or more hidden layers, and one output layer of neurons. The neurons of each layer are connected with all the neurons of the previous layer. The connection weights are the free parameters of a learning process. They are determined by presenting to the network a set of actual input-output values (the training set) and searching the minimum of a suitable error function, which depends on the connections weights.

To ensure a good generalization capability, a cross-validation technique has been used during the training phase, based on monitoring the error on an independent set, called validation set [15].

The validation error will normally decrease during the initial phase of learning, as the training set error does. However, when the network begins to over-fit the data, the error on the validation set will typically begin to rise. When the validation error increases for a specified number of iterations, the learning is stopped, and the weights and biases at the minimum of the validation error are returned.

To correctly estimate the performance of the model and evaluate the generalization capability, a third set (the test set) is used Training, validation and test set are composed of 65, 8, and 8 examples respectively.

The network has been trained using the Levenberg-Marquard learning algorithm.

5.1 Frequency Domain

For each value of k, which represents the amplitude threshold used to select the frequency components, several MLPs have been trained varying the number of the hidden layer nodes. In particular, the growing method has been adopted [4], which consists of progressively increasing the number of hidden neurons, until the performance of the network approaches the desired value.

The value of k giving the best performance in the validation set is chosen to select the best MLP architecture. Best results have been obtained using $k = 0.3$.

The following architecture has been used:

- 92 input nodes;
- one hidden layer with 8 nodes and hyperbolic tangent activation function;
- 3 output nodes: two nodes correspond to the x and y coordinates of the barycentre of the defect; the third one is a flag representing the presence or the absence of the defect.

The output nodes activation functions are linear.

5.2 Time Domain

For each value of h, which corresponds the duration of the considered time window, several MLPs have been trained using the growing method. The value of h giving the best performance in the validation set is chosen to select the best MLP architecture. Best results have been obtained using a temporal window of 17 samples ($h = 13$).

The following architecture has been used:

- 51 input nodes;
- one hidden layer with 10 nodes and hyperbolic tangent activation function;
- 3 output nodes: two nodes correspond to the x and y coordinates of the barycentre of the defect; the third one is a flag representing the presence or the absence of the defect.

The output nodes activation functions are linear.

6 Results

The following results refer to the classification errors evaluated on an independent test set consisting of 8 examples, not used during the training procedure.

Figure 7 shows the actual values of the position of the barycentre and the corresponding values predicted by the neural classifier using the frequency domain analysis.

Figure 8 shows the same parameters for the time domain analysis.

As can be noted, the classifier is able to detect both the defect x and y coordinates with good accuracy in the tested configurations in both frequency and time domain.

The same procedure can be applied in order to estimate the z coordinate of the defect. For this task, the described methodology can be applied positioning emitters and receivers in the faces of the pillar parallel to the $x = 0$ plane and training the neural network to identify the z and y coordinates of the defect. In this way, the position of the barycentre of the defect might be completely identified.

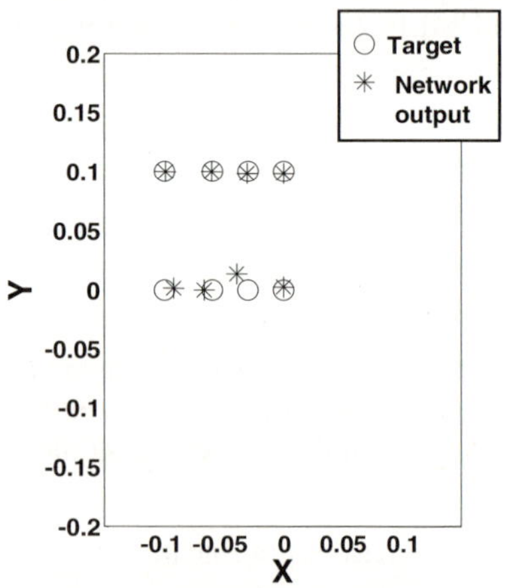

Fig. 7. Position of the barycentre in the XY plane: actual (o) and predicted (*) values for the frequency domain analysis

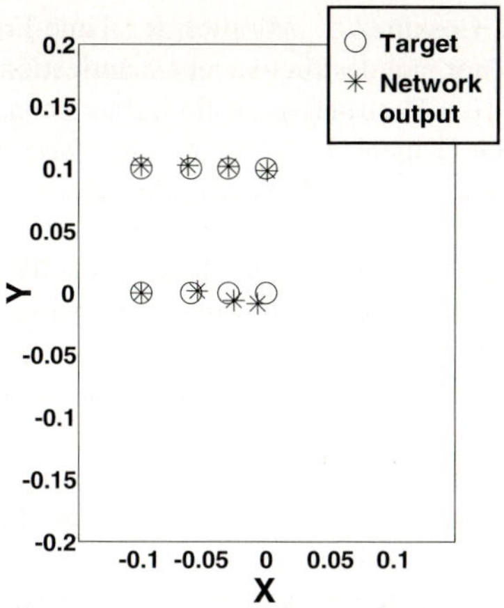

Fig. 8. Position of the barycentre in the XY plane: actual (o) and predicted (*) values for the time domain analysis

7 Conclusions

In the present paper a neural model for non destructive evaluation of concrete's damaging by using ultrasound technology is proposed. The results show that the proposed technique can be efficient for the diagnosis purpose.

Future research will investigate the method applicability using a more realistic model of the pillar including the interaction of the ultrasonic wave with the scatterers. Moreover, tests on a real world pillar will be performed.

Acknowledgments. The work was supported by the Italian Ministry of Education under the PRIN Program.

References

1. Colla, C., Schneider, G., Wöstmann, J., Wiggenhauser, H.: Automated Impact-Echo: 2- and 3-D Imaging of Concrete Elements. In: NDT.net, vol. 4(5) (1999)
2. Berthaud, Y.: Damage measurements in concrete via an ultrasonic technique: Part I. Experiment. NDT and E International 30(4), 264–264 (1997)
3. Popovics, S., Popovics, J.S.: Effect of stresses on the ultrasonic pulse velocity in concrete. Materials and Structures 24, 15–23 (1991)
4. Del Rio, M., Jimenez, A., Lopez, F., Rosa, F.J., Rufo, M.M., Paniagua, J.M.: Characterization and hardening of concrete with ultrasonic testing. Ultrasonics 42(1-9), 527–530 (2004)
5. Asano, M., Kamada, T., Kunieda, M., Rokugo, K.: Impact acoustics methods for defect evaluation in concrete. In: Proceedings of International Symposium on Nondestructive Testing Civil Engineering (2003)
6. Ould Naffa, S., Goueygoua, M., Piwakowskia, B., Buyle-Bodinb, F.: Detection of chemical damage in concrete using ultrasound. Ultrasonics 40(1-8), 247–251 (2002)

7. Fuente, J.V., Vergara, L., Gosalbez, J., Miralles, R.: Time-Frequency analysis of ultrasonics backscattering noise dor nondestructive characterizations on cement pastes. In: 8th European Conference on Non Destructive Testing, Barcelona, June 17-21 (2002)
8. Garnier, V., Corneloup, G., Topani, E., Leygonie, M.: Non destructive evaluation of concrete damage by ultrasounds. In: 15th World Conference on Non-Destructive Testing, Rome (2000)
9. Gaydecki, P.A., Burdekin, F.M., Damaj, W., John, D.G.: The propagation and attenuation of medium-frequency ultrasonic waves in concrete: a signal analytical approach. Meas. Sci. Technol. 3(1), 126–134 (1992)
10. Anugonda, P., Wiehn, J.S., Turner, J.A.: Diffusion of ultrasound in concrete. Ultrasonics 39, 429–435 (2001)
11. Hsu, D.S., Tsai, C.H.: Reinforced concrete structural damage diagnosis by using artificial neural network. In: Proceedings of the 1997 IASTED International Conference on Intelligent Information Systems, p. 149 (1997)
12. Yan, S., Xu, J., Yao, G.: Concrete strength evaluation based on fuzzy neural networks. In: Proceedings of 2004 International Conference on Machine Learning and Cybernetics, vol. 6, pp. 3344–3347 (2004)
13. Ansys Release Note, Version 8.1 (2003)
14. Grosse, C.U., Reinhardt, H.W.: New developments in quality control of concrete using ultrasound. In: Proceedings of the International Symposium on Non-Destructive Testing in Civil Engineering, Berlin (2003)
15. Bishop, C.M.: Neural networks for pattern recognition. Clarendon Press, Oxford (1995)

Innovative Methodology for IR Acquisition

G. Acciani, A. Camposarcone, and S. Vergura

Politecnico di Bari
Dept. of Electrotechnics
St. E. Orabona 4, 70125 Bari, Italy
vergura@deemail.poliba.it

Abstract. This paper propose an innovative and cheap methodology to guarantee a uniform heating of the sample under examination before an IR acquisition. The comparison with the standard heating confirms the goodness of the approach. Moreover, this paper proposes an algorithm for the processing of the IR images, able to detect the weft of the wall. The overall procedure can be also utilized for the defect diagnosis.

1 Introduction

The evaluation of the preservation of the artistic heritage is a very important issue in Italy because of the high number of existing historical buildings. One of the non-destructive tests used for artistic heritage is the InfraRed Thermography (IRT). The drawbacks affecting the IRT are related to the strong influence of the external factors. In order to obtain reliable results it needs to monitor all the variables affecting the system [1]-[10]. The experiments conducted in laboratory, if some variables are not taken into account, give only possible results [3]-[4]-[5].

This paper presents a procedure for IR acquisition, pre processing and processing of images for concrete structures. Particular attention has been devoted to the acquisition methodology and to the images processing. In fact, this paper proposes an innovative procedure to guarantee a uniform heating of the wall under examination. When artificial heating is imposed by means of infrared lamps or heat pump etc., a non-uniform heating is almost always present except for very expensive solutions (eg. a matrix of IR lamps). In order to overcome the issue related to the non-uniform heating, we have designed and realized a structure for conveying uniformly the heat long a desired direction. The system, presented in Section 3, uses the chimney effect.
This paper presents also an algorithm for the processing of the images, called structure, able to give detailed information about the weft of the wall. The procedure can be easily utilized for identification and classification of defects.

Section 2 reports the proposed procedure, Section 3 the pre-acquisition setup, Section 4 the mathematical tools for the pre-processing and processing of the IR images, while Section 5 reports experimental results.

O. Gervasi et al. (Eds.): ICCSA 2008, Part II, LNCS 5073, pp. 617–626, 2008.

2 Proposed Procedure

The proposed procedure for the IR acquisition can be resumed in the following steps:

1. the structure under test - which we will call *sample* from now on – has to be heated by means of a 2-kW heat pump for at least 30 minutes; the heat pump has to be posed in a suitable structure for conveying uniformly the heat onto the surface of the sample (see Section 3).
2. The acquisition frequency has to be at least 1 frame/20 sec.
3. Pre-processing of the thermo grams by means of a Gaussian filter with standard deviation equal to 15 times the standard deviation of the noise, then by applying a median filter with a kernel equal to 4.
4. Processing the thermo grams by means of the algorithm, *Structure*, designed by the authors.

3 Pre-acquisition Set-Up

When heat sources are used to excite the sample, the non-uniform heating happens very often [10]. In this case the thermo-gram shows areas with very different temperatures even if the sample is constituted by the same material; it is evident an area strongly interested by the radiation [3][4].

After a study in depth of several solutions able to guarantee a uniform heating of the sample we have designed and realized a structure (Fig. 1) having sizes 2.0mx1.0mx0.5m (height*width*depth), constituted by a wooden support and covered by a double layer thin, made of aluminium and 3mm-thick expanded polyethylene foam. The top of the structure is open and it behaves like a chimney. The open side of the structure (Fig. 2) is posed in direct contact with the sample.

By means of hinges it is possible to adjust the dimension of the aperture above in order to regulate the heat emission.

A heat source is posed inside the structure; it is a 2-kW heat pump with diffusers.

The heating is made through 30 minutes and then the IR acquisition starts.

Fig. 1. The chimney structure

Fig. 2. Open side with the heat pump

4 Mathematical Tools

Acquired thermo-grams are not able to highlight all the characteristics of the sample, as they are affected by noise. Then a pre-processing of the images has to be utilized before processing them in order to identify the segmentation or to find the desired characteristics.

4.1 Pre-processing

The thermo-gram presents noise deriving from the electronic equipment, the surface reflection of the sample etc. The noise implies an excessive increase of the single pixel intensity; therefore this pixel could become predominant with respect the near pixels and could produce little clean images [9][11]. For this aim local filters are advised because they calculate each point of the resulting image versus the point of both the original image and its neighbourhood [7].

The filtering process or image convolution can be represented as:

$$I_{FILTERED} = \sum_i \sum_j g(i,j) = \sum_i \sum_j f \otimes h = \sum_i \sum_j (\sum_m \sum_n f(i-m, j-n) \cdot h(m,n)) \quad (1)$$

where $I_{FILTERED}$ is the image resulting by the convolution process, h(m,n) is the kernel square matrix. The family of the smoothing filters [7] includes the median filter [7]-[11], very useful in removing both the impulsive noise (salt and pepper) and semantic one.

The *median filter* is nonlinear; the value of each pixel for the resulting image is not algebraically determined, but it is obtained as median of the same pixel and its neighbours in a (2k+1)x(2k+1) around.

Another smoothing filter is the *Gaussian* one; the utilized mask corresponds to a bi-dimensional Gaussian function, having mean value equal to zero and standard deviation obtained by the product of two Gaussian, each one of them oriented along the principal directions, x and y.

The resulting Gaussian will be:

$$G(x,y) = G(x) \cdot G(y) = \frac{1}{2\pi\sigma^2} \cdot e^{\frac{(x^2+y^2)}{2\sigma^2}} \quad (2)$$

The Gaussian has a discretized expression in the space and the mask coincides with a matrix whose elements are calculated in the point of the reticolo del piano dell'immagine. Obtained matrix will have real-value elements decreasing as it allontanandosi dal centro. As the standard deviation value increases it needs to utilize large matrixes in order to not lose the information associated to the Gaussian.

4.2 Processing

There are several techniques to be used for the image processing which differ in basis of the parameter (or parameters) under examination: Color Segmentation, Deblurrig, Enhancement, Morphological Segmentation [7][9][11].

In this paper Edge Detectors algorithms, belonging to the family of Morphological Segmentation, will be used because we are interested to identify and to extract the edge of the images. The edge is a vector dependent on the gradient value of the brightness level and having orthogonal direction with respect to the gradient direction; this last one is oriented long the increase direction of the brightness level. The edge search corresponds with the gradient search of the pixels obtained by means of Canny's method. The Canny's filtering is applied by using a known procedure, called *Non-Maxima Suppression*, by means of a bi-dimensional Gaussian which searches for the local maxima of the first derivative in the gradient direction and realizes the comparison between these local maxima and a prefixed threshold; other values are suppressed.

Even if this approach just gives good enough results, we have implemented a segmentation algorithm, called *Structure* and implemented in Matlab environment, which allows to identify areas with different temperatures in an image with two intensity levels (0,1). The algorithm searches the local minima long all the rows and all the columns of the pre-processed thermo-gram. Two characterizing parameters of the proposed algorithm are *step* and *delta*; they allows to define the desired detail level. The former one, *step*, represents the length of the window, in pixel, in which the search of the local minimum is done. The latter one, *delta*, represents the threshold for which a minimum is considered meaningful; this choice is realized by comparing the delta parameter with the difference between the hypothetical minimum and the value at the window extreme.

5 Experimental Results

Measurement instruments have pointed out the following environmental conditions

Environment temperature	24,5°C
Temperature bifore the heating	24,5°C
Temperatura after the heating	40,5°C
Relative dampness	55%
Initial reflected temperature	23,9°C
Final reflected temperature	35°C
ε_{wall}	0.90

After removing the wooden structure and the heat source, thermo-grams are acquired through 30 minutes with a 20 second sample rate. A Flir Systems ThermoVision^{TM} A40 M has been used. The thermal sensor (24°) has been posed on a table in front of the wall under examination; the table is 1m height while the distance between the table and the wall is 2.0m. In this condition the investigation area corresponds to a 0.63mx0.85m (heigth*length) rectangular. Hence, each pixel represents a 3mm square. Fig 3 reports some thermo-grams during the cooling phase: a), b) and c) are related to an investigation without the proposed wooden structure, while d), e) and f) refer to the investigation with its. In the first three images a concentrated heat is evident, while the last three images are characterized by a uniform heating.

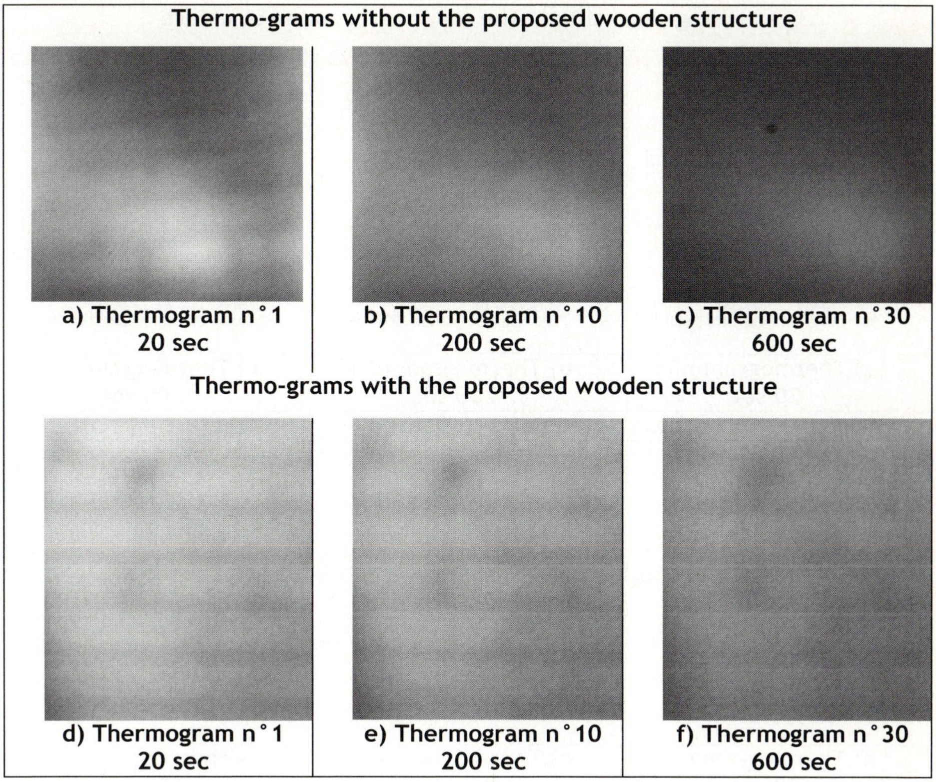

Fig. 3. Comparison between thermograms related to the investigations without and with the proposed wooden structure

The wall weft is just manifest in these images even if none treatment has been conducted for the noise reduction. By evaluating the histogram of the previous thermograms (Fig. 4), a reduction of both the standard deviation and the peak values are present in both the two investigations. In the first investigation the standard deviation of the thermogram c) is strongly reduced with respect the thermo-gram a) (18.115 → 6.151), while in the second one a more modest reduction is present between the thermogram d) and f) (21.079→11.088). The first investigation shows a reduction of 66%, while the second one a reduction of 47%. This parameter shows that the non-uniform heating persists during the overall cooling phase. Obviously, comparing the standard deviation of the respective thermograms [a) with d), b) with e), c) with f)] the non-uniform heating is always present.

5.1 Pre-processing

Analyzing the vertical profiles of thermo-grams it is possible to distinguish the bricks and the lime. The pixels representing the lime show a minimum. After filtering as proposed

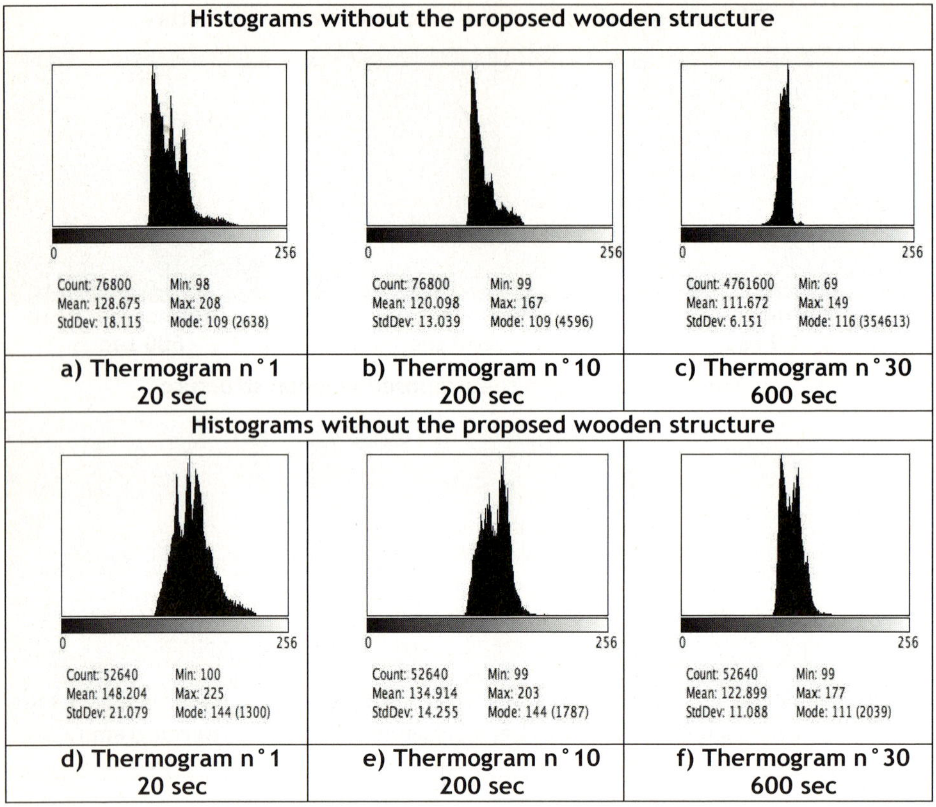

Fig. 4. Hystograms for the thermo-grams n° 1-10-30 for the two investigations

in Secs. 2 and 4.1, the profiles result well-defined. Fig. 5 reports the temperature profiles *without* the wooden structure for several combinations of the pre-processing filters. For analogy, Fig. 6 shows the temperature profiles related to the vertical line reported in the figures *with* the wooden structure for several combinations of the pre-processing filters. It is not possible to affirm that a combination is always superior to the other ones, because the *right* combination strongly depends on the wall under test. Only after a first investigation is possible to affirm what is the *right* combination for the specific wall in those specific conditions.

5.2 Processing

After the pre-processing phase, Canny's algorithm and our algorithm – Structure - have been applied for both the cases (Fig. 7). Although Figs. a) and c) show good enough results in individuating the wall weft, nevertheless Figs. b) and d) – related to the proposed algorithm – highlight a superior quality.

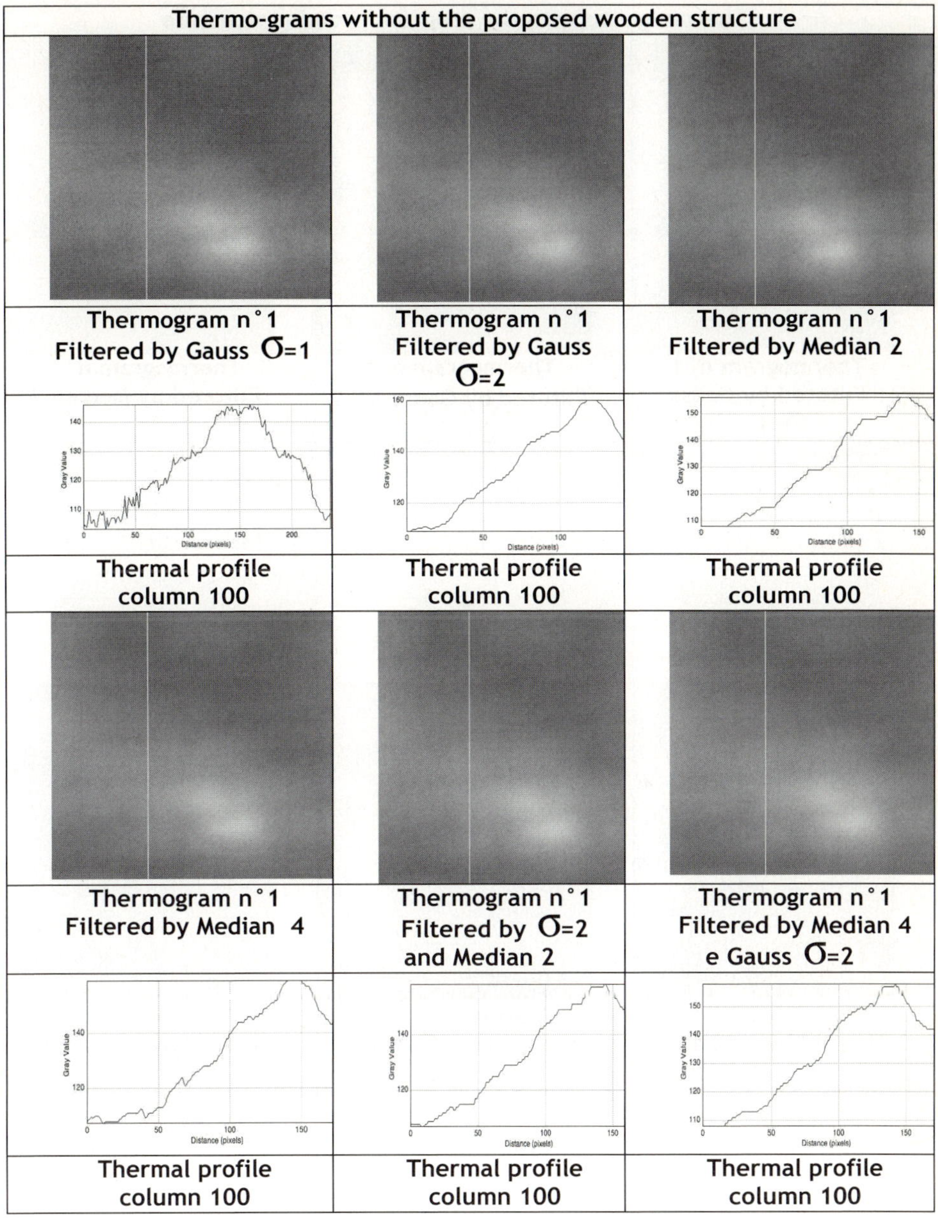

Fig. 5. Filtered thermo-grams e related vertical profile

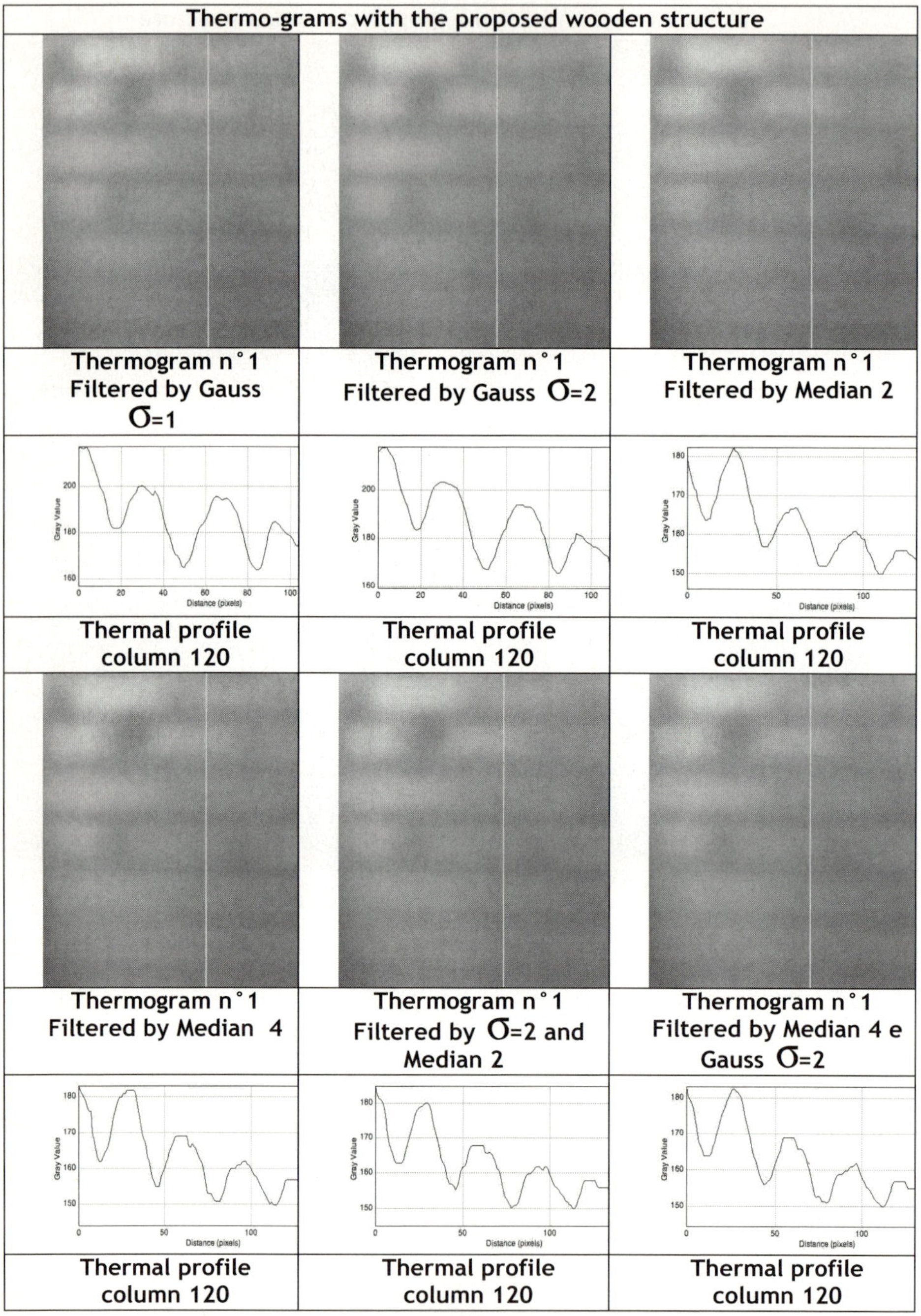

Fig. 6. Filtered thermo-grams e related vertical profile

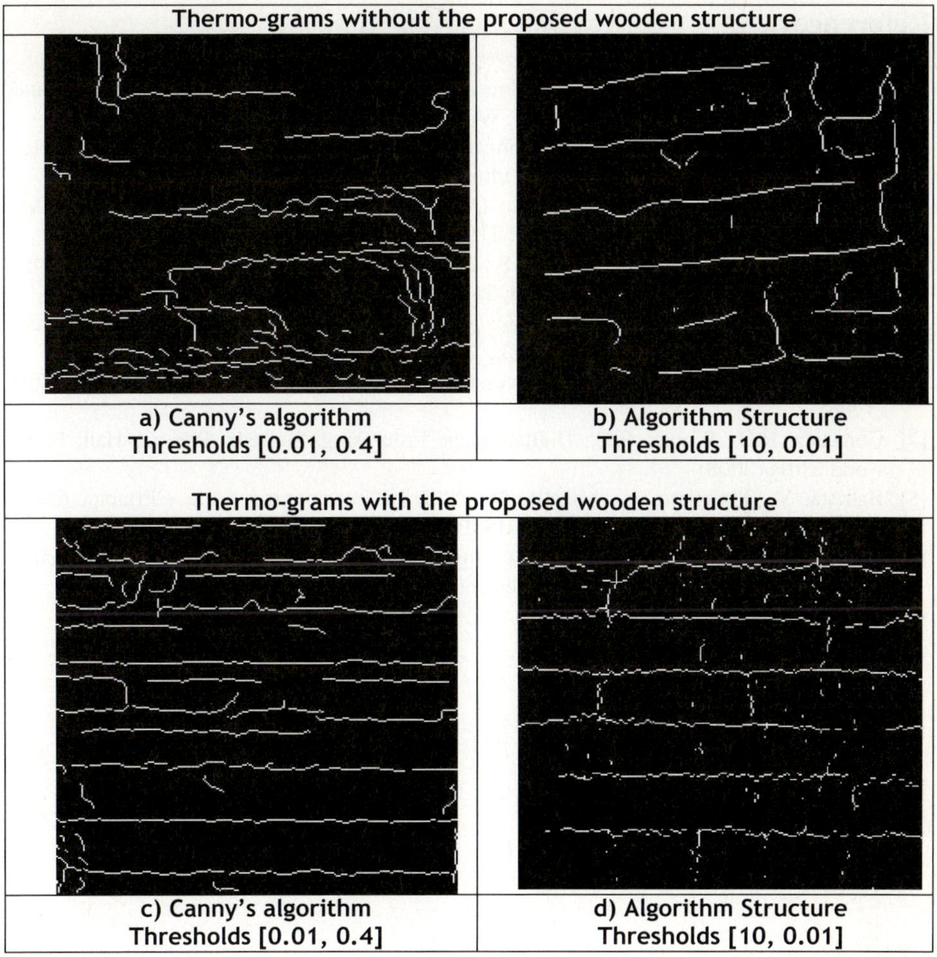

Fig. 7. Processed thermo-grams related to the investigations without and with the proposed wooden structure

6 Conclusions

The results show that the proposed wooden structure as well as the algorithm *Structure* allow to obtain good results in identifying physical and geometrical characteristics of the sample under examination. The overall procedure can be effectively applied also for the defect identification and classification. In fact, at this moment we are just applying this procedure for this aim and the first results are encouraging.

References

[1] Rosina E., Avdelidis N.P., Moropoulou A., Della Torre S.: IRT Monitoring in planned preservatione of built cultural heritage, WCNDT, Monreal, Canada (2004)

[2] Avdelidis, N.P., Moropoulou, A.: Applications of infrared thermography for the investigation of historic structures. Journal of Cultural Heritage 5, 119–127 (2004)

[3] Arndt, R., Maierhofer, C., Röllig, M.: Quantitative Pulse-Phase-Thermography for Masonry and Concrete Structures. In: ECNDT, Berlin, Germany (2006)

[4] Grinzato, E., Bison, P.G., Marinetti, S.: Monitoring of ancient buildings by the thermal method. Journal of Cultural Heritage 3, 21–29 (2002)

[5] Fu, K.S., Gonzalez, R.C., Lee, C.S.G.: Robotics; Control, Sensing, Vision, and Intelligence. McGraw Hill, New York (1987)

[6] Castleman, K.R.: Digital Image Processing, 2nd edn. Prentice Hall, Englewood Cliffs (1996)

[7] Gonzalez, R.C., Woods, R.E.: Digital Image Processing, 2nd edn. Prentice-Hall, Englewood Cliffs (2008)

[8] Battista, V., Maiellaro, N., Micella, N.: 2002 La Termografia Edile - Principi fisici - Applicazioni-Strumentazione. CNR-IRIS, Bari, Italy (1985)

[9] Maldague, X.: Theory and Practice of Infrared Technolgy for Nondestructive Testing. Wiley Interscience Publication, Chichester (2001)

A Splitting Scheme for Large-Scale Atmosphere Dynamics Models

Andrei Bourchtein and Ludmila Bourchtein

Mathematics Department, Pelotas State University, Brazil
burstein@terra.com.br

Abstract. Splitting techniques are applied in the context of the semi-Lagrangian semi-implicit approach in order to design computationally efficient and accurate numerical scheme for large-scale atmosphere dynamics. Description of the constructed numerical algorithm is provided and its properties of accuracy and stability are discussed. Performed numerical experiments with daily gridded atmospheric data showed that the developed scheme is efficient and supplies accurate forecast fields using time steps up to one hour.

1 Introduction

Atmosphere dynamics includes complex processes of different space and time scales providing analogous properties for complete 3D mathematical models of atmosphere (Navier-Stokes or Euler equations). Besides hydrostatic approximation filtering the fastest insignificant acoustic waves and keeping the principal physical processes quasi intact, there is no known differential approximation that contributes to elimination of the fast insignificant waves without distortion of the principal modes [10, 13]. In this way, the nonlinearity and stiffness of mathematical models of atmospheric dynamics have direct effect on the choice of the numerical methods for atmosphere modeling, because the problem of scale separation should be addressed in numerical scheme design in order to achieve accurate and computationally efficient solution.

Generally, explicit schemes are rarely employed because of excessive restriction on time step reflecting presence of the fast acoustic and gravity waves. On the other hand, fully implicit schemes are not used due to complexity of nonlinear systems arising at each time step. In this way, the most popular numerical approach is semi-implicit method with implicit approximation of linear terms responsible for fast waves and explicit approximation of the remaining processes. In such schemes a larger time step can be used and implicit part is reduced to solution of linear systems. In the last two decades, semi-implicit method is usually applied in semi-Lagrangian frame, because it allows to overcome the Courant-Friedrichs-Lewy (CFL) criterion with respect to advection at relatively low cost of solution of trajectory equations [8, 10, 19]. In the last years, a strong preference for two-time-level schemes can be noted in atmospheric community due to their increased accuracy as compared to three-time-level schemes [8, 11, 21]. Thus, design of our scheme follows these general recommendations and splitting techniques are applied in the context of semi-Lagrangian semi-implicit (SLSI) approach.

O. Gervasi et al. (Eds.): ICCSA 2008, Part II, LNCS 5073, pp. 627–640, 2008.

The constructed scheme is aimed for modeling of large-scale (synoptic) atmospheric flows characterized by principal space and time scales of 1000-2000km and 24-48 hours, respectively [13]. Therefore, the choice of mesh size and time step of a scheme should approximately follow the same ratio. Taking also in account that the characteristic wind in the troposphere is about 10-20m/s with elevated values of 20-30m/s in the upper troposphere (included in the prognostic atmospheric layer), one can conclude that for mesh size of 50km it would not be wisely to choose the time step much larger then 30-40 min due to accuracy considerations. This observation is one of the reference points in the determining the scheme design.

The rest of the paper is structured as follows. In section 2 the used splitting techniques are introduced in the case of the linearized model. In section 3 the splitting SLSI algorithm employed for the primitive equations of the hydrostatic atmosphere is described, and accuracy and stability properties of the presented scheme are discussed. Finally, in section 4 the results of numerical experiments are provided to evaluate the forecast skill and computational efficiency of the developed numerical scheme.

2 Design of the Splitting Method

In order to describe the splitting techniques applied to numerical modeling of atmospheric flows, let us consider linearized equations of the atmosphere dynamics under the hydrostatic assumption on a tangent *f*-plane [13, 15]:

$$u_t = f_0 v - RT_0 P_x - \Phi_x, \quad v_t = -f_0 u - RT_0 P_y - \Phi_y, \tag{1}$$

$$\sigma\Phi_\sigma = -RT, \quad T_t = \frac{RT_0}{c_p}\cdot\left(P_t + \frac{\dot{\sigma}}{\sigma}\right), \quad P_t = -u_x - v_y - \dot{\sigma}_\sigma. \tag{2}$$

Here, the first two equations are the horizontal momentum equations, and the remaining are the hydrostatic balance, thermodynamic and continuity equations. The independent variables are the time t and the Cartesian coordinates x, y and $\sigma = p/p_s$, where p and p_s are the pressure and surface pressure, respectively. The unknown functions include the velocity components u, v and $\dot{\sigma} \equiv d\sigma/dt$, the geopotential $\Phi = gz$, where z is the height of the pressure surface, the temperature T, and the surface pressure $P \equiv \ln p_s$. Finally, the following parameters are used: g is the gravitational acceleration, R is the gas constant, c_p is the specific heat at constant pressure, $f_0 = const$ is the characteristic value of the Coriolis parameter, and $T_0 = const$ is the reference temperature profile. Hereinafter, the subscripts with independent variables denote respective partial derivatives, while other sub- and superscripts are used only for indexing.

Multiplying the second equation in (2) by σ, differentiating the result in σ and applying the third equation, one obtains

$$(\sigma T_t)_\sigma = -\frac{RT_0}{c_p} \cdot (u_x + v_y) \, . \tag{3}$$

By substituting Φ for T from the hydrostatic equation, (3) is rewritten as

$$L\Phi_t = RT_0 \cdot (u_x + v_y), \quad L\Phi = \frac{c_p}{R} \left(\sigma^2 \Phi_\sigma\right)_\sigma , \tag{4}$$

where L is the vertical coupling operator. Finally, introducing function $s = \Phi + RT_0 P$ and noticing that the last term does not depend on σ, the equations (1) and (4) can be rewritten in the following closed form:

$$u_t = f_0 v - s_x , \quad v_t = -f_0 u - s_y , \quad Ls_t = RT_0 \cdot (u_x + v_y). \tag{5}$$

Let us introduce uniform time discretization with time step τ and notations φ and φ^τ, $\varphi = u, v, s$ for the function values at the current ($t_n = n\tau$) and next ($t_{n+1} = (n+1)\tau$) time levels, respectively. If one applies forward-backward approximation to the last system

$$\frac{\tilde{u}^\tau - u}{\tau} = f_0 \frac{\tilde{v}^\tau + v}{2} - \tilde{s}_x^\tau , \quad \frac{\tilde{v}^\tau - v}{\tau} = -f_0 \frac{\tilde{u}^\tau + u}{2} - \tilde{s}_y^\tau , \tag{6}$$

$$L\frac{\tilde{s}^\tau - s}{\tau} = RT_0 \cdot (u_x + v_y), \tag{7}$$

then the computation is very simple, because the formulas (6)-(7) are actually explicit in time. However, it can be shown that the stability condition is very rigid:

$$\tau \le \sqrt{2}\, h_g / c_g , \tag{8}$$

where h_g is the mesh size of spatial grid used for gravity terms (s_x, s_y, u_x, v_y) and $c_g \approx 350\, m/s$ is the maximum velocity of gravity waves in the system (5). For example, on unstaggered grid A with the main mesh size $h = 50km$, the minimum gravity mesh size is $h_{gA} = h$ and the maximum allowable time step is about 3 min. On staggered grids B and C (which are frequently used in atmospheric models due to better dispersive properties as compared to grid A), the minimum gravity mesh sizes are $h_{gB} = h/\sqrt{2} \approx 35km$ and $h_{gC} = h/2 = 25km$, and the maximum allowable time steps are about 2 min and 1.5 min, respectively [10, 14]. All these time steps are too small as compared to accuracy requirement of $\tau_{phys} \approx 40\,\text{min}$ (physical time step) relevant for large-scale (synoptic) atmospheric motions.

On the other hand, Crank-Nicholson approximation of (5)

$$\frac{\hat{u}^\tau - u}{\tau} = f_0 \frac{\hat{v}^\tau + v}{2} - \frac{\hat{s}_x^\tau + s_x}{2}, \quad \frac{\hat{v}^\tau - v}{\tau} = -f_0 \frac{\hat{u}^\tau + u}{2} - \frac{\hat{s}_y^\tau + s_y}{2}, \tag{9}$$

$$L\frac{\hat{s}^\tau - s}{\tau} = RT_0 \cdot \left(\frac{\hat{u}_x^\tau + u_x}{2} + \frac{\hat{v}_y^\tau + v_y}{2} \right) \tag{10}$$

provides unconditionally stable scheme, but it requires solution of 3D elliptic equation at each time step that is computationally expensive.

The spectral analysis of the equations (5) shows that they support both fast and slow gravity waves as well the slow inertial oscillations [10, 13]. It is well known, that in the large-scale atmosphere dynamics, the slow inertial processes (advection and Rossby waves) carry the principal part of the atmosphere energy [10, 13]. Therefore, the propagation speed of these slow processes determines the above physical time step. Evidently, one should avoid to use very small time steps (like in forward-backward scheme) due to computational considerations. On the other hand, due to accuracy considerations it is not advisable to use the time steps much greater than the physical one even though these time steps are admissible in numerical scheme (like Crank-Nicholson scheme). Thus, the "optimal" choice would be explicit computationally simple approximation of slow waves and more implicit time differencing for fast waves.

Employing such "optimal" approximation to system (5), one can separate the fast and slow gravity waves, that can be achieved by vertical decoupling using the eigenfunctions of the operator L (or respective eigenvectors in discrete case):

$$\varphi = \sum \Phi_i \psi_i \ , \ L\psi_i = -\lambda_i^{-1} \psi_i \ , \ i = 1,2,... \tag{11}$$

In this case, the equations (5) can be written separately for amplitudes U_i, V_i, S_i of a single vertical mode (from now on the index i is omitted for simplicity):

$$U_t = f_0 V - S_x, \ V_t = -f_0 U - S_y, \ S_t = -c^2 \cdot (U_x + V_y), \tag{12}$$

where $c^2 = \sqrt{\lambda RT_0}$ is the gravity wave velocity for individual vertical mode. It can be shown both in differential and difference formulations [6, 12, 22], that these velocities decrease fast in i and have the following first characteristic values: $c_1 = 342\, m/s$, $c_2 = 203\, m/s$, $c_3 = 121\, m/s$, $c_4 = 78.0\, m/s$, $c_5 = 54.1\, m/s$, $c_6 = 40.6\, m/s$, $c_7 = 32.1\, m/s$, $c_8 = 25.1\, m/s$, $c_9 = 20.2\, m/s$. Besides, the energy contribution also decreases fast in i that makes the first modes the most energy valuable [13, 22]. In this way, the first vertical modes correspond to the fast gravity waves and contain the main part of atmosphere energy. Therefore, these modes should be approximated implicitly and accurately. The remaining modes can be approximated explicitly and with less accuracy because it will not cause strict condition on time step or affect on overall practical accuracy of the numerical solution. Employing two time differencing (6)-(7) and (9)-(10), it means that the "optimal" choice of the numerical scheme for (5) consists of the implicit second-order accuracy Crank-Nicholson differencing

$$\frac{\hat{U}^{\tau}-U}{\tau}=f_0\frac{\hat{V}^{\tau}+V}{2}-\frac{\hat{S}_x^{\tau}+S_x}{2},\quad \frac{\hat{V}^{\tau}-V}{\tau}=-f_0\frac{\hat{U}^{\tau}+U}{2}-\frac{\hat{S}_y^{\tau}+S_y}{2}, \tag{13}$$

$$\frac{\hat{S}^{\tau}-S}{\tau}=-c^2\cdot\left(\frac{\hat{U}_x^{\tau}+U_x}{2}+\frac{\hat{V}_y^{\tau}+V_y}{2}\right), \tag{14}$$

applied to the first vertical modes in (12) and the explicit forward-backward approximation

$$\frac{\tilde{U}^{\tau}-U}{\tau}=f_0\frac{\tilde{V}^{\tau}+V}{2}-\tilde{S}_x^{\tau},\quad \frac{\tilde{V}^{\tau}-V}{\tau}=-f_0\frac{\tilde{U}^{\tau}+U}{2}-\tilde{S}_y^{\tau}, \tag{15}$$

$$\frac{\tilde{S}^{\tau}-S}{\tau}=-c^2\cdot\left(U_x+V_y\right), \tag{16}$$

applied to the remaining vertical modes.

Naturally, the stability condition of such mixed scheme is defined by the gravity waves of the fastest vertical mode treated explicitly, that is,

$$\tau\le\sqrt{2}\,h_g/c_{exp}\ . \tag{17}$$

In particular, if all vertical modes are treated explicitly then $c_{exp}=c_g$ and (8) holds, and if all vertical modes are treated implicitly then the scheme is absolutely stable. The approximation (15)-(16) is of the straightforward solution, and time differencing (13)-(14) leads to 2D elliptic equation separately for each implicit vertical mode, that is much less expensive computational task than 3D equation.

This approach is natural and simple for linearized equations allowing vertical decoupling in differential form, which is not possible for nonlinear atmosphere equations. To make it applicable to nonlinear equations let us reformulate the above algorithm in the following form. First, the forward-backward approximation (15)-(16) is applied for all vertical modes, that is, (6)-(7) is solved. Then the equations for differences between (9)-(10) and (6)-(7) are formed

$$\frac{\hat{u}^{\tau}-\tilde{u}^{\tau}}{\tau}=f_0\frac{\hat{v}^{\tau}-\tilde{v}^{\tau}}{2}-\frac{\hat{s}_x^{\tau}+s_x-2\tilde{s}_x^{\tau}}{2},\quad \frac{\hat{v}^{\tau}-\tilde{v}^{\tau}}{\tau}=-f_0\frac{\hat{u}^{\tau}-\tilde{u}^{\tau}}{2}-\frac{\hat{s}_y^{\tau}+s_y-2\tilde{s}_y^{\tau}}{2}, \tag{18}$$

$$L\frac{\hat{s}^{\tau}-\tilde{s}^{\tau}}{\tau}=\frac{R^2T_0}{c_p}\cdot\left(\frac{\hat{u}_x^{\tau}-u_x}{2}+\frac{\hat{v}_y^{\tau}-v_y}{2}\right), \tag{19}$$

and vertically decoupled using (11):

$$\frac{\hat{U}^\tau - \tilde{U}^\tau}{\tau} = f_0 \frac{\hat{V}^\tau - \tilde{V}^\tau}{2} - \frac{\hat{S}_x^\tau + S_x - 2\tilde{S}_x^\tau}{2}, \quad \frac{\hat{V}^\tau - \tilde{V}^\tau}{\tau} = -f_0 \frac{\hat{U}^\tau - \tilde{U}^\tau}{2} - \frac{\hat{S}_y^\tau + S_y - 2\tilde{S}_y^\tau}{2}, \tag{20}$$

$$\frac{\hat{S}^\tau - \tilde{S}^\tau}{\tau} = -c^2 \cdot \left(\frac{\hat{U}_x^\tau - U_x}{2} + \frac{\hat{V}_y^\tau - V_y}{2} \right). \tag{21}$$

Finally, equations (20)-(21) are solved for corrections $\hat{\varphi}^\tau - \tilde{\varphi}^\tau$, $\varphi = U, V, S$ for the first vertical modes. Evidently, this algorithm (even rather artificial) is equivalent to the splitting (13)-(16) in the case of the linearized model.

This type of vertical splitting was first proposed by Burridge [5] and successfully applied to SLSI schemes in [3, 16]. In the next section it is shown how the splitting method formulated this way can be extended to the nonlinear hydrostatic system.

3 Splitting SLSI Scheme for the Primitive Equations

Keeping the notations of the previous section, the primitive equations of the hydrostatic atmosphere can be written as follows:

$$d_t u = f_0 v - s_x + N_u \ , \quad d_t v = -f_0 u - s_y + N_v \ , \tag{22}$$

$$\sigma \Phi_\sigma = -RT \ , \quad d_t T = \frac{RT_0}{c_p} \cdot \left(d_t P + \frac{\dot{\sigma}}{\sigma} \right) + N_T \ , \quad d_t P = -u_x - v_y - \dot{\sigma}_\sigma \ . \tag{23}$$

Here additionally d_t is individual 3D derivative

$$d_t \varphi = \varphi_t + u\varphi_x + v\varphi_y + \dot{\sigma}\varphi_\sigma \quad , \ \varphi = u, v, P, T \ , \tag{24}$$

and nonlinear and variable coefficient terms are grouped in N_u, N_v, N_T:

$$N_u = (f - f_0)v - R(T - T_0)P_x \ , \ N_v = -(f - f_0)u - R(T - T_0)P_y \ ,$$
$$N_T = -\frac{R(T - T_0)}{c_p} \left(\frac{\dot{\sigma}}{\sigma} - D - \dot{\sigma}_\sigma \right). \tag{25}$$

The case

$$N_u = N_v = N_T \equiv 0, \ \frac{d\varphi}{dt} = \varphi_t \tag{26}$$

corresponds to equations (1)-(2).

According to SLSI method, first the trajectory equations

$$\frac{d\mathbf{r}}{dt} = \mathbf{V} \ , \ \mathbf{r} = (x, y, \sigma) \ , \ \mathbf{V} = (u, v, \dot{\sigma}) \ , \ t \in [t_n, t_{n+1}] \tag{27}$$

are solved backward in time using standard iterative formulas:

$$\mathbf{r}(k+1)=\mathbf{r}^a-\tau\mathbf{V}^{\tau/2}\left(\frac{\mathbf{r}^a+\mathbf{r}(k)}{2}\right), \tag{28}$$

where k is the iteration number, $\mathbf{r}^a=\mathbf{r}(t_{n+1})$ are given coordinates of the arrival points (grid points) and

$$\mathbf{V}^{\tau/2}=\frac{3}{2}\mathbf{V}-\frac{1}{2}\mathbf{V}^{-\tau} \tag{29}$$

is the time extrapolation of velocity components at the intermediate time level $t_{n+1/2}=(n+1/2)\tau$. The use of two iterations with initial guess $\mathbf{r}(0)=\mathbf{r}^a$ ensures computation of the departure points with the second order of accuracy [11, 17, 19]. The sufficient condition for convergence of the iterations (28) can be found in the form

$$\tau\le\frac{2}{3V_d},\quad V_d=\max\left(|u_x|,|u_y|,|u_\sigma|,|v_x|,|v_y|,|v_\sigma|,|\dot\sigma_x|,|\dot\sigma_y|,|\dot\sigma_\sigma|\right), \tag{30}$$

which is a version of the Pudykiewicz-Staniforth-Benoit criterion applied to two-time-level schemes [18, 19]. Using the maximum values of wind component variations $V_d\approx1.5\cdot10^{-4}\,\mathrm{s}^{-1}$ we get the time step $\tau\approx70\,\mathrm{min}$, which is greater than the physical time step.

Then the forward-backward approximation is applied along the found trajectories

$$\frac{\tilde u^\tau-u}{\tau}=f_0\frac{\tilde v^\tau+v}{2}-\tilde s_x^\tau+N_u^{\tau/2},\quad\frac{\tilde v^\tau-v}{\tau}=-f_0\frac{\tilde u^\tau+u}{2}-\tilde s_y^\tau+N_v^{\tau/2}, \tag{31}$$

$$\frac{\tilde P^\tau-P}{\tau}=-u_x-v_y-\dot\sigma_\sigma,\quad\frac{\tilde T^\tau-T}{\tau}=\frac{RT_0}{c_p}\left(\frac{\tilde P^\tau-P}{\tau}+\frac{\dot\sigma}{\sigma}\right)+N_T^{\tau/2}. \tag{32}$$

Here, all the terms at the next time level are evaluated at the arrival points, all the terms at the current level are evaluated at the departure points, and the nonlinear terms are evaluated by time extrapolation formulas of the second order accuracy at the intermediate time level:

$$N^{\tau/2}=\frac{3}{2}N-\frac{1}{2}N^{-\tau},\quad N=N_u,N_v,N_T. \tag{33}$$

These formulas are explicit, but of the first order of accuracy and rigid restriction (8) on the time step.

The Crank-Nicholson SLSI approximation to prognostic equations in (22)-(23) has the form (9)-(10) for the linear terms and the nonlinear ones are treated exactly like in (31)-(32):

$$\frac{\hat{u}^\tau - u}{\tau} = f_0 \frac{\hat{v}^\tau + v}{2} - \frac{\hat{s}_x^\tau + s_x}{2} + N_u^{\tau/2}, \quad \frac{\hat{v}^\tau - v}{\tau} = -f_0 \frac{\hat{u}^\tau + u}{2} - \frac{\hat{s}_y^\tau + s_y}{2} + N_v^{\tau/2}, \tag{34}$$

$$\frac{\hat{P}^\tau - P}{\tau} = -\frac{\hat{u}_x^\tau + \hat{v}_x^\tau + \hat{\dot{\sigma}}_\sigma^\tau + u_x + v_y + \dot{\sigma}_\sigma}{2},$$

$$\frac{\hat{T}^\tau - T}{\tau} = \frac{RT_0}{c_p}\left(\frac{\hat{P}^\tau - P}{\tau} + \frac{\hat{\dot{\sigma}} + \dot{\sigma}}{2\sigma}\right) + N_T^{\tau/2}. \tag{35}$$

Therefore, the equations for the differences between the Crank-Nicholson and forward-backward prognostic values have the same form (18)-(19) as in the linear case. Consequently, they can be vertically decoupled to (20)-(21), which are solved for the first vertical modes. These computations follow the forward-backward step and allow to improve the accuracy and stability for the chosen principal vertical modes.

Let us briefly discuss two points of this algorithm: solution of each of the systems (20)-(21) and choice of the number of the implicit vertical modes. Denoting $\overline{\varphi}^\tau = \hat{\varphi}^\tau - \tilde{\varphi}^\tau$, $\overline{\varphi} = \tilde{\varphi}^\tau - \varphi$, $\varphi = U, V, S$, the system (20)-(21) can be rewritten in the form

$$\frac{\overline{U}^\tau}{\tau} = f_0 \frac{\overline{V}^\tau}{2} - \frac{\overline{S}_x^\tau - \overline{S}_x}{2}, \quad \frac{\overline{V}^\tau}{\tau} = -f_0 \frac{\overline{U}^\tau}{2} - \frac{\overline{S}_y^\tau - \overline{S}_y}{2}, \tag{36}$$

$$\frac{\overline{S}^\tau}{\tau} = -c^2 \cdot \left(\frac{\overline{U}_x^\tau + \overline{U}_x}{2} + \frac{\overline{V}_y^\tau + \overline{V}_y}{2}\right). \tag{37}$$

Usually system (36)-(37) is reduced to 2D elliptic equation for correction $\overline{S}^\tau$:

$$\overline{S}^\tau - \frac{\tau^2 c^2}{4 + \tau^2 f_0^2}\left(\overline{S}_{xx}^\tau + \overline{S}_{yy}^\tau\right) = F, \tag{38}$$

where the right-hand side is a combination of the known functions $\overline{\varphi}$, $\varphi = U, V, S$. Solution of (38) is much more simple than that of its 3D counterpart arising in unsplit semi-implicit model (34)-(35). However, it still requires considerable amount of computations. One option for fast solver is the multigrid method, for example BOXMG algorithm [2, 9], which shows good performance for spatial grids with arbitrary number of the grid points.

Another approach is a slight modification of equation (37) proposed in [20] and applied in hydrostatic model in [4]:

$$\frac{\overline{S}^\tau}{\tau} + \frac{\tau^4 c^4}{\left(4 + \tau^2 f_0^2\right)^2} \frac{\overline{S}_{xxyy}^\tau}{\tau} = -c^2 \cdot \left(\frac{\overline{U}_x^\tau + \overline{U}_x}{2} + \frac{\overline{V}_y^\tau + \overline{V}_y}{2}\right), \tag{39}$$

which leads to the following elliptic equation instead of (38):

$$\bar{S}^{\tau} - \frac{\tau^2 c^2}{4+\tau^2 f_0^2}\left(\bar{S}^{\tau}_{xx} + \bar{S}^{\tau}_{yy}\right) + \frac{\tau^4 c^4}{\left(4+\tau^2 f_0^2\right)^2}\bar{S}_{xxyy} = F\,. \tag{40}$$

The last equation allows further splitting into 1D elliptic systems

$$\left(1 - \frac{\tau^2 c^2}{4+\tau^2 f_0^2}\partial_{xx}\right)\left(1 - \frac{\tau^2 c^2}{4+\tau^2 f_0^2}\partial_{yy}\right)\bar{S}^{\tau} = F\,, \tag{41}$$

each of which is solved by the Thomas algorithm [14].

Numerical experiments show that splitting (41) is about twice as fast as BOXMG and it works well for $\tau \le 40\,\text{min}$. However, for increased time steps the additional splitting term introduced in (39) starts to cause visible deviations from solution of (38). If the time steps larger than 40 min are required, then one of the alternatives is to solve the original elliptic equation (38) for the first implicit modes and the modified equation (41) for others. In fact, since additional term in (39) is proportional to c^2, its influence is greater for the fastest vertical modes. For example, solving (38) for the first vertical mode and (41) for other implicit modes, it is possible to increase the time step up to 60 min without loss of accuracy in forecasting fields.

After finding $\bar{S}^{\tau}$ from (38) or (41), the respective corrections to velocity components $\bar{U}^{\tau}, \bar{V}^{\tau}$ are calculated using the explicit formulas (36).

The choice of the number of implicit modes depends on accuracy, stability and efficiency requirements as well on other scheme parameters. In general, less number of vertical modes treated implicitly means faster solution at a single time step, but the time step should be smaller due to stability condition (17), and if allowable time step is smaller than the physical one, then the number of time steps increases. For example, for modeling the synoptic processes one might not want to use time steps larger than the respective physical time step of 40 min. In this case, for spatial grid *B* with mesh size $h = 50km$ used in the current scheme, implicit treatment of the first eight modes with additional splitting (41) can be an efficient computational option, which ensures accuracy and stability for time steps up to 41 min (fastest gravity waves treated explicitly have propagation velocity $c_9 = 20.2\,m/s$). On the other hand, instead of applying the minimum gravity mesh size $h_{gB} = h/\sqrt{2}$, one can choose a greater mesh size for approximation of linear terms responsible for gravity waves (pressure gradient and divergence) in the explicit modes. For example, choosing $h_{gB} = 3h/\sqrt{2}$ the efficient computational algorithm is obtained with four implicit modes, additional splitting (41) and time steps up to 45 min (fastest gravity waves treated explicitly have propagation velocity $c_5 = 54.1\,m/s$). In the last case, the choice of five implicit modes and unmodified equation (38) for the fastest implicit mode ensures stable and accurate numerical integration with time steps up to 60 min (respective explicit gravity speed is $c_6 = 40.6\, m/s$). Numerical experiments show that a coarser approximation of gravity terms in explicit vertical modes usually does not generate additional inaccuracies in prognostic fields.

4 Numerical Experiments

To evaluate efficiency and accuracy of the splitting techniques within two-time-level SLSI scheme a series of 24-h forecasts were calculated. The initial and boundary conditions were obtained from objective analysis and global forecasts of National Centers for Environmental Prediction (NCEP). The computational domain consists of 100x100 uniform staggered horizontal grid B with mesh size h = 50km and 20 vertical levels, and is centered at Porto Alegre city ($30^0 S, 52^0 W$).

The following scheme configurations described in previous section were used: explicit forward-backward SLSI (S1 in this section), implicit Crank-Nicholson SLSI (S2 in this section), vertical splitting SLSI with eight implicit modes solved through equation (38) (S3 in this section), vertical splitting SLSI with eight implicit modes solved through horizontal splitting equation (41) (S4 in this section), vertical splitting SLSI with first implicit mode solved through equation (38) and remaining seven implicit modes solved through horizontal splitting (41) (S5 in this section), and vertical splitting SLSI with first implicit mode solved through equation (38) and remaining four implicit modes solved through horizontal splitting (41), and with $h_{gB} = 3h/\sqrt{2}$ for explicit modes (S6 in this section). The numerical integrations with different time steps showed that the scheme S1 is practically stable for time steps up to 2 min, the schemes S2 and S6 up to 60 min, and the schemes S3, S4 and S5 up to 40 min.

First, to reveal possible inaccuracy caused by introduced splitting techniques in the schemes with the main time step of 40min, numerical solution of each of schemes S3, S4, S5 was compared with standard SLSI scheme S2. The root-mean-square (rms) differences between geopotential fields at the pressure surfaces of 250hPa, 500hPa, 850hPa and 1000hPa are shown in Table 1. All the schemes were run with the time step $\tau = 40\,\text{min}$. It is seen that these differences are quite small as compared with characteristic 24-h variation of geopotential height which is about 50m.

Table 1. Deviations of splitting solutions S3, S4, S5 from solution S2. δ_{250}, δ_{500}, δ_{850} and δ_{1000} - rms differences in meters between the 24-h forecasts provided by chosen scheme and S2 scheme at the heights 250hPa, 500hPa, 850hPa and 1000hPa, respectively.

scheme	δ_{250}	δ_{500}	δ_{850}	δ_{1000}
S3	3.4	3.2	3.2	3.0
S4	3.7	3.3	3.4	3.1
S5	3.4	3.3	3.2	3.0

Table 2. Mean objective scores of the 24-h forecasts. ε_{G500}, ε_{T850}, ε_{W250} - rms differences between analysis and 24-h forecast of geopotential field at the surface 500hPa (meters), temperature field at the surface 850hPa (degrees), and wind field at the surface 250hPa (m/s); τ - the time step, T_{CPU} - CPU time of a 24-h forecast regarding S1 scheme CPU time

scheme	τ	ε_{G500}	ε_{T850}	ε_{W250}	T_{CPU}
S1	2	20.1	1.72	7.17	1
S2	60	19.4	1.67	6.80	0.19
S3	40	19.1	1.63	6.75	0.20
S4	40	19.3	1.67	6.82	0.18
S5	40	19.1	1.64	6.77	0.18
S6	60	19.5	1.67	6.81	0.11

To assess the accuracy of all the above schemes, the prognostic fields were compared with analysis data available on the date of the forecast. Table 2 shows rms error for the 24-hour forecasts of geopotential field at 500 hPa pressure surface, temperature field at 850 hPa surface, and wind field at 250 hPa surface. These fields are traditionally used for evaluation of forecasting skill of numerical models: the first surface is quite characteristic for evaluation of dynamics of the middle troposphere, the second field is important for definition of vertical convective processes, and the third one for upper troposphere dynamics essential for aviation [1]. Each of the tested schemes was run with the maximum allowed time step showed in the second column. In the last column the CPU time for one forecast is presented in fractions of the CPU time required for S1 scheme. It is seen that forecast accuracy of the splitting schemes is of the same level as that of traditional SLSI scheme S2 and slightly better than forward-backward solution. At the same time, the presented splitting schemes are less computationally expensive.

It is well known that numerical solutions of regional atmospheric models are subject to strong influence of the horizontal boundary conditions after 24-48 hours of integration [1]. For the chosen extension of the computational domain and interior evaluation area of 1500km x 1500km, the maximum prediction period without strong interference of the boundary conditions is about 36 hours. To make systematical evaluation of the forecast skill up to the above maximum integration time and reveal temporal evolution of the statistical scores, the prognostic fields of different meteorological elements were compared against analysis at 12, 24 and 36 hours of numerical integration. The rms scores for five characteristic fields - geopotential field in the middle troposphere, geopotential and temperature fields in the lower troposphere, and geopotential and wind fields in the upper troposphere - are shown in Figs.1-5 as the functions of the integration time. The results for two standard schemes (S1, S2), which can be considered as the reference ones, and two more promising splitting schemes (S5, S6) are compared. One can see a close agreement among forecast fields of all semi-implicit schemes (S2, S5, S6) and a good quality of the forecasts up to 36 hours.

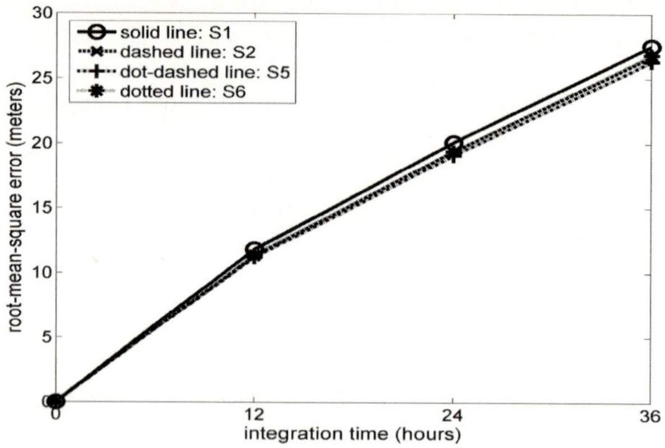

Fig. 1. Evaluation of the S1, S2, S5 and S6 scheme forecasts for geopotential of 500hPa surface

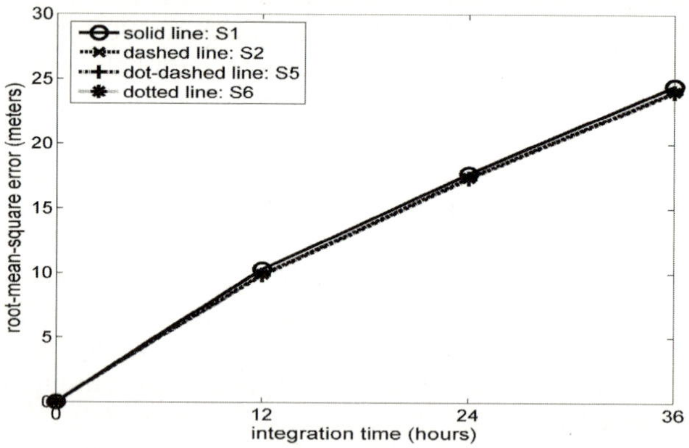

Fig. 2. The same as Fig. 1 but for 850hPa surface

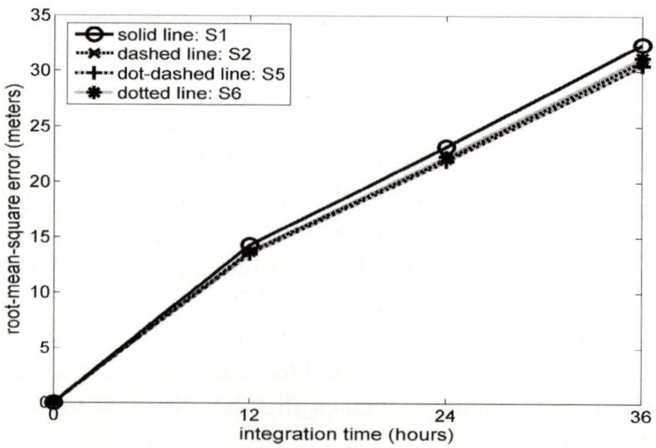

Fig. 3. The same as Fig. 1 but for 250hPa surface

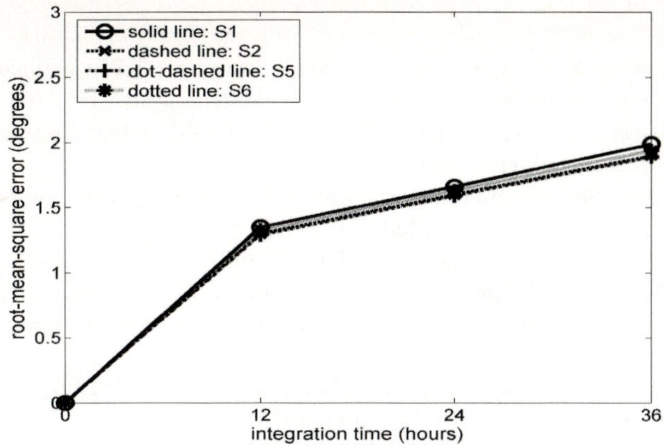

Fig. 4. The same as Fig. 1 but for temperature of 850hPa surface

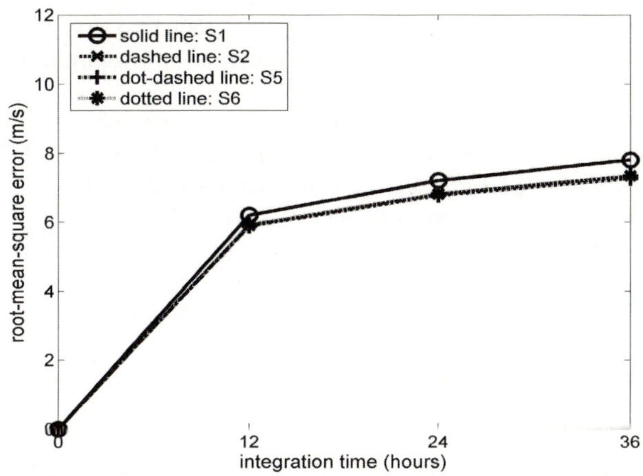

Fig. 5. The same as Fig. 1 but for wind of 250hPa surface

The presented scores are quite typical for the models of this class aimed for dynamical core equations of the atmosphere. These models do not contain the moisture equation and additional physical packages related to subgrid-scale parameterization of the principal diabatic processes, such as vertical convection, condensation/evaporation and turbulence.

Acknowledgements. This research was partially supported by Brazilian science foundation CNPq.

References

1. Anthes, R.A., Kuo, Y.H., Hsie, E.Y., Low-Nam, S., Bettge, T.W.: Estimation of skill and uncertainty in regional numerical models. Q. J. R. Meteorol. Soc. 115, 763–806 (1989)
2. Bandy, V., Sweet, R.: A set of three drivers for BOXMG: a black box multigrid solver. Comm. Appl. Num. Methods 8, 563–571 (1992)
3. Bourchtein, A.: Semi-Lagrangian semi-implicit space splitting regional baroclinic atmospheric model. Appl. Numer. Math. 41, 307–326 (2002)
4. Bourchtein, A.: Semi-Lagrangian semi-implicit fully splitted hydrostatic atmospheric model. In: Sloot, P.M.A., Abramson, D., Bogdanov, A.V., Gorbachev, Y.E., Dongarra, J., Zomaya, A.Y. (eds.) ICCS 2003. LNCS, vol. 2657, pp. 25–34. Springer, Heidelberg (2003)
5. Burridge, D.M.: A split semi-implicit reformulation of the Bushby-Timpson 10 level model. Q. J. R. Meteorol. Soc. 101, 777–792 (1975)
6. Cohn, S.E., Dee, D.P.: An analysis of the vertical structure equation for arbitrary thermal profiles. Q. J. R. Meteorol. Soc. 115, 143–171 (1989)
7. Côté, J., Béland, M., Staniforth, A.: Stability of vertical discretization schemes for semi-implicit primitive equations models: theory and applications. Mon. Wea. Rev. 111, 1189–1207 (1983)
8. Cullen, M.: Modelling atmospheric flows. Acta Numerica 16, 67–154 (2007)
9. Dendy, J.E.: Black box multigrid. J. Comp. Phys. 48, 366–386 (1982)
10. Durran, D.: Numerical Methods for Wave Equations in Geophysical Fluid Dynamics. Springer, New York (1999)
11. Gospodinov, I.G., Spiridonov, V.G., Geleyn, J.-F.: Second-order accuracy of two-time-level semi-Lagrangian schemes. Q. J. R. Meteorol. Soc. 127, 1017–1033 (2001)
12. Gravel, S., Staniforth, A., Cote, J.: A stability analysis of a family of baroclinic semi-Lagrangian forecast models. Mon. Wea. Rev. 121, 815–824 (1993)
13. Holton, J.R.: An Introduction to Dynamic Meteorology. Academic Press, San Diego (1992)
14. Kalnay, E.: Atmospheric Modeling, Data Assimilation and Predictability. Campridge University Press, Cambridge (2002)
15. Kasahara, A.: Various vertical coordinate systems used for numerical weather prediction. Mon. Wea. Rev. 102, 509–522 (1974)
16. Leslie, L.M., Purser, R.J.: Three-dimensional mass-conservating semi-Lagrangian scheme employing forward trajectories. Mon. Wea. Rev. 123, 2551–2566 (1995)
17. McDonald, A.: Accuracy of multiply-upstream semi-Lagrangian advective schemes II. Mon. Wea. Rev. 115, 1446–1450 (1987)
18. Pudykiewicz, J., Benoit, R., Staniforth, A.: Preliminary results from a partial LRTAP model based on an existing meteorological forecast model. Atmos.-Ocean 23, 267–303 (1985)
19. Staniforth, A., Côté, J.: Semi-Lagrangian integration schemes for atmospheric models - A review. Mon. Wea. Rev. 119, 2206–2223 (1991)
20. Tanguay, M., Robert, A.: Elimination of the Helmholtz equation associated with the semi-implicit scheme in a grid point model of the shallow water equations. Mon. Wea. Rev. 114, 2154–2162 (1986)
21. Temperton, C., Hortal, M., Simmons, A.J.: A two-time-level semi-Lagrangian global spectral model. Q. J. R. Meteorol. Soc. 127, 111–126 (2001)
22. Williamson, D.L., Temperton, C.: Normal mode initialization for a multilevel grid-point model. Part II: nonlinear aspects. Mon. Wea. Rev. 109, 744–757 (1981)

A Unique Complexity Metric

Sanjay Misra and Ibrahim Akman

Department of Computer Engineering, Atilim University, Ankara, Turkey
smisra@atilim.edu.tr, akman@atilim.edu.tr

Abstract. Metrics, in general, are defined as "a quantitative measure of the degree to which a system, component, or process possesses a given attribute". Complexity metrics are used to predict critical information about reliability and maintainability of software systems. This paper proposes complexity metric, which includes all major factors responsible for complexity. We validated our metric against the principles of measurement theory since the measurement theory has been proposed and extensively used in the literature as a means to evaluate the software engineering metrics. The scale of the metric is investigated through Extensive structure. It is found that the proposed measure is on ratio scale. The applicability of the proposed measure is tested through test cases and comparative study.

Keywords: Software complexity metric, evaluation criteria, measurement theory, scale, extensive structure.

1 Introduction

The key element of any engineering process is measurement. Engineers use measures to better understand and assess the quality of engineered products or systems that they built. However, absolute measures are uncommon in software engineering [15]. Instead, software engineers attempt to derive a set of indirect measures that lead to metrics that provide an indication of quality of some representation of software. Software engineers plan 'how' an information system should be developed in order to achieve its quality objectives. The quality objectives may be listed as performance, reliability, availability and maintainability [16], and are closely related to software complexity.

Complexity is defined by IEEE [6] as "the degree to which a system or component has a design or implementation that is difficult to understand and verify" Over the years, research on measuring the software complexity has been carried out to understand, what makes software products difficult to develop, maintain, or use. Major complexity measures of software that refer to effort, time and memory expended have been used in the form of different software metrics. Cyclomatic number [9], Halstead programming effort [5], data flow complexity measures [14], cognitive complexity measures [11], [12,] [19], [20] are examples to such metrics. These metrics calculate the complexity of software from the code and measures only specific internal attributes like size, algorithm complexity, control flow structures etc. In all above mentioned complexity metrics, they attempt to quantify the primitives which make software difficult to understand. For many of them, the developer's claim that their

O. Gervasi et al. (Eds.): ICCSA 2008, Part II, LNCS 5073, pp. 641–651, 2008.

complexity metric based on an internal attribute is the most accurate predictor of software quality. However, the authors realize that a single internal attribute is not sufficient for measuring the complexity of the code. For measuring the complexity of a code, one must consider most of the internal attributes responsible for complexity. Therefore, the purpose of this paper is to propose a new complexity metric which calculates complexity of the program code by considering all factors responsible for complexity. For this, first we identified the factors which are responsible for the complexity and then established a metric to reflect a proper relationship between these factors. With these features, it is unique metric of its type. We also evaluated and validated our measure against the principles of measurement theory, since the measurement theory has been proposed and extensively used in the literature as a means to evaluate the software engineering metrics.

In section 2, unique complexity measure is introduced. The quality of the proposed metric has been evaluated in section 3 against the principle of measurement theory. Experimentation and comparative study are given in section 4. The pros cons and future work is discussed in section 5. The last section 6 includes the conclusions drawn.

2 Unique Complexity Metric

Complexity of a code is directly dependent on the understandability of the code and relates to ease of comprehension. All the factors that makes program difficult to understand are responsible for complexity. When we analyze a program code we find that that number of lines (size), total occurrence of operators and operands (size), numbers of control structures (control flow structuredness), function call (coupling) are the factors which directly affect the complexity. In general, these primitives are measured independently by different complexity measures and each one of these is assumed to represent overall complexity of the software. When we look at most of the known complexity measures, we can observe the close relation between number of lines, operator and operand counts, and basic control structures. Consequently, these primitives of software may constitute the components of a unified, comprehensive complexity measure.

In our opinion, the complexity of a software system depends on following factors:

1. Complexity of program depends on the size of the code. Therefore, the complexity due to any single line of code is assumed to be 1 unit.
2. Complexity of program depends on the size of the code. We suggest that the size of the code can be measured by total occurrence of operators and operands. Therefore, the complexity due to i^{th} line of the code can be calculated as

   $$SOO_i = N_{i1} + N_{i2} \text{ Where}$$

   N_{i1}: The total number of occurrences of operators at line i,
   N_{i2}: The total number of occurrences of operands at line i,
3. Complexity of the program is directly proportional to the cognitive weights of Basic Control Structures (BSC). Cognitive weight of software [19] is the extent of difficulty or relative time and effort for comprehending given software modeled by a number of basic control structures (BCS's). BCS's, sequence, branch

and iteration [19] are basic logic building blocks of any software and their weights are one, two and three respectively. These weights are assigned on the classification of cognitive phenomenon as discussed by Wang [19]. He proved and assigned the weights for sub conscious function, meta cognitive function and higher cognitive function as 1, 2 and 3 respectively. In fact, cognitive weights correspond to the number of executed instructions. The details of the weights for different BCS's are given in Table-1, see [19].

Table 1. Basic Control Structures and their weights

Category	Basic Control Structures	Cognitive Weight
Sequence	Sequence	1
Branch	If-Then-Else	2
	Case	3
Iteration	For-do	3
	Repeat-until	3
	While-do	3
Embedded Component	Function Call	2
	Recursion	3

As a result, the cognitive complexity due to i^{th} line of the code, CW_i, can be weighted as in Table-1.

Using the above considerations, we propose the following model to establish a proper relationship among internal attributes of software.

$$UniqueComplexityMetric = \sum_{i=1}^{n} (1 + SOO_i * CW_i) \qquad (1)$$

where unique complexity metric (UCM) of the software code *is* defined as the sum of complexity of its n modules and module i consists of m_i line of code.

The unit of the UCM is given Complexity Units (CU) in terms of positive number. In Equation 1, the SOO_i is multiplied by CW_i because of the structural weight of the line. For a simple program having only basic control structure the "sequence", CW_i will not have any additional contribution. Therefore, for such lines the complexities are only due to SOO_i. Note that in our context, although the concept of cognitive weights is referred in the formulation, the only aspect of cognitive weight idea is used as a multiplier. By this way, the scale compatibility of *SOO* and *CW* are achieved.

For illustrative purpose, an algorithm written in C that finds the sum of n integers corresponding UCM calculations are given in section 4.

3 Theoretical Validation of the Complexity Measure

Practical success of any proposed metric depends on the establishment of (1) its validation, (2) understandability by its users and (3) tight link between the metric and the attribute that it is intended to measure. The establishment of (2) and (3) highly depends on the validation. Therefore, a new metric must be evaluated formally and practically for its validation. In order to make the software discipline more and more mature, tools provided by Measurement Theory (MT) [2] should be used. Several researchers' proposed different validation criteria based on measurement theory [2, 3, 4, 8, 13, 18 and 21]. However, in general all of them suggest that the metric should fulfill some basic requirements based on measurement theory perspective. Amongst available validation criteria, the framework given by Morasca [13] is reported to be more practical. As a consequence, we will define the basics of MT and evaluate the proposed metric formally from MT perspective in section 3.1. We applied admissible transformation and extensive structure to investigate the scale of the UCM in section 3.2.

3.1 UCM and Measurement Theory

The components of the qualified system are (1) Entities whose attributes are targeted for quantification; (2) Empirical binary relations showing the intuitive knowledge about the attributes and (3) Binary operations describing the production of new entities from the existing ones. These components are reflected in the following definitions [13].

Definition 1. (Empirical Relational System-ERS)
For a given attribute, an Empirical Relational System is an ordered tuple $ERS=<E, R_1,...,R_n, o_1,..., o_m>$ where E : the set of entities, R_1, ..., R_n denote n empirical relations such that each R_i has an *arity* n_i, and $R_i \subseteq E^{ni}$. o_1, ...,.. o_m denote m empirical binary operations on the entities that produces new entities from the existing ones, so o_j: $E \times E \rightarrow E$ and the operations are represented with an infix notation, for example, $e_k = e_i\ o_j\ e_l$.

According to this definition, the components of the quantification system are the values representing the decided quantities; the binary relations showing the dependencies among them and the binary operations describing the production of new values from the existing ones.

For UCM, the entities are the program bodies. The only empirical relation is assumed to be *more_or_equal_complex* and the only empirical binary operation is the *concatenation* of program bodies. This can be explained by a solid example. Assume that a program body P is given and a new program body Q is obtained by simply duplicating P. One may easily establish the relation *more_or_equal_complex* between P and Q.

Definition 2. (Numerical Relational System-NRS)
A Numerical Relational System is an ordered tuple $NRS=<V, S_1,...,S_n, p_1,..., p_m>$ where V : the set of values, S_1, ..., S_n denote n relations such that the arity of S_i is equal to the arity of R_i, and $S_i \subseteq V^{ni}$ *and* p_1, ..., p_m denote m numerical binary operations on the values that produce new values from the existing ones, so p_j: $V \times V \rightarrow V$ and the operations are represented with an infix notation, for example, $v_k = v_i\ p_j\ v_l$.

For UCM, V is the set of positive integers, the binary relation is assumed to be $\geq$ and the numerical binary operation is the addition (i.e. +) of two positive integers.

Definition 3. (Measure- *m*)
Measure m is a mapping of entities to the values and it considers neither the empirical nor the numerical knowledge about systems, i.e. $m: E \rightarrow V$.

The measure for UCM complexity is defined by Equation (1). Note that a measure by itself does not provide any mapping between empirical and numerical knowledge.

Definition 4. A measure must satisfy the following two conditions known as *Representation Condition*:

$(1) \forall i \in 1 \ldots\ldots n \forall <e_1, \ldots e_n> \in E_{ni} \ (<e_1, \ldots e_n> \in R_i \Leftrightarrow <m(e_1), \ldots, m(e_n)> \in S_i)$

$(2) \forall j \in 1 \ldots\ldots m \forall <e_1, e_2> \in E \times E \ (m(e_1 o_j e_2 = m(e_1) p_j \ m(e_2))$

The first part of the Representation Condition says that for a given empirically observed relation between entities, there must exist a numerical relation between corresponding measured values and vice versa. In other words, any empirical observation should be measurable and any measurement result should be empirically observable. The second part says a measured value of an entity which is obtained by the application of an empirical binary operation on two entities should be equal to the value obtained by corresponding numerical binary operation executed over individually measured values of entities. In other words, the complexity of the whole should be definable in terms of the complexities of its parts.

For UCM, the representation condition requires that (1) if, for any two program bodies, e_1 and e_2 are in *more_or_equal_complex* relation (i.e. $<e_1, e_2> \in$ *more_-or_equal_complex*) then the measured UCM complexity value of entity e_1 should be greater than the measured complexity value of entity e_2 (i.e. $m(e_1) > m(e_2)$) and vice versa. Considering the program bodies P and Q; if Q is the double of P then the number of BCSs, operators, operands for Q automatically becomes double. Consequently, for part (1) of the condition, it is possible to say that the empirically observed *more_or_equal_complex* relation between two program bodies leads to a numerical binary relation > among those entities or vice versa. However, part (1) is only satisfied if there are such clear empirically observable relations between program bodies for example P and Q.

For part two of the representation condition, we can say that the UCM complexity value of a program body which is obtained by concatenation (i.e. the empirical binary operation) of e_1 and e_2 is equal to the sum (i.e. the numerical binary operation) of their calculated complexity values. Therefore, UCM satisfies the second part of the representation condition. Overall, then, we can say that UCM satisfies the representation condition.

3.2 Evaluation of Measure Based on Scale

In the previous sub-section, we evaluated UCM through the measurement theory. It is identified that the proposed measure is satisfied by the representation condition. Now, there is a need to find out the scale of the proposed measure. There are two ways for getting the scale of the measure: through admissible transformation and extensive structures. Admissible transformation [13] is the simplest way to find the scale of a measure. On the other hand, Zuse [21] has stressed the advantage of using extensive structure because it is one of the most important measurement structures which characterizes empirical conditions of reality, hypothesis of reality and empirical conditions behind the software measure. Therefore, we evaluated UCM by admissible transformation and extensive structure both.

3.2.1 Admissible Transformation

Definition 6. A scale is a triple <ERS, NRS, m>, where ERS is an Empirical Relational System, NRS is a Numerical Relational System, and m is a measure that satisfies the representation condition.

For UCM, we already define ERS, NRS and m in previous section.

Definition 7. Given a scale *<ERS, NRS, m>*, the transformation of a scale f is *admissible* if m\`= f o m (i.e. m\` is the composition of *f* and *m*) and *<ERS, NRS, m'>* is a scale. Based on admissible transformation, four different types of scales can be considered as follows [21]:

Nominal scale: each entity is labeled for categories and there is no ordering relation among them. An example of nominal scale is the labeling of given programs according to the name of their authors.

Ordinal scale: entities are categorized in the form of *total ordering*. The associated values make entities comparable. As an example, program bodies can be assigned degrees from 1 to 5 with comparative meanings (e.g. 1 for least reliable to 5 for the most reliable).

Interval scale: the difference between the assigned numerical values can be quantified in their amount. A new scale m\` from *m* can only be obtained through transformations of the form m\`= a**m* + b where a>0. An example can be the scale Celcius that can be converted into Fahrenheit.

Ratio scale: the ratio between the numerical values associated to the entities is used for quantification. The form of transformation is: m\`= a**m* where a>0. The main difference between interval and ratio scale is the existence of true zero-point in ratio scale. An example of ratio scale is the LOC measure of the size of a program body.

For case of UCM, it can be very easily proved that *<ERS, NRS, m>* for UCM is a ratio scale. Reconsidering the two program bodies P and Q, Q/P =2 and then a=2. This implies that m\`= 2**m*. Therefore, it can be informally stated that the proposed measure UCM is defined on *ratio scale*.

One way of establishing whether a given scale is a ratio scale or not is to investigate whether the scale's Empirical Relation System is an Extensive Structure or not.

3.2.2 Extensive Structure

Definition 8. Let E be a non-empty set, $\bullet\geq$ a binary relation on E, and o a binary operation on E. The relational system (E, $\bullet\geq$, o) is an *Extensive Structure* if and only if the following axioms hold for all $e_1, e_2, e_3, e_4 \in E$. [3, 21].

A1: (E, $\bullet\geq$) is a weak order.
A2: e_1 o (e_2 o e_3) $\approx$ (e_1 o e_2) o e_3 (axiom of *weak associativity*).
A3: e_1 o $e_2 \approx e_2$ o e_1 (axiom of *weak commutativity*).
A4: $e_1 \bullet\geq e_2 \Rightarrow e_1$ o $e_3 \bullet\geq e_2$ o e_3 (axiom of *weak monotonicity*).

A5: If e_1 •> e_2 then for any e_3, e_4 there exists a natural number n, such that n e_1 o e_3 •> n e_2 o e_4 where n E denotes the composition of E within itself n times (*Archimedean Axiom*).

In the above definition, the empirical relation "•>" has the meaning "more complex than", the empirical relation "≈" means "as complex as", and the meaning of the empirical relation "•≥" is "more or equal complex than". In addition, the binary operation "o" between two objects is the composition ("concatenation" for the selected example) of two objects. The validity of the above axioms for the empirical relation system (*ERS=<E,* •>, o>) defined for UCM should be investigated as follows:

A1: There is an order relation between the flowgraph-like entities which is definable for all entities (i.e. complete); any entity is related to itself through the *more_or_equal_complex* relation (i.e. reflexive), and it is clear that if e_1 is *more_or_equal_complex* than e_2 and e_2 is *more_or_equal_complex* than e_3 then e_1 is also *more_or_equal_complex* than e_3 (i.e. transitive). Therefore, (E, •≥) defines a weak order.

A2: The result of a series of concatenations on entities does not depend on the order of execution. Therefore, the concatenation operator of UCM has the weak associativity property and the complexity of UCM will not change.

A3: The UCM complexity of two concatenated entities does not depend on their order. Therefore, this axiom is satisfied by UCM.

A4: For any two entities e_1 and e_2, if there is a *more_or_equal_complex* relation between them (i.e. e_1 •≥ e_2) then the concatenation of a new entity e_3 to both e_1 (i.e. e_1 o e_3) and e_2 (i.e. e_2 o e_3) does not effect the relation between them. Then, this axiom is satisfied.

A5: If entity e_1 is *more_or_equal_complex* than e_2 then for any e_3 and e_4, a new *more_or_equal_complex* relation can established by sufficient number of concatenation; say n times, of e_1 and e_2 to themselves followed by concatenation of e_3 and e_4 with them, respectively. This item is also satisfied.

As a result, the ERS description of the proposed UCM is an extensive structure. Consequently, UCM is concluded to be on ratio scale.

4 Experimentations through Test Cases and Comparative Study

The proposed complexity metric given by equation 1 is demonstrated with the programming example given by the following Table 2.

Empirical studies play an important role in the evaluation of software engineering discipline [1]. We have taken different 'C' programs from Misra and Mishra [10] for the analysis of the UCM approach. We calculated the Complexity Measure (UCM) for each one of those programs (see Table-4). The complexity values of its components and UCM is given in table 3. A plot for UCM and its component is shown in figure 1.

The UCM for all the programs gives higher complexity values. (Figure 1). When we compare UCM with other measures (see table 4), we can conclude that the behavior of UCM is similar to the other complexity metric. However, this is not a surprising result since UCM is a unified complexity metric and its components are mostly overlapping with the other measures like cognitive complexity measure, effort measure and statement counts. When we compare UCM and cognitive functional sixe metric, the missing prospective in cognitive functional size metric is the contribution of sufficient knowledge of information contents. Furthermore, the UCM emerges not only due to cognitive weights but also due to size and number of occurrence of operators and operands. The relatively high values of UCM when compared to cognitive complexity measure (see table 4) for each program are primarily due to the higher number of operators and operands and number of line of code.

Table 2. Calculated complexity values for the example program

N = 10	Sample Algorithm	SOO_i	CW_i	UCM_i
Line 1	`#define N 10`	2	1	3
Line 2	`main ()`	0	1	1
Line 3	`{`	0	1	1
Line 4	`int i, n;`	3	1	4
Line 5	`int sum=0;`	4	1	5
Line 6	printf("enter the number");	0	1	1
Line 7	scanf("%d" , &n);	3	1	4
Line 8	for (i=1;i<=n; i++)	10	3	31
Line 9	printf("the sum is %d" ,sum);	2	1	1
Line 10	getch();}	1	1	2
				54 UCU

Table 3. Calculated complexity values for UCM and its Components

No.	The Number of Lines (NL)	Sum of Operators & Operands (SOO)	Cognitive Weights (CW)	Unique Complexity Metric (UCM)
1	12	20	4	57
2	17	35	3	78
3	18	52	3	90
4	37	58	16	167
5	23	25	10	102
6	15	20	6	70
7	11	10	6	54
8	11	17	9	84

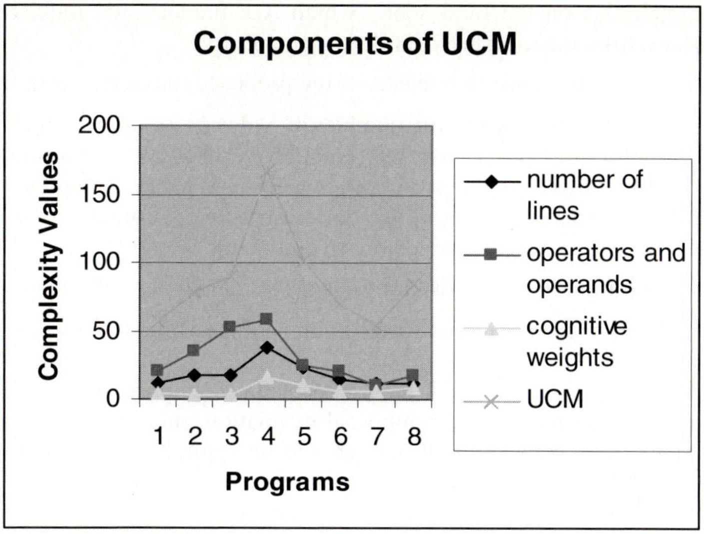

Fig. 1. Graph for UCM and its components

Table-4. Complexity values for statement count, cognitive functional size measure and unique complexity metric

Program No.	1	2	3	4	5	6	7	8
CFS	8	9	9	46	30	14	21	30
Line of Code	12	17	18	37	23	15	11	11
UCM	57	78	90	167	102	70	54	84

5 The Proposed Metric and Future Prospects

It is clear from the above sections that the proposed complexity metric is simple and fulfills the requirements of a good metric since it considers all the parameters responsible for complexity and the nature of the relationship between them. It is reported in the literature that this property is not satisfied for the other complexity metrics. Some of the other features of this metric are:

1. It can be used for the complexity of code and thereby improve the understandability of the code.
2. It can be used to evaluate the efficiency of the design. Low complexity value gives better design information.
3. This metric not only includes the complexity of the code due to one attribute but it also considers the complexity due to other factors. The internal factors give information about the complexity of software before execution and this makes the reliability and maintainability to be easily predicted before the completion of the code. Therefore, it is easy to control the quality of the code at design phase.

4. The metric is on the ratio scale, which is a fundamental requirement for a measure from the measurement theory perspective.

There are, however, also some drawbacks of the proposed measures, as given below.

1. The present method gives the complexity value in numbers, which are generally high for large programs. High complexity values are not desirable.
2. It is difficult to assign the upper and lower bound for the complexity values.
3. It is not possible to identify the underlying source of complexity with the proposed measure since it depends on several factors.

In the light of experience we propose future work to include the following:

1. Assignment of the upper and lower bounds of the complexity values should be investigated.
2. Further analysis is needed for the assessment of complexity.
3. Apart from the preliminary empirical evaluation, more test cases and typical examples (data from the industry) should be applied to the empirical evaluation.
4. Improvement of the proposed metric should be studied for the consideration of the remaining features, if any.
5. Algorithm development to calculate the complexity automatically should be considered.

6 Conclusion

In this paper, we proposed a complexity measure by considering all the internal attributes which directly affect the complexity. We considered the size in terms of number of lines, number of operators and operands. We also considered the compl-exity due to basic control structures since these structures affects the level of the complexity of any code. Furthermore, cognitive complexity is one of the most important factors for increasing overall complexity because it relates to comprehension. Understandability of software is the program comprehension and is a cognitive process. We consider the cognitive complexity in terms of cognitive weights of basic control structures. The cognitive weight also gives the information about the structure of the software, therefore structural complexity. In other terms we proposed a unique complexity measure by using the internal attributes size and (control flow structured ness) structure. The quality of the measures is checked by a practical framework and measurement theory perspective. Further, the applicability of the metric is checked by applying it on different test cases which prove the soundness and robustness of the proposed measure. As a conclusion, we hope that the proposed metric, UCM, will aid the developers and practitioners in evaluating the complexity before and after coding.

References

1. Basili, V.: The Role of Controlled Experiments in Software Engineering Research. In: Basili, V.R., Rombach, H.D., Schneider, K., Kitchenham, B., Pfahl, D., Selby, R.W. (eds.) Empirical Software Engineering Issues. LNCS, vol. 4336, pp. 33–37. Springer, Heidelberg (2007)

2. Brind, L.C., Moraska, S., Basily, V.R.: Property based Software Engineering Measurement. IEEE Transactions on SE 22(1), 68–86 (1996)
3. Briand, L., El Emam, K., Morasca, S.: On the Application of Measurement Theory in Software Engineering. Journal of Empirical Software Engineering 1(1), 61–88 (1996)
4. Fenton, N.: Software measurement: A Necessary Scientific basis. IEEE Trans. on Software Engineering 20(3), 199–206 (1994)
5. Halstead, M.H.: Elements of Software Science. Elsevier North-Holland, New York (1997)
6. IEEE Computer Society: Standard for Software Quality Metrics Methodology. Revision IEEE Standard, 1061–1998 (1998)
7. Kaner, C.: Software Engineering Metrics: What do They Measure and How do We Know? In: Proc. of 10th International Software Metrics Symposium, Metrics (2004)
8. Kitchenham, B., Fenton, N.: Towards a Framework for Software Measurement Validation. IEEE Transactions on SE 21(12), 929–943 (1995)
9. McCabe., T.H.: A Complexity Measure. IEEE Transactions Software Engineering SE-2(6), 308–320 (1976)
10. Misra, S., Misra., A.K.: Evaluating Cognitive Complexity Measure with Weyuker's properties. In: Proc. of IEEE (ICCI 2004), pp. 103–108 (2004)
11. Misra, S.: Cognitive Program Complexity Measure. In: Proceedings of IEEE (ICCI 2007), pp. 120–125 (2007)
12. Misra, S., Akman, I.: A Model for Measuring Cognitive Complexity of Software, Accepted for presentation in KES 2008 (2008)
13. Morasca, S.: Software Measurement. In: Handbook of Software Engineering and Knowledge Engineering, pp. 239–276. World Scientific Pub. Co., Singapore (2001)
14. Oviedo, E.I.: Control flow, Data and Program Complexity. In: Proc. IEEE COMPSAC, Chicago, IL, pp. 146–152 (1980)
15. Pressman, R.S.: Software Engineering: A Practitioner's approach, 5th edn. McGraw Hill, New York (2001)
16. Sommerville, I.: Software Engineering, 6th edn. Addison-Wesley, Reading (2001)
17. Weyuker., E.J.: Evaluating Software Complexity Measure. IEEE Transaction on Software Complexity Measure 14(9), 1357–1365 (1988)
18. Wang, Y.: The Measurement Theory for Software Engineering. In: Proceedings of Canadian Conference on Electrical and Computer Engineering CCECE 2003, pp. 1321–1324 (2003)
19. Wang., Y., Shao, J.: A New Measure of Software Complexity based on Cognitive Weights. Can. J. Elect. Comp. Eng. 28(2), 69–74 (2003)
20. Wang, Y.: On Cognitive Informatics. In: Proceedings of IEEE (ICCI 2002), pp. 34–42 (2002)
21. Zuse, H.A.: Framework of Software Measurement. Walter de Gruyter, Berlin (1998)

Measuring Complexity of Object Oriented Programs

Sanjay Misra and Ibrahim Akman

Department of Computer Engineering, Atilim University, Ankara, Turkey
smisra@atilim.edu.tr, akman@atilim.edu.tr

Abstract. In this paper, a metric for object oriented language is formulated and validated. On the contrary of the other metrics used for object oriented programming (OOPs), the proposed metric calculates the complexity of a class at method level and hence considers the internal architecture of the classes, subclasses and member functions. The proposed metric is evaluated against Weyuker's proposed set of measurement principles through examples and validated through experimentation, case study and comparative study with similar measures. The practical usefulness of the metric is evaluated by a practical framework.

Keywords: Software metrics, method complexity, object oriented programming (OOP), cognitive weights, validation.

1 Introduction

Software metrics have always been important for software engineers to assure software quality. Normally, software engineers attempt to derive a set of indirect measures that lead to metrics that provide an indication of quality of some representation of software. The quality objectives may be listed as performance, reliability, availability and maintainability [20], and are closely related to software complexity. Today, the relevant literature provides a variety of metrics to compute the complexity of software for OO software development. OO software development normally requires different software metrics than conventional ones since it adopts different approaches [15]. We refer the reader to [1], [2], [6], [7], [12], [14], [15],[19] for OO complexity metrics. All the reported complexity metrics are supposed to cover the correctness, effectiveness and clarity of software and to provide good estimate of these parameters. Out of the numerous proposed metrics, selecting a particular complexity metric is again a problem, as every metric has its own advantages and disadvantages. There is an ongoing effort to find such a comprehensive complexity metric, which addresses most of the parameters of software.

Although complexity of methods of OO software development directly effects understandability of the software, complexity metrics based on the method level have not been studied carefully yet. For OO system, traditional metrics are applied to methods that comprise the operation of a class. However, available metrics on the method level like cyclomatic number, line of code etc., do not capture the features of object-oriented system, which differentiate program structure than that of procedural language and are under several criticisms [18], [22], [23],[27]. This may be the underlying reason for the failure of the traditional complexity metrics on method level for OO code.

O. Gervasi et al. (Eds.): ICCSA 2008, Part II, LNCS 5073, pp. 652–667, 2008.

With this backdrop, the present work proposes a metric on the method level. The proposed metric also considers the important feature of OO systems: coupling. An important property of the new metric is that it calculates the complexity of object oriented code in terms of cognitive weights. Cognitive weights are defined as the extent of difficulty or relative time and effort for comprehending given software [24] and measure the complexity of logical structure of software. The complexity metrics based on cognitive weights have already been proved as well-structured [16] for procedural languages. Object orientation requires different approach; therefore, we need the different representation of metric for object-oriented system. In our previous work we presented our metric in ICCI 2007 [17]. In the present work, we extended and modified our previous work and evaluated our metric against Weyuker's properties through examples. We also validated our metric against a practical framework. Empirical validation of the metric has been done through experimentations, case study and comparative study.

In the next section our metric is introduced. Evaluation and validation of the measure has been done in section 3 and 4. We have evaluated our metric against a practical framework in section 4. Case study and comparison with other measures have also been done in same section. The conclusion drawn is given in section 5.

2 Proposed Metric

Any OO system starts by defining a class that contains related attributes and operations (methods). The classes are used as the basis for objects and used to extract subclasses and objects. The complexity metrics developed for object-oriented language are based on these important structures. However, the complexity of the entire code is more important to calculate the complexity of an individual object/class. Therefore, firstly we are interested to calculate the complexity of the method in each object and then the complexity of entire code. For this reason a new approach, is proposed for the complexity of an Object Oriented System on its method level using cognitive weights.

The cognitive weights [24] of software are defined as the extent of difficulty or relative time and effort for comprehending given software modeled by a number of Basic Control Structures (BCS). The basic control structures are the basic building block of software and they measure the complexity of logical structures of the software. The logical structures reside in the method (code) of OOP and are classified as sequence, branch, iteration and call. The corresponding weights of these basic control structures are one, two, three and two. Actually, the weights are assigned on the classification of cognitive phenomenon as discussed by Wang [24]. He proved and assigned the weights for sub conscious function, meta cognitive function and higher cognitive function as 1, 2 and 3 respectively.

The proposed metric first calculates the weight of individual method in a class by associating a number (weight) with each member function (method), and then simply adds all the weights of all methods. This gives the complexity (weight) of a single class/object. If there are several classes then there weights are simply add, which gives the complexity of the entire code.

If there is s number of classes/objects in the object oriented code then the total complexity of the code is calculated by sum of individual class complexities;

$$\text{Code Complexity (CC)} = \sum_{p=1}^{s} ClassComplexity_p \quad (1)$$

Where, the class complexity is the complexity of a class or an object and can be calculated by summing the weights of all methods inside the class/object. If there are y methods in a class/object, then the class complexity is given by;

$$\text{Class Complexity} = \sum_{x=1}^{y} MC_x \quad (2)$$

Where, MC is the method complexity. The weight of a particular method is given by the following equation.

$$MC = \sum_{j=1}^{q} \left[\prod_{k=1}^{m} \sum_{i=1}^{n} W_c(j,k,i) \right] \quad (3)$$

Where W_c is the weight of the concerned basic control structure. The total weight of a single method is defined as the sum of cognitive weights of its q linear blocks composed in individual BCS's. Since each block may consists of m layers of nested BCS's, and each layer with n linear BCS's. Furthermore, the unit of cognitive weight of a class/object is defined as the cognitive weight of the simplest software component (having a single class which includes single method). This corresponds to sequential structure in BCS and hence its unit is taken as 1CWU (Cognitive Weight Unit).

It is important to note that this approach includes the complexity of the object/class due to messages, automatically. In the case of a message between two classes, the complexity of the message is the sum of the weight of the called procedure and the weight due to that call (i.e. two). Furthermore, the proposed metric also provides valuable indication for the level of coupling and cohesion since they are defined as "level of interaction between modules" and "level of interaction within a module" respectively. Clearly, a high complexity value represents the high coupling between classes because of greater number of messages, which is undesirable.

The proposed complexity metric given by equation 1, is demonstrated with the programming example (for complete code see the appendix: Figure 1) in section in 5. This example processes a person (class) of a university system.

3 Evaluation of Proposed Metric against Weyuker's Properties

Any new measure must be validated and evaluated practically and formally both. The purpose of the validation is also to prove the usefulness of attribute, which is measured by the proposed measure. For theoretical validation, several researchers' proposed different criteria [3, 4, 8, 10, 13, 25, 26, 29 and 30] against which the proposed software measure should adhere. We examined our proposed metric against nine well known Weyuker's properties [26] through examples.

Weyuker [26] suggested nine properties to provide a formal approach to determine the effectiveness of various software complexity measures. Although Weyuker has proposed the properties at the time of procedural languages and very much criticized [5, 8, 29], these properties are also valuable to evaluate object-oriented metrics at present and still the topic of research [9], [21], and [28]. Numbers of researchers [7], [12] evaluate the object-oriented metrics by the complexity properties proposed by Weyuker [26]. We also used these properties to evaluate our proposed measure.

Property 1: (∃P) (∃Q) (| P | ≠ | Q |) where P and Q are the two different classes
This property states that a measure should not rank all classes as equally complex. For our proposed measure, there exist numbers of classes whose weights may be different. Therefore, this property is satisfied by our proposed measure. This property can also be proved by taking two examples in fig. 2 and fig. 3 given in the appendix. In the example figure 2 (Class: PERSON-STUDENT), the main class PERSON has only one object/subclass STUDENT. The PERSON class has six methods then

$$\text{Class Complexity} = \sum_{x=1}^{y} MC_x = W_{p1}+W_{p2}+W_{p3}+W_{p4} + W_{p5}+W_{p6} = 1+1+3+1+1+1=8$$

STUDENT object has two methods then

$$\text{Class Complexity} = \sum_{x=1}^{y} MC_x = W_{S1}+W_{S2} = 1+5=6$$

The total complexity of the system is given by Code Complexity (CC);

$$\text{Code Complexity (CC)} = \sum_{p=1}^{s} ClassComplexity_p$$

= Complexity of class PERSON+CC of Class STUDENT
=8 + 6 =14

In the example figure 3 (Class PERSON-EMPLOYEE-FACULTY), the main class PERSON has one object/class EMPLOYEE. The object EMPLOYEE has also sub object FACULTY. The class complexity of each is given by,
PERSON class has six methods then

$$\text{Class Complexity} = \sum_{x=1}^{y} MC_x = W_{p1}+W_{p2}+W_{p3}+ W_{p4} + W_{p5}+W_{p6} = 1+1+3+1+1+1=8$$

EMPLOYEE class/object has three methods then

$$\text{Class Complexity} = \sum_{x=1}^{y} MC_x = W_{E1}+W_{E2}+W_{E3} = 1+1+1=3$$

FACULTY class/object has two methods then

$$\text{Class Complexity} = \sum_{x=1}^{y} MC_x = W_{F1}+W_{F2} = 1+1=2$$

The total complexity of the system is given by Code Complexity (CC);

$$\text{Code Complexity (CC)} = \sum_{p=1}^{s} ClassComplexity_p$$

= Complexity of class PERSON + Complexity of class EMPLOYEE + Complexity of class FACULTY
= 8+3+2 =13.

Clearly, from the above two examples, where code complexities are the different i.e. 14 and 13, this property is satisfied by the proposed measure.

Property 2: Let c be a non-negative number, and then there are only finite number of classes and programs of complexity c

This property states that there is only finite number of classes of the same weight.

All object-oriented languages consist only finite number of cognitive weights of basic control structures. Since, the total complexity is defined as the sum of cognitive weights of all the methods in a program, a possible largest number can be assumed, without harm, to be the upper bound for the number of basic control structures in the program. Therefore, for given number of basic control structures, there are only finitely many programs. Hence, this proposed complexity does hold this property.

Property 3: There are distinct classes P and Q such that | P |=| Q |

This property states that there are multiple classes of the same complexity. In other words, even if there exist different classes, the complexity of these classes may be the same. In our proposed measure, total weights for a class may be same for different classes. Therefore, our measure also satisfies this property. Further, by considering two examples in figures 2 (PERSON–STUDENT) and 4 (PERSON-EMPLOYEE-ADMINSTRATIVE) given in the appendix.

In the example figure 4 (PERSON-EMPLOYEE-ADMINSTRATIVE) the main class PERSON has one object/class EMPLOYEE. The object EMPLOYEE has also sub object ADMISTRATIVE. The class complexity of each is given by,

PERSON class has six methods then

$$\text{Class Complexity}=\sum_{x=1}^{y} MC_x = W_{p1}+W_{p2}+W_{p3}+W_{p4}+W_{p5}+W_{p6}=1+1+3+1+1+1=8$$

EMPLOYEE class/object has three methods then

$$\text{Class Complexity} = \sum_{x=1}^{y} MC_x = W_{E1}+W_{E2}+W_{E3}=1+1+1=3$$

ADMINSTRATIVE class/object has three methods then

$$\text{Class Complexity} = \sum_{x=1}^{y} MC_x = W_{A1}+W_{A2}+W_{A3}=1+1+1=3$$

The total complexity of the system is given by Code Complexity (CC);

$$\text{Code Complexity (CC)} = \sum_{p=1}^{s} ClassComplexity_p$$

=Complexity of class PERSON +Complexity of Class EMPLOYEE + Complexity of ADMISTRATION

=8+3+3 =14.

The code complexity of figure 2 is also 14. Clearly, for two examples in figure 2 and 4, where code complexity are same i.e. 14, this property is satisfied by the proposed measure.

Property 4: (∃P) (∃Q) (P ≡ Q & | P |≠| Q |)

This property states that if there exists classes P and Q such that they produce the same output given the same input. In other words, this property states that even if two classes have same functionality, they are different in details of implementation, and means that even if two programs, which consist of many objects, have the same functionality, they are different in the details of implementations. Since the cognitive weights depend on the internal architecture, therefore cognitive weights for the two classes for the same output may be different. Therefore, this property is also satisfied by the given measure.

Property 5: (∀P) (∀Q) (| P| ≤ | P; Q| & |Q |≤ | P; Q |)

This is the property of monotonic behavior. It is known that, the proposed measure finds the complexity using cognitive weights and therefore the resulting complexity value is always a positive integer. Besides, it is true for all positive integers P and Q that:

$$(P \le P + Q) \text{ and } (Q \le P + Q)$$

This concludes that the above properties extracted from the proposed metric are analogous to Weyuker's property 5. Therefore, property 5 is satisfied by the proposed metric. We can also prove this property by taking the example figures 3 (Class: PERSON-EMPLOYEE-FACULTY (P)), figure 4 (Class: PERSON-EMPLOYEE-ADMINSTRATIVE (Q)) and figure 5 (Class: PERSON-EMPLOYEE-FACULTY-ADMINSTRATIVE (P+Q)) given in the appendix. The CC of figure 3(P) and 4(Q) are 13 and 14 CWU. The CC of combined class for figure 5 (P+Q) is calculated as follows. The main class PERSON has one object EMPLOYEE. The object EMPLOYEE is also a class with two sub objects FACULTY and ADMISTRATIVE. The class complexity of each is calculated as follow.

PERSON class has six methods then

$$\text{Class Complexity}=\sum_{x=1}^{y} MC_x$$

$$=W_{p1}+W_{p2}+W_{p3}+W_{p4}+W_{p5}+W_{p6}=1+1+3+1+1+1=8\text{CWU}$$

EMPLOYEE class/object has three methods then

$$\text{Class Complexity} = \sum_{x=1}^{y} MC_x = W_{E1}+W_{E2}+W_{E3} = 1+1+1=3\text{CWU}$$

FACULTY object has two methods then

$$\text{Class Complexity} = \sum_{x=1}^{y} MC_x = W_{F1}+W_{F2} = 1+1=2\text{CWU}$$

ADMINSTRATIVE class/object has three methods then

$$\text{Class Complexity} = \sum_{x=1}^{y} MC_x = W_{A1}+W_{A2}+ W_{A3}=1+1+1=3\text{CWU}$$

The total complexity of the system is given by Code Complexity (CC);

$$\text{Code Complexity (CC)} = \sum_{p=1}^{s} ClassComplexity_p$$

=Complexity of class PERSON+ Complexity of Class EMPLOYEE +Complexity of Class FACULTY + Complexity of Class ADMINSTRATIVE
=8+3+2+3 =16 CWU

It is clear from the above examples that if we take the two programs in figures 3 and 4 whose Class Complexities are given by 13 and 14, which is less than CC of combined program P+Q in figure 5. i.e. 16. So property five also holds for this complexity measure.

Property 6a: $(\exists P)\ (\exists Q)\ (\exists R)\ (|P|=|Q|).\ \&\ |P;R| \neq |Q;R|)$
6b: $(\exists P)\ (\exists Q)\ (\exists R)\ (|P|=|Q|).\ \&\ |R;P| \neq |R;Q|)$
This property states that if a new class is appended to two classes which have the same class complexity, the class complexities of two new combined classes are different or the interaction between P and R can be different than the interaction between Q and R resulting in different complexity values for P+ R and Q + R. Although the cognitive weights of methods are fixed, the combined complexity depends upon the way of interaction when interacting with other programs. Therefore joining program R with P and Q adds different amount of complexity hence property 6 is also satisfied by our measure.

Property 7: There are program bodies P and Q such that Q is formed by permuting the order of the statements of P, and $(|P| \neq |Q|)$
This property states that permutation is significant. In other words, it states that permutation of elements within the item being measured can change the metric values. The intent is to ensure that metric values change due to permutation of classes. However, in case of object-oriented programming in any class or object, changing the order in which methods or attribute is declared or even in operation the change in order of statement, does not affect the order in which they are executed. Therefore, our proposed measure does not satisfy this property.

Property 8: If P is renaming of Q, then $|P|=|Q|$
This property requires that when the name of the class or object changes it will not affect the complexity of the class. Even if the member function or member data name in the class change, the class complexity should remain unchanged. As in our proposed complexity measure, there is no effect in complexity by renaming, so this property is also satisfied by our measure.

Property 9: $(\exists P)\ (\exists Q)\ (|P|+|Q|) < (|P;Q|)$
This property shows interaction increases complexity. This property states that the class complexity of a new class combined from two classes is greater than the sum of two individual class complexities. For our case, this property is also not satisfied. In general, two classes can have a finite number of identical methods with some cognitive weights. A combination of the two classes into one class would result in one class's version of the identical methods becoming redundant. Therefore, the complexity of the combined class in terms of cognitive weight reduces. Therefore, proposed measure does not satisfy this property. Further if we take the examples figure 2

(Class: PERSON-STUDENT (P)), Figure 5(Class: PERSON-EMPLOYEE-FACULTY-ADMINSTRATIVE (Q)) and combined class in figure 1 (Class: PERSON) of the appendix, where the corresponding CC are 14, 16 and 22 CWU. Clearly, it shows that combined class complexity is less than the sum of class complexity of its components. Therefore, this property is not satisfied by the proposed measure.

4 Practical Evaluations and Case Study

In the previous section, we formally showed that our metric is a valid measure of complexity. An alternative approach to metric validation, which is more practical than the formal approach, is given by Kaner [11]. However, for especially practical approach, the main concern is "how to recognize and describe the attribute in the empirical observation domain in order to relate its values to the proposed metric." We followed the guidelines given by Kaner [11] for our practical evaluation.

When we look our metric from the perspective given in [11], it is a direct metric. It is a function of one component, which contributes to the measurement of software complexity. It should be clear that the proposed metric is neither complete nor unique measure of complexity.

4.1 Evaluation through a Practical Framework

For practical evaluation of our metric, we will apply the metric evaluation framework developed by [11]. This framework and the evaluation of the new metric are as follows:

The purpose of the metric: Two main purposes of the metric are to contribute the judgment about product quality and to provide a self-assessment and improvement for the developer.

Scope of usage of the metric: The proposed metric can be categorized as a technical metric being applicable after coding. Consequently, its scope of use is the software development group.

Identified Attribute to metric: The attributes measured by our metric are the quality of the product and the developer. More complex product makes it less understandable and consequently less maintainable for future development effort. In addition, the developer who can satisfy the user requirements through the usage of less number of branching and looping primitives (implying small time-complexity) is assumed more skillful.

Natural scale of the attribute: The existence of natural scale for the above attributes (but not the metrics) requires the development of a common, non-subjective view about them. We have no knowledge about the natural scale of attributes for our metric.

Definition of metric: The metric has been defined formally in Section 2.

Measuring instrument to perform the measurement: It uses the instrument of *counting* by either human or by machine. The item to be counted is cognitive weights of different BCS's. For automated counting purpose, one can easily develop a token generator and use the string matching algorithms.

Natural scale for the metric: For the natural scale of the metric, we have to go through measurement theory. When we analyze our metric according to Briand and Morasca [3] we find that, it is the ratio scale.

Relationship between the attribute to the metric value: There is a direct relation between the quality of the product and our proposed metric. If the complexity value increases, it is clear that the product quality will decrease since it implies inefficient use of memory and time. Note that the proposed metric is not the unique indicator of product quality. The same argument is also true for the relation between the complexity value and the developer quality attribute.

Natural and foreseeable side effects of using the instrument: Once we automate the complexity calculation, it will not require considerable additional workload of manpower of the company. The only cost will be due to automation.

4.2 Empirical Validations through a Case Study

The proposed complexity metric given by equation 1, is demonstrated with the programming example given by figure 1 (the complete code in the appendix 1). This example processes a person of a university system as given in the following figure.

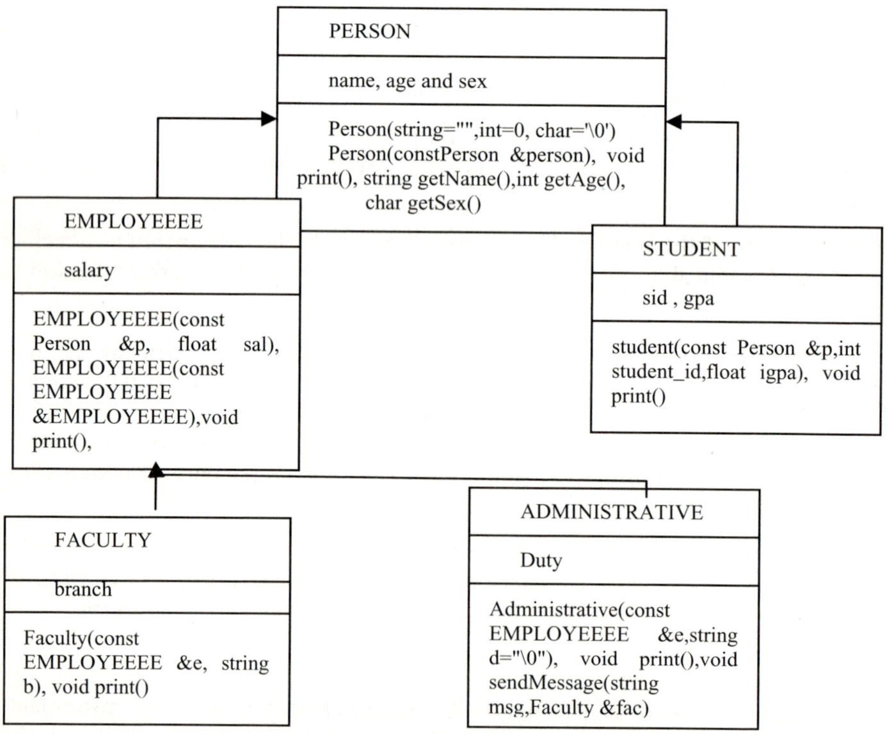

Fig. 1. An Example of Object oriented System

In the above figure the main class PERSON has two objects EMPLOYEE and STUDENTS. The object EMPLOYEE is also a class with two objects FACULTY and ADMISTRATIVE. The class complexity of each is given by,
PERSON class has six methods then

$$\text{Class Complexity} = \sum_{x=1}^{y} MC_x$$

$$= W_{p1}+W_{p2}+W_{p3}+ W_{p4} + W_{p5}+W_{p6} = 1+1+ 3 +1+1+1=8 \text{ CWU}$$

EMPLOYEE class/object has three methods then

$$\text{Class Complexity} = \sum_{x=1}^{y} MC_x = W_{E1}+W_{E2}+W_{E3} = 1+1+1=3 \text{ CWU}$$

STUDENT object has two methods then

$$\text{Class Complexity} = \sum_{x=1}^{y} MC_x = W_{S1}+W_{S2} = 1+5=6 \text{ CWU}$$

FACULTY object has two methods then

$$\text{Class Complexity} = \sum_{x=1}^{y} MC_x = W_{F1}+W_{F2} = 1+1=2 \text{ CWU}$$

ADMINSTRATIVE class/object has three methods then

$$\text{Class Complexity} = \sum_{x=1}^{y} MC_x = W_{A1}+W_{A2}+ W_{A3}=1+1+1=3 \text{ CWU}$$

The total complexity of the system is given by Code Complexity (CC);

$$\text{Code Complexity (CC)} = \sum_{p=1}^{s} ClassCompl\ exity_p$$ = Complexity of class PERSON + Complexity of Class EMPLOYEE+ Complexity of Class FACULTY+ Complexity of Class ADMINSTRATIVE + Complexity of Class STUDENT =8+3+2+3 +6 =22 CWU

We have taken a complete example CLASS PERSON for our case study given in the appendix (Figures 1 to 5). This example processes a person of a university system. It has one main class PERSON and two sub classes EMPLOYEE and STUDENT. Class EMPLOYEE has two sub classes FACULTY and ADMINSTRATION. We demonstrated how we can calculate the class complexity for a big system. We also demonstrated the effects of addition or deletion of the components from the class of the system.

Table 1. Calculated complexities for OO programs 1-5 in appendix 1

Fig. no.	1	2	3	4	5
Name of class	PERSON	PERSON-STUDENT	PERSON-EMPLOYEE-FACULTY	PERSON-EMPLOYEE-ADMISTRATIVE	PERSON-EMPLOYEE-FACULTY-ADMINSTRATIVE
CC	22	14	13	14	16

4.3 A Comparative Study with Other Measures

We compared the proposed measure with other complexity measures in terms of all nine Weyuker's properties.

Table 2. Comparative study with other measures

P.N.	WMC	DIT	NOC	CBO	RFC	LCOM	CMBOE	CC
1	Y	Y	Y	Y	Y	Y	Y	Y
2	Y	Y	Y	Y	Y	Y	Y	Y
3	Y	Y	Y	Y	Y	Y	Y	Y
4	Y	Y	Y	Y	Y	Y	Y	Y
5	Y	Y/N	Y	Y	Y	N	N	Y
6	Y	Y	Y	Y	Y	Y	Y	Y
7	N	N	N	N	N	N	N	N
8	Y	Y	Y	Y	Y	Y	Y	Y
9	N	N	N	N	N	N	Y	N

where WMC= Weighted Methods per Class, DIT= Depth of inheritance tree, NOC = Number of children, CBO= Coupling between objects, RFC= Response for a class, LCOM= Lack of cohesion in Method, CMBOE = Complexity measures for object-Oriented Program Based on Entropy, CC=Code Complexity(Proposed Measure).

We refer the reader [7] and [12] for the details of these metrics. It may be observed from the table 1 that all the measures including the complexity measure based on cognitive weight satisfies properties 1, 2,3,4,6 and 8. This is because of the following facts, as also mentioned by Weyuker.

1. Since not every class can have the same value for a metric so all measures, satisfy this property.
2. Since the universe of discourse deals with at most a finite set of applications, each of which has a finite number of classes, this property will be satisfied by any complexity measure at the class level.
3. Since in every complexity measure, we can find the number of classes, which can have same metric value.
4. The property 4 is also satisfied by every complexity measure given in table 1. This is due to that even though two-class design performs the same function, the details of the design matters in determining the metric for the class.
5. Every complexity measure satisfies the property 6. It is because of the interaction between two classes P and R can be different than interaction between Q and R resulting in different complexity values.
6. As in all complexity measure, there are no effects in complexity by renaming; therefore property eight is also satisfied by all complexity measure.

We have closely examined seven proposed syntactic complexity measures to see which properties, they have in common and which properties distinguish them. When we compare our measure through Weyuker's properties, we found that the proposed measure satisfies seven Weyuker's property out of nine, which established this more measure as well structured one. The proposed measure does not satisfy property seven. It is because of changing the order in which methods or attributes are declared does not affect the order in which they are executed. This property is meaningful in traditional program design where the ordering or if-then-else blocks could alter the program logic and consequent complexity. In OOP, a class is an abstraction of the problem space, and the order of statements within the class definition has no impact on eventual execution or use [7]. This is the reason that no complexity measure does satisfied by this property. Property 9 is also not satisfied by the proposed metric. This property allows the possibility of increased complexity due to interaction. Failing to meet this property implies that complexity metric could increase, rather than reduce, if a class is divided into more classes. Weyuker's Property 9 has received a mixed response regarding its applicability to object oriented software metrics and on the contrary to past beliefs, the relevance of this property to object oriented systems is brought out [21]. The proposed complexity metric is therefore can be considered as suitable for OOP.

5 Conclusion

In this paper, a new complexity measure for object-oriented programming has been presented. On the contrary to the most metrics, the proposed metric consider the internal architecture of methods, classes and subclasses. This measure is evaluated against nine Weyuker's properties. The proposed metric satisfies seven out of the nine Weyuker's Properties. Further we validated our metric for its practical usefulness through practical framework and a case study. The case study and comparison with other similar measures proves its robustness. As such, this metric established itself as a good and comprehensive measure for object-oriented programming.

References

1. Archer, C., Stinson., M.: Object oriented Software Measures, Technical Report CMU/SEI-95-TR-002 (1995)
2. Briand, L., Wust, J.: Modeling Development Effort in Object Oriented System using Design Properties. IEEE TSE 27(11), 963–986 (2001)
3. Briand, L.C., Morasca, S., Basily, V.R.: Property based Software Engineering Measurement. IEEE Transactions on SE 22(1), 68–86 (1996)
4. Basili, V.: The Role of Controlled Experiments in Software Engineering Research. In: Basili, V.R., Rombach, H.D., Schneider, K., Kitchenham, B., Pfahl, D., Selby, R.W. (eds.) Empirical Software Engineering Issues. LNCS, vol. 4336, pp. 33–37. Springer, Heidelberg (2007)
5. Cherniavsky, J.C., Smith, C.H.: On Weyuker's Axioms for Software Complexity Measures. IEEE TSE 17, 636–638 (1991)
6. Costagliola, G., Tortora, G.: Class Points:An Approach for the Size Estimation of Object-Oriented Systems. IEEE TSE 31(1), 52–74 (2005)

7. Chidamber, S.R., Kermer, C.F.: A Metric Suite for Object Oriented Design. IEEE Transactions Software Engineering SE-6, 476–493 (1994)
8. Fenton, N.: Software measurement: A Necessary Scientific Basis. IEEE Trans. Software Engineering SE-20(3), 199–206 (1994)
9. Gursaran,, Roy, G.: On the Applicability of Weyuker Property Nine to Object Oriented Structural Inheritance Complexity Metrics. IEEE TSE 27(4), 361–364 (2001)
10. IEEE Computer Society: Standard for Software Quality Metrics Methodology. Revision IEEE Standard, 1061–1998 (1998)
11. Kaner, C.: Software Engineering Metrics: What do they Measure and How do we know? In: Proc.of 10th Int. Software Metrics Symposium, Metrics (2004)
12. Kim, K., Shin, Y., Wu, C.: Complexity Measures for Object-Oriented Program based on the Entropy. In: Proc. Asia Pacafic SE Conference, pp. 127–136 (1995)
13. Kitchenham, B., Fenton, N.: Towards a framework for software Measurement Validation. IEEE Transactions on SE 21(12), 929–943 (1995)
14. Kim, J., Lerch, J.F.: Cognitive Processes in Logical Design: Comparing Object-Oriented and Traditional Functional Decomposition Software Methodologies. Carnegie Mellon University Graduate School of Industrial Administration Working Paper (1991)
15. Lorenz, M., Kidd, J.: Object-Oriented Software Metrics. Prentice Hall, Englewood Cliffs (1994)
16. Misra, S.: Cognitive Program Complexity Measure. In: Proceedings of IEEE (ICCI 2007), pp. 120–125 (2007)
17. Misra, S.: A Object oriented Metric based on cognitive weights. In: Proceedings of IEEE (ICCI 2007), pp. 134–139 (2007)
18. Prather, R.E.: An Axiomatic Theory of Software Complexity Measures. Computer Journal 27, 340–346 (1984)
19. Pressman, R.S.: Software Engineering: A Practitioner's approach, 5th edn. McGraw Hill, New York (2001)
20. Sommerville, I.: Software Engineering, 6th edn. Addison-Wesley, Reading (2001)
21. Sharma, N., Joshi, P., Joshi, R.K.: Applicability of Weyuker's Property 9 to Object Oriented Metrics. IEEE Trans. on SE 32(3), 209–211 (2006)
22. Vessey, I., Weber, R.: Research on Structured Programming: An Empiricist's Evaluation. IEEE Transactions on Software Engineering SE-10, 394–407 (1984)
23. Wand, Y., Weber, R.: Toward A Theory Of The Deep Structure Of Information Systems. In: Int. Con. on Information Systems, Copenhagen, Denmark, pp. 61–71 (1990)
24. Wang, Y., Shao, J.: A New Measure of Software Complexity based on Cognitive Weights. Can. J. Elec. Comput. Engg., 69–74 (2003)
25. Wang, Y.: The Measurement Theory for Software Engineering. In: Proceedings of Canadian Conference on Electrical and Computer Engineering, pp. 1321–1324 (2003)
26. Weyuker, E.: Evaluating Software Complexity Measures. IEEE Transactions on Software Engineering 14, 1357–1365 (1988)
27. Wilde, N., Huitt, R.: Maintenance Support for Object-Oriented Programs. IEEE Transactions on Software Engineering 18, 1038–1044 (1992)
28. Zhang, L., Xie, D.: Comments on 'On the Applicability of Weyuker Property Nine to Object Oriented Structural Inheritance Complexity Metrics'. IEEE Trans. Software Eng. 28(5), 526–527 (2002)
29. Zuse, H.: Software Complexity Measures and Methods, de Gruyter (1991)
30. Zuse, H.: Properties of Software Measures. Software Quality Journal 1, 225–260 (1992)

Appendix 1

Figure 1 CLASS PERSON: Classes and Subclasses for the Case Study

```
#include <iostream>
#include <string>
using namespace std;
/*Person Class is base class.Student and EMPLOYEE both
  inherits Person Class*/
```

```
/* ******** PERSON CLASS ********** */
class Person{  string name; int age; char sex;
public:Person(string="" ,int=0, char='\0');       // W p1=1
   Person(const Person &person);  //copyconstructorW p2=1
        void print()const;               //Wp3=Wp31+Wp32=2+1=3
        string getName(){                              // Wp4=1
        return name;   }
        int getAge(){return age;}                      //Wp5=1
                char getSex(){return sex; }            //Wp6=1
                  };
              //Person-default constructor
       Person :: Person(string in, int ia, char is)
              { name = in; age = ia; sex = is; }
               //Person-copy constructor
Person :: Person(const Person &p)
{name = p.name; age = p.age; sex = p.sex;   }
void Person :: print()const
{          cout<<"Name\t : "<<name<<'\n' ;             //Wp31=1
           cout<<"Age\t : "<<age<<'\n' ;
      if (sex=='F')cout<<"Sex\t:Female" <<'\n';  //Wp32=2
                  else cout<<"Sex\t : male" <<'\n' ;}
```

```
/* ******** STUDENT CLASS ********** */
class Student: public Person{int sid;float
gpa;public:Student(const Person &p,int student_id,float
igpa): Person(p)                                       //WS1=1
           { sid = student_id;      gpa = igpa; }
           void print()const; }; //WS2=WS21+WS22*WS23=1+2*2=5
         void Student :: print()const
{        Person :: print();
cout<<"S.ID\t:"<<sid<<"\nGPA\t:"<<gpa<<endl;           //WS21=1
        if (gpa>=2.0)                                  //WS22=2
             cout<<" Student is successful"<<endl;
        else {if (gpa>=1.7)                            //WS23=2
               cout<<"Student must improve GPA" <<endl;
        else          cout<<"Student must repeat"
<<endl;         }}
```

```
/* ******** EMPLOYEE CLASS ********** */
class EMPLOYEE: public Person{     float salary;
 public:  EMPLOYEE::EMPLOYEE(const Person &p, float
sal):Person(p) ,salary(sal){}                          //WE1=1
```

```
EMPLOYEE(const EMPLOYEE &EMPLOYEE):Person(EMPLOYEE){
         salary=EMPLOYEE.salary;    }                //WE2=1
          void print()const;  };                     //WE3=1
void EMPLOYEE::print() const{ Person::print();
         cout<<"salary: "<<salary<<endl;  }
```

```
        /* ******** FACULTY CLASS ********** */
class Faculty:public EMPLOYEE{string branch; //Physics,
                                         Math,or Ceng, etc .
public:    Faculty(const EMPLOYEE &e, string        //WF1=1
b):EMPLOYEE(e),branch(b){}void print()const;};      //WF2=1
```

```
     /* ******** ADMINISTRATIVE CLASS ********** */
class Administrative:public EMPLOYEE{string duty;
                                       //Secretary Account.
public: Administrative(const EMPLOYEE &e,string
d="\0"):EMPLOYEE(e){ duty=d;    }                   //WA1=1
       void print() const;};                        //WA2=1
```

```
      /* ********** MAIN *********************** */
int main(void){ Person * per[3];
         per[0]=  new Person ("Aysegul",27,'f');
         per[1]=  new Person ("Remzi",23,'m');
         per[2]=  new Person ("Ali",30,'m');
         EMPLOYEE EMPLOYEE1(* per[0],1000);
         EMPLOYEE1.print();
         Student student1(* per[1],9299,3.5);
         student1.print();
         EMPLOYEE EMPLOYEE2(* per[0],2000);
         Administrative admEMPLOYEE (EMPLOYEE1,
        "Secretary");
         Faculty facEMPLOYEE(EMPLOYEE2,"Computer");
         admEMPLOYEE.sendMessage("Today there is a
        seminar at your university. You are
        invited",facEMPLOYEE);}
```

Figure 2.

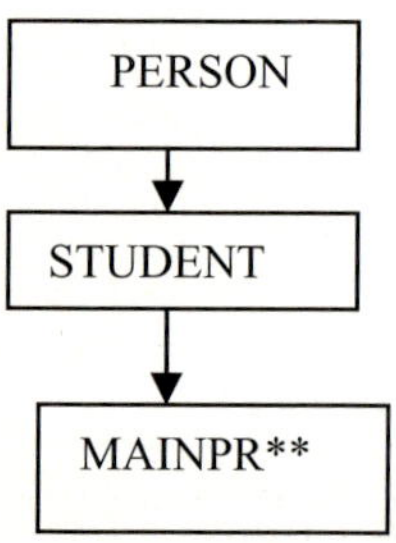

Figure 3.

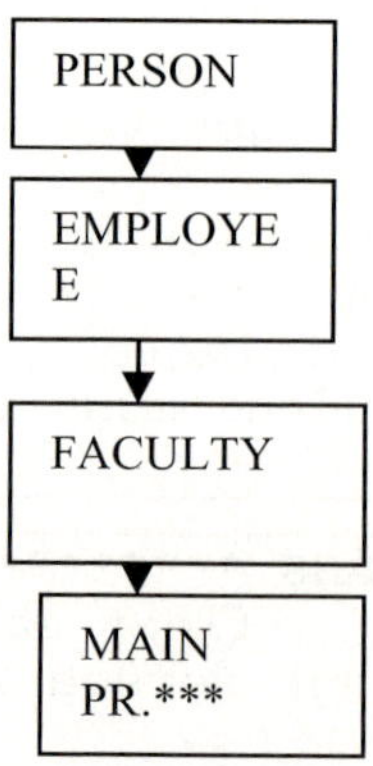

Figure 4.

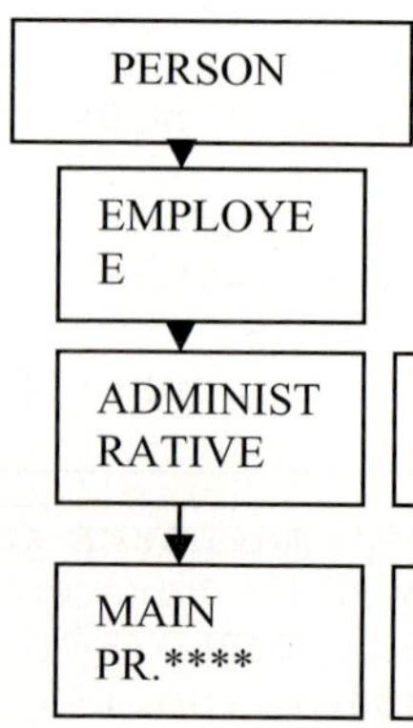

Figure 5.

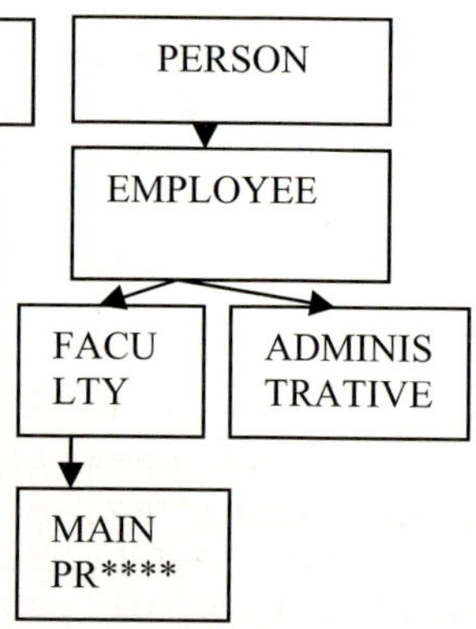

MAIN PROGRAMS(2-5) are given in following figures(2-5).

```
**Main Program for Class PERSON-STUDENT
int main(void) { Person * per[3];
                 per[0]=new Person ("Aysegul",27,'f');
                 per[1]=new Person ("Remzi",23,'m');
                 per[2]=new Person ("Ali",30,'m');
                   Person person1("fatmagul",27,'f');
                 Student student1(* per[1],9299,3.5);
                 student1.print(); }

***Main Programfor Class PERSON-EMPLOYEE-FACULTY
int main(void){Person * per[3];
                per[0]=new Person ("Aysegul",27,'f');
                per[1]=new Person ("Remzi",23,'m');
                per[2]=new Person ("Ali",30,'m');
                 EMPLOYEE EMPLOYEE2(* per[0],2000);
                Faculty
                facEMPLOYEE(EMPLOYEE2,"Computer");}

****Main   Program      for   Class    PERSON-EMPLOYEE-
ADMINISTRATIVE
int main(void) {Person * per[3];
                 per[0]=new Person("Aysegul",27,'f');
                per[1]=new Person ("Remzi",23,'m');
                per[2]=new Person ("Ali",30,'m');
                EMPLOYEE EMPLOYEE1(* per[0],1000);
                EMPLOYEE1.print();
                 AdministrativeadmEMPLOYEE
                (EMPLOYEE1,"Secretary");
                admEMPLOYEE.sendMessage("Today there is
                a seminar at your university.You are
                invited",facEMPLOYEE);}

*****Main Program for PERSON-EMPLOYEE-FACULTY-
ADMINISTRATIVE
int main(void){Person * per[3];
          per[0]=new Person ("Aysegul",27,'f');
          per[1]=new Person ("Remzi",23,'m');
          per[2]=new Person ("Ali",30,'m');
          EMPLOYEE EMPLOYEE1(* per[0],1000);
          EMPLOYEE1.print();
          EMPLOYEE EMPLOYEE2(* per[0],2000);
          AdministrativeadmEMPLOYEE(EMPLOYEE1,"Secretary
           ");
          Faculty facEMPLOYEE(EMPLOYEE2,"Computer");
          admEMPLOYEE.sendMessage("Today there is a
           seminar at youruniversity.You are
           invited",facEMPLOYEE);}
```

Cryptanalytic Attack on Cryptographic Sequence Generators: The Class of Clock-Controlled Shrinking Generators*

A. Fúster-Sabater[1] and P. Caballero-Gil[2]

[1] Institute of Applied Physics, C.S.I.C.
Serrano 144, 28006 Madrid, Spain
amparo@iec.csic.es

[2] Department of Statistics, Operations Research and Computing
Faculty of Mathematics, University of La Laguna, 38271 Tenerife, Spain
pcaballe@ull.es

Abstract. The shrinking generator is a very popular sequence generator with cryptographic applications. Nowadays, it is still considered as a secure keystream generator. In this work, it is shown that the knowledge of only a low number of generated bits is sufficient to break it. Indeed, whereas the linear complexity of the generated sequence (the shrunken sequence) is bounded by $A \cdot 2^{(S-2)} < LC \leq A \cdot 2^{(S-1)}$ (A and S being the lengths of the two component registers), we claim that the generator can be cryptanalyzed with the knowledge of $A \cdot S$ intercepted bits and simple computations. Such a result is proven thanks to the definition of the shrunken sequences as a particular kind of interleaved sequences. A similar attack can be extended to any other generator of the class of clock-controlled shrinking generators. Furthermore, this paper confirms that certain bits of the interleaved sequences have a greater strategic importance than others, which must be considered as a proof of weakness of interleaved sequence generators regarding their use in cryptography.

Keywords: Keystream generator, shrinking generator, interleaved sequence, cryptanalysis.

1 Introduction

Nowadays, stream ciphers are the fastest among the encryption procedures so they are implemented in many practical applications e.g. the algorithms A5 in GSM communications [10], the encryption system E0 in Bluetooth specifications [2] or the algorithm RC4 [15] used in Microsoft Word and Excel. From a short

* This work has been done in the frame of the project HESPERIA (http://www.proyecto-hesperia.org) supported by Centro para el Desarrollo Tecnológico Industrial (CDTI) under programme CENIT and also supported by the companies: Soluziona Consultoría y Tecnología, Unión Fenosa, Tecnobit, Visual-Tools, BrainStorm, SAC and TechnoSafe. The work has also been supported by MEC Project TSI2007-62657.

O. Gervasi et al. (Eds.): ICCSA 2008, Part II, LNCS 5073, pp. 668–679, 2008.

secret key (known only by the two interested parties) and a public algorithm (the sequence generator), stream cipher procedures consist in generating a long sequence of seemingly random bits. Such a sequence is called the keystream sequence. For encryption, the sender realizes the bit-wise (Exclusive-OR) XOR operation among the bits of the original message or plaintext and the keystream sequence. The result is the ciphertext to be sent. For decryption, the receiver generates the same keystream, realizes the same bit-wise XOR operation between the received ciphertext and the keystream sequence and obtains again the original message.

Most keystream generators are based on Linear Feedback Shift Registers (LFSRs) [8]. They are linear structures characterized by their length (the number of memory cells), their characteristic polynomial (the feedback function) and their initial state (the seed or key of the cryptosystem). If the characteristic polynomial is a primitive polynomial [14], then LFSRs generate Pseudo-Noise sequences (PN-sequences) with good characteristics of pseudorandomness. For a survey on recurring sequences, primitive LFSRs, and PN-sequences, the interested reader is referred to [8]. In stream cipher procedures, the PN-sequences are combined by means of nonlinear functions in order to produce keystream sequences of cryptographic application. Combinational generators, nonlinear filters, clock-controlled generators, irregularly decimated generators ... are just some of the most popular nonlinear sequence generators. All of them produce keystreams with high linear complexity, long period and good statistical properties (see [3] and [6]).

Most cryptanalysis on stream ciphers are performed under a known plaintext hypothesis, that is to say, it is assumed that the attacker has direct access to a portion of the keystream sequence (*intercepted sequence*). From the intercepted bits, the attacker has to deduce the cryptosystem key. Once the key is known, as the sequence generator is public, the whole keystream sequence can be reconstructed. The complexity of this attack is always compared with that of the key exhaustive search. If the former complexity is lesser, then the cryptosystem is said to be broken.

This work focuses on a particular kind of stream ciphers based on $LFSRs$: the class of clock-controlled shrinking generators. Among the elements of such a class, shrinking generators are the most popular and well known. They are made out of two LFSRs and an irregular decimation. Shrinking generators have been thoroughly analyzed in the last years and, in fact, different cryptanalytic attacks can be quoted: cryptanalysis based on correlation methods such as [13] and [17] or cryptanalysis based on a linearization procedure by means of cellular automata such as the attack developed in [4]. Nevertheless, in this paper we present an original and efficient cryptanalytic attack requiring much lesser amount of intercepted bits than that of the previous attacks. The basic idea of this cryptanalysis consists in defining the output sequence of a shrinking generator as an interleaved sequence (see [9] and [12]). This definition can be easily extended to the whole class of clock-controlled shrinking generators. The characteristics of the interleaved sequences reveal weaknesses that lead to practical attacks. In addition, these weaknesses can be extended to all interleaved sequence generators with application in cryptography.

The paper is organized as follows: in section 2, the description and characteristics of the shrinking generator and the general class of clock-controlled shrinking generators are introduced. Interleaved configuration and related results are developed in section 3. A cryptanalytic attack against the shrinking generator that exploits the condition of interleaved sequence is presented in section 4, while the generalization of this technique to other cryptographic interleaved generators appears in section 5. Finally, conclusions in section 6 end the paper.

2 The Clock-Controlled Shrinking Generators

First of all, the most popular of the generators of this class, the shrinking generator, is introduced. Next the characteristics of the elements of the general class are presented too.

2.1 The Shrinking Generator

The Shrinking Generator is a pseudorandom number generator based on a non-linear combination of the recurring sequences produced by two $LFSRs$. It was first introduced by Coppersmith, Krawczyk and Mansour at Crypto'93, see [5]. This construction uses two sources of pseudorandom bits to create a third source of potentially better quality than the original ones. Here quality means difficulty of predicting a pseudorandom sequence. We denote by SRA and SRS the first and second LFSR, respectively. The first register SRA has length A, characteristic polynomial $P_A(x) \in GF(2)[x]$ and its output sequence is denoted by $\{a_i\}$ $(i \geq 0)$ with $a_i \in GF(2)$. The selector register SRS has length S, characteristic polynomial $P_S(x) \in GF(2)[x]$ and its output sequence is denoted by $\{s_i\}$ $(i \geq 0)$ with $s_i \in GF(2)$. In addition, the lengths of both registers A, S are relatively prime $(A, S) = 1$, the characteristic polynomials $P_A(x)$, $P_S(x)$ are primitive polynomials in $GF(2)[x]$ and the output sequences $\{a_i\}$, $\{s_i\}$ are PN-sequences of period $(2^A - 1)$ and $(2^S - 1)$, respectively. The output sequence of the shrinking generator, the so-called *shrunken sequence* denoted by $\{z_j\}$ $(j \geq 0)$ with $z_j \in GF(2)$, is a sub-sequence of $\{a_i\}$ whose terms are chosen according to the positions of '1' bits in the sequence $\{s_i\}$. More precisely, the decimation rule is defined such as follows:

1. If $s_i = 1 \Longrightarrow z_j = a_i$
2. If $s_i = 0 \Longrightarrow a_i$ is discarded.

As different pairs of SRA/SRS initial states can generate the same shrunken sequence, in the sequel we assume that the first term of the sequence $\{s_i\}$ equals 1, that is $s_0 = 1$.

According to [5], the period of the shrunken sequence is:

$$T = (2^A - 1) \cdot 2^{(S-1)}, \tag{1}$$

its linear complexity, notated LC, satisfies the following inequality:

$$A \cdot 2^{(S-2)} < LC \leq A \cdot 2^{(S-1)}, \tag{2}$$

and its characteristic polynomial is of the form:

$$P_{sg}(x) = P_D(x)^p \tag{3}$$

where $P_D(x)$ is a unique A-degree primitive polynomial in $GF(2)[x]$ and p is an integer in the interval $2^{(S-2)} < p \leq 2^{(S-1)}$. Moreover, it can be proved [16] that the shrunken sequence has also good distributional statistics. Therefore, this scheme has been traditionally used as keystream sequence generator with application in secret-key cryptography.

2.2 The Clock-Controlled Shrinking Generators

The Clock-Controlled Shrinking Generators (CCSGs) constitute a wide class of sequence generators [13] with applications in cryptography, error correcting codes and digital signature. An CCSG is a sequence generator composed of two LFSRs notated as before SRA and SRS. The parameters of both registers are defined as those of subsection 2.1. At any time t, SRS (the selector register) is clocked normally while the register SRA is clocked a number of times given by an integer decimation function notated X_t. In fact, if $S_0(t)$, $S_1(t)$, ..., $S_{S-1}(t)$ are the binary cell contents of SRS at time t, then X_t is defined as

$$X_t = 1 + 2^0 S_{i_0}(t) + 2^1 S_{i_1}(t) + \ldots + 2^{w-1} S_{i_{w-1}}(t) \tag{4}$$

where $i_0, i_1, \ldots, i_{w-1} \in \{0, 1, \ldots, S-1\}$ and $0 < w \leq S-1$.

In this way, the output sequence of an CCSG is obtained from a double decimation:

1. The output sequence of SRA, $\{a_i\}$, is decimated by means of X_t giving rise to the sequence $\{a'_i\}$.
2. The same decimation rule P, defined in subsection 2.1, is applied to the sequence $\{a'_i\}$.

Remark that if $X_t \equiv 1$ (no cells are selected in SRS), then the proposed generator is just the shrinking generator. Period, linear complexity and statistical properties of the generated sequences by CCSGs have been established in [13]. It can be proved that the characteristic polynomial of a CCSG has the same form as that of the shrinking generator, that is the product of a unique primitive polynomial in $GF(2)[x]$ whose exponent depends on the parameters of the generator [13].

Let us see a simple example of CCSG.

Example 1. For the same LFSRs defined in the previous example and the function $X_t = 1 + 2^0 S_0(t)$ with $w = 1$, the decimated sequence $\{a'_i\}$ is given by:

- $\{a_i\} \rightarrow 1\ \underline{0}\ 0\ 0\ 1\ \underline{0}\ 0\ \underline{1}\ 1\ \underline{0}\ 1\ 0\ \underline{1}\ 1\ 1\ \underline{0}\ 0\ \underline{0}\ 1\ \underline{0}\ 0\ 1\ \underline{1}\ 0\ 1\ 0\ \underline{1}\ 1\ \underline{1}\ 1$ .....
- $X_t \rightarrow 2\ 1\ 1\ 2\ 2\ 2\ 1\ 2\ 1\ 1\ 2\ 2\ 2\ 1\ 2\ 1\ 1\ 2\ 2$ .....
- $\{a'_i\} \rightarrow 1\ 0\ 0\ 1\ 0\ 1\ 1\ 0\ 1\ 1\ 1\ 0\ 1\ 0\ 1\ 0\ 1\ 0\ 1\ 1$ .....

According to the decimation function X_t, the underlined bits $\underline{0}$ or $\underline{1}$ in $\{a_i\}$ are discarded in order to produce the sequence $\{a'_i\}$. Then the output sequence $\{z_j\}$ of the CCSG is given by:

- $\{a_i\} \rightarrow 1\ 0\ 0\ 1\ 1\ 1\ 0\ 1\ 0\ 0\ 1\ 1\ 1\ 0\ 1\ 0\ 0\ 1\ 1\ 1\ 0\ 1$ .....
- $\{a'_i\} \rightarrow 1\ \underline{0}\ \underline{0}\ 1\ 0\ 1\ \underline{1}\ 0\ \underline{1}\ \underline{1}\ 1\ 0\ 1\ \underline{0}\ 1\ \underline{0}\ \underline{1}\ 0\ 1\ 1$ .....
- $\{z_j\} \rightarrow 1\ 1\ 0\ 1\ 0\ 1\ 0\ 1\ 1\ 0\ 1\ 1$ .....

The underlined bits $\underline{0}$ or $\underline{1}$ in $\{a'_i\}$ are discarded.

In brief, the sequence produced by an CCSG is an irregular double decimation of the sequence generated by SRA from the function X_t and the bits of SRS. This construction allows one to generate a large family of different sequences by using the same LFSR initial states and characteristic polynomials but modifying the decimation function.

3 Interleaved Configuration

Although the results of this section are dealing with the shrinking generator, they can be easily generalized to the class of CCSGs. In the previous section, we have considered the form of the characteristic polynomial of these generators, see equation (3). Thus, it seems natural that the output sequence of such generators is related with the *PN*-sequences associated to the unique factor $P(x)_D$. This is the fact that we are going to exploit for the cryptanalysis of this kind of clock-controlled generators.

The $(2^A - 1) \cdot 2^{(S-1)}$ bits of a period of any shrunken sequence $\{z_j\}$ can be arranged into a $(2^A - 1) \cdot 2^{(S-1)}$ matrix that we will call *interleaved configuration* and will denote by IC. In fact,

$$IC = \begin{pmatrix} z_0 & z_1 & z_2 & \cdots & z_{2^{(S-1)}-1} \\ z_{2^{(S-1)}} & z_{2^{(S-1)}+1} & z_{2^{(S-1)}+2} & \cdots & z_{2\cdot 2^{(S-1)}-1} \\ z_{2\cdot 2^{(S-1)}} & z_{2\cdot 2^{(S-1)}+1} & z_{2\cdot 2^{(S-1)}+2} & \cdots & z_{3\cdot 2^{(S-1)}-1} \\ z_{3\cdot 2^{(S-1)}} & z_{3\cdot 2^{(S-1)}+1} & z_{3\cdot 2^{(S-1)}+2} & \cdots & z_{4\cdot 2^{(S-1)}-1} \\ \cdots & \cdots & \cdots & \cdots & \cdots \\ z_{(2^A-2)\cdot 2^{(S-1)}} & z_{(2^A-2)\cdot 2^{(S-1)}+1} & z_{(2^A-2)\cdot 2^{(S-1)}+2} & \cdots & z_{(2^A-1)\cdot 2^{(S-1)}-1} \end{pmatrix}$$

Now the following result allows one to identify each element of the matrix IC with the corresponding term of the sequence $\{a_i\}$.

Theorem 1. *The interleaved configuration matrix IC can be written in terms of the elements of the sequence $\{a_i\}$ such as follows:* $IC =$

$$\begin{pmatrix} a_{o0} & a_{o1} & \cdots & a_{o(2^{(S-1)}-1)} \\ a_{(2^S-1)+o0} & a_{(2^S-1)+o1} & \cdots & a_{(2^S-1)+o(2^{(S-1)}-1)} \\ a_{2\cdot(2^S-1)+o0} & a_{2\cdot(2^S-1)+o1} & \cdots & a_{2\cdot(2^S-1)+o(2^{(S-1)}-1)} \\ a_{3\cdot(2^S-1)+o0} & a_{3\cdot(2^S-1)+o1} & \cdots & a_{3\cdot(2^S-1)+o(2^{(S-1)}-1)} \\ \cdots & \cdots & \cdots & \cdots \\ a_{(2^A-2)\cdot(2^S-1)+o0} & a_{(2^A-2)\cdot(2^S-1)+o1} & \cdots & a_{(2^A-2)\cdot(2^S-1)+o(2^{(S-1)}-1)} \end{pmatrix}$$

where the additive sub-indices oj ($j = 0, 1, \ldots, 2^{(S-1)} - 1$) depend on the bits of the sequence $\{s_i\}$ in the following way: if $s_i = 1$, then the corresponding sub-index oj equals the sub-index i, $oj = i$. All the sub-indices are taken module $2^A - 1$, which is the period of the sequence $\{a_i\}$.

Proof. Since the period of the *PN*-sequence $\{s_i\}$ is $(2^S - 1)$, the number of bits with value '1' in a period is exactly $2^{(S-1)}$, and all the elements of any column of IC come from the same term $s_i = 1$ of the *PN*-sequence, the above expression for the matrix IC in terms of the elements of $\{a_i\}$ is obtained. □

Note that according to the assumption $s_0 = 1$, the sub-index $o0 = 0$. Next, the following result analyzes the characteristics of the columns of the matrix IC.

Theorem 2. *The sequences $\{\boldsymbol{d}_j\} = \{a_{k+oj} : k = 0, (2^S-1), 2\cdot(2^S-1), \ldots, (2^A-2)\cdot(2^S-1)\}$ ($j = 0, 1, \ldots, 2^{(S-1)}-1$) corresponding to the columns of the matrix IC are shifted versions of a unique PN-sequence whose characteristic polynomial is given by:*

$$P_D(x) = (x + \alpha^N)(x + \alpha^{2\cdot N})(x + \alpha^{2^2\cdot N}) \ldots (x + \alpha^{2^{(A-1)}\cdot N}),$$

where N is an integer defined as $N = 2^0 + 2^1 + \ldots + 2^{(S-1)}$ and $\alpha \in GF(2^A)$ a root of the primitive polynomial $P_A(x)$.

Proof. Every sequence $\{\boldsymbol{d}_j\}$ corresponding to the *j-th* column of IC is a regular decimation of the *PN*-sequence $\{a_i\}$. More precisely, such a sequence is obtained by taking one out of $(2^S - 1)$ terms in $\{a_i\}$. The primality of A and S guarantees the primality of $(2^A - 1)$ and $(2^S - 1)$. Thus, the decimated sequence $\{\boldsymbol{d}_j\}$ is also a *PN* sequence. In addition, as every $\{\boldsymbol{d}_j\}$ has been obtained from $\{a_i\}$ with a decimation ratio of value $(2^S - 1)$, then its characteristic polynomial $P_D(x)$ is the polynomial of the cyclotomic coset $(2^S - 1)$ in the Galois Field $GF(2^A)$ generated by the roots of the polynomial $P_A(x)$, see [4]. The starting point of each $\{\boldsymbol{d}_j\}$ is given by the corresponding sub-index oj. □

4 Cryptanalytic Attack on the Shrinking Generator

The cryptanalytic attack consists in the computation of the initial states of both registers SRA and SRS. In fact, from some known bits of the shrunken sequence we have to determine the first A bits $(a_0, a_1, \ldots, a_{A-1})$ of the sequence $\{a_i\}$ as well as the first S bits $(s_0, s_1, \ldots, s_{S-1})$ of the sequence $\{s_i\}$. This attack can be divided into two different steps. In the first one, the computation of the initial state of SRA is carried out. In the second step and based on the SRA initial state, we determine the corresponding initial state of the register SRS.

4.1 Computation of the SRA Initial State

Previously to the computation of the initial state, the following result is introduced.

Lemma 1. *Given A bits of the shrunken sequence corresponding to A successive elements of any column of IC, the remaining bits of such a column can be determined.*

Proof. Theorem 2 defines $P_D(x)$ the characteristic polynomial of the *PN*-sequence corresponding to every column of IC. Thus, knowing A successive bits of any column and its characteristic polynomial, the linear recurrence relationship allows one to compute the remaining bits of such a column. □

Now the computation of the SRA initial state is described in the next result.

Theorem 3. *Given A bits of the shrunken sequence corresponding to A successive elements of the first column of IC, the bits of the initial state of the register SRA can be determined.*

Proof. Lemma 1 shows that the knowledge of A successive elements of the first column of IC allows one to generate the remaining bits of such a column. On the other hand, from Theorem 1 we know that the $(n+1)$-*th* element of the first column of IC corresponds to $a_{n\cdot(2^S-1)}$, that is to say the $(n \cdot (2^S - 1) + 1)$-*th* term of the sequence generated by the register SRA. Consequently, we first solve the following system of modular equations in the unknowns n_i

$$n_i \cdot (2^S - 1) \equiv i \;\; mod\; (2^A - 1) \qquad (i = 0, 1, \ldots, (A-1)),$$

and then we compute successively the (n_i+1)-*th* $(i = 0, 1, \ldots, (A-1))$ elements of the first column of IC in order to obtain $a_0, a_1, \ldots, a_{A-1}$, respectively. □

4.2 Computation of the SRS Initial State

The computation of the SRS initial state is described in the next result.

Theorem 4. *Given A·S bits of the shrunken sequence corresponding to the top-left corner (A x S) sub-matrix of IC, the bits of the initial state of the register SRS can be determined.*

Proof. Firstly, from the knowledge of the $(A$ x $S)$ sub-matrix of IC

$$SUB_{IC} = \begin{pmatrix} a_0 & a_{o1} & \cdots & a_{o(S-1)} \\ a_{2^S-1} & a_{(2^S-1)+o1} & \cdots & a_{(2^S-1)+o(S-1)} \\ a_{2\cdot(2^S-1)} & a_{2\cdot(2^S-1)+o1} & \cdots & a_{2\cdot(2^S-1)+o(S-1)} \\ a_{3\cdot(2^S-1)} & a_{3\cdot(2^S-1)+o1} & \cdots & a_{3\cdot(2^S-1)+o(S-1)} \\ \cdots & \cdots & \cdots & \cdots \\ a_{(A-1)\cdot(2^S-1)} & a_{(A-1)\cdot(2^S-1)+o1} & \cdots & a_{(A-1)\cdot(2^S-1)+o(S-1)} \end{pmatrix}$$

and according to Lemma 1, we can deduce the remaining bits of those S columns. Secondly, the relative shifts among columns may be computed from the comparison between consecutive columns. Since the sequence in every column of IC is exactly the same but starting at different points given by a_{oj}, as soon as a relative shift is found the sub-index oj may be easily computed. In addition, each sub-index oj indicates the position of the $(j+1)$-*th* 1 in the initial state of SRS while the intermediate bits are 0's. Thus, the above procedure can be repeated for $j = 1, 2, \cdots$ till we get $oj \geq (S-1)$. In this way, the initial state of the register SRS is thoroughly determined. □

4.3 An Illustrative Example

Let us consider a shrinking generator characterized by:

1. SRA with length $A = 5$, characteristic polynomial $P_A(x) = x^5 + x^4 + x^3 + x^2 + 1$ and output sequence $\{a_i\}$.
2. SRS with length $S = 4$, characteristic polynomial $P_S(x) = x^4 + x^3 + 1$ and output sequence $\{s_i\}$.
3. The characteristic polynomial of the shrunken sequence is $P_{ss}(x) = P_D(x)^p = (x^5 + x^3 + x^2 + x + 1)^8$.

Given 20 bits of the shrunken sequence corresponding to a (5 x 4) sub-matrix of IC

$$SUB_{IC} = \begin{pmatrix} 1\,0\,1\,1 \\ 1\,0\,0\,1 \\ 0\,1\,0\,1 \\ 0\,1\,1\,1 \\ 0\,0\,0\,1 \end{pmatrix},$$

we can launch a cryptanalytic attack against the shrinking generator in order to obtain the initial states of both LFSRs. Table 1 shows the calculations carried out for cryptanalyzing the above described generator. The most left column represents the indices n_i numbered $(0, 1, \ldots, 2^A - 2 = 30)$. Next column shows from Theorem 1 the position of the terms $(a_0, a_1, \ldots, a_4)$ of the sequence $\{a_i\}$ in the first column $\{\boldsymbol{d}_0\}$ of the matrix IC. The following columns of the Table 1 represent the matrix IC: in boldface the (5 x 4) sub-matrix with the known bits, the remaining bits of $\{\boldsymbol{d}_0\}$ are the bits computed to determine the initial states of SRA and SRS, and the symbol $-$ corresponds to unknown bits of the shrunken sequence.

Computation of the SRA initial state: According to Theorem 3, we compute the positions of the $(n_i + 1)$-*th* elements of the first column of IC by solving the equation system

$$n_i \cdot 15 \equiv i \ mod\ 31 \qquad (i = 0, 1, \ldots, 4),$$

That is, $n_0 = 0$, $n_1 = 29$, $n_2 = 27$, $n_3 = 25$, $n_4 = 23$.

Remark that this computation is carried out in an off-line phase previous to the cryptanalysis. Then, by means of the characteristic polynomial $P_D(x)$ we determine the values of the $(n_i + 1)$-*th* $(i = 0, 1, \ldots, 4)$ elements of the first column $\{\boldsymbol{d}_0\}$ of IC. That just implies the computation of linear recurring equations. That is, $a_0 = 1$, $a_1 = 0$, $a_2 = 0$, $a_3 = 1$, $a_4 = 1$, see Table 1. Therefore, the initial state of the register SRA $(1, 0, 0, 1, 1)$ has been determined.
Computation of the SRS initial state: According to Theorem 4, we compute the relative shifts between consecutive columns in the matrix IC:

- *Computation of* $o1$*:* We know a_1 at the $(29+1)$-*th* position of the first column $\{\boldsymbol{d}_0\}$ and compute its $S - 1 = 4$ successive bits. We compare these 5 bits $(0, 0, 1, 1, 0)$ with the first 5 bits $(0, 0, 1, 1, 0)$ of the second column $\{\boldsymbol{d}_1\}$, see Table 1 . There is coincidence, thus $o1 = 1$.
- *Computation of* $o2$*:* We know a_2 at the $(27 + 1)$-*th* position of the $\{\boldsymbol{d}_0\}$ and compute its 4 successive bits. We compare these 5 bits $(0, 1, 0, 0, 1)$ with the first 5 bits $(1, 0, 0, 1, 0)$ of the third column $\{\boldsymbol{d}_2\}$. There is no coincidence, thus we analyze the following bit a_3. We know a_3 at the $(25+1)$-*th* position of $\{\boldsymbol{d}_0\}$ and compute its 4 successive bits. We compare these 5 bits $(1, 0, 0, 1, 0)$ with the first 5 bits $(1, 0, 0, 1, 0)$ of $\{\boldsymbol{d}_2\}$, see Table 1. There is coincidence, thus $o2 = 3$.

Since $o2 = 3 \geq S - 1$, we have determined the initial state of SRS. In fact, $s_0 = 1$, $o1 = 1$ implies $s_1 = 1$, $o2 = 3$ implies $s_2 = 0$ and $s_3 = 1$. Therefore, the SRS initial state is $(s_0, s_1, s_2, s_3) = (1, 1, 0, 1)$. Remark that only the knowledge of three columns of the sub-matrix has been necessary to identify the initial state of SRS. Indeed, the number of columns needed equals the number of '1' bits in the initial state of the selector register. The maximum number of known bits corresponds to SRS initial state with all bits '1'. In the remaining cases, less bits are sufficient.

In brief, solving A modular equations (Theorem 3), computing S linear recurring relationships and making at most S comparisons is all the computational burden required by this cryptanalytic attack. Actual values for cryptographic parameters can be A, S in the range of 64. So that the proposed attack can be carried out in a range of practical application.

Once the initial states of both register are determined, the whole shrunken sequence that is the keystream sequence can be computed.

5 Generalization of This Technique to Interleaved Sequences

First of all, we introduce the general definition of interleaved sequence [12].

Definition 1. *Let* $f(x)$ *be a polynomial over* $GF(q)$ *of degree* r *and let* m *be a positive integer. For any sequence* $\{u_k\}$ *over* $GF(q)$*, we write* $k = i \cdot m + j$ *with* $(i = 0, 1, \ldots)$ *and* $(j = 0, \ldots, m - 1)$*. If every sub-sequence* $\{\boldsymbol{u}_j\}$ *of* $\{u_k\}$ *defined as* $\{u_{i \cdot m+j}\}$ *is generated by* $f(x)$*, then the sequence* $\{u_k\}$ *is called an interleaved sequence over* $GF(q)$ *of size* m *associated with the polynomial* $f(x)$*.*

Table 1. Matrix IC corresponding to the described shrinking generator

n_i	$\{a_i\}$	d_0	d_1	d_2	d_3	d_4	d_5	d_6	d_7
0	a_0	**1**	**0**	**1**	**1**	–	–	–	–
1		**1**	**0**	**0**	**1**	–	–	–	–
2		**0**	**1**	**0**	**1**	–	–	–	–
3		**0**	**1**	**1**	**1**	–	–	–	–
4		**0**	**0**	**0**	**1**	–	–	–	–
5		–	–	–	–	–	–	–	–
6		–	–	–	–	–	–	–	–
7		–	–	–	–	–	–	–	–
8		–	–	–	–	–	–	–	–
9		–	–	–	–	–	–	–	–
10		–	–	–	–	–	–	–	–
11		–	–	–	–	–	–	–	–
12		–	–	–	–	–	–	–	–
13		–	–	–	–	–	–	–	–
14		–	–	–	–	–	–	–	–
15		–	–	–	–	–	–	–	–
16		–	–	–	–	–	–	–	–
17		–	–	–	–	–	–	–	–
18		–	–	–	–	–	–	–	–
19		–	–	–	–	–	–	–	–
20		–	–	–	–	–	–	–	–
21		–	–	–	–	–	–	–	–
22		–	–	–	–	–	–	–	–
23	a_4	1	–	–	–	–	–	–	–
24		–	–	–	–	–	–	–	–
25	a_3	1	–	–	–	–	–	–	–
26		0	–	–	–	–	–	–	–
27	a_2	0	–	–	–	–	–	–	–
28		1	–	–	–	–	–	–	–
29	a_1	0	–	–	–	–	–	–	–
30		0	–	–	–	–	–	–	–

Table 2 shows the interleaved sequence $\{u_k\}$ over $GF(2)$ associated with the 3-degree characteristic polynomial $f(x) = x^3 + x + 1$ over $GF(2)$ and size $m = 4$. Reading by rows, the interleaved sequence is $\{u_k\} = \{1, 1, 1, 1, 1, 0, 1, 0, 0, 0, 1, 1, 0, 1, 0, 1, 1, 0, 0, 1, 0, 1, 1, 0, 1, 1, 0, 0\}$ while by columns the sequence is made out of $\{\boldsymbol{u}_j\}$ $(j = 0, \ldots, 3)$ four shifted versions of the PN-sequence generated by $f(x)$.

Interleaved sequences are currently used as keystream sequences with application in cryptography. They can be generated in different ways:

1. By a LFSR controlled by another LFSR (which may be the same one) e.g. multiplexed sequences [11], clock-controlled sequences [1], cascaded sequences [7], shrinking generator sequences [5] etc.
2. By one or more than one LFSR and a feed-forward nonlinear function e.g. Gold-sequence family, Kasami (small and large set) sequence families, GMW sequences, Klapper sequences, No sequences etc. See [9] and the references cited therein.

Table 2. Interleaved sequence with 4 shifted versions of the same PN-sequence

u_0	u_1	u_2	u_3
1	1	1	1
1	0	1	0
0	0	1	1
0	1	0	1
1	0	0	1
0	1	1	0
1	1	0	0

In brief, a large number of well-known cryptographic sequences are included in the class of interleaved sequences. Next, the link between interleaved sequences and shrunken sequences is expressed in the following result.

Theorem 5. *Shrunken sequences are interleaved sequences of size* $2^{(S-1)}$.

Proof. Let $\{z_k\}$ be a shrunken sequence with characteristic polynomial $P(x)^p$ where $P(x)$ is an A-degree primitive polynomial and p is an integer in the interval $2^{(S-2)} < p \leq 2^{(S-1)}$. According to the interleaved configuration IC, we may express $\{z_k\}$ in terms of m sequences $\{\boldsymbol{z}_j\}$ where $\{\boldsymbol{z}_j\} = \{z_{i \cdot m+j}\}$ with $i \geq 0$, $m = 2^{(S-1)}$ and $(j = 0, \ldots, m-1)$. Since by Theorem 2 the sequences $\{\boldsymbol{z}_j\}$ are generated by the same characteristic polynomial $P_D(x)$, we get that the shrunken sequence $\{z_k\}$ is an interleaved sequence of size $2^{(S-1)}$ associated with the polynomial $P_D(x)$. □

The previous theorem proves that shrunken sequences are interleaved sequences. Moreover, section 4 shows that the knowledge of a number of bits of the shrunken sequence allows one to launch a cryptanalytic attack against the shrinking generator. As many cryptographic sequence generators produce interleaved sequences, then the previous considerations take us into the following statement:

Given a number of bits corresponding to an initial sub-matrix of the interleaved configuration IC of an interleaved sequence, it is possible to obtain the whole interleaved sequence.

Thus such a statement proves the weakness of interleaved generators for cryptographic purposes.

6 Conclusions

In this work, a new cryptanalytic attack against the class of shrinking generators has been proposed. Actual values for cryptographic parameters can be $A, S \simeq 64$. So that the proposed attack can be carried out in a range of practical application. The amount of intercepted bits necessary to realize such an attack is much lesser than that of other standard cryptanalysis. The basic idea consists in defining the shrunken sequence as an interleaved sequence. Hence the weaknesses inherent to interleaved sequences can be advantageously used in the practical attack. A straight consequence of this technique is its generalization to other interleaved sequence generators of cryptographic purpose. In this way, the security of this kind of generators must be carefully checked.

References

1. Beth, T., Piper, F.: The Stop-and-Go Generator. In: Proceedings of EUROCRYPT 1984. LNCS, vol. 228, pp. 228–238. Springer, Heidelberg (1985)
2. Bluetooth, Specifications of the Bluetooth system, Version 1.1 (February 2001), `http://www.bluetooth.com/`
3. Caballero-Gil, P., Fúster-Sabater, A.: A Wide Family of Nonlinear Filter Functions with a Large Linear Span. Information Sciences 164, 197–207 (2004)
4. Caballero-Gil, P., Fúster-Sabater, A.: Using Linear Hybrid Cellular Automata to Attack the Shrinking Generator. IEICE Transactions on Fundamentals of Electronics Communications and Computer E89-A, 1166–1172 (2006)
5. Coppersmith, D., Krawczyk, H., Mansour, H.: The Shrinking Generator. In: Stinson, D.R. (ed.) CRYPTO 1993. LNCS, vol. 773, pp. 22–39. Springer, Heidelberg (1994)
6. Fúster-Sabater, A.: Run Distribution in Nonlinear Binary Generators. Applied Mathematics Letters 17, 1427–1432 (2004)
7. Gollmann, D., Chambers, W.G.: Clock-Controlled Shift Register. IEEE J. Selected Areas Commun. 7, 525–533 (1989)
8. Golomb, S.: Shift-Register Sequences. Aegean Park Press, Laguna Hill (1982)
9. Gong, G.: Theory and Applications of q-ary Interleaved Sequences. IEEE Trans. Information Theory 41(2), 400–411 (1995)
10. GSM, Global Systems for Mobile Communications, `http://cryptome.org/gsm-a512.htm`
11. Jennings, S.M.: Multiplexed Sequences: Some Properties. In: Proceedings of EUROCRYPT 1983. LNCS, vol. 149, pp. 210–221. Springer, Heidelberg (1983)
12. Jiang, S., Dai, Z., Gong, G.: On interleaved sequences over finite fields. Discrete Maths 252, 161–178 (2002)
13. Kanso, A.: Clock-Controlled Shrinking Generator of Feedback Shift Registers. In: Safavi-Naini, R., Seberry, J. (eds.) ACISP 2003. LNCS, vol. 2727, pp. 443–451. Springer, Heidelberg (2003)
14. Lidl, R., Niederreiter, H.: Introduction to Finite Fields and Their Applications. Cambridge University Press, Cambridge (1986)
15. Rivest, R.L.: RSA Data Security, Inc., March 12 (1998)
16. Shparlinski, I.: On Some Properties of the Shrinking Generator. Designs, Codes and Cryptography 23, 147–156 (2001)
17. Simpsom, L., Golic, J., Dawson, E.: A Probabilistic Correlation Attack on the Shrinking Generator. In: Boyd, C., Dawson, E. (eds.) ACISP 1998. LNCS, vol. 1438, pp. 147–158. Springer, Heidelberg (1998)

Colonial Competitive Algorithm as a Tool for Nash Equilibrium Point Achievement

Ramin Rajabioun, Esmaeil Atashpaz-Gargari, and Caro Lucas

Control and Intelligent Processing Centre of Excellence (CIPCE), School of Electrical and Computer Engineering, University of Tehran, Tehran, Iran
r.rajabioun@ece.ut.ac.ir, e.atashpaz@ece.ut.ac.ir, lucas@ut.ac.ir

Abstract. This paper presents an application of Colonial Competitive Algorithm (CCA) in game theory and multi-objective optimization problems. The recently introduced CCA has proven its excellent capabilities, such as faster convergence and better global optimum achievement. In this paper CCA is used to find Nash Equilibrium points of nonlinear non-cooperative games. The proposed method can also be used as an alternative approach to solve multi-objective optimization problems. The effectiveness of the proposed method, in comparison to Genetic Algorithm, is proven through several static and dynamic example games and also multi-objective problems.

Keywords: Colonial Competitive Algorithm (CCA), game theory, global Nash Equilibrium, evolutionary algorithms.

1 Introduction

Game theory is generally considered to have begun with the publication of "*The Theory of Games and Economic Behaviour*", by von Neumann & Morgenstern in 1944 [1]. It provided the mathematical analysis of the conflict theory and brought the terminology with which one could analyze it. The development of the "Prisoner's Dilemma" by Tucker and J. Nash's papers on the definition and existence of equilibrium in 1950 laid the foundations of the modern non-cooperative game theory [2, 3]. These papers introduced the idea of the decision makers as players in a game and were designed to solve multi-objective optimization problems using game theory. Nash Equilibrium (NE) defined in [2], is a set of strategies such that no player can improve his payoff, given the strategies of all other players in the game, by changing his strategy. Nash proved that all non-cooperative games have a NE point.

Game theory can today be defined as the analysis of rational behaviour under circumstances of strategic interaction, when a player's best strategy depends on that of the others [4]. The players can be any entities that are able to exhibit strategic behaviour.

Economics is the most interesting field where game theory has been studied and applied. In addition to economics, game theory has been applied to biology, mathematics, political science, management science and even philosophy [5].

O. Gervasi et al. (Eds.): ICCSA 2008, Part II, LNCS 5073, pp. 680–695, 2008.

There are different kinds of games resulting in different equilibria points. Some of these equilibria points are Nash, Cournot, Stackelberg, Correlated and Bayesian equilibria. In this paper we will focus on Nash Equilibrium point of nonlinear non-cooperative games.

From a point of view, Games are divided into Dynamic and Static games. *Dynamic Games* mathematically model the interaction among different players, controlling a dynamical system. Instances of such situations happen in armed conflicts and economic competition. These examples are a kind of dynamical systems since the players' actions influence the evolution of the state of a system over time. The difficulty in deciding what should be the behavior of these players relies on the fact that each action a player takes at a given time will influence the reaction of the opponents at later time [6]. A precise mathematical framework to find the NE in linear dynamic games with quadratic cost is introduced in [7]. There have also been made considerable efforts to find the NE Points in nonlinear non-cooperative games. To enhance the poor convergence of simple coevolutionary programming, [8] uses hybrid coevolutionary programming to escape local NE traps and to reach the real NE points. The method used in this paper removes the need for not only such modifications but also the hybrid coevolutionary algorithms themselves. Reference [9] models the behaviour of participants in electricity markets. To study the dynamic behaviour of participants over many trading intervals a coevolutionary approach is developed. Trading agents co-evolve their own populations' strategies using a Genetic Algorithm (GA). [10] proposes a game model based coevolutionary algorithm to solve multi-objective class of problems. It tries to find Evolutionary Stable Strategy (ESS) as a solution to multi-objective problems using game model based coevolutionary algorithm. In [11] efficient approximation algorithms to achieve Nash equilibria in anonymous games, games in which the players' utilities, though different, do not differentiate between other players, are introduced.

In a multi-objective optimization problem including $\boldsymbol{P}$ objectives, there exist $\boldsymbol{P}$ players, each optimizing his own criterion. When a player optimizes his own criterion he knows all other players' actions (strategies) and also knows that these strategies are fixed during his decision making process. This improvement goes on for all players and stops when no player can further improve his criterion. This situation implies that the system has reached to an equilibrium point. This point is called the Nash Equilibrium of the game. If we assume Y_1 as the search space for the first criterion and Y_P the search space for the last criterion. A strategy set $(X_1, X_2, \ldots, X_P) \in Y_1 \times Y_2 \times \ldots \times Y_P$ will be called NE point iff:

$$\begin{aligned} f_{Y1}(X_1, \ldots, X_P) &= \inf_{x_1 \in Y_1} f_{Y1}(X_1, \ldots, X_P) \\ &\vdots \\ f_{Yp}(X_1, \ldots, X_P) &= \inf_{x_p \in Y_p} f_{Yp}(X_1, \ldots, X_P) \end{aligned} \tag{1}$$

As an other definition of NE, if the set $\underline{X} = (X_1, X_2, \ldots, X_p)$ is the strategy set for all players and $J_i(\underline{X})$ is the cost function of player i then $\underline{X}^* = (X^*_1, X^*_2, \ldots, X^*_p)$ will be a NE point iff: $\forall i,\ \forall x_i$

$$J_i(X^*_1, \ldots, X^*_i, \ldots, X^*_P) \le J_i(X_1, \ldots, X_i, \ldots, X_P) \tag{2}$$

Classical approach to find NE points needs the cost (payoff) function to be concave and differentiable. If this condition is satisfied, solving the following set of equations simultaneously gives the NE point of the game:

$$\frac{\partial J_i(X_1, \ldots, X_i, \ldots, X_p)}{\partial X_i} = 0 \quad i = 1, \ldots, p \tag{3}$$

But there are several games in which the payoff function is not differentiable or is so nonlinear that the calculation of differentiation is really challenging. In these cases the classical approach fails to give the NE point and the use of evolutionary algorithms becomes indispensable. Genetic Algorithms are the most preferred evolutionary methods, applied to such problems. [12] uses GA to find NE points. It has constructed the chromosomes in such a way that the actions of all players are included. Each player updates his own part of chromosomes. This way we can obtain NE point of almost all types of games with nonlinearity in cost (payoff) functions. But the main problem with GA is its poor convergence rate and also poor global optima achievement. In this paper, we apply a newly introduced evolutionary algorithm, called CCA, on non-cooperative games with nonlinear cost functions to eliminate mentioned drawbacks.

The recently introduced CCA [13] has been used to solve many optimization problems in different fields. This global search strategy uses the socio-political competition among empires as a source of inspiration. Like other evolutionary strategies that start with an initial population, CCA begins with initial empires. Any individual of an empire is a country. There are two types of countries; colony and imperialist state that collectively form empires. Imperialistic competition among these empires along with modeled assimilation policy forms the basis of this algorithm. During the imperialistic competition, weak empires collapse and powerful ones take possession of their colonies. Imperialistic competition converge to a state in which there exist only one empire and its colonies are in the same position and have the same cost as the imperialist. This evolutionary optimization strategy has shown great performance in both convergence rate and better global optima achievement. Nevertheless, its effectiveness, limitations and applicability in various domains are currently being extensively investigated. In [14] it has been used to design an optimal controller which not only decentralizes but also optimally controls an industrial Multi Input Multi Output (MIMO) Evaporator system. In [15] CCA is used for reverse analysis of an artificial neural network in order to characterize the properties of materials from sharp indentation test. In order to find the optimal priorities for each user in recommender systems, [16] uses CCA in "Prioritized user-profile" approach to recommender systems, trying to implement more personalized recommendation by assigning different priority importance to each feature of the user-profile in different users.

The most interesting aspect of the proposed method in this paper is its ability to find global Nash Equilibrium more reliably and not stopping at local NE traps. Effectiveness of this method will be shown through several static and dynamic games and some multi-objective problems.

The structure of rest of the paper is as follows:

Section 2 studies the idea of Nash Equilibrium search using evolutionary algorithms. In section 3 a brief description of CCA is presented. Simulation results obtained from some examples are presented in section 4. Finally, the conclusion is done in section 5.

2 Evolutionary Algorithms and Nash Equilibrium

In this section we study the method using with, an evolutionary algorithm can be applied to reach Nash Equilibrium. A *P*-objective game corresponding to *P* players each with one action set is considered here. Let each of $\underline{X}_i$s in the potential solution of the game $\underline{\mathbf{X}}=\{\underline{X}_1, \underline{X}_2, \dots, \underline{X}_p\}$ be the action set of players P_1 to P_P, respectively. Each player can optimize and affect only his own set of actions $\underline{X}_i$. But his action might influence the cost function of other players.

At each iteration of the payoff optimization process, player P_i optimizes his cost function ($J_i(\underline{\mathbf{X}})$) knowing that the other players' have chosen and fixed their action sets ($\underline{X}_j$, $j=1,\dots,P$, $j\neq i$). To start the game, each player randomly chooses his action. Then each player optimizes his action with respect to the actions other players have preferred. When all the players have optimized their actions once, the first iteration is over. Players continue the game using the potential solution of iteration *n*, *i. e.*, $\underline{\mathbf{X}}_n^*=\{\underline{X}^*_{1n}, \underline{X}^*_{2n}, \dots, \underline{X}^*_{pn}\}$ as their initial strategies for next iteration. The game stops when no player is able to change his action and so the cost function stops changing. This stop point is the desired Nash Equilibrium point of the game. Fig. 1 shows diagram of the optimization process used in this paper.

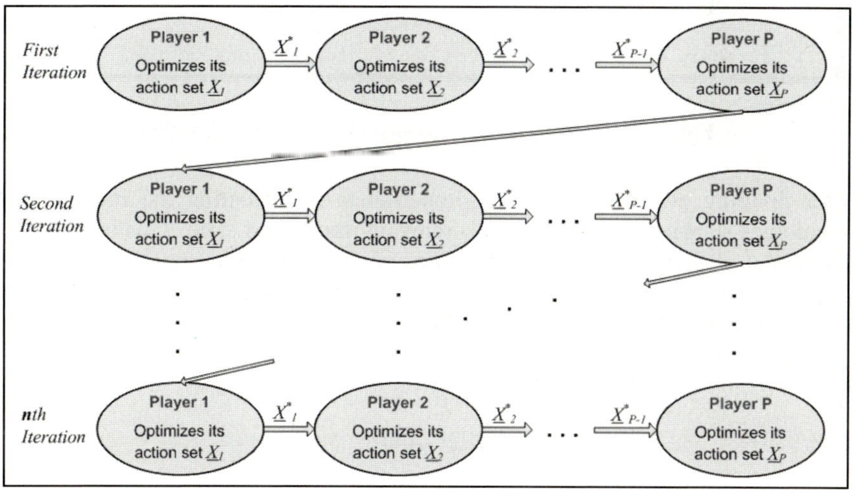

Fig. 1. The strategy scheme used to find Nash Equilibrium

It might seem that this algorithm is similar to that of parallel GA (PGA) introduced in [17]. But it is noteworthy that the principle difference between these two is that PGA uses the same cost function whereas the above mentioned method uses different criteria, resulting in NE point. In this paper we focus on the use of a novel evolutionary algorithm, CCA, to find the NE point of the games.

3 Brief Description of CCA

CCA is a novel global heuristic search method that uses imperialism and imperialistic competition process as a source of inspiration. Fig. 2 shows the pseudo code for this algorithm. This algorithm starts with some initial countries. Some of the best countries are selected to be the *imperialist states* and all the other countries form the *colonies* of these imperialists. The colonies are divided among the mentioned imperialists based on their power.

```
1) Select some random points on the function and initialize the
   empires.
2) Move the colonies toward their relevant imperialist
   (Assimilation).
3) If there is a colony in an empire which has lower cost than
   that of the imperialist, exchange the positions of that
   colony and the imperialist.
4) Compute the total cost of all empires (Related to the power
   of both the imperialist and its colonies).
5) Pick the weakest colony (colonies) from the weakest empires
   and give it (them) to the empire that has the most
   likelihood to possess it (Imperialistic competition).
6) Eliminate the powerless empires.
7) If there is only one empire left, stop, if not go to 2.
```

Fig. 2. Pseudo code of the Colonial Competitive Algorithm

After dividing all colonies among imperialists and creating the initial empires, these colonies start moving toward their relevant imperialist state. This movement is a simple model of assimilation policy that was pursued by some imperialist states [13]. Fig. 2 shows the movement of a colony towards the imperialist. In this movement, θ and *x* are random numbers with uniform distribution as illustrated in (4) and *d* is the distance between colony and the imperialist.

$$x \sim U(0, \beta \times d)$$
$$\theta \sim U(-\gamma, \gamma) \tag{4}$$

where β and γ are arbitrary numbers that modify the area that colonies randomly search around the imperialist. In our implementation β and γ are 2 and 0.5 (rad), respectively.

The total power of an empire depends on both the power of the imperialist country and the power of its colonies. In this algorithm, this fact is modeled by defining the total power of an empire by the power of imperialist state plus a percentage of the mean power of its colonies.

In imperialistic competition, all empires try to take possession of colonies of other empires and control them. This competition gradually brings about a decrease in the power of weak empires and an increase in the power of more powerful ones.

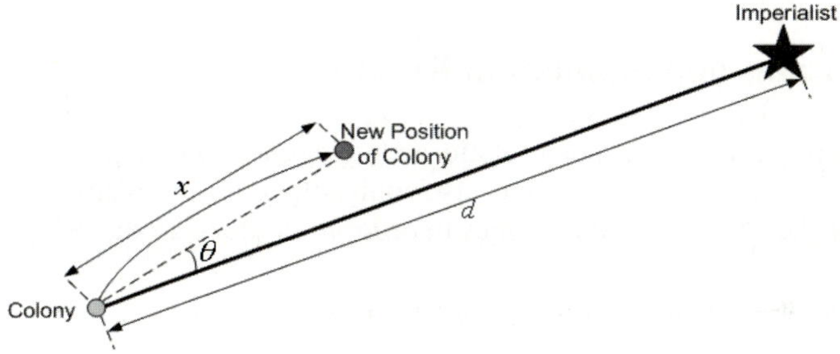

Fig. 3. Motion of colonies toward their relevant imperialist

This competition is modelled by picking some (usually one) of the weakest colonies of the weakest empires and making a competition among all empires to possess these (this) colonies. Fig. 3 shows a big picture of the modelled imperialistic competition. Based on their total power, in this competition, each of empires will have a likelihood of taking possession of the mentioned colonies. The more powerful an empire is, the more likely it will possess these colonies. In other words these colonies will not be certainly possessed by the most powerful empires, but these empires will be more likely to possess them. Any empire that is not able to succeed in imperialist competition and can not increase its power (or at least prevent decreasing its power) will be eliminated.

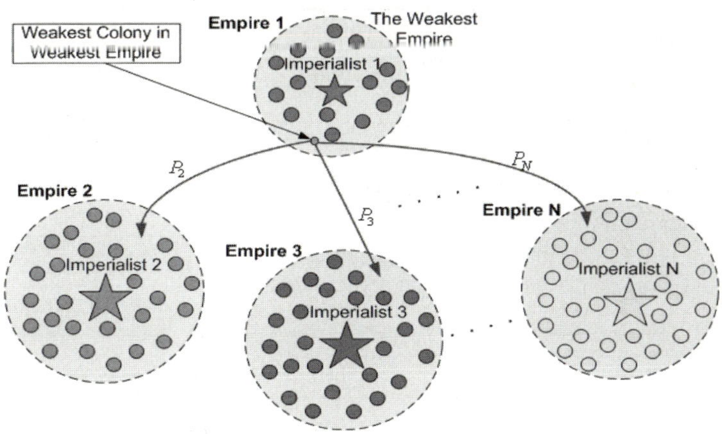

Fig. 4. Imperialistic competition: The more powerful an empire is, the more likely it will possess the weakest colony of the weakest empire

The imperialistic competition will gradually result in an increase in the power of great empires and a decrease in the power of weaker ones. Weak empires will gradually loose their power and ultimately they will collapse.

The movement of colonies toward their relevant imperialists along with competition among empires and also collapse mechanism will hopefully cause all the countries to converge to a state in which there exist just one empire in the world and all the other countries are its colonies. In this ideal new world colonies have the same position and power as the imperialist.

4 Applications and Simulation Results

In this section we focus on some static and dynamic non-cooperative games with nonlinear cost (payoff) functions and also multi-objective problems. Fig. 5 shows a big picture of the CCA applied to games in order to produce global NE points.

4.1 Example of a Simple Nonlinear Static Game

Consider a two player game with the nonlinear cost functions defined as:

$$f_1(x_1,x_2) = (x_1-1)^2 + (x_1-x_2)^2 \tag{5}$$

$$f_2(x_1,x_2) = (x_2-3)^2 + (x_1-x_2)^2 \tag{6}$$

The NE point using analytical method can be obtained as:

$$\begin{cases} \dfrac{\partial f_1(x_1,x_2)}{\partial x_1} = 0 \\ \dfrac{\partial f_2(x_1,x_2)}{\partial x_2} = 0 \end{cases} \Rightarrow \begin{cases} x_2 = 2x_1 - 1 \\ x_2 = \dfrac{x_1+3}{2} \end{cases} \Rightarrow \begin{cases} x_1^* = \dfrac{5}{3} = 1.6667 \\ x_2^* = \dfrac{7}{3} = 2.3333 \end{cases} \tag{7}$$

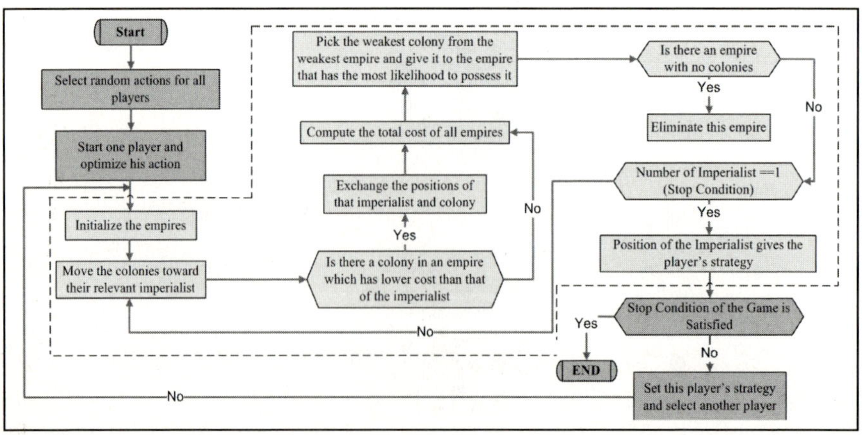

Fig. 5. The flowchart of CCA applied to games for global NE achievement (dashed area is the flowchart of CCA)

So the NE point of this simple game is $(x^*_1, x^*_2) = (1.6667, 2.3333)$ with the corresponding payoff values *(0.88889, 0.88889)*. We will use this value to verify the precision of the answers obtained using both GA and CCA. Fig. 5 and Fig. 6 show the cost function convergence for GA and CCA, respectively.

As it can be seen from Fig. 5 and Fig. 6, even in this simple game CCA has converged in less iterations (5 iterations) compared with GA (8 iterations). Our codes are so that if the payoff values are constant, with the tolerance of 0.0001, for 10 iterations the optimization process stops. In this simple case both the algorithms could achieve the optimal point as computed by analytical method.

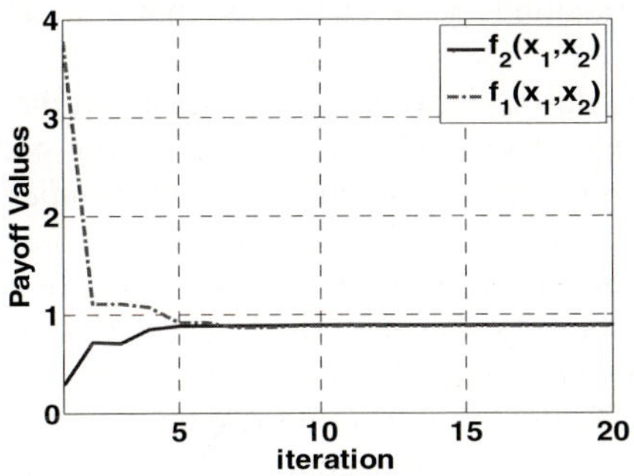

Fig. 6. Payoff convergence for GA

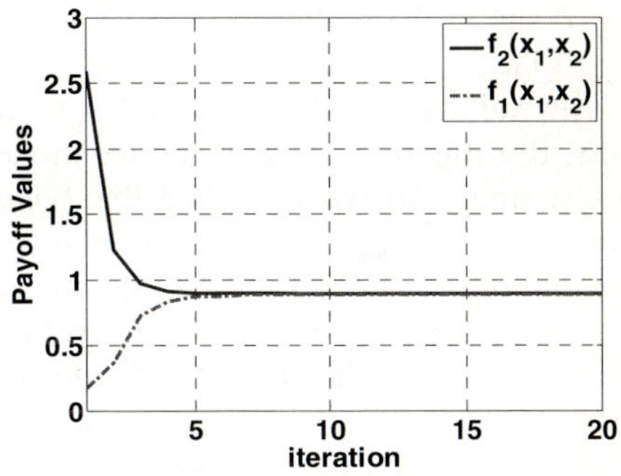

Fig. 7. Payoff convergence for CCA

In this simulation the *Population* size of GA was set to 20 together with the *Mutation* rate of 0.2. On the other hand, *Number of Countries* was set to 20 together with *Number of Imperialists* to 5 for CCA (meaning that 5 out of 20 countries will be chosen as *Imperialists*).

4.2 A Complicated Multi-objective Example

Deb [18] designed some multi objective problems which their optimization was a challenge for GAs. The following example is one of his. Suppose that there are two criterion functions defined as follows:

$$f_1(x_1) = 4x_1 \tag{8}$$

$$f_2(x_1, x_2) = g(x_2).h(f_1(x_1), g(x_2)) \tag{9}$$

where:

$$g(x_2) = \begin{cases} 4 - 3\exp(-(\frac{x_2 - 0.2}{0.02})^2) & \text{if } 0 \le x_2 \le 0.4 \\ 4 - 2\exp(-(\frac{x_2 - 0.7}{0.2})^2) & \text{if } 0.4 \le x_2 \le 1 \end{cases} \tag{10}$$

and

$$h(f_1, g) = \begin{cases} 1 - (\frac{f_1}{g})^{\alpha} & \text{if } f_1 \leq g \\ 0 & \text{otherwise} \end{cases} \quad (11)$$

$$\alpha = 0.25 + 3.75(g(x_2) - 1)$$

We model this optimization problem using game theory and then find its NE point. We suppose that f_1, f_2 are the payoff functions of two players. And also x_1, x_2 are the actions they can take. An investigation of these functions show that for this problem, there is one global Pareto set which is convex and one local Pareto set that is concave. Fig. 8 shows the scatter of (f_1, f_2) for 50000 randomly generated (x_1,x_2) pairs in interval [0,1].

Now we apply both GA and CCA to this multi-objective problem and obtain the results. But this time none of the mentioned algorithms could find the NE using previous settings. So we increased the *Population* size of GA and also the *Number of Countries* and the *Number of imperialists*. But, as seen in Fig. 9, even the increase of *Population* size to 200 and also the increase of its optimization iterations, did not help GA much to find the NE of this problem precisely. On the other hand, CCA could reach the point with *Number of Countries* equal to 35 and *Number of imperialists* to 5. CCA could achieve a NE point on global Pareto front *(0,1)* corresponding to $(x^*_1, x^*_2)=(0, 0.20002)$. Fig. 10 shows the payoff values convergence for CCA.

4.3 An Example of Nonlinear Static Game with Three Players Each with a Single Action to Select

In this example we will focus on a three player static game. There are three market players interconnected with three transmission lines. There are no constraints between transmission lines. The criteria are the profit functions which are going to be maximized. The criteria are defined as:

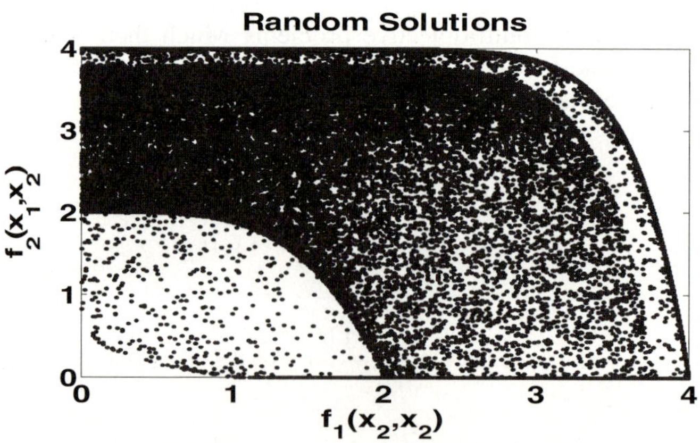

Fig. 8. Randomly generated solutions

$$\pi_i(\underline{x}) = (\theta - \rho(x_1 + x_2 + x_3))x_i - (0.5\phi_i x_i^2 + \gamma_i x_i + \eta_i) \quad i = 1,2,3 \tag{11}$$

This game is in Cournot model format. Connections between the generators are shown in Fig. 9. This system and the market data are quoted from [20]. The transmission lines are assumed to be lossless and have equal reactance [21]. Parameters of the mentioned payoff functions are summarized in table 1. Nash equilibrium is first obtained using analytical method.

$$\frac{\partial \pi_1}{\partial x_1} = 0 \ , \ \frac{\partial \pi_2}{\partial x_2} = 0 \ , \ \frac{\partial \pi_3}{\partial x_3} = 0 \ \Rightarrow \begin{bmatrix} x_1^* \\ x_2^* \\ x_3^* \end{bmatrix} = \begin{bmatrix} 1107.58 \\ 1048.30 \\ 976.73 \end{bmatrix} \tag{12}$$

Now that we've computed NE point of this game we compare it with the results obtained from GA and CCA. Payoff values convergence of GA and CCA are shown in Fig. 12 and Fig. 13, respectively.

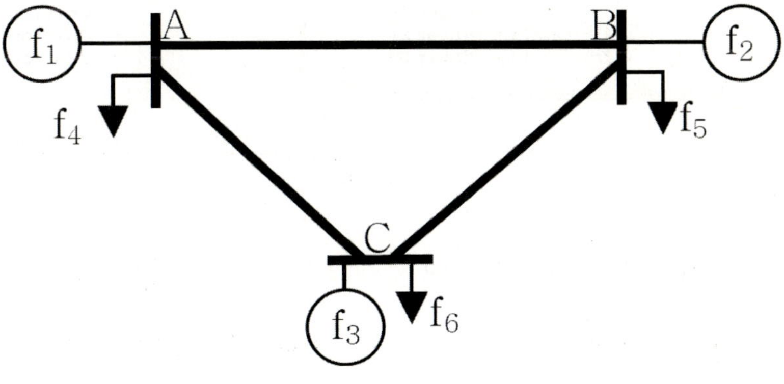

Fig. 9. Sample system for the Cournot model

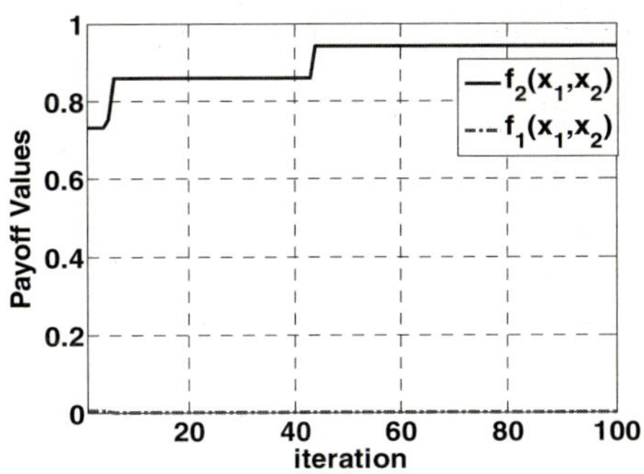

Fig. 10. Payoff convergence for GA

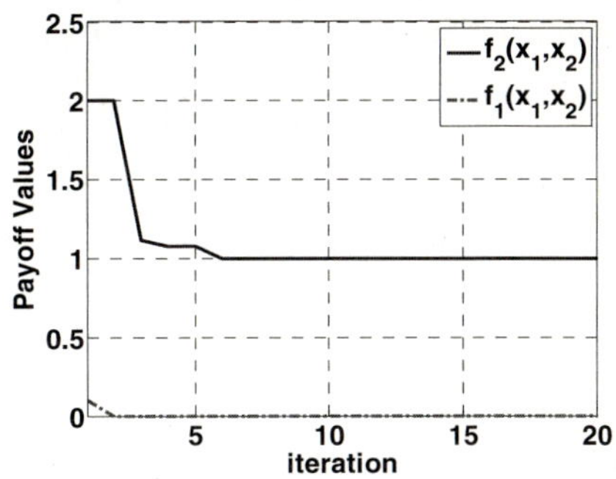

Fig. 11. Payoff convergence for CCA

Table 1. Parameters of the profit functions

	π_1	π_2	π_3
Φ_i	0.015718	0.021052	0.012956
γ_i	1.360575	-2.07807	8.105354
η_i	9490.366	11128.95	6821.482
θ	106.1176	106.1176	106.1176
ρ	0.0206	0.0206	0.0206

Again CCA wins in convergence rate and NE point preciseness, producing the set *(x*1,x*2,x*3)=(1107.5865, 1048.2999, 976.72098)*. This Nash point corresponds to the profit values of *(25421.6369, 23076.6077, 19010.9858)*. GA, in this case, converges to *(x*1,x*2,x*3)=(1104.9563, 1036.1275, 1016.143)* with cost values of *(25256.0385, 22286.6851, 21137.8527)*. As it is seen GA could not find the NE exactly.

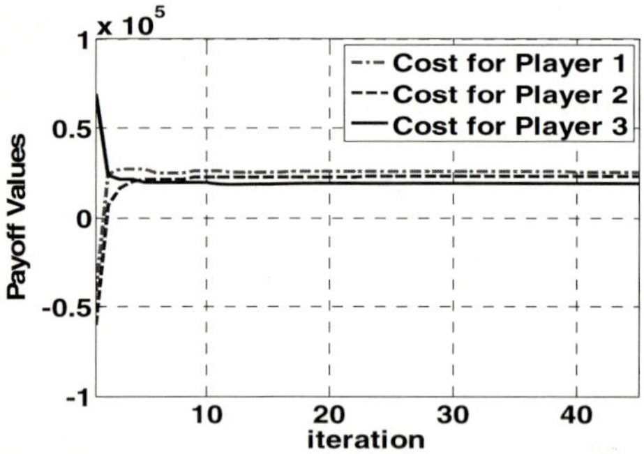

Fig. 12. Payoff convergence for GA

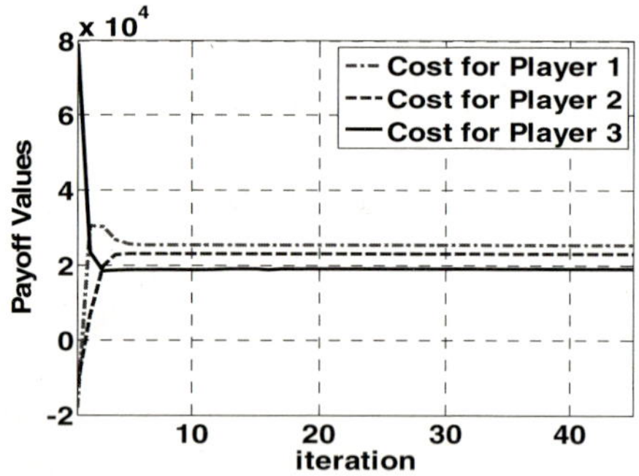

Fig. 13. Payoff convergence for CCA

4.4 An Example of Games with Three Players Each with an Action Set to Select

In this example we extend the players' strategy set so that a three player game each with 2 actions is obtained. Consider the cost functions of the players to be as follows:

$$f_A(x_1, x_2, y_1, y_2, z_1, z_2) = 21 + x \times \sin(5\pi x) + z \times y \times \sin(5\pi yz) \tag{13}$$

$$f_B(x_1, x_2, y_1, y_2, z_1, z_2) = 21 + y \times \sin(5\pi y) + z \times x \times \sin(5\pi xz) \tag{14}$$

$$f_C(x_1, x_2, y_1, y_2, z_1, z_2) = 21 + x \times \sin(5\pi z) + x \times y \times \sin(5\pi yx) \tag{15}$$

where $x = x_1+x_2$, $y = y_1+y_2$ and $z = z_1+z_2$ are the actions of the players P_1 to P_3. Fig. 14 and Fig. 15 depict the payoff function variations for all three players, using GA and CCA. As it can obviously seen from these figures, CCA has converged very faster in just 3 iterations where GA converges after 21 iterations. The payoff values and the NE points obtained by GA and CCA are summarized in table 2.

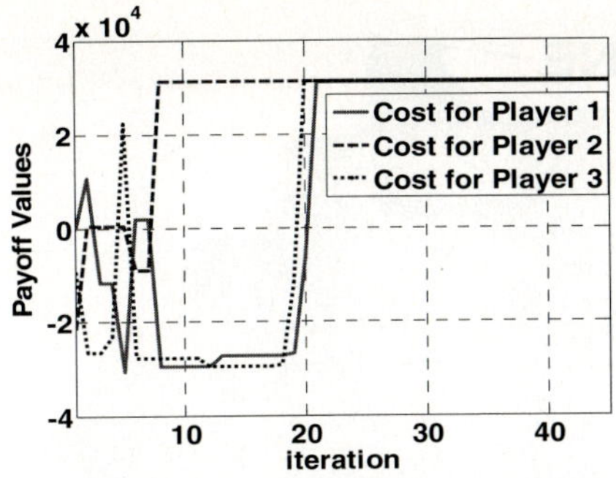

Fig. 14. Payoff convergence for GA

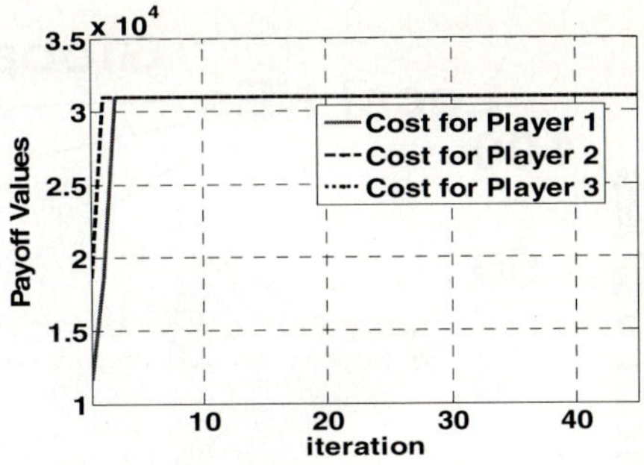

Fig. 15. Payoff convergence for CCA

It can be said that when the players' action sets increase GA losses its convergence speed.

Table 2. NE and payoff values obtained by GA and CCA

	NE obtained by CCA	NE obtained by GA
x^*_1	9.9748	9.9748
x^*_2	9.897	9.897
y^*_1	9.97481	9.9748
y^*_2	9.89703	9.897
z^*_1	9.9748	9.9748
z^*_2	9.897	9.897
Payoff	134.1276	134.4592

4.5 An Example of Static Games with Two Players Each with One Action

In this example a static non-cooperative game is studied. The profit functions of two players are defined as follows:

$$\pi_1 = 21 + x\times\sin(\pi x) + x\times y\times\sin(\pi y) \quad (16)$$

$$\pi_2 = 21 + y\times\sin(\pi y) + y\times x\times\sin(\pi x) \quad (17)$$

Fig. 16 shows these profit functions for different action values that can be chosen by each player. As it can be seen there are several local NE traps and just one global NE in this game.

Fig. 17 and Fig. 18 show the profit functions convergence rates and values for both GA and CCA. It can be seen that CCA converges to NE of *(8.5388, 8.5383)* with corresponding profit values of (x^*, y^*) *=(101.8558, 101.8426)* in just 2 iterations while GA has reached this NE in 9 iterations. The *population* size of GA was set to 30 while the *Number of Countries* and *Number of Imperialists* of CCA were set to 30 and 5, respectively.

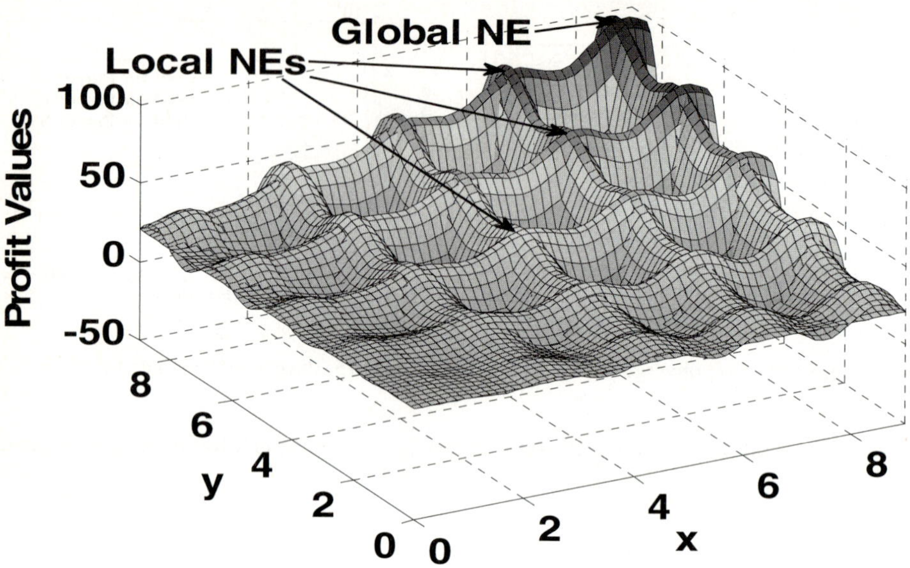

Fig. 16. Profit function values for different actions

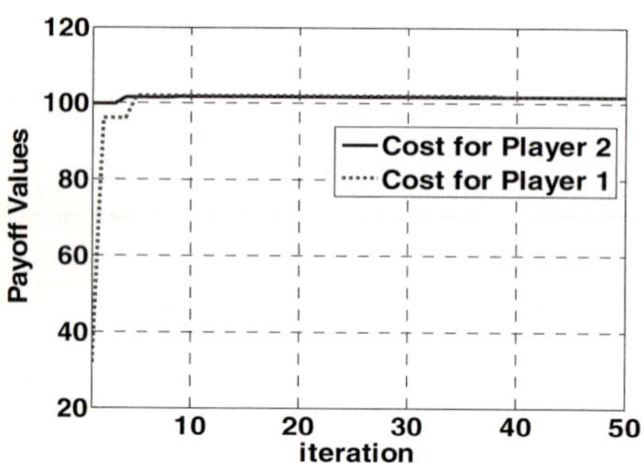

Fig. 17. Profit Values convergence for GA

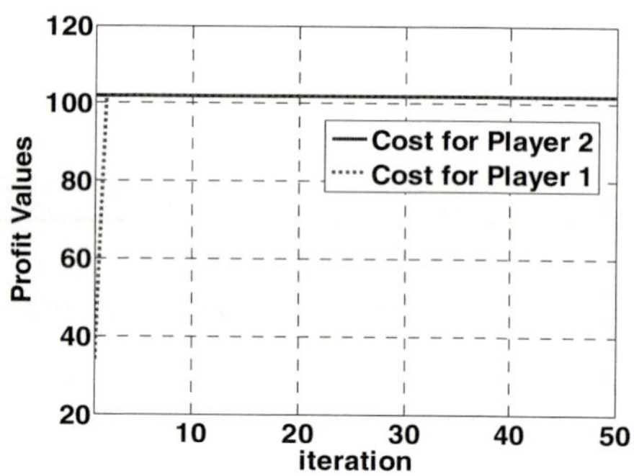

Fig. 18. Profit Values convergence for CCA

4.6 An Example of Games with Three Players Each with an Action Vector of Dimension Three

In this example a game with three players each with three actions is considered and studied. The payoff functions of these three players are as follows:

$$f_A = 21 + x_1 \times \sin(4 \times freq \times x_1) + 1.1 \times x_2 \times \sin(2 \times freq \times x_2) + 1.2 \times x_3 \times \sin(2 \times freq \times x_3) + z \times y \times \sin(\pi \times y \times z) \quad (18)$$

$$f_B = 21 + y_1 \times \sin(4\times freq \times y_1) + 1.1\times y_2 \times \sin(2\times freq \times y_2) + 1.2\times y_3 \times \sin(2\times freq \times y_3) + x\times z\times \sin(\pi \times x\times z) \quad (19)$$

$$f_C = 21 + z_1 \times \sin(4\times freq \times z_1) + 1.1\times z2\times \sin(2\times freq \times z2) + 1.2\times z_3 \times \sin(2\times freq \times z_3) + x\times y\times \sin(\pi \times x\times y) \quad (20)$$

where $x=x_1+x_2+x_3$, $y=y_1+y_2+y_3$ $z=z_1+z_2+z_3$ *and* $freq = 5$. The *freq.* value provides us a tool to make the game more or less complicated. The more *freq.* value is, the more complicated the game will be. Payoffs convergence and precision are depicted in Fig. 19 and Fig. 20. Table 3 shows the achieved NE points together with payoff values obtained for three players. Due to the symmetry in game, looking at payoff functions, a symmetry is expected in payoff values and also in actions. Looking table 3 it can be said that CCA has this symmetry in produced NE point. But GA has not been able to converge to such a symmetry.

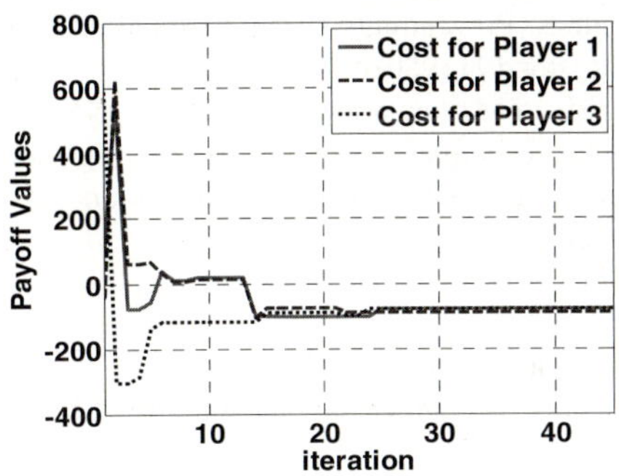

Fig. 19. Payoff Values convergence for GA

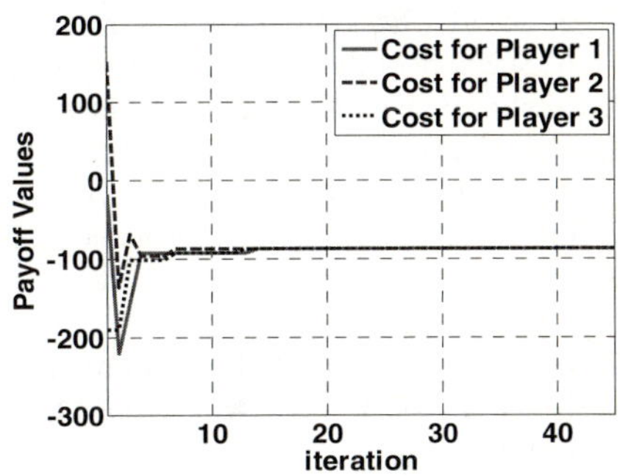

Fig. 20. Payoff Values convergence for CCA

As stated before when the players' action sets increase GA losses its convergence speed. This fact is again seen in Fig. 19 compared with Fig. 20.

Table 3. NE and payoff values obtained by GA and CCA

	NE obtained by CCA	NE obtained by GA
x^*_1	8.7182	8.7182
x^*_2	8.6405	8.6405
x^*_3	8.6405	8.6405
y^*_1	8.7182	8.7187
y^*_2	8.6405	8.6403
y^*_3	8.6405	8.6404
z^*_1	8.7182	8.7182
z^*_2	8.6405	8.6405
z^*_3	8.6405	8.6405
$Payoff_A$	-86.8510	-77.8117
$Payoff_B$	-86.7845	-87.0486
$Payoff_C$	-86.7785	-77.5606

5 Conclusion

In this paper CCA was applied to some multi-objective problems and some non-cooperative games. Better convergence rate and also better optima achievement of CCA, when used in different optimization problems, encouraged us to use it in order to find the NE of non-cooperative games with nonlinear cost functions. Usage of CCA helped to eliminate the drawbacks of GA, such as poor convergence rate and poor optima achievement, while not stopping in local Nash Equilibriums. This method also removed the need for modification of simple coevolutionary algorithms generating hybrid coevolutionary algorithms.

References

[1] Neumann, J., Morgenstern, O.: Theory of Games and Economic Behavior, 60th-Anniversary edn. Princeton University Press, Princeton (2007)
[2] Nash, J.F.: Equilibrium points in n-person games. Proceedings of the National Academy of Sciences of the United States of America 36, 46–49 (1950)
[3] Nash, J.F.: Non-Cooperative Games. The Annals of Mathematics 54, 286–295 (1951)
[4] Varoufakis, Y.: General introduction: Game theory's quest for a single, unifying framework for the social sciences. In: Varoufakis, Y. (ed.) Game Theory: Critical Concepts in the Social Sciences, vol. 1. Routledge, London (2001)
[5] `http://plato.stanford.edu/entries/game-evolutionary` (January 2008)
[6] Haurie, A., Krawczyk, J.: An Introduction to Dynamic Games (2000)
[7] Basar, T., Olsder, G.J.: Noncooperative Game Theory, 2nd edn. Academic Press, London (1995)
[8] Son, Y.S., Baldick, R.: Hybrid Coevolutionary Programming for Nash Equilibrium Search in Games with Local Optima. IEEE Trans. on evolutionary computation 8, 305–315 (2004)
[9] Cau, T.D.H., Anderson, E.J.: A co-evolutionary approach to modeling the behavior of participants in competitive electricity markets. In: Proc. IEEE Power Engineering Soc. Summer Meeting, vol. 3, pp. 1534–1540 (2002)
[10] Gaoping, W., Yongji, W.: A game model based co-evolutionary for constrained multiobjective optimization problems. In: IEEE International Symposium on Communications and Information Technology, vol. 1, pp. 187–190 (2005)
[11] Daskalakis, C., Papadimitriou, C.: Computing Equilibria in Anonymous Games. In: 48th Annual IEEE Symposium on Foundations of Computer Science, pp. 83–93 (2007)
[12] Sefrioui, M., Perlaux, J.: Nash genetic algorithms: examples and applications. In: Proceedings of the 2000 Congress on Evolutionary Computation, vol. 1, pp. 509–516 (2000)
[13] Atashpaz-Gargari, E., Lucas, C.: Imperialist Competitive Algorithm: An algorithm for optimization inspired by imperialistic competition. In: IEEE Congress on Evolutionary Computation, Singapore (2007)
[14] Rajabioun, R., Hashemzadeh, F., Atashpaz-Gargari, E., Mesgari, B., Rajaei Salmasi, F.: Identification of a MIMO Evaporator and Its Decentralized PID Controller Tuning Using Colonial Competitive Algorithm (accepted to be presented in IFAC World Congress 2008)

[15] Biabangard-Oskouyi, A., Atashpaz-Gargari, E., Soltani, N., Lucas, C.: Application of Imperialist Competitive Algorithm for materials property characterization from sharp indentation test. International Journal of Engineering Simulation, under revision (submitted, 2008)

[16] Sepehri Rad, H., Lucas, C.: Application of Imperialistic Competition Algorithm in Recommender Systems. In: 13th Int'l CSI Computer Conference (CSICC 2008), Kish Island, Iran (2008)

[17] Muhlenbein, H., Schomisch, M., Born, J.: The parallel genetic algorithm as function optimizer. Parallel Computing 17(6-7), 619–632 (1991)

[18] Deb, K.: Multi-objective genetic algorithms: Problem difficulties and construction of test problems. Technical report, Department of Computer Science / LS11, University of Dortmund (1998)

[19] Dekel, E., Scotchmer, S.: On the Evolution of Optimizing Behavior. Journal of Economic Theory 57, 392–406 (1992)

[20] Cunningham, L.B., Baldick, R., Baughman, M.L.: An Empirical Study of Applied Game Theory: Transmission Constrained Cournot Behavior. IEEE Transactions on Power Systems 17(1), 166–172 (2002)

[21] Wood, A.J., Wollenberg, B.F.: Power Generation Operation and Control, 2nd edn. Wiley, Chichester (1996)

Enhanced Empirical Mode Decomposition

Denis Donnelly

Department of Physics, Siena College, Loudonville, NY 12211
donnelly@siena.edu
http://www.sos.siena.edu

Abstract. Empirical mode decomposition (EMD) associated with the Hilbert-Huang transform (HHT) deconstructs a time-series signal into a set of monocomponent signals called intrinsic mode functions (IMF). EMD also acts as a filter limiting the frequency range of each IMF. EMD filtering is less than ideal and can lead to misleading results. This difficulty is ameliorated by first subjecting the time-series to bandpass filtration, where the pass-band frequency range is sufficiently narrow that the entire pass-band is captured in a single IMF. A series of such filtrations are required to treat a multicomponent signal. This approach avoids partial representation of those frequencies at the crossover between two successive IMF's. Bandpass enhanced EMD is applied to a bat chirp signal. That the fundamental, the first, second, and part of the third harmonic are expressed, demonstrates the improved sensitivity of this method over the standard HHT approach. The fundamental and harmonics of this chirp have an exponential form with a decay rate proportional to the square root of the time.

Keywords: time-series, empirical mode decomposition, Hilbert spectral analysis, Hilbert-Huang transform, proto-intrinsic mode functions.

1 Introduction

In many experiments, data is captured in the form of a time-series. Of these, many of the datasets are non-stationary, where frequency or statistical correlations, for example, vary with time. As stationarity is an assumption underlying the Fourier transform, it is inappropriate for such series. Further, this transform yields global results devoid of time dependence in the frequency domain. Sliding a time-constraining window across the time-series and taking the Fourier transform of each windowed subset provides an approach to obtaining time-dependent frequencies. A principal disadvantage of this method is the degradation of frequency resolution associated with each of the window-shortened time-series.

The Hilbert-Huang (HHT) transform [1–11] provides a different approach and can be applied directly to non-stationary time-series yielding time dependent amplitude and frequency information. The HHT is carried out in two stages: empirical mode decomposition (EMD) and Hilbert spectral analysis (HSA). Unlike the Fourier transform, this transform is not constrained by a time-frequency uncertainty principle. However, a different constraint limits the applicability of

O. Gervasi et al. (Eds.): ICCSA 2008, Part II, LNCS 5073, pp. 696–706, 2008.

HSA; the method can only be applied to monocomponent signals. Such signals can be expressed in the form $s(t) = A(t)\exp(i\omega(t)t)$, a generalization of the comparable Fourier description $f(t) = A\exp(i\omega t)$. The direct application of HSA to a multicomponent signal may well yield non-physical results, including negative frequencies. The goal of EMD is to generate a set of monocomponent signals called intrinsic mode functions (IMF) that can be individually analyzed using HSA. Extracting an IMF acts as a filter on the time-series; the frequency range spanned by each IMF is limited. The first IMF contains the highest frequencies; each subsequent IMF contains frequencies lower than the previous one. The IMF's are the basis states and are thus specific to the data set being analyzed, unlike the case for the Fourier transform or the wavelet transforms.

One problem associated with the HHT is the limited sharpness of the filtration between successive IMF's. Any given multicomponent signal may have frequency components in the frequency range where successive IMF filtrations overlap. When this happens, those frequencies in the overlap region are partially represented in each of the two IMF's. Analysis for such cases generates misleading results. An example of this signal division is presented in sect ion 5. An enhancement to the HHT process avoids this overlap problem. The enhancement is the addition of a new step which precedes and supplements the standard HHT process. The function of this step is to remove filtration control from the EMD process. EMD is still used but only to render the local mean of the signal to be zero. Filtration, instead of being performed via EMD, is accomplished using a set of non-overlapping bandpass filters. The filter passband ranges are chosen to be sufficiently narrow that the passed signal will be contained entirely in one IMF when EMD is performed. Not only does this enhanced EMD method (bandpass filtration followed by EMD) avoid the overlap problem referred to above, it also increases the sensitivity of HHT analysis significantly. In sections 2 and 3, the standard HHT procedure (EMD followed by HSA) is described very briefly. In section 4, the HHT is applied to a beat signal of unequal amplitude waves. In section 5, the frequency structure of a bat chirp is analyzed using enhanced EMD. When using standard HHT, only the fundamental and first harmonic are seen. The bandpass enhanced method described here shows the fundamental, the first, second, and part of a third harmonic.

2 Empirical Mode Decomposition

EMD generates a set of data specific basis states, IMF's, that are not necessarily orthogonal but which are analogous to the harmonic functions associated with Fourier analysis. Each IMF extracted can be expressed as a moncomponent, $s(t)$, with time dependent amplitude and frequency. A real valued signal $y(t)$ can be represented as a set of such IMF's plus a residual

$$y(t) = \sum_{n=1}^{N} s_n(t) + r_n(t). \tag{1}$$

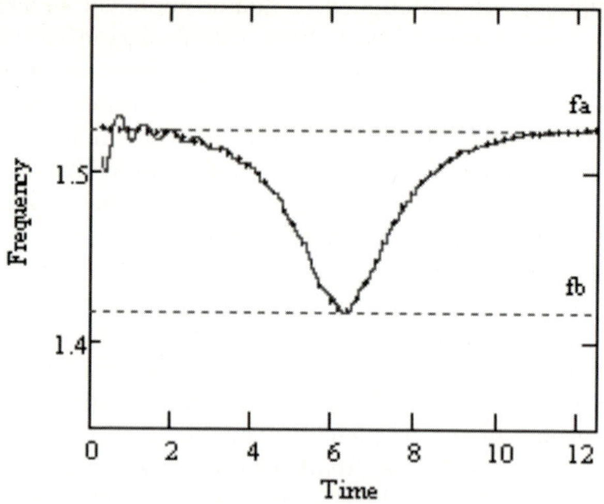

Fig. 1. Frequency vs. time as determined by HSA (solid curve) and from the analytic formulation as per Eqns. 5 and 6 (dotted curve). The signal is the sum of two sine waves with frequencies 1.5 and 1.58 Hz and amplitudes 2 and 1 respectively. The Hilbert spectrum has been smoothed with a moving average filter. The dashed curves, fa and fb, represent analytically determined maximum and minimum frequencies for the signal as per Eqn. 7; fa = 1.527 Hz. fb = 1.42 Hz. The time shown is for one period of the beat envelope.

The signal under analysis is deconstructed via a sifting process designed to generate IMF's, each with a local mean of zero and with the difference between the number of extrema and the number of zero- crossings not exceeding one. EMD comprises several steps. Determine the location of all maxima, $ymax(t)$, and minima $ymin(t)$ of $y(t)$. Fit a cubic spline through the $ymax(t)$ and another through the $ymin(t)$. Calculate the mean of the spline curves at each point $m(t) = [ymax(t) + ymin(t)]/2$. Remove the trend, $m(t)$. Let $d(t) = y(t) - m(t)$. Is $d(t)$ an IMF? If $d(t)$ meets the criteria defining an IMF, store the IMF in a vector, by letting $s_n(t) = d(t)$. Extract the residual $r(t) = y(t) - d(t)$. If $d(t)$ doesn't meet the criteria, further sifting is required; the above steps are repeated, substituting $d(t)$ for $y(t)$. Repeat the process until the last residual no longer contains any useful frequency information.

3 Hilbert Spectral Analysis

HSA extracts the instantaneous amplitude and angular frequency [12] associated with each IMF from the complex number that the IMF and its Hilbert transform define. This complex number has the IMF as its real part and its Hilbert transform as the complex part

$$S_n(t) = s_n(t) + iH\left(s_n(t)\right) \tag{2}$$

where H denotes the Hilbert transform and $i = \sqrt{-1}$. This expression is a generalization of the Euler relation: $\exp(i\omega t) = \cos(\omega t) + i\sin(\omega t)$. The complex part of this signal lags the real part by $\pi/2$. Thus, the imaginary part is equivalent to the Hilbert transform of the real part which shifts the phase by $-\pi/2$ for any positive frequency but does not change the amplitude. With a signal comprising many frequencies, the Hilbert transform shifts the phase of each individual frequency by $-\pi/2$. The instantaneous amplitude of the signal is given by the instantaneous amplitude of this complex number, $S_n(t)$. The instantaneous angular frequency associated with an IMF is extracted from the phase, $\phi_n(t) = \tan^{-1}(\mathrm{Im}(S_n(t))/\mathrm{Re}(S_n(t)))$ by taking the derivative of the unwrapped phase angle. (Unwrapping is necessary as the arctangent function returns the angle within a restricted range.) For the Euler relation example above, clearly, the angular frequency ω is the derivative of the phase. In general, $f_n(t) = (1/2\pi)\phi'_n(t)$.

4 EMD Filtering

Filtering associated with EMD roughly approximates that of a dyadic filter bank. However, with limited stopband attenuation at both the high and low ends of the frequency range associated with a given IMF, frequency response curves of successive IMF's overlap. Consequently, signals with frequencies in the transition zone between two successive IMF's will have those frequencies partially represented in each of the two IMF's. Such frequencies will then be analyzed in different ways in the two IMF's with misleading results. In the example described at the beginning of section 5, undesirable filtering via EMD is demonstrated. Here, comments on beat signals with equal and unequal amplitudes are presented first, as the features of each will help clarify the partial representation example.

The simple multicomponent signal, the sum of two equal amplitude sine waves, can be expressed

$$S(t) = 2A\ cos\left(\frac{\omega_1 - \omega_2}{2}t\right) sin\left(\frac{\omega_1 + \omega_2}{2}t\right) \tag{3}$$

As is well known for the equal amplitude case, the frequency of the carrier wave is constant while the amplitude varies with time. When the amplitudes are unequal, the signal can be expressed as the sum of two distinct equal amplitude beat signals with the same frequency for the carrier wave which is constant and with the same time dependence for the envelope but with different amplitudes and different phases:

$$\begin{aligned} S(t) = {} & (A+B)\ sin\left(\frac{\omega_1 + \omega_2}{2}t\right) cos\left(\frac{\omega_1 - \omega_2}{2}t\right) \\ & + (A-B)\ sin\left(\frac{\omega_1 - \omega_2}{2}t\right) cos\left(\frac{\omega_1 + \omega_2}{2}t\right). \end{aligned} \tag{4}$$

The summing of these constant frequency signals generates a frequency-varying carrier wave. For the equal amplitude case, the carrier frequency is the average

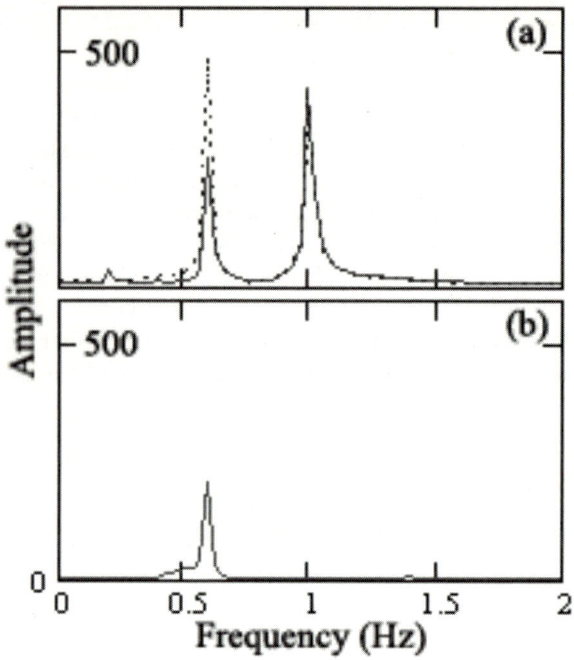

Fig. 2. a) The solid curve represents the FFT of the first IMF. The dotted curve represents the FFT of the original signal. The 0.6 Hz signal is only partially represented in this IMF. For emphasis, only a restricted portion of the frequency range of the FFT is shown. b) An FFT of the second IMF. The 0.6 Hz signal is also partially represented in this IMF. For emphasis only a restricted portion of the frequency range is shown.

frequency, $f(t) = (f_1 + f_2)/2$. With unequal amplitudes, the frequency can be expressed

$$f(t) = \left(\frac{f_1 + f_2}{2}\right) + \left(\frac{(f_2 - f_1)\left(A_2^2 - A_1^2\right)}{2C^2(t)}\right) \tag{5}$$

where

$$C(t) = \sqrt{A_1^2 + A_2^2 + 2A_1A_2\cos(\omega_1 - \omega_2)t}. \tag{6}$$

For this unequal amplitude sum, the maximum and minimum values of the variable frequency occur when the envelope is at its maximum or minimum. The limiting frequencies are

$$f_{Env\ max} = \frac{A_1f_1 + A_2f_2}{A_1 + A_2}$$

$$f_{Env\ min} = \frac{A_1f_1 - A_2f_2}{A_1 - A_2}. \tag{7}$$

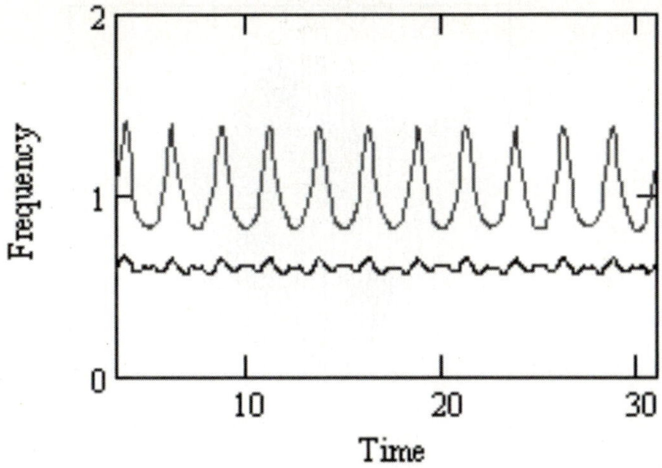

Fig. 3. a) Frequency vs. time as determined by HSA of the first and second IMF's. The upper curve expresses the frequency associated with the first IMF; the lower curve the second IMF. The modulation in the IMF1 frequency is that which occurs when combining unequal amplitude sine waves. That the amplitudes are unequal is a result of the EMD filtering. The frequency of the second IMF is 0.6 Hz. A proper analysis of the entire signal would yield a single constant frequency of 0.8 Hz. Although completely counter to the entire spirit of a frequency vs. time representation, a time averaged average of the two signals does yield 0.8 Hz.

HSA of an equal amplitude beat signal will show the average frequency, although there are artifacts around those times where the envelope passes through zero. If a beat pattern is formed from unequal amplitude waves, the carrier frequency is no longer constant and varies cyclically with the same period as that of the envelope. An example is shown in Fig.1 where the time-frequency information has been extracted from the sum of two sine waves with frequencies 1.5 and 1.58 Hz and with corresponding amplitudes of 2 and 1. The time varying frequency as determined from HHT analysis is in good agreement with the analytic expression for the time-dependent frequency. Both show the same frequency modulation over a period of one envelope cycle.

5 Bandpass Filtering Plus EMD

Consider now a signal comprising two sine waves of equal amplitude and frequencies of 1 and 0.6 Hz. The filtering associated with the generation of the IMF's is such that the 0.6 Hz signal is partially represented in both the first and second IMF as can be seen from the fast Fourier transforms (FFT) of the IMF's (Fig. 2a and 2b). The result of this division (Fig. 3) is that when the first IMF is subjected to HSA, the signal is treated as if there are two unequal amplitude waves

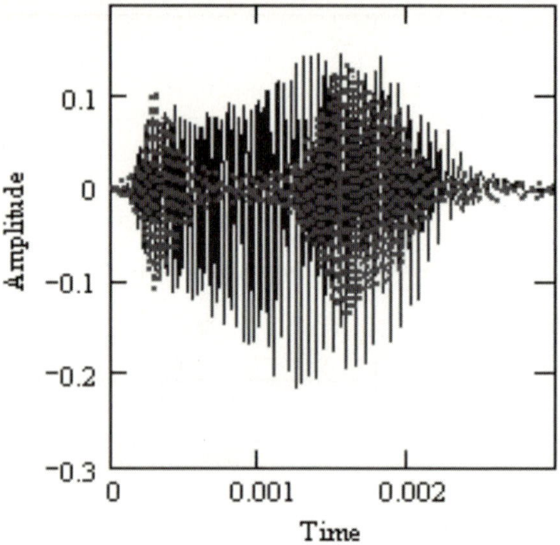

Fig. 4. Chirp of the large brown bat Eptesicus fuscus. The solid curve is the complete chirp. The dotted curve shows the filtered signal with a 30-40 kHz passband.

(see Fig.1). When the second IMF is subjected to HSA, the signal rendered has an essentially constant frequency of 0.6 Hz, but with reduced amplitude.

That there is frequency modulation associated with IMF1 occurs because the 0.6 Hz wave is only partially represented in this IMF. IMF1 shows a modulated frequency similar to the unequal amplitude case of Fig. 1, which is, of course, incorrect here. (The fact that the frequency rises in Fig. 3 rather than dips as it did in Fig. 1 is because in this case the signal with the larger frequency also has the larger amplitude, rather than the reverse as per Fig. 1. This reversal, for a given pair of frequencies, occurs with only a switch of the amplitudes. Such signals are known as Voelcker pairs. [13,14]) HSA applied to the original signal without EMD yields, as it should, a constant frequency signal with the average frequency of 0.8 Hz, although, as mentioned above for beat signals with equal amplitude waves, the signal is corrupted by a series of artifacts that occur in those regions surrounding the point where the envelope approaches zero.

EMD was designed to prepare a multicomponent signal for HSA. Ideally, EMD should create a set of monocomponent signals, each with a limited frequency range and each having a local mean of zero. However, as noted above, the filtration accomplished via EMD is not sufficiently sharp and this limitation can generate misleading results. Signal preparation for HSA is improved by using a different mechanism for filtration, creating proto-IMF's before EMD takes place. Proto-IMF's are frequency limited portions of the time-series, with the frequency range sufficiently narrow that each proto-IMF will appear in a single IMF when EMD is applied. Consequently, with this pre-filtering the effect of

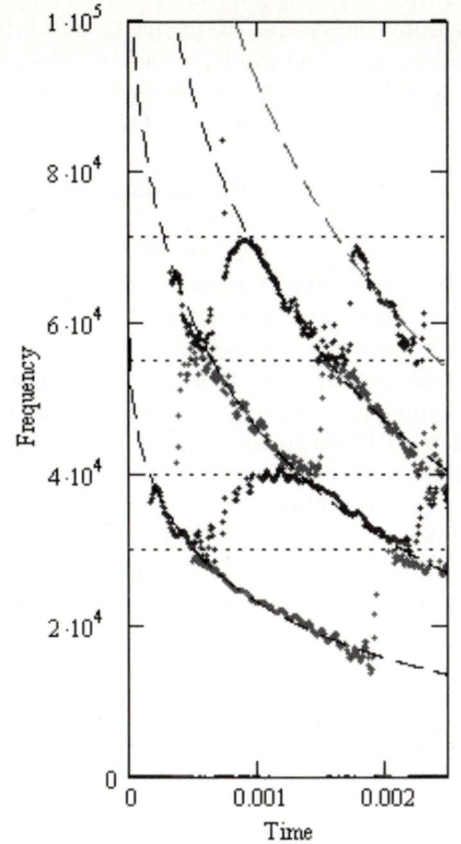

Fig. 5. The Hilbert spectrum from four bandpassed IMF's extracted from the chirp shown in Fig. 4. The horizontal dotted curves show the filtration transitions and the Nyquist frequency. The dashed curves represent an empirical exponential fit.

EMD is to render the local mean of the proto-IMF to be zero, thereby creating monocomponent IMF's.

At the beginning of this section, a signal containing frequency components of f_1 and $0.6f_1$ was subjected to EMD. The lower frequency component was split between two successive IMF's. To avoid partition when EMD is applied, the range of frequencies spanned by a proto-IMF should be smaller than those used in this example. A component with a frequency of $0.7f_1$ is adequately contained in the same IMF as f_1. Thus, as a general procedure, determine the highest relevant frequency of the signal under consideration, f_{max} (which might be much less than the Nyquist frequency but, of course, not greater). Then the range of the first bandpass filter is from f_{max} to $0.7f_{max}$. Each subsequent bandpass filter has as its maximum the minimum frequency of the previous filter and for the new minimum, 70% of the maximum of the current filter. It is not unknown for standard HHT analyses to generate 9 or more IMF's (e.g. there are 11 IMF's in [15]). Nine such 70% filters would cover the range from the Nyquist frequency to 4% of the Nyquist frequency nicely. Often, fewer filters than nine would be needed. Once the signal is filtered, EMD is applied to each of the proto-IMF's. For each, the essential information for that frequency range would be contained in one IMF, and the residual would have only contributions from leakage and noise.

The bandpass filtration can be carried out in the frequency domain. After transforming the time-series data to the frequency domain, use a rectangular window (zero outside , one inside) to select the appropriate frequency range in the frequency domain and transform the windowed frequency data back to the time domain. Once this is performed for each filtration range, apply EMD to each of the proto-IMF's to render the local mean zero, and finally apply HSA to each of the IMF's to extract the time-dependent amplitude and frequency formation. The data is further improved by smoothing and by applying a threshold to help eliminate noise.

Application of enhanced EMD to the chirp of the large brown bat Eptesicus fuscus (Fig. 4) provides a test of the method. The Nyquist frequency associated with this data set is 71.4 kHz. Given that the time-series associated with this chirp contains little information below 20 kHz, four frequency ranges were selected: 0 - 30, 30 - 40, 40 - 55, and > 55 (kHz). For the three filter ranges where $f_{min} \geq 30$ kHz, the ratios of f_{max}/f_{min} were conservative with values ranging from 0.73 to 0.77. The complete chirp and the filtered signal with a passband of 30 - 40 kHz are shown in Fig. 4. (EMD has not yet been performed). While a direct application of standard HHT shows only the fundamental and part of the first harmonic associated with this chirp [16], Fig. 5 based on enhanced EMD shows the fundamental and harmonics in time-frequency form more completely. The horizontal dotted curves (Fig. 5) show the filtration transitions and the Nyquist frequency. The dashed curves are empirical representations of a fundamental and its first three harmonics. These curves are exponentials of the form $f(t) = C \cdot 5.8 \cdot 10^4 \exp\left(-29.25\sqrt{t}\right)$ where C takes on integer values from 1 - 4. Near the filtration edges, where two IMF's represent the same frequencies, e.g. near 30 and 40 kHz an average of the two contributions would be appropriate.

Olhede et al. [17] analyzed this same chirp using wavelet projections. They found that the maximal overlap discrete wavelet packet transform Fejer-Korovkin FK(18) (MODWPT) provided superior results to that of EMD, which it does. Part of their analysis included denoising and thresholding. Their results showed the fundamental, first harmonic, and most of a second. However, the bandpass enhanced EMD shown here is at least comparable if not superior to the filtering provided by MODWPT. Using bandpass enchanced EMD, one sees (Fig. 5) the fundamental, the first and second harmonics and even part of the third which has not been previously noted.

6 Final Comments

The enhanced EMD described here provides a significant improvement over the direct application of EMD. Increased sensitivity renders a more complete representation of the spectrum, and the overlap problem is avoided. For this particular data set, the transitions between harmonics as one signal exits and another enters the bandpass region seems to leave an unavoidable signature. Further testing of enhanced EMD on portions of bird songs is in progress.

Acknowledgments

I wish to thank Curtis Condon, Ken White, and Al Feng of the Beckman Institute of the University of Illinois for the bat data and for permission to use it in this paper.

References

1. Huang, N.E., Shen, Z., Long, S.R., Wu, M.C., Shih, H.H., Zheng, Q., Yen, N.C., Tung, C.C., Liu, H.H.: The empirical mode decomposition and the Hilbert spectrum for nonlinear and nonstationary time series analysis. Proceedings of the Royal Society of London, Series A, Mathematical, Physical and Engineering Sciences 454, 903–995 (1998)
2. Huang, N.E., Wu, M.L., Qu, W.D., Long, S.R., Shen, S.S.P. (eds.): Hilbert-Huang Transform and Its Applications. World Scientific, Singapore (2003)
3. Donnelly, D.: The Fast Fourier and Hilbert-Huang Transforms: A Comparison. International Journal of Computers, Communications & Control I(4), 45–52 (2006)
4. Rilling, G., Flandrin, P., Goncalves, P.: On Empirical Mode Decomposition and its algorithms. In: IEEE-EURASIP Workshop on Nonlinear Signal and Image Processing NSIP 2003, Grado (I) (2003)
5. Phillips, S.C., Gledhill, R.J., Essex, J.W.: Application of the Hilbert-Huang Transform to the Analysis of Molecular Dynamics Simulations. J. Chem. Phys. A 27(24), 4869–4876 (2003)
6. Datig, M., Schlurmann, T.: Performance and limitations of the Hilbert-Huang transformation (HHT) with an application to irregular water waves. Ocean Eng. 31(14-15), 1783–1834 (2004)
7. Flandrin, P., Rilling, G., Goncalves, P.: Empirical mode decomposition as a filter bank. Signal Processing Letters, IEEE, Part 1, 11(2), 112–114 (2004)
8. Huang, N., et al.: A Confidence Limit for the Empirical Mode Decomposition and Hilbert Spectral Analysis. Proc. Royal Soc. A 459(2037), 2317–2345 (2003)
9. Zhang, R.R., Ma, S., Safak, E., Hartzell, S.: Hilbert-Huang Transform Analysis of Dynamic and Earthquake Motion Recordings. J. Engrg. Mech. 129(8), 861–875 (2003)
10. Le Van Quyen, M., et al.: Comparison of Hilbert Transform and Wavelet Methods for the Analysis of Neuronal Ssynchrony. J. Neurosci. Meth. 111, 83–98 (2001)
11. Kerschen, G., et al.: Toward a Fundamental Understanding of the Hilbert-Huang Transform in Nonlinear Structural Dynamics. J. Vibration Control 14(1-2), 77–105 (2008)
12. Loughlin, P.J., Tacer, B.: Comments on the Interpretation of Instantaneous Frequency. IEEE Signal Processing Letters 4(5), 123–125 (1997)
13. Voelcker, H.B.: Toward a Unified Theory of Modulation Part I: Phase-Envelope Relationships. Proc. IEEE 54(3), 340–353 (1966)

14. Voelcker, H.B.: Toward a Unified Theory of Modulation Part II: Zero Manipulation. Proc. IEEE 54(5), 735–755 (1966)
15. Liang, H., Lin, Q.-H., Chen, J.D.Z.: Application of the Empirical Mode Decomposition to the Analysis of Esophageal Manometric Data in Gastroesophageal Reflux Disease. IEEE Transactions of Biomedical Engineering 52(10), 1692–18701 (2005)
16. Donnelly, D.: The Fast Fourier Transform for Experimentalists, Part VI: Chirp of a Bat. Comput. Sci. Eng. 8(2), 72–78 (2006)
17. Olhede, S., Walden, A.T.: The Hilbert Spectrum via Wavelet Projections. Proc. R. Soc. Lond. A 460, 955–975 (2004)

Time-Space FE-PDAS Method for Dynamic Unilateral Contact Problem in Viscoelasticity

Jiří Nedoma[1] and Josef Daněk[1,2]

[1] Institute of Computer Science AS CR, Prague, Czech Republic
[2] Dept. of Mathematics, Faculty of Applied Science, University of West Bohemia, Pilsen, Czech Republic

Abstract. The paper deals with numerical analysis of a class of multibody dynamic unilateral contact problems. The presented problem describes the seismological model problem, representing a new approach in this branch and of studying propagation of seismic waves through broken up upper parts of the Earth. The semi-implicit finite element and the primal-dual active set strategy methods will be discussed.

Keywords: multibody unilateral dynamic contact problem, time-space solution of hyperbolic equation, semi-implicit scheme, FEM, seismology.

1 Introduction

This paper deals with a computational efficient method for multibody dynamic visco-elastic bodies being in contact, describing propagation of seismic waves through the elastic and/or viscoelastic media. The dynamic contact problems with given (Tresca) friction involving linearly (thermo−)elastic and visco-elastic bodies were analyzed in Duvaut, Lions (1976), Hilber et al. (1977), Jarušek (1996), Rivera, Racke (1998), Nedoma (1998), Nedoma (2000), Nedoma (2001), Eck et al. (2005), Nedoma, Daněk (2007), etc.

Analyses of seismic wave propagation play an important role in geodynamic investigations. The propagation of elastic as well as seismic waves through the elastic and the visco-elastic media or the Earth body, characterized by inhomogeneous with elastic or visco-elastic parameters are functions of place were studied theoretically by means of simplified models in several papers and books by use of different techniquies, like the ray method, the wave method, the method of propagators, the finite difference and finite element methods, etc. Theoretical seismograms, constructed on this basis of results obtained by various methods mentioned above, do not yield a complete answer to the question of the structure of the Earths crust, mantle as well as core. Their physical structure is so complicated that the present-day methods cannot give the actual information about their structure. The present-day methods analyse the interior of the Earth under the presumption that the Earths interior can be described by continuous inhomogeneous media. But in the real Earth the crust and the upper mantle are broken up into a great number of 3D plates and geological blocks which are in

O. Gervasi et al. (Eds.): ICCSA 2008, Part II, LNCS 5073, pp. 707–719, 2008.

common contacts. Such problems are up-to-date unsolved. Therefore, the main goal of this paper is to give the new methodology and the optimal algorithm which facilitate to investigate such type of seismological problems. The methodology discussed in the paper can be applied also in the other branches like the building industries, the structural mechanics, etc. These type of problems lead to study the dynamic multibody contact problem in linear viscoelasticity, where the damping term and the Coulombian type of friction acting on the contact boundaries between lithosperic plates or geological blocks being in common contact are also assumed.

2 Formulation of the Dynamic Contact Problem

We introduce a viscoelastic bodies of a Kelvin-Voigt type being in a mutual contact which initially occupy a bounded domains of arbitrary shapes Ω^ι of $\mathbb{R}^2$, such that $\Omega = \cup_{\iota=1}^r \Omega^\iota$. Let several neighboring bodies, say Ω^s and Ω^m, be in a mutual contact and let Γ_c^{sm} be the common contact boundary between them before their deformation. Let the boundary $\partial\Omega = \Gamma_\tau \cup \Gamma_u \cup \Gamma_c$, where $\Gamma_c = \cup_{s,m}\Gamma_c^{sm}$, $\Gamma_c^{sm} = \partial\Omega^s \cap \partial\Omega^m$, $s \neq m$, $s, m \in \{1, \ldots, r\}$ and where Γ_j, $j = u, \tau, c$ are open subsets in $\partial\Omega$. Let $I = (0, t_p)$ be a time interval and let $\Omega_t = I \times \Omega$ denote the time-space domain and $\Gamma_\tau(t) = I \times \Gamma_\tau$, $\Gamma_u(t) = I \times \Gamma_u$, $\Gamma_c(t) = I \times \Gamma_c$ denote the parts of its boundary $\partial\Omega_t = I \times \partial\Omega$.

We assume that in the region Ω will act the damping $\alpha(\mathbf{x}) \geq 0$, defined by such a way that the waves propagated through the areas of a certain thickness D (in which $\alpha(\mathbf{x}) > 0$) near the boundary Γ_u are completely absorbed (for more details see Nedoma (1998)). In the body different nonzero densities ρ are assumed. The body is subjected to volume forces $\mathbf{F}$ and as a result of the effect of damping $\alpha(\mathbf{x})$ zero displacements are imposed on Γ_u and on Γ_τ zero or nonzero tractions $\mathbf{P}$ are imposed. The point sources of seismic waves can be of different types and can be simulated by nonzero volume forces (in remaning points of Ω the volume forces are assumed to be zero), see e.g. Nedoma (1998).

Let $\mathbf{n}$ denote the outer normal vector of the boundary $\partial\Omega$, $u_n = u_i n_i$, $\mathbf{u}_t = \mathbf{u} - u_n\mathbf{n}$, $\tau_n = \tau_{ij}n_j n_i$, $\boldsymbol{\tau}_t = \boldsymbol{\tau} - \tau_n\mathbf{n}$ be the normal and tangential components of the displacement and stress vectors $\mathbf{u} = (u_i)$, $\boldsymbol{\tau} = (\tau_i)$, $\tau_i = \tau_{ij}n_j$, $i, j = 1, 2$. On Γ_c^{sm} the positive direction of the outer normal vector $\mathbf{n}$ is assumed with respect to Ω^s. The respective time derivatives are denoted by " $'$ ".

The stress strain relation in every Ω^ι will be defined by the Hooke's law

$$\tau_{ij}^\iota = \tau_{ij}^\iota(\mathbf{u}^\iota, \mathbf{u}^{\iota\prime}) = c_{ijkl}^{(0)\iota}(\mathbf{x})e_{kl}(\mathbf{u}^\iota) + c_{ijkl}^{(1)\iota}(\mathbf{x})e_{kl}(\mathbf{u}^{\iota\prime}), \quad i, j, k, l = 1, 2,\ \iota = 1, \ldots, r, \tag{1}$$

where $c_{ijkl}^{(n)\iota}(\mathbf{x})$, $n = 0, 1$, are elastic and viscous coefficients and $e_{ij}(\mathbf{u}) = \frac{1}{2}(\frac{\partial u_i}{\partial x_j} + \frac{\partial u_j}{\partial x_i})$ are components of the small strain tensor. For the tensors $c_{ijkl}^{(n)\iota}(\mathbf{x})$, $n = 0, 1$, we assume

$$\begin{gathered} c_{ijkl}^{(n)\iota} \in L^\infty(\Omega^\iota), \quad n = 0, 1, \quad c_{ijkl}^{(n)\iota} = c_{jikl}^{(n)\iota} = c_{klij}^{(n)\iota} = c_{ijlk}^{(n)\iota}, \\ c_{ijkl}^{(n)\iota} e_{ij} e_{kl} \geq c_0^{(n)\iota} e_{ij} e_{ij} \quad \forall e_{ij},\ e_{ij} = e_{ji} \text{ and a.e. } \mathbf{x} \in \Omega^\iota, c_0^{(n)\iota} > 0. \end{gathered} \tag{2}$$

A repeated index implies summation from 1 to 2.

The problem to be solved has the following classical formulation:

Problem 1. Let $s \geq 2$. Find a vector function $\mathbf{u} : I \times \overline{\Omega} \to \mathbb{R}^2$, satisfying

$$\rho^\iota \frac{\partial^2 u_i^\iota}{\partial t^2} + \alpha^\iota \frac{\partial u_i^\iota}{\partial t} = \frac{\partial \tau_{ij}^\iota (\mathbf{u}^\iota, \mathbf{u}^{\iota\prime})}{\partial x_j} + F_i^\iota, \qquad i,j = 1,2,\ \iota = 1,\ldots,r,\ (t,\mathbf{x}) \in \Omega_t^\iota = I \times \Omega^\iota, \tag{3}$$

$$\tau_{ij} n_j = P_i,\ i,j = 1,2,\ (t,\mathbf{x}) \in \Gamma_\tau(t) = I \times \cup_{\iota=1}^r (\Gamma_\tau \cap \partial\Omega^\iota), \tag{4}$$

$$u_i = 0,\ i = 1,2,\ \text{on } \Gamma_u(t) = I \times \cup_{\iota=1}^r (\Gamma_u \cap \partial\Omega^\iota), \tag{5}$$

$$\begin{cases} u_n^s - u_n^m \leq d^{sm},\ \tau_n^s = \tau_n^m \equiv \tau_n^{sm} \leq 0,\ (u_n^s - u_n^m - d^{sm})\, \tau_n^{sm} = 0, \\ \mathbf{u}_t^{\prime s} - \mathbf{u}_t^{\prime m} = 0 \Rightarrow |\boldsymbol{\tau}_t^{sm}| \leq \mathcal{F}_c^{sm}(0)\, |\tau_n^{sm}|, \\ \mathbf{u}_t^{\prime s} - \mathbf{u}_t^{\prime m} \neq 0 \Rightarrow \boldsymbol{\tau}_t^{sm} = -\mathcal{F}_c^{sm}(\mathbf{u}_t^{\prime s} - \mathbf{u}_t^{\prime m})\, |\tau_n^{sm}| \frac{\mathbf{u}_t^{\prime s} - \mathbf{u}_t^{\prime m}}{|\mathbf{u}_t^{\prime s} - \mathbf{u}_t^{\prime m}|}, \\ \qquad (t,\mathbf{x}) \in \Gamma_c(t) = I \times \cup_{s,m} \Gamma_c^{sm}, \end{cases} \tag{6}$$

$$\mathbf{u}(\mathbf{x},0) = \mathbf{u}_0(\mathbf{x}), \quad \mathbf{u}'(\mathbf{x},0) = \mathbf{u}_1(\mathbf{x}), \quad \mathbf{x} \in \Omega, \tag{7}$$

where $\mathbf{u}_0$, $\mathbf{u}_1$ are given functions.

The coefficient of friction $\mathcal{F}_c^{sm} \equiv \mathcal{F}_c^{sm}(\mathbf{x}, \mathbf{u}')$ is globally bounded, non-negative and satisfies the Carathéodory property and has a compact support $\mathcal{SF}_c$, which depends on the space variable $\mathbf{x}$ and since we model also the difference between the coefficients of friction and of stick as well as slip, it depends also on the tangential displacement rate component $\mathbf{u}_t'$. Let us denote by ${}^a\Gamma_c^{sm} \subset \Gamma_c^{sm}$ the actual contact set, i.e. for which $u_n^s - u_n^m = d^{sm}$ on ${}^a\Gamma_c^{sm}$ and $u_n^s - u_n^m < d^{sm}$ on ${}^c\Gamma_c^{sm} = \Gamma_c^{sm} \setminus {}^a\Gamma_c^{sm}$. For the continuous displacement $\mathbf{u}$ the actual contact zone ${}^a\Gamma_c^{sm}$ is well-defined and closed subset of $\cup_{s,m}\, {}^a\Gamma_c^{sm}$.

In the next, the Hilbert and the Sobolev spaces will be used as usual, see e.g. (see Adams (1975)).

The dynamic contact problem with Coulombian friction where the Signorini conditions are formulated in displacements is up-to-date unsolved. In the next, for the numerical solution of the problem discussed, we will assume that the Coulombian friction in every time level depends on its value g_c^{sm} from the previous time levels, where g_c^{sm} is non-negative function and has a meaning of a given friction limit and $-g_c^{sm}$ represents a given frictional force.

The contact problem (3)-(7) has a weak formulation in terms of a variational inequality. Let us introduce the set of virtual displacements and the set of admissible displacements by

$$\mathcal{V} = \{\mathbf{v} \mid \mathbf{v} \in L^2(I; \sqcap_{\iota=1}^r H^{1,2}(\Omega^\iota)), \mathbf{v} = 0 \text{ on } \Gamma_u(t), \quad v_n^s - v_n^m = d^{sm} \text{ a.e. on } I \times \cup_{s,m}\Gamma_c^{sm}\},$$

$$\mathcal{K} = \{\mathbf{v} \mid \mathbf{v} \in L^2(I; \sqcap_{\iota=1}^s H^{1,2}(\Omega^\iota)), \mathbf{v} = \mathbf{0} \text{ on } \Gamma_u(t), \quad v_n^s - v_n^m \leq d^{sm} \text{ a.e. on } I \times \cup_{s,m}\Gamma_c^{sm}\}.$$

Then we have the following problem:

Problem 2. A weak solution of (3)-(7) is a function $\mathbf{u} \in B(I; H^{1,2}(\Omega))$ with $\mathbf{u}(t,.) \in \mathcal{K}$ for a.e. $t \in I$, $\mathbf{u}' \in L^2(I; H^{1,2}(\Omega)) \cap L^\infty(I; L^{2,2}(\Omega))$, $\mathbf{u}'(t_p,.) \in L^{2,2}(\Omega)$ such that for all $\mathbf{v} \in H^{1,2}(\Omega(t))$ with $\mathbf{v}(t,.) \in \mathcal{K}$ a.e. in I the following inequality holds

$$\begin{aligned} &\textstyle\int_I \{(\mathbf{u}''(t), \mathbf{v} - \mathbf{u}(t)) + (\alpha \mathbf{u}'(t), \mathbf{v} - \mathbf{u}(t)) + a^{(0)}(\mathbf{u}(t), \mathbf{v} - \mathbf{u}(t)) + \\ &+ a^{(1)}(\mathbf{u}'(\mathbf{t}), \mathbf{v} - \mathbf{u}(t)) + j(\mathbf{v}) - j(\mathbf{u}(t))\}\, dt \geq \textstyle\int_I (\overline{\mathbf{f}}(t), \mathbf{v} - \mathbf{u}(t))\, dt, \qquad (8) \\ &\mathbf{u}(\mathbf{x}, 0) = \mathbf{u}_0(\mathbf{x}),\ \mathbf{u}'(\mathbf{x}, 0) = \mathbf{u}_1(\mathbf{x}). \end{aligned}$$

where we assume that the initial data $\mathbf{u}_0$, $\mathbf{u}_1$ satisfy the static contact multibody linear elastic problem (see Nedoma (1983), Nedoma (1987), Hlaváček, Nedoma (2002)) and where

$$\begin{aligned} &(\mathbf{u}'', \mathbf{v}) = \textstyle\sum_{\iota=1}^{r} (\mathbf{u}''^{\iota}, \mathbf{v}^{\iota}) = \int_\Omega \rho u_i'' v_i \, d\mathbf{x}, \\ &(\alpha \mathbf{u}', \mathbf{v}) = \textstyle\sum_{\iota=1}^{r} (\alpha^{\iota} \mathbf{u}'^{\iota}, \mathbf{v}^{\iota}) = \int_\Omega \alpha u_i' v_i \, d\mathbf{x}, \\ &a^{(n)}(\mathbf{u}, \mathbf{v}) = \textstyle\sum_{\iota=1}^{r} a^{(n)\iota}(\mathbf{u}^{\iota}, \mathbf{v}^{\iota}) = \int_\Omega c_{ijkl}^{(n)} e_{kl}(\mathbf{u})\, e_{ij}(\mathbf{v})\, d\mathbf{x}, \\ &(\overline{\mathbf{f}}, \mathbf{v}) = \textstyle\sum_{\iota=1}^{r} \left(\overline{\mathbf{f}}^{\iota}, \mathbf{v}^{\iota}\right) = \int_\Omega F_i v_i \, d\mathbf{x} + \int_{\Gamma_\tau} P_i v_i \, ds, \\ &j(\mathbf{v}) = \textstyle\sum_{\iota=1}^{r} j^{\iota}(\mathbf{v}^{\iota}) = \int_{\cup_{s,m} \Gamma_c^{sm}} g_c^{sm} |\mathbf{v}_t^s - \mathbf{v}_t^m| \, ds \equiv \langle g_c^{sm} |\mathbf{v}_t^s - \mathbf{v}_t^m| \rangle_{\Gamma_c}, \end{aligned}$$

where the bilinear forms $a^{(n)}(\mathbf{u}, \mathbf{v})$, $n = 0, 1$, are symmetric in $\mathbf{u}, \mathbf{v}$ and satisfy $a^{(n)}(\mathbf{u}, \mathbf{u}) \geq c_0^{(n)} \|\mathbf{u}\|_{1,2}^2$, $c_0^{(n)} = const. > 0$, $\left|a^{(n)}(\mathbf{u}, \mathbf{v})\right| \leq c_1^{(n)} \|\mathbf{u}\|_{1,2} \|\mathbf{v}\|_{1,2}$, $c_1^{(n)} = const. > 0$, $\mathbf{u}, \mathbf{v} \in V$.

To prove the existence of the solution the decomposion $\mathbf{v} - \mathbf{u} = \mathbf{v} - \mathbf{u} + \mathbf{u}' - \mathbf{u}' = \mathbf{w} - \mathbf{u}'$ can be used. Then the proof is similar of that of Eck et al. (2005), Chap. 4. For the proof the technique of penalization and regularization will be used.

3 Numerical Solution and the Algorithm

Let Ω be approximated by a polygon Ω_h with the boundary $\partial\Omega_h = \Gamma_{\tau h} \cup \Gamma_{uh} \cup \Gamma_{ch}$. Let $I = (0, t_p)$, $t_p > 0$, let $m > 0$ be an integer, then $\triangle t = t_p/m$, $t_i = i\triangle t$, $i = 0, \ldots, m$. Let $\{\mathcal{T}_{h,\Omega_h}\}$ be a regular family of finite element partitions $\mathcal{T}_h$ of $\overline{\Omega}_h$ compatible to the boundary subsets $\overline{\Gamma}_{\tau h}$, $\overline{\Gamma}_{uh}$ and $\overline{\Gamma}_{ch}$. Let $V_h \subset V$ be the finite element space of linear elements corresponding to the partition $\mathcal{T}_h$. Then $K_h = V_h \cap K$ is the set of continuous piecewise linear functions that vanish at the nodes of $\overline{\Gamma}_{uh}$ and whose normal components are non-positive at the nodes on $\cup_{s,m} \Gamma_c^{sm}$. It is evident that K_h is a nonempty, closed, convex subset of $V_h \subset V$. Let $\mathbf{u}_{0h} \in K_h$, $\mathbf{u}_{1h} \in K_h$ be an approximation of $\mathbf{u}_0$ or $\mathbf{u}_1$. Further, we assume that the end points $\overline{\Gamma}_{\tau h} \cup \overline{\Gamma}_{uh}$, $\overline{\Gamma}_{uh} \cup \overline{\Gamma}_{ch}$, $\overline{\Gamma}_{\tau h} \cup \overline{\Gamma}_{ch}$ coincide with the vertices of T_{hi}. Moreover, we will assume that the frictional term is approximated by its value in the previous time levels, so that the frictional term is approximated by a given friction limit. Then in every time level we have the following discrete problem:

Problem 3. Find a displacement field $\mathbf{u}_h : I \to V_h$ with $\mathbf{u}_h(0) = \mathbf{u}_{0h}, \mathbf{u}_h'(0) = \mathbf{u}_{1h}$, such that for a.e. $t \in I$, $\mathbf{u}_h(t) \in K_h$

$$\begin{aligned}(\mathbf{u}_h''(t), \mathbf{v}_h - \mathbf{u}_h(t)) + (\alpha \mathbf{u}_h'(t), \mathbf{v}_h - \mathbf{u}_h(t)) + a^{(0)}(\mathbf{u}_h(t), \mathbf{v}_h - \mathbf{u}_h(t)) +\\ + a^{(1)}(\mathbf{u}_h'(t), \mathbf{v}_h - \mathbf{u}_h(t)) + j(\mathbf{v}_h) - j(\mathbf{u}_h(t)) \geq (\overline{\mathbf{f}}(t), \mathbf{v}_h - \mathbf{u}_h(t))\\ \forall \mathbf{v}_h \in K_h, \text{ a.e. } t \in I,\end{aligned} \tag{9}$$

where

$$\begin{aligned}&(\mathbf{u}_h'', \mathbf{v}_h) = \textstyle\sum_{\iota=1}^{r} (\mathbf{u}_h''^{\iota}, \mathbf{v}_h^{\iota}) = \int_\Omega \rho u_{hi}'' v_{hi}\, d\mathbf{x},\\ &(\alpha \mathbf{u}_h', \mathbf{v}_h) = \textstyle\sum_{\iota=1}^{r} (\alpha^\iota \mathbf{u}_h'^{\iota}, \mathbf{v}_h^{\iota}) = \int_\Omega \alpha u_{hi}' v_{hi}\, d\mathbf{x},\\ &a^{(n)}(\mathbf{u}_h, \mathbf{v}_h) = \textstyle\sum_{\iota=1}^{r} a^{(n)\iota}(\mathbf{u}_h^\iota, \mathbf{v}_h^\iota) = \int_\Omega c_{ijkl}^{(n)} e_{kl}(\mathbf{u}_h) e_{ij}(\mathbf{v}_h)\, d\mathbf{x},\ n = 0, 1,\\ &(\overline{\mathbf{f}}, \mathbf{v}_h) = \textstyle\sum_{\iota=1}^{r} (\overline{\mathbf{f}}^\iota, \mathbf{v}_h^\iota) = \int_\Omega F_i v_{hi}\, d\mathbf{x} + \int_{\Gamma_\tau} P_i v_{hi}\, ds,\\ &j(\mathbf{v}_h) = \textstyle\sum_{\iota=1}^{r} j^\iota(\mathbf{v}_h^\iota) = \int_{\bigcup_{s,m} \Gamma_c^{sm}} g_c^{sm} |\mathbf{v}_t^s - \mathbf{v}_t^m|\, ds \equiv \langle g_c^{sm}, |\mathbf{v}_t^s - \mathbf{v}_t^m| \rangle_{\Gamma_c},\end{aligned}$$

where we also assume that the frictional term is approximated by the given friction functional in the previous time levels.

To prove the existence of discrete solution $\mathbf{u}_h$ the technique similar of that as in the continuous case based on the decomposition $\mathbf{v}_h - \mathbf{u}_h = \mathbf{v}_h - \mathbf{u}_h + \mathbf{u}_h' - \mathbf{u}_h' = \mathbf{w}_h - \mathbf{u}_h'$, the penalty and regularization technique (Eck et al. (2005)) is used. In the next the test function $\mathbf{v}$ will correspond with $\mathbf{v} + \mathbf{u}' - \mathbf{u}$ in (8) and $\mathbf{v}_h$ with $\mathbf{v}_h + \mathbf{u}_h' - \mathbf{u}_h$ in (9) after used decomposition.

3.1 Algorithm

For the algorithm the semi-implicit scheme in time and the finite elements in space will be used. Let $m > 0$ be integer, then $\triangle t = t_p/m$, $t_i = i\triangle t$, $i = 0, 1, \ldots, m$. Then, the derivatives are approximated by the differences, i.e. $\mathbf{u}_h'' = \frac{\mathbf{u}_h^{i+1} - 2\mathbf{u}_h^i + \mathbf{u}_h^{i-1}}{\triangle t^2}$, $\mathbf{u}_h' = \frac{\mathbf{u}_h^{i+1} - \mathbf{u}_h^{i-1}}{\triangle t}$, and after some algebra, in every time level $t = t_{i+1}$ we have to solve the following problem:

$$\begin{aligned}&(\mathbf{u}_h, \mathbf{v}_h - \mathbf{u}_h) + \triangle t(\alpha \mathbf{u}_h, \mathbf{v}_h - \mathbf{u}_h) +\\ &\quad + \triangle t^2 a^{(0)}(\mathbf{u}_h, \mathbf{v}_h - \mathbf{u}_h) + \triangle t a^{(1)}(\mathbf{u}_h, \mathbf{v}_h - \mathbf{u}_h) +\\ &\quad + \triangle t^2 \int_{\bigcup_{s,m} \Gamma_c^{sm}} \tilde{g}_{ch}^{sm} (|\mathbf{v}_{th}^s - \mathbf{v}_{th}^m| - |\mathbf{u}_{th}^s - \mathbf{u}_{th}^m|)\, ds \geq\\ &\quad \geq (\mathbf{f}_h, \mathbf{v}_h - \mathbf{u}_h),\ t = t_{i+1} \in I,\\ &\mathbf{u}_h(0) = \mathbf{u}_{0h},\ \mathbf{u}_h'(0) = \mathbf{u}_{1h},\end{aligned} \tag{10}$$

where we set $\mathbf{u}_h^i = \mathbf{u}_h(t_i)$, $\triangle \mathbf{u}_h^i = \mathbf{u}_h(t_i) - \mathbf{u}_h(t_{i-1})$, $\mathbf{u}_h^{i+1} \equiv \mathbf{u}_h$, $\tilde{g}_{ch}^{sm} = g_{ch}^{sm}(t_i) = \mathcal{F}_c^{sm}(\triangle t^{-1}(\triangle \mathbf{u}_h^{si} - \triangle \mathbf{u}_h^{mi})) |\tau_n^{sm}(\mathbf{u}_h^i, \frac{\triangle \mathbf{u}_h^i}{\triangle t})|$, $(\mathbf{F}(t_{i+1}), \mathbf{v}_h) = \triangle t^2 (\overline{\mathbf{f}}_h(t_{i+1}), \mathbf{v}_h) + (2\mathbf{u}_h^i - \mathbf{u}_h^{i-1}, \mathbf{v}_h) + \triangle t(\alpha \mathbf{u}_h^i, \mathbf{v}_h) + \triangle t a_h^{(1)}(\mathbf{u}_h^i, \mathbf{v}_h)$, $\mathbf{F}(t_{i+1}) \equiv \mathbf{f}_h$.

Let us put

$$\begin{aligned}&A(\mathbf{u}_h, \mathbf{v}_h) = (\mathbf{u}_h, \mathbf{v}_h) + \triangle t(\alpha \mathbf{u}_h, \mathbf{v}_h) + \triangle t^2 a^{(0)}(\mathbf{u}_h, \mathbf{v}_h) + \triangle t a^{(1)}(\mathbf{u}_h, \mathbf{v}_h),\\ &j(\mathbf{v}_h) = \triangle t^2 \int_{\bigcup_{s,m} \Gamma_c^{sm}} \tilde{g}_{ch}^{sm} |\mathbf{v}_{th}^s - \mathbf{v}_{th}^m|\, ds,\end{aligned}$$

where $\tilde{g}_{ch}^{sm}$ is the approximate given frictional limit.

Since we assumed that $\rho \geq \rho_0 > 0$, $\alpha \geq \alpha_0 > 0$ and since bilinear forms $a_h^{(n)}(\mathbf{u}_h, \mathbf{v}_h)$, $n = 0, 1$, are symmetric in $\mathbf{u}_h$ and $\mathbf{v}_h$ and satisfy $a^{(n)}(\mathbf{u}_h, \mathbf{u}_h) \geq c_0^{(n)} \|\mathbf{u}_h\|_{1,2}^2$, $c_0^{(n)} = const. > 0$, $\left|a^{(n)}(\mathbf{u}_h, \mathbf{v}_h)\right| \leq c_0^{(n)} \|\mathbf{u}_h\|_{1,2} \|\mathbf{v}_h\|_{1,2}$, $c_1^{(n)} = const. > 0$, $n = 0, 1$, $\mathbf{u}_h, \mathbf{v}_h \in V_h$, then the bilinear form $A(\mathbf{u}_h, \mathbf{v}_h)$ is also symmetric in $\mathbf{u}_h$ and $\mathbf{v}_h$ and

$$A(\mathbf{u}_h, \mathbf{v}_h) \geq a_0 \|\mathbf{u}_h\|_{1,2}^2, \quad a_0^{(n)} = const. > 0,$$
$$|A(\mathbf{u}_h, \mathbf{v}_h)| \leq a_1 \|\mathbf{u}_h\|_{1,2} \|\mathbf{v}_h\|_{1,2}, \quad a_1 = const. > 0, \quad \mathbf{u}_h, \mathbf{v}_h \in V_h,$$

hold.

Then we have to solve in every time level the equivalent problem:

Problem 4. Find $\mathbf{u}_h \in K_h$, a.e. $t = t_{i+1} \in I$, such that

$$A(\mathbf{u}_h, \mathbf{v}_h - \mathbf{u}_h) + j(\mathbf{v}_h) - j(\mathbf{u}_h) \geq (\mathbf{f}_h, \mathbf{v}_h - \mathbf{u}_h), \ \forall \mathbf{v}_h \in K_h, \ t = t_{i+1} \in I, \quad (11)$$

Then Problem 4 is equivalent with the following problem:

Problem 5. Find $\mathbf{u}_h \in K_h$, a.e. $t = t_{i+1} \in I$, such that

$$\mathcal{L}(\mathbf{u}_h) \leq \mathcal{L}(\mathbf{v}_h), \quad \forall \mathbf{v}_h \in K_h, \ t = t_{i+1} \in I, \quad (12)$$

where the Lagrangian $\mathcal{L}$ is defined as $\mathcal{L}(\mathbf{v}_h) = \mathcal{L}_0(\mathbf{v}_h) + j(\mathbf{v}_h)$, $\mathcal{L}_0(\mathbf{v}_h) = \frac{1}{2}A(\mathbf{v}_h, \mathbf{v}_h) - (\mathbf{f}_h, \mathbf{v}_h)$.

The Mortar Approach - The Non-matching Case. For every polygonal domain Ω^ι, $\iota = 1, \ldots, r$, we introduce triangulations $\mathcal{T}_{h,\Omega^\iota}$ in such a way that on the contact boundaries Γ_c^{sm} the points of Γ_c^s and Γ_c^m are not identical, and therefore, the meshsizes $h_s \neq h_m$.

To give a saddle point formulation, we introduce a Lagrange multiplier space M. We introduce the trace space $W = H^{\frac{1}{2}}(\cup_{s,m}\Gamma_c^{sm})$, being the trace space of V^s restricted to $\cup_{s,m}\Gamma_c^{sm}$, and its dual $W' = H^{-\frac{1}{2}}(\cup_{s,m}\Gamma_c^{sm})$. We introduce the Lagrange multiplier space $M = M_n \times M_t$, where

$$M_n = \{\mu_n \in W'; \ \mu_n \geq 0\},$$
$$M_t = \left\{\mu_t \in L^2(\cup_{s,m}\Gamma_c^{sm}) \mid \|\mu_t\| \leq 1 \ \text{ a.e., } \mu_t = 0 \ \text{ on } \cup_{s,m} \Gamma_c^{sm} \setminus \text{supp } g_c^{sm}\right\}.$$

We define the bilinear form

$$b(\boldsymbol{\mu}, \mathbf{v}) = \langle \mu_n, [\mathbf{v}.\mathbf{n}]^s - d^{sm} \rangle_{\Gamma_c^s} + \int_{\cup_s \Gamma_c^s} \tilde{g}_{ch}^{sm} \mu_t . [\mathbf{v}_t]^s \, ds,$$

where $[\mathbf{v}.\mathbf{n}]^s = v_n^s(\mathbf{x}) - v_n^m(\mathcal{R}^{sm}(\mathbf{x}))$, $[\mathbf{v}_t]^s = \mathbf{v}_t^s(\mathbf{x}) - \mathbf{v}_t^m(\mathcal{R}^{sm}(\mathbf{x}))$, where $\mathcal{R}^{sm} : \Gamma_c^s(t) \mapsto \Gamma_c^m(t)$, at $t \in I$, is a bijective map satisfying $\Gamma_c^m(t) \subset \mathcal{R}^{sm}(\Gamma_c^s(t)), t \in I$ and where $\langle ., . \rangle_{\Gamma_c^s}$ denotes the duality pairing between W and $M = M_n \times M_t$.

We introduce the discrete approximation of the Lagrange multiplier space $M_{hH} = M_{hn} \times M_{Ht}$, where

$$W_{hH}(\Gamma_c^s) = W_{hn}(\Gamma_c^s) \times W_{Ht}(\Gamma_c^s) = \{\boldsymbol{\mu}_{hH}|_{\triangle_r} \in [P_0(\triangle_r)]^2, 0 \leq r \leq n(h^s)\}$$
$$M_{hn} = \{\mu_{hn} \in W_{hn}(\Gamma_c^s), \ \mu_{hn} \geq 0 \ \text{ a.e. on every } \Gamma_c^s\},$$
$$M_{Ht} = \{\mu_{Ht} \in W_{Ht}(\Gamma_c^s), \int_{\Gamma_c^s} \mu_{Ht}\psi_H \, ds - \int_{\Gamma_c^s} \tilde{g}_{ch}^{sm} |\psi_H| \, ds \leq 0, \forall \psi_H \in W_{Ht}(\Gamma_c^s)\},$$

where $\Gamma_c^s = \cup_{p=1}^{\overline{p}} \Gamma_c^{sp}$, $\Gamma_c^{sp} = \cup_r \triangle_r$, $\triangle_r = (q_r, q_r + h_p)$, $r = 0, 1, \ldots, m(h_p) - 1$, where h_p is a step of mesh in the $p - th$ segment of Γ_c^s.

The problem then leads to the following problem:
In every time level find $(\boldsymbol{\lambda}_{hH}, \mathbf{u}_h) \in M_{hH} \times V_h$ satisfying

$$\begin{aligned} &A(\mathbf{u}_h, \mathbf{v}_h) + b(\boldsymbol{\lambda}_{hH}, \mathbf{v}_h) = (\mathbf{f}_h, \mathbf{v}_h) \quad \mathbf{v}_h \in V_h = \sqcap_{\iota=1}^{r} V_h^{\iota}, \\ &b(\boldsymbol{\mu}_{hH} - \boldsymbol{\lambda}_{hH}, \mathbf{v}_h) \leq \langle d^{sm}, \mu_{hn} - \lambda_{hn} \rangle_{\cup \Gamma_c^s} \end{aligned} \tag{13}$$

The existence and uniqueness of $(\boldsymbol{\lambda}_{hH}, \mathbf{u}_h)$ of this saddle-point problem has been stated e.g. in Haslinger et al. (1996), and the discrete Babuška-Brezzi "inf-sup" condition must be satisfied. Thus

Proposition 1. *Let $-\tau_n(\mathbf{u}) \in M_{hn}$. Then the problem (13) has a unique solution $(\boldsymbol{\lambda}_{hH}, \mathbf{u}_h) \in M_{hH} \times V_h$, a.e. $t \in I$. Moreover, we have*

$$\lambda_{hn}^s = -\tau_n^s(\mathbf{u}_h) \quad \textit{and} \quad g_c^s \lambda_{Ht}^s = -\tau_t^s(\mathbf{u}_h),$$

where $\mathbf{u}_h$ is the solution of the discrete primal problem and $g_c^s \equiv \tilde{g}_{ch}^{sm}$.

The above problem is equivalent to that of the Lagrangian formulation:
Find in every time level a pair $(\boldsymbol{\lambda}_{hH}, \mathbf{u}_h) \in (M_{hn} \times M_{Ht}) \times V_h$, such that

$$\begin{aligned} &\mathcal{H}(\boldsymbol{\mu}_{hH}, \mathbf{u}_h) \leq \mathcal{H}(\boldsymbol{\lambda}_{hH}, \mathbf{u}) \leq \mathcal{H}(\boldsymbol{\lambda}_{hH}, \mathbf{v}_h) \\ &\qquad \forall (\boldsymbol{\mu}_{hH}, \mathbf{v}_h) \in (M_{hn} \times M_{Ht}) \times V_h, \ t \in I, \end{aligned} \tag{14}$$

where $\mathcal{H}(\boldsymbol{\mu}_{hH}, \mathbf{v}_h) = \frac{1}{2} A(\mathbf{v}_h, \mathbf{v}_h) + b(\boldsymbol{\mu}_{hH}, \mathbf{v}_h) - (\mathbf{f}_h, \mathbf{v}_h)$.

Matrix Formulation and the Primal-Dual Active Set Strategy Method (PDAS) for the Frictionless Case. Since in the frictionless case the tangential stress component on the contact boundary $\cup \Gamma_c^{sm}$ is zero, then $\lambda_{Ht} = 0$, and therefore, $M_{hH} = M_{hn}$. Wolhmuth and Krause (2003) show that the standard basis of the space $V_h(\Omega^{\iota}, \mathcal{T}_{h,\Omega^{\iota}})$ is not a good choice for these type of problems and they introduced the space with a weak condition on $\cup \Gamma_c^{sm}$

$$\hat{V}_h = \{\mathbf{v} \in \sqcap_{\iota=1}^{r} V_h(\Omega^{\iota}, \mathcal{T}_{h,\Omega^{\iota}}) \mid \int_{\cup \Gamma_c^{sm}} [\mathbf{v}.\mathbf{e}_i]^s \psi \, ds = 0, \quad 1 \leq i \leq 2, \psi \in M_{hn}\},$$

where $\mathbf{e}_i$ denotes the i-th unit vector. The definition of $\hat{V}_h$ yields

$$\int_{\cup \Gamma_c^s} [\mathbf{v}.\mathbf{n}_p]^s \, \psi_s \, ds \leq d_p^s$$

for all vertices on the slide (non-mortar) side. Then the strong form of the non-penetration condition $[\mathbf{u}.\mathbf{n}]^s \leq d^{sm}$ will be replaced by its weak discrete form

$$\int_{\cup \Gamma_c^s} [\mathbf{u}.\mathbf{n}]^s \, \psi_p \, ds \leq \int_{\cup \Gamma_c^s} d_p^s \psi_p \, ds, \quad p \in S, \tag{15}$$

where S is the set of all vertices in the potential contact part on the slave side. The constraints (15) give a coupling between the vertices on the slave side and the master side. We introduce now a basis transformation of the basis of V_h in such a way that the weak non-penetration condition (15) in the new basis (the so-called dual basis) only deals with the vertices on the slave side. Thus the non-penetration condition is satisfied for all elements in $\hat{V}_h$, and $\hat{V}_h$ is a subspace of K_h.

The Lagrange multipliers λ_{hn} are the approximations of the contact forces $-\tau(\mathbf{u}_h).\mathbf{n} = -(n_j n_k \tau_{jk}(\mathbf{u}_h))$ which are necessary to adjust the contact displacements on the contact boundary $\cup\Gamma_c^{sm}$.

Let the space $V_h = span\{\varphi_p,\ p = 1,\dots,n_V\}$, n_V denotes the number of degrees of freedom of the space V_h and $M_{hn} = span\{\psi_q, q = 1,\dots,n_{cn}\}$, where ψ_q denotes the basis function associated with the vertex p and n_{cn} denotes the number of degrees of freedom of the space M_{hn} in each component. φ_p and ψ_q are the scalar basis functions associated with the node p resp. the node q, and satisfying the biorthogonality relation

$$\int_{\cup\Gamma_c^s} \varphi_p \psi_q \, ds = \delta_{pq} \int_{\cup\Gamma_c^s} \varphi_p \, ds, \tag{16}$$

with the Kronecker symbol δ_{pq}, where

$$\delta_{pq} = \begin{cases} 1 & \text{where the structure node } p \text{ coincides with potential node } q, \\ 0 & \text{otherwise.} \end{cases}$$

The discrete mortar formulation of the saddle point problem (13) for every time level is defined as follows:

Problem 6. In every time level find $\mathbf{u}_h \in V_h, \lambda_{hn} \in M_{hn}$, a.e. $t \in I$, satisfying

$$\begin{aligned} &A(\mathbf{u}_h, \mathbf{v}_h) + b(\lambda_{hn}, \mathbf{v}_h) = (\mathbf{f}_h, \mathbf{v}_h) \quad \mathbf{v}_h \in V_h,\ t \in I, \\ &b(\mu_{hn} - \lambda_{hn}, \mathbf{v}_h) \le \langle d^{sm}, (\mu_{hn} - \lambda_{hn})\rangle_{\cup\Gamma_c^{sm}} \quad \forall \mu_{hn} \in M_{hn},\ t \in I, \end{aligned} \tag{17}$$

where

$$M_{hn} = \{\mu_{hn} \in M_n | \langle \mu_{hn}, v_h\rangle_{\cup\Gamma_c^{sm}} \ge 0\ \forall v_h \in \{v_h \in W_h | v_h.\mathbf{n}^s \ge 0\}\},$$
where W_h is the discrete approximation of W

and

$$A(\mathbf{u}_h, \mathbf{v}_h) = (\rho \mathbf{u}_h, \mathbf{v}_h) + \triangle t(\alpha \mathbf{u}_h, \mathbf{v}_h) + \triangle t^2 a^{(0)}(\alpha \mathbf{u}_h, \mathbf{v}_h) + \triangle t a^{(1)}(\alpha \mathbf{u}_h, \mathbf{v}_h),$$

$$b(\mu_{hn}, \mathbf{v}_h) = \int_{\cup\Gamma_c^{sm}} \mu_{hn} [\mathbf{v}_h.\mathbf{n}]^s \, ds = \langle \mu_{hn}, [\mathbf{v}_h.\mathbf{n}]^s\rangle_{\cup\Gamma_c^{sm}} \quad \forall \mathbf{v}_h \in V_h, \mu_{hn} \in M_{hn}.$$

The condition (17*b*) is equivalent with the following conditions

$$\begin{aligned} &b(\lambda_{hn}, \mathbf{u}_h) = \langle d^{sm}, \lambda_{hn}\rangle_{\cup\Gamma_c^{sm}}, \\ &b(\mu_{hn}, \mathbf{u}_h) \le \langle d^{sm}, \mu_{hn}\rangle_{\cup\Gamma_c^{sm}} \quad \forall \mu_{hn} \in M_{hn}. \end{aligned} \tag{18}$$

Since the space M_{hn} is the convex cone, then putting $\mu_{hn} = 0$ and $\mu_{hn} = 2\lambda_{hn}$ in (17b), we have $b(\lambda_{hn}, \mathbf{u}_h) \geq \langle d^{sm}, \lambda_{hn}\rangle_{\cup\Gamma_c^{sm}}$ and $b(\lambda_{hn}, \mathbf{u}_h) \leq \langle d^{sm}, \lambda_{hn}\rangle_{\cup\Gamma_c^{sm}}$, therefore (18$a$) follows immediately. From (17b) using (18a) the condition (18b) follows and it represents the weak nonpenetration condition.

Let us denote the nodal parameters $\mathbf{u}_h$ by $\mathbf{U}$, of λ_{hn} by $\boldsymbol{\lambda}_{hn}$, and since for the frictionless case the tangential stress component on the contact boundary $\cup\Gamma_c^{sm}$ is zero, then $\lambda_{Ht} = 0$, and thus $\boldsymbol{\lambda}_{Ht} = 0$.

The Eq. (17a) for every $t \in I$ in the matrix form is of the form

$$\mathbb{A}_h\mathbf{U} + \mathbb{B}_h\boldsymbol{\lambda}_h = \mathbb{F}_h. \tag{19}$$

To examine the structure of $\mathbb{B}_h$, we introduce three sets of indices $\mathcal{N}$, $\mathcal{M}$, $\mathcal{S}$. We decompose the set of all vertices in every time level into three disjoint parts $\mathcal{N}$, $\mathcal{M}$ and $\mathcal{S}$, where by $\mathcal{S}$ (dim n_V) we denote all vertices on the possible contact part on the slave side, by $\mathcal{M}$ (dim n_{cn}) all vertices of the possible contact part of the master side, and by $\mathcal{N}$ all the other one. Then the strong formulation of the non-penetration condition (i.e. $[\mathbf{u}.\mathbf{n}]^{sm} \leq d^{sm}$ on $\cup\Gamma_c^{sm}$) will be replaced by its weak discrete form

$$\int_{\cup\Gamma_c^{sm}} [\mathbf{U}.\mathbf{n}]^s\,\psi_p\,ds \leq \int_{\cup\Gamma_c^{sm}} d^{sm}\psi_p\,ds, \quad p \in \mathcal{S}. \tag{20}$$

This condition connects points of sets $\mathcal{S}$ and $\mathcal{M}$. We introduce a basis in a new transformation of the basis of the space V_h in such a way that the weak non-penetration condition (20) in the new basis only deals with the vertices on the slave side (see Hüeber, Wolhmuth (2005), Wolhmuth, Krause (2003)).

Let us introduce the transformation $\varphi = (\varphi_{\mathcal{N}}, \varphi_{\mathcal{M}}, \varphi_{\mathcal{S}})^T$. The matrices and vectors in (19) can be decomposed in the sense of decomposition of the set of all vertices into three disjoint parts $\mathcal{N}$, $\mathcal{M}$ and $\mathcal{S}$, i.e.

$$\begin{bmatrix} \mathbb{A}_{\mathcal{NN}} & \mathbb{A}_{\mathcal{NM}} & \mathbb{A}_{\mathcal{NS}} & \mathbb{O} \\ \mathbb{A}_{\mathcal{MN}} & \mathbb{A}_{\mathcal{MM}} & \mathbb{A}_{\mathcal{MS}} & -\mathbb{M}^T \\ \mathbb{A}_{\mathcal{SN}} & \mathbb{A}_{\mathcal{SM}} & \mathbb{A}_{\mathcal{SS}} & \mathbb{D} \end{bmatrix} \begin{bmatrix} \mathbb{U}_{\mathcal{N}} \\ \mathbb{U}_{\mathcal{M}} \\ \mathbb{U}_{\mathcal{S}} \\ \boldsymbol{\lambda}_{\mathcal{S}} \end{bmatrix} = \begin{bmatrix} \mathbb{F}_{\mathcal{N}} \\ \mathbb{F}_{\mathcal{M}} \\ \mathbb{F}_{\mathcal{S}} \end{bmatrix}, \tag{21}$$

where elements of $\mathbb{B}_h$ are defined by $\mathbb{B}_h[p,q] = \int_{\cup\Gamma_c^{sm}} \varphi_p\psi_q\,ds\,\mathbb{I}_2 = \delta_{pq}\int_{\cup\Gamma_c^s} \varphi_p\,ds$, $p = 1,\ldots,n_V$, $q = 1,\ldots,n_{cn}$, and where $\mathbb{I}_2$ denotes the identity matrix in $\mathbb{R}^{2\times 2}$. The biorthogonality of the basis functions satisfies (16).

Further, the matrix $\mathbb{M}$ represents the coupling matrix the trace of the finite element shape functions on the master side "m" and the shape functions for the Lagrange multiplier on the slave side "s", defined by

$$\mathbb{M}[p,q] = \int_{\cup\Gamma_c^{sm}} \varphi_p\psi_q\,ds\,\mathbb{I}_2, \qquad p \in \mathcal{S}, q \in \mathcal{M}. \tag{22}$$

The matrix $\mathbb{M}$ is the block matrix and $\mathbb{D}$ is the block diagonal matrix with

$$\mathbb{D}[p,q] = \delta_{pq}\,\mathbb{I}_2.\int_{\cup\Gamma_c^{sm}} \varphi_p\,ds, \quad p = q \in \mathcal{S}. \tag{23}$$

The structure of the matrix $\mathbb{B}_h$ in every time level, using an approximate node numbering, is thus of the form $\mathbb{B}_h = (\mathbb{O}, -\mathbb{M}^T, \mathbb{D})^T$. The block matrices associated with the basis functions of the free structure nodes (i.e. $\mathcal{N}$), the potential contact nodes of the master side (i.e. $\mathcal{M}$) and the potential contact nodes on the slide side (i.e. $\mathcal{S}$) are denoted by $\mathbb{A}_{k,l}$, $k, l \in \{\mathcal{N}, \mathcal{M}, \mathcal{S}\}$. The entries of vectors $\mathbb{U}$ and $\mathbb{F}$ for $k \in \{\mathcal{N}, \mathcal{M}, \mathcal{S}\}$ are denoted by $\mathbb{U}_k$ and $\mathbb{F}_k$, respectively.

We introduce a new modified basis $\Phi = (\Phi_\mathcal{N}, \Phi_\mathcal{M}, \Phi_\mathcal{S})^T$ instead of the basis $\varphi = (\varphi_\mathcal{N}, \varphi_\mathcal{M}, \varphi_\mathcal{S})^T$ and the matrix $\widehat{\mathbb{M}} = \mathbb{D}^{-1}\mathbb{M}$. Then

$$\Phi = (\Phi_\mathcal{N}, \Phi_\mathcal{M}, \Phi_\mathcal{S})^T = \begin{bmatrix} \mathbb{I}_2 & \mathbb{O} & \mathbb{O} \\ \mathbb{O} & \mathbb{I}_2 & \widehat{\mathbb{M}}^T \\ \mathbb{O} & \mathbb{O} & \mathbb{I}_2 \end{bmatrix} \begin{bmatrix} \varphi_\mathcal{N} \\ \varphi_\mathcal{M} \\ \varphi_\mathcal{S} \end{bmatrix} = Q\varphi. \tag{24}$$

Hence

$$\mathbb{U} = Q^T \widehat{\mathbb{U}},$$

where $\widehat{\mathbb{U}}$ is the vector of coefficients with respect to the transformed basis Φ.

The modified stiffness matrix $\widehat{\mathbb{A}}_h$ associated with the transformed basis Φ is of the form

$$\widehat{\mathbb{A}}_h = \begin{bmatrix} \mathbb{I}_2 & \mathbb{O} & \mathbb{O} \\ \mathbb{O} & \mathbb{I}_2 & \widehat{\mathbb{M}}^T \\ \mathbb{O} & \mathbb{O} & \mathbb{I}_2 \end{bmatrix} \begin{bmatrix} \mathbb{A}_{\mathcal{NN}} & \mathbb{A}_{\mathcal{NM}} & \mathbb{A}_{\mathcal{NS}} \\ \mathbb{A}_{\mathcal{MN}} & \mathbb{A}_{\mathcal{MM}} & \mathbb{A}_{\mathcal{MS}} \\ \mathbb{A}_{\mathcal{SN}} & \mathbb{A}_{\mathcal{SM}} & \mathbb{A}_{\mathcal{SS}} \end{bmatrix} \begin{bmatrix} \mathbb{I}_2 & \mathbb{O} & \mathbb{O} \\ \mathbb{O} & \mathbb{I}_2 & \mathbb{O} \\ \mathbb{O} & \widehat{\mathbb{M}} & \mathbb{I}_2 \end{bmatrix} = Q\mathbb{A}_h Q^T =$$

$$= \begin{bmatrix} \mathbb{A}_{\mathcal{NN}} & \mathbb{A}_{\mathcal{NM}} + \mathbb{A}_{\mathcal{NS}}\widehat{\mathbb{M}} & \mathbb{A}_{\mathcal{NS}} \\ \mathbb{A}_{\mathcal{MN}} + \widehat{\mathbb{M}}^T\mathbb{A}_{\mathcal{SN}} & \mathbb{A}_{\mathcal{MM}} + \mathbb{A}_{\mathcal{MS}}\widehat{\mathbb{M}} + \widehat{\mathbb{M}}^T\mathbb{A}_{\mathcal{SM}} + \widehat{\mathbb{M}}^T\mathbb{A}_{\mathcal{SS}}\widehat{\mathbb{M}} & \mathbb{A}_{\mathcal{MS}} + \widehat{\mathbb{M}}^T\mathbb{A}_{\mathcal{SS}} \\ \mathbb{A}_{\mathcal{SN}} & \mathbb{A}_{\mathcal{SM}} + \mathbb{A}_{\mathcal{SS}}\widehat{\mathbb{M}} & \mathbb{A}_{\mathcal{SS}} \end{bmatrix}$$

and the modified vector of the right hand side $\widehat{\mathbb{F}}_h$ is the following

$$\widehat{\mathbb{F}}_h = Q\mathbb{F}_h = \begin{bmatrix} \mathbb{F}_\mathcal{N} \\ \mathbb{F}_\mathcal{M} + \widehat{\mathbb{M}}^T\mathbb{F}_\mathcal{S} \\ \mathbb{F}_\mathcal{S} \end{bmatrix}.$$

The algebraic representation of the weak non-penetration condition associated with the transformed basis Φ is the following:

$$[\mathbb{U}_n]^s = \textstyle\sum_{p\in\mathcal{S}} \widehat{\mathbb{U}}_p\, [\Phi_{pn}]^s + \sum_{q\in\mathcal{M}} \widehat{\mathbb{U}}_q\, [\Phi_{qn}]^s =$$
$$= (\textstyle\sum_{p\in\mathcal{S}} \widehat{\mathbb{U}}_p \varphi_p).\mathbf{n}_p^s + \sum_{q\in\mathcal{M}} \widehat{\mathbb{U}}_q(-\varphi_q + \sum_{p'\in\mathcal{S}} \widehat{\mathbb{M}}\,[p', q]\, \varphi_{p'}).\mathbf{n}_p^s.$$

By multiplying this equation with ψ_p, $p \in \mathcal{S}$ and integrating the resulting equation over $\cup_s \Gamma_c^s$, we have due to the biorthogonality condition (16) and the definition of the matrix $\widehat{\mathbb{M}}$ and (24)

$$\widehat{\mathbb{U}}_{n,p} \equiv (\mathbf{n}_p^s)^T \mathbb{D}\,[p,p]\, \widehat{\mathbb{U}}_p \le d_p^{sm} \qquad \forall p \in \mathcal{S}, \tag{25}$$

where $d_p^{sm} = \int_{\cup_s \Gamma_c^s} d_h^{sm} \psi_p \, ds$, $p \in \mathcal{S}$, as the coefficients at $\widehat{\mathbb{U}}_q, q \in \mathcal{M}$, are nullified. This basis transformation glues the vertices of the slave (non-master) side and on the master side together. The displacements of the glued vertices are given by $\widehat{\mathbb{U}}_{\mathcal{M}}$ and the relative displacements between the vertices on the slave side and on the master side are given by $\widehat{\mathbb{U}}_{\mathcal{S}}$.

The modified matrix $\widehat{\mathbb{B}}_h$ associated with the transformed basis Φ has the form

$$\widehat{\mathbb{B}}_h = Q\mathbb{B}_h = \begin{bmatrix} \mathbb{I}_2 & \mathbb{O} & \mathbb{O} \\ \mathbb{O} & \mathbb{I}_2 & \mathbb{M}^T \\ \mathbb{O} & \mathbb{O} & \mathbb{I}_2 \end{bmatrix} \begin{bmatrix} \mathbb{O} \\ -\mathbb{M}^T \\ \mathbb{D} \end{bmatrix} = (\mathbb{O}, \mathbb{O}, \mathbb{D})^T .$$

Then, in every time level, we will solve the following problem

$$\begin{aligned} &\widehat{\mathbb{A}}_h\widehat{\mathbb{U}} + \widehat{\mathbb{B}}_h \boldsymbol{\lambda}_{hH} = \widehat{\mathbb{F}}_h, \\ &\widehat{\mathbb{U}}_{n,p} \leq d_p^{sm}, \quad \boldsymbol{\lambda}_{hn,p} \geq 0, \quad (\widehat{\mathbb{U}}_{n,p} - d_p^{sm})\boldsymbol{\Lambda}_{hn,p} = 0, \quad \forall p \in \mathcal{S}, \\ &\boldsymbol{\Lambda}_{Ht,p} = 0, \quad \forall p \in \mathcal{S}, \end{aligned} \tag{26}$$

where in (26) the second line represents the Karush-Kuhn-Tucker conditions of a constrained optimization problem for inequality constraints,

$$\begin{aligned} &\boldsymbol{\Lambda}_{hn,p} = \mathbf{n}_p^{sT}\widehat{\mathbb{D}}\,[p,p]\,\boldsymbol{\Lambda}_{hn}\,(p)\,, \ \boldsymbol{\Lambda}_{hn}\,(p) \in \mathbb{R}^2, \\ &\boldsymbol{\Lambda}_{Ht,p} = \boldsymbol{\Lambda}_{hn}\,(p) - \left(\boldsymbol{\Lambda}_{hn}\,(p)\,.\mathbf{n}_p^s\right)\mathbf{n}_p^s = \left(\boldsymbol{\Lambda}_{hn}\,(p)\,.\mathbf{t}_p^s\right)\mathbf{t}_p^s. \end{aligned}$$

We decompose the set $\mathcal{S}$ as $\mathcal{S} = \mathcal{A} \cup \mathcal{I}$, where $\mathcal{A}$ is the active set and $\mathcal{I}$ is the inactive set. Then the above problem in every time level leads to **the primal-dual active set algorithm**, which is of the form:

Algorithm PDAS:

STEP 1. Initiate the sets $\mathcal{A}_1$ (active set) and $\mathcal{I}_1$ (inactive set), such that $\mathcal{S} = \mathcal{A}_1 \cup \mathcal{I}_1$ and $\mathcal{A}_1 \cap \mathcal{I}_1 = \emptyset$, put the initial value $(\widehat{\mathbb{U}}^0, \boldsymbol{\Lambda}_{hn}^0)$, $c_1 \in \left(10^3, 10^4\right)$ and set $k = 1$.

STEP 2. If $(\widehat{\mathbb{U}}^{k-1}, \boldsymbol{\Lambda}_{hn}^{k-1})$ is known, find the primal-dual pair $(\widehat{\mathbb{U}}^k, \boldsymbol{\Lambda}_{hn}^k)$ such that

$$\begin{aligned} &\widehat{\mathbb{A}}_h\widehat{\mathbb{U}}^k + \widehat{\mathbb{B}}_h\boldsymbol{\Lambda}_{hn}^k = \widehat{\mathbb{F}}_h, \\ &\widehat{\mathbb{U}}_{n,p}^k = d_p^{sm} \quad \text{for all } p \subset \mathcal{A}_k, \\ &\boldsymbol{\Lambda}_{hn,p}^k = 0 \quad \text{for all } p \in \mathcal{I}_k, \\ &\boldsymbol{\Lambda}_{Ht,p}^k = 0 \quad \text{for all } p \in \mathcal{S}. \end{aligned} \tag{27}$$

STEP 3. Set $\mathcal{A}_{k+1}$ and $\mathcal{I}_{k+1}$ to

$$\begin{aligned} \mathcal{A}_{k+1} &= \{p \in \mathcal{S} : \boldsymbol{\Lambda}_{hn,p}^k + c_1(\widehat{\mathbb{U}}_{n,p}^k - d_p^{sm}) > 0\}, \\ \mathcal{I}_{k+1} &:= \{p \in \mathcal{S} : \boldsymbol{\Lambda}_{hn,p}^k + c_1(\widehat{\mathbb{U}}_{n,p}^k - d_p^{sm}) \leq 0\}. \end{aligned}$$

STEP 4. If $\mathcal{A}_{k+1} = \mathcal{A}_k$ and $\mathcal{I}_{k+1} = \mathcal{I}_k$, then STOP else $k = k+1$; goto STEP 2, where the active $(\mathcal{A}_k)$ and inactive $(\mathcal{I}_k)$ sets are associated with the transformed basis Φ.

Remark 1. The system (27) can be rewritten if we decompose the set of vertices $\mathcal{S}$ on the slave side in each step k of the PDAS algorithm into the disjoint active and inactive sets $\mathcal{S} = \mathcal{A}_k \cup \mathcal{I}_k$.

4 Conclusion

With the development of computers, methods of numerical analysis of solving partial differential equations were evolved and were applied to the solution of the propagation of seismic waves in the stratified homogeneous and nonhomegeneous media in the time domain. Such models allow the simulation of propagations of seismic waves inside the Earth closed to the real rheology of the investigated upper parts of the Earth. All these methods were evolved for the continuous Earths crust and the mantle. Since the upper parts of the Earth are broken up into a great number of plates and blocks the up-to-date evolved methods cannot be used and we are forced urgently to investigate new methods and new algorithms permit to simulate and study propagation of seismic ans elastic waves through the broken up media. The different problem is to approximate optimally the contact conditions between geological blocks being in common contact. In our paper the primal-dual active set strategy technique for the 2D seismological model is used. The real Earths body is connected from a great number of 3D bodies being in common contact, and therefore, the methods as well as algorithms must be extended and investigated also in the three dimensions, and moreover, for very complicated contact boundaries between geological blocks, as well as for complicated cracks problems which accompanied the propagation seismic waves through elastic media. It is evident that these problems are enormous difficult and will be problems for very long future time period.

Acknowledgments

The paper was partially supported by the Institutional Research Plan AV0Z10300504 and by the Research Plan MSM 4977751301.

References

Adams, R.A.: Sobolev Spaces. Academic Press, New York (1975)

Duvaut, G., Lions, J.L.: Inequalities in Mechanics and Physics. Springer, Heidelberg (1976)

Eck, C., Jarušek, J., Krbec, J.: Unilateral Contact Problems. Variational Methods and Existence Theorems. Chapman & Hall/CRC, Taylor & Francis Group, Boca Raton, London, New York, Singapore (2005)

Haslinger, J., Hlaváček, I., Nečas, J.: Numerical Methods for Unilateral Problems in Solid Mechanics. In: Handbook of Numerical Analysis, vol. IV, pp. 313–486. Elsevier, Amsterdam (1996)

Hilber, H.M., Hughes, T.J.R., Taylor, R.L.: Improved numerical dissipation for time integration algorithms in structural dynamics. Earthquake Engrg. Struct. Dyn. 5, 283–292 (1977)

Hlaváček, I., Nedoma, J.: On a Solution of a Generalized Semi-Coercive Contact Problem in Thermo-Elasticity. Math. Comput. Simulat. 60, 1–17 (2002)

Hüeber, S., Wolhmuth, B.: A primal-dual active set strategy for non-linear multibody contact problems. Computer Methods in Appl. Mech. and Engrg. 194, 3147–3166 (2005)
Jarušek, J.: Dynamic contact problems with given friction for viscoelastic bodies. Czechoslovak Math. J. 46, 475–487 (1996)
Nedoma, J.: On one Type of Signorini Problem without Friction in Linear Thermo-Elasticity. Appl. Math. 28(6), 393–407 (1983)
Nedoma, J.: On the Signorini Problem with Friction in Linear Thermo-Elasticity: the Quasi-Coupled 2D-Case. Appl. Math. 32(3), 186–199 (1987)
Nedoma, J.: Numerical Modelling in Applied Geodynamics. John Wiley&Sons, New York (1998)
Nedoma, J.: Finite element analysis of elastic (seismic) wave propagation in region with unilateral contacts. In: Arabnia, H.R. (ed.) Proc. Conference PDPTA 2000, Las Vegas, USA (2000)
Nedoma, J.: Dynamic contact problems with friction in elasticity and thermo-elasticity. In: Feistauer., M., Rannacher, R., Kozel, K. (eds.) Proc. of the 4th Conference on Numerical Modelling in Continuum Mechanics, MATFYZPRESS, Prague (2001)
Nedoma, J., Daněk, J.: Dynamic and quasi-static simulations of the function of knee joint and its artificial replacements in 2D and 3D. In: Proc. of the conference ICCBB 2007 (to appear, 2007)
Rivera, J., Racke, R.: Multidimensional contact problems in thermoelasticity. SIAM J. Appl. Math. 58, 1307–1337 (1998)
Wolhmuth, B.I., Krause, R.: Monotone methods on non-matching grids for non-linear contact problems. SIAM J. Scient. Computing 25, 324–347 (2003)

Parallel Implementation of Complexity Reduction Approach to Fourth Order Approximation on 2D Free Space Wave Propagation

Mohammad Khatim Hasan[1], Mohamed Othman[2], Zulkifly Abbas[3], Jumat Sulaiman[4], and Fatimah Ahmad[2]

[1] Faculty of Information Science and Technology, Universiti Kebangsaan Malaysia, 43600 Bangi, Selangor D.E.
[2] Faculty of Computer Science and Info. Tech., Universiti Putra Malaysia, 43400 Serdang, Selangor D.E.
[3] Faculty of Science, Universiti Putra Malaysia, 43400 Serdang, Selangor D.E.
[4] School of Science and Technology, Universiti Malaysia Sabah, Locked Bag 2073, 88999 Kota Kinabalu, Sabah, Malaysia
khatim@ftsm.ukm.my

Abstract. Recently, a new variant of FDTD method known as High Speed Low Order FDTD (HSLO-FDTD) shows to solve 1D electromagnetic problem faster than the standard FDTD method by 67%. Application of parallel strategy to the method for 2D electromagnetic problem gain better saving in computational time to the parallel FDTD method by 85.2%. This method is called Ultra High Speed Low Order FDTD (UHSLO-FDTD). Both method applies the second order discretization with complexity reduction approach. In this paper, fourth order discretization with complexity reduction approach have succeeds to improve the accuracy of UHSLO-FDTD method. However, the fourth order scheme need higher computational time than UHSLO-FDTD method, but still faster than the FDTD method. This fourth order scheme is called Ultra High Speed High Order Finite Difference Time Domain (UHSHO-FDTD) method. In this paper we solve 2D wave propagation problems on a Symmetrical Multiprocessor machine using message-passing interface. We examine the parallelism efficiency of the algorithm by analyzing the simulation time and speedup.

Keywords: finite difference method, complexity reduction approach, free space wave propagation.

1 Introduction

Electromagnetics plays a major role in modern lifestyle of living. Numerical techniques is an alternative method to simulate electromagnetics behavior. Yee proposed a numerical technique which was called Finite difference time domain (FDTD) to solve electromagnetic problem [1]. The method approximates

O. Gervasi et al. (Eds.): ICCSA 2008, Part II, LNCS 5073, pp. 720–735, 2008.

Maxwell equations using a low "discrete" complexity algorithm [2], [3]. Advantages and widely applied problems of FDTD method can be found in [2], [3], [4], [5]. However, FDTD method needs a long processing time to simulate problem [2], [4], [5].

Complexity reduction approach have been actively implemented into several numerical methods to speed-up the computational execution time. For example, Iterative Alternating Decomposition Explicit (IADE) method have been implemented the complexity reduction approach [6]. The method utilized the special case (using $g = \frac{6}{5}$) of the IADE method. The proposed method is called the Reduced Iterative Alternating Decomposition Explicit (RIADE) method. The method have been proven to be faster than the IADE method to solve heat conduction equation.

Beside that Abdullah [7] have proposed another approach of complexity reduction concept. This concept only solve parts of solution domain in the iterative manner until it is converge and then the rest of points will be solve directly. He utilized this concept and proposed a method called Explicit Decouple Group (EDG) method. In the method, only half of the nodes in the solution domain is solved iteratively. Further application of this method can be found in Ibrahim & Abdullah [8], Abdullah & Ali [9], and Yousif & Evans [10]. Othman & Abdullah[11] extend this method an proposed Modified Explicit Group (MEG) method. This method only solve a quarter of its solution domain iteratively. Since then, various new method using half and quarter sweep concept have been proposed [12], [13]. Both EDG and MEG have increase the speed of the original methods by two and four time, respectively.

Another approach which also rapidly becoming dominant theme in all areas of computer science and its applications, is the parallel computation. The approach capable to increase speed of any computation. Their behavior follows Amdahl's or Gustafson's laws [14]. This computing paradigm has been used by several researchers on multiprocessor machine to develop parallel FDTD algorithm [5], [15].

Using the complexity reduction concept, Hasan *et al.* [2], [3] had proposed High Speed Low Order Finite Difference Time Domain (HSLO-FDTD) method to solve 1D and 2D free space wave propagation problem. This method has successfully reduced 67% of 1D FDTD and 85.2% of 2D FDTD computing time (for time step approaching infinity). In this paper we proposed the fourth order approximation with complexity reduction approach to solve 2D problems and named Ultra High Speed High Order Finite Difference Time Domain (UHSHO-FDTD) algorithm.

The remaining of this paper is organized as follows. In Section 2, the formulation of 2D UHSHO-FDTD algorithms is presented. Section 3 analyze the stability and the convergence characteristic of the method. Section 4 demonstrates the results of some numerical test which shows the performance of the algorithm. Finally, conclusions are included in Section 5.

2 Formulation of the Method

Consider a normalized 2D transverse electric (TE) time domain Maxwell equations in linear, homogenous, isotropic lossless medium as given by

$$\frac{\partial H_z}{\partial t} = \frac{1}{\sqrt{\varepsilon_0 \mu_0}} \left(\frac{\partial \widetilde{E}_x}{\partial y} - \frac{\partial \widetilde{E}_y}{\partial x} \right), \tag{1}$$

$$\frac{\partial \widetilde{E}_x}{\partial t} = \frac{1}{\sqrt{\varepsilon_0 \mu_0}} \frac{\partial H_z}{\partial y} \tag{2}$$

and

$$\frac{\partial \widetilde{E}_y}{\partial t} = -\frac{1}{\sqrt{\varepsilon_0 \mu_0}} \frac{\partial H_z}{\partial x} \tag{3}$$

where ε_0 and μ_0 are the electric permittivity and magnetic permeability of free space medium,and **E** and **H** are electric and magnetic field vectors, respectively.

To develop the approximate equations for Maxwell equations in 2D TE mode, let first consider Faraday's law at the space-time point $(m\Delta x, n\Delta y, q\Delta t)$

$$\left.\frac{\partial H_z}{\partial t}\right|_{(m\Delta x, n\Delta y, q\Delta t)} = \frac{1}{\sqrt{\varepsilon_0 \mu_0}} \left(\left.\frac{\partial \widetilde{E}_x}{\partial y}\right|_{(m\Delta x, n\Delta y, q\Delta t)} - \left.\frac{\partial \widetilde{E}_y}{\partial x}\right|_{(m\Delta x, n\Delta y, q\Delta t)} \right). \tag{4}$$

Finite difference approximation for (4) is given by

$$\begin{aligned} &\tfrac{H_z^{q+\frac{1}{2}}[m][n] - H_z^{q-\frac{1}{2}}[m][n]}{\Delta t} = \\ &\frac{1}{\sqrt{\varepsilon_0 \mu_0}} \left(\frac{9}{8} \left(\frac{\widetilde{E}_x^q[m][n+\frac{3}{2}] - \widetilde{E}_x^q[m][n-\frac{3}{2}]}{3\Delta y} - \frac{\widetilde{E}_y^q[m+\frac{3}{2}][n] - \widetilde{E}_y^q[m-\frac{3}{2}][n]}{3\Delta x} \right) \right. \\ &\left. -\frac{1}{24} \left(\frac{\widetilde{E}_x^q[m][n+\frac{9}{2}] - \widetilde{E}_x^q[m][n-\frac{9}{2}]}{3\Delta y} - \frac{\widetilde{E}_y^q[m+\frac{9}{2}][n] - \widetilde{E}_y^q[m-\frac{9}{2}][n]}{3\Delta x} \right) \right). \end{aligned} \tag{5}$$

Rearranging (5), yields

$$\begin{aligned} &H_z^{q+\frac{1}{2}}[m][n] = H_z^{q-\frac{1}{2}}[m][n] \\ &+ \frac{\Delta t}{\sqrt{\varepsilon_0 \mu_0}} \left(\frac{9}{8} \left(\frac{\widetilde{E}_x^q[m][n+\frac{3}{2}] - \widetilde{E}_x^q[m][n-\frac{3}{2}]}{3\Delta y} - \frac{\widetilde{E}_y^q[m+\frac{3}{2}][n] - \widetilde{E}_y^q[m-\frac{3}{2}][n]}{3\Delta x} \right) \right. \\ &\left. -\frac{1}{24} \left(\frac{\widetilde{E}_x^q[m][n+\frac{9}{2}] - \widetilde{E}_x^q[m][n-\frac{9}{2}]}{3\Delta y} - \frac{\widetilde{E}_y^q[m+\frac{9}{2}][n] - \widetilde{E}_y^q[m-\frac{9}{2}][n]}{3\Delta x} \right) \right). \end{aligned} \tag{6}$$

The equation (6) is an explicit equation and known as magnetic field update equation. Now let consider the Ampere's law for E_y at the space-time point $((m + \frac{3}{2})\Delta x, n\Delta y, (q + \frac{1}{2})\Delta t)$

$$\left.\frac{\partial \widetilde{E}_y}{\partial t}\right|_{((m+\frac{3}{2})\Delta x, n\Delta y, (q+\frac{1}{2})\Delta t)} = -\frac{1}{\sqrt{\varepsilon_0 \mu_0}} \left.\frac{\partial H_z}{\partial x}\right|_{((m+\frac{3}{2})\Delta x, n\Delta y, (q+\frac{1}{2})\Delta t)}. \tag{7}$$

Following the same procedure, (7) is approximated by

$$\begin{aligned}&\widetilde{E}_y^{q+1}[m+\tfrac{1}{2}][n]=\widetilde{E}_y^{q}[m+\tfrac{1}{2}][n]\\&-\frac{\Delta t}{3\sqrt{\varepsilon_0\mu_0}\Delta x}\left(\frac{9}{8}\left(H_z^{q+\frac{1}{2}}[m+2][n]-H_z^{q+\frac{1}{2}}[m-1][n]\right)\right.\\&\left.-\frac{1}{24}\left(H_z^{q+\frac{1}{2}}[m+5][n]-H_z^{q+\frac{1}{2}}[m-4][n]\right)\right),\end{aligned}\tag{8}$$

which is also explicit. The equation (8) is known as the E_y update equation.

Next, let consider the Faraday's law for E_x at the space-time point $(m\Delta x,(n+\frac{3}{2})\Delta y,(q+\frac{1}{2})\Delta t)$

$$\left.\frac{\partial\widetilde{E}_x}{\partial t}\right|_{(m\Delta x,(n+\frac{3}{2})\Delta y,(q+\frac{1}{2})\Delta t)}=\left.\frac{1}{\sqrt{\varepsilon_0\mu_0}}\frac{\partial H_z}{\partial y}\right|_{(m\Delta x,(n+\frac{3}{2})\Delta y,(q+\frac{1}{2})\Delta t)}.\tag{9}$$

Following the same approach as before, (9) is approximated by

$$\begin{aligned}&\widetilde{E}_x^{q+1}[m][n+\tfrac{3}{2}]=\widetilde{E}_x^{q}[m][n+\tfrac{3}{2}]\\&+\frac{\Delta t}{3\sqrt{\varepsilon_0\mu_0}\Delta y}\left(\frac{9}{8}\left(H_z^{q+\frac{1}{2}}[m][n+2]-H_z^{q+\frac{1}{2}}[m][n-1]\right)\right.\\&\left.-\frac{1}{24}\left(H_z^{q+\frac{1}{2}}[m][n+5]-H_z^{q+\frac{1}{2}}[m][n-4]\right)\right),\end{aligned}\tag{10}$$

which is also explicit. The equation (10) is known as update equation for the E_x field. Equations (6), (8) and (10) are the main update equations for UHSHO-FDTD in direct domain (DD).

In order to solve the problem simultaneously, the solution domain must be divided into P (number of processor) sub domains. Following [3], [5], [15], we use the 1D topology strategy. The computational domain is divided into P sub domains along the x-direction. Each sub domain is assigned to one processor, as shown in Fig. 1. All fields at the boundaries are set to zero, since we use the perfectly conducting boundary. At the internal cells for UHSHO-FDTD in direct domain (DD) approach, the fields are updated using (6), (8) and (10).

However, updating fields at the sub domain boundaries are not straightforward. The problem occurs since to update electric or magnetic field at sub domain boundary, magnetic or electric field from neighboring processors are needed. This problem can be overcome by developing lattices with overlapping nodes. The overlapping nodes are the same points of the electromagnetic fields on neighboring processors. Additional memories have to be added at both end of sub domains in each processors to hold the sub boundary values from each neighboring processors. These values are then communicated to the neighbors.

MPI_Ssend and MPI_Recv commands are used to exchange data between neighboring processors. These MPI communication commands will assure the pair of send and receive activities completed. We communicate these data in vector form. It should be pointed out that the computation of H_z does not need E_x values from neighboring processor. This is due to the fact that all E_x values

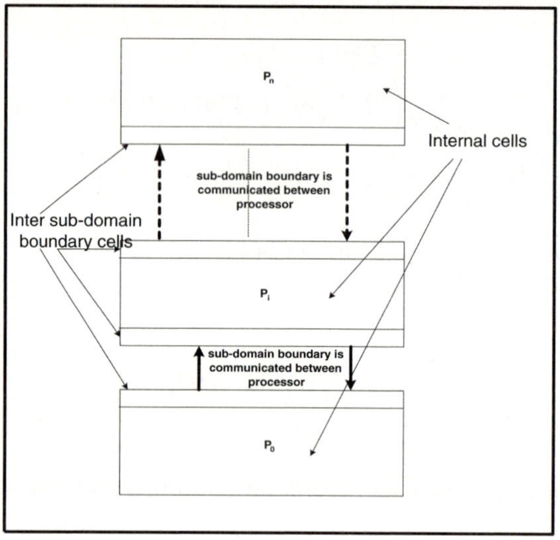

Fig. 1. Decomposition of the solution domain

needed to compute H_z fields are available in the sub domain. Thus, only H_z and E_y sub boundaries value need to be routed. However before the data routing process can be triggered, minimum and maximum valued point's location in each sub domain have to be located.

Following [2], [3], we can develop another variant of the method in temporary domain approach (TD). For temporary domain (TD) approach, equations (6), (8) and (10) have to be modified [2], [3]. The solution domain have to be "compressed" until only black nodes exist (Fig. 2). This "compressed" domain will be called as temporary domain. To suit the temporary domain, equations (6), (8) and (10) becomes

$$\begin{aligned} H_z^{q+\frac{1}{2}}[m][n] &= H_z^{q-\frac{1}{2}}[m][n] \\ &+ \frac{\Delta t}{\sqrt{\varepsilon_0 \mu_0}} \Bigg(\frac{9}{8} \Bigg(\frac{\widetilde{E}_x^q[m][n+\frac{1}{2}] - \widetilde{E}_x^q[m][n-\frac{1}{2}]}{3\Delta y} - \frac{\widetilde{E}_y^q[m+\frac{1}{2}][n] - \widetilde{E}_x^q[m-\frac{1}{2}][n]}{3\Delta x} \Bigg) \\ &- \frac{1}{24} \Bigg(\frac{\widetilde{E}_x^q[m][n+\frac{3}{2}] - \widetilde{E}_x^q[m][n-\frac{3}{2}]}{3\Delta y} - \frac{\widetilde{E}_y^q[m+\frac{3}{2}][n] - \widetilde{E}_x^q[m-\frac{3}{2}][n]}{3\Delta x} \Bigg) \Bigg), \end{aligned} \tag{11}$$

$$\begin{aligned} \widetilde{E}_y^{q+1}[m+\tfrac{1}{2}][n] &= \widetilde{E}_y^q[m+\tfrac{1}{2}][n] \\ &- \frac{\Delta t}{3\sqrt{\varepsilon_0 \mu_0}\Delta x} \Bigg(\frac{9}{8} \Big(H_z^{q+\frac{1}{2}}[m+1][n] - H_z^{q+\frac{1}{2}}[m][n] \Big) \\ &- \frac{1}{24} \Big(H_z^{q+\frac{1}{2}}[m+2][n] - H_z^{q+\frac{1}{2}}[m-1][n] \Big) \Bigg) \end{aligned} \tag{12}$$

and

$$\widetilde{E}_x^{q+1}[m][n+\tfrac{1}{2}] = \widetilde{E}_x^{q}[m][n+\tfrac{1}{2}] + \frac{\Delta t}{3\sqrt{\varepsilon_0\mu_0}\Delta y}\left(\frac{9}{8}\left(H_z^{q+\frac{1}{2}}[m][n+1] - H_z^{q+\frac{1}{2}}[m][n]\right) - \frac{1}{24}\left(H_z^{q+\frac{1}{2}}[m][n+2] - H_z^{q+\frac{1}{2}}[m][n-1]\right)\right). \quad (13)$$

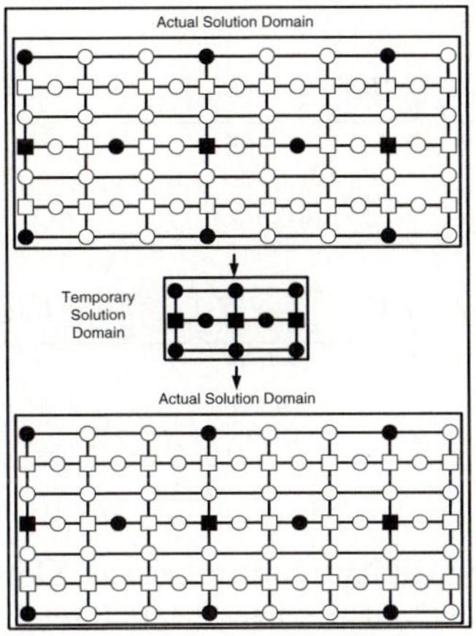

Fig. 2. Transformation of actual solution domain to temporary solution domain

Implementing TD approach simplifies the data routing process. Here, we do not have to find the minimum and maximum valued point's location. We can implement the same data routing process used in parallel FDTD method. At the internal cells for UHSHO-FDTD in temporary domain (TD) approach, the fields are updated using (11), (12) and (13).

Fig. 1 shows data exchange process between neighboring processors. It is clearly shown from Fig. 1 that to update the H_z at the sub domain boundaries, the E_y from the neighboring processor has to be communicated. Similar action has to be done to update E_y, H_z values from neighboring processor has to be communicated.

After the temporary domain is re-transform into actual solution domain, then only we solved all remaining points. The remaining point will be calculated only at pre-specified time step deploying $x - y$ approach [2], [3] using formulation given below.

$$\begin{pmatrix} 2 & -1 \\ -1 & 2 \end{pmatrix}\begin{pmatrix} E_{i+1} \\ E_{i+2} \end{pmatrix} = \begin{pmatrix} E_i \\ E_{i+3} \end{pmatrix}.$$

on x-direction first then only on y-direction.

3 Stability and Convergence Analysis

Consider a 2D time domain Maxwell equations with lossless, isotropic and homogeneous medium. Suppose UHSHO-FDTD method approximate the Ampere's and Faraday's laws in 2D time domain Maxwell equations as follows.

$$\frac{E_x|_{i,j}^{n+\frac{1}{2}} - E_x|_{i,j}^{n-\frac{1}{2}}}{\Delta t} = \frac{1}{\sqrt{\mu_o \epsilon_o}}\left(\frac{9}{8}\frac{H_z|_{i,j+\frac{3}{2}}^{n} - H_z|_{i,j-\frac{3}{2}}^{n}}{3\Delta y} - \frac{1}{24}\frac{H_z|_{i,j+\frac{9}{2}}^{n} - H_z|_{i,j-\frac{9}{2}}^{n}}{3\Delta y}\right), \tag{14}$$

$$\frac{E_y|_{i,j}^{n+\frac{1}{2}} - E_y|_{i,j}^{n-\frac{1}{2}}}{\Delta t} = -\frac{1}{\sqrt{\mu_o \epsilon_o}}\left(\frac{9}{8}\frac{H_z|_{i+\frac{3}{2},j}^{n} - H_z|_{i-\frac{3}{2},j}^{n}}{3\Delta x} - \frac{1}{24}\frac{H_z|_{i+\frac{9}{2},j}^{n} - H_z|_{i-\frac{9}{2},j}^{n}}{3\Delta x}\right) \tag{15}$$

and

$$\frac{H_z|_{i,j}^{n+1} - H_z|_{i,j}^{n}}{\Delta t} = \frac{1}{\sqrt{\mu_o \epsilon_o}}\left(\frac{9}{8}\left(\frac{E_x|_{i,j+\frac{3}{2}}^{n+\frac{1}{2}} - E_x|_{i,j-\frac{3}{2}}^{n+\frac{1}{2}}}{3\Delta y} - \frac{E_y|_{i+\frac{3}{2},j}^{n+\frac{1}{2}} - E_y|_{i-\frac{3}{2},j}^{n+\frac{1}{2}}}{3\Delta x}\right)\right.$$
$$\left. -\frac{1}{24}\left(\frac{E_x|_{i,j+\frac{9}{2}}^{n+\frac{1}{2}} - E_x|_{i,j-\frac{9}{2}}^{n+\frac{1}{2}}}{3\Delta y} - \frac{E_y|_{i+\frac{9}{2},j}^{n+\frac{1}{2}} - E_y|_{i-\frac{9}{2},j}^{n+\frac{1}{2}}}{3\Delta x}\right)\right). \tag{16}$$

In order to analyze the stability characteristic, Courant-Frederich-Levy approach is used [4]. Following Courant-Frederich-Levy, the problem is separated into time and space eigenvalue problem.

Let's first consider the time eigenvalue problem. The time eigenvalue problem for equations (14), (15) and (16) are given by

$$\frac{E_x|_{i,j}^{n+\frac{1}{2}} - E_x|_{i,j}^{n-\frac{1}{2}}}{\Delta t} = \Lambda E_x|_{i,j}^{n}, \tag{17}$$

$$\frac{E_y|_{i,j}^{n+\frac{1}{2}} - E_y|_{i,j}^{n-\frac{1}{2}}}{\Delta t} = \Lambda E_y|_{i,j}^{n} \tag{18}$$

and

$$\frac{H_z|_{i,j}^{n+1} - H_z|_{i,j}^{n}}{\Delta t} = \Lambda H_z|_{i,j}^{n+\frac{1}{2}}, \tag{19}$$

respectively. A generic equation for (17), (18) and (19) can be constructed as follows.

$$\frac{V|_{i,j}^{n+\frac{1}{2}} - V|_{i,j}^{n-\frac{1}{2}}}{\Delta t} = \Lambda V|_{i,j}^{n}. \tag{20}$$

A growth factor can be defined by $q_{i,j} = \frac{V|_{i,j}^{n+\frac{1}{2}}}{V|_{i,j}^{n}} = \frac{V|_{i,j}^{n}}{V|_{i,j}^{n-\frac{1}{2}}}$. Therefore, $V|_{i,j}^{n+\frac{1}{2}} = q_{i,j} V|_{i,j}^{n}$. Thus, $V|_{i,j}^{n-\frac{1}{2}} = \frac{V|_{i,j}^{n}}{q_{i,j}}$. Substituting into equation (20) yield

$$q_{i,j}^2 - \Delta t \Lambda q_{i,j} - 1 = 0. \tag{21}$$

Using quadratic formula to (21) yield

$$q_{i,j} = \frac{\Lambda \Delta t}{2} \pm \sqrt{\left(\frac{\Lambda \Delta t}{2}\right)^2 + 1}. \tag{22}$$

Suppose that $a = \frac{\Lambda \Delta t}{2}$ is purely imaginary, therefore

$$\Re(a) = 0 \rightarrow a = j\Im(a), \tag{23}$$

where $-1 \leq \Im(a) \leq 1$. Therefore (22) can be rewritten as

$$q_{i,j} = j\Im(a) \pm \sqrt{a^2+1} \rightarrow j\Im(a) \pm \sqrt{(j\Im(a))^2+1} \rightarrow j\Im(a) \pm \sqrt{1-(\Im(a))^2}. \tag{24}$$

Taking the $|q_{i,j}|$ of (24) yields one. This means $|q_{i,j}|$ is always equal to one when a is purely imaginary. Thus, the method is stable if it satisfies certain conditions. Let's find the condition that assures the stability of UHSHO-FDTD method. Using information in (22) and (23), an inequality relation can be constructed as follows.

$$-1 \leq \frac{\Delta t \Im(\Lambda)}{2} \leq 1 \rightarrow \frac{-2}{\Delta t} \leq \Im(\Lambda) \leq \frac{2}{\Delta t}. \tag{25}$$

Next, let's consider the space eigenvalue problem for equations (14), (15) and (16). The space eigenvalue problem for equations (14), (15) and (16) are given by

$$\frac{1}{\sqrt{\mu_o \epsilon_o}}\left(\frac{9}{8}\frac{H_z|_{i,j+\frac{3}{2}} - H_z|_{i,j-\frac{3}{2}}}{3\Delta y} - \frac{1}{24}\frac{H_z|_{i,j+\frac{9}{2}} - H_z|_{i,j-\frac{9}{2}}}{3\Delta y}\right) = \Lambda E_x|_{i,j}, \tag{26}$$

$$-\frac{1}{\sqrt{\mu_o \epsilon_o}}\left(\frac{9}{8}\frac{H_z|_{i+\frac{3}{2},j} - H_z|_{i-\frac{3}{2},j}}{3\Delta x} - \frac{1}{24}\frac{H_z|_{i+\frac{9}{2},j} - H_z|_{i-\frac{9}{2},j}}{3\Delta x}\right) = \Lambda E_y|_{i,j} \tag{27}$$

and

$$\begin{aligned}&\tfrac{1}{\sqrt{\mu_o \epsilon_o}}\left(\tfrac{9}{8}\left(\tfrac{E_x|_{i,j+\frac{3}{2}} - E_x|_{i,j-\frac{3}{2}}}{3\Delta y} - \tfrac{E_y|_{i+\frac{3}{2},j} - E_y|_{i-\frac{3}{2},j}}{3\Delta x}\right)\right.\\ &\left.-\frac{1}{24}\left(\frac{E_x|_{i,j+\frac{9}{2}} - E_x|_{i,j-\frac{9}{2}}}{3\Delta y} - \frac{E_y|_{i+\frac{9}{2},j} - E_y|_{i-\frac{9}{2},j}}{3\Delta x}\right)\right) - \Lambda H_z|_i.\end{aligned} \tag{28}$$

Replacing $E_x|_{i,j}$, $E_y|_{i,j}$ and $H_z|_{i,j}$ with $E_{x_o}e^{j(\tilde{k}_x i\Delta x + \tilde{k}_y j \Delta y)}$, $E_{y_o}e^{j(\tilde{k}_x i\Delta x + \tilde{k}_y j \Delta y)}$ and $H_{z_o}e^{j(\tilde{k}_x i\Delta x + \tilde{k}_y j \Delta y)}$, respectively yields

$$\begin{aligned}&-\tfrac{1}{3\sqrt{\mu_o\epsilon_o}\Delta y}\left(\tfrac{9}{8}(H_{z_o}e^{j(\tilde{k}_x i\Delta x+\tilde{k}_y(j+\frac{3}{2})\Delta y)} - H_{z_o}e^{j(\tilde{k}_x i\Delta x+\tilde{k}_y(j-\frac{3}{2})\Delta y)})\right.\\ &\left.-\frac{1}{24}(H_{z_o}e^{j(\tilde{k}_x i\Delta x+\tilde{k}_y(j+\frac{9}{2})\Delta y)} - H_{z_o}e^{j(\tilde{k}_x i\Delta x+\tilde{k}_y(j-\frac{9}{2})\Delta y)})\right) = \Lambda E_{x_o}e^{j(\tilde{k}_x i\Delta x+\tilde{k}_y j\Delta y)},\end{aligned} \tag{29}$$

$$\begin{aligned}&-\tfrac{1}{3\sqrt{\mu_o\epsilon_o}\Delta x}\left(\tfrac{9}{8}(H_{z_o}e^{j(\tilde{k}_x (i+\frac{3}{2})\Delta x+\tilde{k}_y j\Delta y)} - H_{z_o}e^{j(\tilde{k}_x (i-\frac{3}{2})\Delta x+\tilde{k}_y j\Delta y)})\right.\\ &\left.-\frac{1}{24}(H_{z_o}e^{j(\tilde{k}_x (i+\frac{9}{2})\Delta x+\tilde{k}_y j\Delta y)} - H_{z_o}e^{j(\tilde{k}_x (i-\frac{9}{2})\Delta x+\tilde{k}_y j\Delta y)})\right) = \Lambda E_{x_o}e^{j(\tilde{k}_x i\Delta x+\tilde{k}_y j\Delta y)}\end{aligned} \tag{30}$$

and

$$-\frac{1}{\sqrt{\mu_o\epsilon_o}}\Bigg(\frac{9}{8}\Bigg(\frac{1}{3\Delta y}E_{x_o}e^{j(\tilde{k}_x i\Delta x+\tilde{k}_y(j+\frac{3}{2})\Delta y)}-E_{x_o}e^{j(\tilde{k}_x i\Delta x+\tilde{k}_y(j-\frac{3}{2})\Delta y)}$$
$$-\frac{1}{3\Delta x}E_{y_o}e^{j(\tilde{k}_x(i+\frac{3}{2})\Delta x+\tilde{k}_y j\Delta y)}-E_{y_o}e^{j(\tilde{k}_x(i-\frac{3}{2})\Delta x+\tilde{k}_y j\Delta y)}\Bigg)$$
$$\frac{1}{24}\Bigg(\frac{1}{3\Delta y}E_{x_o}e^{j(\tilde{k}_x i\Delta x+\tilde{k}_y(j+\frac{9}{2})\Delta y)}-E_{x_o}e^{j(\tilde{k}_x i\Delta x+\tilde{k}_y(j-\frac{9}{2})\Delta y)}$$
$$-\frac{1}{3\Delta x}E_{y_o}e^{j(\tilde{k}_x(i+\frac{9}{2})\Delta x+\tilde{k}_y j\Delta y)}-E_{y_o}e^{j(\tilde{k}_x(i-\frac{9}{2})\Delta x+\tilde{k}_y j\Delta y)}\Bigg)\Bigg)=\Lambda H_{z_o}e^{j(\tilde{k}_x i\Delta x+\tilde{k}_y j\Delta y)}. \quad (31)$$

Simplifying equation (29) yield

$$-\frac{H_{z_o}}{3\sqrt{\mu_o\epsilon_o}\Delta y}\left(\frac{9}{8}\left(e^{j(\tilde{k}_y(\frac{3}{2})\Delta y}-e^{-j(\tilde{k}_y(\frac{3}{2})\Delta y)}\right)-\frac{1}{24}\left(e^{j(\tilde{k}_y(\frac{9}{2})\Delta y}-e^{-j(\tilde{k}_y(\frac{9}{2})\Delta y)}\right)\right)=\Lambda E_{x_o}. \quad (32)$$

Applying Euler's identity yield

$$\frac{2jH_{z_o}}{3\sqrt{\mu_o\epsilon_o}\Delta y}\left(\frac{9}{8}\sin\left(\frac{3}{2}\tilde{k}_y\Delta y\right)-\frac{1}{24}\sin\left(\frac{9}{2}\tilde{k}_y\Delta y\right)\right)=\Lambda E_{x_o}. \quad (33)$$

Thus,

$$E_{x_o}=\frac{2jH_{z_o}}{3\Lambda\sqrt{\mu_o\epsilon_o}\Delta y}\left(\frac{9}{8}\sin\left(\frac{3}{2}\tilde{k}_y\Delta y\right)-\frac{1}{24}\sin\left(\frac{9}{2}\tilde{k}_y\Delta y\right)\right). \quad (34)$$

Following the same procedure to arrive at equation (30) and (31) yield

$$E_{y_o}=-\frac{2jH_{z_o}}{3\Lambda\sqrt{\mu_o\epsilon_o}\Delta x}\left(\frac{9}{8}\sin\left(\frac{3}{2}\tilde{k}_x\Delta x\right)-\frac{1}{24}\sin\left(\frac{9}{2}\tilde{k}_x\Delta x\right)\right). \quad (35)$$

and

$$H_{z_o}=\frac{2j}{\Lambda\sqrt{\mu_o\epsilon_o}}\left(\frac{9}{8}\left(\frac{E_{x_o}}{3\Delta y}\sin\left(\frac{3}{2}\tilde{k}_y\Delta y\right)-\frac{E_{y_o}}{3\Delta x}\sin\left(\frac{3}{2}\tilde{k}_x\Delta x\right)\right)\right.$$
$$\left.-\frac{1}{24}\left(\frac{E_{x_o}}{3\Delta y}\sin\left(\frac{9}{2}\tilde{k}_y\Delta y\right)-\frac{E_{y_o}}{3\Delta x}\sin\left(\frac{9}{2}\tilde{k}_x\Delta x\right)\right)\right). \quad (36)$$

Substituting equations (34) and (35) into (36) yield

$$H_{z_o}=\frac{2j}{\Lambda\sqrt{\mu_o\epsilon_o}}\left(\frac{2jH_{z_o}}{(3\Delta y)^2\Lambda\sqrt{\mu_o\epsilon_o}}\left(\frac{9}{8}\sin\left(\frac{3}{2}\tilde{k}_y\Delta y\right)-\frac{1}{24}\sin\left(\frac{9}{2}\tilde{k}_y\Delta y\right)\right)^2\right.$$
$$\left.+\frac{2jH_{z_o}}{(3\Delta x)^2\Lambda\sqrt{\mu_o\epsilon_o}}\left(\frac{9}{8}\sin\left(\frac{3}{2}\tilde{k}_x\Delta x\right)-\frac{1}{24}\sin\left(\frac{9}{2}\tilde{k}_x\Delta x\right)\right)^2\right). \quad (37)$$

Simplifying (37) yield

$$\Lambda^2=-\frac{4}{\mu_o\epsilon_o}\left(\frac{1}{(3\Delta y)^2}\left(\frac{9}{8}\sin\left(\frac{3}{2}\tilde{k}_y\Delta y\right)-\frac{1}{24}\sin\left(\frac{9}{2}\tilde{k}_y\Delta y\right)\right)^2\right.$$
$$\left.+\frac{1}{(3\Delta x)^2}\left(\frac{9}{8}\sin\left(\frac{3}{2}\tilde{k}_x\Delta x\right)-\frac{1}{24}\sin\left(\frac{9}{2}\tilde{k}_x\Delta x\right)\right)^2\right). \quad (38)$$

Since sinus function follows the inequality $-1 \leq \sin\left(\frac{M}{2}\tilde{k}_x \Delta x\right) \leq 1$ and $-1 \leq \sin\left(\frac{M}{2}\tilde{k}_y \Delta y\right) \leq 1$, where $M = 3, 9$. Thus implies that Λ^2 negative, hence $\Re(\Lambda) = 0$. Thus

$$-2c\sqrt{\left(\frac{7}{18\Delta x}\right)^2 + \left(\frac{7}{18\Delta y}\right)^2} \leq \Im(\Lambda) \leq 2c\sqrt{\left(\frac{7}{18\Delta x}\right)^2 + \left(\frac{7}{18\Delta y}\right)^2}.$$

Using information in (25), gives $2c\sqrt{\left(\frac{7}{18\Delta x}\right)^2 + \left(\frac{7}{18\Delta y}\right)^2} \leq \frac{2}{\Delta t}$. Thus, $c\Delta t \leq \frac{1}{\sqrt{\left(\frac{7}{18\Delta x}\right)^2 + \left(\frac{7}{18\Delta y}\right)^2}}$. Taking $\Delta x = \Delta y = \Delta$, thus $c\Delta t \leq \frac{18\Delta}{7\sqrt{2}}$, where $c = \frac{1}{\mu_o \epsilon_o}$ is the speed of light in free space. This is the condition that has to be fulfilled in order to maintain UHSHO-FDTD method's stability.

Next, lets analysis the convergence criteria of the 2D UHSHO-FDTD method. Assume that

$$E_x|_i^n = E_{x_o} e^{j(\tilde{k}_x i\Delta x + \tilde{k}_y j\Delta y - wn\Delta t)}, \tag{39}$$

$$E_y|_i^n = E_{y_o} e^{j(\tilde{k}_x i\Delta x + \tilde{k}_y j\Delta y - wn\Delta t)} \tag{40}$$

and

$$H_z|_i^n = H_{z_o} e^{j(\tilde{k}_x i\Delta x + \tilde{k}_y j\Delta y - wn\Delta t)}. \tag{41}$$

Replacing (14) with (39) and (41) gives

$$\begin{aligned}
&\frac{E_{x_o} e^{j(\tilde{k}_x i\Delta x + \tilde{k}_y j\Delta y - w(n+\frac{1}{2})\Delta t)} - E_{x_o} e^{j(\tilde{k}_x i\Delta x + \tilde{k}_y j\Delta y - w(n-\frac{1}{2})\Delta t)}}{\Delta t} = \\
&\frac{1}{\sqrt{\mu_o \epsilon_o}}\left(\frac{9}{8}\left(\frac{H_{z_o} e^{j(\tilde{k}_x i\Delta x + \tilde{k}_y (j+\frac{3}{2})\Delta y - wn\Delta t)} - H_{z_o} e^{j(\tilde{k}_x i\Delta x + \tilde{k}_y (j-\frac{3}{2})\Delta y - wn\Delta t)}}{3\Delta y}\right)\right. \\
&\left. -\frac{1}{24}\left(\frac{H_{z_o} e^{j(\tilde{k}_x i\Delta x + \tilde{k}_y (j+\frac{9}{2})\Delta y - wn\Delta t)} - H_{z_o} e^{j(\tilde{k}_x i\Delta x + \tilde{k}_y (j-\frac{9}{2})\Delta y - wn\Delta t)}}{3\Delta y}\right)\right).
\end{aligned} \tag{42}$$

Simplifying equation (42) yield

$$\begin{aligned}
&\frac{E_{x_o}}{\Delta t}\left(e^{-j(w(\frac{1}{2})\Delta t)} - e^{j(w(\frac{1}{2})\Delta t)}\right) = \frac{H_{z_o}}{3\Delta y\sqrt{\mu_o \epsilon_o}}\left(\frac{9}{8}\left(e^{j(\tilde{k}_y(\frac{3}{2})\Delta y)} - e^{-j(\tilde{k}_y(\frac{3}{2})\Delta y)}\right)\right. \\
&\left. -\frac{1}{24}\left(e^{j(\tilde{k}_y(\frac{9}{2})\Delta y)} - e^{-j(\tilde{k}_y(\frac{9}{2})\Delta y)}\right)\right).
\end{aligned} \tag{43}$$

Thus,

$$E_{x_o} = \frac{\Delta t H_{z_o}}{3\Delta y\sqrt{\mu_o \epsilon_o}}\left(\frac{\frac{9}{8}\left(e^{j(\tilde{k}_y(\frac{3}{2})\Delta y)} - e^{-j(\tilde{k}_y(\frac{3}{2})\Delta y)}\right) - \frac{1}{24}\left(e^{j(\tilde{k}_y(\frac{9}{2})\Delta y)} - e^{-j(\tilde{k}_y(\frac{9}{2})\Delta y)}\right)}{e^{-j(w(\frac{1}{2})\Delta t)} - e^{j(w(\frac{1}{2})\Delta t)}}\right). \tag{44}$$

Using Euler's identity yield

$$E_{x_o} = -\frac{\Delta t H_{z_o}\left(\frac{9}{8}\sin(\frac{3\tilde{k}_y \Delta y}{2}) - \frac{1}{24}\sin(\frac{9\tilde{k}_y \Delta y}{2})\right)}{3\Delta y\sqrt{\mu_o \epsilon_o}\sin(\frac{w\Delta t}{2})}. \tag{45}$$

Following the same procedure to arrive at equations (15) and (16) yield

$$E_{y_o} = \frac{\Delta t H_{z_o}\left(\frac{9}{8}\sin(\frac{3\tilde{k}_x\Delta x}{2}) - \frac{1}{24}\sin(\frac{9\tilde{k}_x\Delta x}{2})\right)}{3\Delta x\sqrt{\mu_o\epsilon_o}\sin(\frac{w\Delta t}{2})} \tag{46}$$

and

$$H_{z_o}\sin(\frac{w\Delta t}{2}) = \frac{\Delta t}{\sqrt{\mu_o\epsilon_o}}\left(\frac{E_{y_o}}{3\Delta x}\left(\frac{9}{8}\sin(\frac{3\tilde{k}_x\Delta x}{2}) - \frac{1}{24}\sin(\frac{9\tilde{k}_x\Delta x}{2})\right)\right.$$
$$\left.-\frac{E_{x_o}}{3\Delta y}\left(\frac{9}{8}\sin(\frac{3\tilde{k}_y\Delta y}{2}) - \frac{1}{24}\sin(\frac{9\tilde{k}_y\Delta y}{2})\right)\right). \tag{47}$$

Substituting (45) and (46) into (47) yield

$$H_{z_o}\sin(\frac{w\Delta t}{2}) = \frac{\Delta t}{\sqrt{\mu_o\epsilon_o}}\left(\frac{\left(\frac{\Delta t H_{z_o}\left(\frac{9}{8}\sin(\frac{3\tilde{k}_x\Delta x}{2}) - \frac{1}{24}\sin(\frac{9\tilde{k}_x\Delta x}{2})\right)}{3\Delta x\sqrt{\mu_o\epsilon_o}\sin(\frac{w\Delta t}{2})}\right)}{3\Delta x}\right)$$
$$\left(\frac{9}{8}\sin\left(\frac{3\tilde{k}_x\Delta x}{2}\right) - \frac{1}{24}\sin\left(\frac{9\tilde{k}_x\Delta x}{2}\right)\right)$$
$$-\frac{\Delta t}{\sqrt{\mu_o\epsilon_o}}\left(\frac{\left(-\frac{\Delta t H_{z_o}\left(\frac{9}{8}\sin(\frac{3\tilde{k}_y\Delta y}{2}) - \frac{1}{24}\sin(\frac{9\tilde{k}_y\Delta y}{2})\right)}{3\Delta y\sqrt{\mu_o\epsilon_o}\sin(\frac{w\Delta t}{2})}\right)}{3\Delta y}\right)$$
$$\left(\frac{9}{8}\sin\left(\frac{3\tilde{k}_y\Delta y}{2}\right) - \frac{1}{24}\sin\left(\frac{9\tilde{k}_y\Delta y}{2}\right)\right). \tag{48}$$

Simplifying (48) yield

$$\left(\frac{1}{c\Delta t}\sin\left(\frac{w\Delta t}{2}\right)\right)^2 = \left(\frac{1}{3\Delta x}\left(\frac{9}{8}\sin\left(\frac{3\tilde{k}_x\Delta x}{2}\right) - \frac{1}{24}\sin\left(\frac{9\tilde{k}_x\Delta x}{2}\right)\right)\right)^2$$
$$+\left(\frac{1}{3\Delta y}\left(\frac{9}{8}\sin\left(\frac{3\tilde{k}_y\Delta y}{2}\right) - \frac{1}{24}\sin\left(\frac{9\tilde{k}_y\Delta y}{2}\right)\right)\right)^2, \tag{49}$$

where $c = \frac{1}{\mu\epsilon}$. Taking $\Delta x = \Delta y = \Delta \to 0$, therefore $\Delta t \to 0$ since $\Delta t \le \frac{18\Delta}{7c\sqrt{2}}$. Hence equation (49) becomes

$$\begin{aligned}\left(\frac{w\Delta t}{2c\Delta t}\right)^2 &= \left(\frac{9}{8(3\Delta x)}\frac{3\tilde{k}_x\Delta x}{2} - \frac{1}{24(3\Delta x)}\frac{9\tilde{k}_x\Delta x}{2}\right)^2 \\ &\quad + \left(\frac{9}{8(3\Delta y)}\frac{3\tilde{k}_y\Delta y}{2} - \frac{1}{24(3\Delta y)}\frac{9\tilde{k}_y\Delta y}{2}\right)^2 \\ \to \left(\frac{w}{2c}\right)^2 &= \left(\frac{\tilde{k}_x}{2}\right)^2 + \left(\frac{\tilde{k}_y}{2}\right)^2. \end{aligned} \tag{50}$$

Suppose that $\tilde{k}, \tilde{k}_x$ and $\tilde{k}_y$ is related by a perpendicular triangle. Thus, $\tilde{k}^2 = \tilde{k}_x^2 + \tilde{k}_y^2$. Using this relationship, (50) can be rewritten as

$$\begin{aligned}\left(\frac{w}{2c}\right)^2 &\approx \left(\frac{\tilde{k}}{2}\right)^2 \\ \tilde{k} &\approx \pm\frac{w}{c} \\ &= k.\end{aligned}$$

which is the exact solution. This completes the analysis.

4 Numerical Experiments

The performance of proposed algorithms were conducted by executing some numerical experiments. The experiments were carried out using a Gaussian source at the middle of solution domain. The source will radiates equally in all directions in 2D solution domain. The sample size considered are 0.09, 0.18, 0.27 and 0.36 millions. The execution time was taken to be 500 time steps. The tests were carried out using a Symmetrical Multi-processor (SMP) machine, Sun Fire V1280 with Ultra SPARC III Cu processors. Each processor has the speed of 900 MHz.

Fig. 3 shows the comparison of elapsed time between methods. The figure shows that although both UHSHO-FDTD methods use fourth order approximation, which is more complex than approximation use in parallel FDTD (PFDTD), implementing the complexity reduction approach resulted in the overall complexity of both UHSHO-FDTD methods less than complexity for PFDTD. As the consequence, both UHSHO-FDTD methods solve problem faster than PFDTD. Applying complexity reduction approach with fourth order approximation save elapsed time between 58% to 70% for UHSHO-FDTD (DD) method and 57% to 69% for UHSHO-FDTD (TD) method compared to PFDTD method (depends on the sample size used and time level, bigger sample size leads to more saving in elapsed time and using time level approaching infinity, the more saving will be gained). However, these elapsed time for both UHSHO-FDTD methods are still higher than both UHSLO-FDTD methods [3].

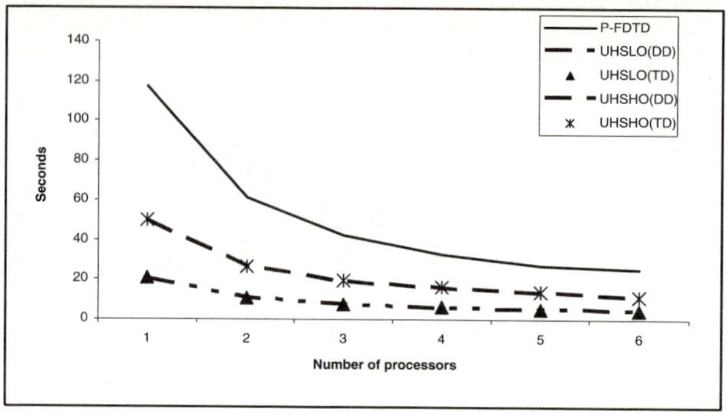

Fig. 3. Comparison of elapsed time

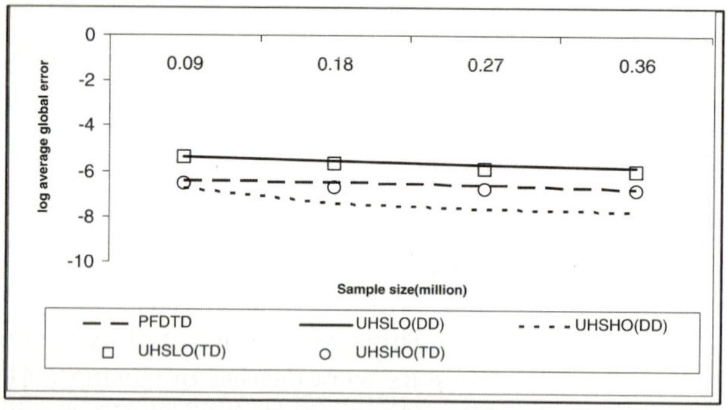

Fig. 4. Comparison of accuracy

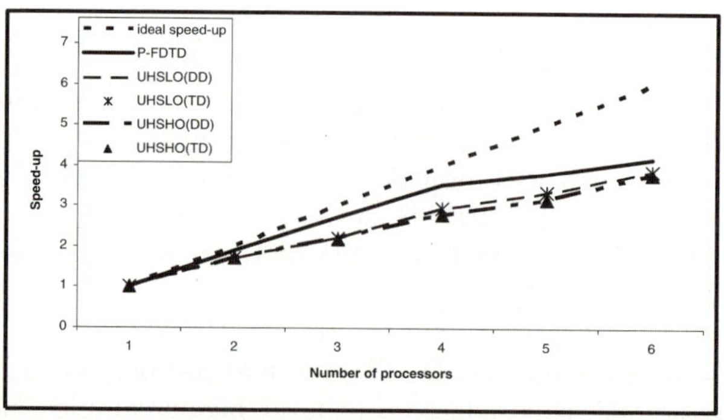

Fig. 5. Comparison of speed-up between methods for sample size 0.18 millions

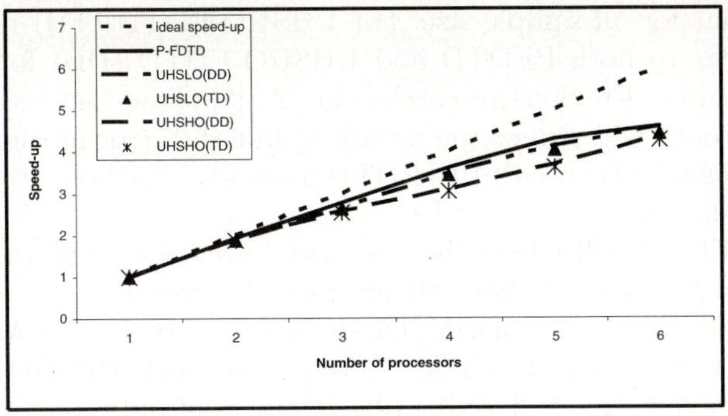

Fig. 6. Comparison of speed-up between methods for sample size 0.36 millions

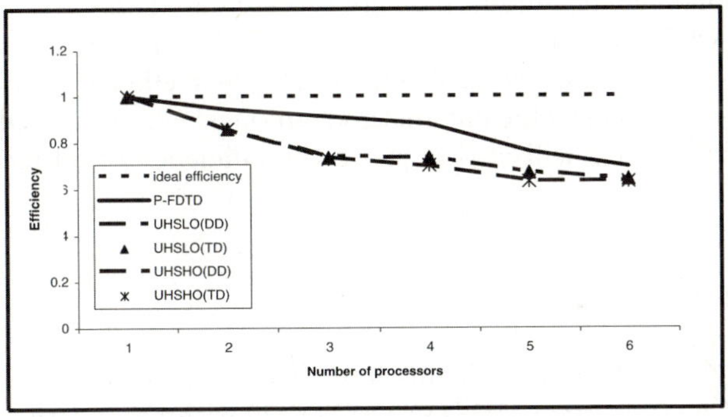

Fig. 7. Comparison of efficiency between methods for sample size 0.18 millions

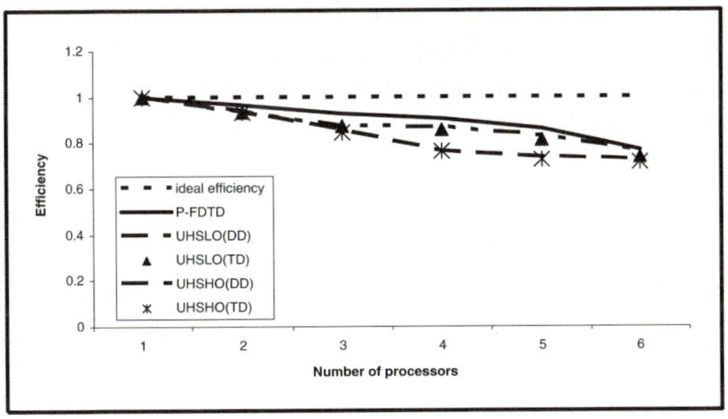

Fig. 8. Comparison of efficiency between methods for sample size 0.36 millions

Fig. 4 shows that the accuracy for both PFDTD and UHSHO-FDTD (DD) is almost similar for all sample size. For UHSHO-FDTD (TD) method, its accuracy is similar to both PFDTD and UHSHO-FDTD (DD) for high number of sample size only. Previous researches in [2], [3] show that using complexity reduction approach will reduce the accuracy, but the error introduce by $3h$ size discretization used in both UHSHO-FDTD methods have been recover by fourth order approximation.

However, PFDTD still posses the best speed-up amongst all compared methods. Increasing the sample size will increase the speed-up for each methods. Figs. 5 and 6 show these scenario. These figures also show that the speed-up of both UHSLO-FDTD methods are better than both UHSHO-FDTD methods. These behavior can be described by numbers of data involved in the data routing processes. Only one vectors of electric and magnetic fields involved in the data routing process in each side of sub-domain for PFDTD and both UHSLO-FDTD methods, while two vectors for both UHSHO-FDTD methods. The scenario of PFDTD have better speed-up than both UHSLO-FDTD methods can be explained by numbers of nodes calculated in each sub domain for PFDTD is much higher than both UHSLO-FDTD methods. Since the speed-up metrics has close relationship to efficiency, their efficiency values also show that PFDTD is the most efficient method amongst all compared methods. Increasing the sample size will also improve the efficiency of all methods (Figs. 7 and 8.)

5 Conclusions

In this paper, five parallel algorithms have been implemented, one is the existing PFDTD algorithm, two UHSLO-FDTD algorithms and the other two are the proposed UHSHO-FDTD algorithms. These algorithms have been implemented for solving the time domain free space Maxwell equations on Sun Fire V1280. Performances for all algorithms are studied using Gaussian source radiating in a 2D domain. From the result, it is observed that the implementations of complexity reduction approach with fourth order approximation to FDTD algorithm provide significant reduction in elapsed time. The elapsed time saving is becoming more significant for large computational domains. Generally, we can conclude that both UHSHO-FDTD algorithms are much faster than the PFDTD algorithm.

Acknowledgements. We are very grateful to the Universiti Kebangsaan Malaysia and the Malaysian government for the first author study leave. Special thanks also for Institute of Mathematical Research, INSPEM UPM for the permission of using the Sun Fire V1280 machine. We are indebted to GUP research grantt UKM-GUP-TMK-07-01-035 for financial support. Last but not least to editors for their comments.

References

1. Yee, K.S.: Numerical Solution of Initial Boundary Value Problems Involving Maxwell's Equations in Isotropic Media. IEEE Transaction on Antennas and Propagation 14, 302–307 (1996)
2. Hasan, M.K., Othman, M., Abbas, Z., Sulaiman, J., Desa, J.: New Algorithms for Time-domain Maxwell Equations. Int. J. Math. and Comp. Sci. 1(3), 271–288 (2006)
3. Hasan, M.K., Othman, M., Abbas, Z., Sulaiman, J., Ahmad, F.: Parallel Solution of High Speed Low Order FDTD on 2D Free Space Wave Propagation. In: Gervasi, O., Gavrilova, M.L. (eds.) ICCSA 2007, Part II. LNCS, vol. 4706, pp. 13–24. Springer, Heidelberg (2007)
4. Taflove, A.: Computational Electrodynamics: The Finite-Difference Time-Domain Method. Artech House, Boston (1995)
5. Ramadan, O., Akaydin, O., Salamah, M., Oztoprak, A.Y.: Parallel Implementation of the Wave-Equation Finite-Difference Time-Domain Method Using the Message Passing Interface. In: Aykanat, C., Dayar, T., Körpeoğlu, İ. (eds.) ISCIS 2004. LNCS, vol. 3280, pp. 810–818. Springer, Heidelberg (2004)
6. Sahimi, M.S., Khatim, M.: The Reduced Iterative Alternating Decomposition Explicit (RIADE) Method for the Diffusion Equation. Pertanika Journal of Science and Technolology 9(1), 13–20 (2001)
7. Abdullah, A.R.: The Four Point Explicit Decoupled Group (EDG) Method: A Fast Poisson Solver. Int. J. of Comp. Math. 38, 61–70 (1991)
8. Ibrahim, A., Abdullah, A.R.: Solving the Two-dimensional Diffusion Equation by the Four Point Explicit Decoupled Group (EDG) Iterative Method. Int. J. Comp. Math. 58, 253–256 (1995)
9. Abdullah, A.R., Ali, N.H.M.: A Comparative Study of Parallel Strategies for the Solution of Elliptic PDE's. Parallel Algorithms and Aplications 10, 93–103 (1996)
10. Yousif, W.S., Evans, D.J.: Explicit De-coupled Group Iterative Methods and Their Implementations. Parallel Algorithms and Applications 7, 53–71 (2001)
11. Othman, M., Abdullah, A.R.: An Efficient Four Point Modified Explicit Group Poisson Solver. Int. J. of Comp. Math. 76, 203–217 (2000)
12. Sulaiman, J., Hasan, M.K., Othman, M.: The Half-Sweep Iterative Alternating Decomposition Explicit (HSIADE) Method for Diffusion Equation. In: Zhang, J., He, J.-H., Fu, Y. (eds.) CIS 2004. LNCS, vol. 3314, pp. 57–63. Springer, Heidelberg (2004)
13. Sulaiman, J., Othman, M., Hasan, M.K.: Quarter-sweep Iterative Alternating Decomposition Explicit Algorithm Applied to Diffusion Equations. Int. J. of Comp. Math. 81(12), 1559–1565 (2004)
14. Roosta, S.H.: Parallel Processing and Parallel Algorithms: Theory and Computation. Springer, Heidelberg (1999)
15. Varadarajan, V., Mittra, R.: Finite-difference Time-domain (FDTD) Analysis using Distributed Computing. IEEE Microwave and Guided Wave Letters 4, 144–145 (1994)

Collocation–Based Two Step Runge–Kutta Methods for Ordinary Differential Equations

R. D'Ambrosio[1], M. Ferro[2], and B. Paternoster[1]

[1] Dipartimento di Matematica e Informatica,
Università degli Studi di Salerno,
84084 Fisciano (Sa) - Italy
{rdambrosio,beapat}@unisa.it
[2] Dipartimento di Matematica e Applicazioni,
Università degli Studi di Napoli - "Federico II",
80126 Napoli - Italy
maria.ferro@dma.unina.it

Abstract. We introduce a general family of collocation based two–step Runge–Kutta methods for the numerical integration of Ordinary Differential Equations depending on the stage values at two consecutive step points. We describe two constructive techniques and analyze the properties of the resulting methods.

1 Introduction

We are concerned with the numerical solution of an Initial Value Problem

$$\begin{cases} y'(x) = f(x, y(x)) \\ y(x_0) = y_0 \end{cases} \tag{1}$$

through multistep collocation methods.

Let us assume that $x \in I := [x_0, X]$, and $f : I \times \Omega \subset \mathbb{R}^d \to \mathbb{R}^d$ is sufficiently smooth. Let $I_h = \{x_h : x_0 < x_1 < ... < x_N = X\}$ be a uniform grid on I, where $h = \frac{X - x_0}{N}$ is the stepsize. The new extension falls into the class of General Two Step Runge–Kutta methods (TSRK), introduced in [7], depending on the stage values at two consecutive step points, that is, in the autonomous case,

$$Y_n^j = u_j y_{n-1} + (1 - u_j) y_n + h \sum_{s=1}^{m} [a_{js} f(Y_{n-1}^s) + b_{js} f(Y_n^s)] \tag{2}$$

$$y_{n+1} = \theta y_{n-1} + (1 - \theta) y_n + h \sum_{j=1}^{m} [v_j f(Y_{n-1}^j) + w_j f(Y_n^j)]. \tag{3}$$

Here θ, v_j, w_j, a_{js}, b_{js}, $j, s, = 1, \ldots, m$ are the coefficients of the methods, which can be represented by the following array:

$$\begin{array}{c|c|c} \mathbf{u} & \mathbf{A} & \mathbf{B} \\ \hline \theta & \mathbf{v}^{\mathbf{T}} & \mathbf{w}^{\mathbf{T}} \end{array} \tag{4}$$

O. Gervasi et al. (Eds.): ICCSA 2008, Part II, LNCS 5073, pp. 736–751, 2008.

These methods belong to the class of General Linear Methods, introduced by Butcher [1] with the aim to provide an unifying approach to analyze the classical subjects of consistency, convergence and stability of numerical methods for Ordinary Differential Equations (ODEs), which represents an active and increasing area of investigation [6].

The reason of interest in methods TSRK (2, 3) lies in the fact that, advancing from x_n to x_{n+1}, we only have to compute Y_n, because Y_{n-1} was already evaluated in the previous step. Therefore the computational cost of the method depends on the matrix B, while the matrix A adds extra degrees of freedom, without need for extra function evaluations. Therefore it is of interest to investigate the class of collocation based methods within the class of TSRK methods, in order to derive continuous methods with higher order of convergence, in comparison with classical collocation methods.

In Section 2 we recall the classical collocation methods, and also the extension to multistep methods already considered in the literature. In Section 3 we extend the idea of multistep collocations methods using two different constructive techniques, considering the two–step case. By adding some extra collocation conditions, the resulting methods depend on the stage values at two consecutive step points and provide a uniform approximation to the solution of order $2m+1$, where m is the number of stages. In Section 4 we analyze the linear stability properties of the methods. In Section 5 we describe some numerical experiments.

2 Classical One Step and Multistep Collocation Methods

The idea of collocation is old and well known in Numerical Analysis [9,4,5]. In order to advance from x_n to x_{n+1}, an algebraic polynomial $P(x)$ is constructed, which interpolates the numerical solution in the step point x_n, and satisfies the ODEs in the points $x_n + c_i h$, where $\{c_1, c_2, ..., c_m\}$ are m real numbers (typically between 0 and 1), that is

$$\begin{cases} P(x_n) = y_n, \\ P'(x_n + c_i h) = f(x_n + c_i h, P(x_n + c_i h)), \quad i = 1, 2, ..., m. \end{cases} \tag{5}$$

The solution in x_{n+1} is given by

$$y_{n+1} = P(x_{n+1}). \tag{6}$$

It is known that one step collocation methods are a subset of implicit Runge-Kutta methods

$$\begin{array}{c|cccc} c_1 & a_{11} & a_{12} & & a_{1m} \\ c_2 & a_{21} & a_{22} & & a_{2m} \\ \\ \dots & \dots & \dots & & \dots \\ \\ c_m & a_{m1} & a_{m2} & & a_{mm} \\ \hline & b_1 & b_2 & & b_m \end{array}$$

where

$$a_{ij} = \int_0^{c_i} L_j(t)dt, \qquad b_j = \int_0^1 L_j(t)dt, \qquad i,j = 1,2,...,m \tag{7}$$

and $L_j(t)$, $j = 1,...,m$, are fundamental Lagrange polynomials. Moreover the maximum attainable order is at most $2m$, and it is obtained by using Gaussian collocation points [9,4]. Guillou and Soulé introduced multistep collocation methods [3], by adding interpolation conditions in the previous k step points, so that the collocation polynomial is defined by

$$\begin{cases} P(x_{n-i}) = y_{n-i} \quad i = 0,1,...,k-1, \\ P'(x_n + c_jh) = f(x_n + c_jh, P(x_n + c_jh)) \quad j = 1,...,m. \end{cases} \tag{8}$$

The numerical solution is then

$$y_{n+1} = P(x_{n+1}) \tag{9}$$

In [5] it is proved that this method is equivalent to a multistep Runge–Kutta method, and the points which guarantee superconvergence are called Radau points. Lie and Norsett analyze the order of the resulting methods [10].

3 General Two–Step Collocation Methods

We now extend the idea of multistep collocation methods, by considering the case of $k = 2$, and by adding some extra collocation conditions, so that the resulting methods depend on the stage values at two consecutive step points. We extend the technique used by Hairer and Wanner [5]. In more detail, the collocation polynomial is then defined by the following conditions:

$$\begin{cases} P(x_{n-1}) = y_{n-1}, \\ P(x_n) = y_n, \\ P'(x_{n-1} + c_ih) = f(x_{n-1} + c_ih, P(x_{n-1} + c_ih)), \quad i = 1,2,...,m, \\ P'(x_n + c_ih) = f(x_n + c_ih, P(x_n + c_ih)), \quad i = 1,2,...,m. \end{cases} \tag{10}$$

The previous problem constitutes a Hermite interpolation problem with incomplete data (because the function values $P(x_{n-1} + c_jh)$, $P(x_n + c_jh)$ are missing). In order to compute the collocation polynomial we introduce the generalized Lagrange basis

$$\{\overline{\phi}_i(x), \overline{\chi}_j(x), \overline{\psi}_j(x), \quad i = 0,1, \quad j = 1,2,...,m\}$$

such that the collocation polynomial is expressed in the following way:

$$P(x) = \overline{\phi}_0(x)y_{n-1} + \overline{\phi}_1(x)y_n + \\ + h\sum_{j=1}^{m}\left[\overline{\chi}_j(x)P'(x_{n-1}+c_jh) + \overline{\psi}_j(x)P'(x_n+c_jh)\right], \tag{11}$$

Introducing the dimensionless coordinate $t = \frac{x-x_n}{h}$, the collocation polynomial takes the form

$$P(x_n+th) = \phi_0(t)y_{n-1} + \phi_1(t)y_n + \\ + h\sum_{j=1}^{m}\left[\chi_j(t)P'(x_{n-1}+c_jh) + \psi_j(t)P'(x_n+c_jh)\right]. \tag{12}$$

We must exhibit the expression of the basis functions $\phi_i(t)$ $(i = 0, 1)$, $\psi_j(t)$ and $\chi_j(t)$ $(j = 1, 2, ...m)$. They can be obtained by applying the interpolation conditions

$$\begin{array}{l} \phi_0(t_0) = 1,\ \phi_0(t_1) = 0,\ \phi_1(t_0) = 0,\ \phi_1(t_1) = 1, \\ \chi_i(t_l) = 0,\ \ \psi_i(t_l) = 0, \end{array} \tag{13}$$

and the collocation ones

$$\begin{array}{l} \phi_0'(c_i-1) = 0,\ \ \ \phi_0'(c_i) = 0,\ \phi_1'(c_i-1) = 0,\ \phi_1'(c_i) = 0, \\ \chi_j'(c_i-1) = \delta_{ij},\ \chi_j'(c_i) = 0,\ \psi_j'(c_i-1) = 0,\ \psi_j'(c_i) = \delta_{ij}, \end{array} \tag{14}$$

where $t_0 = -1$, $t_1 = 0$, $l = 0, 1$, $i, j = 1, 2, \ldots, m$.

Theorem 1. *The method defined by (10) is equivalent to a two–step Runge–Kutta method [7]*

$$\begin{aligned} Y_n^j = u_jy_{n-1} + (1-u_j)y_n &+ h\sum_{s=1}^{m}[a_{js}f(x_{n-1}+c_sh, Y_{n-1}^s) \\ &+ b_{js}f(x_n+c_sh, Y_n^s)],\quad j = 1, ..., m, \\ y_{n+1} = \theta y_{n-1} + (1-\theta)y_n &+ h\sum_{j=1}^{m}[v_jf(x_{n-1}+c_jh, Y_{n-1}^j) \\ &+ w_jf(x_n+c_jh, Y_n^j)]. \end{aligned}$$

where

$$\begin{array}{l} \theta = \phi_0(1),\ \ \ \ v_j = \chi_j(1),\ \ w_j = \psi_j(1), \\ b_{js} = \psi_j(c_s),\ u_j = \phi_0(c_j),\ a_{js} = \chi_j(c_s),\ j, s = 1, ..., m. \end{array}$$

Proof. In order to compute the polynomials $\phi_i(t)$, $\chi_j(t)$, $\psi_j(t)$, we follow the procedure indicated in [5]. We first expand the basis polynomials in the following form

$$\phi_i(t) = \sum_{l=0}^{2m+1} d_l^{(i)} t^l, \qquad i = 1, 2 \tag{15}$$

$$\chi_j(t) = \sum_{l=0}^{2m+1} p_l^{(j)} t^l, \qquad \psi_j(t) = \sum_{l=0}^{2m+1} q_l^{(j)} t^l \qquad j = 1, ..., m \tag{16}$$

Applying (13), (14) in (15), (16) we obtain the following $2m+2$ linear systems:

$$Hd^{(i)} = N_1, \quad i = 1, 2, \tag{17}$$
$$Hp^{(i)} = N_2, \quad Hq^{(i)} = N_3 \quad i = 1, ..., m, \tag{18}$$

where H is the coefficient matrix

$$H = \begin{pmatrix} 1 & t_0 & t_0^2 & ... & ... & t_0^{2m+1} \\ 1 & t_1 & t_1^2 & ... & ... & t_1^{2m+1} \\ 0 & 1 & 2(c_1-1) & 3(c_1-1)^2 & ... & (2m+1)(c_1-1)^{2m} \\ ... & ... & & & & \\ ... & ... & & & & \\ 0 & 1 & 2(c_m-1) & 3(c_m-1)^2 & ... & (2m+1)(c_m-1)^{2m} \\ 0 & 1 & 2c_1 & 3c_1^2 & ... & (2m+1)c_1^{2m} \\ ... & ... & & & & \\ ... & ... & & & & \\ 0 & 1 & 2c_m & 3c_m^2 & ... & (2m+1)c_m^{2m} \end{pmatrix}.$$

N_1, N_2, N_3 are the following vectors

$$N_1 = [\delta_{i1}, \delta_{i2}, 0, ..., 0]^T, \quad N_2 = [0, 0, \delta_{i1}, ..., \delta_{im}, 0, ..., 0]^T$$
$$N_3 = [0, ..., 0, \delta_{i1}, ..., \delta_{im}]^T$$

and $d^{(i)}, p^{(i)}, q^{(i)}$ are the unknowns vectors. These linear systems can be solved (apart from some exceptional values of the collocation abscissa), giving the expressions of the collocation polynomial $P(x)$. □

Each linear system arising in the construction of these methods is nonsingular, because its coefficient matrix is of Vandermonde type. We omit the details of the proof, simply because it uses the well-known technique applied in computing the determinant of the Vandermonde matrix [11].

3.1 Another Constructive Technique

Let us extend the technique used by Lie and Norsett [10] in order to derive general two–step Runge–Kutta methods of collocation type.

Theorem 2. *The method defined by (10) is equivalent to a two-step Runge–Kutta method having the following form:*

$$Y_n^j = \phi_0(c_s)y_{n-1} + \phi_1(c_s)y_n + h\sum_{s=1}^{m}[\chi_j(c_s)f(x_{n-1}+c_sh, Y_{n-1}^s) + \psi_j(c_s)f(x_n+c_sh, Y_n^s)], \quad j=1,\ldots,m, \tag{19}$$

$$y_{n+1} = \phi_0(1)y_{n-1} + \phi_1(1)y_n + h\sum_{j=1}^{m}[\chi_j(1)f(x_{n-1}+c_jh, Y_{n-1}^j) + \psi_j(1)f(x_n+c_jh, Y_n^j)], \tag{20}$$

where

$$\psi_j(t) = \int_0^t l_j(\tau)d\tau - \frac{\int_{-1}^0 l_j(\tau)d\tau}{\int_{-1}^0 M(\tau)d\tau}\int_0^t M(\tau)d\tau, \quad j=1,\ldots,m, \tag{21}$$

$$\chi_j(t) = \int_0^t \tilde{l}_j(\tau)d\tau - \frac{\int_{-1}^0 \tilde{l}_j(\tau)d\tau}{\int_{-1}^0 M(\tau)d\tau}\int_0^t M(\tau)d\tau, \quad j=1,\ldots,m, \tag{22}$$

$$\phi_0(t) = -\frac{\int_0^t M(\tau)d\tau}{\int_{-1}^0 M(\tau)d\tau}, \tag{23}$$

$$\phi_1(t) = 1 + \frac{\int_0^t M(\tau)d\tau}{\int_{-1}^0 M(\tau)d\tau}. \tag{24}$$

with

$$l_i(t) = \prod_{j=1, j\neq i}^{2m}\frac{t-d_j}{d_i-d_j}, \qquad M(t) = \prod_{j=1}^{2m}(t-d_j), \qquad \begin{cases} d_i = c_i \\ d_{m+i} = c_i - 1, \end{cases} \quad i=1,2,\ldots,m \tag{25}$$

$$\tilde{l}_j(t) = \prod_{i=1, i\neq j}^{2m}\frac{t-e_i}{e_j-e_i}, \qquad \begin{cases} e_i = c_i - 1 \\ e_{m+i} = c_i, \end{cases} \qquad = 1,2,\ldots,m$$

Proof. To prove that the general two–step collocation method defined by (10) is equivalent to the TSRK method (19)-(20), we must exhibit the form of the basis polynomials $\phi_i(x)$ $(i=0,1)$, $\psi_j(x)$ and $\chi_j(x)$ $(j=1,2,\ldots m)$. We use the scaled time variable t.

- We first consider $\psi_j(t)$, $j=1,2,\ldots m$.

The interpolation and collocation conditions on ψ_j are

$$\begin{aligned} \psi_j(-r) &= 0, \qquad r = 0, 1 \\ \psi_j'(c_s - 1) &= 0, \qquad s = 1, ..., m \\ \psi_j'(c_s) &= \delta_{js}, \qquad s = 1, ..., m. \end{aligned}$$

We denote the collocation knots in the following way:

$$\begin{cases} d_i = c_i \\ d_{m+i} = c_i - 1, \qquad i = 1, 2, ..., m. \end{cases}$$

Therefore the previous conditions on ψ_j are

$$\psi_j(-r) = 0, \qquad r = 0, 1, \tag{26}$$

$$\psi_j'(d_s) = \Delta_{js}, \qquad s = 1, ..., 2m. \tag{27}$$

where

$$\Delta_{js} = \begin{cases} \delta_{js}, & \text{if } 1 \leq s \leq m \\ 0, & \text{else.} \end{cases} \tag{28}$$

Following [10], the collocation conditions can be satisfied by a polynomial of the form

$$\psi_j'(t) = l_j(t) + \frac{\alpha_0}{a_j} M(t) \tag{29}$$

with

$$l_j(t) = \prod_{i=1, i \neq j}^{2m} \frac{t - d_i}{d_j - d_i}, \qquad M(t) = \prod_{j=1}^{2m} (t - d_j)$$

$$a_j = \prod_{s=1, s \neq j}^{2m} (d_j - d_s), \qquad \alpha_0 \in \mathbb{R}.$$

Setting $\bar{\alpha}_0 = \frac{\alpha_0}{a_j}$, equation (29) becomes

$$\psi_j'(t) = l_j(t) + \bar{\alpha}_0 M(t). \tag{30}$$

Integrating the last equation, we find

$$\psi_j(t) = \int_0^t l_j(\tau) d\tau + \bar{\alpha}_0 \int_0^t M(\tau) d\tau. \tag{31}$$

Imposing the interpolation conditions (26), we compute $\bar{\alpha}_0$ by solving the linear equation

$$\bar{\alpha}_0 \int_{-1}^0 M(\tau) d\tau = - \int_{-1}^0 l_j(\tau) d\tau. \tag{32}$$

For $\psi_j(t)$ we have

$$\psi_j(t) = \int_0^t l_j(\tau) d\tau - \frac{\int_{-1}^0 l_j(\tau) d\tau}{\int_{-1}^0 M(\tau) d\tau} \int_0^t M(\tau) d\tau. \tag{33}$$

- We then consider $\chi_j(t)$, $j = 1, 2, ...m$.

The proof is the same as above, putting

$$\begin{cases} e_i = c_i - 1 \\ e_{m+i} = c_i, \qquad = 1, 2, ..., m \end{cases}$$

instead of d_i, $i = 1, ..., 2m$.

- We now consider $\phi_i(t)$ $(i = 0, 1)$.

Using collocation knots d_j, for ϕ_i we have:

$$\phi_i(-r) = 1 - \delta_{ir} \qquad r = 0, 1, \tag{34}$$

$$\phi_i'(d_j) = 0 \qquad j = 1, ..., m. \tag{35}$$

The previous conditions are then verified by

$$\phi_i(t) = \gamma_0^{(i)} + \gamma_1^{(i)} \int_0^t M(\tau)d\tau,$$

where $M(t) = \prod_{j=1}^{2m}(t - d_j)$ and $\gamma_j^{(i)} \in \mathbb{R}$. First of all we consider ϕ_0:

$$\phi_0(t) = \gamma_0^{(0)} + \gamma_1^{(0)} \int_0^t M(\tau)d\tau.$$

We know that

$$0 = \phi_0(0) = \gamma_0^{(0)} + \gamma_1^{(0)} \int_0^0 M(\tau)d\tau$$

so we have $\gamma_0^{(0)} = 0$. Moreover, it is

$$1 = \phi_0(-1) = \gamma_1^{(0)} \int_0^{-1} M(\tau)d\tau = -\gamma_1^{(0)} \int_{-1}^0 M(\tau)d\tau \tag{36}$$

therefore we obtain

$$\gamma_1^{(0)} = -\frac{1}{\int_{-1}^0 M(\tau)d\tau}. \tag{37}$$

To conclude the proof we must exhibit the basis polynomial $\phi_1(t)$. As for $\phi_0(t)$, we impose the interpolation conditions (34), obtaining

$$1 = \phi_1(0) = \gamma_0^{(1)}$$

while, applying the collocation ones (35), it is

$$0 = \phi_1(-1) = \gamma_0^{(1)} + \gamma_1^{(1)} \int_0^{-1} M(\tau)d\tau = 1 - \gamma_1^{(1)} \int_{-1}^0 M(\tau)d\tau. \tag{38}$$

So, we arrive to the following expression of the coefficients of $\phi_1(t)$:

$$\gamma_0^{(1)} = 1, \qquad \gamma_1^{(1)} = \frac{1}{\int_{-1}^0 M(\tau)d\tau}. \tag{39}$$

□

3.2 Order Results

It is now possible to prove that $P(x)$ provides a uniform approximation to the solution of order $2m+1$ for any choice of the collocation abscissa $\{c_1, c_2, ..., c_m\}$. In particular, the following theorem deals with the order conditions.

Theorem 3. *Let*

$$G_i = det \begin{pmatrix} \int_{-1}^{0} M(\tau)\tau^i d\tau & \int_{-1}^{0} M(\tau)\tau^{i+1} d\tau \\ \int_{0}^{1} M(\tau)\tau^i d\tau & \int_{0}^{1} M(\tau)\tau^{i+1} d\tau \end{pmatrix}. \tag{40}$$

Then the general m stages collocation method has order $2m+\sigma$ if and only if $G_i = 0$ for $i = 0, 1, \ldots, \sigma - 1$.

Proof. The proof uses Alekseev–Gröbner theorem and the knots d_i defined by (25). Technical details are given in [10]. □

The following theorem concerns the possibility to achieve superconvergence.

Theorem 4. *The maximum attainable order of a two–steps and m stages collocation method arising from (10) is $3m$.*

Proof. The set of conditions $G_i = 0$ of theorem 3 gives a nonlinear system in the unknowns $c_1, c_2, ..., c_m$. Let us consider a subset $\Gamma(\zeta)$ of s equations of the system, where $\Gamma(\zeta) : \mathbb{R}^m \to \mathbb{R}^s$, which has unique solution if and only if $s = m$. As a consequence, σ can be at most equal to m and, for this reason, the maximum attainable order is $3m$. □

4 Linear Stability Analysis

From [6], we recall that a TSRK method (2)-(3)

$$Y_i^{[n]} = (1-u_i)y_n + u_i y_{n-1} + h\sum_{j=1}^{m}(a_{ij} f(Y_j^{[n]}) + b_{ij} f(Y_j^{[n-1]}))$$

$$y_{n+1} = (1-\theta)y_n + \theta y_{n-1} + h\sum_{j=1}^{m}(v_j f(Y_j^{[n]}) + w_j f(Y_j^{[n-1]}))$$

is a General Linear Method of the form

$$\left[\begin{array}{c} Y^{[n]} \\ \hline y_{n+1} \\ y_n \\ hf(Y^{[n]}) \end{array}\right] = \left[\begin{array}{c|ccc} A & e-u & u & B \\ \hline v^T & 1-\theta & \theta & w^T \\ 0 & 1 & 0 & \mathbf{0} \\ \mathbf{I} & \mathbf{0} & 0 & \mathbf{0} \end{array}\right] \left[\begin{array}{c} hf(Y^{[n]}) \\ \hline y_n \\ y_{n-1} \\ hf(Y^{[n-1]}) \end{array}\right], \tag{41}$$

where **I** is the identity matrix whose dimension is the number of stages and **0** is the zero matrix or vector of appropriate dimensions. The stability (or amplification) matrix $M(z)$ associated to this method takes the form (compare with [6])

$$M(z) = \begin{bmatrix} M_{11} & M_{12} & M_{13} \\ 1 & 0 & \mathbf{0} \\ zS(z)(\mathbf{e}-u) & zS(z)u & zS(z)B \end{bmatrix} \tag{42}$$

where

$$\begin{aligned} M_{11} &= 1-\theta + zv^T S(z)(\mathbf{e}-u), \\ M_{12} &= \theta + zv^T S(z)u, \\ M_{13} &= w^T + zv^T S(z)B, \end{aligned}$$

$S(z) = (I - zA)^{-1}$, $\mathbf{e}$ is the vector with unitary entries. Let us consider our class of methods for $m = 1$. The matrix representing these methods has the form

$$\left[\begin{array}{c|ccc} a & 1-u & u & b \\ \hline v & 1-\theta & \theta & w \\ 0 & 1 & 0 & 0 \\ 1 & 0 & 0 & 0 \end{array}\right] = \left[\begin{array}{c|ccc} \chi(c) & 1-\phi_0(c) & \phi_0(c) & \psi(c) \\ \hline \chi(1) & 1-\phi_0(1) & \phi_0(1) & \psi(1) \\ 0 & 1 & 0 & 0 \\ 1 & 0 & 0 & 0 \end{array}\right] \tag{43}$$

and the amplification matrix has the form

$$M(z) = \begin{bmatrix} M_{11} & M_{12} & M_{13} \\ 1 & 0 & 0 \\ zS(z)(1-\phi_0(c)) & zS(z)\phi_0(c) & zS(z)\psi(c) \end{bmatrix} \tag{44}$$

where

$$\begin{aligned} M_{11} &= 1-\phi_0(1) + z\chi(1)S(z)(1-\phi_0(c)), \\ M_{12} &= \phi_0(1) + z\chi(1)S(z)\phi_0(c), \\ M_{13} &= \psi(1) + z\chi(1)S(z)\psi(c). \end{aligned}$$

We perform a numerical search to find collocation abscissae to get wide stability regions. For $c = 1$, using the boundary locus technique, we obtain the stability region plotted in fig. 1.

In particular, the stability interval of this method is $[-\frac{79}{20}, 0]$.

Let us look at two–stage methods. These methods are represented by the following array

$$\left[\begin{array}{cc|cccc} a_{11} & a_{12} & 1-u_1 & u_1 & b_{11} & b_{12} \\ a_{21} & a_{22} & 1-u_2 & u_2 & b_{21} & b_{22} \\ v_1 & v_2 & 1-\theta & \theta & w_1 & w_2 \\ \hline 0 & 0 & 1 & 0 & 0 & 0 \\ 1 & 0 & 0 & 0 & 0 & 0 \\ 0 & 1 & 0 & 0 & 0 & 0 \end{array}\right] \tag{45}$$

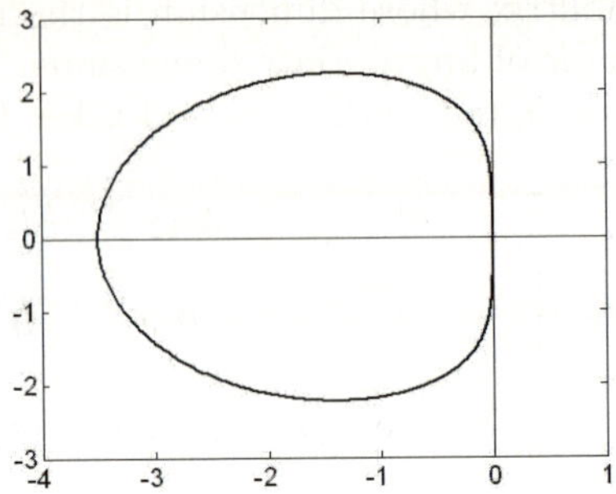

Fig. 1. Stability region of one-stage method for $c = 1$

Fig. 2 shows the stability region of the two–stage method associated to $c_1 = \frac{1}{2}$ and $c_2 = 1$. In this case, the stability interval is $[-\frac{14}{5}, 0]$.

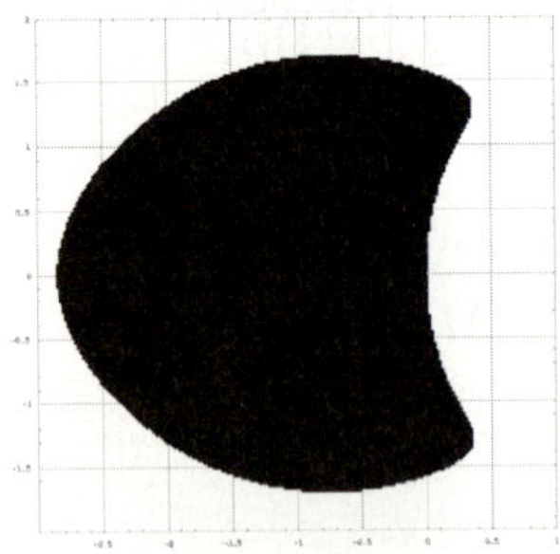

Fig. 2. Stability region of two-stage method for $c_1 = \frac{1}{2}$ and $c_2 = 1$

5 Numerical Experiments

We have tested our class of methods on many non-stiff and stiff problems. We present below a selection of results of numerical experiments with fixed stepsize, designed to compare our methods with respect to classical one step collocation Runge–Kutta methods. Even if our methods have no unbounded stability regions, they can be suitable to integrate also stiff problems, just by adapting the choice of the step size to the amplitude of the interval of stability. Indeed, methods having unbounded stability regions, for example A-stable methods, can integrate stiff problems also with larger step size, but this kind of integration could highten the error committed too much.

The numerical experiments are carried over with a fixed stepsize, without the usage of strategies to save function evaluations.

We first consider the following linear problem (cfr. [9])

$$\begin{cases} y_1'(x) = -2y_1(x) + y_2(x) + 2\sin x \\ y_2'(x) = y_1(x) - 2y_2(x) + 2(\cos x - \sin x) \end{cases} \tag{46}$$

with $x \in [0, 10]$, with the initial condition $y(0) = [2, 3]^T$, whose exact solution is

$$\begin{cases} y_1(x) = 2e^{-x} + \sin x \\ y_2(x) = 2e^{-x} + \cos x \end{cases}. \tag{47}$$

We solve this problem using the general two–step collocation method defined in theorem 1 and in (19)-(20) with $m = 1$ and $c = 1$ (GTSCOLL) and we compare it with the one step Gauss method with one stage (of order 2) and the Radau IIA with 2 stages (of order 3), in order to have a comparison between methods having the same number of stages (Gauss) and the same order (Radau IIA) and also with BDF method of order 3 (BDF), which is usually considered a standard method for stiff problems [8]. The result of the implementation is shown in the following tables, where h is the step size used, fe is the number of function evaluations, cd is the number of correct digits, ge is the global error committed at the end of the integration interval.

GTSCOLL method, m=1, p=3			
h	fe	cd	ge
0.1	2242	4.9435	1.1387e-005
0.05	3522	5.8438	1.4328e-006
0.025	5866	6.7454	1.7968e-007
0.0125	9580	7.6491	2.2430e-008
0.00625	18144	8.5507	2.8133e-009
0.003125	31984	9.4569	3.4917e-010

Gauss method, m=1, p=2			
h	fe	cd	ge
0.1	297	3.0565	8.7792e-004
0.05	597	3.6588	2.1936e-004
0.025	1197	4.2609	5.4835e-005
0.0125	2397	4.8630	1.3708e-005
0.00625	4797	5.4650	3.4270e-006
0.003125	9597	6.0671	8.5676e-007

Radau IIA method, m=2, p=3			
h	fe	cd	ge
0.1	2966	4.7535	1.7637e-005
0.05	4750	5.6481	2.2484e-006
0.025	7904	6.5468	2.8386e-007
0.0125	12972	7.4478	3.5660e-008
0.00625	25036	8.3497	4.4689e-009
0.003125	44684	9.2523	5.5928e-010

BDF method, k=3, p=3			
h	fe	cd	ge
0.1	198	4.0413	9.0920e-005
0.05	398	4.9149	1.2163e-005
0.025	798	5.8036	1.5716e-006
0.0125	1598	6.6996	1.9968e-007
0.00625	3190	7.5992	2.5163e-008
0.003125	3708	8.5006	3.1579e-009

As shown in fig. 3, the method GTSCOLL gives the best accuracy. If we compare the results obtained by our method and the Gauss one, we can see that the GTSCOLL method, using the same number of stages, gives a higher order of accuracy. If compared with the Radau IIA method with same order, but with two stages, our method gives a better accuracy as well. BDF method gives less accurate results, by using less function evaluations.

We next consider the well known Van der Pol's equation [5]. We reduce this second order ODE to the following first order system of two equations

$$\begin{cases} y_1' = y_2 \\ y_2' = \mu(1 - y_1^2)y_2 - y_1 \end{cases} \tag{48}$$

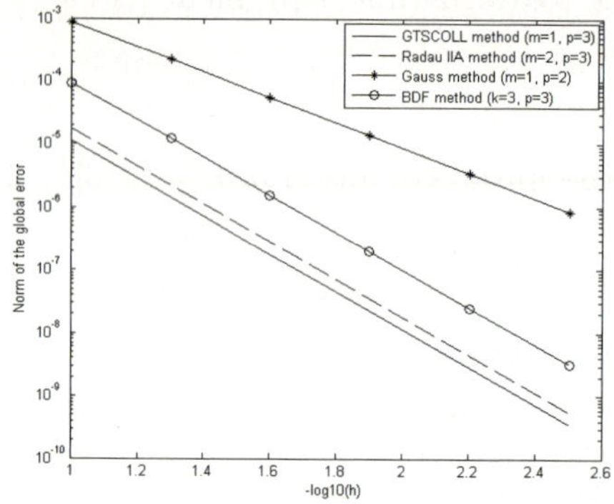

Fig. 3. Comparison between three solvers for the problem (46)

with $x \in [0, 100]$ and with the initial condition $y(0) = [2, 0]^T$. In particular, we consider the case $\mu = 1000$. The parameter $\mu > 0$ hightens the importance of the nonlinear part of the equation. This problem exhibits a particular phenomenon: the problem switches from stiff to nonstiff with a very sharp changing solution. This makes the equation quite challenging for ODEs solvers.

The GTSCOLL method, having a bounded stability region, is able to integrate this nonlinear stiff problem using a stepsize adapted to the amplitude of the stability region: in particular, it shows the same behaviour of the Gauss method and it is better than the Radau and BDF methods:

GTSCOLL method			
h	fe	cd	ge
0.1	1889	6.8483	1.4180e-007
0.05	3098	6.8870	1.2971e-007

Gauss method			
h	fe	cd	ge
0.1	2996	6.9023	1.2520e-007
0.05	5996	6.9023	1.2520e-007

Radau IIa method			
h	fe	cd	ge
0.1	13010	4.1495	7.0869e-005
0.05	24010	4.4498	3.5496e-005

BDF method			
h	fe	cd	ge
0.1	1554	6.9013	1.2509e-007
0.05	3108	6.9029	1.2510e-007

In many cases, it is possible to give an upper bound for the stepsize, in order to integrate a stiff system also with methods having a bounded stability interval. We show an example in which we obtain such an estimation of the stepsize. The following problem is Kramarz's system [8], which is often used in numerical experiments on periodic stiffness (cfr. [14]) on second order ODEs:

$$y''(x) = Ay(x) = \begin{pmatrix} 2498 & 4998 \\ -2499 & -4999 \end{pmatrix} y(x), \qquad (49)$$

where A is the matrix of coefficients in (49), $y(x) = [y_1(x), y_2(x)]^T$, with $y(0) = [2, -1]^T$ and $x \in [0, 2\pi]$. The exact solution of this problem is

$$y(x) = \begin{pmatrix} 2\cos x \\ -\cos x \end{pmatrix}. \tag{50}$$

The eigenvalues of A are $\lambda_1 = -2500$, $\lambda_2 = -1$; then the analytical solution of the system exhibits the two frequencies 1 and $\sqrt{2500}$, but the high frequency component is eliminated by the initial conditions. Notwithstanding this, its presence in the general solution of the system dictates restrictions on the choice of the stepsize, so that the system is stiff.

We transform this problem in a system of 4 ordinary differential equations of the first order and we integrate it by using the GTSCOLL method, whose stability interval is $[-\frac{79}{20}, 0]$. In order to get stable results, the product $h\lambda$ must be in $[-\frac{79}{20}, 0]$, so $h \leq \frac{79}{50000} = 0.00158$. If we use a stepsize $h \leq 0.00158$, the GTSCOLL method integrates the above problem with a bounded error. The following table shows that this actually happens. Then we compare these results with the one obtained by the Gauss method.

GTSCOLL method				Gauss method			
h	fe	cd	ge	h	fe	cd	ge
0.00158	19883	2.6095	2.4575e-3	0.00158	19873	2.6065	2.4745e-3
0.00079	39768	3.1539	7.0154e-4	0.00079	39726	3.1513	7.0579e-4
0.000395	79533	3.1523	7.0418e-4	0.000395	79237	3.1517	7.0523e-4
0.0001975	159068	3.5797	2.6320e-4	0.0001975	156735	3.5793	2.6347e-4
0.00009875	318138	4.3710	4.2559e-5	0.00009875	298833	4.3703	4.2624e-5
0.000049375	636273	4.3706	4.2600e-5	0.000049375	509014	4.3704	4.2616e-5

BDF method			
h	fe	cd	ge
0.00158	6543	2.6070	2.4715e-3
0.00079	9694	3.1517	7.0504e-4
0.000395	15906	3.1517	7.0504e-4
0.0001975	31813	3.5793	2.6342e-4
0.00009875	63627	4.3704	4.2613e-5
0.000049375	127254	4.3706	4.2607e-5

Actually our collocation method shows the same behaviour of the Gauss method and BDF method, despite of its bounded stability region.

6 Conclusions

We presented the family of collocation–based General Two Step Runge–Kutta methods for $y' = f(x, y)$, which exhibit high order of convergence, and high stage order. The knowledge of the collocation polynomial, which provides a uniform approximation in any point of the integration interval, may allow their usage in a variable stepsize implementation, so that they seem promising for future development.

In order to overcome the limit of a limited stability regions, a modification of the collocation approach by relaxing some collocation conditions is under development in [2], to derive A–stable methods of lower order within the same family.

It is also the purpose of our research to investigate on the use of different classes of function basis (cfr. [12]), e.g. trigonometric polynomials, exponentials or mixed basis, in order to obtain continuous methods which are particularly suitable to integrate problems having periodic or oscillatory solutions.

Acknowledgments

We wish to thank Prof. Zdzislaw Jackiewicz for his helpful and valuable discussions and comments along time on various issues in the field.

References

1. Butcher, J.C.: Numerical methods for Ordinary Differential Equations. Wiley, Chichester (2003)
2. D'Ambrosio, R., Ferro, M., Jackiewicz, Z., Paternoster, B.: Almost two-step collocation methods for Ordinary Differential Equations (in preparation)
3. Guillou, A., Soulé, F.L.: La résolution numérique des problèmes differentiels aux conditions par des méthodes de collocation. RAIRO Anal. Numér. Ser. Rouge R-3, 17–44 (1969)
4. Hairer, E., Norsett, S.P., Wanner, G.: Solving Ordinary Differential Equations I – Nonstiff Problems. Springer Series in Computational Mathematics, vol. 8. Springer, Berlin (2000)
5. Hairer, E., Wanner, G.: Solving Ordinary Differential Equations II – Stiff and Differential–Algebraic Problems. Springer Series in Computational Mathematics, vol. 14. Springer, Berlin (2002)
6. Jackiewicz, Z.: General linear methods (in preparation)
7. Jackiewicz, Z., Tracogna, S.: A general class of two–step Runge–Kutta methods for ordinary differential equations. SIAM J. Numer. Anal. 32(5), 1390–1427 (1995)
8. Kramarz, L.: Stability of collocation methods for the numerical solution of $y'' = f(t, y)$. BIT 20, 215–222 (1980)
9. Lambert, J.D.: Numerical methods for ordinary differential systems: The initial value problem. Wiley, Chichester (1991)
10. Lie, I., Norsett, S.P.: Superconvergence for Multistep Collocation. Mathematics of Computation 52(185), 65–79 (1989)
11. Martucci, S., Paternoster, B.: Vandermonde–type matrices in two step collocation methods for special second order Ordinary Differential Equations. In: Bubak, M., van Albada, G.D., Sloot, P.M.A., Dongarra, J. (eds.) ICCS 2004. LNCS, vol. 3039, pp. 418–425. Springer, Heidelberg (2004)

12. Paternoster, B.: General Two-Step Runge-Kutta methods based on algebraic and trigonometric polynomials. Int. J. Appl. Math. 6(4), 347–362 (2001)
13. Paternoster, B.: Two step Runge-Kutta-Nyström methods for $y'' = f(x, y)$ and P–stability. In: Sloot, P.M.A., Tan, C.J.K., Dongarra, J., Hoekstra, A.G. (eds.) ICCS 2002. LNCS, vol. 2331, pp. 459–466. Springer, Heidelberg (2002)
14. Petzold, L.R., Jay, L.O., Yen, J.: Numerical solution of highly oscillatory ordinary differential equations. Acta Numerica, 437–483 (1997)

Analysing the Dependence between Variance and Mean Estimated Values: A Theoretical and Experimental Approach

Adriana Vlad[1,2] and Bogdan Badea[1]

[1] Faculty of Electronics, Telecommunications and Information Technology, POLITEHNICA University of Bucharest, Romania
avlad@racai.ro, adriana_vlad@yahoo.com
[2] The Research Institute for Artificial Intelligence, Romanian Academy
bogdan_badea@yahoo.com

Abstract. The starting point of this paper is a known theoretical result - for normal populations, mean and variance estimators are statistically independent. By using an original statistical independence test method, we enriched the study by experimentally investigating the mean-variance data pairs in case of non-gaussian laws. We experimentally noticed that, for non-gaussian symmetrical distributions, mean and variance estimators are still uncorrelated, but not independent random variables. We further theoretically proved that, for symmetrical distributions, the two estimators, mean and variance, are uncorrelated. This new result provides the basis for an original skewness test method, *i.e.* if mean-variance data pairs are correlated, then the probability density function has an asymmetric shape.

Keywords: sample mean and sample variance, statistical independence, Spearman's rank correlation test, scatter diagrams, testing skewness.

1 Introduction

A known result in statistical mathematics is that variance and mean estimators are independent random variables for a normal population (which is not a straightforward result).

The paper started as a search for an easier proof and visual understanding of the difficult, but very important in practice, mathematical result: the statistical independence existing between variance and mean estimators for normal population. We also carried out this investigation concerning the dependence existing between the sample mean and the sample variance on experimental data provided by non-gaussian laws.

From the experimental results we observed that, for populations with a symmetric probability density function with respect to the theoretical mean (in the paper simply denoted by symmetrical function), the two estimators (mean and variance) are dependent, but uncorrelated random variables. We proved this result both by means of an intuitive and graphical statistical test method, see Section 2 and Appendix, and in analytical form, see Section 3. In fact, we first reached this theoretical result by applying the original statistical independence test [1] on the mean-variance experimental data pairs. The used

O. Gervasi et al. (Eds.): ICCSA 2008, Part II, LNCS 5073, pp. 752–767, 2008.

experimental data set came out from an *i.i.d.* (independently and identically distributed) random variable generator of several statistical laws and then by repeatedly grouping successive data in order to compute the two estimates on each group.

The original statistical test [1] is an independence test method (derived from Spearman's Rank Correlation Test), applicable for all kind of continuous random variables (not restricted to normal distribution and even for an unknown probability law). Note that this test method was strictly necessary here at least for the following reasons: (a) even when referring to the mean and variance estimators for normal (gaussian) population, these two investigated random variables are not jointly gaussian, [2]-[4]; (b) if the investigated population is not normal, the probability law of the variance estimator is unknown, [2]-[4].

In the literature, there are tests for statistical independence, [2]-[5], [8], but applicable only in the case of jointly normal random variables. Otherwise, one can apply Spearman's test, [8], but this test has to be improved in order to decide upon statistical independence, [1]. In the Appendix we briefly present the independence test algorithm used in this paper.

Section 2 presents the experimental study. The test method [1] was first applied to verify the statistical independence between sample mean and sample variance for gaussian populations. The test was further applied for other probability laws represented by symmetrical and non-symmetrical functions. The experimental results brought into evidence that a symmetrical probability density function leads to non-correlation between mean and variance estimators. This could be of some interest in practice, as for example when testing skewness.

Section 3 is our analytical proof that, for symmetrical probability density functions, mean and variance estimators are uncorrelated.

2 Experimental Study and Examples

Be X a continuous random variable providing the sample mean and the sample variance under discussion.

Be $x_1, x_2, x_3, \ldots, x_i, \ldots$ the observations (measured values) of X random variable. We shall consider that these observations $x_1, x_2, x_3, \ldots, x_i, \ldots$ comply with the statistical *i.i.d.* model required for statistical inferences (*i.e.* each measurement of X is performed by an identical procedure and the repeated experiments are independent) .

The experimental data are grouped in blocks of size N. For each and every *i.i.d.* experimental data set of size N (the respective blocks of N observations), we compute the estimated values for the mean and for the variance, usually denoted by $\bar{x}$ and s^2.

So, if $[x_1, x_2, \ldots, x_N]$ is one of these *i.i.d.* data sets of size N, we compute:

$$\bar{x} = \frac{1}{N}\sum_{i=1}^{N} x_i \qquad s^2 = \frac{1}{N-1}\sum_{i=1}^{N}(x_i - \bar{x})^2 \tag{1}$$

For example, for blocks of size $N = 3$ and using (1), we have:

$$\underbrace{x_1, x_2, x_3}_{\left(\bar{x}_1,\, s_1^2\right)},\ \underbrace{x_4, x_5, x_6}_{\left(\bar{x}_2,\, s_2^2\right)},\ \underbrace{x_7, x_8, x_9}_{\left(\bar{x}_3,\, s_3^2\right)}, \ldots$$

In this manner, we obtain a new experimental *i.i.d.* data set of size n (n must be a large sample size), where each element is a pair of mean-variance values:

$$\left[\left(\bar{x}_1, s_1^2\right), \left(\bar{x}_2, s_2^2\right), \left(\bar{x}_3, s_3^2\right), \ldots, \left(\bar{x}_n, s_n^2\right)\right].$$

So, N will denote the sample size used for computing the sample mean and sample variance, and n will denote the sample size submitted to the test in Appendix (to check-up the statistical independence of the two estimators).

This *i.i.d.* data set of pairs of mean-variance type will be submitted to the test presented in Appendix in order to establish whether or not there is statistical independence between the corresponding $\overline{X}$ and S^2 estimators. Equivalently, we submit to the test from Appendix the two *i.i.d.* data sets of size n: $[\bar{x}_1, \bar{x}_2, \bar{x}_3, \ldots, \bar{x}_n]$ and $[s_1^2, s_2^2, s_3^2, \ldots, s_n^2]$. Note that, separately, each of these two data sets is an *i.i.d.* data set; the problem is to find out whether or not the two *i.i.d.* data sets of size n are independent sets.

Now, using the notations $\bar{x}_i \equiv y_i$ and $s_i^2 \equiv z_i$ we can follow the test method steps presented in Appendix. Accordingly, $\overline{X} \equiv Y$ and $S^2 \equiv Z$.

The entire dependence analysis is shifted by two bijective transforms from the yz coordinates to the uv coordinates. Testing independence of Y and Z is done by applying Person's r correlation test on U and V; if U and V are dependent then Y and Z are dependent.

For the test decision concerning non-correlation and also statistical independence, we have to inspect two scatter diagrams: for $y'z'$ coordinates and for $u\,v$ coordinates.

Spearman's rank correlation test is performed on $y'z'$ coordinates because $n\,y_i' = Rank(y_i)$ and $n\,z_i' = Rank(z_i)$. Note that the regression slope in $y'z'$ coordinates is equal to Spearman's rank correlation coefficient r_S. If there is non-correlation between Y and Z random variables, then the regression slope r_S will be practically zero. This regression slope is displayed on the same figure with the $y'z'$ scatter diagram (here denoted as Spearman's ranks scatter diagram).

It is known that the statistical independence of two random variables implies non-correlation, but the converse is not always true. The scatter diagram for $u\,v$ data enables a final visual decision concerning statistical independence between the Y and Z random variables. In case of statistical independence, this $u\,v$ scatter diagram should be like the one in Fig. 11, see Appendix. On this scatter diagram we also display the linear regression slope for the $u\,v$ data. According to the test method, the regression slope for $u\,v$ pairs of data is equal to the r value in (7). r is the experimental correlation coefficient between the U and V random variables.

In conclusion, if $\overline{X}$ and S^2 are statistically independent random variables, then the r and r_S experimental correlation coefficients must be small enough (the test value passes the test) and the $u\,v$ and the $y'z'$ scatter diagrams should be like the ones in Fig. 11 and Fig. 12, respectively. If $\overline{X}$ and S^2 are dependent, but uncorrelated random variables, then the regression slope (r and r_S values) must be practically zero, but $u\,v$ and the $y'z'$ scatter diagrams will differ from the referred scatter diagrams from Fig. 11 and Fig. 12 in Appendix.

We further exemplify the test method for the probability density functions given in Fig. 1.

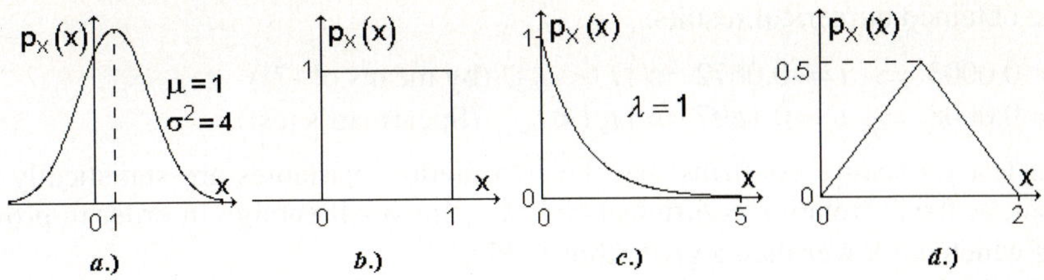

Fig. 1. Simulation used statistical laws: (*a*) normal law; (*b*) uniform law; (*c*) exponential law; (*d*) sum of two *i.i.d.* uniform random variables

The sample mean $\bar{x}$ and the sample variance s^2 are first computed on small samples of size $N=2$ and $N=3$, because in this case the results are rather unexpected. We also considered larger samples of $N=200$ size, if the experimental study asked for. In all cases, the sample size for the statistical independence test presented in Appendix was $n=50000$ and the statistical significance level of the test method was $\alpha=0.05$, thus $t_{\alpha/2}=1.96$.

a. Experimental Study on the Gaussian Law

The X investigated random variable is a normal random variable, of mean 1 and variance 4. In Fig. 2 we display the scatter diagram corresponding to Spearman's ranks, Fig. 2*a*, and the scatter diagram corresponding to the $u\,v$ coordinates, Fig. 2*b*. Mean and variance estimates are computed on $N=2$ sample size.

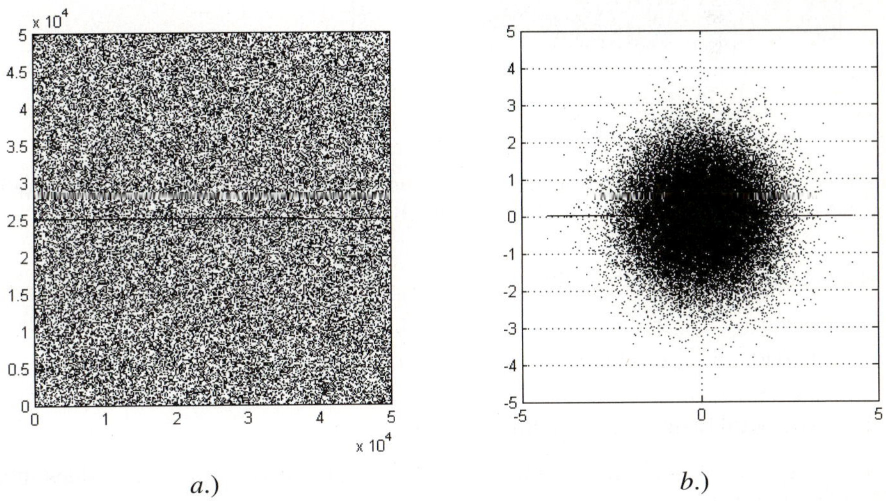

Fig. 2. Scatter diagrams for Spearman's ranks (*a*) and for $u\,v$ coordinates (*b*). Mean and variance computed for: normal law of mean 1 and variance 4; $N=2$.

The obtained numerical results:

$r = -0.0003 \Rightarrow t = -0.0672 \Rightarrow |t| < t_{\alpha/2}$ (by means of (7))
$r_S = 0.0008 \Rightarrow t_S = 0.1897 \Rightarrow |t_S| < t_{\alpha/2}$ (Spearman's test).

The test procedure confirms that the two random variables are statistically independent, as the correlation coefficients r and r_S are small enough in order to provide t values which are lower than a given threshold $t_{\alpha/2}$.

Besides, the $u\,v$ scatter diagram resembles the reference scatter diagram given in Fig. 11, see Appendix.

Similar results indicating that mean and variance estimators are statistically independent random variables are presented in Fig. 3, where we investigated the same normal law, but mean and variance estimated values are computed on $N=3$ sample size.

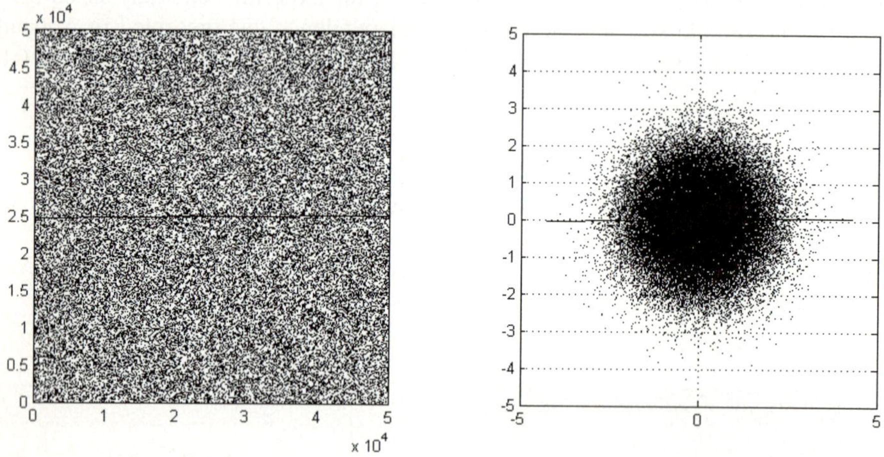

a.) *b.)*

Fig. 3. Scatter diagrams for Spearman's ranks (*a*) and for $u\,v$ coordinates (*b*). Mean and variance computed for: normal law of mean 1 and variance 4; $N=3$.

The obtained numerical results:

$r = 0.0032 \Rightarrow t = 0.7175 \Rightarrow |t| < t_{\alpha/2}$ (by means of (7))
$r_S = 0.0040 \Rightarrow t_S = 0.8958 \Rightarrow |t_S| < t_{\alpha/2}$ (Spearman's test)

b. Experimental Study on the Uniform Law

The X investigated random variable is a uniform random variable, with values in the (0; 1) interval.

Note that the scatter diagram in Fig. 4*a* corresponds to Spearman's ranks and the scatter diagram in Fig. 4*b* is the $u\,v$ scatter diagram. Mean and variance estimates are computed on $N=2$ sample size.

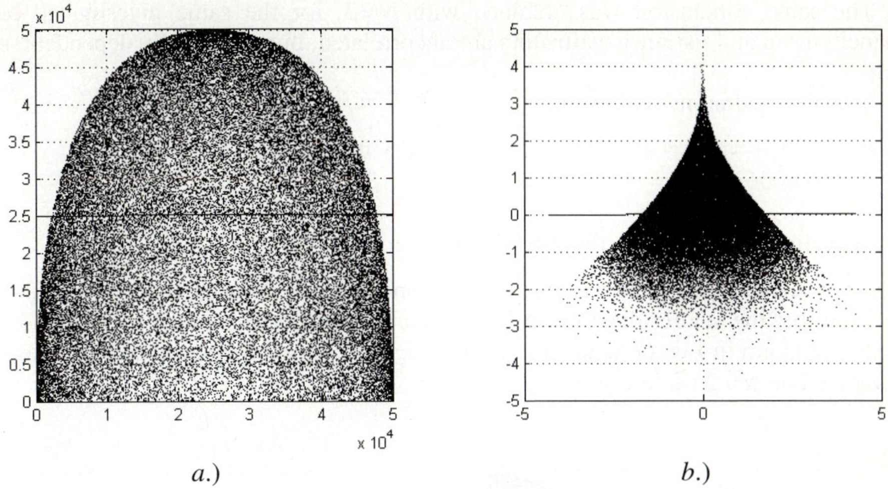

Fig. 4. Scatter diagrams for Spearman's ranks (*a*) and for *u v* coordinates (*b*). Mean and variance computed for: uniform law in (0; 1); $N = 2$.

The obtained numerical results:

$r = 0.0038 \Rightarrow \; t = 0.8559 \Rightarrow | t | < t_{\alpha/2}$ (by means of (7))

$r_S = 0.0047 \Rightarrow \; t_S = 1.0573 \Rightarrow | t_S | < t_{\alpha/2}$ (Spearman's test)

The test procedure shows that the two estimators are not statistically independent. Though the correlation coefficients r and r_S are small enough (see the linear regression slope in both figures, Fig. 4*a* and Fig. 4*b*), scatter diagram in *u v* coordinates is different from the reference scatter diagram given in Fig. 11. So, the mean and variance estimators are uncorrelated, but dependent.

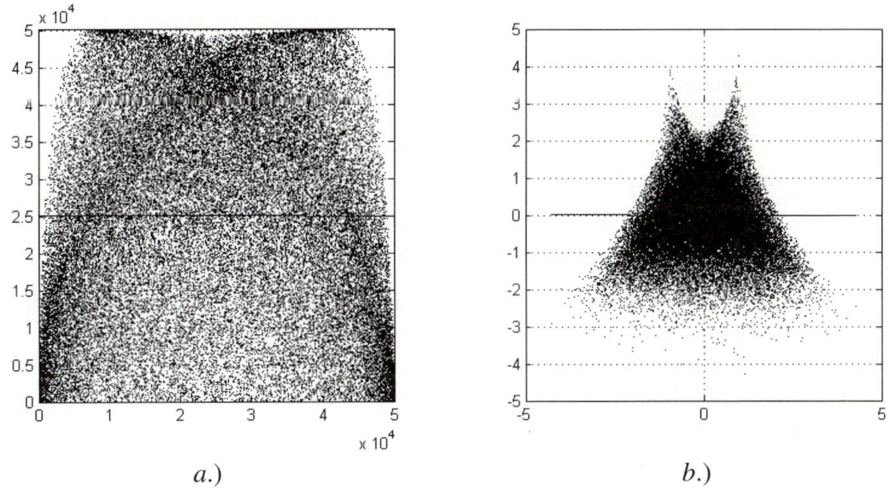

Fig. 5. Scatter diagrams for Spearman's ranks (*a*) and for *u v* coordinates (*b*). Mean and variance computed for: uniform law in (0; 1); $N = 3$.

The same conclusion was obtained with $N=3$, for the same investigated law, namely mean and variance estimators are uncorrelated, but statistically dependent, see Fig. 5.

The obtained numerical results:

$r = -0.0014 \Rightarrow t = -0.3158 \Rightarrow |t| < t_{\alpha/2}$ (by means of (7))
$r_S = -0.0006 \Rightarrow t_S = -0.1379 \Rightarrow |t_S| < t_{\alpha/2}$ (Spearman's test)

c. Experimental Study on the Exponential Law

The X investigated random variable is an exponential law of mean 1.

Note that the scatter diagram in Fig. 6*a* corresponds to Spearman's ranks and the scatter diagram in Fig. 6*b* is the $u\,v$ scatter diagram. Mean and variance estimates are computed on $N=2$ sample size.

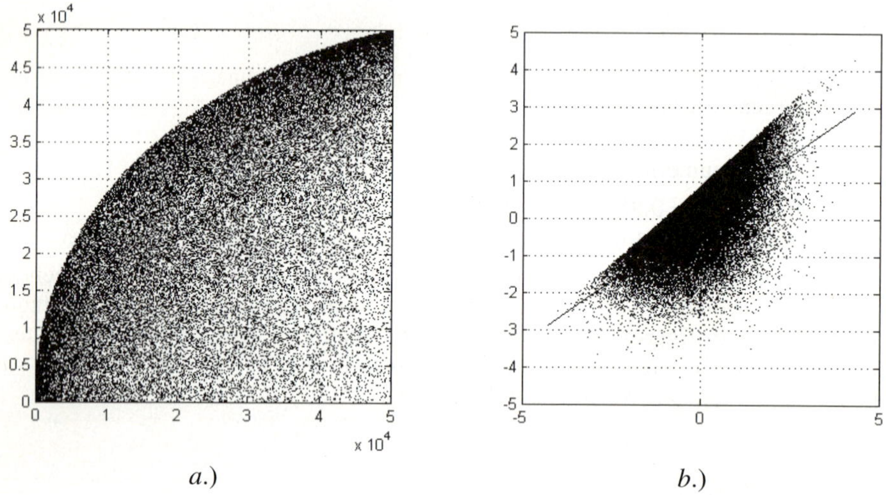

Fig. 6. Scatter diagrams for Spearman's ranks (*a*) and for $u\,v$ coordinates (*b*). Mean and variance computed for: exponential law of mean 1; $N=2$.

The obtained numerical results:

$r = 0.6744 \Rightarrow t = 204.2438 \Rightarrow |t| > t_{\alpha/2}$ (by means of (7))
$r_S = 0.6673 \Rightarrow t_S = 200.3401 \Rightarrow |t_S| > t_{\alpha/2}$ (Spearman's test)

The test procedure indicates that the two random variables are statistically correlated. Scatter diagram in $u\,v$ coordinates is different from the reference scatter diagram; also notice the linear regression slope on both figures, Fig. 6*a* and Fig. 6*b*.

The same conclusion, meaning that the mean and variance estimators are statistically correlated, is obtained for $N = 3$, see Fig. 7.

The obtained numerical results:

$r = 0.7688 \Rightarrow t = 268.7838 \Rightarrow |t| > t_{\alpha/2}$ (by means of (7))
$r_S = 0.7610 \Rightarrow t_S = 262.2747 \Rightarrow |t_S| > t_{\alpha/2}$ (Spearman's test)

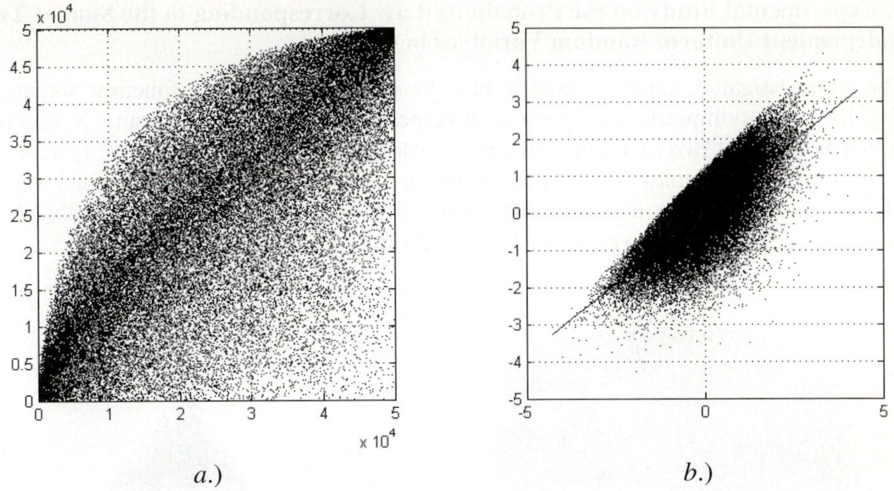

Fig. 7. Scatter diagrams for Spearman's ranks (*a*) and for *u v* coordinates (*b*). Mean and variance computed for: exponential law of mean 1; $N = 3$.

For the exponential law, we also investigated larger sample data of $N = 200$; the conclusion was the same, *i.e.* mean and variance estimators are correlated.

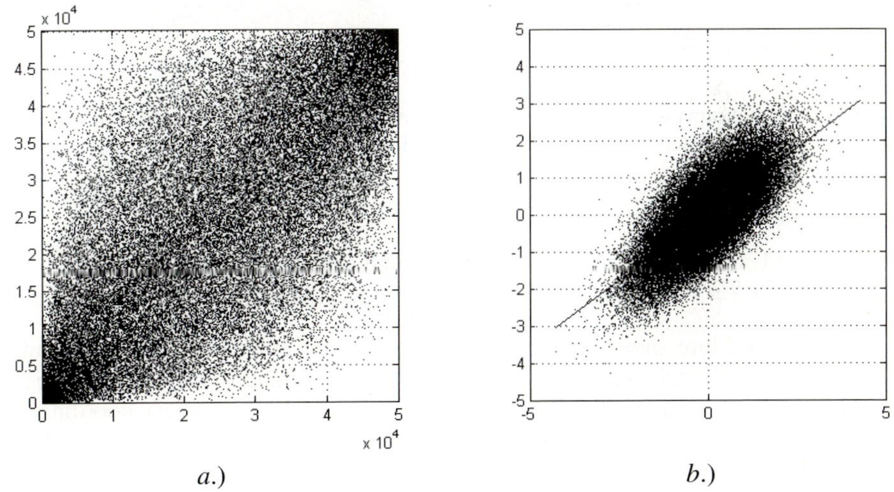

Fig. 8. Scatter diagrams for Spearman's ranks (*a*) and for *u v* coordinates (*b*). Mean and variance computed for: exponential law of mean 1; $N = 200$.

The obtained numerical results:

$r = 0.7178 \Rightarrow \ t = 230.5104 \ \Rightarrow |\, t \,| > t_{\alpha/2}$ (by means of (7))
$r_S = 0.7022 \Rightarrow \ t_S = 220.5209 \ \Rightarrow |\, t_S \,| > t_{\alpha/2}$ (Spearman's test)

d. Experimental Study on the Probability Law Corresponding to the Sum of Two Independent Uniform Random Variables in (0; 1)

The X investigated random variable has the probability density function shown in Fig. 1*d*, *i.e.* a symmetrical function with respect to the theoretical mean. X was obtained as a sum of two *i.i.d.* uniform random variables with values in the (0; 1) interval.

In Fig. 9 we display the scatter diagram corresponding to Spearman's ranks, Fig. 9*a*, and the scatter diagram corresponding to the $u\,v$ coordinates, Fig. 9*b*. Mean and variance estimates are computed on $N = 2$ sample size.

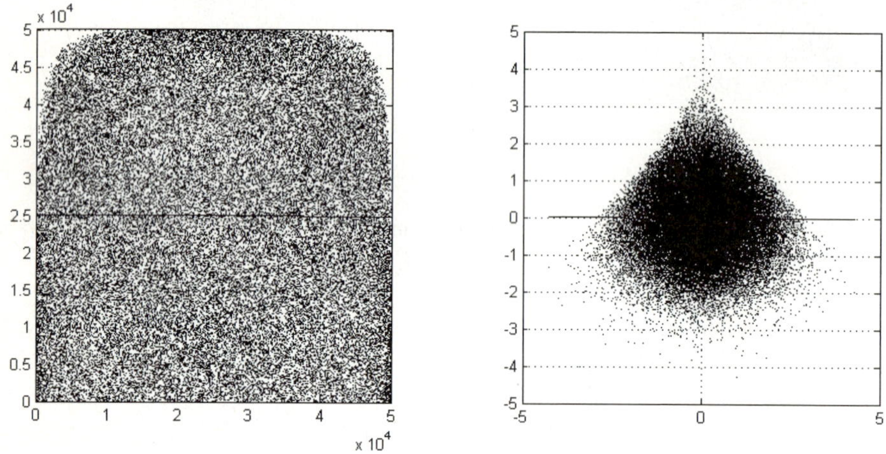

a.) *b.*)

Fig. 9. Scatter diagrams for Spearman's ranks (*a*) and for $u\,v$ coordinates (*b*). Mean and variance computed for: sum of two independent uniform random variables in (0; 1); $N = 2$.

The obtained numerical results:

$r = -0.0011 \Rightarrow t = -0.2569 \Rightarrow |t| < t_{\alpha/2}$ (by means of (7))
$r_S = -0.0018 \Rightarrow t_S = -0.4120 \Rightarrow |t_S| < t_{\alpha/2}$ (Spearman's test)

The test procedure shows that the two estimators are not statistically independent. Though the correlation coefficients r and r_S are small enough (see the linear regression slope in both figures, Fig. 9*a* and Fig. 9*b*), scatter diagram in $u\,v$ coordinates is different from the reference scatter diagram, see Fig. 11 in Appendix. Accordingly, the mean and variance estimators are uncorrelated, but dependent.

The same conclusion was obtained for $N = 3$, namely mean and variance estimators are uncorrelated, but statistically dependent, see Fig. 10.

The obtained numerical results:

$r = 0.0023 \Rightarrow t = 0.5190 \Rightarrow |t| < t_{\alpha/2}$ (by means of (7))
$r_S = 0.0009 \Rightarrow t_S = 0.1905 \Rightarrow |t_S| < t_{\alpha/2}$ (Spearman's test)

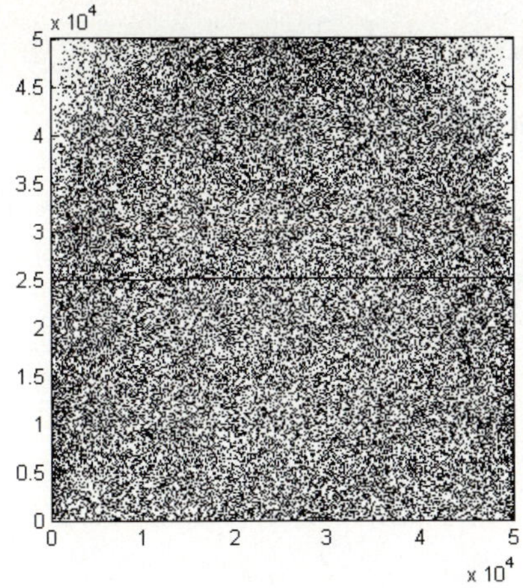

a.)

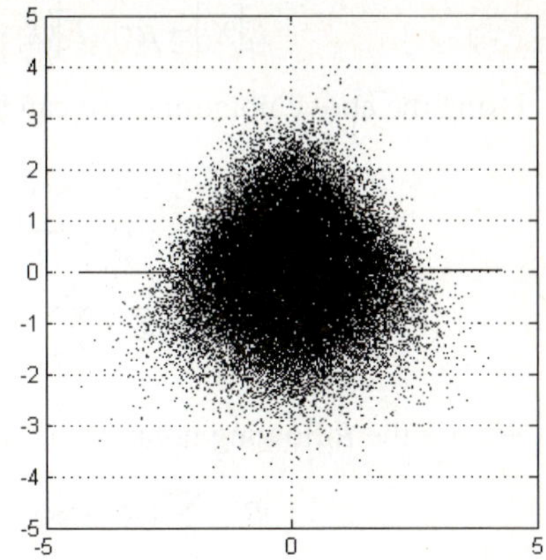

b.)

Fig. 10. Scatter diagrams for Spearman's ranks (*a*) and for *u v* coordinates (*b*). Mean and variance computed for: sum of two independent uniform random variables in (0; 1); $N=3$.

3 A Theoretical Approach Concerning Symmetrical Distributions

In this section we present the mathematical proof of the result we first noticed by means of the experimental study (see Section 2). *That is, if the investigated population has a probability density function symmetric with respect to the theoretical mean, then the mean and variance estimators are uncorrelated random variables.* In Section 3.1 we present the proof of this theorem; in Section 3.2 we provide a test procedure for detecting asymmetry.

3.1 Proof of Theorem

Be $[x_1, x_2, \ldots, x_N]$ the *i.i.d.* experimental data set of size N, on which we compute, according to (1), the mean and the variance estimated values.

Be $[X_1, X_2, \ldots, X_N]$ the corresponding random *i.i.d.* sample. So, $X_1, X_2, \ldots, X_N$ will be statistically independent random variables, all having the same probability density function as the investigated X random variable. We suppose this probability density function to be a symmetric function.

With the usual notations, the two estimators under discussion, $\overline{X}$ and S^2, are:

$$\overline{X} = \frac{1}{N}\sum_{i=1}^{N} X_i \qquad S^2 = \frac{1}{N-1}\sum_{i=1}^{N}\left(X_i - \overline{X}\right)^2 \tag{2}$$

Be μ and σ^2 the theoretical mean and variance of X.

$$\mu = E[X] \qquad \sigma^2 = E[(X-\mu)^2]$$

Using the known theory [2]- [8] we have:

$$E[\overline{X}]=\mu;\quad E\left[(\overline{X}-\mu)^2\right]=\sigma^2/N;\quad E\left[s^2\right]=\sigma^2$$

Using the above notations, we can immediately write (3) and (4):

$$(N-1)\,s^2=\sum_{i=1}^{N}(X_i-\mu)^2-N\left(\overline{X}-\mu\right)^2 \tag{3}$$

$$(N-1)\,s^2\,\overline{X}=\sum_{i=1}^{N}(X_i-\mu)^2\,\overline{X}-N\left(\overline{X}-\mu\right)^2\,\overline{X} \tag{4}$$

We use the following notations concerning (4):

$$B=\sum_{i=1}^{N}(X_i-\mu)^2\,\overline{X}\qquad\qquad C=N\left(\overline{X}-\mu\right)^2\,\overline{X}$$

$$B=\frac{1}{N}\sum_{i=1}^{N}(X_i-\mu)^2\sum_{j=1}^{N}\left(X_j-\mu+\mu\right)=$$

$$=\frac{1}{N}\sum_{i=1}^{N}\sum_{j=1}^{N}(X_i-\mu)^2\,(X_j-\mu)+\frac{1}{N}\sum_{i=1}^{N}\sum_{j=1}^{N}(X_i-\mu)^2\,\mu=$$

$$=\frac{1}{N}\sum_{i=1}^{N}(X_i-\mu)^3+\frac{1}{N}\sum_{\substack{i=1\\ i\neq j}}^{N}\sum_{j=1}^{N}(X_i-\mu)^2\,(X_j-\mu)+\frac{\mu}{N}N\sum_{i=1}^{N}(X_i-\mu)^2$$

The theoretical mean of B is:

$$E(B)=\frac{1}{N}\sum_{i=1}^{N}E\left[(X_i-\mu)^3\right]+\frac{1}{N}\sum_{\substack{i=1\\ i\neq j}}^{N}\sum_{j=1}^{N}E\left[(X_i-\mu)^2\,(X_j-\mu)\right]+\mu\sum_{i=1}^{N}E\left[(X_i-\mu)^2\right]$$

As the probability density function of X is symmetrical with respect to its mean, we have:

$$E\left[(X_i-\mu)^3\right]=0$$

Also, because X_i and X_j are identically distributed (as the random variable X), and statistically independent for $i\neq j$, we have:

$$E\left[(X_i-\mu)^2\,(X_j-\mu)\right]=0\quad \text{for } i\neq j$$

$$E\left[(X_i-\mu)^2\right]=\sigma^2$$

It results:

$$E(B)=0+0+N\mu\sigma^2=N\mu\sigma^2$$

The theoretical mean of C is:

$$E(C)=E\left[N\left(\overline{X}-\mu\right)^2\overline{X}\right]=E\left[N\left(\overline{X}-\mu\right)^2\left(\overline{X}-\mu+\mu\right)\right]=$$

$$=N\,E\left[\left(\overline{X}-\mu\right)^3\right]+\mu N\,E\left[\left(\overline{X}-\mu\right)^2\right]$$

Because the probability density function of $\overline{X}$ is computed as the convolution of the X probability density function, and because symmetry is preserved by convolution, we can write:

$$E[(\overline{X}-\mu)^3]=0$$

It results:

$$E(C)=0+\mu N\frac{\sigma^2}{N}=\mu\sigma^2$$

Finally, the theoretical mean of (4) is:

$$E\left[(N-1)S^2\,\overline{X}\right]=E(B)-E(C)=N\mu\sigma^2-\mu\sigma^2=(N-1)\mu\sigma^2$$

$$(N-1)E\left[S^2\,\overline{X}\right]=(N-1)\mu\sigma^2 \;\Rightarrow\; E\left[S^2\,\overline{X}\right]=\sigma^2\mu \tag{5}$$

As $E\left[S^2\right]=\sigma^2$ and $E\left[\overline{X}\right]=\mu$, (5) becomes:

$$E\left[S^2\,\overline{X}\right]=E\left[S^2\right]E\left[\overline{X}\right] \tag{6}$$

Relation (6) means that the two mean and variance estimators are uncorrelated random variables.

3.2 Detecting Asymmetry of a Distribution

By using both the theoretical result proved in Section 3.1 and the independence test presented in Appendix, we provide a way to reveal the asymmetry of the probability distribution (to detect skewness, [2], [5]).

By applying the test procedure [1], see Appendix, we can detect correlation between sample mean and sample variance. If the independence test from Appendix indicates correlation, then the probability density function has an asymmetric shape (as we proved in Section 3.1). The correlation decision will be taken by computing the correlation coefficients r and r_S; this decision can be also graphically revealed by inspecting the regression slope on the two scatter diagrams (corresponding to Spearman's ranks and to the $u\,v$ data).

4 Conclusions

To sum up, there are basically three aspects of interest presented in this article, as follows.

First, by using the original statistical independence test method [1], see Appendix, we reached an experimental way to verify the known results regarding statistical independence between mean and variance estimators in the case of normal populations. As already mentioned in Section 1, this is a very important theoretical and practical result, but the proof is quite difficult and therefore avoided in most textbooks. Considering this as the starting point of the article, we enriched the study by experimentally investigating the mean-variance data pairs in case of non-gaussian laws. We noticed that, for non-gaussian symmetrical laws, mean and variance are still uncorrelated, but not independent. So, symmetrical distributions seem to produce uncorrelated mean and variance.

Second, we theoretically proved that, for symmetrical distributions, the two estimators, mean and variance, are uncorrelated.

Third, we provide a way to test skewness (a property indicating the asymmetry of the probability distribution).

References

1. Badea, B., Vlad, A.: Revealing statistical independence of two experimental data sets. An improvement on Spearman's algorithm. In: Gavrilova, M.L., Gervasi, O., Kumar, V., Tan, C.J.K., Taniar, D., Laganá, A., Mun, Y., Choo, H. (eds.) ICCSA 2006. LNCS, vol. 3980, pp. 1166–1176. Springer, Heidelberg (2006)
2. Frieden, B.R.: Probability, Statistical Optics and Data Testing. Springer, Heildelberg (1983)
3. Devore, J.: Probability and Statistics for Engineering and the Sciences, 2nd edn. Brooks/Cole Publishing Company, Monterey (1987)
4. Walpole, R.E., Myers, R.H.: Probability and Statistics for Engineers and Scientists, 4th edn. MacMillan Publishing, NY (1989)
5. Mood, A., Graybill, F., Boes, D.: Introduction to the Theory on Statistics, 3rd edn. McGraw-Hill Book Company, New York (1974)
6. Lévine, B.: Fondements Théoriques de la Radiotechnique Statistique. Tome II, Edition Mir, Moscou (1973)
7. Mihoc, Gh., Urseanu, V.: Matematici aplicate în statistică. Ed. Academiei, Bucureşti (1962)
8. Iosifescu, M., Mihoc, Gh., Theodorescu, R.: Teoria probabilităţilor şi statistică matematică. Ed. Tehnică, Bucureşti (1966)

Appendix

A Statistical Independence Test Applicable on Two Continuous Random Variables: An Improvement on Spearman's Algorithm

We briefly resume the test algorithm from [1] to enable both the understanding of the experimental results in Section 2 and the practical steps required in Section 3.2. This algorithm permits to test statistical independence of experimental data pairs coming out from continuous random variables. It is applicable to all kinds of continuous random variables, normal or not, even of unknown probability laws. It also enables a better understanding, from the engineering point of view, of the kind of dependence existing between the two random variables (dependence/independence, correlation/non-correlation).

Note that a statistical correlation test between two continuous random variables is always based on the linear regression analysis. Thus, the correlation of the two random variables is equivalent to the existence of a linear dependence; to test a linear dependence between the two random variables means in fact to reveal that there is a significant data trend (ascending or descending) - *i.e.* correlation exists if there is a non-zero linear regression slope. The linear regression slope can be also inspected graphically; so scatter diagrams are very helpful.

Here, the two random variables under investigation will be denoted Y and Z.

The starting point is the acquisition of the two data samples of n size, corresponding to the two random variables, Y and Z, each of the two data sets complying with the *i.i.d.* statistical model:

$$y_1, y_2, \ldots, y_i, \ldots, y_n \quad \text{and} \quad z_1, z_2, \ldots, z_i, \ldots, z_n .$$

The notation (y_i, z_i) means a pair of values for Y and Z. Thus, there are n experimental data pairs in total.

The n sample size is deduced by means of a known result from the Kolmogorov-Smirnov concordance test (where α is the statistical significance level and Δ_α the corresponding α point value of the Kolmogorov-Smirnov test) [1], [6]:

$$n \geq (ln(2/\alpha))/(2\Delta_\alpha^2) .$$

The two hypotheses of the statistical test [1] are: the null hypothesis, H_0, meaning that random variables Y and Z are statistically independent, and the alternative hypothesis, H_1, meaning that random variables Y and Z are dependent. The decision is based upon a chosen statistical significance level α (of this independence test).

The independence test algorithm consists in the following main steps:

1. Sort y_i and z_i data values in increasing order and obtain the experimental cumulative distribution functions for Y and for Z, $Fe_Y(y)$ and $Fe_Z(z)$.

2. Apply $Fe_Y(y)$ transform on the y_i data values and $Fe_Z(z)$ on the values z_i, $i=1 \ldots n$, and obtain the new data y'_i and z'_i, which are uniformly distributed in (0; 1). By this data transform, Spearman's ranks are revealed: $n\,y'_i = Rank(y_i)$ and $n\,z'_i = Rank(z_i)$.

3. Another random variable transform which is the inverse of the distribution function of the standard normal law, computed by numerical methods, is applied on data y'_i and z'_i, $i=1 \ldots n$, generating the new data u_i and v_i. According to the known theory, both u_i and v_i values are standard normally distributed. Thus, two standard normal random variables U and V are obtained starting from the initial Y and Z. As U and V are normal random variables, one can now benefit from Pearson's independence test on gaussian populations, so test will continue with step 4.

Note the equivalence of testing the independence of Y and Z based on testing the independence of U and V. This is a result of: U is a function only of Y and V is a function only of Z. If Y and Z are independent random variables, it results that U and V are also independent. As a first clear conclusion, if U and V are dependent, then the initial Y and Z random variables are dependent. Conversely, if U and V are independent, that points to independence between Y and Z. Notice that the testing method applied in the continuous case requires a very large amount of data for an accurate estimation of the cumulative distribution functions.

4. Compute r, the correlation coefficient between U and V, and then the t test value:

$$r = \sum_{i=1}^{n} (u_i - \bar{u})(v_i - \bar{v}) \Big/ \sqrt{\sum_{i=1}^{n} (u_i - \bar{u})^2 \sum_{i=1}^{n} (v_i - \bar{v})^2} \; ; \qquad t = \sqrt{\frac{r^2}{1-r^2}(n-2)} \, . \tag{7}$$

If the null hypothesis H_0 is correct, then the t test value should obey the Student law with n–2 degrees of freedom.

It can be easily shown that there is equality between r (from (7)) and b_1 (from (8)), where b_1 is the linear regression slope (V versus U):

$$b_1 = \sum_{i=1}^{n} \left(u_i v_i - \bar{u}\,\bar{v}\right) \Big/ \sum_{i=1}^{n} \left(u_i^2 - \bar{u}^2\right) \tag{8}$$

5. Based on the α significance level of the independence test, compute $t_{\alpha/2}$, the $\alpha/2$ point value of the Student law of n–2 degrees of freedom.

If $|t| \geq t_{\alpha/2}$, the random variables Y and Z are dependent (and correlated) and no further investigation is required. Otherwise, $|t| < t_{\alpha/2}$, then the two random variables Y and Z could be independent. To decide upon independence, continue with step 6.

6. If $|t| < t_{\alpha/2}$, then test if U and V are jointly normally distributed. If U and V are proven jointly normally distributed, then we can say (if n is large enough) that Y and Z are statistically independent. If U and V are not jointly normally distributed, then Y and Z are dependent, but uncorrelated.

In order to decide upon the normal bivariate law, the visual inspection of the (u, v) scatter diagram will be very useful. Thus, if Y and Z are independent random variables, then the (u, v) scatter diagram should be similar to that shown in Fig. 11 (the reference u v scatter diagram). So, in the case of non-correlation decision, further independence investigation can be made by means of this graphical tool.

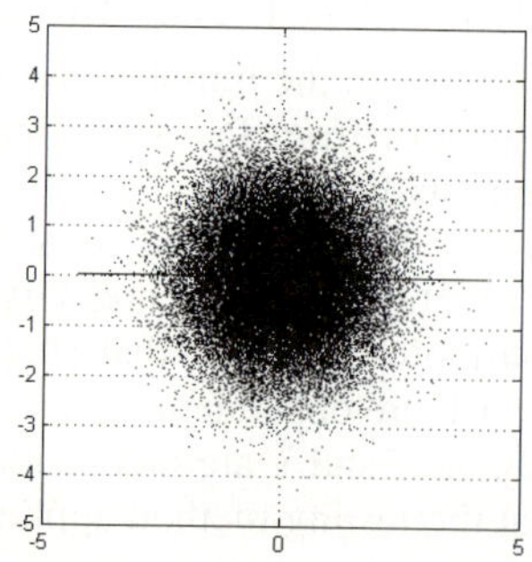

Fig. 11. Reference *u v* scatter diagram

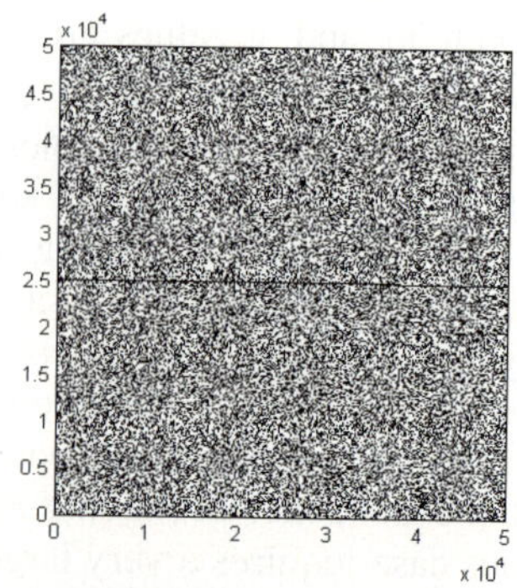

Fig. 12. Reference scatter diagram for Spearman's ranks

The test procedure graphically brings into evidence that the bivariate normal law is essential to decide upon statistical independence (each of the two random variables U and V obeys the standard normal probability law, but not always the jointly normal probability law). The reference $u\,v$ scatter diagram and the linear regression slope are displayed on the same figure (Fig. 11).

In the examples in Section 2, for comparison, the correlation coefficient and the Student test values are computed both for $u\,v$ data and for Spearman's ranks. The Spearman's rank correlation coefficient r_S and the t_S test value are computed as in (7), but instead of $u\,v$ pairs, $y'z'$ are used. It also helps to inspect the scatter diagram for $y'z'$ data (scatter diagram for Spearman's ranks); if Y and Z are independent random variables, then r_S will be practically zero, and Spearman's ranks scatter diagram should resemble the one given in Fig. 12. The respective linear regression slope is also displayed on the same Fig. 12.

Feasibility and Dominance Rules in the Electromagnetism-Like Algorithm for Constrained Global Optimization

Ana Maria A.C. Rocha and Edite M.G.P. Fernandes

Department of Production and Systems
University of Minho
4710-057 Braga, Portugal
{arocha,emgpf}@dps.uminho.pt

Abstract. This paper presents the use of a constraint-handling technique, known as feasibility and dominance rules, in a electromagnetism-like (ELM) mechanism for solving constrained global optimization problems. Since the original ELM algorithm is specifically designed for solving bound constrained problems, only the inequality and equality constraints violation together with the objective function value are used to select points and to progress towards feasibility and optimality. Numerical experiments are presented, including a comparison with other methods recently reported in the literature.

1 Introduction

The problem that is addressed in the paper considers finding a global solution of a nonlinear optimization problem in the following form:

$$\begin{array}{ll} \text{minimize} & f(x) \\ \text{subject to} & g(x) \leq 0,\ h(x) = 0,\ x \in \Omega, \end{array} \tag{1}$$

where $f : \mathbb{R}^n \to \mathbb{R}$, $g : \mathbb{R}^n \to \mathbb{R}^p$ and $h : \mathbb{R}^n \to \mathbb{R}^m$ are nonlinear continuous functions and the closed set Ω is defined by $\Omega = \{x \in \mathbb{R}^n : l \leq x \leq u\}$. We assume that the objective function f is non-convex and possesses many local minima in the feasible region. This class of global optimization problems is very important and frequently encountered in engineering applications.

In the last decades, many algorithms have been proposed to solve problem (1). Probably the most extensively used in practice are stochastic-type algorithms.

In this paper, we are interested in the electromagnetism-like (ELM) algorithm proposed in [2]. This is a population-based algorithm that simulates the electromagnetism theory of physics by considering each point in the population as an electrical charge. The method uses an attraction-repulsion mechanism to move a population of points towards optimality. The ELM algorithm is specifically designed for solving optimization problems with bound constraints [1,2,3]

$$\begin{array}{ll} \text{minimize} & f(x) \\ \text{subject to} & x \in \Omega. \end{array} \tag{2}$$

O. Gervasi et al. (Eds.): ICCSA 2008, Part II, LNCS 5073, pp. 768–783, 2008.

We have been incorporating different strategies into the ELM algorithm in order to improve the accuracy of the results as well as the efficiency of the overall algorithm. A pattern search method was introduced as a local search procedure and compared with the original ELM algorithm [18]. A modification in the total force vector, that is used to move each point in the population, was tested and compared with other well-known heuristics such as the genetic algorithm, the particle swarm optimization (PSO) and an hybrid evolutionary algorithm (EA) [20]. The effect of different charge calculations has been analyzed and compared with PSO, EA and differential evolution methods [19].

In this work, the ELM algorithm is extended for solving constrained optimization problems like (1). A previous extension of the ELM algorithm to constrained problems relies on penalty and barrier methods [1]. In these methods, constrained problems like (1) can be transformed into the bound constrained form (2) so that the ELM algorithm can be directly applied.

The extension herein presented for handling the constraints $g(x) \leq 0$ and $h(x) = 0$ considers a very simple heuristic consisting of three rules, designated by feasibility and dominance (FAD). The reader is referred to [13,26] for details. In [13], FAD is used in an artificial bee colony (ABC) optimization algorithm context, whereas in [26], the authors incorporate the heuristic FAD into a particle evolutionary swarm optimization algorithm, therein denoted by PESO.

The paper is organized as follows. In Sect. 2 we describe the ELM algorithm. Sect. 3 resumes the FAD rules for handling the constraints, and Sect. 4 presents the modifications that were introduced in the ELM algorithm. Sect. 5 contains the results of all the numerical experiments, including comparisons with other methods, and we conclude the paper in Sect. 6.

2 The Electromagnetism-Like Mechanism for Bound Constrained Problems

In this section we describe the ELM algorithm for solving problem (2). The algorithm starts with a population of randomly generated points from the feasible set Ω. Analogous to electromagnetism, each point is a charged particle that is released to the space. The charge of each point is related to the objective function value and determines the magnitude of attraction of the point over the others in the population. The better the objective function value, the higher the magnitude of attraction. The charges are used to find a direction for each point to move in subsequent iterations. The regions that have higher attraction will signal other points to move towards them. In addition, a repulsion mechanism is also introduced to explore new regions for even better solutions. A more detailed explanation follows.

First, a population of p_{size} points is randomly generated from the feasible region Ω. Let x^i be the ith point of the population. Then each coordinate of a point, denoted as x^i_k $(k = 1, \ldots, n)$, is computed by

$$x^i_k = l_k + \lambda(u_k - l_k) \tag{3}$$

where $\lambda \sim U(0,1)$. All points are evaluated (the corresponding objective function values are computed) and compared in order to identify the best point, x^{best}.

Then the total force vector F^i exerted on each point x^i by the other $p_{size}-1$ points is calculated by adding the individual component forces, F_j^i, between any pair of points x^i and x^j of the population. According to the electromagnetism theory, each individual force is inversely proportional to the square of the distance between the two points and directly proportional to the product of their charges.

Since the charge q^i of point x^i determines the power of attraction or repulsion for that point, the charge is computed according to the objective function value by

$$q^i = \exp\left(\frac{-n(f(x^i)-f(x^{best}))}{\sum_{j=1}^{p_{size}}(f(x^j)-f(x^{best}))}\right), i=1,\ldots,p_{size}. \tag{4}$$

In this way the points that have better objective function values possess higher charges. This is a scaled distance of the function value at x^i to the function value of the best point in the population. Different approaches for charge evaluations may be adopted, see for example [6,12,19].

Since the charges (4) are all positive, the direction of a force F_j^i depends on the objective function values at x^i and x^j. If $f(x^j) < f(x^i)$ then the point x^j attracts x^i and the direction of the force should be $\overrightarrow{x^i x^j}$, whereas if $f(x^j) \geq f(x^i)$ then x^j repels x^i and the direction of the force is $\overrightarrow{x^j x^i}$. Thus,

$$F_j^i = \begin{cases} (x^j - x^i)\frac{q^i q^j}{\|x^j - x^i\|^3} & \text{if } f(x^j) < f(x^i) \\ (x^i - x^j)\frac{q^i q^j}{\|x^j - x^i\|^3} & \text{otherwise} \end{cases} \tag{5}$$

for $j \neq i$. The total force vector F^i exerted on each point x^i is then calculated by adding the individual component forces, F_j^i,

$$F^i = \sum_{j \neq i}^{p_{size}} F_j^i, \quad i=1,\ldots,p_{size}.$$

Finally, this total force vector, F^i, is used to move the point x^i in the direction of the force by a random step length $\lambda \sim U(0,1)$ as follows

$$x_k^i = \begin{cases} x_k^i + \lambda\frac{F_k^i}{\|F^i\|}(u_k - x_k^i) & \text{if } F_k^i > 0 \\ x_k^i + \lambda\frac{F_k^i}{\|F^i\|}(x_k^i - l_k) & \text{otherwise} \end{cases} \tag{6}$$

for each k ($k=1,2,\ldots,n$) and for $i=1,\ldots,p_{size}$ and $i \neq best$. The best point, x^{best}, is not moved and is carried out to the subsequent iteration.

We would like to remark that, as shown in (6), feasibility is maintained by using the normalized total force vector and scaling it with the allowed range of movement towards the lower bound l_k, or the upper bound u_k, of the set Ω.

This ELM algorithm was indeed designed to generate and iteratively move points in the population that are feasible in a bound constrained problem context.

Like other hybrid population-based algorithms, the ELM algorithm incorporates a local search procedure in order to improve the accuracy of the solution. Although this local refinement could be applied to all points in the population, it has been shown that when the local procedure is applied only to the best point, the accuracy of the results still improves and the number of function evaluations is not extremely high [2].

Thus, the local refinement is applied coordinate by coordinate to the best point in the population as described below. Based on a parameter δ, the maximum feasible step length $s_{max} = \delta(\max_k(u_k - l_k))$ is computed. Then, for each coordinate k, a random number λ between 0 and 1 is selected as a step length and a new point is calculated along that direction. The new point is compared with the old one. If an improvement is observed, within nit_{max} iterations, the new point replaces the best point and the search continues with the next coordinate.

The ELM algorithm for bound constrained optimization can be described as shown below.

Algorithm 1. *(ELM algorithm)*
Given l, u, p_{size}, nfe_{max}, nit_{max}, δ and set $k = 1$
`step` *1 Randomly generate a population of p_{size} points in the set Ω, using (3)*
`step` *2 Evaluate the points and compare them to select the best point x^{best}*
`step` *3 If termination criterion is satisfied, stop*
`step` *4 Compute the charges of all points by using (4)*
`step` *5 Compute the forces and select their directions based on function values using (5)*
`step` *6 Compute the total force vector for each point in the population*
`step` *7 Move all points, except the best point, in the set Ω according to (6). Evaluate the points and compare them to select the best point*
`step` *8 Perform a local search about the best point to look for an improvement*
`step` *9 Set $k = k + 1$ and go to step 3.*

At this moment there are two issues in the Algorithm 1 that need to be further explained. One involves the evaluation of the points, computing the corresponding objective function values, and their comparison, and the other is concerned with the termination criterion.

We remark here that in *step 2*, *step 5*, *step 7* and *step 8* of the algorithm, the points in the population are compared solely based on function values, in order to select the best point. Thus, in the context of solving problem (2), the rule that is operated is summarized as follows:

– when comparing two points at a time, the one with lowest function value is selected.

Finally, the condition that is used to terminate the algorithm considers a limit on the number of function evaluations, nfe_{max}.

3 Heuristic for Handling Constraints

Most stochastic methods for global optimization are developed primarily for unconstrained problems. Then they are extended to constrained optimization problems by modifying the original procedures or by using penalty functions.

The most common approach to solving constrained problems is based on penalty functions. Penalty terms are added to the objective function to penalize constraints violation. The penalty techniques transform the constrained problem into an unconstrained problem by penalizing f when constraints are violated and then minimizing the penalty function using methods for unconstrained problems. There are two main classes of penalty terms:

i) a stationary penalty term that uses fixed penalty parameters throughout the iterative process and depends on constraints violation [1,4,16];
ii) a non-stationary penalty term that changes dynamically and depends on the iteration counter [14,15,17,22].

Barrier methods use a similar framework although they require initial feasible points and maintain feasibility along the iterative process [1].

Recently, filled function methods have also been used for solving constrained global optimization [25]. Other constraint-handling techniques treat constrained optimization as multi-objective optimization, where constraints violation is regarded as an objective function. Adapted from the multi-objective optimization, a relaxed dominance concept is used in [24] to assess progress towards feasibility and optimality. Another approach that is able to balance dominance between objective and penalty functions in a stochastic manner, known as stochastic ranking, is proposed and tested in [21]. Very simple techniques that aim to preserve feasibility are adopted in [10,11]. A comprehensive survey of the most popular constraint-handling techniques used in evolutionary algorithms is provided in [5].

The approach herein adopted for handling the inequality and equality constraints of problem (1) relies on a simple heuristic consisting of three rules denoted by feasibility and dominance (FAD) rules [13,26].

We assume that all equality constraints are converted into inequality constraints, $|h_j| \leq \epsilon$, $j = 1, \ldots, m$. (For example, $\epsilon = 0.001$.) For simplicity the problem (1) is rewritten as

$$\begin{array}{l} \text{minimize } \ f(x) \\ \text{subject to } c(x) \leq 0, \, x \in \Omega, \end{array} \tag{7}$$

where the vector of inequality constraints of the problem is defined by $c(x) = (g_1(x), \ldots, g_p(x), |h_1(x)| - \epsilon, \ldots, |h_m(x)| - \epsilon)$, and the constraints violation is measured by the l_2 norm of the vector

$$[c_j(x)]^+ = \max\{0, c_j(x)\}, \; j = 1, \ldots, p+m\,.$$

In the sequel, a point x^i with $\|[c(x^i)]^+\|_2 = 0$ is feasible, whereas if $\|[c(x^i)]^+\|_2 > 0$ then the point is infeasible. When using the FAD rules, two points are compared at a time, and

- if both points are feasible, the one with lowest function value is selected;
- if one point is feasible and the other is infeasible, the feasible point is selected;
- if both points are infeasible, the one with lowest constraints violation is selected.

4 FAD-ELM Algorithm for Constrained Problems

The modifications that were introduced in the ELM algorithm for handling the inequality constraints $c(x) \leq 0$ in problem (7) are justified and presented below.

4.1 Evaluation of the Points

The main modification is concerned with the evaluation of the points in the population, so that the best point can be selected. This procedure appears at the beginning of the iterative process, when the population is randomly generated, and when the points are moved according to the total force vector.

In the constrained problem (7) context, when the evaluation of the points is required, the objective function value and the constraints violation are computed for each point of the population. This pair of information $(f(x^i), \|[c(x^i)]^+\|_2)$ defines the point fitness.

4.2 The Direction of Forces

To decide the direction of the individual forces F_j^i, the points x^j and x^i are compared. We say that "x^j is better than x^i" if x^j is the selected point according to the FAD rules and consequently x^i is attracted to x^j. Conversely, x^i is repelled by x^j if x^i is the selected point according to the FAD rules. Hence, the new conditions are summarized as follows (for $j \neq i$):

$$F_j^i = \begin{cases} (x^j - x^i)\frac{q^i q^j}{\|x^j - x^i\|^3} & \text{if "}x^j \text{ is better than } x^i\text{"} \\ (x^i - x^j)\frac{q^i q^j}{\|x^j - x^i\|^3} & \text{otherwise} \end{cases} . \tag{8}$$

4.3 Charges Calculation

Another important issue is related with the charges calculation. Maintaining a similar approach to that of the original ELM algorithm, where only objective function values are compared, then

$$q^i = \exp\left(\frac{-n|f(x^i) - f(x^{best})|}{\sum_{j=1}^{p_{size}} |f(x^j) - f(x^{best})|}\right), i = 1, \ldots, p_{size}. \tag{9}$$

We now use absolute values because there is no guarantee that the best point in the population has the smallest function value. The charges q^i ($i = 1, \ldots, p_{size}$), from (9), remain positive and the direction of each individual force F_j^i relies on x^i and x^j fitness as described above.

Furthermore, when constraints are present in the problem like (7), two objectives have to be achieved in the way to the solution: one is to minimize the objective function f and the other is to minimize constraints violation $\|[c(x)]^+\|_2$. Since a solution must be a feasible point, the minimization of $\|[c(x)]^+\|_2$ is indeed the most important. Thus, the charge of a point could be related to its constraints violation in the sense that feasible points are the only candidates to solutions. The better the constraints violation, the higher the magnitude of attraction. Our proposal here considers the charge calculation as a scaled measure of feasibility

$$q^i = \exp\left(\frac{-n\|[c(x^i)]^+\|_2^2}{\sum_{j=1}^{p_{size}} |f(x^j) - f(x^{best})|} \right), i = 1, \ldots, p_{size}. \tag{10}$$

In this way, the feasible points in the population have unit charges, whereas the infeasible points have smaller charges and the larger the violation the smaller the charge. This means that feasible points have the ability to attract infeasible points.

4.4 The Local Search

When a local refinement is applied coordinate by coordinate to the best point, two points are to be compared at a time to see if an improvement is achieved. The fitness of the points are compared according to the FAD rules.

4.5 The Algorithm

The algorithm for solving constrained global optimization problems that uses the electromagnetism-like mechanism and the feasibility and dominance rules for constraints handling is as follows:

Algorithm 2. *(FAD-ELM algorithm)*
Given l, u, p_{size}, nfe_{max}, nit_{max}, δ and set $k = 1$
`step` *1 Randomly generate a population of p_{size} points in the set Ω, using (3)*
`step` *2 Evaluate the points and apply the* `FAD rules` *to select the best point x^{best}*
`step` *3 If termination criterion is satisfied, stop*
`step` *4 Compute the charges of all points by using (9) or (10)*
`step` *5 Compute the forces and apply the* `FAD rules` *to select the direction of the force using (8)*
`step` *6 Compute the total force vector for each point in the population*
`step` *7 Move all points, except the best point, in the set Ω according to (6). Evaluate the points and apply the* `FAD rules` *to select the best point*
`step` *8 Perform a local search about the best point and apply the* `FAD rules` *to look for an improvement*
`step` *9 Set $k = k + 1$ and go to step 3.*

5 Numerical Experiments

To evaluate the performance of the herein proposed ELM algorithm for constrained problems a set of 13 benchmark problems, described in full detail in the Appendix of [26], is used. The problems are known as g01, g02, ..., g13 [13,21,26]. We remark that g02, g03, g08 and g12 are maximization problems while the others are minimization ones. The algorithm is coded in the C programming language and it contains an interface to connect to AMPL so that the problems coded in AMPL could be easily solved [9]. AMPL is a mathematical programming language that allows the codification of optimization problems in a powerful and easy to learn language. The set of coded problems may be obtained from the first author upon request. The fixed values for the parameters in the ELM algorithm are as follows: $\delta = 0.001$ and $nit_{max} = 10$.

Details of the selected problems are listed in Table 1, where "Prob" identifies the problem, n represents the number of variables, p and m are the number of inequality and equality constraints respectively and n_{act} is the number of active constraints at the solution.

Initially, we study the two versions of the ELM algorithm for constrained optimization based on the FAD rules, which differ in the computation of the charges, see (9) and (10) respectively. We select the best version and report detailed results obtained after 30 independent runs each starting from a random population with different seeds. Here, we consider a population size of 50 points and use a limit of 350000 function evaluations to terminate the algorithm.

We then compare our results using the best and the average objective function values obtained after 30 runs, with other methods recently reported in the literature. For another set of comparisons, we repeat the experiments with different conditions: $p_{size} = 40$ and a limit of 240000 function evaluations.

5.1 Comparison of the Two Versions

The comparison of the two versions of the FAD-ELM algorithm which differ in the computation of the charges, herein designated by FAD-ELM-(9) and FAD-ELM-(10), corresponding to (9) and (10) respectively, is based on the Dolan and Moré's performance profile approach as outline in [7]. The performance profiles give, for every $\tau \geq 1$, the proportion $\rho(\tau)$ of test problems on which each algorithm under comparison has a performance within a factor τ of the best. An explanation of our implementation of this approach follows.

Let $\mathcal{P}$ be the set of all problems and $\mathcal{S}$ the set of solvers used in the comparative study. Let $m_{(p,s)}$ be the performance metric found by solver $s \in \mathcal{S}$ on problem $p \in \mathcal{P}$ after a fixed number of function evaluations. The metric that is used when the performance assessment is on the best result obtained by the solvers is

$$m_{(p,s)} = f_{best\,(p,s)}, \tag{11}$$

where $f_{best\,(p,s)}$ is the best function value found by solver s on problem p, after $nfe_{max} = 350000$ function evaluations, and over 30 runs. However, if the performance assessment is on the average result obtained by the solvers then $f_{avg\,(p,s)}$ should de used in the metric (11).

The performance ratio adopted for this comparative study is

$$r_{(p,s)} = \frac{m_{(p,s)}}{\min\left\{m_{(p,s)} : s \in \mathcal{S}\right\}}$$

and the overall assessment of the performance of a particular solver s is defined by

$$\rho_s(\tau) = \frac{n_{P_\tau}}{n_P}$$

where n_P is the number of problems in the set $\mathcal{P}$ and n_{P_τ} is the number of problems in the set such that the performance ratio $r_{(p,s)}$ is less than or equal to τ for solver s. Hence, $\rho_s(\tau)$ is the probability (for solver $s \in \mathcal{S}$) that the performance ratio $r_{(p,s)}$ is within a factor $\tau \in \mathbb{R}$ of the best possible ratio.

Using the performance profile plot one can compare how well a solver can estimate the optimum relative to the others. The value of $\rho_s(1)$ gives the probability that the solver s will win over the others in the set. However, for large values of τ, the $\rho_s(\tau)$ measures the solver robustness.

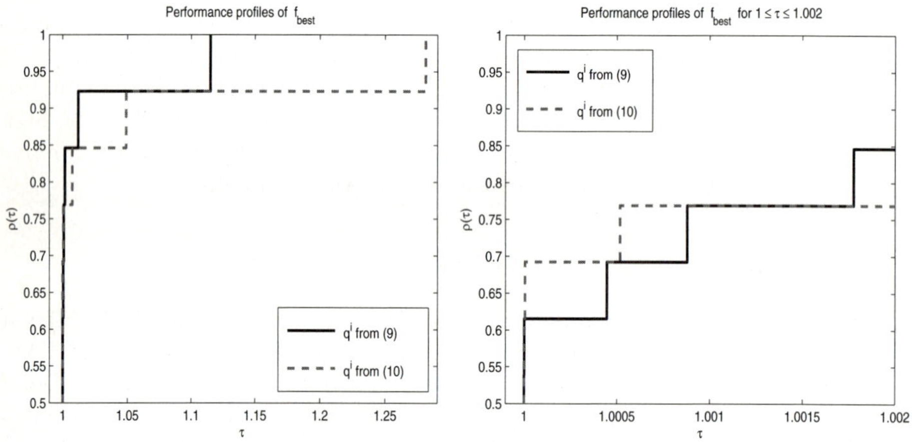

Fig. 1. Performance profiles of f_{best} over 30 runs

Figure 1 contains the performance profiles of f_{best} for the two versions in comparison. The graph on the right contains details of the plots for values of τ near 1 and shows that the version based on the charge calculation (10) is superior on the best performance. For $\tau = 1$, this version has a probability of 70% of winning over the other, whereas the version based on charge (9) is better for values of τ greater than 1.0018. When the average performance is considered, Fig. 2 shows that the version FAD-ELM-(10) outperforms the other version for

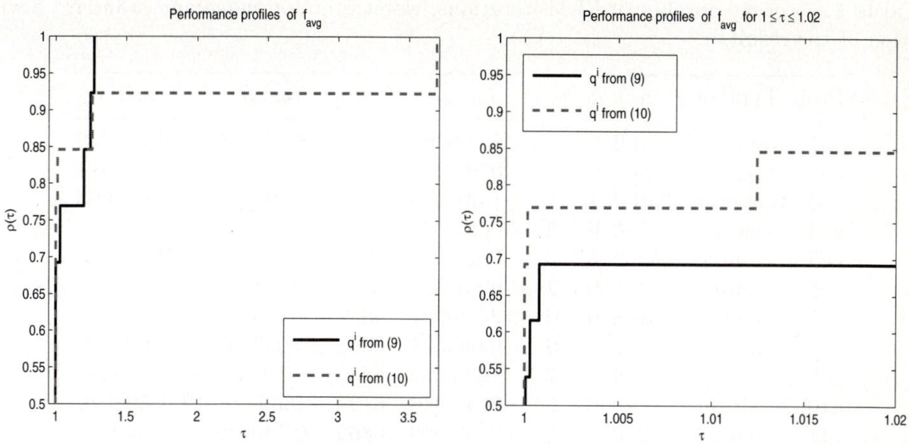

Fig. 2. Performance profiles of f_{avg} over 30 runs

small values of τ (see the graph on the right) and is exceeded by FAD-ELM-(9) for $\tau > 1.3$ (graph on the left).

To compare further the two versions of the FAD-ELM algorithm, we recorded the average number of function evaluations needed for finding a feasible solution satisfying a fixed accuracy level, herein defined by

$$|f(x^{best}) - f_{global}| \leq 0.0001$$

over the 30 runs, where f_{global} is the known global solution. For both versions, the average number of function evaluations was less than 350000 only in 3 problems:

- FAD-ELM-(9) – 323221 (in g03); 270250 (in g11); 529 (in g12);
- FAD-ELM-(10) – 303962 (in g03); 271220 (in g11); 443 (in g12)

Although the differences are not significative, FAD-ELM-(10) seems slightly better.

5.2 Detailed Results of FAD-ELM Algorithm

We now report detailed results of the best version FAD-ELM-(10). The experiments are based on a population size of 50 points and the algorithm is terminated after 350000 function evaluations. Table 1 contains the average number of ELM iterations, n_{ELM}; the worst function value and the standard deviation, *sd*, of the function values obtained over the 30 runs. The best and the average best function values are listed in subsequent tables, when comparisons with other methods are presented. The column headed with f_{global} lists the most recent known global solutions (see http://www.ntu.edu.sg/home/EPNSugan/).

Table 1. Average number of ELM iterations, worst function value and standard deviation of the results

Prob	Type of f	n	p	m	n_{act}	f_{global}	n_{ELM}	f_{worst}	sd
g01	quadratic	13	9	0	6	-15.000000	6103	-12.500515	0.770339
g02	general	20	2	0	1	0.803619	4998	0.377406	0.012574
g03	polynomial	10	0	1	1	1.000500	5928	0.996438	0.001663
g04	quadratic	5	6	0	2	-30665.539	6502	-30521.434	29.849753
g05	cubic	4	2	3	3	5126.497	6626	6117.660	273.027370
g06	cubic	2	2	0	2	-6961.814	6858	-6933.264	4.250238
g07	quadratic	10	8	0	6	24.30621	5852	103.5507	19.236274
g08	general	2	2	0	0	0.095825	6861	0.095825	0.000000
g09	general	7	4	0	2	680.630	6235	701.344	5.104498
g10	linear	8	6	0	6	7049.248	6134	16395.51	1671.593930
g11	quadratic	2	0	1	1	0.749900	6862	0.749346	0.000086
g12	quadratic	3	1	0	0	1.000000	6728	1.000000	0.000000
g13	general	5	0	3	3	0.053942	6476	20.784428	5.010076

5.3 Comparison with Other Heuristics

To demonstrate the performance of the new ELM type algorithm for general constrained global optimization problems, we carried out a comparison with six very well-known and recent methods: the PSO (standard particle swarm optimization), PESO (particle evolutionary swarm optimization), TCPSO (the Toscano and Coello's PSO), SR (stochastic ranking), DE (differential evolution) and ABC (artificial bee colony). To be able to compare our results with those of recent publications, we use the same set of 13 problems. For a fair comparison, we use the population size and the maximum number of objective function evaluations similar to the therein reported.

The results obtained by PSO, PESO and TCPSO are reported in [26] and the ones obtained by SR can be seen in [21,26]. Table 2 reports the best function value, as well as the mean best function values obtained after 30 runs. Our version, denoted by FAD-ELM$_{50}$, considers a population of 50 points, and the algorithm terminates when the number of function evaluations exceeds 350000. It is compared with two specific versions of the particle swarm optimization algorithm: PSO and PESO. From Table 2 we may conclude that FAD-ELM is better in the best results than PSO in problems g03 and g13, is better than PESO only in the problem g13, while in the problems g08, g11 and g12 is competitive with both PSO and PESO. FAD-ELM is also clearly better in the average results than PSO in problems g02, g03 and g12, is competitive in problem g08, but PSO is better in all the remaining problems of the set. PESO is also better than FAD-ELM in all problems, except in problems g08 and g12 that is competitive. We note that FAD-ELM worst values are really far from the best values, giving a poor performance on the average values.

Table 3 compares our average results with those of other algorithms: TCPSO and SR. TCPSO refers to the Toscano and Coello's PSO that incorporates a

Table 2. Comparison of the best and average results, over the 30 runs, with PSO and PESO

Prob	f_{global}	best results FAD-ELM$_{50}$	PSO	PESO	average results FAD-ELM$_{50}$	PSO	PESO
g01	-15.000000	-14.999805	-15.000000	-15.000000	-14.590470	-14.710417	-15.000000
g02	0.803619	0.444939	0.669158	0.792608	0.427705	0.419960	0.721749
g03	1.000500	1.002983	0.993930	1.005010	0.999661	0.764813	1.005006
g04	-30665.539	-30642.600	-30665.539	-30665.539	-30591.897	-30665.539	-30665.539
g05	5126.497	5135.818	5126.484	5126.484	5338.583	5135.973	5129.178
g06	-6961.814	-6953.411	-6961.814	-6961.814	-6942.915	-6961.814	-6961.814
g07	24.30621	27.46692	24.37015	24.30692	53.08965	32.40727	24.37125
g08	0.095825	0.095825	0.095825	0.095825	0.095825	0.095825	0.095825
g09	680.630	681.777	680.630	680.630	689.896	680.630	680.630
g10	7049.248	7187.181	7049.381	7049.459	9492.735	7205.500	7099.101
g11	0.749900	0.749001	0.749000	0.749000	0.749079	0.749000	0.749000
g12	1.000000	1.000000	1.000000	1.000000	1.000000	0.998875	1.000000
g13	0.053942	0.063196	0.085655	0.081498	2.083421	0.569358	0.626881

constraints handling mechanism into the standard PSO algorithm [23], and SR denotes stochastic ranking [21]. This is an evolution strategy enhanced with a stochastic constraints handling mechanism. FAD-ELM average results are better than TCPSO results in problems g05 and g11, and are competitive in g08 and g12. FAD-ELM is better than SR in the problems g06 and g11 and is competitive in problems g08 and g12.

In this part of the paper we compare the FAD-ELM algorithm with DE and ABC algorithms. The results of DE and ABC algorithms are reported in [13]. The best function value and the mean best function values are listed in Table 4. The version under comparison, denoted by FAD-ELM$_{40}$, considers a population of 40 points, and the algorithm terminates when the number of function evaluations exceeds 240000. From Table 4, we note that FAD-ELM best results are better than DE results in problems g02, g03, g11 and g13, and are competitive in problems g08 and g12. When comparing with ABC results, FAD-ELM wins in the problems g03, g11 and g13. Both algorithms are competitive in problems g08 and g12. Although the average results are competitive in problems g08 and g12, FAD-ELM is clearly better only in the problem g11. In the table, "-" means that no feasible solutions were found.

5.4 Comparison with Results in Birbil's Thesis [1]

To further examine the performance of the proposed feasibility and dominance ELM algorithm, we carried out one more comparison. The results reported in [1], where penalty and barrier methods are incorporated into the ELM algorithm for the constraints handling, are compared with FAD-ELM results. The tests involve a set of 5 problems selected from [8]. The results are reported in Table 5. We use

Table 3. Comparison of the average results, over the 30 runs, with TCPSO and SR

Prob	f_{global}	FAD-ELM$_{50}$	TCPSO	SR
g01	-15.000000	-14.590470	-15.000000	-15.000000
g02	0.803619	0.427705	0.790406	0.781975
g03	1.000500	0.999661	1.003814	1.000000
g04	-30665.539	-30591.897	-30665.500	-30665.539
g05	5126.497	5338.583	5461.081	5128.881
g06	-6961.814	-6942.915	-6961.810	-6875.940
g07	24.306201	53.08965	25.35577	24.374
g08	0.095825	0.095825	0.095825	0.095825
g09	680.630	689.896	680.852	680.656
g10	7049.248	9492.735	7560.048	7559.192
g11	0.749900	0.749079	0.750107	0.750000
g12	1.000000	1.000000	1.000000	1.000000
g13	0.053942	2.083421	1.716426	0.057006

Table 4. Comparison of the best and average results, over the 30 runs, with DE and ABC

		best results			average results		
Prob	f_{global}	FAD-ELM$_{40}$	DE	ABC	FAD-ELM$_{40}$	DE	ABC
g01	-15.000000	-14.999830	-15.000	-15.000	-14.503678	-14.555	-15.000
g02	0.803619	0.491136	0.472	0.803598	0.427662	0.665	0.792412
g03	1.000500	1.003318	1.000	1.000	0.999822	1.000	1.000
g04	-30665.539	-30605.032	-30665.539	-30665.539	-30573.439	-30665.539	-30665.539
g05	5126.497	5126.672	5126.484	5126.484	5303.154	5264.270	5185.714
g06	-6961.814	-6953.069	-6954.434	-6961.814	-6936.598	-	-6961.813
g07	24.30621	25.81694	24.306	24.330	66.57473	24.310	24.473
g08	0.095825	0.095825	0.095825	0.095825	0.095825	0.095825	0.095825
g09	680.630	681.792	680.630	680.634	689.423	680.630	680.640
g10	7049.248	7739.708	7049.248	7053.904	9476.079	7147.334	7224.407
g11	0.749900	0.749003	0.752	0.750	0.749074	0.901	0.750
g12	1.000000	1.000000	1.000	1.000	1.000000	1.000	1.000
g13	0.053942	0.095896	0.385	0.760	1.312976	0.872	0.968

"ALG" to identify the algorithm; "P" means penalty method, "B" stands for barrier method, and nfe is the average number of function evaluations computed over the 10 runs. The table also contains the details of the problems (Type of f, n, p and m). In these comparisons the termination criterion in Algorithm 2 limits the number of iterations, herein denoted by $nELM_{max}$, instead of the number of function evaluations.

For a fair comparison we use the conditions reported in [1]: $\delta = 0.01$ in the local procedure of the ELM algorithm; 10 runs were made for each problem; the

Table 5. Comparison with the results in [1]

Prob	Type of f	n	p	m	ALG	f_{global}	f_{best}	f_{avg}	nfe
TP1	quadratic	5	6	0	P [1]	-30665.539	-30563.285	-30374.467	2534
					B [1]		-30596.784	-30447.682	2264
					FAD-ELM		-30399.961	-30168.944	2632
TP2	quadratic	6	6	0	P [1]	-310	-297.7095	-293.9035	4311
					B [1]		-307.2018	-297.8254	3969
					FAD-ELM		-297.2020	-295.190	4577
TP3	linear	2	2	0	P [1]	-5.508013	-5.5036	-5.4256	886
					B [1]		-5.4756	-5.4245	885
					FAD-ELM		-5.500585	-5.451045	1089
TP4	general	3	1	0	P [1]	-83.254	-83.096	-82.5450	1597
					B [1]		-83.1574	-81.9877	1347
					FAD-ELM		-83.063359	-82.481219	1684
TP5	general	4	2	0	P [1]	-5.7398	-5.6450	-5.4857	1092
					B [1]		-5.6346	-5.0523	1251
					FAD-ELM		-5.687241	-5.645319	1812

size of the population and the maximum number of FAD-ELM iterations are problem dependent:

- TP1 – $p_{size} = 30$; $nELM_{max} = 75$;
- TP2 – $p_{size} = 40$; $nELM_{max} = 100$;
- TP3 – $p_{size} = 20$; $nELM_{max} = 50$;
- TP4 – $p_{size} = 30$; $nELM_{max} = 50$;
- TP5 – $p_{size} = 20$; $nELM_{max} = 75$.

Although the herein constraint-handling technique relies on the same set of information that is used in penalty/barrier function methods, namely the objective function value and the constraints violation, to progress towards feasibility and optimality, it avoids the use and updating of the penalty parameter. The FAD rules are less demanding than the usual penalty function approach when a comparison between points is carried out. Nonetheless the obtained results can be considered quite satisfactory.

6 Final Remarks

This paper presents a new ELM type algorithm for solving constrained global optimization problems. The technique used for the constraints handling relies on a very simple heuristic known as the feasibility and dominance rules. These rules are easily incorporated into the ELM algorithm whenever

- the best point in the population has to be selected;
- the direction of individual forces between pair of points in the population has to be selected;

– a local refinement is applied to the best point and an improvement has to be assessed.

The preliminary results reported in the paper seem to show that this approach is competitive with other algorithms, although some modifications are in progress to improve accuracy. Further testing with a different set of constrained problems [4] and engineering problems like the ones reported in [16,24] are required. Since the charges calculation seems to be a crucial issue in the electromagnetism-like mechanism, future developments will focus on a new approach so that a balance between the objective function value and the constraints violation is achieved.

Acknowledgments. The authors are grateful to two anonymous referees for their comments and suggestions.

References

1. Birbil, S.I.: Stochastic global optimization techniques. PhD thesis, North Carolina State University (2002)
2. Birbil, S.I., Fang, S.-C.: An electromagnetism-like mechanism for global optimization. Journal of Global Optimization 25, 263–282 (2003)
3. Birbil, S.I., Fang, S.-C., Sheu, R.-L.: On the convergence of a population-based global optimization algorithm. Journal of Global Optimization 30, 301–318 (2004)
4. Birgin, E.G., Floudas, C.A., Martínez, J.M.: Global minimization using an augmented Lagrangian method with variable lower-level constraints. Report (November 2006)
5. Coello Coello, C.A.: Theoretical and numerical constraint-handling techniques used with evolutionary algorithms: a survey of the state of the art. Computer Methods in Applied Mechanics and Engineering 191, 1245–1287 (2002)
6. Debels, D., DeReyck, B., Leus, R., Vanhoucke, M.: A hybrid scatter search electromagnetism meta-heuristic for project scheduling. European Journal of Operational Research 169, 638–653 (2006)
7. Dolan, E.D., Moré, J.J.: Benchmarking optimization software with performance profiles. Mathematical Programming 91, 201–213 (2002)
8. Floudas, C.A., Pardalos, P.M., Adjiman, C.S., Esposito, W.R., Gumus, Z.H., Harding, S.T., Klepeis, J.L., Meyer, C.A., Schweiger, C.A.: Handbook of test problems in local and global optimization, vol. 33. Kluwer Academic Publishers, Dordrecht (1999)
9. Fourer, R., Gay, D.M., Kernighan, B.W.: A modeling language for mathematical programming. Management Science 36(5), 519–554 (1990)
10. Hu, X., Eberhart, R.: Solving constrained nonlinear optimization problems with particle swarm optimization. In: Proceedings of 6th World Multiconference on Systemics, Cybernetics and Informatics (2002)
11. Hu, X., Eberhart, R.C., Shi, Y.: Engineering optimization with particle swarm. In: Proceedings of 2003 IEEE Swarm Intelligence Symposium, pp. 53–57 (2003)
12. Kaelo, P., Ali, M.M.: Differential evolution algorithms using hybrid mutation. Computational Optimization and Applications 37, 231–246 (2007)

13. Karaboga, D., Basturk, B.: Artificial bee colony (ABC) optimization algorithm for solving constrained optimization problems. In: Melin, P., Castillo, O., Aguilar, L.T., Kacprzyk, J., Pedrycz, W. (eds.) IFSA 2007. LNCS (LNAI), vol. 4529, pp. 789–798. Springer, Heidelberg (2007)
14. Liu, J.-L., Lin, J.-H.: Evolutionary computation of unconstrained and constrained problems using a novel momentum-type particle swarm optimization. Engineering Optimization 39(3), 287–305 (2007)
15. Parsopoulos, K.E., Vrahatis, M.N.: Particle swarm optimization method for constrained optimization problems. In: Intelligent Technologies - Theory and Applications: New Trends in Intelligent Technologies. Frontiers in Artificial Intelligence and Applications series, vol. 76, pp. 214–220 (2002)
16. Parsopoulos, K.E., Vrahatis, M.N.: Unified particle swarm optimization for solving constrained engineering optimization problems. In: Wang, L., Chen, K., S. Ong, Y. (eds.) ICNC 2005. LNCS, vol. 3612, pp. 582–591. Springer, Heidelberg (2005)
17. Petalas, Y.G., Parsopoulos, K.E., Vrahatis, M.N.: Memetic particle swarm optimization. Annals of Operations Research 156, 99–127 (2007)
18. Rocha, A.M.A.C., Fernandes, E.M.G.P.: A modified electromagnetism-like algorithm based on a pattern search method. In: ECC 2007, 8 pages. Springer, Heidelberg (2008)
19. Rocha, A.M.A.C., Fernandes, E.M.G.P.: On charge effects to the electromagnetism-like algorithm. In: Proceedings of EurOPT (to appear, 2008)
20. Rocha, A.M.A.C., Fernandes, E.M.G.P.: Performance profile assessment of electromagnetism-like algorithms for global optimization. In: IeCCS 2007. AIP Conference Proceedings, 4 pages (in press, 2008)
21. Runarsson, T.P., Yao, X.: Stochastic ranking for constrained evolutionary optimization. IEEE Transactions on Evolutionary Computation 4(3), 284–294 (2000)
22. Tahk, M.-J., Woo, H.-W., Park, M.-S.: A hybrid optimization method of evolutionary and gradient search. Engineering Optimization 39(1), 87–104 (2007)
23. Toscano, G., Coello, C.C.: A constraint-handling mechanism for particle swarm optimization. In: Proceedings of the 2004 Congress on Evolutionary Computation, pp. 1396–1403 (2004)
24. Vaz, A.I.F., Fernandes, E.M.G.P.: Optimization of nonlinear constrained particle swarm. Technological and Economic Development of Economy 12(1), 30–36 (2006)
25. Wu, Z.Y., Bai, F.S., Lee, H.W.J., Yang, Y.J.: A filled function method for constrained global optimization. Journal of Global Optimization 39, 495–507 (2007)
26. Zavala, A.E.M., Aguirre, A.H., Diharce, E.R.V.: Constrained optimization via particle evolutionary swarm optimization algorithm (PESO). In: GECCO 2005, pp. 209–216 (2005)

Pseudo-Random Bit Generation Based on 2D Chaotic Maps of Logistic Type and Its Applications in Chaotic Cryptography

C. Pellicer-Lostao and R. López-Ruiz

Department of Computer Science and BIFI,
Universidad de Zaragoza, 50009 - Zaragoza, Spain
carmen.pellicer@unizar.es, rilopez@unizar.es

Abstract. Pseudo-Random Bit Generation (PRBG) is required in many aspects of cryptography as well as in other applications of modern security engineering. In this work, PRBG based on 2D symmetrical chaotic mappings of logistic type is considered. The sequences generated with a chaotic PRBG of this type, are statistically tested and the computational effectiveness of the generator is estimated. Considering this PRBG valid for cryptography, the size of the available key space is also calculated. Different cryptographic applications can be suitable to this PRBG, being a stream cipher probably the most immediate of them.

Keywords: Pseudorandom Bit Generation, Chaotic Cryptography, Security Engineering.

1 Introduction

Many aspects of cryptography and modern security engineering depend upon the generation of pseudo-random numbers [1], [2]. Examples are the use of nonces in authentication protocols, salts in certain signature schemes, generation of keys or the keystreams of stream cyphers. The requirements of randomness in these generators vary according to their application. For example, the generation of master keys normally requires high quality or entropy [2]. But in the case of nonces, uniqueness can be the main requirement for some protocols [1].

Pseudo-Random Bit Generators (PRBG) are implemented by deterministic numeric algorithms and they should pass several statistical tests [2], [3], [4], [5], to prove themselves useful. These tests can be set up to different levels of requirements of randomness depending on the future application of the PRBG. The security of the entire cryptographic protocol or system, relies on the randomness quality of the generator [1].

Over the last two decades, several works have implemented PRBG based on chaotic systems (a complete survey can be found in [6]). Chaotic systems have the property of beeing deterministic in the microscopic space and behave randomly, when observed in a coarse-grained state-space [7]. The sensitivity of chaotic maps to initial conditions make them optimum candidates to relate

O. Gervasi et al. (Eds.): ICCSA 2008, Part II, LNCS 5073, pp. 784–796, 2008.

minimal critical information about the input, in the output sequence [8]. Their iterative nature makes them fast computable [8] and able to produce binary sequences with extremely long cycle lengths [9].

In 2006 Madhekar Suneel proposes in [10], a method for pseudo-random binary sequence generation based on the two-dimensional Hénon map [11]. The author also indicates that the choice of the Hénon map is rather arbitrary and that similar results should also be attainable with other 2D maps.

This paper explore precisely this possibility, and presents a finite automata scheme as the key to achieve that. This comprehensive scheme is then applied to two particular 2D dynamical systems presented in [12], which are formed by two symmetrically coupled logistic maps. The pseudo-random properties of the generators obtained that way, are investigated. The evaluation of the potential size of their key space and the computational cost of the algorithm makes them worth to be considered.

The chaotic PRBG algorithm described in this paper can be used in different ways. One of its applications, and maybe the most immediate, could be the construction of practical stream ciphers. In this way, the chaotic PRBG can expand a short key into a long keystream, which directly exclusive-or'ed with the clear text or message, gives the ciphertext.

The paper is structured as follows: Section 2 introduces chaotic PRBG. Section 3 describes statistical testing of random sequences. Section 4 obtains a PRBG based on a 2D symmetrical chaotic map of logistic type. In Section 5 several binary sequences are obtained and tested. The computational cost of the PRBG algorithm and the size of the key space for cryptographic applications are evaluated in Section 6. Section 7 remarks the final conclusions and discusses further work to be done.

2 Chaotic Random Bit Generation

The inherent properties of chaos, such as ergodicity and sensitivity to initial conditions and control parameters, connect it directly with cryptography characteristics of confusion and diffusion [13]. Additionally chaotic dynamical systems have the advantage of providing simple computable deterministic pseudo-randomness [8].

As a consequence of these observations, several works were presented since 1990s implementing PRBG based on different chaotic systems [8], [9], [10], [14], [15], [16]. In some way, it could be said that chaos has brought into being a novel branch for PRNG in cryptography, called chaotic PRNG.

An N-dimensional deterministic discrete-time dynamical system is an iterative map $f : \Re^N \rightarrow \Re^N$ of the form:

$$X_{k+1} = f(X_k) \tag{1}$$

where $k = 0, 1 \ldots n$. is the discrete time and $X_0, X_1 \ldots X_n$, are the states of the system at different instants of time.

In these systems, the evolution is perfectly determined by the mapping $f : \Re^N \to \Re^N$ and the initial condition X_0. Starting from X_0, the *initial state*, the repeated iteration of (1) gives rise to a fully deterministic series of states known as an *orbit*.

To build a chaotic PRBG is necessary to construct a numerical algorithm that transforms the states under chaotic behaviour of the system into binary numbers. The existing designs of chaotic PRNGs use different techniques to pass from the continuum to the binary world [6]. The most important are:

1. Extracting one or more bits from each chaotic state along chaotic orbits [8], [16].
2. Dividing the phase space into m sub-spaces,and output a binary number $i = 0, 1, , m - 1$ if the chaotic orbit visits the i_{th} subspace [7], [10].
3. Combining the outputs of two or more chaotic systems to generate the pseudo-random numbers [14], [15].

Discrete or digital chaos implemented on computers with finite precision is normally called *pseudo chaos.* In pseudo chaos dynamical degradation of the chaotic properties of the system may appear, as throughout iterations pseudo orbits may depart from the real ones in many different and uncontrolled manners [6], [17], [18]. Even so, the above exposed techniques are able to generate sequences of bits or binary numbers, which appear random-like in many aspects.

One may also consider that using high dimensional chaotic systems could offer additional advantages. While less known, these systems whirl many variables at any calculation and the periodic patterns produced by the finite precision of the computer maybe more difficult to appear than in the low dimensional cases [19].

In this paper the technique of dividing the phase space is followed and applied on two symmetrical two-dimensional (2D) chaotic maps of logistic type.

3 Statistical Tests Suites

In general, randomness cannot be mathematically proved. Alternatively, different statistical batteries of tests are used. Each of these tests evaluates a relevant random property expected in a true random generator. To test a certain randomness property, several output sequences of the generator under test are taken. As one knows a priori the statistical distribution of possible values that true random sequences would be likely to exhibit, a conclusion can be obtained upon the probability of the tested sequences to be random.

Mathematically this is done as follows [3]. For each test, a *statistic variable* X is specified along with its correspondent *theoretical random distribution function* $f(x)$. For non-random sequences, the statistic can be expected to take on larger values, typically far-out in the tails of $f(x)$. A *critical value* x_α is defined for the theoretical distribution so that $P(X > x_\alpha) = \alpha$, which is called the *significance level* of the test. In the same way, theoretically other distribution functions and a β value could be defined to assess non-random properties. But in practice, it is impossible to calculate all the distributions that describe non-randomness, for there are an infinite number of ways that a data stream can be non-random.

When a test is applied, the test statistic value X_s is computed on the sequence being tested. This test statistic value X_s is compared to the critical value x_α. If the test statistic value exceeds the critical value, the hypothesis for randomness is rejected. The rejection is done with a $(100 * \alpha)\%$ probability of having FALSE POSITIVE error (this is called a *TYPE I error*, where the sequence was random and is rejected). Otherwise is not rejected (i.e., the hypothesis is accepted) with a probability of $(100 * \beta)\%$ of error (this is called *TYPE II error* or FALSE NEGATIVE, the sequence was non-random and is accepted). In consequence passing the test merely provides a probabilistic evidence that the generator produces sequences which have certain characteristics of random sequences.

For a given application, the value of α must be selected appropriately. This is because if α is too high TYPE I errors may frequently occur (respectively if α is too low the same will happen for TYPE II errors). For cryptographic applications typical values of α are selected between the interval $\alpha\epsilon[0.001, 0.01]$, which is also referred as a confidence level in the interval $[99.9\%, 99\%]$ for the test. Unlike α, β is not fixed, for it depends on the non-randomness defects of the generator. Nevertheless α, β and the size of the tested sequence (n) are related [3]. Then for a given statistic, a critical value and a minimum n should be selected to minimize the probability of a TYPE II error (β).

There exist different well-known sources of test suites available in the Internet. In the present work, Marsaglia's Diehard test suite (in [4]) and NIST Statistical Test Suite (in [3]) were selected, for they are very accessible and widely used. Table 1 lists the tests comprised in these suites.

In both suites, the test statistic value X_s is used to calculate a p-value that summaries the strength of evidence against the randomness of the PRBG.

Table 1. List of tests in the Diehard and NIST test suites

Number	Diehard test suite	NIST test suite
1	Birthday spacings	Frequency (monobit)
2	Overlapping 5-permutation	Frequency test within a block
3	Binary rank test	Cumulative sums
4	Bitstream	Runs
5	OPSO	Longest run of ones in a block
6	OQSO	Binary matrix rank
7	DNS	Discrete fourier transform
8	Count-the-1's test	Non-overlapping template matching
9	A parking lot	Overlapping template matching
10	Minimum distance	Maurer's universal statistical
11	3D-spheres	Approximate entropy
12	Squeeze	Random excursions
13	Overlapping sums	Random excursions variant
14	Runs	Serial
15	Craps	Linear complexity

4 Pseudo-Random Bit Generation Based on Two-Dimensional Chaotic Maps of Logistic Type

In [12], López-Ruiz and Pérez-García analyze a family of three chaotic systems obtained by coupling two logistic maps. The focus here will be made on models (a) and (b), which will be referred as System A and B:

$$\begin{array}{ll} SYSTEM\ A: & SYSTEM\ B: \\ T_A : [0,1]\times[0,1] \longrightarrow [0,1]\times[0,1] & T_B : [0,1]\times[0,1] \longrightarrow [0,1]\times[0,1] \\ & \\ x_{n+1} = \lambda(3y_n+1)x_n(1-x_n) & x_{n+1} = \lambda(3x_n+1)y_n(1-y_n) \\ y_{n+1} = \lambda(3x_n+1)y_n(1-y_n) & y_{n+1} = \lambda(3y_n+1)x_n(1-x_n) \end{array} \quad (2)$$

Amazingly, these systems show the following symmetry $T_A(x,y) = T_B(y,x)$, which implies that $T_A^2(x,y) = T_B^2(x,y)$. From a geometrical point of view, both present the same chaotic attractor in the interval $\lambda \in [1.032, 1.0843]$. The parameter λ in Eq. 2, could be considered different for the calculus of x and y coordinates. This would give two independent control parameters λ_x and λ_y, which in a similar rank [1.032, 1.0843] maintain the system in a chaotic regime due to its structural stability. The dynamics in this regime is particularly interlaced around the saddle point $P4$, that plays an important role for our proposes:

$$P4 = [P4_x, P4_y], \qquad where \quad P4_x = P4_y = \frac{1}{3}\left(1+\sqrt{4-\frac{3}{\lambda}}\right). \quad (3)$$

To obtain the Symmetric Coupled Logistic Map PRBG, the algorithm presented in [10] is applied on System A. Its functional block structure is represented in Fig.1 and it is explained in the following paragraphs.

The technique of dividing the phase space in four sub-spaces is used here. This is done in the block named as *Sub-space decision* in which the threshold

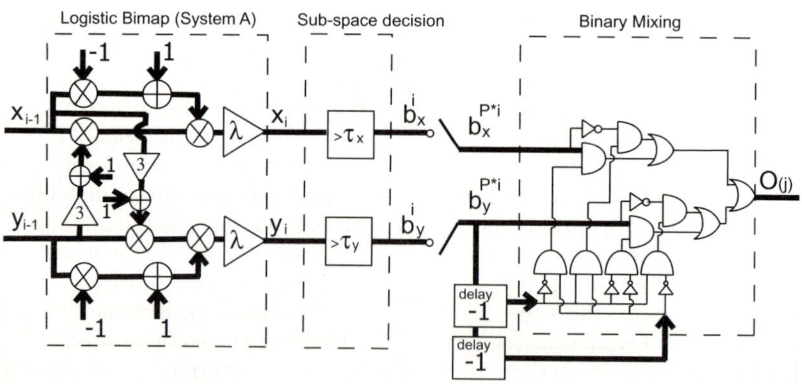

Fig. 1. Functional block structure of the algorithm proposed in [10] for System A

values, τ_x and τ_y, are employed to convert the suite into a binary sequence, by means of the following equations:

$$b_x = \begin{cases} 0 \text{ if } x & \leq & \tau_x \\ 1 \text{ if } x & > & \tau_x \end{cases}; \qquad b_y = \begin{cases} 0 \text{ if } y & \leq & \tau_y \\ 1 \text{ if } y & > & \tau_y \end{cases}. \tag{4}$$

A purely statistical procedure is proposed in [10] to obtain τ_x (or τ_y) as the median of a large T set of x (or y) values. τ_x and τ_y are chosen after the first $T = 1000$ iterations of the system. After obtaining $S_x = \{b_x^i\}_{i=1}^{\infty}$ and $S_y = \{b_y^i\}_{i=1}^{\infty}$, they are sampled with a frequency of $1/P$ (each P iterations) and $B_x = \{b_x^{P*i}\}_{i=1}^{\infty}$ and $B_y = \{b_y^{P*i}\}_{i=1}^{\infty}$ are obtained. The effect of skipping P consecutive values of the orbit is necessary to get a random macroscopic behaviour. With this operation, the correlation existing between consecutive values generated by the chaotic system is eliminated. This is done in a way that over a P_{min}, sequences generated with $P > P_{min}$ will appear macroscopically random. Although P is normally introduced as an additional key parameter in pseudo-random sequences generation [20], it strongly determines the speed of the generation algorithm, so it is recommended to be kept as small as possible.

The output binary pseudorandom sequence $O(j)$ is obtained by a mixing operation of the actual and previous values of the sequence $B(j) = [B_x(j), B_y(j)]$ given by the truth table sketched in Table 2.

Table 2. Truth table generating the binary sequence

	$B_y(j-1)$	
$B_y(j-2)$	0	1
0	$B_x(j)$	Not($B_x(j)$)
1	$B_y(j)$	Not($B_y(j)$)

Unfortunately, the sequences so formed do not pass the minimum requirements of randomness assessed by Diehard test suite. At this point, it is noticed that, to obtain good results, the geometrical characteristics of the system must be taken into account. More precisely, it is found out that the division of the phase space in four sub-spaces must be defined, in a way that the system visits each sub-space according to a particular finite state automata. This automata is inferred from the Hénon map behaviour in [10] and it has a particular pattern of visits of the four sub-spaces. For Systems A and B, this finite automata is depicted in Fig. 2(a) and 2(c).

Let us name the sub-spaces corresponding to $[b_x^i, b_y^i]$ with values $[0,0]$, $[1,0]$, $[0,1]$ and $[1,1]$ as 1,2,3 and 4. Although the four sub-spaces are not visited equally, there exists a symmetry of movements between sub-spaces 1-3 and 2-4, which has a characteristic mixing of 50% and 50%, as long as a predominant (80%) and constant transition between 3 and 2. This leads to a highly variation of binary values in sequences S_x, S_y. In the end, these conditions give the final

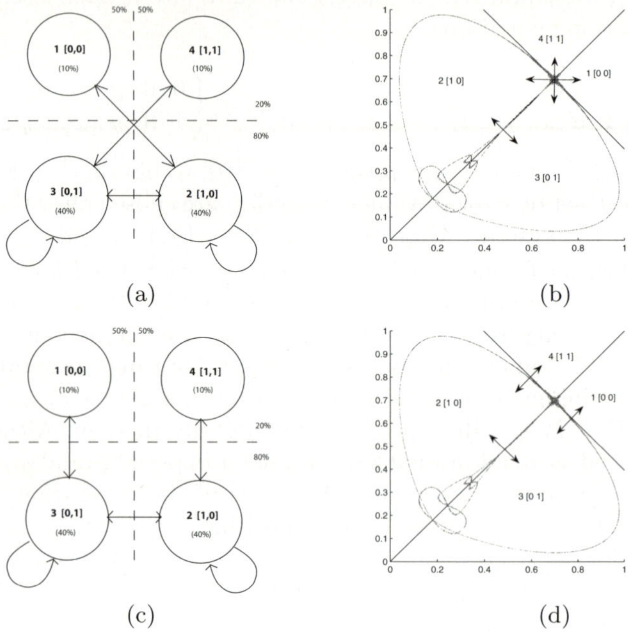

Fig. 2. (a) Finite automata and (b) final sub-space division for System A. (c) Finite automata and (d) final sub-space division for System B. (In both cases, $\lambda = 1.07$).

result of an output sequence $O(j)$ with a proper balance of zeros and ones, or put it in another way, with pseudo-random properties.

To get this automata for the symmetric coupled logistic maps Systems A and B, one should chose the diagonal axis, which divides phase space in two parts, each of which is equally visited (50%). And additional statistical calculus is required to divide these two sub-spaces, in another two with a visiting rate of 40%-10% each one. When this is done, one can observe that this is got by merely selecting P4 and the line perpendicular to the axis in P4 as the other division line. The final sub-space division for each system is presented in Fig. 2(b) and 2(d), along with the indications of the evolution of the visits to each sub-space.

Finally the initial algorithm of Fig. 1 with System A, is modified with the appropriate *Sub-space decision* block. The final PRBG functional scheme is presented in Fig. 3.

As a direct application for cryptography, the PRBG could be used in the construction of a stream cipher. Different initial conditions, x_0 and y_0 and parameters λ and P, can be applied to the input of the system and be used as a key to generate the keystream, or output sequence $O(j)$. The keystream $O(j)$ can be XORed directly with a clear text, obtaining that way the ciphertext.

In next section different sequences are obtained with the system of Fig. 3. Their randomness is assessed and demonstrates them statistically valid for cryptographic applications. This may indicate that the automata scheme presented in

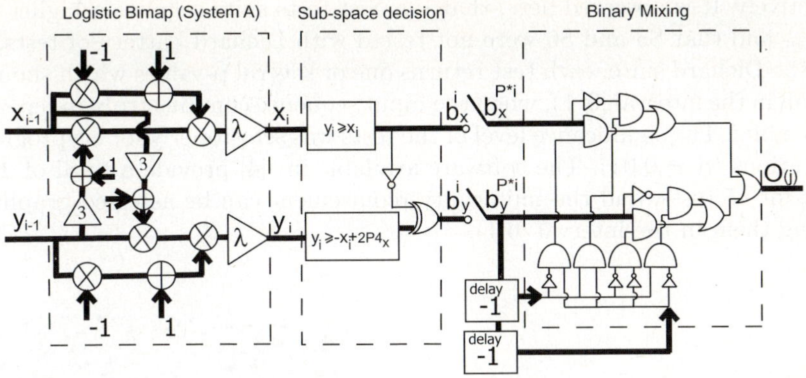

Fig. 3. Functional block structure of the PRBG applied to the symmetric coupled logistic map PRBG with System A

Fig. 2(a) and 2(c) represents a sufficient condition to obtain pseudo-randomness. Consequently, it can represent a systematic scheme to extend the algorithm in [10] to get PRBG on other chaotic maps.

5 Pseudo-Random Sequences Statistical Testing

To assess the randomness of the PRBG obtained in the previous section with Systems A and B, several sequences are obtained and submitted to the Diehard [4] and NIST [3] test suites described in Section 3. Similar results were found for both systems and for simplicity, only those obtained with System A will be presented here after. Ten sequences were generated with six different sets of initial conditions. Their characteristics are described in Table 3.

Six of them (S1,S2,S3,S4,S5 and S6) were tested with Nist tests suite with 200 Mill. of bits and four of them (S1,S2,S3 and S4) were tested with Diehard tests suite with 80 Mill. of bits. Here, the parameters P_{Dmin} and P_{Nmin} are the minimum sampling rate or shift factor, P_{min}, over which, all sequences generated with the same initial conditions and $P > P_{min}$ pass Diehard or Nist tests suites,

Table 3. Parameters P_{Dmin} and P_{Nmin} for different sequences S_i, $i = 1,..,6$, with different initial conditions (x_0, y_0) and map parameter λ

Sequence	S1	S2	S3	S4	S5	S6
x_0	0.989125	0.491335	0.672757	0.726874	0.39565	0.999851
y_0	0.689125	0.691335	0.497757	0.901874	0.49565	0.649851
λ	1.04869	1.05392	1.06961	1.08007	1.06438	1.07489
P_{Dmin}	55	45	35	47	*n.a.*	*n.a.*
P_{Nmin}	83	105	83	83	100	85

respectively. It is observed here, that the Nist tests suite requires a higher value of P_{min} and that S5 and S6 were not tested with Diehard battery of tests.

In the Diehard suite, each test returns one or several p-values which should be uniform in the interval $[0, 1)$, when the input sequence contains truly independent random bits. The significance level of the tests was set properly for cryptographic applications ($\alpha = 0.01$). The software available in [4] provide a total of 218 p-values for 15 tests, and the uniformity requirement can be assessed graphically plotting them in the interval [0,1).

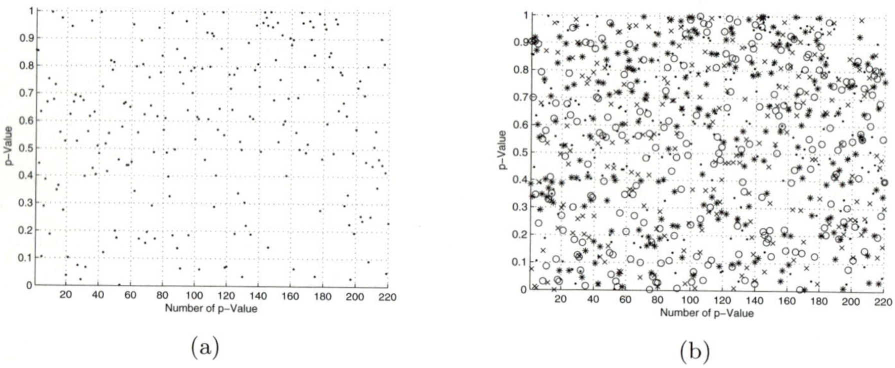

Fig. 4. (a) Diehard test suite p-values obtained with all tests for initial conditions S1 with $P = P_{Dmin} = 55$. (b) p-values obtained for initial conditions S1(•),S2(∘),S3(∗) and S4(×) with $P = P_{Dmin}$ value in Table 3.

Fig. 4(a) shows the uniformity distribution of the p-values over the interval [0,1) obtained for a sequence with initial conditions S1 and $P_{Dmin} = 55$. Sequences with initial values S1 to S4 where proved to pass the Diehard battery of tests. Fig. 4(b) presents a graphical representation of the p-values obtained for each sequence with sampling factor $P = P_{Dmin}$ value in Table 3. It can be observed that some p-values are occasionally near 0 or 1. Although it can not be well appreciated in the figure, it has to be said that those never really reach these values.

In the Nist tests suite [3], one or more p-values are also returned for each sequence under test. The significance level of the tests was set to $\alpha = 0.01$, as in the Diehard case. These tests require a sufficiently high length of sequences and to prove randomness in one test, two conditions should be verified. First, a minimum percentage of sequences should pass the test and second, the p-values of all sequences should also be uniformly distributed in the interval $(0, 1)$.

To perform the tests, each of the six sequences with initial conditions S1 to S6 are arranged in 200 sub-sequences of 1Mill. bits each and submitted to the Nist battery of tests. Sequences S proved to pass all tests over a minimum value of P_{Nmin}, shown in Table 3.

In Fig. 5(a) and 5(b), the results obtained for $S1$ and $S4$ respectively, are graphically presented as an example of what was obtained for each S. The tests in

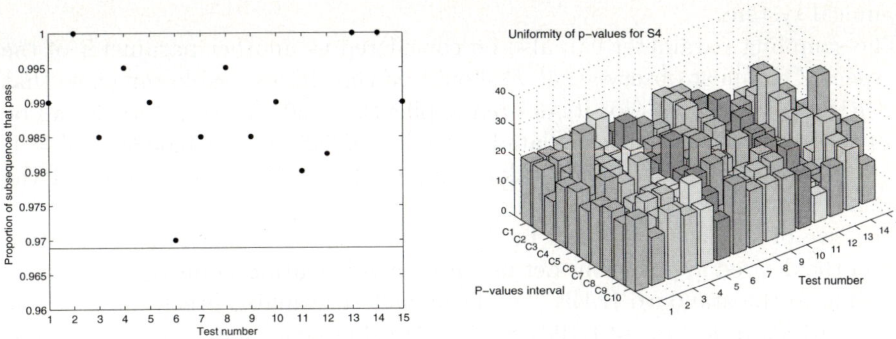

(a) (b)

Fig. 5. In (a), the proportion of sub-sequences S1 that passes each test is displayed. In (b) The distribution of p-values of S4 is examined for each test to ensure uniformity. The interval between 0 and 1 is divided in ten sub-intervals ($C1, C2, ..., C10$), and the p-values that lie within each subinterval are counted and plotted.

the suite are numbered according to Table 1. Fig. 5(a) represents the percentage of the 200 sub-sequences of S1, that pass each of the 15 tests of the suite. These percentages are over the minimum pass rate required of 96.8893% for a sample size = 200 binary sub-sequences. Fig. 5(b) describes the uniformity of the distribution of p-values obtained for the 15 tests of the suite. Here, uniformity is assessed. The interval (0,1) is divided in ten subintervals ($C1, C2, ..., C10$) and the number of p-values that lay in each sub-interval, among a total of 200, are counted and proved to be uniform.

6 Key Space Size and Computational Cost

To establish the complexity, and consequently the speed of the PRBG described in Fig. 3, the principle of invariance is observed. This says that the efficiency of one algorithm in different execution environments differs only in a multiplicative constant, when the values of the parameters of cost are sufficiently high.

In this sense, the asymptotic behaviour of the computational cost of the PRBG is governed by the calculus performed in the chaotic block. This block performs P iterations to obtain an output bit, $O(j)$.

The capital theta notation (Θ) can be used to describe an asymptotic tight bound for the magnitude of cost of the PRBG. And consequently, the 2D symmetric coupled logistic maps have a computational cost or complexity of order $\Theta(P * n)$.

When considered for cryptographic applications, the key space is determined by the interval of the parameter λ and the initial conditions that keep the dynamical system in the chaotic regime. These are $\lambda \in [1.032, 1.0843]$, $x_0 \in (0, 1)$ and $y_0 \in (0, 1)$. As it was said in Section 4, two independent control parameters can be considered, λ_x and λ_y, without altering the chaotic behaviour of the dynamical system.

The sampling parameter can also be considered as another parameter of the key space. One must observe that P should be kept in a suitable range, so that the PRBG is fast enough for its desired application [20]. These intervals can be denoted with brackets and calculated as $[\lambda_x] \approx 0.0523$, $[\lambda_y] \approx 0.0523$, $[x_0] = 1$, $[y_0] = 1$ and $[P] = 890$, when taking $[P] \in [110, 1000]$ as the range of the sampling factor.

The size of the key space will be determined by the numeric representation used in the desired application. Let us consider the floating-point representation according to the standard IEEE 754. The smallest available precision in simple format with 32 bits is $\epsilon_{32} \approx 1.1921 \times 10^{-7}$. For double format this magnitude is $\epsilon_{64} \approx 2.2204 \times 10^{-16}$ for floating-point numbers with 64 bits.

These quantities give us the maximum number of possible values of every parameter in any of the two representations. This is easily computed dividing the intervals by ϵ, as $K_\lambda = [\lambda]/\epsilon$, $K_{x_0} = [x_0]/\epsilon$ and $K_{y_0} = [y_0]/\epsilon$. For the sampling factor $K_P = [P]$ as it takes integer values. The total size of representable parameter values is given by K, calculated as $K = K_{\lambda_x} \times K_{\lambda_y} \times K_{x_0} \times K_{y_0} \times K_P$. K is the size of the available key-space and its logarithm in base 2 gives us the available length of binary keys to produce pseudo-random sequences in the generator.

The values obtained for each number precision, are $K_{32} = 1.21 \times 10^{28}$ with a key length of 93 bits for single precision and $K_{64} = 1.00 \times 10^{63}$, with a key length of 209 bits for double precision.

These results are encouraging for recommending the use of the PRBG in Fig. 3 for cryptographic applications, where a length of keys greater than 100 is considered strong enough against brute force attacks [13].

7 Conclusions

In the present work, a refinement of the algorithm exposed in [10] by M. Suneel is presented. It consists of the introduction of a finite automata that makes possible its application to other chaotic maps. In some way, this finite automata could be said to extend the range of application of this algorithm for other 2D chaotic systems. This is referred in [6] as making the PRBG *chaotic-system-free.*

The fact is that, while systematic, the scheme presented in this paper is not straight-forward. This is because building the finite automata requires necessarily a detailed study of the geometrical properties of the dynamical evolution of the chaotic system.

The authors apply this technique to build two new PRBG using two particular 2D dynamical systems formed by two symmetrically coupled logistic maps. A set of different pseudo-random sequences are generated with one of the PRBG. Statistical testing of these sequences shows fine results of random properties for the PRBG.

The estimation of the PRBG computational cost gives an asymptotic tight bound of $\Theta(P * n)$. The available size of the key space is also calculated and a minimum length of binary keys of 93 and 209 bits is obtained for simple and double precision respectively. These preliminary results indicate a promising quality of the PRBG for cryptographic applications.

Consequently, the chaotic PRBG algorithm could be used for different applications in security engineering. A direct application in cryptography could be the construction of a stream cipher. This can be easily obtained when the output sequences $O(j)$ in Fig. 3 are used as a keystream. Then, a sequence $O(j)$ can be directly XORed with a clear text obtaining that way the ciphertext.

For future work, the authors plan to consider the geometrical properties of the logistic bi-mappings to enhance the performance of the presented PRBG.

References

1. Anderson, R.: Security Engineering, A Guide to Build Dependable Distributed Systems. John Wiley and Sons Inc., New York (2001), `http://www.cl.cam.ac.uk/~rja14/book.html`
2. Menezes, A., van Oorschot, P., Vanstone, S.: Handbook of Applied Cryptography. CRC Press, Florida (1997)
3. NIST Special Publication 800-22: A Statistical Test Suite for the Validation of Random Number Generators and Pseudo Random Number Generators for Cryptographic Applications (2001)
4. Marsaglia, G.: The diehard test suite (1995), `http://stat.fsu.edu/geo/diehard.html`
5. Knuth, D.E.: The Art of Computer Programming, 3rd edn. Seminumerical Algorithms, vol. 2. Addison-Wesley Longman Publishing Co., Massachusetts (1997)
6. Li, S.: Analyses and New Designs of Digital Chaotic Ciphers. PhD thesis, School of Electronic and Information Engineering, Xi'an Jiaotong University (2003)
7. Stojanovski, T., Kocarev, L.: Chaos based Random Number Generators. Part I: Analysis. IEEE Transactions on Circuits and Systems I: Fundamental Theory and Applications 43, 281–288 (2001)
8. Protopopescu, V.A., Santoro, R.T., Tollover, J.S.: Fast secure encryption-decryption method based on chaotic dynamics. US Patent No. 5479513 (1995)
9. Kocarev, L., Jakimoski, G.: Pseudorandom Bits Generated by Chaotic Maps. IEEE Transactions on Circuits and Systems I: Fundamental Theory and Applications 50(1), 123–126 (2003)
10. Madhekar, S.: Cryptographic Pseudo-Random Sequences from the Chaotic Hénon Map, `http://arxiv.org/abs/cs/0604018`
11. Hénon, M.: A two-dimensional mapping with a strange attractor. Communications in Mathematical Physics 50, 69–77 (1976)
12. Lopez-Ruiz, R., Pérez-Garcia, C.: Dynamics of Maps with a Global Multiplicative Coupling. Chaos, Solitons and Fractals 1, 511–528 (1991)
13. Alvarez, G., Li, S.: Some Basic Cryptographic Requirements for Chaos-Based Cryptosystems. International Journal of Bifurcation and Chaos 16, 2129–2151 (2006)
14. Po-Han, L., Soo-Chang, P., Yih-Yuh, C.: Generating chaotic stream ciphers using chaotic systems. Chinese Journal of Physics 41, 559–581 (2003)

15. Li, S., Mou, X., Cai, Y.: Pseudo-random bit generator based on couple chaotic systems and its application in stream-ciphers cryptography. In: Pandu Rangan, C., Ding, C. (eds.) INDOCRYPT 2001. LNCS, vol. 2247, pp. 316–329. Springer, Heidelberg (2001)
16. Bogdan, C., Chargé, P., Fournier-Purnaret, D.: Behaviour of chaotic sequences under a finite representation and its cryptographic applications. In: IEEE Workshop on Nonlinear Maps and Applications (NOMA), Toulouse (2007)
17. Guckenheimer, J.M., Holmes, P.: Nonlinear Oscillations, Dynamical Systems and Bifurcation of Vector Fields. Springer, Heidelberg (1983)
18. Li, S., Chen, G., Mou, X.: On the dynamical degradation of digital piecewise linear chaotic maps. International Journal of Bifurcation and Chaos 15, 3119–3151 (2005)
19. Falcioni, M., Palatella, L., Pigolotti, S., Vulpiani, A.: Properties making a chaotic system a good pseudo random number generator. Physical Review E 016220 (2005)
20. Kocarev, L.: Chaos-based Cryptography: a brief overview. IEEE Circuits and Systems Magazine 1, 6–21 (2001)

A Flow Deviation Algorithm for Joint Optimization of Unicast and Anycast Flows in Connection-Oriented Networks

Krzysztof Walkowiak

Chair of Systems and Computer Networks, Faculty of Electronics, Wroclaw University of Technology, Wybrzeze Wyspianskiego 27, 50-370 Wroclaw, Poland
Krzysztof.Walkowiak@pwr.wroc.pl

Abstract. Our discussion in this article centers around a new heuristic algorithm based on Flow Deviation (FD) method deployed for joint optimization of unicast and anycast flows in connection-oriented networks. The optimization problem is NP-complete. Anycast is a *one-to-one-of-many* technique to deliver a packet to one of many hosts. Current computer networks convey mostly unicast traffic. However, the quantity of anycast traffic will grow in near future, since many new services that can use anycast paradigm are developed, e.g. peer-to-peer networks, content delivery networks, sensor networks, etc. In this work we focus on connection-oriented networks, in view of the fact that many transport networks are based on technologies like MPLS, GMPLS, and DWDM. Results of numerical experiments that compare performance of our approach against other heuristics are presented and discussed.

Keywords: optimization, anycast, unicast.

1 Introduction

The impact of the Internet and other computer networks can be watched in almost all areas of people's life, and in the future its importance will grow. Nowadays computer networks offer mainly two kinds of services: exchanging of data between individual users e.g. VoIP, messengers, access to less popular content and access to content servers e.g. distribution of electronic music, movies, books, software; P2P networks. The former kind of traffic can be modeled as unicast flow; the latter one can be modeled as anycast flow. Currently, a large part of transmissions issued in the Internet are unicast – two individual systems connect *one-to-one* to exchange data. Anycast is a *one-to-one-of-many* technique to deliver a packet to one of many hosts. Significance of anycast transmission will grow with popularity of the Internet and development of new services. One of the most famous technologies that apply anycast traffic is CDN (Content Delivery Network). CDN uses many content servers offering the same content replicated in various locations. User-perceived latency and the other QoS parameters can be easily and inexpensively improved by the CDN [13]. Other applications of anycast paradigm in computer networks are peer-to-peer (P2P) systems, Domain Name Service, Web Service, Distributed Database System, host auto-configuration, overlay network, sensor networks [1], [2], [4], [7], [8].

O. Gervasi et al. (Eds.): ICCSA 2008, Part II, LNCS 5073, pp. 797–807, 2008.

In this work we concentrate on optimization of connection-oriented (c-o) flows in computer networks. Popular network technique using this approach are MPLS (MultiProtocol Label Switching), GMPLS (General MPLS) DWDM (Dense Wavelength Division Multiplexing) and Frame Relay. In c-o networks information between the origin and the destination is transported along an established virtual circuit (connection). C-o techniques gain much attention in recent years, since they can provide controlled traffic engineering for flow of packets for different service/user QoS (Quality of Service) classes in core networks. Flow of c-o networks can be modeled as a non-bifurcated (unsplittable) multicommodity flow [14].

Most of previous work in the field of connection-oriented flow optimization focuses on unicast traffic [3], [5], [6], [11], [12], [14], [16], [18], and [23]. There are very few papers on c-o anycast flow optimization. Only authors of [9], [17] consider the routing and wavelength assignment problem of unicast, anycast and multicast flows in optical networks. In our previous works we proposed heuristics for the anycast flow assignment in c-o networks [19], [21], and [22]. Furthermore, in [20] we formulated the problem of joint optimization of unicast and anycast flows in c-o survivable networks. Our paper [24] discusses the application of anycast flows for effective network restoration. The main contribution of this work is that we propose and evaluate an effective method for joint optimization of unicast and anycast flows in c-o networks, while previous works of the author focus on optimization of unicast or anycast flows. The problem of joint optimization of unicast and anycast flows in c-o networks is a vital problem, since in existing networks these both kinds of traffic are transmitted simultaneously. We present results of numerical experiments comparing our approach against other algorithm proposed previously.

2 Anycasting in Connection-Oriented Networks

Anycast is a *one-to-one-of-many* technique to deliver a packet to one of many hosts. Comparing to unicast transmission, in anycasting one of the endpoints of the transmission must be selected among many possible nodes. Most of previous work on anycasting concentrates on IP networks using connection-less transmission modeled as bifurcated multicommodity flows [4], [7], [13]. Since we consider a connection-oriented network (e.g. MPLS), we model the network flow as non-bifurcated multicommodity flow. Consequently, the anycast request (demand) must consist of two connections: one connecting the client to the server (upstream) and the second one in the opposite direction (downstream). Upstream connection sends user's requests. Downstream connection carries the requested data. Thus, each anycast demand is defined by a following triple: client node, upstream bandwidth requirement and downstream bandwidth requirement. In contrast, a unicast demand is defined by a following triple: origin node, destination node and bandwidth requirement. To establish a unicast demand a path satisfying requested bandwidth and connecting origin and destination nodes must be found. Optimization of anycast demands is more complex. The first step is the server selection process. Next, when the server node is selected, both paths: upstream and downstream can be calculated analogously to unicast approach. However, the main constraint is that both connections associated with a particular anycast demand must connect the same pair of network nodes and one of these nodes must host a server.

3 Problem Formulation

In this section we present the offline optimization problem of unicast and anycast flows in connection-oriented networks. The major novelty of our approach is that we consider joint optimization of unicast and anycast flows. For the remainder of the paper we will refer to the considered problem as NCFA (Non-bifurcated uniCast and anyCast Flow Assignment). To mathematically represent the problem we introduce the following notations.

Sets:

- V Set of n vertices representing the network nodes.
- A Set of m arcs representing network directed links.
- P Set of q connections in the network. A connection can be of two types: unicast and anycast.
- P^{UN} Set of unicast connections in the network defined by a following triple: origin node, destination node and bandwidth requirement
- P^{AN} Set of anycast connections in the network. A connection can be of two types: upstream or downstream. Each anycast connection is defined by a following triple: client node, bandwidth requirement, index of associated connection.
- P^{DS} Set of anycast downstream connections in the network.
- P^{US} Set of anycast upstream connections in the network.
- Π_p The index set of candidate routes (paths) π_p^k for connection p. If p is a unicast connection, route π_p^k connects the origin and destination node of p. If p is an anycast upstream connection, route π_p^k connects the client node and the server. Finally, if p is an anycast downstream connection, candidate routes connect the server and the client node.
- X Set of variables x_p^k, which are equal to one. X determines the unique set of currently selected routes.

Constants:

- δ_{pa}^k Equal to1, if arc a belongs to route k realizing connection p; 0 otherwise
- Q_p Volume (estimated bandwidth requirement) of connection p
- c_a Capacity of arc a
- $\tau(p)$ Index of the connection associated with connection p. If p is a downstream connection $\tau(p)$ must be an upstream connection and vice versa.
- $o(\pi)$ Origin node of route π
- $d(\pi)$ Destination node of route π
- T Total arrive rate of messages in the network

Variables:

- x_p^k 1 if route π_p^k having index $k\in \Pi_p$ is selected for connection p and 0 otherwise.
- f_a Flow of arc a.

The NCFA can be formulated as follows

$$\min_X \quad DEL(X) = \frac{1}{T} \sum_{a \in A} \frac{f_a}{c_a - f_a} \tag{1}$$

subject to

$$\sum_{k \in \Pi_p} x_p^k = 1 \qquad \forall p \in P \tag{2}$$

$$x_p^k \in \{0,1\} \quad \forall p \in P, \forall k \in \Pi_p \tag{3}$$

$$f_a = \sum_{p \in P} \sum_{k \in \Pi_p} \delta_{pa}^k x_p^k Q_p \quad \forall a \in A \tag{4}$$

$$f_a \le c_a \quad \forall a \in A \tag{5}$$

$$\sum_{k \in \Pi_p} x_p^k d(\pi_p^k) = \sum_{k \in \Pi_{\tau(p)}} x_{\tau(p)}^k o(\pi_{\tau(p)}^k) \qquad \forall p \in P^{AN} \tag{6}$$

$$X = \left(\bigcup_{p,k: x_p^k = 1} \left\{ x_p^k \right\} \right) \tag{7}$$

The goal is to minimize the network delay [6], [11], [14]. Condition (2) states that the each connection can use only one route. Constraint (3) assures that decision variables are binary ones. (4) is a definition of an arc flow. Inequality (5) denotes the capacity constraint. Equation (6) guarantees that two routes associated with the same anycast demand connect the same pair of nodes. Notice that constraint (6) must be satisfied only for anycast connections. Finally, (7) is a definition of a set X called a selection that includes all variables x, which are equal to 1. We call a selection *feasible* if for each arc flow calculated according to the selection holds the capacity constraint (5). The NCFA problem given by (1-7) is NP-complete. This follows from the fact that the unicast non-bifurcated flow assignment is NP-complete [14]. NCFA is 0/1 optimization problem. Complexity of NCFA problem is a result of huge solution space..

4 Algorithm

In this section we describe a new algorithm for joint optimization of unicast and anycast flows in connection-oriented networks. We will refer to the algorithms as to CFD_DEL (uniCast and anyCast Flow Deviation for Delay Function). The main idea of our approach is based on the Flow Deviation (FD) algorithm proposed in [6]. The FD method applies a general approach for solving minimization problems with convex objective function and linear constraints. Note that Flow Deviation was applied to a wide range of optimization problems [3], [5], [11], [12], [14], [16], [18], [19], [23].

Let *DEL*(*H*) denote value of the objective function for a feasible selection *H* and l_a^{DEL} denotes a metric of arc *a* calculated according to arc flows given by selection *H*. The common approach is to calculate this metric as partial derivative of the objective function over arc flow [6], i.e.

$$l_a^{DEL} = \frac{c_a}{T(c_a - f_a)^2} \tag{8}$$

Let l_a^{INI} be the value of metric l_a^{DEL} calculated for flow of arc *a* equal to zero

$$l_a^{INI} = \frac{1}{Tc_a} \tag{9}$$

To find a starting feasible solution we apply algorithm CFD_UNI, which is a modification of the initial phase of FD algorithm presented in [6]. The original FD algorithm optimizes only unicast connections. CFD_INI can jointly assign unicast and anycast connections.

Algorithm CFD_INI

Let $\mathbf{Q} = [Q_1, Q_2, \ldots, Q_q]$ denote a vector of connections' volume (estimated bandwidth requirement). We assume that RE_r is a real number. θ, ε – are tolerances applied in the algorithm.

Step 1. Find selection X_1 by assigning all connections to shortest paths calculated under metric l_a^{INI} (9) in the following way. For each unicast connection $i \in P^{UN}$ find the shortest route $k \in \Pi_i$ under metric l_a^{INI}. For each anycast downstream connection $i \in P^{DS}$ find the shortest route $k \in \Pi_i$ under metric l_a^{INI}. Next for each anycast upstream connection $i \in P^{US}$ find the shortest route $k \in \Pi_i$ under metric l_a^{INI}, for which the following condition is satisfied $d(\pi_i^k) = o(\pi_{\tau(i)}^j)$, where $x_{\tau(i)}^j \in X_1$.

Set $RE_1 = 1$, $r = 1$ and go to step 2.

Step 2. Calculate flows in each arc $a \in A$ using formula (4) according to selection X_r and volume vector $\mathbf{Q}_r$. Find

$$\sigma_r = \max_{a \in A} \frac{f_a}{c_a} \tag{10}$$

If $\sigma_r / RE_r < 1$, then stop the algorithm. Otherwise let $RE_{r+1} = RE_r (1 - \varepsilon(1 - \sigma_r)) / \sigma_r$, where ε is a proper tolerance and $0 < \varepsilon < 1$. Let $\mathbf{Q}_{r+1} = (RE_{r+1} / RE_r)\mathbf{Q}_r$. Find $X_{r+1} = \text{CFD\_DEL}(X_r, \mathbf{Q}_{r+1})$ and go to step 3.

Step 3. If

$$|RE_{r+1} - RE_r| < \theta \tag{11}$$

where θ is a proper positive tolerance, stop: the problem is infeasible within tolerance θ. Otherwise set $r = r + 1$ and go to step 2.

In step 2 of CFD_INI we apply algorithm CFD_DEL (see below), which is run for the values of decision variables given by selection X_r and connections' volume given by the vector $\mathbf{Q}_r$.

Algorithm CFD_DEL
Operator *first*(B) returns the index of first connection in set B. F and H are selections.

Step 1. Find feasible selection X_1 using the CFD_INI algorithm. If CFD_INI cannot find a feasible solution stop the algorithm. Otherwise set $r = 1$, and go to step 2.

Step 2. Compute $SR(X_r)$, defined as the set of shortest routes under metric $l_a(X_r)$ in the following way. For each unicast connection $i \in P^{UN}$ find the shortest route $k \in \Pi_i$ under metric $l_a^{DEL}(X_r)$. For each anycast downstream connection $i \in P^{DS}$ find the shortest route $k \in \Pi_i$ under metric $l_a(X_r)$. Next for each anycast upstream connection $i \in P^{US}$ find the shortest route $k \in \Pi_i$ under metric $l_a^{DEL}(X_r)$, for which the following condition is satisfied $d(\pi_i^k) = o(\pi_{\tau(i)}^j)$, where $x_{\tau(i)}^j \in SR(X_r)$.

Step 3. Set $H = X_r$ and $K = P^{UN} \cup P^{DS}$.
a) Find $i = first(K)$. If $i \in P^{UN}$, set $F = \left(H - \{x_i^j\}\right) \cup \{x_i^k\}$, where $x_i^k \in SR(X_r)$ and $x_i^j \in H$. Otherwise, if $i \in P^{AN}$ set $F = \left(H - \{x_i^j\}\right) \cup \{x_i^k\}$, where $x_i^k \in SR(X_r)$, $x_i^j \in H$ and next set $F = \left(F - \{x_{\tau(i)}^{j'}\}\right) \cup \{x_{\tau(i)}^{k'}\}$, where $x_{\tau(i)}^{j'} \in H$, $x_{\tau(i)}^{k'} \in SR(X_r)$.
b) If F is a feasible selection and $DEL(F) < DEL(H)$, let $H = F$.
c) Set $K = K - \{i\}$. If $K = \varnothing$, go to 4. Otherwise, go to 3a.

Step 4. If $H = X_r$, stop. The algorithm cannot improve the solution any further. Otherwise, let $X_{r+1} = H$, $r = r + 1$ and go to 2.

Theorem 1. *Algorithm CFD_DEL converges in a finite number of steps and constitutes a feasible solution.*

The maximum number of iterations is $\left(\left(\prod_{p \in P^{UN}} \zeta_p\right)\left(\prod_{p \in P^{DS}} \left(\sum_{v \in V} (\zeta_p^v \zeta_{\tau(p)}^v)\right)\right) - 1\right)$,

where ζ_p is the number of candidate routes for $p \in P^{UN}$. ζ_p^v is the number of candidate routes connecting the client node and the server node v for $p \in P^{AN}$.

Proof. The main idea of algorithm CFD_DEL is as follows. We start with a feasible selection X_1 (Step1). For each considered selection X_r we calculate $SR(X_r)$ containing the shortest paths according to the metric l_a^{DEL} (Step 2). Next, we try to improve the solution by deviation of one selected connection to other paths (Step 3). Since there is a finite number of non-bifurcated flows, the algorithm converges in a finite number of steps. Repetition of the same flow is impossible due to the stopping condition (Step 4). We assume that the initial selection X_1 is feasible. Next, in Step 3b we check whether the new selection F is feasible. If F is not feasible it is not analyzed further. Therefore, CFD_DEL constitutes a feasible solution.

The number of possible candidate routes of a unicast connection $p\in P^{UN}$ is ζ_p. Therefore, the number of routes' combinations for all unicast connections is $\prod_{p\in P} \zeta_p$. For a pair of associated anycast connections $p\in P^{DS}$ and $\tau(p)\in P^{US}$ the number of feasible (in terms constraint (6)) routes' combinations is $\sum_{v\in V} (\zeta_p^v \zeta_{\tau(p)}^v)$, where ζ_p^v denotes the number of candidate routes connecting the client node and the server node v for $p\in P^{AN}$. Consequently, the number of routes' combinations for all anycast connections is $\left(\prod_{p\in P^{DS}} \left(\sum_{v\in V} (\zeta_p^v \zeta_{\tau(p)}^v)\right)\right)$. If we take into account both unicast and anycast connections, the number of all routes' combinations is $\left(\prod_{p\in P^{UN}} \zeta_p\right)\left(\prod_{p\in P^{DS}} \left(\sum_{v\in V} (\zeta_p^v \zeta_{\tau(p)}^v)\right)\right)$. Since we start with a feasible selection and repetitions of the same flow are impossible, the maximum number of CFD_DEL iterations is $\left(\left(\prod_{p\in P^{UN}} \zeta_p\right)\left(\prod_{p\in P^{DS}} \left(\sum_{v\in V} (\zeta_p^v \zeta_{\tau(p)}^v)\right)\right)-1\right)$. □

Presented algorithm is similar to the FD algorithm for non-bifurcated flows proposed in [6]. The major distinction follows from the fact that CFD_DEL optimizes jointly unicast and anycast connections. While unicast connections are processed analogously as in FD, anycast connections are processed in a different way. This is a consequence of the definition of anycast demand, which consists of two connections: upstream and downstream. Since both associated anycast connection must connect the same pair of nodes (constraint (6)), there have to be considered jointly. However, due to the asymmetry of anycast flows (usually downstream bandwidth is much higher than downstream bandwidth), we first find the shortest route of downstream connection taking into account all replica servers networks. Next, for the upstream connection we take into account only routes between the client node and the server node already selected for downstream connection.

Notice that the CFD_DEL algorithm can be easily adapted to the NCFA problem with another objective function. The only modification is to change the link metric (Step 2) and objective function (Step 3b) used in the algorithm. Besides the network delay function we have developed CFD_LFL and CFD_RCL algorithms with LFL [18] and RCL [23] functions, respectively. Functions LFL and RCL have been defined in context of flow optimization in survivable connection-oriented networks. More information on these functions and unicast FD algorithms applying these functions can be found in [18] and [23].

5 Results

Algorithms CFD_INI, CFD_DEL, CFD_LFL and CFD_RCL were coded in C++. The network on which we run our experiment consists of 36 nodes and 144 directed links (Fig. 1). The bold lines represent links of size 96 units while other lines are links of size 48 units. The link capacities were chosen to model capacity ratio of OC-48 circuits used in optical networks. During simulations, the link capacities were scaled by a factor 100 to enable establishing of thousands of connections. Since we consider an existing backbone network, which is in an operational phase and augmenting the replica servers is constant. However, to verify performance of our approach for various cases in test 12 various servers' location scenarios (Table 1). The number of servers varies from 2 to 4. For each of server location we test 13 different demand patterns consisting of 180 anycast demands (360 anycast connections) and 2500 unicast connections.

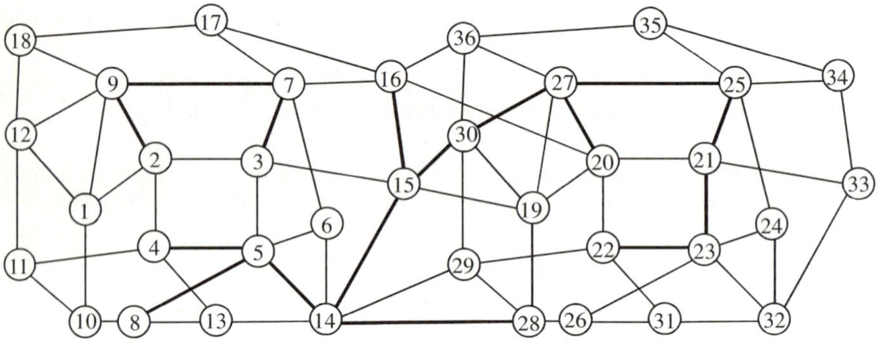

Fig. 1. Topology of tested network

In simulations we use the anycast ratio (AR) parameter defined as the ratio of the anycast demands' bandwidth requirement to the bandwidth requirement of all demands issued in the network. Demand patterns are generated randomly satisfying various values of AR.

Table 1. Location of replica servers – simulation cases

Number of servers	Location of servers	Number of servers	Location of servers	Number of servers	Location of servers
2	5, 23	3	5, 23, 30	4	5, 9, 23, 30
2	5, 25	3	5, 25, 30	4	5, 9, 25, 30
2	7, 23	3	7, 14, 23	4	7, 14, 23, 30
2	7, 25	3	7, 14, 25	4	7, 14, 25, 30

To examine performance of proposed algorithms, we compare CFD with the heuristic algorithm - which we will refer to as UFD - based on the Flow Deviation and suggestions given in [7], [10], and [25]. The main idea of UFD is to assign each anycast client to the closest (in terms of hops) content server. Next, since both end nodes of anycast demand are fixed, the unicast Flow Deviation algorithm is applied for both unicast and anycast demands. The main difference between CFD and UFD is that the former algorithm enables the anycast demand to change the content server during subsequent iterations of the algorithm, while the latter algorithm assigns clients to servers in a static manner.

Fig. 2 shows the performance of CFD_INI against UFD_INI – heuristics developed to find feasible starting solution. We can easily notice that CFD_INI can find a feasible solution for each tested case (52 cases for each number of servers). Recall that each case differs with the location of servers and demand pattern. UFD_INI provides much worse results, only in 69 of 156 tests UFD_INI is able to find a feasible solution.

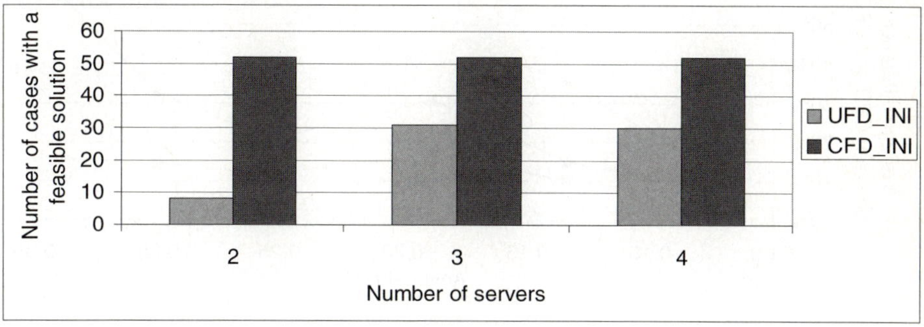

Fig. 2. Comparison of UFD_INI and CFD_INI algorithms

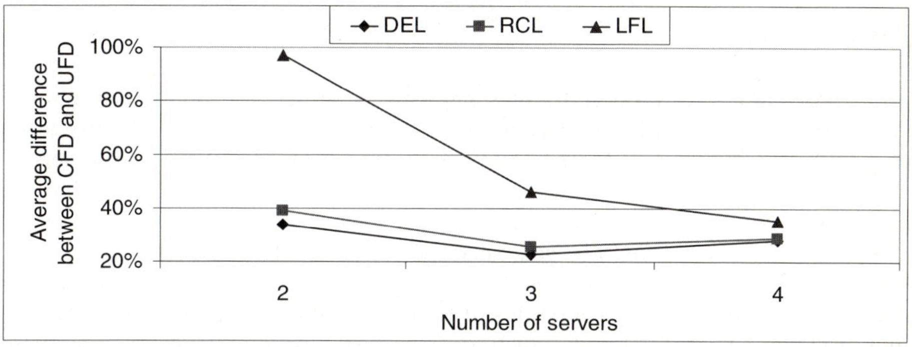

Fig. 3. Average percentage difference between CFD and UFD with various objective functions as a function of server number

In Fig. 3 we compare CFD and UFD algorithms. We present results only for these cases, in which UFD_INI can find a feasible solution. Each curve in the graphs represent the percentage difference between both algorithms obtained for one of the following objective functions: network delay [6], LFL [18] and RCL [23]. Results show that CFD algorithm yields much better performance than UFD. However, detailed observation indicates that the difference depends on the objective function and number of servers.

On Fig. 4 we plot curves presenting difference between algorithms CFD_DEL and UFD_DEL as a function of anycast ratio. Each curve represents results obtained for a particular number of servers placed in the network. Note that in the case of 2 servers only in 2 of all reported tests UFD_DEL finds a feasible solution. We can observe that generally the difference between both algorithm grows with as the value of anycast ratio increases. The experimental data is reasonably well explained by the fact that CFD can deviate anycast requests to various servers while in the case of UFD algorithm the assignment of anycast requests to servers is fixed.

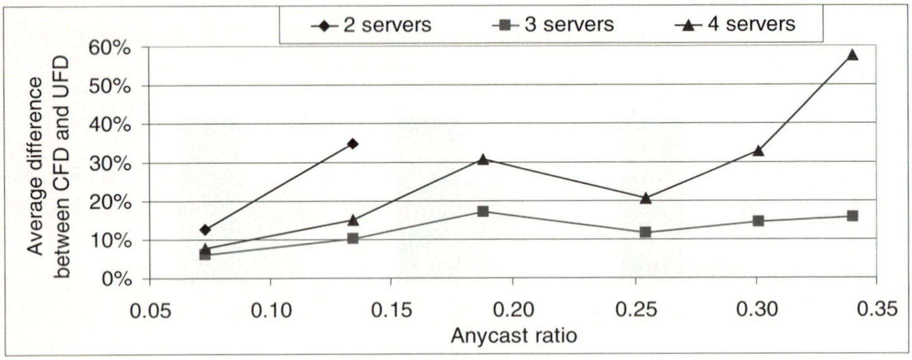

Fig. 4. Percentage difference between CFD_DEL and UFD_DEL as a function of anycast ratio for various numbers of servers

6 Conclusion

In this paper, we have proposed a new heuristic algorithm for joint optimization of unicast and anycast flows in connection-oriented networks. The main novelty of our work is that the CFD algorithm can deviate anycast flows using all content servers, i.e. servers are selected dynamically in subsequent iterations of the algorithm. Although our goal in this paper is biased towards the connection-oriented techniques, we believe that the results should be generally applicable also to connection-less network techniques. We have provided a numerical example to illustrate the proposed approach. Simulations have provided positive results to show that using the CFD algorithm provides substantial improvement over heuristic, which assigns anycast clients statically to the closest replica server. In future work we plan to apply the proposed approach to flow optimization in peer-to-peer networks.

References

1. Awerbuch, B., Brinkmann, A., Scheideler, C.: Anycasting in adversarial systems: routing and admission control. In: Baeten, J.C.M., Lenstra, J.K., Parrow, J., Woeginger, G.J. (eds.) ICALP 2003. LNCS, vol. 2719, pp. 1153–1168. Springer, Heidelberg (2003)
2. Byun, S., Yoo, C.: Minimum DVS gateway deployment in DVS-based overlay streaming. Computer Communications 31, 537–550 (2008)
3. Burns, J., Ott, T., Krzesinski, A., Muller, K.: Path selection and bandwidth allocation in MPLS networks. Performance Evaluation 52, 133–152 (2003)
4. Doi, S., Ata, S., Kitamura, K., Murata, M.: IPv6 anycast for simple and effective service-oriented communications. IEEE Communication Magazine (5), 163–171 (2004)

5. Duhamel, C., Matul, A.: An Augmented Lagrangean Approach for the QoS Constrained Routing Problem. In: Proc. International Network Optimization Conference INOC (2007)
6. Fratta, L., Gerla, M., Kleinrock, L.: The Flow Deviation Method: An Approach to Store-and-Forward Communication Network Design. Networks 3, 97–133 (1973)
7. Hao, F., Zegura, E., Ammar, M.: QoS routing for anycast communications: motivation and an architecture for DiffServ networks. IEEE Comm. Magazine 6, 48–56 (2002)
8. Hou, Y.T., Shi, Y., Sherali, H.D.: Optimal base station selection for anycast routing in wireless sensor networks. IEEE Transactions on Vehicular Technology 3, 813–821 (2006)
9. Hyytia, E.: Heuristic Algorithms for the Generalized Routing and Wavelength Assignment Problem. In: Proc. of Seventeenth Nordic Teletraffic Seminar NTS-17, pp. 373–386 (2004)
10. Jia, W., Xuan, D., Tu, W., Lin, L., Zhao, W.: Distributed admission control for anycast flows. IEEE Transactions on Parallel and Distributed Systems 8, 673–686 (2004)
11. Kasprzak, A.: Designing of Wide Area Networks. Wroclaw Univ. of Tech. Press (2001)
12. Markowski, M., Kasprzak, A.: An Approximate Algorithm for Web Replica Allocation and Topology Assignment Problem in WAN. In: Proc. of 17th IMACS World Congress, Paris, France (2005)
13. Peng, G.: CDN: Content Distribution Networks. Technical Report (2003), `http://sunysb.edu/tr/rpe13.ps.gz`
14. Pióro, M., Medhi, D.: Routing, Flow, and Capacity Design in Communication and Computer Networks. Morgan Kaufman Publishers, San Francisco (2004)
15. Rabinovich, M.: Issues in Web Content Replication. Data Engineering Bulletin 4 (1998)
16. Ryba, P., Kasprzak, A.: The Gateways Location and Capacity Assignment Problem in Hierarchical WANs: An Approximate Algorithm and Computational Results. In: Proc. Of 18th European Meetings on Cybernetics and Systems Research (EMCSR) (2006)
17. Tang, M., Jia, W., Wang, H., Wang, J.: Routing and Wavelength Assignment for Anycast in WDM Networks. In: Proc. of 3rd International Conference on Wireless and Optical Communication WOC, pp. 301–306 (2003)
18. Walkowiak, K.: A New Method of Primary Routes Selection for Local Restoration. In: Mitrou, N.M., Kontovasilis, K., Rouskas, G.N., Iliadis, I., Merakos, L. (eds.) NETWORKING 2004. LNCS, vol. 3042, pp. 1024–1035. Springer, Heidelberg (2004)
19. Walkowiak, K.: Heuristic Algorithm for Anycast Flow Assignment in Connection-Oriented Networks. In: Sunderam, V.S., van Albada, G.D., Sloot, P.M.A., Dongarra, J. (eds.) ICCS 2005. LNCS, vol. 3516, pp. 1092–1095. Springer, Heidelberg (2005)
20. Walkowiak, K.: A Unified Approach to Survivability of Connection-Oriented Networks. In: Yolum, p., Güngör, T., Gürgen, F., Özturan, C. (eds.) ISCIS 2005. LNCS, vol. 3733, pp. 3–12. Springer, Heidelberg (2005)
21. Walkowiak, K.: Lagrangean Heuristic for Anycast Flow Assignment in Connection-Oriented Networks. In: Alexandrov, V.N., van Albada, G.D., Sloot, P.M.A., Dongarra, J. (eds.) ICCS 2006. LNCS, vol. 3991, pp. 618–625. Springer, Heidelberg (2006)
22. Walkowiak, K.: Unsplittable Anycast Flow Problem Formulation and Algorithms. In: Alexandrov, V.N., van Albada, G.D., Sloot, P.M.A., Dongarra, J. (eds.) ICCS 2006. LNCS, vol. 3991, pp. 626–633. Springer, Heidelberg (2006)
23. Walkowiak, K.: A New Function for Optimization of Working Paths in Survivable MPLS Networks. In: Levi, A., Savaş, E., Yenigün, H., Balcısoy, S., Saygın, Y. (eds.) ISCIS 2006. LNCS, vol. 4263, pp. 424–433. Springer, Heidelberg (2006)
24. Walkowiak, K.: Anycast Communication – A New Approach to Survivability of Connection-Oriented Networks. Communications in Computer and Information Science 1, 378–389 (2007)
25. Yamada, H., Okamura, A., Nakamichi, K., Sakai, K., Chugo, A.: QoS Control by Traffic Engineering in Content Delivery Networks. Fujitsu Sci. Tech. J. 2, 244–254 (2003)

Scheduling of Updates in an IP Forwarding Engine with Partitioned Prefix Tables*

Junghwan Kim and Jinsoo Kim**

Department of Computer Science, Konkuk University,
322 Danwol-dong, Chungju-si, Chungbuk 380-701, Korea
{jhkim,jinsoo}@kku.ac.kr

Abstract. Many IP forwarding engines have used TCAM(Ternary Content Addressable Memory) for IP address lookup. While TCAM has the advantage of parallel lookup in a single step, it has a difficulty in updating a prefix table. TCAM requires the prefix table to be ordered for the longest prefix matching, so each update may incur several memory movements to maintain the ordering. For the fast updating of the prefix table, it is needed to eliminate the ordering constraint. If a prefix table is partitioned into several sets each of which contains disjoint prefixes, there is at most one match in each set and the ordering of prefixes is not needed any more for the longest prefix matching. Prefixes of each partitioned table should be kept disjoint after updating as well as in the initial configuration. In this paper we present several algorithms for scheduling of updates to keep prefixes of partitioned tables disjoint. We compare and evaluate our scheduling algorithms by simulation with a real-world routing table and updates. Experiment results show that most of updates can be scheduled to the partitioned tables having disjoint prefixes so good update performance can be achieved.

Keywords: update scheduling, IP address lookup, partitioned prefix tables.

1 Introduction

As the demand for broadband multimedia services and the number of users have continuously increased on the Internet, the amount of Internet traffic has exponentially grown as well. To handle the growing Internet traffic, not only transmission media but routers take a greater role to provide higher bandwidth.

The main function of a router is to forward each incoming packet to the next hop router toward its final destination. For the packet forwarding, the router must determine the output port and the next hop address by means of IP address lookup operation. In the incipient stage of the Internet, IP address lookup could be easily performed by the exact matching operation in the fixed-sized classful addressing. In these days, however, the matching operation has become more complex because the variable-sized network addresses or prefixes have been extensively used under the

---

* This work was supported by Konkuk University.
** Corresponding author.

O. Gervasi et al. (Eds.): ICCSA 2008, Part II, LNCS 5073, pp. 808–819, 2008.

CIDR(Classless Interdomain Routing) to alleviate the deficiency of the address space[1]. In contrast to classful addressing, there are possibly multiple matches for a destination IP address in the CIDR. A router must select the longest matching prefix as the best matching one, which has great impact on the performance of the router. For the development of the high performance routers, many fast IP address lookup schemes have been researched in software or hardware approaches[2, 3].

Most of the hardware approaches for IP address lookup are based on TCAM(Ternary Content Addressable Memory). TCAM is a fully associative memory in which each memory cell can store a "don't care" state in addition to 0's and 1's states[4]. So TCAM can store variable-sized prefixes and find the longest matching prefix in a single cycle using the associative memory. However, it requires the ordering of prefixes and also a priority encoder logic to find the longest matching prefix.

The router has to update its prefix table frequently due to the Internet routing instability[5]. Such frequent updates may consume many computation cycles in an IP forwarding engine and affect the performance of the lookup operation. So the update efficiency has a great impact on the performance of an IP forwarding engine. Many schemes have been proposed for fast update of a prefix table[6-10]. In case of TCAM-based lookup schemes, even a single update may induce several movements of prefix entries, so the update performance is more crucial.

We introduce a partitioning scheme to reduce the number of memory movements during updating. A prefix table is partitioned into several sets so that any two prefixes are *disjoint*, *i.e.*, not the prefix of each other in the same set. A priority encoder is not needed any more because there is at most one matching prefix in each set. However, it is not always possible to insert a new prefix into one of the partitions because the new prefix is possibly not disjoint with one or more than one prefixes in every partition. To handle such a case we introduce a supplementary partition called the *OVF*(OVerFlow). The OVF partition can accommodate any arbitrary prefixes whether those are disjoint with existing prefixes or not. A prefix table is composed of several partitions each of which contains disjoint prefixes, and an OVF partition. We call the former *DISJ*(DISJoint) partitions.

The OVF partition is more complicated than the DISJ partitions in not only an IP address lookup operation but an update operation. It is due to the fact that the OVF partition needs the ordering of prefixes as the conventional TCAM. In this paper we propose and compare several update scheduling algorithms to minimize the size and updates of the OVF partition. If there are several candidates of DISJ partitions for the insertion of a new prefix, it may affect the future OVF partition which DISJ partition is scheduled for the new prefix.

The rest of this paper is organized as follows. Section 2 describes the longest prefix matching in the TCAM and previous works related to the fast update and the table partition. In section 3, we explain our IP address lookup approach based on partitioned prefix tables to reduce the update overhead. Section 4 presents several scheduling algorithms for an update operation to determine which partition will contain a new prefix. Section 5 compares and evaluates the performance of our scheduling algorithms with simulation results. We finally conclude the paper in section 6.

2 Background

2.1 Longest Prefix Matching in TCAM

The longest prefix matching is to select the longest prefix among all the matching prefixes as the best matching one. Fig. 1 shows an example of the longest prefix matching in a TCAM. In the example, the length of an IP address is assumed to be 8 for simplicity. The prefix table in the TCAM contains 8 prefixes 10100*, 1011*, 1110*, 010*, 100*, 101*, 00*, 10* and 11*, as shown in Fig. 1. For a given 8-bit destination IP address 10100100, there are three matching prefixes 10100*, 101* and 10* whose lengths are 5, 3 and 2, respectively. The priority encoder selects the longest prefix 10100* which is located at the topmost among those prefixes.

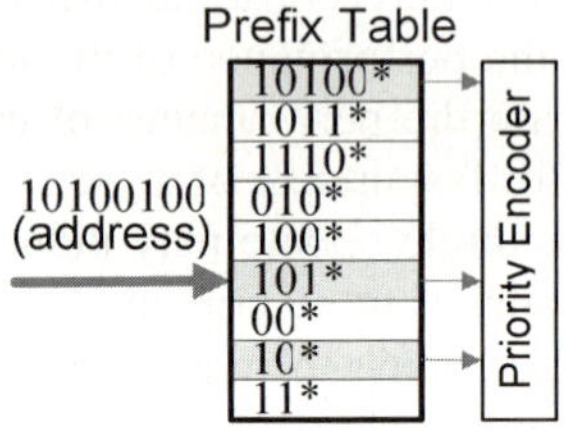

Fig. 1. TCAM-based Longest Prefix Matching

As shown in Fig. 1, all the prefixes are placed by their lengths for the longest prefix matching. The prefixes with the same length can be located in any order because those cannot be simultaneously matched for any destination IP address. Owing to the prefix ordering, the topmost prefix will be the longest matching one. Though the TCAM finds the longest matching prefix in a single step using parallel search and a priority encoder, it has a drawback to maintain the ordering of prefixes while updating.

2.2 Related Works

There are two approaches for the fast update in TCAM. The first type of update schemes defines its own ordering criteria for prefixes so that it reduces memory movements per update. The second type removes the prefix ordering constraint itself by utilizing the hardware modification or characteristics of the forwarding table.

PLO_OPT and CAO_OPT are based on the ordering criteria of the prefix length and the ancestor-descendent relationship on the prefix search trie respectively, in order to reduce the number of memory movements during update[6]. The worst case numbers of memory movements per update in PLO_OPT and CAO_OPT are $L/2$ and $D/2$, respectively, where L is the maximum prefix length and D is the maximum path length from the root to any leaf node in the prefix search trie. L is 32 in case of IPv4 and D is about 8. Wu et al. presented an update algorithm based on the ordering criterion of the level on the prefix search trie[10]. A forwarding table is partitioned and ordered by the level. Each partition includes free space prepared for future update. A new prefix can be easily inserted if its parent and children are in different level partitions. However, the free space may contain several prefixes of different levels within the free space, as updates proceed by means of that algorithm. It will cause the memory movements within the free space as in the CAO_OPT.

Kobayashi et al. modified TCAM by adding vertical OR circuits in the mask column to directly determine the longest mask among the matched entries without the priority encoder logic[11]. Since it does not require any ordering constraint, the fast update can be achieved. However, as the total number of prefixes increases in the table, the actual lookup delay remarkably increases due to the vertical ORing. Ng and Lee partitioned a forwarding table into several TCAM modules so that each module only contains the prefixes with the same output port number[12]. A new prefix can be inserted at any location within the corresponding TCAM module without considering the ordering. However, many updates are just to change the output port number of the existing entry. Such updates lead to excessive memory movements because the prefix must be moved to another module associated with the new output port.

Some IP address lookup schemes[10, 13-15] are similar to our scheme in a sense of partitioning. Those schemes statically partition a prefix table into several sets depending on the depth level or the height level in the prefix search trie. The partitioning techniques in those schemes are conceptually equivalent. They utilize the fact that all the prefixes on the same level are disjoint and can be contained in the same partitioned set. However, the levels of prefixes should be consistently managed. Therefore, the insertion or the deletion of a new prefix causes a series of modifications of the levels for the related prefixes which are ancestors and descendants of the new prefix. Moreover, the set sizes are severely different, since the numbers of prefixes are extraordinarily different among the levels. It is undesirable factor in the lookup performance and the implementation. Although some scheme[13] has tried to make the balance among sets by distributing standalone prefixes whose depth level and height level are 0, it causes the updates to be more complicated.

3 IP Address Lookup with Partitioned Prefix Tables

A conventional TCAM-based IP address lookup scheme needs both a priority encoder and some ordering constraint on prefixes. If a prefix table is partitioned into several sets so that there can be at most one match in each set, then both the requirements can be eliminated. Since at most one match can occur in each partitioned table, we don't need to maintain ordered prefixes for finding the best match. It makes update operations very fast because there are no movements of existing prefixes for the ordering constraint. Lookup operations also become fast by eliminating the delay of the priority encoder. Each partitioned prefix table should be a *disjoint* set to have at most one match on a lookup operation. We define some terms in similar manner to the related works.

Definition 1. Two prefixes p_i and p_j are *disjoint* if neither p_i is the prefix of p_j nor p_j is the prefix of p_i.

For example, p_1=10100* and p_2 = 1011* of Fig. 1 are disjoint because neither p_2 can be a prefix of p_1, nor p_1 can be a prefix of p_2. However, p_8 = 10* and p_2 are not disjoint because p_8 is a prefix of p_2.

Definition 2. A set of prefixes P is the *disjoint set* if any two prefixes, $p_i, p_j \in P$ are disjoint. For a prefix p and a disjoint set P, if $S = P \cup \{p\}$ is a disjoint set as well, p is *adaptable to* the set P. P is called an *adaptable set for* the prefix p.

Fig. 2 shows an example of IP address lookup with partitioned prefix tables. The prefix table of Fig. 1 is partitioned into three tables in Fig. 2(a). Those tables are called *DISJ partitions*. In each DISJ partition, at most one match is possible on any lookup. For example, given an 8-bit destination address, 10100100, there are three matches, 10100*, 101* and 10* each of which resides in a different disjoint partition. Note that each DISJ partition does not need not only prefix ordering but also a priority encoder. However, since there can be several matches on the whole system, a kind of selection logic is needed to select the best one among those matches. The selection logic is similar to the priority encoder in a sense, but its complexity is much less than that of the priority encoder. The time complexity of the selection logic is $O(log\ n)$ where n is the number of partitions, whereas that of the priority encoder is $O(log\ N)$ where N is the number of prefixes total. Certainly, it is true $n << N$.

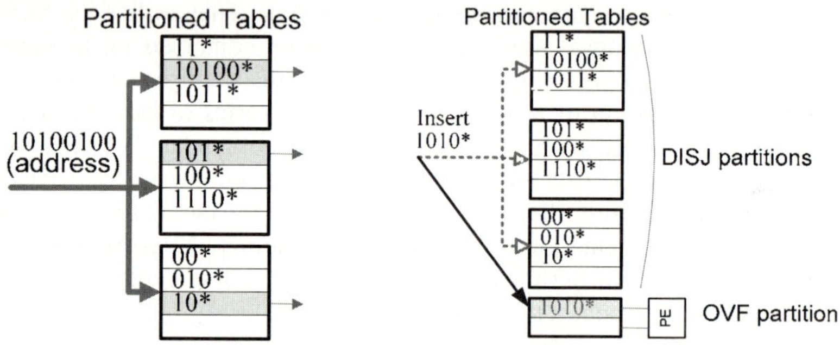

(a) Search with DISJ Partitions (b) Insertion with an OVF Partition

Fig. 2. An Example of IP Address Lookup with Partitioned Prefix Tables

Now let's consider a situation that a new prefix 1010* is being inserted, as shown in Fig. 2(b). There is no way to insert it because there is no adaptable DISJ partition for the prefix 1010*. Fig. 2(b) shows how to resolve it with an additional partitioned table called *OVF partition*. In case that a new prefix is not adaptable to any existing DISJ partitions, the prefix will be inserted into the OVF partition. Undoubtedly, the OVF partition may expose multiple matches, so it requires both a PE(priority encoder) and the maintenance of the ordering. Therefore, the updating cost of the OVF partition is more expensive than that of the DISJ partitions. So, it is crucial to minimize the number of updates on the OVF partition.

4 Update Scheduling Algorithms

4.1 Procedure of Update

There are three types of updates: the insertion of a new prefix, the deletion of an existing prefix, and the modification of the output port number for an existing prefix. In case of the deletion and the modification, the updating partition is determined as the partition which contains the corresponding prefix to be deleted or modified, so it does not need any scheduling of an update to decide the updating partition. In this section we focus on the insertions and the scheduling of them.

The first step of the insertion is to find the available partitioned tables to which a new prefix can be inserted. The available partitions must contain free space and be adaptable for the prefix. If there is no available DISJ partition, the prefix will be inserted into the OVF partition. For example, if a new prefix 01* tries to be inserted in Fig. 2(a), the available partitions are the first and the second disjoint partitions which are adaptable for the prefix 01*. If a new prefix is 1010* as shown in Fig. 2(b), all disjoint partitions are not available and the prefix must be inserted into the OVF partition.

Whether a partition is adaptable for a new prefix can be easily tested by the slight modification of the normal lookup operation in TCAM. Normally, TCAM can find the matching prefix with a given 32-bit IP address. Suppose a new prefix is given to a partition as a search key instead of a full 32-bit IP address. If there is no matching in the partition with respect to the prefix, the partition is clearly adaptable for the prefix.

The second step of the insertion is to schedule the insertion, *i.e.*, to decide one partition among all available partitions. The scheduling will affect the prefix distribution over the partitions and also the size of the OVF partition. Note that the insertion to a DISJ partition doesn't need any memory movement because it does not require any ordering of prefixes. Therefore, the cost of the insertion to a DISJ partition is less than those of the previous works[6, 10]. In case of OVF partition, the previous methods can be applied to the insertion. So, the cost of the insertion to the OVF partition is the same as that of a conventional table. Consequently, the scheduling goal should be to minimize the number of updates to the OVF partition.

4.2 Update Scheduling for the Prefix Insertions

We developed four update scheduling algorithms called *first-fit*, *next-fit*, *random* and *hashing*. These algorithms focus on the minimization of updates at the OVF partition and the fairly even distribution of prefixes over the DISJ partitions.

Fig. 3 shows the first-fit scheduling algorithm. It selects the first partition with the smallest index among all the available DISJ partitions, as shown in steps 1 to 4. If there is no available one among the DISJ partitions, the OVF partition is selected in step 5.

```
First-Fit(np, available)
// np: the number of disjoint partitions + 1
1.   for i ← 0 to np-2 do // for disjoint partitions
2.      if (available[i] = true)
3.          return i
4.   endfor
5.   return np-1 // for OVF partition
```

Fig. 3. First-Fit Scheduling Algorithm

Fig. 4 shows some examples of various update scheduling algorithms. In these examples, partitions 1, 2, 3, and 5 are available. As shown in Fig. 4(a), the first-fit scheduling selects the first available partition 1.

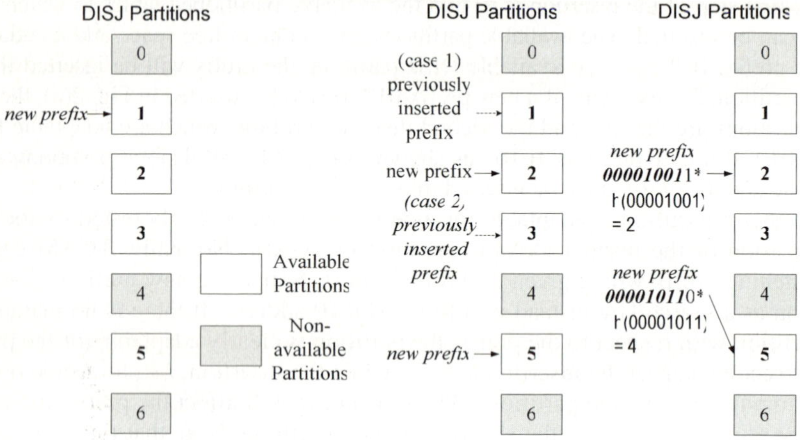

(a) First-Fit (b) Next-Fit (c) Hashing

Fig. 4. Examples of Update Scheduling Algorithms

The first-fit scheduling puts prefixes to the same partition if possible, so it tries to increase the availability of other partitions. For the intuition, suppose that there are only two partitioned tables and three prefixes p_1=1010*, p_2=1011*, and p_3=10* are inserted in sequence. If two prefixes p_1 and p_2 are inserted in different disjoint partitions, then there is no available partition for the insertion of the prefix p_3. However, if the two prefixes are inserted in the same partitions, the prefix p_3 can be inserted into another partition. If the first-fit scheduling is applied, three prefixes can be assigned to two partitions irrespective of the inserting sequence in this example.

The next-fit scheduling and random scheduling algorithms try to distribute the inserting prefixes over all the partitions as evenly as possible, so those prevent early exhaustion of free spaces in specific partitions. The random scheduling selects an available partition as arbitrary. The next-fit scheduling selects the next available partition to the partition that was assigned at the previous insertion. It differs from the first-fit in steps 2 and 3 of Fig. 3 as follows.

```
// prev: the number of previously assigned partition
// mod: modulo operator, mod(a,b) = a % b
2-1.    j ← mod(prev + 1 +  i, np-1)
2-2.    if (available[j] = true)
3.          return j
```

As shown in Fig. 4(b), the next-fit scheduling selects the partition 2 if the previously assigned partition is 1 (case 1). If the previously assigned partition is 3, it selects the partition 5 because the directly next partition 4 is not available (case 2).

The hashing scheduling algorithm has strengths of the first-fit and the next-fit. The hashing scheduling evenly distributes the inserting prefixes over all DISJ partitions using a hash function. Even though it scatters the prefixes, the related prefixes attempt to be mapped into the same partition for the availability like the first-fit. Fig. 5 describes the hashing scheduling algorithm.

```
Hashing(p, np, available)
// p=p0···p7p8···p31: new prefix
// mod(a,b): modular operation a%b
1.    hash ← mod(p0···p7,np-1) // hashing function
2.    for i ← 0 to np-2 do // for disjoint partitions
3.       j ← mod(hash + i, np-1)
2.       if (available[j] = true)
3.          return j
4.    endfor
5.    return np-1 // for OVF partition
```

Fig. 5. Hashing Scheduling Algorithm

As shown in step 1 of Fig. 5, the hashing key K is the MSB 8 bits of the prefix and the hashing function is simply h(K) = mod(K, n) where n is the number of the DISJ partitions. The 8-bit key is adopted from the observation that the prefix length is practically not less than 8. In Fig. 4(c), for a new prefix, 000010011*, the key is 00001001 and h(00001001) = mod(9, 7) = 2. So, the hashing scheduling selects the partition 2. In case that a new prefix is 000010110*, the key is 00001011 and h(00001011) = 4. Unfortunately, the partition 4 is not available, so the algorithm selects the partition 5 which is the next available one.

5 Experiments

5.1 Experiment Environment

For simulation we used a routing table and an update stream which were originally collected from Route Views in July 2007[16]. We converted the routing table into the forwarding table which has 243511 prefixes total. The update stream data of one month were converted to the format on our experiment. Three types of updates which are insertion, deletion and modification were used in the experiment. However, we will summarize only the results of the insertion and the deletion because the modification of an output port number does not incur any memory movements in TCAM and has little performance impact on updates. In this section the 'update' implies only the insertion and the deletion.

The prefixes in the forwarding table were initially partitioned into several sets. Our update scheduling algorithms are compared by performing on the same initial partitions. The initial partitions were obtained with satisfying the criteria: (a) as small OVF partition as possible, (b) as evenly distributed DISJ partitions as possible.

If there is no OVF partition, the number of DISJ partitions should be more than or equal to the maximum height of the prefix search trie, because any two prefixes which have ancestor-descendant relation in the trie cannot be contained in the same DISJ partition. We need an OVF partition for future updating in any case, so the number of DISJ partitions can be smaller than the maximum height of the trie. Table 1 shows the number of prefixes at each height level. Table 2 shows the initial number of prefixes in each partition when the prefix table is divided into 4, 6 and 8 partitions including an OVF partition. From P_0 to P_{np-2} are the DISJ partitions and P_{np-1} is the OVF partition where *np* is the number of partitions. Note that it is inevitable and desirable that the OVF partition contains all the prefixes which reside in the higher or equal level to the OVF partition number. For example, when $np = 6$, the number of prefixes in the OVF partition P_5 in Table 2 is 19 which is the total number of prefixes at heights 5, 6, 7 and 8 in Table 1.

Table 1. Number of Prefixes at Height Levels

Height	0	1	2	3	4	5	6	7	8	Total
Number of Prefixes	222810	17314	2807	465	96	14	3	1	1	243511
Ratio (%)	91.50	7.11	1.15	0.19	0.04	0.01	0.00	0.00	0.00	100.00

Table 2. Initial Prefix Partitions

Number of Partitions	P_0	P_1	P_2	P_3	P_4	P_5	P_6	P_7
$np = 8$	34788	36824	32745	34788	34788	34788	34788	2
$np = 6$	48703	48680	48703	48703	48703	19		
$np = 4$	81171	82170	79590	580				

5.2 Simulation Results

Fig. 6 shows the distribution of updates according to update scheduling algorithms. Fig. 6(a), (b) and (c) represent the distribution where the number of partitions, *np* is 4, 6 and 8, respectively. FF, NF, RAND and HASH represent *first-fit*, *next-fit*, *random* and *hashing* scheduling algorithms respectively. The number of updates in the OVF partition is quite small in either of update scheduling algorithms, but much smaller in HASH than in others. The number of updates in OVF tends to decrease as *np* increases.

Fig. 7 shows the ratio of updates in the OVF partition for 31 days. The ratio represents the accumulated average ratio from the first day to that day, not on that day. In Fig. 7(a), NF and RAND indicate the rapid increment of the average ratio at early stage and then the ratio shows very slow increasing. In FF and HASH, the ratios steadily increase from the early stage. However, the ratio of FF abruptly increases from 18th day. This phenomenon originates from the early exhaustion of the front partitioned table in FF. In FF, since the early exhaustion reduces the number of the available tables, it has negative impact on the update performance especially in case of the small number of partitioned tables. In Fig. 7(b) and (c) every algorithm represents steady increment of ratio, but the ratio of HASH is lower than any others.

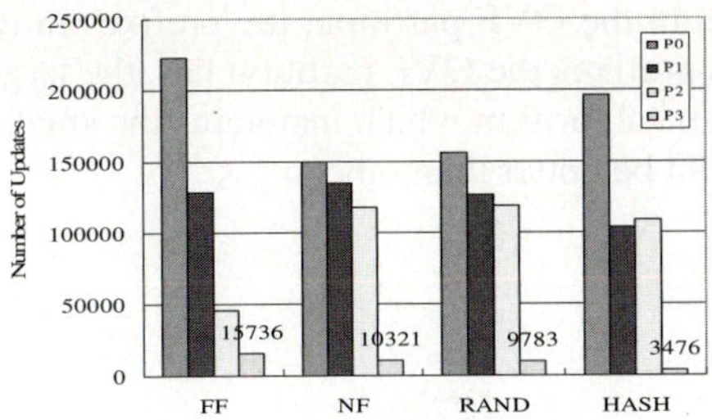

(a) $np = 4$

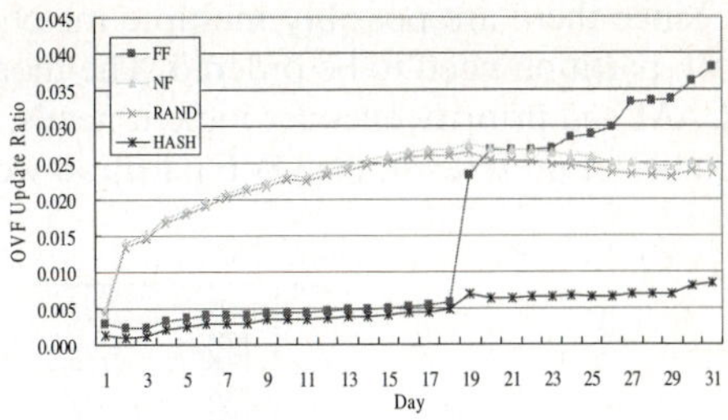

(a) $np = 4$

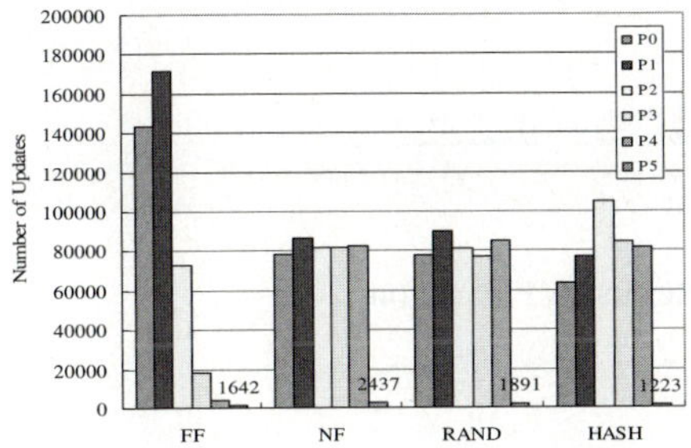

(b) $np = 6$

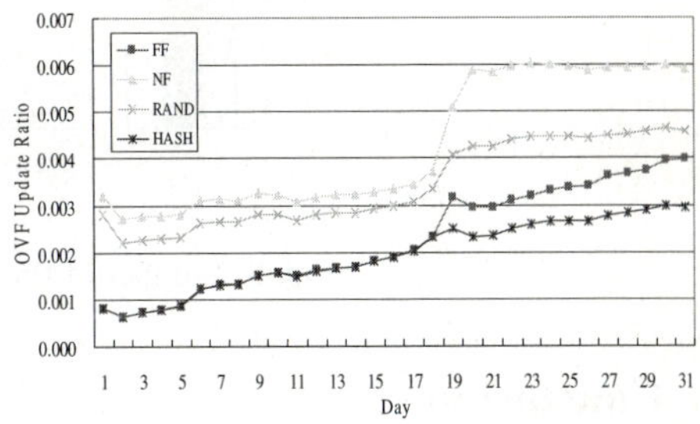

(b) $np = 6$

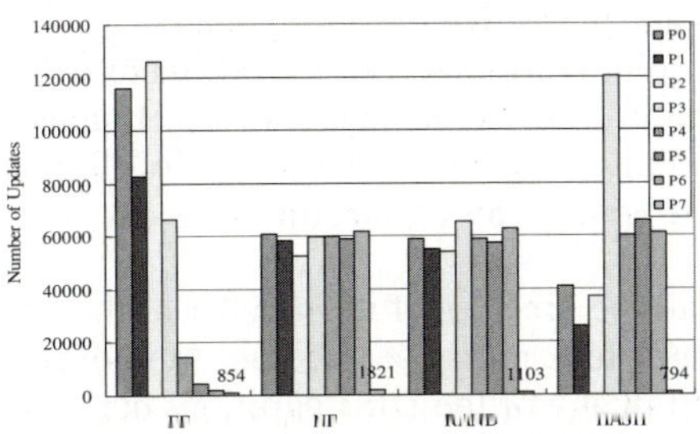

(c) $np = 8$

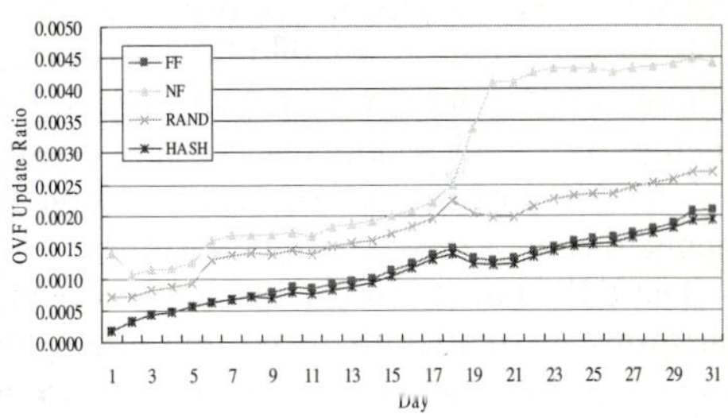

(c) $np = 8$

Fig. 6. Distribution of Updates

Fig. 7. Avg. Update Ratio in the OVF Partition

Since update on the OVF partition is more expensive than those of DISJ partitions, the higher ratio of updates in the OVF partition implies the update performance degradation. As the number of partitions, *np* increases, the absolute level of the ratio becomes lower in either of the algorithms.

Fig. 8 shows the number of prefixes contained in the OVF partition after 31-days updates. Prefixes of the OVF partition are less in order of HASH, RAND, NF and FF. The number of prefixes in the OVF partition tends to decrease rapidly as *np* increases. Note that there is no actual increment of prefixes in HASH for 31 days in case that *np* is 8, comparing with Table 2.

Since there are possibly multiple matches in the OVF partition, the prefixes in the OVF partition need to be ordered. The more prefixes the OVF partition has, the larger TCAM and priority encoder logic it needs. The algorithm which maintains the smaller number of prefixes in the OVF partition would be better than others.

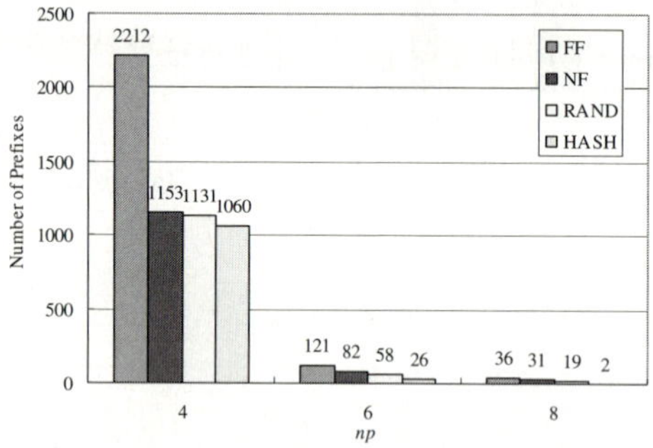

Fig. 8. Number of Prefixes in OVF Partition

6 Conclusion

Conventional TCAM-based IP forwarding engines should have ordered prefixes in the TCAMs. The ordering constraint is due to the fact that there are possibly multiple matches and they need the priority encoder logic to resolve the multiplicity.

If a prefix table is divided into several disjoint partitions called DISJ, there will be at most one match in each partition and the ordering constraint can be eliminated. Such partitioning results in desirable effects in two aspects: fast lookups by eliminating the delay of the priority encoder and fast updates by eliminating memory movements to satisfy the ordering constraint.

The DISJ partitions do not always cover all the prefixes, if the number of the DISJ partitions is smaller than the maximum height in a prefix search trie. Moreover, in some cases, a new prefix cannot be contained in any of the DISJ partitions during the update. We use an additional partition called the OVF to accommodate such prefixes.

The OVF partition is similar to the conventional TCAM, so it requires the ordering constraint to resolve multiple matches and has burden of memory movements on updates. In this paper we proposed four update scheduling algorithms, *first-fit*, *next-fit*, *random*, and *hashing* to minimize the number of updates and prefixes in the OVF partition.

The experiment results from a real-world routing table and an update stream show that the numbers of updates and prefixes in the OVF partition can be quite well controlled by the algorithms. The *next-fit* and the *random* have the characteristics of evenly distributed updates on partitions compared to other algorithms. The *first-fit* has a problem that the number of updates of the OVF partition is radically increased due to the biased prefix distribution among DISJ partitions. The *hashing* algorithm is much better than any others in a point of view that there are fewer updates and prefixes in the OVF partition. It is expected that, in case of the *hashing* algorithm, fewer DISJ partitions can cover a large amount of the prefixes and the size of the OVF partition still remains small even after the updates proceed for a long time period.

References

1. Fuller, V., Li, T., Yu, J., Varadhan, K.: Classless Inter-Domain Routing (CIDR): An Address Assignment and Aggregation Strategy. RFC1519 (1993)
2. Ruiz-Sanchez, M.A., Biersack, E.W., Dabbous, W.: Survey and Taxonomy of IP Address Lookup Algorithms. IEEE Network 15, 8–23 (2001)
3. Varghese, G.: Network Algorithmics: An Interdisciplinary Approach to Designing Fast Networked Devices. Morgan Kaufmann Pub., San Francisco (2005)
4. Pagiamtzis, K., Sheikholeslami, A.: Content-addressable memory (CAM) circuits and architectures: A tutorial and survey. IEEE Journal of Solid-State Circuits 41, 712–727 (2006)
5. Labovitz, C., Malan, G.R., Jahanian, F.: Internet Routing Instability. IEEE/ACM TON 6, 515–528 (1998)
6. Shah, D., Gupta, P.: Fast Updating Algorithms for TCAMs. IEEE Micro. 21, 36–47 (2001)
7. Wang, P., Chan, C., Chen, Y.: A Fast Table Update Scheme for High-Performance IP Forwarding. In: 8th ICPADS, pp. 592–600 (2001)
8. Pao, D., Li, Y.: Enabling Incremental Updates to LC-Trie for Efficient Management of IP Forwarding Tables. IEEE Communications Letters 7, 245–247 (2003)
9. Basu, A., Narlikar, G.J.: Fast Incremental Updates for Pipelined Forwarding Engines. In: IEEE INFOCOM (2003)
10. Wu, W., Shi, B., Wang, F.: Efficient location of free spaces in TCAM to improve router performance. IJCS 18, 363–371 (2005)
11. Kobayashi, M., Murase, T., Kuriyama, A.: A Longest Prefix Match Search Engine for Multi-Gigabit IP Processing. In: 2000 International Conf. on Communications, pp. 1360–1364. IEEE Press, New Orleans (2000)
12. Ng, E., Lee, G.: Eliminating Sorting in IP Lookup Devices using Partitioned Table. In: 16th IEEE International Conf. on Application-Specific Systems, Architecture and Processors (ASAP), pp. 119–126. IEEE Press, Greece (2005)
13. Liang, Z., Xu, K., Wu, J.: A Scalable Parallel Lookup Framework Avoiding Longest Prefix Match. In: Kahng, H.-K., Goto, S. (eds.) ICOIN 2004. LNCS, vol. 3090, pp. 616–625. Springer, Heidelberg (2004)
14. Sun, X., Sahni, S.K., Zhao, Y.Q.: Fast Update Algorithm for IP Forwarding Table Using Independent Sets. In: Mammeri, Z., Lorenz, P. (eds.) HSNMC 2004. LNCS, vol. 3079, pp. 324–335. Springer, Heidelberg (2004)
15. Chang, Y.-K., Lin, Y.-C.: Dynamic Routing Tables Using Simple Balanced Search Trees. In: Chong, I., Kawahara, K. (eds.) ICOIN 2006. LNCS, vol. 3961, pp. 389–398. Springer, Heidelberg (2006)
16. University of Oregon Route Views Project, http://www.routeviews.org/

Path Generation Issues for Survivable Network Design*

Michał Pióro, Tomasz Śliwiński, Michał Zagożdżon,
Mateusz Dzida, and Włodzimierz Ogryczak

Warsaw Univ. of Technology, Faculty of Electronics & Info. Technology, Warsaw, Poland
(tsliwins,wogrycza)@ia.pw.edu.pl,
(mpp,mzagozdz,mdzida)@tele.pw.edu.pl

Abstract. Link dimensioning and routing problems in resilient network design are considered. Reliable network operation is ensured by means of flow restoration which is performed on preselected protection (backup) paths that can absorb traffic overflows from failed primary paths. Backup and primary flows use separated link capacities, and can be split among many paths. In the paper, two restoration models are considered. The first model assumes that once the backup path is assigned it must be used in every state in which the protected primary path fails while the second model allows different protection paths to be used in different network failure states. The problems are formulated as multiple commodity linear programming (LP) models using the link-path (L-P) notation and solved by the column generation technique. Consequent pricing models and algorithms are introduced. Computational efficiency of the presented approaches is analyzed.

1 Introduction

Large capacity of optical transmission systems in telecommunication networks can drastically influence the network performance in the case of network link failures. The implied increasing aggregation level of the traffic flows makes them more and more sensitive to such failures. For this reason, transport network operators are interested in providing reliable services, that thanks to fast recovery mechanisms are able to survive network failures. To provide reliable transport services, the transport network operator must foresee failures that can appear in the network, and accordingly design protection (backup) paths that can absorb the traffic overflows from failed primary paths that are used in the normal, failure-less state of network operation. This is referred to as *flow restoration*. In this paper, we investigate link dimensioning and flow routing design problems for resilient (i.e., robust to failures) networks that assume flow restoration of affected flows against the foreseen failures. Hence, the problem consists in searching for a configuration of paths for each operating state, such that the capacity installation cost resulting from realization of given traffic demands is minimized. Throughout the paper we assume that capacity of the primary flows is separated from the capacity used by the protection flows. There is a strong motivation for this assumption, stemming from practical aspects of the network operation. Having reserved capacity for primary flows

* The research was partially supported by the Dean's grant from the Faculty of Electronics and Information Technology, Warsaw University of Technology.

O. Gervasi et al. (Eds.): ICCSA 2008, Part II, LNCS 5073, pp. 820–835, 2008.

and establishing them, a network operator is in fact able to immediately bring back primary path flows once a failure is removed. Such capacity model is known as the case of *no stub-release*, since it does not allow protection flows to use the basic capacity that could be released due to failing primary paths.

As indicated above, we assume that network survivability is achieved by means of a flow restoration mechanism. We distinguish two problem variants, depending on the assumptions concerning restoration of affected primary flows. The first, called *state-independent flow restoration* assumes that once a backup path is assigned, it is used in every state in which the protected primary path fails. The second model, though, allows different protection paths to be used in different network failure states and is therefore called *state-dependent flow restoration*. Thus, in this paper we assume that each primary flow is protected by its backup paths (state-dependant or not), and bifurcation is permitted for each primary flow as well for its protection flows.

The considered resilient design problems are well recognized in the literature (see [1] and discussion there) and were considered in many research papers, for example in [2,3,4,5,6,7,8,9,10]. The two problems considered in this paper have a property that is common in resilient network design. Both problems are $\mathcal{NP}$-complete and cannot be formulated through LP formulations in a compact way. Still, they can be formulated using the Link-Path (L-P) notation (see [11,1]), but certainly not in a compact way. Hence, the use of L-P formulations is useful only in two situations: when all required paths can be used in the formulation, or when an efficient path (column) generation method can be proposed. Since the former solution is not scalable with growing number of nodes, we focus on the latter method.

Path generation (PG) is an application of the general column generation technique, developed to effectively solve large-scale LP problems of special structure (see [11]). The PG framework can be described as follows. Having given traffic demands and costs of link capacities, we determine preliminary sets of the candidate paths, containing, for each demand, at least one primary path together with its necessary backup paths. Using these sets we formulate an LP problem in the L-P notation. Having the LP problem in hand, we treat it with an LP solver and get the values of the dual variables related to the L-P problem constraints. The dual equalities allow us to verify if the current sets of candidate paths should be extended in order to get an optimal flow distribution in the sense of minimal capacity cost. If the current set of candidate paths for a specific demand does not contain a path (either primary or backup) that due to the dual constraints may be necessary, then the so-called *pricing problem* is formulated and the required path is obtained as a solution of this problem. The procedure consists of re-optimizing the LP problem and introducing new paths generated by the pricing problem, until no new path is necessary.

Application of the PG technique allows one to keep the lists of candidate paths very short with respect to the sets of all possible paths. Thanks to that, the corresponding LP problem can be solved very efficiently. Still, the efficiency of the overall technique depends on the efficiency of the path generation method resolving the pricing problems. It appears that for some network failure scenarios the pricing problem can become $\mathcal{NP}$-complete [6,5,7,8]. One of the investigated approaches is to solve pricing problems formulated as mixed-integer (MIP) programmes. Another approach assumes a special variant of a shortest path algorithm, namely a label-setting algorithm [12], that allows to compute shortest paths with respect to additional resource constraint requirements. In our case these requirements are induced by backup paths used in the failure states in which the generated primary path fails. In any case, exact generation methods are sometimes inefficient, still in such cases it is often sufficient to use path lists produced by a heuristic method. Therefore, we also propose a heuristic for solving the considered PD problems. Its application makes PG quite easy and allows to obtain solutions of good quality for the overall problem. The heuristic is based on precomputed sets of potential primary paths. Then, the pricing problem consists in searching through these sets and selecting the best path for each demand. The presented approach assumes that the sets of candidate primary paths are initially determined as a solution of a K-shortest path algorithm.

The paper is organized as follows. In Section 2 we give mathematical formulations for the problems under consideration. Section 3 introduces the related pricing problems. There are also presented exact and approximate approaches to generate efficiently new columns (paths). The paper is summarized with the discussion on computational experiments (Section 4) and some concluding remarks (Section 5).

2 The Survivability Models

2.1 Network Model

The considered network is modeled with a directed graph $\mathcal{G} = (\mathcal{V}, \mathcal{E})$ composed of a set of nodes $\mathcal{V}$ and a set of (directed) links $\mathcal{E}$ between the nodes. For ease of exposition, we assume that the graph does not contain loops nor parallel links, i.e., $\mathcal{E} \subseteq \mathcal{V}^2 \setminus \{(v,v) : v \in \mathcal{V}\}$. For any link $e \in \mathcal{E}$, its source node is denoted by u_e, whereas the target node is denoted by v_e, so that $e = (u_e, v_e)$. The cost of realizing one unit of demand on link $e \in \mathcal{E}$ is denoted by ξ_e.

The set $\mathcal{D} \subseteq \mathcal{V}^2$ represents (directed) point-to-point demands. For notational convenience, at most one demand between each ordered pair of nodes is assumed. The source and target of a demand $d \in \mathcal{D}$ are denoted by u_d and v_d, respectively, and assumed to be different from each other. The demand value of demand $d \in \mathcal{D}$ is given by $h_d > 0$.

Let $\mathcal{S} \subseteq 2^{\mathcal{E}}$ be the family of all considered failure states (failure scenario) where each state (also called failure situations) $s \in \mathcal{S}$ is identified with the set of its failing links. Family $\mathcal{S}$ is assumed to include the failure-less state $\mathcal{O}$ (formally equal to the empty subset of $\mathcal{E}$) in which all links are operational (state $\mathcal{O}$ is also called the *normal state*). It is assumed that links fail totally. The sets of surviving links in state $s \in \mathcal{S}$ are denoted by $\mathcal{E}^s \subseteq \mathcal{E}$. Similarly, $\bar{\mathcal{E}}^s = \mathcal{E} \setminus \mathcal{E}^s$ denotes the set of failing links in state $s \in \mathcal{S}$. Certainly, $\mathcal{E}^{\mathcal{O}} = \mathcal{E}$.

We observe that the node failures can be easily modeled by link failures. This is achieved by a suitable transformation of the network graph. We simply substitute each node $v \in \mathcal{V}$ that can fail by two nodes v' and v'' connected by a dummy link $e_v = (v', v'')$ with infinite capacity and zero cost. Also, in the transformed graph we terminate all original links $e \in \mathcal{E}$ such that $v_e = v$ at node v', and start all original links $e \in \mathcal{E}$ such that $u_e = v$ at node v''.

Let $\mathcal{P} := \bigcup_{d \in \mathcal{D}} \mathcal{P}_d$ be the set of all candidate (undirected) paths that can be used for realizing demand flows, where $\mathcal{P}_d$ is the set of candidate paths for demand $d \in \mathcal{D}$. We assume that paths do not contain loops so for each demand $d \in \mathcal{D}$ the set $\mathcal{P}_d$ is a subset of all simple paths from u_d to v_d. Note that the sets $\mathcal{P}_d, d \in \mathcal{D}$ are mutually disjoint (since we have assumed at most one demand per node-pair) and in consequence, each path $p \in \mathcal{P}$ can be identified with the set of the links it traverses, so that $p \subseteq \mathcal{E}$.

Notation $\mathcal{P}_d^s \subseteq \mathcal{P}_d$ refers to the set of all those candidate paths for demand $d \in \mathcal{D}$ that are available in state $s \in \mathcal{S}$, i.e., $\mathcal{P}_d^s := \{p \in \mathcal{P}_d \mid p \cap s = \emptyset\}$. Similarly, $\bar{\mathcal{P}}_d^s = \{p \in \mathcal{P}_d \mid p \cap s \neq \emptyset\}$ denotes the set of all candidate paths for demand $d \in \mathcal{D}$ that fail in state s, so that $\mathcal{P}_d^s \cup \bar{\mathcal{P}}_d^s = \mathcal{P}_d$. Furthermore, $\mathcal{P}_e \subseteq \mathcal{P}$ is the set of all paths containing link $e \in \mathcal{E}$, and $\mathcal{P}_e^s := \mathcal{P}_e \cap (\bigcup_{d \in \mathcal{D}} \mathcal{P}_d^s)$ denotes the set of all paths containing link $e \in \mathcal{E}$ that are available in state $s \in \mathcal{S}$.

For convenience, we also introduce the notation $\mathcal{S}_p \subseteq \mathcal{S}$ for the set of all states $s \in \mathcal{S}$ in which path $p \in \mathcal{P}$ is available, i.e., $\mathcal{S}_p = \{s \in \mathcal{S} \mid p \cap s = \emptyset\}$, and $\bar{\mathcal{S}}_p := \mathcal{S} \backslash \mathcal{S}_p$ is the set of all states $s \in \mathcal{S}$ where path $p \in \mathcal{P}$ fails, i.e., $\bar{\mathcal{S}}_p = \{s \in \mathcal{S} \mid p \cap s \neq \emptyset\}$. Similarly, $\mathcal{S}_e = \{s \in \mathcal{S} \mid e \notin s\}$ will denote the set of all states $s \in \mathcal{S}$ in which link $e \in \mathcal{E}$ is available, and $\bar{\mathcal{S}}_e = \mathcal{S} \backslash \mathcal{S}_e$— the set of all failure states in which it is failed.

For the restoration paths (called also protection or backup paths) we will use the following notation. For each demand $d \in \mathcal{D}$ and each path $p \in \mathcal{P}_d$, the set $\mathcal{Q}_p$, $\mathcal{Q}_p \subseteq \mathcal{P}_d$ denotes all candidate backup paths $q \in \mathcal{P}_d$ that never fail together with p. Hence, if $q \in \mathcal{Q}_p$ then for all $s \in \mathcal{S}$, $(p \cap s \neq \emptyset) \Rightarrow (q \cap s = \emptyset)$, i.e., paths p and q never fail simultaneously (and therefore are called *failure-disjoint*). Besides, $\mathcal{Q}_{pe} := \{q \in \mathcal{Q}_p \mid q \ni e\}$ denotes the set of all paths protecting path p that contain a particular link $e \in \mathcal{E}$.

2.2 State-Dependent Restoration – FD

Both restoration models considered in this paper assume protection of traffic against node or link failures using flow restoration. We also adopt a natural assumption that in the case of a failure of one or more links (multiple failure), only the flows affected by this failure are restored, and the flows not containing any of the failed links are not modified. This mechanism is sometimes referred to as "restricted reconfiguration" (see [1]). In the model considered in this section we assume that the failed flows are restored in a failure-dependent fashion, i.e., the flow restoration pattern depends on the particular failure state. We also assume that the failed flows are restored using protection capacity of links that is separated from the basic capacity on the links used by primary flows. As already mentioned, this is known as path restoration "without stub release". Hence, we note that separated protection capacity assumes that each link $e \in \mathcal{E}$ has its "primary" (basic) capacity, denoted by y'_e, used for realizing flows in the normal operating state (i.e., primary flows), and the protection capacity, y''_e, used for restoration of failed primary flows. The (primal) formulation of the state-dependent flow restoration (FD) design problem is as follows (alternative formulations can be found in [1]):

$$\text{minimize } F(\boldsymbol{y}) = \sum_{e \in \mathcal{E}} \xi_e (y'_e + y''_e) \tag{1a}$$

$$\sum_{p \in \mathcal{P}_d} x_p \geq h_d \quad d \in \mathcal{D} \tag{1b}$$

$$\sum_{p \in \mathcal{P}_e} x_p \leq y'_e \quad e \in \mathcal{E} \tag{1c}$$

$$\sum_{p \in \bar{\mathcal{P}}_d^s} x_p \leq \sum_{q \in \mathcal{P}_d^s} z_{qs} \quad d \in \mathcal{D},\ s \in \mathcal{S} \setminus \{\mathcal{O}\} \tag{1d}$$

$$\sum_{q \in \mathcal{P}_e^s} z_{qs} \leq y''_e \quad e \in \mathcal{E},\ s \in \mathcal{S}_e \setminus \{\mathcal{O}\} \tag{1e}$$

$$\boldsymbol{x}, \boldsymbol{y}, \boldsymbol{z} \geq \boldsymbol{0}. \tag{1f}$$

Formulation (1) uses flow variables $x_p \geq 0$ for the primary paths $p \in \mathcal{P}$, and backup path flow variables z_{qs} for restoration paths $\mathcal{P}_d^s, d \in \mathcal{D}, s \in \mathcal{S} \setminus \{\mathcal{O}\}$. Link capacities, both basic and protection to be installed are modeled using variables y'_e and y''_e respectively.

At this point it is worth to mention the *Path Diversity* (PD) problem—a network design problem in which traffic is protected against node or link failures using over-provisioning. The concept behind PD is to route, for each demand $d \in \mathcal{D}$, possibly more traffic than specified by the demand value h_d in the failure-less state, and ensuring that at least a specified fraction of h_d survives any failure state in the considered failure scenario; in effect no flow restoration/rerouting has to be considered. This concept, called also *diversification*, was later extended [13] to develop a new survivability concept called *demand-wise shared protection* (DSP). The DSP was generalized [14] to the model where it is required that at least the amount of flow equal to h_d^s (where $h_d^s \leq h_d, s \in \mathcal{S}$ and $h_d^{\mathcal{O}} = h_d$) should survive in failure state $s \in \mathcal{S}$. The problem formulation is following:

$$\text{minimize} \sum_{e \in \mathcal{E}} \xi_e y_e \tag{2a}$$

$$h_d^s \leq \sum_{p \in \mathcal{P}_d^s} x_p \quad d \in \mathcal{D},\ s \in \mathcal{S} \tag{2b}$$

$$\sum_{p \in \mathcal{P}_e} x_p \leq y_e \quad e \in \mathcal{E}. \tag{2c}$$

As shown in Section 3 the pricing model for the above problem and, hence, the used algorithms, are essentially the same as for the FD problem, in spite of different model origin.

2.3 State-Independent Restoration – FI

Assume that restoration of a failed flow is state-independent, i.e., that restoration of a failed primary flow is performed in the same fashion in every failure state in which the restoration of this particular primary flow is called for. Now the corresponding primal problem FI reads as follows:

$$\text{minimize } F(\boldsymbol{y}) = \sum_{e \in \mathcal{E}} \xi_e (y'_e + y''_e) \tag{3a}$$

$$\sum_{p \in \mathcal{P}_d} x_p \geq h_d \quad d \in \mathcal{D} \tag{3b}$$

$$\sum_{p \in \mathcal{P}_e} x_p \leq y'_e \quad e \in \mathcal{E} \tag{3c}$$

$$x_p \leq \sum_{q \in \mathcal{Q}_p} z_{pq} \quad p \in \mathcal{P} \tag{3d}$$

$$\sum_{d \in \mathcal{D}} \sum_{p \in \bar{\mathcal{P}}_d^s} \sum_{q \in \mathcal{Q}_{pe}} z_{pq} \leq y''_e \quad e \in \mathcal{E},\ s \in \mathcal{S}_e \setminus \{\mathcal{O}\} \tag{3e}$$

$$\boldsymbol{x}, \boldsymbol{y}, \boldsymbol{z} \geq \mathbf{0}, \tag{3f}$$

3 Solution Approach

The problems considered in this paper are formulated as non-compact LP models with the number of path-flow variables growing exponentially with the problem size. Such problems are treated by column (path) generation [15,16], as this does not require all the columns of the constraint matrix to be held in the computer memory (which is impossible for large problem instances). Instead, we only keep a subset of the variables (columns) and this can be seen as an approximation (restriction) of the original problem. The algorithm iteratively modifies this subset by introducing new variables in a way that improves the current optimal solution. At the end, the set contains all the variables (paths) necessary to construct the overall optimal solution which can use all possible paths in graph $\mathcal{G}$. Certainly, to make the calculation efficient we have to be able to effectively generate the paths required during the optimization. At each iteration we are interested in generating paths for which the corresponding dual constraint is most violated, as we can expect that this will improve the current optimal solution to the greatest possible extent. As shown in the next paragraphs, finding such a path corresponds to finding the shortest path in a graph according to specific criteria and is called *path generation*, the *pricing problem*, or *dual separation*.

3.1 Pricing Models

Pricing for FD Models. Let $\lambda_d, \pi_e^s, \sigma_d^s$ be the dual variables corresponding to constraints (1b), (1d) and (1e), respectively (note that the dual variable corresponding to constraint (1c) equals ξ_e in the optimal solution). Then, the dual problem to (1) reads as follows:

$$\text{maximize } W(\boldsymbol{\lambda}) = \sum_{d \in \mathcal{D}} h_d \lambda_d \tag{4a}$$

$$\sum_{s \in \mathcal{S}_e \setminus \{\mathcal{O}\}} \pi_e^s = \xi_e \quad e \in \mathcal{E} \tag{4b}$$

$$\sigma_d^s \leq \sum_{e \in q} \pi_e^s \quad d \in \mathcal{D},\ s \in \mathcal{S} \setminus \{\mathcal{O}\},\ q \in \mathcal{P}_d^s \tag{4c}$$

$$\lambda_d \leq \sum_{e \in p} \xi_e + \sum_{s \in \bar{\mathcal{S}}_p} \sigma_d^s \quad d \in \mathcal{D},\ p \in \mathcal{P}_d \tag{4d}$$

$$\boldsymbol{\lambda}, \boldsymbol{\pi}, \boldsymbol{\sigma} \geq \mathbf{0}. \tag{4e}$$

It follows from formulation (4) that if for the current optimal dual variables $\boldsymbol{\sigma}^*$ and $\boldsymbol{\pi}^*$ we are able to find, for some $d \in \mathcal{D}$ and a failure state $s \in \mathcal{S} \setminus \{\mathcal{O}\}$, a protection path $q_d^s \notin \mathcal{P}_d^s$ from node u_d to node v_d with

$$\sum_{e \in q_d^s} {\pi^*}_e^s < (\sigma_d^s)^* \tag{5}$$

then adding this path to the set $\mathcal{P}_d^s$ can potentially decrease the optimal objective (1a). Note that it is easy to generate a shortest protection paths—we just need to use state-dependent link weights equal to $\pi_e^s, e \in \mathcal{E}$ in the shortest path algorithm. Let q_d^s denote a shortest path for demand $d \in \mathcal{D}$ in situation $s \in \mathcal{S} \setminus \{\mathcal{O}\}$, and its length by r_d^s. Clearly, $r_d^s \leq (\sigma_d^s)^*$. Further, when considering the problem of generating a shortest primary path for demand $d \in \mathcal{D}$ we need to find a path p^* such that the following condition holds

$$\sum_{e \in p^*} \xi_e + \sum_{s \in \bar{\mathcal{S}}_{p^*}} r_d^s < \lambda_d^*. \tag{6}$$

Note that in inequality (6), r_d^s can be substituted with $(\sigma_d^s)^*$ and the procedure will still be valid. As noted above, the pricing problem associated with backup paths is solvable in polynomial time as it can be solved using shortest path algorithm. This is not always the case when the pricing of primary paths is concerned. We comment on the related issues below.

It has been shown in [17,9] (see also [10]) that the path generation problem is solvable in polynomial time in the case of single link (or node) failures. To see this assume only single link failures, i.e., $\mathcal{S} \subseteq \{\{e\} \mid e \in \mathcal{E}\} \cup \{\mathcal{O}\}$, and observe that for demand $d \in \mathcal{D}$ the condition (6) can be rewritten as follows:

$$\sum_{e \in p} (\xi_e + r_d^{\{e\}}) < \lambda_d^*. \tag{7}$$

The right-hand side depends only on the demand, and the left-hand side is nonnegative. Hence, for each demand $d \in \mathcal{D}$, violation of dual constraint (4d) can be tested by searching for a shortest path between the end-nodes of d with respect to the demand-dependent link weights $\gamma_d^e := \xi_e + {r^*}_d^{\{e\}}$ using for example the Dijkstra algorithm, and comparing its length to the value of λ_d^* (if link e does not fail at all we assume $r_d^{\{e\}} = 0$). If the shortest path fulfills condition (7), then the corresponding path variable can be added to the dual formulation (4) (and the corresponding flow variable to the primal formulation (1)) to reduce the domain of the dual problem; otherwise we know that no path for this demand violates its dual constraint for the current set of optimal dual variables.

A general case of a failure state admits multiple link (node) failures; the set of links that fail simultaneously is sometimes called "shared risk link group" (SRLG), and the set of links and nodes that fail simultaneously—"shared risk resource group" (SRRG), see [18]. For this case the path generation problem for FD is difficult.

As for the PD problem, the corresponding dual problem augmented with an auxiliary dual variable Λ_d, $d \in \mathcal{D}$ reads as follows:

$$\text{minimize } W(\boldsymbol{\lambda}) = \sum_{d \in \mathcal{D}} \sum_{s \in \mathcal{S}} h_d^s \lambda_d^s \tag{8a}$$

$$\Lambda_d = \sum_{s \in \mathcal{S}} \lambda_d^s \quad d \in \mathcal{D} \tag{8b}$$

$$\Lambda_d \leq \sum_{e \in p} \xi_e + \sum_{s \in \bar{\mathcal{S}}_p} \lambda_d^s \quad d \in \mathcal{D},\ p \in \mathcal{P}_d, \tag{8c}$$

where λ_d^s is dual variable corresponding to constraint (2b). It follows from formulation (8) that if for current optimal dual variables $\boldsymbol{\lambda}^*$ and $\boldsymbol{\Lambda}^*$ we are able to find for some $d \in \mathcal{D}$ a path $p \notin \mathcal{P}_d$ from node u_d to node v_d with

$$\sum_{e \in p} \xi_e + \sum_{s \in \bar{\mathcal{S}}_p} (\lambda_d^s)^* < \Lambda_d^* \tag{9}$$

then adding this path to set $\mathcal{P}_d$ can decrease the optimal objective (2a). Note that condition (9) has the form equivalent to (6), and in effect the two pricing problems can be considered as essentially the same.

Pricing for FI Models. Let λ_d, π_e^s be the dual variables corresponding to constraints (3b) and (3e), respectively (again, the dual variable corresponding to constraint (3c) equals ξ_e in the optimal solution). Then, the dual problem to (3) reads:

$$\text{maximize } W(\boldsymbol{\lambda}) = \sum_{d \in \mathcal{D}} h_d \lambda_d \tag{10a}$$

$$\sum_{s \in \mathcal{S}_e \setminus \{\mathcal{O}\}} \pi_e^s = \xi_e \quad e \in \mathcal{E} \tag{10b}$$

$$\lambda_d \leq \sum_{e \in p} \xi_e + \sum_{e \in q} \Big(\sum_{s \in \bar{\mathcal{S}}_p} \pi_e^s \Big) \quad d \in \mathcal{D},\ p \in \mathcal{P}_d,\ q \in \mathcal{Q}_p \tag{10c}$$

$$\boldsymbol{\lambda}, \boldsymbol{\pi} \geq \mathbf{0}. \tag{10d}$$

Observe that in this case the pricing problem is always difficult, i.e., also in single link failure scenarios (see [8]). Path generation consists, for each $d \in \mathcal{D}$, in finding a pair of paths $(p, q), p \in \mathcal{P}_d, q \in \mathcal{Q}_p$ with the smallest length defined by the right-hand side of (10c). It follows that the link metrics for calculating the length of the basic path $p \in \mathcal{P}_d$ are equal to the true unit link costs ξ_e, while the link metrics for calculating the length of the backup path $q \in \mathcal{Q}_p$ are given by $\sum_{s \in \bar{\mathcal{S}}_p} \pi_e^s$.

The pricing problem for FI consists in finding, for each demand $d \in \mathcal{D}$, a pair of failure-disjoint paths $r = (p, q)$ from u_d to v_d minimizing

$$\langle r \rangle = \sum_{e \in p} \xi_e + \sum_{e \in q} \Big(\sum_{s \in \bar{\mathcal{S}}_p} \pi_e^s \Big). \tag{11}$$

3.2 Pricing Algorithms

In the following section we analyze three algorithms for the path generation: two exact (MIP and label-setting), and one approximate (primary path precomputation).

MIP Models. For FD problem with multiple link failures, the pricing problem (which is: for each demand $d \in \mathcal{D}$ find a path p from u_d to v_d minimizing $\sum_{e \in p} \xi_e + \sum_{s \in \bar{\mathcal{S}}_p} r^{*\{s\}}_d$) can be formulated as the following mixed-binary problem in binary variables $\boldsymbol{x} = (x_e : e \in \mathcal{E})$ and continuous variables $\boldsymbol{Y} = (Y^s : s \in \mathcal{S})$:

$$\text{minimize } F(\boldsymbol{x}, \boldsymbol{Y}) = \sum_{e \in \mathcal{E}} \xi_e x_e + \sum_{s \in \mathcal{S}} r^s_d Y^s \tag{12a}$$

$$\sum_{e \in \delta^+(v)} x_e - \sum_{e \in \delta^-(v)} x_e = 0 \quad v \in \mathcal{V} \setminus \{u_d, v_d\} \tag{12b}$$

$$\sum_{e \in \delta^+(u_d)} x_e - \sum_{e \in \delta^-(u_d)} x_e = 1 \tag{12c}$$

$$Y^s \geq x_e \quad s \in \mathcal{S},\ e \in \bar{\mathcal{E}}^s \tag{12d}$$

For the FI problem, the pricing problem (minimizing (11)) can be formulated as a mixed-binary problem in binary variables $\boldsymbol{x}' = (x'_e : e \in \mathcal{E})$ and $\boldsymbol{x}'' = (x''_e : e \in \mathcal{E})$, and continuous variables $\boldsymbol{Y} = (Y^s : s \in \mathcal{S})$ and $\bar{\boldsymbol{Z}} = (\bar{Z}^s_e : s \in \mathcal{S}, e \in \mathcal{E})$:

$$\text{minimize } F(\boldsymbol{x}, \bar{\boldsymbol{Z}}) = \sum_{e \in \mathcal{E}} \xi_e x'_e + \sum_{e \in \mathcal{E}} \sum_{s \in \mathcal{S}} \pi^s_e \bar{Z}^s_e \tag{13a}$$

$$\sum_{e \in \delta^+(v)} x'_e - \sum_{e \in \delta^-(v)} x'_e = 0 \quad v \in \mathcal{V} \setminus \{u_d, v_d\} \tag{13b}$$

$$\sum_{e \in \delta^+(u_d)} x'_e - \sum_{e \in \delta^-(u_d)} x'_e = 1 \tag{13c}$$

$$\sum_{e \in \delta^+(v)} x''_e - \sum_{e \in \delta^-(v)} x''_e = 0 \quad v \in \mathcal{V} \setminus \{u_d, v_d\} \tag{13d}$$

$$\sum_{e \in \delta^+(u_d)} x''_e - \sum_{e \in \delta^-(u_d)} x''_e = 1 \tag{13e}$$

$$Y^s \geq x'_e \quad s \in \mathcal{S},\ e \in \bar{\mathcal{E}}^s \tag{13f}$$

$$\bar{Z}^s_e \geq 0,\ \bar{Z}^s_e \geq x''_e + Y^s - 1 \quad s \in \mathcal{S},\ e \in \mathcal{E}. \tag{13g}$$

In the above formulation it is important to assume that $\pi^s_e = +\infty$ for $e \in \mathcal{E}^s$.

Label-Setting. The pricing problems for FD and FI can be approached in a way described in [8]. The idea is as follows. Consider a demand $d \in \mathcal{D}$ and a path $p \in \mathcal{P}_d$ between nodes u_d and v_d. The length of path p with respect to weights ξ_e ($e \in \mathcal{E}$) will be denoted by $\|p\| := \sum_{e \in p} \xi_e$. For fixed dual variables, the notation $\langle p \rangle$ will stand for the dual length of path p, and in case of FD, can be expressed as follows (9):

$$\langle p \rangle = \sum_{e \in p} \xi_e + \sum_{s \in \bar{\mathcal{S}}_p} r^s_d \tag{14}$$

and in case of FI:

$$\langle p \rangle = \sum_{e \in p} \xi_e + \sum_{e \in q} (\sum_{s \in \bar{\mathcal{S}}_p} (\pi_e^s)^*). \tag{15}$$

We say that path $p \in \mathcal{P}_d$ *dominates* another path $p' \in \mathcal{P}_d$ if $\|p\| \le \|p'\|$ and $\bar{\mathcal{S}}_p \subseteq \bar{\mathcal{S}}_p$, i.e., when path p is not longer than path p' and when the set of failure states that affect path p is a subset (not necessarily proper) of the set of failure states affecting path p'. It is obvious that if path p dominates path p' then $\langle p \rangle \le \langle p' \rangle$. Consequently, in order to generate a new path p for demand d (i.e., a path with minimum value of $\langle p \rangle$ with respect to the current optimal dual variables) it is sufficient to generate the set $\mathcal{U}(d)$ of minimal paths with respect to the relation of domination defined above. That is, no path from $\mathcal{U}(d)$ dominates another path from this set.

Such a set $\mathcal{U}(d)$ of all non-dominated paths can be generated by means of a label-setting algorithm for *shortest-path problems with resource constraints* (SPPRC) [12]. This can be done in pseudo-polynomial time, i.e., in time which is polynomial with respect to the size of the network graph, and exponential with respect to the number of failure states $|\mathcal{S}|$ ("resources" are the failure states in this case).

The label-setting algorithm works by creating a set of paths originating at node u_d. Each path is represented by a label placed in its last node. The label contains information about the already visited nodes, as well as the utilized resources. The path can be extended by a link incident to its last node, but only if it does not result in a cycle or in a violation of the resource constraints. In the worst case the number of different paths between the nodes u_d and v_d can grow exponentially with the number of resources. In many cases, however, we are able to apply some dominance rules to decrease the number of the resulting paths. This can be done not only in the target node v_d but also in the intermediate nodes. It is easy to show that if path p' is dominated with respect to the rules defined above, then any extension of p' would be finally dominated in v_d as well.

As an extension of the SPPRC algorithm applied in [8], we have also introduced path length limitation—an important contribution to the reduction of the size of set $\mathcal{U}(d)$. The extension is based on the observation that excessively long paths are useless as they cannot improve the current solution, or the solutions they represent are known to be worse than some already known solutions. For example, a simple path length restriction for FD can be constructed from (6): $\sum_{e \in p} \xi_e < \lambda_d^*$. Also, knowledge of some path p' representing a feasible solution can help to tighten the path length restriction. In such a case, for FD we are only interested in finding a path p satisfying $\sum_{e \in p} \xi_e < \sum_{e \in p'} \xi_e + \sum_{s \in \bar{\mathcal{S}}_{p'}} r_d^s$. Similar constraints can be derived for problem FI as well. Applying path length limitation to pricing in problem FI results in significant (up to 3 orders of magnitude in our experiments) reduction of the pricing time.

Primary Path Precomputation. Usefulness of the pricing algorithms presented before may sometimes be limited if applied for large network instances. Therefore, we consider a heuristic solution based on large predefined sets of allowable primary paths that are kept aside from the current problem formulation (using smaller sets of candidate primary paths $\mathcal{P}_d$). Having sets $\mathcal{R}_d$, $d \in \mathcal{D}$ of the allowable primary paths defined in advance, we can reduce the pricing problem to simple enumeration of the primary paths contained in these (large) sets.

Table 1. Network characteristics

	Network					
	Pdh	Newyork	Ta1	France	Norway	Cost266
Nodes	11	16	24	25	27	37
Links	34	49	51	45	51	57
Demands	24	120	163	300	351	666

For problem FD, we simply compute the lengths of the paths in each $\mathcal{R}_d$ according to the left hand side of formula (6), where r_d^s are the appropriate shortest path lengths computed previously for the current optimal dual variables. For each $d \in \mathcal{D}$ the shortest path in set $\mathcal{R}_d$ should be added to the problem (i.e., to set $\mathcal{P}_d$) if its length is strictly less than λ_d^*.

For problem FI, we enumerate, for each $d \in \mathcal{D}$, the allowable primary paths p in the set $\mathcal{R}_d$, and for each of them we determine the shortest backup path q computed as a shortest path with respect to link weights π_e^s, $s \in \bar{\mathcal{S}}_p$, i.e., for the states in which the considered primary path p fails. Obviously, the resulting backup paths should be state-disjoined with respect to the protected primary paths. We assure this property simply by removing links belonging to the protected primary path p. As a result, a pair $r = (p, q)$ for each $p \in \mathcal{R}_d$ is formed. Finally, a pair with the smallest total length defined by the right-hand side of (10c) is found, and added to the formulation of problem (3), as long as its length is strictly smaller than the value of λ_d^*.

As shown by numerical experiments presented in Section 4, this approach seems to be reasonable, often allowing for the globally optimal solution of the considered problem. Still, the effectiveness of the proposed simplified method heavily depends on the construction of the sets of allowable primary paths $\mathcal{R}_d$. Sets $\mathcal{R}_d$ should contain only "good" candidate primary paths and should be kept as small as possible. Having routing lists that contain "good" paths (sufficient for resolving the problem in hand) the resolution process would be simplified a lot. It seems that a reasonable approach for arriving at lists of paths which have a good chance to be sufficient for solving the problem can be based on the computation of K-shortest paths with respect to unit link costs ξ_e, $e \in \mathcal{E}$. Hence, using these costs as link weights, we can construct the lists of paths that are most likely to be used in the globally optimal solution. This follows from the observation that in the simplified LP dimensioning problems without failure states, an optimal solutions consists in realizing the demands along the shortest paths computed with respect to unit link costs ξ_e, $e \in \mathcal{E}$. Algorithms for finding K-shortest path are well known and can be found in [19,20,21,22,23,24].

4 Computational Experiments

In our computational investigations we conducted a series of performance tests of the discussed methods. For this purpose we selected a set of six network instances (cf. Table (1)). Network topologies and demand volumes were taken from SNDLib library [25]. The unit link costs ξ_e, $e \in \mathcal{E}$ were generated randomly since not all the network

Table 2. FD – double link failures

Network	Path generation	Iterations	Columns	Time [s] master	pricing	total
	MIP	53	1253	1	1	2
Pdh	Label-setting	63	1057	1	0	1
	Precomputed	52	1226	1	0	1
	MIP	88	10101	61	15	76
Newyork	Label-setting	78	7785	49	3	52
	Precomputed	82	9965	56	0	56
	MIP	69	13964	39	19	58
Ta1	Label-setting	62	12427	36	4	40
	Precomputed	60	13124	32	3	29
	MIP	34	12583	32	28	60
France	Label-setting	32	10762	29	4	33
	Precomputed	32	12531	36	0	36
	MIP	39	29277	323	43	366
Norway	Label-setting	34	21741	301	53	354
	Precomputed	40	28363	329	1	330
	MIP	33	44752	394	85	479
Cost266	Label-setting	34	34714	350	104	454
	Precomputed	34	44850	424	4	428

instances contained in SNDlib contained this information. Also, we assumed the presence of a demand requirement between each pair of nodes. Regarding failure states, we tested the two following scenarios: all single link failures, and all double link failures (for FD only). In the second case the number of failure states was equal to the number of states in the single link failure scenario, i.e., equal to the number of links (plus 1 for the normal state). The second scenario was constructed by randomly choosing the second failing links in the considered state and marking it to be failed. Clearly, this operation should be performed carefully as the network graph must be at least two-connected in all failure states.

In our numerical experiments we considered the three following computational scenarios: state-dependant restoration (FD) with single link failures, state-dependant restoration (FD) with double link failures, and state-independent restoration (FI) for single link failures. Initial problems (master problems) were formulated using L-P notation, and solved within our own path generation (PG) framework. Optimal values of dual multipliers were obtained using the LP solver of CPLEX 9.1. In order to resolve the pricing problems, we used all the three presented algorithms: exact MIP pricing, exact pricing based on the shortest path computations using label-setting algorithm, and the heuristic for precomputing the sets of allowable primary paths. (In the primary path precomputation we decided to pre-generate 5 primary shortest paths for each demand computed, with respect to link weights $\xi_e,\ e \in \mathcal{E}$.

In the case of problem FD with single link failures, the pricing problem was solved using a more general version of label-setting algorithm designed for the multiple failures case. Moreover, in both failure scenarios for FD, at each iteration of the PG

Table 3. FD – single link failures

Network	Path generation	Iterations	Columns	Time [s] master	pricing	total
Pdh	MIP	162	1212	1	4	5
	Label-setting	115	1032	1	0	1
	Precomputed	149	1246	1	0	1
Newyork	MIP	87	7706	34	17	51
	Label-setting	74	6144	33	2	35
	Precomputed	75	7612	42	0	42
Ta1	MIP	112	10166	21	30	51
	Label-setting	128	9017	23	6	29
	Precomputed	133	9810	24	1	25
France	MIP	38	8483	23	17	40
	Label-setting	41	7687	20	3	23
	Precomputed	36	8469	22	0	22
Norway	MIP	71	19766	190	33	223
	Label-setting	84	19079	244	17	261
	Precomputed	53	19193	174	2	176
Cost266	MIP	36	31122	278	48	326
	Label-setting	35	28398	334	30	364
	Precomputed	34	31597	266	3	269

Table 4. FI – single link failures

Network	Path generation	Iterations	Columns	Time [s] master	pricing	total
Pdh	MIP	121	682	8	112	120
	Label-setting	124	822	1	0	1
	Precomputed	157	814	1	1	2
Newyork	MIP	62	2887	152	660	812
	Label-setting	57	2867	25	1	26
	Precomputed	66	3079	126	1	127
Ta1	MIP	74	3385	122	1455	1577
	Label-setting	101	3913	22	2	24
	Precomputed	122	4484	95	10	105
France	MIP	14	1957	11	297	308
	Label-setting	16	2294	5	2	7
	Precomputed	18	2696	22	1	23
Norway	MIP	27	4453	420	2044	2464
	Label-setting	29	4419	105	7	112
	Precomputed	30	4665	323	5	328
Cost266	MIP	9	4012	67	900	967
	Label-setting	12	5465	52	5	57
	Precomputed	13	5569	146	2	148

algorithm at most one primary path and at most one backup path for each state and for each demand were generated. In problem FI, at each iteration at most one primary-backup path pair for each demand was generated. All the experiments were performed on a PC computer with a Pentium 4 (2.8GHz) processor.

Tables 2, 3 and 4 present results for the considered network instances, while Table 5 contains objective function values for the three pricing algorithms. First of all, we notice that the pricing problem (path generation) does not seem to affect considerably the overall performance, as the overall time spent on solving the master problems has much bigger impact on the total computation time. When comparing directly the pricing problem solution times, a very good performance of the primary path precomputation method in all the tested cases is observed. What is more important, it gives either optimal solutions or excellent approximations. The label-setting algorithm shows its strength for the FI problem, where it performs similarly to the heuristic algorithm, at the same time assuring optimal solutions. The approach utilizing MIPs for pricing is quite efficient for the FD problem but fails for the FI problems.

Table 5. Objective values

FD – double link failures						
Path gen.	Pdh	Newyork	Ta1	France	Norway	Cost266
MIP	62805.2	16508.6	8.78397e7	4.15572e6	120837	1.78992e7
Label-setting	62805.2	16508.6	8.78397e7	4.15572e6	120837	1.78992e7
Precomputed	62805.2	16508.6	8.78534e7	4.15584e6	120837	1.79049e7
FD – single link failures						
Path gen.	Pdh	Newyork	Ta1	France	Norway	Cost266
MIP	57564.3	15692.4	8.15358e7	3.67687e6	108303	1.56683e7
Label-setting	57564.3	15692.4	8.15358e7	3.67687e6	108303	1.56683e7
Precomputed	57564.3	15692.4	8.15388e7	3.67687e6	108305	1.56683e7
FI – single link failures						
Path gen.	Pdh	Newyork	Ta1	France	Norway	Cost266
MIP	57564.3	15692.4	8.15494e7	3.67687e6	108313	1.56686e7
Label-setting	57564.3	15692.4	8.15494e7	3.67687e6	108313	1.56686e7
Precomputed	57564.3	15692.4	8.15545e7	3.67687e6	108317	1.56686e7

5 Conclusions

The paper discusses the problem of link dimensioning and flow optimization in resilient networks with two different flow restoration schemes to be used in case of network failures, namely state-dependant restoration and state-independent restoration. The two resulting problems are specified in the L-P formulation.

Three appropriate path generation methods, necessary to make the L-P formulation really useful, are discussed: exact approach based on Mixed Integer Programming, an adaptation of the label-setting algorithm, and a heuristic algorithm making use of large precomputed lists of primary paths. As shown by the numerical experiments (Section 4), together the presented algorithms provide an efficient means for paths generation for the considered types of problems. In fact, only application of the MIP pricing for problem FI appears inefficient. (This is probably caused by the complexity of problem (13) incorporating large number of variables and constraints.)

We emphasize that in practice the simplified pricing based on the precomputation of K-shortest primary paths is very efficient and sufficient. In most cases, the values of the optimal objective function are correct. Still, this approach, as compared to the exact label-setting algorithm, does not provide a significant performance gain. This is due to the impact of the solution time of the master problem which is much longer when compared to the time needed to generate new paths. Our analysis shows that the presented path generation algorithms enable efficient resolution of the two considered resilient network design problems for networks of realistic size.

Acknowledgment

The authors are indebted to Dritan Nace, Sebastian Orlowski, and Thomas Stidsen for valuable and enlightening discussions on the paper.

References

1. Pióro, M., Medhi, D.: Routing, Flow, and Capacity Design in Communication and Computer Networks. Morgan Kaufman, San Francisco (2004)
2. Minoux, M., Serrault, J.Y.: Subgradient optimization and large scale programming: an application to optimum multicommodity network synthesis with security constraints. R.A.I.R.O. Operations Research 15(2) (1981)
3. Dahl, G., Stoer, M.: A cutting plane algorithm for multicommodity survivable network design problems. INFORMS Journal on Computing 10, 1–11 (1998)
4. Wessäly, R.: Dimensioning Survivable Capacitated NETworks. PhD thesis, Technische Universität Berlin (2000)
5. Hu, J.: Diverse routing in optical mesh networks. IEEE Trans. Com. 51(3), 489–494 (2003)
6. Maurras, J.F., Vanier, S.: Network synthesis under survivability constraints. 4OR (2), 52–67 (2004)
7. Coudert, D., Datta, P., Perennes, S., Rivano, H., Voge, M.E.: Complexity and approximability issues of shared risk resource group. Technical report, Technical report 5859, INRIA (2006)
8. Stidsen, T., Petersen, B., Rasmussen, K., Spoorendonk, S., Zachariasen, M., Rambach, F., Kiese, M.: Optimal routing with single backup path protection. In: International Network Optimization Conference INOC 2007, Spa, Belgium (2007)
9. Bashllari, A., Nace, D., Rourdin, E., Klopfenstein, O.: The MMF rerouting computation problem. In: International Network Optimization Conference INOC 2007, Spa, Belgium (2007)
10. Orlowski, S., Pióro, M.: On the complexity of column geneartion in survivable network design. Technical report, Zuse Institut Berlin and Warsaw University of Technology (2007)
11. Ahuja, R.K., Magnanti, T.L., Orlin, J.B.: Network Flows: Theory, Algorithms, and Applications. Prentice-Hall, Englewood Cliffs (1993)
12. Irnich, S., Desaulniers, G.: Shortest path problems with resource constraints. In: Desaulniers, G., Desrosier, J., Solomon, M. (eds.) Column Generation, pp. 33–65. Springer, Heidelberg (2005)

13. Koster, A., Zymolka, A., Jäger, M., Hülsermann, R.: Demand-wise shared protection for meshed optical networks. Journal of Network and Systems Management 13(1), 35–55 (2005)
14. Wessäly, R., Orlowski, S., Zymolka, A., Koster, A., Gruber, C.: Demand-wise shared protection revisited: A new model for survivable network design. In: Proc. INOC 2005, Lisbon, pp. 100–105 (2005)
15. Dantzig, G.B., Wolfe, P.: The decomposition algorithm for linear programming. Operations Research 8, 101–111 (1960)
16. Desrosiers, J., Luebbecke, M.: A primer in column generation. In: Desaulniers, G., Desrosier, J., Solomon, M. (eds.) column generation, pp. 1–32. Springer, Heidelberg (2005)
17. Orlowski, S.: Local and global restoration of node and link failures in telecommunication networks. M.sc. thesis, Technische Universität Berlin (2003), `http://www.zib.de/orlowski/`
18. Strand, J., Chiu, A.L., Tkach, R.: Issues for routing in the optical layer. IEEE Communications Magazine, 81–87 (2001)
19. Yen, J.Y.: Finding the k shortest loopless paths in a network. Management Science 17, 712–716 (1971)
20. McCallum, C.J.: An algorithm for finding the k shortest paths in a network. Bell Laboratories Technical Memorandum TM73-1713-9 (1973)
21. Eppstein, D.: Finding the k shortest paths. In: 35th IEEE Symposium on Foundations of Computer Science, pp. 154–165 (1994)
22. Dreyfus, S.E.: An appraisal of some shortest-path algorithms. Operations Research 17, 395–412 (1999)
23. Jiménez, V.M., Marzal, A.: Computing the k shortest paths: A new algorithm and an experimental comparison. In: Proceedings of 3rd Annual Workshop on Algorithmic Engineering, London (1999)
24. Pascoal, M.M.B., Eugénia, M., Captivo, V., Climaco, J.C.N.: An algorithm for ranking quickest simple paths. Computers and Operations Research 32, 509–520 (2005)
25. Orlowski, S., Pióro, M., Tomaszewski, A., Wessäly, R.: SNDlib 1.0–Survivable Network Design Library. In: Proceedings of the 3rd International Network Optimization Conference (INOC 2007), Spa, Belgium (2007), `http://sndlib.zib.de`

Semi-systolic Modular Multiplier over GF(2^m)*

Hyun-Sung Kim[1] and Sung-Woon Lee[2,**]

[1] Kyungil University, School of Computer Engineering,
712-701, Kyungsansi, Kyungbuk Province, Korea
kim@kiu.ac.kr

[2] Tongmyong University, Dept. of Information Security,
608-711, Busan, Korea
staroun@tu.ac.kr

Abstract. The key operations for the public-key cryptosystems are modular inversion, division, and exponentiation. The modular multiplication is considered to be the basic arithmetic for them. This paper proposes a new algorithm and it's semi-systolic array architecture to compute the modular multiplication over GF(2^m). They are based on the standard basis representation and use the property of irreducible all one polynomial as a modulus. The architecture has the critical path with 1-D_{XOR} per cell and has the latency with m+1. These properties are better than the existing multipliers. Since the proposed multiplier has regularity, modularity and concurrency, it is suitable for VLSI implementation and can be easily utilized for the crypto-processor chip design.

1 Introduction

The arithmetic operations in the finite field have several applications in error-correcting codes, cryptography, digital signal processing, and so on [1-4]. Information processing in such areas usually requires performing multiplication, inversion/division, and exponentiation. Among these operations, the modular multiplication is known as the basic operation for public key cryptosystems over GF(2^m) [2- 4]. Thereby, exponentiation can be computed efficiently by the sequences of modular multiplications. Division and inversion also can be regarded as a special case of the exponentiation because $B^{-1} = B^{2^m - 2}$ [5, 6].

In recent years, three types of multipliers for GF(2^m) have been proposed that are easily realized using VLSI techniques. They are normal, dual, and standard basis multipliers, which have their own distinct features. The normal basis architecture is very effective in performing inversion, squaring, and exponentiation operations. The dual basis architecture requires less chip area than the other two types. The standard basis architecture has lower design complexity, and its size is easier to extend to meet various applications due to its simplicity, regularity, and modularity in architecture.

---

* This research was supported by Ministry of Knowledge and Economy, Republic of Korea, under the ITRC(Information Technology Research Center) support program supervised by IITA(Institute for Information Technology Advancement)(IITA-2008-C1090-0801-0004).

** Corresponding author.

O. Gervasi et al. (Eds.): ICCSA 2008, Part II, LNCS 5073, pp. 836–844, 2008.

However, the former two types of finite field multipliers need basis conversion, while the latter type does not. From now on, we are focused on the standard basis multiplier.

Numerous architectures for modular multiplication in GF(2^m) have been proposed in [7-10] over the standard basis. In 1984, Yeh et al. proposed two systolic array architectures with the LSB-first modular multiplication [7]. Wang et al. in [8] proposed two systolic architectures with the MSB-first fashion with less control problems as compared to [6]. Jain et al. proposed semi-systolic multipliers [9]. Its latency is smaller than those of the other standard-basis multipliers. Kim in [10] proposed a bit-level systolic array with a simple hardware complexity with the LSB-first modular multiplication. Thus, further research for efficient circuit for cryptographic applications is needed. To reduced the system complexity, Itoh and Tsujii designed two low-complexity multipliers for the class of GF(2^m), based on the irreducible AOP (All One Polynomial) of degree m and the irreducible ESP (Equally Spaced Polynomial) of degree m [11]. Later, Kim in [5] proposed various AOP architectures based on LFSR(Linear Feedback Shift Register) architecture.

This paper proposes a new algorithm and a parallel-in parallel-out semi-systolic array architecture to compute the modular multiplication over finite field GF(2^m). They are based on the standard basis representation and use the property of irreducible AOP as a modulus. Let D_{AND} and D_{XOR} be the latency of AND and XOR gate, respectively. The architecture has a critical path with 1-D_{XOR} per cell and has a latency of m+1. These characteristics are better than the existing architectures focused on the area and time complexities. Since the proposed architecture has regularity, modularity and concurrency, it is suitable for VLSI implementation and could be used as a basic architecture for the public-key cryptosystems.

The remainder of this paper is organized as follows. Section 2 gives an overview of finite fields and their operations for the basic knowledge of understanding our paper and proposes a new modular multiplication algorithm using the irreducible AOP as a modulus. In section 3, we derive two architectures for a semi-systolic modular multiplier using the new algorithm and a reduction circuit for an additional modular reduction. Section 4 gives a comparison and analysis between the proposed and previous multipliers. Finally, section 5 concludes the paper.

2 Finite Fields and Their Operation

GF(2^m) is the finite extension field of finite field GF(2) [4,5]. First of all, all the nonzero elements in finite field GF(2^m) may be represented by three bases. These are the standard, normal and dual basis. In the standard basis, an arbitrary element A over GF(2^m) can be represented with $\{1, \alpha, \alpha^2, \cdots, \alpha^{m-1}\}$, i.e., $A = A_{m-1}\alpha^{m-1} + A_{m-2}\alpha^{m-2} + \cdots + A_1\alpha + A_0$. For normal basis, an arbitrary element A over GF(2^m) can be represented with $\{\alpha, \alpha^2, \cdots, \alpha^{2m-2}\}$, i.e., $A = A_{m-1}\alpha^{2m-2} + A_{m-2}\alpha^{2m-4} + \cdots + A_1\alpha^2 + A_0$. The dual basis represents an arbitrary element A over GF(2^m) with $\{u_0, u_1, \cdots, u_{m-1}\}$, i.e., $A = A_{m-1}u_{m-1} + A_{m-2}u_{m-2} + \cdots + A_1u_1 + A_0u_0$. The A_i's (i = 0, 1, ⋯, m-1) are elements of GF(2), respectively. From now on, we will only consider the standard basis representation.

A polynomial of the form $f(x) = f_mx^m + f_{m-1}x^{m-1} + \cdots + f_1x + f_0$ is called an irreducible polynomial if and only if a divisor of $f(x)$ is 1 or $f(x)$. Also, a polynomial of the form $f(x) = f_mx^m + f_{m-1}x^{m-1} + \cdots + f_1x + f_0$ is called a primitive polynomial if and only if a greatest common divisor of f_i ($0\le i \le m$) is '1'. Assume that a polynomial of the form $f(x) = f_m x^m + f_{m-1}x^{m-1} + \cdots + f_1 x + f_0$ over GF(2) is called an AOP (All One Polynomial) with degree m if f_i=1 for i=0,1, ⋯ ,m. It has been shown that an AOP is irreducible if and only if m+1 is the prime and 2 is the primitive modulo m+1. Let a set $\{1, \alpha, \alpha^2, \cdots, \alpha^{m-1}\}$ be generated by α which is a root of AOP $f(x)$ and be the standard basis. In the standard basis, an element A over GF(2^m) is presented by $A = A_{m-1}\alpha^{m-1} + A_{m-2}\alpha^{m-2} + \cdots + A_1\alpha + A_0$. A set with $\{1, \alpha, \alpha^2, \cdots, \alpha^{m-1}, \alpha^m\}$ is called an extended basis of $\{1, \alpha, \alpha^2, \cdots, \alpha^{m-1}\}$. In the extended basis, an element a over GF(2^m) is represented by $a = a_m\alpha^m + a_{m-1}\alpha^{m-1} + \cdots + a_1\alpha + a_0$ with $A_i=a_m+a_i$ ($0\le i \le m$-1). Thus, an element over GF(2^m) has two different representations. Let $F(x) = x^m + x^{m-1} + \cdots + x + 1$ be an irreducible AOP of degree m: and let α be a root of $F(x)$. i.e., $F(\alpha) = \alpha^m + \alpha^{m-1} + \cdots + \alpha + 1$. Then, we have $\alpha^m = \alpha^{m-1} + \cdots + \alpha + 1$,

$$\alpha^{m+1}=1 \tag{1}$$

The public-key schemes and other applications described in the previous section are based on modular exponentiation. Let C and M be elements of GF(2^m), the exponentiation of M is then defined as

$$C = M^E,\ 0 \le E \le n, \tag{2}$$

where $n = 2^m-1$. For a special case, $M = \alpha$, the exponent E, which is an integer can be expressed by $E = e_{m-1}2^{m-1} + e_{m-2}2^{m-2} + \ldots + e_1 2^1 + e_0$. The exponent also can be represented with vector representation [$e_{m-1}\ e_{m-2}\ \ldots\ e_1\ e_0$]. A popular algorithm for computing exponentiation is the binary method [5]. There are two ways, the LSB (Least Significant Bit) and MSB (Most Significant Bit) first operations, this can be done. We only focus on the LSB first operation in this paper. Starting from the LSB of the exponent, the exponentiation of M can be expressed as

$$M^E = M^{e_0}(M^{2^1})^{e_1}(M^{2^2})^{e_2}\ldots(M^{2^{m-1}})^{e_{m-1}} \tag{3}$$

Based on the equation 3, an algorithm for computing exponentiations is presented as following algorithm 1. The algorithm shows that the exponentiation can be performed with squaring and multiplication operations. To implement the exponentiation in algorithm 1, either outputs are used as inputs for the same multiplier or separate units for multiplication and squaring are joined. This paper only focuses on the method of using the same multiplier.

[Algorithm 1] LSB first Exponentiation
Input : $A, E, f(x)$
Output : $C=A^E \bmod f(x)$

Step 1 $T=A$
Step 2 if (e_0==1) $C=T$ else $C=\alpha^0$
Step 3 for i=1 to m-1
Step 4 $\quad T=TT \bmod f(x)$
Step 5 $\quad$ if (e_i==1) $C=CT \bmod f(x)$

The modular multiplication is the key operation in the implementation of circuit for cryptosystems. This is because the process of encryption and decryption of message requires the modular exponentiation. The modular exponentiation could be decomposed into repeated multiplications.

The multiplication of elements a and b over GF(2^4) in the extended basis can be performed by ab mod p with $p=\alpha^{m+1}+1$ which applied the property of AOP as a modulus. Let the result of this multiplication, ab mod p, be $r = r_m\alpha^m + r_{m-1}\alpha^{m-1} + \cdots + r_1\alpha + r_0$. The recurrence equation for the MSB first algorithm with the property of AOP is as follows:

$$\begin{aligned} r &= ab \bmod p \\ &= \{\cdots[[ab_m]\alpha \bmod p + ab_{m-1}]\alpha \bmod p + \cdots + ab_1\}\alpha \bmod p + ab_0 \end{aligned} \quad (4)$$

From the equation 4, a new algorithm to compute ab mod p can be derived as following algorithm 2.

[Algorithm 2] Modular Multiplication
Input : $a=(a_m,a_{m-1},\cdots,a_1,a_0)$, $b=(b_m,b_{m-1},\cdots,b_1,b_0)$
Output : $r=ab$ mod p
Initial value : $r^{m+1}=(r_m,r_{m-1},\cdots,r_1,r_0)=(0,0,\cdots,0,0)$

Step 1 for $i=m$ to 1
Step 2 $d^i=r^{i+1}+ab_i$
Step 3 $r^i=Circular\_Left(d^i)$
Step 4 $r=r^1+ab_{i,}$,

where $Circular\_Left(x)$ is the 1-bit-left-circular shift of x and r^i is used to represent the i-th intermediate result for the final result r. In the above algorithm 2, the modular reduction is performed just by using 1-bit-left-circular-shift operation. Specially, all the operations in the for loop can be performed bit by bit in parallel.

Next section presents a parallel-in parallel-out semi-systolic array architecture based on the algorithm 2.

3 Semi-systolic Multiplier

This section proposes a semi-systolic array architecture by following the procedure in [9] and [13]. It is based on the property of all one polynomial as a modulus and is derived from the algorithm 2.

3.1 Modular Multiplier

Let a, b, and b^2 be an elements in GF(2^4). Then a and b with an extended basis $\{1, \alpha, \alpha^2, \alpha^3, \alpha^4\}$ can be represented as follows:

$$a=a_4\alpha^4+a_3\alpha^3+a_2\alpha^2+a_1\alpha+a_0$$
$$b=b_4\alpha^4+b_3\alpha^3+b_2\alpha^2+b_1\alpha+b_0$$

When $p=\alpha^5+1$ is used as a modular in the extended basis, we have

$$r = ab \bmod p$$

$$= a(b_4\alpha^4+b_3\alpha^3+b_2\alpha^2+b_1\alpha+b_0) \bmod p$$

$$= \{\cdots[[ab_4]\alpha \bmod p+ab_3]\alpha \bmod p+\cdots+ab_1\}\alpha \bmod p+ab_0$$

$$= r_4\alpha^4+r_3\alpha^3+r_2\alpha^2+r_1\alpha+r_0$$

Base on the algorithm 2, the bit level operation is to be

$$d^4=r^5+ b_4(a_4\alpha^4+a_3\alpha^3+a_2\alpha^2+a_1\alpha^1+a_0\alpha^0)$$

$$r^4=Circular\_Left(d^4)$$

$$d^3=r^4+ b_3(a_4\alpha^4+a_3\alpha^3+a_2\alpha^2+a_1\alpha^1+a_0\alpha^0)$$

$$r^3=Circular\_Left(d^3)$$

$$d^2=r^3+ b_2(a_4\alpha^4+a_3\alpha^3+a_2\alpha^2+a_1\alpha^1+a_0\alpha^0)$$

$$r^2=Circular\_Left(d^2)$$

$$d^1=r^2+ b_1(a_4\alpha^4+a_3\alpha^3+a_2\alpha^2+a_1\alpha^1+a_0\alpha^0)$$

$$r^1=Circular\_Left(d^1)$$

$$r=r^1+ b_0(a_4\alpha^4+a_3\alpha^3+a_2\alpha^2+a_1\alpha^1+a_0\alpha^0),$$

where r^i ($0\le i \le 4$) represents the intermediate results.

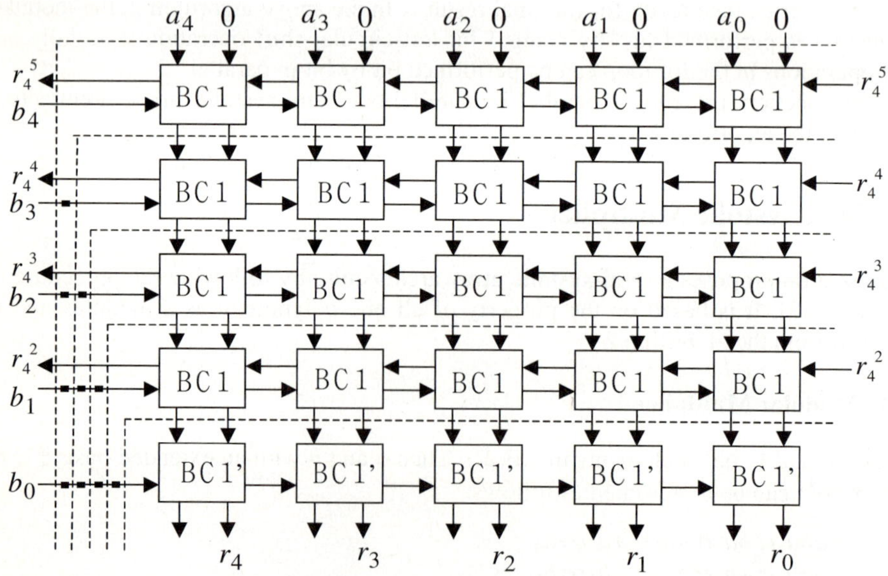

Fig. 1. Dependency Graph over GF(2^4)

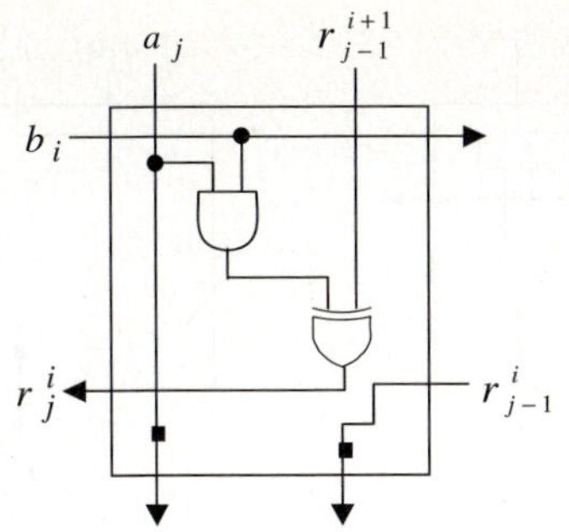

(a) BC1 (b) BC1'

Fig. 2. Basic cell of SMM

Fig.1 shows a dependency graph based on the algorithm 2 over GF(2^4). It is composed of $(m+1)(m+1)$ basic cells. Fig.2 shows the basic cells for the graph. Fig.2 (a) shows the BC (Basic Cell) for the normal cell except the bottom row cells with Fig. 2 (b).

If we can postpone the modular reduction of each cell in the graph from Fig. 1, we could get a better architecture with a low-critical path. Fig. 3 shows the proposed architecture for the modular multiplication over GF(2^4) by postponing the modular reduction of each cell.

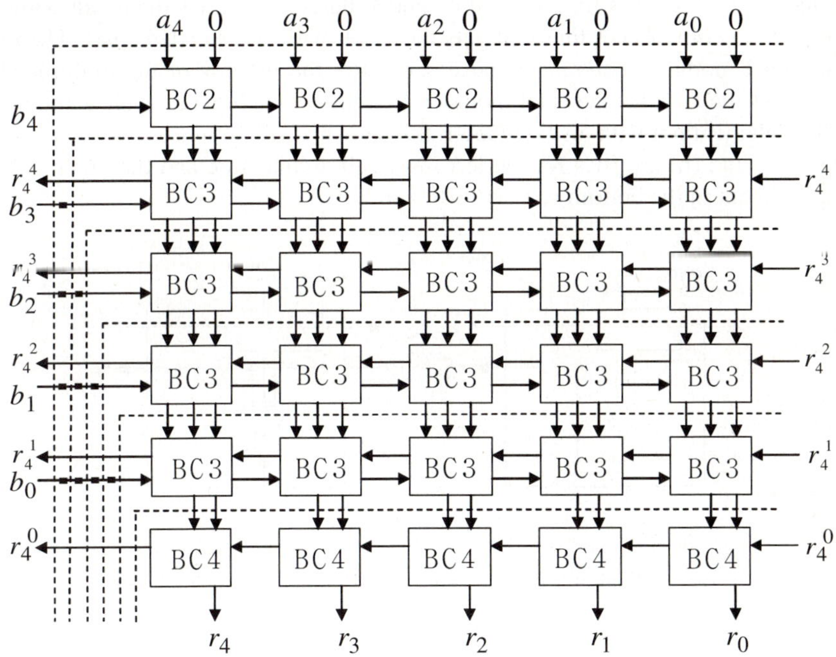

Fig. 3. Proposed Modular Multiplier over GF(2^4)

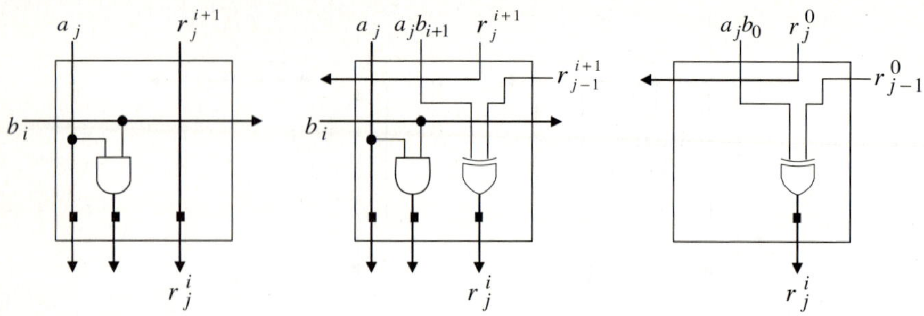

(a) BC2 (b) BC3 (c) BC4

Fig. 4. Basic cells in Fig. 3

It is a parallel architecture which a_i and b_i $(0\leq i \leq m)$ are inputted at the same time. The m-bits of data a are inputted from the top row and transmitted to the adjacent cells following each row. But the data b_i in each row is broadcasted to all cells in the same row at the same time. The multiplier in Fig. 3 has a basic architecture with Fig. 4 (a) except the last row. The last row has the architecture of Fig. 4 (b). The multiplier in Fig. 3 can be generalized for every m as well as m=4.

3.2 Final Reduction

Note that the multiplier is based on an extended basis, which has the result with m+1 bits, not with m bits. Accordingly, it gets the result with an extended basis. Therefore, an additional modular reduction is necessary after the process of the multiplication. Fig. 5 shows the BCs for the final modular reduction, which get a result with an ordinary field element not with an extended field element.

Fig. 5 (a) and (b) are BCs for the left most cell architecture and the other cells for the replacement in the bottom row with BC4 in Fig. 3.

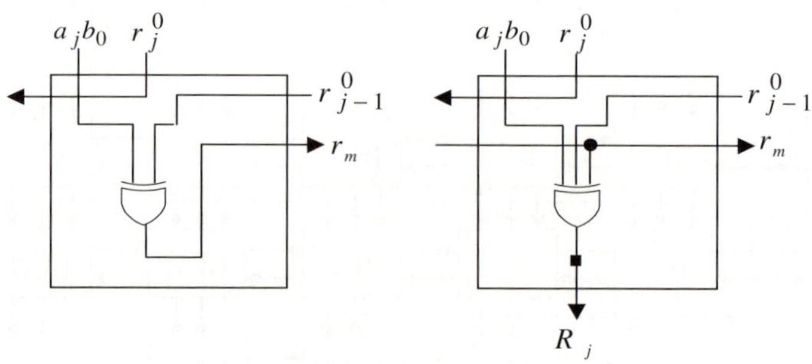

(a) BC-final1 for the multiplier (b) BC-final2 for the multiplier

Fig. 5. BCs for the final modular reduction

Table 1. Comparisons

<table>
<tr><th>Item \ Multiplier</th><th>Jain [9]</th><th>Kim [5]</th><th colspan="2">Proposed multiplier</th></tr>
<tr><td>Basic architecture</td><td>Semi-systolic</td><td>LFSR</td><td colspan="2">Semi-systolic</td></tr>
<tr><td>Irreducible polynomial</td><td>Generalized polynomial</td><td>AOP</td><td colspan="2">AOP</td></tr>
<tr><td>Number of cells</td><td>m^2</td><td>m+1</td><td colspan="2">(m+1)(m+2)</td></tr>
<tr><td rowspan="3">Cell complexity</td><td rowspan="3">2 AND
2 XOR
3 latches</td><td rowspan="3">1 AND
1 XOR
2 REG
1 MUX</td><td>BC2</td><td>1 AND
3 latches</td></tr>
<tr><td>BC3</td><td>1 AND
1 XOR
3 latches</td></tr>
<tr><td>BC4</td><td>1 XOR
1 latch</td></tr>
<tr><td>Latency</td><td>m</td><td>2m+1</td><td colspan="2">m+2</td></tr>
<tr><td>Critical path</td><td>1-D_{AND}
+1-D_{XOR}</td><td>1-D_{AND}+
log m(1-D_{XOR})</td><td colspan="2">1-D_{XOR}</td></tr>
</table>

4 Comparison and Analysis

A comparison and analysis of the proposed multiplier with existing structures were performed focusing on the latency and hardware complexity. We selected a semi-systolic multiplier with a generalized polynomial basis of Jain et al and a LFSR multiplier with a AOP of Kim. Table 1 shows the detailed comparison between the proposed multiplier and the existing two architectures.

For the comparison, it is assumed that AND and XOR represent 2-input AND and XOR gates, respectively, and latch for 1-bit latch. Let D_{AND} and D_{XOR} be the latency of AND and XOR gate, respectively.

The architecture in [9] has the time complexity with m(1-D_{AND}+1-D_{XOR}) because it has a latency of m and the critical path of 1-D_{AND}+1-D_{XOR}. In [5], Kim proposed the architecture with the time complexity with (2m+1)(1-D_{AND}+log m(1-D_{XOR})) due to the latency with 2m+1 and the critical path with 1-D_{AND}+ log m(1-D_{XOR}). However, the proposed multiplier has the time complexity with (m+2)(1-D_{XOR}) because of a latency with m+2 and a critical path with 1-D_{XOR}. As a result, the proposed architecture has the lower time complexity than existing architectures due to the low cell critical path.

Additionally, for the perspective of the area complexity Jain et al's architecture has m^2(2 AND + 2 XOR + 3 latches) and Kim's with (m+1)(1 AND + 1 XOR + 2 REG + 1 MUX). The proposed multiplier has (m+1)(1 AND + 3 latches)(m(m+1)(1 AND + 1 XOR + 3 latches)) (m+1)(1 XOR + 1 latch). The equation to compute the complexity of our multiplier seems like very complex but it is not. It is just because that our multiplier is composed with different three kinds of basic architectures. It could be roughly summarized to $(m+1)(m+3)$(1 AND + 1 XOR + 3 latches). We could gain that the proposed multiplier has the better area complexity than Jain et al's architecture with the same parallel-in parallel-out circuit but not than Kim's architecture with the serial-in serial-out multiplier. However, the disadvantage to Kim's can be compensated with the time complexity. If we consider both of these two complexities, time and area, we could get a result that our proposed multiplier has the best characteristics.

5 Conclusions

This paper proposed a new algorithm and a parallel-in parallel-out semi-systolic array architecture to compute the modular multiplication over finite field GF(2^m). The property of irreducible AOP was used as a modulus to get a better modular multiplier with a good hardware and time complexity. Proposed multiplier has the best characteristics in the terms of the hardware complexity and the time complexity compared to the existing architectures as shown in Table 1. Since it has regularity, modularity and concurrency, it is suitable for VLSI implementation.

References

[1] Peterson, W.W., Weldon, E.J.: Error-correcting codes. MIT Press, MA (1972)

[2] Denning, D.E.R.: Cryptography and data security. Addison-Wesley, Reading (1983)

[3] Menezes, A.: Elliptic Curve Public Key Cryptosystems. Kluwer Academic Publishers, Boston (1993)

[4] Reed, I.S., Truong, T.K.: The use of finite fields to compute convolutions. IEEE Trans. Inform. Theory 21, 208–213 (1975)

[5] Kim, H.S.: Bit-Serial AOP Arithmetic Architecture for Modular Exponentiation, PhD. Thesis, Kyungpook National University (2002)

[6] Wei, S.W.: VLSI architectures for computing exponentiations, multiplicative inverses, and divisions in GF(2<Superscript>m</Superscript>). IEEE Trans. Circuits and Systems 44, 847–855 (1997)

[7] Yeh, C.S., Reed, S., Truong, T.K.: Systolic multipliers for finite fields GF(2^m). IEEE Trans. Comput. C-33, 357–360 (1984)

[8] Wang, C.L., Lin, J.L.: Systolic Array Implementation of Multipliers for Finite Fields GF(2^m). IEEE Trans. Circuits and Systems 38, 796–800 (1991)

[9] Jain, S.K., Song, L.: Efficient Semi-systolic Architectures for finite field Arithmetic. IEEE Trans. on VLSI Systems 6(1) (March 1998)

[10] Kim, H.S.: Efficient Systolic Architecture for Modular Multiplication over GF(2m). IMM-Technical report-2005-09, pp. 104–110 (2005)

[11] Itoh, T., Tsujii, S.: Structure of parallel multipliers for a class of finite fields GF(2^m). Info. Comp. 83, 21–40 (1989)

[12] Kung, S.Y.: VLSI Array Processors. Prentice-Hall, Englewood Cliffs (1987)

[13] Yoo, K.Y.: A Systolic Array Design Methodology for Sequential Loop Algorithms, PhD. thesis, Rensselaer Polytechnic Institute, New York (1992)

Network-Aware Replica Optimization in the SCoPE Grid Infrastructure

Francesco Palmieri and Silvio Pardi

Università degli Studi di Napoli Federico II
Via Cinthia, 5 – 80126, Napoli, Italy
{fpalmier,spardi}@unina.it

Abstract. In a Data Grid, replication of data is critical for maximizing the overall job throughput. Such replication involves the creation of copies of data files at different sites according to specific Replica Optimization strategies that define when and where replicas should be created or deleted on a per-site basis, and which replicas should be used by Grid jobs. To be really effective these strategies have to take into account the available network bandwidth as a primary resource, prior to any consideration about storage or processing power. We present a novel replica management service, integrated within the GlueDomains active network monitoring architecture, designed and implemented within the centralized collective middleware framework of the SCoPE project to provide network-aware replica optimization for data intensive applications.

Keywords: Data Grid, Replica Optimization, Network-aware Grid.

1 Introduction

Many scientific computing experiences involving high energy physics, engineering, bioinformatics, high resolution medical imaging, data mining, and astrophysics now require their operational data from massive, one-of-a-kind, real-time remote equipments, or from immense remote storages, and transfer large portions of them (typically several Terabytes) near to the obtained computational resources for their use. To cope with such problem, many experimental activities have adopted a hierarchical multi-Tier data distribution architecture that is based on the replication of the mostly used datasets in regional computing centres located on a lower layer respect to the above data, starting from the production site (Tier 0). These requirements have led to the development of distributed Grid infrastructures for processing and analyzing these data. A Data Grid is a collection of geographically distributed computation and storage resources which allows transparent sharing of data and processing power in a reliable and consistent way, simultaneously hiding all the unnecessary technical details, so that the users and their applications will only perceive the abstraction of a very large virtual computer. Since the transport network connecting the Grid sites and hardware components is the real "glue" realizing the above abstraction, the optimization of the use of data storage and underlying network resources is essential for applications to be executed efficiently in such a complex distributed execution environment, and data replication may become a crucial tool in achieving this. Replication

O. Gervasi et al. (Eds.): ICCSA 2008, Part II, LNCS 5073, pp. 845–857, 2008.

is the process of creating copies of data files at different sites on the Grid in order to reduce the network traffic and enhance the overall job execution performance. Unfortunately, in current computational Grid deployments, the underlying communication network infrastructure is considered as a pure facility and the middleware components act agnostically with respect to network performance parameters, in particular during data transfer and data replication. In this situation the users are obliged to select by themselves the best storage resources where to put or get their replica of the data they need to analyze. The total absence of any aid or support to the replica management activity from the Grid infrastructure brings the overall system to work globally below the best effort threshold. Thus, the main challenge is to assist the users and application in getting the right data from the right location, at the right time. To achieve this, the location of replicas needs to be controlled by a replica optimization strategy, which can be used to determine which files to copy and which to delete. Such strategy can significantly reduce data access latencies and improve system robustness and scalability by working according to a locality principle. However, dealing with replicas of files adds a number of problems not present when only a single file instance exists. Replicas must be kept consistent and up-to-date, their locations must be traced, their lifetimes need to be managed, etc. Replica management is achieved by coordinating underlying Grid services like file transfer services and protocols (i.e. GridFTP) [1], and metadata catalogs [2]. In this paper we present a simple and effective network-aware replica optimization service, implemented within the Grid framework realized for the SCoPE project (co-funded by the Italian Ministry for Education and Research under the EU PON program) and fully integrated within the GlueDomains [3] active network monitoring architecture, in order to determine the best location of a data to be replicated file among the storage elements available on the Grid. The decisions are obtained by a cost-estimation model, based on a set of very simple and not invasive network measurements. Our prototype implementation, based on the web services paradigm and open source technologies, has been evaluated through a set of real world experiments and performance tests on the SCoPE metropolitan Grid infrastructure and provided promising results.

2 The SCoPE Environment

SCoPE (Italian acronymic for high Performance, Cooperative and distributed System for scientific Elaboration) is a high performance computing research project that aims at developing several applications in the field of fundamental research. The SCoPE architecture has been conceived to provide the unification of all the main computational and storage resources already available in the participating sites, located in the Naples urban area, by creating an open and multidisciplinary distributed computing infrastructure based on the state-of-the-art Grid paradigm. The Grid's connectivity is supported by our metropolitan multi-ring shaped optical network that offers high performance communication facilities to the four main university research sites: Medicine, Engineering, CSI (Center for Information Services) and the Monte S. Angelo Campus as shown in fig. 1.

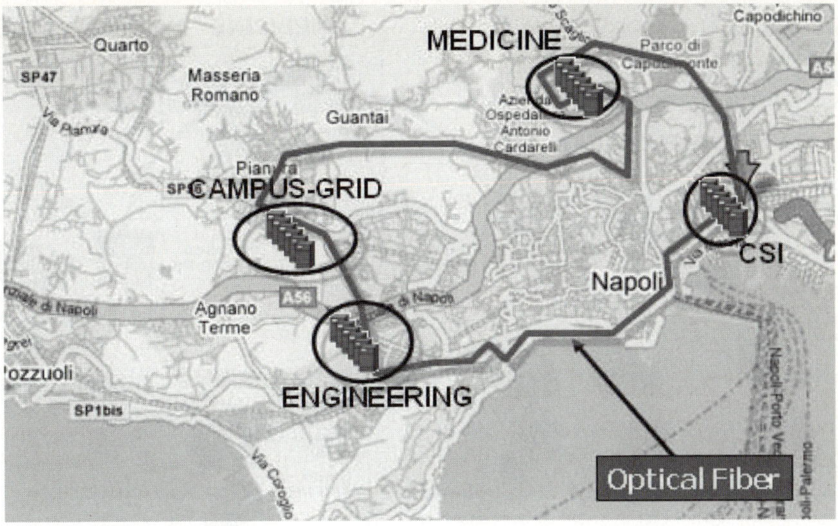

Fig. 1. The SCoPE network layout

The SCoPE Grid solution is based on the INFN-GRID [4] middleware, realizing a structured architecture built on a customization of the LCG storage and computing infrastructure [5] projected for the high-energy physics community using the Large Hadron Collider (LHC) and the gLite middleware [6] developed within the EGEE project context. It offers a computational and storage environment where the Resource Framework Layer (RFL), the Information System Framework (IS) and the Information Data Model encompass semantic and interaction models for all the available resources. The actual implementation provides the creation of a central Virtual Organization (VO) named *scope* managed by a local Virtual Organization Membership Services (VOMS) that collect all the users of the project and that is enabled in each shared resource. Other VOs, are enabled respectively on the resources used in explicit local groups associated to specific experiments. The centralized collective services offer Services Discovery, Resource Brokering, File Catalog, Grid Monitoring and graphical user interface facilities. In the current setup 12 local sites have been activated for a total of more than 330 CPU and 80 TB of storage. The sites are distributed as follow: 10 sites in the MSA Campus, one site in the Medicine pole, another in the Via Claudio Engineering site and the last in the CSI data centre. All the local sites provide one or more traditional Computing Element (CE) batch execution systems, with several associated Worker Nodes (WN) offering computing power, together with one or more Storage Element (SE) servers based on the Disk Pool Manager (DPM). The infrastructure will be integrated in a few months with a new 60 TB storage system located in the bioinformatics research center, and a collective supercomputing center, composed by more than 2000 computing cores and 250 TB of storage located in a new dedicated site in the MSA Campus. The base middleware layer is also powered by the *GlueDomains*

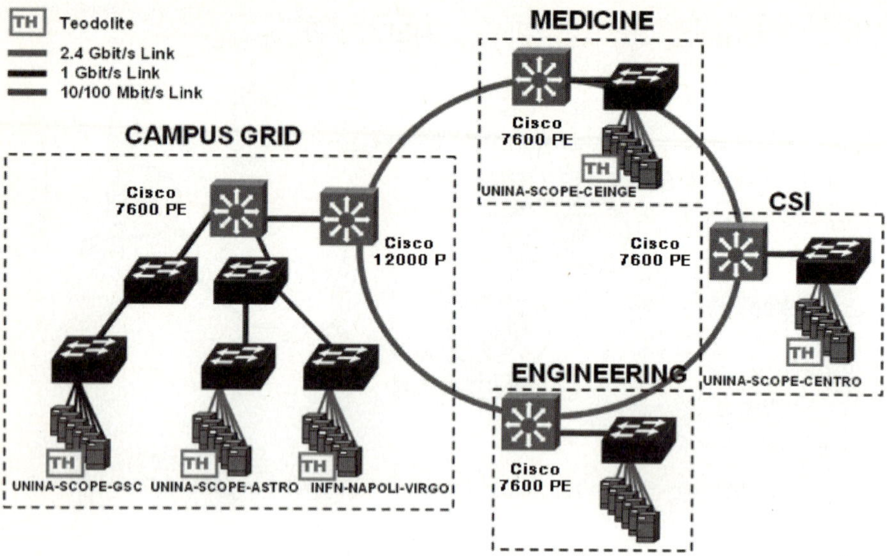

Fig. 2. The GlueDomains-based monitoring architecture

prototype service supporting the domain-based INFN Grid network monitoring activity. The GlueDomains Service is supported by a dedicated central server node, placed in the main MSA Campus Grid site. This server manages the central database used to configure the topology measurement and the network monitoring activity of each "*Theodolite*" agent actually deployed to perform active network monitoring on five sites of the SCoPE infrastructure, as shown in fig. 2.

Theodolite agents are installed within the Storage Elements and configured in a full mesh, making available Round-trip Time, Packet Loss Rate and One-Way Jitter performance measurements for each domain-to-domain path. The data produced by the agents are periodically uploaded and published through the GridICE web interface (see fig. 3 below) which is also used to check theodolite operation.

The SCoPE Grid infrastructure implements storage access via the LCG SRM (Storage Resource Manager) interface. Each file, that can be available on several storage elements, is registered in a central logical file catalogue (LFC - LCG File Catalogue) and the applications access to the physical replicas by querying this service. Such catalogue makes sure that the datasets and associated files and metadata are properly accessible and secured for the end-user application. In the absence of a replica optimization facility, when a user or an application asks for a file, addressed using its *logical file name* (LFN), the LFC catalogue returns a physical location selected randomly among the different replicas that correspond at the same logical name: this way is clearly far from optimal.

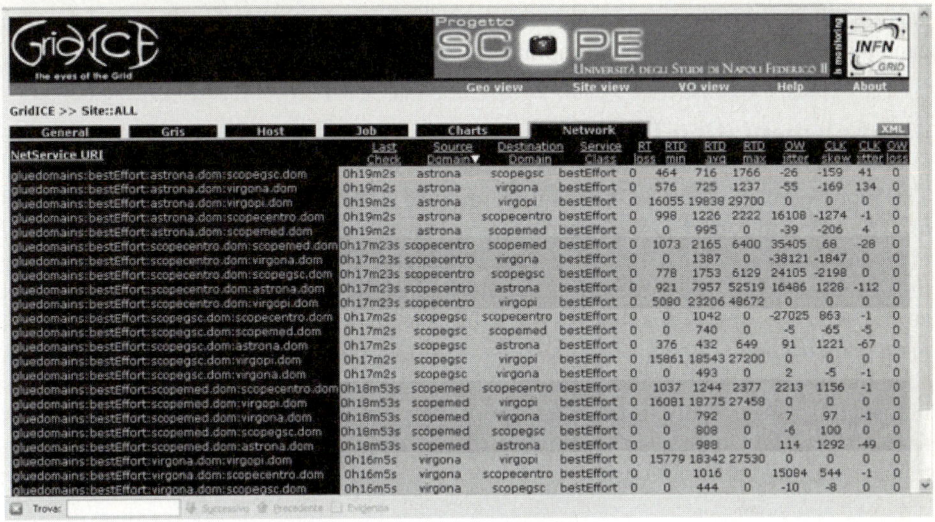

Fig. 3. The GridICE network view

3 Service Objectives and Architecture

Replica Optimization needs to work on two different time scales: short and long-term. In the short term - the hours or minutes that a specific job takes to run - the Replica service must try to minimize the execution cost of each job that is currently running on the Grid. In the Long-term, on the other side, the service main objective is to optimize the overall Grid throughput by putting files in the best places to ensure high performance access for future jobs. All the replication and file replacement issues have to be managed after each job has started running, so that they require short-term optimization actions on each of the involved sites, by simultaneously optimizing the overall performance on the long term. To do this the primary optimization criterion that has to be taken into account is the transport network resource usage. Replicating a file takes time and uses network bandwidth. However, performing no replication will be clearly ineffective compared to even the simplest replication algorithm. Thus, a good balance must be found, where any replication is in the interest of reducing future network traffic.

From the architectural point of view, it is convenient to split the replica optimization process into three main phases, each dependent from the others. The first stage, starting with an initial *data discovery* operation aiming to detect all the copies of the required dataset already existing on the Grid, is that of the *Replication Decision*. Specifically, when a necessary file is not present locally, the Replica Optimization service must decide, starting from the location of the already existing copies, whether to open it remotely or replicate by issuing a data transfer operation and creating a local copy. In the second or *Replica Selection* stage, if the replication decision was positive, the best remote replica must be chosen for data transfer/copying. Finally, if there are no available resources on the local Storage Element - as it most probably will be on an highly loaded Grid - a properly sized file must be selected for deletion to free the necessary space. This is known as *File Replacement* stage. A real replication strategy can be defined by the specific combination of algorithms used for each stage.

4 Implementation Details

The replica optimization service is implemented as a very simple web service that is used to select the best replica respect to a specific Computing or Storage Element, identified from its Site URL, of a dataset which will be needed as input to one or more jobs. Such dataset, identified by its logical file name, is considered to be an atomic unit of data, defined within a VO. The implemented network-aware replica management logic gathers network status information from the GlueDomains Grid's active monitoring service and performs access optimization calculations based on this information. Specifically, it provides optimal replica information on the basis of both faster access and better network performance characteristics to minimize the data transfer time and network costs.

4.1 Network-Aware Replica Management

The replication decision process retrieves available replica's information from the logical file catalogs. It is able to locate datasets as well as individual physical files (depending on the information available on the underlying catalog). Furthermore, a dataset itself can consist of several physical files but the end-user normally only knows the dataset concept. Due to the distributed nature of the Grid, the files may be already replicated at several sites, and the catalog ensures that the user application does not need to know which locations these are. The LFC catalogue is searched in order to find all the locations where the requested dataset is available, obtaining a list of all the already available replicas. The replica decision process then evaluates the network costs for accessing directly each physical replica location in the above list. For this it uses complex estimations of dataset transfer times based on network monitoring statistics and if the resulting network cost is greater than a properly defined tolerability threshold the available copy having the minimum access and transfer cost is selected and a new replica is created on a specified Storage Element. To perform the best replica selection, the previously obtained list of dataset locations is sorted by the datasets "accessibility" which is provided by the network cost and network features such as bandwidth, packet loss etc, and by the "closeness" determined by some network round-trip time or other ICMP-based network measurements. More precisely, the above accessibility and closeness criteria are defined by using a simple model that estimates the network cost using very basic network measurements according to the DIANA approach [7]. The Mathis's formula [8] allows the estimation of maximum TCP throughput by using the historical measurements of the packet loss and RTT with the following equation:

$$MaxBand < \frac{MSS}{RTT} \times \frac{1}{\sqrt{loss}} . \tag{1}$$

Where MSS is the maximum segment size, RTT is the observed Round Trip Time and loss is the packet loss rate during the period of observation. Starting from this approximation we can calculate the network Cost with the following equation:

$$netCost = \frac{Losses}{MaxBand}. \quad (2)$$

Where

$$Losses = w_1 RTT + w_2 Loss + w_3 \,|\, Jitter \,|. \quad (3)$$

The weights w_1, w_2 and w_3 can be chosen based on the different characteristics of the involved data transfer operations. An higher weight indicates the importance of that parameter in the cost calculation so that we can manipulate these weights to prioritize particular parameters in the formula. A higher RTT indicates that a computation site is distant from the storage site where the data resides and therefore the cost to fetch the data would increase. We can increase the significance of this parameter, if required, by assigning a higher value to its associated weight. Moreover, jitter is of less importance for data intensive applications and typically there is no significant impact on the cost due to a higher or lower jitter. We can assign a minimal weight to this parameter but cannot ignore it completely since if it assumes a value higher than an acceptable threshold, then there is some bottleneck involved and this element should then have a higher value to reflect this in the overall cost evaluation criterion. The same is the true for the packet loss: a higher packet loss implies a less reliable network, and we should give less importance to such a connected site when making replica selection. In any case, the value obtained enables the user to estimate the overall performance provided by the different network links in best effort regime. Once the best candidate to replication, in terms of minimum network cost, has been selected from the list of available replicas, a copy operation is issued (lcg-cp, lcg-rep or Gridftp commands) from the involved storage elements and the new replica is registered on the Grid catalog. The current implementation provides replication decision and replica selection stages (interfacing to the network statistics and catalog services) leaving the file replacement activity for future versions.

4.2 Service Interface

Web services technologies are becoming increasingly popular in wide-area distributed computing, since they provide easy and standardized access to distributed services in a multi-lingual, multi-domain environment. The same is true in Grid computing where the upcoming OGSA standard [9] aims at leveraging web services in the Grid context through the Web Service Resource Framework (WSRF) [10]. Accordingly, the top-level service interface of the proposed replica optimization service has been deployed as a centralized web service within the SCoPE collective services framework. Requests and responses conform to Web Services specifications, i.e., they are XML/SOAP messages, carried in HTTP envelopes and transported over TCP/IP connections. The interface has been developed by using the PHP library *nusoap* [11]. Each replication request takes a logical file name (LFN) and the physical file name or SURL of a destination SE as input parameters and chooses the best replica as described above. For the given LFN, all the available physical replicas are identified and the best one in terms of minimal data transfer cost is used as the source file for further

```
<message name="ReplicaRequest">
   <part name="lfn" type="xsd:string"/>
   <part name="dest" type="xsd:string"/>
</message>

<message name="ReplicaResponse">
   <part name="netCost" type="xsd:string"/>
</message>

<portType name="Network Resource ManagerPortType">

<operation name="Replica">
   <input message="tns:ReplicaRequest"/>
   <output message="tns:ReplicaResponse"/>
</operation>
```

Fig. 4. The web service interface

replication. Users may either control the source and destination of the copy or let the system decide on the best resources to use on their behalf. The WSDL representation of the top level interface is reported below.

4.3 Gathering Network Information

Most of the inner components of the presented service are implemented in Perl and work according to a multi-stage process. At the first stage the software wraps the *lcg-utils* command to discover the location of all the available physical replicas of the logical file name that has to be replicated. In the second stage the *gdip-client* Glue-domains tool is used to gather all the theodolite-based performance measurements against all the destination sites containing a replica of the former file. In the following figure we show a dump of the measurement performed between the INFN-NAPOLI-VIRGO and UNINA-SCOPE-ASTRO sites.

For each site containing a valid replica of the required LFN the network cost, based on the previously presented DIANA model, is computed from the network parameters gathered from GlueDomains system against the supplied destination. Finally the list of the available replicas is sorted by network cost, the best replica is successfully selected as the uppermost element and its associated network cost returned to the requesting application.

4.4 Security issues

The Replica Optimization Service security is provided within the SCoPE Grid Security framework, already including versatile security solutions that support X.509v3 certificate-based management of credentials and access/authorization policies. We developed our interfaces basing on the above technologies to ensure that on-demand access to the replica services will be preceded by the necessary identification activities. Each service offered within the replica management framework is able to authenticate Grid certificates via a trust manager, an authentication module which has been

```
 dn:GlueNetworkServiceURI=gluedomains:bestEffort:virgona.dom:astrona.do
m, GlueHostUniqueID=192.135.36.39, mds-vo-name=local, o=grid
 objectclass: GlueTop
 objectclass: GlueNetworkService
 objectclass: GlueKey
 objectclass: GlueRoundTripLoss
 objectclass: GlueOneWayJitter
 objectclass: GlueClockSkew
 objectclass: GlueRoundTripDelay
 GlueNetworkServiceURI: gluedomains:bestEffort:virgona.dom:astrona.dom
 GlueChunkKey: GlueHostUniqueID=192.135.36.39
 GlueRoundTripLossTool: glueping
 GlueRoundTripLossPacketSize: 1024
 GlueRoundTripLossTimestamp: 1202034790
 GlueRoundTripLossValue: 0
 GlueOneWayJitterTool: gluejmeterprobe
 GlueOneWayJitterTimestamp: 1202034757
 GlueOneWayJitterValue: -276
 GlueClockSkewTool: gluejmeterprobe
 GlueClockSkewTimestamp: 1202034757
 GlueClockSkewSkewjitter: 0
 GlueClockSkewValue: 5
 GlueRoundTripDelayTool: gluejmeterprobe
 GlueRoundTripDelayPacketSize: 64
 GlueRoundTripDelayTimestamp: 1202034817
 GlueRoundTripDelayAverage: 1052
```

Fig. 5. GlueDomains measurements

plugged into web service containers. We also have a fine grained authorization mechanism in place that can interpret certificate extensions provided by the VOMS [12], so that each request issued within the context of a Virtual Organization has to be strongly authenticated against the SCoPE Grid PKI infrastructure, and for each replication action to be triggered the associated application or user must use its X.509 certificate provided in the SCoPE environment.

5 Experimental Evaluation

In this section, we report the results of the tests ran for performance analysis sake on five major sites of the SCoPE DataGrid testbed by performing optimized replication of sample files. Specifically the involved sites are:

- UNINA-SCOPE-GSC (MSA Campus Grid)
- UNINA-SCOPE-ASTRO (MSA Campus Grid)
- INFN-NAPOLI-VIRGO (MSA Campus Grid)
- UNINA-SCOPE-CENTRO (Centro Storico)
- UNINA-SCOPE-CEINGE (Medicine)

All the above sites are connected each other through the dark fiber metropolitan area ring described in section 2, where the involved link capacities vary from 2.5 Gbps POS STM-16 to the single or multiple Gigabit Ethernet connections. Each site participating to the testbed is provided with at least a Computing Elements and a Storage Element. In particular, we analyzed the performance of the core optimization functionalities that determines the network cost of each replica. Thus, the goal of the above performance tests was twofold: we verified the consistence of the network cost calculation model and demonstrated that the replica optimization strategy is really able to improve significantly the data access performances. In order to analyze the behaviour of the cost calculation model, we evaluated the *netcost* function between different sites both during a period of low traffic (normal network condition), and under a heavy network load by gathering the information available from their respective theodolites. During an entire night of sampling we observed an almost constant behaviour in the normal traffic scenario (network unloaded or slightly loaded) and, as we expected, a noticeable growth of the netcost value during the heavy network activity. More specifically, regarding the network cost trend between the UNINA-SCOPE-ASTRO and INFN-NAPOLI-VIRGO sites, in 12 hours of observation we found an average netcost value of 6.3 with 0.8 standard deviation, during the normal network activity. On the other side, during the different phases of the heavy network workload we measured a significant growth of the *netcost* value with its maximum in the order of 10^7. The growth of the netcost value when network congestion increases is also strongly related with the measured jitter value. In figure 6 below we showed the netcost behaviour varying in time under normal and heavy load, calculated between the UNINA-SCOPE-ASTRO and INFN-NAPOLI-VIRGO sites' theodolites.

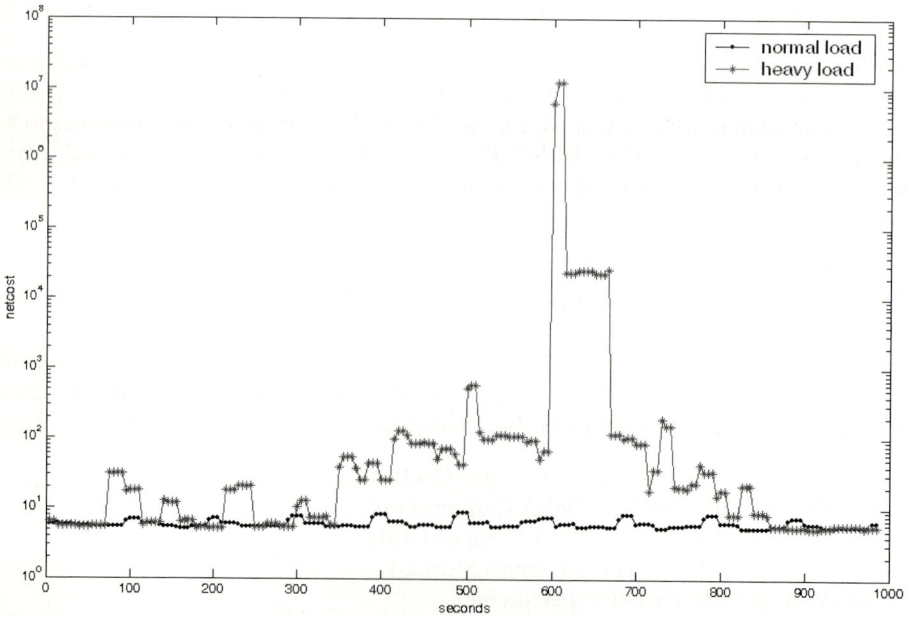

Fig. 6. Netcost behaviour between UNINA-SCOPE-ASTRO and INFN-NAPOLI-VIRGO

As it can be seen from the above trends the network cost function behaviour, allows us to assert the full validity and consistency of the DIANA model in our environment.

The second experiment consists in downloading in the storage element of the INFN-NAPOLI-VIRGO site, a set of 100 1.2 GB sized files that are replicated in the UNINA-SCOPE-ASTRO, UNINA-SCOPE-GSC, UNINA-SCOPE-CEINGE and UNINA-SCOPE-CENTRO sites. The files are registered in the SCoPE logical file catalogue: this allows us to have a single namespace for all the replicas distributed along the different storage elements. The tests have been deliberately run during an intensive synchronization activity among UNINA-SCOPE-GSC, UNINA-SCOPE-CENTRO, UNINA-SCOPE-CEINGE and UNINA-SCOPE-ASTRO sites, and have been split in two phases:

– Replication of 100 files on the INFN-NAPOLI-VIRGO site by using the lcg-utils tools and the logical file name.
– Replication of 100 files on the INFN-NAPOLI-VIRGO site aided by the network-aware replica optimization service

During the first test sequence, the lcg-cp command, after querying the LFC catalog, starts downloading data randomly from the available sites. In the second sequence of tests, data is downloaded using the SURL address returned by the replica optimization service by selecting the "best" replica location based on the network cost measurements gathered through the GlueDomains theodolite interface.

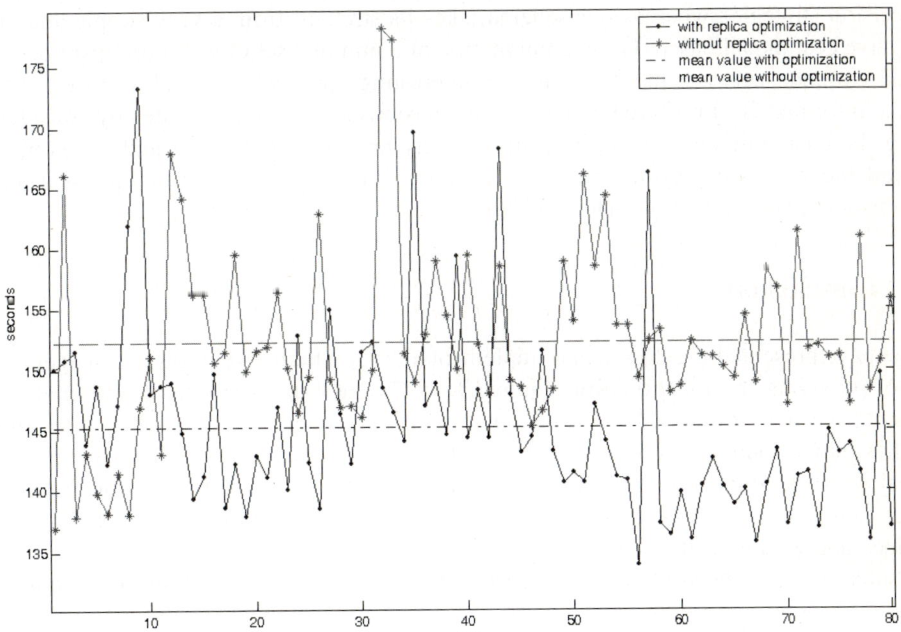

Fig. 7. The data transfer time on a window of 80 file

We measured an average transfer time on 100 files of 154 seconds for each file by using the lcg-utils versus 145 seconds by obtained by using the replica optimization services. In fig. 7 above we show the replica optimization services effects in reducing the data transfer time.

We can note that an average gap of 9 seconds per file is a very good result on a Metropolitan dedicated fiber network, with 1 Gbps or 2.5 Gbps capacities per-link. This result allows us to expect very interesting improvements when applying this model in a geographical scale with significantly slower links.

6 Related Work

A fully functional Replica Optimization System, selecting the best location for replicas based on network and storage costs has been described in [13] and [14] and developed within the EGEE project [15]. However, the system used a "complicated" network monitoring infrastructure that was not further maintained within EGEE and therefore was no longer deployed. Another significant effort has been described in [16] where the high level architecture of a new flexible framework addressing both short-term and long-term replica optimization problems is sketched. The work in [17] considers the replica optimization problem form a different perspective by presenting a fast replica selection algorithm inspired by the Utility Theory in economics that balances the two conflicting demands of increasing parallelism and reducing redundancy. The above works are based, however, on sophisticated and often complex decision strategies whereas our schema takes its strength from a very simple and effective cost model based on the combination of a minimal set of weighting parameters and common network metrics. Another interesting approach (very close to ours) to take informed Grid resource management decisions across multiple sites by considering the changing state of the network, locality and size of the data and the pool of available processing cycles has been proposed in [7] with the Data Intensive and Network Aware (DIANA) meta-scheduler.

7 Conclusion

We have presented the design and implementation of a high level replica management tool providing the functionality required for efficient data access in a Grid environment. We have discussed in detail our optimized replica selection mechanism that takes into account network monitoring performance data in order to identify the optimal network connection with respect to the required data transfer operation. The experimental results show that our approach significantly reduces wide area transfer times and enhances the overall performance of the involved jobs. The encouraging results that have been obtained suggest us to investigate the possibility of acquiring more accurate bandwidth and Quality of Service measurements in order to improve decision accuracy.

References

1. Allcock, B., et al.: Efficient data transport and replica management for high-performance data-intensive computing. In: Proceedings of the 18th IEEE Symposium on Mass Storage Systems and 9th NASA Goddard Conference on Mass Storage Systems and Technologies, S. Diego (2001)
2. Chervenak, A., et al.: Giggle: a framework for constructing scalable replica location services. In: SC 2002, Baltimore (2002)
3. Ciuffoletti, A., et al.: Architecture of monitoring elements for the network element modeling in a grid infrastructure. In: Proc. of Workskop on Computing in High Energy and Nuclear Physics (2003)
4. INFN Grid, `http://grid.infn.it/`
5. LCG middleware, `http://lcg.web.cern.ch/LCG/`
6. gLite middleware, `http://glite.web.cern.ch/glite/`
7. McClatchey, R., et al.: Data Intensive and Network Aware (DIANA) Grid Scheduling. Jouunal of Grid Computing, DOI 10.1007/s10723-006-9059-z
8. Mathis, S., Mahdavi, O.: The macroscopic behavior of the TCP congestion avoidance algorithm. Computer Communications Rev. 27(3), 62–82 (1997)
9. Foster, I., et al.: The Physiology of the Grid: An Open Grid Services Architecture for Distributed Systems Integration, Globus Project (2002), `http://www.globus.org/research/papers/ogsa.pdf`
10. Czajkowski, K., et al.: From Open Grid Services Infrastructure to WS-Resource Framework: Refactoring and Evolution (2004), `http://www.globus.org`
11. NuSoap - SOAP Toolkit for PHP, `http://sourceforge.net/projects/nusoap`
12. Alifieri, R., et al.: An authorization system for virtual organizations. In: European Across Grids Conference, Santiago de Compostela, Spain (2003)
13. Bell, W., Cameron, D., Capozza, L., Millar, A.P., Stockinger, K., Zini, F.: Design of a Replica Optimisation Framework, Technical Report DataGrid-02-TED-021215, Geneva, Switzerland (2002)
14. Cameron, D., Casey, J., Guy, L., Kunszt, P., Lemaitre, S., McCance, G., Stockinger, H., Stockinger, K., et al.: Replica Management in the EU DataGrid Project. International Journal of Grid Computing 2(4), 341–351 (2004)
15. Andreetto, P., et al.: Practical Approaches to Grid Workload & Resource Management in the EGEE Project. In: CHEP 2004, Interlaken, Switzerland (2004)
16. Yin, D., Chen, B., Fang, Y.: A Fast Replica Selection Algorithm for Data Grid, compsac. In: 31st Annual International Computer Software and Applications Conference (COMPSAC 2007), vol. 1, pp. 383–387 (2007)
17. Ciglan, M., Hluchy, L.: Towards Scalable Grid Replica Optimization Framework, ispdc. In: The 4th International Symposium on Parallel and Distributed Computing (ISPDC 2005), pp. 43–50 (2005)

Fault-Tolerant Cluster Management Tool for Supporting Nonstop of Cluster DBMS*

Jae-woo Chang and Jung-ho Um

Dept. of Computer Engineering, Chonbuk National University
Chonju, Chonbuk 561-756, South Korea
{jwchang,jhum}@chonbuk.ac.kr

Abstract. In this paper, we design and implement a fault-tolerant cluster management tool which can monitor the status of the nodes in a cluster DBMS (Database Management System) as well as the status of the various DBMS instances at a given node. Based on the monitoring, our cluster management tool can provide an intelligent fail-over mechanism for dealing with multiple node failures in a cluster DBMS, thus making its fault-tolerance possible when combined with a load balancer. We implement our cluster management tool by using Linux virtual server 0.81 as a load balancer and iBASE/Cluster as a cluster-based DBMS, under a cluster system with four server nodes. It is shown that the sensing and recovery time of our cluster management tool for the iBASE/Cluster is about 2 seconds while that of OCMS for Oracle DBMS is more than 20 seconds. Finally we show that our cluster management tool can support the nonstop and self-managing of cluster DBMS with its intelligent fail-over mechanism even in case of multiple node failures.

Keywords: fault-tolerant, Cluster, Nonstop.

1 Introduction

Much research has been conducted regarding the use of cluster DBMSs which are needed to supports 24 hours nonstop database application services, e.g., map database provision in LBS (Location-based Service) and Telematics. They include Oracle 9i Real Application Server [1], Informix Extended Parallel Server [2] and IBM DB2 Universal Database EEE [3]. As a well-known cluster management tool, OCMS (Oracle Cluster Management System) [4] is included as a part of the Oracle8i Parallel Server product on Linux. To manage the cluster DBMS efficiently, a cluster management tool should be designed by considering a couple of requirements. First, a cluster management tool should provide users with a graphic interface to visualize a cluster system being composed of multiple nodes as a single virtual system, as well as to show the status of all of the nodes in the system and all of the resources (i.e. CPU, memory) at a given node [5, 6]. Next, a cluster management tool should provide a

* This work is financially supported by the Ministry of Education and Human Resources Development (MOE), the Ministry of Commerce, Industry and Energy (MOCIE) and the Ministry of Labor (MOLAB) though the fostering project of the Lab of Excellency.

O. Gervasi et al. (Eds.): ICCSA 2008, Part II, LNCS 5073, pp. 858–870, 2008.

fail-over mechanism to support its fault-tolerance in case of multiple node failures when combined with a load balancer [7, 8]. In this paper, we design and implement a fault-tolerant cluster management tool which can monitor the status of the nodes in a cluster DBMS as well as the status of the various DBMS instances at a given node. Based on the monitoring, our cluster management tool can provide an intelligent fail-over mechanism for dealing with multiple node failures in a cluster DBMS, thus making its fault-tolerance possible. The remainder of this paper is organized as follows. The next section discusses related work on the existing cluster management systems. In section 3, we describe the design of our cluster management tool for a cluster-based DBMS and its graphical user interface. In section 4, we present the implementation and the performance analysis of our cluster management tool. Finally, we draw our conclusions in section 5.

2 Related Work

As related work, we introduce the typical cluster management systems, such as OCMS (Oracle Cluster Management System) [4] and SCMS (SMILE Cluster Management System) [9]. First of all, OCMS is a well known as a cluster management tool for the Oracle8i Parallel Server product on Linux. It provides cluster membership services, a global view of clusters, node monitoring and cluster reconfiguration. OCMS also consists of a watchdog daemon, node monitor and cluster manager. First, the watch-dog daemon offers services to the cluster manager and to the node monitor. It makes use of the standard Linux watchdog timer to monitor selected system resources for the purpose of preventing database corruption. Secondly, the node monitor passes node-level cluster information to the cluster manager. It maintains a consistent view of the cluster and informs the cluster manager of the status of each of its local nodes. The node monitors also cooperates with the watchdog daemon to stop nodes which have an abnormally heavy load. Finally, the cluster manager passes instance-level cluster information to the Oracle instance. It maintains process-level status information for a cluster system. The cluster manager accepts the registration of Oracle instances to the cluster system and provides a consistent view of the Oracle instances. Next, SCMS was developed by Kasetsart university in Thailand as a cluster system management tool for Beowulf clusters. It consists of CMA (Control and Monitoring Agent), SMA (Systems Management Agent) and RMI (Resource Management Interface). First, the CMA runs on each node and collects system statistics continuously. The CMA reads system information through a layer called HAL (Hardware Abstraction Layer). Secondly, the system statistics are collected by a centralized resource management server called SMA. The SMA responds to user queries concerning the system status and sends some commands to the CMA. Finally, the RMI is provided as a set of APIs for system monitoring and logging applications. By using the RMI, the SCMS provides monitoring tool, configuration utilities and parallel Unix commands.

3 Fault-Tolerant Cluster Management Tool

A cluster DBMS consists of multiple server nodes and uses a shared disk. All of the server nodes are connected to each other by means of a high-speed gigabit Ethernet, which is called cluster network. A master node receives a service request from clients through a service network and makes job assignment through the cluster network. Each server is connected to a shared disk by means of SAN (Storage Area Network) for data transmission. The Backup node serves as a new master node when it detects the failure of the master node and uses the cluster network for a fail-over mechanism. Figure 1 shows the overall architecture of a cluster DBMS which are needed to supports 24 hours nonstop database application services, e.g., map database provision.

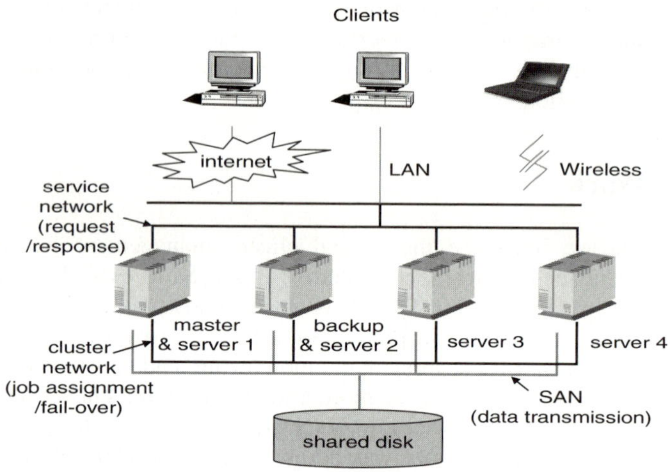

Fig. 1. Overall architecture of a cluster DBMS

By extending the cluster management tool designed for cluster DBMS [10], we proposed a fault-tolerant cluster management tool which can monitor the status of the nodes in a cluster DBMS as well as the status of the various DBMS instances at a given node. Based on the monitoring, our cluster management tool can provide an intelligent fail-over mechanism for dealing with multiple node failures in a cluster DBMS, thus making its fault-tolerance possible when combined with a load balancer. The cluster management tool consists of four components; probe, handler, CM (Cluster Manager) and NM (Node Manager). The probe monitors the status of the each node and generates events according to the status of the system. The handler stores the monitored information into a status table and performs a specific procedure for each type of event. The NM performs network communication with the CM at the master node. If the NM detects an error during the communication with the CM, the NM makes a connection with the backup node and transmits all of the available information to it. If the NM does not receive a response from the CM during a defined interval, the NM considers that the CM has failed and makes a connection with a backup node. The CM, running on the master node, perceives the status of the networks by sending a ping message to each node through both the cluster network and the service network. Based on the status of the networks, the CM manages all of the system resources and services. If the CM detects that the NM has encountered an error, it restarts the NM. If the CM detects an error, it generates an event and transmits it to the corresponding service handler, thus requesting it to perform its recovery procedure.

If the master node fails, our fault-tolerant cluster management tool terminates the master node, and then makes the backup node and one of server nodes be a new master and a new backup one, respectively. If the backup node fails, our management tool terminates the backup node and let one of server nodes be a new backup node. If a server node fails, our management tool just terminates the server node. Figure 2(a) shows the status of the system having four nodes before and after a node failure.

(M: master node, B: backup node, S: database server node)

Before node failure				Failure node	After node failure			
1	2	3	4	Master node	failure	M	B	S
M	B	S	S	Backup node	M	failure	B	S
				Server node	M	B	failure	S

(a) System having four nodes

Before node failure				Failure node	After node failure			
1	2	3	4	Master node	failure	failure	M	B
failure	M	B	S	Backup node	failure	M	failure	B
				Server node	failure	M	B	failure

(b) System having three nodes

Before node failure				Failure node	After node failure			
1	2	3	4	Master node	failure	failure	failure	M
failure	failure	M	B	Backup node	failure	failure	M	failure
				Server node	failure	failure	M	failure

(c) System having two nodes

Fig. 2. Status of the system before and after a node failure

At this time, we assume that the master node has failed, and then the second and the third node have been selected as a new master and a new backup one, respectively. Thus, if the new master node fails continuously, our fault-tolerant cluster management tool terminates the master node, and then assigns the backup node and the remaining database server node to the role of a new master and a new backup one, respectively. If the new backup node fails after the first node failure in Figure 2(a), our management tool terminates the backup node (the third one) and let the remaining database server node (the fourth one) be a new backup node. If a database server node fails after the fist node failure in Figure 2(a), our management tool just terminates the database server node (the fourth one). Figure 2(b) shows the status of the system having three nodes before and after a node failure. At this time, we also assume that the new master node has failed, and then the third and the fourth node have been selected as a new master and a new backup one, respectively. Thus, if the new master node fails continuously, our fault-tolerant cluster management tool terminates the master node and just assigns the backup node to the role of a new master node. This is because there is no running node. If the backup node fails after the first two nodes' failures in Figure 2(b), our management tool just terminates the backup node (the fourth one) because there is no running database server node. Figure 2(c) shows the status of the system having two nodes before and after a node failure. The other cases can be dealt with similarly by our fault-tolerant cluster management tool. Conclusively, it is shown that our cluster management tool is fault-tolerant because it can work well even if one of nodes (database servers) survives in the system.

4 Intelligent Fail-over Mechanism

The server nodes comprising a cluster system can be classified into the master node, the backup node, and the database server nodes. If there is no failure in both the service and the cluster network, the cluster system runs as a normal situation. If the service network fails, a server node can communicate with the other nodes, but it cannot receive user requests or return the results to the user. If the cluster network fails, a given server node cannot communicate with the others and so the master node cannot distribute user requests to any of the database server nodes. If neither the service nor the cluster network work, the situation is considered as node failure since the nodes cannot function anymore. In order to recover the cluster system from the node failure, we provide an intelligent fail-over mechanism which can deal with multiple node failures in a cluster DBMS. The proposed fail-over mechanism is divided into two procedures; node failure detection and failure recovery. First, in order to detect the node failure, the CM and NM sends a ping message to each database server node in the cluster system and analyze the response of each server node. If it fails to receive any response from any of the database server nodes, it regards the situation as its node failure or its cluster network failure. At that time, there can be an ambiguity as shown in the Figure 3. In the Figure 3(a), all of the database server nodes except the first one has failures (node failures or cluster network failures). In the Figure 3(b), only the first server node has a failure. However, both situations can never be distinguished because they have the same ping status. To solve this ambiguity, we propose an intelligent failure detection procedure which uses the history information of the ping status. Once, our cluster management tool detects a failure, it compares the current ping status with the past status. If two database server nodes have already failed due to their node failure or their cluster network failures, this situation can be considered as the failures of all the nodes except one. Otherwise, this situation can be considered as a failure in only one node. In our intelligent failure detection procedure, it is

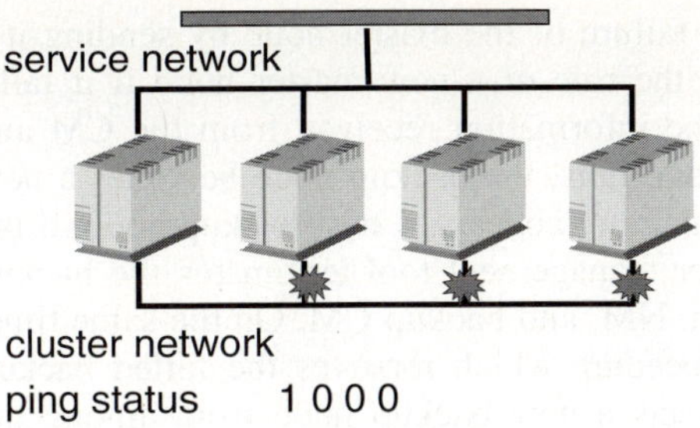

(a) All of the database server nodes except the first one have failures

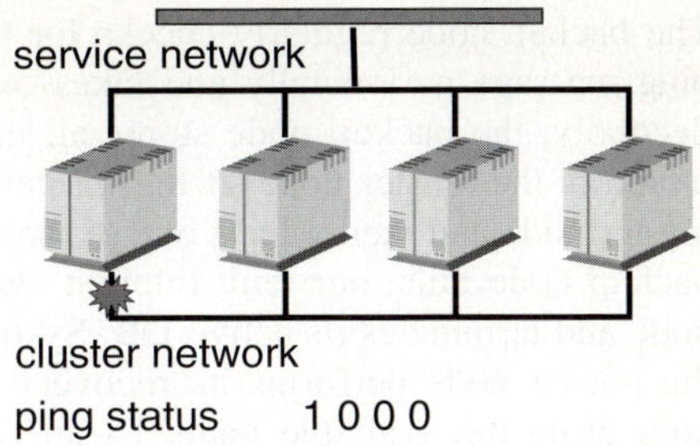

(b) First server node has a failure

Fig. 3. Ambiguity between two failures

```
Failure_detection_procedure(){
  for(i=0; i<NODE_NUM; i++){
    if (service network of node[i] is down based on
        old_node_status)
       then increase sudden_death_servicenet
    if (cluster network of node[i] is down based on
        old_node_status)
       then increase sudden_death_clusternet
    Store status of network into old_node_status }
    if(sudden_death_servicenet >= 2) {
       Generate event ERROR_CLUSTER_NET
       Change mode to TERMINATE;
       Terminate processes in node
    } else if(sudden_death_clusternet >= 2) {
       Generate event ERROR_SERVICE_NET
       Change mode to TERMINATE;
       Terminate processes in node
    } else if(sudden_death_clusternet ==1 && sud-
             den_death_servicenet==1)
       Generate event ERROR_SINGLE } //end of procedure
```

Fig. 4. Intelligent failure detection procedure

assumed that more than one node (or cluster network) never fail at the same time. This assumption is very realistic in a cluster DBMS environment. Figure 4 shows the procedure of our intelligent failure detection.

After we detect node failures (or network failures), we next perform a failure recovery procedure to bring the cluster DBMS back to a normal situation. First, the master node manages the cluster DBMS by distributing user requests to each server node. The master node sends a ping message to each server node in the cluster system and detects the failure of a node by analyzing the response of each node. If the master node fails to receive any response from the nodes, it regards the situation as cluster network failure. Meanwhile, when the backup node cannot communicate with the master node using a ping message, it regards the situation as the master node failure.

The backup node regularly checks for the failure of the master node by sending it a ping message periodically and takes over the role of a new master node if it fails. Secondly, the backup node stores all of the information received from the CM and monitors the master node. If the master node fails, the backup node becomes a new master node and then selects one of the remaining nodes as a new backup node. If the backup node fails, our fault-tolerant cluster management tool terminates the backup node and terminates its active DB, Sysmon, NM, and backup CM. On the same time, the master node performs its recovery procedure which removes the failed backup node from the available node list and selects a new backup node from among the remaining nodes. Finally, the failure of a node can cause the failure of the entire cluster DBMS because the cluster DBMS uses a shared disk. Therefore, our fault-tolerant cluster management tool performs a recovery procedure to preserve the integrity of the data by preventing the failed node from accessing any data in the shared disk. If the service network fails, all of the nodes terminates their transactions and perform their recovery procedure, since they cannot return the results of any user requests to the users. If the cluster network fails, all of the nodes perform the same thing as that described above, because they can neither receive any user requests from the master node, nor communicate with any other nodes. If a server node fails, our fault-tolerant cluster management tool removes this server node from the available node list and informs the other nodes of the failure of this server node. On the same time, the failed node performs its recovery procedure to recovers its transactions and terminate its database. The other nodes recover the data of the failed node. Table 5 shows the failure recovery procedure of the master node due to its importance. The failure recovery procedures of the other nodes can be written similarly.

```
Recovery_procedure_Masternode(int event_num)
  switch(event_num) {
   case ERROR_MASTER :
     Remove VIP from the master node
     Clear the the available server list
     Terminate probe, NM, CM, DBMS
   case ERROR_BACKUP:
     Remove the backupnode from the available server list
     Terminate bcm, DBMS in backupnode
     Select a new backupnode
     break;
   case ERROR_SERVER:
     Remove backupnode from available server list
     Terminate bcm, DBMS in backupnode
   default :
     Print a message of not defining event }} //end of procedure
```

Fig. 5. Failure recovery procedure of the master node

5 Implementation and User Interfaces

We implement our cluster management tool using iBASE/Cluster under a cluster system with four nodes, each being 450 Mhz CPU/HDD 30GB/128MB Memory. We make use of Linux virtual server 0.81 as the load balancer [8], the Redhat Linux 7.1

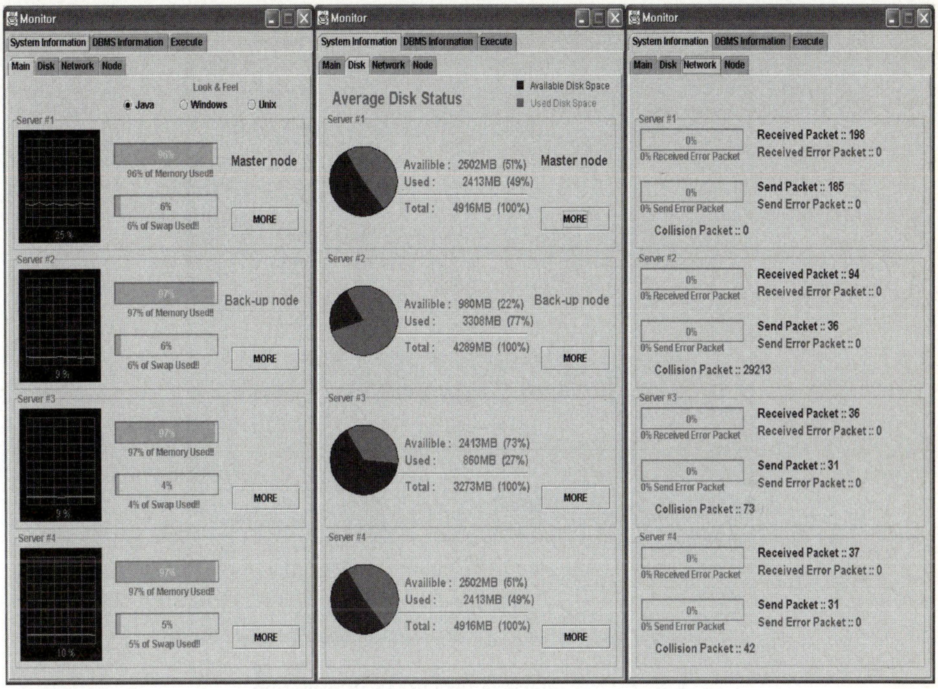

Fig. 6. User interface for system information

operating system and iBASE/Cluster DBMS [11]. We describe the convenient user interface of our fault-tolerant cluster management tool which presents both the status of the system information and the status of the DBMS information being monitored by the system probe at each server node. Figure 6 shows the user interface for system information which consists of the CPU, memory, disk, network and node status.

The left part of Figure 6 shows the user interface for the CPU and memory status part of the system information. It is shown that our cluster system consists of four server nodes and the first node is the master node. To present the status of the disk, we depict the current used disk space, the available disk space, and the total mounted space. The central part of Figure 6 shows the user interface for the disk status part of the system information. It is shown from the master node that the total mounted disk space, available disk space and current used space are 4916 Mbytes, 2502 Mbytes and 2413 Mbytes, respectively. The right part of Figure 6 shows the user interface for the network status part of the system information. For the status of the network, we depict the network information such as the number of received packets, received error packets, sending packets, sending error packets. Figure 7 shows the user interface for the DBMS information. We can observe that there are multiple DBMS processes and node commit is running on the master node. The PID 959 process with S (Sleep) status makes use of 7% of the CPU.

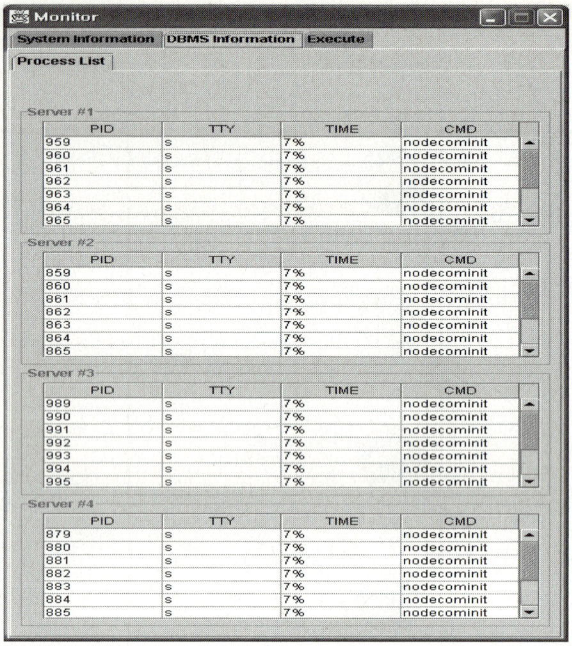

Fig. 7. User interface for DBMS information

To test our cluster management tool, we use a cluster system including four servers, each of which is running on iBASE/Cluster. In the normal situation, the service requests of users are transmitted to the master node. The Linux virtual server on the master node distributes a given user request to one of the available server nodes according to its round-robin algorithm. Figure 8 shows the user interface of our cluster management tool for a normal situation, wherein the four nodes comprising the cluster system are active. The first database server node serves as the master node and the second server node serves as the backup node. It is shown from this figure that the NM, Sys (Sysmon) and iBase (iBASE/Cluster) processes are in the active state and that the Linux virtual server and GLM (global lock manager) of the cluster DBMS are running on the master node. When the master node fails, we disconnect the master node from the cluster network that is used to communicate with the other server nodes. Thus, the second node, which functions as the backup node, takes on the role of a new master node. If the backup node fails, we disconnect the backup node from the cluster network. The second server node, which functions as the backup node, is now in an inactive state, and one of the other server nodes is selected as a new backup node. Figure 9 shows the user interface of our cluster management tool when the master node fails. It is shown that our cluster management tool is fault-tolerant even for a situation when the mater node is inactive.

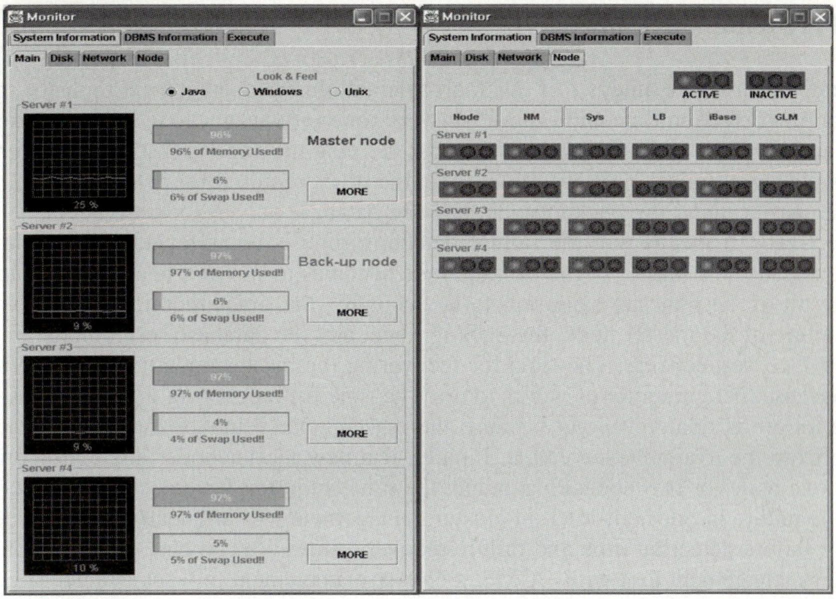

Fig. 8. User interface for a normal situation

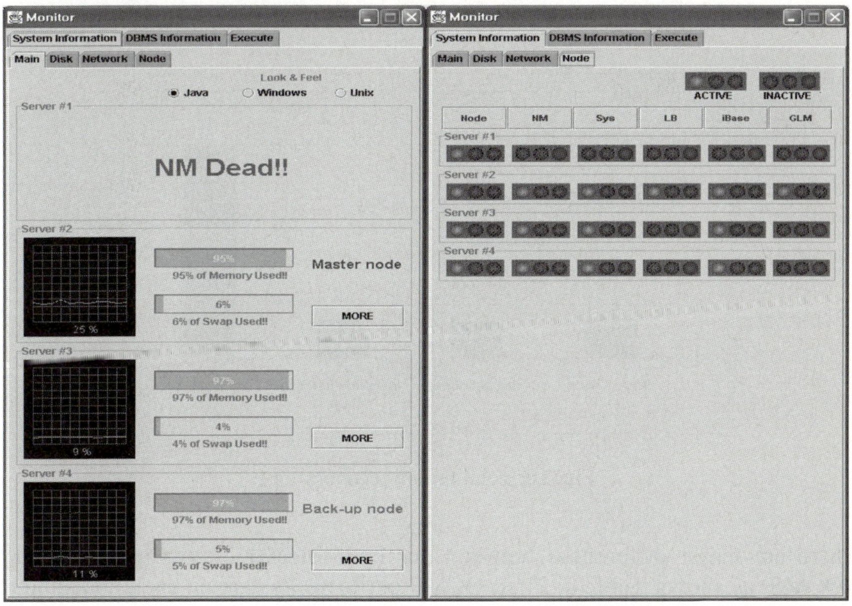

Fig. 9. User interface when the master node fails

6 Performance Analysis

For the performance analysis of our fault-tolerant cluster management tool, we measure failure detection time and failure recovery time for various node failure types. For this, we set the maximum response time for the ping message to 2 seconds. Figure 10 shows the total failure recovery time. First, the times required for detecting the failure of the master node, the backup node, and the database server node are about 0.9 sec, respectively. It means that the failure detection time is not affected by node failure types because the master or the backup node becomes aware of a node failure through the result of the ping message sent to it. Secondly, the times required for recovering the failure of the master node, the backup node, and the database server node are 0.8, 0.5, 0.7sec, respectively. The time for recovering the master node failure is the longest because there are a lot of works to do. The time for recovering the database server node failure is relatively long because the master node removes the database server node from the available server list. Finally, if a new master node fails consecutively after one node (or two nodes) has failed, the times required for detecting and recovering the failure in our fault-tolerant cluster management tool are nearly the same as the above failure detection time and failure recovery time. We compare our fault-tolerant cluster management tool with OCMS, a cluster management tool for Oracle. The total time for failure detection and failure recovery in OCMS is more than 20 sec, according to the Oracle 9i Administrator's Reference [12]. The performance of the OCMS is much worse than our cluster management tool because OCMS has to wait until Oracle instance's reconfiguration.

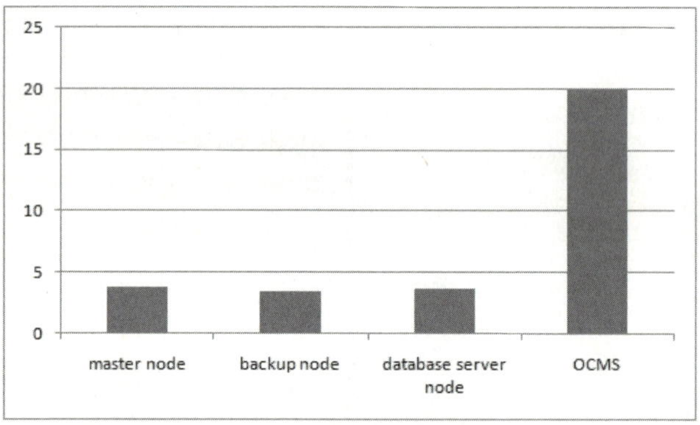

Fig. 10. Total failure recovery time

There are major differences between our fault-tolerant cluster management tool and OCMS, as shown Table 4. First, OCMS performs its task on each node and communicate each other through the private network. Assume that there are N server nodes, the total number of messages for communication in OCMS is (N-1)*N. However, our cluster management tool only needs 2N messages. Secondly, because OCMS uses a quorum partition to recognize the failure of other nodes, more nodes attempt to read the quorum partition, thus resulting in the degradation of recovery performance. However, because our cluster management tool keeps the history status information of every node, it can detect multiple node failures based on the comparison between a current status and a history status. Because being not affected by the number of nodes, our cluster management tool is more efficient than OCMS in terms of scalability.

Table 1. Differences between our cluster management tool and OCMS

	our cluster management tool	OCMS
Network communication	Centralized	Distributed
Failure detection	History status information	Quorum partition

7 Conclusions

In this paper, we designed a fault-tolerant cluster management tool which can monitor the status of the nodes in a cluster DBMS as well as the status of the various DBMS instances at a given node. Based on the monitoring, our cluster management tool can provide an intelligent fail-over mechanism for dealing with multiple node failures in a cluster DBMS, thus making its fault-tolerance possible. In addition, we implemented our cluster management tool by using Linux virtual server 0.81 as a load balancer and iBASE/Cluster as a cluster-based DBMS, under a cluster system with four server nodes. The graphical user interface (GUI) of our cluster management tool enabled users to visualize a single virtual system image by providing them with the status of all of the nodes and resources. Finally, we showed from our performance analysis that the sensing and recovery time of our fault-tolerant cluster management tool for the iBASE/Cluster was about 2 seconds while that of OCMS for Oracle DBMS should be more than 20 seconds. Therefore, we showed that our fault-tolerant cluster management tool could support nonstop database application service, e.g., map database provision, by using its intelligent fail-over mechanism even in case of multiple node failures.

References

1. O. Corporation Oracle 9i database (2002), `http://oracle.com/`
2. Informix Software Inc.: Informix OnLine Dynamic Server administrator's guide (1994)
3. Baru, C.K., Fecteau, G., Goyal, A., Hsiao, H., Jhingran, A., Padmanabhan, S., Copeland, G.P., Wilson, W.G.: DB2 parallel edition. IBM Systems Journal 34(2), 292–322 (1995)
4. Oracle Corporation Oracle 8i Administrator's Reference Release3 (8.1.7) for Linux Intel, ch. 7, Oracle Cluster Management Software (2000)
5. Buyya, R.: High Performance Cluster Computing, vol. 1&2. Prentice Hall PTR, Englewood Cliffs (1999)
6. Gregory,, Pfister, F.: In Search of Clusters, 2nd edn. Prentice-Hall, Englewood Cliffs (1998)
7. Linux Clustering, `http://dpnm.postech.ac.kr/cluster/index.htm`
8. Linux Virtual Server, `http://www.linuxvirtualserver.org`

9. Uthayopas, P., Maneesilp, J., Ingongnam, P.: SCMS: An Integrated Cluster Management Tool for Beowulf Cluster System. In: International Conference on Parallel and Distributed Proceeding Techniques and Applications, pp. 26–28 (2000)
10. Kim, Y.C., Chang, J.W.: HACM: High-Availability Cluster Management Tool for Cluster DBMS. In: International Workshop on Database and Expert Systems Applications (2005)
11. Kim, H.Y., Jin, K.S., Kim, J., Kim, M.J.: iBASE/Cluster: Extending the BADA-IV for a Cluster Environment. In: 18th Korea Information Processing Society Fall Conference, vol. 9(2), pp. 1769–1772 (2002)
12. Oracle Corporation.: Oracle 9i Administrator's Reference Release2(9.2.0.1.0), ch. F, Oracle Cluster Management Software (2002)

Finding Synchronization-Free Slices of Operations in Arbitrarily Nested Loops

Anna Beletska[1], Wlodzimierz Bielecki[2], Krzysztof Siedlecki[2], and Pierluigi San Pietro[1]

[1] Dipartimento di Elettronica e Informazione, Politecnico di Milano, 20122 via Ponzio 34/5, Milano, Italy
{beletska,sanpietr}@elet.polimi.it

[2] Faculty of Computer Science, Technical University of Szczecin, 70210 Zolnierska 49, Szczecin, Poland
{wbielecki,ksiedlecki}@wi.ps.pl

Abstract. This paper presents a new approach for extracting synchronization-free parallelism being represented by dependent statement instances of an arbitrarily nested loop. Presented algorithms can be applied to both uniform and non-uniform loops. The main advantage is that more synchronization-free parallelism may be extracted than that yielded by existing techniques. Our approach, based on operations on relations and sets, requires exact dependence analysis, such as the one by Pugh and Wonnacott, where dependences are found in the form of tuple relations. Results of experiments with the NAS benchmark are presented.

Keywords: loop transformations, arbitrarily nested loops, synchronization-free parallelism, slicing.

1 Introduction

Finding synchronization-free parallelism available in loops is of great importance for parallel and distributed computing, enhancing code locality, and reducing memory requirements. It is important that a compiler is able to find parallelism that does not require synchronization when loops are executed in shared memory multiprocessors or does not require communication when they are executed in distributed systems. The purpose of synchronization-free parallelism, however, is not limited to this, since it may also reduce memory requirements and increase performance on a uniprocessor system by enhancing data locality. Numerous techniques have been developed to extract synchronization-free parallelism available in loops, for example, [1,2,4,13,14,16,18,20,21,28]. However, all well-known techniques have limitations that prevent extracting the entire synchronization-free parallelism available in arbitrarily nested loops.

The goal of this paper is to present an approach which permits us to extract more synchronization-free parallelism available in arbitrarily nested loops than that extracted by well-known techniques as well as to define research problems still to be resolved in order to utilize the entire power of this approach.

O. Gervasi et al. (Eds.): ICCSA 2008, Part II, LNCS 5073, pp. 871–886, 2008.

2 Background

A loop is called *perfectly nested* if all its statements are comprised within the innermost nest. Otherwise the loop is called *imperfectly nested*. An *arbitrarily nested* loop can be both perfectly and imperfectly nested. Given a nest of n loops, the iteration vector I of a particular iteration of the innermost loop is a vector of integers that contains the iteration numbers for each of the loops in order of nesting level [1].

We refer to a particular execution of a statement T for a certain iteration I of the loops, which surround this statement, as an operation T(I) or an instance of statement T at iteration I.

Two operations T1(I) and T2(J) are *dependent* if both access the same memory location and if at least one access is a write. T1(I) and T2(J) are called the *source* and *destination* of a dependence, respectively. An *operation space* is a set of all operations that are executed by a loop.

In this paper, we deal with *affine loop nests*, i.e., loops where 1) for given loop indices, lower and upper bounds as well as array subscripts and conditionals are affine functions of surrounding loop indices and possibly of structure parameters (i.e., parameterized loop bounds), and 2) the loop steps are known positive constants.

The approach presented in this paper requires an exact representation of both loop independent and loop-carried dependences, and consequently an exact dependence analysis which detects a dependence if and only if it actually exists[1]. In this paper, we refer to a dependence to mean a loop-independent or loop-carried dependence. In general, any known technique, extracting exact dependences, can be applied to implement our algorithm. But the description of algorithms and the fulfillment of experiments depend, respectively, on the presentation of exact dependences and on available tools for carrying out a dependence analysis. To describe and carry out experiments with our algorithms, we chose the dependence analysis proposed by Pugh and Wonnacott [23] where dependences are represented with dependence relations comprised of Presburger formulas, which can be built up out of linear constraints over integer variables, logical connectives, and universal and existential quantifiers. This analysis is implemented in Petit, a research tool for doing dependence analysis and program transformations, developed by the Omega Project (http://www.cs.umd.edu/projects/omega).

In this paper, we distinguish between a dependence that is a pair of dependent operations (source and destination) and a dependence *relation* that represents all dependences among operations of a statement (self dependences) or of a pair of statements. To expose all the sources and destinations of dependences yielded with a dependence relation R, we must compute the domain and range of R, respectively.

We distinguish between a dependence graph representing all the dependences among loop operations and a reduced dependence graph being composed by vertices for each statement s_i, $1 \le i \le r$, of the loop and edges joining vertices according to dependence relations $R_{i,j}$, $i,j \in [1,r]$, being exposed by an exact dependence analysis, where r is the number of statements within the loop body.

---

[1] A non-exact representation of dependences is also possible, but this will cause the loss in some parallelism because of the over-approximation of dependences, while this work aims at extracting maximal synchronization-free parallelism.

An *ultimate dependence source(destination)* is a source(destination) that is not the destination(source) of another dependence. Ultimate dependence sources and destinations represented by relation R can be found by means of the following calculations: (domain R – range R) and (range R – domain R), respectively.

The algorithm presented in this paper deals with a strongly connected component (SCC) – a maximal subset of vertices and edges of a reduced dependence graph where for every pair of vertices there exists a directed path.

3 Iteration Space Slicing

Iteration space slicing [23] takes dependence information as input to find all statement instances that must be executed to produce the correct values for the specified array elements. In this paper, we introduce a specific definition of slice as follows.

Definition 1. Given a dependence graph, D, defined by a set of dependence relations, S, a *slice* is a weakly connected component of graph D, i.e., a maximal subgraph of D such that for each pair of vertices in the subgraph there exists a directed or undirected path.

If there exist two or more slices in D, then taking in the account the above definition, we may conclude that all slices are *synchronization-free*, i.e., there is no dependence between them.

Definition 2. The *source(s) of a slice* is the ultimate dependence source(s) that this slice comprises.

We use standard operations on relations and sets, such as intersection ($\cap$), union ($\cup$), difference (–), domain (dom R), range (ran R), composition of relations ($\circ$), relation application (given a relation R and set S, $R(S):=\{ [e'] \mid \exists e \in S, e \rightarrow e' \in R\}$), positive transitive closure (R+), transitive closure ($R^* = R^+ \cup I$, where I is the identity relation). These operations are described in detail in [19].

Input: A set of dependence relations $\{R_{i,j}\}$ describing direct dependences between each pair of statements i,j in an SCC /* $R_{i,j}$ can be empty if a dependence analysis does not extract direct dependences between statements i and j */

```
foreach statement r
    foreach statement p
        foreach statement q
            R_{p,q} = R_{p,q} ∪ R_{r,q} o (R_{r,r})* o R_{p,r} ;
```

$$R_{p,q} = R_{p,q} \cup R_{r,q} \circ (R_{r,r})^* \circ R_{p,r};$$

Output: After running the algorithm, each $R_{i,j}$ (in our algorithm we denote it as $\overline{R}^{+}_{i,j}$) describes all transitive dependences between statements i and j in the SCC.

Fig. 1. Modified Floyd-Warshal algorithm

The algorithm proposed in this paper uses the modified Floyd-Warshal algorithm [17] for calculating all transitive dependences among operations. It starts with the results of an exact dependence analysis, in the form of relations $R_{i,j}$ describing direct dependences between every pair of statements i and j in an SCC. At each iteration of the algorithm, relations $R_{i,j}$ are updated, yielding at the end relations representing all transitive dependences between each pair of statements i and j in the SCC. The algorithm is shown in Figure 1.

4 Extracting Synchronization-Free Slices and Code Generation

In this section, we describe an algorithm to find synchronization-free slices available in an arbitrarily-nested loop, and to generate code to scan slices and iterations of each slice in lexicographical order. Given precomputed dependence relations for a loop, our approach consists of the following steps.

1. Extract a set comprising all ultimate dependence sources.
2. Extract sources of slices from the set comprising ultimate dependence sources.
3. Find operations belonging to each synchronization-free slice.
4. Generate code scanning synchronization-free slices and operations of each slice in lexicographical order.

Our algorithm deals only with one strongly connected component (SCC). Hence, before applying it, a reduced dependence graph has to be transformed into its strongly connected component graph. If there exist two or more SCCs in a loop, then code is first generated independently for each SCC, and next resulting code is formed from the code generated taking into account the topological order of SCCs in the reduced dependence graph.

Algorithm 1, Part 1: Sources extraction

Input:

- a set $S := \{R_{i,j} \mid i, j \in [1,q]\}$ of dependence relations representing an SCC, where values of i, j represent the statement identifiers numbered in the order in which statements appear in the source text; each $R_{i,j}$ denotes the union of all the relations describing dependences between statements i and j; q is the number of vertices in the SCC;
- the dimension of a loop, n.

Output

- set Sources, composed of the (lexicographically minimal) sources of slices.

Method:

begin

1. **foreach** relation $R_{i,j} \in S$ **do**

1.1. Transform relation $R_{i,j}$ so that each its input and output tuples has exactly n elements by inserting the value -1 at the rightmost positions of tuples, e.g., replace a tuple $[e] = [e_1\ e_2\ ..\ e_{n-k}]$, where k is some integer, for the tuple $[e_1\ e_2\ ...\ e_{n-k}\ \underbrace{-1\ \ -1\ ....-1}_{k}]$.

1.2. Extend the input and output tuples of $R_{i,j}$ with additional objects representing identifiers of statements i and j, respectively, i.e., transform $R_{i,j} := \{[e] \rightarrow [e']\}$ into $R_{i,j} := \{[e,i] \rightarrow [e',j]\}$.

2. Find a set, UDS, containing ultimate dependence sources as the difference between the union of domains and the union of ranges of all the relations in S

$$UDS := \bigcup_{R_{i,j} \in S} \text{dom } R_{i,j} - \bigcup_{R_{i,j} \in S} \text{ran } R_{i,j} .$$

3. Calculate the exact transitive closure, R*, representing all the transitive dependences in the SCC, by applying the modified Floyd-Warshal algorithm to calculate relations, $\overline{R}^{+}_{i,j}$, representing all transitive dependences between each pair of statements i, j in the SCC (note: $\overline{R}^{+}_{i,j}$ should not be confused with $R^{+}_{i,j}$ being the exact positive transitive closure of a single dependence relation describing dependences between statements i, j) and then by computing R* as follows:

$$R^* = \bigcup_{1 \le i, j \le q} (\overline{R}^{+}_{i,j}) \cup I ,$$ where I is the identity relation.

4. Form a relation, R_USC, representing all pairs of ultimate dependence sources that are connected (by an undirected path) in the dependence graph formed on the basis of the SCC:

$R\_USC := \{[e] \rightarrow [e'] \mid e, e' \in UDS,\ e' \succ e,\ (R^*(e')) \cap (R^*(e)) \neq \varnothing \}$.

5. Form set, Sources, comprising the (lexicographically minimal) sources of slices

Sources := UDS – ran R_USC.

end

If R*, calculated in step 3 of the above algorithm, is described with affine forms, then set Sources, produced in step 5, is also affine, and the following algorithm can be applied for code generation.

Algorithm 1, Part 2: Code generation

Input:

- Set Sources, represented with affine forms, as produced by Algorithm 1, Part 1.
- Exact transitive closure R*, as produced by step 3 of Algorithm 1, Part 1.
- Relation R_USC, as produces by step 4 of Algorithm 1, Part 1.

Output:

- Code scanning synchronization-free slices and operations of each slice in lexicographical order for the SCC.

Method:

```
begin
   genLoops (in: Sources; out: OuterLoops, L_I);
   foreach I in L_I do
      S_Slice := R* (R_USC* (I)) /* note that if R_USC=∅ , then R_USC*(I) =I */
      genLoops (in: S_Slice; out: InnerLoops, L_J);
         foreach J in L_J do
               genLoopBody (in: OuterLoops, InnerLoops, J; out: LoopBody);
end
```

where

- *genLoops (in: OperSet; out: Loops, VectorList)* generates set, *Loops,* of loop nests to scan operations comprised in set *OperSet*, and returns a list of corresponding parameterized iteration vectors *VectorList* comprising the loops index values. Because set *Sources* is described with affine forms, any well-known technique, such as [3,6,8,25,29], can be applied to implement the function *GenLoops*.
- *genLoopBody (in: OuterLoops, InnerLoops, Iter; out: LoopBody)* generates body *LoopBody* of *InnerLoops* containing the statements of a source loop to be executed at iteration *Iter* and inserts loops *InnerLoops* in the corresponding nest of loops *OuterLoops*. To generate *LoopBody*, the last elements of tuples representing set *InnerLoop* are taken into consideration (they represent the statement identifiers and allow for choosing appropriate statements to be inserted in the loop body).

Note that outermost loops generated by the above algorithm can be executed in parallel because they scan independent sources of synchronization-free slices.

Our paper deals only with extracting dependent operations belonging to slices. In order to scan independent operations, additional code should be added that can be generated as follows. For each statement S_i, $1 \le i \le n$, executed by a source loop, we calculate a set of independent operations, S_IND_i, as the difference between the operation space of this statement, OS_i, and the set containing all the dependent operations yielded by S_i, that is,

$$S\_IND_i := OS_i - \Big(\bigcup_{R_{i,k} \in S} \text{domain } R_{i,k} \cup \bigcup_{R_{k,i} \in S} \text{range } R_{k,i} \Big), \text{ where } 1 \le k \le n.$$

Because all constraints of S_IND_i are affine (for the same reason as for set *Sources*), we can apply any well-known technique to generate loops scanning elements of each set S_IND_i, $1 \le i \le n$. Obviously, these loops can be executed in parallel.

To illustrate Algorithm 1, let us consider the following example.

Example 1

```
for (i=1; i<=n; i++) {
s1:    b(i,i)=a(i-3,i);
       for j=1; j<=n; j++)
s2:          a(i,j)=a(i,j-1)+b(i,i); }
```

Petit extracts the following dependence relations for this example

$R_{1,2} := \{[i] \rightarrow [i,j] : 1 <= i <= n \ \&\& \ 1 <= j <= n\};$

$R_{2,1} := \{[i,i+3] \rightarrow [i+3] : 1 <= i <= n-3\};$

$R_{2,2} := \{[i,j] \rightarrow [i,j+1] : 1 <= i <= n \ \&\& \ 1 <= j < n\}.$

Figure 2 shows the SCC described with relations $R_{1,2}$, $R_{2,1}$, and $R_{2,2}$.

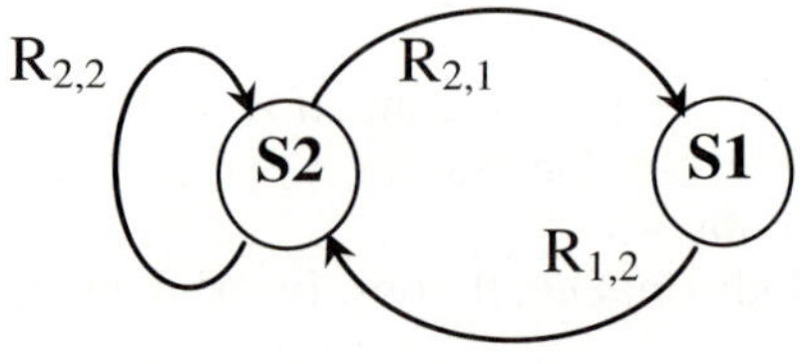

Fig. 2. SCC represented with relations $R_{1,2}$, $R_{2,1}$, $R_{2,2}$.

Figure 3 represents dependences for the loop of Example 1 when n=6. It is obvious from the figure that there exist the three independent slices: slice *One* with the source in s1(1), slice *Two* with the source in s1(2) and slice *Three* with the source in s1(3).

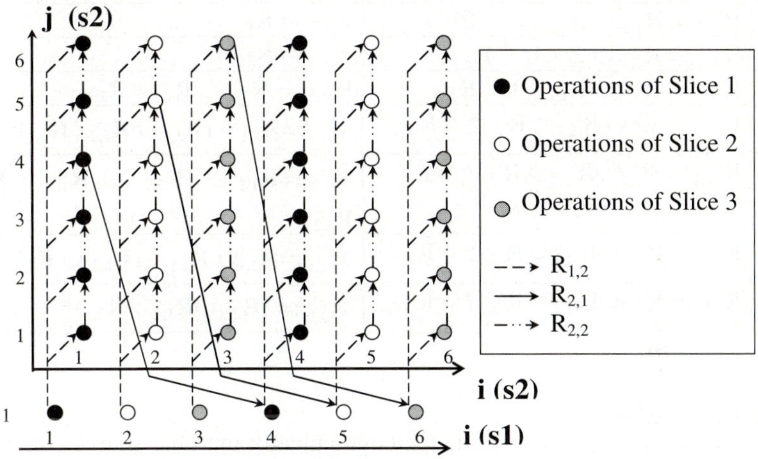

Fig. 3. Dependences for Example 1 when n=6

For this example, there does not exist any affine transformation extracting two or more slices. Our approach can be applied to this example as follows.

<u>**Algorithm 1. Part 1**</u>

1. $R_{1,2}$:= {[i,**-1**,**1**] → [i,j,**2**] : 1 <= i <= n & 1 <= j <= n};
   $R_{2,1}$:= {[i,i+3,**2**] → [i+3,**-1**,**1**] : 1 <= i <= n-3};
   $R_{2,2}$:= {[i,j,**2**] → [i,j+1,**2**] : 1 <= i <= n & 1 <= j < n}.
2. UDS := {[i,-1,1]: 1 <= i <= n, 3}.
3. In order to compute the exact transitive closure R*, we first find relations $\overline{R}^+_{1,1}$, $\overline{R}^+_{1,2}$, $\overline{R}^+_{2,1}$, $\overline{R}^+_{2,2}$ according to the modified Floyd-Warshal algorithm (see Table 1).

Next, we compute R* as follows:

$R^* := \overline{R}^+_{1,1} \cup \overline{R}^+_{1,2} \cup \overline{R}^+_{2,1} \cup \overline{R}^+_{2,2} \cup I$ = {[i,-1,1] →[i',-1,1]: ∃ (alpha: i' = i+3alpha & 1≤i≤i'-3 & i'≤n)} ∪{[i,-1,1] → [i,j',2] : 1≤i≤n & 1≤j'≤n & 4≤n} ∪ {[i,-1,1] → [i',j',2] : ∃ (alpha : i+3alpha = i' & 1≤i≤i'-3 & 1≤j'≤n & i'≤n)} ∪ {[i,-1,1] →[i,j',2] : j', i≤n≤3 & 1≤i & 1≤j'} ∪ {[i,i+3,2] →[i+3,-1,1] : 1≤i≤n-3} ∪ {[i,j,2] →[i+3,-1,1] : 1,j-2≤i≤ n-3 & 1≤j} ∪ {[i,j,2] -> [i,j',2]: 1≤j≤j'≤n & 1≤i≤n} ∪ {[i,j,In_3] →[i,j,In_3] }.

4. R_USC :=∅.
5. Sources := UDS – R_USC = {[i,-1,1]: 1 <= i <= min(n,3)}.

Table 1. Results of iterations of the Floyd-Warshal algorithm

r	p	q	Results of iterations of the Floyd-Warshal algorithm	Simplified results
1	1	1	$R'_{1,1} := \varnothing \cup \varnothing \circ U \circ \varnothing$	$R'_{1,1} := \varnothing$
1	1	2	$R'_{1,2} := R_{1,2} \cup R_{1,2} \circ U \circ \varnothing$	$R'_{1,2} := R_{1,2}$
1	2	1	$R'_{2,1} := R_{2,1} \cup \varnothing \circ U \circ R_{2,1}$	$R'_{2,1} := R_{2,1}$
1	2	2	$R'_{2,2} := R_{2,2} \cup R_{1,2} \circ U \circ R_{2,1}$	$R'_{2,2}' := R_{2,2} \cup R_{1,2} \circ R_{2,1}$
2	1	1	$R''_{1,1} := \varnothing \cup R'_{2,1} \circ R'_{2,2}{}^* \circ R'_{1,2}$	$\mathbf{\overline{R}^+_{1,1} := R_{2,1} \circ (R_{2,2} \cup R_{1,2} \circ R_{2,1})^* \circ R_{1,2}}$
2	1	2	$R''_{1,2} := R'_{,2}' \cup R'_{2,2} \circ R'_{2,2}{}^* \circ R'_{1,2}$	$\mathbf{\overline{R}^+_{1,2} := R_{1,2} \cup (R_{2,2} \cup R_{1,2} \circ R_{2,1})^* \circ R_{1,2}}$
2	2	1	$R''_{2,1} := R'_{2,1} \cup R'_{2,1} \circ R'_{2,2}{}^* \circ R'_{2,2}$	$\mathbf{\overline{R}^+_{2,1} := R_{2,1} \cup R_{2,1} \circ (R_{2,2} \cup R_{1,2} \circ R_{2,1})^*}$
2	2	2	$R''_{2,2} := R'_{2,2} \cup R'_{2,2} \circ R'_{2,2}{}^* \circ R'_{2,2}$	$\mathbf{\overline{R}^+_{2,2} := (R_{2,2} \cup R_{1,2} \circ R_{2,1})^*}$

Algorithm 1. Part 2

- Generate outer loops to scan the lexicographically minimal sources of synchronization-free slices and being represented with operations of statement s1.

```
parfor (t1= 1; t1<= min(n,3); t1++) {
        s1(t1,-1,1); }
```

  List L_I contains vector I equal to (t1,-1,1)'.

- Generate inner loops to enumerate operations belonging to the slice with the source represented by the vector I=(t1,-1,1)' and being contained in the set S_Slice := R* (R_USC* (I)) ={[i,-1,1]: ∃ (alpha : i = t1+3alpha & 1≤t1≤i-3 & i≤n)} ∪ {[t1,j,2]: 1≤t1≤n & 1≤j≤n & 4≤n}∪{[i,j,2]:∃ (alpha : i = t1+ 3alpha & 1≤t1≤i-3 & 1≤j≤n & i≤n)} ∪ {[t1,j,2]: 1≤t1≤n≤3 & 1≤j≤n} ∪ {[i,-1,1]: i = t1}.

```
s(t1,-1,1);
for(t2 = 1; t2 <= n; t2++)
        s(t1,t2,2);
for(t3 = t1+3; t3<= n; t3 += 3) {
        s(t3,-1,1);
        for(t2 = 1; t2 <= n; t2++)
                s(t3,t2,2); }
```

- Generate the body of inner loops containing statements of the source loop body to be executed at iteration J, and insert it as the body of outer loops.

```
parfor (t1 = 1; t1 <= min(n,3); t1++) {
        b(t1,t1)=a(t1-3,t1);
        for(t2 = 1; t2 <= n; t2++)
                a(t1,t2)=a(t1,t2-1)+b(t1,t1);
        for(t3=t1+3; t3<= n; t3+= 3) {
                b(t3,t3)=a(t3-3,t3);
                for(t2 = 1; t2 <= n; t2++)
                        a(t3,t2)=a(t3,t2-1)+b(t3,t3); }}
```

Note that the outermost loops can be run in parallel (this is denoted with the keyword "parfor") because they scan independent sources of synchronization-free slices.

5 Extracting Slices in Inner Nests

When Algorithm 1 extracts only a single slice in the whole loop domain, the following algorithm can be used to seek for slices in inner nests of a loop.

Algorithm 2
Input:

- a source loop;
- set S_1 comprising indices $i_1, i_2, \ldots, i_n$ representing the loop nests, where n is the number of the loop indices.

Output:

- code whose inner loops scan slices (when the algorithm extracts such slices) while outer loops are responsible for synchronization.

Method:
begin

1. $S_2 := \varnothing$, k:=1.
2. Extract dependence relations for the loop whose nests are defined with indices being comprised in set S_1 provided that the indices contained in set S_2 are declared as symbolic constants; form set S containing extracted dependence relations and run Algorithm 1.
3. **if** Algorithm 1 extracts a single slice **then**
   remove index i_k from set S_1 and insert it in set S_2
   if S_1 is the empty set **then** the end, the algorithm extracts a single slice.
   else k=k+1; go to step 2

   else
   form resulting code consisting of the first k-1 outer nests of the source loop (their indices are contained in S_2) and the inner nests representing the code enumerating slices and being generated with Algorithm 1.

end

Note that in a loop generated by Algorithm 2, the first k-1 outer nests are sequential (they enumerate the loop sub-domains where slices are extracted), nest k together with the first k-1 outer nests enumerates sources of slices (it can be run in parallel), and the rest n-k+1 nests enumerate operations of a slice defined with the first k nests. Let us illustrate Algorithm 2 by means of the following loop.

Example 2

```
for(i = 1; i <= n; i++)
      for(j = 1; j <= n; j++)
            a(i,j)=a(i-1,j)+a(i,j-3)+a(i-1,j-1);
```

Petit extracts the following dependence relations for this loop:

$R_1 = \{[i,j] \rightarrow [i+1,j] : 1 \le i < n \ \& \ 1 \le j \le n\}$,
$R_2 = \{[i,j] \rightarrow [i,j+3] : 1 \le i \le n \ \& \ 1 \le j \le n\text{-}3\}$,
$R_3 = \{[i,j] \rightarrow [i+1,j+1] : 1 \le i < n \ \& \ 1 \le j < n\}$.

Figure 4 represents dependences for the loop of Example 2 described with the above relations when n=6.

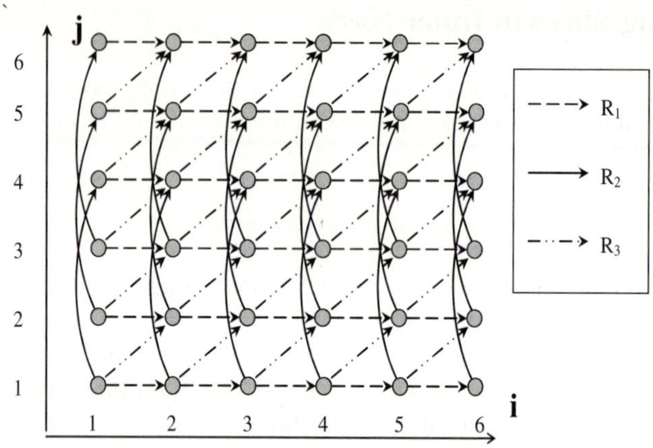

Fig. 4. Dependences for the loop of Example 2 when n=6

For this set of dependence relations, Algorithm 1 extracts a single slice with multiple sources (1,1), (1,2), and (1,3). According to Algorithm 2, we now have to extract dependence relations for the loop

```
for (j=1; j<=n; j++)
  a(i,j)=a(i-1,j)+a(i,j-3)+a(i-1,j-1);
```

where i is declared as a symbolic constant.

Under this constraint, Petit extracts the following dependence relation

$R'_1=\{[j] \rightarrow [j+3] : 1 \leq j \leq n\text{-}3\}$.

Now Algorithm 1 extracts the sources of slices comprised in the set Sources := {[j]: 1≤j≤min(n-3,3)}. After executing step 3 of Algorithm 2, we yield the following nested loop

```
for (i=1; i<=n; i++)
   parfor(j' = 1; j' <= min(n-3,3); j'++)
       for(j = j'; j <= n; j += 3)
           a(i,j)=a(i-1,j)+a(i,j-3)+a(i-1,j-1);
```

For each value of i, the loop above scans 3 slices (if n>=6).

To the best of our knowledge, other well-known techniques fail to extract synchronization-free slices for this example.

6 Experiments

To evaluate the scope of the applicability of the proposed approach, we have developed a tool that uses the Omega Project Software (Petit – for exposing dependence relations and the Omega calculator – for the calculations envisaged by our algorithms and code generation) and implements the algorithms presented in this paper. Using this tool, we have carried out experiments with the NAS Parallel Benchmark (NPB)

Table 2. Results of experiments with some parallel NPB loops

Loop description	**P**	**speedup**	**efficiency**	**upper bounds**
Bt_error, No 2	2	1,573	**0,787**	N1=512
Algorithm 2	4	1,729	**0,432**	N2=512
5 slices if (N3 ≥ 1)	8	1,841	**0,230**	N3=512
	16	1,428	**0,089**	
ft_auxfnct, No 2	2	1,592	**0,796**	N1=512;
Algorithm 1	4	1,848	**0,462**	N2=512;
N1*N2*N3 slices	8	2,362	**0,295**	N3=512;
if (N3 ≥ 1 & N2 ≥ 1)	16	2,208	**0,138**	
LU_HP_l2norm, No 2	2	2,709	**1,355**	N1=512 ;
Algorithm 2	4	3,259	**0,815**	N3=512 ;
5 slices if (N4 ≤ N5−1)	8	3,466	**0,433**	N5=512 ;
	16	2,822	**0,176**	N2=N3=32;
UA_adapt, No 1	2	4,887	**2,444**	N=256;
Algorithm 1	4	9,677	**2,419**	
2*N1*N6*N7+4*N1*N6*N7 sl.	8	19,008	**2,376**	
if (N3,N4,N6,N7≥1;N2,N5≥2)	16	35,201	**2,200**	
UA_diffuse, No 3	2	14,375	**7,187**	N1=512;
Algorithm 2	4	25,571	**6,393**	N2=512;
N1*N3*N4 slices	8	49,832	**6,229**	N3=512;
if(N4≥1&N3≥1&N2≥2)	16	92,865	**5,804**	
UA_precond, No 4	2	0,941	**0,470**	N1=512;
Algorithm 2	4	1,539	**0,385**	N2=512;
N1*(N2-1) slices	8	15,248	**1,906**	N3=512;
if (N3≥2 & N2≥2)	16	29,960	**1,872**	
UA_setup, No 16	2	3,581	**1,790**	N1=512;
Algorithm 1	4	7,662	**1,916**	N2=512;
N1*N2 slices	8	0,201	**0,025**	N3=512;
if (N3≥2 & N2≥1)	16	0,060	**0,004**	
UA_transfer, No 11	2	1,107	**0,553**	N1=512;
Algorithm 1	4	2,148	**0,537**	N2=512;
N1*(N2-1) slices	8	22,959	**2,870**	N3=512;
if (N3≥2 & N2≥2)	16	40,264	**2,517**	
_BT_rhs, No 1	2	5,556	**2,778**	N1=512;
Algorithm 1	4	10,940	**2,735**	N2=512;
(N1+1)(N2+1)(N3+1) slices	8	10,204	**1,275**	N3=512;
if (N3≥0 & N2≥0)	16	5,875	**0,367**	
_SP_ninvr, No 1	2	1,352	**0,676**	N1=512;
Algorithm 1	4	1,397	**0,349**	N2=512;
2* (N1*N2*N3) slices	8	1,386	**0,173**	N3=512;
if (N3≥1 & N2≥1)	16	1,166	**0,073**	

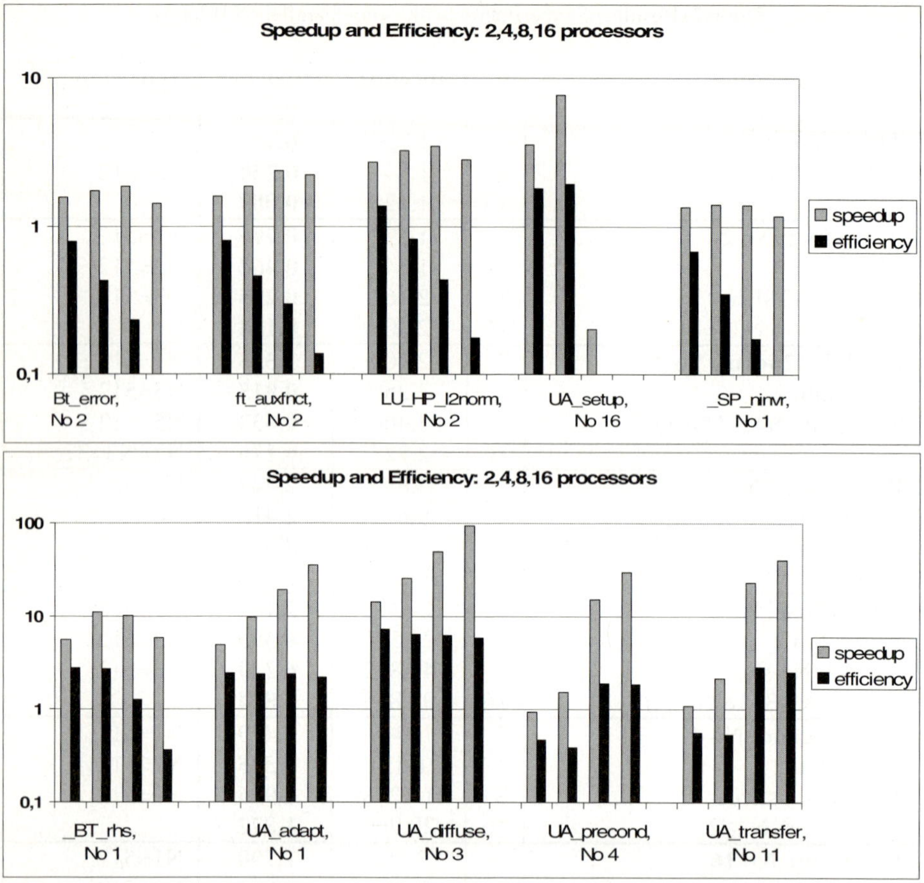

Fig. 5. Speed-up and efficiency of some parallel NPB loops (logarithmic scale)

[31]. The loops of NBP are derived from computational fluid dynamics codes and have gained wide acceptance as a standard indicator of supercomputer performance and are used to study highly parallel processor systems.

From all 431 NPB loops, Petit is able to perform a dependence analysis only for 257 loops that gives about 60% of all NPB loops. After extracting dependence relations, we firstly removed as many as possible redundant dependence relations applying well-known techniques. From those 257 loops, 108 loops did not expose any data flow dependence. From the remaining 149 loops, we were able to extract synchronization-free slices for 59 loops that gives ~14% of all NPB loops, ~23% of the loops for which Petit is able to perform a dependence analysis, and ~40% of the NBP loops after removing redundant dependence relations.

To get parallel code in the OpenMP standard, we inserted appropriate pragmas and clauses into code representing slices extracted for 10 computationally heavy loops from NPB. Next, we measured execution time of the parallel programs for the different number of processors and calculated the speedup and efficiency for each parallel program. Let us recall that if T(P) is the time required to complete the task on P processors, the speedup is the ratio S=T(1)/T(P). Efficiency, on the other hand, is a measure of how much of available processing power is being used: E = S/P = T(1) / (P * T(P)).

Experiments were performed in the following environment:

- parallel machine SGI Altix 3700, 128 x Itanium2 1.5GHz (NUMA), 256 GB RAM, HDD 1TB;
- compiler Intel C/C++ (ver. 9.1).

Table 2 and Figure 5 present the results of experiments. The first column of the table contains the loop description, an algorithm applied for extracting slices, and the number of slices extracted. Let us note that the notation "**Bt_error, No 2**" in the loop description means that we deal with the 2nd loop from the file "error.f" taken from the test "Bt" of NPB. The second column of the table contains the number of processors executing a parallel loop presented in the OpenMP standard. The third and the fourth columns present obtained speedup and efficiency of a parallel program executed on P processors. The fifth column presents values of the loop upper bounds.

Analyzing Table 2 and Figure 5, we can see that for 92.5% of all 40 experiments (10 NBP loops executed on 2, 4, 8, and 16 processors), we achieved speedup greater than one and for 45% of those, efficiency is greater than one, meaning that we deal with super-linear speedup. This is possible through extracting synchronization-free slices permitting for considerable improving program locality.

7 Related Work

The results of the paper are within the iteration space slicing framework (ISSF) introduced by Pugh and Rosser [24]. That paper examines one of possible uses of the slicing framework, namely how to optimize interprocessor communication. In particular, it showed how slicing may enable loop fusion, tolerate message latency and allow message coalescing. However, paper [24] does not show how to construct synchronization-free slices. Our contribution to ISSF consists in demonstrating how to extract sources of slices, how to compute iterations belonging to each slice and how to generate code enumerating slices. We are unaware of any work describing extracting synchronization-free slices with ISSF.

The approach described in our paper [7] deals with extracting synchronization-free slices from perfectly nested loops, but it cannot be applied to arbitrarily-nested loops nor can extract slices in the inner loop nests. Also, the current paper presents new experimental results demonstrating the effectiveness of our approach on real-life codes (NAS benchmark).

Unimodular loop transformations [4,28], permitting the outermost loop in a nest of loops to be parallelized, find synchronization-free parallelism. But unimodular transformations do not allow such transformations as loop fission, fusion, scaling, reindexing, or reordering to be applied to extract synchronization-free parallelism.

Paper [14] describes an approach based on Hamiltonian recurrences, extracting the whole synchronization-free parallelism. But this approach is applicable only to uniform non-parameterized loops.

The affine transformation framework (ATF), considered in many papers, for example, [5,10,11,12,13,20,21] unifies a large number of previously proposed loop transformations. Today, it is one of the most powerful frameworks for loop transformations allowing us to extract synchronization-free parallelism presented in loops with both uniform and affine dependences. But it does not extract all synchronization-free slices in the general case of a loop, for example, for the loops of Examples 1 and 2 in this paper.

We believe that the presented technique extracts more slices than those extracted with ATF. For example, for a loop with the body a(i)=a(i–m), where m is integer, ATF extracts a single slice while the presented technique extracts m slices; for the loops of Examples 1 and 2 of our paper, ATF extracts a single slice, while the presented algorithms do extract all three available slices for Example 1 and three slices in each subdomain of the loop from Example 2 defined with its outer nest.

8 Conclusions and Future Work

In this paper, we described an approach to find coarse grained parallelism being represented with slices of operations of imperfectly nested loops. Such slices can exist in the whole loop domain (extracted with Algorithm 1) or in the loop subdomains (exposed with Algorithm 2).

The results of this paper not only aim at a practical usage, but have also a theoretical character. They permit us to answer, among others, the following questions: how many slices are available in a loop? what is the cost to be paid for extracting slices and generating corresponding code? is this code effective? how many synchronization-free slices are not exposed by means of well-known approaches?

A future research direction is to derive more sophisticated techniques for synchronization of the slices execution. One of possible solutions is partitioning a set of relations into two subsets: one subset allowing for the extraction of multiple synchronization-free slices while the second subset is to be used for implementing synchronization among those slices. The choice of the two subsets and the implementation of synchronization depend on various criteria (e.g., the number of slices, size of slices, number of synchronization-points, etc.).

References

[1] Allen, R., Kennedy, K.: Optimizing Compilers for Modern Architectures, p. 790. Morgan Kaufmann, San Francisco (2001)

[2] Amarasinghe, S.P., Lam, M.S.: Communication optimization and code generation for distributed memory machines. In: Proceedings of the SIGPLAN 1993, pp. 126–138 (1993)

[3] Ancourt, C., Irigoin, F.: Scanning polyhedra with do loops. In: Proc. of the Third ACM/SIGPLAN Symp. on Principles and Practice of Parallel Programming, pp. 39–50. ACM Press, New York (1991)

[4] Banerjee, U.: Unimodular transformations of double loops. In: Proceedings of the Third Workshop on Languages and Compilers for Parallel Computing, pp. 192–219 (1990)

[5] Bastoul, C., Cohen, A., Girbal, S., Sharma, S., Temam, O.: Putting polyhedral loop transformations to work. In: LCPC 16 Intern.l Workshop on Languages and Compilers for Parallel Computing. LNCS, vol. 2958, pp. 209–225. College Station (September 2003)
[6] Bastoul, C.: Code Generation in the Polyhedral Model Is Easier Than You Think. In: Proceedings of the PACT 13 IEEE International Conference on Parallel Architecture and Compilation Techniques, Juan-les-Pins, pp. 7–16 (2004)
[7] Beletska, A., Bielecki, W., San Pietro, P.: Extracting Synchronization-Free Slices of Operations in Perfectly-Nested Loops. In: Proceedings of PDCS 2007 (2007)
[8] Boulet, P., Darte, A., Silber, G.A., Vivien, F.: Loop parallelization algorithms: from parallelism extraction to code generation. Parallel Computing 24, 421–444 (1998)
[9] Cohen, A., Girbal, S., Temam, O.: A polyhedral approach to ease the composition of program transformations. In: Danelutto, M., Vanneschi, M., Laforenza, D. (eds.) Euro-Par 2004. LNCS, vol. 3149, pp. 292–303. Springer, Heidelberg (2004)
[10] Darte, A., Robert, Y., Vivien, F.: Scheduling and Automatic Parallelization. Birkhäuser Boston (2000)
[11] Feautrier, P.: Some efficient solutions to the affine scheduling problem, part i, one dimensional time. International Journal of Parallel Programming 21, 313–348 (1992)
[12] Feautrier, P.: Some efficient solutions to the affine scheduling problem, part ii, multidimensional time. International Journal of Parallel Programming 21, 389–420 (1992)
[13] Feautrier, P.: Toward automatic distribution. Journal of Parallel Processing Letters 4, 233–244 (1994)
[14] Gavaldà, R., Ayguadé, E., Torres, J.: Obtaining Synchronization-Free Code with Maximum Parallelism, Technical Report LSI-96-23-R, Universitat Politècnica de Catalunya (1996)
[15] Griebl, M., Lengauer, C.: Classifying Loops for Space-Time Mapping. In: Proceedings of the Euro-Par. LNCS, pp. 467–474. Springer, Heidelberg (1996)
[16] Huang, C., Sadayappan, P.: Communication-free hyperplane partitioning of nested loops. Journal of Parallel and Distributed Computing 19, 90–102 (1993)
[17] Kelly, W., Pugh, W., Rosser, E., Shpeisman, T.: Transitive Closure of Infinite Graphs and its Applications. International Journal of Parallel Programming 24(6), 579–598 (1996)
[18] Kelly, W., Pugh, W.: Minimizing communication while preserving parallelism. In: Proc. of the 1996 ACM International Conference on Supercomputing, pp. 52–60 (1996)
[19] Kelly, W., Maslov, V., Pugh, W., Rosser, E., Shpeisman, T., Wonnacott, D.: The omega library interface guide, Technical Report CS-TR-3445, University of Maryland (1995)
[20] Lim, W., Lam, M.S.: Communication-free parallelization via affine transformations. In: Proc. of the 7th workshop on languages and compilers for parallel computing, pp. 92–106 (1994)
[21] Lim, W., Cheong, G.I., Lam, M.S.: An affine partitioning algorithm to maximize parallelism and minimize communication. In: Proceedings of the 13th ACM SIGARCH International Conference on Supercomputing (1999)
[22] Lim, W., Liao, S.W., Lam, M.: Blocking and Array Contraction Across Arbitrarily Nested Loops Using Affine Partitioning. In: Proceedings of ACM SIGPLAN Symposium on Principles and Practice of Parallel Programming (2001)
[23] Pugh, W., Wonnacott, D.: Constraint-based array dependence analysis. ACM Trans. on Programming Languages and Systems (1998)
[24] Pugh, W., Rosser, E.: Iteration Space Slicing and Its Application to Communication Optimization. In: Proc. of the International Conf. on Supercomputing, pp. 221–228 (1997)
[25] Quillere, F., Rajopadhye, S., Wilde, D.: Generation of efficient nested loops from polyhedra. International Journal of Parallel Programming 28 (2000)

[26] Weiser, M.: Program slices: formal, psychological, and practical investigations of an automatic program abstraction method, PhD thesis, University of Michigan, Ann Arbor, MI (1979)

[27] Weiser, M.: Program Slicing. IEEE Transactions on Software Engineering SE-10(7), 352–357 (1984)

[28] Wolf, M.E.: Improving locality and parallelism in nested loops, Ph.D. Dissertation CSL-TR-92-538, Stanford University, Dept. Computer Science (1992)

[29] Vasilache, N., Bastoul, C., Cohen, A.: Polyhedral code generation in the real world. In: Proceedings of the International Conference on Compiler Construction (ETAPS CC 2006). LNCS, pp. 185–201. Springer, Vienna (2006)

[30] Netlib Repository at UTK and ORNL, `http://www.netlib.org/benchmark/livermorec`

[31] `http://www.nas.nasa.gov`

Simulation Analysis for the Pricing of Bond Option on Arbitrage-Free Models with Jump

Kisoeb Park[1], Moonseong Kim[2], and Seki Kim[1,⋆]

[1] Department of Mathematics, Sungkyunkwan University, Korea
{kisoeb,skim}@skku.edu
[2] Electrical and Computer Engineering
Michigan State University, USA
mkim@msu.edu

Abstract. We derive a closed-form solutions for bond option pricing on arbitrage-free models with jump and perform Monte Carlo simulation for them. We know that the bond option price on arbitrage-free models with jump is humped. The fact shows that valuations of this model estimate highly precision. In this paper, we simulate the proposed models by the Monte Carlo simulation with scenarious. By simulation analysis, we obtain that bond option prices on arbitrage-free models are larger than bond option prices on arbitrage-free models with jump. Lower value of precision in arbitrage-free models with jump correspond to sharper estimates.

Keyword: Arbitrage-free Models with Jump, Bond Option Price, and Simulation.

1 Introduction

Approaches to modeling the term structure of interest rates in continuous time may be broadly described in terms of either the equilibrium approach or the arbitrage-free approach even though some early models include concepts from both approaches. Current research using jump-diffusion processes relies mostly on two classes of models: the affine jump-diffusion class and the quadratic Gaussian[4].

We introduces a new class of arbitrage-free models of the term structure of interest rates that is supported by an economy with habit formation[6]. We study an analytic solution for European option on discount bonds. This leads to a parabolic partial differential equation for the prices of bonds and to models for bonds and many other interest rate derivative products[3]. The first, arbitrage-free models with jump[7] obtained closed form approximate solutions for bond price by the linearization technique. The second, using this solutions, we study solutions of the zero-coupon bond option. The bond option values that are presented in the paper indicate that there can be substantial differences between

⋆ Corresponding author.

O. Gervasi et al. (Eds.): ICCSA 2008, Part II, LNCS 5073, pp. 887–895, 2008.

the values produced by arbitrage-free models and arbitrage-free models with jump[2].

In addition, we introduce the Monte Carlo simulation. One of the many uses of Monte Carlo simulation by financial engineers is to place a value on financial derivatives. Interest in use of Monte Carlo simulation for zero-coupon bond option pricing is increasing because of the flexibility of the methods in handling complex financial instruments. The Monte Carlo simulation lends itself naturally to the evaluation of the European option of bond pricing models with jump represented as expectations. One measure of the sharpness of the point estimate of the mean is mean standard error(MSE).

The structure of the remainder of this paper is as follows. In Section 2, the European option on arbitrage-free models with jump are presented. In Section 3, calculated numerical solutions using Monte Carlo simulation for bond option price on arbitrage-free models with jump. Conclusions and future research are in Section 4.

2 The European Option on Arbitrage-Free Models with Jump

We drive a closed-form solution (CFS) for European options of bond pricing models on arbitrage-free models with jump. Suppose that the term structure is completely determined by the value of the interest rate $r(t)$ and that $r(t)$ follows a jump diffusion process. Let be $V(r, t, T)$ the price at time t of a **discount bond**. A solution of the form:

$$V(r, t, T) = \exp[-A(t, T)r + B(t, T)] \tag{1}$$

can be guessed. In our framework, the yield curve is the same as **term structure of interest rate**, as we only work with zero-coupon bonds.

Now consider a quite different type of random environment. Suppose $N(t)$ represents the total number of extreme shocks that occur in a financial market until time t[2]. We extend the jump-diffusion version of equilibrium single factor model to reflect this time dependence. This leads to the following model for $r(t)$:

$$dr(t) = [\theta(t) - a(t)r(t)]dt + \sigma(t)r(t)^{\beta}dW(t) + JdN(t), \tag{2}$$

where $\theta(t)$ is a time-dependent drift; $\sigma(t)$ is the volatility factor; $a(t)$ is the reversion rate; $dW(t)$ is the standard Wiener process; $dN(t)$ is the Poisson process with intensity rate h. The probability that only one jump happens during the instantaneous period is hdt. Therefore, $dN(t)$ is a Poisson jump process having the following characteristic:

$$dN(t) = \begin{cases} 0 \quad 1 - hdt & : \quad \textit{no jump} \\ 1 \quad hdt & : \quad \textit{otherwise.} \end{cases}$$

We investigate the $\beta = 0$ case is an extension of Vasicek's jump-diffusion model; the $\beta = 0.5$ case is an extension of CIR jump-diffusion model. Under the process specified in equation (2), $r(t)$ is defined as:

$$r(t) = r_0(t)\{r(0) + \int_0^t {r_0}^{-1}(u)\theta(u)du + \int_0^t {r_0}^{-1}(u)\sigma(u)r(u)^{\beta}dW(u)$$
$$+ \sum_{i=1}^{N(t)} {r_0}^{-1}(T_i)J_i\}, \quad (3)$$

where $r_0(t) = \exp\{-\int_0^t -a(u)du\} = e^{-a(t)t}$, T_i is time that the $j - th$ jump happens, $0 < T_1 < T_2 < \cdots < T_N(t) < t$, and $N(t)$ is the number of jumps happening during the period $[0, t]$. It can be shown that the probability density function for $r(t)$. Let $X(t) = e^{-a(t)t}(r(0)+\int_0^t e^{a(u)u}\theta(u)du+\int_0^t \sigma(u)r(u)^{\beta}dW(u))$ and $Y(t) = e^{-a(t)t}\sum_{i=1}^{N(t)} e^{a(t)(T_i)}J_i$, then $r(t) = X(t) + Y(t)$.

Therefore, the conditional expectation and variance of the jump-diffusion process given the current level are

$$\mathbb{E}[r(t)] = e^{-a(t)t}r(0) + \frac{\theta(t)}{a(t)}(1 - e^{-a(t)t}) + \frac{h\mu}{a(t)}(1 - e^{-a(t)t}) \quad (4)$$

and

$$\mathbb{V}ar[r(t)] = \begin{cases} \frac{e^{-2a(t)(e^{a(t)t}-1)^2\theta(t)\sigma(t)^2}}{2a(t)^2} & : \beta = 0 \\ \frac{e^{-2a(t)(e^{a(t)t}-1)^2\theta(t)\sigma(t)^2}}{2a(t)^2} + \frac{h}{2a(t)}(\mu^2 + \gamma^2)(1 - e^{-2a(t)t}) & : \beta = \frac{1}{2}. \end{cases} \quad (5)$$

2.1 The Pricing of Bond on Arbitrage-Free Models with Jump

We set up a riskless portfolio and the jump-diffusion version of Ito's lemma to functions of r and t [5]. We get the partial differential difference bond pricing equation:

$$[\theta(t) - a(t)r(t) - \lambda(t)\sigma(t)r(t)^{\beta}]V_r + V_t + \frac{1}{2}\sigma(t)^2 V_{rr} - rV$$
$$+ hV[-\mu A(t, T) + \frac{1}{2}(\gamma^2 + \mu^2)A(t, T)^2] = 0. \quad (6)$$

Bond price derivatives can be calculated from (1). We omit the details, but the substitution of this derivatives into (6) and equating powers of $r(t)$ yields the following equations for A and B.

Theorem 1. **(The Equation (6) with** $\beta = 0$**)**

$$-\frac{\partial A}{\partial t} + a(t)A - 1 = 0 \quad (7)$$

and

$$\frac{\partial B}{\partial t} - \phi(t)A + \frac{1}{2}\sigma(t)^2 A^2 + h[-\mu A + \frac{1}{2}(\gamma^2 + \mu^2)A^2] = 0, \tag{8}$$

where $\phi(t) = \theta(t) - \lambda(t)\sigma(t)$ and all coefficients is constants. In order to satisfy the final data that $V(r, T, T) = 1$ we must have $A(T, T) = 0$ and $B(T, T) = 0$.

Theorem 2. **(The Equation (6) with** $\beta = 0.5$)

$$-\frac{\partial A}{\partial t} + \psi(t)A + \frac{1}{2}\sigma(t)^2 A^2 - 1 = 0 \tag{9}$$

and

$$\frac{\partial B}{\partial t} - (\theta(t) + h\mu)A + \frac{1}{2}h[(\gamma^2 + \mu^2)A^2] = 0, \tag{10}$$

where $\psi(t) = a(t) + \lambda(t)\sigma(t)$ and all coefficients is constants. In order to satisfy the final data that $V(r, T, T) = 1$ we must have $A(T, T) = 0$ and $B(T, T) = 0$.

2.2 The Pricing of Bond Option on Arbitrage-Free Models with Jump

We derive a closed-form solution for European bond options when the price of the underlying instantaneous interest rate evolve as a discontinuous process. Since the interest rate and is normally distributed, the discount bond prices are lognormally distributed; and so the value of option on discount bonds can be expressed as a variant of the Black-Scholes formula.

We consider the value of European option on a discount bond equations (1). The price of a call option on the T_V - maturity discount bond with exercise price K and maturity $T < T_V$ is given by

$$C = V(r, t, T)\mathbb{E}[\max(\bar{V} - K, 0)], \tag{11}$$

where $\mathbb{E}$ denotes expected value in a world that is forward risk neutral with respect to a zero coupon bond maturing at time T_V;

$$\mathbb{E}[\bar{V}] = V(r, t, T_V)/V(r, t, T). \tag{12}$$

Assuming the bond price is lognormally with the standard deviation of the logarithm of the bond price equal to σ_V, that is, $\bar{V}$ is lognormally with

$$\mathbb{V}ar[\log \bar{V}] = {\sigma_V}^2, \tag{13}$$

where $\mathbb{V}ar[\log \bar{V}]$ are the $\beta = 0$ case is an extension of Vasicek's jump diffusion model; the $\beta = 0.5$ case is an extension of CIR jump diffusion model. By the Black-Scholes-Morton formula the equation (11) becomes

$$C = V(r,t,T)(\mathbb{E}[\bar{V}]N(d_1) - KN(d_2))$$
$$= V(r,t,T_V)N(d_1) - KV(r,t,T)N(d_2), \tag{14}$$

where

$$d_1 = \frac{\log(\mathbb{E}[\bar{V}]/K) + {\sigma_V}^2/2}{\sigma_V}$$
$$= \frac{\log([V(r,t,T_V)/V(r,t,T)]/K) + {\sigma_V}^2/2}{\sigma_V},$$

$$d_2 = d_1 - \sigma_V,$$

and

$$N(x) = \frac{1}{\sqrt{2\pi}} \int_{-\infty}^{x} e^{-x^2/2} dx.$$

The price of a put option on the discount bond is

$$P = V(t,T)\mathbb{E}[\max(K - \bar{V}, 0)]$$
$$= V(t,T)(KN(-d_2) - \mathbb{E}[\bar{V}]N(-d_1)). \tag{15}$$

The resemblance between the bond option pricing formula and the Black-Scholes and Morton formula is obvious. In both cases, the random variables is lognormal, resulting in similar formulas.

3 Simulation Analysis

The Monte Carlo simulation(MCS) is actually a very general tool, and its applications are by no means restricted to numerical integration. In simple and general terms, Monte Carlo simulation may be described as a numerical method based an random sampling. A Monte Carlo simulation of a stochastic process is a procedure for sampling random outcomes for the process. This uses the risk-neutral valuation result. The bond price can be expressed as:

$$V(r_t, t, T) = E_t^Q \left[e^{-\int_t^T r(s)ds} \right], \tag{16}$$

where E_t^Q is the expectations operator with respect to the equivalent risk-neutral measure. Under the equivalent risk-neutral measure, the local expectation hypothesis holds (that is, $E_t^Q \left[\frac{dV}{V}\right]$).

To execute the Monte Carlo simulation, we discrete the equation (2). We divide the time interval $[t, T]$ into m equal time steps of length Δt each. For small time steps, we are entitled to use the discreted version of the risk-adjusted stochastic differential equation (2):

$$r_j = r_{j-1} + [(\theta \cdot t) - (a \cdot t)r_{j-1} - (\lambda \cdot t)(\sigma \cdot t)(r_{j-1} \cdot t)^{\beta}]\Delta t$$
$$+ (\sigma \cdot t)(r_{j-1} \cdot t)^{\beta} \ \varepsilon_j \sqrt{\Delta t} + J_j N_{\Delta t}, \quad (17)$$

where $j = 1, 2, \cdots, m, \varepsilon_j$ is standard normal variable with $\varepsilon_j \sim N(0,1)$, and $N_{\Delta t}$ is a Poisson random variable with parameter $h\Delta t$.

We know that the $\beta = 0$ case is an extension of Vasicek's jump diffusion model; the $\beta = 0.5$ case is an extension of CIR jump diffusion model. We can investigate the value of the bond by sampling n short rate paths under the discrete process approximation of the risk-adjusted processes of the equation (17). The bond price estimate is given by:

$$V(r_t, t, T) = \frac{1}{n} \sum_{i=1}^{n} \exp\left(- \sum_{j=0}^{m-1} r_{ij} \Delta t \right), \quad (18)$$

where r_{ij} is the value of the short rate under the discrete risk-adjusted process within sample path i at time $t + \Delta t$.

We investigate the bond option price on arbitrage-free model with jump. Let $C(t, T, T_V)$ be the time t-value of a European call option maturing at time T on the zero-coupon bond with maturity T_V, where $0 < t < T < T_V$. The current value of a European call option $C(0, T, T_V)$ can be evaluated, under the risk neutral measure as the expected discounted payoff of the option at the option's maturity

$$C(0, T, T_V) = \mathbb{E}_t^Q \left[\exp\left(-\int_0^T r_s ds \right) \max(V(r, 0, T, T_V) - K) | F_0 \right] \quad (19)$$

For simulation based approaches to bond option pricing, we have found that the use of one or the other probability measure does not seem to provide any significant advantage. Here we report the simulations under the risk neutral probability measure. Given the interest rate dynamics under the risk neutral measure, which is the equation (17), the option price is evaluated using formula (19), using the Euler-Maruyama scheme for the integration, as

$$C(0, T, T_V) = \frac{1}{n} \sum_{i=1}^{n} \exp\left(- \sum_{j=0}^{m-1} r_{ij} \Delta t \right) \max(V_i(r, 0, T, T_V) - K). \quad (20)$$

The bond price $V_i(r, 0, T, T_V)$ is computed by the term structure models with jump. The Monte Carlo simulation is clearly less efficient computationally than the numerical method. The precision of the mean as a point estimate is often defined as the half-width of a 95% confidence interval, which is calculated as

$$Precision = 1.96 \times MSE, \qquad (21)$$

where MSE $= \nu/\sqrt{n}$ and ν^2 is the estimate of the variance of bond option prices as obtained from n sample paths of the short rate:

$$\nu^2 = \frac{\sum_{i=1}^{n}\left[\exp\left(-\sum_{j=0}^{m-1} r_{ij}\Delta t\right)\max(V_i(r,0,T,T_V)-K) - C(0,T,T_V)\right]^2}{n-1}$$

Lower values of Precision in Equation (21) correspond to sharper estimates. Increasing the number of n is a brute force method of obtaining sharper estimates. This reduces the MSE by increasing the value of n. For the purpose of simulation, we conduct three runs of 10,000 trials each and divide the year into 365 time steps.

4 Random Experiment Results

We investigate the valuation of European bond options on arbitrage-free models and arbitrage-free models with jump. Experiment 1, 2 plot estimated bond call option prices on arbitrage-free models and arbitrage-free models with jump.

In Experiment 1 and 2, the parameter values are assumed to be $r = 0.05$, $a = 0.5$, $b = 0.05$, $\theta = 0.025$, $\sigma = 0.08$, $\lambda = -0.5$, $\gamma = 0.01$, $\mu = 0$, $h = 10$, $t = 0.05$, $T = T_V - 0.5$, and $T = 10$.(the $K = 0.9$ case is the call option, the $K = 1.2$ case is the put option.)

Interest in use of Monte Carlo simulation for bond pricing is increasing because of the flexibility of the methods in handling complex financial instruments. One measure of the sharpness of the point estimate of the mean is mean standard error. Numerical methods that are known as Monte Carlo methods can be loosely described as statistical simulation methods, where statistical simulation is defined in quite general terms to be any method that utilizes sequences of random numbers to perform the simulation.

Table 1 and 2 (Experiment 3 and 4) investigate the bond option prices by Monte Carlo simulation. In experiment 3 and 4, the parameter values are assumed to be $r = 0.05$, $a = 0.5$, $b = 0.05$, $\theta = 0.025$, $\sigma = 0.08$, $\lambda = -0.5$, $\Delta t = (T - t)/m$, $m = 365$, $n = 10,000$, $\gamma = 0.01$, $\mu = 0$, $h = 10$, $t = 0.05$, $T = 0.5$, and $T_V = 1$.

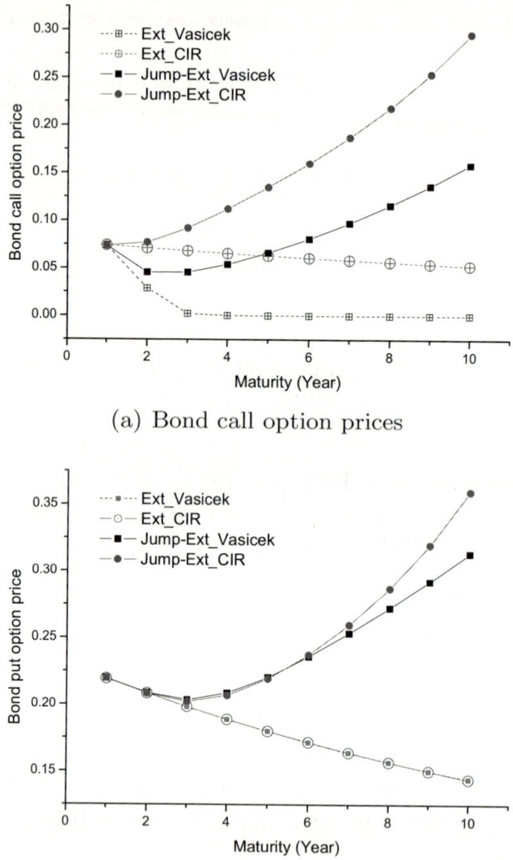

(a) Bond call option prices

(b) Bond put option prices

Fig. 1. [*Experiment 1, 2*] The various bond option prices

Table 1. [*Experiment 3*] Bond call option prices on arbitrage-free models and arbitrage models-free with jump estimated by the Monte Carlo simulation

	Extended Vasicek		Extended CIR	
	No Jump	Jump	No Jump	Jump
CFS	0.0740042	0.0737438	0.0740362	0.0735928
MCS	0.0734676	0.0734272	0.0734537	0.0732046
CFS-MCS	0.000536632	0.000316614	0.000582451	2.610E-07
Variance	2.59E-07	3.38E-09	1.62E-05	1.80E-10
Precision	3.15E-05	3.60E-06	7.61E-04	8.32E-07

Table 2. [*Experiment 4*] Bond put option prices estimated by the Monte Carlo simulation on arbitrage-free models and arbitrage-free models with jump

	Extended Vasicek		Extended CIR	
	No Jump	Jump	No Jump	Jump
CFS	0.219353	0.219604	0.219324	0.219728
MCS	0.219121	0.21899	0.219139	0.218939
CFS-MCS	0.000232267	0.000613734	0.00018485	0.000788546
Variance	2.08E-06	3.21E-08	2.08E-06	1.45E-09
Precision	8.95E-05	1.09E-05	8.94E-05	2.37E-06

5 Conclusions and Future Research

We derive a closed-form solution for European bond option price on arbitrage-free models with jump. This paper develop an approach method to valuing bond options. We study that there is substantial differences between the pricing of bond option produced by arbitrage-free models and arbitrage-free models with jump. In particular, we obtain the European bond option price of arbitrage-free models with jump are increasing as maturity is long. Interest in use of Monte Carlo simulation for bond option pricing is increasing because of the flexibility of the methods in handling complex financial instruments. In Monte Carlo simulation, we know that the bond option prices on arbitrage-free models larger than the bond option prices on arbitrage-free models with jump. Lower values of Precision in arbitrage-free models with jump corresponds to sharper estimates. There are still many problems remaining for further research. For instance, using the Monte Carlo simulation to simulate more complicated the two factor term structure models with jump.

References

1. Ahn, C., Thompson, H.: Jump-Diffusion Processes and the Term Structure of Interest Rates. Journal of Finance 43, 155–174 (1998)
2. Wei, J.Z.: A Simple Approach to Bond Option Pricing. Journal of Futures Markets 17, 131–160 (1997)
3. Duffee, G.R.: Term Premia and Interest Rate Forecasts in Affine Models. Journal of Finance 57, 405–443 (2002)
4. Cheng, P., Scaillet, O.: Linear-Quadratic Jump-Diffusion Modeling. Mathematical Finance 17(4), 575–598 (2007)
5. Dai, Q., Singleton, K.: Expectation Puzzles, Time-varying Risk Premia, and Affine Models of the Term Structure. Journal of Financial Economics 63, 415–441 (2002)
6. Buraschi, A., Jiltsov, A.: Habit Formation and Macroeconomic Models of the Term Structure of Interest Rates. Journal of Finance 62(6), 3009–3063 (2007)
7. Das, S.R., Foresi, S.: Exact solutions for bond and option prices with systematic jump risk. Review of Derivatives Research 1 (2005)

The Two-Criteria Gateways Location and Topology Assignment Problem in Hierarchical WANs: An Exact Algorithm and Computational Results

Przemyslaw Ryba and Andrzej Kasprzak

Wroclaw University of Technology, Chair of Systems and Computer Networks
Wybrzeze Wyspianskiego 27, 50-370 Wroclaw, Poland
{przemyslaw.ryba,andrzej.kasprzak}@pwr.wroc.pl

Abstract. In the paper the problem of designing two-level hierarchical structure of wide area network (WAN) is considered. The goal is to select gateways location, topology of 2nd level network, channel capacities and flow routes in order to minimize linear combination of the total average delay per packet and the total cost of leasing capacities of channels of 2nd level network, subject to budget constraint. An exact algorithm based on branch and bound method is proposed. Some results of computational experiments are reported. Relying on computational experiments, several properties of the considered problem, important from practical point of view, are formulated.

1 Introduction

Nowadays wide-area networks (WAN) and metropolitan-area networks (MAN) support many mission-critical applications requiring high bandwidth, specified level of performance, availability, scalability, affordability, security, and manageability. To meet these needs, difficult network design choices and tradeoffs must be made. Design procedures for huge wide area networks become very costly (from computational point of view) and sometimes infeasible. Thus, they need to be decomposed. Common way to decompose design of huge WAN is designing it as a hierarchical network [1]. In the hierarchical WAN nodes are grouped in clusters on the 1st level of the hierarchy, which in turn are grouped into 2nd level cluster and so on. Communication network of each cluster can be designed separately. In each cluster are selected special communication nodes called "gateways" which provide communication between nodes from different 1st level networks. Traffic between nodes in the same cluster uses paths restricted to local communication network. Traffic between nodes in different 1st level networks is first sent to local gateway, then via 2nd level network of gateways is sent to gateway located in destination 1st level network to finally reach the destination node [1]. Example of the structure of hierarchical WAN is presented in the Figure 1.

In this paper problem of designing two-level hierarchical wide area network interconnecting existing networks in order to minimize the total average delay per packet and total cost of leasing capacities of channels of 2nd level network, subject to budget constraint is considered. The Two-criteria Gateways Location, Capacity and Flow Allocation problem is formulated as follows:

O. Gervasi et al. (Eds.): ICCSA 2008, Part II, LNCS 5073, pp. 896–906, 2008.

given: topology of 1st level networks and 2nd level network, sets of potential gateways locations, set of potential 2nd level network channels and their possible capacities and costs (i.e. discrete cost-capacity function), traffic requirements, budget of the network,

minimize: linear combination of the total average delay per packet and leasing cost of channel capacities of 2nd level network,

over: gateways locations, 2nd level network topology, 2nd level network channel capacities, multicommodity flow (i.e. routing),

subject to: multicommodity flow constraints, channel capacity constraints, budget constraint.

Discrete cost-capacity function considered here is justified for the reason that channels capacities can be chosen from the sequence defined by ITU-T (International Telecommunication Union – Telecommunication Standardization Section) recommendations. Such formulated problem is NP-complete. Short proof of NP-completeness of the problem is presented in the section 3 of the paper.

Some algorithms for hierarchical network design can be found in literature [4, 5, 6]. In the paper [4] problem limited to tree topology of hierarchical network is presented. Algorithm for router location which includes simplified topology assignment without allocating capacities to channels, is presented in the paper [5]. Also, in the paper [6] presenting algorithm for interconnecting two WANs – assigning capacities to channels is not considered. In our earlier papers we presented algorithms for solving problem minimizing average packet delay in hierarchical network [7] and problem with total cost of leasing channel capacities as a criterion [8, 9]. Problem considered in this paper joins these two problems. We consider problem of simultaneously minimizing total network cost and average packet delay to help in de-sign of large WAN satisfying quality demands. We apply linear combination of leasing cost of channel capacities of 2nd level network and total average packet delay in the network as a criterion function. Weight of components of the sum can be adjusted to meet specific design requirements.

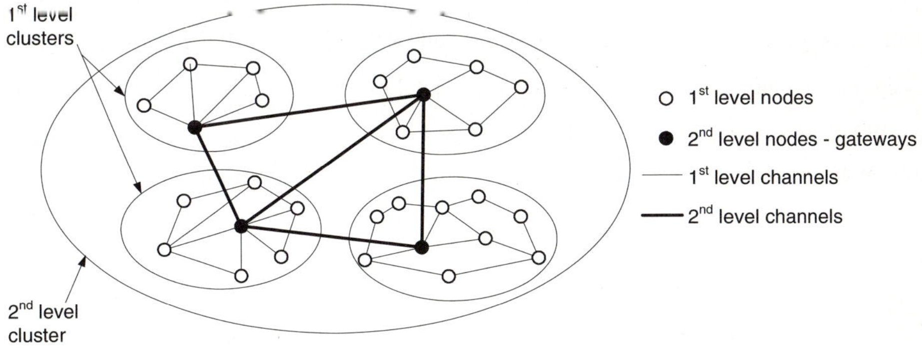

Fig. 1. Structure of two-level hierarchical wide area network [9]

2 Problem Formulation

Consider a network with n nodes grouped in K clusters. The communication network of each cluster (1st level networks) is denoted by S_1^l, $l=1,...,K$. Let N_1^l be set of nodes and L_1^l set of channels in the network S_1^l. In order to set up 2nd level network S_2 connecting 1st level networks, in each network S_1^l given number of gateways must be located. Let y_a^g be the discrete variable for choosing one of available locations for gateway g defined as follows: $y_a^g=1$, if gateway g is located in node a and $y_a^g=0$, otherwise. Each gateway must be placed in exactly one node, thus it is required to satisfy following condition:

$$\sum_{a\in J_g} y_a^g = 1, \quad g\in H^l, \quad l=1,...,K \tag{1}$$

where H^l is a set of gateways to be placed in network S_1^l and J_g is a set of possible locations for gateway g. Let $Y^g=\{y_1^g,...,x_{J_g}^g\}$ be the set of variables y_a^g corresponding to the gateway g.

Let m be number of potential channels between gateways and p total number of channels in hierarchical network. For each potential channel i there is the set $\overline{C}^i=\{c_1^i,...,c_{s(i)-1}^i\}$ of alternative values of capacities from which exactly one must be chosen if the channel was chosen to build the 2nd level network. Let d_k^i be the cost of leasing capacity c_k^i [€/month]. Let $c_{s(i)}^i=0$ for $i=1,...,m$. Then, $C^i=\overline{C}^i\cup\{c_{s(i)}^i\}$ is the set of alternative capacities from among which exactly one must be used to channel i. If the capacity $c_{s(i)}^i$ is chosen, then the channel i is not used to build the network. Let x_k^i be the discrete variable for choosing one of available capacities for channel i defined as follows: $x_k^i=1$, if the capacity c_k^i is assigned to channel i and $x_k^i=0$ otherwise. Let $X^i=\{x_1^i,...,x_{s(i)}^i\}$ be the set of variables x_k^i corresponding to the i-th channel. Since exactly one capacity value from the set C^i must be chosen for channel i, the following condition must be satisfied:

$$\sum_{k=1}^{s(i)} x_k^i = 1 \quad \text{for} \quad i=1,...,m \tag{2}$$

Let d_k^i be the cost of leasing capacity c_k^i for channel i defined as follows:

$$d_k^i = \sum_{a\in J_{g1}} \sum_{b\in J_{g2}} y_a^{g1} y_b^{g2} d_{ab}^{ik} \tag{3}$$

where g_1 and g_2 are gateways adjacent to channel i, d_{ab}^{ik} is cost of leasing capacity c_k^i of channel i between gateway g_1 located in node a and gateway g_2 located in node b. Let $X_r^{'}$ be the permutation of values of variables x_k^i, $i = 1,..., m$ for which the condition (2) is satisfied, and let X_r be the set of variables which are equal to 1 in $X_r^{'}$. Similarly, let $Y_r^{'}$ be the permutation of values of all variables y_a^g for which condition (1) is satisfied and let Y_r be the set of variables which are equal to 1 in $Y_r^{'}$. The pair of sets (X_r, Y_r) is called a selection. Each selection (X_r, Y_r) determines locations of gateways and channels capacities in the 2nd level of hierarchical network. Let $\Re$ be the family of all selections.

Let $d(X_r, Y_r)$ be the total cost of leasing capacities of channels of 2nd level network, in which values of channels capacities are given by X_r and locations of gateways are given by Y_r:

$$d(X_r, Y_r) = \sum_{x_k^i \in X_r} x_k^i d_k^i \tag{4}$$

Let $T(X_r, Y_r)$ denote total average delay per packet in two-level hierarchical network, given by Kleinrock's formula [10] and $Q(X_r, Y_r)$ be linear combination of the total average delay per packet and the cost of leasing capacities of channels of 2nd level network:

$$Q(X_r, Y_r) = \alpha \cdot T(X_r, Y_r) + d(X_r, Y_r) \tag{5}$$

where positive coefficient α converts value of average packets delay to cost expressed in [€/month].

Then, the considered two-criteria gateway location, flow and capacity assignment problem in the two-level hierarchical wide area network can be formulated as follows:

$$\min_{(X_r, Y_r)} Q(X_r, Y_r) \tag{6}$$

subject to:

$$(X_r, Y_r) \in \Re \tag{7}$$

$$d(X_r, Y_r) \leq B \tag{8}$$

where B denotes the budget of the hierarchical wide area network.

3 The Branch and Bound Algorithm

Assuming that locations of all gateways are fixed, i.e. values of variables y_a^g are constant and $C^i = \overline{C}^i$ for $i = 1,...,p$, the problem (6-8) becomes capacity and flow assignment problem (CFA) which is known as a NP-complete [2, 3]. Since the problem (6-8) is more general it is also NP-complete. Then, the branch and bound method can be used to construct the exact algorithm for solving the considered problem. The detailed description of the calculation scheme of the branch and bound method may be found in the papers [11, 12].

Starting with the initial selection (X_1, Y_1) we generate a sequence of selections. The variables from the sets X_1 and Y_1 are called *normal*, other variables are called *reverse*. The generation of new selection involves the choice of the certain normal variable from X_1 or Y_1 for replacement by the reverse variable; the operation is called the branching rules. The choice of the normal and reverse variables is based on local optimization criterion. For each selection (X_r, Y_r) lower bound is calculated to check the possibility of the generation of the selection (X_s, Y_s) with less value of a criterion d than already found. If such (X_s, Y_s) does not exist, (X_r, Y_r) and all its possible successors are abandoned and next backtrack is made to the predecessor from which (X_r, Y_r) was generated. If new selection (X_s, Y_s) is generated from (X_r, Y_r) by replacing the normal variable by the reverse variable, then the reverse variable is *constantly fixed*. This reverse variable cannot be replaced by any other in every possible successor of (X_s, Y_s). After backtrack from (X_s, Y_s) to (X_r, Y_r) by reverse variable of certain normal variable from (X_r, Y_r) this normal variable is momentarily fixed. So, for each (X_r, Y_r), there is a set F_r of *constantly fixed* and set F_r^t. of *momentarily fixed* variables. The reverse variables in F_r are constantly fixed. Each momentarily fixed variable in F_r^t is the reverse variable abandoned during backtracking process. Variables, which do not belong to F_r or F_r^t are called free in (X_r, Y_r). Algorithm terminates, after backtrack from (X_1, Y_1).

The branch and bound method involves constructing two important operations specific for considered problem: branching rules and lower bound. These operations are presented in following sections.

3.1 Branching Rules

The purpose of branching rules is to find the normal variable from the selection (X_r, Y_r) for replacement and to produce a successor (X_s, Y_s) of the selection (X_r, Y_r) with the least possible value of the criterion function (5). We can choose a variable x_k^i or a variable y_a^g. Assuming, that variables y_a^g are fixed, the problem (6-8) may be resolved into classical topology design problem. Then we can use the choice criterion Δ_{kj}^{ir} on variables x_k^i formulated as follows [2, 12]:

$$\Delta_{kj}^{i} = \begin{cases} \frac{\alpha}{\gamma}\left(\frac{f_i}{c_k^i - f_i} - \frac{f_i}{c_j^i - f_i}\right) + \left(d_k^i - d_j^i\right) \text{ if } f_i < c^i \\ \infty \quad \text{otherwise} \end{cases} \tag{9}$$

First term of the sum in expression (9) estimates the change of average packet delay in the network caused by change of capacity of channel i. Second term represents change of the cost of the channel i resulting from change of the channel capacity.

To formulate the choice criterion on variables y_a^g we use the following theorem:

Theorem 1. Let $(X_r, Y_r) \in \Re$. If the selection (X_s, Y_s) is obtained from the selection (X_r, Y_r) by complementing the variable $y_a^g \in Y_r$ by the variable $y_b^g \in X_s$ then $T(X_s, Y_s) \leq T(X_r, Y_r) - \delta_{ab}^{gr}$, where

$$\delta_{ab}^{g} = \begin{cases} \dfrac{\alpha}{\gamma} \displaystyle\sum_{i \in L_1^l} \left(\frac{\tilde{f}_i}{c^i - \tilde{f}_i} - \frac{f_i}{c^i - f_i} \right) + \sum_{i:(g,g_2) \in V^g} \left(d_{eb}^{ij} - d_{ea}^{ij} \right) & \text{for } \tilde{f}_i \leq c^i \\ \infty \quad \text{otherwise} & \end{cases} \tag{10}$$

$\tilde{f}_i = f_{ir} - f_{ia}^{'} + f_{ia}^{''}$: $f_{ia}^{''}$ corresponds to the packets flow between the nodes of network S_1^l and nodes of the remaining 1st level networks via the gateway located at node *a*, $f_{ia}^{'}$ corresponds to the packets exchanged between the nodes of network S_1^l and nodes of the remaining 1st level networks via the gateway located at node *b* after reallocating gateway from node *a* to node *b*.

Let $E_r = (X_r \cup Y_r) - F_r$, and let G_r be the set of all reverse variables of normal variables, which belong to the set E_r, where F_r is the set of variables constantly fixed in the rth iteration of branch and bound algorithm. We want to choose a normal variable the complementing of which generates a successor with the possible least value of total average delay per packet. We should choose such pairs $\{(y_a^g, y_b^g): y_a^g \in E_r, y_b^g \in G_r\}$ or $\{(x_k^i, x_j^i): x_k^i \in E_r, x_j^i \in G_r\}$ for which the value of criterion δ_{ab}^{gr} or Δ_{kj}^{ir} is maximal.

3.2 Lower Bound

The lower bound LB_r of the criterion function (5) for every possible successor (X_s, Y_s) generated from the selection (X_r, Y_r) may be obtained by relaxing some constraints in the problem (6-8). To find the lower bound LB_r we reformulate the problem (6-8) in the following way:

- we assume that the variables x_k^i and y_a^g are continuous variables,
- we approximate the discrete cost-capacity curves (given by the set C^i) with the lower linear envelope [10]. Then, the constraint (8) may be relaxed by the constraint $\Sigma d^i c^i \leq B$, where $d^i = \min\limits_{x_k^i \in X^i} (d_k^i / c_k^i)$ and c^i is the capacity of the channel i (continuous variable).

The solution of such reformulated problem can be found using the method proposed and described in the paper [9].

3.3 Heuristic Algorithm

The presented exact algorithm involves the initial selection $(X_1, Y_1) \in \Re$ for which the constraints (7) and (8) are satisfied [13]. Moreover, the initial selection should be the near-optimal solution of the problem (6-8). To find the initial selection the following heuristic algorithm is proposed. This heuristic algorithm may be also used to design of the hierarchical WAN when the optimal solution is not necessary and for hierarchical networks with large number of 1st level networks. Detailed description of the algorithm calculation scheme can be found in our earlier paper [8].

Step 1. For each gateway g, choose initial location from the set J_g such that cost of 2nd level network is minimal. Next, solve the classical topology assignment problem [4]. If this problem has no solution then the algorithm terminates – the problem (6-8) has no solution. Otherwise, perform $Q^* = Q$, where Q is the value of criterion function obtained by solving the topology assignment problem.

Step 2. Choose the pair of gateway locations with maximal value δ_{ab}^{gr} given by expression (10). Change location of gateway on this indicated by that expression.

Step 3. Solve the topology assignment problem. If the obtained value Q is lower than Q^* then perform $Q^* = Q$ and go to step 2. Otherwise, the algorithm terminates. The feasible solution is found. The network topology and gateways locations associated with the current Q^* is heuristic solution of the problem (6-8).

4 Computational Results

The presented exact algorithm was implemented in C++ code. Extensive numerical experiments have been performed with this algorithm for many different hierarchical network topologies. Number of gateways in 1st level networks was varying from one to six and number of potential location of each gateway was ranging from two to six. These experiments were conducted with two main purposes in mind: first, to examine the impact of various parameters on solutions to find properties of the problem (6-8) important from practical point of view and second, to test the computational efficiency of the algorithms.

We examined the impact of value of coefficient α in the criterion function (5) on the total average packet delay in the network T^{opt} and on the total leasing cost of channel capacities of 2nd level network (network cost) d^{opt} in the hierarchical network obtained as a solution of the problem (6-8). In the Fig. 2. is presented the typical dependence of network cost d^{opt} on the budget of the network B for different values of coefficient α. It can be observed that for higher values of coefficient α, cost of the network raises with increasing budget value. Nearly entire budget is utilized to build the network with total average packet delay in the network T^{opt} as low as possible until network cost reach certain level. It follows from Fig. 2. that the level is higher for greater values of weight coefficient α, while small values of coefficient α result in constant low network cost, not depending on budget value.

The Fig. 3. shows the dependence of average packet delay T^{opt} on the budget of the network for different values of coefficient α obtained during the same experiment. It can be noticed, that for small value of α weight of average packet delay in criterion function is neglected. Increasing budget of the network does not improve performance indices i.e. average packet delay in the network. It follows form Fig 2. and Fig. 3 that

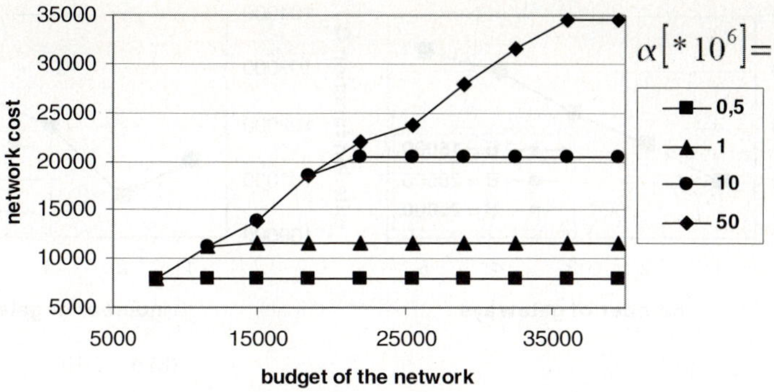

Fig. 2. Dependence of network cost d^{opt} on the budget of the network B

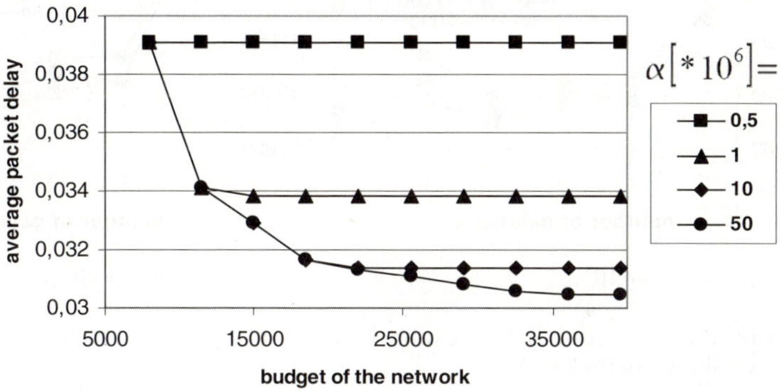

Fig. 3. Dependence of average packet delay in the network T^{opt} on the budget of the network B

value of coefficient α in criterion function (5) has significant impact on dependence of the total average packet delay in the network T^{opt} and of the network cost d^{opt} on the budget of the network B.

In the Fig. 4. plots showing typical dependence of criterion function value Q on the number of gateways located in the 1st level network are presented. Each plot presents results obtained for different value of coefficient α. It can be observed in the Fig. 4(a) and Fig 4(b) that for $\alpha = 10^6$ and for $\alpha = 3{\cdot}10^6$ values of criterion function do not depend on the budget values. Moreover, optimal number of gateways, i.e. number of gateways for which value of criterion function (5) is minimal, depends on the value of coefficient α. It follows from the Fig. 4. that the optimal number of gateways to locate in 1st level network is higher for greater values of weight coefficient α.

To analyze and compare results obtained for many networks of various sizes, topologies, numbers of gateways and sets of capacities of channels we introduce normalized coefficient α and normalized budget of hierarchical network, which values are independent of analyzed networks. Normalized coefficient α is formulated as follows:

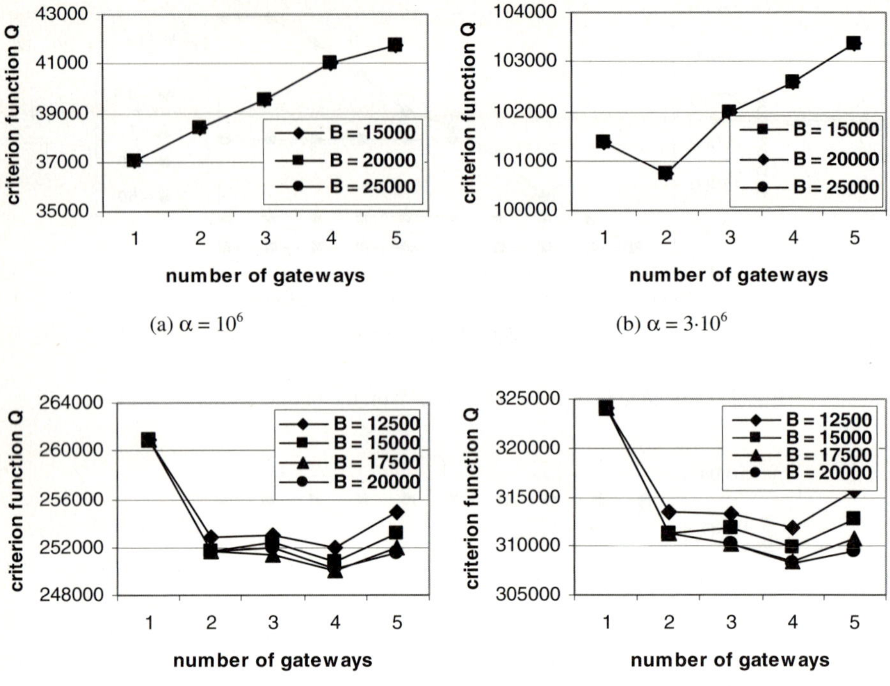

(a) $\alpha = 10^6$ (b) $\alpha = 3 \cdot 10^6$

(c) $\alpha = 8 \cdot 10^6$ (d) $\alpha = 10^7$

Fig. 4. Dependence of criterion function value on number of gateways in the 1st level network for different values of coefficient α

$$\overline{\alpha} = \alpha \cdot T_{\min} / D_{\max} \text{ ,}$$

where $T_{\min}$ is minimal value of average packet delay in considered hierarchical wide area network obtained by solving problem (6-8) without budget constraint and $D_{\max}$ is maximal building cost of the network with given number of gateways, sets of potential gateways locations, sets of potential 2nd level network channels capacities and their costs.

Similarly, let $D_{\min}$ be the minimal building cost of the network with given number of gateways, sets of potential gateways locations, sets of potential 2nd level network channels capacities and their costs. Normalized budget is formulated as follows:

$$\overline{B} = ((B - D_{\min})/(D_{\max} - D_{\min})) \cdot 100\%.$$

Let $\Phi^i(\overline{B}, \overline{\alpha})$ be the number of iterations of the branch and bound algorithm to obtain the optimal value of criterion function Q for normalized budget equal to $\overline{B}$ and normalized coefficient α equal $\overline{\alpha}$ for i-th considered network topology. Let

$$\Phi(u_1, u_2, v_1, v_2) = \frac{1}{Z} \sum_{i=1}^{Z} \left(\sum_{\overline{B} \in [u_1, u_2],\, \overline{\alpha} \in [v_1, v_2]} \Phi^i(\overline{B}, \overline{\alpha}) \Big/ \sum_{\overline{B} \in [0,100],\, \overline{\alpha} \in [0,100]} \Phi^i(\overline{B}, \overline{\alpha}) \right) \cdot 100\%$$

be the arithmetic mean of the relative number of iterations for $\overline{B} \in [u_1, u_2]$ and for $\overline{\alpha} \in [v_1, v_2]$ calculated for all considered network topologies and for different number of gateways, where Z is the number of considered hierarchical WANs.

Fig. 5. shows the dependency of Φ on ranges [0%, 10%), [10%, 20%), ..., [90%, 100%] of normalized budget $\overline{B}$ and ranges [0, 10), [10, 20), ..., [90, 100] of normalized coefficient $\overline{\alpha}$. Highest values of Φ were obtained for both $\overline{\alpha} \in (20, 40]$ and $\overline{B} \in (30\%, 50\%]$. It can be also observed in the Fig 5. that for $\overline{\alpha} \geq 50$ values of Φ are below 1% regardless of budget value, while for $\overline{\alpha} < 50$ low values of Φ are obtained for $\overline{B} \leq 10\%$ and $\overline{B} > 80\%$. Areas with low values Φ are the one for which the presented algorithm is most effective from computational point of view.

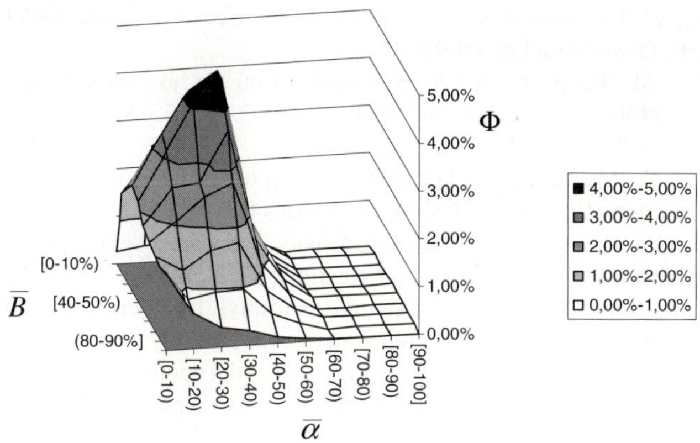

Fig. 5. Performance of presented algorithm for distinct ranges of values of normalized values of coefficient α and budget

5 Conclusions

In the paper the exact algorithm for solving the two-criteria gateway location and network topology assignment problem in two-level hierarchical network to minimize total cost of leasing channel capacities is presented. The considered problem is more general than the problems presented in the literature. It joins two problems considered in our earlier papers: problem minimizing average packet delay in hierarchical network and problem with total cost of leasing channel capacities as a criterion. Linear combination of leasing cost of channel capacities of 2nd level network and total average packet delay in the network is used as a criterion function. Weight of components of the sum can be adjusted to meet requirements specified by network design engineer. From computational experiments presented in the paper follows conclusion that the value of weight coefficient in criterion function has significant impact on dependence of the network cost d^{opt} and of the total average packet delay in the network T^{opt} on the budget of the network B (Fig. 2) and (Fig. 3) and also on optimal number of gateways located in 1st level network (Fig.4).

The high costs of WANs design justify use of computationally costly exact algorithms. Improvement of results obtained by heuristic algorithms, can lead to significant savings in costs of network maintenance. That makes our algorithm important from economical point of view. Moreover, optimal solutions obtained using our algorithm can be used to evaluate results obtained using various heuristic algorithms.

This work was supported by a research project of The Polish State Committee for Scientific Research in 2005-2007.

References

1. Kleinrock, L., Kamoun, F.: Optimal Clustering Structures for Hierarchical Topological Network Design of Large Computer Networks. Networks 10, 221–248 (1980)
2. Kasprzak, A.: Topological Design of the Wide Area Networks. Wroclaw University of Technology Press, Wroclaw (2001)
3. Markowski, M., Kasprzak, A.: An exact algorithm for host allocation, capacity and flow assignment problem in WAN. In: WITASI 2002, pp. 73–82. Kluwer Academic Publ., Dordrecht (2002)
4. Liu, Z., Gu, Y., Medhi, D.: On Optimal Location of Switches/Routers and Interconnection. Technical Report, University of Missouri-Kansas City (1998)
5. Saha, D., Mukherjee, A.: On the multidensity gateway location problem for multilevel high speed network. Computer Communications 20, 576–585 (1997)
6. Liang, S.C., Yee, J.R.: Locating Internet Gateways to Minimize Nonlinear Congestion Costs. IEEE Transactions On Communications 42, 2740–2750 (1994)
7. Ryba, P., Kasprzak, A.: The Gateways Location and Topology Assignment Problem in Hierarchical Wide Area Networks: Algorithms and Computational Results. In: Gavrilova, M.L., Gervasi, O., Kumar, V., Tan, C.J.K., Taniar, D., Laganá, A., Mun, Y., Choo, H. (eds.) ICCSA 2006. LNCS, vol. 3982, pp. 1100–1109. Springer, Heidelberg (2006)
8. Ryba, P., Kasprzak, A.: An Approximate Algorithm for the Minimal Cost Gateways Location, Capacity and Flow Assignment in Two-Level Hierarchical Wide Area Networks. In: Alexandrov, V.N., van Albada, G.D., Sloot, P.M.A., Dongarra, J. (eds.) ICCS 2006. LNCS, vol. 3991. Springer, Heidelberg (2006)
9. Ryba, P., Kasprzak, A.: An Exact Algorithm for The Minimal Cost Gateways Location, Capacity and Flow Assignment Problem. In: Gervasi, O., Gavrilova, M.L. (eds.) ICCSA 2007, Part III. LNCS, vol. 4707, pp. 343–352. Springer, Heidelberg (2007)
10. Fratta, L., Gerla, M., Kleinrock, L.: The Flow Deviation Method: an Approach to Store-and-Forward Communication Network Design. Networks 3, 97–133 (1973)
11. Walkowiak, K.: A New Method of Primary Routes Selection for Local Restoration. In: Mitrou, N.M., Kontovasilis, K., Rouskas, G.N., Iliadis, I., Merakos, L. (eds.) NETWORKING 2004. LNCS, vol. 3042, pp. 1024–1035. Springer, Heidelberg (2004)
12. Markowski, M., Kasprzak, A.: The Web Replica Allocation and Topology Assignment Problem in Wide Area Networks: Algorithms and Computational Results. In: Gervasi, O., Gavrilova, M.L., Kumar, V., Laganá, A., Lee, H.P., Mun, Y., Taniar, D., Tan, C.J.K. (eds.) ICCSA 2005. LNCS, vol. 3483, pp. 772–781. Springer, Heidelberg (2005)
13. Wolsey, L.A.: Integer Programming. Wiley-Interscience, New York (1998)

A Control Packet Minimized Routing Protocol for Ad-Hoc Wireless Networks

Youn-Sik Hong

Department of Computer Science and Eng., Univ. of Incheon,
177 Dowha-dong Nam-gu 402-749 Incheon, Korea
yshong@incheon.ac.kr

Abstract. AODV routing protocol is intended for use by mobile nodes in ad-hoc wireless networks. As the degree of node mobility becomes high, however, the number of RREQ and RREP messages during the route discovery process increase so rapidly. The unexpected increases in the number of control packets cause the destination node to decrease the packet receiving rate. Besides, the overall energy consumption for the network can be increased. Thus, we propose a novel method of adaptively controlling the occurrences of the control packets based on AIAD (additive increase additive decrease) under a consideration of the current network status. We have tested our proposed method with both the conventional AODV and the method using timestamp based on the three performance metrics; i.e., node mobility, node velocity, and node density, to compare their performances.

Keywords: Ad-hoc Network, AODV, Expanding Ring Search, Node Mobility, Node Traversal Time.

1 Introduction

Ad-hoc wireless network architecture is a self-organizing and distributed controlled network formed by a set of stations (called *nodes*) that can freely and dynamically self-configure and organize themselves to set up a temporary wireless network [1]. In the ad-hoc network configuration, a node acts as a mobile terminal as well as a router to forward messages to neighboring nodes if possible. The network topology in an ad hoc network is highly dynamic due to the movement of nodes; hence an on-going session suffers frequent path breaks. A disruption may be occurred either due to the movement of the intermediate nodes in the path or due to the movement of end nodes. Routing protocols for ad-hoc networks must be able to perform efficient and effective mobility management.

Existing ad-hoc routing protocols may generally be categorized as *table-driven* and *source-initiated demand driven* [2]. In table-driven routing protocols, every node maintains the network topology information in the form of routing tables by periodically exchanging routing information. These protocols suffer from excessive control overhead that is proportional to the number of nodes in the network and therefore is not scalable in ad-hoc wireless networks, which have limited bandwidth and whose topologies are highly dynamic.

O. Gervasi et al. (Eds.): ICCSA 2008, Part II, LNCS 5073, pp. 907–919, 2008.

Unlike the table-driven routing protocols, on-demand routing protocols execute the *path-finding* process and exchange routing information only when a path is required by a source node to communicate with a destination. One of the on-demand routing protocols is AODV (*Ad-hoc On-demand Distance Vector*) [5]. In AODV, a source node floods the Route Request (RREQ) packet in the network when a route is not available for the desired destination. When an intermediate node receives a RREQ, it either forwards it or prepares a Route Reply (RREP) if it has a valid route to the destination. All intermediate nodes having valid routes to the destination, or the destination node itself, are allowed to send RREP packets to the source. AODV reduces the need for the system wide broadcasts by localizing the propagation of changes in the network. Even though it performs well in static and low-mobility environments, the performance degrades rapidly with increasing mobility.

As shown in Fig 1, data packet transmission is slightly affected by the node mobility. However, as the node mobility increases the control packets like RREQs and RREPs increase rapidly. Then it causes route discovery latency to increase. Besides, the amount of energy consumption increases corresponding to the increased number of control packets. To the end, the overall performance should be degraded with high node mobility. Our primary concern is to enhance the performance of AODV by reducing the volume of these control packets effectively due to the node mobility.

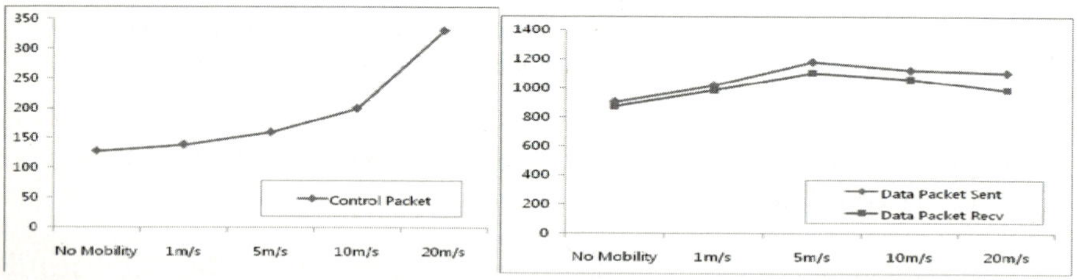

Fig. 1. The number of control packets (*left*) and the number of data packets (*right*) with node mobility

The rest of this paper is organized as follows. In section 2, we describe the expanding ring search algorithm and related works. In section 3, we present a proposed approach based on AIAD method. In section 4, a series of experimental results will be given and finally we conclude our paper in section 5.

2 Expanding Ring Search and Related Works

2.1 Expanding Ring Search

Consider the example depicted in Fig 2. In this figure, the source node 1 initiates a path-finding process by originating a RREQ to be flooded in the network for the destination node 8. When the intermediate nodes 2, 3, and 4 receive the RREQ packet, they check their routes to the destination. In case a route to the destination is not available, they further forward it to their neighbors. If the RREQ reaches the node 8 through path 1-4-6-8 or any other alternate route, it also sends a RREP to the source.

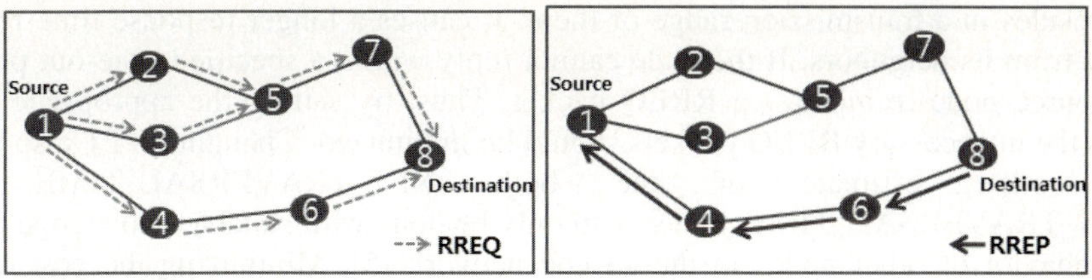

Fig. 2. AODV route discovery; propagation of the RREQ packets (*left*) and a path of the RREP packet to the source (*right*)

To prevent unnecessary network-wide dissemination of RREQs, the source node should use an *expanding ring search* (ERS) technique [6]. Centered on the source node, ERS successively searches larger areas until a node with the information being searched is located. The search begins with a small *time to live* (TTL) value (usually takes as 1) for a query and a time-out period for receiving a reply (proportional to TTL). If there is no reply within the time-out period, the source node increments the TTL by a specified value and re-initiates the query. Thus, by setting the appropriate TTL value in the query, the source node can control the search radius. Table 1 gives default values for some important parameters associated with ERS techniques used in the AODV routing protocol.

Table 1. Default values for AODV parameters

Parameter name	value
NET_DIAMETER	35
NODE_TRAVERSAL_TIME	40 ms
NET_TRAVERSAL_TIME	2 x NODE_TRAVERSAL_TIME x NET_DIAMETER
RING_TRAVERSAL_TIME	2 x NODE_TRAVERSAL_TIME x TTL_VALUE
TTL_START	1
TTL_INCREMENT	2
TTL_THRESHOLD	7

In an expanding ring search, the source node initially uses a TTL = TTL_START in the RREQ packet IP header and sets the timeout for receiving a RREP to RING_TRAVERSAL_TIME milliseconds [5]. If the RREQ times out without a corresponding RREP, the source node broadcasts the RREQ again with the TTL incremented by TTL_INCREMENT. This continues until the TTL set in the RREQ reaches TTL_THRESHOLD, beyond which a TTL = NET_DIAMETER is used for each attempt. Beyond this, the RREQ is not forwarded any further. Each time, the timeout for receiving a RREP is RING_TRAVERSAL_TIME.

NODE_TRAVERSAL_TIME (NTT) is a conservative estimate of the average one hop traversal time for packets and should include queuing delays, interrupt processing times and transfer times [5]. With a fixed value of 40ms for NTT, it does reflect a full dynamic topology with respect to node mobility in ad-hoc wireless networks. For example, one node moves to the right in some distance from its neighbors, while it still resides in a transmission range of them. It causes a longer response time for the query from its neighbors. If the node cannot reply within a specified time-out period, the source node re-initiates a RREQ packet. Thus, by setting the appropriate NTT value the unnecessary RREQ packets should be minimized. Changing NTT also alters the node's estimate of the both NET_TRAVERSAL_TIME and RING_TRAVERSAL_TIME, and so can only be done with suitable knowledge about the behavior of other nodes in the ad hoc network [5]. Minimizing the cost of the initiating and the forwarding RREQ packets is crucial for resource-constraint multi-hop wireless network, which motivates this work.

2.2 Related Works

A number of heuristics for enhancements of the performance of AODV routing protocol has been proposed. *Hassan and Jha* [7] explored to find an optimum TTL threshold *L* that would minimize the expected bandwidth cost of ERS. They give the following experimental results; search threshold L of 3 is optimum for three categories of networks, that is, large networks with small radius, medium networks with large radius and medium networks with small radius. In the AODV, TTL_THRESHOLD is typically set to 7, which gives the search threshold of 3.

Some of the works tries to identify the inefficient elements of ERS. *Lee* et al[9] proposed a method of timestamp to compute NTT; every node adds the current time to the message (i.e., RREQ and RREP) before sending it out, and then the neighbor that receives it computes a measured traversal time (called *M*) using the timestamp. A smoothed NTT is updated by using the similar equation of the round-trip-time estimator:

$$\mathrm{NTT} = \alpha \times \mathrm{NTT} + (1 - \alpha)\,\mathrm{M} \qquad (1)$$

where, α is a smoothing factor with a range of 0 and 1. *Kim* et al [10] presented a technique of estimating NTT that divides a time difference between the source node and the destination node by the number of hops. In addition, *Park* et al [11] proposed a blocking ERS (B-ERS for short). One of the differences from the TTL-ERS is that the B-ERS does not resume its route search procedure from the source node each time a rebroadcast is required. *Tripathi* et al [8] proposed four heuristics; one for them is the utilization of route caches at intermediate nodes for route discovery.

All of the works mentioned above have an emphasis on the adaptive control of AODV routing protocol with respect to node mobility. Since the timestamp based approaches [9] changes NTT too rapidly, the stability of the network becomes relatively low. Besides, with low mobility their performance will be degraded due to increases in the number of RREQ packets initiated by the source node. Even if B-ERS reduces in route discovery latency, it may suffer from dynamic changes in the network topology with high mobility. With the proposed heuristics proposed by *Tripathi* et al, the complexity of maintenance of multiple paths demands a larger memory and greater processing power for every node.

3 A Control Packet Minimized AODV Protocol

TTL based ERS restraints its searching range by giving RREQs a pre-defined TTL number. The TTL number corresponds to the radius of a searching area. Once an intermediate node located at *H* hop distance receives an RREQ, the source node waits for the following waiting time. The *waiting time* WT is calculated as:

$$WT = 2 \times H \times NODE\_TRAVERAL\_TIME \quad (2)$$

Clearly, *WT* is equal to or less than the RING_TRAVERSAL_TIME. In addition, *H* is proportional to the TTL number. With the fixed value for NTT, it can cause to increase the unnecessary RREQs initiated by a source node. For example, if a propagation delay of one link is longer than 40ms, the source node can be received a corresponding RREP after the RREQ times out. In that case, the source node already initiates a new RREQ with the incremented TTL number.

Our proposed method relies on packet based propagation of timestamp information to calculate more realistic NTT by considering the network status. While initiating a route reply, the destination node appends the timestamp information in the RREP packet. When the source node receives the RREP packet, it checks the timestamp *ST* available at the source node and the timestamp *DT* available at the destination node. Let *MT* denotes estimated time taken for a packet to be delivered from one node to one-hop neighboring node. Let the number of hops on the route path established between the source node and the destination node be denoted by H. Then the *MT* is calculated as:

$$MT = (ST - DT) / H \quad (3)$$

A computed estimate *MT* of one-hop traversal time just gives a reference relative to a previous NTT. A new NTT should be adjusted depending on MT. For example, the path 8-6-4-1 in Fig.2 consists of three links. Assume that each of them has a propagation delay of 40ms, 30ms, and 20ms, respectively. In that case, MT is computed as 30ms. However, since the longest propagation delay should be considered, a NTT is not changed and still remains 40ms. If the propagation delays of the all links increase equally in 6ms due to node mobility, MT is 36ms and then the amount of the difference between NTT and MT becomes smaller. Thus, a new NTT should be increased by a certain amount Δ. On the contrary, if the delays of the all links decrease equally in 6ms, MT is 24ms and then the amount of difference between them becomes greater. So a new NTT should be decreased by Δ. The above explanations are summarized as shown in Table 2.

Table 2. The recalculated NTT with respect to the propagation delays of each link (unit: ms)

Propagation delay			MT	previous NTT	NTT – MT	New NTT
Link1	**Link2**	**Link3**				
40	30	20	30	40	10	40
46	36	26	36	40	4	40+Δ
34	24	14	24	40	16	40-Δ

Then the following algorithm, called CP_AODV (*Control Packet minimized* AODV), is used to resolve the number of control packets including RREQ and RREP packets. To minimize the flooding of control packets, a NTT should be adjusted smoothly by AIAD (*additive increase additive decrease*) technique.

```
IF (min_threshold > NTT - MT)
{
     NTT = NTT + increment;
}
ELSE IF (max_threshold < NTT - MT)
{
     NTT = NTT - increment;
     IF (NTT < lower_bound)
     {
            NTT = lower_bound;
     }
}
```

The parameters *min_threshold, max_threshold,* and *increment* are adjustable with respect to the network status. Notice that *max_threshold > min_threshold.* When the amount of the difference between NTT and MT is less than *min_threshold*, the average propagation delay of the links seems to be increased. So a new NTT should be increased by a specific amount *increment.* On the contrary, when the amount of the difference between them is greater than *max_threshold*, the average delay seems to be shorter. Then a new NTT should be decreased by *increment*. The parameter *increment* is controlled adaptively depending on the movement of nodes. Our experimental results show that with less than 40ms of NTT the number of retransmission of the RREQ packets increases so rapidly and thus the overall performance is degraded. Thus, the lower bound on NTT is set to 40ms. While the amount of difference between them is in the range of *min_threshold* and *max_threshold,* the node mobility seems to be relatively low. Then the value of a current NTT remains as before.

4 Performance Evaluations

For the purpose of the simulation, AODV protocol implemented in QualNet 4.0 simulator [12] has been modified to incorporate the changes which have been proposed in this paper. Data payload is of 512 bytes. All the data packets are CBR (*continuous bit rate*) packets. Each node moves with randomly chosen maximum speed up to 10m/s. The simulation parameters are summarized as shown in Table 3. Notice that the control parameters used in the CP_AODV are set through our simulations as the following: *min_threshold* = 5, *max_threshold* = 10, *increment* = 5.

Table 3. Simulation set-up

attributes	value
space	1,500m x 1,500m
bandwidth	2MB
Number of nodes	49
Placement strategy	GRID
Transmission range	250m
Node velocity	0 ~ 10m/s
run time	600 seconds

The performance evaluation is based on the comparison of the three different metrics; node mobility, node velocity and node density. In order to get a realistic insight into the effects of the proposed method (CP_AODV), these metrics also have been evaluated for both the legacy AODV [6] and the modified AODV with the timestamp method [9] (*T-AODV* for short).

4.1 The Experiments for the Node Mobility

After reaching the destination the node begins to move again after pause time. In our experiments we varied it between 0 to 600 seconds. The network between the pause times 0 to 150 have a high mobility, whereas the network beyond 450 have a low mobility. The network between the pause times 150 to 450 have a moderate mobility. The pause time implies the level of the mobility in the network.

We classify RREQ messages into two types; a RREQ initiated by the source node and a RREQ forwarded by an intermediate node. The source node re-initiates the RREQ if there is no route reply within a time-out period. Thus, a possible measurement to evaluate the cost of network-wide flooding is the number of the RREQ packets initiated by the source node. RREQ packets are flooded in the network by intermediate nodes forwarding the requests to their neighbors. So, another measurement is the number of RREQ packets forwarded by them.

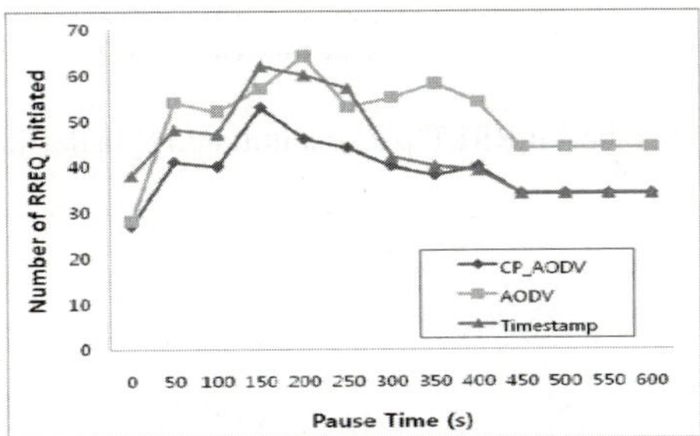

Fig. 3. The number of RREQ packets initiated by a source node

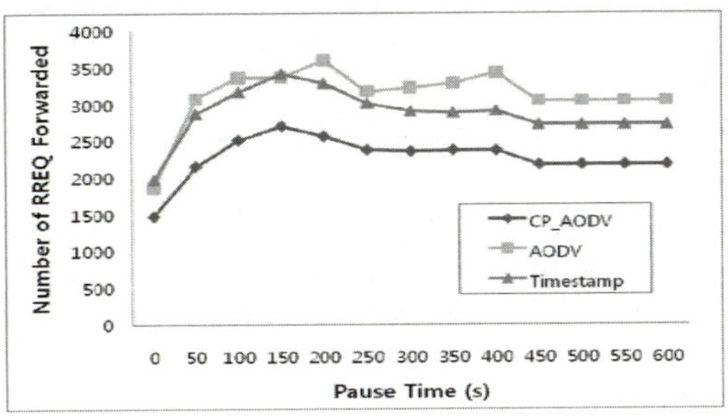

Fig. 4. The number of RREQ forwarded by intermediate nodes

For the number of RREQ packets initiated by the source node, CP_AODV is remarkably less than AODV as shown in Fig 3. In addition, CP_AODV is better than T-AODV except in low mobility. For the number of RREQ packets forwarded by intermediate nodes, CP_AODV has significantly lower average packets than the two protocols. Typically, with CP_AODV the number of RREQ packets forwarded is reduced by 28% and 21% compared to AODV and T-AODV, respectively.

The decreased number of RREQ packets result in the decreased number of both RREP and RERR. For the number of RREP packets initiated by the destination node, CP_AODV has significantly low average packets in all levels of node mobility than both protocols as shown in Fig 5. In addition, with CP_AODV the number of RERR packets is reduced by more than 30% in average than the two protocols. Notice that both curves for RREP (Fig 5) and for RERR (Fig 6) are similar to it for RREQ in Fig 3.

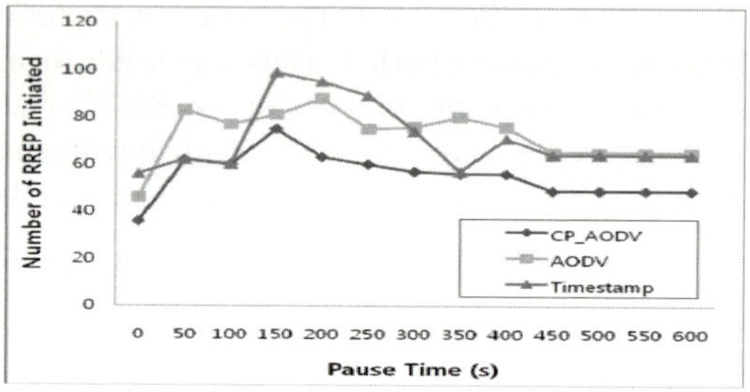

Fig. 5. The number of RREP packets initiated by a destination node

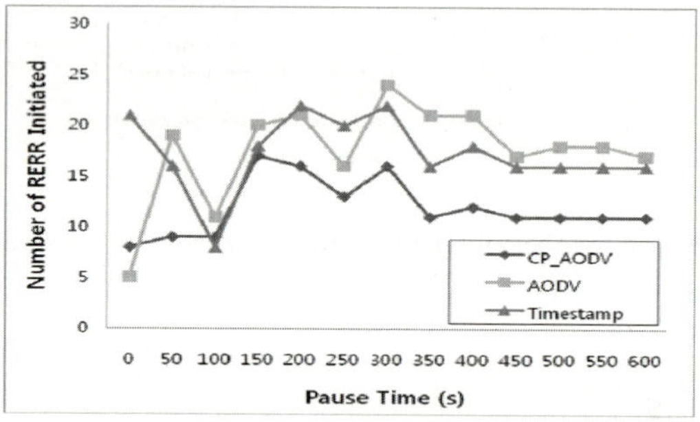

Fig. 6. The number of RERR packets initiated by nodes in the network

The proposed approach can save the amount of energy consumption by reducing the number of control packets including RREQs, RREPs and RERRs without increase in route discovery latency. The amount of energy consumption by CP_AODV, AODV and T-AODV are plotted against the pause time (i.e., *node mobility*) in Fig 7. Notice that the amount of energy consumption in Fig 7 includes both one for the route discovery process and one for data packet transmissions. Thus the shapes of the curves in Fig 7 are quite different from it in Fig 3. In all levels of node mobility, CP_AODV is remarkably less than AODV.

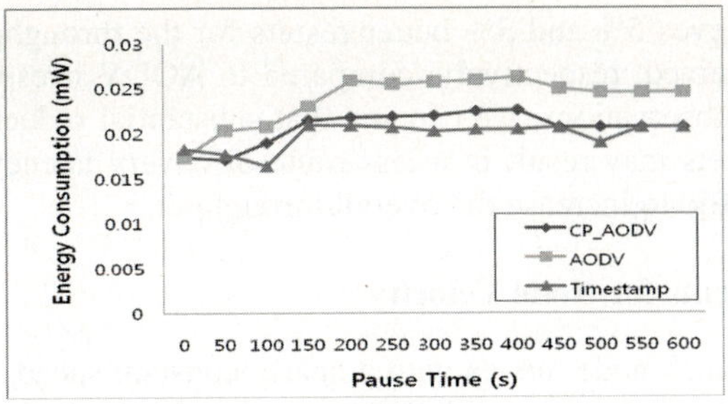

Fig. 7. Energy consumption versus node mobility

Notice that the amount of energy consumption is greater than T-AODV by 6% in average. The reason may be attributed to the fact that with T-AODV the number of data packets received by the destination node is relatively small. To be more precise, few observations can be made from our simulation results (Fig 8 and Fig 9). We see that there is no clear distinction for the throughput versus the pause time in Fig 8 for both CP_AODV and T-AODV. For the number of data packets received by the destination node, CP_AODV, however, is significantly higher than T-AODV by 10% as shown in Fig 9. We can say that T-AODV transmits less data packets compared to CP_AODV and thus its energy consumption becomes low.

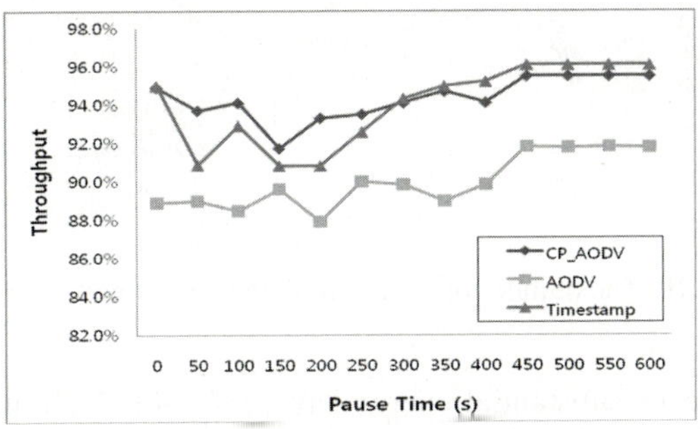

Fig. 8. The throughput versus the pause time

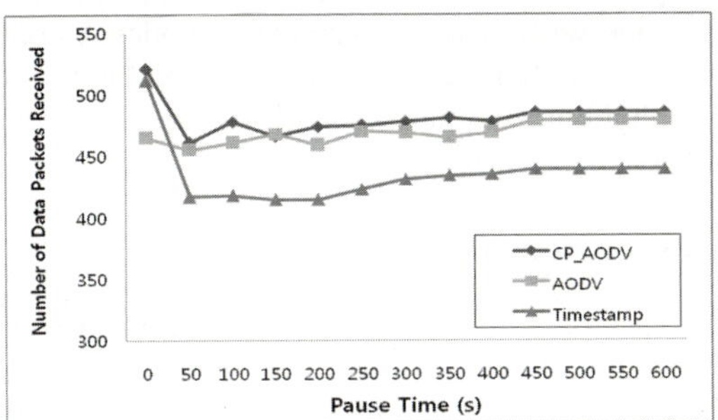

Fig. 9. The number of data packets received

CP_AODV achieves 5% and 3% better results for the throughput and the number of data packets received, respectively, compared to AODV irrespective of node mobility. From these observations, we can say that substantial reduction in the flooding of the control packets may result in a less route discovery latency. In turn, it causes low data packet traffic to increase the overall throughput.

4.2 The Experiments for Node Velocity

In this experiment each node moves with a nearly constant speed. When each node is set to the speed of 5m/s, it moves with the speed of 4~5 m/s. At the experiments in the previous section, it moves with the random speed of 0~10 m/s. Thus, this section focuses on the performance evaluation under the node speed ranges. Through this experiment, the pause time is set to 100 seconds and thus the network has a high mobility. The other parameters are used in Table 3. As you see in Fig 10, the average number of RREQ packets of CP_AODV turns out to be better. More specifically, it of CP_ADOV has been measured 58.7, whereas that of AODV and that of T-AODV is 67.9 and 66.1, respectively. Besides, the standard deviation for the number of RREQs of CP_ADOV is the lowest among the three protocols. Thus, we can say that our proposed approach is more stable under the node speed ranges.

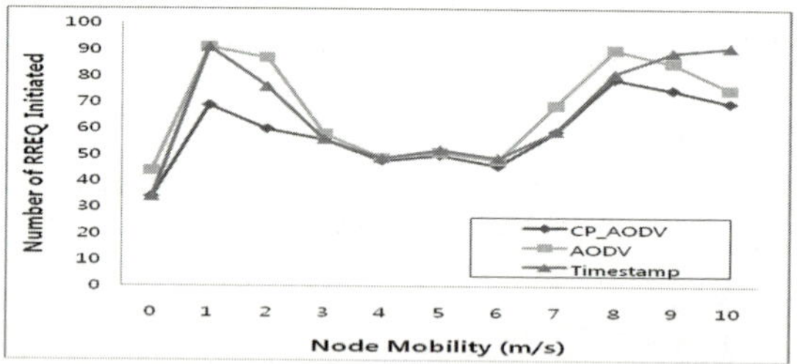

Fig.10. The number of RREQ initiated versus node velocity

CP_AODV achieves substantial energy savings by 11~1.7% in average compared to both AODV and T-AODV (Fig 11). In addition, for the number of data packets delivered to the destination it is better than the two protocols by 14~2.4% in average (Fig 12). It cannot say that a network with slower movement of nodes achieves a better throughput than one with faster movement of nodes. In other words, to reduce the volume of control packets that needs to be exchanged between the source and the destination, all of the neighboring nodes keep the distance to exchange messages each other irrespective of node velocity.

4.3 The Experiments for Node Density

In this experiment we vary the number of nodes starting from 36 up to 121 to evaluate the number of RREQ initiated (Fig 13), the amount of energy consumption (Fig 14) and the number of data packets received by the destination (Fig 15). Since the rectangular field configuration is fixed as 1,500m x 1,500m, the distance between each pair

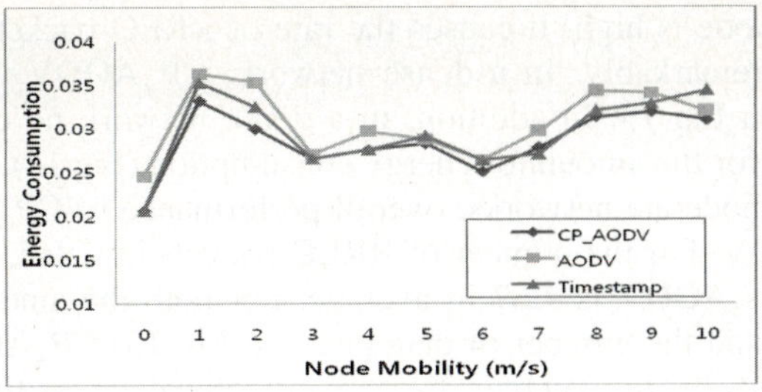

Fig. 11. The energy consumption rate versus node velocity

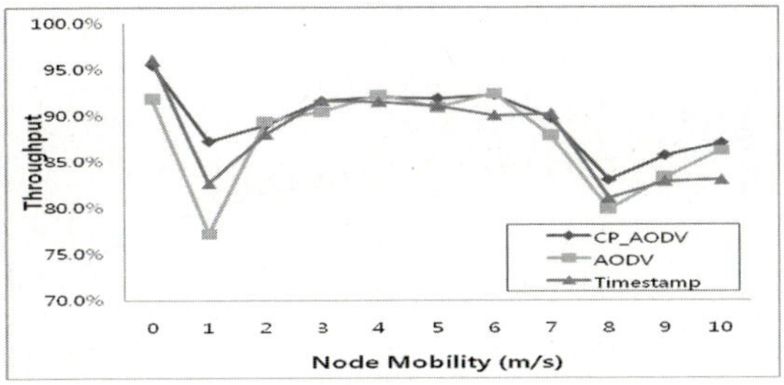

Fig. 12. The throughput versus node velocity

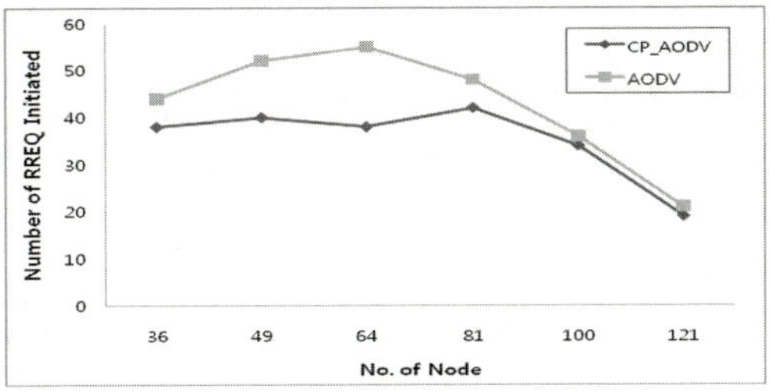

Fig. 13. The number of RREQ initiated versus the number of nodes

of the neighboring nodes becomes shorter with a denser network. Depending on the number of nodes present in the ad-hoc network, we refer to the network as *sparse* which consists of less than 50 nodes. A *dense* network consists of more than 100 nodes. Through this experiment, the pause time is also set to 100 seconds and thus the network has a high mobility. In the following results (Fig 13-15), we omit the results of T-AODV because they are similar to them in the previous sections. Thus, through this subsection, CP_AODV is directly compared to the legacy AODV for simplicity and clarity.

As in a dense network the number of pairs of the two neighboring nodes which locate within a possible transmission range of each other increase, the number of RREQ packets initiated by the source decreases significantly (Fig 13). That means in a dense network the possibility for finding a neighboring node to forward the request packets to the destination node is high. It causes the rate of RREQ packets initiated and forwarded to reduce remarkably. In a dense network, CP_AODV performs similar to AODV as shown in Fig 13. In addition, in a dense network no clear difference between them occurs for the amount of energy consumption (Fig 14).

For sparse and moderate networks, overall performance of CP_AODV is found to be better than AODV. For the number of RREQ packets initiated, CP_AODV is 35.2 in average, whereas AODV is 42.7 in average. For both the amount of energy consumption (Fig 14) and the number of data packets (Fig 15) CP_AODV is better than AODV by 13% and 2% in average, respectively. However, as the node density increases such the differences between them decreases. Thus our proposed method can achieve a better result with a sparse and moderate network than legacy AODV.

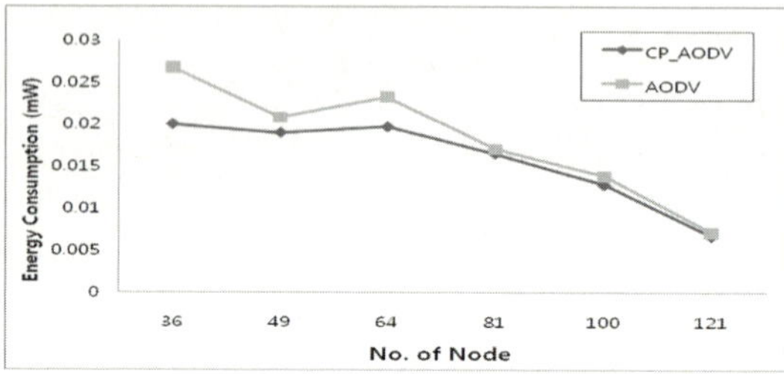

Fig. 14. The energy consumption rate versus the number of nodes

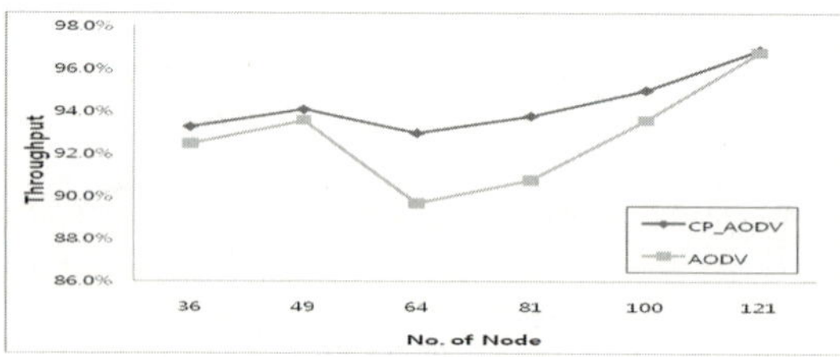

Fig. 15. The throughput versus the number of nodes

5 Concluding Remarks

We proposed a control packet minimized AODV protocol for rapidly changing the network topology in ad-hoc wireless networks due to node mobility. During the route discovery process, the volume of the control packets increases with the changes in the network topology due to the movement of nodes. To reduce them significantly, more realistic estimation of NTT is needed. To minimize the flooding of control packets, a NTT should be adjusted smoothly based on AIAD technique. CP_AODV protocol provides good improvement in terms of the number of control packets, the amount of energy consumption, the throughput and the number of data packets received by the destination, compared to the legacy AODV and the modified AODV with timestamp technique. These results have been obtained under three distinct metrics; node mobility, node velocity and node density. However, in a dense network, CP_AODV performs similar to AODV, whereas it achieves a better result with a sparse and moderate network.

References

1. Corson, S., Macker, J.: Mobile Ad hoc Networking (MANET): Routing Protocol Performance Issues and Evaluation Considerations. RFC 2501 (1999)
2. Royer, E.M., Toh, C.-K.: A review of current routing protocols for ad-hoc mobile wireless networks. IEEE Personal Communications Magazine, 46–55 (1999)
3. Toh, C.K.: Ad-hoc Mobile Wireless Networks: Protocols and Systems. ACM SIGCOMM. Prentice Hall PTR, Englewood Cliffs (2002)
4. Perkins, C.: Performance Comparison of Two On-Demand Routing Protocols for Ad-hoc Networks. In: Proc. of the IEEE Infocom, pp. 3–12 (2000)
5. Ad hoc On-Demand Distance Vector (AODV) Routing (RFC 3561), `http://www.faqs.org/rfcs/rfc3561.html`
6. Carlberg, K., Crowcroft, J.: Building Shared Trees Using a One-to-Many Joining Mechanism. ACM Computer Communication Review 27(1), 5–11 (1997)
7. Hassan, J., Jha, S.: Optimizing expanding ring search for multi-hop wireless networks. In: GLOBECOM, pp. 1061–1065 (2004)
8. Tripathi, S.M., Ghosh, R.K.: An Implementational Study of Certain Heuristics for the Performance Enhancements of AODV. In: Proc. of Euro-Micro Workshop on Parallel, Distributed and Network-based Processing, pp. 395–402 (2002)
9. Lee, J., Park, H., Lee, K., Suk, K.: Performance Improvement of AODV Routing Algorithm Using Timestamp. Journal of Korea Information and Communication Society (domestic) 31(8), 830–836 (2006)
10. Kim, H.-c., Chunng, S.-m.: A study of Optimization using state space survey in ad hoc network. Journal of Korea Institute of Military Science and Technology (domestic) 8(4), 68–76 (2005)
11. Park, I., Pu, I.: Energy Efficient Expanding Ring Search. In: 1st ASIA Int. Conf. on Modeling & Simulation(AMS), pp. 198–199 (2007)
12. `http://www.scalable-networks.com`

A New Paradigm for Prioritizing Multiple Class Services in Optical Burst Switched Networks

Aresh Dadlani[1,2], Ali Rajabi[1], and Ahmad Khonsari[2,1]

[1]IPM School of Computer Science, Tehran, Iran
[2]Department of Electrical and Computer Engineering,
University of Tehran, Tehran, Iran
{a.dadlani,alirajabi,ak}@ipm.ir

Abstract. Quality of service (QoS) differentiation within the optical infrastructure is under constant development due to the direct impact of the ever-rising demand for time-critical and multimedia applications in the Internet. In this paper, we propose an efficient prioritized hybrid model based on two eminent approaches namely, resource-reservation and extra offset-time based QoS that have already been investigated in the literature. We show that this model guarantees better performance with increase in number of traffic classes. We also simulate a complete topologically optical network bearing Poisson traffic to assess the performance of the proposed model as compared with the two aforementioned approaches.

Keywords: Optical burst switching (OBS), quality of service (QoS), resource-reservation based QoS, extra offset-time based QoS, prioritized hybrid model.

1 Introduction

In present day networking, the need for supporting acceptable quality of service (QoS) in emerging multimedia applications such as High Definition Television (HDTV), video conferencing and Internet telephony has become a crucial issue. To satisfy such service demands along with the ever-rising growth of Internet traffic, the surge for higher bandwidth in national as well as international backbones has become inevitable.

In recent years, all-optical WDM networks have been considered as appropriate substitutes for the next generation Internet due to their huge deliverable bandwidth and high data transparency [1, 2]. Among all the QoS provisioning paradigms suggested for the WDM layer, optical burst switching (OBS) [3] has received much attention as it combines the best of the coarse-grained optical circuit switching (OCS) [4, 5] and the fine-grained optical packet switching (OPS) [6, 7] so as to integrate IP over WDM and support diverse services. In [8-10], a novel technique based on extra offset-time has been proposed to provide QoS in OBS networks carrying multiple traffic classes. Resource-reservation based scheme, studied in [11, 12], is another innovation that guarantees differentiation of QoS among traffic classes with little buffering requirements. Although this method has been proposed for an OPS network, the idea can be well applied to an OBS network as well. In [11], the authors offer a

O. Gervasi et al. (Eds.): ICCSA 2008, Part II, LNCS 5073, pp. 920–935, 2008.

network wide QoS management scheme for OPS which is based on the idea that when encountering congestion in intermediate switches, lower priority classes might be directed to routes other than their original one, while higher priority classes always travel through fixed paths, which guarantees in-order delivery of packets at the destination.

In this paper, we propose a more effective QoS provisioning scheme by combining the two prominent QoS management schemes namely, extra-offset-time and resource-reservation based QoS mentioned above. Through simulations, we prove that with increase in number of traffic classes, our proposed hybrid model guarantees improved QoS in terms of latency and burst blocking probability as compared to the individual schemes.

The rest of the paper is structured as follows. In Section 2, an overview of OBS is provided to gain a better understanding of the concepts that follow. The two major schemes devoted to supporting QoS in multiple traffic classes and thus, forming the basis of our proposed model are highlighted in Section 3, followed by our hybrid model in Section 4. Simulation scenarios and results are brought in Section 5. Finally, in Section 6, we present conclusions and ideas for future work.

2 Preliminaries of OBS

In the literature, all OBS network switches are mainly classified into two types: *edge switches* and *core switches*. As shown in Fig. 1, edge switches are the switches that connect the optical network to the outside world, while core switches are the switches that inter-connect the edge switches within the OBS network. Based on their functionality, the edge switches can further be categorized into two types: *ingress* and *egress* edge nodes. An edge switch is said to act as an ingress node whenever it aggregates the incoming packets into optical bursts according to their destination edge switch. An egress node is any edge switch at which data bursts are disassembled into packets. For the sake of clarity, Fig. 1 depicts a scenario in which a burst travels from edge switch 1 (E_1) to edge switch 3 (E_3) via core switches C_1, C_3 and C_4. In this scenario, E_1 acts as the ingress node while E_3 acts as the egress node.

Most of the common burst assembly algorithms can be classified as timer-based, threshold-based, and mixed timer/threshold based algorithms. In the timer-based approach [16], a timer starts at the beginning of each new assembly cycle, determining the transmission time of the burst into the core network. After a fixed time, all the packets that arrived in this time period are assembled into a burst. In the threshold-based approach [17], a threshold is provided as a limiting parameter to determine when to generate and transmit a burst into the optical network. The threshold specifies the number of packets to be aggregated into a burst or the (minimum or maximum) length of the burst. The incoming packets are stored in the prioritized packet queues in the ingress node, until the threshold condition is satisfied. Once the threshold is reached, a burst is created and sent into the optical core. The timeout value for the timer-based schemes should be set carefully. If the value is chosen to be too large, the packet delay at the edge might become intolerable. On the other hand, if the value is too small, too many small bursts will be generated, resulting in higher control overhead. In either approach, there is a control packet assigned to each burst which is sent into the network prior to its respective burst in order to reserve the resources required throughout the transmission. While timer-based schemes might result in undesirable burst lengths, threshold-based assembly algorithms do not guarantee on the assembly delay experienced by the packets. A mixed timer/threshold-based assembly algorithm may provide better performance, especially with self-similar traffic, but may have higher operational complexity [18].

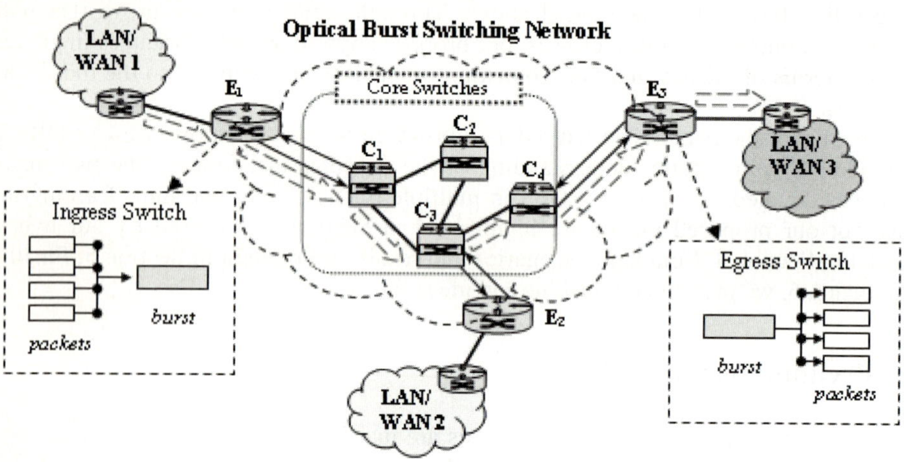

Fig. 1. An OBS network comprising of three edge switches (E_1, E_2 and E_3) and four core switches (C_1, C_2, C_3 and C_4). Edge switches E_1 and E_3 act as ingress and egress nodes respectively, for a burst traveling from LAN/WAN 1 to LAN/WAN 3 via core switches C_1, C_3 and C_4.

A signaling protocol is the procedure through which services such as switching-path establishment, deletion and modification are provisioned. By using a signaling protocol, a control packet can reserve resources for the corresponding data burst by guiding it through a routing path. In an optical network, there are *one-way* and *two-way* reservation signaling protocols. In one-way reservation [13-15], a control packet reserves resources along the path for the corresponding data burst without any acknowledgement from the destination node. On the contrary, in a two-way reservation [5], a control packet collects link and topology information instead of reserving resources for the data burst. The acknowledgement packet from the destination node to the source node reserves resources for the corresponding data burst while traversing along the reverse path. After receiving the acknowledgment packet, the source sends the burst along the reserved path. Since one-way reservation protocols are more flexible, have lower latency, and are more efficient as compared to two-way reservation protocols, they are mainly adopted in OBS networks.

The major responsibilities of the core switches comprise of handling contention resolution and forwarding. When an incoming control packet enters a core switch through one of its input ports, it undergoes an optical to electrical conversion (O/E) so that it can be easily processed electrically. Based on the destination information carried within the control packet, the appropriate output link is determined and the control packet tries to reserve one of the available wavelengths for the time when its corresponding burst arrives at the switch. If any unused wavelength is available, the burst is forwarded to its specified output link and leaves the switch, provided that all the required resources have been reserved beforehand by the control packet. If all the wavelengths are in use, then the burst should wait for one of the wavelength to be freed. Since there is no optical random access-like memory to hold such bursts, they are delayed by employing fiber delay lines (FDLs). Such FDLs serve as optical buffers as they are capable of postponing data burst for a certain amount of time. If a switch is deprived of such FDLs, the burst is immediately dropped out and considered lost. Unlike control packets, data bursts are always transmitted in the optical layer and never undergo any O/E conversion.

2.1 More on Fiber Delay Lines (FDLs)

In this subsection, we examine the structure and functionality of FDLs in more details. Among all proposed structures, we adopt variable-delay fixed-length FDLs, as demonstrated in Fig. 2. As is evident from the figure, each FDL buffer is composed of *D delay lines* (DL), each of which is capable of delaying an optical burst for a total of B time units, where B is given as:

$$B = b\sum_{j=0}^{m} 2^j , \tag{1}$$

in which $m+1$ is the number of stages in a DL and b is the amount of delay a single delay element (specified by circular rings in Fig. 2) can produce. The jth stage of a DL provides a delay of $b2^j$ time units. A burst entering a DL may exit only at the end of any stage, and not in the middle of a stage.

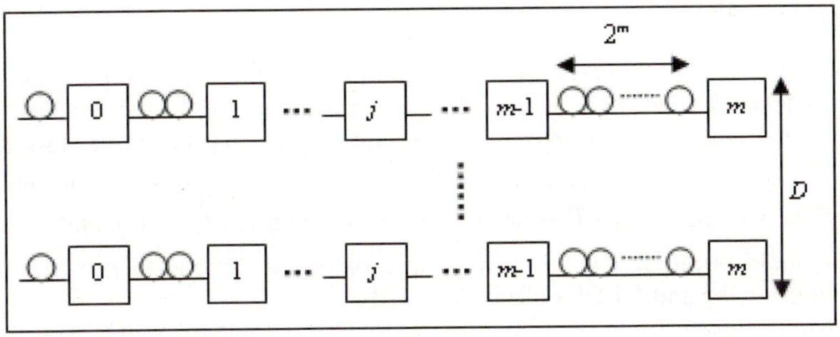

Fig. 2. The structure of a variable-delay fixed-length FDL buffer. The buffer consists of D delay lines, each having m+1 stages. The end of the jth stage is specified by the square box labeled j.

Although FDLs are used to temporarily buffer an optical burst when there is no wavelength available at the instant of burst arrival, they possess specific characteristics that distinguish them thoroughly from typical buffers studied in queuing theory. This difference between two kinds of buffers may result in ambiguities when an optical switch is modeled mathematically using results of the theory. For example, in standard queuing systems, a typical customer may stay in a buffer as long as all servers are busy serving other customers. But the same scenario does not hold in the case of optical buffers. Once entering a DL, a burst can stay only for a limited amount of time (the length of a DL) before it is served. So it appears that ordinary queuing systems (systems with limited waiting space) are not too powerful to model such phenomena. Instead, queues with impatient customers, in which the waiting time in queue is limited in some way, are more realistic tools of modeling.

3 Previous Works on QoS

In this section we investigate, in great details, two previously reported techniques for supporting QoS in OBS. In fact, our proposed model is a hybrid of these two techniques which, as we will show later, combines the individual benefits of the two techniques. The rest of this section is organized as follows: In subsection 3.1, we delineate the *resource-reservation* (or *resource-allocation*) based QoS management technique [7], followed by the novel *extra offset-time* based technique [10] in subsection 3.2. Finally, in subsection 3.3, we explain our motivations for seeking a new paradigm for QoS support in OBS.

3.1 Resource-Allocation Based QoS Management

The first QoS approach we consider is based on resource (wavelength and FDL) allocation and has been reported in [7]. Although the method was originally intended for OPS, it can also be applied to OBS with slight alterations. The main idea is to allocate high priority classes with more resources than those allotted to classes with lower priority. For the sake of simplicity, it is assumed in [7] that only two classes of service, High (H) and Low (L), are available. The scheme may be applied in either of the following ways:

1. *Threshold-based Technique:* In this technique, a burst belonging to the H class may be delayed in an FDL longer than a burst belonging to the L class. In other words, let T_h and T_l be the maximum amount of time that bursts of class H and L, respectively, might be delayed in an FDL when encountering a contention in an intermediate switch. Then, T_h is set to be greater than T_l $(T_h > T_l)$ and T_l is set as the threshold point. In Fig. 3, an example of this technique in presence of three wavelengths (W) and 5 FDLs (B) is depicted.
2. *Wavelength-based Technique:* In this approach, a total of S out of W available wavelengths are reserved to be exclusively utilized by class H bursts, while the other $W - S$ wavelengths are shared between class H and L bursts. Note that W is the total number of wavelengths available in a single optical fiber. This technique is illustrated using a graphical example as shown in Fig. 4 with $W = 4$ and $S = 2$.

3.2 Extra Offset-Time Based QoS Management

In this subsection, we briefly describe the novel extra-offset-time QoS management based approach already reported in [10]. Assume n different classes of service (class 0 to n-1), where class i prioritized over class i-1 ($1 \le i \le n$-1). To each class i, an extra

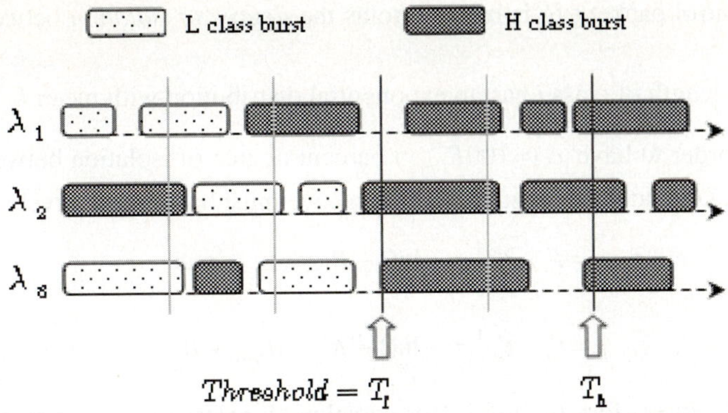

Fig. 3. An example of the threshold-based technique applied to a system having two service classes (L and H) with $W = 3$, $B = 5$ and the threshold value set to T_l

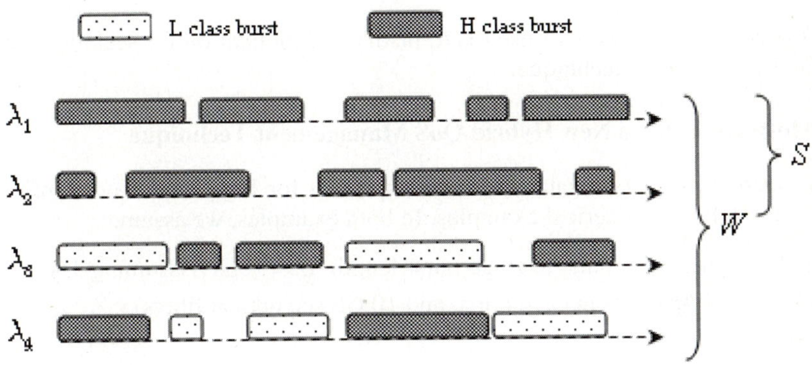

Fig. 4. An example of the wavelength-based technique applied to a system having two service classes (L and H) with $W = 4$ and $S = 2$

offset time t_0^i is assigned (class 0 is assumed to have no extra offset time, i.e. $t_0^0 = 0$). This extra offset time is added to the base offset time t_{base}, such that the expression ($t_{base} + t_0^i$) determines the amount of time between sending a class i control packet and its corresponding data burst. For the sake of simplicity, it is assumed that the base offset time is negligible as compared to the extra offset time, and hence can be ignored in the rest of the discussion. Also, let:

- $t_{ij} = t_0^i + t_0^j$ be the *offset time difference* between classes i and j.
- B be the maximum amount of time that a burst might be delayed in an FDL (following the structure of variable-delay FDLs, described in the previous section, we suppose B is of the form $\left(\sum_{k=0}^{m} 2^k\right) \times b$, for some constant m).
- R_{ij} be the probability that a class i control packet succeeds in reserving resources (wavelength and/or FDL) for its corresponding data burst while competing with a

class j control packet (R_{ij} is referred to as the *degree of isolation* between classes i and j).

- The burst length of class i has an exponential distribution with mean L_i .

Then, in order to have $\alpha\,(=100R_{i,i-1})$ percent degree of isolation between classes i and $i-1$ in FDL and wavelength reservations, we must have, respectively [10],

$$t_{i,i-1} = t_o^i - t_o^{i-1} = -\ln(1-R_{i,i-1})L_{i-1}\text{, and} \tag{2}$$

$$t_{i,i-1} = t_o^i - t_o^{i-1} = -\ln(1-R_{i,i-1})L_{i-1} + B. \tag{3}$$

To derive a direct formula for t_0^i (for wavelength reservation), one may expand (3) as follows:

$$t_o^i = -\sum_{j=1}^{i}\left(\ln(1-R_{j,j-1})L_{j-1}\right) + iB. \tag{4}$$

In the next section, we will use (3) to justify our motivation for seeking a new hybrid QoS management technique.

3.3 Motivations for a New Hybrid QoS Management Technique

In this section, we investigate the behavior of (4) for higher number of prioritized classes through two numerical examples. In both examples, we assume:

- All classes have the same average burst length $L = 12\mu s$ (Assuming 1500 byte IP packets, 10 IP packets in each burst, and 10 Gb/sec optical fibers).
- $B = 31\mu s$ $(=\sum_{k=0}^{4} 2^k)$.
- $R_{i,i-1} = 0.9$ $(1 \le i \le n - 1)$, i.e. we require a 90% degree of isolation between classes i and $i-1$.

We begin with the first example in which the number of classes, n, is taken to be 10. Substituting our assumptions in (4), t_0^9 (extra offset time associated with the most important class) would be approximately equal to 0.5 ms. This upper bound (less than a millisecond) on latency due to extra offset time is tolerable for many applications. However, let us consider the case when the number of classes grows even larger. In the second example, the number of classes increases to 100. Again, using (4), the required extra offset time obtained for the class with the highest priority, t_0^{99}, is approximately equal to 6 ms, which is 12 times that of the highest priority class in the first example.

From the above discussion, it becomes clear that with increase in number of service classes, classes with higher priority suffer from greater network latency (due to extra offset time). This increase in latency might be unacceptable for some real-time applications that require both a high QoS and a response time of less than a few milliseconds. Notice that the latency due to extra offset time is only a fraction of the total latency experienced by a packet through its path in the Internet. Hence, it is desirable to reduce the contribution of extra offset time in total network latency, while still maintaining an effective QoS management scheme. The benefit of this technique, however, is that it imposes no load on core switches. In other words, all QoS management is performed in the ingress switches and core switches are left to their original task of processing headers and switching optical bursts.

On the other hand, resource allocation based QoS does not impose any extra latency to the optical bursts since in this case, bursts are simply transmitted to the network core after a base offset time, which is negligible as compared to the extra offset time, yet the QoS management is left to the switches in the OBS network core. Leaving QoS to the core switches introduces another problem: increase in processing time of burst headers. The more the number of classes is, the higher the processing time of headers becomes. Therefore, this method also seems to be vulnerable to such increase in number of classes. From the above discussions, it is apparent that we are in seek of a QoS technique that concurrently satisfies the following set of goals:

- Supporting a large number of classes.
- Keeping network latency, due to QoS management, as low as possible.
- Imposing little load, due to QoS management, on header processing times at the core network switches.

Based upon the inferences made in this subsection, we propose and explain our hybrid model in the following section.

4 Hybrid Model

From the foregoing discussions, we learnt that neither of the two classical techniques namely, resource reservation and extra offset-time based techniques serve as appropriate traffic differentiation approaches for large number of service classes, as they impose additional processing overload and unacceptable fractional network latency, respectively, to the OBS infrastructure. However, by merging the advantages of these two techniques, we propose a new model that greatly alleviates the processing time and network latency as compared to the two individual basic techniques by classifying the differentiated services, denoted by *n*, as follows:

1. In order to reduce the latency imposed due to the extra offset-time based technique in an ingress edge, we divide the set of *n* service classes into smaller subsets, each having approximately equal number of service classes. Doing so would make it possible for the ingress switch to assign a single extra offset time to every subset rather than every single service class. Such classification would reduce the upper bound on latency and thus, make these services appropriate for time-critical applications. Here onwards, we refer to such subsets of service classes as *inter-service classes* and denote them by $C_{\text{inter}}^{k}(i, i-1, \ldots, j)$, where *i*, *i*-1, …, *j* denote service classes belonging to inter-service class *k* and arranged in the descending order of their priority, i.e. class *i* has a higher priority than class *i*-1, and so on. In Fig. 5, this classification of service classes into inter-service classes at an ingress switch is depicted (as Phase 1), where each inter-service class is composed of two service classes. Rectangular blocks denote service class bursts such that darker blocks represent bursts with higher priority. Also, the curved arrows denote the action of prioritizing classes (in this case, prioritizing inter-service classes using the extra offset-time approach).

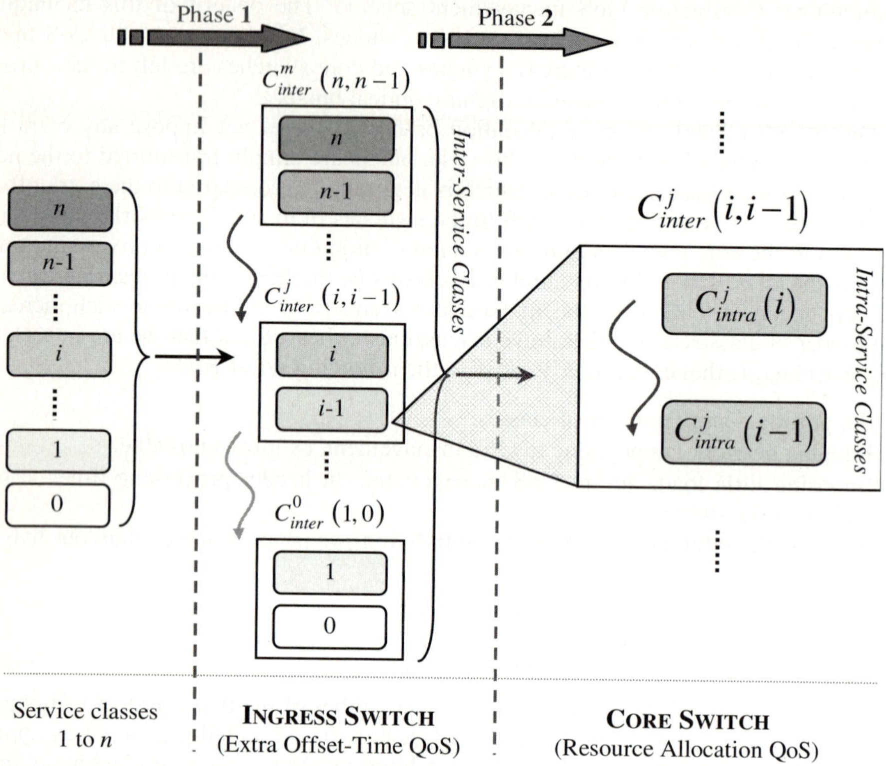

Fig. 5. Graphical representation of the proposed hybrid QoS model for n service classes. Darker rectangular blocks denote bursts of service classes with higher priority. In phase 1, the n service classes are partially prioritized into m inter-service classes ($m < \lceil n/2 \rceil$) at the ingress switch using the extra offset-time approach. In phase 2, the intra-service classes in each inter-service class are prioritized using the resource allocation approach as they reach the core switch.

The new extra offset-time assigned to each inter-service class $C^k_{\mathrm{inter}}(i, i-1, \ldots, j)$ is determined as:

$$t_{k,k-1} = t_o^k - t_o^{k-1} = -\ln(1 - R_{k,k-1})L_{k-1} \,. \tag{5}$$

In (5), the average burst length of any inter-service class k, (L_k), is taken to be a weighted summation of average burst lengths in $C^k_{\mathrm{inter}}(i, i-1, \ldots, j)$, where the weight of average burst length of class $q\,(i \le q \le j)$ is the ratio of the traffic intensity of class q to the total traffic intensity of inter-service class $C^k_{\mathrm{inter}}(i, i-1, \ldots, j)$, i.e.:

$$L_k = \sum_{q=i}^{j} \frac{\rho_q}{\sum_{l=i}^{j} \rho_l} L_q \,, \tag{6}$$

2. Once the service classes are classified into inter-service classes, they reach the core switch to be further prioritized by the resource allocation based technique. Since these arriving inter-service classes have already been partially prioritized at the ingress switches, the overhead existing in such switches is reduced to merely processing and prioritizing service classes within the inter-service classes rather than the entire set of *n* service classes. Here after, we refer to service classes within a specific inter-service class as *intra-service classes* and denote them by $C^k_{\text{int}ra}(i)$, where *k* denotes the inter-service class to which service class *i* belongs. This second phase of intra-service class prioritization is clearly shown in Fig. 5. It should be noticed that the offset time allocated to any $C^k_{\text{int}ra}(i)$ and $C^k_{\text{int}ra}(i-1)$ belonging to the same inter-service class $C^k_{\text{int}er}(i,i-1)$ is the same, and they are prioritized only in the second phase, where $C^k_{\text{int}ra}(i)$ is provided with more resources (wavelength and/or FDL) with respect to $C^k_{\text{int}ra}(i-1)$ using the resource allocation based QoS management techniques (shown as a curved arrow).

5 Simulation Scenarios and Results

In this section, we describe the simulation scenarios and justify our hybrid model by comparing it with the two fundamental QoS management techniques. The performance measures taken into account are burst network latency and blocking probability for the topologically available optical network given in Fig. 6. All scenarios described in this section have been created, simulated and compared using the Ptolemy simulator and in accordance with the assumptions made in subsection 3.3. In all figures through out this section, solid lines denote results obtained using the basic extra offset-time approach, while dotted lines show results obtained through the proposed hybrid model. Hereafter, we use notations *w* and *B* to denote the number of wavelengths and FDLs, respectively.

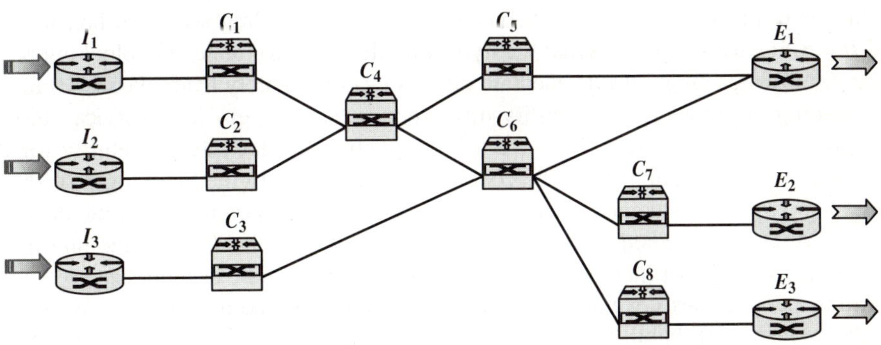

Fig. 6. The optical network considered for simulation. The network comprises of three ingress (I_1, I_2, I_3), eight core (C_1, C_2, C_3, C_4, C_5, C_6, C_7, C_8), and three egress (E_1, E_2, E_3) switches.

Fig. 7 compares the network latency of the given network with 100 service classes for $w = 40$ and $B = 5$. In this scenario, w is kept constant for both models, while the intra-service classes are prioritized in the hybrid model using the 5 FDLs. As shown, for large number of service classes, the network latency suffered by higher prioritized classes in the hybrid model reduces to a great extent. In the extra offset-time model, since the 5 FDLs are shared among all 100 classes, the network latency highly increases. On the contrary, since the 5 FDLs are prioritized among only 10 intra- service classes (after being partially prioritized into inter-service classes using the extra offset-time approach), this latency is greatly reduced.

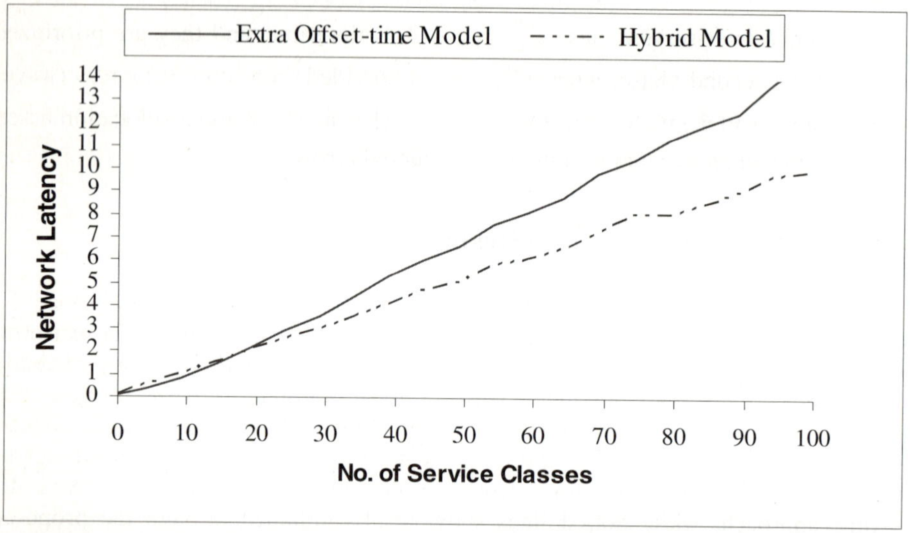

Fig. 7. Latency comparison of the extra offset-time and hybrid models for $w = 40$ and $B = 5$

In Fig. 8, the blocking probability for the large number of service classes has been studied with the same parameters considered in the preceding scenario, i.e., $w = 40$ and $B = 5$. It can be observed that the hybrid model tends to reduce the blocking probability of a wider range of intermediate classes (approximately from classes 10 to 55) by reducing the blocking probability difference between every intra-service class of the same inter-service class. However, this reduction in blocking probability for the intermediate classes happens at the cost of slight increase in the blocking probability of the higher prioritized classes (from 55 to 100). But, when compared to the decrease in network latency studied earlier and reduced blocking probability for the intermediate classes, such minor increase in blocking probability is quite acceptable.

In our next scenario, we study the effect of increasing the number of wavelengths on the blocking probability of the service classes. Due to the capability of WDM technology in supporting large number of wavelengths, we can expect a greater set of wavelengths to be available for prioritizing huge number of service classes. By doing so, the number of wavelengths in each subset assigned to every intra-service class in the resource reservation technique would increase and thus promote lesser blocking of

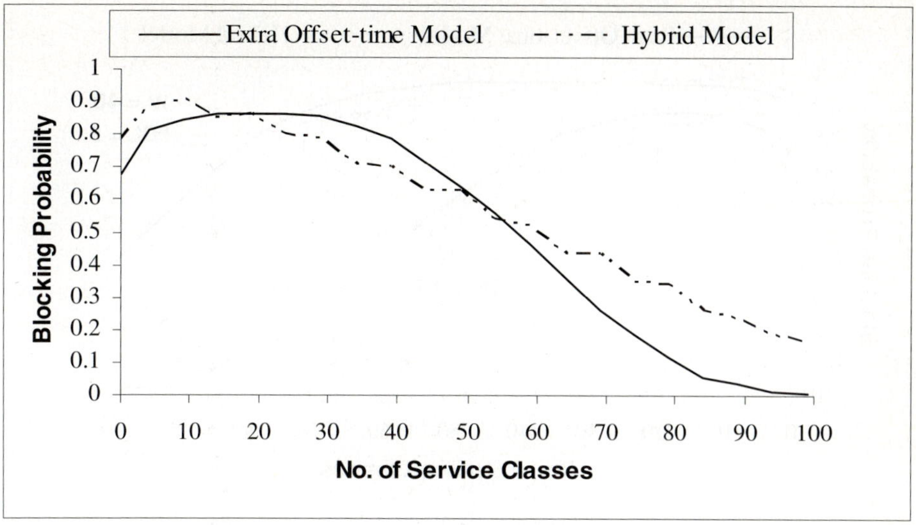

Fig. 8. Blocking probability comparison of the extra offset-time and hybrid models for $w = 40$ and $B = 5$

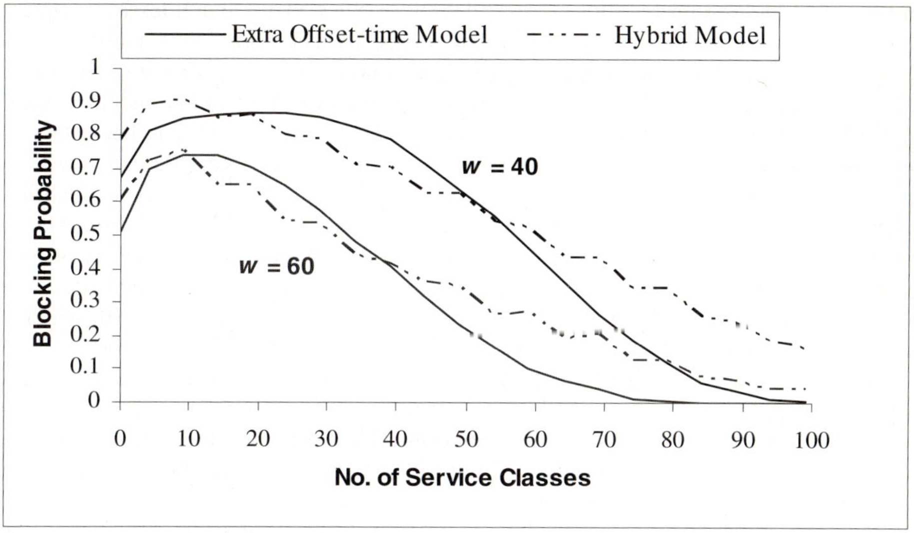

Fig. 9. Blocking probability comparison of the extra offset-time and hybrid models with increase of w from 40 to 60 and $B = 5$

service bursts. This fact is clearly illustrated in Fig. 9, where the number of wavelengths, w, is increased from 40 to 60 and $B = 5$. As can be seen, with increase in w, the high blocking probability of the intermediate classes is greatly reduced to only a small number of low priority classes (approximately classes 5 to 15) in the extra offset-time model. In the hybrid model, as we approach higher prioritized classes, the

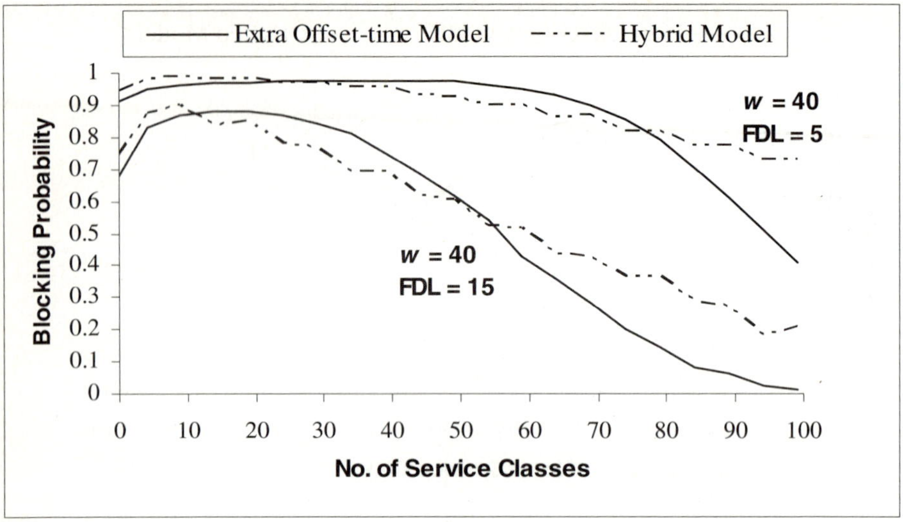

Fig. 10. Blocking probability comparison of the extra offset-time and hybrid models with increase of B from 5 to 15 and $w = 40$

increase in blocking probability of such classes (for $w = 60$) is much lesser compared to the case where $w = 40$. This proves that for the hybrid model, with increase in w, the increase in the blocking probability of higher classes becomes almost negligible.

In Fig. 10, we study effect of increasing the number of FDLs on the blocking probability. We expect to obtain a graph similar to that of Fig. 9. This is because with increase in B, bursts can be delayed for a longer time, thus enabling the core switches to prioritize more number of intra-service classes facing lower blocking probability. In this scenario, w is set to 40, while B increases from 5 to 15. As can be observed, in the case of $B = 15$, the hybrid model pulls down the high blocking probability of most of the intermediate classes (approximately classes 10 to 55), while adding little to the blocking probability of the higher classes (approximately classes 80 to 100), which is nothing compared to the improved network latency obtained (as shown in Fig. 7).

We now study the trend of our performance measures for a specific service class, say class 60 in the following two figures. Extensive validation experiments have been performed for several parameter combinations. However, for the sake of specific illustration, the results are presented for the following cases only. In Fig. 11, we compare the network latency of our hybrid model with that of the extra offset-time model for three different data sets of (w, B), i.e., (10, 5), (20, 10) and (40, 15). These data sets are chosen in such a way so as to illustrate the prominent changes in the trend. We learn that the hybrid model reduces the latency of the entire network as the number of resources (wavelength and/or FDL) increases, while such does not happen in the case of the extra offset-time model.

Similarly, in Fig. 12, we compare the two models on the basis of blocking probability with the same data sets. This figure shows that with increase in resource availability, the blocking probability also increases for our hybrid model. For the data set(40, 15), this probability grows more than obtained for the extra offset-time model.

However, such minor increase in blocking probability is quite tolerable when compared to the improved network latency. Thus, we can conclude that for large number of traffic classes and WDM-based implementation of optical switches, the proposed hybrid QoS management technique provides more improved latency and almost the same blocking probability as the two fundamental techniques namely, extra offset-time based and resource reservation based QoS management techniques.

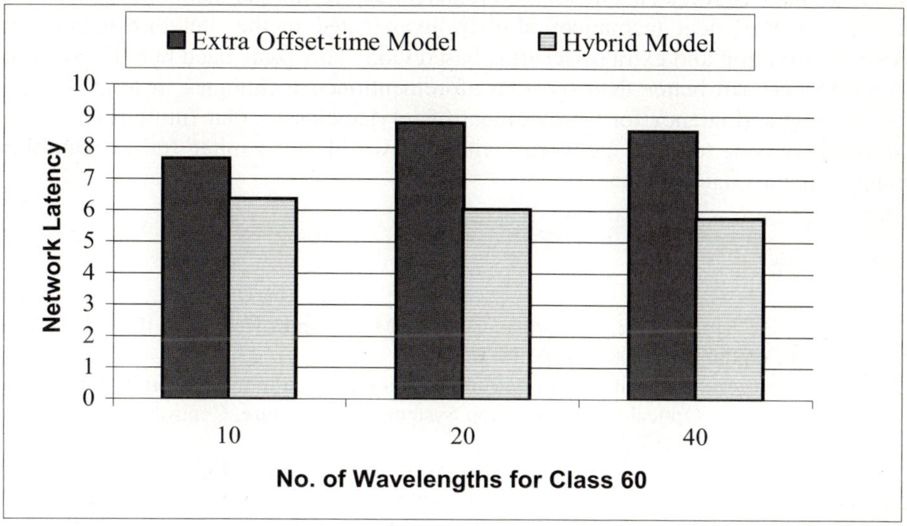

Fig. 11. Network latency comparison of the extra offset-time and hybrid models for service class 60 with $(w, B) = ((10, 5), (20, 10), \text{and } (40, 15))$

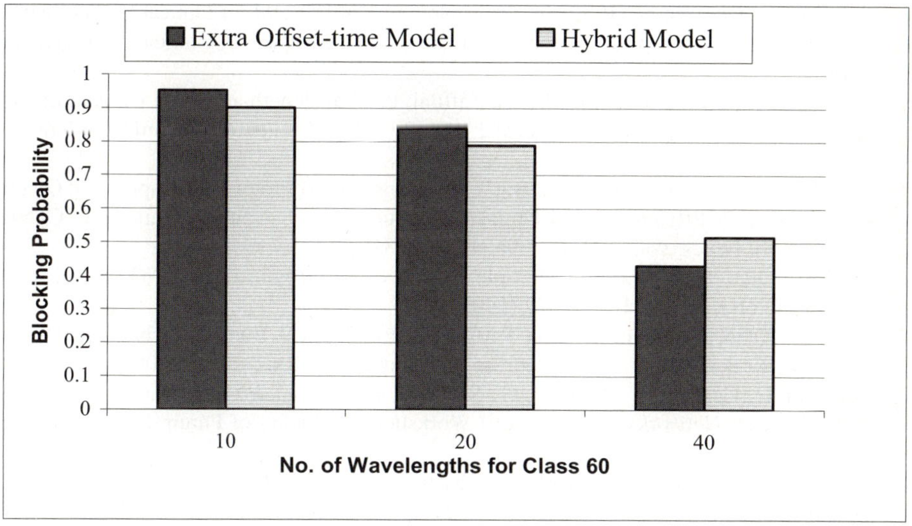

Fig. 12. Blocking probability comparison of the extra offset-time and hybrid models for service class 60 with $(w, B) = ((10, 5), (20, 10), \text{and } (40, 15))$

6 Conclusions

The number of service classes has been growing extensively due to the increase in on-demand multimedia and time-critical applications over the Internet. Several techniques have been suggested and analyzed in the literature so as to improve such quality of service (QoS) differentiation in optical burst switched (OBS) networks. In this paper, we have introduced a novel QoS management technique that combines the best of the two prominent techniques already investigated in the literature namely, resource-reservation and extra offset-time based QoS. This prioritized hybrid model has shown to perform better than the two aforementioned techniques in terms of QoS provisioning and latency for large number of service classes. Our future research will extend this work to OBS networks with real-world self-similar traffic conditions dominating the edge switches.

References

1. Karasan, E., Ayanoglu, E.: Performance of WDM Transport Networks. IEEE J. Selected Areas in Commun. 16, 1081–1096 (1998)
2. Arnaud, B.: Architectural and Engineering Issues for Building an Optical Internet. In: SPIE Proc. - All Optical Communication Systems: Architecture, Control and Network Issues, vol. 3531, pp. 358–377 (1998)
3. Qiao, C., Yoo, M.: Optical Burst Switching (OBS) – A New Paradigm for an Optical Internet. J. High Speed Network 8, 69–84 (1999)
4. Widjaja, I.: Performance Analysis of Burst Admission Control Protocols. IEE Proc. – Commun. 142, 7–14 (1995)
5. Boyer, P.E., Tranchier, D.P.: A Reservation Principle with Applications to the ATM Traffic Control. Computer Networks and ISDN Systems 24, 321–334 (1992)
6. Callegati, F., Corazza, G., Raffaelli, C.: Exploitation of DWDM for Optical Packet Switching with Quality of Service Guarantees. IEEE J. Selected Areas in Commun. 20, 190–201 (2002)
7. Callegati, F., Cerroni, W., Raffaelli, C., Zaffoni, P.: Wavelength and Time Domain Exploitation for QoS Management in Optical Packet Switches. Computer Networks 44, 569–582 (2004)
8. Yoo, M., Qiao, C.: A New Optical Burst Switching (OBS) Protocol for supporting Quality of Service. In: SPIE Proc. – All Optical Communication Systems: Architecture, Control and Network Issues, vol. 3531, pp. 396–405 (1998)
9. Yoo, M., Qiao, C.: Supporting Multiple Classes of Services in IP over WDM Networks. In: IEEE Globecom, pp. 1023–1027 (1999)
10. Yoo, M., Qiao, C., Dixit, S.: QoS Performance of Optical Burst Switching in IP-over-WDM Networks. IEEE J. Selected Areas in Commun. 18, 2062–2071 (2000)
11. Callegati, F., Cerroni, W., Muretto, G., Raffaelli, C., Zaffoni, P.: QoS Routing in DWDM Optical Packet Networks. In: Proc. Int'l Workshop on Quality of Future Internet Services (QofIS) (2004)
12. Callegati, F., Cerroni, W., Raffaelli, C., Zaffoni, P.: DWDM for QoS Management in Optical Packet Switches. In: Proc. Int'l Workshop on Quality of Service in Multiservice IP Networks (QoS-IP), pp. 447–459 (2003)

13. Yoo, M., Qiao, C.: Just-Enough-Time (JET): A High Speed Protocol for Bursty Traffic in Optical Networks. In: IEEE/LEOS Technologies for a Global Information Infrastructure, pp. 26–27 (1997)
14. Yoo, M., Jeong, M., Qiao, C.: A High Speed Protocol for Busty Traffic in Optical Networks. In: SPIE Proc. - All Optical Communication Systems: Architecture, Control and Network Issues, vol. 3230, pp. 79–90 (1997)
15. Hudek, G., Muder, D.: Signaling Analysis for a Multi-Switch All-Optical Network. In: Proc. Int'l Conf. on Commun (ICC), pp. 1206–1210 (1995)
16. Ge, A., Callegati, F., Tamil, L.S.: On Optical Burst Switching and Self-Similar Traffic. IEEE Commun. Lett. 4, 98–100 (2000)
17. Vokkarane, V.M., Haridoss, K., Jue, J.P.: Threshold-based Burst Assembly Policies for QoS Support in Optical Burst-Switched Networks. In: SPIE Proc. OptiComm., vol. 4874, pp. 125–136 (2002)
18. Xiong, Y., Vandenhoute, M., Cankaya, H.C.: Control Architecture in Optical Burst-Switched WDM Networks. IEEE J. Selected Areas in Commun. 18(10), 1838–1851 (2000)

Increasing the Scalability and the Speedup of a Fish School Simulator

Christianne Dalforno, Diego Mostaccio, Remo Suppi, and Emilio Luque

University Autonoma of Barcelona
Computer Architecture and Operating System Department (CAOS)
Barcelona, Spain
chris@caos.uab.es, diego.mostaccio@caos.uab.es,
remo.suppi@uab.es emilio.luque@uab.es

Abstract. In this work we present improvements in the speedup and scalability of a distributed fish school simulator. The results were achieved using a new communication strategy for logical processes (LPs) and changing the algorithm of neighbors' selection that is applied to every fish in each simulation step. In the proposed approach each sender process anticipates future neighbors LP's data needs. The new proposed strategy reduces communication time by limiting the amount of messages exchanged among LPs. The new neighbors' selection algorithm was developed with the aim of avoiding unnecessary work. Diminishing the instructions executed by each fish being simulated in each simulation step caused a great reduction in the simulation time.

1 Introduction

The Individual-oriented Models (IoM) were created to be a solution to the limitations imposed by the models based in a population that use, for example, differential equations, to explain the behavior of a group of individuals.

In IoM the unit of the system is the individual, so the group's heterogeneity can be represented in the differentiated behaviour of each individual. Therefore, "(...) individuals are the building blocks of ecological system. The properties and behavior of individuals determine the properties of the systems they compose" [1].

In [2] many examples of IoM application are given and the authors outline the importance of this kind of model to researches where the individuals' interactions are important to acuratelly represent the group's features. "Organisms get together and form populations and societies, have properties none of which can be explained by properties of the individual organism alone" [3].

We choosed as a case study the simulation of fish schools. Some kinds of fish present the characteristic behavior of being together for a long time maintaining a self-coordinated movement without the presence of leaders[4]. This behavior gained the attention of the biologists interested in investigating this kind of formation.

The issue of April of 2007 of National Geographic [5] presents an overview of the situation of the worldwide fishing. In fact, the fish consumption has been

O. Gervasi et al. (Eds.): ICCSA 2008, Part II, LNCS 5073, pp. 936–949, 2008.

growing and, in around 50 years, the fishing went from about 30 million tons to almost 100 million tons.

One of the species that has been suffering a lot with the growing of its consumption is the blue tuna. The situation of this specie is so critical that was created the *Comisión Internacional para la Conservación del Atún Atlántico* (CICAA), responsible to administrate the reserves of the blue tuna.

It has to be found a balance between commerce interests and preservation of the species. Thus the tendency is the growing of the interests in the investigations about fish behavior, reproduction, development and exploration.

There are some models that represent the behavior of these populations[6]. The one used in this work is an Individual-oriented Model (IoM). This kind of model represents the behavior of a group by applying some rules to a group of individuals interacting among each other.

Our general interest is the application of high-performance computing (HPC) to simulations based on IoM. We beleave that an efficient simulation architecture would permit the simulation of more complex models with greater amount of individuals spending less time.

Furthermore the techniques used in this simulation may be applied to simulations of individuals' displacement in a defined space. It could be more easily done to the ones based on the interaction among the individuals being simulated. Some examples should be: displacement of people in dangerous situations (like presence of fire in a building), dislocation of particles in an environment (considering that an individual can be something different than a biological organism), etc.

In this work we relate a practical experience of improving the scalability and the speedup of a three-dimensional Fish School distributed simulator. The implemented model defines the behavior of a fish as a consequence of its interaction with four selected neighbors [4]. The reaction of a fish to its neighbors is to change its direction. So in each simulation step each fish being simulated will define a new position in the space and a new direction.

This system can simulate a big fish population by dividing the limited simulated space into fixed slices, corresponding to logical processes (LPs), that are each one assigned to a processor in a parallel architecture.

Experimentations done with this application shown a limited scalability because the communication strategy used demands the exchange of a great amount of messages in some situations. The simulator's parallelization caused a great computation time reduction that points to another problem: the sequential simulator's version executes unnecessary operations.

To achieve a better simulator's speedup and scalability, this work presents two main improvements over the previous simulator architecture: (a) It proposes a new communication strategy that permits to control the number of messages exchanged by the logical processes of the system with the aim to increase the system's scalability. (b) It establishes changes in the original algorithm to avoid unnecessary operations reducing the simulation time.

2 The Distributed Simulator

Convetional modelling used to consider the population as the system's base. To do that it is necessary to consider that all of the individuals are similar, presenting the same reactions and affecting in the same way the others individuals and the environment [4][7]. This contrasts with the fact that the individuals have to adapt themselves and its environment to achieve their main objective of reproduction. This adaptation is the result of the interactions of the individual with the environment and other individuals that it has contact and not as a decision based on the conscience about its population. The population's properties will emerge from the actuation of those adaptive individuals [1].

The investigators started to model a population based on simple rules applied to a single individual. The result of this was the developing of the Individual-oriented Models (IoM). This kind of model makes possible the study and understanding of all population's levels based in the relationships between the individuals and between each individual and its environment [8].

The simulator used in this work implements an IoM that represents the movement of a group of fish based in the relationships among its individuals. In each simulation step the new position and the new direction of each fish being simulated will be defined based in the distance between the fish and its four selected neighbors.

Once the model was defined and implemented, the efforts went in the direction of distributing the system. As it is explained in details in section 2.2, the distribution was done dividing the simulated space.

2.1 The Biological Model

The biological model implemented in this work is based on an IoM developed by Huth and Wissel [4] to represent the behavior of a fish school establishing simple rules to be applied to each individual.

This model describes some behavior patterns constituted by four basically reactions to the influence of the individual's neighbors: repulsion, parallel orientation, attraction and look for neighbors.

These reactions are based in the distance between the individual and its nearest neighbors. Some investigations shown that an individual needs to choose only four neighbors to calculate the influences of them in its orientation because more than that will produce the same behavior [4]. To select the best neighbors it is used the front priority algorithm that choose the individuals that are nearest to the direct vision line. So three ranges are defined to select the neighbors and calculate the influence as it can be seen in the figure 1.

The repulsion occurs when the chosen neighbor is in the range I. When two fish are so close, they tend to go away to avoid collision. The neighbor found in the range II influence the fish to go parallel to it. It's a way to maintain a regular distance between them, that is a parallel orientation situation. And when a neighbor in the range III is chosen, it attracts the fish to it. The attraction reduces the distances between the fishes maintaining the group's cohesion. It still

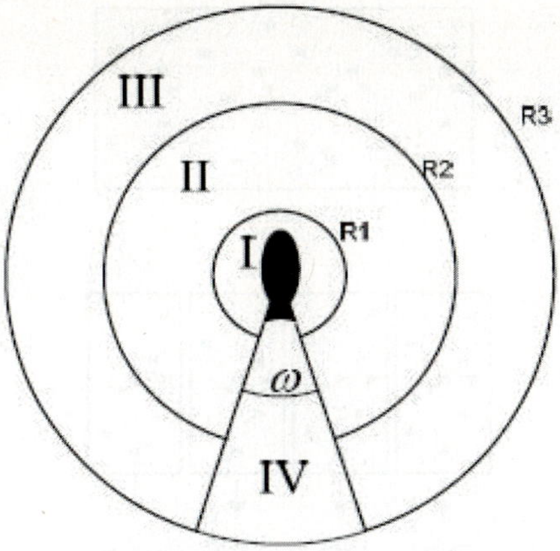

Fig. 1. Ranges of reaction: (I) repulsion area; (II) parallel orientation area; (III) attraction area; (IV) dead area, (ω) dead angle

can be seen in the figure 1, a region called "dead area" (IV), the place behind the individual. The fish can't perceive a neighbor in that area because it is out of its vision area.

The final fish's reaction will be the combination of the reactions to each one of the selected neighbors. When there are no neighbors in those ranges, the fish starts looking for neighbors, swimming in random directions.

2.2 The Computational Model

The first effort of the Mostaccio's work [9] was to extend the Huth and Wissel's model [4] to simulate a three-dimensional space. To do that, it was necessary to revise the mathematical model starting by the definition of a fish that is now represented by the three-dimensional position and velocity vectors. The mathematical model had to be redefined to produce the same behavior in the fish but as a three-dimensional movement.

Once that the mathematical model was established, the challenge was to parallelize the application. In this case, it was decided to split the application based in the space being simulated, i.e. the three-dimensional space in which the fish are being simulated is divided and each one of the resultant logical processes (LPs) are sent to a processor as can be seen in the figure 2 [9].

The main list of fish that contains the identification, the position and the velocity vector of each fish is distributed to the existent process with each fish being assigned to the LP simulating the space that contains its position.

This form of dividing the space causes a situation where an LP will need data from the next LP. This will be produced when one or more fish are next to the border of the LP as can be seen in figure 3 [9]. In this situation, the fish in the border needs data from the neighbor LP that consists in the position and velocity vectors of its possible neighbor fish from outside of its own slice.

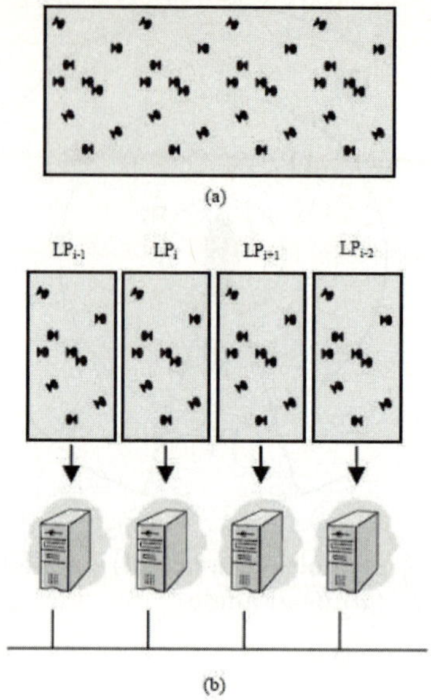

Fig. 2. Model distribution. (a) The whole simulated space. (b) The simulated space divided into slices that will be processed each one by a processor.

Sometimes, it may be necessary to transfer the data of a fish from an LP to another when the fish migrates to a place that is simulated by another processor.

It is evident that a communication strategy is needed to attend the LP's exchanging data requirements. The original communication strategy used guarantees that all exchange data needs are attended but doesn't limit the quantity of messages exchanged.

A simulation step consists of, to each fish been simulated:

- Select the neighbors candidates by calculating the distance between each fish and the other simulated by the same LP.
- Ask for information needed by the fish in the border.
- Send requested data.
- Select four neighbors from the candidates.
- Calculate each fish's new position and velocity vector.

The Original Communication Strategy. As it was mentioned before, there are two basically communication needs:

- Send the migrant fish's data to the appropriate LP.
- Attend the data needs of the fish in the border.

The migrant fish's data are accumulated during the entire simulation step processing to be sent all together in the end of it. In that way, each LP will send a maximum of two messages with migration's data: one to each one of its two neighbor LPs.

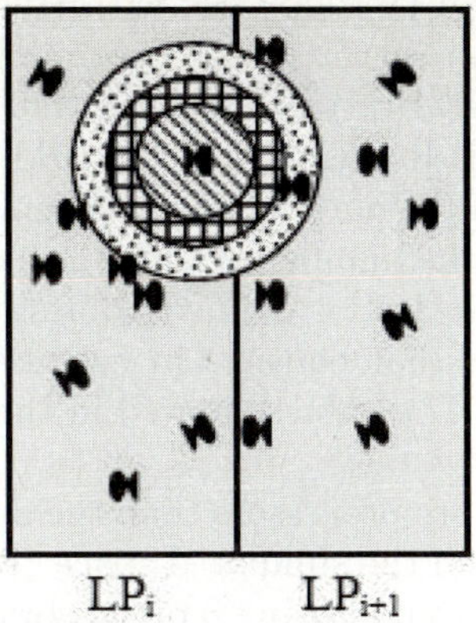

Fig. 3. Fish with a neighbor in the next LP

When it is detected that the fish being calculated is in the border, the LP sends a message to the appropriated neighbor asking for the needed data. This message will be received and processed and the neighbor LP will send an answer message.

The problem with this strategy is that to each fish in the border it will be generated two messages. As more fish in the border as more messages will be interchanged. There is no preoccupation in limiting or controlling the quantity of messages interchanged. Another problem is the fact that two fish in the border can have the same neighbor fish in the neighbor LP. In this case the data of this neighbor fish will be sent more than once.

The Original Neighbor Candidates' Selection Algorithm. In the original neighbor candidates selection algorithm, to select the neighbor candidates of a fish it is necessary firstly to calculate the distance between each fish and the others to select all the ones that are in a distance lower than the third radius excluding the individual in the dead area. [9]

In a sequential version, this algorithm will have the complexity $O(N^2)$. This contrasts with the parallel version that has approximately the complexity of $O((\frac{N}{P})^2)$. [9] What really occurs is that, in the parallel version, the fish are divided by the processors based in its location in space. With this way of distributing the fish it can restrict the neighbor candidates' selection to the group of fish assigned to an LP and its neighbors LPs.

The observation that in the parallel version it wasn't necessary to verify all simulated fish to select the neighbors of each one of them gave us the idea of trying to avoid execution of those unnecessary tasks in the sequential version and after that, apply this solution to each LP of the parallel version.

3 Method

First of all, we did some experiments simulating 640.000 individuals in a space with the dimensions x = 25600, y = 500 and z = 500. We use a Beowulf Cluster with 32 nodes (Intel Pentium IV 3.00 GHz with Hyper-Threading, 1 GB Ram, 80 GB Hdd), an application server (Intel Pentium IV 3.00 GHz with Hyper-Threading, 1 GB Ram, 80 GB Hdd) and a file server (Intel Pentium IV 3.00 GHz with Hyper-Threading, 1 GB Ram, two 200 GB Hdd in RAID-1) interconnected by a 1 Gbit Ethernet switch (3Com). The experiments were done using different numbers of LPs. Each time the number of LPs was incremented, it was done five experiments maintaining the initial defined values. It was measured the time needed to simulate a frame that consists in calculating the new position and velocity vector of each fish. The values showed in the table 1 are the average of these experiments.

The experiments using more processors than the available were done reducing the amount of individuals and the simulated space. To simulate 640.000 individuals that are uniformly distributed using 8 processors means that each processor will simulate 80.000 individuals. To estimate the time spent using 64 processors we did the simulation of 80.000 individuals using 8 processors because simulating 640.000 individuals using 64 processors, each LP is supposed to simulate 10.000 fish. A similar reduction is done with the space dimensions. So what we really do is equivalent to execute only the 8 first slices of the 64 slices that would be simulated.

Analyzing the results presented in table 1, it can be observed that it is not advisable to scale the system with more than 128 LPs. The values in the table show that using more than 128 LPs, the communication spends the best part of the time of simulation. In fact using 64 LPs the communication time represents approximately 25% of computation time what is an undesirable ratio. This can be explained by the fact that the communication strategy used should, in this case, demand the exchange of a great number of messages between neighbor LPs.

When the number of fish being simulated or the number of LPs increases, the number of messages generated can grow in an undesirable way or can stay almost unalterable as can be seen in the table 1. When we used 32, 64 or 128 processors, the communication time measured was around 2.4 seconds in all the cases. Communication time does not decrease in the same way the processing time does, what lowers system scalability.

The great decrease in simulation time achieved with the parallelization of the application can be explained by the change in the complexity of the algorithm. In the sequential version, the complexity is $O(N^2)$, where N is the number of fish

Table 1. Experimental results with the original simulator

LPs	Computation time(seconds)	Communication time (seconds)	Total time (seconds)
1	30785.500	0.000	30785.500
8	489.769	5.176	517.417
16	122.845	3.674	131.548
32	31.293	2.421	36.005
64	8.348	2.492	11.598
128	2.394	2.346	5.255
256	0.758	2.439	3.652

being simulated. This complexity is justified by the fact that to find the neighbors of each fish, its position is compared with all other ones being simulated.

3.1 The New Communication Strategy

As the results' analysis pointed to a limitation of the scalability caused by the communication, we started to analyze the original communication strategy to find a new solution.

The communications were implemented using MPI. Each message sent using MPI has an additional fixed time that is independent of the quantity of data. If the contents of two messages are sent as one, it will spend less time than if it is sent separately.

A way to reduce communication time is to try to reduce the number of messages sent. As it was explained in the case of migrant fish, the minimum amount of messages is exchanged. Therefore, the revision in the communication strategy has to be focused in the messages interchanged to process the fish in the border. The challenge is to reduce the quantity of messages exchanged between the consecutives LPs without avoiding attending the data needs of the fish in the border.

There is a well known technique called halo exchange that suggests a solution to this problem. Bertacchini and Benabén [10] explain this technique with an example that consists in updating a data array in each iteration of a system execution. To change each position of this array it is necessary to use the data of the neighbor positions. If the array is divided to be processed in parallel, as can be seen in figure 4, what happens is that each process needs data that is located at another processor.

Therefore, the boundary cells of each sub-array have to be sent to the neighbors to be used. This data is called hallo cell or ghost cell.

We can apply this solution to the communication problem of the fish school simulator. The idea is instead of waiting for a message asking for the data related to a fish next to the border, it could be sent previously to its consecutive LP all data that could be used to process a fish in that situation. It means that the data about all fish that are in a distance smaller than the third radius needs to be sent to the appropriated consecutive LP.

The idea is to send the data about the fish that are in a distance of the border that can be of interest to a fish in a neighbor LP. Thus, if a fish is next to the border it will not need to send a message asking for data because this data was just received by its LP.

As can be seen in figure 5, the received data will be used as an extension of the slice and thus the neighbor candidates of the fish next to the border will be found locally. This new strategy guarantees that the maximum amount of messages sent by an LP to be used to choose the neighbors of the fish in the border will be two and thus each LP will receive one or two messages with data that extend its simulated subspace to the right and to the left.

Considering the messages with migrant data and simulated space extension data the maximum amount of messages exchanged by one LP is 8. Using the

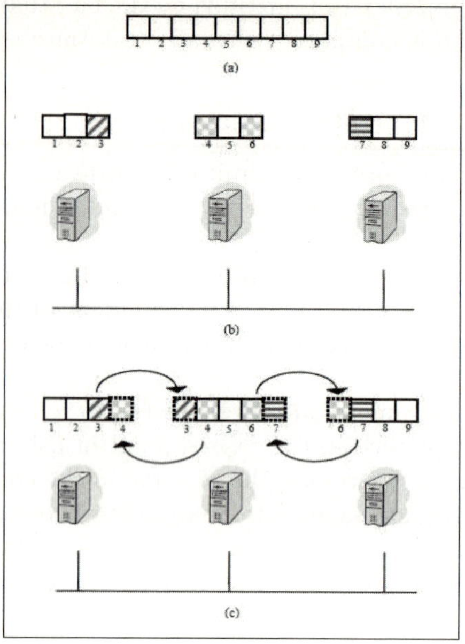

Fig. 4. Example of halo exchange. (a) A vector to be actualized. (b) The same vector distributed into 3 processors. (c) Example of the halo cells exchange.

original strategy, the number of messages increases in function of the quantity of fish in the border of the LP. Thus the maximum amount of messages expected exchanged to be exchanged by an LP is given by:

$$(4 + (8 \cdot \frac{(NR_3)}{X})) \tag{1}$$

where N is the quantity of fish been simulated, R_3 is the third radius and X is the dimension of the simulated space in relation to the x-axis.

3.2 The New Neighbor Candidates Selection Algorithm

To achieve the new propose of avoid consulting all fish to determine the neighbor candidates of one of them, it was necessary to organize the fish's data into the vector to reflect its position in the simulated space.

The strategy used was to sort the fish's data. The ordering is done in function of the x coordinate of the fish's position. To identical values of x, it will be used the y coordinate to order the fish's data and to identical values of y, the z coordinate will be used.

To sort the data into the vector it was used the sort algorithm of the Standard Template Library (STL).

Once the data are organized, a more efficient algorithm of neighbor candidates' selection can be developed. In our new propose we will only avoid to consult the

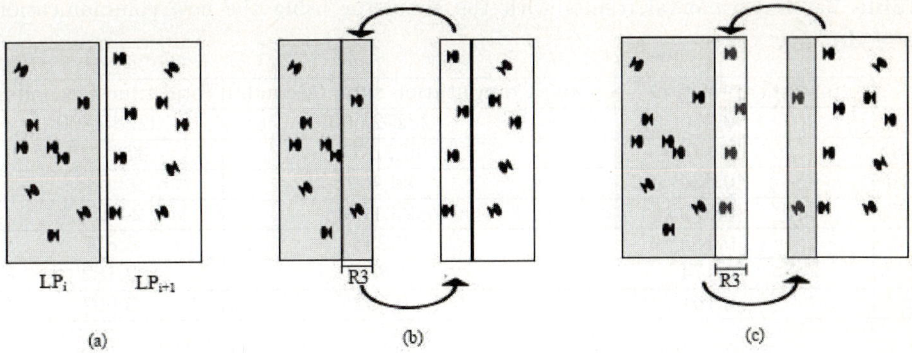

Fig. 5. Example of a slice extension. (a) Two consecutive slices. (b) The area into the slices that contains the fish that can be used by the neighbor slice (i.e. the fish that are in a distance smaller than R_3). This area defines what data will be sent to the neighbor LP. (c) The slices extended with the received data.

fish that are in a distance greater than the third radius in relation to the x coordinate.

Considering that the fish are uniformly distributed in the space, we can assume that the quantity of fish that exists in a space with dimensions x = R_3, y = Y and z = Z (with Y and Z being the dimensions of the total simulated space) is equal to:

$$N_{R_3} = R_3(\frac{N}{X}), \tag{2}$$

where N is the total amount of simulated fish, R_3 is the third radius and X is the x's dimension of the total simulated space.

The new algorithm to choose neighbor candidates has complexity

$$O(\frac{(N^2 2R_3)}{X}), \tag{3}$$

compared to the sequential version. In the parallel version, this algorithm will have, to each LP, complexity

$$O(\frac{(N^2 2R_3)}{XP}), \tag{4}$$

where P is the number of LPs.

4 Results

In a first step, we implemented the new communication strategy. Then we repeated the experiments and the obtained results can be seen in table 2.

Comparing the new data with that presented in table 1, we can see that communication time reduced by the use of the new strategy.

Table 2. Experimental results with the simulator using the new communication strategy

LPs	Communication time(seconds)	Computation time (seconds)	Total time (seconds)
1	0.000	22272.500	22285.800
8	14.401	357.257	374.427
16	0.830	90.469	92.638
32	0.272	23.118	24.191
64	0.185	6.238	6.857
128	0.174	1.805	2.182
256	0.101	0.592	0.807

Table 3. Time spent with communication, computation and in the task of selection of neighbor candidates using the new simulator version

LPS	Communication Time	Computation Time	Candidate Selection Time	Total Time
1	0.000	57.104	35.498	57.104
8	1.150	11.923	9.198	16.213
16	0.726	5.941	4.570	8.388
32	0.210	2.988	2.300	4.130
64	0.398	1.368	1.025	2.293
128	0.117	0.764	0.590	1.201
256	0.089	0.411	0.322	0.634

With the communication problem solved, we implemented the new neighbor candidates' selection algorithm.

In experiments realized using this new algorithm, we found the times showed in table 3.

The results show that the use of the new neighbor candidates selection algorithm reduced a lot the simulation time.

It is important to know that this great reduction is in function of the quantity of individuals and the dimension x of the simulated space. A simulation of a reduced number of individuals would not give so different result between the original algorithm and the new one. In the same way, a space with a small value to the X dimension would not have the same reduction in the time of simulation.

The figure 6 presents the speedup curve of each simulator's versions. It can be observed an improvement achieved by the new solutions.

To calculate the speedup we used the simulation time of the sequential version of the last version of the simulator founded in the table 3, because it is the sequential version with the best simulation time.

Observing the graphic in the figure 6 we can see the speedup's increase achieved by each new solution proposed. If we calculate the tendency line to the original simulator and the last version, we will found the graphic that can be seen in figure 7. We can observe that, after a time, the original simulator's speedup stay almost unalterable while the last version's speedup continues to increase, showing an improvement in the system's scalability.

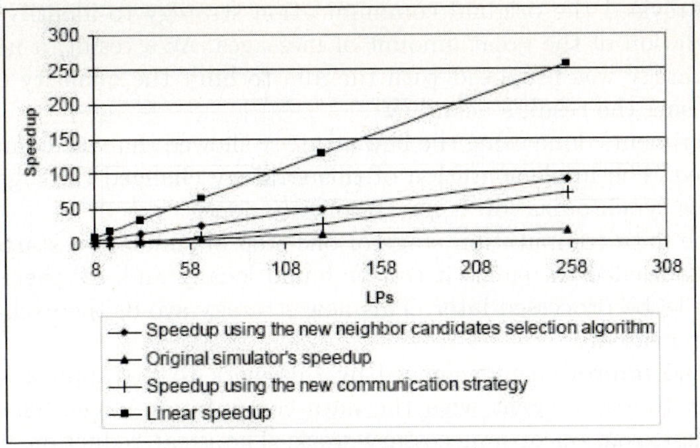

Fig. 6. Speedup

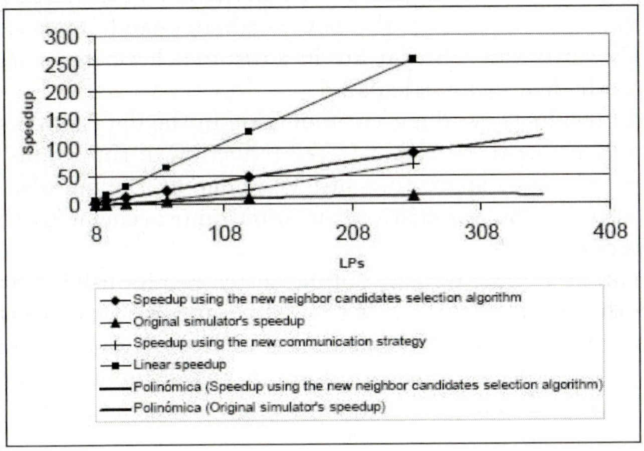

Fig. 7. Speedup with tendency line

5 Conclusion

In this work we did some experimentations with a distributed fish school simulator [9] with the aim to found a way to improve its speedup and scalability.

Experimentations were done with the original simulator to found possible ways to improve it. The results founded with the experiments using the original simulator signalized two points that harms the scalability and the speedup of the system:

- The communication strategy that suggests the exchange of an undesirable number of messages in some situations.
- The execution of unnecessary operations.

First we revised the original communication strategy to identify the causes of the generation of the great amount of messages. As a result, a new communication strategy was proposed with the aim to limit the quantity of messages without loosing the result's accuracy.

The experiments done using the new strategy showed that the communication time reduced. The implementation of this strategy changed the algorithm in a way that the synchronization is now implicitly done.

When the data computation stage of one-step of simulation starts, it means that all data needed to realize it can be found locally and, so, there will be no fish waiting to be processed later. This new strategy avoids the exchange of the same data repeatedly.

The second improvement achieved by this work to the application was the reduction of the time spent with the data computation stage. This was done avoiding the execution of unnecessary work. The great reduction in the complexity of the algorithm pointed to a probably execution of tasks that was not really needed. Analyzing the application, we found that the subdivision of the space, done to parallelize the application, was the key to avoid these tasks.

Based on that, it was developed a new neighbors candidates' selection algorithm that only verify the fish that are in a distance lower than R_3 (in relation to x-axis) from the fish being calculated.

The founded results showed a great reduction in the data processing time. As fewer processors are used, as bigger is the reduction of time. This is explained by the subdivision of the space that just was done by the parallelization. Thus the parallelization of the application suggested improvement to the sequential version in this case.

Other advantage is that we can obtain better results using fewer resources. We spend less time executing the new simulator using 8 processors than the time we spend executing the original simulator using 32 processors. As we mentioned before, with the improvement of the speedup and scalability a more complex model and a greater amount of individuals can be simulated spending a reasonable time.

In future works we aim to restrict the neighbor candidates' selection in relation to the y and z axis objecting to reduce the time avoiding some unnecessary calculus. If the fish are not uniformly distributed a load unbalance situation will be observed. In future works we intend to propose a new distribution of the system.

Acknowledgments. This work has been supported by the MEC-Spain under contracts TIN 2004-03388 and TIN2007-64974.

References

1. Grimm, V.: Individual-based modelling and ecology (2005), `http://jasss.soc.surrey.ac.uk`
2. Breckling, B., Middelhoff, U., Reuter, H.: Individual-based models as tools for ecological theory and application: Understanding the emergence .... Ecological Modelling (2006)

3. Nielsen, S., IMiiller, F.: 11.2. 2 emergent properties of ecosystems. Handbook of Ecosystem Theories and Management (2000)
4. Huth, A., Wissel, C.: The simulation of the movement of fish schools. Journal of theoretical biology (1992)
5. Montaigne, F., Warne, K., Carroll, C.: Crisis mundial de la pesca. National Geographic 20 (2007)
6. Parrish, J., Viscido, S., Grunbaum, D.: Self-organized fish schools: an examination of emergent properties. Biol. Bull. (2002)
7. Huston, M., DeAngelis, D., Post, W.: New computer models unify ecological theory. BioScience (1988)
8. Lorek, H., Sonnenschein, M.: Using parallel computers to simulate individual-oriented models in ecology (1995), `citeseer.ist.psu.edu`
9. Mostaccio, D., Dalforno, C., Suppi, R., Luque, E.: Distributed simulation of large-scale individual oriented models (2006), `journal.info.unlp.edu.ar`
10. Bertacchini, M., Benabén, A.: Reducing mpi communication latency with gamma. Encuentro Científico Internacional (2005)

Hopfield Neural Network for Sea Surface Current Tracking from Tiungsat-1 Data

Maged Marghany, Mazlan Hashim, and Arthur P. Cracknell

Department of Remote Sensing
Faculty of Geoinformation Science and Engineering
Universiti Teknologi Malaysia
81310 UTM, Skudai, Johore Bahru, Malaysia
maged@fksg.utm.my, magedupm@hotmail.com,
mazlan@fksg.utm.my, cracknellarthur@hotmail.com

Abstract. This paper introduces a new approach for neural network application to coastal studies. The method is based on the utilization of the Hopfield neural network to model sea surface current movements from single TiungSAT-1 image. In matching process using Hopfield neural network, identified features have to be mathematically compared to each other in order to build an energy function that will be minimized. In this context, the neuron network has been taken in two dimensions; raw and column in order to match between the similar features of surface pattern. It was required that the two features were extracted from the same location. The Euler method is used to minimized the energy function of neuron equation. The study shows that the surface current features such as structure morphology of water plume can be automatically detected. In TiungSAT-1 data, green and near-infrared bands were competent at sea surface current features detection with high accuracy speed of ±0.14 m/s. It can be said that, Hopfield neural network has highly promised feature enhancement and detection in optical satellite sensor such as TiungSAT-1 image. In conclusion, Hopfield neural network can be used advance computational tool for modeling the pattern movement of sea surface in satellite data.

Keywords: Hopfield neural network, Eluer method, current movement, TiungSAT-1 data.

1 Introduction

Neural networks have been applied successfully to pattern recognition, stereo vision, motion analysis and object tracking problems [4]. A characteristic of neural network is that they may require a long training time but are relatively fast data classifier. In addition, the neural network approach does not require explicit modelling of the data from each source and have shown to approximate class-conditional probabilities for their entire training set in the mean-squared sense[3]. However, in those studies a specific dynamic model of a target system, i.e., a set of individually coupled neuron models, is required for each particular pattern of connectivity among neurons. Therefore, the capabilities of these approaches are restricted since it is usually difficult to

O. Gervasi et al. (Eds.): ICCSA 2008, Part II, LNCS 5073, pp. 950–958, 2008.

find a uniform model through these methods to host all potential dynamical patterns underlying a biological subject[6]. According to Côté and Tatnall [4] promising technique has been used for stereo correspondence and motion understanding problems is based on Hopfield neural networks. In fact, the main advantages of Hopfield neural networks are: the matching problem is coded into any energy function; the approach is used to acquire a minimum of energy function and Hopfield neural networks represent actual physical constraints, heuristics or prior information of the system can be coded into energy function[4] and [6].

In this paper, we address the question of estimating sea surface current pattern using TiungSAT-1 satellite data. This is demonstrated using neural network technique. Hypotheses examined are: (1) there are significant differences between the different bands in detecting ocean current feature patterns, (ii) Hopfield neural network can be applied to single data without needing to include sequential satellite data, (iii) Hopfield neural network can be used as procedures for eliminating inherent noises from TuingSAT-1 satellite data, and (iv) tidal has major impacts on current movements in such water body as the Malacca Straits.

2 Methodology

2.1 Data Set

The TiungSAT-1 satellite data were acquired on 17 March 2001 over the coastline of Penang Island, Malaysia between 100° 09' E to 100° 16' E and 5° 18' N to 5° 26' N(Fig.1). TiungSAT-1 is originally a synchronous orbit at 90° inclination and altitude of 670 km where by the nominal orbital period can be taken as 98 minutes. Furthermore, TiungSAT-1 MSEIS satellite data is a multispectral data with Near Infra-Red (810-890 nm), Red(610-690 nm) and Green(510-590 nm) (Table 1) [8,9,10].

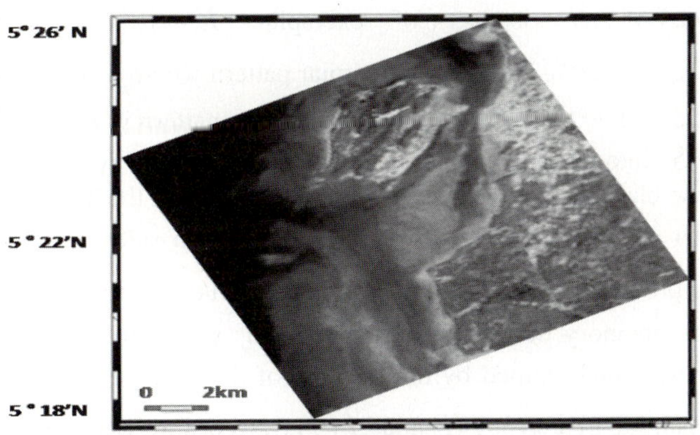

Fig. 1. Study area in Penang coastal waters

Table 1. TiungSAT-1 Satellite Sensor Bands and Wavelengths [1]

Band	Wavelength (nm)	Nominal spectral band
1	510-590	Green
2	610-690	Red
3	810-890	Near-infrared

2.2 Hopfield Neural Network

Neural network techniques are an artificial intelligence procedure that they belong to expert system and knowledge–based approaches to learning [2], [15]. On the other hand, scientists have agreed that neural networks belong to the same classes of techniques as automated pattern recognitions, stereo vision, motion analysis, and object tracking problems [4],[3]. In this context, scientists have defined a Hopfield model as a kind of neural network. The Hopfield network has no special input or output neurons but all are both input and output, and all are connected to all others in both directions (with equal weights in the two directions). Input is applied simultaneously to all neurons which then output to each other and the process continues until a stable state is reached, which represents the network output [11].

According to Côté and Tatnall [4], Hopfield neural networks is considered as a promising method for determining a minimum of energy of function. For instance, motion analysis and object pattern recognitions might be coded into an energy function. Furthermore, the actual physical constraint, heuristics, or prior knowledge of objects and system can be coded into the energy function. A pattern, in the context of the N node Hopfield neural network is an N-dimensional vectors $V=(v_1,v_2,.......v_n)$ and $U=(u_1,u_2,.......u_n)$ from space $S=\{-1,1\}^N$. A special subset of S is set of exemplar $E=\{e^k : 1\leq k\leq K\}$, where, $e^k=(e^k{}_1,e^k{}_2,.......,e^k{}_n)$ and k is exemplar pattern where $1\leq k\leq K$. The Hopfield net associates a vector from S with an exemplar pattern in E. In this context, the net partitions S into classes whose members are in same way of exemplar pattern that presents the classes. For sea surface current features in the TuingSAT-1 satellite image (f), let $f_t(i)\in\{-1,1\}$ to be represent neuron sates, i.e. either -1 or +1, which serve as processing units. Each neuron has a value at time t denotes by $f_t(i)$. The permanent memory of neural set was resided within the interconnections between its neurons which named by the strength of the synapse (w_{ij}) or connection between two pair of neuron $f(i)$ and $f(j)$, According to Nasrabadi and Choo [11], the design specifications for this version of Hopfield net require that $w_{ij}=w_{ji}$ and $w_{ii}=0$. Following, Zhang et al. [14], the propagation rule τ_i which defines how neuron sates and weight combined as input to a neuron can be described by

$$\tau_i = \sum_{j=1}^{N} f_t(j) w_{ij} \tag{1}$$

The Hopfield algorithm has consisted of (i) assign weights to synaptic connections, (ii) initialize the net with unknown pattern, (iii) iterate until convergence and continue features tracking (Fig.2). First step of assign weight w_{ij} to synaptic connection can be achieved as follows:

$$w_{ij} = \begin{cases} \sum_{k=1}^{K} e_i^k e_j^k & \text{if } i \neq j \\ 0 & i = j \end{cases} \tag{2}$$

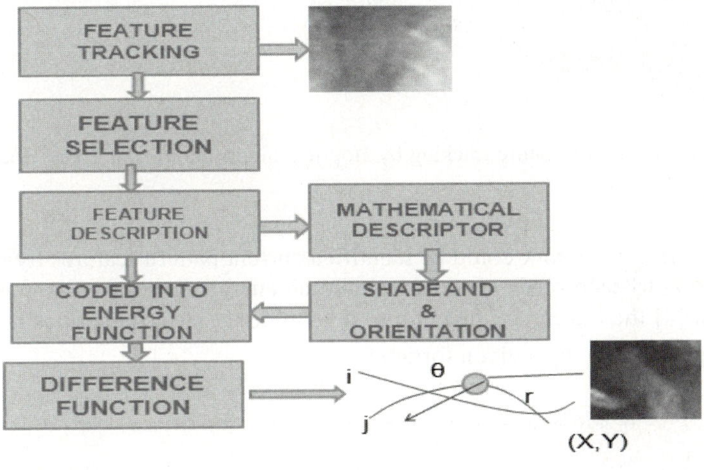

Fig. 2. Block diagram of feature tracking by Hopfield algorithm of step 1

In second step, the pattern that is to be quantified is introduced to the net. If the vectors $V = (v_1, v_2, ...v_n)$ and $U = (u_1, u_2, ...u_n)$ are unknown patterns then, set

$$f_0(i) = v_i \qquad ,1 \leq i \leq N \tag{3}$$

$$f_0(j) = uj \qquad ,1 \leq j \leq N \tag{4}$$

Third step is involved the estimation of next sate values for the neuron in the set using the propagation rule and activation function F. This step can be expressed mathematically by given formula:

$$f_{t+1}(i) = F\left(\sum_{j=1}^{N} f_t(j)w_{ij}, f_t(i)\right) \tag{5}$$

Repetition of second and third steps is involved as fourth step to perform quantification of another sea surface current pattern in TuingSAT-1 image (Fig.3).

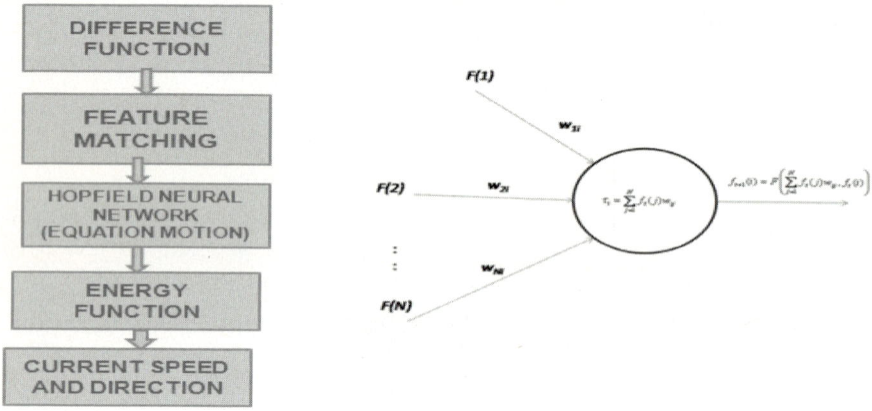

Fig. 3. Block diagram of feature tracking by Hopfield algorithm of step 2 and Hopfield neural network is used in the study

Hopfield neural network could be identified current pattern features by mathematical comparing to each other in order to build an energy function. According to Côté and Tatnall [4] the difference function to determine the discriminations between different features f_i, f_j by a given formula:

$$\begin{aligned} diff\,(f_i, f_j) = {} & G.\max\left|\max\left(\frac{l_i}{l_j}, \frac{l_j}{l_i}\right) - L'', 0\right| + H.\max[\min(\;|\theta_i - \theta_j|, 2\pi - |\theta_i - \theta_j| - \theta'', 0] \\ & + J.\max\,|dis_{ij} - dist'', \theta]| \end{aligned} \tag{6}$$

where, L'' is curvature shape of current feature, dis_{ij} is the distance between sea surface current features f_i and f_j, and G and H and J are constants, and is an angle of orientation of local curve element θ. In addition, $dist''$ and θ'' are the minimum acceptable distance and the maximum acceptable rotation angle, respectively before energy function. In practice, the Hopfield neural network can be quantified by an Euler equation [4] of motion which can be expressed a

$$\frac{dW_i}{dt} = \sum_{j}^{N} \tau_{ij}.\left(F\left(\sum_{j=1}^{N} f_t(j)w_{ij}, f_t(i)\right)\lambda_i W_i\right) + B = \frac{\partial E}{\partial f_{t+1}(i)} \tag{7}$$

where B is the external bias on neuron i, λ is the steepness of the function and E is the network energy, Wang et al.[13]; Juang [5]; Arik [2] which can be defined as

$$E(V,U) = -0.5(v_1, v_2 ...., v_n) \begin{pmatrix} w_{11}\, w_{12} \cdots\cdots\, w_{1k} \\ w_{21}\, w_{22} \cdots\cdots\, w_{2k} \\ . \\ . \\ w_{n1}\, w_{n2} \cdots\cdots\, w_{nk} \end{pmatrix} \begin{pmatrix} u_1 \\ u_2 \\ \vdots \\ \vdots \\ u_k \end{pmatrix} \quad (8)$$

Following Côté and Tatnall [4], the minimization of energy equation 8 can be used for sea surface current features tracking and stereo matching. The contribution of this study is to determine the current velocity in single TuigSAT-1 image. Sequences of point-like current features are selected as candidates for sea surface current movements in TiungSAT-1 image (Fig.2). The adjoining pixels of current features have to full the following criteria: (i) the mean energy of the current pixels must be higher than the mean energy of the surrounding pixels , and (ii) the number of sea current pixels should to be lied between a lower and upper bound. Finally, Hopfield algorithm is implemented on detected features, a displacement vectors, then is estimated by determining the center position of sequences of frames generated over candidate current feature (Fig.3).

3 Result and Discussion

Fig. 4 shows the results of Hopfield neural network for water plume feature in TiungSAT-1 image. It is interesting to find that the Hopfield algorithm provides a details of water plume structure morphology. This result is agreed with Côté and Tatnall [4]. Fig. 5 shows the vector displacement which are appears closed to each other in green band compared to thermal band. Nevertheless, the Hopfield algorithm was tasted in green band and near-infrared band shows that the water vector flows are rotated around Penang island. Conversely, red band shows poor ability for sea current features detection due to its rapid spectra absorption [12]. In fact, Penang Island might be spilt the water flow from Malacca Straits to move parallel to Penang main land coastline and in westward direction along Penang Island. The current flow vectors are produced without noise. In fact, in homogenous flow such as water plume, the pixels are connected continuously which generated neighbor vectors are similar. In this context, the correlation peak corresponds to the true water flow displacement because the best matches between current vectors due to their local similarity. It is also can be noticed that Hopfield algorithm shows current vector variations.

Fig. 5 depicts the similarity of vector displacement in both bands. In fact, the blur band is translated into its pattern representation and introduced to Hopfield net $f_t(i)$ is set equal to v_i , ,$(1 \le i \le N)$. The input pattern was evolved through neuron state changes into the pattern $\hat{V} = (\hat{v}_1, \hat{v}_2, ..., \hat{v}_n)$ of the neuron states at convergence. Furthermore, exemplar pattern was represented the uncorrupted image. This was contributed due to the change of net to an exemplar pattern. This confirms the study of Liang and Wang [6].

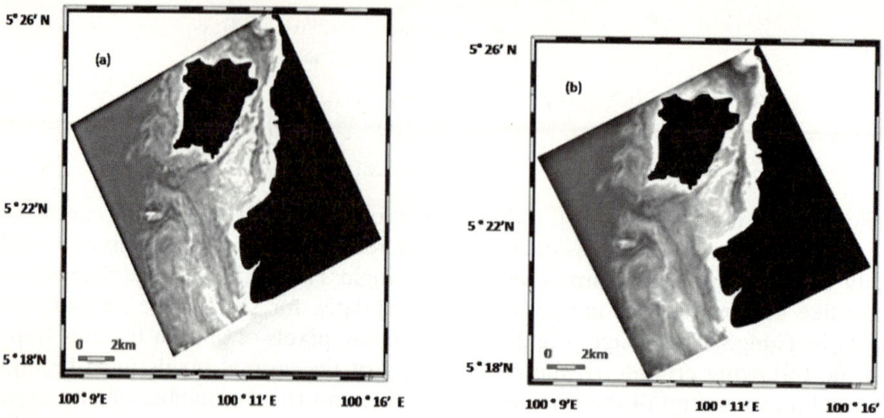

Fig. 4. Output results of Hopfield algorithm for (a) green band and (b) near-infrared Band

The maximum current flow vectors is 0.6 m/s during its rotation around the island. The vectors within 0.24 m/s are appeared to move towards the island and it might be represented the water flow from the south of Malacca Straits. According to Maged and Mazlan [7] the dominant water flow in Malacca Straits in one direction towards the northward. Moreover, it obvious that low current velocity ranged between 0.2 to 0.6 m/s is due to the impact of low tide. This is clear in image due to existence of high sediment accumulations during the acquisition time of the TuingSAT-1 data.

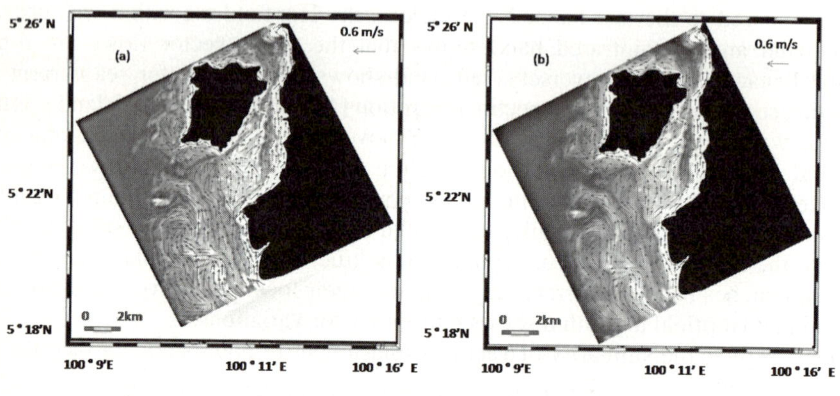

Fig. 5. Vector displacements simulated by Hopfield neural network for (a) green band and (b) near-infrared band

Fig. 6 shows the regression relationship between the sea surface current modeled by Hopfield neural network and one estimated from tidal information. The scatter points are more close to the regression line with r^2 value of 0.989 and accuracy (root

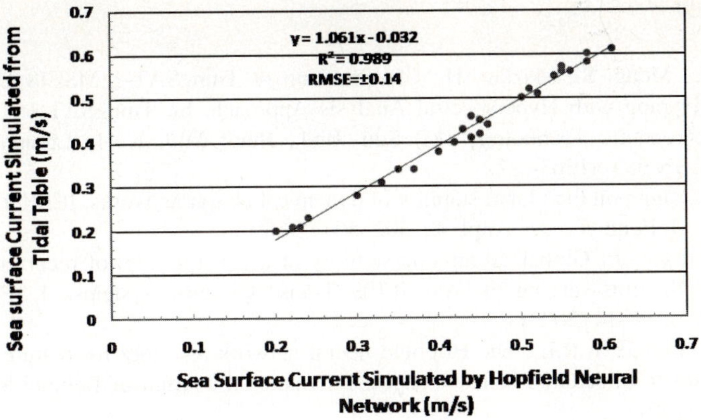

Fig. 6. Regression model for simulated sea surface current by Hopfield neural network and tidal ground data

mean square) of ±0.14 m/s. This confirms that Hopfield neural network provides excellent information on coastal current movements. In fact, Hopfield algorithm parameters are function on the feature size and grey level gradient which could ensure accurate network convergence behavior.

4 Conclusions

This work has demonstrated a new technique for estimating sea surface current from microsatellite data such as TiungSAT-1. The new technique is based on the modification of the Hopfield neural network parameters especially the way of estimation energy function. It is shown that blur image is corrected by utilizing Hopfield neural network. The study confirms that Hopfield is capable algorithm to estimate the sea surface current velocity from green band and near infrared band with high accuracy of ±0.14 m/s. It can be concluded that Hopfield neural network algorithm could be an excellent tool for image reconstruction and sea surface current modeling in single satellite data.

Acknowledgements

We would like to express our appreciation to Universiti Teknologi Malaysia Research Management Center (RMC) for a great cooperation. In fact, RMC is providing excellent advice, plans and management for UTM researchers. We also acknowledge the financial e-science grants have received from the Malaysian Ministry of Science and Technology via its Intensified Research for Priority Areas programme.

References

1. Alvin, L., Meng, S., Mazlan, H.: Classification of TiungSAT-1 MSEIS Data for Land Cover Mapping with Hyperspectral Analysis Approach. In: TiungSAT-1: Data Applications. Astronautic Technology (M) Sdn. Bhd., Bukit Jalil, Kuala Lumpur, pp. 50–60 (2003) ISBN 983-41363-0-7
2. Arik, S.: A note on the global stability of dynamical neural networks. IEEE Trans. Circuits Systems—I: Fund. Theory Appl. 49, 502–504 (2002)
3. Cao, J., Wang, J.: Global asymptotic stability of a general class of recurrent neural networks with time-varying delays. IEEE Trans. Circuits Systems—I: Fund. Theory Appl. 50(1), 34–44 (2003)
4. Côté, S., Tatnall, A.R.L.: The Hopfield neural network as a tool for feature tracking and recognition from satellite sensor images. International Journal of Remote Sensing 18(4), 871–885 (1997)
5. Juang, J.C.: Stability analysis of Hopfield-type neural networks. Neural Networks 10(6), 1365–1373 (1999)
6. Liang, X.B., Wang, J.: Absolute exponential stability of neural networks with a general class of activation functions. IEEE Trans. Circuits Systems—I: Fund. Theory Appl. 47(8), 1258–1263 (2000)
7. Maged, M., Mazlan, H.: Simulation of Oil Slick Trajectory Movements from the RADARSAT-1 SAR. Asian Journal of Geoinformatics 5(4), 17–27 (2005)
8. Mazlan, H., Hazli, H.: Radiometric and geometric information content of TiungSAT-1 MSEIS Data. In: TiungSAT-1: From Inception to Inauguration. Astronautic Technology (M) Sdn. Bhd., Bukit Jalil, Kuala Lumpur, pp. 185–200 (2001) ISBN 983-867-193-2
9. Mohd, I.S., Ming, L.C.: Assessment of the Capability of TiungSAT-1 Satellite Data for Mapping Chlorophyll Distribution. In: TiungSAT-1: Data Applications. Astronautic Technology (M) Sdn. Bhd., Bukit Jalil, Kuala Lumpur, pp. 44–49 (2003) ISBN 983-41363-0-7
10. MatJafri, M.Z., Abdullah, K., Lim., H.S.: Malaysian Tiungsat-1 Imagery for Water Quality Mapping (2003), `http://www.gisdevelopment.net/application/nrm/water/quality/watq0005.htm`
11. Nasrabadi, N.M., Choo, C.Y.: Hopfield network for stereo vision correspondence. IEEE transactions on neural networks 3(1), 5–13 (1992)
12. Robinson, I.S.: Satellite Oceanography An Introduction for Oceanographers and Remote Sensing Scientists, 2nd edn. John Wiley & Sons, New York (1994)
13. Wang, L., Zhang, Y., Zhang, Y.: On absolute stability for a class of nonlinear control systems with delay. Chin. Sci. Bull. 38(16), 1445–1448 (1993)
14. Zhang, Y., Heng., P.A., Fu, A.W.C.: Estimate of exponential convergence rate and exponential stability for neural networks. IEEE Trans. Neural Networks 10(6), 1487–1493 (1999)
15. Zhao, H.: Global asymptotic stability of Hopfield neural network involving distributed delays. Neural Networks 17, 47–53 (2004)

Generator of Synthetic Rainfall Time Series through Markov Hidden States

Héctor Sanvicente-Sánchez and Yolanda Solís-Alvarado

Instituto Mexicano de Tecnología del Agua, Paseo Cuauhnáhuac 8532,
Col. Progreso Jiutepec, Morelos, México. C.P. 62550
{hsanvice,ysolis}@tlaloc.imta.mx

Abstract. This paper presents a method for generating synthetic daily rainfall time-series using Markov hidden states that preserve the temporal, spatial and rainfall quantity correlation in a climatologic network station or in a basin divided into catchments. The method does not require that the rainfall time series must be normally distributed and the seasonal rainfall dependence is managed with a non-homogeneous Markov process. Because the computer implementation does not require a large amount of resources, it is fast and accurately produces synthetic time series. The hydrologic model for simulation and planning developed for the Lerma river basin in Mexico was fed with synthetic rainfall time series generated with this method.

Keywords: Time series, multivariate time series analyst, Markov process, Markov hidden states, hidden Markov model, hydrologic modeling.

1 Introduction

A hydrologic analysis is the first step in the integrated water resources management [1][2]; nowadays this step is carried out using a hydrologic model for simulation and planning that permits determining the basin behavior for different rainfall conditions. This model must be fed with synthetic precipitation time series that preserve the historical statistics and the temporal and spatial correlation.

The most common models used to generate univariate synthetic time series are the autoregressive models AR(p), MA(q), ARMA(p,q) and ARIMA(p,d,q) where p sets the delay, q sets the moving average order and d the differentiation order needed to obtain a stationary process [3][4][5]. The AR(p), MA(q) and ARMA(p,q) models are applied only to stationary processes (processes which are not time-dependent), the ARIMA(p,d,q) model can be applied to processes which are non-stationary in the mean, essentially to processes whose successive differentiation are stationary.

In general, the stationary condition is satisfied only for annual rainfall time series. Periodic rainfall time series (monthly, weekly and daily) always present a seasonal variation, and then statistical parameters (mean, variance, etc) depending on the time interval considered within the year. The autoregressive model application for periodic precipitation time series requires that a year should be divided into stationary time intervals (in general a monthly division is suggested), increasing the statistical parameters to be estimated.

O. Gervasi et al. (Eds.): ICCSA 2008, Part II, LNCS 5073, pp. 959–969, 2008.

Another condition imposed by the autoregressive models is that the process must be normally distributed. To satisfy this condition, generally the precipitation time series are processed taking their logarithm assuming that the rainfall process is log-normally distributed, which not always is true.

The autoregressive model extension for making a multivariate analysis presents the problem that high dimensional statistical matrices occur and that there is a high demand on computational resources to estimate them, as well.

The limitations pointed out before for the autoregressive models make very difficult to get synthetic precipitation time series for a climatological network inside a basin. It is because all spatial, temporal and quantity precipitation correlations must be held into the synthetic precipitation time series to be sure that climatological conditions are according to the historical behavior and that basin system behavior is not an artificial consequence of bad synthetic precipitation time series.

Thus, developing an efficient and accurate method to generate synthetic precipitation time series is important for the hydrologist.

This paper presents a method for generating synthetic rainfall time series using Markov hidden states that preserve the temporal, spatial and rainfall quantity correlation. The method is based in the Robertson, Kirshner and Smyth paper [6] and in the WGEN and ClimGen software [7][8]. The Markov model that was built here represents better the nature because states are defined through the climatological relations that exist inside a basin (precipitation relations). In addition, a Markov model has intrinsic the stochastic behavior of the precipitation.

2 Model for Generating Rainfall Time Series Using Markov Hidden States

Let us consider a climatological network of M stations or a hydrological basin subdivided in M sub-basins (catchments) with rainfall records throughout a given time (see figure 1). From these rainfall records, it is possible to obtain two correlated time series. The first is a series of multivariate stochastic precipitation vector observations along the time $\mathbf{P}_{1:T} = \{\mathbf{P}_t\}$, $\mathbf{P}_t = (P_t^1, \ldots, P_t^M)$, $P_t^m \in \Re^+$, where T sets the series size, t is an index that determines the observation time (for daily observations t sets the consecutive day on which the observation is carried out) and m is the station number or catchment number. The second is a series of hidden climatological states at which the system (the network or the basin) is at time t, $E_{1:T} = \{E_t\}$, $E_t \in \{1, \ldots, K\}$, where K is the maximum number of states in the system. The two time series are bi-directionally related, in such a way that the climatological state E_t is function of the rainfall condition in the system at time t, which is set by the precipitation vector, $\mathbf{P}_t$, in the M dimensional space defined by the climatological stations or catchments. But in the same way, the inverse relation establishes that the precipitation probability, $p(\mathbf{P}_t)$, in the network depends on the climatological state E_t at that time.

$$E_t = f(\mathbf{P}_t) \tag{1}$$

$$p(\mathbf{P}_t) = p(\mathbf{P}_t/E_t) \tag{2}$$

Figure 2 shows a graphical representation for equation 1 which establishes the relationship among the climatological states E_t and the multivariate rainfall observations $\mathbf{P}_t$, in the tridimensional system defined by the catchments in figure 1.

Figure 2 shows how the tridimensional space of rainfall is subdivided into regions that define the climatological states of the system, and how, through a rainfall vector, $\mathbf{P}_t$, it is possible to establish the climatological state, E_t, in the network only by determining the region in which such a point is located.

Equation 2 establishes for $\mathbf{P}_t$ independence for all other variables in the model up to time t, conditioned only by the climatological state E_t at time t, that is,

$$p(\mathbf{P}_t/E_{1:t}, \mathbf{P}_{1:t-1}) = p(\mathbf{P}_t/E_t) \tag{3}$$

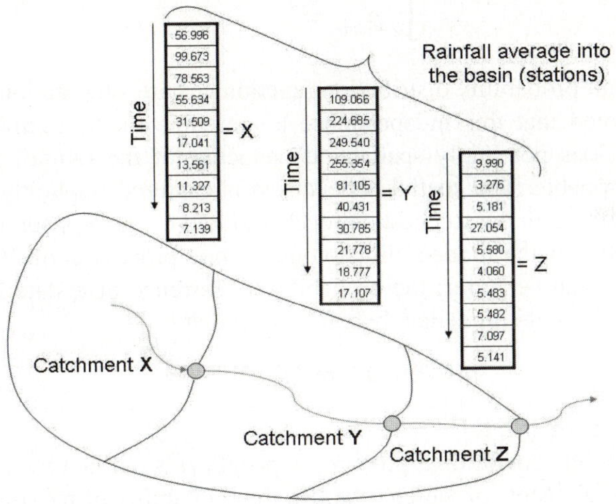

Fig. 1. Precipitation time series for a basin divided into three catchments

X	Y	Z	E
56.996	109.066	9.990	1
99.673	224.685	3.276	2
78.563	249.540	5.181	2
55.634	255.354	27.054	3
31.509	81.105	5.580	1
17.041	40.431	4.060	4
13.561	30.785	5.483	4
11.327	21.778	5.482	5
8.213	18.777	7.097	5
7.139	17.107	5.141	5

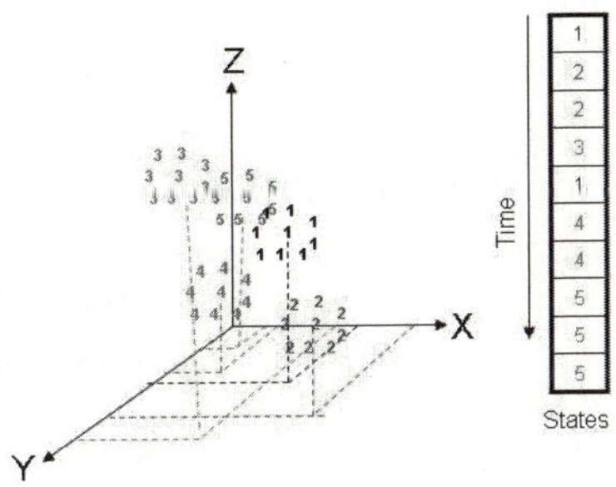

Centres

E	X	Y	Z
1	55.093	253.027	25.000
2	54.810	99.999	8.821
3	17.041	40.431	4.060
4	13.561	30.785	5.483
5	11.327	21.778	5.482

Spatial and temporal correlation

1. Moderate rainfall in all the catchments
2. Strong rainfall in (**X**, **Y**) and scarce in **Z**
3. Strong rainfall in (**Y**, **Z**) and moderate in **X**
4. Scarce rainfall in (**X**, **Z**) and moderate in **Y**
5. Scarce rainfall in (**X**, **Y**) and moderate in **Z**

Fig. 2. Relation among rainfall vectors series $\mathbf{P}_{1:T}$ and states series $E_{T:1}$

To evaluate $p(\mathbf{P}_t/E_t)$, rainfall observation independence at each station at time t must be assumed. That is, the rainfall observation at each station depends only on the hidden climatological state E_t and not on observations at other stations at time t.

$$p(\mathbf{P}_t = \mathbf{r}/E_t = s) = \prod_{m=1}^{M} p(P_t^{m} = r^{m}/E_t = s) = \prod_{m=1}^{M} p_{smr} \tag{4}$$

where $r^m \in \Re^+$ is the precipitation value at station m, $s \in \{1, \ldots, K\}$ is the hidden climatological state of the system and $p_{smr} \in [0, 1]$ is the probability that a rainfall r^m occurred in station m due to the hidden climatological state, s. Then,

$$\int_{m} p_{smr} = 1 \tag{5}$$

defines the rainfall probability distribution function at state s for station m.

It must be noted that this independence assumption on the rainfall observations among stations does not imply spatial independence of the rainfall process, which would be unreasonable. The spatial dependence is captured implicitly by the hidden climatological state E_t in the stochastic process defined by the $E_{1:T}$ series.

Let us assume that the hidden climatological state process defined by $E_{1:T}$, has a first order Markovian behavior; the probability occurrence of a state E_t at the time t depends only on the state immediately before, E_{t-1}; that is,

$$p(E_t/E_{1:t-1}) = p(E_t/E_{t-1}) \tag{6}$$

where E_t and $E_{t-1} \in \{1, \ldots, K\}$.

For equation 6, the conditional probability $p(E_t/E_{t-1})$ is defined by a stochastic matrix, at time t, the elements of which establish the probability of moving from a state i to another state j. This matrix is known as the transition probability matrix of the Markov process at time t and its elements are known as the transition probabilities at time t. These transition probabilities, then, are defined as:

$$p(E_t = j / E_{t-1} = i) = p_{ij}(t)$$

where $p_{ij}(t) \in [0, 1]$. The transition probability matrix is defined as:

$$p(E_t / E_{t-1}) = [p_{ij}(t)] = \begin{bmatrix} p_{11}(t) & p_{12}(t) & \cdots & p_{1K}(t) \\ p_{21}(t) & p_{22}(t) & \cdots & p_{2K}(t) \\ \vdots & \vdots & \ddots & \vdots \\ p_{K1}(t) & p_{K2}(t) & \cdots & p_{KK}(t) \end{bmatrix} \tag{7}$$

where $\forall i$, $\sum_{j=1}^{K} p_{ij}(t) = 1$.

If the transition probabilities among states do not change with time,

$$p_{ij}(t) = p_{ij}$$

so the Markov process defined by $E_{1:T}$ and $\mathbf{P}_{1:T}$ and modeled by equation 6 is homogeneous, but if the transition probabilities changes with time, the process is non-homogeneous.

3 Model Implementation

Notice that the modeling described until now considers real rainfall value vectors. This allows that state definition (figure 2) determines in which catchments or stations it is raining and in which ones it is not; it also allows us to define the type of storms according to the rainfall quantity that the stations record, instead of only defining precipitation occurrence as in Robertson et al [6]. That is, the state makes a correlation among stations in temporal, spatial and rainfall quantity. For this reason we have chosen the terminology "climatological state". The fact that states define the type of storms at each station allows the adjustment of a probability distribution function (pdf) by station and by state, according to equations 4 and 5 and permits calculation of the rainfall quantity in the synthetic series generation process.

According to the modeling process described, it is quite clear that state definition is a fundamental step and such definition does not depend on the time variable in the series; rather it depends on relations that rainfall has among stations or catchments (the precipitation relations into the M-dimensional precipitation space defined by the stations or catchments). This is the reason why, in the present work, we use the traditional algorithm k-means using the minimal distance criteria to cluster the precipitation points and define the states [9].

Although the climatological state definition does not depend on the time variable, rainfall is a seasonal process that varies throughout the year; consequently for periodic series (monthly, weekly, daily) the transition probabilities between states, $p_{ij}(t)$, depend on the time (i.e. they vary throughout the year). Because a state intrinsically determines the quantity of rainfall in the stations, its occurrence probability in the series depends on the season in the year (or the period of the year being considered).

According to the facts described, it can be established that the model given by equation 6 is a homogeneous hidden-state Markov model (HMM) if the transition probabilities between states does not vary with time, this condition is true for stationary series (generally for series of annual precipitation) and this is a non-homogeneous hidden-state Markov model (NHMM) if the transition probabilities change with time, as in periodic series.

For monthly, weekly and daily rainfall series a set of matrices that determine the transition probabilities variation throughout the year must be established. These matrices will be repeated year by year, assuming that the process of annual rainfall is stationary. This can be accomplished by dividing the year into time intervals and determining the transition probabilities for each interval. An appropriate division of the year is to consider each month separately and to obtain its corresponding matrix of transition. In this way, it is possible to generate synthetic series of hidden climatological states starting from an initial climatological state and considering the transition matrix for the month of the year in which it is going to generate the new state (see figure 3) and so on.

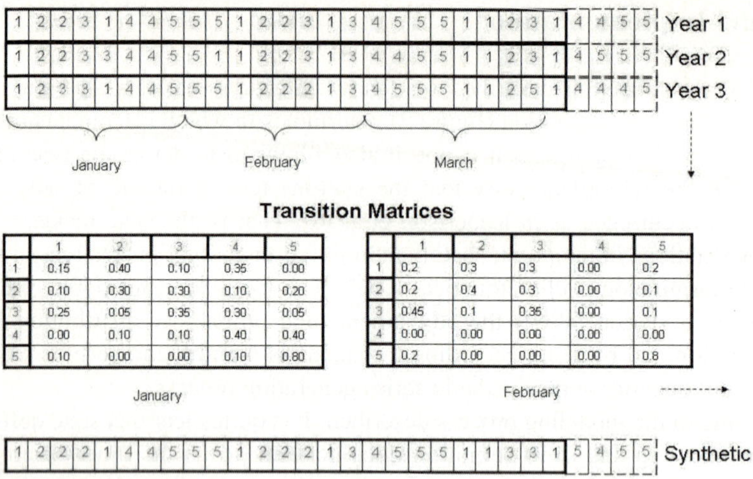

	1	2	3	4	5
1	0.15	0.40	0.10	0.35	0.00
2	0.10	0.30	0.30	0.10	0.20
3	0.25	0.05	0.35	0.30	0.05
4	0.00	0.10	0.10	0.40	0.40
5	0.10	0.00	0.00	0.10	0.80

	1	2	3	4	5
1	0.2	0.3	0.3	0.00	0.2
2	0.2	0.4	0.3	0.00	0.1
3	0.45	0.1	0.35	0.00	0.1
4	0.00	0.00	0.00	0.00	0.00
5	0.2	0.00	0.00	0.00	0.8

Fig. 3. Setting of transition matrices for each month for a non-homogeneous hidden-state Markov model (NHMM) and generation of a synthetic series

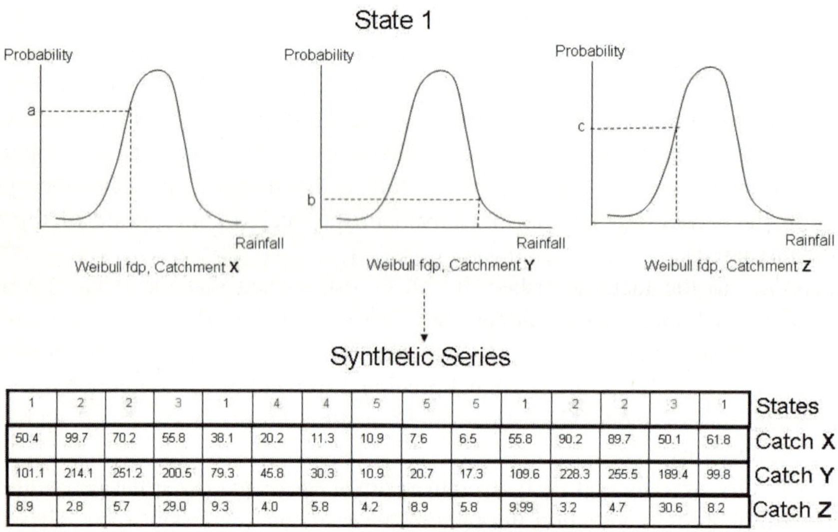

1	2	2	3	1	4	4	5	5	5	1	2	2	3	1	States
50.4	99.7	70.2	55.8	38.1	20.2	11.3	10.9	7.6	6.5	55.8	90.2	89.7	50.1	61.8	Catch X
101.1	214.1	251.2	200.5	79.3	45.8	30.3	10.9	20.7	17.3	109.6	228.3	255.5	189.4	99.8	Catch Y
8.9	2.8	5.7	29.0	9.3	4.0	5.8	4.2	8.9	5.8	9.99	3.2	4.7	30.6	8.2	Catch Z

Fig. 4. Fitting a pdf for each state and each catchment to calculate rainfall quantity for a synthetic hidden climatological state series

As mentioned above, precipitation quantity from a hidden climatological state is carried out using equations 4 and 5, that is, for each hidden climatological state and for each station or catchment, a probability distribution function (pdf) must be fitted from the $\mathbf{P}_{1:T}$ and $E_{1:T}$ series that define how much rainfall is presented (figure 4).

Following the schema given by the WGEN and ClimGen software[7][8], the pdf defined by equations 4 and 5 for each state has the form:

$$p_{smr} = \begin{cases} \pi_0 & si\, r^m = 0 \\ \pi_1 . f(r^m) & si\, r^m > 0 \end{cases} \tag{8}$$

where $r^m \in \Re^+$ is the precipitation value in station m, π_0 is the dry day probability for state s, π_1 is the wet day probability for state s and $f(r^m)$ is a pdf fitted to the wet days in state s, $f(r^m)$ is a gamma function for the WGEN software and is a Weibull function for the ClimGen software. The algorithm developed for this work uses a Weibull pdf function which is suggested by Selker and Haith[10] as well.

From the above description the algorithm to generate the synthetic precipitation time series can be divided into four steps. The first step consists of determining the state regions using the M-dimensional precipitation system given by the climatological stations network or the catchments, and from this definition integrating the historic climatological hidden-state time series, $E_{1:T}$. The second step consists of determining the transition matrix (or matrices for periodic time series) from $E_{1:T}$ and fitting the pdf's for each state and each station that determine the rainfall quantity (equations 5 and 8). The third step is to generate a synthetic climatological hidden-state time series through a MonteCarlo simulation process, given an initial climatological state and the transition matrix (or matrices). The final step is to calculate the precipitation-quantity synthetic time series from the synthetic climatological hidden-state time series and the pdf fitted for each climatological state and station. The rainfall quantities are calculated through MonteCarlo events, using the pdf's calculated in the third step.

4 Results

The non-homogeneous hidden-state Markov model (NHMM) was implemented into a computer program. It has been applied in the Lerma river basin (located in the middle of Mexico) to generate synthetic series for feeding a hydrologic model for simulation and planning. In order to generate the transition matrices set, a 50-year historical daily

Table 1. Correlation coefficients for the historic precipitation time series for the Lerma river catchments

	ALZ	RAM	TEP	TEP	SOL	BEG	AME	PER	YUR	SAL	ADJ	ANG	COR	YUR	DUE	ZUL	CHA
ALZ	1.00																
RAM	0.63	1.00															
TEP	0.60	0.66	1.00														
TEP	0.64	0.85	0.74	1.00													
SOI	0.50	0.52	0.55	0.59	1.00												
BEG	0.32	0.34	0.36	0.40	0.43	1.00											
AME	0.29	0.31	0.31	0.37	0.39	0.64	1.00										
PER	0.38	0.42	0.43	0.50	0.52	0.59	0.51	1.00									
YUR	0.38	0.39	0.41	0.44	0.48	0.42	0.35	0.62	1.00								
SAL	0.43	0.45	0.47	0.52	0.59	0.49	0.42	0.67	0.67	1.00							
ADJ	0.37	0.38	0.38	0.44	0.47	0.56	0.44	0.56	0.47	0.52	1.00						
ANG	0.36	0.33	0.32	0.36	0.38	0.27	0.21	0.33	0.35	0.38	0.41	1.00					
COR	0.42	0.43	0.44	0.50	0.56	0.60	0.49	0.65	0.56	0.62	0.70	0.43	1.00				
YUR	0.40	0.40	0.40	0.45	0.46	0.40	0.34	0.46	0.47	0.51	0.52	0.44	0.59	1.00			
DUE	0.36	0.34	0.34	0.38	0.40	0.30	0.28	0.36	0.34	0.40	0.39	0.44	0.45	0.58	1.00		
ZUL	0.33	0.31	0.33	0.36	0.41	0.53	0.41	0.43	0.39	0.44	0.49	0.35	0.54	0.48	0.39	1.00	
CHA	0.39	0.36	0.35	0.40	0.40	0.35	0.28	0.37	0.38	0.43	0.44	0.45	0.49	0.61	0.59	0.50	1.00

Table 2. Correlation coefficients for the 50-year synthetic precipitation time series for the Lerma river catchments

	ALZ	*RAM*	*TEP*	*TEP*	*SOL*	*BEG*	*AME*	*PER*	*YUR*	*SAL*	*ADJ*	*ANG*	*COR*	*YUR*	*DUE*	*ZUL*	*CHA*
ALZ	1.00																
RAM	0.54	1.00															
TEP	0.52	0.60	1.00														
TEP	0.55	0.61	0.59	1.00													
SOL	0.48	0.53	0.51	0.55	1.00												
BEG	0.35	0.35	0.36	0.38	0.47	1.00											
AME	0.33	0.34	0.33	0.35	0.42	0.52	1.00										
PER	0.40	0.43	0.44	0.47	0.52	0.54	0.48	1.00									
YUR	0.38	0.40	0.42	0.43	0.45	0.46	0.44	0.53	1.00								
SAL	0.40	0.42	0.44	0.46	0.49	0.51	0.44	0.58	0.51	1.00							
ADJ	0.40	0.40	0.43	0.44	0.49	0.52	0.47	0.55	0.51	0.54	1.00						
ANG	0.34	0.36	0.35	0.37	0.39	0.34	0.32	0.38	0.38	0.38	0.42	1.00					
COR	0.44	0.45	0.45	0.49	0.54	0.51	0.47	0.56	0.52	0.54	0.55	0.43	1.00				
YUR	0.40	0.40	0.43	0.44	0.46	0.44	0.43	0.50	0.48	0.50	0.51	0.50	0.51	1.00			
DUE	0.35	0.36	0.37	0.37	0.41	0.36	0.33	0.41	0.38	0.42	0.43	0.56	0.46	0.50	1.00		
ZUL	0.36	0.37	0.38	0.40	0.46	0.55	0.52	0.55	0.50	0.51	0.55	0.40	0.54	0.49	0.42	1.00	
CHA	0.39	0.40	0.39	0.42	0.44	0.40	0.37	0.45	0.43	0.45	0.49	0.52	0.49	0.51	0.53	0.46	1.00

precipitation record was used for each of the 17 catchments located in the case study area. Previous to feeding the hydrologic model for simulation and planning, a statistical analysis was carried out for an ensemble of one 50-year synthetic series and five 60-year synthetic series.

Table 1 shows the correlation coefficient matrix for the 50-year historic precipitation records (the correlation precipitation between catchments) and table 2 shows the correlation coefficient matrix for the 50-year synthetic series (the 60-year synthetic series have similar values). By analyzing tables 1 and 2, it can be noted that precipitation correlation is preserved acceptably in the synthetic time series.

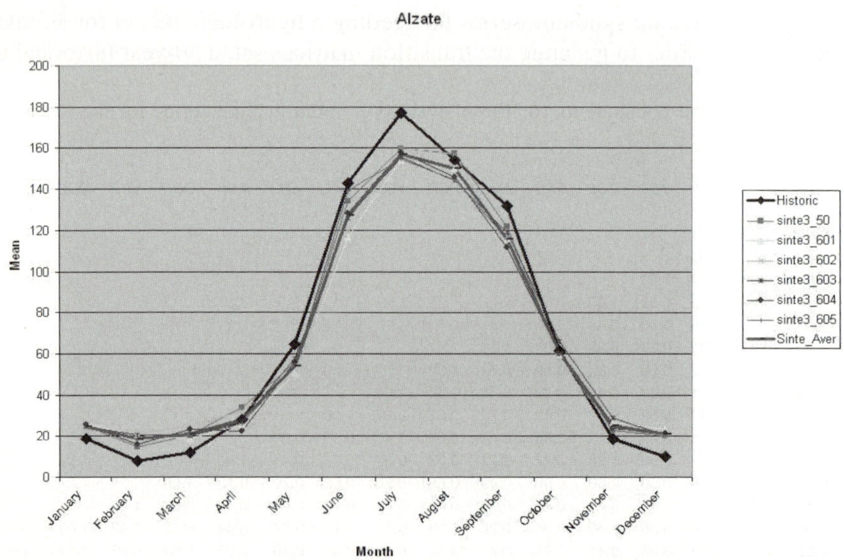

Fig. 5. Monthly mean precipitation for the historic time series and the ensemble of synthetic time series

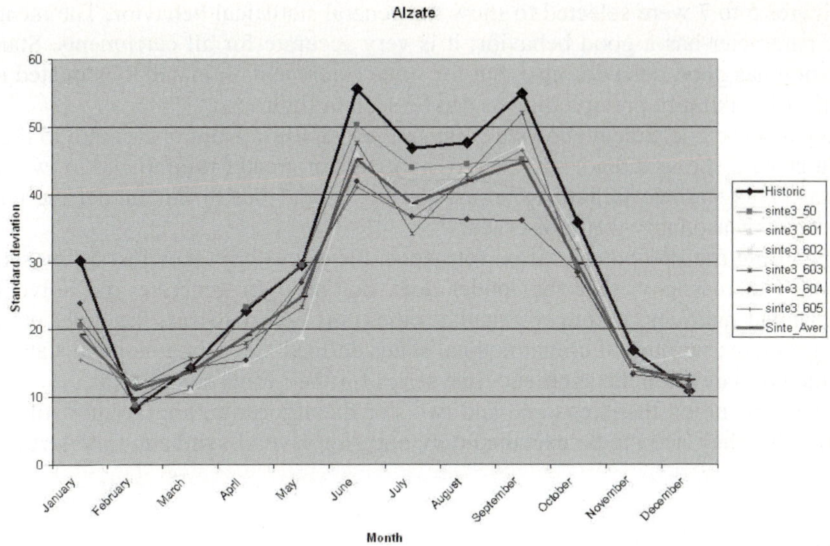

Fig. 6. Monthly standard deviation precipitation for the historic time series and the ensemble of synthetic time series

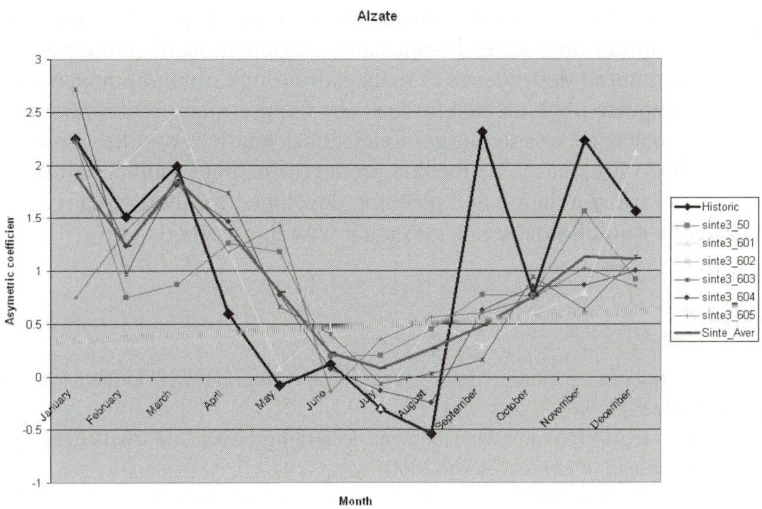

Fig. 7. Monthly asymmetric coefficient for the historic time series and the ensemble of synthetic time series

Other statistic parameters that must be preserved are the mean, the standard deviation and the asymmetric coefficient. Figures 5 to 7 compare the historic statistic with each one of the ensemble statistics and the average ensemble statistic, for the mean, standard deviation and asymmetric coefficient, respectively, for the Alzate catchment (ALZ). A comparison must be made between the historic and the ensemble average.

Figures 5 to 7 were selected to show the general statistical behavior. The mean statistic parameter has a good behavior; it is very accurate for all catchments. Standard deviation has good behavior also, but for some catchments it is under estimated in the months of maximum precipitation as can be seen in figure 6.

Asymmetric coefficient shows a great deal of variation from one month to the next but in general shows a tendency to zero in months of greater rainfall and to two in dry months. This statistic is hard to adjust for any model but in our model the general tendency is reasonably well preserved.

Regarding the computational performance, the algorithm takes just a few seconds to run. The reason is that the model does not need to generate and solve high-dimensional matrices; it simply calculates the transition matrices, the order of which depends on the number of climatological states defined. Six climatological states were considered to develop the synthetic time series for the Lerma river basin.

It must be noted that steps one and two, for the algorithm, are the most time consuming, but they need to be executed only once for several synthetic time series to be generated.

5 Conclusions

A method using Markov hidden states to generate synthetic rainfall time series which preserve the temporal, spatial and rainfall quantity correlation in a climatologic network station or in a basin divided into catchments has been developed. This method does not require that the rainfall time series be normally distributed as in autoregressive models and the seasonal rainfall dependence is managed through a non-homogeneous Markov process. The computer implementation does not require many resources; and it is fast because it does not need to solve high-dimensional matrices and the most consuming steps are executed only once. The method produces accurate synthetic time series. The hydrologic model for simulation and planning developed for the Lerma river basin was fed with synthetic rainfall time series generated with this method.

References

1. Biswas, A.: Integrated Water Management: Some International Dimensions. Journal of Hydrology 51(1-4), 369–379 (1981)
2. Bouwer, H.: Integrated Water Management: Emerging Issues and Challenges. Agricultural Water Management 45(3), 217–228 (2000)
3. Bras, R.L., Rodríguez-Iturbe, I.: Random Functions and Hydrology. Dover Publications, Inc., New York (1993)
4. Salas, J.D., Delleur, J.W., Yevjevich, V., Lane, W.L.: Applied Modeling of Hydrologic, 3rd edn. Time Series. Water Resources Publications, USA (1988)
5. EinsteinNet: Análisis de Series Temporales Univariates, Metodología de Box-Jenkins, Modelo ARIMA (2000-2001) Web Book consulted at, `http://www.einsteinnet.com/econometria/`ARIMA
6. Robertson, A.W., Kirshner, S., Smith, P.: Hidden Markov Models for Modeling Daily Rainfall Ocurrence Over Brasil. Technical Report UCI-ICS 03-27, University of California, Irvine (2003)

7. Richardson, C.W., Wright, D.A.: WGEN: A Model for Generating Daily Weather Variables. USDA-ARS (1984)
8. Nelson, R.: ClimGen – Climatic Data Generator, User's Manual. Washington State University (2002)
9. Jensen, J.R.: Introductory Digital Image Processing. Prentice-Hall, Englewood Cliffs (1986)
10. Selker, J.S., Haither, D.A.: Development and Testing of Single-Parameter Precipitation Distributions. Water Resources Research 26(11), 2733–2740 (1990)

Splitting the Unity, Bisecting a Graph: Applications to Stochastic Boolean Systems*

Luis González

University of Las Palmas de Gran Canaria,
Department of Mathematics, Research Institute IUSIANI,
35017 Las Palmas de Gran Canaria, Spain
luisglez@dma.ulpgc.es

Abstract. This paper deals with the reliability and risk analysis of those complex systems depending on n stochastic Boolean variables. Each one of the 2^n elementary states associated to such a system is given by its corresponding binary n-tuple of 0s and 1s. A symmetric fractal graph on 2^n nodes (the so-called intrinsic order graph I_n) is used for displaying all the binary n-tuples in decreasing order of their occurrence probabilities. The successive bisections of this graph into smaller subgraphs one-to-one correspond to the successive splits of 1 as sum of the occurrence probabilities of those subgraphs. This iterative bisection process satisfies a nice property: if we replace each one of the subgraphs C (obtained after k successive bisections of the original intrinsic order graph I_n) by an unique node weighted by the sum of the probabilities of all vertices lying on C and we sort the new nodes in decreasing order of their weights, then the new condensed graph is exactly the intrinsic order graph I_k. Finally, based on these results, a new algorithm for estimating the unavailability of stochastic Boolean systems is presented.

1 Introduction

A complex stochastic Boolean system (CSBS) is a system depending on an arbitrary number n of random Boolean variables. That is, the n basic system components $x_1, \ldots, x_n$ are assumed to be stochastic (i.e. non-deterministic) and they only take two possible values: 0, 1. These systems can be found in many different knowledge areas, since the assumption "random Boolean variables" is very usual in practice.

Using the statistical terminology, these n random Boolean variables can be considered as Bernoulli variables $x_i = 0, 1$, so that the whole CSBS can be modeled by the n-dimensional Bernoulli distribution. This distribution consists of n random variables $x_1, \ldots, x_n$, which only take two possible values, either 0 or 1, with probabilities

$$\Pr\{x_i = 1\} = p_i, \quad \Pr\{x_i = 0\} = 1 - p_i \quad (1 \leq i \leq n).$$

* Partially supported by MEC (Spain) and FEDER. Grant contracts: CGL2004-06171-C03-02/CLI and CGL2007-65680-C03-01/CLI.

O. Gervasi et al. (Eds.): ICCSA 2008, Part II, LNCS 5073, pp. 970–986, 2008.

Throughout this paper, we assume that the marginal Bernoulli variables $x_1, \ldots, x_n$ of the CSBS are mutually independent. Then the occurrence probability of each binary n-tuple, $u = (u_1, \ldots, u_n) \in \{0,1\}^n$, can be computed as the product

$$\Pr\{u\} = \Pr\{(u_1, \ldots, u_n)\} = \prod_{i=1}^{n} \Pr\{x_i = u_i\} = \prod_{i=1}^{n} p_i^{u_i} (1-p_i)^{1-u_i}, \quad (1)$$

that is, $\Pr\{u\}$ is the product of factors p_i if $u_i = 1$, 1-p_i if $u_i = 0$ $(1 \le i \le n)$.

Example 1. Let $n = 3$, $u = (0,1,1)$ and $v = (1,0,0)$. Using (1) we have

$$p_1 = 0.1,\ p_2 = 0.2,\ p_3 = 0.3: \quad \Pr\{(0,1,1)\} = 0.054 < \Pr\{(1,0,0)\} = 0.056,$$

$$p_1 = 0.2,\ p_2 = 0.3,\ p_3 = 0.4: \quad \Pr\{(0,1,1)\} = 0.096 > \Pr\{(1,0,0)\} = 0.084.$$

The behavior of a given CSBS depends on the ordering between the current values of the 2^n associated binary n-tuple probabilities $\Pr\{u\}$. Of course, for large values of n, computing the 2^n binary n-tuple probabilities (using (1)) and sorting them in decreasing order of their values is not feasible, due to the exponential complexity of the problem. To overcome this obstacle, we have established a simple positional criterion –the so-called intrinsic order criterion– that allows one to compare the occurrence probabilities $\Pr\{u\}, \Pr\{v\}$ of two given binary n-tuples $u, v \in \{0,1\}^n$ without computing them, just looking at the relative positions of their 0s and 1s [3,7]. In this sense, the intrinsic order graph for n variables, denoted by I_n, provides us with an useful representation of these CSBSs, by displaying the 2^n binary n-tuples in decreasing order of their occurrence probabilities [5].

In this paper, we study new graphical properties of the intrinsic order graph that can be applied to the analysis of CSBSs. We present and illustrate a natural partition of the intrinsic order graph into two subgraphs with half the vertices each, i.e. a bisection of I_n. This bisection process is successively iterated, obtaining smaller and smaller subgraphs of I_n, until we get 2^n subgraphs coinciding with the 2^n graph vertices. The recursive bisection of the intrinsic order graph into 2^k subgraphs corresponds to a recursive "split" of the unity (i.e. a partition of 1) into 2^k summands $(1 \le k \le n)$. Each one of these summands gives exactly the sum of the occurrence probabilities of all the binary strings lying on the corresponding subgraph. Continuing with these graphical properties of the intrinsic order graph I_n, our main result states a nice property, which is in accordance with its fractal character (self-similarity property). We prove that if we "concentrate" each one of the 2^k subgraphs C (obtained after k successive bisections of I_n) into an unique node, weighted by the sum of the probabilities of all nodes of the subgraph C and we sort these new nodes in decreasing order of their weights, then the new "condensed" graph on 2^k nodes is precisely the intrinsic order graph I_k.

Finally, based on these theoretical results, we present a new algorithm for estimating the failure probability of those CSBSs that can be described by a stochastic Boolean function depending on its n basic variables x_i, i.e.

$$\Phi : \{0,1\}^n \longrightarrow \{0,1\}$$
$$(x_1,\ldots,x_n) \mapsto \Phi(x_1,\ldots,x_n) .$$

Technical systems or, in general, CSBSs that can be represented by such a stochastic Boolean function, appear in many different scientific or engineering areas (see, e.g. [1,8,9]). Throughout this paper, we assume that $x_i = 1$ if the i-th component fails, $x_i = 0$ otherwise. Assuming also that $\Phi = 1$ if the system fails, $\Phi = 0$ otherwise, then the failure probability of the whole system (also called the system unavailability, or the top event probability) can be evaluated by computing the probability $\Pr\{\Phi = 1\}$. Among the many different techniques proposed in the literature to evaluate the system unavailability (see, e.g. [1,10]), we consider here an algebraic, deterministic method whose efficiency is based on the selection of system elementary states $u = (u_1,\ldots,u_n)$ with occurrence probabilities $\Pr\{u\}$ "as large as possible" [1,9]. This method will be exposed in detail in the next section.

In this context, the main goal of this paper is to determine which is the "optimal" subgraph of I_n among all the 2^k equal-sized subgraphs obtained after k successive bisections of I_n (i.e. after k successive splits of the unity). Here, the term "optimal" means the subgraph that leads to the best accuracy/computational-cost ratio, when evaluating the system unavailability. Using our main theorem, we shall prove that this subgraph is precisely the one placed at the top of I_n. This paper has been organized as follows. In Sect. 2, we present all the required background about the above mentioned deterministic method and about the intrinsic order relation and its graph. Beginning with our new approaches, in Sect. 3, we present a natural iterative bisection of the intrinsic order graph. Section 4 is devoted to the split of the unity associated to this graph bisection process. In Sect. 5, we state our main theorem and apply these previous theoretical results, proposing a new, simple algorithm for evaluating the unavailability of CSBSs. Finally, conclusions are presented in Sect. 6.

2 Some Preliminary Results

2.1 Bounds on System Unavailability

As commented in Sect. 1, our strategy for evaluating the top event probability $\Pr\{\Phi = 1\}$ is based on the selection of system elementary states with large occurrence probabilities. Let us recall the basic idea of this method. It is well known that the probability of a Boolean function Φ taking value 1 can be exactly computed as the sum of the occurrence probabilities of all system elementary states for which $\Phi = 1$ or, alternatively, as the remainder to 1 of the sum of the occurrence probabilities of all system elementary states for which $\Phi = 0$ [9], i.e.

$$\sum_{u\in\{0,1\}^n,\Phi(u)=1} \Pr\{u\} = \Pr\{\Phi=1\} = 1 - \Pr\{\Phi = 0\} = 1 - \sum_{u\in\{0,1\}^n,\Phi(u)=0} \Pr\{u\} . \tag{2}$$

For small values of the number n of system components, (2) unables one to exactly compute the failure probability, $\Pr\{\Phi = 1\}$, of the system. However, for large values of n this procedure is not feasible in practice because of the exponential nature of the problem (there are 2^n n-tuples of 0s and 1s). To overcome this obstacle, we can obtain lower and upper bounds L, U on the top event probability, $\Pr\{\Phi = 1\}$, as follows. Let C_1 and C_0 be two any subsets of the sets of binary n-tuples for which $\Phi = 1$ and $\Phi = 0$, respectively, i.e.

$$C_1 \subseteq \{u \in \{0,1\}^n \mid \Phi(u) = 1\}, \qquad C_0 \subseteq \{u \in \{0,1\}^n \mid \Phi(u) = 0\}.$$

Then using (2) we have

$$\sum_{u \in C_1} \Pr\{u\} \le \sum_{u \in \{0,1\}^n,\, \Phi(u)=1} \Pr\{u\} = \Pr\{\Phi = 1\}, \tag{3}$$

$$\sum_{u \in C_0} \Pr\{u\} \le \sum_{u \in \{0,1\}^n,\, \Phi(u)=0} \Pr\{u\} = \Pr\{\Phi = 0\} \tag{4}$$

and from (3) and (4) we get the following lower and upper bounds L, U

$$L = \sum_{u \in C_1} \Pr\{u\} \le \Pr\{\Phi = 1\} = 1 - \Pr\{\Phi = 0\} \le 1 - \sum_{u \in C_0} \Pr\{u\} = U. \tag{5}$$

Note that, for a fixed set C of binary n-tuples, the above bounds, L, U, depend on the Boolean function Φ, because the same happens with the set partition

$$C = \{u \in C \mid \Phi(u) = 0\} \cup \{u \in C \mid \Phi(u) = 1\} = C_0 \cup C_1. \tag{6}$$

However, the maximum error E in the estimate (difference between upper and lower bounds)

$$E = U - L = \left(1 - \sum_{u \in C_0} \Pr\{u\}\right) - \sum_{u \in C_1} \Pr\{u\} = 1 - \sum_{u \in C_0 \,\cup\, C_1 = C} \Pr\{u\} \tag{7}$$

is completely independent of the logical structure of the system (i.e. it is the same for any Boolean function Φ), and it only depends on the total sum, $\sum_{u \in C} \Pr\{u\}$, of the occurrence probabilities of all the selected binary n-tuples $u \in C$. More precisely, the accuracy in the estimate of the system unavailability, $\Pr\{\Phi = 1\}$, improves at the same time as the error E decreases, i.e. at the same time as this total sum, $\sum_{u \in C} \Pr\{u\}$, increases, as (7) has shown; see [7].

In particular, to estimate the system unavailability $\Pr\{\Phi = 1\}$ with a prespecified maximum error ϵ, according to (7) we must assure that

$$E = 1 - \sum_{u \in C} \Pr\{u\} \le \epsilon \;\Leftrightarrow\; \sum_{u \in C} \Pr\{u\} \ge 1 - \epsilon, \tag{8}$$

that is, the sum of the occurrence probabilities of all selected binary strings $u \in C$ must be greater than or equal to $1 - \epsilon$.

Consequently, the main question is how to select the minor number of system elementary states, $u \in C$, with occurrence probabilities as large as possible, in order to minimize the computational cost –by increasing the sum in (8)– when evaluating the system unavailability by (5).

2.2 Intrinsic Order

The ordering between two given binary n-tuple probabilities $\Pr(u)$ and $\Pr(v)$ clearly depends on the parameters $p_1, \ldots, p_n$ of the Bernoulli distribution, as Example 1 has shown. However, for some pairs of binary strings, the ordering between their occurrence probabilities is independent of the basic probabilities p_i, and it only depends on the relative positions of their 0s and 1s. More precisely, the following theorem [3,7] provides us with an intrinsic order criterion for comparing the occurrence probabilities of two given n-tuples without computing them.

Theorem 1 (The intrinsic order theorem). *Let $n \geq 1$. Let $x_1, \ldots, x_n$ be n mutually independent Bernoulli variables whose parameters $p_i = \Pr\{x_i = 1\}$ satisfy*

$$0 < p_1 \leq p_2 \leq \ldots \leq p_n \leq \frac{1}{2}. \tag{9}$$

Then the probability of the n-tuple $v = (v_1, \ldots, v_n) \in \{0,1\}^n$ is intrinsically less than or equal to the probability of the n-tuple $u = (u_1, \ldots, u_n) \in \{0,1\}^n$ (that is, for all set $\{p_i\}_{i=1}^n$ satisfying (9)) if and only if the matrix

$$M_v^u := \begin{pmatrix} u_1 \ldots u_n \\ v_1 \ldots v_n \end{pmatrix}$$

either has no $\binom{1}{0}$ columns, or for each $\binom{1}{0}$ column in M_v^u there exists (at least) one corresponding preceding $\binom{0}{1}$ column (Intrinsic order criterion).

Remark 1. In the following, we assume that the parameters p_i always satisfy condition (9). Note that this hypothesis is not restrictive for practical applications because, if for some $i : p_i > \frac{1}{2}$, then we only need to consider the variable $\overline{x_i} = 1 - x_i$, instead of x_i. Next, we order the n Bernoulli variables by increasing order of their probabilities.

Remark 2. The $\binom{0}{1}$ column preceding to each $\binom{1}{0}$ column is not required to be necessarily placed at the immediately previous position, but just at previous position.

Remark 3. The term *corresponding,* used in Theorem 1, has the following meaning: for each two $\binom{1}{0}$ columns in matrix M_v^u, there must exist (at least) two *different* $\binom{0}{1}$ columns preceding to each other. In other words: for each $\binom{1}{0}$ column in matrix M_v^u, the number of preceding $\binom{0}{1}$ columns must be strictly greater than the number of preceding $\binom{1}{0}$ columns.

The matrix condition stated by Theorem 1 is called the *intrinsic order criterion,* because it is independent of the basic probabilities p_i and it only depends on the relative positions of the 0s and 1s in the binary n-tuples u, v. Theorem 1 naturally leads to the following partial order relation on the set $\{0,1\}^n$ [3,7]. The so-called intrinsic order will be denoted by "$\preceq$", and we shall write $u \succeq v$ ($u \preceq v$) to indicate that u is intrinsically greater (less) than or equal to v.

Definition 1. *For all* $u, v \in \{0,1\}^n$

$$v \preceq u \text{ if and only if } \Pr\{v\} \leq \Pr\{u\} \text{ for all set } \{p_i\}_{i=1}^n \text{ such that (9)}$$
$$\text{if and only if } M_v^u \text{ satisfies the intrinsic order criterion.}$$

Throughout this paper we denote the partially ordered set (poset, for short), defined by the intrinsic order relation, by $I_n := (\{0,1\}^n, \preceq)$. Many different properties of the intrinsic order can be immediately derived from its simple matrix description stated by Theorem 1; see, e.g. [3,4,6].

Example 2. Neither $(0,1,1) \preceq (1,0,0)$, nor $(1,0,0) \preceq (0,1,1)$ because the matrices

$$\begin{pmatrix} 1\,0\,0 \\ 0\,1\,1 \end{pmatrix} \text{ and } \begin{pmatrix} 0\,1\,1 \\ 1\,0\,0 \end{pmatrix}$$

do not satisfy the intrinsic order criterion (Remark 3). Then $(0,1,1)$ and $(1,0,0)$ are incomparable by intrinsic order, i.e. the ordering between $\Pr\{(0,1,1)\}$ and $\Pr\{(1,0,0)\}$ depends on the basic probabilities p_i, as Example 1 has shown.

Example 3. $(1,1,0,1,0,0) \preceq (0,0,1,1,0,1)$ because

$$\begin{pmatrix} 0\,0\,1\,1\,0\,1 \\ 1\,1\,0\,1\,0\,0 \end{pmatrix}$$

satisfies the intrinsic order criterion (Remark 2). Then

$$\Pr\{(1,1,0,1,0,0)\} \leq \Pr\{(0,0,1,1,0,1)\}, \text{ for all } 0 < p_1 \leq \ldots \leq p_6 \leq 1/2.$$

Throughout this paper, the decimal numbering of a binary string u is denoted by the symbol $u_{(10}$. We use this symbol, instead of the more usual notation u_{10}, to avoid confusion with the 10-th component u_{10} of the binary string u. In the following, we use indistinctly the binary representation or the decimal representation to denote the elements or subsets of $\{0,1\}^n$, i.e.

$$(u_1, \ldots, u_n) = u_{(10} = \sum_{i=1}^{n} 2^{n-i} u_i, \tag{10}$$

e.g. for $n = 3$ we have

$$(1,0,1) = 5, \qquad \{(1,0,0), (1,0,1), (1,1,0), (1,1,1)\} = [4,7].$$

Example 4. For all $n \geq 1$, the binary n-tuples

$$\left(0, \overset{n}{\ldots}, 0\right) = 0 \text{ and } \left(1, \overset{n}{\ldots}, 1\right) = 2^n - 1$$

are the maximum and minimum elements, respectively, in the poset I_n. Indeed, both matrices

$$\begin{pmatrix} 0 & \ldots & 0 \\ u_1 & \ldots & u_n \end{pmatrix} \text{ and } \begin{pmatrix} u_1 & \ldots & u_n \\ 1 & \ldots & 1 \end{pmatrix}$$

satisfy the intrinsic order criterion, since they have no $\binom{1}{0}$ columns. Hence, using Definition 1, we get

$$\left(1, \overset{n}{\ldots}, 1\right) \preceq (u_1, \ldots, u_n) \preceq \left(0, \overset{n}{\ldots}, 0\right), \quad \text{for all } (u_1, \ldots, u_n) \in \{0,1\}^n.$$

2.3 The Intrinsic Order Graph

To finish this section, we present the graphical representation of the poset I_n. The usual representation of a poset is its Hasse diagram (see, e.g. [11] for more details about posets and Hasse diagrams). This is a directed graph whose vertices are the binary n-tuples of 0s and 1s, and whose edges connect each pair (u, v) of binary n-tuples whenever u covers v, that is, whenever u is intrinsically greater than v with no other elements between them, i.e.

$$u \text{ covers } v \text{ if and only if } u \succ v \text{ and there is no } w \in \{0,1\}^n \text{ satisfying } u \succ w \succ v.$$

Moreover, according to the usual convention for Hasse diagrams, if u covers v then u is drawn above v. The Hasse diagram of the poset I_n will be also called the intrinsic order graph for n variables. From now on, looking for a more comfortable and simple notation, we shall denote the vertices (binary n-tuples) in the intrinsic order graph by their decimal numbering (with the only exception of the last figure).

For small values of n, the Hasse diagram of I_n can be constructed by direct application of the intrinsic order criterion. For instance, the Hasse diagram of $I_1 = (\{0,1\}, \preceq)$ is shown in Fig. 1.

0
|
1

Fig. 1. The intrinsic order graph for $n = 1$

Note that 0 covers 1, because the only elements of I_1 are 0 and 1, and $0 \succ 1$ since matrix $\binom{0}{1}$ satisfies the intrinsic order criterion: it has no $\binom{1}{0}$ columns! In other words (see (9)),

$$\Pr\{0\} = 1 - p_1 \geq p_1 = \Pr\{1\}, \text{ since } p_1 \leq 1/2.$$

However, for large values of n we need a more efficient method. For this purpose, in [5] we have developed an algorithm for iteratively building up the directed graph of I_n, for all $n \geq 2$, from the directed graph of I_1 (shown in Fig. 1). The next theorem states this algorithm, which uses the decimal representation of the binary strings. See [5] for the proof and for additional properties of the intrinsic order graph.

Theorem 2 (Iterative construction of I_n from I_1). *For all $n > 1$, the graph of $I_n = \{0, \ldots, 2^n - 1\}$ can be drawn simply by adding to the graph of $I_{n-1} = \{0, \ldots, 2^{n-1} - 1\}$ its isomorphic copy $2^{n-1} + I_{n-1} = \{2^{n-1}, \ldots, 2^n - 1\}$. This addition must be performed placing the powers of 2 at consecutive levels of the Hasse diagram of I_n. Finally, the edges connecting one vertex u of I_{n-1} with the other vertex v of $2^{n-1} + I_{n-1}$ are given by the set of vertex pairs*

$$\left\{(u, v) = \left(u_{(10}, 2^{n-2} + u_{(10)}\right) \mid 2^{n-2} \leq u_{(10} \leq 2^{n-1} - 1\right\}.$$

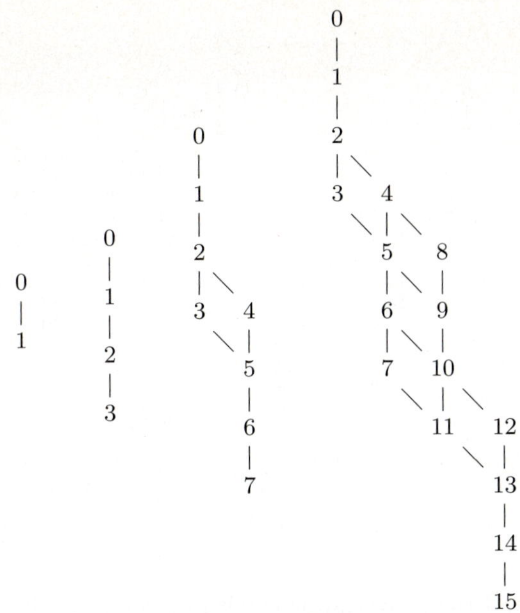

Fig. 2. The intrinsic order graph for $n = 1, 2, 3, 4$

Figure 2 illustrates, for the first few values of n, the iterative construction of the directed graph of I_n from I_1, using Theorem 2 and denoting all the binary n-tuples by their decimal numbering.

3 A Bisection of the Intrinsic Order Graph

For a clearer illustration of the results presented in this and next section, we remove all edges in the intrinsic order graph, keeping the relative position of its nodes. A graph consisting of only isolated nodes with no edges is called an empty or edgeless graph [2]. Of course, according to Theorem 2, the empty graph of $I_n = [0, 2^n - 1]$ can be iteratively built up from the empty graph of I_1, simply by "adding to the graph of $I_{n-1} = \left[0, 2^{n-1} - 1\right]$ its isomorphic copy $2^{n-1} + I_{n-1} = \left[2^{n-1}, 2^n - 1\right]$, placing the powers of two at consecutive levels", as shown in Fig. 3. A bisection of a graph is a partition of its vertex set into two subsets with half the vertices each [2]. Hence, Theorem 2 provides a bisection of the (edgeless) graph I_n into its two isomorphic (edgeless) subgraphs I_{n-1} and $2^{n-1} + I_{n-1}$.

Of course, this bisection process of the edgeless graph I_n can be reiterated by successively partitioning each one of the obtained subgraphs into its top and bottom halves. This iterative bisection process finishes when we have partitioned I_n into 2^n singleton subgraphs (with 1 vertex each), i.e. into its 2^n nodes.

This particular bisection of the intrinsic order graph means that the poset I_n has a "fractal structure": the whole graph has the same "shape" that each one

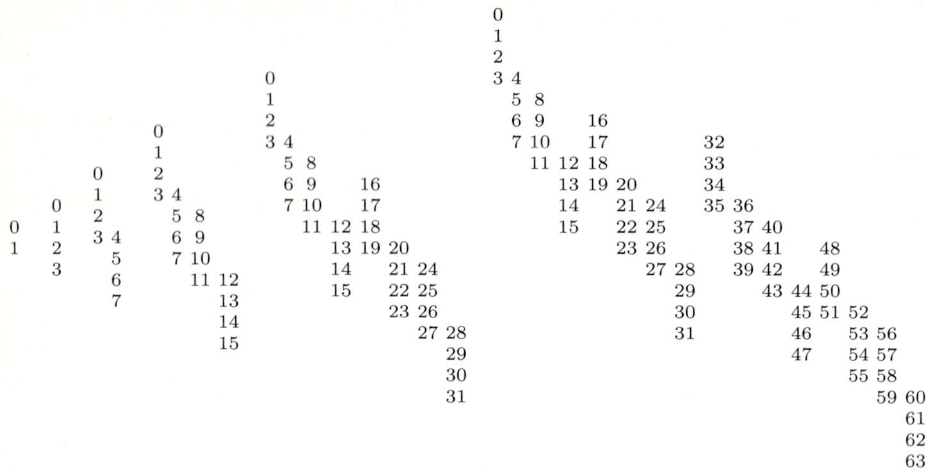

Fig. 3. The edgeless intrinsic order graph for $n = 1, 2, 3, 4, 5, 6$

of its two halves, and the same happens with each one of them, and so on, i.e. the poset I_n has the self-similarity property. Figure 3 illustrates this fact.

Let us set a consistent notation for this iterative bisection process. Recursively bisecting the graph I_n (with 2^n binary n-tuples) is equivalent to recursively bisecting the truth-table for n Boolean variables (with 2^n rows). Since, by construction, the first bit u_1 in all the n-tuples of the first and second half of the truth-table is 0 and 1, respectively, we denote the first and second half of I_n by I_n^0 and I_n^1, respectively. Analogously, since, by construction, the second bit u_2 in all the n-tuples of the first and second half of both halves of the truth table is 0 and 1, respectively, we denote the first and second half of I_n^0 by $I_n^{0,0}$ and $I_n^{0,1}$, respectively; and we denote the first and second half of I_n^1 by $I_n^{1,0}$ and $I_n^{1,1}$, respectively, and so on.

In general, for all $n \geq 1$, for all $1 \leq k \leq n$ and for all k fixed binary digits $\bar{u}_1, \ldots, \bar{u}_k \in \{0, 1\}$, we denote by $I_n^{\bar{u}_1,\ldots,\bar{u}_k}$ the $\bar{u}_k + 1$-th half of the $\bar{u}_{k-1} + 1$-th half ... of the $\bar{u}_1 + 1$-th half of the poset I_n. In other words, $I_n^{\bar{u}_1,\ldots,\bar{u}_k}$ can be graphically obtained after k successive bisections of I_n $(1 \leq k \leq n)$ simply by changing the "0" and "1" bits of the vector $(\bar{u}_1, \ldots, \bar{u}_k)$, by the words "first half" and "second half", respectively. Hence, this is the subset of binary n-tuples whose k first or left-most components are fixed, namely $u_1 = \bar{u}_1, \ldots, u_k = \bar{u}_k$; while the $n - k$ last or right-most components, $u_{k+1}, \ldots, u_n$, take all possible values (0 or 1). More precisely, according to (10), we have

$$
\begin{aligned}
I_n^{\bar{u}_1,\ldots,\bar{u}_k} &= \left\{ (\bar{u}_1, \ldots, \bar{u}_k, u_{k+1}, \ldots, u_n) \;\middle|\; (u_{k+1}, \ldots, u_n) \in \{0,1\}^{n-k} \right\} \\
&= \left[(\bar{u}_1, \ldots, \bar{u}_k, 0, \ldots, 0)_{(10}, (\bar{u}_1, \ldots, \bar{u}_k, 1, \ldots, 1)_{(10} \right] \\
&= \left[2^{n-k}\bar{u}_k + \cdots + 2^{n-1}\bar{u}_1, 2^0 + \cdots + 2^{n-k-1} + 2^{n-k}\bar{u}_k + \cdots + 2^{n-1}\bar{u}_1 \right]
\end{aligned} \tag{11}
$$

and the cardinality of this subposet is

$$\left|I_n^{\bar{u}_1,\ldots,\bar{u}_k}\right| = \left|\{0,1\}^{n-k}\right| = 2^{n-k}. \tag{12}$$

Remark 4. In particular, for $k = n$, the subgraph $I_n^{\bar{u}_1,\ldots,\bar{u}_n}$, obtained after n bisections of I_n, is reduced to a single node of this graph which, according to (11), is precisely

$$I_n^{\bar{u}_1,\ldots,\bar{u}_n} = \{(\bar{u}_1,\ldots,\bar{u}_n)\} \quad (a\ curious\ fact!). \tag{13}$$

With this notation, we can formalize the iterative bisection process as follows

$$\begin{aligned} I_n &= I_n^0 \cup I_n^1 = I_n^{0,0} \cup I_n^{0,1} \cup I_n^{1,0} \cup I_n^{1,1} \\ &= I_n^{0,0,0} \cup I_n^{0,0,1} \cup I_n^{0,1,0} \cup I_n^{0,1,1} \cup I_n^{1,0,0} \cup I_n^{1,0,1} \cup I_n^{1,1,0} \cup I_n^{1,1,1} \\ &= \cdots = \bigcup_{(\bar{u}_1,\ldots,\bar{u}_n)\in\{0,1\}^n} I_n^{\bar{u}_1,\ldots,\bar{u}_n} = \bigcup_{(\bar{u}_1,\ldots,\bar{u}_n)\in\{0,1\}^n} \{(\bar{u}_1,\ldots,\bar{u}_n)\}, \end{aligned} \tag{14}$$

e.g. for the graph of I_3 (the third one from the left in Fig. 3), we have

$$\begin{aligned} I_3 &= [0,7] = [0,3] \cup [4,7] = [0,1] \cup [2,3] \cup [4,5] \cup [6,7] \\ &= \{0\} \cup \{1\} \cup \{2\} \cup \{3\} \cup \{4\} \cup \{5\} \cup \{6\} \cup \{7\}. \end{aligned}$$

Example 5. For $n = 5$, $k = 3$ and $(\bar{u}_1, \bar{u}_2, \bar{u}_3) = (0,1,1)$, using (11), we get the subgraph

$$\begin{aligned} I_5^{0,1,1} &= \left\{(0,1,1,u_4,u_5) \;\middle|\; (u_4,u_5) \in \{0,1\}^2\right\} = \left[2^2+2^3, 2^0+2^1+2^2+2^3\right] \\ &= [12,15] = \{12,13,14,15\} \end{aligned}$$

and looking at the fifth diagram from the left in Fig. 3, we confirm that $[12,15]$ is exactly the second half ($\bar{u}_3 = 1$) of the second half ($\bar{u}_2 = 1$) of the first half ($\bar{u}_1 = 0$) of the poset I_5. In accordance with (12), $I_5^{0,1,1}$ has $2^{5-3} = 4$ elements

Example 6. For $n = 6$, $k = 6$ and $(\bar{u}_1, \bar{u}_2, \bar{u}_3, \bar{u}_4, \bar{u}_5, \bar{u}_6) = (1,0,1,0,1,0)$, using (11) or directly (13) –here $k = n$–, we get the singleton subgraph

$$I_6^{1,0,1,0,1,0} = \{(1,0,1,0,1,0)\} = \left\{2^1+2^3+2^5\right\} = \{42\} \quad (a\ curious\ fact!)$$

and looking at the right-most diagram in Fig. 3, we confirm that $\{42\}$ is exactly the first half ($\bar{u}_6 = 0$) of the second half ($\bar{u}_5 = 1$) of the first half ($\bar{u}_4 = 0$) of the second half ($\bar{u}_3 = 1$) of the first half ($\bar{u}_2 = 0$) of the second half ($\bar{u}_1 = 1$) of the poset I_6. In accordance with (12), $I_6^{1,0,1,0,1,0}$ has $2^{6-6} = 1$ element.

4 A Split of the Unity

In this section, we consider a natural split of the unity, closely related to the bisection of the intrinsic order graph performed in Sect. 3. For this purpose, let us consider a CSBS depending on n mutually independent Bernoulli variables $x_1, \ldots, x_n$ whose parameters $p_i = \Pr\{x_i = 1\}$ satisfy (9). Now, using the obvious fact that $1 = (1 - x) + x$, we can split the unity as follows

$$\begin{aligned} 1 &= (1-p_1) + p_1 = [(1-p_1)+p_1]\,[(1-p_2)+p_2] \\ &= [(1-p_1)+p_1]\,[(1-p_2)+p_2]\,[(1-p_3)+p_3] \\ &= \cdots = [(1-p_1)+p_1]\cdots[(1-p_n)+p_n]\,, \end{aligned}$$

that is, using (1), we get

$$\begin{aligned} 1 &= (1-p_1)+p_1 = (1-p_1)(1-p_2) + (1-p_1)p_2 + p_1(1-p_2) + p_1p_2 \\ &= (1-p_1)(1-p_2)(1-p_3) + (1-p_1)(1-p_2)p_3 + (1-p_1)p_2(1-p_3) \\ &+ (1-p_1)p_2p_3 + p_1(1-p_2)(1-p_3) + p_1(1-p_2)p_3 + p_1p_2(1-p_3) + p_1p_2p_3 \\ &= \cdots = \sum_{(\bar{u}_1,\ldots,\bar{u}_n)\in\{0,1\}^n} \prod_{i=1}^{n} p_i^{\bar{u}_i}(1-p_i)^{1-\bar{u}_i} = \sum_{u\in\{0,1\}^n} \Pr\{u\}\,. \end{aligned} \tag{15}$$

Now, we bridge the gap between the (iterative) bisection (14) of the graph I_n and the (iterative) split (15) of the unity. Observe the similarity between both formulas. Indeed, one can directly obtain (15) from (14), by taking probabilities in this last equation, and by changing the unions by sums because of the additivity probability axiom.

Moreover, each one of the summands $\prod_{i=1}^{k} p_i^{\bar{u}_i}(1-p_i)^{1-\bar{u}_i}$, appearing in the split (15), is exactly the sum of the occurrence probabilities of all the binary n-tuples u lying on the corresponding subgraph $I_n^{\bar{u}_1,\ldots,\bar{u}_k}$, appearing in the bisection (14). This statement is proved in the following theorem, which generalizes (1).

Theorem 3. *Let $n \geq 1$, $1 \leq k \leq n$ and let $\bar{u}_1, \ldots, \bar{u}_k \in \{0,1\}$ be k fixed but arbitrary binary digits. Then*

$$S_n^{\bar{u}_1,\ldots,\bar{u}_k} := \sum_{u\in I_n^{\bar{u}_1,\ldots,\bar{u}_k}} \Pr\{u\} = \Pr\{(\bar{u}_1,\ldots,\bar{u}_k)\} = \prod_{i=1}^{k} p_i^{\bar{u}_i}(1-p_i)^{1-\bar{u}_i}\,, \tag{16}$$

that is, $S_n^{\bar{u}_1,\ldots,\bar{u}_k}$ is the product of factors p_i if $\bar{u}_i = 1$, $1-p_i$ if $\bar{u}_i = 0$ $(1 \leq i \leq k)$.

Proof. Using (11), (15) and (1) we get

$$\begin{aligned} S_n^{\bar{u}_1,\ldots,\bar{u}_k} &= \sum_{(u_{k+1},\ldots,u_n)\in\{0,1\}^{n-k}} \Pr\{(\bar{u}_1,\ldots,\bar{u}_k,u_{k+1},\ldots,u_n)\} \\ &= \Pr\{(\bar{u}_1,\ldots,\bar{u}_k)\} \sum_{(u_{k+1},\ldots,u_n)\in\{0,1\}^{n-k}} \Pr\{(u_{k+1},\ldots,u_n)\} \\ &= \Pr\{(\bar{u}_1,\ldots,\bar{u}_k)\} = \prod_{i=1}^{k} p_i^{\bar{u}_i}(1-p_i)^{1-\bar{u}_i}\,. \end{aligned}$$

□

Corollary 1. *Let $n \geq 1$ and $1 \leq k \leq n$. Then*

$$S_n^{0,\overset{k}{\ldots},0} = \sum_{u \in I_n^{0,\overset{k}{\ldots},0}} \Pr\{u\} = \Pr\left\{\left(0,\overset{k}{\ldots},0\right)\right\} = \prod_{i=1}^{k} (1-p_i)\,. \tag{17}$$

Proof. Using Theorem 3 for $\bar{u}_1 = \cdots = \bar{u}_k = 0$, the proof is straightforward. □

5 Applications to Stochastic Boolean Systems

In this section, we apply the previous results to the reliability analysis of CSBSs. Let us recall that for the algebraic method described in Sect. 2.1, the maximum error E in the evaluation of the system unavailability is given by (7), i.e.

$$E = 1 - \sum_{u \in C} \Pr\{u\}\,, \quad C \subseteq \{0,1\}^n\,, \tag{18}$$

so that, the estimation accuracy improves at the same as the sum of the occurrence probabilities $\Pr\{u\}$ of the selected binary n-tuples $u \in C$ increases.

5.1 The Main Result

Our choice of C is based on the recursive graph bisection process described in Sect. 3. The selected vertex set C will be one of the 2^k subgraphs obtained after k successive bisections of the intrinsic order graph I_n $(1 \leq k \leq n)$, i.e.

$$C = I_n^{\bar{u}_1,\ldots,\bar{u}_k}, \quad \text{for some} \quad (\bar{u}_1,\ldots,\bar{u}_k) \in \{0,1\}^k\,. \tag{19}$$

Now, the question is: for each fixed k, which is the optimal choice of the set C among all the 2^k subgraphs of I_n given in (19)? Here, the term "optimal" means that C must provide the best accuracy/computational-cost ratio. Regarding the computational cost, it only depends on the number of n-tuples in C. Since all the subgraphs (19) have the same cardinality 2^{n-k} (see formula (12)), then the computational cost will be the same for all of them. So, we only need to compare their contributions to the accuracy. According to (18), this is equivalent to comparing the 2^k sums (16) of the occurrence probabilities of all the nodes lying on each one of those 2^k subgraphs. This comparison is performed in the following theorem, the main result of this work, which is illustrated in Fig. 4.

Theorem 4 (The main result). *Let $n \geq 1$ and $1 \leq k \leq n$. Bisect the edgeless graph I_n into its 2^k subgraphs $I_n^{\bar{u}_1,\ldots,\bar{u}_k}$ (i.e. make k successive bisections of I_n). Replace each subgraph $I_n^{\bar{u}_1,\ldots,\bar{u}_k}$ by an unique node labeled by its corresponding vector of upper indices $(\bar{u}_1,\ldots,\bar{u}_k)$ and weighted by the occurrence probability $\Pr\{(\bar{u}_1,\ldots,\bar{u}_k)\}$ of its label. Next, sort these 2^k new nodes in decreasing order of their weights. Then the new "condensed" graph obtained from the intrinsic order graph I_n –with 2^n vertices– by this "bisecting-replacing-sorting" process, is* *precisely the intrinsic order graph I_k –with 2^k vertices. Moreover, this ordering between the 2^k new nodes coincides with the ordering between the 2^k sums of the occurrence probabilities of all the nodes lying on each one of the respective replaced subgraphs.*

Proof. Sorting the 2^k vertices of the new condensed graph in decreasing order of their weights $\Pr\{(\bar{u}_1,\ldots,\bar{u}_k)\}$ is equivalent to ordering the 2^k binary k-tuples $(\bar{u}_1,\ldots,\bar{u}_k) \in \{0,1\}^k$ in decreasing order of their occurrence probabilities. Thus, the new condensed graph is, by definition, the intrinsic order graph I_k. Finally, the last statement immediately follows from the equality

$$\Pr\{(\bar{u}_1,\ldots,\bar{u}_k)\} = \sum_{u\in I_n^{\bar{u}_1,\ldots,\bar{u}_k}} \Pr\{u\},$$

stated by Theorem 3. □

Corollary 2. *Let $n \geq 1$ and $1 \leq k \leq n$. Then the subgraphs $I_n^{0,\overset{k}{\ldots},0}$ and $I_n^{1,\overset{k}{\ldots},1}$ are the best and worst choice, respectively, for the estimation accuracy, among all the 2^k subgraphs $I_n^{\bar{u}_1,\ldots,\bar{u}_k}$ obtained after k successive bisections of I_n.*

Proof. Using (18) and Theorem 4, we see that proving the current theorem is equivalent to proving that, for all $k \geq 1$, the binary k-tuples

$$\left(0,\overset{k}{\ldots},0\right) = 0 \quad \text{and} \quad \left(1,\overset{k}{\ldots},1\right) = 2^k - 1$$

are the maximum and minimum elements, respectively, in the poset I_k. This fact, illustrated by Figs. 2 and 4, has been demonstrated in Example 4. □

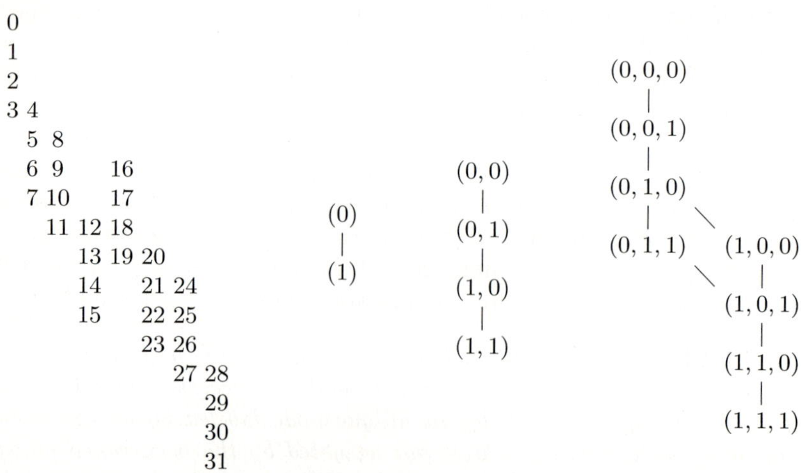

Fig. 4. k successive bisections of I_n lead to I_k ($n = 5$, $k = 1, 2, 3$)

5.2 The Top Subgraph Algorithm

Based on the previous results, we present a new algorithm for estimating the unavailability of a CSBS described by a Boolean function Φ. We call it the *Top Subgraph Algorithm* because the optimal, selected subgraphs $C = I_n^{0,\overset{k}{\ldots},0}$ –see Corollary 2– are placed at the top of the intrinsic order graph I_n, for all $1 \leq k \leq n$. Note that $I_n^{0,\overset{k}{\ldots},0}$ is the first part of the first part of $\overset{k}{\ldots}$ of the first part of I_n.

Our algorithm provides lower and upper bounds L, U on the system unavailability $\Pr\{\Phi = 1\}$ with a prespecified maximum error ϵ. For this purpose, according to (8) and (17), we must assure that

$$S_n^{0,\overset{k}{\ldots},0} = \sum_{u \in C = I_n^{0,\overset{k}{\ldots},0}} \Pr\{u\} = \prod_{i=1}^{k} (1 - p_i) \geq 1 - \epsilon. \tag{20}$$

Remark 5. In order to cover all possible situations, our algorithm includes the singular case $k = 0$: no bisections are made and the chosen top subgraph is the entire graph $C = I_n = \{0,1\}^n$. In this way, $S_n := \sum_{u \in I_n} \Pr\{u\} = 1$ (see (15)) and the Top Subgraph Algorithm even enables one to exactly compute the system unavailability according to (2), i.e. with no error according to (18). Moreover, including the singular case $k = 0$, the Top Subgraph Algorithm theoretically makes sense for all prespecified maximum error $\epsilon \geq 0$.

The Top Subgraph Algorithm:

Step 1. Define $S_n^{0,\overset{0}{\ldots},0} = S_n = 1$ (by convention, see Remark 5).

Step 2. For $k = 0$ to $n - 1$

$$\text{compute } S_n^{0,\overset{k+1}{\ldots},0} = S_n^{0,\overset{k}{\ldots},0} (1 - p_{k+1}) \text{ (see (17))}$$

until $S_n^{0,\overset{k+1}{\ldots},0} < 1 - \epsilon$ (see (20)).

Step 3. For all $n-k$-tuples $(u_{k+1}, \ldots, u_n) \in \{0,1\}^{n-k}$ compute (see (1) & (17))

$$\Pr\{u\} = \Pr\left\{\left(0, \overset{k}{\ldots}, 0, u_{k+1}, \ldots, u_n\right)\right\} = S_n^{0,\overset{k}{\ldots},0} \prod_{i=k+1}^{n} p_i^{u_i} (1 - p_i)^{1-u_i}.$$

Step 4. For all binary n-tuples $u = \left(0, \overset{k}{\ldots}, 0, u_{k+1}, \ldots, u_n\right)$ do (see (5) & (6)):

$$\text{If } \Phi(u) = 1 \text{ then } L = L + \Pr\{u\},$$
$$\text{If } \Phi(u) = 0 \text{ then } V = V + \Pr\{u\}.$$

Step 5. Compute $U = 1 - V$ and write (see (5), (7) & (8))

$$L \leq \Pr\{\Phi = 1\} \leq U, \quad E = U - L \leq \epsilon.$$

Example 7. Recall that a k-out-of-n:F system is an n-component system that fails if and only if at least k of the n components fail [1]. Consider the multiengine system in an airplane, the multipump system in an hydraulic control system, or the accumulator system of a pressured water reactor in a nuclear power plant. Suppose that each one of these technical systems depends on $n = 150$ statistically independent basic components, and that their logical structures correspond to a 5-out-of-150:F system, i.e. all of them are described by the Boolean function $\Phi(x) = 1$ if $\sum_{i=1}^{150} x_i \geq 5$, $\Phi(x) = 0$ otherwise. Let the basic (failure) component probabilities be

$$\{p_i\}_{i=1}^{10} = 10^{-4} < \{p_i\}_{i=11}^{20} = 2 \cdot 10^{-4} < \cdots < \{p_i\}_{i=141}^{150} = 15 \cdot 10^{-4} = 1.5 \cdot 10^{-3}.$$

Let us estimate the system unavailability $\Pr\{\Phi = 1\}$ with a maximum error $\epsilon = 0.1$, using the Top Subgraph Algorithm (with eight-decimal-digit precision).
Steps 1 & 2.

$$S_{150}^{0,\overset{140}{\smile},0} = \prod_{i=1}^{140} (1 - p_i) = 0.90027717 \geq 1 - \epsilon = 0.9,$$
$$S_{150}^{0,\overset{141}{\smile},0} = \prod_{i=1}^{141} (1 - p_i) = 0.89892675 < 1 - \epsilon = 0.9.$$

Step 3. Hence, $k = 140$ and we choose $C = I_{150}^{0,\overset{140}{\smile},0}$, i.e. the top subgraph –on $2^{n-k} = 2^{10} = 1024$ nodes– obtained after $k = 140$ bisections of I_{150}. Then, for all the 1024 binary 10-tuples $(u_{141}, \ldots, u_{150}) \in \{0,1\}^{10}$, we compute

$$\Pr\{u\} = \Pr\left\{\left(0,\overset{140}{\ldots},0,u_{141},\ldots,u_{150}\right)\right\} = S_{150}^{0,\overset{140}{\smile},0} \prod_{i=141}^{150} p_i^{u_i} (1 - p_i)^{1-u_i}.$$

Step 4. $L = \sum_{\substack{u \in C \\ \Phi(u)=1}} \Pr\{u\} = 4 \cdot 10^{-8}$, $\quad V = \sum_{\substack{u \in C \\ \Phi(u)=0}} \Pr\{u\} = 0.90027713.$

Step 5.

$$L = 4 \cdot 10^{-8} \leq \Pr\{\Phi = 1\} \leq 0.09972287 = U,$$
$$E = U - L = 0.09972283 \leq \epsilon = 0.1.$$

6 Summary and Conclusions

In this paper, we have developed new approaches to the reliability analysis of CSBSs. Our starting point has been the intrinsic order graph I_n: a symmetric directed graph, with 2^n vertices, displaying all the bitstrings associated to a given CSBS in decreasing order of their occurrence probabilities.

We have proved that successively bisecting this graph into smaller and smaller subgraphs is equivalent to successively splitting the unity as sum of the occurrence probabilities of those subgraphs. Our main theorem has stated that if we "concentrate" each one of the 2^k subgraphs C of I_n –obtained after k successive bisections of I_n– into an unique node weighted by the sum of the probabilities of all nodes lying on C, and we sort these new nodes in decreasing order of their weights, then the new "condensed" graph is precisely the intrinsic order graph I_k. Based on these results, we have proposed a new algorithm for obtaining lower and upper bounds on the failure probability of those CSBSs that can be described by a Boolean function. The efficiency of the algorithm depends on the proper choice of a subgraph of I_n.

The main idea of this paper is that (in accordance with our main theorem) sorting the 2^k subgraphs of I_n in decreasing order of their contributions to the estimation accuracy is equivalent to ordering the 2^k corresponding binary k-tuples (consisting on the k left-most constant components of the n-tuples of each subgraph) in decreasing order of their probabilities. In particular, the "optimal" subgraph (for the efficiency of the algorithm) of I_n is the subgraph $I_n^{\overset{k}{\overbrace{0,\ldots,0}}}$ placed at the top of I_n. Every time we bisect the graph I_n (i.e. every time we split the unity), and select $I_n^{\overset{k}{\overbrace{0,\ldots,0}}}$ instead of $I_n^{\overset{k-1}{\overbrace{0,\ldots,0}}}$, we reduce the computational cost by a factor of 2, because the number of bitstrings to be evaluated, i.e. the cardinality of $I_n^{\overset{k}{\overbrace{0,\ldots,0}}}$, is 2^{n-k} (see (12)). However, at the same time, we reduce the contribution to the accuracy by a factor of $1 - p_k$, because of (20).

Since the basic probabilities p_i in real technical systems are usually close to 0 [1,8], then the factors $1 - p_i$ in (20) are usually close to 1. Hence, our algorithm provides a good accuracy/computational-cost ratio for most real-world cases. Moreover, it can be applied to many different phenomena –modeled as CSBSs– arising from diverse scientific or engineering fields, e.g. Reliability Engineering, Meteorology, System Analysis, Biology, Economics, etc. The main advantage is that the intrinsic order criterion drastically reduces the complexity of the problem from exponential (sorting 2^n binary n-tuples) to linear (ordering n Boolean variables). Finally, to improve the balance between accuracy and computational cost for CSBSs with a large number of variables with larger probabilities p_i, other related algorithms can be constructed by choosing other subgraphs (19) in addition to the optimal one $I_n^{\overset{k}{\overbrace{0,\ldots,0}}}$. This choice must be performed following the intrinsic order in I_k, as our main result, Theorem 4, has stated.

References

1. Andrews, J., Moss, B.: Reliability and Risk Assessment. Professional Engineering Publishing, London (2002)
2. Diestel, R.: Graph Theory. Graduate Texts in Mathematics, vol. 173. Springer, Heidelberg (2005)
3. González, L.: N-tuples of 0s and 1s: Necessary and sufficient conditions for intrinsic order. LNCS, vol. 2667(1), pp. 937–946. Springer, Heidelberg (2003)
4. González, L.: A new algorithm for complex stochastic Boolean systems. LNCS, vol. 3980(1), pp. 633–643. Springer, Heidelberg (2006)
5. González, L.: A picture for complex stochastic Boolean systems: The intrinsic order graph. LNCS, vol. 3993(3), pp. 305–312. Springer, Heidelberg (2006)

6. González, L.: Algorithm comparing binary string probabilities in complex stochastic Boolean systems using intrinsic order graph. Adv. Complex Syst. 10(1), 111–143 (2007)
7. González, L., García, D., Galván, B.J.: An intrinsic order criterion to evaluate large, complex fault trees. IEEE Trans. Reliab. 53(3), 297–305 (2004)
8. National Aeronautics and Space Administration: Fault Tree Analysis: A Bibliography. Technical report NASA/SP–2000-6111 (2000)
9. Schneeweiss, W.G.: Boolean Functions with Engineering Applications and Computer Programs. Springer, Heidelberg (1989)
10. Singpurwalla, N.D.: Foundational Issues in Reliability and Risk Analysis. SIAM Rev. 30(2), 264–282 (1988)
11. Stanley, R.P.: Enumerative Combinatorics, vol. 1. Cambridge University Press, Cambridge (1997)

Estimating Metabolic Pathways Parameters Using Distributed Monte Carlo Method

Paweł Lebioda[1] and Piotr Bała[1,2]

[1] Faculty of Mathematics and Computer Science,
Nicolaus Copernicus University
Chopina 12/18, 87-100 Toruń, Poland
{killer,bala}@mat.uni.torun.pl

[2] Interdisciplinary Center for Mathematical and Computational Modelling
University of Warsaw
Pawińskiego 5a, 02-106 Warsaw, Poland

Abstract. Metabolic pathways can be represented as a huge set of differential equations with a large number of parameters. The solution of this system is not trivial because of the problem size and because of the limited information on model parameters. Most of the parameters cannot be calculated directly or obtained as direct experimental results. The only way is to estimate models parameters in such way that result of model approximate available experimental data. In this paper we present method of estimation of the parameters based on distributed computing using Monte Carlo method. This method has been applied to the tricarboxylic acid cycle [1] and compared with the existing model.

1 Introduction

Metabolic pathways are one of the most important subjects for biochemistry. The metabolic pathways are set of the chemical reactions where products of some reactions are substrates in the other ones. Therefore the reactions depend each other and are connected forming complex network. The detailed knowledge of the reactions, especially determination of the critical steps is crucial for modern drug design and is of immediate interest.

From the mathematical point of view, every chemical reaction can be modeled by one or more differential equation. Solving this equation one obtains concentration of chemical products over the time. The proposed model is simple and, in principle, can be solved analytically. Unfortunately, differential equations used to describe even simple metabolic reactions are complicated so there is no chance to achieve exact solution. The problem is additionally complicated by the limited information on the constants used to constitute equations. The reaction rate constant is a good example. The experimental data is limited and uncertain since the parameters are not measured directly. In result the treatment of metabolic pathways is difficult and complicated.

The greatest difficulty is design of the set of differential equation. Usually this process can be divide into two steps. First one is collection of the experimental

O. Gervasi et al. (Eds.): ICCSA 2008, Part II, LNCS 5073, pp. 987–999, 2008.

data, obtained for example from *in vitro* experiments. Second step is estimation of the parameters of differential equation based on the acquired data. This strategy has been already presented in the number of papers, for example [1,2,3]. Because of nature of metabolic pathways, this process is very difficult, long and arduous. In this paper we present method of parameters estimation based on the distributed computing using Monte Carlo method. This method simplifies and automatize parameters prediction for the metabolic pathway models.

2 Metabolic Pathways

2.1 Modeling Chemical Reactions

Every chemical reaction can be described by a rate equation which is as a special differential equation. The detailed form of the equation can vary and depends on the reaction type. In the metabolic pathways, chemical reactions are very complicated. Most of them are of catalytic type and to describe that kind of reactions we use the Michaelis-Menten kinetics [4]. Here we consider simple enzymatic reaction:

$$E + S \underset{k_{-1}}{\overset{k_1}{\rightleftarrows}} ES \overset{k_2}{\rightarrow} E + P,$$

where S is a substrate, P product, E enzyme and k_1, k_2, k_{-1} are proper rate constants.

For this reaction we can easily write rate equation for the product and substrat concentration ($[P]$ and $[S]$ respectively):

$$\frac{d[P]}{dt} = V_{max} \frac{[S]}{\frac{k_{-1}+k_2}{k_1} + [S]}.$$

One should note that the more complex reaction is considered, the more complicated differential equations are obtained.

2.2 Tricarboxylic Acid Cycle

In this paper we consider the tricarboxylic acid cycle (TCA), the oxidative phosphorylian and the mitochondrial Ca^{2+} handling. The reaction path is presented in the Figure 1. The analytical solution of the model has been presented in [1].

The tricarboxylic acid cycle can be described by twelve differential equations involving intermediates, the NADH, the ADP, the matrix concentration of CA^{2+} and the mitochondrial membrane potential $\Delta\Psi_m$ (Figure 2).

The tricarboxylate cycle consists of two separate pathways. First, the tricarboxylate pathway consists of several reactions which product following chemical compounds: the oxalacetate (OAA), the citric acid (CIT), the isocitrate (ISOC) and the α-ketoglutarate (αKG). Second, the dicarboxylate is composed of the

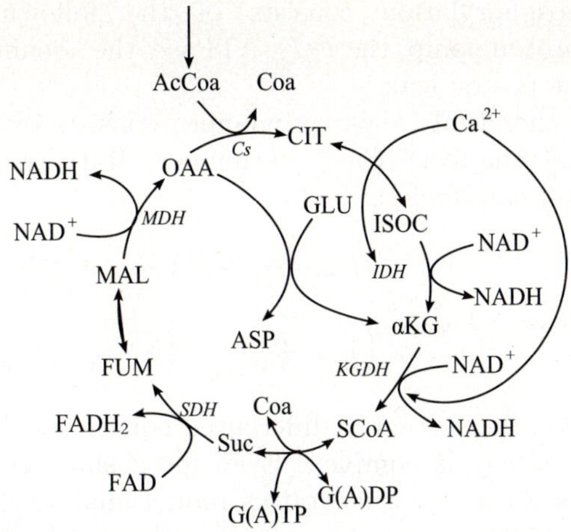

Fig. 1. The schematic graph of the tricarboxylic acid cycle [1]

$[ADP]_m$	$\frac{d[ADP]_m}{dt} = V_{ANT} - V_{ATPase} - V_{SL}$
$[NADH]$	$\frac{d[NADH]}{dt} = V_{IDH} + V_{KGDH} + V_{MDH} - V_{O_2}$
$[ISOC]$	$\frac{d[ISOC]}{dt} = V_{ACO} - V_{IDH}$
$[\alpha KG]$	$\frac{d[\alpha KG]}{dt} = V_{IDH} - V_{KGDH} + V_{AAT}$
$[SCoA]$	$\frac{d[SCoA]}{dt} = V_{KGDH} - V_{SL}$
$[Suc]$	$\frac{d[Suc]}{dt} = V_{SL} - V_{SDH}$
$[FUM]$	$\frac{d[FUM]}{dt} = V_{SDH} - V_{FH}$
$[MAL]$	$\frac{d[MAL]}{dt} = V_{FH} - V_{MDH}$
$[OAA]$	$\frac{d[OAA]}{dt} = V_{MDH} - V_{CS} - V_{AAT}$
$[ASP]$	$\frac{d[ASP]}{dt} = V_{AAT} - V_{C\_ASP}$
$[CA^{2+}]_m$	$\frac{d[CA^{2+}]_m}{dt} = f(V_{uni} - V_{NaCa})$
$\Delta\Psi_m$	$\frac{d\Delta\Psi_m}{dt} = \frac{V_{He}+V_{He(F)}-V_{Hu}-V_{ANT}-V_{HLeak}-V_{NaCa}-2V_{uni}}{C_{mito}}$

Fig. 2. The reactions constituing tricarboxylic cycle [1]

αKG, the succinyl Coa (SCoA), the succinate (Suc), the fumarate (FUM), the malate (MAL) and the OAA. The rate constat for the reaction involving alpha ketoglurate dehydrogenase can be described as:

$$V_{KGDH} = \frac{k_{cat}^{KGDH} E_T^{KGDH}}{1 + \frac{\left(\frac{K_M^{\alpha KG}}{[\alpha KG]}\right)^{n_{aKG}}}{\left(1+\frac{[Mg^{2+}]}{K_D^{Mg^{2+}}}\right)\left(1+\frac{[Ca^{2+}]_m}{K_D^{Ca^{2+}}}\right)} + \frac{\left(\frac{K_M^{NAD}}{[NAD]}\right)^{n_{aKG}}}{\left(1+\frac{[Mg^{2+}]}{K_D^{Mg^{2+}}}\right)\left(1+\frac{[Ca^{2+}]_m}{K_D^{Ca^{2+}}}\right)}}$$

The oxidative phosphorylation consists of the following processes: the respiration-driven proton pomp, the F_1F_0-ATPase, the adenine nucleotide translator (ANT) and the proton leak.

Finally, the mitochondrial calcium dynamics consists of two processes: the Ca^{2+} uniporter and the Na^+/Ca^{2+} exchanger. Rate of the mitochondrial Na^+/Ca^{2+} antiporter can be described as:

$$V_{Naca} = V_{max}^{NaCa} \frac{e^{\left(\frac{bF(\Delta\Psi_m - \Delta\Psi^0}{RT}\right)} e^{\left(\ln \frac{[Ca^{2+}]_i}{[Ca^{2+}]_m}\right)}}{\left(1 + \frac{K_{Na}}{[Na^+]_i}\right)^n \left(1 + \frac{K_{Ca}}{[Ca^{2+}]_m}\right)}$$

As we can see here, the system of differential equations which describes even simple metabolic pathway is complex. Even for a small systems such as tricarboxylic acid cycle we can not solve this model analytically. For the typical metabolic pathways which consist of hundreds or thousands of chemical reactions the problem is much more complicated, and it is difficult to establish proper set of differential equations.

3 Application

3.1 Solving Set of Differential Equations

In order to solve obtained set of the differential equations we need simulation engine which is fast and accurate and allows to solve huge sets of differential equations even with several dozen of thousands of equations. For these purposes we have chosen the SPICE package. Despite it has been design for simulations of analog devices, it is possible to apply it to solve arbitrary set of differential equations. This technique has been described elsewhere and numerical results is almost identical to result achieved with the Matlab package. For the test simulations, the average $PSNR$ (*Peak Signal to Noise Ratio* is approximately 180 dB and the MSE (*Mean Square Error*) is approximately 10^{-10}. Furthermore, the computation time is also comparable. The SPICE package handles very large systems. Additional advantages are small size of package and open source, which is important when we want to install on many computers. The SPICE package has one disadvantage. The source file for this application is description of analog device, but conversion from sets of differential equations to equivalent analog device isn't complicated. With the tools developed previously we are able to get analog device with size growing linearly with the size of the metabolic pathway.

3.2 The Monte Carlo Method

The simulations of the metabolic pathways are limited by the availability of the coefficients which describe individual reactions. As we have mentioned above, these numbers cannot be obtained directly from experiments; in most cases these are obtained indirectly using multiparameter fitting procedure.

One of the methods to use is Monte Carlo [5,6] which simplifies the problem by reduction of the parameter domain to it's finite subset. The reduction process is based on random choice of the representatives of the domain. The stochastic process is performed until good approximation of the experimental data is achieved. The choice of subdomain plays very important role in the whole process. First, two separate simulations for two different datasets of equal size can give different results. Second, the larger dataset the better results we can achieved.

The Monte Carlo is one of the best methods to approximate coefficients of differential equations which reproduce experimentally measured rate constants. This method simplifies to following steps performed as long as similarity to the experimental data reaches acceptable value:

1. randomly choose samples for all examined coefficient (separately and maybe using different distributions)
2. using generated datasets, simulate metabolic pathways
3. compute similarity to the experimental data (measured with $PSNR$ and MSE)

Unfortunately there are some problems. There is variety of distributions which can be used to generate coefficients and the choice is not straightforward.

The another problem is definition of the similarity of the generated solution to the experimental one. In particular case we have tested number of different solutions and one based on the Peak Signal to Noise Ratio ($PSNR$) and Mean Square Error (MSE) was the best solution.

Additionally, source data and results have different sizes and samples are located in the different time domains. The solution we have used is based on the simple linear or cubic interpolation of the samples.

3.3 Approximation Quality

The classification of the results obtained for generated data set is based on the observed kinetics of the chemical reactions. In every iteration we acquire number of curves which are compared to sample ones. For each curve we calculate MSE or $PSNR$. Because $PSNR$ has „logarithmic nature" it is better to use it as a measure for a single curve.

The another problem is comparison of multidimensional data. For these purposes number of values such as average, minimum or maximum can be used. Unfortunately, such approach does not work for our data because there is no relationship between $PSNR$ of different curves. To calculate similarity between two sets of curves we have implemented procedure:

1. For each curve:
   (a) find maximal $PSNR_{max}$.
   (b) normalize all results using $PSNR_{max}$
2. For each iteration:
   (a) find minimal $PSNR$. This value is result of iteration.
3. Result of simulation is maximal number find in 2a.

Practically, as a measure we have used $1 - PSNR_{normalized}$ which is equal to 0 for ideal approximation, and is positive otherwise.

Iteration	Curve 1	Curve 2	Curve 3
1	50 dB	140 dB	100 dB
2	75 dB	180 dB	95 dB
3	40 dB	165 dB	115 dB
4	95 dB	145 dB	90 dB
Maximal value	95 dB	180 dB	115 dB

$\Rightarrow$

Iteration	Curve 1	Curve 2	Curve 3	Result
1	0.52	0.77	0.86	0.52
2	0.78	1.00	0.82	0.78
3	0.42	0.91	1.00	0.42
4	1.00	0.80	0.78	0.78
Maximal value	95 dB	180 dB	115 dB	

Fig. 3. Example procedure used for choosing the best results of simulation. In this simple simulation we have only four iteration and three sample curves. In left table there are results of simulation and already computed maximal $PSNR$ of each curve (step 1a). Right table presents normalized results (step 1b). Last column contains results for each iteration.

3.4 Implementation

The implementation of the presented methodology has to deal with the numerical complexity of the solution. For a simple tricarboxylate cycle the average computation time is slightly less than one second (with minimal output size). In the case of more complex pathways the time rise up to minutes or even more. Using the Monte Carlo method on single computer is very ineffective, e.g. for 1000 samples and average time of solving whole set of differential equations of 1s we can get large computation time. Fortunately, the Monte Carlo method can be easily parallelized. For a given N vectors of samples and M computers we only need to divide dataset to $N' = \frac{N}{M}$ parts, and computations can by done in parallel.

Our implementation is based on a distributed computing concept [7,8]. Created network is of master - slaves type. Master computer assigns jobs, oversees and collects and merge results. Number of workers perform jobs assigned by a master. In this model of network, workers passively wait for jobs from master.

Structure of the application is constrained by a shape of the network. We have divided algorithm to two parts: master and worker.

Master

Application in master role follows steps:

1. Read and parse description of metabolic pathways.
2. Check logical correctness of the description.
3. Read experimental data, which will be used to compare results of simulations.
4. Prepare random dataset.
5. Try to connect to workers (their ip and port are listed in the configuration file)
6. For each successful connected worker:
   (a) Prepare distinct jobs.
   (b) Send model description, experimental data
   (c) Send jobs.
   (d) Wait for results.

7. If some jobs aren't computed yet go to 6a (and skip 6b)
8. Build final version of simulations.

In the step 2 we need to checks description for logical errors such as undefined variables, functions etc. In the step 6a dataset generated in step 4 is divided into equals parts, and for assigned for each worker. Assigned parts are tagged with worker id. In the step 6d three possibilities may occur: job was successful, job failed (but worker condition is ok), worker disconnected for some reason. In the first two cases job is removed from main pool. In the third situation all unfinished jobs of inactive worker are returned to the main pool. These jobs will be reassigned and send again. The worker condition is checked out by the master using hearthbeat mechanism.

Worker
Worker job is a described by a structure which contains: unique job id, identification of assigned worker, some number of randomly generated values of estimated parameters. Worker's role follows steps:

1. Wait for connection.
2. Receive from master model description, data to compare and set of jobs.
3. For each job:
   (a) Simulate model using values of parameters from job description.
   (b) If simulation completed successful compare result to sample data. Compute PSNR.
   (c) Send to the master information about the job.
4. Send to the master information that all assigned jobs are finished.
5. If Server replies that there are additional jobs, receive them and go to the step 3.
6. Close connection with the master and go to the step 1.

In the step 2 worker gets model description in processed form. In the step 3c there are possible two situations. The first is when the simulation finished successfully - worker sends computed value of PSNR and optional result of simulation. Second situation: during simulation an error occurred. In this case worker sends information job which cannot be completed. Situation with an additional jobs processed in the step 5 happens when another worker has lost connection with the master and was not able to perform scheduled work. Master reassigns unfinished jobs to active workers.

3.5 Communication Protocol

The communication between master and workers is based on the TCP/IP. The protocol allows to: send and receive descriptions of metabolic pathways, jobs, data to compare, receive simulations results and error control. Additionally we would like to use the protocol as file format. It has to be ease to implement, extend and has to be portable across different operating systems. XML has such features, but it has also important weakness - the redundancy. This has impact

idSimulation	/U
idJobId	/k56
idJobValues	/u
idName	/nCAM
idValue	/v9.4969e-10
idName	/nDAP
idValue	/v3.964
idEnd	//
idJobPSNR	/z200
idEnd	//

Fig. 4. Sample communicate schema. Worker informs master that Job 56 has been successful completed. The parameter values has been generated and *CAM* is 9.4969^{-10}, *ADP* is 3.9648. For this job obtained PSNR = 200 dB. Right column presents structure of communicate, left column shows this communicate.

on communication speed, which is important especially when large datasets has to be transmitted. For example: the XML description of TCA cycle is few times larger than original one.

In order to solve this problem we have designed dedicated extensible communication protocol which is used to exchange date between master and workers. An example message is presented in the Figure 4.

3.6 Performance and Scaling

The time assigned for connections to the workers, jobs transfer and results collection is proportional to number of workers. Notwithstanding, the time consumed by additional computers is much less that profit gained on computation time reduction. It is obvious, that number of workers computer has significant influence on the computation time. Doubled number of workers computers should reduce

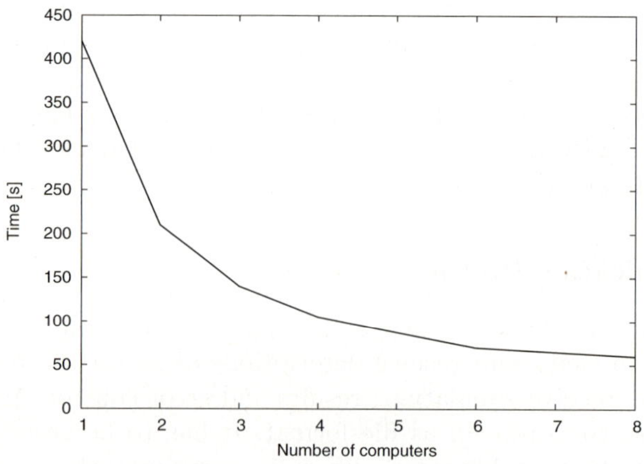

Fig. 5. Dependence between time and number of workers

computation time by half. Practical profit is a bit worse, because some delays occur in the results transfer (specially when transmitting full results) and when handling many workers computers.

4 Results

The presented algorithms has been used for determining reaction parameters for number of different cases. We have started from the single reaction where solution is straightforward and moved to the multiple reactions. In the more complicated examples we have tried to calculate reaction rate coefficient based on the different number of available data. This allows us to test procedure and estimate its quality and efficiency.

4.1 Equilibrium Reaction

First example is simple reaction. We have estimated parameters of equilibrium reaction between two species:

$$A \leftrightarrow B$$

This reaction can by described by a set of two differential equations:

$$\begin{cases} \frac{dA}{dt} = -pA + qB \\ \frac{dB}{dt} = pA - qB \end{cases}$$

The initial concentration of the compound $[A]$ was 5 and $[B]$ was 0. The experimental data was generated for $p = 1$ and $q = 2$. The number of tests was 5000. From the results of simulations presented in the Figure 6 and Figure 7 one can see that original curves are well approximated. Especially further increase of the number of samples do not lead to significant improvement of the found approximation.

4.2 Tricarboxylic Acid Cycle

The tricarboxylic acid cycle is composed of twelve differential equations [1] (Figure 2). Each equation gives us a sample curve, which we use for comparison of the results. In this section we present estimation of the parameters of *citrate synthase*, *succinyl coa lyase* and for the all reactions in the cycle. For each experiment we have performed several simulation, each of them contained 5000 tests. Each estimated parameter has been searched in the interval $\pm 20 - 25\%$ of original value.

Rate Equation for Citrate Synthase
The citrate synthase (CS) can be described by the reaction equation [1]:

$$V_{CS} = \frac{k_{cat}^{CS} E_T^{CS}}{1 + \frac{K_M^{AcCoA}}{[AcCoA]} + \frac{K_M^{OAA}}{[OAA]} + \frac{K_M^{AcCoA}}{[AcCoA]} \frac{K_M^{OAA}}{[OAA]}} \tag{1}$$

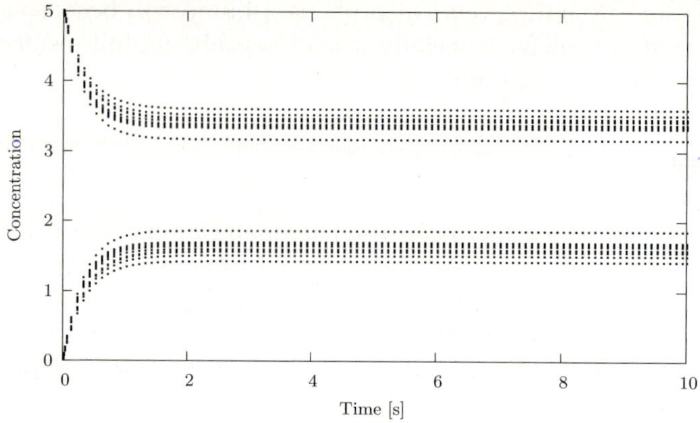

Fig. 6. Results of the simulations for example 4.1. The best 10 approximations of the original curves are presented.

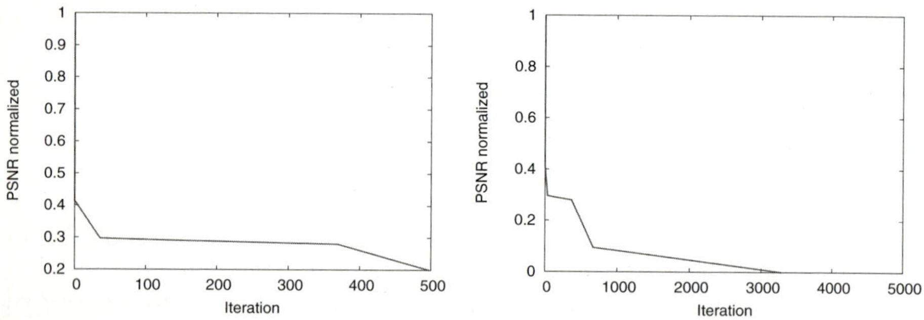

Fig. 7. Results of approximation of equations parameters (eq. 4.1). Left column presents first 500 iterations, right column presents all simulation (5000 iterations).

In this case we have estimated value of k_{cat}^{CS},E_T^{CS}, K_m^{OAA} and K_M^{AcCoA} assuming that all other parameters are known. Three different kind of simulations has been perform. In first group of simulations as experimental data to compare we have used values of OAA (*Case 1*). In second, for comparison we used OAA, MAL and αKG curves (*Case 2*). In the third case we have used all available samples (*Case 3*). Results of simulations are presented in the Figure 8 (left column). Approximation of the parameters are shown in the right column. Average difference between estimated and original parameters is about $4-6\%$.

Rate Equation for Succinyl Coa Lyase
We have tried to estimate parameters of equation describing Succinyl Coa lyase (SL) [1]:

$$V_{SL} = k_f^{SL}\left([SCoA][ADP]_m - \frac{[Suc][ATP]_m[CoA]}{K_E^{SL}}\right) \quad (2)$$

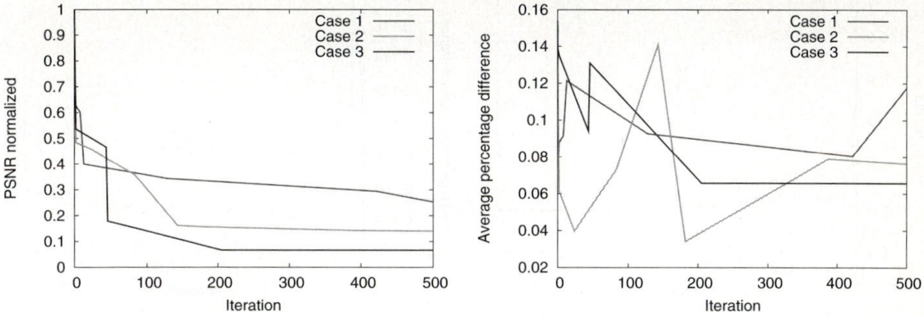

Fig. 8. Results of the approximation of the parameters of the equation 1 (left column). Average difference between estimated and original parameters are presented in the right column. Case 1 - using value of OAA, case 2 - using value OAA, MAL and αKG. Case 3 - using all available samples.

We have estimated value of k_f^{SL} and K_E^{SL}. Two different kind of simulations has been perform. In first group of simulations we have used Suc and $ScoA$ curves as experimental data to compare (*Case 1*). In the second one we have used all available curves (*Case 2*). Results of the simulations are presented in the Figure 9 (left column). Approximation of the parameters is shown in the right column. An average difference between estimated and original parameters is about $3-5\%$.

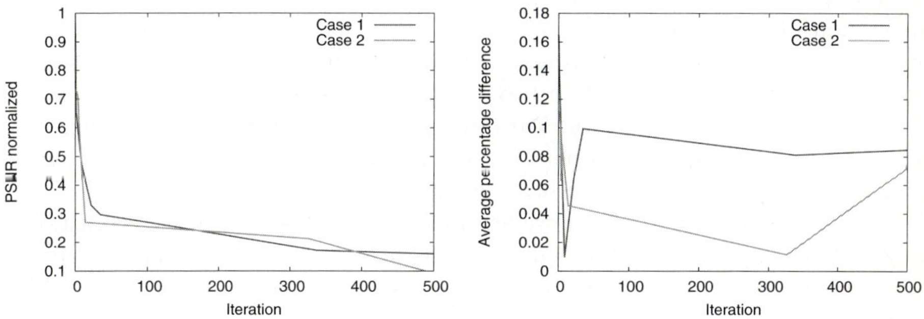

Fig. 9. Results of approximation of the parameters of the equation 2 (left). Average difference between estimated and original parameters equation (right). Case 1 - values of Suc and $SCoA$ are used for comparison, Case 2 - all available samples are used.

Tricarboxylic Acid Cycle

In the last experiment presented here we have tried to estimate all parameters of the TCA cycle [1]. Please note that the number of parameters is almost one hundred. In the simulations we have reduced that number by removing physical

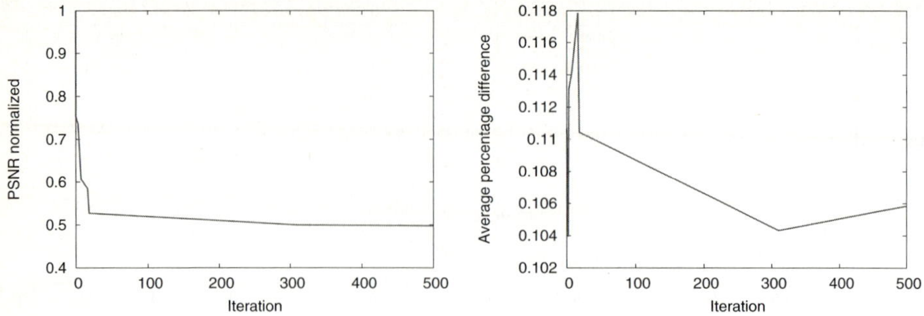

Fig. 10. Results of the estimation of the parameters of the TCA cycle (left). Mean difference between values of estimated and original parameters is presented in the right.

and chemical constants. Results of the experiment are presented in the Figure 10. We have tried to estimate quite large set of parameters with only twelve original curves therefore the results are not so good as in the previous examples and mean difference of estimated parameters is about 10.8%

5 Conclusions

Approximation of the coefficients of reaction rate equations for the metabolic pathways is a challenge. The Monte Carlo method is possible solution to obtained parameters which will allow to reproduce experimental rate curves with the calculated ones. The simulations are long and time consuming therefore parallelization of the process has to be used. We have presented algorithm used for such purposes as well as presented practical implementation. Presented results show that our approach to estimate parameters of reaction equations of the metabolic pathways is effective. We have provided test cases which differ with the number of estimated parameters and with the number of known data. In all cases reasonable approximation has been achieved.

The presented method has however some limitations. In some circumstances not all of parameters and independent and can be precisely estimated. For example in the equation 1 is very hard to approximate coefficients k_{cat}^{CS} and E_T^{CS}. These parameters occur only in one equation and they are link each other. We cannot compute their values separately but if we consider the product, we can get good results. This problem can be solved by replacing both coefficients by a common parameter. The linked parameters can be separated in the another or the same simulation with the knowledge of extra curves which depend on other relationship between both parameters.

In the future we plan to extend our application by automatic choose of the best result and narrow search to smaller domain. For the distributed simulations we plan to introduce some security like authorization which will allow to use our application on distributed resources.

References

1. Cortassa, S., Miguel, A.A., Marban, E., Winslow, R., L., O'Rourke, B.: An integrated model of cardiac mitochondrial energy metabolism and calcium dynamics. Biophys. J. 84, 2734–2755 (2003)
2. Yang, C.R., Shapiro, B.E., Hung, S.p., Mjolsness, E.D., Hatfield, G.W.: A mathematical model for the branched chain amino acid biosynthetic pathways of escherichia coli k12. J. Biol. Chem. 280(12), 11224–11232 (2005)
3. Ishii, N., Suga, Y., Hagiya, A., Watanabe, H., Mori, H., Yoshino, M., Tomita, M.: Dynamic simulation of an in vitro multi-enzyme system. FEBS LettersVolume 581(3), 413–420 (2007)
4. Lehninger, A.L., Nelson, D.L., Cox, M.M.: Principles of Biochemistry, 3rd edn. Worth Publishers (2000)
5. Fishman, G.S.: Monte Carlo: Concepts, Algorithms, and Applications. Springer, Heidelberg (1996)
6. Gentle, J.E.: Random Number Generation and Monte Carlo Methods. Springer, Heidelberg (2003)
7. Ben-Ari, M.: Principles of Concurrent and Distributed Programming. Prentice-Hall, Englewood Cliffs (1990)
8. Garg, V.K.: Elements of distributed computing. John Wiley & Sons, Inc., Chichester (2002)

Parallel Actions and Generalized Multivalued Constraints in Multivalued Planning

M. Baioletti, A. Milani, V. Poggioni, and S. Suriani

Dept. of Mathematics and Computer Science, University of Perugia,
via Vanvitelli 1, 06123, Perugia (Italy)
{baioletti,milani,poggioni,suriani}.dipmat.unipg.it

Abstract. In this work an extension of the model for planning with multivalued fluents and graded actions introduced in [8] is proposed. This model is based on the infinity–valued Łukasiewicz logic, where the fluents can assume truth values in the interval $[0, 1]$ and actions can be executed at different application degrees also varying in $[0, 1]$. Multivalued fluents and graded actions allow to model many real situations where some features of the world are fuzzy and where actions can be executed with varying strength. The main contributions of this paper are given by the introduction of the simultaneous executability of the graded actions and the extension of multivalued constraints to generalized multivalued constraints. An extension of the correct/complete algorithm which solves bounded multivalued planning problems is presented. It allows to solve problems with generalized constraints and simultaneous actions.

1 Introduction

Several logical approaches to AI automated planning have been proposed in addition to planners based on specialized algorithms. These approaches propose to encode the planning problem by a logical theory and to synthesize the plans by means of some general logical procedures. Traditionally, planning has been formalized as deduction. The best-known logical formalization of planning in the deductive view is the Situation Calculus [2]. Dually to the "planning as derivability" approach, the "planning as satisfiability" paradigm was introduced by [3], where planning is reduced to model search in classical propositional logic. More recently, modal logics have been used to model non classical and non-deterministic planning domains (for example [4]), to handle control knowledge of the domain expert and extended goals (for example [5,6]). Finally a fuzzy model for planning is proposed in [7], where fluents are denoted with finite–valued variables and action effects are represented by conditional clauses. A different fuzzy planning model managing multivalued fluents and graded actions has been introduced in [8]. This model proposes to use the Łukasiewicz logic [9] in order to model the planning domains. It allows to represent some crucial features of the actual domains that are not suitably representable by boolean variables.

O. Gervasi et al. (Eds.): ICCSA 2008, Part II, LNCS 5073, pp. 1000–1011, 2008.

In many planning domains disagreement and fuzziness of real world properties need to be modeled. Since classical actions and constraints with crisp boolean values are not adequate to represent these domains, it is necessary to introduce into planning no–boolean fluents and soft constraints that can be partially violated. The idea underlying multivalued planning is to model in a logical framework fuzzy concepts, such as "the door is slightly open" or "the door is wide open", and graded application of actions with additive effects, such as "open the door a bit more" or "open the door as much as possible".

In a typical probabilistic planning model the degrees of uncertainty are linked to the fluents and the actions are considered as a further source of uncertainty of the system. Moreover, it is worth noticing that many fuzzy concepts cannot be represented by numerical resources, because the fuzzy values could not be obtainable by a measurement process or they could be affected by subjective factors. In the multivalued/fuzzy logics scenario the use of a logic based on T-norms is preferred because of its well-founded mathematical aspects. Among this class of logics the Łukasiewicz logic is chosen because the semantics of its operators is more suitable for the intended framework.

The main contributions of this paper are given by the extension of multivalued constraints to generalized multivalued constraints and the introduction of the simultaneous executability of the graded actions. The former extension allows to express constraints by arbitrary Łukasiewicz formulae in the action preconditions and preferences and in the problem goals. The latter introduces a definition of simultaneously action execution that extends the classical one proposed originally in [1] and then adopted by the most planning algorithms.

As proposed in [8], the planning problem is translated into a MIP (Mixed Integer Programming) problem, but due to the new features, some changes had to be made to the algorithm and the translation phase had to be enriched with new constraints. The algorithm was implemented and tested. The results were encouraging even if the implementation needs improvements.

2 The Model

In this section the model of multivalued planning presented in [8] is briefly summarized, and the extensions to parallel actions and generalized multivalued constraints are proposed. The model allows to use boolean fluents and classical actions jointly with two new kinds of fluents and actions, called multivalued fluents and graded actions.

2.1 Multivalued States

Let $\mathcal{F}$ be the finite set of fluents. Given a time–step t, a state is described by a function S_t assigning to each fluent f a truth value belonging to $[0,1]$, denoted by $S_t(f)$. S_t can be seen as a natural extension of the classical state definition in which the range of the state function is $[0,1]$ instead of $\{0,1\}$.

The value at a time–step t of an arbitrary formula in Łukasiewicz logic ϕ is denoted by $S_t(\phi)$ and it is computed in a recursive way. For instance, $S_t(\phi \odot \psi) = S_t(\phi) \odot S_t(\psi)$, and so on.

2.2 Generalized Multivalued Constraints

In [8] *multivalued constraints* over fluents values are introduced. In this work we extend this definition proposing *generalized multivalued constraints* as constraints over an arbitrary formula.

Definition 1. *Generalized Multivalued Constraints*
A generalized multivalued constraint *is represented by a tuple* $(\mathtt{k},\chi,\phi)$, *where* $\mathtt{k} \in \{\mathtt{b},\mathtt{s}\}$ *denotes the kind of the constraint (boolean or soft),* χ *is a value in* $[0,1]$ *representing a threshold and* ϕ *is an arbitrary formula of Łukasiewicz logic. A boolean constraint in the state* S_t *is defined by the classical constraint* $S_t(\phi) \geq \chi$, *while the soft constraint is defined by the Łukasiewicz implication* $\chi \rightarrow S_t(\phi)$.

The expression $\chi \rightarrow S_t(\phi)$ has been chosen to encode a soft constraint because it is used as a fuzzy extension of the crisp relation $S_t(\phi) \geq \chi$. In fact, if $S_t(\phi) \geq \chi$, then the implication assumes truth value 1, while if $S_t(\phi) < \chi$, the implication assumes a value lesser than 1. The higher the difference between $S_t(\phi)$ and χ, the lesser the value of the implication.

Definition 2. *Satisfaction Degree of a Generalized Multivalued Constraint*
Let $\mathcal{S}$ *a set of states,* X *a set of generalized constraints* $mc = (\mathtt{k},\chi,\phi)$, *Let* $S \in \mathcal{S}$, *then the* satisfaction degree *of the constraint in the state* S *is defined by the function* $\mathcal{D} : \mathcal{S} \times X \rightarrow [0,1]$,

$$\mathcal{D}(S, mc) = \begin{cases} 1 & \text{if } k\text{=}b \text{ and } S(\phi) \geq \chi \\ 0 & \text{if } k\text{=}b \text{ and } S(\phi) < \chi \\ \chi \rightarrow S(\phi) & \text{if } k\text{=}s \end{cases} \tag{1}$$

2.3 Graded Actions

In this section we present the extensions to *generalized multivalued constraints* while we recall the concept of *graded action* proposed in [8,10].

An action a is described in terms of a list of preconditions *pre(a)*, a list of preferences *pref(a)* and a list of effects *eff(a)*. A major difference with the classical model is that an action a can be executed with an application degree $\alpha \in [0,1]$, where the degrees 0 and 1 have the usual meaning of *not executed* and *fully executed*. The pair composed by the action a and its application degree α is called simply action and denoted by $a : \alpha$.

Definition 3. *Action Preconditions*
The preconditions of an action a are defined by a list of generalized multivalued constraints $\mathtt{mc}=(\mathtt{k},\chi,\phi)$.

The satisfaction degree $\mathcal{D}(S, mc)$ of each precondition mc=(k,χ,ϕ) of an action a is used to compute how much a can be executed in a given state S.

Definition 4. *Action Executability*
An action $a : \alpha$ is executable in a state S if $\alpha \le \alpha_{max}(a, S)$, where

$$\alpha_{max}(a, S) = \bigwedge_{mc \in pre(a)} \mathcal{D}(S, mc) \tag{2}$$

is the maximum application degree of the action a in the state S.

The maximum application degree is the highest degree at which an action can be executed, but the planner can always choose to apply a with any degree smaller or equal to $\alpha_{max}(a, S)$. Note that the *min*–conjunction $\wedge$ is used instead of the usual Łukasiewicz conjunction $\odot$ because the latter, differently from the former, is not idempotent and if x and y are smaller than 1, $x \odot y$ is smaller than x and y. Moreover note that the violation of any boolean precondition yields to a maximum application degree equal to 0, while the use of soft precondition allows to execute actions even if $S(\phi) \le \chi$, i.e it is possible to have a solution plan that violates soft preconditions.

In this model, effects are intended as modifications of the truth values of fluents. A positive effect provides an increment of the truth value, while a negative effect provides a decrement. The amount of increment is proportional to the application degree of the action and can be tuned by a parameter which models the strength of the effect.

Definition 5. *Action Effects*
The effects of an action a are defined by a list of pairs (γ, e), where e is a literal and γ is a real number in $[0, +\infty)$, called weight of a over e.

Note that positive (negative) literals denote positive (negative) effects.

Definition 6. *Action Execution*
The state resulting from the execution of $a : \alpha$ in a state S is denoted by $\mathcal{R}(S, a : \alpha)$. If $S' = \mathcal{R}(S, a : \alpha)$, then for each $(\gamma, e) \in eff(a)$

$$S'(e) = S(e) \oplus \Delta_{ae}, \tag{3}$$

where $\Delta_{ae} = \min(\gamma \cdot \alpha, 1)$.

The truncated product by γ can be seen as a unary operator added to the standard Łukasiewicz logic. It allows to reduce or to amplify the effect over the fluent e with respect to the application degree. In particular, it reflects the idea of effects which are proportional to the application degree of the action. It is easy to prove that the application of the definition (3) to a negated fluent $\neg e$ yields to $S'(e) = S(e) \ominus \Delta_{ae}$. Therefore, as expected, the action execution decrements the truth value of fluents in its negative effects by the quantity Δ_{ae}, while increments the the truth value of fluents in its positive effects. It is also possible to encode boolean effects that assign 0 or 1 to the truth value of fluents, independently from the execution degree. See [10] for the details. Moreover note that a boolean effect corresponds to a fixed $\Delta_{ae} = 1$, no matter the value of α.

Recently, the AI planning community has driven its attention to the specification of preferences in goal and action precondition conditions [11]. The aim is to specify some conditions that are not necessary for the action executability or the goal satisfiability but that are considered important for the plan quality estimation. In this multivalued model we consider both boolean and soft preferences specifying either a crisp or a fuzzy threshold.

Definition 7. *Preferences.*
The preferences of an action a are defined by a list of generalized multivalued constraints `mc=(k,`χ`,`ϕ`)`*, where* χ *is called* preference threshold *of* ϕ.

In Section 2.6 is shown how the penalties for violations of preferences are used in the metric.

2.4 Parallel Execution of Actions

In this section we define when two or more actions are *simultaneously executable*, i.e. they can be executed in the same state and produce a well defined final state [1]. This concept is important because it allows to use parallel actions, which can be executed in any order, while producing always the same result. The presence of the parallel actions can produce shorter plans with respect to the number of time–steps.

Definition 8. *Several actions* $a_1 : \alpha_1, \ldots, a_n : \alpha_n$ *are simultaneously executable in the state* S *if, for any possible permutation* π *of* $\{1, \ldots, n\}$*, let* $S_0^\pi = S$ *and for each* $i = 1, \ldots, n$ *let* $S_i^\pi = \mathcal{R}(S_{i-1}^\pi, a_{\pi(i)} : \alpha_{\pi(i)})$*, the following two conditions hold:*

1. *Each action* $a_{\pi(i)} : \alpha_{\pi(i)}$ *can be executed in the state* S_{i-1}^π
2. *The final state* S_n^π *does not depend on* π.

Note that in this definition the concept of simultaneous executability of the actions $a_1, \ldots, a_n$ depends also both on the application degrees and on the state in which the actions have to be executed.

In fact, let for instance a an action decreasing the fluent f with $\gamma = 1$, and b an action increasing f with $\gamma = 2$. It is easy to see that $a : 0.3$ and $b : 0.25$ are simultaneously executable in a state S having $S(f) = 0.4$, while with $b : 0.4$, the resulting final states are different. In fact, in the latter case, the execution of the sequence $\langle a : 0.3; b : 0.4\rangle$ yields a state S' in which $S'(f) = 0.9$, while the sequence $\langle b : 0.4, a : 0.3\rangle$ causes the final value $S''(f) = 0.7$.

It is easy also to see that if $S(f) = 0.2$, then $a : 0.3$ and $b : 0.25$ would not be simultaneously executable.

If $\mathcal{B}$ is a set of actions and λ is a literal, $\mathcal{I}(\mathcal{B}, \lambda)$ is the simultaneous "increment" caused to λ by the execution of the actions in $\mathcal{B}$.

$$\mathcal{I}(\mathcal{B}, \lambda) = \sum_{\substack{a\,:\,\alpha \in \mathcal{B} \text{ s.t.} \\ (\gamma, \lambda) \in eff(a)}} \gamma\alpha$$

Actually, $\mathcal{I}(\mathcal{B}, \lambda)$ it is not the effective increment because it is not computed by (3). Anyway it is easy to prove that it does not affect the validity of the Proposition 1 and moreover make easier the computation with respect the effective increment computed by (3).

Proposition 1. *Given a set of actions $\mathcal{B} = \{a_1 : \alpha_1, \ldots, a_n : \alpha_n\}$, in the case that each precondition of every action a_i is a constraints on literal, then the actions in $\mathcal{B}$ are simultaneously executable in the state S if and only if*

1. *for each fluent f, if $\mathcal{I}(\mathcal{B}, f) > 0$ and $\mathcal{I}(\mathcal{B}, \neg f) > 0$ then $S(f) + \mathcal{I}(\mathcal{B}, f) \leq 1$ and $S(f) - \mathcal{I}(\mathcal{B}, \neg f) \geq 0$*
2. *for each precondition (s, χ, f) of the action a_i, $\alpha_i \leq (1-\chi) \oplus (S(f) \ominus \mathcal{I}(\mathcal{B} \setminus \{a_i : \alpha_i\}, \neg f))$*
3. *for each precondition (b, χ, f) of the action a_i, $(S(f) \ominus \mathcal{I}(\mathcal{B} \setminus \{a_i : \alpha_i\}, \neg f)) \geq \chi$,*
4. *for each precondition $(s, \chi, \neg f)$ of the action a_i, $\alpha_i \leq (1 - \chi) \oplus 1 - (S_t(f) \oplus \mathcal{I}(\mathcal{B} \setminus \{a_i : \alpha_i\}, f))$,*
5. *for each precondition $(b, \chi, \neg f)$ of the action a_i, $1 - (S_t(f) \oplus \mathcal{I}(\mathcal{B} \setminus \{a_i : \alpha_i\}, f)) \geq \chi$.*

The rôle played by the hypothesis that all the preconditions are expressed by literals is important. In fact the condition 2 exploits the fact that $S(f) \ominus \mathcal{I}(\mathcal{B} \setminus \{a_i : \alpha_i\}, \neg f)$ is the lowest truth value of the fluent f obtained among all the possible execution orders of all the other actions. A similar property holds for conditions 3, 4, 5. Therefore if an action a_i respects all the conditions 2–5, it will be executable no matters which actions are executed before.

On the other hand, if an action would have an arbitrary propositional formula ϕ in a precondition, it would be hard, in the general case, to find the equivalent of the conditions 2–5. In fact, in order to compute the lowest truth value of ϕ it would be necessary to take into account that the same action can increase a subcomponent of ϕ, while decreasing another component of ϕ, without having a neat impact on ϕ. Therefore it is not clear how to compute the worst case for the executability of a_i, without considering and computing the states coming from many different execution orders.

In conclusion, it is always possible, from the theoretical point of view, to verify the simultaneous executability of actions with general preconditions by using a criterion analogous to the conditions 2–5. However this seems to be infeasible from the computational point of view. Therefore, even if we lose some expressivity in modelling planning domains, in the rest of the paper we will restrict our attention only to actions with preconditions on literals, while allowing to use generalized constraints in preferences and goals.

2.5 Multivalued Planning Problems

Definition 9. *Multivalued Planning Problems*
A multivalued planning problem *is defined by a quadruple $(I, \mathcal{O}, G, \sigma)$ where $I : \mathcal{F} \to [0, 1]$ is a truth assignment over the fluents and denotes the initial state;*

$\mathcal{O}$ is a set of boolean and multivalued actions; G is the goal list and is defined by a list of generalized multivalued constraints mc=(k,χ,ϕ)*; σ is a real number in* $[0, 1]$ *called* global threshold.

A valid plan is an ordered sequence $(\mathcal{A}_0, \mathcal{A}_1, \ldots, \mathcal{A}_{T-1})$ where each $\mathcal{A}_t$, $t = 0, \ldots, T-1$ is a set of k_t actions $\mathcal{A}_t = \{a_{t1} : \alpha_{t1}, \ldots, a_{t,k_t} : \alpha_{t,k_t}\}$.
Moreover, let $S_0 = I$ and, for $t = 0, \ldots, T-1$, $S_{t+1} = \mathcal{R}(\langle a_{t1} : \alpha_{t1}, \ldots, a_{t,k_t} : \alpha_{t,k_t}\rangle, S_t)$, then the actions in $\mathcal{A}_t$ must be simultaneously executable in the corresponding state S_t, for $t = 0, \ldots, T-1$.

Definition 10. *Solution Plan.*
A solution plan is a valid plan where in the final state S_T

$$\bigwedge_{mc \in G} \mathcal{D}(S_T, mc) \geq \sigma \tag{4}$$

Note that a solution plan is a plan which achieves all the goals in a multivalued fashion. Similarly to the preconditions, the problem goals are defined by classical/soft constraints and preferences and their satisfaction degrees are computed by the $\mathcal{D}$ function. Then, σ specifies a global parameter which represents a minimal satisfaction degree desired for the solution plan.

2.6 Plan Quality

The plan quality is evaluated by a metric specified by any arithmetic expression. It allows to involve in the metric violations of preferences or soft preconditions, as well as action application degrees. An optimal plan will minimize (maximize) the metric. We briefly recall how the violations and the action application degrees are computed.

Labeling every generalized multivalued constraint mc by $\mathtt{n_{mc}}$, the total of penalties for the violations of the constraint mc is computed by
$\Lambda(\mathtt{n_{mc}}) = \sum_i \neg\mathcal{D}(S_i, \mathtt{mc})$. This sum is computed considering only the actions that are in the solution plan, having the constraint mc as precondition or preference.

Moreover, the function $\Omega(a) = \sum_{t=0}^{T-1} \alpha_t$ is used to compute the total of application degrees α_t of each occurrence of the action a in the solution plan. It is important to note that it is very interesting to specify action application degrees in the metric for the plan quality, especially in such domains where action application degrees are directly proportional to action costs.

3 An Algorithm for Multivalued Planning

The idea of the algorithm is to solve bounded multivalued planning problems in three steps. At first, a planning graph [1] is constructed until either the upper bound U for time–steps is exceeded or a state verifying the necessary condition for the solution existence is reached. Then the graph representation is translated into a MIP problem (Mixed Integer Programming), and finally a MIP problem solver is called. If the MIP problem has a solution, then also the planning problem has a solution which can be directly extracted from the MIP problem solution. If the MIP problem has no solution, the main loop continues by adding a new level until a solution is found or the bound U is exceeded. It is easy to prove that this algorithm is correct and complete in the class of the U–bounded multivalued planning problems.

3.1 Planning Graph Construction

The planning graph built in our system is similar to the planning graph used in Graphplan and other planning systems. The planning graph is a $(2T+1)$–leveled graph $G = (V, E)$ whose vertex set is $V = F_0 \cup A_0 \cup F_1 \cup A_1 \cup \ldots \cup F_T$, where F_t is the set of *multivalued fluents* at time $t = 0, \ldots, T$ and A_t is the set of *graded actions* at time $t = 0, \ldots, T-1$. Each edge in E links either a fluent $f \in F_t$ to an action $a \in A_t$, such that $(k, \chi, f) \in pre(a)$, or an action $a \in A_t$ to a fluent $f \in F_{t+1}$, such that $(\gamma, f) \in eff(a)$.

In the following, $\alpha_t^*(a)$ denotes the upper bound for the application degree of action a at time t. It is computed by a function depending on the precondition set of the action and on upper and lower bounds of the truth values of the fluents involved in such preconditions. Analogously, $S_t(f)$ is the value of the fluent f at time t.

The values of $S_0(f)$ and $\alpha_0^*(a)$ can be computed in an exact way using the formula (2) because the initial state is completely known. If $t > 0$, since it is not known which actions will be executed in each time–step, the truth value $S_t(f)$ for each fluent $f \in F_t$ is unknown and, for the same reason, it is not possible to compute a realistic value of $\alpha_t^*(a)$, for each action $a \in A_t$.

Anyway, it is possible to compute a lower bound $\underline{S}_t(f)$ and an upper bound $\overline{S}_t(f)$ for the value $S_t(f)$. Given a fluent f,

$$[\underline{S}_t(f), \overline{S}_t(f)] = [\underline{S}_{t-1}(f) \ominus \underline{\Delta}_{t-1}(f), \overline{S}_{t-1}(f) \oplus \overline{\Delta}_{t-1}(f)],$$

where $\underline{\Delta}_{t-1}(f)$ and $\overline{\Delta}_{t-1}(f)$ are respectively the greatest decrement and the greatest increment of f caused by the simultaneous execution of actions at time $t-1$. The quantity $\overline{\Delta}_{t-1}(f)$ should be computed by solving a maximization program in which the objective function is $\mathcal{I}(A_{t-1}, f)$, the variables are the corresponding α's and the constrains correspond to the conditions guaranteeing the simultaneous executability introduced by Proposition 1. A similar program has to be used to find the value of $\underline{\Delta}_{t-1}(f)$. Note that the conditions produce nonlinear constraints.

Since these maximization programs should be performed twice for each fluent at each time–step, it is computationally infeasible to find the exact bounds $\overline{S}_t(f)$ and $\underline{S}_t(f)$.

Therefore, instead of computing $\overline{\Delta}_{t-1}(f)$ and $\underline{\Delta}_{t-1}(f)$, we use respectively $\mathcal{I}^*(A_{t-1}, f)$ and $\mathcal{I}^*(A_{t-1}, \neg f)$, by computing $\mathcal{I}(A_{t-1}, f)$ and $\mathcal{I}(A_{t-1}, \neg f)$ with $\alpha_{t-1}^*(a)$ as the execution degree for each action a. These quantities are much easier to compute and provide an overestimation of $\overline{\Delta}_{t-1}(f)$ and $\underline{\Delta}_{t-1}(f)$, neglecting the interactions among the actions in terms of simultaneous executability.

Then, for each fluent $f \in F_t$ we actually compute

$$[\underline{S}_t(f), \overline{S}_t(f)] = [\underline{S}_{t-1}(f) \ominus \mathcal{I}^*(A_{t-1}, \neg f), \overline{S}_{t-1}(f) \oplus \mathcal{I}^*(A_{t-1}, f)] \tag{5}$$

We can now compute the values of $\alpha_t^*(a)$, for any action $a \in A_t$, by

$$\alpha_t^*(a) = \bigwedge_{(k,\chi,f) \in pre(a)} \mathcal{D}(\overline{S}_t, (k, \chi, f)) \wedge \bigwedge_{(k,\chi,\neg f) \in pre(a)} \mathcal{D}(\underline{S}_t, (k, \chi, \neg f)) \tag{6}$$

The first time–step where a solution can exist is the first t in which the satisfaction degree for the goals, computed by using $\underline{S}_t(f)$ and $\overline{S}_t(f)$, can be greater than the goal threshold σ. Therefore, the graph construction phase ends at a time $t \leq U$, if any, when the goal condition

$$\bigwedge_{(k,\chi,f) \in G} \mathcal{D}(\overline{S}_t, (k, \chi, f)) \wedge \bigwedge_{(k,\chi,\neg f) \in G} \mathcal{D}(\underline{S}_t, (k, \chi, \neg f)) \geq \sigma \tag{7}$$

is verified.

3.2 Constraints Extraction

In this section we will show how to solve a planning problem by a compilation into a Mixed Integer Programming (MIP) problem $\mathcal{M}$. Each constraint expressed by operators of Łukasiewicz logic can be translated into linear constraints on integer/real variables as, for instance, shown in [12]. The details of this translation, fully described in [8,10], are not reported here due to lack of space.

In the following, the truth value of each fluent f and the application degree of each action a at time t are respectively denoted by the real variables x_{ft} and x_{at} in $[0, 1]$. For the boolean cases x_{ft} and x_{at} will be integer variables in {0,1}. A planning problem is encoded through different kinds of constraints, grouped in *action executability rules*, *action execution rules*, *simultaneous executability rules* and *goal satisfiability rules.* Finally if a metric is present in the problem, the MIP system has be transformed into an optimization MIP problem by means of a suitable *objective function.* It is easy to see that the number of constraints and the number of variables in the MIP problem $\mathcal{M}$ will be a polynomial of the number of fluents and actions in the planning graph.

Action Executability Rules. At each time–step t, the actual application degree of an action a must be less (or equal) than the upper bound represented by $\alpha_t^*(a)$. Since the formula (2) requires to compute a minimum, it is easy to separate the contribution of soft preconditions from the contribution of boolean preconditions.

First, we have that for the soft preconditions

$$x_{at} \leq \min(\min_{(s,\chi,f) \in pre(a)} \{\chi \rightarrow x_{ft}\}, \min_{(s,\chi,\neg f) \in pre(a)} \{\chi \rightarrow \neg x_{ft}\}).$$

While, for each boolean precondition (b, χ, f) or $(b, \chi, \neg f)$, we have to use a new 0/1 variable μ_{ft}^a, which is 1 if and only if $x_{ft} \geq \chi$ or $1 - x_{ft} \geq \chi$, respectively, and add the constraint $x_{at} \leq \mu_{ft}^a$.

Action Execution Rules. The truth value of a fluent f at a time–step $t = 1, \ldots, T$ can be affected by the actions executed at the previous time–step $t-1$. Since these action must be simultaneously executable, $x_{f,t+1}$ can be computed by the following equation

$$x_{f,t+1} = x_{ft} \oplus \left(\sum_{(\infty,f)\in eff(a)} \lambda_{at} + \sum_{(\gamma,f)\in eff(a)} \gamma x_{at} \right) \ominus \left(\sum_{(\infty,\neg f)\in eff(a)} \lambda_{at} + \sum_{(\gamma,\neg f)\in eff(a)} \gamma x_{at} \right) \quad (8)$$

where λ_{at} is a suitable 0/1 variable, which is 1 if and only if the action a is executed (i.e. if $x_{at} > 0$) at time t.

Goal Satisfiability Rule. This rule derives directly from the goal condition defined in (7) and requires that each subgoal (k, χ, ϕ) must be fulfilled with a truth value greater or equal to the global threshold.

First, soft goals are encoded as

$$\sigma \leq \min_{(s,\chi,\phi)\in G} \{\chi \rightarrow x_{\phi t}\}$$

where for each goal (s, χ, ϕ) the value of the variable $x_{\phi t}$ is the value of the propositional formula ϕ, computed by the values x_{ft} of the fluents occurring in ϕ.

In particular, $x_{\phi t}$ can replaced with x_{ft}, if ϕ is the fluent f, or with $1 - x_{ft}$, if ϕ is $\neg f$.

The treatment of boolean goals is similar to the treatment of boolean preconditions and requires the use of new 0/1 variables.

Simultaneous Executability Rule. In each time–step t, the actions selected to be executed have to be simultaneously executable. Then, some constraints, equivalent to the conditions of proposition 1, are imposed to the variables x_{at}.

The condition 1 requires that, for each fluent f, if

$$\sum_{(\gamma,f)\in eff(a)} x_{at} > 0, \text{ and } \sum_{(\gamma,\neg f)\in eff(a)} x_{at} > 0$$

then

$$\left(x_{tf} + \sum_{(\gamma,f)\in eff(a)} \gamma x_{at} \right) \leq 1$$

and

$$\left(x_{tf} - \sum_{(\gamma,\neg f)\in eff(a)} \gamma x_{at} \right) \geq 0$$

This constraint, being an implication, needs to be transformed to a suitable form before being translated into MIP inequalities.

The condition 2 requires that, for each precondition (s, χ, f) of the action a,

$$x_{at} \leq (1 - \chi) \oplus \left(x_{tf} \ominus \sum_{\substack{b \neq a \text{ and} \\ (\gamma, \neg f) \in eff(b)}} \gamma x_{bt} \right)$$

Analogous constraints are derived by condition 3–5.

Objective Function. If a metric is present in the planning problem specification, then a suitable objective function ϕ is added to the MIP system $\mathcal{M}$. It is obvious that a MIP solver can only optimize linear functions, therefore only metric expressed as a linear combination of terms involving violations of preferences and soft preconditions and terms related to action execution degrees are allowed.

The objective function is built, as shown in [10], by replacing each term involving violation of a preference/soft preconditions mc with a new variable, whose value is the sum of the violations of mc, and each term involving execution degrees of an action a with a new variable, whose value is the sum of x_{at}, for $t = 0, \ldots, T - 1$.

3.3 Implementation

A prototype implementation of the previously described algorithm has been made in C++, using as a MIP solver the open–source library *lpsolve*, whose performance are not very good, although the preliminary results obtained on small examples are encouraging.

We expect that some further code optimizations and mainly the use of some more efficient linear programming libraries, like CPLEX, will produce good results from the computational point of view.

Another different approach to solve multivalued planning problems without metric could be a compilation into a mixed boolean– real linear system, handled by solvers like MathSAT [14].

4 Conclusions

In this work an extension of the model for "multivalued planning" has been proposed. The model allows to manage multivalued fluents, graded execution of actions, soft constraints. In this paper extensions to generalized multivalued constraints and simultaneously execution of the actions are introduced. These extensions allow both to define more expressive preconditions, preferences and goals and to find shorter solution plans.

An interesting direction for future works is to study the effects of building the planning model on a finite–valued logic. In this case it could be possible to change over the "planning as satisfiability" approach playing on the existence of several multivalued SAT solvers. On the other hand the current algorithm, which uses a translation into a MIP problem, allows to introduce into the model linear objective functions and plan metrics and to solve the corresponding optimization problems. Finally, from an experimental point of view, it is necessary to develop a set of benchmark problems for significant domains involving graded actions in order to test the performance of the implementation.

References

1. Blum, A., Furst, M.: Fast Planning Through Planning Graph Analysis. Artificial Intelligent 90, 279–298 (1997)
2. McCarthy, J., Hayes, P.J.: Some philosophical problems from the standpoint of artificial intelligence. Machine Intelligence 90, 463–502 (1969)
3. Kautz, H., Selman, B.: Planning as Satisfiability. In: 10th European Conference on Artificial Intelligence (ECAI 1992), pp. 360–363 (1992)
4. Cimatti, A., Roveri, M., Traverso, P.: Strong Planning in Non-Deterministic Domains Via Model Checking. Artificial Intelligence Planning Systems, 36–43 (1998)
5. Cialdea, M., Orlandini, A., Balestreri, G., Limongelli, C.: A planner fully based on Linear Time Logic. In: Proc. of AIPS 2000, pp. 347–354 (2000)
6. Bacchus, F., Kabanza, F.: Using Temporal Logics to Express Search Control Knowledge for Planning. Artificial Intelligence 16, 123–191 (2000)
7. Miguel, I., Jarvis, P., Shen, Q.: Efficient Flexible Planning via Dynamic Flexible Constraint Satisfaction. Engineering Applications of AI 14, 301–327 (2001)
8. Baioletti, M., Milani, A., Poggioni, V., Suriani, S.: A Multivalued logic model of planning. In: Proc. of ECAI 2006 (2006)
9. Hajek, P.: Mathematics of Fuzzy Logic, Machine Intelligence. Kluwer Academic Publisher, Dordrecht (1998)
10. Baioletti, M., Milani, A., Poggioni, V., Suriani, S.: A Multivalued planning model for soft constraints and preferences. In: Proc. of ICAPS 2006, WS on Preferences and Soft Constraints (2006)
11. Gerevini, A., Long, D., SPlan Constraints, D.: Preferences in PDDL3, Tech. Rep. of Dept. of Electronics for Automation, University of Brescia, Italy (2005)
12. Hähnle, R.: Some philosophical problems from the standpoint of artificial intelligence. Proof Theory on many–valued logic – linear optimization – logic design: Connections And Interactions 1, 107–119 (1997)
13. Balas, E.: Disjunctive Programming. Annals of Discrete Mathematics 5, 3–51 (1979)
14. Bozzano, M., et al.: The MathSAT 3 System. In: Nieuwenhuis, R. (ed.) CADE 2005. LNCS (LNAI), vol. 3632, Springer, Heidelberg (2005)

An Adaptive Multi Pattern Scheme for Fast Block Motion Estimation

Humaira Nisar and Tae-Sun Choi

Gwangju Institute of Science and Technology,
261 Cheomdan Gwagiro, 1 Oryong-Dong, Buk-Gu, Gwangju, 500712, Korea
{humaira,tschoi}@gist.ac.kr

Abstract. A simple, adaptive, computationally efficient and robust scheme for multi pattern block motion estimation has been proposed. A combination of spatial and temporal predictors has been used for multiple initial search center prediction, determination of magnitude of motion and multiple search pattern selection. The multiple initial predictors help to identify the absolute zero motion blocks and true location of global minimum based on the characteristic of unimodal error surfaces. Hence the final predictive search center selected is closer to the global minimum thus decreases the effect of monotonic error surface assumption and its impact on the motion field. This also increases the computation speed. Further computational speed up has been obtained by considering half stop technique and threshold for minimum distortion point. The computational complexity of the proposed algorithm is drastically decreased (average speedup ~ 43%) whereas the image quality measured in terms of PSNR also shows results close to Full Search algorithm.

Keywords: Full Search, Motion estimation, Block Matching, Spatial Correlation, Temporal Correlation, Video Coding, Compression.

1 Introduction

Block matching motion estimation and compensation is an essential part of several video coding standards such as MPEG-1/2/4 [1], ITU-T H.261/263/263+ [2] and H.264/AVC [3]. A variety of motion estimation algorithms have been developed for video coding. Block matching motion estimation algorithms are the most popular and simplest in concept, design, and implementation. It allows us to exploit temporal correlation, and reduce the redundancy that exists between the frames of video sequences which leads to higher compression. In block based motion estimation (BBME) the current frame is partitioned into square blocks of pixels and the best match of these blocks is found inside the reference frame using a predefined distortion criterion. The best match is then used as a predictor for the block in the current frame, whereas the displacement between the two blocks is usually defined as the motion vector (MV), which is associated with the current block. In the encoder it is only necessary to send the motion vector and the residue block defined as the difference between the current block and the predictor. This requires fewer bits than the direct coding of the original.

O. Gervasi et al. (Eds.): ICCSA 2008, Part II, LNCS 5073, pp. 1012–1024, 2008.

Full Search (FS) is the most straightforward and optimal block matching algorithm (BMA) which searches exhaustively inside the search window to find the motion vector. Despite the very heavy computations required in FS it is still widely used in video coding applications due to its simplicity and ease of hardware implementation. However its large computational requirements have fuelled many research activities and designing the most efficient block matching algorithm remains an open research problem. Several fast block matching motion estimation algorithms [4]-[21] have been proposed so far. These fast algorithms involve approaches like Unimodal Error Surface Assumption (UESA), variable search range instead of fixed one, methods using multi-resolution, spatial and temporal correlation of motion vectors, pixel decimation etc.

Some well known examples are the Three Step Search (TSS) [6], New Three Step Search (NTSS) [7], Four Step Search (FSS) [9], Diamond Search (DS) [13], Simple and Efficient Search (SES) [18], Adaptive Rood Pattern Search (ARPS) [21], etc. Experimental results show that these algorithms reduce the computational requirement significantly by checking only some points inside the search window, while keeping a good error performance when compared with the FS algorithm. However most of these algorithms get trapped in local minimum, yielding a significant loss in motion vector estimation performance. DS is an outstanding algorithm adopted by MPEG-4 verification model (VM) [14] due to its superiority to other methods in the class of fixed search pattern algorithms. However fixed search patterns are unable to constantly match the dynamic motion content and could incur redundant searches, thus not only wasting the computational power but also leading to local minimum matching error trapping and large prediction errors. In low bit rate coding applications such as video telephony the motion field is very structured and is slowly varying. Moreover the motion vectors are usually limited in magnitude. This suggests that significant computational gains can be achieved by taking advantage of strong spatio-temporal dependencies that exist between the motion vectors.

In this paper we present a predictive motion estimation technique employing both spatial and temporal correlation to find the initial search center. There is a possibility that if the prediction goes wrong the initial search point could be misleading. To avoid this we have used the concept of multiple predictors. For example at the first step instead of choosing only one initial search center we will chose multiple initial search centers to start search. The best one having the minimum error will be assumed to be closer to the global minimum. By accurately predicting the location of the best motion vector candidate we can search a relatively small area in the neighborhood of the predicted MV. This is indeed possible especially in low bit rate video applications owing to the large amount of motion field redundancies within the same as well as consecutive frames.

In the proposed algorithm a dynamic search pattern is constructed for each block based on its spatio temporal neighboring motion vectors. This pattern quickly identifies the best center point for refined search to effectively reduce the unnecessary intermediate searches and avoid being trapped in local minimum. The proposed technique reduces the number of computations, produces a smooth motion field and yields images of better quality. The algorithm proposed in this paper can significantly reduce ME computational cost that leads to higher compression ratio and fast processing speed, and achieves the best performance in sequences with low complexity as can have in video surveillance. Since sequential video frames often contain a lot of correlated information and ME based approaches can achieve higher compression ratio. In the following section we will present the proposed algorithm that is followed by the experimental environment, results and discussions on results.

2 Proposed Algorithm

Motion estimation is a multi step process that involves a combination of techniques such as motion starting point, motion search patterns and adaptive control to curb the search, avoidance of search stationary regions and avoidance of local minimum. The collective efficiency of these techniques makes a motion estimation algorithm robust and efficient. The main objective of the proposed algorithm is to decrease the computational burden while keeping a good predicted image quality. Important aspects of the proposed algorithm are 1) Using spatio-temporal neighborhood information that leads to the prediction of initial search center, 2) Multiple, initial starting points selection 3) Multiple Search Patterns 4) Absolute zero motion blocks determination. All these points help in reducing the computational complexity and finding the true minimum error point.

2.1 Motion Vector Prediction

Motion vector prediction is based on spatio-temporal correlation information. The spatio-temporal neighboring blocks are often associated with the same object and have similar MVs as that of the current block. Commonly used regions of support (ROS) for spatio-temporal prediction are shown in Fig. 1. As can be seen normally only spatial neighboring blocks are utilized. However in our proposed algorithm we have used both spatial and temporal neighboring blocks, as shown in Fig. 2. From the spatial domain we have used the left block (relative to the current block). The left block is unavailable for left margin blocks. The majority of MVs are located purely in the horizontal and vertical directions, in other words oblique motion vectors are less common. It has also been observed that MV correlation decreases rapidly if we move away from the current macro block along the spatial axis [16].

In temporal domain the MV temporal correlation is effected not only by the moving speed of the moving objects but also the frame rate. Thus it is less stable than spatial correlation. Furthermore the temporal neighboring vectors are not used in the MV differential coding. The temporal neighboring block used for prediction block is the block at the same location in the previous (reference) frame is. Our preliminary experiments show that this ROS is superior to those composed of other immediate neighboring blocks both in terms of average PSNR and search speed. Thus we have two initial MVs, one each from spatial and temporal neighboring blocks.

The first step of our algorithm utilizes these neighboring MVs for predicting an initial search point which is closer to the global optimum. These motion vectors, $MV_L(n)$ (Left), and $MV_P(n-1)$ (Previous) perform as the candidates of the predicted motion vector P (Xn) for the current block X in frame n. If the predictor accuracy is high the optimal MV (inside a given search window) can be attained faster thus enabling computational savings for fast searches [16]. The commonly used schemes using median X/Y have its problems since its performance can deteriorate dramatically for high/mixed irregular motion sequences. So e will calculate the predicted motion vector by using the simple mean prediction method:

$$\text{PMV(Xn)} = \frac{1}{k}\sum_{k=1}^{K} V_k \ . \tag{1}$$

The x and y components of the mean predicted motion vector are computed independently. As confirmed by our experimental results, only the vectors corresponding to the closest blocks should be used. Including poorly correlated motion vectors in the computation adversely affects prediction accuracy.

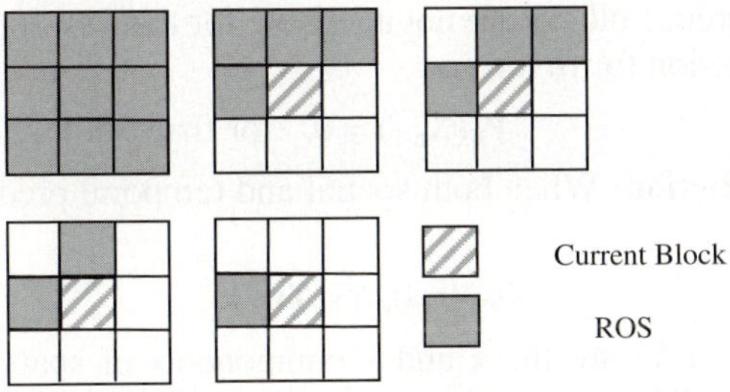

Fig. 1. Commonly used Regions of Support (ROS) for motion prediction (Spatial domain only)

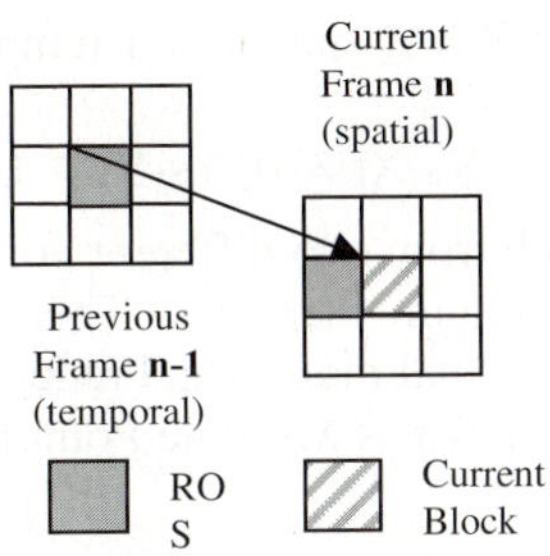

Fig. 2. Proposed Region of Support (ROS) in spatial and temporal domain

2.2 Multiple Initial Search Point Prediction

Most error surfaces encountered in real world video sequences are not truly unimodal. However the overall characteristics of the distortion surfaces are unimodal. Also the neighboring region of the global optimal point can be considered as unimodal distortion surface. Therefore in our proposed algorithm we are using the idea of multiple predictors acting as multiple initial search centers. It reduces the risk of being trapped in the local minimum by choosing the point having smaller minimum value.

In our algorithm we have used left neighboring block and previous block (same block in reference frame) for initial starting point prediction. The two initial point predictors can be obtained as follows:

1 From spatial frame

$$P_S(X_n) = MV_L. \tag{2}$$

2 From temporal frame

$$P_T(X_{n-1}) = MV_P. \tag{3}$$

Where $P_S(X_n)$ and $P_T(X_{n-1})$ are the spatial and temporal predicted motion vectors respectively. MV_L, and MV_P are the motion vectors of the left and temporal reference blocks respectively.

In spatial prediction, the left column for the left block is not available, therefore:

$$P_S(Xn) = 0, \quad \text{for left column blocks.} \tag{4}$$

The temporal reference blocks are not available for the first frame so we will not be using temporal prediction for first frame.

$$P_T(X_{n-1}) = 0, \quad \text{for first frame.} \tag{5}$$

Single Point Prediction: When both spatial and temporal predicted motion vectors are equal, i.e.

$$Xs\text{-}Xt=0,\ Ys\text{-}Yt=0. \tag{6}$$

Where Xs, Ys, Xt and Yt are the x and y components of spatial and temporal predicted motion vectors. We start the final search from only one point, as there is a high probability that the predicted motion vector is in a region that has unimodal characteristics. This is known as Single Point Prediction.

Multiple Point Prediction: When spatial and temporal predicted motion vectors are not equal, i.e.

$$Xs\text{-}Xt \neq 0,\ Ys\text{-}Yt \neq 0. \tag{7}$$

Then we start the final search from two different points. In this way the probability of finding the true global minimum increases. This choice of multiple points decreases the risk of ignoring the actual motion and reduces the chance of being trapped in the local minimum. This is known as Multiple Point Prediction

2.3 Final Search pattern

The distribution of the global minimum point in real world video sequences is centered at the position of zero motion, at the search window center as in TSS, FSS, NTSS etc. Most MVs are found to be enclosed in a circular support within a radius of 2~3 pels centered at the position of zero motion [20]. Using this characteristic only few steps of the search pattern will give the final result. Since the initial search center is already closer to the global minimum point any local search using a small compact search pattern should be fairly efficient. Because searching on a patterns first step search points is unavoidable the minimum necessary computational cost of a search pattern is directly related to the number of their first step search points.

2.3.1 Absolute Zero Motion Blocks

These are those blocks whose mean predicted motion vector is zero. Also both spatial and temporal predictors are zero, i.e.

$$Xs=0,\ Ys=0\ \&\&\ Xt=0,\ Yt=0. \tag{8}$$

To capture any motion the algorithm takes the following steps.

1. If SAD of search center < threshold, then search only one point and the initial search point is taken as the final MV (zero point) as shown in Fig. 3 (a). SAD stands for Sum of Absolute Difference.
2. If SAD of search center > threshold we search five points, i.e. small rood pattern, and then stop the search. Step Size is defined as the horizontal/vertical distance between two pixels as shown in Fig. 3 (b).

2.3.2 Zero Motion Blocks

These are those blocks whose mean predicted motion vector is zero. However all spatial and temporal predictors are not zero. Therefore we consider these blocks separately, as these blocks are not single point prediction blocks but are classified under multiple point prediction blocks. However both spatial and temporal predictors are very closely located and hence there is a great probability that the minimum error point will also be in very close vicinity. The search pattern is as follows:

1. Check the SAD of two initial search centers. Select the minimum one.
2. If SAD of search center < threshold, we will stop the search.
3. Otherwise search a small rood pattern.
4. If the minimum point lies at the center or SAD of minimum point < threshold, stop the search.
5. Otherwise, repeat step 3 and 4 and then stop the search.

This search pattern is shown in Fig. 4.

2.3.3 Single Point Prediction Motion Blocks

These are those blocks for which mean predicted motion vector is not equal to zero. But both the predictors are same. For these blocks we will follow the search pattern exactly same as that of zero motion blocks. However here we have only one initial search center.

2.3.4 Multiple Point Prediction Motion Blocks

These are those blocks for which predicted motion vector is not equal to zero. And both spatial and temporal predictors could be quite different. For these blocks we will follow the search pattern same as that of zero motion blocks. However we will not stop the search after 2 steps, instead we will continue the search till the minimum location is at the search window center or the end of search window. However to speed up the process here also we have used the same threshold for the minimum location as in previous steps.

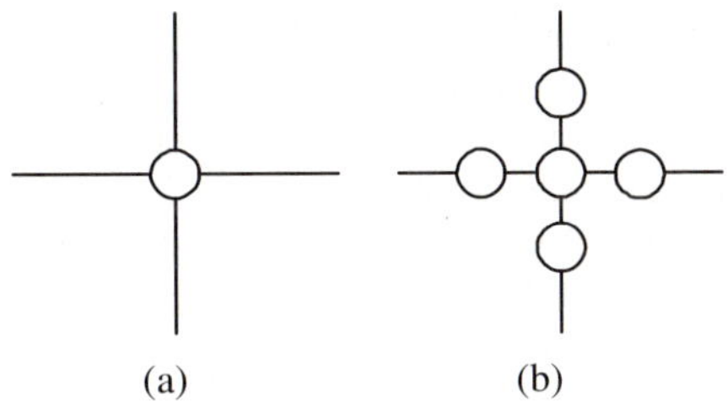

Fig. 3. Search Pattern for absolute zero motion blocks (a) if SAD < Threshold (b) if SAD > Threshold

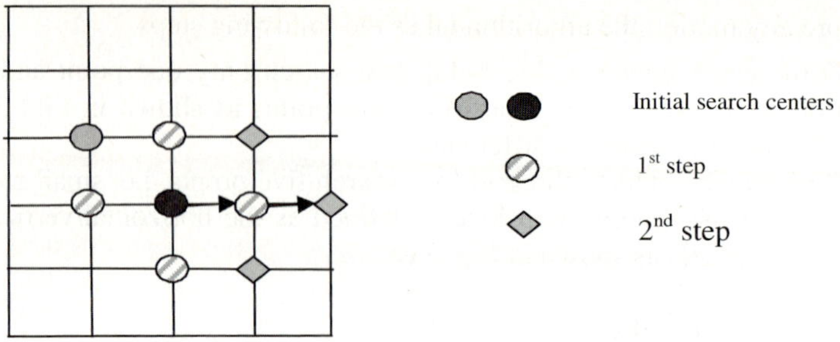

Fig. 4. Multiple Point Prediction Search Pattern

3 Experimental Results

The proposed algorithm is simulated using a number of video sequences that cover a wide range of motion contents from very slow video conferencing sequences to very fast sequences like Football and Table Tennis and various formats including QCIF, CIF and SIF. For example, Miss America is a typical videoconferencing sequence, with movement restricted to the face area of the speaker with a fixed background. Hall Monitor represents a 'security camera' case. The camera shows a corridor with two people entering, walking through it and eventually leaving it again. In Car Phone (a more complex sequence) you can observe movement of hand and some change in the background as well on the right side. Other sequences are Foreman, Coastguard, Salesman, Flower and Mobile.

Each pixel in the image sequences is uniformly quantized to 8 bits. The block size selected is 16x16 pixels and the maximum motion displacement of search area is $\pm$ 7 in both horizontal and vertical directions. Sum of absolute difference (SAD) distortion function is used as the block distortion measure (BDM). We use three measurements to evaluate the objective performance of the proposed algorithm. These are average Peak Signal to Noise Ratio (PSNR) between estimated and original frames, computational speed up obtained by getting the average number of search points (NSP) per motion block required for the MV estimation and the speed up ratio. The speed up ratio is defined as the ratio of total search number of the FS algorithm to the number of search points used by the search algorithm. The proposed algorithm is compared with some well known motion estimation algorithms such as FS, TSS, FSS, DS and ARPS. A fixed threshold equal to 512 is used as stopping criteria at various steps.

Table I and II show a detailed comparison of the PSNR and NSP performance of the proposed algorithm with many fast search algorithms. Although all algorithms yield reliable performance but its quite clear that the proposed algorithm shows definitely better behavior in terms of visual quality as well as computational complexity. Our algorithm employs spatio temporal correlation in a different way from ARPS. Firstly it utilizes the neighborhood information to predict an initial starting point. Secondly it uses the neighborhood information to select multiple rood search patterns based on the motion content with multiple starting points and early termination criteria. The average PSNR of proposed algorithm is same as ARPS and hence is very close to FS as well. But the number of search points for the proposed algorithm is

Table 1. Average PSNR Comparison

Sequence/Format/No.of frames	FS	TSS	FSS	DS	ARPS	**Proposed**
Carphone (QCIF)300	31.4	30.72	30.85	30.72	31.07	***31.1***
Foreman(QCIF)300	31.27	30.66	30.73	30.78	30.87	***31.36***
Coastguard(CIF)300	30.43	26.87	29.73	30.25	30.40	***30.38***
Salesman (CIF)150	35.38	34.63	34.96	34.86	35.32	***35.26***
Ms. America(CIF)150	37.71	36.14	36.62	36.82	37.59	***37.41***
Hall_monitor(CIF)300	34.8	34.51	34.6	34.6	34.74	***34.71***
Flower(SIF)150	23.97	23.29	21.2	23.59	23.78	***23.65***
Table Tennis(SIF)300	27.46	24.00	25.27	26.14	26.9	***26.81***
Football(SIF)210	24.79	23.64	23.69	23.43	24.22	***24.20***
Mobile(SIF)100	22.67	22.4	22.36	22.50	22.65	***22.64***
Average	**29.99**	**28.69**	**29.00**	**29.37**	**29.75**	***29.75***

Table 2. Average Number of Search Points Comparison

Sequence/Format/No.of frames	FS	TSS	FSS	DS	ARPS	Proposed	Speedup w.r.t. FS
Carphone (QCIF) 300	184.56	21.62	16.04	15.89	5.98	***4.01***	***46.02***
Foreman (QCIF)300	184.56	21.68	15.59	16.47	8.22	***4.29***	***43.02***
Coastguard (CIF)300	204.28	23.38	12.64	17.87	8.72	***4.81***	***42.47***
Salesman (CIF)150	204.28	23.21	17.28	15.67	6.35	***4.24***	***48.18***
Ms. America (CIF)150	204.28	23.46	19.3	16.14	8.4	***3.95***	***51.72***
Hall_monitor (CIF) 300	204.28	23.25	16.69	15.62	6.06	***3.45***	***59.21***
Flower (SIF)150	202.408	23.23	14.15	17.79	8.89	***4.73***	***42.79***
Table Tennis (SIF) 300	202.48	23.10	14.4	16.98	8.08	***4.10***	***49.39***
Football (SIF)210	202.408	23.24	16.76	19.77	12.54	***7.93***	***25.52***
Mobile (SIF)100	202.408	23.02	17.82	15.89	7.43	***4.8***	***42.17***
Average	**199.59**	**22.92**	**16.07**	**16.81**	**8.07**	***4.63***	
Speedup	**1**	**8.71**	**12.42**	**11.87**	**24.73**	***43.1***	

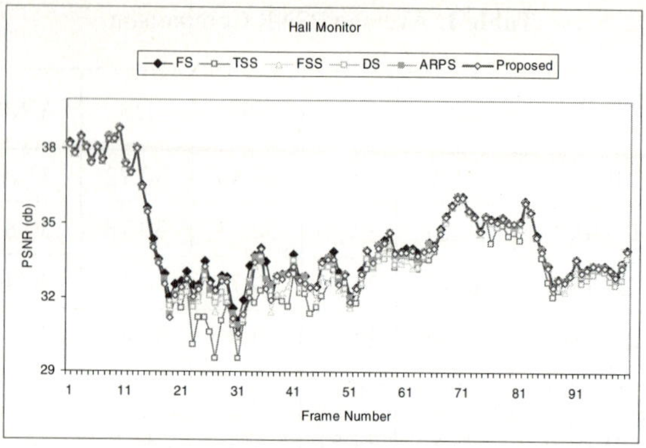

(a)

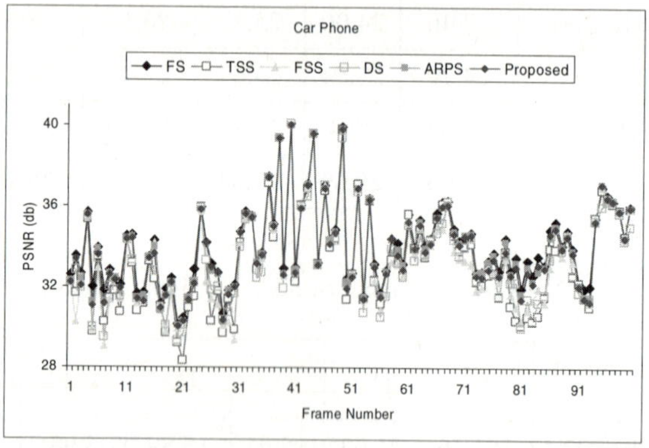

(b)

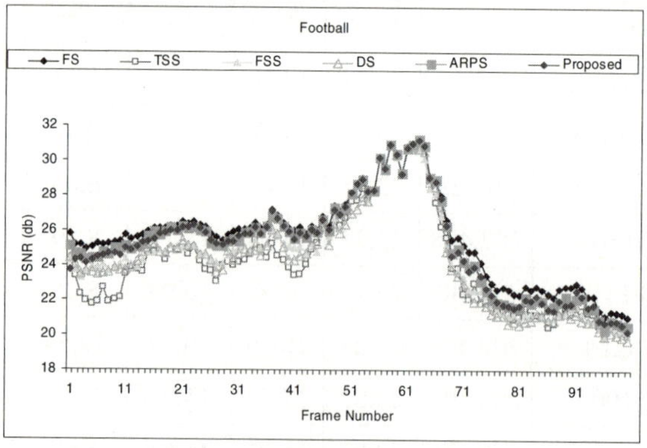

(c)

Fig. 5. PSNR verses frame number for first 100 frames of Hall Monitor, Car Phone and Football video sequence

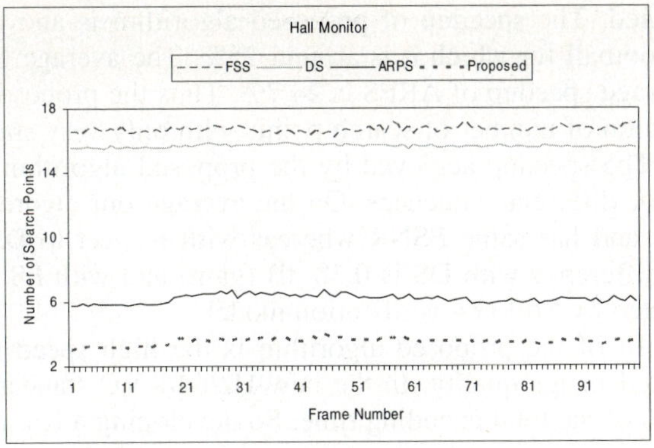

(a)

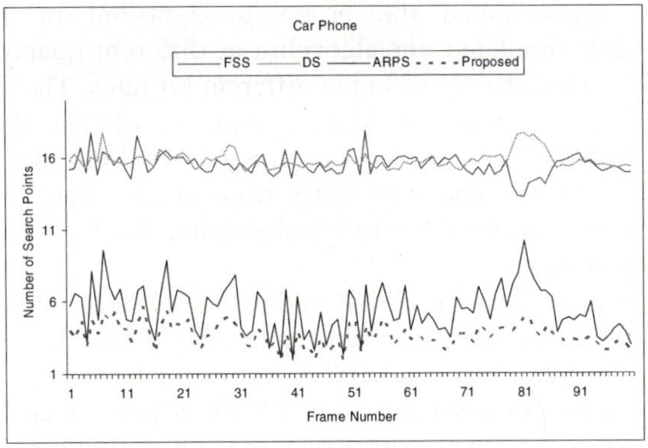

(b)

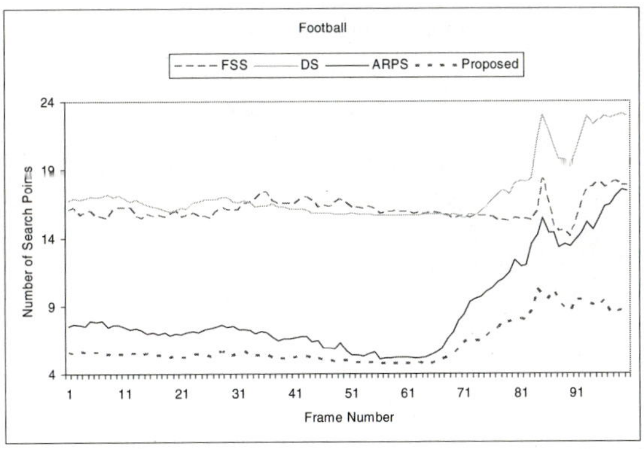

(c)

Fig. 6. Average Search number verses frame number for first 100 frames of Hall Monitor, Car Phone and Football video sequence drastically decreased. The speedup of proposed algorithm is above 40% for all sequences except Football for which it is around 25%. The average NSP is very small i.e. 4.63. The average speedup of ARPS is 24.7%. Thus the proposed algorithm has a tremendous reduction of number of search points with only very small quality loss as compared to FS. The speedup achieved by the proposed algorithm is approximately from 25 ~ 59% for different sequences. On the average our algorithm is 1.74 times faster than ARPS and has same PSNR whereas with respect to DS it is 3.63 times faster. Its PSNR difference with DS is 0.38 dB (gain) and with FS is -0.24dB (loss). DS has been selected by MPEG 4 verification model.

The main benefit of the proposed algorithm is the high speed up achieved with very good predicted image quality. In the new H.264/AVC standard, ME consumes about 70% to 90% of the total encoding time. So developing a fast motion estimation algorithm is an open research problem.

We have also implemented the proposed algorithm in **JM-12.2** [22] of H.264/AVC. We have simulated our algorithm at different quantization parameters (QP=28, 32, 36, 40) to test the algorithm at different bit rates. The average PSNR difference with FS is -.03 dB (loss) for Mobile sequence whereas the average bit rate difference is only .129%. The speedup in terms of Total encoder time and Motion Estimation time is 5.4 times and 9.23 times respectively. Rate Distortion graph in Fig. 7 also shows results comparable to FS algorithm. We have achieved very good results with other sequences as well.

Fig. 5 and Fig. 6 shows average PSNR and average search number comparison of the simulated algorithms for different video sequences. From the results it is obvious that the proposed algorithm has good PSNR values that are quite close to FS, DS, ARPS algorithms. In fact in most cases our PSNR is better than the DS algorithm. The number of search points is very small as compared to all other algorithms.

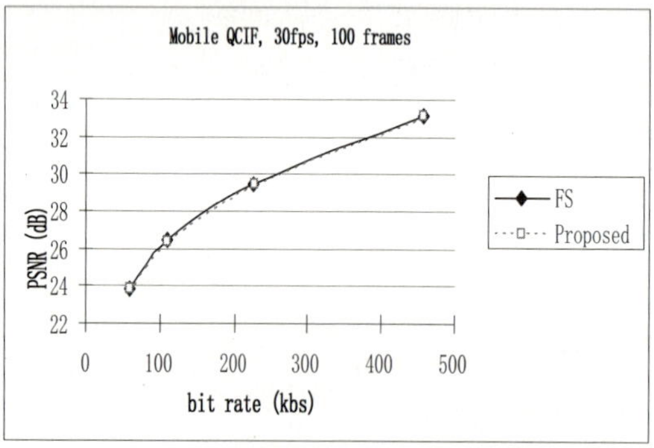

Fig. 7. Rate Distortion Comparison for Mobile Sequence

4 Conclusions

In this paper, a fast motion estimation algorithm has been proposed. The algorithm takes advantage of the spatio-temporal correlation information and utilizes multiple initial point prediction to accurately predict the search center and avoid being trapped in local minimum. Hence the initial search point selected in this manner is closer to the global minimum that helps to speed up the matching process. The final search pattern selected is adaptive, depends on the magnitude of the predicted motion vector and the motion content of the block. Hence we can conclude that the proposed algorithm shows very good predicted image quality (in terms of PSNR and bit rate) and has high speed up (in terms of Total time, ME time). And has less computational complexity as compared to other algorithms in the literature. So it is very suitable for video coding applications such as H.264/AVC as well as can be used in standalone motion estimation applications.

Acknowledgement

This work was supported by the Korea Science and Engineering Foundation (KOSEF) grant funded by the Korean government (MOST) (No. R01-2007-000-20227-0).

References

1. LeGall, D.: MPEG: A video compression standard for multimedia. Commun. ACM 34(4), 47–58 (1991)
2. CCITT SG XV, Recommendation H.261 video codec for audiovisual services at p*64 kbits/sec, Tech. Rep. COMXVR37-E (August 1990)
3. Wiegand, T., Sullivan, G., Luthra, A.: Draft ITU-T Recommendation and Final Draft International Standard of Joint Video Specification (ITU-T Rec. H.264| ISO/IEC 14496-10 AVC) (May 2003)
4. Puri, H.M.H., Schilling, D.L.: An efficient block matching algorithm for motion compensated coding. In: Proc. IEEE Int. Conf. Acoust., Speech, and Signal Proc., pp. 1063–1066 (1987)
5. Nisar, H., Choi, T.-S.: Fast Motion Estimation Algorithm based on spatio-temporal correlation and direction of motion vectors. Electronics Letters 42(24), 1384–1385 (2006)
6. Jain, J.R., Jain, A.K.: Displacement measurement and its application in interframe image coding. IEEE Trans. on Communications COM-29(12), 1799–1808 (1981)
7. Li, R., Zeng, B., Liou, M.L.: A new three step search algorithm for block motion estimation. IEEE Trans. on Circuits and Systems for Video Technology 4(4), 438–442 (1994)
8. Chow, K.H.K., Liou, M.L.: Generic motion search algorithm for video compression. IEEE Trans. on Circuits and Systems for Video Technology 3, 148–157 (1993)
9. Po, L.M., Ma, W.C.: A novel four-step search algorithm for fast block motion estimation. IEEE Trans. on Circuits and Systems for Video Technology 6(3), 313–317 (1996)
10. Koga, T., Iinuma, K., Hirano, A., Iijima, Y., Ishiguro, T.: Motion compensated interframe coding for video conferencing. In: Pro. Nat. Telecommun. Conf., New Orleans, pp.G5.3.1–G3.5 (November 1981)
11. Po, L.M., Ma, W.C.: New center biased search algorithm for block motion estimation. In: Proc. of International Conference on Image Processing, USA, vol. 1, pp. 410–413 (1995)
12. Namuduri, K.R.: Motion estimation using spatio-temporal contextual information. IEEE Transactions On Circuits And Systems For Video Technology 14(8) (August 2004)
13. Zhu, S., Ma, K.K.: A new diamond search algorithm for fast block matching motion estimation. IEEE Trans. Image Process. 9(2), 287–290 (2000)

14. MPEG-4 Video Verification Model (Version 14.0), ISO/IEC JTC1/SC29/WG11 N2932 (October 1999)
15. Nisar, H., Choi, T.-S.: An Adaptive block motion estimation algorithm based on spatio temporal correlation. In: Digest of Technical papers. International Conference on Consumer Electronic, January 7-11, 2006, pp. 393–394 (2006)
16. Lee, Y., Kossentini, F., Smith, M., Ward, R.: Predictive RD Constrained Motion Estimation for very low bit rate video coding. IEEE J. Select Areas Commun. 15, 1752–1763 (1997)
17. Turga, D., Chen, T.: Estimation and Mode Decision for Spatially correlated motion sequences. IEEE transactions on CSVT 11(10), 1098–1107 (2001)
18. Lu, J., Liou, M.L.: A simple and efficient search algorithm for block matching motion estimation. IEEE Trans. on Circuits and Systems for Video Technology 7(2), 429–433 (1997)
19. Chung, P.-C., Huang, C.-L., Chen, E.-L.: A region-based selective optical flow back-projection for genuine motion vector estimation. Pattern Recognition 40(3), 1066–1077 (2007)
20. Kim, B.-G., Song, S.-K., Mah, P.-S.: Enhanced block motion estimation based on distortion-directional search patterns. Pattern Recognition Letters 27(12), 1325–1335 (2006)
21. Nie, Y., Ma, K.-K.: Adaptive rood pattern search for fast block-matching motion estimation. IEEE Trans. on Image Processing 11(12), 1442–1449 (2002)
22. Joint Video Team Reference Software, Version 12.2, `http://iphome.hhi.de/suehring/tml/download`

DCT and PCA Based Method for Shape from Focus

M. Tariq Mahmood, Wook-Jin Choi, and Tae-Sun Choi

School of Information and Mechatronics,
Gwangju Institute of Science and Technology,
Oryong-Dong, Buk-Gu, 500-712, South Korea
tariq@gist.ac.kr, taznux@gist.ac.kr, tschoi@gist.ac.kr

Abstract. Discrete Cosine Transform (DCT) and Principal Component Analysis (PCA) are widely used in computer vision applications. In this paper, we introduce a new SFF method based on DCT and PCA. Contrary to computing focus quality locally by summing all values in a 2D or 3D window obtained after applying a focus measure, a vector consisting of seven neighboring pixels is populated for each pixel in the image volume. PCA is applied on the AC part of the DCT of each vector in the sequence to transform data into eigenspace. Considering the first feature, as it contains maximum variation, and discarding all others, is employed to compute the depth. Though DCT and PCA are both computationally expensive transformations, the reduction in data elements and algorithm iterations have made the new approach efficient. Experimental results are presented to demonstrate the effectiveness of new method by using three different image sequences.

Keywords: 3D Shape Recovery, DCT, Focus Measure, PCA, Shape From Focus.

1 Introduction

Three dimensional shape recovery algorithms based on optical reflective model can be categorized as active methods and passive methods. In active techniques, we are projecting light rays while in passive techniques; we simply capture the reflection of light rays without any projections. The passive methods can further be classified as Shape From X (Stereo, Motion, Shading, Focus etc). Hence, Shape From Focus (SFF) is a passive optical method. Fig.1 represents the basic image focus and defocus geometry in a convex lens. In SFF, the depth map is computed by determining the particular distance of the object point p from the camera for which the Image Detector (ID) is exactly at the point P'. The relation between the distance u of the object point P from the lens having a focal length f and the distance v of its focused image at P' is given in the lens formula (1). Once distances for all points of the object are found, the 3D shape can be easily recovered.

$$\frac{1}{f} = \frac{1}{u} + \frac{1}{v} \tag{1}$$

O. Gervasi et al. (Eds.): ICCSA 2008, Part II, LNCS 5073, pp. 1025–1034, 2008.

In SFF methods, a sequence of images is acquired either by relocating the object of an unknown 3D shape in the direction of the optical axes in small steps or by changing the focal length of the lens in the camera. A sharpness criterion, or a focus measure, is applied for each pixel in every frame to measure the focus quality. The depth information is provided by the sharpest pixel among the sequence. To further enhance the results, an approximation method is followed by the focus measure.

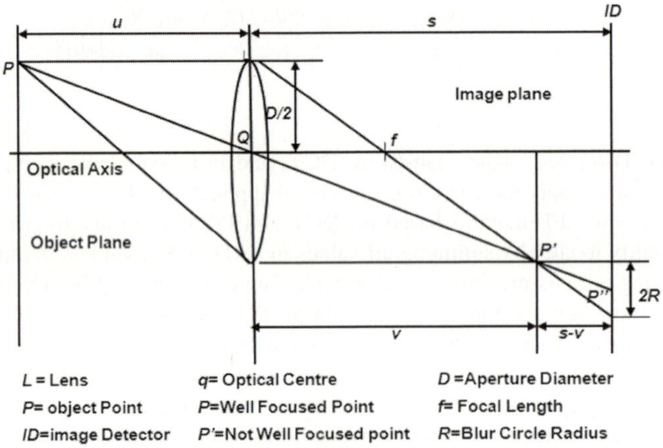

Fig. 1. Basic image formation in convex lens

Conventional focus measures compute the sharpness locally by considering a small 2D window around each pixel, the size of which affects the depth map accuracy and computational complexity [3]. Malik and Choi [4] analyzed the effect of window size on depth map accuracy and concluded that increasing the data elements in the window smoothes the 3D shape but an inaccurate depth map is produced. Bilal and Choi [5]-[6] deduced that the focus measure computed over a 2-D window does not provide an accurate depth map. They suggested that the depth map could be improved by using a 3D window around each pixel. Moreover, these methods compute the focus quality by summing all values obtained by applying a focus measure over the 2D or 3D window. It is mathematically equivalent to the norm of the matrix (window). So, this can be a reasonable approximation of sharpness but it is not an optimal one.

In this paper, we introduce a novel method $F_{DCT-PCA}$ for 3D shape recovery based on Discrete Cosine Transform (DCT) and Principal Component Analysis (PCA). Instead of using a 2D or 3D window, we consider only 6 pixels in the surrounding of each pixel and a vector consisting of seven pixels is then constructed and DCT coefficients are computed. PCA is then applied on the AC part of the DCT of each vector in the sequence. The score for the first principal component or feature is employed to calculate the depth value. The experimental results show that our proposed method is more accurate and robust when compared with previous methods.

In the remainder of this paper, Section 2 describes previously proposed focus measures operators in spatial as well as in DCT domain and approximation techniques. The proposed method is explained in section 3. Experimental results and discussion are presented in section 4 while, section 5 concludes this study.

2 Background

The first step in SFF algorithms is to apply a focus measure operator. In literature, many focus measures have been reported in the spatial as well as in the frequency domain [3]-[8]. Modified Laplacian (ML), Sum Modified Laplacian (SML), Tenengrad (TEN), and Gray Level Variance (GLV) are well known among them and commonly used. The performance of ML focus measure is deteriorated due to the cancellation of derivatives in the opposite direction. Nayar et al [9] proposed Sum Modified Laplacian (SML) to overcome this limitation. In SML the focus value is calculated by summing energy of Laplacian over a small window of size $N \times N$.

$$F_{SML} = \sum_{N} \sum_{N} \left(\frac{\partial^2 g(x,y)}{\partial^2 x} \right)^2 + \left(\frac{\partial^2 g(x,y)}{\partial^2 y} \right)^2 \tag{2}$$

In TEN focus measure, horizontal and vertical Sobel masks are applied and then the focus value is obtained by summing the energy of $G_x(x,y)$ and $G_y(x,y)$ gradients in a small window.

$$F_{TEN} = \sum_{N} \sum_{N} G_x(x,y)^2 + G_y(x,y)^2 \tag{3}$$

GLV focus measure computes the focus value by taking the variance of gray level values within a small window. The difference of each gray level from the mean μ is amplified by the power function.

$$F_{VAR} = \frac{1}{N^2} \sum_{N} \sum_{N} (g(x,y) - \mu)^2 \tag{4}$$

Few focus measures have been proposed in the DCT domain by using the energy compaction property of its coefficients. Baina et al [10] proposed that the energy of the AC part of the DCT is a good approximation of focus quality. If $N \times N$ is a block of image and $F(u,v)$ is its DCT, then the focus measure can be written as

$$F_{DCT\,1} = \sum_{u=1}^{N-1} \sum_{v=1}^{N-1} F(u,v)^2 \tag{5}$$

Shen and Chen [11] suggested that the ratio between energies of AC and DC parts of the DCT is a better choice for focus measure. They observed that its performance had been superior in low contrast images.

$$F_{DCT\,2} = \frac{E_{AC}}{E_{DC}} \tag{6}$$

where E_{AC} and E_{DC} are energies of the AC and DC parts of DCT of an image block. Lee et al [12] showed that DCT coefficients near the centre of 8×8 block have more effect on focus value than others that are near boundaries. They proposed the modified DCT operator by considering the only element $S_{4,4}$ from 8×8 DCT coefficients matrix. The input image is divided into N 8×8 blocks and the focus value is calculated by taking inner of an image block P and modified DCT operator This focus measure is suitable for auto-focusing function in mobile phones due to its low energy consumption.

$$F_{DCT\ 3} = \sum_{k=1}^{N} (P_k . S_{4,4}) \tag{7}$$

Kristan et al [13] proposed another focus measure in the DCT domain. The image sharpness is computed by calculating the entropy of normalized DCT coefficients $\tilde{F}(u, v)$. They used Bayes-sepectral-entropy function and the focus measure is given as

$$F_{DCT\ 4} = 1 - \sum_{u=0}^{N-1} \sum_{v=0}^{N-1} \tilde{F}(u, v)^2 \tag{8}$$

In order to get a refined depth map, a focus measure is followed by an approximation method. In the traditional SFF method proposed by Nayar et al [9], first, the focus value is computed for each pixel of every image by applying SML and then, in the second step, three focus values near the peak are fitted to the Gaussian model. Subbarao and Choi [14] introduced the concept of Focus Image Surface (FIS) and proposed the SFFFIS method to recover an accurate 3D shape. In the SFFFIS method, a rough estimate is taken by applying SML and then the best-focused values lying on FIS are identified in the second step. Yun and Choi [15] proposed an estimation of FIS through a piecewise curved surface approximated through interpolation by using the second order Lagrange polynomial. Asif and Choi [16] applied the neural network to optimize the FIS. GLV is applied to get an initial estimate and then results are provided to the net as input and weights are updated by back propagation algorithm to maximize the focus measure at the output layer. Bilal and Choi [1], [4] proposed a very fast method, SFFDP, by using Dynamic Programming (DP) to get optimal FIS. In this process, a rough depth map is computed through ML in the first step and then optimized by using DP in the second step. Malik and Choi [2] applied the Fuzzy Inference System (FIS) for computing the depth map. These methods provide good results but are computationally expensive. Moreover, their performance relies upon the initial estimate and most commonly used focus measures are gradient based and thus they are noise sensitive.

3 Method

By moving image detector in small steps in the optical axis direction, an image sequence $I(x_i, y_j, z_k)$ of an object is obtained. The sequence consist of Z images each of X×Y pixels, where *i*=1, 2, 3,…,X, *j*=1, 2, 3,…,Y and *k*=1, 2, 3,…,Z, are indices for x, y, and z. Contrary to taking a 2D or 3D window, we consider six neighboring pixels in three directions for each pixel of every frame. Thus a vector consisting of seven pixels is constructed by using the following formula (9).

$$V(i,j,k)=[p(i-1,j,k) \quad p(i,j-1,k) \quad p(i,j,k-1) \quad p(i,j,k) \quad p(i+1,j,k) \quad p(i,j+1,k) \quad p(i,j,k+1)]^T \tag{9}$$

We applied one dimensional DCT on each vector in the image volume. The DCT is a real valued transform, which represents data in the frequency domain. It can be implemented with several variants [17], however equation (10) is one of the most commonly used. For a given vector V_n $0 \le n \le 6$ DCT coefficients $C(u)$ can be computed by

$$C(u) = \omega(u) \sum_{n=0}^{N-1} V_n \cos\left[\frac{\pi(2n+1)u}{2N}\right], \quad 0 \le u \le N-1 \tag{10}$$

where $\omega(u) = \sqrt{1/N}$ *for* $u = 0$ and $\omega(u) = \sqrt{2/N}$ *for* $u \neq 0$. From formula (10) it is clear that the first coefficient, $C(u=0) = \sqrt{\frac{1}{N}} \sum_{n=0}^{N-1} V_n$, is just the average value of the given sequence and it is usually referred to as the DC component. For $C(u \neq 0)$, we obtain AC components for the vector V_n. We considered only the AC part in the proposed method and it is, theoretically, equal to the variance of the given vector V_n [11]. By collecting all AC coefficients $\overset{k}{C}(u)$, where $1 \le u \le N-1$ and $1 \le k \le Z$, for each pixel in the z direction, we get a matrix $\mathbf{M}_{(i,j)} = [m_{ku}]$ having dimensions $Z \times 6$. PCA is applied on this matrix $\mathbf{M}_{(i,j)}$. PCA is a powerful stochastic approach to deal with high dimensional data [18]. It transforms the given data into eigenspace such that the first few components contain more variance than others and noise is concentrated in lower order components or features. The eigenvalues $\boldsymbol{\lambda}$ and their corresponding eigenvectors $\mathbf{E}$ are computed from the covariance matrix of $\mathbf{M}_{(i,j)}$. The transformed data $\mathbf{F}$ in eigenspace is then obtained by multiplying matrix $\mathbf{E}$ with the mean subtracted data. The columns of the transformed matrix $\mathbf{F}$ are known as the principal components or features in eigenspace.

$$\mathbf{F} = \mathbf{E} \times (\mathbf{m}_{ku} - \boldsymbol{\mu}_u) \tag{11}$$

The analysis of the features obtained by PCA shows that the first principal component provides more discriminating power as compared to others and is suitable for calculating the depth map. The depth value for the pixel $P(i, j)$ is computed by using formula (12). It locates the position of the maximum absolute value in the first component.

$$Depth_{(i,j)} = \arg\max_k |\mathbf{f}_{k1}|, \quad 1 \le k \le Z \tag{12}$$

The algorithm iterates XY times to compute the complete depth map for the object. Finally, the median filter is applied on the depth map to reduced the impulse noise due to the possiblity of more than one maxima in the focus curve.

4 Simulation Results

We used three different image sequences for conducting experiments and evaluating the performance of the proposed method. A sequence of 97 images of a simulated cone was generated synthetically by using simulation software. The details about these images can be found in [14]. Images for the other two objects, TFT-LCD color filter and the head of the Lincoln statue on an American one cent coin, were obtained from Microscope Control System (MCS). Details about MCS can be found in [6]. The sample images are shown in Fig. 2. The performance of a focus measure is usually gauged on the basis of unimodality, monotonicity and resolution of the focus curve. Noise sensitivity and computational complexity are also important factors for focus measure evaluation. The resolution metric is used in [19] and is very similar to the Autofocusing Uncertainty Measure (AUM) proposed by Subbarao et al [20]. It provides the deviation of the fitted (Gaussian or quadratic) focus curve from the best focused position. According to the Heiserberg uncertainty principal, the resolution σ^2 for a focus measure function $f(x)$ is defined as

$$\sigma^2 = \frac{1}{\|f\|^2} \sum_{i=1}^{N} (x_i - x_o)^2 |f(x_i)|^2 \tag{13}$$

where x_o is the lens position corresponding to the best focus point among the sequence of N positions and $\|f\|$ is the Euclidian norm of the focus measure function. The sharper curve near the best focus position provides better results so, the smaller the resolution is, the better the performance will be. The resolution was computed using seven positions on each side of the best position (peak) after fitting the focus measure curves to the Gaussian model. Two other statistical metrics, Root Mean Square Error (RMSE) and correlation used by Bilal and Choi [1] and Malik and Choi [2], have also been applied for synthetic image sequence. If $f(x, y)$ is the original image and $g(x, y)$ is the processed image, then the RMSE is defined in equation (14).The higher the correlation and the smaller the RMSE, the better the result will be.

$$RMSE = \sqrt{\frac{1}{XY} \sum_{x=0}^{X-1} \sum_{y=0}^{Y-1} |f(x, y) - g(x, y)|^2} \tag{14}$$

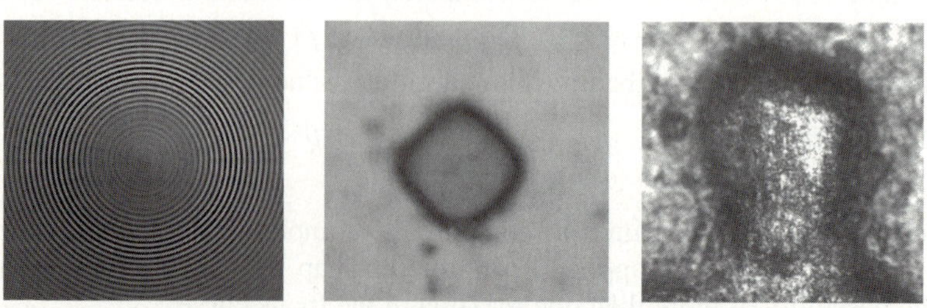

Fig. 2. Sample images (left) simulated cone (middle) TFT-LCD color filter (right) coin

Table 1 and Fig. 3(d) compare the normalized resolution values of the proposed approach with several others using three different image sequences and their focus curves are plotted in Fig. 3(a-c). It can be observed that the resolution value for the proposed method is the lowest among the previously suggested algorithm in the spatial domain and the DCT domain.

Table 1. Computed depth resolution

Focus Measure	Resolution(σ^2)		
	Cone	TFT-LCD	Coin
SML	0.3882	0.2554	0.1807
TEN	0.2971	0.1137	0.1924
VAR	0.0257	0.0512	0.1600
DCT1	0.2134	0.2663	0.1788
DCT2	0.0708	0.3093	0.1810
DCT-PCA	0.0047	0.0041	0.1072

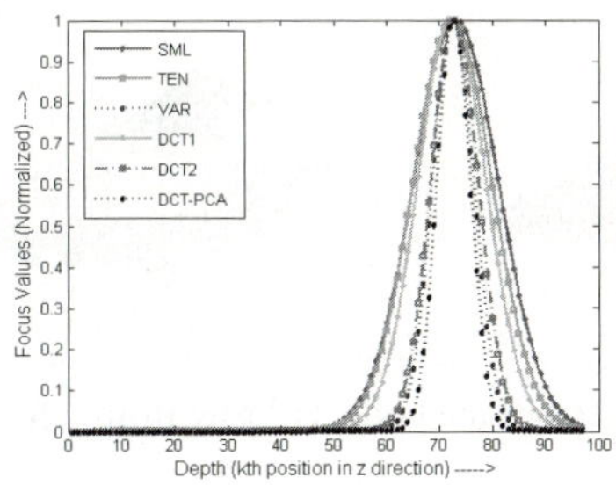

(a) Simulated Cone

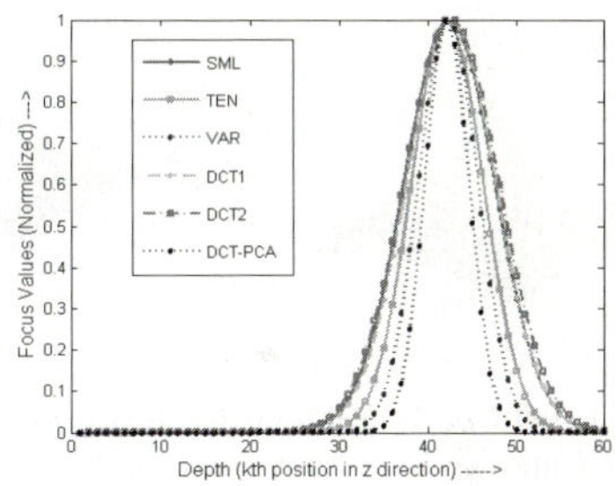

(b) TFT-LCD color filter

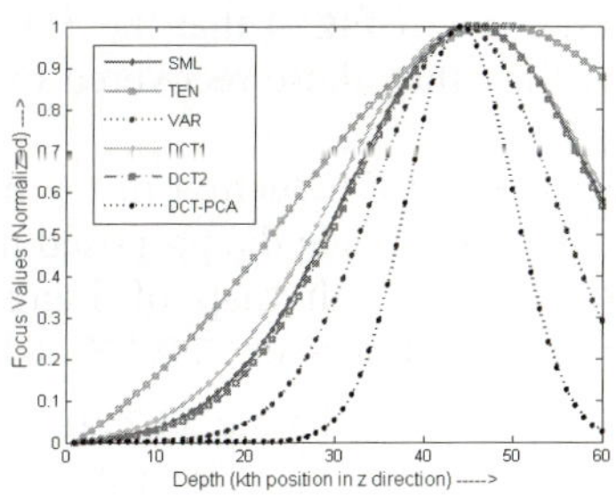

(c) Coin

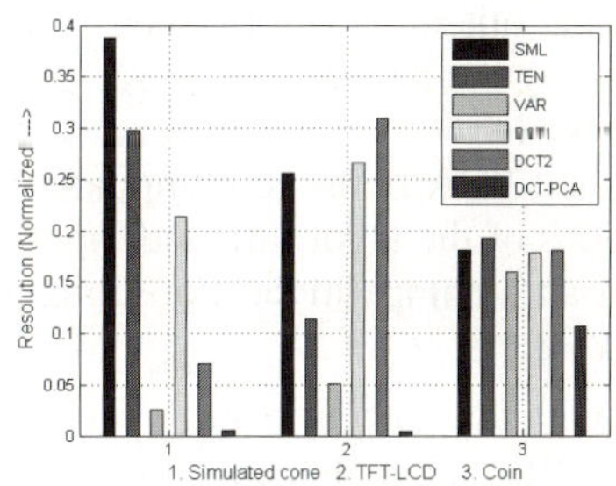

(d)

Fig. 3. (a-c) Fitted focus measure curves to Gaussian model for test objects (d) Resolution comparison

Table 2 shows the comparison of the RMSE and correlation obtained from different focus measure methods using the image sequence of the simulated cone. The correlation value for the proposed method is the highest and the RMSE value is the lowest among the others.

Table 2. Comparisons of focus measures

Focus Measure	RMSE	Correlation
SML	8.6785	0.8957
TEN	8.6599	0.9028
VAR	9.0884	0.8799
DCT1	8.7968	0.9089
DCT2	8.6970	0.9170
DCT-PCA	7.8181	0.9558

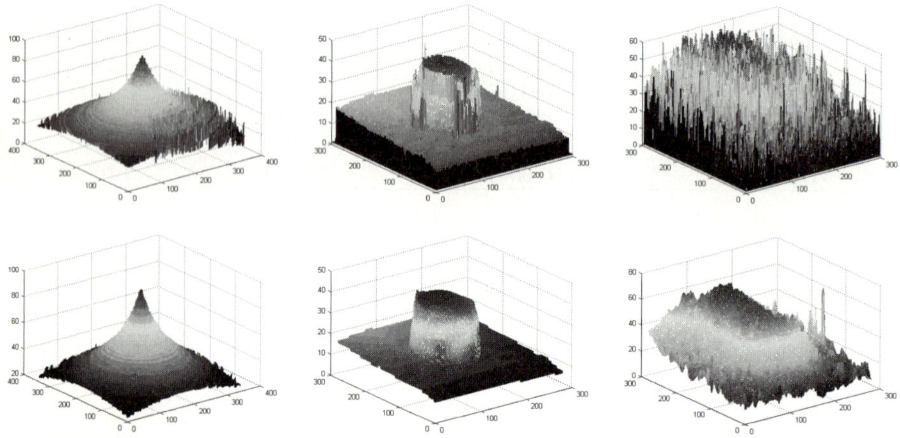

Fig. 4. 3D shapes reconstruted for test objects by different methods. (1st row from left to right) SML, VAR and TEN (2nd row) using DCT-PCA.

Fig. 4 shows the 3D shapes reconstructed for test objects by using proposed method and others. It can be seen from the second row of Fig. 4 that the 3D shapes recovered by the new approach are smoother and finer than those recovered by using other methods.

DCT and PCA both are computationally expensive but a reduction in the number of iterations of the algorithm and input data elements have made the proposed method efficient and competitive. For computing a complete depth map of image volume $I(X \times Y \times Z)$, $F_{DCT-PCA}$ runs XY times and processes $7XYZ$ data elements. On the other hand, window based algorithms iterate XYZ times and $(N \times N)XYZ$ data elements are processed where $N \times N$ is the size of the window. Considering a 7×7 window, $49ZXY$ data elements are required to be processed. In the case of 3D window, even more computations are required. The proposed algorithm has been implemented using MATLAB on a PC, with a 2.8 GHz Pentium IV processor and 512 MB of RAM. It takes approximately 0.0112 s for computing the depth value for a pixel among a sequence of 97 pixels. Whereas the time calculated for F_{DCT1}, F_{DCT2} and F_{DCT4} methods using an 8×8 block of image has been 0.0403 s, 0.0414 s, and 0.0428 s respectively. Thus, it may be said that by reducing the data elements and algorithm iterations, our approach requires fewer computations.

Inherently, noise is collected in lower order features in eigenspace. We have considered only the first feature in the new algorithm, thus the noise effect is reduced by discarding all others noisy features. Therefore, the proposed algorithm performs well even in the presence of noise. The proposed approach has been tested for noisy images and its performance was compared with the previously proposed methods in the DCT domain. Additive white Gaussian noise, with zero mean and nonzero variance, was added to the simulated cone image sequence. Fig. 5 shows the reconstructed 3D shape by SML and DCT-PCA. We can observe that result from SFFDP has degraded significantly while SFFPCA result is still much better.

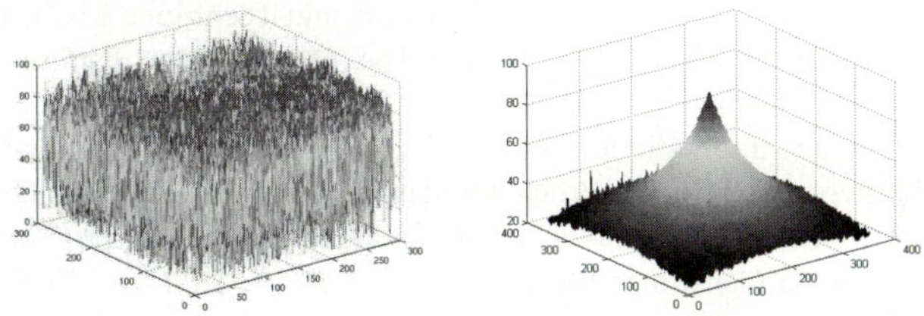

Fig. 5. 3D Shape recovery of simulated cone from images corrupted with noise. (left) SML (right) DCT PCA.

5 Conclusion

In this paper, we introduced a new method for SFF based on discrete cosine transform and PCA. In contrast of using a 2D or 3D window for computing focus value, we considered seven neighboring pixels and applied PCA on the AC part of the DCT. The position of the maximum absolute value from the score of the first component was taken as depth value. The experimental results show that the proposed method yields a more accurate and robust depth map as compared with earlier approaches. The research could be further extended to investigate the application of PCA on wavelet coefficients and results then could be analyzed and compared with DCT based method

Acknowledgements

This work was supported by the Korea Science and Engineering Foundation (KOSEF) grant funded by the Korean government (MOST) (No. R01-2007-000-20227-0).

References

1. Bilal, A.M., Choi, T.S.: A Heuristic approach for finding best focused shape. IEEE Transactions on Circuits and Systems for Video Technology 15(4), 566–574 (2005)
2. Malik, A.S., Choi, T.S.: Application of Passive Techniques for Three Dimensional Cameras. IEEE Trans. Consumer Electronics 53(2) (2007)
3. Krotkov, E.: Focusing. Int. J. Computer Vision 1, 223–237 (1987)

4. Malik, A.S., Choi, T.S.: Consideration of illumination effects and optimization of window size for accurate calculation of depth map for 3D shape recovery. Pattern Recognition 40(1), 154–170 (2007)
5. Ahmad, M.B., Choi, T.S.: Shape from Focus using Optimization Technique. In: Proc. IEEE, Int. Conf. on Acoustics, Speech, and Signal Processing, vol. 2, pp. 493–496 (2006)
6. Ahmad, M.B., Choi, T.S.: Application of Three Dimensional Shape from Image Focus in LCD/TFT Displays Manufacturing. IEEE Trans. Consumer Electronics 53(1), 1–4 (2007)
7. Helmli, F.S., Scherer, S.: Adaptive Shape from Focus with an Error Estimation in Light Microscopy. In: 2nd Int'l Symposium on Image and Signal Processing and Analysis (ISPA 2001) (2001)
8. Sun,, Stefan, Y.D., Nelson Bradley, J.: Autofocusing in computer microscopy: Selecting the optimal focus algorithm. Microscopy Research and Technique 65(3), 139–149 (2004)
9. Nayar, S.K., Nakagawa, Y.: Shape from focus. IEEE Trans. Pattern Anal. Mach. Intell 16(8), 824–831 (1994)
10. Baina, J., Dublet, J.: Automatic focus and iris control for video cameras. In: IEEE Fifth International Conference on Image Processing and its Applications, pp. 232–235 (1995)
11. Chun-Hung, S., Chen, H.H.: Robust focus measure for low-contrast images. In: International Conference on Consumer Electronics ICCE 2006. Digest of Technical Papers, pp. 69–70 (2006)
12. Lee, S.Y., Park, S.S., Kim, C.S., Kumar, Y., Kim, S.W.: Low-power auto focus algorithm using modified DCT for the mobile phones. In: International Conference on Consumer Electronics ICCE 2006. Digest of Technical Papers, pp. 67–68 (2006)
13. Kristan, M., Perse, P.J., Kovacic, S.M.: A Bayes-spectral-entropy-based measure of camera focus using a discrete cosine transform. Pattern Recognition Letters 27(13), 1431–1439 (2006)
14. Subbarao, M., Choi, T.S.: Accurate recovery of three dimensional shape from image focus. IEEE Trans. Pattern Anal. Mach. Intell 17(3), 266–274 (1995)
15. Yun, J., Choi, T.S.: Accurate 3-D shape recovery using curved window focus measure. In: ICIP 1999, vol. 3, pp. 910–914 (1999)
16. Asif, M., Choi, T.S.: Shape from focus using multilayer feedforward neural network. IEEE Trans. Image Process. 10(11), 1670–1675 (2001)
17. Ahmed, N., Natarajan, T., Rao, K.R.: Discrete Cosine Transform. IEEE Transactions on Computers C-23(1), 90–93 (1974)
18. Schölkopf, B., Smola, A.: Learning with Kernels. MIT Press, Cambridge (2002)
19. Yang, G., Nelson, B.J.: Wavelet-based autofocusing and unsupervised segmentation of microscopic images. In: IEEE/RSJ International Conference on Intelligent Robots and Systems (IROS), vol. 3, pp. 2143–2148 (2003)
20. Subbarao, M., Tyan, J.K.: Selecting the optimal focus measure for autofocusing and depth-from-focus. IEEE Trans. Pattern Anal. Mach. Intell. 20(8), 864–870 (1998)

Formula-Dependent Abstraction for CTL Model Checking

Junyan Qian, Lingzhong Zhao, Guoyong Cai, and Tianlong Gu

Department of Computer Science and Technology,
Guilin University of Electronic Technology, Guilin 541004, China
qjy2000@gmail.com

Abstract. We present a state abstraction that is defined with respect to a given CTL formula. Since it does not attempt to preserve all ACTL formula, like simulation does, we can expect to compute coarser abstraction. Specifically, the abstraction is used to reduce the size of each Kripke structure, so that their product will be smaller. When the abstraction is too coarse, we show how refinement can be applied to produce a more precise abstract model. We also extend the notion of formula-dependent abstraction to Kripke structure with fairness, and define the coarsest abstraction that preserves the given CTL formula interpreted with respect to the fair paths. The method is exact and fully automatic, and handles full CTL.

1 Introduction

Model checking [1] is a fully automatic technique of verifying finite-state reactive-system for the correctness of design specifications such as hardware design and protocol verification. Branching time propositional temporal logic has been proven very useful in the automatic verification of concurrent finite-state systems [2]. The systems are modeled using labeled state transition structures called Kripke or temporal structures [3]. The properties that one wishes to verify can be expressed in terms of a branching time temporal logic such as CTL [8]. An efficient search procedure, which was originally developed in 1981 by Clarke and Allen Emerson [4], is used to determine automatically whether the model satisfies the specifications [3].

Oftentimes, the straightforward approach to model checking concurrent systems is to construct the product of the component structure to yield a single structure, and then proceed to model check the single structure. However, the size of the product structure can be exponential in the number of component structures. This is known as the state explosion problem. In order to ameliorating the state explosion problem, Abstraction, where one build up a smaller abstract model in a manner which ensures that the specification holds for the original model if it holds for the abstract model, is a main method for reducing the state space of the checked system by hiding some of the system details that might be irrelevant for the checked property.

Our goal is to develop an algorithm that alleviates the explosion problem by formula-dependent abstraction in each component structure. The abstraction is used to simplify the component before taking their product, thus leading to a smaller product

O. Gervasi et al. (Eds.): ICCSA 2008, Part II, LNCS 5073, pp. 1035–1048, 2008.

structure. It is well known that simulation preorder relation is a coarse abstraction that preserves the truth of all ACTL* formulas. However, in general we are interested in model checking a system with respect to just a few formulas, and hence preserving all ACTL formulas is stronger than needed. Thus, we investigate a formula-dependent abstraction that preserves the truth of a particular formula of interest, but possibly not of other formulas. This leads to a coarser abstraction, and thus to a greater opportunity for simplification.

In many cases, we are only interested in the correctness along fair computation paths. We also extend the notion of formula-dependent abstraction to Kripke structures with fairness. Fairness constraints allow us to reason about only those infinite paths in the Kripke structure with satisfy some fairness specification, which is evaluated over the infinite path. Often, it is more natural to think of the fairness constraints as part of the system specification, instead of part of the property being verified. We will refer to Kripke structures with fairness constraints as fair Kripke structures, and the problem of checking a CTL formula on a fair Kripke structure as the FairCTL model checking.

The remainder of this paper is organized as follows: Section 2 discusses related work, and Section 3 presents some preliminaries. In Section 4 we develop our formula-dependent abstraction in Kripke structure, and in Section 5 we describe parallel composition. Section 6 model checking the abstract model, and Section 7 extends formula-dependent abstraction to fair Kripke structure. Finally, Section 8 gives conclusions.

2 Related Work

As key issue in model checking, Abstraction [16] is a method for reducing the state space of the checked system. The reduction is achieved by hiding (abstracting away) some of the system details that might be irrelevant for the checked property. When model checking using abstraction, the main concern is that the abstractions must be property-preserving. There are two forms of property preservation: *Weak Preservation* and *Strong Preservation*. An abstraction is a weak property preserving if a set of properties true in the abstract system has corresponding properties in the concrete system that are also true, while an abstraction is a strong preserving abstraction if a set of properties with truth values either true or false in the abstract system has corresponding properties in the concrete system with the same truth values.

It is usually difficult and expensive to compute a precise abstraction directly. In order to reduce the complexity, approximation is often used. There are two main forms of approximation abstraction: over-approximation and under-approximation. In over-approximation, more behaviors are added in the abstract system than are present in the concrete system. This approach provides a very popular class of weakly preserving abstractions for universally quantified path properties. However, over-approximation often only works well for safety (or invariant) properties. In under-approximation, the behaviors are removed when going from the concrete to the abstract system. Under-approximation is also often found in the construction of an environment for a system to be checked.

The abstract interpretation framework that Cousot et al. introduced in [6] is based on the use of Galois connections (or equivalently closure operators) to establish the correspondence between the domain of concrete or exact properties and the domain of abstract or approximate properties. This corresponds to a perfect situation, where each concrete property has a unique best abstract approximation.

Abstraction has been widely used to make model checking more efficient for large systems. *Predicate Abstraction* presents firstly by Graf and Saidi [5], which is a special forum of *Abstract Interpretation* [6, 7], where the abstract domain is constructed using a given set of predicates, i.e. a potentially unbound data type is abstracted to a finite set of points. During the last years, predicate abstraction is being used in software model checking such as Bandera [23], Java PathFinder [24], SLAM [25], MAGIC [26] and BLAST [27], where the first two focuses on Java while the last three all deal with C.

The abstraction refinement process has been automated by the *Counterexample Guided Abstraction Refinement* paradigm [9], or CEGAR for short. The notion of CEGAR was originally described by Kurshan [10] for model checking of finite state models. Counterexample guided refinement has even been used with predicate abstraction by Lakhnech et al. [11]. One starts with a coarse abstraction, and if it is found that an error-trace reported by the model checker is not realistic, the error trace is used to automatically refine the abstract model, and the process proceeds until no spurious error traces can be found.

Giacobazzi and Quintarelli [20] first noted that strong preservation is related to completeness in abstract interpretation by studying the relationship between complete abstract interpretations, strongly preserving and spurious counterexamples of abstract model checking. They indicated spurious counterexamples as a lack of completeness in the abstract interpretation sense for the abstract model checking. Then, in order to eliminate spurious counterexamples, they designed a method for minimally making abstract interpretations complete. F. Ranzato and F. Tapparo [21, 22] turns out that this least refinement exist and can be constructively characterized as fixed point solution of abstract domain equations.

The more general abstraction technique to avoid the state explosion problem is known as *Existential Abstraction* [12], resulting in an over-approximation of the set of behaviors of the original program. If a property holds on the abstract structure, then it also holds on the concrete structure.

In [13], Aziz *et al* present a notion of formula-dependent bisimulation equivalence for CTL and interacting finite state machines. Their algorithm calculates the quotient space and uses "pass" and "fail" sets which are also some kind of under- and over-approximations for the satisfaction relation in the concrete model. Dams *et al* [14, 15] represents an abstraction refinement algorithm for ACTL, and deal with abstract models with two transition relations.

Dams *et al* [17] describes model checking temporal logic over abstracted transition systems under fairness constraints. Aziz *et al* [18] extends the notion of bisimulation to Kripke structures with fairness. However, our goal is that a coarser abstraction model with respect to formula-dependent be obtained, and we are interested in techniques that can handle full CTL, not but ACTL. We also extend the notion of formula-dependent to Kripke structures with fairness constraints.

The work of Bensalem *et al* [19] has similar aims as ours. They present a method for computing abstractions of infinite state systems compositionally and automatically. Given a concrete system $S=S_1||\ldots||S_n$ of programs and given an abstraction function α, using our method one can compute an abstract system $S^a = S_1^a||\ldots||S_n^a$ such that S simulates S^a, i.e. every computation of S can be mapped to a computation of S^a.

3 Preliminary

We provide some background on abstraction and model checking.

3.1 Kripke Structure

Kirpke structure is a finite graph in which nodes represent system states and edges represent possible transitions between states. More formally define as follow.

Definition 1. Given a set AP of atomic propositions, a Kripke structure is a four tuple $M=\langle S, R, L\rangle$, where S is a set of states, $R\subseteq S\times S$ a transition relation, and L: $S\rightarrow 2^{AP}$ a labeling function that associates each state in the structure with the set of atomic propositions that hold in that state.

A trace (or simply a path) in M from a state s_0 is an infinite sequence of state $\pi = s_0$, s_1, …, such that s_0 is an initial state and for every $i\geq 0$ there is a transition $\tau\in R$ satisfying $s_{i+1}\in \tau(s_i)$. $P_M(s)$ denote all paths starting from state s in M.

3.2 Syntax and Semantic of CTL

Computation Tree Logic (CTL) [2] is a branching temporal logic, which used originally by Clarke and Emerson in 1981. We will define the syntax first, follow by semantics. CTL formulae are composed of atomic proposition, Boolean connectives, temporal operators and path quantifiers. There are four basic temporal operators, including **G**, **F**, **U** and **X**, and two quantifiers, including **A** and **E**. In CTL, every quantifier must be followed by a temporal operator. However, each of the CTL temporal operators can be expressed in terms of three operators that are **EX**, **EU** and **EG**. The syntax of CTL is defined as follow.

- True, and false, the atomic propositions, and the negations are CTL formulae;
- If ϕ and φ are CTL formulae, then are so $\neg\phi$, $\phi\vee\varphi$, $\mathbf{EX}\ \phi$, $\mathbf{EG}\ \phi$, $\mathbf{E}(\phi\ \mathbf{U}\ \varphi)$.

Let $p\in$ AP is an atomic proposition, $M=(S, R, L)$ is CTL model, $s\in S$, and ϕ, φ be CTL-formulas. The satisfaction relation |= is defined by:

- $s \models p$ iff $p\in L(s)$
- $s \models \neg\phi$ iff $\neg(s \models \phi)$
- $s \models \phi\vee\varphi$ iff $(s \models \phi) \vee (s \models \varphi)$
- $s \models \mathbf{EX}\ \phi$ iff $\exists\pi\in P_M(s)$, $\pi[1] \models \phi$
- $s \models \mathbf{EG}\ \phi$ iff $\exists\pi\in P_M(s)$, $\forall j\geq 0$, $\pi[j] \models \phi$
- $s \models \mathbf{A}[\phi\ \mathbf{U}\ \varphi]$ iff $\forall\pi\in P_M(s)$, $(\exists j\geq 0, \pi[j] \models \varphi \wedge \forall 0\leq k<j, \pi[k] \models \phi))$

3.3 CTL Model Checking

The model checking problem is easy to describe. Given a Kripke structure $M=(S, R, L)$ that represents a finite-state concurrent system and a temporal logic formula ϕ expressing some desired specification, find the set of all state in S that satisfy ϕ. $\{s\in S \mid M, s \models \phi\}$. The system satisfies the specification provided that all of the initial states are in the set.

We define the CTL model checking problem as the problem of determining all states of the system that satisfy the CTL formula ϕ. Suppose we want to determine the CTL formula ϕ is valid in the finite CTL model $M=(S, R, L)$. The basic concept of the model checking algorithm is to "label" each state $s\in S$ with the sub-formula of ϕ that are valid in s.

4 Formula-Dependent Abstraction

Our goal is to define an abstraction on the states of each component that is as coarse as possible with respect to a given CTL formula ϕ, while being efficiently computable. Next section explains how we intend to apply this abstraction to model checking, but the main idea is to minimize the size of each component.

We describe Milner's simulation relation before defining our formula dependent abstraction. Given two Kripke structures $M=(S, R, L)$ and $M'=(S', R', L')$ with $AP \supseteq AP'$, a relation $H\subseteq S\times S'$ is a simulation relation between M and M' iff and only if for all s and s', if $H(s, s')$ then the following conditions hold.

- $L(s)\cap AP'=L'(s')$
- For every state s_1 such that $R(s, s_1)$, there is a state s_1' with the property that $R'(s', s_1')$ and $H(s_1, s_1')$

We say that M' simulates M (denoted by $M \preceq M'$) if there exists a simulation relation H such that for every initial state s_0 in M there is an initial state s_0' in M' for which $H(s_0, s_0')$. We say that paths $\pi'=s_0', s_1',\ldots$ in M' simulates $\pi=s_0, s_1,\ldots$ in M if and only if for every i, $H(s_i, s_i')$. The following lemma is important in establishing the connection between CTL and simulation relation.

Lemma 1. Let s and s' be two states such that $H(s, s')$. Then for every path π starting form s there is a corresponding path π' starting form s'.

Intuitively, state s' simulates s if the corresponding infinite computation trees match. This means that $L(s)\cap AP'=L'(s')$, and the two states have next states whose infinite computation trees again match.

We use the notion of *SAT* and *UNSAT* states to ease the strict requirement of simulation that the infinite computation trees of two states match. If a state is a SAT^{ϕ} state with respect to a CTL formula ϕ, then it satisfies in all environment; likewise, if a state is a $UNSAT^{\phi}$, then it does not satisfy in all environment. Given SAT^{ϕ} and $UNSAT^{\phi}$ states, the modification to simulation we ignore these transitions to $UNSAT^{\phi}$

states from one state. That is the subtrees rooted at $UNSAT^{\phi}$ states need not matched. We say that s' simulates s with respect to a CTL formula ϕ if

- s' simulates s with respect to the immediate subformulas of ϕ, and
- Either they are both SAT^{ϕ} states or $UNSAT^{\phi}$ states, or the infinite computation trees of the two states match up to SAT^{ϕ} states, ignoring all subtrees rooted at $UNSAT^{\phi}$ states.

Before formally defining our abstraction relation, we define the SAT^{ϕ} and $UNSAT^{\phi}$ states. For a given CTL formulas ϕ, SAT^{ϕ} and $UNSAT^{\phi}$ sets are defined as follows.

Definition 2. Let M=(S, R, L) and M'=(S', R', L') be Kripke structures, and let ϕ be a CTL formula. Let $p \in AP$, SAT^{ϕ} and $UNSAT^{\phi}$ for M are subsets of S, as follows:

- If $\phi = p$ then $SAT^{\phi} = \{s \in S \mid p \in L(s)\}$, $UNSAT^{\phi} = S \backslash SAT^{\phi}$
- If $\phi = \neg\varphi$ then $SAT^{\phi} = UNSAT^{\varphi}$, $UNSAT^{\phi} = SAT^{\varphi}$
- If $\phi = \varphi_1 \vee \varphi_2$ then $SAT^{\phi} = SAT^{\varphi_1} \cup SAT^{\varphi_2}$, $UNSAT^{\phi} = UNSAT^{\varphi_1} \cap UNSAT^{\varphi_2}$
- If $\phi = \exists\mathbf{X}\ \varphi$ then $SAT^{\phi} = \{s \in S \mid \exists t,\ R(s, t) \text{ and } t \in SAT^{\varphi}\}$, $UNSAT^{\phi} = \{s_0 \in S \mid$ for every path $s_0, s_1, \ldots, s_1 \in UNSAT^{\varphi}\}$
- If $\phi = \mathbf{EG}\ \varphi$ then great fixed-point of: $S_0 = SAT^{\varphi}$; $S_{i+1} = S_i \cap \{s \in S \mid s \in S$, and $\exists t$, $R(s, t)$ and $t \in S_i\}$, $UNSAT^{\phi} = \{s_0 \in S \mid$ for every path $s_0, s_1, \ldots,$ there exists $i \geq 0$, $s_i \in UNSAT^{\varphi_2}\}$
- If $\phi = \mathbf{E}(\varphi_1 \mathbf{U} \varphi_2)$then least fixed-point of: $S_0 = SAT^{\varphi_2}$; $S_{i+1} = S_i \cup \{s \in S \mid s \in SAT^{\varphi_1}$, and $\exists t$, $R(s, t)$ and $t \in S_i\}$, $UNSAT^{\phi} = \{s_0 \in S \mid$ for every path $s_0, s_1, \ldots,$ either (1) there exists $i \geq 0$, $s_i \in UNSAT^{\varphi_2}$ and $\forall j \leq i$, $s_j \in UNSAT^{\varphi_1}$, or (2) $\forall i \geq 0$, $s_i \in UNSAT^{\varphi_2}\}$

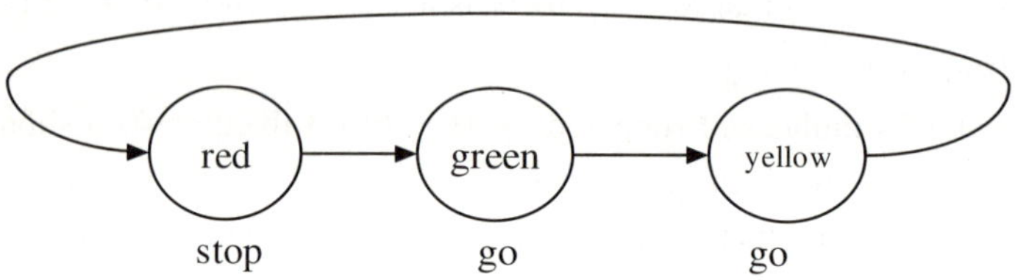

Fig. 1. A Traffic Light

As an example of SAT^{ϕ} and $UNSAT^{\phi}$, consider the Kripke structure in Figure 1. For ϕ=*go,* states *green* and *yellow* lie in SAT^{ϕ} and state *red* lie in $UNSAT^{\phi}$, For ϕ=**EF** *stop*, states *green*, *yellow* and *red* lie in SAT^{ϕ}.

Now we formally define our abstraction relation. Let M=(S, R, L) and M'=(S', R', L') be Kripke structures, and let ϕ be a CTL formula. We define the abstraction relation H^{ϕ} on the states of M and M' as the coarsest abstraction relation satisfying the following: For $s \in S$, $s' \in S'$, $H^{\phi}(s, s')$ iff

- Case $\phi = p$: $p \in AP'$, $s \in SAT^{\phi}$ and $s' \in SAT^{\phi}$, or $s \in UNSAT^{\phi}$ and $s' \in UNSAT^{\phi}$
- Case $\phi = \neg\varphi$: $H^{\varphi}(s, s')$
- Case $\phi = \varphi_1 \vee \varphi_2$: $H^{\varphi_1}(s, s')$ and $H^{\varphi_2}(s, s')$

- Case $\phi = \exists \mathbf{X}\ \varphi$: $H^{\varphi}(s, s')$ and
  1. $s \in SAT^{\phi}$ and $s' \in SAT^{\phi}$, or $s \in UNSAT^{\phi}$ and $s' \in UNSAT^{\phi}$, or
  2. $L(s) \cap AP' = L'(s')$, and for every state $s_1 \notin UNSAT^{\varphi}$ such that $R(s, s_1)$, $\exists s'_1$, $R'(s', s'_1)$ and $H^{\phi}(s_1, s'_1)$
- Case $\phi = \mathbf{EG}\ \varphi$: $H^{\varphi}(s, s')$ and
  1. $s \in SAT^{\phi}$ and $s' \in SAT^{\phi}$, or $s \in UNSAT^{\phi}$ and $s' \in UNSAT^{\phi}$, or
  2. $L(s) \cap AP' = L'(s')$, and for every state $s_1 \notin UNSAT^{\phi}$ such that $R(s, s_1)$, $\exists s'_1$, $R'(s', s'_1)$ and $H^{\phi}(s_1, s'_1)$
- Case $\phi = \mathbf{E}(\varphi_1\ \mathbf{U}\ \varphi_2)$: $H^{\varphi_1}(s, s')$ and $H^{\varphi_2}(s, s')$ and
  1. $s \in SAT^{\phi}$ and $s' \in SAT^{\phi}$, or $s \in UNSAT^{\phi}$ and $s' \in UNSAT^{\phi}$, or
  2. $L(s) \cap AP' = L'(s')$, and for every state $s_1 \notin UNSAT^{\phi}$ such that $R(s, s_1)$, $\exists s'_1$, $R'(s', s'_1)$ and $H^{\phi}(s_1, s'_1)$

We say that M' formula-dependent simulates M (denoted by $M \preceq^{\phi} M'$) if there exists a simulation relation H^{ϕ} such that for every initial state s_0 in M there is an initial state s'_0 in M' for which $H^{\phi}(s_0, s'_0)$. We say that paths $\pi' = s'_0, s'_1, \ldots$ in M' formula-dependent simulates $\pi = s_0, s_1, \ldots$ in M if and only if for every i, $H^{\phi}(s_i, s'_i)$. The following lemma is important in establishing the connection between CTL and formula-dependent simulation relation.

Lemma 2. Let s and s' be two states such that $H^{\phi}(s, s')$. Then for every path π starting form s there is a corresponding path π' starting form s'.

In a manner similar to Milner, using a standard fixed-point computation H^{ϕ} can be computed in polynomial time.

Theorem 1. Let ϕ be a CTL formula, let s be a state of M, and let s' be a state of any M' such that $H^{\phi}(s, s')$, then $s' \models \phi \Rightarrow s \models \phi$.

Proof: According to the definition of H^{ϕ}, we know that $s' \models \phi \Rightarrow s \models \phi$.

Now, we show that $\preceq^{\phi}$ is a preorder, that is a reflexive and transitive relation.

Lemma 3. $\preceq^{\phi}$ is a preorder on the set of structures.

Proof: The relation $H^{\phi} = \{(s, s) \mid s \in S\}$ is a formula-dependent simulation relation between M and M, so is reflexive.

Thus it only remains to show that $\preceq^{\phi}$ is transitive. Assume $M \preceq^{\phi} M'$ and $M' \preceq^{\phi} M''$. Let H_1^{ϕ} be a formula-dependent simulation between M and M', and let H_2^{ϕ} be a formula-dependent simulation between M' and M''. Define H^{ϕ} as the relational product of H_1^{ϕ} and H_2^{ϕ}, that is

$$H^{\phi} = \{(s, s'') \mid \exists s',\ H_1^{\phi}(s, s') \wedge H_2^{\phi}(s', s'')\}$$

If $s_0 \in S_0$, then by the definition of formula-dependent simulation, there exists $s_0' \in S_0'$ such that $H_1^\phi(s_0, s_0')$. Similarly, there exists $s_0'' \in S_0''$ such that $H_2^\phi(s_0', s_0'')$, and hence $H^\phi(s_0, s_0'')$.

Suppose $H^\phi(s, s'')$, and let s' be such that $H_1^\phi(s, s')$ and $H_2^\phi(s', s'')$. By definition of formula-dependent simulation, $L(s) \cap AP' = L'(s')$ and $L'(s') \cap AP'' = L''(s'')$. Then because $AP' \supseteq AP''$, we have $L(s) \cap AP'' = L''(s'')$. Let $R(s, s_1)$ be a transition in M from s. Then there exists a transition $R'(s', s_1')$ in M' such that $H_1^\phi(s_1, s_1')$. Likewise, there exists a transition $R''(s'', s_1'')$ in M'' such that $H_2^\phi(s_1', s_1'')$. Hence $H^\phi(s_1, s_1'')$, and H^ϕ is a formula-dependent simulation between M' and M''. Then $M \preceq^\phi M''$. □

THEOREM 2. Let ϕ be a CTL formula, suppose $M \preceq^\phi M'$. Then for given a CTL formula ϕ, $M' \models \phi \Rightarrow M \models \phi$.

PROOF: By definition, $R(s, s_1)$ implies $R'(s', s_1')$ and for every initial state s_0 in M there is an initial state s_0' in M' for which $H^\phi(s_0, s_0')$. Then $M \preceq^\phi M'$ according to the definition of the formula-dependent simulation relation, that is $M' \models \phi \Rightarrow M \models \phi$.

5 Parallel Composition

When dealing with complex system obtained as the parallel composition of the systems, the application of this method requires the computation of the corresponding global transition relation from which an abstraction can be computed. The question then arises whether it is possible to compute abstractions of complex system as the parallel composition of abstractions of their components in order to avoid building the transition relation associated with the complex system. This is guaranteed if the compositionality property.

$$\frac{M_1 \prec M_1' \text{ and } M_2 \prec M_2'}{(M_1 \parallel M_2) \prec (M_1' \parallel M_2')}$$

holds, where || is a parallel composition operator.

DEFINTION 4. (Parallel composition) The synchronous composition of transition systems $M_i = (S_i, R_i, L_i)$, i=1,2, denoted $M_1 \parallel M_2$, is the transition system $(S_\parallel, R_\parallel, L_\parallel)$, where:

- $S_\parallel = S_1 \otimes S_2$ is the set of states. $s = \langle s_1, s_1 \rangle \in S_\parallel$ iff $s_1 \in S_1$, $s_2 \in S_2$.
- $R_\parallel = R_1 \otimes R_2$, $(s, s') \in R_\parallel$ iff $(s_{|S_i}, s'_{|S_i}) \in R_\parallel$, for i=1,2, where $s_{|S_i}$ denotes the restriction of the state s to S_i.
- $L_\parallel = L_1 \otimes L_2$, $L(s_1, s_2) = L_1(s_1) \cup L_2(s_2)$

THEOREM 3. Let ϕ be a CTL formula, and Let $M_1,\ldots, M_n$ be Kripke structures. Let M_i/ H_i^{φ} be the abstraction of M_i with respect to H_i^{φ}, Then $M_1/ H_1^{\varphi} \| \ldots \| M_n/ H_n^{\varphi} \models \phi \Rightarrow M_1 \| \ldots \| M_n \models \phi$

Then following proposition says that, indeed, if s is in SAT^{ϕ}, then and product state with s as a component stratifies ϕ.

PROPOSITION 1 [13]. Let ϕ be a CTL formula, let s be a state of M_1, and let t be a state of any M_2. if $s\in SAT^{\phi}$, then $M_1\times M_2$, $\langle s, t\rangle \models \phi$. Likewise, if $s\in UNSAT^{\phi}$, then $M_1\times M_2$, $\langle s, t\rangle \not\models \phi$.

By the formula-dependent abstraction, we can all alleviates the explosion problem encountered in compositional model checking. The assumptions are that each component is relatively small and easy to manipulate, and that the full product structure is too large to build and manipulate. The general idea is to minimize each component with respect to H^{ϕ} before computing the full product structure.

Sometimes, in some cases, we may be able to check whether the model satisfy the specification without composing all the component. This is known as early pass/fail detection. Let $s = (s_1, s_2,\ldots, s_n)$ is a product state. By proposition 1, if $s_i\in UNSAT_i^{\phi}$, then the product state s does not satisfy ϕ; and if $s_i\in SAT_i^{\phi}$, then the product state s satisfies ϕ.

6 Model Checking the Abstract Model

Given a CTL formula ϕ, and a Kripke structure M, let M' be the abstract Kripke structure corresponding to the formula-dependent abstraction. We use standard model checking procedures to determine whether M' satisfies the specification ϕ. If it does, then by Theorem 1 we can conclude that the original Kripke structure also satisfies ϕ, and then the verification is successful. Otherwise, assume that the model checker produces a counterexample C corresponding to the abstract model M'.

Because the over-approximation may be not enough to prove original model correctness, resulting in the failure to concretize the abstract counterexample, we need to refine abstract model using the spurious counterexample.

We determine whether counterexample is spurious. Assume that a abstract trace π' of M' is a finite sequence $\pi' = \langle s_1', s_2' \ldots, s_n' \rangle$ such that (i) for $1\leq i\leq n$, $s_1' \in S'$, (ii) s_1' be initial state, and (iii) for $1\leq i\leq n$, $(s_i', s_{i+1}') \in R'$. Assume that h is a abstraction function with respect to formula-dependent simulation H^{ϕ}, that is $h(s)=s'$ iff $H^{\phi}(s, s')$. Given an abstract state s', the set of concrete states s is denoted by $h^{-1}(s')$, i.e., $h^{-1}(s') = \{s \mid h(s)=s'\}$. We extend h^{-1} to sequences in the following way: $h^{-1}(\pi')$ is the set of concrete paths given by the following expression

$$\{\langle s_1, s_2,\ldots, s_n\rangle \mid \bigwedge_{i=1}^{n} h(s_i)= s_i' \wedge \bigwedge_{i=1}^{n-1} R(s_i, s_{i+1})\}$$

Let $S_1 = h^{-1}(s_1')$ and R be the transition relation corresponding to the original Kripke structure M. For 1, we define in the following manner: $S_i = h^{-1}(s_i') \cap Img(S_{i-1}, R)$, where $Img(S_{i-1}, R)$ is the forward image of S_{i-1} with respect to the transition relation R. If i be the largest index such that $S_i \neq \varnothing$, s_i' is called the failure state of the spurious counterexample π'. The algorithm in Figure 2 computes the set of states S_{i-1}.

```
Procedure Find_Spurious
begin
  S = h^-1(s'_1);
  i = 1;
  while (S≠∅ and i<n){
    i = i+1;
    S_prev =S;
    S = Img(S,R)∩h^-1(s'_i); }
  If S≠∅ then return(S,0);
  else return(S_prev,i);
end
```

Fig. 2. Computing the reachable state within π'

The algorithm above is a forward approach. Likewise, the algorithm based on a backward approach can find the failure state. Let the abstract state s_i' was the failure state. There are three types of concrete states in the failure state s_i'.

- The dead-end state is reachable, but there are no outgoing transitions to the next state in the counterexample.
- The bad state is not reachable but outgoing transitions cause the spurious counter-example. The spurious counterexample is caused by the bad state.
- The irrelevant state is neither reachable nor bad.

The goal of the refinement methodology is to refine H^{ϕ} so that the dead-end and bad states do not belong to the same abstract state. Then the spurious counterexample will be eliminated.

If the abstract counterexample π' is not spurious, report the concretion counterexample and stop; otherwise refine abstract model by the spurious counterexample π'. If the abstract counterexample π' is spurious, by combining the forward and backward approaches we can compute a minimal spurious sub-path. In particular, there might be several spurious counterexample paths, and in each spurious path we could split several states. However some paths might share common spurious sub-paths, using the minimal spurious sub-paths to avoid significant redundant computation. Finally, we address the termination properties of the refinement process. The process will terminate if one of the two following conditions is met:

- Model checking the abstract model shows that $M' \models \phi$
- A real counterexample path that is feasible in concrete model is found.

7 Fairness

In many cases, we are only interested in the correctness along fair computation paths. However, the major limitation of CTL is that cannot express correctness under fairness constraints. In order to deal expressed directly in CTL we must modify its semantics slightly. A fairness constraint can be an arbitrary set of states, usually described by a formula of the logic. If fairness constraints are interpreted as sets of states, then a fair path must contain an element of each fairness constraint infinitely often. If fairness constraints are interpreted as CTL formulas, then a path is fair if each constraint is true infinitely often along the path.

Formally, a fair Kripke structure is a 4-tuple $M=\langle S, R, L, F\rangle$, where S, I, R, and L are defined as before and $F\in 2^S$ is a set of fairness constraints. Let $\pi = s_0, s_1, \ldots$ be a path in M. Define $inf(\pi) = \{s \mid s = s_i$ for infinitely many $i\}$. We say that is fair if and only if for every $P\in F$, $inf(\pi)\cap P\neq\emptyset$. The semantics of CTL with respect to a fair Kripke structure is very similar to the semantics of CTL with respect to an ordinary Kripke structure. We will write $M, s \models_F \phi$ to indicate that the formula ϕ is true in state s of the fair Kripke structure M. Only some clauses in the original semantics change.

- $s \models_F p$ iff there exists a fair path π starting from s and $p\in L(s)$
- $s \models_F \mathbf{EX}\ \phi$ iff there exists a fair path π starting from s and $\pi[1] \models_F \phi$
- $s\models_F \mathbf{EG}\ \phi$ iff there exists a fair path π starting from s, and $\forall j\geq 0$, $\pi[j] \models_F \phi$
- $s \models_F \mathbf{A}[\phi\,\mathbf{U}\,\varphi]$ iff for all fair path π starting from s, ($\exists j\geq 0$, $\pi[j] \models_F \varphi \wedge \forall 0\leq k<j$, $\pi[k] \models_F \phi$))

Simulation can be extended to fair structures. Let M and M' be two structures with fairness constraints. Assume that $AP\supseteq AP'$. The relation $H_F\subseteq S\times S'$ is a fair simulation relation between M and M' if and only for all s and s', if $H_F(s, s')$ then the following conditions hold:

- $L(s)\cap AP'=L'(s')$
- For every fair path $\pi = s_0, s_1, \ldots$, from $s= s_0$ in M, there is a fair path $\pi' = s'_0, s'_1$, …from $s'= s'_0$ in M' such that for all i≥0, $H(s_i, s'_i)$.

We write $M \preceq_F M'$ if there exists a fair simulation relation H such that for every initial s_0 in M there is an initial state s'_0 in M' for which $H(s_0, s'_0)$.

In this section we define simulation relation on the states of fair Kripke structures with respect to given a CTL formula. Essentially, our formula-dependent abstraction incorporate fairness constraints by requiring that states be simulation on all fair paths. We show it suffices to examine a restricted class of paths, namely the rational paths defined below.

Definition 5. Let π be a path through a Kripke structure K. Define π to be a rational path if $\exists n, m$ such that $\forall i.(i>n \Rightarrow \pi[i] = \pi[i \bmod m+n])$ and $(i>0 \Rightarrow \pi[i]\notin \mathrm{UNSAT}^{\phi})$.

Thus rational paths are those which end in cycle such as $\pi = (s_0, s_1, \ldots s_n)(s_{n+1},\ldots, s_{n+k})^{\omega}$. Formula-dependent abstraction can be extended to fair structures. Let M and M' be two structures with fairness constraints. Assume that $AP\supseteq AP'$. The relation

$H_F^\phi \subseteq S \times S'$ is a fair simulation relation with respect to formula ϕ if and only for all s and s', if $H_F^\phi(s, s')$ then the following conditions hold:

- $L(s) \cap AP' = L'(s')$
- For every fair rational path $\pi = (s_0, s_1, \ldots s_n)(s_{n+1}, \ldots, s_{n+k})^\omega$, from $s = s_0$ in M, there is a fair rational path π' such that $\forall i(1 \leq i \leq n \rightarrow H^\phi(s_i, \pi'[i])$ and $\forall i > n$, $\pi'[i]$ is H^ϕ-abstraction to some state in $inf(\pi)$.

We write $M \preceq_F^\phi M'$ if there exists a fair simulation relation H_F^ϕ with respect to formula ϕ such that for every initial s_0 in M there is an initial state s_0' in M' for which $H_F^\phi(s_0, s_0')$. It is easy to show that $\preceq_F^\phi$ determines a preoder on fair structures.

Lemma 4. $\preceq_F^\phi$ is a preorder on the set of fair structures.

Every fair behavior of M is a fair behavior of M. Thus, if the semantics of CTL is given with respects to fair paths then we can prove the following theorem.

Theorem 4. If $M \preceq_F^\phi M'$, the for given a CTL formula ϕ interpreted over fair path, $M' \models_F \phi \Rightarrow M \models_F \phi$

8 Conclusion

Model checking is an automatic formal verification technique for a finite state system, where all the states of the system are exhaustively enumerated and the correctness condition checked at each state, and it also yields extremely useful counter examples if model checking fails. Abstraction has been widely used to make model checking more efficient for large systems. We have presented a formula-dependent abstraction that can be used to manage the size of the kripke structure encountered in compositional CTL model checking. Given a CTL formula ϕ, our method firstly computer an abstraction on the states of each component which preserves ϕ. If an abstraction is too coarser, we refine abstraction model using the counterexample guided abstraction refinement paradigm.

We have defined formula-dependent abstraction on Kripke structures that incorporate fairness. Finally we plan to extend our method to fair-CTL model checking.

Acknowledgement

This work is supported by Natural Science Foundation of China (No.60663005, No.60563005), Guangxi Natural Science Foundation of China (No.0542036, No.0728093, No.0728089).

References

1. Clarke, E.M., Grumberg, O., Peled, D.A.: Model Checking. MIT Press, Cambridge (1999)
2. Clarke, E.M., Burch, J.R., Grumberg, O., Long, D.E., McMillan, K.L.: Automatic Verification of Sequential Circuit Designs. Phil. Trans. of the Royal Society of London 339, 105–120 (1992)

3. Emerson, E.A.: Temporal and Model Logic. Formal Models and Semantics. Handbook of theoretical Computer Science, vol. B, pp. 996–1072 (1990)
4. Clarke, E.M., Emerson, E.A.: Synthesis of Synchronization Skeletons for Branching Time Temporal Logic. In: Logic of Programs 1981. LNCS, vol. 131, pp. 52–71. Springer, Heidelberg (1982)
5. Graf, S., Saidi, H.: Construction of Abstract State Graphs with PVS. In: Proceedings of Computer Aided Verification. LNCS, vol. 1254, pp. 72–83. Springer, Heidelberg (1997)
6. Cousot, P., Cousot, R.: Abstract Interpretation: a Unified Lattice Model for Static Analysis of Programs by Construction or Approximation of Fixpoints. In: Proceedings of POPL, pp. 238–252 (1977)
7. Cousot, P., Cousot, R.: Abstract Interpretation Frameworks. Journal of Logic and Computation 2(4), 511–547 (1992)
8. Clarke, E.M., Emerson, E.A., Sistla, A.P.: Automatic Verfication of Finite-state Concurrent Systems Using Temporal Logic Specifications. ACM Transactions on Programming Languages and Systems 8(2), 244–263 (1986)
9. Clarke, E.M., Grumberg, O., Jha, S., Lu, Y., Veith, H.: Counterexample-guided Abstraction Refinement. In: Proceedings of Computer Aided Verification, pp. 154–169 (2000)
10. Alur, R., Itai, A., Kurshan, R.P., Yannakakis, M.: Timing Verification by Successive Approximation. Information and Computation 118(1), 142–157 (1995)
11. Lakhnech, Y., Bensalem, S., Berezin, S., Owre, S.: Incremental Verification by Abstraction. In: Margaria, T., Yi, W. (eds.) ETAPS 2001 and TACAS 2001. LNCS, vol. 2031, pp. 98–112. Springer, Heidelberg (2001)
12. Clarke, E., Grumberg, O., Long, D.E.: Model Checking and Abstraction. In: Proceedings of TOPLAS, pp. 1512–1542 (1994)
13. Aziz, A., Shiple, T.R., Singhal, V., Sangiovanni-Vincentelli, A.L.: Formula-dependent Equivalence for Compositional CTL Model Checking. In: Dill, D.L. (ed.) CAV 1994. LNCS, vol. 818, pp. 324–337. Springer, Heidelberg (1994)
14. Dams, D., Gerth, R., Grumberg, O.: Generation of Reduced Models for Checking Fragments of CTL. In: Courcoubetis, C. (ed.) CAV 1993. LNCS, vol. 697, pp. 479–490. Springer, Heidelberg (1993)
15. Dams, D., Gerth, R., Grumberg, O.: Abstract Interpretation of Reactive Systems. ACM Transactions on Programming Languages and Systems 19(2), 253–291 (1997)
16. Loiseaux, C., Graf, S., Sifakis, J., Bouajjani, A., Bensalem, S.: Property Preserving Abstractions for the Verification of Concurrent Systems. Formal Methods in System Design 6(1), 11–44 (1995)
17. Dams, D., Gerth, R., Grumberg, O.: Fair model Checking of Abstractions (extended abstract). In: Proceedings of VCL, number DSSE-TR-2000-6, University of Southampton (July 2000)
18. Aziz, A., Singhal, V., Shiple, T.R., SangiovanniVincentelli, A.L.,, F.: balarin, and R.K. Brayton. Equivalences for Fair Kripke Structures. In: Shamir, E., Abiteboul, S. (eds.) ICALP 1994. LNCS, vol. 820, pp. 364–375. Springer, Heidelberg (1994)
19. Bensalem, S., Lakhnech, Y., Owre, S.: Computing Abstractions of Infinite State Systems Compositionally and Automatically. In: Proceedings of Computer Aided Verification. LNCS, vol. 1427, pp. 319–331 (1998)
20. Giacobazzi, R., Ranzato, F.: Incompleteness, Counterexamples and Refinements in Abstract Model Checking. In: Cousot, P. (ed.) SAS 2001. LNCS, vol. 2126, pp. 356–373. Springer, Heidelberg (2001)

21. Ranzato, F., Tapparo, F.: Strong Preservation as Completeness in Abstract Interpretation. In: Schmidt, D. (ed.) ESOP 2004. LNCS, vol. 2986, pp. 18–32. Springer, Heidelberg (2004)
22. Ranzato, F., Tapparo, F.: Strong Preservation of Temporal Fixpoint-based Operators by Abstract Interpretation. In: Emerson, E.A., Namjoshi, K.S. (eds.) VMCAI 2006. LNCS, vol. 3855, pp. 332–347. Springer, Heidelberg (2005)
23. Corbett, J.C., Dwyer, M.B., Hatcliff, J., Laubach, S., Pasareanu, C.S., Robby., Zheng, H.: Bandera: Extracting Finite-state Models from Java Source Code. In: Proceedings of ICSE, pp. 439–448 (2000)
24. Havelund, K., Pressburger, T.: Model Checking JAVA Programs using JAVA PathFinder. International Journal on Software Tools for Technology Transfer 2(4), 366–381 (2000)
25. Ball, T., Majumdar, R., Millstein, T.D., Rajamani, S.K.: Automatic Predicate Abstraction of C Programs. In: SIGPLAN Conference on Programming Language Design and Implementation, pp. 203–213 (2001)
26. Chaki, S., Clarke, E.M., Groce, A., Ouaknine, J., Strichman, O., Yorav, K.: Efficient Verification of Sequential and Concurrent C Programs. Formal Methods in System Design 25(2-3), 129–166 (2004)
27. Henzinger, T.A., Jhala, R., Majumdar, R., Sutre, G.: Lazy Abstraction. In: Proceedings of POPL, pp. 58–70 (2002)

Navigating the New Product Process from Strategic Viewpoint

Wang Kung-Jeng[1], Lee Yun-Huei[2], and Wang Sophia[2]

[1] Department of Industrial Management, National Taiwan University of Science and Technology, Taipei, Taiwan, ROC
[2] Department of Business Administration, National Dong Hwa University, Hua-Lien, Taiwan, ROC

Abstract. Resource scarcity is a major difficulty facing firms that engage in new product development (NPD) projects. In order to understand how resource allocation strategies affect NPD performance and which strategy is the best alternative, we construct a research and development (R&D) process model using system dynamics. Moreover, resource allocation strategies are categorized into two types: design-stage-first strategy and manufacturing-stage-first strategy, and several important indicators of performance evaluation are defined. We then use simulation to test the relationships between resource allocation strategy and various NPD performance measures. The most important finding from simulation results is that a firm should allocate its resources into early development stage first in order to obtain superior R&D performance.

Keywords: New product development, System dynamics, Resource allocation strategy.

1 Introduction

It is widely agreed that the development of new products is of increasing importance to profitability and competitiveness in many industries. Over the past twenty years several studies have examined the determinants of new product success and identified many factors that distinguish successful products from unsuccessful ones.

Spivey et al. (1997) suggests that various resources can strongly influence new product development (NPD), but few articles deeply explore how to allocate resources in the NPD process to improve project effectiveness and efficiency. The predominant view about resource allocation in the NPD process focuses on investigating resources that are input into each development activity in successful or unsuccessful projects (Booz-Allen and Hamilton, 1982; Souder, 1987; Cooper and Kleninschmidt, 1988). However, previous research has not yet been able to conclude that more resource put in specific stages will increase the odds of success due to the difficulty in classifying the causality issue. Therefore, which resource allocation strategy is better for the NPD is still unclear. Most importantly, how the resource strategy contributes to NPD performance has not been fully explored.

O. Gervasi et al. (Eds.): ICCSA 2008, Part II, LNCS 5073, pp. 1049–1060, 2008.

Despite the breadth and depth of the product development literature, it must be noted that NPD is a complex and dynamic process that involve external R&D workload and internal R&D defects/failures. The final development results would be affected by the interactions of various variables. To capture this complexity, we used a system dynamics approach instead of a survey approach to construct and assess a simulation model which characterizes an NPD system with highly complex internal and external dynamics. Longitudinal results obtained in the study provide managers with more meaningful guidelines on critical R&D resource-allocation decision than do conventional cross-sectional results.

The main objective of this research is to examine the relationship between resource allocation strategy and NPD performance. We identify two types of resource allocation strategies: design-stage-first and manufacturing-stage-first strategies based on the priority of different development stages in resource allocation. This study makes several contributions. First, we successfully build an R&D model with cause-effect loops using system dynamics. Second, we justify experimentally and quantitatively performance differences of various resource allocation configurations. An important practical implication of the results is that a firm should allocate its resources into early development stage first, which will result in efficient completion rate and rework rate, lower defect rate, and higher project completion rate.

The rest of the article is organized as follows. In section 2 we review the literature on the NPD process, R&D resource types, resource allocation in the NPD process, and NPD performances. Then, we present our proposed NPD system dynamics model in section 3. Followed are findings about the effect of resource allocation strategy on NPD performance. We discuss the limitations and contributions of this study along with its implications for future research and managerial practices in section 5.

2 Literature Review

2.1 New Product Development Process

A number of researchers have proposed an NPD process mainly from technical viewpoint, or marketing viewpoint, or combination of them (Cooper, 1980; Cooper and Kleinschmidt, 1986; Cooper, 1994; Kuczmarski, 1992; Crawford, 1994; Ulrich and Eppinger, 1995; McGrath, 1996; Rochford and Rudelius, 1997; Etzel, et al., 1997; Griffin, 1997; Song and Montoya-Weiss, 1998; Carbonara and Schiuma, 2004; Tzokas et al., 2004; Prasnikar and Skerlj, 2006). For example, Ulrich and Eppinger (1995) presented a generic development process with five phases. These phases are concept development, system-level design, detail design, testing and refinement, and production ramp-up, some primarily stressed on marketing viewpoint. Tzokas et al. (2004) contended that the quest for success in NPD requires management to navigate complex processes. In their study, the development stages of the NPD process include the generation of new product ideas, the development of an initial product concept, an assessment of its business attractiveness, the actual development of the product, testing it within the market, and the actual launch of the product in the marketplace. Rochford and Rudelius (1997) suggested an NPD process that consists of idea generation, initial screening, preliminary market analysis, preliminary technical analysis, preliminary production analysis, preliminary financial analysis, market study, product development, in-house product testing, customer product testing, market testing and pre-commercial financial analysis. Generally speaking, companies most successful at new products have had the same type of development process in place, following two main phases: design and manufacturing. The steps from idea generation to commercialization remain the same.

2.2 R&D Resources

R&D resources are critical to NPD performance when a firm engages in R&D activities (Lai et al., 2004). Ayal and Rothberg (1986) suggested that R&D resource allocation deals not only with the expenditure of funds but also with the allocation of personnel, other support services, equipment, and facilities. And Cooper and Kleinschmidt (1988) mentioned that resource expenditures are measured both in terms of dollars and man-days, so that different types of resource—people and dollars—can be considered. Spivey et al. (1997) indicated that various resources can strongly affect NPD; the resource factors include concerns about information, infrastructure, time, and money are required to improve the NPD process. Di et al. (1999) showed that R&D resources comprise of R&D workmen, expenditure, equipment, information and reward. The allocation of workmen in a multi-project environment poses a problem for many organizations such as software developers, R&D organizations, construction and engineering companies. Given the notion that human knowledge is the most important and most scare resource in R&D projects, efficient human resource allocation in NPD projects has significant implications for NPD performance (Hendriks, et al., 1999).

2.3 Resource Allocation and New Product Development

Mansfield and Rapoport (1975) estimated how innovation costs might be distributed across major stages of new product development. They concluded that about 39% of the total costs of a new product project went to the R&D phase. Similarly, Booz-Allen and Hamilton (1982) provided a breakdown of total company expenditures by major stages in the innovation process, and the proportion spent on successes versus failures. They provided some qualitative evidence that successful firms spent proportionately more on the early stages of the new product process.

Considering two types of innovative approaches of NPD -- needs-generated or technology-pull projects and means-generated or technology-push projects, Souder (1987) investigated resource allocation of NPD across industries and found that in glass, transportation, plastic, machinery and other "fore-load" industries, technical costs in early stages (including exploration, concept development, prototype development, prototype and testing) constituted the largest percentage of the R&D costs. In contrast, in "aft-load" industries such as metals, electronics, chemical, and food industries, most of the money was spend in late phases (including market development, manufacturing, commercialization and service).

Cooper and Kleninschmidt (1988) compared successful and unsuccessful NPD projects in terms of resource allocation in various activities. The variables were measured in divided a NPD process into a number of stages, and probed resource allocation at various activities in terms of dollar and man-days. The research object is industrial firms, the units of analysis are successful and unsuccessful projects, and the variables are measured by absolute values. They found that:

(1) Of the thirteen new product activities studied (initial screening, preliminary market assessment, preliminary technical assessment, detailed market study/market research, business/financial analysis, product development, in-house product testing, customer tests of product, test market/trial sell, trial production, pre-commercialization business analysis, production start-up and market launch), three (product, production start-up, and market launch) accounted for more than three quarters of dollar expenditures. Seventy-eight percent of man-days were expended on technical/production activities. Successful projects spent more than unsuccessful ones in terms of dollar amount on preliminary market / technical assessment, in-house product tests, customer tests of product and production start-up, and more in man-days on preliminary market / technical assessment, pre-commercialization business analysis and market launch than unsuccessful ones.

(2) Of the three product activities studied (pre-development activities, product development and product testing and commercialization), 53.9% of dollars was spent on commercialization. Product development and product testing accounted for 60% of man-days. Successful projects were clearly characterized by more spending on the three major stages, whether measured by dollars or man-days.

These authors concluded that more resources devoted to the first few steps of new product development process were strongly connected to success. However, the R&D workload was assumed to be constant and its dynamic property along time was ignored.

2.4 New Product Development Performance

Previous researchers have employed various NPD performance measures based on their different research purposes. NPD performance has often been measured along financial and non-financial dimensions.

Examples of objective financial performance measures include return on investment, sale and sales growth, internal rate of return, the proportion of sales by products introduced with the last 3/5 years, return on sales, net present value, profit, ROA, ROE, and profitability (Cooper, 1984; Page, 1993; Griffin and Page, 1993; Barczak, 1995; Calantone et al., 1995; Loch et al., 1996; Cooper and Kleinschmidt, 1996; Swink, 2000, Millson and Wileman, 2002). In addition, some studies focus on performance relative to that of competitors or industry peers and firm's historical performance. Examples include unit cost reduction compared to industry, profitability versus competitors and overall success, the degree to which the company's costs stay within budget, and so on (Loch et al., 1996; Cooper and Kleinschmidt, 1996, Song et al., 1998a; Driva et al., 2000; Gruner and Homburg, 2000).

Non-financial performance measures are often subjective variables that can be classified into four categories. The "customer" category includes customer satisfaction as a performance measure (Sbragia, 1984; Page, 1993; Griffin and Page, 1993; Barczak, 1995). The "market" category, by comparison, consists of market share (Cusumano and Nobeoka, 1991; Page, 1993; Barczak, 1995; Calantone et al., 1995) and success rate (Cooper, 1984; Song et al., 1998a). The "product efficiency" category focuses on time-based indicators, such as time to market, speed to market, and time for project completion (Sbragia, 1984; Griffin and Page, 1993; Song et al., 1998a; Driva et al., 2000; Swink, 2000). Finally, the "product effectiveness" category mainly refers to product quality (Griffin and Page, 1993; Olson, et al., 1995; Song et al., 1998a; Kusunoki et al., 1998; Swink, 2000; Gruner and Homburg, 2000). Still, other researchers use overall project completion as a performance measure (Cooper, 1994; Griffin and Page, 1993; Song et al., 1998a).

2.5 Summary

NPD success is crucial for organizational survival in rapidly changing environments characterized by short product life cycle, dramatically technological evolution, and demand fluctuation. Most previous studies on resource allocation in NPD have adopted a static point of view and failed to capture the effect of resource allocation on R&D performance in dynamic environments. To fill up this gap, this study explores the effect of external complexity on resource allocation strategies by emphasizing extra workloads associated with NPD project change.

This study uses the simulation approach to investigate the long-term performance consequences of R&D processes with lumpy workloads. Simulation is preferred because data related to R&D processes are very difficult to get from industry. Also, previous studies have mostly focused on the short-term relation between resource allocation and NPD performance. This study is able to develop a longitudinal framework about resource allocation and performance.

3 Research Methodology

3.1 NPD Process

For the purpose of simplicity, we employ a model structure developed by Repenning (2000) and divide the NPD process into two stages: design stage (including concept development, product development, and testing) and manufacturing stage (comprising trial production, full production, and launch).

A simplified and physical multi-project flow of NPD is presented in Fig. 1. There are four stocks both in the design stage and in the manufacturing stage: tasks not completed, tasks in testing, tasks to be reworked and tasks completed. In the figure, new tasks are introduced into the design stage at the top left and exit from the manufacturing stage at the bottom right. Hollow symbols represent the continuous movement of work between the stocks during the given phase, and at the annual transition of design work to manufacturing work, the all stocks in the design stage are transferred to the corresponding stocks in the manufacturing stage in one time step.

The flows of stocks are the same in each stage. After introducing new tasks into the system, completed tasks accumulate in the stock of tasks testing. Once tasks pass the testing, flowing to the stock of tasks completed. On the other hand, tasks failing testing flow to the stock of tasks to be reworked, and once reworked, tasks flow back to the stock of tasks awaiting testing. Each time a task is completed or reworked there is some probability it is done incorrect and is defective. Those related defects are a portion of the stock of tasks in testing. Similarly, defects failing testing flow to the stock of rework, and after reworking, some tasks pass to the stock of tasks completed while some fail to continuous rework.

We assumed an NPD cycle is 24 weeks: the first 12 week interval is the design stage and the second 12 week interval is the manufacturing stage. Due to the urgency of the firm's reputation and continued viability, every 12 weeks a new product is introduced to the market, and a new project is introduced into the system at 12-week intervals.

3.2 Resource Allocation Strategy

Most previous studies are case studies that show the absolute/relative investigation of resource allocation into each development activity. In this paper, we applied the concept of "fore-load" and "aft-load" to illustrate the allocation of resources. The concept of "fore-load" means that most resource was spend in early stages, including exploration, concept development, prototype development, prototype and testing ; the concept of "aft-load" means that most resource was spend in late phases, including market development, manufacturing, commercialization and service (Souder, 1987). Our resource allocation strategy was divided into two categories: one is design-stage-first strategy (A), in which engineers give priority to the project in early phases; the other is manufacturing-stage-fist strategy (B), in which engineers give priority to the project in late phases. Since a firm usually has only limited resources available, if one of two competing projects gets resources, then the other would lose the exact amount.

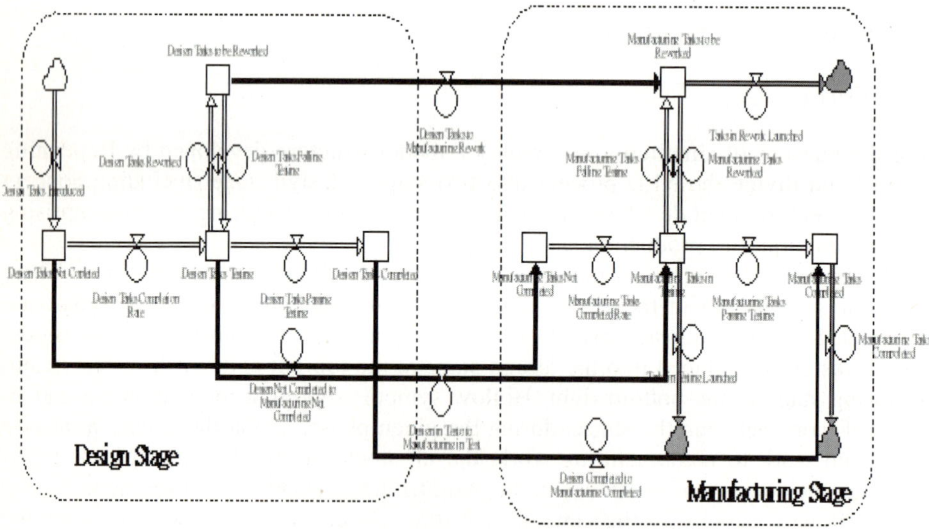

Fig. 1. The physical flow of NPD process

3.3 New Product Development Performance

New product performance is a multidimensional construct. The literatures have suggested several product level measures of NPD performance. We included product efficiency and effectiveness commonly used as dependent dimensions, and developed three indicators based in part on Song et al. (1998) and Swing's (2000) works. Timeless-accurate project completion in each stage is described by the time of completed tasks and remaining-uncompleted tasks. Defect rate in two stages is measured as total defective tasks of flows divided by total tasks of flows. Project completion rate is calculated as completed tasks divided by total tasks.

3.4 System Dynamics

System dynamics, a powerful tool for modeling the behavior of complex systems and understanding change, has been applied to a large number of problems in the social sciences. Because developing really new products is a high risk, difficult venture for firms to undertake (Kleinschmidt and Cooper, 1991; Olson et al., 1995). And NPD involves complexity and uncertainty. Therefore, we use system dynamics to provide a simple method for modeling management decision-making processes embedded in complex programs. Then computer software (powersim) was used to simulate a system dynamics model of the situation being studied. Running "what if" simulations to test certain strategies on such a model can greatly aid in understanding how the system changes over time.

4 Experiments and Results

4.1 Data Analysis

Following figures present the basic model's behaviors under different resource allocation strategies. In all runs the task introduction rate is 1500 per 24 week, constituting a new development project.

4.1.1 The Relationship between Resource Allocation Strategies and Timeless - Accurate Project Completion

Fig. 2 shows that tasks in the design stage with strategy A are completed at a constant rate throughout 12 weeks faster than those with strategy B. In strategy A, there are no new tasks arrived at the manufacturing stage because all tasks are already completed in the design stage, while some numbers (about 50) are sent to the manufacturing stage uncompleted when the firm adopts strategy B.

Moreover, we find that although the level of rework grows during the both stages as tasks done incorrectly are discovered through testing, in strategy A the rate of rework completion is completed in manufacturing stage incrementally over time; in strategy B although the number of rework has been increasing week by week (Fig. 3), the rate of rework completion rises as the product launch date approaches.

4.1.2 The Relationship between Resource Allocation Strategies and Defect Rate

Fig. 4 presents the defect rate in the design and manufacturing stages. In strategy A, the defect fraction in design task begins at approximately 35%, declining gradually over time. The defect fraction (less than 4%) in the manufacturing stage is lower than that in the design stage. On the other hand, in strategy B, the defect fraction reduces

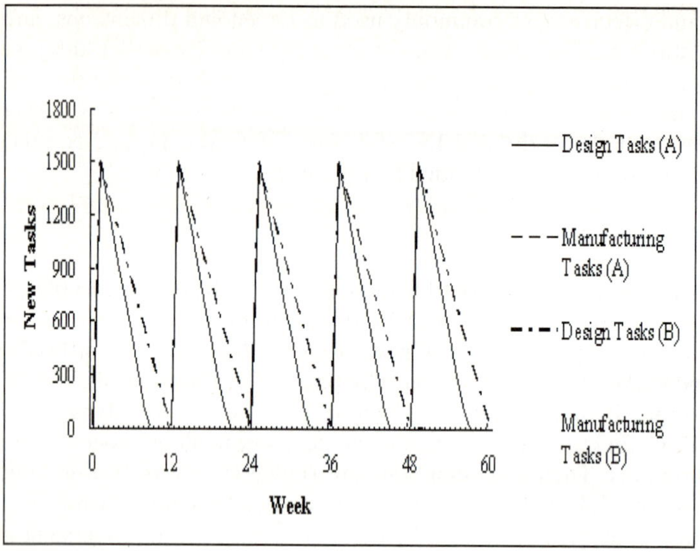

Fig. 2. Timeless-accurate project completion in two stages under different resource allocation strategies

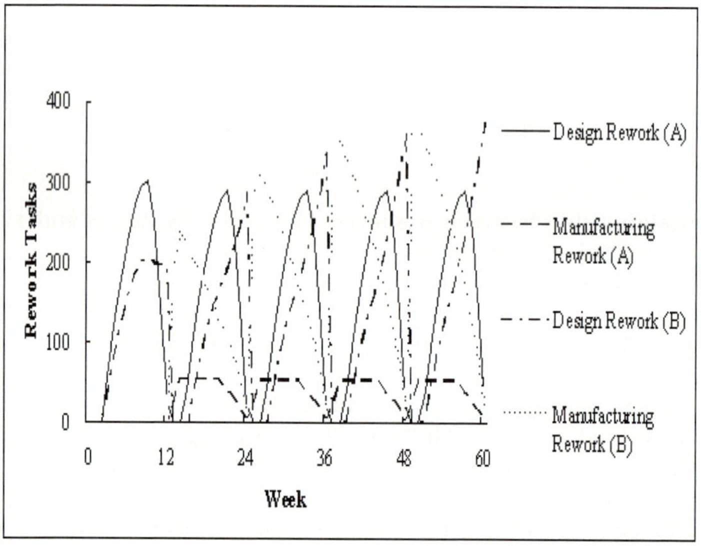

Fig. 3. Timeless-accurate project completion in two stages under different resource allocation strategies

for the first several weeks of 12-weeks, rising then as a result of the high rate of rework completion in the design stage. The defect fraction in the manufacturing stage decreases more dramatic due to multiple iterations through the rework cycle.

4.1.3 The Relationship between Resource Allocation Strategies and Project Completion Rate

A longitudinal simulation result regarding the performance of different resource allocation strategies is presented as following. The Fig. 5 shows as strategy A is applied, it always reaches the high fraction of completed tasks appropriately at 98.7%.

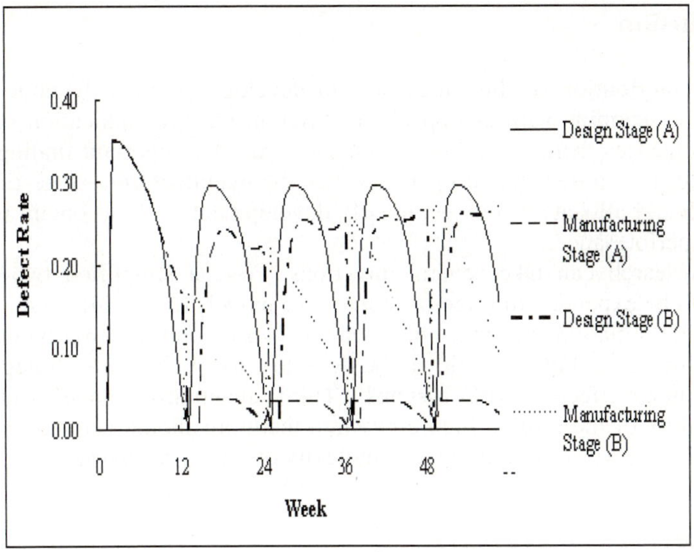

Fig. 4. Defect rate in two stages under different resource allocation strategies

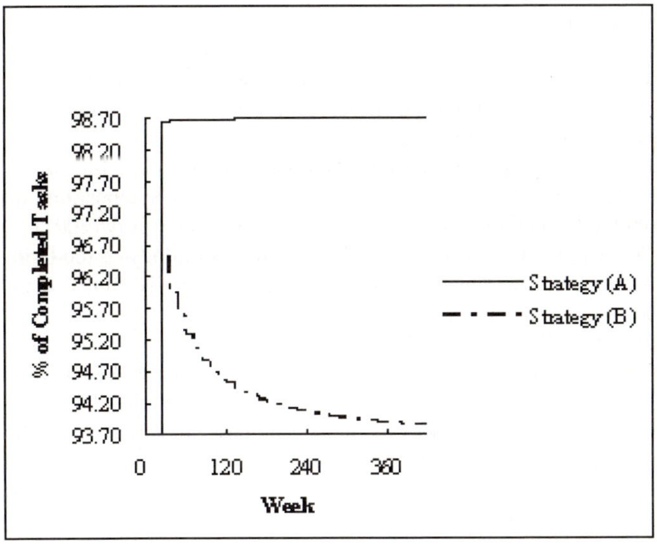

Fig. 5. Project Completion Rate under Different Resource Allocation Strategies

However, the result also indicates that when the resources are given priority to the project in the manufacturing stage, the performance at a high state in a short time, and then the project completion rate is becoming worse over time appropriately from 96.5% to 93.8%. Comparing these two resource allocation strategies, it is concluded that strategy A outperforms strategy B in the NPD process.

5 Conclusion

The main contribution of this study was to develop a system dynamic simulation model and to examine various impacts of different resource allocation strategies on NPD performances from a longitudinal perspective. An important finding is that the design-stage-first strategy performs better than the manufacturing-stage-first strategy. The firm should allocate resources in early development cycle to obtain superior development performance.

Future research can take several directions. First, a simplified two-stage NPD model could be expanded to three-or-more-stage models to investigate more sophisticated real-life situation. Second, our research focuses only on two typical resource allocation strategies. Future work can be done to survey elaborate strategies to form potentially more effective NPD systems. Third, quantitative model could limit the generalizability of the results. Future research therefore should use multiple methods including field surveys to capture the complexity of NPD processes.

References

1. Smith, T.F., Waterman, M.S.: Identification of Common Molecular Subsequences. J. Mol. Biol. 147, 195–197 (1981)
2. May, P., Ehrlich, H.C., Steinke, T.: ZIB Structure Prediction Pipeline: Composing a Complex Biological Workflow through Web Services. In: Nagel, W.E., Walter, W.V., Lehner, W. (eds.) Euro-Par 2006. LNCS, vol. 4128, pp. 1148–1158. Springer, Heidelberg (2006)
3. Foster, I., Kesselman, C.: The Grid: Blueprint for a New Computing Infrastructure. Morgan Kaufmann, San Francisco (1999)
4. Czajkowski, K., Fitzgerald, S., Foster, I., Kesselman, C.: Grid Information Services for Distributed Resource Sharing. In: 10th IEEE International Symposium on High Performance Distributed Computing, pp. 181–184. IEEE Press, New York (2001)
5. Foster, I., Kesselman, C., Nick, J., Tuecke, S.: The Physiology of the Grid: an Open Grid Services Architecture for Distributed Systems Integration. Technical report, Global Grid Forum (2002)
6. National Center for Biotechnology Information, `http://www.ncbi.nlm.nih.gov`
7. Ayal, I., Rothberg, R.: Strategic Control of R&D Resource Allocations in Diversified businesses. Journal of Product Innovation Management 3(4), 238–250 (1986)
8. Barczak, G.: New Product Strategy, Structure, process, and Performance in the Telecommunications Industry. Journal of Product Innovation Management 12(2), 224–234 (1995)
9. Booz-Allen, Hamilton: New Product Management for the 1980s, Booz-Allen & Haniliton, New York (1982)

10. Carbonara, N., Schiuma, G.: The New Product Development Process within Industrial Districts: A Cognitive Approach. International. Journal of Product Development 1(1), 92–107 (2004)
11. Calantone, R., Vickery, S., Droge, C.: Business Performance and Strategic New Product Development Activities: An Empirical Investigation. Journal of Product Innovation management 12(3), 214–223 (1995)
12. Crawford, C.M.: New Product Management, 4th edn. Irwin, Inc., Burr Ridge (1994)
13. Cooper, R.G.: Project Newprod: What Makes a New Product a Winner? Montreal: Centre Quebecois d' Innovation Industrielle (1980)
14. Cooper, R.G.: The Strategy-Performance Link in Product Innovation. R&D Management 14(4), 247–259 (1984)
15. Cooper, R.G., Kleinschmidt, E.J.: An Investigation into the New Product Process: Steps, Deficiencies, and Impact. Journal of Product Innovation Management 3, 71–85 (1986)
16. Cooper, R.G., Kleinschmidt, E.J.: Success Factors in Product Innovation. Industrial marketing management 16(3), 215–223 (1987)
17. Cooper, R.G., Kleinschmidt, E.J.: Resource Allocation in the New Product Process. Industrial Marketing Management 17(3), 249–262 (1988)
18. Cooper, R.G., Kleinschmidt, E.J.: New Products: The Key Factors in Success, American Marketing Association, Chicago (1990)
19. Cooper, R.G.: Perspective: Third-Generation New Product Processes. Journal of Product Innovation Management 11, 3–14 (1994)
20. Cooper, R.G., Kleinschmidt, E.J.: Winning Business in Product Development: The Critical Success Factors. Research Technology Management 39(4), 18–29 (1996)
21. Cusumano, M.A., Nobeoka, K.: Strategy, Structure and Performance in Product Development: Observations from the Auto Industry. Research Policy 21(3), 265–293 (1992)
22. Driva, H., Pawar, K.S., Menon, U.: Measuring Product Development Performance in Manufacturing Organizations. International Journal of Product Economics 63, 147–159 (2000)
23. Di Shi-Yung.: Discipline of R&D Management and Cases of Science Management, Lan De, Taipei (1999)
24. Etzel, M.J., Walker, B.J., Stanton, W.J.: Marketing, 11th edn. McGraw-Hill, New York (1997)
25. Griffin, A.: PDMA Research on New Product Development Practices: Updating Trends and Benchmarking Best Practices. Journal of Product Innovation Management 14(6), 429–458 (1997)
26. Griffin, A., Page, A.L.: An Interim Report on Measuring Product Development Success and Failure. Journal of Product Innovation Management 10(4), 291–308 (1993)
27. Gruner, K.E., Homburg, C.: Does Customer Interaction Enhance New Product Success? Journal of Business Research 49, 1–14 (2000)
28. Hendriks, M.H.A., Voeten, B., Kroep, L.: Human Resource Allocation in a Multi-Project R&D Environment. International Journal of Project Management 17(3), 181–188 (1998)
29. Kuczmarski, T.D.: Managing New Products: The Power of Innovation, 2nd edn. Prentice-Hall, Inc., New Jersey (1992)
30. Kusunoki, K., Nonaka, I., Nagata, A.: Organizational Capabilities in Product Development of Japanese Firms: A Conceptual Framework and Empirical Findings. Organization Science 9(6), 699–718 (1998)
31. Lai, S.B., Chen, S.P., Hsieh, L.F.: Introduction to Science Management. National Open University, Taipei (2004)

32. Loch, C., Stein, L., Terwiesch, C.: Measuring Development Performance in the Electronics Industry. Product Innovation Management 13, 3–20 (1996)
33. Mansfield, E., Rapoport, J.: The Costs of Industrial Product Innovations. Management Science 21(12), 1380–1386 (1975)
34. McGrath, M.E.: Setting the PACE in Product Development, revised edition. Butterworth-Heinemann, MA (1996)
35. Millson, M.R., Wilemon, D.: The Impact of Organizational Integration and Product Development Proficiency on Market Success. Industrial Marketing Management 31, 1–23 (2002)
36. Olson, E.M., Walker Jr., O.C., Ruekert, R.W.: Organizing for Effective New Product Development: The Moderating Role of Product Innovativeness. Journal of Marketing 59(1), 48–62 (1995)
37. Page, A.L.: Assessing New Product Development Practices and Performance: Establishing Crucial Norms. Journal of Innovation Management 10, 290–293 (1993)
38. Prašnikar, J., Škerlj, T.: New Product Development Process and Time-to-Market in the Generic Pharmaceutical Industry. Industrial Marketing Management 35(6), 690–702 (2006)
39. Repenning, N.P.: A Dynamic Model of Resource Allocation in Multi-Project Research and Development Systems. System Dynamics Review 16(3), 173–212 (2000)
40. Rochford, L., Rudelius, W.: New Product Development Process. Industrial Marketing Management 26, 67–84 (1997)
41. Sbragia, R.: Clarity of Management Roles and Performance of R&D Multidisciplinary Projects in Matrix Structures. R&D Management 14(2), 113–126 (1984)
42. Song, X.M., Perry, M.E.: The Determinants of Japanese New Product Successes. Journal of Marketing Research 34(1), 64–76 (1997)
43. Song, X.M., Thieme, R.J., Xie, J.: The Impact of Cross-Functional Joint Involvement across Product Development Stages: An Exploratory Study. Journal of Product Innovation Management 15, 289–303 (1998a)
44. Song, X.M., Montoya-Weiss, M.M.: Critical Development Activities for Really New versus Incremental Products. Journal of Product Innovation management 15(2), 124–135 (1998b)
45. Souder, W.E.: Managing New Product Innovation, Toronto, Massachusetts, pp. 1–6. D.C. Heath and Company/Lexington (1987)
46. Souder, W.E., Sherman, J.D., Cooper, R.D.: Environmental Uncertainty, Organizational Integration, and New Product Development Effectiveness: A Test of Contingency Theory. Journal of Product Innovation Management 15(6), 520–533 (1998)
47. Spivey, W.A., Munson, J.M., Wolcott, J.H.: Improving the New Product Development Process: A Fractal Paradigm for High Technology Products. Journal of Product Innovation Management 14(3), 203–218 (1997)
48. Swink, M.: Technological Innovativeness as a Moderator of New Product Design Integration and Top Management Support. Journal of Product Innovation Management 17(3), 208–220 (2000)
49. Tzokas, N., Hultink, E.J., Hart, S.: Navigating the New Product Development Process. Industrial Marketing Management 33(7), 619–626 (2004)
50. Ulrich, K.T., Eppinger, S.D.: Product Design and Development. McGraw-Hill, Inc., New York (1995)

Using EDP-Polynomials for Network Structure Optimization

Olga Gadyatskaya[1], Alexey Rodionov[2], and Olga Rodionova[3]

[1] Novosibirsk State University
Pirogova st., 2, Novosibirsk, 630090, Russia
twice3@gmail.com

[2] Institute of Computational Mathematics and Mathematical Geophysics
SB RAS, Lavrentieva, 6, Novosibirsk, 630090, Russia
alrod@rav.sscc.ru

[3] Siberian State University of Telecommunications and Informatics
Kirova, 86, Novosibirsk, 630102, Russia
rolkon@mail.ru

Abstract. The analysis of network reliability by the EDP (expectation of a number of disconnected pairs of nodes in the network) criterion is discussed. The difference between results of network optimization by the EDP-criterion and the well-known criterion of the probability of connectivity is shown. Both theoretical statements and results of numerical experiments for network topology optimisation are given.

Keywords: Network reliability, network structure optimization, disconnected nodes, random graphs, reliability polynomial.

1 Introduction

Reliability analysis is an important part of network design and modernization. When random graphs (RG) are used as a network model, the most well-known criterion for network reliability evaluating is probability of connectivity (PC). The PC criterion evaluates the probability for every node to be connected with all the others [1, 4, 3, 2]. The EDP-criterion, which is being discussed in this work, evaluates how many pairs of nodes in RG are disconnected. The more pairs of nodes are disconnected in RG, the less is its reliability by the EDP-criterion. This criterion is mentioned as an important one, for example, in [5]. Earlier only approximate methods (mostly Monte-Carlo method) were used to compute EDP. Exact calculation of EDP is discussed in [6, 7, 8]. Current work develops the ideas, mentioned in these papers. The main part of the paper is devoted to choosing optimal topologies for some special kinds of RG by using EDP-polynomials. Obtaining of such polynomials is discussed also.

The rest of the paper is organized as follows: in Section 2 we list the main definitions and problem specification. In Section 3 we provide basic facts about obtaining EDP-polynomials. In Section 4 we give an important example of the difference between results of graph topology optimization by the PC-criterion and the EDP-criterion. In Section 5 we present some theoretical results about

O. Gervasi et al. (Eds.): ICCSA 2008, Part II, LNCS 5073, pp. 1061–1076, 2008.

choosing the optimal network topologies using EDP-polynomials. In Section 6 we discuss some experimental results concerning choosing the optimal network topologies using exact calculation of EDP-polynomials.

2 Main Definitions and Preliminaries

We consider random graphs with perfectly reliable nodes and unreliable edges, that fail independently. The reliability of each node (operating probability) is assumed to be equal to p. The vertices in the graph have prescribed weights, which are set at start depending on, for example, the number of the terminals connected to these vertices initially or the importances of these vertices for network's functioning. Here we assume that weight is an additive function. When we compute EDP-functional, if two vertices are contracted, their weights are summed up. Consequently, network topology depends not only on the disposition of the edges between the vertices, but on the disposition of the vertices with different weights too.

$G(n,m)$ – non-oriented graph with n vertices and m edges.
w_i – weight of a node v_i, $WT = (w_1, \ldots, w_n)$;
$N(G)$ – EDP-functional of graph G, EDP.
$R(G)$ – PC-functional of graph G, PC.
l_{ij} – a distance between vertices v_i and v_j, a number of the edges in the shortest path between the vertices.

If we consider some special edges, then we usually assign them personal numbers: we use notation e_k (k-th edge) instead of e_{ij} (edge that connects v_i and v_j). As k_{ij} (k_i) we denote multiplicity of the correspondent edge. Degree of an arbitrary polynomial $H(p)$ we denote as $deg(H(p))$ or $deg(H)$.

EDP-functional can be obtained in 2 basic ways:

A. By definition. From [6]:

$$N(G) = \sum_{i=1}^{n-1} \sum_{j=i+1}^{n} f_{ij} w_i w_j, \tag{1}$$

where f_{ij} is the probability that vertices v_i and v_j are disconnected.

B. By Factoring Method. Factoring or branching method is commonly used for obtaining PC (see [4], [9], for instance). We use it for obtaining EDP. In essence this method consists in applying of the formula of total probability, using hypothesis of existence of an arbitrary edge and hypothesis of it's absence as the alternatives:

$$N(G) = pN(G^*(e_{ij})) + (1-p)N(G\backslash\{e_{ij}\}). \tag{2}$$

Here $G^*(e_{ij})$ – graph, where the vertices v_i and v_j are contracted by the edge e_{ij} and the weight of joint vertex v^*_{ij} equals to $w_i + w_j$, and $G\backslash\{e_{ij}\}$ – graph, obtained from G by deleting the edge e_{ij}. Hence, if the reliabilities of all edges in G are equal to p, EDP-functional $N(G)$ is the polynomial function of p. Further we refer to it as EDP-polynomial. Recursive call of (2) goes on until obtaining graph that allows direct calculation of coefficients of its EDP-polynomial (CEDPP). Larger dimension of graph for which we can obtain CEDPP directly contributes to higher performance of the method.

We denote EDP polynomial of graph G as $N(G,p)$. Reliability polynomials can be presented in different forms (see [10, 11], for example) with different meaning of it's coefficients. In current paper the following presentations for EDP-polynomials are used:

$$N(G,p) = \sum_{i=0}^{m} a_i(1-p)^i p^{m-i}, \tag{3}$$

$$N(G,p) = \sum_{i=0}^{m} b_i p^i. \tag{4}$$

In (3) a_i means the total number of disconnected pairs of nodes by all the variants of removing exactly i edges from the original graph, while the classic presentation of polynomials (4) is more convenient for some derivations though the coefficients b_i can't be interpreted. For short we do not discuss mutual transformations between these representations, considering them as simple and well-known task (refer to [11]).

If we have some arbitrary polynomial of the type (3) and degree r, we use capital letters for its name and small letters for coefficients:

$$H(p) = \sum_{i=0}^{r} h_i(1-p)^i p^{r-i}, \tag{5}$$

The vector of polynomial's coefficients we denote by correspondent name in bold letters: $\boldsymbol{H}$; a_i and $\boldsymbol{A}$ are used for EDP-polynomial exclusively.

Subvector of first k elements of vector $\boldsymbol{H}$ is denoted as $\boldsymbol{H}^{(k)}$, while correspondent subvector of its last elements is denoted as $\boldsymbol{H}^{[k]}$. Vector $\boldsymbol{H}$ shifted by k positions to the right (left) is denoted as $\boldsymbol{H}_{k+}$ ($\boldsymbol{H}_{k-}$). $deg(H)$ is the degree of $H(p)$.

As $\boldsymbol{S}_n(m)$, $m < n$, we denote the vector of length $n+1$ in which first $n-m$ elements are zeros while the last $m+1$ are binomial coefficients C_m^i, $i = 0, \ldots, m$. Let $\boldsymbol{X_n(i)}$ be the vector of length $n+1$, in which all elements but i-th ($0 \le i \le n$) are zeros and i-th element is 1. We also define useful for the following derivations function:

$$I(n) \equiv 1^n = [p + (1-p)]^n = \sum_{i=0}^{n} C_n^i p^i (1-p)^{n-i}. \tag{6}$$

As $r(k,p)$ we denote reliability polynomial (RP) of a multi-edge with multiplicity k while $q(k,p) = 1 - r(p)$ is its unreliability polynomial (URP). Obviousely

$$q(k,p) = (1-p)^k, \tag{7}$$

$$r(k,p) = 1 - (1-p)^k = I(k) - (1-p)^k = \sum_{i=0}^{k-1} C_k^i p^{k-i}(1-p)^i. \tag{8}$$

Note that $q(k,p)q(l,p) = q(k+l)$.

3 Obtaining EDP-Polynomial

3.1 Graphs That Allow Direct Calculation of CEDPP

As it was stated above, the factoring process stops when graph achieved allows direct obtaining of EDP-polynomial. Note that the factoring method naturally leads to obtaining EDP-polynomial in representation (3), thus we try using this representation in the current subsection. For short we denote $w_i w_j$ as W_{ij}.

Case of $n = 2$. It is primitive. If e_{12} is the multi-edge with multiplicity k, then

$$N(G(2,k),p) = w_1 w_2 q(k,p) = w_1 w_2 (1-p)^k = W_{12}(1-p)^k. \tag{9}$$

Case of $n = 3$. Without loss of generality, we can assume that $k_1 \le k_2 \le k_3$. Let us consider all possible disconnected sugraphs of $G(3,m)$. Using (6),(7) and (8), we have that

$$\begin{aligned} N(G(3,m),p) &= q(k_1,p)q(k_2,p)q(k_3,p)(W_{12}+W_{13}+W_{23}) + \\ &\quad r(k_1,p)q(k_2,p)q(k_3,p)(W_{13}+W_{23}) + q(k_1,p)r(k_2,p)q(k_3,p)(W_{12} + \\ &\quad W_{13}) + q(k_1,p)q(k_2,p)r(k_3,p)(W_{12}+W_{23}) \\ &= (W_{13}+W_{23})(1-p)^{m-k_1}I(k_1) + (W_{12}+W_{13})(1-p)^{m-k_2}I(k_2) + \\ &\quad (W_{12}+W_{23})(1-p)^{m-k_3}I(k_3) - (W_{12}+W_{13}+W_{23})(1-p)^m \qquad (10) \\ &= (W_{13}+W_{23})\sum_{i=0}^{k_1} C_{k_1}^i p^i (1-p)^{m-i} + (W_{12}+W_{13})\sum_{i=0}^{k_2} C_{k_2}^i p^i (1-p)^{m-i} + \\ &\quad (W_{12}+W_{23})\sum_{i=0}^{k_3} C_{k_3}^i p^i (1-p)^{m-i} - (W_{12}+W_{13}+W_{23})(1-p)^m. \end{aligned}$$

From (10) we obtain the following equations for the CEDPP:

$$a_i = \begin{cases} 0, & \text{if } 0 \le i < m-k_3; \\ C_{k_3}^{m-i}(W_{13}+W_{23}), & \text{if } m-k_3 \le i < m-k_2; \\ C_{k_3}^{m-i}(W_{13}+W_{23}) + C_{k_2}^{m-i}(W_{12}+W_{13}), & \text{if } m-k_2 \le i < m-k_1; \\ C_{k_3}^{m-i}(W_{13}+W_{23}) + C_{k_2}^{m-i}(W_{12}+W_{13}) + & \\ C_{k_1}^{m-i}(W_{12}+W_{23}), & \text{if } m-k_1 \le i < m; \\ W_{12}+W_{13}+W_{23}, & \text{if } i = m. \end{cases} \tag{11}$$

Thus for the 3-node graph we obtain from (11) that

$$\begin{aligned} \boldsymbol{A} &= (W_{13}+W_{23})\boldsymbol{S_m(k_1)} + (W_{12}+W_{13})\boldsymbol{S_m(k_2)} + \qquad (12) \\ &\quad (W_{12}+W_{23})\boldsymbol{S_m(k_3)} - (W_{12}+W_{13}+W_{23})\boldsymbol{X_m(m)}. \end{aligned}$$

In the case when weights of nodes are polynomials of p we have more complex expression (W_{ij} is the polynomial of p). Thus we have that

$$\begin{aligned} \boldsymbol{A} = & \sum_{i=0}^{deg(W_{13}(p)+W_{23}(p))} (\boldsymbol{W_{13}} + \boldsymbol{W_{23}})_i \boldsymbol{S_m}(\boldsymbol{k_1})_{+i} + \\ & \sum_{i=0}^{deg(W_{12}(p)+W_{13}(p))} (\boldsymbol{W_{12}} + \boldsymbol{W_{13}})_i \boldsymbol{S_m}(\boldsymbol{k_2})_{+i} + \\ & \sum_{i=0}^{deg(W_{12}(p)+W_{23}(p))} (\boldsymbol{W_{12}} + \boldsymbol{W_{23}})_i \boldsymbol{S_m}(\boldsymbol{k_3})_{+i} - (\boldsymbol{W_{12}} + \boldsymbol{W_{13}} + \boldsymbol{W_{23}})_{+m}. \end{aligned} \tag{13}$$

Case of $n = 4$. Let multiplicities of e_{12}, e_{23}, e_{34}, e_{14}, e_{13}, and e_{24} be k_1, k_2, k_3, k_4, k_5, and k_6, correspondingly. Similar to the previous case we examine all possible variants of dividing the graph into mutually disconnected components. They correspond to deletion of 3 (incidental to one node) or more edges, all 26 variants are presented in the Fig.1. Using (1) we easily obtain polynomials for

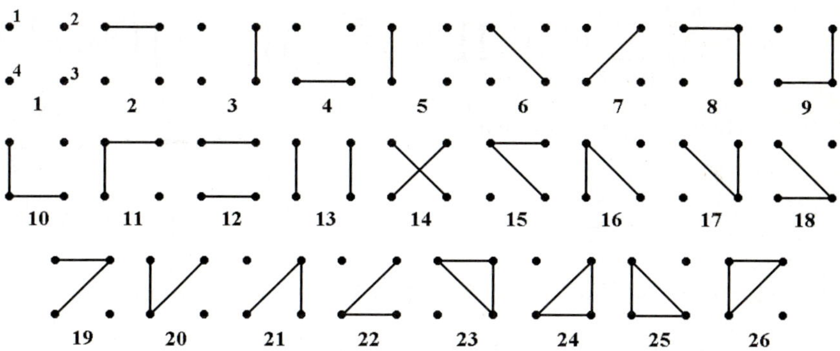

Fig. 1. 4-node disconnected graphs

these subgraphs. We have no space for them all, so we present here some typical cases:

1) $(W_{12} + W_{13} + W_{14} + W_{23} + W_{24} + W_{34})r(m,p)$,
2) $(W_{13} + W_{14} + W_{23} + W_{24} + W_{34})r(k_1,p)q(m - k_1,p)$,
8) $(W_{14} + W_{24} + W_{34})r(k_1,p)r(k_2,p)q(m - k_1 - k_2,p)$,
12) $(W_{13} + W_{14} + W_{23} + W_{24})r(k_1,p)r(k_3,p)q(m - k_1 - k_3,p)$,
23) $(W_{14} + W_{24} + W_{34})r(k_1,p)r(k_2,p)r(k_5,p)q(k_3 + k_4 + k_6,p)$.

From this, using (7), (8) and binomial theorem, we obtain in the vector form that for the 4-node graph

$$\begin{aligned}\boldsymbol{A} = &(W_{12}+W_{13}+W_{14}+W_{23}+W_{24}+W_{34})\boldsymbol{S_m(m)} - \\ &(W_{13}+W_{14}+W_{23}+W_{24}+W_{34})\boldsymbol{S_m(k_1)} - \\ &(W_{12}+W_{13}+W_{14}+W_{24}+W_{34})\boldsymbol{S_m(k_2)} - \\ &(W_{12}+W_{13}+W_{14}+W_{23}+W_{24})\boldsymbol{S_m(k_3)} - \\ &(W_{12}+W_{13}+W_{23}+W_{24}+W_{34})\boldsymbol{S_m(k_4)} - \\ &(W_{12}+W_{14}+W_{23}+W_{24}+W_{34})\boldsymbol{S_m(k_5)} - \\ &(W_{12}+W_{13}+W_{14}+W_{23}+W_{34})\boldsymbol{S_m(k_6)}+(W_{13}+W_{14}+W_{23}+W_{24})\boldsymbol{S_m(k_1+k_3)}+ \\ &(W_{12}+W_{13}+W_{24}+W_{34})\boldsymbol{S_m(k_2+k_4)}+(W_{12}+W_{14}+W_{23}+W_{34})\boldsymbol{S_m(k_5+k_6)}+ \\ &(W_{14}+W_{24}+W_{34})\boldsymbol{S_m(k_1+k_2+k_5)}+(W_{13}+W_{23}+W_{34})\boldsymbol{S_m(k_1+k_4+k_6)}+ \\ &(W_{12}+W_{13}+W_{14})\boldsymbol{S_m(k_2+k_3+k_6)}+(W_{12}+W_{23}+W_{24})\boldsymbol{S_m(k_3+k_4+k_5)}.\end{aligned} \tag{14}$$

Case of a Cycle-Shaped Graph. Let a cycle-shape graph $\mathbb{C}(n,m)$ consists of n multi-edges. We denote the multi-edge that connects v_i with v_{i+1} as e_i, e_n connects v_{n-1} with v_n. Each pair (v_i, v_j) of nodes can be connected by two paths: clockwise and anticlockwise. Thus $f_{ij}(\mathbb{C}(n,m,p))$ is equal to the probability of non-existence of both these paths and so

$$N(\mathbb{C}(n,m)) = \sum_{i=1}^{n-1}\sum_{j=i+1}^{n} W_{ij}\Big[1-\prod_{s=i}^{j-1} r(k_s,p)\Big]\Big[1-\prod_{s=j}^{n} r(k_s,p)\prod_{s=1}^{i-1} r(k_s,p)\Big]. \tag{15}$$

In the case when all $k_i = 1$ we have

$$N(\mathbb{C}(n,n)) = \sum_{i=1}^{n-1}\sum_{j=i+1}^{n} W_{ij}\,(1+p^n) - \sum_{i=1}^{n-1} p^i \sum_{j=0}^{n-1} W_{j,(i+j)\mathrm{mod}(n+1)}. \tag{16}$$

And for this case, if all weights are equal to 1, then we have simple

$$N(\mathbb{C}(n,n)) = C_n^2 - n\sum_{i=1}^{n-1} p^i + C_n^2 p^n. \tag{17}$$

Case of a Tree. From (1) we have that

$$N(\mathbb{T}(n,m)) = \sum_{i=1}^{n-1}\sum_{j=i+1}^{n} W_{ij}\Big[1-\prod_{e_{st}\in Pt_{ij}} r(k_{st},p)\Big], \tag{18}$$

where Pt_{ij} is an unique path between v_i and v_j. Thus in the case of no multi-edges we have

$$N(\mathbb{T}(n)) = \sum_{i=1}^{n-1}\sum_{j=i+1}^{n} W_{ij}\big(1-p^{l_{ij}}\big), \tag{19}$$

For partial cases we can obtain simpler expressions.

Case of a Chain. Let us discuss the case of a chain $Ch(n,m)$ with n nodes and $n-1$ multi-edges e_i. We have $k-1$ pairs with distance 1, $k-2$ pairs with distance 2 and so on. Thus in the general case we have that

$$N(Ch(n,m)) = \sum_{i=1}^{n-1} \sum_{j=i+1}^{n} W_{ij} \left[1 - \prod_{s=i}^{j-1} r(k_s, p)\right]. \tag{20}$$

In the case of no multi-edges ($m = n-1$) we have simpler

$$N(Ch(n, n-1)) = \sum_{i=1}^{n-1} \sum_{j=i+1}^{n} W_{ij}(1 - p^{j-i}), \tag{21}$$

and if all weights are equal to 1, then

$$N(Ch(n, n-1)) = \sum_{i=1}^{n-1} i(1 - p^{n-i}). \tag{22}$$

Case of a Star. Now let us have a star-like tree $S(n,m)$ in which $n-1$ nodes are adjacent to a single node (root). Thus we have some node v_1 with weight w_1 that is adjacent to dangling nodes v_i, $i = 2, \ldots, n$ with weights w_i by multi-edges e_{1i}. Using (19) we obtain:

$$N(S(n,m)) = \sum_{i=2}^{n} W_{1j} q(k_i, p) + \sum_{i=2}^{n-1} \sum_{j=i+1}^{n} W_{ij}[1 - r(k_i, p) r(k_j, p)]. \tag{23}$$

In the simplest case when all weights and multiplicities are equal to 1 we have

$$N(S(n, n-1) = C_n^2 - (n-1)p + C_{n-1}^2 p^2. \tag{24}$$

3.2 Reducing of the Graph Dimension

***k*-Component Graphs.** If a graph G consist of s mutually disconnected subgraphs $G_1, \ldots, G_s$ then

$$N(G) = \sum_{i=1}^{s} N(G_i) + \sum_{i=1}^{s-1} \sum_{j=i+1}^{s} \Omega(G_i)\Omega(G_j). \tag{25}$$

Deleting Dangling Nodes. In [6] we have proved the following statement: *If a graph G consists of subgraph G^* and dangling node v_j with weight w_j, that is connected to some node v_i with weight w_i in G^* by edge with operational probability p_{ij}, then*

$$N(G) = N(G^o) + (1 - p_{ij}) w_i \left[\Omega(G) - w_i\right], \tag{26}$$

where G^o — a graph which has the same structure as G^* and a weight of v_j

$$WT_j(G^o) = w_j + p_{ij}w_i. \tag{27}$$

Thus in our case we have that

$$N(G,p) = N(G^o,p) + (1-p)^{k_{ij}} w_j \left[\Omega(G) - w_j\right], \tag{28}$$

where G^o – a graph which has the same structure as $G \backslash v_j$ and a weight of v_i

$$WT_i(G^o) = w_i + r(k_{ij},p)w_i. \tag{29}$$

Note that after applying these equations we obtain w_i as a polynomial of p. Thus in all equations for EDP-polynomials above w_i is really $w_i(p)$.

3.3 Branching by Chain

Next important corollary from results of [6, 7] is connected with the following theorem first presented in [6] and then improved in [7]:

Theorem 1. *If a connected random graph has a simple chain $Ch = \{e_{sx}, e_{xt}\}$ connecting nodes v_s and v_t through a node v_x with degree 2, then*

$$\begin{aligned} N(G) = {} & [p_{st} + (1-p_{st})p_{sx}p_{xt}]N(G^*) + (1-p_{st})(1-p_{sx}p_{xt})N(G^o) - \\ & f_{st}\frac{(1-p_{st})(1-p_{xt})(1-p_{sx})p_{sx}p_{xt}}{(1-p_{sx}p_{xt})}w_x^2 + (1-p_{sx})(1-p_{xt})w_x\left[\Omega(G) - w_x\right], \end{aligned} \tag{30}$$

where G^* is a graph, obtained from G by contracting v_s and v_t by the chain (let the joint node have number s) and removing e_{st} (if exists), G^o is obtained from G by removing the chain and e_{st},

$$WT_s(G^*) = w_s + w_t + \frac{(p_{sx}p_{st} + p_{sx}p_{xt} + p_{st}p_{xt} - 2p_{sx}p_{st}p_{xt})}{p_{st} + p_{sx}p_{st} - p_{sx}p_{st}p_{xt}}w_x. \tag{31}$$

$$WT_s(G^o) = w_s + w_x\frac{p_{sx}(1-p_{xt})}{1-p_{sx}p_{xt}}; \quad WT_t(G^o) = w_t + w_x\frac{p_{xt}}{1-p_{sx}p_{xt}}. \tag{32}$$

In simple case, equalling all p_i to p we obtain that

$$\begin{aligned} N(G) = {} & p(1+p-p^2)N(G^*,p) + \\ & (1-p)^2\left\{(1+p)N(G^o,p) - f_{st}(p)\frac{p^2}{1+p}w_x^2 + w_x\left[\Omega(G) - w_x\right]\right\}, \end{aligned} \tag{33}$$

$$WT_s(G^*) = w_s + w_t + \frac{3p-2p^2}{1+p-p^2}w_x,$$

$$WT_s(G^o) = w_s + w_x\frac{p}{1+p}, \qquad WT_t(G^o) = w_t + w_x\frac{p}{1-p^2},$$

if e_{st} exists and

$$N(G) = p^2 N(G^*,p)+(1-p)^2 \left\{ N(G^o,p)-f_{st}(p)\frac{p^2}{1-p^2}w_x^2+w_x\left[\Omega(G)-w_x\right]\right\},$$

otherwise. Let us obtain the expression for the case of multi-edges. First we denote $k_{st}+k_{sx}+k_{xt}$ as u and $q_{k_{sx}}(p)+q_{k_{xt}}(p)-q_{k_{sx}+k_{xt}}(p)$ as $Z(p)$. Then, using (7) and (8), we obtain that

$$\begin{aligned} N(G) = & \left[1-(1-p)^{k_{st}}Z(p)\right]N(G^*,p)+ \\ & (1-p)^{k_{st}}Z(p)N(G^o,p)-f_{st}(p)\frac{(1-p)^u\left[1-Z(p)\right]}{Z(p)}w_x^2+ \\ & \left[1-Z(p)\right]w_x\left[\Omega(G)-w_x\right], \end{aligned} \tag{34}$$

$$WT_s(G^*) = w_s+w_t+\frac{\left[1-q(u-k_{xt},p)-q(u-k_{st},p)-q(u-k_{sx},p)\right]w_x}{1-q(k_{st},p)+q(k_{xt},p)-q(u-k_{st},p)-q(u-k_{sx},p)+q(u,p)},$$

$$WT_s(G^o) = w_s+w_x\frac{q(k_{st},p)-q(k_{sx}+k_{st},p)}{Z(p)}, \quad WT_t(G^o) = w_t+w_x\frac{r(k_{xt},p)}{Z(p)}.$$

Now let us consider the value of $f_{st}(p) = f_{st}(G,p)$. There are well-known algorithms for finding 2-terminal probabilistic connectivity (see [4], for example) and a complexity of this task is obviously less then complexity of calculating the EDP of G. Note that though (34) contains fractional polynomial, (1) and fact that $f_{st}(p)$ is a simple polynomial [11] guarantee that at the end we have no fractions.

4 Comparison of Using PC-Criterion and the EDP-Criterion

Reliability polynomials are used for optimal network structures choosing. If EDP-polynomial $N(G_1)$ (PC-polynomial $R(G_1,p)$) overlords EDP-polynomial $N(G_2)$ (PC-polynomial $R(G_2)$) in some interval (p_x,p_y), where $0<p_x<p_y\leq 1$, then graph G_1 is less reliable on (p_x,p_y) than graph G_2 by the EDP-criterion (more reliable on (p_x,p_y) than graph G_2 by the PC-criterion). Intersection of reliability polynomials in the interval $(0,1)$ shows, that both network topologies G_1 and G_2 can be more reliable by the corresponding criterion, depending on the value of edge reliability p.

Lemma 1. *Suppose $G_1(n,n)$ is a cycle-shaped graph and $G_2(n,n-1)$ is a star-shaped graph. Then G_1 is not less reliable by the PC-criterion than G_2.*

Proof. Any tree, star-shaped including, is connected if and only if all its edges are reliable, thus $R(G_2(n,n-1)) = R_2 = p^{n-1}$. A cycle keeps its connection if not more than one of its edges fails. Thus $R(G_1(n,n)) = R_1 = p^n + np^{n-1}(1-p)$ and so $R_2 \geq R_1$ on $[0,1]$. □

Lemma 2. *Adding an edge doesn't decrease graph's reliability by the EDP-criterion.*

Proof. We use equation (1). It is obvious, that for each pair of vertices if we add an edge to the graph, we can't increase a_{ij}. Consequently, after adding an edge EDP-functional can't increase. □

An obvious corollary follows:

Corollary 1. *A cycle-shaped graph* $G_1(n,n)$ *isn't less reliable by the EDP-criterion than a chain-shaped graph* $G_2(n,n-1)$.

Though lemma 2 and it's corollary show that the EDP-criterion for some cases can give similar results to the PC-criterion, next example will show us the difference between how these criteria work.

Statement 1. *EDP-polynomials of a cycle-shaped graph* $G_1(n,n)$ *and a star-shaped graph* $G_2(n,n-1)$ *can intersect.*

Proof. Two tables of prevalence are given below. Stars and cycles with 8 or 10 vertices are considered. In the graphs, which intervals of prevalence are

Table 1. Intervals of prevalence for star and cycle with with equal weight

Graph	Intervals of prevalence
Star-8	$(0.08135, 0.66958)$
Cycle-8	$(0, 0.08135) \cup (0.66958, 1)$
Star-10	$(0.03907, 0.81500)$
Cycle-10	$(0, 0.03807) \cup (0.81500, 1)$

Table 2. Intervals of prevalence for star and cycle with one peculiar vertex

Graph	Intervals of prevalence
Star-8	$(0, 0.80715)$
Cycle-8	$(0.80715, 1)$
Star-10	$(0, 0.87702)$
Cycle-10	$(0.87702, 1)$

illustrated by the table (1), all the vertices are equal and the start weights are assumed as 1. In the graphs, which intervals of prevalence are illustrated by the table (2), one peculiar vertex with start weight 5 is selected, central for star-shaped graphs, all the others' start weights are assumed as 1. The intervals' boundaries were obtained in program Maple 11, using EDP-polynomial coefficients, computed basing on the results of [6, 7] [1].

Consequently, though star-shaped graphs have less edges than cycles of the same number of vertices, star-shaped graphs can be more reliable by the EDP-criterion than cycles, with certain values of edge reliability, and the intervals of star-shape graph prevalence can be significant.

[1] We use the C++ library of our design that allows to compute EDP-polynomials including the case with network's diameter restriction.

5 Special Cases of Choosing an Optimal Network Topology Using the EDP-Polynomials

5.1 Two Chains with One Peculiar Vertex

Lemma 3. *If two chain-shaped graphs G_1 and G_2 with n vertices have one (peculiar) vertex v_x of the weight w_x, all the rest vertices have the weight $w < w_x$ and in G_1 the vertex v_x is situated closer to the center of the chain than in G_2, then $N(G_1) < N(G_2)$.*

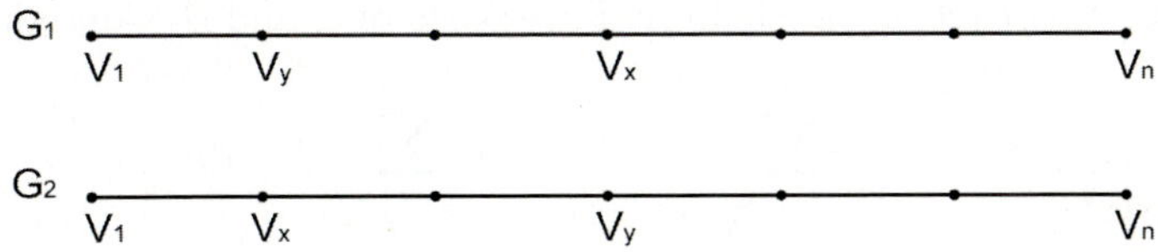

Fig. 2. Two chains with different locations of v_x

Proof. We denote a vertex, that is situated in G_1 on the place of the vertex v_x in G_2 as v_y and a vertex situated in G_2 in the place of v_x in G_1 as v_y. Figure (2) shows the case when both v_x and v_y in both graphs are situated by one side of the chains' centers, but the proof is the same for the case, when the vertices v_x and v_y are situated by the different sides. We number the vertices in the graphs from 1 to n one after another so, that in G_1 vertex v_x has the number greater than v_y, and in G_2 the vertex v_y has the number greater than v_x. Further we will treat x and y like numbers, but they depend on the chain we are analyzing.

Now we can use formula (21). It is obvious, that EDPs for chains G_1 and G_2 are equal, except items that include weight of vertices v_x and v_y. The rest of items contributes in EDP equally and we can denote it as Δ.

$$N(G_1) = \sum_{\substack{i=1 \\ v_i \neq v_x, v_y}}^{n} w w_x (1 - p^{l_{ix}}) + \sum_{\substack{i=1 \\ v_i \neq v_x, v_y}}^{n} w w_y (1 - p^{l_{iy}}) + w_x w_y l_{xy} + \Delta, \quad (35)$$

$$N(G_2) = \sum_{\substack{i=1 \\ v_i \neq v_x, v_y}}^{n} w w_x (1 - p^{l_{ix}}) + \sum_{\substack{i=1 \\ v_i \neq v_x, v_y}}^{n} w w_y (1 - p^{l_{iy}}) + w_x w_y l_{xy} + \Delta. \quad (36)$$

We also can notice here, that the third items in (35) and (36) cancel each other. Also we can divide $N(G_1) - N(G_2)$ by w, as it doesn't influence to the sign. If v_i is a vertex in G_1 and G_2, we can notice that l_{ix} in G_1 equals to l_{iy} in G_2 and vice versa.

We denote

$$\alpha = \sum_{\substack{i=1 \\ v_i \neq v_x, v_y}}^{n} (1 - p^{l_{ix}}), \quad \beta = \sum_{\substack{i=1 \\ v_i \neq v_x, v_y}}^{n} (1 - p^{l_{iy}}), \quad (37)$$

in the chain G_1. Then

$$N(G_1) - N(G_2) = w_x\alpha + w_y\beta - w_y\alpha - w_x\beta \qquad (38)$$
$$= \alpha(w_x - w_y) + \beta(w_y - w_x) = (\alpha - \beta)(w_x - w_y).$$

We remember $w_y = w$, consequently $w_x - w_y > 0$ and we need to find the sign of $\alpha - \beta$. By the data the vertex v_x is closer to the center of the chain G_1 than it is in G_2. A part of the chain G_1 from the vertex v_1 to the vertex v_{x+y-1} is symmetric for the vertices v_x and v_y, hence sums $\sum_{\substack{j=1 \\ v_j \neq v_x, v_y}}^{x+y-1} (1 - p^{l_{jx}})$ and $\sum_{\substack{j=1 \\ v_j \neq v_x, v_y}}^{x+y-1} (1 - p^{l_{jy}})$ are equal for this part of the chain and are parts of α and β. Consequently,

$$\alpha - \beta = \sum_{\substack{j=x+y \\ v_j \neq v_x, v_y}}^{n} (1 - p^{l_{jx}}) - \sum_{\substack{j=x+y \\ v_j \neq v_x, v_y}}^{n} (1 - p^{l_{jy}}). \qquad (39)$$

It is obvious, that for any vertex v_k from a part of the chain G_1 from the vertex v_{x+y} to the vertex v_n $l_{k,x} < l_{k,y}$. Hence, $\alpha - \beta < 0$.

Consequently, $N(G_1) - N(G_2) < 0$. The chain G_1 is more reliable than the chain G_2 by the EDP-criterion. □

5.2 Optimal Addition of an Edge to Some Network Topologies

The best way of elements addition to an existing network to make it more reliable is an important and interesting problem. Here we give some results about adding an edge to some of the most widespread topologies.

Adding an Edge to a Star-Shaped Graph. Suppose $G(n+1, n)$ is a star-shaped graph. All weights are equal to w. It is obvious that there are only two possible ways of adding an edge to a star – doubling of an existing edge or connecting two dangling vertices.

Theorem 2. *Suppose G_1 is a star-shaped graph with $n+1$ vertices and one multi-edge, and G_2 is derived from a star G by connecting of two dangling vertices. Then if $n < 5$ $N(G_1) - N(G_2) > 0$. If $n \geq 5$ G_2 is more reliable by the EDP-criterion for all $p \in (0, \frac{n-4}{2n-4})$ and G_1 is more reliable for all $p \in (\frac{n-4}{2n-4}, 1)$.*

Proof. We number the vertices in G_1 so that the central one is number 0 and the vertex that is connected to the central one by the multi-edge is number n. Also we number the vertices of G_2 so that the central one is number 0 and the dangling vertices that are connected by an edge are numbers $n-1$ and n. We apply the factoring method (2) to the added edge in both G_1 and G_2. Then, with probability $1-p$ we have initial star G and with probability p we have 2 new graphs S_1 and S_2, obtained from G_1 and G_2 correspondingly. S_1 is a star with n vertices, the central vertices has the weight $2w$. S_2 is a star with n vertices, the vertex number $n-1$ weights $2w$ and is connected to the central vertex by the multi-edge. Obviously, $N(G_1) - N(G_2) = N(S_1) - N(S_2)$.

The reliability of the multi-edge in S_2 equals to $p_1 = 1 - (1-p)^2 = 2p - p^2$.

$$N(S_1) = \sum_{i=1}^{n-1} 2w^2(1-p) + \sum_{i=1}^{n-2} \sum_{j=i+1}^{n-1} w^2(1-p^2) =$$

$$= 2(n-1)w^2(1-p) + \frac{(n-1)(n-2)}{2} w^2(1-p^2),$$

$$N(S_2) = \sum_{i=1}^{n-2} w^2(1-p) + (1-p_1)2w^2 + \sum_{i=1}^{n-3} \sum_{j=i+1}^{n-2} w^2(1-p^2) +$$

$$\sum_{i=1}^{n-2}(1-pp_1)2w^2 = (n-2)w^2(1-p) + 2(1-p)^2w^2 +$$

$$\frac{(n-2)(n-3)}{2} w^2(1-p^2) + 2(n-2)w^2(1-2p^2+p^3).$$

Obviously, $1 - 2p^2 + p^3 = (1-p)(1+p-p^2)$, hence

$$N(S_1) - N(S_2) = w^2(1-p)[2(n-1) + (n-2)(1+p) - (n-2) - 2(1-p) - 2(n-2)(1+p-p^2)] = w^2(1-p)p[(2n-4)p - (n-4)].$$

Let us analyze $f(p) = (2n-4)p - (n-4)$. It is clear that $f(p) = 0$ when $p_{root} = \frac{n-4}{2n-4}$. If $n < 5$ there is no root of $f(p)$ on the interval $(0,1)$. If $n \geq 5$ exists $p_{root} \in (0,1)$. If $p \in (0, p_{root})$ and $n \geq 5$ $f(p) < 0$, if $p \in (p_{root}, 1)$ $n \geq 5$ $f(p) > 0$. Hence, if $n < 5$ $N(S_1) - N(S_2) > 0$ for all $p \in (0,1)$. If $n \geq 5$ $N(S_1) - N(S_2) < 0$ for all p in $(0, p_{root})$, $N(S_1) - N(S_2) > 0$ for all p in $(p_{root}, 0)$. □

Adding an Edge to a Chain-Shaped Graph

Theorem 3. *Consider we have a chain $G(n, n-1)$, all weights are equal to w. Then, if we want to add an edge in this chain, the most reliable by the EDP-criterion network topology can be obtained by closing the chain into the cycle*

Proof. At first we will show, that, for all the addings of an edge by doubling of some existing edge, the most reliable by the EDP-criterion is such adding, that the multi-edge is the central in the network (the case of even n) or the multi-edge is incident to the central vertex of the chain (the case of odd n). We call this the closest to the center edge as the central edge. Indeed, if we apply the factoring method (2) to the added edge, then with probability $1-p$ we will obtain the initial chain and with probability p we will obtain a chain of $n-1$ vertices with one vertex of weight $2w$. The more this vertex of maximum weight is closer to the center of the new chain, the more reliable this chain is, according to the lemma (3).

Now we will prove that closing a chain into a cycle produces more reliable graph by the EDP-criterion than doubling of the central edge. We denote the cycle as G_1 and the graph with the central multi-edge as G_2. Applying the factoring method to the added edge, we obtain with probability $1-p$ the initial chain G and with probability p from G_1 cycle G_1^0 of $n-1$ vertices with one vertex of weight $2w$ and from G_2 and chain G_2^0 of $n-1$ vertices with one vertex of weight $2w$ also. Obviously, this new cycle G_1^0 is a chain G_2^0 after adding an edge. Thus using lemma (2) we prove the theorem. □

6 Using EDP-Polynomials for Network Reliability Comparison

Now let us discuss the usage of EDP-polynomials for choosing better graph topology in general case. First we present example of $G(6,6)$. There are 27 non-isomorphic connected $G(6,6)$ multi-graphs, the EDP-polynomial coefficients were computed for all of them. Here the analysis of 5 of them, as the most indicative, is adduced. It is assumed that all the vertices have weight 1.

We will name the chosen graphs as on Fig.(3). The coefficients of the corresponding EDP-polynomials in the form (3) are presented in the table 3.

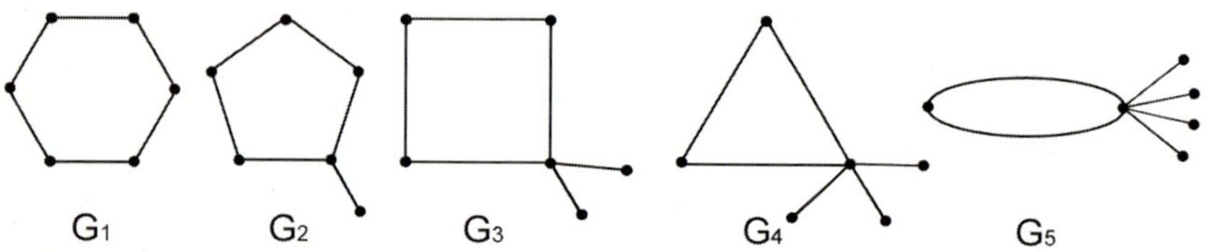

Fig. 3. Multi-graphs with 6 vertices and 6 edges

Table 3. EDP-polynomials' coefficients of the graphs from the Fig.(3)

Graph	Polynomial coefficients
G_1	0 0 105 210 189 84 15
G_2	0 5 95 205 188 84 15
G_3	0 10 89 196 186 84 15
G_4	0 15 90 189 183 84 15
G_5	0 20 99 192 182 84 15

Table 4. EDP-polynomial's points of intersection

	G_1	G_2	G_3	G_4	G_5
G_1	–	0.709544	0.694539	0.651013	0.554942
G_2	0.709544	–	0.680734	0.622905	0.5
G_3	0.694539	0.680734	–	0.559816	0.387425
G_4	0.651013	0.622905	0.559816	–	0.166667
G_5	0.554942	0.5	0.387425	0.166667	–

In the table 4 the root of polynomial $N(G_i)-N(G_j)$ is in the intersection of the column G_i and the row G_j. The roots were computed in program Maple 11. Consequently, the interval $(0,1)$ can be divided into intervals of prevalence of some graph by the EDP-criterion, as in table 5. Hence, having an information

Table 5. Intervals of graphs' optimality

Graph	Interval of optimality
G_1	$(0.709544, 1)$
G_2	$(0.680734, 0.709544)$
G_3	$(0.559816, 0.680734)$
G_4	$(0.166667, 0.559816)$
G_5	$(0, 0.166667)$

of communication circuits' reliability, we can choose optimal network topology by the analysis of EDP-polynomials of possible topologies.

Now let us consider more complex networks and EDP-criterion approach to better topology selection. Let us have two graphs G_1 and G_2 with 10 vertices and 17 edges, which are pictured at Fig.4.

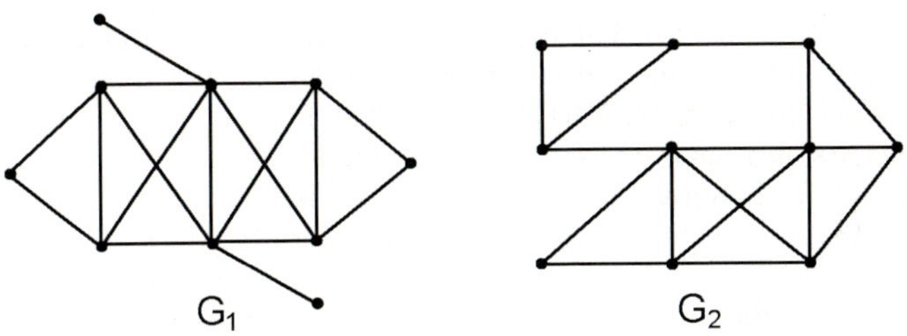

Fig. 4. Graphs with 10 vertices and 17 edges

Using techniques that are described in the section 3, we obtain that

$$N(G_1) = 18p^{16}(1-p)+305p^{15}(1-p)^2+2411p^{14}(1-p)^3+11950p^{13}(1-p)^4+ 41942p^{12}(1-p)^5+110959p^{11}(1-p)^6+228747p^{10}(1-p)^7+ 372645p^9(1-p)^8+479339p^8(1-p)^9+480185p^7(1-p)^{10}+ 367664p^6(1-p)^{11}+211619p^5(1-p)^{12}+90062p^4(1-p)^{13}+ 27605p^3(1-p)^{14}+5792p^2(1-p)^{15}+748p(1-p)^{16}+45(1-p)^{17}$$

$$N(G_2) = 39p^{15}(1-p)^2+677p^{14}(1-p)^3+p^{13}(1-p)^4 5377+26069p^{12}(1-p)^5+ 86762p^{11}(1-p)^6+210214p^{10}(1-p)^7+380438p^9(1-p)^8+515498p^8(1-p)^9+ 518639p^7(1-p)^{10}+389367p^6(1-p)^{11}+219022p^5(1-p)^{12}+91615p^4(1-p)^{13}+ 27792p^3(1-p)^{14}+5802p^2(1-p)^{15}+748p(1-p)^{16}+45(1-p)^{17}.$$

The EDP-polynomials of graphs G_1 and G_2 intersect on the segment (0,1) in the point $p^* = 0.5312978$ approximately (obtained with Maple 11). By EDP-criterion G_1 is more reliable when $p \in (0, p^*)$ and G_2 is more reliable when $p \in (p^*, 1)$. Note that $R(G_2) > R(G_1)$ for all $p \in (0, 1)$, difference in the point p^* is about 0.11.

We can notice, that the last 2 coefficients of EDP-polynomials of graphs with n vertices and m edges are always equal. The last coefficient corresponds total number of graph's pairs of vertices, and the last but one is summarized number of disconnected pairs in graphs with one edge and $n-2$ isolated vertices, which are obtained from the initial graph by deleting all the edges but one.

References

1. Krivoulets, V.G., Polesskii, V.P.: What is the Theory of Bounds for Network Reliability? Inf. Processes 1(2), 199–203 (2001) (in Russian)
2. Rodionova, O., Rodionov, A., Choo, H.: Network Probabilistic Connectivity: Optimal Structures. In: EWSA 2004. LNCS, vol. 3047, pp. 431–440. Springer, Heidelberg (2004)
3. Satyanarayana, A., Chang, M.K.: Network reliability and the factoring theorem. Networks 13, 107–120 (1983)
4. Shooman, A.M.: Algorithms for Network Reliability and Connection Availability Analysis. In: Electro 1995 Int. Professional Program Proc., pp. 309–333 (1995)
5. Palmer, C.R., Siganos, G., Faloutsos, M., Faloutsos, C., Gibbons, P.B.: The connectivity and faulttolerance of the Internet topology. In: Workshop on Network-Related Data Management, NRDM- 2001 (2001), `http://www.cs.ucr.edu/~siganos/papers/nrdm.ps`
6. Rodionov, A., Rodionova, O.: Network Probabilistic Connectivity: Expectation of a Number of Disconnected Pairs of Nodes. In: Gerndt, M., Kranzlmüller, D. (eds.) HPCC 2006. LNCS, vol. 4208, pp. 101–109. Springer, Heidelberg (2006)
7. Rodionov, A., Rodionova, O.K., Choo, H.: On the Expected Value of a Number of Disconnected Pairs of Nodes in Unreliable Network. In: Gervasi, O., Gavrilova, M.L. (eds.) ICCSA 2007, Part III. LNCS, vol. 4707, pp. 534–543. Springer, Heidelberg (2007)
8. Rodionov, A., Gadyatskaya, O.: Usage and Calculation of random graph's EDP-polynomial. In: Proceedings of ICMMG SBRAS (Papers of Third Asian International Workshop on Optimisation Problems of Complex Systems), Novosibirsk, pp. 93–107 (2007) (in Russian)
9. Moore, E.F., Shannon, C.E.: Reliable Circuits Using Less Reliable Relays. J. Franclin Inst. 262(4b), 191–208 (1956)
10. Colbourn, C.J.: Some open problems on reliability polynomials. Congr. Numer. 93, 187–202 (1993)
11. Chari, M., Colbourn, C.J.: Reliability polynomials: a survey. J. Combin. Inform. System Sci. 22, 177–193 (1997)

On Sharp Estimating of Bond Option Prices for Heath-Jarrow-Morton Model Based on Jump

Kisoeb Park[1], Moonseong Kim[2], and Seki Kim[1,⋆]

[1] Department of Mathematics, Sungkyunkwan University, Korea
440-746, Suwon, Korea
{kisoeb,skim}@skku.edu
[2] Electrical and Computer Engineering
Michigan State University, USA
mkim@msu.edu

Abstract. We propose Heath-Jarrow-Morton (HJM) model in which the forward rate follows a jump-diffusion process. We represent HJM model based on jump and give the analytic proof for it. We derive a closed-form solution of the European bond option for HJM model based on jump. We get the fact that bond option price for the proposed model estimate highly precision. Furthermore, we simulate the proposed model by Monte Carlo method with scenarios. Lower value of precision of bond option price for HJM model based on jump means sharper estimates.

Keyword: Heath-Jarrow-Morton Model with Jump, Bond Option Price, and Monte Carlo Method.

1 Introduction

Current research using jump-diffusion processes relies mostly on two classes of models: the affine jump-diffusion class and the quadratic Gaussian[3]. We assume that HJM model based on jump is completely determined by the value of forward rate $f(t,T)$ and $f(t,T)$ follows a Gaussian and Poisson process[1]. The HJM model based on jump uses as driving stochastic dynamic variable forward rates whose evolution is dependent on a specified volatility function[4].

This paper introduces a new class of arbitrage-free models of the term structure of forward rates that is supported by an economy with habit formation[5]. We examines the one-factor HJM model based on jump which we use restrictive condition of RS. We find the existing HJM model based on jump, as this model is currently specified, belong to a more general, arbitrage-free framework, which will be established in this paper. We derive an analytic solution of European bond option on HJM model based on jump. This paper develop the valuation for European bond option[6] on HJM model based on jump.

In addition, we introduce Monte Carlo simulation. One of the many uses of Monte Carlo simulation by financial engineers is to place a value on financial

⋆ Corresponding author.

O. Gervasi et al. (Eds.): ICCSA 2008, Part II, LNCS 5073, pp. 1077–1085, 2008.

derivatives. Interest in use of Monte Carlo simulation for European bond option pricing is increasing because of the flexibility of the methods in handling complex financial instruments. The Monte Carlo simulation lends itself naturally to the evaluation of European options of bond pricing models with jump represented as expectations. One measure of the sharpness of the point estimate of the mean is Mean Standard Error (MSE). Lower value of precision of bond option price for the HJM model based on jump means sharper estimates.

The structure of the remainder of this paper is as follows. In the section 2, investigated bond option values given for HJM model based on jump. In the section 3, calculated numerical solutions using Monte Carlo simulation of bond option pricing for HJM model based on jump. In Section 4, examined bond option values given for HJM model based on jump. This paper is finally concluded in Section 5.

2 Bond Option Pricing for HJM Model Based on Jump

We derive a closed-form solution (CFS) for European bond options when the price of the underlying the forward rate evolve as a discontinuous process. The HJM model consider forward rates rather than bond prices as their basic building blocks. Under the corresponding risk-neutral measure Q, the explicit dependence on the market price of risk can be suppressed, and we obtain the differential form of the equation is given by

$$df(t,T) = {\sigma_f}^*(t,T)dt + \sigma_f(t,T)dW^Q(t) + JdN(t), \tag{1}$$

where ${\sigma_f}^*(t,T) = \sigma_f(t,T)\int_t^T \sigma_f(t,s)ds$, $dW^Q(t)$ is a standard Wiener process generated by the risk-neutral measure Q, and $dN(t)$ is the Poisson process with intensity rate h. The probability that only one jump happens during the instantaneous period is hdt.

Therefore, $dN(t)$ is a Poisson jump process having the following characteristic:

$$dN(t) = \begin{cases} 0 \quad 1 - hdt \;\; : \;\; no\ jump \\ \\ 1 \quad hdt \;\; : \;\; otherwise. \end{cases}$$

At time t, **forward** $f(t,T)$ of discount bond $V(t,T)$ is the continuously compounded rate of return that causes bond price to rise to one at time T

$$V(t,T) = \exp\left(-\int_t^T f(t,s)ds\right). \tag{2}$$

We examines one-factor HJM model based on jump which we use restrictive condition of RS[5]. We introduce relation between short rate process and forward rate process as follows. In this study, jump-diffusion version of Hull and White model to reflect this restriction condition. We know the following model for interest rate r;

$$dr(t) = a(t)[\theta(t)/a(t) - r(t)]dt + \sigma_r(t)r(t)^{\beta}dW^Q(t) + JdN(t), \tag{3}$$

where $\theta(t)$ is a time-dependent drift; $\sigma_r(t)$ is volatility factor; $a(t)$ is reversion rate.

Theorem 1. *Let be the jump-diffusion process in short rate $r(t)$ is the equation (3). Let be the volatility form is*

$$\sigma_f(t,T) = \sigma_r(t)(r(t))^\beta \eta(t,T) \tag{4}$$

with $\eta(t,T) = \exp\left(-\int_t^T a(s)ds\right)$ is deterministic functions. We know the stochastic differential equation for forward rate (1). Then we obtain the equivalent model is

$$\begin{aligned} f(0,T) = r(0)\eta(0,T) + \int_0^T \theta(t)\eta(s,T)ds \\ - \int_0^T \sigma_r^2(s)(r(s)^2)^\beta \eta(s,T) \int_s^T (\eta(s,u)du)ds \end{aligned} \tag{5}$$

that is, all forward rates are normally distributed.

Proof). (i) For the case $\beta = 0$, in equation (1), both sides takes an integral we obtain

$$f(T,T) = f(0,T) + \int_0^T \sigma_r(s)\eta(u,T) \int_t^T \sigma_r(s)\eta(u,s)du\ ds$$

$$+ \int_0^T \sigma_r(s)\eta(s,T)dW^Q(s) + \sum_{j=1}^{N(t)} J_j.$$

Also, we obtain the solution $r(t)$ of the above equation can be written

$$r(T) = r_0(T)\{r(0) + \int_0^T {r_0}^{-1}(s)\theta(s)ds + \int_0^T {r_0}^{-1}(s)\sigma_r(s)dW^Q(s)$$
$$+ \sum_{j=1}^{N(t)} {r_0}^{-1}(T_j)J_j\},$$

where $r_0(T) = \exp\left(-\int_0^T a(s)ds\right)$. From this equation we get

$$\int_0^T \sigma_r(t)\eta(s,T)dW^Q(s) + \sum_{j=1}^{N(t)} J_j = r(T) - r(0)\eta(0,T) - \int_0^T \theta(t)\eta(s,T)ds$$

since, we have $r(T) = f(T,T)$. Therefore,

$$f(0,T) = r(0)\eta(0,T) + \int_0^T \theta(t)\eta(s,T)ds$$

$$- \int_0^T {\sigma_r}^2(s)\eta(s,T)\left(\int_s^T \eta(s,u)du\right)ds$$

(ii) For the case $\beta = 0.5$, in (i) a similar way, we have

$$f(0,T) = r(0)\eta(0,T) + \int_0^T \theta(t)\eta(s,T)ds$$
$$- \int_0^T {\sigma_r}^2(s)r(s)\eta(s,T)\left(\int_s^T \eta(s,u)du\right)ds$$

The forward rates are normally distributed, which means that the bond prices are log-normally distributed. Both the short term rate and the forward rates can become negative. □

By the theroem 1, we derive the relation between short rate and forward rate. Using the equation (5), we obtain bond price as follows;

Corollary 1. *Let be the HJM model based on jump is the stochastic differential equation (SDE) for forward rate $f(t,T)$ is given by*

$$df(t,T) = \sigma_f(t,T)\int_t^T \sigma_f(t,s)dsdt + \sigma_f(t,T)dW^Q(t) + JdN(t), \qquad (6)$$

where dW^Q is a Gaussian process generated by an equivalent martingale measure Q and $\sigma_f(t,T)$ be as given in the equation (4). Then discount bond price $V(t,T)$ for forward rate is given by the formula

$$V(t,T) = \frac{V(0,T)}{V(0,t)}\exp\{-\frac{1}{2}\left(\frac{\int_t^T \sigma_f(t,s)ds}{\sigma_f(t,T)}\right)^2 \int_0^t \sigma_f^2(s,t)ds$$
$$- \frac{\int_t^T \sigma_f(t,s)ds}{\sigma_f(t,T)}[f(0,t) - r(t)]\} \qquad (7)$$

with the equation (5).

Note that we know that $\beta = 0$ case is an extension of Vasicek's jump diffusion model; the $\beta = 0.5$ case is an extension of CIR jump diffusion model.

Suppose $N(t)$ represents the total number of extreme shocks that occur in a financial market until time t. We have the stochastic differential equation (1). In stochastic integral equation form, equation (1) may be written as

$$f(t,T) = f(0,T) + \int_0^t {\sigma_f}^*(s,T)ds + \int_0^t \sigma_f(s,T)dW^Q(s) + \sum_{j=1}^{N(t)} J_j. \qquad (8)$$

It can be shown that the probability density function for $f(t,T)$. Let $X(t,T) = f(0,T) + \int_0^t {\sigma_f}^*(s,T)ds + \int_0^t \sigma_f(s,T)dW^Q(s)$ and $Y(t,T) = \sum_{j=1}^{N(t)} J_j$, then $f(t,T) = X(t,T) + Y(t,T)$.

Therefore, the conditional expectation and variance of jump-diffusion process of the forward rate given the current level are

$$\mathbb{E}[f(t,T)] = f(0,T) + \mathbb{E}\left[\int_0^t \sigma_f{}^*(s,T)ds\right] + \mu ht \tag{9}$$

and

$$\begin{aligned}\mathbb{V}ar[f(t,T)] &= \mathbb{E}[(f(t,T) - \mathbb{E}[f(t,T)])^2] \\ &= \int_0^t \sigma_f{}^2(s,T)ds + \gamma^2 ht + \mu ht,\end{aligned} \tag{10}$$

where the volatility form is the equation (4).

We derive the value of European option on a discount bond equation (6). Since the forward rate is normally distributed, the discount bond prices are lognormally distributed; and so the value of option on discount bonds can be expressed as a variant of the Black-Scholes formula. The price of a call option on the T_V - maturity discount bond with exercise price K and maturity $T < T_V$ is given by

$$C = V(t,T)\mathbb{E}[\max(\bar{V} - K, 0)], \tag{11}$$

where $\mathbb{E}$ denotes expected value in a world that is forward risk neutral with respect to a bond maturing at time T_V;

$$\mathbb{E}[\bar{V}] = V(t,T_V)/V(t,T). \tag{12}$$

Assume that bond price is lognormally with standard deviation of the logarithm of bond price equal to σ_V, that is, $\bar{V}$ is lognormally with

$$\mathbb{V}ar[\log \bar{V}] = \sigma_V{}^2, \tag{13}$$

where $\mathbb{V}ar[\log \bar{V}]$ are the $\beta = 0$ case is an extension of Vasicek's jump diffusion model; the $\beta = 0.5$ case is an extension of CIR jump diffusion model. By the Black-Scholes-Merton formula the equation (11) becomes

$$\begin{aligned}C &= V(t,T)(\mathbb{E}[\bar{V}]N(d_1) - KN(d_2)) \\ &= V(t,T_V)N(d_1) - KV(t,T)N(d_2),\end{aligned} \tag{14}$$

where

$$\begin{aligned}d_1 &= \frac{\log(\mathbb{E}[\bar{V}]/K) + \sigma_V{}^2/2}{\sigma_V} \\ &= \frac{\log([V(t,T_V)/V(t,T)]/K) + \sigma_V{}^2/2}{\sigma_V},\end{aligned}$$

$$\begin{aligned} d_2 &= \frac{\log(\mathbb{E}[\bar{V}]/K) - \sigma_V{}^2/2}{\sigma_V} \\ &= \frac{\log([V(t,T_V)/V(t,T)]/K) - \sigma_V{}^2/2}{\sigma_V} \\ &= d_1 - \sigma_V, \end{aligned}$$

and

$$N(x) = \frac{1}{\sqrt{2\pi}} \int_{-\infty}^{x} e^{-x^2/2} dx.$$

The price of a put option on the discount bond is

$$\begin{aligned} P &= V(t,T)\mathbb{E}[\max(K - \bar{V}, 0)] \\ &= V(t,T)(KN(-d_2) - \mathbb{E}[\bar{V}]N(-d_1)). \end{aligned} \quad (15)$$

The resemblance between the bond option pricing formula and the Black-Scholes and Morton formula is obvious. In both cases, the random variables is lognormal, resulting in similar formulas.

3 Stochastic Simulation

Recent methods of bond pricing do not necessarily exploit partial differential equations (PDEs) implied by risk-neutral portfolios. They rest on converting prices of such assets into martingales. This is done through transforming the underlying probability distribution using the tools provided by the Girsanov's theorem. A risk-neutral measure is any probability measure, equivalent to the market measure P, which makes all discounted bond prices martingales.

We now move on to discuss Monte Carlo simulation (MCS). The Monte Carlo simulation is actually a very general tool, and its applications are by no means restricted to numerical integration. In simple and general terms, Monte Carlo simulation may be described as a numerical method based an random sampling. A Monte Carlo simulation of a stochastic process is a procedure for sampling random outcomes for the process. This uses the risk-neutral valuation result The bond price can be expressed as:

$$V(f_t, t, T) = E_t^Q \left[e^{-\int_t^T f(t,s)ds} \right], \quad (16)$$

where E_t^Q is the expectations operator with respect to the equivalent risk-neutral measure. Under the equivalent risk-neutral measure, the local expectation hypothesis holds (that is, $E_t^Q \left[\frac{dV}{V}\right]$).

To execute the Monte Carlo simulation, we discrete the equation (1). We divide the time interval $[t, T]$ into m equal time steps of length Δt each. For small time steps, we are entitled to use the discreted version of the risk-adjusted stochastic differential equation (1):

$$f_j = f_{j-1} + \left[\sigma_f(t,T)\int_t^T \sigma_f(t,s)dsdt\right]\Delta t + \sigma_f(t,T)\varepsilon_j\sqrt{\Delta t} + J_j N_{\Delta t}, \quad (17)$$

where $\sigma_f(t,T) = \sigma_r(t)(r(t))^\beta \exp\left(-\int_t^T a(s)ds\right)$, $j = 1, 2, \cdots, m$ and ε_j is standard normal variable with $\varepsilon_j \sim N(0,1)$. The $N_{\Delta t}$ is a Poisson random variable with parameter $h\Delta t$.

We can investigate the value of the bond by sampling n short rate paths under the discrete process approximation of the risk-adjusted processes of the equation (17). The bond price estimate is given by:

$$V(f_t,t,T) = \frac{1}{n}\sum_{i=1}^{n}\exp\left(-\sum_{j=0}^{m-1} f_{ij}\Delta t\right), \quad (18)$$

where f_{ij} is the value of forward rate under the discrete risk-adjusted process within sample path i at time $t + \Delta t$.

We now study bond option pricing for HJM model based on jump. Let $C(t,T,T_V)$ be the time t-value of a European call option maturing at time T on the zero-coupon bond with maturity T_V, where $0 < t < T < T_V$. The current value of a European call option $C(0,T,T_V)$ can be evaluated, under the risk neutral measure as the expected discounted payoff of the option at the option's maturity

$$C(0,T,T_V) = \mathbb{E}_t^Q\left[\exp\left(-\int_0^T f(t,s)ds\right)\max(V(f,0,T,T_V) - K)|F_0\right] \quad (19)$$

For simulation based approaches to bond option pricing, we have found that the use of one or the other probability measure does not seem to provide any significant advantage. Here we report the simulations under the risk neutral probability measure. Given the forward rate dynamics under the risk neutral measure, which is the equation (17), the option price is evaluated using formula (14), using the Euler-Maruyama scheme for the integration, as

$$C(0,T,T_V) = \frac{1}{n}\sum_{i=1}^{n}\exp\left(-\sum_{j=0}^{m-1} f_{ij}\Delta t\right)\max(V_i(f,0,T,T_V) - K). \quad (20)$$

The bond price $V_i(f,0,T,T_V)$ is computed by the term structure models with jump. Numerical methods that are known as Monte Carlo methods can be loosely described as statistical simulation methods, where statistical simulation is defined in quite general terms to be any method that utilizes sequences of random numbers to perform the simulation. The Monte Carlo simulation is clearly less efficient computationally than the numerical method. The precision of the mean as a point estimate is often defined as the half-width of a 95% confidence interval, which is calculated as

$$Precision = 1.96 \times MSE, \quad (21)$$

where MSE $= \nu/\sqrt{n}$ and ν^2 is the estimate of the variance of bond option prices as obtained from n sample paths of the short rate:

$$\nu^2 = \frac{\sum_{i=1}^{n}\left[\exp\left(-\sum_{j=0}^{m-1} f_{ij}\Delta t\right)\max(V_i(f,0,T,T_V)-K) - C(0,T,T_V)\right]^2}{n-1}$$

Lower values of precision in equation (21) correspond to sharper estimates. Increasing the number of n is a brute force method of obtaining sharper estimates. This reduces the MSE by increasing the value of n. For the purpose of simulation, we conduct three runs of 10,000 trials each and divide the year into 365 time steps.

4 Estimation Results

We investigate the valuation of European bond options for HJM model based on jump. We know that $\beta = 0$ case is an extension of Vasicek's jump diffusion model (JEV); the $\beta = 0.5$ case is an extension of CIR jump diffusion model (JECIR). In experiment, the parameter values are assumed to be $r = 0.05$, $a = 0.5$, $\theta = 0.025$, $\sigma_r = 0.08$, $\lambda = -0.5$, $\beta = 0$(Vasicek), $\beta = 0.5$(CIR), $\gamma = 0.01$, $\mu = 0$, $h = 10$, $K = 0.9$, $t = 0.05$, $Tc = T - 0.5$ and $T = 10$.(the $K = 0.9$ case is the call option, the $K = 1.2$ case is the put option.)

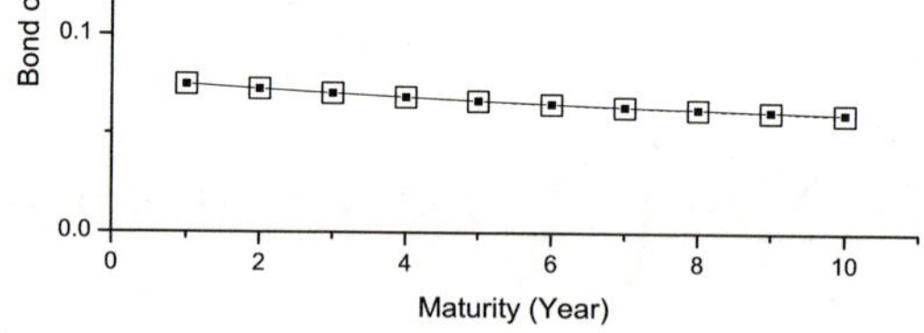

Fig. 1. *Experiment:*Bond option prices for HJM model based on jump

Experiment 2, contrasts bond option prices for HJM model based jump by Monte Carlo simulation. In experiment 2, the parameter values are assumed to be $r[0] = 0.05$, $f[0,t] = 0.0498751$, $a = 0.5$, $\theta = 0.025$, $\sigma_r = 0.08$, $\lambda = -0.5$, $\beta = 0$(Vasicek), $\beta = 0.5$(CIR), $\gamma = 0.01$, $\mu = 0$, $h = 10$, $\Delta t = (T - Tc)/m$, $m = 365$, $n = 10,000$, $t = 0.05$, $Tc = T - 0.5$, and $T = 1$.(the $K = 0.9$ case is the call option, the $K = 1.2$ case is the put option.)

Table 1. [*Experiment 2*] Bond option prices for HJM model based on jump simulated by the Monte Carlo simulation

	Call Option HJM-Jump		Put Option HJM-Jump	
	Ext-Vasicek	Ext-CIR	Ext-Vasicek	Ext-CIR
CFS	0.074610	0.074612	0.218822	0.218820
MCS	0.073558	0.073644	0.218898	0.218955
CFS-MCS	0.001052	0.000969	7.60E-05	1.35E-04
Variance	2.35E-07	2.52E-07	2.19E-06	2.10E-06
Precision	3.01E-05	3.11E-05	9.17E-05	8.95E-05

5 Conclusion and Future Research

In this paper, we derive a closed-form solution for the valuation of European options when the price of the underlying instantaneous forward rate evolves as a discontinuous process. We investigate the valuation of bond option prices for HJM model based on jump. Even though it is hard to achieve the value of bond option prices to term structure models when forward rates follow jump diffusions. We know that this values estimate highly precision. We get the fact that bond option prices for HJM models based on jump by Monte Carlo simulation correspond with a closed-form solutions. Lower values of precision in bond option price for HJM model based on jump signify the sharper estimates. There are still many problems remaining for further research. For instance, using the Monte Carlo simulation to simulate more complicated the two factor HJM model based on jump.

References

1. Ahn, C., Thompson, H.: Jump-Diffusion Processes and the Term Structure of Interest Rates. Journal of Finance 43, 155–174 (1998)
2. Johannes, M.: The Statistical and Economic Role of Jumps in Continuous-Time Interest Rate Models. Journal of Finance 59, 227–260 (2004)
3. Cheng, P., Scaillet, O.: Linear-Quadratic Jump-Diffusion Modeling. Mathematical Finance 17(4), 575–598 (2007)
4. Dai, Q., Singleton, K.: Expectation Puzzles, Time-varying Risk Premia, and Affine Models of the Term Structure. Journal of Financial Economics 63, 415–441 (2002)
5. Buraschi, A., Jiltsov, A.: Habit Formation and Macroeconomic Models of the Term Structure of Interest Rates. Journal of Finance 62(6), 3009–3063 (2007)
6. Das, S.R., Foresi, S.: Exact solutions for bond and option prices with systematic jump risk. Review of Derivatives Research 1 (2005)

Analysis of Interhelical Angle and Sequential Distance in Proteins

Sangyoon Lee

School of Mechanical Engineering, Konkuk University
1 Hwayang-dong, Seoul, South Korea
slee@konkuk.ac.kr

Abstract. This paper presents the results and discussion of statistical analysis of helix-helix interactions in proteins in terms of interhelical angle and sequential distance. The protein data includes 1290 high-resolution globular protein structures with less than 20 percent homology. We first define possible cases of interactions between helices and identify protein pairs that satisfy the conditions for each case. Two cases of helix-helix interactions are investigated using six parameters that can be obtained by considering the interactions as rigid-body motions. The computational results show that orientational preferences in helix packing are dominated by interactions between helices that are proximal in sequence. This bias is described with a univariate probability density function that depends only on the sequential distance between interacting helices. We also compute bivariate probability densities that depend on both interhelical orientation and sequential distance. The dominance of certain preferred orientations and sequential distance values in this bivariate distribution are also observed. Angle preferences persist even when the bivariate distribution is normalized by the univariate distribution. This statistical analysis provides quantitative probability distributions from which statistical potentials for secondary structure assembly can be derived.

Keywords: protein, helix, interhelical angle, sequential distance.

1 Introduction

A protein is a polypeptide chain composed of amino acid residues. For a protein to do its specific biological function, it has to fold into a complex 3-dimensional configuration. Such a process is called the protein folding and it has been described as a hierarchical process in several works [1, 2]. In this hierarchical process, amino acid residues in a polypeptide chain first form local interactions which lead to secondary structures. The remaining freedom in loop regions allows for secondary structures to then make contacts which stabilize the tertiary structure. Secondary structures usually consist of two types: helices and sheets. Helices possess a very regular structure in the arrangement of amino acids, and in particular, interactions between helices are investigated in the statistical way in this paper.

Many researchers have sought to determine principles for predicting the aggregation and contact patterns in alpha-helices and beta-strands [3-7]. Classical works have

O. Gervasi et al. (Eds.): ICCSA 2008, Part II, LNCS 5073, pp. 1086–1095, 2008.

sought to explain helix-helix packing angle preferences in proteins based on models of steric fit and optimal packing of helices around hydrophobic cores. In contrast, recent works have analyzed entries in the protein data bank (PDB) [8] to look for patterns in the way helices interact in globular and membrane proteins [9-11].

Since a large amount of protein data and a fast computation at a less cost are available now, the latter approach becomes useful. For instance, one can examine helix-helix pairs in a very large set of proteins, examine their crossing angle, and presumably make predictions based on these observations. In a recent work, the joint distribution of interhelical distance and angle in proteins has been examined [11]. However, no prior work has investigated the joint distribution of sequential distance and angle in proteins, as is done here.

The emphasis in this paper is how interhelical angle distributions vary with the sequential distance separating the helices. In other words, the goal is to determine how crossing angle preferences change as the sequential distance between helices is varied. This should provide a useful result for the protein secondary structure prediction because patterns of helix interactions can be found from the primary structure of proteins.

2 Methods

The data we examine here includes 1290 protein structures which have been resolved to 2.0 Å or better and possess less than 20 percent of their sequences in common. This data is obtained from [12] and includes 12,025 helices and 89,017 helix-helix pairs. In the previous work [11], we observed three helix-helix interaction cases including two cases in Fig. 1, where a pair of helices is denoted by h_1 and h_2. Since the other case is considered as less significant based on the results of previous work, it is not examined in this work.

In the first case (line-on-line contact case shown in the left of Fig. 1), the connecting line segment of shortest distance interacts both finite helix axes at right angles. The orientational order between helices persists over a wide range of spatial distances between helices. The second case shown in the right of Fig. 1 consists of helices where the end of one finite helix axis interacts with the interior of the other.

The amount of helix pair data can be reduced by imposing a distance cutoff of 15 Å between alpha-carbon atoms in any two helices. It is based on the assumption that helix pairs that are close in space have a higher probability of interaction. When it is used as a criterion to determine their possible interactions, the number of helix pairs in the first case is 5,562 and the number in the second case is 13,839. For each helix pair, we find the number of amino acids in in-between loops and define it as the sequential distance.

In this analysis every helix is considered as a rigid body. Therefore all the possible interactions between helices can be described as rigid-body motions, and the rigid-body motions are expressed in parameters. For example, six parameters $(\alpha, \beta, \gamma, z_1, r, z_2)$ in the case 1 describe the rigid-body motions between the helices h_1 and h_2. The first three parameters α, β, γ are for rotations and the other three are for translations.

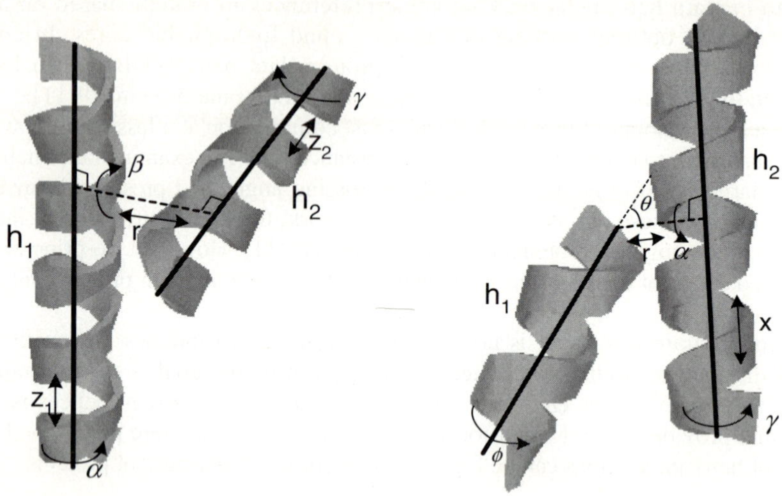

Fig. 1. Case 1(left) consists of helix axes with a common normal (a dotted line) that interacts the interior of both axes; Case 2 (right) consists of helices where the end of one finite helix axis interacts with the interior of the other

In order to consider all the possible interactions between helices, the integration over relative rigid-body motion is employed, and so the parameters are associated with a natural volume element in the integration. For each case, the volume element is derived afterward.

In order to find interhelical angle, we assign a unit vector for each helix along the helix axis. Here we choose the direction that represents the increase in amino acid sequence number. The angle between two helices is considered to be positive if helix 2 is rotated clockwise relative to helix 1 by an angle between 0 and 180 degrees around the unique line segment that points from the axis of helix 1 to that of helix 2 and intersects both axes at right angles.

This convention does not depend on the numbering of the helices after the direction of the helix unit vector is established. Such a line of closest approach will always exist for infinitely long helix axes. In contrast, alpha-helices have finite length, and so a distinction must be made between the axis of the helix and the finite part of the axis which lies inside the helix.

The relative rigid-body motion between two coordinate frames of reference in 3-dimensional space can be described by a 4×4 homogeneous transformation matrix of the form:

$$H = \begin{pmatrix} A & \mathbf{a} \\ \mathbf{0}^{\mathrm{T}} & 1 \end{pmatrix} . \tag{1}$$

Here the 3×3 rotation matrix A and the 3×1 vector $\mathbf{a}$ respectively represent the relative orientation and position of the second coordinate frame of reference relative to the first. The multiplication of two homogeneous transformation matrices can be written as

$$H[A_1, \mathbf{a}_1]\, H[A_2, \mathbf{a}_2] = H[(A_1 A_2, A_1 \mathbf{a}_2 + \mathbf{a}_1)] \ . \tag{2}$$

In the first case, the series of rigid-body motions that result in the frame attached at the base of helix 1 being moved to the base of helix 2 parameterize the homogeneous transformation

$$H = \begin{pmatrix} R_3(\alpha)R_1(\beta)R_3(\gamma) & z_2 R_3(\alpha)R_1(\beta)\mathbf{e}_3 + r\, R_3(\alpha)\mathbf{e}_1 + z_1\, \mathbf{e}_3 \\ \mathbf{0}^{\mathrm{T}} & 1 \end{pmatrix}. \qquad (3)$$

Here $R_i(\phi)$ represents a 3×3 rotation matrix around the natural unit basis vector $\mathbf{e}_i$. Therefore we get $R_1(\phi) = \begin{bmatrix} 1 & 0 & 0 \\ 0 & \cos\phi & -\sin\phi \\ 0 & \sin\phi & \cos\phi \end{bmatrix}$, $R_2(\phi) = \begin{bmatrix} \cos\phi & 0 & \sin\phi \\ 0 & 1 & 0 \\ -\sin\phi & 0 & \cos\phi \end{bmatrix}$,

$R_3(\phi) = \begin{bmatrix} \cos\phi & -\sin\phi & 0 \\ \sin\phi & \cos\phi & 0 \\ 0 & 0 & 1 \end{bmatrix}$, $\mathbf{e}_1 = [1 \;\; 0 \;\; 0]^T$, $\mathbf{e}_2 = [0 \;\; 1 \;\; 0]^T$, and $\mathbf{e}_3 = [0 \;\; 0 \;\; 1]^T$.

Equation (3) is a curvilinear coordinate system which parameterizes the subset of rigid-body motions observed in this case. In this case the Jacobian will have the form

$$\vartheta_R = \begin{pmatrix} J_R & \mathbf{0}_{3\times 3} \\ A^{\mathrm{T}} \dfrac{\partial \mathbf{a}}{\partial \mathbf{q}_1} & A^{\mathrm{T}} \dfrac{\partial \mathbf{a}}{\partial \mathbf{q}_2} \end{pmatrix}. \qquad (4)$$

where the variables $\mathbf{q}_1 = (\alpha, \beta, \gamma)$ and $\mathbf{q}_2 = (z_1, r, z_2)$, and the matrix J_R is obtained in [11]

$$J_R(A) = \begin{pmatrix} \sin\beta\sin\gamma & \cos\gamma & 0 \\ \sin\beta\cos\gamma & -\sin\gamma & 0 \\ \cos\beta & 0 & 1 \end{pmatrix}. \qquad (5)$$

Then the unique volume element for correctly integrating over rigid-body motions is

$$dV = \left|\det \vartheta_R\right| dq_1 \ldots dq_6 = \sin^2 \beta \;\, d\alpha\, d\beta\, d\gamma\, dr\, dz_1\, dz_2 \,. \qquad (6)$$

In the second case of helix-helix interaction, we can get the homogeneous transformation matrix similarly

$$H = \begin{pmatrix} R_3(\phi)R_1(\theta)R_3(\alpha)R_1(\gamma) & r\,R_3(\phi)R_1(\theta)\mathbf{e}_3 + xR_3(\phi)R_1(\theta)R_3(\alpha)\mathbf{e}_1 \\ \mathbf{0}^{\mathrm{T}} & 1 \end{pmatrix}. \tag{7}$$

The corresponding volume element is

$$dV = r\sin\theta\,dr\,dx\,d\phi\,d\theta\,d\alpha\,d\gamma\,. \tag{8}$$

When one can assume the exactness of the measurement, one can use $\sin^2\beta$ for the regularization in the case 1. However our analysis of helix-helix interactions also considers that the measurement of helix-helix angles can contain error from various sources. Given a set of interhelical angles $\{\beta_1,\ldots,\beta_N\}$, one wishes to construct an estimate. If every measurement were exact, and $N \to \infty$, then this distribution is

$$f(\beta) = \frac{1}{N}\sum_{i=1}^{N}\frac{\delta(\beta-\beta_i)}{\sin^2\beta} = \frac{1}{N}\sum_{i=1}^{N}\frac{\delta(\beta-\beta_i)}{\sin^2\beta_i} = \frac{1}{N}\sum_{i=1}^{N}\frac{\delta(\beta-\beta_i)}{\int_{-\pi}^{\pi}\delta(\beta'-\beta_i)\sin^2\beta'd\beta'}\,. \tag{9}$$

where $\delta(\beta)$ is the Dirac delta function.

However, due to uncertainties, one would like to replace the above equation with

$$\tilde{f}(\beta,t) = \frac{1}{N}\sum_{i=1}^{N}\frac{k(\beta-\beta_i,t)}{\int_{-\pi}^{\pi}k(\beta'-\beta_i,t)\sin^2\beta'd\beta'}\,. \tag{10}$$

where $k(\beta,t)$ is a kernel for constructing $\tilde{f}(\beta)$ such that $\|\tilde{f}-f\| \to 0$ as both the accuracy and number of measurements increase. A natural kernel in the current context is obtained by "wrapping" the solution of the heat equation around the circle

$$k(\beta,t) = \sum_{n=-\infty}^{\infty} h(\beta-2\pi n,t)\,. \tag{11}$$

When t is very small (corresponding to high confidence in the measurement) the above equation is accurately approximated when only the $n = 0$ term is retained. In general, the quantity $C(\beta_i,t) = \int_{-\pi}^{\pi}k(\beta'-\beta_i,t)\sin^2\beta'd\beta'$ can be computed in closed form as

$$C(\beta_i,t) = \frac{1}{2}\left(1-e^{-2t}\cos 2\beta_i\right). \tag{12}$$

and for $t > 0$ this provides some regularization which reduces the sensitivity of dividing directly by $\sin^2\beta_i$ when there are uncertainties in the value β_i.

3 Results and Discussion

Among several kinematic parameters shown in Fig. 1, the parameter which shows the strongest patterns is the crossing angle β, which ranges from -180 to +180 degrees. When distributions of β are normalized by the correct volume element in Equation (12), the strongest peaks are close to parallel and anti-parallel stacking of the helices as seen in Fig. 2.

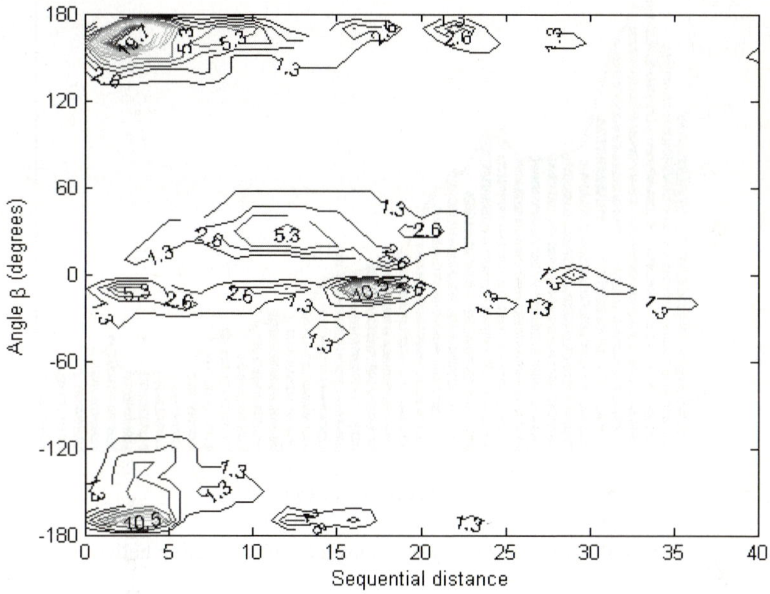

Fig. 2. Angle β preferences broken down by the number of residues in sequence

Note that each occurrence in Fig. 2 is normalized by the quantity in (12) with the variance t = 0.004 before it is binned. The contours in the figure are generated by replacing each normalized data point by a 2-D Gaussian function with the variance of 1.0 for the distance and the variance of 0.004 for the angle.

Interestingly, all three packing preferences persist as sequential distance varies. It is also worth noting is that the parallel packing appears most prominent when the sequential distance between the helices is about 17 residues. Other small peaks are observed, though no significant orientational preferences were observed for helix pairs separated by more than 40 residues. And no kinematic parameters other than the angle β have significant preferred values in this case.

One might argue that since sequentially proximal interactions are more common than sequentially distal interactions (see Fig. 3), it should be no surprise that higher peaks are observed for small values of sequential distance. By normalizing the data in Fig. 2 by the data in Fig. 3, we can verify the validity of the argument.

As seen in Fig. 4, major peaks in Fig. 2 remain even after normalization by the curve in Fig. 3. Compared to the results of Case 1 in [11], the angle locations of peaks in Fig. 4 are nearly identical. We also note that in Fig. 4 the parallel packing seems to exist for larger sequential distances than the anti-parallel one, but in [11] peaks are found in almost the same spatial distance.

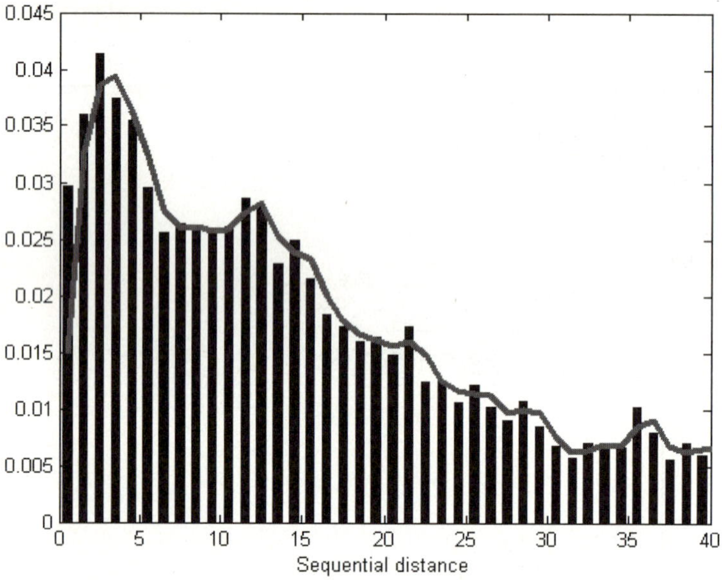

Fig. 3. Relative frequency of occurrence of interacting helices

Very distinctive peaks have been observed in this case of helix interactions. These peaks indicate that nearly parallel or anti-parallel interactions between sequentially proximal helices dominate. No other significant peaks are observed in this case. Parallel packing interactions appear to exist for larger sequential distances than anti-parallel ones.

In this case, helices prefer to turn back on the nearest helix in sequence since higher peaks are observed for small values of sequential distance. Hence, this data appears to support the view that protein folding is hierarchical. In addition, the probability densities constructed from this statistical analysis may be appropriate for use in folding algorithms in which potential functions derived from these probability density functions favor helix-helix interactions at particular angles, with the greatest emphasis on those pairs which occur close in sequence.

In Case 2, two of the parameters show patterns. The strongest peak is seen in the angle θ. As shown in Fig. 5, this occurs in the range from 80 to 90 degrees, and it appears when the sequential distance between the interacting helices is less than 15 residues. Figure 6 illustrates that for a range of values for the angle α (i.e., the angle that rotates helix 1 around the normal to helix 2), peaks appear near 0 and ±180 degrees, which indicates a preference for helices in the case 2 to interact with their helix axes being coplanar. These preferences also decay when the sequential distance is greater than 15 residues.

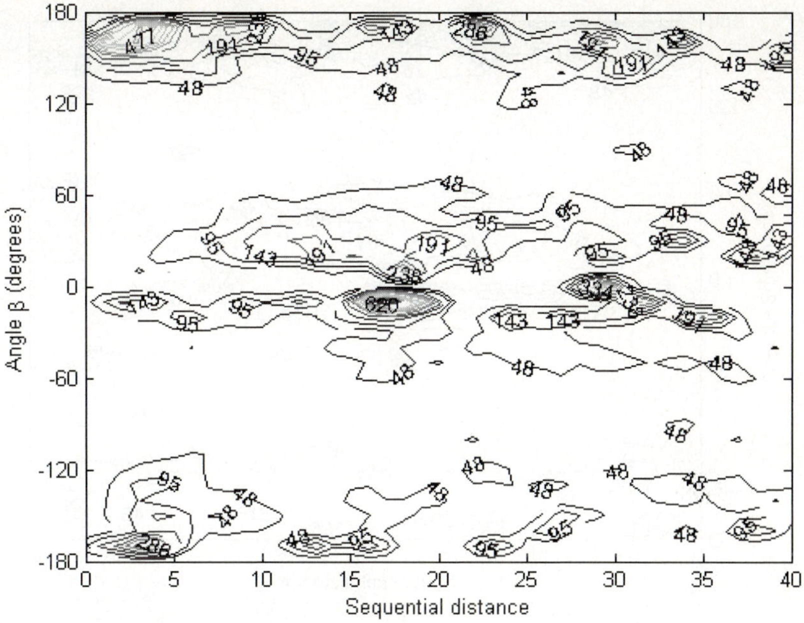

Fig. 4. Result of normalizing Figure 2 by Figure 3: The high anti-parallel peak observed for sequentially proximal pairs remains, and new near-parallel peaks emerge

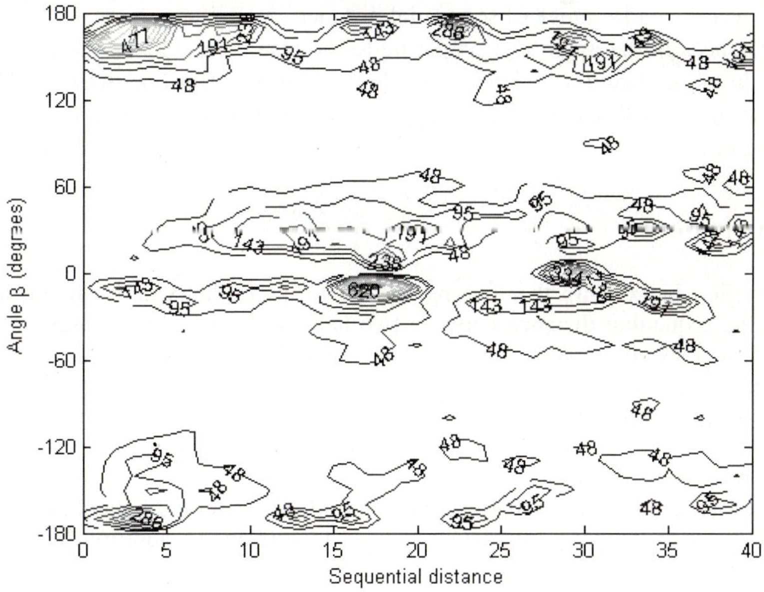

Fig. 5. Angle θ preferences for Case 2

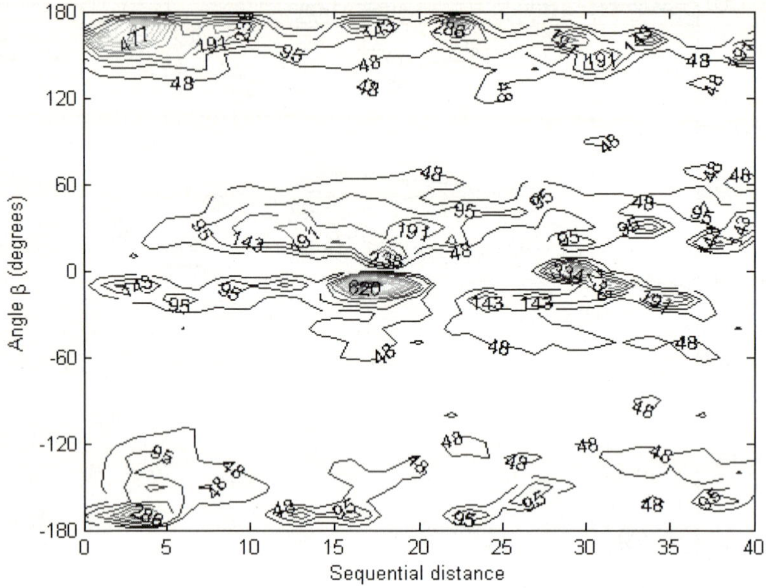

Fig. 6. Angle α preferences for Case 2

However, unlike the results of the case 1, when the data in Fig. 5 and 6 are divided by the probability density of helix pairs, it is observed that the strongest peak in Fig. 5 spreads over the entire range of sequential distance. A similar pattern is also found for the preferences in the angle α. Comparing to the results in [11], we note that similar patterns in angle locations are found for both θ and α. But one difference is noted that the peaks near 0 degree of α exist in larger sequential distances than those near ±180 degrees.

4 Conclusions

We identified helix pairs that belong to two forms of helix-helix interactions and defined the sequential distance between two helices. Then the two cases of helix-to-helix interactions are examined in terms of interhelical angle and sequential distance. The interactions are considered as all possible rigid-body motions, and for each case corresponding parameters are found and an associated volume element for each class is derived. The volume element is used to normalize the relative frequency of occurrence of relative helix orientations.

The bivariate distribution that depends on both the observed helical angles and sequential distance between interacting helices is studied. In addition, each of these bivariate distributions is normalized by the frequency of occurrence of interactions at a particular sequential distance. This analysis shows that very distinctive peaks appear in preferred orientations, and that the relations of peaks to sequential distance are discussed. This analysis supports both the theory that protein folding is hierarchical and the theory that discrete preferences exist in helix-helix interactions. This result should be useful for the protein secondary structure prediction because patterns of helix interactions are found in terms of sequential distance and angle.

Acknowledgements. This paper was supported by Konkuk University in 2006.

References

1. Baldwin, R.L., Rose, G.D.: Is protein folding hierarchic? I. Local structure and peptide folding. Trends in Biochemical Sciences 24(1), 26–33 (1999)
2. Baldwin, R.L., Rose, G.D.: Is protein folding hierarchic? II. Folding intermediates and transition states. Trends in Biochemical Sciences 24(2), 77–83 (1999)
3. Crick, F.: The packing of α-helical: simple coiled coils. Acta Crystallogr 6, 689–697 (1953)
4. Chothia, C., Levitt, M., Richardson, D.: Structure of proteins: Packing of (-helices and pleated sheets. Proc. Natl. Acad. Sci. USA 74(10), 4130–4134 (1977)
5. Chothia, C., Levitt, M., Richardson, D.: Helix to helix packing in proteins. J. Mol. Biol. 145, 215–250 (1981)
6. Finkelstein, A.V., Ptitsyn, O.B.: Why do globular proteins fit the limited set of folding patterns? Prog. Biophys. Molec. Biol. 50, 171–190 (1987)
7. Murzin, A.G., Finkelstein, A.V.: General architecture of the α-helical: globule. J. Mol. Biol. 204, 749–769 (1988)
8. Berman, H.M., Westbrook, J., Feng, Z., Gilliland, G., Bhat, T.N., et al.: The Protein Data Bank. Nucleic Acids Res. 28, 235–242 (2000)
9. Bowie, J.U.: Helix packing angle preferences. Nature Structural Biology 4(11), 915–917 (1997)
10. Walther, D., Springer, C., Cohen, F.E.: Helix-helix packing angle preferences for finite helix axes. Proteins: Structure, Function and Genetics 33, 457–459 (1998)
11. Lee, S., Chirikjian, G.S.: Interhelical Angle and Distance Preferences in Globular Proteins. Biophysical Journal 86, 1105–1117 (2004)
12. Wang, G., Dunbrack, R.L.: PISCES: a Protein Sequence Culling Server. Bioinformatics 19, 1589–1591 (2003)

Classification of Ligase Function Based on Multi-parametric Feature Extracted from Protein Sequence

Bum Ju Lee[1], Heon Gyu Lee[1], Moon Sun Shin[2], and Keun Ho Ryu[1,*]

[1] Database/Bioinformatics Lab., Chungbuk National University, 12 Gaeshindong, Cheongju, Chungbuk, South Korea, 361-763
{bjlee,hglee,khryu}@dblab.chungbuk.ac.kr
[2] Dept. of Computer Science, Konkuk University, 322 Danwol-Dong, Chungju-Si, Chungbuk, South Korea, 380-701
msshin@kku.ac.kr

Abstract. One of the important goals of bioinformatics is to classify and predict the functions of proteins that have no sequence homolog of known functions. The purpose of this paper is to classify protein function by using multi-parametric feature, without sequence similarity. Firstly, we propose a method for generating novel features that present various local information of protein sequence based on positively and negatively charged residues. Then, we introduce a process of making optimal feature subset through combination of traditional and novel features extracted from protein sequence. Finally, we classify ligase enzymes by support vector machine (SVM). In experiment, only 375 out of 483 features were selected by feature selection, and the classification accuracy for 4th sub-classes in Enzyme Commission (EC) number is 98.35%. Our results demonstrate that most of novel features are valuable for specific enzyme function classification.

Keywords: Ligases, Protein function, Classification, Feature, Correlation-based feature selection (CFS), SVM.

1 Introduction

Generally, if unknown gene or protein appears through sequencing in laboratory, researchers carry out sequence similarity search methods, which are BLAST [6], PSI-BLAST [7], and FASTA [21] for finding similar proteins or annotation information in databases. However, proteins that have diverged from a common ancestral gene may have the same function but have very different sequences. Therefore, the sequence similarity based-approach often fails to identify the relevant proteins when sequence similarity is either absent or very low (called twilight zone), or when homology of known proteins in protein databases does not exists [5], [10], [11], [22], [26], [27], [31], [41], [42], and [43]. Also, a structure-based approach such as Deli [24] and MATRAS [23] using structure comparison was often used for identifying similar structure since protein structural information is better conserved than sequence information [8]. But, the same

* Corresponding author.

O. Gervasi et al. (Eds.): ICCSA 2008, Part II, LNCS 5073, pp. 1096–1106, 2008.

functional proteins can have different structures, and moreover structural comparison through structure determination is more difficult task than sequence comparison [9], [26], and [35].

In bygone years, many studies on classification and prediction efficiency of enzyme or protein function independently of traditional sequence or structural similarity approaches have proposed [3], [9], [10], [11], [14], [15], [25], [29], and [40]. The approach based on alternative representation of proteins characterizes the simple sequence and structural properties of each enzyme class. For instance, method to predict enzyme families based on features extracted from protein structure was proposed [10]. In that study, prediction of enzyme class using features such as secondary structure contents, ligands, surface properties, etc. was performed through building one-class versus one-class models by means of the SVM statistical learning method, and obtained 62.16% accuracy on the average. Also, combination method of statistical strength and data mining for enzyme class prediction was suggested [16]. The method used the features from the STING_DB database and applied the naïve bayesian classifier for building model.

Features, one of the most important compositions for successful classification and prediction of protein function, were continuously extracted from various resources of proteins. For instance, for novel features, method to divide 20 amino acids into three different groups based on different physiochemical properties was introduced in previous studies, and predicted enzyme function by SVM [4] and [11]. Also, work onenzyme function prediction using dataset composed of various features extracted from protein sequence, secondary structure, motif as well as annotation information of protein database was recently introduced, and used the mixture model of stochastic decision trees for building model [9]. In the meantime, Prediction of function based on only several meaningful features, which are O-ß-GlcNAc sites, N-linked glycosylation, secondary structure, physicochemical properties, etc. derived from protein sequence, were conducted [28], [29], and [34]. In that study, discriminative factor of each feature was very well presented by correlation coefficient and used neural networks for prediction.

These former studies suggested new classification and prediction methods of enzyme function using the direct relationship between basic protein properties and their function. However, subtle distinction occurring in small part of protein sequence in the most of previous studies was not represented as features. Indeed, although proteins share similar structural organization, biological properties and sequences, small changes of amino acids in protein sequence may derive different functions [9], [35], and [36].

Here, we design a simple method of making four features, which present subtle differences of local regions in protein sequences, using positively/negatively charged residues. Also, after feature subset selection, we find out optimal subset so that the highest accuracy in the given dataset is achieved. Lastly, we obtained 98.35% accuracy in classification for 4th sub-classes of ligase in EC number. In EC classification [2], enzyme families are defined based on the functional role of the enzymes in the cell rather than on common evolutionary ancestry. Therefore, if we classify or predict subordinate enzyme classes, classification task is difficult because accuracy is lower.

2 Dataset and Feature Extraction

2.1 Dataset

Enzymes, which are biological catalysts that mediate almost all chemical reactions and are broadly distributed in all tissues and fluids of the body, are proteins executing a primary role in metabolic pathways.

For protein dataset, we collected 10036 protein sequences annotated with EC numbers (release of 12-Jun-2007) [2] from SWISS-PROT release 53.2 [1]. In preprocessing, we removed proteins that include missing values, and discarded classes of which the number of member proteins is less than 20. Finally, we prepared a total of 9925 proteins and 54 classes.

2.2 Traditional Feature Generation

Features are properties that characterize specific protein (i.e. characterize the simple sequence and structure properties). We extracted a total of 483 features from each protein sequence. The features consist of the traditional features adopted from previous studies [10], [11], [18], and [34] and novel features. Traditional features consist of amino acid composition, amino acid pair ratios, protein length, number of atoms, molecular weight, etc.

Among traditional features, 33 features were extracted by ProtParam tool [17] with SWISS-PROT sequences. And for usage of the physiochemical properties, we selected 17 features (16 properties adopted from [9] and an additional one property), because usage of positively charged residue is two types (i.e. lysine and arginine or histidine, lysine, and arginine). Also, we computed percentage of each pair of amino acids in the protein sequences (400 features).

2.3 Novel Feature Generation

For novel features, we design and compute the ratio of change of charged residues as well as the distribution of charged residues (*PPR, NNR, PNPR*, and *Dist(x, y)* of the ratio value of each *PPR, NNR* and *PNPR*). We noted that the importance of negatively/positively charged residues in protein function has been described in several studies [9], [18], [19], [28], [29], and [30]. Our method for identifying novel features is simple. *NNR* value was calculated by the following formula:

$$PPR = \frac{(C-1)}{N} \times 100, \tag{1}$$

where N is the total number of amino acids in sequence, C is the number of continuous changes from positively charged residue to next positively charged residue. Similar to *PPR*, *NNR* value is calculated in accordance with the following formula:

$$NNR = \frac{(E-1)}{N} \times 100, \tag{2}$$

where E is the total number of continuous changes from negatively charged residue to next negatively charged residue. *PNPR* value is calculated by the following formula:

$$PNPR = \frac{(A-1)}{N} \times 100, \tag{3}$$

where A is the total number of continuous changes from positively charged residue to next negatively charged residue or vice verse. Lastly, *Dist(x, y)* is the distribution of each *PPR*, *NNR*, and *PNPR*. For example, *PPRDist(x,y)* value as distribution value of *PPR* can be defined as following formula:

$$PPRDist(x, y) = \left(\sum_{i=x}^{y} PP_n \right) \times 100, \tag{4}$$

where x is the starting point and y is the ending point in the sequence under the stipulation that x value is smaller than y value, and n is the total number of transitions from positively charged to positively charged in given sequence. This feature will present various local information of proteins sequence according to the values of x and y because regional difference among the similar proteins under the same family exists slightly in sequences. We generated a total of 33 features based on above formulas (exactly, *PPR* (1), *NNR* (1), *PNPR* (1), *PPRDist(x, y)* (10), *NNRDist(x, y)* (10), and *PNPRDist(x, y)* (10), as the sequence length is basically divided into 10 local regions). Specific contents of entire features were described in table 1.

Table 1. Features extracted and computed from protein sequences

	Feature	Description	Number
1	Number of AAs	Number of residues in each protein	1
2	Molecular weight	The molecular weight of the protein	1
3	Theoretical PI	The pH at which the net charge of the protein is zero (isoelectric point)	1
4	Amino acid composition	Percentage of each amino acid in the protein	20
5	Positively charged residues_2	Percentage of positively charged residues in the protein (lysine and arginine)	1
6	Carbon	Atomic composition	1
7	Hydrogen	Atomic composition	1
8	Nitrogen	Atomic composition	1
9	Oxygen	Atomic composition	1
10	Sulfur	Atomic composition	1
11	Number of atoms	Total number of atoms	1
12	Extinction coefficient	How much light a protein absorbs at a certain wavelength (in units of M^{-1} cm^{-1} at 280 nm measured in water)	1
13	Instability Index	The stability of the protein	1
14	Aliphatic Index	The relative volume of the protein occupied by aliphatic side chains	1
15	GRAVY	Grand average of hydropathicity	1
16	PPR	Percentage of continuous changes from positively charged residue to positively charged residue	1

Table 1. (*Continued*)

	Feature	Description	Number
17	NNR	Percentage of continuous changes from negatively charged residue to negatively charged residue	1
18	PNPR	Percentage of continuous changes from negatively charged residue to negatively charged residue or from negatively charged residue to positively charged residue	1
19	NNR(x, y)	Percentage of NNR from x to y (local information)	10
20	PPR(x, y)	Percentage of PPR from x to y	10
21	PNPR(x, y)	Percentage of PNPR from x to y	10
22	Amino acid pair ratio	Percentage of each pair of amino acids in each protein	400
23	Charged	Physiochemical properties	1
24	Positively charged residues_3	Percentage of positively charged residues in the protein (histidine, lysine, and arginine)	1
25	Negatively charged residues	Percentage of negatively charged residues in the protein	1
26	Polar	Physiochemical properties	1
27	Aliphatic	Physiochemical properties	1
28	Aromatic	Physiochemical properties	1
29	Small	Physiochemical properties	1
30	Tiny	Physiochemical properties	1
31	Bulky	Physiochemical properties	1
32	Hydrophobic	Physiochemical properties	1
33	Hydrophobic and Aromatic	Physiochemical properties	1
34	Neutral, weakly and Hydrophobic	Physiochemical properties	1
35	Hydrophilic and Acidic	Physiochemical properties	1
36	Hydrophilic and Basic	Physiochemical properties	1
37	Acidic	Physiochemical properties	1
38	Polar and uncharged	Physiochemical properties	1

3 Feature Subset Selection

One of the challenges of the classification and prediction task is to make the better performance in a given dataset. In order to execute a classification task efficiently, feature selection method is very important because of improving the prediction performance of the predictors and providing more cost-effective predictors [12], [13], [18], and [19]. Additionally, the method selects relevant features and removes redundant and/or irrelevant features. So, we applied correlation-based feature selection (CFS) method to extract optimal feature subset. In feature subset selection, with n features, through the heuristic forward selection or backward elimination, there are n^2 possible subsets [37]. The only way to find the best subset would be to examine the effect of adding every feature to each of the all possible subsets.

Briefly, the method is a filter and uses search algorithm along with a function to evaluate the merit of feature subset on the basis of hypothesis that "Good feature subsets contain features highly correlated with the class, yet uncorrelated with each other" [37] and [38]. Therefore, this method evaluates subsets of features rather than individual features. The core of CFS is subset evaluation heuristic that handles irrelevant features as they will be poor predictor of the classes. Redundant features are distinguished against as they will be highly correlated with one or more of the other features. After calculating the correlation between features, CFS estimates merit of a feature subset. Heuristic formula (Pearson's correlation) is as following:

$$Merit_s = \frac{k\overline{r_{cf}}}{\sqrt{k + k(k-1)\overline{r_{ff}}}}. \quad (5)$$

$Merit_s$ is the heuristic merit of a feature subset S that comprise k features, $\overline{r_{cf}}$ means average correlation of features and class, $\overline{r_{ff}}$ is average inter-correlation of features. Above formula was excerpted from [38] and [39].

Prior to feature subset selection, we tested various binning from 6 to 24 in discretization step in which continuous features are converted to nominal. Number of features and kind of features in selected subset differ according to the number of binning. Table 2 explains detailed information in selected subset in accordance with the number of binning (traditional features is 1~15, novel features is 16~21, amino acid pair ratios is 22, and physiochemical properties is 23~38 shown in table 1).

Through classifications using selected subsets, we finally selected 375 features out of 483 features (specifically, number of binning is 8, and selected feature subset contains traditional features of 17, novel features of 31, amino acid pair ratios of 323, and physiochemical properties of 4).

Even though many features were discarded during feature subset selection procedure, but many of novel features were preserved in every subset selections according to number of binning. So, our features are highly correlated with the class and have low inter-correlation with each other.

4 Experiments and Discussion

The SVM is a basically two-class classifier and can be extended for multi-class classification. In model, each object is mapped to a point in a high dimensional space, each dimension corresponding to feature. The coordinates of the point are the frequencies of the features in the corresponding dimensions. The SVM learns, in the training step, the maximum-margin hyper-planes separating each class. In the testing step, it classifies a new object by mapping it to a point in the same high-dimensional space divided by the hyper-plane learned in the training step.

Implementation of SVM algorithm used here is libSVM [32] and [33] in Weka [16]. In C-SVC type machine, kernel type was used with radial basis function. Also, value of error penalty (cost parameter) is 4, and values of other parameters fixed default. Validation is performed using 10-fold cross validation test (additionally, we classified dataset respectively, the training of 80% and the testing of 20%, but our experiment performance was slightly decreased).

Table 2. Specific results of feature subset according to the number of binning (T_Fs : traditional features, N_Fs : novel features, AAPR : amino acid pair ratio, P_P : physiochemical properties)

Num. of bin	Num. of T_Fs (34)	Num. of N_Fs (33)	Num. of AAPR (400)	Num. of P_Ps (16)	Num. of total features	Merit of selected subset
6	16	30	298	4	348	0.663
8	17	31	323	4	375	0.733
10	17	30	327	3	377	0.765
12	16	32	287	4	339	0.785
14	15	31	301	4	351	0.798
16	15	31	245	4	295	0.8
18	15	32	226	4	277	0.801
20	15	32	219	4	270	0.8
22	16	31	180	5	232	0.794
24	15	32	151	5	203	0.786

In final experiment, our classification achieved an accuracy of 98.35%, in this case the value of MAE (mean absolute error) is 0.0006 and the value of RMSE (Root mean squared error) is 0.0247%. We described detailed results in Table 3.

Table 3. Detailed results of each ligase sub-class (TP : true positive, FP : false negative)

Class	TP	FP	Precision	F-measure	ROC area
6.1.1.1	0.994	0	0.994	0.994	0.997
6.1.1.10	1	0	1	1	1
6.1.1.11	0.995	0	1	0.998	0.998
6.1.1.12	0.997	0	1	0.998	0.998
6.1.1.13	0.909	0	1	0.952	0.955
6.1.1.14	0.995	0	0.995	0.995	0.997
6.1.1.15	0.993	0.001	0.985	0.989	0.996
6.1.1.16	1	0	0.997	0.998	1
6.1.1.17	0.988	0.001	0.98	0.984	0.994
6.1.1.18	0.982	0	1	0.991	0.991
6.1.1.19	1	0	0.991	0.995	1
6.1.1.2	0.987	0	0.994	0.99	0.994
6.1.1.20	1	0.002	0.971	0.985	0.999
6.1.1.21	0.996	0	1	0.998	0.998
6.1.1.22	0.992	0	1	0.996	0.996
6.1.1.3	0.985	0	0.99	0.987	0.992
6.1.1.4	0.995	0	1	0.998	0.998
6.1.1.5	1	0	0.993	0.996	1
6.1.1.6	0.973	0.001	0.968	0.971	0.986
6.1.1.7	1	0	0.996	0.998	1
6.1.1.9	1	0	0.996	0.998	1
6.2.1.1	1	0	1	1	1
6.2.1.26	0.958	0	1	0.979	0.979
6.2.1.3	0.941	0	1	0.97	0.971
6.2.1.5	1	0	1	1	1

Table 3. (*Continued*)

Class	TP	FP	Precision	F-measure	ROC area
6.3.1.1	1	0	1	1	1
6.3.1.2	0.994	0	1	0.997	0.997
6.3.1.5	1	0	0.988	0.994	1
6.3.2.1	1	0	1	1	1
6.3.2.13	0.984	0	0.984	0.984	0.992
6.3.2.19	0.989	0.001	0.935	0.961	0.994
6.3.2.2	0.75	0.001	0.776	0.763	0.874
6.3.2.3	0.803	0.001	0.824	0.813	0.901
6.3.2.4	0.972	0.001	0.958	0.965	0.986
6.3.2.6	0.986	0	0.986	0.986	0.993
6.3.2.8	0.987	0	0.983	0.985	0.993
6.3.2.9	0.993	0	0.986	0.99	0.996
6.3.3.1	0.951	0.001	0.959	0.955	0.975
6.3.3.3	0.969	0	0.984	0.976	0.984
6.3.4.13	0.895	0.001	0.895	0.895	0.947
6.3.4.14	0.1	0.001	0.133	0.114	0.549
6.3.4.2	1	0	1	1	1
6.3.4.3	0.98	0	1	0.99	0.99
6.3.4.4	1	0	1	1	1
6.3.4.5	1	0	1	1	1
6.3.5.2	0.993	0	1	0.996	0.996
6.3.5.3	1	0	0.997	0.999	1
6.3.5.4	0.806	0	1	0.893	0.903
6.3.5.5	1	0	0.996	0.998	1
6.4.1.2	0.92	0.002	0.915	0.917	0.959
6.5.1.1	0.93	0	0.985	0.957	0.965
6.5.1.2	0.935	0	1	0.966	0.967
6.5.1.4	0.927	0	1	0.962	0.963
6.6.1.1	0.933	0	1	0.966	0.967

Precision is the ratio of TP in a given class divided by the total number of instances classified in that class. F-measure is harmonic mean of precision and recall, and receiver operating characteristic (ROC) measures the area under the curve [15] and [20]. Most of instances per each class are well classified except for one class (6.3.4.14). There are reasons leading to such a result. Firstly, the number of member proteins is few. Secondly, discriminatory features for the class were not enough. The class 6.3.4.14 (biotin carboxylase) in Swiss-Prot database retained only 22 proteins that were collected from various organism species.

Like the structure and sequence similarity approach for assignment of function to unknown proteins, feature based approach can be divided into features extracted from structure and features extracted from sequence. Up to the present, data of protein sequences are richer than structure data. The reasons of the use of features extracted from only protein sequences in our experiment are that data of protein sequences are abundant, and it reduces exertion for extraction of structure information, because predictive or experimental work for structure information is both costly and time consuming.

Many studies have been endeavored biologically and computationally to find the meaningful as well as broad features for classification and prediction of more accurate and specific protein function, but features showing obvious propensity for enzyme function was not enough as yet. It alludes to the incipient stage of the discovery of relationship between the feature and function of protein. Accordingly, we need more insight about feature analysis and biological understanding.

5 Conclusions

In this paper, we proposed a method to generate novel features that present subtle differences of local regions in protein sequences, and investigated feature subset selection of features. Also, we obtained good accuracy of classification of 4th subclasses of ligase enzyme.

A trade-off exists between the use of many features extracted from various resources and the use of small features extracted from sequence resource with respect to efficiency, computational time, accuracy and data size. However, our framework not only gives a way to select the protein features, but also gives an alternative approach when the sequence and structure similarity approach fails in enzyme function prediction. Future work of this research would involve more discriminatory features to subset and apply various protein functions.

Acknowledge

This work was supported by the Korea Research Foundation Grant funded by the Government of the Republic of Korea (MOEHRD) (The Regional Research Universities Program/Chungbuk BIT Research-Oriented University Consortium) and supported by the Korea Science and Engineering Foundation (KOSEF) grant funded by the Korea government(MOST) (R01-2007-000-10926-0).

References

1. Bairoch, A., Apweiler, R.: The SWISS-PROT protein sequence data bank and its supplement TrEMBL in 1999. Nucleic Acids Res. 27, 49–54 (1999)
2. Bairoch, A.: The Enzyme Database in 2000. Nucleic Acids Res. 28, 304–305 (2000)
3. Cai, C.Z., Wang, W.L., Sun, L.Z., Chen, Y.Z.: Protein function classification via support vector machine approach. Math. Biosci. 185, 111–122 (2003a)
4. Cai, C.Z., Han, L.Y., Ji, Z.L., Chen, X., Chen, Y.Z.: SVM-Prot: web-based support vector machine software for functional classification of a protein from its primary sequence. Nucleic Acids Res. 31, 3692–3697 (2003b)
5. Wang, X., Schroeder, D., Dobbs, D., Honavar, V.: Automated data-driven discovery of motif-based protein function classifiers. Inf. Sci (ISCI) 155, 1–18 (2003)
6. Altschul, S.F., Gish, W., Miller, W., Myers, E.W., Lipman, D.J.: Basic Local Alignment Search Tool. J. Mol. Biol. 215, 403–410 (1990)

7. Altschul, S.F., Madden, T.L., Schaffer, A.A., Zhang, J., Zhang, Z., Miller, W., Lipman, D.J.: Gapped BLAST and PSI-BLAST: a new generation of protein database search programs. Nucleic Acids Res. 35, 3389–3402 (1997)
8. Eidhammer, I., Jonassen, I., Taylor, W.R.: Protein Structure Comparison and Structure Patterns. J. Comput. Biol. 7, 685–716 (2000)
9. Syed, U., Yona, G.: Enzyme function prediction with interpretable models. In: Methods in Molecular Biology: Computational Systems Biology, pp. 1–33. Humana Press (2007)
10. Dobson, P.D., Doig, A.J.: Predicting Enzyme Class from Protein Structure without Alignments. J. Mol. Biol. 345, 187–199 (2005)
11. Han, L.Y., Cai, C.Z., Ji, Z.L., Cao, Z.W., Cui, J., Chen, Y.Z.: Predicting functional family of novel enzymes irrespective of sequence similarity: a statistical learning approach. Nucleic Acids Res. 32, 6437–6444 (2004)
12. Noble, W.S., Ben-Hur, A.: Integrating Informmation for protein function prediction, Bioinformatics-From Genomes Therapies. In: Lengauer, T. (ed.), WILE-VCH, Weinheim, vol. 3, pp. 1297–1314 (2007)
13. Guyon, I.: An Introduction to Variable and Feature Selection. J. Mach. Learn. Res. 3, 1157–1182 (2003)
14. Chawla, N.V.: C4.5 and Imbalanced Data sets: Investigating the effect of sampling method, probabilistic estimate, and decision tree structure. In: Proc. International Conference on Machine Learning (ICML), Workshop on learning from imbalanced datasets II (2003)
15. Borro, L.C., Oliveira, S.R.M., Yamagishi, M.E.B., Mancini, A.L., Jardine, J.G., Mazoni, I., Santos, E.H.D., Higa, R.H., Kuser, P.R., Neshich, G.: Predicting enzyme class from protein structure using Bayesian classification. Genet. Mol. Res. 5, 193–202 (2006)
16. Ian, H.: Data Mining: Practical machine learning tools and techniques, 2nd edn. Morgan Kaufmann, San Francisco (2005), http://www.cs.waikato.ac.nz/ml/weka/
17. Gasteiger, E., Hoogland, C., Gattiker, A., Duvaud, S., Wilkins, M.R., Appel, R.D., Bairoch, A.: Protein Identification and Analysis Tools on the ExPASy Server. In: John, M.W. (ed.) The Proteomics Protocols Handbook, pp. 571–607. Humana Press (2005)
18. Al-Shahib, A., Breitling, R., Gilbert, D.: Feature Selection and the class imbalance problem in predicting protein function from sequence. Appl. Bioinformatics 4, 195–203 (2005a)
19. Al-Shahib, A., Breitling, R., Gilbert, D.: FRANKSUM: New feature selection method for protein funciton prediction. Int. J. Neural Syst. 15, 259–275 (2005b)
20. Tan, P.N., Steinbach, M., Kumar, V.: Introduction to Data Mining. In: Goldstein, et al. (eds.), pp. 163–298. Addison Wesley, Reading (2006)
21. Pearson, W.R., Lipman, D.J.: Improved tools for biological sequence comparison. Proc. Natl. Acad. Sci. U.S.A. 85, 2444–2448 (1988)
22. Rost, B.: Twilight zone of protein sequence alignments. Protein Eng. 12, 85–94 (1999)
23. Kawabata, T.: MATRAS: a program for protein 3D structure comparison. Nucleic Acids Res. 31, 3367–3369 (2003)
24. Holm, L., Sande, C.: Dali: a network tool for protein structure comparison. Trends Biochem. Sci. 20, 478–480 (1995)
25. Drummond, C., Holte, R.C.: C4.5, class imbalance, and cost sensitivity: why undersampling beats over-sampling. In: Proc. International Conference on Machine Learning (ICML), Workshop on Learning from Imbalanced Datasets II (2003)
26. Lapinsh, M., Gutcaits, A., Prusis, P., Post, C., Lundstedt, T., Wikberg, J.E.S.: Classification of G-protein coupled receptors by alignment-independent extraction of principal chemical properties of primary amino acid sequences. Protein Sci. 11, 795–805 (2002)

27. Claeyssens, M., Henrissat, B.: Specificity mapping of cellulolytic enzymes: Classification into families of structurally related proteins confirmed by biochemical analysis. Protein Sci. 1, 1293–1297 (1992)
28. Jensen, L.J., Gupta, R., Blom, N., Devos, D., Tamames, J., Kesmir, C., Nielsen, H., Stærfeldt, H.H., Rapacki, K., Workman, C., Andersen, C.A.F., Knudsen, S., Krogh, A., Valencia, A., Brunak, S.: Prediction of Human Protein Function from Post-translational Modifications and Localization Features. J. Mol. Biol. 319, 1257–1265 (2002a)
29. Jensen, L.J., Skovgaard, M., Brunak, S.: Prediction of novel archaeal enzymes from sequence-derived features. Protein Sci. 3, 2894–2898 (2002b)
30. Truniger, V., Lazaro, J.M., Esteban, F.J., Blanco, L., Salas, M.: A positively charged residue of φ 29 DNA polymerase, highly conserved in DNA polymerases from families A and B, is involoved in binding the incoming nucleotide. Nucleic Acids Res. 30, 1483–1492 (2002)
31. Pawlowski, K., Jaroszewski, L., Rychlewski, L., Godzik, A.: Sensitive sequence comparison as protein function predictor. In: Proc. pacific Symposium on Biocomputing, vol. 5, pp. 42–53 (2000)
32. Chang, C.C., Lin, C.J.: LIBSVM: a library for support vector machines, Software (2001), http://www.csie.ntu.edu.tw/~cjlin/libsvm
33. Yasser E.L.M.: WLSVM (2005), http://www.cs.iastate.edu/~yasser/wlsvm/
34. Bendtsen, J.D., Jensen, L.J., Blom, N., Heijne, G.V., Brunak, S.: Feature-based prediction of non-classical and leaderless protein secretion. Protein Eng. Des. Sel. 17, 349–356 (2004)
35. Russell, R.B., Saqi, M.A., Bates, P.A., Sayle, R.A., Sternberg, M.J.: Recognition of analogous and homologous protein folds-assessment of prediction success and associated alignment accuracy using empirical substitution matrices. Protein Eng. 11, 1–9 (1998)
36. Todd, A.E., Orengo, C.A., Thornton, J.M.: Evolution of Function in Protein Superfamilies, from a Structural Perspective. J. Mol. Biol. 307, 1113–1143 (2001)
37. Hall, M.A.: Correlation-based feature selection for machine learning, Ph.D. thesis, Department of Computer Science, University of Waikato, Hamilton, New Zealand (1998)
38. Hall, M.A.: Correlation-based feature selection for discrete and numeric class machine learning. In: Proc. of the 17th Int. Conf. on Machine Learning (ICML2000), pp. 359–366. Morgan Kaufmann Publishers Inc., San Francisco (2000)
39. Hall, M.A., Holmes, G.: Benchmarking Feature Selection Techniques for Discrete Class Data Mining. IEEE Transactions on Knowledge and Data Engineering 15, 1–16 (2003)
40. Lee, B.J., Lee, H.G., Lee, J.Y., Ryu, K.H.: Classification of Enzyme Function from Protein Sequence based on Feature Representation. In: Proc. of the 7th IEEE Int. Conf. on Bioinformatics and Bioengineering (BIBE 2007), vol. 2, pp. 741–752 (2007)
41. Lee, B.J., Lee, H.G., Kim, D.S., Ryu, K.H.: Feature Extraction in Spatially-Conserved Regions and Protein Functional Classification. In: Proc. of the 2th Int. Conf. on Frontiers in the Convergence of Bioscience and Information Technologies (FBIT 2007), vol. 1, pp. 165–170 (2007)
42. Kim, S.S., Kang, J.W., Chung, Y.J., Li, J.Y., Ryu, K.H.: Clustering orthologous proteins across phylogenetically distant species. Proteins 71, 1113–1122 (2008)
43. Kim, S.S., Jung, K.S., Ryu, K.H.: Automatic Orthologous-Protein-Clustering from Multiple Complete-Genomes by the Best Reciprocal BLAST Hits. In: Li, J., Yang, Q., Tan, A.-H. (eds.) BioDM 2006. LNCS (LNBI), vol. 3916, pp. 60–70. Springer, Heidelberg (2006)

Thai Sign Language Translation Using Fuzzy C-Means and Scale Invariant Feature Transform

Suwannee Phitakwinai[1], Sansanee Auephanwiriyakul[1], and Nipon Theera-Umpon[2]

[1] Computer Engineering Department, Faculty of Engineering,
Biomedical Engineering Center
Chiang Mai University, Chiang Mai 50200 Thailand
sansanee@ieee.org

[2] Electrical Engineering Department, Faculty of Engineering,
Biomedical Engineering Center
Chiang Mai University, Chiang Mai 50200 Thailand
nipon@ieee.org

Abstract. Visual communication is important for a deft and/or mute person. It is also one of the tools for the communication between human and machines. In this paper, we develop an automatic Thai finger-spelling sign language translation system using Fuzzy C-Means (FCM) and Scale Invariant Feature Transform (SIFT) algorithms. We collect key frames from several subjects at different times of day and for several days. We also collect testing Thai finger-spelling words video from 4 subjects. The system achieves 79.90% and 51.17% correct alphabet translation and the correct word translation, respectively, with the SIFT threshold of 0.7 and 1 nearest neighbor prototype. However, when we change the number of nearest neighbor prototypes to 3, the system yields 82.19% and 55.08% correct alphabet and correct word translation, respectively, at the same SIFT threshold. These results are comparable with the manually-picked Rframe translation system.

Keywords: Shot Clustering, SIFT, FCM, Hand gesture recognition.

1 Introduction

There are several forms of communication among people. One of those is visual communication, e.g., hand gesture, body language, etc. This type of communication is a useful tool to improve a quality of life for deaf or non-vocal persons. Besides being the communication between humans, visual communication becomes one of the important communication channels between human and machines.

Deaf or non-vocal persons communicate using sign language or hand gesture. However, there is a barrier between hearing and deaf persons because most hearing persons cannot understand the language. Hence, the problem of recognizing a sign language becomes an interesting and popular research area. There are several approaches [1–7] that have been proposed in this research area. However, these approaches are based on the idea that the hand gesture data is collected through a cyber-glove. Although, these approaches provide good recognition rates, it is more natural

O. Gervasi et al. (Eds.): ICCSA 2008, Part II, LNCS 5073, pp. 1107–1119, 2008.

to recognize hand gesture without using a special hardware. There are a few researches that propose techniques in recognizing static hand gestures with image caption or dynamic hand gestures with video caption [8–13]. However, these methods need to either automatic or manually segment hand regions before performing any recognition.

In this paper, we build a dynamic hand gesture translation system with video caption without prior hand region segmentation. In particular, we apply the fuzzy C-means algorithm to find a representative frame (Rframe) that represents a set of frames with little significant difference forming a scene or shot [14–16]. We also apply the scale invariant feature transform (SIFT) [17] in the recognition process of the dynamic hand gestures in the Thai finger-spelling sign language for 10 words, i.e., “Far”, “Good”, “Leaf”, “Meet”, “Fire”, “Rich”, “Erase”, “Float”, “Color”, and “Comb”.

2 System Description

Although, this system recognizes only 10 words of the Thai finger-spelling sign language, there are actually 15 Thai finger spelling alphabets [18] shown in figure 1. We collect these alphabets from 11 subjects who are asked to wear a black shirt with long sleeves and stand in front of dark background. Each subject is recorded at different times of day for several days. However, people tend to make slightly difference in each time they perform finger-spelling. We collect several numbers of each Thai finger-spelling alphabet from each person shown in Table 1 to be our signature library. We call each image with the size of 190×190 in the signature library as a key frame. There are 1091 key frames in total. Since, we want to separate ใ (ai) from ไ (ai), we will call “ใ” as “ai1” and “ไ” as “ai2” from now on.

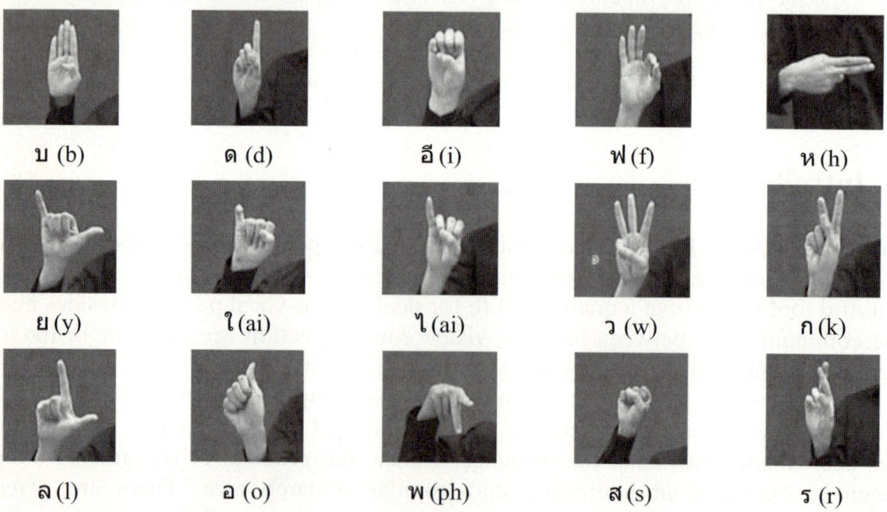

Fig. 1. Hand gestures for 15 Thai finger-spelling alphabets

After we collect all key frames in the signature library, we compute keypoint descriptors of each key frame using the SIFT. We keep all keypoint descriptors in the signature library database. In order to translate hand gesture, we need a system that grabs an image through a camera and analyzes it. Figure 2 shows the structure of the hand gesture translation system.

Table 1. Numbers of each Thai finger-spelling alphabet collected in the signature library

		Subject No.										
		1	2	3	4	5	6	7	8	9	10	11
Thai finger-spelling alphabet	บ (b)	5	8	8	6	6	5	6	5	6	10	6
	ด (d)	5	6	7	6	6	6	6	6	12	11	6
	อี (i)	6	6	5	5	7	8	6	6	5	11	6
	ฟ (f)	5	7	5	5	6	7	7	6	5	10	8
	ห (h)	12	6	7	4	5	6	6	5	6	7	8
	ย (y)	4	5	5	5	7	5	7	4	4	9	6
	ใ (ai)	6	6	5	5	5	5	12	5	5	6	10
	ไ (ai)	7	6	6	6	6	5	6	10	5	10	9
	ว (w)	8	6	8	4	6	6	6	5	6	9	7
	ก (k)	5	6	9	7	6	7	6	10	5	9	4
	ล (l)	4	6	6	6	6	7	6	6	6	8	8
	อ (o)	7	7	6	5	6	5	4	7	10	10	11
	พ (ph)	4	11	6	6	6	9	6	7	9	6	12
	ส (s)	7	11	5	5	7	7	6	8	3	8	5
	ร (r)	7	6	7	6	6	6	6	7	5	9	11

Suppose the size of each image frame grabbed from a video sequence is $J \times K$. The pixel values of the image frame $f(j, k)$, $j = 1, 2, \ldots, J$ and $k = 1, 2, \ldots, K$, can be viewed as a feature vector with the dimension of $J \times K$ representing this frame. However, in our experiment, we use row-column projection. The projection for the nth row is $l_n^r = \sum_{j=1}^{J} f(j,n)$ and the projection for the mth column is $l_m^c = \sum_{k=1}^{K} f(m,k)$. The projection for each row and column gives a feature vector with the dimension of $J+K$.

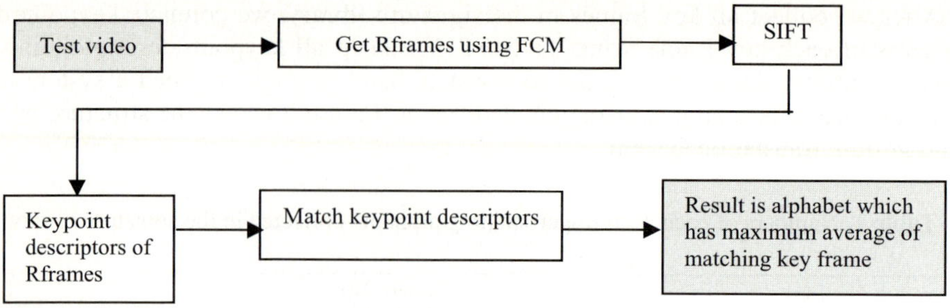

Fig. 2. Hand gestures translation structure

Fig. 3. Rframes from FCM of word "Far"

The features are obtained from the luminance (Y) [19] extracted from a video sequence. The image size from a video sequence is 720×576. Each frame in a video sequence is extracted every 0.04 second.

We use the fuzzy C-means (FCM) algorithm [20] to find Rframes which we briefly review here. Let $\mathbf{X} = \{\mathbf{x}_1, \mathbf{x}_2, \ldots, \mathbf{x}_N\}$ be a set of vectors, where each vector is a p-dimensional vector.

The update equation for FCM is as follows [20]

$$u_{ij} = 1 \Big/ \sum_{k=1}^{C} \left(d^2\left(\mathbf{x}_j, \mathbf{c}_i\right) \Big/ d^2\left(\mathbf{x}_j, \mathbf{c}_k\right) \right)^{1/(m-1)}, \quad (1)$$

$$\mathbf{c}_i = \sum_{j=1}^{N} \left(u_{ij}\right)^m \mathbf{x}_j \Big/ \sum_{j=1}^{N} \left(u_{ij}\right)^m. \quad (2)$$

In the above equations, u_{ij} is the membership value of vector $\mathbf{x}_j$ belonging to cluster i, $\mathbf{c}_i$ is the center of cluster i, and $m \in [1, \infty)$ is the fuzzifier. In our experiment, we set $m = 2$ and we also use the Euclidean distance between feature vectors and the centers. The algorithm for the FCM is:

```
Fix number of clusters C;
Fix the fuzzifier m;
Initialize center;
Do{
   Update membership
   Update centers
}
Until (centers are stable)
```

Since it does not make sense if we utilize the actual center prototypes, we pick the actual frame nearest to it to be our Rframes instead. An example of the Rframes from the FCM of the word "Far" is shown in figure 3.

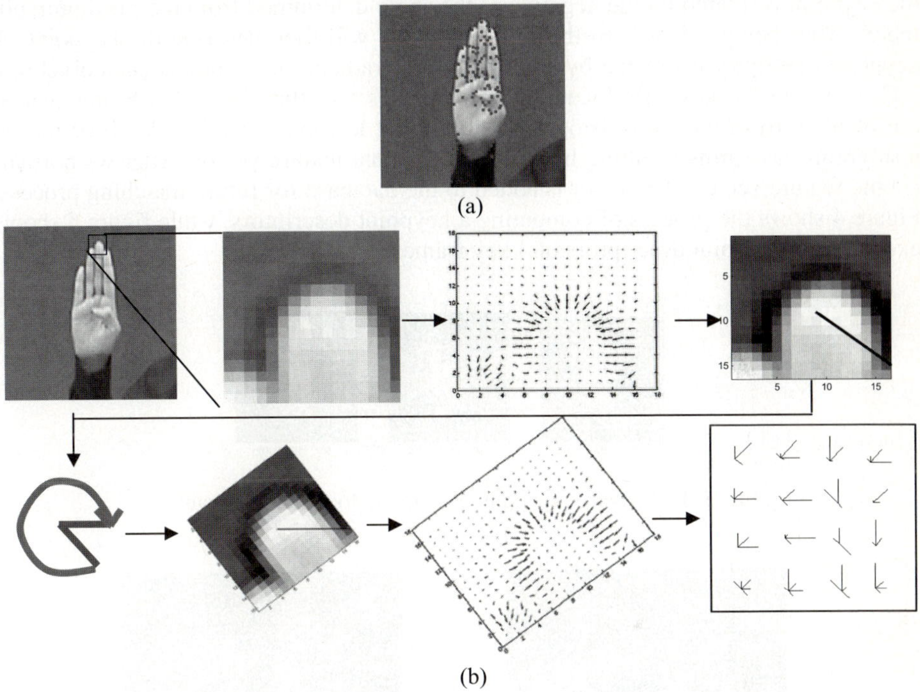

Fig. 4. Keypoint descriptor generation (a) Keypoints found on a keyframe, (b) The description of one keypoint. After a keypoint is found in 4(a), the gradients are determined and a major orientation is found. The area around the keypoint is rotated by this major angle. Then the gradients are added to the histogram bins that will give the description of this keypoint.

Now let us briefly review the Scale Invariant Feature Transform (SIFT) algorithm [17]. The SIFT finds a specific location called the keypoint in the scale space of training image. There are four major stages used to generate a keypoint, i.e., scale-space extrema detection, keypoint localization, orientation assignment, and keypoint descriptor. In scale-space extrema detection, a specific keypoint location is determined by constructing Gaussian pyramid from the original image. Each image in the scale space is subtracted from their nearby scales to produce the difference-of-Gaussian images. Each sample point is compared with eight neighbors in the same scale space and nine neighbors in the scale space above and below. If the pixel is greater than or less than all 26 neighbors, that location is an extremum and will be selected as a keypoint location. In keypoint localization, a detailed fit to the nearby data for location, scale, and ratio of principal curvatures is performed. This information allows candidate keypoints that have low contrast or are poorly localized along an edge to be rejected. In orientation assignment, orientations are assigned to each keypoint location based on the local image gradient direction. An orientation histogram is built from the gradient direction of sample points around the keypoint. The magnitude of each gradient is added to the nearest orientation bin of the histogram. When the largest bin of the histogram is found, we assign orientation to that keypoint. Then the area around the keypoint is rotated by the negative of the angle determined from the maximum bin index. After being oriented to the zero-angle, we will then describe the keypoint. A keypoint descriptor is created by computing the gradient magnitude at each pixel in a region around the keypoint location. These values will then be added to the nearest bin of their 16 orientation histograms around the keypoint location. Each of the 16 histograms has 8 bins resulting in a 128 dimensional feature vector. After we normalize the feature vector, the vector is added to the database for future matching process. Figure 4 shows the process of computing a keypoint descriptors, while figure 5 shows examples of keypoint descriptors of 3 key frames.

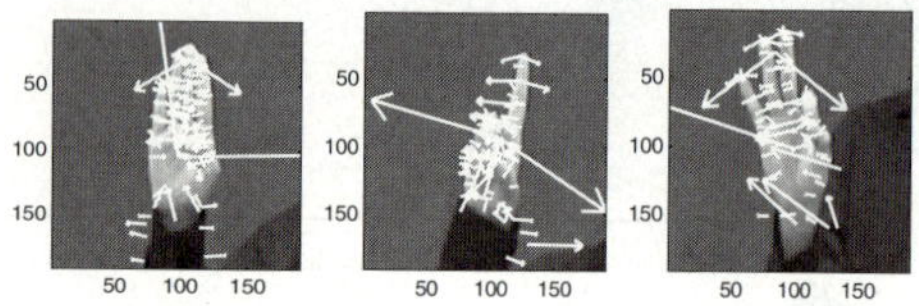

Fig. 5. Examples of keypoint descriptors found on key frames

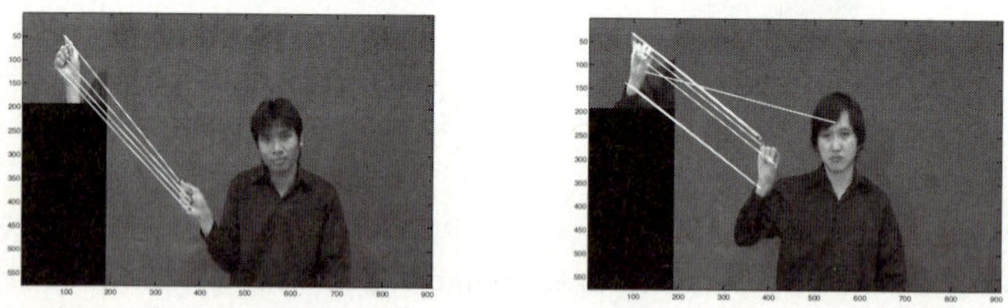

(a) (b)

Fig. 6. Recognition using SIFT (a) อ ("o") alphabet (b) ใ ("ai1") alphabet

Matching is obtained by comparing the Euclidean distances between two nearest neighbor keypoint descriptors in the signature library database to the current keypoint descriptor. If the ratio of the smallest distance to the second smallest one is less than a given threshold, then the two keypoints match. Figure 6 shows matching process between keypoints' descriptors in signature library database and the testing images.

3 Experimental Results

The testing video data sets are recorded at different times of day for several days from 4 subjects. Subject numbers 1 and 2 are the same subjects who are recorded for Thai finger-spelling alphabets. In the testing video, the subjects are asked to wear a black shirt with long sleeves and stand in front of dark background while they are performing the Thai finger-spelling words. The number of times that each subject performs the Thai finger-spelling words is shown in table 2. Hence, the total number of the Thai finger-spelling words is 256 words. The Thai finger-spelling words used in our experiment that are comprised of 2 Thai finger-spelling alphabets are "Good", "Leaf", "Meet", "Fire", "Erase", and "Color", whereas those that consists of 3 Thai finger-spelling alphabets are "Far", "Rich", "Float", and "Comb". Because of the nature of Thai finger-spelling words in this experiment, we choose the number of clusters = 3 for all the words. Hence, there are 831 Thai finger-spelling alphabets needed to be translated. If the translation system provides correct alphabets and the position of each alphabet is in the right place in the case of a 3-alphabet word, we count that as a correct translation. If this is a 2-alphabet word, e.g., "ph" and "b" for "Meet", and the translation system gives "ph", "ph", and "b" or "ph", "b", and "b", this would be considered as a correct translation. Otherwise, it will be considered as a wrong translation.

In this experiment, we set m = 2 and the threshold in matching process are varied from 0.65, 0.7, and 0.75. Since, the key frames for each alphabet in the signature library may have different numbers of keypoints, we compute an average number of matched keypoints per keyframes (*Ave_Match*) of each alphabet as

$$Ave\_Match = \frac{\text{No. of Matched Keypoints of the alphabet}}{\text{No. of Key Frames of that alphabet}}. \qquad (3)$$

Then we pick the matched alphabet by choosing the one that gives maximum *Ave_Match*. An example of the computed *Ave_Match* of the word "Float" is shown in figure 7. This word consists of 3 alphabets, i.e., "l", "o", and "y". The matched alphabet in figures 7(a), 7(b), and 7(c) are "l", "o", and "y" with *Ave_Match* of 3.043, 1.372, and 3.197, respectively.

Table 2. Numbers of each Thai finger-spelling words collected in tesing video data set

No.	ไกล (Far)	ดี (Good)	ใบ (Leaf)	พบ (Meet)	ไฟ (Fire)	รวย (Rich)	ลบ (Erase)	ลอย (Float)	สี (Color)	หวี (Comb)
1	8	15	8	10	8	8	9	15	10	15
2	5	5	5	5	5	5	5	5	5	5
3	5	5	5	5	5	5	5	5	5	5
4	5	5	5	5	5	5	5	5	5	5

We compare the results from the automatic translation system with the results from a manual translation system (Rframes from this system are chosen manually but the translation is still done by the SIFT algorithm). The result when we count the number of correct alphabets from the automatic and manual translation systems are shown in tables 3 and 4. The result when we count the number of correct words from the automatic and manual translation systems are shown in table 5 and 6. We can see that the average of the correct alphabet translation with the thresholding values of 0.65, 0.7, and 0.75 from the automatic translation system are 77.86%, 79.90%, and 79.66%,

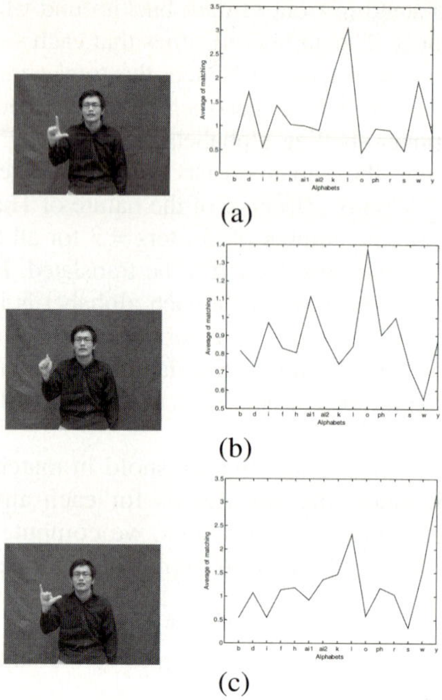

Fig. 7. Rframes and *Ave_Match* of alphabets of the word "Float" (a) matched alphabet is "l", (b) matched alphabet is "o", and (c) matched alphabet is "y"

respectively. In the manual translation system, we know which video is for what words, hence, we choose 3 Rframes for the 3-alphabet words and 2 Rframes for the 2-alphabet words. Therefore, there are only 618 alphabets instead of 831 alphabets. The average of the correct alphabet translation with the thresholding values of 0.65, 0.7, and 0.75 from the manual translation system are 79.93%, 80.58%, and 79.45%, respectively. The average of the correct word translation with the thresholding values of 0.65, 0.7, and 0.75 from the automatic translation system are 48.44%, 51.17%, and 51.18%, respectively, while that from the manual translation system are 59.37%, 61.33%, and 57.81%, respectively. But if we look at each individual subject, we can see that at the SIFT threshold of 0.7, the automatic translation system is able to translate alphabet correctly for subjects 1, 2, 3, and 4 with 90.88%, 85.33%, 62%, and 60.67%, while the manual translation system is able to do that with 93.80%, 85.83%, 60.83%, and 66.67%. At the same SIFT threshold, the correct results on words from the automatic translation system are 74.53%, 62%, 16%, and 26% and that from the manual translation system are 85.85%, 66%, 26%, and 40%. We can see that the results from subjects 3 and 4 are not as good as subject 1 and 2 in the automatic and manual translation systems. The reason is that the keypoints' descriptors generated for the key frames in the signature library are different from the keypoints' descriptors generated in the Rframes from these two subjects. Because the SIFT algorithm is a little sensitive to the shape of hand and fingers. Examples of keypoints' descriptors in a key frame in the signature library of the Thai finger-spelling alphabet "ai1" and the keypoints' descriptors of subjects 3 and 4 of this alphabet are shown in figure 8. The other reason that the automatic translation system cannot translate correctly is because some of the Thai finger-spelling alphabets have similar shapes, e.g., "ï", "s", and "o" as shown in figure 1.

Table 3. Correct translation alphabets from automatic translation system

SIFT threshold	Subject 1	Subject 2	Subject 3	Subject 4	Average
0.65	88.99%	80%	61.33%	62%	77.86%
0.7	90.88%	85.33%	62%	60.67%	79.90%
0.75	89.31%	83.33%	66%	60%	79.66%

Table 4. Correct translation alphabets from manual translation system

SIFT threshold	Subject 1	Subject 2	Subject 3	Subject 4	Average
0.65	91.86%	87.5%	61.67%	65%	79.93%
0.7	93.80%	85.83%	60.83%	66.67%	80.58%
0.75	91.86%	82.5%	63.33%	65.83%	79.45%

Table 5. Correct translation words from automatic translation system

SIFT threshold	Subject 1	Subject 2	Subject 3	Subject 4	Average
0.65	69.81%	52%	20%	28%	48.44%
0.7	74.53%	62%	16%	26%	51.17%
0.75	73.58%	62%	22%	22%	51.18%

Table 6. Correct translation words from manual translation system

SIFT threshold	Subject 1	Subject 2	Subject 3	Subject 4	Average
0.65	80.19%	70%	26%	38%	59.37%
0.7	85.85%	66%	26%	40%	61.33%
0.75	82.07%	60%	26%	36%	57.81%

We perform the FCM with the number of clusters of 3. Hence, when we want to translate a Thai finger-spelling word with 2 alphabets, there may be a chance that the SIFT algorithm will match a wrong alphabet to the nearest neighbor prototype in some clusters. In addition, the selected Rframe may be the image with moving hand or a blur image. From the experiment, the wrong results are almost from the words with one incorrect matched alphabet. Hence, if we choose 3 nearest neighbor prototypes instead of 1 nearest neighbor prototype to be the Rframes of that shot, and then pick the alphabet if two out of three matched key frames pointing to the same alphabet. Otherwise, this cluster is rejected. Examples of 3 nearest neighbor prototypes of the shot that is supposed to be "ï" is shown in figure 9.

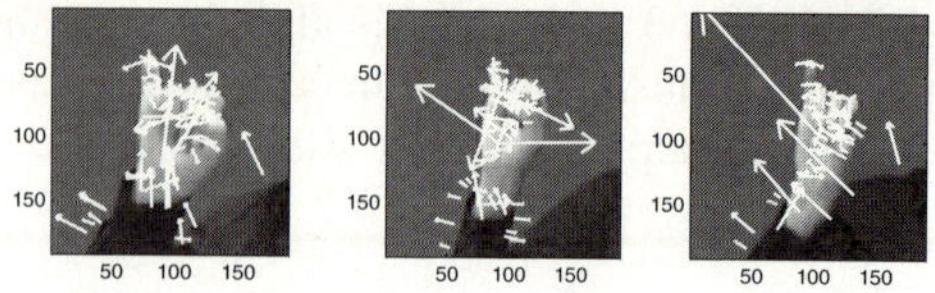

(a) (b) (c)

Fig. 8. Examples of keypoints descriptors generated in a key frame of (a) "ai1" in signature library, (b) "ai1" from subject 3 and (c) "ai1" from subject 4

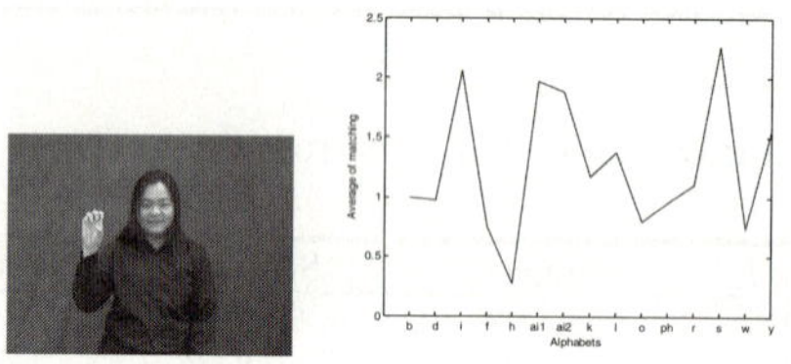

(a)

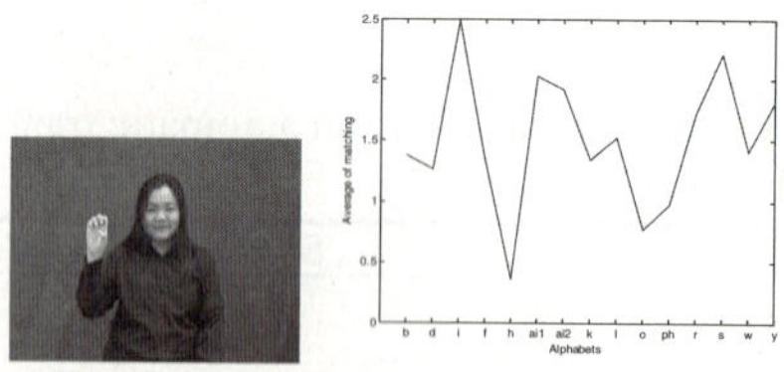

(b)

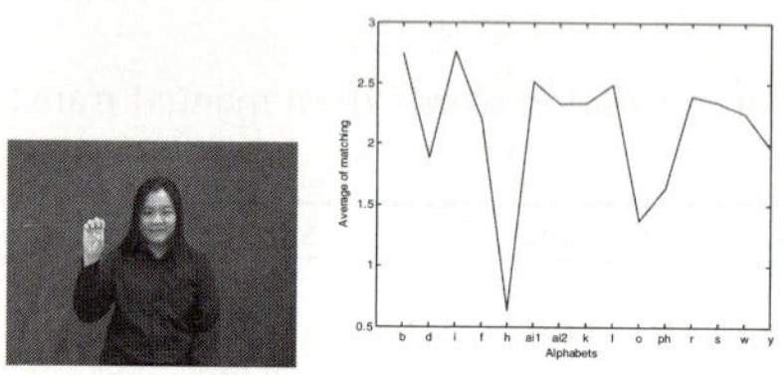

(c)

Fig. 9. An example of 3 nearest prototypes of a shot of alphabet "i" (a) the first nearest prototype and *Ave_Match* of alphabet, (b)the second nearest prototype and *Ave_Match* of alphabet and (c) the third nearest prototype and *Ave_Match* of alphabet

In this figure, the first nearest neighbor prototype is matched with the alphabet "s" with *Ave_Match* = 2.264, while the second and third nearest neighbor prototypes are both matched to the alphabet "i" with *Ave_Match* = 2.479 and 2.761. In this case, the matched alphabet for this shot will be "i". However, in figure 9(c), we can see that the second peak of *Ave_Match* is at the alphabet "b" with the value of 2.746. Now, we want to see if we choose 3 nearest neighbor prototypes from each cluster, would the automatic translation system get improved? Tables 7 and 8 show the results from the automatic translation system with 3 nearest neighbor prototypes from each cluster when counting only correct alphabets and when counting only correct words, respectively. The average of the correct alphabet translation with the SIFT thresholds of 0.65, 0.7, and 0.75 from the automatic translation system with 3 nearest neighbor prototypes are 80.50%, 82.19%, and 82.07%, respectively. While the average of the correct word translation with the SIFT threshold of 0.65, 0.7, and 0.75 from this system are 50.39%, 55.08%, and 55.86%, respectively. However, when we look at the results of each individual subject, we can see that at 0.7 SIFT threshold, the automatic translation system with 3 nearest neighbor prototypes is able to translation alphabets correctly for subjects 1, 2, 3, and 4 with 92.45%, 86.67%, 68%, and 62.67%. Whereas at the same SIFT threshold, the results on words are 79.24%, 64%, 22%, and 28%. We can see that the automatic translation system is improved in overall and individual cases.

Table 7. Correct translation alphabets from automatic translation system with 3 nearest neighbor prototypes from each cluster

SIFT threshold	Subject 1	Subject 2	Subject 3	Subject 4	Average
0.65	89.94%	84%	68%	59.33%	80.50%
0.7	92.45%	86.67%	68%	62.67%	82.19%
0.75	92.14%	86%	68.67%	62.67%	82.07%

Table 8. Correct translation words from manual translation system with 3 nearest neighbor prototypes from each cluster

SIFT threshold	Subject 1	Subject 2	Subject 3	Subject 4	Average
0.65	73.58%	58%	20%	24%	50.39%
0.7	79.24%	64%	22%	28%	55.08%
0.75	80.19%	64%	24%	28%	55.86%

4 Conclusion

In this work, we build an automatic Thai finger-spelling sign language translation system using the fuzzy C-means (FCM) and scale invariant feature transform (SIFT) algorithms. In particular, we pick a representative frame (Rframe) of each shot using the FCM algorithm. We then find a matched alphabet from the keypoints descriptors in the signature library database collected from several subjects at different times of day and for several days using the SIFT algorithm. We found that with 1 nearest neighbor prototype from each cluster, the result is comparable with that when we choose the Rframe manually. At 0.7 SIFT threshold, the average of correct alphabet translation and correct word translation from the automatic translation system with 1 nearest neighbor are 79.90% and 51.17%, respectively. While at the same SIFT threshold, these average are 80.58% and 61.33%. However, the error may occur if the automatically-chosen Rframe is an image with moving hand or blur image. Also, in the case of a 2-alphabet word, one of the Rframes may be matched to a wrong alphabet. This will cause a wrong translation. But if we choose 3 nearest neighbor prototype, the result get improved. The average of correct alphabet translation and correct word translation at 0.7 SIFT threshold are 82.19% and 55.08%. However, the results from subjects 3 and 4 are not as good as that from subjects 1 and 2.

Acknowledgement

The research is supported by the Faculty of Engineering, Chiang Mai University. We would like to thank the teachers and the students at Anusarnsunthorn school for the deaf in Chiang Mai, Thailand for the help and advice in Thai sign language.

References

1. Kramer, J., Leifer, L.: The talking glove: a speaking aid for nonvocal deaf and deaf-bilid individuals. In: Proc. of RESNA 12th Annual Conference, New Orleans, Louisiana, pp. 471–472 (1989)
2. Fels, S.S., Hinton, G.E.: Glove-talk: a neural network interface between a data-glove and a speech synthesizer. IEEE Trans. On Neural Networks 4(1), 2–8 (1993)
3. Su, M.C., Jean, W.F., Chang, H.T.: A static hand gesture recognition system using a composite neural network. In: Proceedings of the Fifth IEEE International Conference on Fuzzy Systems, September 1996, pp. 786–792 (1996)
4. Chen, L.Y., Mizuno, N., Fujimoto, H., Fujimoto, M.: Hand shape recognition using the bending element of fingers in sign language. Transactions of the Japan Society of Mechanical Engineers, Part C 68(12), 3689–3696 (2002)
5. Wu, J.Q., Gao, W., Chen, X.L., Ma, J.Y.: Hierarchical DGMM recognizer for Chinese sign language recognition. Journal of Software 11(11), 1430–1439 (2000)
6. Gao, W., Chen, X.L., Ma, J.Y., Wang, Z.Q.: Building language communication between deaf people and hearing society through multimodal human-computer interface. Chinese Journal of Computers 23(12), 1253–1260 (2000)
7. Wu, J.Q., Gao, W., Song, Y., Liu, W., Pang, B.: Simple sign language recognition system based on data glove. In: International Conference on Signal Processing Proceedings (ICSP), vol. 2, pp. 1257–1260 (1998)
8. Min, B.W., Yoon, H.S., Soh, J., Yang, Y.M., Ejima, T.: Hand gesture recognition using Hidden Markov Models. In: IEEE International Conference on Systems, Man, and Cybernatics (Computational Cybernatics and Simulation), October 1997, pp. 4232–4235 (1997)
9. Huang, C.L., Huang, W.Y.: Sign language recognition using model-based tracking and a 3D Hopfield neural network. Machine Vision and Applications 10(5-6), 292–307 (1998)
10. Kobayashi, T., Haruyama, S.: Partly-Hidden Markov model and its application to gesture recognition. In: IEEE International Conference on Acoustics, Speech and Signal Processing – Proceedings (ICASSP), vol. 4, pp. 3081–3084 (1997)
11. Chen, F.S., Fu, C.M., Huang, C.L.: Hand gesture recognition using a real-time tracking method and hidden Markov models. Image and Vision Computing 21(8), 745–758 (2003)
12. Alon, J., Athisos, V., Yuan, Q., Sclaroff, S.: Simultaneous localization and recognition of dynamic hand gestures. In: Proc. IEEE Workshop on Motion and Video Computing, January 2005, pp. 254–260 (2005)
13. Auephanwiriyakul, S., Chaisatian, P.: Static Hand Gesture Translation Using String Grammar Hard C-Means. In: The Fifth International Conference on Intelligent Technologies, Houston, Texas, USA (December 2004)

14. Jiang, H., Helal, A., Elmagarmid, A.K., Joshi, A.: Scene change detection techniques for video database systems. ACM Multimedia Information Systems (1996)
15. Joshi, A., Auephanwiriyakul, S., Krishnapuram, R.: On Fuzzy Clustering and Content Based Access to Networked Video databases. In: 8th IEEE Workshop on Research Issue in Data Engineering (1998)
16. Auephanwiriyakul, S., Joshi, A., Krishnapuram, R.: Fuzzy Shot Clustering to Support Networked Video Databases. In: IEEE FUZZ-IEEE 1998/WCC I998 (May 1998)
17. Lowe, D.G.: Distinctive Image Features from Scale-Invariant Keypoints. International Journal of Computer Vision 60(2), 91–110 (2004)
18. Office of The Basic Education Commission, Thai Hand Sign Language Handbook under the Initiatives of Her Royal Highness Princess Maha Chakri Sirindhorn, Bangkok, Thailand (1997) (in Thai)
19. Jiang, H., Helal, A., Elmagarmid, A.K., Joshi, A.: Scene change detection techniques for video database systems. ACM Multimedia Information Systems (1996)
20. Bezdek, J.C.: Pattern Recognition with Fuzzy Objective Function Algorithms. Plenum Press, New York (1981)

Statistical Prediction for the Pricing of Bond Using Random Number Generation*

Kisoeb Park[1], Moonseong Kim[2], and Seki Kim[1,]**

[1] Department of Mathematics Sungkyunkwan University
440-746, Suwon, Korea
{kisoeb,skim}@skku.edu
[2] Electrical and Computer Engineering
Michigan State University, USA
mkim@msu.edu

Abstract. In this paper, we propose a dynamic prediction algorithm to predict the interest rate for the bond price using real data set. Our algorithm is based on term structure model of the interest rate, which takes place in financial modelling, such as the standard Wiener process. We use cumulative distribution function (CDF) of real data for the interest rate measurement. The CDF is obtained by the natural cubic spline (CS) method, which is the frequently used numerical methods for interpolation. Most useful when the CDF $F(x)$ has an inverse $F^{-1}(x)$ which is easy to compute. We use the random number generation to calculate the pricing of bond. In empirical computer simulation, we show that the lower values of precision in the proposed prediction algorithm corresponds to sharper estimates. It is very reasonable on prediction.

Keyword: Bond Option Price, Cubic Spline (CS) Interpolation, Random Number Generation, Stochastic Simulation.

1 Introduction

Approaches to modeling the term structure of interest rates in continuous time may be broadly described in terms of either the equilibrium approach or arbitrage-free approach even though some early models include concepts from both approaches. We introduce a model of Vasicek [10], Cox, Ingersoll, and Ross [3], the extended model of the Hull and White [8], and the development of the model is the jump-diffusion model of the Ahn and Thompson [1] and the Baz and Das [2]. Also, Heath, Jarrow and Morton [7] is widely accepted as the most general methodology for term structure models. Furthermore, current research using jump-diffusion processes relies mostly on two classes of models: the affine jump-diffusion class [4][5] and the quadratic Gaussian. In the previous studies, interest rate modelling was performed based on the diffusion process. That is, any variable whose value

* This paper was supported by Faculty Research Fund, Sungkyunkwan University, 2006.

** Corresponding author.

O. Gervasi et al. (Eds.): ICCSA 2008, Part II, LNCS 5073, pp. 1120–1130, 2008.

changes over time in an uncertain way is said to follow a stochastic process. Models of interest rate behavior are usually expressed in terms of what are known as Wiener processes. It has been used in physics to describe the motion of a particle that is subject to a large number of small molecular interest rate and is sometimes referred to as Brownian motion. The basic Wiener process that has been developed so far has a drift rate of zero and a variance rate of 1.0. We present a models which allows the short term interest rate is followed a random walk. This leads to a parabolic partial differential equation (PDE) for the prices of bonds. Using the PDE, we obtain a solution of the bond pricing for the term structure models. This method is difficult of the actual proof analysis of the term structure models. Also, highly precise estimates with the brute force method can take a long time to achieve.

We should be make use the real data of the interest rate for the complexity of the problem. In this paper, we propose a dynamic prediction algorithm to predict the interest rate for the bond price. Our algorithm is based on term structure model of the interest rate, which takes place in financial modelling, such as the standard Wiener process. We use CDF for current the interest rate measurement. The CDF is obtained by the natural CS method [6], which is the frequently used numerical methods for interpolation. We use the inverse-transform technique to calculate the random number R generation. We generate uniform random numbers using the CDF. And then, we investigate the prices of bond by the generated random numbers.

The paper is organized into five sections. In Section 2, the motivations of our work is described. Section 3 presents details of cubic spline interpolation method, which is numerical method used in our prediction algorithm. In Section 4, the proposed algorithms performances are evaluated based on Monte Carlo simulation using random number generation. Then, in Section 5, investigated bond prices by Monte Carlo simulation using random number generation. Finally, Section 6 concludes this paper.

2 Motivations

Bond prices are usually assumed to follow a Markov process. A Markov process is a particular type of stochastic process where only the present value of a variable is relevant for predicting the future. Models of bond prices behavior are usually expressed in term of what are known as Wiener processes. A Wiener process is a particular type of Markov stochastic process. In the same way that we proposed a model for the asset price as a lognormal random walk let us suppose that the interest rate r is governed by a stochastic differential equation (SDE) of the form

$$dr = u(r,t)dt + \sigma(r,t)dW, \tag{1}$$

where $u(r,t)$ is the instantaneous drift and $\sigma(r,t)$ is the instantaneous volatility. The bond price can be expressed as:

$$V(r_t, t, T) = \mathbb{E}^{Q}{}_t \left[e^{-\int_t^T r_s ds} | r(t) \right], \tag{2}$$

where $\mathbb{E}^Q$ is the expectations operator with respect to the equivalent risk-neutral measure. We now move on to discuss Monte Carlo simulation. A Monte Carlo simulation (MCS) of a stochastic process is a procedure for sampling random outcomes for the process. To execute MCS, we discretize the equation (1). We divide the time interval $[t, T]$ into m equal time steps of length Δt each. For small time steps, we are entitled to use the discretized version of the risk-adjusted SDE (1):

$$r_j = r_{j-1} + u(r,t)\Delta t + \sigma(r,t)\varepsilon_j\sqrt{\Delta t}, \tag{3}$$

where $j = 1, 2, \cdots, m$ and ε_j is the standard normal variable with $\varepsilon_j \sim N(0,1)$.

We can investigate the price of the bond by sampling n interest rate paths under the discrete process approximation of the risk-adjusted processes of the equation (2). The bond price estimate is given by:

$$V(r_t, t, T) = \frac{1}{n}\sum_{i=1}^{n} \exp\left(-\sum_{j=0}^{m-1} r_{ij}\Delta t\right), \tag{4}$$

where r_{ij} is the value of the interest rate under the discrete risk-adjusted process within sample path i at time $t + \Delta t$.

Numerical methods that are known as Monte Carlo methods can be loosely described as statistical simulation methods, where statistical simulation is defined in quite general terms to be any method that utilizes sequences of random numbers to perform the simulation. Even though Monte Carlo method is both harder and conceptually more difficult to implement than the other numerical methods, interest in use of Monte Carlo method for bond pricing is getting stronger because of its flexibility in evaluating and handling complicated financial instruments. However, it takes a long time to achieve highly precise estimates with the brute force method. That is, the interest rate follows no specific CDF. Fig. 1 shows histograms for 300 real data set of a two-year, three-year, five-year, ten-year and thirty-year treasury note.

In Fig. 1, (a) is $\mu = 4.8646$ and $\sigma = 0.008951$, (b) is , $\mu = 4.81573$ and $\sigma = 0.01003$, (c) is $\mu = 4.79097$ and $\sigma = 0.01077$, (d) is $\mu = 4.83$ and $\sigma = 0.0118$, and (e) is $\mu = 4.9003$ and $\sigma = 0.0114$ for the real data set. The appearance of the six histograms is very different. Since a ten-year Treasury note has more dynamic motions than a two-year Treasury note. This histogram of it leans to right side rather than that of a two-year Treasury note.

In the previous studies, interest rate modelling was performed based on the Wiener process. In other words, the Wiener process was obtained in advance, and then authors got the statistical properties. However, since the pre-measured interest rate traces already lost the interest rate prediction. Thus, to keep the bond price dynamics, we use CDF for the interest rate measurement. In order to obtain the CDF, we use the natural cubic spline method, which is the frequently used numerical methods for interpolation. In this paper, we propose a dynamic prediction algorithm to predict the interest rate for the bond price.

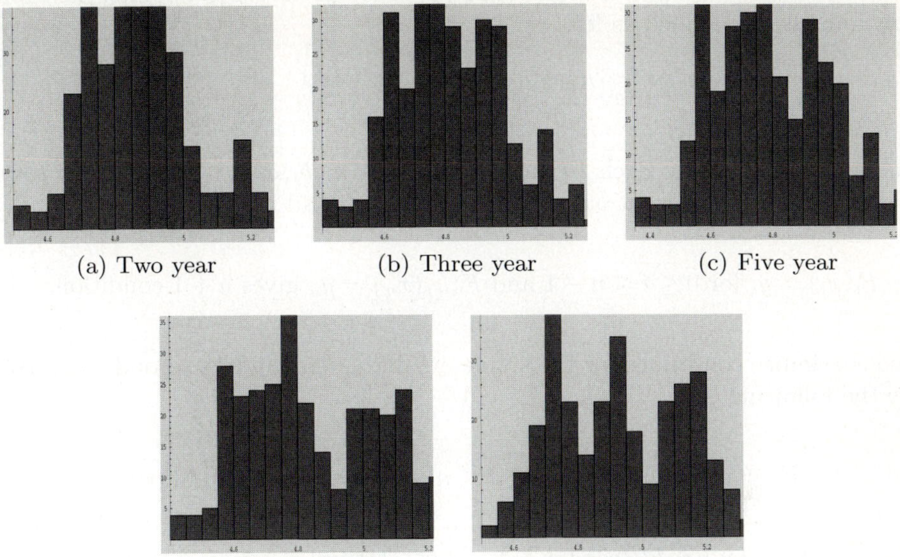

(a) Two year (b) Three year (c) Five year

(d) Ten year (e) Thirty year

Fig. 1. Histogram for treasure note($N = 300$)

3 The CDF Using CS Interpolation Method

We propose an interpolation method to estimate the CDF of real data set for the interest rate measurement. In order to the CDF, we use the CS interpolation method which is widely used in piecewise polynomial approximation [6]. Cubic splines are the most popular spline functions. They are smooth functions with which to fit data, and when used for interpolation, they do not have the oscillatory behavior that is characteristic of high-degree polynomial interpolation (like Lagrange Interpolation, Hermite Interpolation, etc.).

We will now consider cubic splines. Given $n+1$ points (x_0, y_0) to (x_n, y_n), where without loss of generality we assume $x_0 < x_1 < \cdots < x_n$, we seek a cubic spline function $F(x)$ such that $F(x_i) = y_i$ for $0 \leq i \leq n$. As with linear splines, write

$$F(x) = \begin{cases} F_0(x), & x_0 \leq x \leq x_1 \\ F_1(x), & x_1 \leq x \leq x_2 \\ \vdots & \vdots \\ F_{n-1}(x), & x_{n-1} \leq x \leq x_n. \end{cases}$$

To construct the interpolated function $F(x)$ through the control points, the cubic polynomials are defined as follows:

$$F_i(x) = a_i(x - x_i)^3 + b_i(x - x_i)^2 + c_i(x - x_i) + d_i, \tag{5}$$

where $i = 0, \cdots, n-1$. Each Fi has 4 unknowns (a_i, b_i, c_i, d_i). So there is a total of $4n$ unknowns. We want our spline to interpolate all $n+1$ points:

$$F_i(x_i) = y_i \text{ for } 0 \le i \le n-1 \text{ and } F_{n-1}(x_n) = y_n \text{ gives } n+1 \text{ conditions.}$$

The continuity conditions for the spline, its derivative, and its second derivative are the following:

$$F_{i-1}(x_i) = F_i(x_i) \text{ for } 1 \le i \le n-1 \text{ gives } n-1 \text{ conditions,}$$

$$F'_{i-1}(x_i) = F'_i(x_i) \text{ for } 1 \le i \le n-1 \text{ gives } n-1 \text{ conditions,}$$

$$F''_{i-1}(x_i) = F''_i(x_i) \text{ for } 1 \le i \le n-1 \text{ gives } n-1 \text{ conditions.}$$

So there are a total of $4n-2$ conditions. In order to get a unique solution, we must impose 2 extra conditions. A common approach is to impose the following:

$$F''(x_0) = F''(x_n) = 0.$$

The resulting function is called a natural CS. Note that $F'_i(\cdot)$ and $F''_i(\cdot)$ are the first and second derivatives of $F_i(\cdot)$, respectively. As you know, the cubic spline method is very flexible and a good estimator of graph through the control points. Hence, we have to use it in order to prediction of the interest rate, in this paper. To compute the control points, first, we change histogram(cf. Fig 1.) into cumulative distribution that accumulates the occurrences for each of different sizes of yield rate for treasure note. The second, we find calculate the maximum ($max = x_n$, log value) and minimum ($min = x_0$, log value) for frame sizes received. The third we divide the closed interval $[x_0, x_n]$ into n of pieces. The length of subinterval, $[x_{i-1}, x_i]$, is $\triangle x_i$. As a matter of convenience, we assume that the Euclidian distance for each subinterval is all same in this paper(i.e., $\triangle x_i = \triangle x$). After we obtain the frequency β_i for each $[x_{i-1}, x_i]$, the control points are calculated. Let the point $(x_i - x_{i-1})/2$ be α_i. The control points are (α_i, β_i), where $i = 1, \ldots, n$. If the total number of real data set is N, N is $\sum_i \beta_i$. We use the CS interpolation method, and then the function $F(x)$ through the control points is obtained. For example, we can calculate $F(x)$ with the thirty year treasure note($N = 300, n = 6$) in Fig. 2.

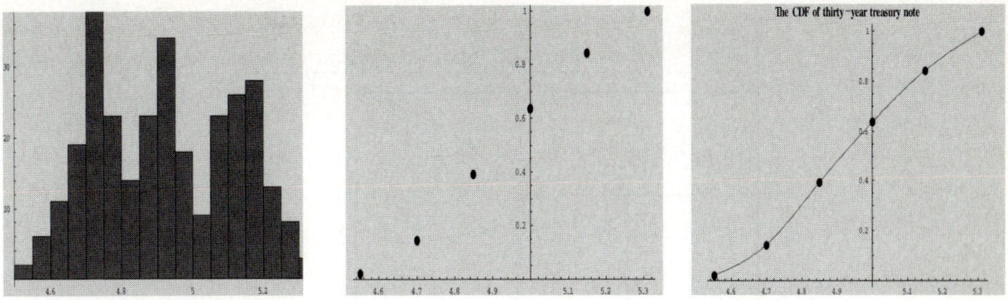

(a) Histogram for thirty year treasure note (b) Listplot for cumulative distribution of 2(a) (c) The CDF for 2(b) using CS method

Fig. 2. A process of generation for CDF of thirty year treasure note($N = 300, n = 6$)

4 Stochastic Simulation Using Random Number Generation

Recent methods of bond pricing do not necessarily exploit partial differential equations implied by risk-neutral portfolios. They rest on converting prices of such assets into martingales. This is done through transforming the underlying probability distribution using the tools provided by the Girsanov's theorem. A risk-neutral measure is any probability measure, equivalent to the market measure P, which makes all discounted bond prices martingales.

Interest in use of MCS for bond pricing is increasing because of the flexibility of the methods in handling complex financial instruments. The MCS is actually a very general tool, and its applications are by no means restricted to numerical integration. In simple and general terms, MCS may be described as a numerical method based an random sampling. This method is difficult of the actual proof analysis of the term structure models. Also, highly precise estimates with the brute force method can take a long time to achieve.

In order to settle this difficulty, we proposed a model for the bond price as a lognormal random walk let us suppose that the interest rate r is obtained by a random number generation. This paper proposes a MCS using random number generation for the pricing of bond. We now move on to discuss MCS using random number generation. To execute the MCS using random number generation, we discrete the equation (2). We divide the time interval $[t, T]$ into m equal time steps of length Δt each. For small time steps, we are entitled to use the discrete version of the interest rate $r_j (j = 1, 2, \cdots, m)$.

We can investigate the value of bond by sampling n interest rate paths. The values of the interest rate find to use the random number generation. We use the inverse-transform technique to calculate the random number generation.

For CDF is $R = F(x)$. Generate R from uniform distribution $[0, 1]$. Find the equation $x = F^{-1}(R)$. This technique can be used in principle for any distribution. Most useful when the CDF $F(x)$ has an inverse $F^{-1}(x)$ which is easy to compute. Required steps as following: (a) Compute the CDF of the

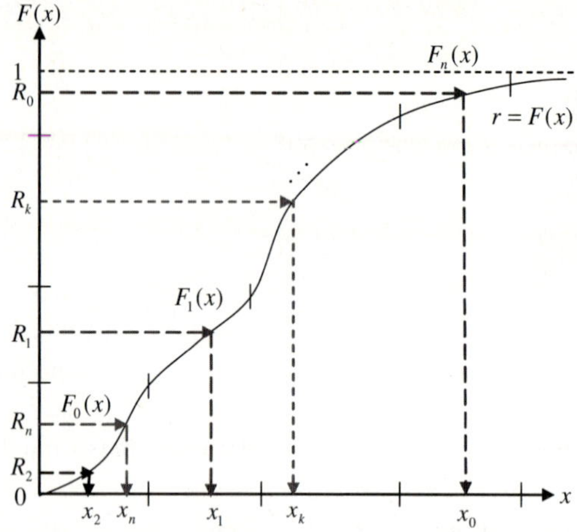

Fig. 3. Process of random number generation

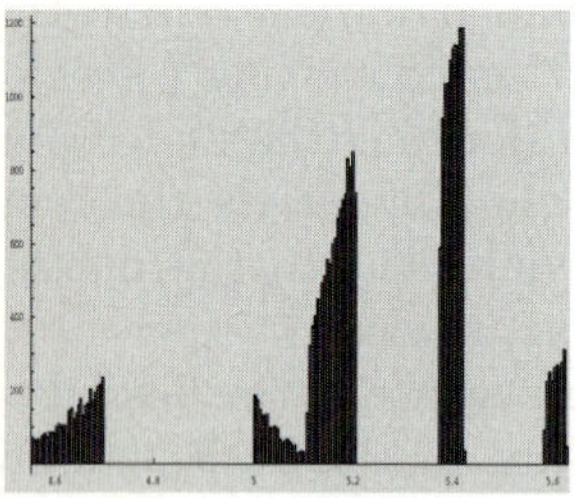

(a) Two year

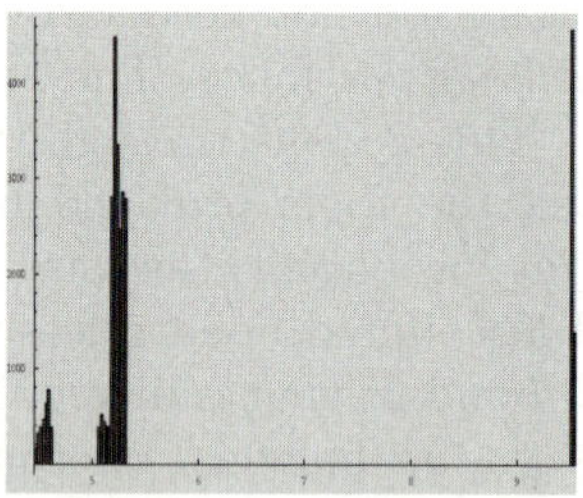

(b) Three year

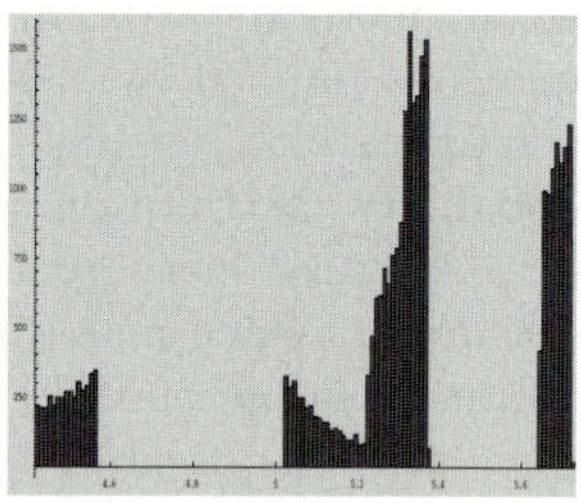

(c) Five year

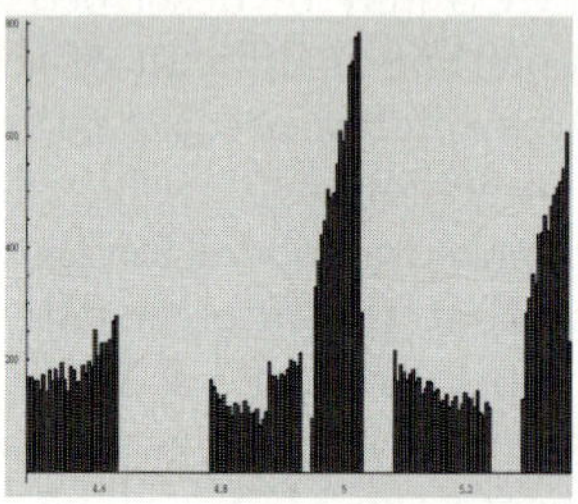

(d) Ten year

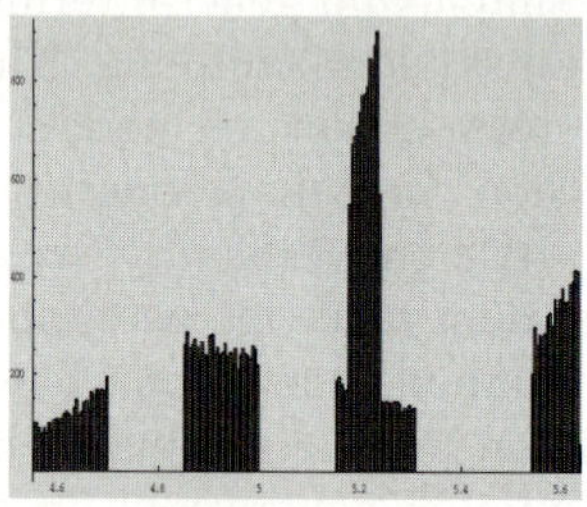

(e) Thirty year

Fig. 4. Histogram for random number generation of the real data set

desired random variable x. (b) Set $F(x) = R$ on the range of x. (c) Solve the equation $F(x) = R$ for x in terms of r. (d) Generation uniform random numbers $R_0, R_1, \cdots, R_n$ and compute the desired random variable by $x_i = F^{-1}(R_i)(i = 0, \cdots, n)$, as shown in Fig. 3.

For example, we calculate $x_i = F_i^{-1}(R_i)(i = 0, \cdots, n)$ with $m = 300, n = 6, N = 300$, and $k(\text{trial}) = 100$ in Fig. 4.

The bond price estimate is given by:

$$V(t,T) = \frac{1}{k}\sum_{i=1}^{k} \exp\left(-\sum_{j=0}^{m-1} r_{ij}\Delta t\right), \tag{6}$$

where r_{ij} means the interest rate. under the discrete risk-adjusted process within sample path i at time $t+\Delta t$. The Monte Carlo simulation is clearly less efficient computationally than the numerical method. One measure of the sharpness of the point estimate of the mean is mean standard error (MSE), defined as

$$MSE = \frac{\nu}{\sqrt{k}} \quad \text{or equivalently} \quad k = \frac{\nu^2}{MSE^2} \tag{7}$$

where ν^2 is the estimate of the variance of bond prices as obtained from k sample paths of the interest rate:

$$\nu^2 = \frac{\sum_{i=1}^{k}\left[\exp\left(-\sum_{j=0}^{m-1} r_{ij}\Delta t\right) - V(t,T)\right]^2}{k-1} \tag{8}$$

The precision (PCS) of the mean as a point estimate is often defined as the half-width of a 95% confidence interval, which is calculated as

$$PCS(\%) = 1.96 \times MSE \times 100. \tag{9}$$

Lower values of PCS in Equation (9) correspond to sharper estimates. Increasing the number of k is a brute force method of obtaining sharper estimates. However, increasing the value of k have little influence on sharper estimates. Hence highly precise estimates with the MSE using random number generation can take a short time to achieve.

5 Estimation Results

In this section, we investigate the pricing of bond by MCS using random number generation. For the purpose of simulation, we conduct three runs of $k = 10, 100$ trials each and divide a year into $m = 300$ time steps. We divide the closed interval [minimum size, maximum size] into $n = 6, 9$ of pieces.The size of real data set for natural CS interpolation method is monotone increasing.

Table 1, 2, 3, and 4 (Experiment 1, 2, 3, and 4) examine the bond prices by MCS and MCS using random number generation. In experiment 1, 2, 3, and 4 are the parameter values are assumed to be u and σ are the mean and standard deviation for the real data set, respectively, $\Delta t = (T-t)/m$, $m = 300$, $t = 0$, and $T = 1$. As a performance analysis, we use relative error (RE) is the ratio between the difference of the actual values. It can be defined by

$$RE(\%) = \frac{OLDS - NEWS}{OLDS} \times 100, \tag{10}$$

Table 1. [*Experiment 1*] Bond prices for real data of treasure note estimated by MCS and MCS using random number generation.$(k = 10, n = 6)$

	2-year	3-year	5-year	10-year	30-year
OLDS	0.951858	0.955632	0.952468	0.950180	0.952982
NEWS	0.949310	0.941683	0.948561	0.951382	0.949762
OLDS-PCS(%)	0.289054	0.321524	0.347103	0.472733	0.259883
NEWS-PCS(%)	0.008994	0.039906	0.010310	0.010777	0.007100
RE(%)	0.267704	1.45965	0.410217	0.137522	0.337897

Table 2. [*Experiment 2*] Bond price for real data of treasure note estimated by MCS and MCS using random number generation.$(k = 10, n = 9)$

	2-year	3-year	5-year	10-year	30-year
OLDS	0.951858	0.955632	0.952468	0.950180	0.952982
NEWS	0.951624	0.950922	0.952030	0.952060	0.951104
OLDS-PCS(%)	0.289054	0.321524	0.347103	0.472733	0.259883
NEWS-PCS(%)	0.006050	0.003900	0.005968	0.006292	0.005369
RE(%)	0.024567	0.492914	0.045943	0.197904	0.197060

Table 3. [*Experiment 3*] Bond prices for real data of treasure note estimated by MCS and MCS using random number generation.$(k = 100, n = 6)$

	2-year	3-year	5-year	10-year	30-year
OLDS	0.954364	0.954303	0.954408	0.954385	0.955926
NEWS	0.949311	0.941657	0.948592	0.951343	0.949790
OLDS-PCS(%)	0.087741	0.112154	0.109330	0.122572	0.130106
NEWS-PCS(%)	0.002951	0.018156	0.004207	0.003029	0.003180
RE(%)	0.529424	1.32510	0.609403	0.318794	0.64193

Table 4. [*Experiment 4*] Bond price for real data of treasure note estimated by MCS and MCS using random number generation.$(k = 100, n = 9)$

	2-year	3-year	5-year	10-year	30-year
OLDS	0.954364	0.954303	0.954408	0.954385	0.955926
NEWS	0.951586	0.950954	0.952042	0.952044	0.951100
OLDS-PCS(%)	0.087741	0.112154	0.109330	0.122572	0.130106
NEWS-PCS(%)	0.001645	0.002561	0.002010	0.001846	0.002507
RE(%)	0.291053	0.350897	0.247897	0.245329	0.504867

where OLDS is solution of the equation (2) for the equation (1) by MCS and NEWS is solution of the equation (2)for real data set by MCS using random number generation.

In empirical computer simulation, we show that the lower values of PCS in the proposed prediction algorithm corresponds to sharper estimates. It is very reasonable on prediction.

6 Conclusion and Future Research

The pricing of bond option is a very important problem encountered in complex financial markets. In this paper, we propose a dynamic prediction algorithm to predict the pricing of bond. In the previous studies, interest rate modelling was performed based on the Wiener process. A Wiener process is a particular type of Markov stochastic process. In other words, the Wiener process was obtained in advance, and then authors got the statistical properties. However, since the pre-measured interest rate traces already lost the interest rate prediction. Thus, to keep the interest rate dynamics, we use CDF for real data of interest rate measurement. In order to obtain the CDF, we use the natural CS interpolation method, which is the frequently used numerical methods for interpolation. We use the inverse-transform technique to calculate the random number R generation. We generate uniform random numbers by the CDF. And then, we investigate the prices of bond by MCS using the random number generation. This paper shows that there can be substantial differences between the bond prices produced by MCS and MCS using the random number generation. The differences are measured by PCS. In empirical computer simulation, we show that the lower values of precision in the proposed prediction algorithm corresponds to sharper estimates. It is very reasonable on prediction. Furthermore, increasing the value of k (trail) have little influence on sharper estimates. Hence highly precise estimates with the proposed prediction algorithm can take a short time to achieve. There are still many problems remaining for further research. Some of them, for instance, are: (i) predict very easily the bond option price by the proposed algorithm; (ii) using the proposed algorithm to simulate more complicated a financial models.

References

1. Ahn, C., Thompson, H.: Jump-Diffusion Processes and the Term Structure of Interest Rates. Journal of Finance 43, 155–174 (1998)
2. Baz, J., Das, S.R.: Analytical Approximations of the Term Structure for Jump-Diffusion Processes: A Numerical Analysis. Journal of Fixed Income 6(1), 78–86 (1996)
3. Cox, J.C., Ingersoll, J., Ross, S.: A Theory of the Term Structure of Interest Rate. Econometrica 53, 385–407 (1985)
4. Dai, Q., Singleton, K.: Expectation Puzzles, Time-varying Risk Premia, and Affine Models of the Term Structure. Journal of Financial Economics 63, 415–441 (2002)

5. Peng, C., Olivier, S.: Linear-Quadratic Jump-Diffusion Modeling. Mathematical Finance 17(4), 575–598 (2007)
6. Gerald, C.F.: Applied Numerical Analysis. Addison-Wesley, Reading (1999)
7. Health, D., Jarrow, R., Morton, A.: Bond Pricing and the Term Structure of Interest Rates. Econometrica 60, 77–105 (1992)
8. Hull, J., White, A.: Pricing Interest Rate Derivative Securities. Review of Financial Studies 3, 573–592 (1990)
9. Longstaff, F.A., Schwartz, S.: Interest Rate Volatility and the Term structure: A Two-Factor General Equilibrium Model. Journal of Finance 47, 1259–1282 (1992)
10. Vasicek, O.A.: An Equilibrium Characterization of the Term Structure. Journal of Financial Economics 5, 177–188 (1977)

Applied Statistical Indicators to the Vehicle Routing Problem with Time Windows for Discriminate Appropriately the Best Algorithm

Jorge A. Ruiz-Vanoye[1], José A. Zárate M.[1], Ocotlán Díaz-Parra[2], and Vanesa Landero N.[1]

[1] Centro Nacional de Investigación y Desarrollo Tecnológico
Interior internado Palmira s/n., Col. Palmira, Cuernavaca, Mexico
{ruizvanoye04c,jazarate,landerov}@cenidet.edu.mx
[2] CIICAp. Universidad Autónoma del Estado de Morelos
Av. Universidad, Col. Chamilpa, 62209, Cuernavaca, México
odiazp@uaem.mx

Abstract. In this paper, we propose indicators based on the position method of the descriptive statistic for the Vehicle Routing Problem with Time Windows (VRPTW). The indicators are based on the calculation of the median; they will serve altogether with the technique of discriminant analysis to select appropriately the algorithm that better solves an instance of the VRPTW.

Keywords: Discriminant analysis, VRPTW, prediction, genetics algorithms.

1 Introduction

The Vehicle Routing Problem (VRP) is an optimization problem combinatorial complex [1]. It is considered naturally like a central problem in the areas of transportation, logistic and distribution. In some sectors of the industry, the transportation means a high percentage of value added to products, for this reason the use of computational methods offers good results to the transportation; it is great to savings from 5% to 20% in the total of costs, as it reports Toth & Vigo [2].

The VRP has diverse variants of the problem, one of them is the VRPTW (Vehicle Routing Problem with Time Windows), consists basically of diminishing the transportation costs fulfilling to time restrictions of each route and vehicle capacity considering the demand of each client [2]. The parameters of the VRPTW instances they are in Table 1.

Table 1. Design of VRPTW instances

VN	*C*					
CN	XCO	YCO	D	RT	DT	ST
0	x_0	Y_0	d_0	RT_0	DT_0	ST_0
…	…	…	…	…	…	…
100	x_{100}	Y_{100}	d_{100}	RT_{100}	DT_{100}	ST_{100}

O. Gervasi et al. (Eds.): ICCSA 2008, Part II, LNCS 5073, pp. 1131–1141, 2008.

Where: *VN*=Vehicle Number, *C*=Capacity, *CN*=Customer Number, *XCO*=X Coord., *YCO*=Y Coord., *D*=Demand, *RT*=Ready time, *DT*=Due date, *ST* =Service Time.

In the VRPTW exist additional parameters: *C*=Clustered data, *R*=Random data, *RC*= Random and clustered data, the value 1 is for small time window and small vehicle capacity, 2 is big time window and big vehicle capacity,.

The main contribution of this paper is propose indicators applied to the VRPTW that is representative, trustworthy, and excellent; in order that it allows to select appropriately the algorithm that better solves a VRPTW instance. An indicator is a relation between quantitative or qualitative variables that allows observing the situation and the tendencies of changes generated in the object or observed situation, in relation to objectives, reach and awaited impacts [3]. Indicators will be generated for the Vehicle Routing Problem with Time Windows (VRPTW) [3].

In order to verify the proposed indicators in this paper, we used the discriminant analysis contained in software SPSS [4] software, like a method of machine learning to find the relation between the characteristics of the problem and the performance of algorithms [5], as well as the execution of 3 variants of the genetic algorithms and an the algorithm Random Search.

The discriminant analysis [6] is a multivariate statistical technique whose purpose is to analyze if exist significant differences between groups of objects with respect to a set of variables measured.

The genetic algorithms, originally developed by Hoolland [7], are heuristic adaptive that simulate processes of optimization with the natural evolution of genes in an organic population. The genetic algorithms to solve problems VRPTW were used pioneering by Blanton, et. al. [8] and Thangiah, et. al.[1]. The general steps that a genetic algorithm makes are: To initialize randomly population of solutions to problem (represented by a suitable structure of data), to evaluate each one of solutions (assign him score or fitness according to solution feasible that finds), to choose of initial population (the one that has a greater score), to apply crossover operator and mutation operator to the different solutions from the selected population, to create a new population, to repeat a determined number of times, or until has been the solution wished.

In this paper, we proposed indicators for the VRPTW, altogether with the technique of discriminant analysis to select appropriately the algorithm that better solves an instance of the problem.

2 Related Works

In this section of the paper are first the related works to this investigation (Table 2), and in the end is mentioned the hypothesis of the investigation.

Exist investigations about the creation of indicators used for self-tuning in diverse areas [9,10,11]. Fink [12] developed an algorithm selection technique for decision problems, which is based on the estimation of the algorithm gain, obtained from the statistical analysis of their previous performance. Pérez et. al. [5,13] proposed a methodology based in automatic learning systematically to develop mathematical models of the algorithm performance. Borghetti [14] developed a method to correlate each instance characteristic to algorithm performance. An important shortcoming of this method is that it does not consider the combined effect of all the characteristics. The METAL group [15] proposed a method to select the most appropriate classification algorithm for a set of similar instances. Most of these investigations it does not use the discriminant analysis like predictive technique, nor until the moment we have either found evidence that discriminant analysis in VRPTW will be used.

Table 2. Related works

Research	Problem characteristics modeling	Prediction of the best algorithm for VRPTW	Performance prediction for one instance	Discriminant Analysis for VRPTW
Fink [12]			√	
Borgethi[14]	√		√	
METAL[15]	√			
Pérez[5,13]	√		√	
This	√	√	√	√

On the basis of the related works arise the investigation hypothesis for this paper. The hypothesis of the investigation (Fig. 1) consists of verifying if it is possible to create indicators applied to problem VRPTW that appropriately helps to discriminate the algorithm that better solves the instances of problem VRPTW.

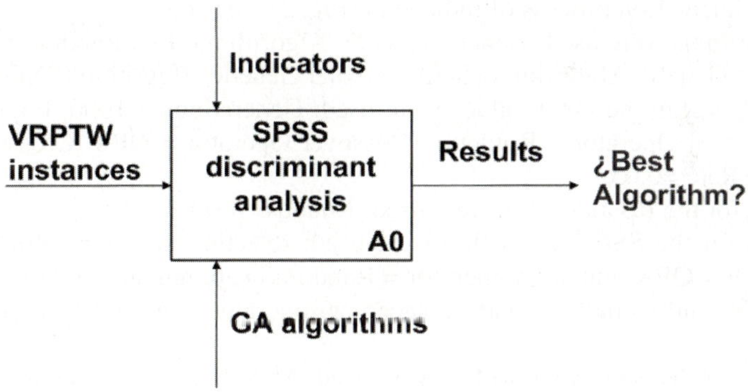

Fig. 1. Hypothesis of the investigation

3 Indicators for VRPTW

The objective of the indicators proposed for VRPTW is to allow being base appropriately to select by machine learning an algorithm that better solves an instance of the problem VRPTW and that is representative, trustworthy and excellent. The indicator *IAREA* (equation 1) is based on a position method of the known descriptive statistics called median. Median is applied to three parameters of the instance Xcoord, Ycoord, and Demand.

$$IAREA=((\mathrm{MEDIAN}(X_0:X_{CN})+\mathrm{MEDIAN}(Y_0:Y_{CN})+\mathrm{MEDIAN}(D_0:D_{CN}))*CN)/100 \quad (1)$$

Where: Indicating *IAREA*=the indicator proposed 1, *X*=*X* Coord., *Y*=*Y* Coord., *D*=Demand, *CN*=Customer Number.

The justification to unite parameters with different units was proposed by Michie [17]: To Transform the attributes usually by applying a monotonic transformation of the power law type. Monotonic transformations do not affect the machine learning methods. We propose to use what Michie says on the VRPTW parameters.

We proposed the indicator *ISIZE* (equation 2) as size of VRPTW instance. Previously in the literature only was based on the value of Customer Number (CN) [2].

$$ISIZE = CN*R*A \quad (2)$$

Where: *ISIZE*= proposed Indicator, *R*= the restrictions of the data on the basis of the value 1 for small time window and small vehicle capacity, 2 big time window and big vehicle capacity, *A*=the randomness of the data with *A*=1 for data type C, *A*=2 for data type RC, *A*=3 for data type R .

The indicator *ITIME* (equation 3) it is calculated on the basis of three time parameters (*RT*, *DT*, *ST*) of the instance of the problem.

$$ITIME=((\mathrm{MEDIAN}(RT_0:RT_n)+\mathrm{MEDIAN}(DT_0:DT_n)+\mathrm{MEDIAN}(ST_0:ST_n))*CN)/100 \quad (3)$$

Where: *ITIME* = proposed indicator, *RT* = Ready time, *DT* = Due date, *ST* = Service Time.

4 Experimentation

In the experimentation process of indicators (Fig. 2) we used:

The algorithms were used: Generic Genetic Algorithm (GA), Random Search (RS), Steady-State Genetic Algorithm (SSGA), Sexual Genetic Algorithm (SXGA) [18,19].

In GA algorithm the input values were used: Generations = 1000, Population Size = 100, Selection Operator = Roulette, Crossover Operator = OPX, Crossover Rate = 1, Mutation Rate =0.05.

The data for the RS algorithm were used: Rounds=1000.

The data for the SSGA algorithm were used: selection operator =Roulette, crossover operator = OPX, mutation operator = RandomSwap, mutation rate=0.05, population size=100, individuals = 1000, replacement operator = Worst, Tournament group size = 2.

The data for the SXGA algorithm were used: Male Selection Operator = Roulette, Female Selection Operator = Roulette, Crossover Operator = OPX, Mutation Operator = RandomSwap, mutation rate = 0.05, Generations = 1000, Population Size = 100, Tournament Group Size = 2.

The experimentation was performed on a computer Acer Travelmate 2330LC Intel Celeron 1.5GHz, 512 MB, 80 GB hard disk, 15.0” XGA.

The instances of problem VRPTW were obtained from Solomon [20] benchmark, this instances was used in HeuristicLab [18] software with random values in the options: OverloadPenalty, TardinessPenalty, RouteTimePenalty, TravelTimeExcessPenalty, and DistancePenalty. In addition was used the technique of discriminant analysis contained in software SPSS [4] software.

The results are given in the run time of the algorithm on the given instance, and the theoretical ratio (equation 4) that means that so close it was the algorithm of the solution that is reported in computer science literature.

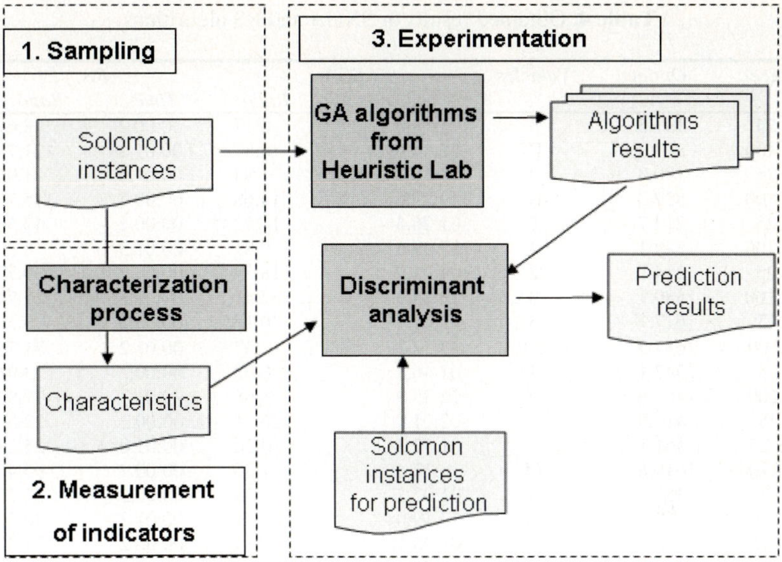

Fig. 2. Experimentation of indicators

$$Theoretical\ ratio = (Sol/NVsol) - (OptimalKnow\ /NV) \quad (4)$$

Where: *OptimalKnow* = it is the best reported optimal solution, *NV* = Number of reported optimal Vehicles, *Sol* = Found solution, *NVSol* = Number of found vehicles.

Table 3. Obtained results of GA and SSGA algorithms

Instances	*Optimal*	*Vehicles*	*GA*		*SSGA*	
			Time	*Ratio*	*Time*	*Ratio*
c101.25	191.3	3	01:59.0	0.17	00:04.7	0.17
c101.100	827.3	10	25:46.7	13.91	00:54.6	24.38
c102.25	190.3	3	01:23.2	1.94	00:00.9	0.50
c102.100	827.3	10	20:49.2	27.08	00:11.0	38.09
c201.25	214.7	2	01:30.0	11.22	00:02.1	11.22
c201.100	589.1	3	46:56.7	0.81	00:23.3	0.81
c202.25	214.7	2	01:39.2	118.23	00:00.9	118.23
c202.100	589.1	3	43:19.6	32.63	00:21.5	14.52
r101.25	617.1	8	01:01.0	0.15	00:00.7	1.84
r101.100	1637.7	20	20:59.9	1.53	00:09.8	4.6784
r102.25	547.1	7	01:06.9	1.94	00:00.7	4.59
r102.100	1466.6	18	17:41.7	6.17	00:09.9	10.37
r202.25	410.5	4	01:42.8	127.38	00:01.0	44.41
rc101.25	461.1	4	01:10.0	0.26	00:02.0	1.01
rc101.100	1619.8	15	17:31.1	4.79	00:00.1	8.85
rc102.25	351.8	3	01:06.3	0.38	00:00.8	2.01
rc102.100	1457.4	14	18:36.6	4.76	00:10.3	16.28
rc201.25	360.2	3	01:38.0	0.34	00:00.9	0.34
rc201.100	1261.8	9	35:02.5	75.82	00:19.8	134.90
rc202.25	338.0	3	01:42.0	0.27	00:02.1	2.52
rc202.100	1092.3	8	39:49.2	76.74	00:19.7	96.60

Table 4. Obtained results of SXGA and RS algorithms

Instances	*Optimal*	*Vehicles*	*SXGA*		*RS*	
			Time	*Ratio*	*Time*	*Ratio*
c101.25	191.3	3	01:18.0	0.17	00:00.2	8.67
c101.100	827.3	10	20:54.0	6.87	00:01.2	121.78
c102.25	190.3	3	01:19.1	1.94	00:01.0	1.06
c102.100	827.3	10	19:27.6	14.68	00:00.2	115.76
c201.25	214.7	2	01:26.4	11.22	00:00.2	43.58
c201.100	589.1	3	37:49.6	0.81	00:01.3	8.20
c202.25	214.7	2	01:41.9	118.23	00:00.2	39.77
c202.100	589.1	3	39:25.9	37.41	00:01.3	25.74
r101.25	617.1	8	01:01.7	0.15	00:00.2	9.38
r101.100	1637.7	20	17:37.2	2.37	00:01.2	71.75
r102.25	547.1	7	01:46.1	0.68	00:00.2	3.91
r102.100	1466.6	18	22:25.9	5.68	00:01.3	77.93
r202.25	410.5	4	02:01.9	126.71	00:00.2	22.09
rc101.25	461.1	4	01:12.3	0.26	00:01.0	25.24
rc101.100	1619.8	15	18:12.4	1.36	00:00.2	95.45
rc102.25	351.8	3	01:07.2	0.38	00:00.3	22.77
rc102.100	1457.4	14	17:19.0	11.44	00:01.3	90.21
rc201.25	360.2	3	01:25.1	0.34	00:00.2	24.32
rc201.100	1261.8	9	32:55.5	105.14	00:03.3	90.31
rc202.25	338.0	3	01:24.8	0.52	00:00.2	5.65
rc202.100	1092.3	8	34:19.1	75.90	00:01.2	91.57

Table 5. Obtained results from the indicators of the problem instances

Instances	*CN*	*ISIZE*	*IAREA*	*ITIME*	*Best Algorithm*
c101.25	25	25	28.25	231.87	GA (1)
c101.100	100	100	100.00	981.00	SXGA (3)
c102.25	25	25	28.25	305.00	SSGA (2)
c102.100	100	100	100.00	956.00	SXGA (3)
c201.25	25	50	28.00	22.50	GA (1)
c201.100	100	200	100	981.00	GA (1)
c202.25	25	50	28.00	1039.12	GA (1)
c202.100	100	50	95.00	3231.00	SSGA (2)
r101.25	25	75	19.25	51.75	GA (1)
r101.100	100	75	79.00	205.00	GA (1)
r102.25	25	75	19.25	58.25	SXGA (3)
r102.100	100	300	79.00	220.00	SXGA (3)
r202.25	25	150	19.25	253.5	SSGA (2)
rc101.25	25	50	20.00	55.00	GA (1)
rc101.100	100	200	98.00	214.00	GA (1)
rc102.25	25	50	20.00	60.87	GA (1)
rc102.100	100	50	98.00	231.00	GA (1)
rc201.25	25	100	20.00	212.50	GA (1)
rc201.100	100	400	98.00	831.00	GA (1)
rc202.25	25	100	20.00	243.75	GA (1)
rc202.100	100	400	98.00	907.00	SXGA (3)

In table 3 are the obtained results of GA and SSGA algorithms with the instances of VRPTW contained in HeuristicLab software. In Table 4 are the obtained results of SXGA and RS algorithms with the instances of VRPTW contained in HeuristicLab software. Table 5 shows the list of better algorithms for the instances of the sampling (obtained of table 3 and table 4), as well as the calculation of the proposed indicators in each one of the instances. The values of table 5 were used like input values of the discriminant analysis.

For each instance the list of algorithms was determined that better solve it, defined by the following criteria of evaluation: the run time divided between the obtained theoretical radius, and the smaller value is the algorithm champion with better performance for the instance. In order to obtain the first criterion of algorithms classification, we used two indicators (CN and ISIZE) which they were used like independent variables, and the number of the best algorithm like dependent variable. The discriminant classification was trained with VRPTW instances from Solomon benchmark.

In Table 6 are the results of the discriminant analysis (SPSS software) with the training instances (Table 3 and Table 4), showing 38.1 % of the original group classified correctly.

Table 6. Results of the discriminant analysis 1

Group Origin	***Predicted Group Membership***				*TOTAL*
	GA	*SSGA*	*SXGA*	*RS*	
GA	2	7	4	0	13
SSGA	1	2	0	0	3
SXGA	0	1	4	0	5
RS	0	0	0	0	0

Where: Predicted Group Membership = these are the independent variables, also called predictors. Group Origin = this is the dependent variable or it is the object of classification efforts, also called the grouping variable in SPSS.

In order to obtain a second criterion of algorithms classification we used two indicators (IAREA and ITIME) which they were used like independent variables, and the number of the best algorithm like dependent variable. The discriminant classification was trained with VRPTW instances from Solomon benchmark. In Table 7 are to the results of the discriminant analysis with the training instances, showing 57.1% of the original group classified correctly.

Table 7. Results of the discriminant analysis 2

Group Origin	***Predicted***	***Group***	***Membership***		
	GA	*SSGA*	*SXGA*	*RS*	*TOTAL*
GA	7	1	5	0	13
SSGA	2	1	0	0	3
SXGA	1	0	4	0	5
RS	0	0	0	0	0

Where: Predicted Group Membership = these are the independent variables, also called predictors. Group Origin = this is the dependent variable or it is the object of classification efforts, also called the grouping variable in SPSS.

In order to obtain the third criterion of algorithms classification, the three indicators (IAREA, ISIZE, ITIME) that was used like independent variables, and the number of the best algorithm was used like dependent variable. The discriminant classification was trained with VRPTW instances from Solomon benchmark.

In Table 8 and Table 9 are to the results of the discriminant analysis with the training instances, showing 66.7 % of the original group classified correctly.

Table 8. Results of the discriminant analysis 3

Group Origin	***Predicted Group Membership***				
	GA	*SSGA*	*SXGA*	*RS*	*TOTAL*
GA	9	1	3	0	13
SSGA	2	1	0	0	3
SXGA	1	0	4	0	5
RS	0	0	0	0	0

Where: Predicted Group Membership = these are the independent variables, also called predictors. Group Origin = this is the dependent variable or it is the object of classification efforts, also called the grouping variable in SPSS.

Table 9. Results of the obtained functions of the discriminant analysis

	Wilks Lambda	*Chi-square*
Function 1	.590	8.977
Function 2	.866	2.446

The Fig. 3 shows the results of the obtained functions of the discriminant analysis.

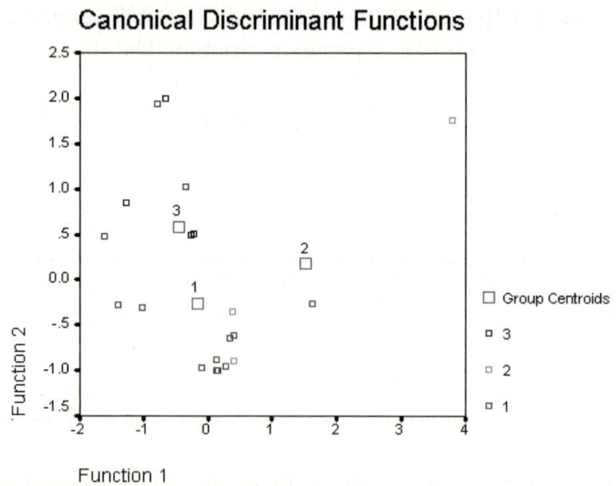

Fig. 3. Experimentation of indicators

The first function maximizes the differences between the values of the dependent variable. The second function is orthogonal to it (uncorrelated with it) and maximizes the differences between values of the dependent variable, controlling for the first factor. Though mathematically different, each discriminant function is a dimension which differentiates a case into categories of the dependent based on its values on the independents. The first function will be the most powerful differentiating dimension, but later functions may also represent additional significant dimensions of differentiation.

If the functions are effective for the training sample of VRPTW instances, the percentage of the new observations classified correctly is an indicator of the effectiveness of the discriminant functions. In order to validate the effectiveness of the discriminant classification, we were considered other instances of problem VRPTW.

Table 10 represents a fraction of the instances where is the result of the proposed indicators and the prediction of the best algorithm than solves the given instance. The prediction of the obtained classification of the discriminant analysis was of 66.7%.

Table 10. Example of instances with its characteristics and the best algorithm predicted

Instances	*CN*	*ISIZE*	*IAREA*	*ITIME*	*Best Algorithm predicted*
C105.25	25	25	28.25	232.12	GA(1)
C105.100	100	100	100.00	983.00	SXGA (3)
R210.25	25	150	19.25	234.25	GA(1)
R210.100	100	600	79.00	945.00	SXGA (3)
RC208.25	25	100	20.00	217.37	GA(1)
RC208.100	100	400	98.00	841.00	SXGA (3)

Where: CN = Customer Number, *ISIZE* = the size indicator of VRPTW instance, *IAREA* = The indicator is based on a position method of the known descriptive statistics called median, *ITIME* = The indicator *ITIME* is calculated on the basis of three time parameters (*RT*, *DT*, *ST*) of the instance of the problem.

We can see in that example that the GA and SXGA algorithm correspond 1 to 1, but is not that they correspond of that way, can vary best algorithm predicted in all the instance sample of the Solomon benchmark.

5 Conclusions

The main contribution is the proposal of indicators generated using a contained method of position in the descriptive statistics. By means of the experimentation we concludes that if it is possible to create indicating applied to VRPTW that help appropriately to predict the algorithm that better solves the instances of the VRPTW.

The prediction of the classification obtained of the discriminant analysis was 38.1% (we used two indicators: *CN* and *ISIZE*).

The prediction of the classification obtained of the discriminant analysis was 57.1% (we used two indicators: *IAREA* and *ITIME*).

The prediction of the classification obtained of the discriminant analysis was of 66.7% (with three indicators: *IAREA*, *ISIZE*, *ITIME*).

It is a good percentage of prediction taking into account that this based in techniques statistics and not in data-mining techniques.

As future works, we proposed continue with the creation of indicators to improve the percentage of the algorithm prediction that better solves a given instance of the VRPTW, as well as to use other classification to improve the prediction.

References

1. Thangiah, S.R.: Genetic Algorithms, Tabu Search and Simulated Annealing Methods for Vehicle Routing Problems with Time Windows. In: Chambers, L. (ed.) Practical Handbook of Genetic Algorithms: Complex Structures. CRC Press, Boca Raton (1998)
2. Toth, P., Vigo, D.: The Vehicle Routing Problem. Monographs on Discrete Mathematics and Applications. SIAM, Philadelphia (2001)
3. Rascon, C., Octacio, A.: Introducción a la estadística descriptiva (in Spanish). Editorial UNAM (1983)
4. SPSS, Inc. Headquarters, Chicago, Illinois, `http://www.spss.com/es/`
5. Pérez, J., Pazos, R.A., Frausto, J., Rodriguez, G., Romero, D., Cruz, L.: A Statical Approach for Algorithm Selection. In: Ribeiro, C.C., Martins, S.L. (eds.) WEA 2004. LNCS, vol. 3059. Springer, Heidelberg (2004)
6. Huberty, C.J.: Applied Discriminant Analysis. Wiley Interscience, Chichester (1994)
7. Holland, J.H.: Adaptation in Natural and Artificial Systems. University of Michigan Press, Ann Arbor (1975)
8. Blanton, J., Wainwright, R.L.: Multiple Vehicle Routing with Time and Capacity Constrainst using Genetic Algorithms. In: Proceedings of the Fifth International Conference on Genetic Algorithms (1993)
9. Pérez, J., Pazos, R.A., Frausto, J., Rodriguez, G., Cruz, L.: Self-Tuning Mechanism for Genetic Algorithms Parameters an Application to Data-Object Allocation in the Web. In: Laganá, A., Gavrilova, M.L., Kumar, V., Mun, Y., Tan, C.J.K., Gervasi, O. (eds.) ICCSA 2004. LNCS, vol. 3046. Springer, Heidelberg (2004)
10. Pérez, J., Pazos, R.A., Fraire, H., Cruz, L., Pecero, J.: Adaptive Allocation of Data-Objects in the Web using Neural Networks. LNCS, vol. 2829. Springer, Heidelberg (2003)
11. Pérez, J., Pazos, R.A., Fraire, H., Cruz, L., Santiago, E., García, N.E.: A Machine Learning Approach for Modeling Algorithm Performance Predictors. In: Torra, V., Narukawa, Y. (eds.) MDAI 2004. LNCS (LNAI), vol. 3131. Springer, Heidelberg (2004)
12. Fink, E.: How to Solve it Automatically, Selection among Problem-solving Methods. In: Proceedings of the Fourth International Conference on AI Planning Systems AIPS 1998, pp. 128–136 (1998)
13. Pérez, J., Pazos, R.A., Frausto, J., Rodriguez, G., Cruz, L.: Comparison and Selection of Exact and Heuristic Algorithm. In: Laganá, A., Gavrilova, M.L., Kumar, V., Mun, Y., Tan, C.J.K., Gervasi, O. (eds.) ICCSA 2004. LNCS, vol. 3045. Springer, Heidelberg (2004)
14. Borghetti, B.J.: Inference algorithm Performance and Selection under Constrained Resources. MS Thesis. AFIT/GCS/ENG/96D-05 (1996)
15. Soares, C., Brazdil, P.: Zoomed Ranking, Selection of Classification Algorithms Based on Relevant Performance Information. In: Zighed, A.D.A., Komorowski, J., Żytkow, J.M. (eds.) PKDD 2000. LNCS (LNAI), vol. 1910, pp. 126–135. Springer, Heidelberg (2000)
16. Michie, D., Spiegelhalter, D.J., Taylor, C.C.: Machine Learning, Neural and Statistical Classification. Ellis Horwood (1994) ISBN: 978-0131063600

17. Wagner, S., Affenzeller, M.: HeuristicLab: A Generic and Extensible Optimization Environment. Adaptive and Natural Computing Algorithms. Springer Computer Science, pp. 538–541 (2005)
18. Affenzeller, M.: New Generic Hybrids Based Upon Genetic Algorithms. Advances in Artificial Intelligence. In: Garijo, F.J., Riquelme, J.-C., Toro, M. (eds.) IBERAMIA 2002. LNCS (LNAI), vol. 2527, pp. 329–339. Springer, Heidelberg (2002)
19. Solomon, M.M.: Algorithms for Vehicle Routing and Scheduling Problems with Time Window Constrains. Operations Research 35 (1987)

The ASRS_RL – A Research Platform for Spoken Language Recognition and Understanding Experiments

Inge Gavat and Corneliu Octavian Dumitru

Faculty of Electronics Telecommunications and Information Technology,
Politehnica University Bucharest, 313 Splaiul Independetei, Bucharest Romania
{igavat,odumitru}@lpsv.pub.ro

Abstract. In the present paper are described the component stages implemented on the Automatic Speech Recognition System for Romanian Language, ASRS_RL and the speech recognition experiments made with the system. The discussed stages are the feature extraction based on cepstral analysis and linear prediction and the learning strategies implemented on the system for acoustical modelling. The first goal of our research was to develop and experiment a system for continuous speech recognition and understanding in the statistical framework of hidden Markov models (HMM). A second goal was to asses other learning strategies like support vector machines (SVM), artificial neural networks (ANN), and some hybrid structures as alternatives for the HMM. The made experiments in isolated words (digits) and vowels recognition gave good results and proved the competitively of the applied new structures.

Keywords: Speech recognition, feature extraction, hidden Markov model, artificial neural network, MFCC, PLP, LPC coefficients.

1 Introduction

Speech is the specific, most used, and most comfortable communication means for humans. To communicate with machines, man still need today sophisticated languages, difficult to be learned and used. To overcome this difficulty, a simple solution could be a communication in spoken language that becomes possible if the machine could mimic the human production and understanding of speech. To realize that, nowadays was developed and had a spectacular evolution a new technical domain, called speech technology, having as important component spoken language recognition and understanding. The history of this new domain is extended over approximately one half of a century, time for a huge progress from the first laboratory models to the commercial products available today. But despite this unbelievable development, a problem remained: the speech recognition machines have a recognition rate that can not reach that of humans, a gap existing between the two performance levels. Therefore, spoken language recognition and understanding is still a challenging domain, in which considerable efforts in research have to be done. Of course, development of electronics and computing techniques will play further an important role, but also important is to study the human's mechanisms in recognizing and understanding speech. A better knowledge in this direction could lead to results diminishing the today gap between human and machine performance.

O. Gervasi et al. (Eds.): ICCSA 2008, Part II, LNCS 5073, pp. 1142–1157, 2008.

Speech is a communication modality, so how the communication is accomplished is a question of high interest. Traditionally [1] it can be admitted that the speech communication chain is organized in four stages: detection of acoustic-phonetic cues to form words, syntactic and grammatical analysis to build sentences, semantic evaluation to determine the possible meanings of a sentence and pragmatic evaluation in order to select the convenient meaning. A machine that performs all this four stages can converse with a human, but the whole knowledge necessary to entirely realize this task is not completely acquired yet.

Today, technologies for speech recognition and understanding are based on two strategies. The first assumes that a phrase is structured in a sequence of linguistically meaningful words that has first to be decoded from the acoustical wave as accurately as possible, and from this sequence meaning and intention expressed in the sentence will be derived. In this strategy, a first step is to optimize the automatic speech recognition part and the second step is to optimize the understanding part in order to derive the desired meaning. The second strategy makes no explicit assumption on the linguistically structure, trying to capture the intended meaning, useful for the desired action, the accuracy being the result of a global optimization procedure.

Our proposed system follows the first strategy; the paper will be focused on our research and results especially in the automatic speech recognition part, as important element in the success of the whole understanding process of the spoken utterance.

The block diagram of the system, based on the pattern recognition paradigm, and applied in a continuous speech recognition task is presented in Fig. 1. The speech signal is analysed and sequence of feature vectors are calculated, and grouped in linguistic unit patterns. Each obtained pattern is compared with reference patterns, pretrained and stored with class identities. These pre-trained patterns, obtained in a learning process are in our system the acoustical models for phonemes, with or without context, and represent a first knowledge source for the word sequence decoder.

Further, based on the second knowledge source, namely the dictionary and the language model, the recognized phonemes flow will be organized in a sequence of words that represents the goal of the automatic speech recognition task.

Its performance is characterized by the word recognition rate. Like it is represented in Fig 1, after the ASR part follows the understanding part, supported by a grammar and other knowledge sources.

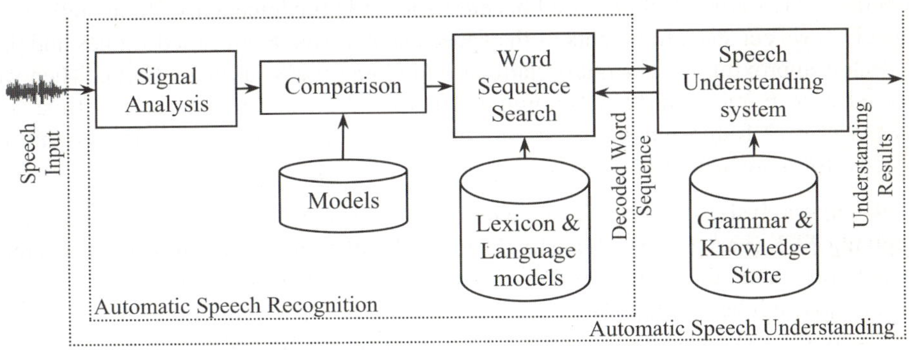

Fig. 1. Block diagram for an automatic speech recognition and understanding system

Of course the system can solve not only continuous speech recognition tasks, but also simpler tasks, like recognition of isolated words or phonemes. In our research, all this tasks helped to evaluate the obtainable performance in a great variety of possible choices concerning the two principal steps of automatic speech recognition, namely the speech analysis to extract features and the learning structures to build models.

The remainder of this paper will be structured as follows. Chapter two will briefly describe the techniques applied in our system for extraction of features in form of formants, cepstral coefficients (CC) and linear prediction coefficients LPC, or perceptual features like mel-frequency cepstral coefficients (MFCC) and perceptual linear prediction (PLP) coefficients. In Chapter three are described the learning structures applied in our system to build acoustical models. The basic learning strategies are: the hidden Markov models (HMM), the artificial neural networks (ANN), and the support vector machines (SVM). Some enhanced models like the HMM – ANN hybrid or fuzzy variant of the multilayer perceptron (MLP) are also considered. The fourth Chapter is dedicated to the databases in Romanian language constructed for our experiments, namely continuous speech database, digit database and vowel database. The fifth Chapter presents a large variety of conditions into which can be carried out the basic experiments with our Automatic Speech Recognition System for Romanian Language (ASRS_RL) in tasks for continuous speech recognition, digit recognition, and vowel recognition. The paper will end with conclusions extracted from the done experiments for our future work in understanding of spoken language.

2 Speech Analysis

First step in all recognition tasks is speech analysis, where the speech signal is processed in order to obtain important characteristics, further called features. By features extraction data compression is done, so that the amount of data used for comparisons is greatly reduced and thus, less computation and less time is needed for comparisons [9].

Our features extraction is based on perceptual cepstral coding and perceptual linear predictive coding, methods that will be presented further.

The block scheme of the acoustical processor is shown in Fig. 2. Few blocks are common in both linear prediction and cepstral coding [2]. The first block in the scheme is the frame blocking, used because speech is fundamentally a non-stationary signal, so we cut short fragments of the speech signal, which are called frames and the speech is approximated as a quasi-stationary random process during a frame. Then we passed each frame through a Hamming window. We can compute at this time the energy of each frame and we can use the energy set of coefficients in the recognition process for more accuracy.

Cepstral Analysis is a very reliable method to speech analysis and it can be realized applying FFT (Fast Fourier Transform) to the blocked and windowed time discrete signal $s(n)$. After that, the modulus of the signal is calculated and the logarithm is taken, the result being proportional in fact to the power spectrum of the speech signal. Applying IFFT (Inverse Fast Fourier Transform) leads to the real cepstrum.

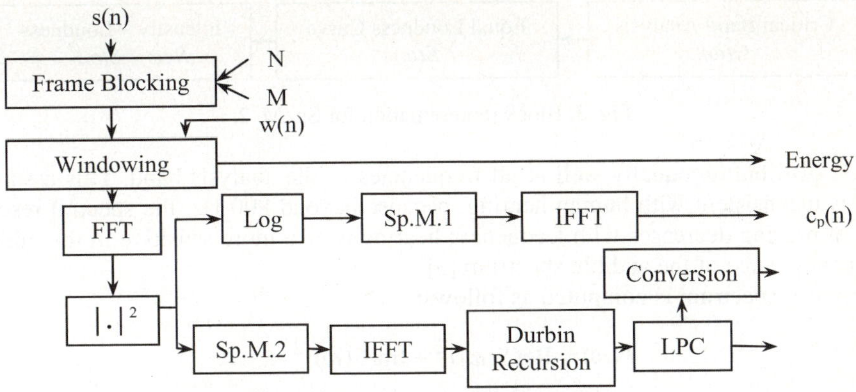

Fig. 2. The block scheme of the acoustical processor

Cepstral Coefficients, are obtained by re-sampling of the real cepstrum; alone or in addition with the energy *E*, and/or the first and second order differences constitutes a feature vector successfully applied in speech recognition

Mel-frequency Cepstral Coefficients are calculated from the power spectrum after a spectral manipulation Sp.M.1 in form of a filter bank that is a model for the critical band perception of the human cochlea. Often, in noisy environments the first (D – Delta) and second (A – Acceleration) order variations of the MFCC are applied to enhance the feature vector.

Formants can be calculated from the smoothed spectrum, resulted applying FFT to the windowed cepstrum near the origin.

Linear Predictive coding requires computation of the autocorrelation coefficients possible to be obtained according to the Wiener-Hinchin theorem, by applying IFFT to the power spectrum.

Linear Prediction coefficients are obtained by low order all-pole modelling. The Levinson – Durbin recursive algorithm is used to solve the Yule – Walker equations.

$$\begin{bmatrix} R(1) & R(2) & \dots & R(N) \\ R(2) & R(3) & \dots & R(N-1) \\ \dots & \dots & \dots & \dots \\ R(N) & R(N-1) & \dots & R(1) \end{bmatrix} \times \begin{bmatrix} A(1) \\ A(2) \\ \dots \\ A(N) \end{bmatrix} = \begin{bmatrix} -R(2) \\ -R(3) \\ \dots \\ -R(N+1) \end{bmatrix} \qquad (1)$$

where *R(n)* are the autocorrelation coefficients, and *A(n)* are the all-pole model coefficients (the predictor), and *A(1)* =1; through the conversion block the lasts can be converted in other types of coefficients like the reflection coefficients or the LPC cepstral coefficients.

Perceptual Linear Prediction (PLP) coefficients are obtained by spectral manipulation Sp.M.2, displayed in Fig.3. The PLP analysis method [10] is more adapted to human hearing, in comparison to the classic Linear Prediction Coding because of the critical band processing of the analysed signal. The LP all-pole model approximates

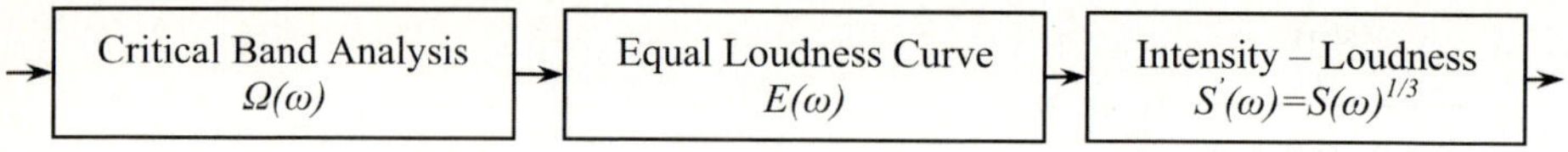

Fig. 3. Block representation for Sp. M. 2.

power distribution equally well at all frequencies of the analysis band. This assumption is inconsistent with human hearing, because beyond 800 Hz, the spectral resolution of hearing decreases with frequency; hearing is also more sensitive in the middle frequency range of the audible spectrum [3].
The power spectrum is computed as follows:

$$P(\omega) = \mathrm{Re}(S(\omega))^2 + \mathrm{Im}(S(\omega))^2 \tag{2}$$

The first step is a conversion from frequency to bark, which is a better representation of the human hearing resolution in frequency. The bark frequency corresponding to an audio frequency is:

$$\Omega(\omega) = 6\ln\left(\frac{\omega}{1200\pi} + \left(\left(\frac{\omega}{1200\pi}\right)^2 + 1\right)^{0.5}\right) \tag{3}$$

The resulting warped spectrum is convoluted with the power spectrum of the critical band-masking curve, which act like a bank of filters centred on Ω_i.
The spectrum is pre-emphasized by an equal loudness curve, which is an approximation to the non-equal sensitivity of human hearing at different frequencies, at about 40dB level. A filter having the following transfer function gives the curve:

$$E(\omega) = \frac{(\omega^2 + 56.8\times10^6)\omega^4}{(\omega^2 + 6.3\times10^6)^2 \times (\omega^2 + 0.38\times10^9)} \tag{4}$$

The last operation prior to the all-pole modelling is the cubic-root amplitude compression (Intensity – Loudness Conversion), which simulates the non-linear relation between the intensity of sound and its perceived loudness. Together with the psychophysical equal-loudness pre-emphasis, this operation also reduces the spectral amplitude variation of the critical-band spectrum.

Concluding, in this chapter are presented two methods for features extraction, the first based on cepstral analysis, the second on the linear prediction. Through cepstral analysis we can compute as features cepstral coefficients, formants and mel-frequency cepstral coefficients as perceptual variant. Through linear prediction, we can compute the linear prediction coefficients and the perceptual linear prediction coefficients. Because they mimic better the human perception, in our speech recognition experiments we used especially the perceptual coding variant of features in form of MFCC or PLP coefficients. As reference we kept also the LPC; in some experiments we used the formants as features. The features corresponding to a frame are organized as features sequences (vectors) in the recognition phase and also in the training phase giving the models.

3 Learning Strategies

In the same way into which a human learns to perceive speech from examples by listening, the automatic speech recognizer does apply a learning strategy in order to decode the pronounced word sequence from a sequence of elementary speech units like phonemes, with or without context. By learning are created models for each elementary speech unit, serving in the comparisons to make a decision about the uttered speech unit. We will describe further the learning strategies implemented in our research platform: hidden Markov models (HMM), support vector machines (SVM), artificial neural networks (ANN) in form of multilayer perceptrons (MLP) or Kohonen maps (KM). Some hybrid learning strategies are also implemented in form of a HMM-ANN combination and a fuzzy perceptron.

Hidden Markov Models with Context – depending Modelling

HMMs are finite automata, with a given number of states; passing from one state to another is made instantaneously at equally spaced time moments. At every pass from one state to another, the system generates observations, two processes taking place: the transparent one represented by the observations string (features sequence), and the hidden one, which cannot be observed, represented by the state string [4].

In context independent phoneme modelling each word results as a concatenation of the component phonemes; therefore a model is necessary for each phoneme. In Romanian language, like phoneticians claimed, there are 34 phonemes, requiring 34 different models.

In real speech, the words are not simple strings of independent phonemes: as effect of co-articulation, the immediate neighbour – phonemes, for instance the preceding and the following one, affect each phoneme in the word. This immediate neighbour – phonemes are called respectively the left and the right context; a phoneme constitutes with the left and right context is a triphone. For each such a triphone a model must be trained: in Romanian it would give a number which equals $34^3=39304$ models, which is totally unacceptable for a real world system. In our speech recognition task we have modelled only internal – word triphones and the adopted state tying procedure has conducted to a controllable situation.

If triphones are used in place of monophones, the number of needed models increases and the problem of insufficient training data may occur. To solve this problem, tying of acoustically similar states of the models built for triphones corresponding to each context - there is an efficient solution. For example, in Fig. 4a four models are represented for different contexts of the phoneme “a”, namely the triphones “k – a + S”, “g – a + z”, “n – a + j”, “m – a + j”. In Fig. 4b the clusters formed with acoustically similar states of the corresponding HMMs are represented.

The choice of the states and the clustering in phonetic classes are achieved by mean of phonetic decision trees. A phonetic decision tree built as a binary tree and has in the root node all the training frames to be tied, in other words all the contexts of a phoneme. To each node of the tree, beginning with the parent – nodes, a question is associated concerning the contexts of the phoneme [6]. Depending on the answer, yes or no, child nodes are created and the frames are placed in them. New questions are further made for the child nodes, and the frames are divided again.

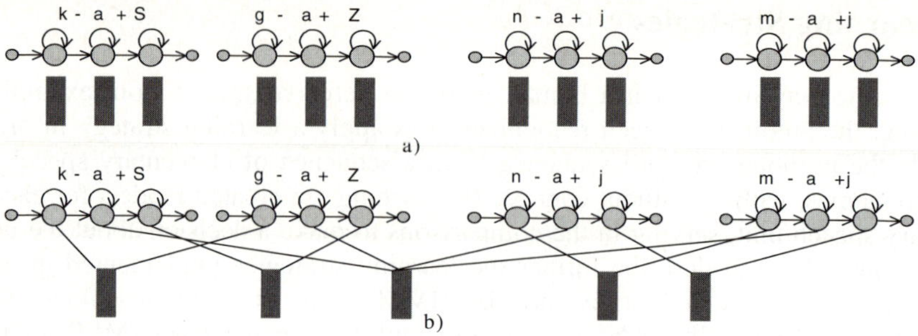

Fig. 4. a) Different models for triphones, b) Tying of acoustically similar states

The questions are chosen in order to increase the log likelihood of the data after splitting. Splitting is stopped when increasing in log likelihood is less than an imposed threshold, resulting a leaf node. In such leaf nodes are concentrated all states having the same answer to the question made along the path from the root node and therefore states reaching the same leaf node can be tied as regarded acoustically similar. For each leaf node pair the occupancy must be calculated in order to merge insufficient occupied leaf nodes [7].

A decision tree is built for each state of each phoneme. The sequential top down construction of the decision trees was realized automatically, with an algorithm selecting the questions to be answered from a large set of 130 questions, established after knowledge about phonetic rules for Romanian language [8].

Support Vector Machine

The foundations of Support Vector Machines have been developed by Vapnik [11].

The formulation is based on Structural Risk Minimization (SRM) principle, which minimizes an upper bound on the generalization error, as opposed to Empirical Risk Minimization (ERM) which minimizes the error on the training data.

SVM is a statistical algorithm with a great potential to generalize, that can successfully be used in pattern recognition and information retrieval tasks. The main idea in training a SVM system is finding a hyperplane as a decision boundary between two classes. Considering the general case of non-separable patterns, Fig. 5 is representative.

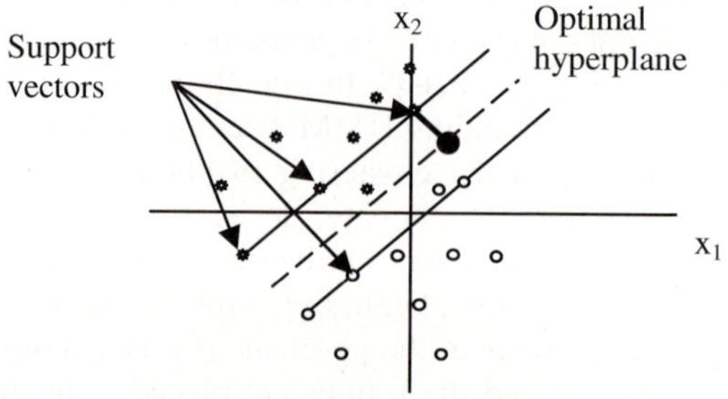

Fig. 5. Non-separable patterns

The equation that is verified for each data sample is:

$$d_i(w^T x_i + b) \geq 1 - \xi_i \tag{5}$$

where ξ_i represents the number of data samples left inside the decision area, giving the number of training errors. The support vectors are the data samples for which the equation (5) is verified with the equal sign. After the training process only the support vector will be kept from all data.

The problem of finding the optimal hyperplane becomes a problem of minimizing the cost function described by equation (6):

$$\Phi(w,\xi) = \frac{1}{2} w^T w + C \sum_{i=1}^{N} \xi_i \tag{6}$$

where minimizing the first term means maximizing the distance between the two classes and minimizing the second term means reducing the number of training errors. Under those circumstances, parameter C becomes a balance between a smaller training error and a bigger distance between classes.

Fundamentally, SVM is a binary decision method, but there are several algorithms that allow its use in classification tasks with more than 2 classes. The most common algorithms use combinations of binary SVMs: *"one against one"* method, *"one against all"* method and DAGSVM (Directly Acyclic Graph SVM) method. We used the DAGSVM method, which consist in building *k(k-1)* binary SVMs if we have k classes, each class is trained with another class, until we trained each class with all the other classes, then we form a tree. In the testing phase, we start at the top of the tree and if the decision is made for the i class, then we follow to the left path. If not, we go to the right path, continuing until the end of the tree where we will have the final decision.

Classical and Fuzzy Variant of Multilayer Perceptron

MLP is the most common ANN architecture used for speech recognition. Typically, the classical MLPs have a layered feed–forward architecture, with an input layer, one or more intermediate (hidden) layers, and one output layer. The structure without hidden layer is called Boolean network and is a simple perceptron.

Each layer computes a set of linear discriminative functions, followed by a non-linear function, which is often a sigmoid function.

The numbers of neurons in the hidden layer was experimentally determined, trying to achieve an optimum between the following two opposite requirements: (a) lower computing volume and more rapid process of convergence in the learning period; (b) better performances from the correct classification of input patterns percentage.

In the learning phase are determined the optimum values for weights [12], [14] connecting the pairs of neurons from the adjoint layers in the input–output direction using the Back-Propagation algorithm.

Introducing a fuzzy processing of the input layer of the MLP is a solution to improve the MLP performances.

The structure of the fuzzy neural network (Fig. 6) [14], like the classical one, is composed from an input layer, a hidden layer and an output layer.

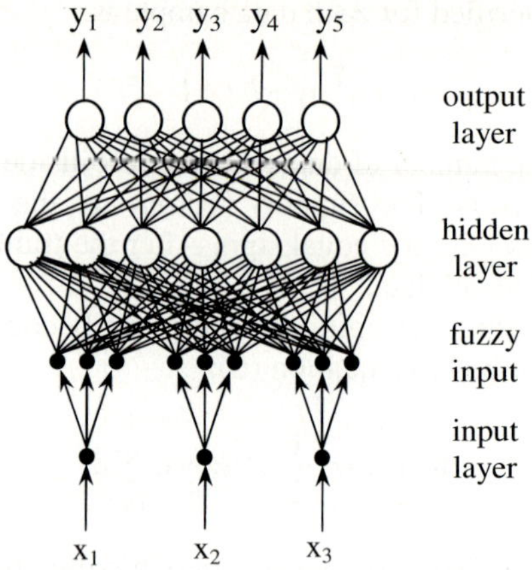

Fig. 6. Schematic diagram of fuzzy MLP

In learning phase, the Back-Propagation algorithm is used to determine the weights which minimized the mean square error between the real output and the desired one.

Kohonen maps

Kohonen maps are competitive neural networks with topological character. The setting up of the winner neurons at output is done with keeping the topological relations between the input vectors. That is the reason for which this neural network is successfully used in pattern recognition.

It was applied the following structures to recognize vowels: (1) the input layer with 3 neurons, corresponding to the three formant frequencies; (2) three case for the output layer: one-dimensional with 25 neurons, bidimensional with 4×4 neurons, 5×5 neurons, 6×6 neurons, and toroidal with 25 neurons.

In the learning phase [15] the structures are trained and the weights of the networks are established in two steps: (a) the determination of the winner neurons; (b) the adaptation of the weights for the winner neurons and for the neurons existing in a certain neighbourhood. In this step important are: the neighbourhood dimension $r(t)$ decreasing during the training, and the learning rate $\eta(t)$ following in our experiments one of the laws:

$$\eta(t) = t^{-1} \text{ or } \eta(t) = t^{-1/2} \tag{7}$$

Hybrid System

The HMM-based speech recognition methods make use of a probability estimator, in order to approximate emission probabilities $p(x_n/q_k)$, where x_n represents the observed data feature, and q_k is the hypothesized HMM state [13]. These probabilities are used by the basic HMM equations, and because the HMM is based on a strict formalism, when the HMM is modified, there is a great risk of losing the theoretical foundations or the efficiency of the training and recognition algorithms. Fortunately, a proper use of the MLPs can lead to obtain probabilities that are related with the HMM emission probabilities [14].

In particular, MLPs can be trained to produce the *a posteriori* probability $p(x_n / q_k)$, that is, the *a posteriori* probability of the HMM state given the acoustic data, when each MLP output is associated with a specific HMM state. Many authors have shown that the outputs of an ANN used as described above can be interpreted as estimates of *a posteriori* probabilities of output classes conditioned by the input, so we will not insist on this matter, but we will mention an important condition, useful for finding an acceptable connectionist probability estimator: the system must contain enough parameters to be trained to a good approximation of the mapping function between the input and the output classes.

Thus, the *a posteriori* probabilities that are estimated by MLPs can be converted in emission probabilities by applying Bayes' rule (8) to the MLP outputs:

$$\frac{p(x_n / q_k)}{p(x_n)} = \frac{p(q_k / x_n)}{p(q_k)} \tag{8}$$

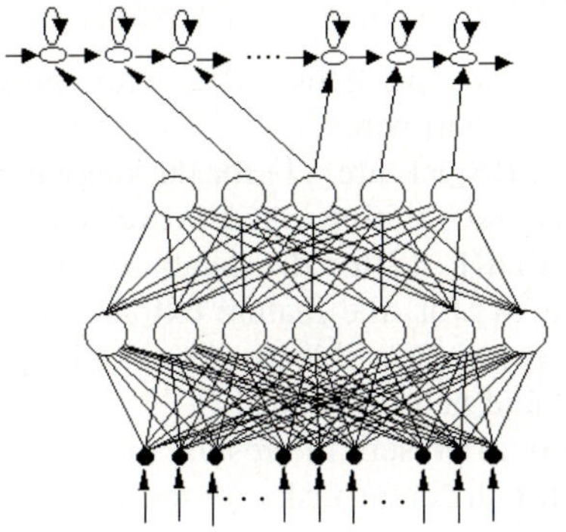

Fig. 7. Schematic diagram of a hybrid system

That is, the emission probabilities are obtained by dividing the *a posteriori* estimations from the MLP outputs by estimations of the frequencies of each class, while the scaling factor $p(x_n)$ is considered a constant for all classes, and will not modify the classification.

This was the idea that leads to hybrid neuro-statistical methods, that is, hybrid MLP - HMM methods (Fig. 7), applied for solving the speech recognition problem.

4 Databases

The databases created for our experiments are: CDRL (Continuous Database for Romanian Language), DDRL (Digit Database for Romanian Language) and VDRL (Vowel Database for Romanian Language). The data are sampled by 16 kHz, quantified with 16 bits, and recorded in a laboratory environment [16].

CDRL: for *continuous speech recognition*, the database is constituted for training by 3300 phrases, spoken by 11 speakers (7 males and 4 females), each speaker reading 300 phrases, and for testing by 880 phrases spoken by the same speakers, each of them reading 80 phrases. The training database contains over 3200 distinct words, while the testing database contains 1500 distinct words.

In order to query out our experiments the database was reorganized as follows: one database for male speakers (MS), one database for female speakers (FS) and one mixed database for male and female speakers (MS and FS). In all cases we have excluded one MS and one FS from the training and we used for testing.

DDRL: for *digit recognition*, the database contained speech data from 9 speakers (6 males and 3 females) each speakers reading 9 digits (*unu, doi, trei, patru, cinci, şase, şapte, opt, nouă*).

VDRL: for *vowel recognition*, the database contained speech data from 19 speakers (9 males and 10 females) each reading the same 5 vowels (*a, e, i, o, u*).

The database based on formant contained for the training 500 formant vectors, 100 for each vowel and for the testing 250 formant vectors, 50 for each vowel.

5 Experimental Results

The experiment results are made by ASRS_RL, with multiple options for a large variety of speech recognition experiments.

The possible options in ASRS_RL are: (1) for the recognition application: continuous speech recognition, digit recognition , and vowel recognition; (2) for the databases: CDRL, DDRL, and VDRL; (3)for the database configuration for training and testing: MS, FS, MS and FS (4) for the feature extraction method: MFCC, PLP and LPC; (5) for the modelling techniques: statistical modelling, neural modelling, hybrid modelling, (neuro-statistical and neuro-fuzzy)[17].

The goal of this Chapter is to present the results of experiments realized by applying the learning strategies for different tasks of speech recognition, like continuous speech recognition, digit and vowel recognition in order to evaluate the obtained performance.

Continuous Speech Recognition Based on Statistical Method

To assess the performance of our ASR system we initiated comparative tests on the CDRL database for the word recognition rate (WRR) and the phrase recognition rate (PRR) under different conditions.

The conditions for feature extraction are: perceptive cepstral analysis giving a 36 - dimensional vector (MFCC_D_A) having as components 12 MFCCs with the corresponding first and second order variations, perceptual linear prediction giving a 5 - dimensional feature vector having as components 5 PLP coefficients, and linear prediction, giving a 12 - dimensional feature vector having as components the LPC.

For WRR the comparison is made for the following situations: (1) MFCC, LPC and PLP coefficients, (2) triphone modelling/monophone modelling; (3) gender based training/mixed training; [18].

The results obtained in the experiments realized under these conditions are summarized in Table 1. The WRR is displayed in this table for the case of testing with MS and FS and training with MS, FS and mixed database).

Trying to "understand" the spoken phrases, we introduced some restrictions in the word sequence in form of a simple grammar. We used a *loop grammar* which means

Table 1. WRR (%) for training with all three databases and testing with MS or FS

	Test	Type	MFCC_D_A	LPC	PLP
Train MS	MS	Monophone	56.33%	30.85%	34.02%
		Triphone	81.02%	49.73%	68.10%
	FS	Monophone	40.98%	23.23%	25.12%
		Triphone	72.86%	47.68%	59.00%
Train FS	MS	Monophone	53.56%	26.72%	23.78%
		Triphone	69.23%	49.73%	53.02%
	FS	Monophone	56.67%	31.11%	34.22%
		Triphone	78.43%	61.15%	58.55%
Train MS&FS	MS	Monophone	57.44%	26.10%	47.00%
		Triphone	78.24%	51.50%	70.11%
	FS	Monophone	49.89%	24.06%	41.22%
		Triphone	74.95%	50.49%	69.65%

Table 2. PRR (%) for training with all three databases and testing with MS or FS

Train	Test	MFCC_D_A	LPC	PLP
MS	MS	66.25%	11.25%	37.50%
	FS	48.75%	12.50%	32.50%
FS	MS	37.50%	11.25%	21.25%
	FS	57.50%	33.75%	30.00%
MS&	MS	60.00%	10.00%	36.25%
FS	FS	55.00%	11.25%	46.25%

any word from the dictionary can appear after any other word with the same probability.The results are displayed in Table 2 in the case of triphone modelling for phrase recognition rate (PRR) applying a *loop grammar*. The results are weaker for PRR as for WRR and that is a direction in that it remains to do a hard work.

Digit Recognition Based on Statistical and Hybrid Methods

In the first digit recognition experiment, we compare the performance of two kinds of classifiers, namely the SVM and HMM, the performance being appreciated by their recognition rate and by their generalization capacity. For that, we performed two types of tests on the DDRL database: first, with enrolled speakers, which mean that the speakers were involved both in training and testing. We used five speakers for training and four for testing. For the second type of tests, with unenrolled speakers we used the *leave-one-out* method: for each word, we trained the classifier with 8 speakers and tested with the 9th repeating the procedure for each speaker.

The conditions for feature extraction are: perceptive cepstral analysis giving a 13 - dimensional vector having as components 13 MFCCs and perceptual linear prediction giving a 5 - dimensional feature vector having as components 5 PLP coefficients [19].

In the second digit recognition experiment, we evaluate the performance obtained with HMM and with the hybrid neuro-statistical system (HMM - MLP) for unenrolled and enrolled speaker [16].

The digit parameters were extracted by cepstral analysis, in form of 12 mel-frequency cepstral coefficients.

a) HMM: for each digit we constructed a left-right hidden Markov model with 3 states;

b) HMM - MLP: the system consists of 9 hybrid models corresponding to 9 digits. Each hybrid model is made of 5 states, each state being associated with one output node of the MLP. The MLP has one hidden layer (100 nodes), and the input layer consisting of 12 nodes [14].

The results for digit recognition are presented in Table 3.

Table 3. WRR (%) for digit recognition using SVM, HMM and HMM-MLP

Type	Features extraction	Enrolled	Unenrolled
SVM	MFCC	97.70%	91.70%
HMM	MFCC	98.00%	97.50%
HMM-MLP	MFCC	98.50%	98.30%
SVM	PLP	91.70%	84.70%
HMM	PLP	95.10%	94.20%

The performance of the HMM, SVM and hybrid system are validate using other databases, for example Advanced Multimedia Laboratory from the Carnegie Mellon University. The results show the same behaviour, but the WRRs are slightly higher for the English database, recorded in a studio with specialized speakers.

Vowel Recognition Based on Statistical, Neural and Neuro-fuzzy Methods

The learning strategies applied in our recognition experiments are: the unsupervised KM, the MLP, the fuzzy-MLP and the HMM.

In the first case, the vowel recognition rate obtained using the formants as features are depicted in Table 4. The learning structures applied for this investigation are:

(1) KM was tested in different variants: 1-dimensional, 2-dimensional and toroidal.
(2) MLP contain 3 layers organized as it follows: (a) the input layer with 3 neurons, corresponding to the three formant frequencies; (b) the hidden layer with 0 or 4 neurons, (c) the output layer with 5 neurons corresponding each to a processed class (in our case the vowel *a, e, i, o, u*) [19].
(3) Fuzzy-MLP the learning rate (during the training) is gradually decreased in discrete steps {1, 0.5, 0.3, 0.1, 0.05, 0.03, 0.01}, until the network converges to a minimum error solution. The network was trained in 7500 steps and it was obtained a 0.021 mean square error per vector's component.

Table 4. Vowel recognition rate (%) in the case of format

Vowel	KM 1-dim	KM 2-dim	KM toroidal	MLP	Fuzzy MLP
a	97.40%	98.80%	98.00%	100.00%	98.50%
e	96.80%	97.60%	97.60%	97.50%	99.00%
i	97.80%	98.40%	98.80%	99.00%	99.50%
o	98.20%	98.80%	98.80%	99.00%	99.00%
u	98.80%	97.90%	98.20%	99.00%	100.00%
Mean	***97.60%***	***98.30%***	***98.28%***	***98.90%***	***99.20%***

In the second case, the vowel recognition rate obtained using the MFCC coefficients (in form of 12 mel-frequency cepstral coefficients) is presented in Table 5. The following learning structures are considered:

(1) HMM: for each vowel on the VDRL database is constructed a hidden Markov model with 3 state left-right.

(2) MLP is a two-layer perceptron trained with Back-Propagation algorithm, having in the output layer 5 nodes corresponding to the 5 vowels to be classified and 100 nodes in the hidden layer (experimentally chosen). The number of the input nodes is equal to the number of features (12) used to describe the vowels.

For the experiments, in the VDRL were excluded one MS and one FS for the training and used for testing. The results presented in this table are in the case of training only MS. Similarly results were obtained for the training with FS [17].

Table 5. Vowel recognition rate (%) in the case of the MFCC and the training with MS

Vowel	HMM		MLP	
	Testing MS	Testing FS	Testing MS	Testing FS
a	100.00%	80.71%	100.00%	82.28%
e	85.81%	43.25%	94.82%	50.67%
i	85.71%	85.71%	95.15%	92.41%
o	90.90%	51.33%	97.00%	52.78%
u	88.88%	71.42%	94.83%	77.85%
Mean	***91.26%***	***66.48%***	***96.36%***	***71.20%***

6 Conclusions

The following conclusions can be reported after the made experiments:

A. Continuous speech recognition

(1) Concerning the speech parameterization, there are different behaviours. In the case of PLP coefficients, the best WRR are obtained on the database combined - trained with MS and FS for both cases of tests with MS or FS, this proving a high generalization capacity of the combined system. In the case of LPC and MFCC coefficients, the combined - trained database is not so efficient, the best results being obtained if the tests are made on the same type of database used in the training process.

(2) The triphone modelling is effective, conducting to increasing in WRR between 15% and 30% versus the monophone modelling. The maximal enhancement exceeds 30% for training MS and testing FS for MFCC_D_A.

(3) A gender based training conduct to good result for test made with speakers from the same gender (training MS/testing MS: 81.02%, testing FS: 72.86%; training FS/testing FS: 78.43%, testing MS: 69.23%); changing gender in testing versus training leads to a decrease in WRR around 10%. For a mixed trained data base changing gender determines only variations around 5% in WRR.

(4) The phrase recognition rate variation follows the word recognition rate variation, which was expected because nothing was especially done to enhance these, but the results (in %) are drastically lower when we compare with word recognition rate.

B. Digit recognition

(1) It is to seen that SVM performance is slightly after that of the HMM, but is really promising, taking into account that the HMM has the benefit of a so long refinement time.

(2) Trying to reduce these limitation effects of HMMs (the model training is not discriminative), chosen an alternative approaches can be a good solution. This approaches combine the HMM with MLP into a hybrid system. The results for HMM are lower than the results for the hybrid system.

C. Vowel recognition

(1) The KM 2-dimensional structures and the toroidal have the same performance, weaker is the performance of the 1-dimensional structure. The best balanced situation corresponds to the 2-dimensional map 5x5, in which all neurons are associated to a vowel to be recognized.
Using fuzzy multilayer structure it is an improvement with a mean value of 0.30% comparative with the non-fuzzy structure.

(2) The recognition rates in the case of HMM are lower than in the case of MLP. A possible explanation can be the fact that the model training is discriminative, while in the case of HMM the training is not discriminative, which represents a disadvantage of HMM utilization.

The ASRS_RL platform proved to be a valuable tool in realizing a large variety of speech recognition experiments.

We could develop a system for continuous speech recognition in Romanian language with good results in WRR, but weaker in PRR. In both directions improvements are necessary: higher quality and larger databases for better training the models could improve WRR; PRR can be improved by introducing more severe restrictions by higher level grammars or improved language models. Of course semantic and pragmatic knowledge would be a next step in improving the system.

We could also assess some new structures for speech recognition that we hope to can implement in further experiments.

References

1. Juanhg, B.H., Furui, S.: Automatic Recognition and Understanding of Spoken Language–A First Step Toward Natural Human–Machine Communication. Proc. IEEE 88(8), 1142–1165 (2000)
2. Gavat, I., Dumitru, C.O., Costache, G., Militaru, D.: Continuous Speech Recognition Based on Statistical Methods, Speech Technology and Human-Computer-Dialog (SPED 2003), Bucharest, pp. 115–126 (2003)
3. Dumitru, C.O., Gavat, I.: A Comparative Study of Features Extraction Methods Applied for Continuous Speech Recognition in Romanian Language. In: The 48th International Symposium ELMAR 2006, Zadar, pp. 115–118 (2006)
4. Gavat, I., et al.: Elemente de sinteza si recunoasterea vorbirii, Ed.Printech, Bucharest (2000)
5. SAMPA, `http://www.phon.ucl.ac.uk/home/sampa/home.htm`
6. Young, S.J.: The general use of tying in phoneme-based HMM speech recognizers. In: Proc. ICASSP 1992, San Francisco, vol. 1, pp. 569–572 (1992)

7. Young, S.J., Odell, J.J., Woodland, P.C.: Tree based state tying for high accuracy modelling. In: ARPA Workshop on Human Language Technology, Princeton (1994)
8. Rosetti, Al., Introducere in fonetica, Ed. Stiintifica, Bucharest (1967)
9. Huang, X., Acero, A., Hon, H.W.: Spoken Language Processing – A Guide to Theory, Algorithm, and System Development. Prentice Hall, Englewood Cliffs (2001)
10. Hermansky, H.: Perceptual Linear Predictive (PLP) Analysis of Speech. J. Acoustic Soc. America 87(4), 1738–1752 (1990)
11. Vapnik, V.: An overview of statistical learning theory. IEEE Trans. on Neural Networks 10(5), 988–1000 (1999)
12. Goronzy, S.: Robust Adaptation to Non-Native Accents in Automatic Speech Recognition. Springer, Berlin (2002)
13. Rabiner, L.R.: A Tutorial on Hidden Markov Models and Selected Applications in Speech Recognition. Proceedings of the IEEE 77(8), 257–286 (1989)
14. Valsan, Z., Gavat, I., Sabac, B., Cula, O., Grigore, O., Militaru, D., Dumitru, C.O.: Statistical and Hybrid Methods for Speech Recognition in Romanian. International Journal of Speech Technology 5(3), 259–268 (2002)
15. Gold, B., Morgan, N.: Speech and audio signal processing. John Wiley&Sons, N. Y (2002)
16. Dumitru, C.O., Gavat, I.: Vowel, Digit and Continuous Speech Recognition Based on Statistical, Neural and Hybrid Modelling by Using ASRS_RL. In: EUROCON 2007, Warsaw, pp. 856–863 (2007)
17. Dumitru, O.: Modele neurale si statistice pentru recunoasterea vorbirii, Ph.D. thesis (2006)
18. Dumitru, O., Gavat, I., Militaru, D.: Progresses In Continuous Speech Recognition Based On Statistical Modelling For Romanian Language. In: ICINCO 2007, Angers, pp. 262–267 (2007)
19. Gavat, I., Dumitru, O., Iancu, C., Costache, G.: Learning Strategies in Speech Recognition. In: The 47th International Symposium ELMAR 2005, Zadar, pp. 237–240 (2005)

A Method for Automatic Text Categorization Using Word Sense Disambiguation

Azucena Montes Rendon, Rocio Vargas A., Hugo Estrada Esquivel,
Juan G. Gonzalez Serna, and Jose Ruiz Ascencio

Centro Nacional de Investigación y Desarrollo Tecnológico
Interior Internado Palmira s/n, col. Palmira
Cuernavaca, Morelos, México. C.P. 62490
{amr,rvargas,hestrada,gabriel,josera}@cenidet.edu.mx

Abstract. At present time, Information plays a relevant role in current societies. In this context, Internet is one of the most extended mechanisms to communicate and distribute information around the word. Today, due to the extremely large number of information sources, automatic mechanisms are needed to filter the information that could be useful for each user. However, one of the problems that the usual techniques of automatic text categorization have not been able to handle is polysemy (words with two o more senses). In this paper, we have faced this problem by proposing a semantic analyzer for the automatic categorization of texts in Spanish. Context exploration techniques were used as a key mechanism for guiding the disambiguation process. A specific lexical database and its existing semantic relations fulfilled the objective of appropriately categorizing the analyzed text. To validate this analyzer, a tool was developed that classifies web pages by semantic sense. We present performance results for this classifier. Finally, a comparison with four other classification tools is reported.

1 Introduction

At present, most information could be accessible through electronics means. Several techniques have been proposed to improve the performance in appropriately accessing and using this information. In this context, it is clear that mechanisms are needed to filter the extremely large volume of information that a user obtains when looking for a specific content in the Web. In this sense, automatic text classification is one of the most successful approaches today to address the information filtering problem.

Machine learning and information retrieval are traditional techniques to carry out the classification process. However, there are certain issues that current techniques have not been able to manage. One of these unsolved problems is the polysemy (words with two o more senses) that exists in almost all languages. The lack of means to handle this problem induces the search mechanisms to obtain irrelevant information and thus, an incorrect classification of texts is obtained as a result. In this study, we present a semantic analyzer which analyzes text from the point of view of a polysemic word and which performs a classification according to the different meanings of the word. Therefore, given a polysemic word, the analyzer disambiguates the meaning of

O. Gervasi et al. (Eds.): ICCSA 2008, Part II, LNCS 5073, pp. 1158–1169, 2008.

the word according to the context in which it is found and classifies it in the corresponding domain.

This analyzer disambiguates the polysemic word using contextual exploration [1, 2] and the semantic relationships of the words in lexical databases: WordNet (WN), EuroWordNet (EWN) [3]. The classification is based on WordNet Domains Hierarchy (WNDH) [4].

Several studies have tried to solve the problem of polysemy in automatic text classification. Gómez et al [5] present an interesting study in which, starting from a thesaurus and statistical approaches they manage the disambiguation of sentences. In addition to the use of statistical approaches, we think that it is also important to consider the linguistic markers that (make explicit) underscore relevant parts of the text. In this paper, a semantic analyzer that considers linguistic approaches is presented as the main contribution of our research. Since we work in Spanish we have used EWN and WNDH to determine the classes or domains for the selected documents.

To validate the semantic analyzer a tool was developed that classifies web pages. This tool starts using a standard search engine that gives as result a set of web pages. The tool takes these pages and classifies them according to the different meanings of the searched word. A state of the art on web page classifiers in Spanish was made in order to compare our approach with others of the same class.

The paper is structured as follows. Section 2 is an overview of the proposed approach. Section 3 explains the filtering tasks: filtering and preprocessing. Section 4 describes the tasks of the semantic analyzer: paragraph selector, disambiguation and classification. Section 5 presents the validations and results. Finally, the last section is dedicated to the conclusions of this work.

2 Overview of the Proposed Method

The proposed method is composed of four keys steps (Figure 1): first, a filtering process of a Web query is applied in order to obtain a set of pages to be analyzed. To do the search we use standard Web searchers. Following, in preprocessing process the HTML code is extracted from the selected pages generated by the query. The idea is using the structure of the HTML to obtain meta-data and information needed for the semantic analysis. As a result of this process, a document that contains the relevant paragraphs is generated. In semantic analysis, the relevant segments of the document are classified using EWN and WNDH to determine the classes or domains for the selected words. As a result of the proposed approach, the page can be classified according to their corresponding context.

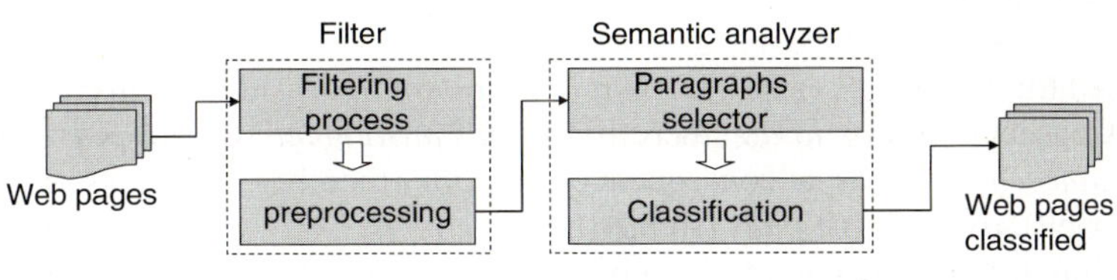

Fig. 1. Architecture of the proposed method

3 Information Filtering Process

In the first process of the proposed approach, it is necessary to obtain a set of target pages that contain the text to be analyzed. The filter obtains the pages from a query performed in a Web search. In order to do this, we have performed an analysis of the petition chains to the most popular searchers: Yahoo, AltaVista, Google, AllTheWeb and Clusy. Table 1 shows the petition chains for most popular Web searchers. In order to manage the most complicated case of analysis, the filter performs the analysis according to the viewpoint of an isolated polysemic word.

Table 1. Simplified petition chains

Searcher	Petition chains
Yahoo!	http://mx.search.yahoo.com/search?vp=planta&vf=html&fl=1&vl=lang_es&n= 10
Altavista	http://es.altavista.com/web/results?q=planta&filetype=html&nbq=10
AllTheWeb	http://www.alltheweb.com/search?q=planta&l=es &doctype=html &hits=25
Clusty	http://clusty.com/search?query=planta&language=spanish&num=200 &filetype=html

Using the petition chains, the system performs queries according the specific search tool. As a result, a set of hyperlinks are obtained that represent the pages where the term specified in the query has been found. An analysis is also performed in this stage in order to validate that links returned by the Web search contain the search word. Then, the filter selects only the documents where the search word is found.

The second step of the method is the preprocessing. In this step the system analyzed the complete text of the Web pages in order to find the relevant text. The use of linguistic markers is a key factor in the preprocessing stage.

The markers are linguistic units that establish the order of the discourse and guide the reader to the text segments emphasized by the author. In Prada's work [6, 7] the markers have been categorized into four groups (information structures, connectors, reformulation markers and argumentative operators). In each group, a more detailed set of associated concepts have been defined. Following, a classification of these markers proposed by Prada, that we have adopted, is presented.

a) Information structurers:

Commentators: *pues* (well), *pues bien* (well), *así las cosas* (as things stand)
Ordainers: *en primer lugar / en segundo lugar* (in first place/ in second place), *por una parte / por otra parte* (on the one hand / on the other hand), *de un lado / de otro lado* (on one side, on the other side).
Digressors: *por cierto* (by the way), *a todo esto* (meanwhile), *a propósito* (talking of).

b) Connectors:

Additive: *además* (also), *encima* (on top), *aparte* (aside), *incluso* (even).
Consecutive: *por tanto* (therefore), *por consiguiente* (thus), *por ende* so, *en consecuencia* (as a consequence), *de ahí* (thereby), *entonces* (then), *pues* (well), *así* (so), *así pues* (so then).
Counter-argumentative: *en cambio* (on the other hand), *por el contrario* (on the contrary), *antes bien* (notwithstanding), *sin embargo* (however), *no obstante* (nonetheless), *con todo* (despite).

c) Reformulation markers:

Explanatory: *o sea* (or), *es decir* (in other words), *esto es* (that is), *a saber* (namely).

Rectification: *mejor dicho* (more accurately), *mejor aún* (better yet), *más bien* (rather)

Rift: *en cualquier caso* (in any case), *de todos modos* (anyway).

Recapitulation: *en suma* (to sum up), *en conclusión* (in conclusion), *en definitiva* (definitely), *en fin* (in short), *al fin y al cabo* (at last).

d) Argumentative operators:

Argumentative reinforcement: *en realidad* (In reality), *en el fondo* (deep down), *de hecho* (as a matter of fact).

Realization: *por ejemplo* (for example), *en particular* (in particular)

We have used linguistic markers as a basis for our automatic approach because we have found that, in practice, a very close relation exists between the text where a marker is used and the general topic of the document. We perform several practical tests that confirm this hypothesis. We also perform several tests to determine the segments of text that are useful for our analysis. We found that the following sections are relevant for our analysis purposes: title, metadata and the plain text of the Web page. As a result of the filtering process, a new representation of the Web page is generated that contains the relevant sections before mentioned. This new page representation is the input for the semantic analysis process.

4 Semantic Analysis of the Text

The objective of the semantic analyzer is to appropriately classify the pages according to contextual information of the page. The semantic analyzer is made up of the following parts: paragraph selection, text disambiguation and text classification.

4.1 Paragraph Selection

In the first phase of the semantic analysis, the analyzer selects the relevant text fragments based on two subtasks: a) the localization and selection of segments where the polysemic word appears, and b) the selection of text segment where discourse markers appear. As commented before, we have concluded that linguistic markers offer indications of the relevance of the analyzed text. As a result of preprocessing, a new representation of the page is created that contains only relevant text fragments. The generated page is then the input to the disambiguation process, which makes a contextual exploration over the page text.

4.2 The Disambiguation Process

The objective of disambiguation is giving an appropriate meaning to a specific polysemic word in a certain context. In order to carry out the disambiguation phase, we use the lexical knowledge contained in WordNet (WN) and EuroWordNet (EWN) [3] to determine the existing semantic relationships among words. We carried out a context exploration starting from the term to disambiguate. The EWN data were complemented with a simple table that contains nouns related to their root, the nouns can be found either in singular or plural (inflexion).

The next step in the process consists in performing a new search with the purpose to locate the set of homonym synsets (first vicinity synsets) of the searched word (figure 2).

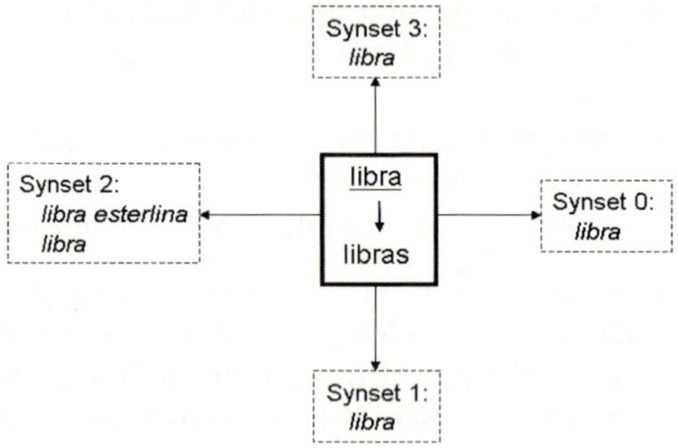

Fig. 2. Homonym synsets of the word *libra* and its inflexion, *libras*

The words contained in each of the homonym synsets make up the meanings that the word to disambiguate may take. Each of these groups of words makes up the set of indicators that later will be complemented with the indexes to carry out the exploration context. The indicators are formally defined in the following way:

$$I_i = \{p \mid p \in S_i, i = 0, 1, 2, \dots, numSyn\}$$

where:

S_i is the first vicinity synset of the word to disambiguate
numSyn is the number of first vicinity synsets

The indexes are the sets of words semantically related to the indicators that allow the determination of the meaning of the searched word given a context. The sets of indexes are built from the following rules:

$$J_i = \left\{ \begin{array}{c} R_i \neq \phi \rightarrow \{p \mid p \in R_i\} \\ R_i = \phi \rightarrow \{p \mid p \in V_i\} \\ R_i \cup V_i = \phi \rightarrow \{p \mid p \in T_i\} \end{array} \right\}$$

where:

$R_i = \{r \mid r \text{ is a descendant of } S_i\}$
$V_i = \{h \cup m \mid h \text{ is a holonym of } S_i, m \text{ is a meronym of } h\}$
$T_i = \{n \mid n \text{ is a hyperonym of } S_i\}$

It is important to point out that first classification experiments showed that the minimum depth of exploration recommended is 3, in the others words, it must descend at least to the third level of each of the branches (or to the deepest level if the branch does not reach level 3) in order to obtain the set of indexes that complement the previously defined indicators. For example, Figure 3 shows the indexed for the word *libra* (pound).

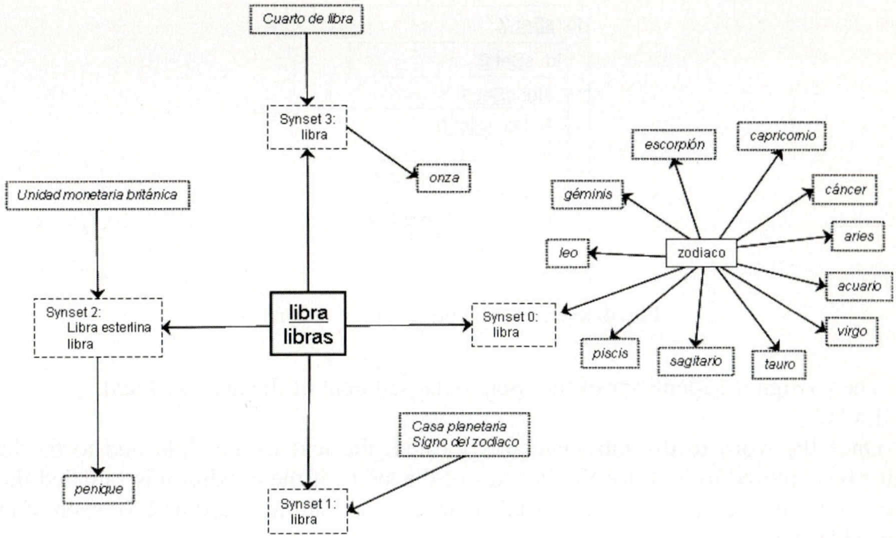

Fig. 3. Enquiry of indicators and indexes for the searched word *libra* (pound) and its inflexion *libras* (pounds)

Following the example of the word *libra* (pound) as the word to disambiguate, it has 4 possible meanings from which the sets of indicators are made up; each of these possible meanings are semantically related to other words that make up the set of indexes.

I_o = { *libra* (libra)}
J_o = { *zodiaco* (zodiac), *piscis* (pisces), *sagitario* (sagittarius), *tauro* (taurus), virgo (virgo), *acuario* (aquarius), *aries* (aries), *cáncer* (cancer), *capricornio* (capricorn), *escorpión* (scorpio), *géminis* (gemini), *leo* (leo)}

I_1 = { *libra* (libra)}
J_1 = {*casa* (house), *planetaria* (planetary), *signo del zodiaco* (sign of the zodiac)}

I_2 = {*libra, libra esterlina* (pound, pound sterling)}
J_2 = {*penique* (penny), *unidad monetaria británica* (British monetary unit)}

I_3 = { libra (pound)}
J_3 = {*onza* (ounce), *cuarto de libra* (a fourth of a pound)}

Once the indicators as well as the indexes have been identified, a new representation of the original first vicinity synsets is created. Such representation is called the sSet and it is made up of: an identifier of the sSet, which corresponds to the key field of the original synset, a set of indicators and a set of indexes (figure 4).

Starting from the identified sSets, the set of rules is dynamically created:

Indicators = I_i

Indexes: J_i

Rule:

If in the context of some element of the set I_i it appears any element belonging to the set J_i

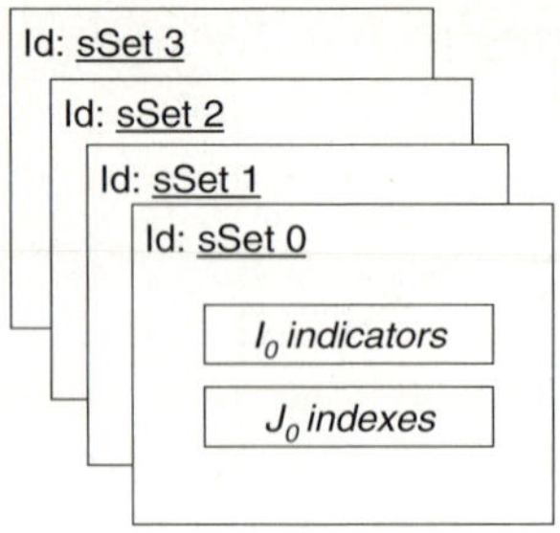

Fig. 4. sSet, a new representation of synsets

Then assign the identifier of the $sSet_i$ to the segment of the analyzed text.
End if

Once the word to disambiguate was located, the text to the right and to the left must be explored to look for the indexes of the $sSet_0$, if the condition is satisfied then the rule is applied. The next sSet is taken and the operation is repeated for each of the available sSets.

Once the context exploration has concluded, the results are quantified and two heuristics are applied in order to assign the category to the page. In the next subsection such heuristics are detailed.

4.3 The Information Classification

The classification consists on grouping the Web pages according to the specific domains represented in WordNet.

To carry out the classification of text, two heuristics are applied. The first heuristic assigns a category. In first place, it is necessary to determine the number of text segments (*# seg*) that were dedicated to each of the sSets (possible meanings). In second place, it is necessary to determine how many indexes of each sSet were found (*#ocurr*) in order to obtain a general total. From this information, a decision table is created (table 2).

Table 2. Decision table to determine the category to which a page belongs

sSet	# seg	#ocurr
$sSet_0$	3	4
$sSet_1$	0	0
$sSet_2$	3	6
$sSet_3$	3	4

The meaning of the winning sSet is associated to the text, to begin with, evaluating the number of segments that were dedicated to such meaning.

When there is a tie in the number of paragraphs (*# parr*), then a second heuristic is applied. This breaks the tie of the winning sSets of the first heuristic, and in this case proceeds to select the sSet with the higher number of occurrences of indexes (*#ocurr*) per page (table 3).

Table 3. Tied sSets

sSet	# parr	#ocurr
$sSet_0$	3	4
$sSet_2$	3	6
$sSet_3$	3	4

If the tie persists, then the winning set is the first (of those tied) that appears on the decision list. Once we have a winning sSet, the identifier is used to associate the page with a meaning label.

At the end of the process, the analyzed text has two labels, one corresponding to the meaning of the word to disambiguate in one context, and another corresponding to whether there is a relationship or not to the word to disambiguate.

To carry out the association of each possible meaning with its corresponding domain in Spanish, it was necessary to make a translation of the labels of the WNDH because this resource is only available in English.

Originally, the association between meanings and domains is done by WN as shown in figure 5. However, during experimentation it was not possible to find a direct correspondence between the synset indexes of EWN in Spanish and the synset indices of WN, therefore, it was necessary to build a domain index in which the synset indexes of EWN in Spanish are related to their corresponding domain index (translated) of WDH (figure 6).

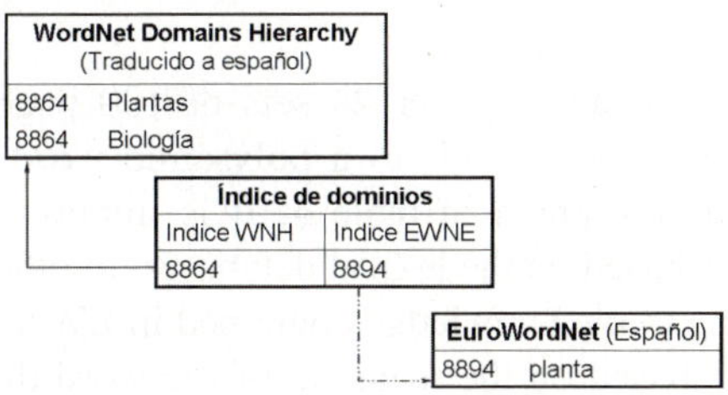

Fig. 5. Original association between meanings and domains

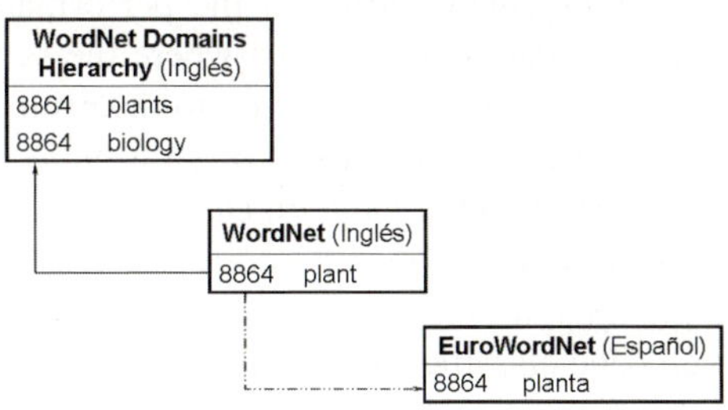

Fig. 6. New association between meanings and domains by using the proposed domain index

5 Validating and Comparing the Proposed Method

In order to validate the semantic analyzer we implemented a tool CLASSE that permits us to classify a set of web pages. For practical reasons it was decided to work only with html pages obtained from a simple search. In figure 7 are shown the elements integrated in the validation model of the semantic analyzer.

A set of hyperlinks is obtained as a result of the search on a web finder. The links are filtered in order to have access to HTML pages, excluding other page formats. The semantic analyzer carries out an analysis of the contents of the pages in order to determine the category to which such page belongs to. The knowledge base (lexical database) has a direct relationship with the semantic analyzer, because the knowledge for the classification process is extracted from it. Finally, an interface is integrated for the presentation of results.

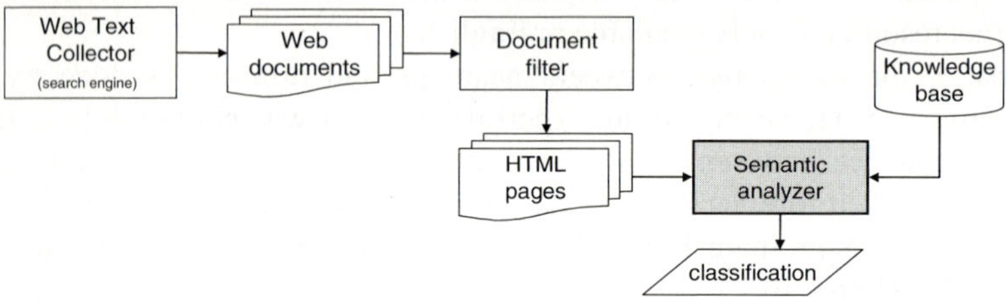

Fig. 7. Model for the validation of the semantic analyzer

The experiment was carried out over 24 sets of 100 pages stored in a local resource; each of these sets corresponds to a polysemic word selected at random. A compilation of results is presented in table 4. It is important to point out that the evaluation of results was based on the lexical databases, in other words, the evaluation was performed according to the knowledge contained in EWN.

For the classification based on the meaning of the word disambiguated, results of up to 95% were correctly obtained. Unfortunately, certain aspects such as spelling and the structure of the page affected the performance of the system during the classification. Another factor that undoubtedly affected the performance of the tool was the incompleteness of the lexical database. In cases where the sets related to each meaning of the searched word have a useful population to determine the sense of the word, the results we found have levels of correctness of up to 90% (see *guía* (guide), *rosca* (roll), and *plano* (plane)). On the other hand, in cases where the population was insufficient or non-useful, results were below 50% correct (see *cubo* (cube) and *carta* (letter) in Table 4.

A state of the art web page classifier in Spanish allowed us to decide to validate the semantic analyzer in this field. We present a comparison of our tool CLASSE with some existing classifiers.

Table 4. Evaluation results of the 24 test cases

Word	Classification by meaning	
	% success	% fault
banco (bank)	86%	14%
banda (band)	62%	38%
campo (field)	57%	43%
carta (letter)	36%	67%
cola (tail)	75%	25%
cubo (cube)	7%	93%
guía (guide)	95%	5%
hoja (sheet)	77%	23%
lengua (tongue)	67%	33%
libra (pound)	90%	10%
liga (link)	75%	25%
lima (lime)	88%	12%
marco (frame)	59%	41%
metro (meter)	78%	22%
mina (mine)	85%	15%
palma (palm)	85%	15%
papa (potato)	59%	41%
peso (weight)	68%	32%
plano (plane)	94%	6%
planta (plant)	58%	42%
pluma (pen)	77%	23%
red (net)	72%	28%
rosca (roll)	93%	7%
sierra (saw)	91%	9%
average	**72%**	**28%**

Activity	Theseus [8, 9]	Clusty [10, 11]	Iboogie [12]	Poesia [13]	CLASSE
Automatic classification of web pages in Spanish	✓	✓	✓	✓	✓
Exploitation of the semantic context of the text	×	×	×	×	✓
Exploitation of the content of the web document	×	×	×	✓	✓
Consideration of the phenomenon of polysemy	×	×	×	×	✓
Utilization of linguistic methods	×	×	✓	×	✓

Exploration of the context in which the words are being used	×	×	×	×	✓
Resulting categorization in Spanish	×	×	×	✓	✓
Independence of the domain	✓	✓	✓	×	✓
Taxonomy of flexible and non-predefined classification	✓	×	×	✓	✓

6 Conclusions

In order to propose a solution to the problem of polysemy, which has adversely affected conventional techniques of automatic text classification, we have presented a method which combines statistical approaches as well as context exploration techniques. The proposed method uses linguistic markers that provide us with information about the most relevant parts of the text. In addition to this, we rely on WNDH to determine the classes or domains of the classification. The proposed method was evaluated by an analyzer that implements automatic classification and text categorization in Web pages. It is important to point out that a bibliographical study has revealed that other conventional tools (web page classifiers in Spanish) do not take into account the polysemy problem.

We argue that this tool is an alternative for users who wish to have more structured information of their search requests. Nevertheless, our tool, CLASSE, yielded very good results as well as bad ones. There are factors that could have affected these results such as bad spelling or pages that were not properly downloaded. However; we believe that the information contained in the lexical database and in its relations is very important for determining a good or poor classification.

References

1. Desclés, J.P., Cartier, E., Jackiewicz, A., Minel, J.L.: Textual Processing and Contextual Exploration Method, Context 97, Rio de Janeiro (February 1997)
2. Couto, J.: Los sistemas de exploración contextual de cara al usuario, Tesis de Maestría, InCo, Facultad de Ingeniería, Universidad de la República, Uruguay (2002)
3. Amsterdam University: EuroWordNet: Building a multilingual database with wordnets for several European languages (March 2006), http://www.illc.uva.nl/EuroWordNet/
4. Bentivogli, L., Forner, P., Magnini, B., Pianta, E.: Revising WordNet Domains Hierarchy Semantics, Coverage, and Balancing. In: COLING 2004 Workshop on Multilingual Linguistic Resources, Geneva, Switzerland, August 28, 2004, pp. 101–108 (2004)
5. Gómez, J., de Buenaga, M., Cortizo, J.: The role of word sense disambiguation in automated text categorization. In: Montoyo, A., Muńoz, R., Métais, E. (eds.) NLDB 2005. LNCS, vol. 3513, pp. 298–309. Springer, Heidelberg (2005)
6. Prada, J.J., Moncecchi, G.: Reconocimiento eficiente de marcadores del discurso en español. In: VIII Simposio Internacional de Comunicación Social, Santiago de Cuba, Cuba (January 2003)

7. Prada, J.J.: Marcadores del discurso en español: Análisis y Representación, Tesis de Maestría, InCo, Facultad de Ingeniería, Universidad de la República, Uruguay (2001)
8. Gulli, A., Ferragina, P.: The anatomy of a hierarchical clustering engine for web-page, news and book snippets. In: Fourth IEEE International Conference on Data Mining, ICDM 2004, Brighton, UK (2004)
9. Gulli, A.: SnakeT. Pisa University, Italy, `http://www.snaket.com`
10. Vivísimo, Inc.: How the Vivísimo clustering engine works (2003), `http://www.vivisimo.com`
11. Vivísimo, Inc.: Clusty (August 2006), `http://www.clusty.com`
12. `http://www.iboogie.com` (November 2006)
13. Puertas, E., Gómez, J., Carrero, F., Buenaza, M.: Filtrado de contenidos Web en español dentro del proyecto POESIA. In: Proceedings of Ibero-America Conference www/Internet (2004)

On Regularization Parameters Estimation in Edge–Preserving Image Reconstruction

Ivan Gerace and Francesca Martinelli

Dipartimento di Matematica e Informatica
Università degli Studi di Perugia
Via Vanvitelli, 1, I-06123 Perugia, Italy
{gerace,martinelli}@dipmat.unipg.it

Abstract. The image restoration problem is a well known ill-posed inverse problem. Thus, to solve it some regularization techniques are necessary. By these techniques the solution of the problem is defined as the minimum of an energy function. This function is given by the sum of two terms: the first one is due to the data consistency condition, the second one is related to the a priori smoothness condition on the solution. A regularization parameter regulate the degree of strength of the two terms. A right estimation of this parameter is very important to obtain a correct image reconstruction.

Dealing with an edge–preserving image reconstruction, another free parameter is introduced. This parameter controls the number of discontinuities presenting in the restored image. In this paper we deal with joint estimation of these two parameters, proposing a novel technique for a correct second order reconstruction.

1 Introduction

Images are a fundamental tool of investigation in many branches of science, as biology, medicine, astronomy, robotic and so on. However, sometimes the bad quality of the available images does not permit the immediate use. To improve the cleanness of the observed images it is necessary to eliminate the noise and effects of the blur. The problem of restoring images deals with estimating the original images, by starting from the observed image and the characteristics of the blur and the noise. The image restoration problem is ill-posed in the sense of Hadamard [10,26], that is in some case, the solution nether exists, nor is unique, nor can be stable in presence of noise. Thus, regularization techniques [6,10,13,17,23,26] are useful in order to transform this problem in a well-posed one. By these techniques the solution is defined as the minimum of a suitable energy function that is given by the sum of two terms. The former measures the data consistency and later the regularity proprieties of the solution. The second term is multiplied by a regularization parameter that reflects the confidence that we have into the data. The estimation of such a parameter is very important to have a correct image reconstruction. To this aim, many authors plot the values of the consistency data term versus the smoothness ones [17,18,22,23]. The shape

O. Gervasi et al. (Eds.): ICCSA 2008, Part II, LNCS 5073, pp. 1170–1183, 2008.

of such a graph is like an L-curve. The corner of the L-curve is often considered corresponding to the best value of the regularization parameter.

Exploiting the common observation that most real scenes are piecewise smooth, the energy function can be opportunely modified, obtaining an edge–preserving reconstruction. In this case is necessary to estimate another free parameter appearing in the dual energy function, that allows us to avoid the presence of too many discontinuities in the restored image. One of the common open problem in visual restoration is the joint estimation of the two parameters [12,14,27].

Many authors consider just first order constraints imposing that the solution has to be piecewise constant. In this way the restored image loses many characteristics of the originals scene. These characteristics could be better recovered imposing second order smoothness constrains. In [16] Geman and Reynolds propose a model with second order constraints. Starting at the data, they use the first order model to generate a starting point for the second order model. Moreover, the minimization of the energy function is performed by a stochastic relaxation algorithm. In [19] Li propose to model the image using planar surfaces. In this approach the smoothness constraints are imposed on the parameters related to the surfaces, in such a way to obtain local planar results. The minimization of the associated energy function is performed by a simulated annealing algorithm incorporated into an iterative deterministic descent process. In [9] Cricco et al. propose the use of second order model in which the minimization of the energy function is performed by a *Graduated Non–Convexity* (GNC) algorithm [4,6,21].

In this paper, for solving the problem of the free parameters estimation, we extend the novel approach proposed in [14] to permit a second order reconstruction. In particular, for a correct edge–preserving parameters estimation, we ask for the two following requires: the estimated noise should have variance equal to the one of the effective noise; the estimated image has to be piecewise planar and has not to present triple adjacent parallel lines. The inhibition of closed triple parallel discontinuities permits to obtain cleaner images. Note that, differently from the first order reconstruction, in this case double parallel edges have not to be inhibit. In fact, in many cases, a double second order edge corresponds just to a simple first order one. Thus, we define an appropriate *target function* whose zeros are in correspondence of parameters that have the above described characteristics. To find the roots of this function we use a secant method. The proposed technique is similar to a classic shooting method for solving boundary value differential equation problems [25]. Comparing the results obtained by first and second order models, it is easy to note the best quality of the reconstructions obtained by the proposed algorithm. The paper is organized as follow: in §2 we introduce the classical regularization techniques; in §3 regularization techniques are extended for an edge–preserving reconstruction; in §4 we deal with color edge–preserving reconstruction; in §5 a new technique for the estimation of the free parameters in the image reconstruction is proposed; in §6 experimental results are analyzed. Finally in §7 some conclusion remarks are given.

2 Image Restoration Problem

Let us consider the following model:

$$\mathbf{y} = A\mathbf{x} + \mathbf{n}, \tag{1}$$

where $\mathbf{y}$, $\mathbf{x}$, $\mathbf{n} \in \mathbb{R}^{N^2}$ are taken in lexicographic order and $A \in \mathbb{R}^{N^2 \times N^2}$. The vectors $\mathbf{y}$, $\mathbf{x}$ and $\mathbf{n}$ represent the observed image, the original image and an additive noise respectively while A is a linear blur operator. In this paper we assume that the noise is Gaussian, independent, white with zero mean and known variance σ^2. The matrix A can be easily recovered from a given convolution operator $M \in \mathbb{R}^{(2z+1)\times(2z+1)}$, $z \in \mathbb{N}^+$ [14].

The *image restoration problem* consists in estimating the original image $\mathbf{x}$, given the degraded image $\mathbf{y}$, the blur operator A and the variance of the noise σ^2. It is well-known that this problem is ill-posed in the sense of Hadamard [10,17]. In order to solve visual ill-posed problems, several approaches have been investigated, like regularization [10,17] and Bayesian approaches [13]. Regularization techniques impose some constraints of smoothness in order to restrict the number of feasible solutions. Many authors consider just constrains of first order imposing that the solution has be locally constant. In such a way the restored image loses many characteristics of the original scene. These characteristics could be better recovered imposing smoothness constrains of second order obtaining images locally planar [9,16,19]. Thus, in the following we consider just a second order restoration. Let C_2 be the set of the second order cliques, that is

$$\begin{aligned} C_2 = \{c = \{(i,j),(h,l),(r,q)\} \ : \ & i = h = r, j = l + 1 = q + 2, \\ \text{or } \ & i = h + 1 = r + 2, j = l = q\} \end{aligned}$$

then the second order finite difference operator of the vector $\mathbf{x}$ on the clique c is

$$D_c^2\mathbf{x} = x_{i,j} - 2x_{h,l} + x_{r,q}.$$

Using classical regularization techniques [1,17,23,26], the solution of the image restoration problem is defined as the minimum of the following *energy function*

$$E_\lambda(\mathbf{x}) = T(\mathbf{x},\mathbf{y}) + \lambda^2 S(\mathbf{x}), \tag{2}$$

where

$$T(\mathbf{x},\mathbf{y}) = \|\mathbf{y} - A\mathbf{x}\|^2 \tag{3}$$

is a term which measures faithfulness of the solution to the data and

$$S(\mathbf{x}) = \sum_{c \in C_2} (D_c^2\mathbf{x})^2 \tag{4}$$

is a regularization term which imposes the smoothness condition on $\mathbf{x}$. The *regularization parameter* λ in (2) permits to regulate the degree of smoothness of the solution. An open problem consists in estimating the best regularization parameter. Given the parameter λ, using the *Generalized Singular Value Decomposition* (GSVD), it is possible to compute a direct solution [1]

$$\mathbf{x}(\lambda) = \arg\min E_\lambda(\mathbf{x}).$$

Lawson and Hanson in [20] proposed a interesting way to understand the properties of λ by plotting the function $T(\mathbf{x}(\lambda), \mathbf{y})$ in (3) versus the function $S(\mathbf{x}(\lambda))$ in (4). The generated curve has a characteristic "L" shape, so it is used referred as *L-curve.* Many authors suggest to choose the regularization parameter λ obtained in correspondence of the L-curve corner [17,18,22,23].

According to the classical approach, based on the Lagrange multipliers theory, the solution of the image restoration problem can be obtained iteratively by solving several minimization problems of very large dimensions. This involves enormous computation costs. Thus, in this context, Bedini et al. propose an alternative method, based on the primal-dual theory of convex optimization [2]. The original problem in [2] is reformulated in an equivalent form so that the related dual problem can be solved through a single unconstrained maximization. The solution of the original problem is then related to the solution of the dual problem via a simple model which depends on the particular cost functional adopted.

3 Edge–Preserving Image Restoration

The image restoration often deals with the problem of reconstructing a piecewise smooth solution from noisy and blurred observation. In this reconstruction it is important not only to interpolate smooth patches over uniform regions, but to locate and preserve the discontinuities that bound this regions, since they may represents object boundaries in vision problem [3,11,13,24,27]. Indeed, discontinuities very often are the most important part of images. To deal with these considerations, some auxiliary variables associated to the discontinuities of the image $\mathbf{x}$ are introduced. A good estimate of discontinuities improves the quality of restored image [6,12,13]. Such variables have the role of eliminating the regularity constraints, where discontinuities should appear. We indicate with b_c the boolean line variable associated to clique c; in particular, the one value corresponds to a discontinuity of the involved image in c. The vector $\mathbf{b}$ is the set of all line variables b_c. In this case the solution of the image restoration problem can be defined as the minimum of the following energy function:

$$E_{\lambda,\alpha}(\mathbf{x}, \mathbf{b}) = T(\mathbf{x}, \mathbf{y}) + \lambda^2 \overline{S}_\alpha(\mathbf{x}, \mathbf{b}), \tag{5}$$

where $T(\mathbf{x}, \mathbf{y})$ is as in (3) and

$$\overline{S}_\alpha(\mathbf{x}, \mathbf{b}) = \sum_{c \in C_2} [(D_c^2 \mathbf{x})^2 (1 - b_c) + \alpha b_c] \tag{6}$$

is the smoothness term. The positive parameter α is used to regulate the presence of discontinuities in the restored image. Small values of α correspond to reconstructions presenting a lot of discontinuities. Thus, now the problem consists in a joint estimation of the pair of parameters (λ, α). The restored image has to present both the "right" degree of smoothness and the "right" number of discontinuities. In [12], Figuerido and Leitão propose an estimation technique for these hyper-parameters using a stochastic approach. They assume the variance σ^2 of the noise unknown and fix α using the *Minimum Description Length Principle* (MDL). They use an iterative method that, at each time, computes one of the variables $\mathbf{x}$, $\mathbf{b}$, λ^2 or σ^2 fixing the other ones. This procedure terminates when it finds a fixed points.

Using a Bayesian approach, with a *Markov Random Field* (MRF) model, Bedini et al. in [5] deal with the same energy function given in (5). They proposed a *Mixed–Annealing* algorithm for the minimization of the energy with respect to the image field, periodically interrupted to compute, via *Maximum Likelihood* (ML) estimation, a new set of parameters. They first consider the weak membrane model and show that, by adopting a saddle point approximation [7] for the partition function, these new parameters are defined as those that maximize the conditional prior distribution of the lines given the intensities, evaluated on the current estimate of the whole image field. In this way the computation of the expectations involved in the ML estimation can be performed by analytical summation over the binary line elements alone, with a strong reduction of the computational cost. This approach can be extended to the general case of self–interacting line models, by substituting the analytical computations with a binary, shot–range Gibbs sampler.

In [14] the authors fix the properties that the restored image should present. Thus, starting from these properties, they propose an innovative approach to estimate the hyper-parameters λ and α.

4 Edge–Preserving Color Image Restoration

A color image $\boldsymbol{z} \in \mathbb{R}^{3N^2}$ is composed by three vectors $\boldsymbol{z}_1$,$\boldsymbol{z}_2$ and $\boldsymbol{z}_3$ that represent the red, green and blue components respectively. Considering a color image reconstruction, the variables in (1) have the following dimensions: $\mathbf{x}$, $\mathbf{y}$, $\mathbf{n} \in \mathbb{R}^{3N^2}$ and $A \in \mathbb{R}^{3N^2 \times 3N^2}$. Using regularization techniques [15] for solving the color image reconstruction problem, the term $\overline{S}_\alpha(\mathbf{x}, \mathbf{b})$ in (5) turns to be

$$\overline{S}_\alpha(\mathbf{x}, \mathbf{b}) = \sum_{c \in C_2} \{[(D_c^2 \mathbf{x}_1)^2 + (D_c^2 \mathbf{x}_2)^2 + (D_c^2 \mathbf{x}_3)^2](1 - b_c) + \alpha b_c\}$$

Note that a discontinuity in the clique c corresponds to a discontinuity in each channel.

5 Proposed Technique

In this paper we propose a new technique with the aim to afford a second order reconstruction using the approach introduced in [14]. By means of a second order edge–preserving model, the restored image has to be piecewise planar.

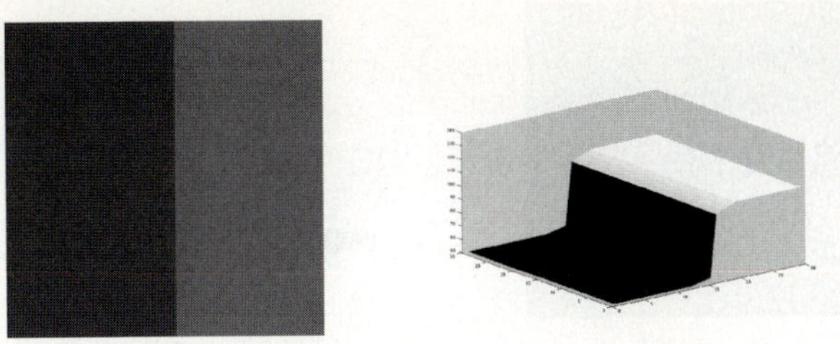

(a) (b)

Fig. 1. First sample image: (a) gray-level representation; (b) surface representation

Let consider the following order between second order cliques:

$$\{(i,j),(i-1,j),(i-2,j)\} \preceq \{(h,j),(h-1,j),(h-2,j)\} \quad \Leftrightarrow i \le h,$$
$$\{(i,j),(i,j-1),(i,j-2)\} \preceq \{(i,h),(i,h-1),(i,h-2)\} \quad \Leftrightarrow j \le h,$$

We denote by $c-1$ and $c+1$ the clique that precedes and follows c in the order $\preceq$ respectively. Let consider the following detail of the image in Figure 1 (a) where the bold numbers locate a clique c.

50	51	52	53	100	102	104	106
50	51	52	53	100	102	104	106
50	51	52	53	100	102	104	106
50	51	52	**53**	**100**	**102**	104	106.
50	51	52	53	100	102	104	106
50	51	52	53	100	102	104	106
50	51	52	53	100	102	104	106

As shown in Figure 1 (a)-(b), the image is piecewise planar and does not appear corrupted by blur. We consider that in c there is a discontinuity because $D_c^2\mathbf{x} = -45$, while the greater part of the other cliques have second order derivative equal to 0. There is a discontinuities also in $c-1$ being $D_{c-1}^2\mathbf{x} = 46$, while $c+1$ does not present discontinuities as $D_{c+1}^2\mathbf{x} = 0$. In this case we say that the image has *double edges* but not *triple edges*. This is the typical situation dealing with clear piecewise planar images. Consider now the following detail of the image in Figure 2 (a):

50	50	50	80	100	120	120	120
50	50	50	80	100	120	120	120
50	50	50	80	100	120	120	120
50	50	50	**80**	**100**	**120**	120	120.
50	50	50	80	100	120	120	120
50	50	50	80	100	120	120	120
50	50	50	80	100	120	120	120

As it is possible to notice in Figure 2 (a)-(b), the image turn to be blurred. Assuming that the clique c is the one setted in bold, we have that $D_c^2\mathbf{x} = 100$, $D_{c-1}^2\mathbf{x} = 70$, $D_{c+1}^2\mathbf{x} = -20$. Thus, a blurred image can presents triple edges.

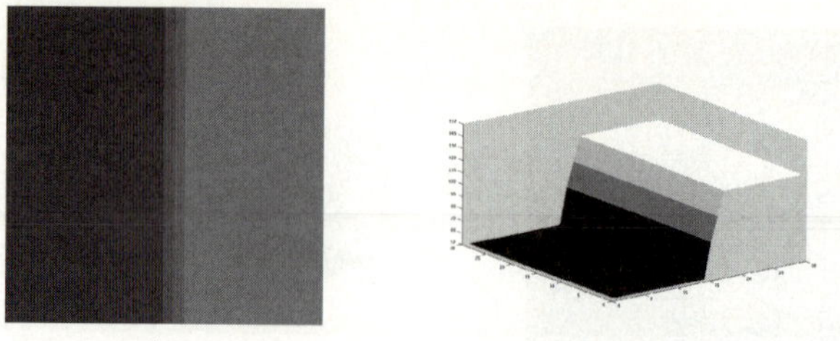

(a) (b)

Fig. 2. Second sample image: (a) gray-level representation; (b) surface representation

According with these considerations we propose to estimate (λ, α) in such way that:

1. the restored image should be piecewise linear and without any triple edges;
2. the estimated noise should have variance equal to σ^2.

In order to satisfy the above condition 1, we define the function

$$\Gamma(\lambda, \alpha) = \sum_{c \in C_2} g_{\lambda,\alpha} \left(D_c^2 \mathbf{x}(\lambda, \alpha), D_{c-1}^2 \mathbf{x}(\lambda, \alpha), D_{c+1}^2 \mathbf{x}(\lambda, \alpha) \right), \tag{7}$$

where

$$g_{\lambda,\alpha}(t_1, t_2, t_3) = \begin{cases} t_1^2 \,, \text{ if } & |t_1| < \tau \\ \quad\quad \text{ or } & |t_2| > \tau \\ \quad\quad \text{ or } & |t_3| > \tau \\ & \\ 0 \ , \text{ otherwise.} & \end{cases}$$

where τ is a threshold, which determines the presence of a discontinuity. Note that the function $\Gamma(\lambda, \alpha)$ is positive and is equal to zero if and only if the restored image $\mathbf{x}(\lambda, \alpha)$ is piecewise planar and without triple discontinuities.

To the goal to satisfy the condition 2, a *dual energy function* can be defined by the explicit minimization of the energy function in (5) respect to the line variables b_c [6,16] with the aim to reduce computational costs. Namely

$$\overline{E}_{\lambda,\alpha}(\mathbf{x}) = \inf_{\mathbf{b}} E_{\lambda,\alpha}(\mathbf{x}) = T(\mathbf{x}, \mathbf{y}) + \inf_{\mathbf{b}} \overline{S}_\alpha(\mathbf{x}, \mathbf{b}) = T(\mathbf{x}, \mathbf{y}) + \overline{\overline{S}}_\alpha(\mathbf{x}).$$

Dual energy function is not convex. In order to find

$$\mathbf{x}(\lambda, \alpha) = \arg \min_{\mathbf{x}} \bar{E}_{\lambda,\alpha}(\mathbf{x})$$

we use a classical *Graduated Non-Convexity* (GNC) technique [4,6,21]. In particular, for gray–level images, we use the algorithm proposed in [21] while for color images, the one proposed in [15]. Let $\sigma^2(\lambda, \alpha)$ be the variance of the estimated noise $\mathbf{n}(\lambda, \alpha) = \mathbf{y} - A\mathbf{x}(\lambda, \alpha)$.

Hence, we select the free parameters (λ, α) as the roots of the *target function* Υ defined as follows

$$\Upsilon(\lambda, \alpha) = (\sigma^2(\lambda, \alpha) - \sigma^2)\, \Gamma(\lambda, \alpha). \tag{8}$$

Note that, to compute the roots of $\Upsilon(\lambda, \alpha)$ a numerical method is applied. In this way values of (λ, α) close to real roots are found satisfying both requires 1 and 2. In particular, a secant method is chosen. Namely, we apply the following algorithm:

initialize $\lambda_0, \lambda_1, \alpha_0$ and α_1 in such a way that $\Upsilon(\lambda_0, \alpha_0) \cdot \Upsilon(\lambda_1, \alpha_1) < 0$;
$F_0 = \Upsilon(\lambda_0, \alpha_0)$ and $F = F_1 = \Upsilon(\lambda_1, \alpha_1)$;
while $F \neq 0$ do
$\quad \lambda = \lambda_1 - \dfrac{F_1 \cdot (\lambda_1 - \lambda_0)}{F_1 - F_0}$;
$\quad \alpha = \alpha_1 - \dfrac{F_1 \cdot (\alpha_1 - \alpha_0)}{F_1 - F_0}$;
$\quad F = \Upsilon(\lambda, \alpha)$;
$\quad$ if $F \cdot F_0 > 0$, $\lambda_0 = \lambda_1$ and $\alpha_0 = \alpha_1$, then
$\quad \lambda_1 = \lambda$, $\alpha_1 = \alpha$ and $F_1 = F$;

6 Experimental Results

Our experimental results can be classified in two categories. The first one, where we verify if the L-curve theory could be extended to the edge–preserving reconstruction; the second one in which we analyze the results of the proposed techniques for a joint estimation of (λ, α) both for gray-levels and color images.

Differently from which happens in classical image reconstruction, in the edge–preserving one the number of parameters to be estimated turn to be two. Thus, to analyze the L-curve theory, we have fixed one parameter and plotted $T(\mathbf{x}(\lambda, \alpha), \mathbf{y})$ versus $\overline{\overline{S}}_\alpha(\mathbf{x}(\lambda, \alpha))$. Figure 5 reports the graphs obtained fixing α, while Figure 6 the ones obtained fixing λ. Watching at the Figure 6, it is immediate to observe that the L-curve theory could have some applications just for the λ estimation. Analyzing Figure 5, it should be taken into consideration that in the edge–preserving reconstruction it is not possible to determine directly the solution by the GSVD. This solution, according with the GNC technique, has to be computed by a descendent algorithm [4,6,21]. The use of such an algorithm amplifies the regularization effects [18]. This consideration can explain why the graphs in Figure 5 have not a shape L-corner.

Let now consider the experimental results of proposed algorithm in Section §5. All the data images are blurred, corrupted by noise of variance $\sigma^2 = 25$ and of dimension 256×256 . First experimental result is reported to show the behavior of the method. Data image is presenting in Figure 3(a). Assuming $\lambda_0 = 6$ and $\alpha_0 = 7000$ we obtain an oversmoothed reconstruction of the original image (see Figure 3(b)). On the other hand, assuming $\lambda_1 = 0.1$ and $\alpha_1 = 15$ we obtain a quite noisy image (see Figure 3(c)). The proposed method estimates as free parameters $\lambda = 0.417$ and $\alpha = 61.777$. The corresponding reconstructed image is presented in Figure 3(d). In Table 1 we report the *Mean Squared Error* (MSE) between the ideal image and the restored image given by these different values of free parameters. It is possible to notice that the reconstruction closer

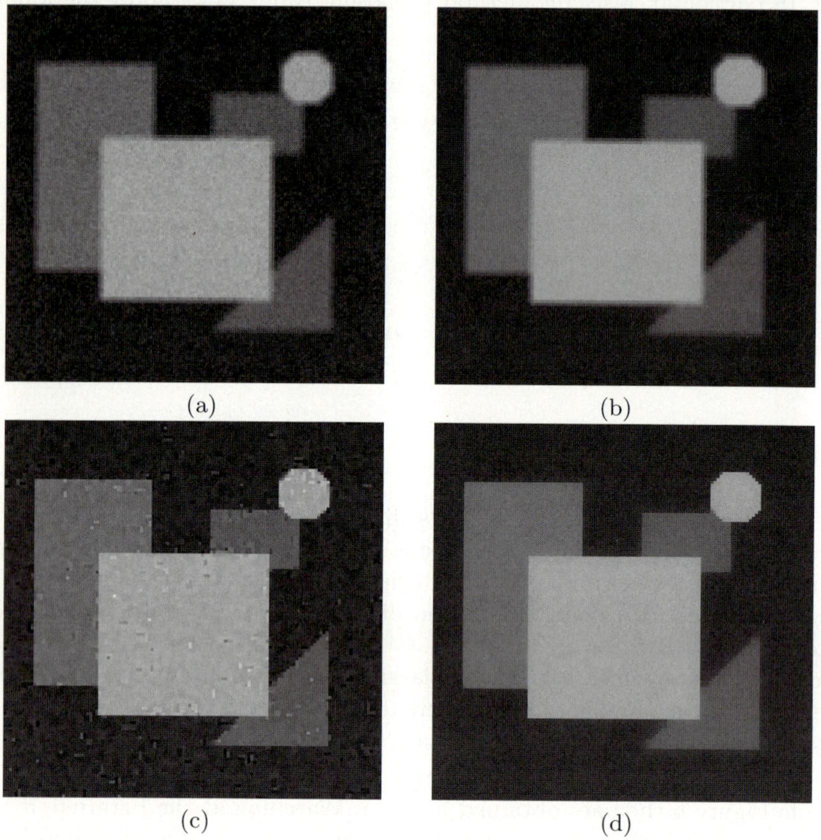

Fig. 3. (a) Data image; (b) restored image with $\lambda_0 = 6$ and $\alpha_0 = 7000$; (c) restored image with $\lambda_0 = 0.01$ and $\alpha_0 = 0.1$; (d) restoration by the proposed method (estimated parameters $\lambda_0 = 0.417$ and $\alpha_0 = 61.777$)

Table 1. MSEs between the ideal image and the restored images in Figure 3(b), 3(c) e 3(d) respectively

Figure	λ	α	MSE
3(b)	6	7000	23.7725
3(c)	0.1	15	19.5537
3(d)	0.494	208.16	17.6702

Fig. 4. (a) Data image; (b) restoration by the proposed method (estimated parameters $\lambda = 0.385$ and $\alpha = 150.918$)

Table 2. MSEs between the restored images, using the proposed method and the ones in [14] and [5], and the ideal ones

Figure	proposed method	method in [14]	method in [5]
3	17.67	19.24	21.97
4	8.83	11.04	11.84
7	11.64	13.22	12.41
9	8.72	11.69	11.72
8	9.49	10.57	10.91

to the ideal image is the one obtained by the proposed algorithm. In the following we present a sequence of results that can be obtained by the proposed algorithm. Considering as data the image in Figure 7(a), the method estimates $\lambda = 0.624$ and $\alpha = 277.403$, obtaining the image in Figure 7(b). The Figure 4(a) shows observed distorted data, while in Figure 4(b) is presented the image obtained by the estimated parameters $\lambda = 0.385$ and $\alpha = 150.918$. Let now consider results on RGB images. The first data color image is presented in Figure 8(a), the estimated parameters are $\lambda = 0.403$ and $\alpha = 160.534$ and the corresponding restored image is shown in Figure 8(b). Finally, the image appearing in Figure 9(b) is the one obtained restoring the image in Figure 9(a) with the estimated parameters $\lambda = 0.587$ and $\alpha = 256.992$. In order to compare the proposed method with other methods using the same energy function as in (5), that is the ones in [14] and [5], we report in table 2 the MSEs between ideal images and restored ones. The better performance of the proposed technique is confirmed by the fact that the MSEs of its results are the smallest in every case.

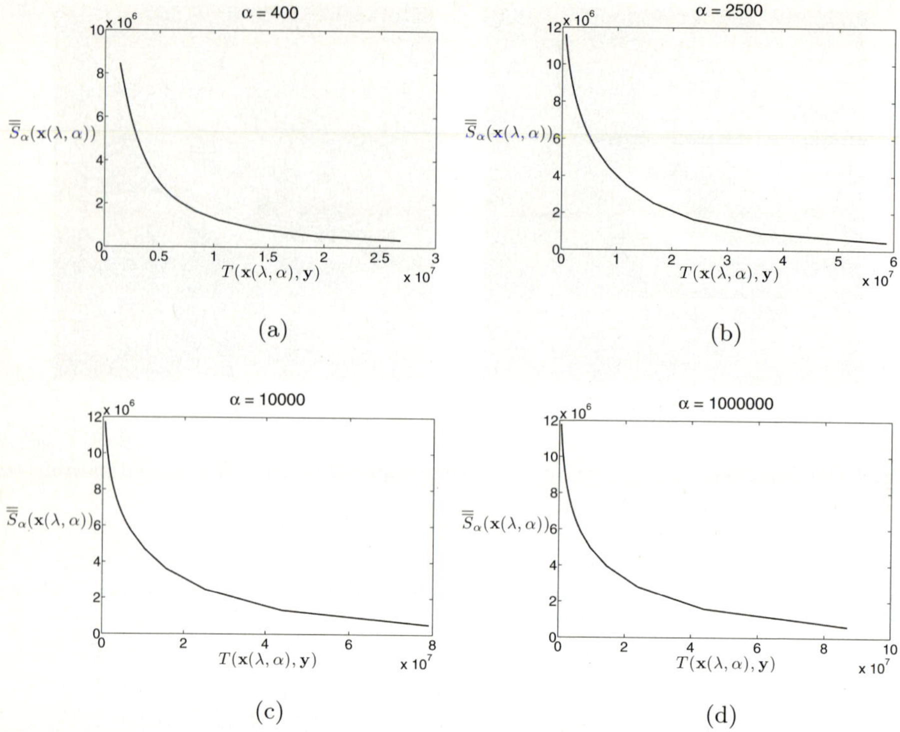

Fig. 5. Graphs of $T(\mathbf{x}(\lambda,\alpha),\mathbf{y})$ versus $\overline{\overline{S}}_\alpha(\mathbf{x}(\lambda,\alpha)$, with α fixed, obtained in the reconstruction of the image in Figure (4) in which: (a) $\alpha = 400$; (b) $\alpha = 2500$; (c) $\alpha = 10000$; (d) $\alpha = 100000$

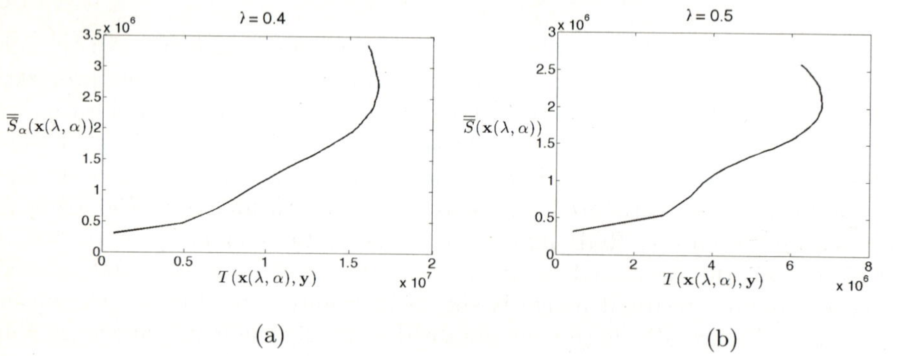

Fig. 6. Graphs of $T(\mathbf{x}(\lambda,\alpha),\mathbf{y})$ versus $\overline{\overline{S}}_\alpha(\mathbf{x}(\lambda,\alpha))$ with λ fixed, obtained in the reconstruction of the image in Figure (4) in which: (a) $\lambda = 0.4$; (b) $\lambda = 0.5$.

Fig. 7. (a) Data image; (b) restoration by the proposed method (estimated parameters $\lambda = 0.489$ and $\alpha = 205.786$)

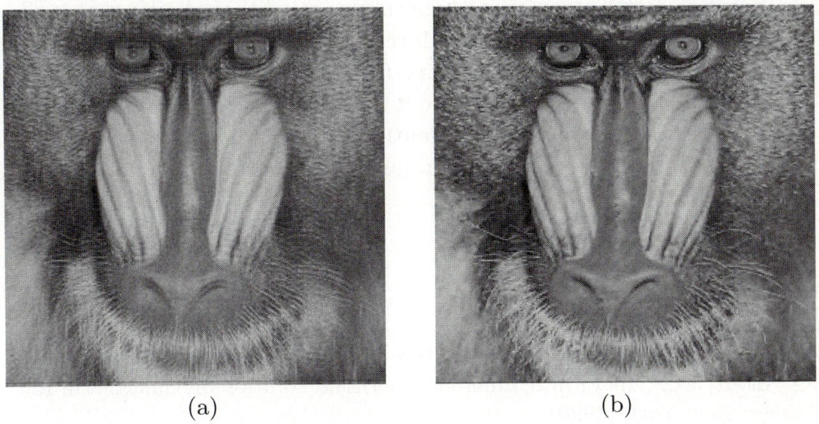

Fig. 8. (a) Data image; (b) restoration by the proposed method (estimated parameters $\lambda = 0.403$ and $\alpha = 160.534$

7 Conclusion

In this paper we have dealt with the problem of the estimation of the free parameters in an edge-preserving image reconstruction. We have proposed an original approach to this problem to the aim to restore images that are piecewise planar and without triple discontinuities. Indeed, we have shown as this condition is essential to obtain a cleaner reconstruction. Thus, we have defined a target function that is equal to zero just when the result verifies the above condition and the variance of the estimated noise is equal to the known one. We have proposed a secant method to find a root of the target function. By our experimental results

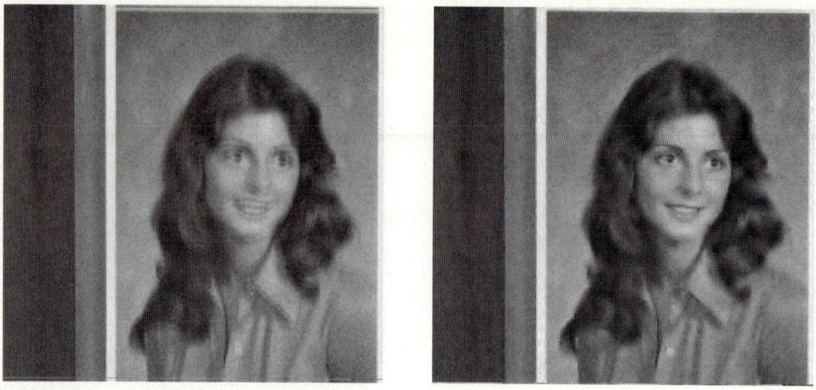

(a) (b)

Fig. 9. (a) Data image; (b) restoration by the proposed method (estimated parameters $\lambda = 0.489$ and $\alpha = 205.786$)

we have proved that better reconstructions are obtained using second order operators instant of first order ones. Moreover, we have analyzed the possibility of extending the classical L-curve theory to the edge–preserving reconstruction. Anyway, in this case the use of a descent algorithm for the minimization of the dual energy, do not permit a clear identification of the L-corner. Indeed, it is well known how iterative methods for minimizing a function can increase the degree of regularization.

References

1. Björck, Å.: Least Squares methods. Handbook of numerical analysis. In: Ciarlet, P.G., Lions, J.L. (eds.) Finite difference methods–solution of equations in $\mathbb{R}^n$, vol. 1, Elsevier, New York (1990)
2. Bedini, L., Fantini, E., Tonazzini, A.: A dual approach to regularization in image restoration. Pattern Recogn. Lett. 12, 687–692 (1991)
3. Bedini, L., Gerace, I., Salerno, E., Tonazzini, A.: Models and algorithms for edge-preserving image reconstruction. In: Hawkes, P.W. (ed.) Advances in Imaging and Electron Physics, vol. 97, pp. 86–189 (1996)
4. Bedini, L., Gerace, I., Tonazzini, A.: A deterministic algorithm for reconstruction images with interacting discontinuities. CVGIP: Graphical Models Image Process 56, 109–123 (1994)
5. Bedini, L., Tonazzini, A., Minutoli, S.: Unsupervised Edge–Preserving Image Restoration via a Saddle Point Approximation. Image and Vision Computing 17, 779–793 (1999)
6. Blake, A., Zisserman, A.: Visual Reconstruction. MIT Press, Cambridge (1987)
7. Chandler, D.: Introduction to Modern Statistical Mechanics. Oxford University Press, Oxford (1987)
8. Chen, W., Chen, M.J., Zhou, J.: Adaptively regularized constrained total least-squares image restoration. IEEE Trans. Image Process 9, 588–596 (2000)

9. Cricco, F., De Santi, G., Gerace, I.: A deterministic algorithm for deblurring considering higher order smoothness constrains. In: Proc. of the 2004 International TICSP Workshop on Spectral Methods and Multirate Signal Processing SMMSP 2004, pp. 325–332 (2004)
10. Demoment, G.: Image reconstruction and restoration: overview of common estimation structures and problems. IEEE Trans. Acoust., Speech, and Signal processing 37, 2024–2036 (1989)
11. Discepoli, M., Gerace, I., Pandolfi, R.: An edge-preserving high-resolution image reconstruction. In: 4th EURASIP Conference focused on Video/Image processing and multimedia communications, 2003, vol. 1, pp. 77–82 (2003)
12. Figueiredo, M.A.T., Leitão, M.N.: Unsupervised image restoration and edge location using compound Gauss-Markov random fields and the MDL principle. IEEE Trans. Image Process. 6, 1089–1102 (1997)
13. Geman, S., Geman, D.: Stochastic relaxation, Gibbs distributions, and the Bayesian restoration of images. IEEE Trans. Pattern Anal. Machine Intelll 6, 721–740 (1984)
14. Gerace, I., Martinelli, F., Sanchini, G.: Estimation of the free parameters in the problem of edge-preserving image reconstraction by a shooting method. In: Proc. of SMMSP 2006, pp. 205–212 (2006)
15. Gerace, I., Pandolfi, R.: A Color Image Restoration with Adjacent Parallel Lines Inhibition. In: Proc. of the 12th International Conference on Image Analysis and Processing, pp. 391–396 (2003)
16. Geman, D., Reynolds, G.: Constrained Restoration and the Recovery of Discontinuities. IEEE Trans. Pattern Anal. Machine Intell 14, 367–383 (1992)
17. Hansen, P.C.: Analysis of discrete Ill-posed problems by means of the L-Curve. SIAM Review 34, 561–580 (1992)
18. Hansen, P.C., O' Leary, D.P.: The use of the L-Curve in the regularization of discrete ill–posed problems. SIAM J. on Sci. Comput 14, 1487–1503 (1993)
19. Li, S.Z.: Roof–edge preserving image smoothing based on MRFs. IEEE Trans. Image Process 9, 1134–1138 (2000)
20. Lawson, C.L., Hanson, R.J.: Solving least squares problems. Prentice-Hall, Englewood Cliffs (1974)
21. Nikolova, M.: Markovian reconstruction using a GNC approach. IEEE Trans. Image Process 8, 1204–1220 (1999)
22. Oraintara, S., Karl, W.C., Castanon, D.A., Nguyen, T.Q.: A method for choosing the regularization paramotor in generalized Tikhonov regularized linear inverseproblems. In: Proc. of IEEE Int. Conf. on Image Processing, vol. 1, pp. 93–96 (2000)
23. Reginska, T.: A regularization parameter in discrete Ill-Posed problems. SIAM J. Sci. Comput. 17, 740–749 (1996)
24. Raghavan, S., Gupta, N., Kanal, L.: Computing discontinuity–preserved image flow. In: Proc. of the 11th International Conference on Pattern Recognition, ICPR 1992, pp. 764–767 (1992)
25. Stoer, J., Bulirsch, R.: Introduction to numerical analysis. Springer, New York (2002)
26. Tikhonov, A.N., Arsenin, V.Y.: Solutions of ill-posed problems, Winston, Washington, DC (1977)
27. Tonazzini, A., Bedini, L.: Degradation identification and model parameter estimation in discontinuity-adaptive visual reconstruction. In: Hawkes, P.W. (ed.) Advances in Imaging and Electron Physics, vol. 120, pp. 193–284. Academic Press, London (2002)

Table Based Single Pass Algorithm for Clustering News Articles in NewsPage.com

Gwyduk Yeom[1], Taeho Jo[2,1], and Youngik Yeom[1]

[1] School of Information and Communication Engineering
Sungkyunkwan University
300 Cheoncheon Dong, Suwon, Gyeonggi 440-746, Korea
[2] School of Information and Information Engineering, Inha University
{kdyeom,yieom}@ece.skku.ac.kr, tjo018@inha.ac.kr

Abstract. This research proposes a modified version of single pass algorithm specialized for text clustering. Encoding documents into numerical vectors for using the traditional version of single pass algorithm causes the two main problems: huge dimensionality and sparse distribution. Therefore, in order to address the two problems, this research modifies the single pass algorithm into its version where documents are encoded into other forms than numerical vectors. In the proposed version, documents are mapped into tables and an operation on two tables is defined for using the single pass algorithm. The goal of this research is to improve the performance of single pass algorithm for text clustering by modifying it.

1 Introduction

Text clustering refers to the process of segmenting a particular group of documents into subgroups each of which contains content-based similar documents. A collection or group of documents is given as the input of the task. Several smaller groups of content-based similar documents are generated from the task as its output. Although there are many heuristic approaches to the task, unsupervised learning algorithms have been used as state of the art approaches to it. As an instance of text mining, text clustering is necessary for organizing documents automatically.

Even if there are many state of the art strategies of text clustering, as the scope of exploration on previous research, we will mention only three most popular approaches: Kohonen Networks, single pass algorithm, and EM algorithm. Concerning Kohonen Networks, in 1998, a text clustering system, called WEBSOM, was developed by Kaski et al 3, and in 2000, the system was modified into faster version by Kohonen et al 4. In 2000, single pass algorithm was used for text clustering by Hatzivassiloglou et al 5. Concerning EM algorithm, Vinokourov and Girolami proposed five versions of EM algorithm for hierarchical text clustering in 2000 6, and Banerjee et al proposed two variants of EM algorithm for soft text clustering in 2003 7. In order to use a traditional approach including each of the three representative ones, documents should be encoded into numerical vectors.

[1] Taeho Jo is the corresponding author who submitted the copy right.

O. Gervasi et al. (Eds.): ICCSA 2008, Part II, LNCS 5073, pp. 1184–1193, 2008.

Encoding documents into numerical vectors causes the two main problems. The first problem is huge dimensionality where documents must be encoded into very large dimensional numerical vectors for avoiding information loss. In general, documents must be encoded at least into several hundreds dimensional numerical vectors in previous literatures after reducing dimensions by a feature selection method 45. The problem causes very expensive cost for processing each numerical vector representing a document in terms of time and system resources. Furthermore, much more training examples are required proportionally to the dimension for avoiding over-fitting.

The second problem is sparse distribution where each numerical vector has zero values dominantly. In other words, more than 90% of its elements are zero values in each numerical vector. This phenomenon degrades the discrimination among numerical vectors. Ultimately, it may cause poor performance of text categorization or text clustering. In order to improve performance of both tasks, the two problems should be addressed.

There was a previous attempt to solve the two problems without encoding documents so. In 2002, Lodhi et al proposed a string kernel for applying Support Vector Machine to text categorization [1]. The string kernel function proposed by them is the kernel function of SVM where two raw texts are given as the input. The output value of the kernel function is computed based on syntactic similarity between two raw texts, and it was applicable to any text independently of a natural language. However, their proposed version of SVM took very much time for the computation and failed to be better than the traditional version of SVM [1].

Figure 1 illustrates an original document or documents and its or their surrogate given as a table. The table consists of entries of words and their weights indicating their content based importance in the original document or corpus. This research modifies the single pass algorithm into its version which accepts the tables as its input. A semantic similarity between two documents is computed based on words shared by both tables. The computation will be described in detail in section 4.

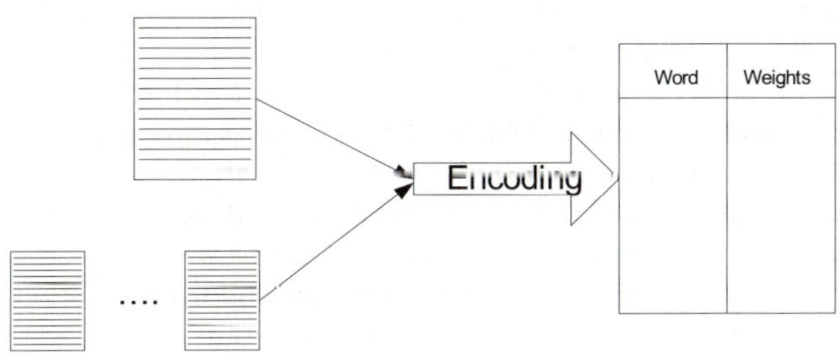

Fig. 1. Original Document or Documents and its or their Table as a Surrogate

As mentioned above, this research proposes another version of single pass algorithm which addresses the two problems. By doing that, this research provides three contributions for developing text clustering systems. As mentioned several times, the first contribution is to avoid the two problems, huge dimensionality and sparse distribution, by encoding documents into another form which is completely different from numerical vectors. The second contribution is to open a way of developing an alternative class of approaches to text clustering to machine learning based ones. The third contribution is to provide an easy way of tracing why a particular document should belong to a cluster, because the table is closer to symbolic data rather than numerical data.

This paper consists of five sections including this section. In section 2, we will describe in detail the process of encoding a document into a table. In section 3, we will describe the proposed version of single pass algorithm. In section 4, the performance of the proposed version is validated by comparing the version with the traditional version. In section 5, we will mention the significance of this research as the conclusion of this paper.

2 Document Encoding

This section concerns the process of encoding a document and documents into a table. The process is regarded as preprocessing step for text categorization which is described in section 3. As illustrated in figure 1, the process which is covered in this section consists of the three steps: tokenization, stemming, and stopword removal. A document or documents are given as the input of the process, and a list of words and their frequencies is generated as the output. The list given as the output becomes a table which is a surrogate of a document or documents.

The first step of the process is tokenization as shown in figure 1. In this step, if more than two documents are given as the input, their full texts are concatenated into a full text. The integrated full text, itself, becomes the target for the tokenization. This step segments the full text into tokens by a white space or a punctuation mark. Here, a list of tokens is generated as the output.

The output of the first step is transferred to the next step, 'stemming and exception handling' to as its input. In this step, each token is converted into its root form. Before doing that, rules of stemming and exception handling are saved into a file. When the process of encoding documents is triggered, all rules saved as a file are loaded into memory and the corresponding one among them is applied to each token. The output of this step is a list of tokens in their root forms.

The last step of the process is to remove stop words as illustrated in figure 1. Here, stop words are defined as words which perform only grammatical functions irrelevantly to content of their document; articles (a, an, or the), prepositions (in, on, into, or at), pronoun (he, she, I, or me), and conjunctions (and, or, but, and so on) belong to this kind of words. It is necessary to remove the kind of words for efficient document processing, whether the given task belong to text mining or information retrieval. This step is carried out by loading a list of stop words from a file. Therefore, counting frequencies of words except of stop words and a list of words and their frequencies is generated as a table representing a document or documents.

Although there are various schemes of assigning weights to words, we will mention only three representative schemes. First, we can assign zero or one to words as their weights, depending on their frequencies. Second, the frequencies of words themselves may be used as their weights. For third, we can assign values computed by equation (1) to words as their weights,

$$weight_i(w_k) = tf_i(w_k)(\log_2 D - \log_2 df(w_k) + 1) \quad (1)$$

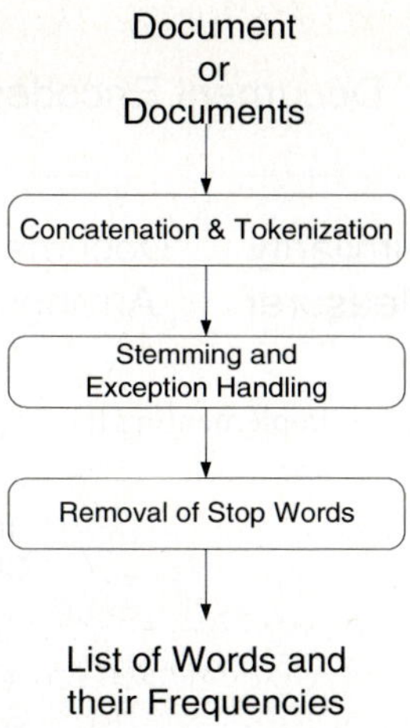

Fig. 2. The Process of Extracting a List of Words and their Frequencies

where $weight_i(w_k)$ indicates a weight of the word, w_k, which indicates its content based importance in the document, i, $tf_i(w_k)$ indicates the frequency of the word, w_k in the document, i, $df(w_k)$ is the number of documents including the word, w_k, and D is the total number of documents in a given corpus. In this research, we adopted the third one as the scheme for assigning weights to words.

3 Table Based Single Pass Algorithm

This section concerns the proposed version of single pass algorithm and the text clustering system which adopts the proposed version. Figure 3 illustrates the modules involved in implementing the proposed text clustering system. The first module is document encoder given as the interface and its role is to encode documents into tables. The second module is similarity measurer and its role is to compute a semantic similarity between two documents. The third module is document arranger and its role is to arrange documents into their content based corresponding clusters or creates a new cluster.

Figure 4 illustrates the initialization of the single pass algorithm as its first step applicable to the first document. The initialization refers to the process of creating the first cluster and containing the first document in the cluster. The first document is given as the input of the step. The first document contained in the cluster becomes its prototype which represents it[2]. Therefore, from the initialization, a cluster containing a single document is generated as the output as illustrated in figure 4.

[2] In other literatures on single pass algorithm, an average over similarities of a document with contained ones in a cluster is used as a similarity between the document and the cluster. However, in this research, a similarity between a document and the first document in the cluster is used as the similarity between the document and the cluster for fast clustering.

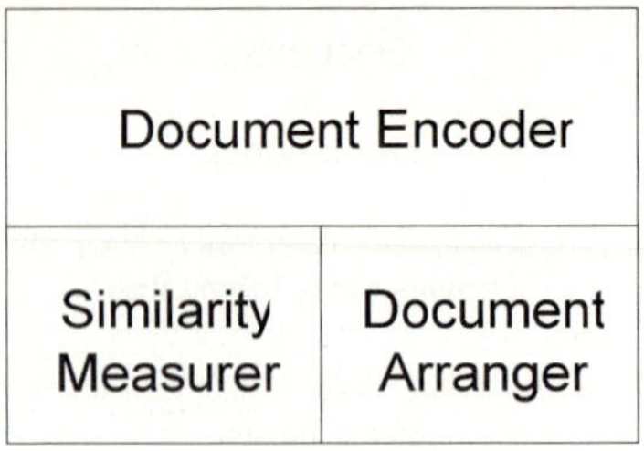

Fig. 3. The Modules inolved in Implementing the Proposed Text Clustering System

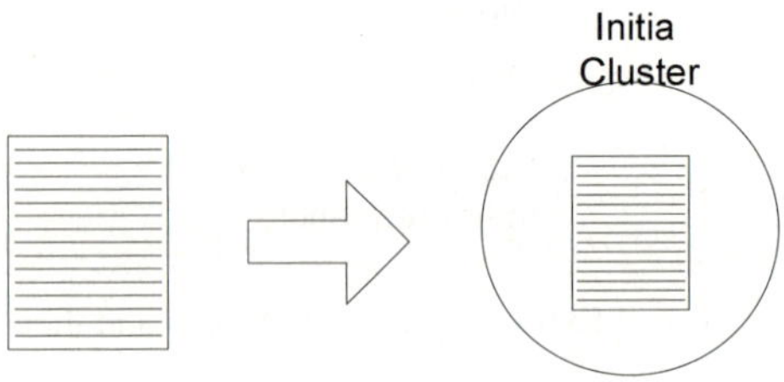

Fig. 4. The Initialization of the Single Pass Algorithm

Figure 5 illustrates the process of generating a similarity between two documents given as a normalized value. The role of document encoder was already mentioned above. The process of encoding documents into tables was already described in detail in section 3. The module, similarity measurer, computes a similarity between two tables based on words shared by both tables. The process illustrated in figure 5 generates a normalized real value as the output.

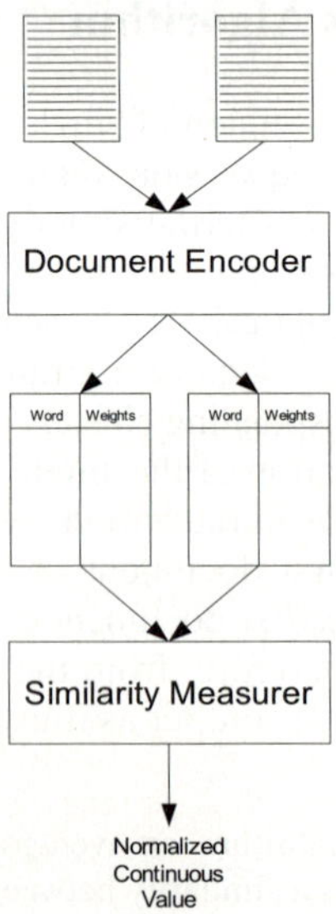

Fig. 5. The Process of Generating a Similarity between Two Documents

Figure 6 illustrates the process of generating an output table from two input tables for computing a semantic similarity between two tables, as the internal structure of the module, 'similarity measurer'. Let the two input tables be 'Table 1' and 'Table 2'. By getting words shared by Table 1 and Table 2, the output table, Table 3, is built, and each word in Table 3 has its two kinds of weights: one from Table 1 and the other from Table 2. From the three tables, we can define four sums of weights as follows.

- Sum_Weight 1: The sum of weights of words contained in Table 1
- Sum_Weight 2: The sum of weights of words contained in Table 2
- Sum_Weight 3: The sum of weights from table 1 of words contained in Table 3
- Sum_Weight 4: The sum of weights from table 2 of words contained in Table 3

Therefore, the similarity between Table 1 and Table 2 is computed using equation (2).

$$similarity = \frac{Sum\_Weight3 + Sum\_Weight4}{Sum\_Weight1 + Sum\_Weight2} \quad (2)$$

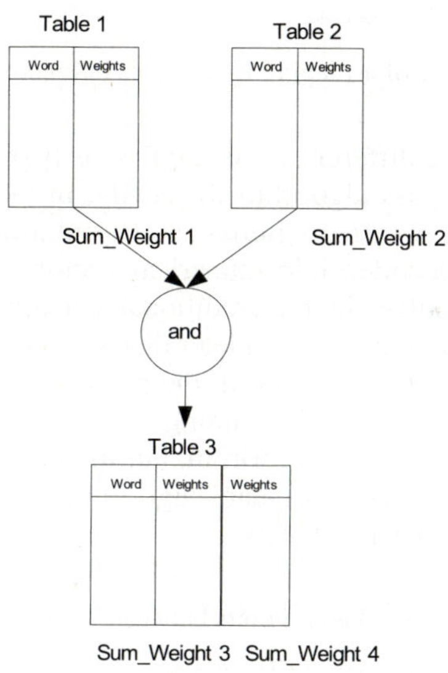

Fig. 6. The Process of computing a Similarity between Two Tables

Figure 7 illustrates the process of arranging documents or creating one more cluster. The threshold of similarity is given as the parameter of the single pass algorithm. For each successive document, its similarities with prototypes of clusters are computed using equation (2). If its maximum similarity is less than the threshold, one more cluster is created and it is contained in the cluster. Otherwise, the document is included in the cluster corresponding to the maximum similarity.

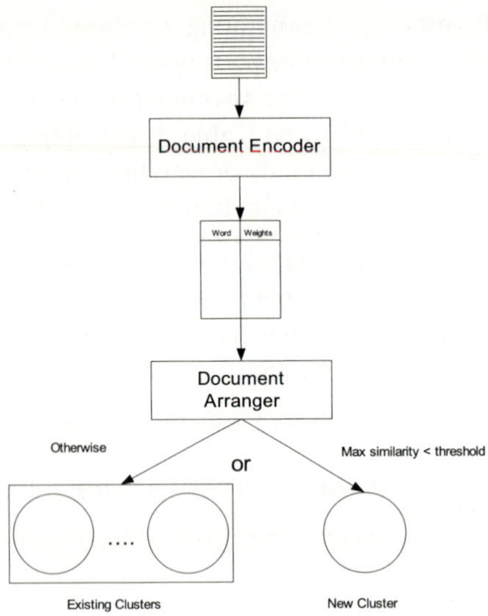

Fig. 7. The Process of arranging Documents or creating one more Cluster

Table 1 summarizes the differences and commons between the traditional and proposed versions of single pass algorithm. Basically, in the both versions, single pass algorithm consists of the two steps: initialization and arrangement. In the traditional version, documents are encoded into numerical vectors, while in the proposed one, they are encoded into tables. In the traditional version, a similarity between two documents is computed based on Euclidean distance or inner product between their corresponding numerical vectors, while in the proposed version it is computed based on words shared by their two corresponding tables using equation (2). Single pass algorithm is a very fast clustering algorithm, but its reliability is relatively poor compared with Kohonen Networks or k means algorithm, since prototypes of clusters are constantly fixed while clustering objects.

Table 1. The Different and Shared Points between Traditional and Proposed Version

	Traditional Version	Modified Version
Clustering Process	Initialization and Arrangement	
Encoding Documents	Numerical Vectors	Tables
Semantic Similarity	Inner Product Cosine Similarity Euclidean Distance	Equation (2)

4 Experiments and Results

This section concerns a set of experiments for comparing the two versions of single pass algorithm on a particular test bed. The test bed used in this set of experiments is Newspage.com which is a collection of electronic news articles. The two versions of single pass algorithm are compared with each other; two versions are summarized in table 1. We use the measure, called clustering index, for evaluating the two versions; the measure is described in detail in the literature 2. The goal of this set of experiments is to observe the results of comparing the two versions on the test bed.

Table 2 illustrates the number of news articles in each category in the test bed, NewsPage.com. There are totally 1,200 news articles which are exclusively labeled with one of five categories: 'business', 'health', 'law', 'internet', and 'sports'. The source of this test bed is from the web site, www.newspage.com; the test bed is named after the URL address. We made the test bed as text files by copying and pasting full texts of news articles. In this test bed, each news article is given as an ASCII text file.

Table 2. NewsPage.com

Category Name	#Document
Business	400
Health	200
Law	100
Internet	300
Sports	200
Total	1200

Table 3 illustrates the five subgroups of news articles of this test bed for evaluating approaches to text clustering. Each subgroup consists of 500 news articles (100 news articles per category). A file name of each ASCII text file consists of its category name and a sequential number. For example, if a particular news article belongs to the category, 'health' and the fifth one, its ASCII file name is assigned to it as 'health005'. As shown in table 3, five subgroups are exclusive with each other.

Table 3. Five Sub-collections of NewsPage.com

Category Name	Subgroup 1	Subgrou p 2	Subgroup 3	Subgroup 4	Subgroup 5
Business	100 1 ~ 100	100 51 ~ 150	100 101 ~ 200	100 151 ~ 250	100 201 ~ 300
Health	100 1 ~ 100	100 26 ~ 125	100 51 ~ 150	100 76 ~ 175	100 101 ~ 200
Law	100 1 ~ 100	100 1 ~ 100	100 1 ~ 100	100 1 ~ 100	100 1 ~ 100
Internet	100 1 ~ 100	100 51 ~ 150	100 101 ~ 200	100 151 ~ 250	100 201 ~ 300
Sports	100 1 ~ 100	100 26 ~ 125	100 51 ~ 150	100 76 ~ 175	100 101 ~ 200
Total	500	500	500	500	500

Differently from k means algorithm and Kohonen Networks, the similarity threshold is given as the parameter of the single pass algorithm, instead of the number of clusters. In the clustering algorithm, the number and the size of clusters are determined automatically depending on a value of the similarity threshold. The parameter is given as a continuous normalized value between zero and one, and if it is close to zero, the small number of large clusters is resulted in. If it is close to one, the large number of small clusters is resulted in. Since the test bed has a small number of target categories, the parameter is set as 10^{-6} close to zero.

Table 4 illustrates the three groups by input size, and within each group the two versions of single pass algorithm are compared with each other. In the first group, documents are encoded into 100 dimensional numerical vectors in the traditional version, and they are encoded into 10 entries tables in the proposed version. In the second group, they are encoded into 250 dimensional numerical vectors in the traditional version, while they are encoded into 25 entries tables in the proposed version. In the third group, they are encoded into 500 dimensional numerical vectors and 50 entries tables in the traditional and proposed version, respectively. Note that we set the dimensions of numerical vectors, following previous dimensions in the previous literatures.

Table 4. Input Sizes for Comparison of the two Versions of Single Pass Algorithm

Groups of Input Sizes	Traditional	Proposed
Small Input Sizes	100 dimensional numerical vectors	10 entries table
Medium Input Sizes	250 dimensional numerical vectors	25 entries table
Large Input Sizes	500 dimensional numerical vectors	50 entries table

Figure 8 illustrates the results of comparing the two versions of the single pass algorithm on this test bed. The x-axis of figure 1 contains the three groups of bars by input size as shown in table 3. Within each group, the black bar indicates the performance of the previous version, while the white bar does that of the proposed version. The y-axis indicates the logarithmic clustering index computed by equation (9),

$$\frac{1}{-\log_{10} CI} \tag{3}$$

where the base of the logarithm is ten. The detail description of clustering index, CI, was described in detail in the literature, [2]. The reason of rescaling the clustering index logarithmically is that the raw value of CI is too tiny to display with a bar-graph.

The difference between the two versions is outstanding by the only logarithmic scale. In the original scale, the proposed version is better 1000 times than the traditional version. In the logarithmic scale, the proposed version is better almost three times than the traditional version. In the traditional version, its clustering performances are not influenced by the dimension of numerical vectors; they are almost identical as illustrated in figure 8. In the proposed version, its performance is highest when documents are encoded into 50 entries tables.

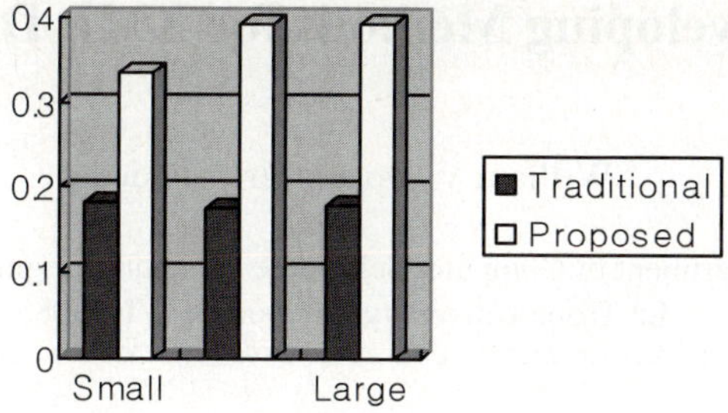

Fig. 8. The Results of Comparing Two Versions on Subgroups of NewsPage.com

5 Conclusion

The significance of this research is to modify the single pass algorithm to be more suitable for text clustering, solving the two problems completely. We used a more suitable measure which was described in the literature [2] for evaluating approaches to text clustering, rather than F1 measure. In section 4, the proposed version worked better than the traditional version through the two sets of experiments. The reason of the better performance of the proposed version is that the two main problems were addressed by encoding documents into another form different from numerical vectors. From the empirical validation in section 5, we may conclude that the proposed version is more desirable than the traditional version.

References

1. Lodhi, H., Saunders, C., Shawe-Taylor, J., Cristianini, N., Watkins, C.: Text Classification with String Kernels. Journal of Machine Learning Research 2(2), 419–444 (2002)
2. Jo, T., Lee, M.: The Evaluation Measure of Text Clustering for the Variable Number of Clusters. In: Liu, D., Fei, S., Hou, Z., Zhang, H., Sun, C. (eds.) ISNN 2007. LNCS, vol. 4492, pp. 871–879. Springer, Heidelberg (2007)
3. Kaski, S., Honkela, T., Lagus, K., Kohonen, T.: WEBSOM-Self Organizing Maps of Document Collections. Neurocomputing 21, 101–117 (1998)
4. Kohonen, T., Kaski, S., Lagus, K., Salojarvi, J., Paatero, V., Saarela, A.: Self Organization of a Massive Document Collection. IEEE Transaction on Neural Networks 11(3), 574–585 (2000)
5. Hatzivassiloglou, V., Gravano, L., Maganti, A.: An Investigation of Linguistic Features and Clustering Algorithms for Topical Document Clustering. In: The Proceedings of 23rd SIGIR, pp. 224–231 (2000)
6. Vinokourov, A., Girolami, M.: A Probabilistic Hierarchical Clustering Method for Organizing Collections of Text Documents. In: The Proceedings of 15th International Conference on Pattern Recognition, pp. 182–185 (2000)
7. Banerjee, A., Dhillon, I., Ghosh, J., Sra, S.: Generative model-based clustering of directional data. In: The Proceedings of the 9th ACM SIGKDD International Conference on Knowledge Discovery and Data Mining, pp. 19–28 (2003)

On Developing Methods for XML Databases

William V. Do and Eric Pardede

Department of Computer Science & Computer Engineering,
La Trobe University, Melbourne, VIC 3083
{wvdo@students,e.pardede@}latrobe.edu.au

Abstract. Although procedures and methods have been long established in Relational and Object-Relational DBMS, they have yet to make an appearance in XML Database. Currently no form of methods exists on XML DBMS, along with any syntactical approach to the problem. The direct translation of methods from relational tables to hierarchical data form is not possible. This paper addresses the issues of building XML Methods, their capability and limitations. For implementation, we use an open-source Native XML DBMS. A preliminary analysis is also provided to show the methods performance.

1 Introduction

Method is one established feature in Relational or Object-Relational Databases. It enables database users to perform data retrieval and manipulation without continuously feeding in ad-hoc queries into the database. On the other hand, XML Database does not support this feature. This fact has made the maintenance of XML-based applications is increasingly difficult.

In a sample scenario, a user wishes to query from a database container that consists of multiple documents with multiple records with the structure as seen in Fig. 1. An ad-hoc query can be built and executed to perform any database operation. A sample of XQuery [5] to retrieve the site place element is shown below.

```
doc('container.dbxml/movies')/directorfilms/director
/dirname/siteplace'
```

The length of the query would mean that it is prone to human error. In addition, if the query is to be re-executed in its current form or under different granularity, retyping would be a consuming process. A method in XML database would solve this by allowing subtle changes to the query to be re-executed with minimal time wasted.

The immaturity of many XML-related standards have contributed to the inexistence of XML Method. Research in this area is also in its initial stage. XML Trigger research [2, 3, 4] provided the closest insight into the fundamentals of XML Methods. Like in XML Trigger, in this paper we will base our XML Method using standard XML query language. General syntaxes are devised to base the development of the method.

O. Gervasi et al. (Eds.): ICCSA 2008, Part II, LNCS 5073, pp. 1194–1203, 2008.

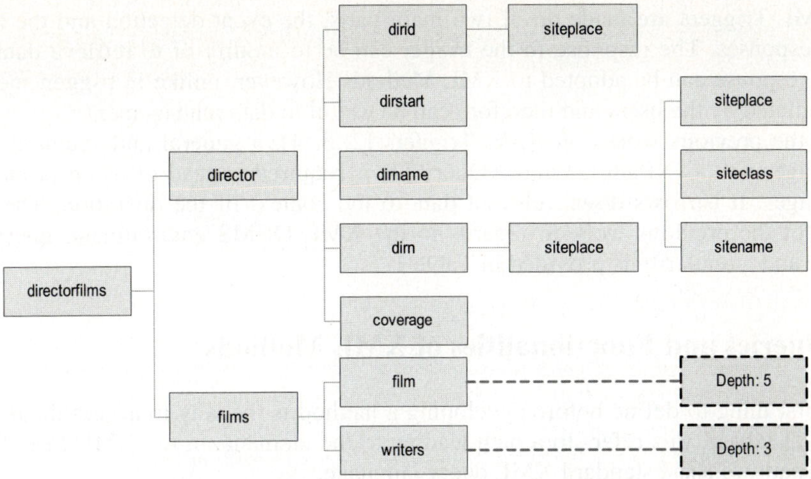

Fig. 1. Sample Deep XML Document "movies.xml"

Since there is no research or support for the syntax format of building a method in XML DBMS shell interface, other means of building and encapsulating methods' functionalities needs to be investigated. In this paper, we encapsulate the functionalities in a programmed wrapper application. The XML DBMS used is an open-source Native XML Database.

The rest of this paper is structured as follows. In section 2, we provide brief description of the background knowledge and the existing works related to this research. Section 3 provides the development of the XML Methods including the queries and functionalities. Section 4 describes the architecture of the implementation and shows some examples. In section 5, we show an evaluation of our proposal. Finally, our paper will be concluded in section 6.

2 Background and Related Works

A lack of research and development into the area of XML Methods provided insufficient guidance, insight and starting point for the research. However, the general outline obtained from the previous works in XML Triggers and Relational Database method implementation gave direction to look into the new area of XML Methods

Table 1. Comparison of Trigger Approaches

Approach	DBMS	Query Language	Granularity
Trigger Support for XML Database	Sedna	XPath, XQuery	Node level
Triggers over XML Views of Relational Data	IBM D2	XPath, XQuery	(No specific granularity due to flat table)
Extending XML Triggers with Path-Granularity	Berkeley DB XML	XPath	Node level

XML Triggers are made up of two main parts, the event detection and the operative responses. The response in the trigger can be to modify or to retrieve data. The same response can be adopted to XML Methods However, unlike in trigger, methods are initiated by the users and therefore can be useful in data management.

In the previous works on XML Triggers [2, 3, 4], a general and standard query language such as XPath [5] and XQuery [1] is required instead of using proprietary languages. It narrows down relevant data to the context of the operation. The summary of the previous work in regards to the XML DBMS environment, query language and granularity is provided in Table 1.

3 Queries and Functionalities of XML Methods

The first thing to define before developing a method is the way to access the data. In XML Database, this refers to a path leading to the elements in the XML tree. To define a path we use a standard XML query language.

Input parameters are also used in the development of the methods. The effects on the executed methods on the data is also taken into consideration. For this reason, the method will incorporate schema checking mechanism to maintain the data integrity of the database. Classifications of the operations that the XML Method would be capable of, along with the method functionalities are proposed.

Table 2 shows possible XML paths that are incorporated into the XML Method. The last three paths show the XQuery 1.0 expressions that take into consideration the predicates such as element and attribute. The classification of these queries shows the limitation and the capability that the XML Method will have in querying and narrowing down data for operations such as adding, removing or modifying nodes.

Table 2. Queries Path in XML Methods

Query	Description
`/part[parent-part]`	Will select all part records containing parent-part nodes in their document structure
`/part/parent-part`	Will display the element of parent-part
`/part/parent-part/string()`	Will only display the data of the parent-part element
`/part[not (parent-part)]`	Will display the documents that do not have parent-part elements
`/part[parent-part=1]`	Will select all parts that have a parent-part as a child and also have a parent-part value of 1
`/part[@number=1070 or @number=1032]`	Searching on the value of an attribute on part with attribute value 1070 or 1032
`/part[@number>100 and @number<105]`	Searching on the value of an attribute on part with attribute value that is greater than 100 and less than 105

Table 3 shows the functional overview of the XML Methods. It consists of the classification of the operations inside an XML Method, the operation signature, its functionality and the granularity level that the operation can support.

Table 3. XML Method Functional Overview

Operation	Signature Syntax	Functionality	Granularity	
			Node	Doc
Querying	XMethodQuery <document> <XQuery-expression>	Query of specific documents or whole containers	✓	✓
Rename Attribute	XMethodOperations <document> <XQuery-expression> <data> <operation-id> <XMethod-connection>	Rename an attribute function that accepts a string of data as the new attribute content	✓	✓
Add Node	XMethodOperations <document> <XQuery-expression> <data> <operation-id> <XMethod-connection>	Adds a Child node to the designated query path with data string consisting of "node-name:node-content"	✓	✓
Remove Node	XMethodOperations <document> <XQuery-expression> <data> <operation-id> <XMethod-connection>	Removes a node which is defined in the query path	✓	✓
Replace Attribute Content	XMethodOperations <document> <XQuery-expression> <data> <operation-id> <XMethod-connection>	Replaces attribute content at the location path with a new data string of content	✓	✓
Display Document	XMethodOperations <document> <XQuery-expression> <data> <operation-id> <XMethod-connection>	Accepts a document name and builds a document retrieval query	×	✓

The method signature as seen above is determined by the general syntax and parameters that are required for the method function. The method signature shows the general method calls for the operations.

The querying call is a single call that can handle both document level and node level granularity. The operative calls contain much more functionality due to the similarities between adding, removing and modifying nodes. This can be seen in the fact that the querying would return a certain result, whereas the operative calls would use the query capability to narrow down data or a path of which data manipulation could then be done.

The granularity of the XML Method are categorized into document and node level (except for the display document function which directly accesses a file on document level). The methods can work on a document level by going through the database container and can access elements inside the data in the container. This is possible due to the nature of the XML Method and its lack of knowledge required on the dataset before runtime. Therefore, the knowledge of the data is provided in the user input during runtime.

4 XML Method Implementation

In this section, we implement the XML Method in an XML Database. It is important to emphasize that the XML Methods described in section 3 is generic and can be implemented in any XML DBMS. Figure 2 shows that the XML Methods will only interact with the XML Data no matter where it is actually stored. Although the concept of XML Methods is capable and implementable in both Native and XML Enabled DBMS, the implementation in this paper has been done in a Native XML DBMS, Berkeley DB XML [6].

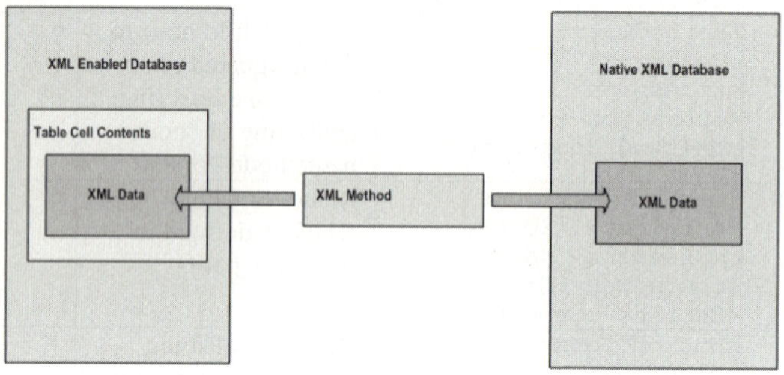

Fig. 2. XML Methods for Various XML DBMS Types

4.1 System Architecture

The XML Methods are implemented on a basic level to highlight different functionalities. To do this implementation, a high level system architecture was devised which identified the major components to the XML Method.

Figure 3 represents the high level system architecture for the implementation of the XML Method. The package includes the main components for executing queries and operations. The XML Method User Interface is the link between the user and the operations within the method. The Berkeley DB XML Connection component deals with the connecting to the Berkeley database container file with its methods that allow for direct access to the database. The XMLQuery Processor component processes any query with the database container or with a document inside the container. The XMLOperations Processor component handles certain operations to the database contents such as renaming node attribute names, adding a node, removing a node, replace attribute content and display document.

An emulated Berkeley Shell User Interface was developed with standard Berkeley DB XML function calls and support for XML Methods. This demonstrated the possibility on XML DBMS expansion to support java plug-ins for XML Methods. This Native XML Database can be accessed by the Berkeley DB XML Connection component which then allows the XMLQuery Processor and XMLOperations Processor to execute their functions.

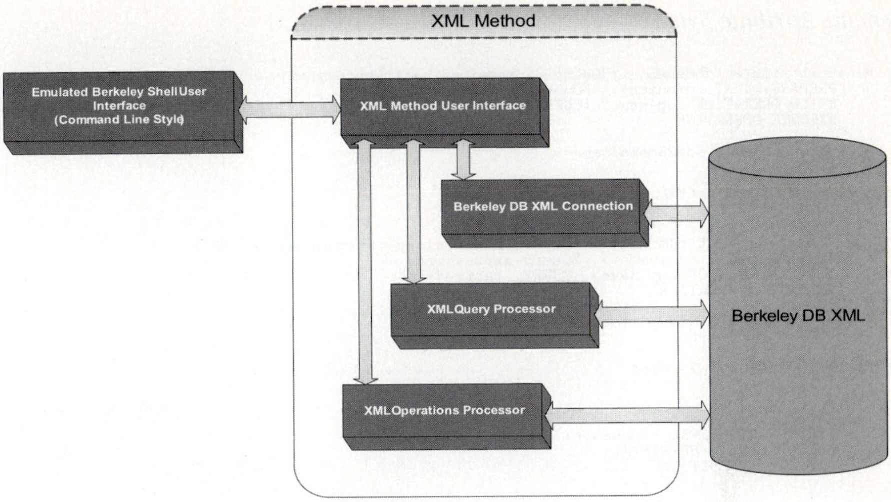

Fig. 3. XML Method High Level System Architecture

4.2 XML Methods Syntaxes

In addition to this high level design, a low level guide for XML Method operations functionality was devised for the XMLQuery Processor and XMLOperations Processor in the form of general syntaxes. These general syntaxes provide an outline and basis on what would be required in the programmatic building of an XML Method operation, and the parameters that would be required when calling a certain operation. Eg. The *XMethodQuery* function as seen below would require a *<document>* parameter and a *<XQuery-expression>* parameter for its operation.

General Query Syntax

```
XMethodQuery <document> <XQuery-expression>

    IF <document> == null THEN
        BUILD QUERY [collection('container-alias') + <XQuery-expression>]
    ELSE
        BUILD QUERY [doc(' + <container-alias> + '/' + <document> + ') + <XQuery-expression>]

    EXECUTE QUERY
```

Add Node Syntax

```
AddNode <document> <XQuery-expression> <node-name> <node-data>
    PREPARE QUERY <document> <XQuery-expression>
    BUILD OPERATION <op-name> <QUERY> <XmlModify.Element> <node-name> <node-data>
    EXECUTE OPERATION

where <op-name> = addAppendStep
```

Remove Node Syntax

```
RemoveNode <document> <XQuery-expression>
    PREPARE QUERY <document> <XQuery-expression>
    BUILD OPERATION <op-name> <QUERY>
    EXECUTE OPERATION

where <op-name> = addRemoveStep
```

Rename Attribute Syntax

```
RenameAttribute <document> <XQuery-expression> <attribute-name>
    PREPARE QUERY <document> <XQuery-expression>
    BUILD OPERATION <op-name> <QUERY> <attribute-name>
    EXECUTE OPERATION

where <op-name> = addRenameStep
```

Replace Attribute Content Syntax

```
ReplaceAttrContent <document> <XQuery-expression> <attr-data>
    PREPARE QUERY <document> <XQuery-expression>
    BUILD OPERATION <op-name> <QUERY> <attr-data>
    EXECUTE <OPERATION>

where <op-name> = addUpdateStep
```

Retrieve document Syntax

```
retrieveDoc <document>
    PREPARE RETRIEVAL <document>
    BUILD RESULT <RETRIEVAL>
    DISPLAY <RESULT>
```

4.3 Implemented XML Methods

The prototype for the XML Method is implemented using JAVA SE 6.0 in conjunction with Berkeley DB XML API under the Eclipse Europa 3.2 IDE.

All of the functionalities were implemented into an encapsulated JAVA wrapper program. The Emulated Berkeley Shell UI component is also developed to support all Berkeley DB XML function calls along with support for XML Methods.

By devising the XML Method syntax and implementing using JAVA and the Berkeley DB XML API, the XML Methods have been developed and functioned to a basic level of complexity. The implementation can solve the issue of removing the repetitive ad-hoc queries by remembering the input parameters of XQuery-expression and granularity used. For slight variation to the operation, parts of the query could be changed with ease such as the granularity of the operation (from node level to document level or vice versa).

The methods allow for user input parameters and XQuery expression to be applied to any document inside the database container. The prototype XML Method operated under an emulated Berkeley DB XML shell interface to demonstrate possible integration of XML Methods with XML DBMS. However, the method also had the capability to run as a stand-alone java application. This demonstrated the portability of the developed XML Method and could provide possible avenues for future XML DBMS with pluggable java application support. Figure 4 shows a sample execution of the XML Methods.

5 Evaluation

In this section, we run some performance testing of various types of queries on different datasets (see Table 4). Since the operative functions of the XML Methods rely on the XQuery querying capability, the tests were conducted on the querying performance in the XML Method and not the specific operations such as adding, removing or modifying nodes or attributes. As a note, the testing results were recorded after 10 warm up simulation runs to avoid any possible cold start problems. The range of queries that were used included: query using element as a condition, query using attribute as a condition, query using both element and attribute as conditions, shallow querying, deep querying and medium depth query.

Two files were used in the datasets. The shallow file that was used in dataset DS1 and DS3. The deeper file used in the datasets had a depth of 8 levels and consisted of 5000 records. The files contain data on Movies detail as it is shown in Fig.1.

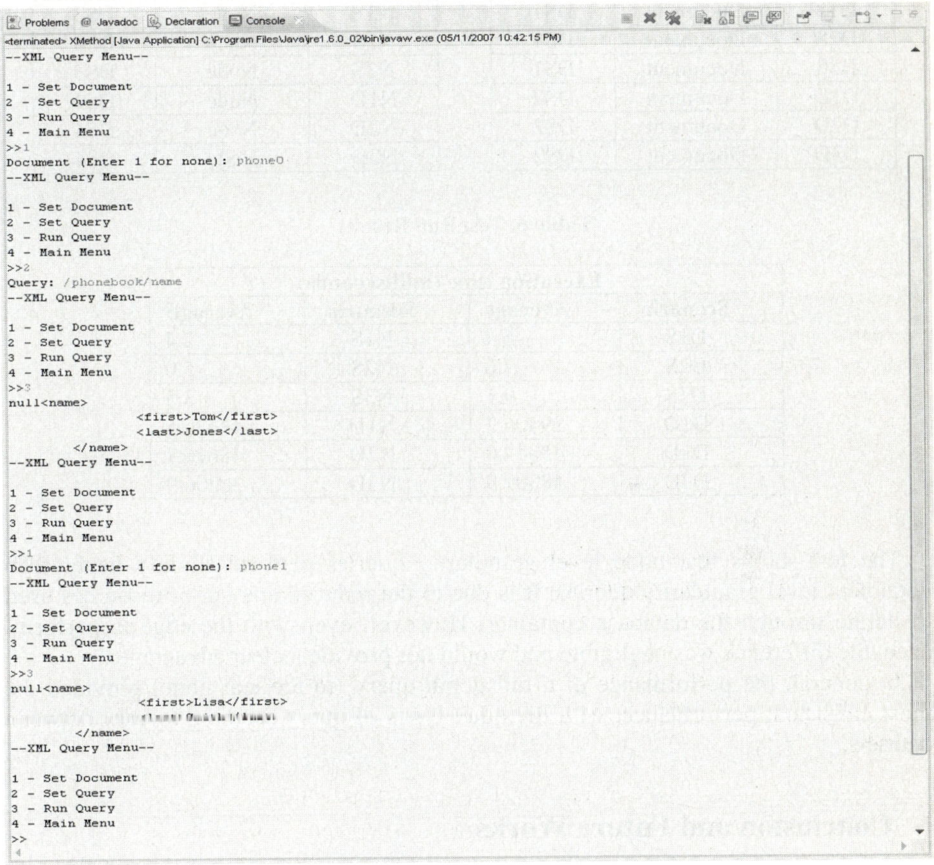

Fig. 4. Sample Execution of XML Methods

Table 4. Dataset Specifications

Dataset	Document Count	Document Specification
DS1	10	Depth: 3
DS2	11	For 10 of the documents: Depth: 3 For 1 of the documents: Depth: 8
DS3	1	Depth: 3
DS4	1	Depth: 8

Based on different query types and dataset specifications, a range of test scenarios were devised to determine and show any difference if any, in the querying performance inside the XML Method. The scenarios were done to test the query granularity (document level or node level) and document type (shallow or deep) with various queries as represented in Table 5. Table 6 shows the experimental results.

Table 5. Test Scenario Plan

Scenario	Granularity	Dataset
D1S	Document	DS1
D2S	Document	DS1
D3S	Document	DS1
D1D	Document	DS2
D2D	Document	DS2
D3D	Document	DS2

Scenario	Granularity	Dataset
N1S	Node	DS3
N2S	Node	DS3
N3S	Node	DS3
N1D	Node	DS4
N2D	Node	DS4
N3D	Node	DS4

Table 6. Test Run Results

Execution time (milliseconds)			
Scenario	**Average**	**Scenario**	**Average**
D1S	6.4	N1S	2.4
D2S	7.6	N2S	5.0
D3S	9.1	N3S	3.5
D1D	29406.7	N1D	43330.5
D2D	45844.0	N2D	48616.7
D3D	48309.9	N3D	30206.9

The test shows that node level granularity queries perform slightly better than document level granularity queries. It is due to the reduced amount of resources used to iterate through the database container. However, even with the edge in performance, the difference was negligible and would not provide a clear advantage.

In general, the performance of a full depth query (to the leaf node) provided the most consistent performance, with the other types of query seeming to vary between datasets.

6 Conclusion and Future Works

In this paper, some functionalities for XML Methods are devised and implemented. At the present time, implementation was limited to the available functionality and capability of XML DBMS, but shows the portability of the XML Method as an encapsulated style of wrapper program. Real world applications for XML Methods in the encapsulated format could include in web applications which would fundamentally operate on a programmatic function call basis.

Future work in this area can involve the optimization of the query function inside the method that are used by various database operations. This optimization can be done by incorporating this work with many indexing algorithms for XML tree.

Research can also be done to investigate methodologies to determine the best type of indexing to apply to a document on-the-fly during execution.

In addition, investigation into the idea of stored methods would also be beneficial. This would be similar to the wrapper program mentioned above but would be done all inside a DBMS. This would involve devising a way to store XML Methods, which would mean the method would always reside in the DBMS and be executed by calling the name with passing parameters.

References

1. Boag, S., Chamberlain, D., Fernández, M.F., Florescu, D., Robie, J., Siméon, J. (eds.): XQuery 1.0: An XML Query Language, W3C Recommendation (January 23, 2007), `http://www.w3.org/TR/xquery`
2. Grinev, M.R.: Introducing Trigger Support for XML Database Systems. SYRCODIS (2005)
3. Shao, F.N.A., Shanmugasundaram, J.: Triggers over XML views of relational data. In: 21st International Conference on Data Engineering, California, United States, pp. 483–484. IEEE CS, Los Alamitos (2005)
4. Landberg, A.H.R., W.J., Pardede, E.: Extending XML Triggers with Path-Granularity. In: Benatallah, B., Casati, F., Georgakopoulos, D., Bartolini, C., Sadiq, W., Godart, C. (eds.) WISE 2007. LNCS, vol. 4831, pp. 410–422. Springer, Heidelberg (2007)
5. Clark, J., DeRose, S. (eds.): W3C, XML Path Language (XPath) 1.0, W3C Recommendation (November 16, 1999), `http://www.w3.org/TR/xpath`
6. Oracle, Oracle Berkeley DB XML (2007), `http://www.oracle.com/database/berkeley-db/xml/index.html`

A Mixed Approach for Data Warehouse Conceptual Design with MDA

Leopoldo Zepeda[1], Matilde Celma[2], and Ramon Zatarain[1]

[1] Department of Information Systems and Computation
Instituto Tecnológico de Culiacán, Mexico
[2] Technical University of Valencia, Spain
leopoldo@correo.ccs.net.mx

Abstract. This paper is developed in the context of conceptual design of Data Warehoses (DWs) with Model Driven Architecture (MDA) (a methodology for DW conceptual design). The method is made up of a set of transformation rules as a mechanism to extract multidimensional schemas from the conceptual description of the operational database. Next the generation of multidimensional schemas, we apply user requirements to guide the selection of the multidimensional schema that is supported by the operational database and most likely satisfy user requirements.

1 Introduction

Standard database development consists of steps for requirements analysis, logical design, and physical design. In contrast, for DWs development there is not yet a standard method. Following the orientation used for the development of DWs current approaches are usually classified in two categories: process and data-driven.

The first one starts with an early requirements step that focuses on modeling user requirements and deriving from these functional requirements, whereas the second one start with an in-depth analysis of the operational data sources in order to directly derive a conceptual multidimensional schema. Often claimed benefits of process-driven approaches are the principal drawbacks of data-driven approaches and viceversa.

We believe that the definition of the conceptual schema is one of the most critical steps in the overall DW development process and there are several steps that must be correctly followed in order to build a "good quality" multidimensional schema (a schema influenced by user requirements and the availability structures of the data in operational systems).

In this paper, we present a mixed-driven method for DW conceptual design. This method starts from the conceptual description (data-driven) of the operational data source, captures the DW user requirements (process-driven) and ends with a detailed conceptual multidimensional schema of the data to be stored.

The method proposed consists of three phases. The first phase of the method is devoted to examining the conceptual description (ER schema) of the operational databases, finding out candidate multidimensional schemas for the DW. The solution is tackled within the context of MDA, for this we define a set of transformation rules between the

O. Gervasi et al. (Eds.): ICCSA 2008, Part II, LNCS 5073, pp. 1204–1217, 2008.

ER metamodel and the OLAP metamodel. In the second phase, user requirements are collected. The purpose of this phase is to gather information from business analysts and/or managers about the company goals and needs. For the basis of our elicitation method, we adopt a goal model. The final phase integrates these two viewpoints and thus generates a feasible solution (supported by the existing data sources) that best reflects the user's requirements. The contributions of this paper include:

1. The adaptation of a goal model for consideration and representation of multiple users requirements.
2. A set of transformation rules between the ER and the OLAP metamodel.
3. A new design method, which is not only initialized from user's requirements, but also from the conceptual description of the operational database schema.
4. An uniform and standard way for DW conceptual design.

The rest of the paper is organized in four sections. Section 2, relates the proposed approach to the state of the art and terminology. In section 3, the design process is presented. The construction of the conceptual schema is summarized in section 4. Finally, in section 5 we present our conclusions.

2 Related Work and Terminology

Our method of DW design is influenced by MDA [1], which define a new way to develop software by transforming an input model into an output model. MDA is organized and aligned in three viewpoints: 1) The Computation Independent Model (CIM) describes the system without showing details about how it is constructed. 2) The Platform Independent Model (PIM) reflects the functionalities, structure and behavior of a system and contains no information specific to the platform or the technology that is used in the realization of it on a specific platform. 3) The Platform Specific Model (PSM) is more implementation oriented and corresponds to a first binding phase of a given PIM to a given execution platform. Using a series of transformations, also called model transformations, the software system is developed from a PIM to source code. The key technologies of MDA are Unified Modeling Language (UML), Meta-Object Facility (MOF), XML and Common Warehouse Metamodel (CWM). CWM is a language created specifically to model database applications [2][3]. CWM has a set of metamodels for relational, OLAP, XML, ER, etc. Although, to our best knowledge, there is no existing mixed approaches (which incorporate MDA in the design process) devoted specifically to DW conceptual design, some approaches do the conceptual or logical design of a DW. For example, in [4] a methodology to build a conceptual multidimensional schema in terms of the user requirements and the conceptual description of the operational database is proposed.

Its main purpose is to adapt the i^* framework for elicit user requirements and then referred them to concepts of the multidimensional model. The conceptual multidimensional schemas are identified using the algorithmic procedure presented in [5]. The approach presented in [6] starts from the goals of decision makers, goals are analyzed using abstractions sheets and general considerations about how they relate to the organization activities are given in natural language. Next, a set of multidimensional schemas are identified using an algorithmic procedure.

The guidelines and techniques presented in both approaches for deriving user requirements from goals are considered as a suitable starting point for our work. However, the principal lack in these approaches are: 1) the use of a proprietary notation that difficult the communication of the requirements between developers and end users. 2) Data sources are analyzed with conventional algorithmic procedures that turn these methodologies in particular and isolate solutions.

Starting with the notion of goals in conceptual DW design in not a new idea. See for example [7] where the first step is to define a set of strategic goals, which describe the high-level objectives of the business process, from these objectives decision goals are identified and finally the information goals are defined. The requirements are used to build a conceptual model in a fully process-driven without any feedback from the data sources. In our approach, goals play a central role in the analysis of initial data requirements, but we use the requirements during an integration process that produce the multidimensional schema.

Database researches have proposed *Data-driven methodologies*, for example, [5] defines a method for developing multidimensional schemas. The design method starts from an existing ER schema, derives a multidimensional schema, and provides implementations in terms of relational tables as well as multidimensional arrays. As another example, [8], the design is carried out by producing a fact schema, which is derived from an ER schema using an algorithmic procedure. This method represents the fact as the tree's root and the dimension as the tree's descendents. Both approaches are concerned with the study of DW conceptual design starting from the conceptual description of the operational schemas. Still, they do not integrate this phase with the analysis of user requirements.

Finally, in [9] an MDA framework for the logical design of DW is presented. The authors present a set of transformation rules between the multidimensional and CWM Relational metamodels. This approaches shares some similarities with ours, since it is based on MDA. However, the multidimensional metamodel used is too generic to represent all peculiarities of multidimensional modeling in a conceptual model. For example, there is not possible to represents a dimension with multiples hierarchies.

3 Overview of Our Method

An overview of the proposed method is shown in fig. 1. It covers both the analysis of initial requirements and the transformation process. Goals requirements analysis starts with a list of users and their height-level goals, which are refined to produce a goal model. From the goal model user requirement are identified. A transformation process on the other hand, starts with the conceptual description of the operational data sources (ER) and produces a set of multidimensional schemas (OLAP). A final multidimensional schema is obtained through a integration process. The final schema represents the semantics of the user requirements that is supported by the existing data in the operational data sources. As we mentioned in the introduction, our method consists of three phases. In the next sections, we describe the three phases of the method.

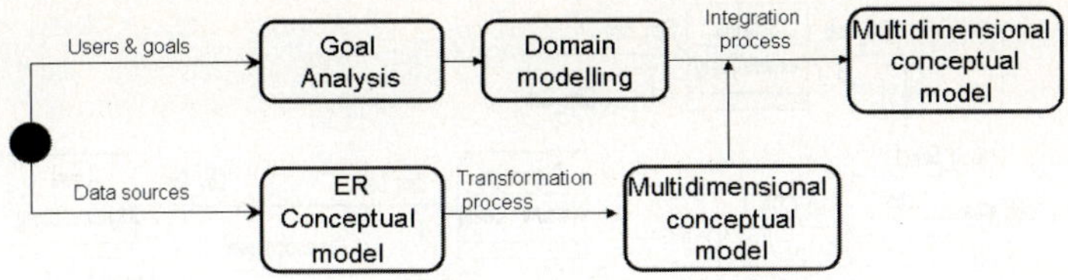

Fig. 1. Method for conceptual DW design

3.1 Phase 1. From ER Schemas to OLAP Schemas

Starting from an ER schema, this phase performs an analysis in order to discover the candidate multidimensional (OLAP) schemas. This is achieved through a set of transformation rules between the ER and OLAP PIMs. For the construction of the OLAP schema, we require a number of previous activities, like the reorganization of the ER schema; we will not consider these activities here, since they are outside the scope of the paper. In this section, we first describe the metamodels. Next, we describe how an ER schema is mapped to an OLAP schema. Following, we show the transformation rules. Finally, we show how an ER schema is transformed into an OLAP schema.

3.1.1 PIMs Description

In this section we summarize the ER and the OLAP metamodel. However, for a more detailed explanation, we refer the reader to [2, 3].The source and target PIMs are modeled using the ER and OLAP metamodels from CWM (Common Warehouse Metamodel). The CWM ER [2] metamodel (fig. 2), contains all the elements of an ER model. In the ER metamodel each Entity may own Attributes. A RelationShip represents associations between two or more Entities. Each end of a RelationShip is a RelationShipEnd, which may optionally have name and multiplicity.

The CWM OLAP [3] metamodel (fig 3) is structured into a schema class that owns all elements of an OLAP model. In the OLAP metamodel, each Dimension is a collection of Members. Cubes are used to store Measures and they are related to the Dimensions through the CubeDimensionAssociation class. Dimensions can contain multiple and diverse hierarchical arrangements of Members including a specialized hierarchies that support ordering Members by hierarchy levels (HierarchyLevelAssociation).

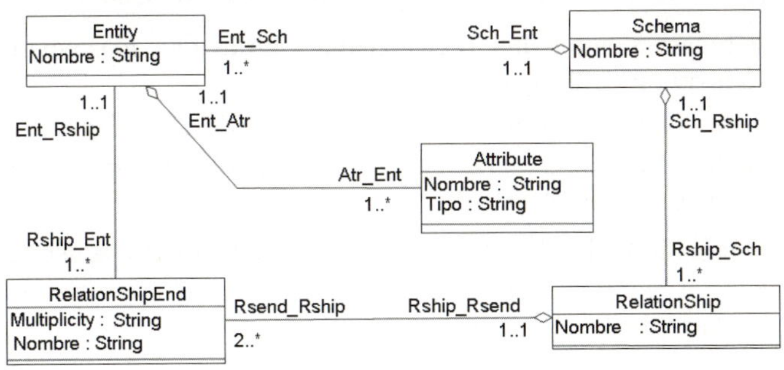

Fig. 2. ER metamodel

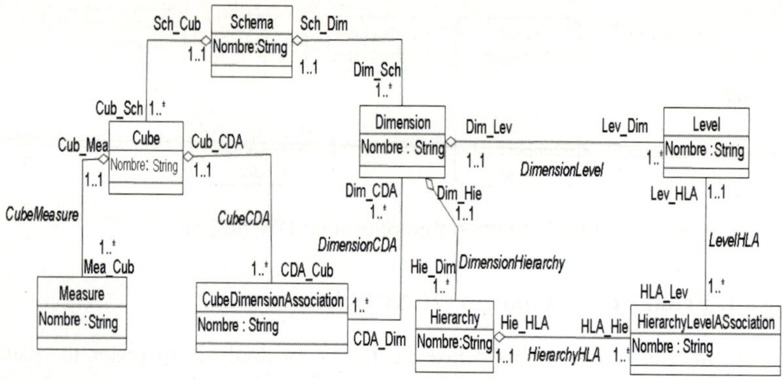

Fig. 3. OLAP metamodel

3.1.2 Mapping an ER Schema into an OLAP Schema

In this subsection, we describe how an ER schema is mapped to an OLAP schema.

Definition 1: A candidate cube entity *H* in ER is an entity such that *H* has references to other entities but it is not referenced by any Entity[1].

Definition 2: A path c in ER, is a set of entities c=(H, E1,…, En) such that exists a sequence of binary relationships R1(H, E1), R2(E1, E2),…, Rn(En-1, En), where H is a candidate cube entity.

The OLAP schema is constructed from the *ER* as follows:

1. Each candidate cube entity *H* is mapped to a cube in the OLAP schema.
2. Each numeric attribute from *H* is mapped to a measure of the cube in the OLAP schema.

- Let *R* be a binary relationship between the entities *(H, E)*, where *R* has multiplicity *H(1,1), E(0,N)*, *H* is a candidate cube entity and *h, e* are, respectively the roles (RelationShipEnd) of *H* and *E* then:

– *R* is mapped to a dimension *D* in the OLAP schema.
– *h* is mapped to a CubeDimensionAssociattion (CDA), between the candidate cube entity *H* and the dimension *D* in the OLAP schema.
– *E* is mapped to a level *L* of the dimension *D* in the OLAP schema.

- Let *R* be a binary relationship between the entities (E_j, E_k), where E_j has been mapped to a level of the dimension *D* then:

– E_k is mapped to a level L_k of the dimension *D* in the OLAP schema.

- Let *c* be a path in ER and *j, k* are, respectively the roles(RelationShipEnd) of E_j, E_k in one binary relationship $R_{jk}(E_j, E_k)$ of *c* then:

– *c* is mapped to a hierarchy *J* of the dimension *D* in the OLAP schema.
– *k* is mapped to a hierarchy-level association class (HLA) between the hierarchy *J* and the level E_k in the OLAP schema.

---

[1] An entity E_i has references to other entity E_j, if exists a binary relationship R_{ij} between E_i and E_j with multiplicity (1,1) on the side of E_i and (0,N) on the side of E_j.

3.1.3 Transformation Rules

In this section, we briefly describe the transformations rules using the diagrammatic notation of the declarative approach of QVT [10]. Each transformation contains the following elements:

- Domains: each domain identifies one candidate model, and a set of elements to match in that model by means of patterns.
- A relation domain: it specifies the kind of relation between domains, since it can be marked as checkonly (labeled as C) or as enforced (labeled as E). A checkonly domain is checked to see if there exists a valid match in the model that satisfies the relationship; whereas for a domain that is enforced, when the domain patterns do not match, model elements may be created, deleted or modified in the target model. For each domain the name of its underlying metamodel is specified.
- When clause: it specifies the pre-conditions that must be satisfied to carry out the transformation.
- Where clause: it specifies the post-condition that must be satisfied by all model elements participating in the relationship.

Main rule.- This rule, is the starting point of the transformation process. On the left side of the fig. 4, we can see the part of the ER PIM (labeled as er:ENTREL) that has to match with the part of the OLAP-PIM (labeled as md:OLAP) on the right side. In this case, an ER schema matches with an OLAP schema having the same name of the ER schema but prefixed with 'OLAP'. The goal of the function IsCandidate is identify the candidate cube entities of the source schema. Each candidate cube entity is matching with a Cube through the transformation rule *EntityToCube*. The Relationships with multiplicity many to one, that are linked to a candidate cube entity are matching with a Dimension through the transformation rule *RelationShipToDim.*

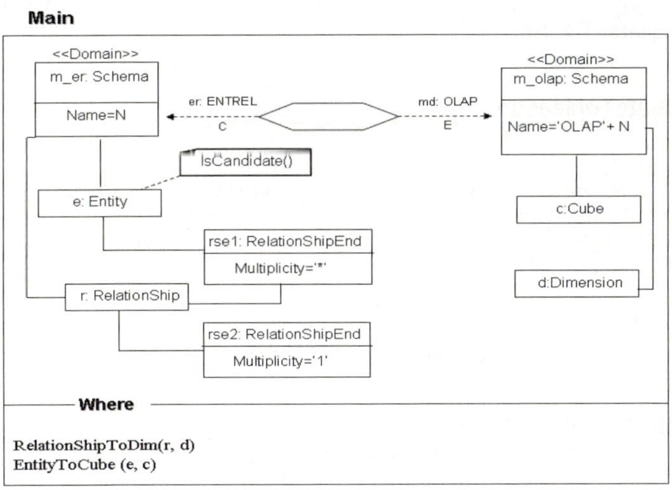

Fig. 4. Main Rule

EntityToCube. According to the transformation rule (fig. 5), a candidate cube entity gets transformed to a corresponding *Cube*, having the same name of the Entity but prefixed with a "C". Once this transformation is done, the transformation rules *RelationShipEndToCDA* and *AttributeToMeasure* must be done. The check domain (C arrow) determines the transformation in the following way: 1) each role (*RelationShipEnd*) with multiplicity equal to many is matching with a *CubeDimensionAssosiation* through the transformation rule *RelationShipEndToCDA*. 2) The numeric attributes are matching with a *Measure* through the transformation rule *AttributeToMeasure*.

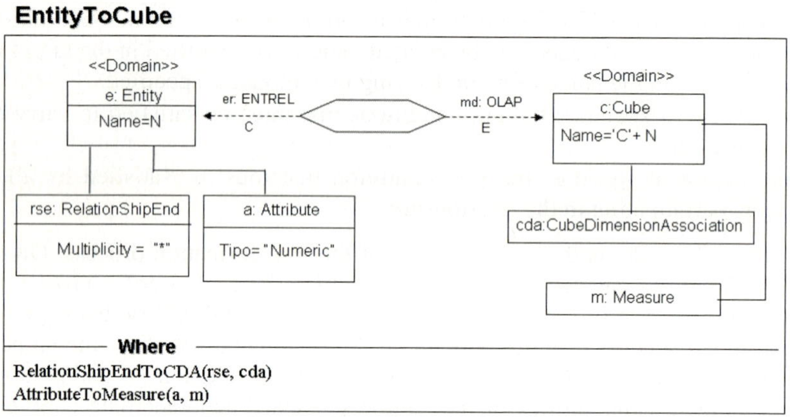

Fig. 5. EntityToCube rule

AttributeToMeasure. The numeric attributes of the candidate cube entity, gets transformed to a corresponding Measure of the Cube through the transformation rule AttributeToMeasure (fig. 6).

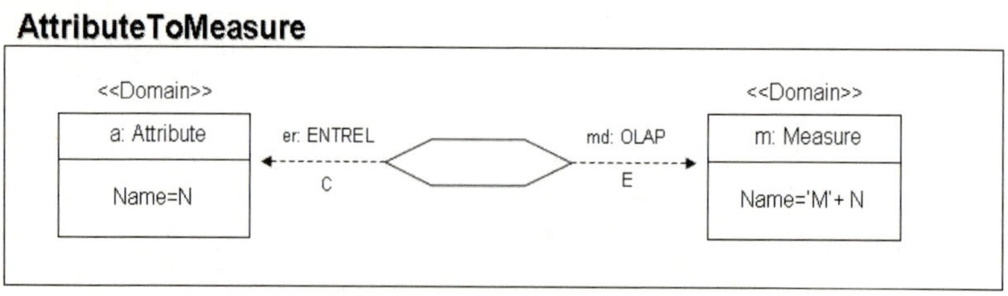

Fig. 6. AttributeToMeasure rule

RelationShipToDimension. In this rule (fig. 7), a RelationShip gets converted to a corresponding Dimension, having the same name as the RelationShip, but prefixed with a "D". Once this transformation is done, the transformation rules EntityToLevel and *HierarchyAuxToHierarchy* must be done. The check domain (C arrow) determines the transformation in the following way: 1) each path identified is matching with a Hierarchy through the transformation rule *HierarchyAuxToHierarchy*. 2) The goal of the function *Get_Hierarchies* is identify each path starting from the relationship. 3) Each entity in the path is matching with a level through the transformation rule *EntityToLevel*. 4) The goal of the function *Get_levels* is identify each entity in the path. We wish to point out that due to space constrains, we only show the transformation rule EntityToLevel.

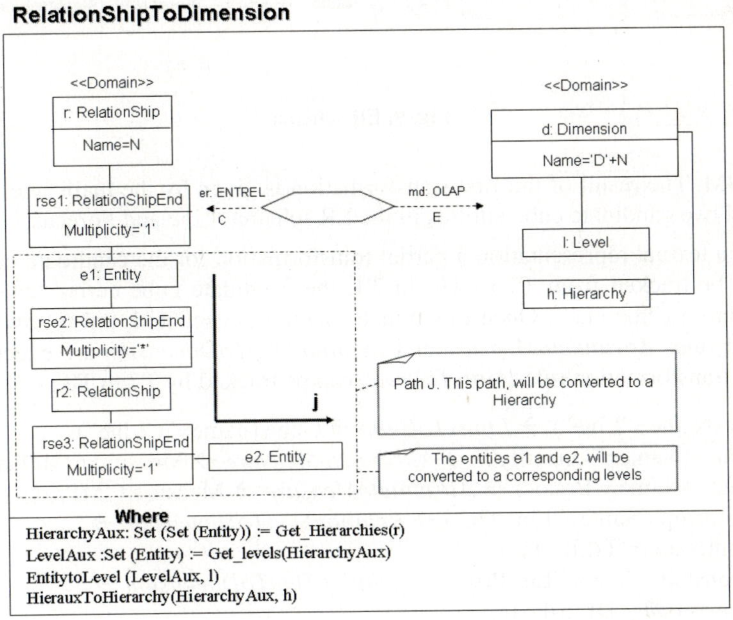

Fig. 7. RelationShipToDimension rule

EntityToLevel. Each *Entity* that is linked to a candidate cube entity is matching with a *Level*. In this rule (fig. 8), an *Entity* gets converted to a corresponding *Level*, having the same name as the *Entity*, but prefixed with a "L".

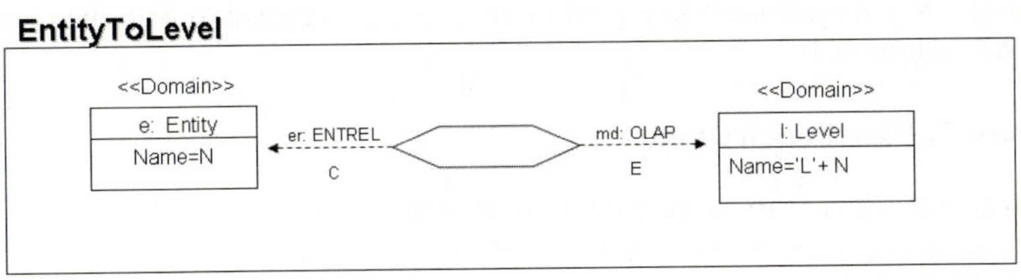

Fig. 8. EntityToLevel rule

3.1.4 Example

In this section, we provide an hypothetical example to show how to apply the presented transformation rules to an ER PIM (fig. 9) in order to obtain the corresponding

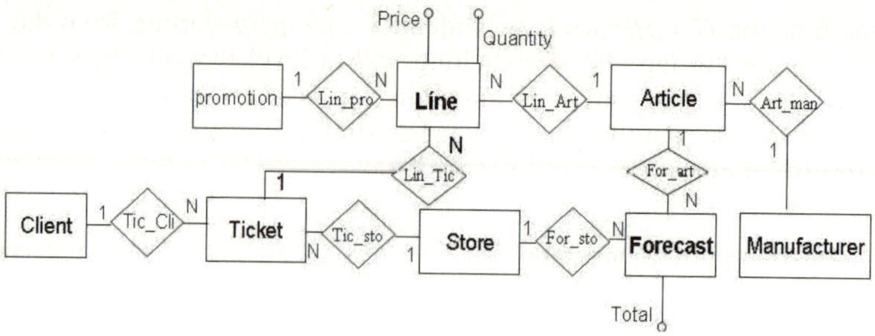

Fig. 9. ER schema

OLAP PIM. The result of the first transformation is given by the main rule. This rule identified two candidate cube entities in the ER schema: Line and Forecast.

Using a textual representation a partial transformation for the candidate cube entity *Line* can be tracked from T1 to T9. In T1, the candidate cube entity *Line* is transformed into a Cube class. Once this transformation is executed, the following transformation rules *AttributetoMeasure* and *RelationShipToDimension* are executed (T2-T7). The transformation rule *EntityToLevel,* can be tracked by T7 to T8.

T1:Entity (Name="Line") → *EntityToCube* → Cube (name="CLine").
T2:Attribute (Name="Quantity") → *AttributetoMeasure* → Measure ("MQuantity")
T3:Attribute (Name="Price") → *AttributetoMeasure* → Measure ("MPrice")
T4:Relationship (Name="Lin_Tic") → *RelationShipToDimension* →
Dimension(Name="DLin_Tic)
T5:Relationship (Name="Lin-Pro") → *RelationShipToDimension* →
Dimension(Name="DLin-Pro)
T6:Relationship (Name="Lin-Art") → *RelationShipToDimension* →
Dimension (Name="DLin-Art)
T7:Entity (Name="Ticket") → *EntityToLevel* → Level (Name="LTicket")
T8: Entity (Name="Client")→ *EntityToLevel* → Level (Name="LClient")

The candidate multidimensional schemas obtained from the ER schema are shown in fig. 10. As we can see, the elements of the schemas are stereotyped with the cube, dimension, level and measure keyword to indicate its relationship with the concepts of the OLAP metamodel.

3.2 Phase 2. User Requirements

The aim of this phase is to capture the information requirements to be kept in the DW. This phase deals with the identification of the strategic goals of the organization, decisions that can be taken to achieve these goals and the information requirements needed for decision making. Our approach proposes the construction of a goal model. It uses the concepts of goal and task of the i* conceptual framework [11]. In agreement with goal orientation philosophy, our goal model is built from the strategic goals and tasks that users must be able to accomplish when interacting with the DW. Afterwards, the information requirements will be discovered from these tasks using

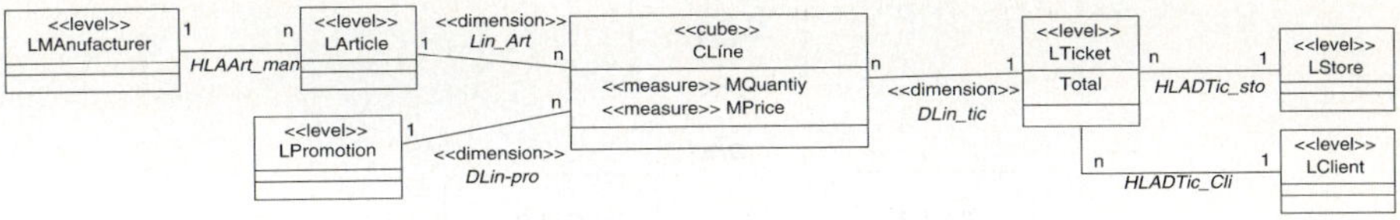

a). CLíne

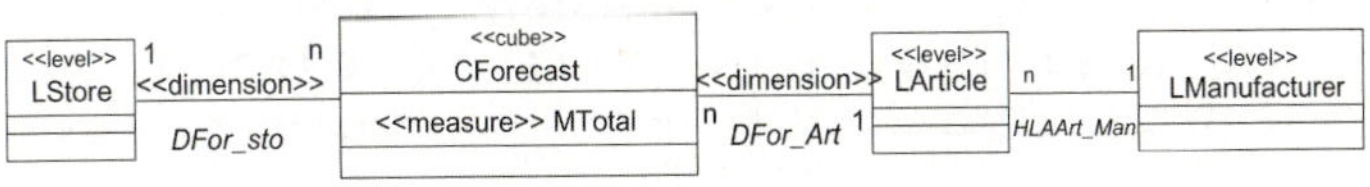

b). CForecast

Fig. 10. Candidate multidimensional schemas

UML activity diagram. Finally, the information requirements will be used for the selection and refinement of the conceptual multidimensional schema.

We propose three steps to define the task model: 1) Goal model definition, 2) Task description and 3) Identify domain notation.

3.2.1 Step 1. Goal Model Definition

We specify the information requirements of a DW system by means of a goal model. This model is build from the strategic goals that the stakeholders of a DW are interested in analyzing. In a DW environment, strategic goals represent the main objectives of the organization. These objectives deal with the business process to be analyzed but usually they lack details. A Goal Refinement Tree (GRT), can be used to refine these strategic goals. For the construction of the GRT, we take as the starting point, a strategic goal. From this strategic goal, goals are obtained following structural refinements. The refinement, consists of decomposing goals into sub-goals through an OR/AND relationship. This refinement of goals can continue until we have tasks that are tangible. In Fig. 11, we show the elements used for the construction of the GRT.

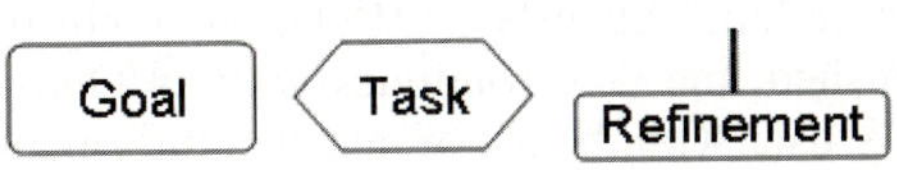

Fig. 11. Elements used for the construction of the GRT

Example (Continued).- In our example, one main domain stakeholders are identified: sales manager and offer manager. The strategic goals of the sales manager are: *G1.- Increase return on investment* and *G2.- Increase customer fidelity*. For instance the strategic goal Increase return on investment may be AND decomposed into *G.1.1.- Increase sales volume* and *G1.2.- Increase sales profit*. Likewise, increase sales volume might be OR decomposed into *G.1.1.1.- Increase consumer appeal* or *G1.1.2.- Expand market*. In our example, at least two well-established tasks can be to Increase sales profit: *G.1.2.1.- Increase sales price* or *G.1.2.2.- Lower production costs*. The partial representation of the goal model is shown in fig. 12.

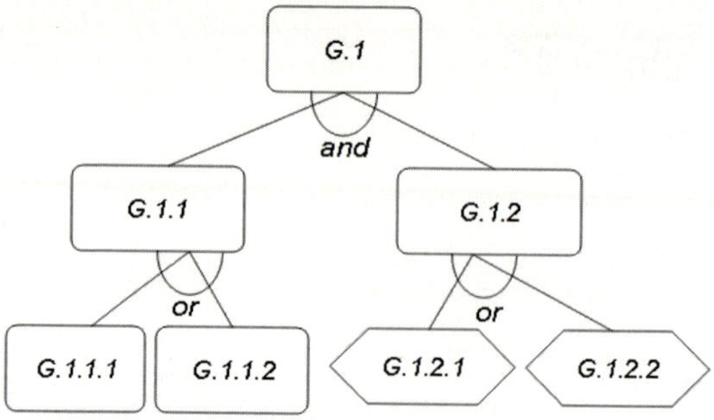

Fig. 12. A partial goal model

3.2.2 Step 2. Task Description

During this step, each task of the GRT is related to the actions that stakeholders consider necessary in order to satisfy each task. These actions are formulated in terms of the information required by a task to be achieved.

Similarly to [12] for tasks descriptions, we use UML Activity Diagrams [13]. In these diagrams, we show the actions performed to obtain some task, indicating the roles that are in charge of each activity, and the data required and produced by each activity. Data appear as objects that flow between activities. We refer to these objects as Data Objects (DO). We distinguish two different types of DOs. 1) Output DO: the system provides actors with information about data. 2) Input DO: the system is waiting for the user to introduce some data. This information is taken by the system to correctly perform a specific action. Fig. 13, shows an activity diagram for the description of the *increase sales price* task. This task is related with two actions: analyze the *margin profit* and *the quantity sold.* The activity diagram starts with the selection of an individual action. So, for instance if the selected action is *quantity sold*, this action, will search information that matches with the information provided by the DW user through an Input DO (*year, promotion and store*). In order to perform its purpose, this action needs access to the data-source *sales.* (An Entity is considered a data source, if an Entity-Relationship schema models the operational database. If a Relational schema models the operational database, a Relation is considered a data source.). Once this action is finished, the task continues with an Output DO, where the DW system provides the DW user with the list of matched information.

3.2.3 Step 3. Identify Domain Notation

In this step, we extract initial information requirements from the task description. The extracted information is called *domain notation.* Domain notation represents potential data to be stored in the DW. Well known heuristics for DW design can be applied here: within the multidimensional model a domain notation usually corresponds to factors that are supposed to influence the values of the measures (dimensions), data descriptors (measures) and data to be analyzed (facts). In particular, the DOs are considered as measures and dimensions in the multidimensional schema. Then, the Input DO defines the variables that may cause changes to measures (dimension analysis) and each Output DO contributes a measure.

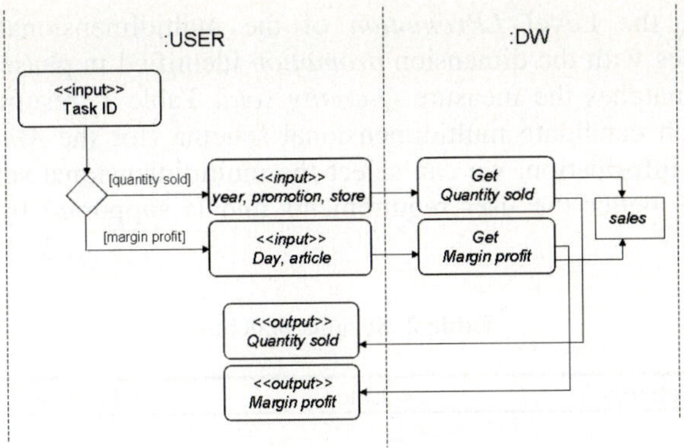

Fig. 13. Task description

Example (Continued)- The set of domain notation corresponding to the UML activity diagram of the fig. 13, is shown in table 1. According to the information showed in table I, the information that the DW must store about the *increase sales price* task is: *promotions, year, day, store, Quantity sold and Margin profit.* This information is related with the data source *sales.*

Table 1. Domain notation

Task	Domain notation	Data Object
T1	Promotions	Input
T1	Year	Input
T1	Store	Input
T1	Quantity sold	Output
T1	Margin profit	Output
T1	Day	Input
Data source	Sales	-------

3.3 Phase 3. Integration

In this section, the requirements identified in phase 2 are mapped with the elements of the candidate multidimensional schemas to generate the conceptual schema for DW. Two steps are involved: 1) selection, where a candidate multidimensional schema is selected; and 2) refinement, where the selected multidimensional schema is manually modified with the goal of represent the user expectations.

Selection.- Given user requirements and candidate multidimensional schemas, the problem is to find semantic mappings between their elements. For this, we use a *structure-level* matching technique [14]. This technique only considers the properties of the elements, such as name, description, data type and not instances data.

For example, the Level *LPromotion* of the multidimensional schema *CLine* (Fig.. 3a), matches with the dimension *promotion* identified in phase 2 and the measure *MQuantity* matches the measure *Quantity sold*. Table 1, resume the number of properties of each candidate multidimensional schema (for the *Analyze promotions* task). From this information, we can select the multidimensional schema *CLine*, because it captures better the user requirements and is supported by the operational database.

Table 2. Schema matching

Property\Schema	CLíne	CForecast
Measures	1	0
Dimensions	2	1

Refinement.-This step is aimed at reorganizing the selected multidimensional schema in order to better fit them to the users´ requirements. During this process the operations that can be carried out are: remove unnecessary attributes, remove dimension's levels, add measures, add summarization constraints and define time granularity (Due space constraints, we only describe the process to remove a dimension level).

-*Remove a dimension level*: probably, not all of the levels represented in the selected multidimensional schema, are interesting for DW. Thus, the level must be removed. In general, removing a level correspond a change in the granularity of cube instances, if the level eliminated has two or more descendents levels, leads to increasing the number of dimensions in the multidimensional schema [4]. For instance, in the final multidimensional schema (fig. 12), we assume that the user want to classify the information directly by LClient and LStore levels without considering the LTicket Level (of the selected schema) generating two new dimensions: DLin_Cli and DLin_Sto. For this example, we also assume that the user is interested in day granularity (Time dimension) and that the measure Quantity sold is computed with the aggregation operator Sum.

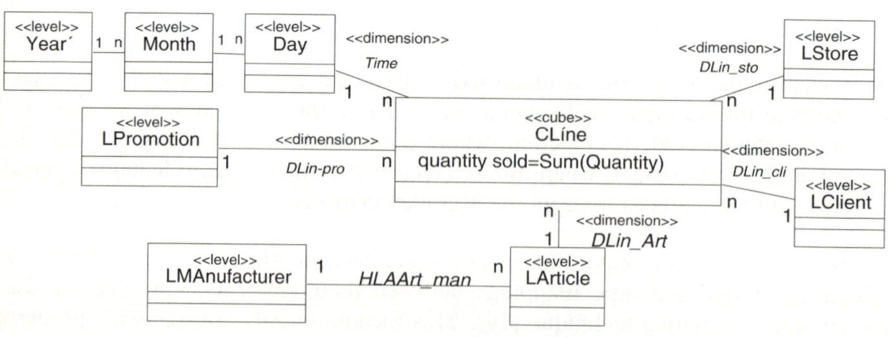

Fig. 14. Final multidimensional schema

4 Conclusions and Future Works

In this paper, we have introduced our MDA method for DW conceptual design. In conclusion, the experience we have gathered by applying the proposed method in our case study is encouraging. The method can be essential to direct the designer toward a solution that is both efficient to implement and consistent with users' requirements. The choice of MDA as a guiding methodology phase is justified by the need of exploiting an ER schema using standardized means. We think, that the description of each metamodel and its transformations is a good example about the use of MDA in conceptual design of DWs, at the same time the final DW schema is strongly rooted to the operational database which makes the design of ETL simpler. The transformation rules were programmed with QVT operational mapping language obtaining the expected results. We plan for future work to extend the approach presented in this paper by considering richer criteria for the third phase.

References

1. Kleppe, A., Warmer, J., Bast, W.: MDA Explained. The Practice and Promise of the Model Driven Architecture. Addison Wesley, Reading (2003)
2. Object Management Group (OMG), Common Warehouse Metamodel (CWM) specification (Extensions), http://www.omg.org/cgi-bin/doc?ad/2001-02-02
3. OMG, CWM specification 1.0.1, http://www.omg.org/cgi-bin/doc?formal/03-03-02
4. Giorgine, P., Rizzi, S., Garzetti, M.: Goal-oriented requirements analysis for data warehouses design. In: Intl. Workshop on Data warehousing and OLAP (DOLAP), pp. 47–56. ACM Press, Bremen (2005)
5. Golfarelli, M., Rizzi, S.: A methodological Framework for Data Warehouse Design. In: ACM 1st Workshop on Data warehousing and OLAP (DOLAP), pp. 3–9. ACM Press, Washington (1998)
6. Bonifati, S., Ceri, A., Fuggetta, S., Paraboschi.: Designing data marts for data warehouses. ACM Trans. Softw. Eng. Method 0, 452–483 (2001)
7. Mazon, J.-N., Trujillo, J., Serrano, M., Piattini, M.: Designing data warehouses: From business requirement analysis to multidimensional modeling. In: Proc. 1st Int. Workshop on Requirements Engineering for Business Need and IT Alignment, Paris (2005)
8. Cabibbo, L., Torlone, R.: A Logical Approach to Multidimensional Databases. In: Schek, H.-J., Saltor, F., Ramos, I., Alonso, G. (eds.) EDBT 1998. LNCS, vol. 1377, pp. 183–197. Springer, Heidelberg (1998)
9. Mazón, J.-N., Trujillo, J., Serrano, M., Piattini, M.: Applying MDA to the Development of Data Warehouses. In: ACM 8th International Workshop on Data Warehousing and OLAP (DOLAP), pp. 57–66. ACM Press, Bremen (2005)
10. OMG 2nd Revised Submission: MOF 2.0 Query/Views/Transformations, http://www.omg.org/cgi-bin/doc?ad/05-03-02
11. Lamsweerde, A.: Goal-Oriented Requirements Engineering: A Guided Tour. In: 5 IEEE International Symposium on Requirements Engineering. IEEE press, Toronto (2001)
12. Valderas, P., Fons, J., Pelechano, V.: Using Task Descriptions for the Specification of Web Application Requirements. In: Workshop en ingeniería de requisitos, Porto Portugal (2005)
13. UML Specification Version 2.0 Final Adopted Specification, http://www.omg.org
14. Rahm, E., Bernstein, P.: A survey of approaches to automatic schema matching. In: Very Large Databases (VLDB), vol. 10, pp. 334–350. ACM Press, New York (2001)

A New Bitmap Index and a New Data Cube Compression Technology

Jianqing Xi[1], Fuqiang Chen[1,2,*], and Pingjian Zhang[1]

[1] School of Computer Science & Engineering, South China University of Technology, Guangzhou, P.R. China
[2] Guangdong Polytechnic Normal University, Guangzhou, P.R. China
csjqxi@scut.edu.cn, chenfuqiang@163.com, pjzhang@scut.edu.cn

Abstract. This paper introduces a new kind of bitmap index. A tuple in the data cube is mapped to a sequential key (seqkey) determined by its value in each dimension. Furthermore, the quotient bit sequence is constructed according to whether the corresponding cell of a seqkey exists in the cover quotient cube or not, and the cover quotient cube is indexed by this quotient bit sequence (qcbit index). A compression method is presented for the seqkey cover quotient cube, which compress the cover quotient cube via omitting dimension attributes for all cells. To improve the storage and query of data cubes, based on these index and compression methods, algorithms are proposed to query the cover quotient cube and seqkey cover quotient cube. Experimental results on the dataset weather show that the volume of the qcbit index file is only 11% of the value-list index file, and the volume of seqkey cover quotient cube is only 27.75% of the original cover quotient cube.

Keywords: qcbit index, seqkey cover quotient cube, bitmap index, data cube, quotient cube, cover quotient cube.

1 Introduction and Related Works

Data cubes [1] in warehouse usually occupy huge storage space, and the time cost to query a data cube is always high. Laks et al proposed quotient cube [2] and cover quotient cube [3] to shrink the volume of data cube, in a cover quotient cube only upper bounds are contained. They also proposed the idea of QC-Tree to compress cover quotient cube, however, QC-Tree cannot be directly applied to data cubes based on RDBMS. Zhao Y. proposed the sorted-list index and adapt the value-list index [4] to improve the efficiency to query cover quotient cube, however, the required storage space and retrieval time for index files are proportional to the size of base tables, which does not scale well for large dataset. Li S.E. et al proposed the concept of closed data cube [5], where upper bounds are grouped into layers according to the number of their definite dimension values. Nevertheless, no index is adopted for closed data cubes, hence upper bounds may not be located directly when querying a data cube. The incremental maintenance of the corresponding data cubes are also handled in [5]. Li C.P.

* Corresponding author.

O. Gervasi et al. (Eds.): ICCSA 2008, Part II, LNCS 5073, pp. 1218–1228, 2008.

et al studied the incremental maintenance for quotient cube for median [6]. Li H.S. et al discussed the online clustering algorithm based on QC-Tree [7].

As we know there is only a little work on quotient cube index mechanism, there are only sorted-list index and value-list index are used to index cover quotient cube[4], and no published work to further compress the cover quotient cube based on RDBMS. In this paper we explore new methods to store and query cover quotient cubes more efficiently. The new methods include: (1) a new kind of bitmap index, (2) corresponding query algorithm to answer point query and (3) a new data cube compression technology.

Table 1. (a). Base table *Sale* and the tuples

D_1(Location)	D_2(Product)	D_3(Time)	M_1(Sales)
van	b	d1	9
van	f	d2	3
tor	b	d2	6

(b). Values of each dimension and their CARDKEY

Dimension	Values and corresponding CARDKEYs
D_1(Location)	(*,0),(tor,1),(van,2)
D_2(Product)	(*,0),(b,1),(f,2)
D_3(Time)	(*,0),(d1,1),(d2,2)

2 Preliminaries

Given base table T ($D_1,\ldots,D_i,\ldots,D_n, M_1,\ldots,M_j,\ldots,M_m$), where D_i 's are dimension attributes, $1\le i\le n$, the cardinality of D_i is denoted by cdn_i , $M_1,\ldots,M_j,\ldots,M_m$ are measure attributes, $1\le j\le m$. Denote by * the special value *All*. Let D_T be the data cube of T, then, the number of cells in D_T is given by $ce=\prod_{i=1}^{n}(cdn_i+1)$ .Denote by CQ_T the cover quotient cube of T, $CQ_T=\{ub|ub\in D_T$, ub is an upper bound$\}$, let cu denote the number of upper bounds in CQ_T.

For $c1,c2 \in D_T$, if $c1.D_i = c2.D_i$ holds true for all $1\le i\le n$, then $c1$ and $c2$ are said to be equal, denoted by $c1=c2$; otherwise, $c1$ and $c2$ are said to be not equal, denoted by $c1\neq c2$. For $c \in D_T$, denote by UB_c the upper bound of the class to which c belongs.

Definition 1. CARDKEY Each value of dimension D_i ($1\le i\le n$) in base table T can be uniquely mapped into a non negative integer k not greater than cdn_i, k is defined to be the cardinality key(CARDKEY) for this dimension value. Specifically, the CARDKEY for * is defined to be 0. Denote by ck_{ij} the CARDKEY of j-th value in dimension D_i.

As an example, consider the base table *Sale* with three dimension attributes D_1, D_2, D_3 as given in table 1(a). A possible CARDKEY encoding is given for all dimensions, as shown in table 1(b). Hereafter, every cell in D_{Sale} has two representations: one in dimension values, the other in CARDKEY. (van,b,d1,9) and (2,1,1,9), for example, represent the same cell.

Definition 2. Sequential Key For $c \in D_T$ with CARDKEY ck_i in dimension $D_i, 1 \le i \le n$, the sequential key of cell c is defined to be *seqkey*(c) $= \sum_{i=1}^{n-1} (ck_i * \prod_{j=i+1}^{n} (cdn_i + 1)) + ck_n + 1$. For example, let c be the cell (van, b, d1,9) in D_{sale}, then *seqkey*(c) = 2 * 3 * 3 +1 * 3 + 1 + 1=23, as shown in table 2.

Definition 3. Quotient Bit and Quotient Bit String Associated with each data cube D_T, there is a bit string of length *ce*, denoted by QBS_T, each bit in QBS_T corresponds to a cell c in D_T and is set to be 1 if c is a upper bound, otherwise, the bit is set to be 0. This bit is called the quotient bit of cell c and QBS_T is called the Quotient Bit String of T. For example , given the base table *Sale*, if seqkey is defined as shown in table 2 below, then, QBS_{Sale}="101100000000001000100010001".

Definition 4. Ancestor Let $c \in D_T$, $t \in T$, if for all i, $1 \le i \le n$, such that $c.D_i \ne *$, one has $t.D_i = c.D_i$, then c is said to cover t. The cover set of c in T is defined to be $BTS(c)=\{t| t \in T, c \in D_T, c \text{ covers } t\}$. For example, *BTS*(*, b,*,15)={(van,b,d1,9),(tor,b,d2,6)}.

Zhao Y. also introduced the concept of ancestor [4] . Let $c \in D_T$, $c' \in D_T$, $c' \ne c$, if for all i, $1 \le i \le n$, such that $c.D_i \ne *$, one has $c'.D_i = c.D_i$, then, c' is said to be an ancestor of c, denoted by $c' \Delta c$. For example, in table 2, cell (*,b,d1,9) is an ancestor of (*,b,*,15).

Definition 5. Set of Upper Bound Ancestors Let $c \in D_T$, the set of upper bound ancestors of c is defined to be $SAU(c)=\{ub | ub \in CQ_T, ub = c \text{ or } ub \Delta c\}$. For example, *SAU*(*,b,*,15) = {(*,b,*,15), (tor,b,d2,6), (van,b,d1,9)}.

Definition 6. Set of Seqkey's of Ancestors Let $c \in D_T$, the set of Seqkeys' of ancestors of c is defined to be $KSAC(c)=\{i | i= seqkey(c'), c' = c \text{ or } c' \Delta c\}$. The set of Seqkeys' of upper bound ancestors of c is defined to be $KSAU(c) = \{i | i \in KSAC(c), QBS_T(i)=1\}$. For example, *KSAC*(*,b,*,15) = {4,5,6,13,14,15,22,23,24}, *KSAU*(*,b,*,15) = {4,15,23}.

Definition 7. Cubeseqrow Assume that all upper bounds in CQ_T are ordered by ascending seqkey, if a upper bound *ub* is in the *k*-th place, then, the Cubeseqrow of *ub* is defined to be *k*, denoted by *CSR*(*ub*)=*k*. For example, *CSR*(van,b,d1,9)=6.

Definition 8. Bit Cubeseqrow If *b* is the *k*-th 1 in QBS_T, then, the bit cubeseqrow of *b* is defined to be *k*, denoted by *BCSR*(*b*) = *k* . Obviously, for a upper bound *ub*, $BCSR(QB_{ub}) = CSR(ub)$.

Lemma 1: Let $c \in D_T$, the following properties hold true :

(A) For $c' \in D_T$, if $c' \Delta c$, then, $seqkey(c') > seqkey(c)$.
(B) $seqkey(UB_c) \ge seqkey(c)$.

Proof : (A) Since $c'\Delta c$, c' is different from c in some dimensions where c is * and c' is not, the corresponding CARDKEY is 0 for c and positive integer for c', hence, $seqkey(c')>seqkey(c)$.

(B) If $UB_c=c$, then, $seqkey(UB_c)=seqkey(c)$. Otherwise, $UB_c \neq c$, since $UB_c\Delta c$,we have $seqkey(UB_c)>seqkey(c)$ by (A).

Theorem 1: Let $c \in D_T$, then for all $ub\in SAU(c)$ such that $ub\neq UB_c$, one has $seqkey(ub)> seqkey(UB_c)$.

Proof : Since $c \in D_T$, $UB_c \in CQ_T$. Note that $ub \neq c$. In fact, if $c = ub\in SAU(c)$, $c \in CQ_T$, since upper bound for each class is unique, we have $UB_c = c$, thus, $ub=UB_c$, a contradiction with $ub\neq UB_c$. From $ub\in SAU(c)$, it follows that either $ub= c$ or $ub\Delta c$, since $ub \neq c$, $ub\Delta c$, hence $BTS(ub) \subset BTS(c) =BTS(UB_c)$. For $1\leq i\leq n$ such that $UB_c.D_i\neq *$, one has $c.D_i= UB_c.D_i$, since $ub\Delta c$, $ub.D_i= c.D_i =UB_c.D_i$, thus, $ub\Delta UB_c$ by definition. Since $ub\neq UB_c$, Lemma 1(A) yields $seqkey(ub)> seqkey(UB_c)$.

Lemma 2 : Let $c \in D_T$, $x\in KSAU(c)$, if there is no $x'\in KSAU(c)$ such that $x' < x$, then, $x = seqkey(UB_c)$, i.e., $QBS_T(x)$ is the quotient bit of UB_c.

Proof : Since $c \in D_T$, we have $UB_c \in CQ_T$. By defintion 6, there must exist $y \in KSAU(c)$ such that $y = seqkey(UB_c)$. Suppose $x\neq seqkey(UB_c)$, then, there exists $x''\neq x$, $x'' \in KSAU(c)$, such that $x'' = seqkey(UB_c)$. Since $x\in KSAU(c)$, there exists $ub\in SAU(c)$ such that $ub\neq UB_c$ and $x = seqkey(ub)$, then, by theorem 1, we have $seqkey(ub) > seqkey(UB_c)$, or $x'' < x$, a contradiction to the assumption, hence, the lemma is proved.

Lemma 3 : Let $c \in D_T$, $x\in KSAC(c)$, $QBS_T(x) = 1$, if there is no $x'\in KSAC(c)$ such that $x' < x$ and $QBS_T(x')= 1$, then, $x = seqkey(UB_c)$ i.e., $QBS_T(x)$ is the quotient bit of UB_c.

Proof : It is straight forward from definition 6 and lemma 2.

Let $c \in D_T$,from lemma 3 we know that if we check the quotient bits of the ancestors of c in QBS_T, beginning from the quotient bit of c, then the first 1 will be the quotient bit of UB_c, and the corresponding upper bound in CQ_T will be UB_c . From the definition 7 and definition 8 we know that by calculating the bit cubeseqrow of the quotient bit of UB_c, we get the cubeseqrow of UB_c . The cubeseqrow depicts the physical location of UB_c. Based on these, we designed the index mechanism named qcbit index for cover quotient cube.

3 Qcbit Index

3.1 The Concept of Qcbit Index

Let F_T be the disk file that contains CQ_T, I_T be the disk file that contains QBS_T. The disk file M_T is used to store the number of 1 in each block of I_T. Suppose a point query is used to find out the aggregate value in a specific cell c. To answer this query, one needs to locate UB_c in F_T, then, fetch the aggregate value. To locate UB_c efficiently, we propose a new index called qcbit index and design the APQ algorithm to answer the point query. In the description below, all upper bounds in F_T are assumed to be ordered by ascending seqkey .

Table 2. Seqkey and quotient bit for base table *Sale*

Seqkey	D_1	D_2	D_3	ck_1	ck_2	ck_3	Sum(M_I)	Quotient bit	Cubeseqrow
1	*	*	*	0	0	0	18	1	1
2	*	*	d1	0	0	1	9	0	
3	*	*	d2	0	0	2	9	1	2
4	*	b	*	0	1	0	15	1	3
5	*	b	d1	0	1	1	9	0	
6	*	b	d2	0	1	2	6	0	
7	*	f	*	0	2	0	3	0	
8	*	f	d1	0	2	1	false	0	
9	*	f	d2	0	2	2	3	0	
10	tor	*	*	1	0	0	6	0	
11	tor	*	d1	1	0	1	false	0	
12	tor	*	d2	1	0	2	6	0	
13	tor	b	*	1	1	0	6	0	
14	tor	b	d1	1	1	1	false	0	
15	tor	b	d2	1	1	2	6	1	4
16	tor	f	*	1	2	0	false	0	
17	tor	f	d1	1	2	1	false	0	
18	tor	f	d2	1	2	2	false	0	
19	van	*	*	2	0	0	12	1	5
20	van	*	d1	2	0	1	9	0	
21	van	*	d2	2	0	2	3	0	
22	van	b	*	2	1	0	9	0	
23	van	b	d1	2	1	1	9	1	6
24	van	b	d2	2	1	2	false	0	
25	van	f	*	2	2	0	3	0	
26	van	f	d1	2	2	1	false	0	
27	van	f	d2	2	2	2	3	1	7

```
Algorithm APQ. Answering point query using qcbit index.
    Input: q=(......, ck1 ,......,cki ,......,ckj ,......), IT , FT
    Output: the aggregate value in q
    Procedure: //disk block size is assumed to be 8K bytes;
    1. ikey = sk(q); if(ikey==0) then return false;
      // find the SEQKEY of UBq
    2. ibit1 = sum of first ikey/64K integers in MT;
      ibit2 = number of 1 in the first ikey%64K bits of the
  ikey/64K-th block in IT;
      k = ibit1 + ibit2;      // k is the bit cubeseqrow of
  QBST(ikey);
    3. fetch the k/rc-th block of FT, and get its k%rc-th
  tuple;
      // rc is the number of records in a block;
    4. return the aggregate value.
    Function sk(cArg)
    {  int ck[n+1],l[n+1],cdn[n+1],begin[n+1],i;
       //n is the number of dimension attributes in q
```

```
    for(i=1;i≤n;i++)
        if(ck[i]==0) begin[i]=0; else begin[i]=cdn[i];
        //ck[i] is equal to ck_i
    for(l[1]=begin[1]; l[1]≤cdn[1]; l[1]++)
     ......
         for(l[i]=begin[i]; l[i]≤cdn[i]; l[i]++)
     ......
         for(l[n]=begin[n]; l[n]≤cdn[n]; l[n]++)
             {  x=seqkey(cArg) ;
                if(the x%64K-th bit in the x/64K-th block
  of I_T is 1)  return x ;
             }
    return 0 ; }
// end of algorithm APQ
```

3.2 Cost Analysis for the Algorithm APQ

In the first step of algorithm APQ, one needs to locate the quotient bit of q to find out the seqkey of UB_q, then find the quotient bit of UB_q, the average cost is $COST_q = \frac{1}{n}\sum_{i=0}^{n}\frac{n-i}{2} = \frac{n}{4}$, where n is the length of QBS_T; for step 2, since M_T is usually very small and can reside in memory, and also notice that the *ikey*/64K-th block of I_T is already fetched into memory in step 1, there is no need to access disk during this stage; step 3 results in disk access only once. Considerable savings can be achieved compared to HQ algorim[5], whose average cost is $n/2$.

3.3 Creation and Incremental Maintenance

I_T can be created at the same time F_T is generated. For example, if one calculates the seqkeys for all the upper bounds when a Depth-first Search Algorithm is adapted[2], and update the corresponding QBS_T, one will get QBS_T which is the content of I_T.

Qcbit index is suitable for data environment where the most operations on data is reading. Adding, deleting or updating tuples in the base table T may cause dimension values to be added, deleted or updated for all dimensions of T and the CARDKEYs may also change accordingly, resulting changes in the seqkey of cells of the data cube. A reasonable solution to incrementally maintain qcbit index is to regenerate the cover quotient cube and qcbit index when there is a low workload on the database server.

4 Seqkey Cover Quotient Cube

Definition 9: In the cover quotient cube CQ_T, order all upper bounds by ascending seqkey, then remove all dimensions except the measures, the resulted data cube is called the seqkey cover quotient cube of T, denoted by SQ_T.

For example, the column SUM(M_1) in table 3 forms the SQ_{Sale} for base table *Sale* in table 1(a), with measures ordered by ascending seqkey. Obviously, the volume of SQ_{Sale} is by far smaller than that of CQ_{Sale}.

The algorithm APQ is still valid if F_T contains SQ_T instead of CQ_T.

Table 3. The Seqkey Cover Quotient Cube

D_1	D_2	D_3	SEQKEY	Cubeseqrow	SUM(M_1)
*	*	*	1	1	18
*	*	d2	3	2	9
*	b	*	4	3	15
tor	b	d2	15	4	6
van	*	*	19	5	12
van	b	d1	23	6	9
van	f	d2	27	7	3

5 Experimental Results

Experiments are conducted on a 2.4GHz Intel PC with 512MB main memory, running Microsoft Windows 2000 Server. We used the *weather* dataset [8], provided by the Carbon Dioxide Information Analysis Center, The dataset contains many data files that record weather conditions at thousands of weather stations on land. Each data file that contains records for a month can be represented by a relation scheme which includes many dimension attributes and measure attributes. One of the data files, SEP85L.DAT, which contains records for September 1985, has been recently used in related works of [2,4,5]. Tuples randomly selected from files SEP85L.DAT, OCT91L.DAT, MAY86L.DAT, MAR84L.DAT, JUN87L.DAT, JUL88L.DAT, JAN82L.DAT, FEB83L.DAT, DEC81L.DAT, AUG89L.DAT are extracted for the experiment. The results are summarized in figure 1 to figure 5.

Figure 1 and figure 2 show that the volume of qcbit index file becomes much smaller than that of value-list index file as base tuples count increases. Figure 3 shows that when the count of the base tuples is less than some critical value, the volume of qcbit index file is bigger than that of value-list index file; the situation reverses when the count of the tuples exceeds the critical value. Figure 4 shows that the volume ratio of an compressed qcbit index file/compressed value-list file is always less than 1, and the ratio decreases as base tuples count increases, the ratio reaches 11% when base tuples count is 10M, here both the two kinds of index files are compressed using run-length encoding approach. The experimental results demonstrate that the qcbit index has advantages for applications involving large dataset. When both of the two kinds of index files are compressed using run-length encoding approach, the advantages become more obvious.

It is also observed that when the base tuples count approaches to some extent, the cardinalities of the dimensions increase slowly or even stop growing, as shown in figure 5. There are many such examples in reality. Surveying all-year student scores for a major, for example, show that the increase in number of courses is by far behind the increase in number of students. Due to the slow growing or even not growing of cardinalities, the volume of qcbit index grows slowly or even stop growing.

We calculated the aggregate function *count*() on dimension attributes *solaraltitude*, *longitude* and *presentweather* for data file SEP85L.dat. It turns out that the volume of seqkey cover quotient cube is only 27.75% of that of cover quotient cube.

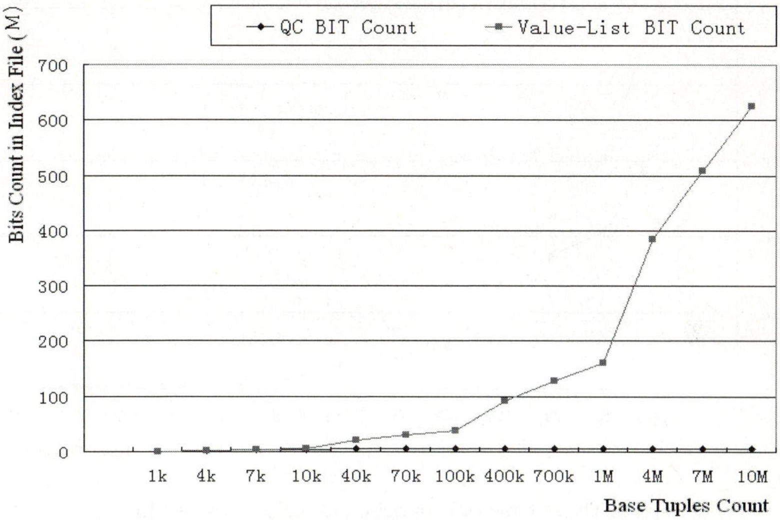

Fig. 1. The volume of qcbit index file (*solaraltitude*, *longitude* and *presentweather* are selected as dimension attributes)

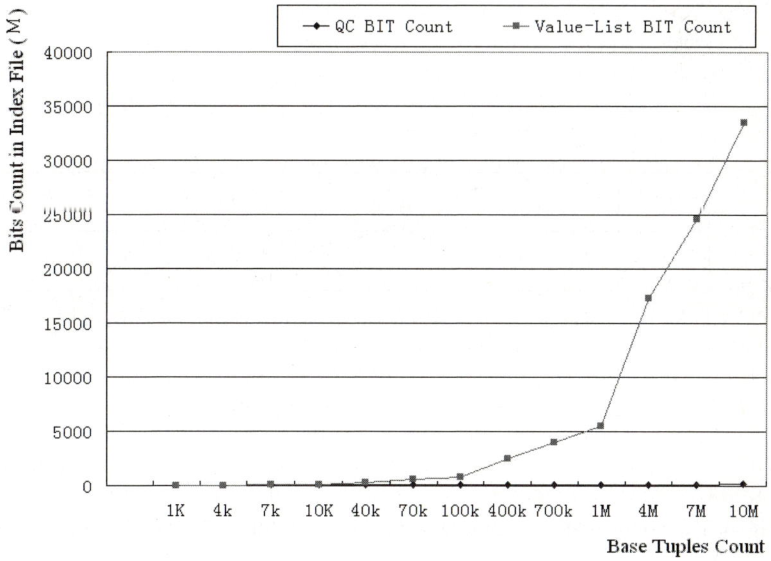

Fig. 2. The volume of qcbit index file (*solaraltitude*, *stationid* and *presentweather* are selected as dimension attributes

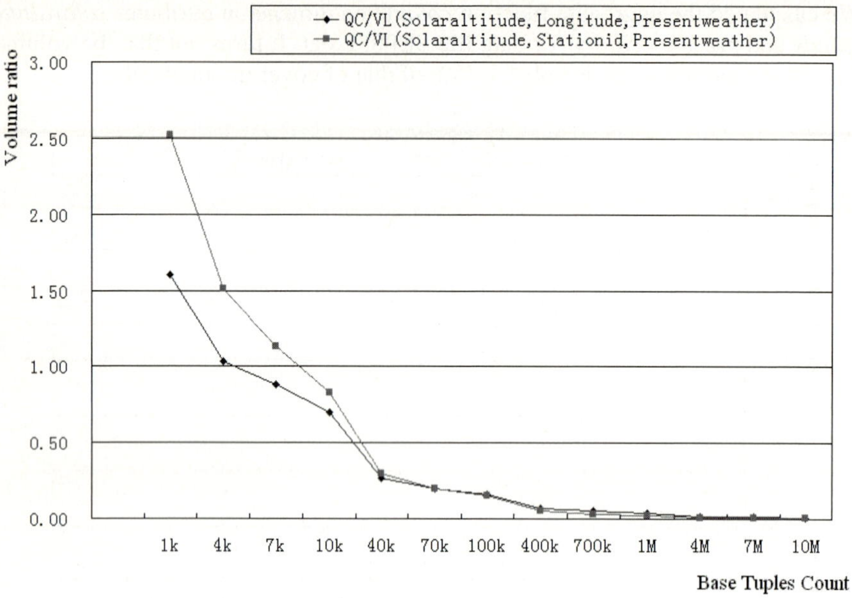

Fig. 3. The volume ratio of qcbit index file/value-list file

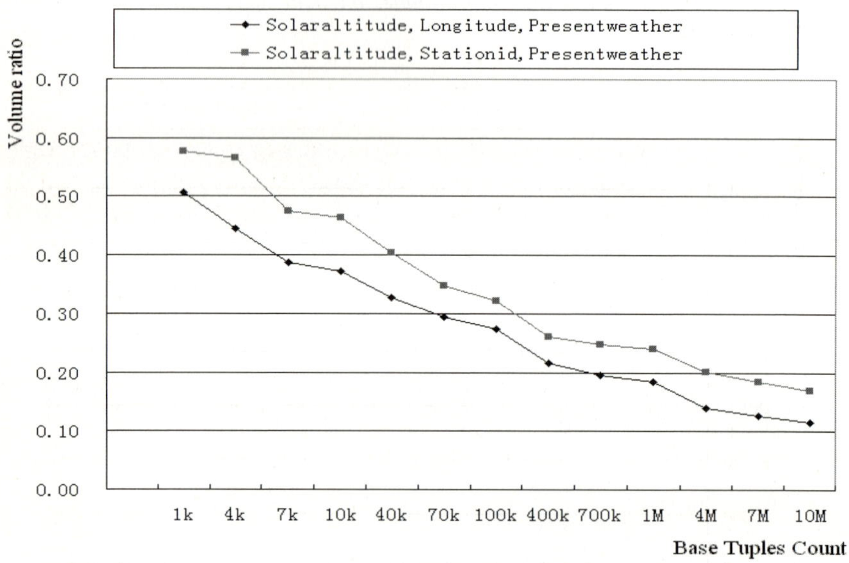

Fig. 4. The volume ratio of compressed qcbit index file/compressed value-list file

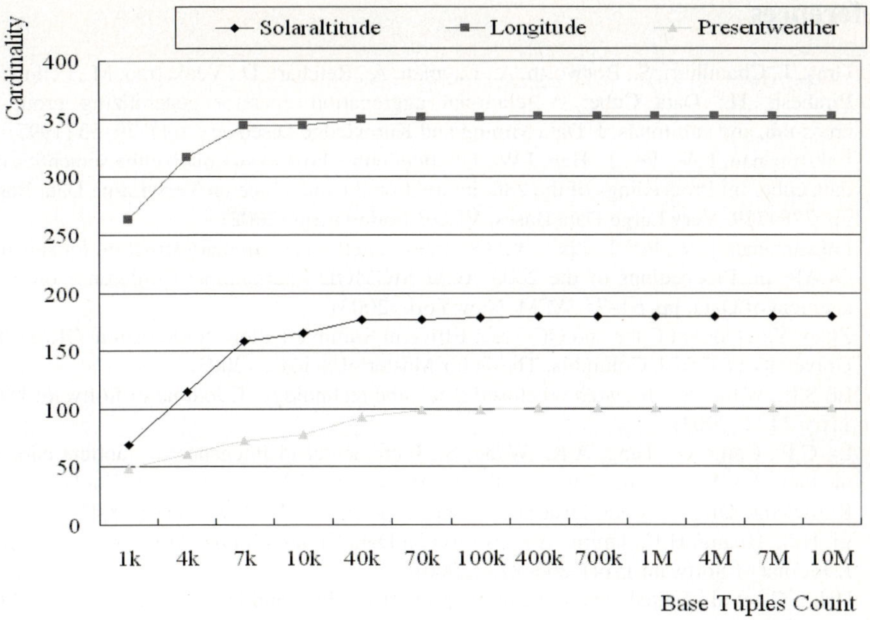

Fig. 5. The cardinality increasing trend

6 Conclusions and Further Works

The concept of qcbit index is presented in the paper. Algorithm for point query are proposed. Studies show that the volume of qcbit index file is by far smaller than that of value-list index file for large datasets, and the volume of seqkey quotient cube is much smaller than that of cover quotient cube.

When querying data cubes, the method based on qcbit index requires less disk I/O's, outperforming the value-list index method. For large base table, the advantage of qcbit index and seqkey cover quotient cube is even more evident.

Further works includes to find efficient incremental maintenance method for qcbit index, and to apply qcbit index to base tables.

Acknowledgements

This work is supported by Science & Technology program of Guangdong Province, China (2006B11301001, 2006B80407001), and International Science &Technology Cooperation Program of Guangdong Province, China (2007A050100026).

References

[1] Gray, J., Chaudhuri, S., Bosworth, A., Layman, A., Reichart, D., Venkatrao, M., Pellow, F., Pirahesh, H.: Data Cube: A relational aggregation operator generalizing group-by, cross-tab, and sub-totals. J. Data Mining and Knowledge Discovery 1(1), 29–53 (1997)

[2] Lakshmanan, L.V., Pei, J., Han, J.W.: Quotient cube: how to summarize the semantics of a data cube. In: Proceedings of the 28th international Conference on Very Large Data Bases, pp. 778–789. Very Large Data Bases. VLDB Endowment (2002)

[3] Lakshmanan, L.V., Pei, J., Zhao, Y.: QC-trees: an efficient summary structure for semantic OLAP. In: Proceedings of the 2003 ACM SIGMOD international Conference on Management of Data, pp. 64–75. ACM, New York (2003)

[4] Zhao, Y.: Quotient Cube and QC-Tree: Efficient Summarizations for Semantic OLAP, The University of British Columbia, Thesis for Master of Science (2003)

[5] Li, S.E., Wang, S.: Research on closed data cube technology. J. Journal of Software 15(8), 1165–1171 (2004)

[6] Li, C.P., Cong, G., Tung, A.K., Wang, S.: Incremental maintenance of quotient cube for median. In: Proceedings of the Tenth ACM SIGKDD international Conference on Knowledge Discovery and Data Mining, pp. 226–235. ACM, New York (2004)

[7] Li, H.S., Huang, H.K.: Online Aggregation on Data Cubes Without Auxiliary Information. J. Journal of Software 17(4), 806–813 (2006)

[8] Hahn, C., et al.: Edited synoptic cloud reports from ships and land stations over the globe, 1982-1991, http://cdiac.ornl.gov/ftp/ndp026b/

Network Stability Assessment Using the Number of Tree Adjacent to an Articulation Node

Yonghyoun Kim[1], Hyuncheol Kim[2], Jin-wook Chung[1], Kwang-jong Cho[3], and Ki-sung Yu[3]

[1] Dept. of Electrical and Computer Engineering, Sungkyunkwan Univ., Suwon, Korea
{yhkim,jwchung}@songgang.skku.ac.kr
[2] Dept. of Computer Science, Namseoul Univ., Cheonan, Korea
hckim@nsu.ac.kr
[3] Korea Institute of Science and Technology Information, Daejeon, Korea
{kjcho,ksyu}@kisti.re.kr

Abstract. In this study, we examine the stability assessment of the current network by using the number of trees generated when an articulation node within the network becomes functional. As a consequence, we will be able to design networks with more stability using the same amount of resources. Previous studies assessed the stability of a network by using the number of nodes adjacent to a broken node but they were inapplicable to complex networks with more than two broken nodes. The network stability assessment method proposed in this study will be applicable to complex networks with two or more broken nodes and will offer a more detailed assessment by considering the distribution of network risk factors.

Keywords: articulation point, biconnected component, network stability, network design, standard deviation.

1 Introduction

The 1.25 Internet crises occurred when an Internet worm found a weakness in the KT server, which, in 2003, occupied 54% of Korea's Internet traffic, to generate a network traffic surge and transfer large volumes of network packets thereby overloading and paralyzing the DNS server or an attack known as DDOS (Distributed Denial Of Service). This has become the background to Korea's network environment in which Internet traffics are handled in bulks thereby allowing hackers to bring down network functions by overloading with traffics [1].

A network is consisted of nodes and edges where a large scope has to be maintained. Thus, some of the core elements of the network must be fully maintained in order for an efficient maintenance. This paper introduces a network component, which is the only connection between networks and a path through which all traffics must pass before reaching another network. We will denote this component an articulation node [1].

While increasing the number of links used when designing a network will cause the number of articulation nodes to drop and stabilize the network, an appreciation in

O. Gervasi et al. (Eds.): ICCSA 2008, Part II, LNCS 5073, pp. 1229–1241, 2008.

the network management cost is unavoidable. Therefore, we can say that the number of links we can use to establish a network is limited. As a consequence, an accurate stability analysis of a network is necessary in order to create a stable network with a smaller number of links that connect the same nodes. We propose a method that assesses the stability of an entire network based on the number of adjacent trees to an articulation node.

This paper is organized as follows. Chapter 2 introduces the general notion of articulation nodes. Chapter 3 discusses the network stability assessment method using the existing articulation nodes. In chapter 4, we apply the network stability assessment method showed in chapter 3 with many kinds of topologies composed of same resources and find network stability about node distribution. Finally, Chapter 5 proposes future study topics and the conclusion.

2 Related Studies

2.1 Articulation Point

In order to search for articulation points, we say G=<V, E> is a given connection undirected graph and we generate a depth first search spanning tree by applying the depth first search for G that has root node or start node $r \in V$ and perform a depth first numbering from the root node [3, 4, 5, 6, 7, 8, 9, 10, 11, 12].

V has n number of nodes and E has e number of edges. If deleting a node in G along with all affiliated edges, node $i \in V$ that generates graph G with at least two connection elements is called the articulation point [3, 10].

(1) Conditions for an Articulation Point

① The root of a breadth-first spanning tree that has more than two children.

② Node i that has at least one child node k, which is a node of which the ancestors cannot be reached using a path consisted of only one white edge, node k, and its descendants.

③ If node i is the root of a spanning tree that has more than two children or if it has child k (child of i) that satisfies low(k) ≥ DFN(i).

④ low(i) is the minimum depth first number reached from I using at most one white edge and the descendants of I as the path.

low(i) = min {
dfn(i),
min{low(k)|k is the children node of i},
min(dfn(j)|(i, j) is the DFN of a back edge node}

⑤ Node I is an articulation point if it is the root of a spanning tree that has more than children or if it has child k that satisfies low(k) ≥ dfn(i).

B, E, G, I nodes are articulation points as shown in [Figure 1].

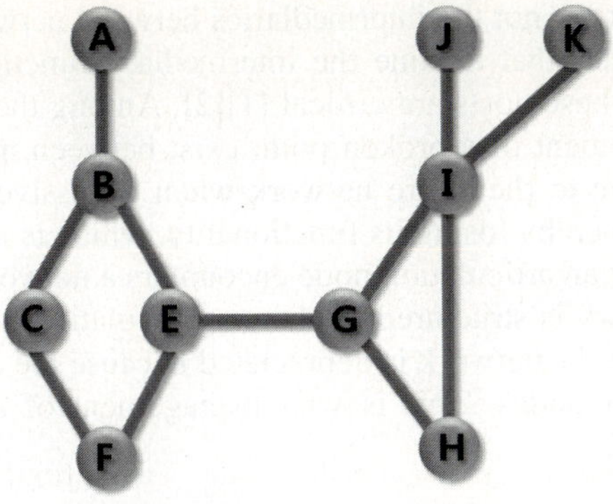

Fig. 1. Articulation point of a connection undirected graph

2.2 Biconnected Component

If connection undirected graph is G = <V, E> and if G-{i} is connected for all I ∈ V, graph G is a biconneted graph [1]. As illustrated in [Figure 2], connection undirected graph with no articulation point entails biconnected component and (A,B), (B,C,F,E), (E,G), (G,H,I), (I,K), (I,J) are oriented as biconnected component [3, 4, 5, 6, 7, 8, 9].

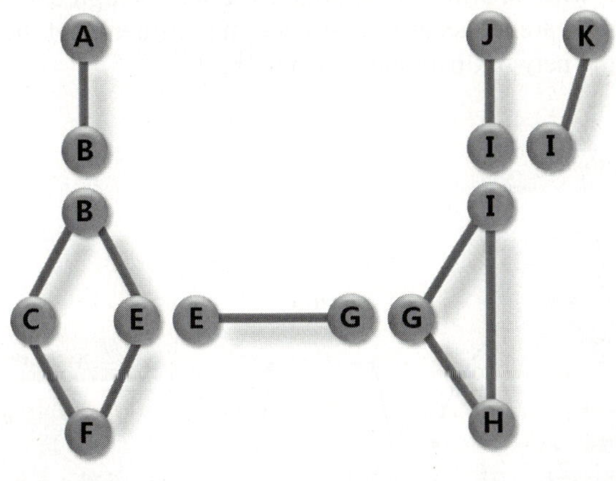

Fig. 2. Biconnected component of a connection undirected graph

2.3 Articulation Node in Network

The structure of a network can be expressed with a connection undirected graph. A node can indicate a hub, a switch, or a router and an edge can represent a communication line. Also, the factors of a broken point can be applied to some nodes where they connect biconnected components. In other words, since this is the only path that connects two or more networks and the only node that can share data, it must handle more traffic than other nodes [1][2].

The many hosts that are not the intermediaries between networks are serviced communication by the nodes that assume the intermediary functions. The roles of each node that administers these hosts are critical [1][2]. Among these intermediary nodes, those that have the element of a broken point exist between networks and will bring about a critical damage to the entire network when excessive traffic or an attack is taken on those nodes thereby losing its functionality, which is much more critical than when a node that is not an articulation node encounters a network problem [1][2].

Also, when a network is structured so that an articulation node has many adjacent nodes, the reliability of the network is depreciated because the articulation node forms connections with other nodes. This is why management of articulation nodes in a network is important [2].

3 Network Stability Formation through the Relationship between an Articulation Node and Its Adjacent Nodes

Network stability is determined by the relationships between the many elements that form the network. The network stability of a network can be figured out by using the number and the location of the nodes.

As per the location, a node can be classified as an articulation node or an adjacent node when the node has the elements of a broken point.

When a network with an articulation node experiences high external traffic volume and when many nodes are adjacent to that node, the influence of the articulation node on the stability of the network becomes greater [1][2].

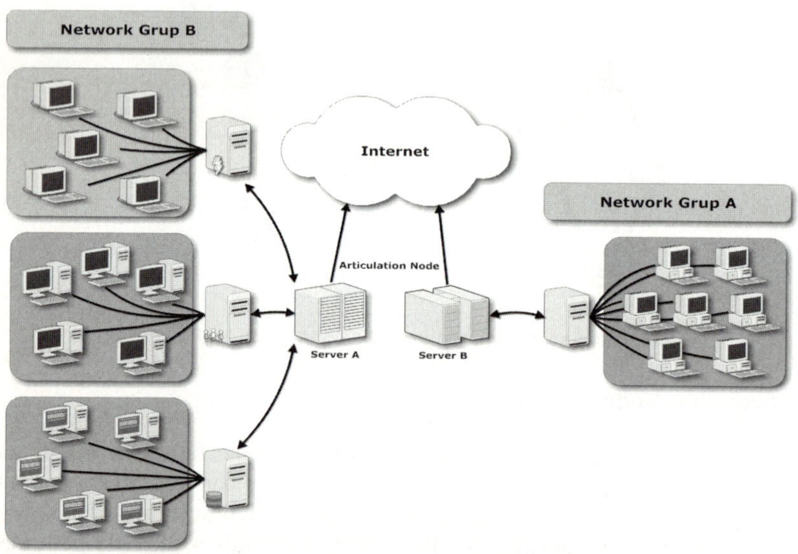

Fig. 3. Number of adjacent nodes of an articulation node

An articulation node having many adjacent nodes indicates that the articulation node is heavily depended upon [2]. An increase in the number of adjacent nodes with high external traffic usage would naturally increase the traffic volume the articulation node has to handle and this eventually leads to more risk factors [1].

There are two routers in [Figure 3]. 16 hosts and 8 hosts are connected to Server A and B, respectively. Assuming that each host consumes an equal amount of network resources, Server A would have to handle twice as much traffic volume as Server B because it simply has twice as many hosts. Therefore, Server A has a greater influence on the network than does Server B [2].

3.1 The Method of Assessing Network Stability Based on the Number of Adjacent Node to an Articulation Node

In previous research, Network stability analysis method use to one Articulation node and many adjacent nodes can not apply to the network environment that Articulation node many existed [2]. Therefore, we suggested that analysis method can apply to network environments had more than one Articulation node. The following is formula (1) for network stability measurement using the number of nodes adjacent to an articulation node [2].

$$N_{stability} = \begin{cases} \dfrac{a}{\sum_{i=1}^{a} D_i} & (\text{if } a \geq 1) \\ 1 & (\text{if } a = 0) \end{cases} \tag{1}$$

$N_{\text{stability}}$: Network stability
a : Number of Articulation node ($a \geq 0$)
D_i : Number of Adjacent node

The following algorithm is that formula (1) transformed.

Algorithm. Network Stability Measurement (ref ar_node_info <object array>)
Conditions before
- The number of nodes and edges of the network is the same.
- The volume of generated traffic and the importance of every host are the same.
- The network has one or more articulation nodes.
- An adjacent node must have an edge and be connected to an articulation node.
- If two trees that include one or more nodes are connected by one or more edges, the two trees are then adjacent.
- The number of articulation nodes is or grater.
- There must be two or more adjacent nodes to from an articulation node. (including the articulation node)

Conditions after
- Evaluate $N_{stability}$ divide number of Articulation node of Articulation node's adjacent node.

```
 - Nstability == 1 : Articulation node does not exist to network.
 - 0 < Nstability < 1 : return Nstability
 - Nstability == -1 : Cannot measure network's stability.

Return conditions
 - Nstability <float>

i = 0
net_ad_node_num = 0;
if ( ar_node_info.length == 0 )
   Nstability = 1;
else if ( ar_node_info.length > 0 )
  loop ( i < ar_node_info.length )
    net_ad_node_num = net_ad_node_num + ar_node_info[i++].ad_node_num
  end loop
  if ( net_ad_node_num < ar_node_info.length )
    Nstability = ar_node_info.length / net_ad_node_num
    return Nstability
  end if
else
  Nstability = -1
end if
return Nstability
end Network Stability Measurement
```

3.2 The Method of Assessing Network Stability Based on the Number of Adjacent Trees to an Articulation Node

A method measuring network's stability using adjacent node which Articulation node has $N_{stability}$ is the same though having same network resource and consisting another topology. So a finding method safer than topology having the same $N_{stability}$ is suggested. We can measure the network stability calculating the standard deviation using the number of articulation node, the number of adjacent tree and $N_{stability}$.

There are networks with structures as shown in [Figure 4], [Figure 5], [Figure 6].

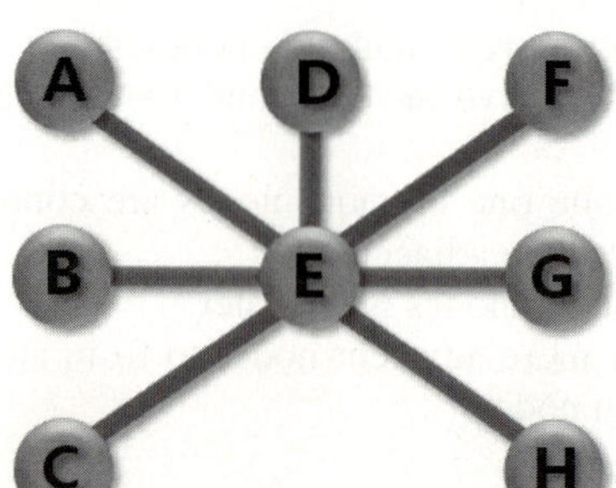

Fig. 4. Network Topology 1

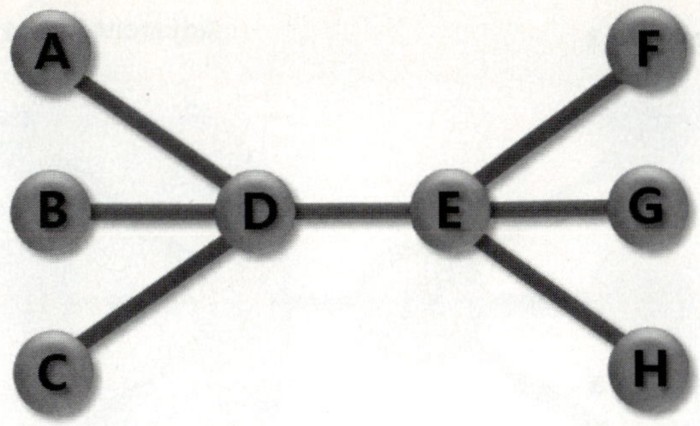

Fig. 5. Network Topology 2

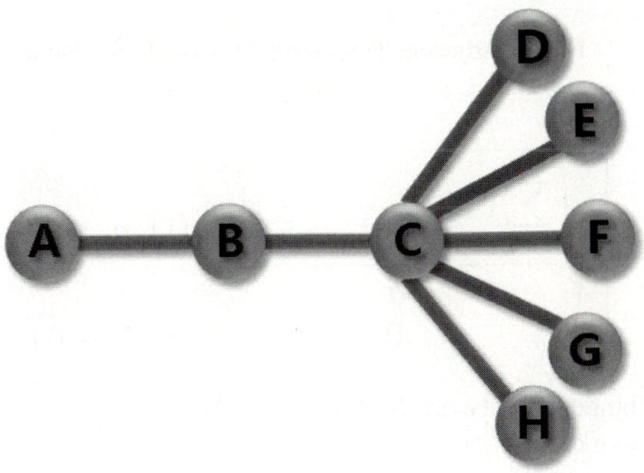

Fig. 6. Network Topology 3

The following is an analysis of each network's articulation nodes.

1) In [Figure 4], there is one articulation node and that is Node E.
2) In [Figure 5], there are two articulation nodes and they are Nodes D, E.
3) In [Figure 6], there are two articulation nodes and they are Nodes B, C.

Prior to applying the network stability assessment method proposed in this paper, we should go over the notion of adjacent trees. In [Figure 5], the adjacent tree created when removing Node D, which is the articulation node, is as shown in [Figure 7]. Simply put, the adjacent tree of an articulation node is a partition of the network generated by removing the articulation node.

If the value of $N_{stabiltiy}$ is the same, we can state that the network with equally scattered risk factors of an articulation node is more stable. This can be computed by using the following formula (2) [13].

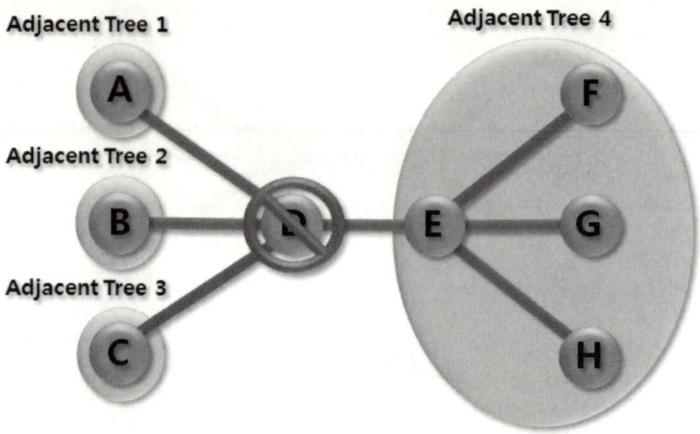

Fig. 7. Adjacent Tree of the D Articulation Node

$$N_{variance} = \begin{cases} \sqrt{\dfrac{\sum_{i=1}^{a}(T_i - \frac{1}{N_{stability}})^2}{a}} & (\text{if } a \geq 1) \\ 0 & (\text{if } a = 0) \end{cases} \tag{2}$$

$N_{variance}$: Variance of Network Stability
$N_{stability}$: Network Stability
a : Number of Articulation Node
T_i : Number of Adjacent Tree of the Articulation Node

Here, a network is more stable when the value of $N_{variance}$ is smaller. Because the smaller value indicates that the likelihood that a critical situation will occur when an articulation node has been broken is lower.

The following algorithm is that formula (2) transformed.

Algorithm. Network Stability Measurement (ref ar_node_info <object array>)
Conditions before
- The number of nodes and edges of the network is the same.
- The volume of generated traffic and the importance of every host are the same.
- The network has one or more articulation nodes.
- An adjacent node must have an edge and be connected to an articulation node.
- If two trees that include one or more nodes are connected by one or more edges, the two trees are then adjacent.
- The number of articulation nodes is 0 or greater.
- There must be two or more adjacent nodes to form an articulation node. (including the articulation node)

Conditions after
- Evaluate $N_{stability}$ divide number of Articulation node of Articulation node's adjacent node.
- $N_{stability}$ == 1 : Articulation node does not exist to network.
- 0 〈 $N_{stability}$ 〈 1 : return $N_{stability}$
- $N_{stability}$ == -1 : Cannot measure network's stability.

Return Conditions
- $N_{variance}$ <float>

i = 0
j = 0
net_ad_tree_num = 0;
net_stDev_E1 = 0;
net_stDev_E2 = 0;

if (ar_node_info.length == 0)
  $N_{variance}$ = 0
else if (ar_node_info.length ≥ 1)
  loop (i < ar_node_info.length)
    net_ad_tree_num = net_ad_tree_num + ar_node_info[i++].ad_tree_num
  if (net_ad_tree_num 〈 ar_node_info.length)
    $N_{stability}$ = ar_node_info.length / net_ad_tree_num
  end if
  loop (j < ar_node_info.length)
    net_stDev_E1 = net_stDev_E1 + **square** (ar_node_info[j++].ad_tree_num - $N_{stability}$)
  end loop
  net_stDev_E2 = net_stDev_E1 / ar_node_length
  $N_{variance}$ = **sqare root** (net_stDev_E2)
else
  $N_{variance}$ = -1
end if
return $N_{variance}$
end Network Stability Measurement

4 Experiment and Analysis

There are many topologies in the network. So, we have to find the method which can make the network stable by using the same network resources.

[Figure 8], [Figure 9] making a comparison between two topologies, $N_{stability}$ is the same because the number of articulation nodes composing the network is same as the number of adjacent nodes composing the network. But, as each two topologies are composed of different structures, their stabilities are different.

So, we seek more stable network structure comparing $N_{variance}$ of two networks by way to use articulation node and adjacent tree.

There are networks with structures as shown in [Figure 8], [Figure 9].

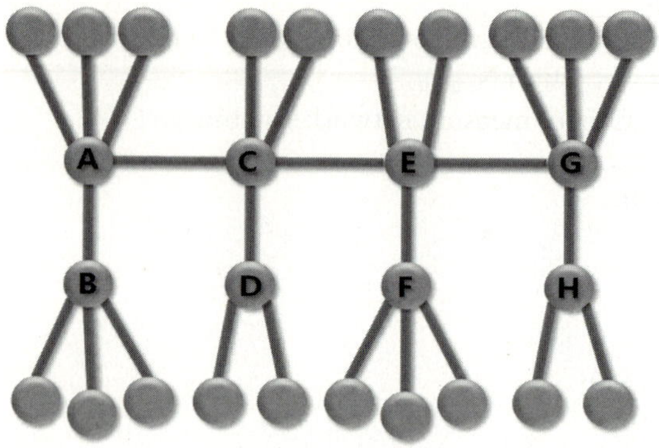

Fig. 8. Network Topology 1

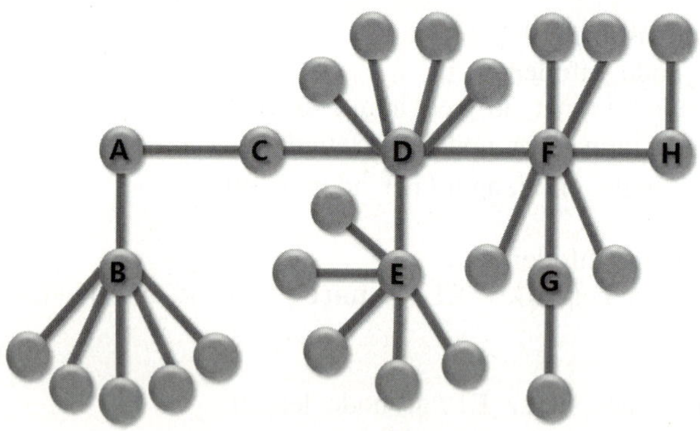

Fig. 9. Network Topology 2

The $N_{stability}$ The value of [Figure 8] obtained by using the proposed network stability assessment method is as follows.

$$N_{stability} = \frac{8}{5+4+5+3+5+4+5+3} = \frac{8}{34} \tag{3}$$

The $N_{stability}$ The value of [Figure 9] obtained by using the proposed network stability assessment method is as follows.

$$N_{stability} = \frac{8}{2+6+2+7+6+7+2+2} = \frac{8}{34} \tag{4}$$

The $N_{stability}$ Since the values of [Figure 8] and [Figure 9] are the same when applying the proposed network stability assessment method, we must compare the $N_{variance}$ values.

The $N_{variance}$ value of [Figure 8] is as follows.

$$N_{variance} = \sqrt{\frac{0.56+0.06+0.56+1.56+0.56+0.06+0.56+1.56}{8}} = 0.83 \tag{5}$$

The $N_{variance}$ value of in [Figure 9] is as follows.

$$N_{variance} = \sqrt{\frac{5.06+3.06+5.06+7.56+3.06+7.56+5.06+5.06}{8}} = 2.28 \tag{6}$$

As a result, we can say that [Figure 8] is the most stable network followed by [Figure 9]. When we are given a certain amount expense to create a network, it would be fairly economical and stable to create one as shown in [Figure 8].

The proposed network stability assessment method is applicable to not only networks with simpler structures as mentioned earlier but also to those networks with complex and more number of nodes.

You can see many kind of topologies composed of eight articulation nodes and twenty non articulation nodes on Table 1. And we find $N_{variance}$ to show each topology's stabilities.

Table 1. Variance of Network Stability results by distributions of articulation node's adjacent tree

Topology		**A**	**B**	**C**	**D**	**E**	**F**	**G**	**H**	$N_{variance}$
		Articulation Node								
1	**Adjacent Tree**	4	4	4	4	4	4	5	5	0.4330
2		3	3	4	4	4	4	5	7	1.1989
3		3	3	3	4	4	4	4	9	1.8540
4		2	2	2	3	4	5	7	9	2.4366
5		2	2	2	2	4	4	7	11	3.0310
6		2	2	2	2	3	3	7	13	3.6657
7		2	2	2	2	2	3	6	15	4.2646
8		2	2	2	2	2	3	4	17	4.8669
9		2	2	2	2	2	2	3	19	5.5845
10		2	2	2	2	2	2	2	20	5.9529

You can find network variance changes using distributions of articulation node's adjacent tree in [Figure 10].

In case of Topology 1 had minimum $N_{variance}$, each articulation node has balanced adjacent trees. Therefore, in case of articulation node broken, networks face minimum damages, so it is stable. And in case of topology 10 had maximum $N_{variance}$ adjacent trees are concentrated on H articulation node while other articulation nodes have minimum adjacent trees. When H articulation node is broken, networks face fatal damages, so it is very instable.

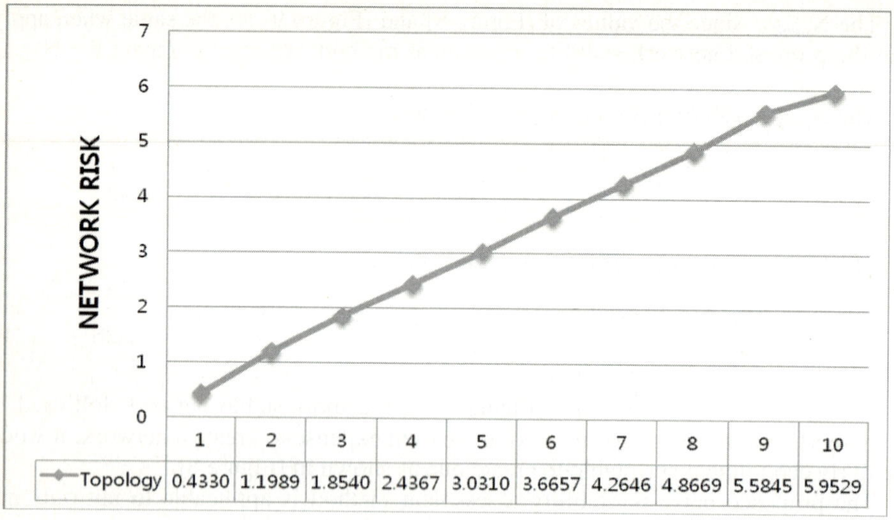

Fig. 10. Variance of Network Stability changes by distributions of Articulation node's adjacent tree

5 Conclusion

The network stability assessment method proposed in this research is applicable to complex networks with many nodes and its ability to reflect the restrictions of a network is appropriate for stability assessment of network designs. It is also economic and efficient in that it can automate network assessments.

However, it is rather distant from actual networks because it does not take into consideration the importance of each node or the traffic generated by each node. It will also need to implement cases for two or more edges.

We will therefore continue to examine and study ways that reflect the importance of each node, the traffic generated by nodes, and if a node is a network device. We will also focus on various network issues such as when two or more articulation nodes have been created or when two or more edges have been lost thereby a node that was not an articulation node becomes an articulation node.

References

1. Kim, Y., Ahn, S., Chung, J.-w.: Importance of network Articulation node of traffic base analysis techniques. In: Proceedings of the Digital Contents Society, December 2007, pp. 113–117 (2007)
2. Kim, Y.h., Kim, K., Lim, C.S., Ahn, S., Chung, J.-w.: Technique of Analysis of Network's Importance through Analysis on the Relationship between an Articulation Node and its Adjacent Nodes. In: GESTS, pp. 1–8 (January 2008)
3. Horowitz, E., Sahni, S., Anderson-Freed, S.: Fundamentals of Data Structure in C, 2nd edn., pp. 265–292. Silicon Press (August 2007)

4. Aho, A.V., Hopcroft, J.E., Ullman, J.D.: Data Structures and Algorithms. Addison-Wesley, Reading (1983)
5. Cormen, T.H., Leiserson, C.E., Rivest, R.L.: Introduction to Algorithms. McGraw-Hill, New York (1990)
6. Gabow, H.N.: Path-based depth-first search for strong and biconnected components, Tech. Report CU-CS-890-99, revised version, Dept. of Computer Sciencs, University of Colorado at Boulder (2000)
7. Weiss, M.A.: Data Structures and Algorithm Analysis in C++. Addison-Wesley, Reading (1999)
8. Huang, S.T.: A new distributed algorithm for the biconnectivity problem. In: Proc. of International Conf. on Parallel Processing, vol. 3, pp. 106–113 (1989)
9. Tarjan, R.E., Vishkin, U.: An efficient parallel biconnectivity algorithm, TR-69, Ultracomputer Note-51, Courant Institute of Mathematical Sciences, NY (1983)
10. Chaudhuri, P.: An optimal distributed algorithm for finding articulation points in a network. Computer Communications 21(18), 1707–1715 (1998)
11. Liang, Y.D., Rhee, C.: Finding biconnected components in O(n) time for a class of graphs. Information Processing Letters 60(3), 159–163 (1996)
12. Woo, J., Sahni, S.: Computing biconnected components on a hypercube. In: Second IEEE Workshop on Future Trends of 1990 Distributed Computing Systems, pp. 277–283 (1990)
13. Ziemer, R.E (translator)., Kim, J.-k.: Elements of Engineering Probability and Statistics, November 2003, pp. 90–92. HongRueng pub. (2003)

Discovering Knowledge from Local Patterns with Global Constraints

Bruno Crémilleux[1] and Arnaud Soulet[2]

[1] GREYC-CNRS, Université de Caen
Campus Côte de Nacre
F-14032 Caen Cédex France
bruno.cremilleux@info.unicaen.fr
[2] LI, Université François Rabelais de Tours
3 place Jean Jaurès
F-41029 Blois France
arnaud.soulet@univ-tours.fr

Abstract. It is well known that local patterns are at the core of a lot of knowledge which may be discovered from data. Nevertheless, use of local patterns is limited by their huge number and computational costs. Several approaches (e.g., condensed representations, pattern set discovery) aim at selecting or grouping local patterns to provide a global view of the data. In this paper, we propose the idea of global constraints to write queries addressing global patterns as sets of local patterns. Usefulness of global constraints is to take into account relationships between local patterns, such relations expressing a user bias according to its expectation (e.g., search of exceptions, top-k patterns). We think that global constraints are a powerful way to get meaningful patterns. We propose the generic Approximate-and-Push approach to mine patterns under global constraints and we give a method for the case of the top-k patterns w.r.t. any measure. Experiments show its efficiency since it was not feasible to mine such patterns beforehand.

Keywords: data mining, local patterns, constraint-based paradigm, global constraints, top-k patterns.

1 Introduction

In current scientific, industrial or business areas, the critical need is not to generate data, but to derive knowledge from huge datasets produced at high throughput. Extracting or mining knowledge from large amounts of data is at the core of the Knowledge Discovery from Data task, often also named "data mining". This involves different challenges, such as designing efficient tools to tackle data and the discovery of patterns of a potential user's interest. The constraint-based pattern mining framework is a powerful paradigm to discover new highly valuable knowledge [16]. Constraints provide a focus on the most promising knowledge by reducing the number of extracted patterns to those of a potential interest given by the user. There are now generic approaches to discover *local patterns*

O. Gervasi et al. (Eds.): ICCSA 2008, Part II, LNCS 5073, pp. 1242–1257, 2008.

under constraints [7, 20]ăand this issue is rather well-mastered, at least for the data described by items (i.e., boolean attributes). A Dagstuhl seminar has been devoted on the topic of local patterns in order to provide a clearer view of this new field [15] and a Definition of local patterns is given in Section 2.1.

Nevertheless, even if the number of produced local patterns is reduced thanks to the constraint, the output still remains too large for an individual and global analysis performed by the end-user. The most significant patterns are lost among too many trivial, noisy and redundant information. Many works propose methods to reduce the collection of patterns, like the so-called condensed representations [5] or the compression of the dataset by exploiting Minimum Description Length Principle [19] but most of them are dedicated to the particular case of frequent patterns (a pattern X is said *frequent* if the number of examples in the database supporting X exceeds a given threshold). Recent approaches - pattern teams [12], constraint-based pattern set mining [8] and selecting patterns according to the added value of a new pattern given the currently selected patterns [2] - aim at reducing the redundancy by selecting patterns from the initial large set of patterns on the basis of their usefulness in the context of the other selected patterns. Even if these approaches explicitly compare patterns between them (as the global constraints, see Section 3), they are mainly based on the reduction of the redundancy and they cannot take into account in a flexible way a bias given by the user to direct the final set of patterns toward a specific aim such as finding the best patterns according to an interestingness measure or the search of exceptions. In this paper, we call *global pattern* a set of local patterns satisfying a property involving several local patterns (cf. Section 2).

On the other hand, it should be a pity to consider the summarization of local patterns only from the point of view of the redundancy. We think that local patterns are also suitable to revisit classical data exploration tasks as for example classification or clustering and more generally to produce global patterns as models. Assuming that the user would like to produce a classifier based on rules. A complete and correct constraint-based data mining method ensures to produce a proper set of classification rules. It differs for instance from the decision trees technique where the attribute selection criterion ensures to pick the best attribute at each node but it does not guarantee the best whole tree [4]. Obviously, having a proper set of classification rules does not straightforwardly provide a relevant classifier. More generally, it is clear that moving from local patterns to global patterns as models like classifiers is still a challenge. Studying carefully the complementarity of the various pattern domains (e.g., clustering and local pattern discovery, local pattern discovery as feature construction for supervised classification) is needed.

This paper addresses this general issue by using the constraint paradigm. Our first contribution is to propose the notion of *global constraints* as a flexible and declarative way to design global patterns as sets of local patterns. This approach enables the user to express a bias and discover richer global patterns. Second, we provide the generic Approximate-and-Push approach to mine patterns satisfying global constraints. We give a method for mining the top-k patterns w.r.t. any measure and we show its efficiency in practice.

This paper is organized as follows. Section 2 outlines basic definitions and several examples of global patterns as set of local patterns. We will see that the current techniques are specific to the kind of targeted global patterns. In Section 3, we propose the notion of global constraints and we show how ităenables us to address a lot of global patterns. We present in Section 4 our generic Approximate-and-Push approach to mine patterns under global constraints and the case of the top-k patterns w.r.t. any measure. In Section 5, we discuss of the usefulness of our approach to design global patterns as models. Finally, Section 6 concludes.

2 Basic Definitions and Examples of Global Patterns

We give below basic definitions used among this paper. Then we sketch examples of methods providing global patterns coming from local patterns.

2.1 Basic Definitions

Let $\mathcal{I}$ be a set of distinct literals called items, an itemset (also called *pattern*) corresponds to a non-null subset of $\mathcal{I}$. The language of patterns corresponds to $\mathcal{L}_{\mathcal{I}} = 2^{\mathcal{I}} \backslash \emptyset$. A transactional dataset $\mathcal{D}$ is a multi-set of patterns of $\mathcal{L}_{\mathcal{I}}$ usually called transactions. For instance, Table 1 gives a transactional dataset $\mathcal{D}$ where 9 transactions $t_1, \ldots, t_9$ are described by 6 items $A, \ldots, C_2$.

Table 1. Example of a transactional context $\mathcal{D}$

$\mathcal{D}$

Trans.	Items					
t_1	A	B			C_1	
t_2	A	B			C_1	
t_3			C		C_1	
t_4			C		C_1	
t_5			C		C_1	
t_6	A	B	C	D		C_2
t_7			C	D		C_2
t_8			C			C_2
t_9				D		C_2

Local patterns are regularities that hold for a particular part of the data. A local pattern is of special interest if it exhibits a deviating behavior w.r.t. the underlying global model of the data [11] because we are seeking for surprising knowledge which deviates from the already known background model. But local patterns can also be considered as fragmented and incomplete knowledge conveying some aspect of the data. As introduced in Section 1, a challenge is then to gather the pieces of the puzzle to turn *global patterns*, i.e., sets of patterns satisfying a property involving several local patterns. Such global patterns provide summarizations of the data.

Let X be a pattern. The constraint-based pattern mining framework aims at discovering all the patterns of $\mathcal{L}_\mathcal{I}$ satisfying a given predicate q, named *constraint*, and occurring in $\mathcal{D}$. A well-known example is the frequency constraint focusing on patterns having a frequency in the database exceeding a given minimal threshold $minfr > 0$: $freq(X) \geq minfr$. For instance, AB is a frequent pattern with $minfr = 3$ because $freq(AB) = 3$. There are a lot of constraints to evaluate the relevance of local patterns. Examples of various and powerful constraints can be found in [16, 20]. We call *local constraint* such constraints because they concern only one pattern contrary to the global constraints that we propose in Section 3. The constraint paradigm also uses interestingness measures (the frequency is an example) to select local patterns. Another example of measure is the area of a pattern $area(X)$: it corresponds to the frequency of the pattern times its length (i.e., $area(X) = freq(X) \times count(X)$ where $count(X)$ denotes the cardinality of X). Previous examples are local constraints and Section 3 depicts examples of global constraints. We will come-back on the area as an example of an interestingness measure that can be used in the top-k global constraint (see Section 3).

Mining patterns under constraints requires the exploration of the search space depicted by $\mathcal{L}_\mathcal{I}$. Unfortunately, this space is generally huge and thus, pruning conditions are necessary. The property of monotonicity is well-used because pruning conditions are straightforwardly deduced [14]. A constraint q is anti-monotone w.r.t. the item specialization (resp. monotone) iff for all $X \in \mathcal{L}_\mathcal{I}$ satisfying q, any subset (resp. superset) of X also satisfies q. For instance, the minimal support constraint is anti-monotone. Whenever a pattern X is frequent, any subset of X is also frequent.

2.2 Global Patterns as Sets of Local Patterns

This section gives examples of methods producing the sets of all patterns satisfying a property involving several local patterns. Have a look on characterization rules which are an important issue in a lot of applications. The usefulness of the simplest rules (i.e., rules having minimal premises) for this task is shown in [6]. Basically, given a bounded number of exceptions δ, a rule has a minimal premise if any rule with a proper subset of its premise and less than δ exceptions does not enable to conclude on the same class value (Section 3 gives a formal definition highlighting the comparison between two local patterns). Such rules are inferred from a specific subset of patterns satisfying a property of δ-freeness [3]. Prototypes extracting the condensed representations of the δ-free patterns can be updated to extract the correct and complete collection of characterization rules as defined above [6]. Unfortunately, this approach cannot be extended if we are interested in other kinds of characterization rules.

The discovery of exception rules from a data set without domain-specific information is also of a great interest [22]. An exception rule is defined as a deviational pattern to a strong rule and the interest of a rule is evaluated according to another rule. Again, the comparison between rules means that these exception rules are not local patterns. Suzuki proposes a method based on sound pruning and probabilistic estimation [22] to extract all the exception rules. But, as previously, this approach is devoted to this kind of patterns.

Another example is the top-k patterns [10]. These patterns are the k patterns optimizing an interestingness measure m (e.g., frequency, area). The collection of the top-k patterns is concise and gathers the most relevant patterns according to m. It can be seen as a global pattern. Discovering the top-k patterns optimizing a measure m becomes difficult as soon as m has no suitable property like anti-monotonicity. Naive post-processing approaches extract in a first step all patterns whose the value of m exceeds a threshold, then the k patterns optimizing m are selected. But, in practice, the choice of the threshold is too subtle. A too high threshold may lead to miss patterns satisfying the top-k constraint, a too small may generate thousands of patterns giving intractable computations. For instance, a naive method for mining the top-k patterns w.r.t. area requires to extract all present patterns because even patterns whose frequency value is 1 may satisfy the area constraint. A branch-and-bound algorithm for top-k patterns is proposed in [8], but only few measures (stemming from upper-boundable constraints) are tackled.

In the next section, we show that all of these global patterns (and many others) can be easily expressed with global constraints. In Section 4.2, we provide a method belonging to our generic Approximate-and-Push approach which efficiently mines the top-k patterns w.r.t. any measure (whereas the works on top-k patterns in the literature are limited to the frequency measure [10] and specific forms of the patterns).

3 Global Constraints for Mining Global Patterns

We propose in this section the notion of *global constraint* which formalizes the building of the global patterns previously introduced. Global constraints consider simultaneously at least two patterns contrary to local constraints which are checked individually on each pattern (see Section 2.1). By comparing several patterns, global characteristics from the database can be revealed.

Definition 1 (Global constraint). *A constraint q is said global if several patterns have to be compared to check if q is satisfied or not.*

A global constraint satisfaction is based on information coming from at least two patterns. This point is highlighted in the examples given below by the two patterns X and Y. For instance, several constraints (e.g., freeness, characterization rules, exception rules) require to check properties of the subsets Y of X. Peak constraint compares two neighbor patterns and top-k any pair of patterns. This definition of global constraints enables to write properties both minimizing the redundancy (e.g., freeness, characterization rules) and directed toward a bias given by the user (e.g., minimal premise with the characterization rules, unexpected information with the exception rules, exceptional behavior with the peak constraint).

We present now a non-exhaustive list of global constraints illustrating Definition 1 in the context of itemset mining. All the examples given in Section 2.2 are written with the formalism of the global constraints. Some of them are very well-known, like freeness or closedness. We also propose the peak constraint which highlights an exceptional behavior of a pattern with respect to its neighbors.

- **Condensed representations (freeness and closedness):** A free (resp. closed) pattern is a minimal (resp. the maximal) pattern of its equivalence class of frequency [3, 17] (the frequency has been introduced in Section 2.1). This kind of patterns is useful to design condensed representations or to derive association rules. Freeness and closedness require to check the frequency of X with respect to the frequencies of its subsets.

$$freeness(X) \equiv \begin{cases} true & \text{if } \forall Y \in \mathcal{L}_\mathcal{I} \text{ such that } Y \subset X, \text{ one have } \mathit{freq}(X) < \mathit{freq}(Y) \\ false & \text{otherwise} \end{cases}$$

$$closedness(X) \equiv \begin{cases} true & \text{if } \forall Y \in \mathcal{L}_\mathcal{I} \text{ such that } X \subset Y, \text{ one have } \mathit{freq}(X) > \mathit{freq}(Y) \\ false & \text{otherwise} \end{cases}$$

  For instance, AC satisfies the *freeness* constraint in dataset $\mathcal{D}$ given by Table 1 but not ABC since $\mathit{freq}(ABC) = \mathit{freq}(AB)$. Dually, $ABCD$ satisfies the *closedness* constraint but not ABC ($\mathit{freq}(ABC) = \mathit{freq}(ABCD)$).
- **Characterization rules:** A formulation of characterization rules is given in [6]. It can be defined by the following global constraint:

$$characterization(X \rightarrow c) \equiv \begin{cases} true & \text{if } \forall Y \in \mathcal{L}_\mathcal{I} \text{ such that } Y \subset X, \text{ one have} \\ & \qquad X \rightarrow c \wedge \neg(Y \rightarrow c) \\ false & \text{otherwise} \end{cases}$$

  where $X \rightarrow c$ denotes a rule with $\mathit{freq}(X \rightarrow c) \geq \gamma$ and $\mathit{freq}(X \rightarrow \neg c) \leq \delta$. It means that $X \rightarrow c$ is a rule with a minimal premise to conclude to c with lower than δ exceptions. For instance, with $\gamma = 1$ and $\delta = 1$, the rule $A \rightarrow C_1$ is minimal and has only one exception in dataset $\mathcal{D}$ (see Table 1). Then, $A \rightarrow C_1$ satisfies the constraint *characterization*.
- **Exceptions:** Suzuki [22] defines exception rules which isolate unexpected information as the following set or rules (I is an item):

$$exception(X \rightarrow \neg I) \equiv \begin{cases} true & \text{if } \exists Y \in \mathcal{L}_\mathcal{I} \text{ such that } Y \subset X, \text{ one have} \\ & \qquad X \backslash Y \rightarrow I \wedge X \rightarrow \neg I \wedge Y \not\rightarrow \neg I \\ false & \text{otherwise} \end{cases}$$

  The rule $X \rightarrow \neg I$ is an exception rule since usually if $X \backslash Y$ then I and if Y then not frequently $\neg I$. Let us illustrate this notion by assuming that a rule $X \rightarrow Y$ holds iff at least 2/3 of transactions containing X also contains Y. Following our running example, $AC \rightarrow \neg C_1$ is an exception rule because we jointly have $A \rightarrow C_1$ and $C \not\rightarrow \neg C_1$.

- **Top-k constraint:** Let $k > 0$ be an integer and $m : \mathcal{L}_\mathcal{I} \rightarrow \Re$ be a measure (e.g., frequency [10]). The top-k constraint w.r.t m are the k best patterns according to m and it equals to:

$$top_{k,m}(X) \equiv |\{Y \in \mathcal{L}_\mathcal{I} | Y \neq X \wedge m(Y) > m(X)\}| < k$$

  Typically, by ignoring class values C_1 and C_2 in dataset $\mathcal{D}$, the top-2 patterns according to the area measure are C ($area(C) = 6$) and AB ($area(AB) = 6$). Indeed, all the other patterns have an area lower than 6.
- **Peak constraint:** A peak pattern is a pattern whose neighbors have a value for a measure m lower than a threshold. It means that a peak pattern has an exceptional behavior compared to its neighbors. Let $m : \mathcal{L}_\mathcal{I} \rightarrow \Re$ be a measure, d be a distance (e.g., $d(X,Y) = |X \backslash Y| + |Y \backslash X|$), δ be an integer and ρ be a real, one have:

$$peak(X) \equiv \begin{cases} true & \text{if } \forall Y \in \mathcal{L}_\mathcal{I} \text{ such that } d(X,Y) < \delta, \text{ one have } m(X) \geq \rho \times m(Y) \\ false & \text{otherwise} \end{cases}$$

  For instance, according to Table 1 and not considering C_1 and C_2, the pattern AB is a peak with $\delta = 1$ and $\rho = 1.5$. More precisely, the area of AB ($= 6$) exeeds those of all its 4 neighbors (i.e., $area(A) = 3$, $area(B) = 3$, $area(ABC) = 3$ and $area(ABD) = 3$ and $3 \times \rho \leq 6$).

Among others, all examples of global patterns given in Section 2.2 can be straightforwardly expressed with global constraints. We think that this unified approach brought by the global constraints improves the understandability of the global patterns. Furthermore, this approach enables to easily design new global patterns like peaks, but many other examples can be addressed.

Global constraints provide a reduced collection of patterns forming a coverage or gathering the best ones according to a bias or a criterion expressing the interest of the user. The *covering constraints* synthesize the whole set of local patterns by deleting redundancies: freeness, closedness, characterization rules. The *optimizing constraints* optimize a given criterion over some patterns (i.e., exceptions, peak constraint) or over all the patterns (i.e., the $top_{k,m}$ constraint). There is no strict border between these two categories: for instance, a characterization rule also satisfies an optimizing constraint, i.e., a maximal number of exceptions.

4 The Approximate-and-Push Approach and the Case of the Top-k Patterns

This section proposes the generic Approximate-and-Push approach for mining patterns satisfying global constraints. We start by giving the key ideas of the approach and then we show how it performs in the case of the top-k patterns.

4.1 Overview of the Approximate-and-Push Approach

Basically, the idea is to automatically deduce local constraints from a global constraint. These local constraints are relaxations of the global one and can be pushed in the mining step. They are dynamically refined during the process to be more and more efficient. As suggested by its name, the Approximate-and-Push approach is divided in two steps which are iteratively repeated.

Approximate. This step provides a collection of candidate patterns which is an approximation of the final collection of patterns satisfying the global constraint. The initialization of the method has to guarantee that this approximation ensures to mine all patterns satisfying the global constraint. In the case of the top-k patterns, we will see in the next section that preserving *the* k best already extracted patterns is enough. Then, at each approximate step, a pattern X is tested to know if it has to be included in the collection of candidate patterns. X is added in the collection if it may satisfy the global constraint with regard to the other patterns already present in the collection. Candidate patterns of the collection are removed when a new tested pattern ensures that they cannot satisfy the global constraint.

Push. The aim of this step is to deduce from the global constraint, pruning conditions of the search space which benefit from the characteristics of the approximation, i.e., the collection of candidate patterns. Pruning conditions may be relaxations to get suitable properties like (anti-)monotonicity which are efficiently pushed by usual pattern mining algorithms. The next section details this process in the case of the top-k patterns. Moreover, when new patterns enter in the collection of candidate patterns, these pruning conditions become more efficient. This is clearly the main interest of the iterative process of the Approximate-and-Push approach.

The Approximate-and-Push approach can be performed with any pattern mining algorithm, provided that the pruning conditions which are inferred are useful for the selected algorithm. Furthermore, the Approximate-and-Push approach is easily extendable to other languages (sequences, trees, etc). In the next section, we show how an APRIORI-like algorithm can be efficiently reused by the Approximate-and-Push approach to mine the top-k patterns w.r.t. any measure.

4.2 Mining the Top-k Patterns w.r.t. a Measure

This section shows how to push the $top_{k,m}$ constraint into the mining step thanks to the Approximate-and-Push approach. Besides its efficiency, one of the original features of our method is to deal with measures without properties of (anti)-monotonicity (i.e., such measures do not provide efficient pruning conditions for usual algorithms to mine patterns satisfying the constraint). To the best of our knowledge, our method is the first one to mine the top-k patterns w.r.t. any measure.

Principles of the Method. The definition of $top_{k,m}$ given in Section 3 does not straightforwardly provide a local constraint which can be pushed by an usual pattern mining algorithm. We reformulate the $top_{k,m}$ constraint to get a relevant local constraint. First, we introduce the threshold $\rho_{k,m}$ corresponding to $\min\{m(X)|X \in \mathcal{L}_\mathcal{I} \wedge top_{k,m}(X)\}$. Then, we can rewrite the $top_{k,m}$ constraint as $top_{k,m}(X) \equiv m(X) \geq \rho_{k,m}$. This last constraint is still a global one because the threshold $\rho_{k,m}$ implicitly implies comparisons between patterns. But, fixing this threshold makes possible the definition of relevant local constraints as we will see below. The key idea is to start by fixing an arbitrary threshold, then it will be refined to improve the efficiency of the pruning conditions coming from the local constraint thanks to the iterative process of the Approximate-and-Push approach.

Approximating the top-k patterns. As already said, this step preserves the approximation which is the k patterns *Cand* optimizing the measure m from the patterns already mined. This ensures the correctness and the completeness of the result at the end of the scanning of the search space. We define an *adding threshold* ρ to decide if an applicant pattern has to be added or not in the collection *Cand* (the following definition assumes that m has to be maximized, if m has to be minimized, it is enough to change the sign of m).

$$\rho = \begin{cases} -\infty, & \text{if } |Cand| < k \\ \min_{X \in Cand} m(X), & \text{otherwise} \end{cases}$$

An applicant pattern is added to *Cand* if and only if its measure m is greater than or equal to the adding threshold ρ. At the beginning, there is a filling phase (i.e., $\rho = -\infty$) and all the mined patterns are added until *Cand* has k candidate patterns. Thereafter, the addition of an applicant pattern depends on ρ which provides the minimum value of m of the k best current patterns. A pattern X is deleted from *Cand* when k other patterns of *Cand* have a better measure than $m(X)$. The great interest of this approach is that the adding threshold will increase ($\rho_{k,m}$ is its optimal value) according to the updating of the *Cand* leading to more and more powerful pruning conditions in the push step.

Table 2. The top-2 pattern w.r.t the area with APRIORI algorithm

Level 1

Pattern	*Cand*	ρ
A (3)	*A*	$-\infty$
B (3)	*A*, *B*	3
C (6)	*C*, *A*, *B*	3
D (3)	*C*, *A*, *B*, *D*	3

Level 2

Pattern	*Cand*	ρ
AB (6)	*C*, *AB*	6
AC (2)	*C*, *AB*	6
AD (2)	*C*, *AB*	6
BC (2)	*C*, *AB*	6
BD (2)	*C*, *AB*	6
CD (4)	*C*, *AB*	6

Table 2 depicts the modifications of candidate patterns $Cand$ during the mining of the constraint $top_{2,area}$ with a levelwise approach (as performed by APRIORI [1]) in the transactional context provided by Table 1 (ignoring class values C_1 and C_2). The levelwise algorithm generates two successive sets of applicant patterns. For each level, patterns whose area is greater than ρ (bold patterns) enter in the collection of candidate patterns. The value of area is specified between brackets in the left column and the candidate patterns are gathered in the central column. As explained above, the threshold ρ (the right column) is progressively adjusted. While the number of patterns of $Cand$ is less than k, the threshold ρ remains $-\infty$. After, ρ corresponds to the minimal area satisfying by at least one pattern of $Cand$. The pattern A is not excluded when B enters in the collection because its area equals ρ. On the contrary, A, B and D are deleted when the pattern AB arrives in $Cand$ because then $\rho = 6$. At the end of the last level, $Cand$ contains exactly C and AB which are the two patterns satisfying $top_{2,area}$ (i.e., the two patterns having the best area).

Pushing the approximation. Suitable pruning conditions can be inferred from the collection of candidate patterns $Cand$ and the adding threshold. As $Cand$ preserves the $top_{k,m}$ constraint and any pattern in $Cand$ has a value of m greater than or equal to ρ, we can deduce that a pattern X must satisfy the local constraint $m(X) \geq \rho$ to be introduced into $Cand$. Unfortunately, this constraint does not straightforwardly provide suitable pruning conditions. For instance, if we come back to the area measure, $area(X) \geq \rho$ does not satisfy (anti-)monotone properties. For that purpose, we propose to relax the constraint $m(X) \geq \rho$ by a *relaxation* $m'(X) \geq \rho$ satisfying the two following conditions: $m'(X) \geq \rho$ is anti-monotone (condition $C1$) and $\forall X, m(X) \geq \rho \Rightarrow m'(X) \geq \rho$ (condition $C2$). Condition $C2$ ensures the completeness of the mining process. Finding an anti-monotone relaxation $m'(X) \geq \rho$ is not a trivial task. An automated and generic method taking into account any measure, which can be decomposed according to primitives, is given in [21]. Let us illustrate this step w.r.t. the area measure. If the k candidate patterns in the approximation have an area higher or equal to A and the size of the longest transaction in the database is L (so a pattern cannot be longer than L), we can deduce that the frequency of any top-k pattern X w.r.t. the area measure must be higher or equal to $\frac{A}{L}$ because $area(X) = freq(X) \times count(X)$. The key point is that we get a pruning condition (i.e., $freq(X) < \frac{A}{L}$) which corresponds to an anti-monotone constraint and thus can be efficiently pushed [16]. In other words, we have deduced from the global constraint a local constraint (here, $freq(X) \geq \frac{A}{L}$) leading to a pruning condition for the usual levelwise mining algorithms. Moreover, when new patterns enter in the collection of candidate patterns, the adding threshold can only increase (following our example, L is fixed and A can only increase) and this pruning condition becomes more and more efficient. Experiments described below demonstrate the practical efficiency of this process.

Let us come back on our running example of top-2 patterns w.r.t area given by Table 2. The used anti-monotone relaxation is $freq(X) \geq \rho/L$ (here $L = 4$ because the size of the longest transaction is 4). At the end of the first level,

$\rho = 3$ and the relaxation $freq(X) \geq 3/4$ eliminates no pattern. So, all the patterns containing two items and present in the context $\mathcal{D}$ are generated. On the contrary, at the end of the second level, $\rho = 6$ and the relaxation becomes $freq(X) \geq 6/4$. For the level 3, the generated patterns must have a frequency greater than or equal to 2 (none pattern has a sufficient frequency). The mining process stops at level 3 because no more pattern can be generated. Finally, the Approximate-and-Push approach avoids the generation of 4 patterns having a length of 3 and 1 pattern of length 4.

4.3 Experiments

The aim of the experiments is to show and quantify the efficiency of our method. Recalling that mining $top_{k,m}$ patterns w.r.t. any measure was not feasible beforehand, except by using a rather naive post-processing technique ($top_{k,m}$ patterns are filtered in a second time). Furthermore, as explained below, $top_{k,m}$ patterns may be missed by this technique!

In the following, **Approximate-and-Push** means our method described above and using an Apriori-like algorithm to push the anti-monotone constraint automatically deduced. As there is no other generic approaches in the literature, we think that demonstrating the feasibility of mining the top-k patterns w.r.t. measures like area (which do not check good properties like (anti-)monotonicty) is itself a first interesting result of these experiments. Nevertheless, to go further and better understand the behavior and the efficiency of Approximate-and-Push, we compare it with two other specific strategies also based on Apriori algorithm:

- **Optimal:** this strategy exploits the *optimal* anti-monotone relaxation of $m(X) \geq \rho_{k,m}$. We fix for the threshold $\rho_{k,m}$ the value optimizing the pruning condition without loss of top-k patterns. This is ideal but unfeasible in practice because this threshold cannot be known by the user. Nevertheless, the interest of this experiment is this strategy enables us to compare Approximate-and-Push w.r.t. the optimal situation pushing the constraint.
- **Post-processing:** until now, this is the single strategy to mine the top-k patterns w.r.t. measures like area, even if such an approach is not safe as we show below. In a first step, all patterns having a frequency higher than 10% are mined. Then, the k best patterns w.r.t. the measure are selected. The frequency threshold (here, 10%) is fixed by the user. In practice, the user makes a trade-off between tractability and completeness (avoid missing top-k patterns) according to his intuition. But, this method *cannot* be used safely. For instance, it can happen that a long pattern having a frequency below the frequency threshold (consequently not mining) has an area higher than a shorter and frequent pattern. Let us recall that Approximate-and-Push is sound and complete.

We use two datasets (`mushroom` and `letter` [1]) and two measures, the frequency and the area. We also performed experiments with other measures and

[1] `www.ics.uci.edu/~mlearn/MLRepository.html`

datasets, these experiments gave similar results. For each dataset, we mine the top-k patterns w.r.t these measures. All the experiments are done with the same APRIORI implementation and we can compare the different running-times. Experiments were conducted on an Xeon 2.2 GHz with 3GB of RAM memory running the Linux system. Figure 1 reports the extraction running times according to k, i.e., the number of desired patterns.

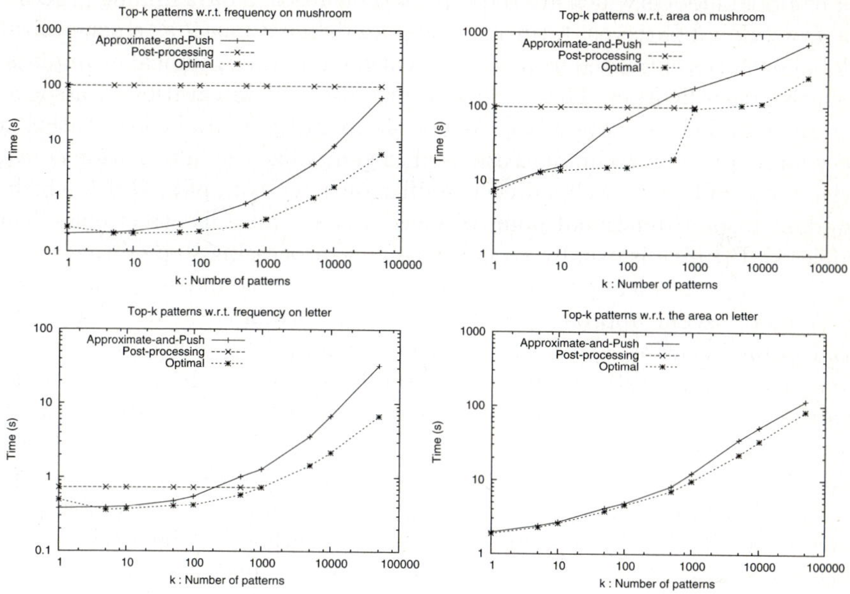

Fig. 1. Running-times for mining the top-k patterns

The Post-processing strategy has a specific behavior because its running-time is a constant which does not depend on the value k for a experiment. It suffers from two main drawbacks. When k is small, its running time is much longer than the other ones. More serious, this strategy fails for some values of k because it misses top-k patterns. In these cases, curves in Figure 1 are interrupted. With the area measure on `letter`, this strategy systematically returns false top-k patterns for any k! That it is why the curve for the Post-processing strategy is not plotted in this figure. These results recall that the Post-processing strategy is not safe.

Figure 1 indicates that the shapes of the Approximate-and-Push and Optimal curves are similar. The higher k is, the longer the running-time. Obviously, the Optimal strategy is better than the Approximate-and-Push strategy. But recalling that the Optimal strategy cannot be used in practice because it is impossible to guess the suitable threshold which has to be used to achieve the optimality. A major lesson is that the Approximate-and-Push curve is not so far from the Optimal curve (see for instance results on `letter`). Approximate-and-Push is especially efficient (and very close to the optimality) when k is small. This is due to a fast filling phase of candidate patterns which provides quickly an efficient local constraint. This result is important because the user often fixes low values for k in order to obtain a small and analyzable output.

5 Future Issue: Designing Global Patterns as Models

A lot of global models which are the expected results of a data mining process can come from global patterns. Intuitively, we can define a *model* as a generalization of global patterns extracted from an application domain. A typical example is the associative classification. This technique proposes the integration of association rule mining and classification [13]. As the number of potential classification rules is very large, pruning techniques are used to generalize the information conveyed by the rules and get a well-suited classifier on new examples. Rules which are redundant from a functional point of view or may cause incorrect classification are deleted. Pruning is usually applied as a post-processing step on the extracted rules by using statistical parameters such as support, confidence and chi-square test. A more recent approach finds the best k correlated association rules for classification by using a measure which has suitable pruning properties [24]. A common characteristic of all these methods is to use heuristics techniques to select the rules from a complete collection of local patterns to produce a classifier.

Associations, like the frequent patterns, are also at the core of clustering works [23]. For instance, ECCLAT [9] is based on frequent closed patterns and has the originality to enable a slight overlapping between clusters. A global pattern produced by ECCLAT is performed by a greedy method in which the interest of a cluster is evaluated according to an interestingness measure.

Interestingly, global patterns can capture the joint effect of local patterns. The co-classification is a way of conceptual clustering and provides a limited collection of bi-clusters. These bi-clusters are linked for both objects and attribute-value pairs. In [18], the authors propose a generic framework for co-classification. Its great interest is that the bi-partition comes from a reconstruction of the objects and attributes defining the local patterns: the bi-clusters of the final bi-partition (i.e., a global pattern) are not necessary elements of the initial set of the local patterns. Nevertheless, a distance between the bi-sets which are at the origin of the bi-clusters has to be chosen.

We see that there are several techniques to infer global patterns and models from local patterns. Unfortunately, combinations of local patterns are ad hoc to a specific goal and often use heuristics or parameters which have to be set by the user. That is why the question of how to turn large collections of local patterns into global models deserves attention. We think that global constraints enable us to significantly go further in this direction by providing appropriate sets of patterns easier to manage than the local patterns currently used. Indeed, patterns addressing by global constraints reveal relationships and a global structure inside the data, they are noticeably less numerous and they can be oriented toward a bias given by the user. We think that global constraints enable us to get a sound basis of patterns to design models.

6 Conclusion

We think that designing and discovering global patterns and models are crucial goals for the end-users. In this paper, we have proposed the notion of global constraints to address this general issue. By comparing local patterns between them, global constraints take into account relationships in the data. We have shown that global constraints are a flexible and declarative way to define a lot of global patterns (e.g., condensed representations, exceptions, top-k patterns, characterization rules). This approach also enables the user to express a bias and discover meaningful global patterns. In the literature, only few global patterns can be mined and only by using dedicated algorithms mainly relying on anti-monotone pruning. In this paper, we have proposed the generic Approximate-and-Push approach and we have given an efficient method for mining the top-k patterns w.r.t. a measure.

In the future, we would like to perform further experiments with other global constraints using the Approximate-and-Push approach. It would also be interesting to evaluate the quality of such new discovered patterns in real-world applications. Furthermore, a promising open avenue is to design high-level constraints to directly build global models, it would avoid the choice of heuristics, as it is the case in the current methods.

Acknowledgments. The authors would like to thank the participants of the Dagstuhl seminar on Parallel Universes and Local Patterns for enlightening on the subjects of this paper. Special thanks to Johannes Fürnkranz, Arno J. Knobbe and Martin Scholz for the fruitful discussions on the use of local patterns to design models. This work is partly supported by the ANR (French Research National Agency) funded project Bingo2 ANR-07-MDCO-014, which is a follow-up of the first Bingo project (2004-2007).

References

[1] Agrawal, R., Mannila, H., Srikant, R., Toivonen, H., Verkamo, A.I.: Fast discovery of association rules. In: Advances in Knowledge Discovery and Data Mining, ch.12, AAAI/MIT Press (1996)

[2] Bringmann, B., Zimmermann, A.: The chosen few: On identifying valuable patterns. In: Proceedings of the 12th IEEE International Conference on Data Mining (ICDM 2007), pp. 63–72 (2007)

[3] Boulicaut, J.-F., Bykowski, A., Rigotti, C.: Free-sets: A condensed representation of boolean data for the approximation of frequency queries. Data Min. Knowl. Discov. 7(1), 5–22 (2003)

[4] Breiman, L., Friedman, J.H., Olshen, R.A., Stone, C.J.: Classification and regression trees. Statistics probability series. Wadsworth, Belmont (1984)

[5] Calders, T., Rigotti, C., Boulicaut, J.-F.: A survey on condensed representations for frequent sets. In: Boulicaut, J.-F., De Raedt, L., Mannila, H. (eds.) Constraint-Based Mining and Inductive Databases. LNCS (LNAI), vol. 3848, pp. 64–80. Springer, Heidelberg (2006)

[6] Crémilleux, B., Boulicaut, J.-F.: Simplest rules characterizing classes generated by delta-free sets. In: 22nd Int. Conf. on Knowledge Based Systems and Applied Artificial Intelligence (ES 2002), Cambridge, UK, December 2002, pp. 33–46. Springer, Heidelberg (2002)

[7] De Raedt, L., Jäger, M., Lee, S.D., Mannila, H.: A theory of inductive query answering. In: Proceedings of the IEEE Conference on Data Mining (ICDM 2002), Maebashi, Japan, 2002, pp. 123–130 (2002)

[8] De Raedt, L., Zimmermann, A.: Constraint-based pattern set mining. In: Proceedings of the Seventh SIAM International Conference on Data Mining, Minneapolis, Minnesota, USA, April 2007, SIAM, Philadelphia (2007)

[9] Durand, N., Crémilleux, B.: ECCLAT: a New Approach of Clusters Discovery in Categorical Data. In: 22nd Int. Conf. on Knowledge Based Systems and Applied Artificial Intelligence (ES 2002), Cambridge, UK, December 2002, pp. 177–190. Springer, Heidelberg (2002)

[10] Fu, A., R.W., Kwong, W., Tang, J.: Mining n-most interesting itemsets. In: Ohsuga, S., Raś, Z.W. (eds.) ISMIS 2000. LNCS (LNAI), vol. 1932, pp. 59–67. Springer, Heidelberg (2000)

[11] Hand, D.J.: ESF exploratory workshop on Pattern Detection and Discovery in Data Mining. In: Pattern Detection and Discovery. LNCS, vol. 2447, pp. 1–12. Springer, Heidelberg (2002)

[12] Knobbe, A., Ho, E.: Pattern teams. In: Fürnkranz, J., Scheffer, T., Spiliopoulou, M. (eds.) PKDD 2006. LNCS (LNAI), vol. 4213, pp. 577–584. Springer, Heidelberg (2006)

[13] Liu, B., Hsu, W., Ma, Y.: Integrating classification and association rules mining. In: Proceedings of Fourth International Conference on Knowledge Discovery & Data Mining (KDD 1998), New York, August 1998, pp. 80–86. AAAI Press, Menlo Park (1998)

[14] Mannila, H., Toivonen, H.: Levelwise search and borders of theories in knowledge discovery. Data Mining and Knowledge Discovery 1(3), 241–258 (1997)

[15] Morik, K., Boulicaut, J.-F., Siebes, A. (eds.): Local Pattern Detection. LNCS (LNAI), vol. 3539. Springer, Heidelberg (2005)

[16] Ng, R.T., Lakshmanan, V.S., Han, J., Pang, A.: Exploratory mining and pruning optimizations of constrained associations rules. In: Proceedings of ACM SIGMOD 1998, pp. 13–24. ACM Press, New York (1998)

[17] Pasquier, N., Bastide, Y., Taouil, R., Lakhal, L.: Discovering frequent closed itemsets for association rules. In: Beeri, C., Bruneman, P. (eds.) ICDT 1999. LNCS, vol. 1540, pp. 398–416. Springer, Heidelberg (1998)

[18] Pensa, R., Robardet, C., Boulicaut, J.-F.: A bi-clustering framework for categorical data. In: Jorge, A.M., Torgo, L., Brazdil, P.B., Camacho, R., Gama, J. (eds.) PKDD 2005. LNCS (LNAI), vol. 3721, pp. 643–650. Springer, Heidelberg (2005)

[19] Siebes, A., Vreeken, J., van Leeuwen, M.: Item sets that compress. In: Proceedings of the Sixth SIAM International Conference on Data Mining, Bethesda, MD, USA, April 2006, SIAM, Philadelphia (2006)

[20] Soulet, A., Crémilleux, B.: An efficient framework for mining flexible constraints. In: Ho, T.-B., Cheung, D., Liu, H. (eds.) PAKDD 2005. LNCS (LNAI), vol. 3518, pp. 661–671. Springer, Heidelberg (2005)

[21] Soulet, A., Crémilleux, B.: Mining constraint-based patterns using automatic relaxation. Intelligent Data Analysis 13(1) (to appear). IOS Press

[22] Suzuki, E.: Undirected discovery of interesting exception rules. International Journal of Pattern Recognition and Artificial Intelligence 16(8), 1065–1086 (2002)
[23] Wang, K., Chu, X., Liu, B.: Clustering transactions using large items. In: Proceedings of ACM CIKM 1999 (1999)
[24] Zimmermann, A., De Raedt, L.: Corclass: correlated association rule mining for classification. In: Suzuki, E., Arikawa, S. (eds.) DS 2004. LNCS (LNAI), vol. 3245, pp. 60–72. Springer, Heidelberg (2004)

SQL/XML Performance Analysis of Parent/Ancestor Queries

Eric Pardede[1], J. Wenny Rahayu[1], David Taniar[2], and Ramanpreet Kaur Aujla[1]

[1] Department of Computer Science and Computer Engineeering, LaTrobe University, Bundoora 3086, Australia
{E.Pardede,W.Rahayu}@latrobe.edu.au
[2] Clayton School of IT, Monash University, Clayton 3800, Australia
David.Taniar@infotech.monash.edu.au

Abstract. The fact that XML data can be stored in different repositories creates a layer of complexity in efficient XML query processing. Most existing works discussed the query analysis for Native XML Database using XML Query Language such as XQuery and XPath. However, it is inevitable that many XML users still use established database repositories, such as relational and object-relational database, for their XML storage (called XML-Enabled Database), and an extended SQL, such as SQL/XML, for their query processing. Query analysis using SQL/XML on the XML-Enabled Database System family has not been well-researched. In this paper, we present our research on parent/ancestor XML queries using the SQL/XML standard.

1 Introduction

The reasons for storing XML in a database system are the same as for relational data: persistent storage, transactional consistency, recoverability, high availability, security, efficient search and update operations, and scalability [1]. For this purpose, two databases are used, *Native XML Database* (NXD) and *XML Enabled Database* (XED).

Despite the claims of their powerful features, NXD is still in the development stage. Many people are reluctant to use this database family. On the other hand, XED is extended from established databases such as relational, object-oriented and object-relational [2, 3], for XML data management purposes. Therefore, this database family can be used to store traditional data, such as relational data as well as XML data [3]. It gives them an advantage over NXD.

Since XED has mainly been developed from a relational or object-relational database management system, they support SQL with some extension of XML query languages (e.g. XPath and XQuery). This extended SQL is widely known as SQL/XML.

In this paper, we study XML queries focusing on parent/ancestor relationships. The purpose is to compare and analyze the performance of different querying techniques and to propose a recommendation of which query writing is suitable for *Parent/Ancestor Query*. This work is motivated by the fact that current XEDs deal with considerably large number of queries and thus, the most efficient process is required.

O. Gervasi et al. (Eds.): ICCSA 2008, Part II, LNCS 5073, pp. 1258–1273, 2008.

The analysis has to consider the nature of XML tree and also the relational-based performance measurement. The study considers the structure of the XML data in two aspects, the path and the level of hierarchy, where the path analysis is a based on the way of writing XML tree path, and the level of hierarchy analysis is based on the placing selection and the projection at different levels of hierarchy. In the experimental testing, we used a major XED product, Oracle 10g Release 2 [4].

2 Background

We start by briefly explain Parent/Ancestor XML queries, followed by different query structures in XED.

2.1 Parent/Ancestor XML Queries

Based on the retrieval target, we classify XML Query into three types: *Child/Descendant Queries*, *Parent Queries* and *Ancestor Queries* (see figure 1). In this paper, we focus on the last two query types. They are further classified into *Parent Element(s) Queries, Parent Attribute Queries, Ancestor Element(s) Queries* and *Ancestor Attribute Queries*. The analysis on the Child/Descendent queries has been reported in our previous paper [5].

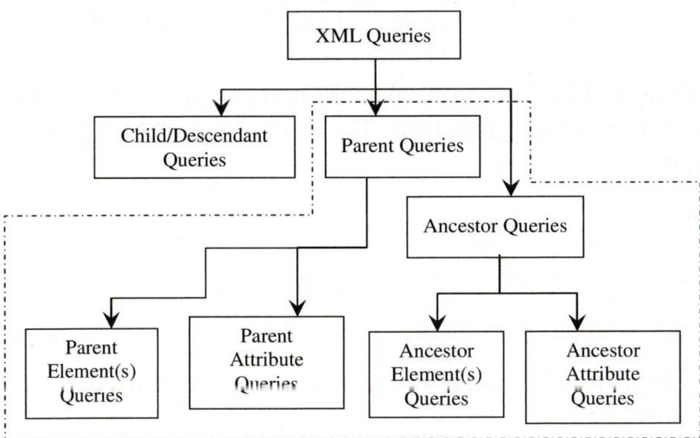

Fig. 1. XML-Enabled Database Query Classification

The query implementation is dependent on the database type. Even in one database family, such as XED, the implementation can be different from one product to another. We discuss the variations in the next section.

2.2 Queries in XED

This section will show how an XED, in this case Oracle DBMS, retrieves XML documents. XML documents can be stored in Oracle using either *CLOB*, *XMLType tables* or tables with *XMLType columns* [4]. The first type is used to store non-schema based XML documents, while the next two can be used to store schema-based XML documents.

Different ways of storing XML documents in XEDs introduce different query types. In [4], XML queries can be broadly described as: *Queries having extract, Queries having XMLTABLE, Queries having XMLQuery* and *Queries having XMLSequence*. The basic difference between these types is the structure used for projection and selection condition.

XPath Extract Query. It takes XPath expression as the projection and the selection elements. The function *existsNode* checks the XPath expression and returns true or false depending upon whether or not the XML document contains an XPath expression node. If the function returns true, the XML are retrieved using the extract function.

```
SELECT extract(<alias>.OBJECT_VALUE, '//<projection_element>')
FROM <table_name> <alias>
WHERE existsNode(<alias>.OBJECT_VALUE, '//<selection_condition>') = 1;
```

XMLTABLE XQuery. This query takes XQuery expression, containing the *FLOWR* expressions, as an input. The "xtab" serves as a pointer which stores the results of the XQuery expressions and the results are displayed using "xtab.COLUMN_VALUE".

```
SELECT xtab.COLUMN_VALUE
FROM <table_name>, XMLTABLE('<XQuery Expression using FLOWR Expressions>
   for $a in //<element_node>
    where <selection_predicate [conditional_expression]>
   return $a/<projection_element>' PASSING OBJECT_VALUE)xtab;
```

XML Query XQuery. This function takes XQuery expressions as an input. There are two ways of using XMLQuery function. The syntax variation introduces the variation in the query execution.

```
SELECT XMLQuery('<XQuery Expression using FLOWR Expressions>
  for $a in //<element_node>
   where <selection_predicate [conditional_expression]>
   return $a/<projection_element>'
  PASSING OBJECT_VALUE RETURNING CONTENT) FROM <table_name>;

SELECT XMLQuery('<XQuery Expression using FLOR Expressions>
  for $a in //<element_node>
   return $a/<projection_element>'
  PASSING OBJECT_VALUE RETURNING CONTENT) FROM <table_name>
WHERE existsNode(<alias>.OBJECT_VALUE, '//<selection_condition>') = 1;
```

XPath XMLSequence Query. This query takes XPath expression as an input. The *existsNode* function checks the XPath expression and returns true or false depending upon whether XML document contains an XPath expression node or not. XMLSequence function creates a collection of all these XML documents. The table function un-nests the collection and stores them in virtual table. Then *extractValue* function is used to extract the scalar values for projected element. If *existsNode* function returns false, then no rows are projected.

```
SELECT extractValue(value(x), '/<projection_element>')
FROM <table_name> <alias>,table( XMLSe-
  quence(extract(<alias>.OBJECT_VALUE,'//<projection_element>')))x
WHERE existsNode(<alias>.OBJECT_VALUE,'//<selection_condition>')= 1;
```

3 XML Query Optimization

As it is mentioned earlier, there are two major repositories for XML Database, NXD and XED. For NXD, there are works that optimized the XML query by extending the XML data model representation [6], by proposing new algorithms for query tree processing [7, 8, 9], and by using optimized path expressions [10, 11]. There are rooms for more research for this storage option. However, in this paper we focus on the query optimization in the other XML storage, XED.

3.1 Existing Work

Some of the existing works used approaches that are based on how the XML is stored in XED. For example, [12] investigated the query optimization using loosely-coupled information retrieval engine and using DBMS tables with query execution engines. It shows the effect of join algorithm and hardware cache utilization towards containment queries performance. Unfortunately, there is no detail discussion on how to improve the performance, so that, the containment queries can be efficiently processed. In this paper, we use different approach. We will only concern on the usage of DBMS tables and query execution engine for query optimization. However, we will show how to improve the query performance for different query types. The approach has been attempted in pure Relational Database as it is seen in [13, 14].

Other works focused on investigating translation from XML standard to XED language. [15] proposed an algebraic approach for order-sensitive XQuery processing over relational databases, which is called XQuery-to-SQL Order-sensitive Translation (XSOT). This work has more focus on direct rewriting from one query family to another query family. [16] investigated the algorithm using XML Algebra Tree (XAT) for explicitly represents the semantics of XQuery. Unfortunately, it does not clearly describe how it is applied in XED language such as SQL or SQL/XML. A similar approach is taken by [17], where the authors proposed a technique to rewrite XQuery by using an XML element constructor. It provided XQuery samples and their optimized result. However, there is no justification for the performance evaluation.

Table 1 summarizes the optimization techniques in XEDs with comments on their limitations. Another observation, these works focus on algorithm level and thus, do not give end users ability to perform query optimization.

Table 1. Optimization Techniques in XEDs

Techniques	Refs	Comments
Join Algorithm and Cache Utilization	[11]	Missing discussion on performance improvement
XAT Query Rewriting and Cleanup	[12]	Missing description on implementation
Required Kinds of Query Rewriting	[3]	Missing performance evaluation
XSOT Approach	[10]	Missing description on implementation

3.2 Current Trends

Many XEDs are developed with relational database as the underlying DBMS. Among the products are Oracle, DB2, SQL Server, Cache and Matisse. In regards to query optimization, most of the products have been working in the following area.

- **XML Parsing Optimization.** It eliminates the need to keep re-parsing the XML documents. Any changes to the contents of a schema or Document Object Model (DOM) results in just writing the updated data back to the disk.
- **Node Searching Optimization.** A more optimum way to search a particular node can be investigated, instead of just using string comparisons. For example, Oracle [6] uses dynamic hash tables rather than string comparisons to perform node searches.
- **XML Schema Optimization.** Once the schema has been registered with the database, no additional schema parsing or validation is required before using it. If the DBMS requires information contained in the schema, it can access the required information directly from the cache.
- **Optimization of SQL functions that modify XML.** SQL/XML provides some functionality to update XML data, such as *updateXML*, *insertChildXML*, etc.
- **Rewriting of XPath expression.** XPath expressions can be rewritten to purely relational SQL statements which results in efficient processing. This rewritten SQL statement is processed by a database optimizer in the same manner as other SQL statements and the execution plan is based on a conventional relational algebra.

All these techniques, except for the last point, are based on hardware implementation and there has not been much work done for query optimization. As evident, query optimization in XED, is in infancy and has not been fully explored. One probable reason is the immaturity of XED standard by ISO/ANSI and W3C.

4 Parent and Ancestor Queries Classification

Parent and Ancestor Queries are further classified based on the projection target, into Element(s) Queries and Attribute Queries. Element(s) queries retrieve the whole element node including child elements and attributes associated with it, while Attribute queries retrieve just the attribute associated with the parent/ancestor element.

4.1 Parent Element(s) Queries

These are the queries which involve the retrieval of the Parent element(s) based on the selection on the child element. For example, Figure 2 shows a tree representation for element retrieval (node *E*), having selection on the child element (node *H*). From the figure, we can see that this query type retrieve node *E* and its sub-elements. All these queries represented as tree structure can be written using XPath extract Query, XMLTABLE XQuery and XMLQuery XQuery syntax.

4.2 Parent Attribute Queries

These queries involve with the projection of attributes associated with the parent element based on the selection on the child element. For example, Figure 3 shows a Parent Attribute query, which contains the selection predicate on element *F* (child element) and projection predicate on element *C* (parent element). The output of this query is attribute associated the element *C*.

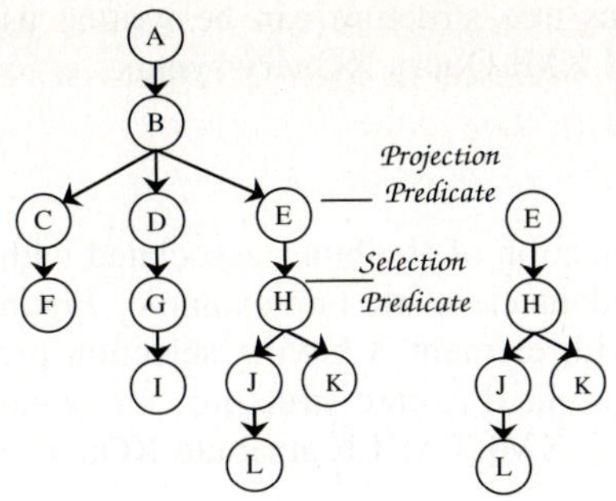

(a) Input Tree (b) Output Tree

Fig. 2. Tree Repr. for *Parent Element* Queries

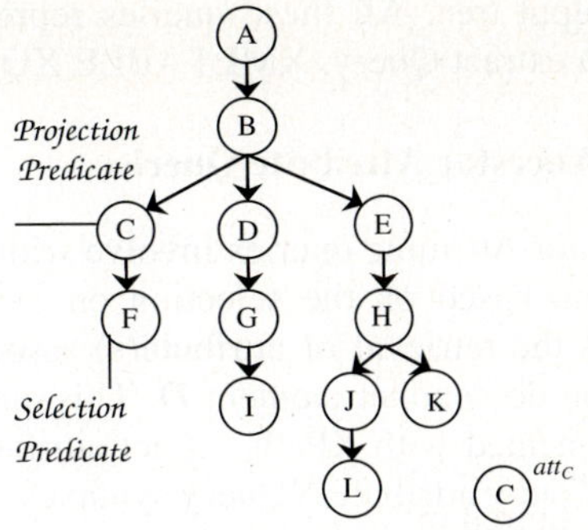

(a) Input Tree (b) Output Tree

Fig. 3. Tree Repr. for *Parent Attribute* Queries

Attribute queries use XPath extract Query, XMLTABLE XQuery and XMLQuery XQuery with a little modification. The changed syntax of the above mentioned queries in order to retrieve attribute associated with Parent element is given below.

XPath extract attribute Query:

```
SELECT extract(<alias>.OBJECT_VALUE,
   '/<element_having_attribute>/@<projection_attribute>')
FROM <table_name> <alias>
WHERE existsNode(<alias>.OBJECT_VALUE, '//<selection_condition>');
```

XMLTABLE attribute XQuery:

```
SELECT xtab.COLUMN_VALUE
FROM <table_name>, XMLTABLE('<XQuery Expression using FLOWR Expressions>
   for $a in //<element_node> where <selec-
   tion_predicate[conditional_expression]>
   RETURN $a/<element_having_attribute>/@<projection_attribute>'
   PASSING OBJECT_VALUE)xtab;
```

XMLQuery attribute XQuery:

```
SELECT XMLQuery('<XQuery Expression using FLOWR Expressions>
   for $a in //<element_node> where <selec-
   tion_predicate[conditional_expression]>
   RETURN $a/<element_having_attribute>/@<projection_attribute>'
   PASSING OBJECT_VALUE RETURNING CONTENT)
FROM <table_name>;
```

4.3 Ancestor Element(s) Queries

Ancestor Element(s) queries involve with the retrieval of the Ancestor element(s) based on the selection on the descendant element. For example, Figure 4 shows the tree representation for element retrieval (node *A*), having selection on the descendant element (node *D*). From the figure, we can see that element *A* is on the top hierarchy. If the descendant node satisfies the selection condition, then the whole input tree will be output tree. All these queries represented as tree structure can be written using XPath extract Query, XMLTABLE XQuery and XMLQuery XQuery syntax.

4.4 Ancestor Attribute Queries

Ancestor Attribute queries involve with the projection of attributes associated with an element based on the selection on the descendant element. For example, Figure 5 shows the retrieval of attribute(s) associated with element *A* having selection predicate on descendant element *D*. This query represented as tree structure can be easily implemented with XPath extract attribute Query, XMLTABLE attribute XQuery and XMLQuery attribute XQuery syntaxes.

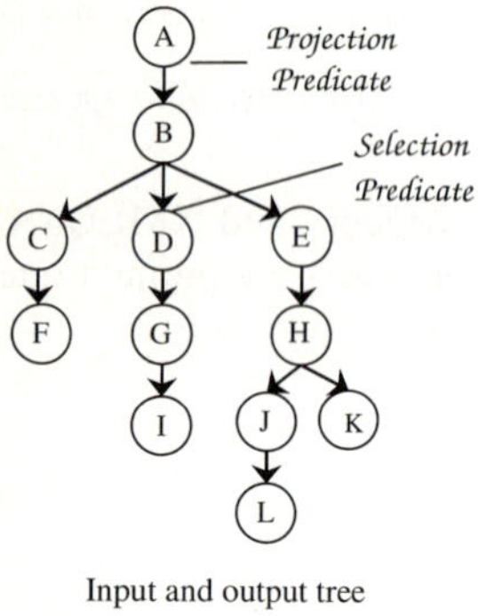

Fig. 4. Tree Repr. for *Ancestor Element* Queries

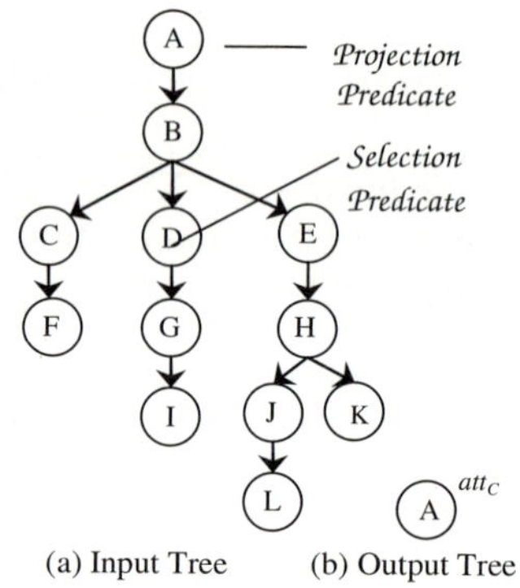

Fig. 5. Tree Repr. *for Ancestor Attribute Querie*

5 Queries Performance Experiments

5.1 Experimental Setup and Tool

For all practical purposes, Oracle 10g Release 2 [4] was used for this paper. The experiments were performed on a Windows workstation having 40 GB of disk storage and 504MB of memory. XML Schema was created based on a publicly available XML document (http://www.dia.uniroma3.it/Araneus/Sigmod/Record/Sigmod Record/SigmodRecord.xml)

The size of the data used for the experiment was 1.01MB. The table was created based on the XML Schema. The XML documents stored contain not only element nodes but also attribute and text nodes. TKPROF utility and Timing utility [4] were used to do performance analysis. TKPROF utility [18], also known as SQL Trace, reports information about each SQL statement executed with the resources it has used, the number of times it was called, and the number of rows which it processed.

Performance analysis of the queries involves the comparison of the statistical information of the queries. The information is based on the following parameters.

- **CPU Time.** It refers to time taken to parse, execute, or fetch calls for the statement. Query performance is shown by Execute and Fetch phases.
- **TKPROF Elapsed Time.** It refers to the time required for all parsing, executing, or fetching calls for the statement. This is the total time from start to finish of the execution of the query.
- **Timing Elapsed Time.** It is the total execution time of the query obtained by timing utility of SQL*Plus.

5.2 Experimentation

Parent Element(s) Queries Experimentation. *"Retrieve Parent Element when the selection is on Child Element."* The query for this problem can be categorized into nine, based on different combinations of projections and selections in the hierarchy (see table 2). The projection predicates are on four different parent elements namely *issuesTuple*, *articles*, *articlesTuple* and *authors*. The selection predicate is located at different levels of hierarchy.

Figure 6 represents the selection and the projection predicates for category 9. In this case, we "*retrieve authors element when its AuthorPosition attribute is located at the lowest level of the tree hierarchy*". Table 3 represents the possible structures for this query.

Table 2. Parent Element(s) Query Categories

Cat	Description
1	Selection Predicate at attribute associated with Child element and Projection Predicate present at top level of hierarchy Category
2	Selection Predicate at Child element and Projection Predicate present at top level of hierarchy Category
3	Selection Predicate at attribute associated with projected element and Projection Predicate present at top level of hierarchy Category
4	Selection Predicate at attribute associated with Child element and Projection Predicate present at middle level of hierarchy Category
5	Selection Predicate at attribute associated with projected element and Projection Predicate present at middle level of hierarchy Category
6	Selection Predicate at Child element and Projection Predicate present at one level down to middle level of hierarchy Category
7	Selection Predicate at attribute associated with projected element and Projection Predicate present at one level down to middle level of hierarchy Category
8	Selection Predicate at Child element and Projection Predicate present at bottom-most level of hierarchy Category
9	Selection Predicate at attribute associated with projected element and Projection Predicate present at bottom-most level of hierarchy Category.

This query can be written in four different ways based on sub-section 2.2. Queries 1 and 4 are written using XPath extract Query. Query 4 is written using the same query structure as Query 1. However, the way it is written is different. Query 2 is written using XMLTABLE XQuery. Query 3 is written using XMLQuery XQuery.

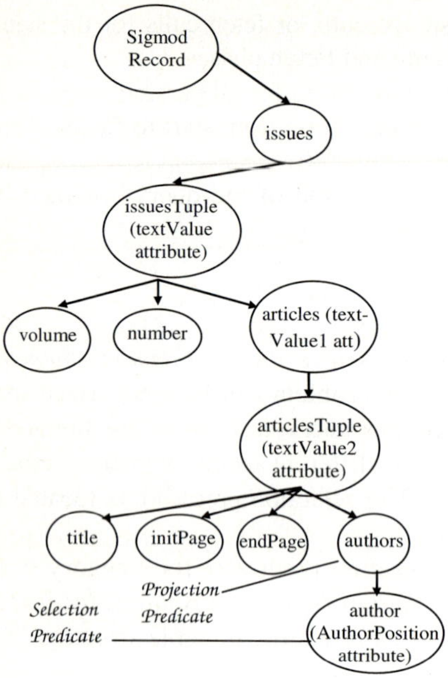

Fig. 6. *Parent Elem.* Query Tree Representation

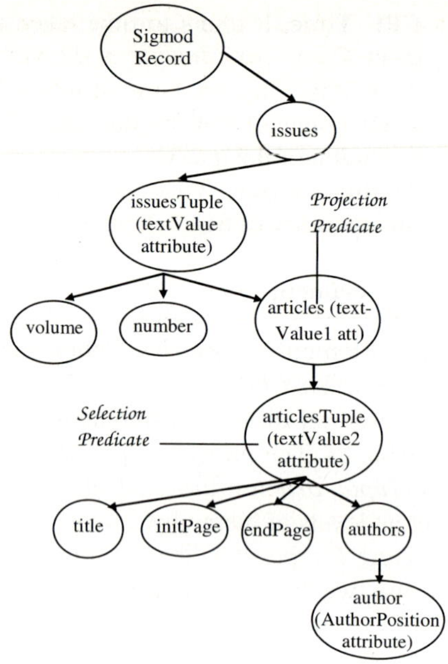

Fig. 7. *Parent Att.* Query Tree Representation

Table 3. Parent Element(s) Query Structures

Query	Query Structures associated with Parent Element(s) Query Category 9

Q1	`SELECT extract(s.OBJECT_VALUE, '//authors')"AUTHORS"`
	`FROM SIGMOD s`
	`WHERE existsNode(s.OBJECT_VALUE,'//authors[(author/@AuthorPosition="89" and author/@AuthorPosition="67") or author/@AuthorPosition="100"]')= 1;`

Q2	`SELECT xtab.COLUMN_VALUE "AUTHORS"`
	`FROM SIGMOD, XMLTABLE ('for $a in //authors`
	`where ($a/author/@AuthorPosition="89" and`
	`$a/author/@AuthorPosition="67") or`
	`$a/author/@AuthorPosition="100"`
	`return $a'`
	`PASSING OBJECT_VALUE)xtab;`

Q3	`SELECT XMLQuery('for $a in //authors`
	`return $a'`
	`PASSING OBJECT_VALUE`
	`RETURNING CONTENT)"AUTHORS"`
	`FROM SIGMOD s`
	`WHERE existsNode(s.OBJECT_VALUE,'//authors[(author/@AuthorPosition="89" and author/@AuthorPosition="67") or author/@AuthorPosition="100"]')= 1;`

Q4	`SELECT extract(s.OBJECT_VALUE, '/SigmodRecord/*//authors') "AUTHORS"`
	`FROM SIGMOD s`
	`WHERE existsNode(s.OBJECT_VALUE,'/SigmodRecord/*//authors`
	`[(author/@AuthorPosition="89" and author/@AuthorPosition="67") or`
	`author/@AuthorPosition="100"]')= 1;`

Parent Attribute Queries Experimentation. *"Retrieve Attribute associated with Parent Element when the selection is on Child Element."* Parent Attribute Queries are based on projection of three different Parent Attributes: *textValue*, *textValue1* and *textValue2*. Based on the projection and the selection nodes, we can classify this query into 4 categories shown in table 4. There are 16 different ways of writing queries for these categories.

Figure 7 represents selection and projection predicate for query category 3. In this case, we *retrieve textValue1 attribute while the selection predicate is textValue2 attribute.* Table 5 represents the possible query structures for this category. The queries include those written using XPath attribute query (Query 5 and 8), using XMLTABLE (Query 6) and using XMLQuery (Query 7).

Table 4. Parent Attribute Query Categories

Cat	Description
1	Selection Predicate at attribute associated with Child element and Projection Predicate present at top level of hierarchy Category
2	Selection Predicate at Child element and Projection Predicate present at top level of hierarchy Category
3	Selection Predicate at attribute associated with Child element and Projection Predicate present at middle level of hierarchy Category
4	Selection Predicate at Child element and Projection Predicate present at middle level of hierarchy Category

Table 5. Parent Attribute Query Structures

Query	Query Structures associated with Parent Element(s) Query Category 3
Q5	SELECT extract(s.OBJECT_VALUE, '//articles/@textValue1')"TEXTVALUE1" FROM SIGMOD s WHERE exists- Node(s.OBJECT_VALUE,'//articles[articlesTuple/@textValue2="51" and articlesTuple/@textValue2="52"]')= 1;
Q6	SELECT xtab.COLUMN_VALUE "TEXTVALUE1" FROM SIGMOD, XMLTABLE ('for $a in //articles where $a/articlesTuple/@textValue2="51" and $a/articlesTuple/@textValue2="52" return $a/@textValue1' PASSING OBJECT_VALUE)xtab;
Q7	SELECT XMLQuery('for $a in //articles return $a/@textValue1' PASSING OBJECT_VALUE RETURNING CONTENT)"TEXTVALUE1" FROM SIGMOD s WHERE exists- Node(s.OBJECT_VALUE,'//articles[articlesTuple/@textValue2="51" and articlesTuple/@textValue2="52"]')= 1;
Q8	SELECT extract(s.OBJECT_VALUE, '/SigmodRecord/*//articles/@textValue1') "TEXTVALUE1" FROM SIGMOD s WHERE existsNode(s.OBJECT_VALUE,'/SigmodRecord/*//articles [articlesTuple/@textValue2="51" and articlesTuple/@textValue2="52"]')= 1;

Ancestor Element(s) Queries Experimentation. *"Retrieve Ancestor Element when the selection is on Descendant Element."* The query for this problem can be categorized into nine, based on different combinations of projections and selections in the hierarchy (see table 6).

Table 6. Ancestor Element(s) Query Categories

Cat	Description
1	Selection Predicate at attribute associated with first level Descendant element and Projection Predicate present at top level of hierarchy Category
2	Selection Predicate at second level Descendant element and Projection Predicate present at top level of hierarchy Category
3	Selection Predicate at bottom-most Descendant attribute and Projection Predicate present at top level of hierarchy Category
4	Predicate at bottom-most Descendant element and Projection Predicate present at top level of hierarchy Category
5	Selection Predicate at Descendant element and Projection Predicate present at middle level of hierarchy Category
6	Selection Predicate at bottom-most Descendant attribute and Projection Predicate present at middle level of hierarchy Category
7	Selection Predicate at bottom-most Descendant element and Projection Predicate present at middle level of hierarchy Category
8	Selection Predicate at bottom-most Descendant attribute and Projection Predicate present at one level down to the middle level of hierarchy Category
9	Selection Predicate at bottom-most Descendant element and Projection Predicate present at one level down to the middle level of hierarchy Category

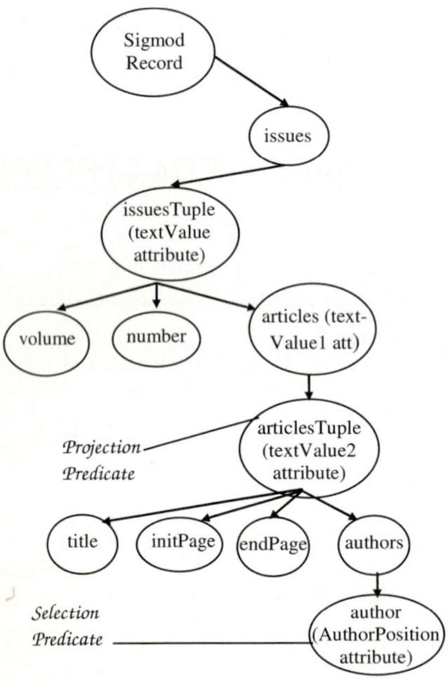

Fig. 8. *Anc. Elem.* Query Tree Representation

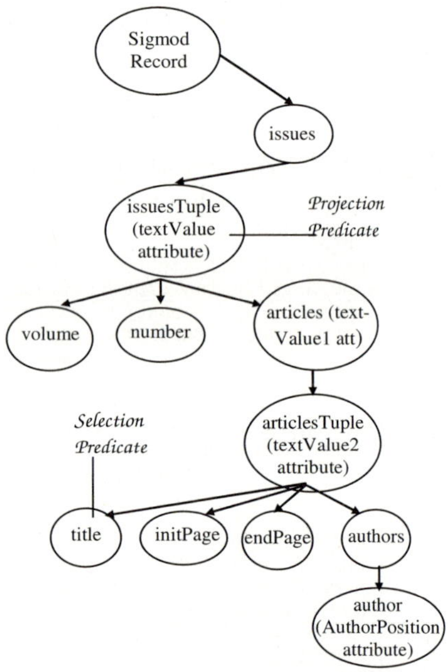

Fig. 9. *Anc Att.* Query Tree Representation

Figure 8 represents selection and projection predicate for query category 9. In this case, we *retrieve textValue2 attribute while the selection predicate is AuthorPosition attribute.* Table 7 represents the possible query structures for this category. Like in Parent Element(s) Query, there are four ways to writes this query, using XPath attribute Query, XMLTABLE attribute XQuery and XMLQuery attribute XQuery.

Table 7. Ancestor Element(s) Query Structures

Query	Query Structures associated with Parent Element(s) Query Category 9

Q9	`SELECT extract(s.OBJECT_VALUE, '//articlesTuple') "ARTICLESTUPLE"`
	`FROM SIGMOD s`
	`WHERE existsNode(s.OBJECT_VALUE,'//articlesTuple[(.//@AuthorPosition= "67" and .//@AuthorPosition="89") or .//@AuthorPosition="100"]')= 1;`

Q10	`SELECT xtab.COLUMN_VALUE "ARTICLESTUPLE"`
	`FROM SIGMOD, XMLTABLE ('for $a in //articlesTuple`
	`where (($a/authors/author/@AuthorPosition = "67" and`
	`$a/authors/author/@AuthorPosition = "89") or`
	`$a/authors/author/@AuthorPosition = "100" )`
	`return $a'`
	`PASSING OBJECT_VALUE) xtab;`

Q11	`SELECT XMLQuery('for $a in //articlesTuple`
	`return $a'`
	`PASSING OBJECT_VALUE`
	`RETURNING CONTENT) "ARTICLESTUPLE"`
	`FROM SIGMOD s`
	`WHERE existsNode(s.OBJECT_VALUE,'//articlesTuple[(.//@AuthorPosition= "67" and .//@AuthorPosition="89") or .//@AuthorPosition="100"]')= 1;`

Q12	`SELECT extract(s.OBJECT_VALUE, '/SigmodRecord/*//articlesTuple') "ARTI-CLESTUPLE"`
	`FROM SIGMOD s`
	`WHERE existsNode(s.OBJECT_VALUE,'/SigmodRecord/*//articlesTuple [(.//@AuthorPosition="67" and .//@AuthorPosition="89") or .//@AuthorPosition="100"]')= 1;`

Ancestor Attribute Queries Experimentation. *"Retrieve Attribute associated with Ancestor Element when the selection is on Descendant Element." The* Ancestor Attribute Queries are based on projection of three different Ancestor Attributes: *textValue*, *textValue1* and *textValue2*. The query for this problem can be categorized into nine, based on different combinations of projections and selections in the hierarchy (see table 8).

Table 8. Ancestor Attribute Query Categories

Cat	Description
1	Selection Predicate at attribute associated with first level Descendant element and Projection Predicate present at top level of hierarchy Category
2	Selection Predicate at bottom-most Descendant attribute and Projection Predicate present at top level of hierarchy Category
3	Selection Predicate at bottom-most Descendant attribute and Projection Predicate present at top level of hierarchy Category
4	Selection Predicate at bottom-most Descendant element and Projection Predicate present at top level of hierarchy Category
5	Selection Predicate at Descendant element and Projection Predicate present at middle level of hierarchy Category
6	Selection Predicate at bottom-most Descendant attribute and Projection Predicate present at middle level of hierarchy Category
7	Selection Predicate at bottom-most Descendant element and Projection Predicate present at middle level of hierarchy Category
8	Selection Predicate at bottom-most Descendant attribute and Projection Predicate present at one level down to the middle level of hierarchy Category
9	Selection Predicate at bottom-most Descendant element and Projection Predicate present at one level down to the middle level of hierarchy Category

Table 9. Ancestor Attribute Query Structures

Query	Query Structures associated with Parent Element(s) Query Category 2
Q13	SELECT extract(s.OBJECT_VALUE, '//issuesTuple/@textValue') "TEXTVALUE" FROM SIGMOD s WHERE existsNode(s.OBJECT_VALUE,'//issuesTuple[.//title="XML QUERIES" and .//title="Interim Report: ANSI/X3/SPARC Study Group on Data Base Management Systems 75-02-08."]')= 1;
Q14	SELECT xtab.COLUMN_VALUE "TEXTVALUE" FROM SIGMOD, XMLTABLE ('for $a in //issuesTuple where $a//articlesTuple/title = "XML QUERIES" and $a//articlesTuple/title = "Interim Report: ANSI/X3/SPARC Study Group on Data Base Management Systems 75-02-08." return $a/@textValue' PASSING OBJECT_VALUE) xtab;
Q15	SELECT XMLQuery('for $a in //issuesTuple return $a/@textValue' PASSING OBJECT_VALUE RETURNING CONTENT) "TEXTVALUE" FROM SIGMOD s WHERE existsNode(s.OBJECT_VALUE,'//issuesTuple[.//title="XML QUERIES" and .//title="Interim Report: ANSI/X3/SPARC Study Group on Data Base Management Systems 75-02-08."]')= 1;
Q16	SELECT extract(s.OBJECT_VALUE, '/SigmodRecord/*//issuesTuple/@textValue') "TEXTVALUE" FROM SIGMOD s WHERE existsNode(s.OBJECT_VALUE,'/SigmodRecord/*//issuesTuple[.//title= "XML QUERIES" and .//title="Interim Report: ANSI/X3/SPARC Study Group on Data Base Management Systems 75-02-08."]')= 1;

Figure 9 represents selection and projection predicate for query category 2. In this case, we *retrieve tissuesTuple attribute while the selection predicate is title element.* Table 9 represents the possible query structures for this category. While Query 13 and 16 are written using XPath attribute Query, Query 14 is written using XMLTABLE attribute XQuery and Query 15 is written using XMLQuery attribute XQuery.

5.3 Experimental Results

We provide four different categories from four different query structures. They are category 9, category 3, category 9 and category 2 respectively for Parent Element(s), Parent Attribute, Ancestor Element(s) and Ancestor Attribute Query.

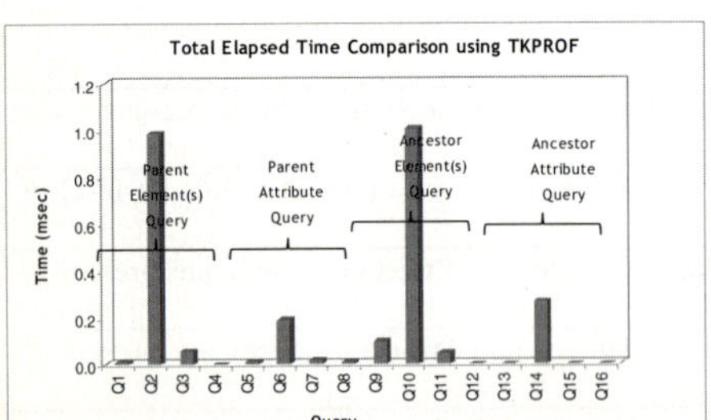

Fig. 10. TKPROF Total Elapsed Time

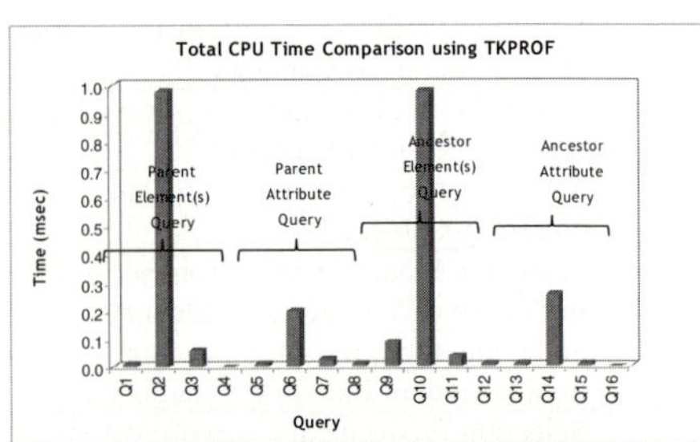

Fig. 11. TKPROF Total CPU Time

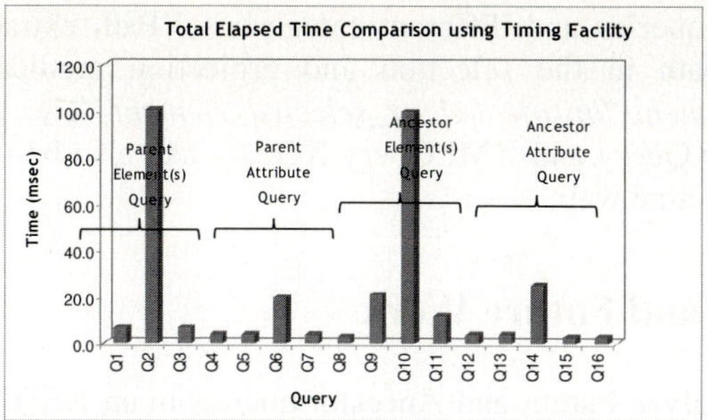

Fig. 12. Total Elapsed Time Comparison using Timing Facility

Each of the categories can be written in different ways. There are four ways of writing queries in each respective category as it is shown in tables 3, 5, 7 and 9. The analysis of the Execution Plan of the Queries is explained below:

- **Type of operation.** All the execution plans consist of full table access. Execution plan for queries 4, 8, 12 and 16 are a subset of all other queries
- **Number of Steps.** The execution plan of queries 4, 8, 12 and 16 involve only one step, fewer than for any other execution plan
- **Number of rows.** In queries 4, 8, 12 and 16, only one row is being accessed and retrieved in the execution plan. It is less when compared with execution plans of other queries. The execution plan of these three queries involves full table access, whereas in others, several steps had to be performed in order to obtain the required output.

From the above analysis of the execution plan, it can be concluded that the execution plan for queries 4, 8, 12 and 16 are the most cost effective for its respective query type and category. A comparative analysis of total elapsed time using TKPROF, Total CPU time and total elapsed time using DBMS timing property is graphically represented in figures 10 to 12. We can see how queries 4, 8, 12 and 16 incur the smallest cost for each respected category in each comparative term.

5.4 Experimental Summary

Based on the performance experimentations, we can derive some analysis for writing Parent and Ancestor XML Queries in XED.

- From the graphs, we can see XQuery written in terms of XPath Expressions using *extract* SQL function is more cost effective than XQuery written in terms of FLOWR Expressions using *XMLTABLE* and *XMLQuery* SQL Functions. XPath extract query yields good performance even when the selection is at bottom of hierarchy
- In all three query types, having the selection predicate at the top level of the hierarchy performs better than having the predicate at the bottom. The performance for the queries having selection at top or bottom can be performed by writing the path //element as */root_node_element/*//element* in both selection and projection

- For all of the queries and different categories, XPath extract query performed better when path in the selection and projection predicates are written as */root_node_element/*//projected_or_selected_element*. The performance of the XMLTABLE XQuery and XMLQuery XQuery can also be improved by writing the path as the same way.

6 Conclusions and Future Work

In this paper we analyze Parent and Ancestor queries in an XED. Based on the result of the performance analysis, each type of query can be written in a most optimized query structure. We use four queries structures, namely XPath extract query, XMLTABLE XQuery, XMLQuery XQuery and XPath XMLSequence query. The performance is then analyzed by using the statistics generated by the XED. The most optimized query structure was identified in terms of cost associated with the execution of the query.

For future work, a benchmark study of different XED can be performed to help database users determine the storage option for their XML data. We will also look into the data consistency [19] for XML transactions.

References

1. Lapis, G.: XML and Relational Storage-Are they mutually exclusive. IBM Corporation (2005)
2. Taniar, D., Rahayu, W.: Object-Oriented Collection Join Queries. In: Proceedings of International Conference on TOOLS Pacific 1996, pp. 115–125 (1996)
3. Pardede, E., Rahayu, J.W., Taniar, D.: Object-relational complex structures for XML storage. Information & Software Technology 48(6), 370–384 (2006)
4. Oracle, Introduction to Oracle XML DB. Oracle XML DB Developer's Guide, 10g Release 2 (2006), `http://www.downloadwest.oracle.com/docs/cd/B19306_01/appdev.102/b14259/xdb01int.htm#i1047170`
5. Pardede, E., Rahayu, J.W., Aujla, R.K., Taniar, D.: Performance Analysis of Child/Descendent Queries in an XML-Enabled Database. In: Gervasi, O., Gavrilova, M.L. (eds.) ICCSA 2007, Part III. LNCS, vol. 4707, pp. 749–762. Springer, Heidelberg (2007)
6. Jagadish, H.V., Lakshmanan, L.V.S., Scannapieco, M., Srivastava, D., Wiwatwattana, N.: Colorful XML: One Hierarchy Isn't Enough. In: SIGMOD 2004, ACM Press, New York (2004)
7. Wang, G., Liu, M., Yu, J.X., Sun, B., Yu, G., Lv, J., Lu, H.: Effective schema-based XML query optimization techniques. In: IDEAS 2003, IEEE-CS (2003)
8. Paparizos, S., Wu, Y., Lakhsmanan, L.V.S., Jagadish, H.V.: Tree Logical Classes for Efficient Evaluation of XQuery. In: SIGMOD 2004. ACM Press, New York (2004)
9. Geng, K., Dobbie, G.: An XML document generator for semantic query optimization experimentation. International Journal of Web Information Systems Emerald 3(1/2) (2007)
10. Bruno, N., Koudas, N., Srivastava, D.: Holistic twig joins: optimal XML pattern matching. In: SIGMOD 2002. ACM Press, New York (2002)

11. Le, D.X.T., Bressan, S., Taniar, D., Rahayu, J.W.: Semantic XPath Query Transformation: Opportunities and Performance. In: Kotagiri, R., Radha Krishna, P., Mohania, M., Nantajeewarawat, E. (eds.) DASFAA 2007. LNCS, vol. 4443. Springer, Heidelberg (2007)
12. Zhang, C., Naughton, J., Dewitt, D., Luo., Q., Lohman, G.: OHMAN. On Supporting Containment Queries in Relational Database Management Systems. In: SIGMOD 2001. ACM Press, New York (2001)
13. Taniar, D., Khaw, H.Y., Tjioe, H.C., Rahayu, J.W.: The use of hints in object-relational query optimization. Computer Systems: Science and Engineering 19(6) (2004) CRL Publishing
14. Taniar, D., Khaw, H.Y., Tjioe, H.C., Pardede, E.: The use of Hints in SQL-Nested query optimization. Information Sciences 177(12), 3110–3128 (2007)
15. Wang, L., Wang, S., Murphy, B., Rundensteiner, E.: Order-sensitive XML Query Processing over Relational Sources: An Algebraic Approach. In: IDEAS 2005. IEEE-CS, Los Alamitos (2005)
16. Zhang, X., Pielech, B., Rundesnteiner, E.: Honey, I Shrunk the XQuery! - An XML Algebra Optimization Approach. In: WIDM 2002. ACM Press, New York (2002)
17. Grinev, M., Pleshachkov, P.: Rewriting-based Optimization for XQuery Transformational Queries. In: IDEAS 2005, IEEE-CS, Los Alamitos (2005)
18. Srivastava, P.G.: Performance Evaluation Tools on Oracle., Department of Computer Science and Computer Engineering, La Trobe University (2002)
19. Goel, S., Sharda, H., Taniar, D.: Preserving Data Consistency in Grid Databases with Multiple Transactions. In: Li, M., Sun, X.-H., Deng, Q.-n., Ni, J. (eds.) GCC 2003. LNCS, vol. 3033, pp. 847–854. Springer, Heidelberg (2004)

Author Index

Printing: Mercedes-Druck, Berlin
Binding: Stein+Lehmann, Berlin

Lecture Notes in Computer Science

Sublibrary 1: Theoretical Computer Science and General Issues

For information about Vols. 1–4770
please contact your bookseller or Springer

Vol. 4972: J. van Hemert, C. Cotta (Eds.), Evolutionary Computation in Combinatorial Optimization. XII, 289 pages. 2008.

Vol. 4971: M. O'Neill, L. Vanneschi, S. Gustafson, A.I. Esparcia Alcázar, I. De Falco, A. Della Cioppa, E. Tarantino (Eds.), Genetic Programming. XI, 375 pages. 2008.

Vol. 4967: R. Wyrzykowski, J. Dongarra, K. Karczewski, J. Wasniewski (Eds.), Parallel Processing and Applied Mathematics. XXIII, 1414 pages. 2008.

Vol. 4963: C.R. Ramakrishnan, J. Rehof (Eds.), Tools and Algorithms for the Construction and Analysis of Systems. XVI, 518 pages. 2008.

Vol. 4962: R. Amadio (Ed.), Foundations of Software Science and Computational Structures. XV, 505 pages. 2008.

Vol. 4961: J.L. Fiadeiro, P. Inverardi (Eds.), Fundamental Approaches to Software Engineering. XIII, 430 pages. 2008.

Vol. 4960: S. Drossopoulou (Ed.), Programming Languages and Systems. XIII, 399 pages. 2008.

Vol. 4959: L. Hendren (Ed.), Compiler Construction. XII, 307 pages. 2008.

Vol. 4957: E.S. Laber, C. Bornstein, L.T. Nogueira, L. Faria (Eds.), LATIN 2008: Theoretical Informatics. XVII, 794 pages. 2008.

Vol. 4943: R. Woods, K. Compton, C. Bouganis, P.C. Diniz (Eds.), Reconfigurable Computing: Architectures, Tools and Applications. XIV, 344 pages. 2008.

Vol. 4942: E. Frachtenberg, U. Schwiegelshohn (Eds.), Job Scheduling Strategies for Parallel Processing. VII, 189 pages. 2008.

Vol. 4941: M. Miculan, I. Scagnetto, F. Honsell (Eds.), Types for Proofs and Programs. VII, 203 pages. 2008.

Vol. 4935: B. Chapman, W. Zheng, G.R. Gao, M. Sato, E. Ayguadé, D. Wang (Eds.), A Practical Programming Model for the Multi-Core Era. VI, 208 pages. 2008.

Vol. 4934: U. Brinkschulte, T. Ungerer, C. Hochberger, R.G. Spallek (Eds.), Architecture of Computing Systems - ARCS 2008. XI, 287 pages. 2008.

Vol. 4927: C. Kaklamanis, M. Skutella (Eds.), Approximation and Online Algorithms. X, 289 pages. 2008.

Vol. 4926: N. Monmarché, E.-G. Talbi, P. Collet, M. Schoenauer, E. Lutton (Eds.), Artificial Evolution. XIII, 327 pages. 2008.

Vol. 4921: S.-i. Nakano, M.. S. Rahman (Eds.), WALCOM: Algorithms and Computation. XII, 241 pages. 2008.

Vol. 4919: A. Gelbukh (Ed.), Computational Linguistics and Intelligent Text Processing. XVIII, 666 pages. 2008.

Vol. 4917: P. Stenström, M. Dubois, M. Katevenis, R. Gupta, T. Ungerer (Eds.), High Performance Embedded Architectures and Compilers. XIII, 400 pages. 2008.

Vol. 4915: A. King (Ed.), Logic-Based Program Synthesis and Transformation. X, 219 pages. 2008.

Vol. 4912: G. Barthe, C. Fournet (Eds.), Trustworthy Global Computing. XI, 401 pages. 2008.

Vol. 4910: V. Geffert, J. Karhumäki, A. Bertoni, B. Preneel, P. Návrat, M. Bieliková (Eds.), SOFSEM 2008: Theory and Practice of Computer Science. XV, 792 pages. 2008.

Vol. 4905: F. Logozzo, D.A. Peled, L.D. Zuck (Eds.), Verification, Model Checking, and Abstract Interpretation. X, 325 pages. 2008.

Vol. 4904: S. Rao, M. Chatterjee, P. Jayanti, C.S.R. Murthy, S.K. Saha (Eds.), Distributed Computing and Networking. XVIII, 588 pages. 2007.

Vol. 4878: E. Tovar, P. Tsigas, H. Fouchal (Eds.), Principles of Distributed Systems. XIII, 457 pages. 2007.

Vol. 4875: S.-H. Hong, T. Nishizeki, W. Quan (Eds.), Graph Drawing. XIII, 402 pages. 2008.

Vol. 4873: S. Aluru, M. Parashar, R. Badrinath, V.K. Prasanna (Eds.), High Performance Computing – HiPC 2007. XXIV, 663 pages. 2007.

Vol. 4863: A. Bonato, F.R.K. Chung (Eds.), Algorithms and Models for the Web-Graph. X, 217 pages. 2007.

Vol. 4860: G. Eleftherakis, P. Kefalas, G. Păun, G. Rozenberg, A. Salomaa (Eds.), Membrane Computing. IX, 453 pages. 2007.

Vol. 4855: V. Arvind, S. Prasad (Eds.), FSTTCS 2007: Foundations of Software Technology and Theoretical Computer Science. XIV, 558 pages. 2007.

Vol. 4854: L. Bougé, M. Forsell, J.L. Träff, A. Streit, W. Ziegler, M. Alexander, S. Childs (Eds.), Euro-Par 2007 Workshops: Parallel Processing. XVII, 236 pages. 2008.

Vol. 4851: S. Boztaş, H.-F.(F.) Lu (Eds.), Applied Algebra, Algebraic Algorithms and Error-Correcting Codes. XII, 368 pages. 2007.

Vol. 4848: M.H. Garzon, H. Yan (Eds.), DNA Computing. XI, 292 pages. 2008.

Vol. 4847: M. Xu, Y. Zhan, J. Cao, Y. Liu (Eds.), Advanced Parallel Processing Technologies. XIX, 767 pages. 2007.

Vol. 4846: I. Cervesato (Ed.), Advances in Computer Science – ASIAN 2007. XI, 313 pages. 2007.

Vol. 4838: T. Masuzawa, S. Tixeuil (Eds.), Stabilization, Safety, and Security of Distributed Systems. XIII, 409 pages. 2007.

Vol. 4835: T. Tokuyama (Ed.), Algorithms and Computation. XVII, 929 pages. 2007.

Vol. 4818: I. Lirkov, S. Margenov, J. Waśniewski (Eds.), Large-Scale Scientific Computing. XIV, 755 pages. 2008.

Vol. 4800: A. Avron, N. Dershowitz, A. Rabinovich (Eds.), Pillars of Computer Science. XXI, 683 pages. 2008.

Vol. 4783: J. Holub, J. Žďárek (Eds.), Implementation and Application of Automata. XIII, 324 pages. 2007.

Vol. 4782: R. Perrott, B.M. Chapman, J. Subhlok, R.F. de Mello, L.T. Yang (Eds.), High Performance Computing and Communications. XIX, 823 pages. 2007.

Vol. 4771: T. Bartz-Beielstein, M.J. Blesa Aguilera, C. Blum, B. Naujoks, A. Roli, G. Rudolph, M. Sampels (Eds.), Hybrid Metaheuristics. X, 202 pages. 2007.